COMPUTER TOOLS promote flexibility and meet ABET

- *PSpice* is introduced in Chapter 3 and appears in special sections throughout as a tutorial on *PSpice* for Windows for readers not familiar with its use. The special sections contain examples and practice problems using *PSpice*. Additional homework problems at the end of each chapter also provide an opportunity to use *PSpice*.
- *MATLAB*® is introduced through a tutorial in Appendix E to show its usage in circuit analysis. A number of examples and practice problems are presented throughout the book in a manner that will allow the student to develop a facility with this powerful tool. A number of end-of-chapter problems will aid in understanding how to effectively use *MATLAB*.
- *KCIDE for Circuits* is a working software environment developed at Cleveland State University. It is designed to help the student work through circuit problems in an organized manner following the process on problem-solving discussed in Section 1.8. Appendix F contains a description of how to use the software. Additional examples can be found at the web site, http://kcide.fennresearch.org/. The actual software package can be downloaded for free from this site. One of the best benefits from using this package is that it automatically generates a Word document and/or a PowerPoint presentation.

CAREERS AND HISTORY of electrical engineering pioneers

Since a course in circuit analysis may be a student's first exposure to electrical engineering, each chapter opens with discussions about how to enhance skills that contribute to successful problem-solving or career-oriented talks on a sub-discipline of electrical engineering. The chapter openers are intended to help students grasp the scope of electrical engineering and give thought to the various careers available to EE graduates. The opening boxes include information on careers in electronics, instrumentation, electromagnetics, control systems, engineering education, and the importance of good communication skills. Historicals throughout the text provide brief biological sketches of such engineering pioneers as Faraday, Ampere, Edison, Henry, Fourier, Volta, and Bell.

OUR COMMITMENT TO ACCURACY

You have a right to expect an accurate textbook, and McGraw-Hill Engineering invests considerable time and effort to ensure that we deliver one. Listed below are the many steps we take in this process.

OUR ACCURACY VERIFICATION PROCESS

First Round

Step 1: Numerous **college engineering instructors** review the manuscript and report errors to the editorial team. The authors review their comments and make the necessary corrections in their manuscript.

Second Round

Step 2: An **expert in the field** works through every example and exercise in the final manuscript to verify the accuracy of the examples, exercises, and solutions. The authors review any resulting corrections and incorporate them into the final manuscript and solutions manual.

Step 3: The manuscript goes to a **copyeditor,** who reviews the pages for grammatical and stylistic considerations. At the same time, the expert in the field begins a second accuracy check. All corrections are submitted simultaneously to the **authors,** who review and integrate the editing, and then submit the manuscript pages for typesetting.

Third Round

Step 4: The **authors** review their page proofs for a dual purpose: 1) to make certain that any previous corrections were properly made, and 2) to look for any errors they might have missed.

Step 5: A **proofreader** is assigned to the project to examine the new page proofs, double check the authors' work, and add a fresh, critical eye to the book. Revisions are incorporated into a new batch of pages which the authors check again.

Fourth Round

Step 6: The **author team** submits the solutions manual to the **expert in the field,** who checks text pages against the solutions manual as a final review.

Step 7: The **project manager, editorial team,** and **author team** review the pages for a final accuracy check.

The resulting engineering textbook has gone through several layers of quality assurance and is verified to be as accurate and error-free as possible. Our authors and publishing staff are confident that through this process we deliver textbooks that are industry leaders in their correctness and technical integrity.

Fundamentals of

Electric Circuits

fourth edition

Fundamentals of
Electric Circuits

Charles K. Alexander

Department of Electrical and
Computer Engineering

Cleveland State University

Matthew N. O. Sadiku

Department of
Electrical Engineering

Prairie View A&M University

Boston Burr Ridge, IL Dubuque, IA New York San Francisco St. Louis
Bangkok Bogotá Caracas Kuala Lumpur Lisbon London Madrid Mexico City
Milan Montreal New Delhi Santiago Seoul Singapore Sydney Taipei Toronto

The McGraw-Hill Companies

FUNDAMENTALS OF ELECTRIC CIRCUITS, FOURTH EDITION

Published by McGraw-Hill, a business unit of The McGraw-Hill Companies, Inc., 1221 Avenue of the Americas, New York, NY 10020.

This book is printed on acid-free paper.

2 3 4 5 6 7 8 9 0 VNH/VNH 0 9

ISBN 978–0–07–352955–4
MHID 0–07–352955–9

Global Publisher: *Raghothaman Srinivasan*
Director of Development: *Kristine Tibbetts*
Developmental Editor: *Lora Neyens*
Senior Marketing Manager: *Curt Reynolds*
Project Manager: *Joyce Watters*
Senior Production Supervisor: *Sherry L. Kane*
Lead Media Project Manager: *Stacy A. Patch*
Associate Design Coordinator: *Brenda A. Rolwes*
Cover Designer: *Studio Montage, St. Louis, Missouri*
(USE) Cover Image: *Astronauts Repairing Spacecraft: © StockTrek/Getty Images; Printed Circuit Board: Photodisc Collection/Getty Images*
Lead Photo Research Coordinator: *Carrie K. Burger*
Compositor: *ICC Macmillan Inc.*
Typeface: *10/12 Times Roman*
Printer: *R. R. Donnelley, Jefferson City, MO*

Library of Congress Cataloging-in-Publication Data

Alexander, Charles K.
Fundamentals of electric circuits / Charles K. Alexander, Matthew N. O. Sadiku. — 4th ed.
p. cm.
Includes index.
ISBN 978–0–07–352955–4 — ISBN 0–07–352955–9 (hard copy : alk. paper) 1. Electric circuits.
I. Sadiku, Matthew N. O. II. Title.

TK454.A452 2009
621.319'24—dc22 2008023020

www.mhhe.com

Dedicated to our wives, Kikelomo and Hannah, whose understanding and support have truly made this book possible.

Matthew

and

Chuck

Contents

Preface

You may be wondering why we chose a photo of astronauts working in space on the Space Station for the cover. We actually chose it for several reasons. Obviously, it is very exciting; in fact, space represents the most exciting frontier for the entire world! In addition, much of the station itself consists of all kinds of circuits! One of the most significant circuits within the station is its power distribution system. It is a complete and self contained, modern power generation and distribution system. That is why NASA (especially NASA-Glenn) continues to be at the forefront of both theoretical as well as applied power system research and development. The technology that has gone into the development of space exploration continues to find itself impacting terrestrial technology in many important ways. For some of you, this will be an important career path.

FEATURES

New to This Edition

A course in circuit analysis is perhaps the first exposure students have to electrical engineering. This is also a place where we can enhance some of the skills that they will later need as they learn how to design.

In the fourth edition, we have included a very significant new feature to help students enhance skills that are an important part of the design process. We call this new feature, ***design a problem.***

We know it is not possible to fully develop a student's design skills in a fundamental course like circuits. To fully develop design skills a student needs a design experience normally reserved for their senior year. This does not mean that some of those skills cannot be developed and exercised in a circuits course. The text already included open-ended questions that help students use creativity, which is an important part of learning how to design. We already have some questions that are open desired to add much more into our text in this important area and have developed an approach to do just that. When we develop problems for the student to solve our goal is that in solving the problem the student learn more about the theory and the problem solving process. Why not have the students design problems like we do? That is exactly what we will do in each chapter. Within the normal problem set, we have a set of problems where we ask the student to design a problem. This will have two very important results. The first will be a better understanding of the basic theory and the second will be the enhancement of some of the student's basic design skills.

We are making effective use of the principle of learning by teaching. Essentially we all learn better when we teach a subject. Designing effective problems is a key part of the teaching process. Students should also be encouraged to develop problems, when appropriate, which have nice numbers and do not necessarily overemphasize complicated mathematical manipulations.

Additionally we have changed almost 40% of the Practice Problems with the idea to better reflect more real component values and to help the student better understand the problem and have added 121 *design a problem* problems. We have also changed and added a total of 357 end-of-chapter problems (this number contains the new *design a problem* problems). This brings up a very important advantage to our textbook, we have a total of 2404 Examples, Practice Problems, Review Questions, and end-of-chapter problems!

Retained from Previous Editions

The main objective of the fourth edition of this book remains the same as the previous editions—to present circuit analysis in a manner that is clearer, more interesting, and easier to understand than other circuit text, and to assist the student in beginning to see the "fun" in engineering. This objective is achieved in the following ways:

- **Chapter Openers and Summaries**
  Each chapter opens with a discussion about how to enhance skills which contribute to successful problem solving as well as successful careers or a career-oriented talk on a sub-discipline of electrical engineering. This is followed by an introduction that links the chapter with the previous chapters and states the chapter objectives. The chapter ends with a summary of key points and formulas.
- **Problem Solving Methodology**
  Chapter 1 introduces a six-step method for solving circuit problems which is used consistently throughout the book and media supplements to promote best-practice problem-solving procedures.
- **Student Friendly Writing Style**
  All principles are presented in a lucid, logical, step-by-step manner. As much as possible, we avoid wordiness and giving too much detail that could hide concepts and impede overall understanding of the material.
- **Boxed Formulas and Key Terms**
  Important formulas are boxed as a means of helping students sort out what is essential from what is not. Also, to ensure that students clearly understand the key elements of the subject matter, key terms are defined and highlighted.
- **Margin Notes**
  Marginal notes are used as a pedagogical aid. They serve multiple uses such as hints, cross-references, more exposition, warnings, reminders not to make some particular common mistakes, and problem-solving insights.
- **Worked Examples**
  Thoroughly worked examples are liberally given at the end of every section. The examples are regarded as a part of the text and are clearly explained without asking the reader to fill in missing steps. Thoroughly worked examples give students a good understanding of the solution process and the confidence to solve problems

themselves. Some of the problems are solved in two or three different ways to facilitate a substantial comprehension of the subject material as well as a comparison of different approaches.

- **Practice Problems**
  To give students practice opportunity, each illustrative example is immediately followed by a practice problem with the answer. The student can follow the example step by step to aid in the solution of the practice problem without flipping pages or looking at the end of the book for answers. The practice problem is also intended to test a student's understanding of the preceding example. It will reinforce their grasp of the material before the student can move on to the next section. Complete solutions to the practice problems are available to students on ARIS.
- **Application Sections**
  The last section in each chapter is devoted to practical application aspects of the concepts covered in the chapter. The material covered in the chapter is applied to at least one or two practical problems or devices. This helps students see how the concepts are applied to real-life situations.
- **Review Questions**
  Ten review questions in the form of multiple-choice objective items are provided at the end of each chapter with answers. The review questions are intended to cover the little "tricks" that the examples and end-of-chapter problems may not cover. They serve as a self-test device and help students determine how well they have mastered the chapter.
- **Computer Tools**
  In recognition of the requirements by ABET® on integrating computer tools, the use of *PSpice, MATLAB, KCIDE for Circuits,* and developing design skills are encouraged in a student-friendly manner. *PSpice* is covered early on in the text so that students can become familiar and use it throughout the text. Appendix D serves as a tutorial on *PSpice* for Windows. *MATLAB* is also introduced early in the book with a tutorial available in Appendix E. *KCIDE for Circuits* is a brand new, state-of-the-art software system designed to help the students maximize their chance of success in problem solving. It is introduced in Appendix F. Finally, *design a problem* problems have been introduced, for the first time. These are meant to help the student develop skills that will be needed in the design process.
- **Historical Tidbits**
  Historical sketches throughout the text provide profiles of important pioneers and events relevant to the study of electrical engineering.
- **Early Op Amp Discussion**
  The operational amplifier (op amp) as a basic element is introduced early in the text.
- **Fourier and Laplace Transforms Coverage**
  To ease the transition between the circuit course and signals and systems courses, Fourier and Laplace transforms are covered lucidly and thoroughly. The chapters are developed in a manner that the interested instructor can go from solutions of first-order

circuits to Chapter 15. This then allows a very natural progression from Laplace to Fourier to AC.

- **Four Color Art Program**
  A completely redesigned interior design and four color art program bring circuit drawings to life and enhance key pedagogical elements throughout the text.
- **Extended Examples**
  Examples worked in detail according to the six-step problem solving method provide a roadmap for students to solve problems in a consistent fashion. At least one example in each chapter is developed in this manner.
- **EC 2000 Chapter Openers**
  Based on ABET's new skill-based CRITERION 3, these chapter openers are devoted to discussions as to how students can acquire the skills that will lead to a significantly enhanced career as an engineer. Because these skills are so very important to the student while in college as well as in their career, we will use the heading, *"Enhancing your Skills and your Career."*
- **Homework Problems**
  There are 358 new or changed end-of-chapter problems which will provide students with plenty of practice as well as reinforce key concepts.
- **Homework Problem Icons**
  Icons are used to highlight problems that relate to engineering design as well as problems that can be solved using *PSpice* or *MATLAB*.
- ***KCIDE for Circuits* Appendix F**
  A new Appendix F provides a tutorial on the Knowledge Capturing Integrated Design Environment (*KCIDE for Circuits*) software, available on ARIS.

Organization

This book was written for a two-semester or three-quarter course in linear circuit analysis. The book may also be used for a one-semester course by a proper selection of chapters and sections by the instructor. It is broadly divided into three parts.

- Part 1, consisting of Chapters 1 to 8, is devoted to dc circuits. It covers the fundamental laws and theorems, circuits techniques, and passive and active elements.
- Part 2, which contains Chapter 9 to 14, deals with ac circuits. It introduces phasors, sinusoidal steady-state analysis, ac power, rms values, three-phase systems, and frequency response.
- Part 3, consisting of Chapters 15 to 19, is devoted to advanced techniques for network analysis. It provides students with a solid introduction to the Laplace transform, Fourier series, Fourier transform, and two-port network analysis.

The material in three parts is more than sufficient for a two-semester course, so the instructor must select which chapters or sections to cover. Sections marked with the dagger sign (†) may be skipped, explained briefly, or assigned as homework. They can be omitted without loss of continuity. Each chapter has plenty of problems grouped according to the sections of the related material and diverse enough that the instructor can choose some as examples and assign some as homework. As stated earlier, we are using three icons with this edition. We are using ps to denote problems that either require *PSpice* in the solution process, where the circuit complexity is such that *PSpice* would make the solution process easier, and where *PSpice* makes a good check to see if the problem has been solved correctly. We are using ML to denote problems where *MATLAB* is required in the solution process, where *MATLAB* makes sense because of the problem makeup and its complexity, and where *MATLAB* makes a good check to see if the problem has been solved correctly. Finally, we use e d to identify problems that help the student develop skills that are needed for engineering design. More difficult problems are marked with an asterisk (*). Comprehensive problems follow the end-of-chapter problems. They are mostly applications problems that require skills learned from that particular chapter.

Prerequisites

As with most introductory circuit courses, the main prerequisites, for a course using the text, are physics and calculus. Although familiarity with complex numbers is helpful in the later part of the book, it is not required. A very important asset of this text is that ALL the mathematical equations and fundamentals of physics needed by the student, are included in the text.

Supplements

McGraw-Hill's ARIS—Assessment, Review, and Instruction System is a complete, online tutorial, electronic homework, and course management system, designed for greater ease of use than any other system available. Available on adoption, instructors can create and share course materials and assignments with other instructors, edit questions and algorithms, import their own content, and create announcements and due dates for assignments. ARIS has automatic grading and reporting of easy-to-assign algorithmically-generated homework, quizzing, and testing. Once a student is registered in the course, all student activity within McGraw-Hill's ARIS is automatically recorded and available to the instructor through a fully integrated grade book that can be downloaded to Excel. Also included on ARIS are a solutions manual, text image files, transition guides to instructors, and Network Analysis Tutorials, software downloads, complete solutions to text practice problems, FE Exam questions, flashcards, and web links to students. Visit www.mhhe.com/alexander.

Knowledge Capturing Integrated Design Environment for Circuits (*KCIDE for Circuits*) This new software, developed at Cleveland State University and funded by NASA, is designed to help the student work through a circuits problem in an organized manner using the six-step problem-solving methodology in the text. *KCIDE for Circuits* allows students to work a circuit problem in *PSpice* and *MATLAB*, track the evolution of their solution, and save a record of their process for future reference. In addition, the software automatically generates a Word document and/or a PowerPoint presentation. Appendix F contains a description of how to use the software. Additional examples can be found at the web site, http://kcide.fennresearch.org/, which is linked from ARIS. The software package can be downloaded for free.

Problem Solving Made *Almost* Easy, a companion workbook to *Fundamentals of Electric Circuits,* is available on ARIS for students who wish to practice their problem-solving techniques. The workbook contains a discussion of problem-solving strategies and 150 additional problems with complete solutions provided.

C.O.S.M.O.S This CD, available to instructors only, is a powerful solutions manual tool to help instructors streamline the creation of assignments, quizzes, and tests by using problems and solutions from the textbook, as well as their own custom material. Instructors can edit textbook end-of-chapter problems as well as track which problems have been assigned.

Although the textbook is meant to be self-explanatory and act as a tutor for the student, the personal contact in teaching is not forgotten. It is hoped that the book and supplemental materials supply the instructor with all the pedagogical tools necessary to effectively present the material.

Acknowledgements

We would like to express our appreciation for the loving support we have received from our wives (Hannah and Kikelomo), daughters (Christina, Tamara, Jennifer, Motunrayo, Ann, and Joyce), son (Baixi), and our extended family members.

At McGraw-Hill, we would like to thank the following editorial and production staff: Raghu Srinivasan, publisher and senior sponsoring editor; Lora Kalb-Neyens, developmental editors; Joyce Watters, project manager; Carrie Burger, photo researcher; and Brenda Rolwes, designer. Also, we appreciate the hard work of Tom Hartley at the University of Akron for his very detailed evaluation of various elements of the text and his many valued suggestions for continued improvement of this textbook.

We wish to thank Yongjian Fu and his outstanding team of students, Bramarambha Elka and Saravaran Chinniah, for their efforts in the development of *KCIDE for Circuits.* Their efforts to help us continue to improve this software are also appreciated.

The fourth edition has benefited greatly from the many outstanding reviewers and symposium attendees who contributed to the success of the first three editions! In addition, the following have made important contributions to the fourth edition (in alphabetical order):

Tom Brewer, Georgia Tech
Andy Chan, City University of Hong Kong
Alan Tan Wee Chiat, Multimedia University
Norman Cox, University of Missouri-Rolla
Walter L. Green, University of Tennessee

Dr. Gordon K. Lee, San Diego State University
Gary Perks, Cal Poly State University, San Luis Obispo
Dr. Raghu K. Settaluri, Oregon State University
Ramakant Srivastava, University of Florida
John Watkins, Wichita State University
Yik-Chung Wu, The University of Hong Kong
Xiao-Bang Xu, Clemson University

Finally, we appreciate the feedback received from instructors and students who used the previous editions. We want this to continue, so please keep sending us emails or direct them to the publisher. We can be reached at c.alexander@ieee.org for Charles Alexander and sadiku@ieee.org for Matthew Sadiku.

C. K. Alexander and M.N.O. Sadiku

GUIDED TOUR

The main objective of this book is to present circuit analysis in a manner that is clearer, more interesting, and easier to understand than other texts. For you, the student, here are some features to help you study and be successful in this course.

The four color art program brings circuit drawings to life and enhances key concepts throughout the text.

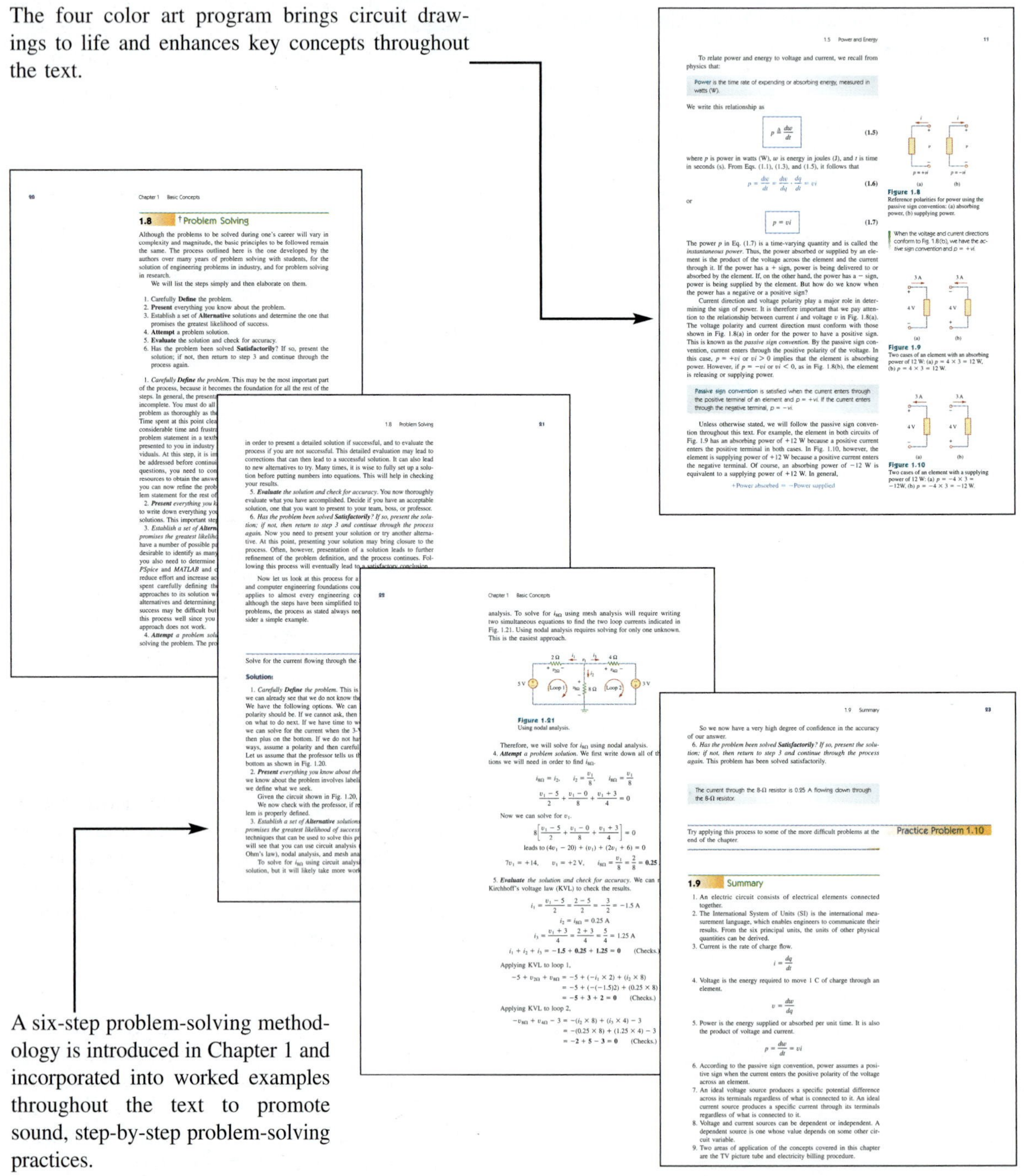

A six-step problem-solving methodology is introduced in Chapter 1 and incorporated into worked examples throughout the text to promote sound, step-by-step problem-solving practices.

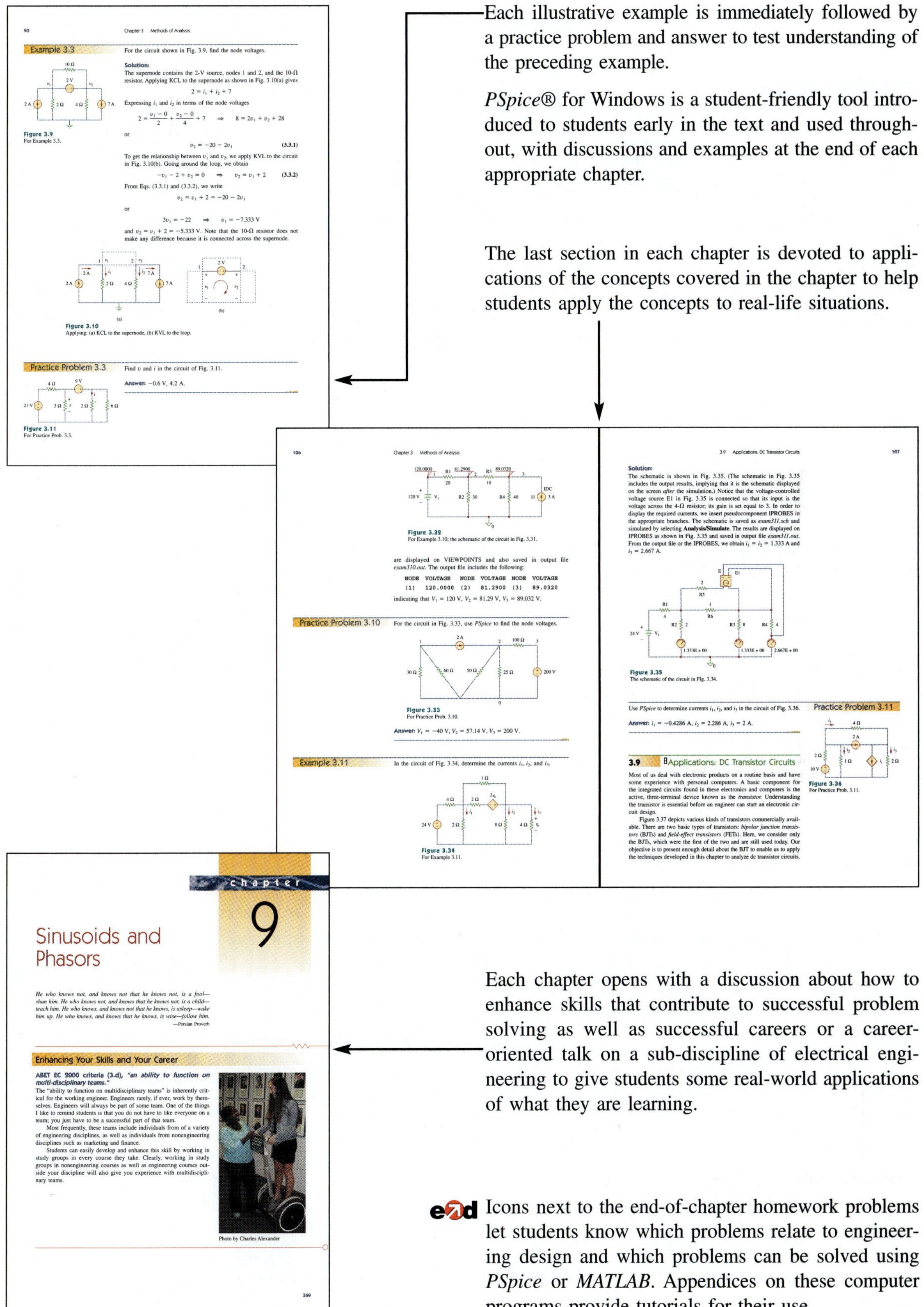

Each illustrative example is immediately followed by a practice problem and answer to test understanding of the preceding example.

PSpice® for Windows is a student-friendly tool introduced to students early in the text and used throughout, with discussions and examples at the end of each appropriate chapter.

The last section in each chapter is devoted to applications of the concepts covered in the chapter to help students apply the concepts to real-life situations.

Each chapter opens with a discussion about how to enhance skills that contribute to successful problem solving as well as successful careers or a career-oriented talk on a sub-discipline of electrical engineering to give students some real-world applications of what they are learning.

Icons next to the end-of-chapter homework problems let students know which problems relate to engineering design and which problems can be solved using *PSpice* or *MATLAB*. Appendices on these computer programs provide tutorials for their use.

Supplements for Students and Instructors

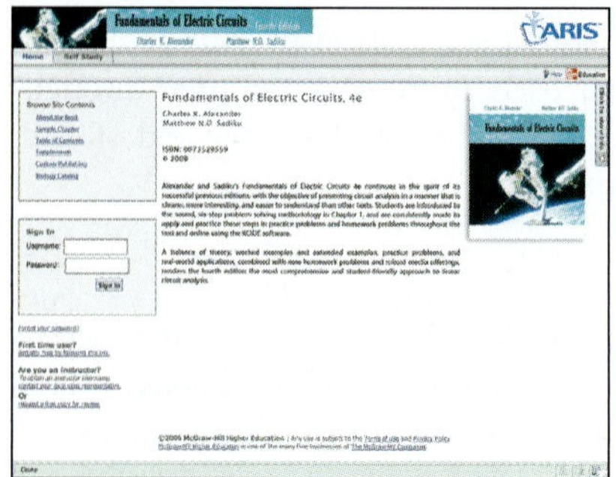

McGraw-Hill's ARIS—Assessment, Review, and Instruction System is a complete, online tutorial, electronic homework, and course management system, designed for greater ease of use than any other system available. With ARIS, instructors can create and share course materials and assignments with other instructors, edit questions and algorithms, import their own content, and create announcements and due dates for assignments. ARIS has automatic grading and reporting of easy-to-assign algorithmically-generated homework, quizzing, and testing. Once a student is registered in the course, all student activity within McGraw-Hill's ARIS is automatically recorded and available to the instructor through a fully integrated grade book that can be downloaded to Excel.

www.mhhe.com/alexander

Knowledge Capturing Integrated Design Environment for Circuits (*KCIDE for Circuits*) software, linked from ARIS, enhances student understanding of the six-step problem-solving methodology in the book. *KCIDE for Circuits* allows students to work a circuit problem in *PSpice* and *MATLAB*, track the evolution of their solution, and save a record of their process for future reference. Appendix F walks the user through this program.

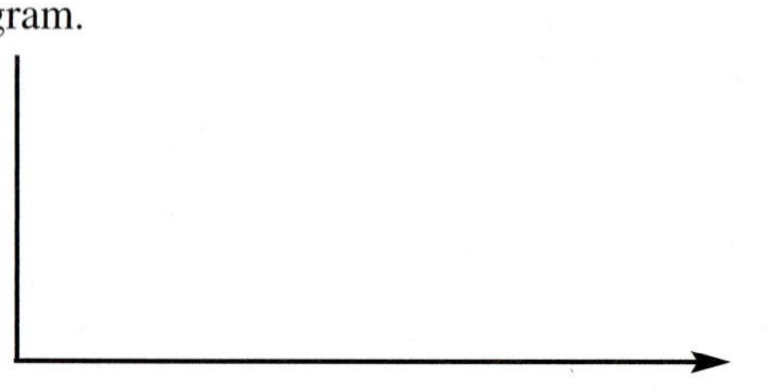

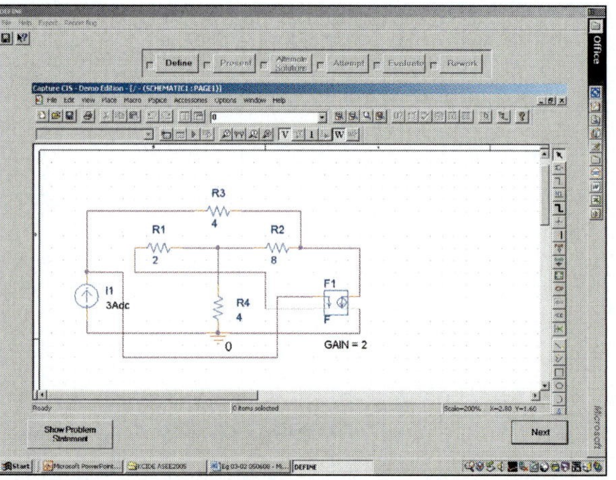

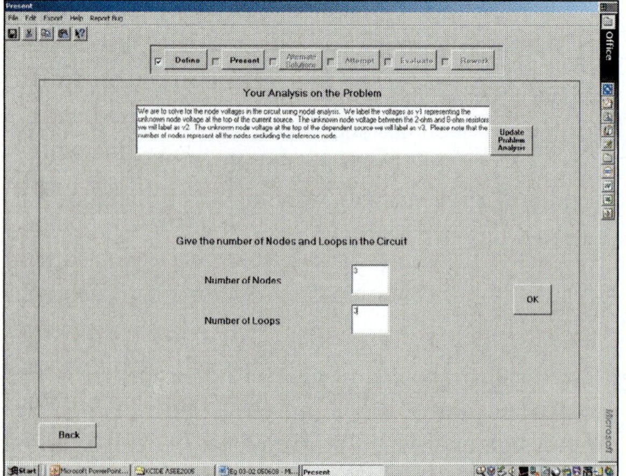

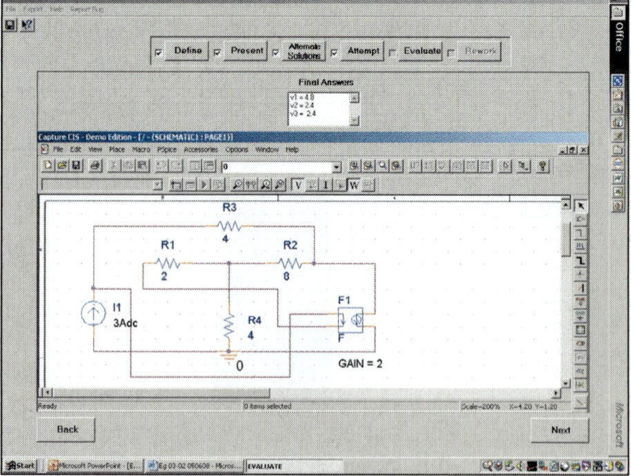

Other resources provided on ARIS.

For Students:

— Network Analysis Tutorials—a series of interactive quizzes to help students practice fundamental concepts in circuits.
— FE Exam Interactive Review Quizzes—chapter based self-quizzes provide hints for solutions and correct solution methods, and help students prepare for the NCEES Fundamentals of Engineering Examination.
— Problem Solving Made *Almost* Easy—a companion workbook to the text, featuring 150 additional problems with complete solutions.
— Complete solutions to Practice Problems in the text
— Flashcards of key terms
— Web links

For Instructors:

— Image Sets—electronic files of text figures for easy integration into your course presentations, exams, and assignments.
— Transition Guides—compare coverage of the third edition to other popular circuits books at the section level to aid transition to teaching from our text.

A Note to the Student

This may be your first course in electrical engineering. Although electrical engineering is an exciting and challenging discipline, the course may intimidate you. This book was written to prevent that. A good textbook and a good professor are an advantage—but you are the one who does the learning. If you keep the following ideas in mind, you will do very well in this course.

- This course is the foundation on which most other courses in the electrical engineering curriculum rest. For this reason, put in as much effort as you can. Study the course regularly.
- Problem solving is an essential part of the learning process. Solve as many problems as you can. Begin by solving the practice problem following each example, and then proceed to the end-of-chapter problems. The best way to learn is to solve a lot of problems. An asterisk in front of a problem indicates a challenging problem.
- *Spice*, a computer circuit analysis program, is used throughout the textbook. *PSpice*, the personal computer version of *Spice*, is the popular standard circuit analysis program at most universities. *PSpice for Windows* is described in Appendix D. Make an effort to learn *PSpice*, because you can check any circuit problem with *PSpice* and be sure you are handing in a correct problem solution.
- *MATLAB* is another software that is very useful in circuit analysis and other courses you will be taking. A brief tutorial on *MATLAB* is given in Appendix E to get you started. The best way to learn *MATLAB* is to start working with it once you know a few commands.
- Each chapter ends with a section on how the material covered in the chapter can be applied to real-life situations. The concepts in this section may be new and advanced to you. No doubt, you will learn more of the details in other courses. We are mainly interested in gaining a general familiarity with these ideas.
- Attempt the review questions at the end of each chapter. They will help you discover some "tricks" not revealed in class or in the textbook.
- Clearly a lot of effort has gone into making the technical details in this book easy to understand. It also contains all the mathematics and physics necessary to understand the theory and will be very useful in your other engineering courses. However, we have also focused on creating a reference for you to use both in school as well as when working in industry or seeking a graduate degree.
- It is very tempting to sell your book after you have completed your classroom experience; however, our advice to you is *DO NOT SELL YOUR ENGINEERING BOOKS!* Books have always been expensive, however, the cost of this book is virtually the same as I paid for my circuits text back in the early 60s in terms of real dollars. In fact, it is actually cheaper. In addition, engineering books of the past are no where near as complete as what is available now.

When I was a student, I did not sell any of my engineering textbooks and was very glad I did not! I found that I needed most of them throughout my career.

A short review on finding determinants is covered in Appendix A, complex numbers in Appendix B, and mathematical formulas in Appendix C. Answers to odd-numbered problems are given in Appendix G.

Have fun!

C. K. A. and M. N. O. S.

About the Authors

Charles K. Alexander

Charles K. Alexander is professor of electrical and computer engineering of the Fenn College of Engineering at Cleveland State University, Cleveland, Ohio. He is also the Director of The Center for Research in Electronics and Aerospace Technology (CREATE), and is the Managing Director of the Wright Center for Sensor Systems (WCSSE). From 2002 until 2006 he was Dean of the Fenn College of Engineering. From 2004 until 2007, he was Director of Ohio ICE, a research center in instrumentation, controls, electronics, and sensors (a coalition of CSU, Case, the University of Akron, and a number of Ohio industries). From 1998 until 2002, he was interim director (2000 and 2001) of the Institute for Corrosion and Multiphase Technologies and Stocker Visiting Professor of electrical engineering and computer science at Ohio University. From 1994–1996 he was dean of engineering and computer science at California State University, Northridge.

From 1989–1994 he was acting dean of the college of engineering at Temple University, and from 1986–1989 he was professor and chairman of the department of electrical engineering at Temple. From 1980–1986 he held the same positions at Tennessee Technological University. He was an associate professor and a professor of electrical engineering at Youngstown State University from 1972–1980, where he was named Distinguished Professor in 1977 in recognition of "outstanding teaching and research." He was assistant professor of electrical engineering at Ohio University in 1971–1972. He received the Ph.D. (1971) and M.S.E.E. (1967) from Ohio University and the B.S.E.E. (1965) from Ohio Northern University.

Dr. Alexander has been a consultant to 23 companies and governmental organizations, including the Air Force and Navy and several law firms. He has received over $85 million in research and development funds for projects ranging from solar energy to software engineering. He has authored 40 publications, including a workbook and a videotape lecture series, and is coauthor of *Fundamentals of Electric Circuits, Problem Solving Made* Almost *Easy,* and the fifth edition of the *Standard Handbook of Electronic Engineering,* with McGraw-Hill. He has made more than 500 paper, professional, and technical presentations.

Dr. Alexander is a fellow of the IEEE and served as its president and CEO in 1997. In 1993 and 1994 he was IEEE vice president, professional activities, and chair of the United States Activities Board (USAB). In 1991–1992 he was region 2 director, serving on the Regional Activities Board (RAB) and USAB. He has also been a member of the Educational Activities Board. He served as chair of the USAB Member Activities Council and vice chair of the USAB Professional Activities Council for Engineers, and he chaired the RAB Student Activities Committee and the USAB Student Professional Awareness Committee.

In 1998 he received the Distinguished Engineering Education Achievement Award from the Engineering Council, and in 1996 he received the Distinguished Engineering Education Leadership Award from the same group. When he became a fellow of the IEEE in 1994, the citation read "for leadership in the field of engineering education and the professional development of engineering students." In 1984 he received the IEEE Centennial Medal, and in 1983 he received the IEEE/RAB Innovation Award, given to the IEEE member who best contributes to RAB's goals and objectives.

Matthew N. O. Sadiku

Matthew N. O. Sadiku is presently a professor at Prairie View A&M University. Prior to joining Prairie View, he taught at Florida Atlantic University, Boca Raton, and Temple University, Philadelphia. He has also worked for Lucent/Avaya and Boeing Satellite Systems.

Dr. Sadiku is the author of over 170 professional papers and almost 30 books including Elements of Electromagnetics (Oxford University Press, 3rd ed., 2001), Numerical Techniques in Electromagnetics (2nd ed., CRC Press, 2000), Simulation of Local Area Networks (with M. Ilyas, CRC Press, 1994), Metropolitan Area Networks (CRC Press, 1994), and Fundamentals of Electric Circuits (with C. K. Alexander, McGraw-Hill, 3rd ed., 2007). His books are used worldwide, and some of them have been translated into Korean, Chinese, Italian, and Spanish. He was the recipient of the 2000 McGraw-Hill/Jacob Millman Award for outstanding contributions in the field of electrical engineering. He was the IEEE region 2 Student Activities Committee chairman and is an associate editor for IEEE "Transactions on Education." He received his Ph.D. at Tennessee Technological University, Cookeville.

Fundamentals of
Electric Circuits

PART ONE

DC Circuits

OUTLINE

chapter

1

Basic Concepts

Some books are to be tasted, others to be swallowed, and some few to be chewed and digested.

—Francis Bacon

Enhancing Your Skills and Your Career

ABET EC 2000 criteria (3.a), *"an ability to apply knowledge of mathematics, science, and engineering."*

As students, you are required to study mathematics, science, and engineering with the purpose of being able to apply that knowledge to the solution of engineering problems. The skill here is the ability to apply the fundamentals of these areas in the solution of a problem. So, how do you develop and enhance this skill?

The best approach is to work as many problems as possible in all of your courses. However, if you are really going to be successful with this, you must spend time analyzing where and when and why you have difficulty in easily arriving at successful solutions. You may be surprised to learn that most of your problem-solving problems are with mathematics rather than your understanding of theory. You may also learn that you start working the problem too soon. Taking time to think about the problem and how you should solve it will always save you time and frustration in the end.

What I have found that works best for me is to apply our six-step problem-solving technique. Then I carefully identify the areas where I have difficulty solving the problem. Many times, my actual deficiencies are in my understanding and ability to use correctly certain mathematical principles. I then return to my fundamental math texts and carefully review the appropriate sections and in some cases work some example problems in that text. This brings me to another important thing you should always do: Keep nearby all your basic mathematics, science, and engineering textbooks.

This process of continually looking up material you thought you had acquired in earlier courses may seem very tedious at first; however, as your skills develop and your knowledge increases, this process will become easier and easier. On a personal note, it is this very process that led me from being a much less than average student to someone who could earn a Ph.D. and become a successful researcher.

Photo by Charles Alexander

1.1 Introduction

Electric circuit theory and electromagnetic theory are the two fundamental theories upon which all branches of electrical engineering are built. Many branches of electrical engineering, such as power, electric machines, control, electronics, communications, and instrumentation, are based on electric circuit theory. Therefore, the basic electric circuit theory course is the most important course for an electrical engineering student, and always an excellent starting point for a beginning student in electrical engineering education. Circuit theory is also valuable to students specializing in other branches of the physical sciences because circuits are a good model for the study of energy systems in general, and because of the applied mathematics, physics, and topology involved.

In electrical engineering, we are often interested in communicating or transferring energy from one point to another. To do this requires an interconnection of electrical devices. Such interconnection is referred to as an *electric circuit*, and each component of the circuit is known as an *element.*

An **electric circuit** is an interconnection of electrical elements.

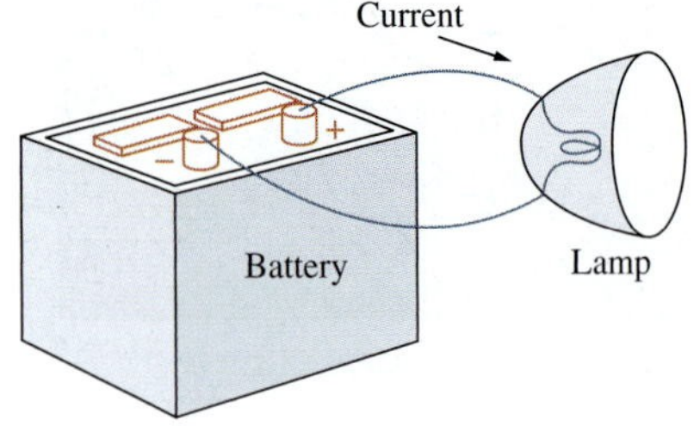

Figure 1.1
A simple electric circuit.

A simple electric circuit is shown in Fig. 1.1. It consists of three basic elements: a battery, a lamp, and connecting wires. Such a simple circuit can exist by itself; it has several applications, such as a flashlight, a search light, and so forth.

A complicated real circuit is displayed in Fig. 1.2, representing the schematic diagram for a radio receiver. Although it seems complicated, this circuit can be analyzed using the techniques we cover in this book. Our goal in this text is to learn various analytical techniques and computer software applications for describing the behavior of a circuit like this.

Electric circuits are used in numerous electrical systems to accomplish different tasks. Our objective in this book is not the study of various uses and applications of circuits. Rather our major concern is the analysis of the circuits. By the analysis of a circuit, we mean a study of the behavior of the circuit: How does it respond to a given input? How do the interconnected elements and devices in the circuit interact?

We commence our study by defining some basic concepts. These concepts include charge, current, voltage, circuit elements, power, and energy. Before defining these concepts, we must first establish a system of units that we will use throughout the text.

1.2 Systems of Units

As electrical engineers, we deal with measurable quantities. Our measurement, however, must be communicated in a standard language that virtually all professionals can understand, irrespective of the country where the measurement is conducted. Such an international measurement

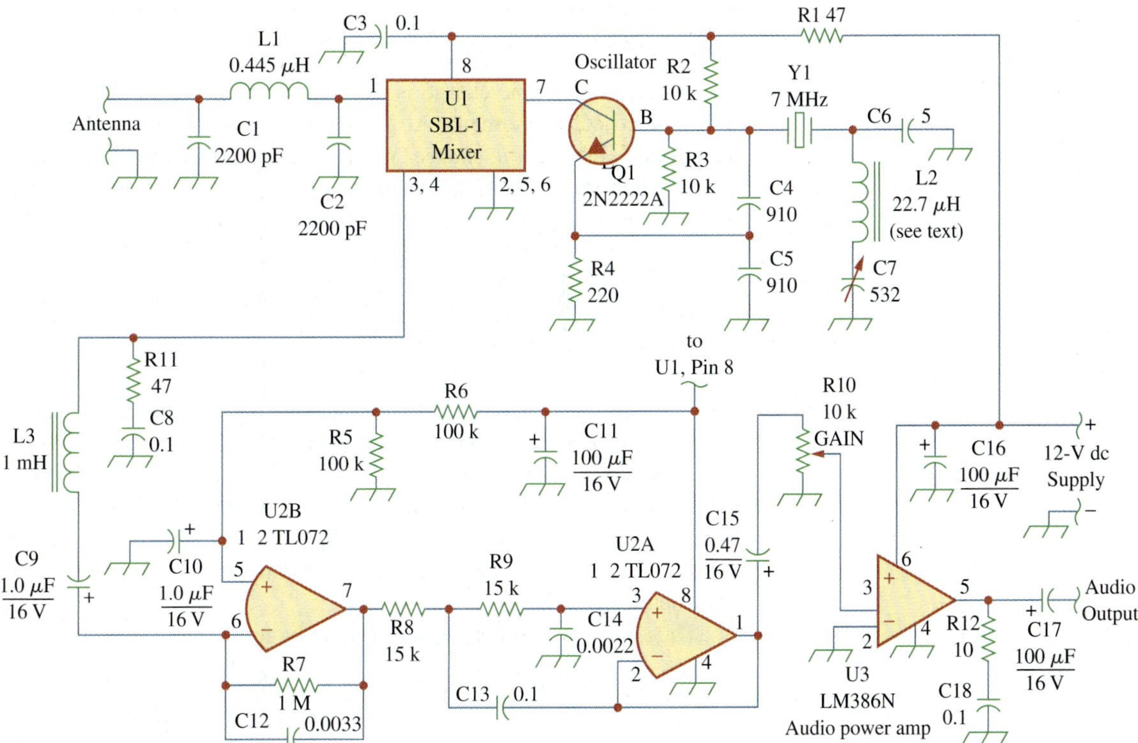

Figure 1.2
Electric circuit of a radio receiver.
Reproduced with permission from QST, August 1995, p. 23.

language is the International System of Units (SI), adopted by the General Conference on Weights and Measures in 1960. In this system, there are six principal units from which the units of all other physical quantities can be derived. Table 1.1 shows the six units, their symbols, and the physical quantities they represent. The SI units are used throughout this text.

One great advantage of the SI unit is that it uses prefixes based on the power of 10 to relate larger and smaller units to the basic unit. Table 1.2 shows the SI prefixes and their symbols. For example, the following are expressions of the same distance in meters (m):

600,000,000 mm   600,000 m   600 km

TABLE 1.1

Six basic SI units and one derived unit relevant to this text.

Quantity	Basic unit	Symbol
Length	meter	m
Mass	kilogram	kg
Time	second	s
Electric current	ampere	A
Thermodynamic temperature	kelvin	K
Luminous intensity	candela	cd
Charge	coulomb	C

TABLE 1.2

The SI prefixes.

Multiplier	Prefix	Symbol
10^{18}	exa	E
10^{15}	peta	P
10^{12}	tera	T
10^{9}	giga	G
10^{6}	mega	M
10^{3}	kilo	k
10^{2}	hecto	h
10	deka	da
10^{-1}	deci	d
10^{-2}	centi	c
10^{-3}	milli	m
10^{-6}	micro	μ
10^{-9}	nano	n
10^{-12}	pico	p
10^{-15}	femto	f
10^{-18}	atto	a

1.3 Charge and Current

The concept of electric charge is the underlying principle for explaining all electrical phenomena. Also, the most basic quantity in an electric circuit is the *electric charge*. We all experience the effect of electric charge when we try to remove our wool sweater and have it stick to our body or walk across a carpet and receive a shock.

> **Charge** is an electrical property of the atomic particles of which matter consists, measured in coulombs (C).

We know from elementary physics that all matter is made of fundamental building blocks known as atoms and that each atom consists of electrons, protons, and neutrons. We also know that the charge e on an electron is negative and equal in magnitude to 1.602×10^{-19} C, while a proton carries a positive charge of the same magnitude as the electron. The presence of equal numbers of protons and electrons leaves an atom neutrally charged.

The following points should be noted about electric charge:

1. The coulomb is a large unit for charges. In 1 C of charge, there are $1/(1.602 \times 10^{-19}) = 6.24 \times 10^{18}$ electrons. Thus realistic or laboratory values of charges are on the order of pC, nC, or μC.[1]
2. According to experimental observations, the only charges that occur in nature are integral multiples of the electronic charge $e = -1.602 \times 10^{-19}$ C.
3. The *law of conservation of charge* states that charge can neither be created nor destroyed, only transferred. Thus the algebraic sum of the electric charges in a system does not change.

We now consider the flow of electric charges. A unique feature of electric charge or electricity is the fact that it is mobile; that is, it can be transferred from one place to another, where it can be converted to another form of energy.

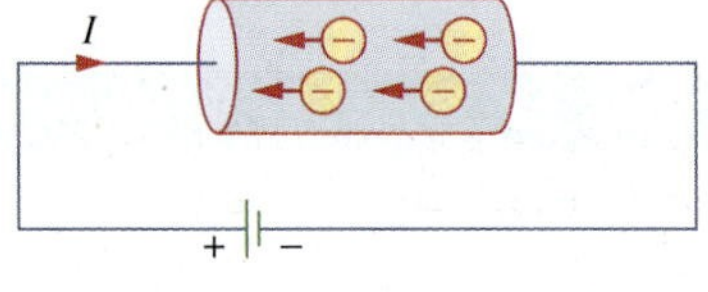

Figure 1.3
Electric current due to flow of electronic charge in a conductor.

A convention is a standard way of describing something so that others in the profession can understand what we mean. We will be using IEEE conventions throughout this book.

When a conducting wire (consisting of several atoms) is connected to a battery (a source of electromotive force), the charges are compelled to move; positive charges move in one direction while negative charges move in the opposite direction. This motion of charges creates electric current. It is conventional to take the current flow as the movement of positive charges. That is, opposite to the flow of negative charges, as Fig. 1.3 illustrates. This convention was introduced by Benjamin Franklin (1706–1790), the American scientist and inventor. Although we now know that current in metallic conductors is due to negatively charged electrons, we will follow the universally accepted convention that current is the net flow of positive charges. Thus,

> **Electric current** is the time rate of change of charge, measured in amperes (A).

[1] However, a large power supply capacitor can store up to 0.5 C of charge.

Historical

Andre-Marie Ampere (1775–1836), a French mathematician and physicist, laid the foundation of electrodynamics. He defined the electric current and developed a way to measure it in the 1820s.

Born in Lyons, France, Ampere at age 12 mastered Latin in a few weeks, as he was intensely interested in mathematics and many of the best mathematical works were in Latin. He was a brilliant scientist and a prolific writer. He formulated the laws of electromagnetics. He invented the electromagnet and the ammeter. The unit of electric current, the ampere, was named after him.

The Burndy Library Collection at The Huntington Library, San Marino, California.

Mathematically, the relationship between current i, charge q, and time t is

$$i \triangleq \frac{dq}{dt} \tag{1.1}$$

where current is measured in amperes (A), and

1 ampere = 1 coulomb/second

The charge transferred between time t_0 and t is obtained by integrating both sides of Eq. (1.1). We obtain

$$Q \triangleq \int_{t_0}^{t} i\,dt \tag{1.2}$$

The way we define current as i in Eq. (1.1) suggests that current need not be a constant-valued function. As many of the examples and problems in this chapter and subsequent chapters suggest, there can be several types of current; that is, charge can vary with time in several ways.

If the current does not change with time, but remains constant, we call it a *direct current* (dc).

A **direct current** (dc) is a current that remains constant with time.

By convention the symbol I is used to represent such a constant current.

A time-varying current is represented by the symbol i. A common form of time-varying current is the sinusoidal current or *alternating current* (ac).

An **alternating current** (ac) is a current that varies sinusoidally with time.

Such current is used in your household, to run the air conditioner, refrigerator, washing machine, and other electric appliances. Figure 1.4

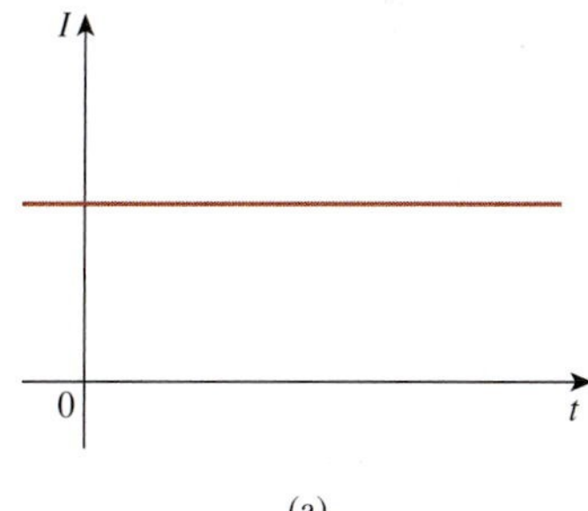

(a)

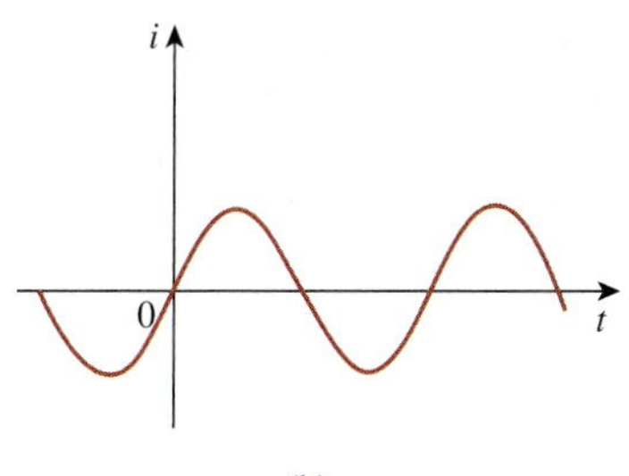

(b)

Figure 1.4
Two common types of current: (a) direct current (dc), (b) alternating current (ac).

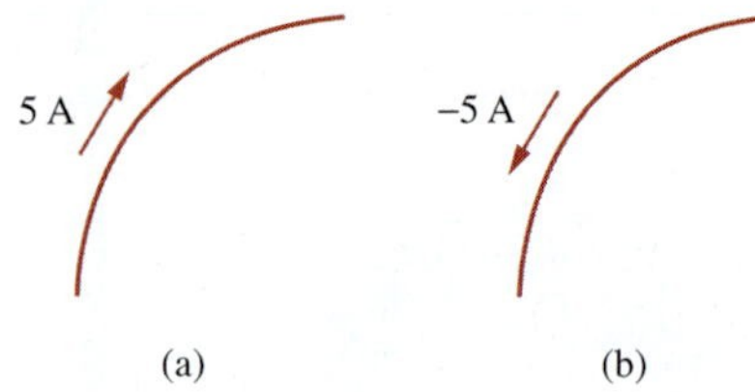

Figure 1.5
Conventional current flow: (a) positive current flow, (b) negative current flow.

shows direct current and alternating current; these are the two most common types of current. We will consider other types later in the book.

Once we define current as the movement of charge, we expect current to have an associated direction of flow. As mentioned earlier, the direction of current flow is conventionally taken as the direction of positive charge movement. Based on this convention, a current of 5 A may be represented positively or negatively as shown in Fig. 1.5. In other words, a negative current of −5 A flowing in one direction as shown in Fig. 1.5(b) is the same as a current of +5 A flowing in the opposite direction.

Example 1.1

How much charge is represented by 4,600 electrons?

Solution:
Each electron has -1.602×10^{-19} C. Hence 4,600 electrons will have

$$-1.602 \times 10^{-19} \text{ C/electron} \times 4{,}600 \text{ electrons} = -7.369 \times 10^{-16} \text{ C}$$

Practice Problem 1.1

Calculate the amount of charge represented by four million protons.

Answer: $+6.408 \times 10^{-13}$ C.

Example 1.2

The total charge entering a terminal is given by $q = 5t \sin 4\pi t$ mC. Calculate the current at $t = 0.5$ s.

Solution:

$$i = \frac{dq}{dt} = \frac{d}{dt}(5t \sin 4\pi t) \text{ mC/s} = (5 \sin 4\pi t + 20\pi t \cos 4\pi t) \text{ mA}$$

At $t = 0.5$,

$$i = 5 \sin 2\pi + 10\pi \cos 2\pi = 0 + 10\pi = 31.42 \text{ mA}$$

Practice Problem 1.2

If in Example 1.2, $q = (10 - 10e^{-2t})$ mC, find the current at $t = 0.5$ s.

Answer: 7.36 mA.

Example 1.3

Determine the total charge entering a terminal between $t = 1$ s and $t = 2$ s if the current passing the terminal is $i = (3t^2 - t)$ A.

Solution:

$$Q = \int_{t=1}^{2} i\,dt = \int_{1}^{2} (3t^2 - t)dt$$
$$= \left(t^3 - \frac{t^2}{2}\right)\Bigg|_1^2 = (8 - 2) - \left(1 - \frac{1}{2}\right) = 5.5 \text{ C}$$

Practice Problem 1.3

The current flowing through an element is

$$i = \begin{cases} 2 \text{ A}, & 0 < t < 1 \\ 2t^2 \text{ A}, & t > 1 \end{cases}$$

Calculate the charge entering the element from $t = 0$ to $t = 2$ s.

Answer: 6.667 C.

1.4 Voltage

As explained briefly in the previous section, to move the electron in a conductor in a particular direction requires some work or energy transfer. This work is performed by an external electromotive force (emf), typically represented by the battery in Fig. 1.3. This emf is also known as *voltage* or *potential difference*. The voltage v_{ab} between two points a and b in an electric circuit is the energy (or work) needed to move a unit charge from a to b; mathematically,

$$v_{ab} \triangleq \frac{dw}{dq} \tag{1.3}$$

where w is energy in joules (J) and q is charge in coulombs (C). The voltage v_{ab} or simply v is measured in volts (V), named in honor of the Italian physicist Alessandro Antonio Volta (1745–1827), who invented the first voltaic battery. From Eq. (1.3), it is evident that

$$1 \text{ volt} = 1 \text{ joule/coulomb} = 1 \text{ newton-meter/coulomb}$$

Thus,

> **Voltage** (or **potential difference**) is the energy required to move a unit charge through an element, measured in volts (V).

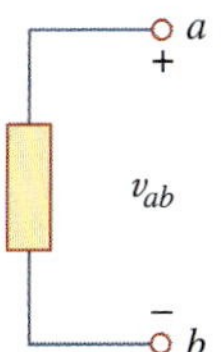

Figure 1.6
Polarity of voltage v_{ab}.

Figure 1.6 shows the voltage across an element (represented by a rectangular block) connected to points a and b. The plus (+) and minus (−) signs are used to define reference direction or voltage polarity. The v_{ab} can be interpreted in two ways: (1) point a is at a potential of v_{ab}

Historical

The Burndy Library Collection at The Huntington Library, San Marino, California.

Alessandro Antonio Volta (1745–1827), an Italian physicist, invented the electric battery—which provided the first continuous flow of electricity—and the capacitor.

Born into a noble family in Como, Italy, Volta was performing electrical experiments at age 18. His invention of the battery in 1796 revolutionized the use of electricity. The publication of his work in 1800 marked the beginning of electric circuit theory. Volta received many honors during his lifetime. The unit of voltage or potential difference, the volt, was named in his honor.

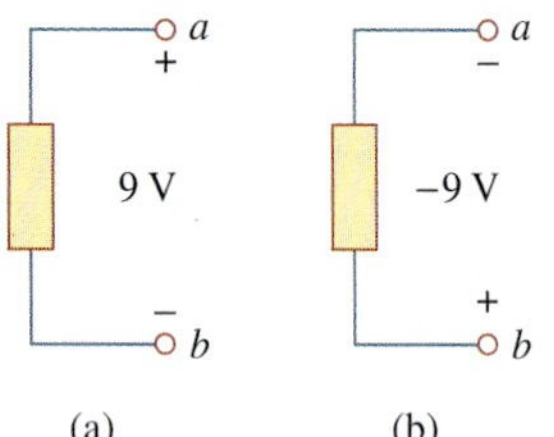

Figure 1.7
Two equivalent representations of the same voltage v_{ab}: (a) point a is 9 V above point b, (b) point b is -9 V above point a.

Keep in mind that electric current is always *through* an element and that electric voltage is always *across* the element or between two points.

volts higher than point b, or (2) the potential at point a with respect to point b is v_{ab}. It follows logically that in general

$$v_{ab} = -v_{ba} \tag{1.4}$$

For example, in Fig. 1.7, we have two representations of the same voltage. In Fig. 1.7(a), point a is $+9$ V above point b; in Fig. 1.7(b), point b is -9 V above point a. We may say that in Fig. 1.7(a), there is a 9-V *voltage drop* from a to b or equivalently a 9-V *voltage rise* from b to a. In other words, a voltage drop from a to b is equivalent to a voltage rise from b to a.

Current and voltage are the two basic variables in electric circuits. The common term *signal* is used for an electric quantity such as a current or a voltage (or even electromagnetic wave) when it is used for conveying information. Engineers prefer to call such variables signals rather than mathematical functions of time because of their importance in communications and other disciplines. Like electric current, a constant voltage is called a *dc voltage* and is represented by V, whereas a sinusoidally time-varying voltage is called an *ac voltage* and is represented by v. A dc voltage is commonly produced by a battery; ac voltage is produced by an electric generator.

1.5 Power and Energy

Although current and voltage are the two basic variables in an electric circuit, they are not sufficient by themselves. For practical purposes, we need to know how much *power* an electric device can handle. We all know from experience that a 100-watt bulb gives more light than a 60-watt bulb. We also know that when we pay our bills to the electric utility companies, we are paying for the electric *energy* consumed over a certain period of time. Thus, power and energy calculations are important in circuit analysis.

To relate power and energy to voltage and current, we recall from physics that:

Power is the time rate of expending or absorbing energy, measured in watts (W).

We write this relationship as

$$p \triangleq \frac{dw}{dt} \tag{1.5}$$

where p is power in watts (W), w is energy in joules (J), and t is time in seconds (s). From Eqs. (1.1), (1.3), and (1.5), it follows that

$$p = \frac{dw}{dt} = \frac{dw}{dq} \cdot \frac{dq}{dt} = vi \tag{1.6}$$

or

$$p = vi \tag{1.7}$$

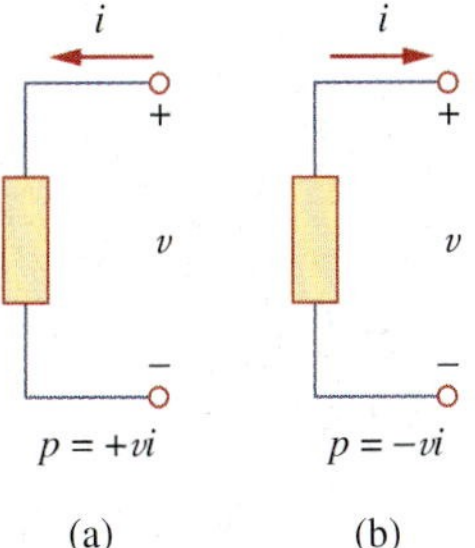

Figure 1.8
Reference polarities for power using the passive sign convention: (a) absorbing power, (b) supplying power.

When the voltage and current directions conform to Fig. 1.8(b), we have the *active sign convention* and $p = +vi$.

The power p in Eq. (1.7) is a time-varying quantity and is called the *instantaneous power*. Thus, the power absorbed or supplied by an element is the product of the voltage across the element and the current through it. If the power has a + sign, power is being delivered to or absorbed by the element. If, on the other hand, the power has a − sign, power is being supplied by the element. But how do we know when the power has a negative or a positive sign?

Current direction and voltage polarity play a major role in determining the sign of power. It is therefore important that we pay attention to the relationship between current i and voltage v in Fig. 1.8(a). The voltage polarity and current direction must conform with those shown in Fig. 1.8(a) in order for the power to have a positive sign. This is known as the *passive sign convention*. By the passive sign convention, current enters through the positive polarity of the voltage. In this case, $p = +vi$ or $vi > 0$ implies that the element is absorbing power. However, if $p = -vi$ or $vi < 0$, as in Fig. 1.8(b), the element is releasing or supplying power.

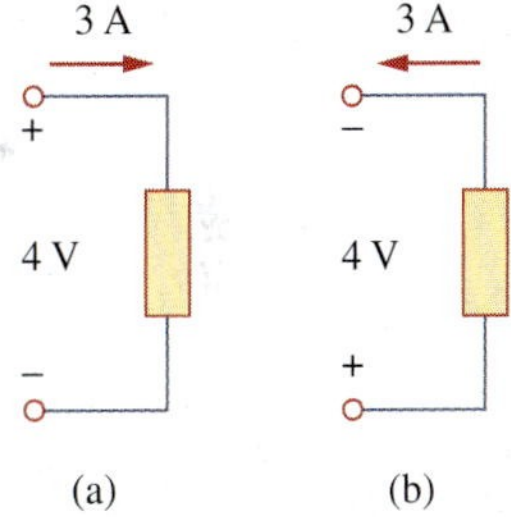

Figure 1.9
Two cases of an element with an absorbing power of 12 W: (a) $p = 4 \times 3 = 12$ W, (b) $p = 4 \times 3 = 12$ W.

Passive sign convention is satisfied when the current enters through the positive terminal of an element and $p = +vi$. If the current enters through the negative terminal, $p = -vi$.

Unless otherwise stated, we will follow the passive sign convention throughout this text. For example, the element in both circuits of Fig. 1.9 has an absorbing power of +12 W because a positive current enters the positive terminal in both cases. In Fig. 1.10, however, the element is supplying power of +12 W because a positive current enters the negative terminal. Of course, an absorbing power of −12 W is equivalent to a supplying power of +12 W. In general,

$$+\text{Power absorbed} = -\text{Power supplied}$$

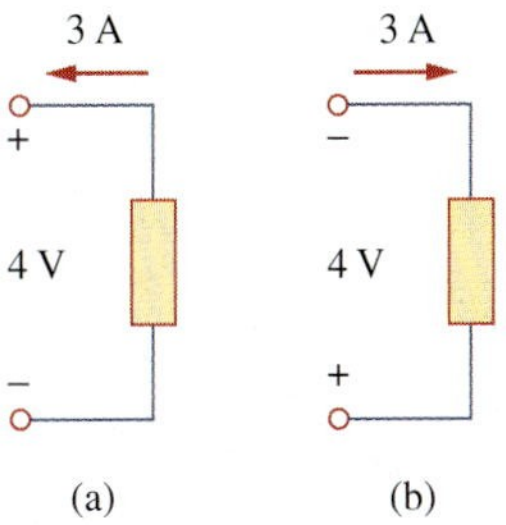

Figure 1.10
Two cases of an element with a supplying power of 12 W: (a) $p = -4 \times 3 = -12$W, (b) $p = -4 \times 3 = -12$ W.

In fact, the *law of conservation of energy* must be obeyed in any electric circuit. For this reason, the algebraic sum of power in a circuit, at any instant of time, must be zero:

$$\sum p = 0 \tag{1.8}$$

This again confirms the fact that the total power supplied to the circuit must balance the total power absorbed.

From Eq. (1.6), the energy absorbed or supplied by an element from time t_0 to time t is

$$w = \int_{t_0}^{t} p\, dt = \int_{t_0}^{t} vi\, dt \tag{1.9}$$

Energy is the capacity to do work, measured in joules (J).

The electric power utility companies measure energy in watt-hours (Wh), where

$$1 \text{ Wh} = 3{,}600 \text{ J}$$

Example 1.4

An energy source forces a constant current of 2 A for 10 s to flow through a lightbulb. If 2.3 kJ is given off in the form of light and heat energy, calculate the voltage drop across the bulb.

Solution:
The total charge is

$$\Delta q = i\,\Delta t = 2 \times 10 = 20 \text{ C}$$

The voltage drop is

$$v = \frac{\Delta w}{\Delta q} = \frac{2.3 \times 10^3}{20} = 115 \text{ V}$$

Practice Problem 1.4

To move charge q from point a to point b requires -30 J. Find the voltage drop v_{ab} if: (a) $q = 2$ C, (b) $q = -6$ C.

Answer: (a) -15 V, (b) 5 V.

Example 1.5

Find the power delivered to an element at $t = 3$ ms if the current entering its positive terminal is

$$i = 5 \cos 60\pi t \text{ A}$$

and the voltage is: (a) $v = 3i$, (b) $v = 3\, di/dt$.

Solution:

(a) The voltage is $v = 3i = 15 \cos 60\pi t$; hence, the power is

$$p = vi = 75 \cos^2 60\pi t \text{ W}$$

At $t = 3$ ms,

$$p = 75 \cos^2 (60\pi \times 3 \times 10^{-3}) = 75 \cos^2 0.18\pi = 53.48 \text{ W}$$

(b) We find the voltage and the power as

$$v = 3\frac{di}{dt} = 3(-60\pi)5 \sin 60\pi t = -900\pi \sin 60\pi t \text{ V}$$

$$p = vi = -4500\pi \sin 60\pi t \cos 60\pi t \text{ W}$$

At $t = 3$ ms,

$$\begin{aligned} p &= -4500\pi \sin 0.18\pi \cos 0.18\pi \text{ W} \\ &= -14137.167 \sin 32.4^\circ \cos 32.4^\circ = -6.396 \text{ kW} \end{aligned}$$

Practice Problem 1.5

Find the power delivered to the element in Example 1.5 at $t = 5$ ms if the current remains the same but the voltage is: (a) $v = 2i$ V, (b) $v = \left(10 + 5\int_0^t i\,dt\right)$ V.

Answer: (a) 17.27 W, (b) 29.7 W.

Example 1.6

How much energy does a 100-W electric bulb consume in two hours?

Solution:

$$\begin{aligned} w = pt &= 100 \text{ (W)} \times 2 \text{ (h)} \times 60 \text{ (min/h)} \times 60 \text{ (s/min)} \\ &= 720{,}000 \text{ J} = 720 \text{ kJ} \end{aligned}$$

This is the same as

$$w = pt = 100 \text{ W} \times 2 \text{ h} = 200 \text{ Wh}$$

Practice Problem 1.6

A stove element draws 15 A when connected to a 240-V line. How long does it take to consume 60 kJ?

Answer: 16.667 s.

Historical

1884 Exhibition In the United States, nothing promoted the future of electricity like the 1884 International Electrical Exhibition. Just imagine a world without electricity, a world illuminated by candles and gaslights, a world where the most common transportation was by walking and riding on horseback or by horse-drawn carriage. Into this world an exhibition was created that highlighted Thomas Edison and reflected his highly developed ability to promote his inventions and products. His exhibit featured spectacular lighting displays powered by an impressive 100-kW "Jumbo" generator.

Edward Weston's dynamos and lamps were featured in the United States Electric Lighting Company's display. Weston's well known collection of scientific instruments was also shown.

Other prominent exhibitors included Frank Sprague, Elihu Thompson, and the Brush Electric Company of Cleveland. The American Institute of Electrical Engineers (AIEE) held its first technical meeting on October 7–8 at the Franklin Institute during the exhibit. AIEE merged with the Institute of Radio Engineers (IRE) in 1964 to form the Institute of Electrical and Electronics Engineers (IEEE).

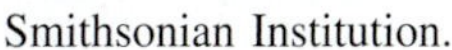

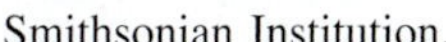
Smithsonian Institution.

1.6 Circuit Elements

As we discussed in Section 1.1, an element is the basic building block of a circuit. An electric circuit is simply an interconnection of the elements. Circuit analysis is the process of determining voltages across (or the currents through) the elements of the circuit.

There are two types of elements found in electric circuits: *passive* elements and *active* elements. An active element is capable of generating energy while a passive element is not. Examples of passive elements are resistors, capacitors, and inductors. Typical active elements include generators, batteries, and operational amplifiers. Our aim in this section is to gain familiarity with some important active elements.

The most important active elements are voltage or current sources that generally deliver power to the circuit connected to them. There are two kinds of sources: independent and dependent sources.

> An **ideal independent source** is an active element that provides a specified voltage or current that is completely independent of other circuit elements.

In other words, an ideal independent voltage source delivers to the circuit whatever current is necessary to maintain its terminal voltage. Physical sources such as batteries and generators may be regarded as approximations to ideal voltage sources. Figure 1.11 shows the symbols for independent voltage sources. Notice that both symbols in Fig. 1.11(a) and (b) can be used to represent a dc voltage source, but only the symbol in Fig. 1.11(a) can be used for a time-varying voltage source. Similarly, an ideal independent current source is an active element that provides a specified current completely independent of the voltage across the source. That is, the current source delivers to the circuit whatever voltage is necessary to maintain the designated current. The symbol for an independent current source is displayed in Fig. 1.12, where the arrow indicates the direction of current i.

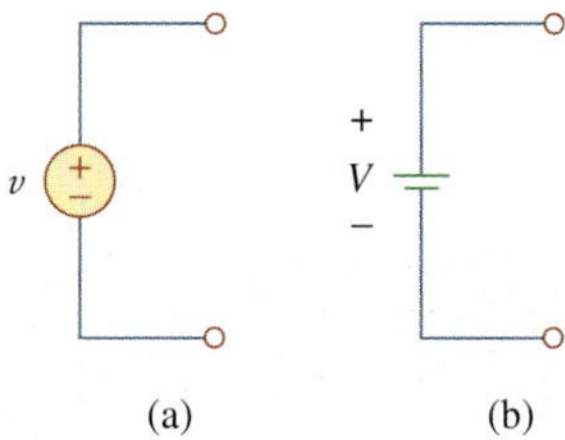

Figure 1.11
Symbols for independent voltage sources: (a) used for constant or time-varying voltage, (b) used for constant voltage (dc).

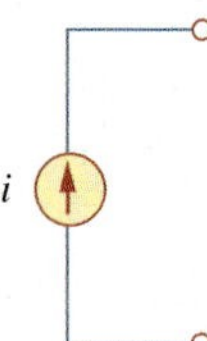

Figure 1.12
Symbol for independent current source.

> An **ideal dependent** (or **controlled**) **source** is an active element in which the source quantity is controlled by another voltage or current.

Dependent sources are usually designated by diamond-shaped symbols, as shown in Fig. 1.13. Since the control of the dependent source is achieved by a voltage or current of some other element in the circuit, and the source can be voltage or current, it follows that there are four possible types of dependent sources, namely:

1. A voltage-controlled voltage source (VCVS).
2. A current-controlled voltage source (CCVS).
3. A voltage-controlled current source (VCCS).
4. A current-controlled current source (CCCS).

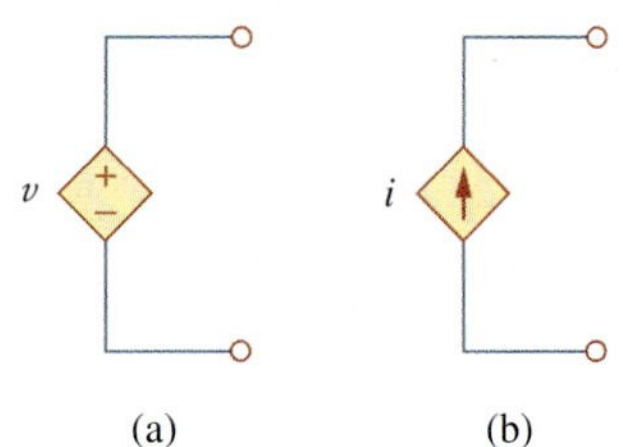

Figure 1.13
Symbols for: (a) dependent voltage source, (b) dependent current source.

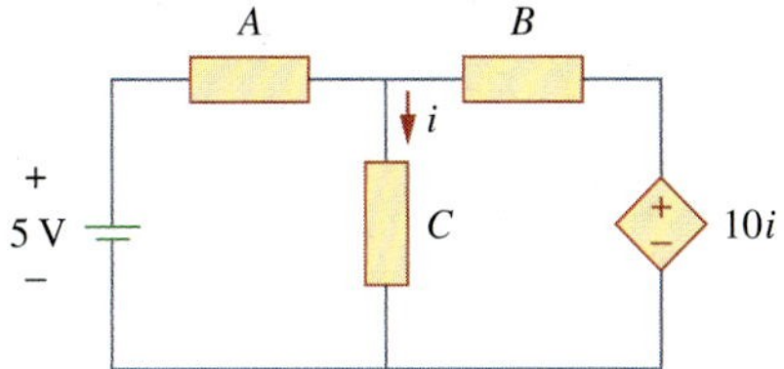

Figure 1.14
The source on the right-hand side is a current-controlled voltage source.

Dependent sources are useful in modeling elements such as transistors, operational amplifiers, and integrated circuits. An example of a current-controlled voltage source is shown on the right-hand side of Fig. 1.14, where the voltage $10i$ of the voltage source depends on the current i through element C. Students might be surprised that the value of the dependent voltage source is $10i$ V (and not $10i$ A) because it is a voltage source. The key idea to keep in mind is that a voltage source comes with polarities (+ −) in its symbol, while a current source comes with an arrow, irrespective of what it depends on.

It should be noted that an ideal voltage source (dependent or independent) will produce any current required to ensure that the terminal voltage is as stated, whereas an ideal current source will produce the necessary voltage to ensure the stated current flow. Thus, an ideal source could in theory supply an infinite amount of energy. It should also be noted that not only do sources supply power to a circuit, they can absorb power from a circuit too. For a voltage source, we know the voltage but not the current supplied or drawn by it. By the same token, we know the current supplied by a current source but not the voltage across it.

Example 1.7

Calculate the power supplied or absorbed by each element in Fig. 1.15.

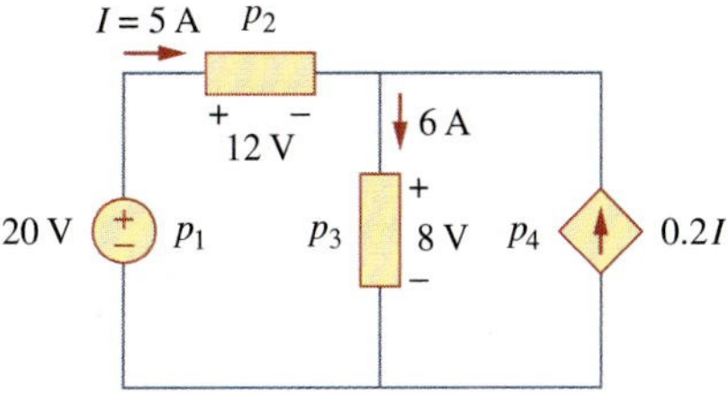

Figure 1.15
For Example 1.7.

Solution:

We apply the sign convention for power shown in Figs. 1.8 and 1.9. For p_1, the 5-A current is out of the positive terminal (or into the negative terminal); hence,

$$p_1 = 20(-5) = -100 \text{ W} \qquad \text{Supplied power}$$

For p_2 and p_3, the current flows into the positive terminal of the element in each case.

$$p_2 = 12(5) = 60 \text{ W} \qquad \text{Absorbed power}$$

$$p_3 = 8(6) = 48 \text{ W} \qquad \text{Absorbed power}$$

For p_4, we should note that the voltage is 8 V (positive at the top), the same as the voltage for p_3, since both the passive element and the dependent source are connected to the same terminals. (Remember that voltage is always measured across an element in a circuit.) Since the current flows out of the positive terminal,

$$p_4 = 8(-0.2I) = 8(-0.2 \times 5) = -8 \text{ W} \qquad \text{Supplied power}$$

We should observe that the 20-V independent voltage source and $0.2I$ dependent current source are supplying power to the rest of the network, while the two passive elements are absorbing power. Also,

$$p_1 + p_2 + p_3 + p_4 = -100 + 60 + 48 - 8 = 0$$

In agreement with Eq. (1.8), the total power supplied equals the total power absorbed.

Practice Problem 1.7

Compute the power absorbed or supplied by each component of the circuit in Fig. 1.16.

Answer: $p_1 = -40$ W, $p_2 = 16$ W, $p_3 = 9$ W, $p_4 = 15$ W.

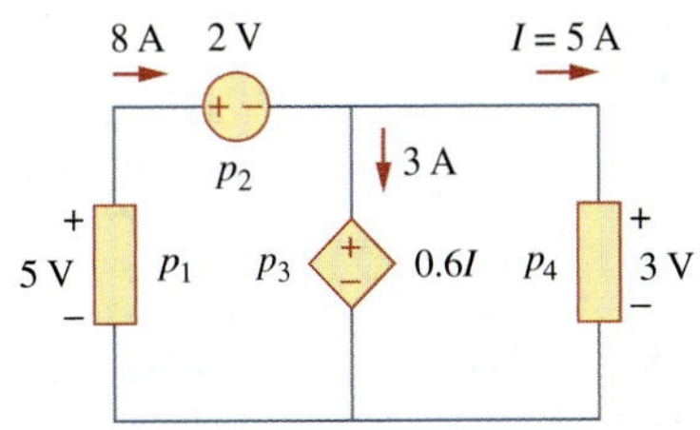

Figure 1.16
For Practice Prob. 1.7.

1.7 †Applications[2]

In this section, we will consider two practical applications of the concepts developed in this chapter. The first one deals with the TV picture tube and the other with how electric utilities determine your electric bill.

1.7.1 TV Picture Tube

One important application of the motion of electrons is found in both the transmission and reception of TV signals. At the transmission end, a TV camera reduces a scene from an optical image to an electrical signal. Scanning is accomplished with a thin beam of electrons in an iconoscope camera tube.

At the receiving end, the image is reconstructed by using a cathode-ray tube (CRT) located in the TV receiver.[3] The CRT is depicted in Fig. 1.17. Unlike the iconoscope tube, which produces an electron beam of constant intensity, the CRT beam varies in intensity according to the incoming signal. The electron gun, maintained at a high potential, fires the electron beam. The beam passes through two sets of plates for vertical and horizontal deflections so that the spot on the screen where the beam strikes can move right and left and up and down. When the electron beam strikes the fluorescent screen, it gives off light at that spot. Thus, the beam can be made to "paint" a picture on the TV screen.

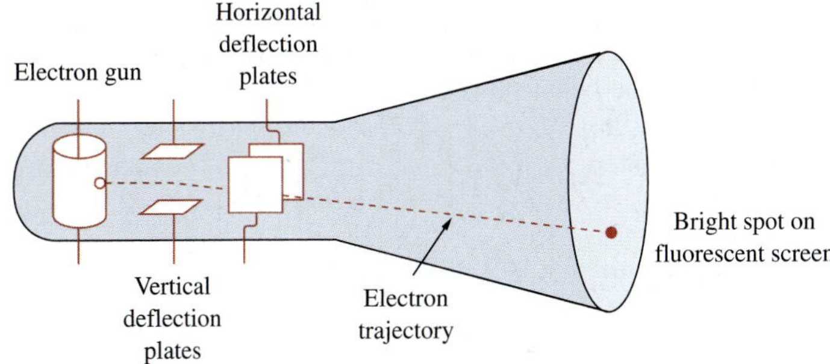

Figure 1.17
Cathode-ray tube.
D. E. Tilley, *Contemporary College Physics* Menlo Park, CA: Benjamin/Cummings, 1979, p. 319.

---

[2] The dagger sign preceding a section heading indicates the section that may be skipped, explained briefly, or assigned as homework.

[3] Modern TV tubes use a different technology.

Historical

Zworykin with an iconoscope.

Karl Ferdinand Braun and Vladimir K. Zworykin

Karl Ferdinand Braun (1850–1918), of the University of Strasbourg, invented the Braun cathode-ray tube in 1879. This then became the basis for the picture tube used for so many years for televisions. It is still the most economical device today, although the price of flat-screen systems is rapidly becoming competitive. Before the Braun tube could be used in television, it took the inventiveness of **Vladimir K. Zworykin** (1889–1982) to develop the iconoscope so that the modern television would become a reality. The iconoscope developed into the orthicon and the image orthicon, which allowed images to be captured and converted into signals that could be sent to the television receiver. Thus, the television camera was born.

Example 1.8

The electron beam in a TV picture tube carries 10^{15} electrons per second. As a design engineer, determine the voltage V_o needed to accelerate the electron beam to achieve 4 W.

Solution:

The charge on an electron is

$$e = -1.6 \times 10^{-19}\text{ C}$$

If the number of electrons is n, then $q = ne$ and

$$i = \frac{dq}{dt} = e\frac{dn}{dt} = (-1.6 \times 10^{-19})(10^{15}) = -1.6 \times 10^{-4}\text{ A}$$

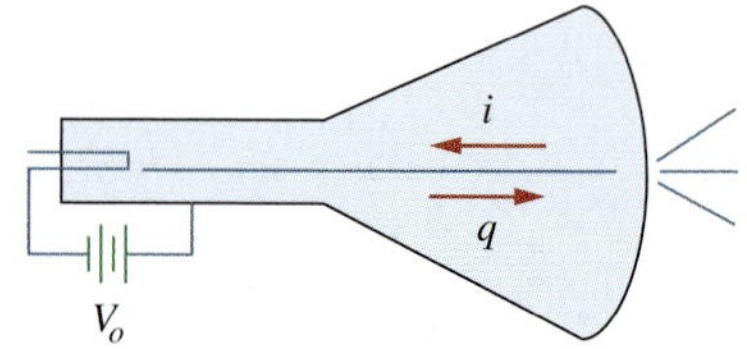

Figure 1.18
A simplified diagram of the cathode-ray tube; for Example 1.8.

The negative sign indicates that the current flows in a direction opposite to electron flow as shown in Fig. 1.18, which is a simplified diagram of the CRT for the case when the vertical deflection plates carry no charge. The beam power is

$$p = V_o i \qquad \text{or} \qquad V_o = \frac{p}{i} = \frac{4}{1.6 \times 10^{-4}} = 25{,}000\text{ V}$$

Thus, the required voltage is 25 kV.

Practice Problem 1.8

If an electron beam in a TV picture tube carries 10^{13} electrons/second and is passing through plates maintained at a potential difference of 30 kV, calculate the power in the beam.

Answer: 48 mW.

TABLE 1.3

Typical average monthly consumption of household appliances.

Appliance	kWh consumed	Appliance	kWh consumed
Water heater	500	Washing machine	120
Freezer	100	Stove	100
Lighting	100	Dryer	80
Dishwasher	35	Microwave oven	25
Electric iron	15	Personal computer	12
TV	10	Radio	8
Toaster	4	Clock	2

1.7.2 Electricity Bills

The second application deals with how an electric utility company charges their customers. The cost of electricity depends upon the amount of energy consumed in kilowatt-hours (kWh). (Other factors that affect the cost include demand and power factors; we will ignore these for now.) However, even if a consumer uses no energy at all, there is a minimum service charge the customer must pay because it costs money to stay connected to the power line. As energy consumption increases, the cost per kWh drops. It is interesting to note the average monthly consumption of household appliances for a family of five, shown in Table 1.3.

Example 1.9

A homeowner consumes 700 kWh in January. Determine the electricity bill for the month using the following residential rate schedule:

Base monthly charge of \$12.00.

First 100 kWh per month at 16 cents/kWh.

Next 200 kWh per month at 10 cents/kWh.

Over 300 kWh per month at 6 cents/kWh.

Solution:
We calculate the electricity bill as follows.

$$\begin{aligned}
\text{Base monthly charge} &= \$12.00 \\
\text{First 100 kWh @ \$0.16/kWh} &= \$16.00 \\
\text{Next 200 kWh @ \$0.10/kWh} &= \$20.00 \\
\text{Remaining 400 kWh @ \$0.06/kWh} &= \$24.00 \\
\text{Total charge} &= \$72.00
\end{aligned}$$

$$\text{Average cost} = \frac{\$72}{100 + 200 + 400} = 10.2 \text{ cents/kWh}$$

Practice Problem 1.9

Referring to the residential rate schedule in Example 1.9, calculate the average cost per kWh if only 400 kWh are consumed in July when the family is on vacation most of the time.

Answer: 13.5 cents/kWh.

1.8 †Problem Solving

Although the problems to be solved during one's career will vary in complexity and magnitude, the basic principles to be followed remain the same. The process outlined here is the one developed by the authors over many years of problem solving with students, for the solution of engineering problems in industry, and for problem solving in research.

We will list the steps simply and then elaborate on them.

1. Carefully **define** the problem.
2. **Present** everything you know about the problem.
3. Establish a set of **alternative** solutions and determine the one that promises the greatest likelihood of success.
4. **Attempt** a problem solution.
5. **Evaluate** the solution and check for accuracy.
6. Has the problem been solved **satisfactorily**? If so, present the solution; if not, then return to step 3 and continue through the process again.

1. *Carefully **define** the problem.* This may be the most important part of the process, because it becomes the foundation for all the rest of the steps. In general, the presentation of engineering problems is somewhat incomplete. You must do all you can to make sure you understand the problem as thoroughly as the presenter of the problem understands it. Time spent at this point clearly identifying the problem will save you considerable time and frustration later. As a student, you can clarify a problem statement in a textbook by asking your professor. A problem presented to you in industry may require that you consult several individuals. At this step, it is important to develop questions that need to be addressed before continuing the solution process. If you have such questions, you need to consult with the appropriate individuals or resources to obtain the answers to those questions. With those answers, you can now refine the problem, and use that refinement as the problem statement for the rest of the solution process.

2. ***Present** everything you know about the problem.* You are now ready to write down everything you know about the problem and its possible solutions. This important step will save you time and frustration later.

3. *Establish a set of **alternative** solutions and determine the one that promises the greatest likelihood of success.* Almost every problem will have a number of possible paths that can lead to a solution. It is highly desirable to identify as many of those paths as possible. At this point, you also need to determine what tools are available to you, such as *PSpice* and *MATLAB* and other software packages that can greatly reduce effort and increase accuracy. Again, we want to stress that time spent carefully defining the problem and investigating alternative approaches to its solution will pay big dividends later. Evaluating the alternatives and determining which promises the greatest likelihood of success may be difficult but will be well worth the effort. Document this process well since you will want to come back to it if the first approach does not work.

4. ***Attempt** a problem solution.* Now is the time to actually begin solving the problem. The process you follow must be well documented

in order to present a detailed solution if successful, and to evaluate the process if you are not successful. This detailed evaluation may lead to corrections that can then lead to a successful solution. It can also lead to new alternatives to try. Many times, it is wise to fully set up a solution before putting numbers into equations. This will help in checking your results.

5. ***Evaluate** the solution and check for accuracy.* You now thoroughly evaluate what you have accomplished. Decide if you have an acceptable solution, one that you want to present to your team, boss, or professor.

6. *Has the problem been solved **satisfactorily**? If so, present the solution; if not, then return to step 3 and continue through the process again.* Now you need to present your solution or try another alternative. At this point, presenting your solution may bring closure to the process. Often, however, presentation of a solution leads to further refinement of the problem definition, and the process continues. Following this process will eventually lead to a satisfactory conclusion.

Now let us look at this process for a student taking an electrical and computer engineering foundations course. (The basic process also applies to almost every engineering course.) Keep in mind that although the steps have been simplified to apply to academic types of problems, the process as stated always needs to be followed. We consider a simple example.

Example 1.10

Solve for the current flowing through the 8-Ω resistor in Fig. 1.19.

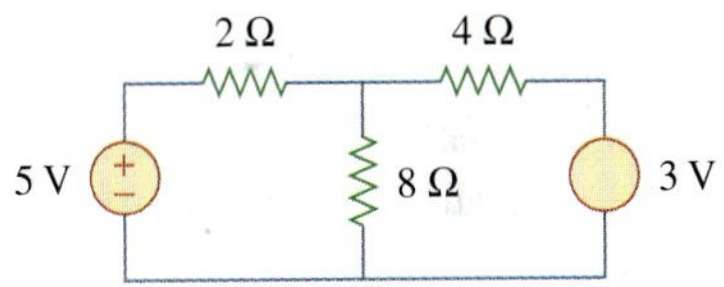

Figure 1.19
Illustrative example.

Solution:

1. *Carefully **define** the problem.* This is only a simple example, but we can already see that we do not know the polarity on the 3-V source. We have the following options. We can ask the professor what the polarity should be. If we cannot ask, then we need to make a decision on what to do next. If we have time to work the problem both ways, we can solve for the current when the 3-V source is plus on top and then plus on the bottom. If we do not have the time to work it both ways, assume a polarity and then carefully document your decision. Let us assume that the professor tells us that the source is plus on the bottom as shown in Fig. 1.20.

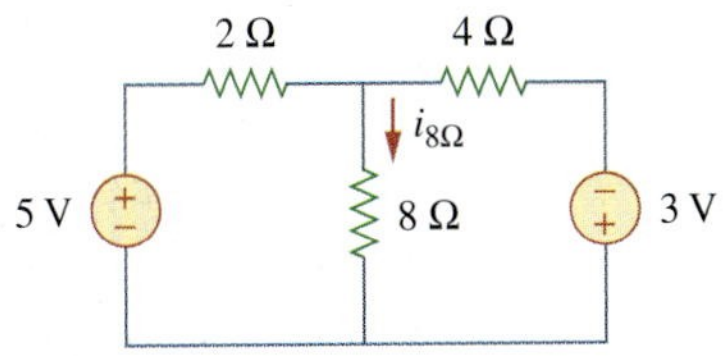

Figure 1.20
Problem defintion.

2. ***Present** everything you know about the problem.* Presenting all that we know about the problem involves labeling the circuit clearly so that we define what we seek.

Given the circuit shown in Fig. 1.20, solve for $i_{8\Omega}$.

We now check with the professor, if reasonable, to see if the problem is properly defined.

3. *Establish a set of **alternative** solutions and determine the one that promises the greatest likelihood of success.* There are essentially three techniques that can be used to solve this problem. Later in the text you will see that you can use circuit analysis (using Kirchhoff's laws and Ohm's law), nodal analysis, and mesh analysis.

To solve for $i_{8\Omega}$ using circuit analysis will eventually lead to a solution, but it will likely take more work than either nodal or mesh analysis. To solve for $i_{8\Omega}$ using mesh analysis will require writing two simultaneous equations to find the two loop currents indicated in Fig. 1.21. Using nodal analysis requires solving for only one unknown. This is the easiest approach.

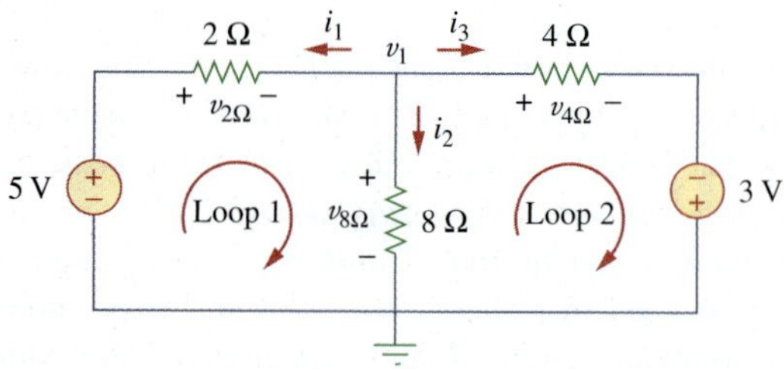

Figure 1.21
Using nodal analysis.

Therefore, we will solve for $i_{8\Omega}$ using nodal analysis.

4. ***Attempt** a problem solution.* We first write down all of the equations we will need in order to find $i_{8\Omega}$.

$$i_{8\Omega} = i_2, \qquad i_2 = \frac{v_1}{8}, \qquad i_{8\Omega} = \frac{v_1}{8}$$

$$\frac{v_1 - 5}{2} + \frac{v_1 - 0}{8} + \frac{v_1 + 3}{4} = 0$$

Now we can solve for v_1.

$$8\left[\frac{v_1 - 5}{2} + \frac{v_1 - 0}{8} + \frac{v_1 + 3}{4}\right] = 0$$

$$\text{leads to } (4v_1 - 20) + (v_1) + (2v_1 + 6) = 0$$

$$7v_1 = +14, \qquad v_1 = +2\text{ V}, \qquad i_{8\Omega} = \frac{v_1}{8} = \frac{2}{8} = \mathbf{0.25\text{ A}}$$

5. ***Evaluate** the solution and check for accuracy.* We can now use Kirchhoff's voltage law (KVL) to check the results.

$$i_1 = \frac{v_1 - 5}{2} = \frac{2 - 5}{2} = -\frac{3}{2} = -1.5\text{ A}$$

$$i_2 = i_{8\Omega} = 0.25\text{ A}$$

$$i_3 = \frac{v_1 + 3}{4} = \frac{2 + 3}{4} = \frac{5}{4} = 1.25\text{ A}$$

$$i_1 + i_2 + i_3 = \mathbf{-1.5 + 0.25 + 1.25 = 0} \qquad \text{(Checks.)}$$

Applying KVL to loop 1,

$$\begin{aligned} -5 + v_{2\Omega} + v_{8\Omega} &= -5 + (-i_1 \times 2) + (i_2 \times 8) \\ &= -5 + [-(-1.5)2] + (0.25 \times 8) \\ &= \mathbf{-5 + 3 + 2 = 0} \qquad \text{(Checks.)} \end{aligned}$$

Applying KVL to loop 2,

$$\begin{aligned} -v_{8\Omega} + v_{4\Omega} - 3 &= -(i_2 \times 8) + (i_3 \times 4) - 3 \\ &= -(0.25 \times 8) + (1.25 \times 4) - 3 \\ &= \mathbf{-2 + 5 - 3 = 0} \qquad \text{(Checks.)} \end{aligned}$$

So we now have a very high degree of confidence in the accuracy of our answer.

6. *Has the problem been solved* ***satisfactorily****? If so, present the solution; if not, then return to step 3 and continue through the process again.* This problem has been solved satisfactorily.

The current through the 8-Ω resistor is 0.25 A flowing down through the 8-Ω resistor.

Practice Problem 1.10

Try applying this process to some of the more difficult problems at the end of the chapter.

1.9 Summary

1. An electric circuit consists of electrical elements connected together.
2. The International System of Units (SI) is the international measurement language, which enables engineers to communicate their results. From the six principal units, the units of other physical quantities can be derived.
3. Current is the rate of charge flow.

$$i = \frac{dq}{dt}$$

4. Voltage is the energy required to move 1 C of charge through an element.

$$v = \frac{dw}{dq}$$

5. Power is the energy supplied or absorbed per unit time. It is also the product of voltage and current.

$$p = \frac{dw}{dt} = vi$$

6. According to the passive sign convention, power assumes a positive sign when the current enters the positive polarity of the voltage across an element.
7. An ideal voltage source produces a specific potential difference across its terminals regardless of what is connected to it. An ideal current source produces a specific current through its terminals regardless of what is connected to it.
8. Voltage and current sources can be dependent or independent. A dependent source is one whose value depends on some other circuit variable.
9. Two areas of application of the concepts covered in this chapter are the TV picture tube and electricity billing procedure.

Review Questions

1.1 One millivolt is one millionth of a volt.

(a) True (b) False

1.2 The prefix *micro* stands for:

(a) 10^6 (b) 10^3 (c) 10^{-3} (d) 10^{-6}

1.3 The voltage 2,000,000 V can be expressed in powers of 10 as:

(a) 2 mV (b) 2 kV (c) 2 MV (d) 2 GV

1.4 A charge of 2 C flowing past a given point each second is a current of 2 A.

(a) True (b) False

1.5 The unit of current is:

(a) coulomb (b) ampere
(c) volt (d) joule

1.6 Voltage is measured in:

(a) watts (b) amperes
(c) volts (d) joules per second

1.7 A 4-A current charging a dielectric material will accumulate a charge of 24 C after 6 s.

(a) True (b) False

1.8 The voltage across a 1.1-kW toaster that produces a current of 10 A is:

(a) 11 kV (b) 1100 V (c) 110 V (d) 11 V

1.9 Which of these is not an electrical quantity?

(a) charge (b) time (c) voltage
(d) current (e) power

1.10 The dependent source in Fig. 1.22 is:

(a) voltage-controlled current source
(b) voltage-controlled voltage source
(c) current-controlled voltage source
(d) current-controlled current source

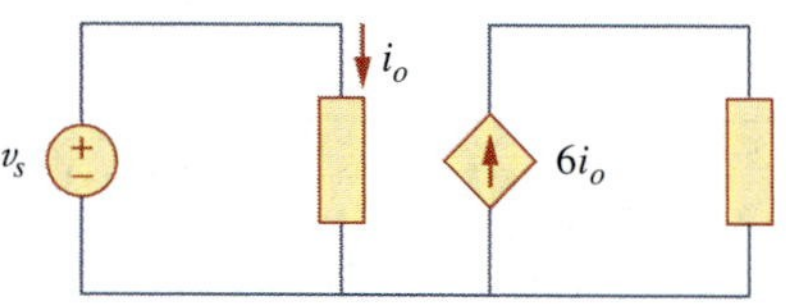

Figure 1.22
For Review Question 1.10.

Answers: 1.1b, 1.2d, 1.3c, 1.4a, 1.5b, 1.6c, 1.7a, 1.8c, 1.9b, 1.10d.

Problems

Section 1.3 Charge and Current

1.1 How many coulombs are represented by these amounts of electrons?

(a) 6.482×10^{17} (b) 1.24×10^{18}
(c) 2.46×10^{19} (d) 1.628×10^{20}

1.2 Determine the current flowing through an element if the charge flow is given by

(a) $q(t) = (3t + 8)$ mC
(b) $q(t) = (8t^2 + 4t - 2)$ C
(c) $q(t) = (3e^{-t} - 5e^{-2t})$ nC
(d) $q(t) = 10 \sin 120\pi t$ pC
(e) $q(t) = 20e^{-4t} \cos 50t\ \mu$C

1.3 Find the charge $q(t)$ flowing through a device if the current is:

(a) $i(t) = 3$ A, $q(0) = 1$ C
(b) $i(t) = (2t + 5)$ mA, $q(0) = 0$
(c) $i(t) = 20 \cos(10t + \pi/6)\ \mu$A, $q(0) = 2\ \mu$C
(d) $i(t) = 10e^{-30t} \sin 40t$ A, $q(0) = 0$

1.4 A current of 3.2 A flows through a conductor. Calculate how much charge passes through any cross-section of the conductor in 20 s.

1.5 Determine the total charge transferred over the time interval of $0 \le t \le 10$ s when $i(t) = \frac{1}{2}t$ A.

1.6 The charge entering a certain element is shown in Fig. 1.23. Find the current at:

(a) $t = 1$ ms (b) $t = 6$ ms (c) $t = 10$ ms

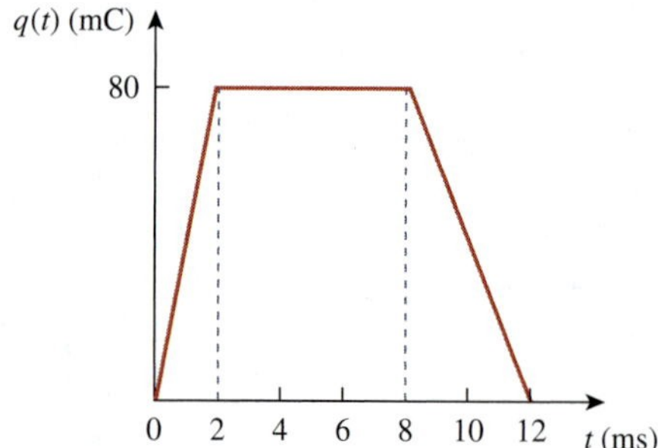

Figure 1.23
For Prob. 1.6.

1.7 The charge flowing in a wire is plotted in Fig. 1.24. Sketch the corresponding current.

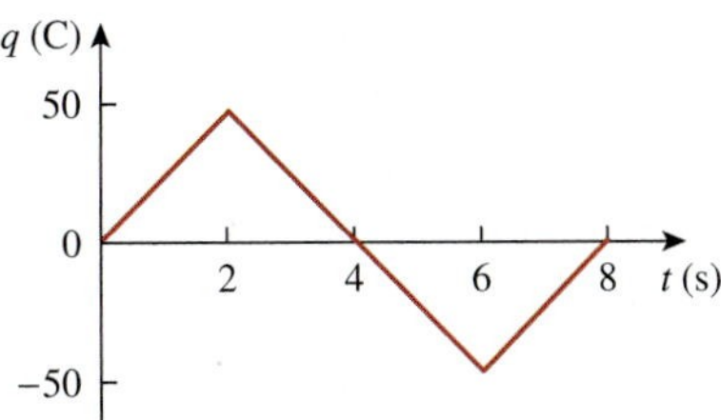

Figure 1.24
For Prob. 1.7.

1.8 The current flowing past a point in a device is shown in Fig. 1.25. Calculate the total charge through the point.

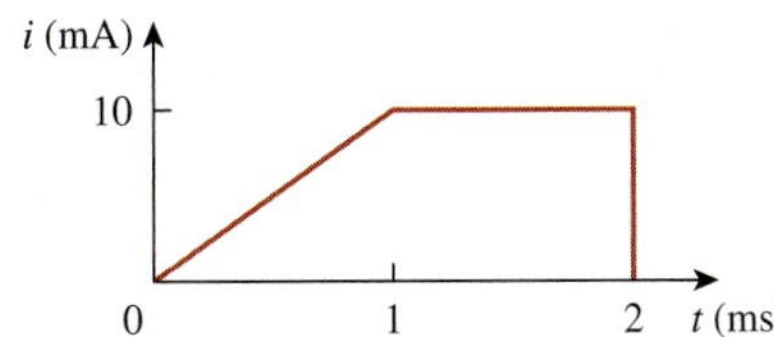

Figure 1.25
For Prob. 1.8.

1.9 The current through an element is shown in Fig. 1.26. Determine the total charge that passed through the element at:

(a) $t = 1$ s (b) $t = 3$ s (c) $t = 5$ s

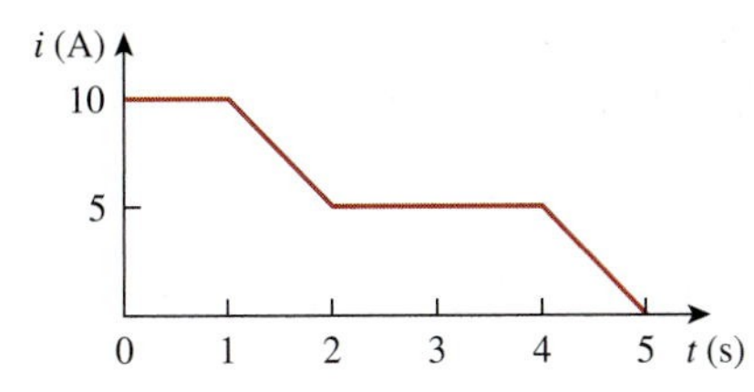

Figure 1.26
For Prob. 1.9.

Sections 1.4 and 1.5 Voltage, Power, and Energy

1.10 A lightning bolt with 8 kA strikes an object for 15 μs. How much charge is deposited on the object?

1.11 A rechargeable flashlight battery is capable of delivering 85 mA for about 12 h. How much charge can it release at that rate? If its terminal voltage is 1.2 V, how much energy can the battery deliver?

1.12 If the current flowing through an element is given by

$$i(t) = \begin{cases} 3t\text{A}, & 0 \leqq t < 6 \text{ s} \\ 18\text{A}, & 6 \leqq t < 10 \text{ s} \\ -12\text{A}, & 10 \leqq t < 15 \text{ s} \\ 0, & t \geqq 15 \text{ s} \end{cases}$$

Plot the charge stored in the element over $0 < t < 20$ s.

1.13 The charge entering the positive terminal of an element is

$$q = 10 \sin 4\pi t \text{ mC}$$

while the voltage across the element (plus to minus) is

$$v = 2 \cos 4\pi t \text{ V}$$

(a) Find the power delivered to the element at $t = 0.3$ s.

(b) Calculate the energy delivered to the element between 0 and 0.6 s.

1.14 The voltage v across a device and the current i through it are

$$v(t) = 5 \cos 2t \text{ V}, \qquad i(t) = 10(1 - e^{-0.5t}) \text{ A}$$

Calculate:

(a) the total charge in the device at $t = 1$ s

(b) the power consumed by the device at $t = 1$ s.

1.15 The current entering the positive terminal of a device is $i(t) = 3e^{-2t}$ A and the voltage across the device is $v(t) = 5di/dt$ V.

(a) Find the charge delivered to the device between $t = 0$ and $t = 2$ s.

(b) Calculate the power absorbed.

(c) Determine the energy absorbed in 3 s.

Section 1.6 Circuit Elements

1.16 Find the power absorbed by each element in Fig. 1.27.

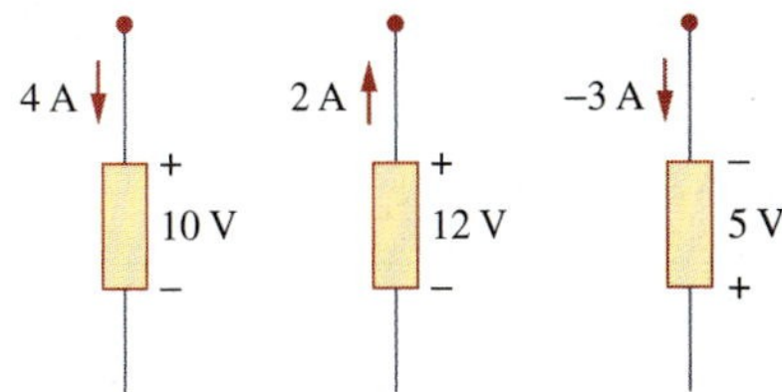

Figure 1.27
For Prob. 1.16.

1.17 Figure 1.28 shows a circuit with five elements. If $p_1 = -205$ W, $p_2 = 60$ W, $p_4 = 45$ W, $p_5 = 30$ W, calculate the power p_3 received or delivered by element 3.

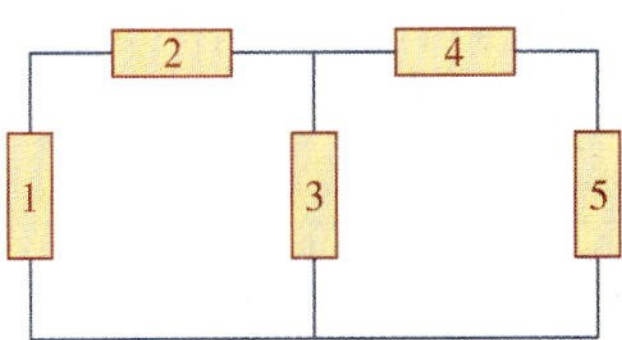

Figure 1.28
For Prob. 1.17.

1.18 Calculate the power absorbed or supplied by each element in Fig. 1.29.

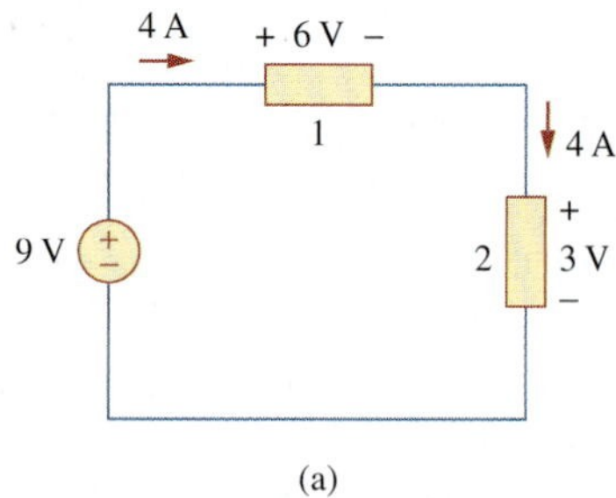

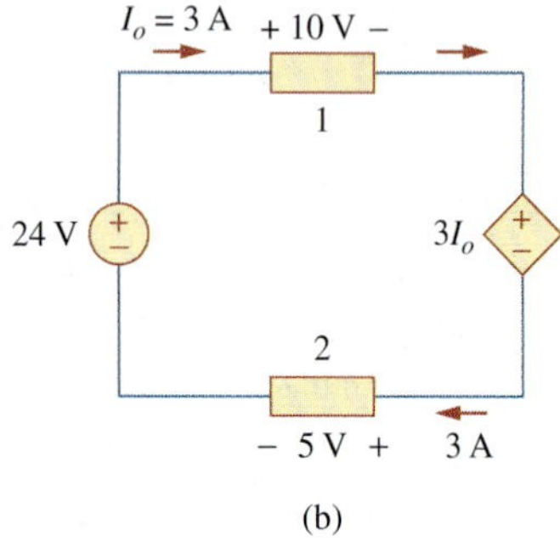

Figure 1.29
For Prob. 1.18.

1.19 Find I in the network of Fig. 1.30.

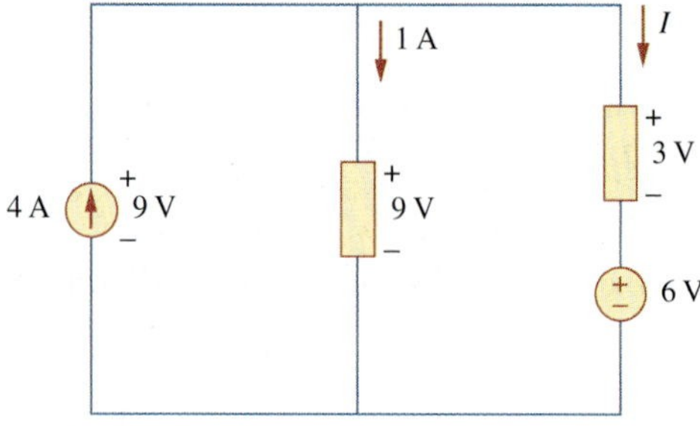

Figure 1.30
For Prob. 1.19.

1.20 Find V_o in the circuit of Fig. 1.31.

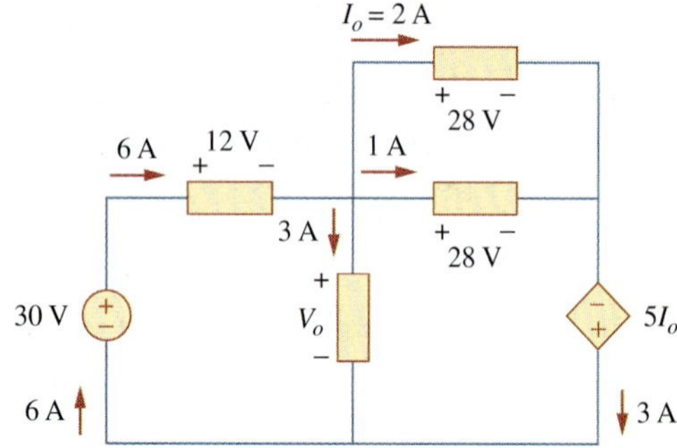

Figure 1.31
For Prob. 1.20.

Section 1.7 Applications

1.21 A 60-W incandescent bulb operates at 120 V. How many electrons and coulombs flow through the bulb in one day?

1.22 A lightning bolt strikes an airplane with 30 kA for 2 ms. How many coulombs of charge are deposited on the plane?

1.23 A 1.8-kW electric heater takes 15 min to boil a quantity of water. If this is done once a day and power costs 10 cents/kWh, what is the cost of its operation for 30 days?

1.24 A utility company charges 8.5 cents/kWh. If a consumer operates a 40-W light bulb continuously for one day, how much is the consumer charged?

1.25 A 1.2-kW toaster takes roughly 4 minutes to heat four slices of bread. Find the cost of operating the toaster once per day for 1 month (30 days). Assume energy costs 9 cents/kWh.

1.26 A 12-V car battery supported a current of 150 mA to a bulb. Calculate:

(a) the power absorbed by the bulb,

(b) the energy absorbed by the bulb over an interval of 20 minutes.

1.27 A constant current of 3 A for 4 hours is required to charge an automotive battery. If the terminal voltage is $10 + t/2$ V, where t is in hours,

(a) how much charge is transported as a result of the charging?

(b) how much energy is expended?

(c) how much does the charging cost? Assume electricity costs 9 cents/kWh.

1.28 A 30-W incandescent lamp is connected to a 120-V source and is left burning continuously in an otherwise dark staircase. Determine:

(a) the current through the lamp.

(b) the cost of operating the light for one non-leap year if electricity costs 12 cents per kWh.

1.29 An electric stove with four burners and an oven is used in preparing a meal as follows.

Burner 1: 20 minutes Burner 2: 40 minutes
Burner 3: 15 minutes Burner 4: 45 minutes
Oven: 30 minutes

If each burner is rated at 1.2 kW and the oven at 1.8 kW, and electricity costs 12 cents per kWh, calculate the cost of electricity used in preparing the meal.

1.30 Reliant Energy (the electric company in Houston, Texas) charges customers as follows:

Monthly charge \$6

First 250 kWh @ \$0.02/kWh

All additional kWh @ \$0.07/kWh

If a customer uses 1,218 kWh in one month, how much will Reliant Energy charge?

1.31 In a household, a 120-W personal computer (PC) is run for 4 h/day, while a 60-W bulb runs for 8 h/day. If the utility company charges \$0.12/kWh, calculate how much the household pays per year on the PC and the bulb.

Comprehensive Problems

1.32 A telephone wire has a current of 20 μA flowing through it. How long does it take for a charge of 15 C to pass through the wire?

1.33 A lightning bolt carried a current of 2 kA and lasted for 3 ms. How many coulombs of charge were contained in the lightning bolt?

1.34 Figure 1.32 shows the power consumption of a certain household in 1 day. Calculate:

(a) the total energy consumed in kWh,

(b) the average power per hour.

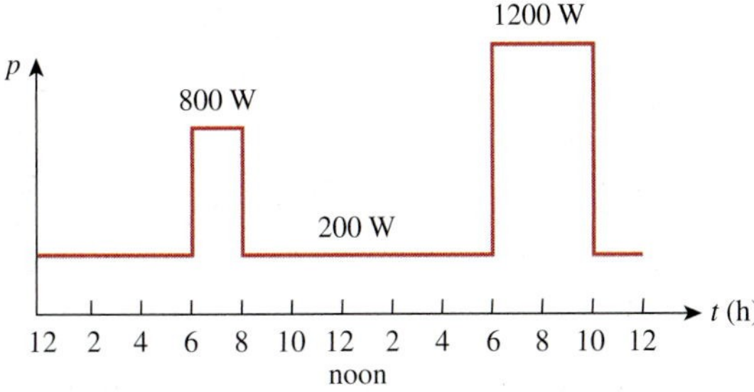

Figure 1.32
For Prob. 1.34.

1.35 The graph in Fig. 1.33 represents the power drawn by an industrial plant between 8:00 and 8:30 A.M. Calculate the total energy in MWh consumed by the plant.

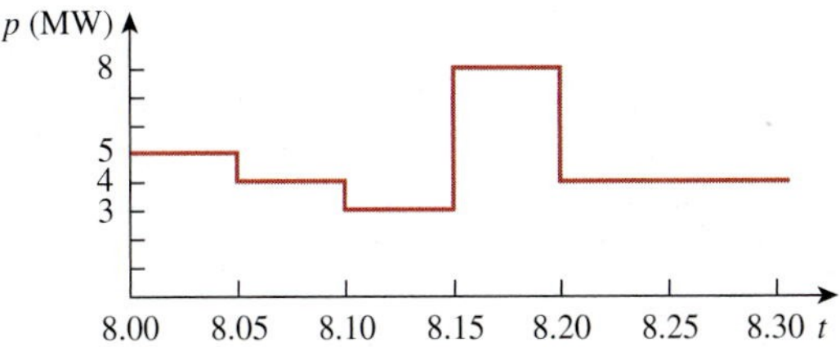

Figure 1.33
For Prob. 1.35.

1.36 A battery may be rated in ampere-hours (Ah). A lead-acid battery is rated at 160 Ah.

(a) What is the maximum current it can supply for 40 h?

(b) How many days will it last if it is discharged at 1 mA?

1.37 A unit of power often used for electric motors is the horsepower (hp), which equals 746 W. A small electric car is equipped with a 40-hp electric motor. How much energy does the motor deliver in one hour, assuming the motor is operating at maximum power for the whole time?

1.38 How much energy does a 10-hp motor deliver in 30 minutes? Assume that 1 horsepower = 746 W.

1.39 A 600-W TV receiver is turned on for 4 h with nobody watching it. If electricity costs 10 cents/kWh, how much money is wasted?

chapter

2

Basic Laws

There are too many people praying for mountains of difficulty to be removed, when what they really need is the courage to climb them!

—Unknown

Enhancing Your Skills and Your Career

Photo by Charles Alexander

ABET EC 2000 criteria (3.b), *"an ability to design and conduct experiments, as well as to analyze and interpret data.* Engineers must be able to design and conduct experiments, as well as analyze and interpret data. Most students have spent many hours performing experiments in high school and in college. During this time, you have been asked to analyze the data and to interpret the data. Therefore, you should already be skilled in these two activities. My recommendation is that, in the process of performing experiments in the future, you spend more time in analyzing and interpreting the data in the context of the experiment. What does this mean?

If you are looking at a plot of voltage versus resistance or current versus resistance or power versus resistance, what do you actually see? Does the curve make sense? Does it agree with what the theory tells you? Does it differ from expectation, and, if so, why? Clearly, practice with analyzing and interpreting data will enhance this skill.

Since most, if not all, the experiments you are required to do as a student involve little or no practice in designing the experiment, how can you develop and enhance this skill?

Actually, developing this skill under this constraint is not as difficult as it seems. What you need to do is to take the experiment and analyze it. Just break it down into its simplest parts, reconstruct it trying to understand why each element is there, and finally, determine what the author of the experiment is trying to teach you. Even though it may not always seem so, every experiment you do was designed by someone who was sincerely motivated to teach you something.

2.1 Introduction

Chapter 1 introduced basic concepts such as current, voltage, and power in an electric circuit. To actually determine the values of these variables in a given circuit requires that we understand some fundamental laws that govern electric circuits. These laws, known as Ohm's law and Kirchhoff's laws, form the foundation upon which electric circuit analysis is built.

In this chapter, in addition to these laws, we shall discuss some techniques commonly applied in circuit design and analysis. These techniques include combining resistors in series or parallel, voltage division, current division, and delta-to-wye and wye-to-delta transformations. The application of these laws and techniques will be restricted to resistive circuits in this chapter. We will finally apply the laws and techniques to real-life problems of electrical lighting and the design of dc meters.

2.2 Ohm's Law

Materials in general have a characteristic behavior of resisting the flow of electric charge. This physical property, or ability to resist current, is known as *resistance* and is represented by the symbol R. The resistance of any material with a uniform cross-sectional area A depends on A and its length ℓ, as shown in Fig. 2.1(a). We can represent resistance (as measured in the laboratory), in mathematical form,

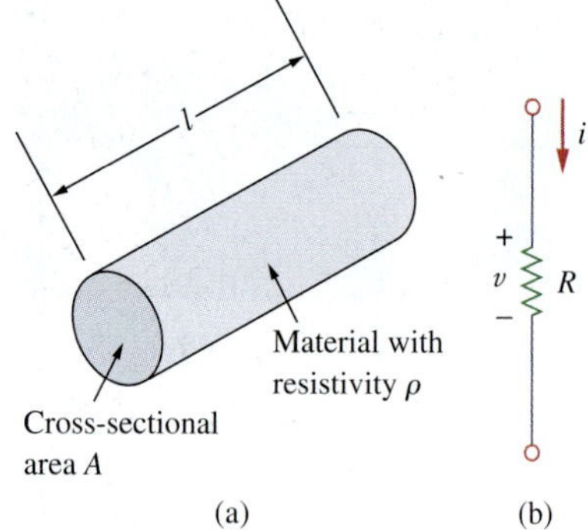

Figure 2.1
(a) Resistor, (b) Circuit symbol for resistance.

$$R = \rho\frac{\ell}{A} \tag{2.1}$$

where ρ is known as the *resistivity* of the material in ohm-meters. Good conductors, such as copper and aluminum, have low resistivities, while insulators, such as mica and paper, have high resistivities. Table 2.1 presents the values of ρ for some common materials and shows which materials are used for conductors, insulators, and semiconductors.

The circuit element used to model the current-resisting behavior of a material is the *resistor*. For the purpose of constructing circuits, resistors are usually made from metallic alloys and carbon compounds. The circuit

TABLE 2.1

Resistivities of common materials.

Material	Resistivity (Ω·m)	Usage
Silver	1.64×10^{-8}	Conductor
Copper	1.72×10^{-8}	Conductor
Aluminum	2.8×10^{-8}	Conductor
Gold	2.45×10^{-8}	Conductor
Carbon	4×10^{-5}	Semiconductor
Germanium	47×10^{-2}	Semiconductor
Silicon	6.4×10^{2}	Semiconductor
Paper	10^{10}	Insulator
Mica	5×10^{11}	Insulator
Glass	10^{12}	Insulator
Teflon	3×10^{12}	Insulator

symbol for the resistor is shown in Fig. 2.1(b), where R stands for the resistance of the resistor. The resistor is the simplest passive element.

Georg Simon Ohm (1787–1854), a German physicist, is credited with finding the relationship between current and voltage for a resistor. This relationship is known as *Ohm's law*.

> **Ohm's law** states that the voltage v across a resistor is directly proportional to the current i flowing through the resistor.

That is,

$$v \propto i \tag{2.2}$$

Ohm defined the constant of proportionality for a resistor to be the resistance, R. (The resistance is a material property which can change if the internal or external conditions of the element are altered, e.g., if there are changes in the temperature.) Thus, Eq. (2.2) becomes

$$\boxed{v = iR} \tag{2.3}$$

which is the mathematical form of Ohm's law. R in Eq. (2.3) is measured in the unit of ohms, designated Ω. Thus,

> The **resistance** R of an element denotes its ability to resist the flow of electric current; it is measured in ohms (Ω).

We may deduce from Eq. (2.3) that

$$R = \frac{v}{i} \tag{2.4}$$

so that

$$1\ \Omega = 1\ \text{V/A}$$

To apply Ohm's law as stated in Eq. (2.3), we must pay careful attention to the current direction and voltage polarity. The direction of current i and the polarity of voltage v must conform with the passive

Historical

Georg Simon Ohm (1787–1854), a German physicist, in 1826 experimentally determined the most basic law relating voltage and current for a resistor. Ohm's work was initially denied by critics.

Born of humble beginnings in Erlangen, Bavaria, Ohm threw himself into electrical research. His efforts resulted in his famous law. He was awarded the Copley Medal in 1841 by the Royal Society of London. In 1849, he was given the Professor of Physics chair by the University of Munich. To honor him, the unit of resistance was named the ohm.

sign convention, as shown in Fig. 2.1(b). This implies that current flows from a higher potential to a lower potential in order for $v = iR$. If current flows from a lower potential to a higher potential, $v = -iR$.

Since the value of R can range from zero to infinity, it is important that we consider the two extreme possible values of R. An element with $R = 0$ is called a *short circuit*, as shown in Fig. 2.2(a). For a short circuit,

$$v = iR = 0 \tag{2.5}$$

showing that the voltage is zero but the current could be anything. In practice, a short circuit is usually a connecting wire assumed to be a perfect conductor. Thus,

A **short circuit** is a circuit element with resistance approaching zero.

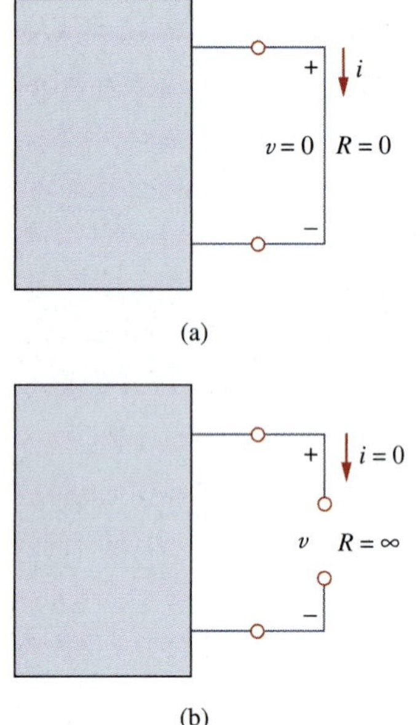

Figure 2.2
(a) Short circuit ($R = 0$), (b) Open circuit ($R = \infty$).

Similarly, an element with $R = \infty$ is known as an *open circuit*, as shown in Fig. 2.2(b). For an open circuit,

$$i = \lim_{R \to \infty} \frac{v}{R} = 0 \tag{2.6}$$

indicating that the current is zero though the voltage could be anything. Thus,

An **open circuit** is a circuit element with resistance approaching infinity.

A resistor is either fixed or variable. Most resistors are of the fixed type, meaning their resistance remains constant. The two common types of fixed resistors (wirewound and composition) are shown in Fig. 2.3. The composition resistors are used when large resistance is needed. The circuit symbol in Fig. 2.1(b) is for a fixed resistor. Variable resistors have adjustable resistance. The symbol for a variable resistor is shown in Fig. 2.4(a). A common variable resistor is known as a *potentiometer* or *pot* for short, with the symbol shown in Fig. 2.4(b). The pot is a three-terminal element with a sliding contact or wiper. By sliding the wiper, the resistances between the wiper terminal and the fixed terminals vary. Like fixed resistors, variable resistors can be of either wirewound or composition type, as shown in Fig. 2.5. Although resistors like those in Figs. 2.3 and 2.5 are used in circuit designs, today most

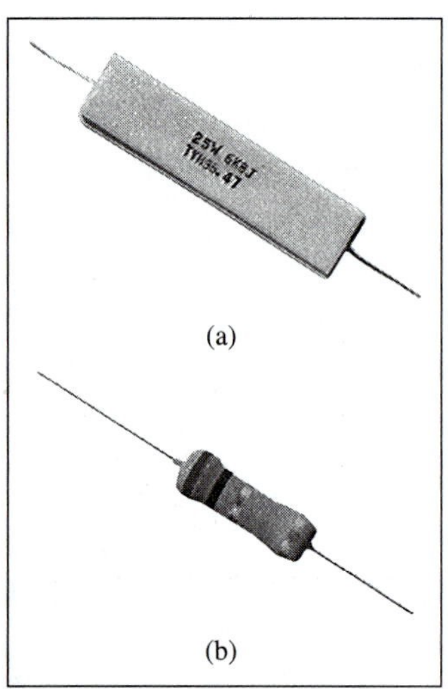

Figure 2.3
Fixed resistors: (a) wirewound type, (b) carbon film type.
Courtesy of Tech America.

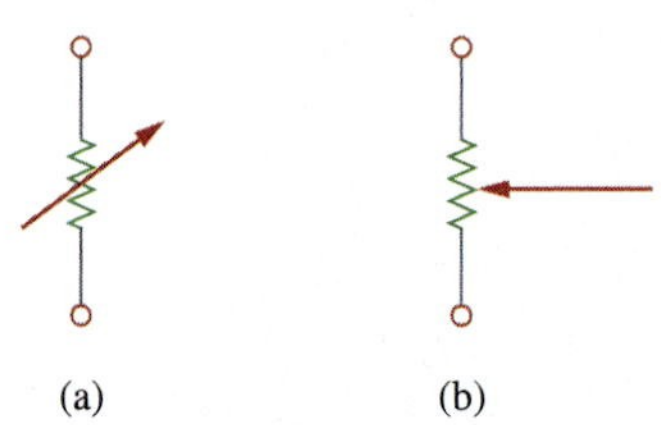

Figure 2.4
Circuit symbol for: (a) a variable resistor in general, (b) a potentiometer.

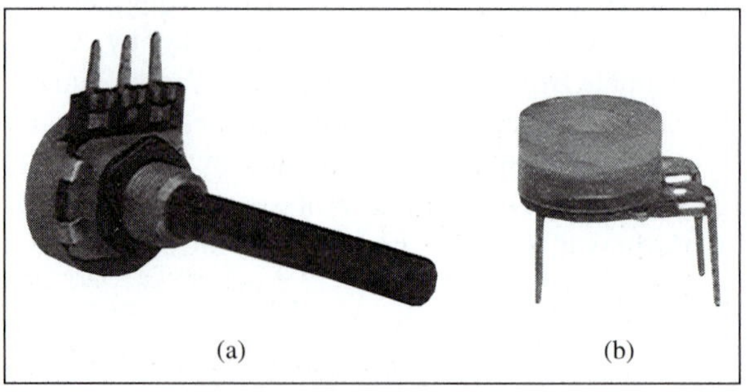

Figure 2.5
Variable resistors: (a) composition type, (b) slider pot.
Courtesy of Tech America.

circuit components including resistors are either surface mounted or integrated, as typically shown in Fig. 2.6.

It should be pointed out that not all resistors obey Ohm's law. A resistor that obeys Ohm's law is known as a *linear* resistor. It has a constant resistance and thus its current-voltage characteristic is as illustrated in Fig. 2.7(a): its i-v graph is a straight line passing through the origin. A *nonlinear* resistor does not obey Ohm's law. Its resistance varies with current and its i-v characteristic is typically shown in Fig. 2.7(b). Examples of devices with nonlinear resistance are the lightbulb and the diode. Although all practical resistors may exhibit nonlinear behavior under certain conditions, we will assume in this book that all elements actually designated as resistors are linear.

A useful quantity in circuit analysis is the reciprocal of resistance R, known as *conductance* and denoted by G:

$$G = \frac{1}{R} = \frac{i}{v} \tag{2.7}$$

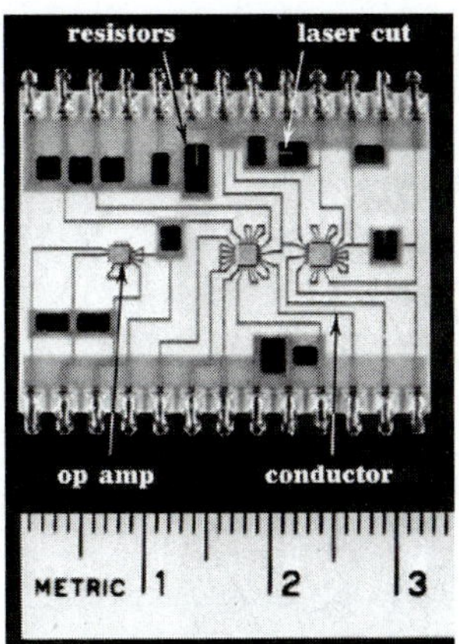

Figure 2.6
Resistors in a thick-film circuit.
G. Daryanani, *Principles of Active Network Synthesis and Design* (New York: John Wiley, 1976), p. 461c.

The conductance is a measure of how well an element will conduct electric current. The unit of conductance is the *mho* (ohm spelled backward) or reciprocal ohm, with symbol ℧, the inverted omega. Although engineers often use the mho, in this book we prefer to use the siemens (S), the SI unit of conductance:

$$1\text{ S} = 1\ ℧ = 1\text{ A/V} \tag{2.8}$$

Thus,

> **Conductance** is the ability of an element to conduct electric current; it is measured in mhos (℧) or siemens (S).

The same resistance can be expressed in ohms or siemens. For example, 10 Ω is the same as 0.1 S. From Eq. (2.7), we may write

$$i = Gv \tag{2.9}$$

The power dissipated by a resistor can be expressed in terms of R. Using Eqs. (1.7) and (2.3),

$$p = vi = i^2R = \frac{v^2}{R} \tag{2.10}$$

The power dissipated by a resistor may also be expressed in terms of G as

$$p = vi = v^2G = \frac{i^2}{G} \tag{2.11}$$

We should note two things from Eqs. (2.10) and (2.11):

1. The power dissipated in a resistor is a nonlinear function of either current or voltage.
2. Since R and G are positive quantities, the power dissipated in a resistor is always positive. Thus, a resistor always absorbs power from the circuit. This confirms the idea that a resistor is a passive element, incapable of generating energy.

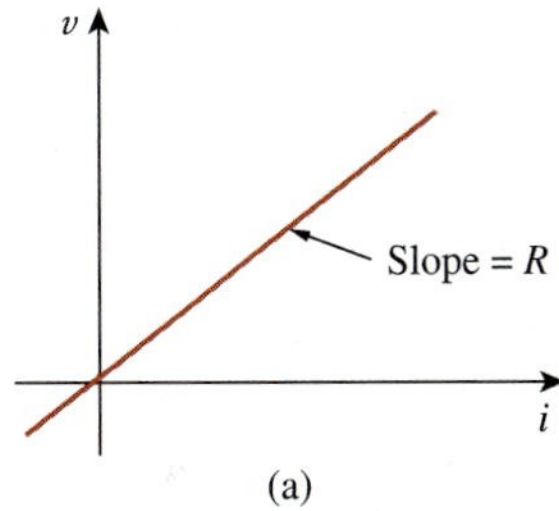

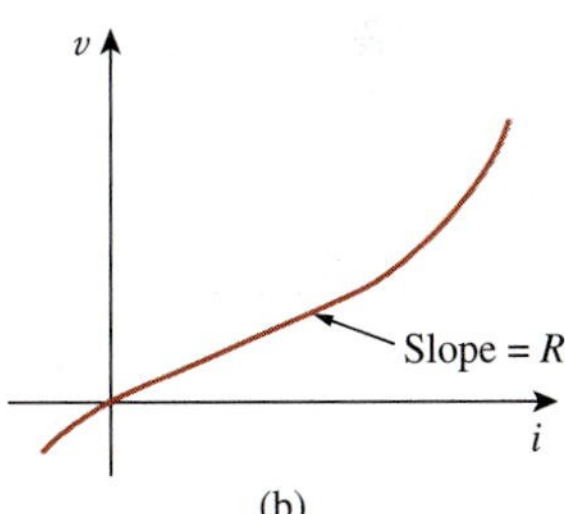

Figure 2.7
The i-v characteristic of: (a) a linear resistor, (b) a nonlinear resistor.

Example 2.1

An electric iron draws 2 A at 120 V. Find its resistance.

Solution:
From Ohm's law,

$$R = \frac{v}{i} = \frac{120}{2} = 60\ \Omega$$

Practice Problem 2.1

The essential component of a toaster is an electrical element (a resistor) that converts electrical energy to heat energy. How much current is drawn by a toaster with resistance 10 Ω at 110 V?

Answer: 11 A.

Example 2.2

In the circuit shown in Fig. 2.8, calculate the current i, the conductance G, and the power p.

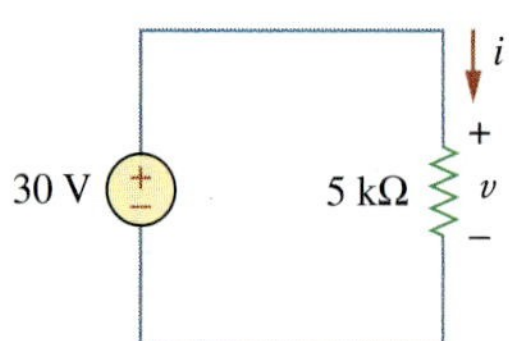

Figure 2.8
For Example 2.2.

Solution:
The voltage across the resistor is the same as the source voltage (30 V) because the resistor and the voltage source are connected to the same pair of terminals. Hence, the current is

$$i = \frac{v}{R} = \frac{30}{5 \times 10^3} = 6\text{ mA}$$

The conductance is

$$G = \frac{1}{R} = \frac{1}{5 \times 10^3} = 0.2\text{ mS}$$

We can calculate the power in various ways using either Eqs. (1.7), (2.10), or (2.11).

$$p = vi = 30(6 \times 10^{-3}) = 180\text{ mW}$$

or

$$p = i^2R = (6 \times 10^{-3})^2 5 \times 10^3 = 180\text{ mW}$$

or

$$p = v^2G = (30)^2 0.2 \times 10^{-3} = 180\text{ mW}$$

Practice Problem 2.2

For the circuit shown in Fig. 2.9, calculate the voltage v, the conductance G, and the power p.

Answer: 20 V, 100 μS, 40 mW.

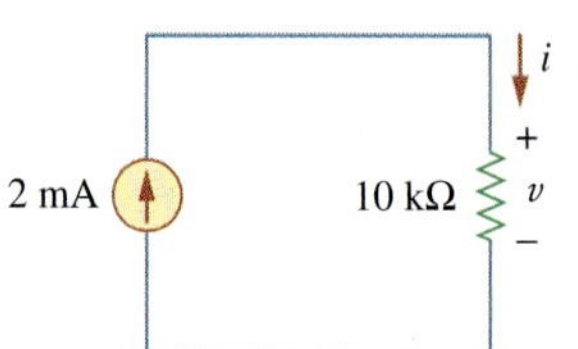

Figure 2.9
For Practice Prob. 2.2

Example 2.3

A voltage source of $20 \sin \pi t$ V is connected across a 5-kΩ resistor. Find the current through the resistor and the power dissipated.

Solution:

$$i = \frac{v}{R} = \frac{20 \sin \pi t}{5 \times 10^3} = 4 \sin \pi t \text{ mA}$$

Hence,

$$p = vi = 80 \sin^2 \pi t \text{ mW}$$

Practice Problem 2.3

A resistor absorbs an instantaneous power of $20 \cos^2 t$ mW when connected to a voltage source $v = 10 \cos t$ V. Find i and R.

Answer: $2 \cos t$ mA, 5 kΩ.

2.3 †Nodes, Branches, and Loops

Since the elements of an electric circuit can be interconnected in several ways, we need to understand some basic concepts of network topology. To differentiate between a circuit and a network, we may regard a network as an interconnection of elements or devices, whereas a circuit is a network providing one or more closed paths. The convention, when addressing network topology, is to use the word network rather than circuit. We do this even though the words network and circuit mean the same thing when used in this context. In network topology, we study the properties relating to the placement of elements in the network and the geometric configuration of the network. Such elements include branches, nodes, and loops.

> A **branch** represents a single element such as a voltage source or a resistor.

In other words, a branch represents any two-terminal element. The circuit in Fig. 2.10 has five branches, namely, the 10-V voltage source, the 2-A current source, and the three resistors.

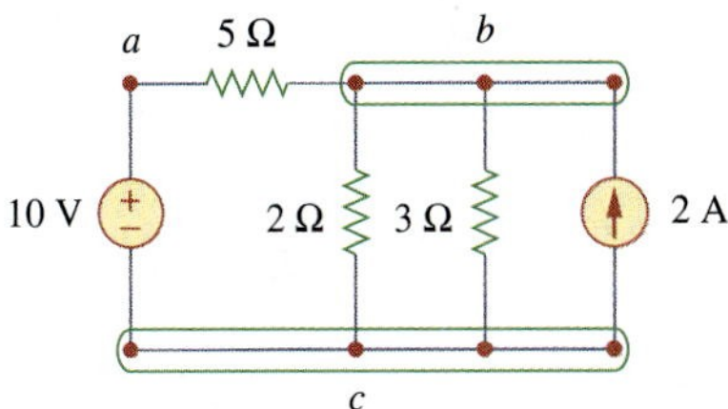

Figure 2.10
Nodes, branches, and loops.

> A **node** is the point of connection between two or more branches.

A node is usually indicated by a dot in a circuit. If a short circuit (a connecting wire) connects two nodes, the two nodes constitute a single node. The circuit in Fig. 2.10 has three nodes a, b, and c. Notice that the three points that form node b are connected by perfectly conducting wires and therefore constitute a single point. The same is true of the four points forming node c. We demonstrate that the circuit in Fig. 2.10 has only three nodes by redrawing the circuit in Fig. 2.11. The two circuits in Figs. 2.10 and 2.11 are identical. However, for the sake of clarity, nodes b and c are spread out with perfect conductors as in Fig. 2.10.

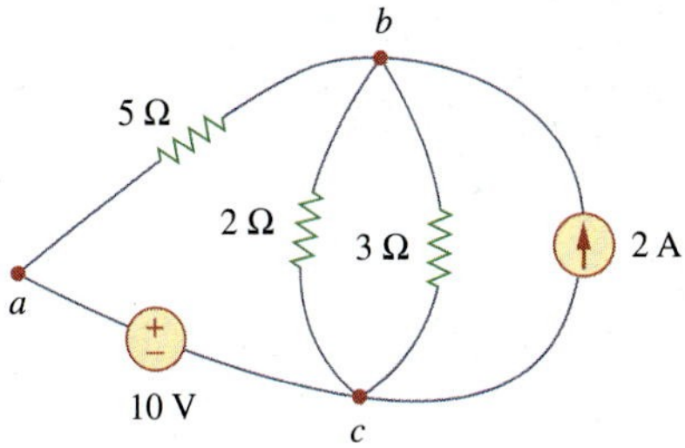

Figure 2.11
The three-node circuit of Fig. 2.10 is redrawn.

A **loop** is any closed path in a circuit.

A loop is a closed path formed by starting at a node, passing through a set of nodes, and returning to the starting node without passing through any node more than once. A loop is said to be *independent* if it contains at least one branch which is not a part of any other independent loop. Independent loops or paths result in independent sets of equations.

It is possible to form an independent set of loops where one of the loops does not contain such a branch. In Fig. 2.11, *abca* with the 2Ω resistor is independent. A second loop with the 3Ω resistor and the current source is independent. The third loop could be the one with the 2Ω resistor in parallel with the 3Ω resistor. This does form an independent set of loops.

A network with b branches, n nodes, and l independent loops will satisfy the fundamental theorem of network topology:

$$b = l + n - 1 \tag{2.12}$$

As the next two definitions show, circuit topology is of great value to the study of voltages and currents in an electric circuit.

Two or more elements are in **series** if they exclusively share a single node and consequently carry the same current.
Two or more elements are in **parallel** if they are connected to the same two nodes and consequently have the same voltage across them.

Elements are in series when they are chain-connected or connected sequentially, end to end. For example, two elements are in series if they share one common node and no other element is connected to that common node. Elements in parallel are connected to the same pair of terminals. Elements may be connected in a way that they are neither in series nor in parallel. In the circuit shown in Fig. 2.10, the voltage source and the 5-Ω resistor are in series because the same current will flow through them. The 2-Ω resistor, the 3-Ω resistor, and the current source are in parallel because they are connected to the same two nodes b and c and consequently have the same voltage across them. The 5-Ω and 2-Ω resistors are neither in series nor in parallel with each other.

Example 2.4

Determine the number of branches and nodes in the circuit shown in Fig. 2.12. Identify which elements are in series and which are in parallel.

Solution:

Since there are four elements in the circuit, the circuit has four branches: 10 V, 5 Ω, 6 Ω, and 2 A. The circuit has three nodes as identified in Fig. 2.13. The 5-Ω resistor is in series with the 10-V voltage source because the same current would flow in both. The 6-Ω resistor is in parallel with the 2-A current source because both are connected to the same nodes 2 and 3.

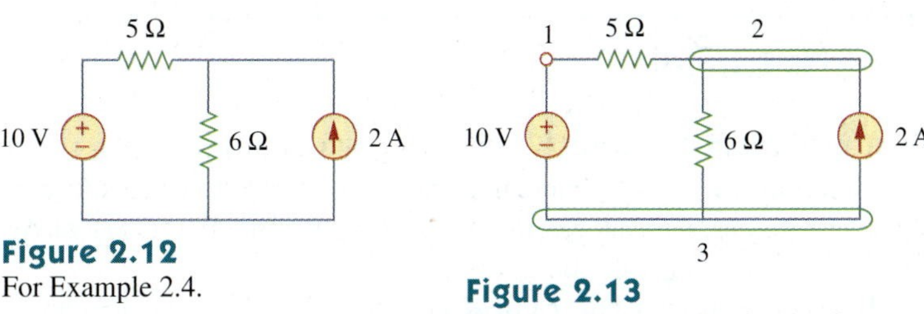

Figure 2.12
For Example 2.4.

Figure 2.13
The three nodes in the circuit of Fig. 2.12.

Practice Problem 2.4

How many branches and nodes does the circuit in Fig. 2.14 have? Identify the elements that are in series and in parallel.

Answer: Five branches and three nodes are identified in Fig. 2.15. The 1-Ω and 2-Ω resistors are in parallel. The 4-Ω resistor and 10-V source are also in parallel.

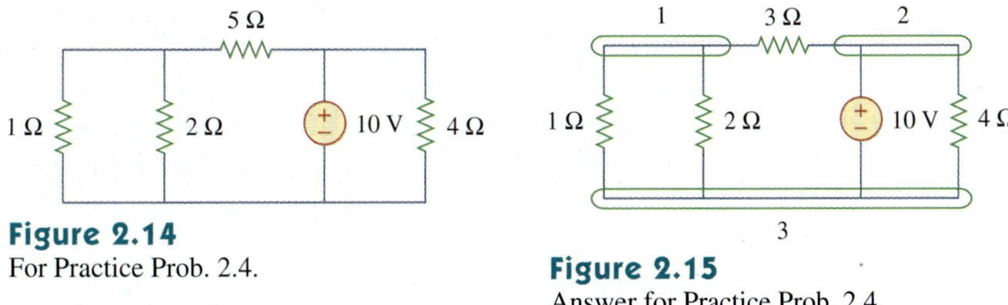

Figure 2.14
For Practice Prob. 2.4.

Figure 2.15
Answer for Practice Prob. 2.4.

2.4 Kirchhoff's Laws

Ohm's law by itself is not sufficient to analyze circuits. However, when it is coupled with Kirchhoff's two laws, we have a sufficient, powerful set of tools for analyzing a large variety of electric circuits. Kirchhoff's laws were first introduced in 1847 by the German physicist Gustav Robert Kirchhoff (1824–1887). These laws are formally known as Kirchhoff's current law (KCL) and Kirchhoff's voltage law (KVL).

Kirchhoff's first law is based on the law of conservation of charge, which requires that the algebraic sum of charges within a system cannot change.

> **Kirchhoff's current law (KCL)** states that the algebraic sum of currents entering a node (or a closed boundary) is zero.

Mathematically, KCL implies that

$$\sum_{n=1}^{N} i_n = 0 \tag{2.13}$$

where N is the number of branches connected to the node and i_n is the nth current entering (or leaving) the node. By this law, currents

Historical

Gustav Robert Kirchhoff (1824–1887), a German physicist, stated two basic laws in 1847 concerning the relationship between the currents and voltages in an electrical network. Kirchhoff's laws, along with Ohm's law, form the basis of circuit theory.

Born the son of a lawyer in Konigsberg, East Prussia, Kirchhoff entered the University of Konigsberg at age 18 and later became a lecturer in Berlin. His collaborative work in spectroscopy with German chemist Robert Bunsen led to the discovery of cesium in 1860 and rubidium in 1861. Kirchhoff was also credited with the Kirchhoff law of radiation. Thus Kirchhoff is famous among engineers, chemists, and physicists.

entering a node may be regarded as positive, while currents leaving the node may be taken as negative or vice versa.

To prove KCL, assume a set of currents $i_k(t)$, $k = 1, 2, \ldots$, flow into a node. The algebraic sum of currents at the node is

$$i_T(t) = i_1(t) + i_2(t) + i_3(t) + \cdots \tag{2.14}$$

Integrating both sides of Eq. (2.14) gives

$$q_T(t) = q_1(t) + q_2(t) + q_3(t) + \cdots \tag{2.15}$$

where $q_k(t) = \int i_k(t)dt$ and $q_T(t) = \int i_T(t)dt$. But the law of conservation of electric charge requires that the algebraic sum of electric charges at the node must not change; that is, the node stores no net charge. Thus $q_T(t) = 0 \rightarrow i_T(t) = 0$, confirming the validity of KCL.

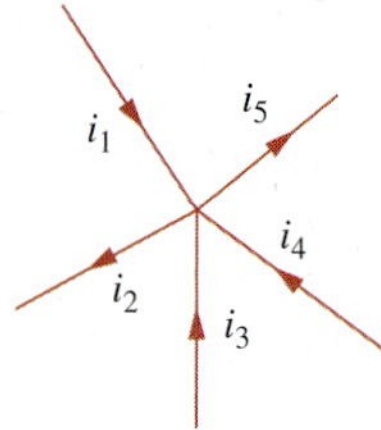

Figure 2.16
Currents at a node illustrating KCL.

Consider the node in Fig. 2.16. Applying KCL gives

$$i_1 + (-i_2) + i_3 + i_4 + (-i_5) = 0 \tag{2.16}$$

since currents i_1, i_3, and i_4 are entering the node, while currents i_2 and i_5 are leaving it. By rearranging the terms, we get

$$i_1 + i_3 + i_4 = i_2 + i_5 \tag{2.17}$$

Equation (2.17) is an alternative form of KCL:

> The sum of the currents entering a node is equal to the sum of the currents leaving the node.

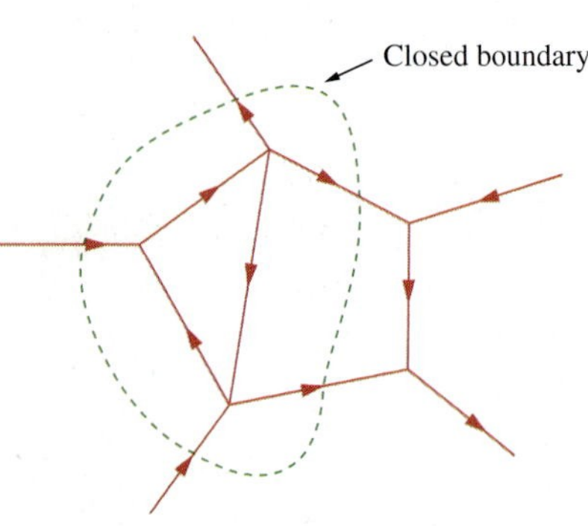

Figure 2.17
Applying KCL to a closed boundary.

Note that KCL also applies to a closed boundary. This may be regarded as a generalized case, because a node may be regarded as a closed surface shrunk to a point. In two dimensions, a closed boundary is the same as a closed path. As typically illustrated in the circuit of Fig. 2.17, the total current entering the closed surface is equal to the total current leaving the surface.

> Two sources (or circuits in general) are said to be equivalent if they have the same *i-v* relationship at a pair of terminals.

A simple application of KCL is combining current sources in parallel. The combined current is the algebraic sum of the current supplied by the individual sources. For example, the current sources shown in

Fig. 2.18(a) can be combined as in Fig. 2.18(b). The combined or equivalent current source can be found by applying KCL to node a.

$$I_T + I_2 = I_1 + I_3$$

or

$$I_T = I_1 - I_2 + I_3 \tag{2.18}$$

A circuit cannot contain two different currents, I_1 and I_2, in series, unless $I_1 = I_2$; otherwise KCL will be violated.

Kirchhoff's second law is based on the principle of conservation of energy:

Kirchhoff's voltage law (KVL) states that the algebraic sum of all voltages around a closed path (or loop) is zero.

Expressed mathematically, KVL states that

$$\sum_{m=1}^{M} v_m = 0 \tag{2.19}$$

where M is the number of voltages in the loop (or the number of branches in the loop) and v_m is the mth voltage.

To illustrate KVL, consider the circuit in Fig. 2.19. The sign on each voltage is the polarity of the terminal encountered first as we travel around the loop. We can start with any branch and go around the loop either clockwise or counterclockwise. Suppose we start with the voltage source and go clockwise around the loop as shown; then voltages would be $-v_1$, $+v_2$, $+v_3$, $-v_4$, and $+v_5$, in that order. For example, as we reach branch 3, the positive terminal is met first; hence, we have $+v_3$. For branch 4, we reach the negative terminal first; hence, $-v_4$. Thus, KVL yields

$$-v_1 + v_2 + v_3 - v_4 + v_5 = 0 \tag{2.20}$$

Rearranging terms gives

$$v_2 + v_3 + v_5 = v_1 + v_4 \tag{2.21}$$

which may be interpreted as

$$\text{Sum of voltage drops} = \text{Sum of voltage rises} \tag{2.22}$$

This is an alternative form of KVL. Notice that if we had traveled counterclockwise, the result would have been $+v_1$, $-v_5$, $+v_4$, $-v_3$, and $-v_2$, which is the same as before except that the signs are reversed. Hence, Eqs. (2.20) and (2.21) remain the same.

When voltage sources are connected in series, KVL can be applied to obtain the total voltage. The combined voltage is the algebraic sum of the voltages of the individual sources. For example, for the voltage sources shown in Fig. 2.20(a), the combined or equivalent voltage source in Fig. 2.20(b) is obtained by applying KVL.

$$-V_{ab} + V_1 + V_2 - V_3 = 0$$

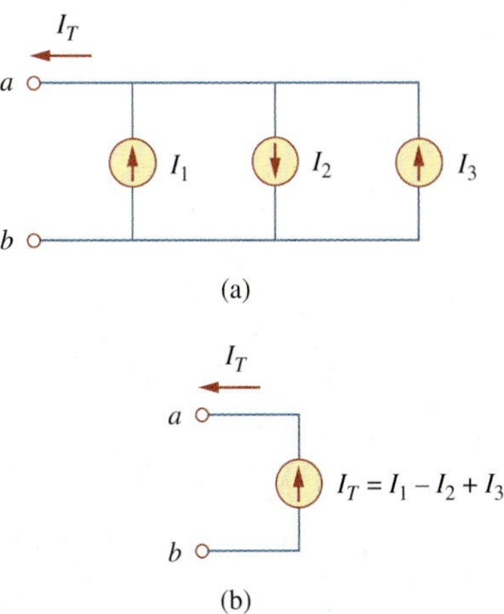

Figure 2.18
Current sources in parallel: (a) original circuit, (b) equivalent circuit.

KVL can be applied in two ways: by taking either a clockwise or a counterclockwise trip around the loop. Either way, the algebraic sum of voltages around the loop is zero.

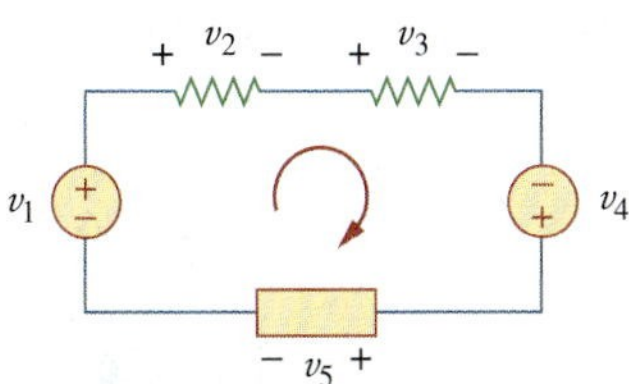

Figure 2.19
A single-loop circuit illustrating KVL.

or

$$V_{ab} = V_1 + V_2 - V_3 \tag{2.23}$$

To avoid violating KVL, a circuit cannot contain two different voltages V_1 and V_2 in parallel unless $V_1 = V_2$.

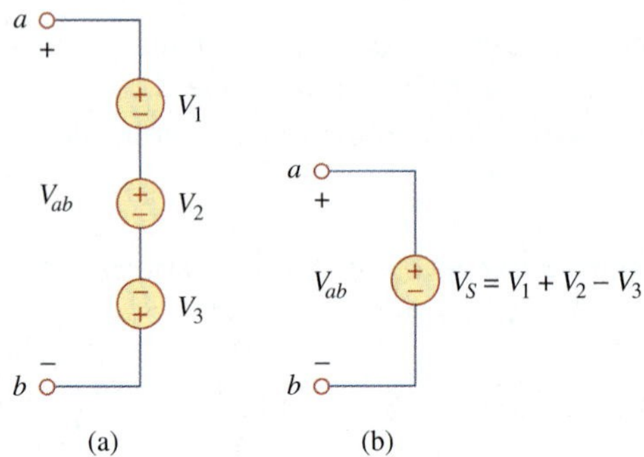

Figure 2.20
Voltage sources in series: (a) original circuit, (b) equivalent circuit.

Example 2.5

For the circuit in Fig. 2.21(a), find voltages v_1 and v_2.

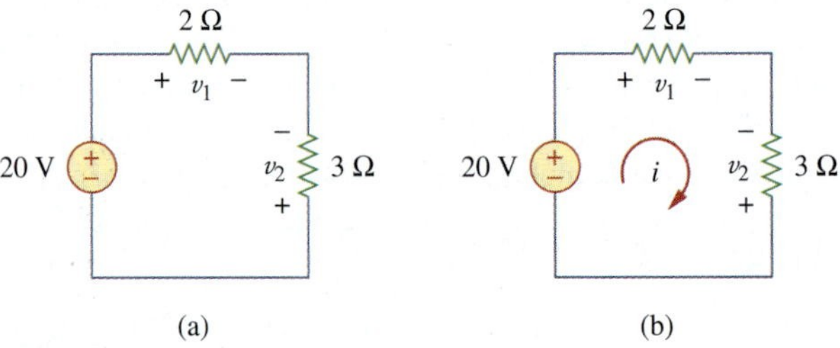

Figure 2.21
For Example 2.5.

Solution:
To find v_1 and v_2, we apply Ohm's law and Kirchhoff's voltage law. Assume that current i flows through the loop as shown in Fig. 2.21(b). From Ohm's law,

$$v_1 = 2i, \qquad v_2 = -3i \tag{2.5.1}$$

Applying KVL around the loop gives

$$-20 + v_1 - v_2 = 0 \tag{2.5.2}$$

Substituting Eq. (2.5.1) into Eq. (2.5.2), we obtain

$$-20 + 2i + 3i = 0 \quad \text{or} \quad 5i = 20 \quad \Rightarrow \quad i = 4 \text{ A}$$

Substituting i in Eq. (2.5.1) finally gives

$$v_1 = 8 \text{ V}, \qquad v_2 = -12 \text{ V}$$

Practice Problem 2.5

Find v_1 and v_2 in the circuit of Fig. 2.22.

Answer: 12 V, −6 V.

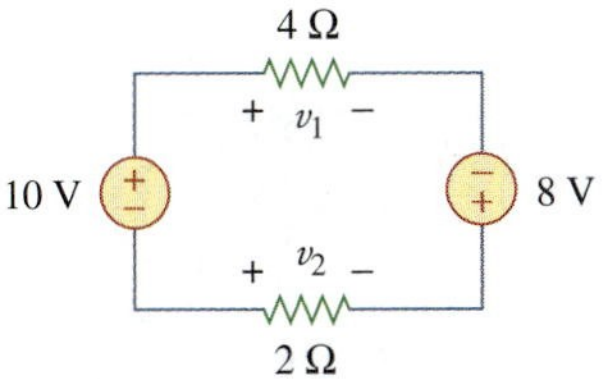

Figure 2.22
For Practice Prob. 2.5.

Example 2.6

Determine v_o and i in the circuit shown in Fig. 2.23(a).

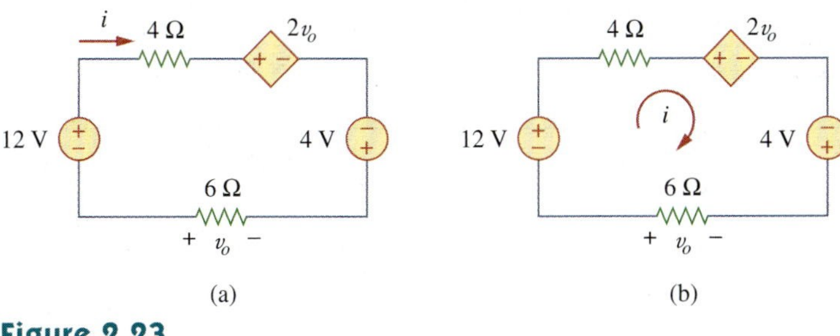

Figure 2.23
For Example 2.6.

Solution:
We apply KVL around the loop as shown in Fig. 2.23(b). The result is

$$-12 + 4i + 2v_o - 4 + 6i = 0 \tag{2.6.1}$$

Applying Ohm's law to the 6-Ω resistor gives

$$v_o = -6i \tag{2.6.2}$$

Substituting Eq. (2.6.2) into Eq. (2.6.1) yields

$$-16 + 10i - 12i = 0 \quad \Rightarrow \quad i = -8 \text{ A}$$

and $v_o = 48$ V.

Practice Problem 2.6

Find v_x and v_o in the circuit of Fig. 2.24.

Answer: 10 V, −5 V.

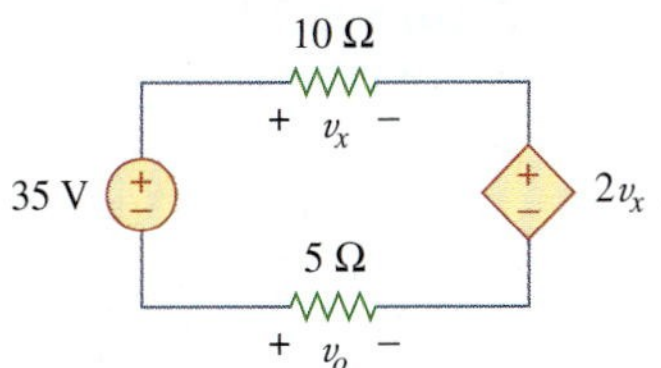

Figure 2.24
For Practice Prob. 2.6.

Example 2.7

Find current i_o and voltage v_o in the circuit shown in Fig. 2.25.

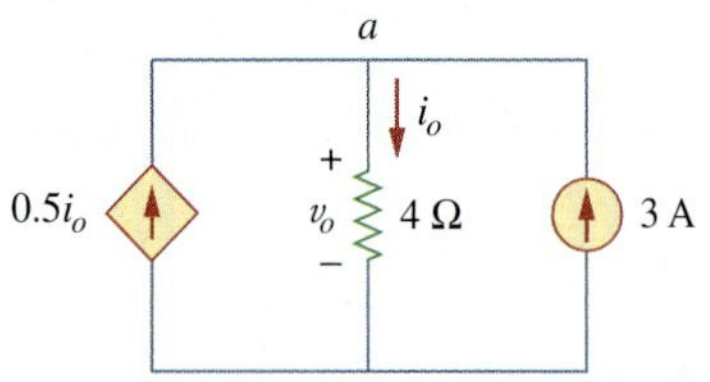

Figure 2.25
For Example 2.7.

Solution:

Applying KCL to node a, we obtain

$$3 + 0.5i_o = i_o \quad \Rightarrow \quad i_o = 6 \text{ A}$$

For the 4-Ω resistor, Ohm's law gives

$$v_o = 4i_o = 24 \text{ V}$$

Practice Problem 2.7

Find v_o and i_o in the circuit of Fig. 2.26.

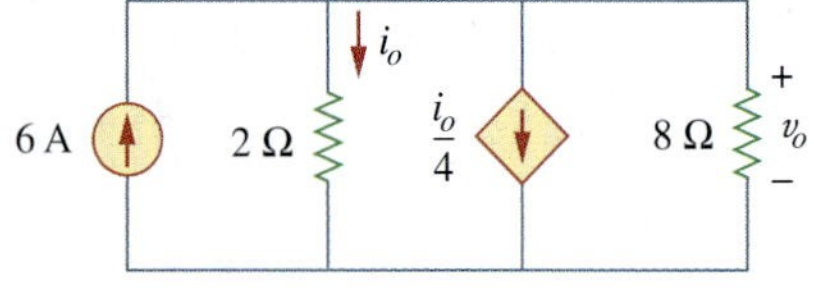

Figure 2.26
For Practice Prob. 2.7.

Answer: 8 V, 4 A.

Example 2.8

Find currents and voltages in the circuit shown in Fig. 2.27(a).

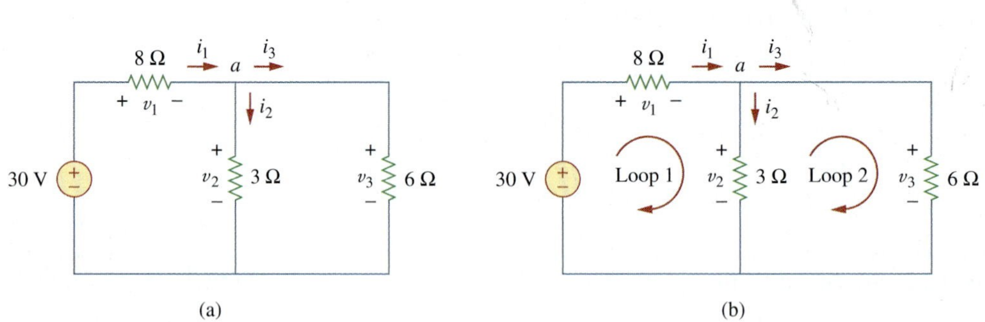

Figure 2.27
For Example 2.8.

Solution:

We apply Ohm's law and Kirchhoff's laws. By Ohm's law,

$$v_1 = 8i_1, \qquad v_2 = 3i_2, \qquad v_3 = 6i_3 \tag{2.8.1}$$

Since the voltage and current of each resistor are related by Ohm's law as shown, we are really looking for three things: (v_1, v_2, v_3) or (i_1, i_2, i_3). At node a, KCL gives

$$i_1 - i_2 - i_3 = 0 \tag{2.8.2}$$

Applying KVL to loop 1 as in Fig. 2.27(b),

$$-30 + v_1 + v_2 = 0$$

We express this in terms of i_1 and i_2 as in Eq. (2.8.1) to obtain

$$-30 + 8i_1 + 3i_2 = 0$$

or

$$i_1 = \frac{(30 - 3i_2)}{8} \tag{2.8.3}$$

Applying KVL to loop 2,

$$-v_2 + v_3 = 0 \qquad \Rightarrow \qquad v_3 = v_2 \tag{2.8.4}$$

as expected since the two resistors are in parallel. We express v_1 and v_2 in terms of i_1 and i_2 as in Eq. (2.8.1). Equation (2.8.4) becomes

$$6i_3 = 3i_2 \qquad \Rightarrow \qquad i_3 = \frac{i_2}{2} \tag{2.8.5}$$

Substituting Eqs. (2.8.3) and (2.8.5) into (2.8.2) gives

$$\frac{30 - 3i_2}{8} - i_2 - \frac{i_2}{2} = 0$$

or $i_2 = 2$ A. From the value of i_2, we now use Eqs. (2.8.1) to (2.8.5) to obtain

$$i_1 = 3\text{ A}, \qquad i_3 = 1\text{ A}, \qquad v_1 = 24\text{ V}, \qquad v_2 = 6\text{ V}, \qquad v_3 = 6\text{ V}$$

Practice Problem 2.8

Find the currents and voltages in the circuit shown in Fig. 2.28.

Answer: $v_1 = 3$ V, $v_2 = 2$ V, $v_3 = 5$ V, $i_1 = 1.5$ A, $i_2 = 0.25$ A, $i_3 = 1.25$ A.

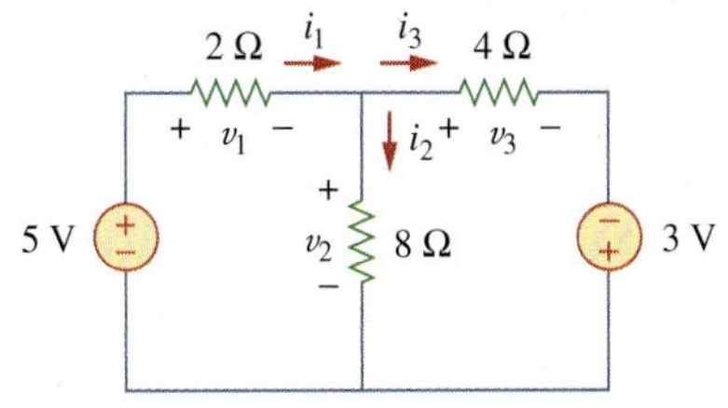

Figure 2.28
For Practice Prob. 2.8.

2.5 Series Resistors and Voltage Division

The need to combine resistors in series or in parallel occurs so frequently that it warrants special attention. The process of combining the resistors is facilitated by combining two of them at a time. With this in mind, consider the single-loop circuit of Fig. 2.29. The two resistors

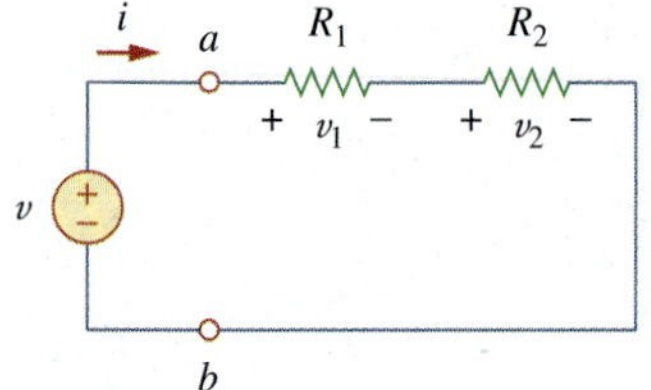

Figure 2.29
A single-loop circuit with two resistors in series.

are in series, since the same current i flows in both of them. Applying Ohm's law to each of the resistors, we obtain

$$v_1 = iR_1, \qquad v_2 = iR_2 \tag{2.24}$$

If we apply KVL to the loop (moving in the clockwise direction), we have

$$-v + v_1 + v_2 = 0 \tag{2.25}$$

Combining Eqs. (2.24) and (2.25), we get

$$v = v_1 + v_2 = i(R_1 + R_2) \tag{2.26}$$

or

$$i = \frac{v}{R_1 + R_2} \tag{2.27}$$

Notice that Eq. (2.26) can be written as

$$v = iR_{eq} \tag{2.28}$$

implying that the two resistors can be replaced by an equivalent resistor R_{eq}; that is,

$$R_{eq} = R_1 + R_2 \tag{2.29}$$

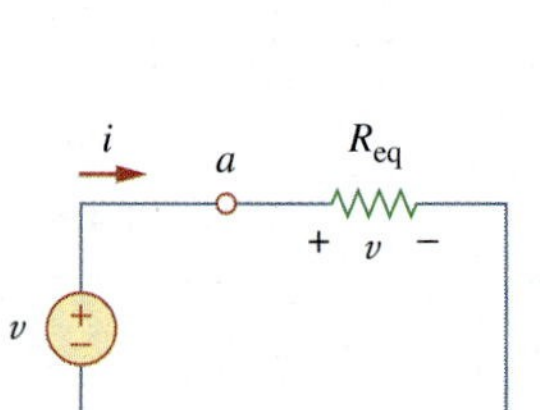

Figure 2.30
Equivalent circuit of the Fig. 2.29 circuit.

Thus, Fig. 2.29 can be replaced by the equivalent circuit in Fig. 2.30. The two circuits in Figs. 2.29 and 2.30 are equivalent because they exhibit the same voltage-current relationships at the terminals *a-b*. An equivalent circuit such as the one in Fig. 2.30 is useful in simplifying the analysis of a circuit. In general,

> The **equivalent resistance** of any number of resistors connected in series is the sum of the individual resistances.

Resistors in series behave as a single resistor whose resistance is equal to the sum of the resistances of the individual resistors.

For N resistors in series then,

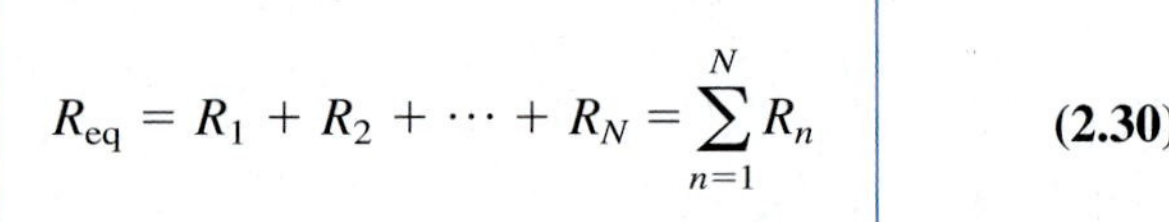

$$R_{eq} = R_1 + R_2 + \cdots + R_N = \sum_{n=1}^{N} R_n \tag{2.30}$$

To determine the voltage across each resistor in Fig. 2.29, we substitute Eq. (2.26) into Eq. (2.24) and obtain

$$v_1 = \frac{R_1}{R_1 + R_2} v, \qquad v_2 = \frac{R_2}{R_1 + R_2} v \tag{2.31}$$

Notice that the source voltage v is divided among the resistors in direct proportion to their resistances; the larger the resistance, the larger the voltage drop. This is called the *principle of voltage division*, and the circuit in Fig. 2.29 is called a *voltage divider*. In general, if a voltage divider has N resistors ($R_1, R_2, \ldots, R_N$) in series with the source voltage v, the nth resistor (R_n) will have a voltage drop of

$$v_n = \frac{R_n}{R_1 + R_2 + \cdots + R_N} v \tag{2.32}$$

2.6 Parallel Resistors and Current Division

Consider the circuit in Fig. 2.31, where two resistors are connected in parallel and therefore have the same voltage across them. From Ohm's law,

$$v = i_1 R_1 = i_2 R_2$$

or

$$i_1 = \frac{v}{R_1}, \qquad i_2 = \frac{v}{R_2} \tag{2.33}$$

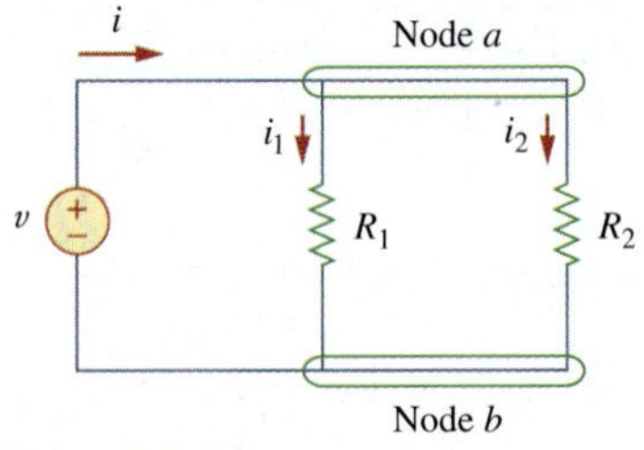

Figure 2.31
Two resistors in parallel.

Applying KCL at node a gives the total current i as

$$i = i_1 + i_2 \tag{2.34}$$

Substituting Eq. (2.33) into Eq. (2.34), we get

$$i = \frac{v}{R_1} + \frac{v}{R_2} = v\left(\frac{1}{R_1} + \frac{1}{R_2}\right) = \frac{v}{R_{eq}} \tag{2.35}$$

where R_{eq} is the equivalent resistance of the resistors in parallel:

$$\frac{1}{R_{eq}} = \frac{1}{R_1} + \frac{1}{R_2} \tag{2.36}$$

or

$$\frac{1}{R_{eq}} = \frac{R_1 + R_2}{R_1 R_2}$$

or

$$\boxed{R_{eq} = \frac{R_1 R_2}{R_1 + R_2}} \tag{2.37}$$

Thus,

> The **equivalent resistance** of two parallel resistors is equal to the product of their resistances divided by their sum.

It must be emphasized that this applies only to two resistors in parallel. From Eq. (2.37), if $R_1 = R_2$, then $R_{eq} = R_1/2$.

We can extend the result in Eq. (2.36) to the general case of a circuit with N resistors in parallel. The equivalent resistance is

$$\boxed{\frac{1}{R_{eq}} = \frac{1}{R_1} + \frac{1}{R_2} + \cdots + \frac{1}{R_N}} \tag{2.38}$$

Note that R_{eq} is always smaller than the resistance of the smallest resistor in the parallel combination. If $R_1 = R_2 = \cdots = R_N = R$, then

$$R_{eq} = \frac{R}{N} \tag{2.39}$$

For example, if four 100-Ω resistors are connected in parallel, their equivalent resistance is 25 Ω.

Conductances in parallel behave as a single conductance whose value is equal to the sum of the individual conductances.

It is often more convenient to use conductance rather than resistance when dealing with resistors in parallel. From Eq. (2.38), the equivalent conductance for N resistors in parallel is

$$G_{eq} = G_1 + G_2 + G_3 + \cdots + G_N \tag{2.40}$$

where $G_{eq} = 1/R_{eq}$, $G_1 = 1/R_1$, $G_2 = 1/R_2$, $G_3 = 1/R_3, \ldots, G_N = 1/R_N$. Equation (2.40) states:

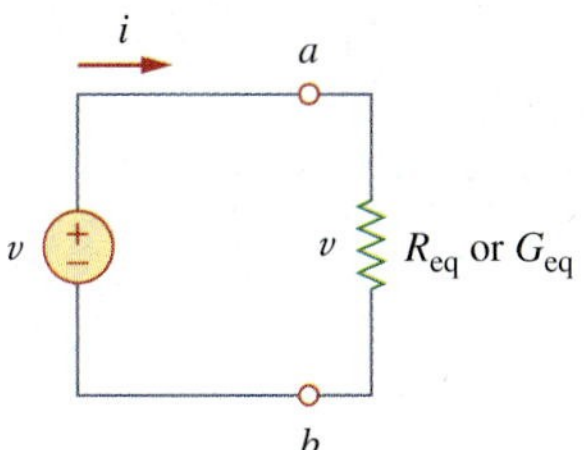

Figure 2.32
Equivalent circuit to Fig. 2.31.

> The **equivalent conductance** of resistors connected in parallel is the sum of their individual conductances.

This means that we may replace the circuit in Fig. 2.31 with that in Fig. 2.32. Notice the similarity between Eqs. (2.30) and (2.40). The equivalent conductance of parallel resistors is obtained the same way as the equivalent resistance of series resistors. In the same manner, the equivalent conductance of resistors in series is obtained just the same way as the resistance of resistors in parallel. Thus the equivalent conductance G_{eq} of N resistors in series (such as shown in Fig. 2.29) is

$$\frac{1}{G_{eq}} = \frac{1}{G_1} + \frac{1}{G_2} + \frac{1}{G_3} + \cdots + \frac{1}{G_N} \tag{2.41}$$

Given the total current i entering node a in Fig. 2.31, how do we obtain current i_1 and i_2? We know that the equivalent resistor has the same voltage, or

$$v = iR_{eq} = \frac{iR_1R_2}{R_1 + R_2} \tag{2.42}$$

Combining Eqs. (2.33) and (2.42) results in

$$i_1 = \frac{R_2\, i}{R_1 + R_2}, \qquad i_2 = \frac{R_1\, i}{R_1 + R_2} \tag{2.43}$$

which shows that the total current i is shared by the resistors in inverse proportion to their resistances. This is known as the *principle of current division*, and the circuit in Fig. 2.31 is known as a *current divider*. Notice that the larger current flows through the smaller resistance.

As an extreme case, suppose one of the resistors in Fig. 2.31 is zero, say $R_2 = 0$; that is, R_2 is a short circuit, as shown in Fig. 2.33(a). From Eq. (2.43), $R_2 = 0$ implies that $i_1 = 0$, $i_2 = i$. This means that the entire current i bypasses R_1 and flows through the short circuit $R_2 = 0$, the path of least resistance. Thus when a circuit

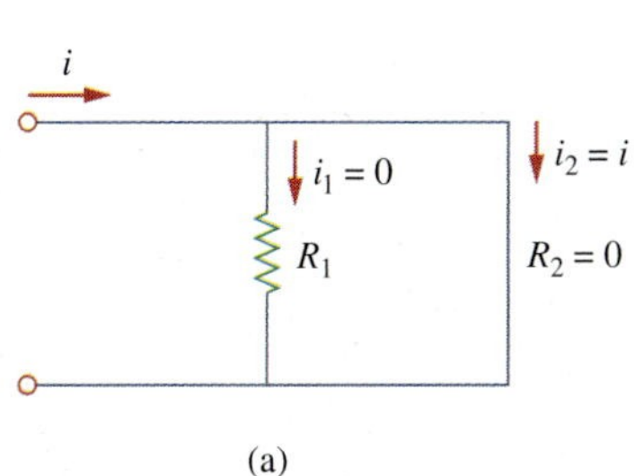

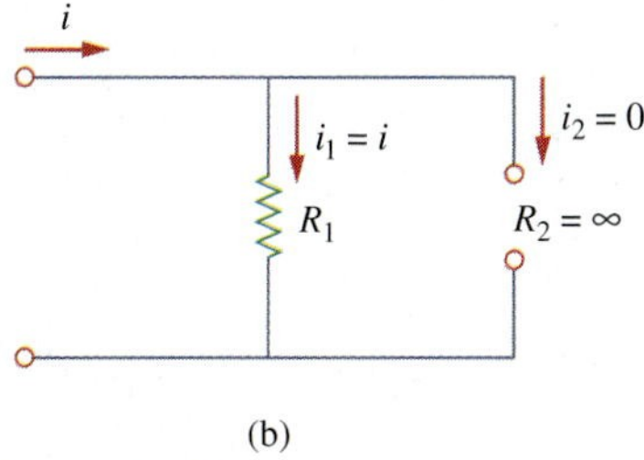

Figure 2.33
(a) A shorted circuit, (b) an open circuit.

is short circuited, as shown in Fig. 2.33(a), two things should be kept in mind:

1. The equivalent resistance $R_{eq} = 0$. [See what happens when $R_2 = 0$ in Eq. (2.37).]
2. The entire current flows through the short circuit.

As another extreme case, suppose $R_2 = \infty$, that is, R_2 is an open circuit, as shown in Fig. 2.33(b). The current still flows through the path of least resistance, R_1. By taking the limit of Eq. (2.37) as $R_2 \to \infty$, we obtain $R_{eq} = R_1$ in this case.

If we divide both the numerator and denominator by R_1R_2, Eq. (2.43) becomes

$$i_1 = \frac{G_1}{G_1 + G_2} i \tag{2.44a}$$

$$i_2 = \frac{G_2}{G_1 + G_2} i \tag{2.44b}$$

Thus, in general, if a current divider has N conductors ($G_1, G_2, \dots, G_N$) in parallel with the source current i, the nth conductor (G_n) will have current

$$i_n = \frac{G_n}{G_1 + G_2 + \cdots + G_N} i \tag{2.45}$$

In general, it is often convenient and possible to combine resistors in series and parallel and reduce a resistive network to a single *equivalent resistance* R_{eq}. Such an equivalent resistance is the resistance between the designated terminals of the network and must exhibit the same *i-v* characteristics as the original network at the terminals.

Example 2.9

Find R_{eq} for the circuit shown in Fig. 2.34.

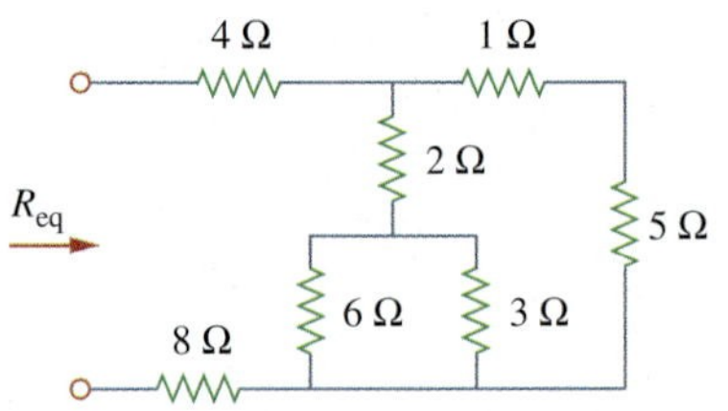

Figure 2.34
For Example 2.9.

Solution:

To get R_{eq}, we combine resistors in series and in parallel. The 6-Ω and 3-Ω resistors are in parallel, so their equivalent resistance is

$$6\ \Omega \parallel 3\Omega = \frac{6 \times 3}{6 + 3} = 2\ \Omega$$

(The symbol $\parallel$ is used to indicate a parallel combination.) Also, the 1-Ω and 5-Ω resistors are in series; hence their equivalent resistance is

$$1\ \Omega + 5\ \Omega = 6\ \Omega$$

Thus the circuit in Fig. 2.34 is reduced to that in Fig. 2.35(a). In Fig. 2.35(a), we notice that the two 2-Ω resistors are in series, so the equivalent resistance is

$$2\ \Omega + 2\ \Omega = 4\ \Omega$$

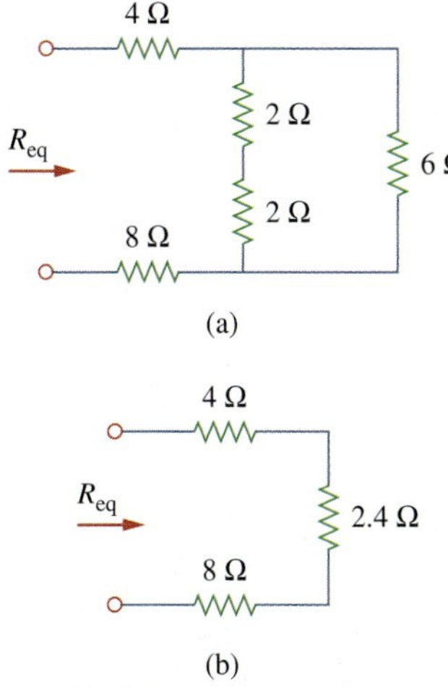

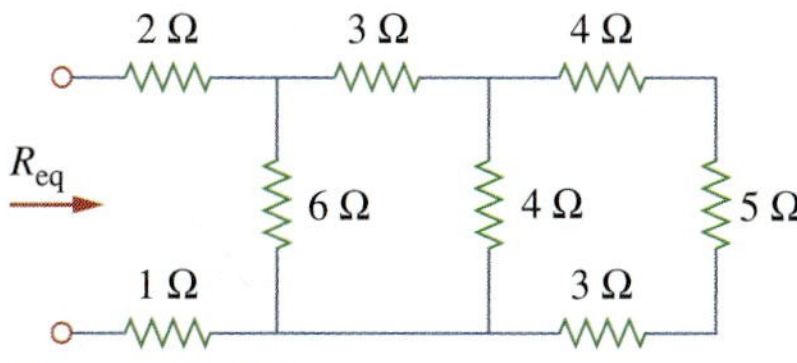

Figure 2.35
Equivalent circuits for Example 2.9.

This 4-Ω resistor is now in parallel with the 6-Ω resistor in Fig. 2.35(a); their equivalent resistance is

$$4\ \Omega \parallel 6\ \Omega = \frac{4 \times 6}{4 + 6} = 2.4\ \Omega$$

The circuit in Fig. 2.35(a) is now replaced with that in Fig. 2.35(b). In Fig. 2.35(b), the three resistors are in series. Hence, the equivalent resistance for the circuit is

$$R_{eq} = 4\ \Omega + 2.4\ \Omega + 8\ \Omega = 14.4\ \Omega$$

Practice Problem 2.9

By combining the resistors in Fig. 2.36, find R_{eq}.

Answer: 6 Ω.

Figure 2.36
For Practice Prob. 2.9.

Example 2.10

Calculate the equivalent resistance R_{ab} in the circuit in Fig. 2.37.

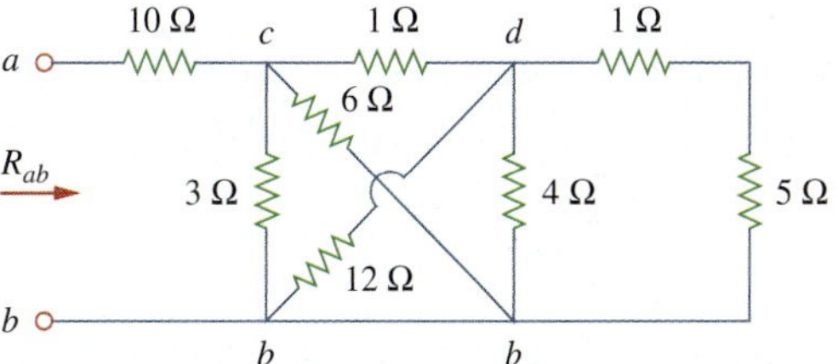

Figure 2.37
For Example 2.10.

Solution:

The 3-Ω and 6-Ω resistors are in parallel because they are connected to the same two nodes c and b. Their combined resistance is

$$3\ \Omega \parallel 6\ \Omega = \frac{3 \times 6}{3 + 6} = 2\ \Omega \qquad \textbf{(2.10.1)}$$

Similarly, the 12-Ω and 4-Ω resistors are in parallel since they are connected to the same two nodes d and b. Hence

$$12\ \Omega \parallel 4\ \Omega = \frac{12 \times 4}{12 + 4} = 3\ \Omega \qquad \textbf{(2.10.2)}$$

Also the 1-Ω and 5-Ω resistors are in series; hence, their equivalent resistance is

$$1\ \Omega + 5\ \Omega = 6\ \Omega \qquad \textbf{(2.10.3)}$$

With these three combinations, we can replace the circuit in Fig. 2.37 with that in Fig. 2.38(a). In Fig. 2.38(a), 3-Ω in parallel with 6-Ω gives 2-Ω, as calculated in Eq. (2.10.1). This 2-Ω equivalent resistance is now in series with the 1-Ω resistance to give a combined resistance of $1\ \Omega + 2\ \Omega = 3\ \Omega$. Thus, we replace the circuit in Fig. 2.38(a) with that in Fig. 2.38(b). In Fig. 2.38(b), we combine the 2-Ω and 3-Ω resistors in parallel to get

$$2\ \Omega \parallel 3\ \Omega = \frac{2 \times 3}{2 + 3} = 1.2\ \Omega$$

This 1.2-Ω resistor is in series with the 10-Ω resistor, so that

$$R_{ab} = 10 + 1.2 = 11.2\ \Omega$$

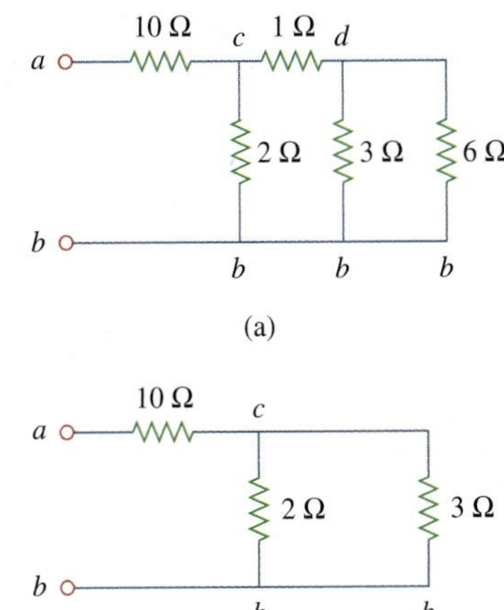

Figure 2.38
Equivalent circuits for Example 2.10.

Practice Problem 2.10

Find R_{ab} for the circuit in Fig. 2.39.

Answer: 11 Ω.

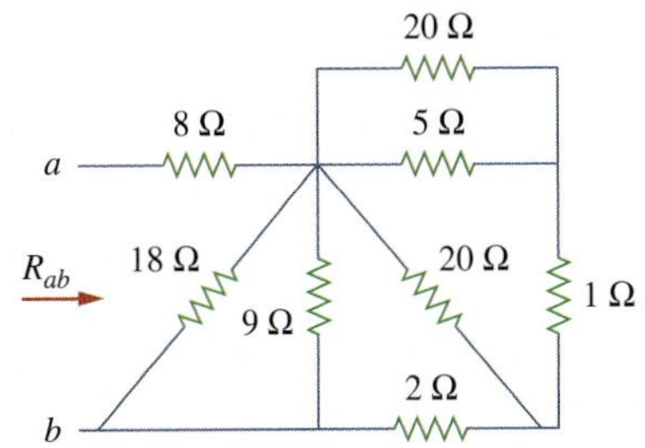

Figure 2.39
For Practice Prob. 2.10.

Example 2.11

Find the equivalent conductance G_{eq} for the circuit in Fig. 2.40(a).

Solution:

The 8-S and 12-S resistors are in parallel, so their conductance is

$$8\ S + 12\ S = 20\ S$$

This 20-S resistor is now in series with 5 S as shown in Fig. 2.40(b) so that the combined conductance is

$$\frac{20 \times 5}{20 + 5} = 4\ S$$

This is in parallel with the 6-S resistor. Hence,

$$G_{eq} = 6 + 4 = 10\ S$$

We should note that the circuit in Fig. 2.40(a) is the same as that in Fig. 2.40(c). While the resistors in Fig. 2.40(a) are expressed in

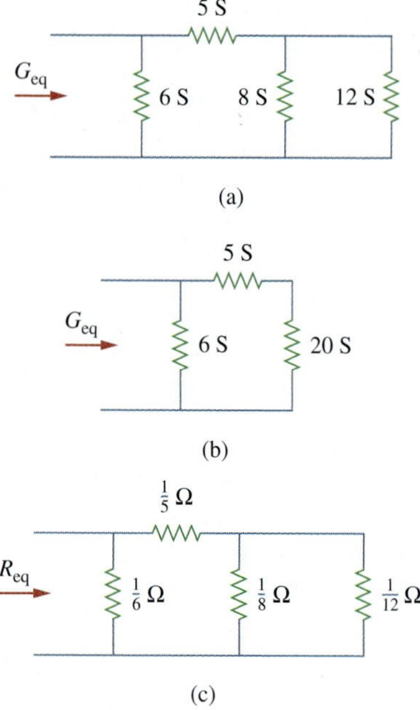

Figure 2.40
For Example 2.11: (a) original circuit, (b) its equivalent circuit, (c) same circuit as in (a) but resistors are expressed in ohms.

siemens, those in Fig. 2.40(c) are expressed in ohms. To show that the circuits are the same, we find R_{eq} for the circuit in Fig. 2.40(c).

$$R_{eq} = \frac{1}{6} \left\| \left(\frac{1}{5} + \frac{1}{8} \left\| \frac{1}{12} \right. \right) = \frac{1}{6} \right\| \left(\frac{1}{5} + \frac{1}{20} \right) = \frac{1}{6} \left\| \frac{1}{4} \right.$$

$$= \frac{\frac{1}{6} \times \frac{1}{4}}{\frac{1}{6} + \frac{1}{4}} = \frac{1}{10}\ \Omega$$

$$G_{eq} = \frac{1}{R_{eq}} = 10\text{ S}$$

This is the same as we obtained previously.

Practice Problem 2.11

Calculate G_{eq} in the circuit of Fig. 2.41.

Answer: 4 S.

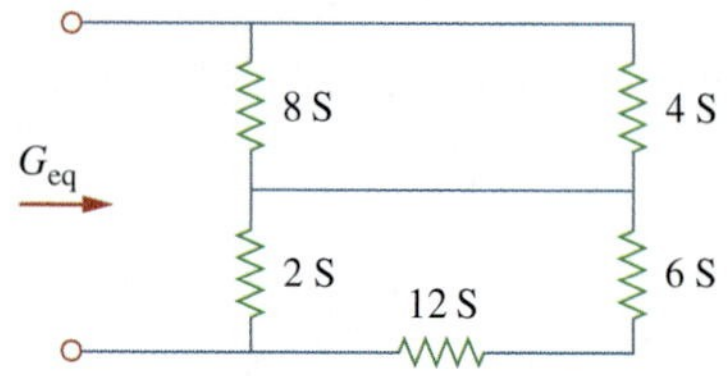

Figure 2.41
For Practice Prob. 2.11.

Example 2.12

Find i_o and v_o in the circuit shown in Fig. 2.42(a). Calculate the power dissipated in the 3-Ω resistor.

Solution:
The 6-Ω and 3-Ω resistors are in parallel, so their combined resistance is

$$6\ \Omega \parallel 3\ \Omega = \frac{6 \times 3}{6 + 3} = 2\ \Omega$$

Thus our circuit reduces to that shown in Fig. 2.42(b). Notice that v_o is not affected by the combination of the resistors because the resistors are in parallel and therefore have the same voltage v_o. From Fig. 2.42(b), we can obtain v_o in two ways. One way is to apply Ohm's law to get

$$i = \frac{12}{4 + 2} = 2\text{ A}$$

and hence, $v_o = 2i = 2 \times 2 = 4$ V. Another way is to apply voltage division, since the 12 V in Fig. 2.42(b) is divided between the 4-Ω and 2-Ω resistors. Hence,

$$v_o = \frac{2}{2+4}(12 \text{ V}) = 4 \text{ V}$$

Similarly, i_o can be obtained in two ways. One approach is to apply Ohm's law to the 3-Ω resistor in Fig. 2.42(a) now that we know v_o; thus,

$$v_o = 3i_o = 4 \quad \Rightarrow \quad i_o = \frac{4}{3} \text{ A}$$

Another approach is to apply current division to the circuit in Fig. 2.42(a) now that we know i, by writing

$$i_o = \frac{6}{6+3}i = \frac{2}{3}(2 \text{ A}) = \frac{4}{3} \text{ A}$$

The power dissipated in the 3-Ω resistor is

$$p_o = v_o i_o = 4\left(\frac{4}{3}\right) = 5.333 \text{ W}$$

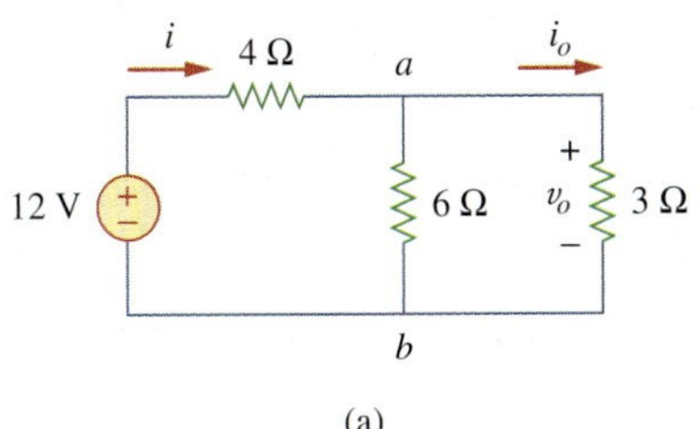

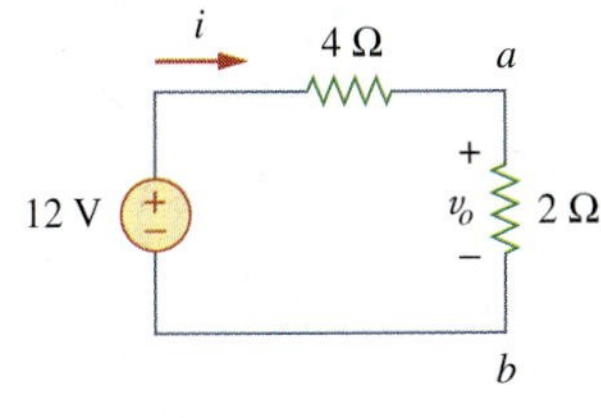

Figure 2.42
For Example 2.12: (a) original circuit, (b) its equivalent circuit.

Practice Problem 2.12

Find v_1 and v_2 in the circuit shown in Fig. 2.43. Also calculate i_1 and i_2 and the power dissipated in the 12-Ω and 40-Ω resistors.

Answer: $v_1 = 5$ V, $i_1 = 416.7$ mA, $p_1 = 2.083$ W, $v_2 = 10$ V, $i_2 = 250$ mA, $p_2 = 2.5$ W.

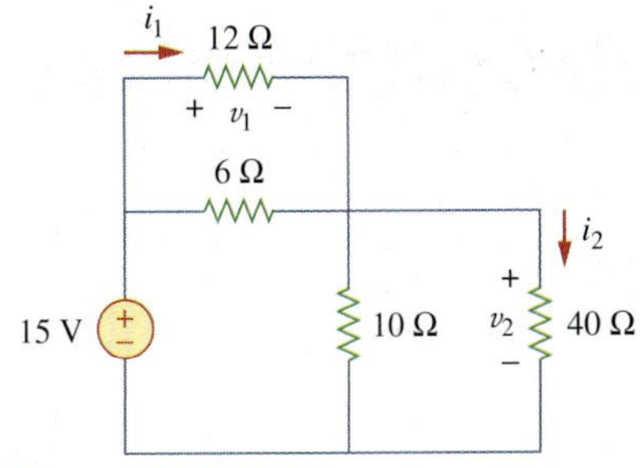

Figure 2.43
For Practice Prob. 2.12.

Example 2.13

For the circuit shown in Fig. 2.44(a), determine: (a) the voltage v_o, (b) the power supplied by the current source, (c) the power absorbed by each resistor.

Solution:

(a) The 6-kΩ and 12-kΩ resistors are in series so that their combined value is 6 + 12 = 18 kΩ. Thus the circuit in Fig. 2.44(a) reduces to that shown in Fig. 2.44(b). We now apply the current division technique to find i_1 and i_2.

$$i_1 = \frac{18{,}000}{9{,}000 + 18{,}000}(30 \text{ mA}) = 20 \text{ mA}$$

$$i_2 = \frac{9{,}000}{9{,}000 + 18{,}000}(30 \text{ mA}) = 10 \text{ mA}$$

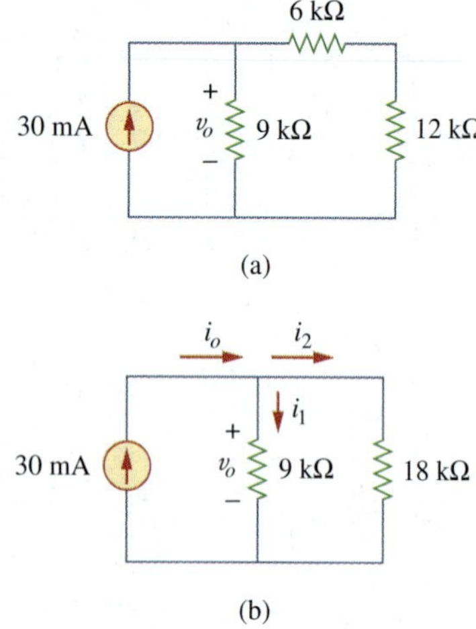

Figure 2.44
For Example 2.13: (a) original circuit, (b) its equivalent circuit.

Notice that the voltage across the 9-kΩ and 18-kΩ resistors is the same, and $v_o = 9{,}000 i_1 = 18{,}000 i_2 = 180$ V, as expected.

(b) Power supplied by the source is

$$p_o = v_o i_o = 180(30) \text{ mW} = 5.4 \text{ W}$$

(c) Power absorbed by the 12-kΩ resistor is

$$p = iv = i_2(i_2 R) = i_2^2 R = (10 \times 10^{-3})^2 (12{,}000) = 1.2 \text{ W}$$

Power absorbed by the 6-kΩ resistor is

$$p = i_2^2 R = (10 \times 10^{-3})^2 (6{,}000) = 0.6 \text{ W}$$

Power absorbed by the 9-kΩ resistor is

$$p = \frac{v_o^2}{R} = \frac{(180)^2}{9{,}000} = 3.6 \text{ W}$$

or

$$p = v_o i_1 = 180(20) \text{ mW} = 3.6 \text{ W}$$

Notice that the power supplied (5.4 W) equals the power absorbed (1.2 + 0.6 + 3.6 = 5.4 W). This is one way of checking results.

Practice Problem 2.13

For the circuit shown in Fig. 2.45, find: (a) v_1 and v_2, (b) the power dissipated in the 3-kΩ and 20-kΩ resistors, and (c) the power supplied by the current source.

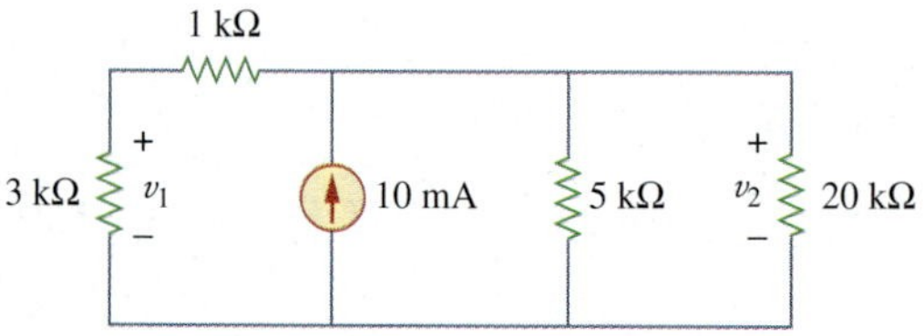

Figure 2.45
For Practice Prob. 2.13.

Answer: (a) 15 V, 20 V, (b) 75 mW, 20 mW, (c) 200 mW.

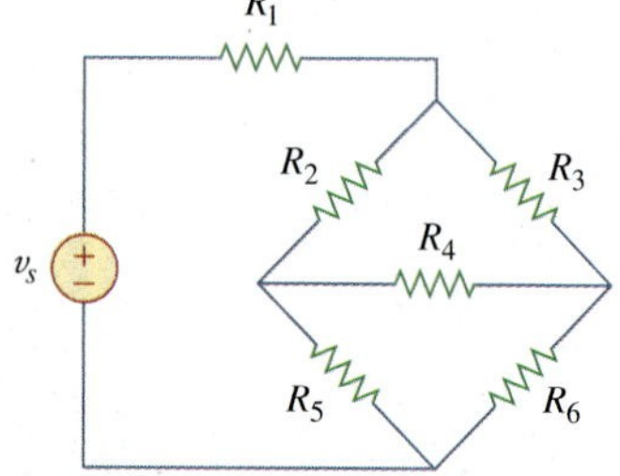

Figure 2.46
The bridge network.

2.7 †Wye-Delta Transformations

Situations often arise in circuit analysis when the resistors are neither in parallel nor in series. For example, consider the bridge circuit in Fig. 2.46. How do we combine resistors R_1 through R_6 when the resistors are neither in series nor in parallel? Many circuits of the type shown in Fig. 2.46 can be simplified by using three-terminal equivalent networks. These are the wye (Y) or tee (T) network shown in Fig. 2.47 and the delta (Δ) or pi (Π) network shown in Fig. 2.48. These networks occur by themselves or as part of a larger network. They are used in three-phase networks, electrical filters, and matching networks. Our main interest here is in how to identify them when they occur as part of a network and how to apply wye-delta transformation in the analysis of that network.

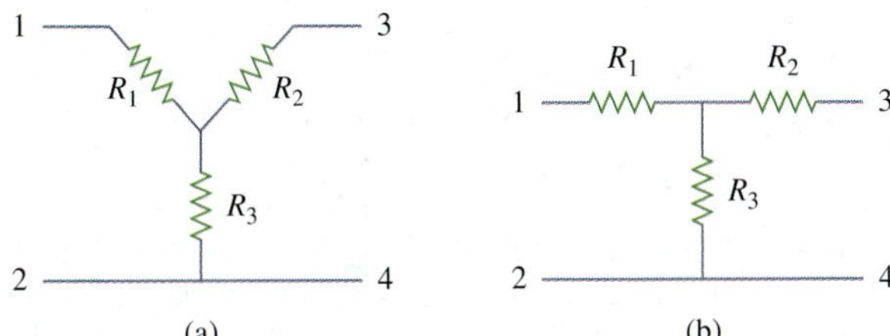

Figure 2.47
Two forms of the same network: (a) Y, (b) T.

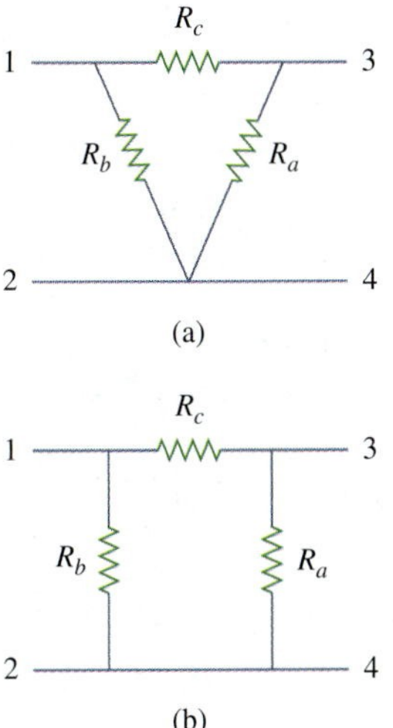

Figure 2.48
Two forms of the same network: (a) Δ, (b) Π.

Delta to Wye Conversion

Suppose it is more convenient to work with a wye network in a place where the circuit contains a delta configuration. We superimpose a wye network on the existing delta network and find the equivalent resistances in the wye network. To obtain the equivalent resistances in the wye network, we compare the two networks and make sure that the resistance between each pair of nodes in the Δ (or Π) network is the same as the resistance between the same pair of nodes in the Y (or T) network. For terminals 1 and 2 in Figs. 2.47 and 2.48, for example,

$$R_{12}(\text{Y}) = R_1 + R_3 \tag{2.46}$$

$$R_{12}(\Delta) = R_b \parallel (R_a + R_c)$$

Setting $R_{12}(\text{Y}) = R_{12}(\Delta)$ gives

$$R_{12} = R_1 + R_3 = \frac{R_b(R_a + R_c)}{R_a + R_b + R_c} \tag{2.47a}$$

Similarly,

$$R_{13} = R_1 + R_2 = \frac{R_c(R_a + R_b)}{R_a + R_b + R_c} \tag{2.47b}$$

$$R_{34} = R_2 + R_3 = \frac{R_a(R_b + R_c)}{R_a + R_b + R_c} \tag{2.47c}$$

Subtracting Eq. (2.47c) from Eq. (2.47a), we get

$$R_1 - R_2 = \frac{R_c(R_b - R_a)}{R_a + R_b + R_c} \tag{2.48}$$

Adding Eqs. (2.47b) and (2.48) gives

$$\boxed{R_1 = \frac{R_b R_c}{R_a + R_b + R_c}} \tag{2.49}$$

and subtracting Eq. (2.48) from Eq. (2.47b) yields

$$R_2 = \frac{R_c R_a}{R_a + R_b + R_c} \tag{2.50}$$

Subtracting Eq. (2.49) from Eq. (2.47a), we obtain

$$R_3 = \frac{R_a R_b}{R_a + R_b + R_c} \tag{2.51}$$

We do not need to memorize Eqs. (2.49) to (2.51). To transform a Δ network to Y, we create an extra node n as shown in Fig. 2.49 and follow this conversion rule:

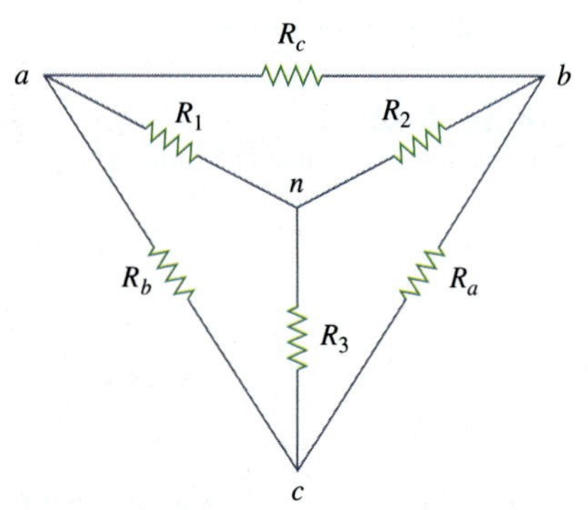

Figure 2.49
Superposition of Y and Δ networks as an aid in transforming one to the other.

> Each resistor in the Y network is the product of the resistors in the two adjacent Δ branches, divided by the sum of the three Δ resistors.

One can follow this rule and obtain Eqs. (2.49) to (2.51) from Fig. 2.49.

Wye to Delta Conversion

To obtain the conversion formulas for transforming a wye network to an equivalent delta network, we note from Eqs. (2.49) to (2.51) that

$$\begin{aligned} R_1 R_2 + R_2 R_3 + R_3 R_1 &= \frac{R_a R_b R_c (R_a + R_b + R_c)}{(R_a + R_b + R_c)^2} \\ &= \frac{R_a R_b R_c}{R_a + R_b + R_c} \end{aligned} \tag{2.52}$$

Dividing Eq. (2.52) by each of Eqs. (2.49) to (2.51) leads to the following equations:

$$R_a = \frac{R_1 R_2 + R_2 R_3 + R_3 R_1}{R_1} \tag{2.53}$$

$$R_b = \frac{R_1 R_2 + R_2 R_3 + R_3 R_1}{R_2} \tag{2.54}$$

$$R_c = \frac{R_1 R_2 + R_2 R_3 + R_3 R_1}{R_3} \tag{2.55}$$

From Eqs. (2.53) to (2.55) and Fig. 2.49, the conversion rule for Y to Δ is as follows:

> Each resistor in the Δ network is the sum of all possible products of Y resistors taken two at a time, divided by the opposite Y resistor.

The Y and Δ networks are said to be *balanced* when

$$R_1 = R_2 = R_3 = R_Y, \qquad R_a = R_b = R_c = R_\Delta \tag{2.56}$$

Under these conditions, conversion formulas become

$$R_Y = \frac{R_\Delta}{3} \qquad \text{or} \qquad R_\Delta = 3R_Y \tag{2.57}$$

One may wonder why R_Y is less than R_Δ. Well, we notice that the Y-connection is like a "series" connection while the Δ-connection is like a "parallel" connection.

Note that in making the transformation, we do not take anything out of the circuit or put in anything new. We are merely substituting different but mathematically equivalent three-terminal network patterns to create a circuit in which resistors are either in series or in parallel, allowing us to calculate R_{eq} if necessary.

Example 2.14

Convert the Δ network in Fig. 2.50(a) to an equivalent Y network.

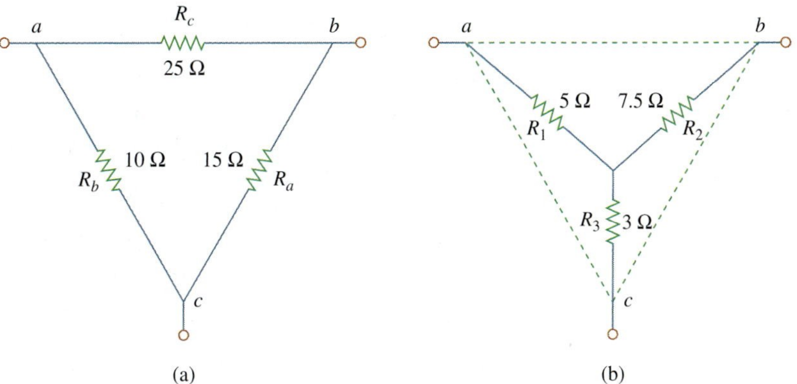

Figure 2.50
For Example 2.14: (a) original Δ network, (b) Y equivalent network.

Solution:
Using Eqs. (2.49) to (2.51), we obtain

$$R_1 = \frac{R_b R_c}{R_a + R_b + R_c} = \frac{10 \times 25}{15 + 10 + 25} = \frac{250}{50} = 5\ \Omega$$

$$R_2 = \frac{R_c R_a}{R_a + R_b + R_c} = \frac{25 \times 15}{50} = 7.5\ \Omega$$

$$R_3 = \frac{R_a R_b}{R_a + R_b + R_c} = \frac{15 \times 10}{50} = 3\ \Omega$$

The equivalent Y network is shown in Fig. 2.50(b).

Practice Problem 2.14

Transform the wye network in Fig. 2.51 to a delta network.

Answer: $R_a = 140\ \Omega$, $R_b = 70\ \Omega$, $R_c = 35\ \Omega$.

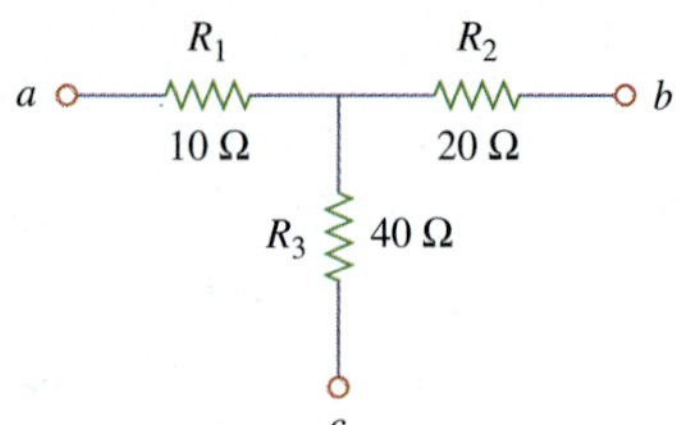

Figure 2.51
For Practice Prob. 2.14.

Example 2.15

Obtain the equivalent resistance R_{ab} for the circuit in Fig. 2.52 and use it to find current i.

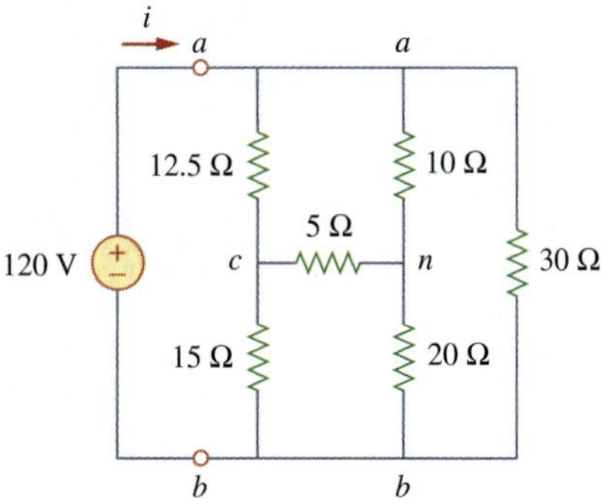

Figure 2.52
For Example 2.15.

Solution:

1. **Define.** The problem is clearly defined. Please note, this part normally will deservedly take much more time.
2. **Present.** Clearly, when we remove the voltage source, we end up with a purely resistive circuit. Since it is composed of deltas and wyes, we have a more complex process of combining the elements together. We can use wye-delta transformations as one approach to find a solution. It is useful to locate the wyes (there are two of them, one at n and the other at c) and the deltas (there are three: *can, abn, cnb*).
3. **Alternative.** There are different approaches that can be used to solve this problem. Since the focus of Sec. 2.7 is the wye-delta transformation, this should be the technique to use. Another approach would be to solve for the equivalent resistance by injecting one amp into the circuit and finding the voltage between a and b; we will learn about this approach in Chap. 4.

   The approach we can apply here as a check would be to use a wye-delta transformation as the first solution to the problem. Later we can check the solution by starting with a delta-wye transformation.
4. **Attempt.** In this circuit, there are two Y networks and three Δ networks. Transforming just one of these will simplify the circuit. If we convert the Y network comprising the 5-Ω, 10-Ω, and 20-Ω resistors, we may select

$$R_1 = 10\ \Omega, \qquad R_2 = 20\ \Omega, \qquad R_3 = 5\ \Omega$$

Thus from Eqs. (2.53) to (2.55) we have

$$R_a = \frac{R_1R_2 + R_2R_3 + R_3R_1}{R_1} = \frac{10 \times 20 + 20 \times 5 + 5 \times 10}{10}$$

$$= \frac{350}{10} = 35\ \Omega$$

$$R_b = \frac{R_1R_2 + R_2R_3 + R_3R_1}{R_2} = \frac{350}{20} = 17.5\ \Omega$$

$$R_c = \frac{R_1R_2 + R_2R_3 + R_3R_1}{R_3} = \frac{350}{5} = 70\ \Omega$$

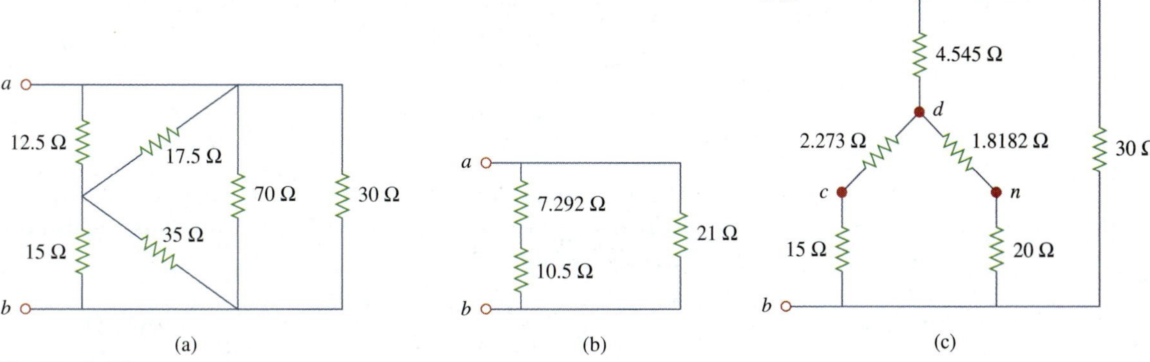

Figure 2.53
Equivalent circuits to Fig. 2.52, with the voltage source removed.

With the Y converted to Δ, the equivalent circuit (with the voltage source removed for now) is shown in Fig. 2.53(a). Combining the three pairs of resistors in parallel, we obtain

$$70 \parallel 30 = \frac{70 \times 30}{70 + 30} = 21\ \Omega$$

$$12.5 \parallel 17.5 = \frac{12.5 \times 17.5}{12.5 + 17.5} = 7.292\ \Omega$$

$$15 \parallel 35 = \frac{15 \times 35}{15 + 35} = 10.5\ \Omega$$

so that the equivalent circuit is shown in Fig. 2.53(b). Hence, we find

$$R_{ab} = (7.292 + 10.5) \parallel 21 = \frac{17.792 \times 21}{17.792 + 21} = \mathbf{9.632\ \Omega}$$

Then

$$i = \frac{v_s}{R_{ab}} = \frac{120}{9.632} = \mathbf{12.458\ A}$$

We observe that we have successfully solved the problem. Now we must evaluate the solution.

5. **Evaluate.** Now we must determine if the answer is correct and then evaluate the final solution.

   It is relatively easy to check the answer; we do this by solving the problem starting with a delta-wye transformation. Let us transform the delta, *can*, into a wye.

   Let $R_c = 10\ \Omega$, $R_a = 5\ \Omega$, and $R_n = 12.5\ \Omega$. This will lead to (let d represent the middle of the wye):

$$R_{ad} = \frac{R_c R_n}{R_a + R_c + R_n} = \frac{10 \times 12.5}{5 + 10 + 12.5} = 4.545\ \Omega$$

$$R_{cd} = \frac{R_a R_n}{27.5} = \frac{5 \times 12.5}{27.5} = 2.273\ \Omega$$

$$R_{nd} = \frac{R_a R_c}{27.5} = \frac{5 \times 10}{27.5} = 1.8182\ \Omega$$

This now leads to the circuit shown in Figure 2.53(c). Looking at the resistance between d and b, we have two series combination in parallel, giving us

$$R_{db} = \frac{(2.273 + 15)(1.8182 + 20)}{2.273 + 15 + 1.8182 + 20} = \frac{376.9}{39.09} = 9.642\ \Omega$$

This is in series with the 4.545-Ω resistor, both of which are in parallel with the 30-Ω resistor. This then gives us the equivalent resistance of the circuit.

$$R_{ab} = \frac{(9.642 + 4.545)30}{9.642 + 4.545 + 30} = \frac{425.6}{44.19} = \mathbf{9.631\ \Omega}$$

This now leads to

$$i = \frac{v_s}{R_{ab}} = \frac{120}{9.631} = \mathbf{12.46\ A}$$

We note that using two variations on the wye-delta transformation leads to the same results. This represents a very good check.

6. **Satisfactory?** Since we have found the desired answer by determining the equivalent resistance of the circuit first and the answer checks, then we clearly have a satisfactory solution. This represents what can be presented to the individual assigning the problem.

Practice Problem 2.15

For the bridge network in Fig. 2.54, find R_{ab} and i.

Answer: 40 Ω, 2.5 A.

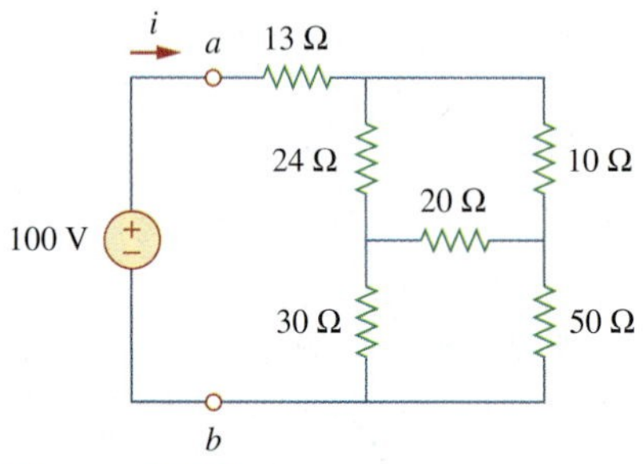

Figure 2.54
For Practice Prob. 2.15.

2.8 †Applications

Resistors are often used to model devices that convert electrical energy into heat or other forms of energy. Such devices include conducting wire, lightbulbs, electric heaters, stoves, ovens, and loudspeakers. In this section, we will consider two real-life problems that apply the concepts developed in this chapter: electrical lighting systems and design of dc meters.

2.8.1. Lighting Systems

So far, we have assumed that connecting wires are perfect conductors (i.e., conductors of zero resistance). In real physical systems, however, the resistance of the connecting wire may be appreciably large, and the modeling of the system must include that resistance.

Lighting systems, such as in a house or on a Christmas tree, often consist of N lamps connected either in parallel or in series, as shown in Fig. 2.55. Each lamp is modeled as a resistor. Assuming that all the lamps are identical and V_o is the power-line voltage, the voltage across each lamp is V_o for the parallel connection and V_o/N for the series connection. The series connection is easy to manufacture but is seldom used in practice, for at least two reasons. First, it is less reliable; when a lamp fails, all the lamps go out. Second, it is harder to maintain; when a lamp is bad, one must test all the lamps one by one to detect the faulty one.

Historical

Thomas Alva Edison (1847–1931) was perhaps the greatest American inventor. He patented 1093 inventions, including such history-making inventions as the incandescent electric bulb, the phonograph, and the first commercial motion pictures.

Born in Milan, Ohio, the youngest of seven children, Edison received only three months of formal education because he hated school. He was home-schooled by his mother and quickly began to read on his own. In 1868, Edison read one of Faraday's books and found his calling. He moved to Menlo Park, New Jersey, in 1876, where he managed a well-staffed research laboratory. Most of his inventions came out of this laboratory. His laboratory served as a model for modern research organizations. Because of his diverse interests and the overwhelming number of his inventions and patents, Edison began to establish manufacturing companies for making the devices he invented. He designed the first electric power station to supply electric light. Formal electrical engineering education began in the mid-1880s with Edison as a role model and leader.

Library of Congress

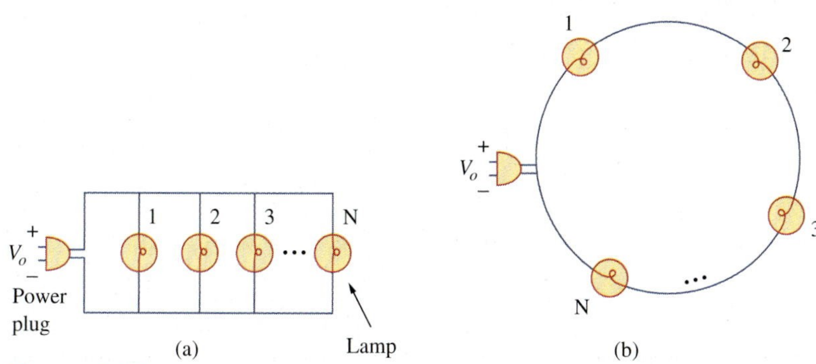

Figure 2.55
(a) Parallel connection of lightbulbs, (b) series connection of lightbulbs.

Example 2.16

Three lightbulbs are connected to a 9-V battery as shown in Fig. 2.56(a). Calculate: (a) the total current supplied by the battery, (b) the current through each bulb, (c) the resistance of each bulb.

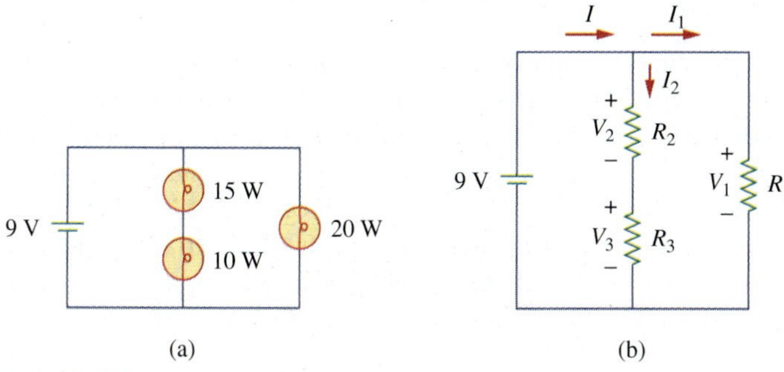

Figure 2.56
(a) Lighting system with three bulbs, (b) resistive circuit equivalent model.

Solution:

(a) The total power supplied by the battery is equal to the total power absorbed by the bulbs; that is,

$$p = 15 + 10 + 20 = 45 \text{ W}$$

Since $p = VI$, then the total current supplied by the battery is

$$I = \frac{p}{V} = \frac{45}{9} = 5 \text{ A}$$

(b) The bulbs can be modeled as resistors as shown in Fig. 2.56(b). Since R_1 (20-W bulb) is in parallel with the battery as well as the series combination of R_2 and R_3,

$$V_1 = V_2 + V_3 = 9 \text{ V}$$

The current through R_1 is

$$I_1 = \frac{p_1}{V_1} = \frac{20}{9} = 2.222 \text{ A}$$

By KCL, the current through the series combination of R_2 and R_3 is

$$I_2 = I - I_1 = 5 - 2.222 = 2.778 \text{ A}$$

(c) Since $p = I^2R$,

$$R_1 = \frac{p_1}{I_1^2} = \frac{20}{2.222^2} = 4.05\ \Omega$$

$$R_2 = \frac{p_2}{I_2^2} = \frac{15}{2.777^2} = 1.945\ \Omega$$

$$R_3 = \frac{p_3}{I_3^2} = \frac{10}{2.777^2} = 1.297\ \Omega$$

Practice Problem 2.16

Refer to Fig. 2.55 and assume there are 10 lightbulbs that can be connected in parallel and 10 lightbulbs that can be connected in series, each with a power rating of 40 W. If the voltage at the plug is 110 V for the parallel and series connections, calculate the current through each bulb for both cases.

Answer: 0.364 A (parallel), 3.64 A (series).

2.8.2 Design of DC Meters

By their nature, resistors are used to control the flow of current. We take advantage of this property in several applications, such as in a potentiometer (Fig. 2.57). The word *potentiometer*, derived from the words *potential* and *meter*, implies that potential can be metered out. The potentiometer (or pot for short) is a three-terminal device that operates on the principle of voltage division. It is essentially an adjustable voltage divider. As a voltage regulator, it is used as a volume or level control on radios, TVs, and other devices. In Fig. 2.57,

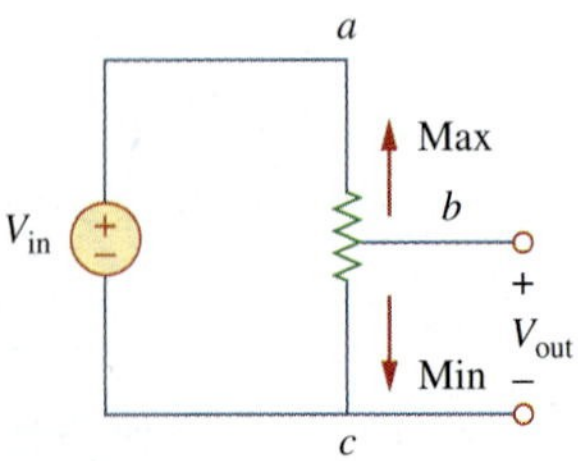

Figure 2.57
The potentiometer controlling potential levels.

$$V_{out} = V_{bc} = \frac{R_{bc}}{R_{ac}} V_{in} \tag{2.58}$$

where $R_{ac} = R_{ab} + R_{bc}$. Thus, V_{out} decreases or increases as the sliding contact of the pot moves toward c or a, respectively.

Another application where resistors are used to control current flow is in the analog dc meters—the ammeter, voltmeter, and ohmmeter, which measure current, voltage, and resistance, respectively. Each of these meters employs the d'Arsonval meter movement, shown in Fig. 2.58. The movement consists essentially of a movable iron-core coil mounted on a pivot between the poles of a permanent magnet. When current flows through the coil, it creates a torque which causes the pointer to deflect. The amount of current through the coil determines the deflection of the pointer, which is registered on a scale attached to the meter movement. For example, if the meter movement is rated 1 mA, 50 Ω, it would take 1 mA to cause a full-scale deflection of the meter movement. By introducing additional circuitry to the d'Arsonval meter movement, an ammeter, voltmeter, or ohmmeter can be constructed.

An instrument capable of measuring voltage, current, and resistance is called a *multimeter* or a *volt-ohm meter* (VOM).

Consider Fig. 2.59, where an analog voltmeter and ammeter are connected to an element. The voltmeter measures the voltage across a *load* and is therefore connected in parallel with the element. As shown

A load is a component that is receiving energy (an energy sink), as opposed to a generator supplying energy (an energy source). More about loading will be discussed in Section 4.9.1.

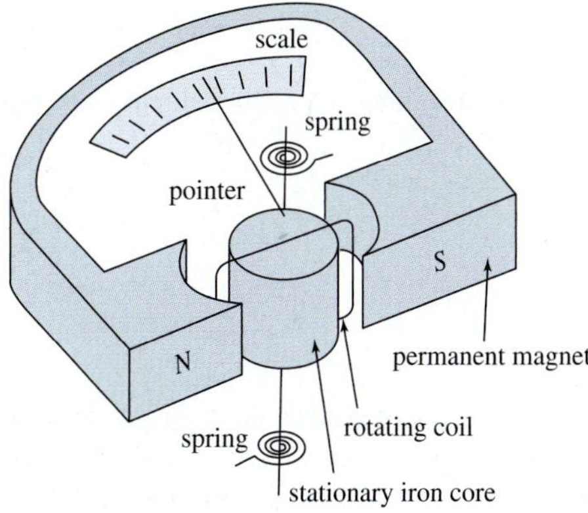

Figure 2.58
A d'Arsonval meter movement.

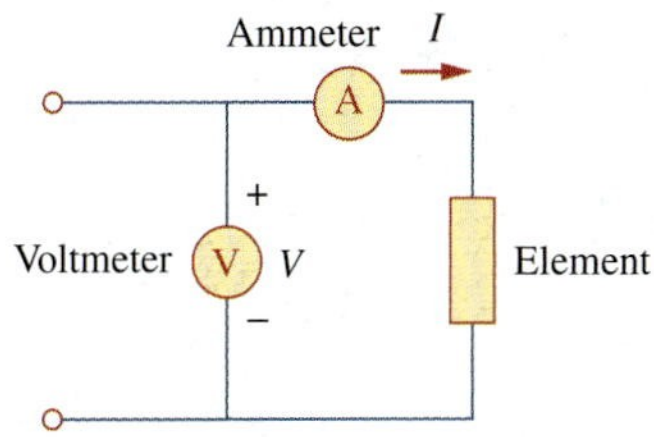

Figure 2.59
Connection of a voltmeter and an ammeter to an element.

in Fig. 2.60(a), the voltmeter consists of a d'Arsonval movement in series with a resistor whose resistance R_m is deliberately made very large (theoretically, infinite), to minimize the current drawn from the circuit. To extend the range of voltage that the meter can measure, series multiplier resistors are often connected with the voltmeters, as shown in Fig. 2.60(b). The multiple-range voltmeter in Fig. 2.60(b) can measure voltage from 0 to 1 V, 0 to 10 V, or 0 to 100 V, depending on whether the switch is connected to R_1, R_2, or R_3, respectively.

Let us calculate the multiplier resistor R_n for the single-range voltmeter in Fig. 2.60(a), or $R_n = R_1, R_2$, or R_3 for the multiple-range voltmeter in Fig. 2.60(b). We need to determine the value of R_n to be connected in series with the internal resistance R_m of the voltmeter. In any design, we consider the worst-case condition. In this case, the worst case occurs when the full-scale current $I_{\text{fs}} = I_m$ flows through the meter. This should also correspond to the maximum voltage reading or the full-scale voltage V_{fs}. Since the multiplier resistance R_n is in series with the internal resistance R_m,

$$V_{\text{fs}} = I_{\text{fs}}(R_n + R_m) \tag{2.59}$$

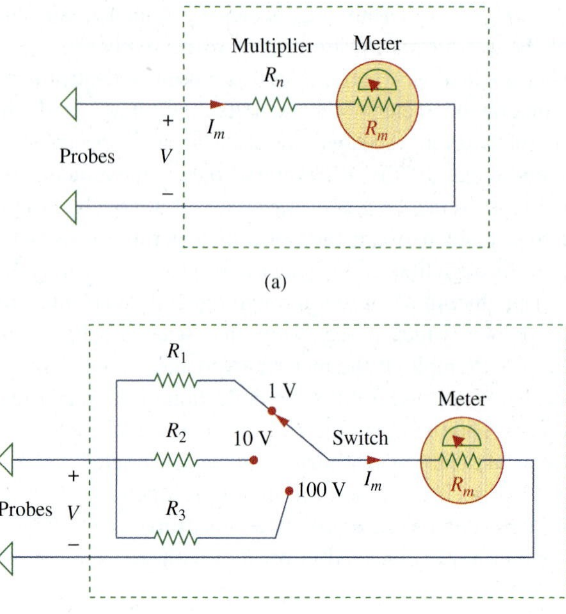

Figure 2.60
Voltmeters: (a) single-range type, (b) multiple-range type.

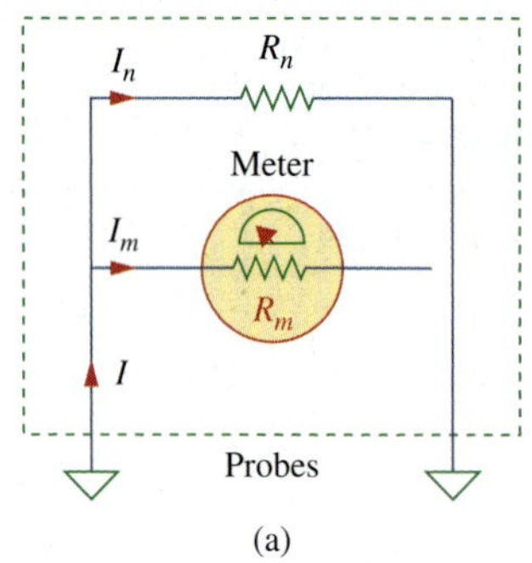

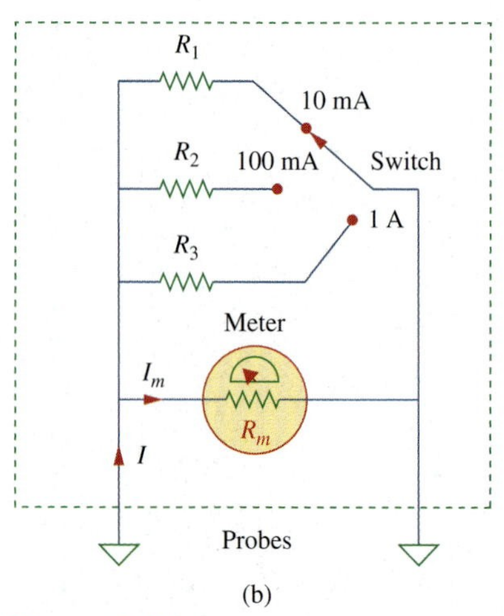

Figure 2.61
Ammeters: (a) single-range type, (b) multiple-range type.

From this, we obtain

$$R_n = \frac{V_{\text{fs}}}{I_{\text{fs}}} - R_m \tag{2.60}$$

Similarly, the ammeter measures the current through the load and is connected in series with it. As shown in Fig. 2.61(a), the ammeter consists of a d'Arsonval movement in parallel with a resistor whose resistance R_m is deliberately made very small (theoretically, zero) to minimize the voltage drop across it. To allow multiple ranges, shunt resistors are often connected in parallel with R_m as shown in Fig. 2.61(b). The shunt resistors allow the meter to measure in the range 0–10 mA, 0–100 mA, or 0–1 A, depending on whether the switch is connected to R_1, R_2, or R_3, respectively.

Now our objective is to obtain the multiplier shunt R_n for the single-range ammeter in Fig. 2.61(a), or $R_n = R_1, R_2$, or R_3 for the multiple-range ammeter in Fig. 2.61(b). We notice that R_m and R_n are in parallel and that at full-scale reading $I = I_{\text{fs}} = I_m + I_n$, where I_n is the current through the shunt resistor R_n. Applying the current division principle yields

$$I_m = \frac{R_n}{R_n + R_m} I_{\text{fs}}$$

or

$$R_n = \frac{I_m}{I_{\text{fs}} - I_m} R_m \tag{2.61}$$

The resistance R_x of a linear resistor can be measured in two ways. An indirect way is to measure the current I that flows through it by connecting an ammeter in series with it and the voltage V across it by connecting a voltmeter in parallel with it, as shown in Fig. 2.62(a). Then

$$R_x = \frac{V}{I} \tag{2.62}$$

The direct method of measuring resistance is to use an ohmmeter. An ohmmeter consists basically of a d'Arsonval movement, a variable resistor or potentiometer, and a battery, as shown in Fig. 2.62(b). Applying KVL to the circuit in Fig. 2.62(b) gives

$$E = (R + R_m + R_x)I_m$$

or

$$R_x = \frac{E}{I_m} - (R + R_m) \tag{2.63}$$

The resistor R is selected such that the meter gives a full-scale deflection; that is, $I_m = I_{fs}$ when $R_x = 0$. This implies that

$$E = (R + R_m)I_{fs} \tag{2.64}$$

Substituting Eq. (2.64) into Eq. (2.63) leads to

$$R_x = \left(\frac{I_{fs}}{I_m} - 1\right)(R + R_m) \tag{2.65}$$

As mentioned, the types of meters we have discussed are known as *analog* meters and are based on the d'Arsonval meter movement. Another type of meter, called a *digital meter*, is based on active circuit elements such as op amps. For example, a digital multimeter displays measurements of dc or ac voltage, current, and resistance as discrete numbers, instead of using a pointer deflection on a continuous scale as in an analog multimeter. Digital meters are what you would most likely use in a modern lab. However, the design of digital meters is beyond the scope of this book.

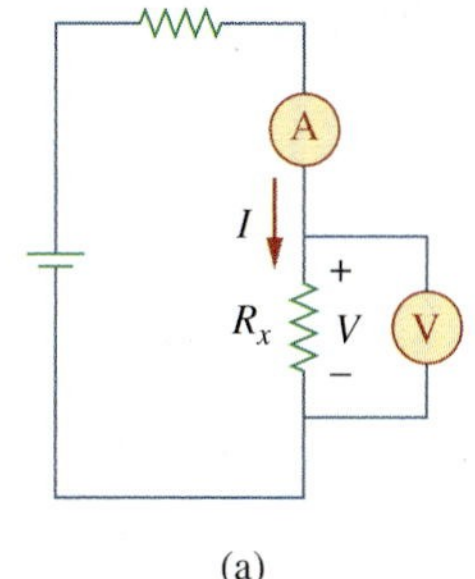

(a)

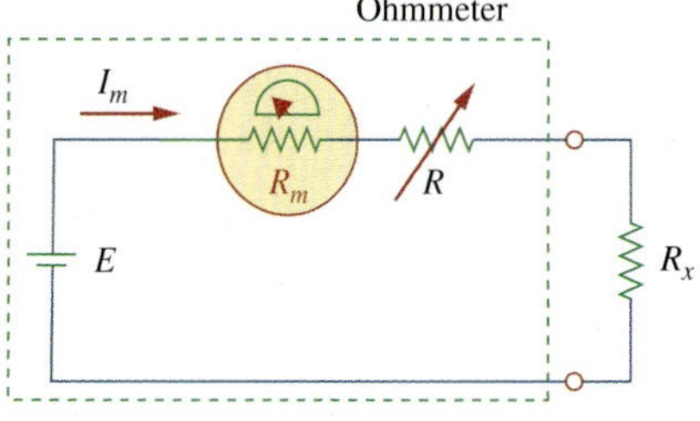

(b)

Figure 2.62
Two ways of measuring resistance: (a) using an ammeter and a voltmeter, (b) using an ohmmeter.

Historical

Samuel F. B. Morse (1791–1872), an American painter, invented the telegraph, the first practical, commercialized application of electricity.

Morse was born in Charlestown, Massachusetts and studied at Yale and the Royal Academy of Arts in London to become an artist. In the 1830s, he became intrigued with developing a telegraph. He had a working model by 1836 and applied for a patent in 1838. The U.S. Senate appropriated funds for Morse to construct a telegraph line between Baltimore and Washington, D.C. On May 24, 1844, he sent the famous first message: "What hath God wrought!" Morse also developed a code of dots and dashes for letters and numbers, for sending messages on the telegraph. The development of the telegraph led to the invention of the telephone.

Library of Congress

Example 2.17

Following the voltmeter setup of Fig. 2.60, design a voltmeter for the following multiple ranges:

(a) 0–1 V (b) 0–5 V (c) 0–50 V (d) 0–100 V

Assume that the internal resistance $R_m = 2\ \text{k}\Omega$ and the full-scale current $I_{fs} = 100\ \mu\text{A}$.

Solution:

We apply Eq. (2.60) and assume that R_1, R_2, R_3, and R_4 correspond with ranges 0–1 V, 0–5 V, 0–50 V, and 0–100 V, respectively.

(a) For range 0–1 V,

$$R_1 = \frac{1}{100 \times 10^{-6}} - 2000 = 10{,}000 - 2000 = 8\ \text{k}\Omega$$

(b) For range 0–5 V,

$$R_2 = \frac{5}{100 \times 10^{-6}} - 2000 = 50{,}000 - 2000 = 48\ \text{k}\Omega$$

(c) For range 0–50 V,

$$R_3 = \frac{50}{100 \times 10^{-6}} - 2000 = 500{,}000 - 2000 = 498\ \text{k}\Omega$$

(d) For range 0–100 V,

$$R_4 = \frac{100\ \text{V}}{100 \times 10^{-6}} - 2000 = 1{,}000{,}000 - 2000 = 998\ \text{k}\Omega$$

Note that the ratio of the total resistance $(R_n + R_m)$ to the full-scale voltage V_{fs} is constant and equal to $1/I_{fs}$ for the four ranges. This ratio (given in ohms per volt, or Ω/V) is known as the *sensitivity* of the voltmeter. The larger the sensitivity, the better the voltmeter.

Practice Problem 2.17

Following the ammeter setup of Fig. 2.61, design an ammeter for the following multiple ranges:

(a) 0–1 A (b) 0–100 mA (c) 0–10 mA

Take the full-scale meter current as $I_m = 1$ mA and the internal resistance of the ammeter as $R_m = 50\ \Omega$.

Answer: Shunt resistors: 0.05 Ω, 0.505 Ω, 5.556 Ω.

2.9 Summary

1. A resistor is a passive element in which the voltage v across it is directly proportional to the current i through it. That is, a resistor is a device that obeys Ohm's law,

   $$v = iR$$

   where R is the resistance of the resistor.

2. A short circuit is a resistor (a perfectly conducting wire) with zero resistance ($R = 0$). An open circuit is a resistor with infinite resistance ($R = \infty$).
3. The conductance G of a resistor is the reciprocal of its resistance:

$$G = \frac{1}{R}$$

4. A branch is a single two-terminal element in an electric circuit. A node is the point of connection between two or more branches. A loop is a closed path in a circuit. The number of branches b, the number of nodes n, and the number of independent loops l in a network are related as

$$b = l + n - 1$$

5. Kirchhoff's current law (KCL) states that the currents at any node algebraically sum to zero. In other words, the sum of the currents entering a node equals the sum of currents leaving the node.
6. Kirchhoff's voltage law (KVL) states that the voltages around a closed path algebraically sum to zero. In other words, the sum of voltage rises equals the sum of voltage drops.
7. Two elements are in series when they are connected sequentially, end to end. When elements are in series, the same current flows through them ($i_1 = i_2$). They are in parallel if they are connected to the same two nodes. Elements in parallel always have the same voltage across them ($v_1 = v_2$).
8. When two resistors $R_1\,(=1/G_1)$ and $R_2\,(=1/G_2)$ are in series, their equivalent resistance R_{eq} and equivalent conductance G_{eq} are

$$R_{eq} = R_1 + R_2, \qquad G_{eq} = \frac{G_1 G_2}{G_1 + G_2}$$

9. When two resistors $R_1\,(=1/G_1)$ and $R_2\,(=1/G_2)$ are in parallel, their equivalent resistance R_{eq} and equivalent conductance G_{eq} are

$$R_{eq} = \frac{R_1 R_2}{R_1 + R_2}, \qquad G_{eq} = G_1 + G_2$$

10. The voltage division principle for two resistors in series is

$$v_1 = \frac{R_1}{R_1 + R_2} v, \qquad v_2 = \frac{R_2}{R_1 + R_2} v$$

11. The current division principle for two resistors in parallel is

$$i_1 = \frac{R_2}{R_1 + R_2} i, \qquad i_2 = \frac{R_1}{R_1 + R_2} i$$

12. The formulas for a delta-to-wye transformation are

$$R_1 = \frac{R_b R_c}{R_a + R_b + R_c}, \qquad R_2 = \frac{R_c R_a}{R_a + R_b + R_c}$$

$$R_3 = \frac{R_a R_b}{R_a + R_b + R_c}$$

13. The formulas for a wye-to-delta transformation are

$$R_a = \frac{R_1R_2 + R_2R_3 + R_3R_1}{R_1}, \qquad R_b = \frac{R_1R_2 + R_2R_3 + R_3R_1}{R_2}$$

$$R_c = \frac{R_1R_2 + R_2R_3 + R_3R_1}{R_3}$$

14. The basic laws covered in this chapter can be applied to the problems of electrical lighting and design of dc meters.

Review Questions

2.1 The reciprocal of resistance is:

(a) voltage (b) current
(c) conductance (d) coulombs

2.2 An electric heater draws 10 A from a 120-V line. The resistance of the heater is:

(a) 1200 Ω (b) 120 Ω
(c) 12 Ω (d) 1.2 Ω

2.3 The voltage drop across a 1.5-kW toaster that draws 12 A of current is:

(a) 18 kV (b) 125 V
(c) 120 V (d) 10.42 V

2.4 The maximum current that a 2W, 80 kΩ resistor can safely conduct is:

(a) 160 kA (b) 40 kA
(c) 5 mA (d) 25 μA

2.5 A network has 12 branches and 8 independent loops. How many nodes are there in the network?

(a) 19 (b) 17 (c) 5 (d) 4

2.6 The current I in the circuit of Fig. 2.63 is:

(a) −0.8 A (b) −0.2 A
(c) 0.2 A (d) 0.8 A

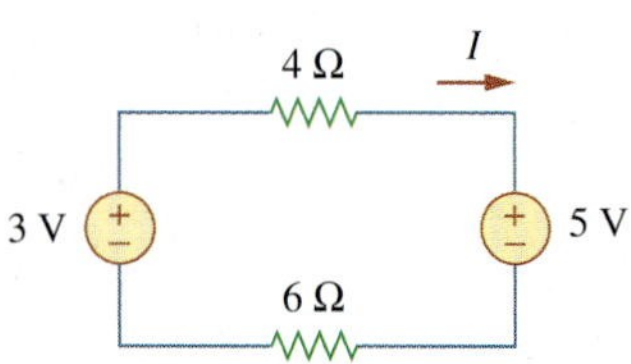

Figure 2.63
For Review Question 2.6.

2.7 The current I_o of Fig. 2.64 is:

(a) −4 A (b) −2 A (c) 4 A (d) 16 A

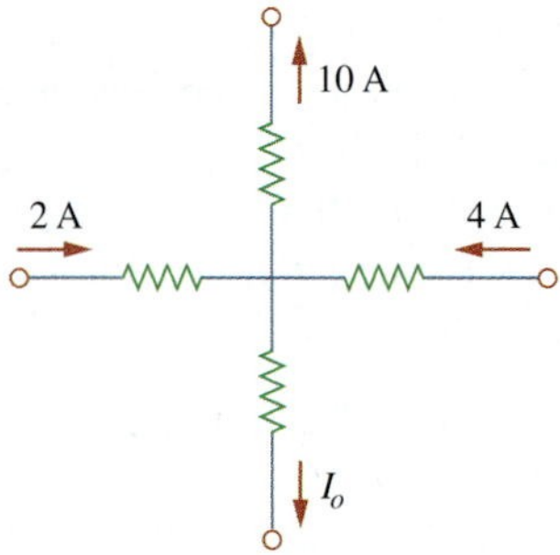

Figure 2.64
For Review Question 2.7.

2.8 In the circuit in Fig. 2.65, V is:

(a) 30 V (b) 14 V (c) 10 V (d) 6 V

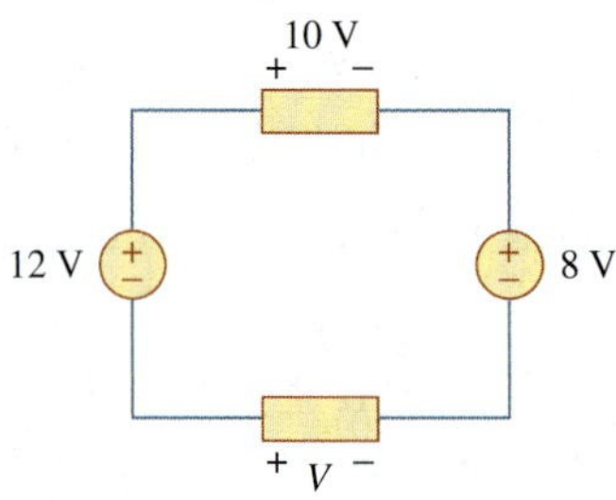

Figure 2.65
For Review Question 2.8.

2.9 Which of the circuits in Fig. 2.66 will give you $V_{ab} = 7$ V?

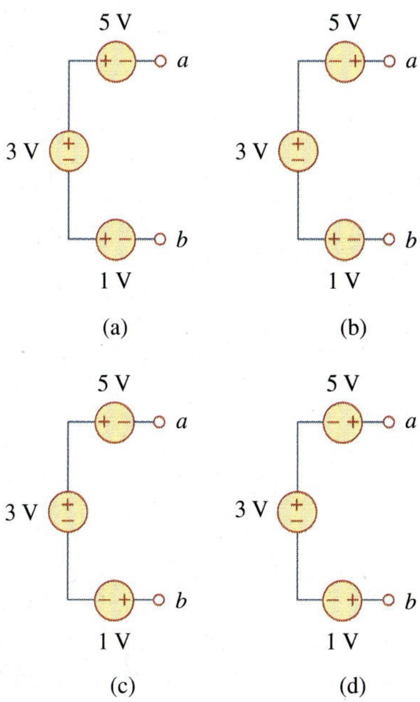

Figure 2.66
For Review Question 2.9.

2.10 In the circuit of Fig. 2.67, a decrease in R_3 leads to a decrease of:

(a) current through R_3

(b) voltage across R_3

(c) voltage across R_1

(d) power dissipated in R_2

(e) none of the above

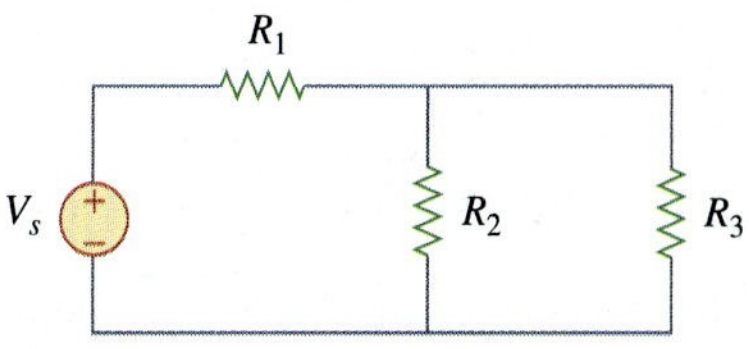

Figure 2.67
For Review Question 2.10.

Answers: 2.1c, 2.2c, 2.3b, 2.4c, 2.5c, 2.6b, 2.7a, 2.8d, 2.9d, 2.10b, d.

Problems

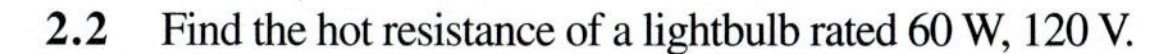

Section 2.2 Ohm's Law

2.1 Design a problem, complete with a solution, to help students to better understand Ohm's Law. Use at least two resistors and one voltage source. Hint, you could use both resistors at once or one at a time, it is up to you. Be creative.

2.2 Find the hot resistance of a lightbulb rated 60 W, 120 V.

2.3 A bar of silicon is 4 cm long with a circular cross section. If the resistance of the bar is 240 Ω at room temperature, what is the cross-sectional radius of the bar?

2.4 (a) Calculate current i in Fig. 2.68 when the switch is in position 1.

(b) Find the current when the switch is in position 2.

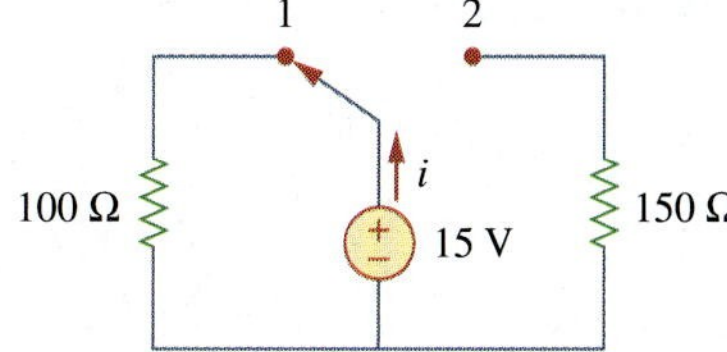

Figure 2.68
For Prob. 2.4.

Section 2.3 Nodes, Branches, and Loops

2.5 For the network graph in Fig. 2.69, find the number of nodes, branches, and loops.

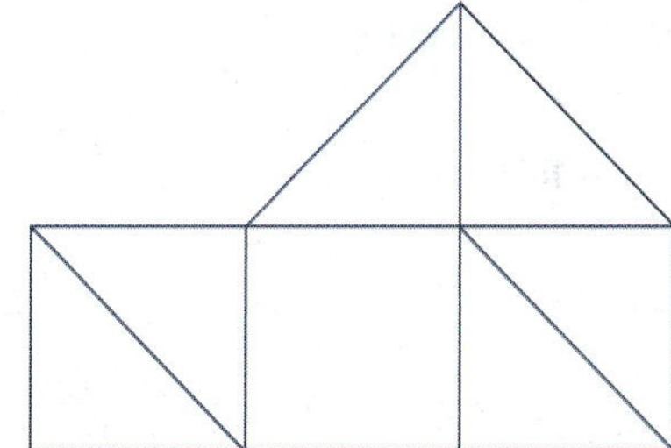

Figure 2.69
For Prob. 2.5.

2.6 In the network graph shown in Fig. 2.70, determine the number of branches and nodes.

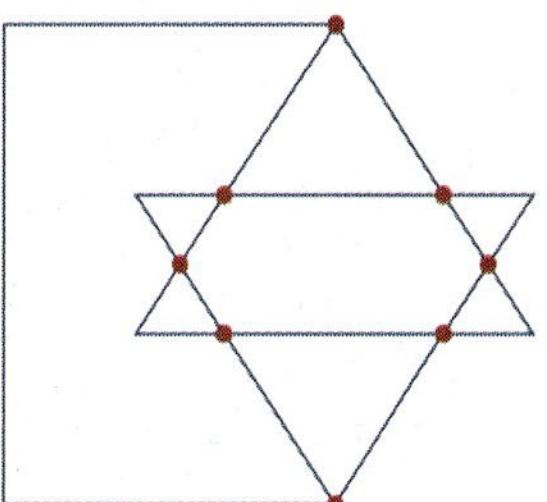

Figure 2.70
For Prob. 2.6.

2.7 Find the number of branches and nodes in each of the circuits of Fig. 2.71.

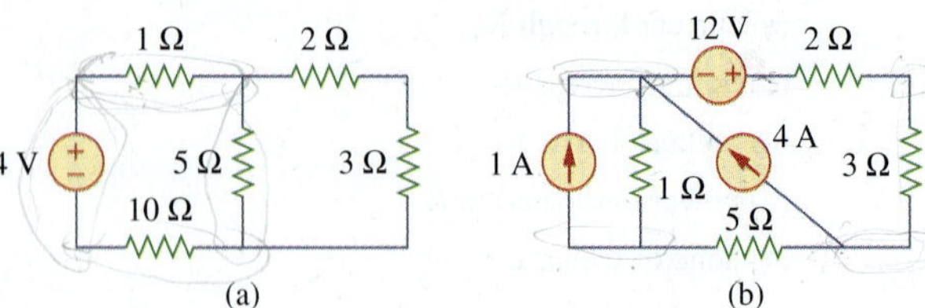

Figure 2.71
For Prob. 2.7.

Section 2.4 Kirchhoff's Laws

2.8 Design a problem, complete with a solution, to help other students better understand Kirchhoff's Current Law. Design the problem by specifying values of i_a, i_b, and i_c, shown in Fig. 2.72, and asking them to solve for values of i_1, i_2, and i_3. Be careful specify realistic currents.

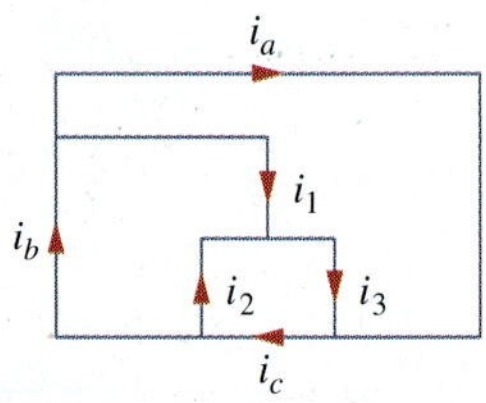

Figure 2.72
For Prob. 2.8.

2.9 Find i_1, i_2, and i_3 in Fig. 2.73.

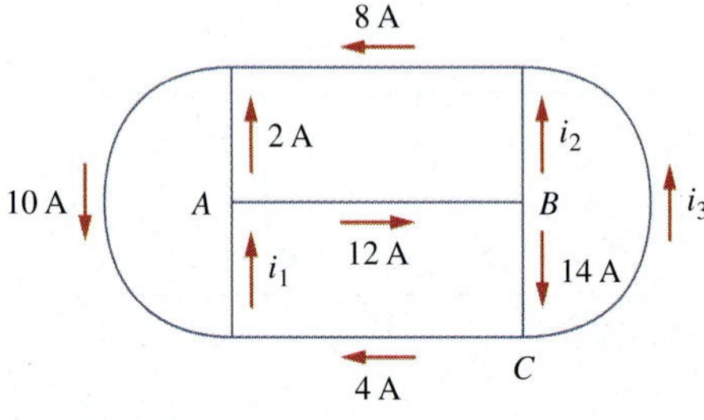

Figure 2.73
For Prob. 2.9.

2.10 Determine i_1 and i_2 in the circuit of Fig. 2.74.

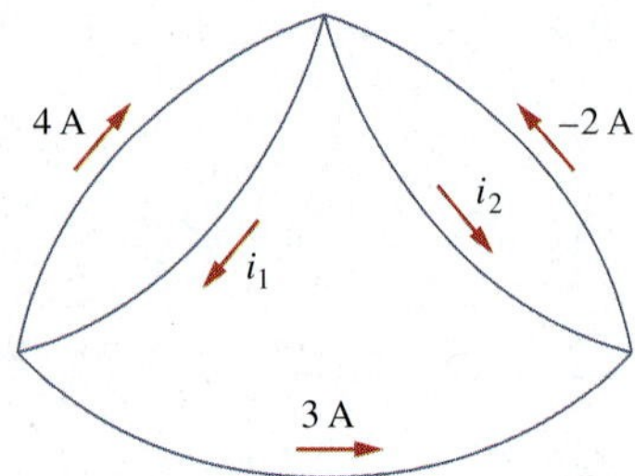

Figure 2.74
For Prob. 2.10.

2.11 In the circuit of Fig. 2.75, calculate V_1 and V_2.

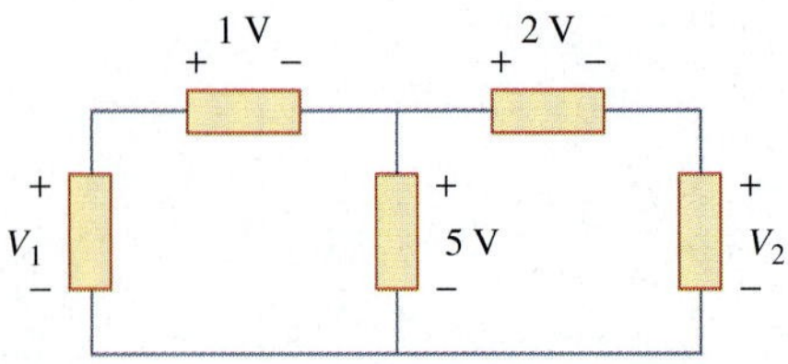

Figure 2.75
For Prob. 2.11.

2.12 In the circuit of Fig. 2.76, obtain v_1, v_2, and v_3.

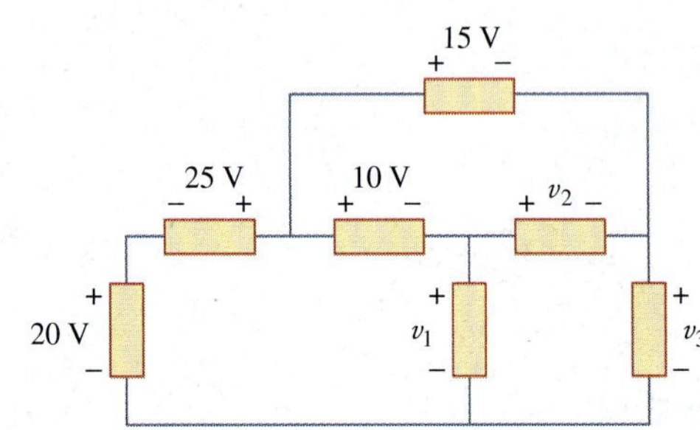

Figure 2.76
For Prob. 2.12.

2.13 For the circuit in Fig. 2.77, use KCL to find the branch currents I_1 to I_4.

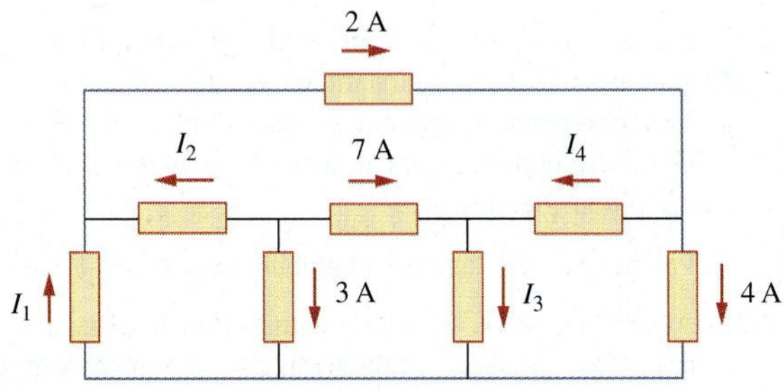

Figure 2.77
For Prob. 2.13.

2.14 Given the circuit in Fig. 2.78, use KVL to find the branch voltages V_1 to V_4.

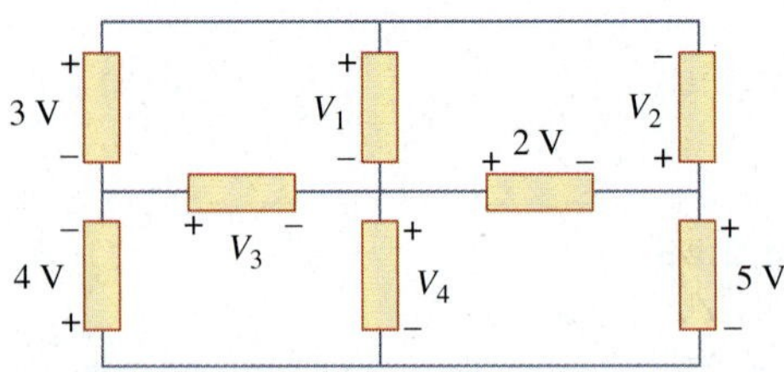

Figure 2.78
For Prob. 2.14.

2.15 Calculate v and i_x in the circuit of Fig. 2.79.

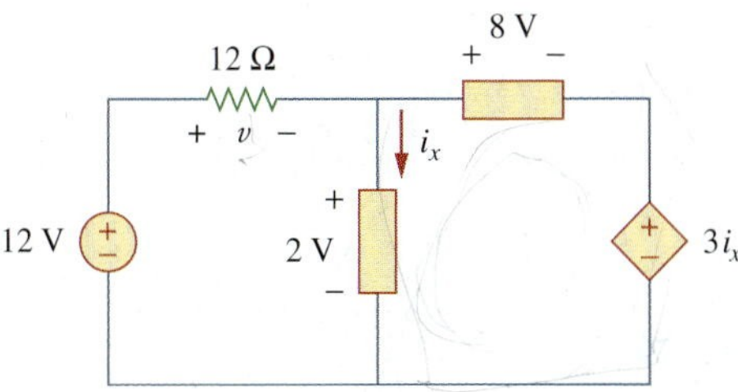

Figure 2.79
For Prob. 2.15.

2.16 Determine V_o in the circuit of Fig. 2.80.

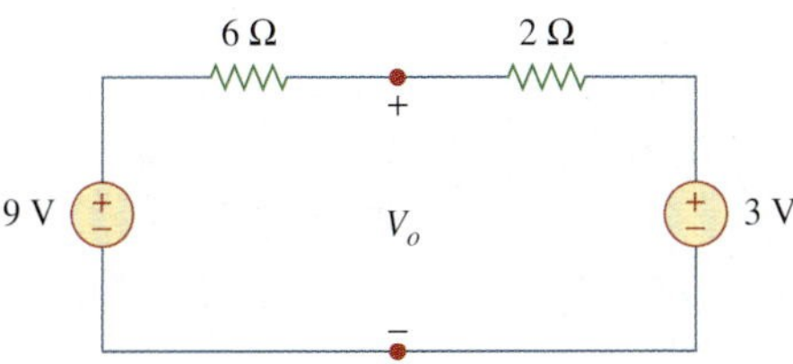

Figure 2.80
For Prob. 2.16.

2.17 Obtain v_1 through v_3 in the circuit of Fig. 2.81.

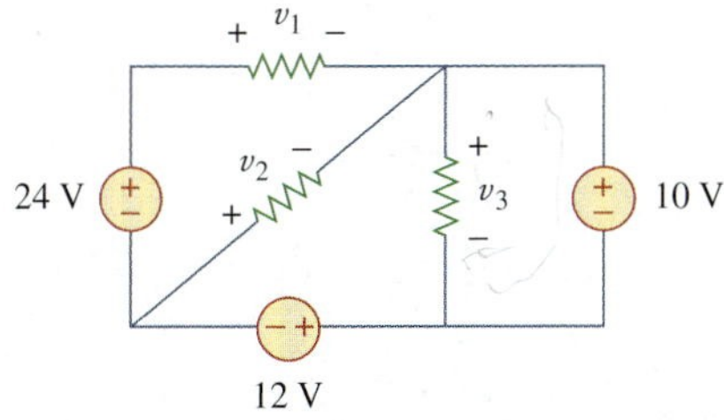

Figure 2.81
For Prob. 2.17.

2.18 Find I and V_{ab} in the circuit of Fig. 2.82.

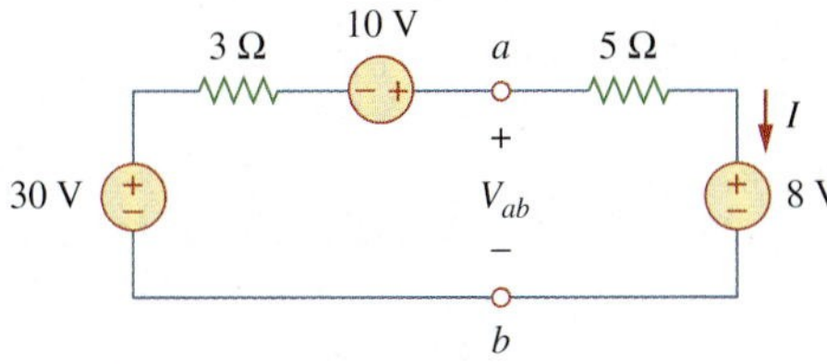

Figure 2.82
For Prob. 2.18.

2.19 From the circuit in Fig. 2.83, find I, the power dissipated by the resistor, and the power absorbed by each source.

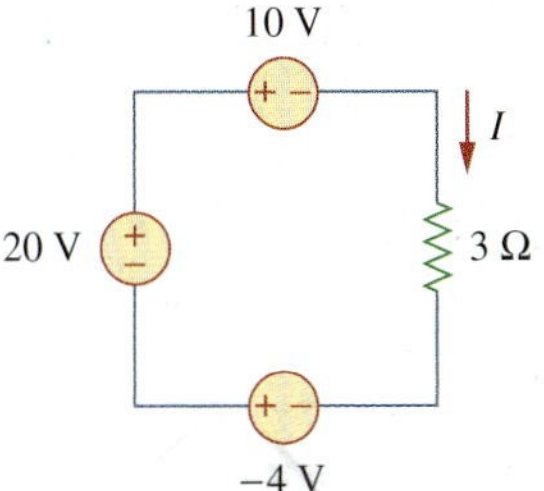

Figure 2.83
For Prob. 2.19.

2.20 Determine i_o in the circuit of Fig. 2.84.

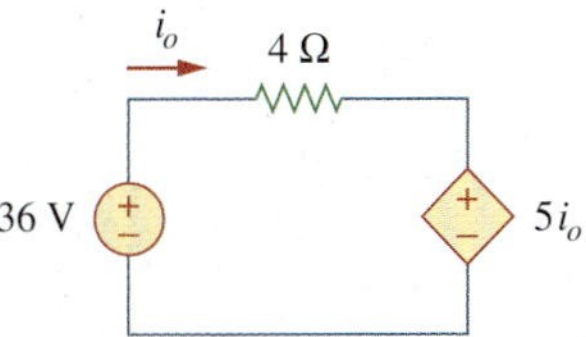

Figure 2.84
For Prob. 2.20.

2.21 Find V_x in the circuit of Fig. 2.85.

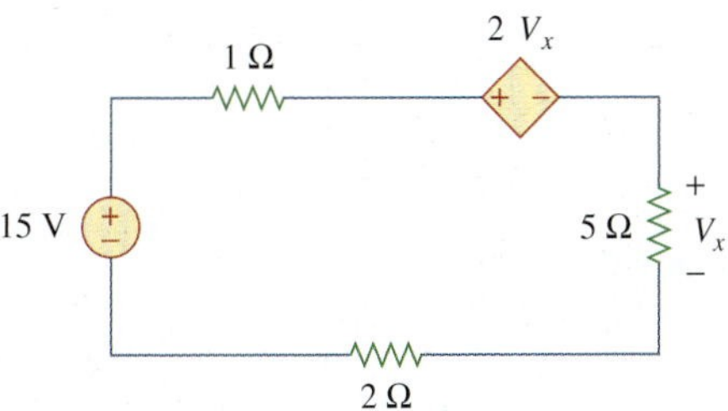

Figure 2.85
For Prob. 2.21.

2.22 Find V_o in the circuit of Fig. 2.86 and the power dissipated by the controlled source.

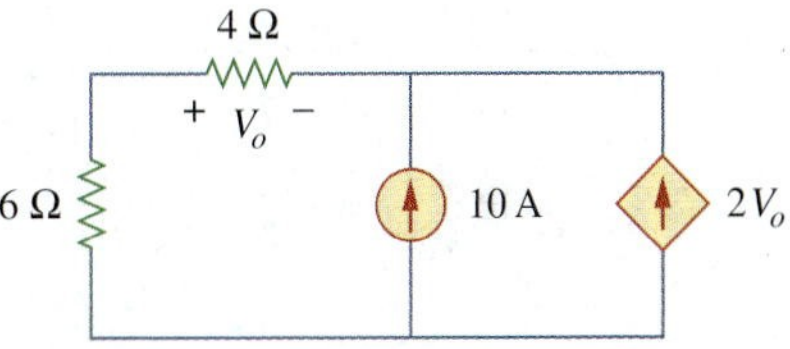

Figure 2.86
For Prob. 2.22.

2.23 In the circuit shown in Fig. 2.87, determine v_x and the power absorbed by the 12-Ω resistor.

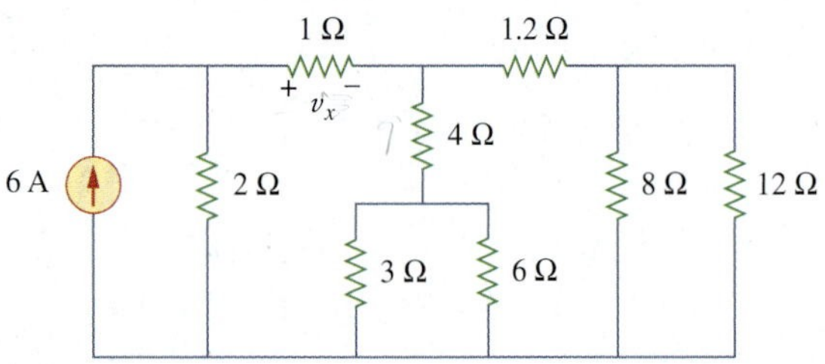

Figure 2.87
For Prob. 2.23.

2.24 For the circuit in Fig. 2.88, find V_o/V_s in terms of α, R_1, R_2, R_3, and R_4. If $R_1 = R_2 = R_3 = R_4$, what value of α will produce $|V_o/V_s| = 10$?

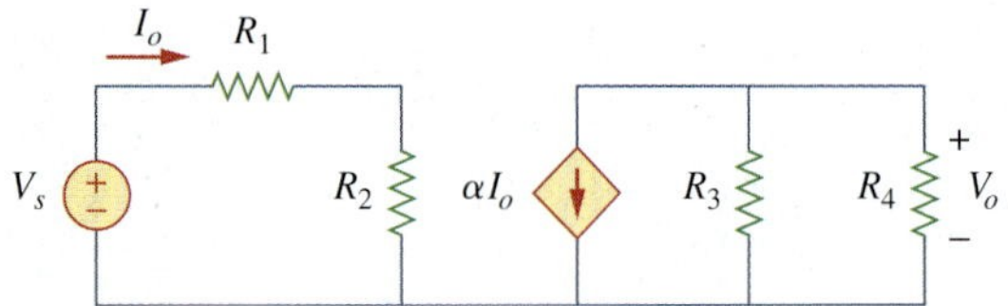

Figure 2.88
For Prob. 2.24.

2.25 For the network in Fig. 2.89, find the current, voltage, and power associated with the 20-kΩ resistor.

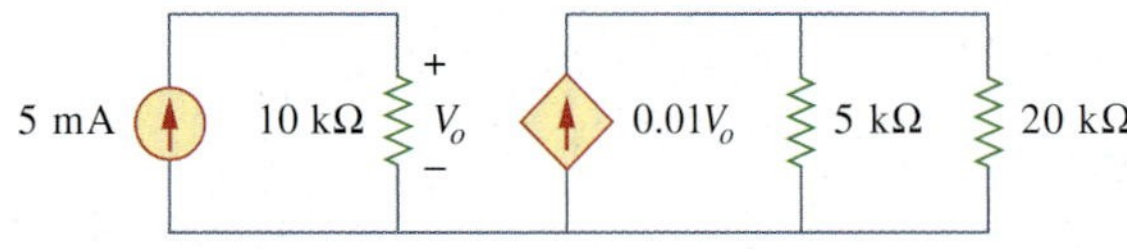

Figure 2.89
For Prob. 2.25.

Sections 2.5 and 2.6 Series and Parallel Resistors

2.26 For the circuit in Fig. 2.90, $i_o = 2$ A. Calculate i_x and the total power dissipated by the circuit.

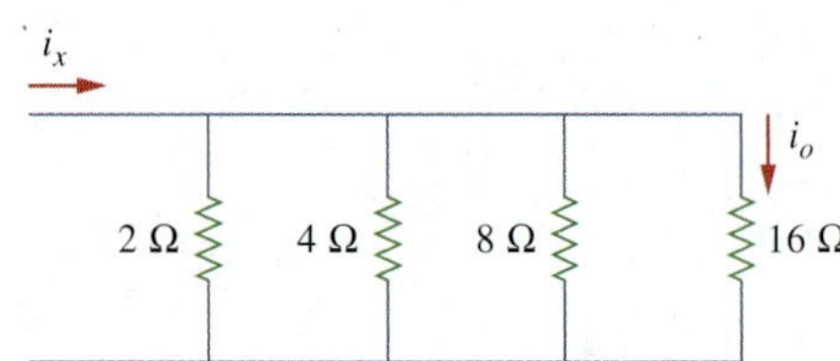

Figure 2.90
For Prob. 2.26.

2.27 Calculate V_o in the circuit of Fig. 2.91.

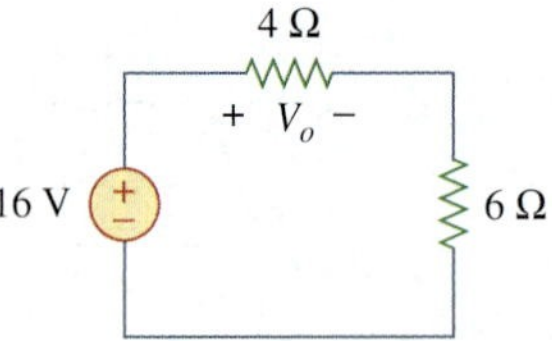

Figure 2.91
For Prob. 2.27.

2.28 Design a problem, using Fig. 2.92, to help other students better understand series and parallel circuits.

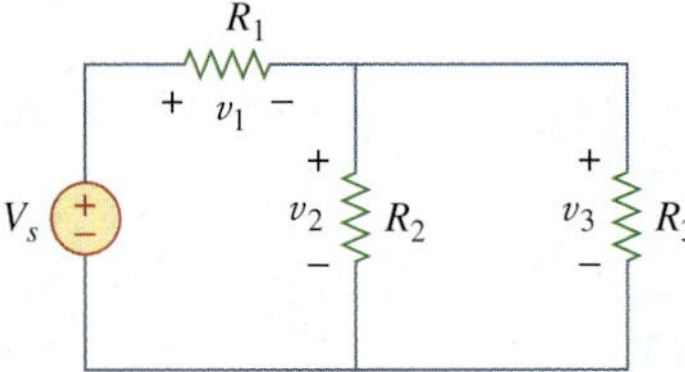

Figure 2.92
For Prob. 2.28.

2.29 All resistors in Fig. 2.93 are 1 Ω each. Find R_{eq}.

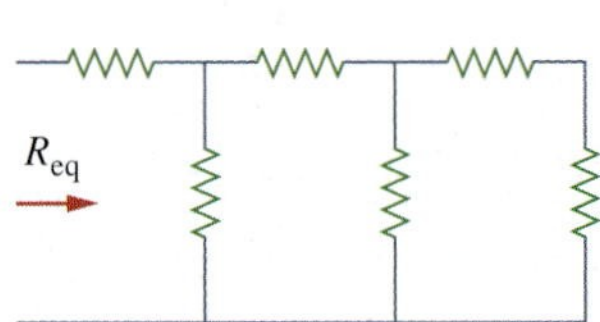

Figure 2.93
For Prob. 2.29.

2.30 Find R_{eq} for the circuit of Fig. 2.94.

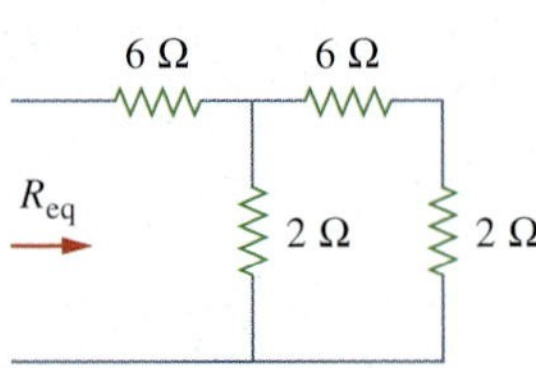

Figure 2.94
For Prob. 2.30.

2.31 For the circuit in Fig. 2.95, determine i_1 to i_5.

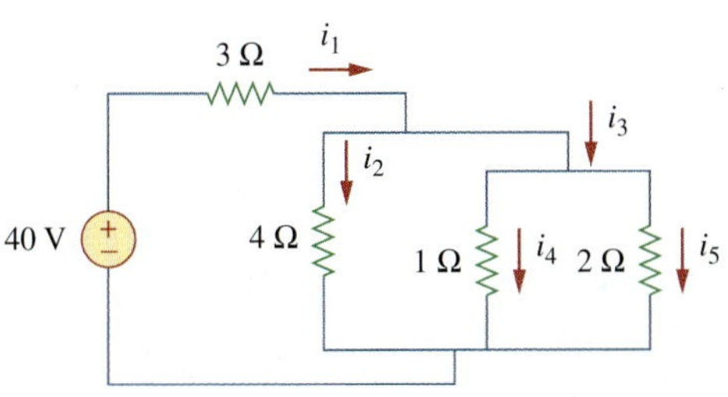

Figure 2.95
For Prob. 2.31.

2.32 Find i_1 through i_4 in the circuit of Fig. 2.96.

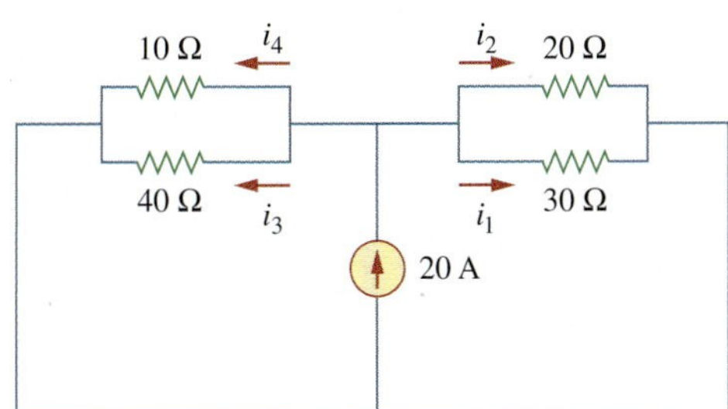

Figure 2.96
For Prob. 2.32.

2.33 Obtain v and i in the circuit of Fig. 2.97.

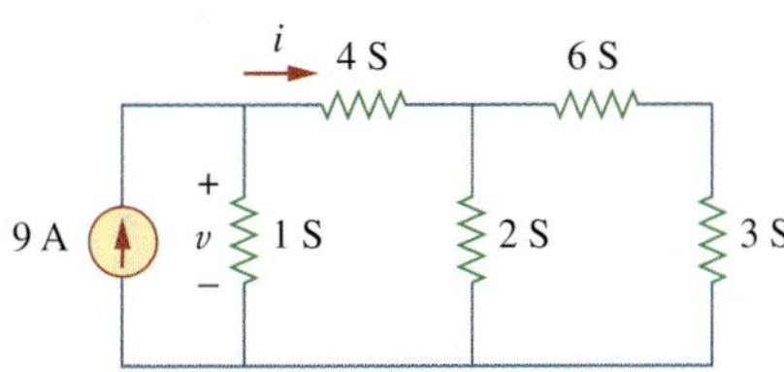

Figure 2.97
For Prob. 2.33.

2.34 Using series/parallel resistance combination, find the equivalent resistance seen by the source in the circuit of Fig. 2.98. Find the overall dissipated power.

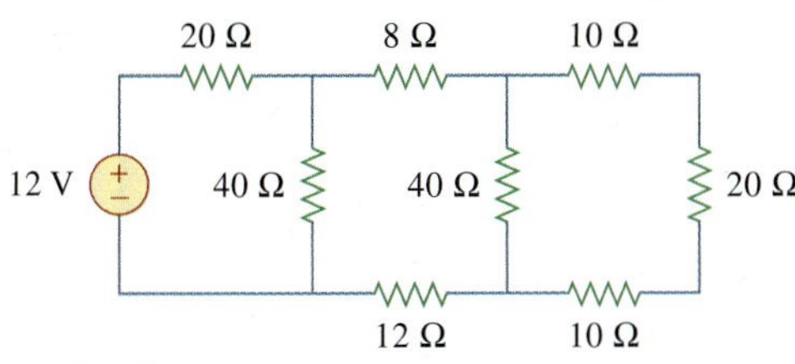

Figure 2.98
For Prob. 2.34.

2.35 Calculate V_o and I_o in the circuit of Fig. 2.99.

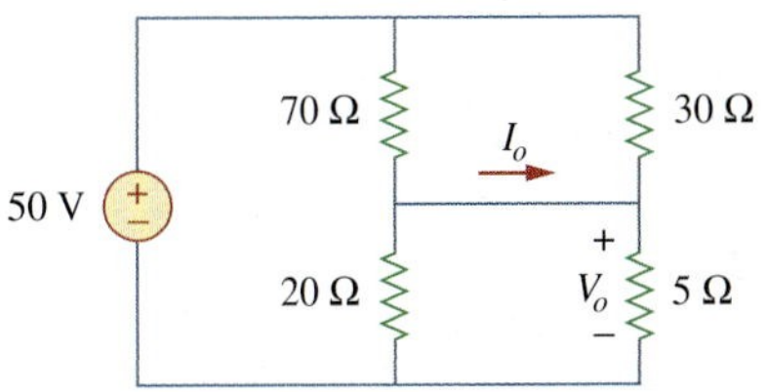

Figure 2.99
For Prob. 2.35.

2.36 Find i and V_o in the circuit of Fig. 2.100.

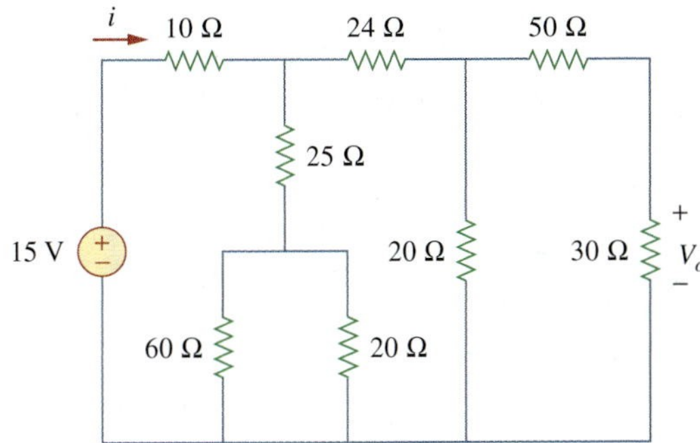

Figure 2.100
For Prob. 2.36.

2.37 Find R for the circuit in Fig. 2.101.

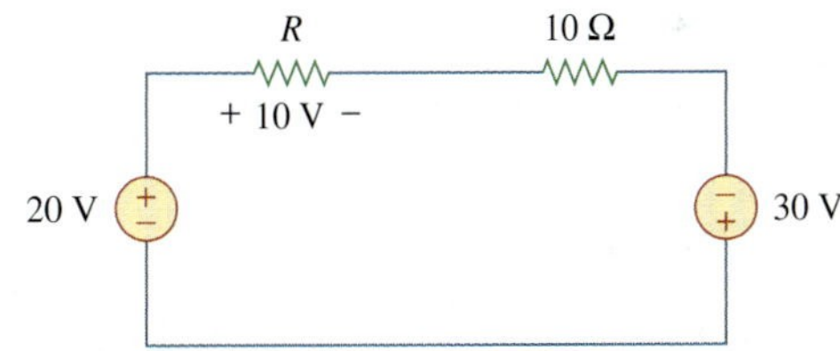

Figure 2.101
For Prob. 2.37.

2.38 Find R_{eq} and i_o in the circuit of Fig. 2.102.

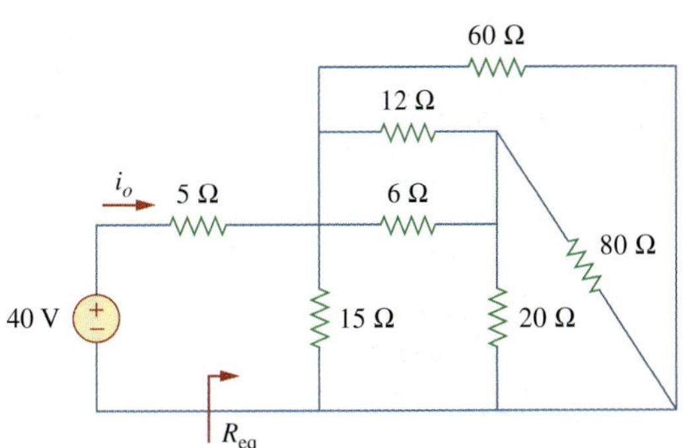

Figure 2.102
For Prob. 2.38.

2.39 Evaluate R_{eq} for each of the circuits shown in Fig. 2.103.

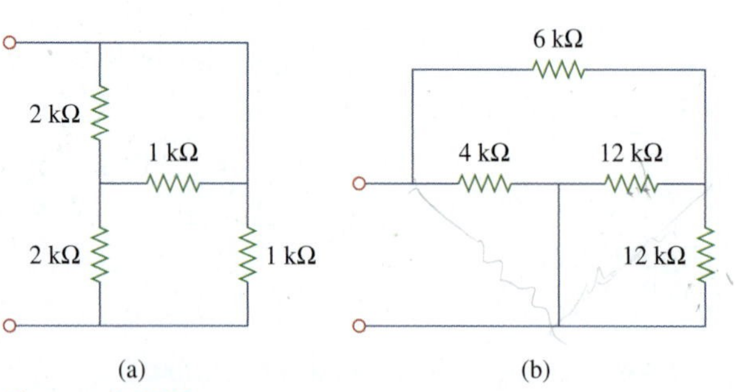

Figure 2.103
For Prob. 2.39.

2.40 For the ladder network in Fig. 2.104, find I and R_{eq}.

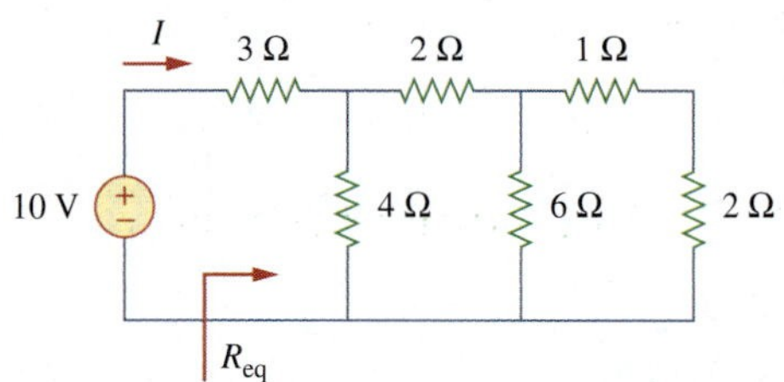

Figure 2.104
For Prob. 2.40.

2.41 If $R_{eq} = 50\ \Omega$ in the circuit of Fig. 2.105, find R.

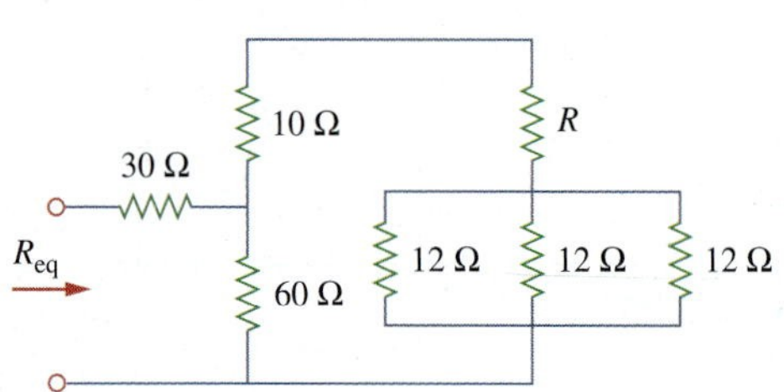

Figure 2.105
For Prob. 2.41.

2.42 Reduce each of the circuits in Fig. 2.106 to a single resistor at terminals a-b.

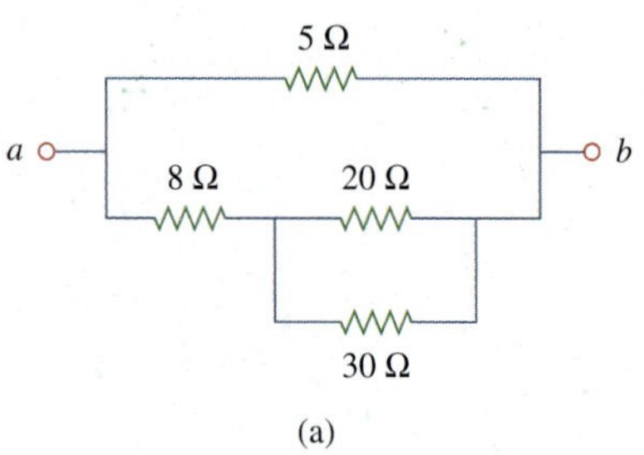

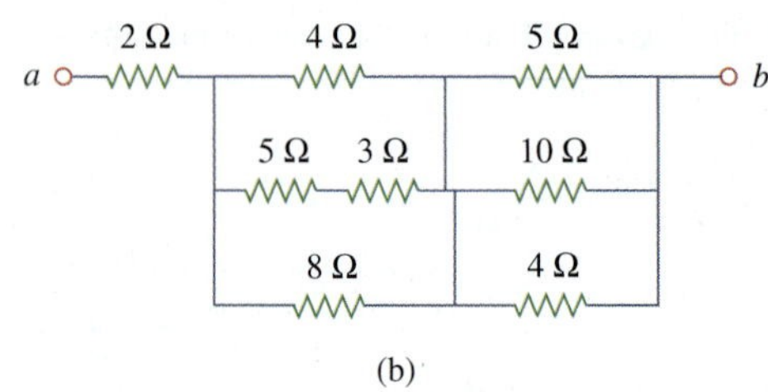

Figure 2.106
For Prob. 2.42.

2.43 Calculate the equivalent resistance R_{ab} at terminals a-b for each of the circuits in Fig. 2.107.

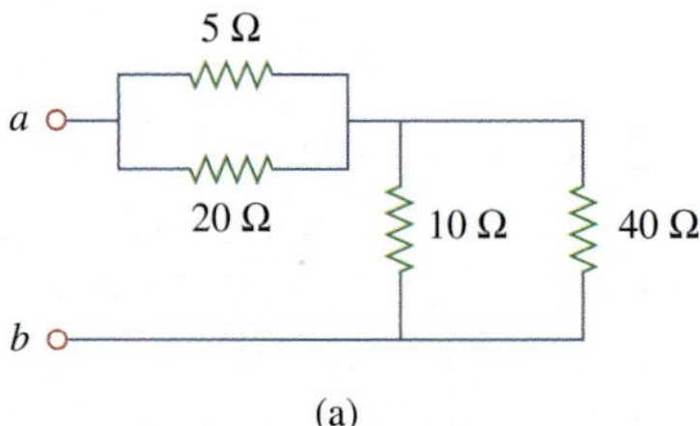

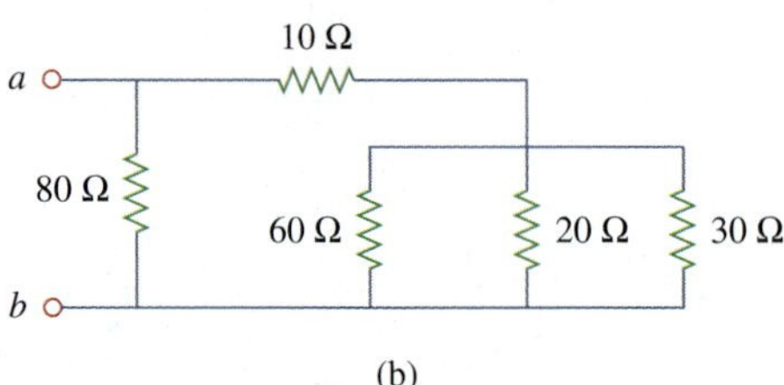

Figure 2.107
For Prob. 2.43.

2.44 For the circuit in Fig. 2.108, obtain the equivalent resistance at terminals a-b.

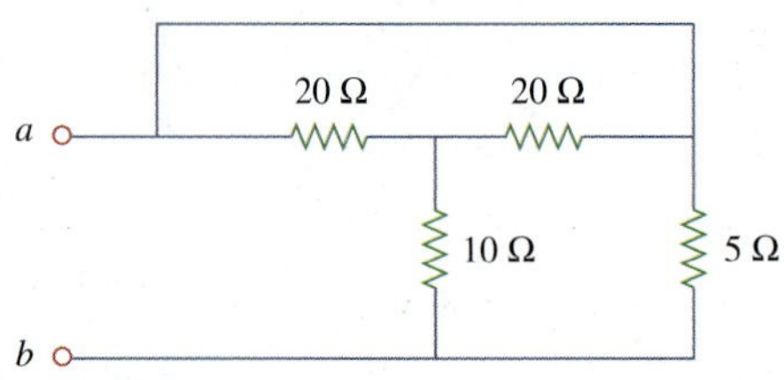

Figure 2.108
For Prob. 2.44.

2.45 Find the equivalent resistance at terminals *a*-*b* of each circuit in Fig. 2.109.

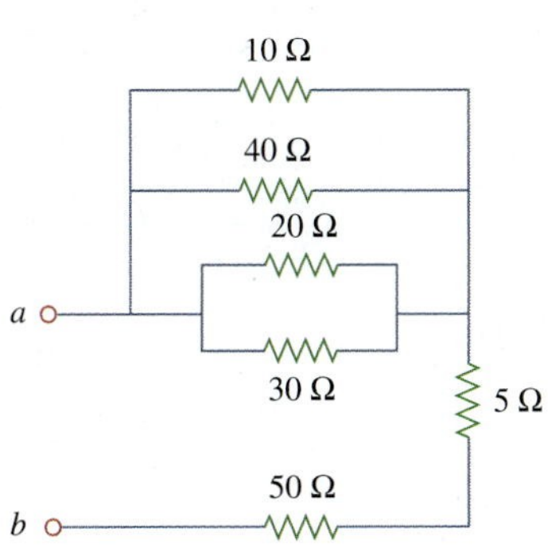

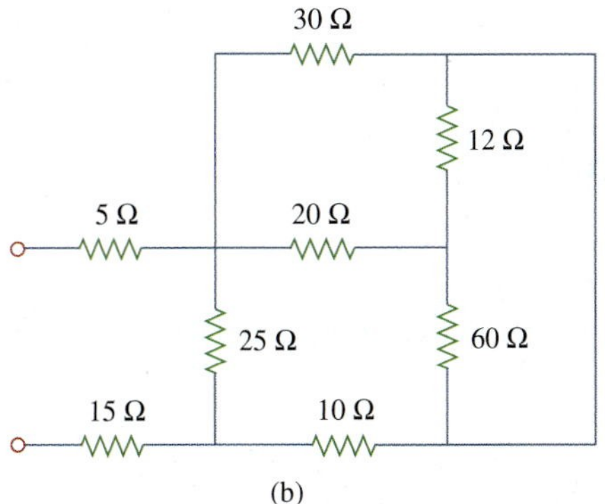

Figure 2.109
For Prob. 2.45.

2.46 Find I in the circuit of Fig. 2.110.

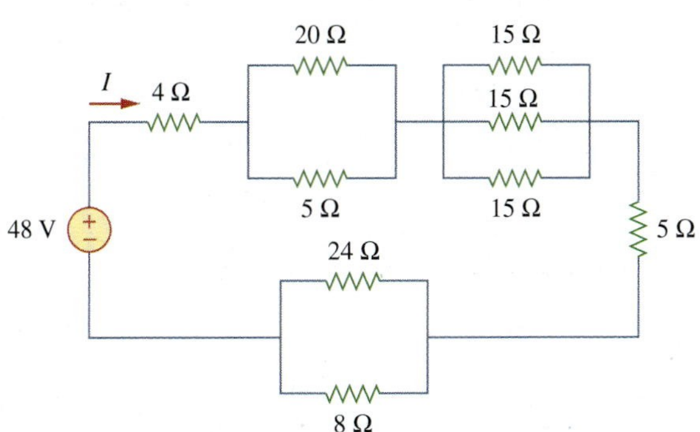

Figure 2.110
For Prob. 2.46.

2.47 Find the equivalent resistance R_{ab} in the circuit of Fig. 2.111.

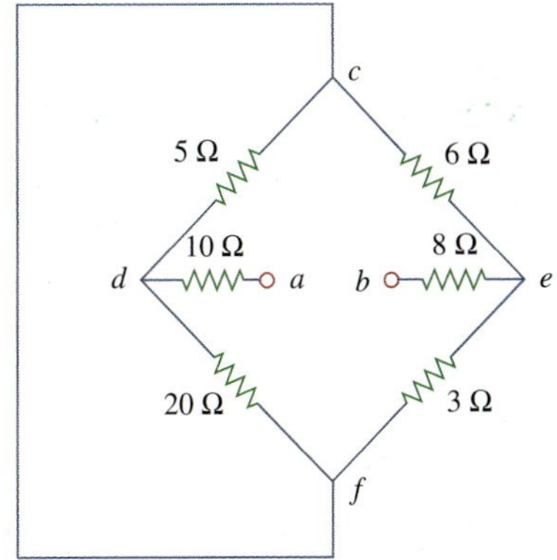

Figure 2.111
For Prob. 2.47.

Section 2.7 Wye-Delta Transformations

2.48 Convert the circuits in Fig. 2.112 from Y to Δ.

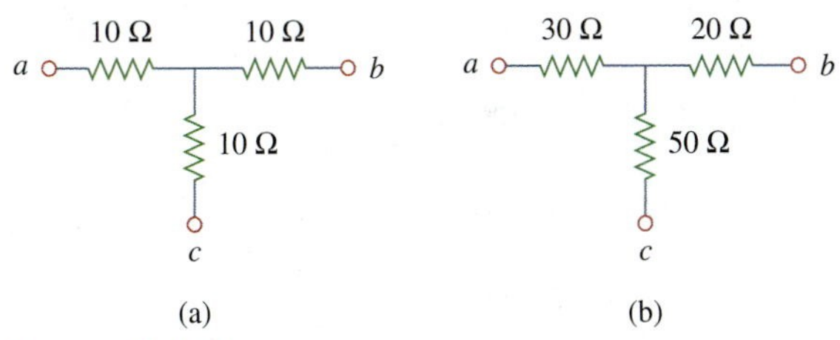

Figure 2.112
For Prob. 2.48.

2.49 Transform the circuits in Fig. 2.113 from Δ to Y.

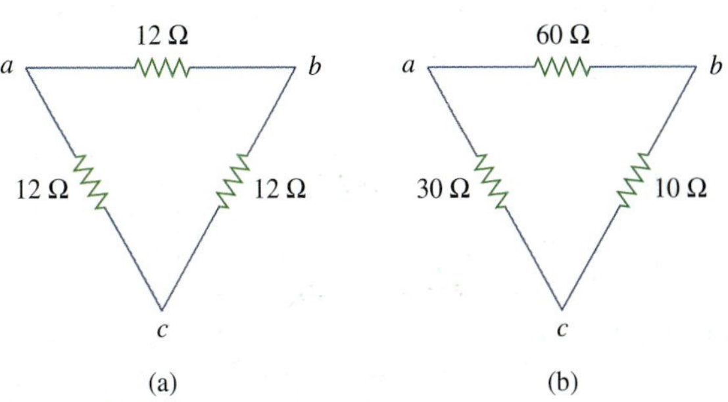

Figure 2.113
For Prob. 2.49.

2.50 Design a problem to help other students better understand wye-delta transformations using Fig. 2.114.

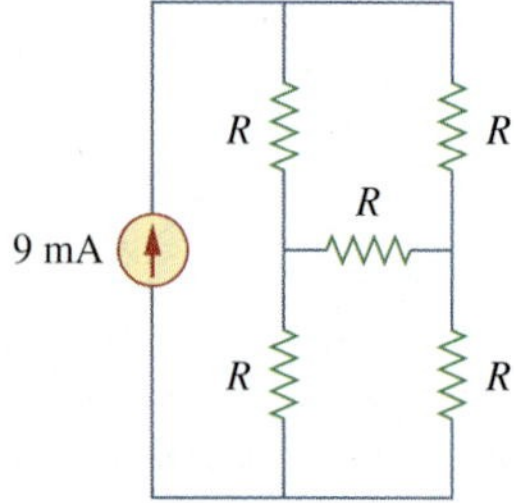

Figure 2.114
For Prob. 2.50.

2.51 Obtain the equivalent resistance at the terminals *a-b* for each of the circuits in Fig. 2.115.

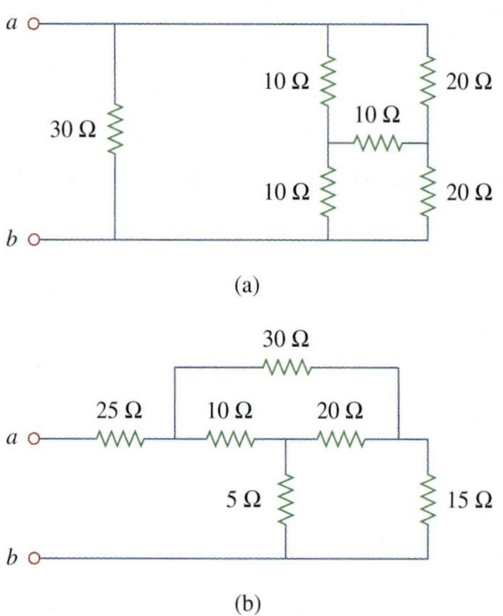

Figure 2.115
For Prob. 2.51.

*__2.52__ For the circuit shown in Fig. 2.116, find the equivalent resistance. All resistors are 1 Ω.

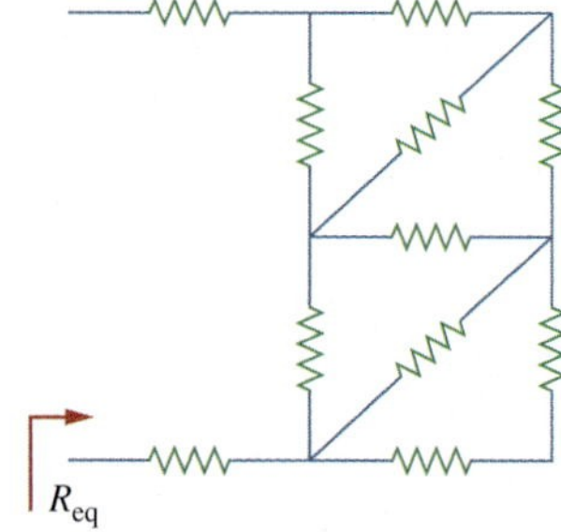

Figure 2.116
For Prob. 2.52.

* An asterisk indicates a challenging problem.

*__2.53__ Obtain the equivalent resistance R_{ab} in each of the circuits of Fig. 2.117. In (b), all resistors have a value of 30 Ω.

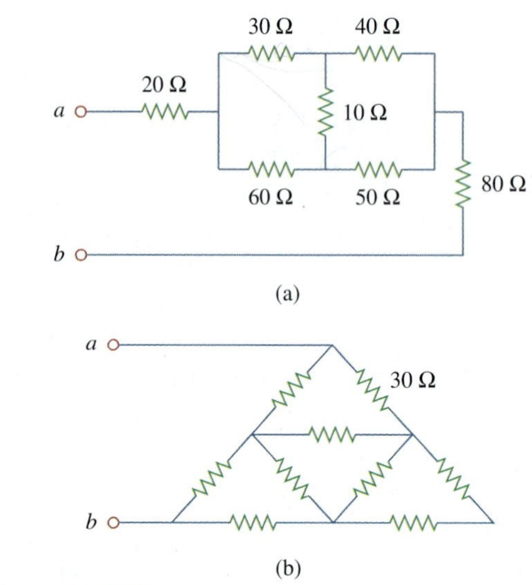

Figure 2.117
For Prob. 2.53.

2.54 Consider the circuit in Fig. 2.118. Find the equivalent resistance at terminals: (a) *a-b*, (b) *c-d*.

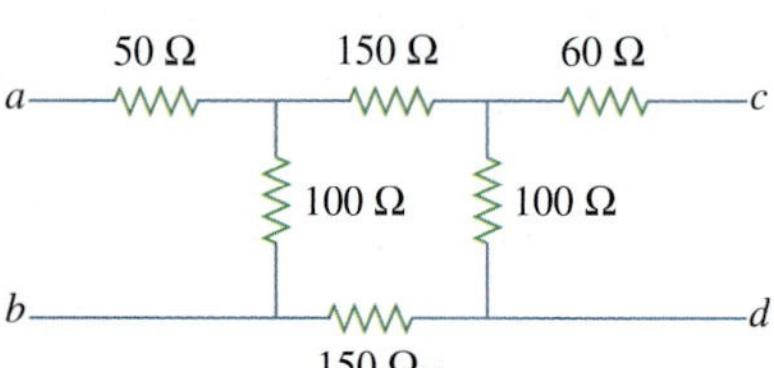

Figure 2.118
For Prob. 2.54.

2.55 Calculate I_o in the circuit of Fig. 2.119.

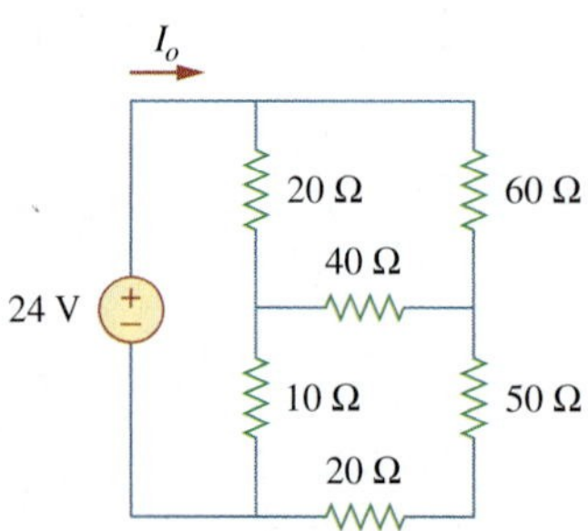

Figure 2.119
For Prob. 2.55.

2.56 Determine V in the circuit of Fig. 2.120.

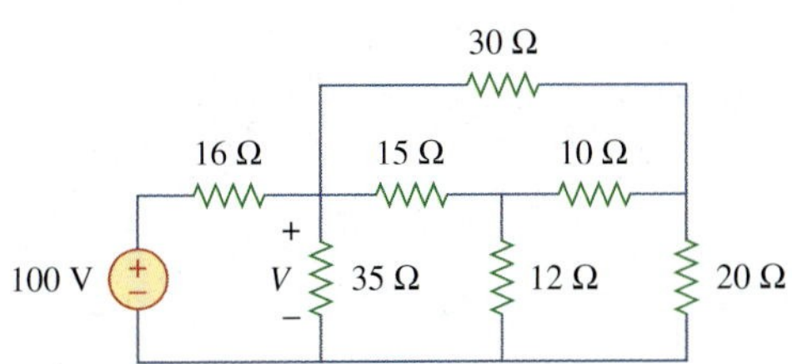

Figure 2.120
For Prob. 2.56.

***2.57** Find R_{eq} and I in the circuit of Fig. 2.121.

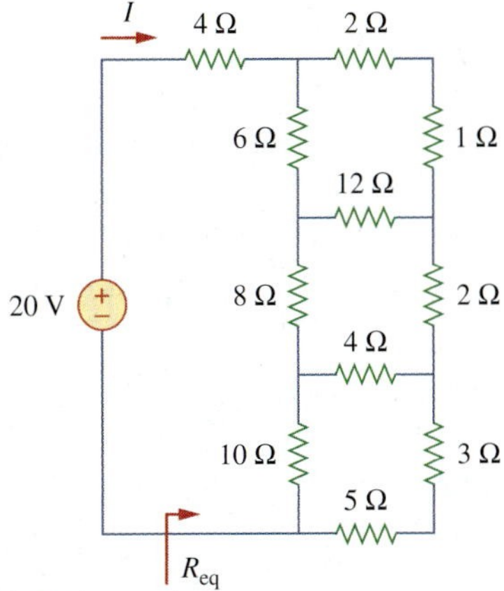

Figure 2.121
For Prob. 2.57.

Section 2.8 Applications

2.58 The lightbulb in Fig. 2.122 is rated 120 V, 0.75 A. Calculate V_s to make the lightbulb operate at the rated conditions.

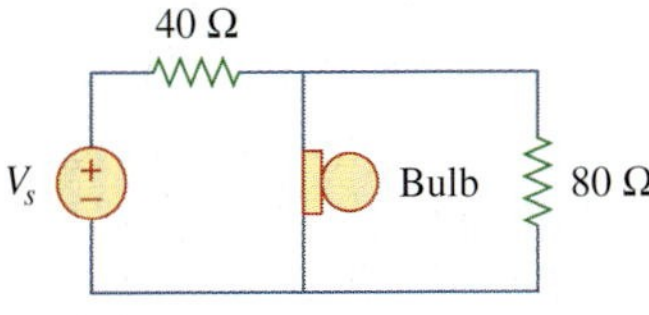

Figure 2.122
For Prob. 2.58.

2.59 Three lightbulbs are connected in series to a 100-V battery as shown in Fig. 2.123. Find the current I through the bulbs.

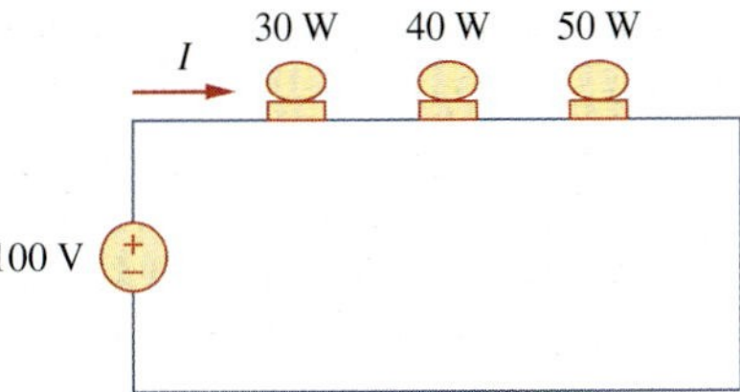

Figure 2.123
For Prob. 2.59.

2.60 If the three bulbs of Prob. 2.59 are connected in parallel to the 100-V battery, calculate the current through each bulb.

2.61 ed As a design engineer, you are asked to design a lighting system consisting of a 70-W power supply and two lightbulbs as shown in Fig. 2.124. You must select the two bulbs from the following three available bulbs.

$R_1 = 80\ \Omega$, cost = \$0.60 (standard size)
$R_2 = 90\ \Omega$, cost = \$0.90 (standard size)
$R_3 = 100\ \Omega$, cost = \$0.75 (nonstandard size)

The system should be designed for minimum cost such that lies within the range $I = 1.2\text{ A} \pm 5$ percent.

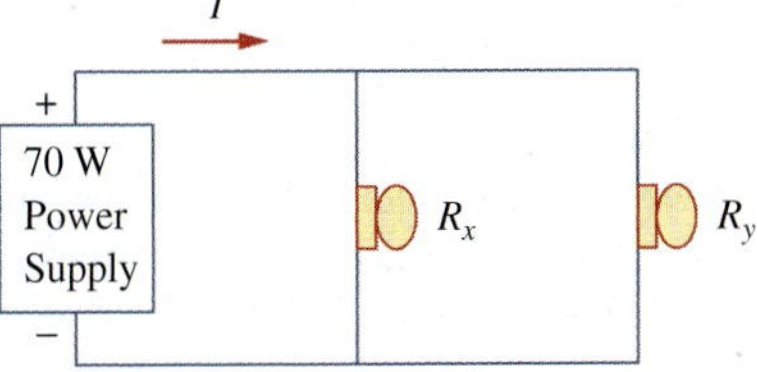

Figure 2.124
For Prob. 2.61.

2.62 A three-wire system supplies two loads A and B as shown in Fig. 2.125. Load A consists of a motor drawing a current of 8 A, while load B is a PC drawing 2 A. Assuming 10 h/day of use for 365 days and 6 cents/kWh, calculate the annual energy cost of the system.

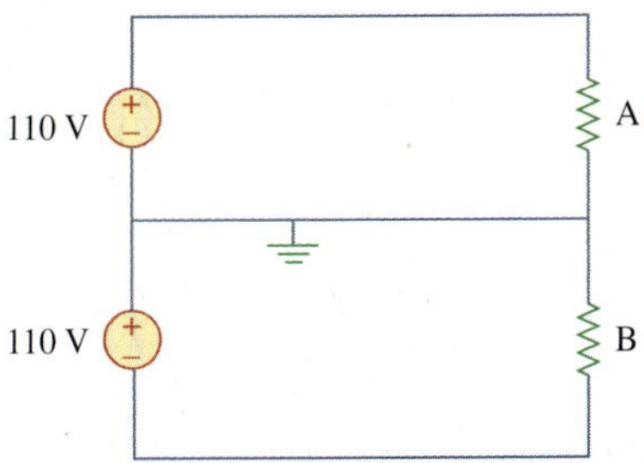

Figure 2.125
For Prob. 2.62.

2.63 If an ammeter with an internal resistance of 100 Ω and a current capacity of 2 mA is to measure 5 A, determine the value of the resistance needed.

Calculate the power dissipated in the shunt resistor.

2.64 The potentiometer (adjustable resistor) R_x in Fig. 2.126 is to be designed to adjust current i_x from 1 A to 10 A. Calculate the values of R and R_x to achieve this.

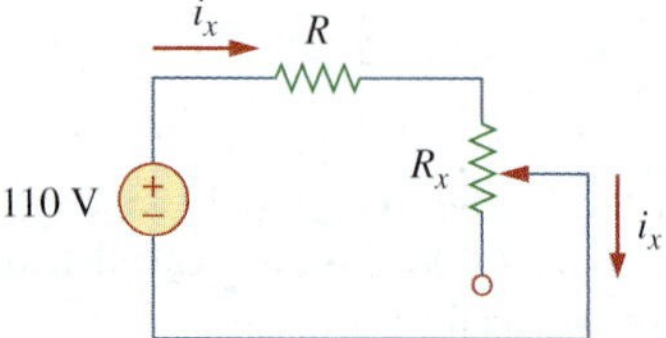

Figure 2.126
For Prob. 2.64.

2.65 A d'Arsonval meter with an internal resistance of 1 kΩ requires 10 mA to produce full-scale deflection. Calculate the value of a series resistance needed to measure 50 V of full scale.

2.66 A 20-kΩ/V voltmeter reads 10 V full scale.

(a) What series resistance is required to make the meter read 50 V full scale?

(b) What power will the series resistor dissipate when the meter reads full scale?

2.67 (a) Obtain the voltage V_o in the circuit of Fig. 2.127(a).

(b) Determine the voltage V_o' measured when a voltmeter with 6-kΩ internal resistance is connected as shown in Fig. 2.127(b).

(c) The finite resistance of the meter introduces an error into the measurement. Calculate the percent error as

$$\left|\frac{V_o - V_o'}{V_o}\right| \times 100\%$$

(d) Find the percent error if the internal resistance were 36 kΩ.

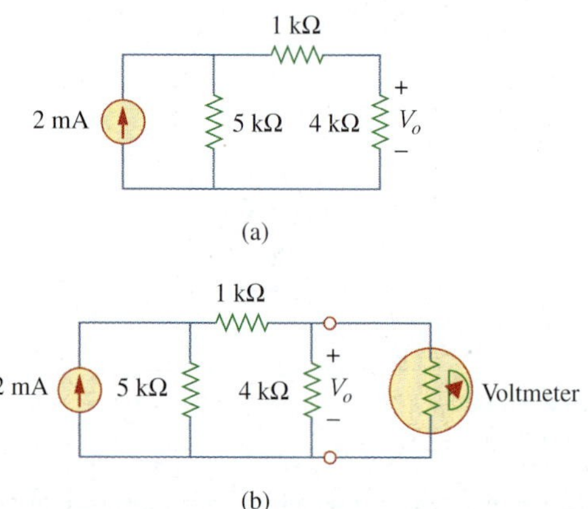

Figure 2.127
For Prob. 2.67.

2.68 (a) Find the current I in the circuit of Fig. 2.128(a).

(b) An ammeter with an internal resistance of 1 Ω is inserted in the network to measure I' as shown in Fig. 2.128(b). What is I'?

(c) Calculate the percent error introduced by the meter as

$$\left|\frac{I - I'}{I}\right| \times 100\%$$

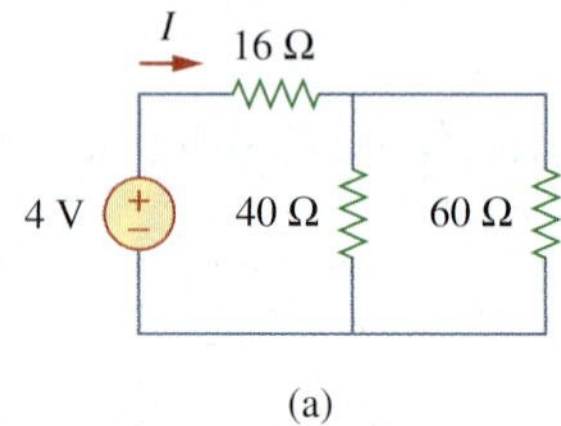

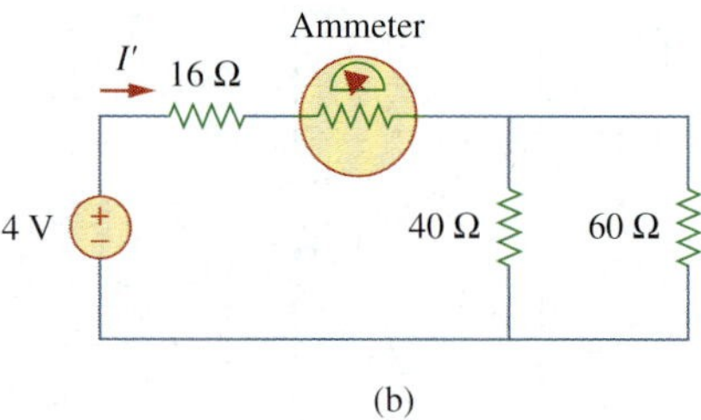

Figure 2.128
For Prob. 2.68.

2.69 A voltmeter is used to measure V_o in the circuit in Fig. 2.129. The voltmeter model consists of an ideal voltmeter in parallel with a 100-kΩ resistor. Let $V_s = 40$ V, $R_s = 10$ kΩ, and $R_1 = 20$ kΩ. Calculate V_o with and without the voltmeter when

(a) $R_2 = 1$ kΩ (b) $R_2 = 10$ kΩ

(c) $R_2 = 100$ kΩ

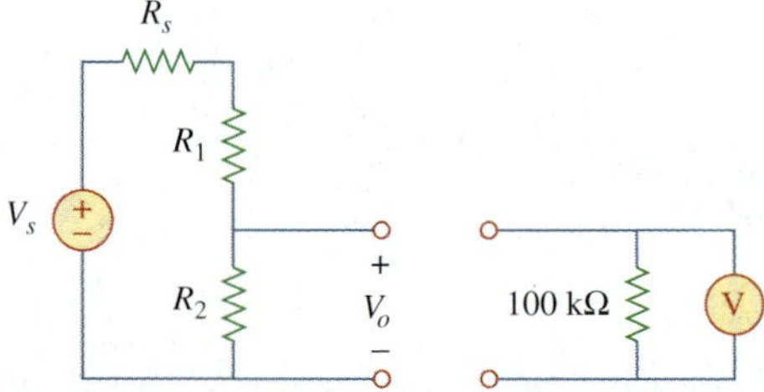

Figure 2.129
For Prob. 2.69.

2.70 (a) Consider the Wheatstone bridge shown in Fig. 2.130. Calculate v_a, v_b, and v_{ab}.

(b) Rework part (a) if the ground is placed at a instead of o.

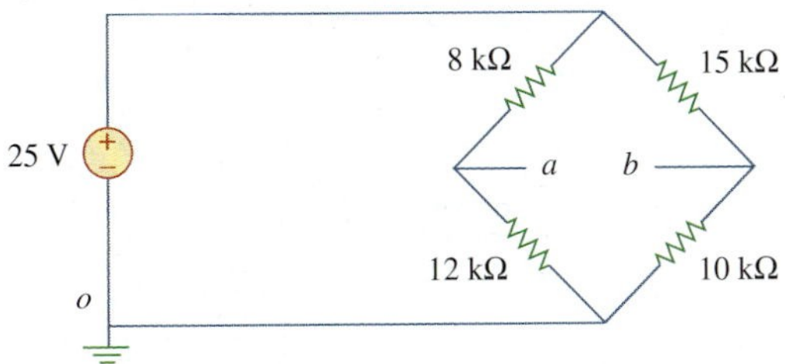

Figure 2.130
For Prob. 2.70.

2.71 Figure 2.131 represents a model of a solar photovoltaic panel. Given that $V_s = 30$ V, $R_1 = 20\ \Omega$, and $i_L = 1$ A, find R_L.

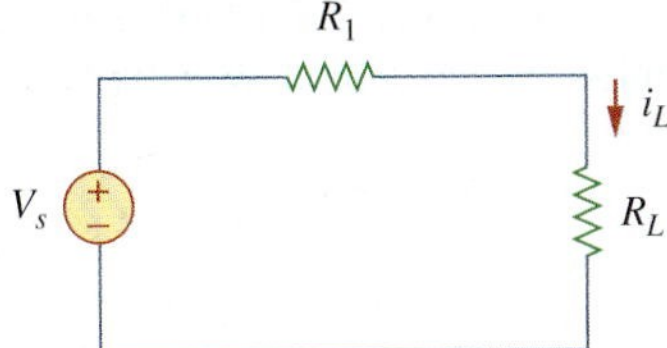

Figure 2.131
For Prob. 2.71.

2.72 Find V_o in the two-way power divider circuit in Fig. 2.132.

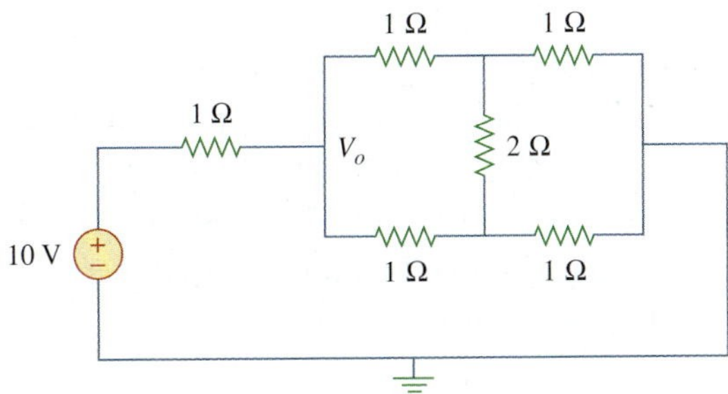

Figure 2.132
For Prob. 2.72.

2.73 An ammeter model consists of an ideal ammeter in series with a 20-Ω resistor. It is connected with a current source and an unknown resistor R_x as shown in Fig. 2.133. The ammeter reading is noted. When a potentiometer R is added and adjusted until the ammeter reading drops to one half its previous reading, then $R = 65\ \Omega$. What is the value of R_x?

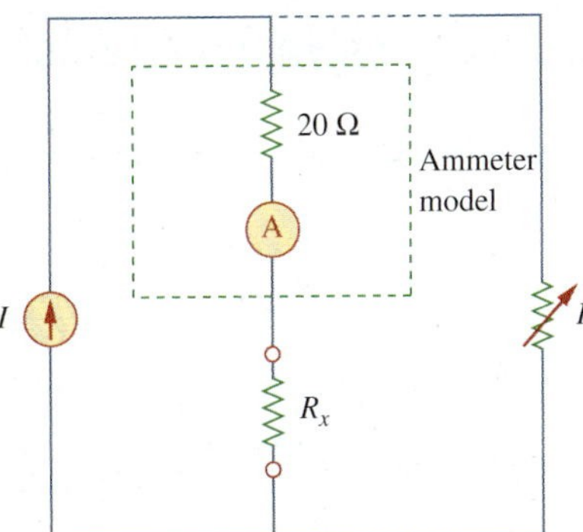

Figure 2.133
For Prob. 2.73.

2.74 The circuit in Fig. 2.134 is to control the speed of a motor such that the motor draws currents 5 A, 3 A, and 1 A when the switch is at high, medium, and low positions, respectively. The motor can be modeled as a load resistance of 20 mΩ. Determine the series dropping resistances R_1, R_2, and R_3.

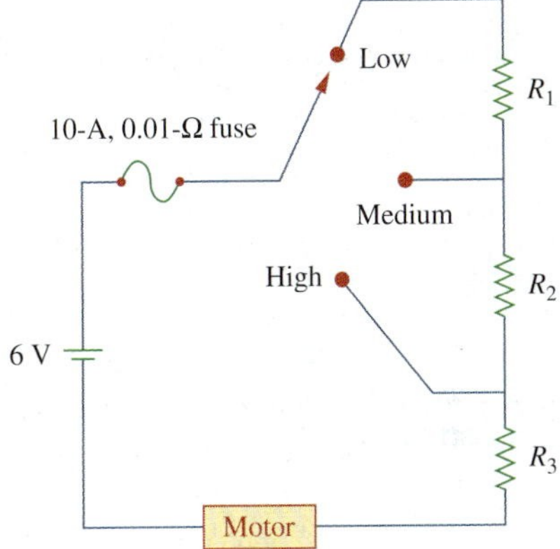

Figure 2.134
For Prob. 2.74.

2.75 Find R_{ab} in the four-way power divider circuit in Fig. 2.135. Assume each element is 1 Ω.

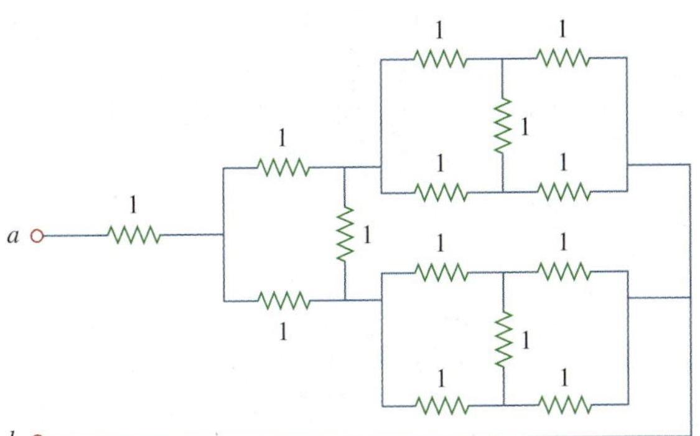

Figure 2.135
For Prob. 2.75.

Comprehensive Problems

2.76 Repeat Prob. 2.75 for the eight-way divider shown in Fig. 2.136.

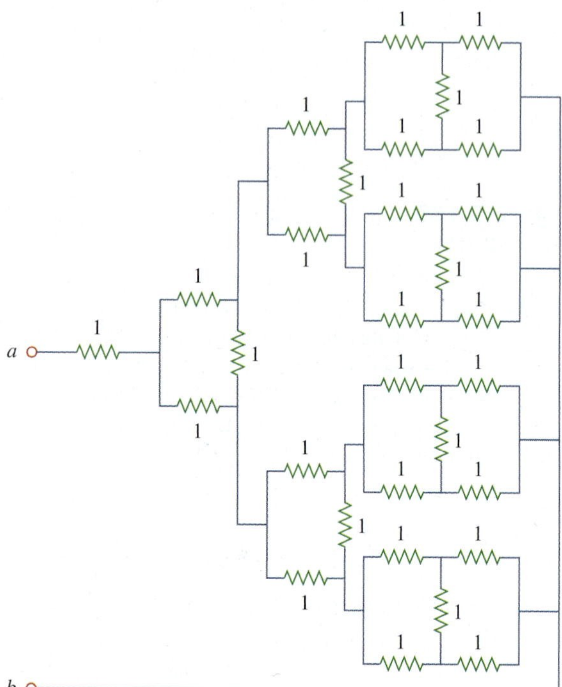

Figure 2.136
For Prob. 2.76.

2.77 Suppose your circuit laboratory has the following standard commercially available resistors in large quantities:

1.8 Ω 20 Ω 300 Ω 24 kΩ 56 kΩ

Using series and parallel combinations and a minimum number of available resistors, how would you obtain the following resistances for an electronic circuit design?

(a) 5 Ω (b) 311.8 Ω
(c) 40 kΩ (d) 52.32 kΩ

2.78 In the circuit in Fig. 2.137, the wiper divides the potentiometer resistance between αR and $(1 - \alpha)R$, $0 \leq \alpha \leq 1$. Find v_o/v_s.

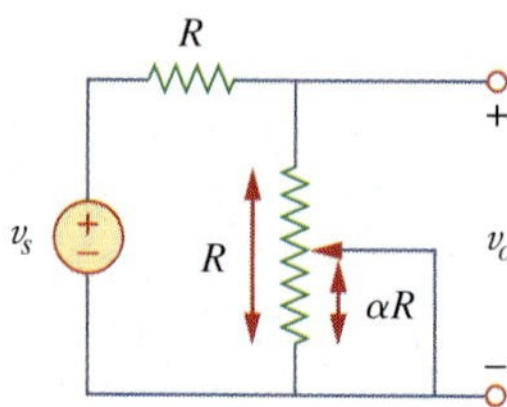

Figure 2.137
For Prob. 2.78.

2.79 An electric pencil sharpener rated 240 mW, 6 V is connected to a 9-V battery as shown in Fig. 2.138. Calculate the value of the series-dropping resistor R_x needed to power the sharpener.

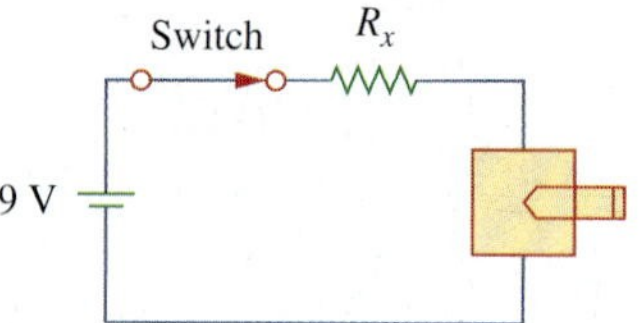

Figure 2.138
For Prob. 2.79.

2.80 A loudspeaker is connected to an amplifier as shown in Fig. 2.139. If a 10-Ω loudspeaker draws the maximum power of 12 W from the amplifier, determine the maximum power a 4-Ω loudspeaker will draw.

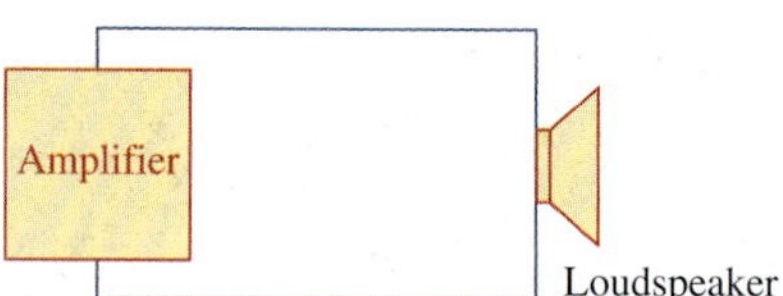

Figure 2.139
For Prob. 2.80.

2.81 In a certain application, the circuit in Fig. 2.140 must be designed to meet these two criteria:

(a) $V_o/V_s = 0.05$ (b) $R_{eq} = 40$ kΩ

If the load resistor 5 kΩ is fixed, find R_1 and R_2 to meet the criteria.

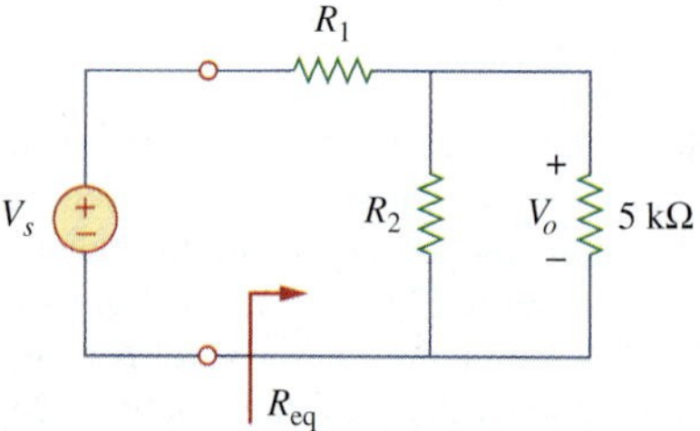

Figure 2.140
For Prob. 2.81.

2.82 The pin diagram of a resistance array is shown in Fig. 2.141. Find the equivalent resistance between the following:

(a) 1 and 2

(b) 1 and 3

(c) 1 and 4

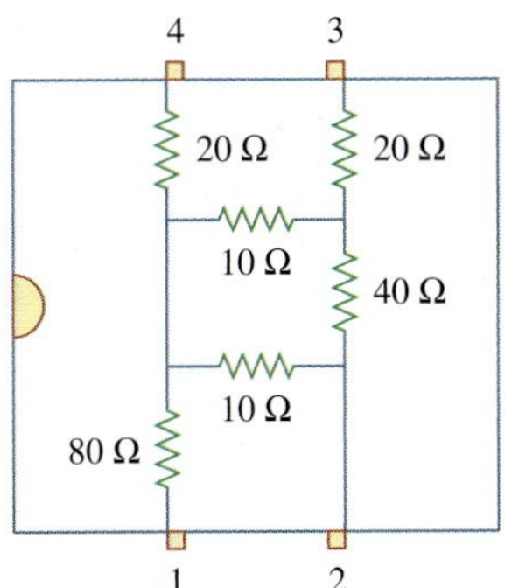

Figure 2.141
For Prob. 2.82.

2.83 Two delicate devices are rated as shown in Fig. 2.142. Find the values of the resistors R_1 and R_2 needed to power the devices using a 24-V battery.

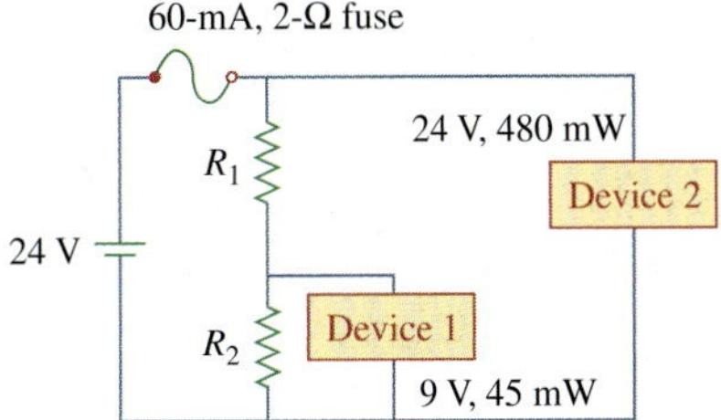

Figure 2.142
For Prob. 2.83.

3

Methods of Analysis

No great work is ever done in a hurry. To develop a great scientific discovery, to print a great picture, to write an immortal poem, to become a minister, or a famous general—to do anything great requires time, patience, and perseverance. These things are done by degrees, "little by little."

—W. J. Wilmont Buxton

Enhancing Your Career

Career in Electronics

One area of application for electric circuit analysis is electronics. The term *electronics* was originally used to distinguish circuits of very low current levels. This distinction no longer holds, as power semiconductor devices operate at high levels of current. Today, electronics is regarded as the science of the motion of charges in a gas, vacuum, or semiconductor. Modern electronics involves transistors and transistor circuits. The earlier electronic circuits were assembled from components. Many electronic circuits are now produced as integrated circuits, fabricated in a semiconductor substrate or chip.

Electronic circuits find applications in many areas, such as automation, broadcasting, computers, and instrumentation. The range of devices that use electronic circuits is enormous and is limited only by our imagination. Radio, television, computers, and stereo systems are but a few.

An electrical engineer usually performs diverse functions and is likely to use, design, or construct systems that incorporate some form of electronic circuits. Therefore, an understanding of the operation and analysis of electronics is essential to the electrical engineer. Electronics has become a specialty distinct from other disciplines within electrical engineering. Because the field of electronics is ever advancing, an electronics engineer must update his/her knowledge from time to time. The best way to do this is by being a member of a professional organization such as the Institute of Electrical and Electronics Engineers (IEEE). With a membership of over 300,000, the IEEE is the largest professional organization in the world. Members benefit immensely from the numerous magazines, journals, transactions, and conference/symposium proceedings published yearly by IEEE. You should consider becoming an IEEE member.

Troubleshooting an electronic circuit board.
© BrandX Pictures/Punchstock

3.1 Introduction

Having understood the fundamental laws of circuit theory (Ohm's law and Kirchhoff's laws), we are now prepared to apply these laws to develop two powerful techniques for circuit analysis: nodal analysis, which is based on a systematic application of Kirchhoff's current law (KCL), and mesh analysis, which is based on a systematic application of Kirchhoff's voltage law (KVL). The two techniques are so important that this chapter should be regarded as the most important in the book. Students are therefore encouraged to pay careful attention.

With the two techniques to be developed in this chapter, we can analyze any linear circuit by obtaining a set of simultaneous equations that are then solved to obtain the required values of current or voltage. One method of solving simultaneous equations involves Cramer's rule, which allows us to calculate circuit variables as a quotient of determinants. The examples in the chapter will illustrate this method; Appendix A also briefly summarizes the essentials the reader needs to know for applying Cramer's rule. Another method of solving simultaneous equations is to use *MATLAB*, a computer software discussed in Appendix E.

Also in this chapter, we introduce the use of *PSpice for Windows*, a circuit simulation computer software program that we will use throughout the text. Finally, we apply the techniques learned in this chapter to analyze transistor circuits.

3.2 Nodal Analysis

Nodal analysis is also known as the *node-voltage method.*

Nodal analysis provides a general procedure for analyzing circuits using node voltages as the circuit variables. Choosing node voltages instead of element voltages as circuit variables is convenient and reduces the number of equations one must solve simultaneously.

To simplify matters, we shall assume in this section that circuits do not contain voltage sources. Circuits that contain voltage sources will be analyzed in the next section.

In *nodal analysis*, we are interested in finding the node voltages. Given a circuit with n nodes without voltage sources, the nodal analysis of the circuit involves taking the following three steps.

Steps to Determine Node Voltages:

1. Select a node as the reference node. Assign voltages v_1, $v_2, \dots, v_{n-1}$ to the remaining $n - 1$ nodes. The voltages are referenced with respect to the reference node.
2. Apply KCL to each of the $n - 1$ nonreference nodes. Use Ohm's law to express the branch currents in terms of node voltages.
3. Solve the resulting simultaneous equations to obtain the unknown node voltages.

We shall now explain and apply these three steps.

The first step in nodal analysis is selecting a node as the *reference* or *datum node*. The reference node is commonly called the *ground* since it is assumed to have zero potential. A reference node is indicated by any of the three symbols in Fig. 3.1. The type of ground in Fig. 3.1(c) is called a *chassis ground* and is used in devices where the case, enclosure, or chassis acts as a reference point for all circuits. When the potential of the earth is used as reference, we use the *earth ground* in Fig. 3.1(a) or (b). We shall always use the symbol in Fig. 3.1(b).

The number of nonreference nodes is equal to the number of independent equations that we will derive.

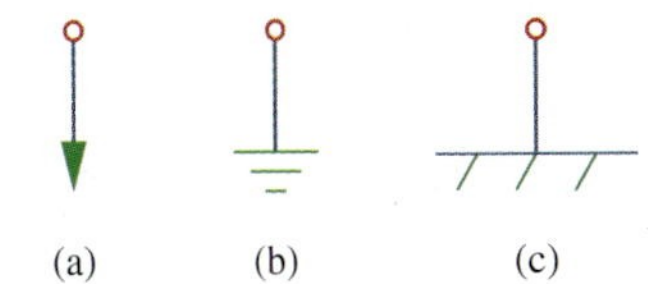

Figure 3.1
Common symbols for indicating a reference node, (a) common ground, (b) ground, (c) chassis ground.

Once we have selected a reference node, we assign voltage designations to nonreference nodes. Consider, for example, the circuit in Fig. 3.2(a). Node 0 is the reference node ($v = 0$), while nodes 1 and 2 are assigned voltages v_1 and v_2, respectively. Keep in mind that the node voltages are defined with respect to the reference node. As illustrated in Fig. 3.2(a), each node voltage is the voltage rise from the reference node to the corresponding nonreference node or simply the voltage of that node with respect to the reference node.

As the second step, we apply KCL to each nonreference node in the circuit. To avoid putting too much information on the same circuit, the circuit in Fig. 3.2(a) is redrawn in Fig. 3.2(b), where we now add i_1, i_2, and i_3 as the currents through resistors R_1, R_2, and R_3, respectively. At node 1, applying KCL gives

$$I_1 = I_2 + i_1 + i_2 \tag{3.1}$$

At node 2,

$$I_2 + i_2 = i_3 \tag{3.2}$$

We now apply Ohm's law to express the unknown currents i_1, i_2, and i_3 in terms of node voltages. The key idea to bear in mind is that, since resistance is a passive element, by the passive sign convention, current must always flow from a higher potential to a lower potential.

Current flows from **a higher** potential to **a lower** potential in a resistor.

We can express this principle as

$$i = \frac{v_{\text{higher}} - v_{\text{lower}}}{R} \tag{3.3}$$

Note that this principle is in agreement with the way we defined resistance in Chapter 2 (see Fig. 2.1). With this in mind, we obtain from Fig. 3.2(b),

$$\begin{aligned} i_1 &= \frac{v_1 - 0}{R_1} \quad \text{or} \quad i_1 = G_1 v_1 \\ i_2 &= \frac{v_1 - v_2}{R_2} \quad \text{or} \quad i_2 = G_2(v_1 - v_2) \\ i_3 &= \frac{v_2 - 0}{R_3} \quad \text{or} \quad i_3 = G_3 v_2 \end{aligned} \tag{3.4}$$

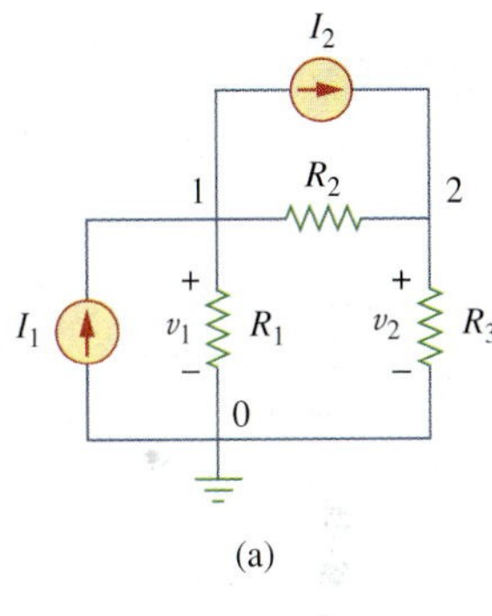

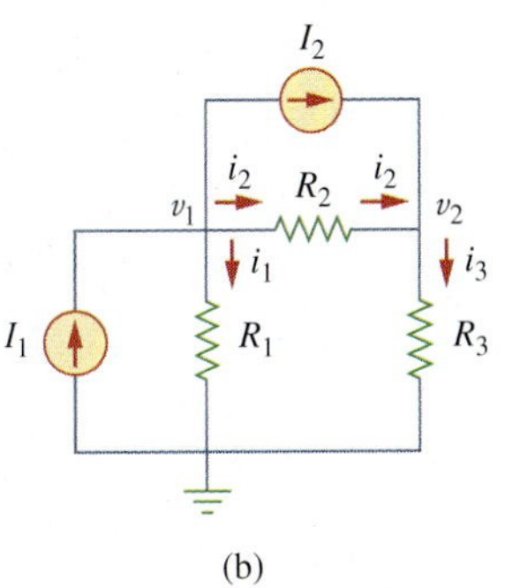

Figure 3.2
Typical circuit for nodal analysis.

Substituting Eq. (3.4) in Eqs. (3.1) and (3.2) results, respectively, in

$$I_1 = I_2 + \frac{v_1}{R_1} + \frac{v_1 - v_2}{R_2} \tag{3.5}$$

$$I_2 + \frac{v_1 - v_2}{R_2} = \frac{v_2}{R_3} \tag{3.6}$$

In terms of the conductances, Eqs. (3.5) and (3.6) become

$$I_1 = I_2 + G_1 v_1 + G_2(v_1 - v_2) \tag{3.7}$$

$$I_2 + G_2(v_1 - v_2) = G_3 v_2 \tag{3.8}$$

The third step in nodal analysis is to solve for the node voltages. If we apply KCL to $n - 1$ nonreference nodes, we obtain $n - 1$ simultaneous equations such as Eqs. (3.5) and (3.6) or (3.7) and (3.8). For the circuit of Fig. 3.2, we solve Eqs. (3.5) and (3.6) or (3.7) and (3.8) to obtain the node voltages v_1 and v_2 using any standard method, such as the substitution method, the elimination method, Cramer's rule, or matrix inversion. To use either of the last two methods, one must cast the simultaneous equations in matrix form. For example, Eqs. (3.7) and (3.8) can be cast in matrix form as

Appendix A discusses how to use Cramer's rule.

$$\begin{bmatrix} G_1 + G_2 & -G_2 \\ -G_2 & G_2 + G_3 \end{bmatrix} \begin{bmatrix} v_1 \\ v_2 \end{bmatrix} = \begin{bmatrix} I_1 - I_2 \\ I_2 \end{bmatrix} \tag{3.9}$$

which can be solved to get v_1 and v_2. Equation 3.9 will be generalized in Section 3.6. The simultaneous equations may also be solved using calculators or with software packages such as *MATLAB*, *Mathcad*, *Maple*, and *Quattro Pro*.

Example 3.1

Calculate the node voltages in the circuit shown in Fig. 3.3(a).

Solution:

Consider Fig. 3.3(b), where the circuit in Fig. 3.3(a) has been prepared for nodal analysis. Notice how the currents are selected for the application of KCL. Except for the branches with current sources, the labeling of the currents is arbitrary but consistent. (By consistent, we mean that if, for example, we assume that i_2 enters the 4-Ω resistor from the left-hand side, i_2 must leave the resistor from the right-hand side.) The reference node is selected, and the node voltages v_1 and v_2 are now to be determined.

At node 1, applying KCL and Ohm's law gives

$$i_1 = i_2 + i_3 \quad \Rightarrow \quad 5 = \frac{v_1 - v_2}{4} + \frac{v_1 - 0}{2}$$

Multiplying each term in the last equation by 4, we obtain

$$20 = v_1 - v_2 + 2v_1$$

or

$$3v_1 - v_2 = 20 \tag{3.1.1}$$

At node 2, we do the same thing and get

$$i_2 + i_4 = i_1 + i_5 \quad \Rightarrow \quad \frac{v_1 - v_2}{4} + 10 = 5 + \frac{v_2 - 0}{6}$$

Multiplying each term by 12 results in

$$3v_1 - 3v_2 + 120 = 60 + 2v_2$$

or

$$-3v_1 + 5v_2 = 60 \tag{3.1.2}$$

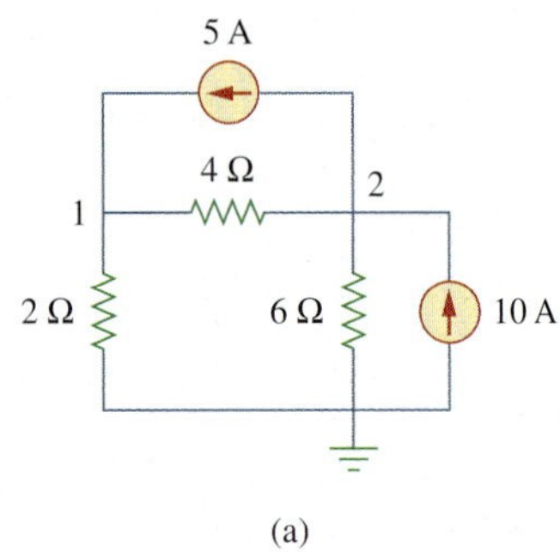

(a)

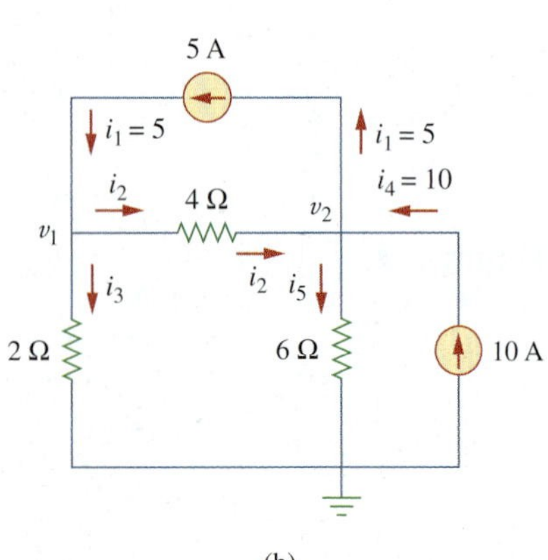

(b)

Figure 3.3
For Example 3.1: (a) original circuit, (b) circuit for analysis.

Now we have two simultaneous Eqs. (3.1.1) and (3.1.2). We can solve the equations using any method and obtain the values of v_1 and v_2.

METHOD 1 Using the elimination technique, we add Eqs. (3.1.1) and (3.1.2).

$$4v_2 = 80 \quad \Rightarrow \quad v_2 = 20 \text{ V}$$

Substituting $v_2 = 20$ in Eq. (3.1.1) gives

$$3v_1 - 20 = 20 \quad \Rightarrow \quad v_1 = \frac{40}{3} = 13.333 \text{ V}$$

METHOD 2 To use Cramer's rule, we need to put Eqs. (3.1.1) and (3.1.2) in matrix form as

$$\begin{bmatrix} 3 & -1 \\ -3 & 5 \end{bmatrix} \begin{bmatrix} v_1 \\ v_2 \end{bmatrix} = \begin{bmatrix} 20 \\ 60 \end{bmatrix} \tag{3.1.3}$$

The determinant of the matrix is

$$\Delta = \begin{vmatrix} 3 & -1 \\ -3 & 5 \end{vmatrix} = 15 - 3 = 12$$

We now obtain v_1 and v_2 as

$$v_1 = \frac{\Delta_1}{\Delta} = \frac{\begin{vmatrix} 20 & -1 \\ 60 & 5 \end{vmatrix}}{\Delta} = \frac{100 + 60}{12} = 13.333 \text{ V}$$

$$v_2 = \frac{\Delta_2}{\Delta} = \frac{\begin{vmatrix} 3 & 20 \\ -3 & 60 \end{vmatrix}}{\Delta} = \frac{180 + 60}{12} = 20 \text{ V}$$

giving us the same result as did the elimination method.

If we need the currents, we can easily calculate them from the values of the nodal voltages.

$$i_1 = 5 \text{ A}, \qquad i_2 = \frac{v_1 - v_2}{4} = -1.6668 \text{ A}, \qquad i_3 = \frac{v_1}{2} = 6.666 \text{ A}$$

$$i_4 = 10 \text{ A}, \qquad i_5 = \frac{v_2}{6} = 3.333 \text{ A}$$

The fact that i_2 is negative shows that the current flows in the direction opposite to the one assumed.

Practice Problem 3.1

Obtain the node voltages in the circuit of Fig. 3.4.

Answer: $v_1 = -2$ V, $v_2 = -14$ V.

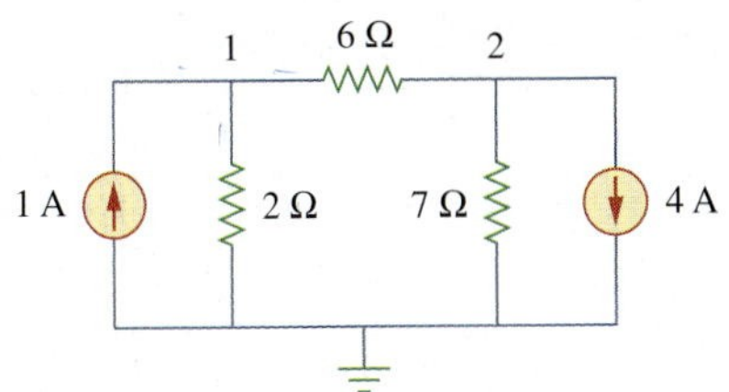

Figure 3.4
For Practice Prob. 3.1.

Example 3.2

Determine the voltages at the nodes in Fig. 3.5(a).

Solution:

The circuit in this example has three nonreference nodes, unlike the previous example which has two nonreference nodes. We assign voltages to the three nodes as shown in Fig. 3.5(b) and label the currents.

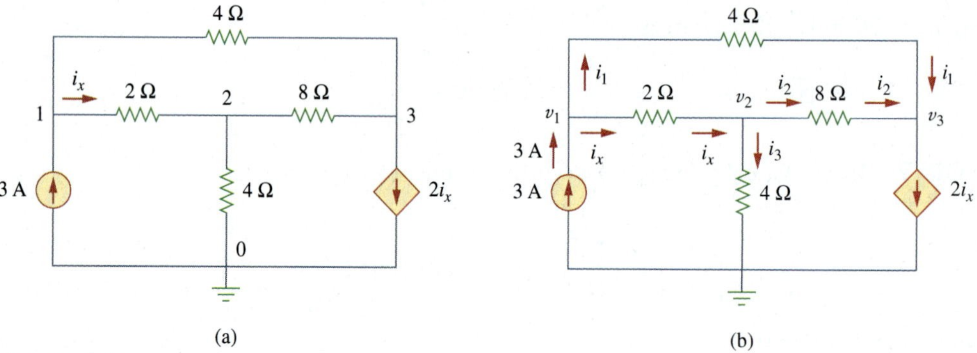

Figure 3.5
For Example 3.2: (a) original circuit, (b) circuit for analysis.

At node 1,

$$3 = i_1 + i_x \quad \Rightarrow \quad 3 = \frac{v_1 - v_3}{4} + \frac{v_1 - v_2}{2}$$

Multiplying by 4 and rearranging terms, we get

$$3v_1 - 2v_2 - v_3 = 12 \tag{3.2.1}$$

At node 2,

$$i_x = i_2 + i_3 \quad \Rightarrow \quad \frac{v_1 - v_2}{2} = \frac{v_2 - v_3}{8} + \frac{v_2 - 0}{4}$$

Multiplying by 8 and rearranging terms, we get

$$-4v_1 + 7v_2 - v_3 = 0 \tag{3.2.2}$$

At node 3,

$$i_1 + i_2 = 2i_x \quad \Rightarrow \quad \frac{v_1 - v_3}{4} + \frac{v_2 - v_3}{8} = \frac{2(v_1 - v_2)}{2}$$

Multiplying by 8, rearranging terms, and dividing by 3, we get

$$2v_1 - 3v_2 + v_3 = 0 \tag{3.2.3}$$

We have three simultaneous equations to solve to get the node voltages v_1, v_2, and v_3. We shall solve the equations in three ways.

■ **METHOD 1** Using the elimination technique, we add Eqs. (3.2.1) and (3.2.3).

$$5v_1 - 5v_2 = 12$$

or

$$v_1 - v_2 = \frac{12}{5} = 2.4 \tag{3.2.4}$$

Adding Eqs. (3.2.2) and (3.2.3) gives

$$-2v_1 + 4v_2 = 0 \quad \Rightarrow \quad v_1 = 2v_2 \tag{3.2.5}$$

Substituting Eq. (3.2.5) into Eq. (3.2.4) yields

$$2v_2 - v_2 = 2.4 \quad \Rightarrow \quad v_2 = 2.4, \qquad v_1 = 2v_2 = 4.8 \text{ V}$$

From Eq. (3.2.3), we get

$$v_3 = 3v_2 - 2v_1 = 3v_2 - 4v_2 = -v_2 = -2.4 \text{ V}$$

Thus,

$$v_1 = 4.8 \text{ V}, \qquad v_2 = 2.4 \text{ V}, \qquad v_3 = -2.4 \text{ V}$$

METHOD 2 To use Cramer's rule, we put Eqs. (3.2.1) to (3.2.3) in matrix form.

$$\begin{bmatrix} 3 & -2 & -1 \\ -4 & 7 & -1 \\ 2 & -3 & 1 \end{bmatrix} \begin{bmatrix} v_1 \\ v_2 \\ v_3 \end{bmatrix} = \begin{bmatrix} 12 \\ 0 \\ 0 \end{bmatrix} \tag{3.2.6}$$

From this, we obtain

$$v_1 = \frac{\Delta_1}{\Delta}, \qquad v_2 = \frac{\Delta_2}{\Delta}, \qquad v_3 = \frac{\Delta_3}{\Delta}$$

where Δ, Δ_1, Δ_2, and Δ_3 are the determinants to be calculated as follows. As explained in Appendix A, to calculate the determinant of a 3 by 3 matrix, we repeat the first two rows and cross multiply.

$$\Delta = \begin{vmatrix} 3 & -2 & -1 \\ -4 & 7 & -1 \\ 2 & -3 & 1 \end{vmatrix} = \begin{vmatrix} 3 & -2 & -1 \\ -4 & 7 & -1 \\ 2 & -3 & 1 \\ 3 & -2 & -1 \\ -4 & 7 & -1 \end{vmatrix}$$

$$= 21 - 12 + 4 + 14 - 9 - 8 = 10$$

Similarly, we obtain

$$\Delta_1 = \begin{vmatrix} 12 & -2 & -1 \\ 0 & 7 & -1 \\ 0 & -3 & 1 \\ 12 & -2 & -1 \\ 0 & 7 & -1 \end{vmatrix} = 84 + 0 + 0 - 0 - 36 - 0 = 48$$

$$\Delta_2 = \begin{vmatrix} 3 & 12 & -1 \\ -4 & 0 & -1 \\ 2 & 0 & 1 \\ 3 & 12 & -1 \\ -4 & 0 & -1 \end{vmatrix} = 0 + 0 - 24 - 0 - 0 + 48 = 24$$

$$\Delta_3 = \begin{vmatrix} 3 & -2 & 12 \\ -4 & 7 & 0 \\ 2 & -3 & 0 \\ 3 & -2 & 12 \\ -4 & 7 & 0 \end{vmatrix} = 0 + 144 + 0 - 168 - 0 - 0 = -24$$

Thus, we find

$$v_1 = \frac{\Delta_1}{\Delta} = \frac{48}{10} = 4.8 \text{ V}, \qquad v_2 = \frac{\Delta_2}{\Delta} = \frac{24}{10} = 2.4 \text{ V}$$

$$v_3 = \frac{\Delta_3}{\Delta} = \frac{-24}{10} = -2.4 \text{ V}$$

as we obtained with Method 1.

■ **METHOD 3** We now use *MATLAB* to solve the matrix. Equation (3.2.6) can be written as

$$\mathbf{AV} = \mathbf{B} \quad \Rightarrow \quad \mathbf{V} = \mathbf{A}^{-1}\mathbf{B}$$

where **A** is the 3 by 3 square matrix, **B** is the column vector, and **V** is a column vector comprised of v_1, v_2, and v_3 that we want to determine. We use *MATLAB* to determine **V** as follows:

```
>>A = [3  -2  -1;  -4  7  -1;  2  -3  1];
>>B = [12  0  0]';
>>V = inv(A) * B
        4.8000
   V =  2.4000
       -2.4000
```

Thus, $v_1 = 4.8$ V, $v_2 = 2.4$ V, and $v_3 = -2.4$ V, as obtained previously.

Practice Problem 3.2

Find the voltages at the three nonreference nodes in the circuit of Fig. 3.6.

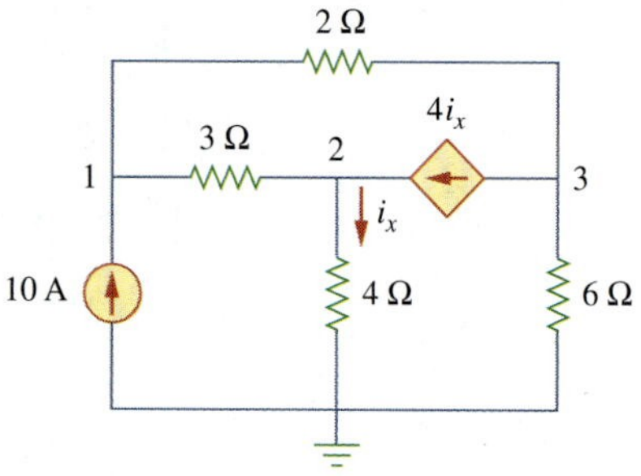

Answer: $v_1 = 80$ V, $v_2 = -64$ V, $v_3 = 156$ V.

2 Ω, 3 Ω, 4i_x, 1, 2, 3, i_x, 10 A, 4 Ω, 6 Ω

Figure 3.6
For Practice Prob. 3.2.

3.3 Nodal Analysis with Voltage Sources

We now consider how voltage sources affect nodal analysis. We use the circuit in Fig. 3.7 for illustration. Consider the following two possibilities.

■ **CASE 1** If a voltage source is connected between the reference node and a nonreference node, we simply set the voltage at the nonreference node equal to the voltage of the voltage source. In Fig. 3.7, for example,

$$v_1 = 10 \text{ V} \qquad \textbf{(3.10)}$$

Thus, our analysis is somewhat simplified by this knowledge of the voltage at this node.

■ **CASE 2** If the voltage source (dependent or independent) is connected between two nonreference nodes, the two nonreference nodes

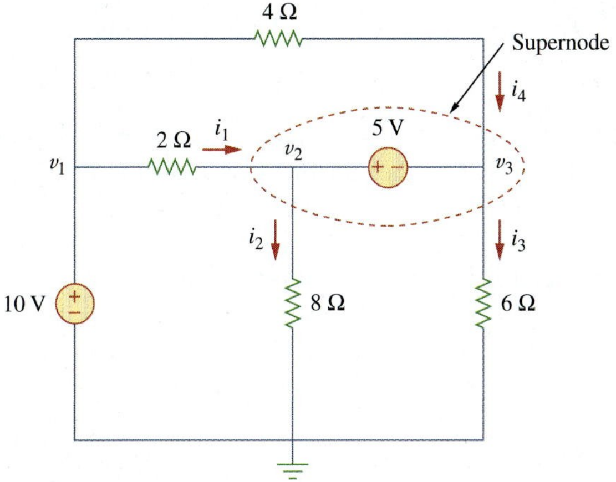

Figure 3.7
A circuit with a supernode.

form a *generalized node* or *supernode*; we apply both KCL and KVL to determine the node voltages.

A supernode may be regarded as a closed surface enclosing the voltage source and its two nodes.

> A **supernode** is formed by enclosing a (dependent or independent) voltage source connected between two nonreference nodes and any elements connected in parallel with it.

In Fig. 3.7, nodes 2 and 3 form a supernode. (We could have more than two nodes forming a single supernode. For example, see the circuit in Fig. 3.14.) We analyze a circuit with supernodes using the same three steps mentioned in the previous section except that the supernodes are treated differently. Why? Because an essential component of nodal analysis is applying KCL, which requires knowing the current through each element. There is no way of knowing the current through a voltage source in advance. However, KCL must be satisfied at a supernode like any other node. Hence, at the supernode in Fig. 3.7,

$$i_1 + i_4 = i_2 + i_3 \tag{3.11a}$$

or

$$\frac{v_1 - v_2}{2} + \frac{v_1 - v_3}{4} = \frac{v_2 - 0}{8} + \frac{v_3 - 0}{6} \tag{3.11b}$$

To apply Kirchhoff's voltage law to the supernode in Fig. 3.7, we redraw the circuit as shown in Fig. 3.8. Going around the loop in the clockwise direction gives

$$-v_2 + 5 + v_3 = 0 \quad \Rightarrow \quad v_2 - v_3 = 5 \tag{3.12}$$

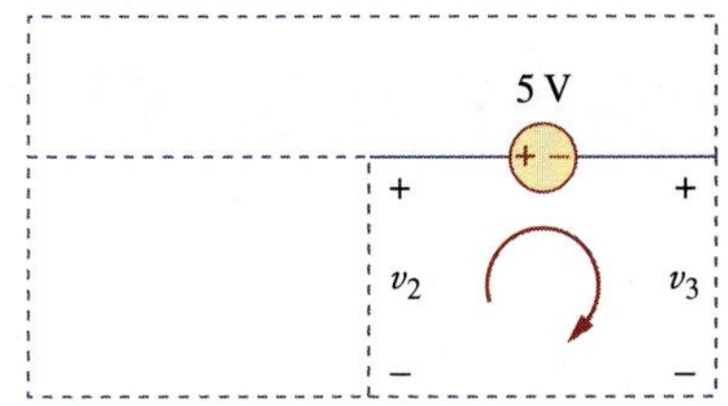

Figure 3.8
Applying KVL to a supernode.

From Eqs. (3.10), (3.11b), and (3.12), we obtain the node voltages.

Note the following properties of a supernode:

1. The voltage source inside the supernode provides a constraint equation needed to solve for the node voltages.
2. A supernode has no voltage of its own.
3. A supernode requires the application of both KCL and KVL.

Example 3.3

For the circuit shown in Fig. 3.9, find the node voltages.

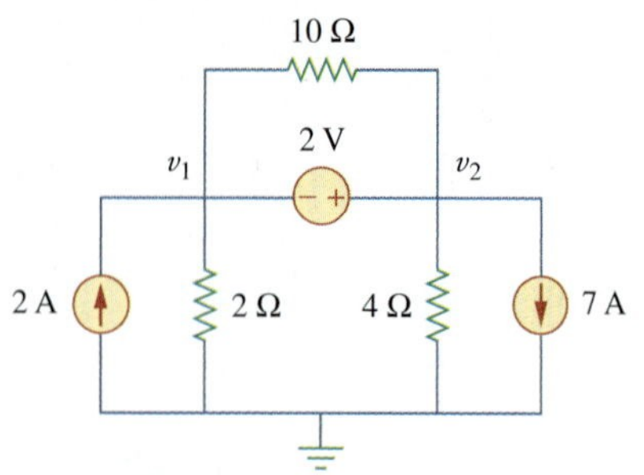

Figure 3.9
For Example 3.3.

Solution:
The supernode contains the 2-V source, nodes 1 and 2, and the 10-Ω resistor. Applying KCL to the supernode as shown in Fig. 3.10(a) gives

$$2 = i_1 + i_2 + 7$$

Expressing i_1 and i_2 in terms of the node voltages

$$2 = \frac{v_1 - 0}{2} + \frac{v_2 - 0}{4} + 7 \quad \Rightarrow \quad 8 = 2v_1 + v_2 + 28$$

or

$$v_2 = -20 - 2v_1 \tag{3.3.1}$$

To get the relationship between v_1 and v_2, we apply KVL to the circuit in Fig. 3.10(b). Going around the loop, we obtain

$$-v_1 - 2 + v_2 = 0 \quad \Rightarrow \quad v_2 = v_1 + 2 \tag{3.3.2}$$

From Eqs. (3.3.1) and (3.3.2), we write

$$v_2 = v_1 + 2 = -20 - 2v_1$$

or

$$3v_1 = -22 \quad \Rightarrow \quad v_1 = -7.333 \text{ V}$$

and $v_2 = v_1 + 2 = -5.333$ V. Note that the 10-Ω resistor does not make any difference because it is connected across the supernode.

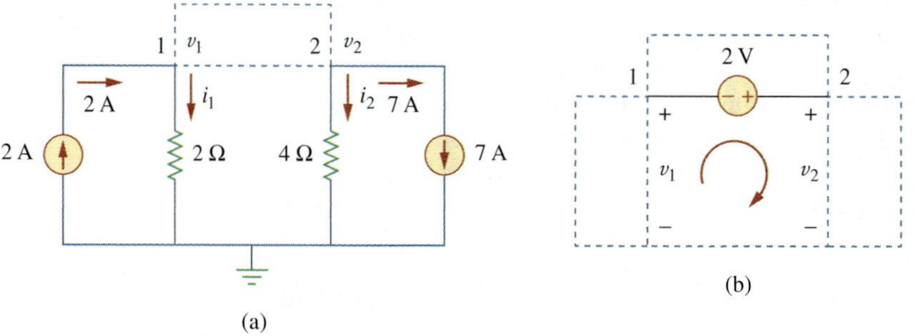

Figure 3.10
Applying: (a) KCL to the supernode, (b) KVL to the loop.

Practice Problem 3.3

Find v and i in the circuit of Fig. 3.11.

Answer: −0.6 V, 4.2 A.

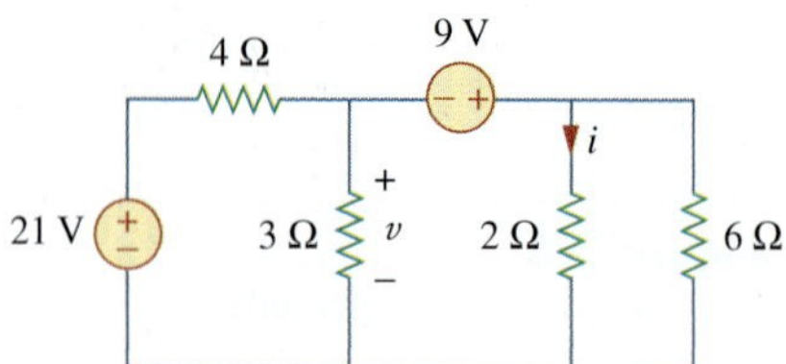

Figure 3.11
For Practice Prob. 3.3.

Example 3.4

Find the node voltages in the circuit of Fig. 3.12.

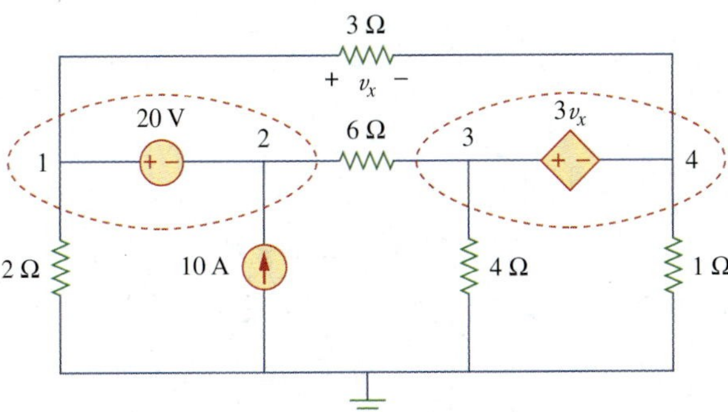

Figure 3.12
For Example 3.4.

Solution:
Nodes 1 and 2 form a supernode; so do nodes 3 and 4. We apply KCL to the two supernodes as in Fig. 3.13(a). At supernode 1-2,

$$i_3 + 10 = i_1 + i_2$$

Expressing this in terms of the node voltages,

$$\frac{v_3 - v_2}{6} + 10 = \frac{v_1 - v_4}{3} + \frac{v_1}{2}$$

or

$$5v_1 + v_2 - v_3 - 2v_4 = 60 \qquad \textbf{(3.4.1)}$$

At supernode 3-4,

$$i_1 = i_3 + i_4 + i_5 \quad \Rightarrow \quad \frac{v_1 - v_4}{3} = \frac{v_3 - v_2}{6} + \frac{v_4}{1} + \frac{v_3}{4}$$

or

$$4v_1 + 2v_2 - 5v_3 - 16v_4 = 0 \qquad \textbf{(3.4.2)}$$

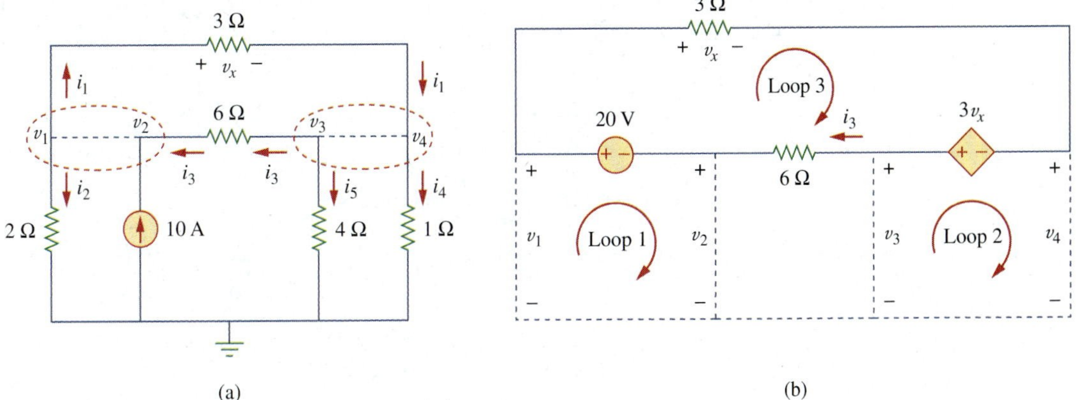

Figure 3.13
Applying: (a) KCL to the two supernodes, (b) KVL to the loops.

We now apply KVL to the branches involving the voltage sources as shown in Fig. 3.13(b). For loop 1,

$$-v_1 + 20 + v_2 = 0 \quad \Rightarrow \quad v_1 - v_2 = 20 \tag{3.4.3}$$

For loop 2,

$$-v_3 + 3v_x + v_4 = 0$$

But $v_x = v_1 - v_4$ so that

$$3v_1 - v_3 - 2v_4 = 0 \tag{3.4.4}$$

For loop 3,

$$v_x - 3v_x + 6i_3 - 20 = 0$$

But $6i_3 = v_3 - v_2$ and $v_x = v_1 - v_4$. Hence,

$$-2v_1 - v_2 + v_3 + 2v_4 = 20 \tag{3.4.5}$$

We need four node voltages, v_1, v_2, v_3, and v_4, and it requires only four out of the five Eqs. (3.4.1) to (3.4.5) to find them. Although the fifth equation is redundant, it can be used to check results. We can solve Eqs. (3.4.1) to (3.4.4) directly using *MATLAB*. We can eliminate one node voltage so that we solve three simultaneous equations instead of four. From Eq. (3.4.3), $v_2 = v_1 - 20$. Substituting this into Eqs. (3.4.1) and (3.4.2), respectively, gives

$$6v_1 - v_3 - 2v_4 = 80 \tag{3.4.6}$$

and

$$6v_1 - 5v_3 - 16v_4 = 40 \tag{3.4.7}$$

Equations (3.4.4), (3.4.6), and (3.4.7) can be cast in matrix form as

$$\begin{bmatrix} 3 & -1 & -2 \\ 6 & -1 & -2 \\ 6 & -5 & -16 \end{bmatrix} \begin{bmatrix} v_1 \\ v_3 \\ v_4 \end{bmatrix} = \begin{bmatrix} 0 \\ 80 \\ 40 \end{bmatrix}$$

Using Cramer's rule gives

$$\Delta = \begin{vmatrix} 3 & -1 & -2 \\ 6 & -1 & -2 \\ 6 & -5 & -16 \end{vmatrix} = -18, \qquad \Delta_1 = \begin{vmatrix} 0 & -1 & -2 \\ 80 & -1 & -2 \\ 40 & -5 & -16 \end{vmatrix} = -480,$$

$$\Delta_3 = \begin{vmatrix} 3 & 0 & -2 \\ 6 & 80 & -2 \\ 6 & 40 & -16 \end{vmatrix} = -3120, \qquad \Delta_4 = \begin{vmatrix} 3 & -1 & 0 \\ 6 & -1 & 80 \\ 6 & -5 & 40 \end{vmatrix} = 840$$

Thus, we arrive at the node voltages as

$$v_1 = \frac{\Delta_1}{\Delta} = \frac{-480}{-18} = 26.67 \text{ V}, \qquad v_3 = \frac{\Delta_3}{\Delta} = \frac{-3120}{-18} = 173.33 \text{ V},$$

$$v_4 = \frac{\Delta_4}{\Delta} = \frac{840}{-18} = -46.67 \text{ V}$$

and $v_2 = v_1 - 20 = 6.667$ V. We have not used Eq. (3.4.5); it can be used to cross check results.

Practice Problem 3.4

Find v_1, v_2, and v_3 in the circuit of Fig. 3.14 using nodal analysis.

Answer: $v_1 = 3.043$ V, $v_2 = -6.956$ V, $v_3 = 0.6522$ V.

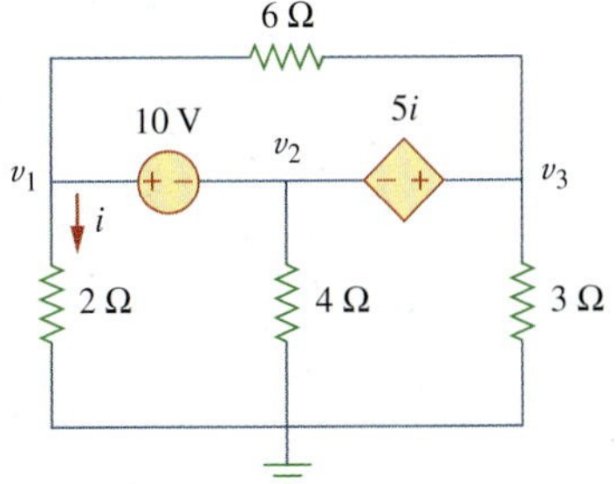

Figure 3.14
For Practice Prob. 3.4.

3.4 Mesh Analysis

Mesh analysis provides another general procedure for analyzing circuits, using mesh currents as the circuit variables. Using mesh currents instead of element currents as circuit variables is convenient and reduces the number of equations that must be solved simultaneously. Recall that a loop is a closed path with no node passed more than once. A mesh is a loop that does not contain any other loop within it.

Mesh analysis is also known as *loop analysis* or the *mesh-current method.*

Nodal analysis applies KCL to find unknown voltages in a given circuit, while mesh analysis applies KVL to find unknown currents. Mesh analysis is not quite as general as nodal analysis because it is only applicable to a circuit that is *planar*. A planar circuit is one that can be drawn in a plane with no branches crossing one another; otherwise it is *nonplanar.* A circuit may have crossing branches and still be planar if it can be redrawn such that it has no crossing branches. For example, the circuit in Fig. 3.15(a) has two crossing branches, but it can be redrawn as in Fig. 3.15(b). Hence, the circuit in Fig. 3.15(a) is planar. However, the circuit in Fig. 3.16 is nonplanar, because there is no way to redraw it and avoid the branches crossing. Nonplanar circuits can be handled using nodal analysis, but they will not be considered in this text.

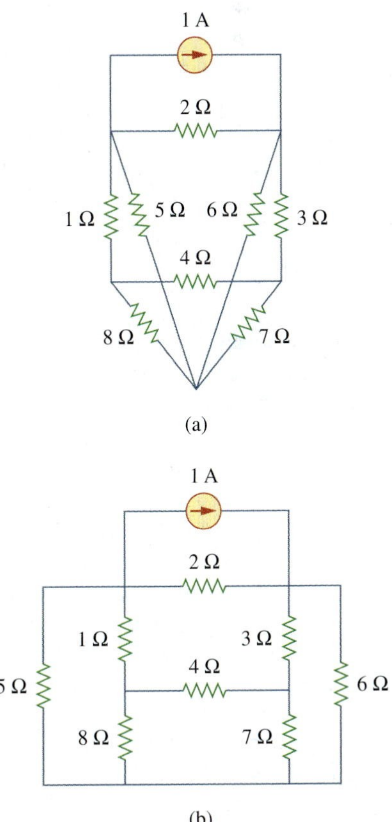

Figure 3.15
(a) A planar circuit with crossing branches, (b) the same circuit redrawn with no crossing branches.

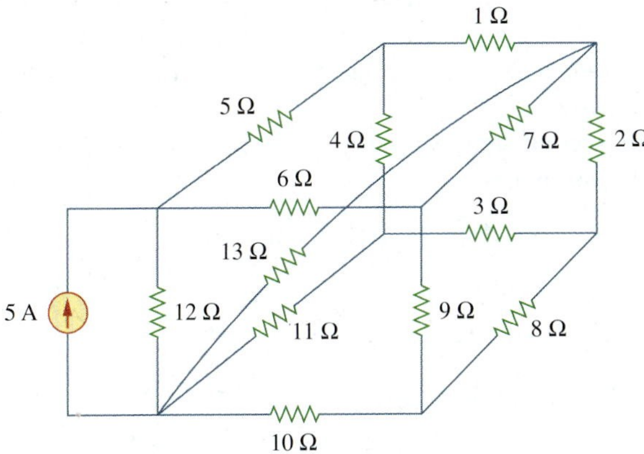

Figure 3.16
A nonplanar circuit.

To understand mesh analysis, we should first explain more about what we mean by a mesh.

A **mesh** is a loop which does not contain any other loops within it.

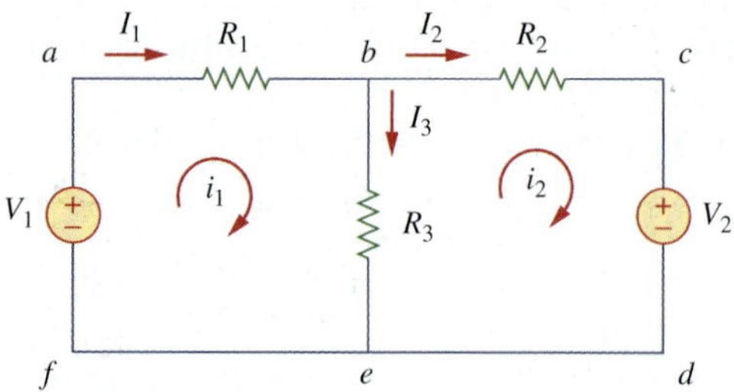

Figure 3.17
A circuit with two meshes.

In Fig. 3.17, for example, paths *abefa* and *bcdeb* are meshes, but path *abcdefa* is not a mesh. The current through a mesh is known as *mesh current*. In mesh analysis, we are interested in applying KVL to find the mesh currents in a given circuit.

Although path *abcdefa* is a loop and not a mesh, KVL still holds. This is the reason for loosely using the terms *loop analysis* and *mesh analysis* to mean the same thing.

In this section, we will apply mesh analysis to planar circuits that do not contain current sources. In the next section, we will consider circuits with current sources. In the mesh analysis of a circuit with n meshes, we take the following three steps.

Steps to Determine Mesh Currents:

1. Assign mesh currents $i_1, i_2, \ldots, i_n$ to the n meshes.
2. Apply KVL to each of the n meshes. Use Ohm's law to express the voltages in terms of the mesh currents.
3. Solve the resulting n simultaneous equations to get the mesh currents.

The direction of the mesh current is arbitrary—(clockwise or counterclockwise)—and does not affect the validity of the solution.

To illustrate the steps, consider the circuit in Fig. 3.17. The first step requires that mesh currents i_1 and i_2 are assigned to meshes 1 and 2. Although a mesh current may be assigned to each mesh in an arbitrary direction, it is conventional to assume that each mesh current flows clockwise.

As the second step, we apply KVL to each mesh. Applying KVL to mesh 1, we obtain

$$-V_1 + R_1 i_1 + R_3(i_1 - i_2) = 0$$

or

$$(R_1 + R_3)i_1 - R_3 i_2 = V_1 \tag{3.13}$$

For mesh 2, applying KVL gives

$$R_2 i_2 + V_2 + R_3(i_2 - i_1) = 0$$

or

$$-R_3 i_1 + (R_2 + R_3)i_2 = -V_2 \tag{3.14}$$

Note in Eq. (3.13) that the coefficient of i_1 is the sum of the resistances in the first mesh, while the coefficient of i_2 is the negative of the resistance common to meshes 1 and 2. Now observe that the same is true in Eq. (3.14). This can serve as a shortcut way of writing the mesh equations. We will exploit this idea in Section 3.6.

The shortcut way will not apply if one mesh current is assumed clockwise and the other assumed counterclockwise, although this is permissible.

The third step is to solve for the mesh currents. Putting Eqs. (3.13) and (3.14) in matrix form yields

$$\begin{bmatrix} R_1 + R_3 & -R_3 \\ -R_3 & R_2 + R_3 \end{bmatrix} \begin{bmatrix} i_1 \\ i_2 \end{bmatrix} = \begin{bmatrix} V_1 \\ -V_2 \end{bmatrix} \quad \textbf{(3.15)}$$

which can be solved to obtain the mesh currents i_1 and i_2. We are at liberty to use any technique for solving the simultaneous equations. According to Eq. (2.12), if a circuit has n nodes, b branches, and l independent loops or meshes, then $l = b - n + 1$. Hence, l independent simultaneous equations are required to solve the circuit using mesh analysis.

Notice that the branch currents are different from the mesh currents unless the mesh is isolated. To distinguish between the two types of currents, we use i for a mesh current and I for a branch current. The current elements I_1, I_2, and I_3 are algebraic sums of the mesh currents. It is evident from Fig. 3.17 that

$$I_1 = i_1, \qquad I_2 = i_2, \qquad I_3 = i_1 - i_2 \quad \textbf{(3.16)}$$

Example 3.5

For the circuit in Fig. 3.18, find the branch currents I_1, I_2, and I_3 using mesh analysis.

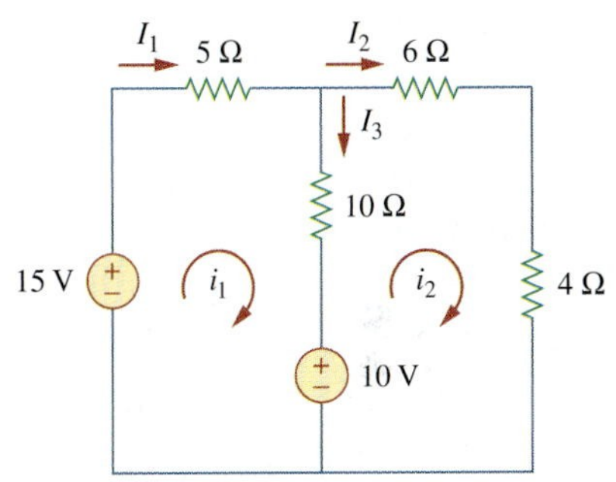

Figure 3.18
For Example 3.5.

Solution:
We first obtain the mesh currents using KVL. For mesh 1,

$$-15 + 5i_1 + 10(i_1 - i_2) + 10 = 0$$

or

$$3i_1 - 2i_2 = 1 \quad \textbf{(3.5.1)}$$

For mesh 2,

$$6i_2 + 4i_2 + 10(i_2 - i_1) - 10 = 0$$

or

$$i_1 = 2i_2 - 1 \quad \textbf{(3.5.2)}$$

■ METHOD 1 Using the substitution method, we substitute Eq. (3.5.2) into Eq. (3.5.1), and write

$$6i_2 - 3 - 2i_2 = 1 \quad \Rightarrow \quad i_2 = 1 \text{ A}$$

From Eq. (3.5.2), $i_1 = 2i_2 - 1 = 2 - 1 = 1$ A. Thus,

$$I_1 = i_1 = 1 \text{ A}, \qquad I_2 = i_2 = 1 \text{ A}, \qquad I_3 = i_1 - i_2 = 0$$

■ METHOD 2 To use Cramer's rule, we cast Eqs. (3.5.1) and (3.5.2) in matrix form as

$$\begin{bmatrix} 3 & -2 \\ -1 & 2 \end{bmatrix} \begin{bmatrix} i_1 \\ i_2 \end{bmatrix} = \begin{bmatrix} 1 \\ 1 \end{bmatrix}$$

We obtain the determinants

$$\Delta = \begin{vmatrix} 3 & -2 \\ -1 & 2 \end{vmatrix} = 6 - 2 = 4$$

$$\Delta_1 = \begin{vmatrix} 1 & -2 \\ 1 & 2 \end{vmatrix} = 2 + 2 = 4, \qquad \Delta_2 = \begin{vmatrix} 3 & 1 \\ -1 & 1 \end{vmatrix} = 3 + 1 = 4$$

Thus,

$$i_1 = \frac{\Delta_1}{\Delta} = 1 \text{ A}, \qquad i_2 = \frac{\Delta_2}{\Delta} = 1 \text{ A}$$

as before.

Practice Problem 3.5

Calculate the mesh currents i_1 and i_2 of the circuit of Fig. 3.19.

Answer: $i_1 = 2$ A, $i_2 = 0$ A.

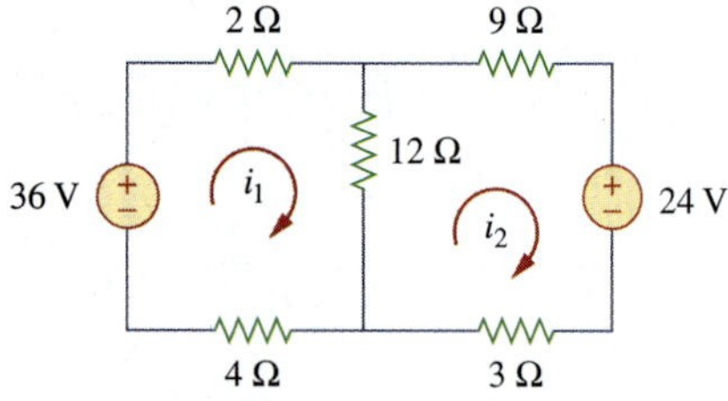

Figure 3.19
For Practice Prob. 3.5.

Example 3.6

Use mesh analysis to find the current I_o in the circuit of Fig. 3.20.

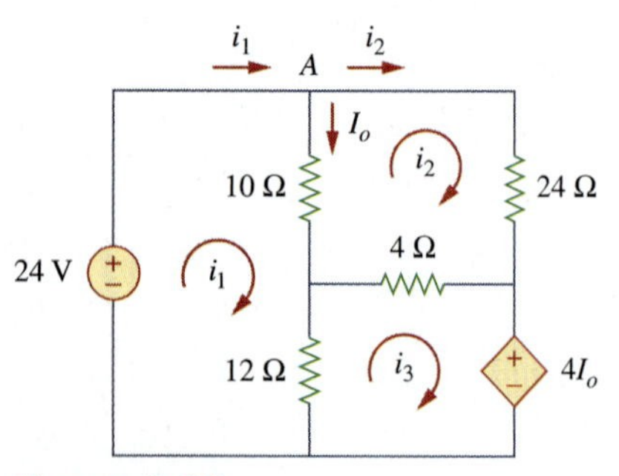

Figure 3.20
For Example 3.6.

Solution:
We apply KVL to the three meshes in turn. For mesh 1,

$$-24 + 10(i_1 - i_2) + 12(i_1 - i_3) = 0$$

or

$$11i_1 - 5i_2 - 6i_3 = 12 \tag{3.6.1}$$

For mesh 2,

$$24i_2 + 4(i_2 - i_3) + 10(i_2 - i_1) = 0$$

or

$$-5i_1 + 19i_2 - 2i_3 = 0 \tag{3.6.2}$$

For mesh 3,

$$4I_o + 12(i_3 - i_1) + 4(i_3 - i_2) = 0$$

But at node A, $I_o = i_1 - i_2$, so that

$$4(i_1 - i_2) + 12(i_3 - i_1) + 4(i_3 - i_2) = 0$$

or

$$-i_1 - i_2 + 2i_3 = 0 \tag{3.6.3}$$

In matrix form, Eqs. (3.6.1) to (3.6.3) become

$$\begin{bmatrix} 11 & -5 & -6 \\ -5 & 19 & -2 \\ -1 & -1 & 2 \end{bmatrix} \begin{bmatrix} i_1 \\ i_2 \\ i_3 \end{bmatrix} = \begin{bmatrix} 12 \\ 0 \\ 0 \end{bmatrix}$$

We obtain the determinants as

$$\Delta = \begin{vmatrix} 11 & -5 & -6 \\ -5 & 19 & -2 \\ -1 & -1 & 2 \\ 11 & -5 & -6 \\ -5 & 19 & -2 \end{vmatrix}$$

$$= 418 - 30 - 10 - 114 - 22 - 50 = 192$$

$$\Delta_1 = \begin{vmatrix} 12 & -5 & -6 \\ 0 & 19 & -2 \\ 0 & -1 & 2 \\ 12 & -5 & -6 \\ 0 & 19 & -2 \end{vmatrix} = 456 - 24 = 432$$

$$\Delta_2 = \begin{vmatrix} 11 & 12 & -6 \\ -5 & 0 & -2 \\ -1 & 0 & 2 \\ 11 & 12 & -6 \\ -5 & 0 & -2 \end{vmatrix} = 24 + 120 = 144$$

$$\Delta_3 = \begin{vmatrix} 11 & -5 & 12 \\ -5 & 19 & 0 \\ -1 & -1 & 0 \\ 11 & -5 & 12 \\ -5 & 19 & 0 \end{vmatrix} = 60 + 228 = 288$$

We calculate the mesh currents using Cramer's rule as

$$i_1 = \frac{\Delta_1}{\Delta} = \frac{432}{192} = 2.25 \text{ A}, \qquad i_2 = \frac{\Delta_2}{\Delta} = \frac{144}{192} = 0.75 \text{ A},$$

$$i_3 = \frac{\Delta_3}{\Delta} = \frac{288}{192} = 1.5 \text{ A}$$

Thus, $I_o = i_1 - i_2 = 1.5$ A.

Practice Problem 3.6

Using mesh analysis, find I_o in the circuit of Fig. 3.21.

Answer: -5 A.

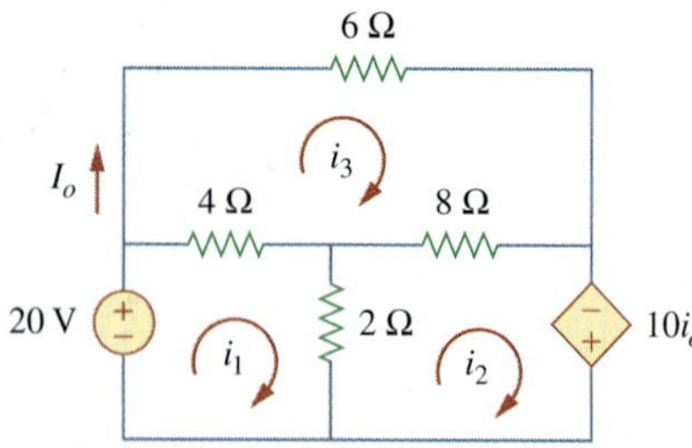

Figure 3.21
For Practice Prob. 3.6.

3.5 Mesh Analysis with Current Sources

Applying mesh analysis to circuits containing current sources (dependent or independent) may appear complicated. But it is actually much easier than what we encountered in the previous section, because the presence of the current sources reduces the number of equations. Consider the following two possible cases.

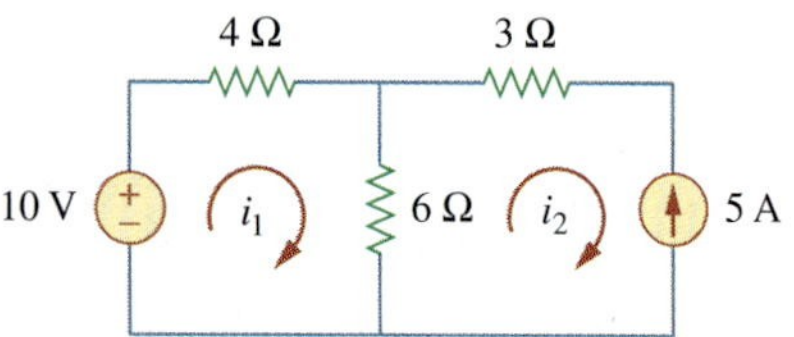

Figure 3.22
A circuit with a current source.

■ **CASE 1** When a current source exists only in one mesh: Consider the circuit in Fig. 3.22, for example. We set $i_2 = -5$ A and write a mesh equation for the other mesh in the usual way; that is,

$$-10 + 4i_1 + 6(i_1 - i_2) = 0 \quad \Rightarrow \quad i_1 = -2 \text{ A} \tag{3.17}$$

■ **CASE 2** When a current source exists between two meshes: Consider the circuit in Fig. 3.23(a), for example. We create a *supermesh* by excluding the current source and any elements connected in series with it, as shown in Fig. 3.23(b). Thus,

A **supermesh** results when two meshes have a (dependent or independent) current source in common.

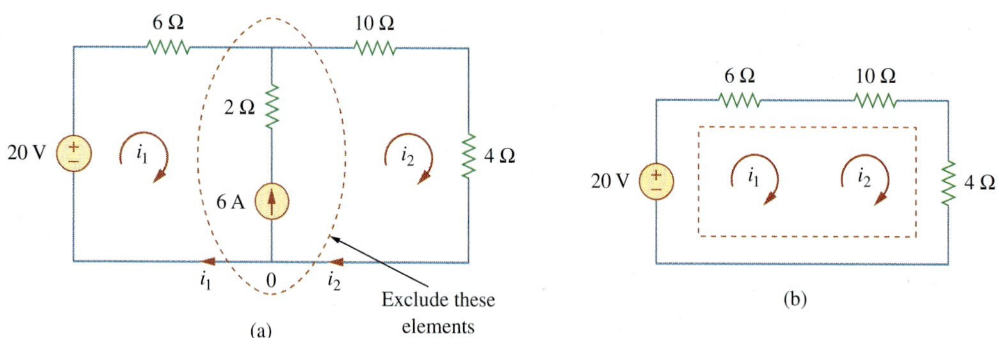

Figure 3.23
(a) Two meshes having a current source in common, (b) a supermesh, created by excluding the current source.

As shown in Fig. 3.23(b), we create a supermesh as the periphery of the two meshes and treat it differently. (If a circuit has two or more supermeshes that intersect, they should be combined to form a larger supermesh.) Why treat the supermesh differently? Because mesh analysis applies KVL—which requires that we know the voltage across each branch—and we do not know the voltage across a current source in advance. However, a supermesh must satisfy KVL like any other mesh. Therefore, applying KVL to the supermesh in Fig. 3.23(b) gives

$$-20 + 6i_1 + 10i_2 + 4i_2 = 0$$

or

$$6i_1 + 14i_2 = 20 \tag{3.18}$$

We apply KCL to a node in the branch where the two meshes intersect. Applying KCL to node 0 in Fig. 3.23(a) gives

$$i_2 = i_1 + 6 \tag{3.19}$$

Solving Eqs. (3.18) and (3.19), we get

$$i_1 = -3.2 \text{ A}, \qquad i_2 = 2.8 \text{ A} \tag{3.20}$$

Note the following properties of a supermesh:

1. The current source in the supermesh provides the constraint equation necessary to solve for the mesh currents.
2. A supermesh has no current of its own.
3. A supermesh requires the application of both KVL and KCL.

Example 3.7

For the circuit in Fig. 3.24, find i_1 to i_4 using mesh analysis.

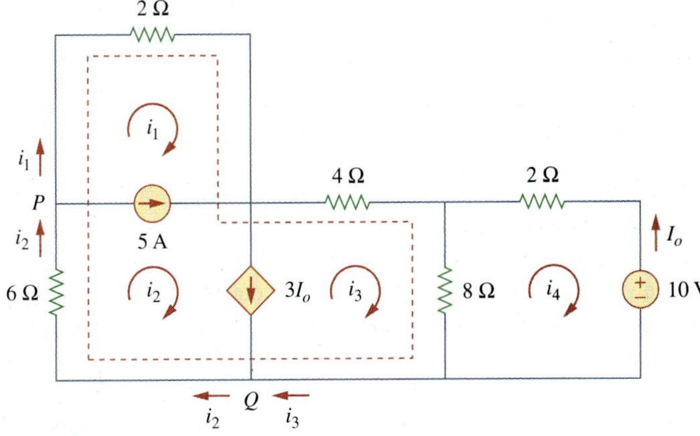

Figure 3.24
For Example 3.7.

Solution:
Note that meshes 1 and 2 form a supermesh since they have an independent current source in common. Also, meshes 2 and 3 form another supermesh because they have a dependent current source in common. The two supermeshes intersect and form a larger supermesh as shown. Applying KVL to the larger supermesh,

$$2i_1 + 4i_3 + 8(i_3 - i_4) + 6i_2 = 0$$

or

$$i_1 + 3i_2 + 6i_3 - 4i_4 = 0 \tag{3.7.1}$$

For the independent current source, we apply KCL to node P:

$$i_2 = i_1 + 5 \tag{3.7.2}$$

For the dependent current source, we apply KCL to node Q:

$$i_2 = i_3 + 3I_o$$

But $I_o = -i_4$, hence,

$$i_2 = i_3 - 3i_4 \tag{3.7.3}$$

Applying KVL in mesh 4,

$$2i_4 + 8(i_4 - i_3) + 10 = 0$$

or

$$5i_4 - 4i_3 = -5 \tag{3.7.4}$$

From Eqs. (3.7.1) to (3.7.4),

$$i_1 = -7.5 \text{ A}, \quad i_2 = -2.5 \text{ A}, \quad i_3 = 3.93 \text{ A}, \quad i_4 = 2.143 \text{ A}$$

Practice Problem 3.7

Use mesh analysis to determine i_1, i_2, and i_3 in Fig. 3.25.

Answer: $i_1 = 3.474$ A, $i_2 = 0.4737$ A, $i_3 = 1.1052$ A.

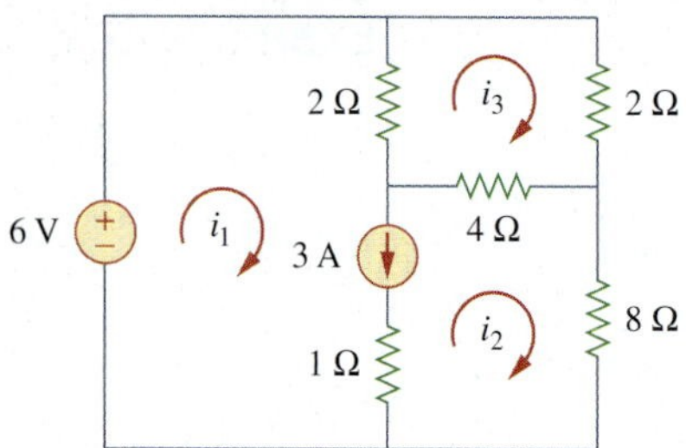

Figure 3.25
For Practice Prob. 3.7.

3.6 †Nodal and Mesh Analyses by Inspection

This section presents a generalized procedure for nodal or mesh analysis. It is a shortcut approach based on mere inspection of a circuit.

When all sources in a circuit are independent current sources, we do not need to apply KCL to each node to obtain the node-voltage equations as we did in Section 3.2. We can obtain the equations by mere inspection of the circuit. As an example, let us reexamine the circuit in Fig. 3.2, shown again in Fig. 3.26(a) for convenience. The circuit has two nonreference nodes and the node equations were derived in Section 3.2 as

$$\begin{bmatrix} G_1 + G_2 & -G_2 \\ -G_2 & G_2 + G_3 \end{bmatrix} \begin{bmatrix} v_1 \\ v_2 \end{bmatrix} = \begin{bmatrix} I_1 - I_2 \\ I_2 \end{bmatrix} \tag{3.21}$$

Observe that each of the diagonal terms is the sum of the conductances connected directly to node 1 or 2, while the off-diagonal terms are the negatives of the conductances connected between the nodes. Also, each term on the right-hand side of Eq. (3.21) is the algebraic sum of the currents entering the node.

In general, if a circuit with independent current sources has N nonreference nodes, the node-voltage equations can be written in terms of the conductances as

$$\begin{bmatrix} G_{11} & G_{12} & \dots & G_{1N} \\ G_{21} & G_{22} & \dots & G_{2N} \\ \vdots & \vdots & \vdots & \vdots \\ G_{N1} & G_{N2} & \dots & G_{NN} \end{bmatrix} \begin{bmatrix} v_1 \\ v_2 \\ \vdots \\ v_N \end{bmatrix} = \begin{bmatrix} i_1 \\ i_2 \\ \vdots \\ i_N \end{bmatrix} \tag{3.22}$$

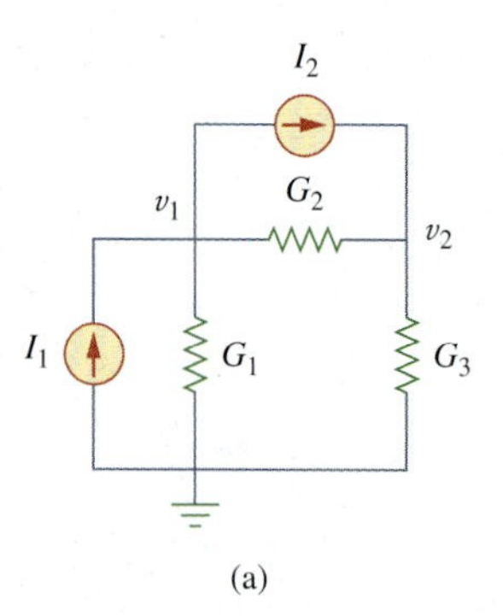

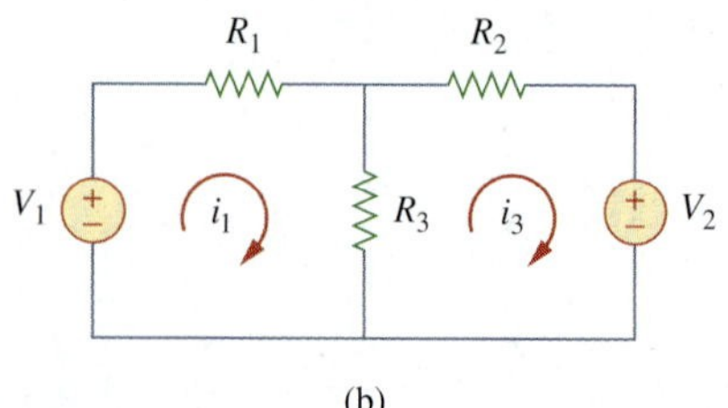

Figure 3.26
(a) The circuit in Fig. 3.2, (b) the circuit in Fig. 3.17.

or simply

$$\mathbf{Gv} = \mathbf{i} \tag{3.23}$$

where

G_{kk} = Sum of the conductances connected to node k

$G_{kj} = G_{jk}$ = Negative of the sum of the conductances directly connecting nodes k and j, $k \neq j$

v_k = Unknown voltage at node k

i_k = Sum of all independent current sources directly connected to node k, with currents entering the node treated as positive

G is called the *conductance matrix*; **v** is the output vector; and **i** is the input vector. Equation (3.22) can be solved to obtain the unknown node voltages. Keep in mind that this is valid for circuits with only independent current sources and linear resistors.

Similarly, we can obtain mesh-current equations by inspection when a linear resistive circuit has only independent voltage sources. Consider the circuit in Fig. 3.17, shown again in Fig. 3.26(b) for convenience. The circuit has two nonreference nodes and the node equations were derived in Section 3.4 as

$$\begin{bmatrix} R_1 + R_3 & -R_3 \\ -R_3 & R_2 + R_3 \end{bmatrix} \begin{bmatrix} i_1 \\ i_2 \end{bmatrix} = \begin{bmatrix} v_1 \\ -v_2 \end{bmatrix} \tag{3.24}$$

We notice that each of the diagonal terms is the sum of the resistances in the related mesh, while each of the off-diagonal terms is the negative of the resistance common to meshes 1 and 2. Each term on the right-hand side of Eq. (3.24) is the algebraic sum taken clockwise of all independent voltage sources in the related mesh.

In general, if the circuit has N meshes, the mesh-current equations can be expressed in terms of the resistances as

$$\begin{bmatrix} R_{11} & R_{12} & \dots & R_{1N} \\ R_{21} & R_{22} & \dots & R_{2N} \\ \vdots & \vdots & \vdots & \vdots \\ R_{N1} & R_{N2} & \dots & R_{NN} \end{bmatrix} \begin{bmatrix} i_1 \\ i_2 \\ \vdots \\ i_N \end{bmatrix} = \begin{bmatrix} v_1 \\ v_2 \\ \vdots \\ v_N \end{bmatrix} \tag{3.25}$$

or simply

$$\mathbf{Ri} = \mathbf{v} \tag{3.26}$$

where

R_{kk} = Sum of the resistances in mesh k

$R_{kj} = R_{jk}$ = Negative of the sum of the resistances in common with meshes k and j, $k \neq j$

i_k = Unknown mesh current for mesh k in the clockwise direction

v_k = Sum taken clockwise of all independent voltage sources in mesh k, with voltage rise treated as positive

R is called the *resistance matrix*; **i** is the output vector; and **v** is the input vector. We can solve Eq. (3.25) to obtain the unknown mesh currents.

Example 3.8

Write the node-voltage matrix equations for the circuit in Fig. 3.27 by inspection.

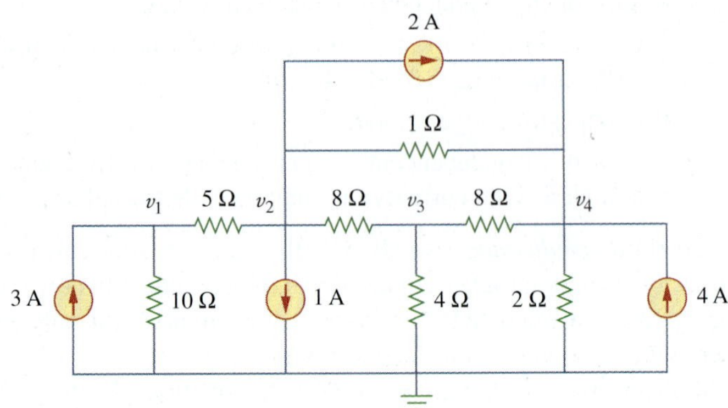

Figure 3.27
For Example 3.8.

Solution:

The circuit in Fig. 3.27 has four nonreference nodes, so we need four node equations. This implies that the size of the conductance matrix **G**, is 4 by 4. The diagonal terms of **G**, in siemens, are

$$G_{11} = \frac{1}{5} + \frac{1}{10} = 0.3, \qquad G_{22} = \frac{1}{5} + \frac{1}{8} + \frac{1}{1} = 1.325$$

$$G_{33} = \frac{1}{8} + \frac{1}{8} + \frac{1}{4} = 0.5, \qquad G_{44} = \frac{1}{8} + \frac{1}{2} + \frac{1}{1} = 1.625$$

The off-diagonal terms are

$$G_{12} = -\frac{1}{5} = -0.2, \qquad G_{13} = G_{14} = 0$$

$$G_{21} = -0.2, \qquad G_{23} = -\frac{1}{8} = -0.125, \qquad G_{24} = -\frac{1}{1} = -1$$

$$G_{31} = 0, \qquad G_{32} = -0.125, \qquad G_{34} = -\frac{1}{8} = -0.125$$

$$G_{41} = 0, \qquad G_{42} = -1, \qquad G_{43} = -0.125$$

The input current vector **i** has the following terms, in amperes:

$$i_1 = 3, \qquad i_2 = -1 - 2 = -3, \qquad i_3 = 0, \qquad i_4 = 2 + 4 = 6$$

Thus the node-voltage equations are

$$\begin{bmatrix} 0.3 & -0.2 & 0 & 0 \\ -0.2 & 1.325 & -0.125 & -1 \\ 0 & -0.125 & 0.5 & -0.125 \\ 0 & -1 & -0.125 & 1.625 \end{bmatrix} \begin{bmatrix} v_1 \\ v_2 \\ v_3 \\ v_4 \end{bmatrix} = \begin{bmatrix} 3 \\ -3 \\ 0 \\ 6 \end{bmatrix}$$

which can be solved using *MATLAB* to obtain the node voltages v_1, v_2, v_3, and v_4.

Practice Problem 3.8

By inspection, obtain the node-voltage equations for the circuit in Fig. 3.28.

Answer:

$$\begin{bmatrix} 1.3 & -0.2 & -1 & 0 \\ -0.2 & 0.2 & 0 & 0 \\ -1 & 0 & 1.25 & -0.25 \\ 0 & 0 & -0.25 & 0.75 \end{bmatrix} \begin{bmatrix} v_1 \\ v_2 \\ v_3 \\ v_4 \end{bmatrix} = \begin{bmatrix} 0 \\ 3 \\ -1 \\ 3 \end{bmatrix}$$

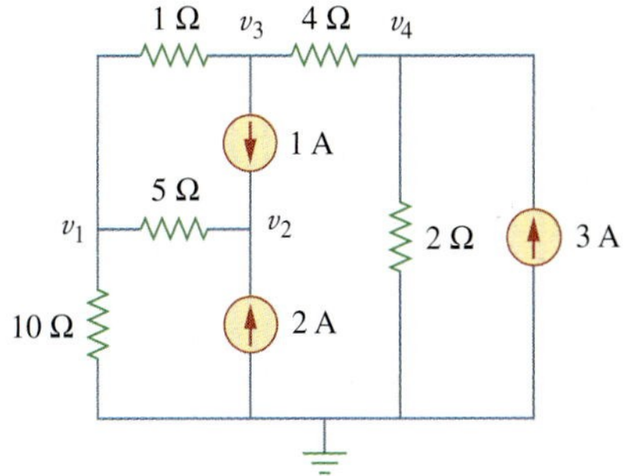

Figure 3.28
For Practice Prob. 3.8.

Example 3.9

By inspection, write the mesh-current equations for the circuit in Fig. 3.29.

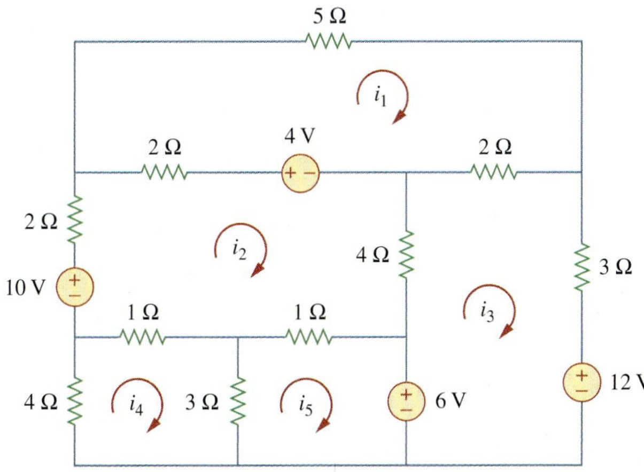

Figure 3.29
For Example 3.9.

Solution:

We have five meshes, so the resistance matrix is 5 by 5. The diagonal terms, in ohms, are:

$$R_{11} = 5 + 2 + 2 = 9, \qquad R_{22} = 2 + 4 + 1 + 1 + 2 = 10,$$
$$R_{33} = 2 + 3 + 4 = 9, \qquad R_{44} = 1 + 3 + 4 = 8, \qquad R_{55} = 1 + 3 = 4$$

The off-diagonal terms are:

$$R_{12} = -2, \qquad R_{13} = -2, \qquad R_{14} = 0 = R_{15},$$
$$R_{21} = -2, \qquad R_{23} = -4, \qquad R_{24} = -1, \qquad R_{25} = -1,$$
$$R_{31} = -2, \qquad R_{32} = -4, \qquad R_{34} = 0 = R_{35},$$
$$R_{41} = 0, \qquad R_{42} = -1, \qquad R_{43} = 0, \qquad R_{45} = -3,$$
$$R_{51} = 0, \qquad R_{52} = -1, \qquad R_{53} = 0, \qquad R_{54} = -3$$

The input voltage vector **v** has the following terms in volts:

$$v_1 = 4, \qquad v_2 = 10 - 4 = 6,$$
$$v_3 = -12 + 6 = -6, \qquad v_4 = 0, \qquad v_5 = -6$$

Thus, the mesh-current equations are:

$$\begin{bmatrix} 9 & -2 & -2 & 0 & 0 \\ -2 & 10 & -4 & -1 & -1 \\ -2 & -4 & 9 & 0 & 0 \\ 0 & -1 & 0 & 8 & -3 \\ 0 & -1 & 0 & -3 & 4 \end{bmatrix} \begin{bmatrix} i_1 \\ i_2 \\ i_3 \\ i_4 \\ i_5 \end{bmatrix} = \begin{bmatrix} 4 \\ 6 \\ -6 \\ 0 \\ -6 \end{bmatrix}$$

From this, we can use *MATLAB* to obtain mesh currents i_1, i_2, i_3, i_4, and i_5.

Practice Problem 3.9

By inspection, obtain the mesh-current equations for the circuit in Fig. 3.30.

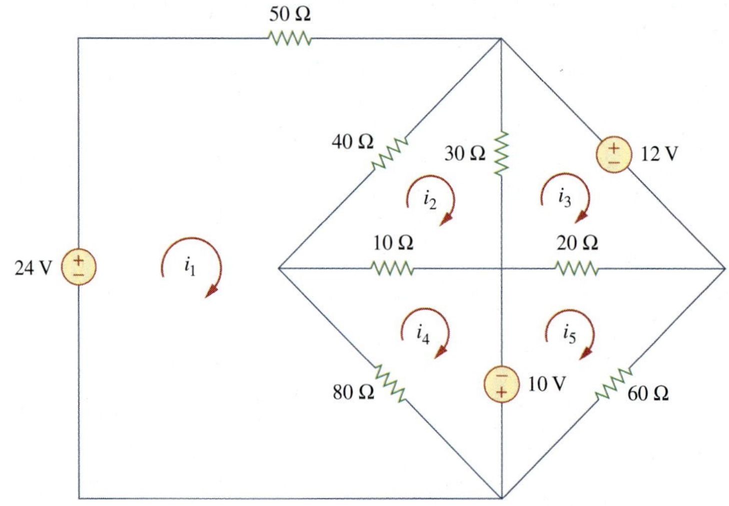

Figure 3.30
For Practice Prob. 3.9.

Answer:

$$\begin{bmatrix} 170 & -40 & 0 & -80 & 0 \\ -40 & 80 & -30 & -10 & 0 \\ 0 & -30 & 50 & 0 & -20 \\ -80 & -10 & 0 & 90 & 0 \\ 0 & 0 & -20 & 0 & 80 \end{bmatrix} \begin{bmatrix} i_1 \\ i_2 \\ i_3 \\ i_4 \\ i_5 \end{bmatrix} = \begin{bmatrix} 24 \\ 0 \\ -12 \\ 10 \\ -10 \end{bmatrix}$$

3.7 Nodal Versus Mesh Analysis

Both nodal and mesh analyses provide a systematic way of analyzing a complex network. Someone may ask: Given a network to be analyzed, how do we know which method is better or more efficient? The choice of the better method is dictated by two factors.

The first factor is the nature of the particular network. Networks that contain many series-connected elements, voltage sources, or supermeshes are more suitable for mesh analysis, whereas networks with parallel-connected elements, current sources, or supernodes are more suitable for nodal analysis. Also, a circuit with fewer nodes than meshes is better analyzed using nodal analysis, while a circuit with fewer meshes than nodes is better analyzed using mesh analysis. The key is to select the method that results in the smaller number of equations.

The second factor is the information required. If node voltages are required, it may be expedient to apply nodal analysis. If branch or mesh currents are required, it may be better to use mesh analysis.

It is helpful to be familiar with both methods of analysis, for at least two reasons. First, one method can be used to check the results from the other method, if possible. Second, since each method has its limitations, only one method may be suitable for a particular problem. For example, mesh analysis is the only method to use in analyzing transistor circuits, as we shall see in Section 3.9. But mesh analysis cannot easily be used to solve an op amp circuit, as we shall see in Chapter 5, because there is no direct way to obtain the voltage across the op amp itself. For nonplanar networks, nodal analysis is the only option, because mesh analysis only applies to planar networks. Also, nodal analysis is more amenable to solution by computer, as it is easy to program. This allows one to analyze complicated circuits that defy hand calculation. A computer software package based on nodal analysis is introduced next.

3.8 Circuit Analysis with *PSpice*

PSpice is a computer software circuit analysis program that we will gradually learn to use throughout the course of this text. This section illustrates how to use *PSpice for Windows* to analyze the dc circuits we have studied so far.

Appendix D provides a tutorial on using *PSpice for Windows*.

The reader is expected to review Sections D.1 through D.3 of Appendix D before proceeding in this section. It should be noted that *PSpice* is only helpful in determining branch voltages and currents when the numerical values of all the circuit components are known.

Example 3.10

Use *PSpice* to find the node voltages in the circuit of Fig. 3.31.

Solution:

The first step is to draw the given circuit using Schematics. If one follows the instructions given in Appendix sections D.2 and D.3, the schematic in Fig. 3.32 is produced. Since this is a dc analysis, we use voltage source VDC and current source IDC. The pseudocomponent VIEWPOINTS are added to display the required node voltages. Once the circuit is drawn and saved as *exam310.sch*, we run *PSpice* by selecting **Analysis/Simulate**. The circuit is simulated and the results

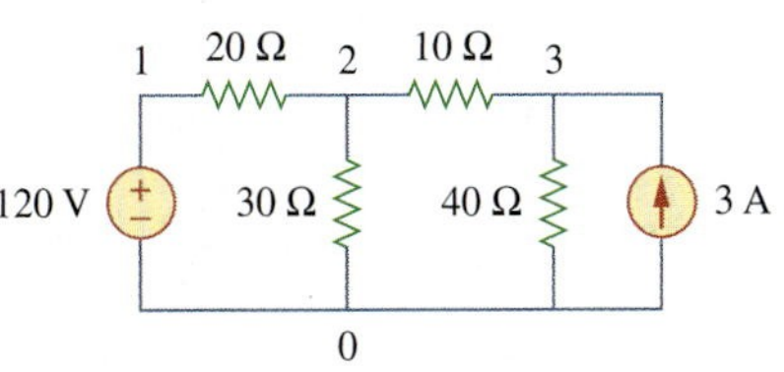

Figure 3.31
For Example 3.10.

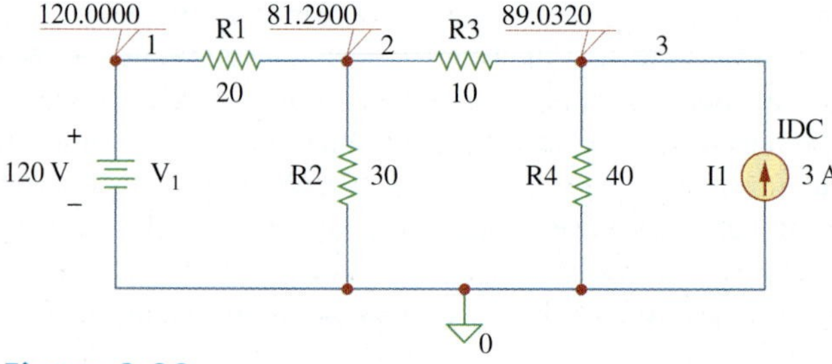

Figure 3.32
For Example 3.10; the schematic of the circuit in Fig. 3.31.

are displayed on VIEWPOINTS and also saved in output file *exam310.out*. The output file includes the following:

```
NODE  VOLTAGE    NODE  VOLTAGE    NODE  VOLTAGE
(1)   120.0000   (2)   81.2900    (3)   89.0320
```

indicating that $V_1 = 120$ V, $V_2 = 81.29$ V, $V_3 = 89.032$ V.

Practice Problem 3.10

For the circuit in Fig. 3.33, use *PSpice* to find the node voltages.

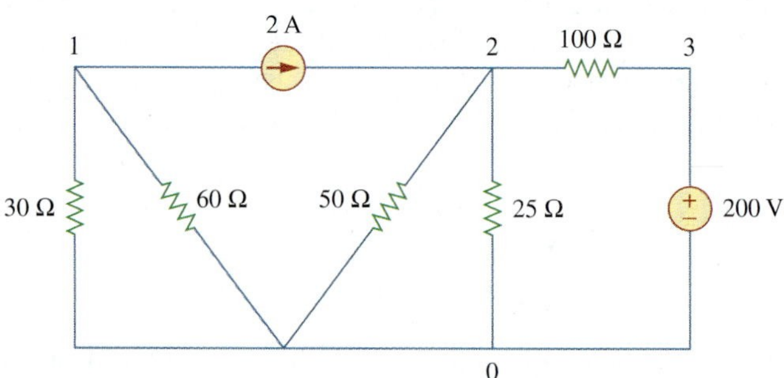

Figure 3.33
For Practice Prob. 3.10.

Answer: $V_1 = -40$ V, $V_2 = 57.14$ V, $V_3 = 200$ V.

Example 3.11

In the circuit of Fig. 3.34, determine the currents i_1, i_2, and i_3.

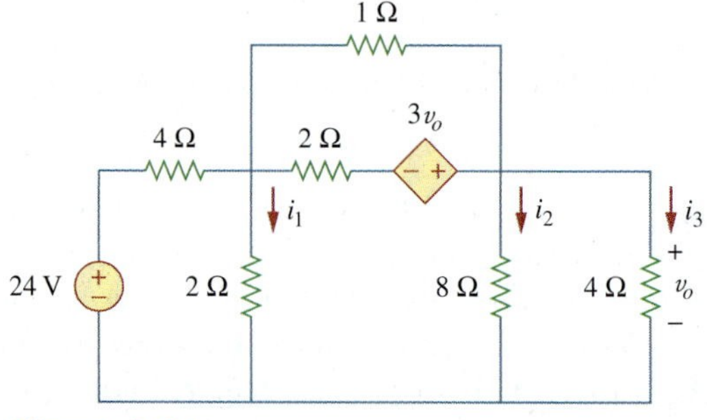

Figure 3.34
For Example 3.11.

Solution:

The schematic is shown in Fig. 3.35. (The schematic in Fig. 3.35 includes the output results, implying that it is the schematic displayed on the screen *after* the simulation.) Notice that the voltage-controlled voltage source E1 in Fig. 3.35 is connected so that its input is the voltage across the 4-Ω resistor; its gain is set equal to 3. In order to display the required currents, we insert pseudocomponent IPROBES in the appropriate branches. The schematic is saved as *exam311.sch* and simulated by selecting **Analysis/Simulate**. The results are displayed on IPROBES as shown in Fig. 3.35 and saved in output file *exam311.out*. From the output file or the IPROBES, we obtain $i_1 = i_2 = 1.333$ A and $i_3 = 2.667$ A.

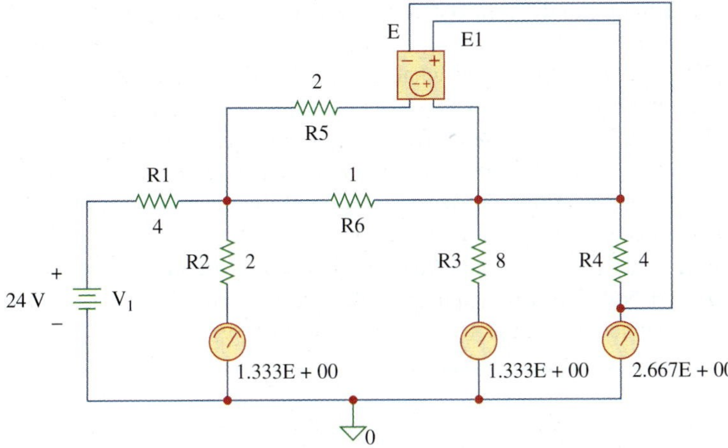

Figure 3.35
The schematic of the circuit in Fig. 3.34.

Practice Problem 3.11

Use *PSpice* to determine currents i_1, i_2, and i_3 in the circuit of Fig. 3.36.

Answer: $i_1 = -0.4286$ A, $i_2 = 2.286$ A, $i_3 = 2$ A.

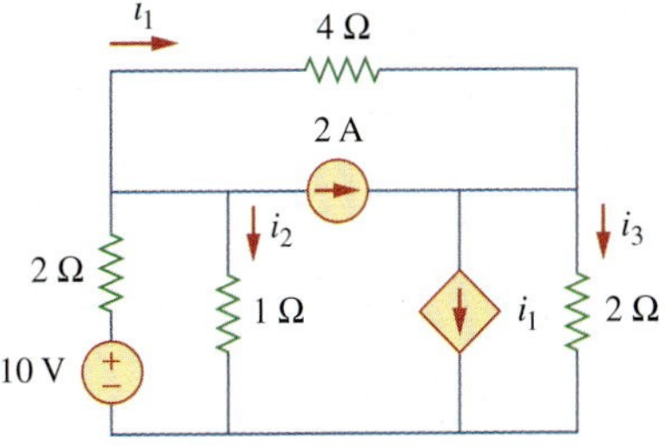

Figure 3.36
For Practice Prob. 3.11.

3.9 †Applications: DC Transistor Circuits

Most of us deal with electronic products on a routine basis and have some experience with personal computers. A basic component for the integrated circuits found in these electronics and computers is the active, three-terminal device known as the *transistor*. Understanding the transistor is essential before an engineer can start an electronic circuit design.

Figure 3.37 depicts various kinds of transistors commercially available. There are two basic types of transistors: *bipolar junction transistors* (BJTs) and *field-effect transistors* (FETs). Here, we consider only the BJTs, which were the first of the two and are still used today. Our objective is to present enough detail about the BJT to enable us to apply the techniques developed in this chapter to analyze dc transistor circuits.

Historical

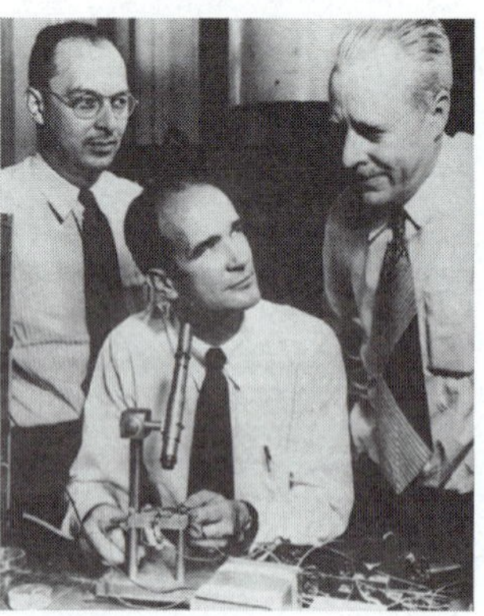
Courtesy of Lucent Technologies/Bell Labs

William Schockley (1910–1989), **John Bardeen** (1908–1991), and **Walter Brattain** (1902–1987) co-invented the transistor.

Nothing has had a greater impact on the transition from the "Industrial Age" to the "Age of the Engineer" than the transistor. I am sure that Dr. Shockley, Dr. Bardeen, and Dr. Brattain had no idea they would have this incredible effect on our history. While working at Bell Laboratories, they successfully demonstrated the point-contact transistor, invented by Bardeen and Brattain in 1947, and the junction transistor, which Shockley conceived in 1948 and successfully produced in 1951.

It is interesting to note that the idea of the field-effect transistor, the most commonly used one today, was first conceived in 1925–1928 by J. E. Lilienfeld, a German immigrant to the United States. This is evident from his patents of what appears to be a field-effect transistor. Unfortunately, the technology to realize this device had to wait until 1954 when Shockley's field-effect transistor became a reality. Just think what today would be like if we had this transistor 30 years earlier!

For their contributions to the creation of the transistor, Dr. Shockley, Dr. Bardeen, and Dr. Brattain received, in 1956, the Nobel Prize in physics. It should be noted that Dr. Bardeen is the only individual to win two Nobel prizes in physics; the second came later for work in superconductivity at the University of Illinois.

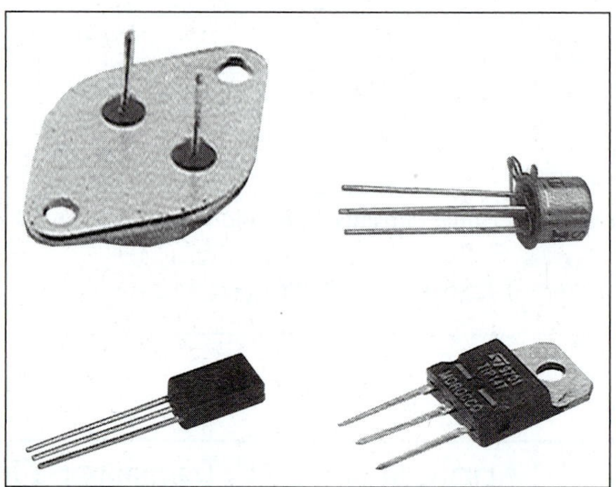
Figure 3.37
Various types of transistors.
(Courtesy of Tech America.)

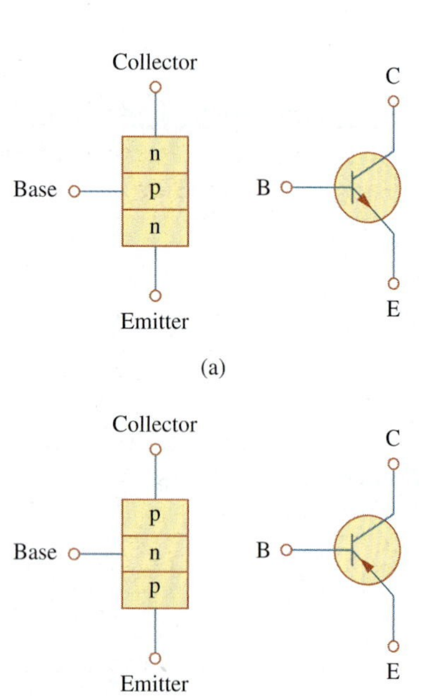

Figure 3.38
Two types of BJTs and their circuit symbols: (a) *npn*, (b) *pnp*.

There are two types of BJTs: *npn* and *pnp*, with their circuit symbols as shown in Fig. 3.38. Each type has three terminals, designated as emitter (E), base (B), and collector (C). For the *npn* transistor, the currents and voltages of the transistor are specified as in Fig. 3.39. Applying KCL to Fig. 3.39(a) gives

$$I_E = I_B + I_C \tag{3.27}$$

where I_E, I_C, and I_B are emitter, collector, and base currents, respectively. Similarly, applying KVL to Fig. 3.39(b) gives

$$V_{CE} + V_{EB} + V_{BC} = 0 \tag{3.28}$$

where V_{CE}, V_{EB}, and V_{BC} are collector-emitter, emitter-base, and base-collector voltages. The BJT can operate in one of three modes: active, cutoff, and saturation. When transistors operate in the active mode, typically $V_{BE} \simeq 0.7$ V,

$$I_C = \alpha I_E \tag{3.29}$$

where α is called the *common-base current gain.* In Eq. (3.29), α denotes the fraction of electrons injected by the emitter that are collected by the collector. Also,

$$\boxed{I_C = \beta I_B} \tag{3.30}$$

where β is known as the *common-emitter current gain.* The α and β are characteristic properties of a given transistor and assume constant values for that transistor. Typically, α takes values in the range of 0.98 to 0.999, while β takes values in the range of 50 to 1000. From Eqs. (3.27) to (3.30), it is evident that

$$I_E = (1 + \beta)I_B \tag{3.31}$$

and

$$\beta = \frac{\alpha}{1 - \alpha} \tag{3.32}$$

These equations show that, in the active mode, the BJT can be modeled as a dependent current-controlled current source. Thus, in circuit analysis, the dc equivalent model in Fig. 3.40(b) may be used to replace the *npn* transistor in Fig. 3.40(a). Since β in Eq. (3.32) is large, a small base current controls large currents in the output circuit. Consequently, the bipolar transistor can serve as an amplifier, producing both current gain and voltage gain. Such amplifiers can be used to furnish a considerable amount of power to transducers such as loudspeakers or control motors.

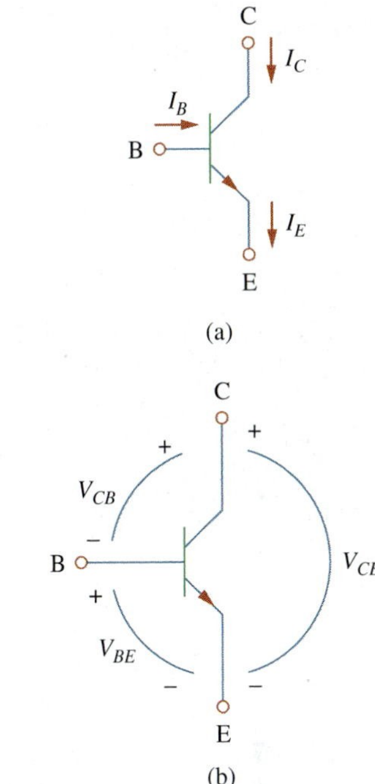

Figure 3.39
The terminal variables of an *npn* transistor: (a) currents, (b) voltages.

In fact, transistor circuits provide motivation to study dependent sources.

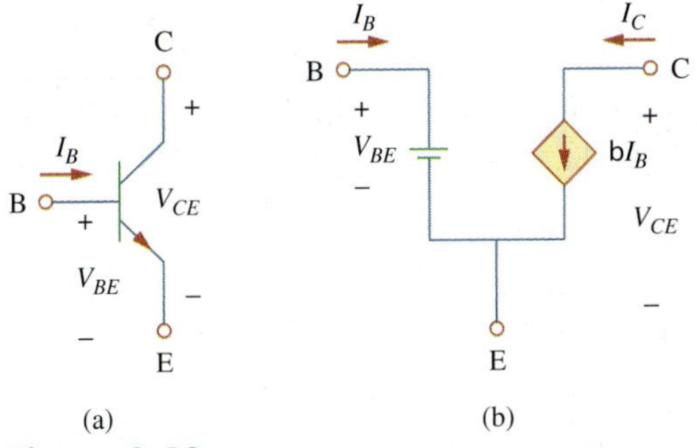

Figure 3.40
(a) An *npn* transistor, (b) its dc equivalent model.

It should be observed in the following examples that one cannot directly analyze transistor circuits using nodal analysis because of the potential difference between the terminals of the transistor. Only when the transistor is replaced by its equivalent model can we apply nodal analysis.

Example 3.12

Find I_B, I_C, and v_o in the transistor circuit of Fig. 3.41. Assume that the transistor operates in the active mode and that $\beta = 50$.

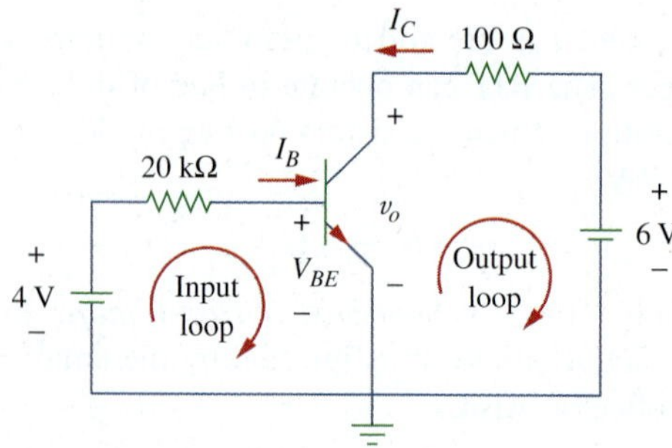

Figure 3.41
For Example 3.12.

Solution:

For the input loop, KVL gives

$$-4 + I_B(20 \times 10^3) + V_{BE} = 0$$

Since $V_{BE} = 0.7$ V in the active mode,

$$I_B = \frac{4 - 0.7}{20 \times 10^3} = 165\ \mu\text{A}$$

But

$$I_C = \beta I_B = 50 \times 165\ \mu\text{A} = 8.25\ \text{mA}$$

For the output loop, KVL gives

$$-v_o - 100I_C + 6 = 0$$

or

$$v_o = 6 - 100I_C = 6 - 0.825 = 5.175\ \text{V}$$

Note that $v_o = V_{CE}$ in this case.

Practice Problem 3.12

For the transistor circuit in Fig. 3.42, let $\beta = 100$ and $V_{BE} = 0.7$ V. Determine v_o and V_{CE}.

Answer: 2.876 V, 1.984 V.

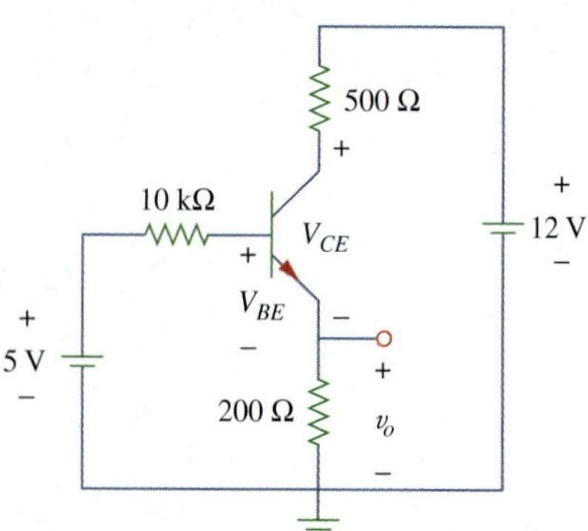

Figure 3.42
For Practice Prob. 3.12.

Example 3.13

For the BJT circuit in Fig. 3.43, $\beta = 150$ and $V_{BE} = 0.7$ V. Find v_o.

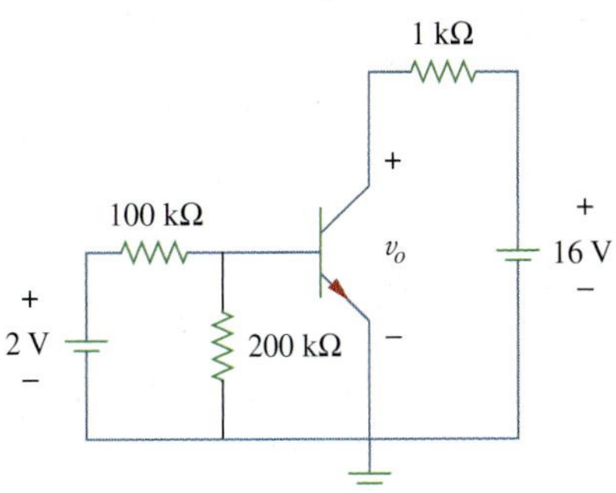

Figure 3.43
For Example 3.13.

Solution:

1. **Define.** The circuit is clearly defined and the problem is clearly stated. There appear to be no additional questions that need to be asked.
2. **Present.** We are to determine the output voltage of the circuit shown in Fig. 3.43. The circuit contains an ideal transistor with $\beta = 150$ and $V_{BE} = 0.7$ V.
3. **Alternative.** We can use mesh analysis to solve for v_o. We can replace the transistor with its equivalent circuit and use nodal analysis. We can try both approaches and use them to check each other. As a third check, we can use the equivalent circuit and solve it using *PSpice*.
4. **Attempt.**

■ **METHOD 1** Working with Fig. 3.44(a), we start with the first loop.

$$-2 + 100kI_1 + 200k(I_1 - I_2) = 0 \qquad \text{or} \qquad 3I_1 - 2I_2 = 2 \times 10^{-5} \tag{3.13.1}$$

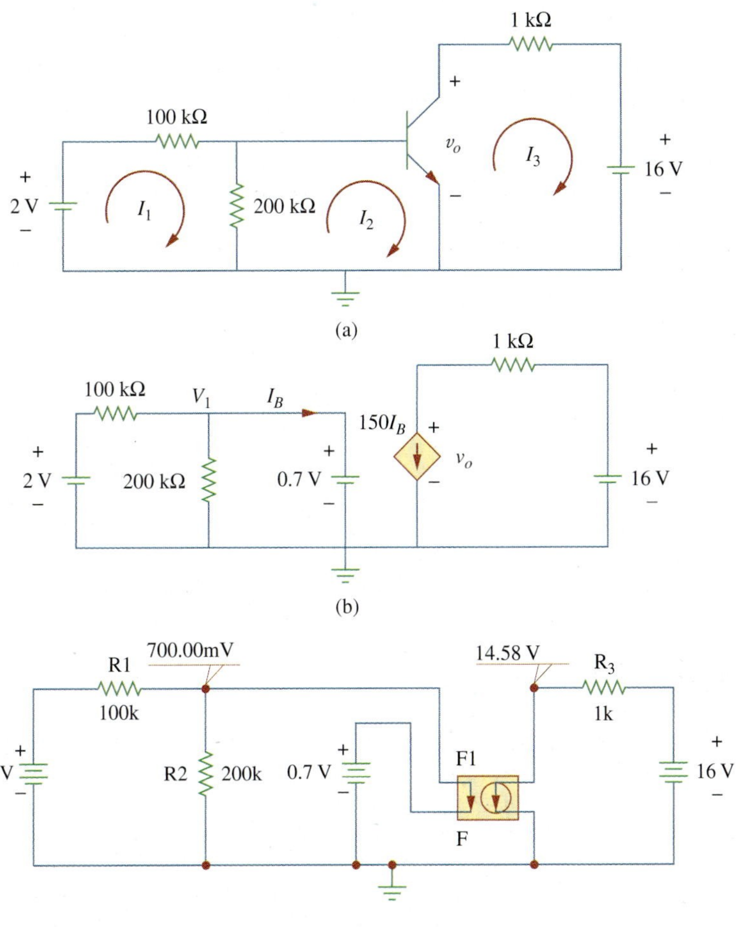

Figure 3.44
Solution of the problem in Example 3.13: (a) Method 1, (b) Method 2, (c) Method 3.

Now for loop 2.

$$200\text{k}(I_2 - I_1) + V_{BE} = 0 \qquad \text{or} \qquad -2I_1 + 2I_2 = -0.7 \times 10^{-5} \qquad \textbf{(3.13.2)}$$

Since we have two equations and two unknowns, we can solve for I_1 and I_2. Adding Eq. (3.13.1) to (3.13.2) we get;

$$I_1 = 1.3 \times 10^{-5}\text{A} \qquad \text{and} \qquad I_2 = (-0.7 + 2.6)10^{-5}/2 = 9.5\ \mu\text{A}$$

Since $I_3 = -150I_2 = -1.425$ mA, we can now solve for v_o using loop 3:

$$-v_o + {}^1\text{k}I_3 + 16 = 0 \qquad \text{or} \qquad v_o = -1.425 + 16 = \mathbf{14.575\ V}$$

■ **METHOD 2** Replacing the transistor with its equivalent circuit produces the circuit shown in Fig. 3.44(b). We can now use nodal analysis to solve for v_o.

At node number 1: $V_1 = 0.7$ V

$$(0.7 - 2)/100\text{k} + 0.7/200\text{k} + I_B = 0 \qquad \text{or} \qquad I_B = 9.5\ \mu\text{A}$$

At node number 2 we have:

$$150I_B + (v_o - 16)/1\text{k} = 0 \qquad \text{or}$$
$$v_o = 16 - 150 \times 10^3 \times 9.5 \times 10^{-6} = \mathbf{14.575\ V}$$

5. **Evaluate.** The answers check, but to further check we can use *PSpice* (Method 3), which gives us the solution shown in Fig. 3.44(c).
6. **Satisfactory?** Clearly, we have obtained the desired answer with a very high confidence level. We can now present our work as a solution to the problem.

Practice Problem 3.13

The transistor circuit in Fig. 3.45 has $\beta = 80$ and $V_{BE} = 0.7$ V. Find v_o and I_o.

Answer: 3 V, 150 μA.

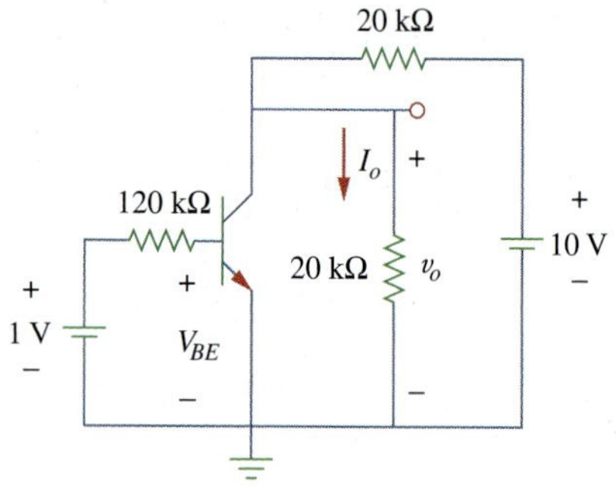

Figure 3.45
For Practice Prob. 3.13.

3.10 Summary

1. Nodal analysis is the application of Kirchhoff's current law at the nonreference nodes. (It is applicable to both planar and nonplanar circuits.) We express the result in terms of the node voltages. Solving the simultaneous equations yields the node voltages.
2. A supernode consists of two nonreference nodes connected by a (dependent or independent) voltage source.
3. Mesh analysis is the application of Kirchhoff's voltage law around meshes in a planar circuit. We express the result in terms of mesh currents. Solving the simultaneous equations yields the mesh currents.

4. A supermesh consists of two meshes that have a (dependent or independent) current source in common.
5. Nodal analysis is normally used when a circuit has fewer node equations than mesh equations. Mesh analysis is normally used when a circuit has fewer mesh equations than node equations.
6. Circuit analysis can be carried out using *PSpice*.
7. DC transistor circuits can be analyzed using the techniques covered in this chapter.

Review Questions

3.1 At node 1 in the circuit of Fig. 3.46, applying KCL gives:

(a) $2 + \dfrac{12 - v_1}{3} = \dfrac{v_1}{6} + \dfrac{v_1 - v_2}{4}$

(b) $2 + \dfrac{v_1 - 12}{3} = \dfrac{v_1}{6} + \dfrac{v_2 - v_1}{4}$

(c) $2 + \dfrac{12 - v_1}{3} = \dfrac{0 - v_1}{6} + \dfrac{v_1 - v_2}{4}$

(d) $2 + \dfrac{v_1 - 12}{3} = \dfrac{0 - v_1}{6} + \dfrac{v_2 - v_1}{4}$

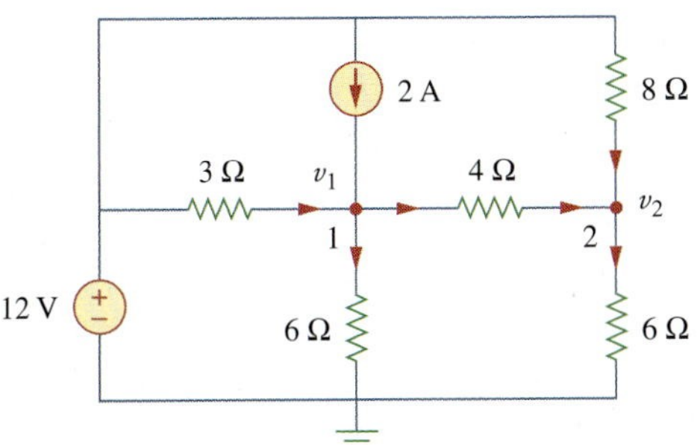

Figure 3.46
For Review Questions 3.1 and 3.2.

3.2 In the circuit of Fig. 3.46, applying KCL at node 2 gives:

(a) $\dfrac{v_2 - v_1}{4} + \dfrac{v_2}{8} = \dfrac{v_2}{6}$

(b) $\dfrac{v_1 - v_2}{4} + \dfrac{v_2}{8} = \dfrac{v_2}{6}$

(c) $\dfrac{v_1 - v_2}{4} + \dfrac{12 - v_2}{8} = \dfrac{v_2}{6}$

(d) $\dfrac{v_2 - v_1}{4} + \dfrac{v_2 - 12}{8} = \dfrac{v_2}{6}$

3.3 For the circuit in Fig. 3.47, v_1 and v_2 are related as:

(a) $v_1 = 6i + 8 + v_2$ (b) $v_1 = 6i - 8 + v_2$
(c) $v_1 = -6i + 8 + v_2$ (d) $v_1 = -6i - 8 + v_2$

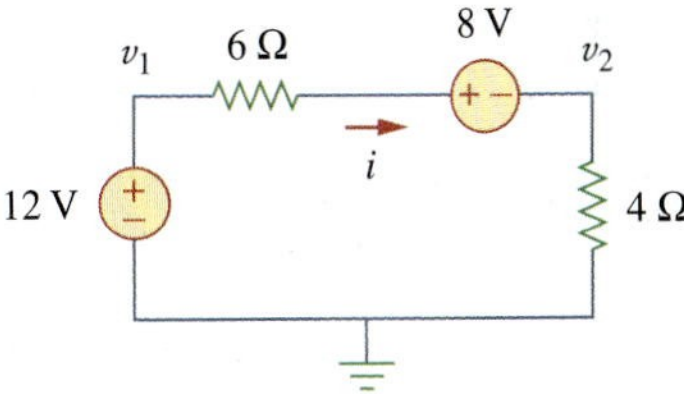

Figure 3.47
For Review Questions 3.3 and 3.4.

3.4 In the circuit of Fig. 3.47, the voltage v_2 is:

(a) −8 V (b) −1.6 V
(c) 1.6 V (d) 8 V

3.5 The current i in the circuit of Fig. 3.48 is:

(a) −2.667 A (b) −0.667 A
(c) 0.667 A (d) 2.667 A

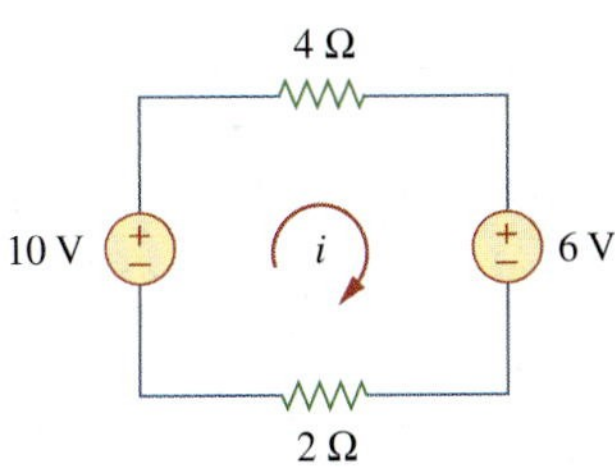

Figure 3.48
For Review Questions 3.5 and 3.6.

3.6 The loop equation for the circuit in Fig. 3.48 is:

(a) $-10 + 4i + 6 + 2i = 0$
(b) $10 + 4i + 6 + 2i = 0$
(c) $10 + 4i - 6 + 2i = 0$
(d) $-10 + 4i - 6 + 2i = 0$

3.7 In the circuit of Fig. 3.49, current i_1 is:

(a) 4 A (b) 3 A (c) 2 A (d) 1 A

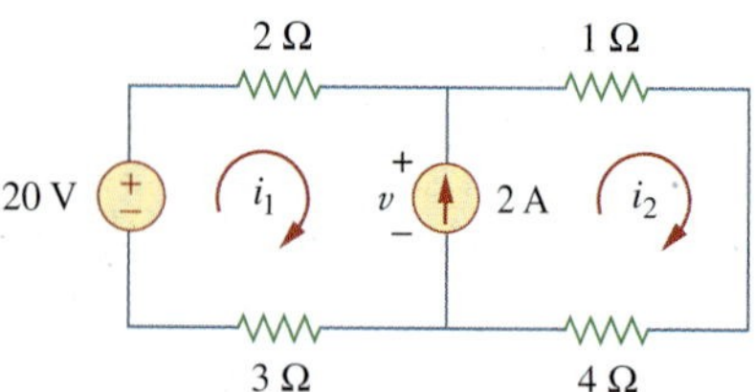

Figure 3.49
For Review Questions 3.7 and 3.8.

3.8 The voltage v across the current source in the circuit of Fig. 3.49 is:

(a) 20 V (b) 15 V (c) 10 V (d) 5 V

3.9 The *PSpice* part name for a current-controlled voltage source is:

(a) EX (b) FX (c) HX (d) GX

3.10 Which of the following statements are not true of the pseudocomponent IPROBE:

(a) It must be connected in series.

(b) It plots the branch current.

(c) It displays the current through the branch in which it is connected.

(d) It can be used to display voltage by connecting it in parallel.

(e) It is used only for dc analysis.

(f) It does not correspond to a particular circuit element.

Answers: 3.1a, 3.2c, 3.3a, 3.4c, 3.5c, 3.6a, 3.7d, 3.8b, 3.9c, 3.10b,d.

Problems

Sections 3.2 and 3.3 Nodal Analysis

3.1 Using Fig. 3.50, design a problem to help other students better understand nodal analysis.
e2d

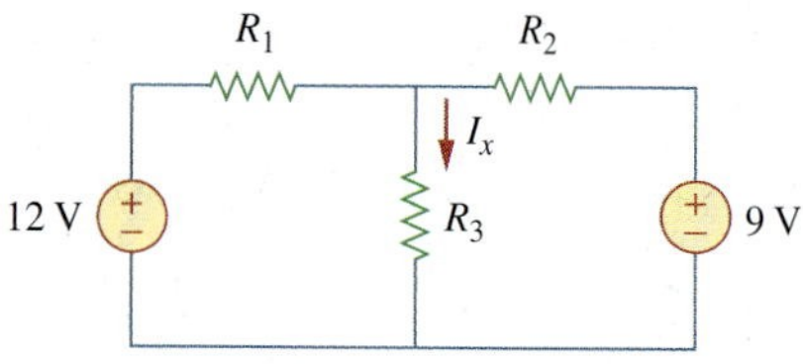

Figure 3.50
For Prob. 3.1.

3.2 For the circuit in Fig. 3.51, obtain v_1 and v_2.

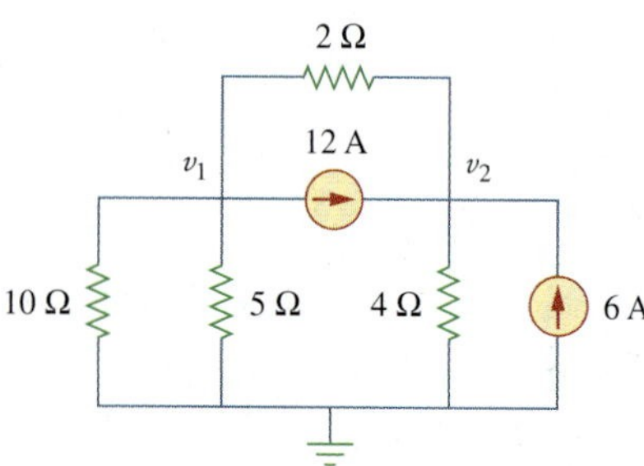

Figure 3.51
For Prob. 3.2.

3.3 Find the currents I_1 through I_4 and the voltage v_o in the circuit of Fig. 3.52.

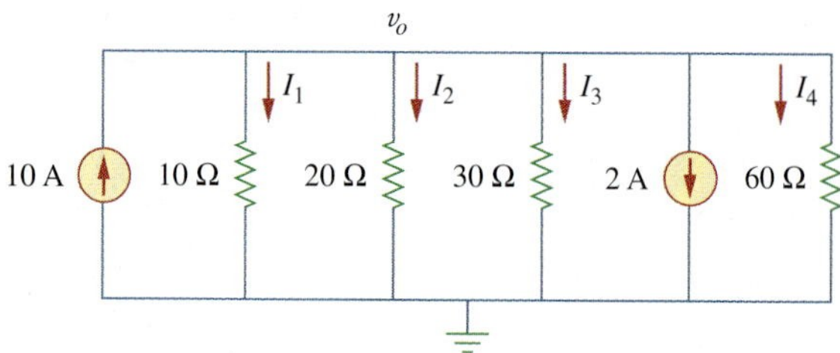

Figure 3.52
For Prob. 3.3.

3.4 Given the circuit in Fig. 3.53, calculate the currents I_1 through I_4.

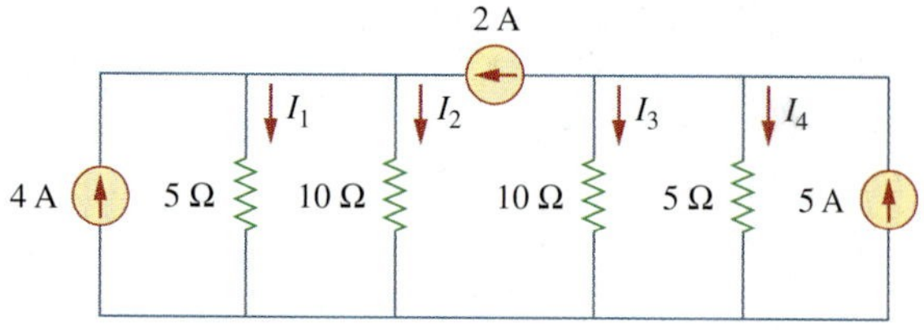

Figure 3.53
For Prob. 3.4.

3.5 Obtain v_o in the circuit of Fig. 3.54.

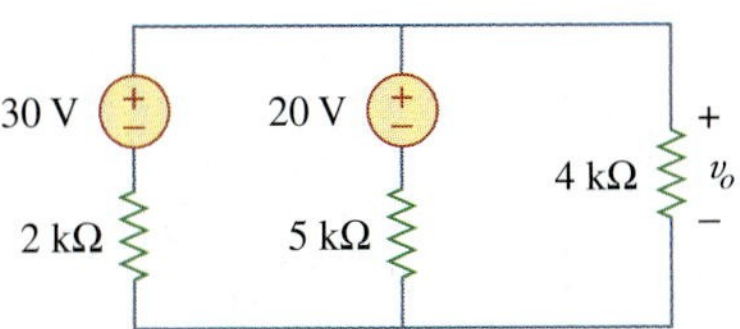

Figure 3.54
For Prob. 3.5.

3.6 Use nodal analysis to obtain v_o in the circuit of Fig. 3.55.

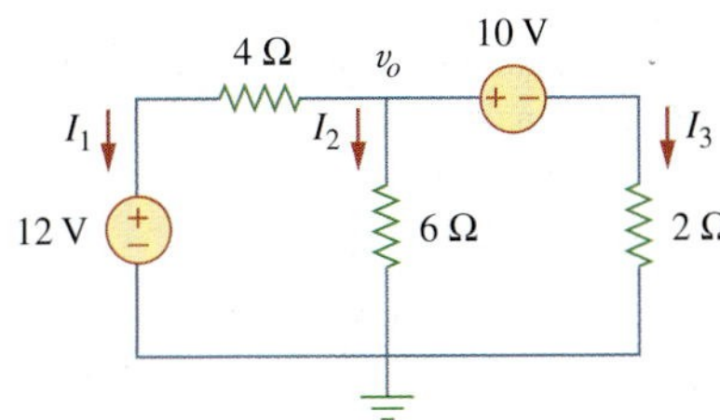

Figure 3.55
For Prob. 3.6.

3.7 Apply nodal analysis to solve for V_x in the circuit of Fig. 3.56.

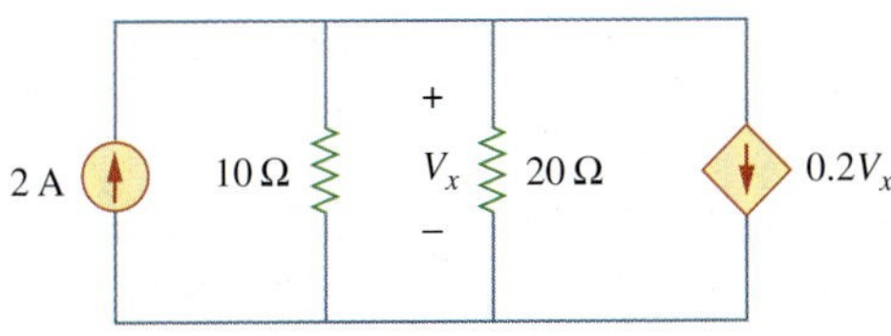

Figure 3.56
For Prob. 3.7.

3.8 Using nodal analysis, find v_o in the circuit of Fig. 3.57.

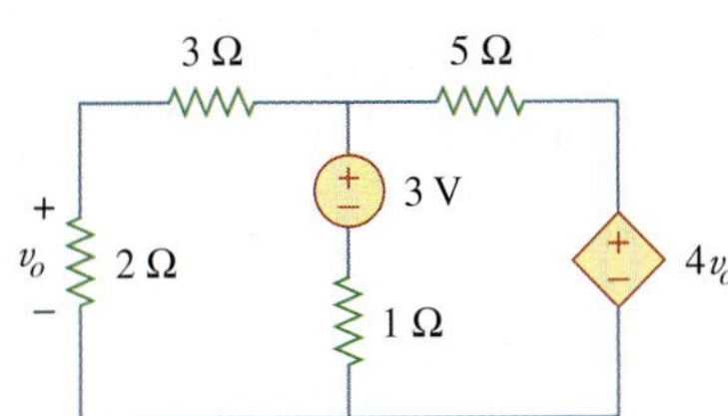

Figure 3.57
For Prob. 3.8.

3.9 Determine I_b in the circuit of Fig. 3.58 using nodal analysis.

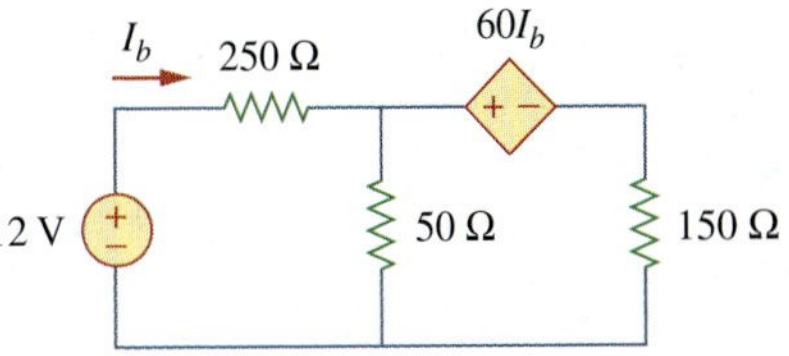

Figure 3.58
For Prob. 3.9.

3.10 Find I_o in the circuit of Fig. 3.59.

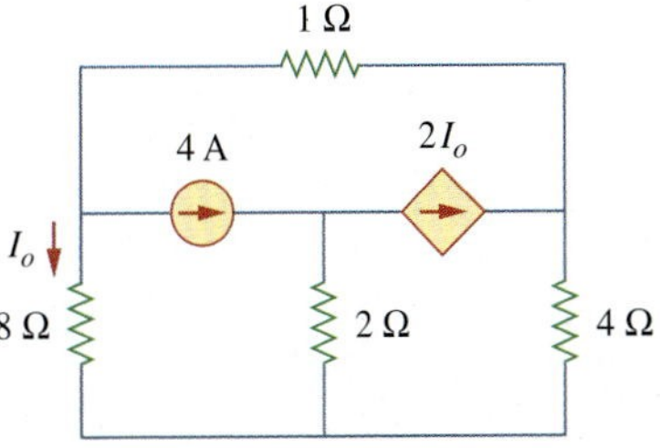

Figure 3.59
For Prob. 3.10.

3.11 Find V_o and the power dissipated in all the resistors in the circuit of Fig. 3.60.

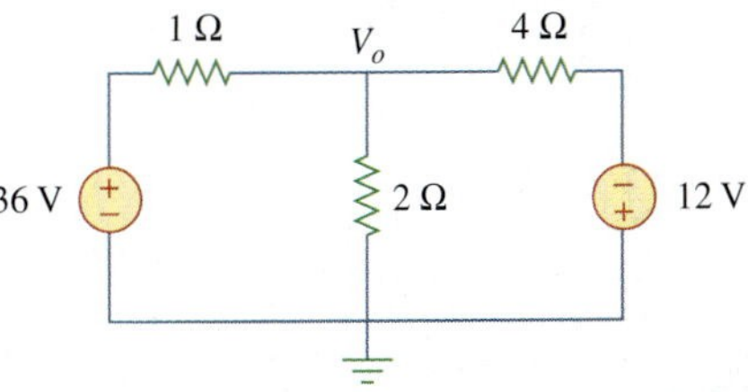

Figure 3.60
For Prob. 3.11.

3.12 Using nodal analysis, determine V_o in the circuit in Fig. 3.61.

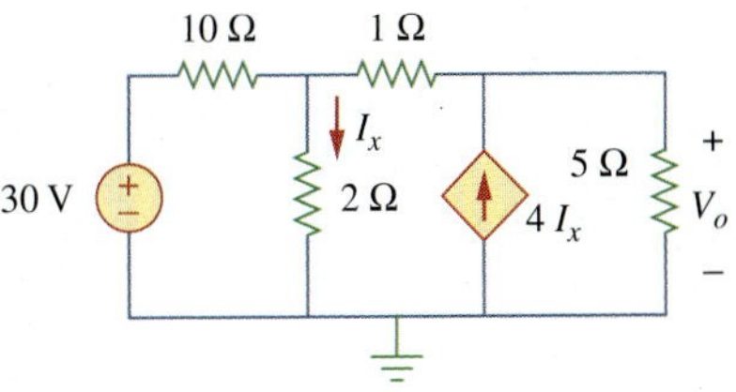

Figure 3.61
For Prob. 3.12.

3.13 Calculate v_1 and v_2 in the circuit of Fig. 3.62 using nodal analysis.

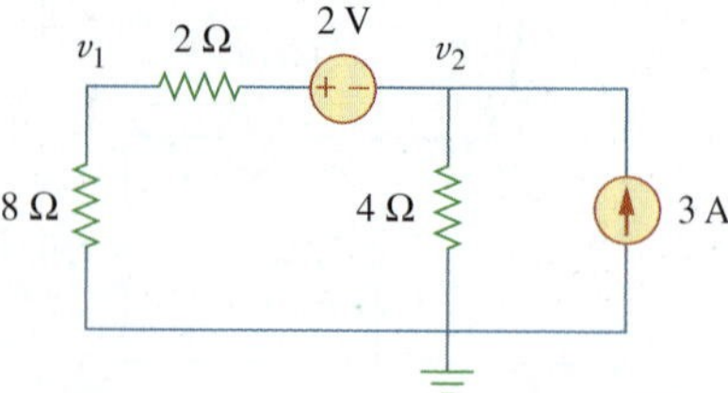

Figure 3.62
For Prob. 3.13.

3.14 Using nodal analysis, find v_o in the circuit of Fig. 3.63.

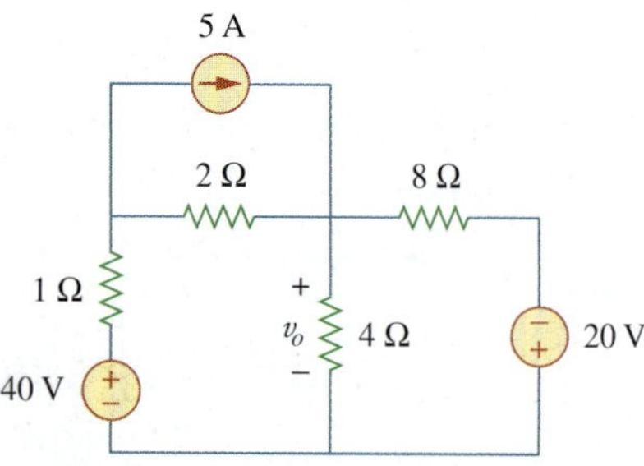

Figure 3.63
For Prob. 3.14.

3.15 Apply nodal analysis to find i_o and the power dissipated in each resistor in the circuit of Fig. 3.64.

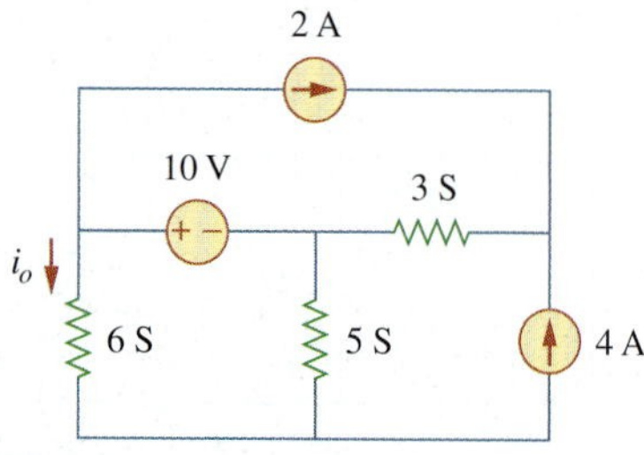

Figure 3.64
For Prob. 3.15.

3.16 Determine voltages v_1 through v_3 in the circuit of Fig. 3.65 using nodal analysis.

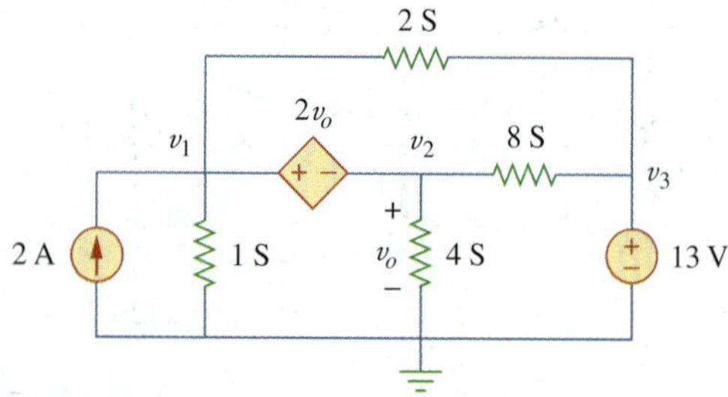

Figure 3.65
For Prob. 3.16.

3.17 Using nodal analysis, find current i_o in the circuit of Fig. 3.66.

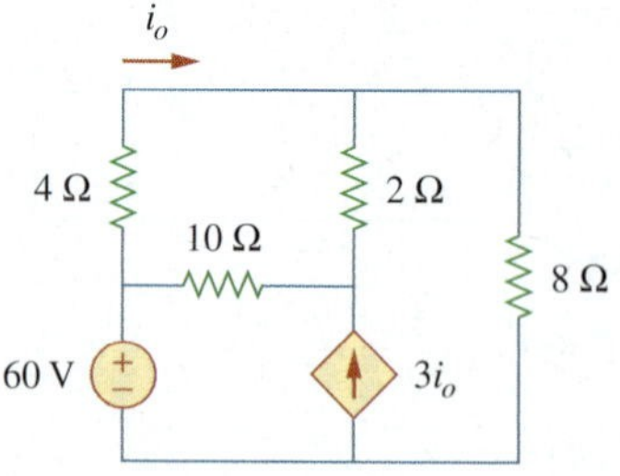

Figure 3.66
For Prob. 3.17.

3.18 Determine the node voltages in the circuit of Fig. 3.67 using nodal analysis.

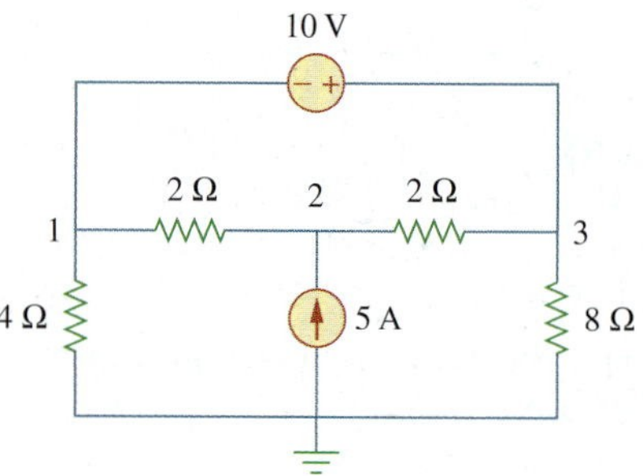

Figure 3.67
For Prob. 3.18.

3.19 Use nodal analysis to find v_1, v_2, and v_3 in the circuit of Fig. 3.68.

ML

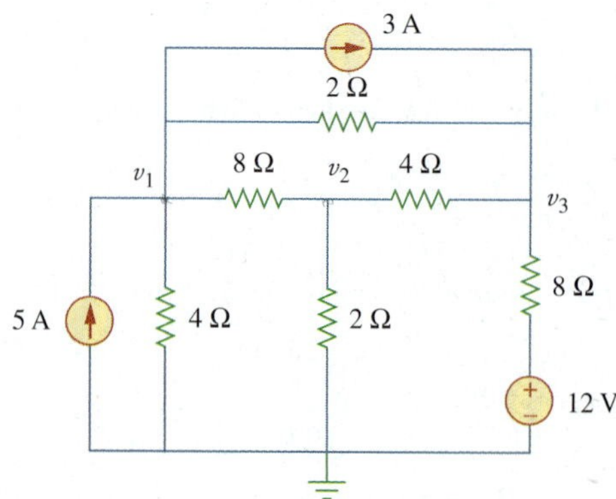

Figure 3.68
For Prob. 3.19.

3.20 For the circuit in Fig. 3.69, find v_1, v_2, and v_3 using nodal analysis.

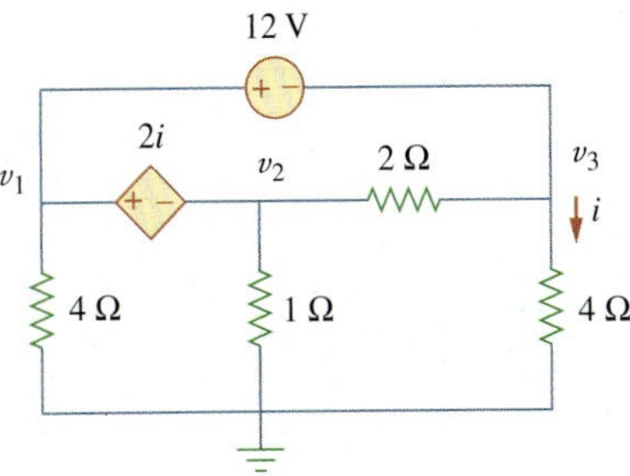

Figure 3.69
For Prob. 3.20.

3.21 For the circuit in Fig. 3.70, find v_1 and v_2 using nodal analysis.

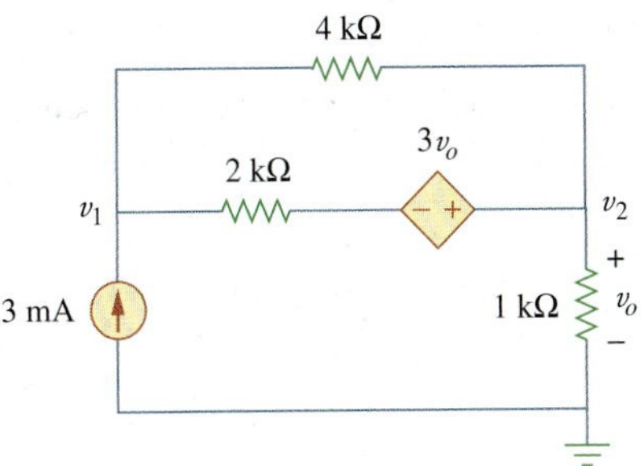

Figure 3.70
For Prob. 3.21.

3.22 Determine v_1 and v_2 in the circuit of Fig. 3.71.

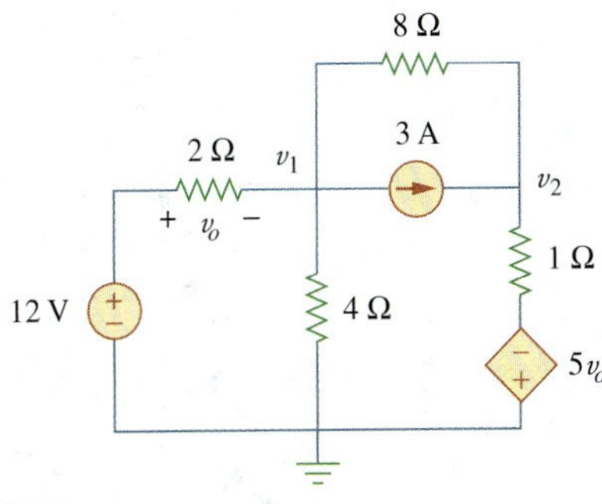

Figure 3.71
For Prob. 3.22.

3.23 Use nodal analysis to find V_o in the circuit of Fig. 3.72.

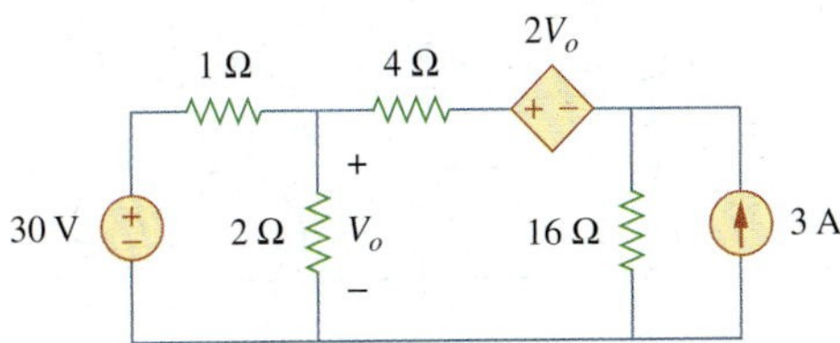

Figure 3.72
For Prob. 3.23.

3.24 Use nodal analysis and *MATLAB* to find V_o in the circuit of Fig. 3.73.
ML

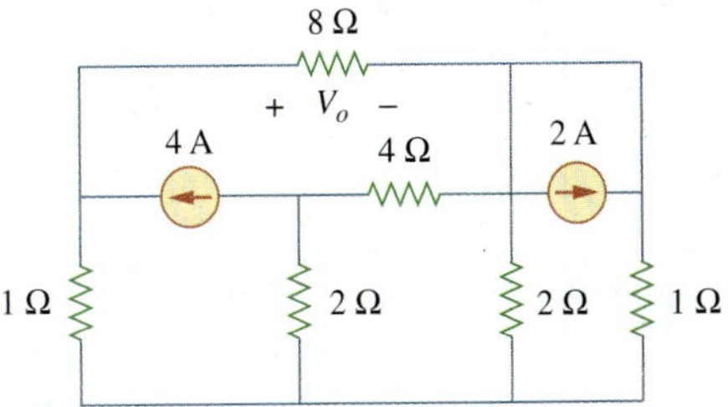

Figure 3.73
For Prob. 3.24.

3.25 Use nodal analysis along with *MATLAB* to determine the node voltages in Fig. 3.74.
ML

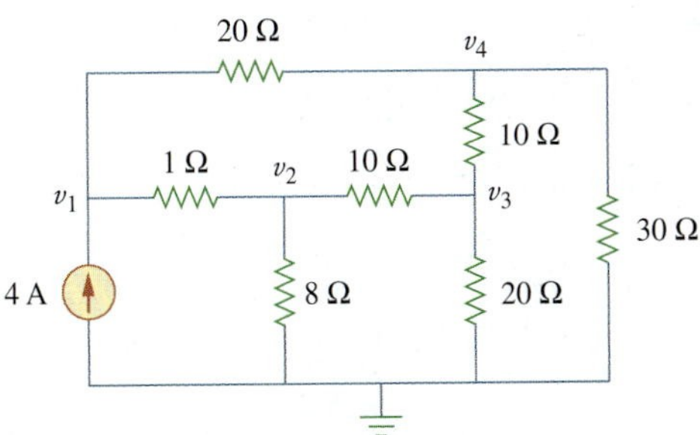

Figure 3.74
For Prob. 3.25.

3.26 Calculate the node voltages v_1, v_2, and v_3 in the circuit of Fig. 3.75.
ML

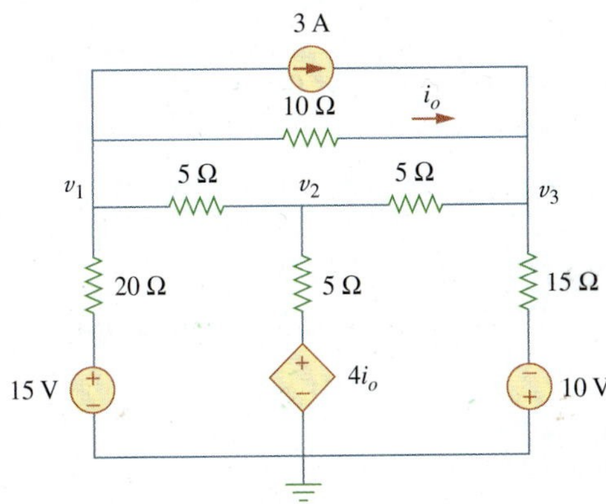

Figure 3.75
For Prob. 3.26.

***3.27** Use nodal analysis to determine voltages v_1, v_2, and v_3 in the circuit of Fig. 3.76.

ML

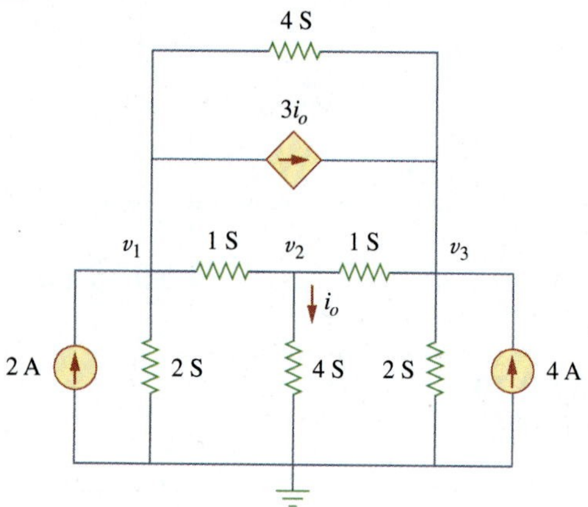

Figure 3.76
For Prob. 3.27.

***3.28** Use *MATLAB* to find the voltages at nodes *a*, *b*, *c*, and *d* in the circuit of Fig. 3.77.

ML

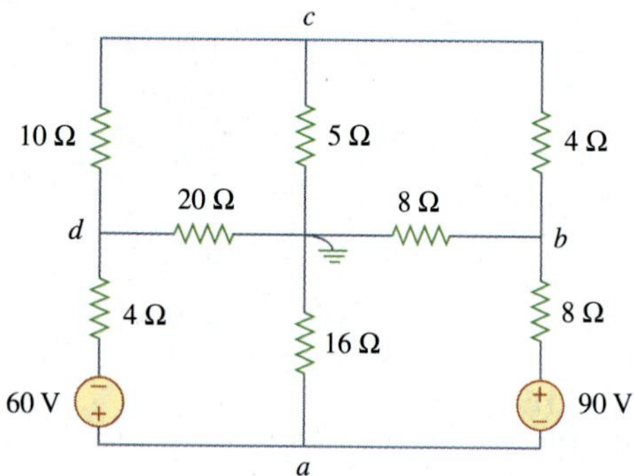

Figure 3.77
For Prob. 3.28.

3.29 Use *MATLAB* to solve for the node voltages in the circuit of Fig. 3.78.

ML

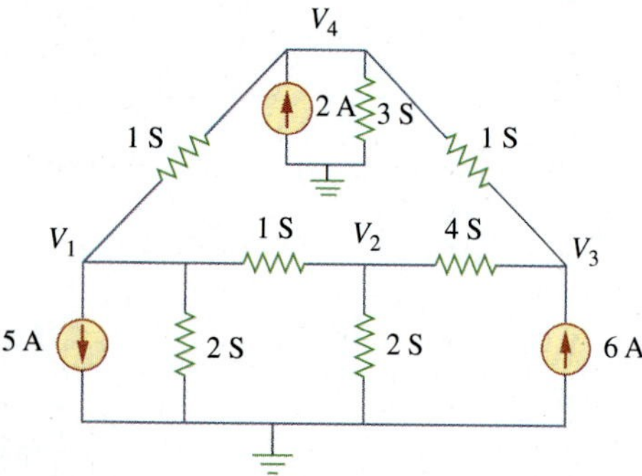

Figure 3.78
For Prob. 3.29.

* An asterisk indicates a challenging problem.

3.30 Using nodal analysis, find v_o and I_o in the circuit of Fig. 3.79.

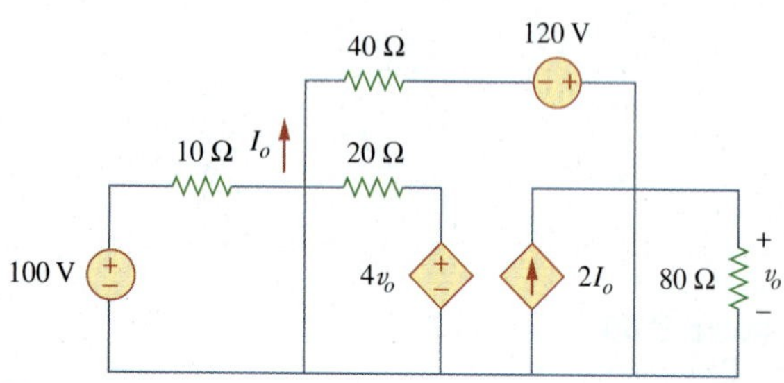

Figure 3.79
For Prob. 3.30.

3.31 Find the node voltages for the circuit in Fig. 3.80.

ML

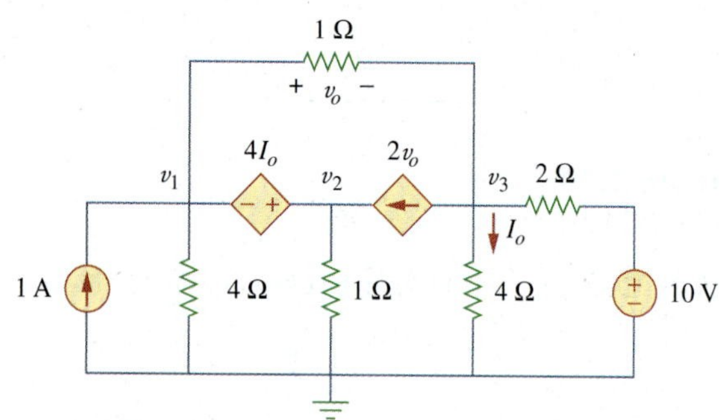

Figure 3.80
For Prob. 3.31.

***3.32** Obtain the node voltages v_1, v_2, and v_3 in the circuit of Fig. 3.81.

ML

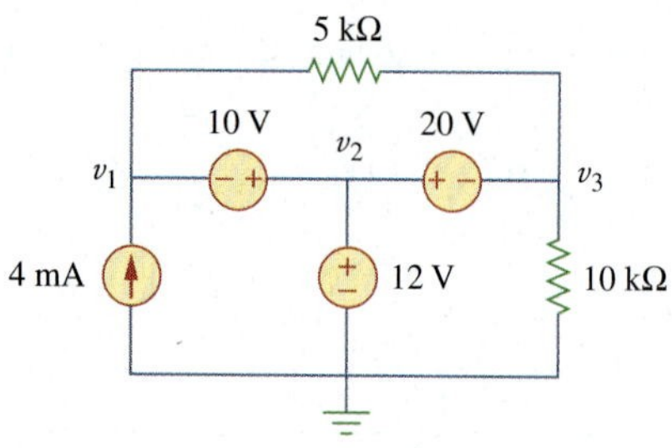

Figure 3.81
For Prob. 3.32.

Sections 3.4 and 3.5 Mesh Analysis

3.33 Which of the circuits in Fig. 3.82 is planar? For the planar circuit, redraw the circuits with no crossing branches.

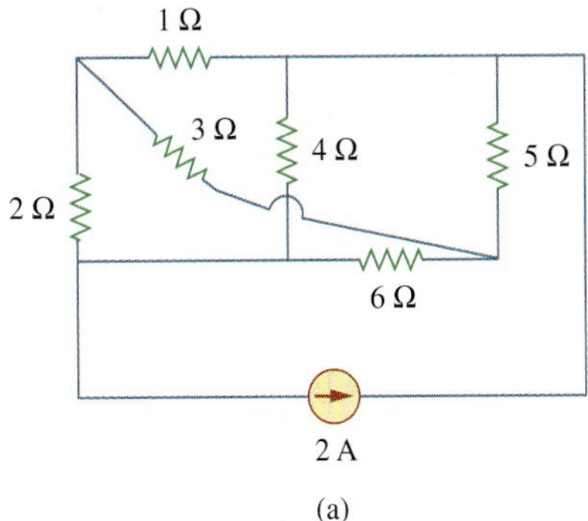

(a)

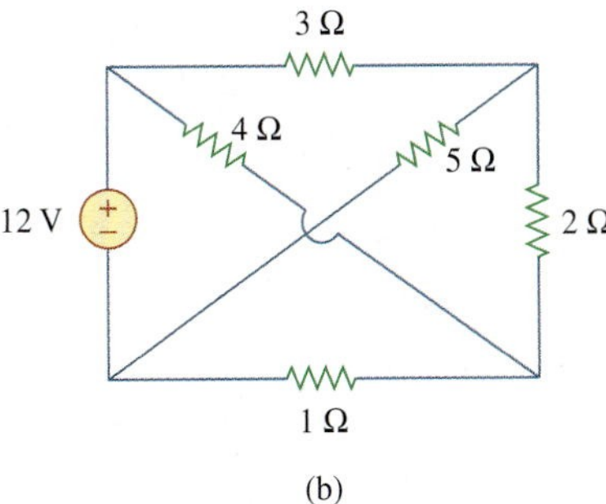

(b)

Figure 3.82
For Prob. 3.33.

3.34 Determine which of the circuits in Fig. 3.83 is planar and redraw it with no crossing branches.

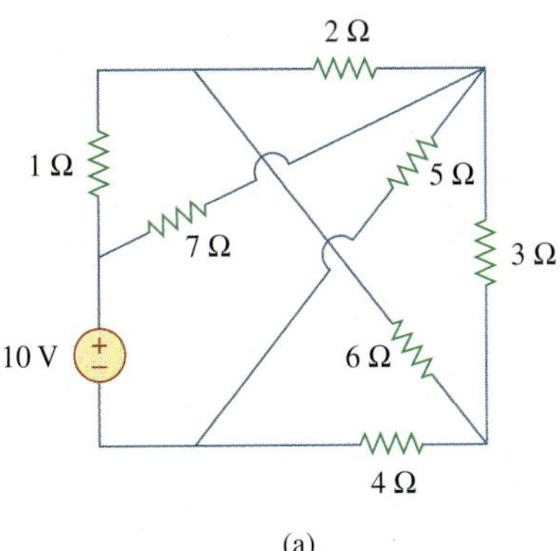

(a)

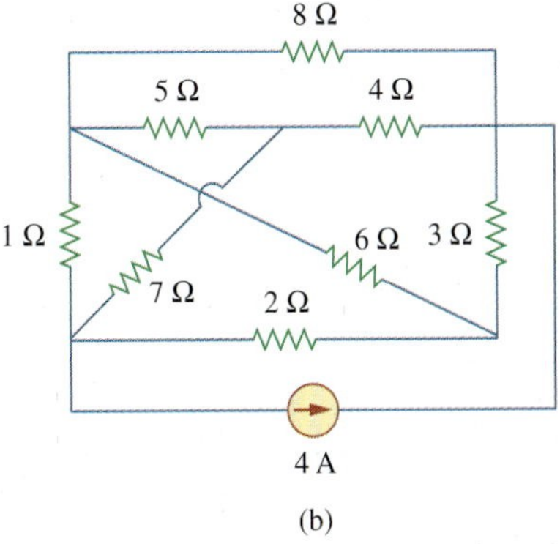

(b)

Figure 3.83
For Prob. 3.34.

3.35 Rework Prob. 3.5 using mesh analysis.

3.36 Rework Prob. 3.6 using mesh analysis.

3.37 Solve Prob. 3.8 using mesh analysis.

3.38 Apply mesh analysis to the circuit in Fig. 3.84 and obtain I_o.

ML

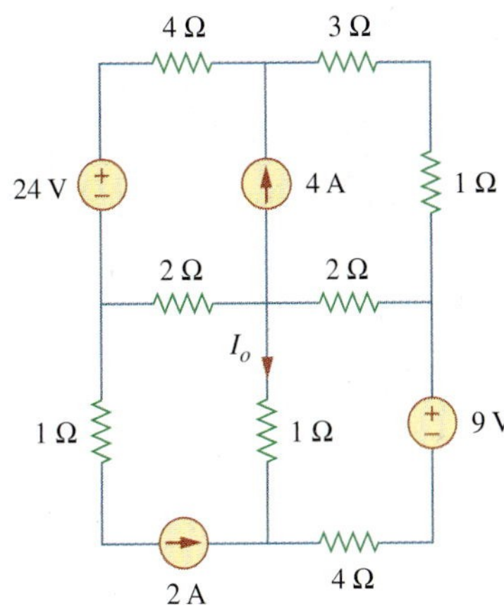

Figure 3.84
For Prob. 3.38.

3.39 Determine the mesh currents i_1 and i_2 in the circuit shown in Fig. 3.85.

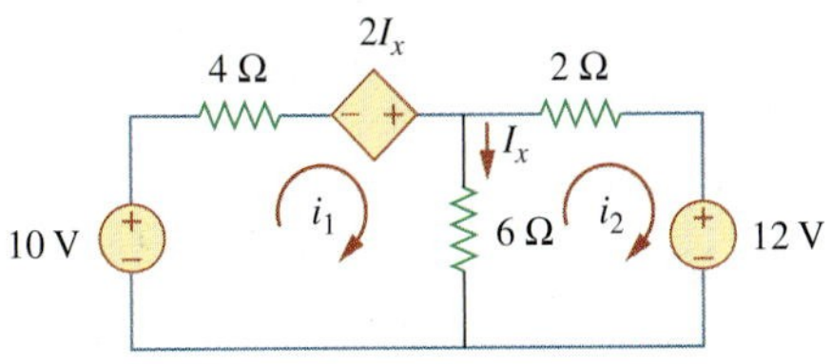

Figure 3.85
For Prob. 3.39.

3.40 For the bridge network in Fig. 3.86, find i_o using mesh analysis.

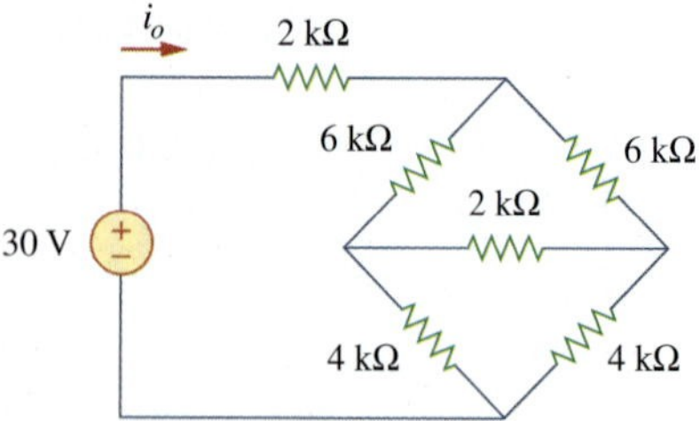

Figure 3.86
For Prob. 3.40.

3.41 Apply mesh analysis to find i in Fig. 3.87.

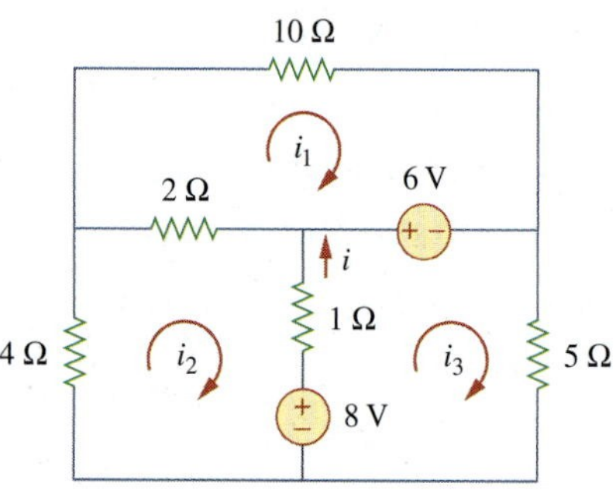

Figure 3.87
For Prob. 3.41.

3.42 Using Fig. 3.88, design a problem to help students better understand mesh analysis using matrices.

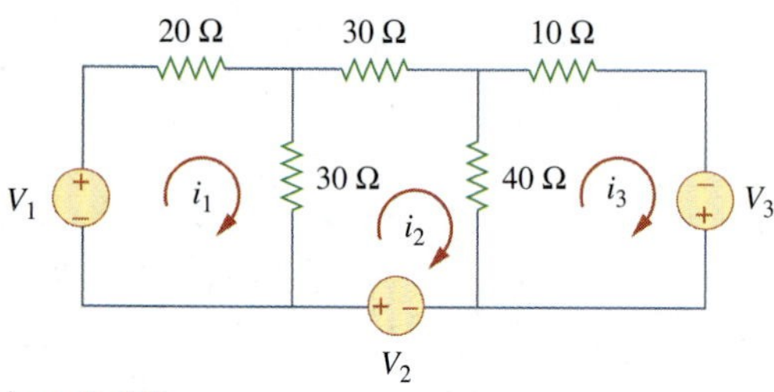

Figure 3.88
For Prob. 3.42.

3.43 Use mesh analysis to find v_{ab} and i_o in the circuit of Fig. 3.89.

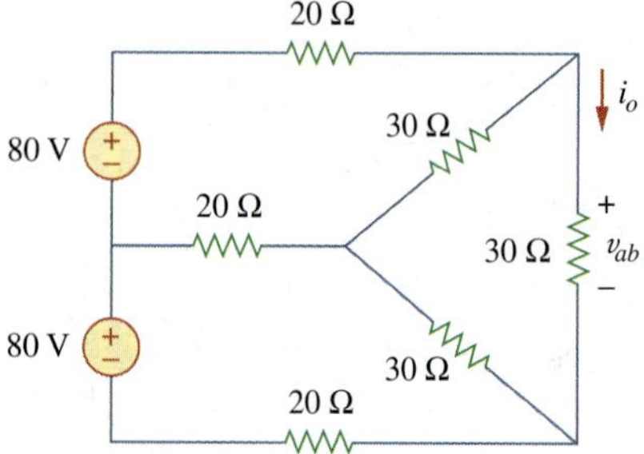

Figure 3.89
For Prob. 3.43.

3.44 Use mesh analysis to obtain i_o in the circuit of Fig. 3.90.

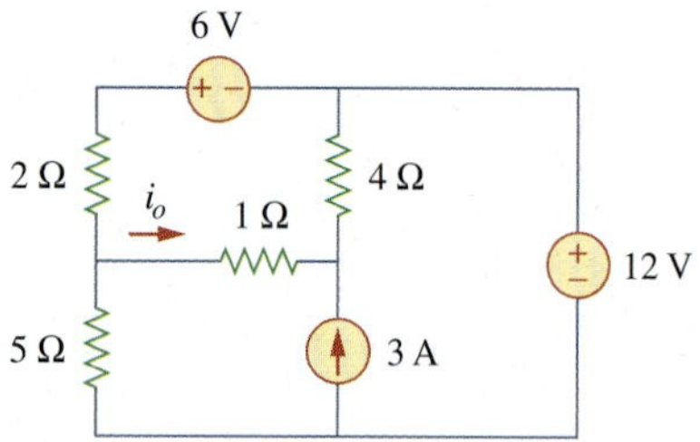

Figure 3.90
For Prob. 3.44.

3.45 Find current i in the circuit of Fig. 3.91.

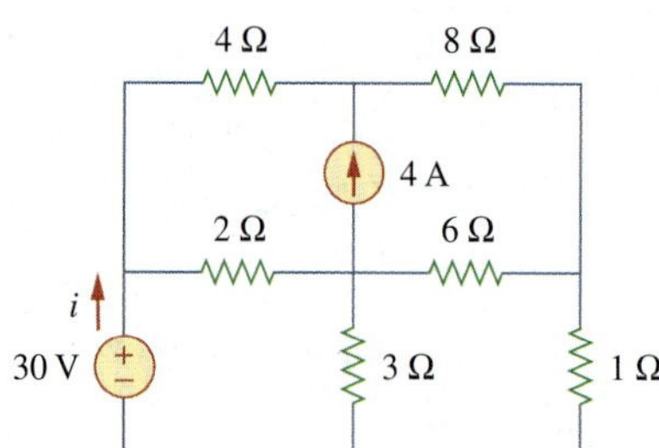

Figure 3.91
For Prob. 3.45.

3.46 Solve for the mesh currents in Fig. 3.92.

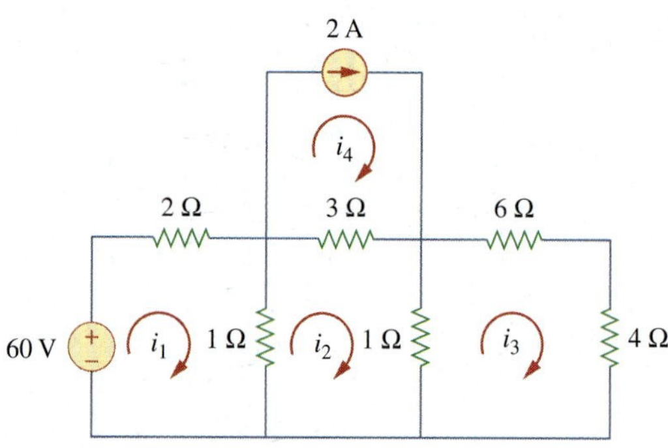

Figure 3.92
For Prob. 3.46.

3.47 Rework Prob. 3.19 using mesh analysis.

3.48 Determine the current through the 10-kΩ resistor in the circuit of Fig. 3.93 using mesh analysis.

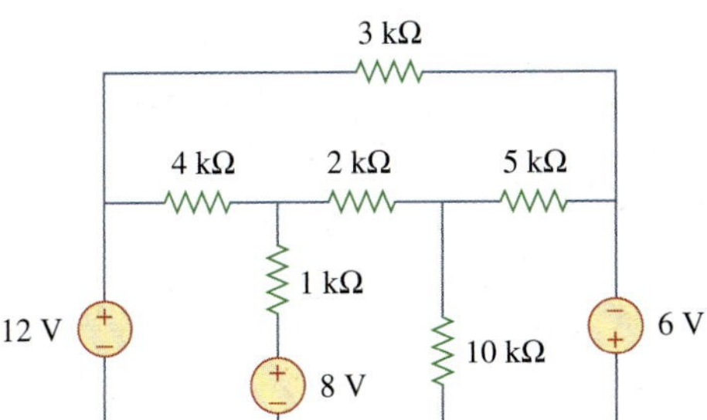

Figure 3.93
For Prob. 3.48.

3.49 Find v_o and i_o in the circuit of Fig. 3.94.

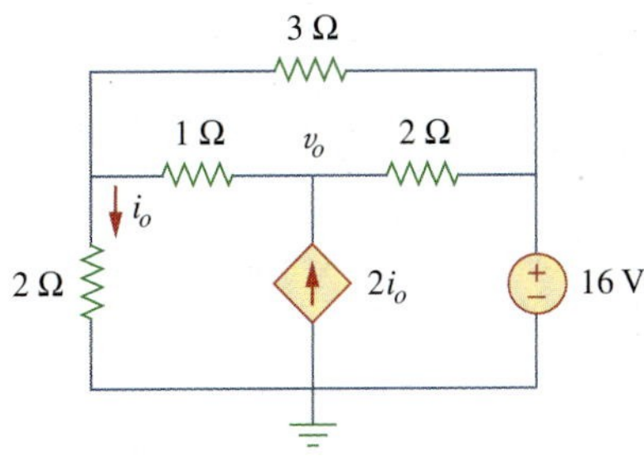

Figure 3.94
For Prob. 3.49.

3.50 Use mesh analysis to find the current i_o in the circuit of Fig. 3.95.

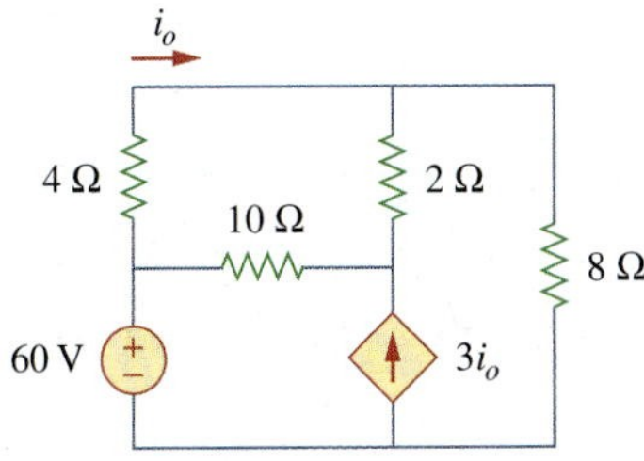

Figure 3.95
For Prob. 3.50.

3.51 Apply mesh analysis to find v_o in the circuit of Fig. 3.96.

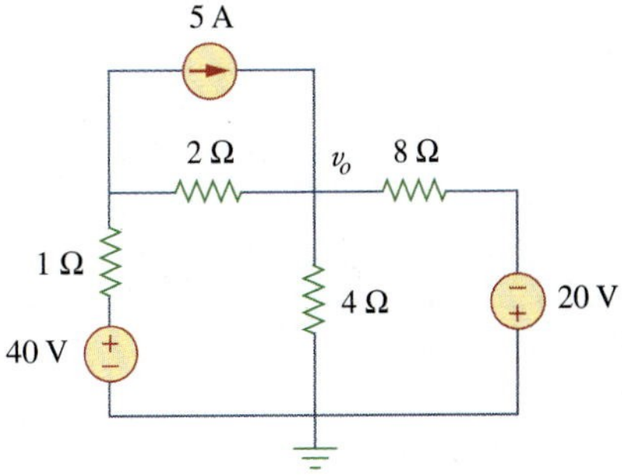

Figure 3.96
For Prob. 3.51.

3.52 Use mesh analysis to find i_1, i_2, and i_3 in the circuit of Fig. 3.97.

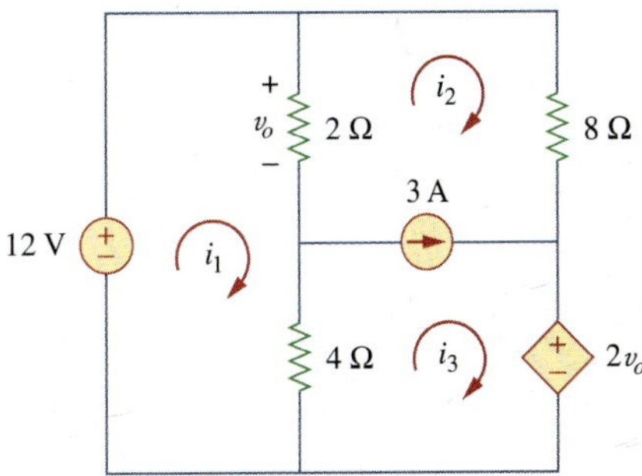

Figure 3.97
For Prob. 3.52.

3.53 Find the mesh currents in the circuit of Fig. 3.98 using *MATLAB*.

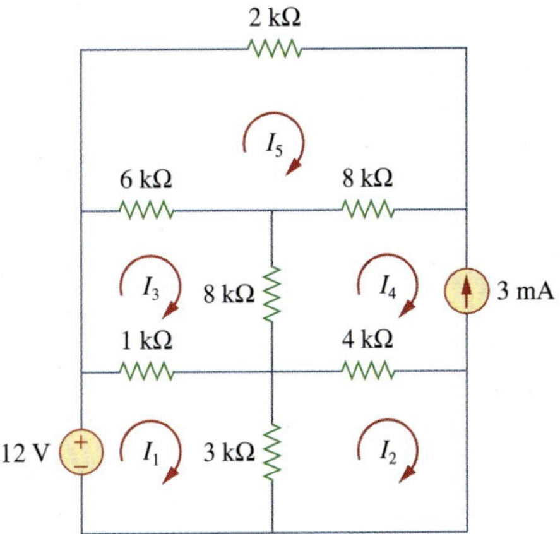

Figure 3.98
For Prob. 3.53.

***3.54** Given the circuit Fig. 3.99, use mesh analysis to find the mesh currents.

ML

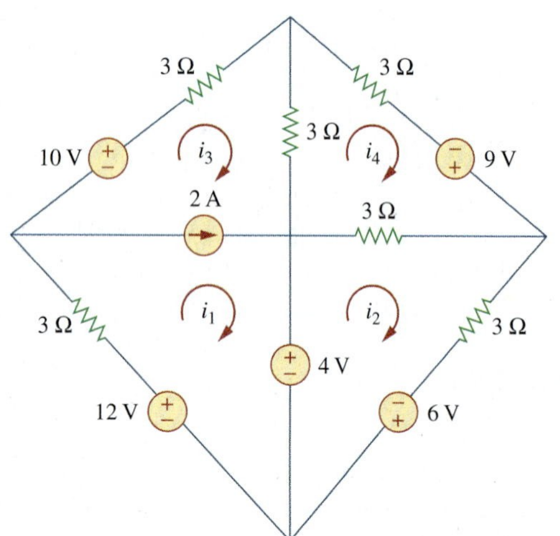

Figure 3.99
For Prob. 3.54.

***3.55** In the circuit of Fig. 3.100, solve for I_1, I_2, and I_3.

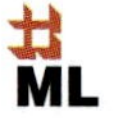

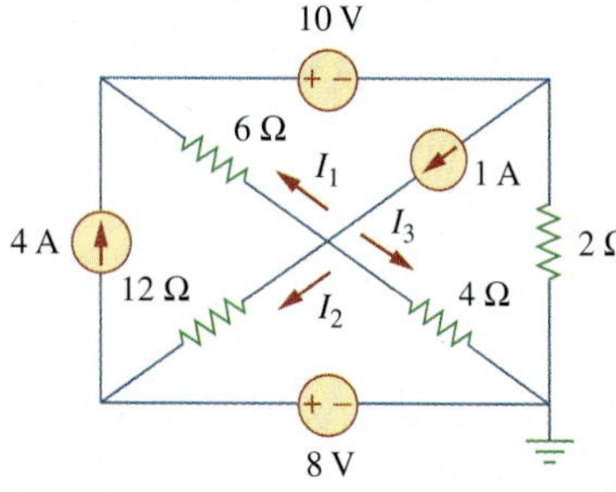

Figure 3.100
For Prob. 3.55.

3.56 Determine v_1 and v_2 in the circuit of Fig. 3.101.

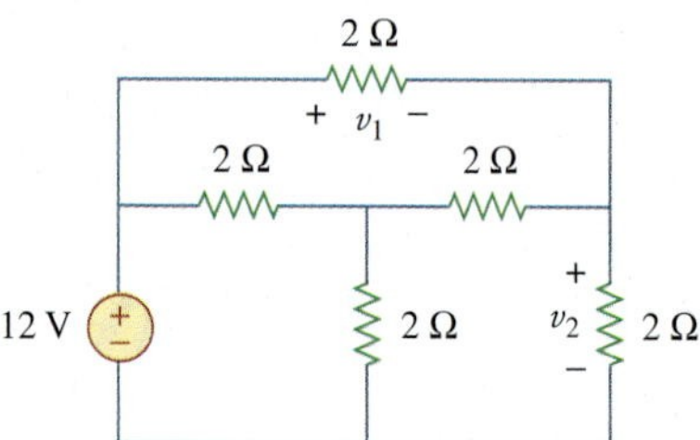

Figure 3.101
For Prob. 3.56.

3.57 In the circuit of Fig. 3.102, find the values of R, V_1, and V_2 given that $i_o = 18$ mA.

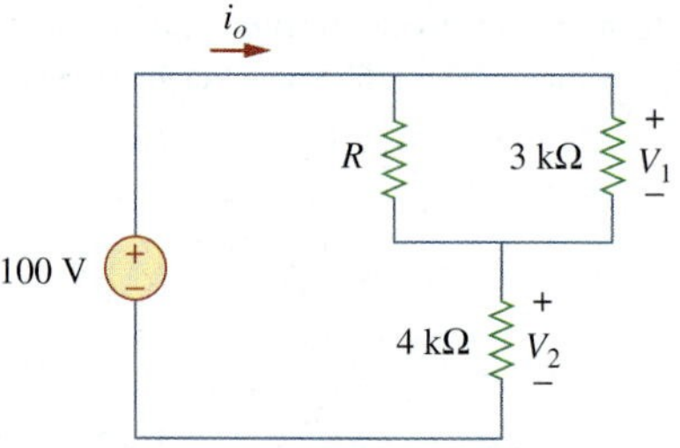

Figure 3.102
For Prob. 3.57.

3.58 Find i_1, i_2, and i_3 in the circuit of Fig. 3.103.

ML

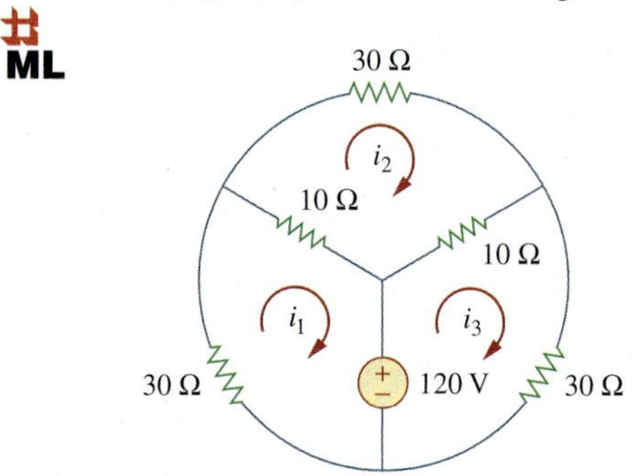

Figure 3.103
For Prob. 3.58.

3.59 Rework Prob. 3.30 using mesh analysis.

3.60 Calculate the power dissipated in each resistor in the circuit of Fig. 3.104.

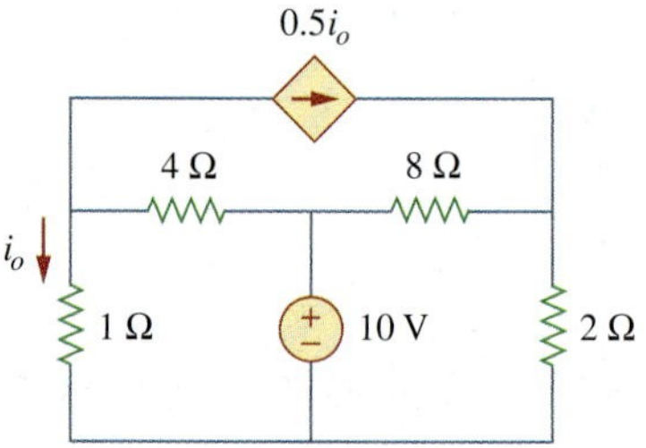

Figure 3.104
For Prob. 3.60.

3.61 Calculate the current gain i_o/i_s in the circuit of Fig. 3.105.

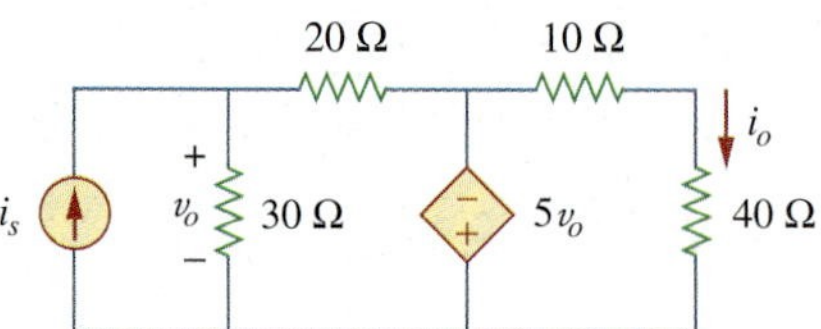

Figure 3.105
For Prob. 3.61.

3.62 Find the mesh currents i_1, i_2, and i_3 in the network of Fig. 3.106.

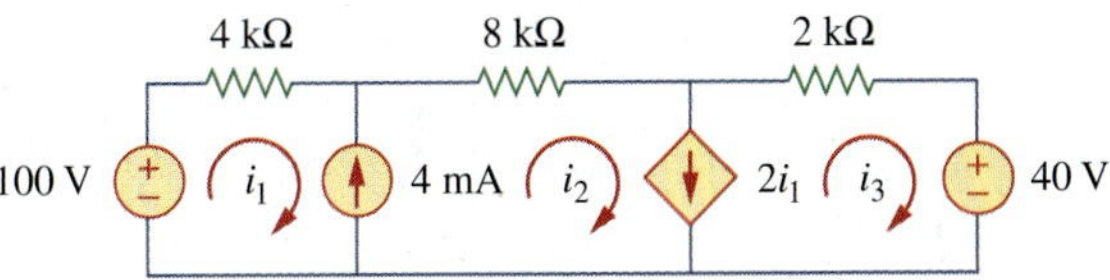

Figure 3.106
For Prob. 3.62.

3.63 Find v_x and i_x in the circuit shown in Fig. 3.107.

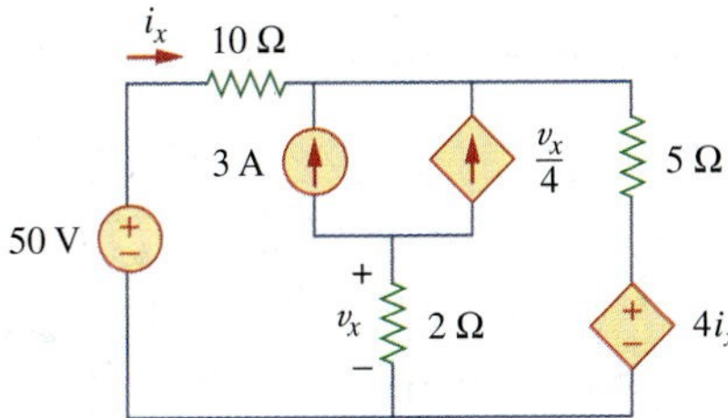

Figure 3.107
For Prob. 3.63.

3.64 Find v_o and i_o in the circuit of Fig. 3.108.

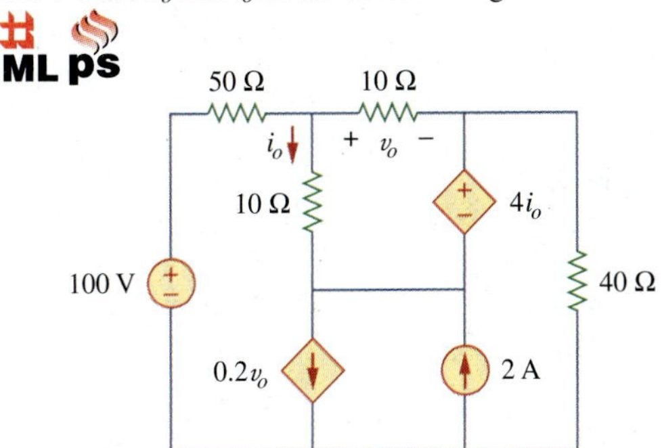

Figure 3.108
For Prob. 3.64.

3.65 Use *MATLAB* to solve for the mesh currents in the circuit of Fig. 3.109.

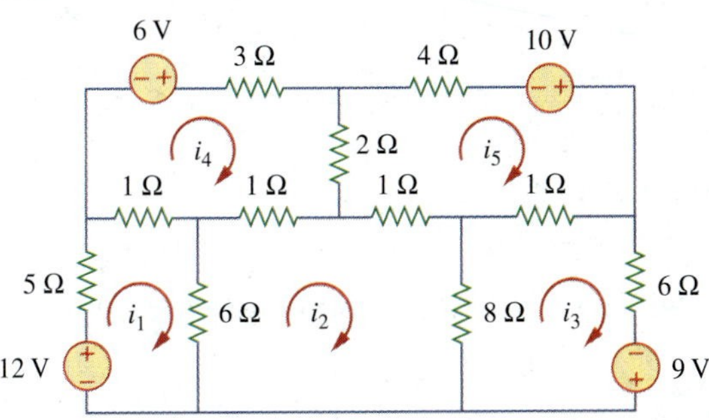

Figure 3.109
For Prob. 3.65.

3.66 Write a set of mesh equations for the circuit in Fig. 3.110. Use *MATLAB* to determine the mesh currents.

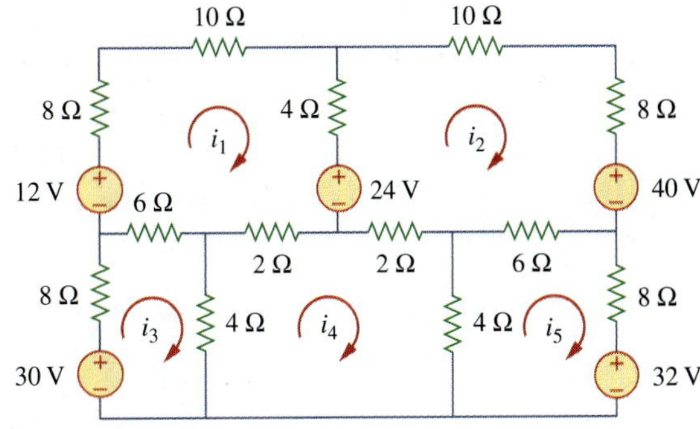

Figure 3.110
For Prob. 3.66.

Section 3.6 Nodal and Mesh Analyses by Inspection

3.67 Obtain the node-voltage equations for the circuit in Fig. 3.111 by inspection. Then solve for V_o.

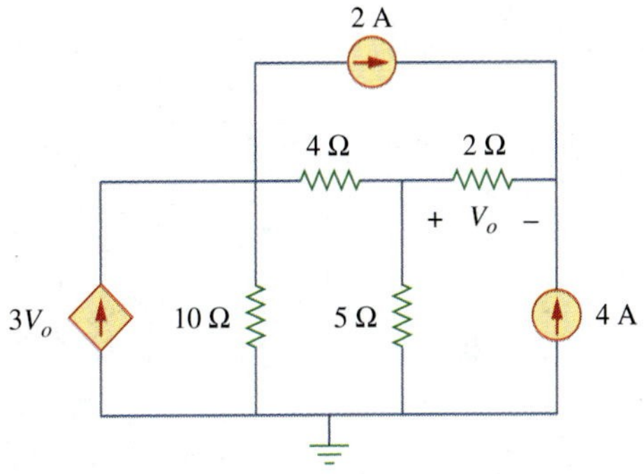

Figure 3.111
For Prob. 3.67.

3.68 Using Fig. 3.112, design a problem, to solve for V_o, to help other students better understand nodal analysis. Try your best to come up with values to make the calculations easier.

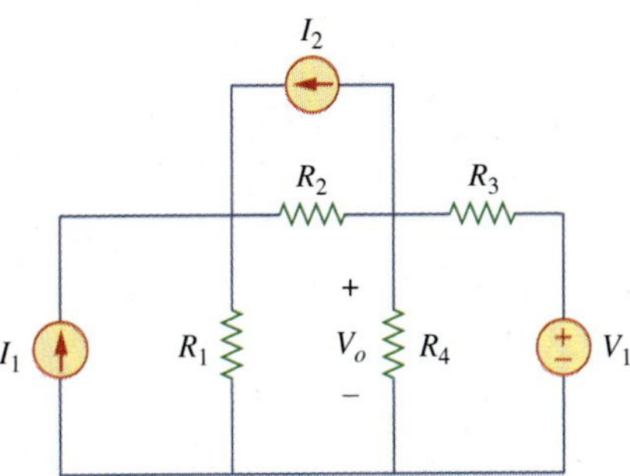

Figure 3.112
For Prob. 3.68.

3.69 For the circuit shown in Fig. 3.113, write the node-voltage equations by inspection.

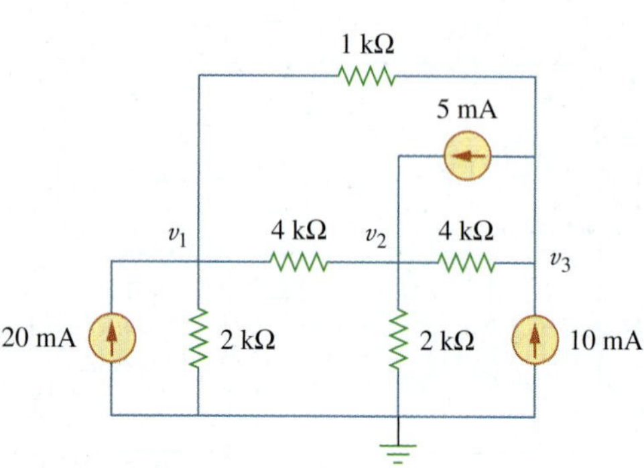

Figure 3.113
For Prob. 3.69.

3.70 Write the node-voltage equations by inspection and then determine values of V_1 and V_2 in the circuit of Fig. 3.114.

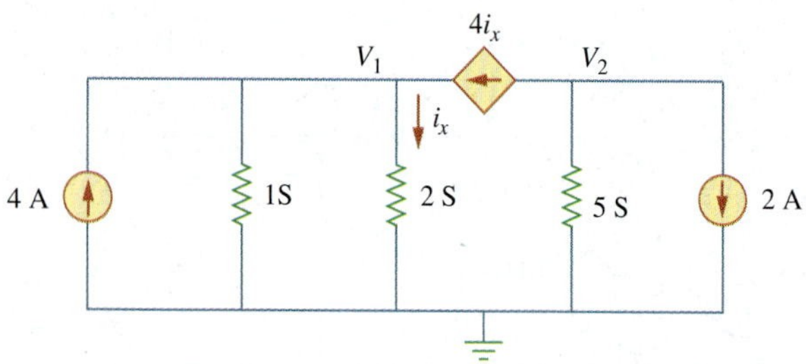

Figure 3.114
For Prob. 3.70.

3.71 Write the mesh-current equations for the circuit in Fig. 3.115. Next, determine the values of i_1, i_2, and i_3.

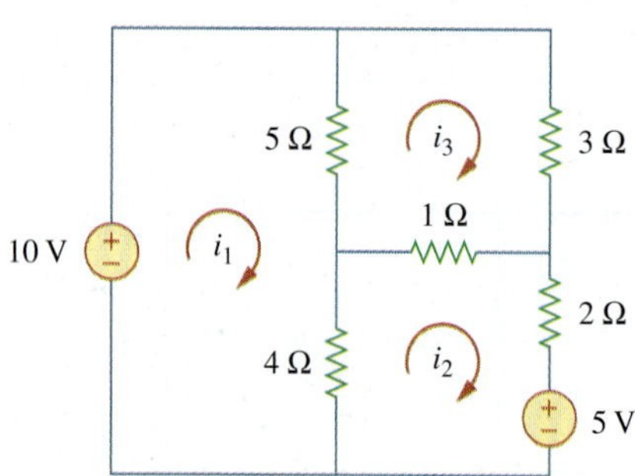

Figure 3.115
For Prob. 3.71.

3.72 By inspection, write the mesh-current equations for the circuit in Fig. 3.116.

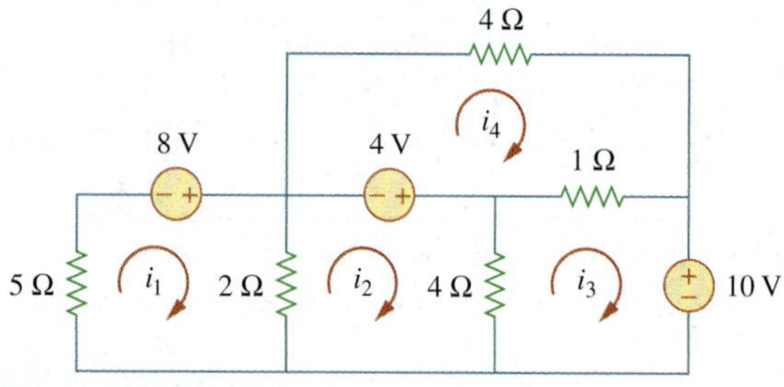

Figure 3.116
For Prob. 3.72.

3.73 Write the mesh-current equations for the circuit in Fig. 3.117.

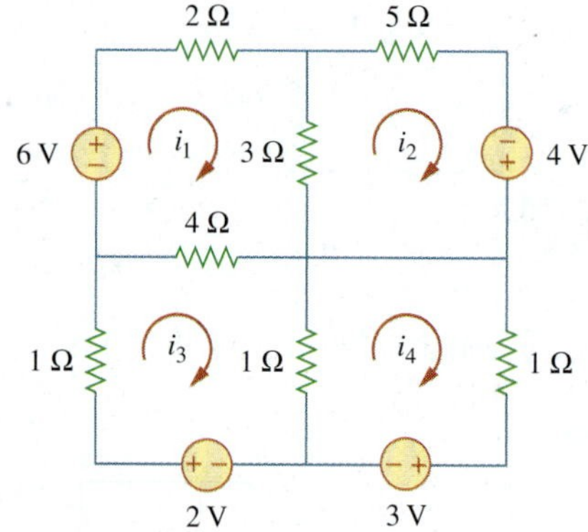

Figure 3.117
For Prob. 3.73.

3.74 By inspection, obtain the mesh-current equations for the circuit in Fig. 3.118.

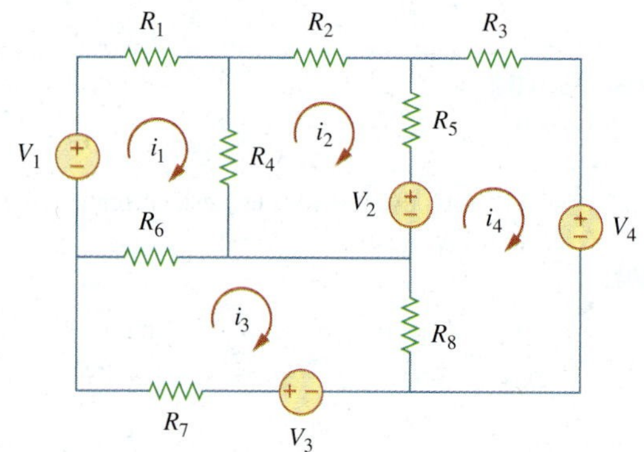

Figure 3.118
For Prob. 3.74.

Section 3.8 Circuit Analysis with *PSpice*

3.75 Use *PSpice* to solve Prob. 3.58.

3.76 Use *PSpice* to solve Prob. 3.27.

3.77 Solve for V_1 and V_2 in the circuit of Fig. 3.119 using *PSpice*.

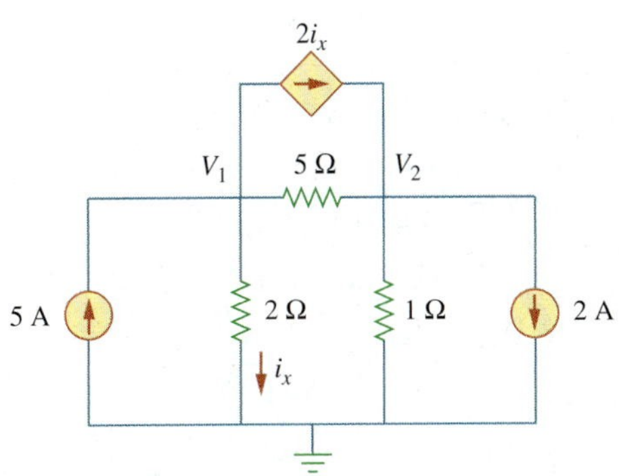

Figure 3.119
For Prob. 3.77.

3.78 Solve Prob. 3.20 using *PSpice*.

3.79 Rework Prob. 3.28 using *PSpice*.

3.80 Find the nodal voltages v_1 through v_4 in the circuit of Fig. 3.120 using *PSpice*.

ML

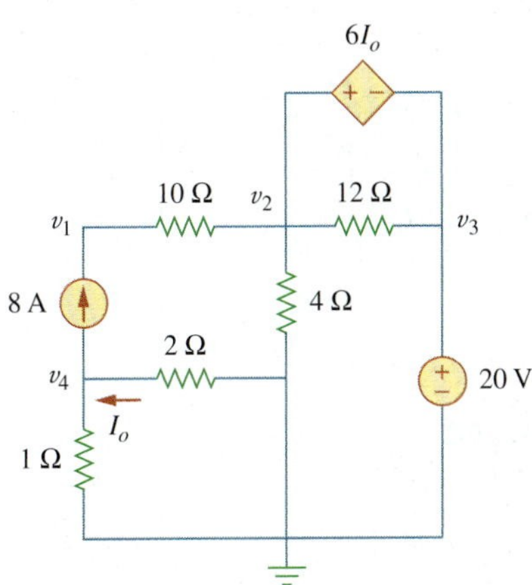

Figure 3.120
For Prob. 3.80.

3.81 Use *PSpice* to solve the problem in Example 3.4.

3.82 If the Schematics Netlist for a network is as follows, draw the network.

```
R_R1   1  2  2K
R_R2   2  0  4K
R_R3   3  0  8K
R_R4   3  4  6K
R_R5   1  3  3K
V_VS   4  0  DC     100
I_IS   0  1  DC     4
F_F1   1  3  VF_F1  2
VF_F1  5  0  0V
E_E1   3  2  1      3    3
```

3.83 The following program is the Schematics Netlist of a particular circuit. Draw the circuit and determine the voltage at node 2.

```
R_R1  1  2  20
R_R2  2  0  50
R_R3  2  3  70
R_R4  3  0  30
V_VS  1  0  20V
I_IS  2  0  DC   2A
```

Section 3.9 Applications

3.84 Calculate v_o and I_o in the circuit of Fig. 3.121.

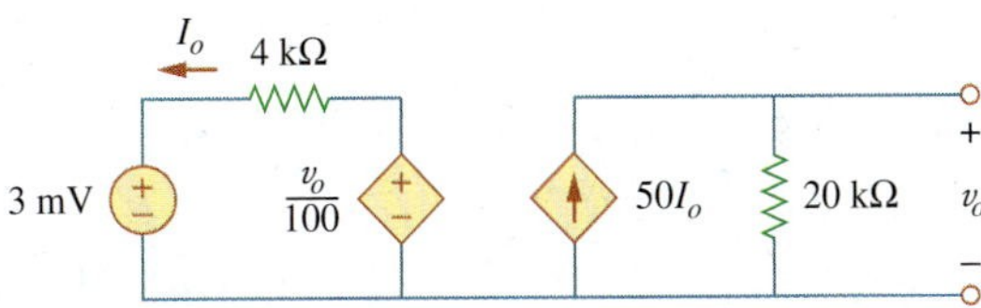

Figure 3.121
For Prob. 3.84.

3.85 An audio amplifier with a resistance of 9 Ω supplies power to a speaker. What should be the resistance of the speaker for maximum power to be delivered?

3.86 For the simplified transistor circuit of Fig. 3.122, calculate the voltage v_o.

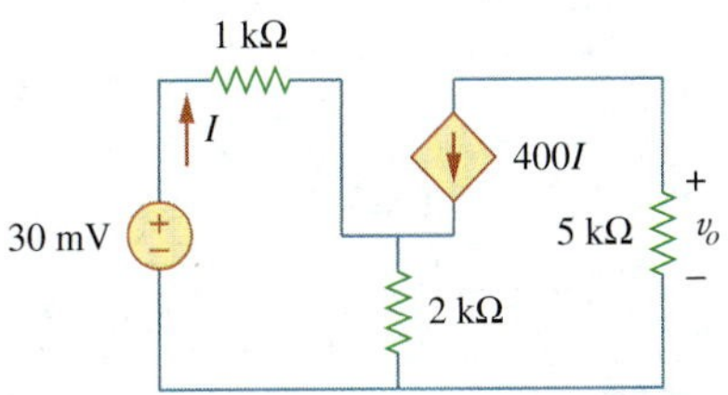

Figure 3.122
For Prob. 3.86.

3.87 For the circuit in Fig. 3.123, find the gain v_o/v_s.

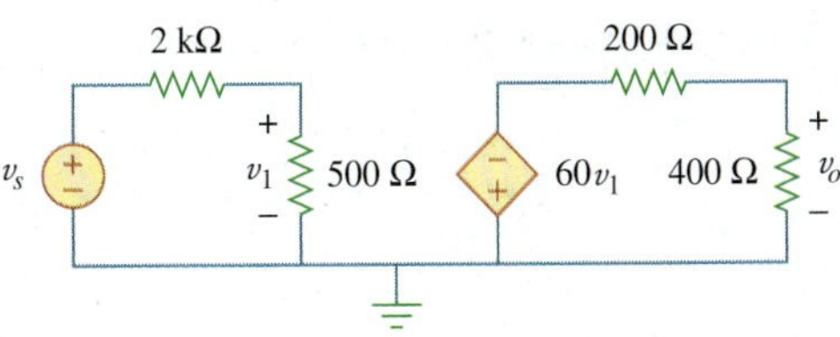

Figure 3.123
For Prob. 3.87.

***3.88** Determine the gain v_o/v_s of the transistor amplifier circuit in Fig. 3.124.

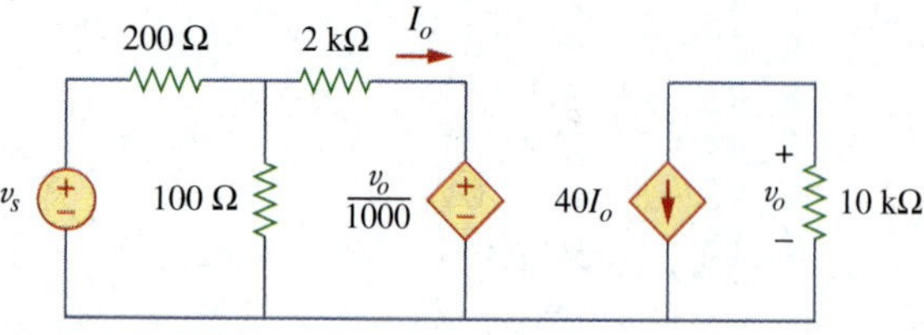

Figure 3.124
For Prob. 3.88.

3.89 For the transistor circuit shown in Fig. 3.125, find I_B and V_{CE}. Let $\beta = 100$, and $V_{BE} = 0.7$ V.

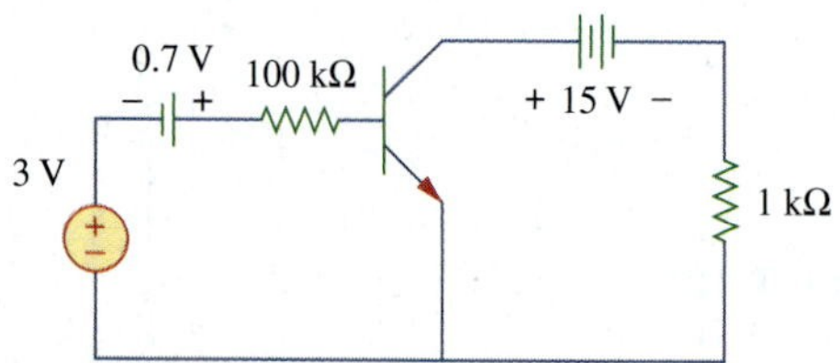

Figure 3.125
For Prob. 3.89.

3.90 Calculate v_s for the transistor in Fig. 3.126 given that $v_o = 4$ V, $\beta = 150$, $V_{BE} = 0.7$ V.

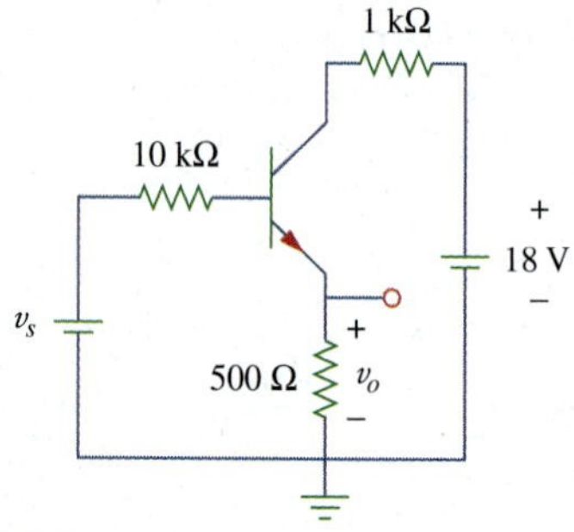

Figure 3.126
For Prob. 3.90.

3.91 For the transistor circuit of Fig. 3.127, find I_B, V_{CE}, and v_o. Take $\beta = 200$, $V_{BE} = 0.7$ V.

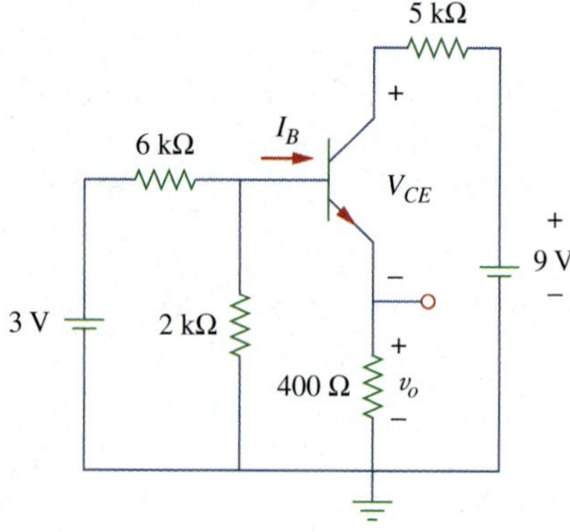

Figure 3.127
For Prob. 3.91.

3.92 Using Fig. 3.128, design a problem to help other students better understand transistors. Make sure you use reasonable numbers!

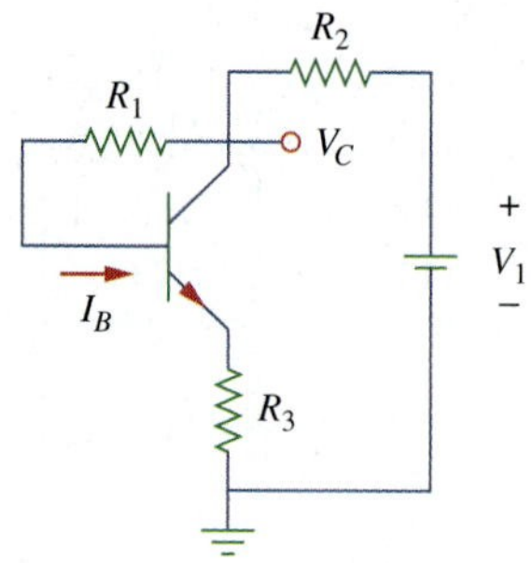

Figure 3.128
For Prob. 3.92.

Comprehensive Problem

***3.93** Rework Example 3.11 with hand calculation.

chapter

4

Circuit Theorems

Your success as an engineer will be directly proportional to your ability to communicate!

—Charles K. Alexander

Enhancing Your Skills and Your Career

Enhancing Your Communication Skills

Taking a course in circuit analysis is one step in preparing yourself for a career in electrical engineering. Enhancing your communication skills while in school should also be part of that preparation, as a large part of your time will be spent communicating.

People in industry have complained again and again that graduating engineers are ill-prepared in written and oral communication. An engineer who communicates effectively becomes a valuable asset.

You can probably speak or write easily and quickly. But how *effectively* do you communicate? The art of effective communication is of the utmost importance to your success as an engineer.

For engineers in industry, communication is key to promotability. Consider the result of a survey of U.S. corporations that asked what factors influence managerial promotion. The survey includes a listing of 22 personal qualities and their importance in advancement. You may be surprised to note that "technical skill based on experience" placed fourth from the bottom. Attributes such as self-confidence, ambition, flexibility, maturity, ability to make sound decisions, getting things done with and through people, and capacity for hard work all ranked higher. At the top of the list was "ability to communicate." The higher your professional career progresses, the more you will need to communicate. Therefore, you should regard effective communication as an important tool in your engineering tool chest.

Learning to communicate effectively is a lifelong task you should always work toward. The best time to begin is while still in school. Continually look for opportunities to develop and strengthen your reading, writing, listening, and speaking skills. You can do this through classroom presentations, team projects, active participation in student organizations, and enrollment in communication courses. The risks are less now than later in the workplace.

Ability to communicate effectively is regarded by many as the most important step to an executive promotion.
© IT Stock/Punchstock

4.1 Introduction

A major advantage of analyzing circuits using Kirchhoff's laws as we did in Chapter 3 is that we can analyze a circuit without tampering with its original configuration. A major disadvantage of this approach is that, for a large, complex circuit, tedious computation is involved.

The growth in areas of application of electric circuits has led to an evolution from simple to complex circuits. To handle the complexity, engineers over the years have developed some theorems to simplify circuit analysis. Such theorems include Thevenin's and Norton's theorems. Since these theorems are applicable to *linear* circuits, we first discuss the concept of circuit linearity. In addition to circuit theorems, we discuss the concepts of superposition, source transformation, and maximum power transfer in this chapter. The concepts we develop are applied in the last section to source modeling and resistance measurement.

4.2 Linearity Property

Linearity is the property of an element describing a linear relationship between cause and effect. Although the property applies to many circuit elements, we shall limit its applicability to resistors in this chapter. The property is a combination of both the homogeneity (scaling) property and the additivity property.

The homogeneity property requires that if the input (also called the *excitation*) is multiplied by a constant, then the output (also called the *response*) is multiplied by the same constant. For a resistor, for example, Ohm's law relates the input i to the output v,

$$v = iR \tag{4.1}$$

If the current is increased by a constant k, then the voltage increases correspondingly by k; that is,

$$\boxed{kiR = kv} \tag{4.2}$$

The additivity property requires that the response to a sum of inputs is the sum of the responses to each input applied separately. Using the voltage-current relationship of a resistor, if

$$v_1 = i_1R \tag{4.3a}$$

and

$$v_2 = i_2R \tag{4.3b}$$

then applying $(i_1 + i_2)$ gives

$$\boxed{v = (i_1 + i_2)R = i_1R + i_2R = v_1 + v_2} \tag{4.4}$$

We say that a resistor is a linear element because the voltage-current relationship satisfies both the homogeneity and the additivity properties.

In general, a circuit is linear if it is both additive and homogeneous. A linear circuit consists of only linear elements, linear dependent sources, and independent sources.

A **linear circuit** is one whose output is linearly related (or directly proportional) to its input.

Throughout this book we consider only linear circuits. Note that since $p = i^2R = v^2/R$ (making it a quadratic function rather than a linear one), the relationship between power and voltage (or current) is nonlinear. Therefore, the theorems covered in this chapter are not applicable to power.

To illustrate the linearity principle, consider the linear circuit shown in Fig. 4.1. The linear circuit has no independent sources inside it. It is excited by a voltage source v_s, which serves as the input. The circuit is terminated by a load R. We may take the current i through R as the output. Suppose $v_s = 10$ V gives $i = 2$ A. According to the linearity principle, $v_s = 1$ V will give $i = 0.2$ A. By the same token, $i = 1$ mA must be due to $v_s = 5$ mV.

For example, when current i_1 flows through resistor R, the power is $p_1 = Ri_1^2$, and when current i_2 flows through R, the power is $p_2 = Ri_2^2$. If current $i_1 + i_2$ flows through R, the power absorbed is $p_3 = R(i_1 + i_2)^2 = Ri_1^2 + Ri_2^2 + 2Ri_1i_2 \neq p_1 + p_2$. Thus, the power relation is nonlinear.

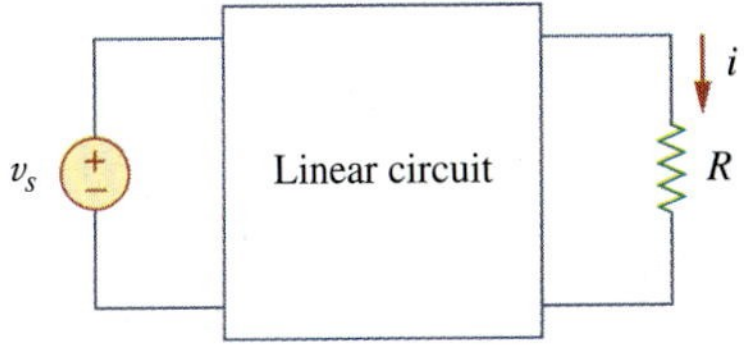

Figure 4.1
A linear circuit with input v_s and output i.

Example 4.1

For the circuit in Fig. 4.2, find I_o when $v_s = 12$ V and $v_s = 24$ V.

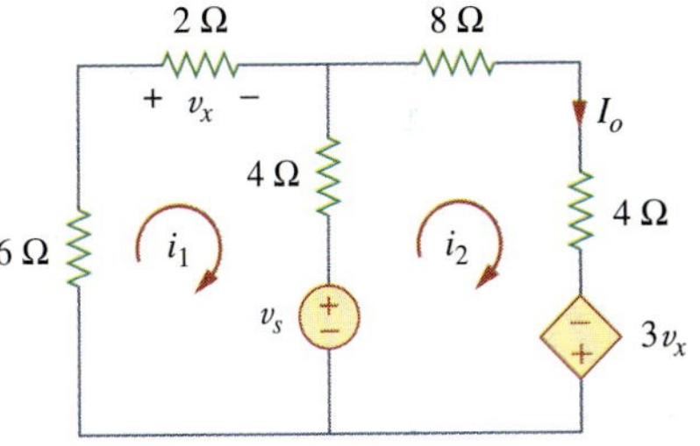

Figure 4.2
For Example 4.1.

Solution:
Applying KVL to the two loops, we obtain

$$12i_1 - 4i_2 + v_s = 0 \tag{4.1.1}$$

$$-4i_1 + 16i_2 - 3v_x - v_s = 0 \tag{4.1.2}$$

But $v_x = 2i_1$. Equation (4.1.2) becomes

$$-10i_1 + 16i_2 - v_s = 0 \tag{4.1.3}$$

Adding Eqs. (4.1.1) and (4.1.3) yields

$$2i_1 + 12i_2 = 0 \quad \Rightarrow \quad i_1 = -6i_2$$

Substituting this in Eq. (4.1.1), we get

$$-76i_2 + v_s = 0 \quad \Rightarrow \quad i_2 = \frac{v_s}{76}$$

When $v_s = 12$ V,

$$I_o = i_2 = \frac{12}{76} \text{ A}$$

When $v_s = 24$ V,

$$I_o = i_2 = \frac{24}{76} \text{ A}$$

showing that when the source value is doubled, I_o doubles.

Practice Problem 4.1

For the circuit in Fig. 4.3, find v_o when $i_s = 15$ and $i_s = 30$ A.

Answer: 20 V, 40 V.

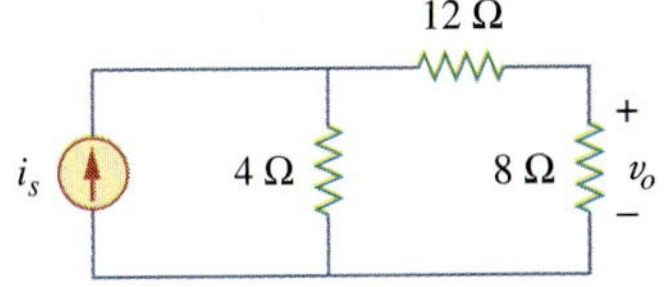

Figure 4.3
For Practice Prob. 4.1.

Example 4.2

Assume $I_o = 1$ A and use linearity to find the actual value of I_o in the circuit of Fig. 4.4.

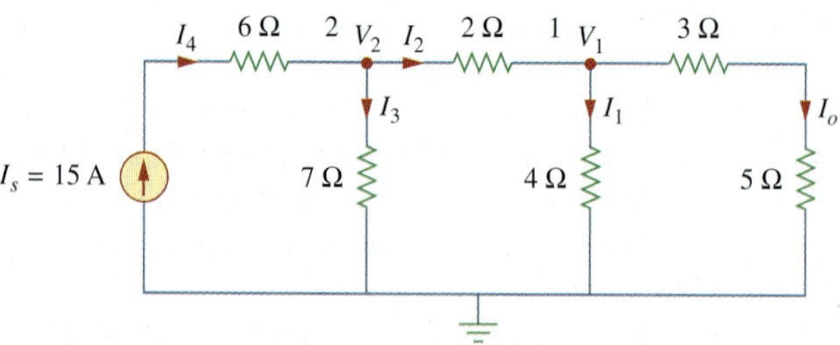

Figure 4.4
For Example 4.2.

Solution:
If $I_o = 1$ A, then $V_1 = (3 + 5)I_o = 8$ V and $I_1 = V_1/4 = 2$ A. Applying KCL at node 1 gives

$$I_2 = I_1 + I_o = 3 \text{ A}$$

$$V_2 = V_1 + 2I_2 = 8 + 6 = 14 \text{ V}, \qquad I_3 = \frac{V_2}{7} = 2 \text{ A}$$

Applying KCL at node 2 gives

$$I_4 = I_3 + I_2 = 5 \text{ A}$$

Therefore, $I_s = 5$ A. This shows that assuming $I_o = 1$ gives $I_s = 5$ A, the actual source current of 15 A will give $I_o = 3$ A as the actual value.

Practice Problem 4.2

Assume that $V_o = 1$ V and use linearity to calculate the actual value of V_o in the circuit of Fig. 4.5.

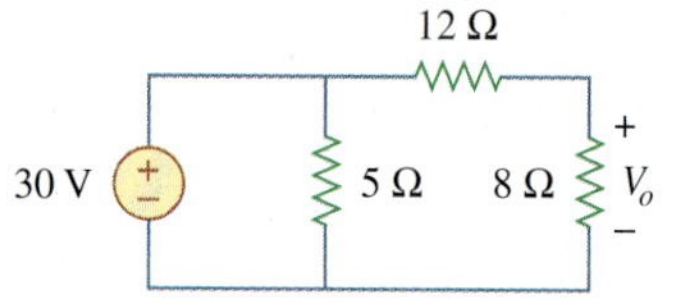

Figure 4.5
For Practice Prob. 4.2.

Answer: 12 V.

4.3 Superposition

If a circuit has two or more independent sources, one way to determine the value of a specific variable (voltage or current) is to use nodal or mesh analysis as in Chapter 3. Another way is to determine the contribution of each independent source to the variable and then add them up. The latter approach is known as the *superposition.*

The idea of superposition rests on the linearity property.

Superposition is not limited to circuit analysis but is applicable in many fields where cause and effect bear a linear relationship to one another.

> The **superposition** principle states that the voltage across (or current through) an element in a linear circuit is the algebraic sum of the voltages across (or currents through) that element due to each independent source acting alone.

The principle of superposition helps us to analyze a linear circuit with more than one independent source by calculating the contribution of each independent source separately. However, to apply the superposition principle, we must keep two things in mind:

1. We consider one independent source at a time while all other independent sources are *turned off*. This implies that we replace every voltage source by 0 V (or a short circuit), and every current source by 0 A (or an open circuit). This way we obtain a simpler and more manageable circuit.
2. Dependent sources are left intact because they are controlled by circuit variables.

Other terms such as *killed, made inactive, deadened,* or *set equal to zero* are often used to convey the same idea.

With these in mind, we apply the superposition principle in three steps:

Steps to Apply Superposition Principle:

1. Turn off all independent sources except one source. Find the output (voltage or current) due to that active source using the techniques covered in Chapters 2 and 3.
2. Repeat step 1 for each of the other independent sources.
3. Find the total contribution by adding algebraically all the contributions due to the independent sources.

Analyzing a circuit using superposition has one major disadvantage: it may very likely involve more work. If the circuit has three independent sources, we may have to analyze three simpler circuits each providing the contribution due to the respective individual source. However, superposition does help reduce a complex circuit to simpler circuits through replacement of voltage sources by short circuits and of current sources by open circuits.

Keep in mind that superposition is based on linearity. For this reason, it is not applicable to the effect on power due to each source, because the power absorbed by a resistor depends on the square of the voltage or current. If the power value is needed, the current through (or voltage across) the element must be calculated first using superposition.

Example 4.3

Use the superposition theorem to find v in the circuit of Fig. 4.6.

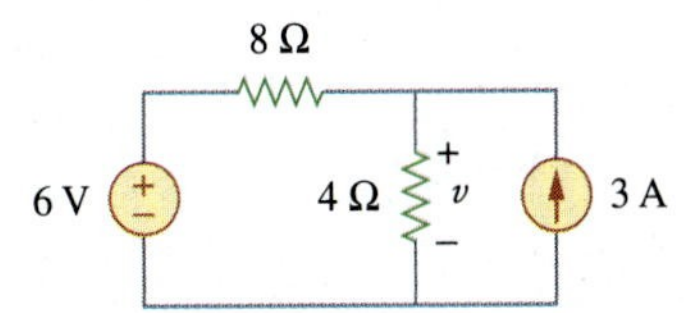

Figure 4.6
For Example 4.3.

Solution:
Since there are two sources, let

$$v = v_1 + v_2$$

where v_1 and v_2 are the contributions due to the 6-V voltage source and the 3-A current source, respectively. To obtain v_1, we set the current source to zero, as shown in Fig. 4.7(a). Applying KVL to the loop in Fig. 4.7(a) gives

$$12i_1 - 6 = 0 \quad \Rightarrow \quad i_1 = 0.5 \text{ A}$$

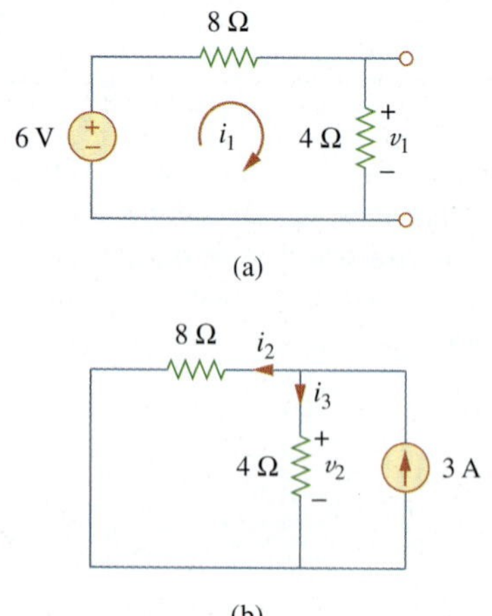

Figure 4.7
For Example 4.3: (a) calculating v_1, (b) calculating v_2.

Thus,

$$v_1 = 4i_1 = 2 \text{ V}$$

We may also use voltage division to get v_1 by writing

$$v_1 = \frac{4}{4 + 8}(6) = 2 \text{ V}$$

To get v_2, we set the voltage source to zero, as in Fig. 4.7(b). Using current division,

$$i_3 = \frac{8}{4 + 8}(3) = 2 \text{ A}$$

Hence,

$$v_2 = 4i_3 = 8 \text{ V}$$

And we find

$$v = v_1 + v_2 = 2 + 8 = 10 \text{ V}$$

Practice Problem 4.3

Using the superposition theorem, find v_o in the circuit of Fig. 4.8.

Answer: 6 V.

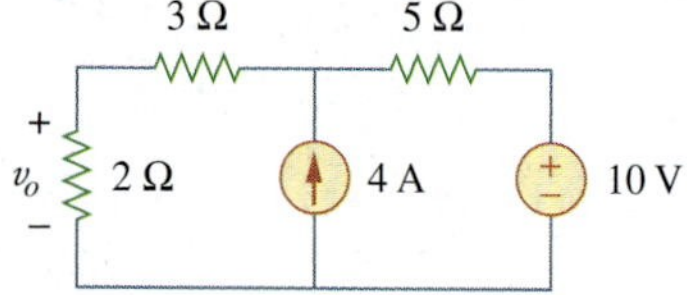

Figure 4.8
For Practice Prob. 4.3.

Example 4.4

Find i_o in the circuit of Fig. 4.9 using superposition.

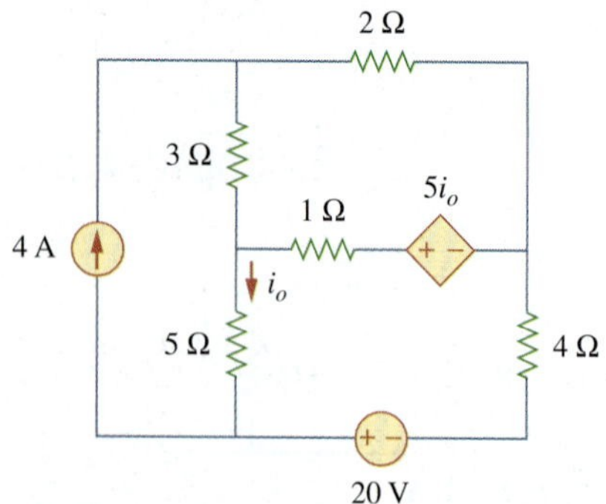

Figure 4.9
For Example 4.4.

Solution:
The circuit in Fig. 4.9 involves a dependent source, which must be left intact. We let

$$i_o = i'_o + i''_o \tag{4.4.1}$$

where i'_o and i''_o are due to the 4-A current source and 20-V voltage source respectively. To obtain i'_o, we turn off the 20-V source so that we have the circuit in Fig. 4.10(a). We apply mesh analysis in order to obtain i'_o. For loop 1,

$$i_1 = 4 \text{ A} \tag{4.4.2}$$

For loop 2,

$$-3i_1 + 6i_2 - 1i_3 - 5i'_o = 0 \tag{4.4.3}$$

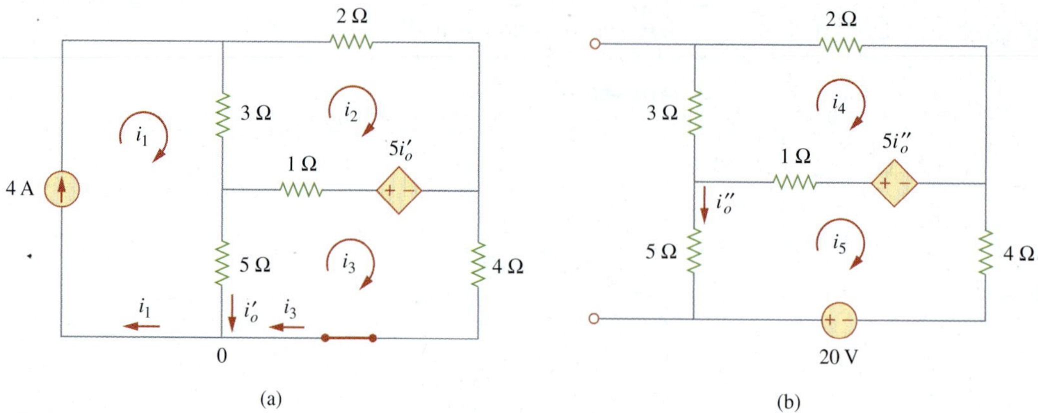

Figure 4.10
For Example 4.4: Applying superposition to (a) obtain i_o', (b) obtain i_o''.

For loop 3,

$$-5i_1 - 1i_2 + 10i_3 + 5i_o' = 0 \quad \textbf{(4.4.4)}$$

But at node 0,

$$i_3 = i_1 - i_o' = 4 - i_o' \quad \textbf{(4.4.5)}$$

Substituting Eqs. (4.4.2) and (4.4.5) into Eqs. (4.4.3) and (4.4.4) gives two simultaneous equations

$$3i_2 - 2i_o' = 8 \quad \textbf{(4.4.6)}$$

$$i_2 + 5i_o' = 20 \quad \textbf{(4.4.7)}$$

which can be solved to get

$$i_o' = \frac{52}{17} \text{ A} \quad \textbf{(4.4.8)}$$

To obtain i_o'', we turn off the 4-A current source so that the circuit becomes that shown in Fig. 4.10(b). For loop 4, KVL gives

$$6i_4 - i_5 - 5i_o'' = 0 \quad \textbf{(4.4.9)}$$

and for loop 5,

$$-i_4 + 10i_5 - 20 + 5i_o'' = 0 \quad \textbf{(4.4.10)}$$

But $i_5 = -i_o''$. Substituting this in Eqs. (4.4.9) and (4.4.10) gives

$$6i_4 - 4i_o'' = 0 \quad \textbf{(4.4.11)}$$

$$i_4 + 5i_o'' = -20 \quad \textbf{(4.4.12)}$$

which we solve to get

$$i_o'' = -\frac{60}{17} \text{ A} \quad \textbf{(4.4.13)}$$

Now substituting Eqs. (4.4.8) and (4.4.13) into Eq. (4.4.1) gives

$$i_o = -\frac{8}{17} = -0.4706 \text{ A}$$

Practice Problem 4.4

Use superposition to find v_x in the circuit of Fig. 4.11.

Answer: $v_x = 25$ V.

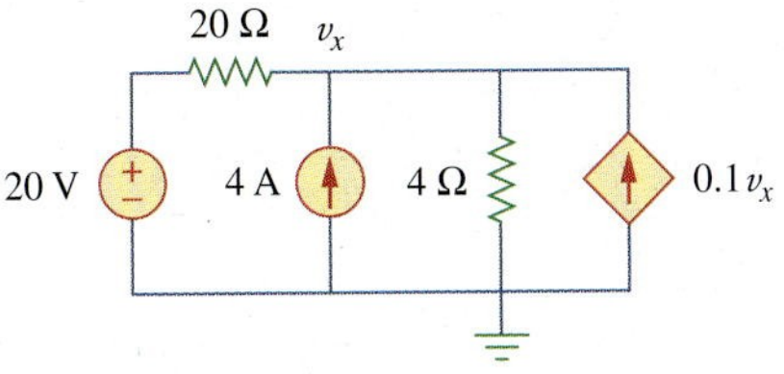

Figure 4.11
For Practice Prob. 4.4.

Example 4.5

For the circuit in Fig. 4.12, use the superposition theorem to find i.

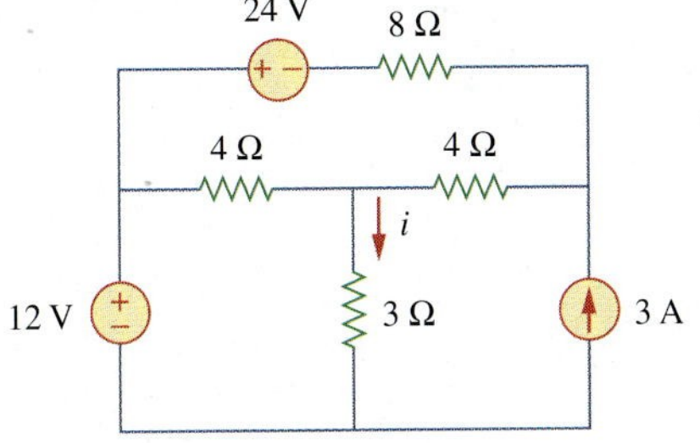

Figure 4.12
For Example 4.5.

Solution:
In this case, we have three sources. Let

$$i = i_1 + i_2 + i_3$$

where i_1, i_2, and i_3 are due to the 12-V, 24-V, and 3-A sources respectively. To get i_1, consider the circuit in Fig. 4.13(a). Combining 4 Ω (on the right-hand side) in series with 8 Ω gives 12 Ω. The 12 Ω in parallel with 4 Ω gives 12 × 4/16 = 3 Ω. Thus,

$$i_1 = \frac{12}{6} = 2 \text{ A}$$

To get i_2, consider the circuit in Fig. 4.13(b). Applying mesh analysis gives

$$16i_a - 4i_b + 24 = 0 \quad \Rightarrow \quad 4i_a - i_b = -6 \tag{4.5.1}$$

$$7i_b - 4i_a = 0 \quad \Rightarrow \quad i_a = \frac{7}{4}i_b \tag{4.5.2}$$

Substituting Eq. (4.5.2) into Eq. (4.5.1) gives

$$i_2 = i_b = -1$$

To get i_3, consider the circuit in Fig. 4.13(c). Using nodal analysis gives

$$3 = \frac{v_2}{8} + \frac{v_2 - v_1}{4} \quad \Rightarrow \quad 24 = 3v_2 - 2v_1 \tag{4.5.3}$$

$$\frac{v_2 - v_1}{4} = \frac{v_1}{4} + \frac{v_1}{3} \quad \Rightarrow \quad v_2 = \frac{10}{3}v_1 \tag{4.5.4}$$

Substituting Eq. (4.5.4) into Eq. (4.5.3) leads to $v_1 = 3$ and

$$i_3 = \frac{v_1}{3} = 1 \text{ A}$$

Thus,

$$i = i_1 + i_2 + i_3 = 2 - 1 + 1 = 2 \text{ A}$$

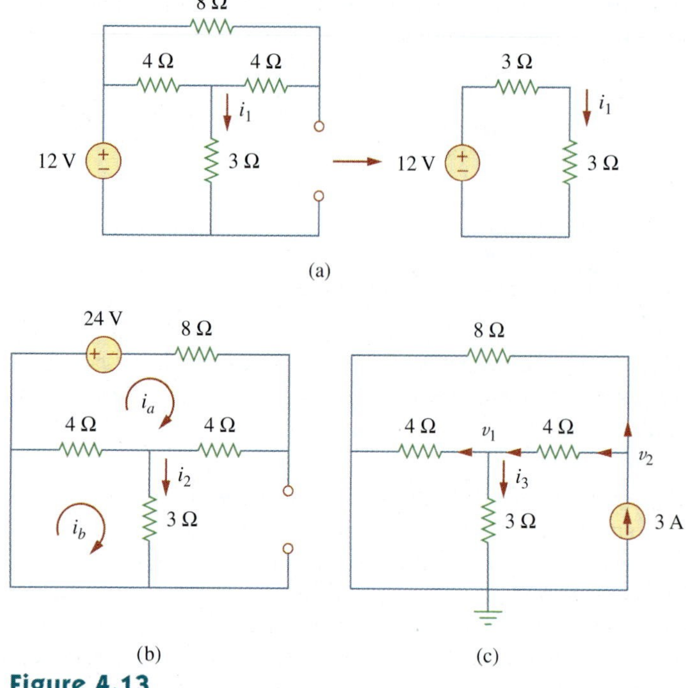

Figure 4.13
For Example 4.5.

Practice Problem 4.5

Find I in the circuit of Fig. 4.14 using the superposition principle.

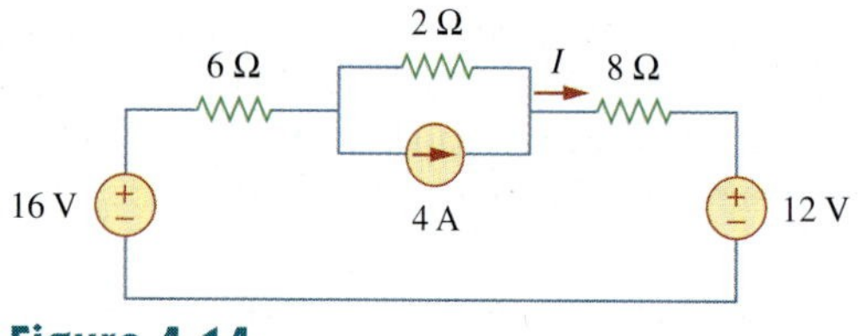

Figure 4.14
For Practice Prob. 4.5.

Answer: 0.75 A.

4.4 Source Transformation

We have noticed that series-parallel combination and wye-delta transformation help simplify circuits. *Source transformation* is another tool for simplifying circuits. Basic to these tools is the concept of *equivalence*. We recall that an equivalent circuit is one whose v-i characteristics are identical with the original circuit.

In Section 3.6, we saw that node-voltage (or mesh-current) equations can be obtained by mere inspection of a circuit when the sources are all independent current (or all independent voltage) sources. It is therefore expedient in circuit analysis to be able to substitute a voltage source in series with a resistor for a current source in parallel with a

resistor, or vice versa, as shown in Fig. 4.15. Either substitution is known as a *source transformation*.

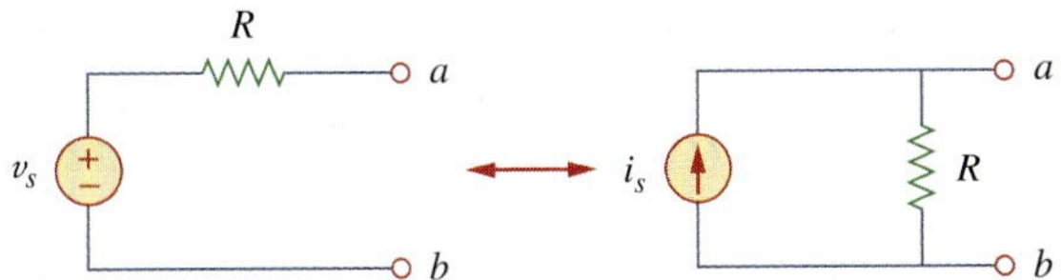

Figure 4.15
Transformation of independent sources.

A **source transformation** is the process of replacing a voltage source v_s in series with a resistor R by a current source i_s in parallel with a resistor R, or vice versa.

The two circuits in Fig. 4.15 are equivalent—provided they have the same voltage-current relation at terminals *a-b*. It is easy to show that they are indeed equivalent. If the sources are turned off, the equivalent resistance at terminals *a-b* in both circuits is R. Also, when terminals *a-b* are short-circuited, the short-circuit current flowing from a to b is $i_{sc} = v_s/R$ in the circuit on the left-hand side and $i_{sc} = i_s$ for the circuit on the right-hand side. Thus, $v_s/R = i_s$ in order for the two circuits to be equivalent. Hence, source transformation requires that

$$v_s = i_s R \qquad \text{or} \qquad i_s = \frac{v_s}{R} \tag{4.5}$$

Source transformation also applies to dependent sources, provided we carefully handle the dependent variable. As shown in Fig. 4.16, a dependent voltage source in series with a resistor can be transformed to a dependent current source in parallel with the resistor or vice versa where we make sure that Eq. (4.5) is satisfied.

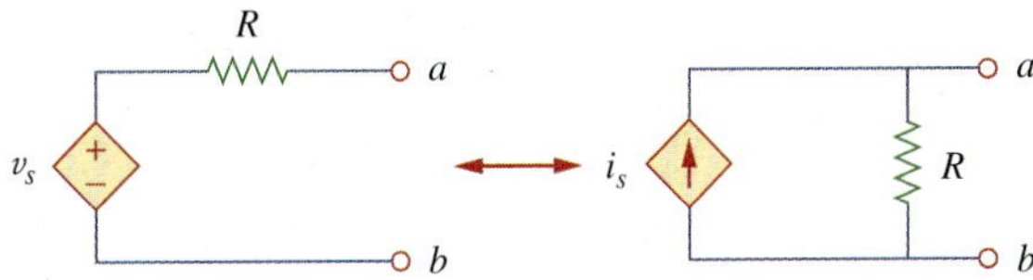

Figure 4.16
Transformation of dependent sources.

Like the wye-delta transformation we studied in Chapter 2, a source transformation does not affect the remaining part of the circuit. When applicable, source transformation is a powerful tool that allows circuit manipulations to ease circuit analysis. However, we should keep the following points in mind when dealing with source transformation.

1. Note from Fig. 4.15 (or Fig. 4.16) that the arrow of the current source is directed toward the positive terminal of the voltage source.
2. Note from Eq. (4.5) that source transformation is not possible when $R = 0$, which is the case with an ideal voltage source. However, for a practical, nonideal voltage source, $R \neq 0$. Similarly, an ideal current source with $R = \infty$ cannot be replaced by a finite voltage source. More will be said on ideal and nonideal sources in Section 4.10.1.

Example 4.6

Use source transformation to find v_o in the circuit of Fig. 4.17.

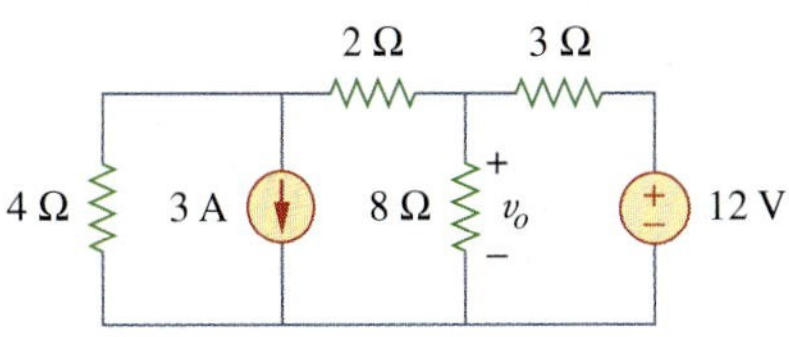

Figure 4.17
For Example 4.6.

Solution:
We first transform the current and voltage sources to obtain the circuit in Fig. 4.18(a). Combining the 4-Ω and 2-Ω resistors in series and transforming the 12-V voltage source gives us Fig. 4.18(b). We now combine the 3-Ω and 6-Ω resistors in parallel to get 2-Ω. We also combine the 2-A and 4-A current sources to get a 2-A source. Thus, by repeatedly applying source transformations, we obtain the circuit in Fig. 4.18(c).

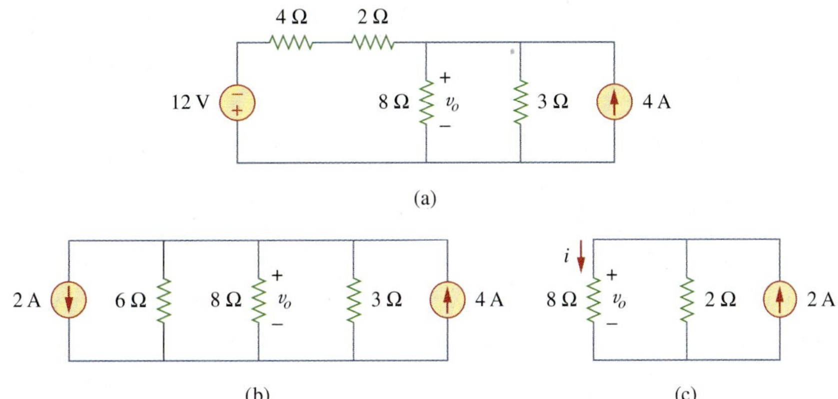

Figure 4.18
For Example 4.6.

We use current division in Fig. 4.18(c) to get

$$i = \frac{2}{2 + 8}(2) = 0.4 \text{ A}$$

and

$$v_o = 8i = 8(0.4) = 3.2 \text{ V}$$

Alternatively, since the 8-Ω and 2-Ω resistors in Fig. 4.18(c) are in parallel, they have the same voltage v_o across them. Hence,

$$v_o = (8 \parallel 2)(2 \text{ A}) = \frac{8 \times 2}{10}(2) = 3.2 \text{ V}$$

Practice Problem 4.6

Find i_o in the circuit of Fig. 4.19 using source transformation.

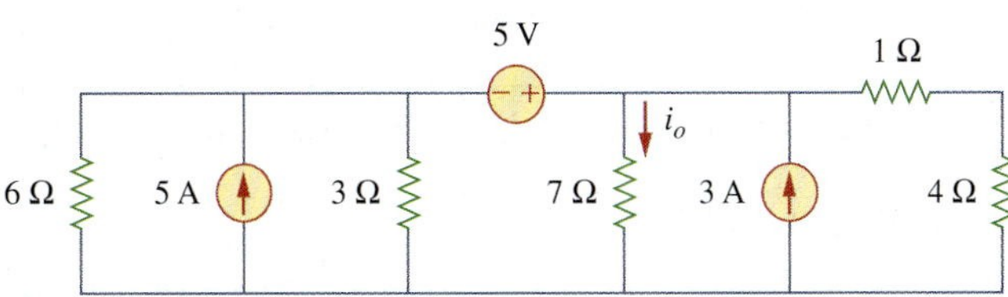

Figure 4.19
For Practice Prob. 4.6.

Answer: 1.78 A.

Example 4.7

Find v_x in Fig. 4.20 using source transformation.

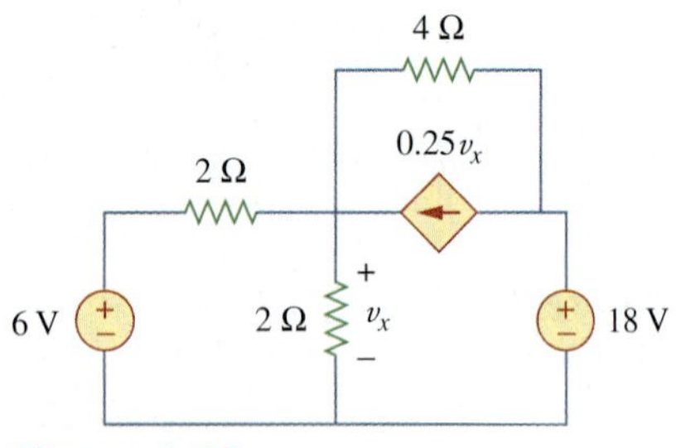

Figure 4.20
For Example 4.7.

Solution:

The circuit in Fig. 4.20 involves a voltage-controlled dependent current source. We transform this dependent current source as well as the 6-V independent voltage source as shown in Fig. 4.21(a). The 18-V voltage source is not transformed because it is not connected in series with any resistor. The two 2-Ω resistors in parallel combine to give a 1-Ω resistor, which is in parallel with the 3-A current source. The current source is transformed to a voltage source as shown in Fig. 4.21(b). Notice that the terminals for v_x are intact. Applying KVL around the loop in Fig. 4.21(b) gives

$$-3 + 5i + v_x + 18 = 0 \tag{4.7.1}$$

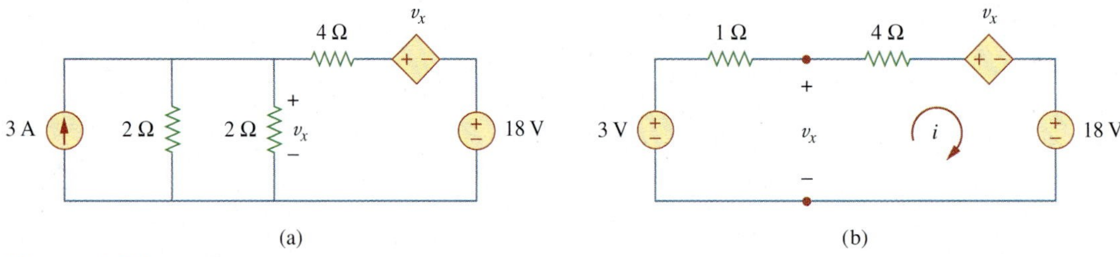

Figure 4.21
For Example 4.7: Applying source transformation to the circuit in Fig. 4.20.

Applying KVL to the loop containing only the 3-V voltage source, the 1-Ω resistor, and v_x yields

$$-3 + 1i + v_x = 0 \quad \Rightarrow \quad v_x = 3 - i \tag{4.7.2}$$

Substituting this into Eq. (4.7.1), we obtain

$$15 + 5i + 3 - i = 0 \quad \Rightarrow \quad i = -4.5 \text{ A}$$

Alternatively, we may apply KVL to the loop containing v_x, the 4-Ω resistor, the voltage-controlled dependent voltage source, and the 18-V voltage source in Fig. 4.21(b). We obtain

$$-v_x + 4i + v_x + 18 = 0 \quad \Rightarrow \quad i = -4.5 \text{ A}$$

Thus, $v_x = 3 - i = 7.5$ V.

Practice Problem 4.7

Use source transformation to find i_x in the circuit shown in Fig. 4.22.

Answer: 7.056 mA.

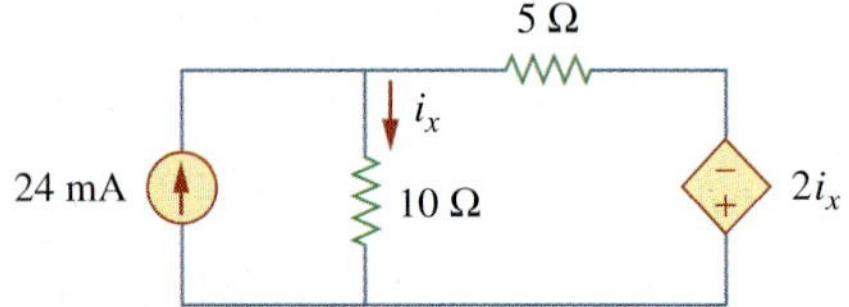

Figure 4.22
For Practice Prob. 4.7.

4.5 Thevenin's Theorem

It often occurs in practice that a particular element in a circuit is variable (usually called the *load*) while other elements are fixed. As a typical example, a household outlet terminal may be connected to different appliances constituting a variable load. Each time the variable element is changed, the entire circuit has to be analyzed all over again. To avoid this problem, Thevenin's theorem provides a technique by which the fixed part of the circuit is replaced by an equivalent circuit.

According to Thevenin's theorem, the linear circuit in Fig. 4.23(a) can be replaced by that in Fig. 4.23(b). (The load in Fig. 4.23 may be a single resistor or another circuit.) The circuit to the left of the terminals *a-b* in Fig. 4.23(b) is known as the *Thevenin equivalent circuit*; it was developed in 1883 by M. Leon Thevenin (1857–1926), a French telegraph engineer.

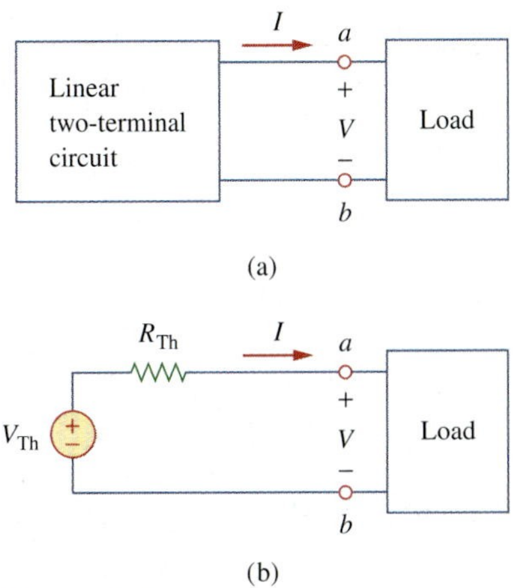

Figure 4.23
Replacing a linear two-terminal circuit by its Thevenin equivalent: (a) original circuit, (b) the Thevenin equivalent circuit.

> **Thevenin's theorem** states that a linear two-terminal circuit can be replaced by an equivalent circuit consisting of a voltage source V_{Th} in series with a resistor R_{Th}, where V_{Th} is the open-circuit voltage at the terminals and R_{Th} is the input or equivalent resistance at the terminals when the independent sources are turned off.

The proof of the theorem will be given later, in Section 4.7. Our major concern right now is how to find the Thevenin equivalent voltage V_{Th} and resistance R_{Th}. To do so, suppose the two circuits in Fig. 4.23 are equivalent. Two circuits are said to be *equivalent* if they have the same voltage-current relation at their terminals. Let us find out what will make the two circuits in Fig. 4.23 equivalent. If the terminals *a-b* are made open-circuited (by removing the load), no current flows, so that the open-circuit voltage across the terminals *a-b* in Fig. 4.23(a) must be equal to the voltage source V_{Th} in Fig. 4.23(b), since the two circuits are equivalent. Thus V_{Th} is the open-circuit voltage across the terminals as shown in Fig. 4.24(a); that is,

$$V_{Th} = v_{oc} \tag{4.6}$$

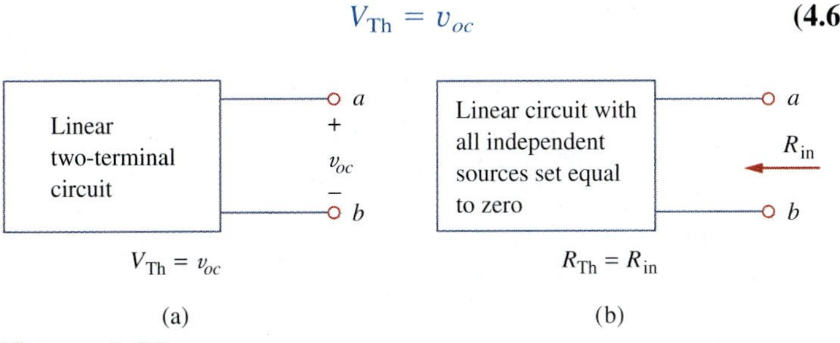

Figure 4.24
Finding V_{Th} and R_{Th}.

Again, with the load disconnected and terminals *a-b* open-circuited, we turn off all independent sources. The input resistance (or equivalent resistance) of the dead circuit at the terminals *a-b* in Fig. 4.23(a) must be equal to R_{Th} in Fig. 4.23(b) because the two circuits are equivalent. Thus, R_{Th} is the input resistance at the terminals when the independent sources are turned off, as shown in Fig. 4.24(b); that is,

$$R_{Th} = R_{in} \tag{4.7}$$

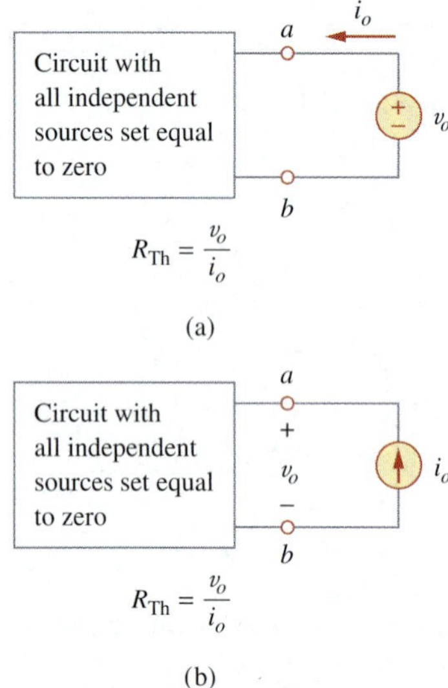

Figure 4.25
Finding R_{Th} when circuit has dependent sources.

To apply this idea in finding the Thevenin resistance R_{Th}, we need to consider two cases.

■ **CASE 1** If the network has no dependent sources, we turn off all independent sources. R_{Th} is the input resistance of the network looking between terminals a and b, as shown in Fig. 4.24(b).

■ **CASE 2** If the network has dependent sources, we turn off all independent sources. As with superposition, dependent sources are not to be turned off because they are controlled by circuit variables. We apply a voltage source v_o at terminals a and b and determine the resulting current i_o. Then $R_{Th} = v_o/i_o$, as shown in Fig. 4.25(a). Alternatively, we may insert a current source i_o at terminals a-b as shown in Fig. 4.25(b) and find the terminal voltage v_o. Again $R_{Th} = v_o/i_o$. Either of the two approaches will give the same result. In either approach we may assume any value of v_o and i_o. For example, we may use $v_o = 1$ V or $i_o = 1$ A, or even use unspecified values of v_o or i_o.

Later we will see that an alternative way of finding R_{Th} is $R_{Th} = v_{oc}/i_{sc}$.

It often occurs that R_{Th} takes a negative value. In this case, the negative resistance ($v = -iR$) implies that the circuit is supplying power. This is possible in a circuit with dependent sources; Example 4.10 will illustrate this.

Thevenin's theorem is very important in circuit analysis. It helps simplify a circuit. A large circuit may be replaced by a single independent voltage source and a single resistor. This replacement technique is a powerful tool in circuit design.

As mentioned earlier, a linear circuit with a variable load can be replaced by the Thevenin equivalent, exclusive of the load. The equivalent network behaves the same way externally as the original circuit. Consider a linear circuit terminated by a load R_L, as shown in Fig. 4.26(a). The current I_L through the load and the voltage V_L across the load are easily determined once the Thevenin equivalent of the circuit at the load's terminals is obtained, as shown in Fig. 4.26(b). From Fig. 4.26(b), we obtain

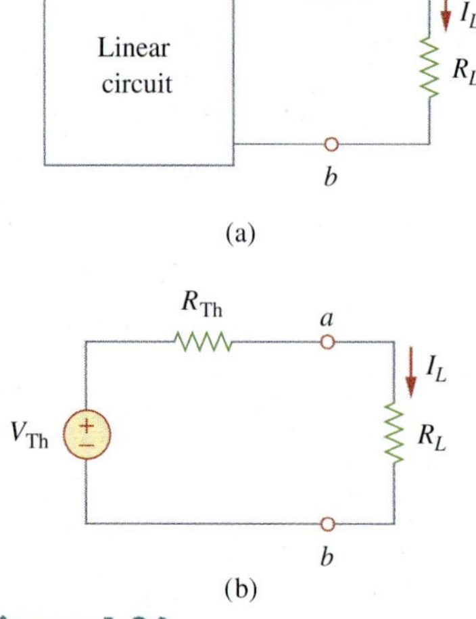

Figure 4.26
A circuit with a load: (a) original circuit, (b) Thevenin equivalent.

$$I_L = \frac{V_{Th}}{R_{Th} + R_L} \tag{4.8a}$$

$$V_L = R_L I_L = \frac{R_L}{R_{Th} + R_L} V_{Th} \tag{4.8b}$$

Note from Fig. 4.26(b) that the Thevenin equivalent is a simple voltage divider, yielding V_L by mere inspection.

Example 4.8

Find the Thevenin equivalent circuit of the circuit shown in Fig. 4.27, to the left of the terminals a-b. Then find the current through $R_L = 6$, 16, and 36 Ω.

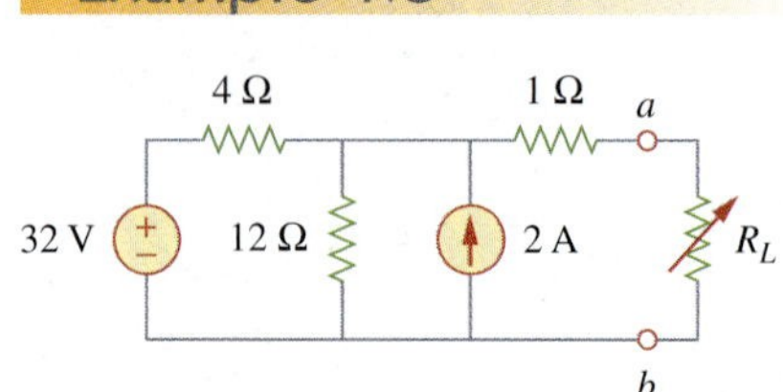

Figure 4.27
For Example 4.8.

Solution:
We find R_{Th} by turning off the 32-V voltage source (replacing it with a short circuit) and the 2-A current source (replacing it with an

open circuit). The circuit becomes what is shown in Fig. 4.28(a). Thus,

$$R_{\text{Th}} = 4 \parallel 12 + 1 = \frac{4 \times 12}{16} + 1 = 4\ \Omega$$

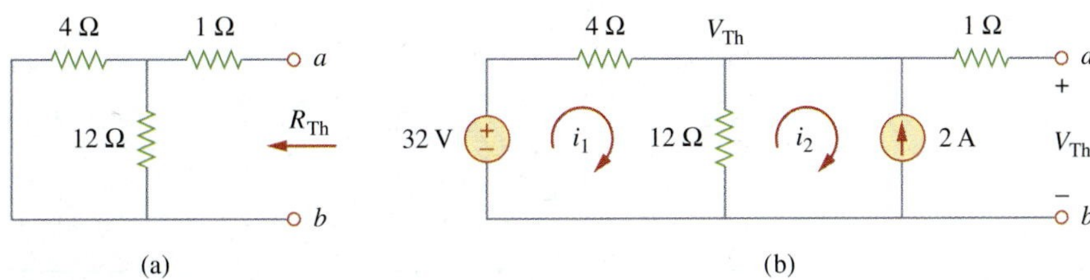

Figure 4.28
For Example 4.8: (a) finding R_{Th}, (b) finding V_{Th}.

To find V_{Th}, consider the circuit in Fig. 4.28(b). Applying mesh analysis to the two loops, we obtain

$$-32 + 4i_1 + 12(i_1 - i_2) = 0, \qquad i_2 = -2\ \text{A}$$

Solving for i_1, we get $i_1 = 0.5$ A. Thus,

$$V_{\text{Th}} = 12(i_1 - i_2) = 12(0.5 + 2.0) = 30\ \text{V}$$

Alternatively, it is even easier to use nodal analysis. We ignore the 1-Ω resistor since no current flows through it. At the top node, KCL gives

$$\frac{32 - V_{\text{Th}}}{4} + 2 = \frac{V_{\text{Th}}}{12}$$

or

$$96 - 3V_{\text{Th}} + 24 = V_{\text{Th}} \quad \Rightarrow \quad V_{\text{Th}} = 30\ \text{V}$$

as obtained before. We could also use source transformation to find V_{Th}.

The Thevenin equivalent circuit is shown in Fig. 4.29. The current through R_L is

$$I_L = \frac{V_{\text{Th}}}{R_{\text{Th}} + R_L} = \frac{30}{4 + R_L}$$

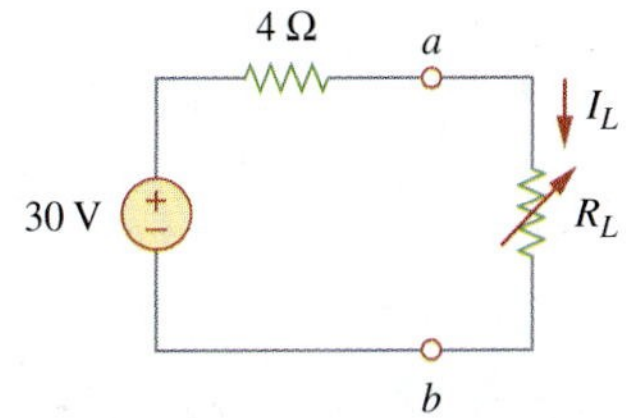

Figure 4.29
The Thevenin equivalent circuit for Example 4.8.

When $R_L = 6$,

$$I_L = \frac{30}{10} = 3\ \text{A}$$

When $R_L = 16$,

$$I_L = \frac{30}{20} = 1.5\ \text{A}$$

When $R_L = 36$,

$$I_L = \frac{30}{40} = 0.75\ \text{A}$$

Practice Problem 4.8

Using Thevenin's theorem, find the equivalent circuit to the left of the terminals in the circuit of Fig. 4.30. Then find I.

Answer: $V_{\text{Th}} = 9$ V, $R_{\text{Th}} = 3\ \Omega$, $I = 2.25$ A.

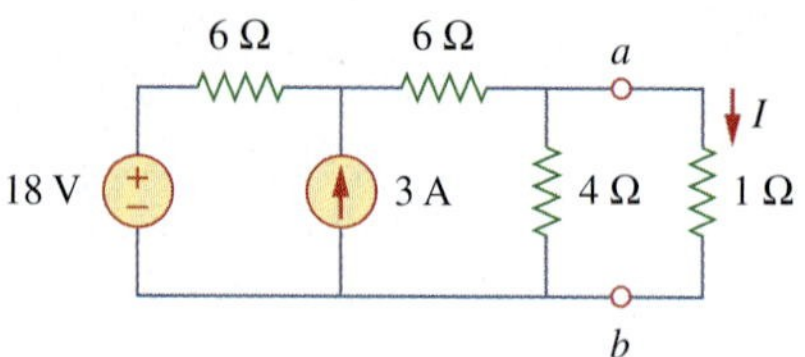

Figure 4.30
For Practice Prob. 4.8.

Example 4.9

Find the Thevenin equivalent of the circuit in Fig. 4.31 at terminals a-b.

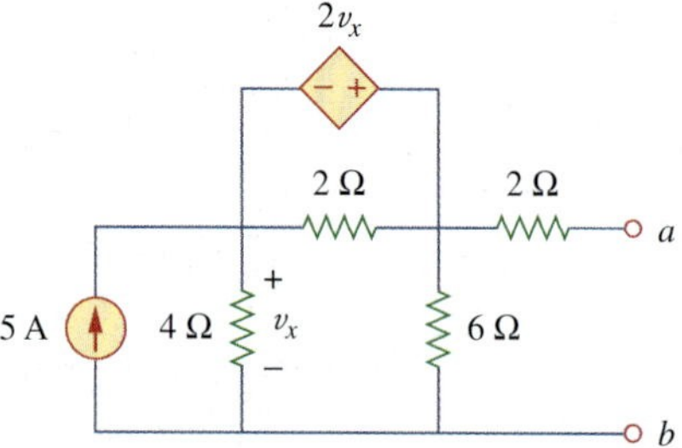

Figure 4.31
For Example 4.9.

Solution:
This circuit contains a dependent source, unlike the circuit in the previous example. To find R_{Th}, we set the independent source equal to zero but leave the dependent source alone. Because of the presence of the dependent source, however, we excite the network with a voltage source v_o connected to the terminals as indicated in Fig. 4.32(a). We may set $v_o = 1$ V to ease calculation, since the circuit is linear. Our goal is to find the current i_o through the terminals, and then obtain $R_{\text{Th}} = 1/i_o$. (Alternatively, we may insert a 1-A current source, find the corresponding voltage v_o, and obtain $R_{\text{Th}} = v_o/1$.)

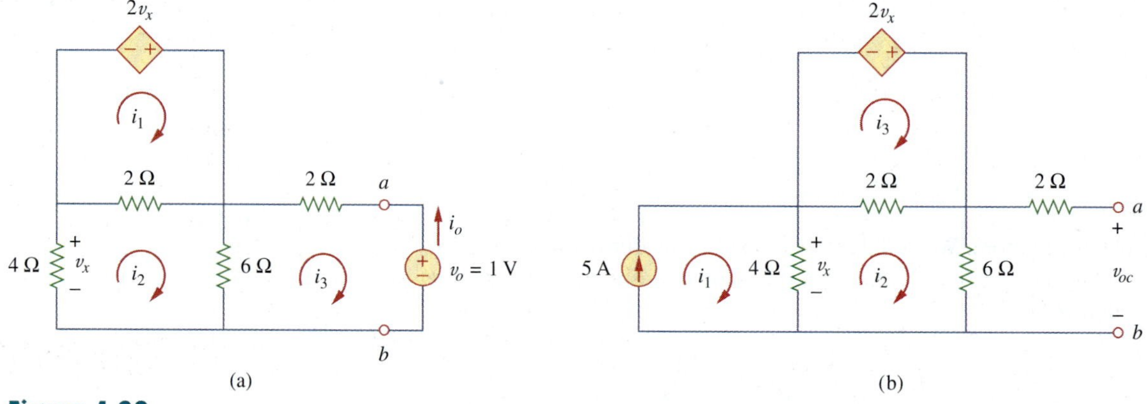

Figure 4.32
Finding R_{Th} and V_{Th} for Example 4.9.

Applying mesh analysis to loop 1 in the circuit of Fig. 4.32(a) results in

$$-2v_x + 2(i_1 - i_2) = 0 \qquad \text{or} \qquad v_x = i_1 - i_2$$

But $-4i_2 = v_x = i_1 - i_2$; hence,

$$i_1 = -3i_2 \tag{4.9.1}$$

For loops 2 and 3, applying KVL produces

$$4i_2 + 2(i_2 - i_1) + 6(i_2 - i_3) = 0 \tag{4.9.2}$$

$$6(i_3 - i_2) + 2i_3 + 1 = 0 \tag{4.9.3}$$

Solving these equations gives

$$i_3 = -\frac{1}{6} \text{ A}$$

But $i_o = -i_3 = 1/6$ A. Hence,

$$R_{\text{Th}} = \frac{1 \text{ V}}{i_o} = 6 \ \Omega$$

To get V_{Th}, we find v_{oc} in the circuit of Fig. 4.32(b). Applying mesh analysis, we get

$$i_1 = 5 \tag{4.9.4}$$

$$-2v_x + 2(i_3 - i_2) = 0 \quad \Rightarrow \quad v_x = i_3 - i_2 \tag{4.9.5}$$

$$4(i_2 - i_1) + 2(i_2 - i_3) + 6i_2 = 0$$

or

$$12i_2 - 4i_1 - 2i_3 = 0 \tag{4.9.6}$$

But $4(i_1 - i_2) = v_x$. Solving these equations leads to $i_2 = 10/3$. Hence,

$$V_{\text{Th}} = v_{oc} = 6i_2 = 20 \text{ V}$$

The Thevenin equivalent is as shown in Fig. 4.33.

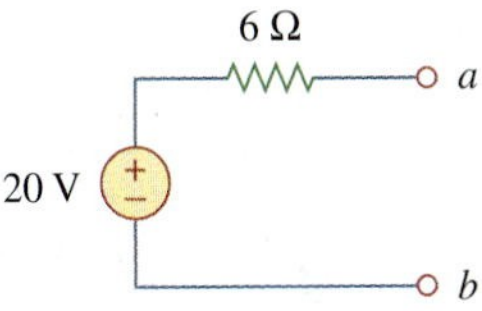

Figure 4.33
The Thevenin equivalent of the circuit in Fig. 4.31.

Practice Problem 4.9

Find the Thevenin equivalent circuit of the circuit in Fig. 4.34 to the left of the terminals.

Answer: $V_{\text{Th}} = 5.33$ V, $R_{\text{Th}} = 0.44 \ \Omega$.

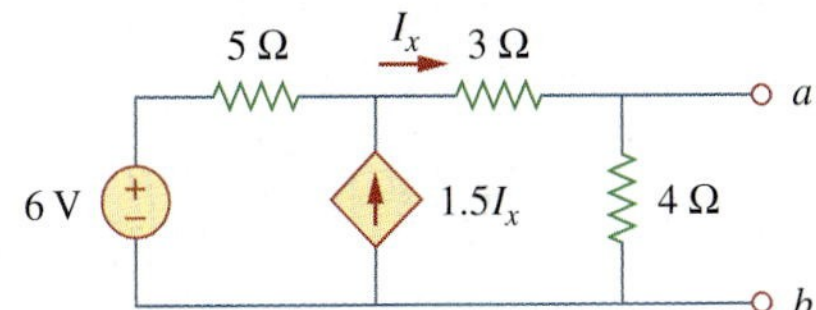

Figure 4.34
For Practice Prob. 4.9.

Example 4.10

Determine the Thevenin equivalent of the circuit in Fig. 4.35(a) at terminals *a-b*.

Solution:

1. **Define.** The problem is clearly defined; we are to determine the Thevenin equivalent of the circuit shown in Fig. 4.35(a).
2. **Present.** The circuit contains a 2-Ω resistor in parallel with a 4-Ω resistor. These are, in turn, in parallel with a dependent current source. It is important to note that there are no independent sources.
3. **Alternative.** The first thing to consider is that, since we have no independent sources in this circuit, we must excite the circuit externally. In addition, when you have no independent sources you will not have a value for V_{Th}; you will only have to find R_{Th}.

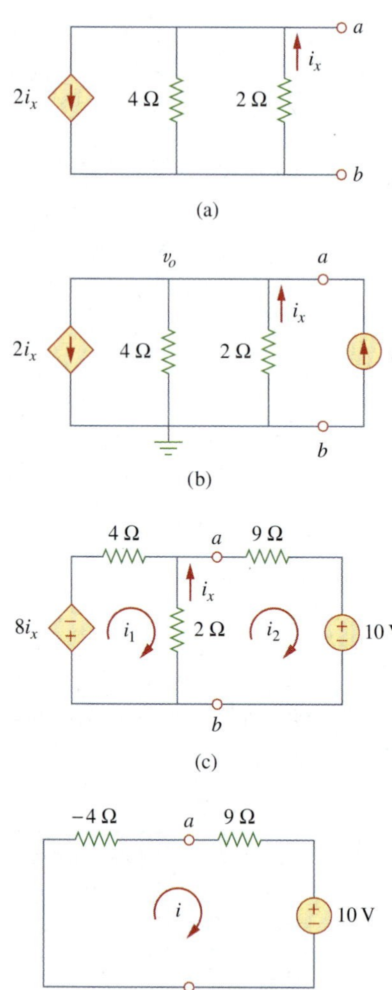

Figure 4.35
For Example 4.10.

The simplest approach is to excite the circuit with either a 1-V voltage source or a 1-A current source. Since we will end up with an equivalent resistance (either positive or negative), I prefer to use the current source and nodal analysis which will yield a voltage at the output terminals equal to the resistance (with 1 A flowing in, v_o is equal to 1 times the equivalent resistance).

As an alternative, the circuit could also be excited by a 1-V voltage source and mesh analysis could be used to find the equivalent resistance.

4. **Attempt.** We start by writing the nodal equation at a in Fig. 4.35(b) assuming $i_o = 1$ A.

$$2i_x + (v_o - 0)/4 + (v_o - 0)/2 + (-1) = 0 \qquad \textbf{(4.10.1)}$$

Since we have two unknowns and only one equation, we will need a constraint equation.

$$i_x = (0 - v_o)/2 = -v_o/2 \qquad \textbf{(4.10.2)}$$

Substituting Eq. (4.10.2) into Eq. (4.10.1) yields

$$2(-v_o/2) + (v_o - 0)/4 + (v_o - 0)/2 + (-1) = 0$$
$$= (-1 + \tfrac{1}{4} + \tfrac{1}{2})v_o - 1 \qquad \text{or} \qquad v_o = -4 \text{ V}$$

Since $v_o = 1 \times R_{\text{Th}}$, then $R_{\text{Th}} = v_o/1 = \mathbf{-4\ \Omega}$.

The negative value of the resistance tells us that, according to the passive sign convention, the circuit in Fig. 4.35(a) is supplying power. Of course, the resistors in Fig. 4.35(a) cannot supply power (they absorb power); it is the dependent source that supplies the power. This is an example of how a dependent source and resistors could be used to simulate negative resistance.

5. **Evaluate.** First of all, we note that the answer has a negative value. We know this is not possible in a passive circuit, but in this circuit we do have an active device (the dependent current source). Thus, the equivalent circuit is essentially an active circuit that can supply power.

Now we must evaluate the solution. The best way to do this is to perform a check, using a different approach, and see if we obtain the same solution. Let us try connecting a 9-Ω resistor in series with a 10-V voltage source across the output terminals of the original circuit and then the Thevenin equivalent. To make the circuit easier to solve, we can take and change the parallel current source and 4-Ω resistor to a series voltage source and 4-Ω resistor by using source transformation. This, with the new load, gives us the circuit shown in Fig. 4.35(c).

We can now write two mesh equations.

$$8i_x + 4i_1 + 2(i_1 - i_2) = 0$$
$$2(i_2 - i_1) + 9i_2 + 10 = 0$$

Note, we only have two equations but have 3 unknowns, so we need a constraint equation. We can use

$$i_x = i_2 - i_1$$

This leads to a new equation for loop 1. Simplifying leads to

$$(4 + 2 - 8)i_1 + (-2 + 8)i_2 = 0$$

or

$$-2i_1 + 6i_2 = 0 \qquad \text{or} \qquad i_1 = 3i_2$$
$$-2i_1 + 11i_2 = -10$$

Substituting the first equation into the second gives

$$-6i_2 + 11i_2 = -10 \qquad \text{or} \qquad i_2 = -10/5 = \mathbf{-2\ A}$$

Using the Thevenin equivalent is quite easy since we have only one loop, as shown in Fig. 4.35(d).

$$-4i + 9i + 10 = 0 \qquad \text{or} \qquad i = -10/5 = \mathbf{-2\ A}$$

6. **Satisfactory?** Clearly we have found the value of the equivalent circuit as required by the problem statement. Checking does validate that solution (we compared the answer we obtained by using the equivalent circuit with one obtained by using the load with the original circuit). We can present all this as a solution to the problem.

Practice Problem 4.10

Obtain the Thevenin equivalent of the circuit in Fig. 4.36.

Answer: $V_{Th} = 0$ V, $R_{Th} = -7.5\ \Omega$.

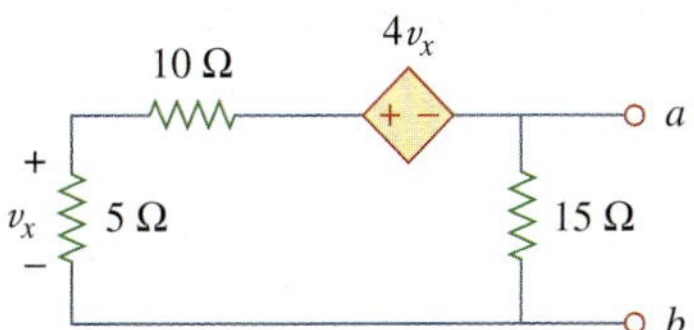

Figure 4.36
For Practice Prob. 4.10.

4.6 Norton's Theorem

In 1926, about 43 years after Thevenin published his theorem, E. L. Norton, an American engineer at Bell Telephone Laboratories, proposed a similar theorem.

> **Norton's theorem** states that a linear two-terminal circuit can be replaced by an equivalent circuit consisting of a current source I_N in parallel with a resistor R_N, where I_N is the short-circuit current through the terminals and R_N is the input or equivalent resistance at the terminals when the independent sources are turned off.

Thus, the circuit in Fig. 4.37(a) can be replaced by the one in Fig. 4.37(b).

The proof of Norton's theorem will be given in the next section. For now, we are mainly concerned with how to get R_N and I_N. We find R_N in the same way we find R_{Th}. In fact, from what we know about source transformation, the Thevenin and Norton resistances are equal; that is,

$$R_N = R_{Th} \tag{4.9}$$

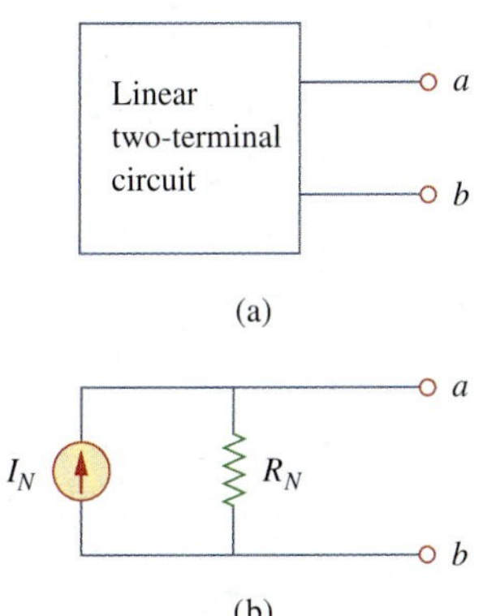

Figure 4.37
(a) Original circuit, (b) Norton equivalent circuit.

To find the Norton current I_N, we determine the short-circuit current flowing from terminal a to b in both circuits in Fig. 4.37. It is evident

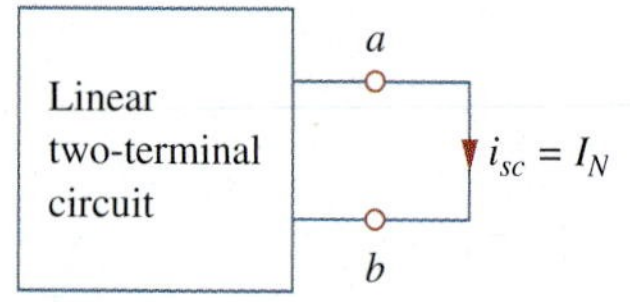

Figure 4.38
Finding Norton current I_N.

that the short-circuit current in Fig. 4.37(b) is I_N. This must be the same short-circuit current from terminal a to b in Fig. 4.37(a), since the two circuits are equivalent. Thus,

$$I_N = i_{sc} \tag{4.10}$$

shown in Fig. 4.38. Dependent and independent sources are treated the same way as in Thevenin's theorem.

Observe the close relationship between Norton's and Thevenin's theorems: $R_N = R_{Th}$ as in Eq. (4.9), and

$$I_N = \frac{V_{Th}}{R_{Th}} \tag{4.11}$$

The Thevenin and Norton equivalent circuits are related by a source transformation.

This is essentially source transformation. For this reason, source transformation is often called Thevenin-Norton transformation.

Since V_{Th}, I_N, and R_{Th} are related according to Eq. (4.11), to determine the Thevenin or Norton equivalent circuit requires that we find:

- The open-circuit voltage v_{oc} across terminals a and b.
- The short-circuit current i_{sc} at terminals a and b.
- The equivalent or input resistance R_{in} at terminals a and b when all independent sources are turned off.

We can calculate any two of the three using the method that takes the least effort and use them to get the third using Ohm's law. Example 4.11 will illustrate this. Also, since

$$V_{Th} = v_{oc} \tag{4.12a}$$

$$I_N = i_{sc} \tag{4.12b}$$

$$R_{Th} = \frac{v_{oc}}{i_{sc}} = R_N \tag{4.12c}$$

the open-circuit and short-circuit tests are sufficient to find any Thevenin or Norton equivalent, of a circuit which contains at least one independent source.

Example 4.11

Find the Norton equivalent circuit of the circuit in Fig. 4.39 at terminals a-b.

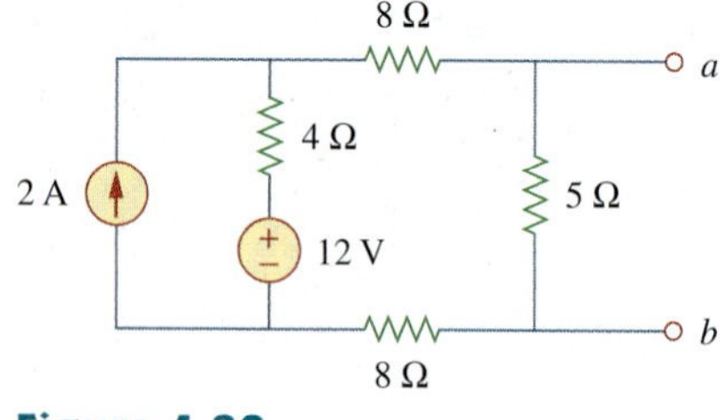

Figure 4.39
For Example 4.11.

Solution:

We find R_N in the same way we find R_{Th} in the Thevenin equivalent circuit. Set the independent sources equal to zero. This leads to the circuit in Fig. 4.40(a), from which we find R_N. Thus,

$$R_N = 5 \parallel (8 + 4 + 8) = 5 \parallel 20 = \frac{20 \times 5}{25} = 4\ \Omega$$

To find I_N, we short-circuit terminals a and b, as shown in Fig. 4.40(b). We ignore the 5-Ω resistor because it has been short-circuited. Applying mesh analysis, we obtain

$$i_1 = 2\text{ A}, \qquad 20i_2 - 4i_1 - 12 = 0$$

From these equations, we obtain

$$i_2 = 1\text{ A} = i_{sc} = I_N$$

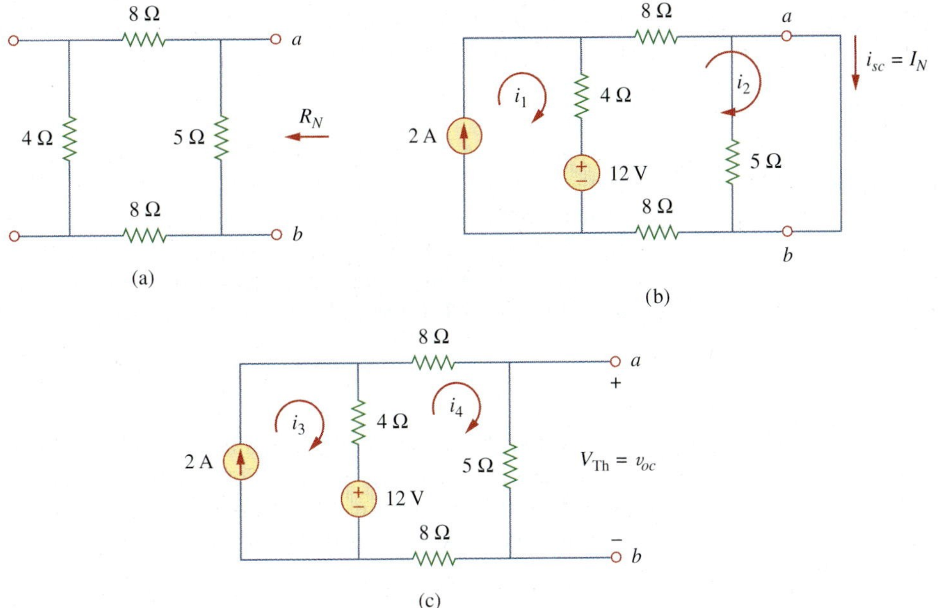

Figure 4.40
For Example 4.11; finding: (a) R_N, (b) $I_N = i_{sc}$, (c) $V_{Th} = v_{oc}$.

Alternatively, we may determine I_N from V_{Th}/R_{Th}. We obtain V_{Th} as the open-circuit voltage across terminals a and b in Fig. 4.40(c). Using mesh analysis, we obtain

$$i_3 = 2 \text{ A}$$

$$25i_4 - 4i_3 - 12 = 0 \quad \Rightarrow \quad i_4 = 0.8 \text{ A}$$

and

$$v_{oc} = V_{Th} = 5i_4 = 4 \text{ V}$$

Hence,

$$I_N = \frac{V_{Th}}{R_{Th}} = \frac{4}{4} = 1 \text{ A}$$

as obtained previously. This also serves to confirm Eq. (4.12c) that $R_{Th} = v_{oc}/i_{sc} = 4/1 = 4\ \Omega$. Thus, the Norton equivalent circuit is as shown in Fig. 4.41.

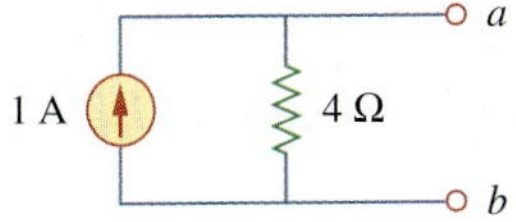

Figure 4.41
Norton equivalent of the circuit in Fig. 4.39.

Practice Problem 4.11

Find the Norton equivalent circuit for the circuit in Fig. 4.42, at terminals *a-b*.

Answer: $R_N = 3\ \Omega$, $I_N = 4.5$ A.

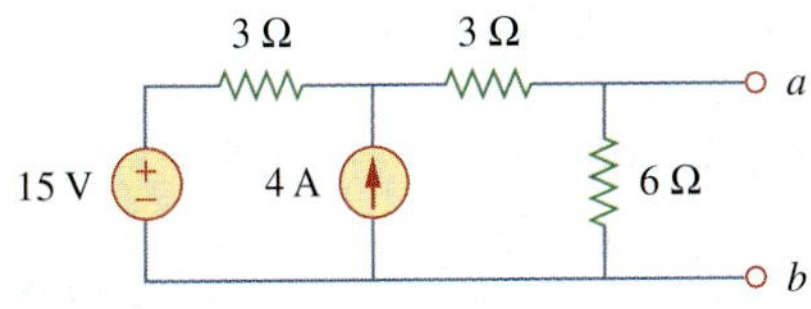

Figure 4.42
For Practice Prob. 4.11.

Example 4.12

Using Norton's theorem, find R_N and I_N of the circuit in Fig. 4.43 at terminals *a-b*.

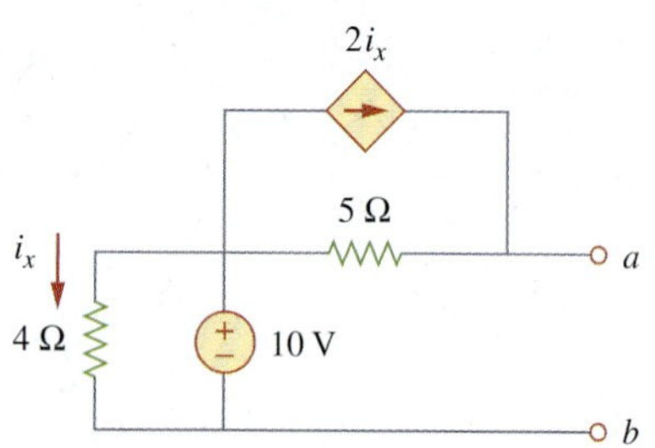

Figure 4.43
For Example 4.12.

Solution:

To find R_N, we set the independent voltage source equal to zero and connect a voltage source of $v_o = 1$ V (or any unspecified voltage v_o) to the terminals. We obtain the circuit in Fig. 4.44(a). We ignore the 4-Ω resistor because it is short-circuited. Also due to the short circuit, the 5-Ω resistor, the voltage source, and the dependent current source are all in parallel. Hence, $i_x = 0$. At node a, $i_o = \frac{1v}{5\Omega} = 0.2$ A, and

$$R_N = \frac{v_o}{i_o} = \frac{1}{0.2} = 5\ \Omega$$

To find I_N, we short-circuit terminals a and b and find the current i_{sc}, as indicated in Fig. 4.44(b). Note from this figure that the 4-Ω resistor, the 10-V voltage source, the 5-Ω resistor, and the dependent current source are all in parallel. Hence,

$$i_x = \frac{10}{4} = 2.5\ \text{A}$$

At node a, KCL gives

$$i_{sc} = \frac{10}{5} + 2i_x = 2 + 2(2.5) = 7\ \text{A}$$

Thus,

$$I_N = 7\ \text{A}$$

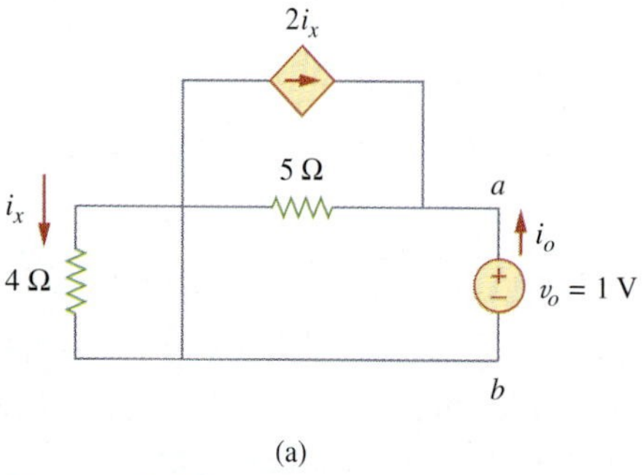

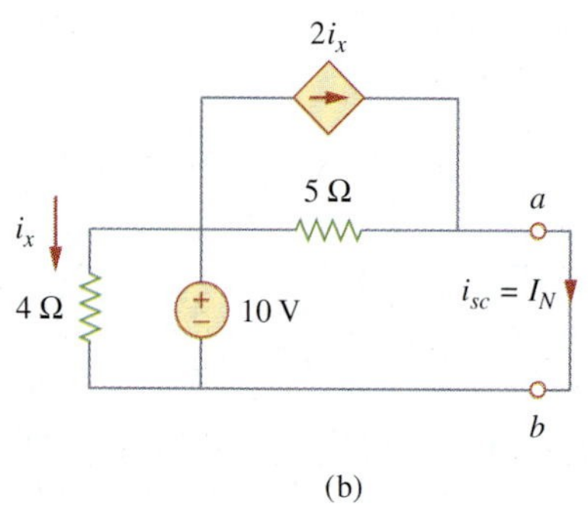

Figure 4.44
For Example 4.12: (a) finding R_N, (b) finding I_N.

Practice Problem 4.12

Find the Norton equivalent circuit of the circuit in Fig. 4.45 at terminals *a-b*.

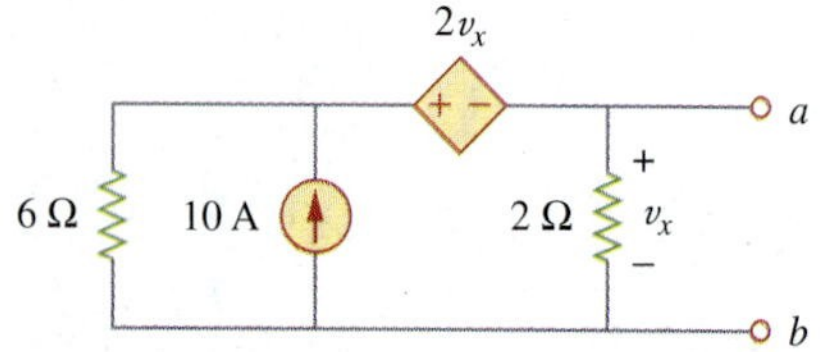

Figure 4.45
For Practice Prob. 4.12.

Answer: $R_N = 1\ \Omega$, $I_N = 10$ A.

4.7 †Derivations of Thevenin's and Norton's Theorems

In this section, we will prove Thevenin's and Norton's theorems using the superposition principle.

Consider the linear circuit in Fig. 4.46(a). It is assumed that the circuit contains resistors, and dependent and independent sources. We have access to the circuit via terminals a and b, through which current from an external source is applied. Our objective is to ensure that the voltage-current relation at terminals a and b is identical to that of the Thevenin equivalent in Fig. 4.46(b). For the sake of simplicity, suppose the linear circuit in Fig. 4.46(a) contains two independent voltage sources v_{s1} and v_{s2} and two independent current sources i_{s1} and i_{s2}. We may obtain any circuit variable, such as the terminal voltage v, by applying superposition. That is, we consider the contribution due to each independent source including the external source i. By superposition, the terminal voltage v is

$$v = A_0 i + A_1 v_{s1} + A_2 v_{s2} + A_3 i_{s1} + A_4 i_{s2} \quad \textbf{(4.13)}$$

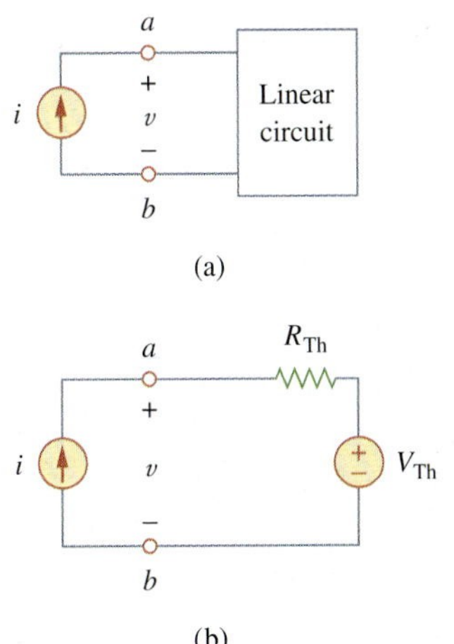

Figure 4.46
Derivation of Thevenin equivalent: (a) a current-driven circuit, (b) its Thevenin equivalent.

where A_0, A_1, A_2, A_3, and A_4 are constants. Each term on the right-hand side of Eq. (4.13) is the contribution of the related independent source; that is, $A_0 i$ is the contribution to v due to the external current source i, $A_1 v_{s1}$ is the contribution due to the voltage source v_{s1}, and so on. We may collect terms for the internal independent sources together as B_0, so that Eq. (4.13) becomes

$$v = A_0 i + B_0 \quad \textbf{(4.14)}$$

where $B_0 = A_1 v_{s1} + A_2 v_{s2} + A_3 i_{s1} + A_4 i_{s2}$. We now want to evaluate the values of constants A_0 and B_0. When the terminals a and b are open-circuited, $i = 0$ and $v = B_0$. Thus, B_0 is the open-circuit voltage v_{oc}, which is the same as V_{Th}, so

$$B_0 = V_{Th} \quad \textbf{(4.15)}$$

When all the internal sources are turned off, $B_0 = 0$. The circuit can then be replaced by an equivalent resistance R_{eq}, which is the same as R_{Th}, and Eq. (4.14) becomes

$$v = A_0 i = R_{Th} i \quad \Rightarrow \quad A_0 = R_{Th} \quad \textbf{(4.16)}$$

Substituting the values of A_0 and B_0 in Eq. (4.14) gives

$$v = R_{Th} i + V_{Th} \quad \textbf{(4.17)}$$

which expresses the voltage-current relation at terminals a and b of the circuit in Fig. 4.46(b). Thus, the two circuits in Fig. 4.46(a) and 4.46(b) are equivalent.

When the same linear circuit is driven by a voltage source v as shown in Fig. 4.47(a), the current flowing into the circuit can be obtained by superposition as

$$i = C_0 v + D_0 \quad \textbf{(4.18)}$$

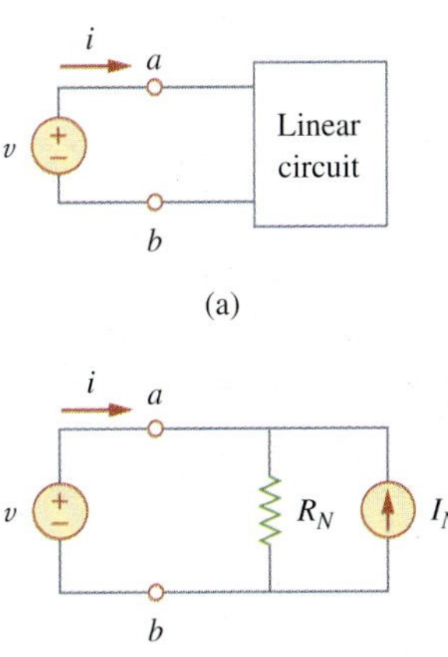

Figure 4.47
Derivation of Norton equivalent: (a) a voltage-driven circuit, (b) its Norton equivalent.

where $C_0 v$ is the contribution to i due to the external voltage source v and D_0 contains the contributions to i due to all internal independent sources. When the terminals a-b are short-circuited, $v = 0$ so that

$i = D_0 = -i_{sc}$, where i_{sc} is the short-circuit current flowing out of terminal a, which is the same as the Norton current I_N, i.e.,

$$D_0 = -I_N \tag{4.19}$$

When all the internal independent sources are turned off, $D_0 = 0$ and the circuit can be replaced by an equivalent resistance R_{eq} (or an equivalent conductance $G_{eq} = 1/R_{eq}$), which is the same as R_{Th} or R_N. Thus Eq. (4.19) becomes

$$i = \frac{v}{R_{Th}} - I_N \tag{4.20}$$

This expresses the voltage-current relation at terminals a-b of the circuit in Fig. 4.47(b), confirming that the two circuits in Fig. 4.47(a) and 4.47(b) are equivalent.

4.8 Maximum Power Transfer

In many practical situations, a circuit is designed to provide power to a load. There are applications in areas such as communications where it is desirable to maximize the power delivered to a load. We now address the problem of delivering the maximum power to a load when given a system with known internal losses. It should be noted that this will result in significant internal losses greater than or equal to the power delivered to the load.

The Thevenin equivalent is useful in finding the maximum power a linear circuit can deliver to a load. We assume that we can adjust the load resistance R_L. If the entire circuit is replaced by its Thevenin equivalent except for the load, as shown in Fig. 4.48, the power delivered to the load is

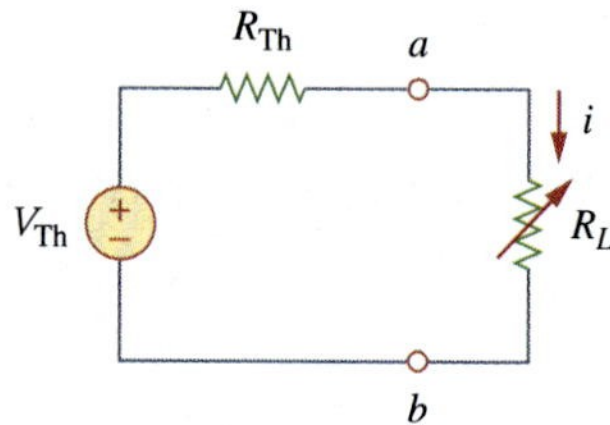

Figure 4.48
The circuit used for maximum power transfer.

$$p = i^2 R_L = \left(\frac{V_{Th}}{R_{Th} + R_L}\right)^2 R_L \tag{4.21}$$

For a given circuit, V_{Th} and R_{Th} are fixed. By varying the load resistance R_L, the power delivered to the load varies as sketched in Fig. 4.49. We notice from Fig. 4.49 that the power is small for small or large values of R_L but maximum for some value of R_L between 0 and ∞. We now want to show that this maximum power occurs when R_L is equal to R_{Th}. This is known as the *maximum power theorem.*

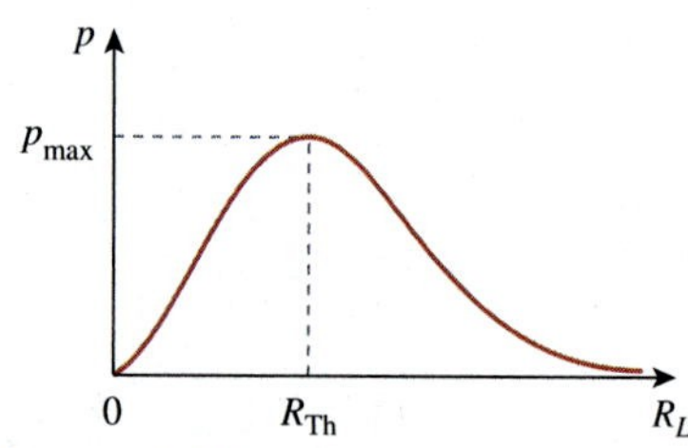

Figure 4.49
Power delivered to the load as a function of R_L.

> **Maximum power** is transferred to the load when the load resistance equals the Thevenin resistance as seen from the load ($R_L = R_{Th}$).

To prove the maximum power transfer theorem, we differentiate p in Eq. (4.21) with respect to R_L and set the result equal to zero. We obtain

$$\begin{aligned}\frac{dp}{dR_L} &= V_{Th}^2\left[\frac{(R_{Th} + R_L)^2 - 2R_L(R_{Th} + R_L)}{(R_{Th} + R_L)^4}\right] \\ &= V_{Th}^2\left[\frac{(R_{Th} + R_L - 2R_L)}{(R_{Th} + R_L)^3}\right] = 0\end{aligned}$$

This implies that

$$0 = (R_{Th} + R_L - 2R_L) = (R_{Th} - R_L) \tag{4.22}$$

which yields

$$R_L = R_{Th} \tag{4.23}$$

showing that the maximum power transfer takes place when the load resistance R_L equals the Thevenin resistance R_{Th}. We can readily confirm that Eq. (4.23) gives the maximum power by showing that $d^2p/dR_L^2 < 0$.

The source and load are said to be *matched* when $R_L = R_{Th}$.

The maximum power transferred is obtained by substituting Eq. (4.23) into Eq. (4.21), for

$$p_{max} = \frac{V_{Th}^2}{4R_{Th}} \tag{4.24}$$

Equation (4.24) applies only when $R_L = R_{Th}$. When $R_L \neq R_{Th}$, we compute the power delivered to the load using Eq. (4.21).

Example 4.13

Find the value of R_L for maximum power transfer in the circuit of Fig. 4.50. Find the maximum power.

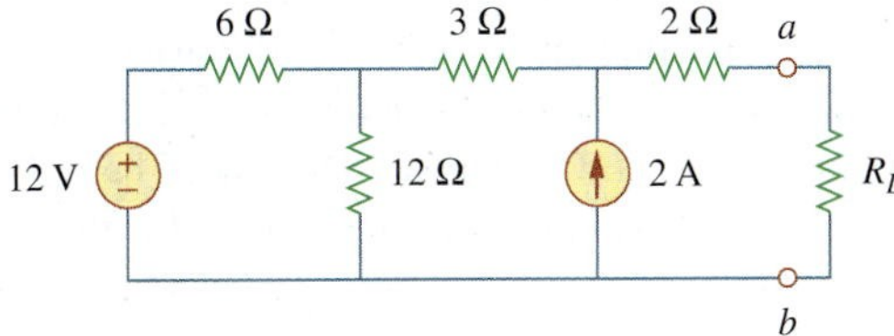

Figure 4.50
For Example 4.13.

Solution:

We need to find the Thevenin resistance R_{Th} and the Thevenin voltage V_{Th} across the terminals *a-b*. To get R_{Th}, we use the circuit in Fig. 4.51(a) and obtain

$$R_{Th} = 2 + 3 + 6 \| 12 = 5 + \frac{6 \times 12}{18} = 9\ \Omega$$

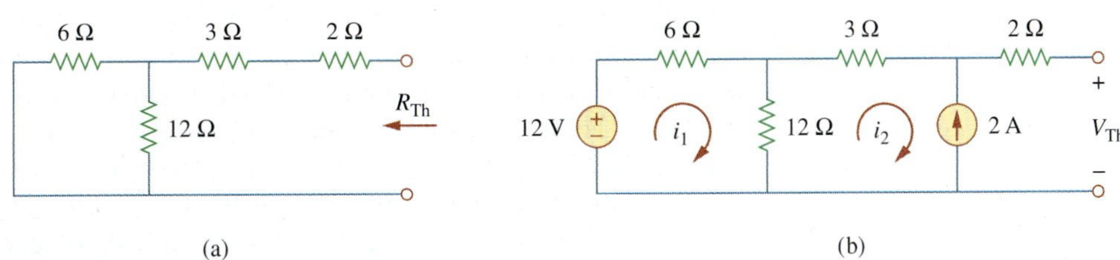

Figure 4.51
For Example 4.13: (a) finding R_{Th}, (b) finding V_{Th}.

To get V_{Th}, we consider the circuit in Fig. 4.51(b). Applying mesh analysis gives

$$-12 + 18i_1 - 12i_2 = 0, \qquad i_2 = -2\text{ A}$$

Solving for i_1, we get $i_1 = -2/3$. Applying KVL around the outer loop to get V_{Th} across terminals *a-b*, we obtain

$$-12 + 6i_1 + 3i_2 + 2(0) + V_{\text{Th}} = 0 \qquad \Rightarrow \qquad V_{\text{Th}} = 22\text{ V}$$

For maximum power transfer,

$$R_L = R_{\text{Th}} = 9\ \Omega$$

and the maximum power is

$$p_{\max} = \frac{V_{\text{Th}}^2}{4R_L} = \frac{22^2}{4 \times 9} = 13.44\text{ W}$$

Practice Problem 4.13

Determine the value of R_L that will draw the maximum power from the rest of the circuit in Fig. 4.52. Calculate the maximum power.

Answer: 4.22 Ω, 2.901 W.

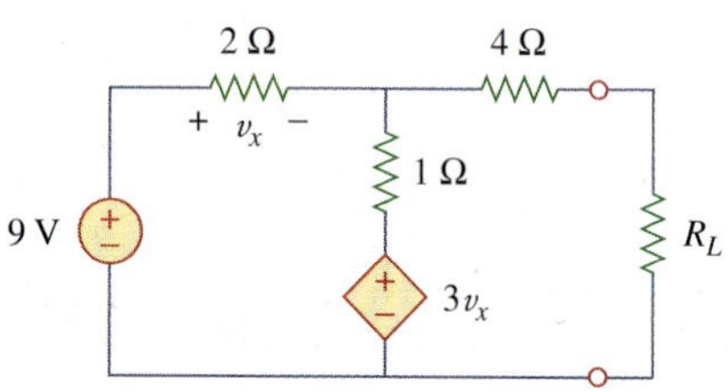

Figure 4.52
For Practice Prob. 4.13.

4.9 Verifying Circuit Theorems with *PSpice*

In this section, we learn how to use *PSpice* to verify the theorems covered in this chapter. Specifically, we will consider using DC Sweep analysis to find the Thevenin or Norton equivalent at any pair of nodes in a circuit and the maximum power transfer to a load. The reader is advised to read Section D.3 of Appendix D in preparation for this section.

To find the Thevenin equivalent of a circuit at a pair of open terminals using *PSpice*, we use the schematic editor to draw the circuit and insert an independent probing current source, say, Ip, at the terminals. The probing current source must have a part name ISRC. We then perform a DC Sweep on Ip, as discussed in Section D.3. Typically, we may let the current through Ip vary from 0 to 1 A in 0.1-A increments. After saving and simulating the circuit, we use Probe to display a plot of the voltage across Ip versus the current through Ip. The zero intercept of the plot gives us the Thevenin equivalent voltage, while the slope of the plot is equal to the Thevenin resistance.

To find the Norton equivalent involves similar steps except that we insert a probing independent voltage source (with a part name VSRC), say, Vp, at the terminals. We perform a DC Sweep on Vp and let Vp vary from 0 to 1 V in 0.1-V increments. A plot of the current through Vp versus the voltage across Vp is obtained using the Probe menu after simulation. The zero intercept is equal to the Norton current, while the slope of the plot is equal to the Norton conductance.

To find the maximum power transfer to a load using *PSpice* involves performing a DC parametric Sweep on the component value of R_L in Fig. 4.48 and plotting the power delivered to the load as a function of R_L. According to Fig. 4.49, the maximum power occurs when $R_L = R_{\text{Th}}$. This is best illustrated with an example, and Example 4.15 provides one.

We use VSRC and ISRC as part names for the independent voltage and current sources, respectively.

Example 4.14

Consider the circuit in Fig. 4.31 (see Example 4.9). Use *PSpice* to find the Thevenin and Norton equivalent circuits.

Solution:

(a) To find the Thevenin resistance R_{Th} and Thevenin voltage V_{Th} at the terminals *a-b* in the circuit in Fig. 4.31, we first use Schematics to draw the circuit as shown in Fig. 4.53(a). Notice that a probing current source I2 is inserted at the terminals. Under **Analysis/Setput**, we select DC Sweep. In the DC Sweep dialog box, we select Linear for the *Sweep Type* and Current Source for the *Sweep Var. Type.* We enter I2 under the *Name* box, 0 as *Start Value*, 1 as *End Value*, and 0.1 as *Increment.* After simulation, we add trace V(I2:–) from the *PSpice* A/D window and obtain the plot shown in Fig. 4.53(b). From the plot, we obtain

$$V_{\text{Th}} = \text{Zero intercept} = 20 \text{ V}, \qquad R_{\text{Th}} = \text{Slope} = \frac{26 - 20}{1} = 6 \ \Omega$$

These agree with what we got analytically in Example 4.9.

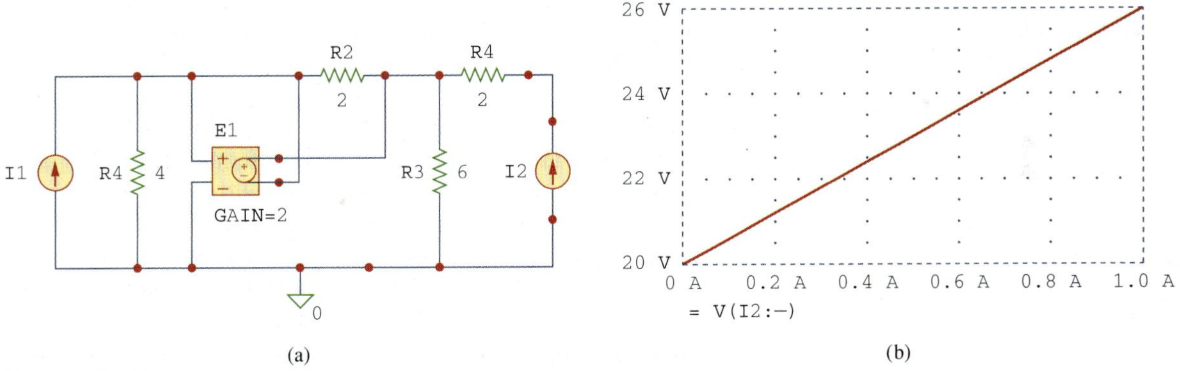

Figure 4.53
For Example 4.14: (a) schematic and (b) plot for finding R_{Th} and V_{Th}.

(b) To find the Norton equivalent, we modify the schematic in Fig. 4.53(a) by replaying the probing current source with a probing voltage source V1. The result is the schematic in Fig. 4.54(a). Again, in the DC Sweep dialog box, we select Linear for the *Sweep Type* and Voltage Source for the *Sweep Var. Type.* We enter V1 under *Name* box, 0 as *Start Value*, 1 as *End Value*, and 0.1 as *Increment.* Under the *PSpice* A/D Window, we add trace I (V1) and obtain the plot in Fig. 4.54(b). From the plot, we obtain

$$I_N = \text{Zero intercept} = 3.335 \text{ A}$$

$$G_N = \text{Slope} = \frac{3.335 - 3.165}{1} = 0.17 \text{ S}$$

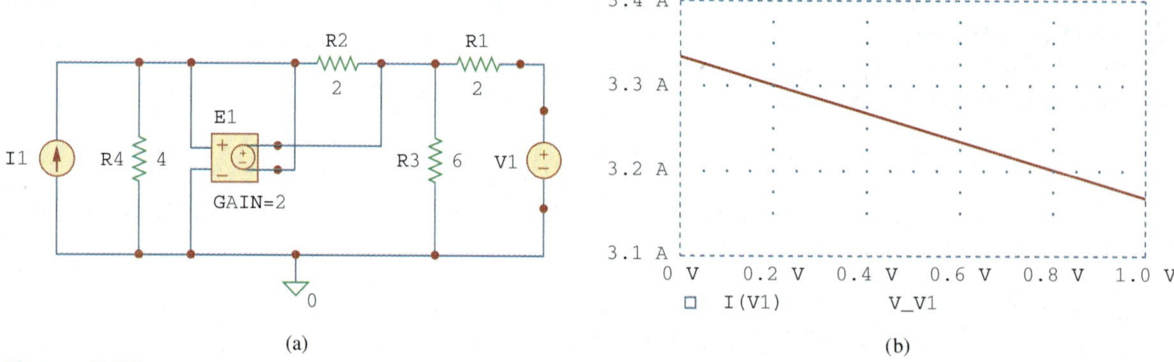

Figure 4.54
For Example 4.14: (a) schematic and (b) plot for finding G_N and I_N.

Practice Problem 4.14

Rework Practice Prob. 4.9 using *PSpice.*

Answer: $V_{\text{Th}} = 5.33$ V, $R_{\text{Th}} = 0.44\ \Omega$.

Example 4.15

Refer to the circuit in Fig. 4.55. Use *PSpice* to find the maximum power transfer to R_L.

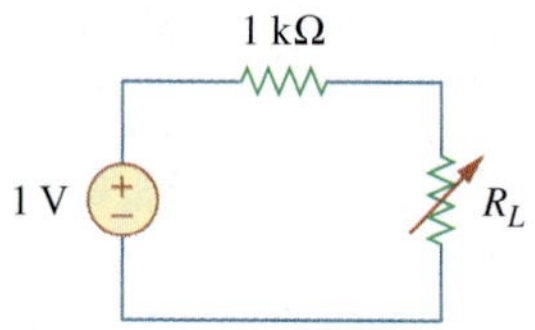

Figure 4.55
For Example 4.15.

Solution:

We need to perform a DC Sweep on R_L to determine when the power across it is maximum. We first draw the circuit using Schematics as shown in Fig. 4.56. Once the circuit is drawn, we take the following three steps to further prepare the circuit for a DC Sweep.

The first step involves defining the value of R_L as a parameter, since we want to vary it. To do this:

1. **DCLICKL** the value 1k of R2 (representing R_L) to open up the *Set Attribute Value* dialog box.
2. Replace 1k with {RL} and click **OK** to accept the change.

Note that the curly brackets are necessary.

The second step is to define parameter. To achieve this:

1. Select **Draw/Get New Part/Libraries** ···**/special.slb**.
2. Type PARAM in the *PartName* box and click **OK**.
3. **DRAG** the box to any position near the circuit.
4. **CLICKL** to end placement mode.
5. **DCLICKL** to open up the *PartName: PARAM* dialog box.
6. **CLICKL** on *NAME1* = and enter RL (with no curly brackets) in the *Value* box, and **CLICKL Save Attr** to accept change.
7. **CLICKL** on *VALUE1* = and enter 2k in the *Value* box, and **CLICKL Save Attr** to accept change.
8. Click **OK**.

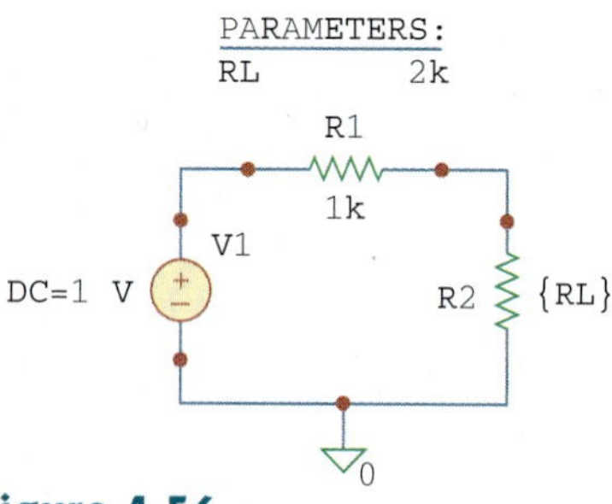

Figure 4.56
Schematic for the circuit in Fig. 4.55.

The value 2k in item 7 is necessary for a bias point calculation; it cannot be left blank.

The third step is to set up the DC Sweep to sweep the parameter. To do this:

1. Select **Analysis/Setput** to bring up the DC Sweep dialog box.
2. For the *Sweep Type*, select Linear (or Octave for a wide range of R_L).
3. For the *Sweep Var. Type*, select Global Parameter.
4. Under the *Name* box, enter RL.
5. In the *Start Value* box, enter 100.
6. In the *End Value* box, enter 5k.
7. In the *Increment* box, enter 100.
8. Click **OK** and **Close** to accept the parameters.

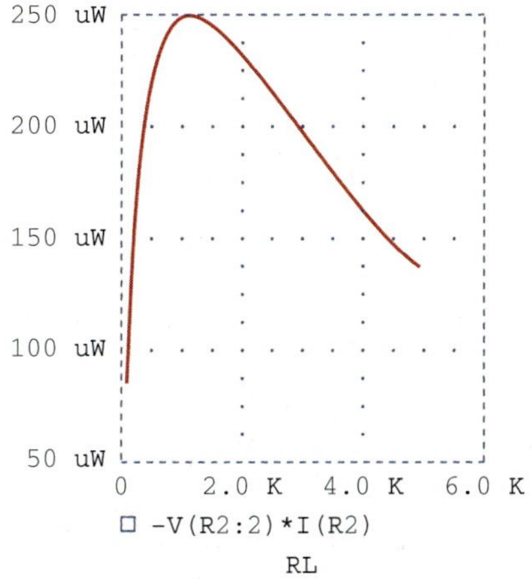

Figure 4.57
For Example 4.15: the plot of power across R_L.

After taking these steps and saving the circuit, we are ready to simulate. Select **Analysis/Simulate**. If there are no errors, we select **Add Trace** in the *PSpice* A/D window and type −V(R2:2)*I(R2) in the *Trace Command* box. [The negative sign is needed since I(R2) is negative.] This gives the plot of the power delivered to R_L as R_L varies from 100 Ω to 5 kΩ. We can also obtain the power absorbed by R_L by typing V(R2:2)*V(R2:2)/RL in the *Trace Command* box. Either way, we obtain the plot in Fig. 4.57. It is evident from the plot that the maximum power is 250 μW. Notice that the maximum occurs when $R_L = 1$ kΩ, as expected analytically.

Practice Problem 4.15

Find the maximum power transferred to R_L if the 1-kΩ resistor in Fig. 4.55 is replaced by a 2-kΩ resistor.

Answer: 125 μW.

4.10 †Applications

In this section we will discuss two important practical applications of the concepts covered in this chapter: source modeling and resistance measurement.

4.10.1 Source Modeling

Source modeling provides an example of the usefulness of the Thevenin or the Norton equivalent. An active source such as a battery is often characterized by its Thevenin or Norton equivalent circuit. An ideal voltage source provides a constant voltage irrespective of the current drawn by the load, while an ideal current source supplies a constant current regardless of the load voltage. As Fig. 4.58 shows, practical voltage and current sources are not ideal, due to their *internal resistances* or *source resistances* R_s and R_p. They become ideal as $R_s \rightarrow 0$ and $R_p \rightarrow \infty$. To show that this is the case, consider the effect

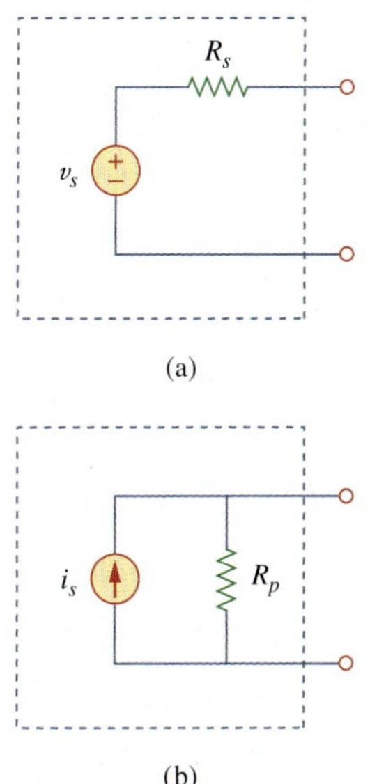

Figure 4.58
(a) Practical voltage source, (b) practical current source.

of the load on voltage sources, as shown in Fig. 4.59(a). By the voltage division principle, the load voltage is

$$v_L = \frac{R_L}{R_s + R_L} v_s \tag{4.25}$$

As R_L increases, the load voltage approaches a source voltage v_s, as illustrated in Fig. 4.59(b). From Eq. (4.25), we should note that:

1. The load voltage will be constant if the internal resistance R_s of the source is zero or, at least, $R_s \ll R_L$. In other words, the smaller R_s is compared with R_L, the closer the voltage source is to being ideal.

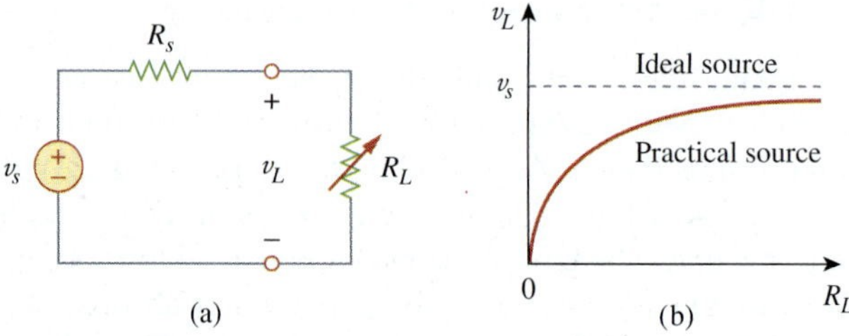

Figure 4.59
(a) Practical voltage source connected to a load R_L, (b) load voltage decreases as R_L decreases.

2. When the load is disconnected (i.e., the source is open-circuited so that $R_L \to \infty$), $v_{oc} = v_s$. Thus, v_s may be regarded as the *unloaded source* voltage. The connection of the load causes the terminal voltage to drop in magnitude; this is known as the *loading effect.*

The same argument can be made for a practical current source when connected to a load as shown in Fig. 4.60(a). By the current division principle,

$$i_L = \frac{R_p}{R_p + R_L} i_s \tag{4.26}$$

Figure 4.60(b) shows the variation in the load current as the load resistance increases. Again, we notice a drop in current due to the load (loading effect), and load current is constant (ideal current source) when the internal resistance is very large (i.e., $R_p \to \infty$ or, at least, $R_p \gg R_L$).

Sometimes, we need to know the unloaded source voltage v_s and the internal resistance R_s of a voltage source. To find v_s and R_s, we follow the procedure illustrated in Fig. 4.61. First, we measure the open-circuit voltage v_{oc} as in Fig. 4.61(a) and set

$$v_s = v_{oc} \tag{4.27}$$

Then, we connect a variable load R_L across the terminals as in Fig. 4.61(b). We adjust the resistance R_L until we measure a load voltage of exactly one-half of the open-circuit voltage, $v_L = v_{oc}/2$, because now $R_L = R_{\text{Th}} = R_s$. At that point, we disconnect R_L and measure it. We set

$$R_s = R_L \tag{4.28}$$

For example, a car battery may have $v_s = 12$ V and $R_s = 0.05\ \Omega$.

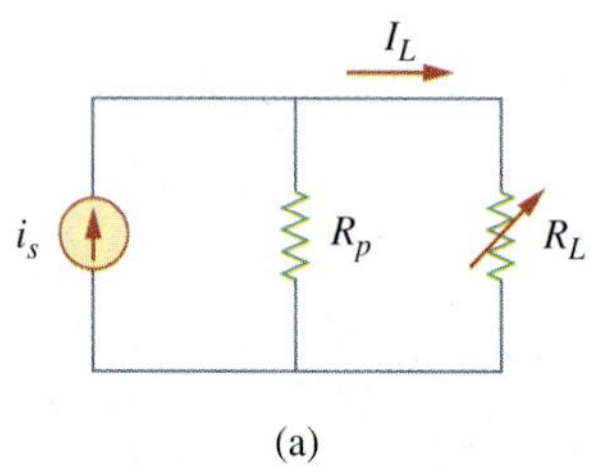

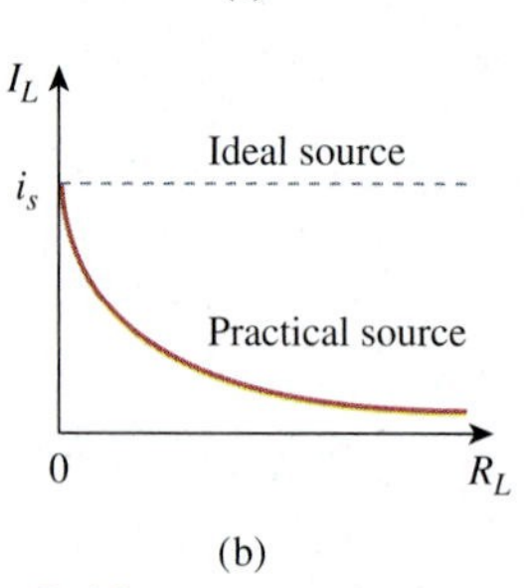

Figure 4.60
(a) Practical current source connected to a load R_L, (b) load current decreases as R_L increases.

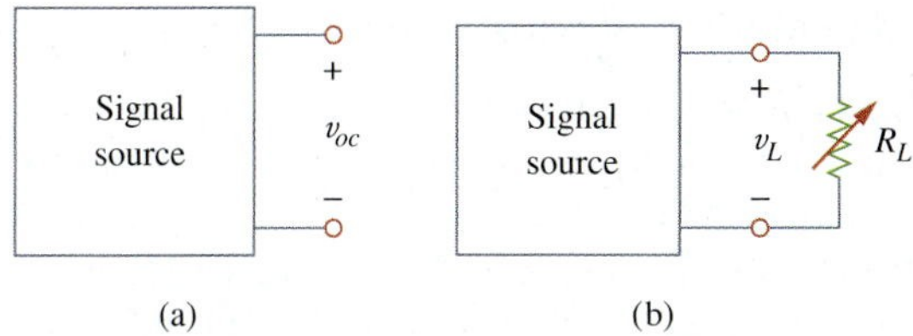

Figure 4.61
(a) Measuring v_{oc}, (b) measuring v_L.

Example 4.16

The terminal voltage of a voltage source is 12 V when connected to a 2-W load. When the load is disconnected, the terminal voltage rises to 12.4 V. (a) Calculate the source voltage v_s and internal resistance R_s. (b) Determine the voltage when an 8-Ω load is connected to the source.

Solution:

(a) We replace the source by its Thevenin equivalent. The terminal voltage when the load is disconnected is the open-circuit voltage,

$$v_s = v_{oc} = 12.4 \text{ V}$$

When the load is connected, as shown in Fig. 4.62(a), $v_L = 12$ V and $p_L = 2$ W. Hence,

$$p_L = \frac{v_L^2}{R_L} \quad \Rightarrow \quad R_L = \frac{v_L^2}{p_L} = \frac{12^2}{2} = 72 \ \Omega$$

The load current is

$$i_L = \frac{v_L}{R_L} = \frac{12}{72} = \frac{1}{6} \text{ A}$$

The voltage across R_s is the difference between the source voltage v_s and the load voltage v_L, or

$$12.4 - 12 = 0.4 = R_s i_L, \qquad R_s = \frac{0.4}{I_L} = 2.4 \ \Omega$$

(b) Now that we have the Thevenin equivalent of the source, we connect the 8-Ω load across the Thevenin equivalent as shown in Fig. 4.62(b). Using voltage division, we obtain

$$v = \frac{8}{8 + 2.4}(12.4) = 9.538 \text{ V}$$

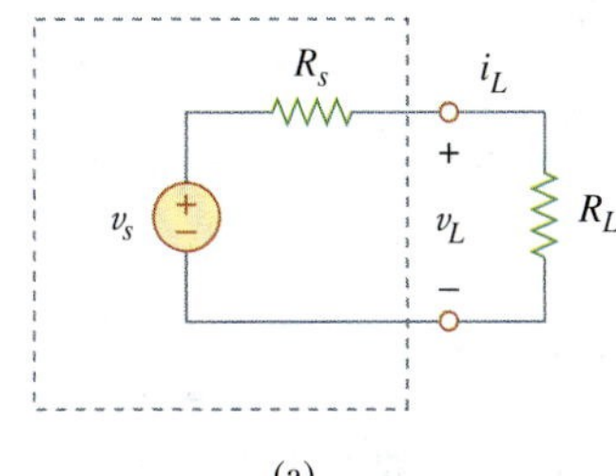

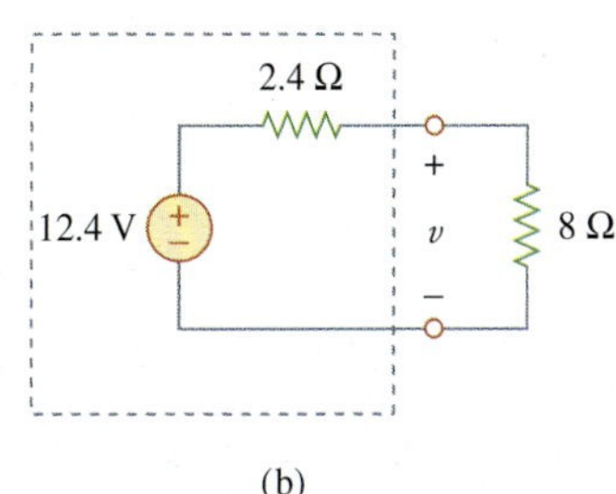

Figure 4.62
For Example 4.16.

Practice Problem 4.16

The measured open-circuit voltage across a certain amplifier is 9 V. The voltage drops to 8 V when a 20-Ω loudspeaker is connected to the amplifier. Calculate the voltage when a 10-Ω loudspeaker is used instead.

Answer: 7.2 V.

4.10.2 Resistance Measurement

Although the ohmmeter method provides the simplest way to measure resistance, more accurate measurement may be obtained using the Wheatstone bridge. While ohmmeters are designed to measure resistance in low, mid, or high range, a Wheatstone bridge is used to measure resistance in the mid range, say, between 1 Ω and 1 MΩ. Very low values of resistances are measured with a *milliohmmeter*, while very high values are measured with a *Megger tester.*

Historical note: The bridge was invented by Charles Wheatstone (1802–1875), a British professor who also invented the telegraph, as Samuel Morse did independently in the United States.

The Wheatstone bridge (or resistance bridge) circuit is used in a number of applications. Here we will use it to measure an unknown resistance. The unknown resistance R_x is connected to the bridge as shown in Fig. 4.63. The variable resistance is adjusted until no current flows through the galvanometer, which is essentially a d'Arsonval movement operating as a sensitive current-indicating device like an ammeter in the microamp range. Under this condition $v_1 = v_2$, and the bridge is said to be *balanced.* Since no current flows through the galvanometer, R_1 and R_2 behave as though they were in series; so do R_3 and R_x. The fact that no current flows through the galvanometer also implies that $v_1 = v_2$. Applying the voltage division principle,

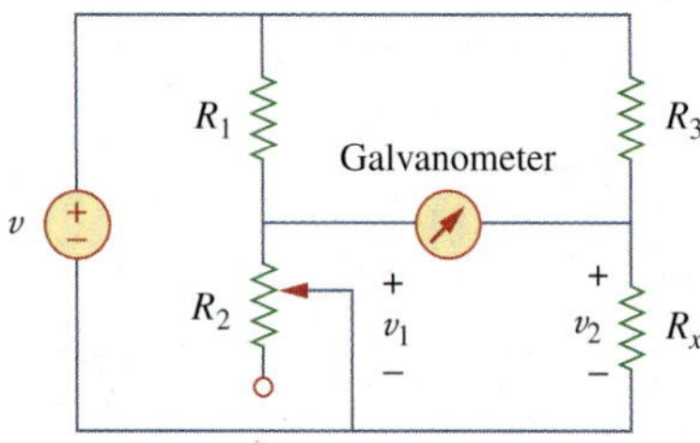

Figure 4.63
The Wheatstone bridge; R_x is the resistance to be measured.

$$v_1 = \frac{R_2}{R_1 + R_2}v = v_2 = \frac{R_x}{R_3 + R_x}v \tag{4.29}$$

Hence, no current flows through the galvanometer when

$$\frac{R_2}{R_1 + R_2} = \frac{R_x}{R_3 + R_x} \quad \Rightarrow \quad R_2R_3 = R_1R_x$$

or

$$\boxed{R_x = \frac{R_3}{R_1}R_2} \tag{4.30}$$

If $R_1 = R_3$, and R_2 is adjusted until no current flows through the galvanometer, then $R_x = R_2$.

How do we find the current through the galvanometer when the Wheatstone bridge is *unbalanced*? We find the Thevenin equivalent (V_{Th} and R_{Th}) with respect to the galvanometer terminals. If R_m is the resistance of the galvanometer, the current through it under the unbalanced condition is

$$I = \frac{V_{\text{Th}}}{R_{\text{Th}} + R_m} \tag{4.31}$$

Example 4.18 will illustrate this.

Example 4.17

In Fig. 4.63, $R_1 = 500\ \Omega$ and $R_3 = 200\ \Omega$. The bridge is balanced when R_2 is adjusted to be 125 Ω. Determine the unknown resistance R_x.

Solution:

Using Eq. (4.30) gives

$$R_x = \frac{R_3}{R_1}R_2 = \frac{200}{500}125 = 50\ \Omega$$

Practice Problem 4.17

A Wheatstone bridge has $R_1 = R_3 = 1\ \text{k}\Omega$. R_2 is adjusted until no current flows through the galvanometer. At that point, $R_2 = 3.2\ \text{k}\Omega$. What is the value of the unknown resistance?

Answer: 3.2 kΩ.

Example 4.18

The circuit in Fig. 4.64 represents an unbalanced bridge. If the galvanometer has a resistance of 40 Ω, find the current through the galvanometer.

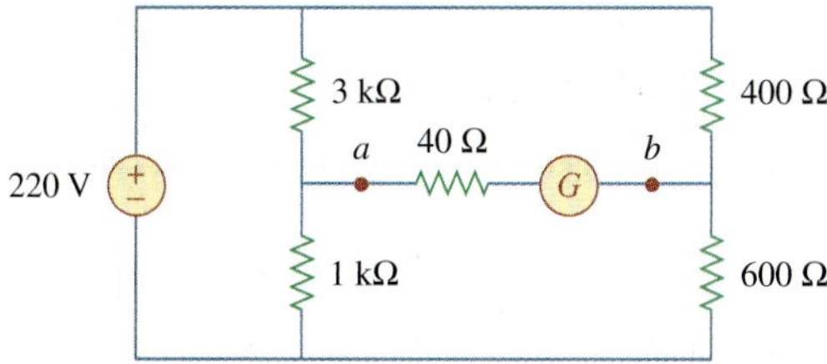

Figure 4.64
Unbalanced bridge of Example 4.18.

Solution:
We first need to replace the circuit by its Thevenin equivalent at terminals a and b. The Thevenin resistance is found using the circuit in Fig. 4.65(a). Notice that the 3-kΩ and 1-kΩ resistors are in parallel; so are the 400-Ω and 600-Ω resistors. The two parallel combinations form a series combination with respect to terminals a and b. Hence,

$$R_{\text{Th}} = 3000 \,\|\, 1000 + 400 \,\|\, 600$$
$$= \frac{3000 \times 1000}{3000 + 1000} + \frac{400 \times 600}{400 + 600} = 750 + 240 = 990\ \Omega$$

To find the Thevenin voltage, we consider the circuit in Fig. 4.65(b). Using the voltage division principle gives

$$v_1 = \frac{1000}{1000 + 3000}(220) = 55\text{ V}, \qquad v_2 = \frac{600}{600 + 400}(220) = 132\text{ V}$$

Applying KVL around loop ab gives

$$-v_1 + V_{\text{Th}} + v_2 = 0 \qquad \text{or} \qquad V_{\text{Th}} = v_1 - v_2 = 55 - 132 = -77\text{ V}$$

Having determined the Thevenin equivalent, we find the current through the galvanometer using Fig. 4.65(c).

$$I_G = \frac{V_{\text{Th}}}{R_{\text{Th}} + R_m} = \frac{-77}{990 + 40} = -74.76\text{ mA}$$

The negative sign indicates that the current flows in the direction opposite to the one assumed, that is, from terminal b to terminal a.

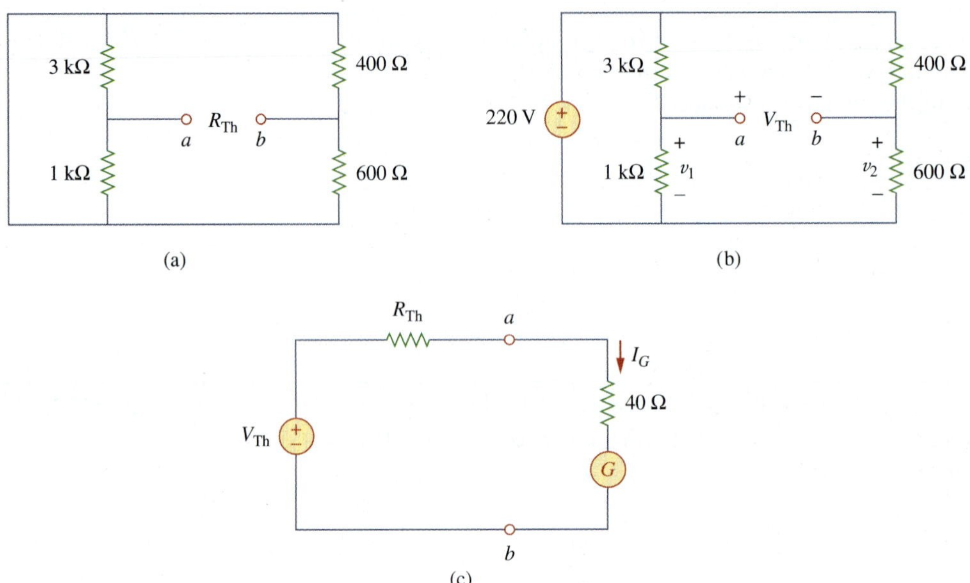

Figure 4.65
For Example 4.18: (a) Finding R_{Th}, (b) finding V_{Th}, (c) determining the current through the galvanometer.

Practice Problem 4.18

Obtain the current through the galvanometer, having a resistance of 14 Ω, in the Wheatstone bridge shown in Fig. 4.66.

Answer: 64 mA.

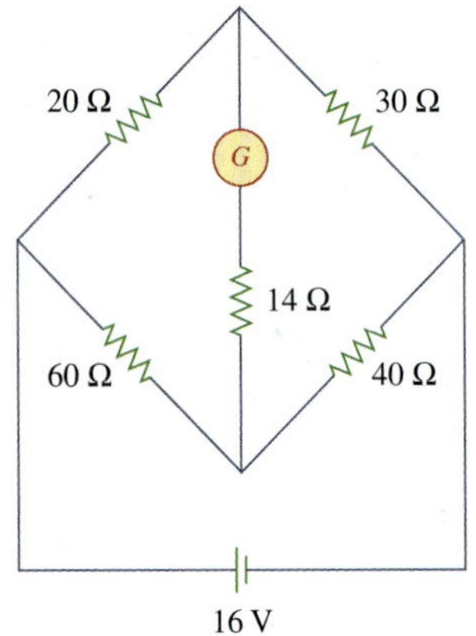

Figure 4.66
For Practice Prob. 4.18.

4.11 Summary

1. A linear network consists of linear elements, linear dependent sources, and linear independent sources.
2. Network theorems are used to reduce a complex circuit to a simpler one, thereby making circuit analysis much simpler.
3. The superposition principle states that for a circuit having multiple independent sources, the voltage across (or current through) an element is equal to the algebraic sum of all the individual voltages (or currents) due to each independent source acting one at a time.
4. Source transformation is a procedure for transforming a voltage source in series with a resistor to a current source in parallel with a resistor, or vice versa.
5. Thevenin's and Norton's theorems allow us to isolate a portion of a network while the remaining portion of the network is replaced by an equivalent network. The Thevenin equivalent consists of a voltage source V_{Th} in series with a resistor R_{Th}, while the Norton equivalent consists of a current source I_N in parallel with a resistor R_N. The two theorems are related by source transformation.

$$R_N = R_{Th}, \qquad I_N = \frac{V_{Th}}{R_{Th}}$$

6. For a given Thevenin equivalent circuit, maximum power transfer occurs when $R_L = R_{Th}$; that is, when the load resistance is equal to the Thevenin resistance.
7. The maximum power transfer theorem states that the maximum power is delivered by a source to the load R_L when R_L is equal to R_{Th}, the Thevenin resistance at the terminals of the load.
8. *PSpice* can be used to verify the circuit theorems covered in this chapter.
9. Source modeling and resistance measurement using the Wheatstone bridge provide applications for Thevenin's theorem.

Review Questions

4.1 The current through a branch in a linear network is 2 A when the input source voltage is 10 V. If the voltage is reduced to 1 V and the polarity is reversed, the current through the branch is:

(a) −2 A (b) −0.2 A (c) 0.2 A
(d) 2 A (e) 20 A

4.2 For superposition, it is not required that only one independent source be considered at a time; any number of independent sources may be considered simultaneously.

(a) True (b) False

4.3 The superposition principle applies to power calculation.

(a) True (b) False

4.4 Refer to Fig. 4.67. The Thevenin resistance at terminals *a* and *b* is:

(a) 25 Ω (b) 20 Ω
(c) 5 Ω (d) 4 Ω

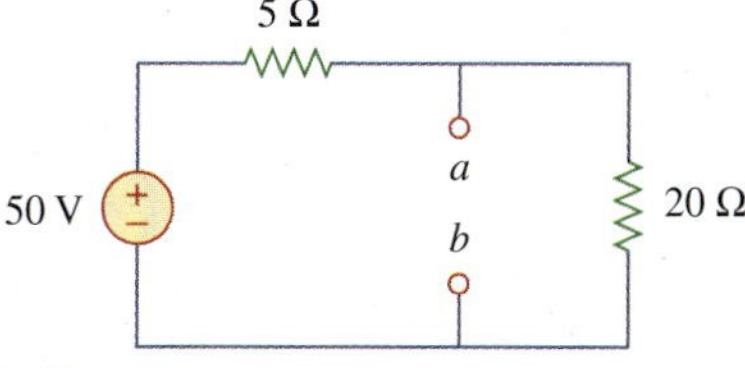

Figure 4.67
For Review Questions 4.4 to 4.6.

4.5 The Thevenin voltage across terminals *a* and *b* of the circuit in Fig. 4.67 is:

(a) 50 V (b) 40 V
(c) 20 V (d) 10 V

4.6 The Norton current at terminals *a* and *b* of the circuit in Fig. 4.67 is:

(a) 10 A (b) 2.5 A
(c) 2 A (d) 0 A

4.7 The Norton resistance R_N is exactly equal to the Thevenin resistance R_{Th}.

(a) True (b) False

4.8 Which pair of circuits in Fig. 4.68 are equivalent?

(a) a and b (b) b and d
(c) a and c (d) c and d

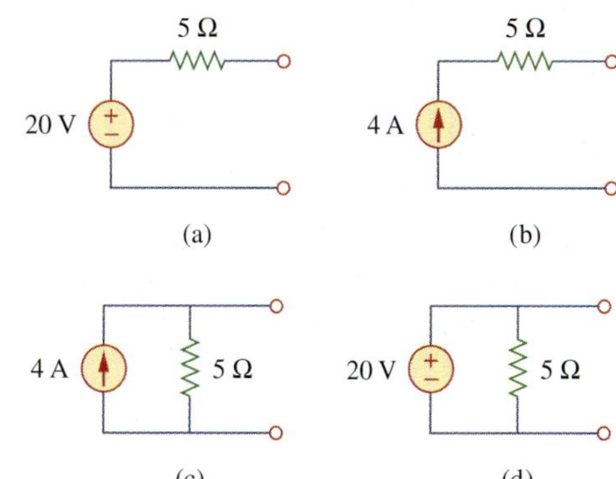

Figure 4.68
For Review Question 4.8.

4.9 A load is connected to a network. At the terminals to which the load is connected, $R_{Th} = 10\ \Omega$ and $V_{Th} = 40$ V. The maximum possible power supplied to the load is:

(a) 160 W (b) 80 W
(c) 40 W (d) 1 W

4.10 The source is supplying the maximum power to the load when the load resistance equals the source resistance.

(a) True (b) False

Answers: 4.1b, 4.2a, 4.3b, 4.4d, 4.5b, 4.6a, 4.7a, 4.8c, 4.9c, 4.10a.

Problems

Section 4.2 Linearity Property

4.1 Calculate the current i_o in the current of Fig. 4.69. What does this current become when the input voltage is raised to 10 V?

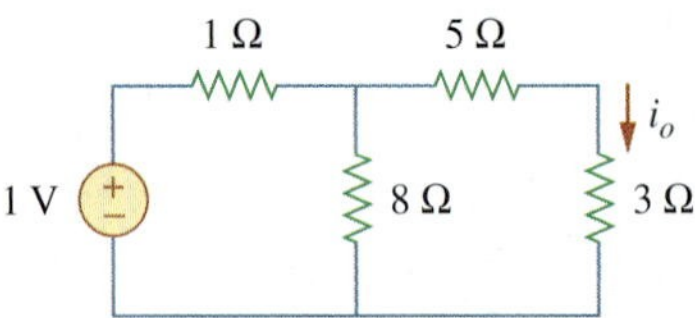

Figure 4.69
For Prob. 4.1.

4.2 Using Fig. 4.70, design a problem to help other students better understand linearity.

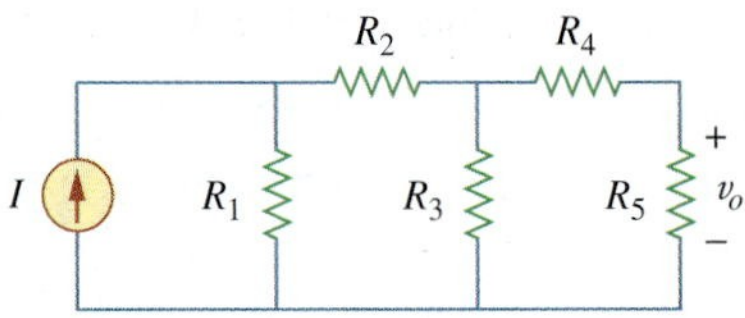

Figure 4.70
For Prob. 4.2.

4.3 (a) In the circuit of Fig. 4.71, calculate v_o and i_o when $v_s = 1$ V.

(b) Find v_o and i_o when $v_s = 10$ V.

(c) What are v_o and i_o when each of the 1-Ω resistors is replaced by a 10-Ω resistor and $v_s = 10$ V?

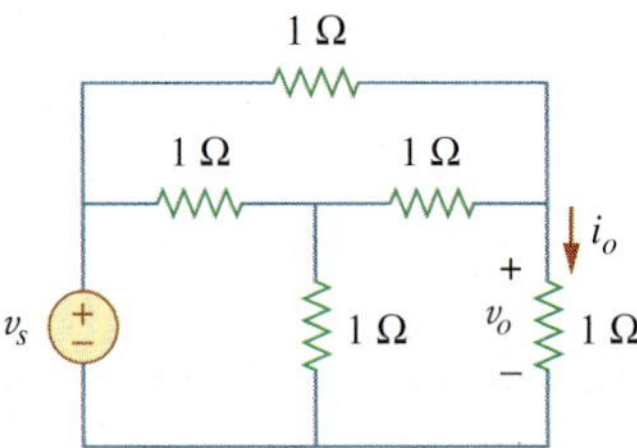

Figure 4.71
For Prob. 4.3.

4.4 Use linearity to determine i_o in the circuit of Fig. 4.72.

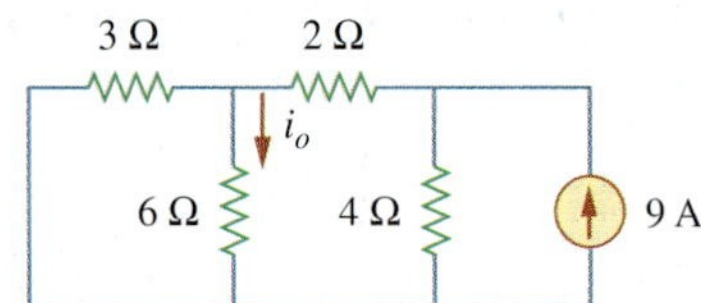

Figure 4.72
For Prob. 4.4.

4.5 For the circuit in Fig. 4.73, assume $v_o = 1$ V, and use linearity to find the actual value of v_o.

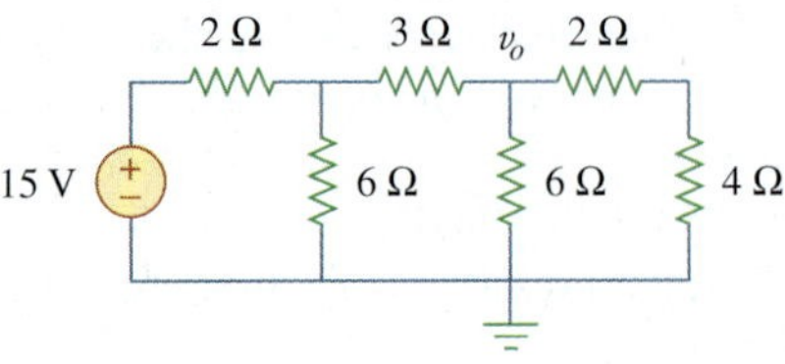

Figure 4.73
For Prob. 4.5.

4.6 For the linear circuit shown in Fig. 4.74, use linearity to complete the following table.

Experiment	V_s	V_o
1	12 V	4 V
2		16 V
3	1 V	
4		−2 V

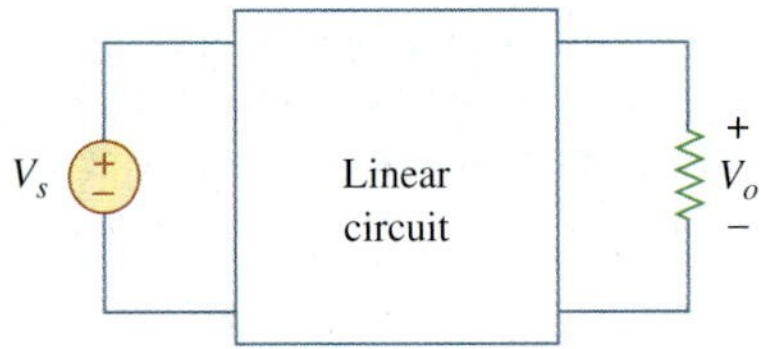

Figure 4.74
For Prob. 4.6.

4.7 Use linearity and the assumption that $V_o = 1$ V to find the actual value of V_o in Fig. 4.75.

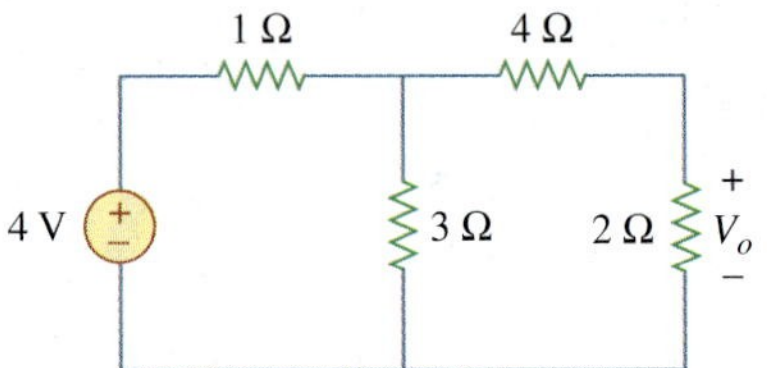

Figure 4.75
For Prob. 4.7.

Section 4.3 Superposition

4.8 Using superposition, find V_o in the circuit of Fig. 4.76. Check with *PSpice*.

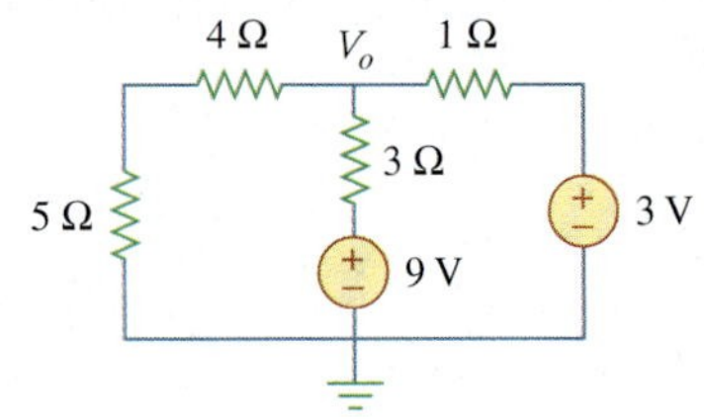

Figure 4.76
For Prob. 4.8.

4.9 Use superposition to find v_o in the circuit of Fig. 4.77.

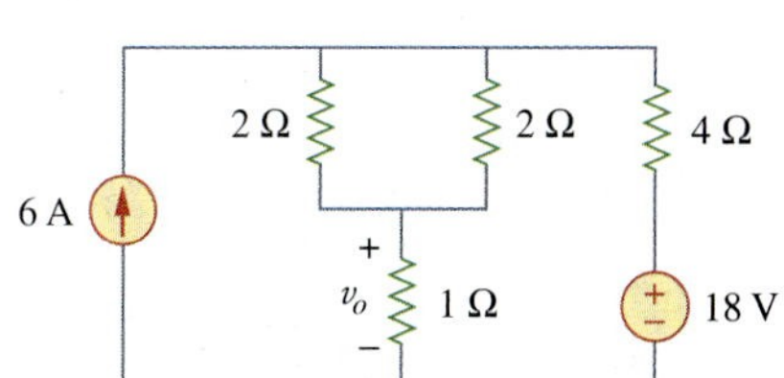

Figure 4.77
For Prob. 4.9.

4.10 Using Fig. 4.78, design a problem to help other students better understand superposition. Note, the letter k is a gain you can specify to make the problem easier to solve but must not be zero.

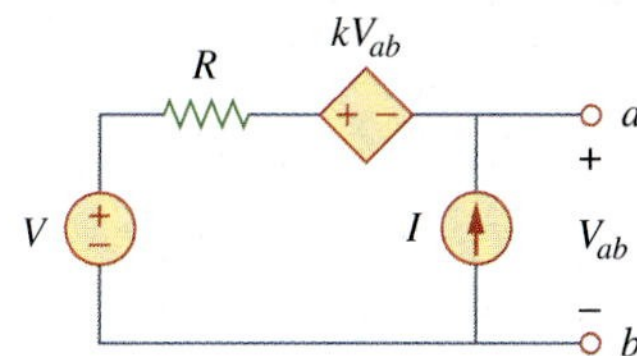

Figure 4.78
For Prob. 4.10.

4.11 Use the superposition principle to find i_o and v_o in the circuit of Fig. 4.79.

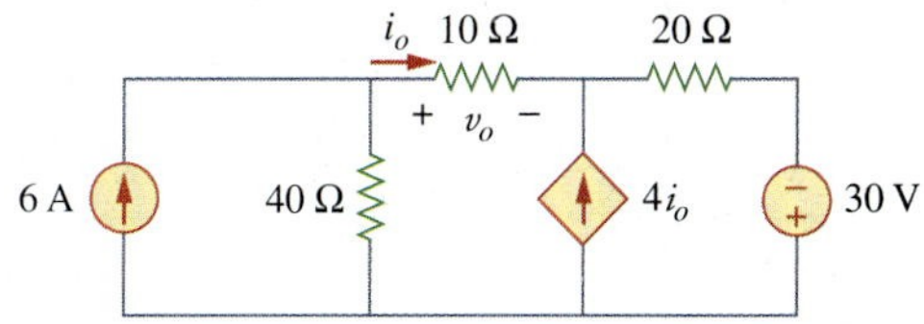

Figure 4.79
For Prob. 4.11.

4.12 Determine v_o in the circuit of Fig. 4.80 using the superposition principle.

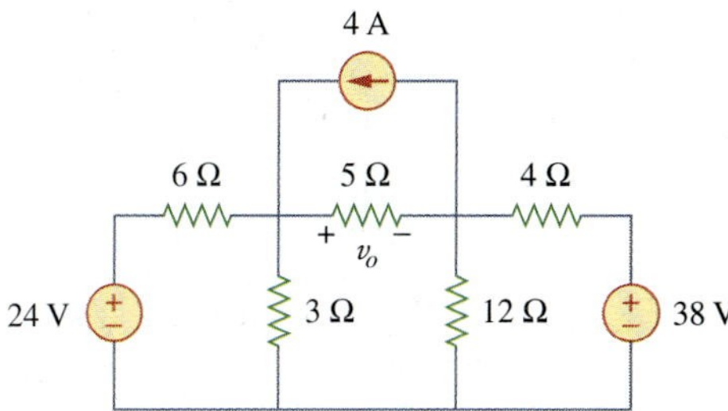

Figure 4.80
For Probs. 4.12 and 4.35.

4.13 Use superposition to find v_o in the circuit of Fig. 4.81.

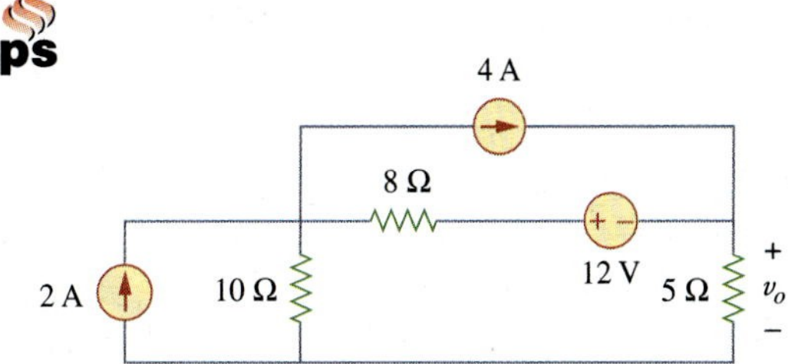

Figure 4.81
For Prob. 4.13.

4.14 Apply the superposition principle to find v_o in the circuit of Fig. 4.82.

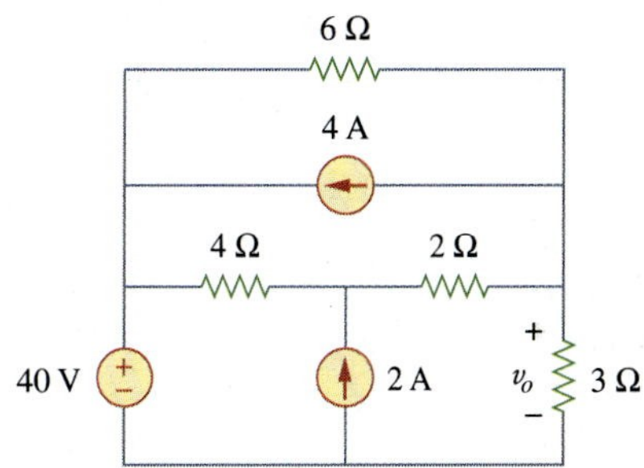

Figure 4.82
For Prob. 4.14.

4.15 For the circuit in Fig. 4.83, use superposition to find i. Calculate the power delivered to the 3-Ω resistor.

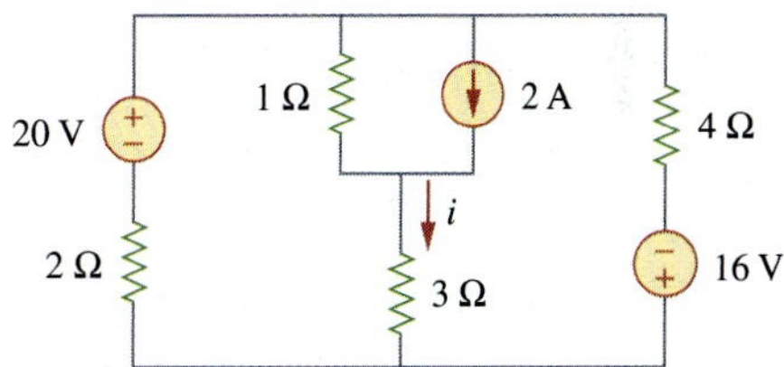

Figure 4.83
For Probs. 4.15 and 4.56.

4.16 Given the circuit in Fig. 4.84, use superposition to get i_o.

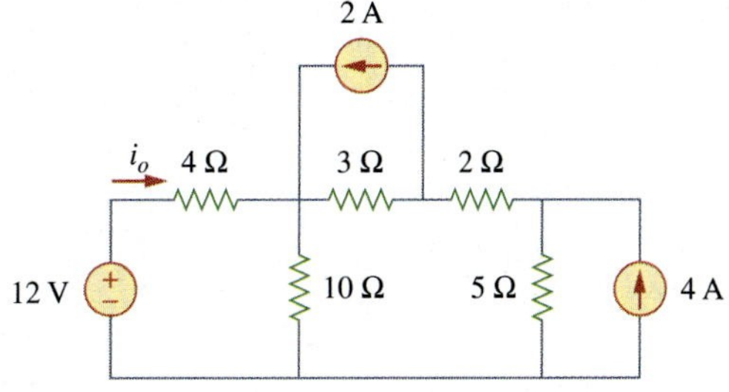

Figure 4.84
For Prob. 4.16.

4.17 Use superposition to obtain v_x in the circuit of Fig. 4.85. Check your result using *PSpice*.

ps ML

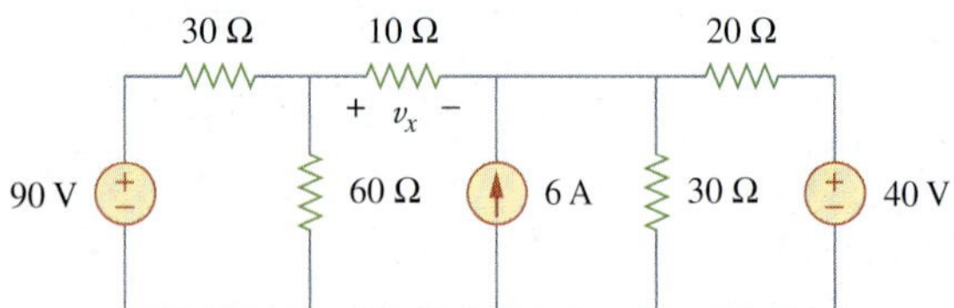

Figure 4.85
For Prob. 4.17.

4.18 Use superposition to find V_o in the circuit of Fig. 4.86.

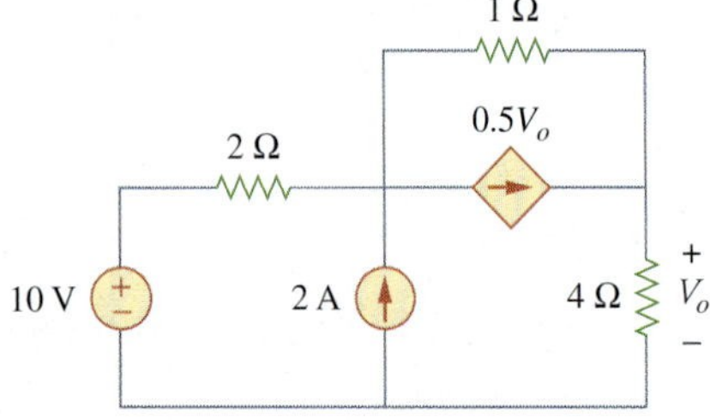

Figure 4.86
For Prob. 4.18.

4.19 Use superposition to solve for v_x in the circuit of Fig. 4.87.

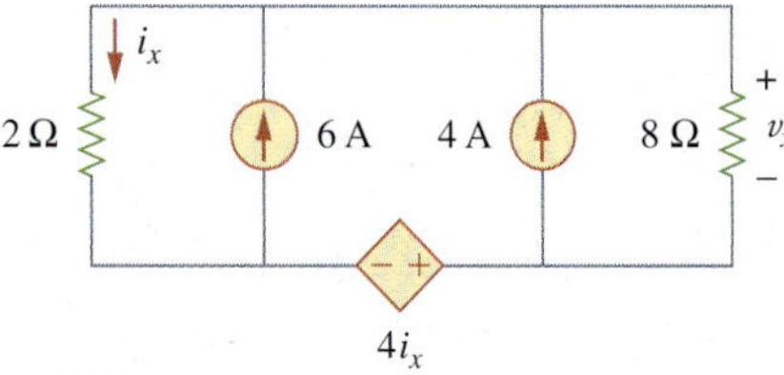

Figure 4.87
For Prob. 4.19.

Section 4.4 Source Transformation

4.20 Use source transformations to reduce the circuit in Fig. 4.88 to a single voltage source in series with a single resistor.

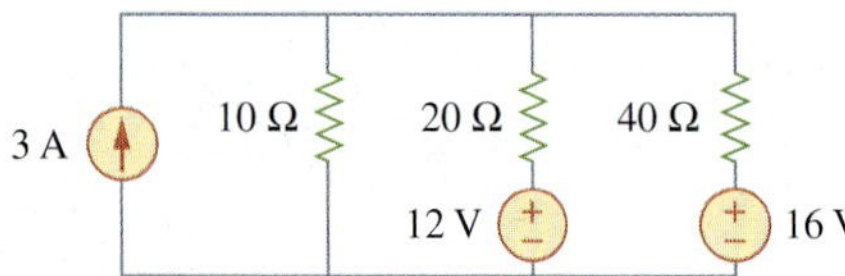

Figure 4.88
For Prob. 4.20.

4.21 Using Fig. 4.89, design a problem to help other students better understand source transformation.

ed

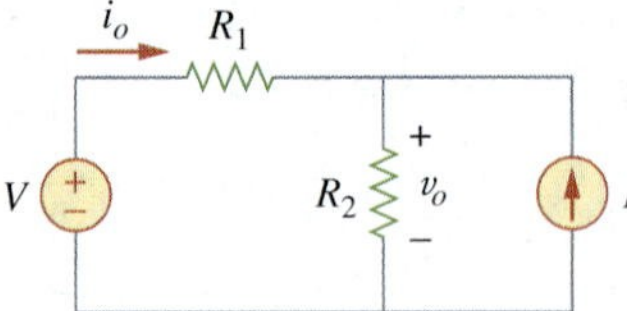

Figure 4.89
For Prob. 4.21.

4.22 For the circuit in Fig. 4.90, use source transformation to find i.

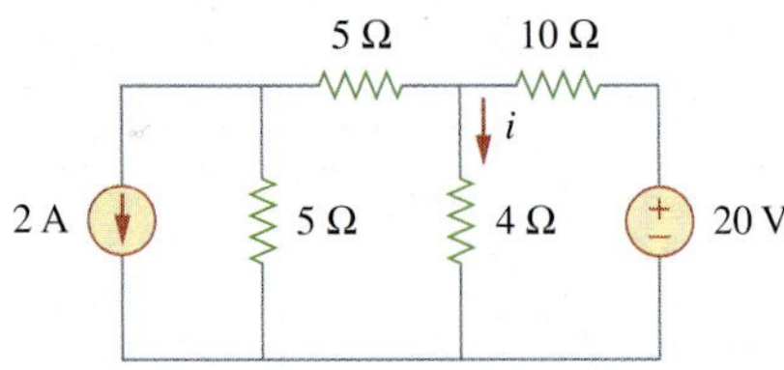

Figure 4.90
For Prob. 4.22.

4.23 Referring to Fig. 4.91, use source transformation to determine the current and power in the 8-Ω resistor.

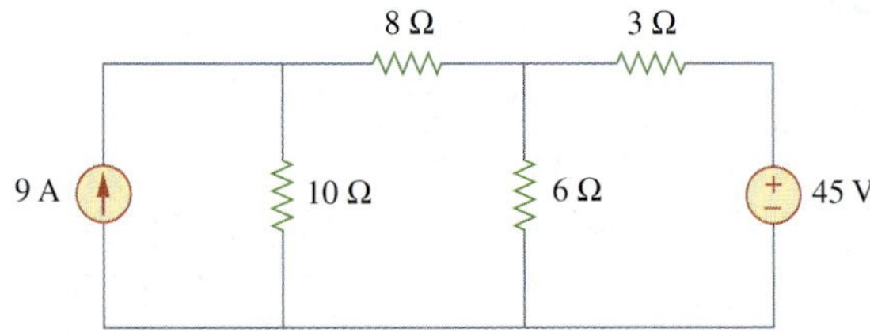

Figure 4.91
For Prob. 4.23.

4.24 Use source transformation to find the voltage V_x in the circuit of Fig. 4.92.

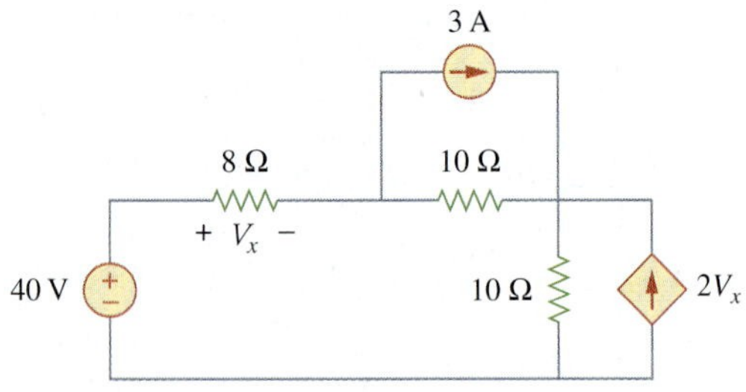

Figure 4.92
For Prob. 4.24.

4.25 Obtain v_o in the circuit of Fig. 4.93 using source transformation. Check your result using *PSpice*.

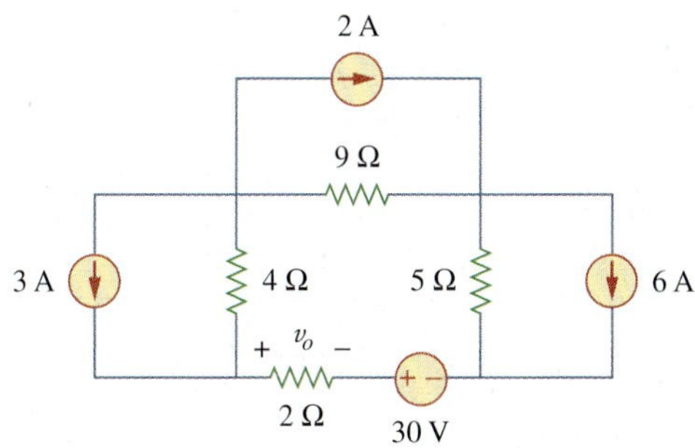

Figure 4.93
For Prob. 4.25.

4.26 Use source transformation to find i_o in the circuit of Fig. 4.94.

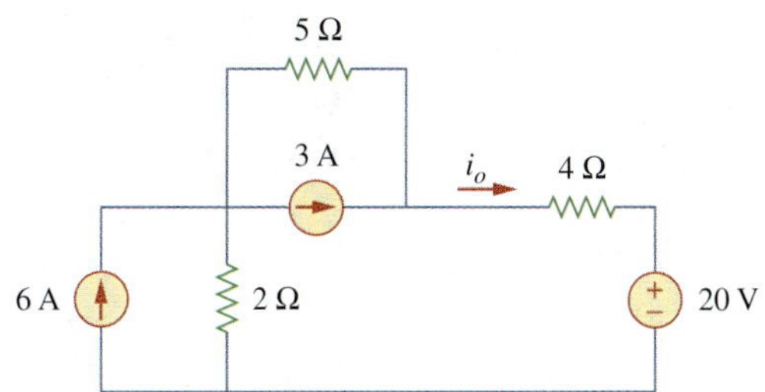

Figure 4.94
For Prob. 4.26.

4.27 Apply source transformation to find v_x in the circuit of Fig. 4.95.

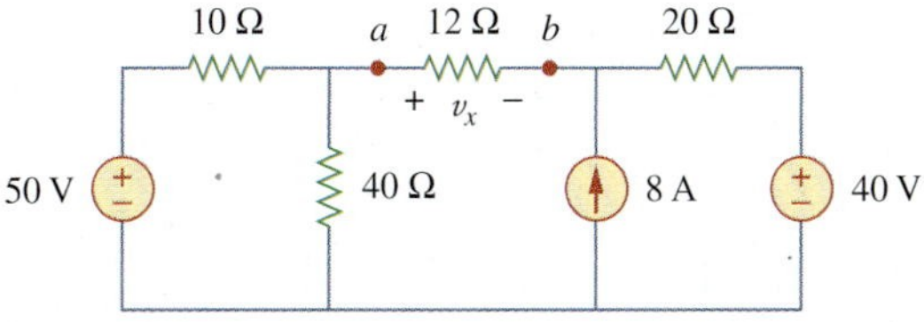

Figure 4.95
For Probs. 4.27 and 4.40.

4.28 Use source transformation to find I_o in Fig. 4.96.

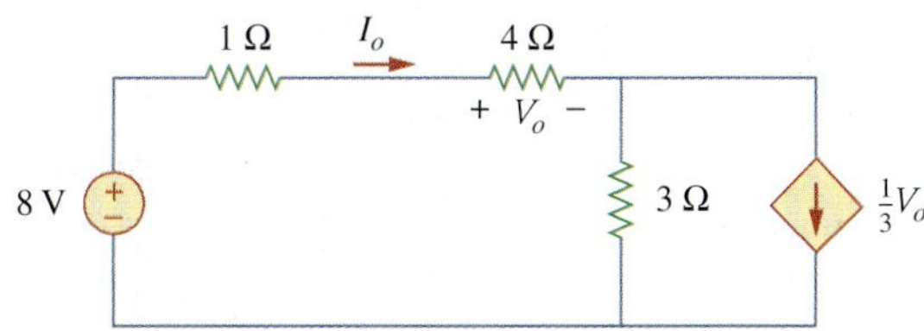

Figure 4.96
For Prob. 4.28.

4.29 Use source transformation to find v_o in the circuit of Fig. 4.97.

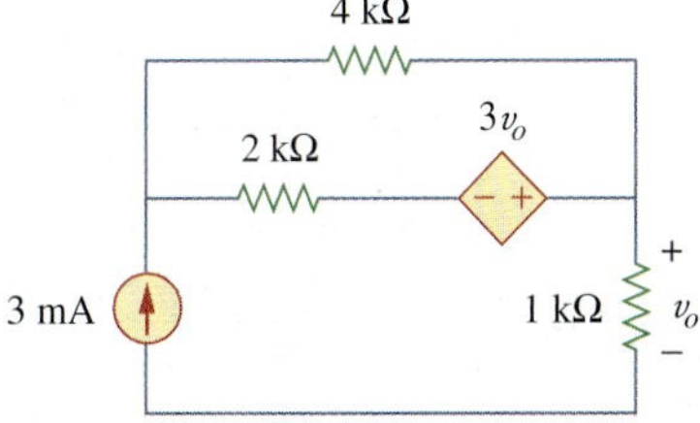

Figure 4.97
For Prob. 4.29.

4.30 Use source transformation on the circuit shown in Fig 4.98 to find i_x.

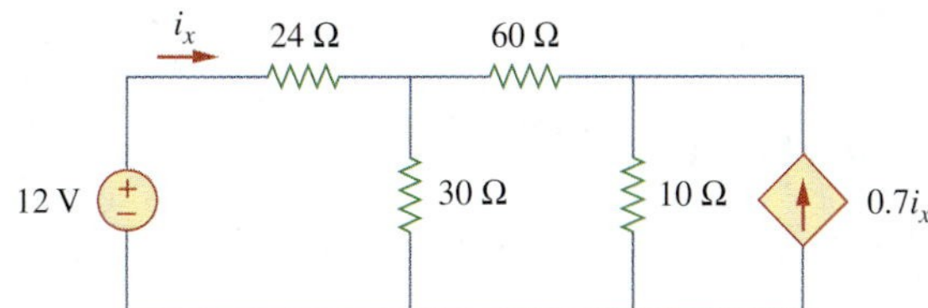

Figure 4.98
For Prob. 4.30.

4.31 Determine v_x in the circuit of Fig. 4.99 using source transformation.

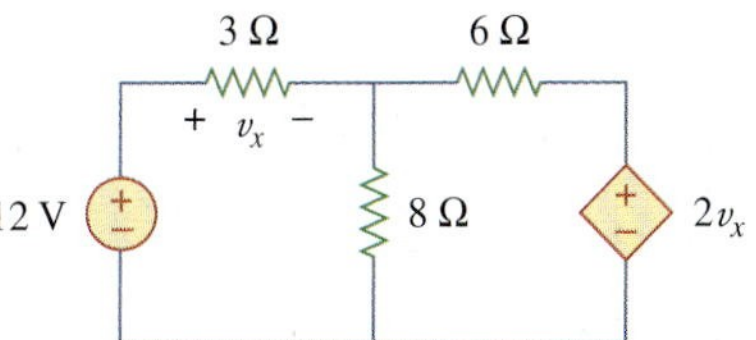

Figure 4.99
For Prob. 4.31.

4.32 Use source transformation to find i_x in the circuit of Fig. 4.100.

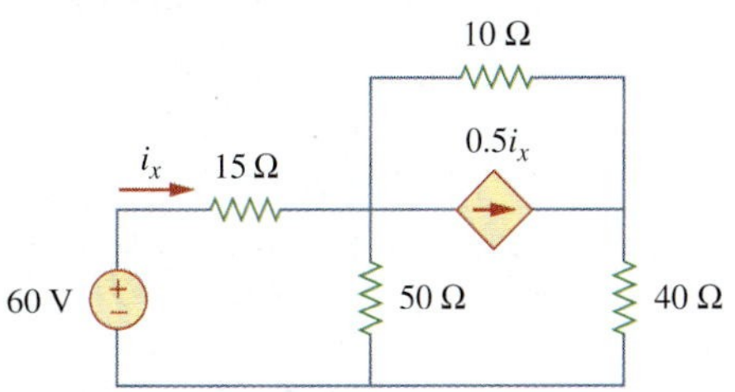

Figure 4.100
For Prob. 4.32.

Sections 4.5 and 4.6 Thevenin's and Norton's Theorems

4.33 Determine R_{Th} and V_{Th} at terminals 1-2 of each of the circuits in Fig. 4.101.

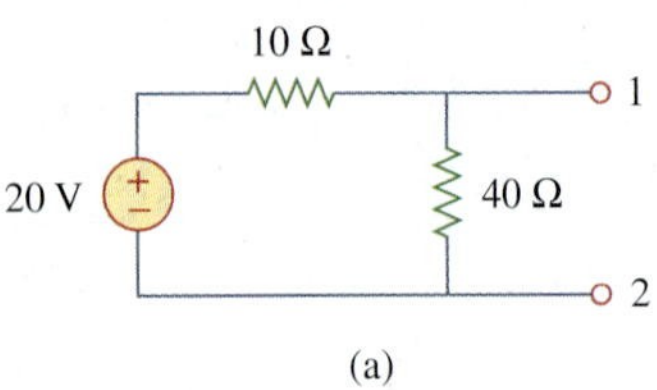

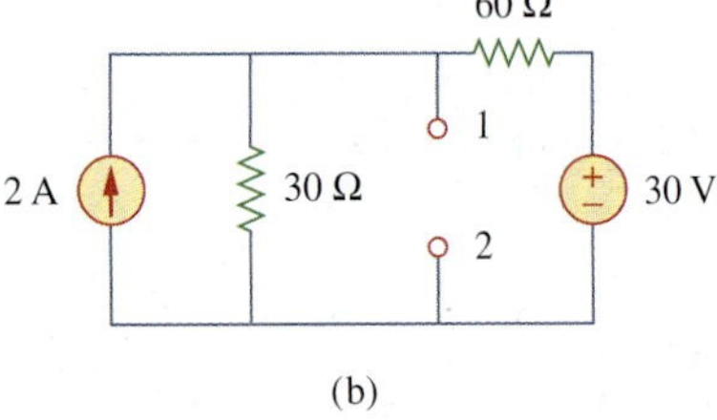

Figure 4.101
For Probs. 4.33 and 4.46.

4.34 Using Fig. 4.102, design a problem that will help other students better understand Thevenin equivalent circuits.

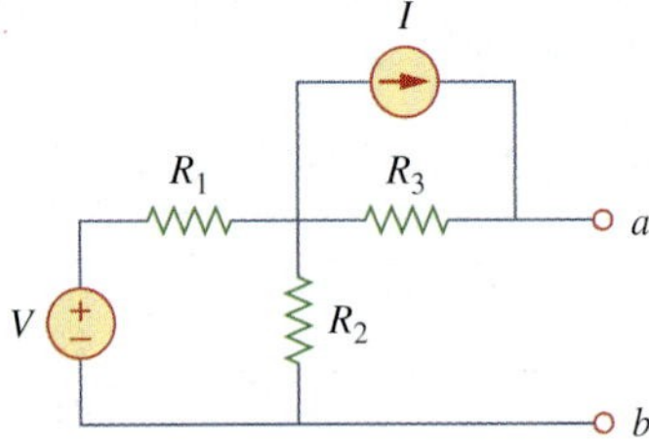

Figure 4.102
For Probs. 4.34 and 4.49.

4.35 Use Thevenin's theorem to find v_o in Prob. 4.12.

4.36 Solve for the current i in the circuit of Fig. 4.103 using Thevenin's theorem. (*Hint:* Find the Thevenin equivalent seen by the 12-Ω resistor.)

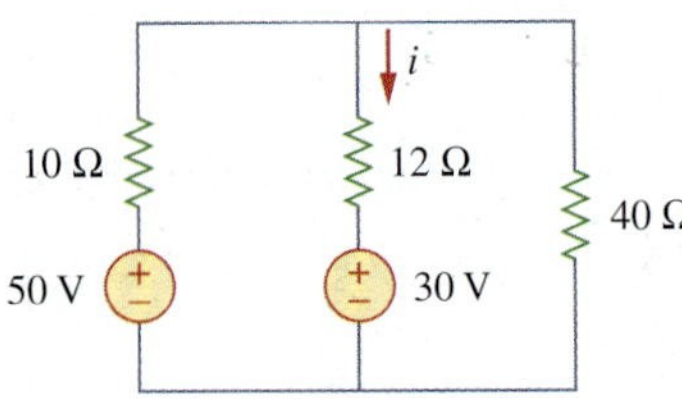

Figure 4.103
For Prob. 4.36.

4.37 Find the Norton equivalent with respect to terminals a-b in the circuit shown in Fig. 4.104.

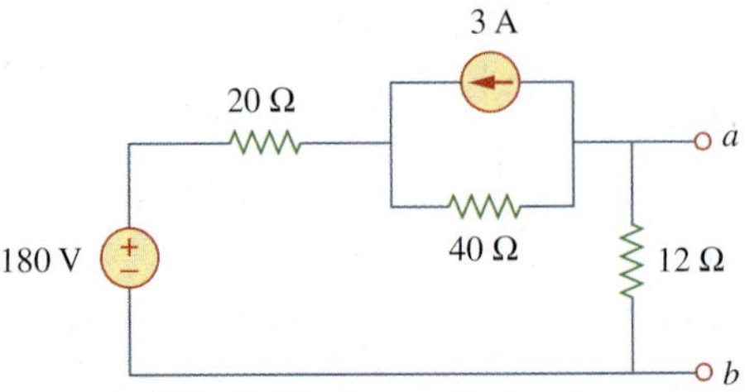

Figure 4.104
For Prob. 4.37.

4.38 Apply Thevenin's theorem to find V_o in the circuit of Fig. 4.105.

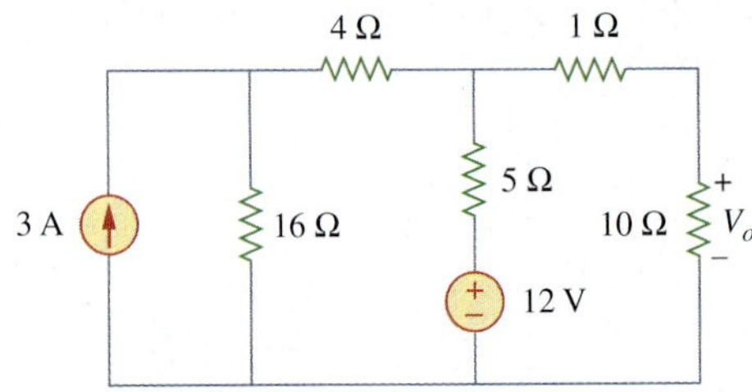

Figure 4.105
For Prob. 4.38.

4.39 Obtain the Thevenin equivalent at terminals a-b of the circuit in Fig. 4.106.

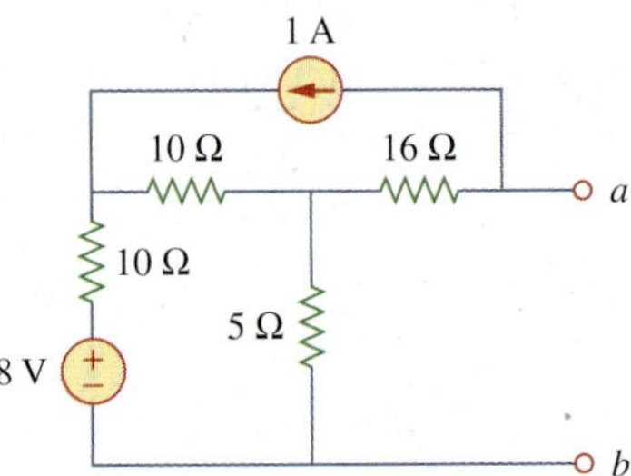

Figure 4.106
For Prob. 4.39.

4.40 Find the Thevenin equivalent at terminals a-b of the circuit in Fig. 4.107.

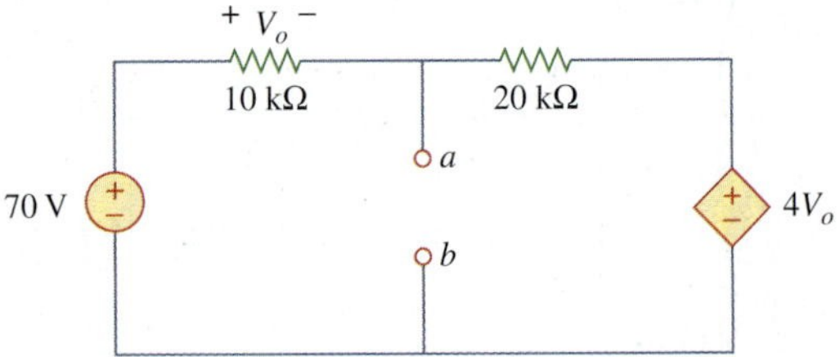

Figure 4.107
For Prob. 4.40.

4.41 Find the Thevenin and Norton equivalents at terminals *a-b* of the circuit shown in Fig. 4.108.

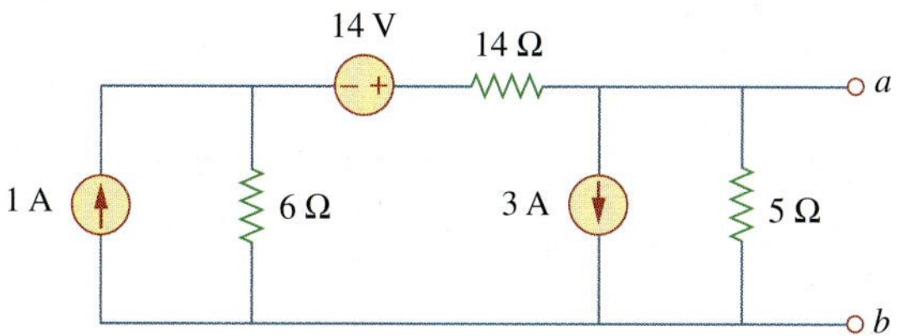

Figure 4.108
For Prob. 4.41.

***4.42** For the circuit in Fig. 4.109, find the Thevenin equivalent between terminals *a* and *b*.

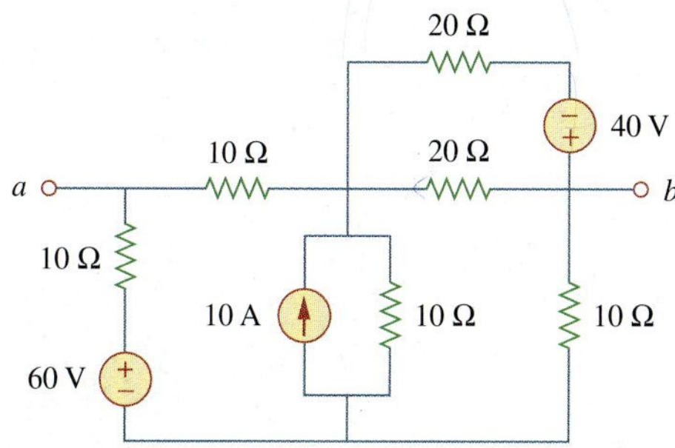

Figure 4.109
For Prob. 4.42.

4.43 Find the Thevenin equivalent looking into terminals *a-b* of the circuit in Fig. 4.110 and solve for i_x.

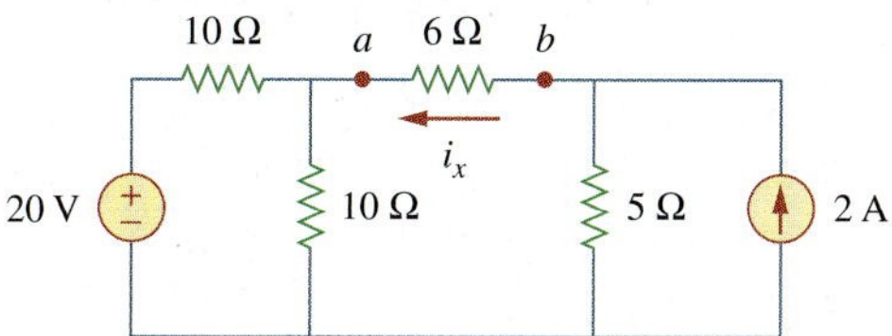

Figure 4.110
For Prob. 4.43.

4.44 For the circuit in Fig. 4.111, obtain the Thevenin equivalent as seen from terminals:

(a) *a-b* (b) *b-c*

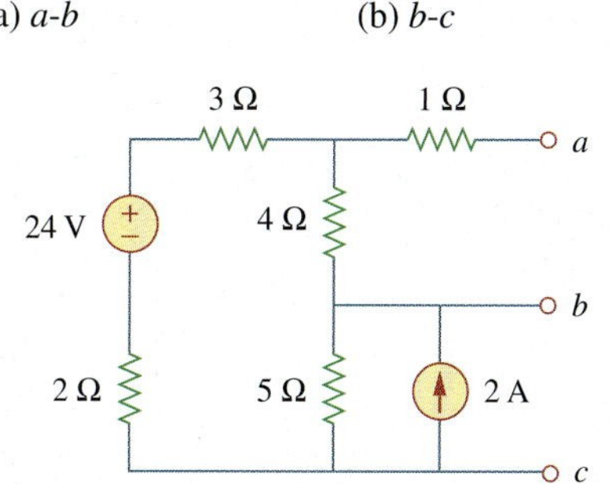

Figure 4.111
For Prob. 4.44.

* An asterisk indicates a challenging problem.

4.45 Find the Norton equivalent of the circuit in Fig. 4.112.

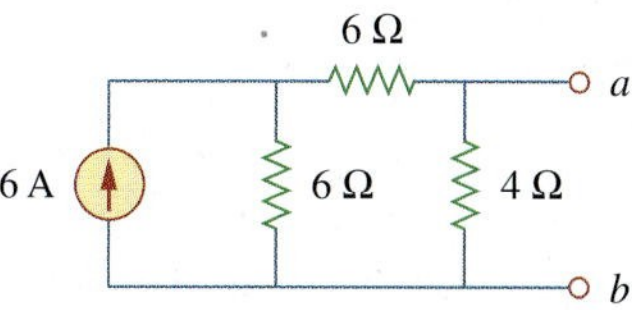

Figure 4.112
For Prob. 4.45.

4.46 Using Fig. 4.113, design a problem to help other students better understand Norton equivalent circuits.

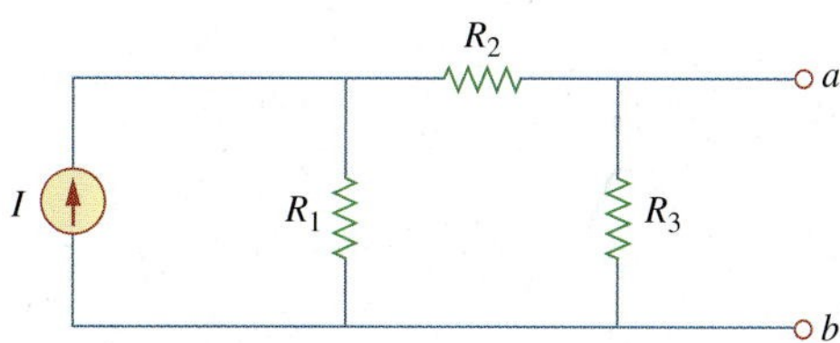

Figure 4.113
For Prob. 4.46.

4.47 Obtain the Thevenin and Norton equivalent circuits of the circuit in Fig. 4.114 with respect to terminals *a* and *b*.

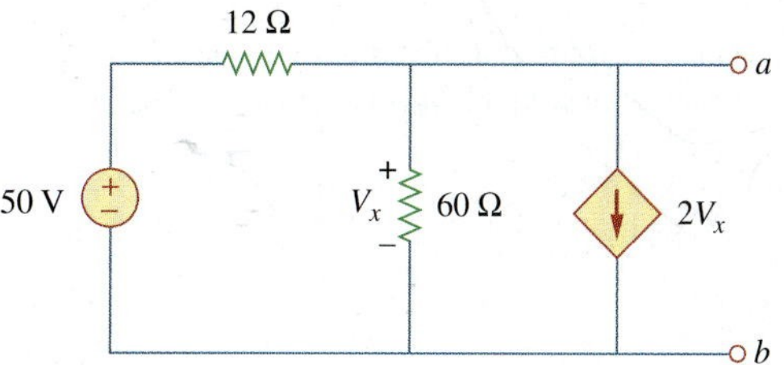

Figure 4.114
For Prob. 4.47.

4.48 Determine the Norton equivalent at terminals *a-b* for the circuit in Fig. 4.115.

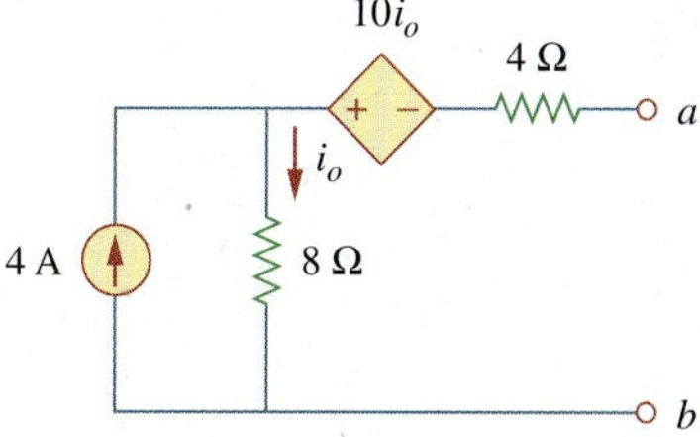

Figure 4.115
For Prob. 4.48.

4.49 Find the Norton equivalent looking into terminals *a-b* of the circuit in Fig. 4.102.

4.50 Obtain the Norton equivalent of the circuit in Fig. 4.116 to the left of terminals *a-b*. Use the result to find current *i*.

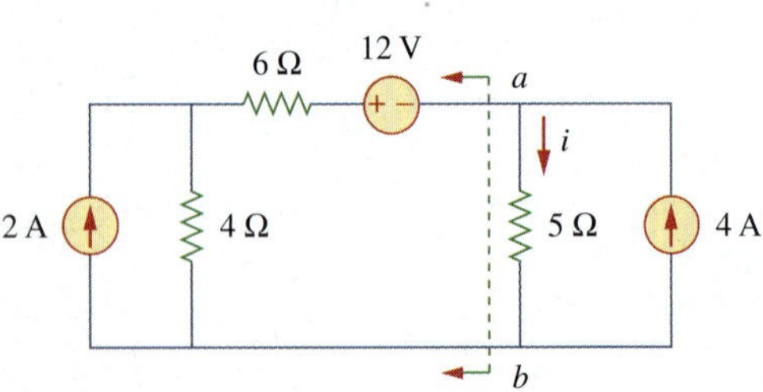

Figure 4.116
For Prob. 4.50.

4.51 Given the circuit in Fig. 4.117, obtain the Norton equivalent as viewed from terminals:

(a) *a-b* (b) *c-d*

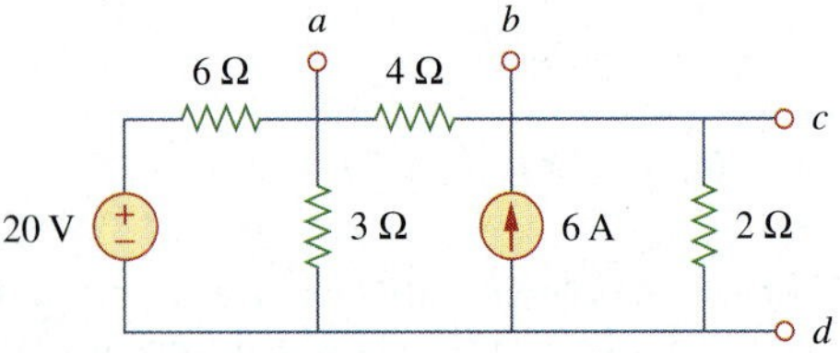

Figure 4.117
For Prob. 4.51.

4.52 For the transistor model in Fig. 4.118, obtain the Thevenin equivalent at terminals *a-b*.

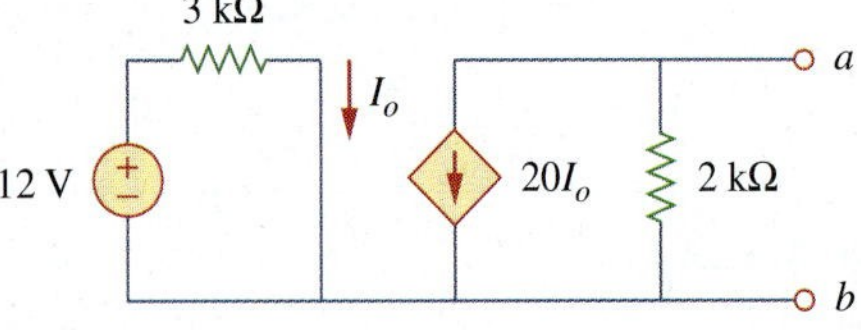

Figure 4.118
For Prob. 4.52.

4.53 Find the Norton equivalent at terminals *a-b* of the circuit in Fig. 4.119.

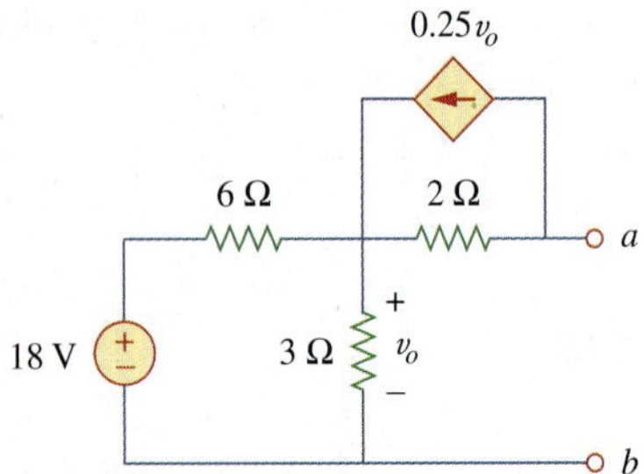

Figure 4.119
For Prob. 4.53.

4.54 Find the Thevenin equivalent between terminals *a-b* of the circuit in Fig. 4.120.

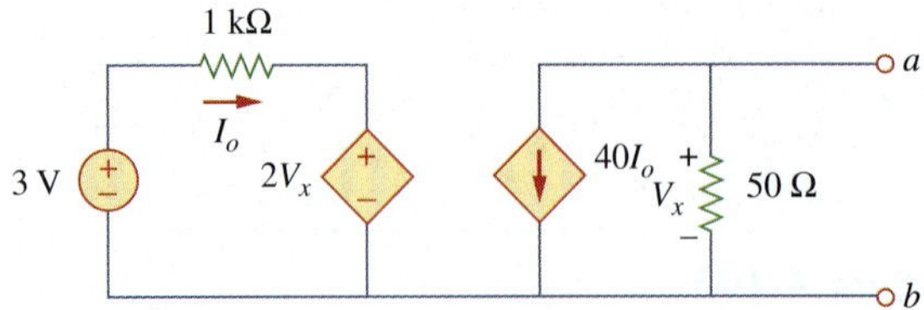

Figure 4.120
For Prob. 4.54.

***4.55** Obtain the Norton equivalent at terminals *a-b* of the circuit in Fig. 4.121.

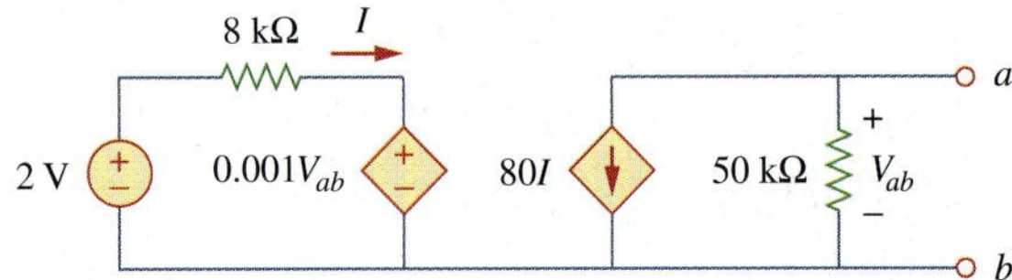

Figure 4.121
For Prob. 4.55.

4.56 Use Norton's theorem to find V_o in the circuit of Fig. 4.122.

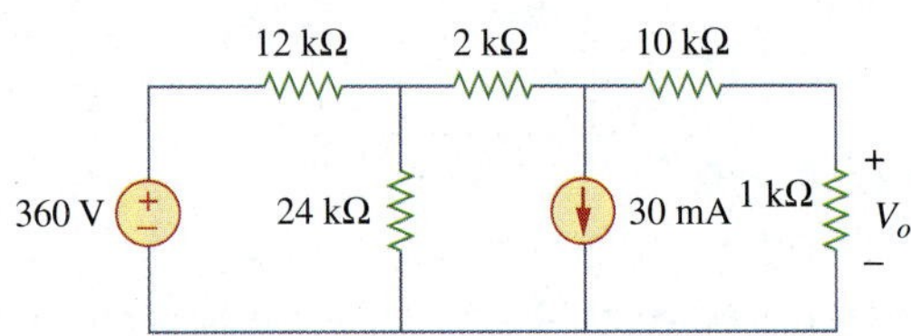

Figure 4.122
For Prob. 4.56.

4.57 Obtain the Thevenin and Norton equivalent circuits at terminals *a-b* for the circuit in Fig. 4.123.

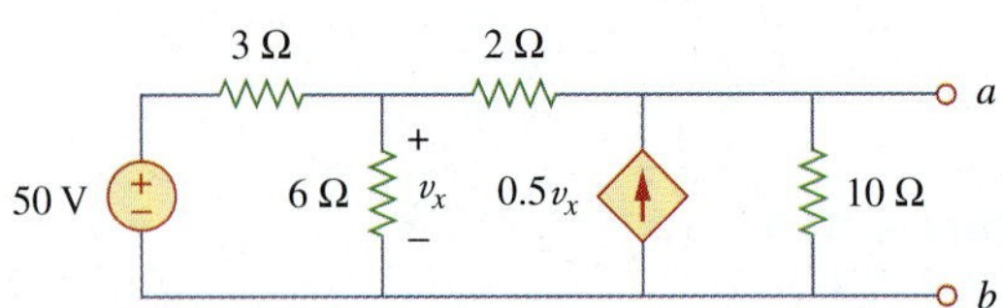

Figure 4.123
For Probs. 4.57 and 4.79.

4.58 The network in Fig. 4.124 models a bipolar transistor common-emitter amplifier connected to a load. Find the Thevenin resistance seen by the load.

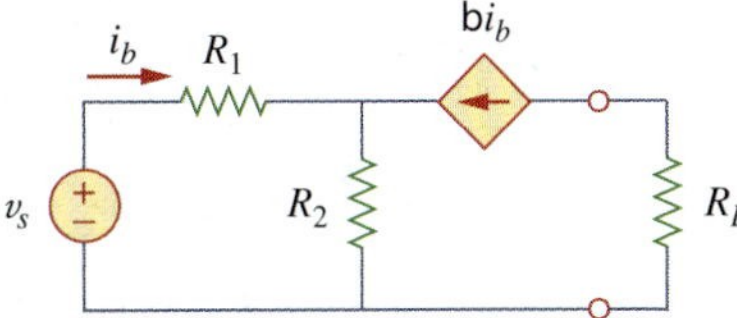

Figure 4.124
For Prob. 4.58.

4.59 Determine the Thevenin and Norton equivalents at terminals *a-b* of the circuit in Fig. 4.125.

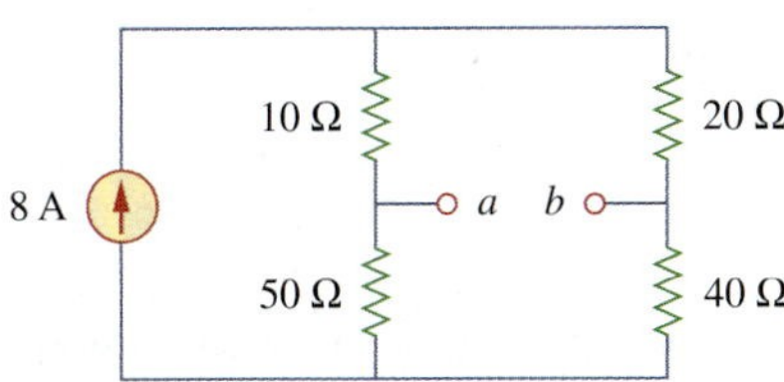

Figure 4.125
For Probs. 4.59 and 4.80.

***4.60** For the circuit in Fig. 4.126, find the Thevenin and Norton equivalent circuits at terminals *a-b*.

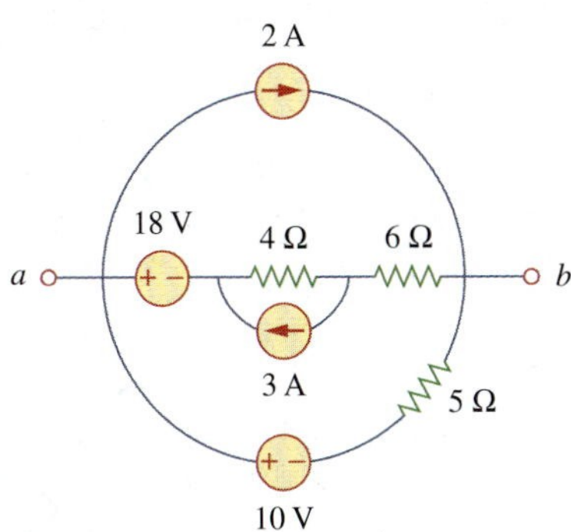

Figure 4.126
For Probs. 4.60 and 4.81.

***4.61** Obtain the Thevenin and Norton equivalent circuits at terminals *a-b* of the circuit in Fig. 4.127.
ML

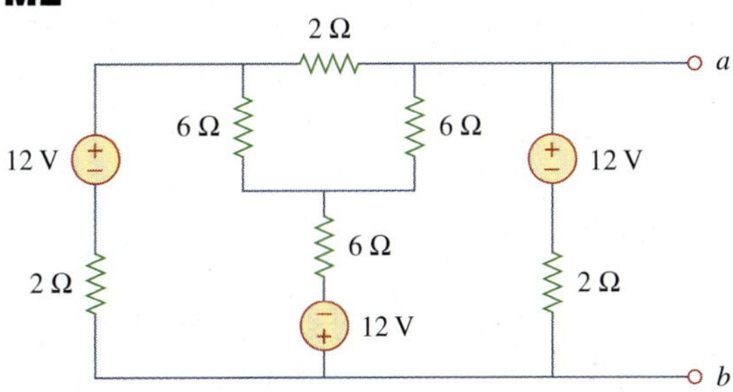

Figure 4.127
For Prob. 4.61.

***4.62** Find the Thevenin equivalent of the circuit in Fig. 4.128.
ML

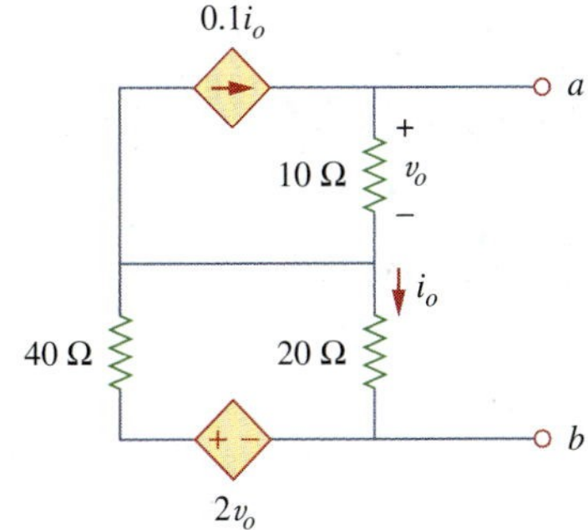

Figure 4.128
For Prob. 4.62.

4.63 Find the Norton equivalent for the circuit in Fig. 4.129.

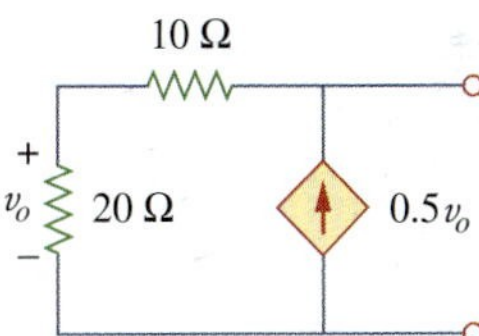

Figure 4.129
For Prob. 4.63.

4.64 Obtain the Thevenin equivalent seen at terminals *a-b* of the circuit in Fig. 4.130.

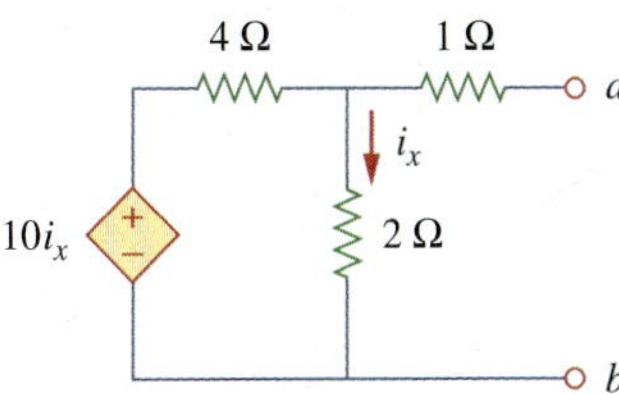

Figure 4.130
For Prob. 4.64.

4.65 For the circuit shown in Fig. 4.131, determine the relationship between V_o and I_o.

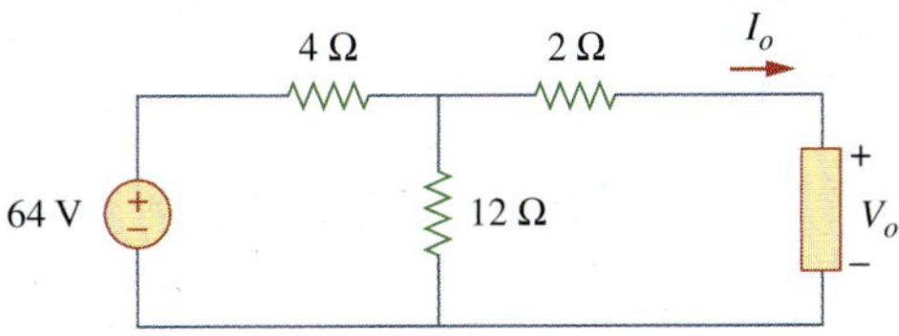

Figure 4.131
For Prob. 4.65.

Section 4.8 Maximum Power Transfer

4.66 Find the maximum power that can be delivered to the resistor R in the circuit of Fig. 4.132.

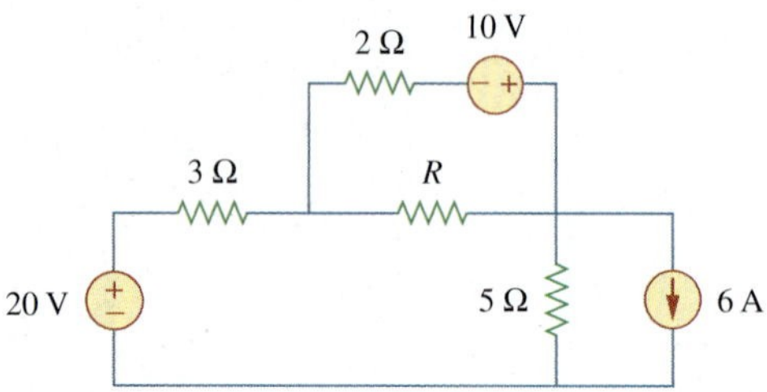

Figure 4.132
For Prob. 4.66.

4.67 The variable resistor R in Fig. 4.133 is adjusted until it absorbs the maximum power from the circuit.
(a) Calculate the value of R for maximum power.
(b) Determine the maximum power absorbed by R.

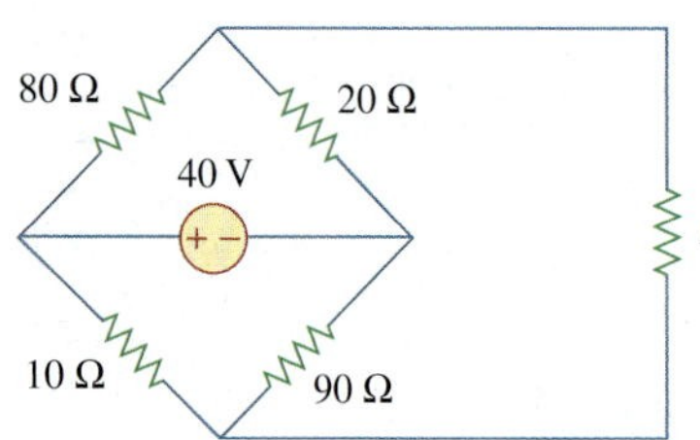

Figure 4.133
For Prob. 4.67.

***4.68** Compute the value of R that results in maximum power transfer to the 10-Ω resistor in Fig. 4.134. Find the maximum power.

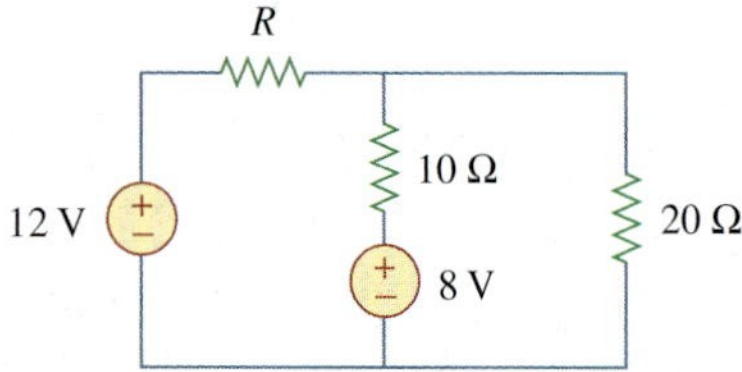

Figure 4.134
For Prob. 4.68.

4.69 Find the maximum power transferred to resistor R in the circuit of Fig. 4.135.

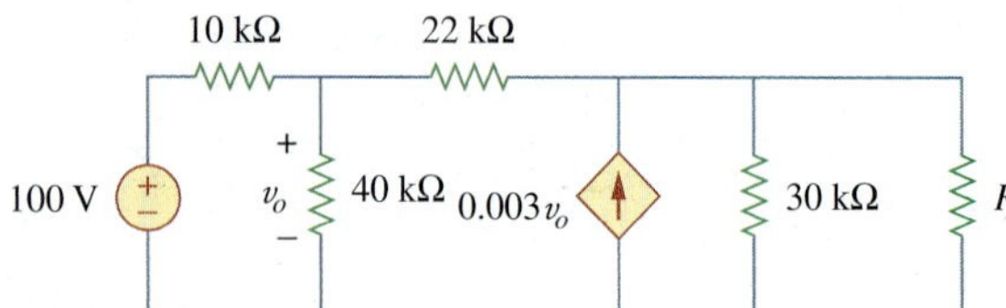

Figure 4.135
For Prob. 4.69.

4.70 Determine the maximum power delivered to the variable resistor R shown in the circuit of Fig. 4.136.

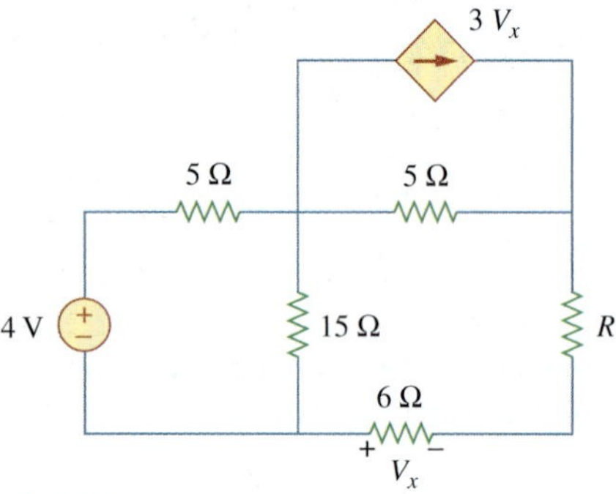

Figure 4.136
For Prob. 4.70.

4.71 For the circuit in Fig. 4.137, what resistor connected across terminals a-b will absorb maximum power from the circuit? What is that power?

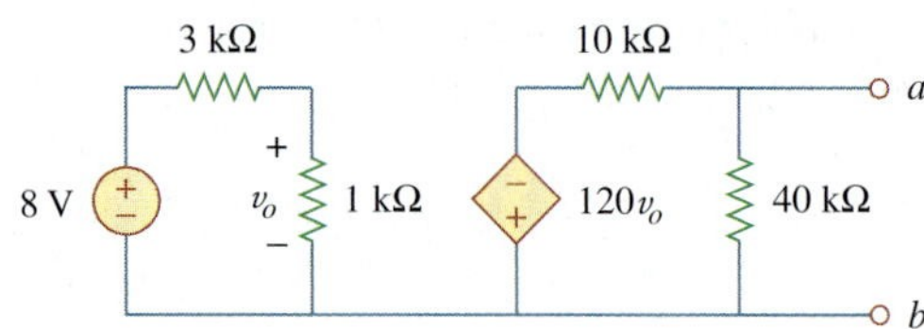

Figure 4.137
For Prob. 4.71.

4.72 (a) For the circuit in Fig. 4.138, obtain the Thevenin equivalent at terminals a-b.
(b) Calculate the current in $R_L = 8\ \Omega$.
(c) Find R_L for maximum power deliverable to R_L.
(d) Determine that maximum power.

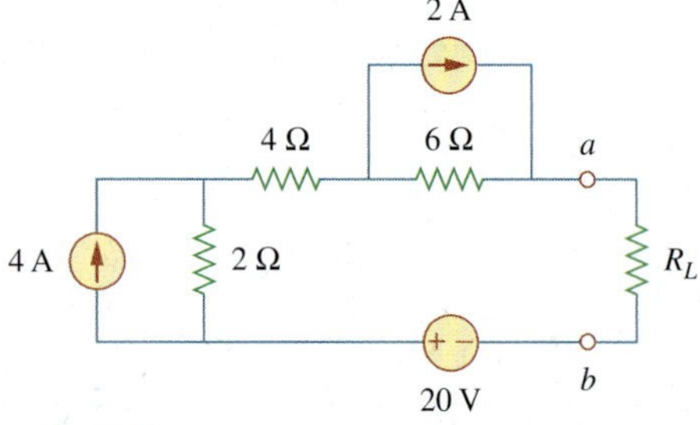

Figure 4.138
For Prob. 4.72.

4.73 Determine the maximum power that can be delivered to the variable resistor R in the circuit of Fig. 4.139.

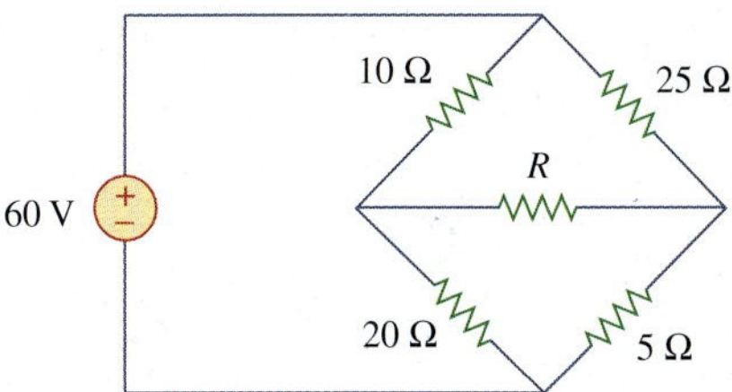

Figure 4.139
For Prob. 4.73.

4.74 For the bridge circuit shown in Fig. 4.140, find the load R_L for maximum power transfer and the maximum power absorbed by the load.

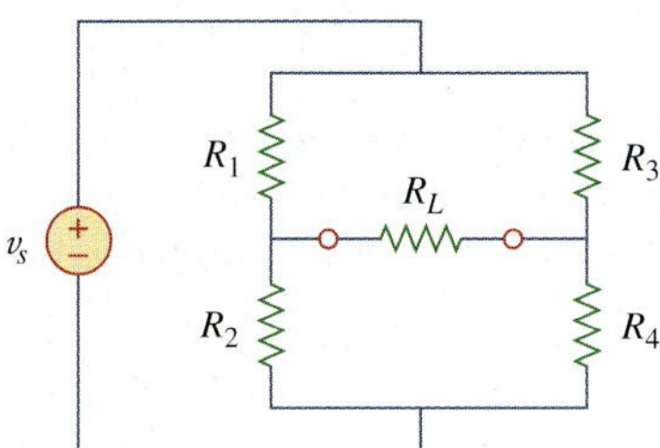

Figure 4.140
For Prob. 4.74.

***4.75** Looking into terminals of the circuit shown in Fig. 4.141, from the right (the R_L side), determine the Thevenin equivalent circuit. What value of R_L produces maximum power to R_L?

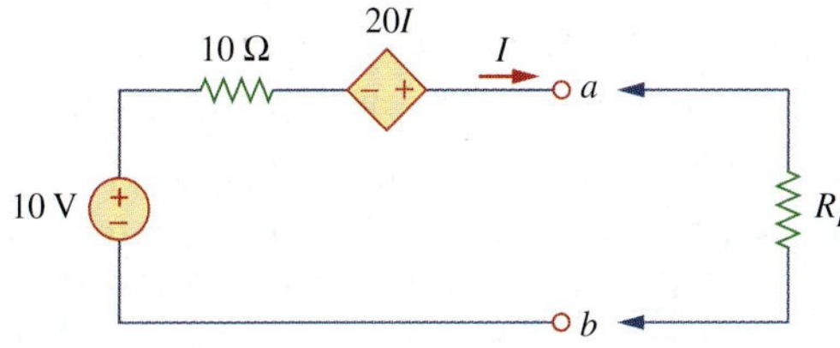

Figure 4.141
For Prob. 4.75.

Section 4.9 Verifying Circuit Theorems with *PSpice*

4.76 Solve Prob. 4.34 using *PSpice*.

4.77 Use *PSpice* to solve Prob. 4.44.

4.78 Use *PSpice* to solve Prob. 4.52.

4.79 Obtain the Thevenin equivalent of the circuit in Fig. 4.123 using *PSpice*.

4.80 Use *PSpice* to find the Thevenin equivalent circuit at terminals *a-b* of the circuit in Fig. 4.125.

4.81 For the circuit in Fig. 4.126, use *PSpice* to find the Thevenin equivalent at terminals *a-b*.

Section 4.10 Applications

4.82 A battery has a short-circuit current of 20 A and an open-circuit voltage of 12 V. If the battery is connected to an electric bulb of resistance 2 Ω, calculate the power dissipated by the bulb.

4.83 The following results were obtained from measurements taken between the two terminals of a resistive network.

Terminal Voltage	12 V	0 V
Terminal Current	0 A	1.5 A

Find the Thevenin equivalent of the network.

4.84 When connected to a 4-Ω resistor, a battery has a terminal voltage of 10.8 V but produces 12 V on an open circuit. Determine the Thevenin equivalent circuit for the battery.

4.85 The Thevenin equivalent at terminals *a-b* of the linear network shown in Fig. 4.142 is to be determined by measurement. When a 10-kΩ resistor is connected to terminals *a-b*, the voltage V_{ab} is measured as 6 V. When a 30-kΩ resistor is connected to the terminals, V_{ab} is measured as 12 V. Determine: (a) the Thevenin equivalent at terminals *a-b*, (b) V_{ab} when a 20-kΩ resistor is connected to terminals *a-b*.

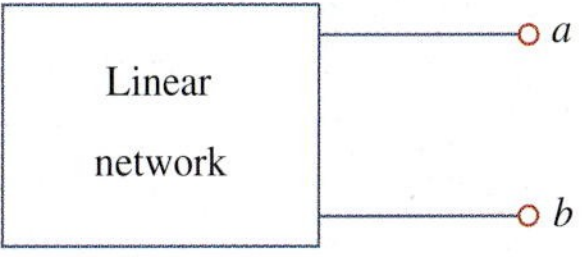

Figure 4.142
For Prob. 4.85.

4.86 A black box with a circuit in it is connected to a variable resistor. An ideal ammeter (with zero resistance) and an ideal voltmeter (with infinite resistance) are used to measure current and voltage as shown in Fig. 4.143. The results are shown in the table on the next page.

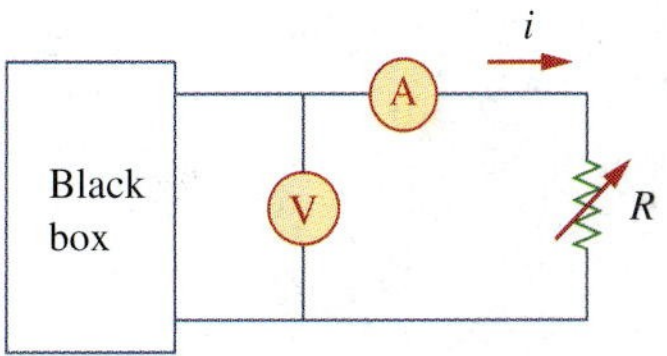

Figure 4.143
For Prob. 4.86.

(a) Find i when $R = 4\ \Omega$.

(b) Determine the maximum power from the box.

$R(\Omega)$	V(V)	i(A)
2	3	1.5
8	8	1.0
14	10.5	0.75

4.87 A transducer is modeled with a current source I_s and a parallel resistance R_s. The current at the terminals of the source is measured to be 9.975 mA when an ammeter with an internal resistance of 20 Ω is used.

(a) If adding a 2-kΩ resistor across the source terminals causes the ammeter reading to fall to 9.876 mA, calculate I_s and R_s.

(b) What will the ammeter reading be if the resistance between the source terminals is changed to 4 kΩ?

4.88 Consider the circuit in Fig. 4.144. An ammeter with internal resistance R_i is inserted between A and B to measure I_o. Determine the reading of the ammeter if: (a) $R_i = 500\ \Omega$, (b) $R_i = 0\ \Omega$. (*Hint*: Find the Thevenin equivalent circuit at terminals a-b.)

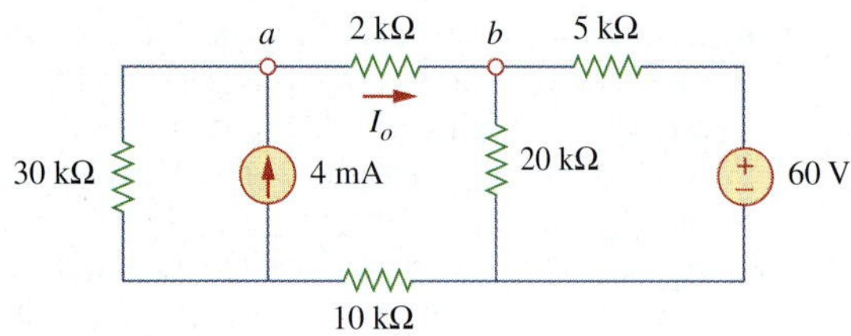

Figure 4.144
For Prob. 4.88.

4.89 Consider the circuit in Fig. 4.145. (a) Replace the resistor R_L by a zero resistance ammeter and determine the ammeter reading. (b) To verify the reciprocity theorem, interchange the ammeter and the 12-V source and determine the ammeter reading again.

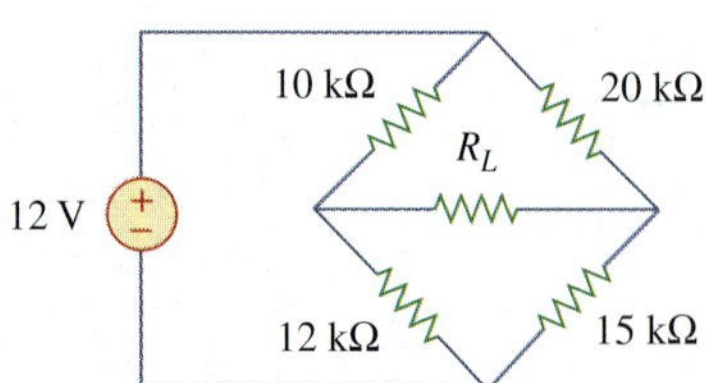

Figure 4.145
For Prob. 4.89.

4.90 The Wheatstone bridge circuit shown in Fig. 4.146 is used to measure the resistance of a strain gauge. The adjustable resistor has a linear taper with a maximum value of 100 Ω. If the resistance of the strain gauge is found to be 42.6 Ω, what fraction of the full slider travel is the slider when the bridge is balanced?

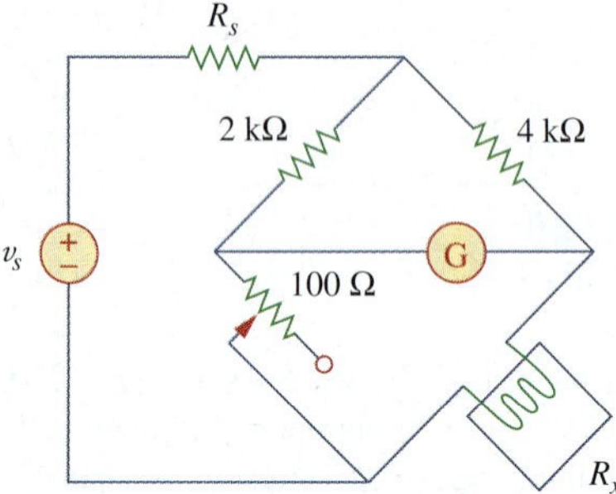

Figure 4.146
For Prob. 4.90.

4.91 (a) In the Wheatstone bridge circuit of Fig. 4.147, select the values of R_1 and R_3 such that the bridge can measure R_x in the range of 0–10 Ω.

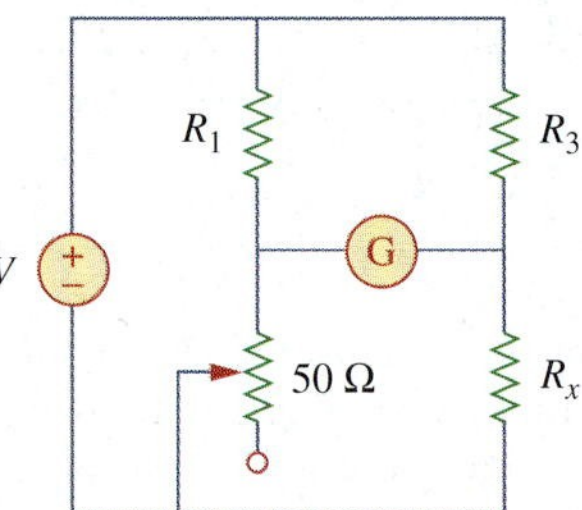

Figure 4.147
For Prob. 4.91.

(b) Repeat for the range of 0–100 Ω.

***4.92** Consider the bridge circuit of Fig. 4.148. Is the bridge balanced? If the 10-kΩ resistor is replaced by an 18-kΩ resistor, what resistor connected between terminals a-b absorbs the maximum power? What is this power?

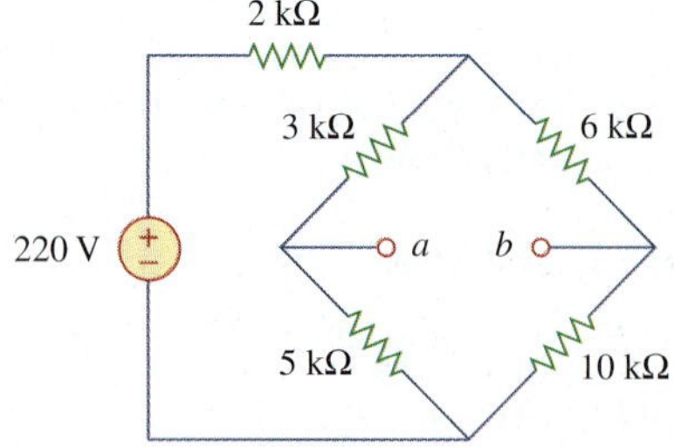

Figure 4.148
For Prob. 4.92.

Comprehensive Problems

4.93 The circuit in Fig. 4.149 models a common-emitter transistor amplifier. Find i_x using source transformation.

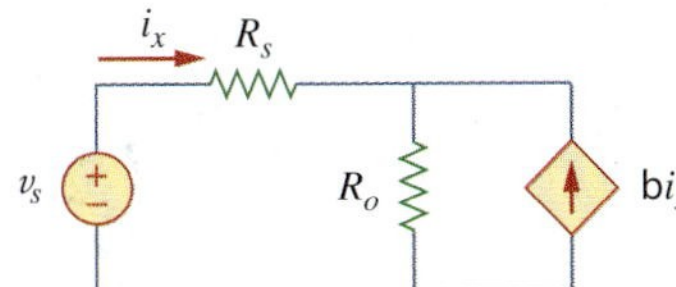

Figure 4.149
For Prob. 4.93.

4.94 An attenuator is an interface circuit that reduces the voltage level without changing the output resistance.

(a) By specifying R_s and R_p of the interface circuit in Fig. 4.150, design an attenuator that will meet the following requirements:

$$\frac{V_o}{V_g} = 0.125, \qquad R_{eq} = R_{Th} = R_g = 100\ \Omega$$

(b) Using the interface designed in part (a), calculate the current through a load of $R_L = 50\ \Omega$ when $V_g = 12$ V.

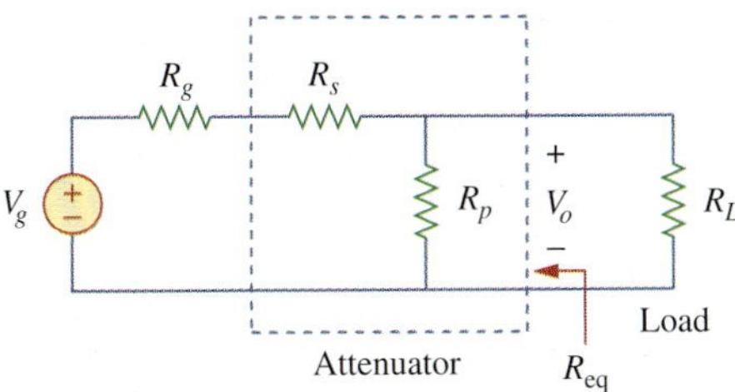

Figure 4.150
For Prob. 4.94.

***4.95** A dc voltmeter with a sensitivity of 20 kΩ/V is used to find the Thevenin equivalent of a linear network. Readings on two scales are as follows:

(a) 0–10 V scale: 4 V (b) 0–50 V scale: 5 V

Obtain the Thevenin voltage and the Thevenin resistance of the network.

***4.96** A resistance array is connected to a load resistor R and a 9-V battery as shown in Fig. 4.151.

(a) Find the value of R such that $V_o = 1.8$ V.

(b) Calculate the value of R that will draw the maximum current. What is the maximum current?

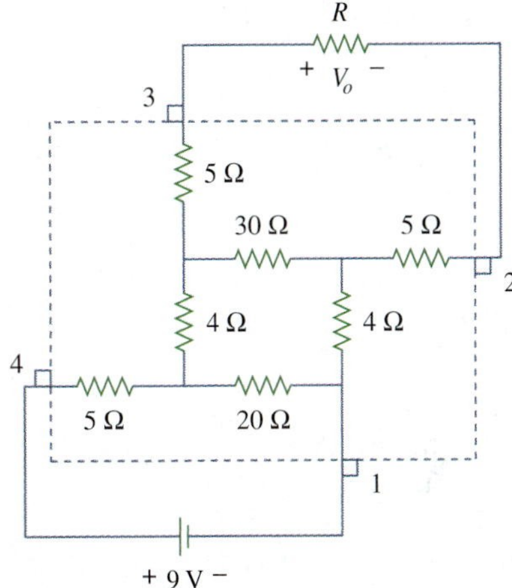

Figure 4.151
For Prob. 4.96.

4.97 A common-emitter amplifier circuit is shown in Fig. 4.152. Obtain the Thevenin equivalent to the left of points B and E.

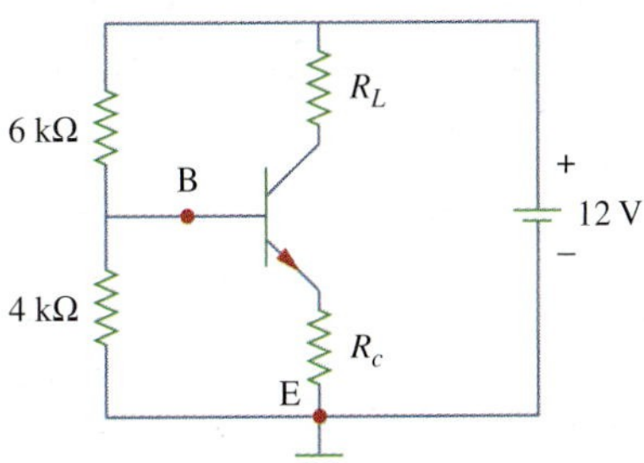

Figure 4.152
For Prob. 4.97.

***4.98** For Practice Prob. 4.18, determine the current through the 40-Ω resistor and the power dissipated by the resistor.

chapter

5

Operational Amplifiers

He who will not reason is a bigot; he who cannot is a fool; and he who dares not is a slave

—Lord Byron

Enhancing Your Career

Career in Electronic Instrumentation

Engineering involves applying physical principles to design devices for the benefit of humanity. But physical principles cannot be understood without measurement. In fact, physicists often say that physics is the science that measures reality. Just as measurements are a tool for understanding the physical world, instruments are tools for measurement. The operational amplifier introduced in this chapter is a building block of modern electronic instrumentation. Therefore, mastery of operational amplifier fundamentals is paramount to any practical application of electronic circuits.

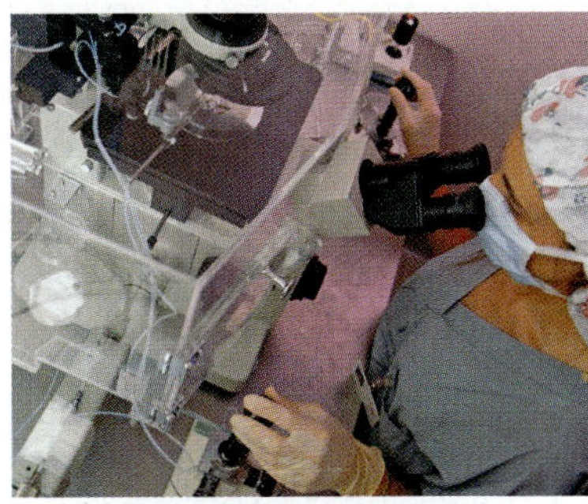

Electronic Instrumentation used in medical research.
© Royalty-Free/Corbis

Electronic instruments are used in all fields of science and engineering. They have proliferated in science and technology to the extent that it would be ridiculous to have a scientific or technical education without exposure to electronic instruments. For example, physicists, physiologists, chemists, and biologists must learn to use electronic instruments. For electrical engineering students in particular, the skill in operating digital and analog electronic instruments is crucial. Such instruments include ammeters, voltmeters, ohmmeters, oscilloscopes, spectrum analyzers, and signal generators.

Beyond developing the skill for operating the instruments, some electrical engineers specialize in designing and constructing electronic instruments. These engineers derive pleasure in building their own instruments. Most of them invent and patent their inventions. Specialists in electronic instruments find employment in medical schools, hospitals, research laboratories, aircraft industries, and thousands of other industries where electronic instruments are routinely used.

5.1 Introduction

The term operational amplifier was introduced in 1947 by John Ragazzini and his colleagues, in their work on analog computers for the National Defense Research Council after World War II. The first op amps used vacuum tubes rather than transistors.

Having learned the basic laws and theorems for circuit analysis, we are now ready to study an active circuit element of paramount importance: the *operational amplifier*, or *op amp* for short. The op amp is a versatile circuit building block.

> The **op amp** is an electronic unit that behaves like a voltage-controlled voltage source.

An op amp may also be regarded as a voltage amplifier with very high gain.

It can also be used in making a voltage- or current-controlled current source. An op amp can sum signals, amplify a signal, integrate it, or differentiate it. The ability of the op amp to perform these mathematical operations is the reason it is called an *operational amplifier.* It is also the reason for the widespread use of op amps in analog design. Op amps are popular in practical circuit designs because they are versatile, inexpensive, easy to use, and fun to work with.

We begin by discussing the ideal op amp and later consider the nonideal op amp. Using nodal analysis as a tool, we consider ideal op amp circuits such as the inverter, voltage follower, summer, and difference amplifier. We will also analyze op amp circuits with *PSpice.* Finally, we learn how an op amp is used in digital-to-analog converters and instrumentation amplifiers.

5.2 Operational Amplifiers

An operational amplifier is designed so that it performs some mathematical operations when external components, such as resistors and capacitors, are connected to its terminals. Thus,

> An **op amp** is an active circuit element designed to perform mathematical operations of addition, subtraction, multiplication, division, differentiation, and integration.

Figure 5.1
A typical operational amplifier.
Courtesy of Tech America.

The op amp is an electronic device consisting of a complex arrangement of resistors, transistors, capacitors, and diodes. A full discussion of what is inside the op amp is beyond the scope of this book. It will suffice to treat the op amp as a circuit building block and simply study what takes place at its terminals.

The pin diagram in Fig. 5.2(a) corresponds to the 741 general-purpose op amp made by Fairchild Semiconductor.

Op amps are commercially available in integrated circuit packages in several forms. Figure 5.1 shows a typical op amp package. A typical one is the eight-pin dual in-line package (or DIP), shown in Fig. 5.2(a). Pin or terminal 8 is unused, and terminals 1 and 5 are of little concern to us. The five important terminals are:

1. The inverting input, pin 2.
2. The noninverting input, pin 3.
3. The output, pin 6.
4. The positive power supply V^+, pin 7.
5. The negative power supply V^-, pin 4.

The circuit symbol for the op amp is the triangle in Fig. 5.2(b); as shown, the op amp has two inputs and one output. The inputs are

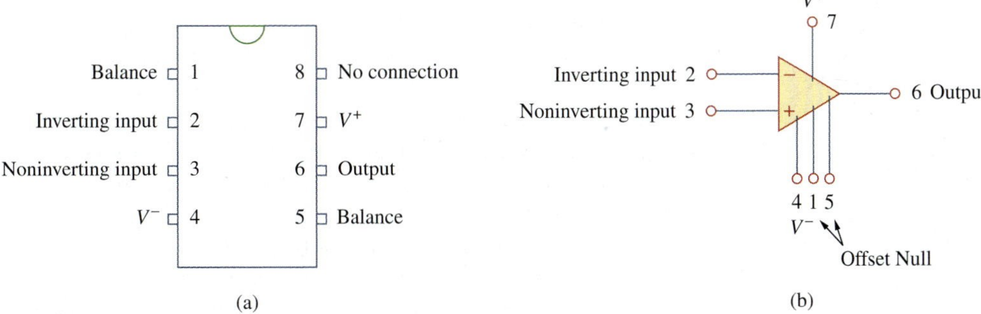

Figure 5.2
A typical op amp: (a) pin configuration, (b) circuit symbol.

marked with minus (−) and plus (+) to specify *inverting* and *noninverting* inputs, respectively. An input applied to the noninverting terminal will appear with the same polarity at the output, while an input applied to the inverting terminal will appear inverted at the output.

As an active element, the op amp must be powered by a voltage supply as typically shown in Fig. 5.3. Although the power supplies are often ignored in op amp circuit diagrams for the sake of simplicity, the power supply currents must not be overlooked. By KCL,

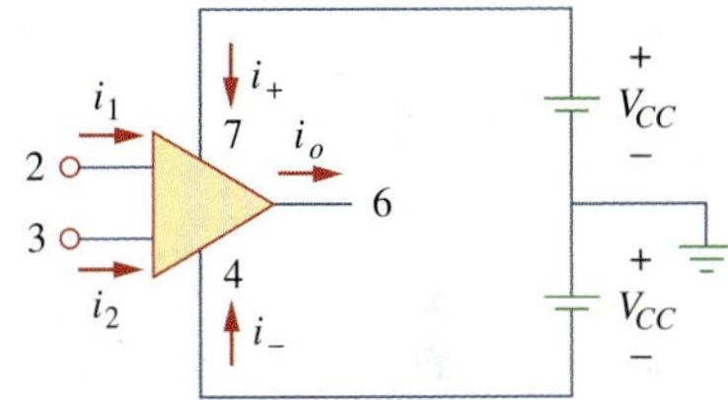

Figure 5.3
Powering the op amp.

$$i_o = i_1 + i_2 + i_+ + i_- \tag{5.1}$$

The equivalent circuit model of an op amp is shown in Fig. 5.4. The output section consists of a voltage-controlled source in series with the output resistance R_o. It is evident from Fig. 5.4 that the input resistance R_i is the Thevenin equivalent resistance seen at the input terminals, while the output resistance R_o is the Thevenin equivalent resistance seen at the output. The differential input voltage v_d is given by

$$v_d = v_2 - v_1 \tag{5.2}$$

where v_1 is the voltage between the inverting terminal and ground and v_2 is the voltage between the noninverting terminal and ground. The op amp senses the difference between the two inputs, multiplies it by the gain A, and causes the resulting voltage to appear at the output. Thus, the output v_o is given by

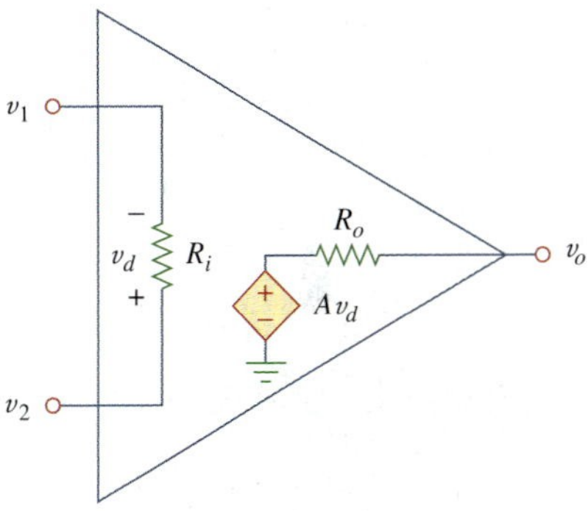

Figure 5.4
The equivalent circuit of the nonideal op amp.

$$\boxed{v_o = Av_d = A(v_2 - v_1)} \tag{5.3}$$

A is called the *open-loop voltage gain* because it is the gain of the op amp without any external feedback from output to input. Table 5.1

Sometimes, voltage gain is expressed in decibels (dB), as discussed in Chapter 14.

$$A \text{ dB} = 20 \log_{10} A$$

Typical ranges for op amp parameters.

Parameter	Typical range	Ideal values
Open-loop gain, A	10^5 to 10^8	∞
Input resistance, R_i	10^5 to 10^{13} Ω	∞ Ω
Output resistance, R_o	10 to 100 Ω	0 Ω
Supply voltage, V_{CC}	5 to 24 V	

shows typical values of voltage gain A, input resistance R_i, output resistance R_o, and supply voltage V_{CC}.

The concept of feedback is crucial to our understanding of op amp circuits. A negative feedback is achieved when the output is fed back to the inverting terminal of the op amp. As Example 5.1 shows, when there is a feedback path from output to input, the ratio of the output voltage to the input voltage is called the *closed-loop gain*. As a result of the negative feedback, it can be shown that the closed-loop gain is almost insensitive to the open-loop gain A of the op amp. For this reason, op amps are used in circuits with feedback paths.

A practical limitation of the op amp is that the magnitude of its output voltage cannot exceed $|V_{CC}|$. In other words, the output voltage is dependent on and is limited by the power supply voltage. Figure 5.5 illustrates that the op amp can operate in three modes, depending on the differential input voltage v_d:

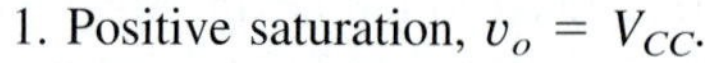

1. Positive saturation, $v_o = V_{CC}$.
2. Linear region, $-V_{CC} \leq v_o = Av_d \leq V_{CC}$.
3. Negative saturation, $v_o = -V_{CC}$.

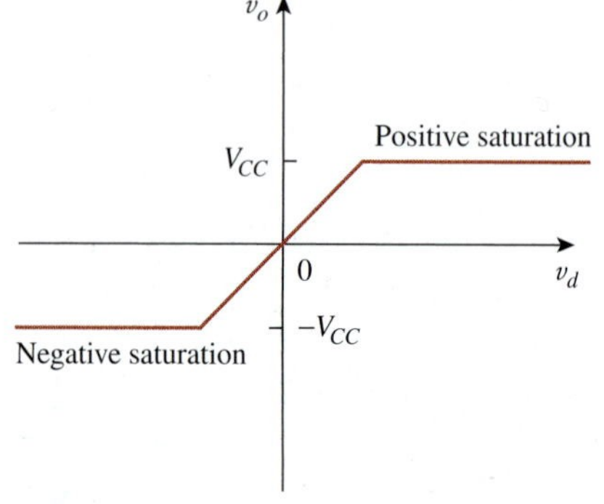

Figure 5.5
Op amp output voltage v_o as a function of the differential input voltage v_d.

If we attempt to increase v_d beyond the linear range, the op amp becomes saturated and yields $v_o = V_{CC}$ or $v_o = -V_{CC}$. Throughout this book, we will assume that our op amps operate in the linear mode. This means that the output voltage is restricted by

$$-V_{CC} \leq v_o \leq V_{CC} \tag{5.4}$$

Although we shall always operate the op amp in the linear region, the possibility of saturation must be borne in mind when one designs with op amps, to avoid designing op amp circuits that will not work in the laboratory.

Throughout this book, we assume that an op amp operates in the linear range. Keep in mind the voltage constraint on the op amp in this mode.

Example 5.1

A 741 op amp has an open-loop voltage gain of 2×10^5, input resistance of 2 MΩ, and output resistance of 50 Ω. The op amp is used in the circuit of Fig. 5.6(a). Find the closed-loop gain v_o/v_s. Determine current i when $v_s = 2$ V.

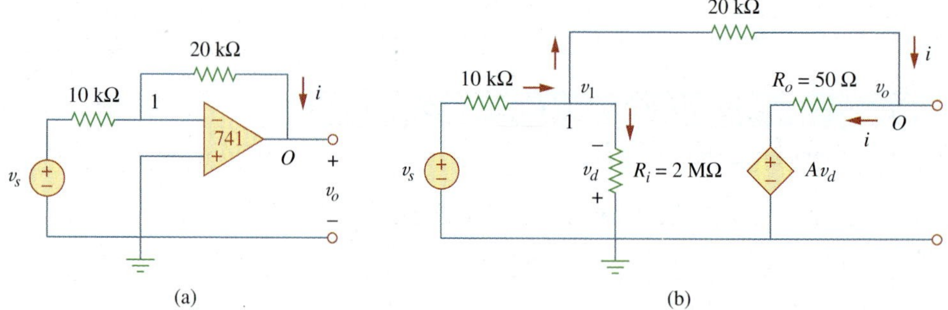

Figure 5.6
For Example 5.1: (a) original circuit, (b) the equivalent circuit.

Solution:

Using the op amp model in Fig. 5.4, we obtain the equivalent circuit of Fig. 5.6(a) as shown in Fig. 5.6(b). We now solve the circuit in Fig. 5.6(b) by using nodal analysis. At node 1, KCL gives

$$\frac{v_s - v_1}{10 \times 10^3} = \frac{v_1}{2000 \times 10^3} + \frac{v_1 - v_o}{20 \times 10^3}$$

Multiplying through by 2000×10^3, we obtain

$$200v_s = 301v_1 - 100v_o$$

or

$$2v_s \simeq 3v_1 - v_o \quad \Rightarrow \quad v_1 = \frac{2v_s + v_o}{3} \tag{5.1.1}$$

At node O,

$$\frac{v_1 - v_o}{20 \times 10^3} = \frac{v_o - Av_d}{50}$$

But $v_d = -v_1$ and $A = 200{,}000$. Then

$$v_1 - v_o = 400(v_o + 200{,}000v_1) \tag{5.1.2}$$

Substituting v_1 from Eq. (5.1.1) into Eq. (5.1.2) gives

$$0 \simeq 26{,}667{,}067v_o + 53{,}333{,}333v_s \quad \Rightarrow \quad \frac{v_o}{v_s} = -1.9999699$$

This is closed-loop gain, because the 20-kΩ feedback resistor closes the loop between the output and input terminals. When $v_s = 2$ V, $v_o = -3.9999398$ V. From Eq. (5.1.1), we obtain $v_1 = 20.066667$ μV. Thus,

$$i = \frac{v_1 - v_o}{20 \times 10^3} = 0.19999 \text{ mA}$$

It is evident that working with a nonideal op amp is tedious, as we are dealing with very large numbers.

Practice Problem 5.1

If the same 741 op amp in Example 5.1 is used in the circuit of Fig. 5.7, calculate the closed-loop gain v_o/v_s. Find i_o when $v_s = 1$ V.

Answer: 9.00041, 0.657 mA.

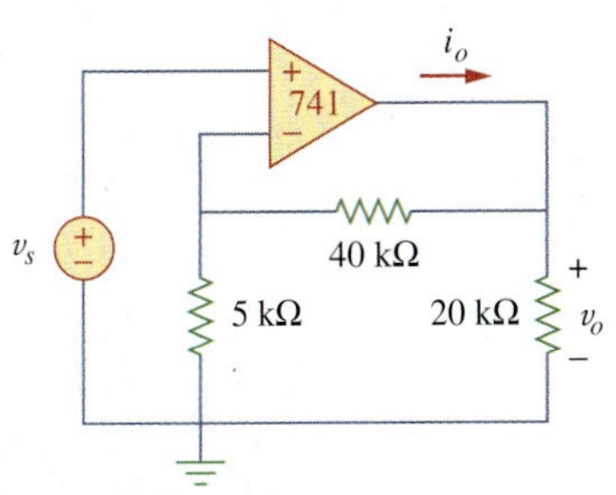

Figure 5.7
For Practice Prob. 5.1.

5.3 Ideal Op Amp

To facilitate the understanding of op amp circuits, we will assume ideal op amps. An op amp is ideal if it has the following characteristics:

1. Infinite open-loop gain, $A \simeq \infty$.
2. Infinite input resistance, $R_i \simeq \infty$.
3. Zero output resistance, $R_o \simeq 0$.

An **ideal op amp** is an amplifier with infinite open-loop gain, infinite input resistance, and zero output resistance.

Although assuming an ideal op amp provides only an approximate analysis, most modern amplifiers have such large gains and input impedances that the approximate analysis is a good one. Unless stated otherwise, we will assume from now on that every op amp is ideal.

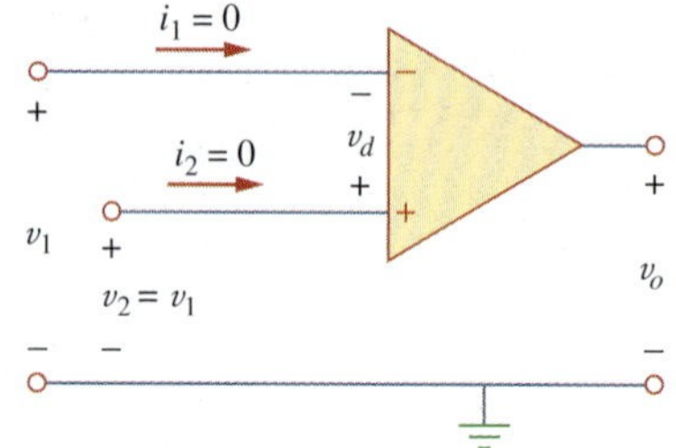

Figure 5.8
Ideal op amp model.

For circuit analysis, the ideal op amp is illustrated in Fig. 5.8, which is derived from the nonideal model in Fig. 5.4. Two important characteristics of the ideal op amp are:

1. The currents into both input terminals are zero:

$$i_1 = 0, \qquad i_2 = 0 \tag{5.5}$$

This is due to infinite input resistance. An infinite resistance between the input terminals implies that an open circuit exists there and current cannot enter the op amp. But the output current is not necessarily zero according to Eq. (5.1).

2. The voltage across the input terminals is equal to zero; i.e.,

$$v_d = v_2 - v_1 = 0 \tag{5.6}$$

or

$$v_1 = v_2 \tag{5.7}$$

The two characteristics can be exploited by noting that for voltage calculations the input port behaves as a short circuit, while for current calculations the input port behaves as an open circuit.

Thus, an ideal op amp has zero current into its two input terminals and the voltage between the two input terminals is equal to zero. Equations (5.5) and (5.7) are extremely important and should be regarded as the key handles to analyzing op amp circuits.

Example 5.2

Rework Practice Prob. 5.1 using the ideal op amp model.

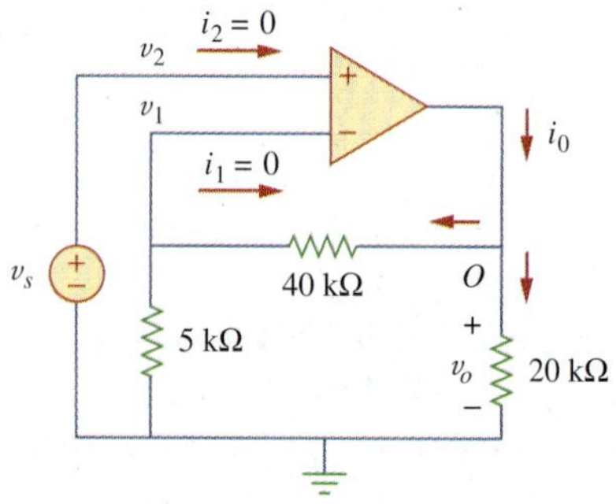

Figure 5.9
For Example 5.2.

Solution:
We may replace the op amp in Fig. 5.7 by its equivalent model in Fig. 5.9 as we did in Example 5.1. But we do not really need to do this. We just need to keep Eqs. (5.5) and (5.7) in mind as we analyze the circuit in Fig. 5.7. Thus, the Fig. 5.7 circuit is presented as in Fig. 5.9. Notice that

$$v_2 = v_s \tag{5.2.1}$$

Since $i_1 = 0$, the 40-kΩ and 5-kΩ resistors are in series; the same current flows through them. v_1 is the voltage across the 5-kΩ resistor. Hence, using the voltage division principle,

$$v_1 = \frac{5}{5 + 40}v_o = \frac{v_o}{9} \tag{5.2.2}$$

According to Eq. (5.7),

$$v_2 = v_1 \tag{5.2.3}$$

Substituting Eqs. (5.2.1) and (5.2.2) into Eq. (5.2.3) yields the closed-loop gain,

$$v_s = \frac{v_o}{9} \quad \Rightarrow \quad \frac{v_o}{v_s} = 9 \tag{5.2.4}$$

which is very close to the value of 9.00041 obtained with the nonideal model in Practice Prob. 5.1. This shows that negligibly small error results from assuming ideal op amp characteristics.

At node O,

$$i_o = \frac{v_o}{40 + 5} + \frac{v_o}{20}\text{mA} \tag{5.2.5}$$

From Eq. (5.2.4), when $v_s = 1$ V, $v_o = 9$ V. Substituting for $v_o = 9$ V in Eq. (5.2.5) produces

$$i_o = 0.2 + 0.45 = 0.65 \text{ mA}$$

This, again, is close to the value of 0.657 mA obtained in Practice Prob. 5.1 with the nonideal model.

Practice Problem 5.2

Repeat Example 5.1 using the ideal op amp model.

Answer: −2, 0.2 mA.

5.4 Inverting Amplifier

In this and the following sections, we consider some useful op amp circuits that often serve as modules for designing more complex circuits. The first of such op amp circuits is the inverting amplifier shown in Fig. 5.10. In this circuit, the noninverting input is grounded, v_i is connected to the inverting input through R_1, and the feedback resistor R_f is connected between the inverting input and output. Our goal is to obtain the relationship between the input voltage v_i and the output voltage v_o. Applying KCL at node 1,

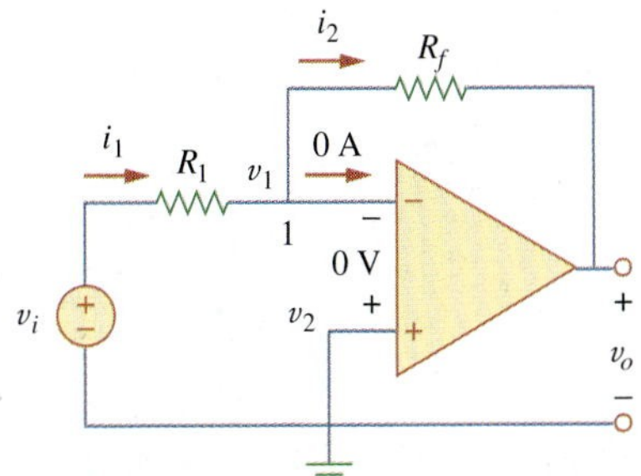

Figure 5.10
The inverting amplifier.

A key feature of the inverting amplifier is that both the input signal and the feedback are applied at the inverting terminal of the op amp.

$$i_1 = i_2 \quad \Rightarrow \quad \frac{v_i - v_1}{R_1} = \frac{v_1 - v_o}{R_f} \tag{5.8}$$

But $v_1 = v_2 = 0$ for an ideal op amp, since the noninverting terminal is grounded. Hence,

$$\frac{v_i}{R_1} = -\frac{v_o}{R_f}$$

or

$$v_o = -\frac{R_f}{R_1} v_i \tag{5.9}$$

Note there are two types of gains: the one here is the closed-loop voltage gain A_v, while the op amp itself has an open-loop voltage gain A.

The voltage gain is $A_v = v_o/v_i = -R_f/R_1$. The designation of the circuit in Fig. 5.10 as an *inverter* arises from the negative sign. Thus,

An inverting amplifier reverses the polarity of the input signal while amplifying it.

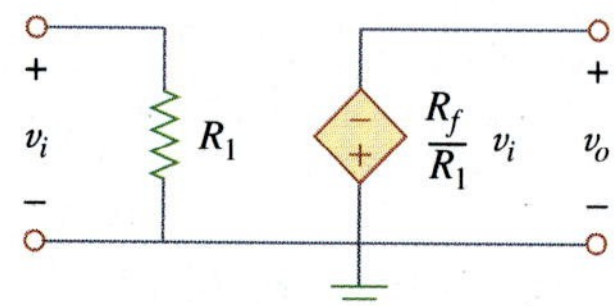

Figure 5.11
An equivalent circuit for the inverter in Fig. 5.10.

Notice that the gain is the feedback resistance divided by the input resistance which means that the gain depends only on the external elements connected to the op amp. In view of Eq. (5.9), an equivalent circuit for the inverting amplifier is shown in Fig. 5.11. The inverting amplifier is used, for example, in a current-to-voltage converter.

Example 5.3

Refer to the op amp in Fig. 5.12. If $v_i = 0.5$ V, calculate: (a) the output voltage v_o, and (b) the current in the 10-kΩ resistor.

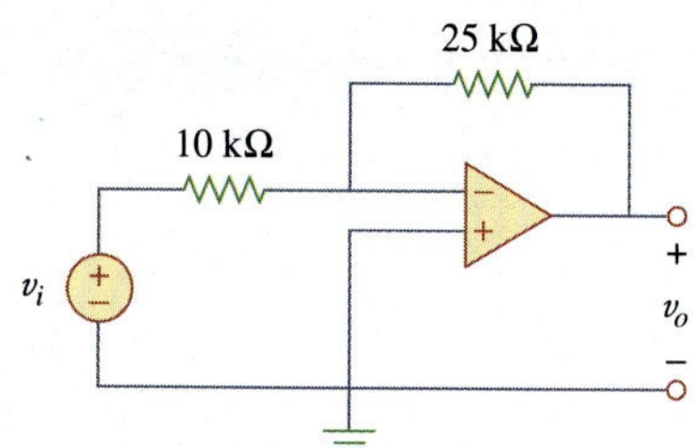

Figure 5.12
For Example 5.3.

Solution:

(a) Using Eq. (5.9),

$$\frac{v_o}{v_i} = -\frac{R_f}{R_1} = -\frac{25}{10} = -2.5$$

$$v_o = -2.5v_i = -2.5(0.5) = -1.25 \text{ V}$$

(b) The current through the 10-kΩ resistor is

$$i = \frac{v_i - 0}{R_1} = \frac{0.5 - 0}{10 \times 10^3} = 50 \ \mu\text{A}$$

Practice Problem 5.3

Find the output of the op amp circuit shown in Fig. 5.13. Calculate the current through the feedback resistor.

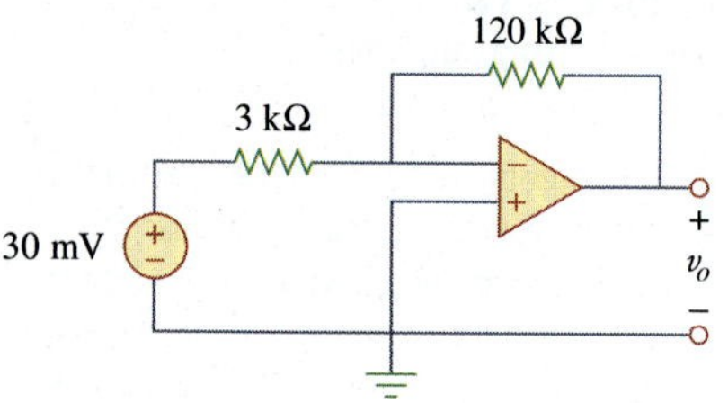

Figure 5.13
For Practice Prob. 5.3.

Answer: −1.2 V, 10 μA.

Example 5.4

Determine v_o in the op amp circuit shown in Fig. 5.14.

Solution:

Applying KCL at node a,

$$\frac{v_a - v_o}{40\ \text{k}\Omega} = \frac{6 - v_a}{20\ \text{k}\Omega}$$

$$v_a - v_o = 12 - 2v_a \quad \Rightarrow \quad v_o = 3v_a - 12$$

But $v_a = v_b = 2$ V for an ideal op amp, because of the zero voltage drop across the input terminals of the op amp. Hence,

$$v_o = 6 - 12 = -6\ \text{V}$$

Notice that if $v_b = 0 = v_a$, then $v_o = -12$, as expected from Eq. (5.9).

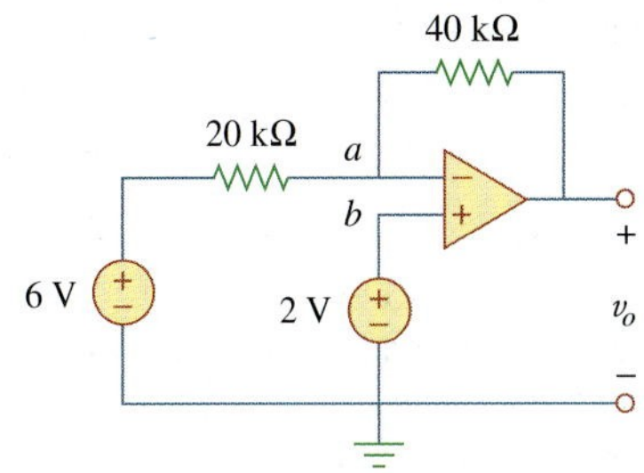

Figure 5.14
For Example 5.4.

Practice Problem 5.4

Two kinds of current-to-voltage converters (also known as *transresistance amplifiers*) are shown in Fig. 5.15.

(a) Show that for the converter in Fig. 5.15(a),

$$\frac{v_o}{i_s} = -R$$

(b) Show that for the converter in Fig. 5.15(b),

$$\frac{v_o}{i_s} = -R_1\left(1 + \frac{R_3}{R_1} + \frac{R_3}{R_2}\right)$$

Answer: Proof.

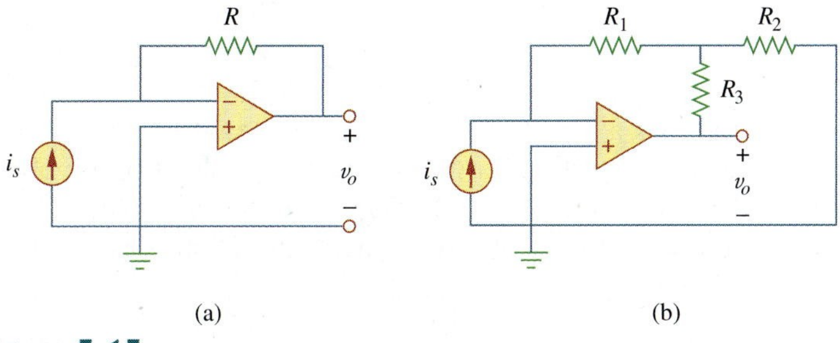

Figure 5.15
For Practice Prob. 5.4.

5.5 Noninverting Amplifier

Another important application of the op amp is the noninverting amplifier shown in Fig. 5.16. In this case, the input voltage v_i is applied directly at the noninverting input terminal, and resistor R_1 is connected

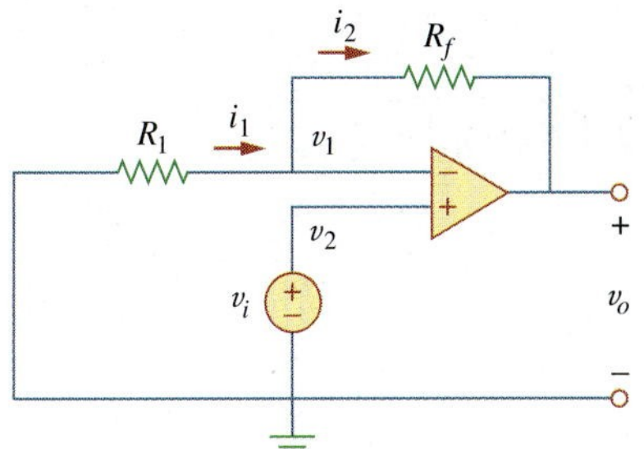

Figure 5.16
The noninverting amplifier.

between the ground and the inverting terminal. We are interested in the output voltage and the voltage gain. Application of KCL at the inverting terminal gives

$$i_1 = i_2 \quad \Rightarrow \quad \frac{0 - v_1}{R_1} = \frac{v_1 - v_o}{R_f} \tag{5.10}$$

But $v_1 = v_2 = v_i$. Equation (5.10) becomes

$$\frac{-v_i}{R_1} = \frac{v_i - v_o}{R_f}$$

or

$$\boxed{v_o = \left(1 + \frac{R_f}{R_1}\right)v_i} \tag{5.11}$$

The voltage gain is $A_v = v_o/v_i = 1 + R_f/R_1$, which does not have a negative sign. Thus, the output has the same polarity as the input.

A noninverting amplifier is an op amp circuit designed to provide a positive voltage gain.

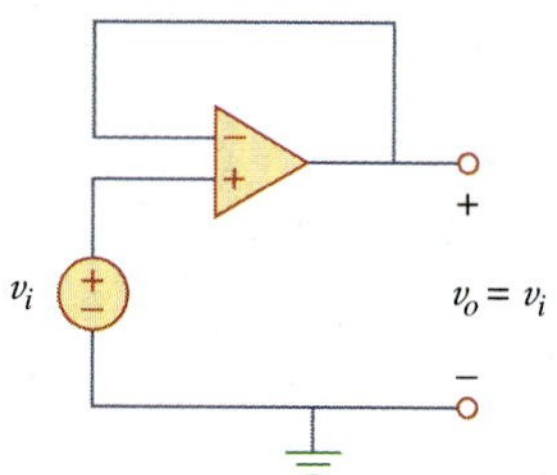

Figure 5.17
The voltage follower.

Again we notice that the gain depends only on the external resistors.

Notice that if feedback resistor $R_f = 0$ (short circuit) or $R_1 = \infty$ (open circuit) or both, the gain becomes 1. Under these conditions ($R_f = 0$ and $R_1 = \infty$), the circuit in Fig. 5.16 becomes that shown in Fig. 5.17, which is called a *voltage follower* (or *unity gain amplifier*) because the output follows the input. Thus, for a voltage follower

$$\boxed{v_o = v_i} \tag{5.12}$$

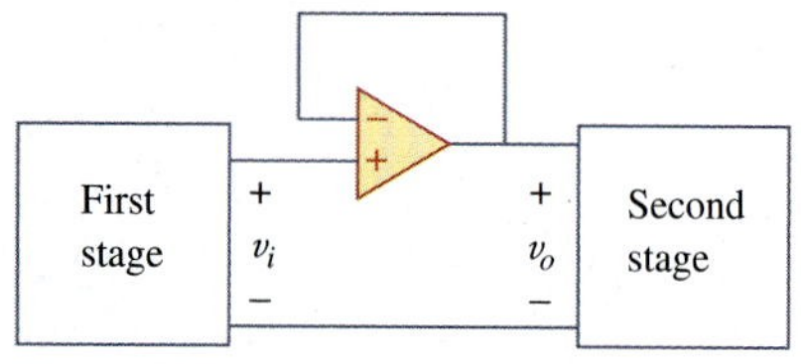

Figure 5.18
A voltage follower used to isolate two cascaded stages of a circuit.

Such a circuit has a very high input impedance and is therefore useful as an intermediate-stage (or buffer) amplifier to isolate one circuit from another, as portrayed in Fig. 5.18. The voltage follower minimizes interaction between the two stages and eliminates interstage loading.

Example 5.5

For the op amp circuit in Fig. 5.19, calculate the output voltage v_o.

Solution:
We may solve this in two ways: using superposition and using nodal analysis.

■ **METHOD 1** Using superposition, we let

$$v_o = v_{o1} + v_{o2}$$

where v_{o1} is due to the 6-V voltage source, and v_{o2} is due to the 4-V input. To get v_{o1}, we set the 4-V source equal to zero. Under this condition, the circuit becomes an inverter. Hence Eq. (5.9) gives

$$v_{o1} = -\frac{10}{4}(6) = -15 \text{ V}$$

To get v_{o2}, we set the 6-V source equal to zero. The circuit becomes a noninverting amplifier so that Eq. (5.11) applies.

$$v_{o2} = \left(1 + \frac{10}{4}\right)4 = 14 \text{ V}$$

Thus,

$$v_o = v_{o1} + v_{o2} = -15 + 14 = -1 \text{ V}$$

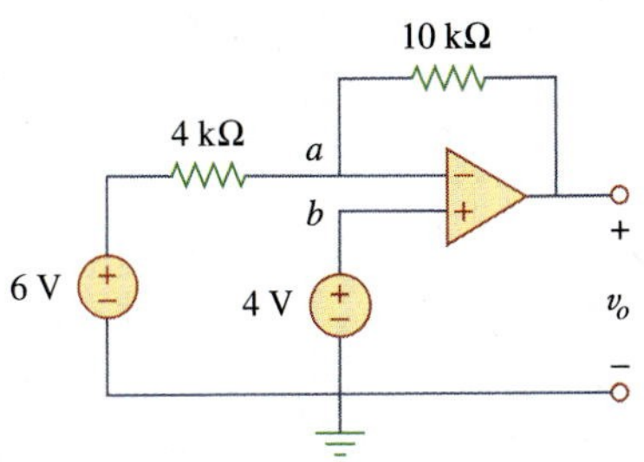

Figure 5.19
For Example 5.5.

METHOD 2 Applying KCL at node a,

$$\frac{6 - v_a}{4} = \frac{v_a - v_o}{10}$$

But $v_a = v_b = 4$, and so

$$\frac{6 - 4}{4} = \frac{4 - v_o}{10} \quad \Rightarrow \quad 5 = 4 - v_o$$

or $v_o = -1$ V, as before.

Practice Problem 5.5

Calculate v_o in the circuit of Fig. 5.20.

Answer: 7 V.

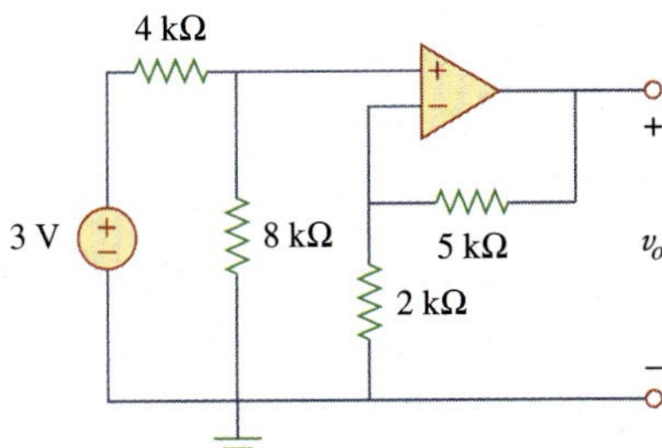

Figure 5.20
For Practice Prob. 5.5.

5.6 Summing Amplifier

Besides amplification, the op amp can perform addition and subtraction. The addition is performed by the summing amplifier covered in this section; the subtraction is performed by the difference amplifier covered in the next section.

> A summing amplifier is an op amp circuit that combines several inputs and produces an output that is the weighted sum of the inputs.

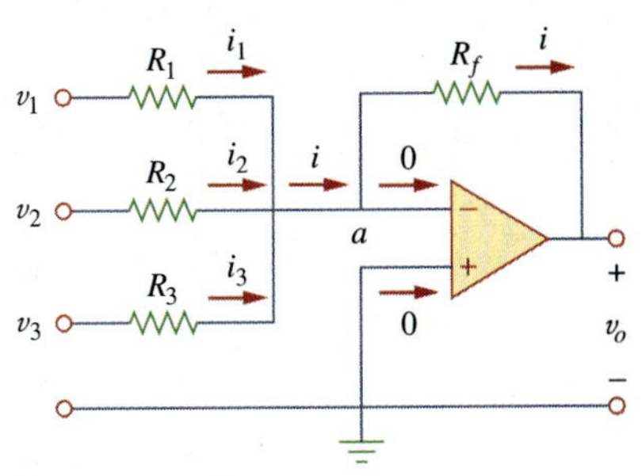

Figure 5.21
The summing amplifier.

The summing amplifier, shown in Fig. 5.21, is a variation of the inverting amplifier. It takes advantage of the fact that the inverting configuration can handle many inputs at the same time. We keep in mind that the current entering each op amp input is zero. Applying KCL at node a gives

$$i = i_1 + i_2 + i_3 \tag{5.13}$$

But

$$i_1 = \frac{v_1 - v_a}{R_1}, \quad i_2 = \frac{v_2 - v_a}{R_2}$$
$$i_3 = \frac{v_3 - v_a}{R_3}, \quad i = \frac{v_a - v_o}{R_f} \tag{5.14}$$

We note that $v_a = 0$ and substitute Eq. (5.14) into Eq. (5.13). We get

$$v_o = -\left(\frac{R_f}{R_1}v_1 + \frac{R_f}{R_2}v_2 + \frac{R_f}{R_3}v_3\right) \tag{5.15}$$

indicating that the output voltage is a weighted sum of the inputs. For this reason, the circuit in Fig. 5.21 is called a *summer*. Needless to say, the summer can have more than three inputs.

Example 5.6

Calculate v_o and i_o in the op amp circuit in Fig. 5.22.

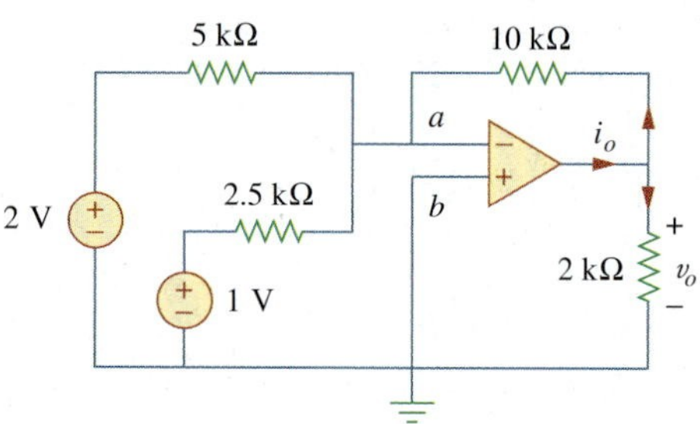

Figure 5.22
For Example 5.6.

Solution:
This is a summer with two inputs. Using Eq. (5.15) gives

$$v_o = -\left[\frac{10}{5}(2) + \frac{10}{2.5}(1)\right] = -(4 + 4) = -8 \text{ V}$$

The current i_o is the sum of the currents through the 10-kΩ and 2-kΩ resistors. Both of these resistors have voltage $v_o = -8$ V across them, since $v_a = v_b = 0$. Hence,

$$i_o = \frac{v_o - 0}{10} + \frac{v_o - 0}{2} \text{ mA} = -0.8 - 4 = -4.8 \text{ mA}$$

Practice Problem 5.6

Find v_o and i_o in the op amp circuit shown in Fig. 5.23.

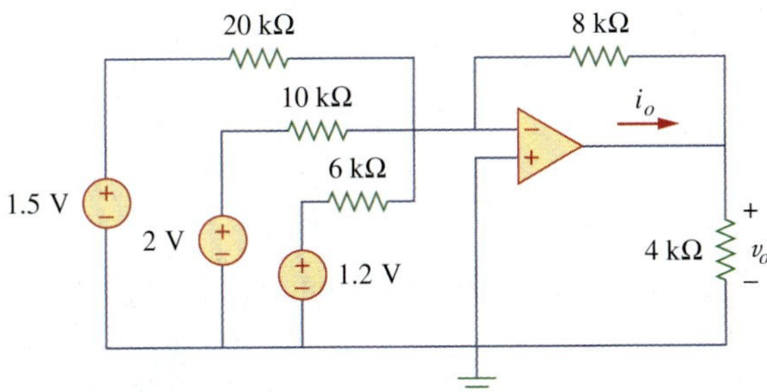

Figure 5.23
For Practice Prob. 5.6.

Answer: −3.8 V, −1.425 mA.

5.7 Difference Amplifier

Difference (or differential) amplifiers are used in various applications where there is need to amplify the difference between two input signals. They are first cousins of the *instrumentation amplifier*, the most useful and popular amplifier, which we will discuss in Section 5.10.

The difference amplifier is also known as the *subtractor*, for reasons to be shown later.

> A difference amplifier is a device that amplifies the difference between two inputs but rejects any signals common to the two inputs.

Consider the op amp circuit shown in Fig. 5.24. Keep in mind that zero currents enter the op amp terminals. Applying KCL to node a,

$$\frac{v_1 - v_a}{R_1} = \frac{v_a - v_o}{R_2}$$

or

$$v_o = \left(\frac{R_2}{R_1} + 1\right)v_a - \frac{R_2}{R_1}v_1 \tag{5.16}$$

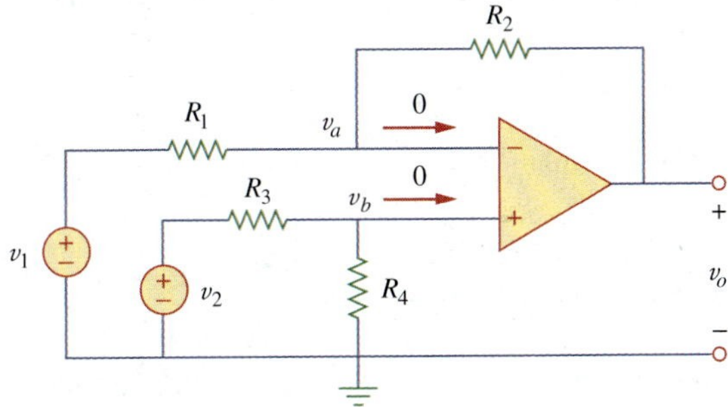

Figure 5.24
Difference amplifier.

Applying KCL to node b,

$$\frac{v_2 - v_b}{R_3} = \frac{v_b - 0}{R_4}$$

or

$$v_b = \frac{R_4}{R_3 + R_4} v_2 \tag{5.17}$$

But $v_a = v_b$. Substituting Eq. (5.17) into Eq. (5.16) yields

$$v_o = \left(\frac{R_2}{R_1} + 1\right)\frac{R_4}{R_3 + R_4} v_2 - \frac{R_2}{R_1} v_1$$

or

$$v_o = \frac{R_2(1 + R_1/R_2)}{R_1(1 + R_3/R_4)} v_2 - \frac{R_2}{R_1} v_1 \tag{5.18}$$

Since a difference amplifier must reject a signal common to the two inputs, the amplifier must have the property that $v_o = 0$ when $v_1 = v_2$. This property exists when

$$\frac{R_1}{R_2} = \frac{R_3}{R_4} \tag{5.19}$$

Thus, when the op amp circuit is a difference amplifier, Eq. (5.18) becomes

$$v_o = \frac{R_2}{R_1}(v_2 - v_1) \tag{5.20}$$

If $R_2 = R_1$ and $R_3 = R_4$, the difference amplifier becomes a *subtractor*, with the output

$$v_o = v_2 - v_1 \tag{5.21}$$

Example 5.7

Design an op amp circuit with inputs v_1 and v_2 such that $v_o = -5v_1 + 3v_2$.

Solution:
The circuit requires that

$$v_o = 3v_2 - 5v_1 \tag{5.7.1}$$

This circuit can be realized in two ways.

Design 1 If we desire to use only one op amp, we can use the op amp circuit of Fig. 5.24. Comparing Eq. (5.7.1) with Eq. (5.18), we see

$$\frac{R_2}{R_1} = 5 \quad \Rightarrow \quad R_2 = 5R_1 \tag{5.7.2}$$

Also,

$$5\frac{(1 + R_1/R_2)}{(1 + R_3/R_4)} = 3 \quad \Rightarrow \quad \frac{\frac{6}{5}}{1 + R_3/R_4} = \frac{3}{5}$$

or

$$2 = 1 + \frac{R_3}{R_4} \quad \Rightarrow \quad R_3 = R_4 \qquad \textbf{(5.7.3)}$$

If we choose $R_1 = 10$ kΩ and $R_3 = 20$ kΩ, then $R_2 = 50$ kΩ and $R_4 = 20$ kΩ.

Design 2 If we desire to use more than one op amp, we may cascade an inverting amplifier and a two-input inverting summer, as shown in Fig. 5.25. For the summer,

$$v_o = -v_a - 5v_1 \qquad \textbf{(5.7.4)}$$

and for the inverter,

$$v_a = -3v_2 \qquad \textbf{(5.7.5)}$$

Combining Eqs. (5.7.4) and (5.7.5) gives

$$v_o = 3v_2 - 5v_1$$

which is the desired result. In Fig. 5.25, we may select $R_1 = 10$ kΩ and $R_3 = 20$ kΩ or $R_1 = R_3 = 10$ kΩ.

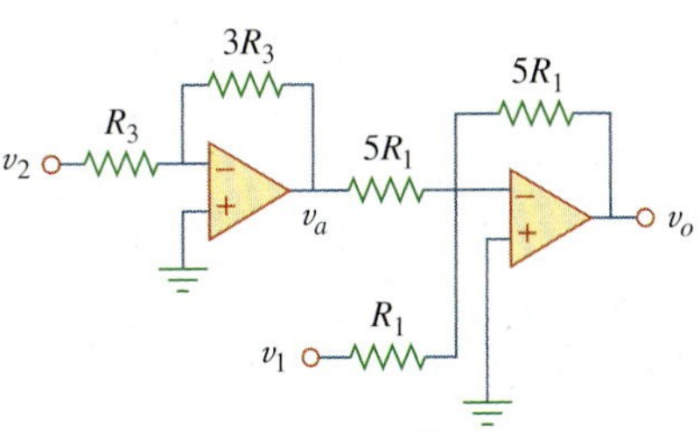

Figure 5.25
For Example 5.7.

Practice Problem 5.7

Design a difference amplifier with gain 5.

Answer: Typical: $R_1 = R_3 = 10$kΩ, $R_2 = R_4 = 50$ kΩ.

Example 5.8

An *instrumentation amplifier* shown in Fig. 5.26 is an amplifier of low-level signals used in process control or measurement applications and commercially available in single-package units. Show that

$$v_o = \frac{R_2}{R_1}\left(1 + \frac{2R_3}{R_4}\right)(v_2 - v_1)$$

Solution:

We recognize that the amplifier A_3 in Fig. 5.26 is a difference amplifier. Thus, from Eq. (5.20),

$$v_o = \frac{R_2}{R_1}(v_{o2} - v_{o1}) \qquad \textbf{(5.8.1)}$$

Since the op amps A_1 and A_2 draw no current, current i flows through the three resistors as though they were in series. Hence,

$$v_{o1} - v_{o2} = i(R_3 + R_4 + R_3) = i(2R_3 + R_4) \qquad \textbf{(5.8.2)}$$

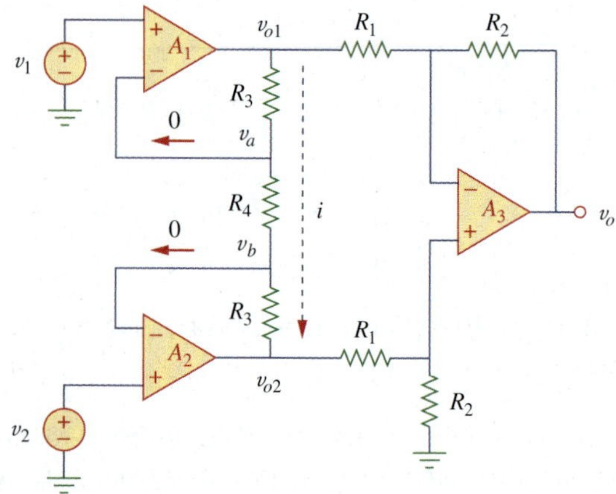

Figure 5.26
Instrumentation amplifier; for Example 5.8.

But

$$i = \frac{v_a - v_b}{R_4}$$

and $v_a = v_1$, $v_b = v_2$. Therefore,

$$i = \frac{v_1 - v_2}{R_4} \tag{5.8.3}$$

Inserting Eqs. (5.8.2) and (5.8.3) into Eq. (5.8.1) gives

$$v_o = \frac{R_2}{R_1}\left(1 + \frac{2R_3}{R_4}\right)(v_2 - v_1)$$

as required. We will discuss the instrumentation amplifier in detail in Section 5.10.

Practice Problem 5.8

Obtain i_o in the instrumentation amplifier circuit of Fig. 5.27.

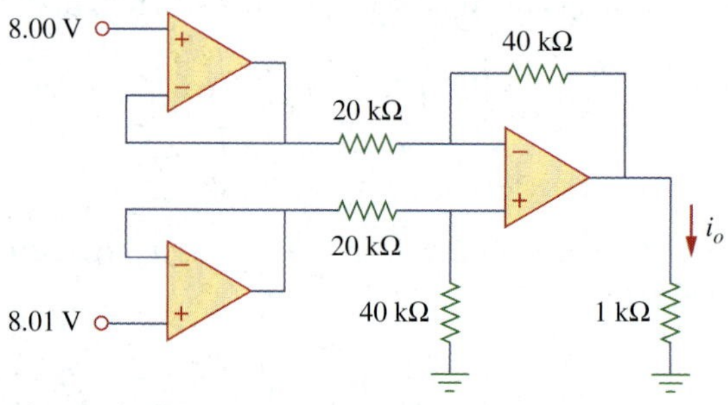

Figure 5.27
Instrumentation amplifier; for Practice Prob. 5.8.

Answer: 20 μA.

5.8 Cascaded Op Amp Circuits

As we know, op amp circuits are modules or building blocks for designing complex circuits. It is often necessary in practical applications to connect op amp circuits in cascade (i.e., head to tail) to achieve a large overall gain. In general, two circuits are cascaded when they are connected in tandem, one behind another in a single file.

A cascade connection is a head-to-tail arrangement of two or more op amp circuits such that the output of one is the input of the next.

When op amp circuits are cascaded, each circuit in the string is called a *stage*; the original input signal is increased by the gain of the individual stage. Op amp circuits have the advantage that they can be cascaded without changing their input-output relationships. This is due to the fact that each (ideal) op amp circuit has infinite input resistance and zero output resistance. Figure 5.28 displays a block diagram representation of three op amp circuits in cascade. Since the output of one stage is the input to the next stage, the overall gain of the cascade connection is the product of the gains of the individual op amp circuits, or

$$A = A_1 A_2 A_3 \tag{5.22}$$

Although the cascade connection does not affect the op amp input-output relationships, care must be exercised in the design of an actual op amp circuit to ensure that the load due to the next stage in the cascade does not saturate the op amp.

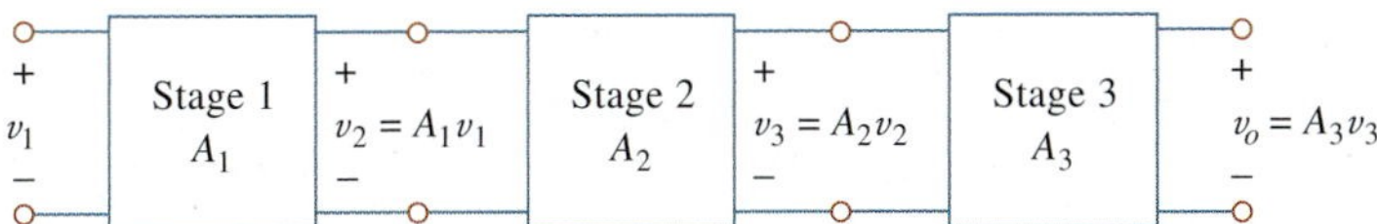

Figure 5.28
A three-stage cascaded connection.

Example 5.9

Find v_o and i_o in the circuit in Fig. 5.29.

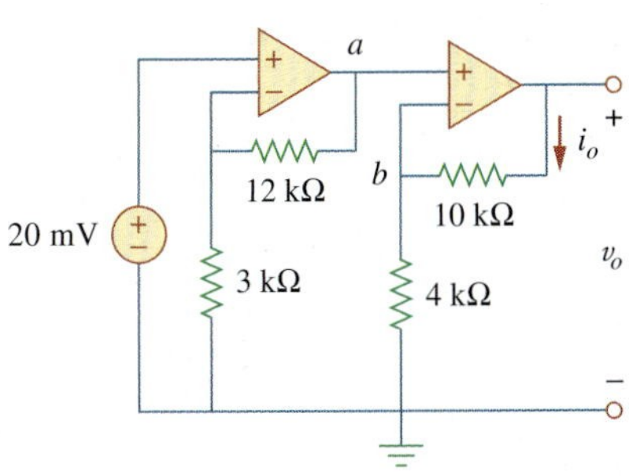

Figure 5.29
For Example 5.9.

Solution:
This circuit consists of two noninverting amplifiers cascaded. At the output of the first op amp,

$$v_a = \left(1 + \frac{12}{3}\right)(20) = 100 \text{ mV}$$

At the output of the second op amp,

$$v_o = \left(1 + \frac{10}{4}\right)v_a = (1 + 2.5)100 = 350 \text{ mV}$$

The required current i_o is the current through the 10-kΩ resistor.

$$i_o = \frac{v_o - v_b}{10} \text{ mA}$$

But $v_b = v_a = 100$ mV. Hence,

$$i_o = \frac{(350 - 100) \times 10^{-3}}{10 \times 10^3} = 25\ \mu\text{A}$$

Practice Problem 5.9

Determine v_o and i_o in the op amp circuit in Fig. 5.30.

Answer: 24 V, 2 mA.

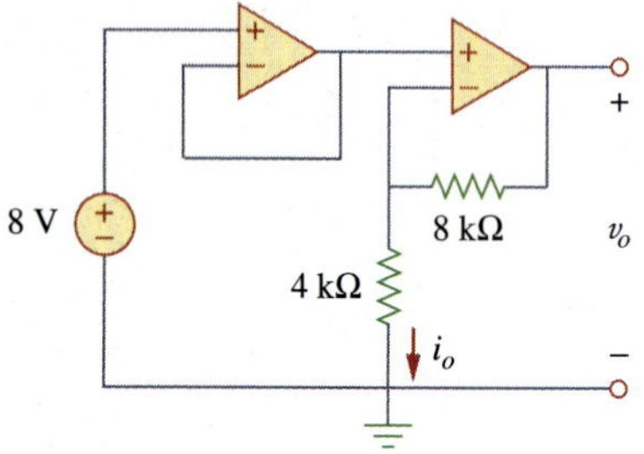

Figure 5.30
For Practice Prob. 5.9.

Example 5.10

If $v_1 = 1$ V and $v_2 = 2$ V, find v_o in the op amp circuit of Fig. 5.31.

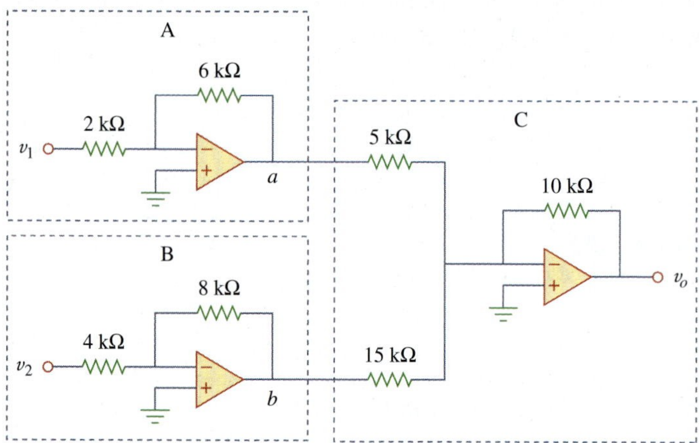

Figure 5.31
For Example 5.10.

Solution:

1. **Define.** The problem is clearly defined.
2. **Present.** With an input of v_1 of 1 V and of v_2 of 2 V, determine the output voltage of the circuit shown in Figure 5.31. The op amp circuit is actually composed of three circuits. The first circuit acts as an amplifier of gain $-3(-6\ \text{k}\Omega/2\ \text{k}\Omega)$ for v_1 and the second functions as an amplifier of gain $-2(-8\ \text{k}\Omega/4\ \text{k}\Omega)$ for v_2. The last circuit serves as a summer of two different gains for the output of the other two circuits.
3. **Alternative.** There are different ways of working with this circuit. Since it involves ideal op amps, then a purely mathematical

approach will work quite easily. A second approach would be to use *PSpice* as a confirmation of the math.

4. **Attempt.** Let the output of the first op amp circuit be designated as v_{11} and the output of the second op amp circuit be designated as v_{22}. Then we get

$$v_{11} = -3v_1 = -3 \times 1 = -3\text{ V},$$
$$v_{22} = -2v_2 = -2 \times 2 = -4\text{ V}$$

In the third circuit we have

$$\begin{aligned} v_o &= -(10\text{ k}\Omega/5\text{ k}\Omega)v_{11} + [-(10\text{ k}\Omega/15\text{ k}\Omega)v_{22}] \\ &= -2(-3) - (2/3)(-4) \\ &= 6 + 2.667 = \mathbf{8.667\text{ V}} \end{aligned}$$

5. **Evaluate.** In order to properly evaluate our solution, we need to identify a reasonable check. Here we can easily use *PSpice* to provide that check.

   Now we can simulate this in *PSpice*. We see the results are shown in Fig. 5.32.

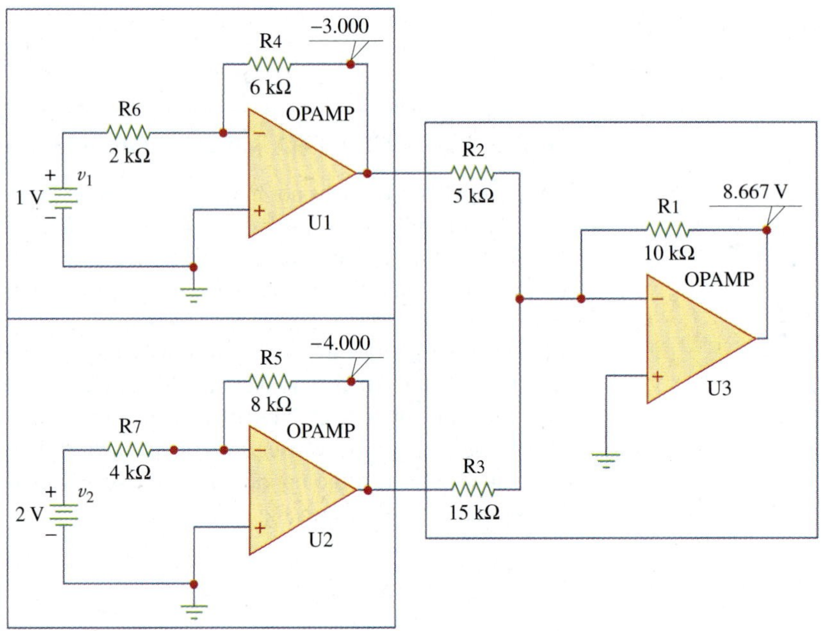

Figure 5.32
For Example 5.10.

We note that we obtain the same results using two entirely different techniques (the first is to treat the op amp circuits as just gains and a summer and the second is to use circuit analysis with *PSpice*). This is a very good method of assuring that we have the correct answer.

6. **Satisfactory?** We are satisfied we have obtained the asked for results. We can now present our work as a solution to the problem.

Practice Problem 5.10

If $v_1 = 4$ V and $v_2 = 3$ V, find v_o in the op amp circuit of Fig. 5.33.

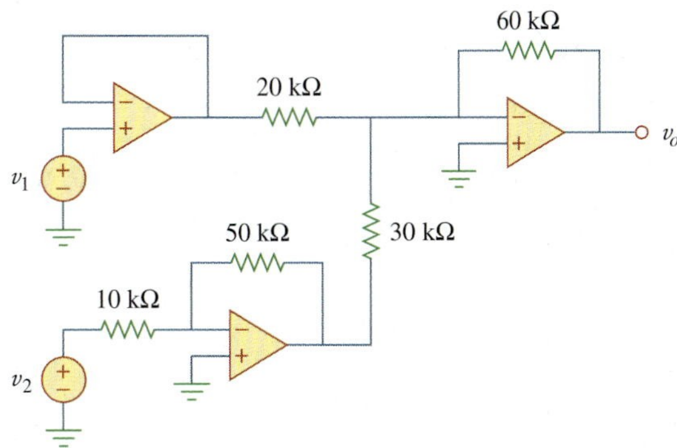

Figure 5.33
For Practice Prob. 5.10.

Answer: 18 V.

5.9 Op Amp Circuit Analysis with *PSpice*

PSpice for Windows does not have a model for an ideal op amp, although one may create one as a subcircuit using the *Create Subcircuit* line in the *Tools* menu. Rather than creating an ideal op amp, we will use one of the four nonideal, commercially available op amps supplied in the *PSpice* library *eval.slb*. The op amp models have the part names LF411, LM111, LM324, and uA741, as shown in Fig. 5.34. Each of them can be obtained from **Draw/Get New Part/libraries · · · /eval.lib** or by simply selecting **Draw/Get New Part** and typing the part name in the *PartName* dialog box, as usual. Note that each of them requires dc supplies, without which the op amp will not work. The dc supplies should be connected as shown in Fig. 5.3.

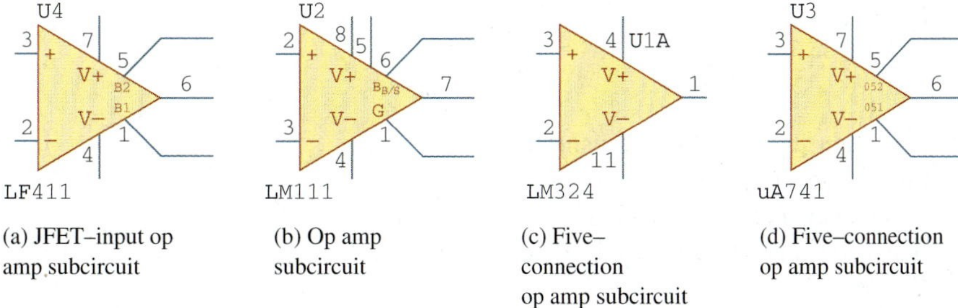

(a) JFET–input op amp subcircuit

(b) Op amp subcircuit

(c) Five–connection op amp subcircuit

(d) Five–connection op amp subcircuit

Figure 5.34
Nonideal op amp model available in *PSpice*.

Example 5.11

Use *PSpice* to solve the op amp circuit for Example 5.1.

Solution:

Using Schematics, we draw the circuit in Fig. 5.6(a) as shown in Fig. 5.35. Notice that the positive terminal of the voltage source v_s is connected to the inverting terminal (pin 2) via the 10-kΩ resistor, while the noninverting terminal (pin 3) is grounded as required in Fig. 5.6(a). Also, notice how the op amp is powered; the positive power supply terminal V+ (pin 7) is connected to a 15-V dc voltage source, while the negative power supply terminal V− (pin 4) is connected to −15 V. Pins 1 and 5 are left floating because they are used for offset null adjustment, which does not concern us in this chapter. Besides adding the dc power supplies to the original circuit in Fig. 5.6(a), we have also added pseudocomponents VIEWPOINT and IPROBE to respectively measure the output voltage v_o at pin 6 and the required current i through the 20-kΩ resistor.

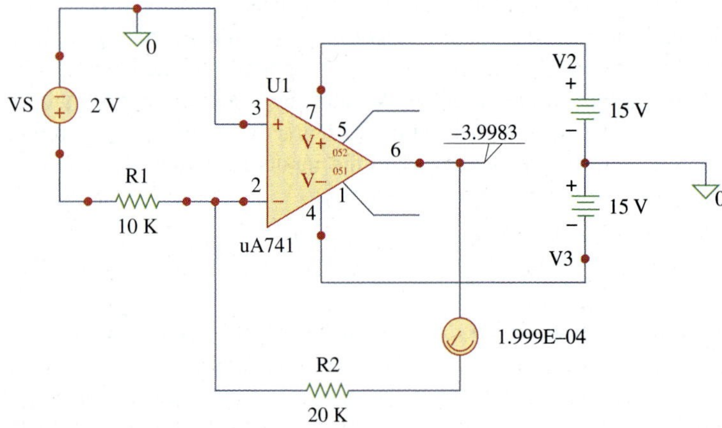

Figure 5.35
Schematic for Example 5.11.

After saving the schematic, we simulate the circuit by selecting **Analysis/Simulate** and have the results displayed on VIEWPOINT and IPROBE. From the results, the closed-loop gain is

$$\frac{v_o}{v_s} = \frac{-3.9983}{2} = -1.99915$$

and $i = 0.1999$ mA, in agreement with the results obtained analytically in Example 5.1.

Practice Problem 5.11

Rework Practice Prob. 5.1 using *PSpice*.

Answer: 9.0027, 0.6502 mA.

5.10 †Applications

The op amp is a fundamental building block in modern electronic instrumentation. It is used extensively in many devices, along with resistors and other passive elements. Its numerous practical applications include instrumentation amplifiers, digital-to-analog converters, analog computers, level shifters, filters, calibration circuits, inverters, summers, integrators, differentiators, subtractors, logarithmic amplifiers, comparators, gyrators, oscillators, rectifiers, regulators, voltage-to-current converters, current-to-voltage converters, and clippers. Some of these we have already considered. We will consider two more applications here: the digital-to-analog converter and the instrumentation amplifier.

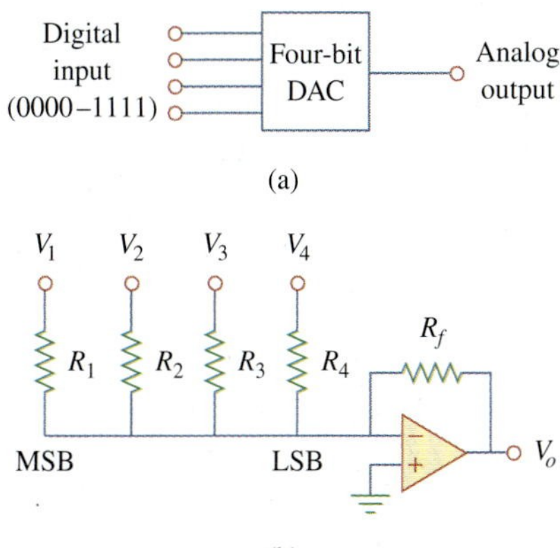

Figure 5.36
Four-bit DAC: (a) block diagram, (b) binary weighted ladder type.

5.10.1 Digital-to-Analog Converter

The digital-to-analog converter (DAC) transforms digital signals into analog form. A typical example of a four-bit DAC is illustrated in Fig. 5.36(a). The four-bit DAC can be realized in many ways. A simple realization is the *binary weighted ladder*, shown in Fig. 5.36(b). The bits are weights according to the magnitude of their place value, by descending value of R_f/R_n so that each lesser bit has half the weight of the next higher. This is obviously an inverting summing amplifier. The output is related to the inputs as shown in Eq. (5.15). Thus,

$$-V_o = \frac{R_f}{R_1}V_1 + \frac{R_f}{R_2}V_2 + \frac{R_f}{R_3}V_3 + \frac{R_f}{R_4}V_4 \tag{5.23}$$

Input V_1 is called the *most significant bit* (MSB), while input V_4 is the *least significant bit* (LSB). Each of the four binary inputs $V_1, \ldots, V_4$ can assume only two voltage levels: 0 or 1 V. By using the proper input and feedback resistor values, the DAC provides a single output that is proportional to the inputs.

In practice, the voltage levels may be typically 0 and ± 5 V.

Example 5.12

In the op amp circuit of Fig. 5.36(b), let $R_f = 10$ kΩ, $R_1 = 10$ kΩ, $R_2 = 20$ kΩ, $R_3 = 40$ kΩ, and $R_4 = 80$ kΩ. Obtain the analog output for binary inputs [0000], [0001], [0010], . . . , [1111].

Solution:
Substituting the given values of the input and feedback resistors in Eq. (5.23) gives

$$-V_o = \frac{R_f}{R_1}V_1 + \frac{R_f}{R_2}V_2 + \frac{R_f}{R_3}V_3 + \frac{R_f}{R_4}V_4$$
$$= V_1 + 0.5V_2 + 0.25V_3 + 0.125V_4$$

Using this equation, a digital input $[V_1V_2V_3V_4] = [0000]$ produces an analog output of $-V_o = 0$ V; $[V_1V_2V_3V_4] = [0001]$ gives $-V_o = 0.125$ V.

Similarly,

$$[V_1\,V_2\,V_3\,V_4] = [0010] \quad \Rightarrow \quad -V_o = 0.25 \text{ V}$$
$$[V_1\,V_2\,V_3\,V_4] = [0011] \quad \Rightarrow \quad -V_o = 0.25 + 0.125 = 0.375 \text{ V}$$
$$[V_1\,V_2\,V_3\,V_4] = [0100] \quad \Rightarrow \quad -V_o = 0.5 \text{ V}$$
$$\vdots$$
$$[V_1\,V_2\,V_3\,V_4] = [1111] \quad \Rightarrow \quad -V_o = 1 + 0.5 + 0.25 + 0.125 = 1.875 \text{ V}$$

Table 5.2 summarizes the result of the digital-to-analog conversion. Note that we have assumed that each bit has a value of 0.125 V. Thus, in this system, we cannot represent a voltage between 1.000 and 1.125, for example. This lack of resolution is a major limitation of digital-to-analog conversions. For greater accuracy, a word representation with a greater number of bits is required. Even then a digital representation of an analog voltage is never exact. In spite of this inexact representation, digital representation has been used to accomplish remarkable things such as audio CDs and digital photography.

TABLE 5.2

Input and output values of the four-bit DAC.

Binary input $[V_1V_2V_3V_4]$	Decimal value	Output $-V_o$
0000	0	0
0001	1	0.125
0010	2	0.25
0011	3	0.375
0100	4	0.5
0101	5	0.625
0110	6	0.75
0111	7	0.875
1000	8	1.0
1001	9	1.125
1010	10	1.25
1011	11	1.375
1100	12	1.5
1101	13	1.625
1110	14	1.75
1111	15	1.875

Practice Problem 5.12

A three-bit DAC is shown in Fig. 5.37.

(a) Determine $|V_o|$ for $[V_1V_2V_3] = [010]$.
(b) Find $|V_o|$ if $[V_1V_2V_3] = [110]$.
(c) If $|V_o| = 1.25$ V is desired, what should be $[V_1V_2V_3]$?
(d) To get $|V_o| = 1.75$ V, what should be $[V_1V_2V_3]$?

Answer: 0.5 V, 1.5 V, [101], [111].

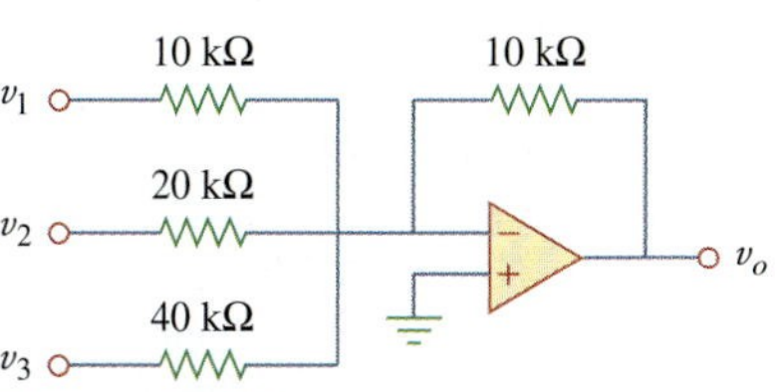

Figure 5.37
Three-bit DAC; for Practice Prob. 5.12.

5.10.2 Instrumentation Amplifiers

One of the most useful and versatile op amp circuits for precision measurement and process control is the *instrumentation amplifier* (IA), so called because of its widespread use in measurement systems. Typical applications of IAs include isolation amplifiers, thermocouple amplifiers, and data acquisition systems.

The instrumentation amplifier is an extension of the difference amplifier in that it amplifies the difference between its input signals. As shown in Fig. 5.26 (see Example 5.8), an instrumentation amplifier typically consists of three op amps and seven resistors. For convenience, the amplifier is shown again in Fig. 5.38(a), where the resistors are made equal except for the external gain-setting resistor R_G, connected between the gain set terminals. Figure 5.38(b) shows its schematic symbol. Example 5.8 showed that

$$v_o = A_v(v_2 - v_1) \tag{5.24}$$

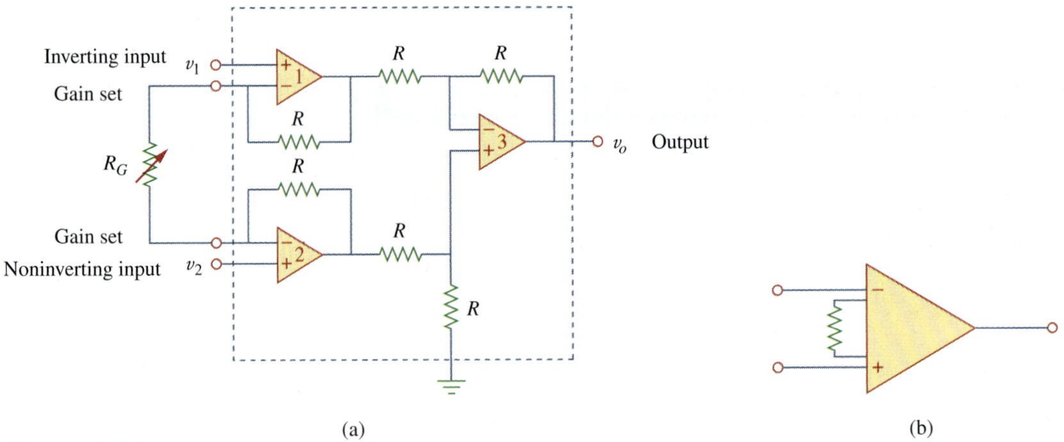

Figure 5.38
(a) The instrumentation amplifier with an external resistance to adjust the gain, (b) schematic diagram.

where the voltage gain is

$$A_v = 1 + \frac{2R}{R_G} \tag{5.25}$$

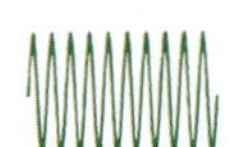

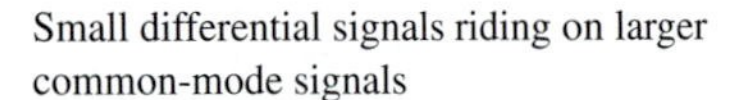

As shown in Fig. 5.39, the instrumentation amplifier amplifies small differential signal voltages superimposed on larger common-mode

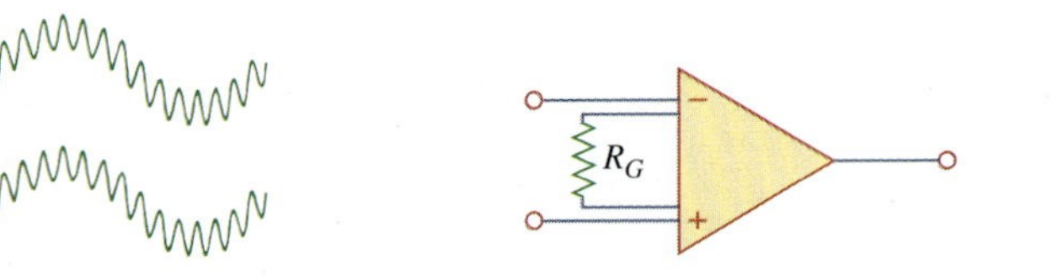

Figure 5.39
The IA rejects common voltages but amplifies small signal voltages.
T. L. Floyd, *Electronic Devices, 2nd ed., Englewood Cliffs,* NJ: Prentice Hall, 1996, p. 795.

voltages. Since the common-mode voltages are equal, they cancel each other.

The IA has three major characteristics:

1. The voltage gain is adjusted by *one* external resistor R_G.
2. The input impedance of both inputs is very high and does not vary as the gain is adjusted.
3. The output v_o depends on the difference between the inputs v_1 and v_2, not on the voltage common to them (common-mode voltage).

Due to the widespread use of IAs, manufacturers have developed these amplifiers on single-package units. A typical example is the LH0036, developed by National Semiconductor. The gain can be varied from 1 to 1,000 by an external resistor whose value may vary from 100 Ω to 10 kΩ.

Example 5.13

In Fig. 5.38, let $R = 10\text{ k}\Omega$, $v_1 = 2.011$ V, and $v_2 = 2.017$ V. If R_G is adjusted to 500 Ω, determine: (a) the voltage gain, (b) the output voltage v_o.

Solution:

(a) The voltage gain is

$$A_v = 1 + \frac{2R}{R_G} = 1 + \frac{2 \times 10{,}000}{500} = 41$$

(b) The output voltage is

$$v_o = A_v(v_2 - v_1) = 41(2.017 - 2.011) = 41(6)\text{ mV} = 246\text{ mV}$$

Practice Problem 5.13

Determine the value of the external gain-setting resistor R_G required for the IA in Fig. 5.38 to produce a gain of 142 when $R = 25\text{ k}\Omega$.

Answer: 354.6 Ω.

5.11 Summary

1. The op amp is a high-gain amplifier that has high input resistance and low output resistance.
2. Table 5.3 summarizes the op amp circuits considered in this chapter. The expression for the gain of each amplifier circuit holds whether the inputs are dc, ac, or time-varying in general.

TABLE 5.3

Summary of basic op amp circuits.

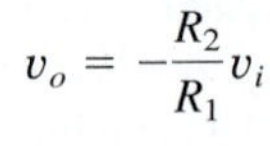

Op amp circuit	Name/output-input relationship
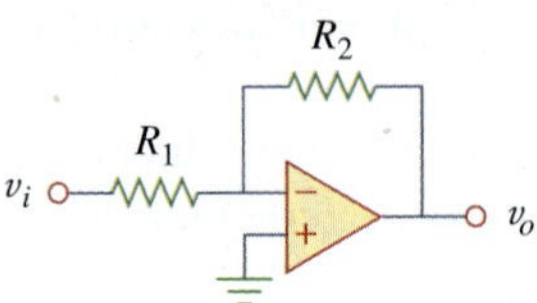	Inverting amplifier $v_o = -\frac{R_2}{R_1} v_i$
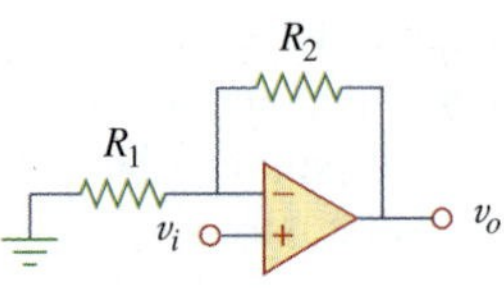	Noninverting amplifier $v_o = \left(1 + \frac{R_2}{R_1}\right) v_i$
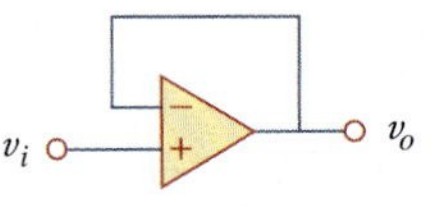	Voltage follower $v_o = v_i$
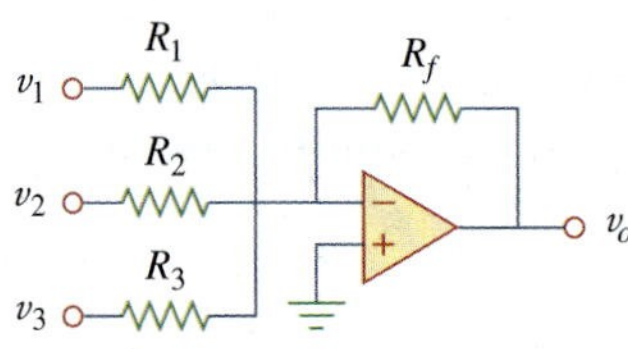	Summer $v_o = -\left(\frac{R_f}{R_1} v_1 + \frac{R_f}{R_2} v_2 + \frac{R_f}{R_3} v_3\right)$
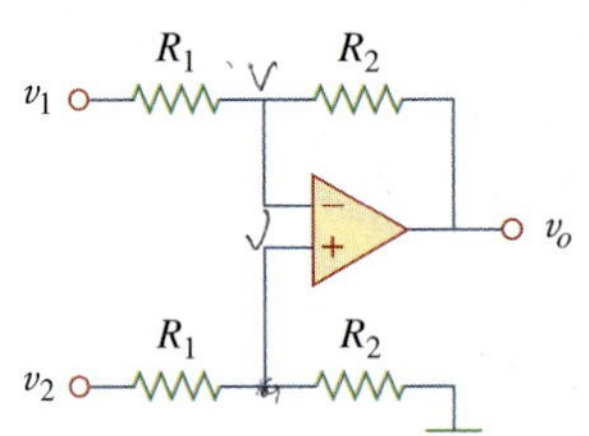	Difference amplifier $v_o = \frac{R_2}{R_1}(v_2 - v_1)$

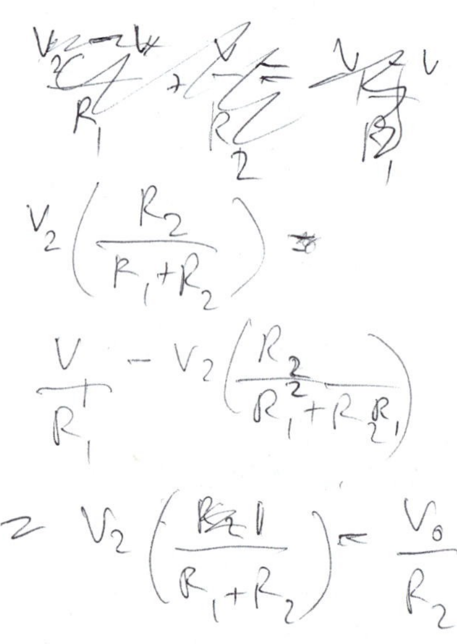

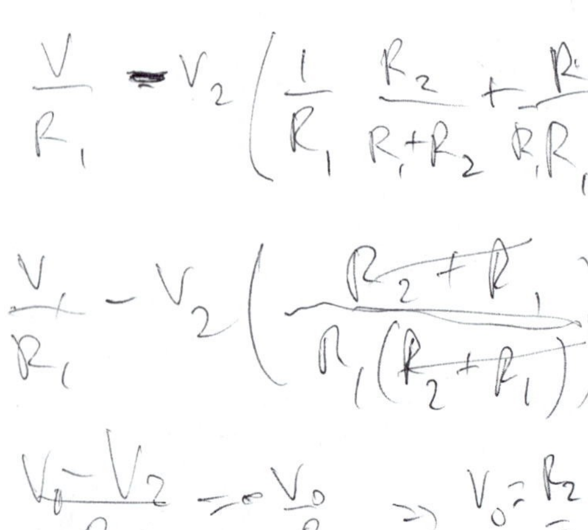

3. An ideal op amp has an infinite input resistance, a zero output resistance, and an infinite gain.
4. For an ideal op amp, the current into each of its two input terminals is zero, and the voltage across its input terminals is negligibly small.
5. In an inverting amplifier, the output voltage is a negative multiple of the input.
6. In a noninverting amplifier, the output is a positive multiple of the input.
7. In a voltage follower, the output follows the input.
8. In a summing amplifier, the output is the weighted sum of the inputs.
9. In a difference amplifier, the output is proportional to the difference of the two inputs.
10. Op amp circuits may be cascaded without changing their input-output relationships.
11. *PSpice* can be used to analyze an op amp circuit.
12. Typical applications of the op amp considered in this chapter include the digital-to-analog converter and the instrumentation amplifier.

Review Questions

5.1 The two input terminals of an op amp are labeled as:

(a) high and low.

(b) positive and negative.

(c) inverting and noninverting.

(d) differential and nondifferential.

5.2 For an ideal op amp, which of the following statements are not true?

(a) The differential voltage across the input terminals is zero.

(b) The current into the input terminals is zero.

(c) The current from the output terminal is zero.

(d) The input resistance is zero.

(e) The output resistance is zero.

5.3 For the circuit in Fig. 5.40, voltage v_o is:

(a) -6 V (b) -5 V

(c) -1.2 V (d) -0.2 V

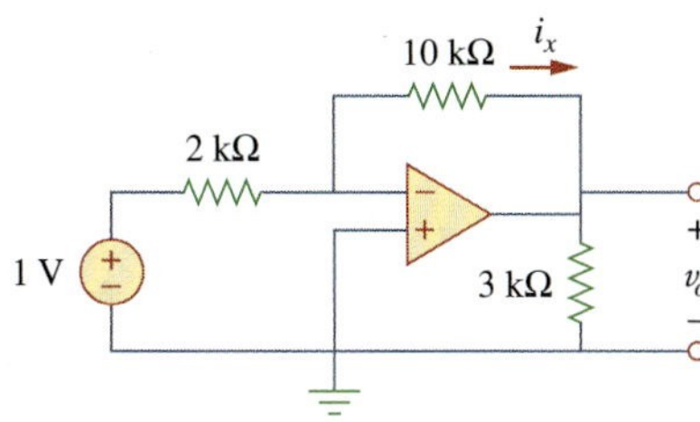

Figure 5.40
For Review Questions 5.3 and 5.4.

5.4 For the circuit in Fig. 5.40, current i_x is:

(a) 0.6 mA (b) 0.5 mA

(c) 0.2 mA (d) 1/12 mA

5.5 If $v_s = 0$ in the circuit of Fig. 5.41, current i_o is:

(a) -10 mA (b) -2.5 mA

(c) 10/12 mA (d) 10/14 mA

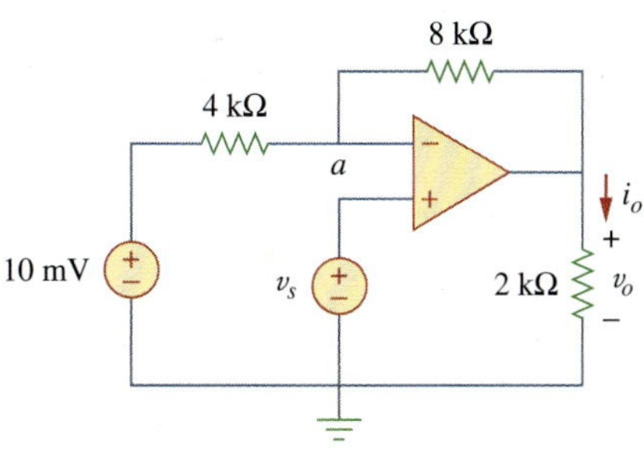

Figure 5.41
For Review Questions 5.5, 5.6, and 5.7.

5.6 If $v_s = 8$ mV in the circuit of Fig. 5.41, the output voltage is:

(a) -44 mV (b) -8 mV

(c) 4 mV (d) 7 mV

5.7 Refer to Fig. 5.41. If $v_s = 8$ mV voltage v_a is:

(a) -8 mV (b) 0 mV

(c) 10/3 mV (d) 8 mV

5.8 The power absorbed by the 4-kΩ resistor in Fig. 5.42 is:

(a) 9 mW (b) 4 mW

(c) 2 mW (d) 1 mW

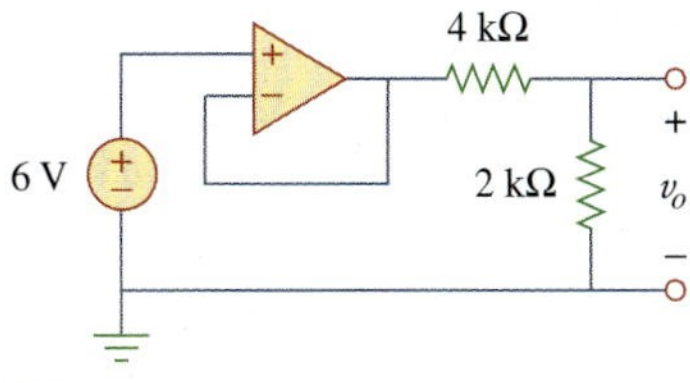

Figure 5.42
For Review Questions 5.8.

5.9 Which of these amplifiers is used in a digital-to-analog converter?

(a) noninverter

(b) voltage follower

(c) summer

(d) difference amplifier

5.10 Difference amplifiers are used in:

(a) instrumentation amplifiers

(b) voltage followers

(c) voltage regulators

(d) buffers

(e) summing amplifiers

(f) subtracting amplifiers

Answers: 5.1c, 5.2c,d, 5.3b, 5.4b, 5.5a, 5.6c, 5.7d, 5.8b, 5.9c, 5.10a,f.

Problems

Section 5.2 Operational Amplifiers

5.1 The equivalent model of a certain op amp is shown in Fig. 5.43. Determine:

(a) the input resistance

(b) the output resistance

(c) the voltage gain in dB.

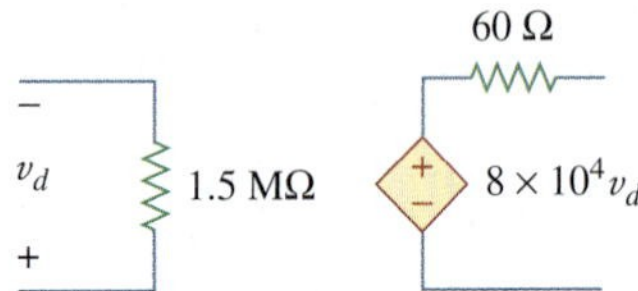

Figure 5.43
For Prob. 5.1.

5.2 The open-loop gain of an op amp is 100,000. Calculate the output voltage when there are inputs of $+10\ \mu$V on the inverting terminal and $+20\ \mu$V on the noninverting terminal.

5.3 Determine the output voltage when $-20\ \mu$V is applied to the inverting terminal of an op amp and $+30\ \mu$V to its noninverting terminal. Assume that the op amp has an open-loop gain of 200,000.

5.4 The output voltage of an op amp is -4 V when the noninverting input is 1 mV. If the open-loop gain of the op amp is 2×10^6, what is the inverting input?

5.5 For the op amp circuit of Fig. 5.44, the op amp has an open-loop gain of 100,000, an input resistance of 10 kΩ, and an output resistance of 100 Ω. Find the voltage gain v_o/v_i using the nonideal model of the op amp.

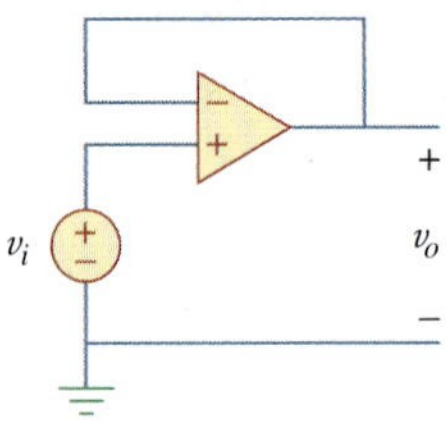

Figure 5.44
For Prob. 5.5.

5.6 Using the same parameters for the 741 op amp in Example 5.1, find v_o in the op amp circuit of Fig. 5.45.

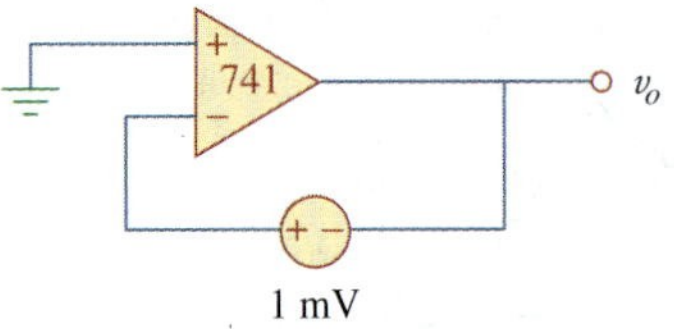

Figure 5.45
For Prob. 5.6.

5.7 The op amp in Fig. 5.46 has $R_i = 100$ kΩ, $R_o = 100\ \Omega$, $A = 100{,}000$. Find the differential voltage v_d and the output voltage v_o.

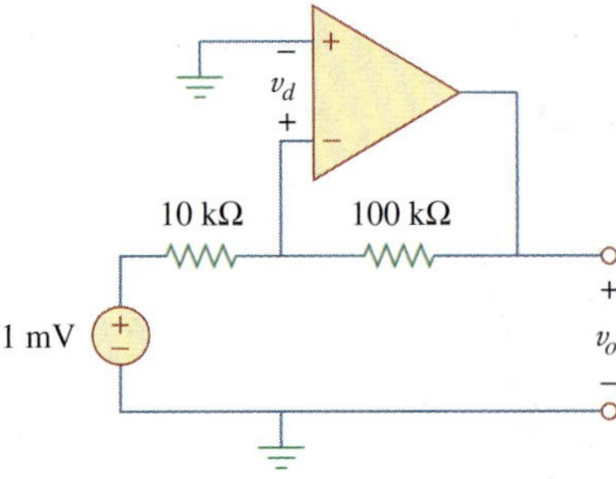

Figure 5.46
For Prob. 5.7.

Section 5.3 Ideal Op Amp

5.8 Obtain v_o for the op amp circuit in Fig. 5.47.

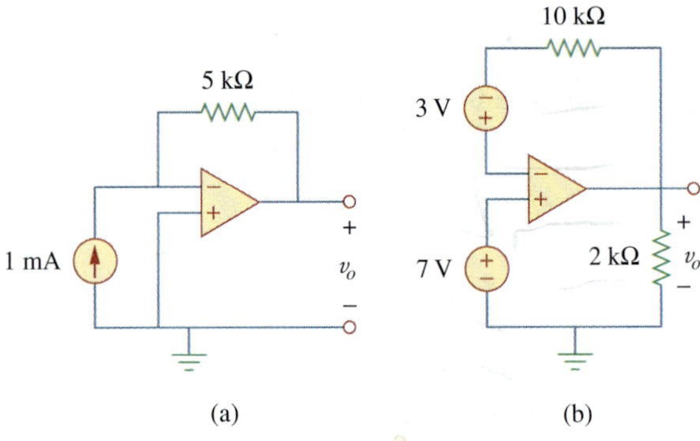

Figure 5.47
For Prob. 5.8.

5.9 Determine v_o for each of the op amp circuits in Fig. 5.48.

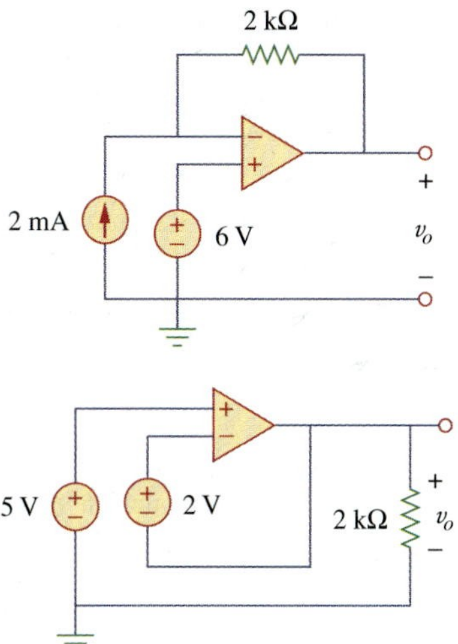

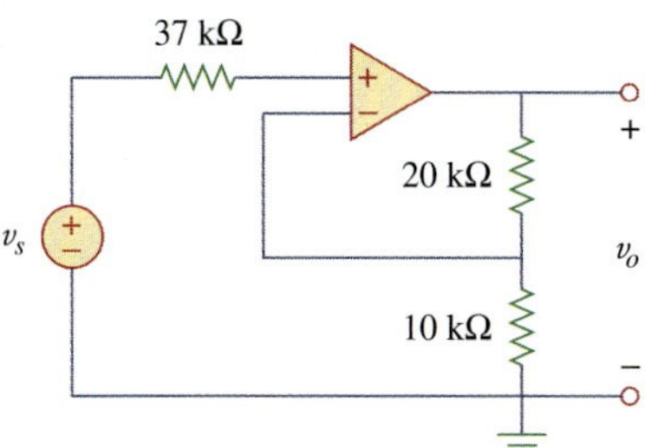

Figure 5.48
For Prob. 5.9.

5.10 Find the gain v_o/v_s of the circuit in Fig. 5.49.

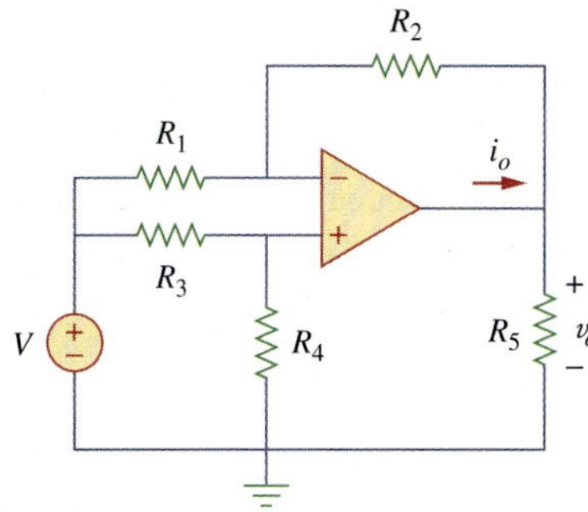

Figure 5.49
For Prob. 5.10.

5.11 Using Fig. 5.50, design a problem to help other students better understand how ideal op amps work.

Figure 5.50
For Prob. 5.11.

5.12 Calculate the voltage ratio v_o/v_s for the op amp circuit of Fig. 5.51. Assume that the op amp is ideal.

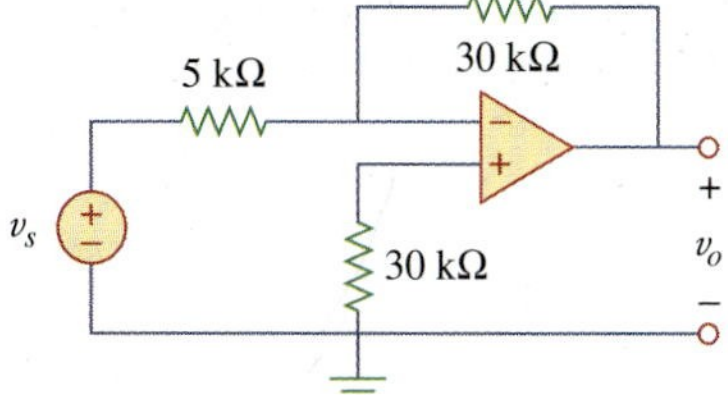

Figure 5.51
For Prob. 5.12.

5.13 Find v_o and i_o in the circuit of Fig. 5.52.

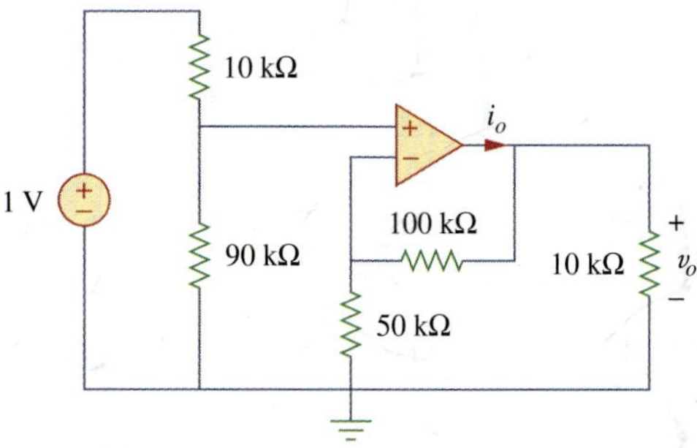

Figure 5.52
For Prob. 5.13.

5.14 Determine the output voltage v_o in the circuit of Fig. 5.53.

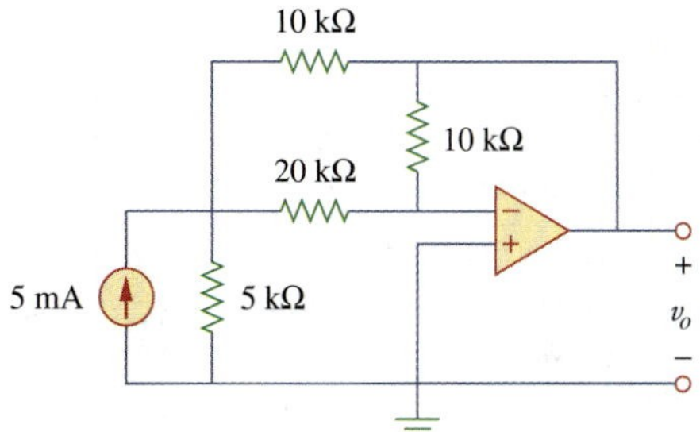

Figure 5.53
For Prob. 5.14.

Section 5.4 Inverting Amplifier

5.15 (a) Determine the ratio v_o/i_s in the op amp circuit of Fig. 5.54.

(b) Evaluate the ratio for $R_1 = 20$ kΩ, $R_2 = 25$ kΩ, $R_3 = 40$ kΩ.

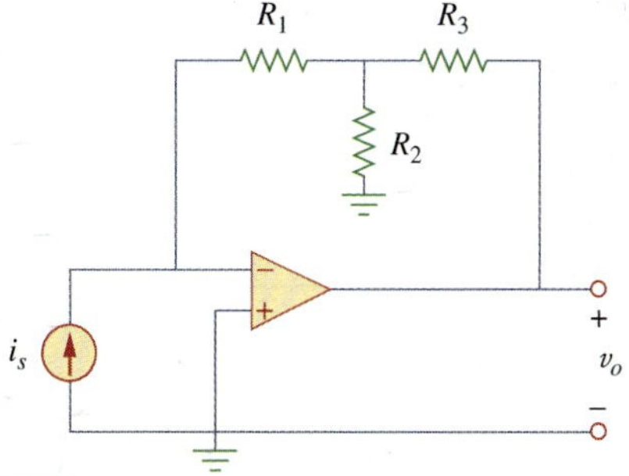

Figure 5.54
For Prob. 5.15.

5.16 Using Fig. 5.55, design a problem to help students better understand inverting op amps.

e◪d

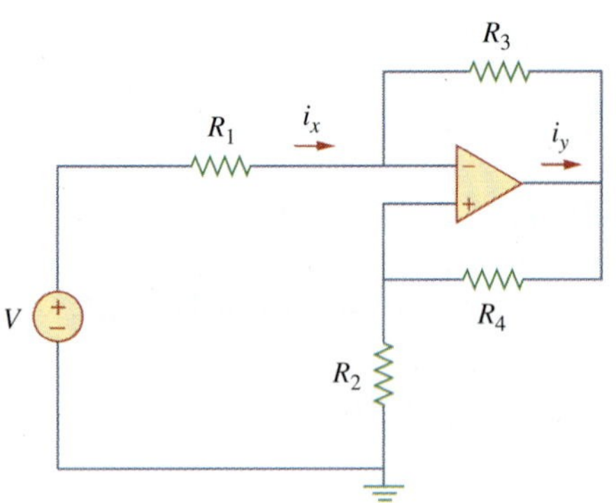

Figure 5.55
For Prob. 5.16.

5.17 Calculate the gain v_o/v_i when the switch in Fig. 5.56 is in:

(a) position 1 (b) position 2 (c) position 3

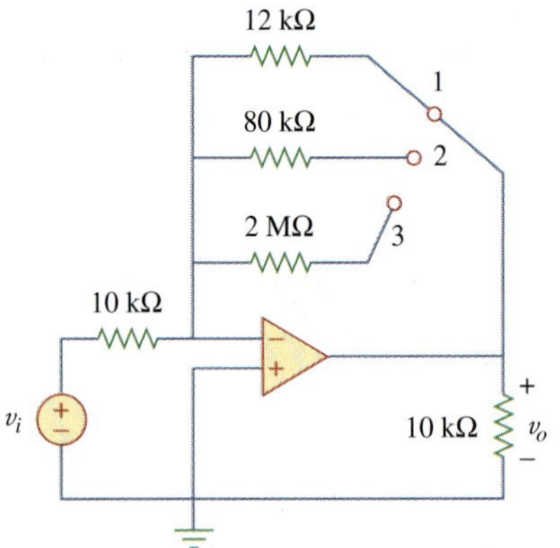

Figure 5.56
For Prob. 5.17.

***5.18** For the circuit shown in Figure 5.57, solve for the Thevenin equivalent circuit looking into terminals A and B.

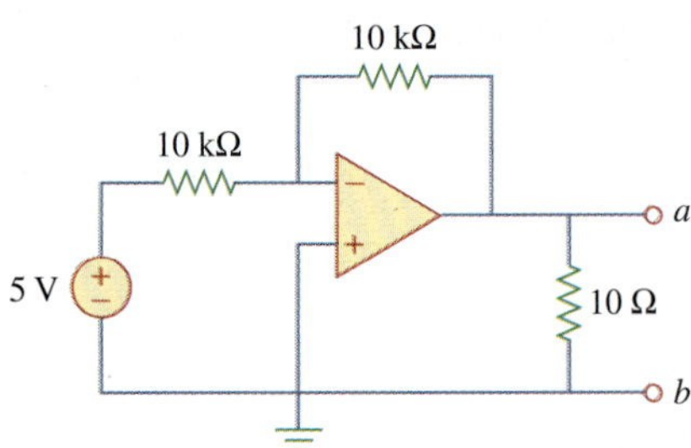

Figure 5.57
For Prob. 5.18.

5.19 Determine i_o in the circuit of Fig. 5.58.

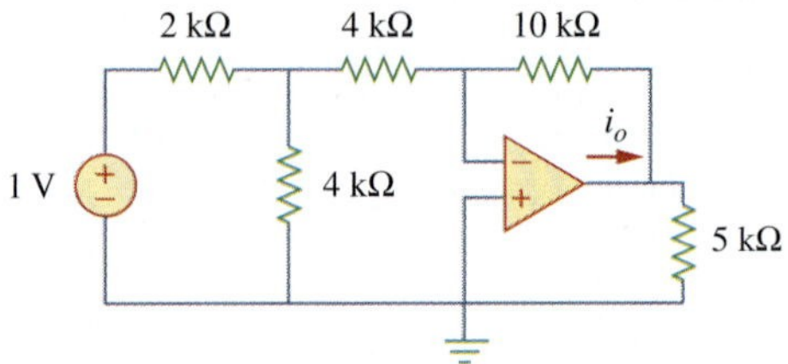

Figure 5.58
For Prob. 5.19.

5.20 In the circuit of Fig. 5.59, calculate v_o of $v_s = 0$.

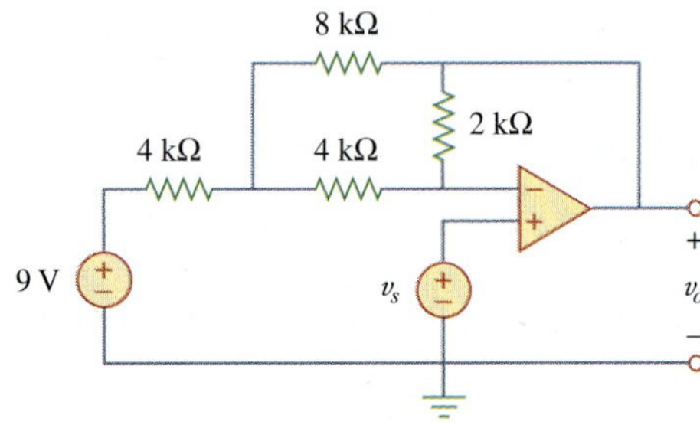

Figure 5.59
For Prob. 5.20.

5.21 Calculate v_o in the op amp circuit of Fig. 5.60.

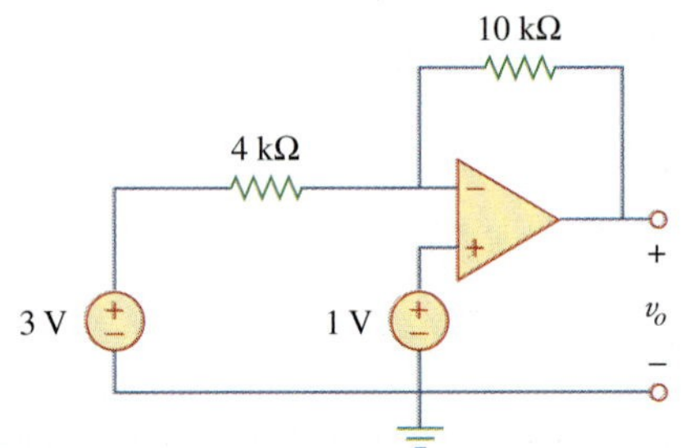

Figure 5.60
For Prob. 5.21.

5.22 Design an inverting amplifier with a gain of −15.

e◪d

5.23 For the op amp circuit in Fig. 5.61, find the voltage gain v_o/v_s.

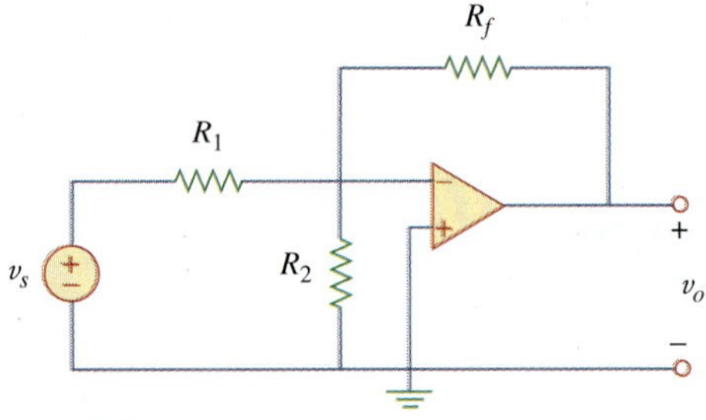

Figure 5.61
For Prob. 5.23.

* An asterisk indicates a challenging problem.

5.24 In the circuit shown in Fig. 5.62, find k in the voltage transfer function $v_o = kv_s$.

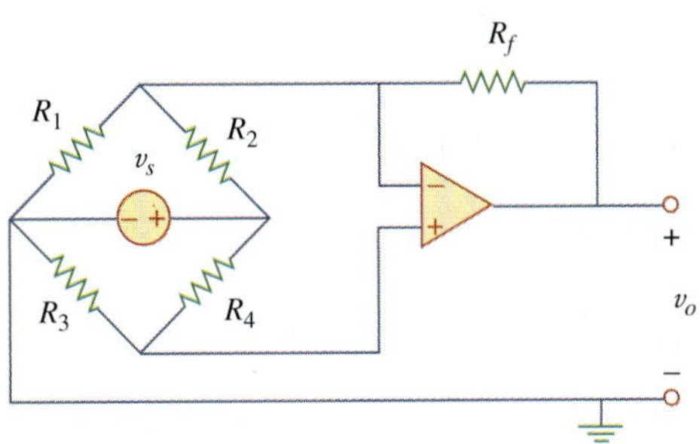

Figure 5.62
For Prob. 5.24.

Section 5.5 Noninverting Amplifier

5.25 Calculate v_o in the op amp circuit of Fig. 5.63.

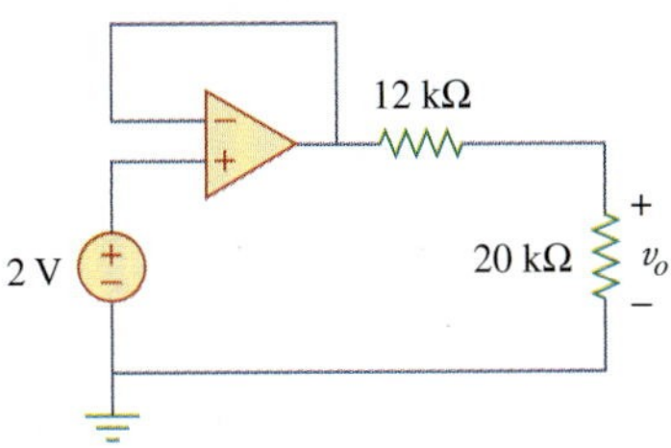

Figure 5.63
For Prob. 5.25.

5.26 Using Fig. 5.64, design a problem to help other students better understand noninverting op amps.

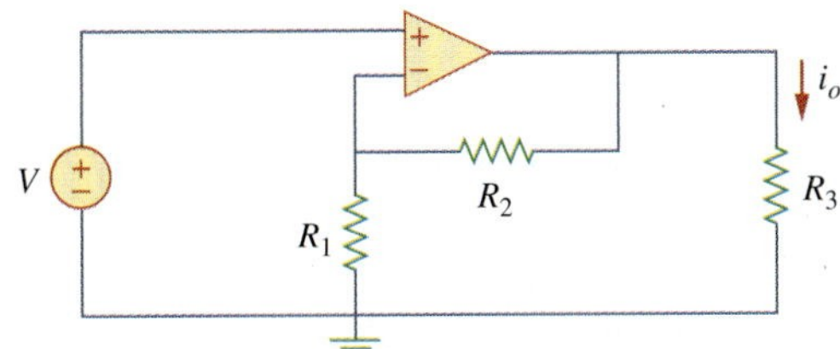

Figure 5.64
For Prob. 5.26.

5.27 Find v_o in the op amp circuit of Fig. 5.65.

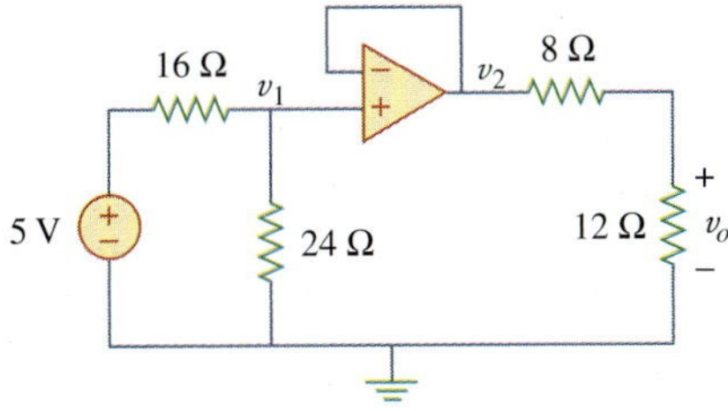

Figure 5.65
For Prob. 5.27.

5.28 Find i_o in the op amp circuit of Fig. 5.66.

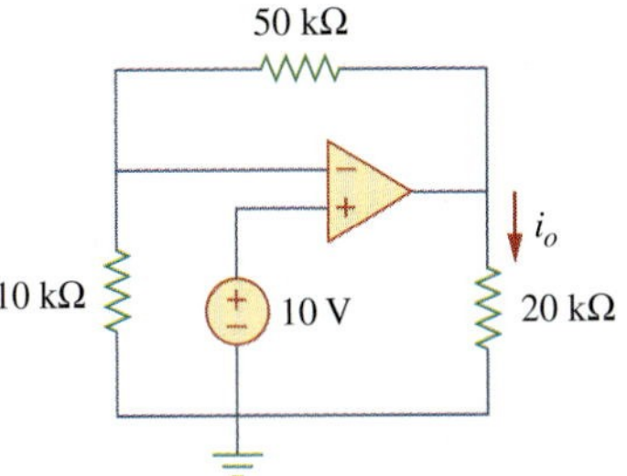

Figure 5.66
For Prob. 5.28.

5.29 Determine the voltage gain v_o/v_i of the op amp circuit in Fig. 5.67.

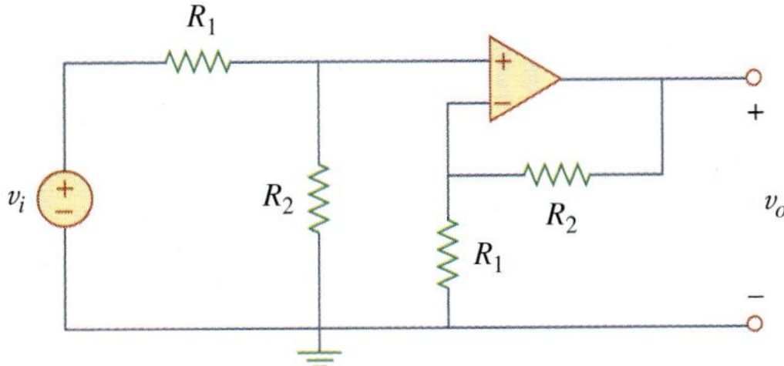

Figure 5.67
For Prob. 5.29.

5.30 In the circuit shown in Fig. 5.68, find i_x and the power absorbed by the 20-kΩ resistor.

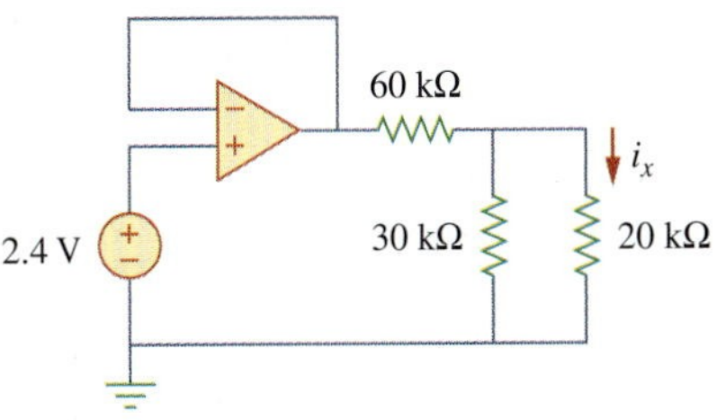

Figure 5.68
For Prob. 5.30.

5.31 For the circuit in Fig. 5.69, find i_x.

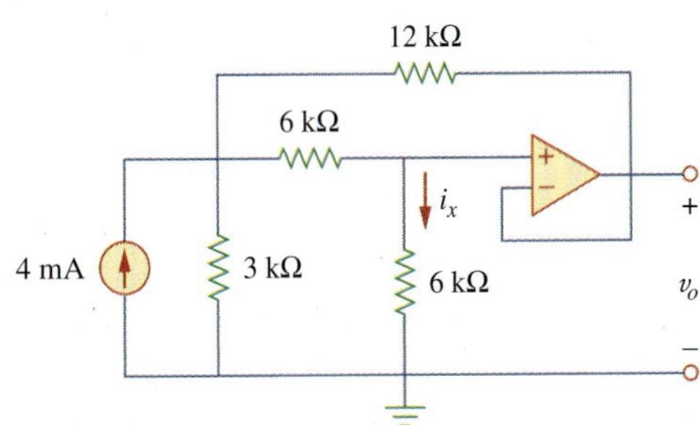

Figure 5.69
For Prob. 5.31.

5.32 Calculate i_x and v_o in the circuit of Fig. 5.70. Find the power dissipated by the 30-kΩ resistor.

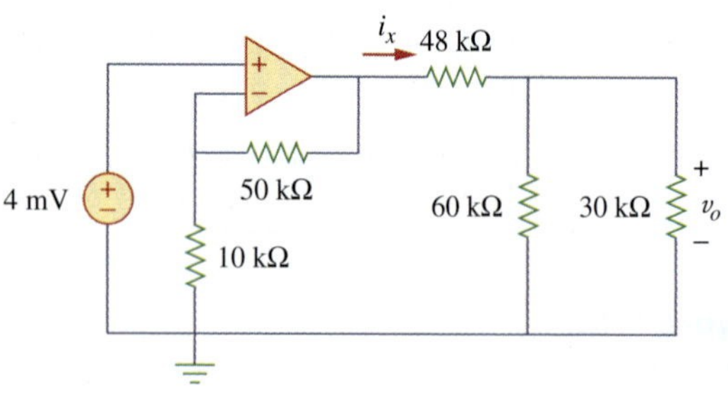

Figure 5.70
For Prob. 5.32.

5.33 Refer to the op amp circuit in Fig. 5.71. Calculate i_x and the power dissipated by the 3-kΩ resistor.

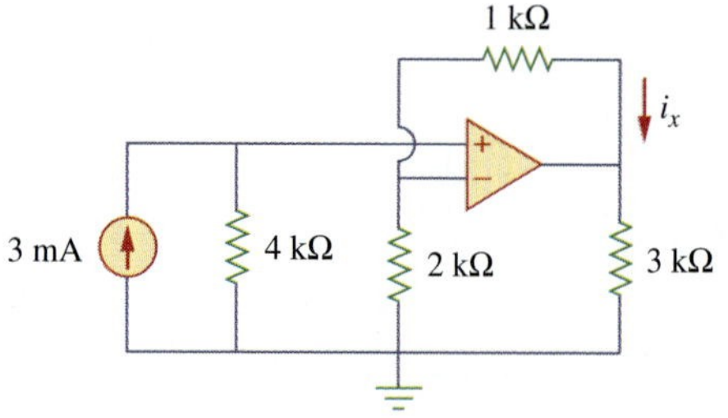

Figure 5.71
For Prob. 5.33.

5.34 Given the op amp circuit shown in Fig. 5.72, express v_o in terms of v_1 and v_2.

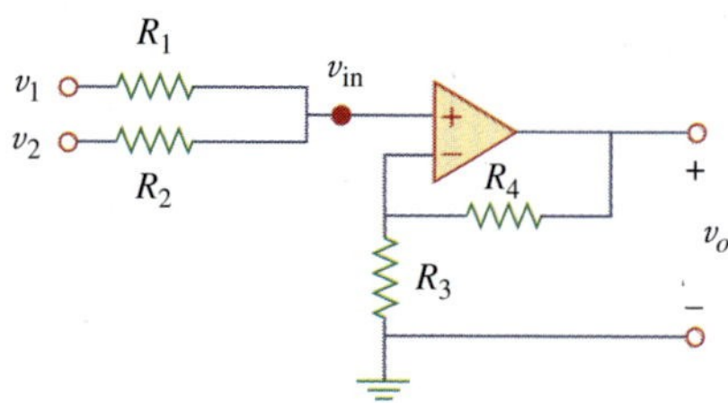

Figure 5.72
For Prob. 5.34.

5.35 Design a noninverting amplifier with a gain of 10.
e⃞d

5.36 For the circuit shown in Fig. 5.73, find the Thevenin equivalent at terminals *a-b*. (*Hint:* To find R_{Th}, apply a current source i_o and calculate v_o.)

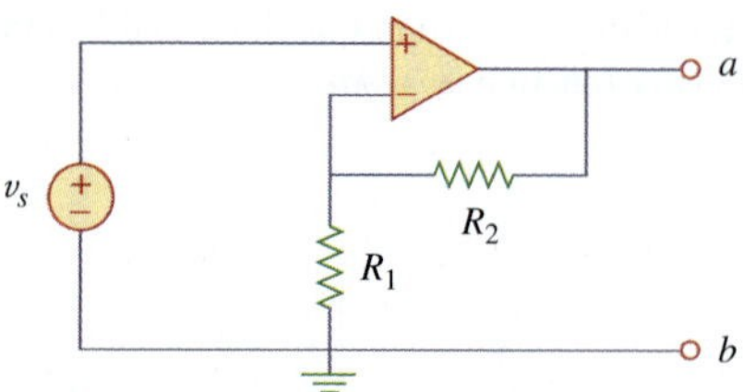

Figure 5.73
For Prob. 5.36.

Section 5.6 Summing Amplifier

5.37 Determine the output of the summing amplifier in Fig. 5.74.

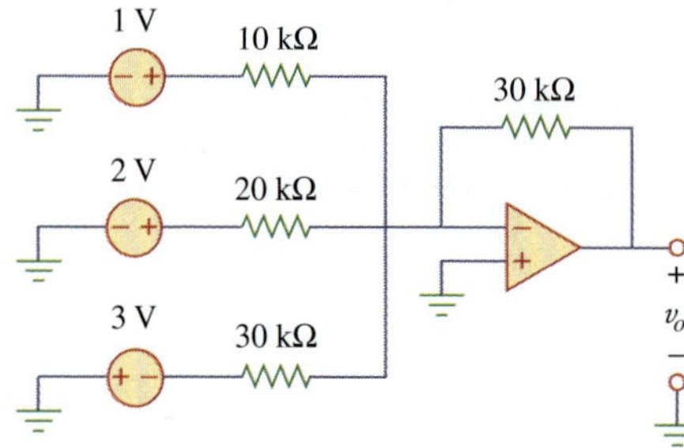

Figure 5.74
For Prob. 5.37.

5.38 Using Fig. 5.75, design a problem to help other students better understand summing amplifiers.
e⃞d

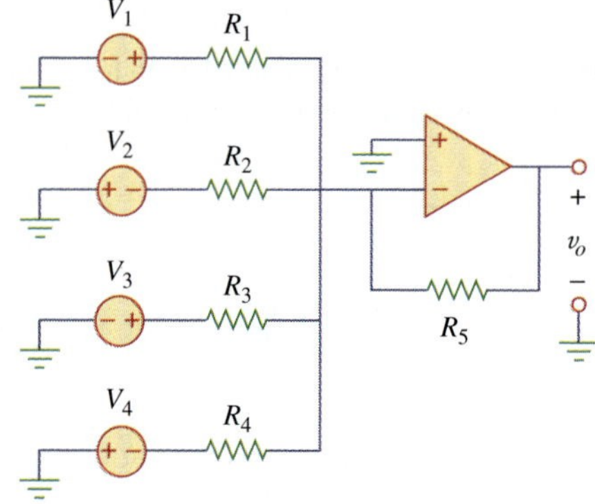

Figure 5.75
For Prob. 5.38.

5.39 For the op amp circuit in Fig. 5.76, determine the value of v_2 in order to make $v_o = -16.5$ V.

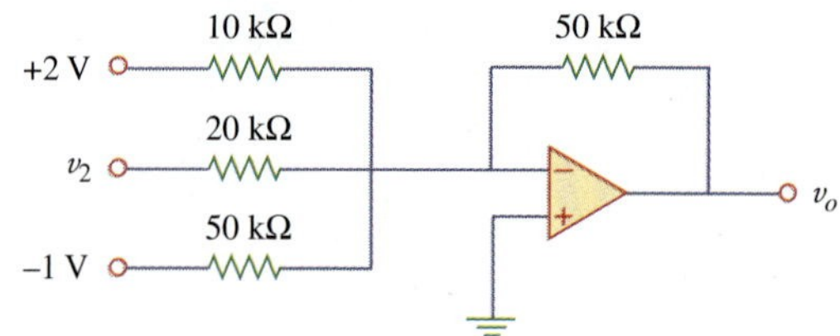

Figure 5.76
For Prob. 5.39.

5.40 Find v_o in terms of v_1, v_2, and v_3 in the circuit of Fig. 5.77.

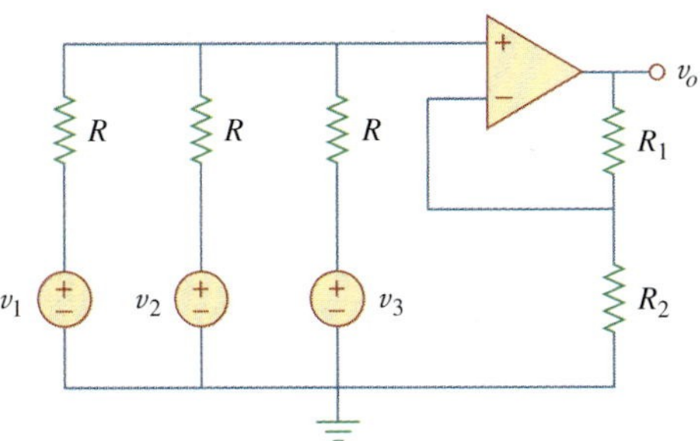

Figure 5.77
For Prob. 5.40.

5.41 An *averaging amplifier* is a summer that provides an output equal to the average of the inputs. By using proper input and feedback resistor values, one can get

$$-v_{out} = \tfrac{1}{4}(v_1 + v_2 + v_3 + v_4)$$

Using a feedback resistor of 10 kΩ design an averaging amplifier with four inputs.

5.42 A three-input summing amplifier has input resistors with $R_1 = R_2 = R_3 = 30$ kΩ. To produce an averaging amplifier, what value of feedback resistor is needed?

5.43 A four-input summing amplifier has $R_1 = R_2 = R_3 = R_4 = 12$ kΩ. What value of feedback resistor is needed to make it an averaging amplifier?

5.44 Show that the output voltage v_o of the circuit in Fig. 5.78 is

$$v_o = \frac{(R_3 + R_4)}{R_3(R_1 + R_2)}(R_2 v_1 + R_1 v_2)$$

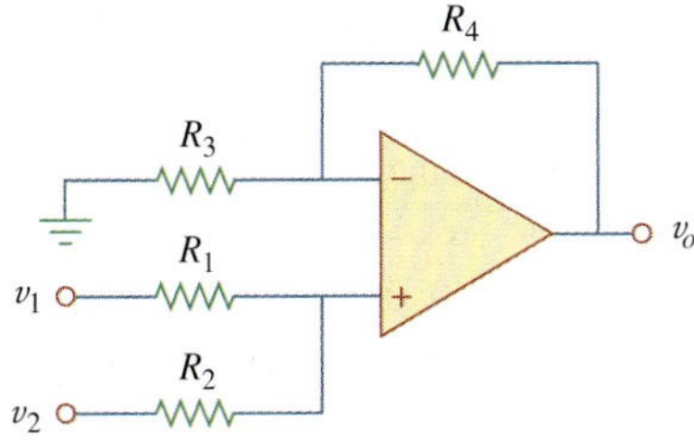

Figure 5.78
For Prob. 5.44.

5.45 Design an op amp circuit to perform the following operation:

$$v_o = 3v_1 - 2v_2$$

All resistances must be ≤ 100 kΩ.

5.46 Using only two op amps, design a circuit to solve

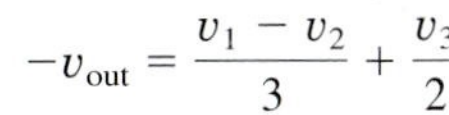

$$-v_{out} = \frac{v_1 - v_2}{3} + \frac{v_3}{2}$$

Section 5.7 Difference Amplifier

5.47 The circuit in Fig. 5.79 is for a difference amplifier. Find v_o given that $v_1 = 1$ V and $v_2 = 2$ V.

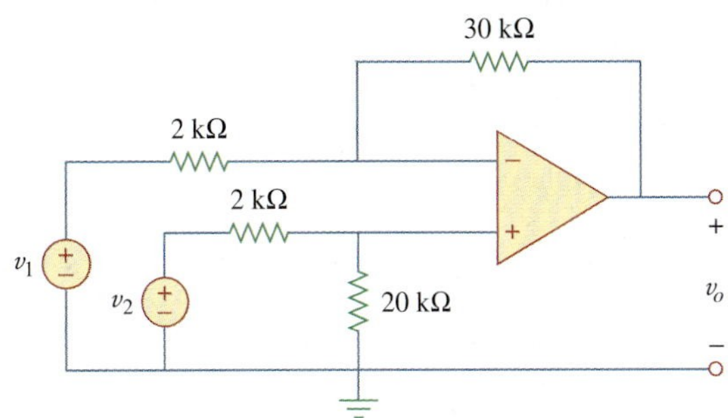

Figure 5.79
For Prob. 5.47.

5.48 The circuit in Fig. 5.80 is a differential amplifier driven by a brige. Find v_o.

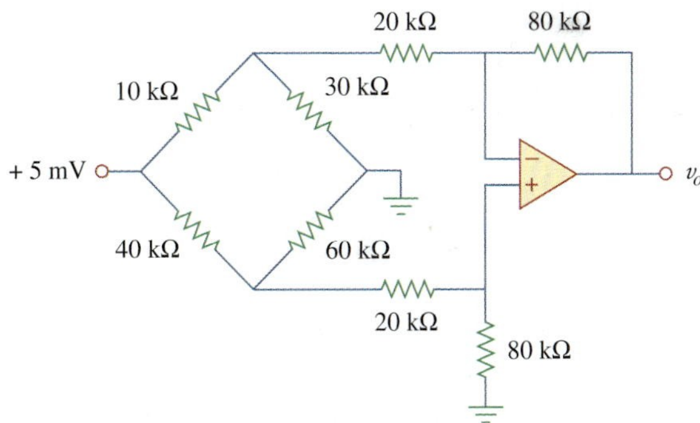

Figure 5.80
For Prob. 5.48.

5.49 Design a difference amplifier to have a gain of 2 and a common-mode input resistance of 10 kΩ at each input.

5.50 Design a circuit to amplify the difference between two inputs by 2.

(a) Use only one op amp.

(b) Use two op amps.

5.51 Using two op amps, design a subtractor.

***5.52** Design an op amp circuit such that

$$v_o = -2v_1 + 4v_2 - 5v_3 - v_4$$

Let all the resistors be in the range of 5 to 100 kΩ.

***5.53** The ordinary difference amplifier for fixed-gain operation is shown in Fig. 5.81(a). It is simple and reliable unless gain is made variable. One way of providing gain adjustment without losing simplicity and accuracy is to use the circuit in Fig. 5.81(b). Another way is to use the circuit in Fig. 5.81(c). Show that:

(a) for the circuit in Fig. 5.81(a),

$$\frac{v_o}{v_i} = \frac{R_2}{R_1}$$

(b) for the circuit in Fig. 5.81(b),

$$\frac{v_o}{v_i} = \frac{R_2}{R_1} \frac{1}{1 + \frac{R_1}{2R_G}}$$

(c) for the circuit in Fig. 5.81(c),

$$\frac{v_o}{v_i} = \frac{R_2}{R_1}\left(1 + \frac{R_2}{2R_G}\right)$$

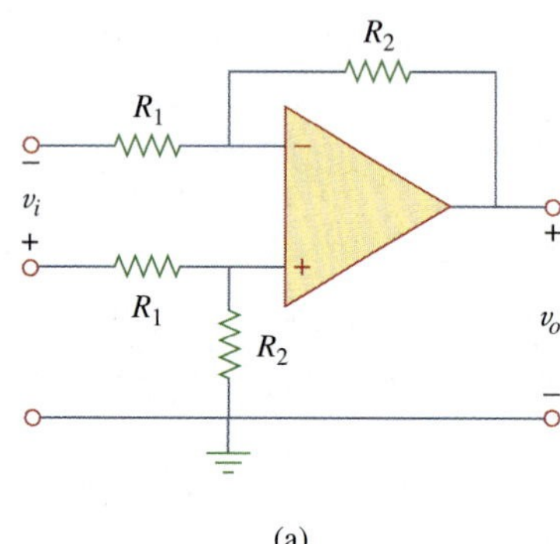

(a)

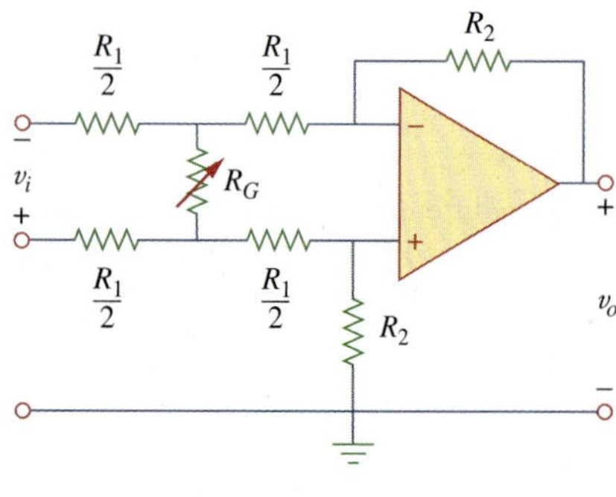

(b)

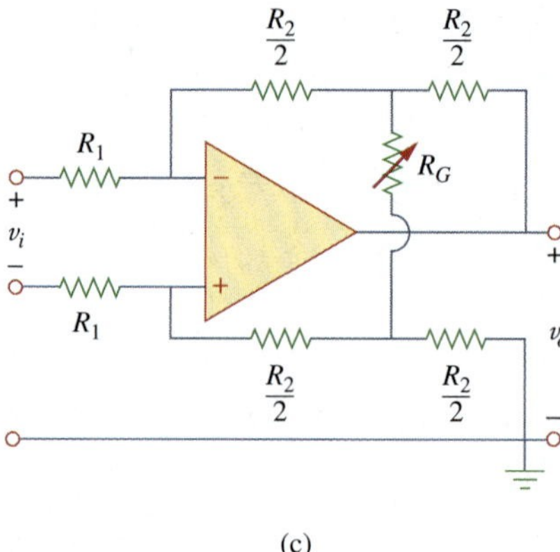

(c)

Figure 5.81
For Prob. 5.53.

Section 5.8 Cascaded Op Amp Circuits

5.54 Determine the voltage transfer ratio v_o/v_s in the op amp circuit of Fig. 5.82, where R = 10 kΩ.

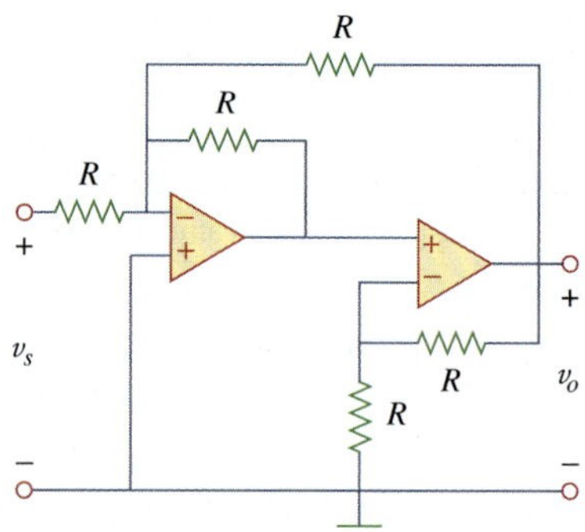

Figure 5.82
For Prob. 5.54.

5.55 In a certain electronic device, a three-stage amplifier is desired, whose overall voltage gain is 42 dB. The individual voltage gains of the first two stages are to be equal, while the gain of the third is to be one-fourth of each of the first two. Calculate the voltage gain of each.

5.56 Using Fig. 5.83, design a problem to help other students better understand cascaded op amps.

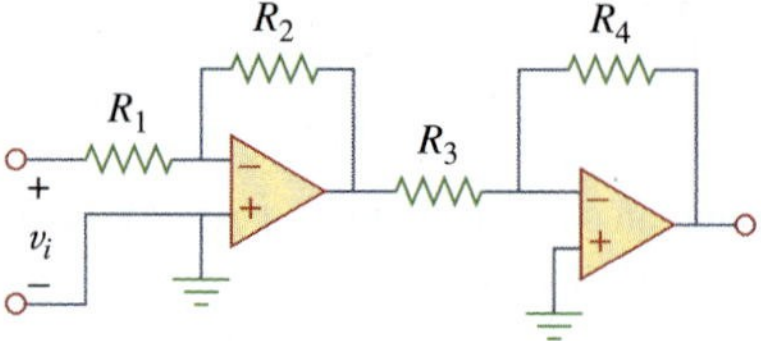

Figure 5.83
For Prob. 5.56.

5.57 Find v_o in the op amp circuit of Fig. 5.84.

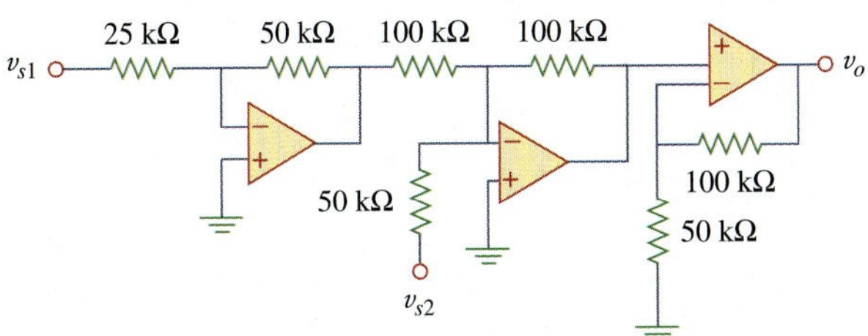

Figure 5.84
For Prob. 5.57.

5.58 Calculate i_o in the op amp circuit of Fig. 5.85.

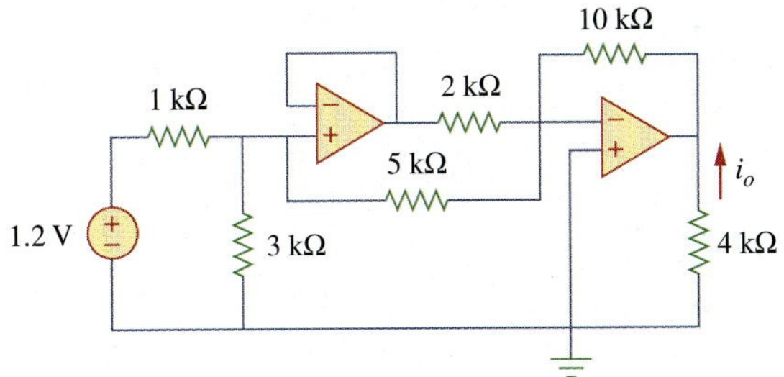

Figure 5.85
For Prob. 5.58.

5.59 In the op amp circuit of Fig. 5.86, determine the voltage gain v_o/v_s. Take $R = 20$ kΩ.

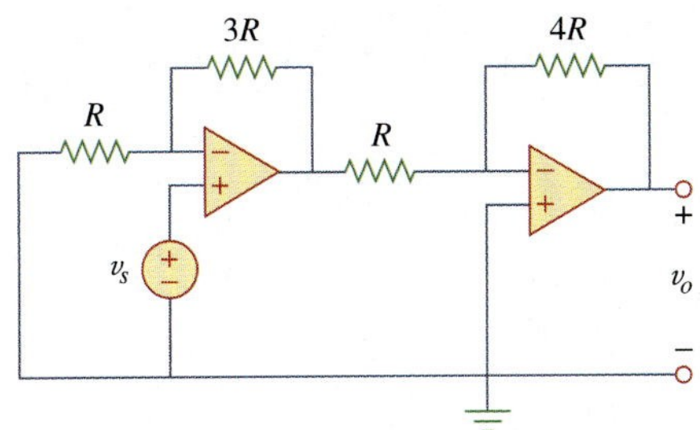

Figure 5.86
For Prob. 5.59.

5.60 Calculate v_o/v_i in the op amp circuit of Fig. 5.87.

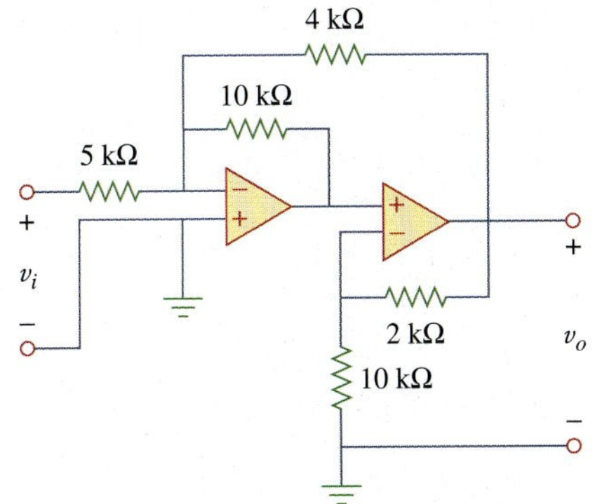

Figure 5.87
For Prob. 5.60.

5.61 Determine v_o in the circuit of Fig. 5.88.

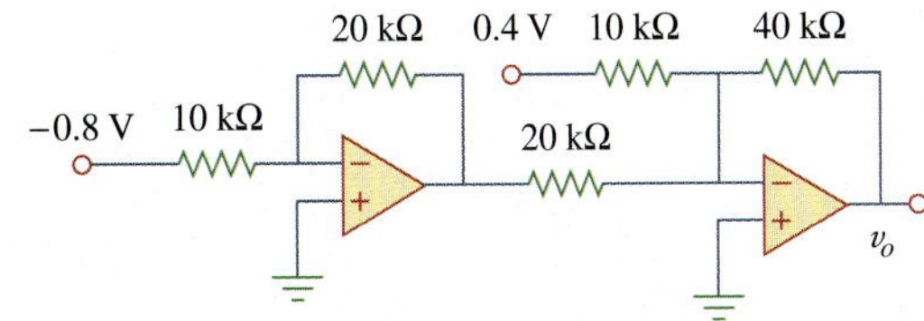

Figure 5.88
For Prob. 5.61.

5.62 Obtain the closed-loop voltage gain v_o/v_i of the circuit in Fig. 5.89.

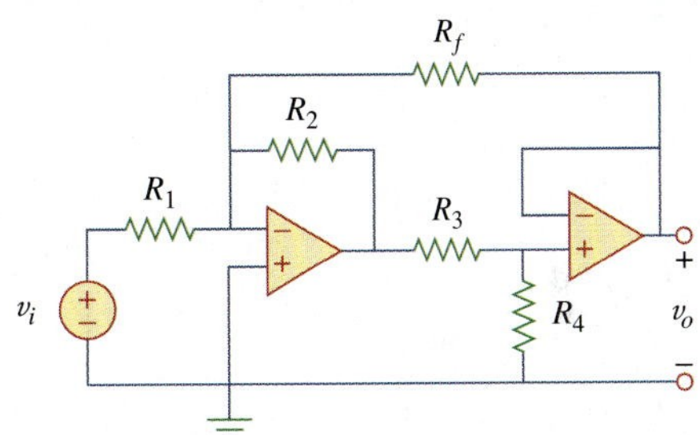

Figure 5.89
For Prob. 5.62.

5.63 Determine the gain v_o/v_i of the circuit in Fig. 5.90.

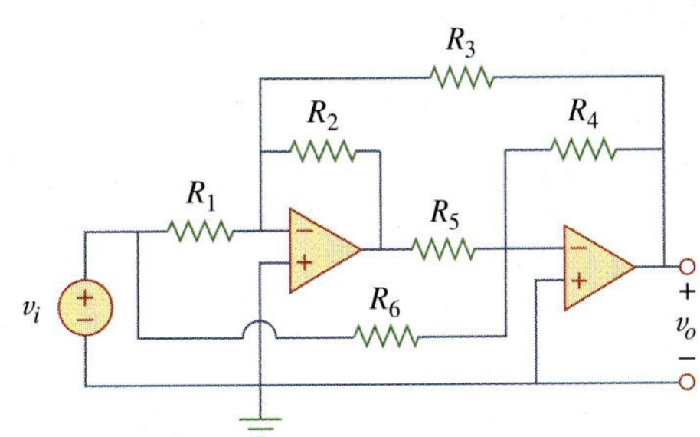

Figure 5.90
For Prob. 5.63.

5.64 For the op amp circuit shown in Fig. 5.91, find v_o/v_s.

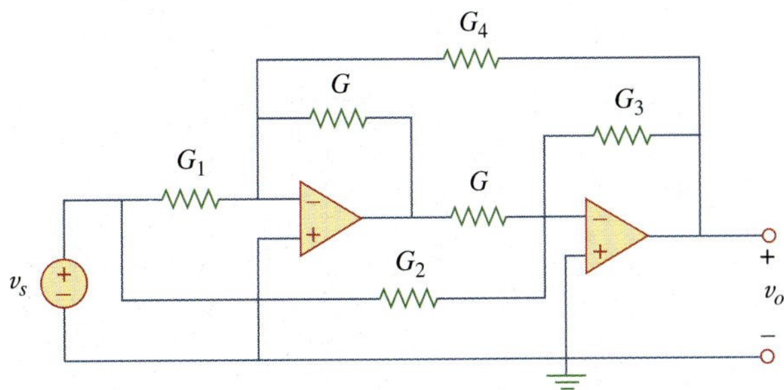

Figure 5.91
For Prob. 5.64.

5.65 Find v_o in the op amp circuit of Fig. 5.92.

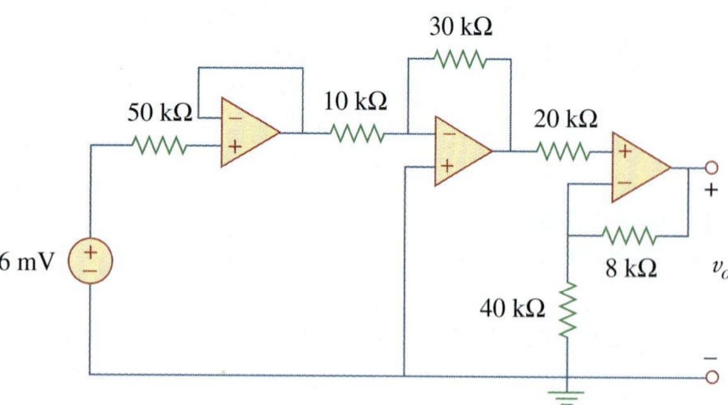

Figure 5.92
For Prob. 5.65.

5.66 For the circuit in Fig. 5.93, find v_o.

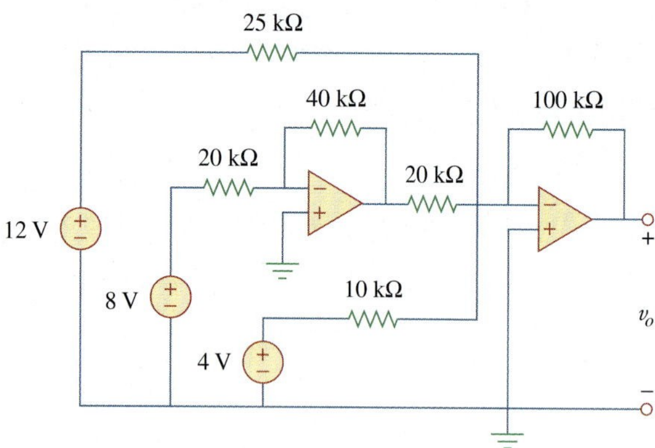

Figure 5.93
For Prob. 5.66.

5.67 Obtain the output v_o in the circuit of Fig. 5.94.

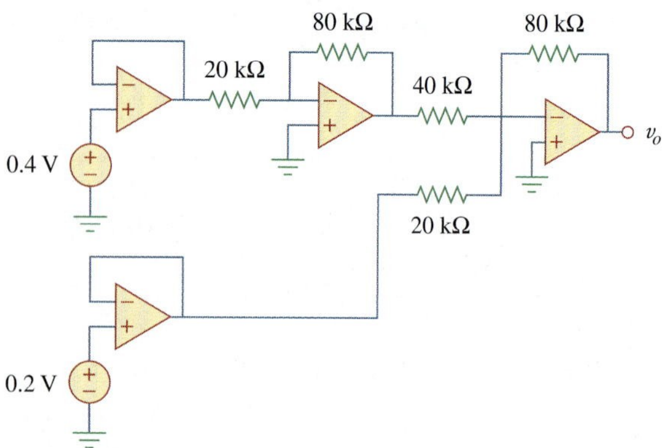

Figure 5.94
For Prob. 5.67.

5.68 Find v_o in the circuit of Fig. 5.95, assuming that $R_f = \infty$ (open circuit).

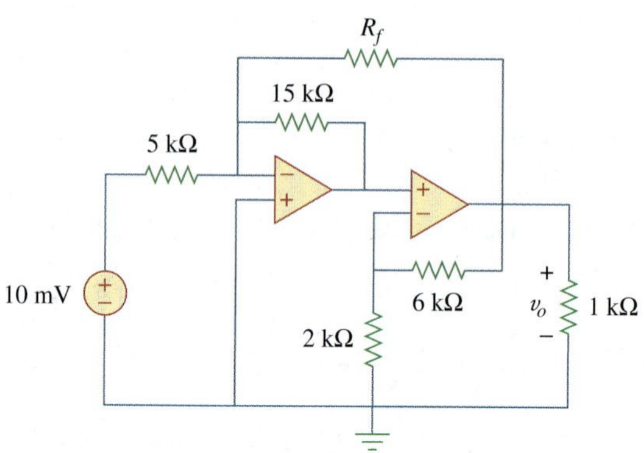

Figure 5.95
For Prob. 5.68 and 5.69.

5.69 Repeat the previous problem if $R_f = 10$ kΩ.

5.70 Determine v_o in the op amp circuit of Fig. 5.96.

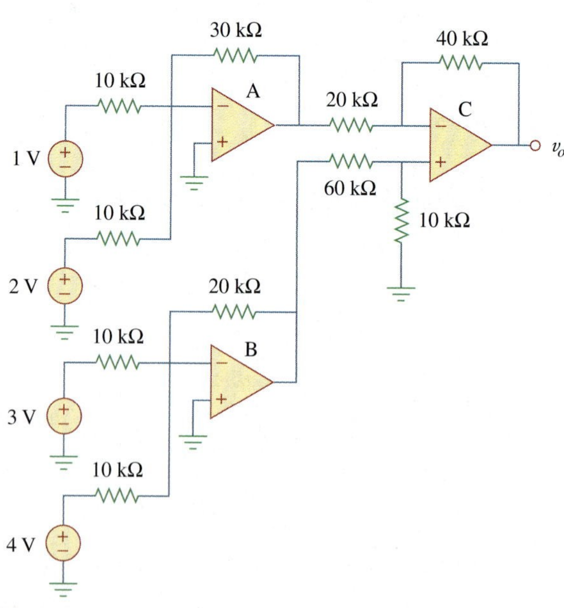

Figure 5.96
For Prob. 5.70.

5.71 Determine v_o in the op amp circuit of Fig. 5.97.

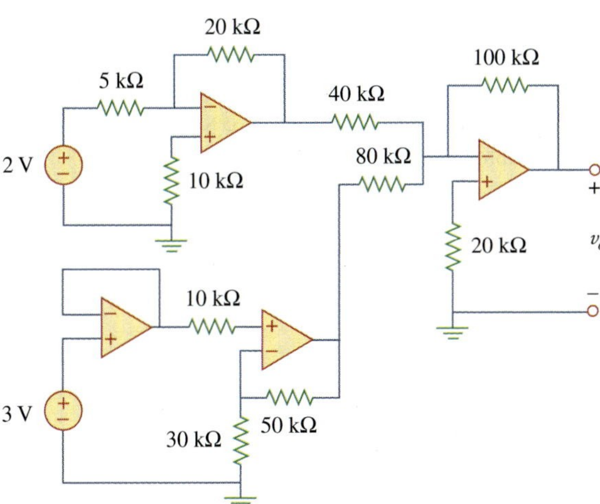

Figure 5.97
For Prob. 5.71.

5.72 Find the load voltage v_L in the circuit of Fig. 5.98.

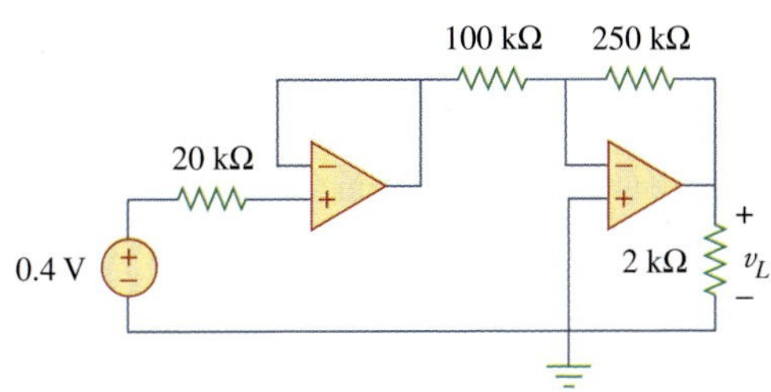

Figure 5.98
For Prob. 5.72.

5.73 Determine the load voltage v_L in the circuit of Fig. 5.99.

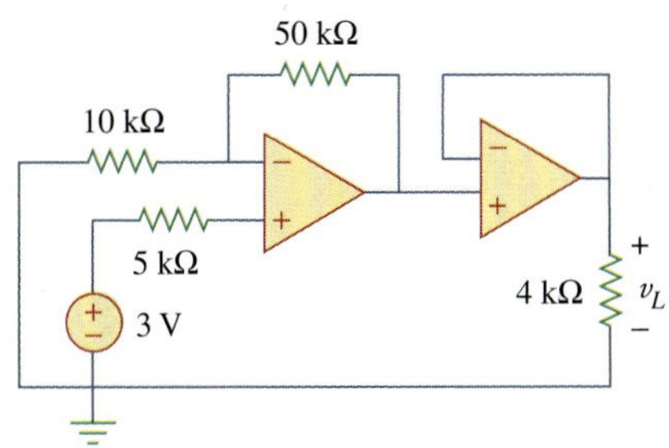

Figure 5.99
For Prob. 5.73.

5.74 Find i_o in the op amp circuit of Fig. 5.100.

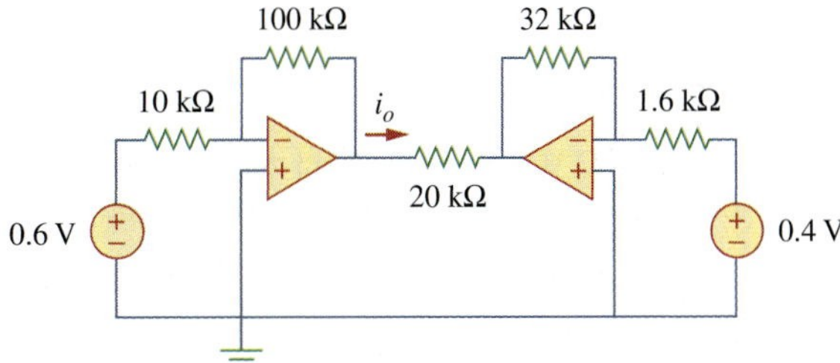

Figure 5.100
For Prob. 5.74.

Section 5.9 Op Amp Circuit Analysis with *PSpice*

5.75 Rework Example 5.11 using the nonideal op amp LM324 instead of uA741.

5.76 Solve Prob. 5.19 using *PSpice* and op amp uA741.

5.77 Solve Prob. 5.48 using *PSpice* and op amp LM324.

5.78 Use *PSpice* to obtain v_o in the circuit of Fig. 5.101.

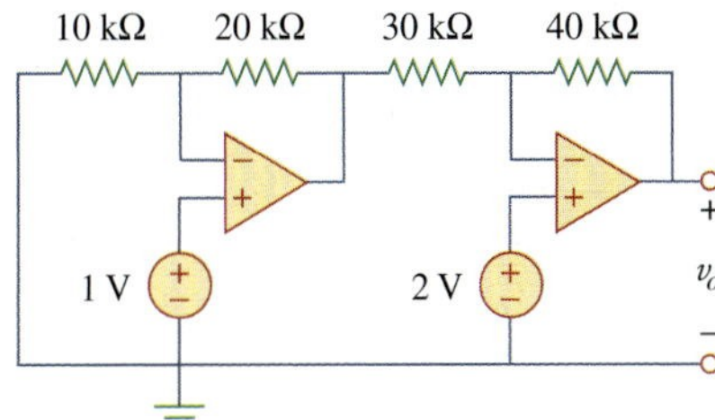

Figure 5.101
For Prob. 5.78.

5.79 Determine v_o in the op amp circuit of Fig. 5.102, using *PSpice*.

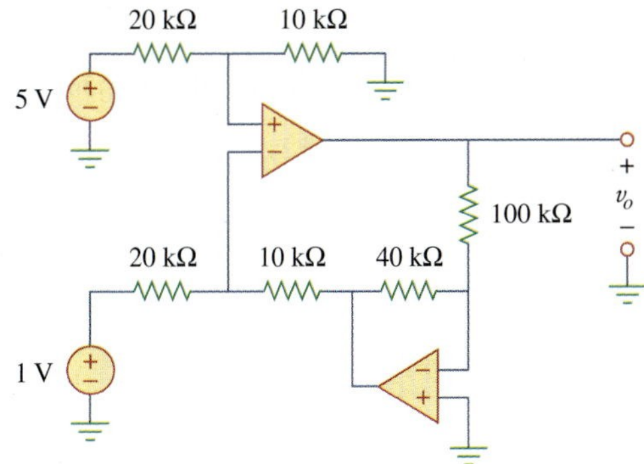

Figure 5.102
For Prob. 5.79.

5.80 Use *PSpice* to solve Prob. 5.70.

5.81 Use *PSpice* to verify the results in Example 5.9. Assume nonideal op amps LM324.

Section 5.10 Applications

5.82 A five-bit DAC covers a voltage range of 0 to 7.75 V. Calculate how much voltage each bit is worth.

5.83 Design a six-bit digital-to-analog converter.

(a) If $|V_o| = 1.1875$ V is desired, what should $[V_1V_2V_3V_4V_5V_6]$ be?

(b) Calculate $|V_o|$ if $[V_1V_2V_3V_4V_5V_6] = [011011]$.

(c) What is the maximum value $|V_o|$ can assume?

***5.84** A four-bit *R-2R ladder* DAC is presented in Fig. 5.103.

(a) Show that the output voltage is given by

$$-V_o = R_f\left(\frac{V_1}{2R} + \frac{V_2}{4R} + \frac{V_3}{8R} + \frac{V_4}{16R}\right)$$

(b) If $R_f = 12$ kΩ and $R = 10$ kΩ, find $|V_o|$ for $[V_1V_2V_3V_4] = [1011]$ and $[V_1V_2V_3V_4] = [0101]$.

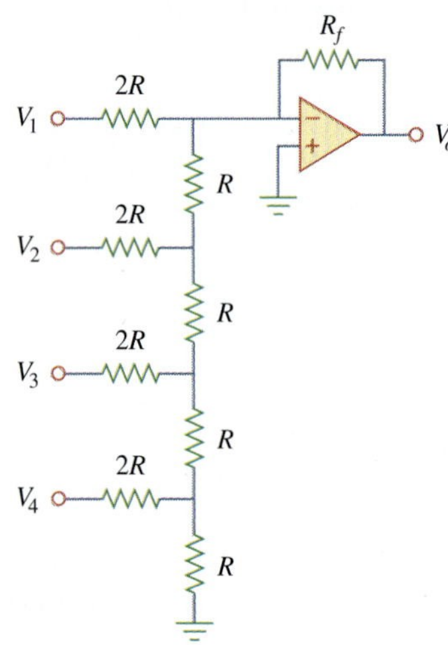

Figure 5.103
For Prob. 5.84.

5.85 In the op amp circuit of Fig. 5.104, find the value of R so that the power absorbed by the 10-kΩ resistor is 10 mW. Take $v_s = 2$ V.

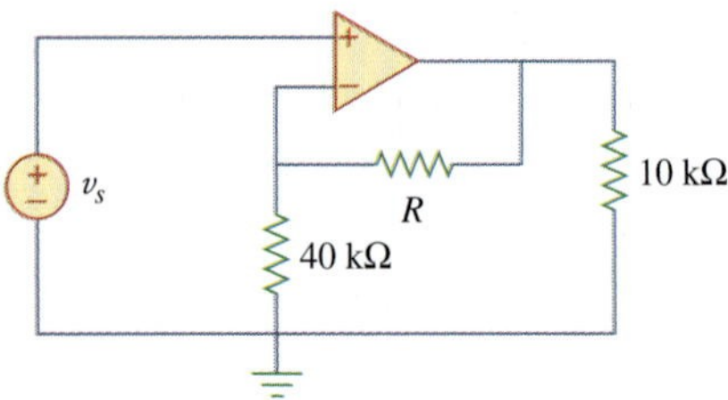

Figure 5.104
For Prob. 5.85.

5.86 Design a voltage controlled ideal current source (within the operating limits of the op amp) where the output current is equal to $200\,v_s(t)$ μA..

5.87 Figure 5.105 displays a two-op-amp instrumentation amplifier. Derive an expression for v_o in terms of v_1 and v_2. How can this amplifier be used as a subtractor?

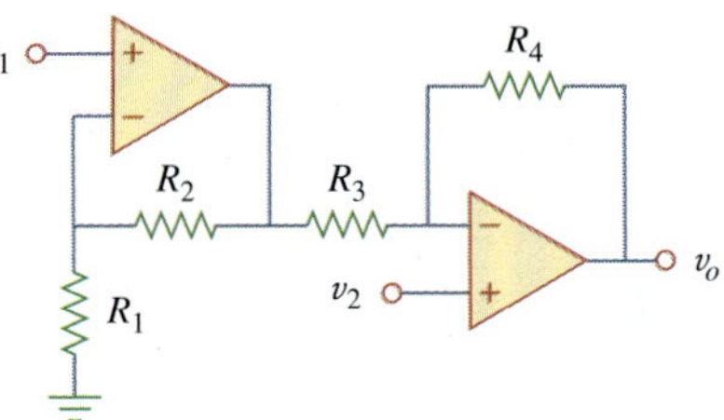

Figure 5.105
For Prob. 5.87.

***5.88** Figure 5.106 shows an instrumentation amplifier driven by a bridge. Obtain the gain v_o/v_i of the amplifier.

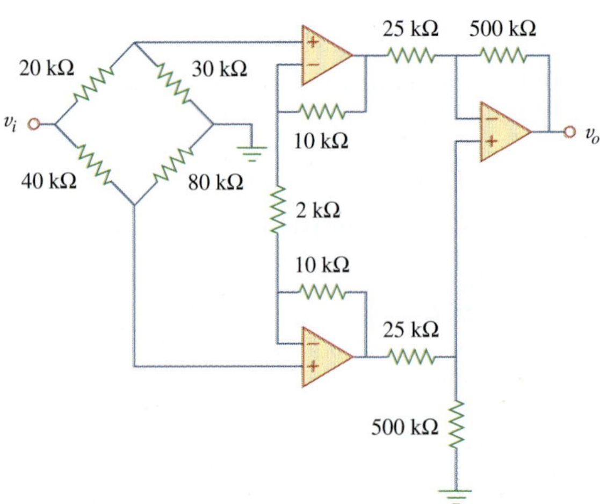

Figure 5.106
For Prob. 5.88.

Comprehensive Problems

5.89 Design a circuit that provides a relationship between output voltage v_o and input voltage v_s such that $v_o = 12v_s - 10$. Two op amps, a 6-V battery, and several resistors are available.

5.90 The op amp circuit in Fig. 5.107 is a *current amplifier*. Find the current gain i_o/i_s of the amplifier.

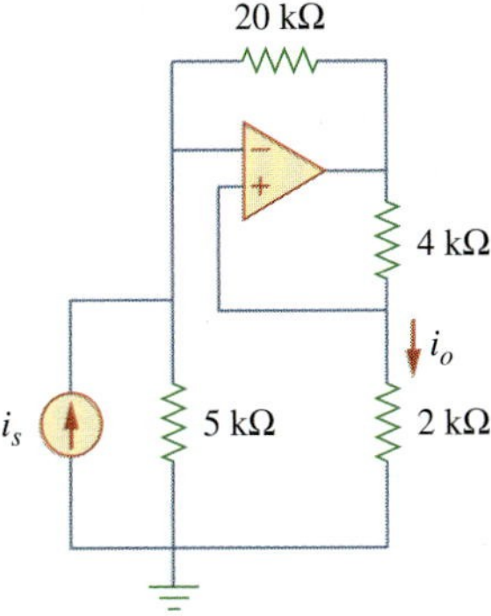

Figure 5.107
For Prob. 5.90.

5.91 A noninverting current amplifier is portrayed in Fig. 5.108. Calculate the gain i_o/i_s. Take $R_1 = 8$ kΩ and $R_2 = 1$ kΩ.

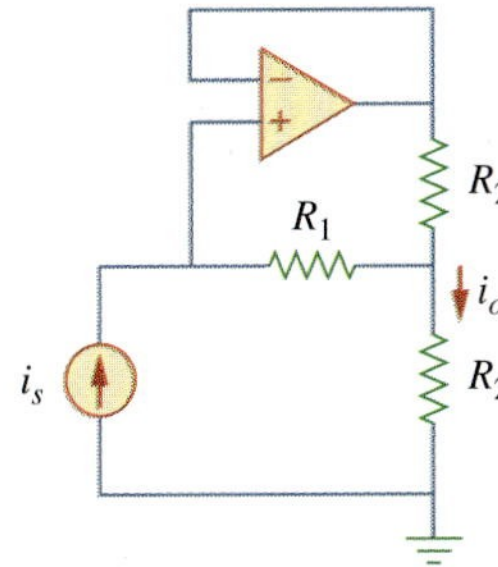

Figure 5.108
For Prob. 5.91.

5.92 Refer to the *bridge amplifier* shown in Fig. 5.109. Determine the voltage gain v_o/v_i.

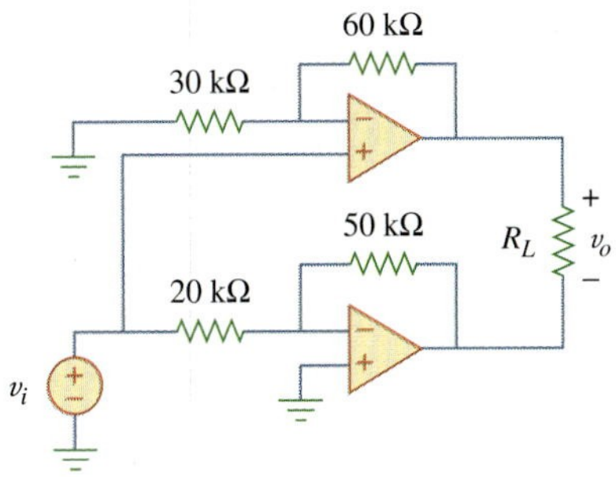

Figure 5.109
For Prob. 5.92.

***5.93** A voltage-to-current converter is shown in Fig. 5.110, which means that $i_L = Av_i$ if $R_1R_2 = R_3R_4$. Find the constant term A.

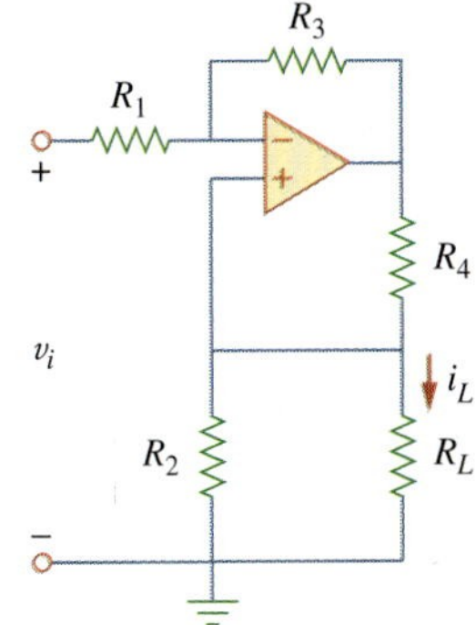

Figure 5.110
For Prob. 5.93.

chapter

6

Capacitors and Inductors

But in science the credit goes to the man who convinces the world, not to the man whom the idea first occurs.

—Francis Darwin

Enhancing Your Skills and Your Career

ABET EC 2000 criteria (3.c), *"an ability to design a system, component, or process to meet desired needs."*

The "ability to design a system, component, or process to meet desired needs" is why engineers are hired. That is why this is the most important *technical* skill that an engineer has. Interestingly, your success as an engineer is directly proportional to your ability to communicate but your being able to design is why you will be hired in the first place.

Design takes place when you have what is termed an open-ended problem that eventually is defined by the solution. Within the context of this course or textbook, we can only explore some of the elements of design. Pursuing all of the steps of our problem-solving technique teaches you several of the most important elements of the design process.

Probably the most important part of design is clearly defining what the system, component, process, or, in our case, problem is. Rarely is an engineer given a perfectly clear assignment. Therefore, as a student, you can develop and enhance this skill by asking yourself, your colleagues, or your professors questions designed to clarify the problem statement.

Exploring alternative solutions is another important part of the design process. Again, as a student, you can practice this part of the design process on almost every problem you work.

Evaluating your solutions is critical to any engineering assignment. Again, this is a skill that you as a student can practice on every problem you work.

Photo by Charles Alexander

6.1 Introduction

So far we have limited our study to resistive circuits. In this chapter, we shall introduce two new and important passive linear circuit elements: the capacitor and the inductor. Unlike resistors, which dissipate energy, capacitors and inductors do not dissipate but store energy, which can be retrieved at a later time. For this reason, capacitors and inductors are called *storage* elements.

In contrast to a resistor, which spends or dissipates energy irreversibly, an inductor or capacitor stores or releases energy (i.e., has a memory).

The application of resistive circuits is quite limited. With the introduction of capacitors and inductors in this chapter, we will be able to analyze more important and practical circuits. Be assured that the circuit analysis techniques covered in Chapters 3 and 4 are equally applicable to circuits with capacitors and inductors.

We begin by introducing capacitors and describing how to combine them in series or in parallel. Later, we do the same for inductors. As typical applications, we explore how capacitors are combined with op amps to form integrators, differentiators, and analog computers.

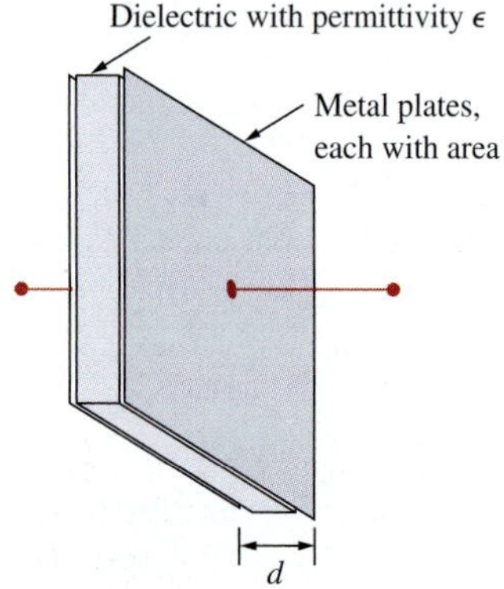

Figure 6.1
A typical capacitor.

6.2 Capacitors

A capacitor is a passive element designed to store energy in its electric field. Besides resistors, capacitors are the most common electrical components. Capacitors are used extensively in electronics, communications, computers, and power systems. For example, they are used in the tuning circuits of radio receivers and as dynamic memory elements in computer systems.

A capacitor is typically constructed as depicted in Fig. 6.1.

> A **capacitor** consists of two conducting plates separated by an insulator (or dielectric).

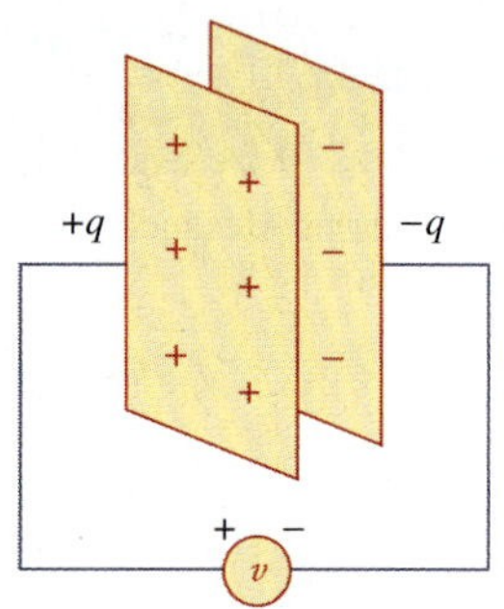

Figure 6.2
A capacitor with applied voltage v.

In many practical applications, the plates may be aluminum foil while the dielectric may be air, ceramic, paper, or mica.

When a voltage source v is connected to the capacitor, as in Fig. 6.2, the source deposits a positive charge q on one plate and a negative charge $-q$ on the other. The capacitor is said to store the electric charge. The amount of charge stored, represented by q, is directly proportional to the applied voltage v so that

$$q = Cv \tag{6.1}$$

where C, the constant of proportionality, is known as the *capacitance* of the capacitor. The unit of capacitance is the farad (F), in honor of the English physicist Michael Faraday (1791–1867). From Eq. (6.1), we may derive the following definition.

Alternatively, capacitance is the amount of charge stored per plate for a unit voltage difference in a capacitor.

> **Capacitance** is the ratio of the charge on one plate of a capacitor to the voltage difference between the two plates, measured in farads (F).

Note from Eq. (6.1) that 1 farad = 1 coulomb/volt.

Historical

Michael Faraday (1791–1867), an English chemist and physicist, was probably the greatest experimentalist who ever lived.

Born near London, Faraday realized his boyhood dream by working with the great chemist Sir Humphry Davy at the Royal Institution, where he worked for 54 years. He made several contributions in all areas of physical science and coined such words as electrolysis, anode, and cathode. His discovery of electromagnetic induction in 1831 was a major breakthrough in engineering because it provided a way of generating electricity. The electric motor and generator operate on this principle. The unit of capacitance, the farad, was named in his honor.

The Burndy Library Collection at The Huntington Library, San Marino, California.

Although the capacitance C of a capacitor is the ratio of the charge q per plate to the applied voltage v, it does not depend on q or v. It depends on the physical dimensions of the capacitor. For example, for the parallel-plate capacitor shown in Fig. 6.1, the capacitance is given by

$$C = \frac{\epsilon A}{d} \tag{6.2}$$

Capacitor voltage rating and capacitance are typically inversely rated due to the relationships in Eqs. (6.1) and (6.2). Arcing occurs if d is small and V is high.

where A is the surface area of each plate, d is the distance between the plates, and ϵ is the permittivity of the dielectric material between the plates. Although Eq. (6.2) applies to only parallel-plate capacitors, we may infer from it that, in general, three factors determine the value of the capacitance:

1. The surface area of the plates—the larger the area, the greater the capacitance.
2. The spacing between the plates—the smaller the spacing, the greater the capacitance.
3. The permittivity of the material—the higher the permittivity, the greater the capacitance.

Capacitors are commercially available in different values and types. Typically, capacitors have values in the picofarad (pF) to microfarad (μF) range. They are described by the dielectric material they are made of and by whether they are of fixed or variable type. Figure 6.3 shows the circuit symbols for fixed and variable capacitors. Note that according to the passive sign convention, if $v > 0$ and $i > 0$ or if $v < 0$ and $i < 0$, the capacitor is being charged, and if $v \cdot i < 0$, the capacitor is discharging.

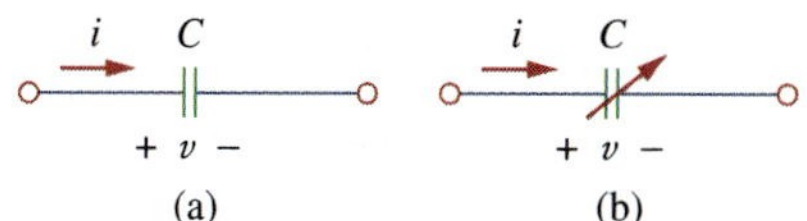

Figure 6.3
Circuit symbols for capacitors: (a) fixed capacitor, (b) variable capacitor.

Figure 6.4 shows common types of fixed-value capacitors. Polyester capacitors are light in weight, stable, and their change with temperature is predictable. Instead of polyester, other dielectric materials such as mica and polystyrene may be used. Film capacitors are rolled and housed in metal or plastic films. Electrolytic capacitors produce very high capacitance. Figure 6.5 shows the most common types of variable capacitors. The capacitance of a trimmer (or padder) capacitor

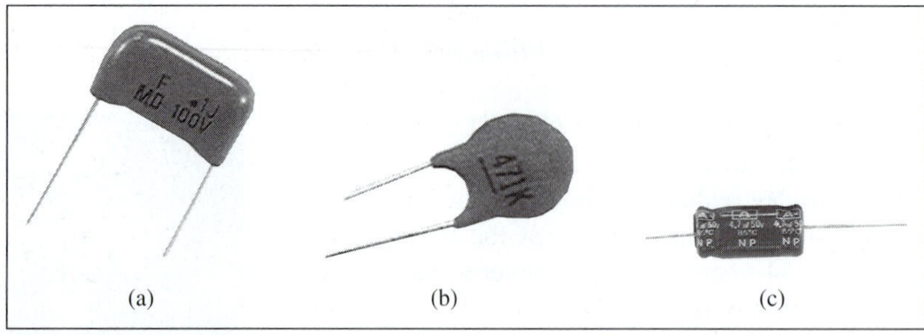

Figure 6.4
Fixed capacitors: (a) polyester capacitor, (b) ceramic capacitor, (c) electrolytic capacitor.
Courtesy of Tech America.

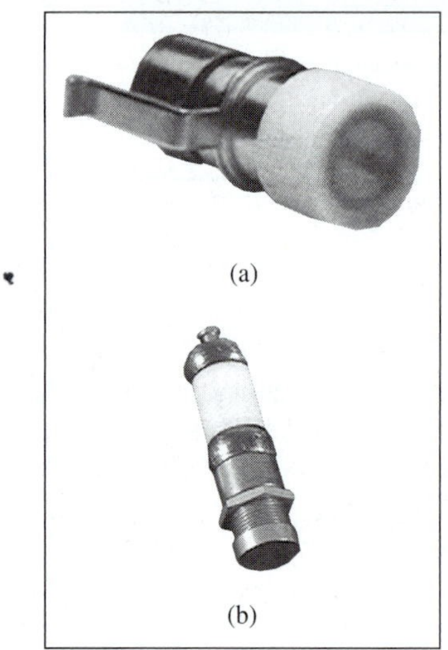

Figure 6.5
Variable capacitors: (a) trimmer capacitor, (b) filmtrim capacitor.
Courtesy of Johanson.

is often placed in parallel with another capacitor so that the equivalent capacitance can be varied slightly. The capacitance of the variable air capacitor (meshed plates) is varied by turning the shaft. Variable capacitors are used in radio receivers allowing one to tune to various stations. In addition, capacitors are used to block dc, pass ac, shift phase, store energy, start motors, and suppress noise.

To obtain the current-voltage relationship of the capacitor, we take the derivative of both sides of Eq. (6.1). Since

$$i = \frac{dq}{dt} \tag{6.3}$$

differentiating both sides of Eq. (6.1) gives

$$i = C\frac{dv}{dt} \tag{6.4}$$

This is the current-voltage relationship for a capacitor, assuming the passive sign convention. The relationship is illustrated in Fig. 6.6 for a capacitor whose capacitance is independent of voltage. Capacitors that satisfy Eq. (6.4) are said to be *linear.* For a *nonlinear capacitor*, the plot of the current-voltage relationship is not a straight line. Although some capacitors are nonlinear, most are linear. We will assume linear capacitors in this book.

According to Eq. (6.4), for a capacitor to carry current, its voltage must vary with time. Hence, for constant voltage, $i = 0$.

The voltage-current relation of the capacitor can be obtained by integrating both sides of Eq. (6.4). We get

$$v = \frac{1}{C}\int_{-\infty}^{t} i\,dt \tag{6.5}$$

or

$$v = \frac{1}{C}\int_{t_0}^{t} i\,dt + v(t_0) \tag{6.6}$$

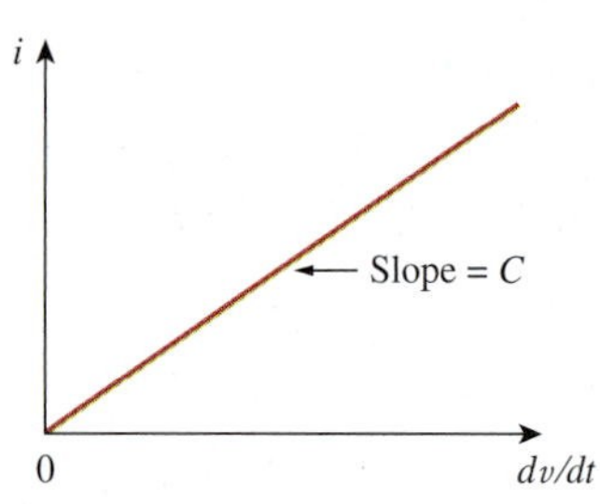

Figure 6.6
Current-voltage relationship of a capacitor.

where $v(t_0) = q(t_0)/C$ is the voltage across the capacitor at time t_0. Equation (6.6) shows that capacitor voltage depends on the past history of the capacitor current. Hence, the capacitor has memory—a property that is often exploited.

The instantaneous power delivered to the capacitor is

$$p = vi = Cv\frac{dv}{dt} \tag{6.7}$$

The energy stored in the capacitor is therefore

$$w = \int_{-\infty}^{t} p\,dt = C\int_{-\infty}^{t} v\frac{dv}{dt}dt = C\int_{v(-\infty)}^{v(t)} v\,dv = \frac{1}{2}Cv^2\Big|_{v(-\infty)}^{v(t)} \tag{6.8}$$

We note that $v(-\infty) = 0$, because the capacitor was uncharged at $t = -\infty$. Thus,

$$\boxed{w = \frac{1}{2}Cv^2} \tag{6.9}$$

Using Eq. (6.1), we may rewrite Eq. (6.9) as

$$w = \frac{q^2}{2C} \tag{6.10}$$

Equation (6.9) or (6.10) represents the energy stored in the electric field that exists between the plates of the capacitor. This energy can be retrieved, since an ideal capacitor cannot dissipate energy. In fact, the word *capacitor* is derived from this element's capacity to store energy in an electric field.

We should note the following important properties of a capacitor:

1. Note from Eq. (6.4) that when the voltage across a capacitor is not changing with time (i.e., dc voltage), the current through the capacitor is zero. Thus,

> A capacitor is an open circuit to dc.

However, if a battery (dc voltage) is connected across a capacitor, the capacitor charges.

2. The voltage on the capacitor must be continuous.

> The voltage on a capacitor cannot change abruptly.

The capacitor resists an abrupt change in the voltage across it. According to Eq. (6.4), a discontinuous change in voltage requires an infinite current, which is physically impossible. For example, the voltage across a capacitor may take the form shown in Fig. 6.7(a), whereas it is not physically possible for the capacitor voltage to take the form shown in Fig. 6.7(b) because of the abrupt changes. Conversely, the current through a capacitor can change instantaneously.

3. The ideal capacitor does not dissipate energy. It takes power from the circuit when storing energy in its field and returns previously stored energy when delivering power to the circuit.
4. A real, nonideal capacitor has a parallel-model leakage resistance, as shown in Fig. 6.8. The leakage resistance may be as high as

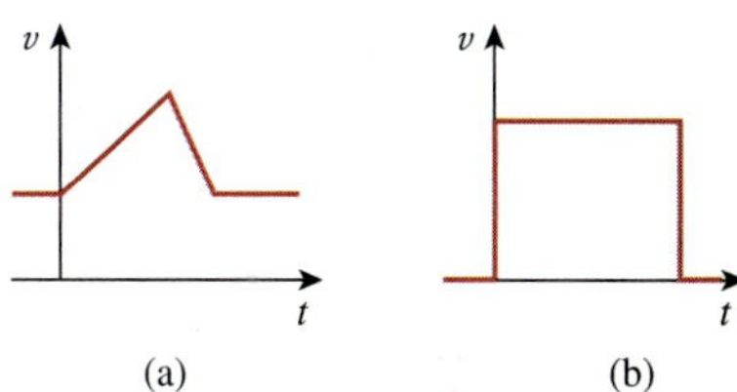

Figure 6.7
Voltage across a capacitor: (a) allowed, (b) not allowable; an abrupt change is not possible.

> An alternative way of looking at this is using Eq. (6.9), which indicates that energy is proportional to voltage squared. Since injecting or extracting energy can only be done over some finite time, voltage cannot change instantaneously across a capacitor.

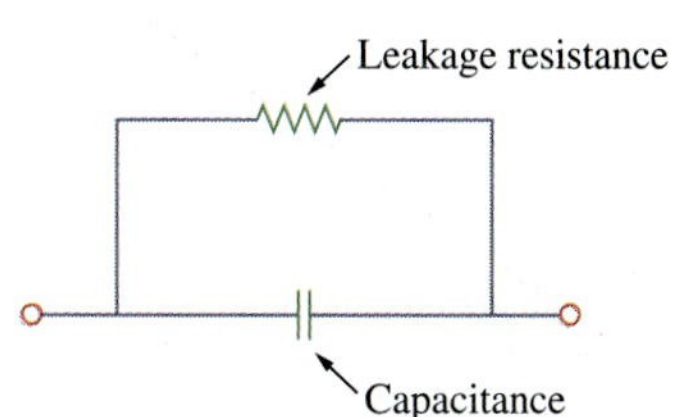

Figure 6.8
Circuit model of a nonideal capacitor.

100 MΩ and can be neglected for most practical applications. For this reason, we will assume ideal capacitors in this book.

Example 6.1

(a) Calculate the charge stored on a 3-pF capacitor with 20 V across it.
(b) Find the energy stored in the capacitor.

Solution:

(a) Since $q = Cv$,

$$q = 3 \times 10^{-12} \times 20 = 60 \text{ pC}$$

(b) The energy stored is

$$w = \frac{1}{2}Cv^2 = \frac{1}{2} \times 3 \times 10^{-12} \times 400 = 600 \text{ pJ}$$

Practice Problem 6.1

What is the voltage across a 3-μF capacitor if the charge on one plate is 0.12 mC? How much energy is stored?

Answer: 40 V, 2.4 mJ.

Example 6.2

The voltage across a 5-μF capacitor is

$$v(t) = 10 \cos 6000t \text{ V}$$

Calculate the current through it.

Solution:
By definition, the current is

$$\begin{aligned} i(t) &= C\frac{dv}{dt} = 5 \times 10^{-6}\frac{d}{dt}(10 \cos 6000t) \\ &= -5 \times 10^{-6} \times 6000 \times 10 \sin 6000t = -0.3 \sin 6000t \text{ A} \end{aligned}$$

Practice Problem 6.2

If a 10-μF capacitor is connected to a voltage source with

$$v(t) = 50 \sin 2000t \text{ V}$$

determine the current through the capacitor.

Answer: cos 2000t A.

Example 6.3

Determine the voltage across a 2-μF capacitor if the current through it is

$$i(t) = 6e^{-3000t} \text{ mA}$$

Assume that the initial capacitor voltage is zero.

Solution:

Since $v = \dfrac{1}{C}\displaystyle\int_0^t i\,dt + v(0)$ and $v(0) = 0$,

$$v = \frac{1}{2 \times 10^{-6}} \int_0^t 6e^{-3000t}\, dt \cdot 10^{-3}$$

$$= \frac{3 \times 10^3}{-3000} e^{-3000t} \Big|_0^t = (1 - e^{-3000t}) \text{ V}$$

Practice Problem 6.3

The current through a 100-μF capacitor is $i(t) = 50 \sin 120\pi t$ mA. Calculate the voltage across it at $t = 1$ ms and $t = 5$ ms. Take $v(0) = 0$.

Answer: 93.14 mV, 1.736 V.

Example 6.4

Determine the current through a 200-μF capacitor whose voltage is shown in Fig. 6.9.

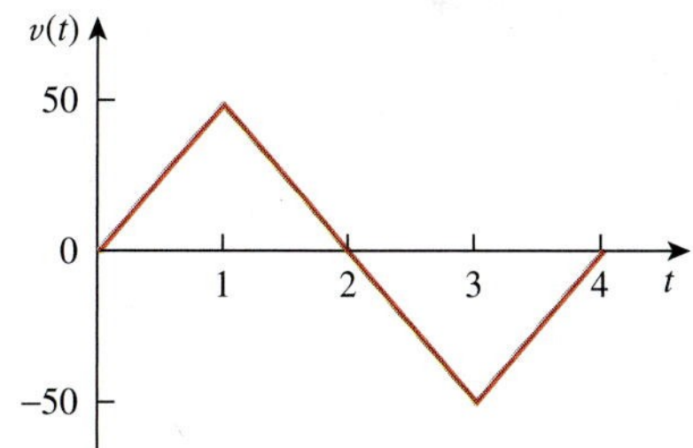

Figure 6.9
For Example 6.4.

Solution:
The voltage waveform can be described mathematically as

$$v(t) = \begin{cases} 50t \text{ V} & 0 < t < 1 \\ 100 - 50t \text{ V} & 1 < t < 3 \\ -200 + 50t \text{ V} & 3 < t < 4 \\ 0 & \text{otherwise} \end{cases}$$

Since $i = C\,dv/dt$ and $C = 200\ \mu$F, we take the derivative of v to obtain

$$i(t) = 200 \times 10^{-6} \times \begin{cases} 50 & 0 < t < 1 \\ -50 & 1 < t < 3 \\ 50 & 3 < t < 4 \\ 0 & \text{otherwise} \end{cases}$$

$$= \begin{cases} 10 \text{ mA} & 0 < t < 1 \\ -10 \text{ mA} & 1 < t < 3 \\ 10 \text{ mA} & 3 < t < 4 \\ 0 & \text{otherwise} \end{cases}$$

Thus the current waveform is as shown in Fig. 6.10.

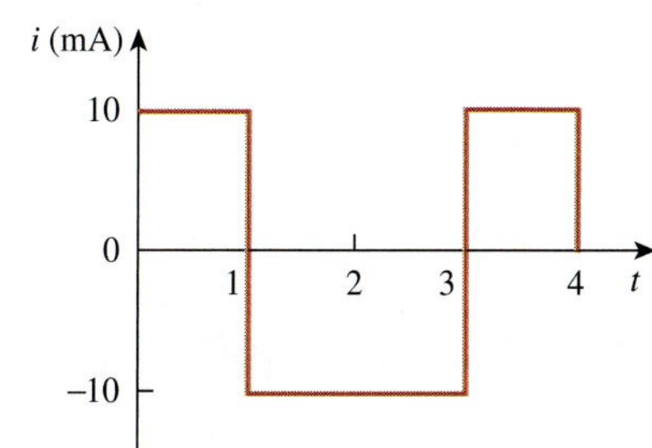

Figure 6.10
For Example 6.4.

Practice Problem 6.4

An initially uncharged 1-mF capacitor has the current shown in Fig. 6.11 across it. Calculate the voltage across it at $t = 2$ ms and $t = 5$ ms.

Answer: 100 mV, 400 mV.

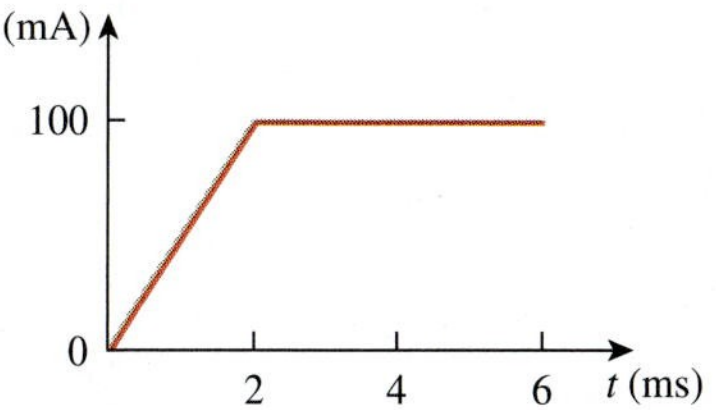

Figure 6.11
For Practice Prob. 6.4.

Example 6.5

Obtain the energy stored in each capacitor in Fig. 6.12(a) under dc conditions.

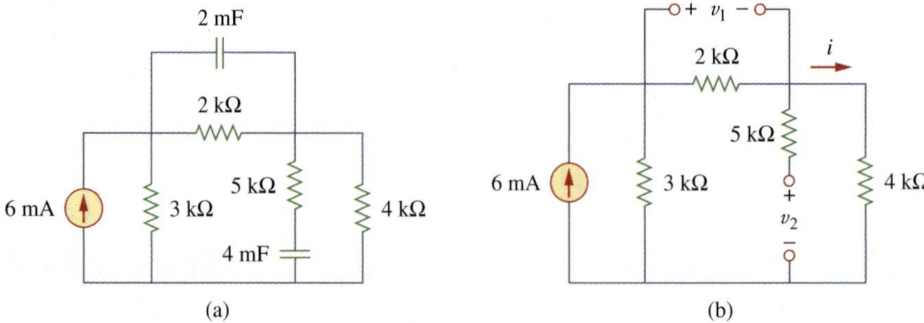

Figure 6.12
For Example 6.5.

Solution:
Under dc conditions, we replace each capacitor with an open circuit, as shown in Fig. 6.12(b). The current through the series combination of the 2-kΩ and 4-kΩ resistors is obtained by current division as

$$i = \frac{3}{3 + 2 + 4}(6 \text{ mA}) = 2 \text{ mA}$$

Hence, the voltages v_1 and v_2 across the capacitors are

$$v_1 = 2000i = 4 \text{ V} \qquad v_2 = 4000i = 8 \text{ V}$$

and the energies stored in them are

$$w_1 = \frac{1}{2}C_1 v_1^2 = \frac{1}{2}(2 \times 10^{-3})(4)^2 = 16 \text{ mJ}$$

$$w_2 = \frac{1}{2}C_2 v_2^2 = \frac{1}{2}(4 \times 10^{-3})(8)^2 = 128 \text{ mJ}$$

Practice Problem 6.5

Under dc conditions, find the energy stored in the capacitors in Fig. 6.13.

Answer: 810 μJ, 135 μJ.

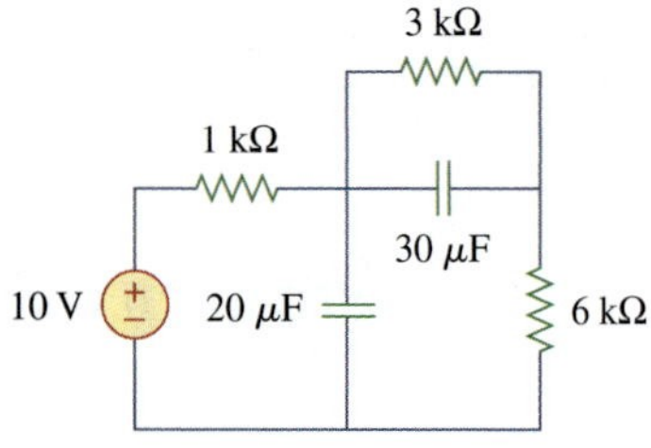

Figure 6.13
For Practice Prob. 6.5.

6.3 Series and Parallel Capacitors

We know from resistive circuits that the series-parallel combination is a powerful tool for reducing circuits. This technique can be extended to series-parallel connections of capacitors, which are sometimes encountered. We desire to replace these capacitors by a single equivalent capacitor C_{eq}.

In order to obtain the equivalent capacitor C_{eq} of N capacitors in parallel, consider the circuit in Fig. 6.14(a). The equivalent circuit is

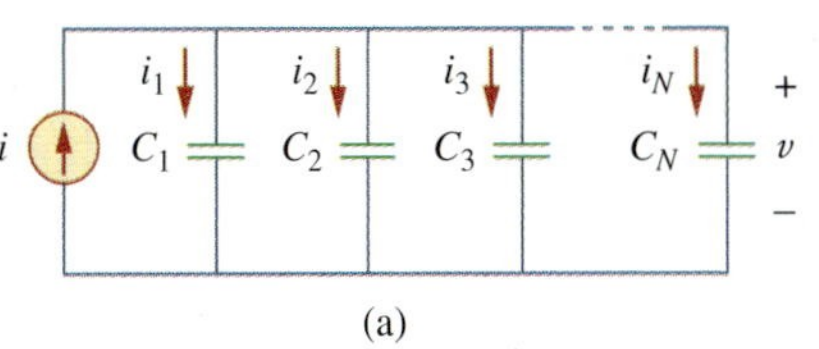

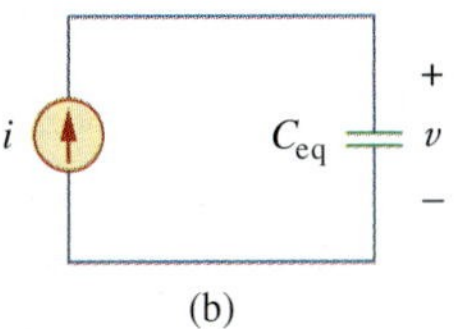

Figure 6.14
(a) Parallel-connected N capacitors, (b) equivalent circuit for the parallel capacitors.

in Fig. 6.14(b). Note that the capacitors have the same voltage v across them. Applying KCL to Fig. 6.14(a),

$$i = i_1 + i_2 + i_3 + \cdots + i_N \tag{6.11}$$

But $i_k = C_k \, dv/dt$. Hence,

$$\begin{aligned} i &= C_1 \frac{dv}{dt} + C_2 \frac{dv}{dt} + C_3 \frac{dv}{dt} + \cdots + C_N \frac{dv}{dt} \\ &= \left(\sum_{k=1}^{N} C_k \right) \frac{dv}{dt} = C_{eq} \frac{dv}{dt} \end{aligned} \tag{6.12}$$

where

$$C_{eq} = C_1 + C_2 + C_3 + \cdots + C_N \tag{6.13}$$

The **equivalent capacitance** of N parallel-connected capacitors is the sum of the individual capacitances.

We observe that capacitors in parallel combine in the same manner as resistors in series.

We now obtain C_{eq} of N capacitors connected in series by comparing the circuit in Fig. 6.15(a) with the equivalent circuit in Fig. 6.15(b). Note that the same current i flows (and consequently the same charge) through the capacitors. Applying KVL to the loop in Fig. 6.15(a),

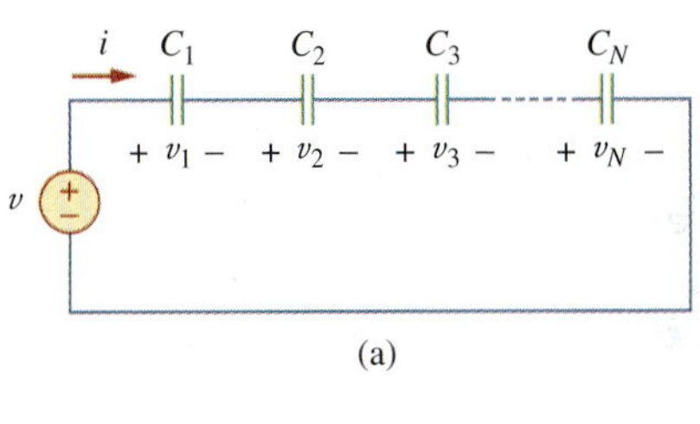

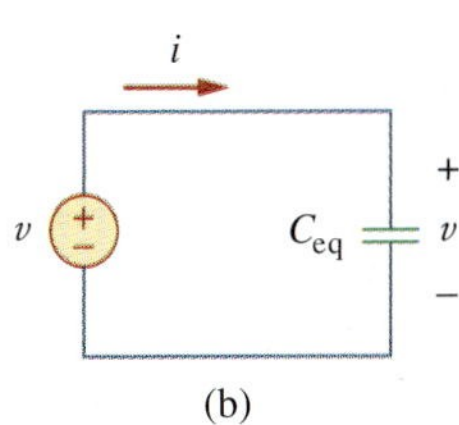

Figure 6.15
(a) Series-connected N capacitors, (b) equivalent circuit for the series capacitor.

$$v = v_1 + v_2 + v_3 + \cdots + v_N \tag{6.14}$$

But $v_k = \dfrac{1}{C_k} \displaystyle\int_{t_0}^{t} i(t)\, dt + v_k(t_0)$. Therefore,

$$\begin{aligned} v &= \frac{1}{C_1} \int_{t_0}^{t} i(t)\, dt + v_1(t_0) + \frac{1}{C_2} \int_{t_0}^{t} i(t)\, dt + v_2(t_0) \\ &\quad + \cdots + \frac{1}{C_N} \int_{t_0}^{t} i(t)\, dt + v_N(t_0) \\ &= \left(\frac{1}{C_1} + \frac{1}{C_2} + \cdots + \frac{1}{C_N} \right) \int_{t_0}^{t} i(t)\, dt + v_1(t_0) + v_2(t_0) \\ &\quad + \cdots + v_N(t_0) \\ &= \frac{1}{C_{eq}} \int_{t_0}^{t} i(t)\, dt + v(t_0) \end{aligned} \tag{6.15}$$

where

$$\frac{1}{C_{eq}} = \frac{1}{C_1} + \frac{1}{C_2} + \frac{1}{C_3} + \cdots + \frac{1}{C_N} \tag{6.16}$$

The initial voltage $v(t_0)$ across C_{eq} is required by KVL to be the sum of the capacitor voltages at t_0. Or according to Eq. (6.15),

$$v(t_0) = v_1(t_0) + v_2(t_0) + \cdots + v_N(t_0)$$

Thus, according to Eq. (6.16),

> The **equivalent capacitance** of series-connected capacitors is the reciprocal of the sum of the reciprocals of the individual capacitances.

Note that capacitors in series combine in the same manner as resistors in parallel. For $N = 2$ (i.e., two capacitors in series), Eq. (6.16) becomes

$$\frac{1}{C_{eq}} = \frac{1}{C_1} + \frac{1}{C_2}$$

or

$$C_{eq} = \frac{C_1 C_2}{C_1 + C_2} \tag{6.17}$$

Example 6.6

Find the equivalent capacitance seen between terminals a and b of the circuit in Fig. 6.16.

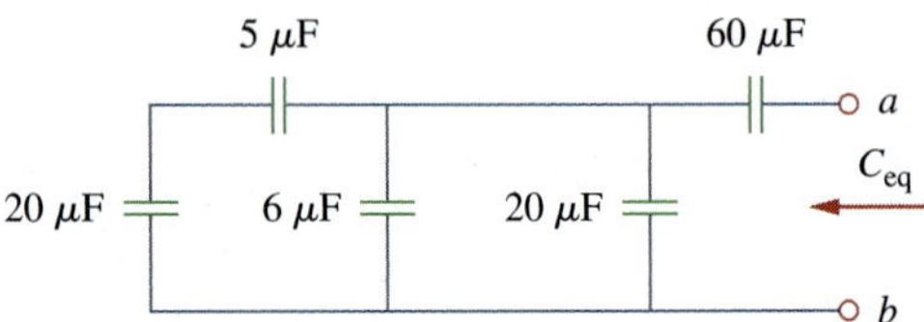

Figure 6.16
For Example 6.6.

Solution:

The 20-μF and 5-μF capacitors are in series; their equivalent capacitance is

$$\frac{20 \times 5}{20 + 5} = 4\ \mu\text{F}$$

This 4-μF capacitor is in parallel with the 6-μF and 20-μF capacitors; their combined capacitance is

$$4 + 6 + 20 = 30\ \mu\text{F}$$

This 30-μF capacitor is in series with the 60-μF capacitor. Hence, the equivalent capacitance for the entire circuit is

$$C_{eq} = \frac{30 \times 60}{30 + 60} = 20\ \mu\text{F}$$

Practice Problem 6.6

Find the equivalent capacitance seen at the terminals of the circuit in Fig. 6.17.

Answer: 40 μF.

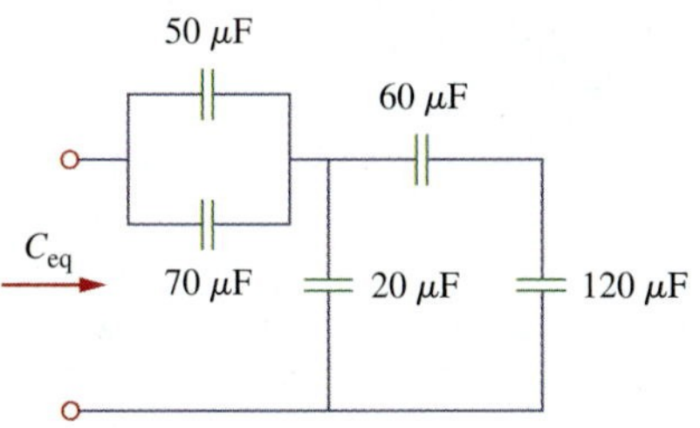

Figure 6.17
For Practice Prob. 6.6.

Example 6.7

For the circuit in Fig. 6.18, find the voltage across each capacitor.

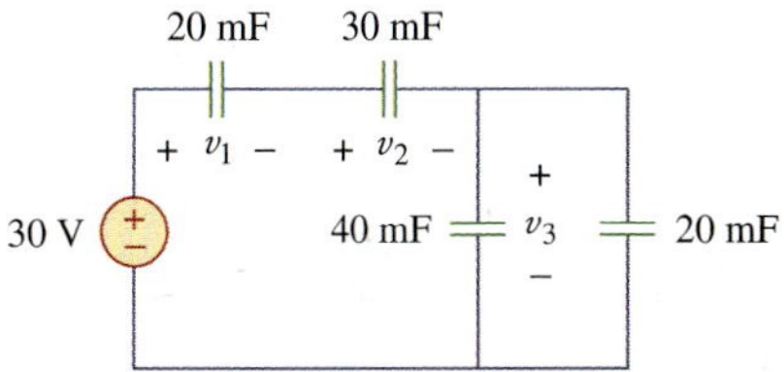

Figure 6.18
For Example 6.7.

Solution:
We first find the equivalent capacitance C_{eq}, shown in Fig. 6.19. The two parallel capacitors in Fig. 6.18 can be combined to get $40 + 20 = 60$ mF. This 60-mF capacitor is in series with the 20-mF and 30-mF capacitors. Thus,

$$C_{eq} = \frac{1}{\frac{1}{60} + \frac{1}{30} + \frac{1}{20}} \text{ mF} = 10 \text{ mF}$$

The total charge is

$$q = C_{eq}v = 10 \times 10^{-3} \times 30 = 0.3 \text{ C}$$

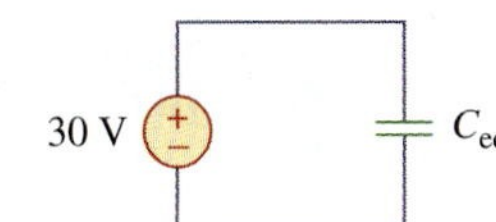

Figure 6.19
Equivalent circuit for Fig. 6.18.

This is the charge on the 20-mF and 30-mF capacitors, because they are in series with the 30-V source. (A crude way to see this is to imagine that charge acts like current, since $i = dq/dt$.) Therefore,

$$v_1 = \frac{q}{C_1} = \frac{0.3}{20 \times 10^{-3}} = 15 \text{ V} \qquad v_2 = \frac{q}{C_2} = \frac{0.3}{30 \times 10^{-3}} = 10 \text{ V}$$

Having determined v_1 and v_2, we now use KVL to determine v_3 by

$$v_3 = 30 - v_1 - v_2 = 5 \text{ V}$$

Alternatively, since the 40-mF and 20-mF capacitors are in parallel, they have the same voltage v_3 and their combined capacitance is $40 + 20 = 60$ mF. This combined capacitance is in series with the 20-mF and 30-mF capacitors and consequently has the same charge on it. Hence,

$$v_3 = \frac{q}{60 \text{ mF}} = \frac{0.3}{60 \times 10^{-3}} = 5 \text{ V}$$

Practice Problem 6.7

Find the voltage across each of the capacitors in Fig. 6.20.

Answer: $v_1 = 30$ V, $v_2 = 30$ V, $v_3 = 10$ V, $v_4 = 20$ V.

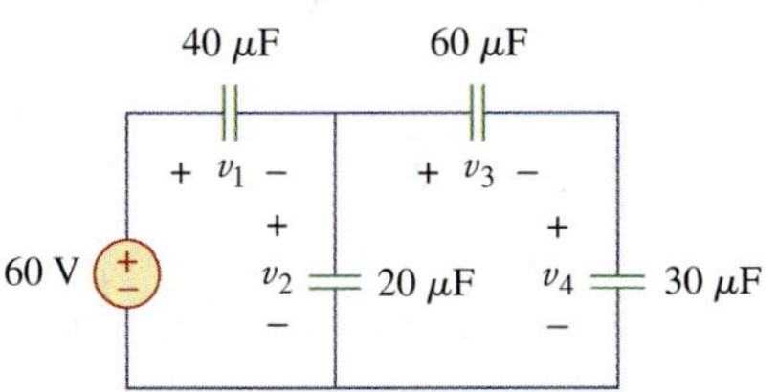

Figure 6.20
For Practice Prob. 6.7.

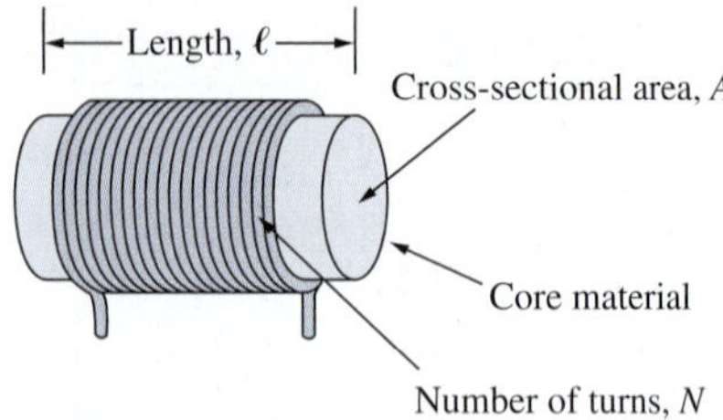

Figure 6.21
Typical form of an inductor.

6.4 Inductors

An inductor is a passive element designed to store energy in its magnetic field. Inductors find numerous applications in electronic and power systems. They are used in power supplies, transformers, radios, TVs, radars, and electric motors.

Any conductor of electric current has inductive properties and may be regarded as an inductor. But in order to enhance the inductive effect, a practical inductor is usually formed into a cylindrical coil with many turns of conducting wire, as shown in Fig. 6.21.

> An **inductor** consists of a coil of conducting wire.

If current is allowed to pass through an inductor, it is found that the voltage across the inductor is directly proportional to the time rate of change of the current. Using the passive sign convention,

$$v = L\frac{di}{dt} \tag{6.18}$$

> In view of Eq. (6.18), for an inductor to have voltage across its terminals, its current must vary with time. Hence, $v = 0$ for constant current through the inductor.

where L is the constant of proportionality called the *inductance* of the inductor. The unit of inductance is the henry (H), named in honor of the American inventor Joseph Henry (1797–1878). It is clear from Eq. (6.18) that 1 henry equals 1 volt-second per ampere.

> **Inductance** is the property whereby an inductor exhibits opposition to the change of current flowing through it, measured in henrys (H).

The inductance of an inductor depends on its physical dimension and construction. Formulas for calculating the inductance of inductors of different shapes are derived from electromagnetic theory and can be found in standard electrical engineering handbooks. For example, for the inductor, (solenoid) shown in Fig. 6.21,

$$L = \frac{N^2\mu A}{\ell} \tag{6.19}$$

where N is the number of turns, ℓ is the length, A is the cross-sectional area, and μ is the permeability of the core. We can see from Eq. (6.19) that inductance can be increased by increasing the number of turns of coil, using material with higher permeability as the core, increasing the cross-sectional area, or reducing the length of the coil.

Like capacitors, commercially available inductors come in different values and types. Typical practical inductors have inductance values ranging from a few microhenrys (μH), as in communication systems, to tens of henrys (H) as in power systems. Inductors may be fixed or variable. The core may be made of iron, steel, plastic, or air. The terms *coil* and *choke* are also used for inductors. Common inductors are shown in Fig. 6.22. The circuit symbols for inductors are shown in Fig. 6.23, following the passive sign convention.

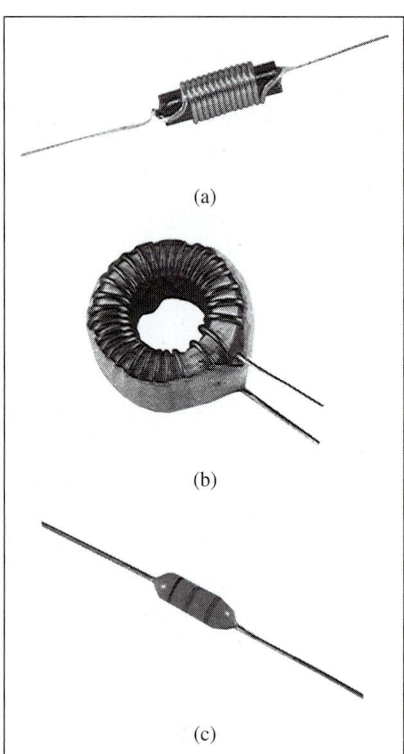

Figure 6.22
Various types of inductors: (a) solenoidal wound inductor, (b) toroidal inductor, (c) chip inductor.
Courtesy of Tech America.

Equation (6.18) is the voltage-current relationship for an inductor. Figure 6.24 shows this relationship graphically for an inductor whose

Historical

Joseph Henry (1797–1878), an American physicist, discovered inductance and constructed an electric motor.

Born in Albany, New York, Henry graduated from Albany Academy and taught philosophy at Princeton University from 1832 to 1846. He was the first secretary of the Smithsonian Institution. He conducted several experiments on electromagnetism and developed powerful electromagnets that could lift objects weighing thousands of pounds. Interestingly, Joseph Henry discovered electromagnetic induction before Faraday but failed to publish his findings. The unit of inductance, the henry, was named after him.

inductance is independent of current. Such an inductor is known as a *linear inductor*. For a *nonlinear inductor*, the plot of Eq. (6.18) will not be a straight line because its inductance varies with current. We will assume linear inductors in this textbook unless stated otherwise.

The current-voltage relationship is obtained from Eq. (6.18) as

$$di = \frac{1}{L} v\, dt$$

Integrating gives

$$i = \frac{1}{L}\int_{-\infty}^{t} v(t)\, dt \tag{6.20}$$

or

$$i = \frac{1}{L}\int_{t_0}^{t} v(t)\, dt + i(t_0) \tag{6.21}$$

where $i(t_0)$ is the total current for $-\infty < t < t_0$ and $i(-\infty) = 0$. The idea of making $i(-\infty) = 0$ is practical and reasonable, because there must be a time in the past when there was no current in the inductor.

The inductor is designed to store energy in its magnetic field. The energy stored can be obtained from Eq. (6.18). The power delivered to the inductor is

$$p = vi = \left(L\frac{di}{dt}\right)i \tag{6.22}$$

The energy stored is

$$\begin{aligned} w &= \int_{-\infty}^{t} p\, dt = \int_{-\infty}^{t} \left(L\frac{di}{dt}\right) i\, dt \\ &= L\int_{-\infty}^{t} i\, di = \frac{1}{2}Li^2(t) - \frac{1}{2}Li^2(-\infty) \end{aligned} \tag{6.23}$$

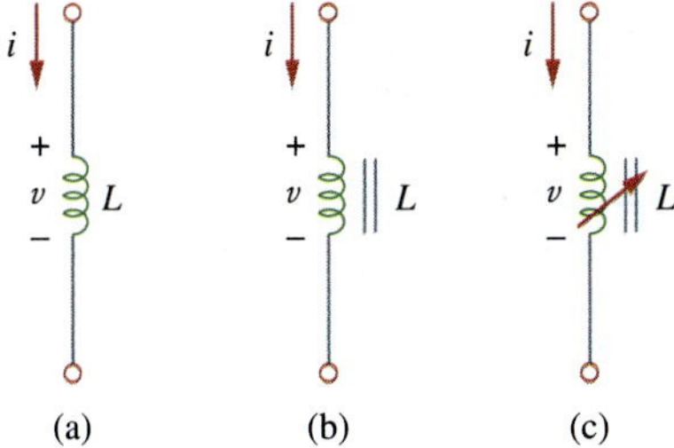

Figure 6.23
Circuit symbols for inductors: (a) air-core, (b) iron-core, (c) variable iron-core.

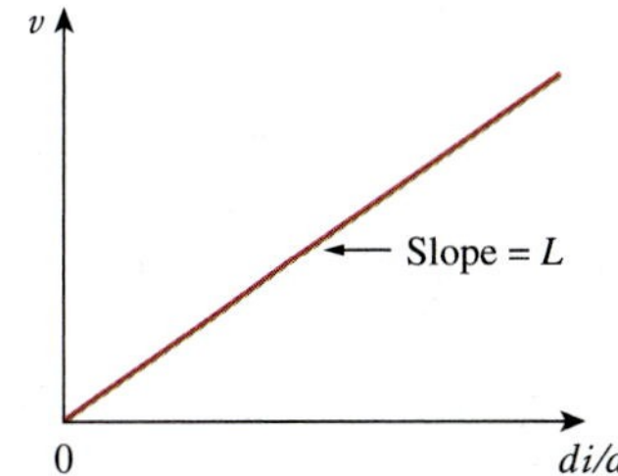

Figure 6.24
Voltage-current relationship of an inductor.

Since $i(-\infty) = 0$,

$$w = \frac{1}{2}Li^2 \tag{6.24}$$

We should note the following important properties of an inductor.

1. Note from Eq. (6.18) that the voltage across an inductor is zero when the current is constant. Thus,

> An inductor acts like a short circuit to dc.

2. An important property of the inductor is its opposition to the change in current flowing through it.

> The current through an inductor cannot change instantaneously.

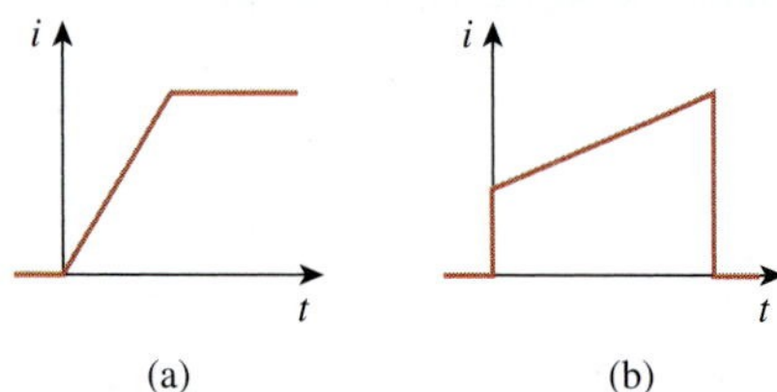

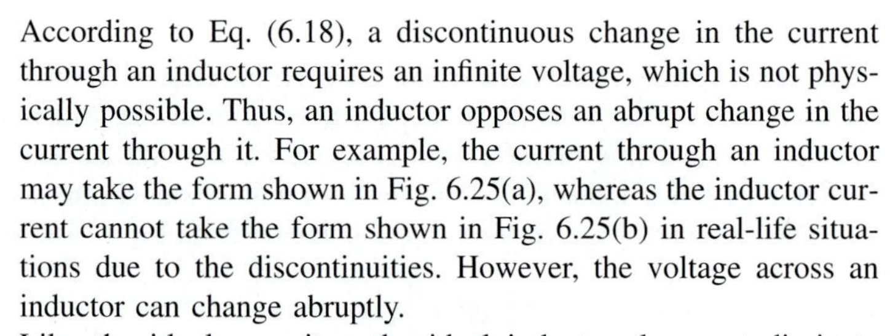

Figure 6.25
Current through an inductor: (a) allowed, (b) not allowable; an abrupt change is not possible.

According to Eq. (6.18), a discontinuous change in the current through an inductor requires an infinite voltage, which is not physically possible. Thus, an inductor opposes an abrupt change in the current through it. For example, the current through an inductor may take the form shown in Fig. 6.25(a), whereas the inductor current cannot take the form shown in Fig. 6.25(b) in real-life situations due to the discontinuities. However, the voltage across an inductor can change abruptly.

3. Like the ideal capacitor, the ideal inductor does not dissipate energy. The energy stored in it can be retrieved at a later time. The inductor takes power from the circuit when storing energy and delivers power to the circuit when returning previously stored energy.

Since an inductor is often made of a highly conducting wire, it has a very small resistance.

4. A practical, nonideal inductor has a significant resistive component, as shown in Fig. 6.26. This is due to the fact that the inductor is made of a conducting material such as copper, which has some resistance. This resistance is called the *winding resistance* R_w, and it appears in series with the inductance of the inductor. The presence of R_w makes it both an energy storage device and an energy dissipation device. Since R_w is usually very small, it is ignored in most cases. The nonideal inductor also has a *winding capacitance* C_w due to the capacitive coupling between the conducting coils. C_w is very small and can be ignored in most cases, except at high frequencies. We will assume ideal inductors in this book.

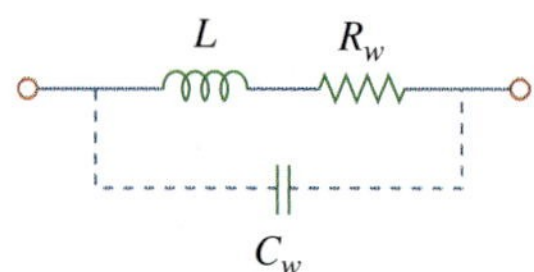

Figure 6.26
Circuit model for a practical inductor.

Example 6.8

The current through a 0.1-H inductor is $i(t) = 10te^{-5t}$ A. Find the voltage across the inductor and the energy stored in it.

Solution:

Since $v = L\,di/dt$ and $L = 0.1$ H,

$$v = 0.1\frac{d}{dt}(10te^{-5t}) = e^{-5t} + t(-5)e^{-5t} = e^{-5t}(1 - 5t) \text{ V}$$

The energy stored is

$$w = \frac{1}{2}Li^2 = \frac{1}{2}(0.1)100t^2e^{-10t} = 5t^2e^{-10t}\text{ J}$$

Practice Problem 6.8

If the current through a 1-mH inductor is $i(t) = 20\cos 100t$ mA, find the terminal voltage and the energy stored.

Answer: $-2\sin 100t$ mV, $0.2\cos^2 100t\ \mu$J.

Example 6.9

Find the current through a 5-H inductor if the voltage across it is

$$v(t) = \begin{cases} 30t^2, & t > 0 \\ 0, & t < 0 \end{cases}$$

Also, find the energy stored at $t = 5$ s. Assume $i(v) > 0$.

Solution:

Since $i = \frac{1}{L}\int_{t_0}^{t} v(t)\,dt + i(t_0)$ and $L = 5$ H,

$$i = \frac{1}{5}\int_0^t 30t^2\,dt + 0 = 6 \times \frac{t^3}{3} = 2t^3\text{ A}$$

The power $p = vi = 60t^5$, and the energy stored is then

$$w = \int p\,dt = \int_0^5 60t^5\,dt = 60\frac{t^6}{6}\bigg|_0^5 = 156.25\text{ kJ}$$

Alternatively, we can obtain the energy stored using Eq. (6.24), by writing

$$w\big|_0^5 = \frac{1}{2}Li^2(5) - \frac{1}{2}Li(0) = \frac{1}{2}(5)(2 \times 5^3)^2 - 0 = 156.25\text{ kJ}$$

as obtained before.

Practice Problem 6.9

The terminal voltage of a 2-H inductor is $v = 10(1 - t)$ V. Find the current flowing through it at $t = 4$ s and the energy stored in it at $t = 4$ s. Assume $i(0) = 2$ A.

Answer: -18 A, 320 J.

Example 6.10

Consider the circuit in Fig. 6.27(a). Under dc conditions, find: (a) i, v_C, and i_L, (b) the energy stored in the capacitor and inductor.

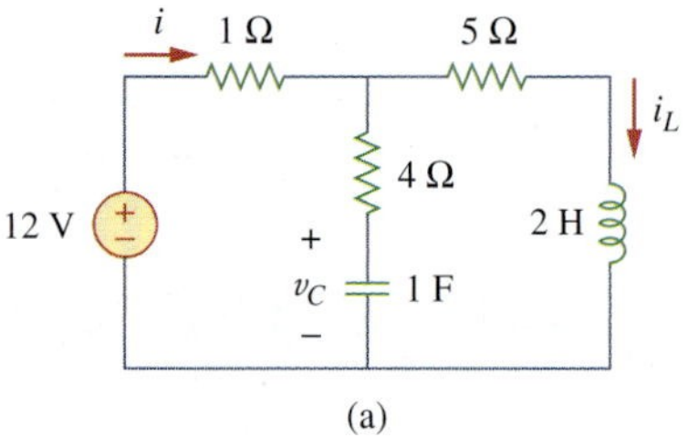

(a)

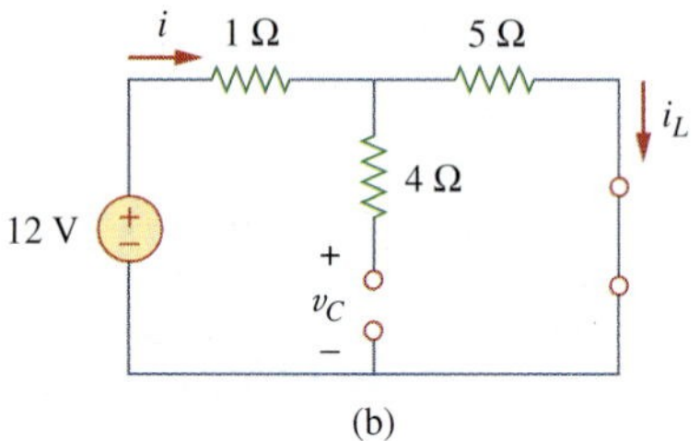

(b)

Figure 6.27
For Example 6.10.

Solution:

(a) Under dc conditions, we replace the capacitor with an open circuit and the inductor with a short circuit, as in Fig. 6.27(b). It is evident from Fig. 6.27(b) that

$$i = i_L = \frac{12}{1 + 5} = 2 \text{ A}$$

The voltage v_C is the same as the voltage across the 5-Ω resistor. Hence,

$$v_C = 5i = 10 \text{ V}$$

(b) The energy in the capacitor is

$$w_C = \frac{1}{2}Cv_C^2 = \frac{1}{2}(1)(10^2) = 50 \text{ J}$$

and that in the inductor is

$$w_L = \frac{1}{2}Li_L^2 = \frac{1}{2}(2)(2^2) = 4 \text{ J}$$

Practice Problem 6.10

Determine v_C, i_L, and the energy stored in the capacitor and inductor in the circuit of Fig. 6.28 under dc conditions.

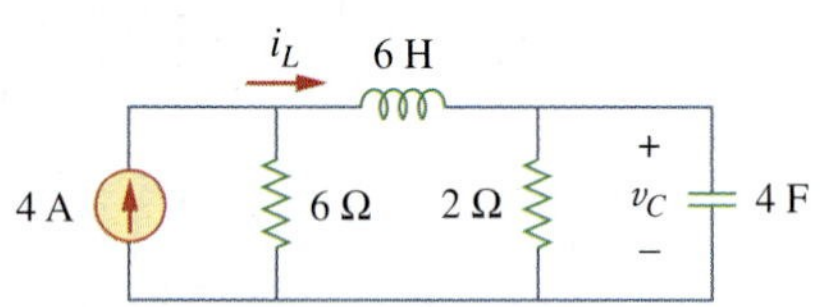

Figure 6.28
For Practice Prob. 6.10.

Answer: 6 V, 3 A, 72 J, 27 J.

6.5 Series and Parallel Inductors

Now that the inductor has been added to our list of passive elements, it is necessary to extend the powerful tool of series-parallel combination. We need to know how to find the equivalent inductance of a series-connected or parallel-connected set of inductors found in practical circuits.

Consider a series connection of N inductors, as shown in Fig. 6.29(a), with the equivalent circuit shown in Fig. 6.29(b). The inductors have the same current through them. Applying KVL to the loop,

$$v = v_1 + v_2 + v_3 + \cdots + v_N \tag{6.25}$$

Substituting $v_k = L_k\, di/dt$ results in

$$\begin{aligned} v &= L_1\frac{di}{dt} + L_2\frac{di}{dt} + L_3\frac{di}{dt} + \cdots + L_N\frac{di}{dt} \\ &= (L_1 + L_2 + L_3 + \cdots + L_N)\frac{di}{dt} \\ &= \left(\sum_{k=1}^{N} L_k\right)\frac{di}{dt} = L_{eq}\frac{di}{dt} \end{aligned} \tag{6.26}$$

(a)

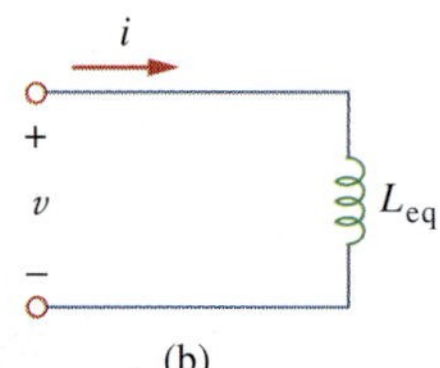

(b)

Figure 6.29
(a) A series connection of N inductors, (b) equivalent circuit for the series inductors.

where

$$L_{eq} = L_1 + L_2 + L_3 + \cdots + L_N \tag{6.27}$$

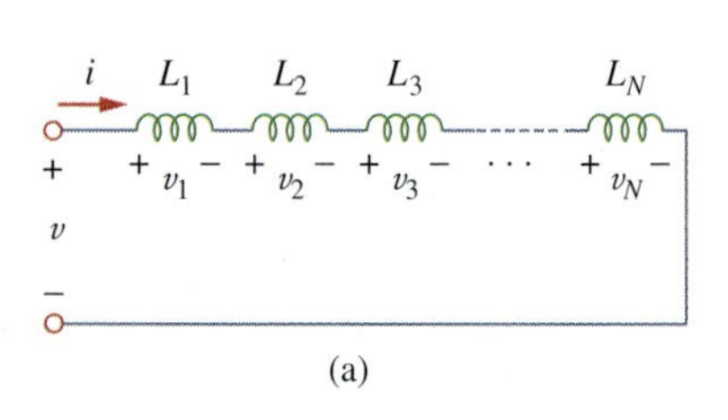

Thus,

> The **equivalent inductance** of series-connected inductors is the sum of the individual inductances.

Inductors in series are combined in exactly the same way as resistors in series.

We now consider a parallel connection of N inductors, as shown in Fig. 6.30(a), with the equivalent circuit in Fig. 6.30(b). The inductors have the same voltage across them. Using KCL,

$$i = i_1 + i_2 + i_3 + \cdots + i_N \tag{6.28}$$

But $i_k = \frac{1}{L_k}\int_{t_0}^{t} v\,dt + i_k(t_0)$; hence,

$$\begin{aligned} i &= \frac{1}{L_1}\int_{t_0}^{t} v\,dt + i_1(t_0) + \frac{1}{L_2}\int_{t_0}^{t} v\,dt + i_2(t_0) \\ &\quad + \cdots + \frac{1}{L_N}\int_{t_0}^{t} v\,dt + i_N(t_0) \\ &= \left(\frac{1}{L_1} + \frac{1}{L_2} + \cdots + \frac{1}{L_N}\right)\int_{t_0}^{t} v\,dt + i_1(t_0) + i_2(t_0) \\ &\quad + \cdots + i_N(t_0) \\ &= \left(\sum_{k=1}^{N}\frac{1}{L_k}\right)\int_{t_0}^{t} v\,dt + \sum_{k=1}^{N} i_k(t_0) = \frac{1}{L_{\text{eq}}}\int_{t_0}^{t} v\,dt + i(t_0) \end{aligned} \tag{6.29}$$

where

$$\frac{1}{L_{\text{eq}}} = \frac{1}{L_1} + \frac{1}{L_2} + \frac{1}{L_3} + \cdots + \frac{1}{L_N} \tag{6.30}$$

The initial current $i(t_0)$ through L_{eq} at $t = t_0$ is expected by KCL to be the sum of the inductor currents at t_0. Thus, according to Eq. (6.29),

$$i(t_0) = i_1(t_0) + i_2(t_0) + \cdots + i_N(t_0)$$

According to Eq. (6.30),

> The **equivalent inductance** of parallel inductors is the reciprocal of the sum of the reciprocals of the individual inductances.

Note that the inductors in parallel are combined in the same way as resistors in parallel.

For two inductors in parallel ($N = 2$), Eq. (6.30) becomes

$$\frac{1}{L_{\text{eq}}} = \frac{1}{L_1} + \frac{1}{L_2} \quad \text{or} \quad L_{\text{eq}} = \frac{L_1 L_2}{L_1 + L_2} \tag{6.31}$$

As long as all the elements are of the same type, the Δ-Y transformations for resistors discussed in Section 2.7 can be extended to capacitors and inductors.

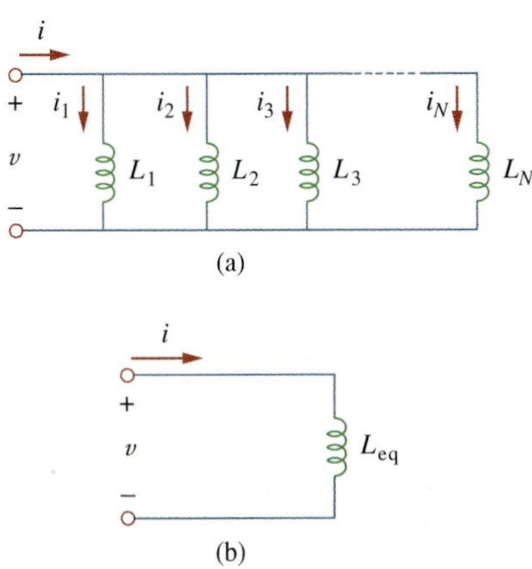

Figure 6.30
(a) A parallel connection of N inductors, (b) equivalent circuit for the parallel inductors.

TABLE 6.1

Important characteristics of the basic elements.†

Relation	Resistor (R)	Capacitor (C)	Inductor (L)
v-i:	$v = iR$	$v = \frac{1}{C}\int_{t_0}^{t} i\,dt + v(t_0)$	$v = L\frac{di}{dt}$
i-v:	$i = v/R$	$i = C\frac{dv}{dt}$	$i = \frac{1}{L}\int_{t_0}^{t} v\,dt + i(t_0)$
p or w:	$p = i^2R = \frac{v^2}{R}$	$w = \frac{1}{2}Cv^2$	$w = \frac{1}{2}Li^2$
Series:	$R_{eq} = R_1 + R_2$	$C_{eq} = \frac{C_1C_2}{C_1 + C_2}$	$L_{eq} = L_1 + L_2$
Parallel:	$R_{eq} = \frac{R_1R_2}{R_1 + R_2}$	$C_{eq} = C_1 + C_2$	$L_{eq} = \frac{L_1L_2}{L_1 + L_2}$
At dc:	Same	Open circuit	Short circuit
Circuit variable that cannot change abruptly:	Not applicable	v	i

† *Passive sign convention is assumed.*

It is appropriate at this point to summarize the most important characteristics of the three basic circuit elements we have studied. The summary is given in Table 6.1.

The wye-delta transformation discussed in Section 2.7 for resistors can be extended to capacitors and inductors.

Example 6.11

Find the equivalent inductance of the circuit shown in Fig. 6.31.

Figure 6.31
For Example 6.11.

Solution:
The 10-H, 12-H, and 20-H inductors are in series; thus, combining them gives a 42-H inductance. This 42-H inductor is in parallel with the 7-H inductor so that they are combined, to give

$$\frac{7 \times 42}{7 + 42} = 6\text{ H}$$

This 6-H inductor is in series with the 4-H and 8-H inductors. Hence,

$$L_{eq} = 4 + 6 + 8 = 18\text{ H}$$

Practice Problem 6.11

Calculate the equivalent inductance for the inductive ladder network in Fig. 6.32.

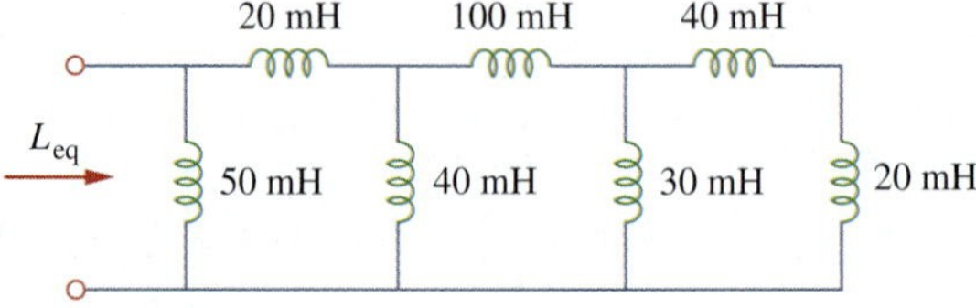

Figure 6.32
For Practice Prob. 6.11.

Answer: 25 mH.

Example 6.12

For the circuit in Fig. 6.33, $i(t) = 4(2 - e^{-10t})$ mA. If $i_2(0) = -1$ mA, find: (a) $i_1(0)$; (b) $v(t)$, $v_1(t)$, and $v_2(t)$; (c) $i_1(t)$ and $i_2(t)$.

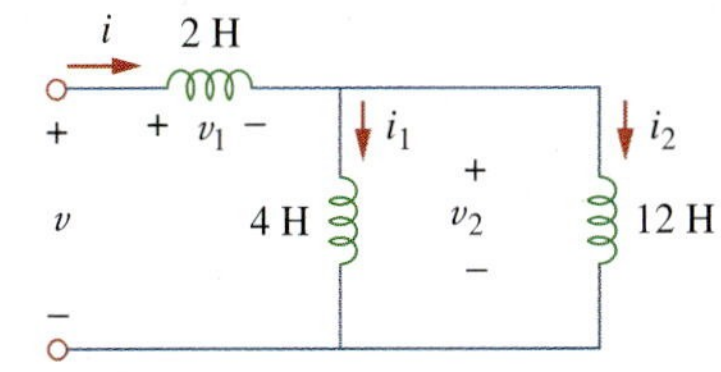

Figure 6.33
For Example 6.12.

Solution:

(a) From $i(t) = 4(2 - e^{-10t})$ mA, $i(0) = 4(2 - 1) = 4$ mA. Since $i = i_1 + i_2$,

$$i_1(0) = i(0) - i_2(0) = 4 - (-1) = 5 \text{ mA}$$

(b) The equivalent inductance is

$$L_{eq} = 2 + 4 \parallel 12 = 2 + 3 = 5 \text{ H}$$

Thus,

$$v(t) = L_{eq}\frac{di}{dt} = 5(4)(-1)(-10)e^{-10t} \text{ mV} = 200e^{-10t} \text{ mV}$$

and

$$v_1(t) = 2\frac{di}{dt} = 2(-4)(-10)e^{-10t} \text{ mV} = 80e^{-10t} \text{ mV}$$

Since $v = v_1 + v_2$,

$$v_2(t) = v(t) - v_1(t) = 120e^{-10t} \text{ mV}$$

(c) The current i_1 is obtained as

$$\begin{aligned} i_1(t) &= \frac{1}{4}\int_0^t v_2\,dt + i_1(0) = \frac{120}{4}\int_0^t e^{-10t}\,dt + 5 \text{ mA} \\ &= -3e^{-10t}\Big|_0^t + 5 \text{ mA} = -3e^{-10t} + 3 + 5 = 8 - 3e^{-10t} \text{ mA} \end{aligned}$$

Similarly,

$$\begin{aligned} i_2(t) &= \frac{1}{12}\int_0^t v_2\,dt + i_2(0) = \frac{120}{12}\int_0^t e^{-10t}\,dt - 1 \text{ mA} \\ &= -e^{-10t}\Big|_0^t - 1 \text{ mA} = -e^{-10t} + 1 - 1 = -e^{-10t} \text{ mA} \end{aligned}$$

Note that $i_1(t) + i_2(t) = i(t)$.

Practice Problem 6.12

In the circuit of Fig. 6.34, $i_1(t) = 0.6e^{-2t}$ A. If $i(0) = 1.4$ A, find: (a) $i_2(0)$; (b) $i_2(t)$ and $i(t)$; (c) $v_1(t)$, $v_2(t)$, and $v(t)$.

Answer: (a) 0.8 A, (b) $(-0.4 + 1.2e^{-2t})$ A, $(-0.4 + 1.8e^{-2t})$ A, (c) $-36e^{-2t}$ V, $-7.2e^{-2t}$ V, $-28.8e^{-2t}$ V.

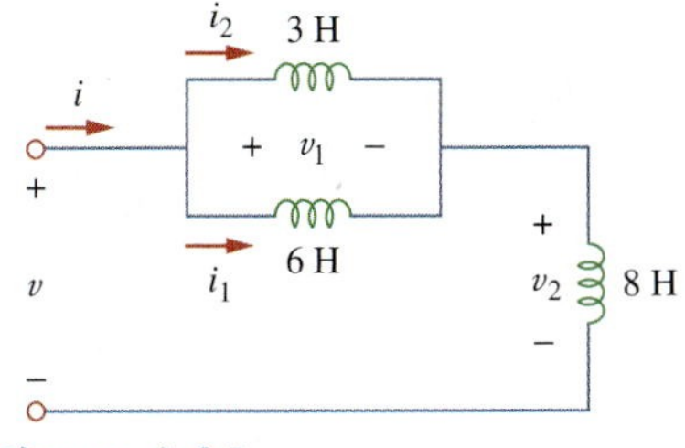

Figure 6.34
For Practice Prob. 6.12.

6.6 †Applications

Circuit elements such as resistors and capacitors are commercially available in either discrete form or integrated-circuit (IC) form. Unlike capacitors and resistors, inductors with appreciable inductance are difficult to produce on IC substrates. Therefore, inductors (coils) usually come in discrete form and tend to be more bulky and expensive. For this reason, inductors are not as versatile as capacitors and resistors, and they are more limited in applications. However, there are several applications in which inductors have no practical substitute. They are routinely used in relays, delays, sensing devices, pick-up heads, telephone circuits, radio and TV receivers, power supplies, electric motors, microphones, and loudspeakers, to mention a few.

Capacitors and inductors possess the following three special properties that make them very useful in electric circuits:

1. The capacity to store energy makes them useful as temporary voltage or current sources. Thus, they can be used for generating a large amount of current or voltage for a short period of time.
2. Capacitors oppose any abrupt change in voltage, while inductors oppose any abrupt change in current. This property makes inductors useful for spark or arc suppression and for converting pulsating dc voltage into relatively smooth dc voltage.
3. Capacitors and inductors are frequency sensitive. This property makes them useful for frequency discrimination.

The first two properties are put to use in dc circuits, while the third one is taken advantage of in ac circuits. We will see how useful these properties are in later chapters. For now, consider three applications involving capacitors and op amps: integrator, differentiator, and analog computer.

6.6.1 Integrator

Important op amp circuits that use energy-storage elements include integrators and differentiators. These op amp circuits often involve resistors and capacitors; inductors (coils) tend to be more bulky and expensive.

The op amp integrator is used in numerous applications, especially in analog computers, to be discussed in Section 6.6.3.

> An **integrator** is an op amp circuit whose output is proportional to the integral of the input signal.

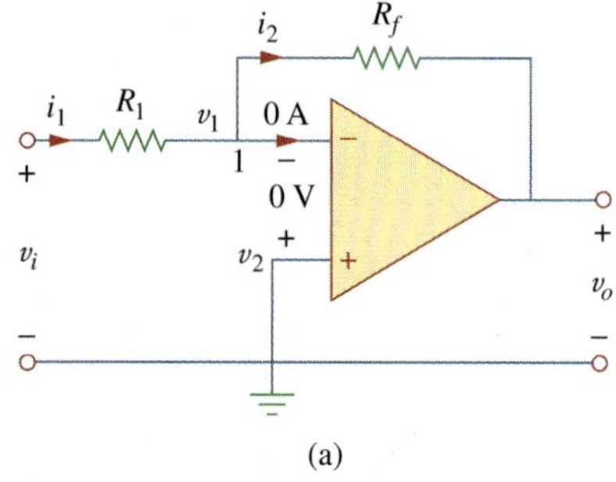

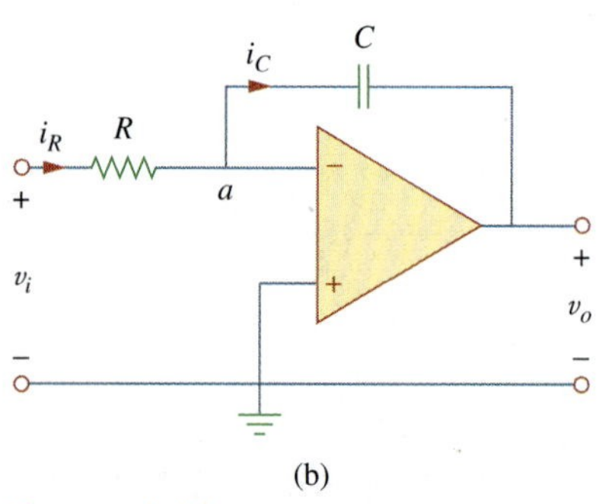

Figure 6.35
Replacing the feedback resistor in the inverting amplifier in (a) produces an integrator in (b).

If the feedback resistor R_f in the familiar inverting amplifier of Fig. 6.35(a) is replaced by a capacitor, we obtain an ideal integrator, as shown in Fig. 6.35(b). It is interesting that we can obtain a mathematical representation of integration this way. At node a in Fig. 6.35(b),

$$i_R = i_C \tag{6.32}$$

But

$$i_R = \frac{v_i}{R}, \qquad i_C = -C\frac{dv_o}{dt}$$

Substituting these in Eq. (6.32), we obtain

$$\frac{v_i}{R} = -C\frac{dv_o}{dt} \tag{6.33a}$$

$$dv_o = -\frac{1}{RC}v_i\, dt \tag{6.33b}$$

Integrating both sides gives

$$v_o(t) - v_o(0) = -\frac{1}{RC}\int_0^t v_i(t)\,dt \tag{6.34}$$

To ensure that $v_o(0) = 0$, it is always necessary to discharge the integrator's capacitor prior to the application of a signal. Assuming $v_o(0) = 0$,

$$v_o = -\frac{1}{RC}\int_0^t v_i(t)\,dt \tag{6.35}$$

which shows that the circuit in Fig. 6.35(b) provides an output voltage proportional to the integral of the input. In practice, the op amp integrator requires a feedback resistor to reduce dc gain and prevent saturation. Care must be taken that the op amp operates within the linear range so that it does not saturate.

Example 6.13

If $v_1 = 10 \cos 2t$ mV and $v_2 = 0.5t$ mV, find v_o in the op amp circuit in Fig. 6.36. Assume that the voltage across the capacitor is initially zero.

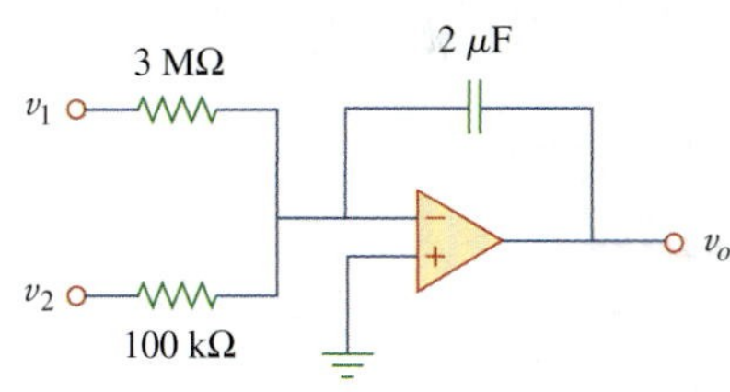

Figure 6.36
For Example 6.13.

Solution:
This is a summing integrator, and

$$\begin{aligned} v_o &= -\frac{1}{R_1C}\int v_1\,dt - \frac{1}{R_2C}\int v_2\,dt \\ &= -\frac{1}{3\times 10^6 \times 2 \times 10^{-6}}\int_0^t 10\cos 2t\,dt \\ &\quad -\frac{1}{100\times 10^3 \times 2\times 10^{-6}}\int_0^t 0.5t\,dt \\ &= -\frac{1}{6}\frac{10}{2}\sin 2t - \frac{1}{0.2}\frac{0.5t^2}{2} = -0.833\sin 2t - 1.25t^2 \text{ mV} \end{aligned}$$

Practice Problem 6.13

The integrator in Fig. 6.35(b) has $R = 100$ kΩ, $C = 20\ \mu$F. Determine the output voltage when a dc voltage of 10 mV is applied at $t = 0$. Assume that the op amp is initially nulled.

Answer: $-5t$ mV.

6.6.2 Differentiator

A **differentiator** is an op amp circuit whose output is proportional to the rate of change of the input signal.

In Fig. 6.35(a), if the input resistor is replaced by a capacitor, the resulting circuit is a differentiator, shown in Fig. 6.37. Applying KCL at node a,

$$i_R = i_C \tag{6.36}$$

But

$$i_R = -\frac{v_o}{R}, \qquad i_C = C\frac{dv_i}{dt}$$

Substituting these in Eq. (6.36) yields

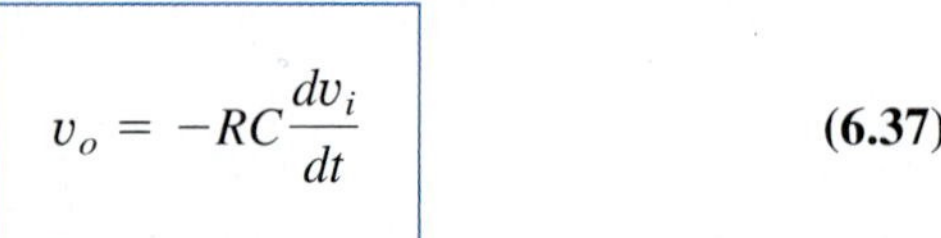

$$v_o = -RC\frac{dv_i}{dt} \tag{6.37}$$

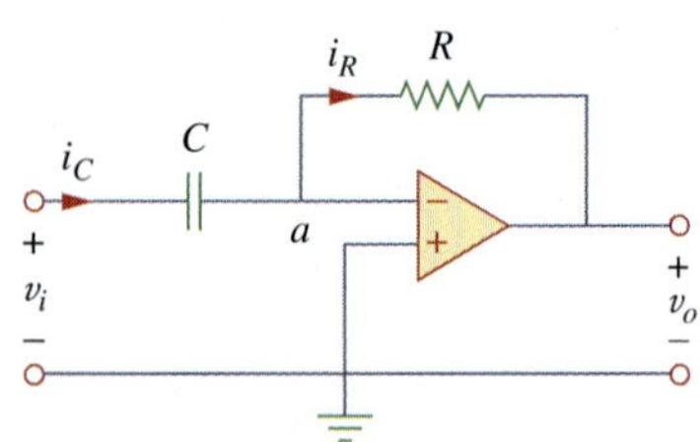

Figure 6.37
An op amp differentiator.

showing that the output is the derivative of the input. Differentiator circuits are electronically unstable because any electrical noise within the circuit is exaggerated by the differentiator. For this reason, the differentiator circuit in Fig. 6.37 is not as useful and popular as the integrator. It is seldom used in practice.

Example 6.14

Sketch the output voltage for the circuit in Fig. 6.38(a), given the input voltage in Fig. 6.38(b). Take $v_o = 0$ at $t = 0$.

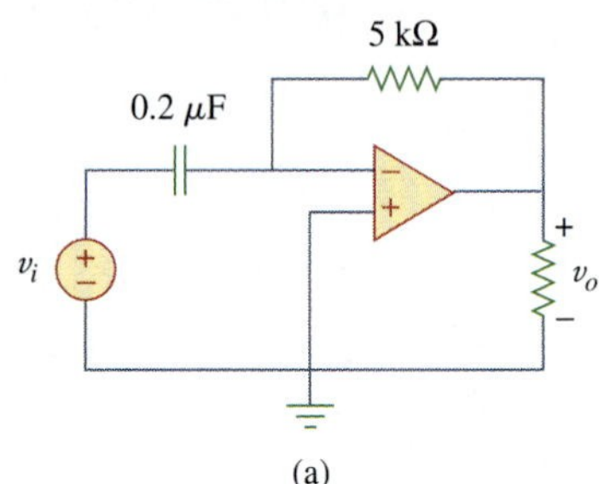

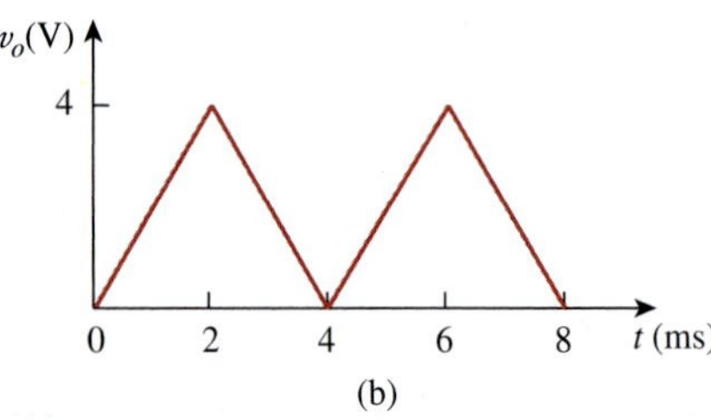

Figure 6.38
For Example 6.14.

Solution:
This is a differentiator with

$$RC = 5 \times 10^3 \times 0.2 \times 10^{-6} = 10^{-3}\text{ s}$$

For $0 < t < 4$ ms, we can express the input voltage in Fig. 6.38(b) as

$$v_i = \begin{cases} 2000t & 0 < t < 2\text{ ms} \\ 8 - 2000t & 2 < t < 4\text{ ms} \end{cases}$$

This is repeated for $4 < t < 8$ ms. Using Eq. (6.37), the output is obtained as

$$v_o = -RC\frac{dv_i}{dt} = \begin{cases} -2\text{ V} & 0 < t < 2\text{ ms} \\ 2\text{ V} & 2 < t < 4\text{ ms} \end{cases}$$

Thus, the output is as sketched in Fig. 6.39.

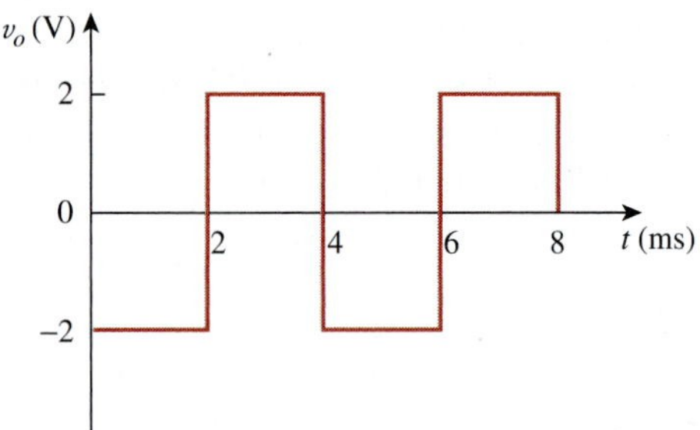

Figure 6.39
Output of the circuit in Fig. 6.38(a).

Practice Problem 6.14

The differentiator in Fig. 6.37 has $R = 100$ kΩ and $C = 0.1$ μF. Given that $v_i = 3t$ V, determine the output v_o.

Answer: −30 mV.

6.6.3 Analog Computer

Op amps were initially developed for electronic analog computers. Analog computers can be programmed to solve mathematical models of mechanical or electrical systems. These models are usually expressed in terms of differential equations.

To solve simple differential equations using the analog computer requires cascading three types of op amp circuits: integrator circuits, summing amplifiers, and inverting/noninverting amplifiers for negative/positive scaling. The best way to illustrate how an analog computer solves a differential equation is with an example.

Suppose we desire the solution $x(t)$ of the equation

$$a\frac{d^2x}{dt^2} + b\frac{dx}{dt} + cx = f(t), \qquad t > 0 \tag{6.38}$$

where a, b, and c are constants, and $f(t)$ is an arbitrary forcing function. The solution is obtained by first solving the highest-order derivative term. Solving for d^2x/dt^2 yields

$$\frac{d^2x}{dt^2} = \frac{f(t)}{a} - \frac{b}{a}\frac{dx}{dt} - \frac{c}{a}x \tag{6.39}$$

To obtain dx/dt, the d^2x/dt^2 term is integrated and inverted. Finally, to obtain x, the dx/dt term is integrated and inverted. The forcing function is injected at the proper point. Thus, the analog computer for solving Eq. (6.38) is implemented by connecting the necessary summers, inverters, and integrators. A plotter or oscilloscope may be used to view the output x, or dx/dt, or d^2x/dt^2, depending on where it is connected in the system.

Although the above example is on a second-order differential equation, any differential equation can be simulated by an analog computer comprising integrators, inverters, and inverting summers. But care must be exercised in selecting the values of the resistors and capacitors, to ensure that the op amps do not saturate during the solution time interval.

The analog computers with vacuum tubes were built in the 1950s and 1960s. Recently their use has declined. They have been superseded by modern digital computers. However, we still study analog computers for two reasons. First, the availability of integrated op amps has made it possible to build analog computers easily and cheaply. Second, understanding analog computers helps with the appreciation of the digital computers.

Example 6.15

Design an analog computer circuit to solve the differential equation:

$$\frac{d^2v_o}{dt^2} + 2\frac{dv_o}{dt} + v_o = 10\sin 4t, \qquad t > 0$$

subject to $v_o(0) = -4$, $v'_o(0) = 1$, where the prime refers to the time derivative.

Solution:

1. **Define.** We have a clearly defined problem and expected solution. I might remind the student that many times the problem is not so well defined and this portion of the problem-solving process could

require much more effort. If this is so, then you should always keep in mind that time spent here will result in much less effort later and most likely save you a lot of frustration in the process.

2. **Present.** Clearly, using the devices developed in Section 6.6.3 will allow us to create the desired analog computer circuit. We will need the integrator circuits (possibly combined with a summing capability) and one or more inverter circuits.
3. **Alternative.** The approach for solving this problem is straightforward. We will need to pick the correct values of resistances and capacitors to allow us to realize the equation we are representing. The final output of the circuit will give the desired result.
4. **Attempt.** There are an infinite number of possibilities for picking the resistors and capacitors, many of which will result in correct solutions. Extreme values of resistors and capacitors will result in incorrect outputs. For example, low values of resistors will overload the electronics. Picking values of resistors that are too large will cause the op amps to stop functioning as ideal devices. The limits can be determined from the characteristics of the real op amp.

   We first solve for the second derivative as

$$\frac{d^2v_o}{dt^2} = 10 \sin 4t - 2\frac{dv_o}{dt} - v_o \qquad \textbf{(6.15.1)}$$

Solving this requires some mathematical operations, including summing, scaling, and integration. Integrating both sides of Eq. (6.15.1) gives

$$\frac{dv_o}{dt} = -\int_0^t \left(-10 \sin 4t + 2\frac{dv_o}{dt} + v_o\right)dt + v_o'(0) \qquad \textbf{(6.15.2)}$$

where $v_o'(0) = 1$. We implement Eq. (6.15.2) using the summing integrator shown in Fig. 6.40(a). The values of the resistors and capacitors have been chosen so that $RC = 1$ for the term

$$-\frac{1}{RC}\int_0^t v_o\,dt$$

Other terms in the summing integrator of Eq. (6.15.2) are implemented accordingly. The initial condition $dv_o(0)/dt = 1$ is implemented by connecting a 1-V battery with a switch across the capacitor as shown in Fig. 6.40(a).

The next step is to obtain v_o by integrating dv_o/dt and inverting the result,

$$v_o = -\int_0^t \left(-\frac{dv_o}{dt}\right)dt + v(0) \qquad \textbf{(6.15.3)}$$

This is implemented with the circuit in Fig. 6.40(b) with the battery giving the initial condition of −4 V. We now combine the two circuits in Fig. 6.40(a) and (b) to obtain the complete circuit shown in Fig. 6.40(c). When the input signal 10 sin 4t is applied, we open the switches at $t = 0$ to obtain the output waveform v_o, which may be viewed on an oscilloscope.

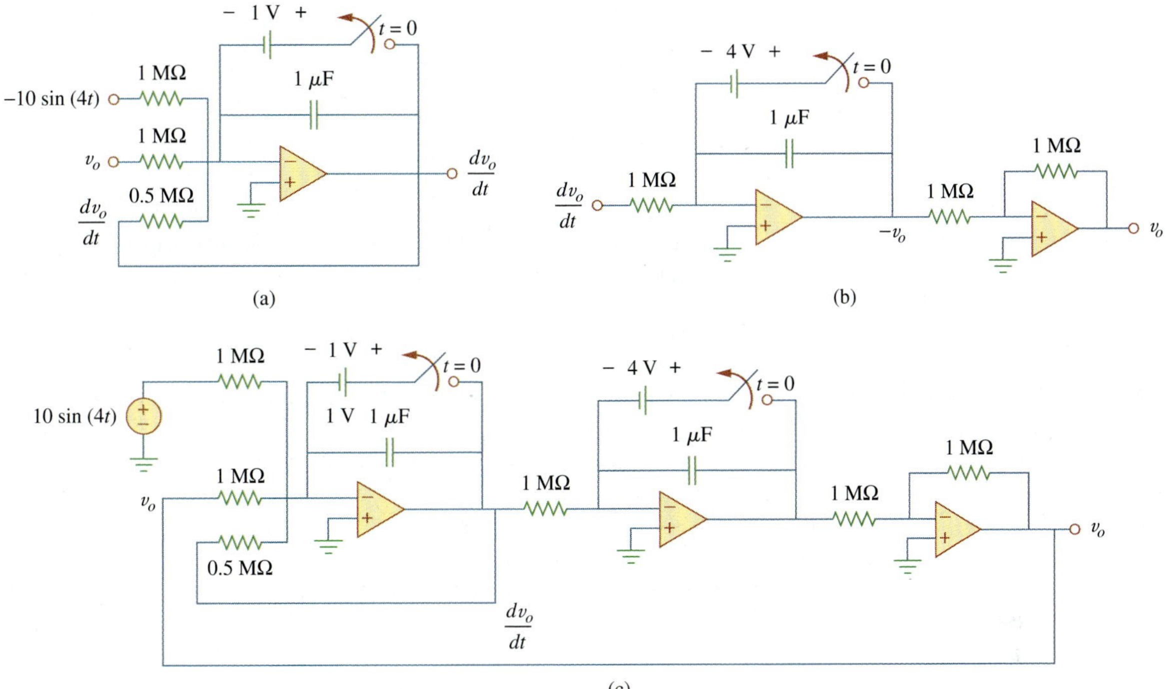

Figure 6.40
For Example 6.15.

5. **Evaluate.** The answer looks correct, but is it? If an actual solution for v_o is desired, then a good check would be to first find the solution by realizing the circuit in *PSpice*. This result could then be compared with a solution using the differential solution capability of *MATLAB*.

Since all we need to do is check the circuit and confirm that it represents the equation, we have an easier technique to use. We just go through the circuit and see if it generates the desired equation.

However, we still have choices to make. We could go through the circuit from left to right but that would involve differentiating the result to obtain the original equation. An easier approach would be to go from right to left. This is the approach we will use to check the answer.

Starting with the output, v_o, we see that the right-hand op amp is nothing more than an inverter with a gain of one. This means that the output of the middle circuit is $-v_o$. The following represents the action of the middle circuit.

$$-v_o = -\left(\int_0^t \frac{dv_o}{dt}\,dt + v_o(0)\right) = -\left(v_o\Big|_0^t + v_o(0)\right)$$
$$= -(v_o(t) - v_o(0) + v_o(0))$$

where $v_o(0) = -4$ V is the initial voltage across the capacitor.

We check the circuit on the left the same way.

$$\frac{dv_o}{dt} = -\left(\int_0^t -\frac{d^2v_o}{dt^2}dt - v_o'(0)\right) = -\left(-\frac{dv_o}{dt} + v_o'(0) - v_o'(0)\right)$$

Now all we need to verify is that the input to the first op amp is $-d^2v_o/dt^2$.

Looking at the input we see that it is equal to

$$-10\sin(4t) + v_o + \frac{1/10^{-6}}{0.5\text{ M}\Omega}\frac{dv_o}{dt} = -10\sin(4t) + v_o + 2\frac{dv_o}{dt}$$

which does produce $-d^2v_o/dt^2$ from the original equation.

6. **Satisfactory?** The solution we have obtained is satisfactory. We can now present this work as a solution to the problem.

Practice Problem 6.15

Design an analog computer circuit to solve the differential equation:

$$\frac{d^2v_o}{dt^2} + 3\frac{dv_o}{dt} + 2v_o = 4\cos 10t, \qquad t > 0$$

subject to $v_o(0) = 2$, $v'_o(0) = 0$.

Answer: See Fig. 6.41, where $RC = 1$ s.

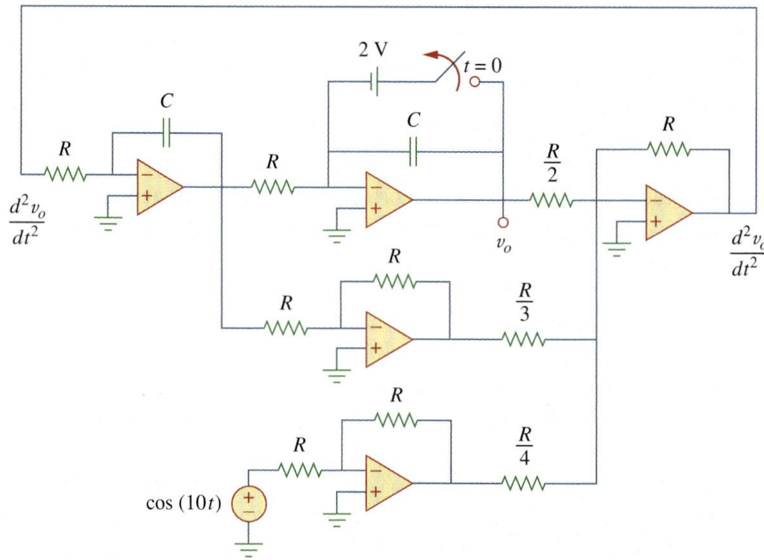

Figure 6.41
For Practice Prob. 6.15.

6.7 Summary

1. The current through a capacitor is directly proportional to the time rate of change of the voltage across it.

$$i = C\frac{dv}{dt}$$

The current through a capacitor is zero unless the voltage is changing. Thus, a capacitor acts like an open circuit to a dc source.

2. The voltage across a capacitor is directly proportional to the time integral of the current through it.

$$v = \frac{1}{C}\int_{-\infty}^{t} i\,dt = \frac{1}{C}\int_{t_0}^{t} i\,dt + v(t_0)$$

The voltage across a capacitor cannot change instantly.
3. Capacitors in series and in parallel are combined in the same way as conductances.
4. The voltage across an inductor is directly proportional to the time rate of change of the current through it.

$$v = L\frac{di}{dt}$$

The voltage across the inductor is zero unless the current is changing. Thus, an inductor acts like a short circuit to a dc source.
5. The current through an inductor is directly proportional to the time integral of the voltage across it.

$$i = \frac{1}{L}\int_{-\infty}^{t} v\,dt = \frac{1}{L}\int_{t_0}^{t} v\,dt + i(t_0)$$

The current through an inductor cannot change instantly.
6. Inductors in series and in parallel are combined in the same way resistors in series and in parallel are combined.
7. At any given time t, the energy stored in a capacitor is $\frac{1}{2}Cv^2$, while the energy stored in an inductor is $\frac{1}{2}Li^2$.
8. Three application circuits, the integrator, the differentiator, and the analog computer, can be realized using resistors, capacitors, and op amps.

Review Questions

6.1 What charge is on a 5-F capacitor when it is connected across a 120-V source?

(a) 600 C (b) 300 C
(c) 24 C (d) 12 C

6.2 Capacitance is measured in:

(a) coulombs (b) joules
(c) henrys (d) farads

6.3 When the total charge in a capacitor is doubled, the energy stored:

(a) remains the same (b) is halved
(c) is doubled (d) is quadrupled

6.4 Can the voltage waveform in Fig. 6.42 be associated with a real capacitor?

(a) Yes (b) No

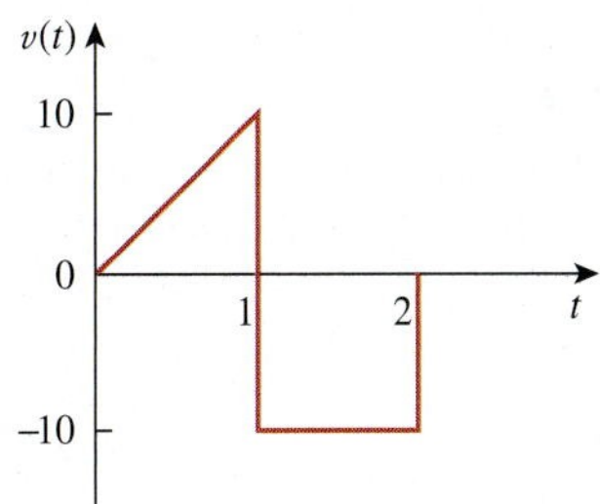

Figure 6.42
For Review Question 6.4.

6.5 The total capacitance of two 40-mF series-connected capacitors in parallel with a 4-mF capacitor is:

(a) 3.8 mF (b) 5 mF (c) 24 mF
(d) 44 mF (e) 84 mF

6.6 In Fig. 6.43, if $i = \cos 4t$ and $v = \sin 4t$, the element is:

(a) a resistor (b) a capacitor (c) an inductor

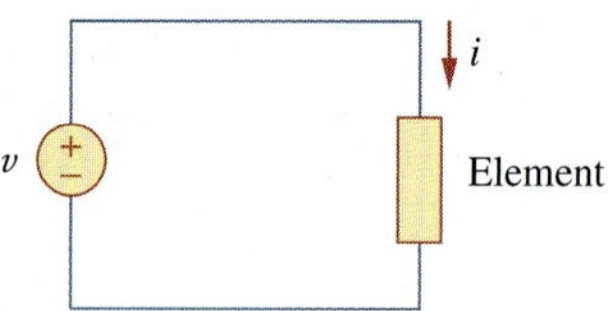

Figure 6.43
For Review Question 6.6.

6.7 A 5-H inductor changes its current by 3 A in 0.2 s. The voltage produced at the terminals of the inductor is:

(a) 75 V (b) 8.888 V
(c) 3 V (d) 1.2 V

6.8 If the current through a 10-mH inductor increases from zero to 2 A, how much energy is stored in the inductor?

(a) 40 mJ (b) 20 mJ
(c) 10 mJ (d) 5 mJ

6.9 Inductors in parallel can be combined just like resistors in parallel.

(a) True (b) False

6.10 For the circuit in Fig. 6.44, the voltage divider formula is:

(a) $v_1 = \dfrac{L_1 + L_2}{L_1} v_s$ (b) $v_1 = \dfrac{L_1 + L_2}{L_2} v_s$

(c) $v_1 = \dfrac{L_2}{L_1 + L_2} v_s$ (d) $v_1 = \dfrac{L_1}{L_1 + L_2} v_s$

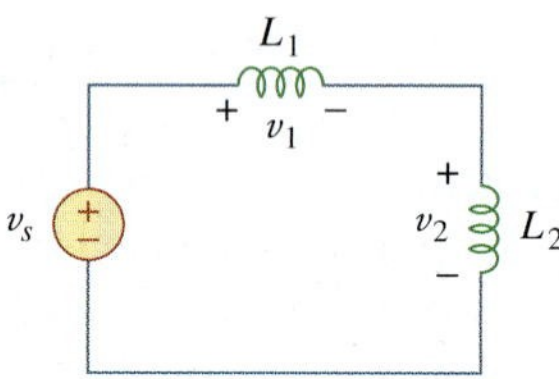

Figure 6.44
For Review Question 6.10.

Answers: 6.1a, 6.2d, 6.3d, 6.4b, 6.5c, 6.6b, 6.7a, 6.8b, 6.9a, 6.10d.

Problems

Section 6.2 Capacitors

6.1 If the voltage across a 5-F capacitor is $2te^{-3t}$ V, find the current and the power.

6.2 A 20-μF capacitor has energy $w(t) = 10 \cos^2 377t$ J. Determine the current through the capacitor.

6.3 Design a problem to help other students better understand how capacitors work.

6.4 A current of $6 \sin 4t$ A flows through a 2-F capacitor. Find the voltage $v(t)$ across the capacitor given that $v(0) = 1$ V.

6.5 The voltage across a 4-μF capacitor is shown in Fig. 6.45. Find the current waveform.

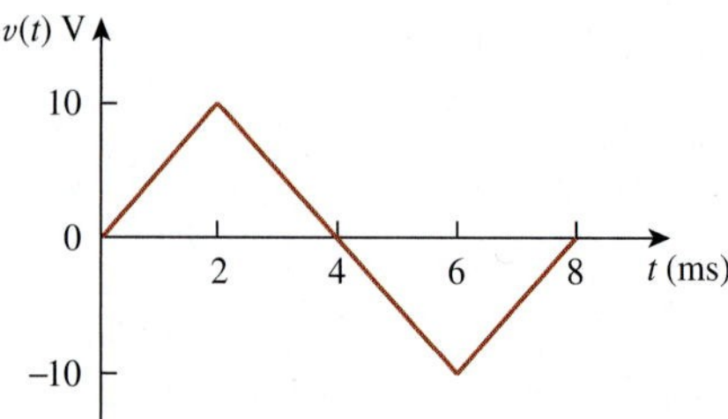

Figure 6.45
For Prob. 6.5.

6.6 The voltage waveform in Fig. 6.46 is applied across a 30-μF capacitor. Draw the current waveform through it.

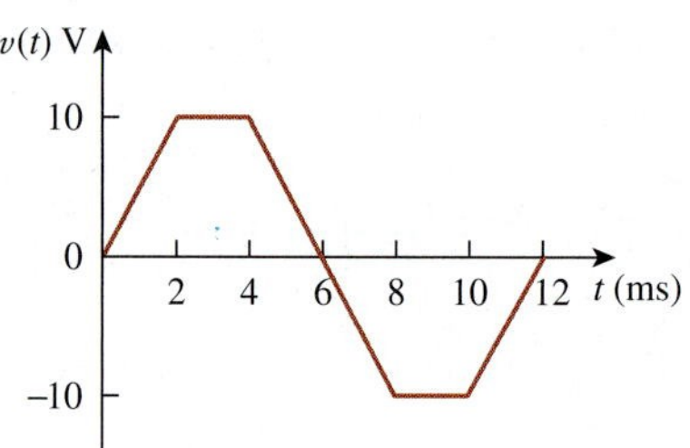

Figure 6.46
For Prob. 6.6.

6.7 At $t = 0$, the voltage across a 50-mF capacitor is 10 V. Calculate the voltage across the capacitor for $t > 0$ when current $4t$ mA flows through it.

6.8 A 4-mF capacitor has the terminal voltage

$$v = \begin{cases} 50 \text{ V}, & t \leq 0 \\ Ae^{-100t} + Be^{-600t} \text{ V}, & t \geq 0 \end{cases}$$

If the capacitor has an initial current of 2 A, find:

(a) the constants A and B,

(b) the energy stored in the capacitor at $t = 0$,

(c) the capacitor current for $t > 0$.

6.9 The current through a 0.5-F capacitor is $6(1 - e^{-t})$ A. Determine the voltage and power at $t = 2$ s. Assume $v(0) = 0$.

6.10 The voltage across a 2-mF capacitor is shown in Fig. 6.47. Determine the current through the capacitor.

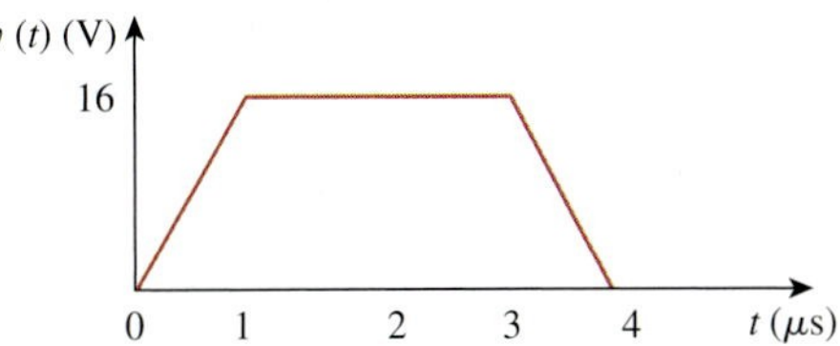

Figure 6.47
For Prob. 6.10.

6.11 A 4-mF capacitor has the current waveform shown in Fig. 6.48. Assuming that $v(0) = 10$ V, sketch the voltage waveform $v(t)$.

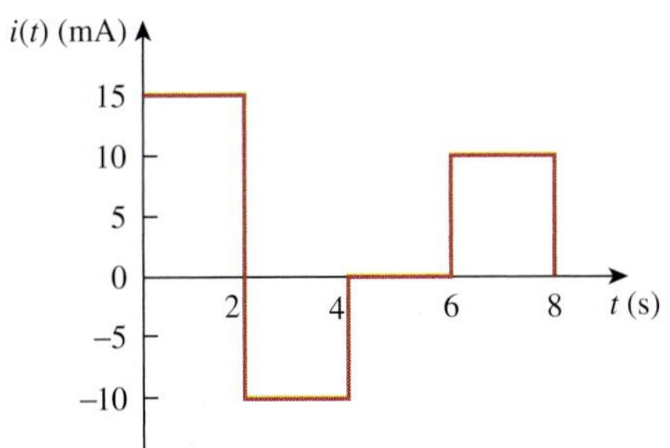

Figure 6.48
For Prob. 6.11.

6.12 A voltage of $6e^{-2000t}$ V appears across a parallel combination of a 100-mF capacitor and a 12-Ω resistor. Calculate the power absorbed by the parallel combination.

6.13 Find the voltage across the capacitors in the circuit of Fig. 6.49 under dc conditions.

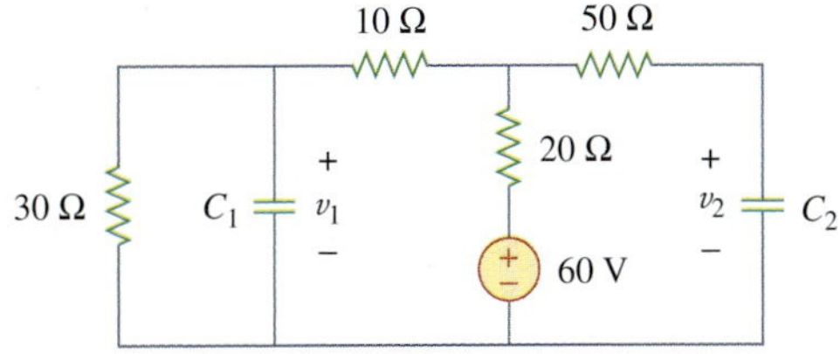

Figure 6.49
For Prob. 6.13.

Section 6.3 Series and Parallel Capacitors

6.14 Series-connected 20-pF and 60-pF capacitors are placed in parallel with series-connected 30-pF and 70-pF capacitors. Determine the equivalent capacitance.

6.15 Two capacitors (20 μF and 30 μF) are connected to a 100-V source. Find the energy stored in each capacitor if they are connected in:

(a) parallel (b) series

6.16 The equivalent capacitance at terminals *a-b* in the circuit of Fig. 6.50 is 30 μF. Calculate the value of C.

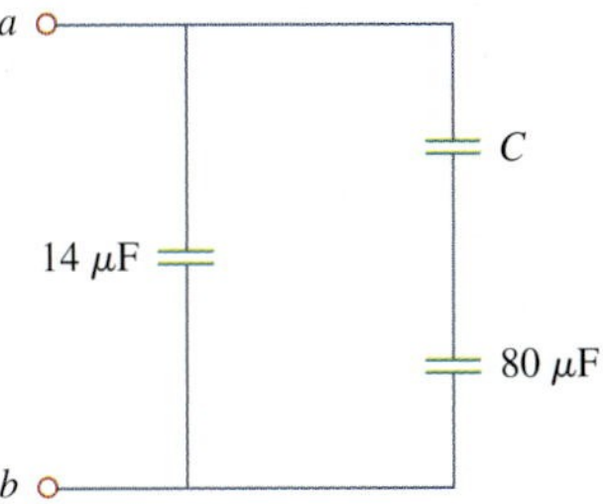

Figure 6.50
For Prob. 6.16.

6.17 Determine the equivalent capacitance for each of the circuits of Fig. 6.51.

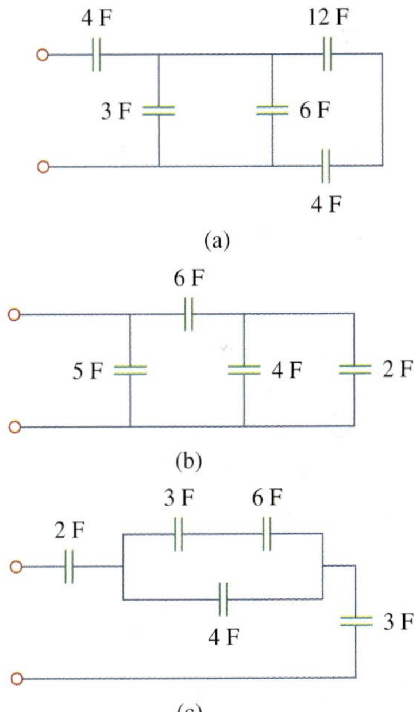

Figure 6.51
For Prob. 6.17.

6.18 Find C_{eq} in the circuit of Fig. 6.52 if all capacitors are 4 μF.

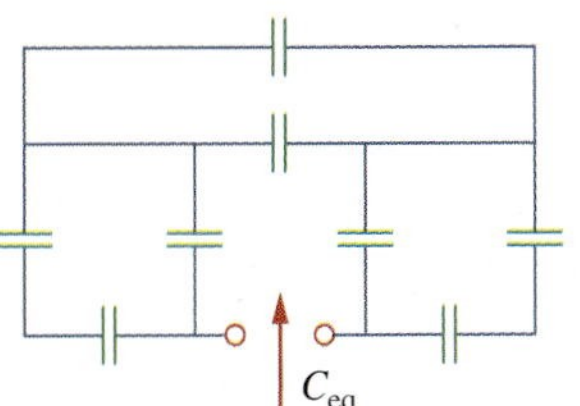

Figure 6.52
For Prob. 6.18.

6.19 Find the equivalent capacitance between terminals a and b in the circuit of Fig. 6.53. All capacitances are in μF.

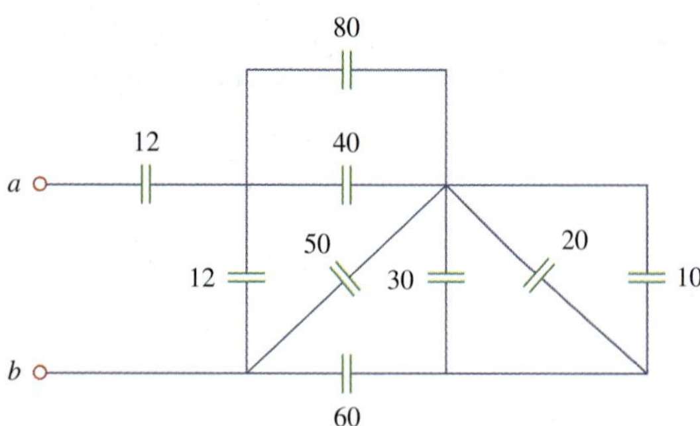

Figure 6.53
For Prob. 6.19.

6.20 Find the equivalent capacitance at terminals a-b of the circuit in Fig. 6.54.

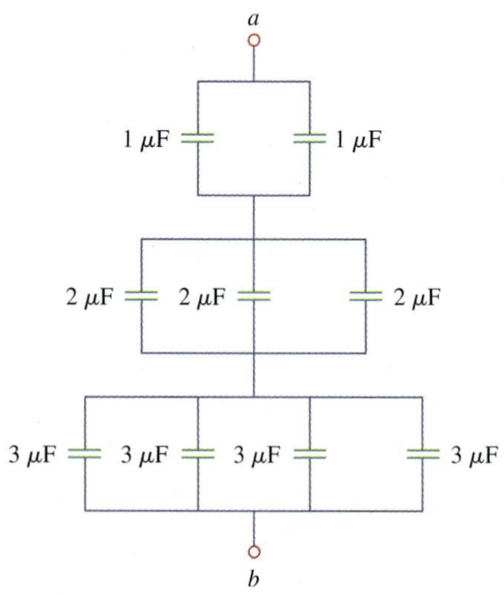

Figure 6.54
For Prob. 6.20.

6.21 Determine the equivalent capacitance at terminals a-b of the circuit in Fig. 6.55.

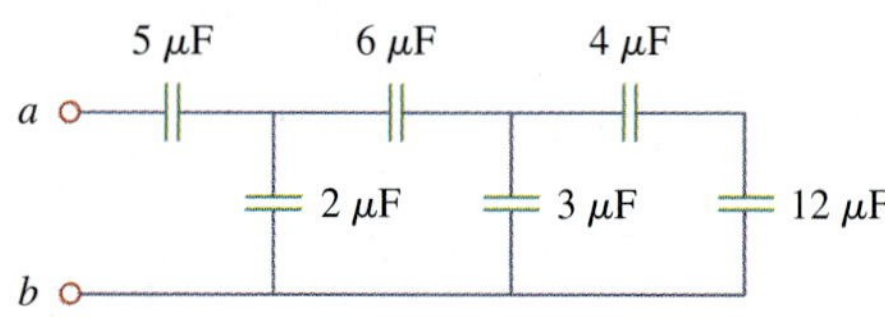

Figure 6.55
For Prob. 6.21.

6.22 Obtain the equivalent capacitance of the circuit in Fig. 6.56.

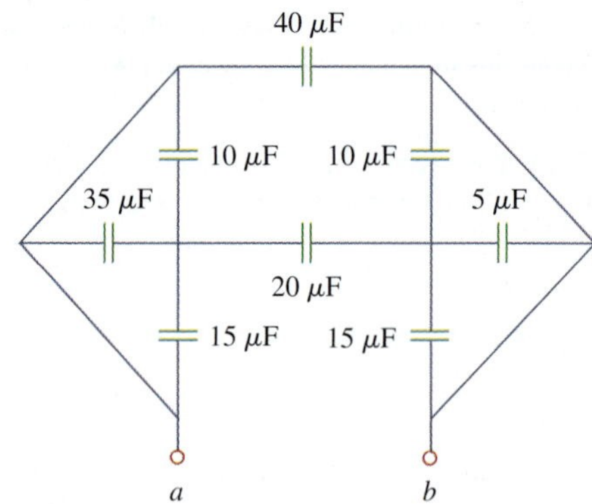

Figure 6.56
For Prob. 6.22.

6.23 Using Fig. 6.57, design a problem that will help other students better understand how capacitors work together when connected in series and in parallel.

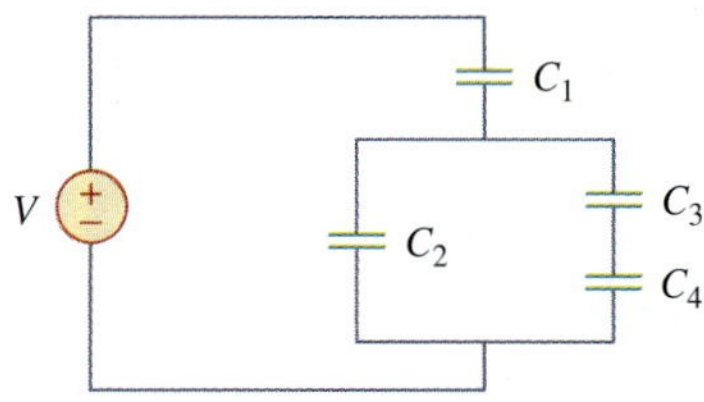

Figure 6.57
For Prob. 6.23.

6.24 Repeat Prob. 6.23 for the circuit of Fig. 6.58.

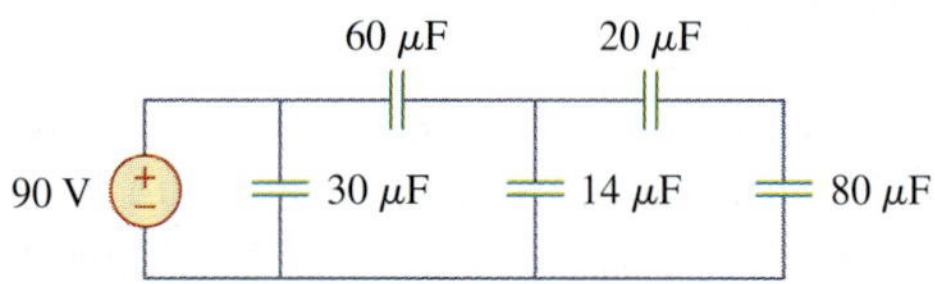

Figure 6.58
For Prob. 6.24.

6.25 (a) Show that the voltage-division rule for two capacitors in series as in Fig. 6.59(a) is

$$v_1 = \frac{C_2}{C_1 + C_2} v_s, \qquad v_2 = \frac{C_1}{C_1 + C_2} v_s$$

assuming that the initial conditions are zero.

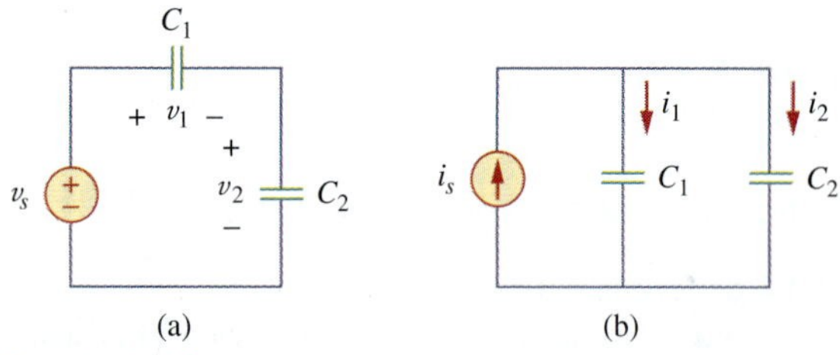

Figure 6.59
For Prob. 6.25.

(b) For two capacitors in parallel as in Fig. 6.59(b), show that the current-division rule is

$$i_1 = \frac{C_1}{C_1 + C_2} i_s, \qquad i_2 = \frac{C_2}{C_1 + C_2} i_s$$

assuming that the initial conditions are zero.

6.26 Three capacitors, $C_1 = 5\ \mu F$, $C_2 = 10\ \mu F$, and $C_3 = 20\ \mu F$, are connected in parallel across a 150-V source. Determine:

(a) the total capacitance,

(b) the charge on each capacitor,

(c) the total energy stored in the parallel combination.

6.27 Given that four 4-μF capacitors can be connected in series and in parallel, find the minimum and maximum values that can be obtained by such series/parallel combinations.

***6.28** Obtain the equivalent capacitance of the network shown in Fig. 6.60.

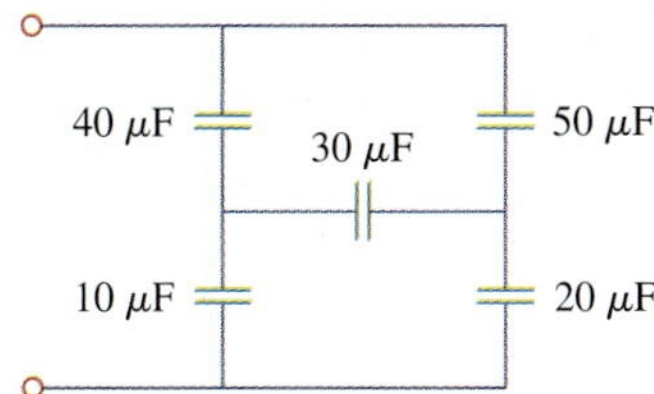

Figure 6.60
For Prob. 6.28.

6.29 Determine C_{eq} for each circuit in Fig. 6.61.

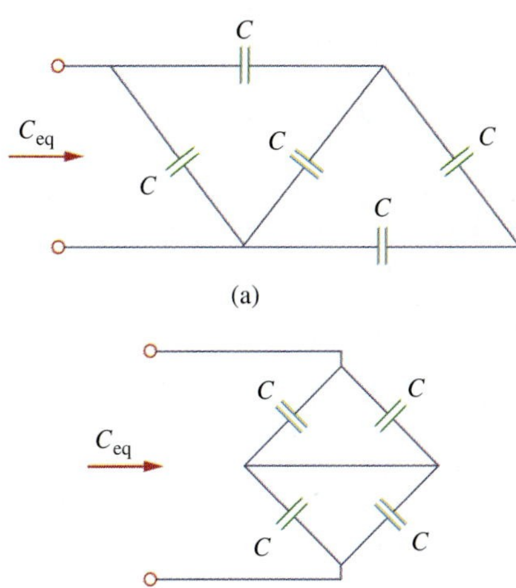

Figure 6.61
For Prob. 6.29.

* An asterisk indicates a challenging problem.

6.30 Assuming that the capacitors are initially uncharged, find $v_o(t)$ in the circuit of Fig. 6.62.

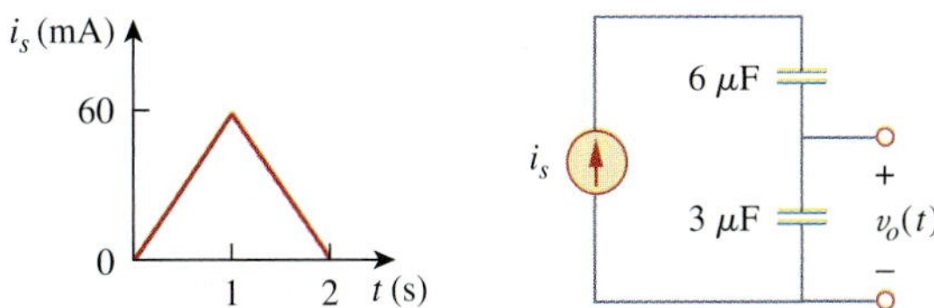

Figure 6.62
For Prob. 6.30.

6.31 If $v(0) = 0$, find $v(t)$, $i_1(t)$, and $i_2(t)$ in the circuit of Fig. 6.63.

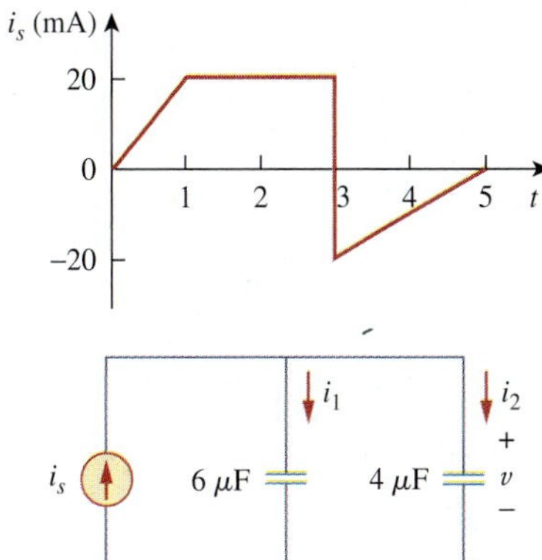

Figure 6.63
For Prob. 6.31.

6.32 In the circuit of Fig. 6.64, let $i_s = 30e^{-2t}$ mA and $v_1(0) = 50$ V, $v_2(0) = 20$ V. Determine: (a) $v_1(t)$ and $v_2(t)$, (b) the energy in each capacitor at $t = 0.5$ s.

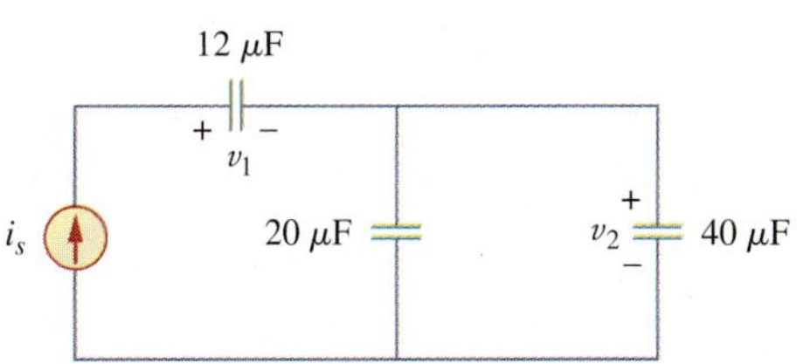

Figure 6.64
For Prob. 6.32.

6.33 Obtain the Thevenin equivalent at the terminals, *a-b*, of the circuit shown in Fig. 6.65. Please note that Thevenin equivalent circuits do not generally exist for circuits involving capacitors and resistors. This is a special case where the Thevenin equivalent circuit does exist.

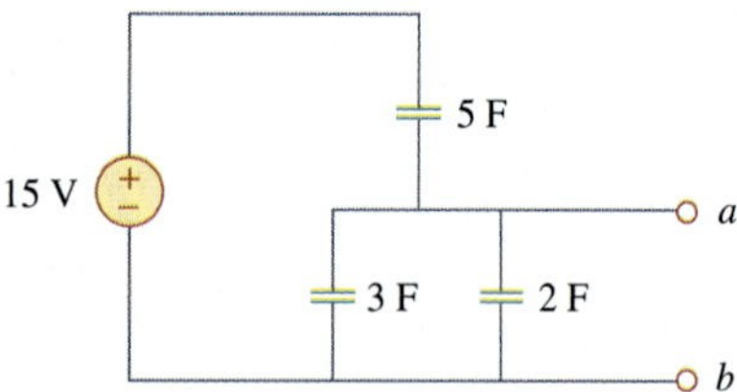

Figure 6.65
For Prob. 6.33.

Section 6.4 Inductors

6.34 The current through a 10-mH inductor is $6e^{-t/2}$ A. Find the voltage and the power at $t = 3$ s.

6.35 An inductor has a linear change in current from 50 mA to 100 mA in 2 ms and induces a voltage of 160 mV. Calculate the value of the inductor.

6.36 Design a problem to help other students better understand how inductors work.

6.37 The current through a 12-mH inductor is 4 sin 100*t* A. Find the voltage, across the inductor for $0 < t < \pi/200$ s, and the energy stored at $t = \frac{\pi}{200}$ s.

6.38 The current through a 40-mH inductor is

$$i(t) = \begin{cases} 0, & t < 0 \\ te^{-2t}\text{ A}, & t > 0 \end{cases}$$

Find the voltage $v(t)$.

6.39 The voltage across a 200-mH inductor is given by

$$v(t) = 3t^2 + 2t + 4 \text{ V} \qquad \text{for } t > 0.$$

Determine the current $i(t)$ through the inductor. Assume that $i(0) = 1$ A.

6.40 The current through a 10-mH inductor is shown in Fig. 6.66. Determine the voltage across the inductor at $t = 1$, 3, and 5 ms.

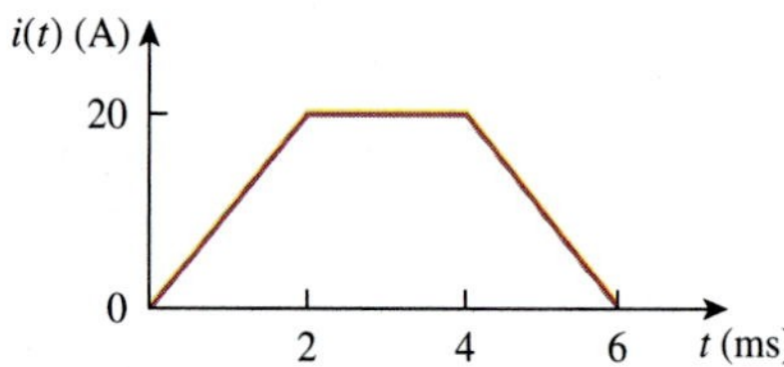

Figure 6.66
For Prob. 6.40.

6.41 The voltage across a 2-H inductor is $20(1 - e^{-2t})$ V. If the initial current through the inductor is 0.3 A, find the current and the energy stored in the inductor at $t = 1$ s.

6.42 If the voltage waveform in Fig. 6.67 is applied across the terminals of a 10-H inductor, calculate the current through the inductor. Assume $i(0) = -1$ A.

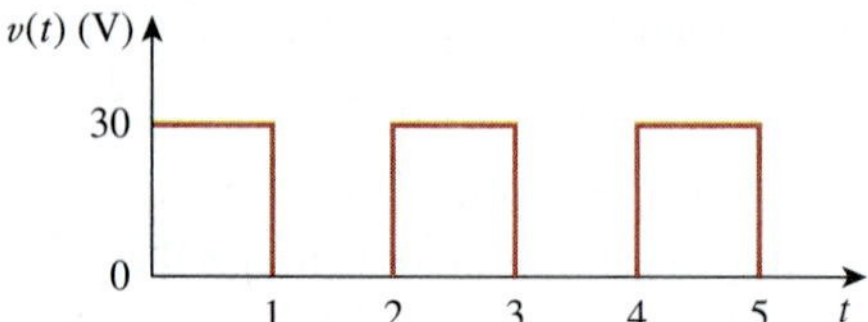

Figure 6.67
For Prob. 6.42.

6.43 The current in an 80-mH inductor increases from 0 to 60 mA. How much energy is stored in the inductor?

***6.44** A 100-mH inductor is connected in parallel with a 2-kΩ resistor. The current through the inductor is $i(t) = 50e^{-400t}$ mA. (a) Find the voltage v_L across the inductor. (b) Find the voltage v_R across the resistor. (c) Does $v_R(t) + v_L(t) = 0$? (d) Calculate the energy in the inductor at $t = 0$.

6.45 If the voltage waveform in Fig. 6.68 is applied to a 50-mH inductor, find the inductor current $i(t)$. Assume $i(0) = 0$.

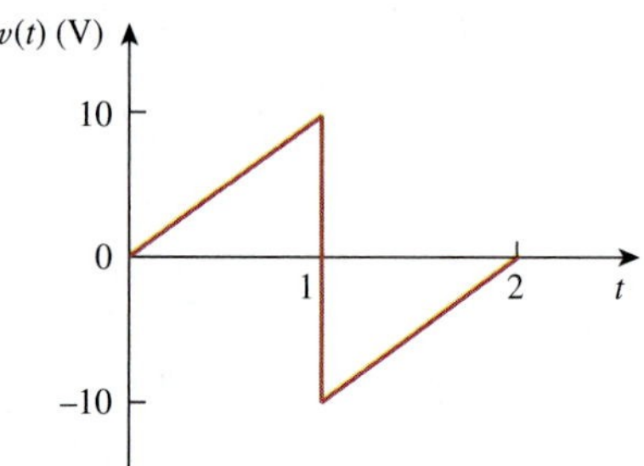

Figure 6.68
For Prob. 6.45.

6.46 Find v_C, i_L, and the energy stored in the capacitor and inductor in the circuit of Fig. 6.69 under dc conditions.

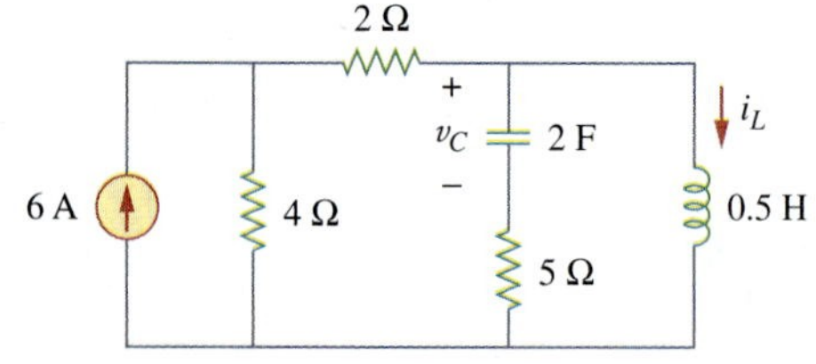

Figure 6.69
For Prob. 6.46.

6.47 For the circuit in Fig. 6.70, calculate the value of R that will make the energy stored in the capacitor the same as that stored in the inductor under dc conditions.

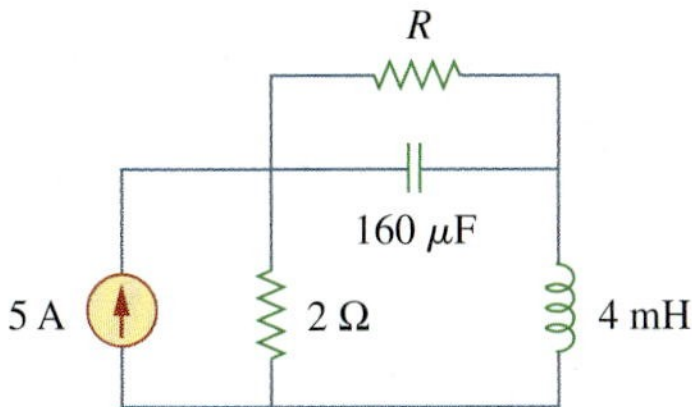

Figure 6.70
For Prob. 6.47.

6.48 Under steady-state dc conditions, find i and v in the circuit of Fig. 6.71.

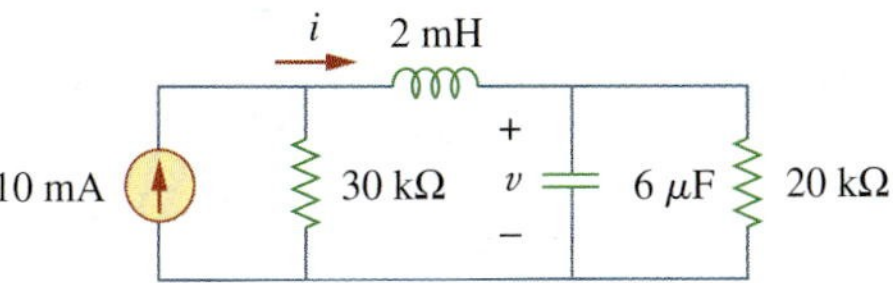

Figure 6.71
For Prob. 6.48.

Section 6.5 Series and Parallel Inductors

6.49 Find the equivalent inductance of the circuit in Fig. 6.72. Assume all inductors are 10 mH.

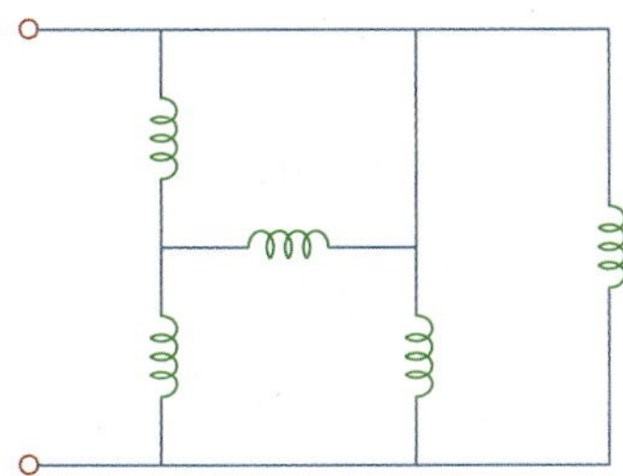

Figure 6.72
For Prob. 6.49.

6.50 An energy-storage network consists of series-connected 16-mH and 14-mH inductors in parallel with series-connected 24-mH and 36-mH inductors. Calculate the equivalent inductance.

6.51 Determine L_{eq} at terminals a-b of the circuit in Fig. 6.73.

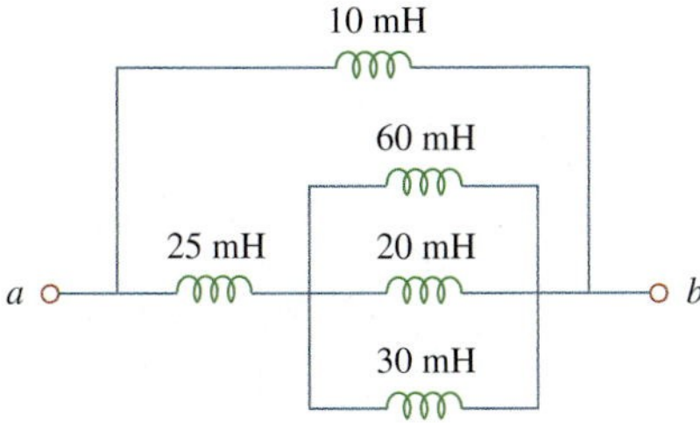

Figure 6.73
For Prob. 6.51.

6.52 Using Fig. 6.74, design a problem to help other students better understand how inductors behave when connected in series and when connected in parallel.

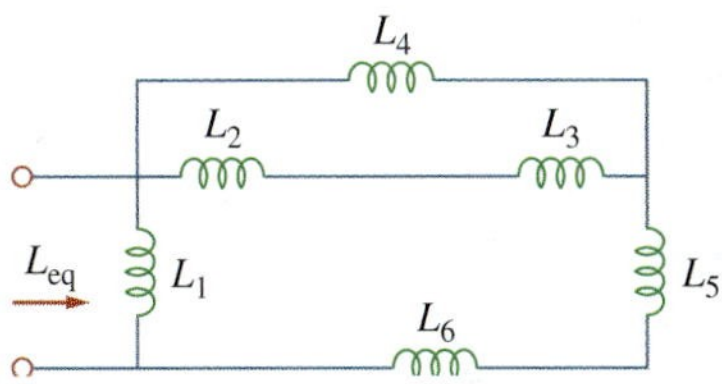

Figure 6.74
For Prob. 6.52.

6.53 Find L_{eq} at the terminals of the circuit in Fig. 6.75.

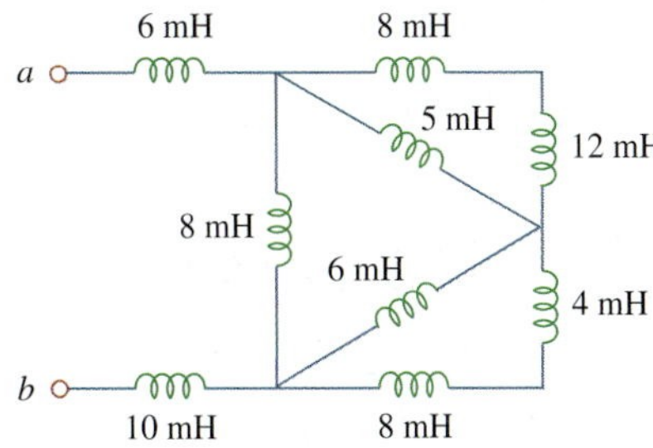

Figure 6.75
For Prob. 6.53.

6.54 Find the equivalent inductance looking into the terminals of the circuit in Fig. 6.76.

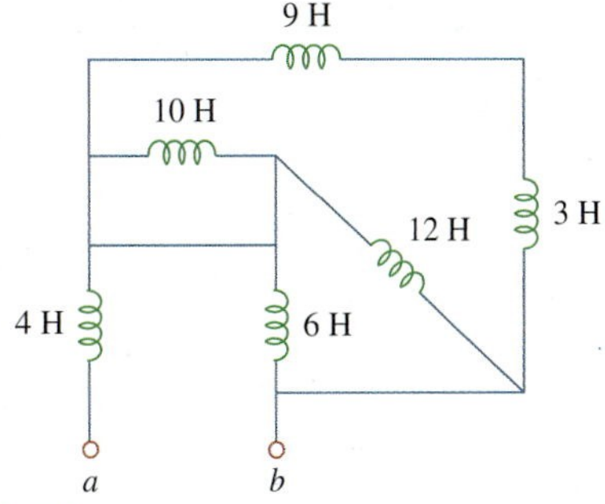

Figure 6.76
For Prob. 6.54.

6.55 Find L_{eq} in each of the circuits in Fig. 6.77.

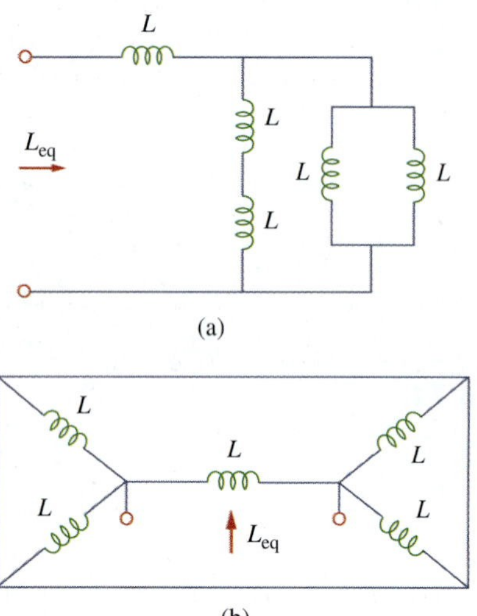

Figure 6.77
For Prob. 6.55.

6.56 Find L_{eq} in the circuit of Fig. 6.78.

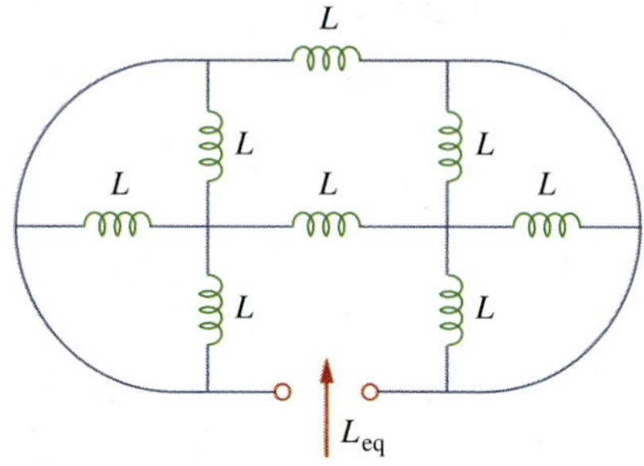

Figure 6.78
For Prob. 6.56.

***6.57** Determine L_{eq} that may be used to represent the inductive network of Fig. 6.79 at the terminals.

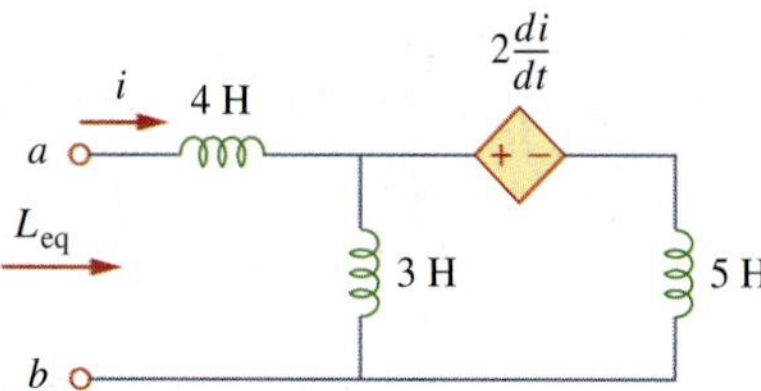

Figure 6.79
For Prob. 6.57.

6.58 The current waveform in Fig. 6.80 flows through a 3-H inductor. Sketch the voltage across the inductor over the interval $0 < t < 6$ s.

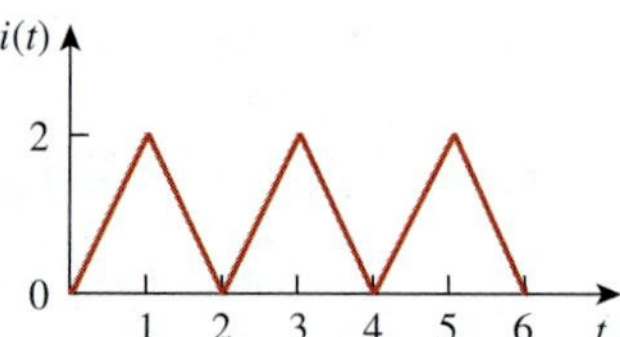

Figure 6.80
For Prob. 6.58.

6.59 (a) For two inductors in series as in Fig. 6.81(a), show that the voltage division principle is

$$v_1 = \frac{L_1}{L_1 + L_2}v_s, \qquad v_2 = \frac{L_2}{L_1 + L_2}v_s$$

assuming that the initial conditions are zero.

(b) For two inductors in parallel as in Fig. 6.81(b), show that the current-division principle is

$$i_1 = \frac{L_2}{L_1 + L_2}i_s, \qquad i_2 = \frac{L_1}{L_1 + L_2}i_s$$

assuming that the initial conditions are zero.

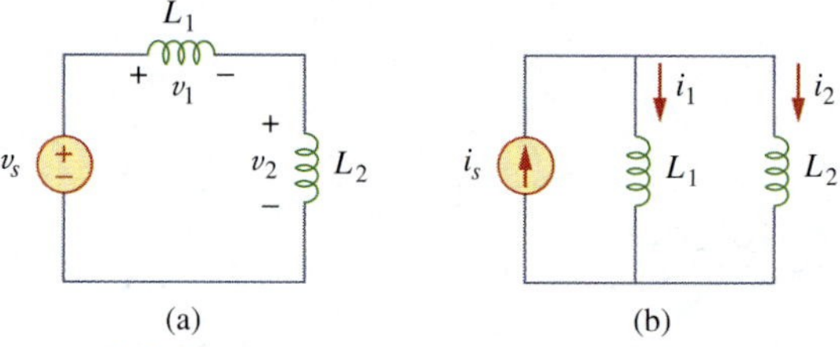

Figure 6.81
For Prob. 6.59.

6.60 In the circuit of Fig. 6.82, $i_o(0) = 2$ A. Determine $i_o(t)$ and $v_o(t)$ for $t > 0$.

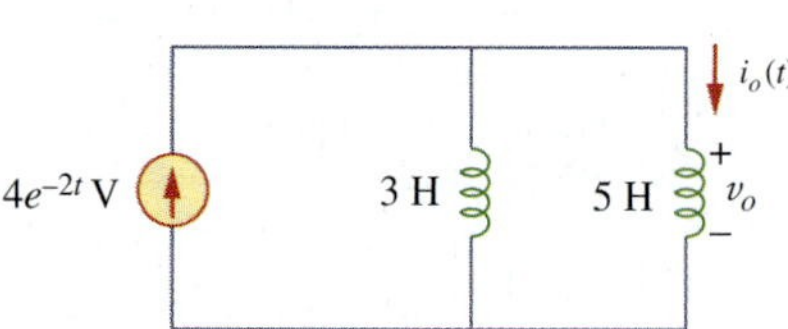

Figure 6.82
For Prob. 6.60.

6.61 Consider the circuit in Fig. 6.83. Find: (a) L_{eq}, $i_1(t)$, and $i_2(t)$ if $i_s = 3e^{-t}$ mA, (b) $v_o(t)$, (c) energy stored in the 20-mH inductor at $t = 1$ s.

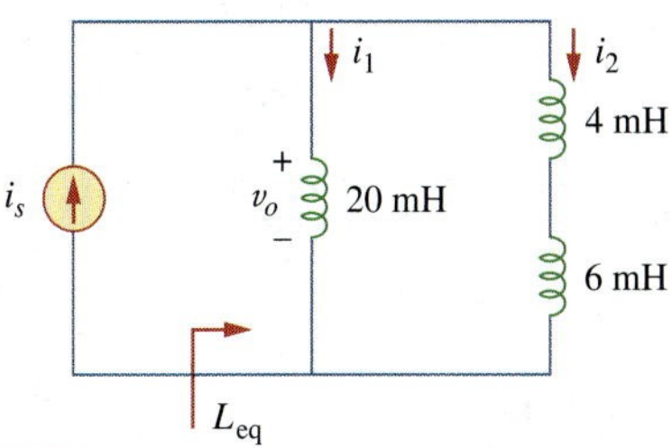

Figure 6.83
For Prob. 6.61.

6.62 Consider the circuit in Fig. 6.84. Given that $v(t) = 12e^{-3t}$ mV for $t > 0$ and $i_1(0) = -10$ mA, find: (a) $i_2(0)$, (b) $i_1(t)$ and $i_2(t)$.

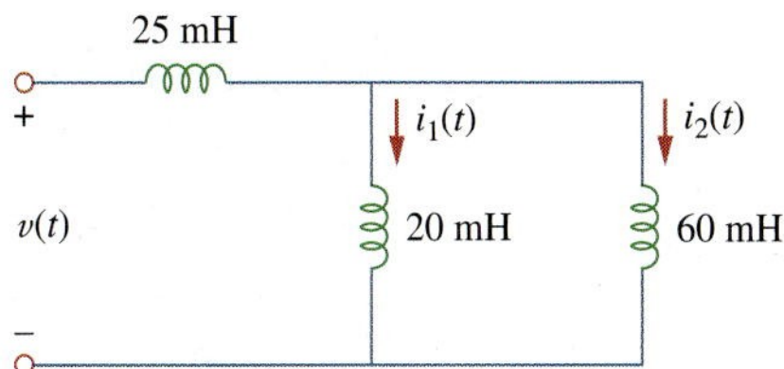

Figure 6.84
For Prob. 6.62.

6.63 In the circuit of Fig. 6.85, sketch v_o.

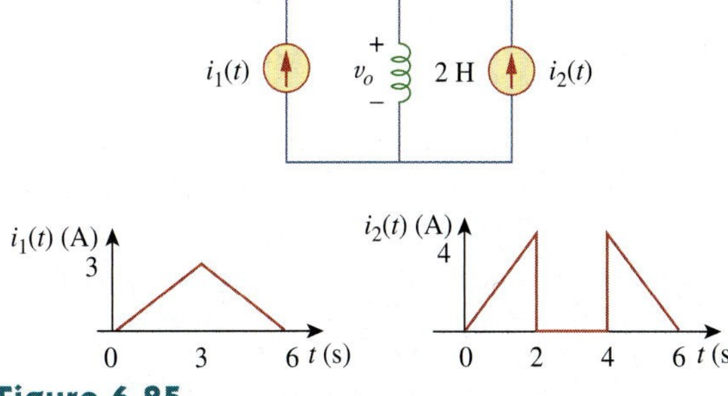

Figure 6.85
For Prob. 6.63.

6.64 The switch in Fig. 6.86 has been in position *A* for a long time. At $t = 0$, the switch moves from position *A* to *B*. The switch is a make-before-break type so that there is no interruption in the inductor current. Find: (a) $i(t)$ for $t < 0$, (b) v *just after* the switch has been moved to position *B*, (c) $v(t)$ long after the switch is in position *B*.

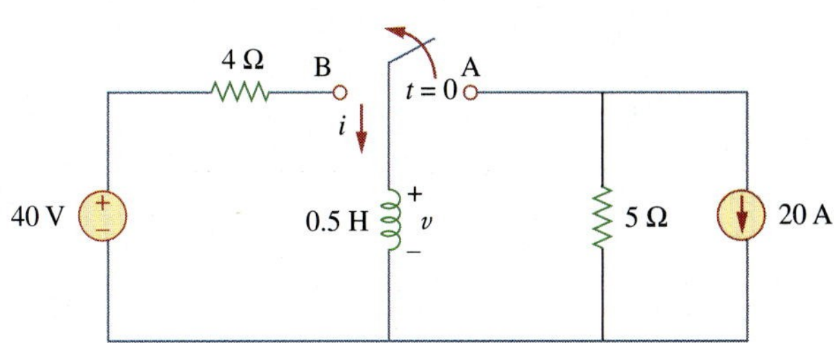

Figure 6.86
For Prob. 6.64.

6.65 The inductors in Fig. 6.87 are initially charged and are connected to the black box at $t = 0$. If $i_1(0) = 4$ A, $i_2(0) = -2$ A, and $v(t) = 50e^{-200t}$ mV, $t \geq 0$, find:

(a) the energy initially stored in each inductor,

(b) the total energy delivered to the black box from $t = 0$ to $t = \infty$,

(c) $i_1(t)$ and $i_2(t)$, $t \geq 0$,

(d) $i(t)$, $t \geq 0$.

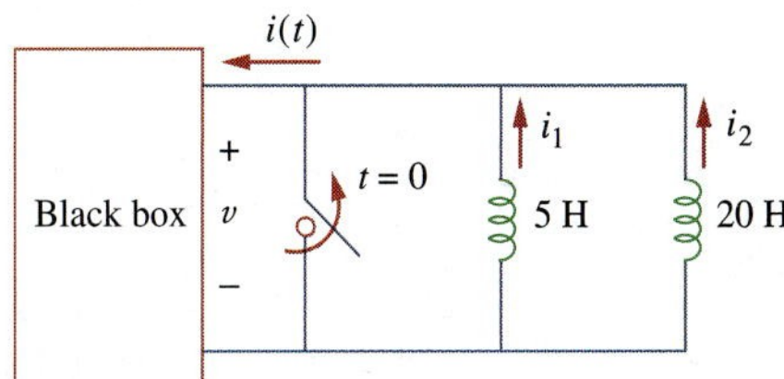

Figure 6.87
For Prob. 6.65.

6.66 The current $i(t)$ through a 40-mH inductor is equal, in magnitude, to the voltage across it for all values of time. If $i(0) = 5$ A, find $i(t)$.

Section 6.6 Applications

6.67 An op amp integrator has $R = 100$ kΩ and $C = 0.01$ μF. If the input voltage is $v_i = 10 \sin 50t$ mV, obtain the output voltage.

6.68 A 10-V dc voltage is applied to an integrator with $R = 50\ \text{k}\Omega$, $C = 100\ \mu\text{F}$ at $t = 0$. How long will it take for the op amp to saturate if the saturation voltages are $+12$ V and -12 V? Assume that the initial capacitor voltage was zero.

6.69 An op amp integrator with $R = 4\ \text{M}\Omega$ and $C = 1\ \mu\text{F}$ has the input waveform shown in Fig. 6.88. Plot the output waveform.

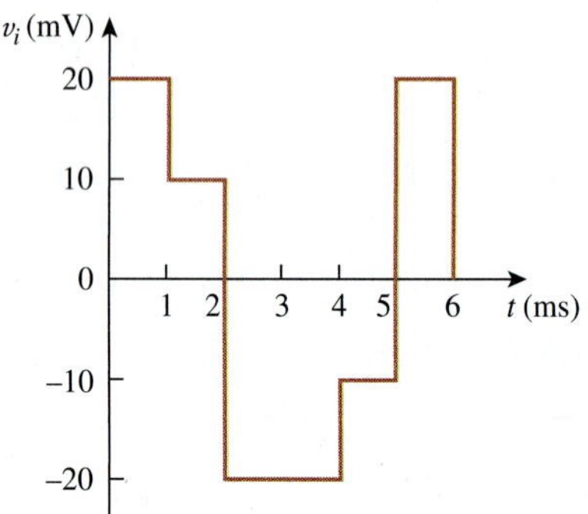

Figure 6.88
For Prob. 6.69.

6.70 Using a single op amp, a capacitor, and resistors of 100 kΩ or less, design a circuit to implement

$$v_o = -50 \int_0^t v_i(t)\, dt$$

Assume $v_o = 0$ at $t = 0$.

6.71 Show how you would use a single op amp to generate

$$v_o = -\int_0^t (v_1 + 4v_2 + 10v_3)\, dt$$

If the integrating capacitor is $C = 2\ \mu\text{F}$, obtain the other component values.

6.72 At $t = 1.5$ ms, calculate v_o due to the cascaded integrators in Fig. 6.89. Assume that the integrators are reset to 0 V at $t = 0$.

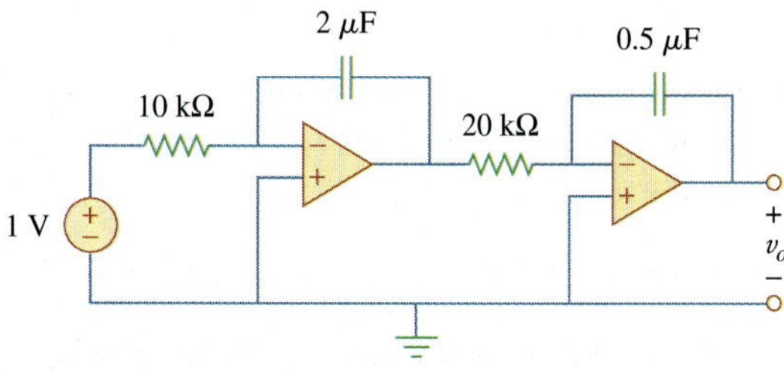

Figure 6.89
For Prob. 6.72.

6.73 Show that the circuit in Fig. 6.90 is a noninverting integrator.

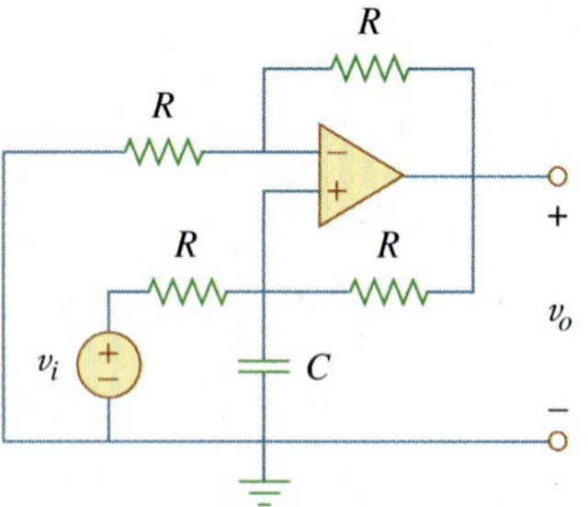

Figure 6.90
For Prob. 6.73.

6.74 The triangular waveform in Fig. 6.91(a) is applied to the input of the op amp differentiator in Fig. 6.91(b). Plot the output.

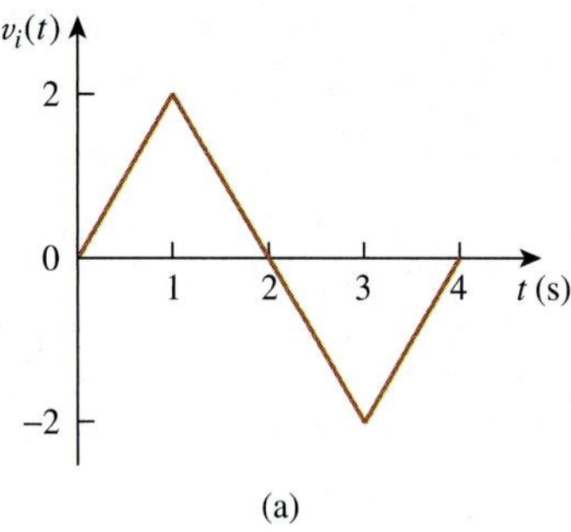

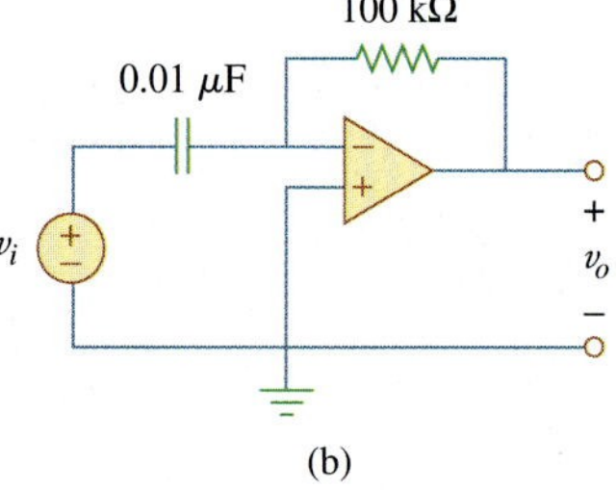

Figure 6.91
For Prob. 6.74.

6.75 An op amp differentiator has $R = 250\ \text{k}\Omega$ and $C = 10\ \mu\text{F}$. The input voltage is a ramp $r(t) = 12t$ mV. Find the output voltage.

6.76 A voltage waveform has the following characteristics: a positive slope of 20 V/s for 5 ms followed by a negative slope of 10 V/s for 10 ms. If the waveform is applied to a differentiator with $R = 50\ \text{k}\Omega$, $C = 10\ \mu\text{F}$, sketch the output voltage waveform.

***6.77** The output v_o of the op amp circuit in Fig. 6.92(a) is shown in Fig. 6.92(b). Let $R_i = R_f = 1\ \text{M}\Omega$ and $C = 1\ \mu\text{F}$. Determine the input voltage waveform and sketch it.

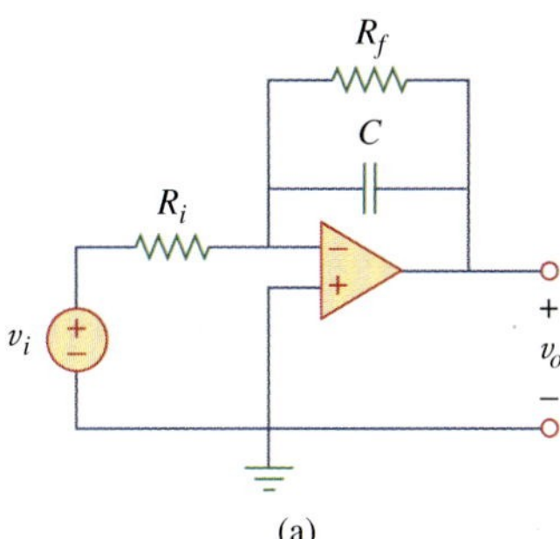

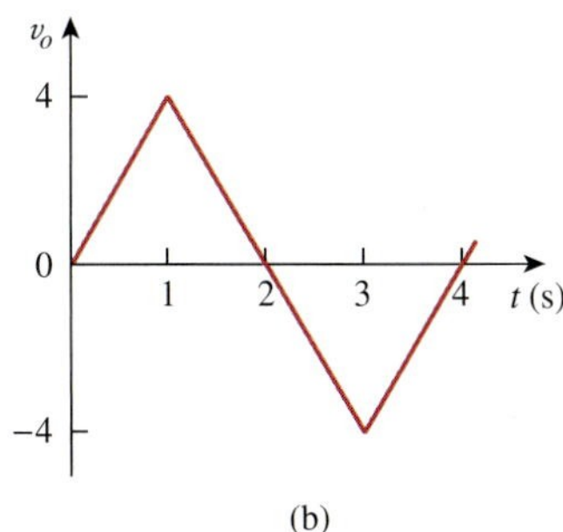

Figure 6.92
For Prob. 6.77.

6.78 Design an analog computer to simulate

e2d

$$\frac{d^2v_o}{dt^2} + 2\frac{dv_o}{dt} + v_o = 10 \sin 2t$$

where $v_0(0) = 2$ and $v_0'(0) = 0$.

6.79 Design an analog computer circuit to solve the following ordinary differential equation.

e2d

$$\frac{dy(t)}{dt} + 4y(t) = f(t)$$

where $y(0) = 1$ V.

6.80 Figure 6.93 presents an analog computer designed to solve a differential equation. Assuming $f(t)$ is known, set up the equation for $f(t)$.

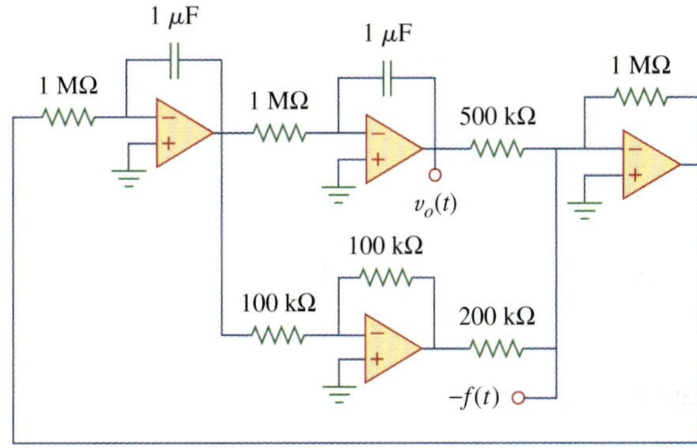

Figure 6.93
For Prob. 6.80.

6.81 Design an analog computer to simulate the following equation:

e2d

$$\frac{d^2v}{dt^2} + 5v = -2f(t)$$

6.82 Design an op amp circuit such that

e2d

$$v_o = 10v_s + 2\int v_s dt$$

where v_s and v_o are the input voltage and output voltage, respectively.

Comprehensive Problems

6.83 Your laboratory has available a large number of 10-μF capacitors rated at 300 V. To design a capacitor bank of 40 μF rated at 600 V, how many 10-μF capacitors are needed and how would you connect them?

e2d

6.84 An 8-mH inductor is used in a fusion power experiment. If the current through the inductor is $i(t) = 5 \sin^2 \pi t$ mA, $t > 0$, find the power being delivered to the inductor and the energy stored in it at $t = 0.5$ s.

6.85 A square-wave generator produces the voltage waveform shown in Fig. 6.94(a). What kind of a circuit component is needed to convert the voltage waveform to the triangular current waveform shown in Fig. 6.94(b)? Calculate the value of the component, assuming that it is initially uncharged.

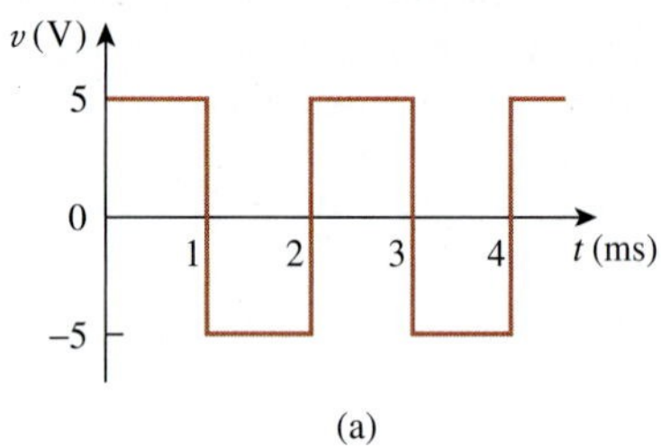

(a)

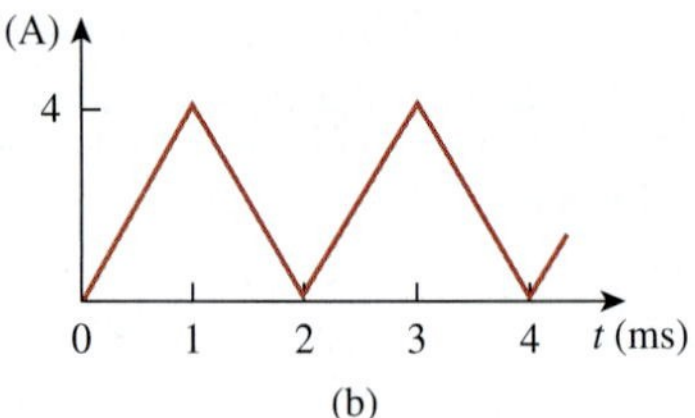

(b)

Figure 6.94
For Prob. 6.85.

6.86 An electric motor can be modeled as a series combination of a 12-Ω resistor and 200-mH inductor. If a current $i(t) = 2te^{-10t}$ A flows through the series combination, find the voltage across the combination.

chapter

7

First-Order Circuits

We live in deeds, not years; in thoughts, not breaths; in feelings, not in figures on a dial. We should count time in heart-throbs. He most lives who thinks most, feels the noblest, acts the best.

—F. J. Bailey

Enhancing Your Career

Careers in Computer Engineering

Electrical engineering education has gone through drastic changes in recent decades. Most departments have come to be known as Department of Electrical and Computer Engineering, emphasizing the rapid changes due to computers. Computers occupy a prominent place in modern society and education. They have become commonplace and are helping to change the face of research, development, production, business, and entertainment. The scientist, engineer, doctor, attorney, teacher, airline pilot, businessperson—almost anyone benefits from a computer's abilities to store large amounts of information and to process that information in very short periods of time. The internet, a computer communication network, is essential in business, education, and library science. Computer usage continues to grow by leaps and bounds.

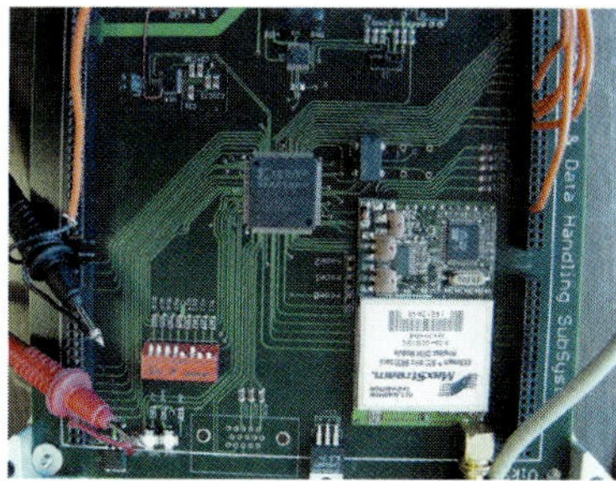

Computer design of very large scale integrated (VLSI) circuits.
Courtesy Brian Fast, Cleveland State University

An education in computer engineering should provide breadth in software, hardware design, and basic modeling techniques. It should include courses in data structures, digital systems, computer architecture, microprocessors, interfacing, software engineering, and operating systems.

Electrical engineers who specialize in computer engineering find jobs in computer industries and in numerous fields where computers are being used. Companies that produce software are growing rapidly in number and size and providing employment for those who are skilled in programming. An excellent way to advance one's knowledge of computers is to join the IEEE Computer Society, which sponsors diverse magazines, journals, and conferences.

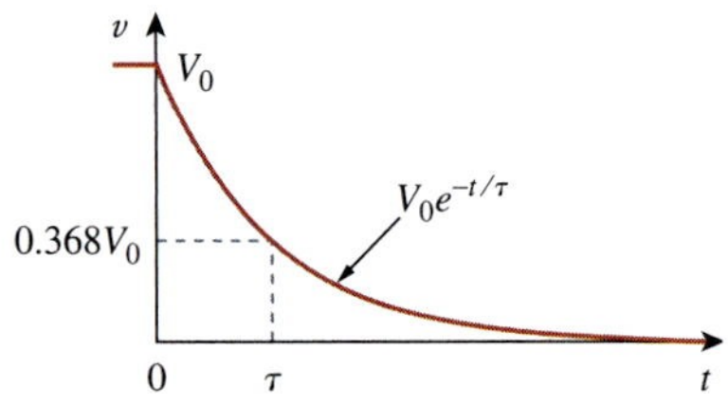

Figure 7.2
The voltage response of the *RC* circuit.

the voltage decreases is expressed in terms of the *time constant*, denoted by τ, the lowercase Greek letter tau.

> The **time constant** of a circuit is the time required for the response to decay to a factor of $1/e$ or 36.8 percent of its initial value.[1]

This implies that at $t = \tau$, Eq. (7.7) becomes

$$V_0 e^{-\tau/RC} = V_0 e^{-1} = 0.368V_0$$

or

$$\boxed{\tau = RC} \tag{7.8}$$

In terms of the time constant, Eq. (7.7) can be written as

$$\boxed{v(t) = V_0 e^{-t/\tau}} \tag{7.9}$$

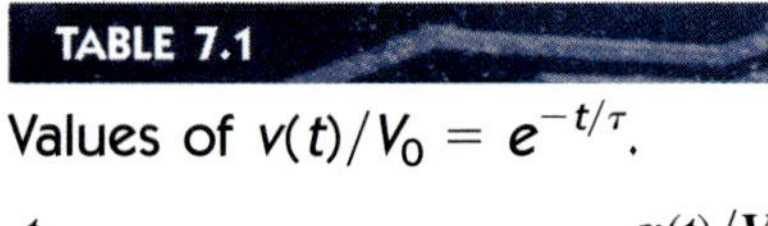

TABLE 7.1

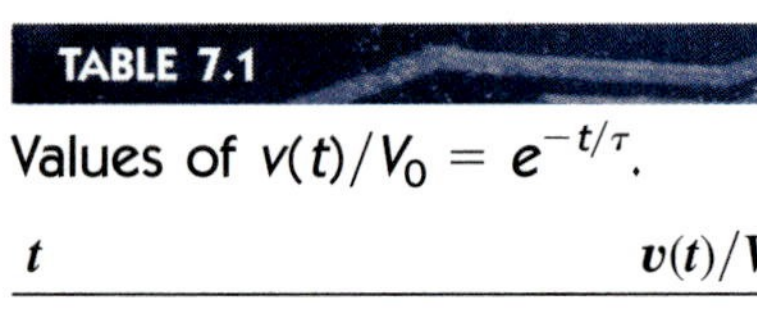

Values of $v(t)/V_0 = e^{-t/\tau}$.

t	$v(t)/V_0$
τ	0.36788
2τ	0.13534
3τ	0.04979
4τ	0.01832
5τ	0.00674

With a calculator it is easy to show that the value of $v(t)/V_0$ is as shown in Table 7.1. It is evident from Table 7.1 that the voltage $v(t)$ is less than 1 percent of V_0 after 5τ (five time constants). Thus, it is customary to assume that the capacitor is fully discharged (or charged) after five time constants. In other words, it takes 5τ for the circuit to reach its final state or steady state when no changes take place with time. Notice that for every time interval of τ, the voltage is reduced by 36.8 percent of its previous value, $v(t + \tau) = v(t)/e = 0.368v(t)$, regardless of the value of t.

Observe from Eq. (7.8) that the smaller the time constant, the more rapidly the voltage decreases, that is, the faster the response. This is illustrated in Fig. 7.4. A circuit with a small time constant gives a fast response in that it reaches the steady state (or final state) quickly due to quick dissipation of energy stored, whereas a circuit with a large time constant gives a slow response because it takes longer to reach steady state. At any rate, whether the time constant is small or large, the circuit reaches steady state in five time constants.

With the voltage $v(t)$ in Eq. (7.9), we can find the current $i_R(t)$,

$$i_R(t) = \frac{v(t)}{R} = \frac{V_0}{R} e^{-t/\tau} \tag{7.10}$$

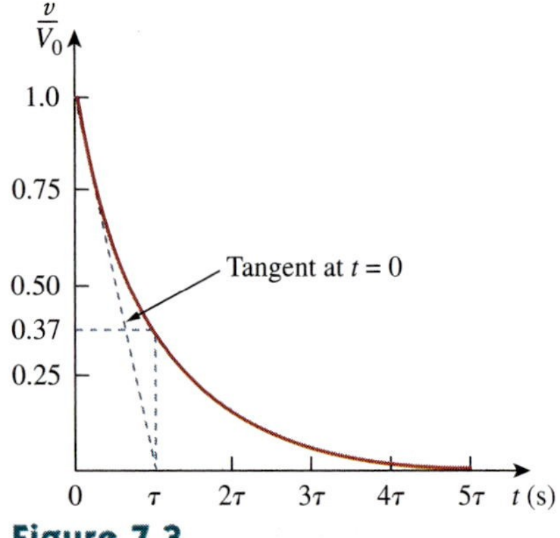

Figure 7.3
Graphical determination of the time constant τ from the response curve.

[1] The time constant may be viewed from another perspective. Evaluating the derivative of $v(t)$ in Eq. (7.7) at $t = 0$, we obtain

$$\left.\frac{d}{dt}\left(\frac{v}{V_0}\right)\right|_{t=0} = \left.-\frac{1}{\tau}e^{-t/\tau}\right|_{t=0} = -\frac{1}{\tau}$$

Thus, the time constant is the initial rate of decay, or the time taken for v/V_0 to decay from unity to zero, assuming a constant rate of decay. This initial slope interpretation of the time constant is often used in the laboratory to find τ graphically from the response curve displayed on an oscilloscope. To find τ from the response curve, draw the tangent to the curve at $t = 0$, as shown in Fig. 7.3. The tangent intercepts with the time axis at $t = \tau$.

chapter

7

First-Order Circuits

We live in deeds, not years; in thoughts, not breaths; in feelings, not in figures on a dial. We should count time in heart-throbs. He most lives who thinks most, feels the noblest, acts the best.

—F. J. Bailey

Enhancing Your Career

Careers in Computer Engineering

Electrical engineering education has gone through drastic changes in recent decades. Most departments have come to be known as Department of Electrical and Computer Engineering, emphasizing the rapid changes due to computers. Computers occupy a prominent place in modern society and education. They have become commonplace and are helping to change the face of research, development, production, business, and entertainment. The scientist, engineer, doctor, attorney, teacher, airline pilot, businessperson—almost anyone benefits from a computer's abilities to store large amounts of information and to process that information in very short periods of time. The internet, a computer communication network, is essential in business, education, and library science. Computer usage continues to grow by leaps and bounds.

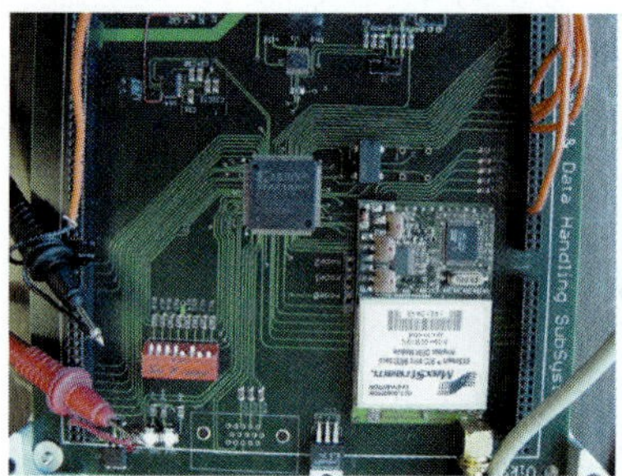

Computer design of very large scale integrated (VLSI) circuits.
Courtesy Brian Fast, Cleveland State University

An education in computer engineering should provide breadth in software, hardware design, and basic modeling techniques. It should include courses in data structures, digital systems, computer architecture, microprocessors, interfacing, software engineering, and operating systems.

Electrical engineers who specialize in computer engineering find jobs in computer industries and in numerous fields where computers are being used. Companies that produce software are growing rapidly in number and size and providing employment for those who are skilled in programming. An excellent way to advance one's knowledge of computers is to join the IEEE Computer Society, which sponsors diverse magazines, journals, and conferences.

7.1 Introduction

Now that we have considered the three passive elements (resistors, capacitors, and inductors) and one active element (the op amp) individually, we are prepared to consider circuits that contain various combinations of two or three of the passive elements. In this chapter, we shall examine two types of simple circuits: a circuit comprising a resistor and capacitor and a circuit comprising a resistor and an inductor. These are called *RC* and *RL* circuits, respectively. As simple as these circuits are, they find continual applications in electronics, communications, and control systems, as we shall see.

We carry out the analysis of *RC* and *RL* circuits by applying Kirchhoff's laws, as we did for resistive circuits. The only difference is that applying Kirchhoff's laws to purely resistive circuits results in algebraic equations, while applying the laws to *RC* and *RL* circuits produces differential equations, which are more difficult to solve than algebraic equations. The differential equations resulting from analyzing *RC* and *RL* circuits are of the first order. Hence, the circuits are collectively known as *first-order* circuits.

> A **first-order** circuit is characterized by a first-order differential equation.

In addition to there being two types of first-order circuits (*RC* and *RL*), there are two ways to excite the circuits. The first way is by initial conditions of the storage elements in the circuits. In these so-called *source-free circuits*, we assume that energy is initially stored in the capacitive or inductive element. The energy causes current to flow in the circuit and is gradually dissipated in the resistors. Although source-free circuits are by definition free of independent sources, they may have dependent sources. The second way of exciting first-order circuits is by independent sources. In this chapter, the independent sources we will consider are dc sources. (In later chapters, we shall consider sinusoidal and exponential sources.) The two types of first-order circuits and the two ways of exciting them add up to the four possible situations we will study in this chapter.

Finally, we consider four typical applications of *RC* and *RL* circuits: delay and relay circuits, a photoflash unit, and an automobile ignition circuit.

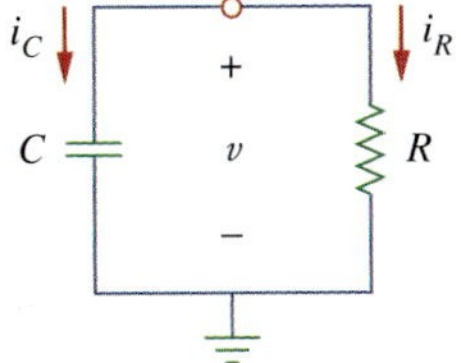

Figure 7.1
A source-free *RC* circuit.

> A circuit response is the manner in which the circuit reacts to an excitation.

7.2 The Source-Free *RC* Circuit

A source-free *RC* circuit occurs when its dc source is suddenly disconnected. The energy already stored in the capacitor is released to the resistors.

Consider a series combination of a resistor and an initially charged capacitor, as shown in Fig. 7.1. (The resistor and capacitor may be the equivalent resistance and equivalent capacitance of combinations of resistors and capacitors.) Our objective is to determine the circuit response, which, for pedagogic reasons, we assume to be the voltage

$v(t)$ across the capacitor. Since the capacitor is initially charged, we can assume that at time $t = 0$, the initial voltage is

$$v(0) = V_0 \tag{7.1}$$

with the corresponding value of the energy stored as

$$w(0) = \frac{1}{2}CV_0^2 \tag{7.2}$$

Applying KCL at the top node of the circuit in Fig. 7.1 yields

$$i_C + i_R = 0 \tag{7.3}$$

By definition, $i_C = C\,dv/dt$ and $i_R = v/R$. Thus,

$$C\frac{dv}{dt} + \frac{v}{R} = 0 \tag{7.4a}$$

or

$$\frac{dv}{dt} + \frac{v}{RC} = 0 \tag{7.4b}$$

This is a *first-order differential equation*, since only the first derivative of v is involved. To solve it, we rearrange the terms as

$$\frac{dv}{v} = -\frac{1}{RC}dt \tag{7.5}$$

Integrating both sides, we get

$$\ln v = -\frac{t}{RC} + \ln A$$

where $\ln A$ is the integration constant. Thus,

$$\ln\frac{v}{A} = -\frac{t}{RC} \tag{7.6}$$

Taking powers of e produces

$$v(t) = Ae^{-t/RC}$$

But from the initial conditions, $v(0) = A = V_0$. Hence,

$$v(t) = V_0e^{-t/RC} \tag{7.7}$$

This shows that the voltage response of the *RC* circuit is an exponential decay of the initial voltage. Since the response is due to the initial energy stored and the physical characteristics of the circuit and not due to some external voltage or current source, it is called the *natural response* of the circuit.

The **natural response** of a circuit refers to the behavior (in terms of voltages and currents) of the circuit itself, with no external sources of excitation.

The natural response depends on the nature of the circuit alone, with no external sources. In fact, the circuit has a response only because of the energy initially stored in the capacitor.

The natural response is illustrated graphically in Fig. 7.2. Note that at $t = 0$, we have the correct initial condition as in Eq. (7.1). As t increases, the voltage decreases toward zero. The rapidity with which

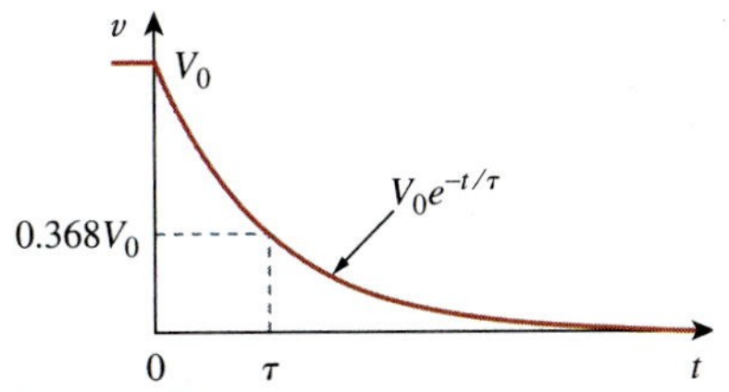

Figure 7.2
The voltage response of the *RC* circuit.

the voltage decreases is expressed in terms of the *time constant*, denoted by τ, the lowercase Greek letter tau.

> The **time constant** of a circuit is the time required for the response to decay to a factor of $1/e$ or 36.8 percent of its initial value.[1]

This implies that at $t = \tau$, Eq. (7.7) becomes

$$V_0e^{-\tau/RC} = V_0e^{-1} = 0.368V_0$$

or

$$\tau = RC \tag{7.8}$$

In terms of the time constant, Eq. (7.7) can be written as

$$v(t) = V_0e^{-t/\tau} \tag{7.9}$$

TABLE 7.1

Values of $v(t)/V_0 = e^{-t/\tau}$.

t	$v(t)/V_0$
τ	0.36788
2τ	0.13534
3τ	0.04979
4τ	0.01832
5τ	0.00674

With a calculator it is easy to show that the value of $v(t)/V_0$ is as shown in Table 7.1. It is evident from Table 7.1 that the voltage $v(t)$ is less than 1 percent of V_0 after 5τ (five time constants). Thus, it is customary to assume that the capacitor is fully discharged (or charged) after five time constants. In other words, it takes 5τ for the circuit to reach its final state or steady state when no changes take place with time. Notice that for every time interval of τ, the voltage is reduced by 36.8 percent of its previous value, $v(t + \tau) = v(t)/e = 0.368v(t)$, regardless of the value of t.

Observe from Eq. (7.8) that the smaller the time constant, the more rapidly the voltage decreases, that is, the faster the response. This is illustrated in Fig. 7.4. A circuit with a small time constant gives a fast response in that it reaches the steady state (or final state) quickly due to quick dissipation of energy stored, whereas a circuit with a large time constant gives a slow response because it takes longer to reach steady state. At any rate, whether the time constant is small or large, the circuit reaches steady state in five time constants.

With the voltage $v(t)$ in Eq. (7.9), we can find the current $i_R(t)$,

$$i_R(t) = \frac{v(t)}{R} = \frac{V_0}{R}e^{-t/\tau} \tag{7.10}$$

Figure 7.3
Graphical determination of the time constant τ from the response curve.

[1] The time constant may be viewed from another perspective. Evaluating the derivative of $v(t)$ in Eq. (7.7) at $t = 0$, we obtain

$$\frac{d}{dt}\left(\frac{v}{V_0}\right)\bigg|_{t=0} = -\frac{1}{\tau}e^{-t/\tau}\bigg|_{t=0} = -\frac{1}{\tau}$$

Thus, the time constant is the initial rate of decay, or the time taken for v/V_0 to decay from unity to zero, assuming a constant rate of decay. This initial slope interpretation of the time constant is often used in the laboratory to find τ graphically from the response curve displayed on an oscilloscope. To find τ from the response curve, draw the tangent to the curve at $t = 0$, as shown in Fig. 7.3. The tangent intercepts with the time axis at $t = \tau$.

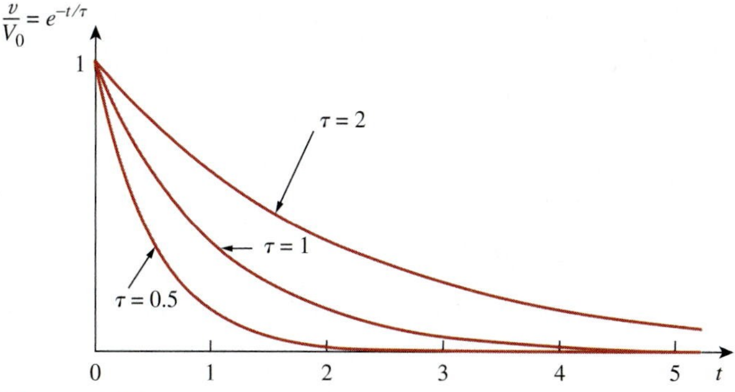

Figure 7.4
Plot of $v/V_0 = e^{-t/\tau}$ for various values of the time constant.

The power dissipated in the resistor is

$$p(t) = v i_R = \frac{V_0^2}{R} e^{-2t/\tau} \tag{7.11}$$

The energy absorbed by the resistor up to time t is

$$\begin{aligned} w_R(t) &= \int_0^t p\,dt = \int_0^t \frac{V_0^2}{R} e^{-2t/\tau}\,dt \\ &= -\frac{\tau V_0^2}{2R} e^{-2t/\tau}\Big|_0^t = \frac{1}{2} C V_0^2 (1 - e^{-2t/\tau}), \qquad \tau = RC \end{aligned} \tag{7.12}$$

Notice that as $t \to \infty$, $w_R(\infty) \to \frac{1}{2} C V_0^2$, which is the same as $w_C(0)$, the energy initially stored in the capacitor. The energy that was initially stored in the capacitor is eventually dissipated in the resistor.

In summary:

The Key to Working with a Source-free *RC* Circuit Is Finding:

1. The initial voltage $v(0) = V_0$ across the capacitor.
2. The time constant τ.

The time constant is the same regardless of what the output is defined to be.

With these two items, we obtain the response as the capacitor voltage $v_C(t) = v(t) = v(0)e^{-t/\tau}$. Once the capacitor voltage is first obtained, other variables (capacitor current i_C, resistor voltage v_R, and resistor current i_R) can be determined. In finding the time constant $\tau = RC$, R is often the Thevenin equivalent resistance at the terminals of the capacitor; that is, we take out the capacitor C and find $R = R_{Th}$ at its terminals.

When a circuit contains a single capacitor and several resistors and dependent sources, the Thevenin equivalent can be found at the terminals of the capacitor to form a simple *RC* circuit. Also, one can use Thevenin's theorem when several capacitors can be combined to form a single equivalent capacitor.

Example 7.1

In Fig. 7.5, let $v_C(0) = 15$ V. Find v_C, v_x, and i_x for $t > 0$.

Solution:

We first need to make the circuit in Fig. 7.5 conform with the standard *RC* circuit in Fig. 7.1. We find the equivalent resistance or the Thevenin

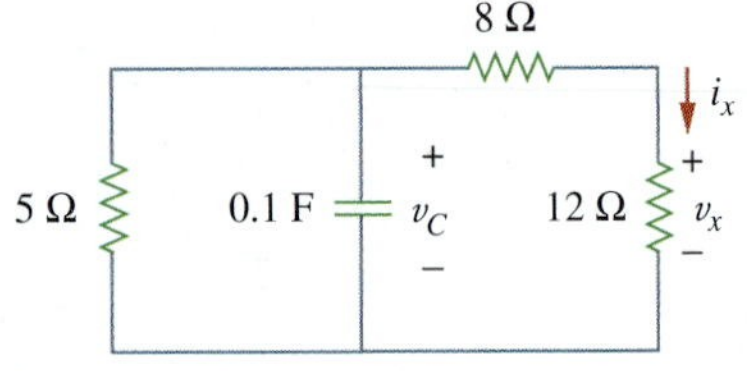

Figure 7.5
For Example 7.1.

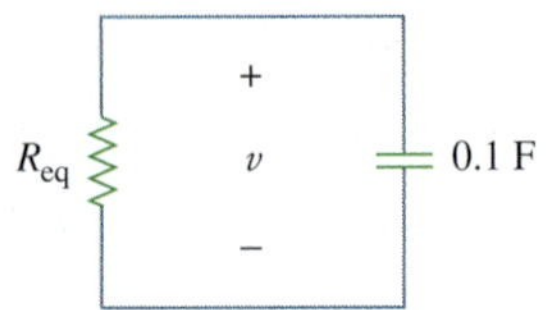

Figure 7.6
Equivalent circuit for the circuit in Fig. 7.5.

resistance at the capacitor terminals. Our objective is always to first obtain capacitor voltage v_C. From this, we can determine v_x and i_x.

The 8-Ω and 12-Ω resistors in series can be combined to give a 20-Ω resistor. This 20-Ω resistor in parallel with the 5-Ω resistor can be combined so that the equivalent resistance is

$$R_{eq} = \frac{20 \times 5}{20 + 5} = 4\ \Omega$$

Hence, the equivalent circuit is as shown in Fig. 7.6, which is analogous to Fig. 7.1. The time constant is

$$\tau = R_{eq}C = 4(0.1) = 0.4\ \text{s}$$

Thus,

$$v = v(0)e^{-t/\tau} = 15e^{-t/0.4}\ \text{V}, \qquad v_C = v = 15e^{-2.5t}\ \text{V}$$

From Fig. 7.5, we can use voltage division to get v_x; so

$$v_x = \frac{12}{12 + 8}v = 0.6(15e^{-2.5t}) = 9e^{-2.5t}\ \text{V}$$

Finally,

$$i_x = \frac{v_x}{12} = 0.75e^{-2.5t}\ \text{A}$$

Practice Problem 7.1

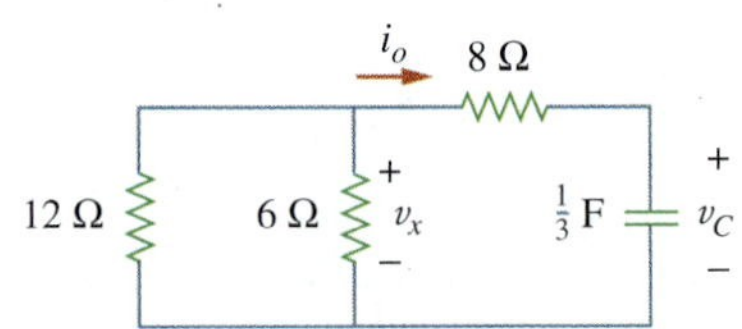

Figure 7.7
For Practice Prob. 7.1.

Refer to the circuit in Fig. 7.7. Let $v_C(0) = 45$ V. Determine v_C, v_x, and i_o for $t \geq 0$.

Answer: $45e^{-0.25t}$ V, $15e^{-0.25t}$ V, $-3.75e^{-0.25t}$ A.

Example 7.2

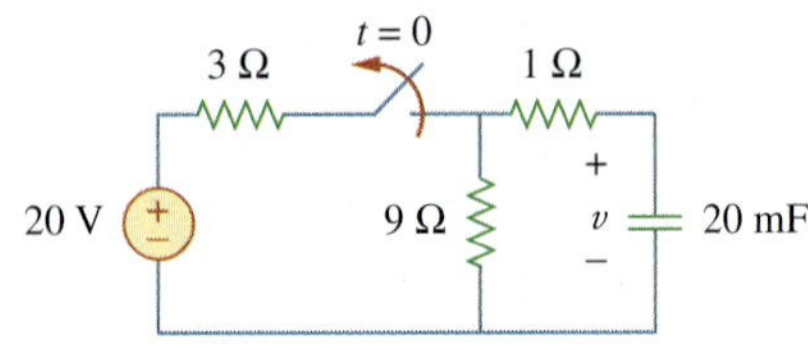

Figure 7.8
For Example 7.2.

The switch in the circuit in Fig. 7.8 has been closed for a long time, and it is opened at $t = 0$. Find $v(t)$ for $t \geq 0$. Calculate the initial energy stored in the capacitor.

Solution:
For $t < 0$, the switch is closed; the capacitor is an open circuit to dc, as represented in Fig. 7.9(a). Using voltage division

$$v_C(t) = \frac{9}{9 + 3}(20) = 15\ \text{V}, \qquad t < 0$$

Since the voltage across a capacitor cannot change instantaneously, the voltage across the capacitor at $t = 0^-$ is the same at $t = 0$, or

$$v_C(0) = V_0 = 15\ \text{V}$$

For $t > 0$, the switch is opened, and we have the *RC* circuit shown in Fig. 7.9(b). [Notice that the *RC* circuit in Fig. 7.9(b) is source free; the independent source in Fig. 7.8 is needed to provide V_0 or the initial energy in the capacitor.] The 1-Ω and 9-Ω resistors in series give

$$R_{eq} = 1 + 9 = 10\ \Omega$$

The time constant is

$$\tau = R_{eq}C = 10 \times 20 \times 10^{-3} = 0.2\text{ s}$$

Thus, the voltage across the capacitor for $t \geq 0$ is

$$v(t) = v_C(0)e^{-t/\tau} = 15e^{-t/0.2}\text{ V}$$

or

$$v(t) = 15e^{-5t}\text{ V}$$

The initial energy stored in the capacitor is

$$w_C(0) = \frac{1}{2}Cv_C^2(0) = \frac{1}{2} \times 20 \times 10^{-3} \times 15^2 = 2.25\text{ J}$$

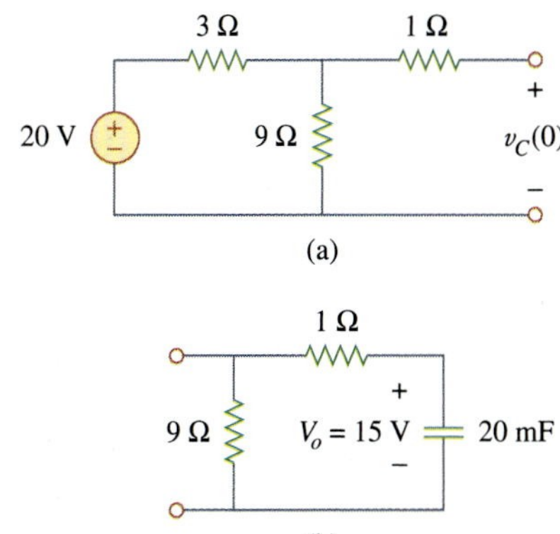

Figure 7.9
For Example 7.2: (a) $t < 0$, (b) $t > 0$.

Practice Problem 7.2

If the switch in Fig. 7.10 opens at $t = 0$, find $v(t)$ for $t \geq 0$ and $w_C(0)$.

Answer: $8e^{-2t}$ V, 5.33 J.

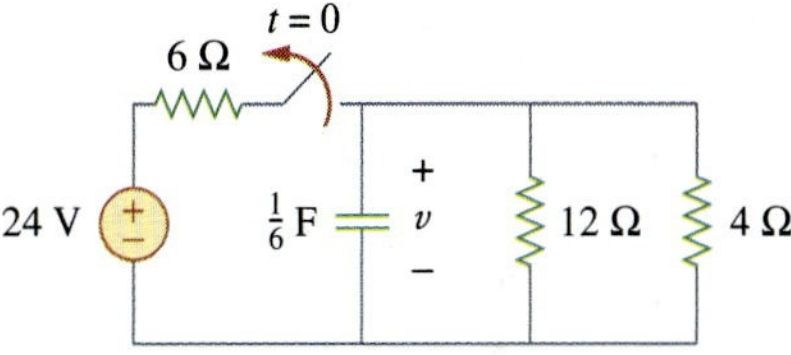

Figure 7.10
For Practice Prob. 7.2.

7.3 The Source-Free *RL* Circuit

Consider the series connection of a resistor and an inductor, as shown in Fig. 7.11. Our goal is to determine the circuit response, which we will assume to be the current $i(t)$ through the inductor. We select the inductor current as the response in order to take advantage of the idea that the inductor current cannot change instantaneously. At $t = 0$, we assume that the inductor has an initial current I_0, or

$$i(0) = I_0 \tag{7.13}$$

with the corresponding energy stored in the inductor as

$$w(0) = \frac{1}{2}L\,I_0^2 \tag{7.14}$$

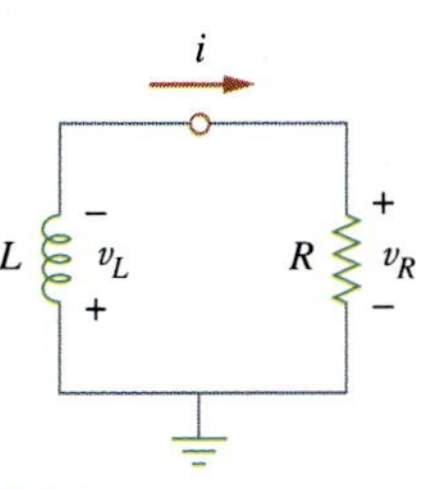

Figure 7.11
A source-free *RL* circuit.

Applying KVL around the loop in Fig. 7.11,

$$v_L + v_R = 0 \tag{7.15}$$

But $v_L = L\,di/dt$ and $v_R = iR$. Thus,

$$L\frac{di}{dt} + Ri = 0$$

or

$$\frac{di}{dt} + \frac{R}{L}i = 0 \tag{7.16}$$

Rearranging terms and integrating gives

$$\int_{I_0}^{i(t)} \frac{di}{i} = -\int_0^t \frac{R}{L}\,dt$$

$$\ln i\Big|_{I_0}^{i(t)} = -\frac{Rt}{L}\Big|_0^t \quad \Rightarrow \quad \ln i(t) - \ln I_0 = -\frac{Rt}{L} + 0$$

or

$$\ln \frac{i(t)}{I_0} = -\frac{Rt}{L} \tag{7.17}$$

Taking the powers of e, we have

$$i(t) = I_0 e^{-Rt/L} \tag{7.18}$$

This shows that the natural response of the RL circuit is an exponential decay of the initial current. The current response is shown in Fig. 7.12. It is evident from Eq. (7.18) that the time constant for the RL circuit is

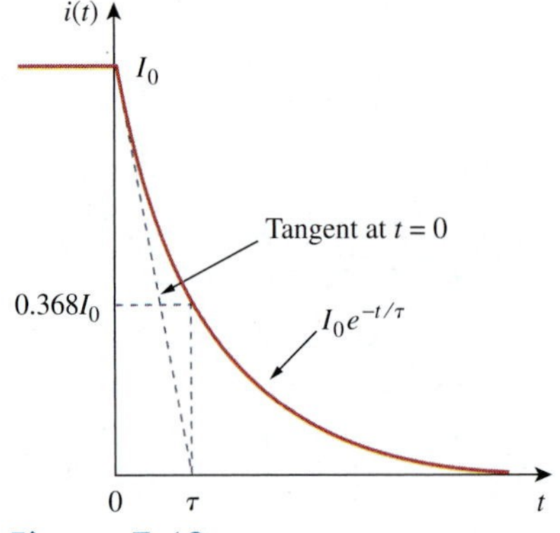

Figure 7.12
The current response of the RL circuit.

$$\boxed{\tau = \frac{L}{R}} \tag{7.19}$$

with τ again having the unit of seconds. Thus, Eq. (7.18) may be written as

$$\boxed{i(t) = I_0 e^{-t/\tau}} \tag{7.20}$$

With the current in Eq. (7.20), we can find the voltage across the resistor as

The smaller the time constant τ of a circuit, the faster the rate of decay of the response. The larger the time constant, the slower the rate of decay of the response. At any rate, the response decays to less than 1 percent of its initial value (i.e., reaches steady state) after 5τ.

$$v_R(t) = iR = I_0 R e^{-t/\tau} \tag{7.21}$$

The power dissipated in the resistor is

$$p = v_R i = I_0^2 R e^{-2t/\tau} \tag{7.22}$$

The energy absorbed by the resistor is

$$w_R(t) = \int_0^t p\,dt = \int_0^t I_0^2 R e^{-2t/\tau}\,dt = -\frac{1}{2}\tau I_0^2 R e^{-2t/\tau}\Big|_0^t, \qquad \tau = \frac{L}{R}$$

or

$$w_R(t) = \frac{1}{2} L I_0^2 (1 - e^{-2t/\tau}) \tag{7.23}$$

Note that as $t \to \infty$, $w_R(\infty) \to \frac{1}{2} L I_0^2$, which is the same as $w_L(0)$, the initial energy stored in the inductor as in Eq. (7.14). Again, the energy initially stored in the inductor is eventually dissipated in the resistor.

Figure 7.12 shows an initial slope interpretation may be given to τ.

In summary:

The Key to Working with a Source-free *RL* Circuit Is to Find:

1. The initial current $i(0) = I_0$ through the inductor.
2. The time constant τ of the circuit.

With the two items, we obtain the response as the inductor current $i_L(t) = i(t) = i(0)e^{-t/\tau}$. Once we determine the inductor current i_L, other variables (inductor voltage v_L, resistor voltage v_R, and resistor current i_R) can be obtained. Note that in general, R in Eq. (7.19) is the Thevenin resistance at the terminals of the inductor.

When a circuit has a single inductor and several resistors and dependent sources, the Thevenin equivalent can be found at the terminals of the inductor to form a simple *RL* circuit. Also, one can use Thevenin's theorem when several inductors can be combined to form a single equivalent inductor.

Example 7.3

Assuming that $i(0) = 10$ A, calculate $i(t)$ and $i_x(t)$ in the circuit of Fig. 7.13.

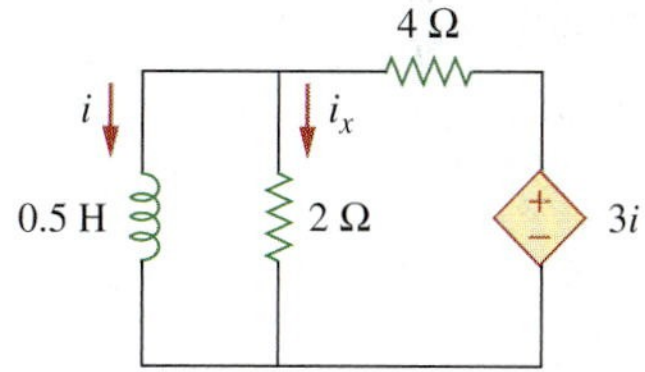

Figure 7.13
For Example 7.3.

Solution:
There are two ways we can solve this problem. One way is to obtain the equivalent resistance at the inductor terminals and then use Eq. (7.20). The other way is to start from scratch by using Kirchhoff's voltage law. Whichever approach is taken, it is always better to first obtain the inductor current.

■ **METHOD 1** The equivalent resistance is the same as the Thevenin resistance at the inductor terminals. Because of the dependent source, we insert a voltage source with $v_o = 1$ V at the inductor terminals *a-b*, as in Fig. 7.14(a). (We could also insert a 1-A current source at the terminals.) Applying KVL to the two loops results in

$$2(i_1 - i_2) + 1 = 0 \quad \Rightarrow \quad i_1 - i_2 = -\frac{1}{2} \tag{7.3.1}$$

$$6i_2 - 2i_1 - 3i_1 = 0 \quad \Rightarrow \quad i_2 = \frac{5}{6}i_1 \tag{7.3.2}$$

Substituting Eq. (7.3.2) into Eq. (7.3.1) gives

$$i_1 = -3 \text{ A}, \qquad i_o = -i_1 = 3 \text{ A}$$

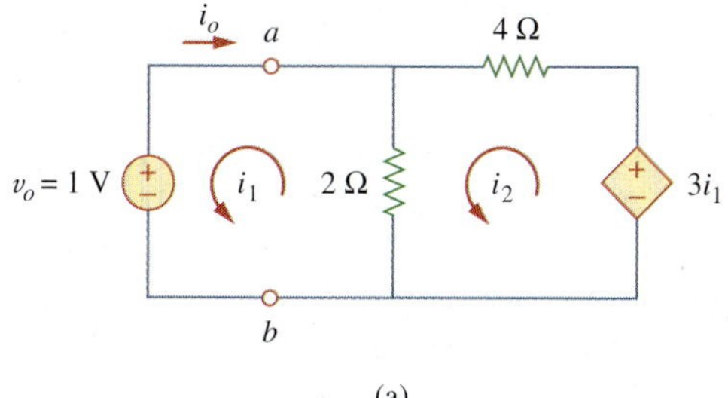

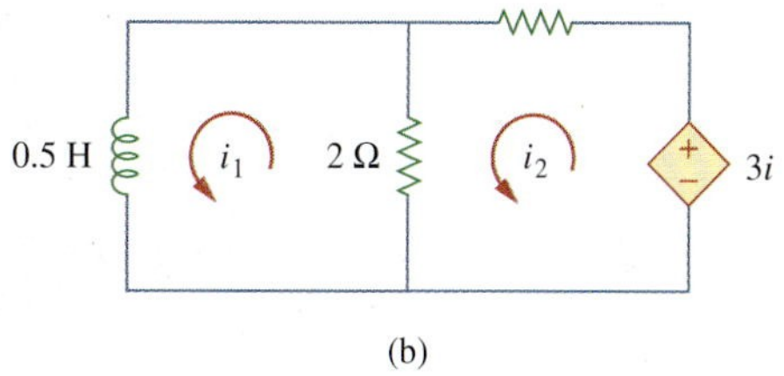

Figure 7.14
Solving the circuit in Fig. 7.13.

Hence,

$$R_{eq} = R_{Th} = \frac{v_o}{i_o} = \frac{1}{3}\,\Omega$$

The time constant is

$$\tau = \frac{L}{R_{eq}} = \frac{\frac{1}{2}}{\frac{1}{3}} = \frac{3}{2}\text{ s}$$

Thus, the current through the inductor is

$$i(t) = i(0)e^{-t/\tau} = 10e^{-(2/3)t}\text{ A}, \qquad t > 0$$

METHOD 2 We may directly apply KVL to the circuit as in Fig. 7.14(b). For loop 1,

$$\frac{1}{2}\frac{di_1}{dt} + 2(i_1 - i_2) = 0$$

or

$$\frac{di_1}{dt} + 4i_1 - 4i_2 = 0 \tag{7.3.3}$$

For loop 2,

$$6i_2 - 2i_1 - 3i_1 = 0 \quad \Rightarrow \quad i_2 = \frac{5}{6}i_1 \tag{7.3.4}$$

Substituting Eq. (7.3.4) into Eq. (7.3.3) gives

$$\frac{di_1}{dt} + \frac{2}{3}i_1 = 0$$

Rearranging terms,

$$\frac{di_1}{i_1} = -\frac{2}{3}dt$$

Since $i_1 = i$, we may replace i_1 with i and integrate:

$$\ln i\Big|_{i(0)}^{i(t)} = -\frac{2}{3}t\Big|_0^t$$

or

$$\ln\frac{i(t)}{i(0)} = -\frac{2}{3}t$$

Taking the powers of e, we finally obtain

$$i(t) = i(0)e^{-(2/3)t} = 10e^{-(2/3)t}\text{ A}, \qquad t > 0$$

which is the same as by Method 1.

The voltage across the inductor is

$$v = L\frac{di}{dt} = 0.5(10)\left(-\frac{2}{3}\right)e^{-(2/3)t} = -\frac{10}{3}e^{-(2/3)t}\text{ V}$$

Since the inductor and the 2-Ω resistor are in parallel,

$$i_x(t) = \frac{v}{2} = -1.6667e^{-(2/3)t}\text{ A}, \qquad t > 0$$

Practice Problem 7.3

Find i and v_x in the circuit of Fig. 7.15. Let $i(0) = 5$ A.

Answer: $5e^{-4t}$ V, $-20e^{-4t}$ V.

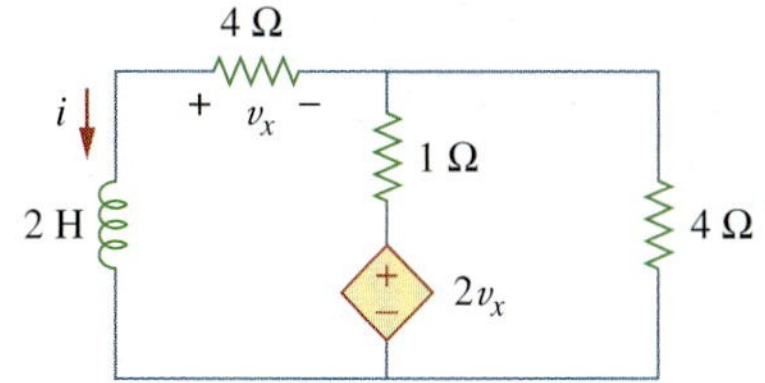

Figure 7.15
For Practice Prob. 7.3.

Example 7.4

The switch in the circuit of Fig. 7.16 has been closed for a long time. At $t = 0$, the switch is opened. Calculate $i(t)$ for $t > 0$.

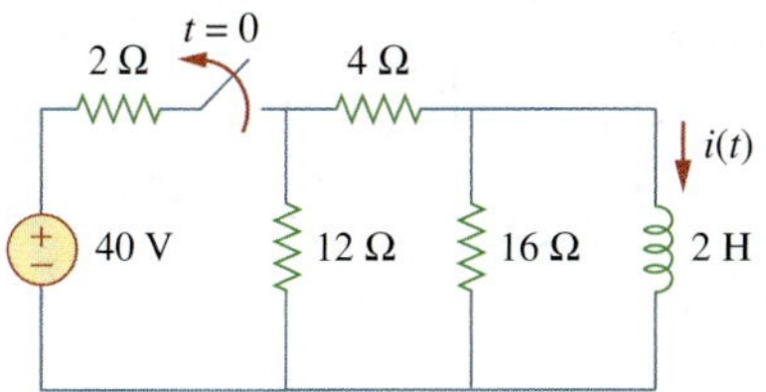

Figure 7.16
For Example 7.4.

Solution:
When $t < 0$, the switch is closed, and the inductor acts as a short circuit to dc. The 16-Ω resistor is short-circuited; the resulting circuit is shown in Fig. 7.17(a). To get i_1 in Fig. 7.17(a), we combine the 4-Ω and 12-Ω resistors in parallel to get

$$\frac{4 \times 12}{4 + 12} = 3\ \Omega$$

Hence,

$$i_1 = \frac{40}{2 + 3} = 8\text{ A}$$

We obtain $i(t)$ from i_1 in Fig. 7.17(a) using current division, by writing

$$i(t) = \frac{12}{12 + 4}i_1 = 6\text{ A}, \qquad t < 0$$

Since the current through an inductor cannot change instantaneously,

$$i(0) = i(0^-) = 6\text{ A}$$

When $t > 0$, the switch is open and the voltage source is disconnected. We now have the source-free *RL* circuit in Fig. 7.17(b). Combining the resistors, we have

$$R_{eq} = (12 + 4) \parallel 16 = 8\ \Omega$$

The time constant is

$$\tau = \frac{L}{R_{eq}} = \frac{2}{8} = \frac{1}{4}\text{ s}$$

Thus,

$$i(t) = i(0)e^{-t/\tau} = 6e^{-4t}\text{ A}$$

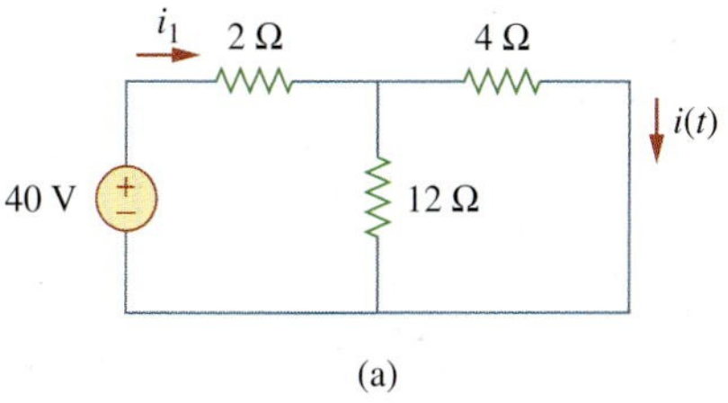

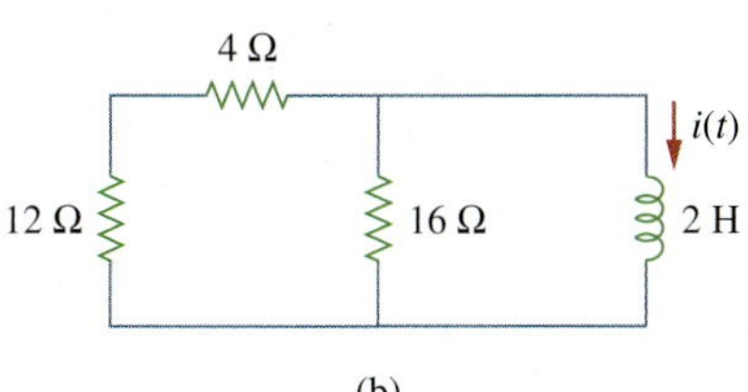

Figure 7.17
Solving the circuit of Fig. 7.16: (a) for $t < 0$, (b) for $t > 0$.

Practice Problem 7.4

For the circuit in Fig. 7.18, find $i(t)$ for $t > 0$.

Answer: $2e^{-2t}$ A, $t > 0$.

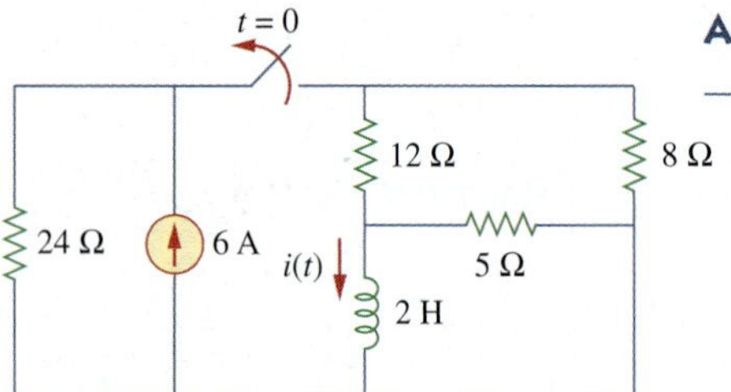

Figure 7.18
For Practice Prob. 7.4.

Example 7.5

In the circuit shown in Fig. 7.19, find i_o, v_o, and i for all time, assuming that the switch was open for a long time.

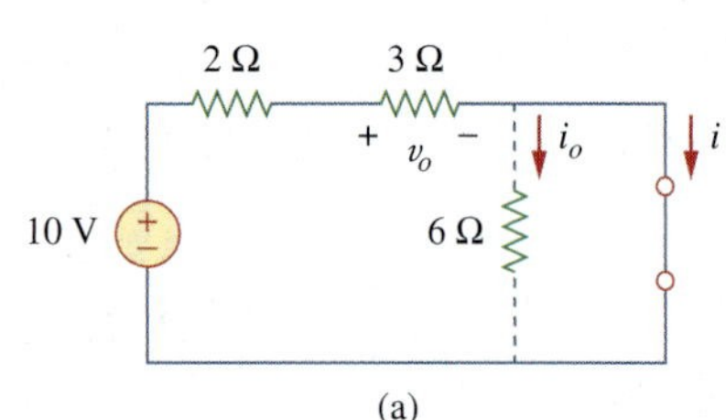

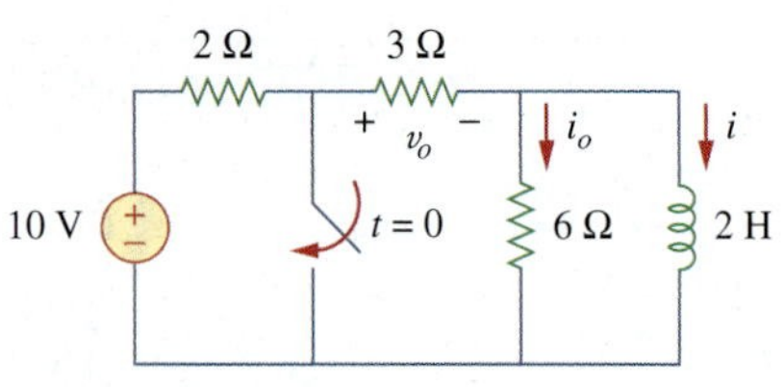

Figure 7.19
For Example 7.5.

Solution:

It is better to first find the inductor current i and then obtain other quantities from it.

For $t < 0$, the switch is open. Since the inductor acts like a short circuit to dc, the 6-Ω resistor is short-circuited, so that we have the circuit shown in Fig. 7.20(a). Hence, $i_o = 0$, and

$$i(t) = \frac{10}{2 + 3} = 2 \text{ A}, \qquad t < 0$$

$$v_o(t) = 3i(t) = 6 \text{ V}, \qquad t < 0$$

Thus, $i(0) = 2$.

For $t > 0$, the switch is closed, so that the voltage source is short-circuited. We now have a source-free RL circuit as shown in Fig. 7.20(b). At the inductor terminals,

$$R_{\text{Th}} = 3 \parallel 6 = 2 \text{ Ω}$$

so that the time constant is

$$\tau = \frac{L}{R_{\text{Th}}} = 1 \text{ s}$$

Hence,

$$i(t) = i(0)e^{-t/\tau} = 2e^{-t} \text{ A}, \qquad t > 0$$

Since the inductor is in parallel with the 6-Ω and 3-Ω resistors,

$$v_o(t) = -v_L = -L\frac{di}{dt} = -2(-2e^{-t}) = 4e^{-t} \text{ V}, \qquad t > 0$$

and

$$i_o(t) = \frac{v_L}{6} = -\frac{2}{3}e^{-t} \text{ A}, \qquad t > 0$$

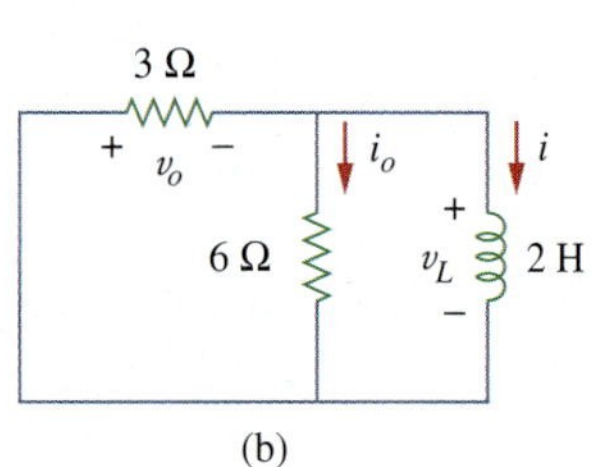

Figure 7.20
The circuit in Fig. 7.19 for: (a) $t < 0$, (b) $t > 0$.

Thus, for all time,

$$i_o(t) = \begin{cases} 0\text{ A}, & t < 0 \\ -\dfrac{2}{3}e^{-t}\text{ A}, & t > 0 \end{cases}, \qquad v_o(t) = \begin{cases} 6\text{ V}, & t < 0 \\ 4e^{-t}\text{ V}, & t > 0 \end{cases}$$

$$i(t) = \begin{cases} 2\text{ A}, & t < 0 \\ 2e^{-t}\text{ A}, & t \geq 0 \end{cases}$$

We notice that the inductor current is continuous at $t = 0$, while the current through the 6-Ω resistor drops from 0 to $-2/3$ at $t = 0$, and the voltage across the 3-Ω resistor drops from 6 to 4 at $t = 0$. We also notice that the time constant is the same regardless of what the output is defined to be. Figure 7.21 plots i and i_o.

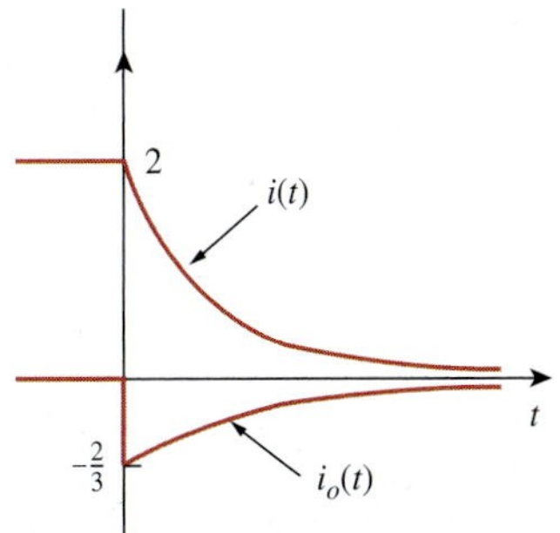

Figure 7.21
A plot of i and i_o.

Practice Problem 7.5

Determine i, i_o, and v_o for all t in the circuit shown in Fig. 7.22. Assume that the switch was closed for a long time. It should be noted that opening a switch in series with an ideal current source creates an infinite voltage at the current source terminals. Clearly this is impossible. For the purposes of problem solving, we can place a shunt resistor in parallel with the source (which now makes it a voltage source in series with a resistor). In more practical circuits, devices that act like current sources are, for the most part, electronic circuits. These circuits will allow the source to act like an ideal current source over its operating range but voltage-limit it when the load resistor becomes too large (as in an open circuit).

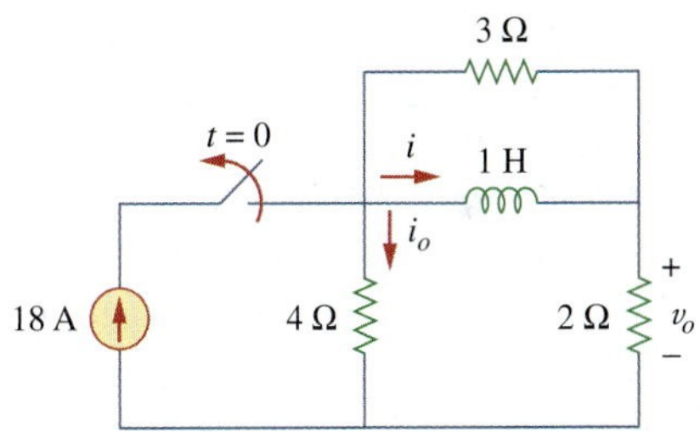

Figure 7.22
For Practice Prob. 7.5.

Answer:

$$i = \begin{cases} 12\text{ A}, & t < 0 \\ 12e^{-2t}\text{ A}, & t \geq 0 \end{cases}, \qquad i_o = \begin{cases} 6\text{ A}, & t < 0 \\ -4e^{-2t}\text{ A}, & t > 0 \end{cases},$$

$$v_o = \begin{cases} 24\text{ V}, & t < 0 \\ 8e^{-2t}\text{ V}, & t > 0 \end{cases}$$

7.4 Singularity Functions

Before going on with the second half of this chapter, we need to digress and consider some mathematical concepts that will aid our understanding of transient analysis. A basic understanding of singularity functions will help us make sense of the response of first-order circuits to a sudden application of an independent dc voltage or current source.

Singularity functions (also called *switching functions*) are very useful in circuit analysis. They serve as good approximations to the switching signals that arise in circuits with switching operations. They are helpful in the neat, compact description of some circuit phenomena, especially the step response of *RC* or *RL* circuits to be discussed in the next sections. By definition,

> **Singularity functions** are functions that either are discontinuous or have discontinuous derivatives.

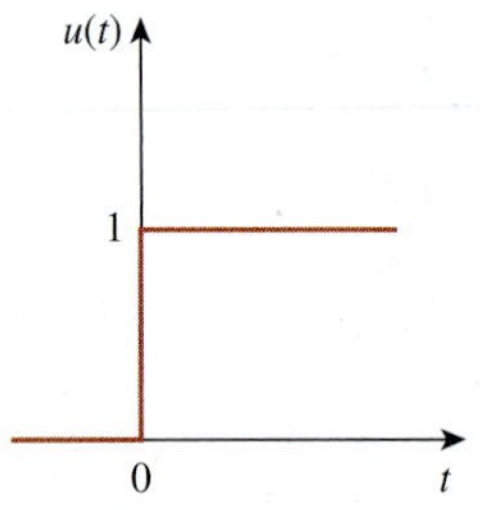

Figure 7.23
The unit step function.

The three most widely used singularity functions in circuit analysis are the *unit step*, the *unit impulse*, and the *unit ramp* functions.

> The **unit step function** $u(t)$ is 0 for negative values of t and 1 for positive values of t.

In mathematical terms,

$$u(t) = \begin{cases} 0, & t < 0 \\ 1, & t > 0 \end{cases} \tag{7.24}$$

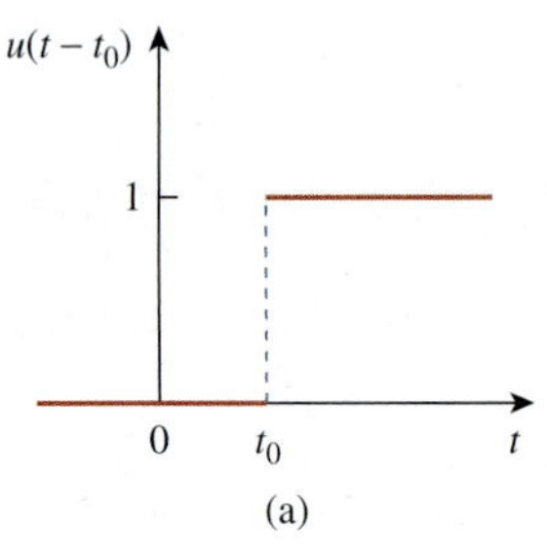

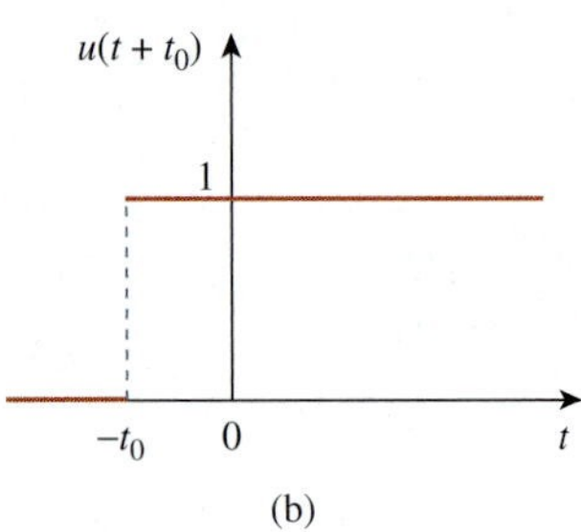

Figure 7.24
(a) The unit step function delayed by t_0, (b) the unit step advanced by t_0.

The unit step function is undefined at $t = 0$, where it changes abruptly from 0 to 1. It is dimensionless, like other mathematical functions such as sine and cosine. Figure 7.23 depicts the unit step function. If the abrupt change occurs at $t = t_0$ (where $t_0 > 0$) instead of $t = 0$, the unit step function becomes

$$u(t - t_0) = \begin{cases} 0, & t < t_0 \\ 1, & t > t_0 \end{cases} \tag{7.25}$$

which is the same as saying that $u(t)$ is delayed by t_0 seconds, as shown in Fig. 7.24(a). To get Eq. (7.25) from Eq. (7.24), we simply replace every t by $t - t_0$. If the change is at $t = -t_0$, the unit step function becomes

$$u(t + t_0) = \begin{cases} 0, & t < -t_0 \\ 1, & t > -t_0 \end{cases} \tag{7.26}$$

meaning that $u(t)$ is advanced by t_0 seconds, as shown in Fig. 7.24(b).

We use the step function to represent an abrupt change in voltage or current, like the changes that occur in the circuits of control systems and digital computers. For example, the voltage

$$v(t) = \begin{cases} 0, & t < t_0 \\ V_0, & t > t_0 \end{cases} \tag{7.27}$$

may be expressed in terms of the unit step function as

$$v(t) = V_0 u(t - t_0) \tag{7.28}$$

> Alternatively, we may derive Eqs. (7.25) and (7.26) from Eq. (7.24) by writing $u[f(t)] = 1$, $f(t) > 0$, where $f(t)$ may be $t - t_0$ or $t + t_0$.

If we let $t_0 = 0$, then $v(t)$ is simply the step voltage $V_0 u(t)$. A voltage source of $V_0 u(t)$ is shown in Fig. 7.25(a); its equivalent circuit is shown in Fig. 7.25(b). It is evident in Fig. 7.25(b) that terminals *a-b* are short-circuited ($v = 0$) for $t < 0$ and that $v = V_0$ appears at the terminals

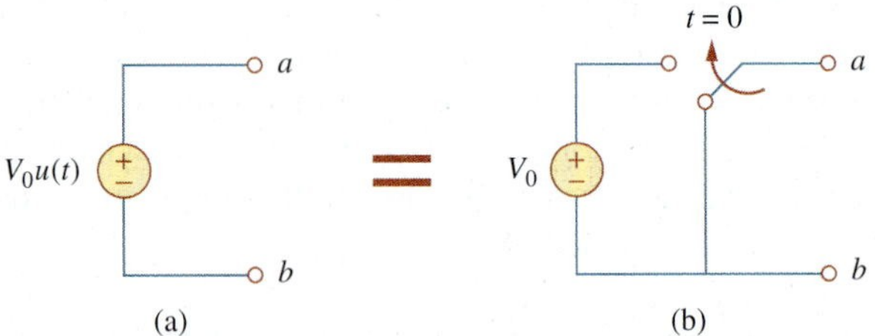

Figure 7.25
(a) Voltage source of $V_0u(t)$, (b) its equivalent circuit.

for $t > 0$. Similarly, a current source of $I_0u(t)$ is shown in Fig. 7.26(a), while its equivalent circuit is in Fig. 7.26(b). Notice that for $t < 0$, there is an open circuit ($i = 0$), and that $i = I_0$ flows for $t > 0$.

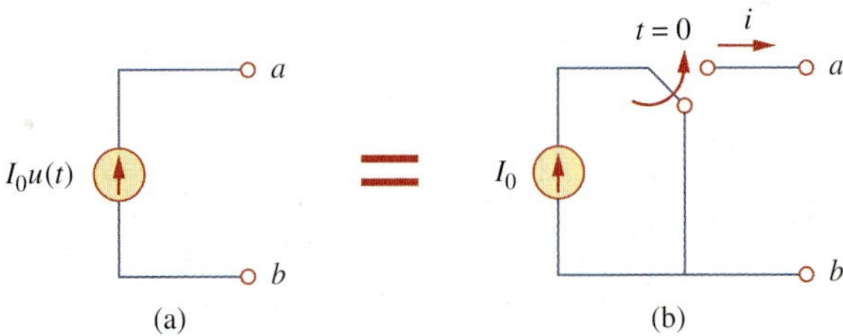

Figure 7.26
(a) Current source of $I_0u(t)$, (b) its equivalent circuit.

The derivative of the unit step function $u(t)$ is the *unit impulse function* $\delta(t)$, which we write as

$$\delta(t) = \frac{d}{dt}u(t) = \begin{cases} 0, & t < 0 \\ \text{Undefined}, & t = 0 \\ 0, & t > 0 \end{cases} \tag{7.29}$$

The unit impulse function—also known as the *delta* function—is shown in Fig. 7.27.

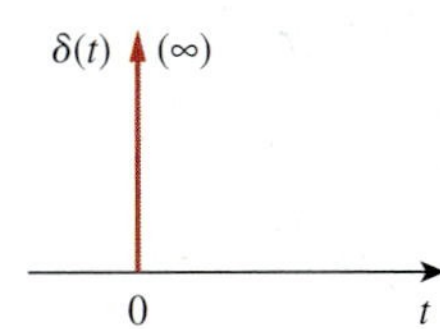

Figure 7.27
The unit impulse function.

The **unit impulse function** $\delta(t)$ is zero everywhere except at $t = 0$, where it is undefined.

Impulsive currents and voltages occur in electric circuits as a result of switching operations or impulsive sources. Although the unit impulse function is not physically realizable (just like ideal sources, ideal resistors, etc.), it is a very useful mathematical tool.

The unit impulse may be regarded as an applied or resulting shock. It may be visualized as a very short duration pulse of unit area. This may be expressed mathematically as

$$\int_{0^-}^{0^+} \delta(t)\, dt = 1 \tag{7.30}$$

where $t = 0^-$ denotes the time just before $t = 0$ and $t = 0^+$ is the time just after $t = 0$. For this reason, it is customary to write 1 (denoting unit area) beside the arrow that is used to symbolize the unit impulse function, as in Fig. 7.27. The unit area is known as the *strength* of the impulse function. When an impulse function has a strength other than unity, the area of the impulse is equal to its strength. For example, an impulse function $10\delta(t)$ has an area of 10. Figure 7.28 shows the impulse functions $5\delta(t + 2)$, $10\delta(t)$, and $-4\delta(t - 3)$.

To illustrate how the impulse function affects other functions, let us evaluate the integral

$$\int_a^b f(t)\delta(t - t_0)\, dt \tag{7.31}$$

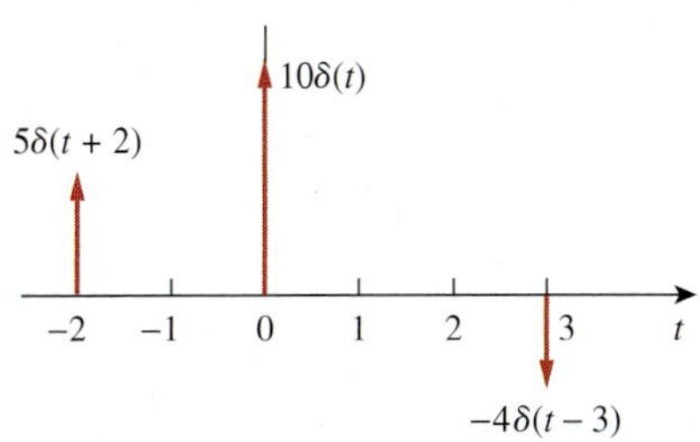

Figure 7.28
Three impulse functions.

where $a < t_0 < b$. Since $\delta(t - t_0) = 0$ except at $t = t_0$, the integrand is zero except at t_0. Thus,

$$\int_a^b f(t)\delta(t - t_0)\,dt = \int_a^b f(t_0)\delta(t - t_0)\,dt$$
$$= f(t_0)\int_a^b \delta(t - t_0)\,dt = f(t_0)$$

or

$$\boxed{\int_a^b f(t)\delta(t - t_0)\,dt = f(t_0)} \tag{7.32}$$

This shows that when a function is integrated with the impulse function, we obtain the value of the function at the point where the impulse occurs. This is a highly useful property of the impulse function known as the *sampling* or *sifting* property. The special case of Eq. (7.31) is for $t_0 = 0$. Then Eq. (7.32) becomes

$$\int_{0^-}^{0^+} f(t)\,\delta(t)\,dt = f(0) \tag{7.33}$$

Integrating the unit step function $u(t)$ results in the *unit ramp function* $r(t)$; we write

$$r(t) = \int_{-\infty}^{t} u(t)\,dt = tu(t) \tag{7.34}$$

or

$$\boxed{r(t) = \begin{cases} 0, & t \leq 0 \\ t, & t \geq 0 \end{cases}} \tag{7.35}$$

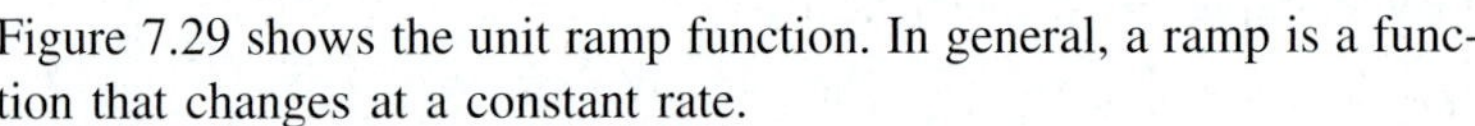
The **unit ramp function** is zero for negative values of t and has a unit slope for positive values of t.

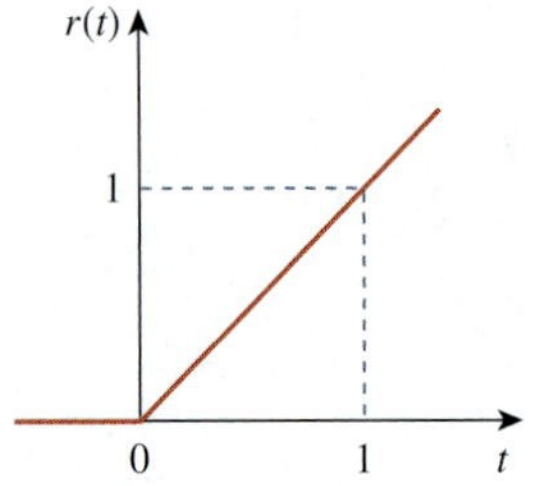

Figure 7.29
The unit ramp function.

Figure 7.29 shows the unit ramp function. In general, a ramp is a function that changes at a constant rate.

The unit ramp function may be delayed or advanced as shown in Fig. 7.30. For the delayed unit ramp function,

$$r(t - t_0) = \begin{cases} 0, & t \leq t_0 \\ t - t_0, & t \geq t_0 \end{cases} \tag{7.36}$$

and for the advanced unit ramp function,

$$r(t + t_0) = \begin{cases} 0, & t \leq -t_0 \\ t + t_0, & t \geq -t_0 \end{cases} \tag{7.37}$$

We should keep in mind that the three singularity functions (impulse, step, and ramp) are related by differentiation as

$$\delta(t) = \frac{du(t)}{dt}, \qquad u(t) = \frac{dr(t)}{dt} \tag{7.38}$$

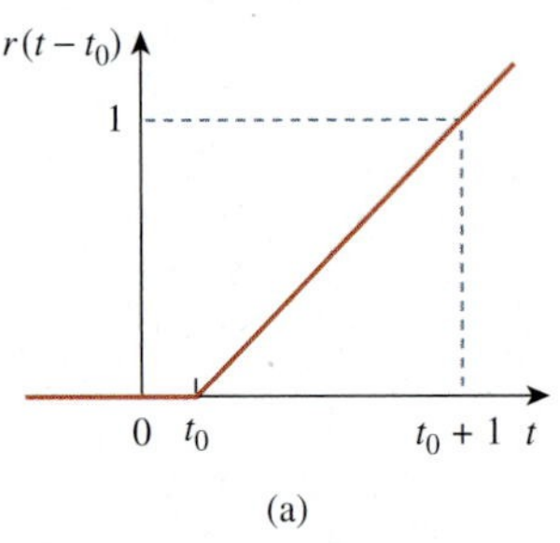

(a)

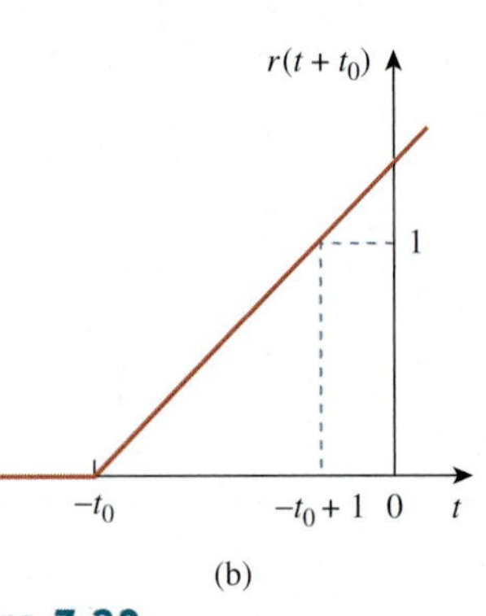

(b)

Figure 7.30
The unit ramp function: (a) delayed by t_0, (b) advanced by t_0.

or by integration as

$$u(t) = \int_{-\infty}^{t} \delta(t)\,dt, \qquad r(t) = \int_{-\infty}^{t} u(t)\,dt \qquad \textbf{(7.39)}$$

Although there are many more singularity functions, we are only interested in these three (the impulse function, the unit step function, and the ramp function) at this point.

Example 7.6

Express the voltage pulse in Fig. 7.31 in terms of the unit step. Calculate its derivative and sketch it.

Gate functions are used along with switches to pass or block another signal.

Solution:

The type of pulse in Fig. 7.31 is called the *gate function*. It may be regarded as a step function that switches on at one value of t and switches off at another value of t. The gate function shown in Fig. 7.31 switches on at $t = 2$ s and switches off at $t = 5$ s. It consists of the sum of two unit step functions as shown in Fig. 7.32(a). From the figure, it is evident that

$$v(t) = 10u(t-2) - 10u(t-5) = 10[u(t-2) - u(t-5)]$$

Taking the derivative of this gives

$$\frac{dv}{dt} = 10[\delta(t-2) - \delta(t-5)]$$

which is shown in Fig. 7.32(b). We can obtain Fig. 7.32(b) directly from Fig. 7.31 by simply observing that there is a sudden increase by 10 V at $t = 2$ s leading to $10\delta(t-2)$. At $t = 5$ s, there is a sudden decrease by 10 V leading to $-10\text{ V }\delta(t-5)$.

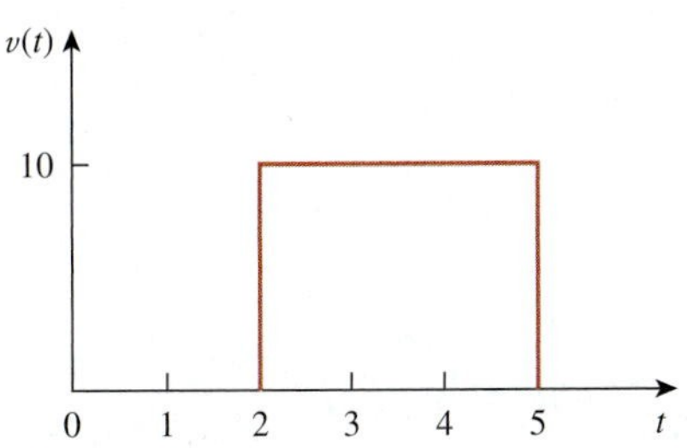

Figure 7.31
For Example 7.6.

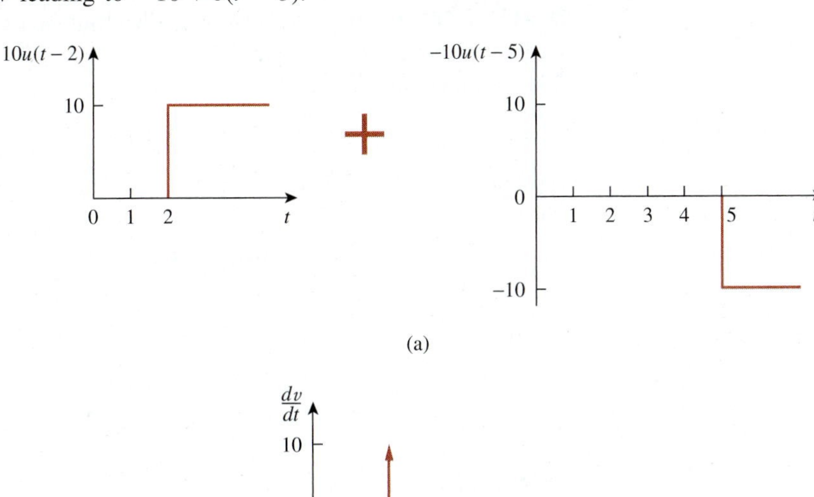

Figure 7.32
(a) Decomposition of the pulse in Fig. 7.31, (b) derivative of the pulse in Fig. 7.31.

Practice Problem 7.6

Express the current pulse in Fig. 7.33 in terms of the unit step. Find its integral and sketch it.

Answer: $10[u(t) - 2u(t-2) + u(t-4)]$, $10[r(t) - 2r(t-2) + r(t-4)]$. See Fig. 7.34.

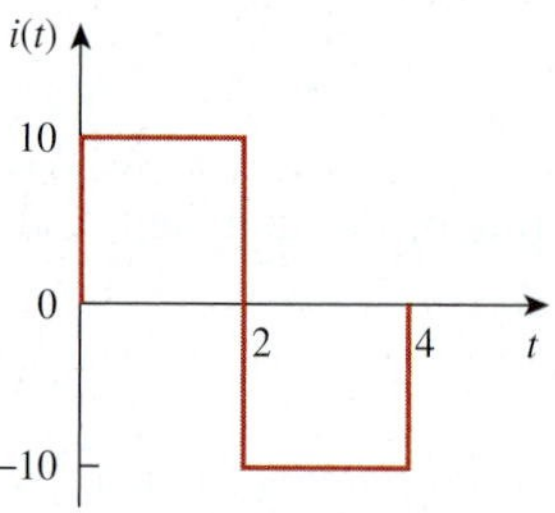

Figure 7.33
For Practice Prob. 7.6.

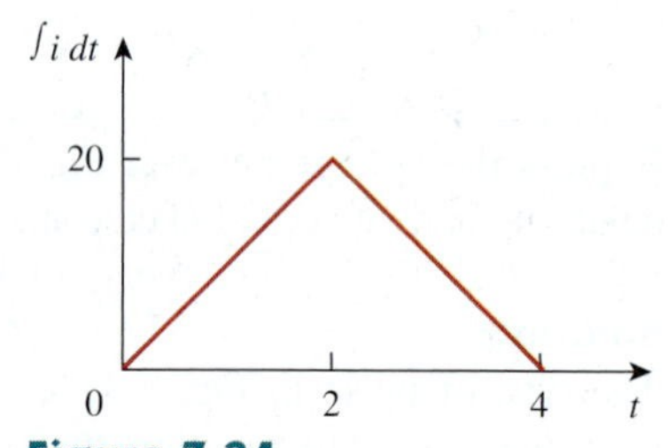

Figure 7.34
Integral of $i(t)$ in Fig. 7.33.

Example 7.7

Express the *sawtooth* function shown in Fig. 7.35 in terms of singularity functions.

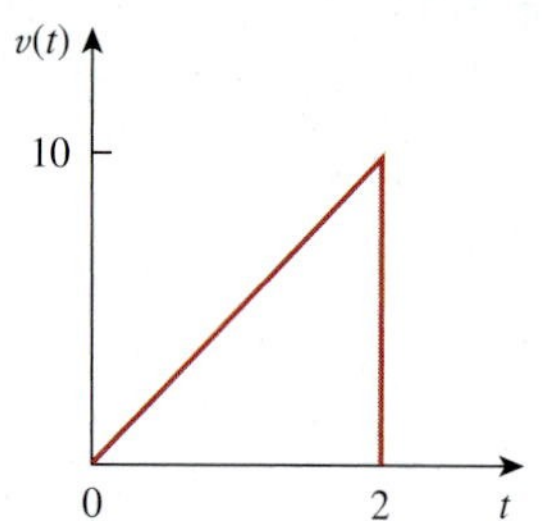

Figure 7.35
For Example 7.7.

Solution:
There are three ways of solving this problem. The first method is by mere observation of the given function, while the other methods involve some graphical manipulations of the function.

■ **METHOD 1** By looking at the sketch of $v(t)$ in Fig. 7.35, it is not hard to notice that the given function $v(t)$ is a combination of singularity functions. So we let

$$v(t) = v_1(t) + v_2(t) + \cdots \tag{7.7.1}$$

The function $v_1(t)$ is the ramp function of slope 5, shown in Fig. 7.36(a); that is,

$$v_1(t) = 5r(t) \tag{7.7.2}$$

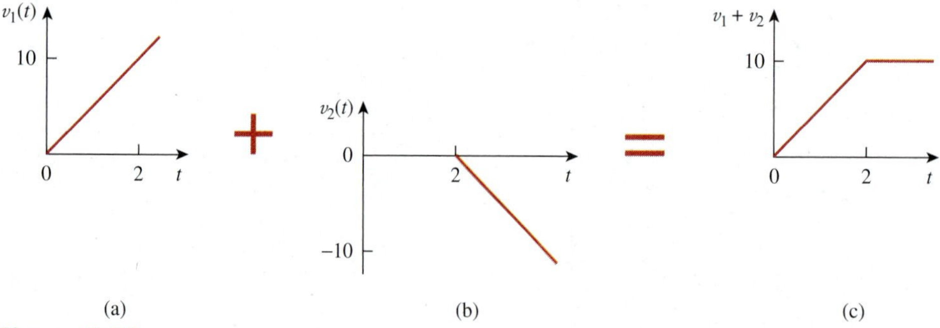

Figure 7.36
Partial decomposition of $v(t)$ in Fig. 7.35.

Since $v_1(t)$ goes to infinity, we need another function at $t = 2$ s in order to get $v(t)$. We let this function be v_2, which is a ramp function of slope -5, as shown in Fig. 7.36(b); that is,

$$v_2(t) = -5r(t - 2) \qquad \textbf{(7.7.3)}$$

Adding v_1 and v_2 gives us the signal in Fig. 7.36(c). Obviously, this is not the same as $v(t)$ in Fig. 7.35. But the difference is simply a constant 10 units for $t > 2$ s. By adding a third signal v_3, where

$$v_3 = -10u(t - 2) \qquad \textbf{(7.7.4)}$$

we get $v(t)$, as shown in Fig. 7.37. Substituting Eqs. (7.7.2) through (7.7.4) into Eq. (7.7.1) gives

$$v(t) = 5r(t) - 5r(t - 2) - 10u(t - 2)$$

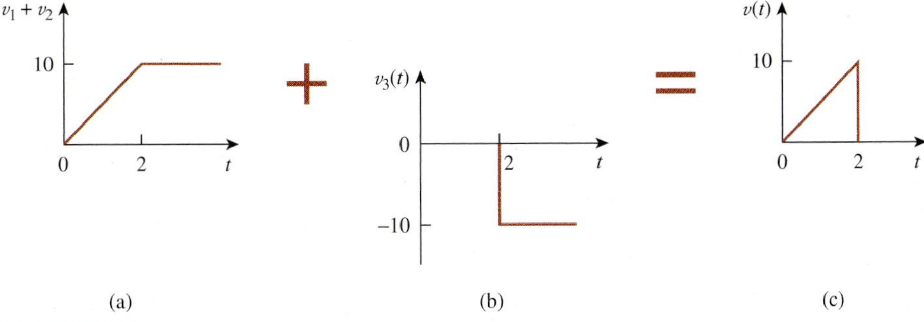

Figure 7.37
Complete decomposition of $v(t)$ in Fig. 7.35.

■ METHOD 2 A close observation of Fig. 7.35 reveals that $v(t)$ is a multiplication of two functions: a ramp function and a gate function. Thus,

$$\begin{aligned} v(t) &= 5t[u(t) - u(t - 2)] \\ &= 5tu(t) - 5tu(t - 2) \\ &= 5r(t) - 5(t - 2 + 2)u(t - 2) \\ &= 5r(t) - 5(t - 2)u(t - 2) - 10u(t - 2) \\ &= 5r(t) - 5r(t - 2) - 10u(t - 2) \end{aligned}$$

the same as before.

■ METHOD 3 This method is similar to Method 2. We observe from Fig. 7.35 that $v(t)$ is a multiplication of a ramp function and a unit step function, as shown in Fig. 7.38. Thus,

$$v(t) = 5r(t)u(-t + 2)$$

If we replace $u(-t)$ by $1 - u(t)$, then we can replace $u(-t + 2)$ by $1 - u(t - 2)$. Hence,

$$v(t) = 5r(t)[1 - u(t - 2)]$$

which can be simplified as in Method 2 to get the same result.

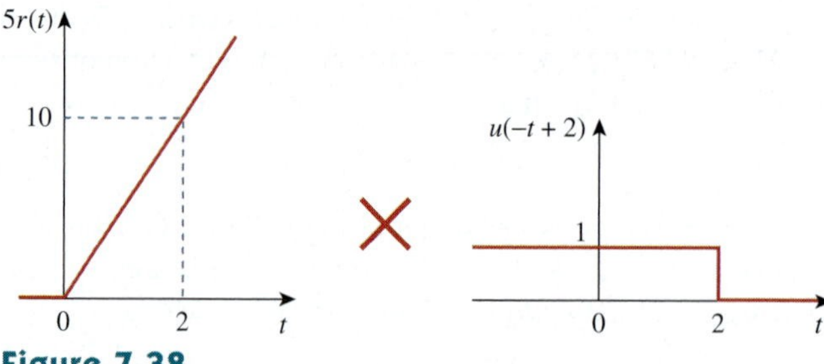

Figure 7.38
Decomposition of $v(t)$ in Fig. 7.35.

Practice Problem 7.7

Refer to Fig. 7.39. Express $i(t)$ in terms of singularity functions.

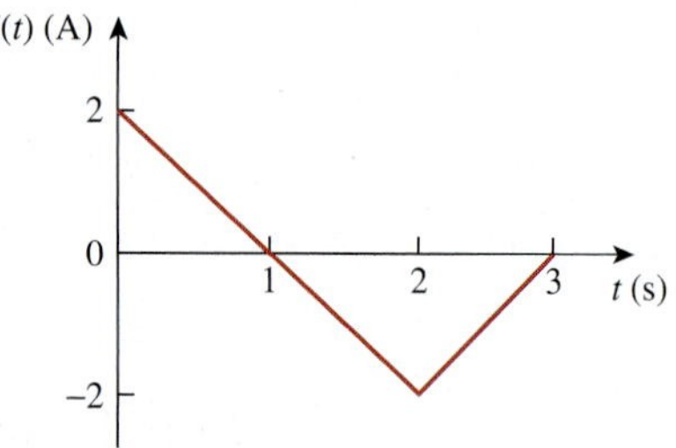

Figure 7.39
For Practice Prob. 7.7.

Answer: $2u(t) - 2r(t) + 4r(t-2) - 2r(t-3)$.

Example 7.8

Given the signal

$$g(t) = \begin{cases} 3, & t < 0 \\ -2, & 0 < t < 1 \\ 2t - 4, & t > 1 \end{cases}$$

express $g(t)$ in terms of step and ramp functions.

Solution:

The signal $g(t)$ may be regarded as the sum of three functions specified within the three intervals $t < 0$, $0 < t < 1$, and $t > 1$.

For $t < 0$, $g(t)$ may be regarded as 3 multiplied by $u(-t)$, where $u(-t) = 1$ for $t < 0$ and 0 for $t > 0$. Within the time interval $0 < t < 1$, the function may be considered as -2 multiplied by a gated function $[u(t) - u(t-1)]$. For $t > 1$, the function may be regarded as $2t - 4$ multiplied by the unit step function $u(t-1)$. Thus,

$$\begin{aligned} g(t) &= 3u(-t) - 2[u(t) - u(t-1)] + (2t-4)u(t-1) \\ &= 3u(-t) - 2u(t) + (2t - 4 + 2)u(t-1) \\ &= 3u(-t) - 2u(t) + 2(t-1)u(t-1) \\ &= 3u(-t) - 2u(t) + 2r(t-1) \end{aligned}$$

One may avoid the trouble of using $u(-t)$ by replacing it with $1 - u(t)$. Then

$$g(t) = 3[1 - u(t)] - 2u(t) + 2r(t-1) = 3 - 5u(t) + 2r(t-1)$$

Alternatively, we may plot $g(t)$ and apply Method 1 from Example 7.7.

Practice Problem 7.8

If

$$h(t) = \begin{cases} 0, & t < 0 \\ 8, & 0 < t < 2 \\ 2t + 6, & 2 < t < 6 \\ 0, & t > 6 \end{cases}$$

express $h(t)$ in terms of the singularity functions.

Answer: $8u(t) + 2u(t - 2) + 2r(t - 2) - 18u(t - 6) - 2r(t - 6)$.

Example 7.9

Evaluate the following integrals involving the impulse function:

$$\int_0^{10} (t^2 + 4t - 2)\delta(t - 2)\,dt$$

$$\int_{-\infty}^{\infty} [\delta(t - 1)e^{-t}\cos t + \delta(t + 1)e^{-t}\sin t]dt$$

Solution:
For the first integral, we apply the sifting property in Eq. (7.32).

$$\int_0^{10} (t^2 + 4t - 2)\delta(t - 2)\,dt = (t^2 + 4t - 2)|_{t=2} = 4 + 8 - 2 = 10$$

Similarly, for the second integral,

$$\begin{aligned} &\int_{-\infty}^{\infty} [\delta(t - 1)e^{-t}\cos t + \delta(t + 1)e^{-t}\sin t]\,dt \\ &\quad = e^{-t}\cos t|_{t=1} + e^{-t}\sin t|_{t=-1} \\ &\quad = e^{-1}\cos 1 + e^{1}\sin(-1) = 0.1988 - 2.2873 = -2.0885 \end{aligned}$$

Practice Problem 7.9

Evaluate the following integrals:

$$\int_{-\infty}^{\infty} (t^3 + 5t^2 + 10)\delta(t + 3)\,dt, \qquad \int_0^{10} \delta(t - \pi)\cos 3t\,dt$$

Answer: 28, -1.

7.5 Step Response of an *RC* Circuit

When the dc source of an *RC* circuit is suddenly applied, the voltage or current source can be modeled as a step function, and the response is known as a *step response*.

The **step response** of a circuit is its behavior when the excitation is the step function, which may be a voltage or a current source.

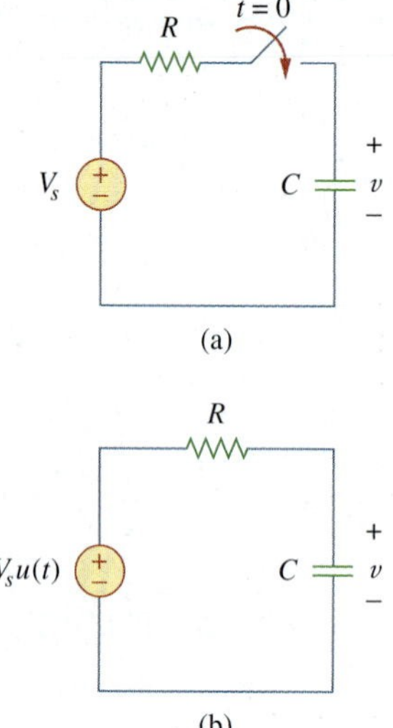

Figure 7.40
An RC circuit with voltage step input.

The step response is the response of the circuit due to a sudden application of a dc voltage or current source.

Consider the RC circuit in Fig. 7.40(a) which can be replaced by the circuit in Fig. 7.40(b), where V_s is a constant dc voltage source. Again, we select the capacitor voltage as the circuit response to be determined. We assume an initial voltage V_0 on the capacitor, although this is not necessary for the step response. Since the voltage of a capacitor cannot change instantaneously,

$$v(0^-) = v(0^+) = V_0 \tag{7.40}$$

where $v(0^-)$ is the voltage across the capacitor just before switching and $v(0^+)$ is its voltage immediately after switching. Applying KCL, we have

$$C\frac{dv}{dt} + \frac{v - V_s u(t)}{R} = 0$$

or

$$\frac{dv}{dt} + \frac{v}{RC} = \frac{V_s}{RC}u(t) \tag{7.41}$$

where v is the voltage across the capacitor. For $t > 0$, Eq. (7.41) becomes

$$\frac{dv}{dt} + \frac{v}{RC} = \frac{V_s}{RC} \tag{7.42}$$

Rearranging terms gives

$$\frac{dv}{dt} = -\frac{v - V_s}{RC}$$

or

$$\frac{dv}{v - V_s} = -\frac{dt}{RC} \tag{7.43}$$

Integrating both sides and introducing the initial conditions,

$$\ln(v - V_s)\Big|_{V_0}^{v(t)} = -\frac{t}{RC}\Big|_0^t$$

$$\ln(v(t) - V_s) - \ln(V_0 - V_s) = -\frac{t}{RC} + 0$$

or

$$\ln\frac{v - V_s}{V_0 - V_s} = -\frac{t}{RC} \tag{7.44}$$

Taking the exponential of both sides

$$\frac{v - V_s}{V_0 - V_s} = e^{-t/\tau}, \qquad \tau = RC$$

$$v - V_s = (V_0 - V_s)e^{-t/\tau}$$

or

$$v(t) = V_s + (V_0 - V_s)e^{-t/\tau}, \qquad t > 0 \tag{7.45}$$

Thus,

$$v(t) = \begin{cases} V_0, & t < 0 \\ V_s + (V_0 - V_s)e^{-t/\tau}, & t > 0 \end{cases} \tag{7.46}$$

This is known as the *complete response* (or total response) of the *RC* circuit to a sudden application of a dc voltage source, assuming the capacitor is initially charged. The reason for the term "complete" will become evident a little later. Assuming that $V_s > V_0$, a plot of $v(t)$ is shown in Fig. 7.41.

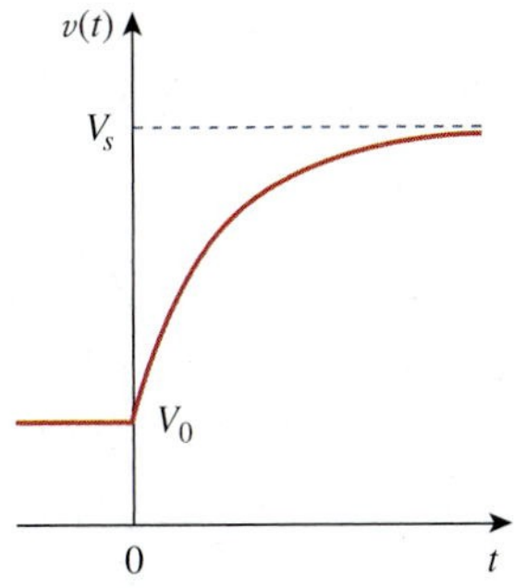

Figure 7.41
Response of an *RC* circuit with initially charged capacitor.

If we assume that the capacitor is uncharged initially, we set $V_0 = 0$ in Eq. (7.46) so that

$$v(t) = \begin{cases} 0, & t < 0 \\ V_s(1 - e^{-t/\tau}), & t > 0 \end{cases} \tag{7.47}$$

which can be written alternatively as

$$v(t) = V_s(1 - e^{-t/\tau})u(t) \tag{7.48}$$

This is the complete step response of the *RC* circuit when the capacitor is initially uncharged. The current through the capacitor is obtained from Eq. (7.47) using $i(t) = C dv/dt$. We get

$$i(t) = C\frac{dv}{dt} = \frac{C}{\tau}V_s e^{-t/\tau}, \qquad \tau = RC, \qquad t > 0$$

or

$$i(t) = \frac{V_s}{R}e^{-t/\tau}u(t) \tag{7.49}$$

Figure 7.42 shows the plots of capacitor voltage $v(t)$ and capacitor current $i(t)$.

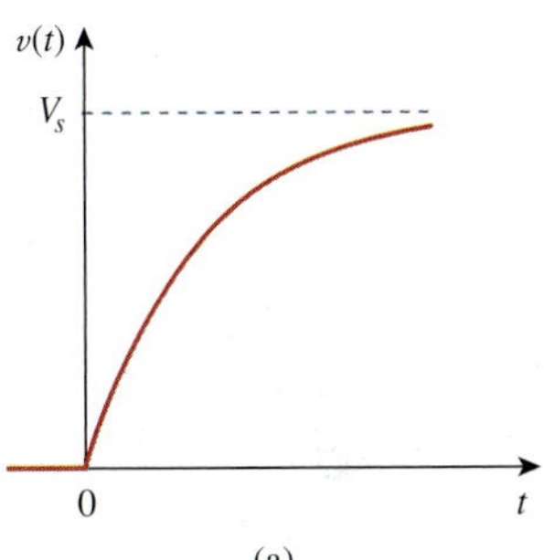

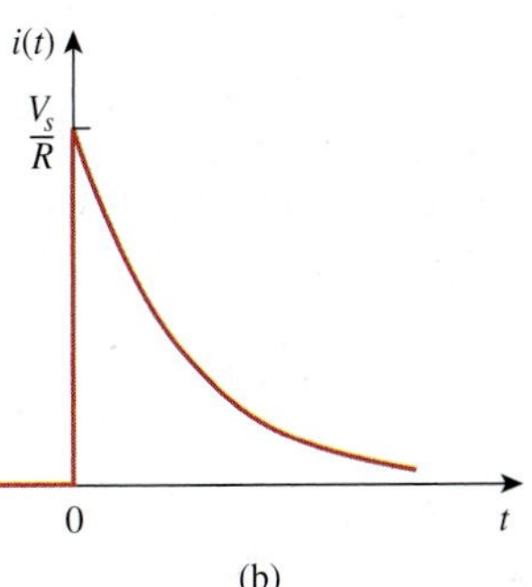

Figure 7.42
Step response of an *RC* circuit with initially uncharged capacitor: (a) voltage response, (b) current response.

Rather than going through the derivations above, there is a systematic approach—or rather, a short-cut method—for finding the step response of an *RC* or *RL* circuit. Let us reexamine Eq. (7.45), which is more general than Eq. (7.48). It is evident that $v(t)$ has two components. Classically there are two ways of decomposing this into two components. The first is to break it into a "natural response and a forced response" and the second is to break it into a "transient response and a steady-state response." Starting with the natural response and forced response, we write the total or complete response as

$$\text{Complete response} = \underset{\text{stored energy}}{\text{natural response}} + \underset{\text{independent source}}{\text{forced response}}$$

or

$$v = v_n + v_f \tag{7.50}$$

where

$$v_n = V_o e^{-t/\tau}$$

and

$$v_f = V_s(1 - e^{-t/\tau})$$

We are familiar with the natural response v_n of the circuit, as discussed in Section 7.2. v_f is known as the *forced* response because it is produced by the circuit when an external "force" (a voltage source in this case) is applied. It represents what the circuit is forced to do by the input excitation. The natural response eventually dies out along with the transient component of the forced response, leaving only the steady-state component of the forced response.

Another way of looking at the complete response is to break into two components—one temporary and the other permanent, i.e.,

$$\underbrace{\text{Complete response}}_{} = \underbrace{\text{transient response}}_{\text{temporary part}} + \underbrace{\text{steady-state response}}_{\text{permanent part}}$$

or

$$v = v_t + v_{ss} \tag{7.51}$$

where

$$v_t = (V_o - V_s)e^{-t/\tau} \tag{7.52a}$$

and

$$v_{ss} = V_s \tag{7.52b}$$

The *transient response* v_t is temporary; it is the portion of the complete response that decays to zero as time approaches infinity. Thus,

The **transient response** is the circuit's temporary response that will die out with time.

The *steady-state response* v_{ss} is the portion of the complete response that remains after the transient reponse has died out. Thus,

The **steady-state response** is the behavior of the circuit a long time after an external excitation is applied.

The first decomposition of the complete response is in terms of the source of the responses, while the second decomposition is in terms of the permanency of the responses. Under certain conditions, the natural response and transient response are the same. The same can be said about the forced response and steady-state response.

Whichever way we look at it, the complete response in Eq. (7.45) may be written as

This is the same as saying that the complete response is the sum of the transient response and the steady-state response.

$$v(t) = v(\infty) + [v(0) - v(\infty)]e^{-t/\tau} \tag{7.53}$$

where $v(0)$ is the initial voltage at $t = 0^+$ and $v(\infty)$ is the final or steady-state value. Thus, to find the step response of an *RC* circuit requires three things:

1. The initial capacitor voltage $v(0)$.
2. The final capacitor voltage $v(\infty)$.
3. The time constant τ.

Once we know $x(0)$, $x(\infty)$, and τ, almost all the circuit problems in this chapter can be solved using the formula

$$x(t) = x(\infty) + [x(0) - x(\infty)]e^{-t/\tau}$$

We obtain item 1 from the given circuit for $t < 0$ and items 2 and 3 from the circuit for $t > 0$. Once these items are determined, we obtain the response using Eq. (7.53). This technique equally applies to *RL* circuits, as we shall see in the next section.

Note that if the switch changes position at time $t = t_0$ instead of at $t = 0$, there is a time delay in the response so that Eq. (7.53) becomes

$$v(t) = v(\infty) + [v(t_0) - v(\infty)]e^{-(t-t_0)/\tau} \tag{7.54}$$

where $v(t_0)$ is the initial value at $t = t_0^+$. Keep in mind that Eq. (7.53) or (7.54) applies only to step responses, that is, when the input excitation is constant.

Example 7.10

The switch in Fig. 7.43 has been in position *A* for a long time. At $t = 0$, the switch moves to *B*. Determine $v(t)$ for $t > 0$ and calculate its value at $t = 1$ s and 4 s.

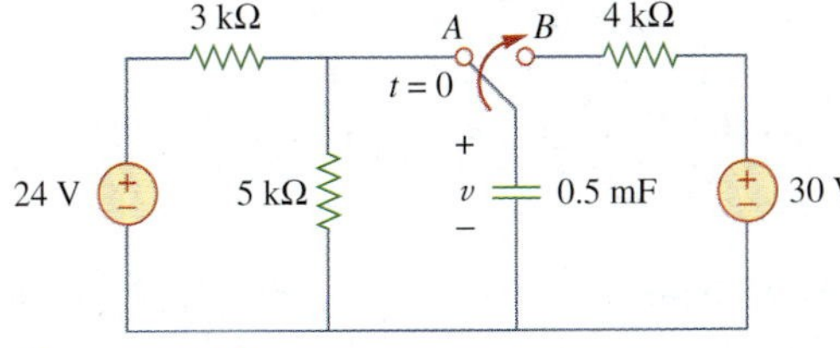

Figure 7.43
For Example 7.10.

Solution:
For $t < 0$, the switch is at position *A*. The capacitor acts like an open circuit to dc, but v is the same as the voltage across the 5-kΩ resistor. Hence, the voltage across the capacitor just before $t = 0$ is obtained by voltage division as

$$v(0^-) = \frac{5}{5+3}(24) = 15 \text{ V}$$

Using the fact that the capacitor voltage cannot change instantaneously,

$$v(0) = v(0^-) = v(0^+) = 15 \text{ V}$$

For $t > 0$, the switch is in position *B*. The Thevenin resistance connected to the capacitor is $R_{Th} = 4$ kΩ, and the time constant is

$$\tau = R_{Th}C = 4 \times 10^3 \times 0.5 \times 10^{-3} = 2 \text{ s}$$

Since the capacitor acts like an open circuit to dc at steady state, $v(\infty) = 30$ V. Thus,

$$\begin{aligned} v(t) &= v(\infty) + [v(0) - v(\infty)]e^{-t/\tau} \\ &= 30 + (15 - 30)e^{-t/2} = (30 - 15e^{-0.5t}) \text{ V} \end{aligned}$$

At $t = 1$,

$$v(1) = 30 - 15e^{-0.5} = 20.9 \text{ V}$$

At $t = 4$,

$$v(4) = 30 - 15e^{-2} = 27.97 \text{ V}$$

Practice Problem 7.10

Find $v(t)$ for $t > 0$ in the circuit of Fig. 7.44. Assume the switch has been open for a long time and is closed at $t = 0$. Calculate $v(t)$ at $t = 0.5$.

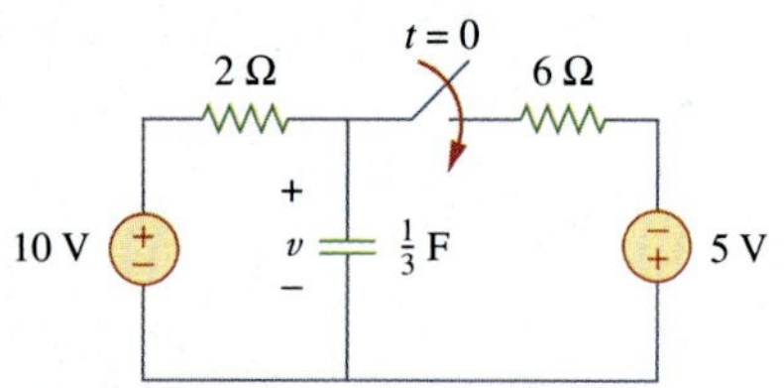

Figure 7.44
For Practice Prob. 7.10.

Answer: $(6.25 + 3.75e^{-2t})$ V for all $t > 0$, 7.63 V.

Example 7.11

In Fig. 7.45, the switch has been closed for a long time and is opened at $t = 0$. Find i and v for all time.

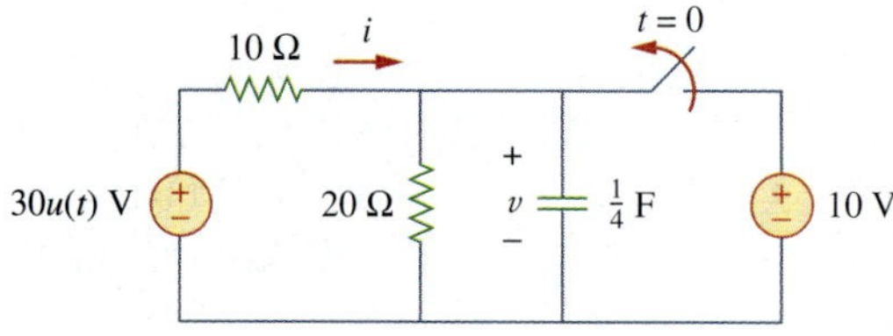

Figure 7.45
For Example 7.11.

Solution:
The resistor current i can be discontinuous at $t = 0$, while the capacitor voltage v cannot. Hence, it is always better to find v and then obtain i from v.

By definition of the unit step function,

$$30u(t) = \begin{cases} 0, & t < 0 \\ 30, & t > 0 \end{cases}$$

For $t < 0$, the switch is closed and $30u(t) = 0$, so that the $30u(t)$ voltage source is replaced by a short circuit and should be regarded as contributing nothing to v. Since the switch has been closed for a long time, the capacitor voltage has reached steady state and the capacitor acts like an open circuit. Hence, the circuit becomes that shown in Fig. 7.46(a) for $t < 0$. From this circuit we obtain

$$v = 10 \text{ V}, \qquad i = -\frac{v}{10} = -1 \text{ A}$$

Since the capacitor voltage cannot change instantaneously,

$$v(0) = v(0^-) = 10 \text{ V}$$

For $t > 0$, the switch is opened and the 10-V voltage source is disconnected from the circuit. The $30u(t)$ voltage source is now operative, so the circuit becomes that shown in Fig. 7.46(b). After a long time, the circuit reaches steady state and the capacitor acts like an open circuit again. We obtain $v(\infty)$ by using voltage division, writing

$$v(\infty) = \frac{20}{20 + 10}(30) = 20 \text{ V}$$

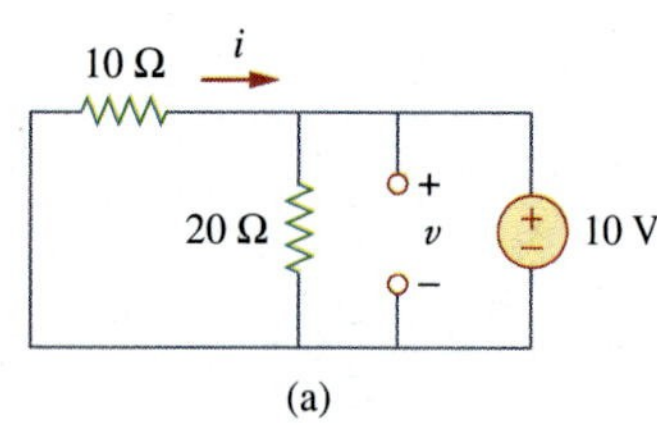

(a)

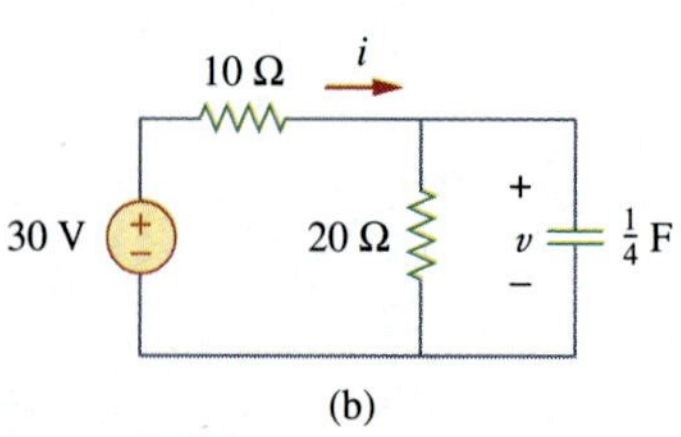

(b)

Figure 7.46
Solution of Example 7.11: (a) for $t < 0$, (b) for $t > 0$.

The Thevenin resistance at the capacitor terminals is

$$R_{Th} = 10 \parallel 20 = \frac{10 \times 20}{30} = \frac{20}{3}\ \Omega$$

and the time constant is

$$\tau = R_{Th}C = \frac{20}{3} \cdot \frac{1}{4} = \frac{5}{3}\ \text{s}$$

Thus,

$$\begin{aligned} v(t) &= v(\infty) + [v(0) - v(\infty)]e^{-t/\tau} \\ &= 20 + (10 - 20)e^{-(3/5)t} = (20 - 10e^{-0.6t})\ \text{V} \end{aligned}$$

To obtain i, we notice from Fig. 7.46(b) that i is the sum of the currents through the 20-Ω resistor and the capacitor; that is,

$$\begin{aligned} i &= \frac{v}{20} + C\frac{dv}{dt} \\ &= 1 - 0.5e^{-0.6t} + 0.25(-0.6)(-10)e^{-0.6t} = (1 + e^{-0.6t})\ \text{A} \end{aligned}$$

Notice from Fig. 7.46(b) that $v + 10i = 30$ is satisfied, as expected. Hence,

$$v = \begin{cases} 10\ \text{V}, & t < 0 \\ (20 - 10e^{-0.6t})\ \text{V}, & t \geq 0 \end{cases}$$

$$i = \begin{cases} -1\ \text{A}, & t < 0 \\ (1 + e^{-0.6t})\ \text{A}, & t > 0 \end{cases}$$

Notice that the capacitor voltage is continuous while the resistor current is not.

Practice Problem 7.11

The switch in Fig. 7.47 is closed at $t = 0$. Find $i(t)$ and $v(t)$ for all time. Note that $u(-t) = 1$ for $t < 0$ and 0 for $t > 0$. Also, $u(-t) = 1 - u(t)$.

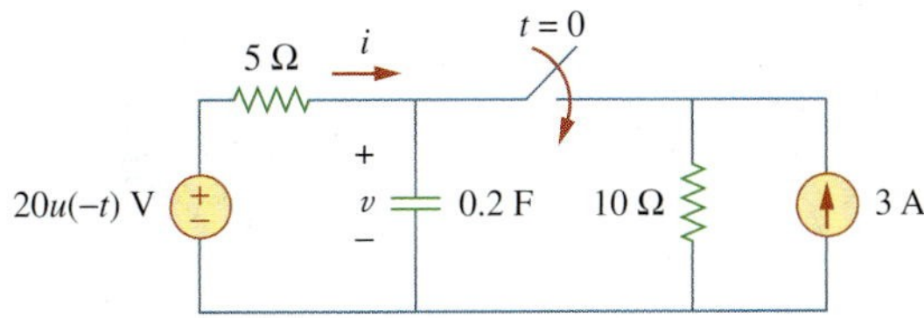

Figure 7.47
For Practice Prob. 7.11.

Answer: $i(t) = \begin{cases} 0, & t < 0 \\ -2(1 + e^{-1.5t})\ \text{A}, & t > 0 \end{cases}$,

$$v = \begin{cases} 20\ \text{V}, & t < 0 \\ 10(1 + e^{-1.5t})\ \text{V}, & t > 0 \end{cases}$$

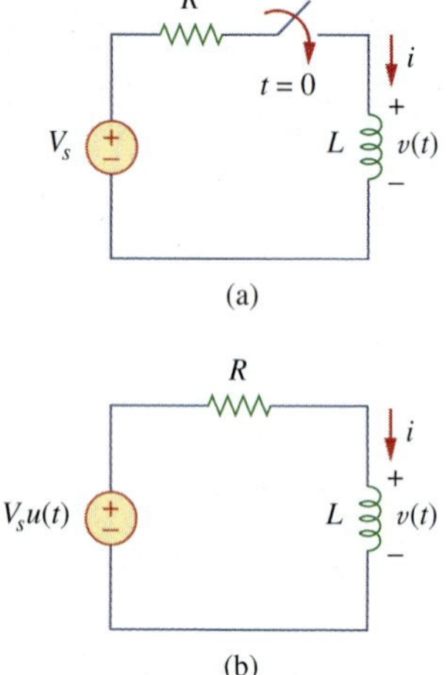

Figure 7.48
An *RL* circuit with a step input voltage.

7.6 Step Response of an *RL* Circuit

Consider the *RL* circuit in Fig. 7.48(a), which may be replaced by the circuit in Fig. 7.48(b). Again, our goal is to find the inductor current i as the circuit response. Rather than apply Kirchhoff's laws, we will use the simple technique in Eqs. (7.50) through (7.53). Let the response be the sum of the transient response and the steady-state response,

$$i = i_t + i_{ss} \tag{7.55}$$

We know that the transient response is always a decaying exponential, that is,

$$i_t = Ae^{-t/\tau}, \qquad \tau = \frac{L}{R} \tag{7.56}$$

where A is a constant to be determined.

The steady-state response is the value of the current a long time after the switch in Fig. 7.48(a) is closed. We know that the transient response essentially dies out after five time constants. At that time, the inductor becomes a short circuit, and the voltage across it is zero. The entire source voltage V_s appears across R. Thus, the steady-state response is

$$i_{ss} = \frac{V_s}{R} \tag{7.57}$$

Substituting Eqs. (7.56) and (7.57) into Eq. (7.55) gives

$$i = Ae^{-t/\tau} + \frac{V_s}{R} \tag{7.58}$$

We now determine the constant A from the initial value of i. Let I_0 be the initial current through the inductor, which may come from a source other than V_s. Since the current through the inductor cannot change instantaneously,

$$i(0^+) = i(0^-) = I_0 \tag{7.59}$$

Thus, at $t = 0$, Eq. (7.58) becomes

$$I_0 = A + \frac{V_s}{R}$$

From this, we obtain A as

$$A = I_0 - \frac{V_s}{R}$$

Substituting for A in Eq. (7.58), we get

$$i(t) = \frac{V_s}{R} + \left(I_0 - \frac{V_s}{R}\right)e^{-t/\tau} \tag{7.60}$$

This is the complete response of the *RL* circuit. It is illustrated in Fig. 7.49. The response in Eq. (7.60) may be written as

$$\boxed{i(t) = i(\infty) + [i(0) - i(\infty)]e^{-t/\tau}} \tag{7.61}$$

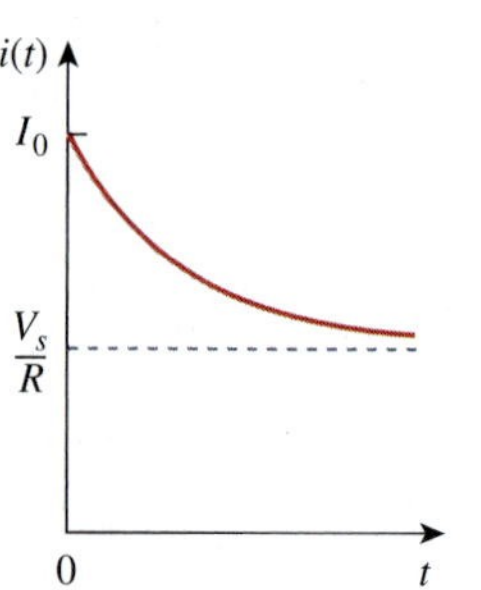

Figure 7.49
Total response of the *RL* circuit with initial inductor current I_0.

where $i(0)$ and $i(\infty)$ are the initial and final values of i, respectively. Thus, to find the step response of an *RL* circuit requires three things:

1. The initial inductor current $i(0)$ at $t = 0$.
2. The final inductor current $i(\infty)$.
3. The time constant τ.

We obtain item 1 from the given circuit for $t < 0$ and items 2 and 3 from the circuit for $t > 0$. Once these items are determined, we obtain the response using Eq. (7.61). Keep in mind that this technique applies only for step responses.

Again, if the switching takes place at time $t = t_0$ instead of $t = 0$, Eq. (7.61) becomes

$$i(t) = i(\infty) + [i(t_0) - i(\infty)]e^{-(t-t_0)/\tau} \tag{7.62}$$

If $I_0 = 0$, then

$$i(t) = \begin{cases} 0, & t < 0 \\ \dfrac{V_s}{R}(1 - e^{-t/\tau}), & t > 0 \end{cases} \tag{7.63a}$$

or

$$i(t) = \frac{V_s}{R}(1 - e^{-t/\tau})u(t) \tag{7.63b}$$

This is the step response of the *RL* circuit with no initial inductor current. The voltage across the inductor is obtained from Eq. (7.63) using $v = L\,di/dt$. We get

$$v(t) = L\frac{di}{dt} = V_s\frac{L}{\tau R}e^{-t/\tau}, \qquad \tau = \frac{L}{R}, \qquad t > 0$$

or

$$v(t) = V_s e^{-t/\tau}u(t) \tag{7.64}$$

Figure 7.50 shows the step responses in Eqs. (7.63) and (7.64).

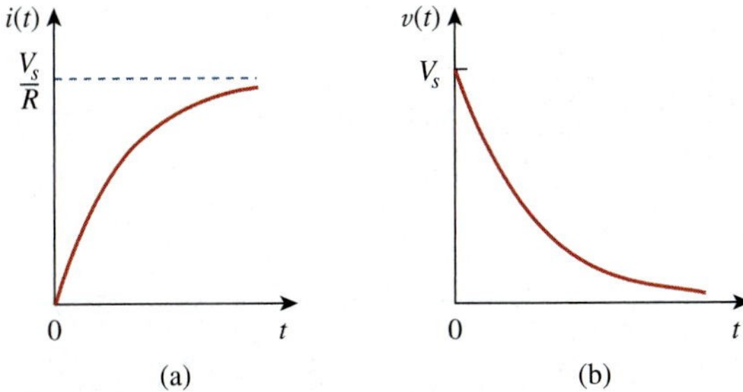

Figure 7.50
Step responses of an *RL* circuit with no initial inductor current: (a) current response, (b) voltage response.

Example 7.12

Find $i(t)$ in the circuit of Fig. 7.51 for $t > 0$. Assume that the switch has been closed for a long time.

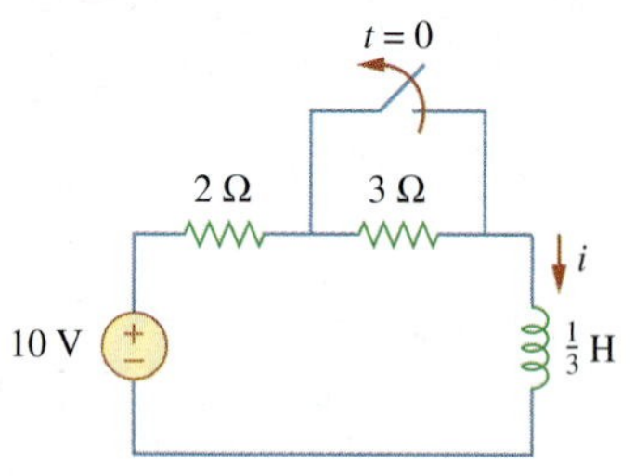

Figure 7.51
For Example 7.12.

Solution:

When $t < 0$, the 3-Ω resistor is short-circuited, and the inductor acts like a short circuit. The current through the inductor at $t = 0^-$ (i.e., just before $t = 0$) is

$$i(0^-) = \frac{10}{2} = 5 \text{ A}$$

Since the inductor current cannot change instantaneously,

$$i(0) = i(0^+) = i(0^-) = 5 \text{ A}$$

When $t > 0$, the switch is open. The 2-Ω and 3-Ω resistors are in series, so that

$$i(\infty) = \frac{10}{2 + 3} = 2 \text{ A}$$

The Thevenin resistance across the inductor terminals is

$$R_{\text{Th}} = 2 + 3 = 5\ \Omega$$

For the time constant,

$$\tau = \frac{L}{R_{\text{Th}}} = \frac{\frac{1}{3}}{5} = \frac{1}{15} \text{ s}$$

Thus,

$$\begin{aligned} i(t) &= i(\infty) + [i(0) - i(\infty)]e^{-t/\tau} \\ &= 2 + (5 - 2)e^{-15t} = 2 + 3e^{-15t} \text{ A}, \qquad t > 0 \end{aligned}$$

Check: In Fig. 7.51, for $t > 0$, KVL must be satisfied; that is,

$$10 = 5i + L\frac{di}{dt}$$

$$5i + L\frac{di}{dt} = [10 + 15e^{-15t}] + \left[\frac{1}{3}(3)(-15)e^{-15t}\right] = 10$$

This confirms the result.

Practice Problem 7.12

The switch in Fig. 7.52 has been closed for a long time. It opens at $t = 0$. Find $i(t)$ for $t > 0$.

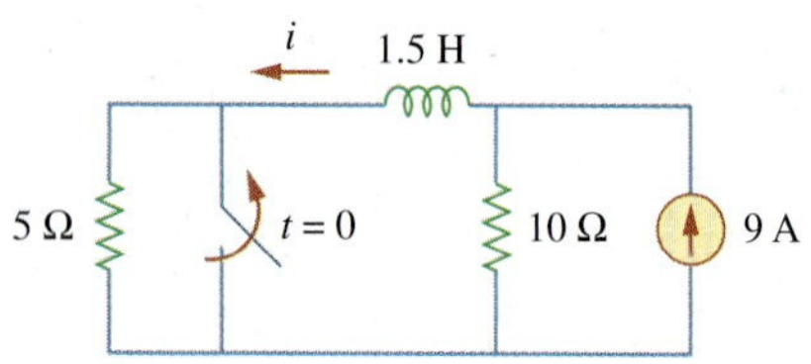

Figure 7.52
For Practice Prob. 7.12.

Answer: $(6 + 3e^{-10t})$ A for all $t > 0$.

Example 7.13

At $t = 0$, switch 1 in Fig. 7.53 is closed, and switch 2 is closed 4 s later. Find $i(t)$ for $t > 0$. Calculate i for $t = 2$ s and $t = 5$ s.

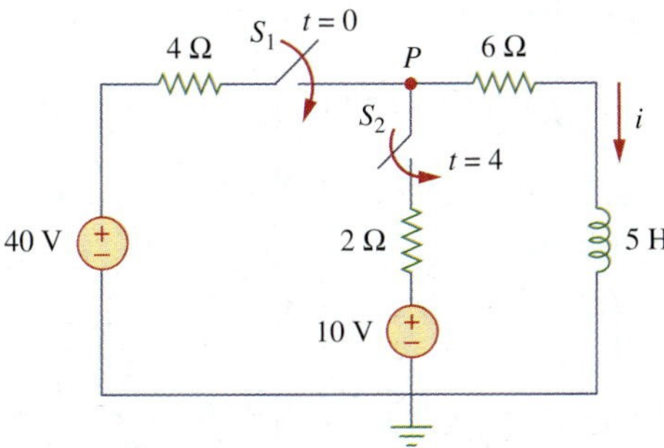

Figure 7.53
For Example 7.13.

Solution:
We need to consider the three time intervals $t \leq 0, 0 \leq t \leq 4$, and $t \geq 4$ separately. For $t < 0$, switches S_1 and S_2 are open so that $i = 0$. Since the inductor current cannot change instantly,

$$i(0^-) = i(0) = i(0^+) = 0$$

For $0 \leq t \leq 4$, S_1 is closed so that the 4-Ω and 6-Ω resistors are in series. (Remember, at this time, S_2 is still open.) Hence, assuming for now that S_1 is closed forever,

$$i(\infty) = \frac{40}{4+6} = 4 \text{ A}, \qquad R_{\text{Th}} = 4 + 6 = 10\ \Omega$$

$$\tau = \frac{L}{R_{\text{Th}}} = \frac{5}{10} = \frac{1}{2} \text{ s}$$

Thus,

$$\begin{aligned} i(t) &= i(\infty) + [i(0) - i(\infty)]e^{-t/\tau} \\ &= 4 + (0 - 4)e^{-2t} = 4(1 - e^{-2t}) \text{ A}, \qquad 0 \leq t \leq 4 \end{aligned}$$

For $t \geq 4$, S_2 is closed; the 10-V voltage source is connected, and the circuit changes. This sudden change does not affect the inductor current because the current cannot change abruptly. Thus, the initial current is

$$i(4) = i(4^-) = 4(1 - e^{-8}) \simeq 4 \text{ A}$$

To find $i(\infty)$, let v be the voltage at node P in Fig. 7.53. Using KCL,

$$\frac{40 - v}{4} + \frac{10 - v}{2} = \frac{v}{6} \quad \Rightarrow \quad v = \frac{180}{11} \text{ V}$$

$$i(\infty) = \frac{v}{6} = \frac{30}{11} = 2.727 \text{ A}$$

The Thevenin resistance at the inductor terminals is

$$R_{\text{Th}} = 4 \parallel 2 + 6 = \frac{4 \times 2}{6} + 6 = \frac{22}{3}\ \Omega$$

and

$$\tau = \frac{L}{R_{\text{Th}}} = \frac{5}{\frac{22}{3}} = \frac{15}{22} \text{ s}$$

Hence,

$$i(t) = i(\infty) + [i(4) - i(\infty)]e^{-(t-4)/\tau}, \qquad t \geq 4$$

We need $(t - 4)$ in the exponential because of the time delay. Thus,

$$i(t) = 2.727 + (4 - 2.727)e^{-(t-4)/\tau}, \qquad \tau = \frac{15}{22}$$

$$= 2.727 + 1.273e^{-1.4667(t-4)}, \qquad t \geq 4$$

Putting all this together,

$$i(t) = \begin{cases} 0, & t \leq 0 \\ 4(1 - e^{-2t}), & 0 \leq t \leq 4 \\ 2.727 + 1.273e^{-1.4667(t-4)}, & t \geq 4 \end{cases}$$

At $t = 2$,

$$i(2) = 4(1 - e^{-4}) = 3.93 \text{ A}$$

At $t = 5$,

$$i(5) = 2.727 + 1.273e^{-1.4667} = 3.02 \text{ A}$$

Practice Problem 7.13

Switch S_1 in Fig. 7.54 is closed at $t = 0$, and switch S_2 is closed at $t = 2$ s. Calculate $i(t)$ for all t. Find $i(1)$ and $i(3)$.

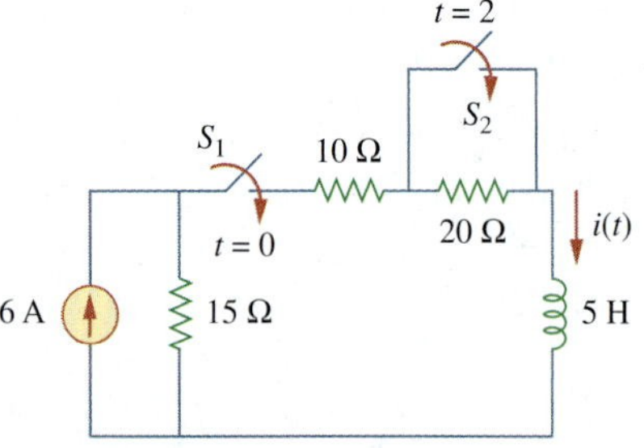

Figure 7.54
For Practice Prob. 7.13.

Answer:

$$i(t) = \begin{cases} 0, & t < 0 \\ 2(1 - e^{-9t}), & 0 < t < 2 \\ 3.6 - 1.6e^{-5(t-2)}, & t > 2 \end{cases}$$

$i(1) = 1.9997$ A, $i(3) = 3.589$ A.

7.7 † First-Order Op Amp Circuits

An op amp circuit containing a storage element will exhibit first-order behavior. Differentiators and integrators treated in Section 6.6 are examples of first-order op amp circuits. Again, for practical reasons, inductors are hardly ever used in op amp circuits; therefore, the op amp circuits we consider here are of the *RC* type.

As usual, we analyze op amp circuits using nodal analysis. Sometimes, the Thevenin equivalent circuit is used to reduce the op amp circuit to one that we can easily handle. The following three examples illustrate the concepts. The first one deals with a source-free op amp circuit, while the other two involve step responses. The three examples have been carefully selected to cover all possible *RC* types of op amp circuits, depending on the location of the capacitor with respect to the op amp; that is, the capacitor can be located in the input, the output, or the feedback loop.

Example 7.14

For the op amp circuit in Fig. 7.55(a), find v_o for $t > 0$, given that $v(0) = 3$ V. Let $R_f = 80$ kΩ, $R_1 = 20$ kΩ, and $C = 5\ \mu$F.

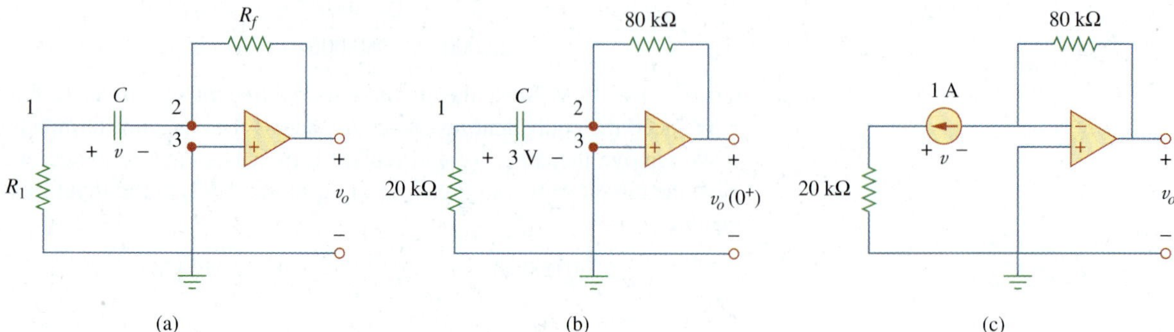

Figure 7.55
For Example 7.14.

Solution:
This problem can be solved in two ways:

■ **METHOD 1** Consider the circuit in Fig. 7.55(a). Let us derive the appropriate differential equation using nodal analysis. If v_1 is the voltage at node 1, at that node, KCL gives

$$\frac{0 - v_1}{R_1} = C\frac{dv}{dt} \tag{7.14.1}$$

Since nodes 2 and 3 must be at the same potential, the potential at node 2 is zero. Thus, $v_1 - 0 = v$ or $v_1 = v$ and Eq. (7.14.1) becomes

$$\frac{dv}{dt} + \frac{v}{CR_1} = 0 \tag{7.14.2}$$

This is similar to Eq. (7.4b) so that the solution is obtained the same way as in Section 7.2, i.e.,

$$v(t) = V_0 e^{-t/\tau}, \qquad \tau = R_1 C \tag{7.14.3}$$

where V_0 is the initial voltage across the capacitor. But $v(0) = 3 = V_0$ and $\tau = 20 \times 10^3 \times 5 \times 10^{-6} = 0.1$. Hence,

$$v(t) = 3e^{-10t} \tag{7.14.4}$$

Applying KCL at node 2 gives

$$C\frac{dv}{dt} = \frac{0 - v_o}{R_f}$$

or

$$v_o = -R_f C\frac{dv}{dt} \tag{7.14.5}$$

Now we can find v_0 as

$$v_o = -80 \times 10^3 \times 5 \times 10^{-6}(-30e^{-10t}) = 12e^{-10t}\ \text{V}, \qquad t > 0$$

■ **METHOD 2** Let us apply the short-cut method from Eq. (7.53). We need to find $v_o(0^+)$, $v_o(\infty)$, and τ. Since $v(0^+) = v(0^-) = 3$ V, we apply KCL at node 2 in the circuit of Fig. 7.55(b) to obtain

$$\frac{3}{20{,}000} + \frac{0 - v_o(0^+)}{80{,}000} = 0$$

or $v_o(0^+) = 12$ V. Since the circuit is source free, $v(\infty) = 0$ V. To find τ, we need the equivalent resistance R_{eq} across the capacitor terminals. If we remove the capacitor and replace it by a 1-A current source, we have the circuit shown in Fig. 7.55(c). Applying KVL to the input loop yields

$$20{,}000(1) - v = 0 \qquad \Rightarrow \qquad v = 20 \text{ kV}$$

Then

$$R_{\text{eq}} = \frac{v}{1} = 20 \text{ k}\Omega$$

and $\tau = R_{\text{eq}}C = 0.1$. Thus,

$$\begin{aligned} v_o(t) &= v_o(\infty) + [v_o(0) - v_o(\infty)]e^{-t/\tau} \\ &= 0 + (12 - 0)e^{-10t} = 12e^{-10t} \text{ V}, \qquad t > 0 \end{aligned}$$

as before.

Practice Problem 7.14

For the op amp circuit in Fig. 7.56, find v_o for $t > 0$ if $v(0) = 4$ V. Assume that $R_f = 50$ kΩ, $R_1 = 10$ kΩ, and $C = 10$ μF.

Answer: $-4e^{-2t}$ V, $t > 0$.

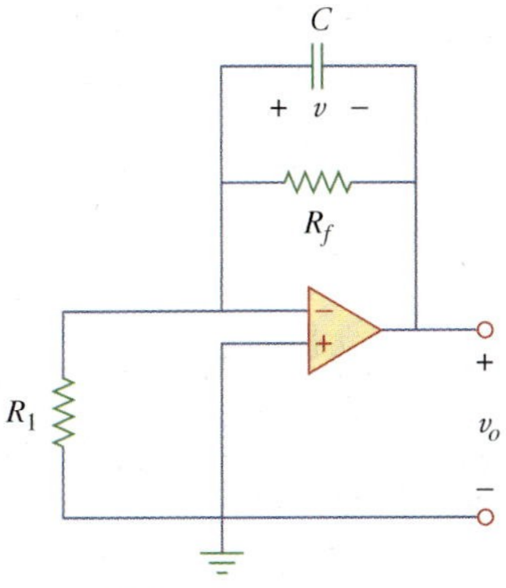

Figure 7.56
For Practice Prob. 7.14.

Example 7.15

Determine $v(t)$ and $v_o(t)$ in the circuit of Fig. 7.57.

Solution:
This problem can be solved in two ways, just like the previous example. However, we will apply only the second method. Since what we are looking for is the step response, we can apply Eq. (7.53) and write

$$v(t) = v(\infty) + [v(0) - v(\infty)]e^{-t/\tau}, \qquad t > 0 \qquad \textbf{(7.15.1)}$$

where we need only find the time constant τ, the initial value $v(0)$, and the final value $v(\infty)$. Notice that this applies strictly to the capacitor voltage due a step input. Since no current enters the input terminals of the op amp, the elements on the feedback loop of the op amp constitute an *RC* circuit, with

$$\tau = RC = 50 \times 10^3 \times 10^{-6} = 0.05 \quad \textbf{(7.15.2)}$$

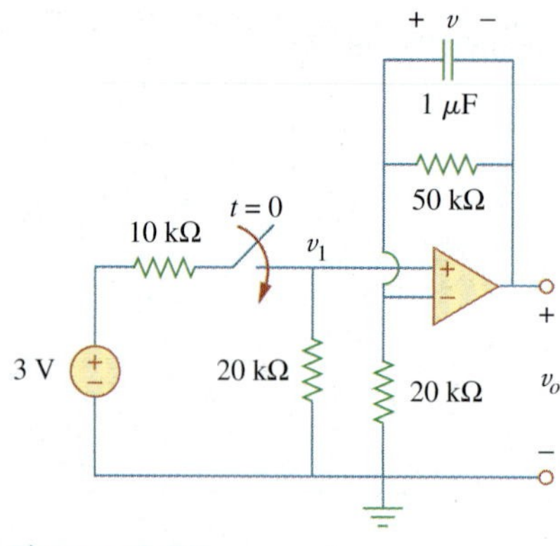

Figure 7.57
For Example 7.15.

For $t < 0$, the switch is open and there is no voltage across the capacitor. Hence, $v(0) = 0$. For $t > 0$, we obtain the voltage at node 1 by voltage division as

$$v_1 = \frac{20}{20 + 10} 3 = 2 \text{ V} \quad \textbf{(7.15.3)}$$

Since there is no storage element in the input loop, v_1 remains constant for all t. At steady state, the capacitor acts like an open circuit so that the op amp circuit is a noninverting amplifier. Thus,

$$v_o(\infty) = \left(1 + \frac{50}{20}\right) v_1 = 3.5 \times 2 = 7 \text{ V} \quad \textbf{(7.15.4)}$$

But

$$v_1 - v_o = v \quad \textbf{(7.15.5)}$$

so that

$$v(\infty) = 2 - 7 = -5 \text{ V}$$

Substituting τ, $v(0)$, and $v(\infty)$ into Eq. (7.15.1) gives

$$v(t) = -5 + [0 - (-5)]e^{-20t} = 5(e^{-20t} - 1) \text{ V}, \qquad t > 0 \quad \textbf{(7.15.6)}$$

From Eqs. (7.15.3), (7.15.5), and (7.15.6), we obtain

$$v_o(t) = v_1(t) - v(t) = 7 - 5e^{-20t} \text{ V}, \qquad t > 0 \quad \textbf{(7.15.7)}$$

Practice Problem 7.15

Find $v(t)$ and $v_o(t)$ in the op amp circuit of Fig. 7.58.

Answer: (Note, the voltage across the capacitor and the output voltage must be both equal to zero, for $t < 0$, since the input was zero for all $t < 0$.) $40(1 - e^{-10t})\, u(t)$ mV, $40(e^{-10t} - 1)\, u(t)$ mV.

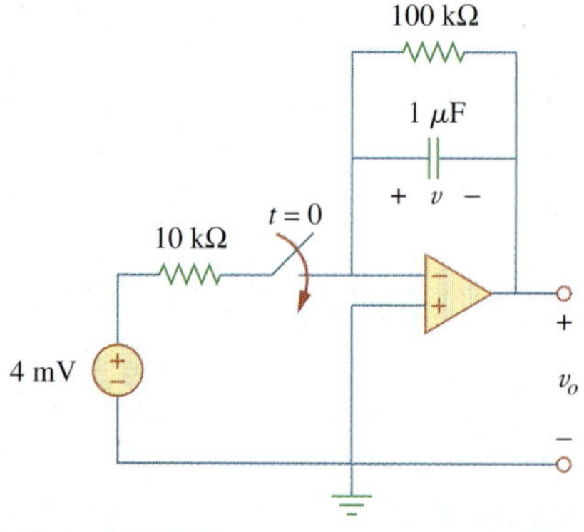

Figure 7.58
For Practice Prob. 7.15.

Example 7.16

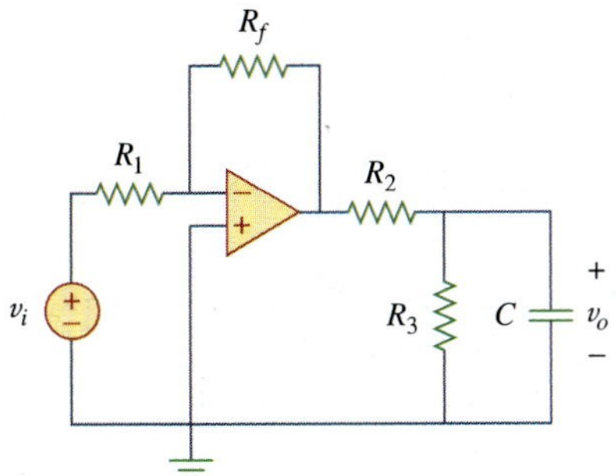

Figure 7.59
For Example 7.16.

Find the step response $v_o(t)$ for $t > 0$ in the op amp circuit of Fig. 7.59. Let $v_i = 2u(t)$ V, $R_1 = 20$ kΩ, $R_f = 50$ kΩ, $R_2 = R_3 = 10$ kΩ, $C = 2\ \mu$F.

Solution:

Notice that the capacitor in Example 7.14 is located in the input loop, while the capacitor in Example 7.15 is located in the feedback loop. In this example, the capacitor is located in the output of the op amp. Again, we can solve this problem directly using nodal analysis. However, using the Thevenin equivalent circuit may simplify the problem.

We temporarily remove the capacitor and find the Thevenin equivalent at its terminals. To obtain V_{Th}, consider the circuit in Fig. 7.60(a). Since the circuit is an inverting amplifier,

$$V_{ab} = -\frac{R_f}{R_1} v_i$$

By voltage division,

$$V_{Th} = \frac{R_3}{R_2 + R_3} V_{ab} = -\frac{R_3}{R_2 + R_3}\frac{R_f}{R_1} v_i$$

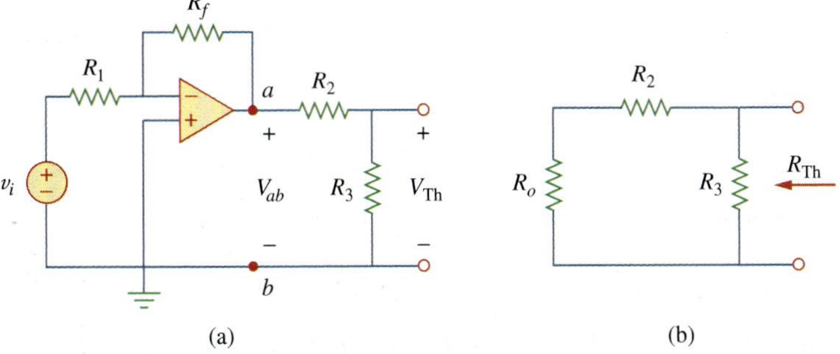

Figure 7.60
Obtaining V_{Th} and R_{Th} across the capacitor in Fig. 7.59.

To obtain R_{Th}, consider the circuit in Fig. 7.60(b), where R_o is the output resistance of the op amp. Since we are assuming an ideal op amp, $R_o = 0$, and

$$R_{Th} = R_2 \| R_3 = \frac{R_2 R_3}{R_2 + R_3}$$

Substituting the given numerical values,

$$V_{Th} = -\frac{R_3}{R_2 + R_3}\frac{R_f}{R_1} v_i = -\frac{10}{20}\frac{50}{20} 2u(t) = -2.5u(t)$$

$$R_{Th} = \frac{R_2 R_3}{R_2 + R_3} = 5\text{ k}\Omega$$

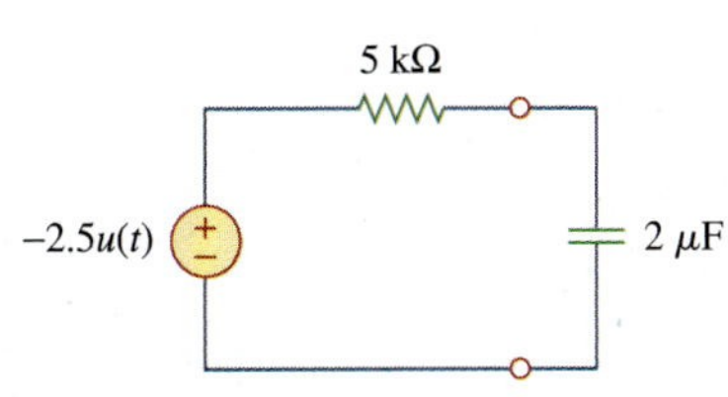

Figure 7.61
Thevenin equivalent circuit of the circuit in Fig. 7.59.

The Thevenin equivalent circuit is shown in Fig. 7.61, which is similar to Fig. 7.40. Hence, the solution is similar to that in Eq. (7.48); that is,

$$v_o(t) = -2.5(1 - e^{-t/\tau})u(t)$$

where $\tau = R_{Th}C = 5 \times 10^3 \times 2 \times 10^{-6} = 0.01$. Thus, the step response for $t > 0$ is

$$v_o(t) = 2.5(e^{-100t} - 1)u(t)\text{ V}$$

Practice Problem 7.16

Obtain the step response $v_o(t)$ for the circuit in Fig. 7.62. Let $v_i = 3u(t)$ V, $R_1 = 20$ kΩ, $R_f = 40$ kΩ, $R_2 = R_3 = 10$ kΩ, $C = 2$ μF.

Answer: $9(1 - e^{-50t})u(t)$ V.

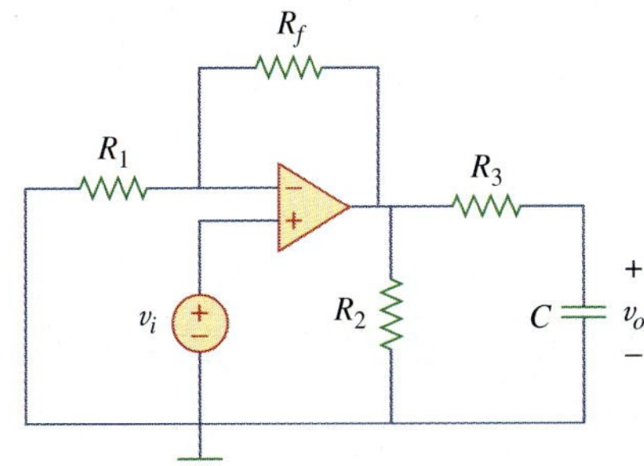

Figure 7.62
For Practice Prob. 7.16.

7.8 Transient Analysis with *PSpice*

As we discussed in Section 7.5, the transient response is the temporary response of the circuit that soon disappears. *PSpice* can be used to obtain the transient response of a circuit with storage elements. Section D.4 in Appendix D provides a review of transient analysis using *PSpice for Windows*. It is recommended that you read Section D.4 before continuing with this section.

If necessary, dc *PSpice* analysis is first carried out to determine the initial conditions. Then the initial conditions are used in the transient *PSpice* analysis to obtain the transient responses. It is recommended but not necessary that during this dc analysis, all capacitors should be open-circuited while all inductors should be short-circuited.

> *PSpice* uses "transient" to mean "function of time." Therefore, the transient response in *PSpice* may not actually die out as expected.

Example 7.17

Use *PSpice* to find the response $i(t)$ for $t > 0$ in the circuit of Fig. 7.63.

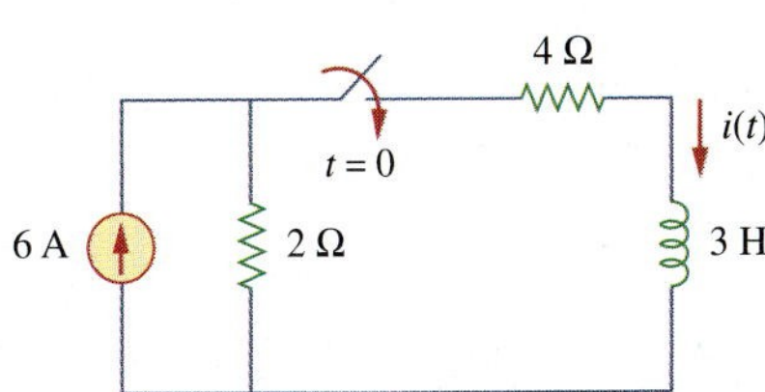

Figure 7.63
For Example 7.17.

Solution:
Solving this problem by hand gives $i(0) = 0$, $i(\infty) = 2$ A, $R_{\text{Th}} = 6$, $\tau = 3/6 = 0.5$ s, so that

$$i(t) = i(\infty) + [i(0) - i(\infty)]e^{-t/\tau} = 2(1 - e^{-2t}), \qquad t > 0$$

To use *PSpice*, we first draw the schematic as shown in Fig. 7.64. We recall from Appendix D that the part name for a closed switch is Sw_tclose. We do not need to specify the initial condition of the inductor because *PSpice* will determine that from the circuit. By selecting **Analysis/Setup/Transient**, we set *Print Step* to 25 ms and *Final Step* to $5\tau = 2.5$ s. After saving the circuit, we simulate by selecting **Analysis/Simulate**. In the *PSpice* A/D window, we select **Trace/Add** and display –I(L1) as the current through the inductor. Figure 7.65 shows the plot of $i(t)$, which agrees with that obtained by hand calculation.

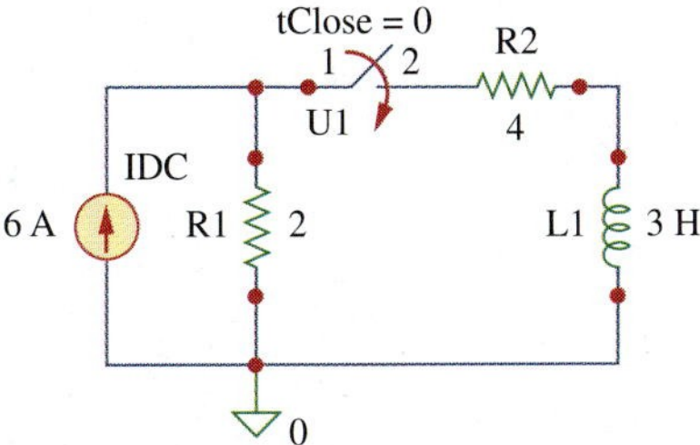

Figure 7.64
The schematic of the circuit in Fig. 7.63.

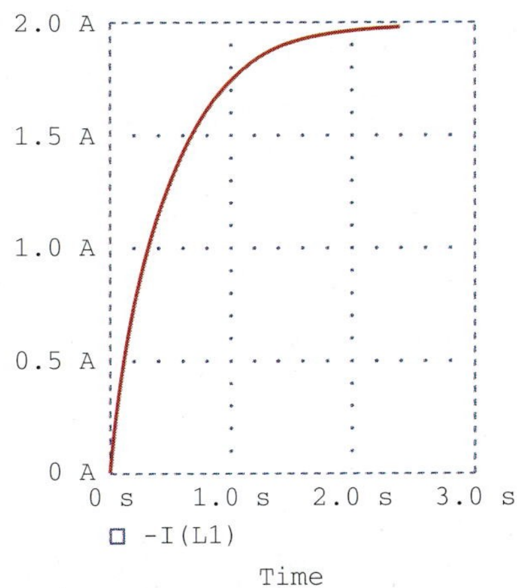

Figure 7.65
For Example 7.17; the response of the circuit in Fig. 7.63.

Note that the negative sign on I(L1) is needed because the current enters through the upper terminal of the inductor, which happens to be the negative terminal after one counterclockwise rotation. A way to avoid the negative sign is to ensure that current enters pin 1 of the inductor. To obtain this desired direction of positive current flow, the initially horizontal inductor symbol should be rotated counterclockwise 270° and placed in the desired location.

Practice Problem 7.17

For the circuit in Fig. 7.66, use *Pspice* to find $v(t)$ for $t > 0$.

Answer: $v(t) = 8(1 - e^{-t})$ V, $t > 0$. The response is similar in shape to that in Fig. 7.65.

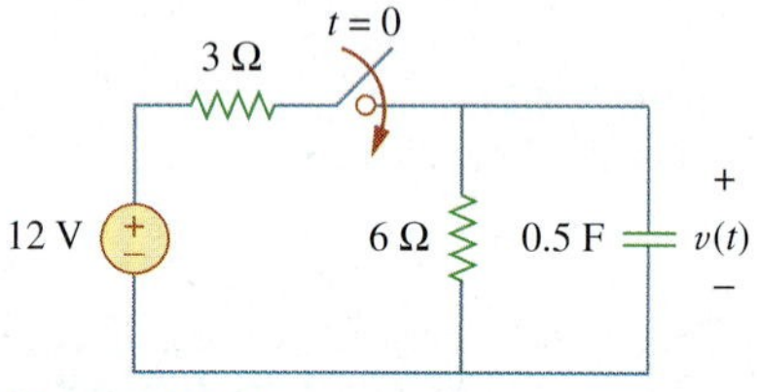

Figure 7.66
For Practice Prob. 7.17.

Example 7.18

In the circuit of Fig. 7.67(a), determine the response $v(t)$.

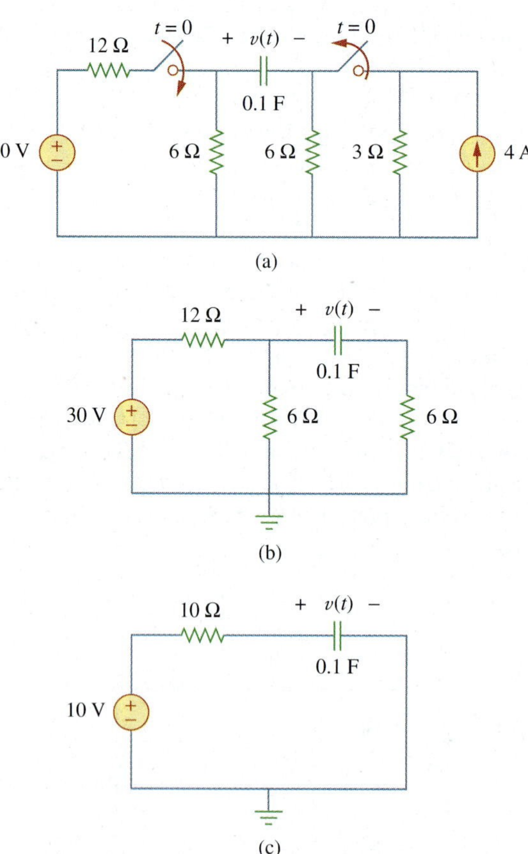

Figure 7.67
For Example 7.18. Original circuit (a), circuit for $t > 0$ (b), and reduced circuit for $t > 0$ (c).

Solution:

1. **Define.** The problem is clearly stated and the circuit is clearly labeled.
2. **Present.** Given the circuit shown in Fig. 7.67(a), determine the response $v(t)$.
3. **Alternative.** We can solve this circuit using circuit analysis techniques, nodal analysis, mesh analysis, or *PSpice*. Let us solve the problem using circuit analysis techniques (this time Thevenin equivalent circuits) and then check the answer using two methods of *PSpice*.
4. **Attempt.** For time <0, the switch on the left is open and the switch on the right is closed. Assume that the switch on the right has been closed long enough for the circuit to reach steady state; then the capacitor acts like an open circuit and the current from the 4-A source flows through the parallel combination of the 6-Ω and 3-Ω resistors ($6 \parallel 3 = 18/9 = 2$), producing a voltage equal to $2 \times 4 = 8\text{ V} = -v(0)$.

   At $t = 0$, the switch on the left closes and the switch on the right opens, producing the circuit shown in Fig. 7.67(b).

   The easiest way to complete the solution is to find the Thevenin equivalent circuit as seen by the capacitor. The open-circuit voltage (with the capacitor removed) is equal to the voltage drop across the 6-Ω resistor on the left, or 10 V (the voltage drops uniformly across the 12-Ω resistor, 20 V, and across the 6-Ω resistor, 10 V). This is V_{Th}. The resistance looking in where the capacitor was is equal to $12 \parallel 6 + 6 = 72/18 + 6 = 10\ \Omega$, which is R_{eq}. This produces the Thevenin equivalent circuit shown in Fig. 7.67(c). Matching up the boundary conditions ($v(0) = -8$ V and $v(\infty) = 10$ V) and $\tau = RC = 1$, we get

$$v(t) = \mathbf{10 - 18e^{-t}\ V}$$

5. **Evaluate.** There are two ways of solving the problem using *PSpice*.

■ **METHOD 1** One way is to first do the dc *PSpice* analysis to determine the initial capacitor voltage. The schematic of the revelant circuit is in Fig. 7.68(a). Two pseudocomponent VIEWPOINTs are inserted to measure the voltages at nodes 1 and 2. When the circuit is simulated, we obtain the displayed values in Fig. 7.68(a) as $V_1 = 0$ V and $V_2 = 8$ V. Thus, the initial capacitor voltage is $v(0) = V_1 - V_2 = -8$ V. The *PSpice* transient analysis uses this value along with the schematic in Fig. 7.68(b). Once the circuit in Fig. 7.68(b) is drawn, we insert the capacitor initial voltage as IC = −8. We select **Analysis/Setup/Transient** and set *Print Step* to 0.1 s and *Final Step* to $4\tau = 4$ s. After saving the circuit, we select **Analysis/Simulate** to simulate the circuit. In the *PSpice* A/D window, we select **Trace/Add** and display V(R2:2) − V(R3:2) or V(C1:1) − V(C1:2) as the capacitor voltage $v(t)$. The plot of $v(t)$ is shown in Fig. 7.69. This agrees with the result obtained by hand calculation, $v(t) = 10 - 18\,e^{-t}$ V.

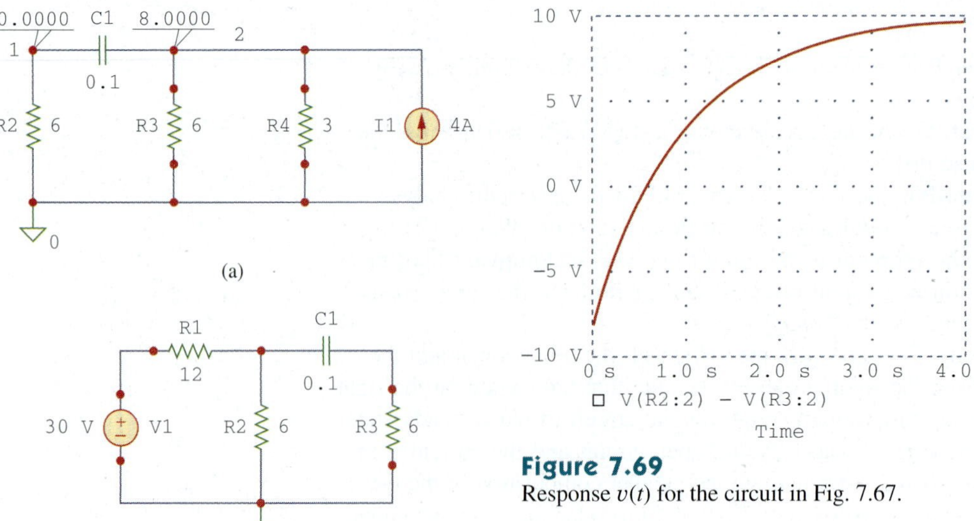

Figure 7.68
(a) Schematic for dc analysis to get $v(0)$, (b) schematic for transient analysis used in getting the response $v(t)$.

Figure 7.69
Response $v(t)$ for the circuit in Fig. 7.67.

■ METHOD 2 We can simulate the circuit in Fig. 7.67 directly, since *PSpice* can handle the open and closed switches and determine the initial conditions automatically. Using this approach, the schematic is drawn as shown in Fig. 7.70. After drawing the circuit, we select **Analysis/Setup/Transient** and set *Print Step* to 0.1 s and *Final Step* to $4\tau = 4$ s. We save the circuit, then select **Analysis/Simulate** to simulate the circuit. In the *PSpice* A/D window, we select **Trace/Add** and display V(R2:2) − V(R3:2) as the capacitor voltage $v(t)$. The plot of $v(t)$ is the same as that shown in Fig. 7.69.

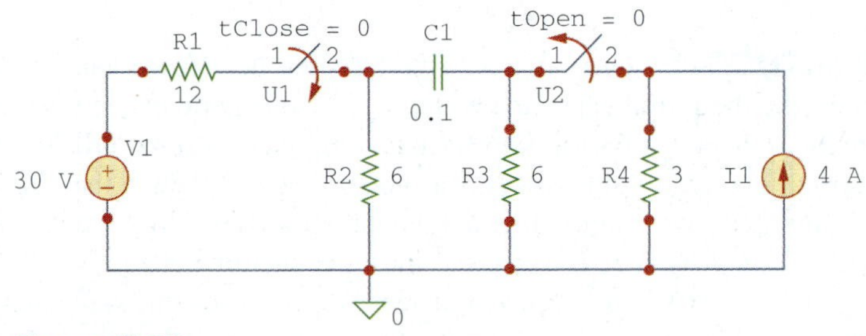

Figure 7.70
For Example 7.18.

6. **Satisfactory?** Clearly, we have found the value of the output response $v(t)$, as required by the problem statement. Checking does validate that solution. We can present all this as a complete solution to the problem.

Practice Problem 7.18

The switch in Fig. 7.71 was open for a long time but closed at $t = 0$. If $i(0) = 10$ A, find $i(t)$ for $t > 0$ by hand and also by *PSpice*.

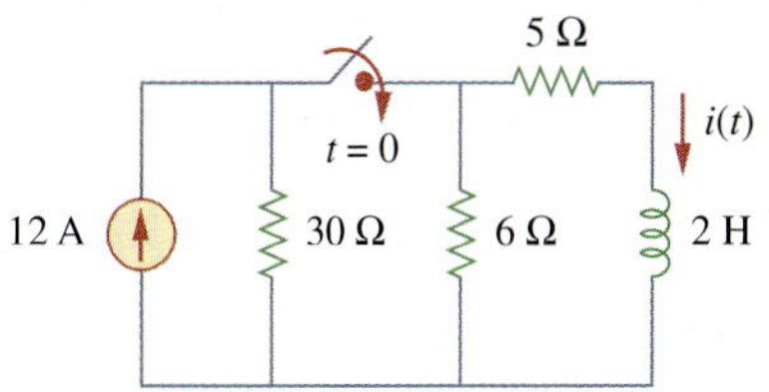

Figure 7.71
For Practice Prob. 7.18.

Answer: $i(t) = 6 + 4e^{-5t}$ A. The plot of $i(t)$ obtained by *PSpice* analysis is shown in Fig. 7.72.

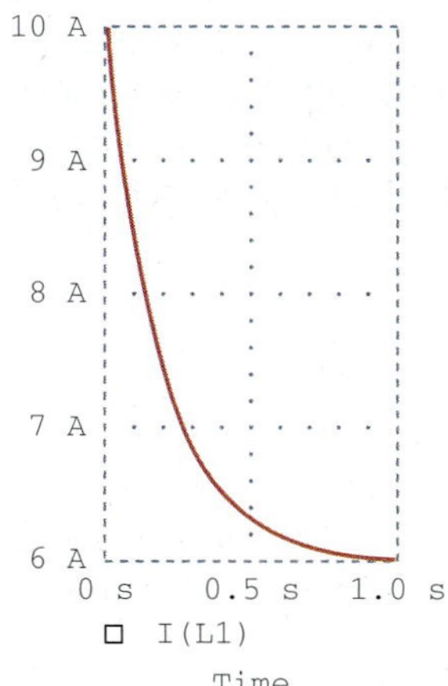

Figure 7.72
For Practice Prob. 7.18.

7.9 †Applications

The various devices in which *RC* and *RL* circuits find applications include filtering in dc power supplies, smoothing circuits in digital communications, differentiators, integrators, delay circuits, and relay circuits. Some of these applications take advantage of the short or long time constants of the *RC* or *RL* circuits. We will consider four simple applications here. The first two are *RC* circuits, the last two are *RL* circuits.

7.9.1 Delay Circuits

An *RC* circuit can be used to provide various time delays. Figure 7.73 shows such a circuit. It basically consists of an *RC* circuit with the capacitor connected in parallel with a neon lamp. The voltage source can provide enough voltage to fire the lamp. When the switch is closed, the capacitor voltage increases gradually toward 110 V at a rate determined by the circuit's time constant, $(R_1 + R_2)C$. The lamp will act as an open

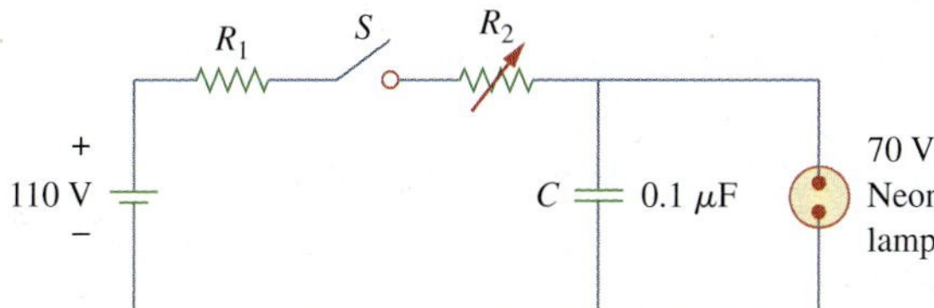

Figure 7.73
An *RC* delay circuit.

circuit and not emit light until the voltage across it exceeds a particular level, say 70 V. When the voltage level is reached, the lamp fires (goes on), and the capacitor discharges through it. Due to the low resistance of the lamp when on, the capacitor voltage drops fast and the lamp turns off. The lamp acts again as an open circuit and the capacitor recharges. By adjusting R_2, we can introduce either short or long time delays into the circuit and make the lamp fire, recharge, and fire repeatedly every time constant $\tau = (R_1 + R_2)C$, because it takes a time period τ to get the capacitor voltage high enough to fire or low enough to turn off.

The warning blinkers commonly found on road construction sites are one example of the usefulness of such an RC delay circuit.

Example 7.19

Consider the circuit in Fig. 7.73, and assume that $R_1 = 1.5\ \text{M}\Omega$, $0 < R_2 < 2.5\ \text{M}\Omega$. (a) Calculate the extreme limits of the time constant of the circuit. (b) How long does it take for the lamp to glow for the first time after the switch is closed? Let R_2 assume its largest value.

Solution:

(a) The smallest value for R_2 is 0 Ω, and the corresponding time constant for the circuit is

$$\tau = (R_1 + R_2)C = (1.5 \times 10^6 + 0) \times 0.1 \times 10^{-6} = 0.15\ \text{s}$$

The largest value for R_2 is 2.5 MΩ, and the corresponding time constant for the circuit is

$$\tau = (R_1 + R_2)C = (1.5 + 2.5) \times 10^6 \times 0.1 \times 10^{-6} = 0.4\ \text{s}$$

Thus, by proper circuit design, the time constant can be adjusted to introduce a proper time delay in the circuit.

(b) Assuming that the capacitor is initially uncharged, $v_C(0) = 0$, while $v_C(\infty) = 110$. But

$$v_C(t) = v_C(\infty) + [v_C(0) - v_C(\infty)]e^{-t/\tau} = 110[1 - e^{-t/\tau}]$$

where $\tau = 0.4$ s, as calculated in part (a). The lamp glows when $v_C = 70$ V. If $v_C(t) = 70$ V at $t = t_0$, then

$$70 = 110[1 - e^{-t_0/\tau}] \quad \Rightarrow \quad \frac{7}{11} = 1 - e^{-t_0/\tau}$$

or

$$e^{-t_0/\tau} = \frac{4}{11} \quad \Rightarrow \quad e^{t_0/\tau} = \frac{11}{4}$$

Taking the natural logarithm of both sides gives

$$t_0 = \tau \ln\frac{11}{4} = 0.4 \ln 2.75 = 0.4046\ \text{s}$$

A more general formula for finding t_0 is

$$t_0 = \tau \ln\frac{-v(\infty)}{v(t_0) - v(\infty)}$$

The lamp will fire repeatedly every t_0 seconds if and only if $v(t_0) < v(\infty)$.

Practice Problem 7.19

The RC circuit in Fig. 7.74 is designed to operate an alarm which activates when the current through it exceeds $120\ \mu A$. If $0 \leq R \leq 6\ k\Omega$, find the range of the time delay that the variable resistor can create.

Answer: Between 47.23 ms and 124 ms.

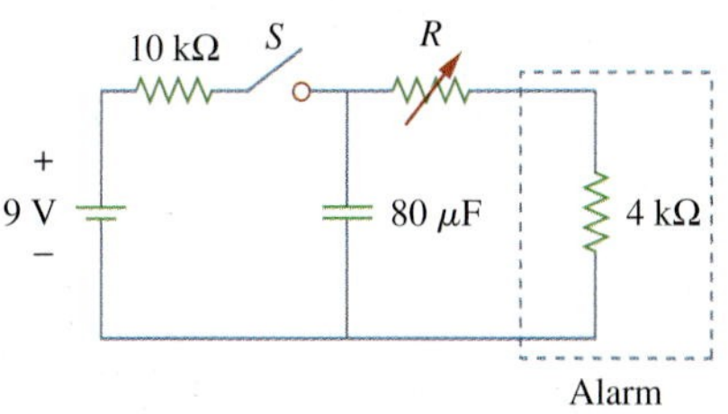

Figure 7.74
For Practice Prob. 7.19.

7.9.2 Photoflash Unit

An electronic flash unit provides a common example of an RC circuit. This application exploits the ability of the capacitor to oppose any abrupt change in voltage. Figure 7.75 shows a simplified circuit. It consists essentially of a high-voltage dc supply, a current-limiting large resistor R_1, and a capacitor C in parallel with the flashlamp of low resistance R_2. When the switch is in position 1, the capacitor charges slowly due to the large time constant ($\tau_1 = R_1 C$). As shown in Fig. 7.76(a), the capacitor voltage rises gradually from zero to V_s, while its current decreases gradually from $I_1 = V_s/R_1$ to zero. The charging time is approximately five times the time constant,

$$t_{\text{charge}} = 5R_1C \tag{7.65}$$

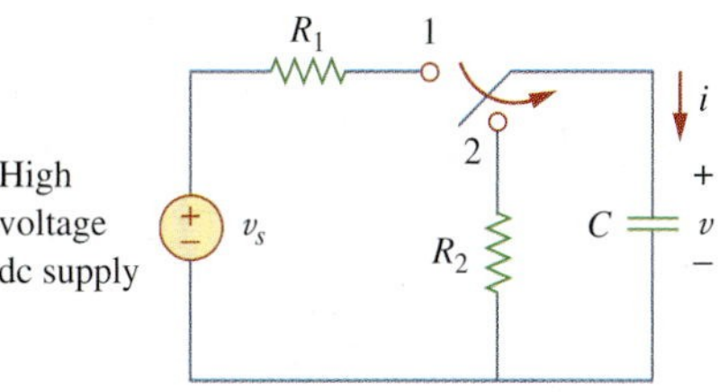

Figure 7.75
Circuit for a flash unit providing slow charge in position 1 and fast discharge in position 2.

With the switch in position 2, the capacitor voltage is discharged. The low resistance R_2 of the photolamp permits a high discharge current with peak $I_2 = V_s/R_2$ in a short duration, as depicted in Fig. 7.76(b). Discharging takes place in approximately five times the time constant,

$$t_{\text{discharge}} = 5R_2C \tag{7.66}$$

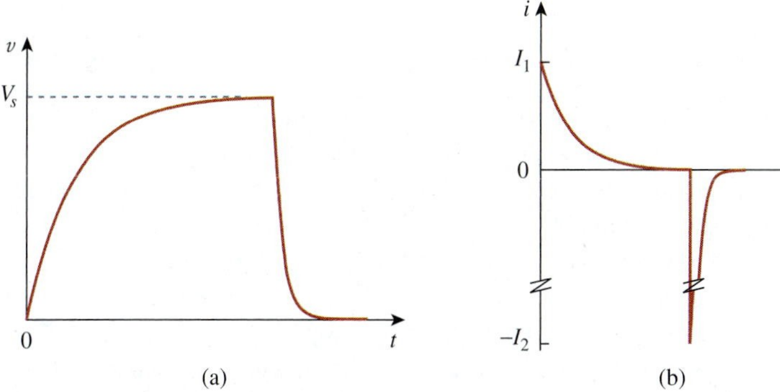

Figures 7.76
(a) Capacitor voltage showing slow charge and fast discharge, (b) capacitor current showing low charging current $I_1 = V_s/R_1$ and high discharge current $I_2 = V_s/R_2$.

Thus, the simple RC circuit of Fig. 7.75 provides a short-duration, high-current pulse. Such a circuit also finds applications in electric spot welding and the radar transmitter tube.

Example 7.20

An electronic flashgun has a current-limiting 6-kΩ resistor and 2000-μF electrolytic capacitor charged to 240 V. If the lamp resistance is 12 Ω, find: (a) the peak charging current, (b) the time required for the capacitor to fully charge, (c) the peak discharging current, (d) the total energy stored in the capacitor, and (e) the average power dissipated by the lamp.

Solution:

(a) The peak charging current is

$$I_1 = \frac{V_s}{R_1} = \frac{240}{6 \times 10^3} = 40 \text{ mA}$$

(b) From Eq. (7.65),

$$t_{\text{charge}} = 5R_1C = 5 \times 6 \times 10^3 \times 2000 \times 10^{-6} = 60 \text{ s} = 1 \text{ minute}$$

(c) The peak discharging current is

$$I_2 = \frac{V_s}{R_2} = \frac{240}{12} = 20 \text{ A}$$

(d) The energy stored is

$$W = \frac{1}{2}CV_s^2 = \frac{1}{2} \times 2000 \times 10^{-6} \times 240^2 = 57.6 \text{ J}$$

(e) The energy stored in the capacitor is dissipated across the lamp during the discharging period. From Eq. (7.66),

$$t_{\text{discharge}} = 5R_2C = 5 \times 12 \times 2000 \times 10^{-6} = 0.12 \text{ s}$$

Thus, the average power dissipated is

$$p = \frac{W}{t_{\text{discharge}}} = \frac{57.6}{0.12} = 480 \text{ watts}$$

Practice Problem 7.20

The flash unit of a camera has a 2-mF capacitor charged to 80 V.

(a) How much charge is on the capacitor?
(b) What is the energy stored in the capacitor?
(c) If the flash fires in 0.8 ms, what is the average current through the flashtube?
(d) How much power is delivered to the flashtube?
(e) After a picture has been taken, the capacitor needs to be recharged by a power unit that supplies a maximum of 5 mA. How much time does it take to charge the capacitor?

Answer: (a) 0.16 C, (b) 6.4 J, (c) 200 A, (d) 8 kW, (e) 32 s.

7.9.3 Relay Circuits

A magnetically controlled switch is called a *relay*. A relay is essentially an electromagnetic device used to open or close a switch that controls another circuit. Figure 7.77(a) shows a typical relay circuit.

The coil circuit is an *RL* circuit like that in Fig. 7.77(b), where R and L are the resistance and inductance of the coil. When switch S_1 in Fig. 7.77(a) is closed, the coil circuit is energized. The coil current gradually increases and produces a magnetic field. Eventually the magnetic field is sufficiently strong to pull the movable contact in the other circuit and close switch S_2. At this point, the relay is said to be *pulled in*. The time interval t_d between the closure of switches S_1 and S_2 is called the *relay delay time*.

Relays were used in the earliest digital circuits and are still used for switching high-power circuits.

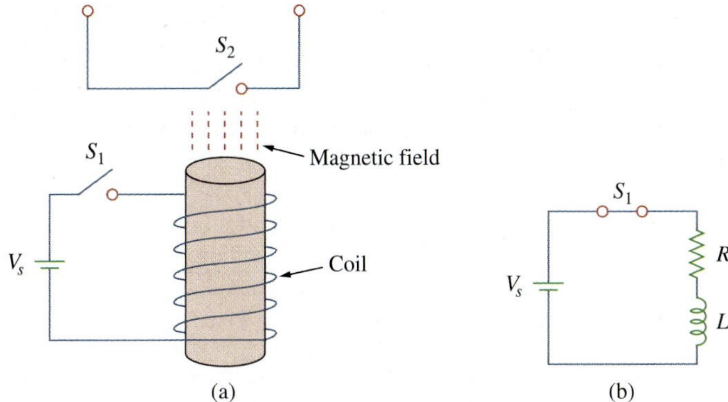

Figure 7.77
A relay circuit.

Example 7.21

The coil of a certain relay is operated by a 12-V battery. If the coil has a resistance of 150 Ω and an inductance of 30 mH and the current needed to pull in is 50 mA, calculate the relay delay time.

Solution:
The current through the coil is given by

$$i(t) = i(\infty) + [i(0) - i(\infty)]e^{-t/\tau}$$

where

$$i(0) = 0, \qquad i(\infty) = \frac{12}{150} = 80 \text{ mA}$$

$$\tau = \frac{L}{R} = \frac{30 \times 10^{-3}}{150} = 0.2 \text{ ms}$$

Thus,

$$i(t) = 80[1 - e^{-t/\tau}] \text{ mA}$$

If $i(t_d) = 50$ mA, then

$$50 = 80[1 - e^{-t_d/\tau}] \quad \Rightarrow \quad \frac{5}{8} = 1 - e^{-t_d/\tau}$$

or

$$e^{-t_d/\tau} = \frac{3}{8} \quad \Rightarrow \quad e^{t_d/\tau} = \frac{8}{3}$$

By taking the natural logarithm of both sides, we get

$$t_d = \tau \ln\frac{8}{3} = 0.2 \ln\frac{8}{3} \text{ ms} = 0.1962 \text{ ms}$$

Alternatively, we may find t_d using

$$t_d = \tau \ln\frac{i(0) - i(\infty)}{i(t_d) - i(\infty)}$$

Practice Problem 7.21

A relay has a resistance of 200 Ω and an inductance of 500 mH. The relay contacts close when the current through the coil reaches 350 mA. What time elapses between the application of 110 V to the coil and contact closure?

Answer: 2.529 ms.

7.9.4 Automobile Ignition Circuit

The ability of inductors to oppose rapid change in current makes them useful for arc or spark generation. An automobile ignition system takes advantage of this feature.

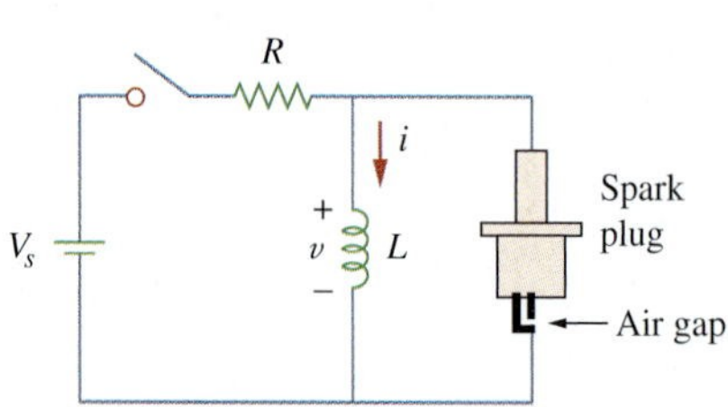

Figure 7.78
Circuit for an automobile ignition system.

The gasoline engine of an automobile requires that the fuel-air mixture in each cylinder be ignited at proper times. This is achieved by means of a spark plug (Fig. 7.78), which essentially consists of a pair of electrodes separated by an air gap. By creating a large voltage (thousands of volts) between the electrodes, a spark is formed across the air gap, thereby igniting the fuel. But how can such a large voltage be obtained from the car battery, which supplies only 12 V? This is achieved by means of an inductor (the spark coil) L. Since the voltage across the inductor is $v = L\,di/dt$, we can make di/dt large by creating a large change in current in a very short time. When the ignition switch in Fig. 7.78 is closed, the current through the inductor increases gradually and reaches the final value of $i = V_s/R$, where $V_s = 12$ V. Again, the time taken for the inductor to charge is five times the *time constant* of the circuit ($\tau = L/R$),

$$t_{\text{charge}} = 5\frac{L}{R} \tag{7.67}$$

Since at steady state, i is constant, $di/dt = 0$ and the inductor voltage $v = 0$. When the switch suddenly opens, a large voltage is developed across the inductor (due to the rapidly collapsing field) causing a spark or arc in the air gap. The spark continues until the energy stored in the inductor is dissipated in the spark discharge. In laboratories, when one is working with inductive circuits, this same effect causes a very nasty shock, and one must exercise caution.

Example 7.22

A solenoid with resistance 4 Ω and inductance 6 mH is used in an automobile ignition circuit similar to that in Fig. 7.78. If the battery supplies 12 V, determine: the final current through the solenoid when the switch is closed, the energy stored in the coil, and the voltage across the air gap, assuming that the switch takes 1 μs to open.

Solution:

The final current through the coil is

$$I = \frac{V_s}{R} = \frac{12}{4} = 3 \text{ A}$$

The energy stored in the coil is

$$W = \frac{1}{2}L\,I^2 = \frac{1}{2} \times 6 \times 10^{-3} \times 3^2 = 27 \text{ mJ}$$

The voltage across the gap is

$$V = L\frac{\Delta I}{\Delta t} = 6 \times 10^{-3} \times \frac{3}{1 \times 10^{-6}} = 18 \text{ kV}$$

Practice Problem 7.22

The spark coil of an automobile ignition system has a 20-mH inductance and a 5-Ω resistance. With a supply voltage of 12 V, calculate: the time needed for the coil to fully charge, the energy stored in the coil, and the voltage developed at the spark gap if the switch opens in 2 μs.

Answer: 20 ms, 57.6 mJ, and 24 kV.

7.10 Summary

1. The analysis in this chapter is applicable to any circuit that can be reduced to an equivalent circuit comprising a resistor and a single energy-storage element (inductor or capacitor). Such a circuit is first-order because its behavior is described by a first-order differential equation. When analyzing *RC* and *RL* circuits, one must always keep in mind that the capacitor is an open circuit to steady-state dc conditions while the inductor is a short circuit to steady-state dc conditions.
2. The natural response is obtained when no independent source is present. It has the general form

$$x(t) = x(0)e^{-t/\tau}$$

   where x represents current through (or voltage across) a resistor, a capacitor, or an inductor, and $x(0)$ is the initial value of x. Because most practical resistors, capacitors, and inductors always have losses, the natural response is a transient response, i.e. it dies out with time.
3. The time constant τ is the time required for a response to decay to $1/e$ of its initial value. For *RC* circuits, $\tau = RC$ and for *RL* circuits, $\tau = L/R$.

4. The singularity functions include the unit step, the unit ramp function, and the unit impulse functions. The unit step function $u(t)$ is

$$u(t) = \begin{cases} 0, & t < 0 \\ 1, & t > 0 \end{cases}$$

The unit impulse function is

$$\delta(t) = \begin{cases} 0, & t < 0 \\ \text{Undefined}, & t = 0 \\ 0, & t > 0 \end{cases}$$

The unit ramp function is

$$r(t) = \begin{cases} 0, & t \leq 0 \\ t, & t \geq 0 \end{cases}$$

5. The steady-state response is the behavior of the circuit after an independent source has been applied for a long time. The transient response is the component of the complete response that dies out with time.
6. The total or complete response consists of the steady-state response and the transient response.
7. The step response is the response of the circuit to a sudden application of a dc current or voltage. Finding the step response of a first-order circuit requires the initial value $x(0^+)$, the final value $x(\infty)$, and the time constant τ. With these three items, we obtain the step response as

$$x(t) = x(\infty) + [x(0^+) - x(\infty)]e^{-t/\tau}$$

A more general form of this equation is

$$x(t) = x(\infty) + [x(t_0^+) - x(\infty)]e^{-(t-t_0)/\tau}$$

Or we may write it as

$$\text{Instantaneous value} = \text{Final} + [\text{Initial} - \text{Final}]e^{-(t-t_0)/\tau}$$

8. *PSpice* is very useful for obtaining the transient response of a circuit.
9. Four practical applications of *RC* and *RL* circuits are: a delay circuit, a photoflash unit, a relay circuit, and an automobile ignition circuit.

Review Questions

7.1 An *RC* circuit has $R = 2\ \Omega$ and $C = 4$ F. The time constant is:

(a) 0.5 s (b) 2 s (c) 4 s
(d) 8 s (e) 15 s

7.2 The time constant for an *RL* circuit with $R = 2\ \Omega$ and $L = 4$ H is:

(a) 0.5 s (b) 2 s (c) 4 s
(d) 8 s (e) 15 s

7.3 A capacitor in an *RC* circuit with $R = 2\ \Omega$ and $C = 4$ F is being charged. The time required for the capacitor voltage to reach 63.2 percent of its steady-state value is:

(a) 2 s (b) 4 s (c) 8 s
(d) 16 s (e) none of the above

7.4 An *RL* circuit has $R = 2\ \Omega$ and $L = 4$ H. The time needed for the inductor current to reach 40 percent of its steady-state value is:

(a) 0.5 s (b) 1 s (c) 2 s
(d) 4 s (e) none of the above

7.5 In the circuit of Fig. 7.79, the capacitor voltage just before $t = 0$ is:

(a) 10 V (b) 7 V (c) 6 V
(d) 4 V (e) 0 V

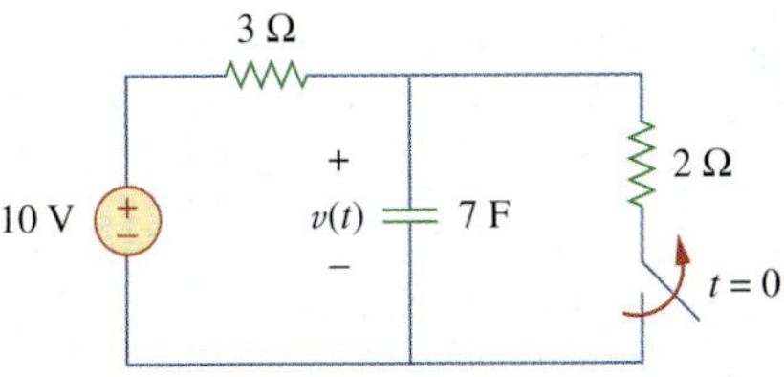

Figure 7.79
For Review Questions 7.5 and 7.6.

7.6 In the circuit in Fig. 7.79, $v(\infty)$ is:

(a) 10 V (b) 7 V (c) 6 V
(d) 4 V (e) 0 V

7.7 For the circuit in Fig. 7.80, the inductor current just before $t = 0$ is:

(a) 8 A (b) 6 A (c) 4 A
(d) 2 A (e) 0 A

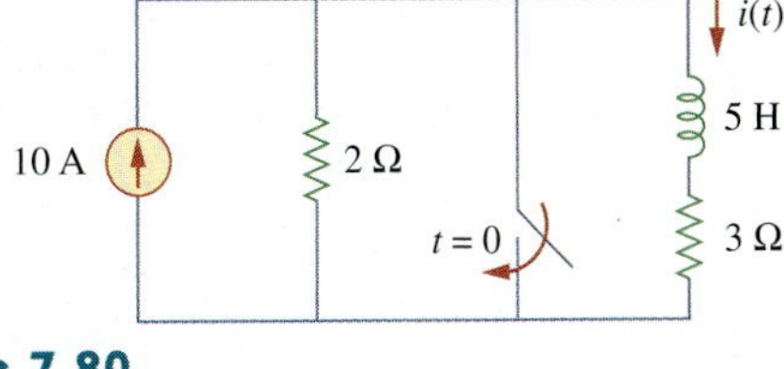

Figure 7.80
For Review Questions 7.7 and 7.8.

7.8 In the circuit of Fig. 7.80, $i(\infty)$ is:

(a) 10 A (b) 6 A (c) 4 A
(d) 2 A (e) 0 A

7.9 If v_s changes from 2 V to 4 V at $t = 0$, we may express v_s as:

(a) $\delta(t)$ V (b) $2u(t)$ V
(c) $2u(-t) + 4u(t)$ V (d) $2 + 2u(t)$ V
(e) $4u(t) - 2$ V

7.10 The pulse in Fig. 7.116(a) can be expressed in terms of singularity functions as:

(a) $2u(t) + 2u(t-1)$ V (b) $2u(t) - 2u(t-1)$ V
(c) $2u(t) - 4u(t-1)$ V (d) $2u(t) + 4u(t-1)$ V

Answers: 7.1d, 7.2b, 7.3c, 7.4b, 7.5d, 7.6a, 7.7c, 7.8e, 7.9c,d, 7.10b.

Problems

Section 7.2 The Source-Free *RC* Circuit

7.1 In the circuit shown in Fig. 7.81

$$v(t) = 56e^{-200t} \text{ V}, \quad t > 0$$
$$i(t) = 8e^{-200t} \text{ mA}, \quad t > 0$$

(a) Find the values of R and C.

(b) Calculate the time constant τ.

(c) Determine the time required for the voltage to decay half its initial value at $t = 0$.

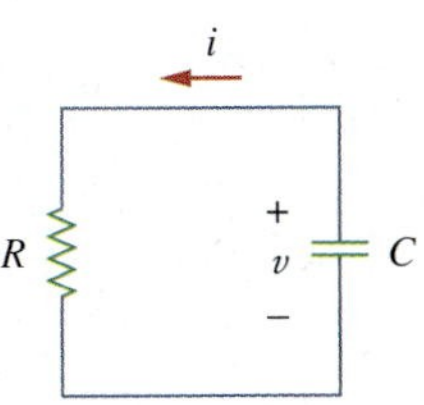

Figure 7.81
For Prob. 7.1.

7.2 Find the time constant for the *RC* circuit in Fig. 7.82.

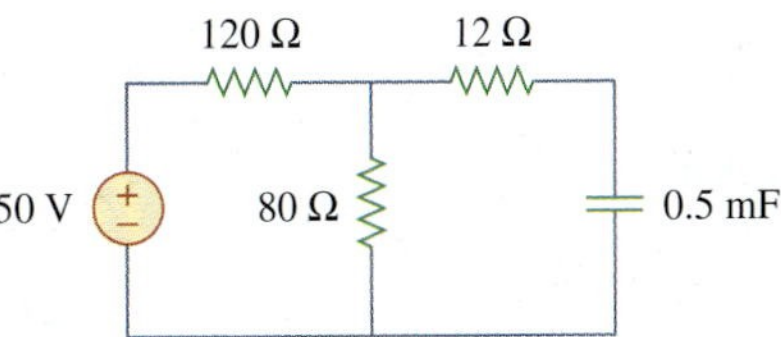

Figure 7.82
For Prob. 7.2.

7.3 Determine the time constant for the circuit in Fig. 7.83.

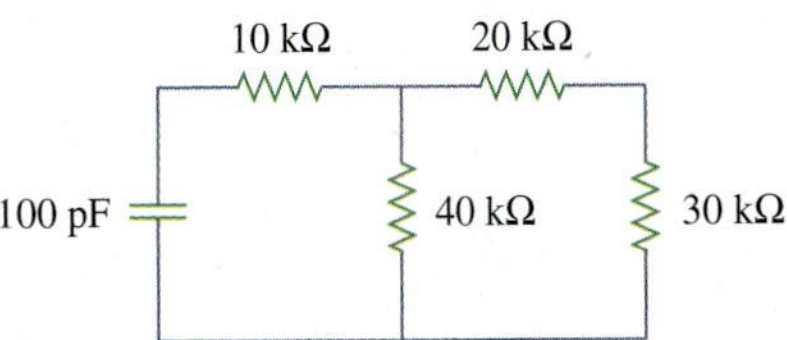

Figure 7.83
For Prob. 7.3.

7.4 The switch in Fig. 7.84 has been in position A for a long time. Assume the switch moves instantaneously from A to B at $t = 0$. Find v for $t > 0$.

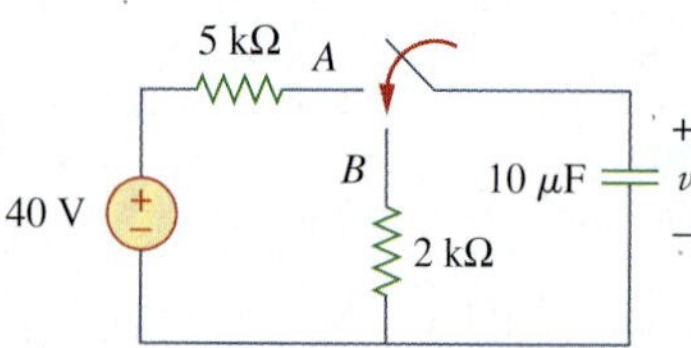

Figure 7.84
For Prob. 7.4.

7.5 Using Fig. 7.85, design a problem to help other students better understand source-free RC circuits.

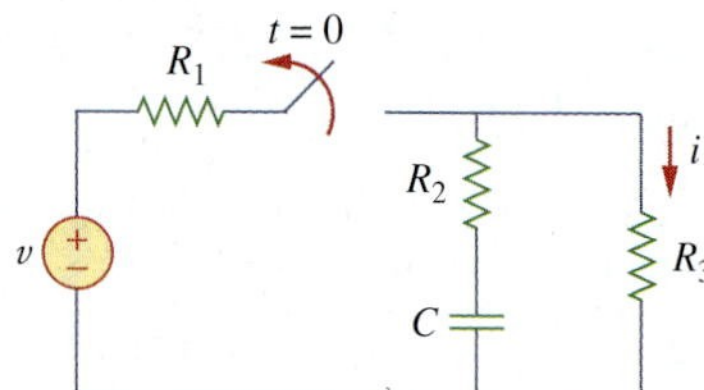

Figure 7.85
For Prob. 7.5.

7.6 The switch in Fig. 7.86 has been closed for a long time, and it opens at $t = 0$. Find $v(t)$ for $t \geq 0$.

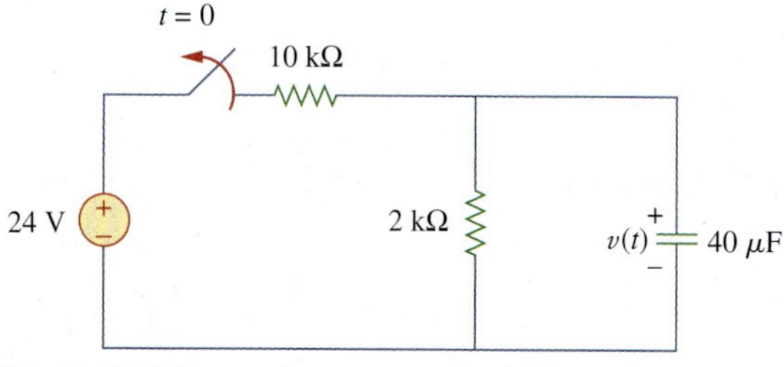

Figure 7.86
For Prob. 7.6.

7.7 Assuming that the switch in Fig. 7.87 has been in position A for a long time and is moved to position B at $t = 0$, find $v_o(t)$ for $t \geq 0$.

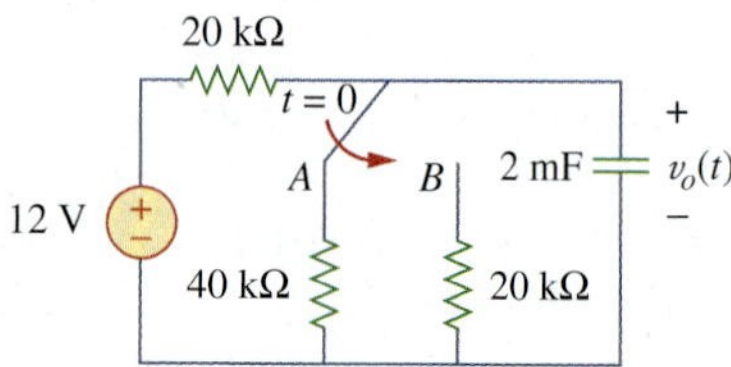

Figure 7.87
For Prob. 7.7.

7.8 For the circuit in Fig. 7.88, if

$$v = 10e^{-4t} \text{ V} \quad \text{and} \quad i = 0.2\, e^{-4t} \text{ A}, \quad t > 0$$

(a) Find R and C.

(b) Determine the time constant.

(c) Calculate the initial energy in the capacitor.

(d) Obtain the time it takes to dissipate 50 percent of the initial energy.

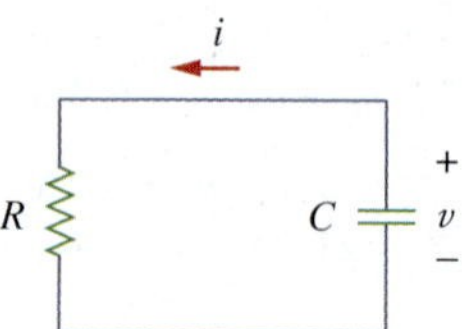

Figure 7.88
For Prob. 7.8.

7.9 The switch in Fig. 7.89 opens at $t = 0$. Find v_o for $t > 0$.

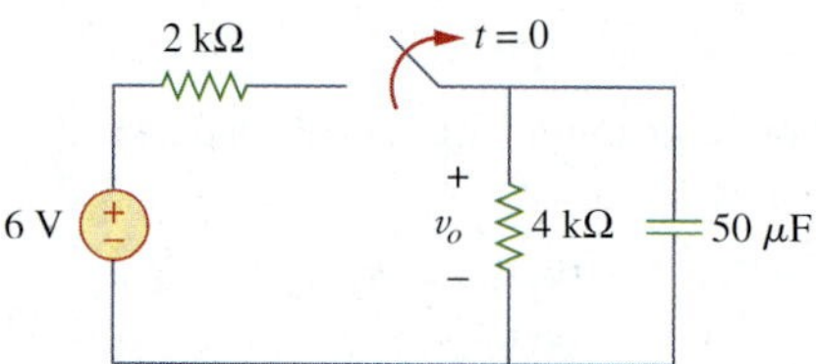

Figure 7.89
For Prob. 7.9.

7.10 For the circuit in Fig. 7.90, find $v_o(t)$ for $t > 0$. Determine the time necessary for the capacitor voltage to decay to one-third of its value at $t = 0$.

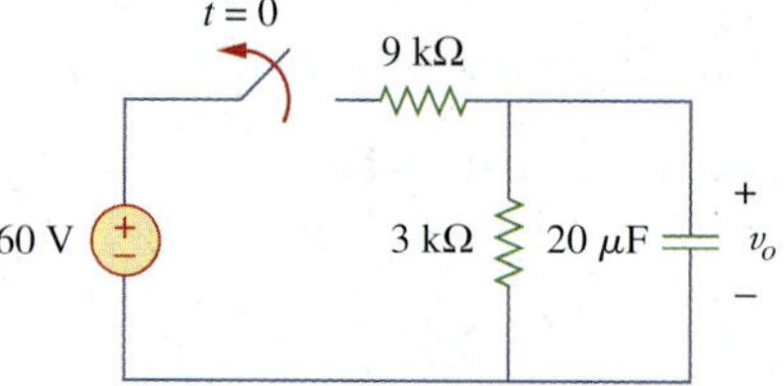

Figure 7.90
For Prob. 7.10.

Section 7.3 The Source-Free *RL* Circuit

7.11 For the circuit in Fig. 7.91, find i_o for $t > 0$.

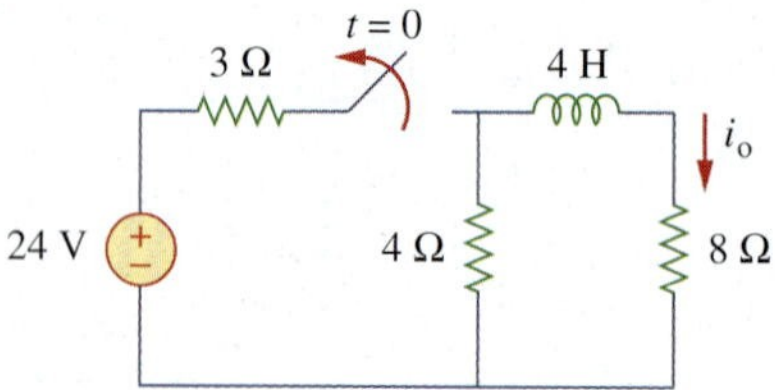

Figure 7.91
For Prob. 7.11.

7.12 Using Fig. 7.92, design a problem to help other students better understand source-free RL circuits.

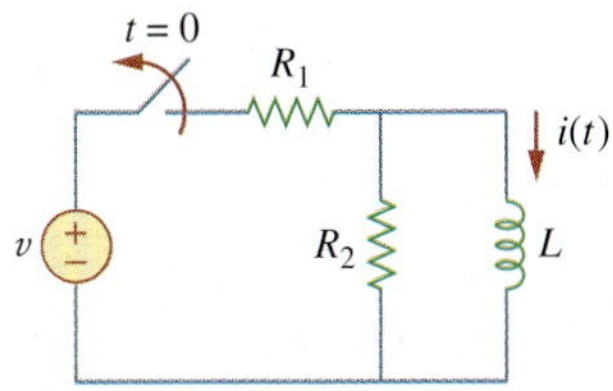

Figure 7.92
For Prob. 7.12.

7.13 In the circuit of Fig. 7.93,

$$v(t) = 20e^{-10^3 t}\ \text{V}, \quad t > 0$$
$$i(t) = 4e^{-10^3 t}\ \text{mA}, \quad t > 0$$

(a) Find R, L, and τ.

(b) Calculate the energy dissipated in the resistance for $0 < t < 0.5$ ms.

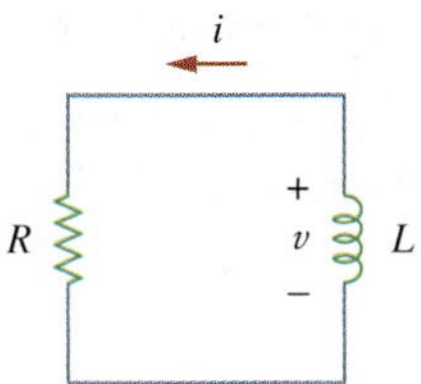

Figure 7.93
For Prob. 7.13.

7.14 Calculate the time constant of the circuit in Fig. 7.94.

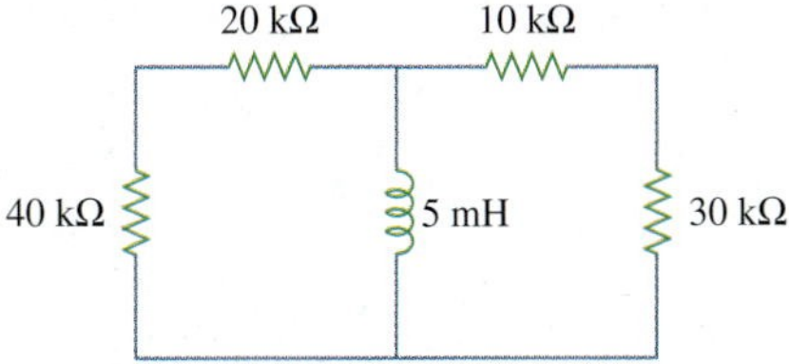

Figure 7.94
For Prob. 7.14.

7.15 Find the time constant for each of the circuits in Fig. 7.95.

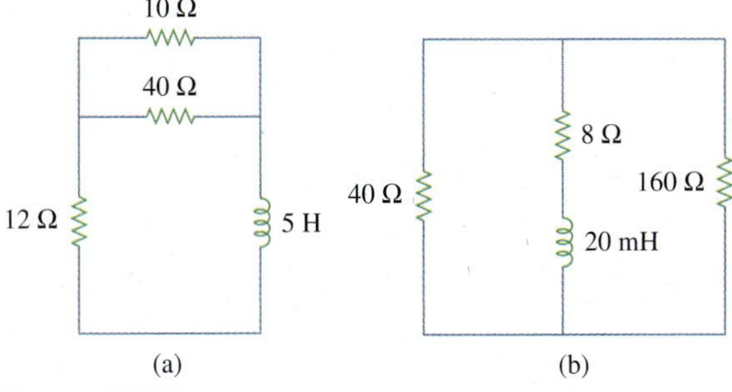

Figure 7.95
For Prob. 7.15.

7.16 Determine the time constant for each of the circuits in Fig. 7.96.

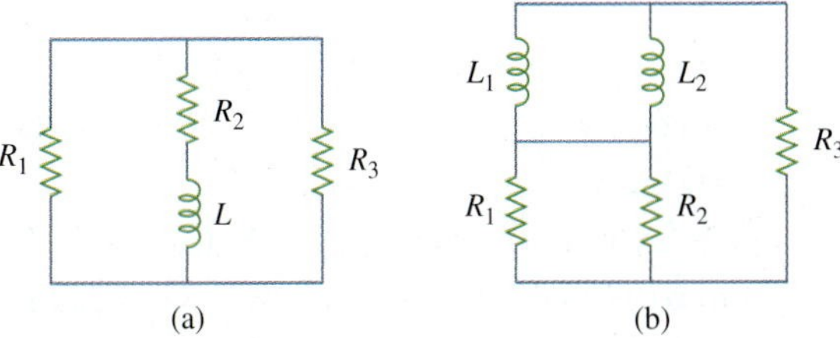

Figure 7.96
For Prob. 7.16.

7.17 Consider the circuit of Fig. 7.97. Find $v_o(t)$ if $i(0) = 2$ A and $v(t) = 0$.

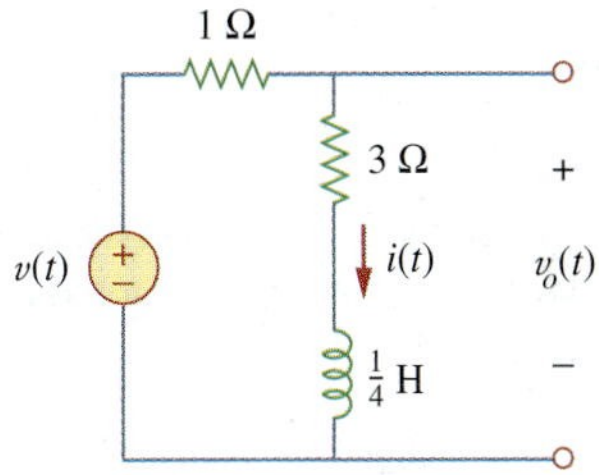

Figure 7.97
For Prob. 7.17.

7.18 For the circuit in Fig. 7.98, determine $v_o(t)$ when $i(0) = 1$ A and $v(t) = 0$.

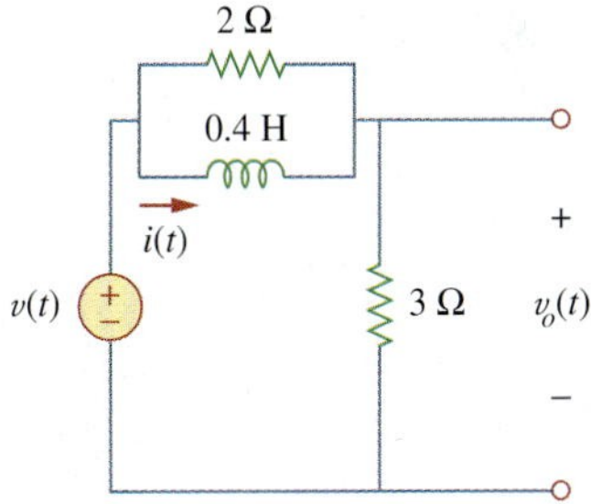

Figure 7.98
For Prob. 7.18.

7.19 In the circuit of Fig. 7.99, find $i(t)$ for $t > 0$ if $i(0) = 2$ A.

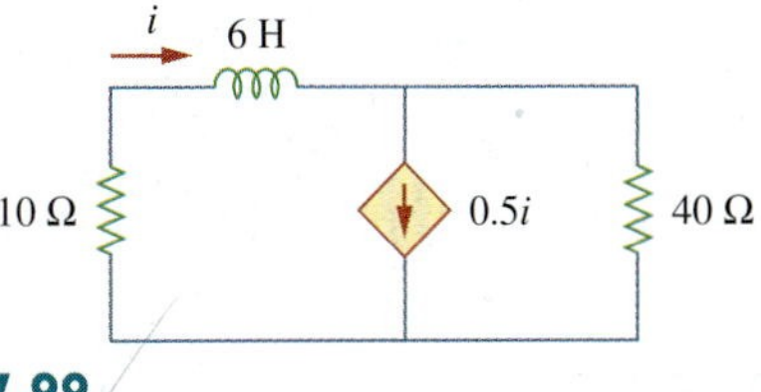

Figure 7.99
For Prob. 7.19.

7.20 For the circuit in Fig. 7.100,

$$v = 150e^{-50t} \text{ V}$$

and

$$i = 30e^{-50t} \text{ A}, \qquad t > 0$$

(a) Find L and R.

(b) Determine the time constant.

(c) Calculate the initial energy in the inductor.

(d) What fraction of the initial energy is dissipated in 10 ms?

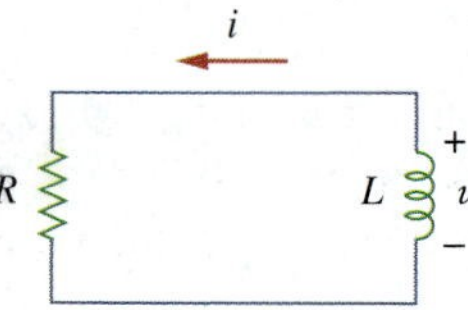

Figure 7.100
For Prob. 7.20.

7.21 In the circuit of Fig. 7.101, find the value of R for which the steady-state energy stored in the inductor will be 0.25 J.

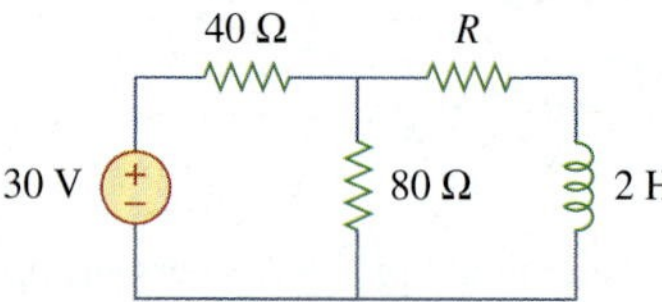

Figure 7.101
For Prob. 7.21.

7.22 Find $i(t)$ and $v(t)$ for $t > 0$ in the circuit of Fig. 7.102 if $i(0) = 20$ A.

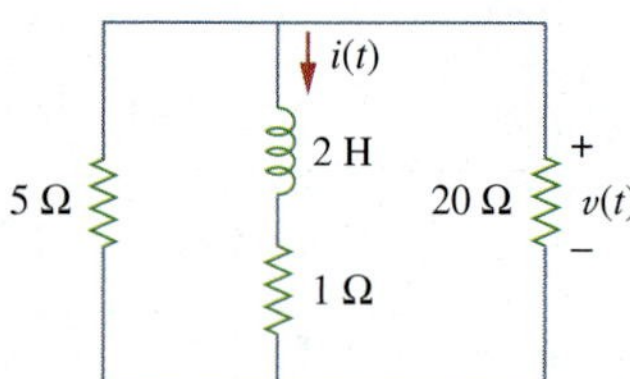

Figure 7.102
For Prob. 7.22.

7.23 Consider the circuit in Fig. 7.103. Given that $v_o(0) = 2$ V, find v_o and v_x for $t > 0$.

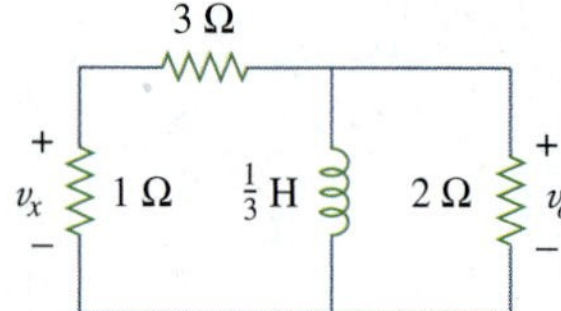

Figure 7.103
For Prob. 7.23.

Section 7.4 Singularity Functions

7.24 Express the following signals in terms of singularity functions.

(a) $v(t) = \begin{cases} 0, & t < 0 \\ -5, & t > 0 \end{cases}$

(b) $i(t) = \begin{cases} 0, & t < 1 \\ -10, & 1 < t < 3 \\ 10, & 3 < t < 5 \\ 0, & t > 5 \end{cases}$

(c) $x(t) = \begin{cases} t - 1, & 1 < t < 2 \\ 1, & 2 < t < 3 \\ 4 - t, & 3 < t < 4 \\ 0, & \text{Otherwise} \end{cases}$

(d) $y(t) = \begin{cases} 2, & t < 0 \\ -5, & 0 < t < 1 \\ 0, & t > 1 \end{cases}$

7.25 Design a problem to help other students better understand singularity functions.

7.26 Express the signals in Fig. 7.104 in terms of singularity functions.

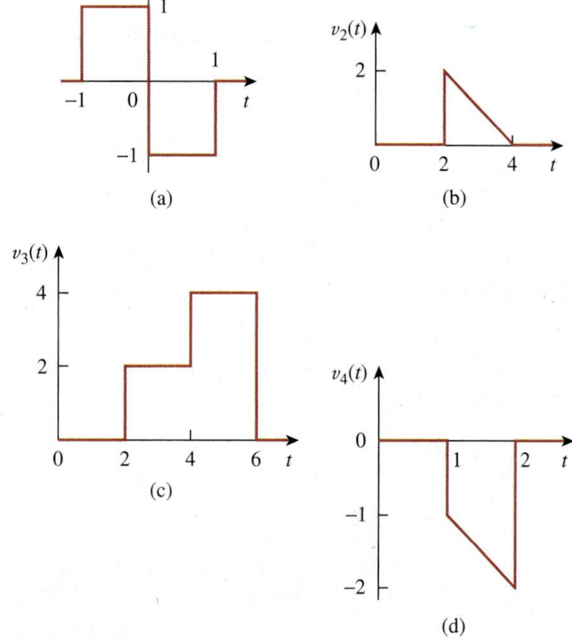

Figure 7.104
For Prob. 7.26.

7.27 Express $v(t)$ in Fig. 7.105 in terms of step functions.

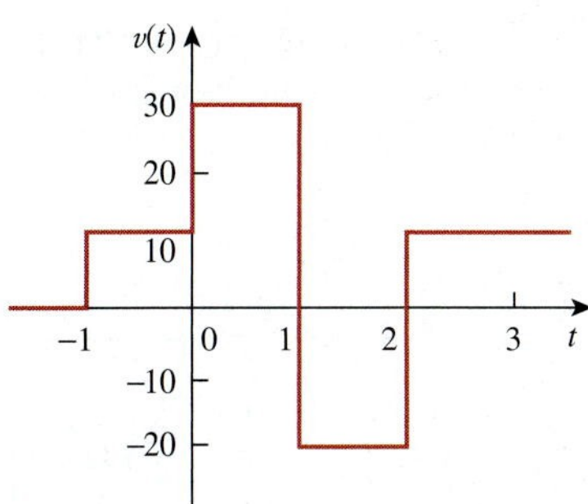

Figure 7.105
For Prob. 7.27.

7.28 Sketch the waveform represented by

$$i(t) = r(t) - r(t-1) - u(t-2) - r(t-2) + r(t-3) + u(t-4)$$

7.29 Sketch the following functions:

(a) $x(t) = 5e^{-t}u(t-1)$

(b) $y(t) = 20e^{-(t-1)}u(t)$

(c) $z(t) = 5\cos 4t\delta(t-1)$

7.30 Evaluate the following integrals involving the impulse functions:

(a) $\int_{-\infty}^{\infty} 4t^2\delta(t-1)\,dt$

(b) $\int_{-\infty}^{\infty} 4t^2\cos 2\pi t\delta(t-0.5)\,dt$

7.31 Evaluate the following integrals:

(a) $\int_{-\infty}^{\infty} e^{-4t^2}\delta(t-2)\,dt$

(b) $\int_{-\infty}^{\infty} [5\delta(t) + e^{-t}\delta(t) + \cos 2\pi t\delta(t)]\,dt$

7.32 Evaluate the following integrals:

(a) $\int_{1}^{t} u(\lambda)\,d\lambda$

(b) $\int_{0}^{4} r(t-1)\,dt$

(c) $\int_{1}^{5} (t-6)^2\delta(t-2)\,dt$

7.33 The voltage across a 10-mH inductor is $20\delta(t-2)$ mV. Find the inductor current, assuming that the inductor is initially uncharged.

7.34 Evaluate the following derivatives:

(a) $\frac{d}{dt}[u(t-1)u(t+1)]$

(b) $\frac{d}{dt}[r(t-6)u(t-2)]$

(c) $\frac{d}{dt}[\sin 4tu(t-3)]$

7.35 Find the solution to the following differential equations:

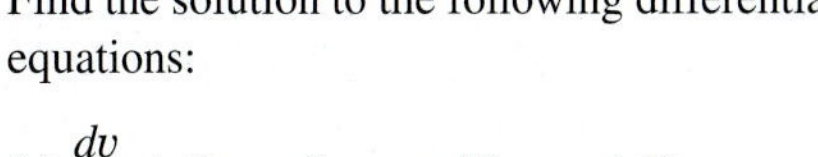

(a) $\frac{dv}{dt} + 2v = 0, \qquad v(0) = -1$ V

(b) $2\frac{di}{dt} - 3i = 0, \qquad i(0) = 2$

7.36 Solve for v in the following differential equations, subject to the stated initial condition.

(a) $dv/dt + v = u(t), \qquad v(0) = 0$

(b) $2\,dv/dt - v = 3u(t), \qquad v(0) = -6$

7.37 A circuit is described by

$$4\frac{dv}{dt} + v = 10$$

(a) What is the time constant of the circuit?

(b) What is $v(\infty)$, the final value of v?

(c) If $v(0) = 2$, find $v(t)$ for $t \geq 0$.

7.38 A circuit is described by

$$\frac{di}{dt} + 3i = 2u(t)$$

Find $i(t)$ for $t > 0$ given that $i(0) = 0$.

Section 7.5 Step Response of an *RC* Circuit

7.39 Calculate the capacitor voltage for $t < 0$ and $t > 0$ for each of the circuits in Fig. 7.106.

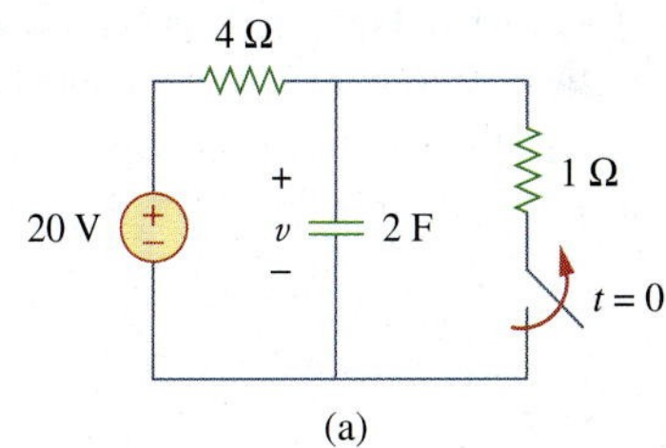

(a)

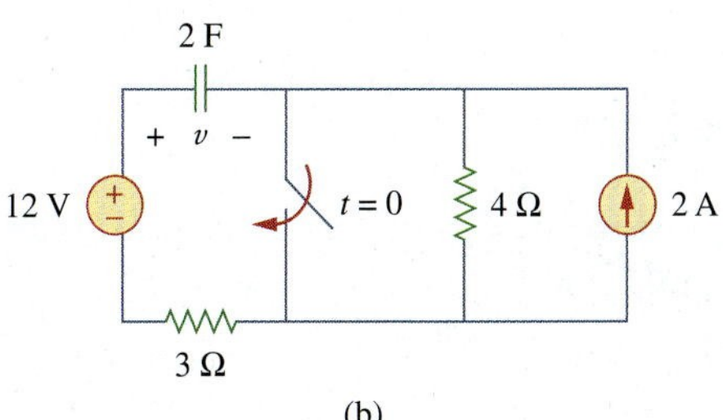

(b)

Figure 7.106
For Prob. 7.39.

7.40 Find the capacitor voltage for $t < 0$ and $t > 0$ for each of the circuits in Fig. 7.107.

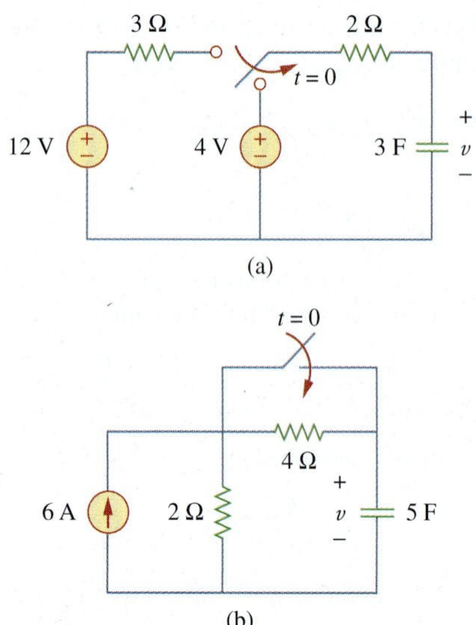

Figure 7.107
For Prob. 7.40.

7.41 Using Fig. 7.108, design a problem to help other students better understand the step response of an *RC* circuit.

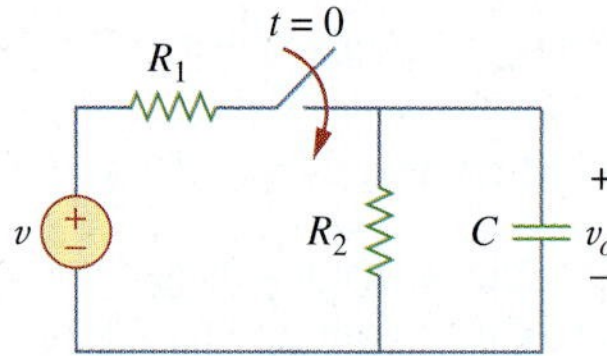

Figure 7.108
For Prob. 7.41.

7.42 (a) If the switch in Fig. 7.109 has been open for a long time and is closed at $t = 0$, find $v_o(t)$.

(b) Suppose that the switch has been closed for a long time and is opened at $t = 0$. Find $v_o(t)$.

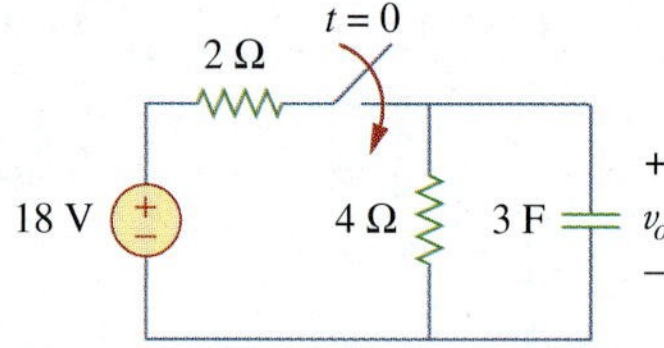

Figure 7.109
For Prob. 7.42.

7.43 Consider the circuit in Fig. 7.110. Find $i(t)$ for $t < 0$ and $t > 0$.

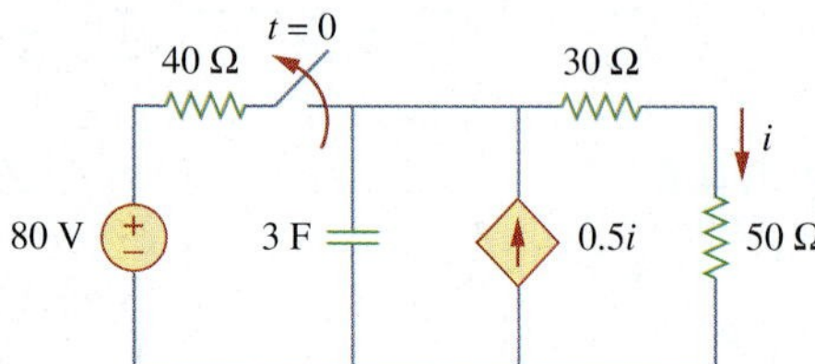

Figure 7.110
For Prob. 7.43.

7.44 The switch in Fig. 7.111 has been in position *a* for a long time. At $t = 0$, it moves to position *b*. Calculate $i(t)$ for all $t > 0$.

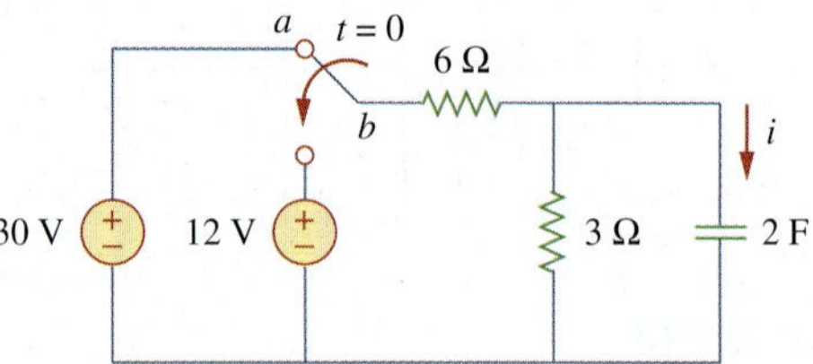

Figure 7.111
For Prob. 7.44.

7.45 Find v_o in the circuit of Fig. 7.112 when $v_s = 6u(t)$. Assume that $v_o(0) = 1$ V.

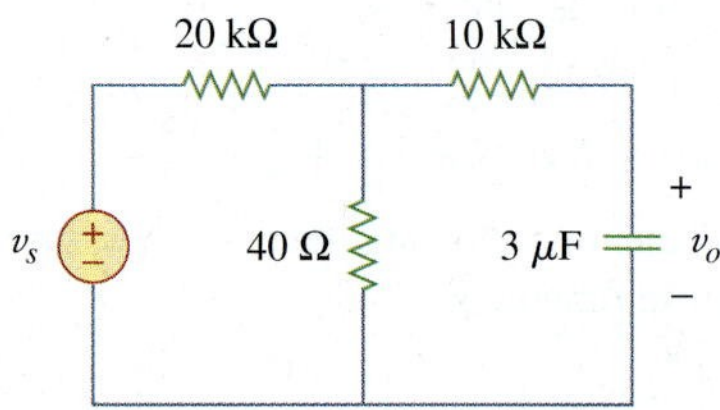

Figure 7.112
For Prob. 7.45.

7.46 For the circuit in Fig. 7.113, $i_s(t) = 5u(t)$. Find $v(t)$.

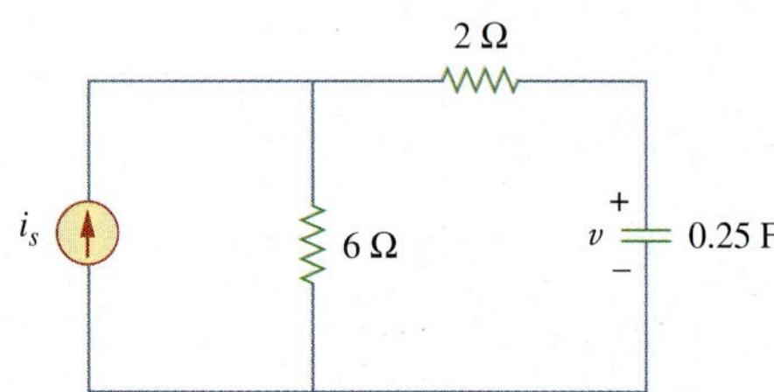

Figure 7.113
For Prob. 7.46.

7.47 Determine $v(t)$ for $t > 0$ in the circuit of Fig. 7.114 if $v(0) = 0$.

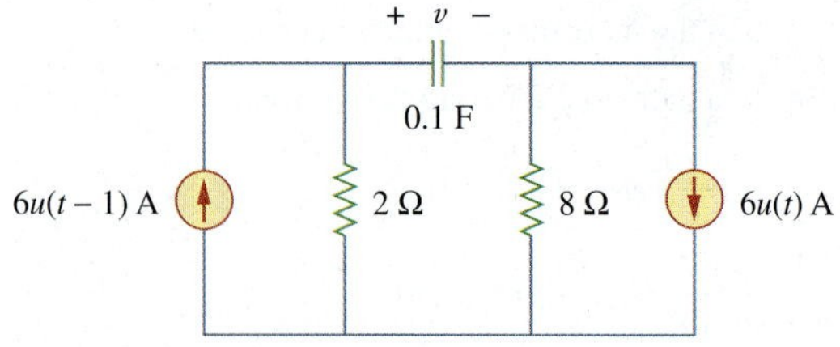

Figure 7.114
For Prob. 7.47.

7.48 Find $v(t)$ and $i(t)$ in the circuit of Fig. 7.115.

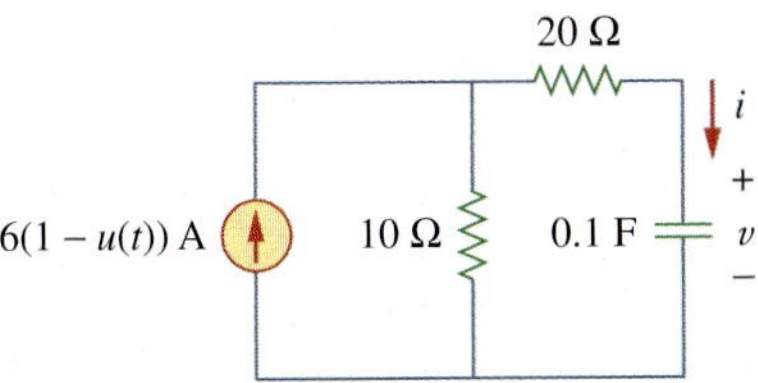

Figure 7.115
For Prob. 7.48.

7.49 If the waveform in Fig. 7.116(a) is applied to the circuit of Fig. 7.116(b), find $v(t)$. Assume $v(0) = 0$.

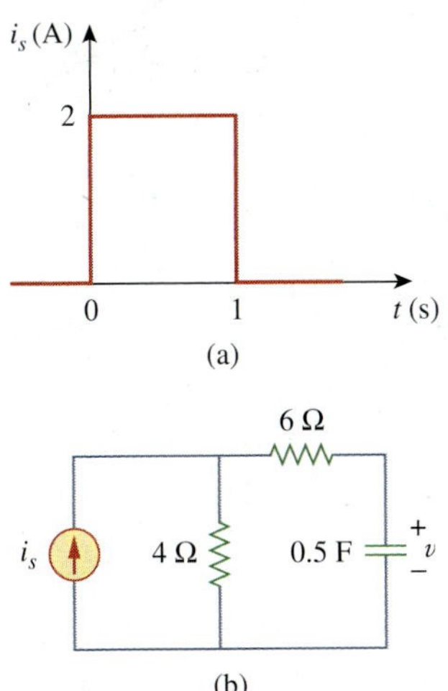

Figure 7.116
For Prob. 7.49 and Review Question 7.10.

***7.50** In the circuit of Fig. 7.117, find i_x for $t > 0$. Let $R_1 = R_2 = 1\text{ k}\Omega$, $R_3 = 2\text{ k}\Omega$, and $C = 0.25$ mF.

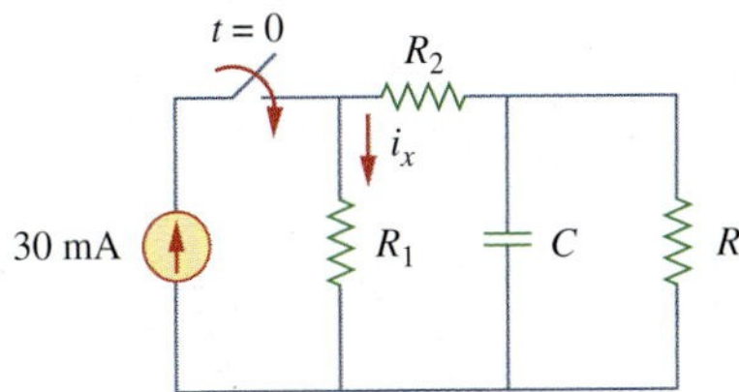

Figure 7.117
For Prob. 7.50.

Section 7.6 Step Response of an *RL* Circuit

7.51 Rather than applying the short-cut technique used in Section 7.6, use KVL to obtain Eq. (7.60).

7.52 Using Fig. 7.118, design a problem to help other students better understand the step response of an *RL* circuit.

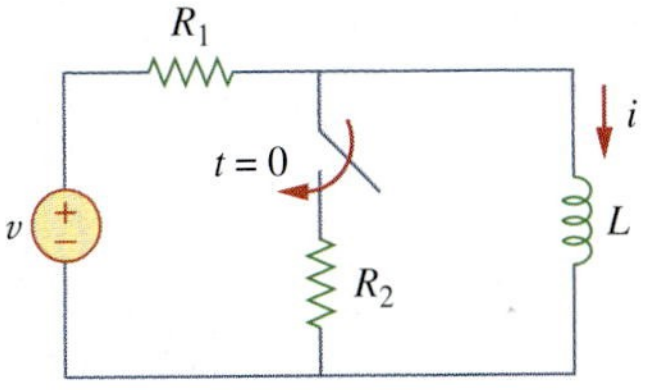

Figure 7.118
For Prob. 7.52.

7.53 Determine the inductor current $i(t)$ for both $t < 0$ and $t > 0$ for each of the circuits in Fig. 7.119.

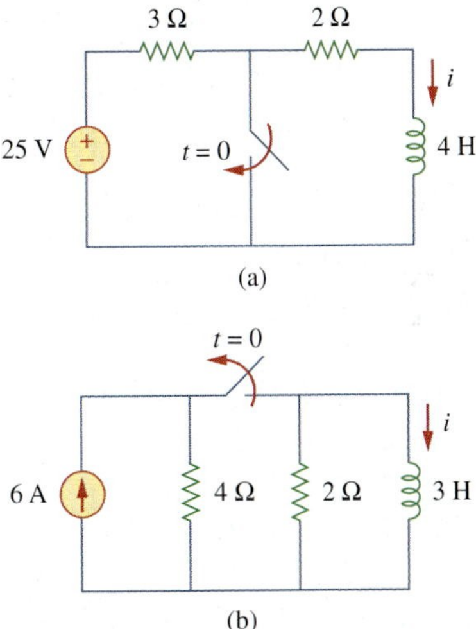

Figure 7.119
For Prob. 7.53.

7.54 Obtain the inductor current for both $t < 0$ and $t > 0$ in each of the circuits in Fig. 7.120.

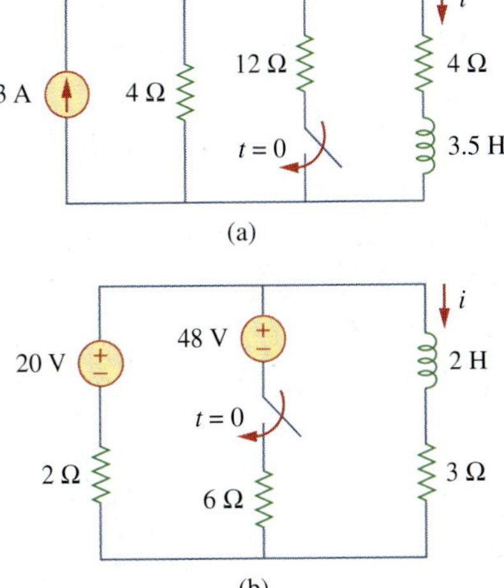

Figure 7.120
For Prob. 7.54.

* An asterisk indicates a challenging problem.

7.55 Find $v(t)$ for $t < 0$ and $t > 0$ in the circuit of Fig. 7.121.

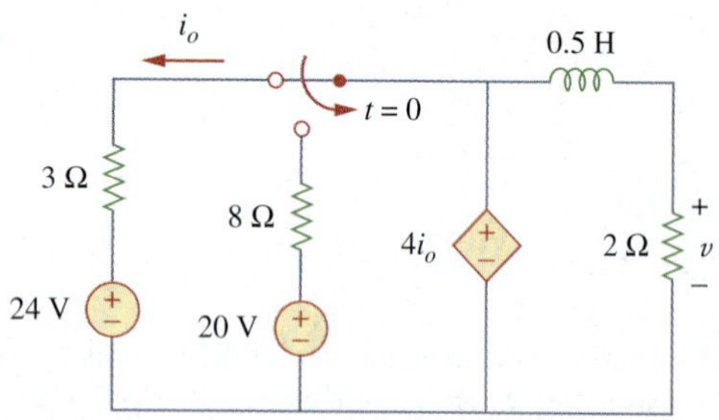

Figure 7.121
For Prob. 7.55.

7.56 For the network shown in Fig. 7.122, find $v(t)$ for $t > 0$.

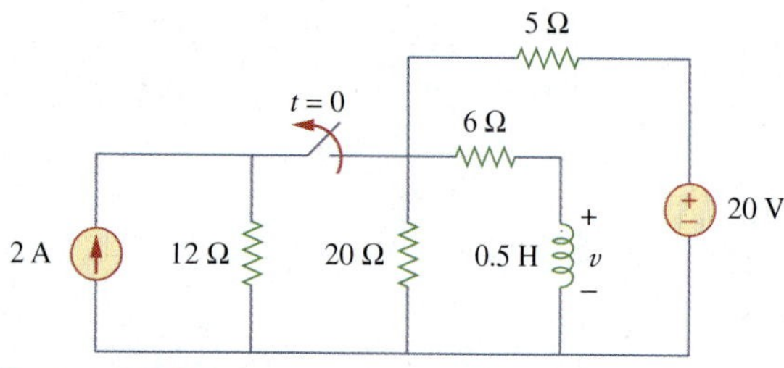

Figure 7.122
For Prob. 7.56.

***7.57** Find $i_1(t)$ and $i_2(t)$ for $t > 0$ in the circuit of Fig. 7.123.

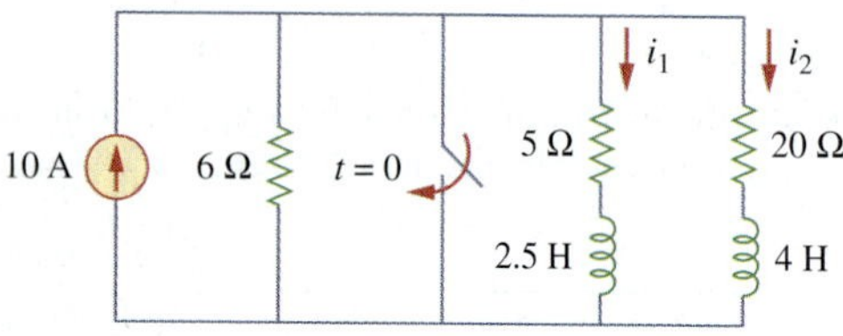

Figure 7.123
For Prob. 7.57.

7.58 Rework Prob. 7.17 if $i(0) = 10$ A and $v(t) = 20u(t)$ V.

7.59 Determine the step response $v_o(t)$ to $v_s = 9u(t)$ V in the circuit of Fig. 7.124.

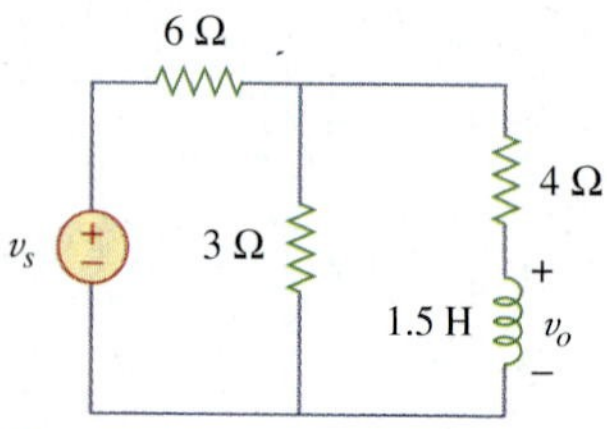

Figure 7.124
For Prob. 7.59.

7.60 Find $v(t)$ for $t > 0$ in the circuit of Fig. 7.125 if the initial current in the inductor is zero.

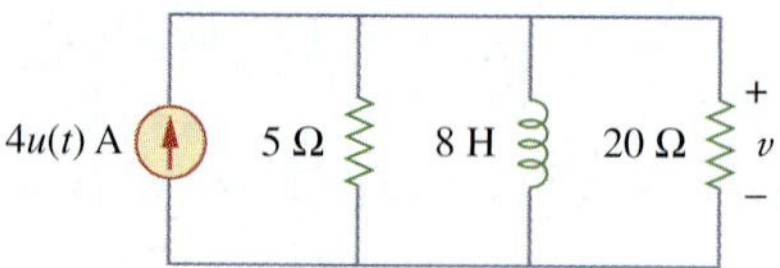

Figure 7.125
For Prob. 7.60.

7.61 In the circuit of Fig. 7.126, i_s changes from 5 A to 10 A at $t = 0$; that is, $i_s = (5 + 5u(t))$ A. Find v and i.

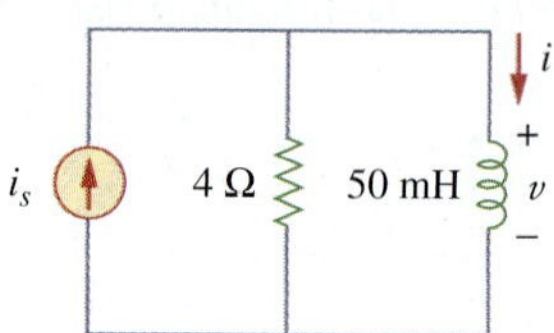

Figure 7.126
For Prob. 7.61.

7.62 For the circuit in Fig. 7.127, calculate $i(t)$ if $i(0) = 0$.

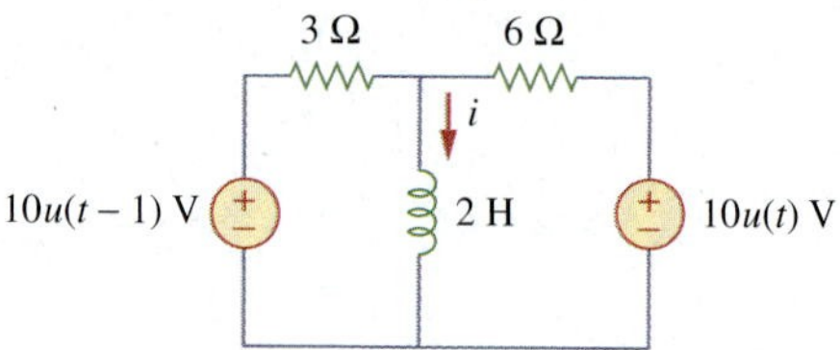

Figure 7.127
For Prob. 7.62.

7.63 Obtain $v(t)$ and $i(t)$ in the circuit of Fig. 7.128.

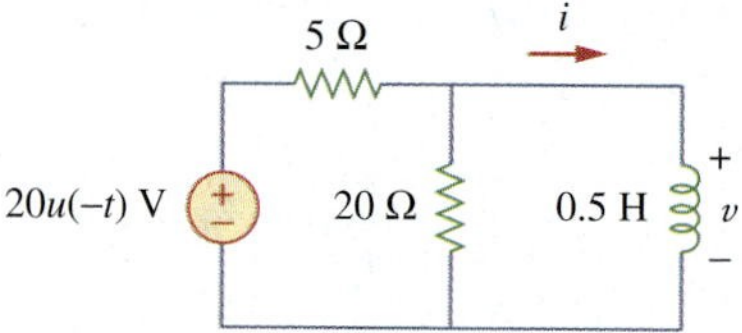

Figure 7.128
For Prob. 7.63.

7.64 Find $v_o(t)$ for $t > 0$ in the circuit of Fig. 7.129.

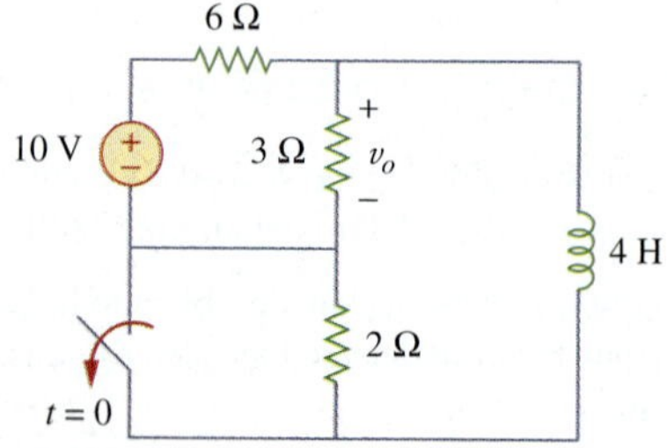

Figure 7.129
For Prob. 7.64.

7.65 If the input pulse in Fig. 7.130(a) is applied to the circuit in Fig. 7.130(b), determine the response $i(t)$.

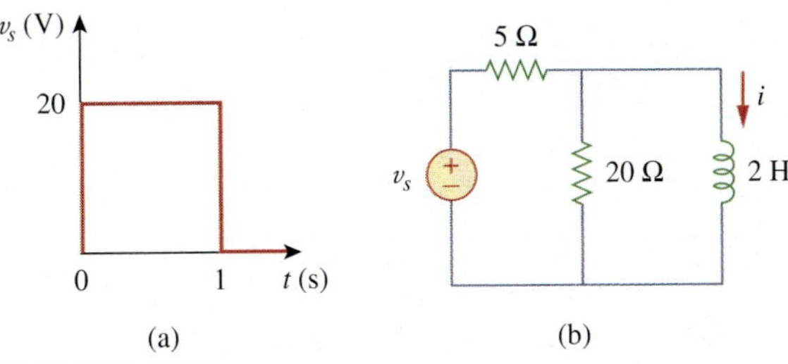

Figure 7.130
For Prob. 7.65.

Section 7.7 First-order Op Amp Circuits

7.66 Using Fig. 7.131, design a problem to help other students better understand first-order op amp circuits.

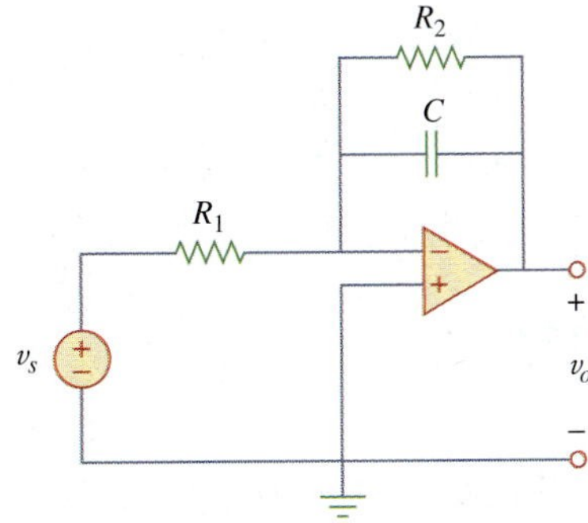

Figure 7.131
For Prob. 7.66.

7.67 If $v(0) = 10$ V, find $v_o(t)$ for $t > 0$ in the op amp circuit of Fig. 7.132. Let $R = 10$ kΩ and $C = 1$ μF.

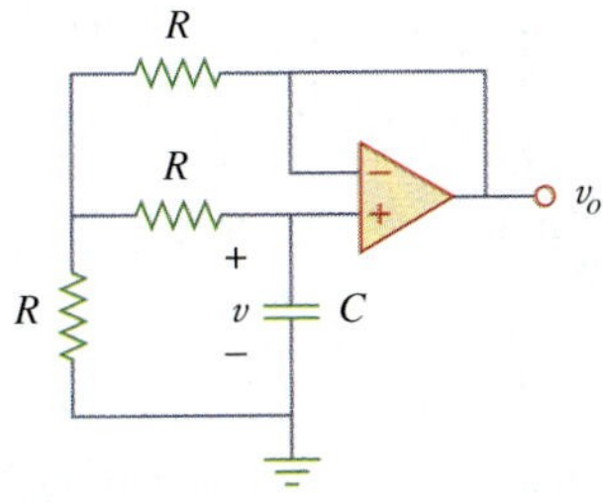

Figure 7.132
For Prob. 7.67.

7.68 Obtain v_o for $t > 0$ in the circuit of Fig. 7.133.

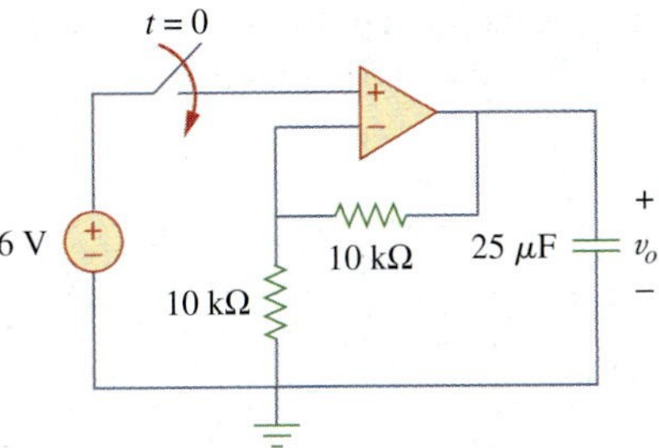

Figure 7.133
For Prob. 7.68.

7.69 For the op amp circuit in Fig. 7.134, find $v_o(t)$ for $t > 0$.

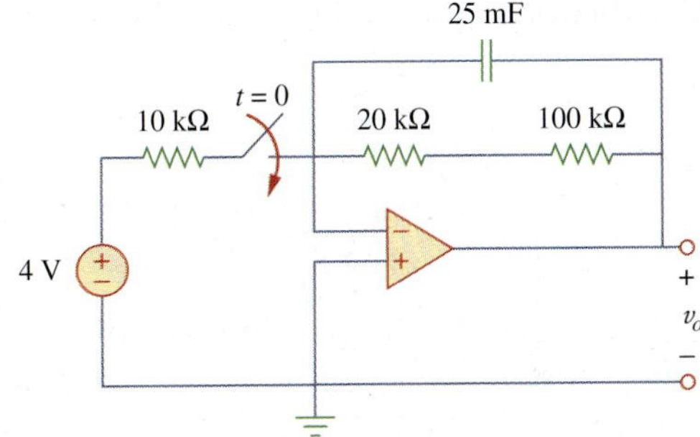

Figure 7.134
For Prob. 7.69.

7.70 Determine v_o for $t > 0$ when $v_s = 20$ mV in the op amp circuit of Fig. 7.135.

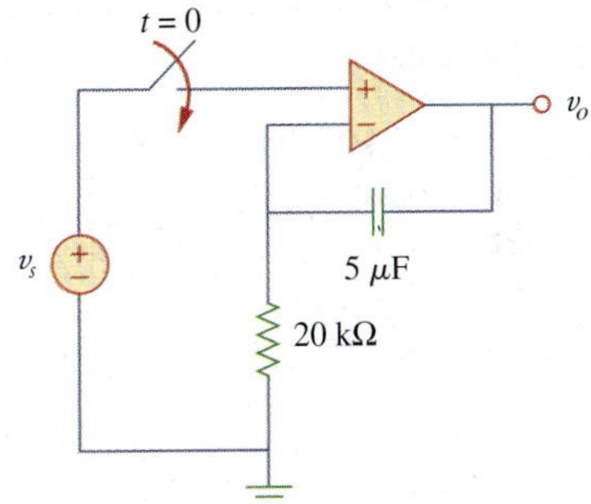

Figure 7.135
For Prob. 7.70.

7.71 For the op amp circuit in Fig. 7.136, suppose $v_0 = 0$ and $v_s = 3$ V. Find $v(t)$ for $t > 0$.

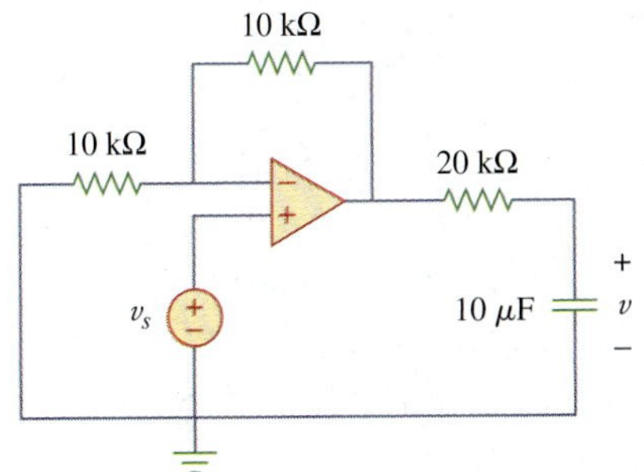

Figure 7.136
For Prob. 7.71.

7.72 Find i_o in the op amp circuit in Fig. 7.137. Assume that $v(0) = -2$ V, $R = 10$ kΩ, and $C = 10$ μF.

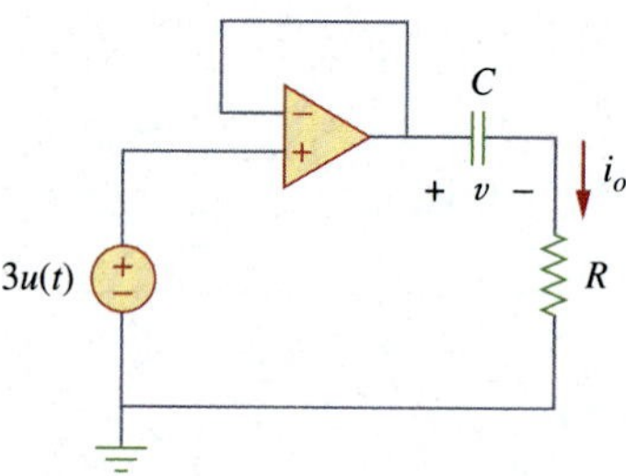

Figure 7.137
For Prob. 7.72.

7.73 For the circuit shown in Fig. 7.138, solve for $i_o(t)$.

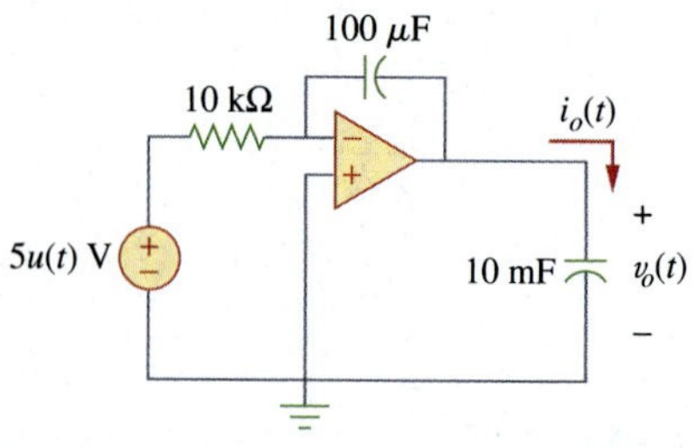

Figure 7.138
For Prob. 7.73.

7.74 Determine $v_o(t)$ for $t > 0$ in the circuit of Fig. 7.139. Let $i_s = 10u(t)$ μA and assume that the capacitor is initially uncharged.

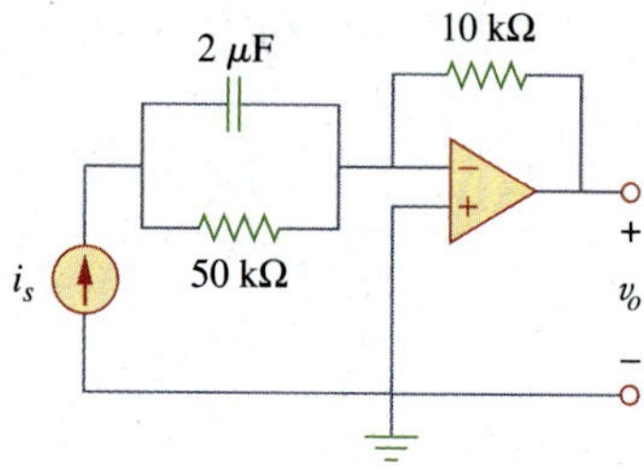

Figure 7.139
For Prob. 7.74.

7.75 In the circuit of Fig. 7.140, find v_o and i_o, given that $v_s = 4u(t)$ V and $v(0) = 1$ V.

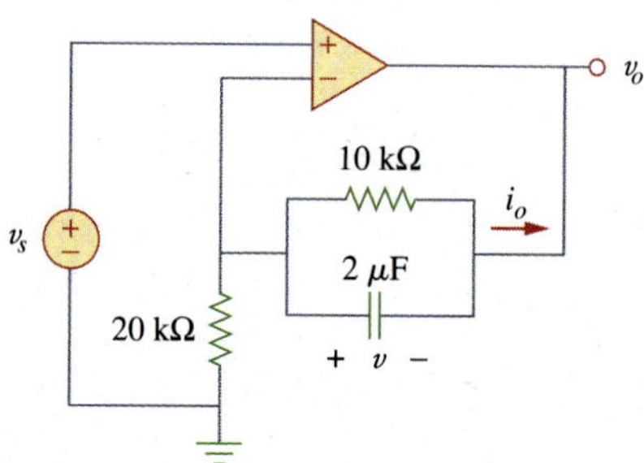

Figure 7.140
For Prob. 7.75.

Section 7.8 Transient Analysis with *PSpice*

7.76 Repeat Prob. 7.49 using *PSpice*.

7.77 The switch in Fig. 7.141 opens at $t = 0$. Use *PSpice* to determine $v(t)$ for $t > 0$.

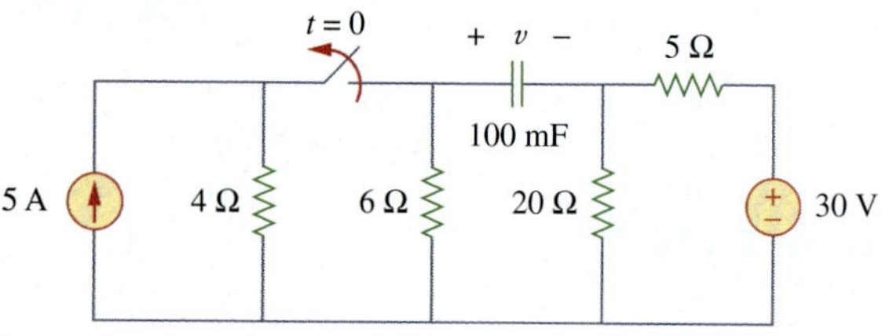

Figure 7.141
For Prob. 7.77.

7.78 The switch in Fig. 7.142 moves from position a to b at $t = 0$. Use *PSpice* to find $i(t)$ for $t > 0$.

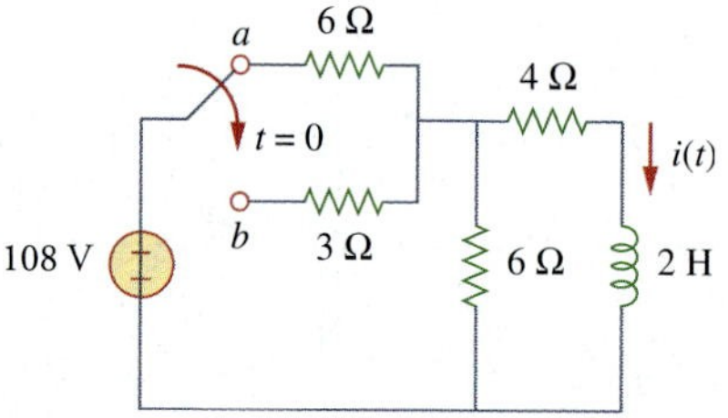

Figure 7.142
For Prob. 7.78.

7.79 In the circuit of Fig. 7.143, the switch has been in position a for a long time but moves instantaneously to position b at $t = 0$. Determine $i_o(t)$.

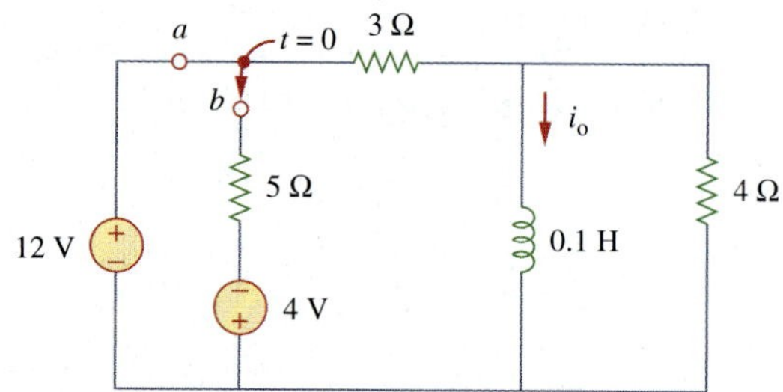

Figure 7.143
For Prob. 7.79.

7.80 In the circuit of Fig. 7.144, assume that the switch has been in position a for a long time, find:

(a) $i_1(0)$, $i_2(0)$, and $v_o(0)$

(b) $i_L(t)$

(c) $i_1(\infty)$, $i_2(\infty)$, and $v_o(\infty)$.

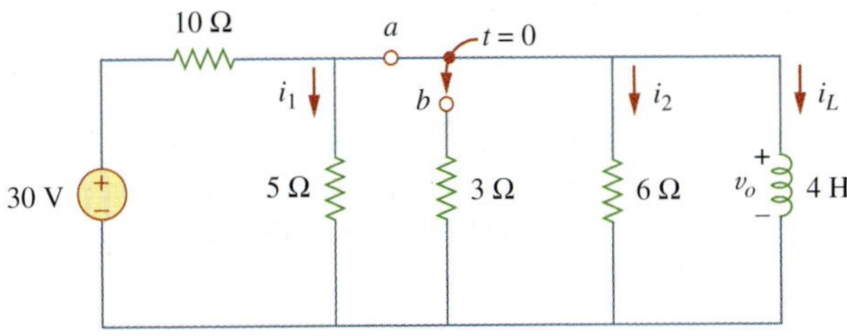

Figure 7.144
For Prob. 7.80.

7.81 Repeat Prob. 7.65 using *PSpice*.

Section 7.9 Applications

7.82 In designing a signal-switching circuit, it was found that a 100-μF capacitor was needed for a time constant of 3 ms. What value resistor is necessary for the circuit?

7.83 An *RC* circuit consists of a series connection of a 120-V source, a switch, a 34-MΩ resistor, and a 15-μF capacitor. The circuit is used in estimating the speed of a horse running a 4-km racetrack. The switch closes when the horse begins and opens when the horse crosses the finish line. Assuming that the capacitor charges to 85.6 V, calculate the speed of the horse.

7.84 The resistance of a 160-mH coil is 8 Ω. Find the time required for the current to build up to 60 percent of its final value when voltage is applied to the coil.

7.85 A simple relaxation oscillator circuit is shown in Fig. 7.145. The neon lamp fires when its voltage reaches 75 V and turns off when its voltage drops to 30 V. Its resistance is 120 Ω when on and infinitely high when off.

(a) For how long is the lamp on each time the capacitor discharges?

(b) What is the time interval between light flashes?

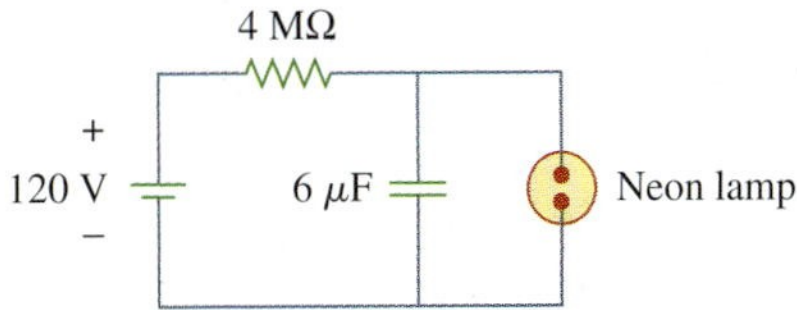

Figure 7.145
For Prob. 7.85.

7.86 Figure 7.146 shows a circuit for setting the length of time voltage is applied to the electrodes of a welding machine. The time is taken as how long it takes the capacitor to charge from 0 to 8 V. What is the time range covered by the variable resistor?

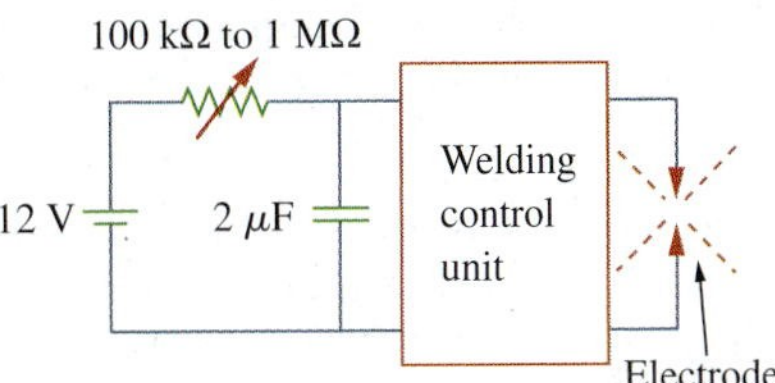

Figure 7.146
For Prob. 7.86.

7.87 A 120-V dc generator energizes a motor whose coil has an inductance of 50 H and a resistance of 100 Ω. A field discharge resistor of 400 Ω is connected in parallel with the motor to avoid damage to the motor, as shown in Fig. 7.147. The system is at steady state. Find the current through the discharge resistor 100 ms after the breaker is tripped.

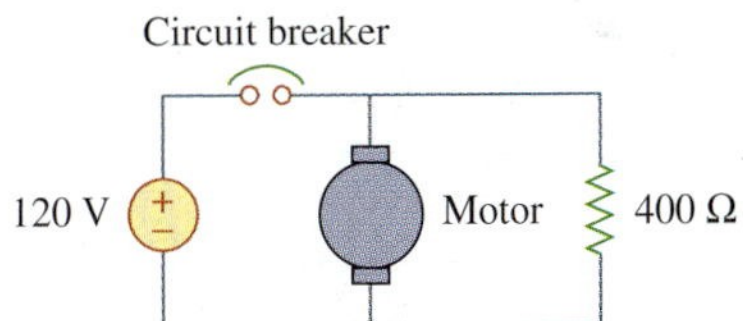

Figure 7.147
For Prob. 7.87.

Comprehensive Problems

7.88 The circuit in Fig. 7.148(a) can be designed as an approximate differentiator or an integrator, depending on whether the output is taken across the resistor or the capacitor, and also on the time constant $\tau = RC$ of the circuit and the width T of the input pulse in Fig. 7.148(b). The circuit is a differentiator if $\tau \ll T$, say $\tau < 0.1T$, or an integrator if $\tau \gg T$, say $\tau > 10T$.

(a) What is the minimum pulse width that will allow a differentiator output to appear across the capacitor?

(b) If the output is to be an integrated form of the input, what is the maximum value the pulse width can assume?

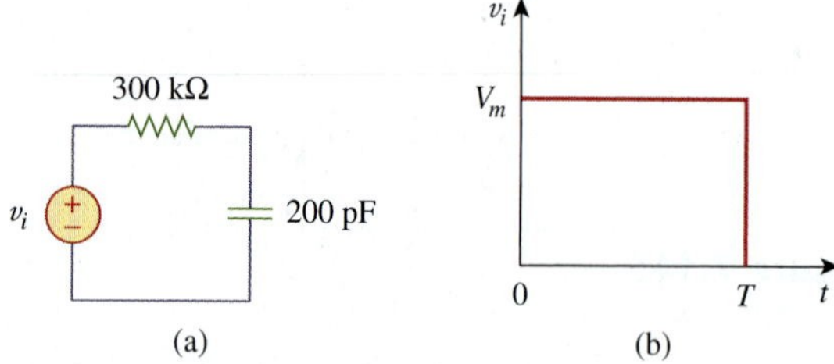

Figure 7.148
For Prob. 7.88.

7.89 An *RL* circuit may be used as a differentiator if the output is taken across the inductor and $\tau \ll T$ (say $\tau < 0.1T$), where T is the width of the input pulse. If R is fixed at 200 kΩ, determine the maximum value of L required to differentiate a pulse with $T = 10\ \mu$s.

7.90 An attenuator probe employed with oscilloscopes was designed to reduce the magnitude of the input voltage v_i by a factor of 10. As shown in Fig. 7.149, the oscilloscope has internal resistance R_s and capacitance C_s, while the probe has an internal resistance R_p. If R_p is fixed at 6 MΩ, find R_s and C_s for the circuit to have a time constant of 15 μs.

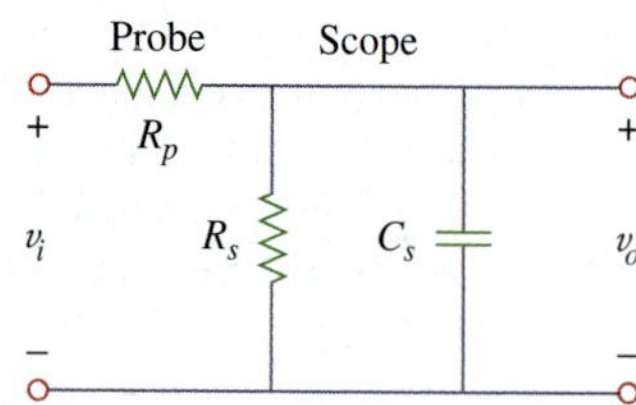

Figure 7.149
For Prob. 7.90.

7.91 The circuit in Fig. 7.150 is used by a biology student to study "frog kick." She noticed that the frog kicked a little when the switch was closed but kicked violently for 5 s when the switch was opened. Model the frog as a resistor and calculate its resistance. Assume that it takes 10 mA for the frog to kick violently.

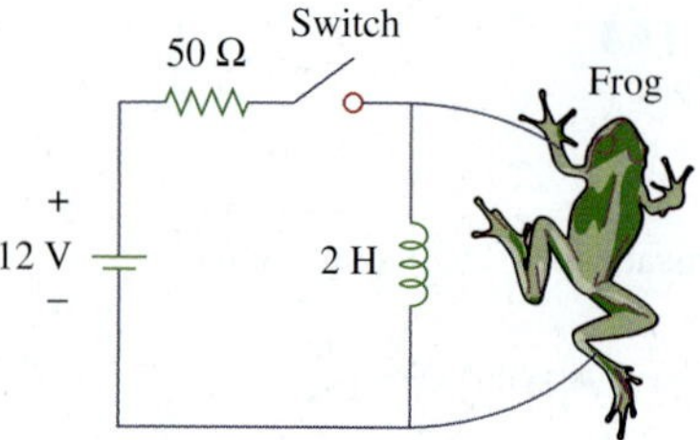

Figure 7.150
For Prob. 7.91.

7.92 To move a spot of a cathode-ray tube across the screen requires a linear increase in the voltage across the deflection plates, as shown in Fig. 7.151. Given that the capacitance of the plates is 4 nF, sketch the current flowing through the plates.

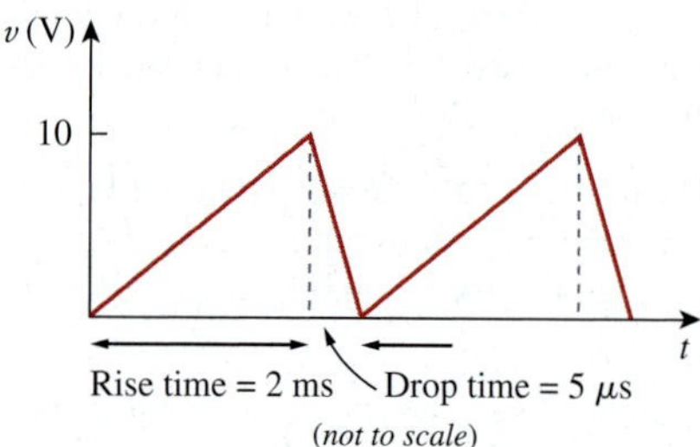

Figure 7.151
For Prob. 7.92.

chapter

8

Second-Order Circuits

Everyone who can earn a masters degree in engineering must earn a masters degree in engineering in order to maximize the success of their career! If you want to do research, state-of-the-art engineering, teach in a university, or start your own business, you really need to earn a doctoral degree!

—Charles K. Alexander

Enhancing Your Career

To increase your engineering career opportunities after graduation, develop a strong fundamental understanding in a broad set of engineering areas. When possible, this might best be accomplished by working toward a graduate degree immediately upon receiving your undergraduate degree.

Each degree in engineering represents certain skills the student acquires. At the Bachelor degree level, you learn the language of engineering and the fundamentals of engineering and design. At the Master's level, you acquire the ability to do advanced engineering projects and to communicate your work effectively both orally and in writing. The Ph.D. represents a thorough understanding of the fundamentals of electrical engineering and a mastery of the skills necessary both for working at the frontiers of an engineering area and for communicating one's effort to others.

If you have no idea what career you should pursue after graduation, a graduate degree program will enhance your ability to explore career options. Since your undergraduate degree will only provide you with the fundamentals of engineering, a Master's degree in engineering supplemented by business courses benefits more engineering students than does getting a Master's of Business Administration (MBA). The best time to get your MBA is after you have been a practicing engineer for some years and decide your career path would be enhanced by strengthening your business skills.

Engineers should constantly educate themselves, formally and informally, taking advantage of all means of education. Perhaps there is no better way to enhance your career than to join a professional society such as IEEE and be an active member.

Enhancing your career involves understanding your goals, adapting to changes, anticipating opportunities, and planning your own niche.
© 2005 Institute of Electrical and Electronics Engineers (IEEE).

8.1 Introduction

In the previous chapter we considered circuits with a single storage element (a capacitor or an inductor). Such circuits are first-order because the differential equations describing them are first-order. In this chapter we will consider circuits containing two storage elements. These are known as *second-order* circuits because their responses are described by differential equations that contain second derivatives.

Typical examples of second-order circuits are *RLC* circuits, in which the three kinds of passive elements are present. Examples of such circuits are shown in Fig. 8.1(a) and (b). Other examples are *RL* and *RC* circuits, as shown in Fig. 8.1(c) and (d). It is apparent from Fig. 8.1 that a second-order circuit may have two storage elements of different type or the same type (provided elements of the same type cannot be represented by an equivalent single element). An op amp circuit with two storage elements may also be a second-order circuit. As with first-order circuits, a second-order circuit may contain several resistors and dependent and independent sources.

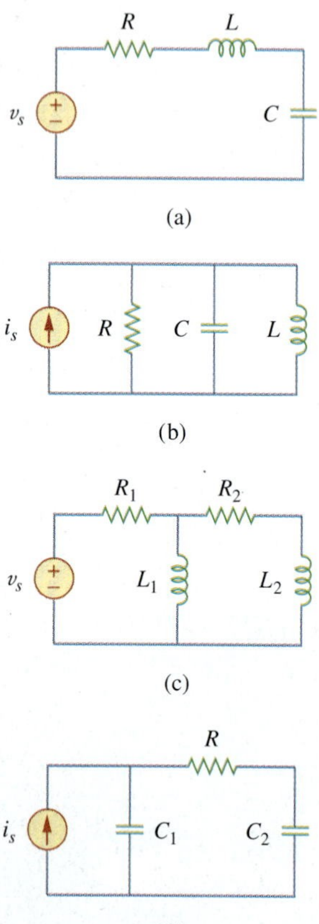

Figure 8.1
Typical examples of second-order circuits: (a) series *RLC* circuit, (b) parallel *RLC* circuit, (c) *RL* circuit, (d) *RC* circuit.

> A **second-order circuit** is characterized by a second-order differential equation. It consists of resistors and the equivalent of two energy storage elements.

Our analysis of second-order circuits will be similar to that used for first-order. We will first consider circuits that are excited by the initial conditions of the storage elements. Although these circuits may contain dependent sources, they are free of independent sources. These source-free circuits will give natural responses as expected. Later we will consider circuits that are excited by independent sources. These circuits will give both the transient response and the steady-state response. We consider only dc independent sources in this chapter. The case of sinusoidal and exponential sources is deferred to later chapters.

We begin by learning how to obtain the initial conditions for the circuit variables and their derivatives, as this is crucial to analyzing second-order circuits. Then we consider series and parallel *RLC* circuits such as shown in Fig. 8.1 for the two cases of excitation: by initial conditions of the energy storage elements and by step inputs. Later we examine other types of second-order circuits, including op amp circuits. We will consider *PSpice* analysis of second-order circuits. Finally, we will consider the automobile ignition system and smoothing circuits as typical applications of the circuits treated in this chapter. Other applications such as resonant circuits and filters will be covered in Chapter 14.

8.2 Finding Initial and Final Values

Perhaps the major problem students face in handling second-order circuits is finding the initial and final conditions on circuit variables. Students are usually comfortable getting the initial and final values of v and i but often have difficulty finding the initial values of their derivatives: dv/dt and di/dt. For this reason, this section is explicitly devoted to the subtleties of getting $v(0)$, $i(0)$, $dv(0)/dt$, $di(0)/dt$, $i(\infty)$, and $v(\infty)$. Unless otherwise stated in this chapter, v denotes capacitor voltage, while i is the inductor current.

There are two key points to keep in mind in determining the initial conditions.

First—as always in circuit analysis—we must carefully handle the polarity of voltage $v(t)$ across the capacitor and the direction of the current $i(t)$ through the inductor. Keep in mind that v and i are defined strictly according to the passive sign convention (see Figs. 6.3 and 6.23). One should carefully observe how these are defined and apply them accordingly.

Second, keep in mind that the capacitor voltage is always continuous so that

$$v(0^+) = v(0^-) \tag{8.1a}$$

and the inductor current is always continuous so that

$$i(0^+) = i(0^-) \tag{8.1b}$$

where $t = 0^-$ denotes the time just before a switching event and $t = 0^+$ is the time just after the switching event, assuming that the switching event takes place at $t = 0$.

Thus, in finding initial conditions, we first focus on those variables that cannot change abruptly, capacitor voltage and inductor current, by applying Eq. (8.1). The following examples illustrate these ideas.

Example 8.1

The switch in Fig. 8.2 has been closed for a long time. It is open at $t = 0$. Find: (a) $i(0^+)$, $v(0^+)$, (b) $di(0^+)/dt$, $dv(0^+)/dt$, (c) $i(\infty)$, $v(\infty)$.

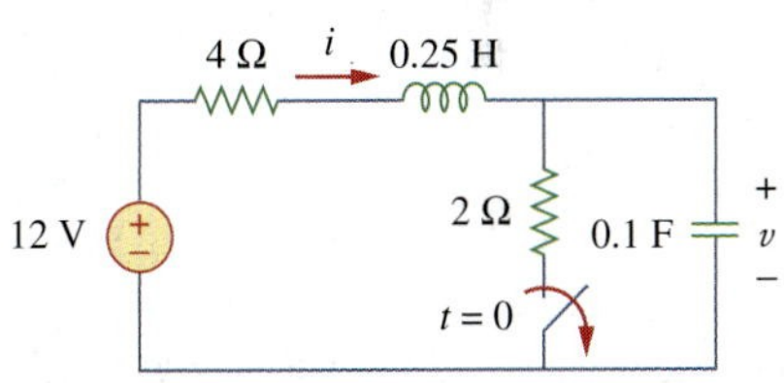

Figure 8.2
For Example 8.1.

Solution:

(a) If the switch is closed a long time before $t = 0$, it means that the circuit has reached dc steady state at $t = 0$. At dc steady state, the inductor acts like a short circuit, while the capacitor acts like an open circuit, so we have the circuit in Fig. 8.3(a) at $t = 0^-$. Thus,

$$i(0^-) = \frac{12}{4+2} = 2 \text{ A}, \qquad v(0^-) = 2i(0^-) = 4 \text{ V}$$

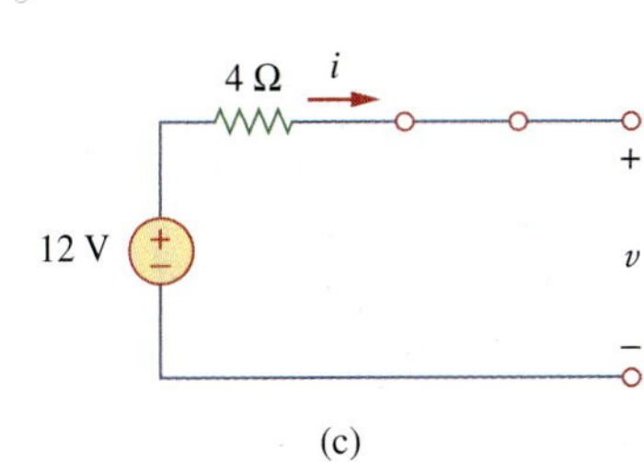

Figure 8.3
Equivalent circuit of that in Fig. 8.2 for: (a) $t = 0^-$, (b) $t = 0^+$, (c) $t \to \infty$.

As the inductor current and the capacitor voltage cannot change abruptly,

$$i(0^+) = i(0^-) = 2 \text{ A}, \qquad v(0^+) = v(0^-) = 4 \text{ V}$$

(b) At $t = 0^+$, the switch is open; the equivalent circuit is as shown in Fig. 8.3(b). The same current flows through both the inductor and capacitor. Hence,

$$i_C(0^+) = i(0^+) = 2 \text{ A}$$

Since $C\,dv/dt = i_C$, $dv/dt = i_C/C$, and

$$\frac{dv(0^+)}{dt} = \frac{i_C(0^+)}{C} = \frac{2}{0.1} = 20 \text{ V/s}$$

Similarly, since $L\,di/dt = v_L$, $di/dt = v_L/L$. We now obtain v_L by applying KVL to the loop in Fig. 8.3(b). The result is

$$-12 + 4i(0^+) + v_L(0^+) + v(0^+) = 0$$

or

$$v_L(0^+) = 12 - 8 - 4 = 0$$

Thus,

$$\frac{di(0^+)}{dt} = \frac{v_L(0^+)}{L} = \frac{0}{0.25} = 0 \text{ A/s}$$

(c) For $t > 0$, the circuit undergoes transience. But as $t \to \infty$, the circuit reaches steady state again. The inductor acts like a short circuit and the capacitor like an open circuit, so that the circuit in Fig. 8.3(b) becomes that shown in Fig. 8.3(c), from which we have

$$i(\infty) = 0 \text{ A}, \qquad v(\infty) = 12 \text{ V}$$

Practice Problem 8.1

The switch in Fig. 8.4 was open for a long time but closed at $t = 0$. Determine: (a) $i(0^+)$, $v(0^+)$, (b) $di(0^+)/dt$, $dv(0^+)/dt$, (c) $i(\infty)$, $v(\infty)$.

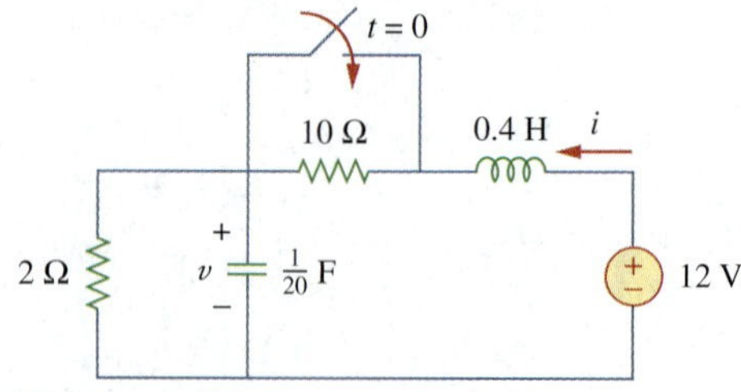

Figure 8.4
For Practice Prob. 8.1.

Answer: (a) 1 A, 2 V, (b) 25 A/s, 0 V/s, (c) 6 A, 12 V.

Example 8.2

In the circuit of Fig. 8.5, calculate: (a) $i_L(0^+)$, $v_C(0^+)$, $v_R(0^+)$, (b) $di_L(0^+)/dt$, $dv_C(0^+)/dt$, $dv_R(0^+)/dt$, (c) $i_L(\infty)$, $v_C(\infty)$, $v_R(\infty)$.

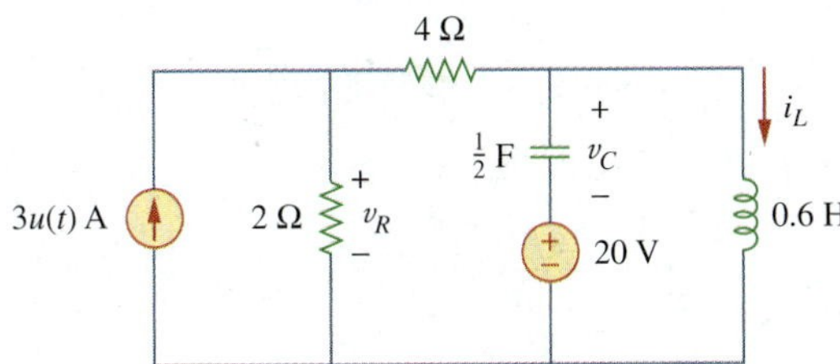

Figure 8.5
For Example 8.2.

Solution:

(a) For $t < 0$, $3u(t) = 0$. At $t = 0^-$, since the circuit has reached steady state, the inductor can be replaced by a short circuit, while the capacitor is replaced by an open circuit as shown in Fig. 8.6(a). From this figure we obtain

$$i_L(0^-) = 0, \qquad v_R(0^-) = 0, \qquad v_C(0^-) = -20 \text{ V} \tag{8.2.1}$$

Although the derivatives of these quantities at $t = 0^-$ are not required, it is evident that they are all zero, since the circuit has reached steady state and nothing changes.

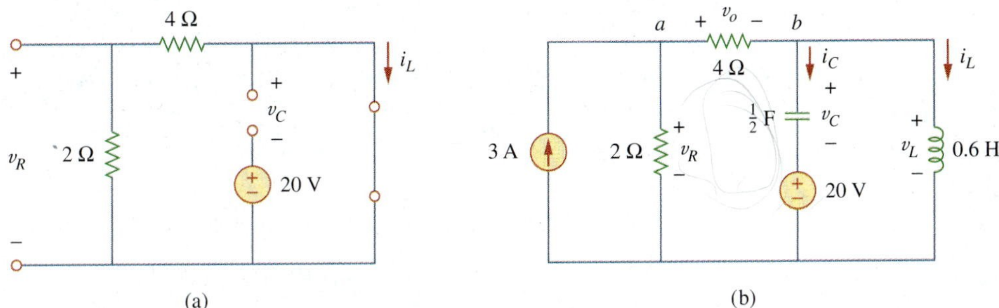

Figure 8.6
The circuit in Fig. 8.5 for: (a) $t = 0^-$, (b) $t = 0^+$.

For $t > 0$, $3u(t) = 3$, so that the circuit is now equivalent to that in Fig. 8.6(b). Since the inductor current and capacitor voltage cannot change abruptly,

$$i_L(0^+) = i_L(0^-) = 0, \qquad v_C(0^+) = v_C(0^-) = -20 \text{ V} \tag{8.2.2}$$

Although the voltage across the 4-Ω resistor is not required, we will use it to apply KVL and KCL; let it be called v_o. Applying KCL at node a in Fig. 8.6(b) gives

$$3 = \frac{v_R(0^+)}{2} + \frac{v_o(0^+)}{4} \tag{8.2.3}$$

Applying KVL to the middle mesh in Fig. 8.6(b) yields

$$-v_R(0^+) + v_o(0^+) + v_C(0^+) + 20 = 0 \tag{8.2.4}$$

Since $v_C(0^+) = -20$ V from Eq. (8.2.2), Eq. (8.2.4) implies that

$$v_R(0^+) = v_o(0^+) \tag{8.2.5}$$

From Eqs. (8.2.3) and (8.2.5), we obtain

$$v_R(0^+) = v_o(0^+) = 4 \text{ V} \tag{8.2.6}$$

(b) Since $L\,di_L/dt = v_L$,

$$\frac{di_L(0^+)}{dt} = \frac{v_L(0^+)}{L}$$

But applying KVL to the right mesh in Fig. 8.6(b) gives

$$v_L(0^+) = v_C(0^+) + 20 = 0$$

Hence,

$$\frac{di_L(0^+)}{dt} = 0 \tag{8.2.7}$$

Similarly, since $C\,dv_C/dt = i_C$, then $dv_C/dt = i_C/C$. We apply KCL at node b in Fig. 8.6(b) to get i_C:

$$\frac{v_o(0^+)}{4} = i_C(0^+) + i_L(0^+) \tag{8.2.8}$$

Since $v_o(0^+) = 4$ and $i_L(0^+) = 0$, $i_C(0^+) = 4/4 = 1$ A. Then

$$\frac{dv_C(0^+)}{dt} = \frac{i_C(0^+)}{C} = \frac{1}{0.5} = 2 \text{ V/s} \tag{8.2.9}$$

To get $dv_R(0^+)/dt$, we apply KCL to node a and obtain

$$3 = \frac{v_R}{2} + \frac{v_o}{4}$$

Taking the derivative of each term and setting $t = 0^+$ gives

$$0 = 2\frac{dv_R(0^+)}{dt} + \frac{dv_o(0^+)}{dt} \tag{8.2.10}$$

We also apply KVL to the middle mesh in Fig. 8.6(b) and obtain

$$-v_R + v_C + 20 + v_o = 0$$

Again, taking the derivative of each term and setting $t = 0^+$ yields

$$-\frac{dv_R(0^+)}{dt} + \frac{dv_C(0^+)}{dt} + \frac{dv_o(0^+)}{dt} = 0$$

Substituting for $dv_C(0^+)/dt = 2$ gives

$$\frac{dv_R(0^+)}{dt} = 2 + \frac{dv_o(0^+)}{dt} \tag{8.2.11}$$

From Eqs. (8.2.10) and (8.2.11), we get

$$\frac{dv_R(0^+)}{dt} = \frac{2}{3} \text{ V/s}$$

We can find $di_R(0^+)/dt$ although it is not required. Since $v_R = 5i_R$,

$$\frac{di_R(0^+)}{dt} = \frac{1}{5}\frac{dv_R(0^+)}{dt} = \frac{1}{5}\frac{2}{3} = \frac{2}{15} \text{ A/s}$$

(c) As $t \to \infty$, the circuit reaches steady state. We have the equivalent circuit in Fig. 8.6(a) except that the 3-A current source is now operative. By current division principle,

$$i_L(\infty) = \frac{2}{2+4}3 \text{ A} = 1 \text{ A}$$

$$v_R(\infty) = \frac{4}{2+4}3 \text{ A} \times 2 = 4 \text{ V}, \qquad v_C(\infty) = -20 \text{ V} \tag{8.2.12}$$

Practice Problem 8.2

For the circuit in Fig. 8.7, find: (a) $i_L(0^+), v_C(0^+), v_R(0^+)$, (b) $di_L(0^+)/dt, dv_C(0^+)/dt, dv_R(0^+)/dt$, (c) $i_L(\infty), v_C(\infty), v_R(\infty)$.

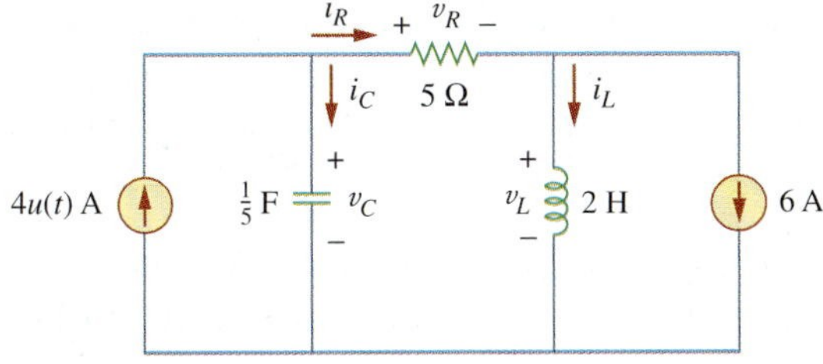

Figure 8.7
For Practice Prob. 8.2.

Answer: (a) −6 A, 0, 0, (b) 0, 20 V/s, 0, (c) −2 A, 20 V, 20 V.

8.3 The Source-Free Series *RLC* Circuit

An understanding of the natural response of the series *RLC* circuit is a necessary background for future studies in filter design and communications networks.

Consider the series *RLC* circuit shown in Fig. 8.8. The circuit is being excited by the energy initially stored in the capacitor and inductor. The energy is represented by the initial capacitor voltage V_0 and initial inductor current I_0. Thus, at $t = 0$,

$$v(0) = \frac{1}{C}\int_{-\infty}^{0} i \, dt = V_0 \tag{8.2a}$$

$$i(0) = I_0 \tag{8.2b}$$

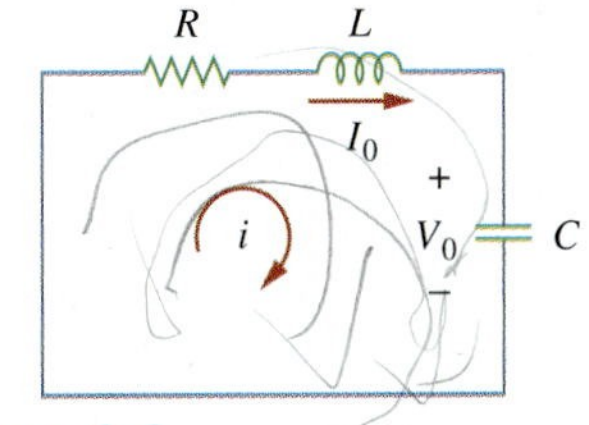

Figure 8.8
A source-free series *RLC* circuit.

Applying KVL around the loop in Fig. 8.8,

$$Ri + L\frac{di}{dt} + \frac{1}{C}\int_{-\infty}^{t} i \, dt = 0 \tag{8.3}$$

To eliminate the integral, we differentiate with respect to t and rearrange terms. We get

$$\frac{d^2i}{dt^2} + \frac{R}{L}\frac{di}{dt} + \frac{i}{LC} = 0 \qquad \textbf{(8.4)}$$

This is a *second-order differential equation* and is the reason for calling the *RLC* circuits in this chapter second-order circuits. Our goal is to solve Eq. (8.4). To solve such a second-order differential equation requires that we have two initial conditions, such as the initial value of i and its first derivative or initial values of some i and v. The initial value of i is given in Eq. (8.2b). We get the initial value of the derivative of i from Eqs. (8.2a) and (8.3); that is,

$$Ri(0) + L\frac{di(0)}{dt} + V_0 = 0$$

or

$$\frac{di(0)}{dt} = -\frac{1}{L}(RI_0 + V_0) \qquad \textbf{(8.5)}$$

With the two initial conditions in Eqs. (8.2b) and (8.5), we can now solve Eq. (8.4). Our experience in the preceding chapter on first-order circuits suggests that the solution is of exponential form. So we let

$$i = Ae^{st} \qquad \textbf{(8.6)}$$

where A and s are constants to be determined. Substituting Eq. (8.6) into Eq. (8.4) and carrying out the necessary differentiations, we obtain

$$As^2e^{st} + \frac{AR}{L}se^{st} + \frac{A}{LC}e^{st} = 0$$

or

$$Ae^{st}\left(s^2 + \frac{R}{L}s + \frac{1}{LC}\right) = 0 \qquad \textbf{(8.7)}$$

Since $i = Ae^{st}$ is the assumed solution we are trying to find, only the expression in parentheses can be zero:

$$s^2 + \frac{R}{L}s + \frac{1}{LC} = 0 \qquad \textbf{(8.8)}$$

See Appendix C.1 for the formula to find the roots of a quadratic equation.

This quadratic equation is known as the *characteristic equation* of the differential Eq. (8.4), since the roots of the equation dictate the character of i. The two roots of Eq. (8.8) are

$$s_1 = -\frac{R}{2L} + \sqrt{\left(\frac{R}{2L}\right)^2 - \frac{1}{LC}} \qquad \textbf{(8.9a)}$$

$$s_2 = -\frac{R}{2L} - \sqrt{\left(\frac{R}{2L}\right)^2 - \frac{1}{LC}} \qquad \textbf{(8.9b)}$$

A more compact way of expressing the roots is

$$s_1 = -\alpha + \sqrt{\alpha^2 - \omega_0^2}, \qquad s_2 = -\alpha - \sqrt{\alpha^2 - \omega_0^2} \qquad \textbf{(8.10)}$$

where

$$\alpha = \frac{R}{2L}, \qquad \omega_0 = \frac{1}{\sqrt{LC}} \tag{8.11}$$

The roots s_1 and s_2 are called *natural frequencies*, measured in nepers per second (Np/s), because they are associated with the natural response of the circuit; ω_0 is known as the *resonant frequency* or strictly as the *undamped natural frequency*, expressed in radians per second (rad/s); and α is the *neper frequency* or the *damping factor*, expressed in nepers per second. In terms of α and ω_0, Eq. (8.8) can be written as

The *neper* (Np) is a dimensionless unit named after John Napier (1550–1617), a Scottish mathematician.

$$s^2 + 2\alpha s + \omega_0^2 = 0 \tag{8.8a}$$

The variables s and ω_0 are important quantities we will be discussing throughout the rest of the text.

The ratio α/ω_0 is known as the *damping ratio* ζ.

The two values of s in Eq. (8.10) indicate that there are two possible solutions for i, each of which is of the form of the assumed solution in Eq. (8.6); that is,

$$i_1 = A_1 e^{s_1 t}, \qquad i_2 = A_2 e^{s_2 t} \tag{8.12}$$

Since Eq. (8.4) is a linear equation, any linear combination of the two distinct solutions i_1 and i_2 is also a solution of Eq. (8.4). A complete or total solution of Eq. (8.4) would therefore require a linear combination of i_1 and i_2. Thus, the natural response of the series *RLC* circuit is

$$i(t) = A_1 e^{s_1 t} + A_2 e^{s_2 t} \tag{8.13}$$

where the constants A_1 and A_2 are determined from the initial values $i(0)$ and $di(0)/dt$ in Eqs. (8.2b) and (8.5).

From Eq. (8.10), we can infer that there are three types of solutions:

1. If $\alpha > \omega_0$, we have the *overdamped* case.
2. If $\alpha = \omega_0$, we have the *critically damped* case.
3. If $\alpha < \omega_0$, we have the *underdamped* case.

We will consider each of these cases separately.

The response is *overdamped* when the roots of the circuit's characteristic equation are unequal and real, *critically damped* when the roots are equal and real, and *underdamped* when the roots are complex.

Overdamped Case ($\alpha > \omega_0$)

From Eqs. (8.9) and (8.10), $\alpha > \omega_0$ implies $C > 4L/R^2$. When this happens, both roots s_1 and s_2 are negative and real. The response is

$$i(t) = A_1 e^{s_1 t} + A_2 e^{s_2 t} \tag{8.14}$$

which decays and approaches zero as t increases. Figure 8.9(a) illustrates a typical overdamped response.

Critically Damped Case ($\alpha = \omega_0$)

When $\alpha = \omega_0$, $C = 4L/R^2$ and

$$s_1 = s_2 = -\alpha = -\frac{R}{2L} \tag{8.15}$$

For this case, Eq. (8.13) yields

$$i(t) = A_1 e^{-\alpha t} + A_2 e^{-\alpha t} = A_3 e^{-\alpha t}$$

where $A_3 = A_1 + A_2$. This cannot be the solution, because the two initial conditions cannot be satisfied with the single constant A_3. What then could be wrong? Our assumption of an exponential solution is incorrect for the special case of critical damping. Let us go back to Eq. (8.4). When $\alpha = \omega_0 = R/2L$, Eq. (8.4) becomes

$$\frac{d^2 i}{dt^2} + 2\alpha \frac{di}{dt} + \alpha^2 i = 0$$

or

$$\frac{d}{dt}\left(\frac{di}{dt} + \alpha i\right) + \alpha\left(\frac{di}{dt} + \alpha i\right) = 0 \tag{8.16}$$

If we let

$$f = \frac{di}{dt} + \alpha i \tag{8.17}$$

then Eq. (8.16) becomes

$$\frac{df}{dt} + \alpha f = 0$$

which is a first-order differential equation with solution $f = A_1 e^{-\alpha t}$, where A_1 is a constant. Equation (8.17) then becomes

$$\frac{di}{dt} + \alpha i = A_1 e^{-\alpha t}$$

or

$$e^{\alpha t}\frac{di}{dt} + e^{\alpha t}\alpha i = A_1 \tag{8.18}$$

This can be written as

$$\frac{d}{dt}(e^{\alpha t} i) = A_1 \tag{8.19}$$

Integrating both sides yields

$$e^{\alpha t} i = A_1 t + A_2$$

or

$$i = (A_1 t + A_2)e^{-\alpha t} \tag{8.20}$$

where A_2 is another constant. Hence, the natural response of the critically damped circuit is a sum of two terms: a negative exponential and a negative exponential multiplied by a linear term, or

$$\boxed{i(t) = (A_2 + A_1 t)e^{-\alpha t}} \tag{8.21}$$

A typical critically damped response is shown in Fig. 8.9(b). In fact, Fig. 8.9(b) is a sketch of $i(t) = te^{-\alpha t}$, which reaches a maximum value of e^{-1}/α at $t = 1/\alpha$, one time constant, and then decays all the way to zero.

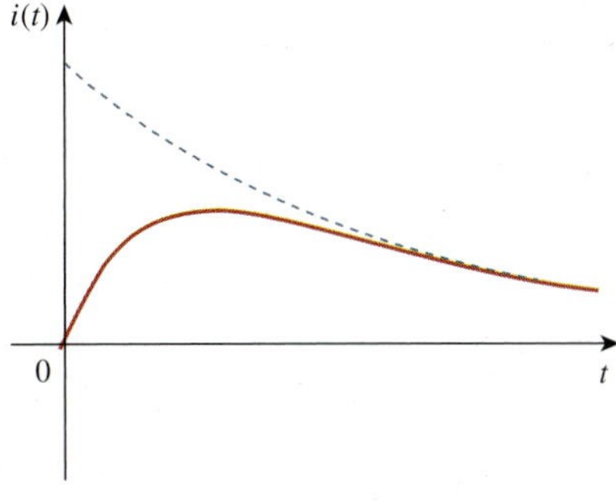

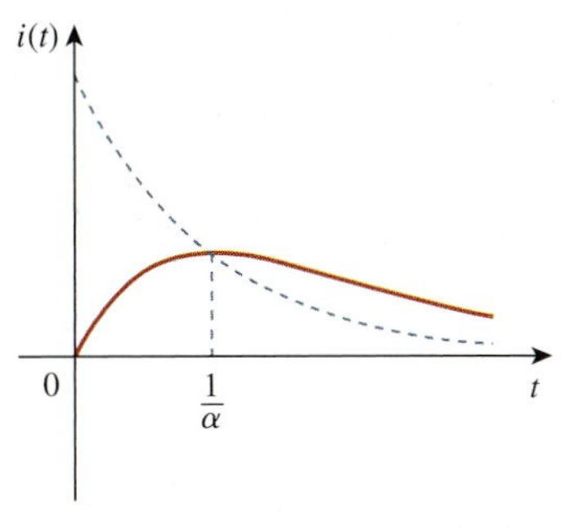

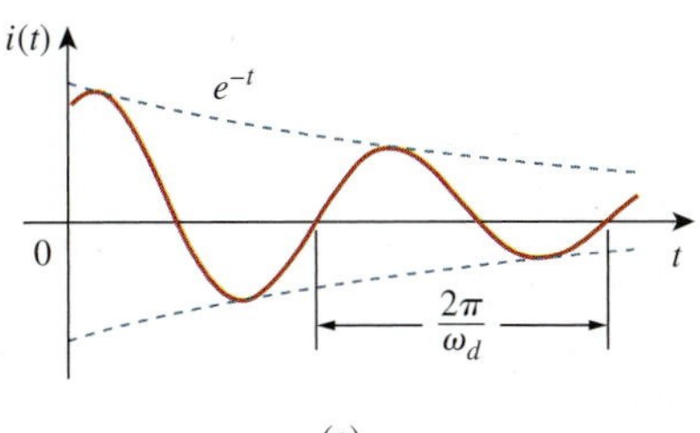

Figure 8.9
(a) Overdamped response, (b) critically damped response, (c) underdamped response.

Underdamped Case ($\alpha < \omega_0$)

For $\alpha < \omega_0$, $C < 4L/R^2$. The roots may be written as

$$s_1 = -\alpha + \sqrt{-(\omega_0^2 - \alpha^2)} = -\alpha + j\omega_d \tag{8.22a}$$

$$s_2 = -\alpha - \sqrt{-(\omega_0^2 - \alpha^2)} = -\alpha - j\omega_d \tag{8.22b}$$

where $j = \sqrt{-1}$ and $\omega_d = \sqrt{\omega_0^2 - \alpha^2}$, which is called the *damping frequency*. Both ω_0 and ω_d are natural frequencies because they help determine the natural response; while ω_0 is often called the *undamped natural frequency*, ω_d is called the *damped natural frequency*. The natural response is

$$\begin{aligned} i(t) &= A_1 e^{-(\alpha - j\omega_d)t} + A_2 e^{-(\alpha + j\omega_d)t} \\ &= e^{-\alpha t}(A_1 e^{j\omega_d t} + A_2 e^{-j\omega_d t}) \end{aligned} \tag{8.23}$$

Using Euler's identities,

$$e^{j\theta} = \cos\theta + j\sin\theta, \qquad e^{-j\theta} = \cos\theta - j\sin\theta \tag{8.24}$$

we get

$$\begin{aligned} i(t) &= e^{-\alpha t}[A_1(\cos\omega_d t + j\sin\omega_d t) + A_2(\cos\omega_d t - j\sin\omega_d t)] \\ &= e^{-\alpha t}[(A_1 + A_2)\cos\omega_d t + j(A_1 - A_2)\sin\omega_d t] \end{aligned} \tag{8.25}$$

Replacing constants $(A_1 + A_2)$ and $j(A_1 - A_2)$ with constants B_1 and B_2, we write

$$i(t) = e^{-\alpha t}(B_1 \cos\omega_d t + B_2 \sin\omega_d t) \tag{8.26}$$

With the presence of sine and cosine functions, it is clear that the natural response for this case is exponentially damped and oscillatory in nature. The response has a time constant of $1/\alpha$ and a period of $T = 2\pi/\omega_d$. Figure 8.9(c) depicts a typical underdamped response. [Figure 8.9 assumes for each case that $i(0) = 0$.]

Once the inductor current $i(t)$ is found for the *RLC* series circuit as shown above, other circuit quantities such as individual element voltages can easily be found. For example, the resistor voltage is $v_R = Ri$, and the inductor voltage is $v_L = L\,di/dt$. The inductor current $i(t)$ is selected as the key variable to be determined first in order to take advantage of Eq. (8.1b).

We conclude this section by noting the following interesting, peculiar properties of an *RLC* network:

1. The behavior of such a network is captured by the idea of *damping*, which is the gradual loss of the initial stored energy, as evidenced by the continuous decrease in the amplitude of the response. The damping effect is due to the presence of resistance R. The damping factor α determines the rate at which the response is damped. If $R = 0$, then $\alpha = 0$, and we have an *LC* circuit with $1/\sqrt{LC}$ as the undamped natural frequency. Since $\alpha < \omega_0$ in this case, the response is not only undamped but also oscillatory. The circuit is said to be *loss-less*, because the dissipating or damping element (R) is absent. By adjusting the value of R, the response may be made undamped, overdamped, critically damped, or underdamped.
2. Oscillatory response is possible due to the presence of the two types of storage elements. Having both L and C allows the flow of

$R = 0$ produces a perfectly sinusoidal response. This response cannot be practically accomplished with L and C because of the inherent losses in them. See Figs 6.8 and 6.26. An electronic device called an *oscillator* can produce a perfectly sinusoidal response.

Examples 8.5 and 8.7 demonstrate the effect of varying R.

The response of a second-order circuit with two storage elements of the same type, as in Fig. 8.1(c) and (d), cannot be oscillatory.

energy back and forth between the two. The damped oscillation exhibited by the underdamped response is known as *ringing*. It stems from the ability of the storage elements L and C to transfer energy back and forth between them.

3. Observe from Fig. 8.9 that the waveforms of the responses differ. In general, it is difficult to tell from the waveforms the difference between the overdamped and critically damped responses. The critically damped case is the borderline between the underdamped and overdamped cases and it decays the fastest. With the same initial conditions, the overdamped case has the longest settling time, because it takes the longest time to dissipate the initial stored energy. If we desire the response that approaches the final value most rapidly without oscillation or ringing, the critically damped circuit is the right choice.

What this means in most practical circuits is that we seek an overdamped circuit that is as close as possible to a critically damped circuit.

Example 8.3

In Fig. 8.8, $R = 40\ \Omega$, $L = 4$ H, and $C = 1/4$ F. Calculate the characteristic roots of the circuit. Is the natural response overdamped, underdamped, or critically damped?

Solution:

We first calculate

$$\alpha = \frac{R}{2L} = \frac{40}{2(4)} = 5, \qquad \omega_0 = \frac{1}{\sqrt{LC}} = \frac{1}{\sqrt{4 \times \frac{1}{4}}} = 1$$

The roots are

$$s_{1,2} = -\alpha \pm \sqrt{\alpha^2 - \omega_0^2} = -5 \pm \sqrt{25 - 1}$$

or

$$s_1 = -0.101, \qquad s_2 = -9.899$$

Since $\alpha > \omega_0$, we conclude that the response is overdamped. This is also evident from the fact that the roots are real and negative.

Practice Problem 8.3

If $R = 10\ \Omega$, $L = 5$ H, and $C = 2$ mF in Fig. 8.8, find α, ω_0, s_1, and s_2. What type of natural response will the circuit have?

Answer: 1, 10, $-1 \pm j9.95$, underdamped.

Example 8.4

Find $i(t)$ in the circuit of Fig. 8.10. Assume that the circuit has reached steady state at $t = 0^-$.

Solution:

For $t < 0$, the switch is closed. The capacitor acts like an open circuit while the inductor acts like a shunted circuit. The equivalent circuit is shown in Fig. 8.11(a). Thus, at $t = 0$,

$$i(0) = \frac{10}{4 + 6} = 1\text{ A}, \qquad v(0) = 6i(0) = 6\text{ V}$$

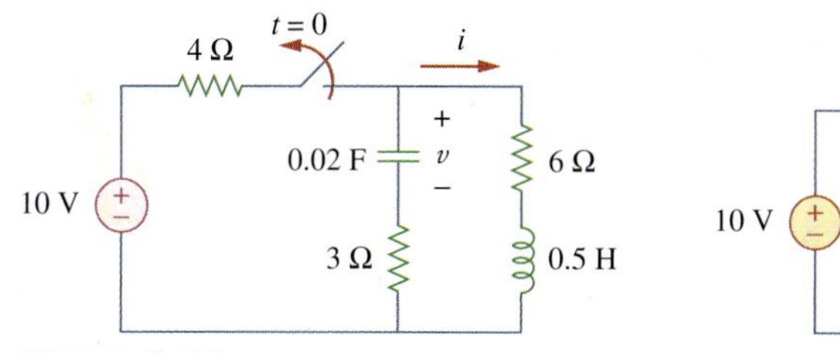

Figure 8.10
For Example 8.4.

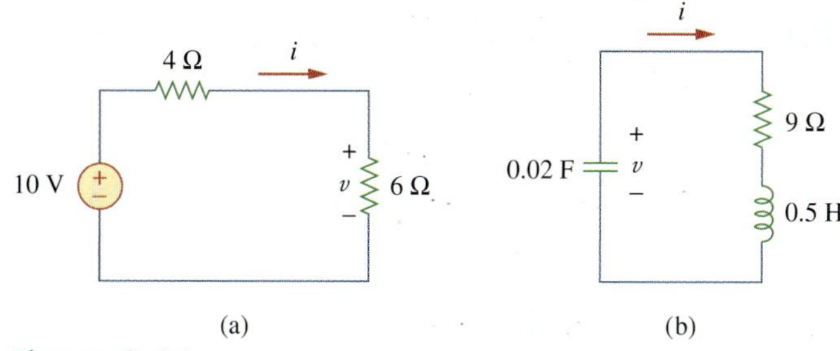

Figure 8.11
The circuit in Fig. 8.10: (a) for $t < 0$, (b) for $t > 0$.

where $i(0)$ is the initial current through the inductor and $v(0)$ is the initial voltage across the capacitor.

For $t > 0$, the switch is opened and the voltage source is disconnected. The equivalent circuit is shown in Fig. 8.11(b), which is a source-free series *RLC* circuit. Notice that the 3-Ω and 6-Ω resistors, which are in series in Fig. 8.10 when the switch is opened, have been combined to give $R = 9\ \Omega$ in Fig. 8.11(b). The roots are calculated as follows:

$$\alpha = \frac{R}{2L} = \frac{9}{2(\frac{1}{2})} = 9, \qquad \omega_0 = \frac{1}{\sqrt{LC}} = \frac{1}{\sqrt{\frac{1}{2} \times \frac{1}{50}}} = 10$$

$$s_{1,2} = -\alpha \pm \sqrt{\alpha^2 - \omega_0^2} = -9 \pm \sqrt{81 - 100}$$

or

$$s_{1,2} = -9 \pm j4.359$$

Hence, the response is underdamped ($\alpha < \omega$); that is,

$$i(t) = e^{-9t}(A_1 \cos 4.359t + A_2 \sin 4.359t) \qquad \textbf{(8.4.1)}$$

We now obtain A_1 and A_2 using the initial conditions. At $t = 0$,

$$i(0) = 1 = A_1 \qquad \textbf{(8.4.2)}$$

From Eq. (8.5),

$$\left.\frac{di}{dt}\right|_{t=0} = -\frac{1}{L}[Ri(0) + v(0)] = -2[9(1) - 6] = -6 \text{ A/s} \qquad \textbf{(8.4.3)}$$

Note that $v(0) = V_0 = -6$ V is used, because the polarity of v in Fig. 8.11(b) is opposite that in Fig. 8.8. Taking the derivative of $i(t)$ in Eq. (8.4.1),

$$\frac{di}{dt} = -9e^{-9t}(A_1 \cos 4.359t + A_2 \sin 4.359t)$$
$$+ e^{-9t}(4.359)(-A_1 \sin 4.359t + A_2 \cos 4.359t)$$

Imposing the condition in Eq. (8.4.3) at $t = 0$ gives

$$-6 = -9(A_1 + 0) + 4.359(-0 + A_2)$$

But $A_1 = 1$ from Eq. (8.4.2). Then

$$-6 = -9 + 4.359A_2 \qquad \Rightarrow \qquad A_2 = 0.6882$$

Substituting the values of A_1 and A_2 in Eq. (8.4.1) yields the complete solution as

$$i(t) = e^{-9t}(\cos 4.359t + 0.6882 \sin 4.359t) \text{ A}$$

Practice Problem 8.4

The circuit in Fig. 8.12 has reached steady state at $t = 0^-$. If the make-before-break switch moves to position b at $t = 0$, calculate $i(t)$ for $t > 0$.

Answer: $e^{-2.5t}(5\cos 1.6583t - 7.5378\sin 1.6583t)$ A.

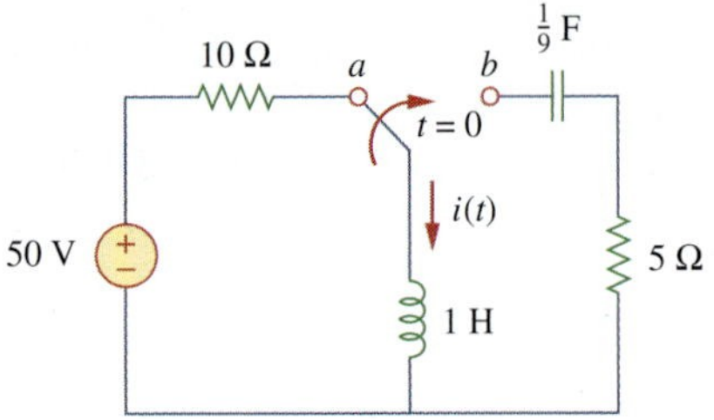

Figure 8.12
For Practice Prob. 8.4.

8.4 The Source-Free Parallel *RLC* Circuit

Parallel *RLC* circuits find many practical applications, notably in communications networks and filter designs.

Consider the parallel *RLC* circuit shown in Fig. 8.13. Assume initial inductor current I_0 and initial capacitor voltage V_0,

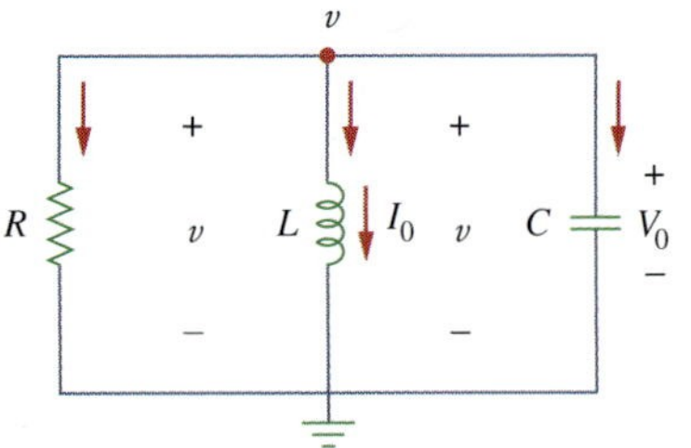

Figure 8.13
A source-free parallel *RLC* circuit.

$$i(0) = I_0 = \frac{1}{L}\int_{\infty}^{0} v(t)\,dt \tag{8.27a}$$

$$v(0) = V_0 \tag{8.27b}$$

Since the three elements are in parallel, they have the same voltage v across them. According to passive sign convention, the current is entering each element; that is, the current through each element is leaving the top node. Thus, applying KCL at the top node gives

$$\frac{v}{R} + \frac{1}{L}\int_{-\infty}^{t} v\,dt + C\frac{dv}{dt} = 0 \tag{8.28}$$

Taking the derivative with respect to t and dividing by C results in

$$\frac{d^2v}{dt^2} + \frac{1}{RC}\frac{dv}{dt} + \frac{1}{LC}v = 0 \tag{8.29}$$

We obtain the characteristic equation by replacing the first derivative by s and the second derivative by s^2. By following the same reasoning used in establishing Eqs. (8.4) through (8.8), the characteristic equation is obtained as

$$s^2 + \frac{1}{RC}s + \frac{1}{LC} = 0 \tag{8.30}$$

The roots of the characteristic equation are

$$s_{1,2} = -\frac{1}{2RC} \pm \sqrt{\left(\frac{1}{2RC}\right)^2 - \frac{1}{LC}}$$

or

$$s_{1,2} = -\alpha \pm \sqrt{\alpha^2 - \omega_0^2} \tag{8.31}$$

where

$$\alpha = \frac{1}{2RC}, \qquad \omega_0 = \frac{1}{\sqrt{LC}} \tag{8.32}$$

The names of these terms remain the same as in the preceding section, as they play the same role in the solution. Again, there are three possible solutions, depending on whether $\alpha > \omega_0$, $\alpha = \omega_0$, or $\alpha < \omega_0$. Let us consider these cases separately.

Overdamped Case ($\alpha > \omega_0$)

From Eq. (8.32), $\alpha > \omega_0$ when $L > 4R^2C$. The roots of the characteristic equation are real and negative. The response is

$$v(t) = A_1 e^{s_1 t} + A_2 e^{s_2 t} \tag{8.33}$$

Critically Damped Case ($\alpha = \omega_0$)

For $\alpha = \omega_0$, $L = 4R^2C$. The roots are real and equal so that the response is

$$v(t) = (A_1 + A_2 t)e^{-\alpha t} \tag{8.34}$$

Underdamped Case ($\alpha < \omega_0$)

When $\alpha < \omega_0$, $L < 4R^2C$. In this case the roots are complex and may be expressed as

$$s_{1,2} = -\alpha \pm j\omega_d \tag{8.35}$$

where

$$\omega_d = \sqrt{\omega_0^2 - \alpha^2} \tag{8.36}$$

The response is

$$v(t) = e^{-\alpha t}(A_1 \cos \omega_d t + A_2 \sin \omega_d t) \tag{8.37}$$

The constants A_1 and A_2 in each case can be determined from the initial conditions. We need $v(0)$ and $dv(0)/dt$. The first term is known from Eq. (8.27b). We find the second term by combining Eqs. (8.27) and (8.28), as

$$\frac{V_0}{R} + I_0 + C\frac{dv(0)}{dt} = 0$$

or

$$\frac{dv(0)}{dt} = -\frac{(V_0 + RI_0)}{RC} \tag{8.38}$$

The voltage waveforms are similar to those shown in Fig. 8.9 and will depend on whether the circuit is overdamped, underdamped, or critically damped.

Having found the capacitor voltage $v(t)$ for the parallel *RLC* circuit as shown above, we can readily obtain other circuit quantities such as individual element currents. For example, the resistor current is $i_R = v/R$ and the capacitor voltage is $v_C = C\,dv/dt$. We have selected the capacitor voltage $v(t)$ as the key variable to be determined first in order to take advantage of Eq. (8.1a). Notice that we first found the inductor current $i(t)$ for the *RLC* series circuit, whereas we first found the capacitor voltage $v(t)$ for the parallel *RLC* circuit.

Example 8.5

In the parallel circuit of Fig. 8.13, find $v(t)$ for $t > 0$, assuming $v(0) = 5$ V, $i(0) = 0$, $L = 1$ H, and $C = 10$ mF. Consider these cases: $R = 1.923\ \Omega$, $R = 5\ \Omega$, and $R = 6.25\ \Omega$.

Solution:

CASE 1 If $R = 1.923\ \Omega$,

$$\alpha = \frac{1}{2RC} = \frac{1}{2 \times 1.923 \times 10 \times 10^{-3}} = 26$$

$$\omega_0 = \frac{1}{\sqrt{LC}} = \frac{1}{\sqrt{1 \times 10 \times 10^{-3}}} = 10$$

Since $\alpha > \omega_0$ in this case, the response is overdamped. The roots of the characteristic equation are

$$s_{1,2} = -\alpha \pm \sqrt{\alpha^2 - \omega_0^2} = -2, -50$$

and the corresponding response is

$$v(t) = A_1 e^{-2t} + A_2 e^{-50t} \tag{8.5.1}$$

We now apply the initial conditions to get A_1 and A_2.

$$v(0) = 5 = A_1 + A_2 \tag{8.5.2}$$

$$\frac{dv(0)}{dt} = -\frac{v(0) + Ri(0)}{RC} = -\frac{5 + 0}{1.923 \times 10 \times 10^{-3}} = -260$$

But differentiating Eq. (8.5.1),

$$\frac{dv}{dt} = -2A_1 e^{-2t} - 50A_2 e^{-50t}$$

At $t = 0$,

$$-260 = -2A_1 - 50A_2 \tag{8.5.3}$$

From Eqs. (8.5.2) and (8.5.3), we obtain $A_1 = -0.2083$ and $A_2 = 5.208$. Substituting A_1 and A_2 in Eq. (8.5.1) yields

$$v(t) = -0.2083e^{-2t} + 5.208e^{-50t} \tag{8.5.4}$$

CASE 2 When $R = 5\ \Omega$,

$$\alpha = \frac{1}{2RC} = \frac{1}{2 \times 5 \times 10 \times 10^{-3}} = 10$$

while $\omega_0 = 10$ remains the same. Since $\alpha = \omega_0 = 10$, the response is critically damped. Hence, $s_1 = s_2 = -10$, and

$$v(t) = (A_1 + A_2 t)e^{-10t} \tag{8.5.5}$$

To get A_1 and A_2, we apply the initial conditions

$$v(0) = 5 = A_1 \tag{8.5.6}$$

$$\frac{dv(0)}{dt} = -\frac{v(0) + Ri(0)}{RC} = -\frac{5 + 0}{5 \times 10 \times 10^{-3}} = -100$$

But differentiating Eq. (8.5.5),

$$\frac{dv}{dt} = (-10A_1 - 10A_2 t + A_2)e^{-10t}$$

At $t = 0$,

$$-100 = -10A_1 + A_2 \tag{8.5.7}$$

From Eqs. (8.5.6) and (8.5.7), $A_1 = 5$ and $A_2 = -50$. Thus,

$$v(t) = (5 - 50t)e^{-10t} \text{ V} \tag{8.5.8}$$

CASE 3 When $R = 6.25\ \Omega$,

$$\alpha = \frac{1}{2RC} = \frac{1}{2 \times 6.25 \times 10 \times 10^{-3}} = 8$$

while $\omega_0 = 10$ remains the same. As $\alpha < \omega_0$ in this case, the response is underdamped. The roots of the characteristic equation are

$$s_{1,2} = -\alpha \pm \sqrt{\alpha^2 - \omega_0^2} = -8 \pm j6$$

Hence,

$$v(t) = (A_1 \cos 6t + A_2 \sin 6t)e^{-8t} \tag{8.5.9}$$

We now obtain A_1 and A_2, as

$$v(0) = 5 = A_1 \tag{8.5.10}$$

$$\frac{dv(0)}{dt} = -\frac{v(0) + Ri(0)}{RC} = -\frac{5 + 0}{6.25 \times 10 \times 10^{-3}} = -80$$

But differentiating Eq. (8.5.9),

$$\frac{dv}{dt} = (-8A_1 \cos 6t - 8A_2 \sin 6t - 6A_1 \sin 6t + 6A_2 \cos 6t)e^{-8t}$$

At $t = 0$,

$$-80 = -8A_1 + 6A_2 \tag{8.5.11}$$

From Eqs. (8.5.10) and (8.5.11), $A_1 = 5$ and $A_2 = -6.667$. Thus,

$$v(t) = (5 \cos 6t - 6.667 \sin 6t)e^{-8t} \tag{8.5.12}$$

Notice that by increasing the value of R, the degree of damping decreases and the responses differ. Figure 8.14 plots the three cases.

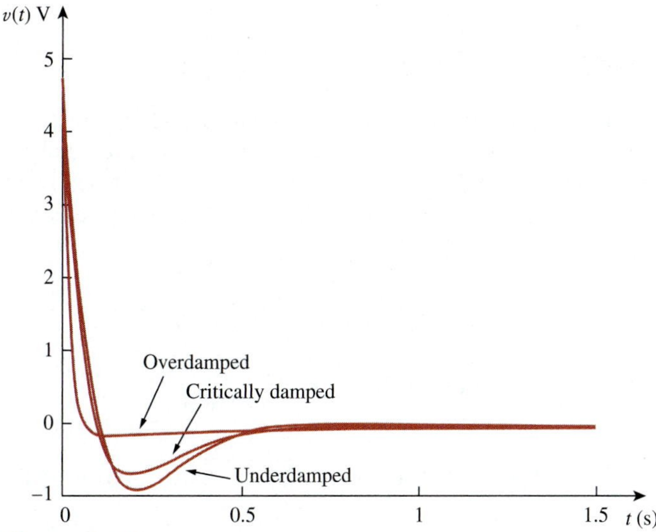

Figure 8.14
For Example 8.5: responses for three degrees of damping.

Practice Problem 8.5

In Fig. 8.13, let $R = 2\ \Omega$, $L = 0.4$ H, $C = 25$ mF, $v(0) = 0$, $i(0) = 10$ mA. Find $v(t)$ for $t > 0$.

Answer: $-400te^{-10t}\,u(t)$ mV.

Example 8.6

Find $v(t)$ for $t > 0$ in the *RLC* circuit of Fig. 8.15.

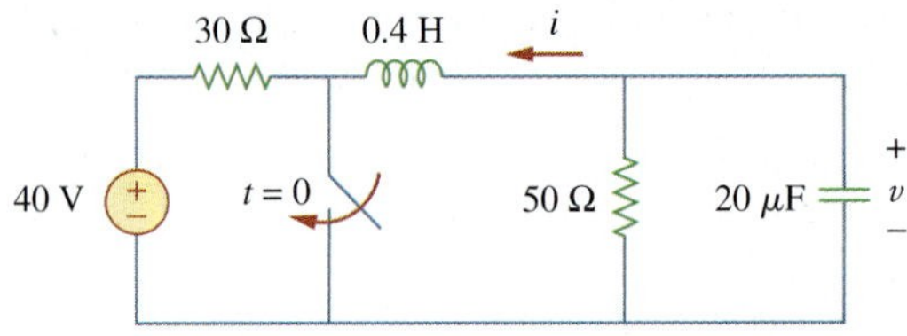

Figure 8.15
For Example 8.6.

Solution:
When $t < 0$, the switch is open; the inductor acts like a short circuit while the capacitor behaves like an open circuit. The initial voltage across the capacitor is the same as the voltage across the 50-Ω resistor; that is,

$$v(0) = \frac{50}{30 + 50}(40) = \frac{5}{8} \times 40 = 25 \text{ V} \tag{8.6.1}$$

The initial current through the inductor is

$$i(0) = -\frac{40}{30 + 50} = -0.5 \text{ A}$$

The direction of i is as indicated in Fig. 8.15 to conform with the direction of I_0 in Fig. 8.13, which is in agreement with the convention that current flows into the positive terminal of an inductor (see Fig. 6.23). We need to express this in terms of dv/dt, since we are looking for v.

$$\frac{dv(0)}{dt} = -\frac{v(0) + Ri(0)}{RC} = -\frac{25 - 50 \times 0.5}{50 \times 20 \times 10^{-6}} = 0 \tag{8.6.2}$$

When $t > 0$, the switch is closed. The voltage source along with the 30-Ω resistor is separated from the rest of the circuit. The parallel *RLC* circuit acts independently of the voltage source, as illustrated in Fig. 8.16. Next, we determine that the roots of the characteristic equation are

$$\alpha = \frac{1}{2RC} = \frac{1}{2 \times 50 \times 20 \times 10^{-6}} = 500$$

$$\omega_0 = \frac{1}{\sqrt{LC}} = \frac{1}{\sqrt{0.4 \times 20 \times 10^{-6}}} = 354$$

$$\begin{aligned} s_{1,2} &= -\alpha \pm \sqrt{\alpha^2 - \omega_0^2} \\ &= -500 \pm \sqrt{250{,}000 - 124{,}997.6} = -500 \pm 354 \end{aligned}$$

or

$$s_1 = -854, \qquad s_2 = -146$$

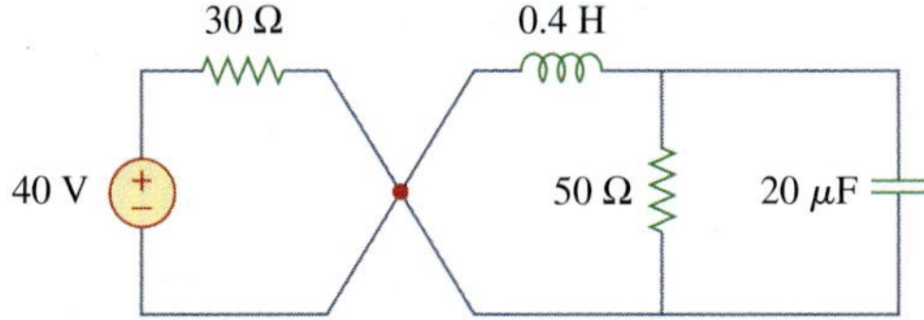

Figure 8.16
The circuit in Fig. 8.15 when $t > 0$. The parallel *RLC* circuit on the right-hand side acts independently of the circuit on the left-hand side of the junction.

Since $\alpha > \omega_0$, we have the overdamped response

$$v(t) = A_1 e^{-854t} + A_2 e^{-146t} \tag{8.6.3}$$

At $t = 0$, we impose the condition in Eq. (8.6.1),

$$v(0) = 25 = A_1 + A_2 \quad \Rightarrow \quad A_2 = 25 - A_1 \tag{8.6.4}$$

Taking the derivative of $v(t)$ in Eq. (8.6.3),

$$\frac{dv}{dt} = -854 A_1 e^{-854t} - 146 A_2 e^{-146t}$$

Imposing the condition in Eq. (8.6.2),

$$\frac{dv(0)}{dt} = 0 = -854 A_1 - 146 A_2$$

or

$$0 = 854 A_1 + 146 A_2 \tag{8.6.5}$$

Solving Eqs. (8.6.4) and (8.6.5) gives

$$A_1 = -5.156, \qquad A_2 = 30.16$$

Thus, the complete solution in Eq. (8.6.3) becomes

$$v(t) = -5.156 e^{-854t} + 30.16 e^{-146t} \text{ V}$$

Practice Problem 8.6

Refer to the circuit in Fig. 8.17. Find $v(t)$ for $t > 0$.

Answer: $100(e^{-10t} - e^{-2.5t})$ V.

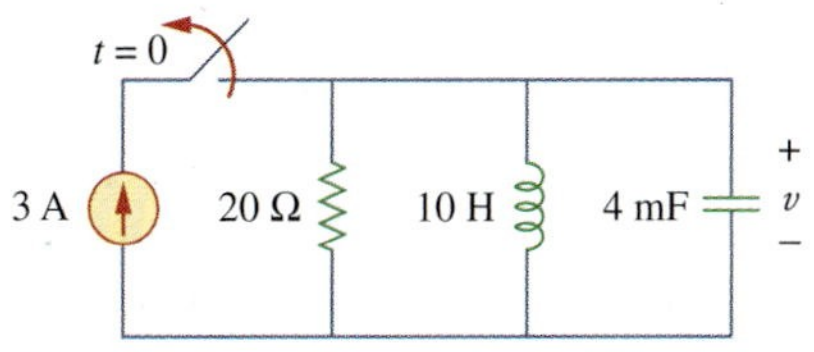

Figure 8.17
For Practice Prob. 8.6.

8.5 Step Response of a Series *RLC* Circuit

As we learned in the preceding chapter, the step response is obtained by the sudden application of a dc source. Consider the series *RLC* circuit shown in Fig. 8.18. Applying KVL around the loop for $t > 0$,

$$L\frac{di}{dt} + Ri + v = V_s \tag{8.39}$$

But

$$i = C\frac{dv}{dt}$$

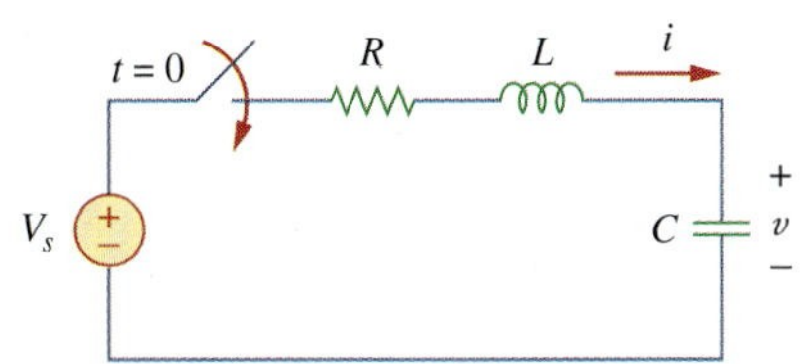

Figure 8.18
Step voltage applied to a series *RLC* circuit.

Substituting for i in Eq. (8.39) and rearranging terms,

$$\frac{d^2v}{dt^2} + \frac{R}{L}\frac{dv}{dt} + \frac{v}{LC} = \frac{V_s}{LC} \tag{8.40}$$

which has the same form as Eq. (8.4). More specifically, the coefficients are the same (and that is important in determining the frequency parameters) but the variable is different. (Likewise, see Eq. (8.47).) Hence, the characteristic equation for the series RLC circuit is not affected by the presence of the dc source.

The solution to Eq. (8.40) has two components: the transient response $v_t(t)$ and the steady-state response $v_{ss}(t)$; that is,

$$v(t) = v_t(t) + v_{ss}(t) \tag{8.41}$$

The transient response $v_t(t)$ is the component of the total response that dies out with time. The form of the transient response is the same as the form of the solution obtained in Section 8.3 for the source-free circuit, given by Eqs. (8.14), (8.21), and (8.26). Therefore, the transient repsonse $v_t(t)$ for the overdamped, underdamped, and critically damped cases are:

$$v_t(t) = A_1e^{s_1t} + A_2e^{s_2t} \quad \text{(Overdamped)} \tag{8.42a}$$

$$v_t(t) = (A_1 + A_2t)e^{-\alpha t} \quad \text{(Critically damped)} \tag{8.42b}$$

$$v_t(t) = (A_1 \cos \omega_d t + A_2 \sin \omega_d t)e^{-\alpha t} \quad \text{(Underdamped)} \tag{8.42c}$$

The steady-state response is the final value of $v(t)$. In the circuit in Fig. 8.18, the final value of the capacitor voltage is the same as the source voltage V_s. Hence,

$$v_{ss}(t) = v(\infty) = V_s \tag{8.43}$$

Thus, the complete solutions for the overdamped, underdamped, and critically damped cases are:

$$v(t) = V_s + A_1e^{s_1t} + A_2e^{s_2t} \quad \text{(Overdamped)} \tag{8.44a}$$

$$v(t) = V_s + (A_1 + A_2t)e^{-\alpha t} \quad \text{(Critically damped)} \tag{8.44b}$$

$$v(t) = V_s + (A_1 \cos \omega_d t + A_2 \sin \omega_d t)e^{-\alpha t} \quad \text{(Underdamped)} \tag{8.44c}$$

The values of the constants A_1 and A_2 are obtained from the initial conditions: $v(0)$ and $dv(0)/dt$. Keep in mind that v and i are, respectively, the voltage across the capacitor and the current through the inductor. Therefore, Eq. (8.44) only applies for finding v. But once the capacitor voltage $v_C = v$ is known, we can determine $i = C\,dv/dt$, which is the same current through the capacitor, inductor, and resistor. Hence, the voltage across the resistor is $v_R = iR$, while the inductor voltage is $v_L = L\,di/dt$.

Alternatively, the complete response for any variable $x(t)$ can be found directly, because it has the general form

$$x(t) = x_{ss}(t) + x_t(t) \tag{8.45}$$

where the $x_{ss} = x(\infty)$ is the final value and $x_t(t)$ is the transient response. The final value is found as in Section 8.2. The transient response has the same form as in Eq. (8.42), and the associated constants are determined from Eq. (8.44) based on the values of $x(0)$ and $dx(0)/dt$.

Example 8.7

For the circuit in Fig. 8.19, find $v(t)$ and $i(t)$ for $t > 0$. Consider these cases: $R = 5\ \Omega$, $R = 4\ \Omega$, and $R = 1\ \Omega$.

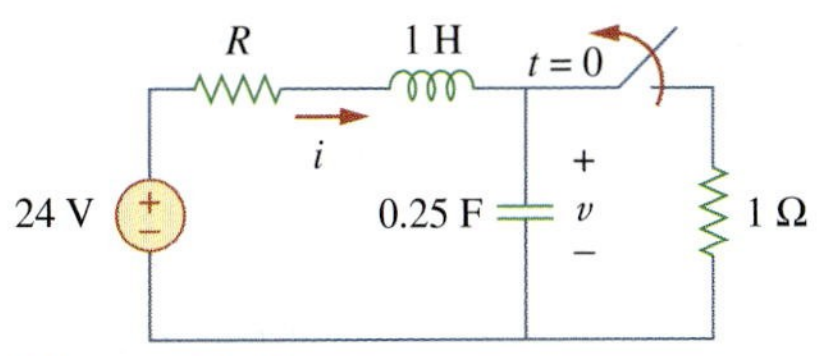

Figure 8.19
For Example 8.7.

Solution:

CASE 1 When $R = 5\ \Omega$. For $t < 0$, the switch is closed for a long time. The capacitor behaves like an open circuit while the inductor acts like a short circuit. The initial current through the inductor is

$$i(0) = \frac{24}{5+1} = 4\text{ A}$$

and the initial voltage across the capacitor is the same as the voltage across the 1-Ω resistor; that is,

$$v(0) = 1i(0) = 4\text{ V}$$

For $t > 0$, the switch is opened, so that we have the 1-Ω resistor disconnected. What remains is the series *RLC* circuit with the voltage source. The characteristic roots are determined as follows:

$$\alpha = \frac{R}{2L} = \frac{5}{2\times 1} = 2.5, \qquad \omega_0 = \frac{1}{\sqrt{LC}} = \frac{1}{\sqrt{1\times 0.25}} = 2$$

$$s_{1,2} = -\alpha \pm \sqrt{\alpha^2 - \omega_0^2} = -1, -4$$

Since $\alpha > \omega_0$, we have the overdamped natural response. The total response is therefore

$$v(t) = v_{ss} + (A_1 e^{-t} + A_2 e^{-4t})$$

where v_{ss} is the steady-state response. It is the final value of the capacitor voltage. In Fig. 8.19, $v_f = 24$ V. Thus,

$$v(t) = 24 + (A_1 e^{-t} + A_2 e^{-4t}) \tag{8.7.1}$$

We now need to find A_1 and A_2 using the initial conditions.

$$v(0) = 4 = 24 + A_1 + A_2$$

or

$$-20 = A_1 + A_2 \tag{8.7.2}$$

The current through the inductor cannot change abruptly and is the same current through the capacitor at $t = 0^+$ because the inductor and capacitor are now in series. Hence,

$$i(0) = C\frac{dv(0)}{dt} = 4 \quad \Rightarrow \quad \frac{dv(0)}{dt} = \frac{4}{C} = \frac{4}{0.25} = 16$$

Before we use this condition, we need to take the derivative of v in Eq. (8.7.1).

$$\frac{dv}{dt} = -A_1 e^{-t} - 4A_2 e^{-4t} \tag{8.7.3}$$

At $t = 0$,

$$\frac{dv(0)}{dt} = 16 = -A_1 - 4A_2 \tag{8.7.4}$$

From Eqs. (8.7.2) and (8.7.4), $A_1 = -64/3$ and $A_2 = 4/3$. Substituting A_1 and A_2 in Eq. (8.7.1), we get

$$v(t) = 24 + \frac{4}{3}(-16e^{-t} + e^{-4t}) \text{ V} \tag{8.7.5}$$

Since the inductor and capacitor are in series for $t > 0$, the inductor current is the same as the capacitor current. Hence,

$$i(t) = C\frac{dv}{dt}$$

Multiplying Eq. (8.7.3) by $C = 0.25$ and substituting the values of A_1 and A_2 gives

$$i(t) = \frac{4}{3}(4e^{-t} - e^{-4t}) \text{ A} \tag{8.7.6}$$

Note that $i(0) = 4$ A, as expected.

CASE 2 When $R = 4\ \Omega$. Again, the initial current through the inductor is

$$i(0) = \frac{24}{4 + 1} = 4.8 \text{ A}$$

and the initial capacitor voltage is

$$v(0) = 1i(0) = 4.8 \text{ V}$$

For the characteristic roots,

$$\alpha = \frac{R}{2L} = \frac{4}{2 \times 1} = 2$$

while $\omega_0 = 2$ remains the same. In this case, $s_1 = s_2 = -\alpha = -2$, and we have the critically damped natural response. The total response is therefore

$$v(t) = v_{ss} + (A_1 + A_2t)e^{-2t}$$

and, as before $v_{ss} = 24$ V,

$$v(t) = 24 + (A_1 + A_2t)e^{-2t} \tag{8.7.7}$$

To find A_1 and A_2, we use the initial conditions. We write

$$v(0) = 4.8 = 24 + A_1 \quad \Rightarrow \quad A_1 = -19.2 \tag{8.7.8}$$

Since $i(0) = C\,dv(0)/dt = 4.8$ or

$$\frac{dv(0)}{dt} = \frac{4.8}{C} = 19.2$$

From Eq. (8.7.7),

$$\frac{dv}{dt} = (-2A_1 - 2tA_2 + A_2)e^{-2t} \tag{8.7.9}$$

At $t = 0$,

$$\frac{dv(0)}{dt} = 19.2 = -2A_1 + A_2 \tag{8.7.10}$$

From Eqs. (8.7.8) and (8.7.10), $A_1 = -19.2$ and $A_2 = -19.2$. Thus, Eq. (8.7.7) becomes

$$v(t) = 24 - 19.2(1 + t)e^{-2t} \text{ V} \tag{8.7.11}$$

The inductor current is the same as the capacitor current; that is,

$$i(t) = C\frac{dv}{dt}$$

Multiplying Eq. (8.7.9) by $C = 0.25$ and substituting the values of A_1 and A_2 gives

$$i(t) = (4.8 + 9.6t)e^{-2t} \text{ A} \tag{8.7.12}$$

Note that $i(0) = 4.8$ A, as expected.

CASE 3 When $R = 1\ \Omega$. The initial inductor current is

$$i(0) = \frac{24}{1 + 1} = 12 \text{ A}$$

and the initial voltage across the capacitor is the same as the voltage across the 1-Ω resistor,

$$v(0) = 1i(0) = 12 \text{ V}$$

$$\alpha = \frac{R}{2L} = \frac{1}{2 \times 1} = 0.5$$

Since $\alpha = 0.5 < \omega_0 = 2$, we have the underdamped response

$$s_{1,2} = -\alpha \pm \sqrt{\alpha^2 - \omega_0^2} = -0.5 \pm j1.936$$

The total response is therefore

$$v(t) = 24 + (A_1 \cos 1.936t + A_2 \sin 1.936t)e^{-0.5t} \tag{8.7.13}$$

We now determine A_1 and A_2. We write

$$v(0) = 12 = 24 + A_1 \quad \Rightarrow \quad A_1 = -12 \tag{8.7.14}$$

Since $i(0) = C\,dv(0)/dt = 12$,

$$\frac{dv(0)}{dt} = \frac{12}{C} = 48 \tag{8.7.15}$$

But

$$\begin{aligned}\frac{dv}{dt} &= e^{-0.5t}(-1.936A_1 \sin 1.936t + 1.936A_2 \cos 1.936t) \\ &\quad - 0.5e^{-0.5t}(A_1 \cos 1.936t + A_2 \sin 1.936t)\end{aligned} \tag{8.7.16}$$

At $t = 0$,

$$\frac{dv(0)}{dt} = 48 = (-0 + 1.936A_2) - 0.5(A_1 + 0)$$

Substituting $A_1 = -12$ gives $A_2 = 21.694$, and Eq. (8.7.13) becomes

$$v(t) = 24 + (21.694 \sin 1.936t - 12 \cos 1.936t)e^{-0.5t} \text{ V} \tag{8.7.17}$$

The inductor current is

$$i(t) = C\frac{dv}{dt}$$

Multiplying Eq. (8.7.16) by $C = 0.25$ and substituting the values of A_1 and A_2 gives

$$i(t) = (3.1 \sin 1.936t + 12 \cos 1.936t)e^{-0.5t} \text{ A} \quad \textbf{(8.7.18)}$$

Note that $i(0) = 12$ A, as expected.

Figure 8.20 plots the responses for the three cases. From this figure, we observe that the critically damped response approaches the step input of 24 V the fastest.

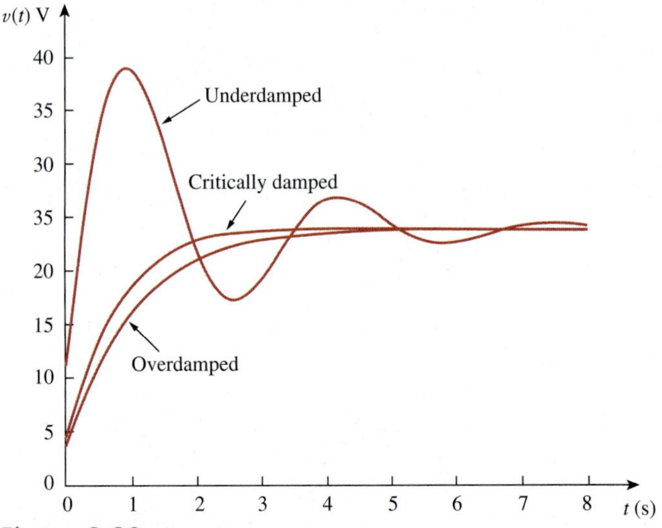

Figure 8.20
For Example 8.7: response for three degrees of damping.

Practice Problem 8.7

Having been in position a for a long time, the switch in Fig. 8.21 is moved to position b at $t = 0$. Find $v(t)$ and $v_R(t)$ for $t > 0$.

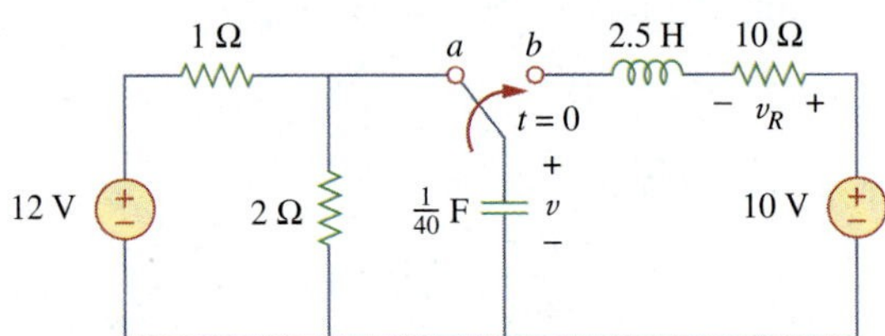

Figure 8.21
For Practice Prob. 8.7.

Answer: $10 - (1.1547 \sin 3.464t + 2 \cos 3.464t)e^{-2t}$ V, $2.31e^{-2t} \sin 3.464t$ V.

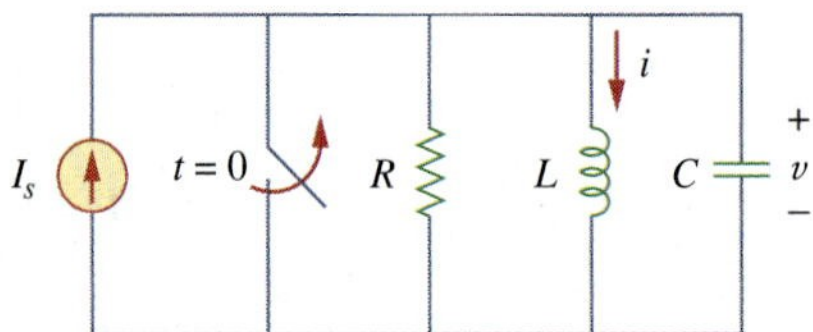

Figure 8.22
Parallel *RLC* circuit with an applied current.

8.6 Step Response of a Parallel *RLC* Circuit

Consider the parallel *RLC* circuit shown in Fig. 8.22. We want to find i due to a sudden application of a dc current. Applying KCL at the top node for $t > 0$,

$$\frac{v}{R} + i + C\frac{dv}{dt} = I_s \quad \textbf{(8.46)}$$

But

$$v = L\frac{di}{dt}$$

Substituting for v in Eq. (8.46) and dividing by LC, we get

$$\frac{d^2i}{dt^2} + \frac{1}{RC}\frac{di}{dt} + \frac{i}{LC} = \frac{I_s}{LC} \tag{8.47}$$

which has the same characteristic equation as Eq. (8.29).

The complete solution to Eq. (8.47) consists of the transient response $i_t(t)$ and the steady-state response i_{ss}; that is,

$$i(t) = i_t(t) + i_{ss}(t) \tag{8.48}$$

The transient response is the same as what we had in Section 8.4. The steady-state response is the final value of i. In the circuit in Fig. 8.22, the final value of the current through the inductor is the same as the source current I_s. Thus,

$$\begin{aligned} i(t) &= I_s + A_1e^{s_1t} + A_2e^{s_2t} \quad \text{(Overdamped)} \\ i(t) &= I_s + (A_1 + A_2t)e^{-\alpha t} \quad \text{(Critically damped)} \\ i(t) &= I_s + (A_1\cos\omega_dt + A_2\sin\omega_dt)e^{-\alpha t} \quad \text{(Underdamped)} \end{aligned} \tag{8.49}$$

The constants A_1 and A_2 in each case can be determined from the initial conditions for i and di/dt. Again, we should keep in mind that Eq. (8.49) only applies for finding the inductor current i. But once the inductor current $i_L = i$ is known, we can find $v = L\,di/dt$, which is the same voltage across inductor, capacitor, and resistor. Hence, the current through the resistor is $i_R = v/R$, while the capacitor current is $i_C = C\,dv/dt$. Alternatively, the complete response for any variable $x(t)$ may be found directly, using

$$x(t) = x_{ss}(t) + x_t(t) \tag{8.50}$$

where x_{ss} and x_t are its final value and transient response, respectively.

Example 8.8

In the circuit of Fig. 8.23, find $i(t)$ and $i_R(t)$ for $t > 0$.

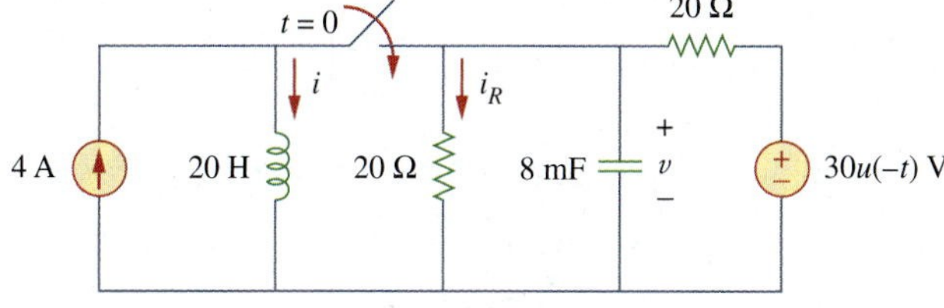

Figure 8.23
For Example 8.8.

Solution:
For $t < 0$, the switch is open, and the circuit is partitioned into two independent subcircuits. The 4-A current flows through the inductor, so that

$$i(0) = 4 \text{ A}$$

Since $30u(-t) = 30$ when $t < 0$ and 0 when $t > 0$, the voltage source is operative for $t < 0$. The capacitor acts like an open circuit and the voltage across it is the same as the voltage across the 20-Ω resistor connected in parallel with it. By voltage division, the initial capacitor voltage is

$$v(0) = \frac{20}{20 + 20}(30) = 15 \text{ V}$$

For $t > 0$, the switch is closed, and we have a parallel *RLC* circuit with a current source. The voltage source is zero which means it acts like a short-circuit. The two 20-Ω resistors are now in parallel. They are combined to give $R = 20 \| 20 = 10\ \Omega$. The characteristic roots are determined as follows:

$$\alpha = \frac{1}{2RC} = \frac{1}{2 \times 10 \times 8 \times 10^{-3}} = 6.25$$

$$\omega_0 = \frac{1}{\sqrt{LC}} = \frac{1}{\sqrt{20 \times 8 \times 10^{-3}}} = 2.5$$

$$s_{1,2} = -\alpha \pm \sqrt{\alpha^2 - \omega_0^2} = -6.25 \pm \sqrt{39.0625 - 6.25}$$
$$= -6.25 \pm 5.7282$$

or

$$s_1 = -11.978, \qquad s_2 = -0.5218$$

Since $\alpha > \omega_0$, we have the overdamped case. Hence,

$$i(t) = I_s + A_1 e^{-11.978t} + A_2 e^{-0.5218t} \qquad \textbf{(8.8.1)}$$

where $I_s = 4$ is the final value of $i(t)$. We now use the initial conditions to determine A_1 and A_2. At $t = 0$,

$$i(0) = 4 = 4 + A_1 + A_2 \quad \Rightarrow \quad A_2 = -A_1 \qquad \textbf{(8.8.2)}$$

Taking the derivative of $i(t)$ in Eq. (8.8.1),

$$\frac{di}{dt} = -11.978A_1e^{-11.978t} - 0.5218A_2e^{-0.5218t}$$

so that at $t = 0$,

$$\frac{di(0)}{dt} = -11.978A_1 - 0.5218A_2 \qquad \textbf{(8.8.3)}$$

But

$$L\frac{di(0)}{dt} = v(0) = 15 \quad \Rightarrow \quad \frac{di(0)}{dt} = \frac{15}{L} = \frac{15}{20} = 0.75$$

Substituting this into Eq. (8.8.3) and incorporating Eq. (8.8.2), we get

$$0.75 = (11.978 - 0.5218)A_2 \quad \Rightarrow \quad A_2 = 0.0655$$

Thus, $A_1 = -0.0655$ and $A_2 = 0.0655$. Inserting A_1 and A_2 in Eq. (8.8.1) gives the complete solution as

$$i(t) = 4 + 0.0655(e^{-0.5218t} - e^{-11.978t}) \text{ A}$$

From $i(t)$, we obtain $v(t) = L\, di/dt$ and

$$i_R(t) = \frac{v(t)}{20} = \frac{L}{20}\frac{di}{dt} = 0.785e^{-11.978t} - 0.0342e^{-0.5218t} \text{ A}$$

Practice Problem 8.8

Find $i(t)$ and $v(t)$ for $t > 0$ in the circuit of Fig. 8.24.

Answer: $12(1 - \cos t)$ A, $60 \sin t$ V.

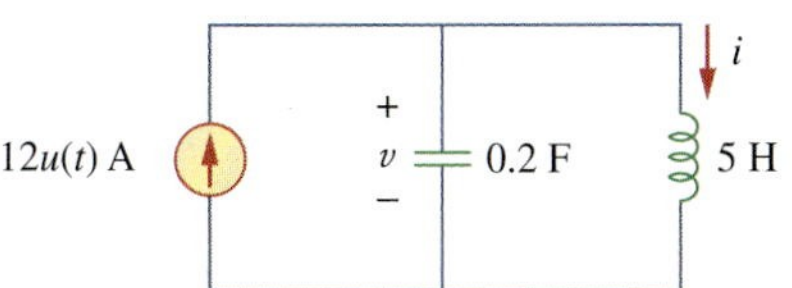

Figure 8.24
For Practice Prob. 8.8.

8.7 General Second-Order Circuits

Now that we have mastered series and parallel *RLC* circuits, we are prepared to apply the ideas to any second-order circuit having one or more independent sources with constant values. Although the series and parallel *RLC* circuits are the second-order circuits of greatest interest, other second-order circuits including op amps are also useful. Given a second-order circuit, we determine its step response $x(t)$ (which may be voltage or current) by taking the following four steps:

A circuit may look complicated at first. But once the sources are turned off in an attempt to find the form of the transient response, it may be reducible to a first-order circuit, when the storage elements can be combined, or to a parallel/series *RLC* circuit. If it is reducible to a first-order circuit, the solution becomes simply what we had in Chapter 7. If it is reducible to a parallel or series *RLC* circuit, we apply the techniques of previous sections in this chapter.

1. We first determine the initial conditions $x(0)$ and $dx(0)/dt$ and the final value $x(\infty)$, as discussed in Section 8.2.
2. We turn off the independent sources and find the form of the transient response $x_t(t)$ by applying KCL and KVL. Once a second-order differential equation is obtained, we determine its characteristic roots. Depending on whether the response is overdamped, critically damped, or underdamped, we obtain $x_t(t)$ with two unknown constants as we did in the previous sections.
3. We obtain the steady-state response as

$$x_{ss}(t) = x(\infty) \tag{8.51}$$

where $x(\infty)$ is the final value of x, obtained in step 1.
4. The total response is now found as the sum of the transient response and steady-state response

$$x(t) = x_t(t) + x_{ss}(t) \tag{8.52}$$

We finally determine the constants associated with the transient response by imposing the initial conditions $x(0)$ and $dx(0)/dt$, determined in step 1.

Problems in this chapter can also be solved by using Laplace transforms, which are covered in Chapters 15 and 16.

We can apply this general procedure to find the step response of any second-order circuit, including those with op amps. The following examples illustrate the four steps.

Example 8.9

Find the complete response v and then i for $t > 0$ in the circuit of Fig. 8.25.

Solution:
We first find the initial and final values. At $t = 0^-$, the circuit is at steady state. The switch is open; the equivalent circuit is shown in Fig. 8.26(a). It is evident from the figure that

$$v(0^-) = 12 \text{ V}, \qquad i(0^-) = 0$$

At $t = 0^+$, the switch is closed; the equivalent circuit is in Fig. 8.26(b). By the continuity of capacitor voltage and inductor current, we know that

$$v(0^+) = v(0^-) = 12 \text{ V}, \qquad i(0^+) = i(0^-) = 0 \tag{8.9.1}$$

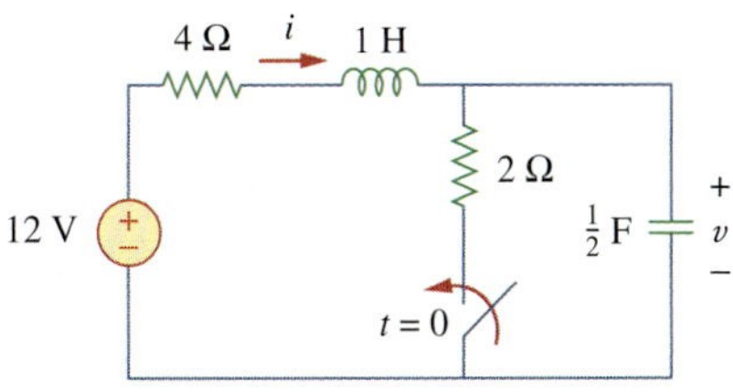

Figure 8.25
For Example 8.9.

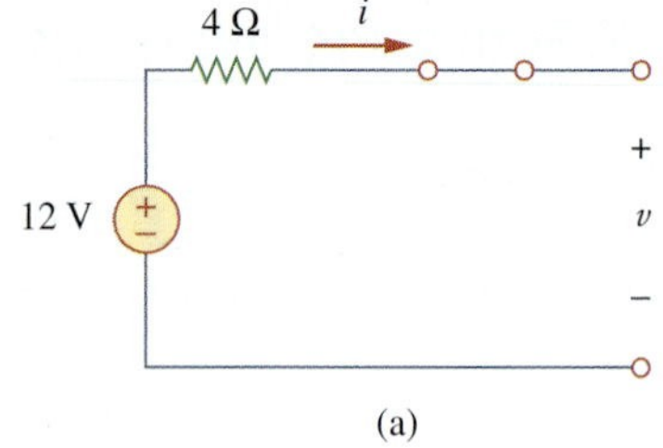

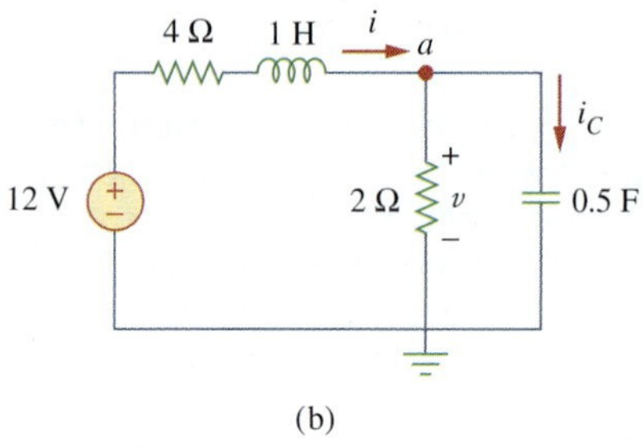

Figure 8.26
Equivalent circuit of the circuit in Fig. 8.25 for: (a) $t < 0$, (b) $t > 0$.

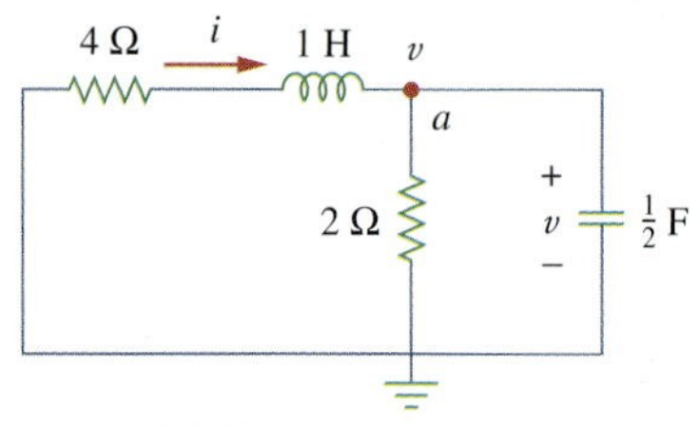

Figure 8.27
Obtaining the form of the transient response for Example 8.9.

To get $dv(0^+)/dt$, we use $C\,dv/dt = i_C$ or $dv/dt = i_C/C$. Applying KCL at node a in Fig. 8.26(b),

$$i(0^+) = i_C(0^+) + \frac{v(0^+)}{2}$$

$$0 = i_C(0^+) + \frac{12}{2} \quad \Rightarrow \quad i_C(0^+) = -6 \text{ A}$$

Hence,

$$\frac{dv(0^+)}{dt} = \frac{-6}{0.5} = -12 \text{ V/s} \tag{8.9.2}$$

The final values are obtained when the inductor is replaced by a short circuit and the capacitor by an open circuit in Fig. 8.26(b), giving

$$i(\infty) = \frac{12}{4+2} = 2 \text{ A}, \qquad v(\infty) = 2i(\infty) = 4 \text{ V} \tag{8.9.3}$$

Next, we obtain the form of the transient response for $t > 0$. By turning off the 12-V voltage source, we have the circuit in Fig. 8.27. Applying KCL at node a in Fig. 8.27 gives

$$i = \frac{v}{2} + \frac{1}{2}\frac{dv}{dt} \tag{8.9.4}$$

Applying KVL to the left mesh results in

$$4i + 1\frac{di}{dt} + v = 0 \tag{8.9.5}$$

Since we are interested in v for the moment, we substitute i from Eq. (8.9.4) into Eq. (8.9.5). We obtain

$$2v + 2\frac{dv}{dt} + \frac{1}{2}\frac{dv}{dt} + \frac{1}{2}\frac{d^2v}{dt^2} + v = 0$$

or

$$\frac{d^2v}{dt^2} + 5\frac{dv}{dt} + 6v = 0$$

From this, we obtain the characteristic equation as

$$s^2 + 5s + 6 = 0$$

with roots $s = -2$ and $s = -3$. Thus, the natural response is

$$v_n(t) = Ae^{-2t} + Be^{-3t} \tag{8.9.6}$$

where A and B are unknown constants to be determined later. The steady-state response is

$$v_{ss}(t) = v(\infty) = 4 \tag{8.9.7}$$

The complete response is

$$v(t) = v_t + v_{ss} = 4 + Ae^{-2t} + Be^{-3t} \tag{8.9.8}$$

We now determine A and B using the initial values. From Eq. (8.9.1), $v(0) = 12$. Substituting this into Eq. (8.9.8) at $t = 0$ gives

$$12 = 4 + A + B \quad \Rightarrow \quad A + B = 8 \tag{8.9.9}$$

Taking the derivative of v in Eq. (8.9.8),

$$\frac{dv}{dt} = -2Ae^{-2t} - 3Be^{-3t} \qquad \textbf{(8.9.10)}$$

Substituting Eq. (8.9.2) into Eq. (8.9.10) at $t = 0$ gives

$$-12 = -2A - 3B \qquad \Rightarrow \qquad 2A + 3B = 12 \qquad \textbf{(8.9.11)}$$

From Eqs. (8.9.9) and (8.9.11), we obtain

$$A = 12, \qquad B = -4$$

so that Eq. (8.9.8) becomes

$$v(t) = 4 + 12e^{-2t} - 4e^{-3t}\text{ V}, \qquad t > 0 \qquad \textbf{(8.9.12)}$$

From v, we can obtain other quantities of interest by referring to Fig. 8.26(b). To obtain i, for example,

$$\begin{aligned} i = \frac{v}{2} + \frac{1}{2}\frac{dv}{dt} &= 2 + 6e^{-2t} - 2e^{-3t} - 12e^{-2t} + 6e^{-3t} \\ &= 2 - 6e^{-2t} + 4e^{-3t}\text{ A}, \qquad t > 0 \end{aligned} \qquad \textbf{(8.9.13)}$$

Notice that $i(0) = 0$, in agreement with Eq. (8.9.1).

Practice Problem 8.9

Determine v and i for $t > 0$ in the circuit of Fig. 8.28. (See comments about current sources in Practice Prob. 7.5.)

Answer: $12(1 - e^{-5t})$ V, $3(1 - e^{-5t})$ A.

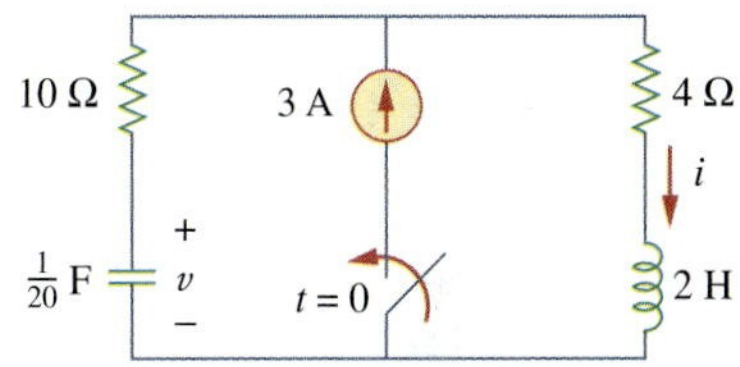

Figure 8.28
For Practice Prob. 8.9.

Example 8.10

Find $v_o(t)$ for $t > 0$ in the circuit of Fig. 8.29.

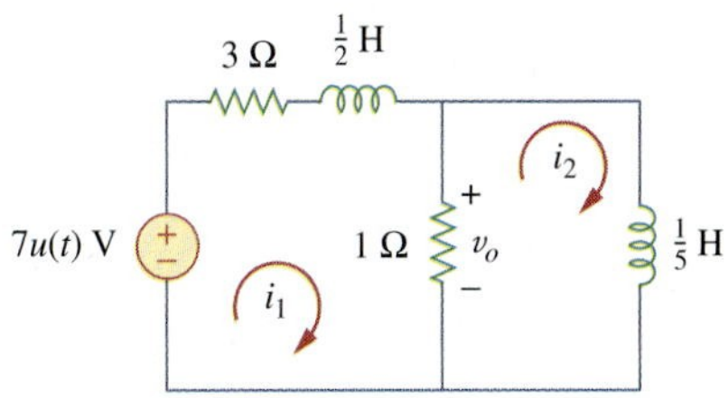

Figure 8.29
For Example 8.10.

Solution:
This is an example of a second-order circuit with two inductors. We first obtain the mesh currents i_1 and i_2, which happen to be the currents through the inductors. We need to obtain the initial and final values of these currents.

For $t < 0$, $7u(t) = 0$, so that $i_1(0^-) = 0 = i_2(0^-)$. For $t > 0$, $7u(t) = 7$, so that the equivalent circuit is as shown in Fig. 8.30(a). Due to the continuity of inductor current,

$$i_1(0^+) = i_1(0^-) = 0, \qquad i_2(0^+) = i_2(0^-) = 0 \qquad \textbf{(8.10.1)}$$

$$v_{L_2}(0^+) = v_o(0^+) = 1[(i_1(0^+) - i_2(0^+)] = 0 \qquad \textbf{(8.10.2)}$$

Applying KVL to the left loop in Fig. 8.30(a) at $t = 0^+$,

$$7 = 3i_1(0^+) + v_{L_1}(0^+) + v_o(0^+)$$

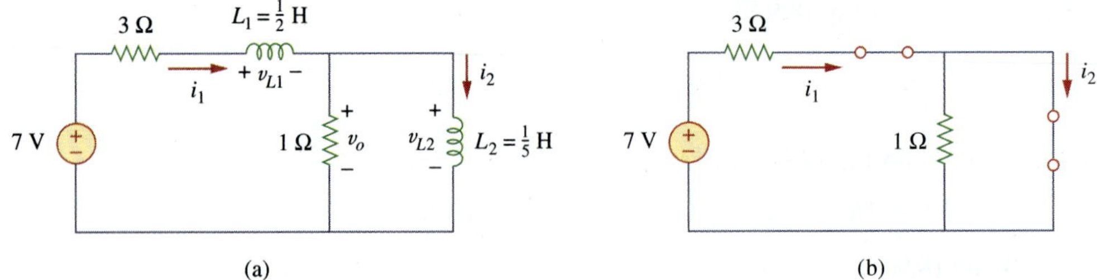

Figure 8.30
Equivalent circuit of that in Fig. 8.29 for: (a) $t > 0$, (b) $t \to \infty$.

or

$$v_{L_1}(0^+) = 7 \text{ V}$$

Since $L_1\, di_1/dt = v_{L_1}$,

$$\frac{di_1(0^+)}{dt} = \frac{v_{L_1}}{L_1} = \frac{7}{\frac{1}{2}} = 14 \text{ V/s} \tag{8.10.3}$$

Similarly, since $L_2\, di_2/dt = v_{L_2}$,

$$\frac{di_2(0^+)}{dt} = \frac{v_{L_2}}{L_2} = 0 \tag{8.10.4}$$

As $t \to \infty$, the circuit reaches steady state, and the inductors can be replaced by short circuits, as shown in Fig. 8.30(b). From this figure,

$$i_1(\infty) = i_2(\infty) = \frac{7}{3} \text{ A} \tag{8.10.5}$$

Next, we obtain the form of the transient responses by removing the voltage source, as shown in Fig. 8.31. Applying KVL to the two meshes yields

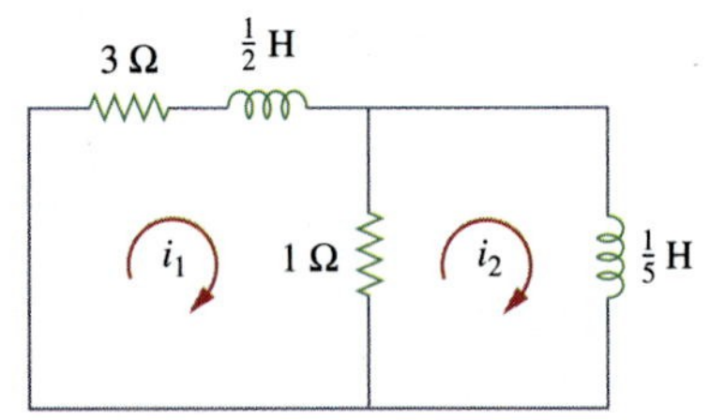

Figure 8.31
Obtaining the form of the transient response for Example 8.10.

$$4i_1 - i_2 + \frac{1}{2}\frac{di_1}{dt} = 0 \tag{8.10.6}$$

and

$$i_2 + \frac{1}{5}\frac{di_2}{dt} - i_1 = 0 \tag{8.10.7}$$

From Eq. (8.10.6),

$$i_2 = 4i_1 + \frac{1}{2}\frac{di_1}{dt} \tag{8.10.8}$$

Substituting Eq. (8.10.8) into Eq. (8.10.7) gives

$$4i_1 + \frac{1}{2}\frac{di_1}{dt} + \frac{4}{5}\frac{di_1}{dt} + \frac{1}{10}\frac{d^2i_1}{dt^2} - i_1 = 0$$

$$\frac{d^2i_1}{dt^2} + 13\frac{di_1}{dt} + 30i_1 = 0$$

From this we obtain the characteristic equation as

$$s^2 + 13s + 30 = 0$$

which has roots $s = -3$ and $s = -10$. Hence, the form of the transient response is

$$i_{1n} = Ae^{-3t} + Be^{-10t} \tag{8.10.9}$$

where A and B are constants. The steady-state response is

$$i_{1ss} = i_1(\infty) = \frac{7}{3}\text{ A} \tag{8.10.10}$$

From Eqs. (8.10.9) and (8.10.10), we obtain the complete response as

$$i_1(t) = \frac{7}{3} + Ae^{-3t} + Be^{-10t} \tag{8.10.11}$$

We finally obtain A and B from the initial values. From Eqs. (8.10.1) and (8.10.11),

$$0 = \frac{7}{3} + A + B \tag{8.10.12}$$

Taking the derivative of Eq. (8.10.11), setting $t = 0$ in the derivative, and enforcing Eq. (8.10.3), we obtain

$$14 = -3A - 10B \tag{8.10.13}$$

From Eqs. (8.10.12) and (8.10.13), $A = -4/3$ and $B = -1$. Thus,

$$i_1(t) = \frac{7}{3} - \frac{4}{3}e^{-3t} - e^{-10t} \tag{8.10.14}$$

We now obtain i_2 from i_1. Applying KVL to the left loop in Fig. 8.30(a) gives

$$7 = 4i_1 - i_2 + \frac{1}{2}\frac{di_1}{dt} \quad \Rightarrow \quad i_2 = -7 + 4i_1 + \frac{1}{2}\frac{di_1}{dt}$$

Substituting for i_1 in Eq. (8.10.14) gives

$$\begin{aligned} i_2(t) &= -7 + \frac{28}{3} - \frac{16}{3}e^{-3t} - 4e^{-10t} + 2e^{-3t} + 5e^{-10t} \\ &= \frac{7}{3} - \frac{10}{3}e^{-3t} + e^{-10t} \end{aligned} \tag{8.10.15}$$

From Fig. 8.29,

$$v_o(t) = 1[i_1(t) - i_2(t)] \tag{8.10.16}$$

Substituting Eqs. (8.10.14) and (8.10.15) into Eq. (8.10.16) yields

$$v_o(t) = 2(e^{-3t} - e^{-10t}) \tag{8.10.17}$$

Note that $v_o(0) = 0$, as expected from Eq. (8.10.2).

Practice Problem 8.10

For $t > 0$, obtain $v_o(t)$ in the circuit of Fig. 8.32. (*Hint:* First find v_1 and v_2.)

Answer: $8(e^{-t} - e^{-6t})$ V, $t > 0$.

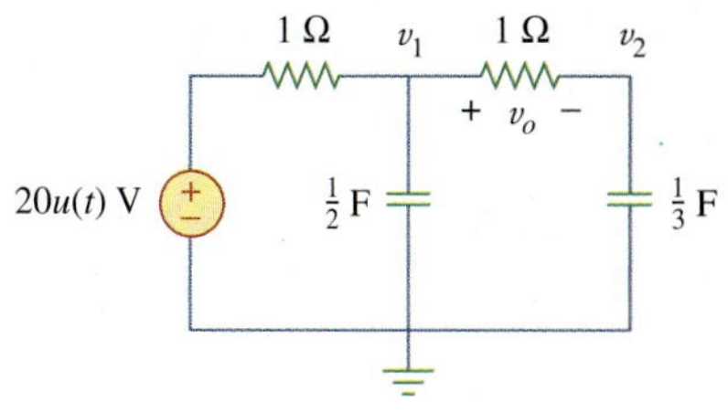

Figure 8.32
For Practice Prob. 8.10.

8.8 Second-Order Op Amp Circuits

The use of op amps in second-order circuits avoids the use of inductors, which are undesirable in some applications.

An op amp circuit with two storage elements that cannot be combined into a single equivalent element is second-order. Because inductors are bulky and heavy, they are rarely used in practical op amp circuits. For this reason, we will only consider *RC* second-order op amp circuits here. Such circuits find a wide range of applications in devices such as filters and oscillators.

The analysis of a second-order op amp circuit follows the same four steps given and demonstrated in the previous section.

Example 8.11

In the op amp circuit of Fig. 8.33, find $v_o(t)$ for $t > 0$ when $v_s = 10u(t)$ mV. Let $R_1 = R_2 = 10$ kΩ, $C_1 = 20$ μF, and $C_2 = 100$ μF.

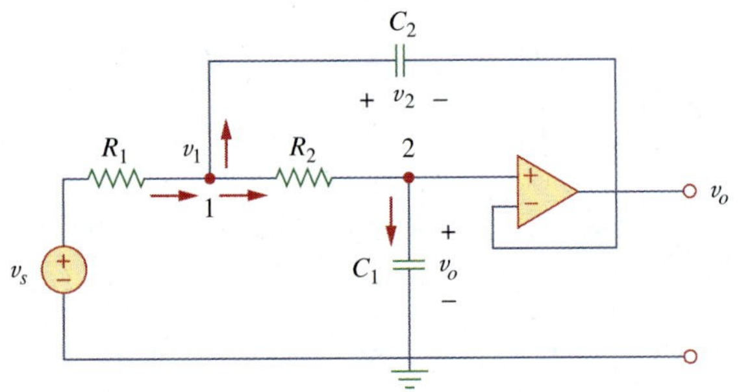

Figure 8.33
For Example 8.11.

Solution:

Although we could follow the same four steps given in the previous section to solve this problem, we will solve it a little differently. Due to the voltage follower configuration, the voltage across C_1 is v_o. Applying KCL at node 1,

$$\frac{v_s - v_1}{R_1} = C_2\frac{dv_2}{dt} + \frac{v_1 - v_o}{R_2} \tag{8.11.1}$$

At node 2, KCL gives

$$\frac{v_1 - v_o}{R_2} = C_1\frac{dv_o}{dt} \tag{8.11.2}$$

But

$$v_2 = v_1 - v_o \tag{8.11.3}$$

We now try to eliminate v_1 and v_2 in Eqs. (8.11.1) to (8.11.3). Substituting Eqs. (8.11.2) and (8.11.3) into Eq. (8.11.1) yields

$$\frac{v_s - v_1}{R_1} = C_2\frac{dv_1}{dt} - C_2\frac{dv_o}{dt} + C_1\frac{dv_o}{dt} \tag{8.11.4}$$

From Eq. (8.11.2),

$$v_1 = v_o + R_2C_1\frac{dv_o}{dt} \tag{8.11.5}$$

Substituting Eq. (8.11.5) into Eq. (8.11.4), we obtain

$$\frac{v_s}{R_1} = \frac{v_o}{R_1} + \frac{R_2C_1}{R_1}\frac{dv_o}{dt} + C_2\frac{dv_o}{dt} + R_2C_1C_2\frac{d^2v_o}{dt^2} - C_2\frac{dv_o}{dt} + C_1\frac{dv_o}{dt}$$

or

$$\frac{d^2v_o}{dt^2} + \left(\frac{1}{R_1C_2} + \frac{1}{R_2C_2}\right)\frac{dv_o}{dt} + \frac{v_o}{R_1R_2C_1C_2} = \frac{v_s}{R_1R_2C_1C_2} \tag{8.11.6}$$

With the given values of R_1, R_2, C_1, and C_2, Eq. (8.11.6) becomes

$$\frac{d^2v_o}{dt^2} + 2\frac{dv_o}{dt} + 5v_o = 5v_s \tag{8.11.7}$$

To obtain the form of the transient response, set $v_s = 0$ in Eq. (8.11.7), which is the same as turning off the source. The characteristic equation is

$$s^2 + 2s + 5 = 0$$

which has complex roots $s_{1,2} = -1 \pm j2$. Hence, the form of the transient response is

$$v_{ot} = e^{-t}(A\cos 2t + B\sin 2t) \tag{8.11.8}$$

where A and B are unknown constants to be determined.

As $t \to \infty$, the circuit reaches the steady-state condition, and the capacitors can be replaced by open circuits. Since no current flows through C_1 and C_2 under steady-state conditions and no current can enter the input terminals of the ideal op amp, current does not flow through R_1 and R_2. Thus,

$$v_o(\infty) = v_1(\infty) = v_s$$

The steady-state response is then

$$v_{oss} = v_o(\infty) = v_s = 10 \text{ mV}, \qquad t > 0 \tag{8.11.9}$$

The complete response is

$$v_o(t) = v_{ot} + v_{oss} = 10 + e^{-t}(A\cos 2t + B\sin 2t) \text{ mV} \tag{8.11.10}$$

To determine A and B, we need the initial conditions. For $t < 0$, $v_s = 0$, so that

$$v_o(0^-) = v_2(0^-) = 0$$

For $t > 0$, the source is operative. However, due to capacitor voltage continuity,

$$v_o(0^+) = v_2(0^+) = 0 \tag{8.11.11}$$

From Eq. (8.11.3),

$$v_1(0^+) = v_2(0^+) + v_o(0^+) = 0$$

and hence, from Eq. (8.11.2),

$$\frac{dv_o(0^+)}{dt} = \frac{v_1 - v_o}{R_2C_1} = 0 \tag{8.11.12}$$

We now impose Eq. (8.11.11) on the complete response in Eq. (8.11.10) at $t = 0$, for

$$0 = 10 + A \quad \Rightarrow \quad A = -10 \tag{8.11.13}$$

Taking the derivative of Eq. (8.11.10),

$$\frac{dv_o}{dt} = e^{-t}(-A\cos 2t - B\sin 2t - 2A\sin 2t + 2B\cos 2t)$$

Setting $t = 0$ and incorporating Eq. (8.11.12), we obtain

$$0 = -A + 2B \qquad \textbf{(8.11.14)}$$

From Eqs. (8.11.13) and (8.11.14), $A = -10$ and $B = -5$. Thus, the step response becomes

$$v_o(t) = 10 - e^{-t}(10\cos 2t + 5\sin 2t)\text{ mV}, \qquad t > 0$$

Practice Problem 8.11

In the op amp circuit shown in Fig. 8.34, $v_s = 10u(t)$ V, find $v_o(t)$ for $t > 0$. Assume that $R_1 = R_2 = 10\text{ k}\Omega$, $C_1 = 20\ \mu\text{F}$, and $C_2 = 100\ \mu\text{F}$.

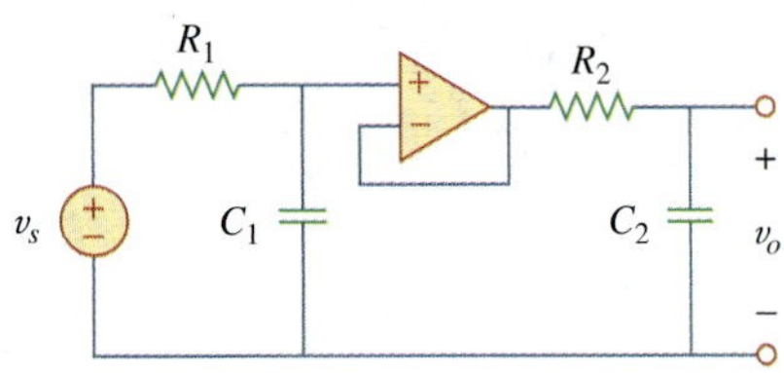

Figure 8.34
For Practice Prob. 8.11.

Answer: $(10 - 12.5e^{-t} + 2.5e^{-5t})$ V, $t > 0$.

8.9 *PSpice* Analysis of *RLC* Circuits

RLC circuits can be analyzed with great ease using *PSpice*, just like the *RC* or *RL* circuits of Chapter 7. The following two examples will illustrate this. The reader may review Section D.4 in Appendix D on *PSpice* for transient analysis.

Example 8.12

The input voltage in Fig. 8.35(a) is applied to the circuit in Fig. 8.35(b). Use *PSpice* to plot $v(t)$ for $0 < t < 4$ s.

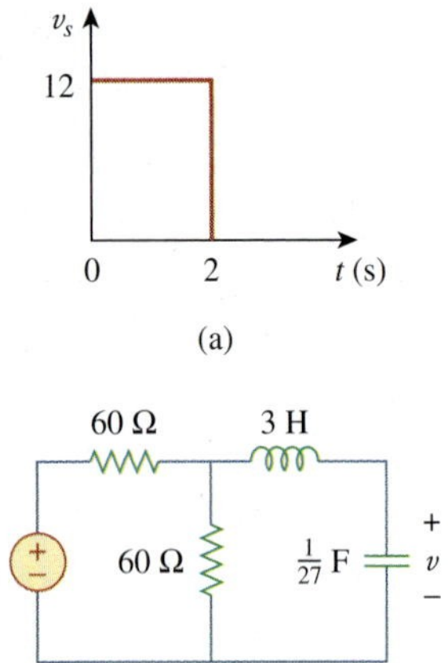

Figure 8.35
For Example 8.12.

Solution:

1. **Define.** As true with most textbook problems, the problem is clearly defined.
2. **Present.** The input is equal to a single square wave of amplitude 12 V with a period of 2 s. We are asked to plot the output, using *PSpice*.
3. **Alternative.** Since we are required to use *PSpice*, that is the only alternative for a solution. However, we can check it using the technique illustrated in Section 8.5 (a step response for a series *RLC* circuit).
4. **Attempt.** The given circuit is drawn using Schematics as in Fig. 8.36. The pulse is specified using VPWL voltage source, but VPULSE could be used instead. Using the piecewise linear function, we set the attributes of VPWL as T1 = 0, V1 = 0, T2 = 0.001, V2 = 12, and so forth, as shown in Fig. 8.36. Two voltage markers are inserted to plot the input and output voltages. Once the circuit is drawn and the attributes are set, we select **Analysis/Setup/Transient** to open up the *Transient Analysis* dialog box. As a parallel *RLC* circuit, the roots of the characteristic equation are −1 and −9. Thus, we may set *Final Time* as 4 s (four times the magnitude of the lower root). When

the schematic is saved, we select **Analysis/Simulate** and obtain the plots for the input and output voltages under the *PSpice* A/D window as shown in Fig. 8.37.

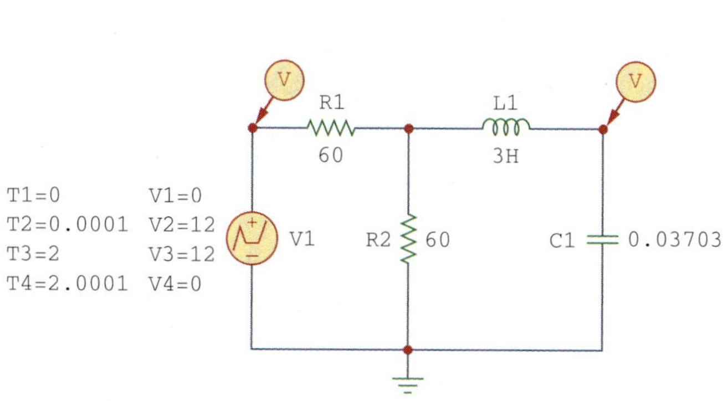

Figure 8.36
Schematic for the circuit in Fig. 8.35(b).

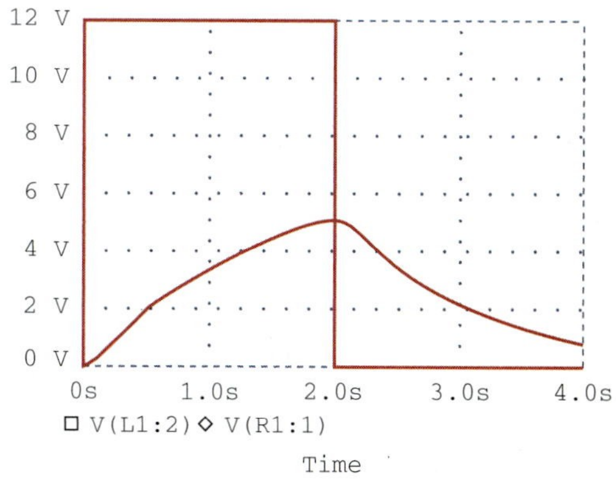

Figure 8.37
For Example 8.12: input and output.

Now we check using the technique from Section 8.5. We can start by realizing the Thevenin equivalent for the resistor-source combination is $V_{\text{Th}} = 12/2$ (the open circuit voltage divides equally across both resistors) $= 6$ V. The equivalent resistance is 30 Ω (60 ∥ 60). Thus, we can now solve for the response using $R = 30\ \Omega$, $L = 3$ H, and $C = (1/27)$ F.

We first need to solve for α and ω_0:

$$\alpha = R/(2L) = 30/6 = 5 \qquad \text{and} \qquad \omega_0 = \frac{1}{\sqrt{3\frac{1}{27}}} = 3$$

Since 5 is greater than 3, we have the overdamped case

$$s_{1,2} = -5 \pm \sqrt{5^2 - 9} = -1, -9, \qquad v(0) = 0, \qquad v(\infty) = 6\text{ V}, \qquad i(0) = 0$$

$$i(t) = C\frac{dv(t)}{dt},$$

where

$$v(t) = A_1e^{-t} + A_2e^{-9t} + 6$$
$$v(0) = 0 = A_1 + A_2 + 6$$
$$i(0) = 0 = C(-A_1 - 9A_2)$$

which yields $A_1 = -9A_2$. Substituting this into the above, we get $0 = 9A_2 - A_2 + 6$, or $A_2 = 0.75$ and $A_1 = -6.75$.

$v(t) = (\mathbf{-6.75}e^{-t} + \mathbf{0.75}e^{-9t} + \mathbf{6})u(t)$ **V** for all $0 < t < 2$ s.

At $t = 1$ s, $v(1) = -6.75e^{-1} + 0.75e^{-9} = -2.483 + 0.0001 + 6 = -3.552$ V. At $t = 2$ s, $v(2) = -6.75e^{-2} + 0 + 6 = 5.086$ V.

Note that from $2 < t < 4$ s, $V_{\text{Th}} = 0$, which implies that $v(\infty) = 0$. Therefore, $v(t) = (A_3e^{-(t-2)} + A_4e^{-9(t-2)})u(t-2)$ V. At $t = 2$ s, $A_3 + A_4 = 5.086$.

$$i(t) = \frac{(-A_3e^{-(t-2)} - 9A_4e^{-9(t-2)})}{27}$$

and

$$i(2) = \frac{(6.75e^{-2} - 6.75e^{-18})}{27} = 33.83 \text{ mA}$$

Therefore, $-A_3 - 9A_4 = 0.9135$.

Combining the two equations, we get $-A_3 - 9(5.086 - A_3) = 0.9135$, which leads to $A_3 = 5.835$ and $A_4 = -0.749$.

$$v(t) = (\mathbf{5.835}e^{-(t-2)} - \mathbf{0.749}e^{-9(t-2)})\, u(t-2) \text{ V}$$

At $t = 3$ s, $v(3) = (2.147 - 0) = 2.147$ V. At $t = 4$ s, $v(4) = 0.7897$ V.

5. **Evaluate.** A check between the values calculated above and the plot shown in Figure 8.37 shows good agreement within the obvious level of accuracy.
6. **Satisfactory?** Yes, we have agreement and the results can be presented as a solution to the problem.

Practice Problem 8.12

Find $i(t)$ using *PSpice* for $0 < t < 4$ s if the pulse voltage in Fig. 8.35(a) is applied to the circuit in Fig. 8.38.

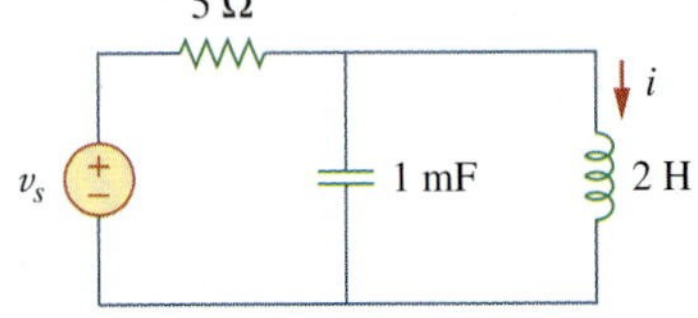

Figure 8.38
For Practice Prob. 8.12.

Answer: See Fig. 8.39.

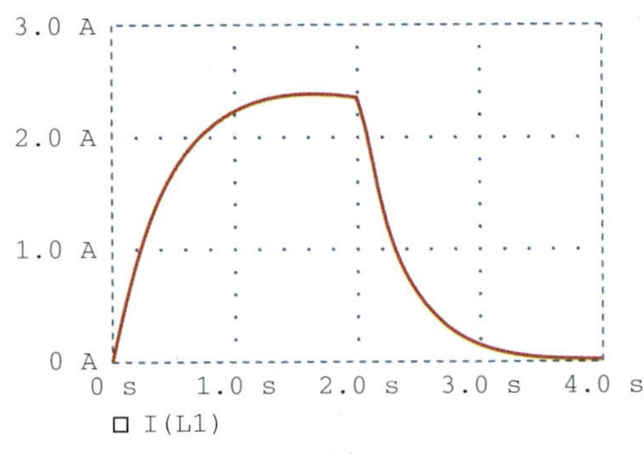

Figure 8.39
Plot of $i(t)$ for Practice Prob. 8.12.

Example 8.13

For the circuit in Fig. 8.40, use *PSpice* to obtain $i(t)$ for $0 < t < 3$ s.

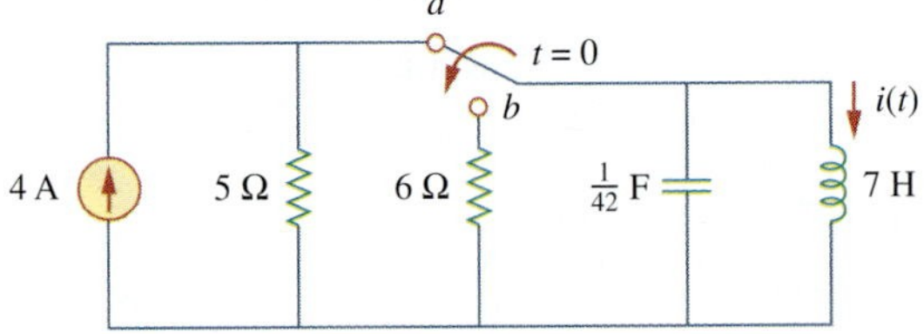

Figure 8.40
For Example 8.13.

Solution:

When the switch is in position a, the 6-Ω resistor is redundant. The schematic for this case is shown in Fig. 8.41(a). To ensure that current

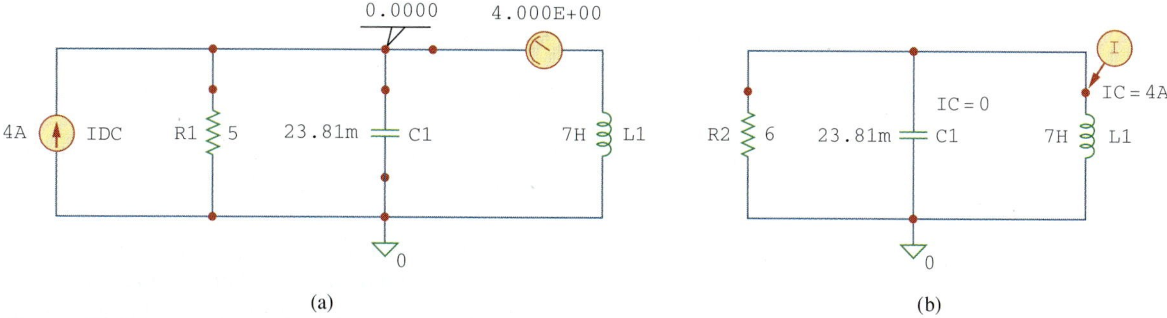

Figure 8.41
For Example 8.13: (a) for dc analysis, (b) for transient analysis.

$i(t)$ enters pin 1, the inductor is rotated three times before it is placed in the circuit. The same applies for the capacitor. We insert pseudo-components VIEWPOINT and IPROBE to determine the initial capacitor voltage and initial inductor current. We carry out a dc *PSpice* analysis by selecting **Analysis/Simulate**. As shown in Fig. 8.41(a), we obtain the initial capacitor voltage as 0 V and the initial inductor current $i(0)$ as 4 A from the dc analysis. These initial values will be used in the transient analysis.

When the switch is moved to position b, the circuit becomes a source-free parallel *RLC* circuit with the schematic in Fig. 8.41(b). We set the initial condition IC $= 0$ for the capacitor and IC $= 4$ A for the inductor. A current marker is inserted at pin 1 of the inductor. We select **Analysis/Setup/Transient** to open up the *Transient Analysis* dialog box and set *Final Time* to 3 s. After saving the schematic, we select **Analysis/Transient**. Figure 8.42 shows the plot of $i(t)$. The plot agrees with $i(t) = 4.8e^{-t} - 0.8e^{-6t}$ A, which is the solution by hand calculation.

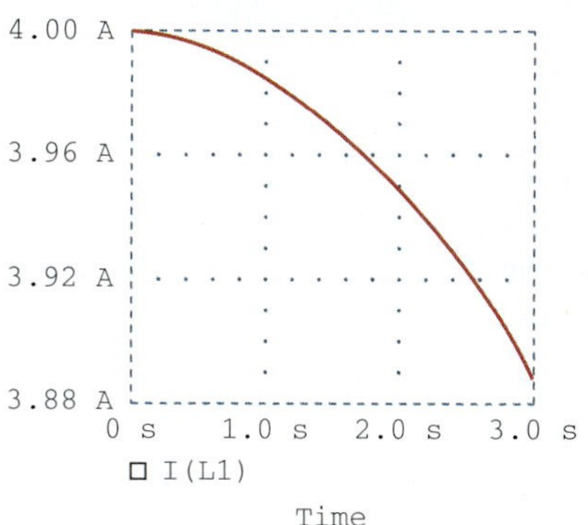

Figure 8.42
Plot of $i(t)$ for Example 8.13.

Practice Problem 8.13

Refer to the circuit in Fig. 8.21 (see Practice Prob. 8.7). Use *PSpice* to obtain $v(t)$ for $0 < t < 2$.

Answer: See Fig. 8.43.

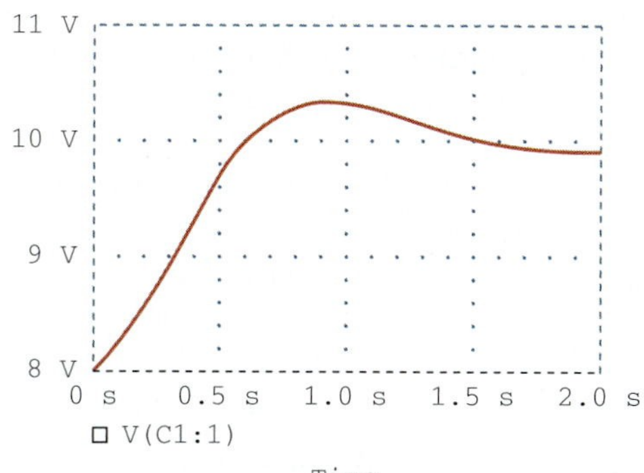

Figure 8.43
Plot of $v(t)$ for Practice Prob. 8.13.

8.10 †Duality

The concept of duality is a time-saving, effort-effective measure of solving circuit problems. Consider the similarity between Eq. (8.4) and Eq. (8.29). The two equations are the same, except that we must interchange the following quantities: (1) voltage and current, (2) resistance and conductance, (3) capacitance and inductance. Thus, it sometimes occurs in circuit analysis that two different circuits have the same equations and solutions, except that the roles of certain complementary elements are interchanged. This interchangeability is known as the principle of *duality*.

> The duality principle asserts a parallelism between pairs of characterizing equations and theorems of electric circuits.

TABLE 8.1

Dual pairs.

Resistance R	Conductance G
Inductance L	Capacitance C
Voltage v	Current i
Voltage source	Current source
Node	Mesh
Series path	Parallel path
Open circuit	Short circuit
KVL	KCL
Thevenin	Norton

> Even when the principle of linearity applies, a circuit element or variable may not have a dual. For example, mutual inductance (to be covered in Chapter 13) has no dual.

Dual pairs are shown in Table 8.1. Note that power does not appear in Table 8.1, because power has no dual. The reason for this is the principle of linearity; since power is not linear, duality does not apply. Also notice from Table 8.1 that the principle of duality extends to circuit elements, configurations, and theorems.

Two circuits that are described by equations of the same form, but in which the variables are interchanged, are said to be dual to each other.

> Two circuits are said to be duals of one another if they are described by the same characterizing equations with dual quantities interchanged.

The usefulness of the duality principle is self-evident. Once we know the solution to one circuit, we automatically have the solution for the dual circuit. It is obvious that the circuits in Figs. 8.8 and 8.13 are dual. Consequently, the result in Eq. (8.32) is the dual of that in Eq. (8.11). We must keep in mind that the principle of duality is limited to planar circuits. Nonplanar circuits have no duals, as they cannot be described by a system of mesh equations.

To find the dual of a given circuit, we do not need to write down the mesh or node equations. We can use a graphical technique. Given a planar circuit, we construct the dual circuit by taking the following three steps:

1. Place a node at the center of each mesh of the given circuit. Place the reference node (the ground) of the dual circuit outside the given circuit.
2. Draw lines between the nodes such that each line crosses an element. Replace that element by its dual (see Table 8.1).
3. To determine the polarity of voltage sources and direction of current sources, follow this rule: A voltage source that produces a positive (clockwise) mesh current has as its dual a current source whose reference direction is from the ground to the nonreference node.

In case of doubt, one may verify the dual circuit by writing the nodal or mesh equations. The mesh (or nodal) equations of the original circuit are similar to the nodal (or mesh) equations of the dual circuit. The duality principle is illustrated with the following two examples.

Example 8.14

Construct the dual of the circuit in Fig. 8.44.

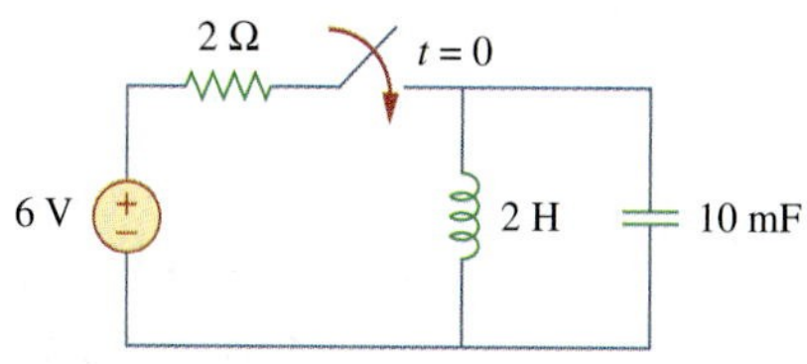

Figure 8.44
For Example 8.14.

Solution:

As shown in Fig. 8.45(a), we first locate nodes 1 and 2 in the two meshes and also the ground node 0 for the dual circuit. We draw a line between one node and another crossing an element. We replace the line joining the nodes by the duals of the elements which it crosses. For example, a line between nodes 1 and 2 crosses a 2-H inductor, and we place a 2-F capacitor (an inductor's dual) on the line. A line between nodes 1 and 0 crossing the 6-V voltage source will contain a 6-A current source. By drawing lines crossing all the elements, we construct the dual circuit on the given circuit as in Fig. 8.45(a). The dual circuit is redrawn in Fig. 8.45(b) for clarity.

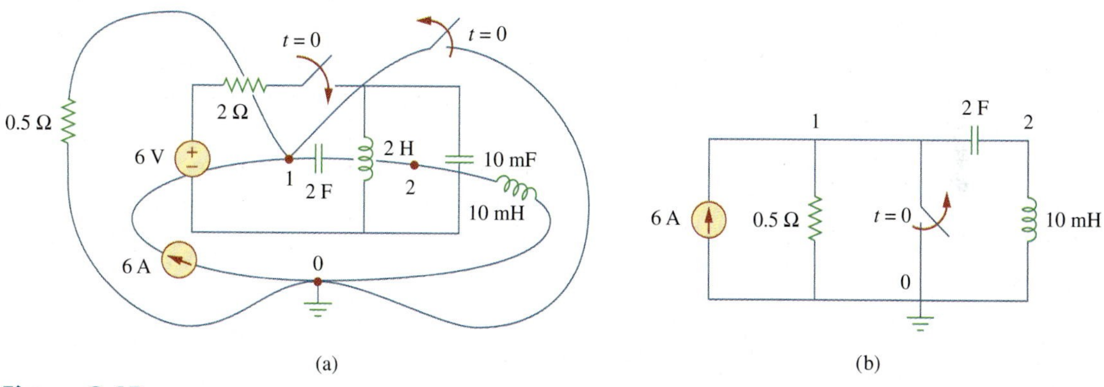

Figure 8.45
(a) Construction of the dual circuit of Fig. 8.44, (b) dual circuit redrawn.

Practice Problem 8.14

Draw the dual circuit of the one in Fig. 8.46.

Answer: See Fig. 8.47.

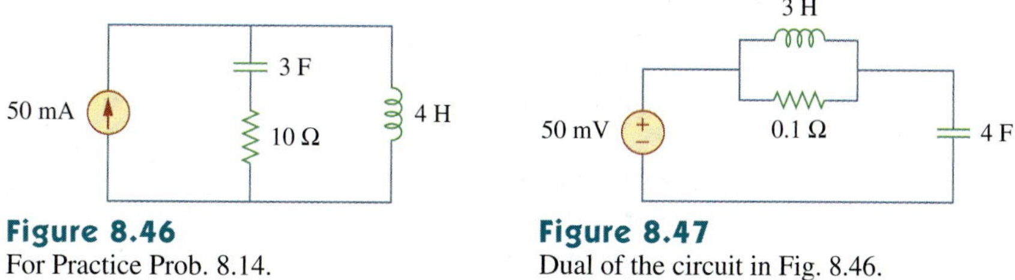

Figure 8.46
For Practice Prob. 8.14.

Figure 8.47
Dual of the circuit in Fig. 8.46.

Example 8.15

Obtain the dual of the circuit in Fig. 8.48.

Solution:

The dual circuit is constructed on the original circuit as in Fig. 8.49(a). We first locate nodes 1 to 3 and the reference node 0. Joining nodes 1 and 2, we cross the 2-F capacitor, which is replaced by a 2-H inductor.

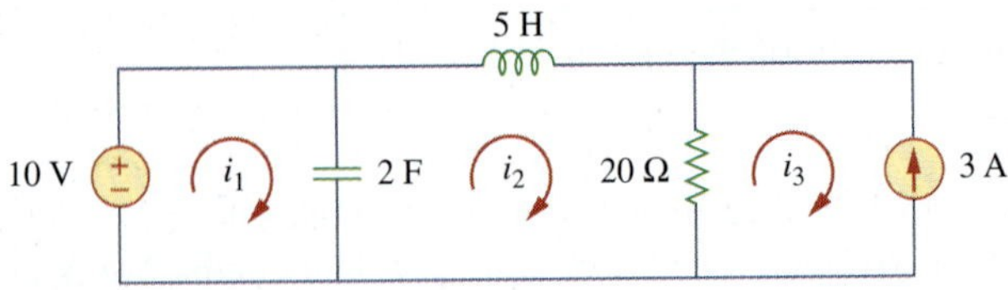

Figure 8.48
For Example 8.15.

Joining nodes 2 and 3, we cross the 20-Ω resistor, which is replaced by a $\frac{1}{20}$-Ω resistor. We keep doing this until all the elements are crossed. The result is in Fig. 8.49(a). The dual circuit is redrawn in Fig. 8.49(b).

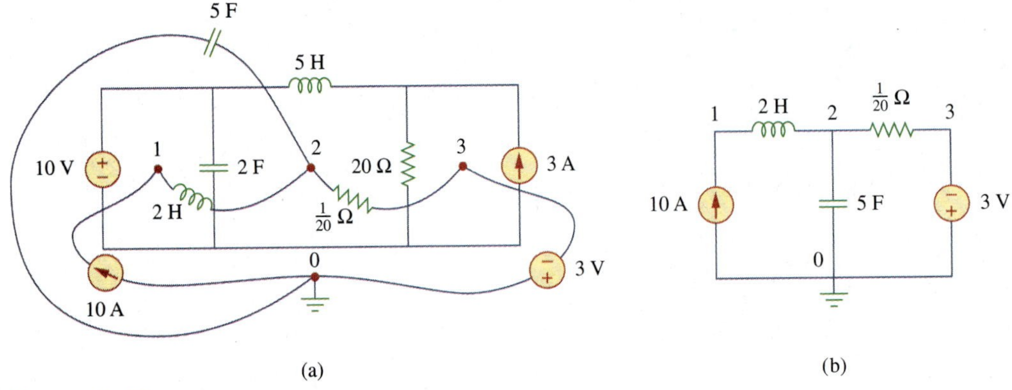

Figure 8.49
For Example 8.15: (a) construction of the dual circuit of Fig. 8.48, (b) dual circuit redrawn.

To verify the polarity of the voltage source and the direction of the current source, we may apply mesh currents i_1, i_2, and i_3 (all in the clockwise direction) in the original circuit in Fig. 8.48. The 10-V voltage source produces positive mesh current i_1, so that its dual is a 10-A current source directed from 0 to 1. Also, $i_3 = -3$ A in Fig. 8.48 has as its dual $v_3 = -3$ V in Fig. 8.49(b).

Practice Problem 8.15

For the circuit in Fig. 8.50, obtain the dual circuit.

Answer: See Fig. 8.51.

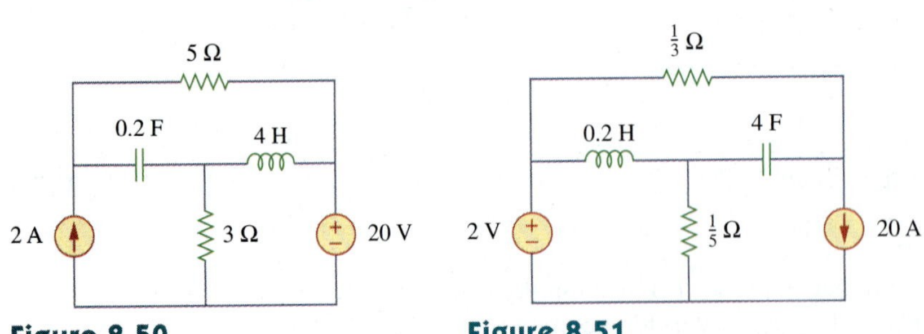

Figure 8.50
For Practice Prob. 8.15.

Figure 8.51
Dual of the circuit in Fig. 8.50.

8.11 †Applications

Practical applications of *RLC* circuits are found in control and communications circuits such as ringing circuits, peaking circuits, resonant circuits, smoothing circuits, and filters. Most of these circuits cannot be covered until we treat ac sources. For now, we will limit ourselves to two simple applications: automobile ignition and smoothing circuits.

8.11.1 Automobile Ignition System

In Section 7.9.4, we considered the automobile ignition system as a charging system. That was only a part of the system. Here, we consider another part—the voltage generating system. The system is modeled by the circuit shown in Fig. 8.52. The 12-V source is due to the battery and alternator. The 4-Ω resistor represents the resistance of the wiring. The ignition coil is modeled by the 8-mH inductor. The 1-μF capacitor (known as the *condenser* to automechanics) is in parallel with the switch (known as the *breaking points* or *electronic ignition*). In the following example, we determine how the *RLC* circuit in Fig. 8.52 is used in generating high voltage.

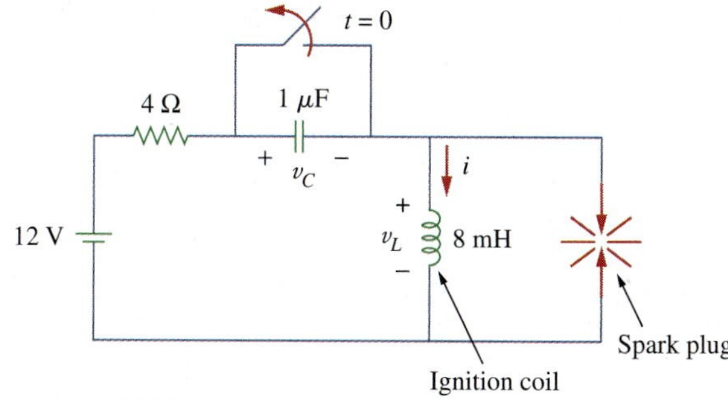

Figure 8.52
Automobile ignition circuit.

Example 8.16

Assuming that the switch in Fig. 8.52 is closed prior to $t = 0^-$, find the inductor voltage v_L for $t > 0$.

Solution:
If the switch is closed prior to $t = 0^-$ and the circuit is in steady state, then

$$i(0^-) = \frac{12}{4} = 3 \text{ A}, \qquad v_C(0^-) = 0$$

At $t = 0^+$, the switch is opened. The continuity conditions require that

$$i(0^+) = 3 \text{ A}, \qquad v_C(0^+) = 0 \tag{8.16.1}$$

We obtain $di(0^+)/dt$ from $v_L(0^+)$. Applying KVL to the mesh at $t = 0^+$ yields

$$-12 + 4i(0^+) + v_L(0^+) + v_C(0^+) = 0$$

$$-12 + 4 \times 3 + v_L(0^+) + 0 = 0 \quad \Rightarrow \quad v_L(0^+) = 0$$

Hence,

$$\frac{di(0^+)}{dt} = \frac{v_L(0^+)}{L} = 0 \qquad \textbf{(8.16.2)}$$

As $t \rightarrow \infty$, the system reaches steady state, so that the capacitor acts like an open circuit. Then

$$i(\infty) = 0 \qquad \textbf{(8.16.3)}$$

If we apply KVL to the mesh for $t > 0$, we obtain

$$12 = Ri + L\frac{di}{dt} + \frac{1}{C}\int_0^t i\,dt + v_C(0)$$

Taking the derivative of each term yields

$$\frac{d^2i}{dt^2} + \frac{R}{L}\frac{di}{dt} + \frac{i}{LC} = 0 \qquad \textbf{(8.16.4)}$$

We obtain the form of the transient response by following the procedure in Section 8.3. Substituting $R = 4\ \Omega$, $L = 8$ mH, and $C = 1\ \mu$F, we get

$$\alpha = \frac{R}{2L} = 250, \qquad \omega_0 = \frac{1}{\sqrt{LC}} = 1.118 \times 10^4$$

Since $\alpha < \omega_0$, the response is underdamped. The damped natural frequency is

$$\omega_d = \sqrt{\omega_0^2 - \alpha^2} \simeq \omega_0 = 1.118 \times 10^4$$

The form of the transient response is

$$i_t(t) = e^{-\alpha}(A\cos\omega_d t + B\sin\omega_d t) \qquad \textbf{(8.16.5)}$$

where A and B are constants. The steady-state response is

$$i_{ss}(t) = i(\infty) = 0 \qquad \textbf{(8.16.6)}$$

so that the complete response is

$$i(t) = i_t(t) + i_{ss}(t) = e^{-250t}(A\cos 11{,}180t + B\sin 11{,}180t) \qquad \textbf{(8.16.7)}$$

We now determine A and B.

$$i(0) = 3 = A + 0 \qquad \Rightarrow \qquad A = 3$$

Taking the derivative of Eq. (8.16.7),

$$\begin{aligned}\frac{di}{dt} &= -250e^{-250t}(A\cos 11{,}180t + B\sin 11{,}180t)\\ &\quad + e^{-250t}(-11{,}180A\sin 11{,}180t + 11{,}180B\cos 11{,}180t)\end{aligned}$$

Setting $t = 0$ and incorporating Eq. (8.16.2),

$$0 = -250A + 11{,}180B \qquad \Rightarrow \qquad B = 0.0671$$

Thus,

$$i(t) = e^{-250t}(3\cos 11{,}180t + 0.0671\sin 11{,}180t) \qquad \textbf{(8.16.8)}$$

The voltage across the inductor is then

$$v_L(t) = L\frac{di}{dt} = -268e^{-250t}\sin 11{,}180t \qquad \textbf{(8.16.9)}$$

This has a maximum value when sine is unity, that is, at $11{,}180t_0 = \pi/2$ or $t_0 = 140.5\ \mu s$. At time $= t_0$, the inductor voltage reaches its peak, which is

$$v_L(t_0) = -268e^{-250t_0} = -259 \text{ V} \qquad \textbf{(8.16.10)}$$

Although this is far less than the voltage range of 6000 to 10,000 V required to fire the spark plug in a typical automobile, a device known as a *transformer* (to be discussed in Chapter 13) is used to step up the inductor voltage to the required level.

Practice Problem 8.16

In Fig. 8.52, find the capacitor voltage v_C for $t > 0$.

Answer: $12 - 12e^{-250t}\cos 11{,}180t + 267.7e^{-250t}\sin 11{,}180t$ V.

8.11.2 Smoothing Circuits

In a typical digital communication system, the signal to be transmitted is first sampled. Sampling refers to the procedure of selecting samples of a signal for processing, as opposed to processing the entire signal. Each sample is converted into a binary number represented by a series of pulses. The pulses are transmitted by a transmission line such as a coaxial cable, twisted pair, or optical fiber. At the receiving end, the signal is applied to a digital-to-analog (D/A) converter whose output is a "staircase" function, that is, constant at each time interval. In order to recover the transmitted analog signal, the output is smoothed by letting it pass through a "smoothing" circuit, as illustrated in Fig. 8.53. An *RLC* circuit may be used as the smoothing circuit.

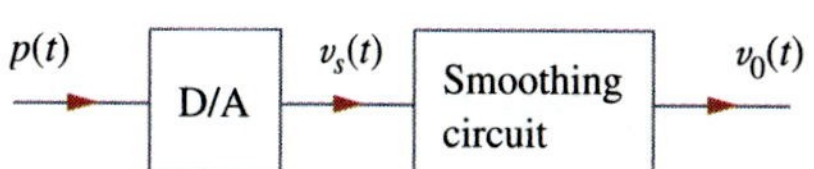

Figure 8.53
A series of pulses is applied to the digital-to-analog (D/A) converter, whose output is applied to the smoothing circuit.

Example 8.17

The output of a D/A converter is shown in Fig. 8.54(a). If the *RLC* circuit in Fig. 8.54(b) is used as the smoothing circuit, determine the output voltage $v_o(t)$.

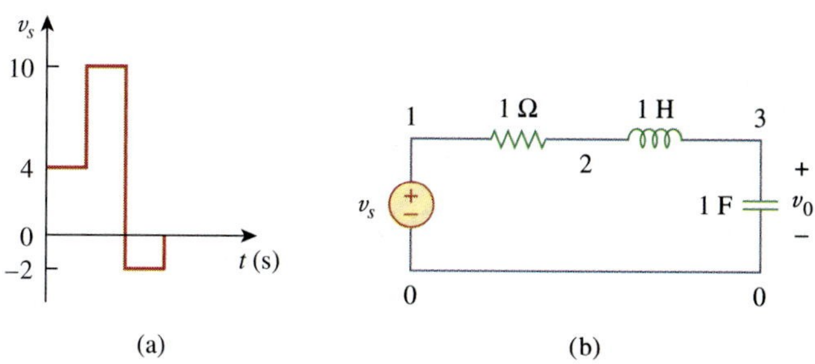

Figure 8.54
For Example 8.17: (a) output of a D/A converter, (b) an *RLC* smoothing circuit.

Solution:
This problem is best solved using *PSpice*. The schematic is shown in Fig. 8.55(a). The pulse in Fig. 8.54(a) is specified using the piecewise

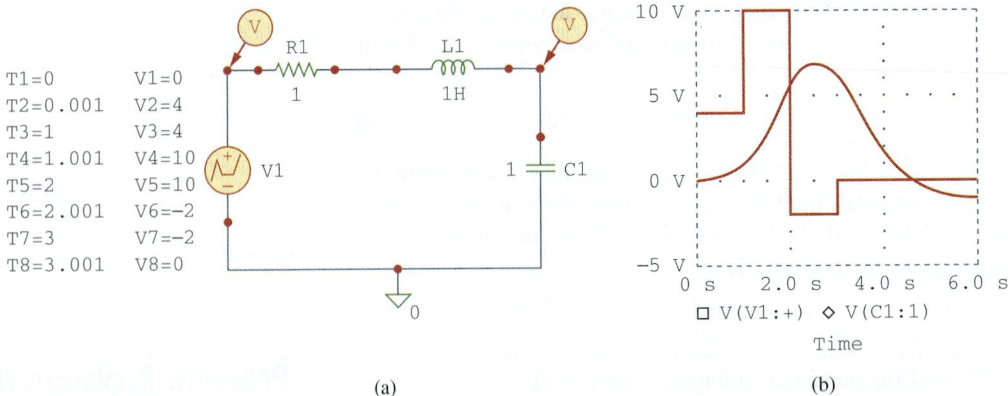

Figure 8.55
For Example 8.17: (a) schematic, (b) input and output voltages.

linear function. The attributes of V1 are set as T1 = 0, V1 = 0, T2 = 0.001, V2 = 4, T3 = 1, V3 = 4, and so on. To be able to plot both input and output voltages, we insert two voltage markers as shown. We select **Analysis/Setup/Transient** to open up the *Transient Analysis* dialog box and set *Final Time* as 6 s. Once the schematic is saved, we select **Analysis/Simulate** to run and obtain the plots shown in Fig. 8.55(b).

Practice Problem 8.17

Rework Example 8.17 if the output of the D/A converter is as shown in Fig. 8.56.

Answer: See Fig. 8.57.

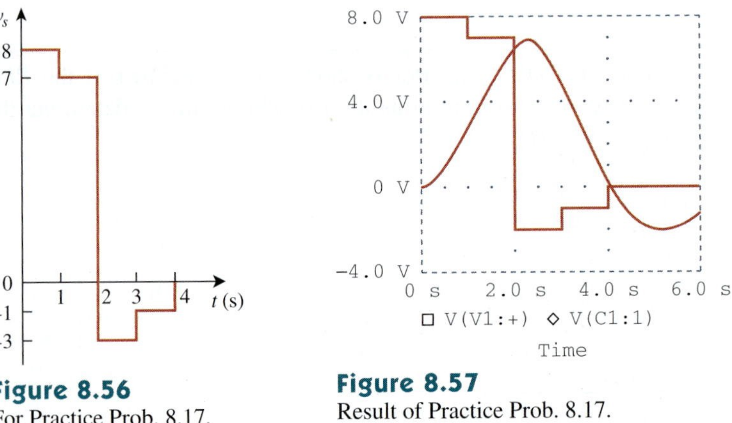

Figure 8.56
For Practice Prob. 8.17.

Figure 8.57
Result of Practice Prob. 8.17.

8.12 Summary

1. The determination of the initial values $x(0)$ and $dx(0)/dt$ and final value $x(\infty)$ is crucial to analyzing second-order circuits.
2. The *RLC* circuit is second-order because it is described by a second-order differential equation. Its characteristic equation is

$s^2 + 2\alpha s + \omega_0^2 = 0$, where α is the damping factor and ω_0 is the undamped natural frequency. For a series circuit, $\alpha = R/2L$, for a parallel circuit $\alpha = 1/2RC$, and for both cases $\omega_0 = 1/0\sqrt{LC}$.

3. If there are no independent sources in the circuit after switching (or sudden change), we regard the circuit as source-free. The complete solution is the natural response.
4. The natural response of an *RLC* circuit is overdamped, underdamped, or critically dampèd, depending on the roots of the characteristic equation. The response is critically damped when the roots are equal ($s_1 = s_2$ or $\alpha = \omega_0$), overdamped when the roots are real and unequal ($s_1 \neq s_2$ or $\alpha > \omega_0$), or underdamped when the roots are complex conjugate ($s_1 = s_2^*$ or $\alpha < \omega_0$).
5. If independent sources are present in the circuit after switching, the complete response is the sum of the transient response and the steady-state response.
6. *PSpice* is used to analyze *RLC* circuits in the same way as for *RC* or *RL* circuits.
7. Two circuits are dual if the mesh equations that describe one circuit have the same form as the nodal equations that describe the other. The analysis of one circuit gives the analysis of its dual circuit.
8. The automobile ignition circuit and the smoothing circuit are typical applications of the material covered in this chapter.

Review Questions

8.1 For the circuit in Fig. 8.58, the capacitor voltage at $t = 0^-$ (just before the switch is closed) is:

(a) 0 V (b) 4 V (c) 8 V (d) 12 V

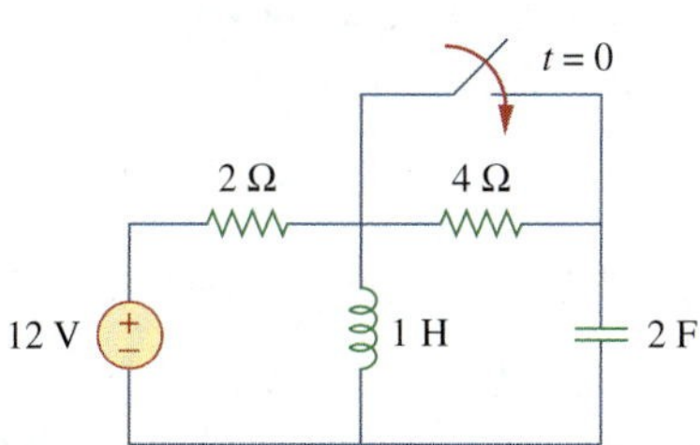

Figure 8.58
For Review Questions 8.1 and 8.2.

8.2 For the circuit in Fig. 8.58, the initial inductor current (at $t = 0$) is:

(a) 0 A (b) 2 A (c) 6 A (d) 12 A

8.3 When a step input is applied to a second-order circuit, the final values of the circuit variables are found by:

(a) Replacing capacitors with closed circuits and inductors with open circuits.

(b) Replacing capacitors with open circuits and inductors with closed circuits.

(c) Doing neither of the above.

8.4 If the roots of the characteristic equation of an *RLC* circuit are -2 and -3, the response is:

(a) $(A \cos 2t + B \sin 2t)e^{-3t}$

(b) $(A + 2Bt)e^{-3t}$

(c) $Ae^{-2t} + Bte^{-3t}$

(d) $Ae^{-2t} + Be^{-3t}$

where A and B are constants.

8.5 In a series *RLC* circuit, setting $R = 0$ will produce:

(a) an overdamped response

(b) a critically damped response

(c) an underdamped response

(d) an undamped response

(e) none of the above

8.6 A parallel *RLC* circuit has $L = 2$ H and $C = 0.25$ F. The value of R that will produce unity damping factor is:

(a) 0.5 Ω (b) 1 Ω (c) 2 Ω (d) 4 Ω

8.7 Refer to the series *RLC* circuit in Fig. 8.59. What kind of response will it produce?

(a) overdamped

(b) underdamped

(c) critically damped

(d) none of the above

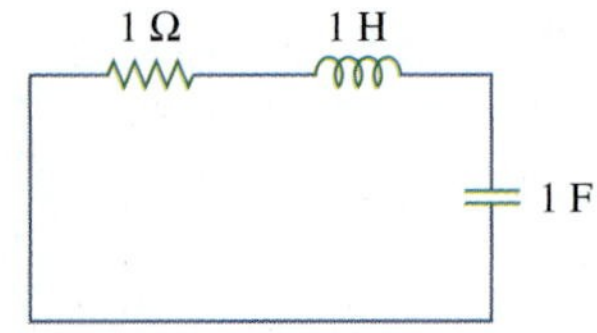

Figure 8.59
For Review Question 8.7.

8.8 Consider the parallel *RLC* circuit in Fig. 8.60. What type of response will it produce?

(a) overdamped

(b) underdamped

(c) critically damped

(d) none of the above

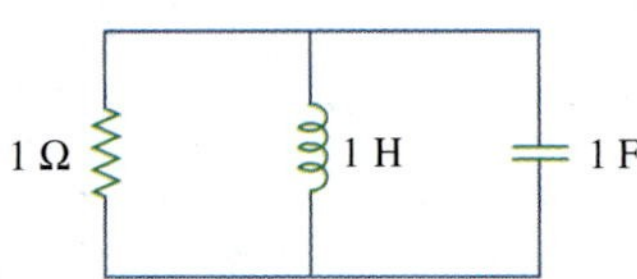

Figure 8.60
For Review Question 8.8.

8.9 Match the circuits in Fig. 8.61 with the following items:

(i) first-order circuit

(ii) second-order series circuit

(iii) second-order parallel circuit

(iv) none of the above

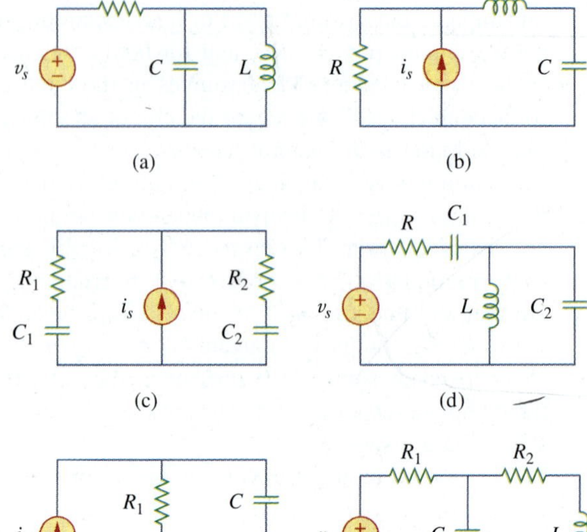

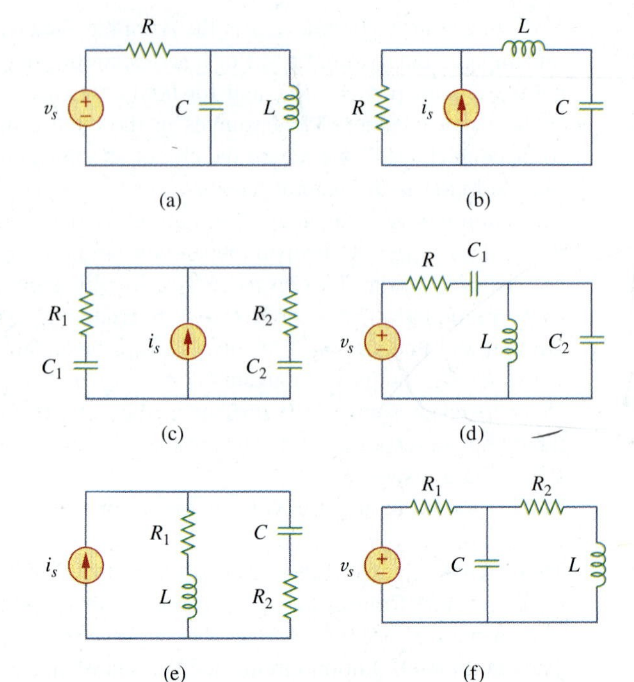

Figure 8.61
For Review Question 8.9.

8.10 In an electric circuit, the dual of resistance is:

(a) conductance (b) inductance

(c) capacitance (d) open circuit

(e) short circuit

Answers: 8.1a, 8.2c, 8.3b, 8.4d, 8.5d, 8.6c, 8.7b, 8.8b, 8.9 (i)-c, (ii)-b, e, (iii)-a, (iv)-d, f, 8.10a.

Problems

Section 8.2 Finding Initial and Final Values

8.1 For the circuit in Fig. 8.62, find:

(a) $i(0^+)$ and $v(0^+)$,

(b) $di(0^+)/dt$ and $dv(0^+)/dt$,

(c) $i(\infty)$ and $v(\infty)$.

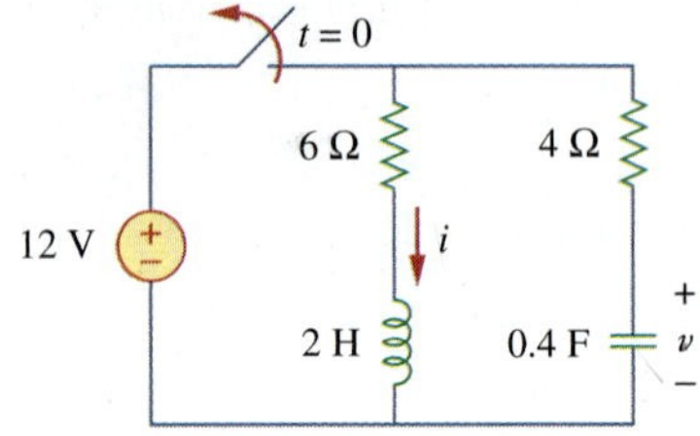

Figure 8.62
For Prob. 8.1.

8.2 Using Fig. 8.63, design a problem to help other students better understand finding initial and final values.

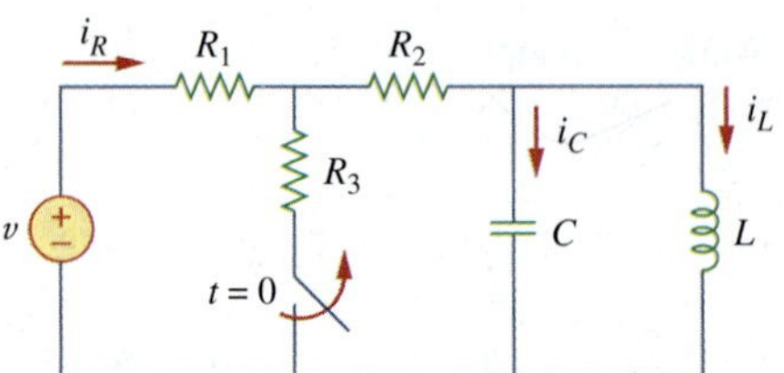

Figure 8.63
For Prob. 8.2.

8.3 Refer to the circuit shown in Fig. 8.64. Calculate:

(a) $i_L(0^+)$, $v_C(0^+)$, and $v_R(0^+)$,

(b) $di_L(0^+)/dt$, $dv_C(0^+)/dt$, and $dv_R(0^+)/dt$,

(c) $i_L(\infty)$, $v_C(\infty)$, and $v_R(\infty)$.

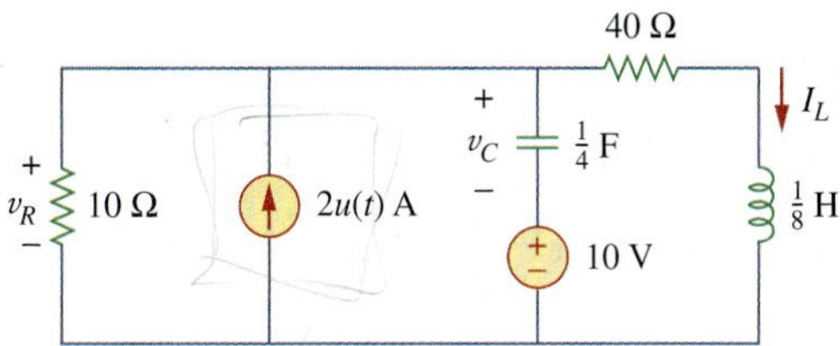

Figure 8.64
For Prob. 8.3.

8.4 In the circuit of Fig. 8.65, find:

(a) $v(0^+)$ and $i(0^+)$,

(b) $dv(0^+)/dt$ and $di(0^+)/dt$,

(c) $v(\infty)$ and $i(\infty)$.

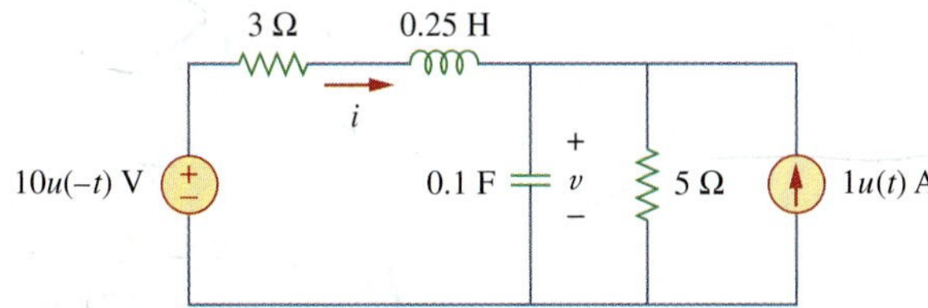

Figure 8.65
For Prob. 8.4.

8.5 Refer to the circuit in Fig. 8.66. Determine:

(a) $i(0^+)$ and $v(0^+)$,

(b) $di(0^+)/dt$ and $dv(0^+)/dt$,

(c) $i(\infty)$ and $v(\infty)$.

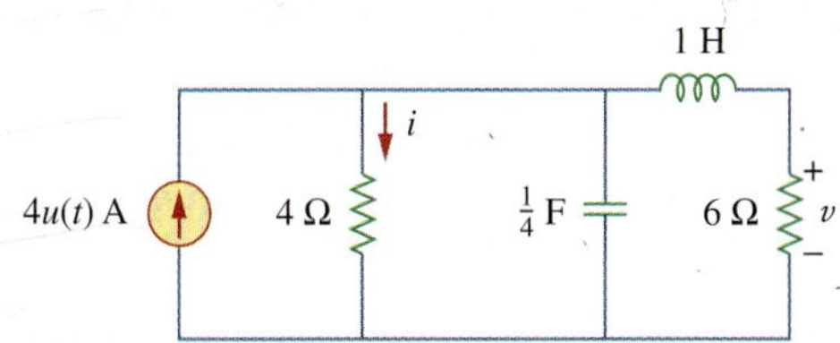

Figure 8.66
For Prob. 8.5.

8.6 In the circuit of Fig. 8.67, find:

(a) $v_R(0^+)$ and $v_L(0^+)$,

(b) $dv_R(0^+)/dt$ and $dv_L(0^+)/dt$,

(c) $v_R(\infty)$ and $v_L(\infty)$.

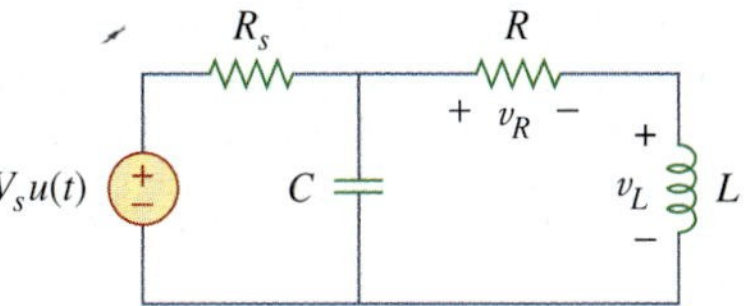

Figure 8.67
For Prob. 8.6.

Section 8.3 Source-Free Series *RLC* Circuit

8.7 A series *RLC* circuit has $R = 10$ kΩ, $L = 0.1$ mH, and $C = 10$ μF. What type of damping is exhibited by the circuit?

8.8 **e2d** Design a problem to help other students better understand source-free *RLC* circuits.

8.9 The current in an *RLC* circuit is described by

$$\frac{d^2i}{dt^2} + 10\frac{di}{dt} + 25i = 0$$

If $i(0) = 2$ A and $di(0)/dt = 0$, find $i(t)$ for $t > 0$.

8.10 The differential equation that describes the voltage in an *RLC* network is

$$\frac{d^2v}{dt^2} + 5\frac{dv}{dt} + 4v = 0$$

Given that $v(0) = 0$, $dv(0)/dt = 5$ V/s, obtain $v(t)$.

8.11 The natural response of an *RLC* circuit is described by the differential equation

$$\frac{d^2v}{dt^2} + 2\frac{dv}{dt} + v = 0$$

for which the initial conditions are $v(0) = 20$ V and $dv(0)/dt = 0$. Solve for $v(t)$.

8.12 If $R = 20$ Ω, $L = 0.6$ H, what value of C will make an *RLC* series circuit:

(a) overdamped,

(b) critically damped,

(c) underdamped?

8.13 For the circuit in Fig. 8.68, calculate the value of R needed to have a critically damped response.

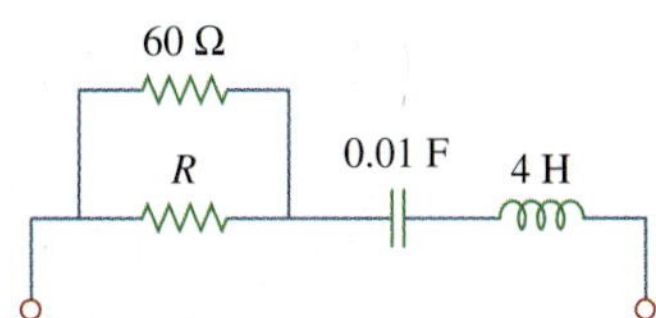

Figure 8.68
For Prob. 8.13.

8.14 The switch in Fig. 8.69 moves from position A to position B at $t = 0$ (please note that the switch must connect to point B before it breaks the connection at A, a make-before-break switch). Find $v(t)$ for $t > 0$.

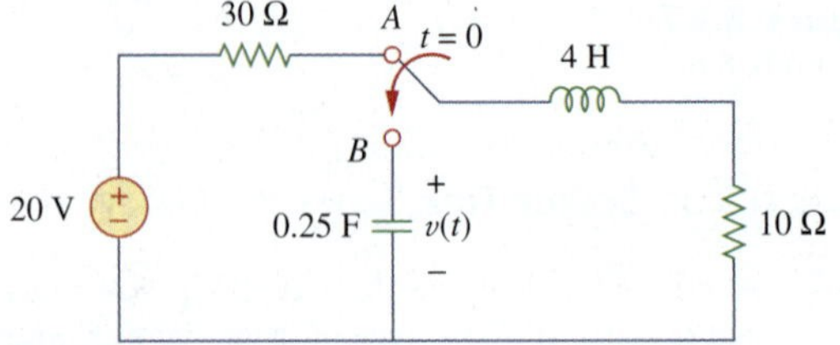

Figure 8.69
For Prob. 8.14.

8.15 The responses of a series RLC circuit are

$$v_C(t) = 30 - 10e^{-20t} + 30e^{-10t}\ \text{V}$$
$$i_L(t) = 40e^{-20t} - 60e^{-10t}\ \text{mA}$$

where v_C and i_L are the capacitor voltage and inductor current, respectively. Determine the values of R, L, and C.

8.16 Find $i(t)$ for $t > 0$ in the circuit of Fig. 8.70.

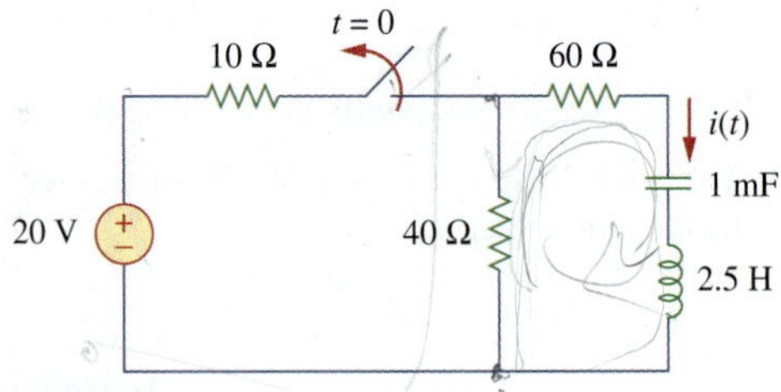

Figure 8.70
For Prob. 8.16.

8.17 In the circuit of Fig. 8.71, the switch instantaneously moves from position A to B at $t = 0$. Find $v(t)$ for all $t \geq 0$.

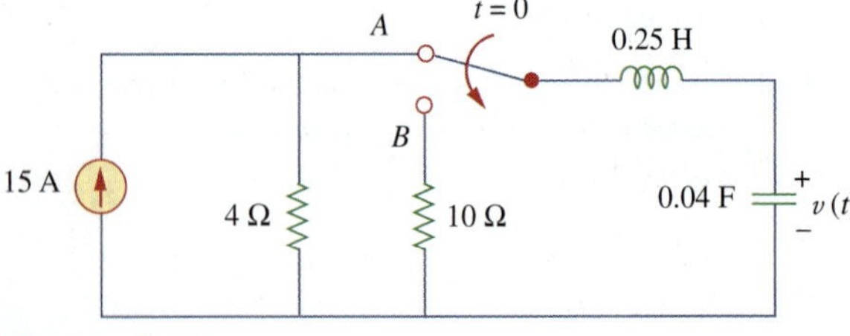

Figure 8.71
For Prob. 8.17.

8.18 Find the voltage across the capacitor as a function of time for $t > 0$ for the circuit in Fig. 8.72. Assume steady-state conditions exist at $t = 0^-$.

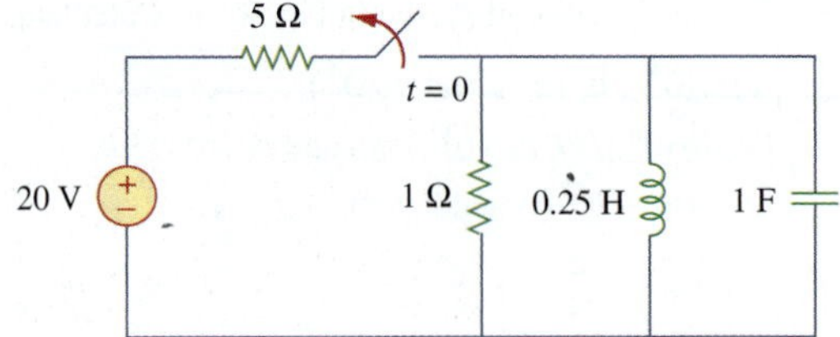

Figure 8.72
For Prob. 8.18.

8.19 Obtain $v(t)$ for $t > 0$ in the circuit of Fig. 8.73.

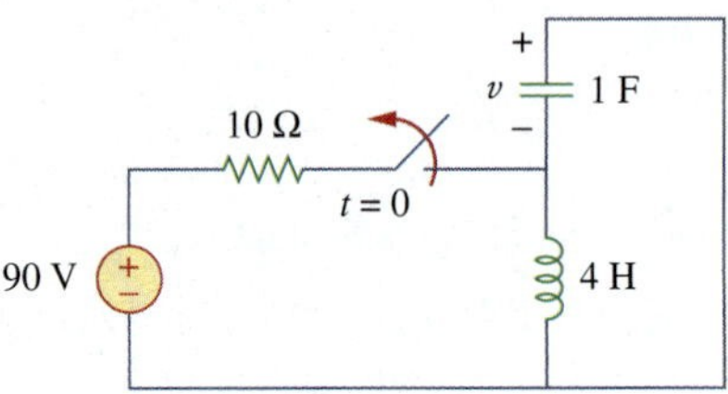

Figure 8.73
For Prob. 8.19.

8.20 The switch in the circuit of Fig. 8.74 has been closed for a long time but is opened at $t = 0$. Determine $i(t)$ for $t > 0$.

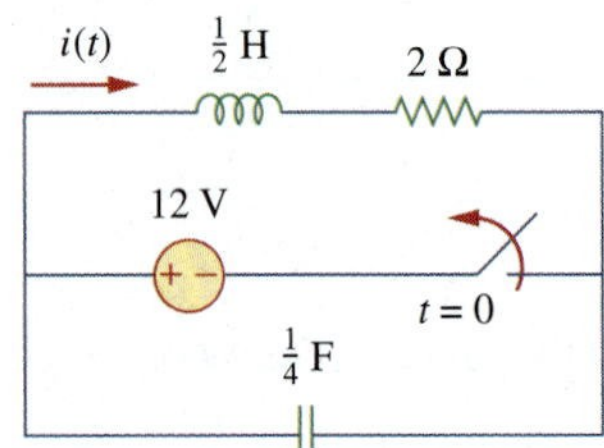

Figure 8.74
For Prob. 8.20.

***8.21** Calculate $v(t)$ for $t > 0$ in the circuit of Fig. 8.75.

ps

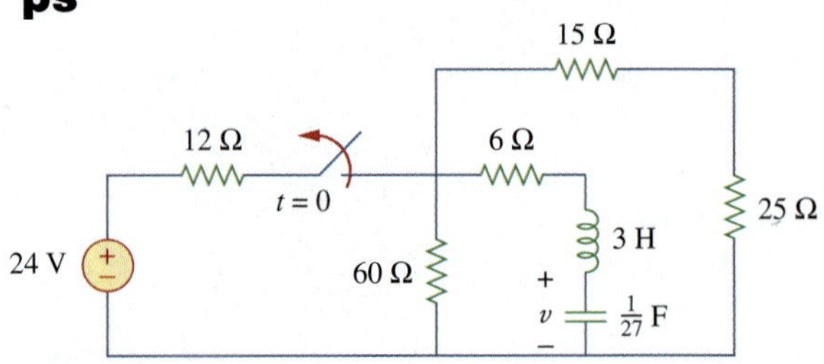

Figure 8.75
For Prob. 8.21.

* An asterisk indicates a challenging problem.

Section 8.4 Source-Free Parallel *RLC* Circuit

8.22 Assuming $R = 2\ \text{k}\Omega$, design a parallel *RLC* circuit that has the characteristic equation

$$s^2 + 100s + 10^6 = 0.$$

8.23 For the network in Fig. 8.76, what value of C is needed to make the response underdamped with unity damping factor ($\alpha = 1$)?

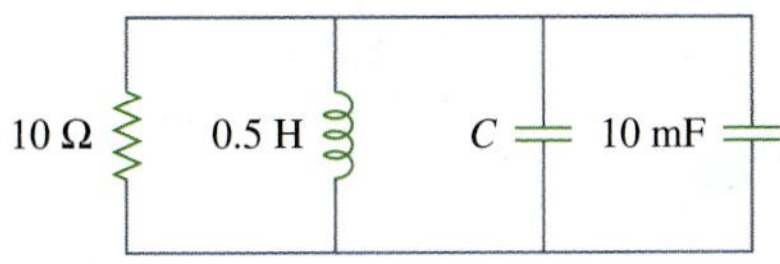

Figure 8.76
For Prob. 8.23.

8.24 The switch in Fig. 8.77 moves from position A to position B at $t = 0$ (please note that the switch must connect to point B before it breaks the connection at A, a make-before-break switch). Determine $i(t)$ for $t > 0$.

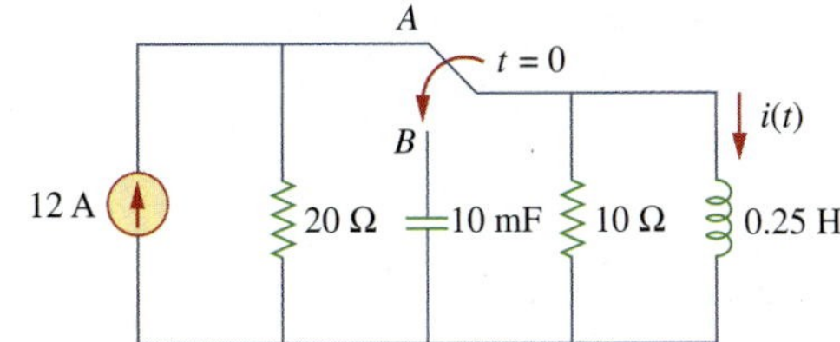

Figure 8.77
For Prob. 8.24.

8.25 Using Fig. 8.78, design a problem to help other students better understand source-free *RLC* circuits.

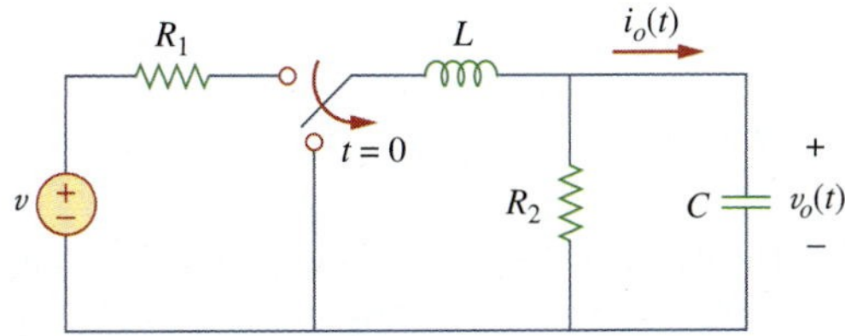

Figure 8.78
For Prob. 8.25.

Section 8.5 Step Response of a Series *RLC* Circuit

8.26 The step response of an *RLC* circuit is described by

$$\frac{d^2i}{dt^2} + 2\frac{di}{dt} + 5i = 10$$

Given that $i(0) = 6$ A and $di(0)/dt = 12$ A/s, solve for $i(t)$.

8.27 A branch voltage in an *RLC* circuit is described by

$$\frac{d^2v}{dt^2} + 4\frac{dv}{dt} + 8v = 48$$

If the initial conditions are $v(0) = 0 = dv(0)/dt$, find $v(t)$.

8.28 A series *RLC* circuit is described by

$$L\frac{d^2i}{dt^2} + R\frac{di}{dt} + \frac{i}{C} = 2$$

Find the response when $L = 0.5$ H, $R = 4\ \Omega$, and $C = 0.2$ F. Let $i(0) = 1$, $di(0)/dt = 0$.

8.29 Solve the following differential equations subject to the specified initial conditions

(a) $d^2v/dt^2 + 4v = 12$, $v(0) = 0$, $dv(0)/dt = 2$

(b) $d^2i/dt^2 + 5\,di/dt + 4i = 8$, $i(0) = -1$, $di(0)/dt = 0$

(c) $d^2v/dt^2 + 2\,dv/dt + v = 3$, $v(0) = 5$, $dv(0)/dt = 1$

(d) $d^2i/dt^2 + 2\,di/dt + 5i = 10$, $i(0) = 4$, $di(0)/dt = -2$

8.30 The step responses of a series *RLC* circuit are

$$v_C = 40 - 10e^{-2000t} - 10e^{-4000t}\ \text{V}, \quad t > 0$$
$$i_L(t) = 3e^{-2000t} + 6e^{-4000t}\ \text{mA}, \quad t > 0$$

(a) Find C. (b) Determine what type of damping is exhibited by the circuit.

8.31 Consider the circuit in Fig. 8.79. Find $v_L(0^+)$ and $v_C(0^+)$.

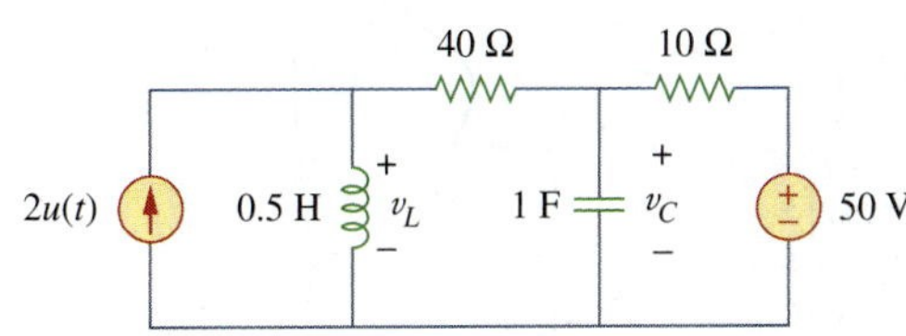

Figure 8.79
For Prob. 8.31.

8.32 For the circuit in Fig. 8.80, find $v(t)$ for $t > 0$.

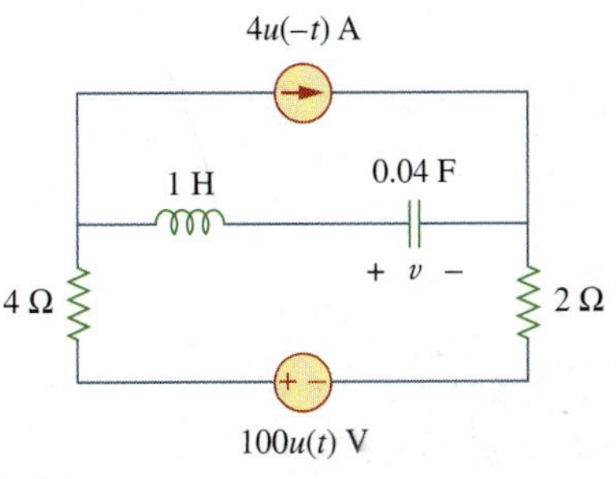

Figure 8.80
For Prob. 8.32.

8.33 Find $v(t)$ for $t > 0$ in the circuit of Fig. 8.81.

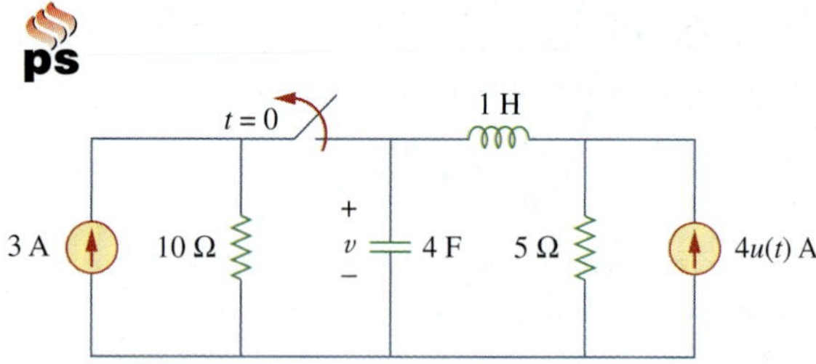

Figure 8.81
For Prob. 8.33.

8.34 Calculate $i(t)$ for $t > 0$ in the circuit of Fig. 8.82.

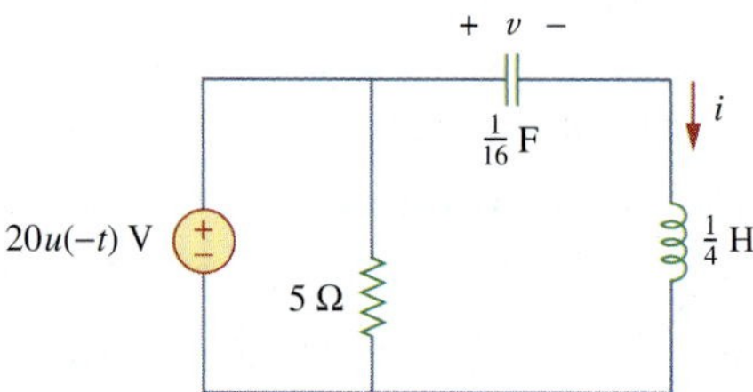

Figure 8.82
For Prob. 8.34.

8.35 Using Fig. 8.83, design a problem to help other students better understand the step response of series *RLC* circuits.

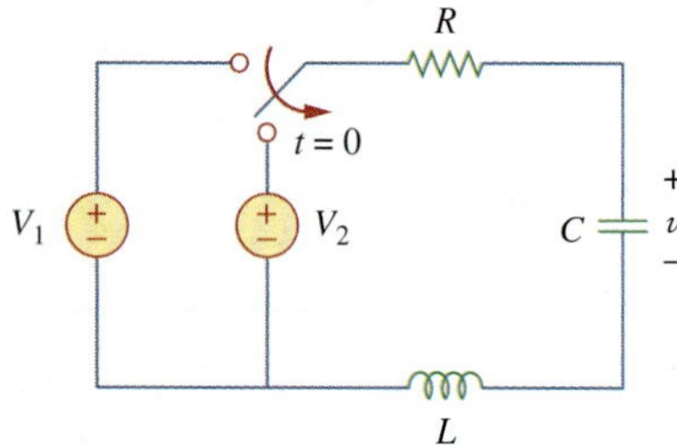

Figure 8.83
For Prob. 8.35.

8.36 Obtain $v(t)$ and $i(t)$ for $t > 0$ in the circuit of Fig. 8.84.

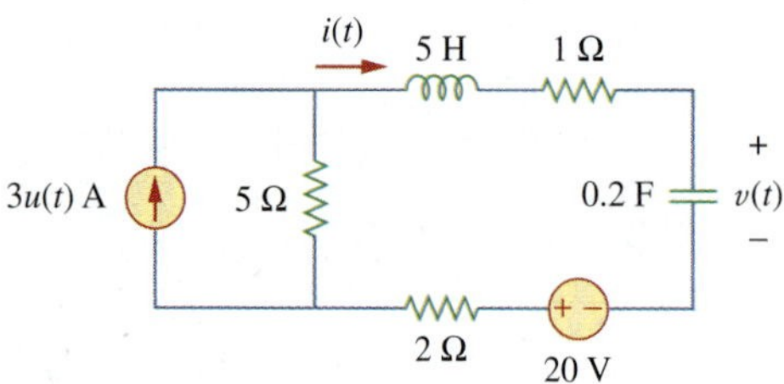

Figure 8.84
For Prob. 8.36.

***8.37** For the network in Fig. 8.85, solve for $i(t)$ for $t > 0$.

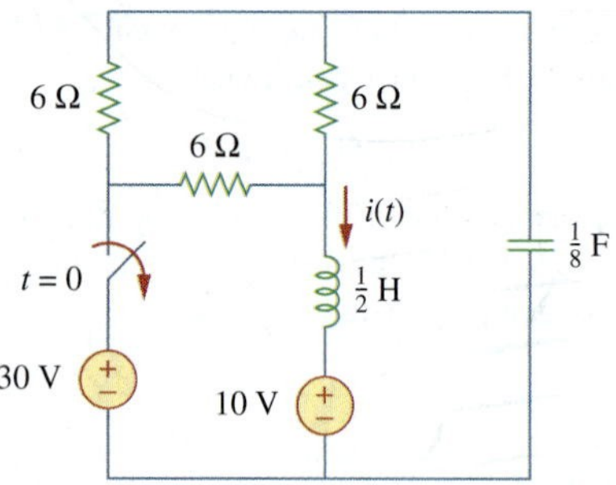

Figure 8.85
For Prob. 8.37.

8.38 Refer to the circuit in Fig. 8.86. Calculate $i(t)$ for $t > 0$.

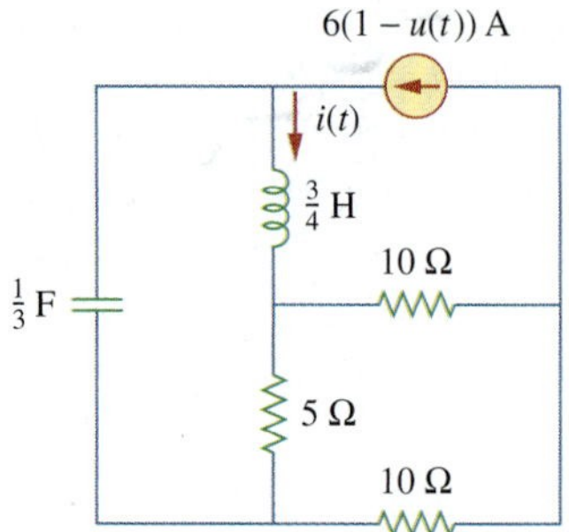

Figure 8.86
For Prob. 8.38.

8.39 Determine $v(t)$ for $t > 0$ in the circuit of Fig. 8.87.

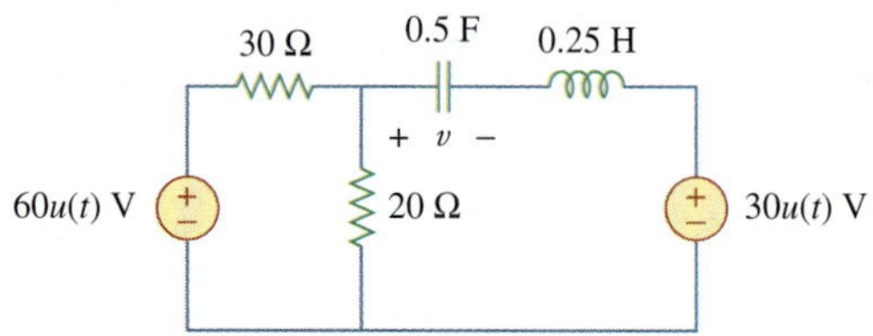

Figure 8.87
For Prob. 8.39.

8.40 The switch in the circuit of Fig. 8.88 is moved from position a to b at $t = 0$. Determine $i(t)$ for $t > 0$.

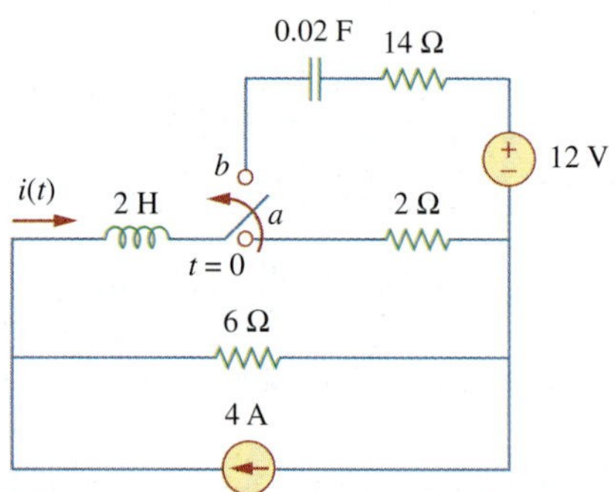

Figure 8.88
For Prob. 8.40.

***8.41** For the network in Fig. 8.89, find $i(t)$ for $t > 0$.

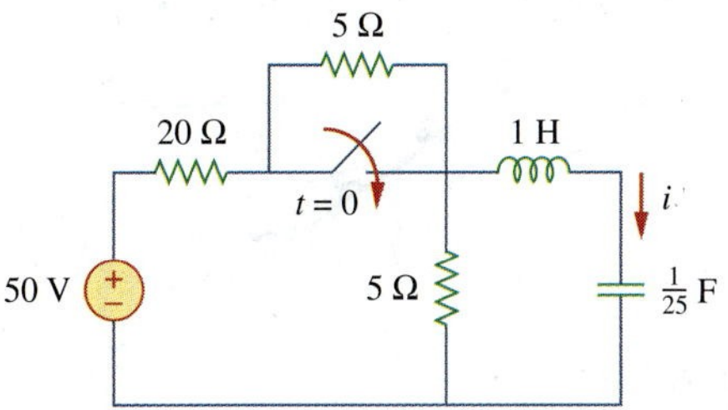

Figure 8.89
For Prob. 8.41.

***8.42** Given the network in Fig. 8.90, find $v(t)$ for $t > 0$.

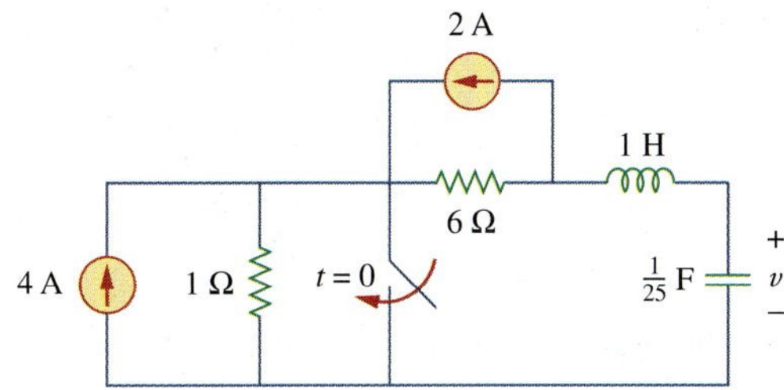

Figure 8.90
For Prob. 8.42.

8.43 The switch in Fig. 8.91 is opened at $t = 0$ after the circuit has reached steady state. Choose R and C such that $\alpha = 8$ Np/s and $\omega_d = 30$ rad/s.

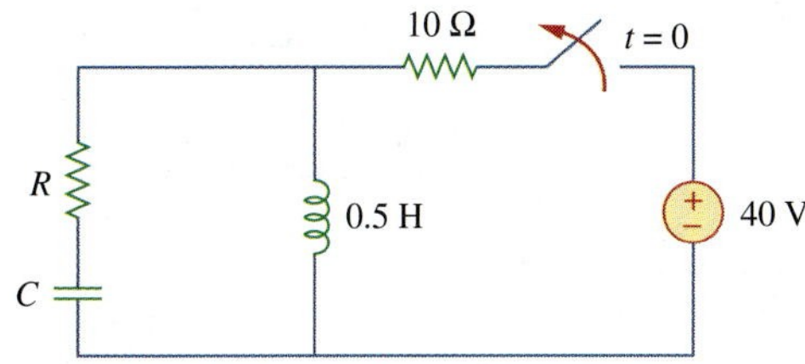

Figure 8.91
For Prob. 8.43.

8.44 A series RLC circuit has the following parameters: $R = 1$ kΩ, $L = 1$ H, and $C = 10$ nF. What type of damping does this circuit exhibit?

Section 8.6 Step Response of a Parallel *RLC* Circuit

8.45 In the circuit of Fig. 8.92, find $v(t)$ and $i(t)$ for $t > 0$. Assume $v(0) = 0$ V and $i(0) = 1$ A.

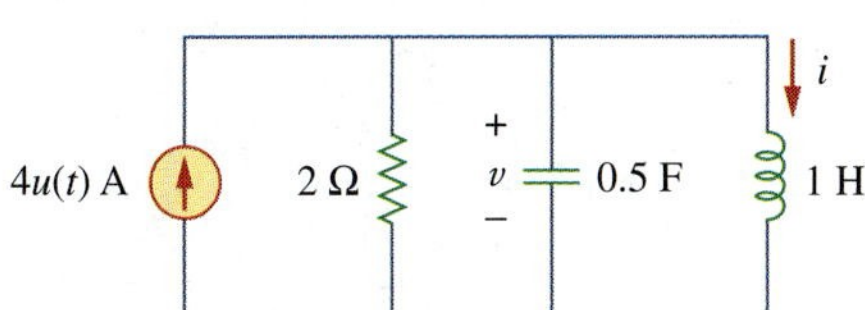

Figure 8.92
For Prob. 8.45.

8.46 Using Fig. 8.93, design a problem to help other students better understand the step response of a parallel RLC circuit.

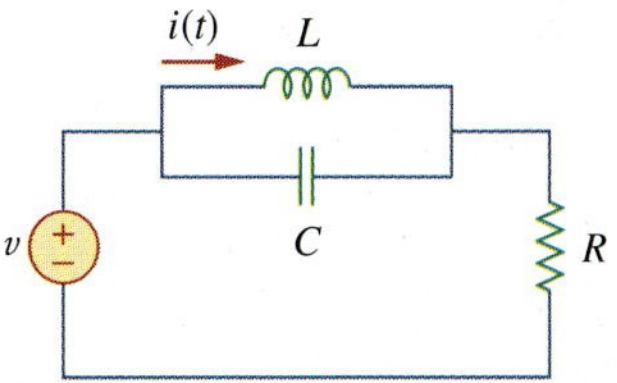

Figure 8.93
For Prob. 8.46.

8.47 Find the output voltage $v_o(t)$ in the circuit of Fig. 8.94.

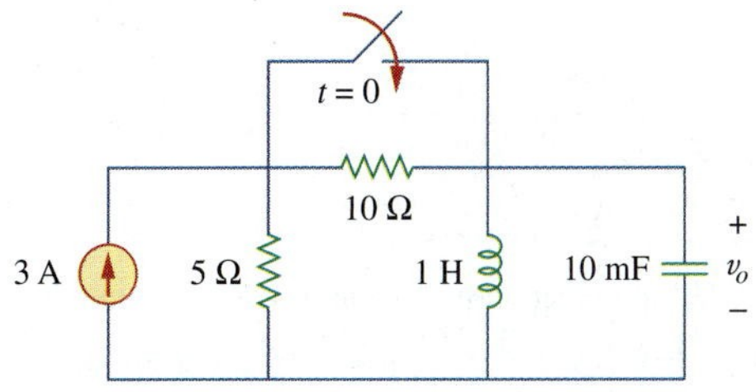

Figure 8.94
For Prob. 8.47.

8.48 Given the circuit in Fig. 8.95, find $i(t)$ and $v(t)$ for $t > 0$.

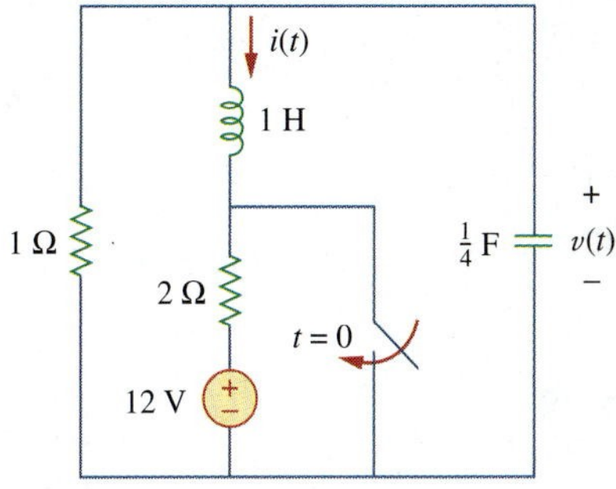

Figure 8.95
For Prob. 8.48.

8.49 Determine $i(t)$ for $t > 0$ in the circuit of Fig. 8.96.

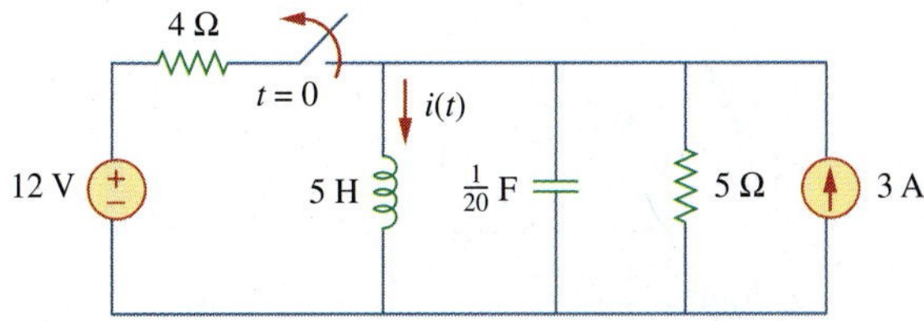

Figure 8.96
For Prob. 8.49.

8.50 For the circuit in Fig. 8.97, find $i(t)$ for $t > 0$.

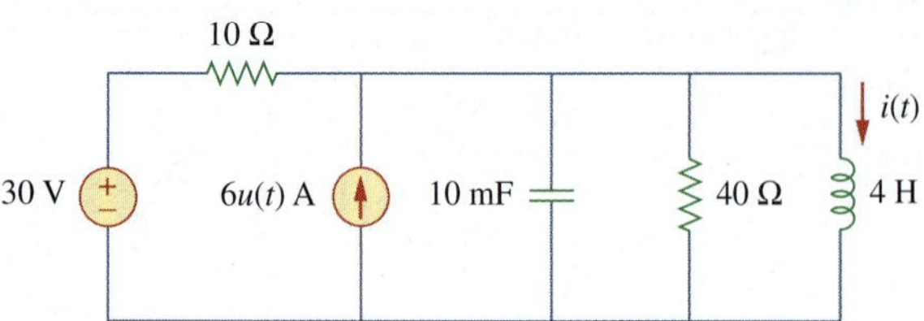

Figure 8.97
For Prob. 8.50.

8.51 Find $v(t)$ for $t > 0$ in the circuit of Fig. 8.98.

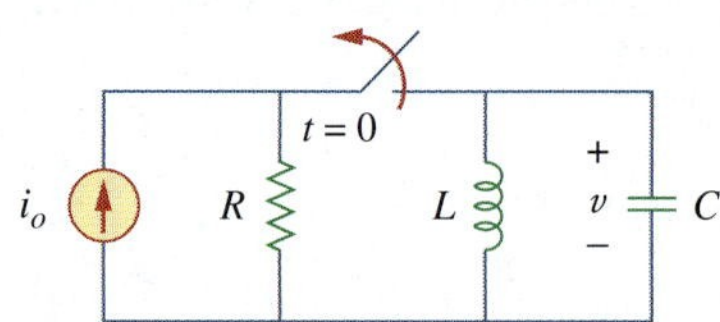

Figure 8.98
For Prob. 8.51.

8.52 The step response of a parallel *RLC* circuit is

$$v = 10 + 20e^{-300t}(\cos 400t - 2 \sin 400t) \text{ V}, \qquad t \geq 0$$

when the inductor is 50 mH. Find R and C.

Section 8.7 General Second-Order Circuits

8.53 After being open for a day, the switch in the circuit of Fig. 8.99 is closed at $t = 0$. Find the differential equation describing $i(t)$, $t > 0$.

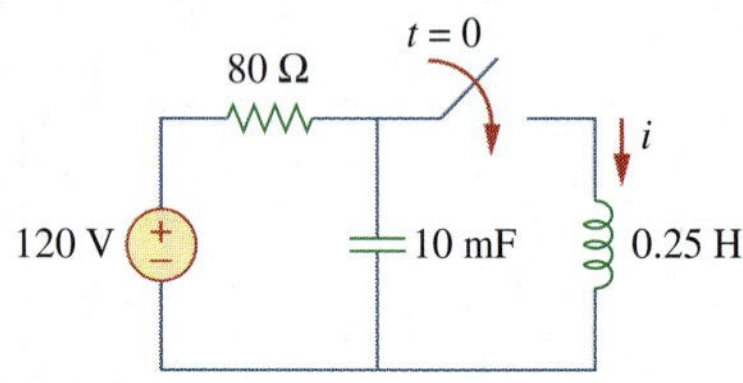

Figure 8.99
For Prob. 8.53.

8.54 Using Fig. 8.100, design a problem to help other students better understand general second-order circuits.

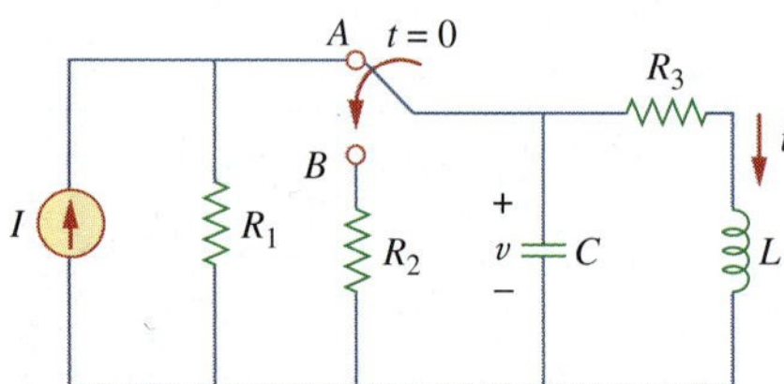

Figure 8.100
For Prob. 8.54.

8.55 For the circuit in Fig. 8.101, find $v(t)$ for $t > 0$. Assume that $v(0^+) = 4$ V and $i(0^+) = 2$ A.

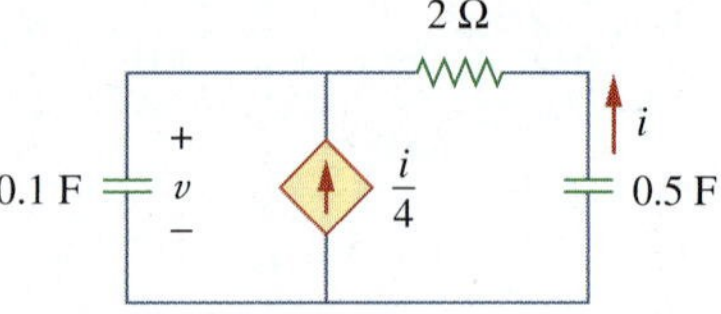

Figure 8.101
For Prob. 8.55.

8.56 In the circuit of Fig. 8.102, find $i(t)$ for $t > 0$.

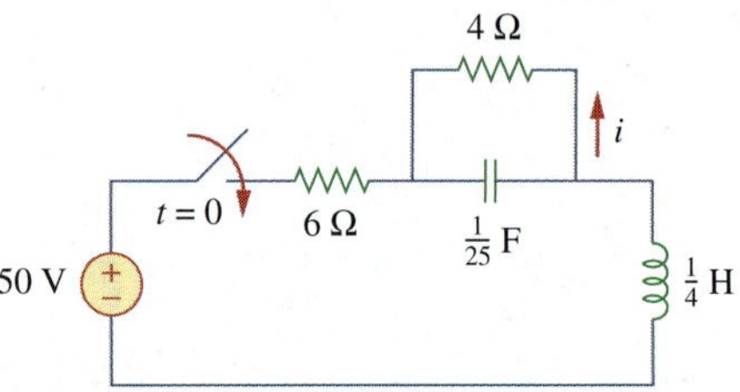

Figure 8.102
For Prob. 8.56.

8.57 If the switch in Fig. 8.103 has been closed for a long time before $t = 0$ but is opened at $t = 0$, determine:

(a) the characteristic equation of the circuit,

(b) i_x and v_R for $t > 0$.

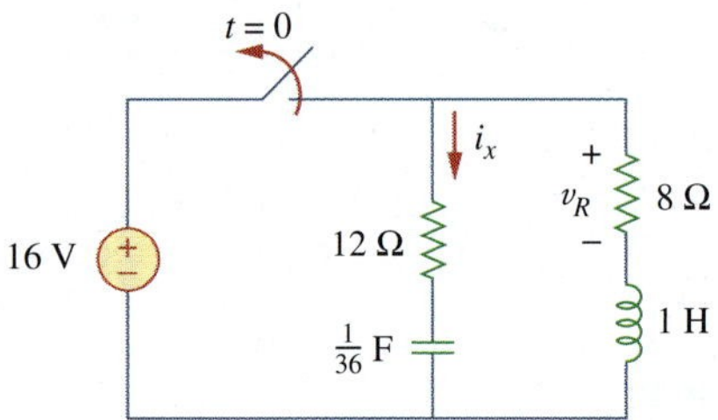

Figure 8.103
For Prob. 8.57.

8.58 In the circuit of Fig. 8.104, the switch has been in position 1 for a long time but moved to position 2 at $t = 0$. Find:

(a) $v(0^+)$, $dv(0^+)/dt$

(b) $v(t)$ for $t \geq 0$.

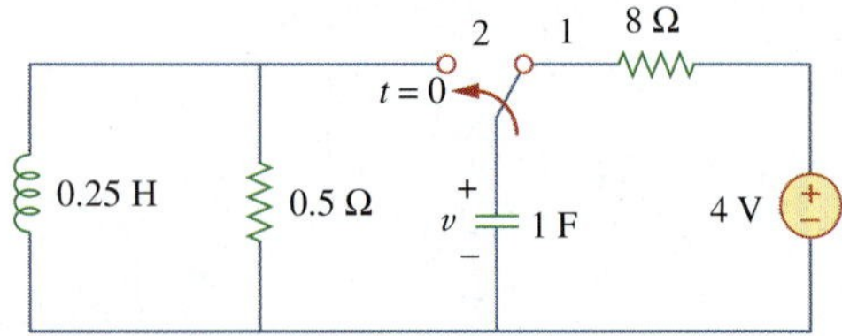

Figure 8.104
For Prob. 8.58.

8.59 The make before break switch in Fig. 8.105 has been in position 1 for $t < 0$. At $t = 0$, it is moved instantaneously to position 2. Determine $v(t)$.

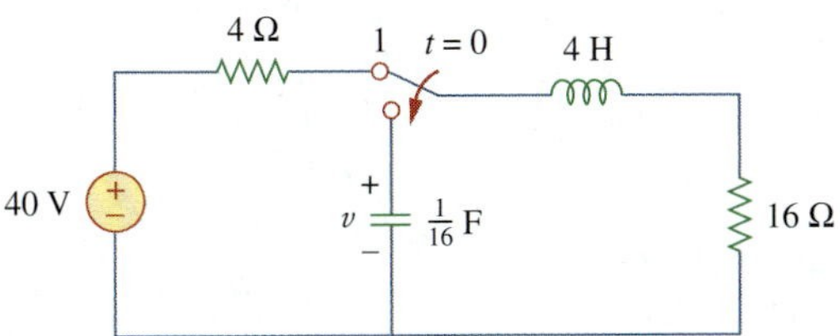

Figure 8.105
For Prob. 8.59.

8.60 Obtain i_1 and i_2 for $t > 0$ in the circuit of Fig. 8.106.

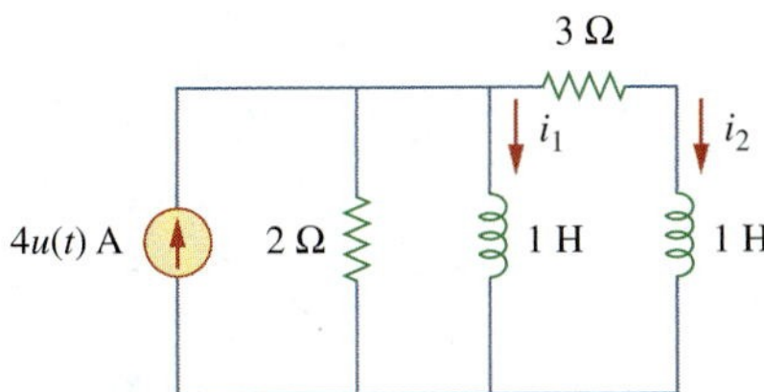

Figure 8.106
For Prob. 8.60.

8.61 For the circuit in Prob. 8.5, find i and v for $t > 0$.

8.62 Find the response $v_R(t)$ for $t > 0$ in the circuit of Fig. 8.107. Let $R = 3\ \Omega$, $L = 2$ H, and $C = 1/18$ F.

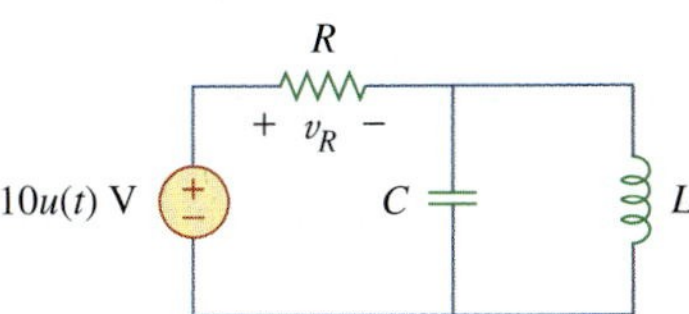

Figure 8.107
For Prob. 8.62.

Section 8.8 Second-Order Op Amp Circuits

8.63 For the op amp circuit in Fig. 8.108, find the differential equation for $i(t)$.

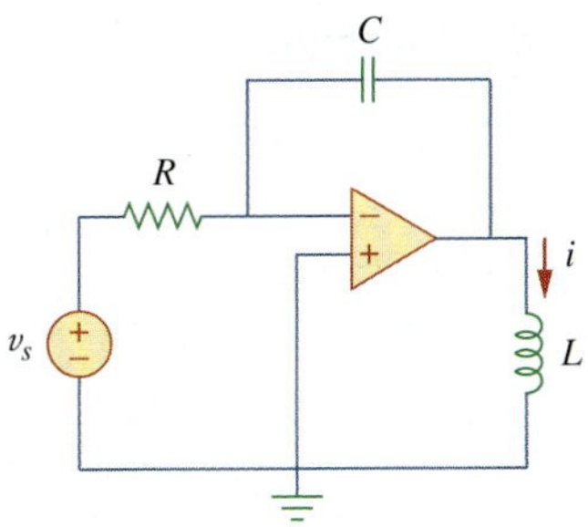

Figure 8.108
For Prob. 8.63.

8.64 Using Fig. 8.109, design a problem to help other students better understand second-order op amp circuits.

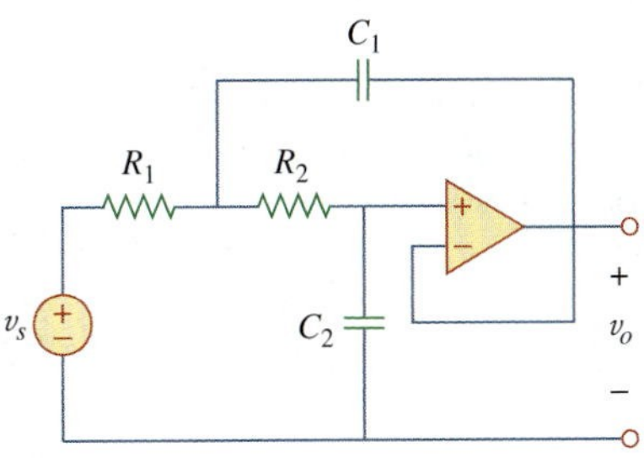

Figure 8.109
For Prob. 8.64.

8.65 Determine the differential equation for the op amp circuit in Fig. 8.110. If $v_1(0^+) = 2$ V and $v_2(0^+) = 0$ V, find v_o for $t > 0$. Let $R = 100\ \text{k}\Omega$ and $C = 1\ \mu\text{F}$.

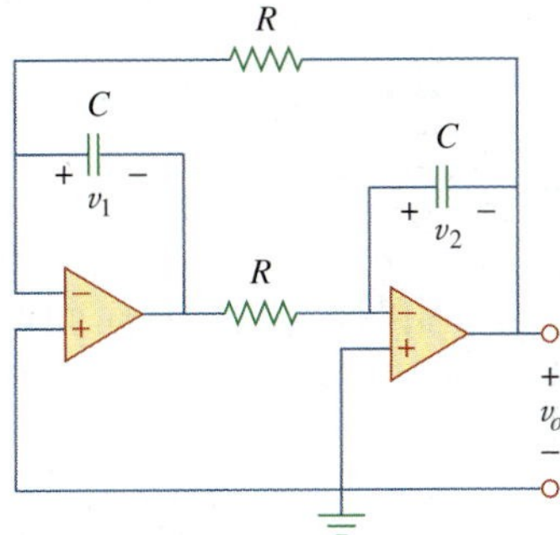

Figure 8.110
For Prob. 8.65.

8.66 Obtain the differential equations for $v_o(t)$ in the op amp circuit of Fig. 8.111.

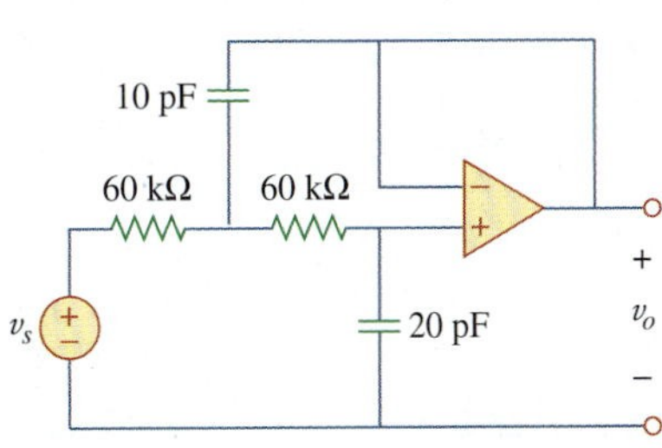

Figure 8.111
For Prob. 8.66.

***8.67** In the op amp circuit of Fig. 8.112, determine $v_o(t)$ for $t > 0$. Let $v_{in} = u(t)$ V, $R_1 = R_2 = 10$ kΩ, $C_1 = C_2 = 100$ μF.

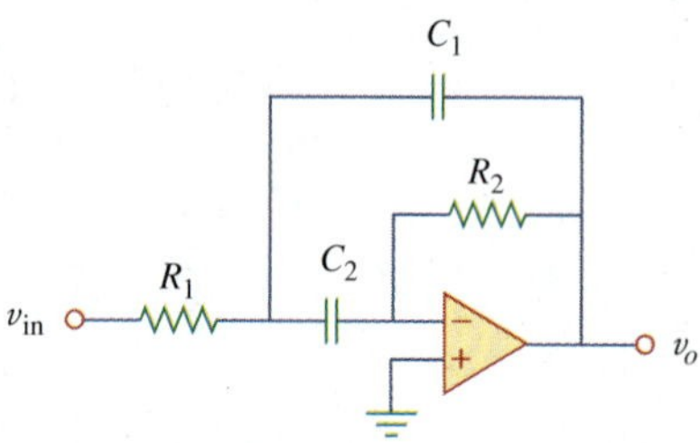

Figure 8.112
For Prob. 8.67.

Section 8.9 *PSpice* Analysis of *RLC* Circuit

8.68 For the step function $v_s = u(t)$, use *PSpice* to find the response $v(t)$ for $0 < t < 6$ s in the circuit of Fig. 8.113.

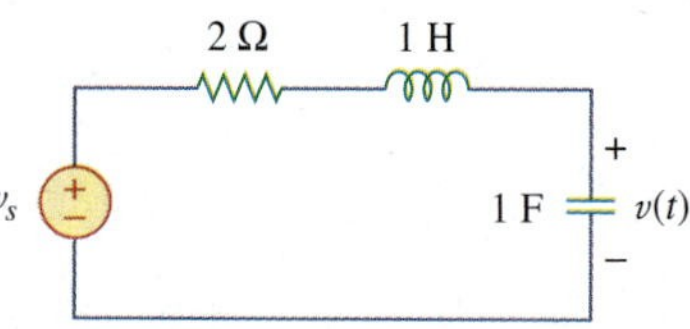

Figure 8.113
For Prob. 8.68.

8.69 Given the source-free circuit in Fig. 8.114, use *PSpice* to get $i(t)$ for $0 < t < 20$ s. Take $v(0) = 30$ V and $i(0) = 2$ A.

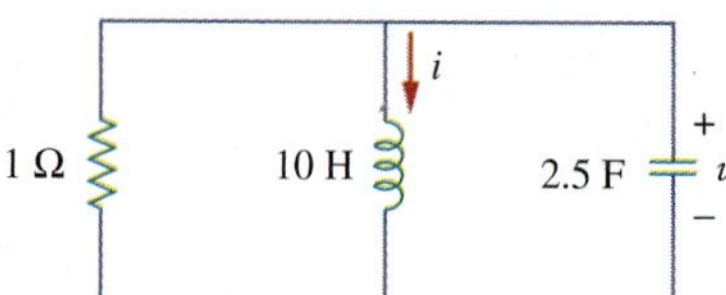

Figure 8.114
For Prob. 8.69.

8.70 For the circuit in Fig. 8.115, use *PSpice* to obtain $v(t)$ for $0 < t < 4$ s. Assume that the capacitor voltage and inductor current at $t = 0$ are both zero.

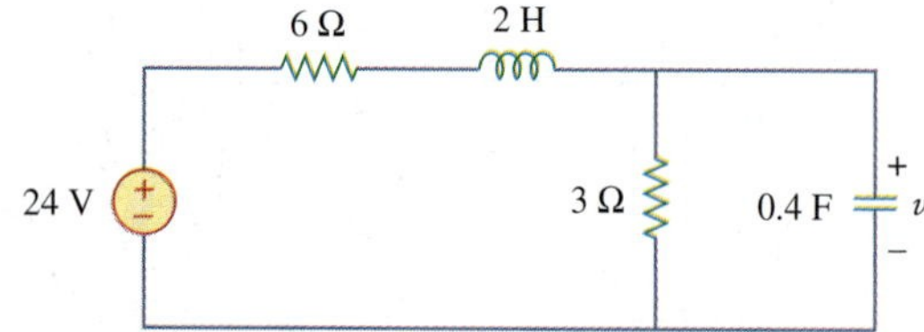

Figure 8.115
For Prob. 8.70.

8.71 Obtain $v(t)$ for $0 < t < 4$ s in the circuit of Fig. 8.116 using *PSpice*.

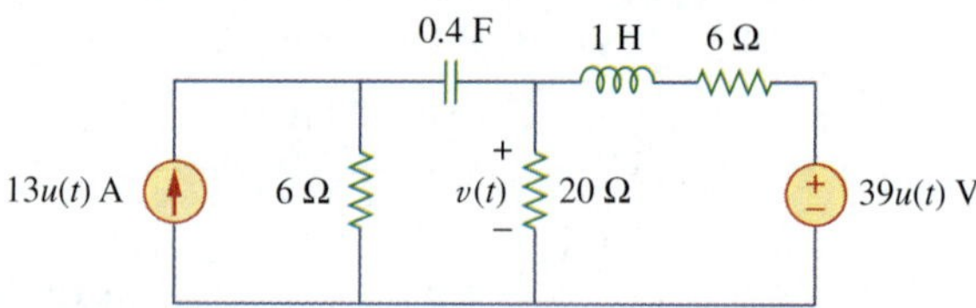

Figure 8.116
For Prob. 8.71.

8.72 The switch in Fig. 8.117 has been in position 1 for a long time. At $t = 0$, it is switched to position 2. Use *PSpice* to find $i(t)$ for $0 < t < 0.2$ s.

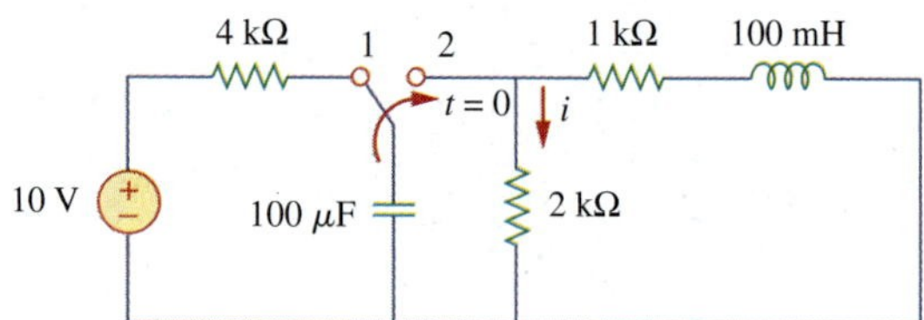

Figure 8.117
For Prob. 8.72.

8.73 Design a problem, to be solved using PSpice, to help other students better understand source-free *RLC* circuits.

Section 8.10 Duality

8.74 Draw the dual of the circuit shown in Fig. 8.118.

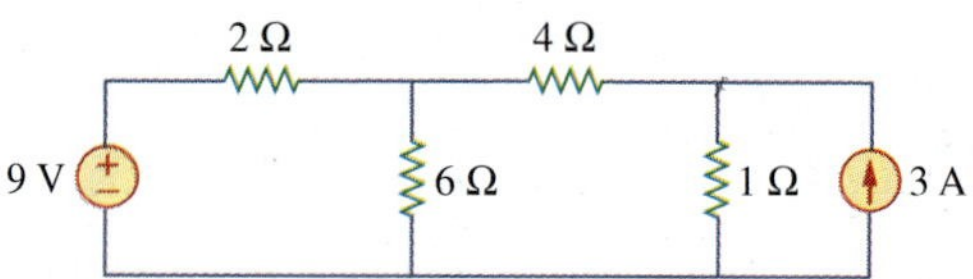

Figure 8.118
For Prob. 8.74.

8.75 Obtain the dual of the circuit in Fig. 8.119.

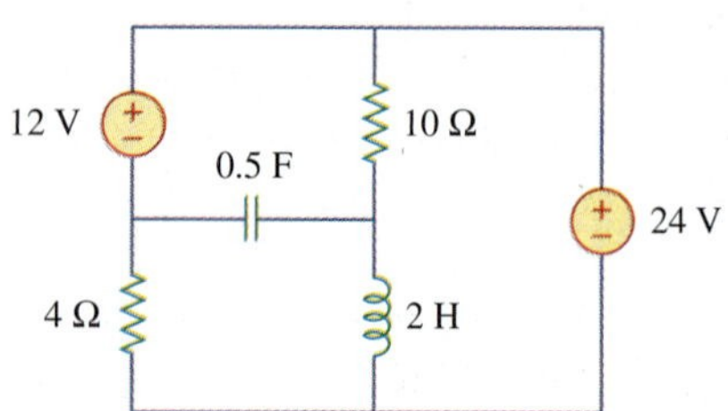

Figure 8.119
For Prob. 8.75.

8.76 Find the dual of the circuit in Fig. 8.120.

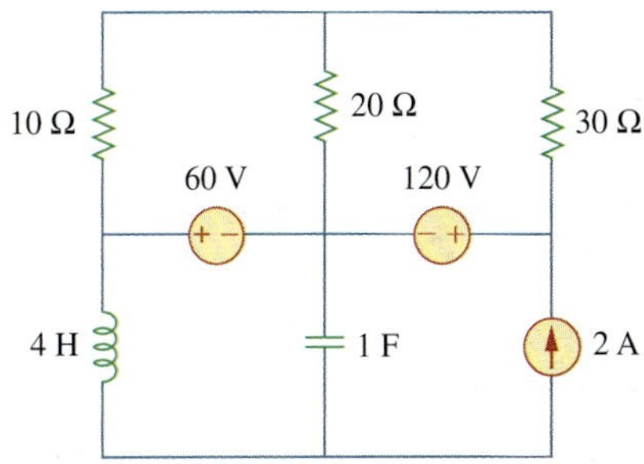

Figure 8.120
For Prob. 8.76.

8.77 Draw the dual of the circuit in Fig. 8.121.

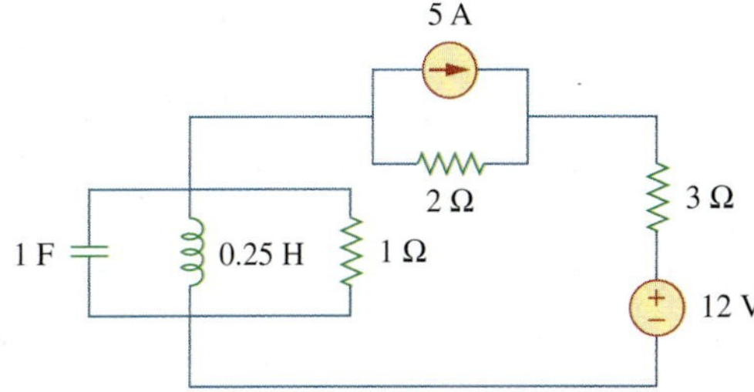

Figure 8.121
For Prob. 8.77.

Section 8.11 Applications

8.78 An automobile airbag igniter is modeled by the circuit in Fig. 8.122. Determine the time it takes the voltage across the igniter to reach its first peak after switching from A to B. Let $R = 3\ \Omega$, $C = 1/30$ F, and $L = 60$ mH.

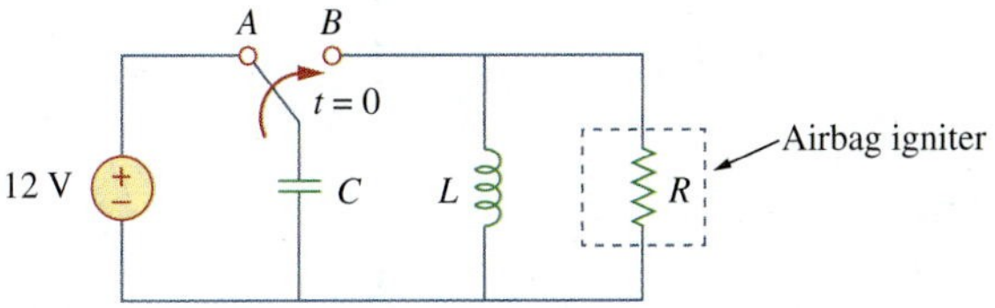

Figure 8.122
For Prob. 8.78.

8.79 A load is modeled as a 250-mH inductor in parallel with a 12-Ω resistor. A capacitor is needed to be connected to the load so that the network is critically damped at 60 Hz. Calculate the size of the capacitor.

Comprehensive Problems

8.80 A mechanical system is modeled by a series *RLC* circuit. It is desired to produce an overdamped response with time constants 0.1 ms and 0.5 ms. If a series 50-kΩ resistor is used, find the values of L and C.

8.81 An oscillogram can be adequately modeled by a second-order system in the form of a parallel *RLC* circuit. It is desired to give an underdamped voltage across a 200-Ω resistor. If the damping frequency is 4 kHz and the time constant of the envelope is 0.25 s, find the necessary values of L and C.

8.82 The circuit in Fig. 8.123 is the electrical analog of body functions used in medical schools to study convulsions. The analog is as follows:

C_1 = Volume of fluid in a drug

C_2 = Volume of blood stream in a specified region

R_1 = Resistance in the passage of the drug from the input to the blood stream

R_2 = Resistance of the excretion mechanism, such as kidney, etc.

v_0 = Initial concentration of the drug dosage

$v(t)$ = Percentage of the drug in the blood stream

Find $v(t)$ for $t > 0$ given that $C_1 = 0.5\ \mu$F, $C_2 = 5\ \mu$F, $R_1 = 5$ MΩ, $R_2 = 2.5$ MΩ, and $v_0 = 60u(t)$ V.

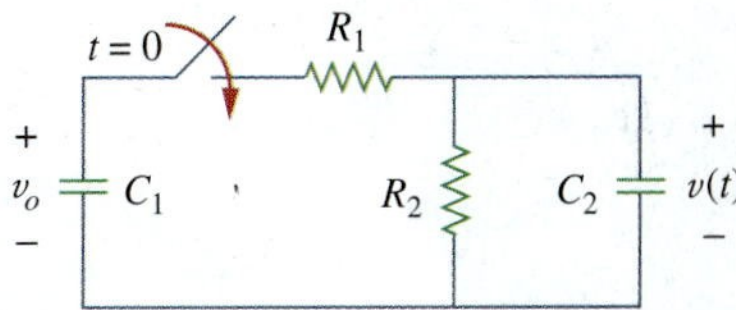

Figure 8.123
For Prob. 8.82.

8.83 Figure 8.124 shows a typical tunnel-diode oscillator circuit. The diode is modeled as a nonlinear resistor with $i_D = f(v_D)$, i.e., the diode current is a nonlinear function of the voltage across the diode. Derive the differential equation for the circuit in terms of v and i_D.

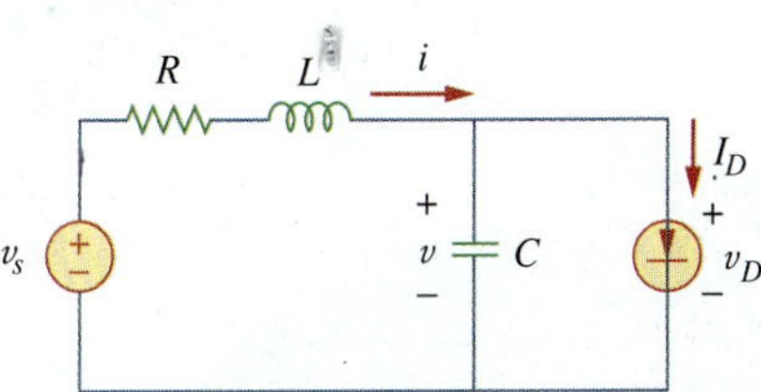

Figure 8.124
For Prob. 8.83.

AC Circuits

OUTLINE

chapter

9

Sinusoids and Phasors

He who knows not, and knows not that he knows not, is a fool—shun him. He who knows not, and knows that he knows not, is a child—teach him. He who knows, and knows not that he knows, is asleep—wake him up. He who knows, and knows that he knows, is wise—follow him.

—Persian Proverb

Enhancing Your Skills and Your Career

ABET EC 2000 criteria (3.d), *"an ability to function on multi-disciplinary teams."*

The "ability to function on multidisciplinary teams" is inherently critical for the working engineer. Engineers rarely, if ever, work by themselves. Engineers will always be part of some team. One of the things I like to remind students is that you do not have to like everyone on a team; you just have to be a successful part of that team.

Most frequently, these teams include individuals from of a variety of engineering disciplines, as well as individuals from nonengineering disciplines such as marketing and finance.

Students can easily develop and enhance this skill by working in study groups in every course they take. Clearly, working in study groups in nonengineering courses as well as engineering courses outside your discipline will also give you experience with multidisciplinary teams.

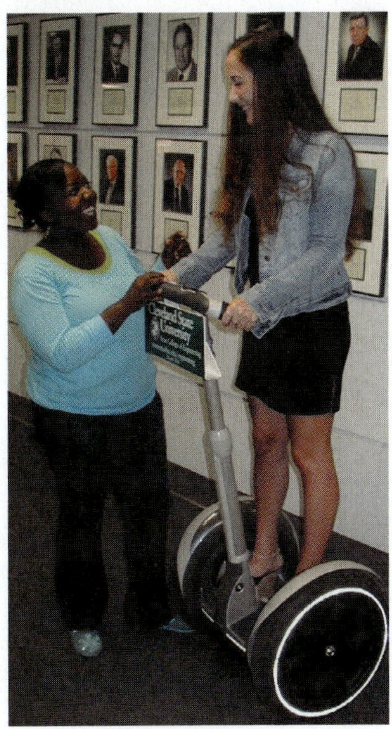

Photo by Charles Alexander

Historical

George Westinghouse. Photo © Bettmann/Corbis

Nikola Tesla (1856–1943) and **George Westinghouse** (1846–1914) helped establish alternating current as the primary mode of electricity transmission and distribution.

Today it is obvious that ac generation is well established as the form of electric power that makes widespread distribution of electric power efficient and economical. However, at the end of the 19th century, which was the better—ac or dc—was hotly debated and had extremely outspoken supporters on both sides. The dc side was lead by Thomas Edison, who had earned a lot of respect for his many contributions. Power generation using ac really began to build after the successful contributions of Tesla. The real commercial success in ac came from George Westinghouse and the outstanding team, including Tesla, he assembled. In addition, two other big names were C. F. Scott and B. G. Lamme.

The most significant contribution to the early success of ac was the patenting of the polyphase ac motor by Tesla in 1888. The induction motor and polyphase generation and distribution systems doomed the use of dc as the prime energy source.

9.1 Introduction

Thus far our analysis has been limited for the most part to dc circuits: those circuits excited by constant or time-invariant sources. We have restricted the forcing function to dc sources for the sake of simplicity, for pedagogic reasons, and also for historic reasons. Historically, dc sources were the main means of providing electric power up until the late 1800s. At the end of that century, the battle of direct current versus alternating current began. Both had their advocates among the electrical engineers of the time. Because ac is more efficient and economical to transmit over long distances, ac systems ended up the winner. Thus, it is in keeping with the historical sequence of events that we considered dc sources first.

We now begin the analysis of circuits in which the source voltage or current is time-varying. In this chapter, we are particularly interested in sinusoidally time-varying excitation, or simply, excitation by a *sinusoid.*

A **sinusoid** is a signal that has the form of the sine or cosine function.

A sinusoidal current is usually referred to as *alternating current* (*ac*). Such a current reverses at regular time intervals and has alternately positive and negative values. Circuits driven by sinusoidal current or voltage sources are called *ac circuits.*

We are interested in sinusoids for a number of reasons. First, nature itself is characteristically sinusoidal. We experience sinusoidal variation in the motion of a pendulum, the vibration of a string, the ripples on the ocean surface, and the natural response of underdamped second-order systems, to mention but a few. Second, a sinusoidal signal is easy to generate and transmit. It is the form of voltage generated throughout the world and supplied to homes, factories, laboratories, and so on. It is the dominant form of signal in the communications and electric power industries. Third, through Fourier analysis, any practical periodic signal can be represented by a sum of sinusoids. Sinusoids, therefore, play an important role in the analysis of periodic signals. Lastly, a sinusoid is easy to handle mathematically. The derivative and integral of a sinusoid are themselves sinusoids. For these and other reasons, the sinusoid is an extremely important function in circuit analysis.

A sinusoidal forcing function produces both a transient response and a steady-state response, much like the step function, which we studied in Chapters 7 and 8. The transient response dies out with time so that only the steady-state response remains. When the transient response has become negligibly small compared with the steady-state response, we say that the circuit is operating at sinusoidal steady state. It is this *sinusoidal steady-state response* that is of main interest to us in this chapter.

We begin with a basic discussion of sinusoids and phasors. We then introduce the concepts of impedance and admittance. The basic circuit laws, Kirchhoff's and Ohm's, introduced for dc circuits, will be applied to ac circuits. Finally, we consider applications of ac circuits in phase-shifters and bridges.

9.2 Sinusoids

Consider the sinusoidal voltage

$$v(t) = V_m \sin \omega t \tag{9.1}$$

where

V_m = the *amplitude* of the sinusoid
ω = the *angular frequency* in radians/s
ωt = the *argument* of the sinusoid

The sinusoid is shown in Fig. 9.1(a) as a function of its argument and in Fig. 9.1(b) as a function of time. It is evident that the sinusoid repeats itself every T seconds; thus, T is called the *period* of the sinusoid. From the two plots in Fig. 9.1, we observe that $\omega T = 2\pi$,

$$\boxed{T = \frac{2\pi}{\omega}} \tag{9.2}$$

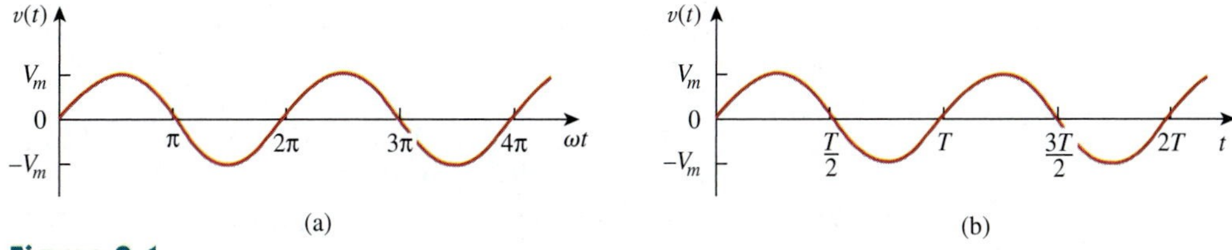

Figure 9.1
A sketch of $V_m \sin \omega t$: (a) as a function of ωt, (b) as a function of t.

Historical

The Burndy Library Collection at The Huntington Library, San Marino, California.

Heinrich Rudorf Hertz (1857–1894), a German experimental physicist, demonstrated that electromagnetic waves obey the same fundamental laws as light. His work confirmed James Clerk Maxwell's celebrated 1864 theory and prediction that such waves existed.

Hertz was born into a prosperous family in Hamburg, Germany. He attended the University of Berlin and did his doctorate under the prominent physicist Hermann von Helmholtz. He became a professor at Karlsruhe, where he began his quest for electromagnetic waves. Hertz successfully generated and detected electromagnetic waves; he was the first to show that light is electromagnetic energy. In 1887, Hertz noted for the first time the photoelectric effect of electrons in a molecular structure. Although Hertz only lived to the age of 37, his discovery of electromagnetic waves paved the way for the practical use of such waves in radio, television, and other communication systems. The unit of frequency, the hertz, bears his name.

The fact that $v(t)$ repeats itself every T seconds is shown by replacing t by $t + T$ in Eq. (9.1). We get

$$\begin{aligned} v(t+T) &= V_m \sin\omega(t+T) = V_m \sin\omega\left(t + \frac{2\pi}{\omega}\right) \\ &= V_m \sin(\omega t + 2\pi) = V_m \sin\omega t = v(t) \end{aligned} \tag{9.3}$$

Hence,

$$\boxed{v(t+T) = v(t)} \tag{9.4}$$

that is, v has the same value at $t + T$ as it does at t and $v(t)$ is said to be *periodic*. In general,

> A **periodic function** is one that satisfies $f(t) = f(t + nT)$, for all t and for all integers n.

As mentioned, the *period* T of the periodic function is the time of one complete cycle or the number of seconds per cycle. The reciprocal of this quantity is the number of cycles per second, known as the *cyclic frequency* f of the sinusoid. Thus,

$$\boxed{f = \frac{1}{T}} \tag{9.5}$$

From Eqs. (9.2) and (9.5), it is clear that

The unit of f is named after the German physicist Heinrich R. Hertz (1857–1894).

$$\omega = 2\pi f \tag{9.6}$$

While ω is in radians per second (rad/s), f is in hertz (Hz).

Let us now consider a more general expression for the sinusoid,

$$v(t) = V_m \sin(\omega t + \phi) \tag{9.7}$$

where $(\omega t + \phi)$ is the argument and ϕ is the *phase*. Both argument and phase can be in radians or degrees.

Let us examine the two sinusoids

$$v_1(t) = V_m \sin \omega t \quad \text{and} \quad v_2(t) = V_m \sin(\omega t + \phi) \tag{9.8}$$

shown in Fig. 9.2. The starting point of v_2 in Fig. 9.2 occurs first in time. Therefore, we say that v_2 *leads* v_1 by ϕ or that v_1 *lags* v_2 by ϕ. If $\phi \neq 0$, we also say that v_1 and v_2 are *out of phase*. If $\phi = 0$, then v_1 and v_2 are said to be *in phase*; they reach their minima and maxima at exactly the same time. We can compare v_1 and v_2 in this manner because they operate at the same frequency; they do not need to have the same amplitude.

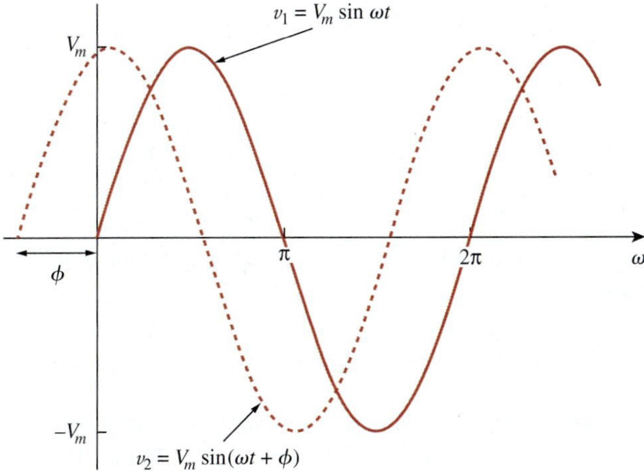

Figure 9.2
Two sinusoids with different phases.

A sinusoid can be expressed in either sine or cosine form. When comparing two sinusoids, it is expedient to express both as either sine or cosine with positive amplitudes. This is achieved by using the following trigonometric identities:

$$\begin{aligned} \sin(A \pm B) &= \sin A \cos B \pm \cos A \sin B \\ \cos(A \pm B) &= \cos A \cos B \mp \sin A \sin B \end{aligned} \tag{9.9}$$

With these identities, it is easy to show that

$$\begin{aligned} \sin(\omega t \pm 180^\circ) &= -\sin \omega t \\ \cos(\omega t \pm 180^\circ) &= -\cos \omega t \\ \sin(\omega t \pm 90^\circ) &= \pm \cos \omega t \\ \cos(\omega t \pm 90^\circ) &= \mp \sin \omega t \end{aligned} \tag{9.10}$$

Using these relationships, we can transform a sinusoid from sine form to cosine form or vice versa.

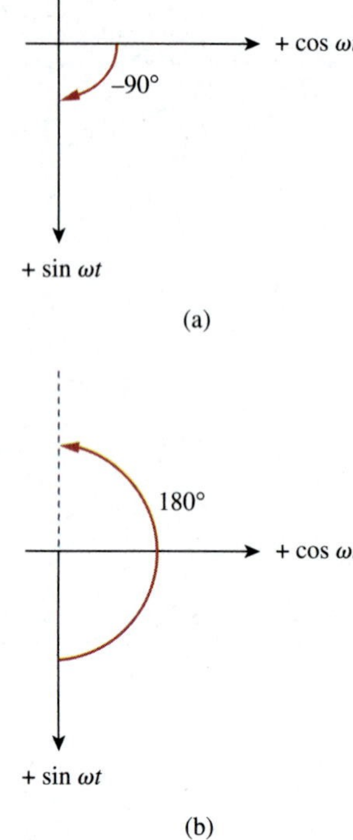

Figure 9.3
A graphical means of relating cosine and sine: (a) $\cos(\omega t - 90^\circ) = \sin\omega t$, (b) $\sin(\omega t + 180^\circ) = -\sin\omega t$.

A graphical approach may be used to relate or compare sinusoids as an alternative to using the trigonometric identities in Eqs. (9.9) and (9.10). Consider the set of axes shown in Fig. 9.3(a). The horizontal axis represents the magnitude of cosine, while the vertical axis (pointing down) denotes the magnitude of sine. Angles are measured positively counterclockwise from the horizontal, as usual in polar coordinates. This graphical technique can be used to relate two sinusoids. For example, we see in Fig. 9.3(a) that subtracting 90° from the argument of $\cos\omega t$ gives $\sin\omega t$, or $\cos(\omega t - 90^\circ) = \sin\omega t$. Similarly, adding 180° to the argument of $\sin\omega t$ gives $-\sin\omega t$, or $\sin(\omega t + 180^\circ) = -\sin\omega t$, as shown in Fig. 9.3(b).

The graphical technique can also be used to add two sinusoids of the same frequency when one is in sine form and the other is in cosine form. To add $A\cos\omega t$ and $B\sin\omega t$, we note that A is the magnitude of $\cos\omega t$ while B is the magnitude of $\sin\omega t$, as shown in Fig. 9.4(a). The magnitude and argument of the resultant sinusoid in cosine form is readily obtained from the triangle. Thus,

$$A\cos\omega t + B\sin\omega t = C\cos(\omega t - \theta) \tag{9.11}$$

where

$$C = \sqrt{A^2 + B^2}, \qquad \theta = \tan^{-1}\frac{B}{A} \tag{9.12}$$

For example, we may add $3\cos\omega t$ and $-4\sin\omega t$ as shown in Fig. 9.4(b) and obtain

$$3\cos\omega t - 4\sin\omega t = 5\cos(\omega t + 53.1^\circ) \tag{9.13}$$

Compared with the trigonometric identities in Eqs. (9.9) and (9.10), the graphical approach eliminates memorization. However, we must not confuse the sine and cosine axes with the axes for complex numbers to be discussed in the next section. Something else to note in Figs. 9.3 and 9.4 is that although the natural tendency is to have the vertical axis point up, the positive direction of the sine function is down in the present case.

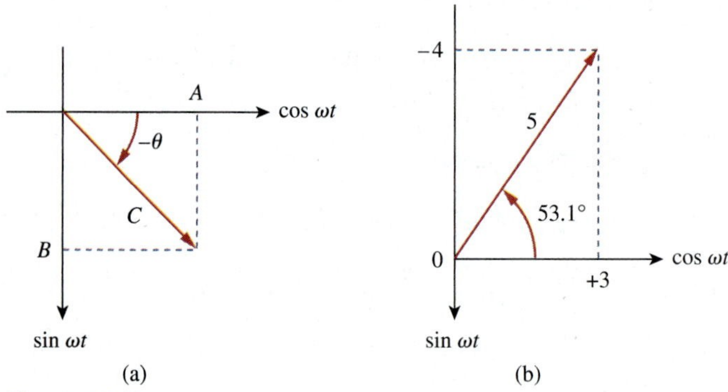

Figure 9.4
(a) Adding $A\cos\omega t$ and $B\sin\omega t$, (b) adding $3\cos\omega t$ and $-4\sin\omega t$.

Example 9.1

Find the amplitude, phase, period, and frequency of the sinusoid

$$v(t) = 12\cos(50t + 10^\circ)$$

Solution:

The amplitude is $V_m = 12$ V.
The phase is $\phi = 10^\circ$.
The angular frequency is $\omega = 50$ rad/s.
The period $T = \dfrac{2\pi}{\omega} = \dfrac{2\pi}{50} = 0.1257$ s.
The frequency is $f = \dfrac{1}{T} = 7.958$ Hz.

Practice Problem 9.1

Given the sinusoid $5\sin(4\pi t - 60^\circ)$, calculate its amplitude, phase, angular frequency, period, and frequency.

Answer: 5, -60°, 12.57 rad/s, 0.5 s, 2 Hz.

Example 9.2

Calculate the phase angle between $v_1 = -10\cos(\omega t + 50^\circ)$ and $v_2 = 12\sin(\omega t - 10^\circ)$. State which sinusoid is leading.

Solution:
Let us calculate the phase in three ways. The first two methods use trigonometric identities, while the third method uses the graphical approach.

■ METHOD 1 In order to compare v_1 and v_2, we must express them in the same form. If we express them in cosine form with positive amplitudes,

$$v_1 = -10\cos(\omega t + 50^\circ) = 10\cos(\omega t + 50^\circ - 180^\circ)$$

$$v_1 = 10\cos(\omega t - 130^\circ) \quad \text{or} \quad v_1 = 10\cos(\omega t + 230^\circ) \tag{9.2.1}$$

and

$$v_2 = 12\sin(\omega t - 10^\circ) = 12\cos(\omega t - 10^\circ - 90^\circ)$$

$$v_2 = 12\cos(\omega t - 100^\circ) \tag{9.2.2}$$

It can be deduced from Eqs. (9.2.1) and (9.2.2) that the phase difference between v_1 and v_2 is 30°. We can write v_2 as

$$v_2 = 12\cos(\omega t - 130^\circ + 30^\circ) \quad \text{or} \quad v_2 = 12\cos(\omega t + 260^\circ) \tag{9.2.3}$$

Comparing Eqs. (9.2.1) and (9.2.3) shows clearly that v_2 leads v_1 by 30°.

■ METHOD 2 Alternatively, we may express v_1 in sine form:

$$v_1 = -10\cos(\omega t + 50^\circ) = 10\sin(\omega t + 50^\circ - 90^\circ)$$
$$= 10\sin(\omega t - 40^\circ) = 10\sin(\omega t - 10^\circ - 30^\circ)$$

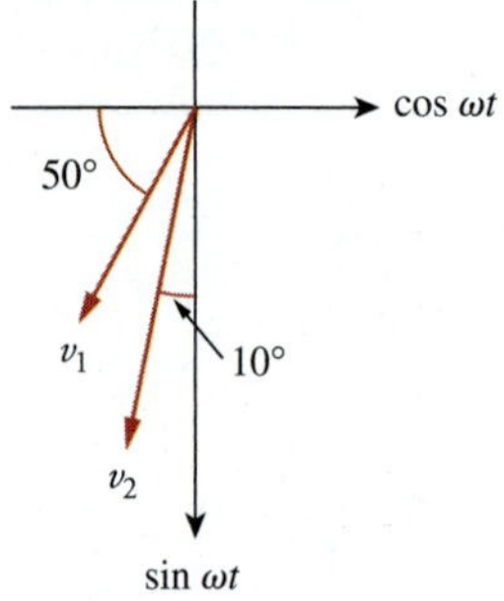

Figure 9.5
For Example 9.2.

But $v_2 = 12 \sin(\omega t - 10°)$. Comparing the two shows that v_1 lags v_2 by 30°. This is the same as saying that v_2 leads v_1 by 30°.

■ **METHOD 3** We may regard v_1 as simply $-10 \cos\omega t$ with a phase shift of $+50°$. Hence, v_1 is as shown in Fig. 9.5. Similarly, v_2 is $12 \sin\omega t$ with a phase shift of $-10°$, as shown in Fig. 9.5. It is easy to see from Fig. 9.5 that v_2 leads v_1 by 30°, that is, $90° - 50° - 10°$.

Practice Problem 9.2

Find the phase angle between

$$i_1 = -4 \sin(377t + 25°) \quad \text{and} \quad i_2 = 5 \cos(377t - 40°)$$

Does i_1 lead or lag i_2?

Answer: 155°, i_1 leads i_2.

9.3 Phasors

Sinusoids are easily expressed in terms of phasors, which are more convenient to work with than sine and cosine functions.

> A **phasor** is a complex number that represents the amplitude and phase of a sinusoid.

Charles Proteus Steinmetz (1865–1923) was a German-Austrian mathematician and electrical engineer.

Appendix B presents a short tutorial on complex numbers.

Phasors provide a simple means of analyzing linear circuits excited by sinusoidal sources; solutions of such circuits would be intractable otherwise. The notion of solving ac circuits using phasors was first introduced by Charles Steinmetz in 1893. Before we completely define phasors and apply them to circuit analysis, we need to be thoroughly familiar with complex numbers.

A complex number z can be written in rectangular form as

$$z = x + jy \tag{9.14a}$$

where $j = \sqrt{-1}$; x is the real part of z; y is the imaginary part of z. In this context, the variables x and y do not represent a location as in two-dimensional vector analysis but rather the real and imaginary parts of z in the complex plane. Nevertheless, we note that there are some resemblances between manipulating complex numbers and manipulating two-dimensional vectors.

The complex number z can also be written in polar or exponential form as

$$z = r\underline{/\phi} = re^{j\phi} \tag{9.14b}$$

Historical

Charles Proteus Steinmetz (1865–1923), a German-Austrian mathematician and engineer, introduced the phasor method (covered in this chapter) in ac circuit analysis. He is also noted for his work on the theory of hysteresis.

Steinmetz was born in Breslau, Germany, and lost his mother at the age of one. As a youth, he was forced to leave Germany because of his political activities just as he was about to complete his doctoral dissertation in mathematics at the University of Breslau. He migrated to Switzerland and later to the United States, where he was employed by General Electric in 1893. That same year, he published a paper in which complex numbers were used to analyze ac circuits for the first time. This led to one of his many textbooks, *Theory and Calculation of ac Phenomena*, published by McGraw-Hill in 1897. In 1901, he became the president of the American Institute of Electrical Engineers, which later became the IEEE.

where r is the magnitude of z, and ϕ is the phase of z. We notice that z can be represented in three ways:

$$\begin{aligned} z &= x + jy && \text{Rectangular form} \\ z &= r\angle\phi && \text{Polar form} \\ z &= re^{j\phi} && \text{Exponential form} \end{aligned} \tag{9.15}$$

The relationship between the rectangular form and the polar form is shown in Fig. 9.6, where the x axis represents the real part and the y axis represents the imaginary part of a complex number. Given x and y, we can get r and ϕ as

$$r = \sqrt{x^2 + y^2}, \qquad \phi = \tan^{-1}\frac{y}{x} \tag{9.16a}$$

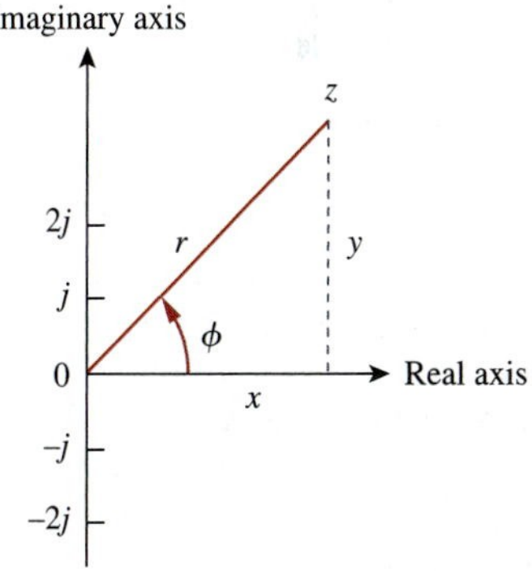

Figure 9.6
Representation of a complex number $z = x + jy = r\angle\phi$.

On the other hand, if we know r and ϕ, we can obtain x and y as

$$x = r\cos\phi, \qquad y = r\sin\phi \tag{9.16b}$$

Thus, z may be written as

$$z = x + jy = r\angle\phi = r(\cos\phi + j\sin\phi) \tag{9.17}$$

Addition and subtraction of complex numbers are better performed in rectangular form; multiplication and division are better done in polar form. Given the complex numbers

$$z = x + jy = r\angle\phi, \qquad z_1 = x_1 + jy_1 = r_1\angle\phi_1$$
$$z_2 = x_2 + jy_2 = r_2\angle\phi_2$$

the following operations are important.

Addition:

$$z_1 + z_2 = (x_1 + x_2) + j(y_1 + y_2) \tag{9.18a}$$

Subtraction:

$$z_1 - z_2 = (x_1 - x_2) + j(y_1 - y_2) \tag{9.18b}$$

Multiplication:

$$z_1 z_2 = r_1 r_2 \underline{/\phi_1 + \phi_2} \tag{9.18c}$$

Division:

$$\frac{z_1}{z_2} = \frac{r_1}{r_2} \underline{/\phi_1 - \phi_2} \tag{9.18d}$$

Reciprocal:

$$\frac{1}{z} = \frac{1}{r} \underline{/-\phi} \tag{9.18e}$$

Square Root:

$$\sqrt{z} = \sqrt{r}\, \underline{/\phi/2} \tag{9.18f}$$

Complex Conjugate:

$$z^* = x - jy = r\underline{/-\phi} = re^{-j\phi} \tag{9.18g}$$

Note that from Eq. (9.18e),

$$\frac{1}{j} = -j \tag{9.18h}$$

These are the basic properties of complex numbers we need. Other properties of complex numbers can be found in Appendix B.

The idea of phasor representation is based on Euler's identity. In general,

$$\boxed{e^{\pm j\phi} = \cos\phi \pm j\sin\phi} \tag{9.19}$$

which shows that we may regard $\cos\phi$ and $\sin\phi$ as the real and imaginary parts of $e^{j\phi}$; we may write

$$\cos\phi = \text{Re}(e^{j\phi}) \tag{9.20a}$$

$$\sin\phi = \text{Im}(e^{j\phi}) \tag{9.20b}$$

where Re and Im stand for the *real part of* and the *imaginary part of*. Given a sinusoid $v(t) = V_m \cos(\omega t + \phi)$, we use Eq. (9.20a) to express $v(t)$ as

$$v(t) = V_m \cos(\omega t + \phi) = \text{Re}(V_m e^{j(\omega t + \phi)}) \tag{9.21}$$

or

$$v(t) = \text{Re}(V_m e^{j\phi} e^{j\omega t}) \tag{9.22}$$

Thus,

$$\boxed{v(t) = \text{Re}(\mathbf{V}e^{j\omega t})} \tag{9.23}$$

where

$$\mathbf{V} = V_m e^{j\phi} = V_m \underline{/\phi} \tag{9.24}$$

V is thus the *phasor representation* of the sinusoid $v(t)$, as we said earlier. In other words, a phasor is a complex representation of the magnitude and phase of a sinusoid. Either Eq. (9.20a) or Eq. (9.20b) can be used to develop the phasor, but the standard convention is to use Eq. (9.20a).

A phasor may be regarded as a mathematical equivalent of a sinusoid with the time dependence dropped.

One way of looking at Eqs. (9.23) and (9.24) is to consider the plot of the *sinor* $\mathbf{V}e^{j\omega t} = V_m e^{j(\omega t+\phi)}$ on the complex plane. As time increases, the sinor rotates on a circle of radius V_m at an angular velocity ω in the counterclockwise direction, as shown in Fig. 9.7(a). We may regard $v(t)$ as the projection of the sinor $\mathbf{V}e^{j\omega t}$ on the real axis, as shown in Fig. 9.7(b). The value of the sinor at time $t = 0$ is the phasor **V** of the sinusoid $v(t)$. The sinor may be regarded as a rotating phasor. Thus, whenever a sinusoid is expressed as a phasor, the term $e^{j\omega t}$ is implicitly present. It is therefore important, when dealing with phasors, to keep in mind the frequency ω of the phasor; otherwise we can make serious mistakes.

If we use sine for the phasor instead of cosine, then $v(t) = V_m \sin(\omega t + \phi) = \text{Im}(V_m e^{j(\omega t+\phi)})$ and the corresponding phasor is the same as that in Eq. (9.24).

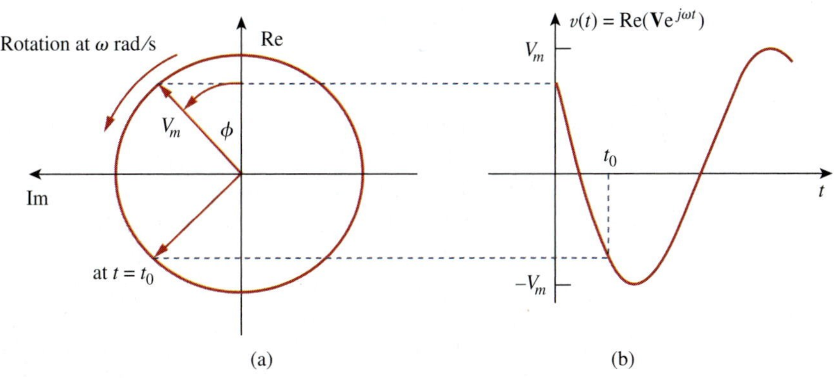

Figure 9.7
Representation of $\mathbf{V}e^{j\omega t}$: (a) sinor rotating counterclockwise, (b) its projection on the real axis, as a function of time.

Equation (9.23) states that to obtain the sinusoid corresponding to a given phasor **V**, multiply the phasor by the time factor $e^{j\omega t}$ and take the real part. As a complex quantity, a phasor may be expressed in rectangular form, polar form, or exponential form. Since a phasor has magnitude and phase ("direction"), it behaves as a vector and is printed in boldface. For example, phasors $\mathbf{V} = V_m \underline{/\phi}$ and $\mathbf{I} = I_m \underline{/-\theta}$ are graphically represented in Fig. 9.8. Such a graphical representation of phasors is known as a *phasor diagram.*

We use lightface italic letters such as *z* to represent complex numbers but boldface letters such as **V** to represent phasors, because phasors are vectorlike quantities.

Equations (9.21) through (9.23) reveal that to get the phasor corresponding to a sinusoid, we first express the sinusoid in the cosine form so that the sinusoid can be written as the real part of a complex number. Then we take out the time factor $e^{j\omega t}$, and whatever is left is the phasor corresponding to the sinusoid. By suppressing the time factor, we transform the sinusoid from the time domain to the phasor domain. This transformation is summarized as follows:

$$v(t) = V_m \cos(\omega t + \phi) \quad \Leftrightarrow \quad \mathbf{V} = V_m \underline{/\phi} \tag{9.25}$$

(Time-domain representation) (Phasor-domain representation)

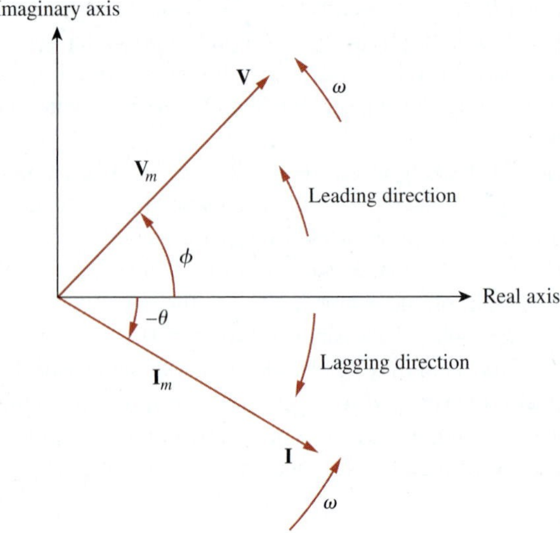

Figure 9.8
A phasor diagram showing $\mathbf{V} = V_m \underline{/\phi}$ and $\mathbf{I} = I_m \underline{/-\theta}$.

Given a sinusoid $v(t) = V_m \cos(\omega t + \phi)$, we obtain the corresponding phasor as $\mathbf{V} = V_m \underline{/\phi}$. Equation (9.25) is also demonstrated in Table 9.1, where the sine function is considered in addition to the cosine function. From Eq. (9.25), we see that to get the phasor representation of a sinusoid, we express it in cosine form and take the magnitude and phase. Given a phasor, we obtain the time domain representation as the cosine function with the same magnitude as the phasor and the argument as ωt plus the phase of the phasor. The idea of expressing information in alternate domains is fundamental to all areas of engineering.

TABLE 9.1

Sinusoid-phasor transformation.

Time domain representation	Phasor domain representation
$V_m \cos(\omega t + \phi)$	$V_m \underline{/\phi}$
$V_m \sin(\omega t + \phi)$	$V_m \underline{/\phi - 90^\circ}$
$I_m \cos(\omega t + \theta)$	$I_m \underline{/\theta}$
$I_m \sin(\omega t + \theta)$	$I_m \underline{/\theta - 90^\circ}$

Note that in Eq. (9.25) the frequency (or time) factor $e^{j\omega t}$ is suppressed, and the frequency is not explicitly shown in the phasor domain representation because ω is constant. However, the response depends on ω. For this reason, the phasor domain is also known as the *frequency domain.*

From Eqs. (9.23) and (9.24), $v(t) = \text{Re}(\mathbf{V}e^{j\omega t}) = V_m \cos(\omega t + \phi)$, so that

$$\begin{aligned}\frac{dv}{dt} &= -\omega V_m \sin(\omega t + \phi) = \omega V_m \cos(\omega t + \phi + 90^\circ) \\ &= \text{Re}(\omega V_m e^{j\omega t} e^{j\phi} e^{j90^\circ}) = \text{Re}(j\omega \mathbf{V} e^{j\omega t})\end{aligned} \quad (9.26)$$

This shows that the derivative $v(t)$ is transformed to the phasor domain as $j\omega\mathbf{V}$

$$\underset{\text{(Time domain)}}{\frac{dv}{dt}} \quad \Leftrightarrow \quad \underset{\text{(Phasor domain)}}{j\omega\mathbf{V}} \tag{9.27}$$

Differentiating a sinusoid is equivalent to multiplying its corresponding phasor by $j\omega$.

Similarly, the integral of $v(t)$ is transformed to the phasor domain as $\mathbf{V}/j\omega$

$$\underset{\text{(Time domain)}}{\int v\, dt} \quad \Leftrightarrow \quad \underset{\text{(Phasor domain)}}{\frac{\mathbf{V}}{j\omega}} \tag{9.28}$$

Integrating a sinusoid is equivalent to dividing its corresponding phasor by $j\omega$.

Equation (9.27) allows the replacement of a derivative with respect to time with multiplication of $j\omega$ in the phasor domain, whereas Eq. (9.28) allows the replacement of an integral with respect to time with division by $j\omega$ in the phasor domain. Equations (9.27) and (9.28) are useful in finding the steady-state solution, which does not require knowing the initial values of the variable involved. This is one of the important applications of phasors.

Besides time differentiation and integration, another important use of phasors is found in summing sinusoids of the same frequency. This is best illustrated with an example, and Example 9.6 provides one.

Adding sinusoids of the same frequency is equivalent to adding their corresponding phasors.

The differences between $v(t)$ and $\mathbf{V}$ should be emphasized:

1. $v(t)$ is the *instantaneous or time domain* representation, while $\mathbf{V}$ is the *frequency or phasor domain* representation.
2. $v(t)$ is time dependent, while $\mathbf{V}$ is not. (This fact is often forgotten by students.)
3. $v(t)$ is always real with no complex term, while $\mathbf{V}$ is generally complex.

Finally, we should bear in mind that phasor analysis applies only when frequency is constant; it applies in manipulating two or more sinusoidal signals only if they are of the same frequency.

Example 9.3

Evaluate these complex numbers:

(a) $(40\underline{/50^\circ} + 20\underline{/-30^\circ})^{1/2}$

(b) $\dfrac{10\underline{/-30^\circ} + (3 - j4)}{(2 + j4)(3 - j5)^*}$

Solution:

(a) Using polar to rectangular transformation,

$$40\underline{/50^\circ} = 40(\cos 50^\circ + j\sin 50^\circ) = 25.71 + j30.64$$
$$20\underline{/-30^\circ} = 20[\cos(-30^\circ) + j\sin(-30^\circ)] = 17.32 - j10$$

Adding them up gives

$$40\underline{/50^\circ} + 20\underline{/-30^\circ} = 43.03 + j20.64 = 47.72\underline{/25.63^\circ}$$

Taking the square root of this,

$$(40\underline{/50^\circ} + 20\underline{/-30^\circ})^{1/2} = 6.91\underline{/12.81^\circ}$$

(b) Using polar-rectangular transformation, addition, multiplication, and division,

$$\frac{10\underline{/-30^\circ} + (3 - j4)}{(2 + j4)(3 - j5)^*} = \frac{8.66 - j5 + (3 - j4)}{(2 + j4)(3 + j5)}$$

$$= \frac{11.66 - j9}{-14 + j22} = \frac{14.73\underline{/-37.66^\circ}}{26.08\underline{/122.47^\circ}}$$

$$= 0.565\underline{/-160.13^\circ}$$

Practice Problem 9.3

Evaluate the following complex numbers:

(a) $[(5 + j2)(-1 + j4) - 5\underline{/60^\circ}]^*$

(b) $\dfrac{10 + j5 + 3\underline{/40^\circ}}{-3 + j4} + 10\underline{/30^\circ} + j5$

Answer: (a) $-15.5 - j13.67$, (b) $8.293 + j7.2$.

Example 9.4

Transform these sinusoids to phasors:

(a) $i = 6\cos(50t - 40^\circ)$ A
(b) $v = -4\sin(30t + 50^\circ)$ V

Solution:

(a) $i = 6\cos(50t - 40^\circ)$ has the phasor

$$\mathbf{I} = 6\underline{/-40^\circ}\text{ A}$$

(b) Since $-\sin A = \cos(A + 90^\circ)$,

$$v = -4\sin(30t + 50^\circ) = 4\cos(30t + 50^\circ + 90^\circ)$$
$$= 4\cos(30t + 140^\circ)\text{ V}$$

The phasor form of v is

$$\mathbf{V} = 4\underline{/140^\circ}\text{ V}$$

Practice Problem 9.4

Express these sinusoids as phasors:

(a) $v = 7\cos(2t + 40^\circ)$ V
(b) $i = -4\sin(10t + 10^\circ)$ A

Answer: (a) $\mathbf{V} = 7\underline{/40^\circ}$ V, (b) $\mathbf{I} = 4\underline{/100^\circ}$ A.

Example 9.5

Find the sinusoids represented by these phasors:

(a) $\mathbf{I} = -3 + j4$ A
(b) $\mathbf{V} = j8e^{-j20^\circ}$ V

Solution:

(a) $\mathbf{I} = -3 + j4 = 5\underline{/126.87^\circ}$. Transforming this to the time domain gives

$$i(t) = 5\cos(\omega t + 126.87^\circ) \text{ A}$$

(b) Since $j = 1\underline{/90^\circ}$,

$$\begin{aligned} \mathbf{V} = j8\underline{/-20^\circ} &= (1\underline{/90^\circ})(8\underline{/-20^\circ}) \\ &= 8\underline{/90^\circ - 20^\circ} = 8\underline{/70^\circ} \text{ V} \end{aligned}$$

Converting this to the time domain gives

$$v(t) = 8\cos(\omega t + 70^\circ) \text{ V}$$

Practice Problem 9.5

Find the sinusoids corresponding to these phasors:

(a) $\mathbf{V} = -10\underline{/30^\circ}$ V
(b) $\mathbf{I} = j(5 - j12)$ A

Answer: (a) $v(t) = 10\cos(\omega t + 210^\circ)$ V or $10\cos(\omega t - 150^\circ)$ V, (b) $i(t) = 13\cos(\omega t + 22.62^\circ)$ A.

Example 9.6

Given $i_1(t) = 4\cos(\omega t + 30^\circ)$ A and $i_2(t) = 5\sin(\omega t - 20^\circ)$ A, find their sum.

Solution:

Here is an important use of phasors—for summing sinusoids of the same frequency. Current $i_1(t)$ is in the standard form. Its phasor is

$$\mathbf{I}_1 = 4\underline{/30^\circ}$$

We need to express $i_2(t)$ in cosine form. The rule for converting sine to cosine is to subtract 90°. Hence,

$$i_2 = 5\cos(\omega t - 20^\circ - 90^\circ) = 5\cos(\omega t - 110^\circ)$$

and its phasor is

$$\mathbf{I}_2 = 5\underline{/-110^\circ}$$

If we let $i = i_1 + i_2$, then

$$\begin{aligned} \mathbf{I} = \mathbf{I}_1 + \mathbf{I}_2 &= 4\underline{/30^\circ} + 5\underline{/-110^\circ} \\ &= 3.464 + j2 - 1.71 - j4.698 = 1.754 - j2.698 \\ &= 3.218\underline{/-56.97^\circ} \text{ A} \end{aligned}$$

Transforming this to the time domain, we get

$$i(t) = 3.218 \cos(\omega t - 56.97^\circ) \text{ A}$$

Of course, we can find $i_1 + i_2$ using Eq. (9.9), but that is the hard way.

Practice Problem 9.6

If $v_1 = -10 \sin(\omega t - 30^\circ)$ V and $v_2 = 20 \cos(\omega t + 45^\circ)$ V, find $v = v_1 + v_2$.

Answer: $v(t) = 12.158 \cos(\omega t + 55.95^\circ)$ V.

Example 9.7

Using the phasor approach, determine the current $i(t)$ in a circuit described by the integrodifferential equation

$$4i + 8\int i\,dt - 3\frac{di}{dt} = 50 \cos(2t + 75^\circ)$$

Solution:
We transform each term in the equation from time domain to phasor domain. Keeping Eqs. (9.27) and (9.28) in mind, we obtain the phasor form of the given equation as

$$4\mathbf{I} + \frac{8\mathbf{I}}{j\omega} - 3j\omega\mathbf{I} = 50\underline{/75^\circ}$$

But $\omega = 2$, so

$$\mathbf{I}(4 - j4 - j6) = 50\underline{/75^\circ}$$

$$\mathbf{I} = \frac{50\underline{/75^\circ}}{4 - j10} = \frac{50\underline{/75^\circ}}{10.77\underline{/-68.2^\circ}} = 4.642\underline{/143.2^\circ} \text{ A}$$

Converting this to the time domain,

$$i(t) = 4.642 \cos(2t + 143.2^\circ) \text{ A}$$

Keep in mind that this is only the steady-state solution, and it does not require knowing the initial values.

Practice Problem 9.7

Find the voltage $v(t)$ in a circuit described by the integrodifferential equation

$$2\frac{dv}{dt} + 5v + 10\int v\,dt = 50 \cos(5t - 30^\circ)$$

using the phasor approach.

Answer: $v(t) = 5.3 \cos(5t - 88^\circ)$ V.

9.4 Phasor Relationships for Circuit Elements

Now that we know how to represent a voltage or current in the phasor or frequency domain, one may legitimately ask how we apply this to circuits involving the passive elements R, L, and C. What we need to do is to transform the voltage-current relationship from the time domain to the frequency domain for each element. Again, we will assume the passive sign convention.

We begin with the resistor. If the current through a resistor R is $i = I_m \cos(\omega t + \phi)$, the voltage across it is given by Ohm's law as

$$v = iR = RI_m \cos(\omega t + \phi) \tag{9.29}$$

The phasor form of this voltage is

$$\mathbf{V} = RI_m \underline{/\phi} \tag{9.30}$$

But the phasor representation of the current is $\mathbf{I} = I_m \underline{/\phi}$. Hence,

$$\mathbf{V} = R\mathbf{I} \tag{9.31}$$

showing that the voltage-current relation for the resistor in the phasor domain continues to be Ohm's law, as in the time domain. Figure 9.9 illustrates the voltage-current relations of a resistor. We should note from Eq. (9.31) that voltage and current are in phase, as illustrated in the phasor diagram in Fig. 9.10.

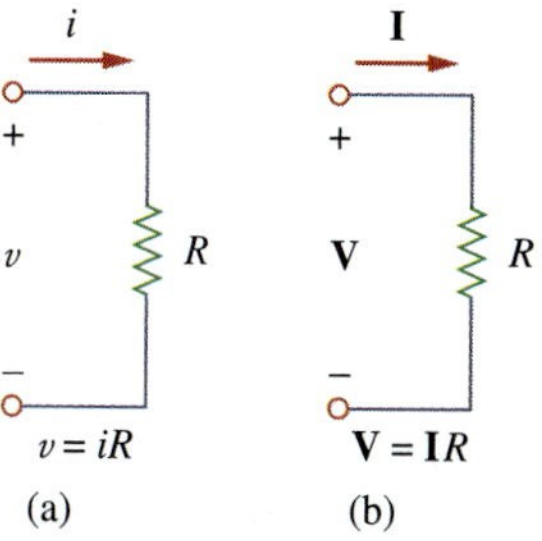

Figure 9.9
Voltage-current relations for a resistor in the: (a) time domain, (b) frequency domain.

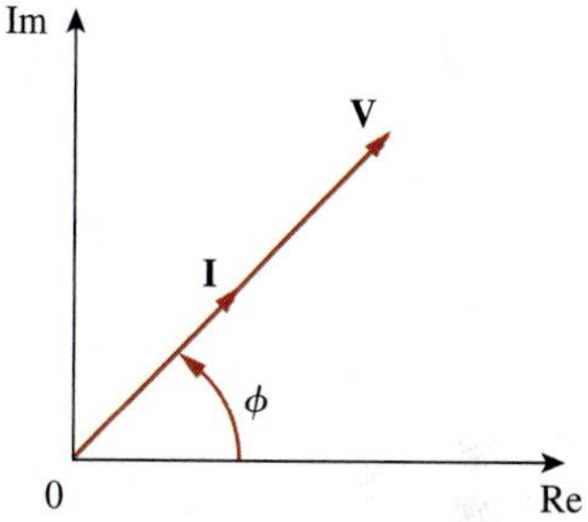

Figure 9.10
Phasor diagram for the resistor.

For the inductor L, assume the current through it is $i = I_m \cos(\omega t + \phi)$. The voltage across the inductor is

$$v = L\frac{di}{dt} = -\omega L I_m \sin(\omega t + \phi) \tag{9.32}$$

Recall from Eq. (9.10) that $-\sin A = \cos(A + 90^\circ)$. We can write the voltage as

$$v = \omega L I_m \cos(\omega t + \phi + 90^\circ) \tag{9.33}$$

which transforms to the phasor

$$\mathbf{V} = \omega L I_m e^{j(\phi+90^\circ)} = \omega L I_m e^{j\phi} e^{j90^\circ} = \omega L I_m \underline{/\phi + 90^\circ} \tag{9.34}$$

But $I_m \underline{/\phi} = \mathbf{I}$, and from Eq. (9.19), $e^{j90^\circ} = j$. Thus,

$$\mathbf{V} = j\omega L\mathbf{I} \tag{9.35}$$

showing that the voltage has a magnitude of $\omega L I_m$ and a phase of $\phi + 90^\circ$. The voltage and current are 90° out of phase. Specifically, the current lags the voltage by 90°. Figure 9.11 shows the voltage-current relations for the inductor. Figure 9.12 shows the phasor diagram.

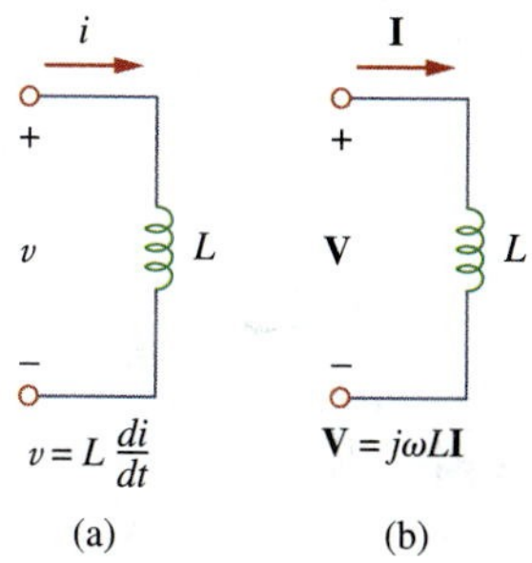

Figure 9.11
Voltage-current relations for an inductor in the: (a) time domain, (b) frequency domain.

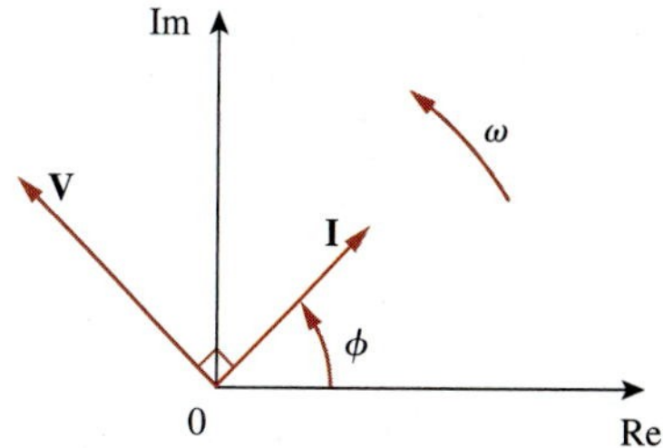

Figure 9.12
Phasor diagram for the inductor; **I** lags **V**.

Although it is equally correct to say that the inductor voltage leads the current by 90°, convention gives the current phase relative to the voltage.

For the capacitor C, assume the voltage across it is $v = V_m \cos(\omega t + \phi)$. The current through the capacitor is

$$i = C\frac{dv}{dt} \tag{9.36}$$

By following the same steps as we took for the inductor or by applying Eq. (9.27) on Eq. (9.36), we obtain

$$\mathbf{I} = j\omega C\mathbf{V} \quad \Rightarrow \quad \mathbf{V} = \frac{\mathbf{I}}{j\omega C} \tag{9.37}$$

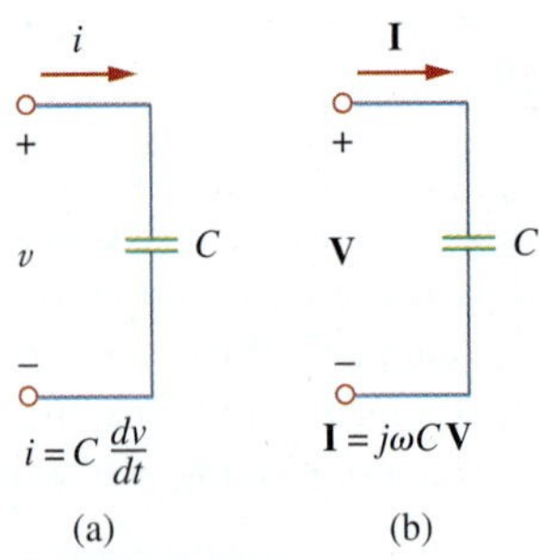

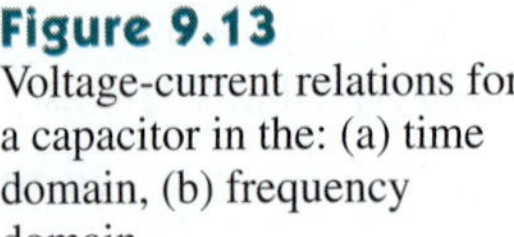

Figure 9.13 Voltage-current relations for a capacitor in the: (a) time domain, (b) frequency domain.

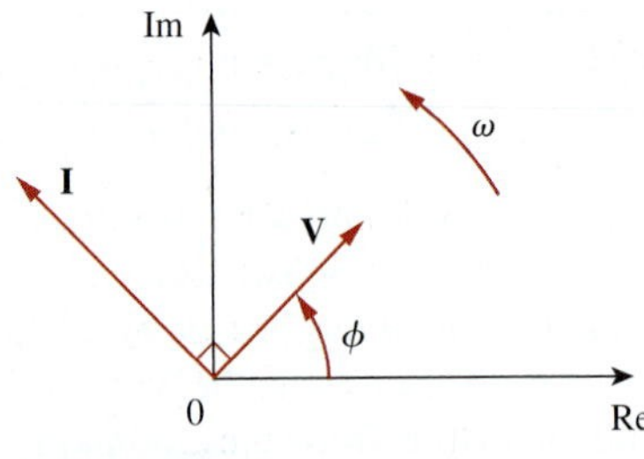

Figure 9.14 Phasor diagram for the capacitor; **I** leads **V**.

showing that the current and voltage are 90° out of phase. To be specific, the current leads the voltage by 90°. Figure 9.13 shows the voltage-current relations for the capacitor; Fig. 9.14 gives the phasor diagram. Table 9.2 summarizes the time domain and phasor domain representations of the circuit elements.

TABLE 9.2

Summary of voltage-current relationships.

Element	Time domain	Frequency domain
R	$v = Ri$	$\mathbf{V} = R\mathbf{I}$
L	$v = L\frac{di}{dt}$	$\mathbf{V} = j\omega L\mathbf{I}$
C	$i = C\frac{dv}{dt}$	$\mathbf{V} = \frac{\mathbf{I}}{j\omega C}$

Example 9.8

The voltage $v = 12\cos(60t + 45°)$ is applied to a 0.1-H inductor. Find the steady-state current through the inductor.

Solution:

For the inductor, $\mathbf{V} = j\omega L\mathbf{I}$, where $\omega = 60$ rad/s and $\mathbf{V} = 12\underline{/45°}$ V. Hence,

$$\mathbf{I} = \frac{\mathbf{V}}{j\omega L} = \frac{12\underline{/45°}}{j60 \times 0.1} = \frac{12\underline{/45°}}{6\underline{/90°}} = 2\underline{/-45°} \text{ A}$$

Converting this to the time domain,

$$i(t) = 2\cos(60t - 45°) \text{ A}$$

Practice Problem 9.8

If voltage $v = 10\cos(100t + 30°)$ is applied to a 50 μF capacitor, calculate the current through the capacitor.

Answer: $50\cos(100t + 120°)$ mA.

9.5 Impedance and Admittance

In the preceding section, we obtained the voltage-current relations for the three passive elements as

$$\mathbf{V} = R\mathbf{I}, \qquad \mathbf{V} = j\omega L\mathbf{I}, \qquad \mathbf{V} = \frac{\mathbf{I}}{j\omega C} \tag{9.38}$$

These equations may be written in terms of the ratio of the phasor voltage to the phasor current as

$$\frac{\mathbf{V}}{\mathbf{I}} = R, \qquad \frac{\mathbf{V}}{\mathbf{I}} = j\omega L, \qquad \frac{\mathbf{V}}{\mathbf{I}} = \frac{1}{j\omega C} \tag{9.39}$$

From these three expressions, we obtain Ohm's law in phasor form for any type of element as

$$\mathbf{Z} = \frac{\mathbf{V}}{\mathbf{I}} \quad \text{or} \quad \mathbf{V} = \mathbf{Z}\mathbf{I} \tag{9.40}$$

where $\mathbf{Z}$ is a frequency-dependent quantity known as *impedance*, measured in ohms.

> The **impedance Z** of a circuit is the ratio of the phasor voltage **V** to the phasor current **I**, measured in ohms (Ω).

The impedance represents the opposition that the circuit exhibits to the flow of sinusoidal current. Although the impedance is the ratio of two phasors, it is not a phasor, because it does not correspond to a sinusoidally varying quantity.

The impedances of resistors, inductors, and capacitors can be readily obtained from Eq. (9.39). Table 9.3 summarizes their impedances. From the table we notice that $\mathbf{Z}_L = j\omega L$ and $\mathbf{Z}_C = -j/\omega C$. Consider two extreme cases of angular frequency. When $\omega = 0$ (i.e., for dc sources), $\mathbf{Z}_L = 0$ and $\mathbf{Z}_C \to \infty$, confirming what we already know—that the inductor acts like a short circuit, while the capacitor acts like an open circuit. When $\omega \to \infty$ (i.e., for high frequencies), $\mathbf{Z}_L \to \infty$ and $\mathbf{Z}_C = 0$, indicating that the inductor is an open circuit to high frequencies, while the capacitor is a short circuit. Figure 9.15 illustrates this.

TABLE 9.3 Impedances and admittances of passive elements.

Element	Impedance	Admittance
R	$\mathbf{Z} = R$	$\mathbf{Y} = \frac{1}{R}$
L	$\mathbf{Z} = j\omega L$	$\mathbf{Y} = \frac{1}{j\omega L}$
C	$\mathbf{Z} = \frac{1}{j\omega C}$	$\mathbf{Y} = j\omega C$

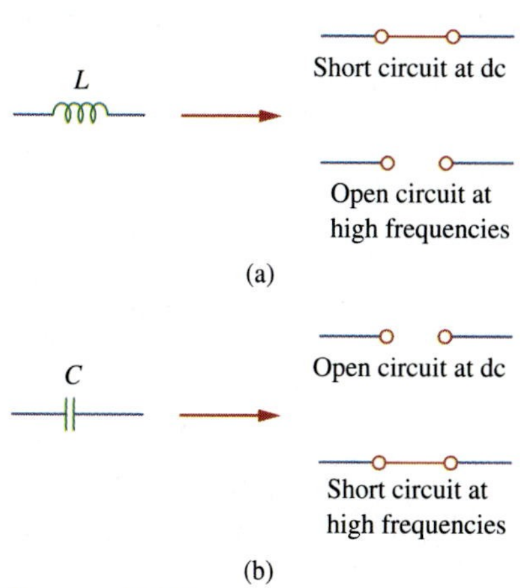

Figure 9.15 Equivalent circuits at dc and high frequencies: (a) inductor, (b) capacitor.

As a complex quantity, the impedence may be expressed in rectangular form as

$$\mathbf{Z} = R + jX \tag{9.41}$$

where $R = \text{Re}\,\mathbf{Z}$ is the *resistance* and $X = \text{Im}\,\mathbf{Z}$ is the *reactance*. The reactance X may be positive or negative. We say that the impedance is inductive when X is positive or capacitive when X is negative. Thus, impedance $\mathbf{Z} = R + jX$ is said to be *inductive* or lagging since current lags voltage, while impedance $\mathbf{Z} = R - jX$ is capacitive or leading because current leads voltage. The impedance, resistance, and reactance are all measured in ohms. The impedance may also be expressed in polar form as

$$\mathbf{Z} = |\mathbf{Z}| \underline{/\theta} \tag{9.42}$$

Comparing Eqs. (9.41) and (9.42), we infer that

$$\mathbf{Z} = R + jX = |\mathbf{Z}|\angle\theta \tag{9.43}$$

where

$$|\mathbf{Z}| = \sqrt{R^2 + X^2}, \qquad \theta = \tan^{-1}\frac{X}{R} \tag{9.44}$$

and

$$R = |\mathbf{Z}|\cos\theta, \qquad X = |\mathbf{Z}|\sin\theta \tag{9.45}$$

It is sometimes convenient to work with the reciprocal of impedance, known as *admittance*.

> The **admittance Y** is the reciprocal of impedance, measured in siemens (S).

The admittance **Y** of an element (or a circuit) is the ratio of the phasor current through it to the phasor voltage across it, or

$$\mathbf{Y} = \frac{1}{\mathbf{Z}} = \frac{\mathbf{I}}{\mathbf{V}} \tag{9.46}$$

The admittances of resistors, inductors, and capacitors can be obtained from Eq. (9.39). They are also summarized in Table 9.3.

As a complex quantity, we may write **Y** as

$$\mathbf{Y} = G + jB \tag{9.47}$$

where $G = \text{Re}\,\mathbf{Y}$ is called the *conductance* and $B = \text{Im}\,\mathbf{Y}$ is called the *susceptance*. Admittance, conductance, and susceptance are all expressed in the unit of siemens (or mhos). From Eqs. (9.41) and (9.47),

$$G + jB = \frac{1}{R + jX} \tag{9.48}$$

By rationalization,

$$G + jB = \frac{1}{R + jX} \cdot \frac{R - jX}{R - jX} = \frac{R - jX}{R^2 + X^2} \tag{9.49}$$

Equating the real and imaginary parts gives

$$G = \frac{R}{R^2 + X^2}, \qquad B = -\frac{X}{R^2 + X^2} \tag{9.50}$$

showing that $G \neq 1/R$ as it is in resistive circuits. Of course, if $X = 0$, then $G = 1/R$.

Example 9.9

Find $v(t)$ and $i(t)$ in the circuit shown in Fig. 9.16.

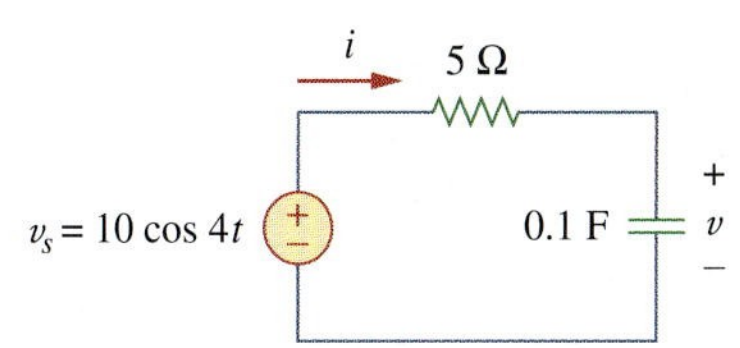

Figure 9.16
For Example 9.9.

Solution:
From the voltage source 10 cos 4t, $\omega = 4$,

$$\mathbf{V}_s = 10\underline{/0^\circ} \text{ V}$$

The impedance is

$$\mathbf{Z} = 5 + \frac{1}{j\omega C} = 5 + \frac{1}{j4 \times 0.1} = 5 - j2.5\ \Omega$$

Hence the current

$$\begin{aligned}\mathbf{I} = \frac{\mathbf{V}_s}{\mathbf{Z}} &= \frac{10\underline{/0^\circ}}{5 - j2.5} = \frac{10(5 + j2.5)}{5^2 + 2.5^2} \\ &= 1.6 + j0.8 = 1.789\underline{/26.57^\circ} \text{ A}\end{aligned} \quad \textbf{(9.9.1)}$$

The voltage across the capacitor is

$$\begin{aligned}\mathbf{V} = \mathbf{I}\mathbf{Z}_C = \frac{\mathbf{I}}{j\omega C} &= \frac{1.789\underline{/26.57^\circ}}{j4 \times 0.1} \\ &= \frac{1.789\underline{/26.57^\circ}}{0.4\underline{/90^\circ}} = 4.47\underline{/-63.43^\circ} \text{ V}\end{aligned} \quad \textbf{(9.9.2)}$$

Converting **I** and **V** in Eqs. (9.9.1) and (9.9.2) to the time domain, we get

$$\begin{aligned} i(t) &= 1.789 \cos(4t + 26.57^\circ) \text{ A} \\ v(t) &= 4.47 \cos(4t - 63.43^\circ) \text{ V}\end{aligned}$$

Notice that $i(t)$ leads $v(t)$ by 90° as expected.

Practice Problem 9.9

Refer to Fig. 9.17. Determine $v(t)$ and $i(t)$.

Answer: $8.944 \sin(10t + 93.43^\circ)$ V, $4.472 \sin(10t + 3.43^\circ)$ A.

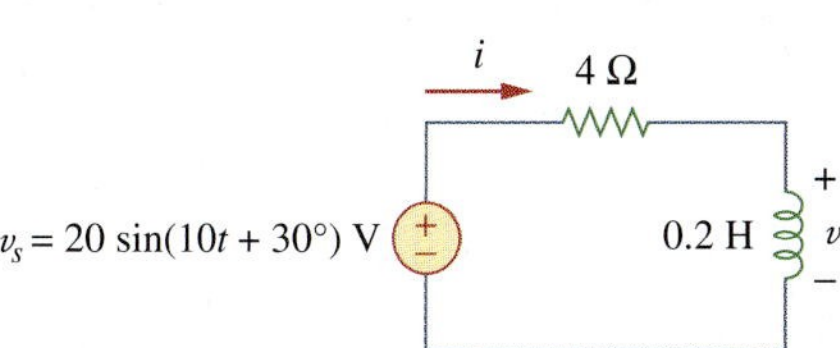

Figure 9.17
For Practice Prob. 9.9.

9.6 †Kirchhoff's Laws in the Frequency Domain

We cannot do circuit analysis in the frequency domain without Kirchhoff's current and voltage laws. Therefore, we need to express them in the frequency domain.

For KVL, let $v_1, v_2, \ldots, v_n$ be the voltages around a closed loop. Then

$$v_1 + v_2 + \cdots + v_n = 0 \quad \textbf{(9.51)}$$

In the sinusoidal steady state, each voltage may be written in cosine form, so that Eq. (9.51) becomes

$$\begin{aligned} &V_{m1} \cos(\omega t + \theta_1) + V_{m2} \cos(\omega t + \theta_2) \\ &\quad + \cdots + V_{mn} \cos(\omega t + \theta_n) = 0 \end{aligned} \quad \textbf{(9.52)}$$

This can be written as

$$\text{Re}(V_{m1}e^{j\theta_1}e^{j\omega t}) + \text{Re}(V_{m2}e^{j\theta_2}e^{j\omega t}) + \cdots + \text{Re}(V_{mn}e^{j\theta_n}e^{j\omega t}) = 0$$

or

$$\text{Re}[(V_{m1}e^{j\theta_1} + V_{m2}e^{j\theta_2} + \cdots + V_{mn}e^{j\theta_n})e^{j\omega t}] = 0 \qquad (9.53)$$

If we let $\mathbf{V}_k = V_{mk}e^{j\theta_k}$, then

$$\text{Re}[(\mathbf{V}_1 + \mathbf{V}_2 + \cdots + \mathbf{V}_n)e^{j\omega t}] = 0 \qquad (9.54)$$

Since $e^{j\omega t} \neq 0$,

$$\mathbf{V}_1 + \mathbf{V}_2 + \cdots + \mathbf{V}_n = 0 \qquad (9.55)$$

indicating that Kirchhoff's voltage law holds for phasors.

By following a similar procedure, we can show that Kirchhoff's current law holds for phasors. If we let $i_1, i_2, \ldots, i_n$ be the current leaving or entering a closed surface in a network at time t, then

$$i_1 + i_2 + \cdots + i_n = 0 \qquad (9.56)$$

If $\mathbf{I}_1, \mathbf{I}_2, \ldots, \mathbf{I}_n$ are the phasor forms of the sinusoids $i_1, i_2, \ldots, i_n$, then

$$\mathbf{I}_1 + \mathbf{I}_2 + \cdots + \mathbf{I}_n = 0 \qquad (9.57)$$

which is Kirchhoff's current law in the frequency domain.

Once we have shown that both KVL and KCL hold in the frequency domain, it is easy to do many things, such as impedance combination, nodal and mesh analyses, superposition, and source transformation.

9.7 Impedance Combinations

Consider the N series-connected impedances shown in Fig. 9.18. The same current $\mathbf{I}$ flows through the impedances. Applying KVL around the loop gives

$$\mathbf{V} = \mathbf{V}_1 + \mathbf{V}_2 + \cdots + \mathbf{V}_N = \mathbf{I}(\mathbf{Z}_1 + \mathbf{Z}_2 + \cdots + \mathbf{Z}_N) \qquad (9.58)$$

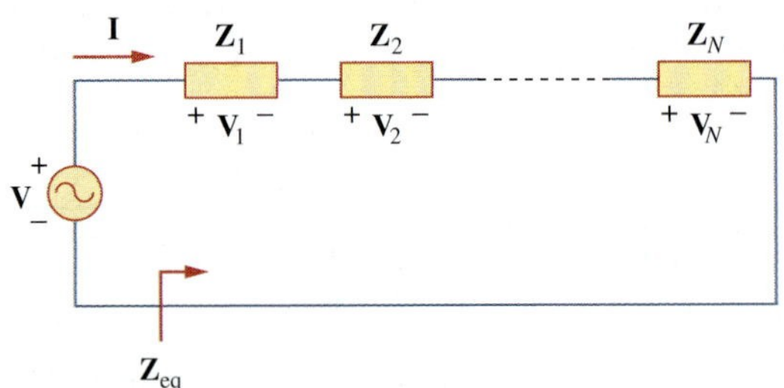

Figure 9.18
N impedances in series.

The equivalent impedance at the input terminals is

$$\mathbf{Z}_{eq} = \frac{\mathbf{V}}{\mathbf{I}} = \mathbf{Z}_1 + \mathbf{Z}_2 + \cdots + \mathbf{Z}_N$$

or

$$\boxed{\mathbf{Z}_{eq} = \mathbf{Z}_1 + \mathbf{Z}_2 + \cdots + \mathbf{Z}_N} \qquad (9.59)$$

showing that the total or equivalent impedance of series-connected impedances is the sum of the individual impedances. This is similar to the series connection of resistances.

If $N = 2$, as shown in Fig. 9.19, the current through the impedances is

$$\mathbf{I} = \frac{\mathbf{V}}{\mathbf{Z}_1 + \mathbf{Z}_2} \tag{9.60}$$

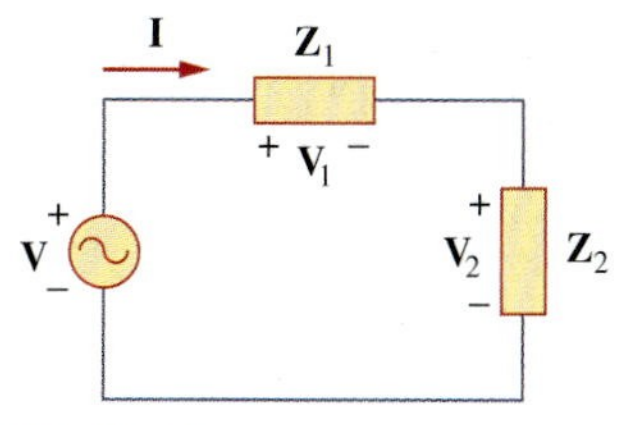

Figure 9.19
Voltage division.

Since $\mathbf{V}_1 = \mathbf{Z}_1\mathbf{I}$ and $\mathbf{V}_2 = \mathbf{Z}_2\mathbf{I}$, then

$$\mathbf{V}_1 = \frac{\mathbf{Z}_1}{\mathbf{Z}_1 + \mathbf{Z}_2}\mathbf{V}, \qquad \mathbf{V}_2 = \frac{\mathbf{Z}_2}{\mathbf{Z}_1 + \mathbf{Z}_2}\mathbf{V} \tag{9.61}$$

which is the *voltage-division* relationship.

In the same manner, we can obtain the equivalent impedance or admittance of the N parallel-connected impedances shown in Fig. 9.20. The voltage across each impedance is the same. Applying KCL at the top node,

$$\mathbf{I} = \mathbf{I}_1 + \mathbf{I}_2 + \cdots + \mathbf{I}_N = \mathbf{V}\left(\frac{1}{\mathbf{Z}_1} + \frac{1}{\mathbf{Z}_2} + \cdots + \frac{1}{\mathbf{Z}_N}\right) \tag{9.62}$$

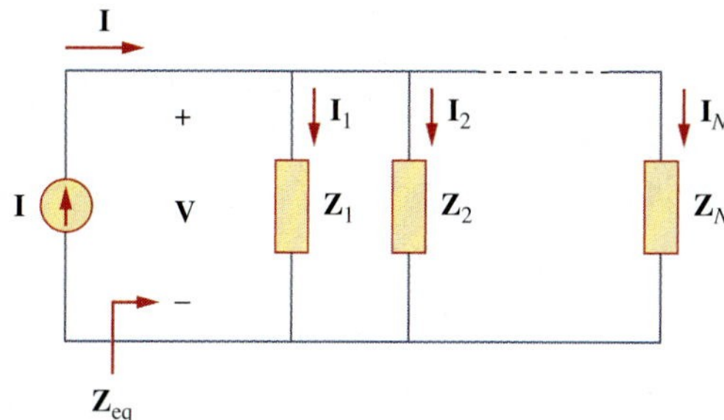

Figure 9.20
N impedances in parallel.

The equivalent impedance is

$$\frac{1}{\mathbf{Z}_{eq}} = \frac{\mathbf{I}}{\mathbf{V}} = \frac{1}{\mathbf{Z}_1} + \frac{1}{\mathbf{Z}_2} + \cdots + \frac{1}{\mathbf{Z}_N} \tag{9.63}$$

and the equivalent admittance is

$$\mathbf{Y}_{eq} = \mathbf{Y}_1 + \mathbf{Y}_2 + \cdots + \mathbf{Y}_N \tag{9.64}$$

This indicates that the equivalent admittance of a parallel connection of admittances is the sum of the individual admittances.

When $N = 2$, as shown in Fig. 9.21, the equivalent impedance becomes

$$\mathbf{Z}_{eq} = \frac{1}{\mathbf{Y}_{eq}} = \frac{1}{\mathbf{Y}_1 + \mathbf{Y}_2} = \frac{1}{1/\mathbf{Z}_1 + 1/\mathbf{Z}_2} = \frac{\mathbf{Z}_1\mathbf{Z}_2}{\mathbf{Z}_1 + \mathbf{Z}_2} \tag{9.65}$$

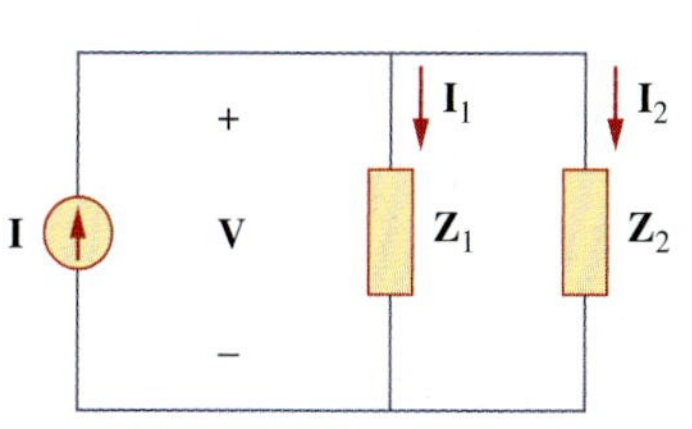

Figure 9.21
Current division.

Also, since

$$\mathbf{V} = \mathbf{I}\mathbf{Z}_{eq} = \mathbf{I}_1\mathbf{Z}_1 = \mathbf{I}_2\mathbf{Z}_2$$

the currents in the impedances are

$$\mathbf{I}_1 = \frac{\mathbf{Z}_2}{\mathbf{Z}_1 + \mathbf{Z}_2}\mathbf{I}, \qquad \mathbf{I}_2 = \frac{\mathbf{Z}_1}{\mathbf{Z}_1 + \mathbf{Z}_2}\mathbf{I} \tag{9.66}$$

which is the *current-division* principle.

The delta-to-wye and wye-to-delta transformations that we applied to resistive circuits are also valid for impedances. With reference to Fig. 9.22, the conversion formulas are as follows.

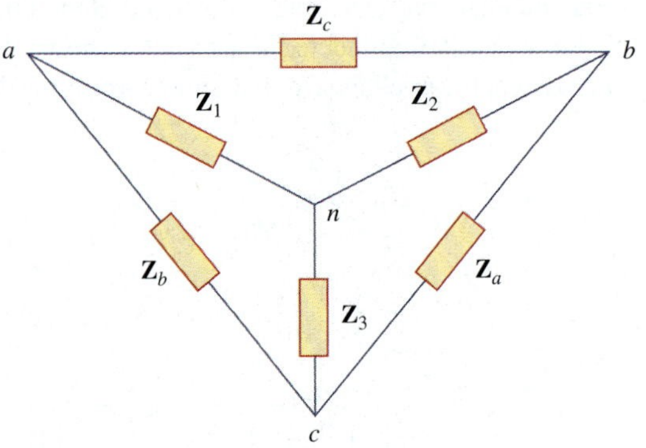

Figure 9.22
Superimposed Y and Δ networks.

Y-Δ Conversion:

$$\begin{aligned}
\mathbf{Z}_a &= \frac{\mathbf{Z}_1\mathbf{Z}_2 + \mathbf{Z}_2\mathbf{Z}_3 + \mathbf{Z}_3\mathbf{Z}_1}{\mathbf{Z}_1} \\
\mathbf{Z}_b &= \frac{\mathbf{Z}_1\mathbf{Z}_2 + \mathbf{Z}_2\mathbf{Z}_3 + \mathbf{Z}_3\mathbf{Z}_1}{\mathbf{Z}_2} \\
\mathbf{Z}_c &= \frac{\mathbf{Z}_1\mathbf{Z}_2 + \mathbf{Z}_2\mathbf{Z}_3 + \mathbf{Z}_3\mathbf{Z}_1}{\mathbf{Z}_3}
\end{aligned} \tag{9.67}$$

Δ-Y Conversion:

$$\begin{aligned}
\mathbf{Z}_1 &= \frac{\mathbf{Z}_b\mathbf{Z}_c}{\mathbf{Z}_a + \mathbf{Z}_b + \mathbf{Z}_c} \\
\mathbf{Z}_2 &= \frac{\mathbf{Z}_c\mathbf{Z}_a}{\mathbf{Z}_a + \mathbf{Z}_b + \mathbf{Z}_c} \\
\mathbf{Z}_3 &= \frac{\mathbf{Z}_a\mathbf{Z}_b}{\mathbf{Z}_a + \mathbf{Z}_b + \mathbf{Z}_c}
\end{aligned} \tag{9.68}$$

A delta or wye circuit is said to be **balanced** if it has equal impedances in all three branches.

When a Δ-Y circuit is balanced, Eqs. (9.67) and (9.68) become

$$\mathbf{Z}_\Delta = 3\mathbf{Z}_Y \qquad \text{or} \qquad \mathbf{Z}_Y = \frac{1}{3}\mathbf{Z}_\Delta \tag{9.69}$$

where $\mathbf{Z}_Y = \mathbf{Z}_1 = \mathbf{Z}_2 = \mathbf{Z}_3$ and $\mathbf{Z}_\Delta = \mathbf{Z}_a = \mathbf{Z}_b = \mathbf{Z}_c$.

As you see in this section, the principles of voltage division, current division, circuit reduction, impedance equivalence, and Y-Δ transformation all apply to ac circuits. Chapter 10 will show that other circuit techniques—such as superposition, nodal analysis, mesh analysis, source transformation, the Thevenin theorem, and the Norton theorem—are all applied to ac circuits in a manner similar to their application in dc circuits.

Example 9.10

Find the input impedance of the circuit in Fig. 9.23. Assume that the circuit operates at $\omega = 50$ rad/s.

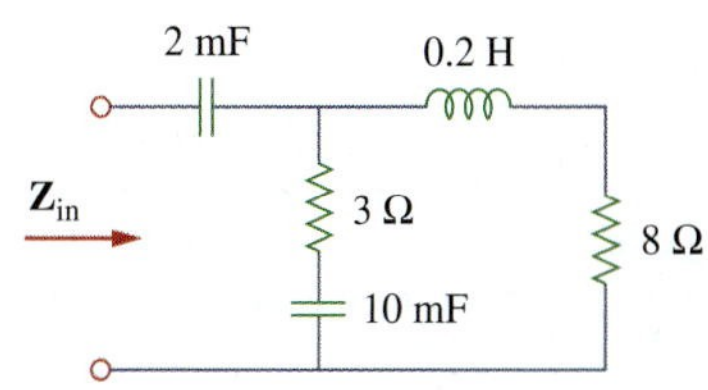

Figure 9.23
For Example 9.10.

Solution:
Let

$\mathbf{Z}_1$ = Impedance of the 2-mF capacitor

$\mathbf{Z}_2$ = Impedance of the 3-Ω resistor in series with the10-mF capacitor

$\mathbf{Z}_3$ = Impedance of the 0.2-H inductor in series with the 8-Ω resistor

Then

$$\mathbf{Z}_1 = \frac{1}{j\omega C} = \frac{1}{j50 \times 2 \times 10^{-3}} = -j10\ \Omega$$

$$\mathbf{Z}_2 = 3 + \frac{1}{j\omega C} = 3 + \frac{1}{j50 \times 10 \times 10^{-3}} = (3 - j2)\ \Omega$$

$$\mathbf{Z}_3 = 8 + j\omega L = 8 + j50 \times 0.2 = (8 + j10)\ \Omega$$

The input impedance is

$$\mathbf{Z}_{\text{in}} = \mathbf{Z}_1 + \mathbf{Z}_2 \parallel \mathbf{Z}_3 = -j10 + \frac{(3 - j2)(8 + j10)}{11 + j8}$$

$$= -j10 + \frac{(44 + j14)(11 - j8)}{11^2 + 8^2} = -j10 + 3.22 - j1.07\ \Omega$$

Thus,

$$\mathbf{Z}_{\text{in}} = 3.22 - j11.07\ \Omega$$

Practice Problem 9.10

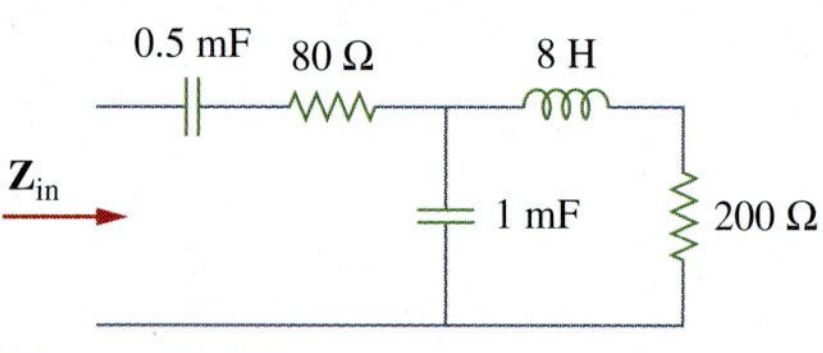

Figure 9.24
For Practice Prob. 9.10.

Determine the input impedance of the circuit in Fig. 9.24 at $\omega = 10$ rad/s.

Answer: $(129.52 - j295)$

Example 9.11

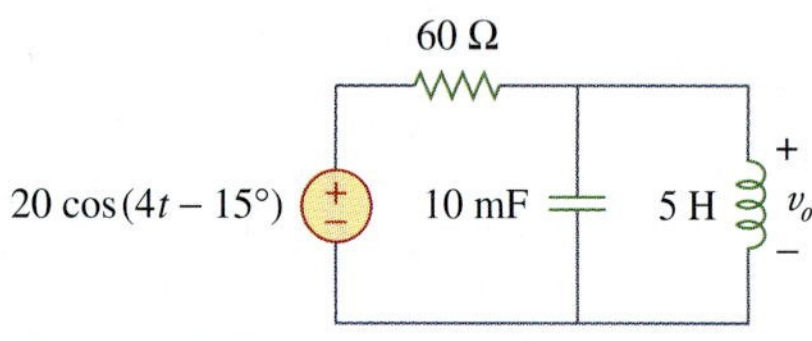

Figure 9.25
For Example 9.11.

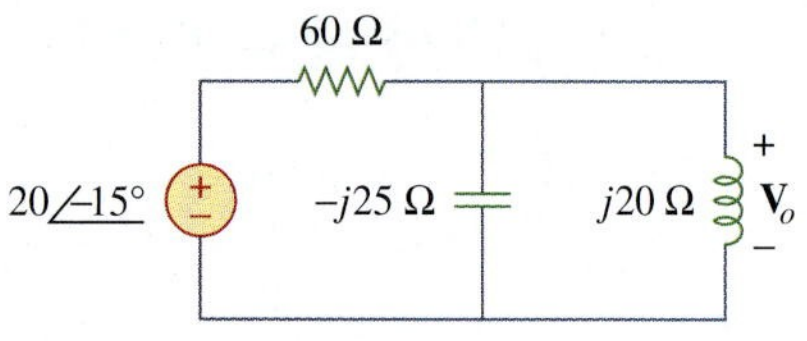

Figure 9.26
The frequency domain equivalent of the circuit in Fig. 9.25.

Determine $v_o(t)$ in the circuit of Fig. 9.25.

Solution:

To do the analysis in the frequency domain, we must first transform the time domain circuit in Fig. 9.25 to the phasor domain equivalent in Fig. 9.26. The transformation produces

$$v_s = 20\cos(4t - 15^\circ) \quad \Rightarrow \quad \mathbf{V}_s = 20\underline{/-15^\circ}\text{ V}, \qquad \omega = 4$$

$$10\text{ mF} \quad \Rightarrow \quad \frac{1}{j\omega C} = \frac{1}{j4 \times 10 \times 10^{-3}} = -j25\ \Omega$$

$$5\text{ H} \quad \Rightarrow \quad j\omega L = j4 \times 5 = j20\ \Omega$$

Let

$\mathbf{Z}_1$ = Impedance of the 60-Ω resistor

$\mathbf{Z}_2$ = Impedance of the parallel combination of the 10-mF capacitor and the 5-H inductor

Then $\mathbf{Z}_1 = 60\ \Omega$ and

$$\mathbf{Z}_2 = -j25 \parallel j20 = \frac{-j25 \times j20}{-j25 + j20} = j100\ \Omega$$

By the voltage-division principle,

$$\mathbf{V}_o = \frac{\mathbf{Z}_2}{\mathbf{Z}_1 + \mathbf{Z}_2}\mathbf{V}_s = \frac{j100}{60 + j100}(20\underline{/-15^\circ})$$
$$= (0.8575\underline{/30.96^\circ})(20\underline{/-15^\circ}) = 17.15\underline{/15.96^\circ}\text{ V}$$

We convert this to the time domain and obtain

$$v_o(t) = 17.15\cos(4t + 15.96^\circ)\text{ V}$$

Practice Problem 9.11

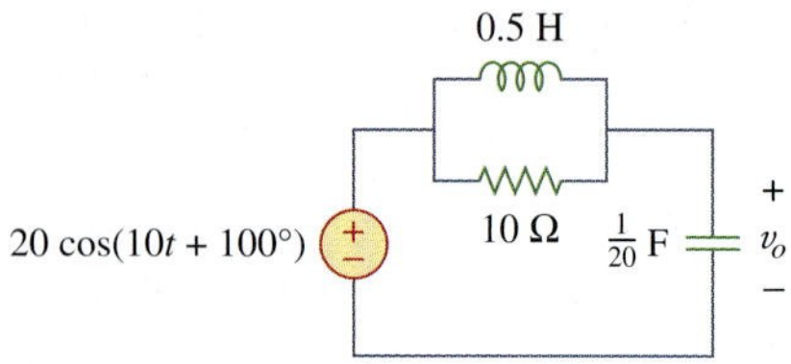

Figure 9.27
For Practice Prob. 9.11.

Calculate v_o in the circuit of Fig. 9.27.

Answer: $v_o(t) = 14.142\cos(10t - 35^\circ)$ V.

Example 9.12

Find current **I** in the circuit of Fig. 9.28.

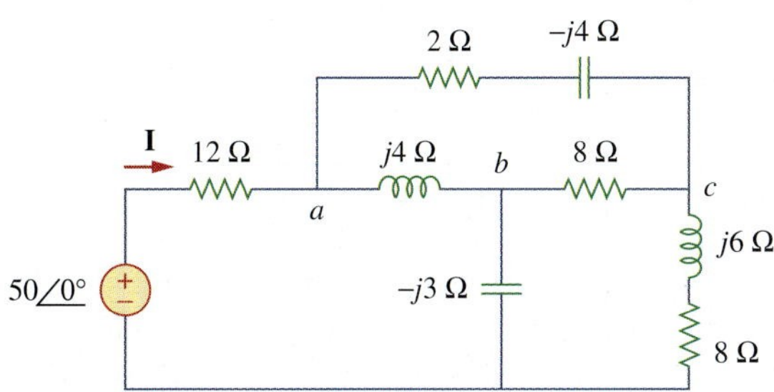

Figure 9.28
For Example 9.12.

Solution:

The delta network connected to nodes a, b, and c can be converted to the Y network of Fig. 9.29. We obtain the Y impedances as follows using Eq. (9.68):

$$\mathbf{Z}_{an} = \frac{j4(2 - j4)}{j4 + 2 - j4 + 8} = \frac{4(4 + j2)}{10} = (1.6 + j0.8)\ \Omega$$

$$\mathbf{Z}_{bn} = \frac{j4(8)}{10} = j3.2\ \Omega, \qquad \mathbf{Z}_{cn} = \frac{8(2 - j4)}{10} = (1.6 - j3.2)\ \Omega$$

The total impedance at the source terminals is

$$\begin{aligned}\mathbf{Z} &= 12 + \mathbf{Z}_{an} + (\mathbf{Z}_{bn} - j3) \parallel (\mathbf{Z}_{cn} + j6 + 8)\\ &= 12 + 1.6 + j0.8 + (j0.2) \parallel (9.6 + j2.8)\\ &= 13.6 + j0.8 + \frac{j0.2(9.6 + j2.8)}{9.6 + j3}\\ &= 13.6 + j1 = 13.64\underline{/4.204^\circ}\ \Omega\end{aligned}$$

The desired current is

$$\mathbf{I} = \frac{\mathbf{V}}{\mathbf{Z}} = \frac{50\underline{/0^\circ}}{13.64\underline{/4.204^\circ}} = 3.666\underline{/-4.204^\circ}\ \text{A}$$

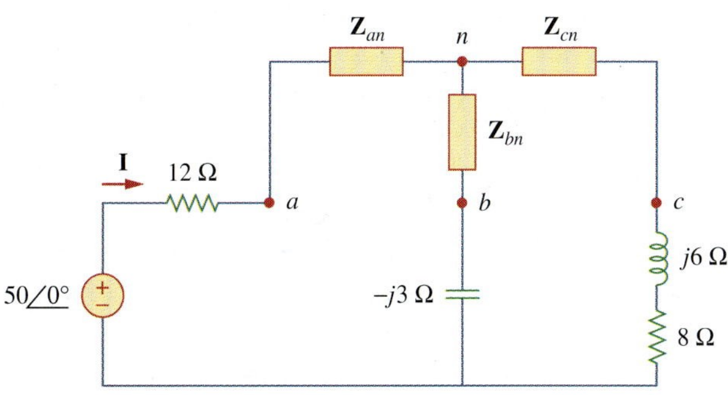

Figure 9.29
The circuit in Fig. 9.28 after delta-to-wye transformation.

Practice Problem 9.12

Find **I** in the circuit of Fig. 9.30.

Answer: $6.364\angle 3.8^\circ$ A.

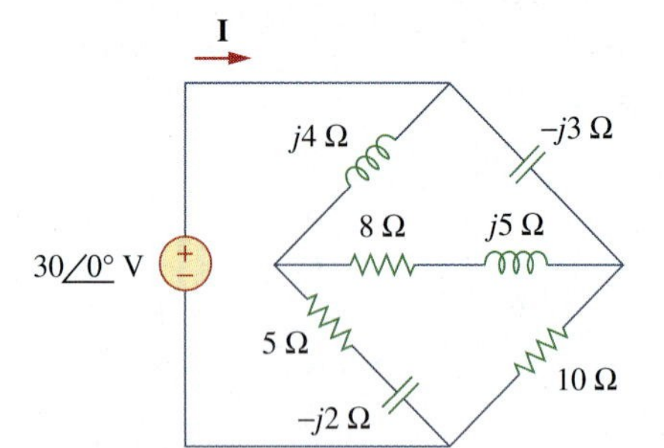

Figure 9.30
For Practice Prob. 9.12.

9.8 †Applications

In Chapters 7 and 8, we saw certain uses of *RC*, *RL*, and *RLC* circuits in dc applications. These circuits also have ac applications; among them are coupling circuits, phase-shifting circuits, filters, resonant circuits, ac bridge circuits, and transformers. This list of applications is inexhaustive. We will consider some of them later. It will suffice here to observe two simple ones: *RC* phase-shifting circuits, and ac bridge circuits.

9.8.1 Phase-Shifters

A phase-shifting circuit is often employed to correct an undesirable phase shift already present in a circuit or to produce special desired effects. An *RC* circuit is suitable for this purpose because its capacitor causes the circuit current to lead the applied voltage. Two commonly used *RC* circuits are shown in Fig. 9.31. (*RL* circuits or any reactive circuits could also serve the same purpose.)

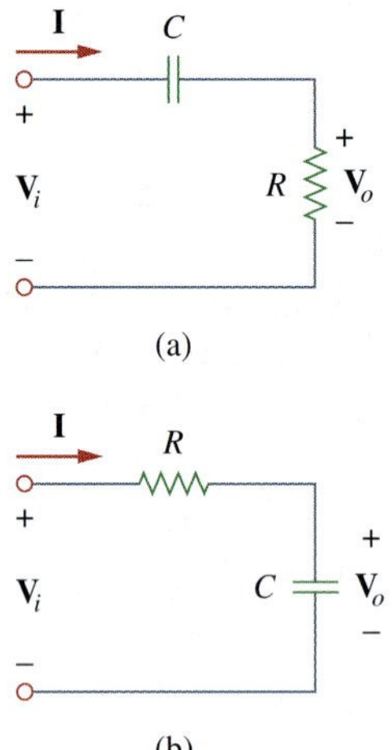

Figure 9.31
Series *RC* shift circuits: (a) leading output, (b) lagging output.

In Fig. 9.31(a), the circuit current **I** leads the applied voltage $\mathbf{V}_i$ by some phase angle θ, where $0 < \theta < 90^\circ$, depending on the values of *R* and *C*. If $X_C = -1/\omega C$, then the total impedance is $\mathbf{Z} = R + jX_C$, and the phase shift is given by

$$\theta = \tan^{-1}\frac{X_C}{R} \tag{9.70}$$

This shows that the amount of phase shift depends on the values of *R*, *C*, and the operating frequency. Since the output voltage $\mathbf{V}_o$ across the resistor is in phase with the current, $\mathbf{V}_o$ leads (positive phase shift) $\mathbf{V}_i$ as shown in Fig. 9.32(a).

In Fig. 9.31(b), the output is taken across the capacitor. The current **I** leads the input voltage $\mathbf{V}_i$ by θ, but the output voltage $v_o(t)$ across the capacitor lags (negative phase shift) the input voltage $v_i(t)$ as illustrated in Fig. 9.32(b).

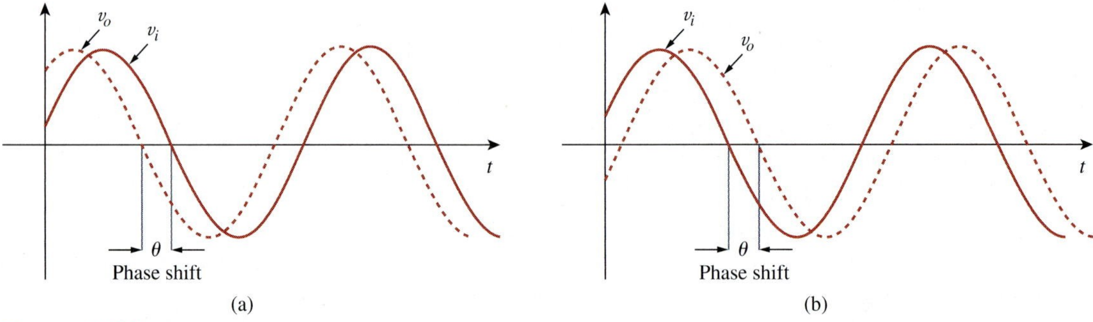

Figure 9.32
Phase shift in *RC* circuits: (a) leading output, (b) lagging output.

We should keep in mind that the simple *RC* circuits in Fig. 9.31 also act as voltage dividers. Therefore, as the phase shift θ approaches 90°, the output voltage $\mathbf{V}_o$ approaches zero. For this reason, these simple *RC* circuits are used only when small amounts of phase shift are required. If it is desired to have phase shifts greater than 60°, simple *RC* networks are cascaded, thereby providing a total phase shift equal to the sum of the individual phase shifts. In practice, the phase shifts due to the stages are not equal, because the succeeding stages load down the earlier stages unless op amps are used to separate the stages.

Example 9.13

Design an *RC* circuit to provide a phase of 90° leading.

Solution:

If we select circuit components of equal ohmic value, say $R = |X_C| = 20\ \Omega$, at a particular frequency, according to Eq. (9.70), the phase shift is exactly 45°. By cascading two similar *RC* circuits in Fig. 9.31(a), we obtain the circuit in Fig. 9.33, providing a positive or leading phase shift of 90°, as we shall soon show. Using the series-parallel combination technique, **Z** in Fig. 9.33 is obtained as

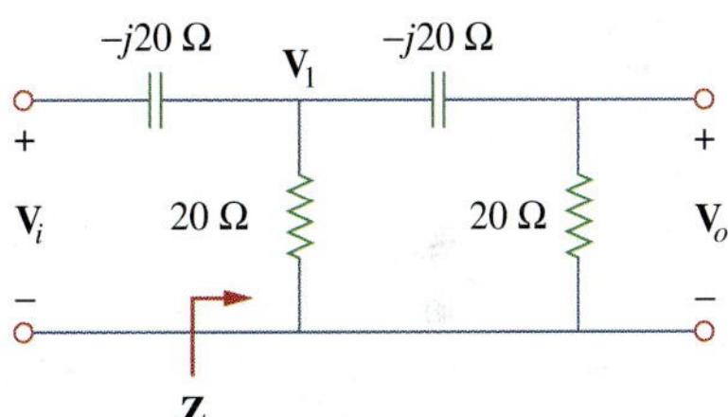

Figure 9.33
An *RC* phase shift circuit with 90° leading phase shift; for Example 9.13.

$$\mathbf{Z} = 20 \parallel (20 - j20) = \frac{20(20 - j20)}{40 - j20} = 12 - j4\ \Omega \qquad \textbf{(9.13.1)}$$

Using voltage division,

$$\mathbf{V}_1 = \frac{\mathbf{Z}}{\mathbf{Z} - j20}\mathbf{V}_i = \frac{12 - j4}{12 - j24}\mathbf{V}_i = \frac{\sqrt{2}}{3}\underline{/45^\circ}\ \mathbf{V}_i \qquad \textbf{(9.13.2)}$$

and

$$\mathbf{V}_o = \frac{20}{20 - j20}\mathbf{V}_1 = \frac{\sqrt{2}}{2}\underline{/45^\circ}\ \mathbf{V}_1 \qquad \textbf{(9.13.3)}$$

Substituting Eq. (9.13.2) into Eq. (9.13.3) yields

$$\mathbf{V}_o = \left(\frac{\sqrt{2}}{2}\underline{/45^\circ}\right)\left(\frac{\sqrt{2}}{3}\underline{/45^\circ}\ \mathbf{V}_i\right) = \frac{1}{3}\underline{/90^\circ}\ \mathbf{V}_i$$

Thus, the output leads the input by 90° but its magnitude is only about 33 percent of the input.

Practice Problem 9.13

Design an *RC* circuit to provide a 90° lagging phase shift of the output voltage relative to the input voltage. If an ac voltage of 10 V rms is applied, what is the output voltage?

Answer: Figure 9.34 shows a typical design; 3.33 V rms.

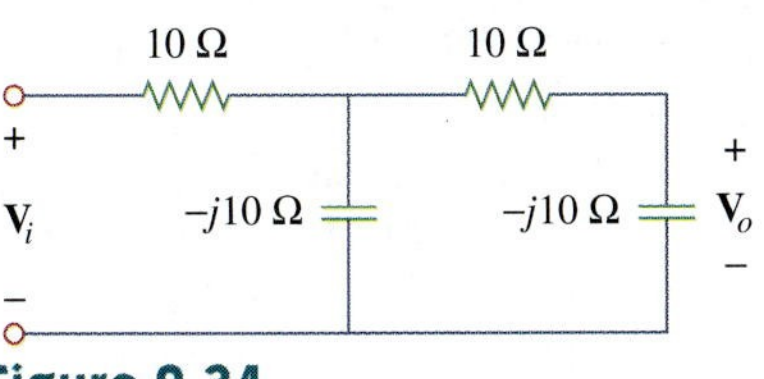

Figure 9.34
For Practice Prob. 9.13.

Example 9.14

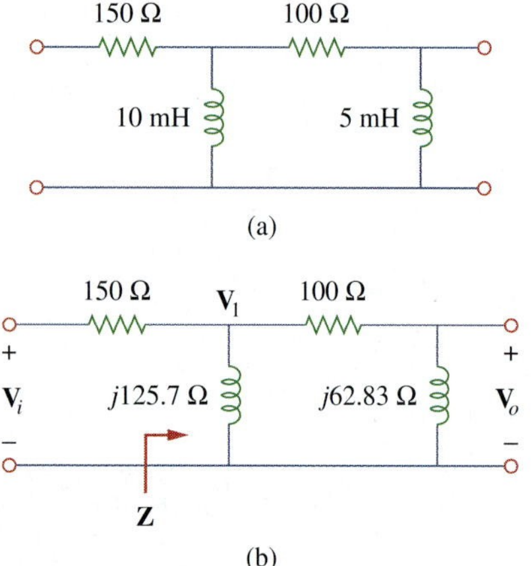

Figure 9.35
For Example 9.14.

For the *RL* circuit shown in Fig. 9.35(a), calculate the amount of phase shift produced at 2 kHz.

Solution:

At 2 kHz, we transform the 10-mH and 5-mH inductances to the corresponding impedances.

$$\begin{aligned} 10\text{ mH} \quad &\Rightarrow \quad X_L = \omega L = 2\pi \times 2 \times 10^3 \times 10 \times 10^{-3} \\ &\qquad\qquad = 40\pi = 125.7\ \Omega \\ 5\text{ mH} \quad &\Rightarrow \quad X_L = \omega L = 2\pi \times 2 \times 10^3 \times 5 \times 10^{-3} \\ &\qquad\qquad = 20\pi = 62.83\ \Omega \end{aligned}$$

Consider the circuit in Fig. 9.35(b). The impedance $\mathbf{Z}$ is the parallel combination of $j125.7\ \Omega$ and $100 + j62.83\ \Omega$. Hence,

$$\begin{aligned} \mathbf{Z} &= j125.7 \parallel (100 + j62.83) \\ &= \frac{j125.7(100 + j62.83)}{100 + j188.5} = 69.56\underline{/60.1^\circ}\ \Omega \end{aligned} \tag{9.14.1}$$

Using voltage division,

$$\begin{aligned} \mathbf{V}_1 &= \frac{\mathbf{Z}}{\mathbf{Z} + 150}\mathbf{V}_i = \frac{69.56\underline{/60.1^\circ}}{184.7 + j60.3}\mathbf{V}_i \\ &= 0.3582\underline{/42.02^\circ}\ \mathbf{V}_i \end{aligned} \tag{9.14.2}$$

and

$$\mathbf{V}_o = \frac{j62.832}{100 + j62.832}\mathbf{V}_1 = 0.532\underline{/57.86^\circ}\ \mathbf{V}_1 \tag{9.14.3}$$

Combining Eqs. (9.14.2) and (9.14.3),

$$\mathbf{V}_o = (0.532\underline{/57.86^\circ})(0.3582\underline{/42.02^\circ})\ \mathbf{V}_i = 0.1906\underline{/100^\circ}\ \mathbf{V}_i$$

showing that the output is about 19 percent of the input in magnitude but leading the input by 100°. If the circuit is terminated by a load, the load will affect the phase shift.

Practice Problem 9.14

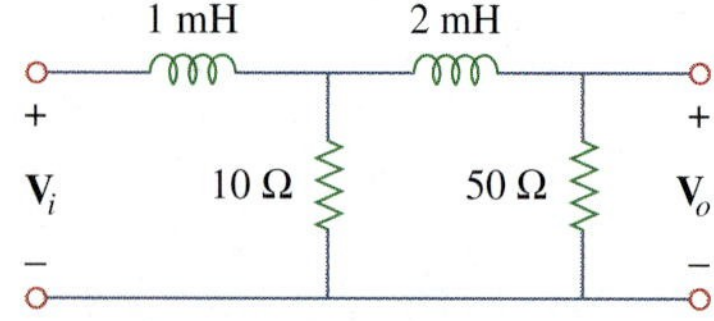

Figure 9.36
For Practice Prob. 9.14.

Refer to the *RL* circuit in Fig. 9.36. If 1 V is applied, find the magnitude and the phase shift produced at 5 kHz. Specify whether the phase shift is leading or lagging.

Answer: 0.172, 120.4°, lagging.

9.8.2 AC Bridges

An ac bridge circuit is used in measuring the inductance L of an inductor or the capacitance C of a capacitor. It is similar in form to the Wheatstone bridge for measuring an unknown resistance (discussed in Section 4.10) and follows the same principle. To measure L and C, however, an ac source is needed as well as an ac meter instead of the galvanometer. The ac meter may be a sensitive ac ammeter or voltmeter.

Consider the general ac bridge circuit displayed in Fig. 9.37. The bridge is *balanced* when no current flows through the meter. This means that $\mathbf{V}_1 = \mathbf{V}_2$. Applying the voltage division principle,

$$\mathbf{V}_1 = \frac{\mathbf{Z}_2}{\mathbf{Z}_1 + \mathbf{Z}_2}\mathbf{V}_s = \mathbf{V}_2 = \frac{\mathbf{Z}_x}{\mathbf{Z}_3 + \mathbf{Z}_x}\mathbf{V}_s \tag{9.71}$$

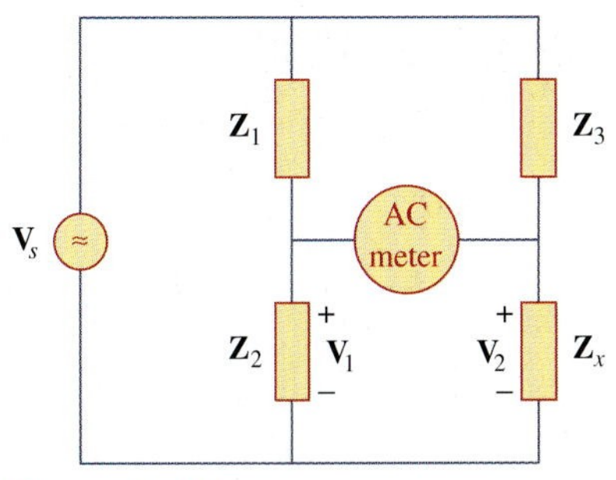

Figure 9.37
A general ac bridge.

Thus,

$$\frac{\mathbf{Z}_2}{\mathbf{Z}_1 + \mathbf{Z}_2} = \frac{\mathbf{Z}_x}{\mathbf{Z}_3 + \mathbf{Z}_x} \quad \Rightarrow \quad \mathbf{Z}_2\mathbf{Z}_3 = \mathbf{Z}_1\mathbf{Z}_x \tag{9.72}$$

or

$$\boxed{\mathbf{Z}_x = \frac{\mathbf{Z}_3}{\mathbf{Z}_1}\mathbf{Z}_2} \tag{9.73}$$

This is the balanced equation for the ac bridge and is similar to Eq. (4.30) for the resistance bridge except that the R's are replaced by $\mathbf{Z}$'s.

Specific ac bridges for measuring L and C are shown in Fig. 9.38, where L_x and C_x are the unknown inductance and capacitance to be measured while L_s and C_s are a standard inductance and capacitance (the values of which are known to great precision). In each case, two resistors, R_1 and R_2, are varied until the ac meter reads zero. Then the bridge is balanced. From Eq. (9.73), we obtain

$$L_x = \frac{R_2}{R_1}L_s \tag{9.74}$$

and

$$C_x = \frac{R_1}{R_2}C_s \tag{9.75}$$

Notice that the balancing of the ac bridges in Fig. 9.38 does not depend on the frequency f of the ac source, since f does not appear in the relationships in Eqs. (9.74) and (9.75).

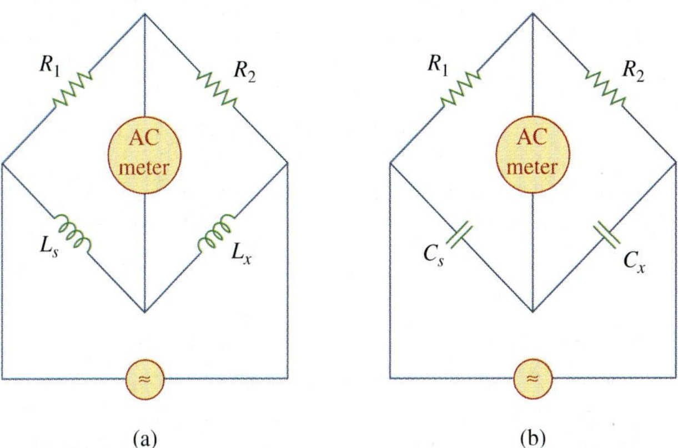

Figure 9.38
Specific ac bridges: (a) for measuring L, (b) for measuring C.

Example 9.15

The ac bridge circuit of Fig. 9.37 balances when $\mathbf{Z}_1$ is a 1-kΩ resistor, $\mathbf{Z}_2$ is a 4.2-kΩ resistor, $\mathbf{Z}_3$ is a parallel combination of a 1.5-MΩ resistor and a 12-pF capacitor, and $f = 2$ kHz. Find: (a) the series components that make up $\mathbf{Z}_x$, and (b) the parallel components that make up $\mathbf{Z}_x$.

Solution:

1. **Define.** The problem is clearly stated.
2. **Present.** We are to determine the unknown components subject to the fact that they balance the given quantities. Since a parallel and series equivalent exists for this circuit, we need to find both.
3. **Alternative.** Although there are alternative techniques that can be used to find the unknown values, a straightforward equality works best. Once we have answers, we can check them by using hand techniques such as nodal analysis or just using *PSpice*.
4. **Attempt.** From Eq. (9.73),

$$\mathbf{Z}_x = \frac{\mathbf{Z}_3}{\mathbf{Z}_1}\mathbf{Z}_2 \tag{9.15.1}$$

where $\mathbf{Z}_x = R_x + jX_x$,

$$\mathbf{Z}_1 = 1000\ \Omega, \qquad \mathbf{Z}_2 = 4200\ \Omega \tag{9.15.2}$$

and

$$\mathbf{Z}_3 = R_3 \parallel \frac{1}{j\omega C_3} = \frac{\dfrac{R_3}{j\omega C_3}}{R_3 + 1/j\omega C_3} = \frac{R_3}{1 + j\omega R_3 C_3}$$

Since $R_3 = 1.5$ MΩ and $C_3 = 12$ pF,

$$\mathbf{Z}_3 = \frac{1.5 \times 10^6}{1 + j2\pi \times 2 \times 10^3 \times 1.5 \times 10^6 \times 12 \times 10^{-12}} = \frac{1.5 \times 10^6}{1 + j0.2262}$$

or

$$\mathbf{Z}_3 = 1.427 - j0.3228\ \text{M}\Omega \tag{9.15.3}$$

(a) Assuming that $\mathbf{Z}_x$ is made up of series components, we substitute Eqs. (9.15.2) and (9.15.3) in Eq. (9.15.1) and obtain

$$\begin{aligned} R_x + jX_x &= \frac{4200}{1000}(1.427 - j0.3228) \times 10^6 \\ &= (5.993 - j1.356)\ \text{M}\Omega \end{aligned} \tag{9.15.4}$$

Equating the real and imaginary parts yields $R_x = 5.993$ MΩ and a capacitive reactance

$$X_x = \frac{1}{\omega C} = 1.356 \times 10^6$$

or

$$C = \frac{1}{\omega X_x} = \frac{1}{2\pi \times 2 \times 10^3 \times 1.356 \times 10^6} = 58.69\ \text{pF}$$

(b) $\mathbf{Z}_x$ remains the same as in Eq. (9.15.4) but R_x and X_x are in parallel. Assuming an *RC* parallel combination,

$$\mathbf{Z}_x = (5.993 - j1.356)\ \text{M}\Omega$$
$$= R_x \parallel \frac{1}{j\omega C_x} = \frac{R_x}{1 + j\omega R_x C_x}$$

By equating the real and imaginary parts, we obtain

$$R_x = \frac{\text{Real}(\mathbf{Z}_x)^2 + \text{Imag}(\mathbf{Z}_x)^2}{\text{Real}(\mathbf{Z}_x)} = \frac{5.993^2 + 1.356^2}{5.993} = \mathbf{6.3\ M\Omega}$$

$$C_x = -\frac{\text{Imag}(\mathbf{Z}_x)}{\omega[\text{Real}(\mathbf{Z}_x)^2 + \text{Imag}(\mathbf{Z}_x)^2]}$$
$$= -\frac{-1.356}{2\pi(2000)(5.917^2 + 1.356^2)} = \mathbf{2.852\ \mu F}$$

We have assumed a parallel *RC* combination which works in this case.

5. **Evaluate.** Let us now use *PSpice* to see if we indeed have the correct equalities. Running *PSpice* with the equivalent circuits, an open circuit between the "bridge" portion of the circuit, and a 10-volt input voltage yields the following voltages at the ends of the "bridge" relative to a reference at the bottom of the circuit:

```
FREQ        VM($N_0002)  VP($N_0002)
2.000E+03   9.993E+00   -8.634E-03
2.000E+03   9.993E+00   -8.637E-03
```

Since the voltages are essentially the same, then no measurable current can flow through the "bridge" portion of the circuit for any element that connects the two points together and we have a balanced bridge, which is to be expected. This indicates we have properly determined the unknowns.

There is a very important problem with what we have done! Do you know what that is? We have what can be called an ideal, "theoretical" answer, but one that really is not very good in the real world. The difference between the magnitudes of the upper impedances and the lower impedances is much too large and would never be accepted in a real bridge circuit. For greatest accuracy, the overall magnitude of the impedances must at least be within the same relative order. To increase the accuracy of the solution of this problem, I would recommend increasing the magnitude of the top impedances to be in the range of 500 kΩ to 1.5 MΩ. One additional real-world comment: the size of these impedances also creates serious problems in making actual measurements, so the appropriate instruments must be used in order to minimize their loading (which would change the actual voltage readings) on the circuit.

6. **Satisfactory?** Since we solved for the unknown terms and then tested to see if they woked, we validated the results. They can now be presented as a solution to the problem.

Practice Problem 9.15

In the ac bridge circuit of Fig. 9.37, suppose that balance is achieved when $\mathbf{Z}_1$ is a 4.8-kΩ resistor, $\mathbf{Z}_2$ is a 10-Ω resistor in series with a 0.25-μH inductor, $\mathbf{Z}_3$ is a 12-kΩ resistor, and $f = 6$ MHz. Determine the series components that make up $\mathbf{Z}_x$.

Answer: A 25-Ω resistor in series with a 0.625-μH inductor.

9.9 Summary

1. A sinusoid is a signal in the form of the sine or cosine function. It has the general form

$$v(t) = V_m \cos(\omega t + \phi)$$

where V_m is the amplitude, $\omega = 2\pi f$ is the angular frequency, $(\omega t + \phi)$ is the argument, and ϕ is the phase.

2. A phasor is a complex quantity that represents both the magnitude and the phase of a sinusoid. Given the sinusoid $v(t) = V_m \cos(\omega t + \phi)$, its phasor $\mathbf{V}$ is

$$\mathbf{V} = V_m \underline{/\phi}$$

3. In ac circuits, voltage and current phasors always have a fixed relation to one another at any moment of time. If $v(t) = V_m \cos(\omega t + \phi_v)$ represents the voltage through an element and $i(t) = I_m \cos(\omega t + \phi_i)$ represents the current through the element, then $\phi_i = \phi_v$ if the element is a resistor, ϕ_i leads ϕ_v by 90° if the element is a capacitor, and ϕ_i lags ϕ_v by 90° if the element is an inductor.

4. The impedance $\mathbf{Z}$ of a circuit is the ratio of the phasor voltage across it to the phasor current through it:

$$\mathbf{Z} = \frac{\mathbf{V}}{\mathbf{I}} = R(\omega) + jX(\omega)$$

The admittance $\mathbf{Y}$ is the reciprocal of impedance:

$$\mathbf{Y} = \frac{1}{\mathbf{Z}} = G(\omega) + jB(\omega)$$

Impedances are combined in series or in parallel the same way as resistances in series or parallel; that is, impedances in series add while admittances in parallel add.

5. For a resistor $\mathbf{Z} = R$, for an inductor $\mathbf{Z} = jX = j\omega L$, and for a capacitor $\mathbf{Z} = -jX = 1/j\omega C$.

6. Basic circuit laws (Ohm's and Kirchhoff's) apply to ac circuits in the same manner as they do for dc circuits; that is,

$$\mathbf{V} = \mathbf{ZI}$$
$$\Sigma\, \mathbf{I}_k = 0 \quad \text{(KCL)}$$
$$\Sigma\, \mathbf{V}_k = 0 \quad \text{(KVL)}$$

7. The techniques of voltage/current division, series/parallel combination of impedance/admittance, circuit reduction, and Y-Δ transformation all apply to ac circuit analysis.
8. AC circuits are applied in phase-shifters and bridges.

Review Questions

9.1 Which of the following is *not* a right way to express the sinusoid $A \cos \omega t$?

(a) $A \cos 2\pi ft$ (b) $A \cos(2\pi t/T)$
(c) $A \cos\omega(t - T)$ (d) $A \sin(\omega t - 90°)$

9.2 A function that repeats itself after fixed intervals is said to be:

(a) a phasor (b) harmonic
(c) periodic (d) reactive

9.3 Which of these frequencies has the shorter period?

(a) 1 krad/s (b) 1 kHz

9.4 If $v_1 = 30 \sin(\omega t + 10°)$ and $v_2 = 20 \sin(\omega t + 50°)$, which of these statements are true?

(a) v_1 leads v_2 (b) v_2 leads v_1
(c) v_2 lags v_1 (d) v_1 lags v_2
(e) v_1 and v_2 are in phase

9.5 The voltage across an inductor leads the current through it by 90°.

(a) True (b) False

9.6 The imaginary part of impedance is called:

(a) resistance (b) admittance
(c) susceptance (d) conductance
(e) reactance

9.7 The impedance of a capacitor increases with increasing frequency.

(a) True (b) False

9.8 At what frequency will the output voltage $v_o(t)$ in Fig. 9.39 be equal to the input voltage $v(t)$?

(a) 0 rad/s (b) 1 rad/s (c) 4 rad/s
(d) ∞ rad/s (e) none of the above

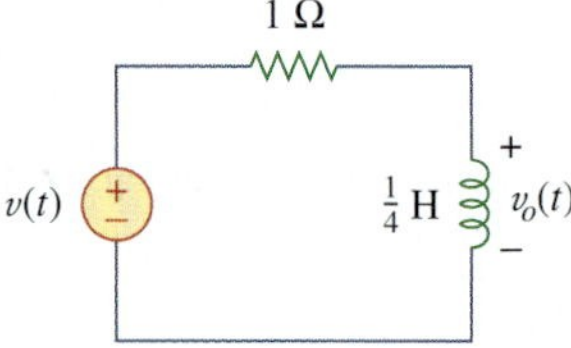

Figure 9.39
For Review Question 9.8.

9.9 A series RC circuit has $|V_R| = 12$ V and $|V_C| = 5$ V. The magnitude of the supply voltage is:

(a) -7 V (b) 7 V (c) 13 V (d) 17 V

9.10 A series RCL circuit has $R = 30\ \Omega$, $X_C = 50\ \Omega$, and $X_L = 90\ \Omega$. The impedance of the circuit is:

(a) $30 + j140\ \Omega$ (b) $30 + j40\ \Omega$
(c) $30 - j40\ \Omega$ (d) $-30 - j40\ \Omega$
(e) $-30 + j40\ \Omega$

Answers: 9.1d, 9.2c, 9.3b, 9.4b,d, 9.5a, 9.6e, 9.7b, 9.8d, 9.9c, 9.10b.

Problems

Section 9.2 Sinusoids

9.1 Given the sinusoidal voltage $v(t) = 50 \cos(30t + 10°)$ V, find: (a) the amplitude V_m, (b) the period T, (c) the frequency f, and (d) $v(t)$ at $t = 10$ ms.

9.2 A current source in a linear circuit has

$$i_s = 8 \cos(500\pi t - 25°) \text{ A}$$

(a) What is the amplitude of the current?
(b) What is the angular frequency?
(c) Find the frequency of the current.
(d) Calculate i_s at $t = 2$ ms.

9.3 Express the following functions in cosine form:

(a) $4 \sin(\omega t - 30°)$ (b) $-2 \sin 6t$
(c) $-10 \sin(\omega t + 20°)$

9.4 Design a problem to help other students better understand sinusoids.
ed

9.5 Given $v_1 = 20 \sin(\omega t + 60^\circ)$ and $v_2 = 60 \cos(\omega t - 10^\circ)$, determine the phase angle between the two sinusoids and which one lags the other.

9.6 For the following pairs of sinusoids, determine which one leads and by how much.

(a) $v(t) = 10 \cos(4t - 60^\circ)$ and $i(t) = 4 \sin(4t + 50^\circ)$

(b) $v_1(t) = 4 \cos(377t + 10^\circ)$ and $v_2(t) = -20 \cos 377t$

(c) $x(t) = 13 \cos 2t + 5 \sin 2t$ and $y(t) = 15 \cos(2t - 11.8^\circ)$

Section 9.3 Phasors

9.7 If $f(\phi) = \cos\phi + j \sin\phi$, show that $f(\phi) = e^{j\phi}$.

9.8 Calculate these complex numbers and express your results in rectangular form:

(a) $\dfrac{15\angle 45^\circ}{3 - j4} + j2$

(b) $\dfrac{8\angle -20^\circ}{(2 + j)(3 - j4)} + \dfrac{10}{-5 + j12}$

(c) $10 + (8\angle 50^\circ)(5 - j12)$

9.9 Evaluate the following complex numbers and leave your results in polar form:

(a) $5\angle 30^\circ \left(6 - j8 + \dfrac{3\angle 60^\circ}{2 + j}\right)$

(b) $\dfrac{(10\angle 60^\circ)(35\angle -50^\circ)}{(2 + j6) - (5 + j)}$

9.10 Design a problem to help other students better understand phasors.
ed

9.11 Find the phasors corresponding to the following signals:

(a) $v(t) = 21 \cos(4t - 15^\circ)$ V

(b) $i(t) = -8 \sin(10t + 70^\circ)$ mA

(c) $v(t) = 120 \sin(10t - 50^\circ)$ V

(d) $i(t) = -60 \cos(30t + 10^\circ)$ mA

9.12 Let $\mathbf{X} = 8\angle 40^\circ$ and $\mathbf{Y} = 10\angle -30^\circ$. Evaluate the following quantities and express your results in polar form:

(a) $(\mathbf{X} + \mathbf{Y})\mathbf{X}^*$

(b) $(\mathbf{X} - \mathbf{Y})^*$

(c) $(\mathbf{X} + \mathbf{Y})/\mathbf{X}$

9.13 Evaluate the following complex numbers:

(a) $\dfrac{2 + j3}{1 - j6} + \dfrac{7 - j8}{-5 + j11}$

(b) $\dfrac{(5\angle 10^\circ)(10\angle -40^\circ)}{(4\angle -80^\circ)(-6\angle 50^\circ)}$

(c) $\begin{vmatrix} 2 + j3 & -j2 \\ -j2 & 8 - j5 \end{vmatrix}$

9.14 Simplify the following expressions:

(a) $\dfrac{(5 - j6) - (2 + j8)}{(-3 + j4)(5 - j) + (4 - j6)}$

(b) $\dfrac{(240\angle 75^\circ + 160\angle -30^\circ)(60 - j80)}{(67 + j84)(20\angle 32^\circ)}$

(c) $\left(\dfrac{10 + j20}{3 + j4}\right)^2 \sqrt{(10 + j5)(16 - j20)}$

9.15 Evaluate these determinants:

(a) $\begin{vmatrix} 10 + j6 & 2 - j3 \\ -5 & -1 + j \end{vmatrix}$

(b) $\begin{vmatrix} 20\angle -30^\circ & -4\angle -10^\circ \\ 16\angle 0^\circ & 3\angle 45^\circ \end{vmatrix}$

(c) $\begin{vmatrix} 1 - j & -j & 0 \\ j & 1 & -j \\ 1 & j & 1 + j \end{vmatrix}$

9.16 Transform the following sinusoids to phasors:

(a) $-10 \cos(4t + 75^\circ)$ (b) $5 \sin(20t - 10^\circ)$

(c) $4 \cos 2t + 3 \sin 2t$

9.17 Two voltages v_1 and v_2 appear in series so that their sum is $v = v_1 + v_2$. If $v_1 = 10 \cos(50t - \pi/3)$ V and $v_2 = 12 \cos(50t + 30^\circ)$ V, find v.

9.18 Obtain the sinusoids corresponding to each of the following phasors:

(a) $\mathbf{V}_1 = 60\angle 15^\circ$ V, $\omega = 1$

(b) $\mathbf{V}_2 = 6 + j8$ V, $\omega = 40$

(c) $\mathbf{I}_1 = 2.8e^{-j\pi/3}$ A, $\omega = 377$

(d) $\mathbf{I}_2 = -0.5 - j1.2$ A, $\omega = 10^3$

9.19 Using phasors, find:

(a) $3 \cos(20t + 10^\circ) - 5 \cos(20t - 30^\circ)$

(b) $40 \sin 50t + 30 \cos(50t - 45^\circ)$

(c) $20 \sin 400t + 10 \cos(400t + 60^\circ) - 5 \sin(400t - 20^\circ)$

9.20 A linear network has a current input $4 \cos(\omega t + 20^\circ)$ A and a voltage output $10 \cos(\omega t + 110^\circ)$ V. Determine the associated impedance.

9.21 Simplify the following:

(a) $f(t) = 5\cos(2t + 15°) - 4\sin(2t - 30°)$

(b) $g(t) = 8\sin t + 4\cos(t + 50°)$

(c) $h(t) = \int_0^t (10\cos 40t + 50\sin 40t)\,dt$

9.22 An alternating voltage is given by $v(t) = 20\cos(5t - 30°)$ V. Use phasors to find

$$10v(t) + 4\frac{dv}{dt} - 2\int_{-\infty}^{t} v(t)\,dt$$

Assume that the value of the integral is zero at $t = -\infty$.

9.23 Apply phasor analysis to evaluate the following.

(a) $v = 50\cos(\omega t + 30°) + 30\cos(\omega t - 90°)$ V

(b) $i = 15\cos(\omega t + 45°) - 10\sin(\omega t + 45°)$ A

9.24 Find $v(t)$ in the following integrodifferential equations using the phasor approach:

(a) $v(t) + \int v\,dt = 5\cos(t + 45°)$ V

(b) $\frac{dv}{dt} + 5v(t) + 4\int v\,dt = 20\sin(4t + 10°)$ V

9.25 Using phasors, determine $i(t)$ in the following equations:

(a) $2\frac{di}{dt} + 3i(t) = 4\cos(2t - 45°)$

(b) $10\int i\,dt + \frac{di}{dt} + 6i(t) = 5\cos(5t + 22°)$ A

9.26 The loop equation for a series *RLC* circuit gives

$$\frac{di}{dt} + 2i + \int_{-\infty}^{t} i\,dt = \cos 2t \text{ A}$$

Assuming that the value of the integral at $t = -\infty$ is zero, find $i(t)$ using the phasor method.

9.27 A parallel *RLC* circuit has the node equation

$$\frac{dv}{dt} + 50v + 100\int v\,dt = 110\cos(377t - 10°) \text{ V}$$

Determine $v(t)$ using the phasor method. You may assume that the value of the integral at $t = -\infty$ is zero.

Section 9.4 Phasor Relationships for Circuit Elements

9.28 Determine the current that flows through an 8-Ω resistor connected to a voltage source $v_s = 110\cos 377t$ V.

9.29 What is the instantaneous voltage across a 2-μF capacitor when the current through it is $i = 4\sin(10^6 t + 25°)$ A?

9.30 A voltage $v(t) = 100\cos(60t + 20°)$ V is applied to a parallel combination of a 40-kΩ resistor and a 50-μF capacitor. Find the steady-state currents through the resistor and the capacitor.

9.31 A series *RLC* circuit has $R = 80\ \Omega$, $L = 240$ mH, and $C = 5$ mF. If the input voltage is $v(t) = 10\cos 2t$, find the currrent flowing through the circuit.

9.32 ed Using Fig. 9.40, design a problem to help other students better understand phasor relationships for circuit elements.

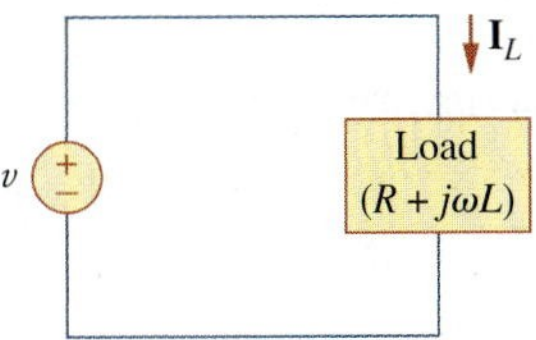

Figure 9.40
For Prob. 9.32.

9.33 A series *RL* circuit is connected to a 110-V ac source. If the voltage across the resistor is 85 V, find the voltage across the inductor.

9.34 What value of ω will cause the forced response v_o in Fig. 9.41 to be zero?

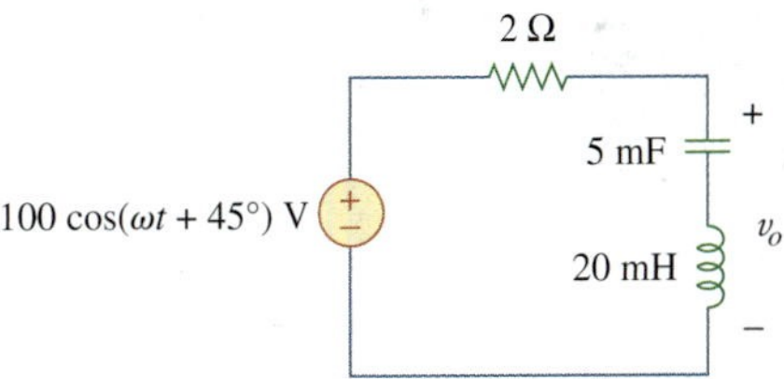

Figure 9.41
For Prob. 9.34.

Section 9.5 Impedance and Admittance

9.35 Find current i in the circuit of Fig. 9.42, when $v_s(t) = 50\cos 200t$ V.

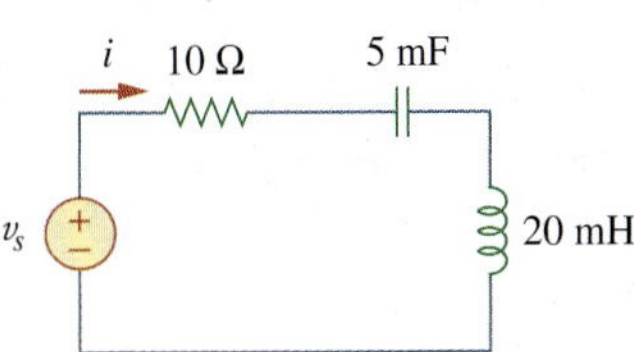

Figure 9.42
For Prob. 9.35.

9.36 Using Fig. 9.43, design a problem to help other students better understand impedance.

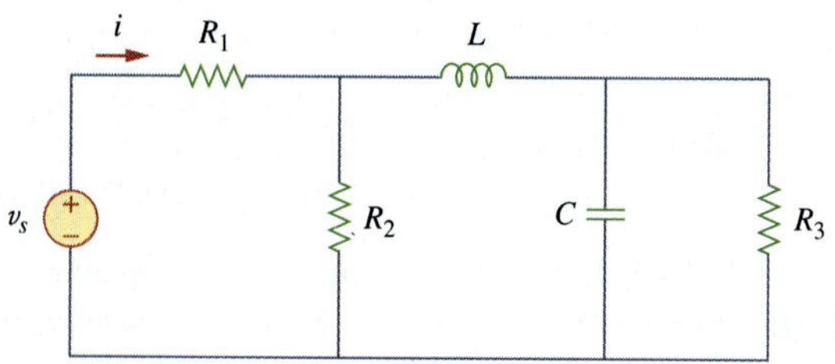

Figure 9.43
For Prob. 9.36.

9.37 Determine the admittance **Y** for the circuit in Fig. 9.44.

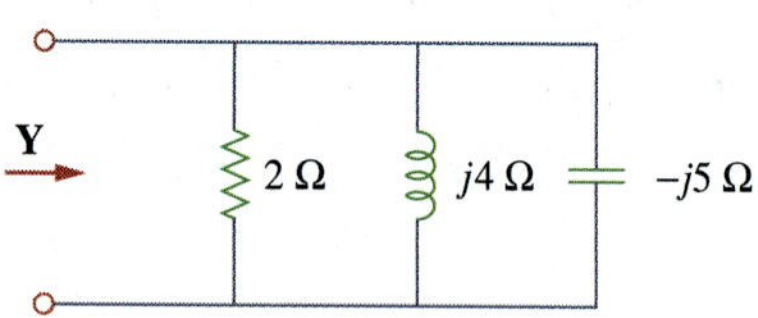

Figure 9.44
For Prob. 9.37.

9.38 Using Fig. 9.45, design a problem to help other students better understand admittance.

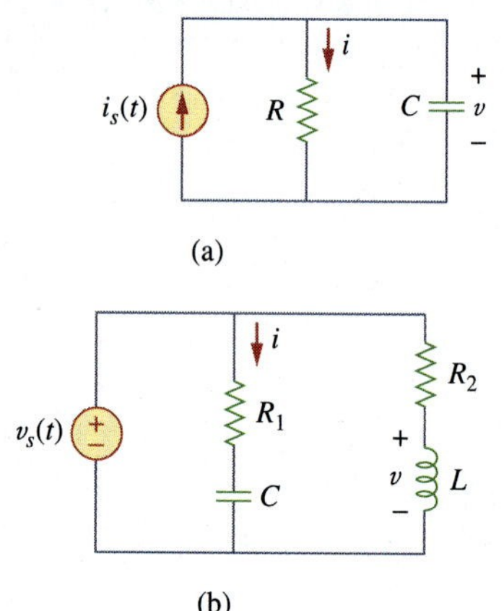

Figure 9.45
For Prob. 9.38.

9.39 For the circuit shown in Fig. 9.46, find Z_{eq} and use that to find current **I**. Let $\omega = 10$ rad/s.

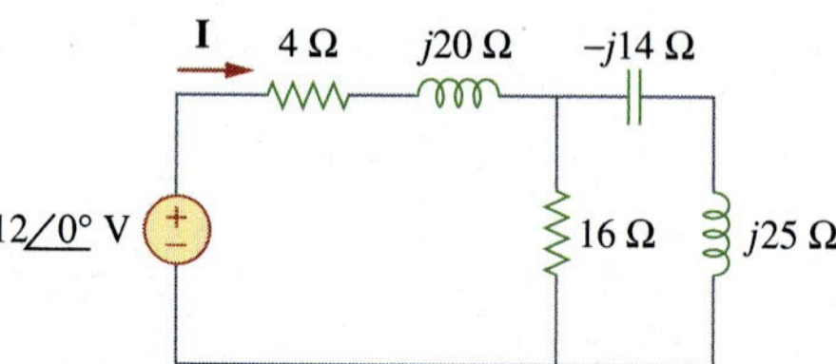

Figure 9.46
For Prob. 9.39.

9.40 In the circuit of Fig. 9.47, find i_o when:

(a) $\omega = 1$ rad/s (b) $\omega = 5$ rad/s
(c) $\omega = 10$ rad/s

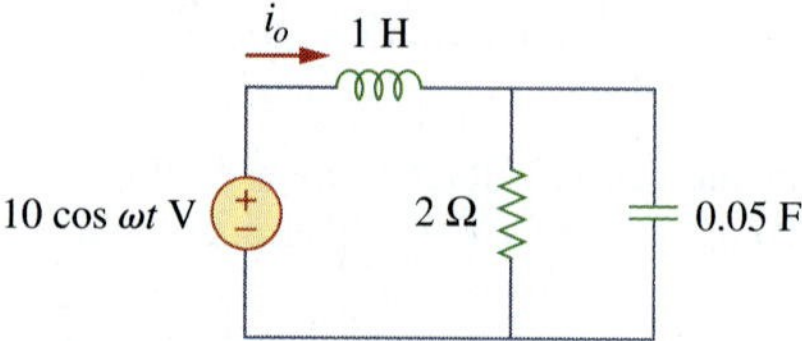

Figure 9.47
For Prob. 9.40.

9.41 Find $v(t)$ in the *RLC* circuit of Fig. 9.48.

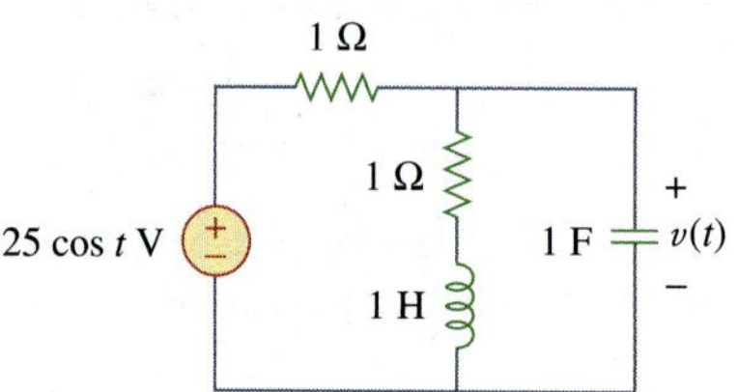

Figure 9.48
For Prob. 9.41.

9.42 Calculate $v_o(t)$ in the circuit of Fig. 9.49.

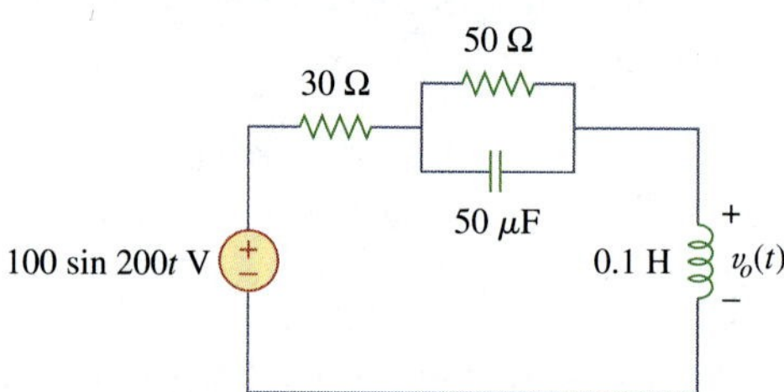

Figure 9.49
For Prob. 9.42.

9.43 Find current $\mathbf{I}_o$ in the circuit shown in Fig. 9.50.

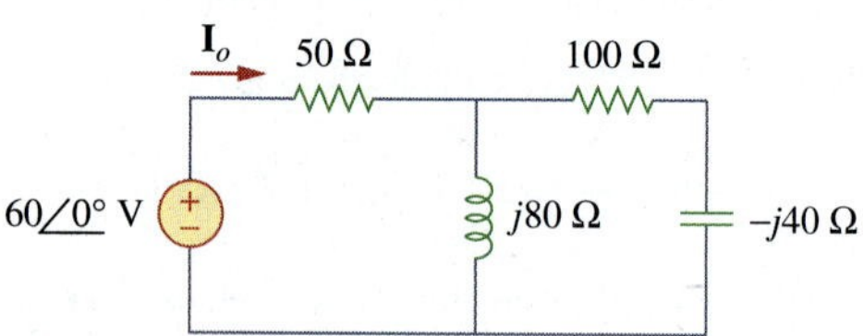

Figure 9.50
For Prob. 9.43.

9.44 Calculate $i(t)$ in the circuit of Fig. 9.51.

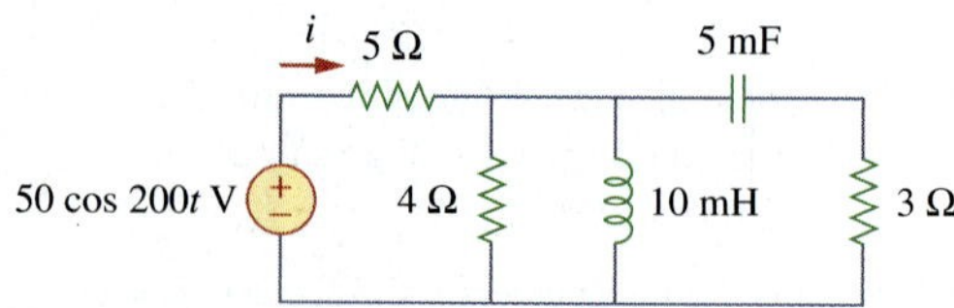

Figure 9.51
For prob. 9.44.

9.45 Find current $\mathbf{I}_o$ in the network of Fig. 9.52.

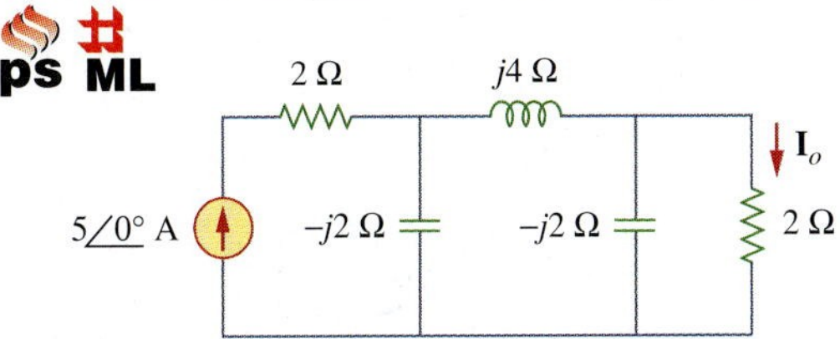

Figure 9.52
For Prob. 9.45.

9.46 If $i_s = 20 \cos(10t + 15°)$ A in the circuit of Fig. 9.53, find i_o.

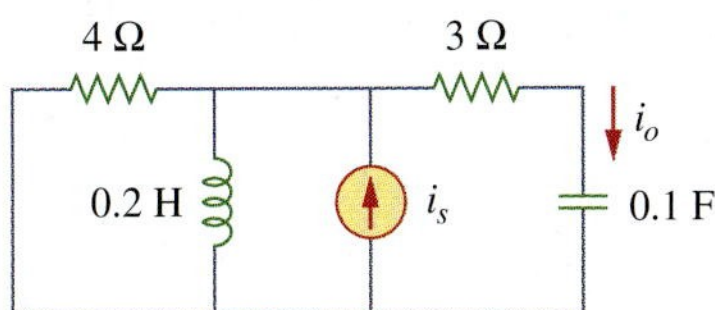

Figure 9.53
For Prob. 9.46.

9.47 In the circuit of Fig. 9.54, determine the value of $i_s(t)$.

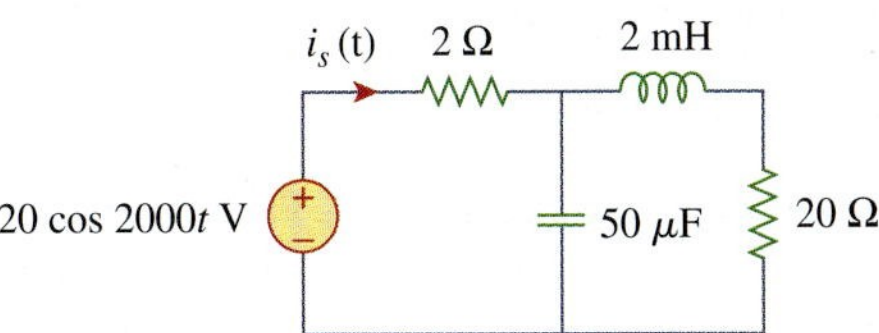

Figure 9.54
For Prob. 9.47.

9.48 Given that $v_s(t) = 20 \sin(100t - 40°)$ in Fig. 9.55, determine $i_x(t)$.

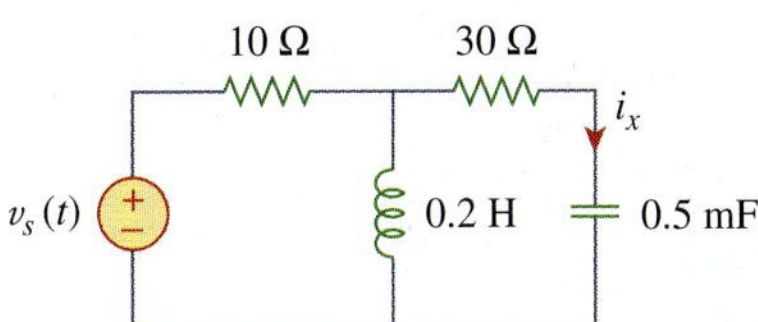

Figure 9.55
For Prob. 9.48.

9.49 Find $v_s(t)$ in the circuit of Fig. 9.56 if the current i_x through the 1-Ω resistor is $0.5 \sin 200t$ A.

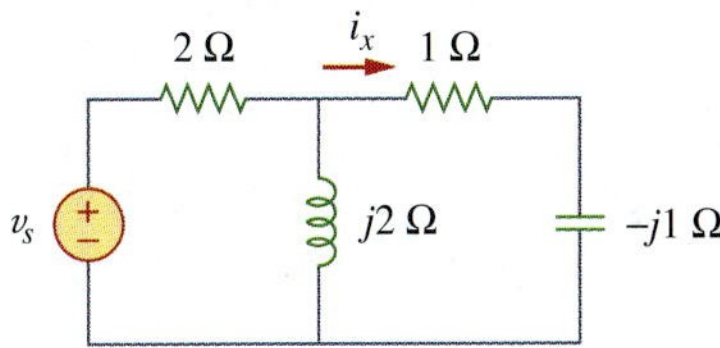

Figure 9.56
For Prob. 9.49.

9.50 Determine v_x in the circuit of Fig. 9.57. Let $i_s(t) = 5 \cos(100t + 40°)$ A.

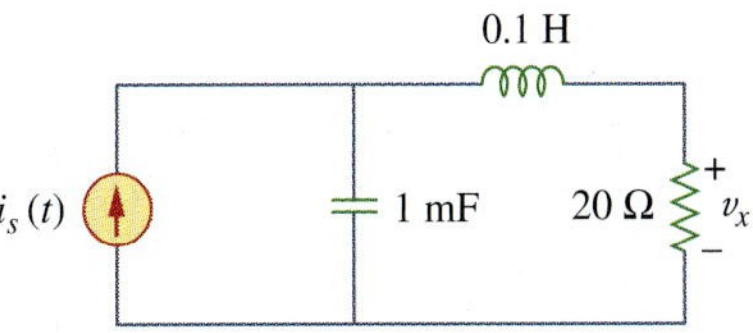

Figure 9.57
For Prob. 9.50.

9.51 If the voltage v_o across the 2-Ω resistor in the circuit of Fig. 9.58 is $-5 \cos 2t$ V, obtain i_s.

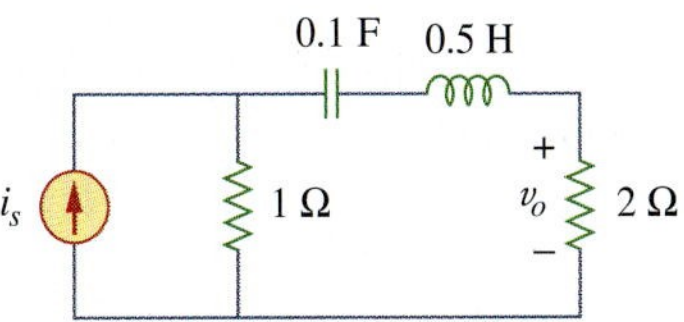

Figure 9.58
For Prob. 9.51.

9.52 If $\mathbf{V}_o = 20\underline{/45°}$ V in the circuit of Fig. 9.59, find $\mathbf{I}_s$.

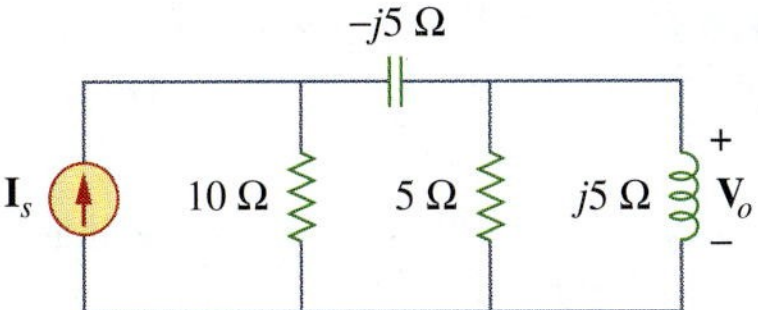

Figure 9.59
For Prob. 9.52.

9.53 Find $\mathbf{I}_o$ in the circuit of Fig. 9.60.

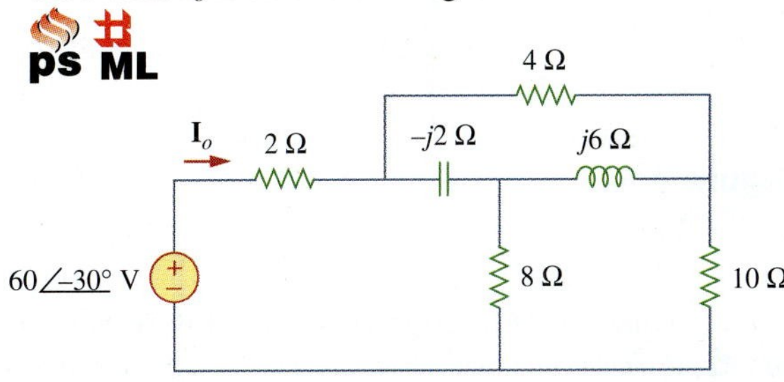

Figure 9.60
For Prob. 9.53.

9.54 In the circuit of Fig. 9.61, find $\mathbf{V}_s$ if $\mathbf{I}_o = 2\underline{/0°}$ A.

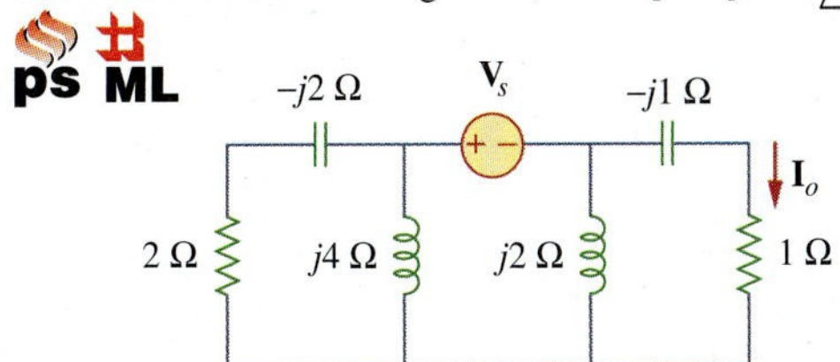

Figure 9.61
For Prob. 9.54.

***9.55** Find **Z** in the network of Fig. 9.62, given that $\mathbf{V}_o = 8\underline{/0°}$ V.

ML

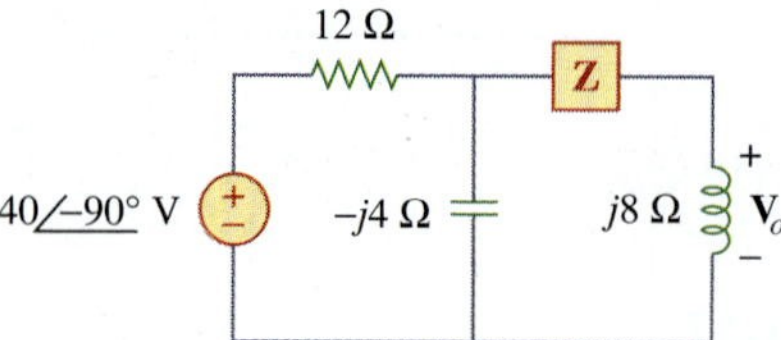

Figure 9.62
For Prob. 9.55.

Section 9.7 Impedance Combinations

9.56 At $\omega = 377$ rad/s, find the input impedance of the circuit shown in Fig. 9.63.

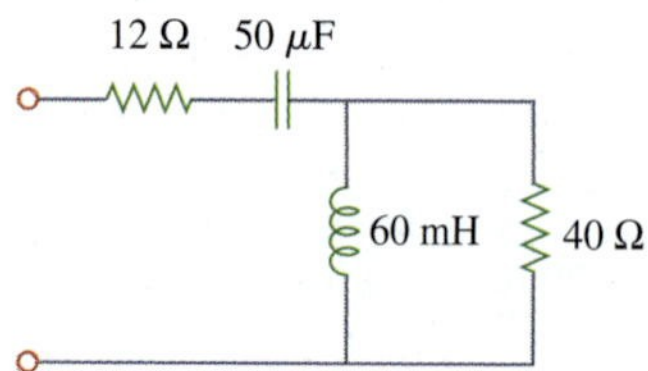

Figure 9.63
For Prob. 9.56.

9.57 At $\omega = 1$ rad/s, obtain the input admittance in the circuit of Fig. 9.64.

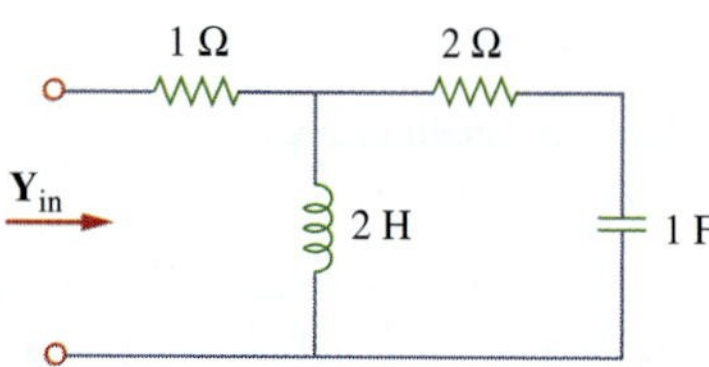

Figure 9.64
For Prob. 9.57.

9.58 Using Fig. 9.65, design a problem to help other students better understand impedance combinations.

e2d

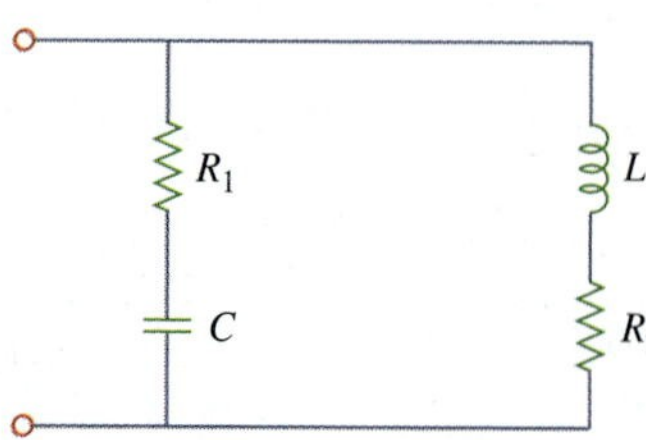

Figure 9.65
For Prob. 9.58.

9.59 For the network in Fig. 9.66, find $\mathbf{Z}_{in}$. Let $\omega = 10$ rad/s.

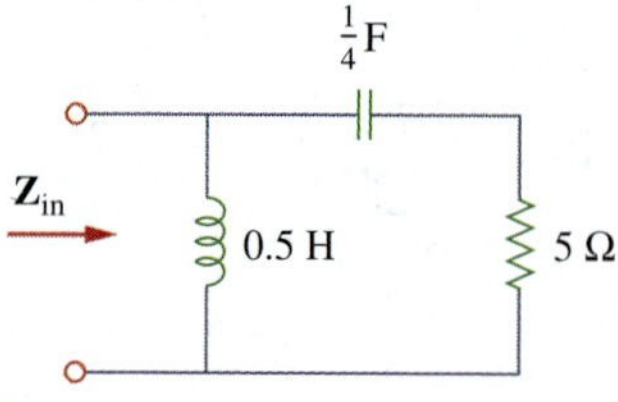

Figure 9.66
For Prob. 9.59.

9.60 Obtain $\mathbf{Z}_{in}$ for the circuit in Fig. 9.67.

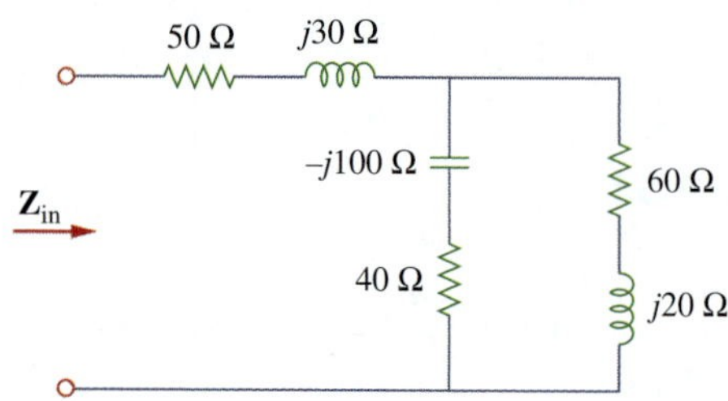

Figure 9.67
For Prob. 9.60.

9.61 Find $\mathbf{Z}_{eq}$ in the circuit of Fig. 9.68.

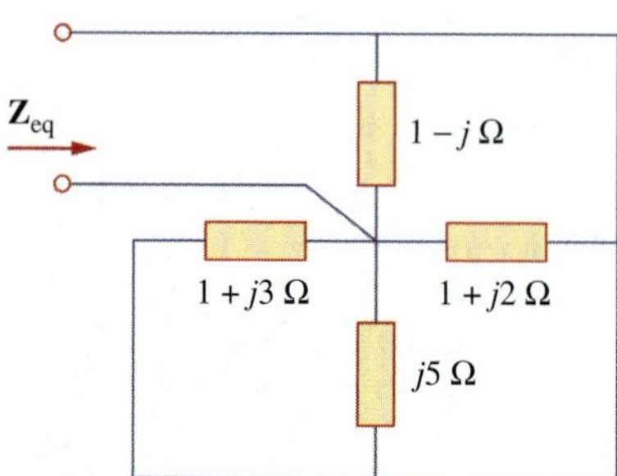

Figure 9.68
For Prob. 9.61.

9.62 For the circuit in Fig. 9.69, find the input impedance $\mathbf{Z}_{in}$ at 10 krad/s.

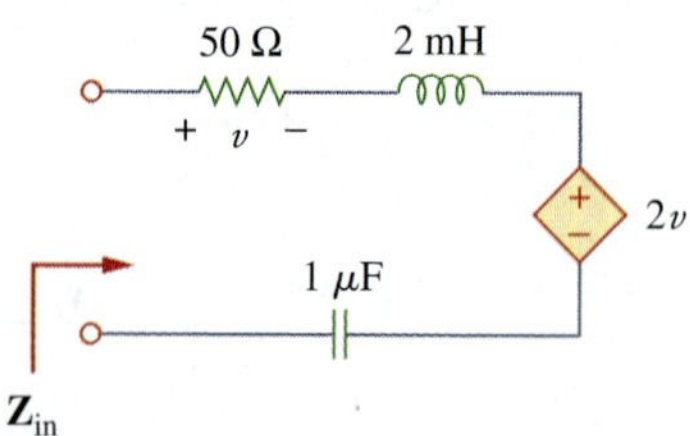

Figure 9.69
For Prob. 9.62.

* An asterisk indicates a challenging problem.

9.63 For the circuit in Fig. 9.70, find the value of $\mathbf{Z}_T$.

ML

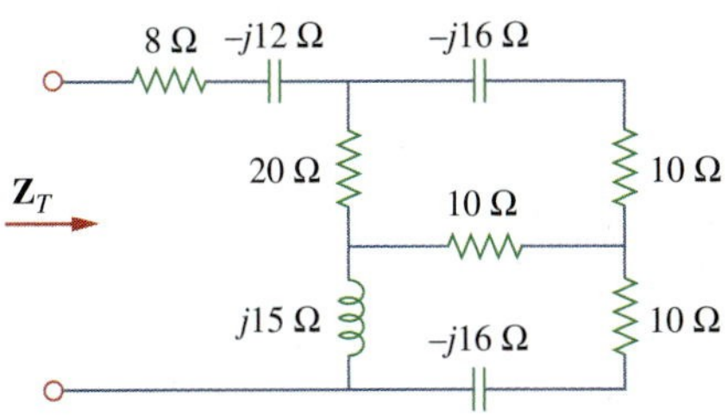

Figure 9.70
For Prob. 9.63.

9.64 Find $\mathbf{Z}_T$ and $\mathbf{I}$ in the circuit of Fig. 9.71.

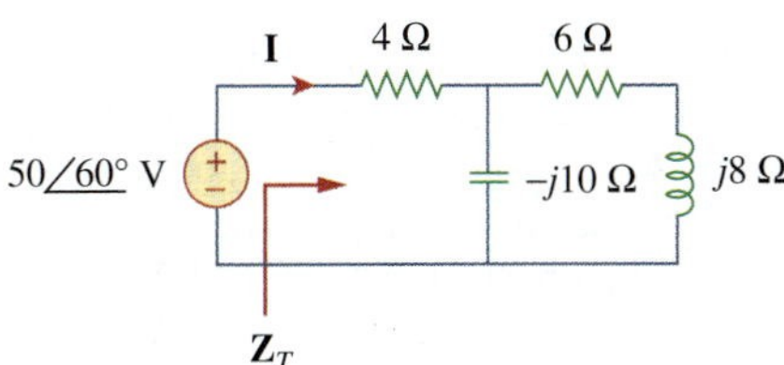

Figure 9.71
For Prob. 9.64.

9.65 Determine $\mathbf{Z}_T$ and $\mathbf{I}$ for the circuit in Fig. 9.72.

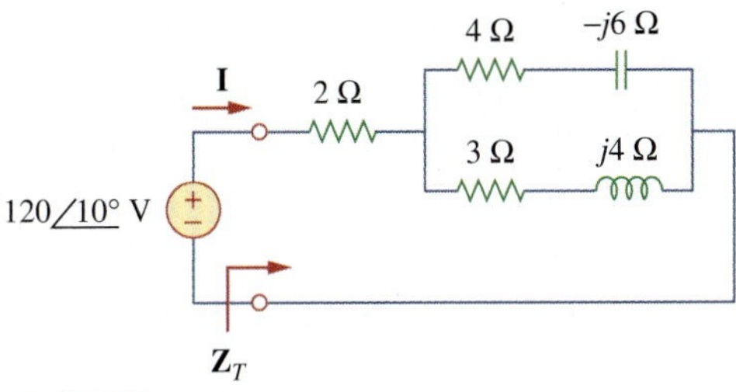

Figure 9.72
For Prob. 9.65.

9.66 For the circuit in Fig. 9.73, calculate $\mathbf{Z}_T$ and $\mathbf{V}_{ab}$.

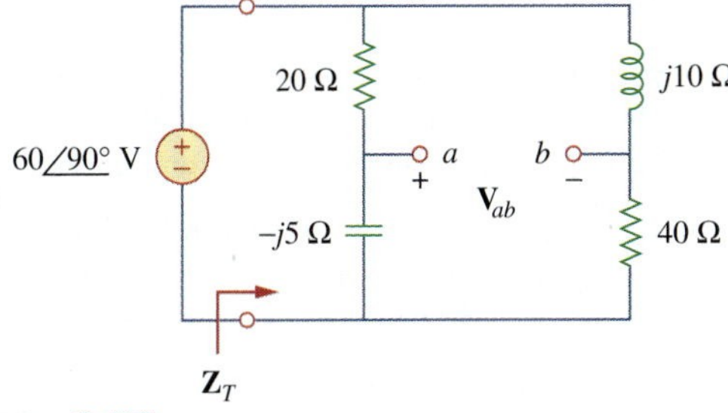

Figure 9.73
For Prob. 9.66.

9.67 At $\omega = 10^3$ rad/s, find the input admittance of each of the circuits in Fig. 9.74.

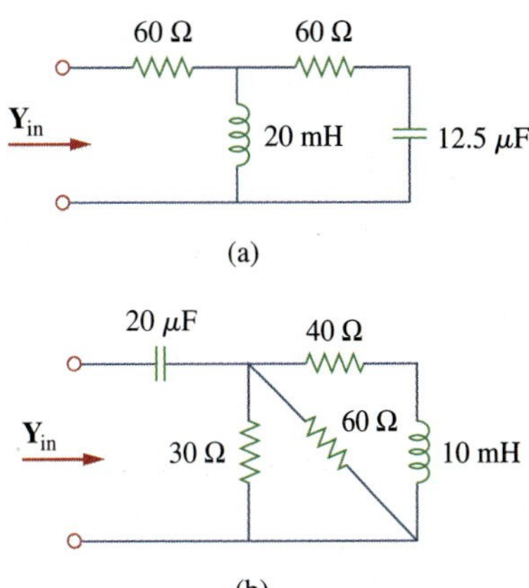

Figure 9.74
For Prob. 9.67.

9.68 Determine $\mathbf{Y}_{eq}$ for the circuit in Fig. 9.75.

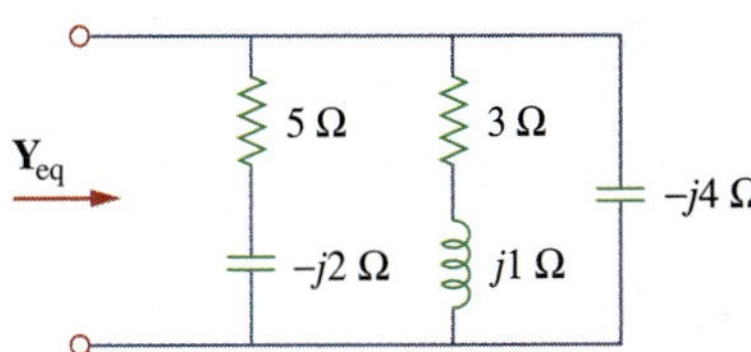

Figure 9.75
For Prob. 9.68.

9.69 Find the equivalent admittance $\mathbf{Y}_{eq}$ of the circuit in Fig. 9.76.

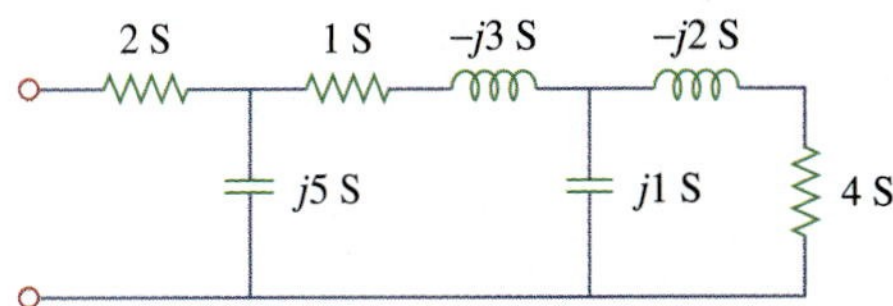

Figure 9.76
For Prob. 9.69.

9.70 Find the equivalent impedance of the circuit in Fig. 9.77.

ML

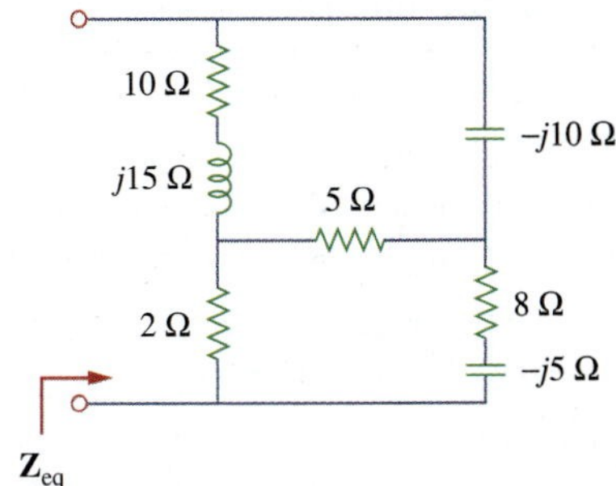

Figure 9.77
For Prob. 9.70.

9.71 Obtain the equivalent impedance of the circuit in Fig. 9.78.

ML

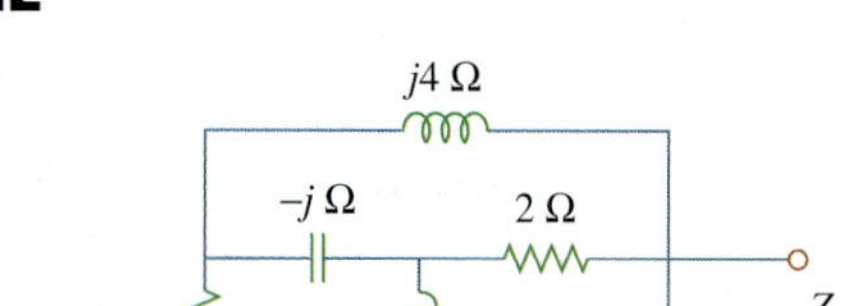

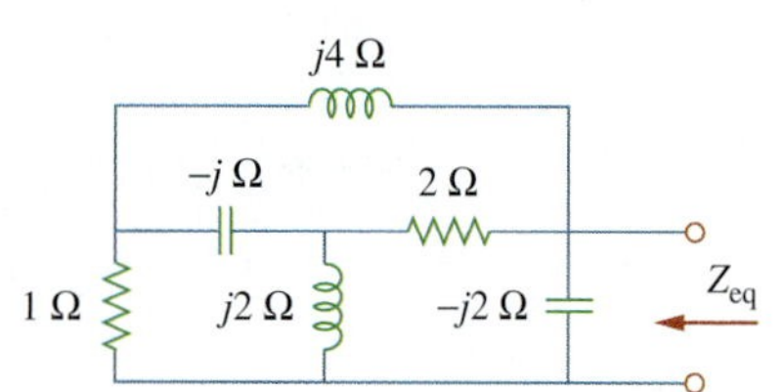

Figure 9.78
For Prob. 9.71.

9.72 Calculate the value of $\mathbf{Z}_{ab}$ in the network of Fig. 9.79.

ML

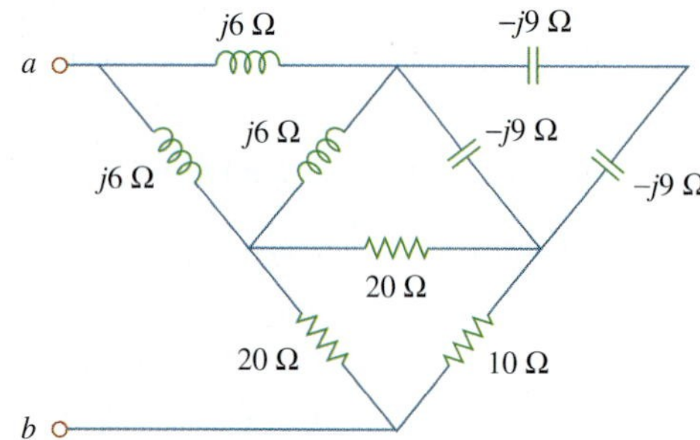

Figure 9.79
For Prob. 9.72.

9.73 Determine the equivalent impedance of the circuit in Fig. 9.80.

ML

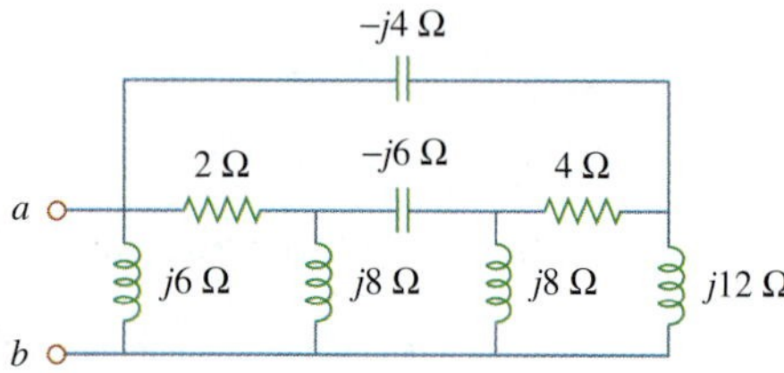

Figure 9.80
For Prob. 9.73.

Section 9.8 Applications

9.74 Design an *RL* circuit to provide a 90° leading phase shift.

9.75 Design a circuit that will transform a sinusoidal voltage input to a cosinusoidal voltage output.

9.76 For the following pairs of signals, determine if v_1 leads or lags v_2 and by how much.

(a) $v_1 = 10\cos(5t - 20°)$, $v_2 = 8\sin 5t$

(b) $v_1 = 19\cos(2t + 90°)$, $v_2 = 6\sin 2t$

(c) $v_1 = -4\cos 10t$, $v_2 = 15\sin 10t$

9.77 Refer to the *RC* circuit in Fig. 9.81.

(a) Calculate the phase shift at 2 MHz.

(b) Find the frequency where the phase shift is 45°.

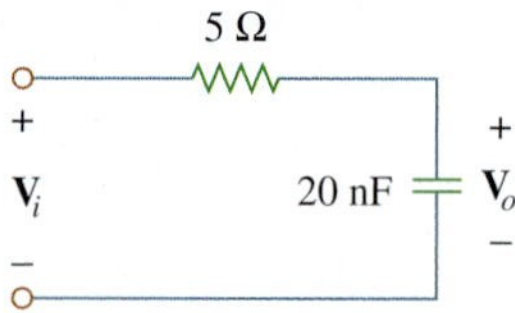

Figure 9.81
For Prob. 9.77.

9.78 A coil with impedance $8 + j6\ \Omega$ is connected in series with a capacitive reactance X. The series combination is connected in parallel with a resistor R. Given that the equivalent impedance of the resulting circuit is $5\angle 0°\ \Omega$, find the value of R and X.

9.79 (a) Calculate the phase shift of the circuit in Fig. 9.82.

(b) State whether the phase shift is leading or lagging (output with respect to input).

(c) Determine the magnitude of the output when the input is 120 V.

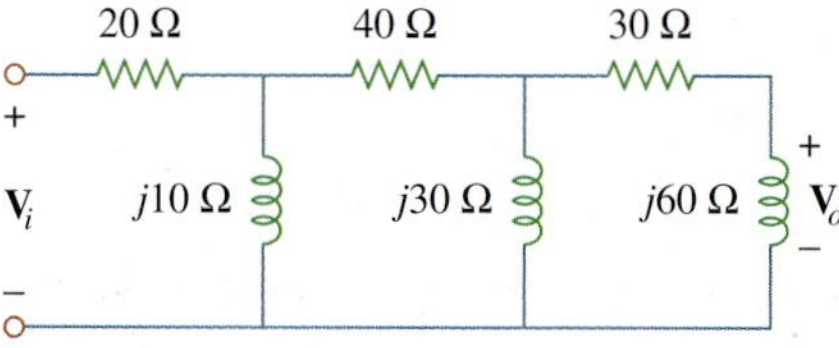

Figure 9.82
For Prob. 9.79.

9.80 Consider the phase-shifting circuit in Fig. 9.83. Let $\mathbf{V}_i = 120$ V operating at 60 Hz. Find:

(a) $\mathbf{V}_o$ when R is maximum

(b) $\mathbf{V}_o$ when R is minimum

(c) the value of R that will produce a phase shift of 45°

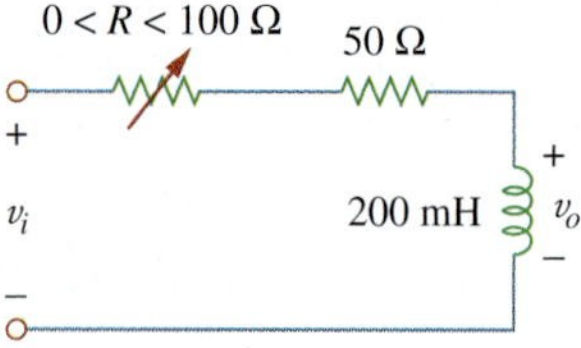

Figure 9.83
For Prob. 9.80.

9.81 The ac bridge in Fig. 9.37 is balanced when $R_1 = 400\ \Omega$, $R_2 = 600\ \Omega$, $R_3 = 1.2\ \text{k}\Omega$, and $C_2 = 0.3\ \mu\text{F}$. Find R_x and C_x. Assume R_2 and C_2 are in series.

9.82 A capacitance bridge balances when $R_1 = 100\ \Omega$, $R_2 = 2\ \text{k}\Omega$, and $C_s = 40\ \mu\text{F}$. What is C_x, the capacitance of the capacitor under test?

9.83 An inductive bridge balances when $R_1 = 1.2\ \text{k}\Omega$, $R_2 = 500\ \Omega$, and $L_s = 250$ mH. What is the value of L_x, the inductance of the inductor under test?

9.84 The ac bridge shown in Fig. 9.84 is known as a *Maxwell bridge* and is used for accurate measurement of inductance and resistance of a coil in terms of a standard capacitance C_s. Show that when the bridge is balanced,

$$L_x = R_2 R_3 C_s \quad \text{and} \quad R_x = \frac{R_2}{R_1} R_3$$

Find L_x and R_x for $R_1 = 40\ \text{k}\Omega$, $R_2 = 1.6\ \text{k}\Omega$, $R_3 = 4\ \text{k}\Omega$, and $C_s = 0.45\ \mu\text{F}$.

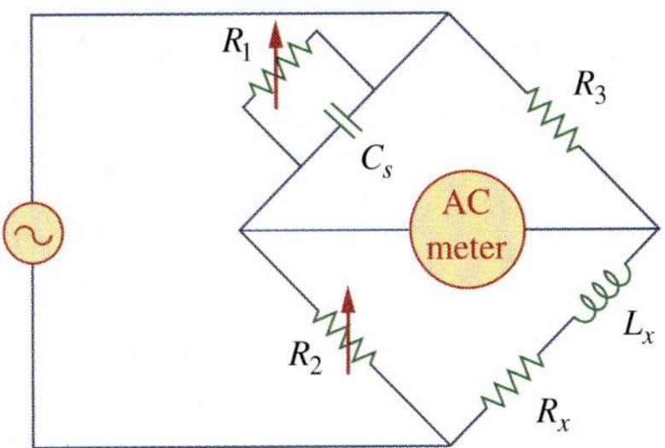

Figure 9.84
Maxwell bridge; For Prob. 9.84.

9.85 The ac bridge circuit of Fig. 9.85 is called a *Wien bridge*. It is used for measuring the frequency of a source. Show that when the bridge is balanced,

$$f = \frac{1}{2\pi\sqrt{R_2 R_4 C_2 C_4}}$$

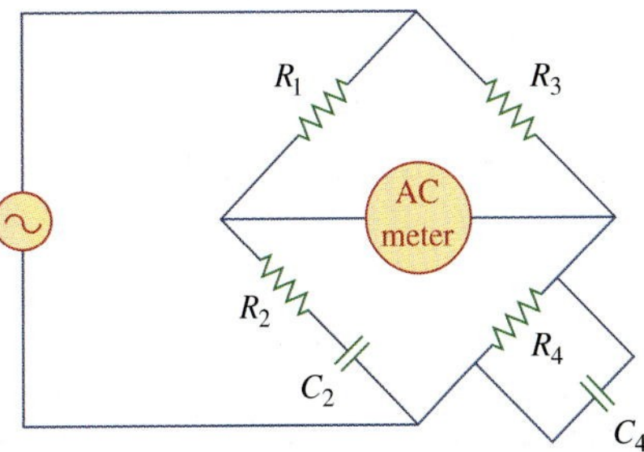

Figure 9.85
Wein bridge; For Prob. 9.85.

Comprehensive Problems

9.86 The circuit shown in Fig. 9.86 is used in a television receiver. What is the total impedance of this circuit?

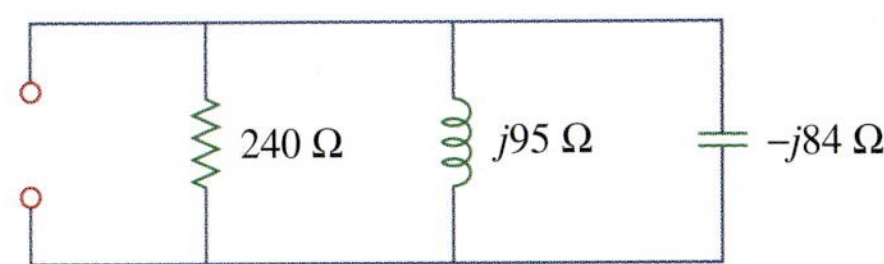

Figure 9.86
For Prob. 9.86.

9.87 The network in Fig. 9.87 is part of the schematic describing an industrial electronic sensing device. What is the total impedance of the circuit at 2 kHz?

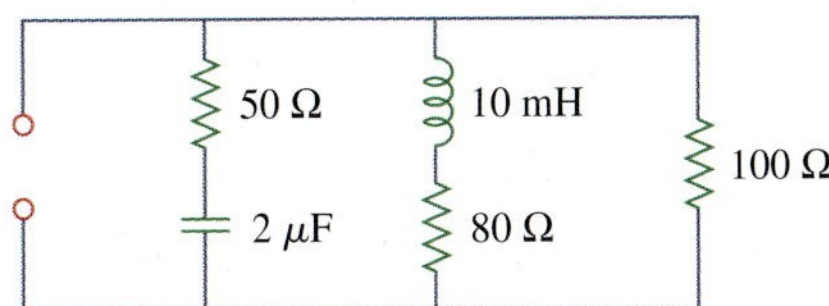

Figure 9.87
For Prob. 9.87.

9.88 A series audio circuit is shown in Fig. 9.88.

(a) What is the impedance of the circuit?

(b) If the frequency were halved, what would be the impedance of the circuit?

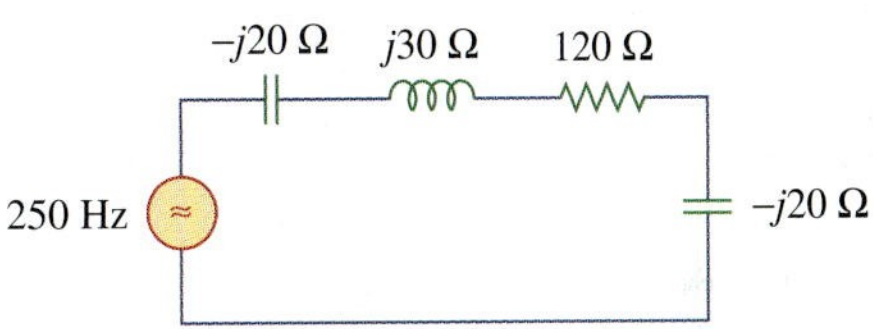

Figure 9.88
For Prob. 9.88.

9.89 An industrial load is modeled as a series combination of a capacitance and a resistance as shown in Fig. 9.89. Calculate the value of an inductance L across the series combination so that the net impedance is resistive at a frequency of 50 kHz.

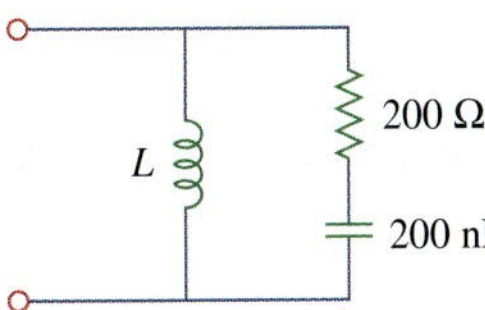

Figure 9.89
For Prob. 9.89.

9.90 An industrial coil is modeled as a series combination of an inductance L and resistance R, as shown in Fig. 9.90. Since an ac voltmeter measures only the magnitude of a sinusoid, the following measurements are taken at 60 Hz when the circuit operates in the steady state:

$|\mathbf{V}_s| = 145$ V, $|\mathbf{V}_1| = 50$ V, $|\mathbf{V}_o| = 110$ V

Use these measurements to determine the values of L and R.

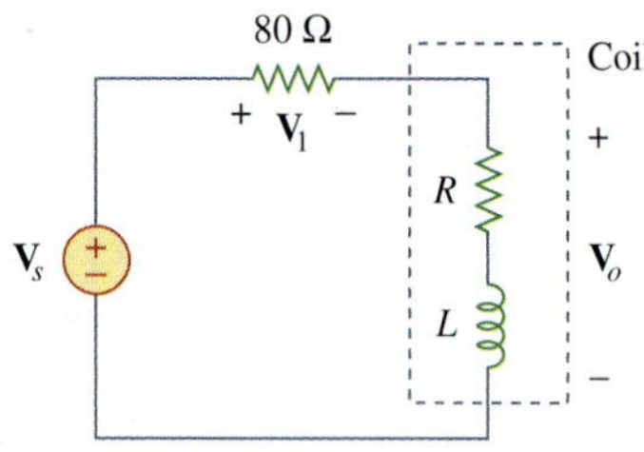

Figure 9.90
For Prob. 9.90.

9.91 Figure 9.91 shows a parallel combination of an inductance and a resistance. If it is desired to connect a capacitor in series with the parallel combination such that the net impedance is resistive at 10 MHz, what is the required value of C?

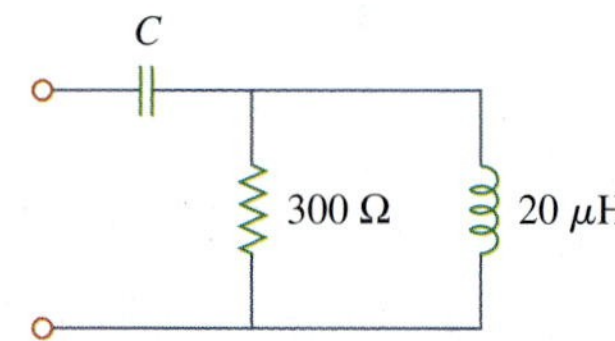

Figure 9.91
For Prob. 9.91.

9.92 A transmission line has a series impedance of $\mathbf{Z} = 100\underline{/75^\circ}$ Ω and a shunt admittance of $\mathbf{Y} = 450\underline{/48^\circ}$ μS. Find: (a) the characteristic impedance $\mathbf{Z}_o = \sqrt{\mathbf{Z}/\mathbf{Y}}$, (b) the propagation constant $\gamma = \sqrt{\mathbf{ZY}}$.

9.93 A power transmission system is modeled as shown in Fig. 9.92. Given the following;

Source voltage $\mathbf{V}_s = 115\underline{/0^\circ}$ V,
Source impedance $\mathbf{Z}_s = (2 + j)$ Ω,
Line impedance $\mathbf{Z}_\ell = (0.8 + j0.6)$ Ω,
Load impedance $\mathbf{Z}_L = (46.4 + j37.8)$ Ω,
Find the load current $\mathbf{I}_L$.

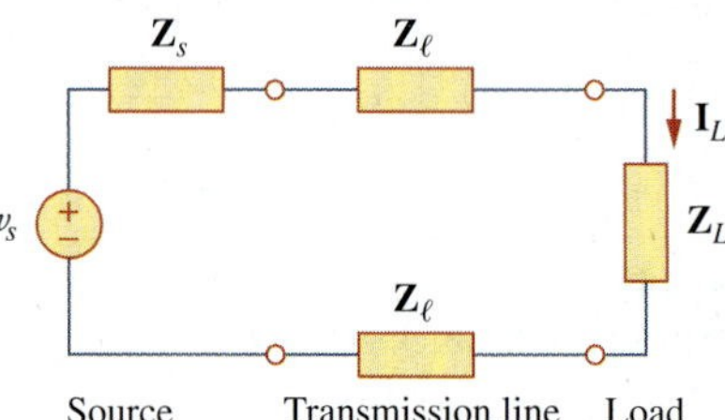

Figure 9.92
For Prob. 9.93.

chapter

10

Sinusoidal Steady-State Analysis

Three men are my friends—he that loves me, he that hates me, he that is indifferent to me. Who loves me, teaches me tenderness; who hates me, teaches me caution; who is indifferent to me, teaches me self-reliance.

—J. E. Dinger

Enhancing Your Career

Career in Software Engineering

Software engineering is that aspect of engineering that deals with the practical application of scientific knowledge in the design, construction, and validation of computer programs and the associated documentation required to develop, operate, and maintain them. It is a branch of electrical engineering that is becoming increasingly important as more and more disciplines require one form of software package or another to perform routine tasks and as programmable microelectronic systems are used in more and more applications.

The role of a software engineer should not be confused with that of a computer scientist; the software engineer is a practitioner, not a theoretician. A software engineer should have good computer-programming skills and be familiar with programming languages, in particular C^{++}, which is becoming increasingly popular. Because hardware and software are closely interlinked, it is essential that a software engineer have a thorough understanding of hardware design. Most important, the software engineer should have some specialized knowledge of the area in which the software development skill is to be applied.

All in all, the field of software engineering offers a great career to those who enjoy programming and developing software packages. The higher rewards will go to those having the best preparation, with the most interesting and challenging opportunities going to those with graduate education.

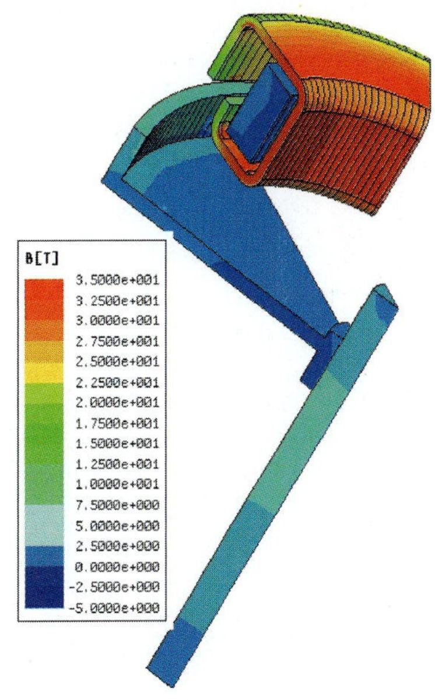

Output of a modeling software.
Courtesy Ansoft

10.1 Introduction

In Chapter 9, we learned that the forced or steady-state response of circuits to sinusoidal inputs can be obtained by using phasors. We also know that Ohm's and Kirchhoff's laws are applicable to ac circuits. In this chapter, we want to see how nodal analysis, mesh analysis, Thevenin's theorem, Norton's theorem, superposition, and source transformations are applied in analyzing ac circuits. Since these techniques were already introduced for dc circuits, our major effort here will be to illustrate with examples.

Analyzing ac circuits usually requires three steps.

Steps to Analyze AC Circuits:

1. Transform the circuit to the phasor or frequency domain.
2. Solve the problem using circuit techniques (nodal analysis, mesh analysis, superposition, etc.).
3. Transform the resulting phasor to the time domain.

Step 1 is not necessary if the problem is specified in the frequency domain. In step 2, the analysis is performed in the same manner as dc circuit analysis except that complex numbers are involved. Having read Chapter 9, we are adept at handling step 3.

Frequency domain analysis of an ac circuit via phasors is much easier than analysis of the circuit in the time domain.

Toward the end of the chapter, we learn how to apply *PSpice* in solving ac circuit problems. We finally apply ac circuit analysis to two practical ac circuits: oscillators and ac transistor circuits.

10.2 Nodal Analysis

The basis of nodal analysis is Kirchhoff's current law. Since KCL is valid for phasors, as demonstrated in Section 9.6, we can analyze ac circuits by nodal analysis. The following examples illustrate this.

Example 10.1

Find i_x in the circuit of Fig. 10.1 using nodal analysis.

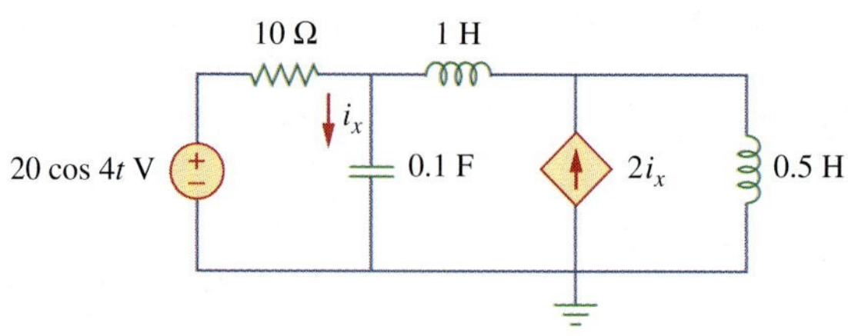

Figure 10.1
For Example 10.1.

Solution:
We first convert the circuit to the frequency domain:

$$\begin{aligned}
20\cos 4t \quad &\Rightarrow \quad 20\underline{/0^\circ}, \qquad \omega = 4 \text{ rad/s}\\
1 \text{ H} \quad &\Rightarrow \quad j\omega L = j4\\
0.5 \text{ H} \quad &\Rightarrow \quad j\omega L = j2\\
0.1 \text{ F} \quad &\Rightarrow \quad \frac{1}{j\omega C} = -j2.5
\end{aligned}$$

Thus, the frequency domain equivalent circuit is as shown in Fig. 10.2.

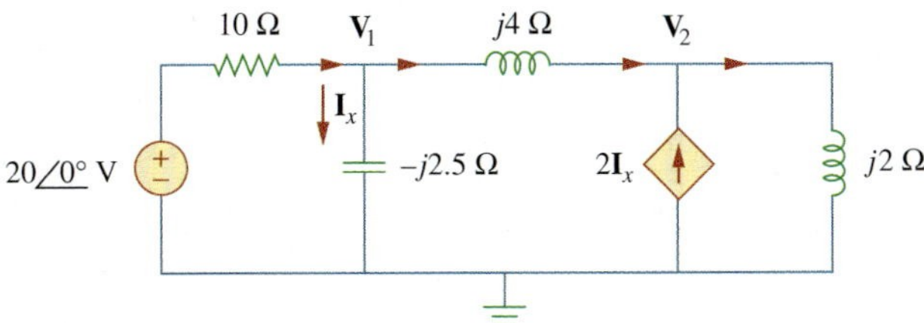

Figure 10.2
Frequency domain equivalent of the circuit in Fig. 10.1.

Applying KCL at node 1,

$$\frac{20 - \mathbf{V}_1}{10} = \frac{\mathbf{V}_1}{-j2.5} + \frac{\mathbf{V}_1 - \mathbf{V}_2}{j4}$$

or

$$(1 + j1.5)\mathbf{V}_1 + j2.5\mathbf{V}_2 = 20 \tag{10.1.1}$$

At node 2,

$$2\mathbf{I}_x + \frac{\mathbf{V}_1 - \mathbf{V}_2}{j4} = \frac{\mathbf{V}_2}{j2}$$

But $\mathbf{I}_x = \mathbf{V}_1/-j2.5$. Substituting this gives

$$\frac{2\mathbf{V}_1}{-j2.5} + \frac{\mathbf{V}_1 - \mathbf{V}_2}{j4} = \frac{\mathbf{V}_2}{j2}$$

By simplifying, we get

$$11\mathbf{V}_1 + 15\mathbf{V}_2 = 0 \tag{10.1.2}$$

Equations (10.1.1) and (10.1.2) can be put in matrix form as

$$\begin{bmatrix} 1 + j1.5 & j2.5 \\ 11 & 15 \end{bmatrix} \begin{bmatrix} \mathbf{V}_1 \\ \mathbf{V}_2 \end{bmatrix} = \begin{bmatrix} 20 \\ 0 \end{bmatrix}$$

We obtain the determinants as

$$\Delta = \begin{vmatrix} 1 + j1.5 & j2.5 \\ 11 & 15 \end{vmatrix} = 15 - j5$$

$$\Delta_1 = \begin{vmatrix} 20 & j2.5 \\ 0 & 15 \end{vmatrix} = 300, \qquad \Delta_2 = \begin{vmatrix} 1 + j1.5 & 20 \\ 11 & 0 \end{vmatrix} = -220$$

$$\mathbf{V}_1 = \frac{\Delta_1}{\Delta} = \frac{300}{15 - j5} = 18.97\underline{/18.43^\circ} \text{ V}$$

$$\mathbf{V}_2 = \frac{\Delta_2}{\Delta} = \frac{-220}{15 - j5} = 13.91\underline{/198.3^\circ} \text{ V}$$

The current $\mathbf{I}_x$ is given by

$$\mathbf{I}_x = \frac{\mathbf{V}_1}{-j2.5} = \frac{18.97\angle 18.43^\circ}{2.5\angle -90^\circ} = 7.59\angle 108.4^\circ \text{ A}$$

Transforming this to the time domain,

$$i_x = 7.59 \cos(4t + 108.4^\circ) \text{ A}$$

Practice Problem 10.1

Using nodal analysis, find v_1 and v_2 in the circuit of Fig. 10.3.

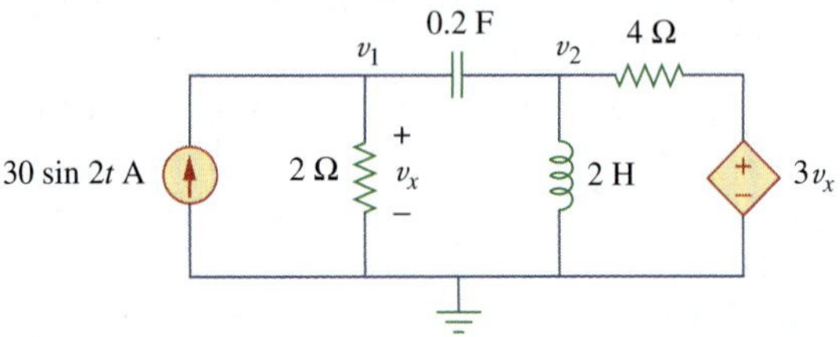

Figure 10.3
For Practice Prob. 10.1.

Answer: $v_1(t) = 33.96 \sin(2t + 60.01^\circ)$ V, $v_2(t) = 99.06 \sin(2t + 57.12^\circ)$ V.

Example 10.2

Compute $\mathbf{V}_1$ and $\mathbf{V}_2$ in the circuit of Fig. 10.4.

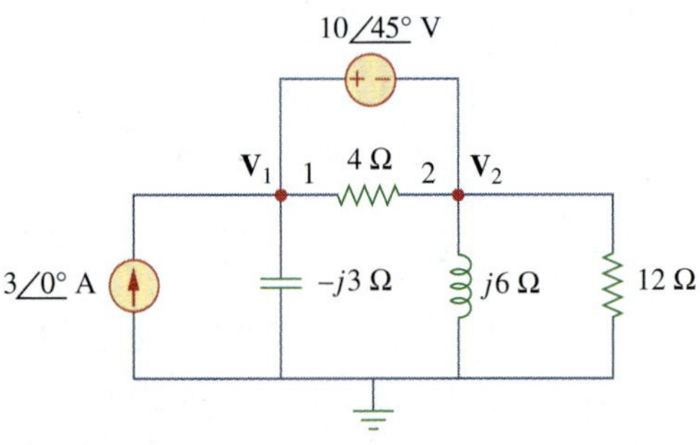

Figure 10.4
For Example 10.2.

Solution:

Nodes 1 and 2 form a supernode as shown in Fig. 10.5. Applying KCL at the supernode gives

$$3 = \frac{\mathbf{V}_1}{-j3} + \frac{\mathbf{V}_2}{j6} + \frac{\mathbf{V}_2}{12}$$

or

$$36 = j4\mathbf{V}_1 + (1 - j2)\mathbf{V}_2 \tag{10.2.1}$$

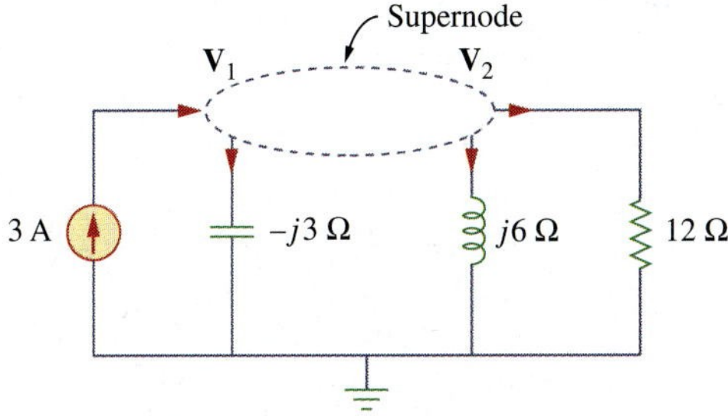

Figure 10.5
A supernode in the circuit of Fig. 10.4.

But a voltage source is connected between nodes 1 and 2, so that

$$\mathbf{V}_1 = \mathbf{V}_2 + 10\angle 45^\circ \tag{10.2.2}$$

Substituting Eq. (10.2.2) in Eq. (10.2.1) results in

$$36 - 40\angle 135^\circ = (1 + j2)\mathbf{V}_2 \quad \Rightarrow \quad \mathbf{V}_2 = 31.41\angle -87.18^\circ \text{ V}$$

From Eq. (10.2.2),

$$\mathbf{V}_1 = \mathbf{V}_2 + 10\angle 45^\circ = 25.78\angle -70.48^\circ \text{ V}$$

Practice Problem 10.2

Calculate $\mathbf{V}_1$ and $\mathbf{V}_2$ in the circuit shown in Fig. 10.6.

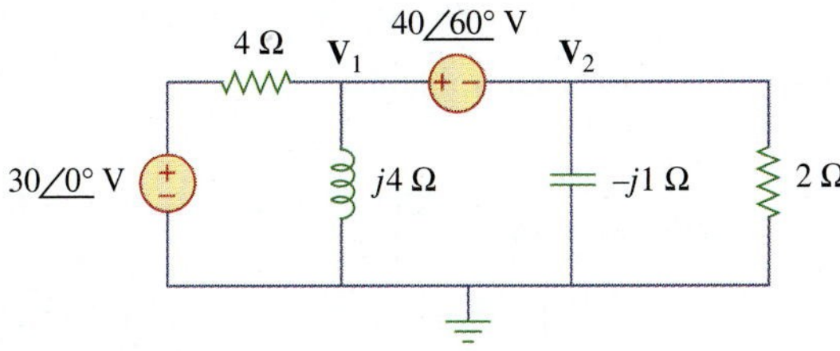

Figure 10.6
For Practice Prob. 10.2.

Answer: $\mathbf{V}_1 = 38.72\angle 69.67^\circ$ V, $\mathbf{V}_2 = 6.752\angle 165.7^\circ$ V.

10.3 Mesh Analysis

Kirchhoff's voltage law (KVL) forms the basis of mesh analysis. The validity of KVL for ac circuits was shown in Section 9.6 and is illustrated in the following examples. Keep in mind that the very nature of using mesh analysis is that it is to be applied to planar circuits.

Example 10.3

Determine current $\mathbf{I}_o$ in the circuit of Fig. 10.7 using mesh analysis.

Solution:

Applying KVL to mesh 1, we obtain

$$(8 + j10 - j2)\mathbf{I}_1 - (-j2)\mathbf{I}_2 - j10\mathbf{I}_3 = 0 \tag{10.3.1}$$

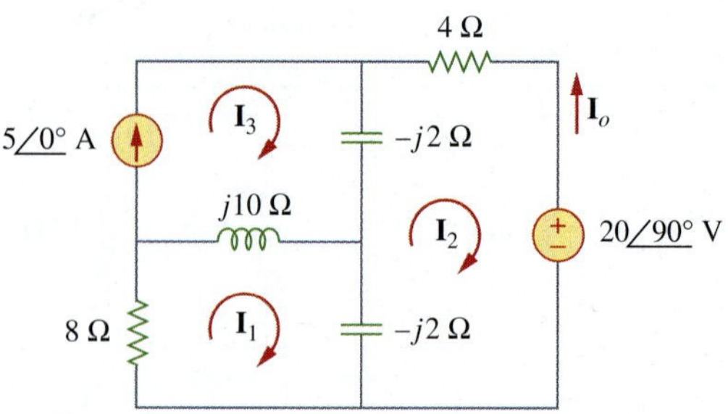

Figure 10.7
For Example 10.3.

For mesh 2,

$$(4 - j2 - j2)\mathbf{I}_2 - (-j2)\mathbf{I}_1 - (-j2)\mathbf{I}_3 + 20\underline{/90^\circ} = 0 \tag{10.3.2}$$

For mesh 3, $\mathbf{I}_3 = 5$. Substituting this in Eqs. (10.3.1) and (10.3.2), we get

$$(8 + j8)\mathbf{I}_1 + j2\mathbf{I}_2 = j50 \tag{10.3.3}$$

$$j2\mathbf{I}_1 + (4 - j4)\mathbf{I}_2 = -j20 - j10 \tag{10.3.4}$$

Equations (10.3.3) and (10.3.4) can be put in matrix form as

$$\begin{bmatrix} 8 + j8 & j2 \\ j2 & 4 - j4 \end{bmatrix} \begin{bmatrix} \mathbf{I}_1 \\ \mathbf{I}_2 \end{bmatrix} = \begin{bmatrix} j50 \\ -j30 \end{bmatrix}$$

from which we obtain the determinants

$$\Delta = \begin{vmatrix} 8 + j8 & j2 \\ j2 & 4 - j4 \end{vmatrix} = 32(1 + j)(1 - j) + 4 = 68$$

$$\Delta_2 = \begin{vmatrix} 8 + j8 & j50 \\ j2 & -j30 \end{vmatrix} = 340 - j240 = 416.17\underline{/-35.22^\circ}$$

$$\mathbf{I}_2 = \frac{\Delta_2}{\Delta} = \frac{416.17\underline{/-35.22^\circ}}{68} = 6.12\underline{/-35.22^\circ} \text{ A}$$

The desired current is

$$\mathbf{I}_o = -\mathbf{I}_2 = 6.12\underline{/144.78^\circ} \text{ A}$$

Practice Problem 10.3

Find $\mathbf{I}_o$ in Fig. 10.8 using mesh analysis.

Answer: $3.582\underline{/65.45^\circ}$ A.

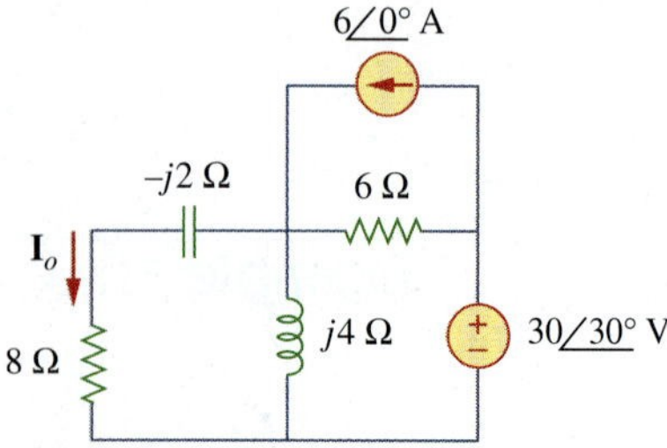

Figure 10.8
For Practice Prob. 10.3.

Example 10.4

Solve for $\mathbf{V}_o$ in the circuit of Fig. 10.9 using mesh analysis.

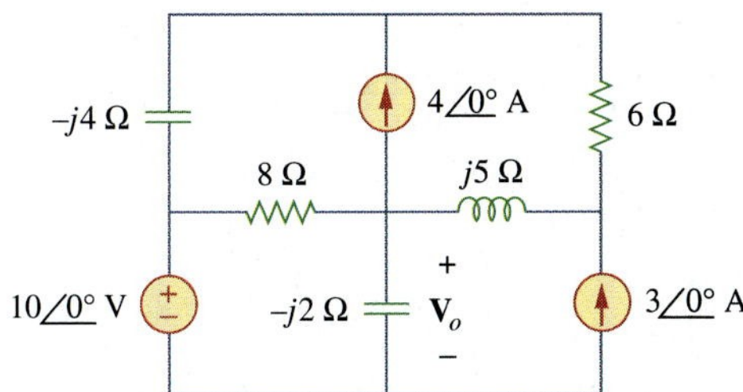

Figure 10.9
For Example 10.4.

Solution:
As shown in Fig. 10.10, meshes 3 and 4 form a supermesh due to the current source between the meshes. For mesh 1, KVL gives

$$-10 + (8 - j2)\mathbf{I}_1 - (-j2)\mathbf{I}_2 - 8\mathbf{I}_3 = 0$$

or

$$(8 - j2)\mathbf{I}_1 + j2\mathbf{I}_2 - 8\mathbf{I}_3 = 10 \qquad \textbf{(10.4.1)}$$

For mesh 2,

$$\mathbf{I}_2 = -3 \qquad \textbf{(10.4.2)}$$

For the supermesh,

$$(8 - j4)\mathbf{I}_3 - 8\mathbf{I}_1 + (6 + j5)\mathbf{I}_4 - j5\mathbf{I}_2 = 0 \qquad \textbf{(10.4.3)}$$

Due to the current source between meshes 3 and 4, at node A,

$$\mathbf{I}_4 = \mathbf{I}_3 + 4 \qquad \textbf{(10.4.4)}$$

■ METHOD 1 Instead of solving the above four equations, we reduce them to two by elimination.

Combining Eqs. (10.4.1) and (10.4.2),

$$(8 - j2)\mathbf{I}_1 - 8\mathbf{I}_3 = 10 + j6 \qquad \textbf{(10.4.5)}$$

Combining Eqs. (10.4.2) to (10.4.4),

$$-8\mathbf{I}_1 + (14 + j)\mathbf{I}_3 = -24 - j35 \qquad \textbf{(10.4.6)}$$

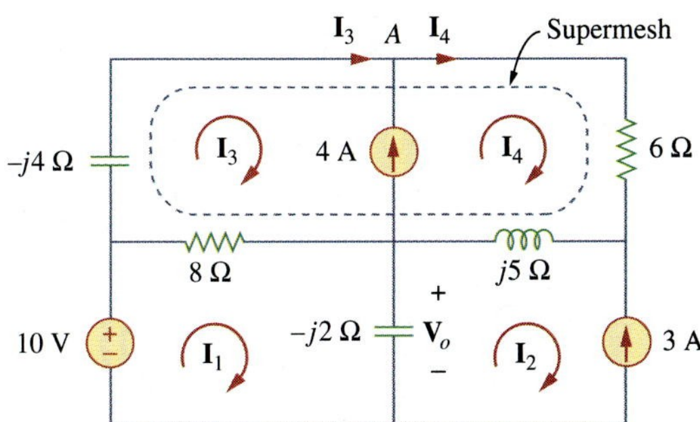

Figure 10.10
Analysis of the circuit in Fig. 10.9.

From Eqs. (10.4.5) and (10.4.6), we obtain the matrix equation

$$\begin{bmatrix} 8 - j2 & -8 \\ -8 & 14 + j \end{bmatrix} \begin{bmatrix} \mathbf{I}_1 \\ \mathbf{I}_3 \end{bmatrix} = \begin{bmatrix} 10 + j6 \\ -24 - j35 \end{bmatrix}$$

We obtain the following determinants

$$\Delta = \begin{vmatrix} 8 - j2 & -8 \\ -8 & 14 + j \end{vmatrix} = 112 + j8 - j28 + 2 - 64 = 50 - j20$$

$$\Delta_1 = \begin{vmatrix} 10 + j6 & -8 \\ -24 - j35 & 14 + j \end{vmatrix} = 140 + j10 + j84 - 6 - 192 - j280$$

$$= -58 - j186$$

Current $\mathbf{I}_1$ is obtained as

$$\mathbf{I}_1 = \frac{\Delta_1}{\Delta} = \frac{-58 - j186}{50 - j20} = 3.618\underline{/274.5^\circ} \text{ A}$$

The required voltage $\mathbf{V}_0$ is

$$\mathbf{V}_o = -j2(\mathbf{I}_1 - \mathbf{I}_2) = -j2(3.618\underline{/274.5^\circ} + 3)$$

$$= -7.2134 - j6.568 = 9.756\underline{/222.32^\circ} \text{ V}$$

■ METHOD 2 We can use *MATLAB* to solve Eqs. (10.4.1) to (10.4.4). We first cast the equations as

$$\begin{bmatrix} 8 - j2 & j2 & -8 & 0 \\ 0 & 1 & 0 & 0 \\ -8 & -j5 & 8 - j4 & 6 + j5 \\ 0 & 0 & -1 & 1 \end{bmatrix} \begin{bmatrix} \mathbf{I}_1 \\ \mathbf{I}_2 \\ \mathbf{I}_3 \\ \mathbf{I}_4 \end{bmatrix} = \begin{bmatrix} 10 \\ -3 \\ 0 \\ 4 \end{bmatrix} \qquad \textbf{(10.4.7a)}$$

or

$$\mathbf{AI} = \mathbf{B}$$

By inverting **A**, we can obtain **I** as

$$\mathbf{I} = \mathbf{A}^{-1}\mathbf{B} \qquad \textbf{(10.4.7b)}$$

We now apply *MATLAB* as follows:

```
>> A = [(8-j*2) j*2   -8       0;
        0       1     0        0;
        -8      -j*5  (8-j*4)  (6+j*5);
        0       0     -1       1];
>> B = [10 -3 0 4]';
>> I = inv(A)*B

I =
  0.2828 - 3.6069i
  -3.0000
  -1.8690 - 4.4276i
   2.1310 - 4.4276i
>> Vo = -2*j*(I(1) - I(2))

Vo =
  -7.2138 - 6.5655i
```

as obtained previously.

Practice Problem 10.4

Calculate current $\mathbf{I}_o$ in the circuit of Fig. 10.11.

Answer: $2.538\underline{/5.943^\circ}$ A.

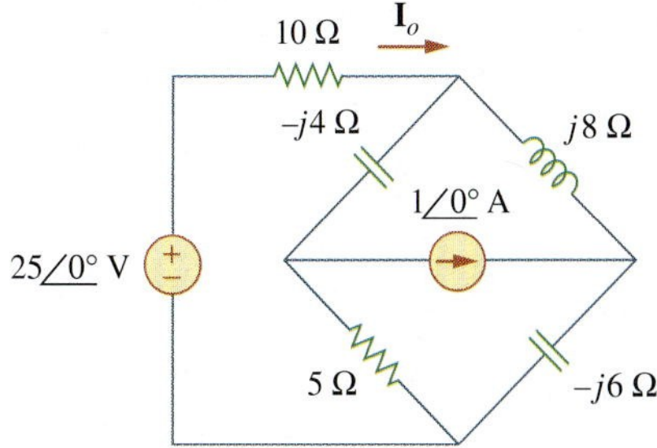

Figure 10.11
For Practice Prob. 10.4.

10.4 Superposition Theorem

Since ac circuits are linear, the superposition theorem applies to ac circuits the same way it applies to dc circuits. The theorem becomes important if the circuit has sources operating at *different* frequencies. In this case, since the impedances depend on frequency, we must have a different frequency domain circuit for each frequency. The total response must be obtained by adding the individual responses in the *time* domain. It is incorrect to try to add the responses in the phasor or frequency domain. Why? Because the exponential factor $e^{j\omega t}$ is implicit in sinusoidal analysis, and that factor would change for every angular frequency ω. It would therefore not make sense to add responses at different frequencies in the phasor domain. Thus, when a circuit has sources operating at different frequencies, one must add the responses due to the individual frequencies in the time domain.

Example 10.5

Use the superposition theorem to find $\mathbf{I}_o$ in the circuit in Fig. 10.7.

Solution:
Let

$$\mathbf{I}_o = \mathbf{I}'_o + \mathbf{I}''_o \tag{10.5.1}$$

where $\mathbf{I}'_o$ and $\mathbf{I}''_o$ are due to the voltage and current sources, respectively. To find $\mathbf{I}'_o$, consider the circuit in Fig. 10.12(a). If we let $\mathbf{Z}$ be the parallel combination of $-j2$ and $8 + j10$, then

$$\mathbf{Z} = \frac{-j2(8 + j10)}{-2j + 8 + j10} = 0.25 - j2.25$$

and current $\mathbf{I}'_o$ is

$$\mathbf{I}'_o = \frac{j20}{4 - j2 + \mathbf{Z}} = \frac{j20}{4.25 - j4.25}$$

or

$$\mathbf{I}'_o = -2.353 + j2.353 \tag{10.5.2}$$

To get $\mathbf{I}''_o$, consider the circuit in Fig. 10.12(b). For mesh 1,

$$(8 + j8)\mathbf{I}_1 - j10\mathbf{I}_3 + j2\mathbf{I}_2 = 0 \tag{10.5.3}$$

For mesh 2,

$$(4 - j4)\mathbf{I}_2 + j2\mathbf{I}_1 + j2\mathbf{I}_3 = 0 \tag{10.5.4}$$

For mesh 3,

$$\mathbf{I}_3 = 5 \tag{10.5.5}$$

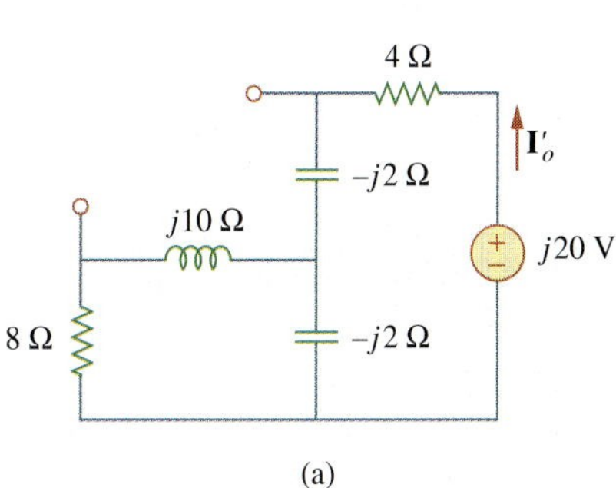

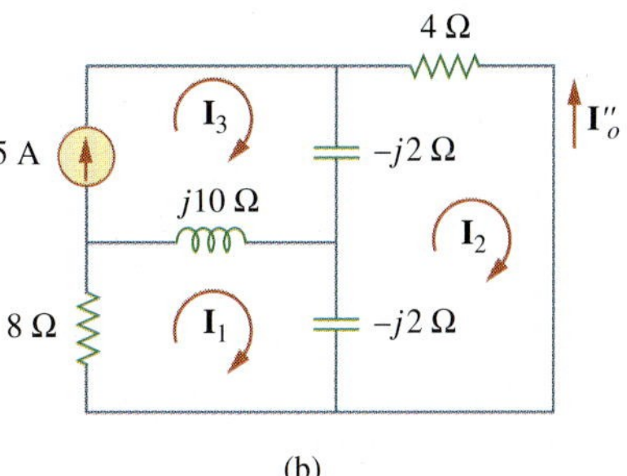

Figure 10.12
Solution of Example 10.5.

From Eqs. (10.5.4) and (10.5.5),

$$(4 - j4)\mathbf{I}_2 + j2\mathbf{I}_1 + j10 = 0$$

Expressing $\mathbf{I}_1$ in terms of $\mathbf{I}_2$ gives

$$\mathbf{I}_1 = (2 + j2)\mathbf{I}_2 - 5 \qquad \textbf{(10.5.6)}$$

Substituting Eqs. (10.5.5) and (10.5.6) into Eq. (10.5.3), we get

$$(8 + j8)[(2 + j2)\mathbf{I}_2 - 5] - j50 + j2\mathbf{I}_2 = 0$$

or

$$\mathbf{I}_2 = \frac{90 - j40}{34} = 2.647 - j1.176$$

Current $\mathbf{I}_o''$ is obtained as

$$\mathbf{I}_o'' = -\mathbf{I}_2 = -2.647 + j1.176 \qquad \textbf{(10.5.7)}$$

From Eqs. (10.5.2) and (10.5.7), we write

$$\mathbf{I}_o = \mathbf{I}_o' + \mathbf{I}_o'' = -5 + j3.529 = 6.12\underline{/144.78^\circ} \text{ A}$$

which agrees with what we got in Example 10.3. It should be noted that applying the superposition theorem is not the best way to solve this problem. It seems that we have made the problem twice as hard as the original one by using superposition. However, in Example 10.6, superposition is clearly the easiest approach.

Practice Problem 10.5

Find current $\mathbf{I}_o$ in the circuit of Fig. 10.8 using the superposition theorem.

Answer: $3.582\underline{/65.45^\circ}$ A.

Example 10.6

Find v_o of the circuit of Fig. 10.13 using the superposition theorem.

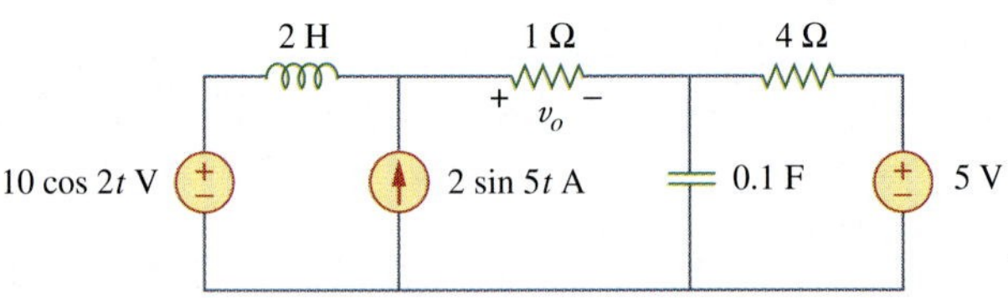

Figure 10.13
For Example 10.6.

Solution:
Since the circuit operates at three different frequencies ($\omega = 0$ for the dc voltage source), one way to obtain a solution is to use superposition, which breaks the problem into single-frequency problems. So we let

$$v_o = v_1 + v_2 + v_3 \qquad \textbf{(10.6.1)}$$

where v_1 is due to the 5-V dc voltage source, v_2 is due to the $10 \cos 2t$ V voltage source, and v_3 is due to the $2 \sin 5t$ A current source.

To find v_1, we set to zero all sources except the 5-V dc source. We recall that at steady state, a capacitor is an open circuit to dc while an inductor is a short circuit to dc. There is an alternative way of looking at this. Since $\omega = 0$, $j\omega L = 0$, $1/j\omega C = \infty$. Either way, the equivalent circuit is as shown in Fig. 10.14(a). By voltage division,

$$-v_1 = \frac{1}{1+4}(5) = 1 \text{ V} \tag{10.6.2}$$

To find v_2, we set to zero both the 5-V source and the $2 \sin 5t$ current source and transform the circuit to the frequency domain.

$$\begin{aligned} 10 \cos 2t \quad &\Rightarrow \quad 10\underline{/0^\circ}, \qquad \omega = 2 \text{ rad/s} \\ 2 \text{ H} \quad &\Rightarrow \quad j\omega L = j4\ \Omega \\ 0.1 \text{ F} \quad &\Rightarrow \quad \frac{1}{j\omega C} = -j5\ \Omega \end{aligned}$$

The equivalent circuit is now as shown in Fig. 10.14(b). Let

$$\mathbf{Z} = -j5 \parallel 4 = \frac{-j5 \times 4}{4 - j5} = 2.439 - j1.951$$

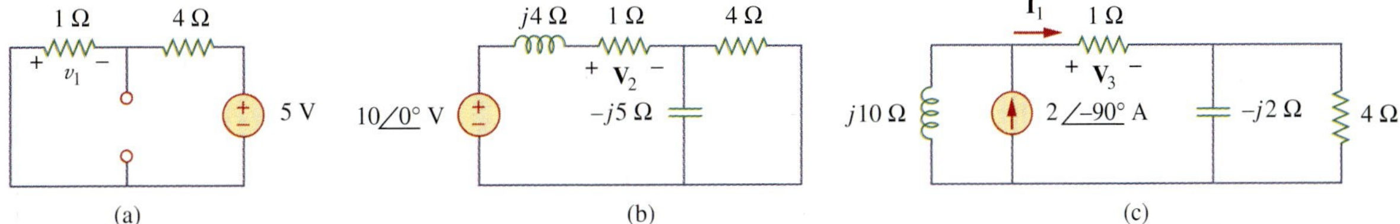

Figure 10.14
Solution of Example 10.6: (a) setting all sources to zero except the 5-V dc source, (b) setting all sources to zero except the ac voltage source, (c) setting all sources to zero except the ac current source.

By voltage division,

$$\mathbf{V}_2 = \frac{1}{1 + j4 + \mathbf{Z}}(10\underline{/0^\circ}) = \frac{10}{3.439 + j2.049} = 2.498\underline{/-30.79^\circ}$$

In the time domain,

$$v_2 = 2.498 \cos(2t - 30.79^\circ) \tag{10.6.3}$$

To obtain v_3, we set the voltage sources to zero and transform what is left to the frequency domain.

$$\begin{aligned} 2 \sin 5t \quad &\Rightarrow \quad 2\underline{/-90^\circ}, \qquad \omega = 5 \text{ rad/s} \\ 2 \text{ H} \quad &\Rightarrow \quad j\omega L = j10\ \Omega \\ 0.1 \text{ F} \quad &\Rightarrow \quad \frac{1}{j\omega C} = -j2\ \Omega \end{aligned}$$

The equivalent circuit is in Fig. 10.14(c). Let

$$\mathbf{Z}_1 = -j2 \,\|\, 4 = \frac{-j2 \times 4}{4 - j2} = 0.8 - j1.6\ \Omega$$

By current division,

$$\mathbf{I}_1 = \frac{j10}{j10 + 1 + \mathbf{Z}_1}(2\underline{/-90^\circ})\ \text{A}$$

$$\mathbf{V}_3 = \mathbf{I}_1 \times 1 = \frac{j10}{1.8 + j8.4}(-j2) = 2.328\underline{/-80^\circ}\ \text{V}$$

In the time domain,

$$v_3 = 2.33\cos(5t - 80^\circ) = 2.33\sin(5t + 10^\circ)\ \text{V} \qquad \textbf{(10.6.4)}$$

Substituting Eqs. (10.6.2) to (10.6.4) into Eq. (10.6.1), we have

$$v_o(t) = -1 + 2.498\cos(2t - 30.79^\circ) + 2.33\sin(5t + 10^\circ)\ \text{V}$$

Practice Problem 10.6

Calculate v_o in the circuit of Fig. 10.15 using the superposition theorem.

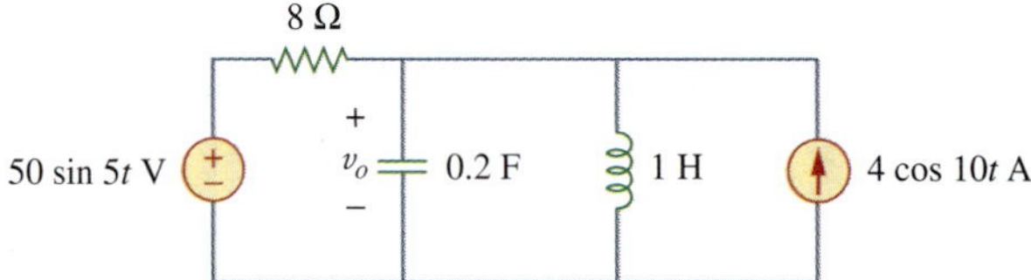

Figure 10.15
For Practice Prob. 10.6.

Answer: $7.718\sin(5t - 81.12^\circ) + 2.102\cos(10t - 86.24^\circ)$ V.

10.5 Source Transformation

As Fig. 10.16 shows, source transformation in the frequency domain involves transforming a voltage source in series with an impedance to a current source in parallel with an impedance, or vice versa. As we go from one source type to another, we must keep the following relationship in mind:

$$\mathbf{V}_s = \mathbf{Z}_s\mathbf{I}_s \quad \Leftrightarrow \quad \mathbf{I}_s = \frac{\mathbf{V}_s}{\mathbf{Z}_s} \qquad \textbf{(10.1)}$$

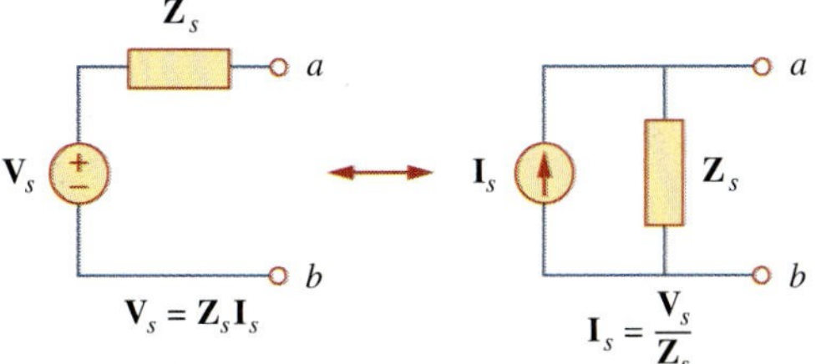

Figure 10.16
Source transformation.

Example 10.7

Calculate $\mathbf{V}_x$ in the circuit of Fig. 10.17 using the method of source transformation.

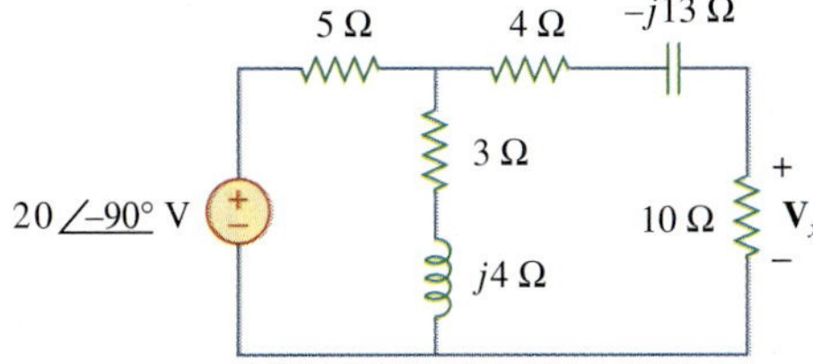

Figure 10.17
For Example 10.7.

Solution:

We transform the voltage source to a current source and obtain the circuit in Fig. 10.18(a), where

$$\mathbf{I}_s = \frac{20\underline{/-90^\circ}}{5} = 4\underline{/-90^\circ} = -j4 \text{ A}$$

The parallel combination of 5-Ω resistance and $(3 + j4)$ impedance gives

$$\mathbf{Z}_1 = \frac{5(3 + j4)}{8 + j4} = 2.5 + j1.25\ \Omega$$

Converting the current source to a voltage source yields the circuit in Fig. 10.18(b), where

$$\mathbf{V}_s = \mathbf{I}_s\mathbf{Z}_1 = -j4(2.5 + j1.25) = 5 - j10 \text{ V}$$

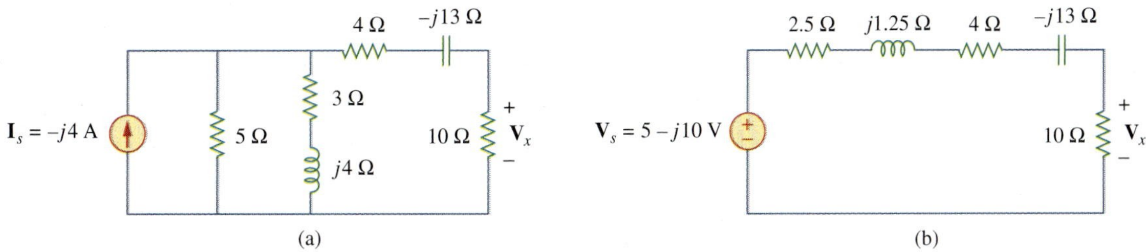

Figure 10.18
Solution of the circuit in Fig. 10.17.

By voltage division,

$$\mathbf{V}_x = \frac{10}{10 + 2.5 + j1.25 + 4 - j13}(5 - j10) = 5.519\underline{/-28^\circ} \text{ V}$$

Practice Problem 10.7

Find $\mathbf{I}_o$ in the circuit of Fig. 10.19 using the concept of source transformation.

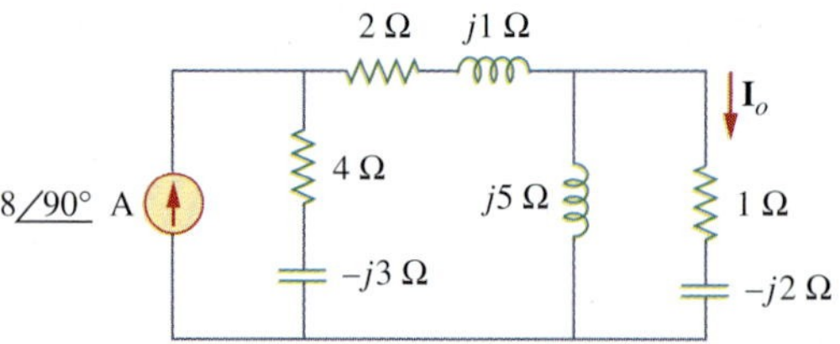

Figure 10.19
For Practice Prob. 10.7.

Answer: $6.576\underline{/99.46^\circ}$ A.

10.6 Thevenin and Norton Equivalent Circuits

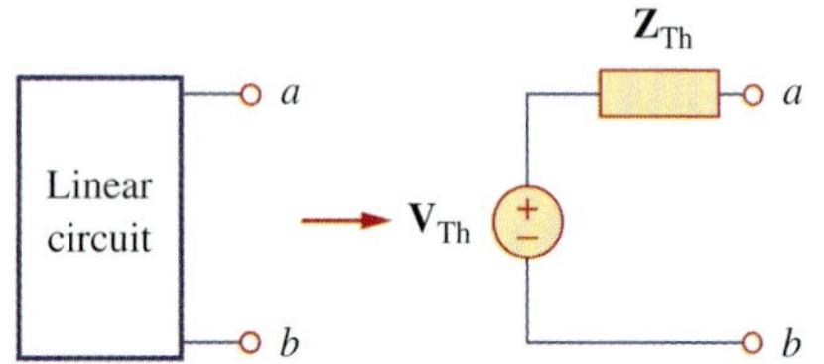

Figure 10.20
Thevenin equivalent.

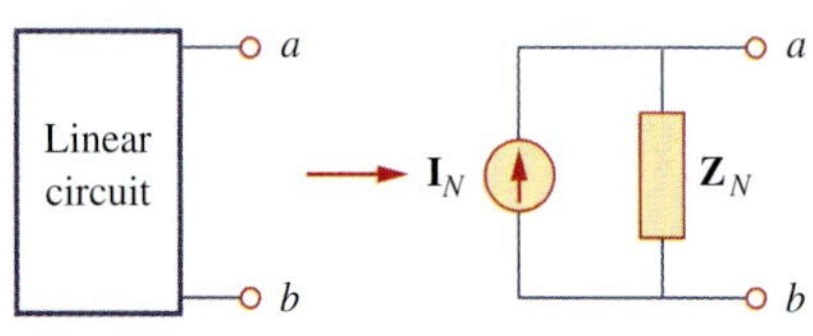

Figure 10.21
Norton equivalent.

Thevenin's and Norton's theorems are applied to ac circuits in the same way as they are to dc circuits. The only additional effort arises from the need to manipulate complex numbers. The frequency domain version of a Thevenin equivalent circuit is depicted in Fig. 10.20, where a linear circuit is replaced by a voltage source in series with an impedance. The Norton equivalent circuit is illustrated in Fig. 10.21, where a linear circuit is replaced by a current source in parallel with an impedance. Keep in mind that the two equivalent circuits are related as

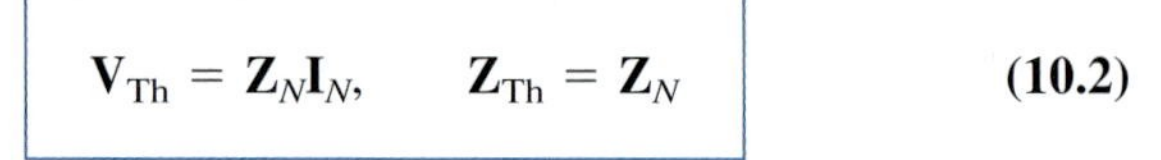

$$\mathbf{V}_{\text{Th}} = \mathbf{Z}_N \mathbf{I}_N, \qquad \mathbf{Z}_{\text{Th}} = \mathbf{Z}_N \tag{10.2}$$

just as in source transformation. $\mathbf{V}_{\text{Th}}$ is the open-circuit voltage while $\mathbf{I}_N$ is the short-circuit current.

If the circuit has sources operating at different frequencies (see Example 10.6, for example), the Thevenin or Norton equivalent circuit must be determined at each frequency. This leads to entirely different equivalent circuits, one for each frequency, not one equivalent circuit with equivalent sources and equivalent impedances.

Example 10.8

Obtain the Thevenin equivalent at terminals *a*-*b* of the circuit in Fig. 10.22.

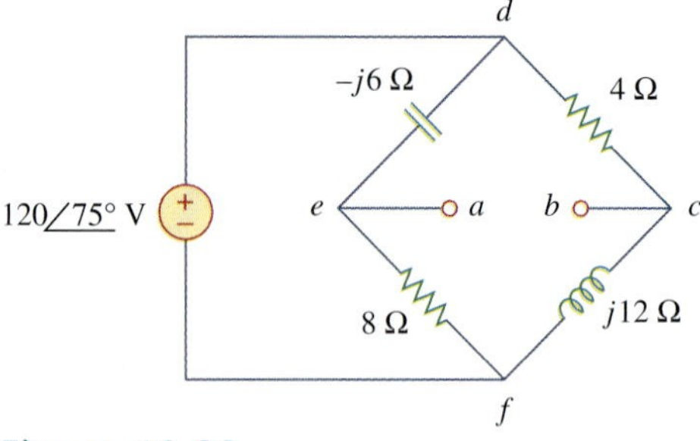

Figure 10.22
For Example 10.8.

Solution:

We find $\mathbf{Z}_{\text{Th}}$ by setting the voltage source to zero. As shown in Fig. 10.23(a), the 8-Ω resistance is now in parallel with the $-j6$ reactance, so that their combination gives

$$\mathbf{Z}_1 = -j6 \parallel 8 = \frac{-j6 \times 8}{8 - j6} = 2.88 - j3.84 \ \Omega$$

Similarly, the 4-Ω resistance is in parallel with the $j12$ reactance, and their combination gives

$$\mathbf{Z}_2 = 4 \parallel j12 = \frac{j12 \times 4}{4 + j12} = 3.6 + j1.2 \ \Omega$$

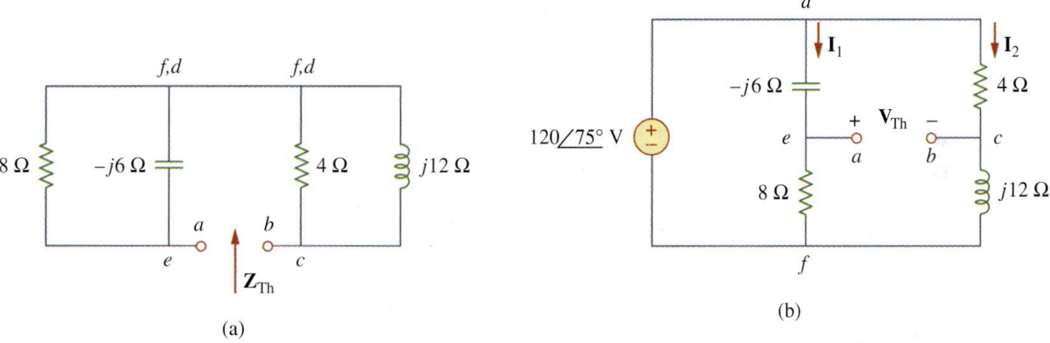

Figure 10.23
Solution of the circuit in Fig. 10.22: (a) finding $\mathbf{Z}_{\text{Th}}$, (b) finding $\mathbf{V}_{\text{Th}}$.

The Thevenin impedance is the series combination of $\mathbf{Z}_1$ and $\mathbf{Z}_2$; that is,

$$\mathbf{Z}_{\text{Th}} = \mathbf{Z}_1 + \mathbf{Z}_2 = 6.48 - j2.64 \ \Omega$$

To find $\mathbf{V}_{\text{Th}}$, consider the circuit in Fig. 10.23(b). Currents $\mathbf{I}_1$ and $\mathbf{I}_2$ are obtained as

$$\mathbf{I}_1 = \frac{120\underline{/75^\circ}}{8 - j6} \ \text{A}, \qquad \mathbf{I}_2 = \frac{120\underline{/75^\circ}}{4 + j12} \ \text{A}$$

Applying KVL around loop *bcdeab* in Fig. 10.23(b) gives

$$\mathbf{V}_{\text{Th}} - 4\mathbf{I}_2 + (-j6)\mathbf{I}_1 = 0$$

or

$$\begin{aligned} \mathbf{V}_{\text{Th}} = 4\mathbf{I}_2 + j6\mathbf{I}_1 &= \frac{480\underline{/75^\circ}}{4 + j12} + \frac{720\underline{/75^\circ + 90^\circ}}{8 - j6} \\ &= 37.95\underline{/3.43^\circ} + 72\underline{/201.87^\circ} \\ &= -28.936 - j24.55 = 37.95\underline{/220.31^\circ} \ \text{V} \end{aligned}$$

Practice Problem 10.8

Find the Thevenin equivalent at terminals *a-b* of the circuit in Fig. 10.24.

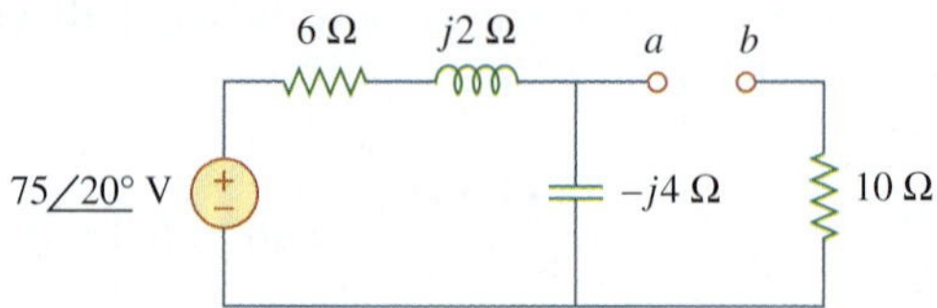

Figure 10.24
For Practice Prob. 10.8.

Answer: $\mathbf{Z}_{\text{Th}} = 12.4 - j3.2\ \Omega$, $\mathbf{V}_{\text{Th}} = 47.42\underline{/-51.57^\circ}$ V.

Example 10.9

Find the Thevenin equivalent of the circuit in Fig. 10.25 as seen from terminals *a-b*.

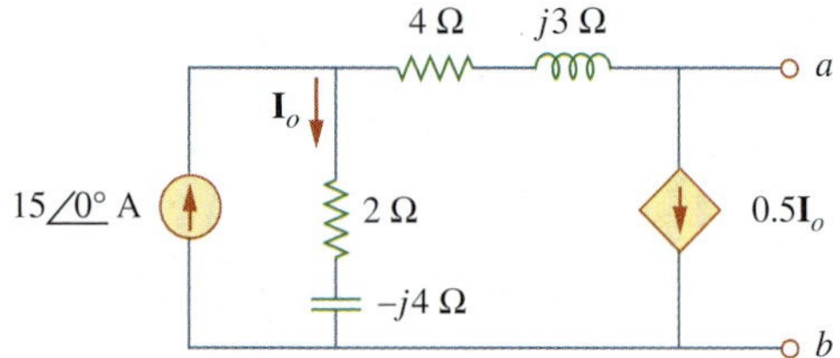

Figure 10.25
For Example 10.9.

Solution:
To find $\mathbf{V}_{\text{Th}}$, we apply KCL at node 1 in Fig. 10.26(a).

$$15 = \mathbf{I}_o + 0.5\mathbf{I}_o \qquad \Rightarrow \qquad \mathbf{I}_o = 10\ \text{A}$$

Applying KVL to the loop on the right-hand side in Fig. 10.26(a), we obtain

$$-\mathbf{I}_o(2 - j4) + 0.5\mathbf{I}_o(4 + j3) + \mathbf{V}_{\text{Th}} = 0$$

or

$$\mathbf{V}_{\text{Th}} = 10(2 - j4) - 5(4 + j3) = -j55$$

Thus, the Thevenin voltage is

$$\mathbf{V}_{\text{Th}} = 55\underline{/-90^\circ}\ \text{V}$$

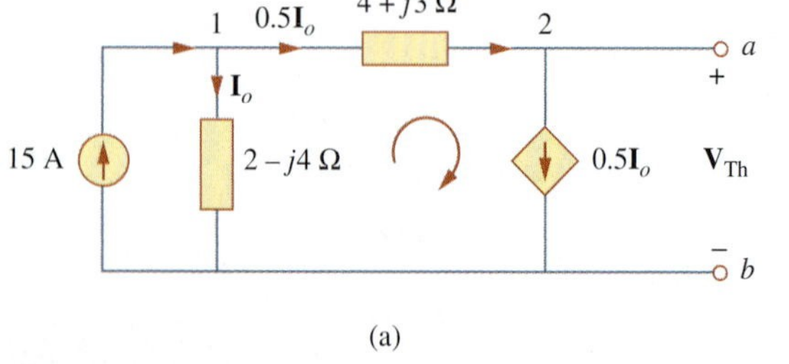

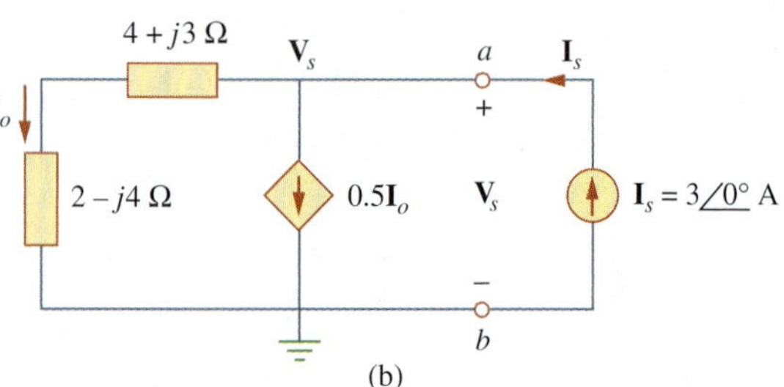

Figure 10.26
Solution of the problem in Fig. 10.25: (a) finding $\mathbf{V}_{\text{Th}}$, (b) finding $\mathbf{Z}_{\text{Th}}$.

To obtain $\mathbf{Z}_{\text{Th}}$, we remove the independent source. Due to the presence of the dependent current source, we connect a 3-A current source (3 is an arbitrary value chosen for convenience here, a number divisible by the sum of currents leaving the node) to terminals *a-b* as shown in Fig. 10.26(b). At the node, KCL gives

$$3 = \mathbf{I}_o + 0.5\mathbf{I}_o \qquad \Rightarrow \qquad \mathbf{I}_o = 2 \text{ A}$$

Applying KVL to the outer loop in Fig. 10.26(b) gives

$$\mathbf{V}_s = \mathbf{I}_o(4 + j3 + 2 - j4) = 2(6 - j)$$

The Thevenin impedance is

$$\mathbf{Z}_{\text{Th}} = \frac{\mathbf{V}_s}{\mathbf{I}_s} = \frac{2(6 - j)}{3} = 4 - j0.6667\ \Omega$$

Practice Problem 10.9

Determine the Thevenin equivalent of the circuit in Fig. 10.27 as seen from the terminals *a-b*.

Answer: $\mathbf{Z}_{\text{Th}} = 4.473\underline{/-7.64^\circ}\ \Omega$, $\mathbf{V}_{\text{Th}} = 29.4\underline{/72.9^\circ}$ V.

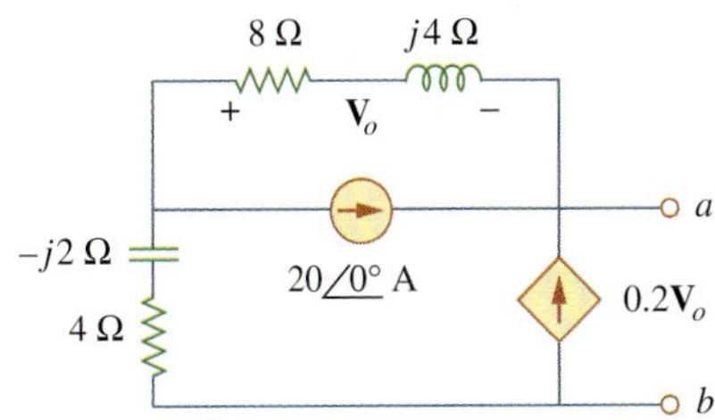

Figure 10.27
For Practice Prob. 10.9.

Example 10.10

Obtain current $\mathbf{I}_o$ in Fig. 10.28 using Norton's theorem.

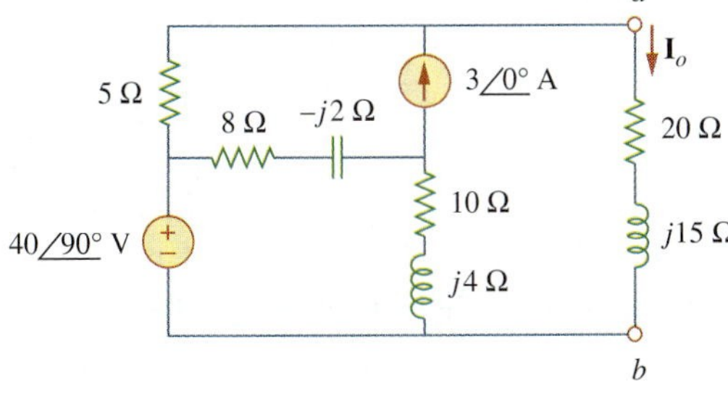

Figure 10.28
For Example 10.10.

Solution:

Our first objective is to find the Norton equivalent at terminals *a-b*. $\mathbf{Z}_N$ is found in the same way as $\mathbf{Z}_{\text{Th}}$. We set the sources to zero as shown in Fig. 10.29(a). As evident from the figure, the $(8 - j2)$ and $(10 + j4)$ impedances are short-circuited, so that

$$\mathbf{Z}_N = 5\ \Omega$$

To get $\mathbf{I}_N$, we short-circuit terminals *a-b* as in Fig. 10.29(b) and apply mesh analysis. Notice that meshes 2 and 3 form a supermesh because of the current source linking them. For mesh 1,

$$-j40 + (18 + j2)\mathbf{I}_1 - (8 - j2)\mathbf{I}_2 - (10 + j4)\mathbf{I}_3 = 0 \qquad \textbf{(10.10.1)}$$

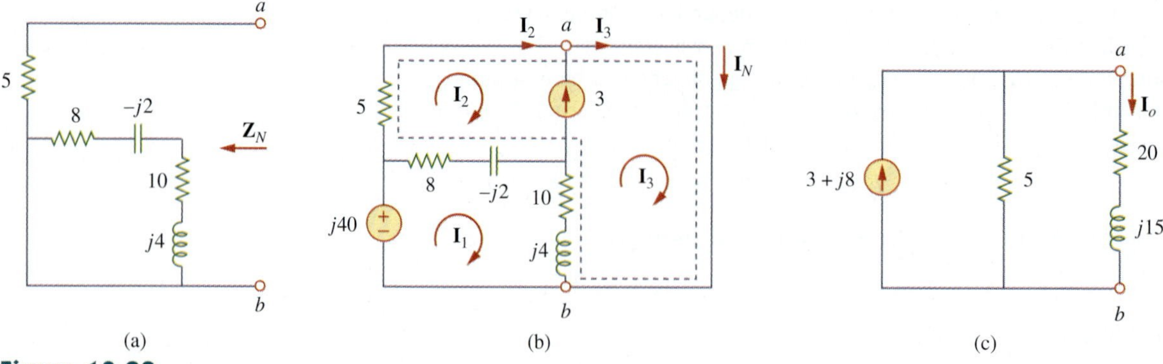

Figure 10.29
Solution of the circuit in Fig. 10.28: (a) finding $\mathbf{Z}_N$, (b) finding $\mathbf{V}_N$, (c) calculating $\mathbf{I}_o$.

For the supermesh,

$$(13 - j2)\mathbf{I}_2 + (10 + j4)\mathbf{I}_3 - (18 + j2)\mathbf{I}_1 = 0 \tag{10.10.2}$$

At node a, due to the current source between meshes 2 and 3,

$$\mathbf{I}_3 = \mathbf{I}_2 + 3 \tag{10.10.3}$$

Adding Eqs. (10.10.1) and (10.10.2) gives

$$-j40 + 5\mathbf{I}_2 = 0 \quad \Rightarrow \quad \mathbf{I}_2 = j8$$

From Eq. (10.10.3),

$$\mathbf{I}_3 = \mathbf{I}_2 + 3 = 3 + j8$$

The Norton current is

$$\mathbf{I}_N = \mathbf{I}_3 = (3 + j8)\text{ A}$$

Figure 10.29(c) shows the Norton equivalent circuit along with the impedance at terminals a-b. By current division,

$$\mathbf{I}_o = \frac{5}{5 + 20 + j15}\mathbf{I}_N = \frac{3 + j8}{5 + j3} = 1.465\underline{/38.48^\circ}\text{ A}$$

Practice Problem 10.10

Determine the Norton equivalent of the circuit in Fig. 10.30 as seen from terminals a-b. Use the equivalent to find $\mathbf{I}_o$.

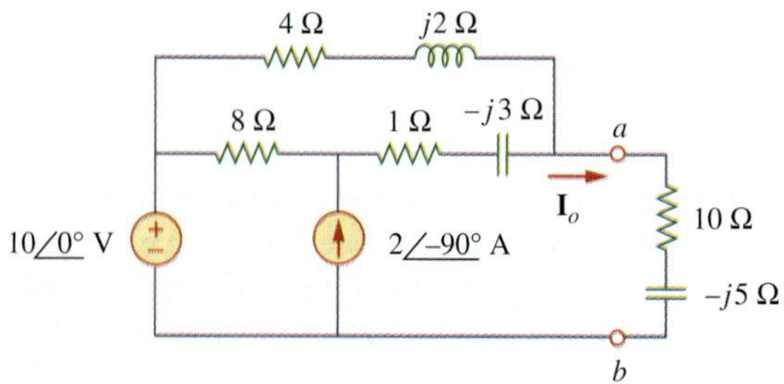

Figure 10.30
For Practice Prob. 10.10.

Answer: $\mathbf{Z}_N = 3.176 + j0.706\ \Omega$, $\mathbf{I}_N = 4.198\underline{/-32.68^\circ}$ A, $\mathbf{I}_o = 985.5\underline{/-2.101^\circ}$ mA.

10.7 Op Amp AC Circuits

The three steps stated in Section 10.1 also apply to op amp circuits, as long as the op amp is operating in the linear region. As usual, we will assume ideal op amps. (See Section 5.2.) As discussed in Chapter 5, the key to analyzing op amp circuits is to keep two important properties of an ideal op amp in mind:

1. No current enters either of its input terminals.
2. The voltage across its input terminals is zero.

The following examples will illustrate these ideas.

Example 10.11

Determine $v_o(t)$ for the op amp circuit in Fig. 10.31(a) if $v_s = 3 \cos 1000t$ V.

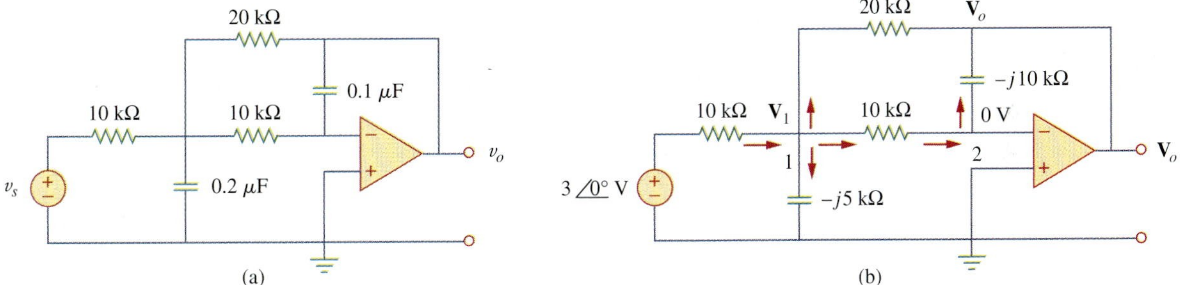

Figure 10.31
For Example 10.11: (a) the original circuit in the time domain, (b) its frequency domain equivalent.

Solution:

We first transform the circuit to the frequency domain, as shown in Fig. 10.31(b), where $\mathbf{V}_s = 3\underline{/0^\circ}$, $\omega = 1000$ rad/s. Applying KCL at node 1, we obtain

$$\frac{3\underline{/0^\circ} - \mathbf{V}_1}{10} = \frac{\mathbf{V}_1}{-j5} + \frac{\mathbf{V}_1 - 0}{10} + \frac{\mathbf{V}_1 - \mathbf{V}_o}{20}$$

or

$$6 = (5 + j4)\mathbf{V}_1 - \mathbf{V}_o \tag{10.11.1}$$

At node 2, KCL gives

$$\frac{\mathbf{V}_1 - 0}{10} = \frac{0 - \mathbf{V}_o}{-j10}$$

which leads to

$$\mathbf{V}_1 = -j\mathbf{V}_o \tag{10.11.2}$$

Substituting Eq. (10.11.2) into Eq. (10.11.1) yields

$$6 = -j(5 + j4)\mathbf{V}_o - \mathbf{V}_o = (3 - j5)\mathbf{V}_o$$

$$\mathbf{V}_o = \frac{6}{3 - j5} = 1.029\underline{/59.04^\circ}$$

Hence,

$$v_o(t) = 1.029 \cos(1000t + 59.04^\circ) \text{ V}$$

Practice Problem 10.11

Find v_o and i_o in the op amp circuit of Fig. 10.32. Let $v_s = 4 \cos 5000t$ V.

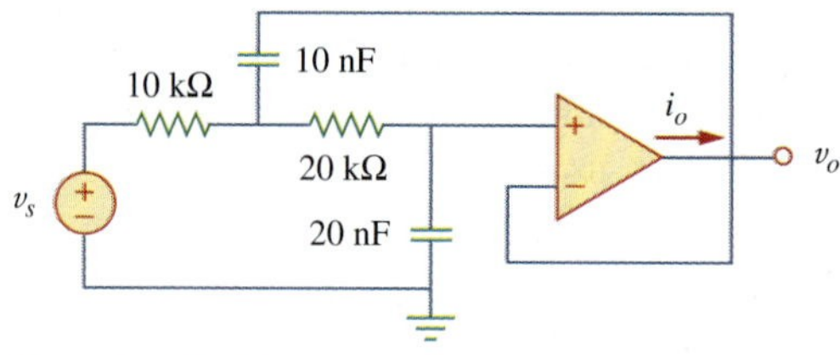

Figure 10.32
For Practice Prob. 10.11.

Answer: 1.3333 sin 5000t V, 133.33 sin 5000t μA.

Example 10.12

Compute the closed-loop gain and phase shift for the circuit in Fig. 10.33. Assume that $R_1 = R_2 = 10$ kΩ, $C_1 = 2$ μF, $C_2 = 1$ μF, and $\omega = 200$ rad/s.

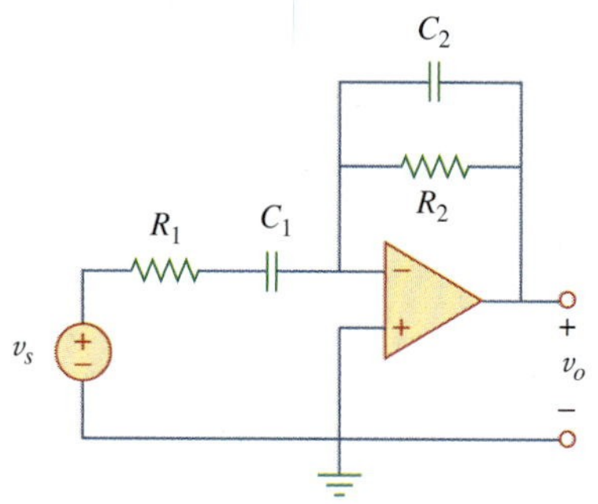

Figure 10.33
For Example 10.12.

Solution:
The feedback and input impedances are calculated as

$$\mathbf{Z}_f = R_2 \Bigg\| \frac{1}{j\omega C_2} = \frac{R_2}{1 + j\omega R_2 C_2}$$

$$\mathbf{Z}_i = R_1 + \frac{1}{j\omega C_1} = \frac{1 + j\omega R_1 C_1}{j\omega C_1}$$

Since the circuit in Fig. 10.33 is an inverting amplifier, the closed-loop gain is given by

$$\mathbf{G} = \frac{\mathbf{V}_o}{\mathbf{V}_s} = -\frac{\mathbf{Z}_f}{\mathbf{Z}_i} = \frac{-j\omega C_1 R_2}{(1 + j\omega R_1 C_1)(1 + j\omega R_2 C_2)}$$

Substituting the given values of R_1, R_2, C_1, C_2, and ω, we obtain

$$\mathbf{G} = \frac{-j4}{(1 + j4)(1 + j2)} = 0.434\underline{/130.6^\circ}$$

Thus, the closed-loop gain is 0.434 and the phase shift is 130.6°.

Practice Problem 10.12

Obtain the closed-loop gain and phase shift for the circuit in Fig. 10.34. Let $R = 10$ kΩ, $C = 1$ μF, and $\omega = 1000$ rad/s.

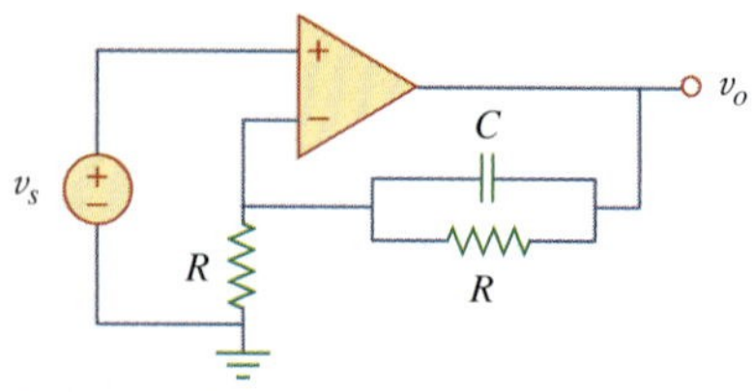

Figure 10.34
For Practice Prob. 10.12.

Answer: 1.015, −5.6°.

10.8 AC Analysis Using *PSpice*

PSpice affords a big relief from the tedious task of manipulating complex numbers in ac circuit analysis. The procedure for using *PSpice* for ac analysis is quite similar to that required for dc analysis. The reader should read Section D.5 in Appendix D for a review of *PSpice* concepts for ac analysis. AC circuit analysis is done in the phasor or frequency domain, and all sources must have the same frequency. Although ac analysis with *PSpice* involves using AC Sweep, our analysis in this chapter requires a single frequency $f = \omega/2\pi$. The output file of *PSpice* contains voltage and current phasors. If necessary, the impedances can be calculated using the voltages and currents in the output file.

Example 10.13

Obtain v_o and i_o in the circuit of Fig. 10.35 using *PSpice*.

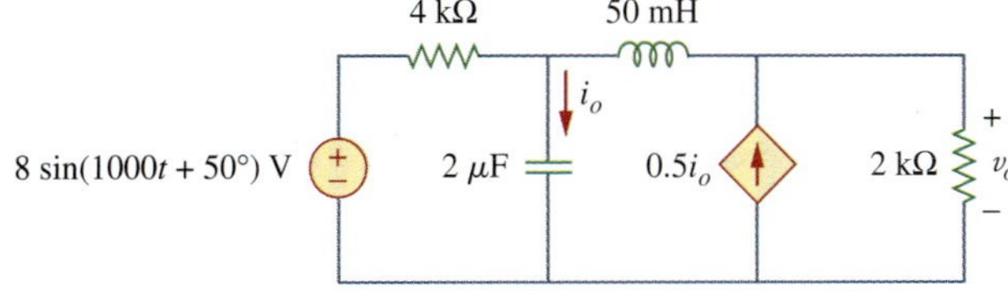

Figure 10.35
For Example 10.13.

Solution:

We first convert the sine function to cosine.

$$\begin{aligned} 8\sin(1000t + 50^\circ) &= 8\cos(1000t + 50^\circ - 90^\circ) \\ &= 8\cos(1000t - 40^\circ) \end{aligned}$$

The frequency f is obtained from ω as

$$f = \frac{\omega}{2\pi} = \frac{1000}{2\pi} = 159.155 \text{ Hz}$$

The schematic for the circuit is shown in Fig. 10.36. Notice that the current-controlled current source F1 is connected such that its current flows from node 0 to node 3 in conformity with the original circuit in Fig. 10.35. Since we only want the magnitude and phase of v_o and i_o, we set the attributes of IPRINT and VPRINT1 each to *AC* = *yes*, *MAG* = *yes*, *PHASE* = *yes*. As a single-frequency analysis, we select **Analysis/Setup/AC Sweep** and enter *Total Pts* = 1, *Start Freq* = 159.155, and *Final Freq* = 159.155. After saving the schematic, we simulate it by selecting **Analysis/Simulate.** The output file includes the source frequency in addition to the attributes checked for the pseudocomponents IPRINT and VPRINT1,

```
FREQ         IM(V_PRINT3)   IP(V_PRINT3)
1.592E+02    3.264E-03      -3.743E+01

FREQ         VM(3)          VP(3)
1.592E+02    1.550E+00      -9.518E+01
```

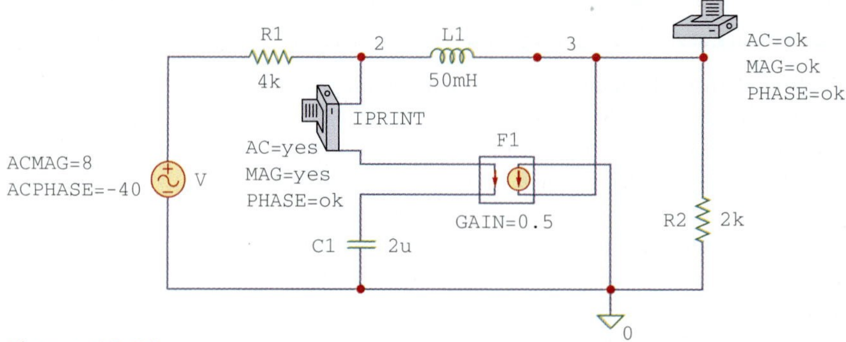

Figure 10.36
The schematic of the circuit in Fig. 10.35.

From this output file, we obtain

$$\mathbf{V}_o = 1.55\underline{/-95.18^\circ}\text{ V}, \qquad \mathbf{I}_o = 3.264\underline{/-37.43^\circ}\text{ mA}$$

which are the phasors for

$$v_o = 1.55\cos(1000t - 95.18^\circ) = 1.55\sin(1000t - 5.18^\circ)\text{ V}$$

and

$$i_o = 3.264\cos(1000t - 37.43^\circ)\text{ mA}$$

Practice Problem 10.13

Use *PSpice* to obtain v_o and i_o in the circuit of Fig. 10.37.

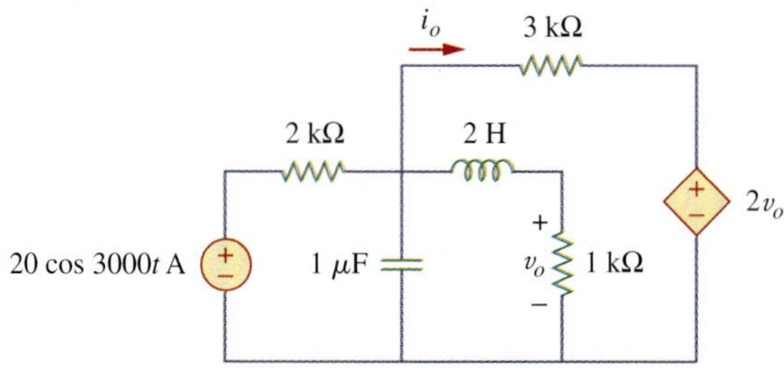

Figure 10.37
For Practice Prob. 10.13.

Answer: $536.4\cos(3000t - 154.6^\circ)$ mV, $1.088\cos(3000t - 55.12^\circ)$ mA.

Example 10.14

Find $\mathbf{V}_1$ and $\mathbf{V}_2$ in the circuit of Fig. 10.38.

Solution:

1. **Define.** In its present form, the problem is clearly stated. Again, we must emphasize that time spent here will save lots of time and expense later on! One thing that might have created a problem for you is that, if the reference was missing for this problem, you would then need to ask the individual assigning

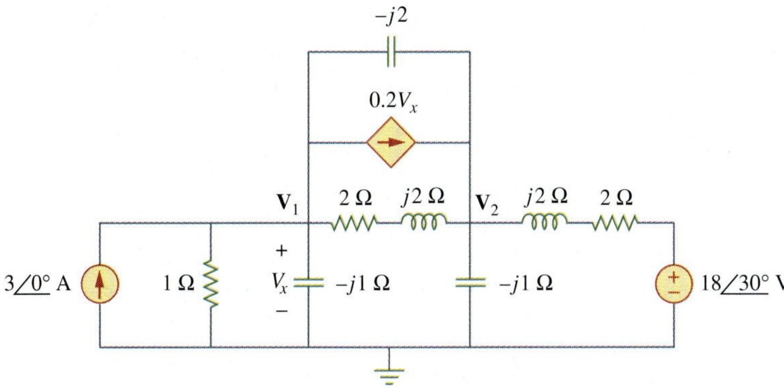

Figure 10.38
For Example 10.14.

the problem where it is to be located. If you could not do that, then you would need to assume where it should be and then clearly state what you did and why you did it.

2. **Present.** The given circuit is a frequency domain circuit and the unknown node voltages $\mathbf{V}_1$ and $\mathbf{V}_2$ are also frequency domain values. Clearly, we need a process to solve for these unknowns in the frequency domain.
3. **Alternative.** We have two direct alternative solution techniques that we can easily use. We can do a straightforward nodal analysis approach or use *PSpice*. Since this example is in a section dedicated to using *PSpice* to solve problems, we will use *PSpice* to find $\mathbf{V}_1$ and $\mathbf{V}_2$. We can then use nodal analysis to check the answer.
4. **Attempt.** The circuit in Fig. 10.35 is in the time domain, whereas the one in Fig. 10.38 is in the frequency domain. Since we are not given a particular frequency and *PSpice* requires one, we select any frequency consistent with the given impedances. For example, if we select $\omega = 1$ rad/s, the corresponding frequency is $f = \omega/2\pi = 0.15916$ Hz. We obtain the values of the capacitance ($C = 1/\omega X_C$) and inductances ($L = X_L/\omega$). Making these changes results in the schematic in Fig. 10.39. To ease wiring, we have

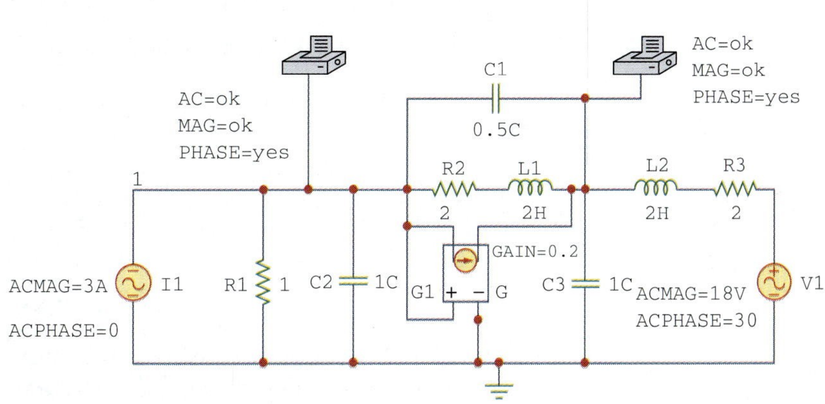

Figure 10.39
Schematic for circuit in the Fig. 10.38.

exchanged the positions of the voltage-controlled current source G1 and the $2 + j2\ \Omega$ impedance. Notice that the current of G1 flows from node 1 to node 3, while the controlling voltage is across the capacitor C2, as required in Fig. 10.38. The attributes of pseudocomponents VPRINT1 are set as shown. As a single-frequency analysis, we select **Analysis/Setup/AC Sweep** and enter *Total Pts* = 1, *Start Freq* = 0.15916, and *Final Freq* = 0.15916. After saving the schematic, we select **Analysis/Simulate** to simulate the circuit. When this is done, the output file includes

```
FREQ        VM(1)       VP(1)
1.592E-01   2.708E+00   -5.673E+01

FREQ        VM(3)       VP(3)
1.592E-01   4.468E+00   -1.026E+02
```

from which we obtain,

$$\mathbf{V}_1 = \mathbf{2.708\underline{/-56.74^\circ}\ V} \qquad \text{and} \qquad \mathbf{V}_2 = \mathbf{6.911\underline{/-80.72^\circ}\ V}$$

5. **Evaluate.** One of the most important lessons to be learned is that when using programs such as *PSpice* you still need to validate the answer. There are many opportunities for making a mistake, including coming across an unknown "bug" in *PSpice* that yields incorrect results.

   So, how can we validate this solution? Obviously, we can rework the entire problem with nodal analysis, and perhaps using *MATLAB*, to see if we obtain the same results. There is another way we will use here: write the nodal equations and substitute the answers obtained in the *PSpice* solution, and see if the nodal equations are satisfied.

   The nodal equations for this circuit are given below. Note we have substituted $\mathbf{V}_1 = \mathbf{V}_x$ into the dependent source.

$$-3 + \frac{\mathbf{V}_1 - 0}{1} + \frac{\mathbf{V}_1 - 0}{-j1} + \frac{\mathbf{V}_1 - \mathbf{V}_2}{2 + j2} + 0.2\mathbf{V}_1 + \frac{\mathbf{V}_1 - \mathbf{V}_2}{-j2} = 0$$

$$(1 + j + 0.25 - j0.25 + 0.2 + j0.5)\mathbf{V}_1$$
$$- (0.25 - j0.25 + j0.5)\mathbf{V}_2 = 3$$
$$(1.45 + j1.25)\mathbf{V}_1 - (0.25 + j0.25)\mathbf{V}_2 = 3$$
$$1.9144\underline{/40.76^\circ}\,\mathbf{V}_1 - 0.3536\underline{/45^\circ}\,\mathbf{V}_2 = 3$$

Now, to check the answer, we substitute the *PSpice* answers into this.

$$1.9144\underline{/40.76^\circ} \times 2.708\underline{/-56.74^\circ} - 0.3536\underline{/45^\circ} \times 6.911\underline{/-80.72^\circ}$$
$$= 5.184\underline{/-15.98^\circ} - 2.444\underline{/-35.72^\circ}$$
$$= 4.984 - j1.4272 - 1.9842 + j1.4269$$
$$= 3 - j0.0003 \qquad \text{[Answer checks]}$$

6. **Satisfactory?** Although we used only the equation from node 1 to check the answer, this is more than satisfactory to validate the answer from the *PSpice* solution. We can now present our work as a solution to the problem.

Practice Problem 10.14

Obtain $\mathbf{V}_x$ and $\mathbf{I}_x$ in the circuit depicted in Fig. 10.40.

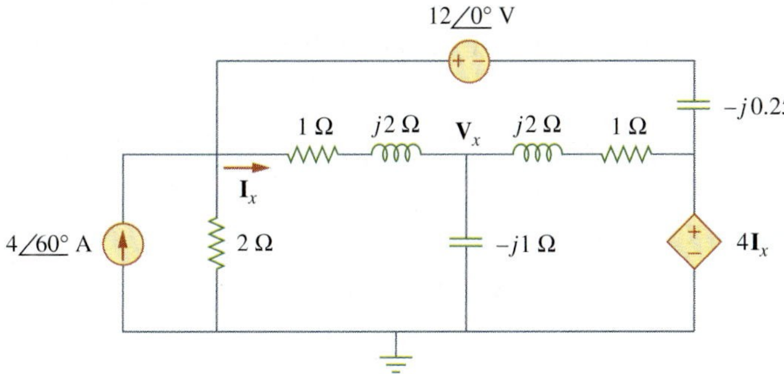

Figure 10.40
For Practice Prob. 10.14.

Answer: $9.842\angle 44.78^\circ$ V, $2.584\angle 158^\circ$ A.

10.9 †Applications

The concepts learned in this chapter will be applied in later chapters to calculate electric power and determine frequency response. The concepts are also used in analyzing coupled circuits, three-phase circuits, ac transistor circuits, filters, oscillators, and other ac circuits. In this section, we apply the concepts to develop two practical ac circuits: the capacitance multiplier and the sine wave oscillators.

10.9.1 Capacitance Multiplier

The op amp circuit in Fig. 10.41 is known as a *capacitance multiplier*, for reasons that will become obvious. Such a circuit is used in integrated-circuit technology to produce a multiple of a small physical capacitance C when a large capacitance is needed. The circuit in Fig. 10.41 can be used to multiply capacitance values by a factor up to 1000. For example, a 10-pF capacitor can be made to behave like a 100-nF capacitor.

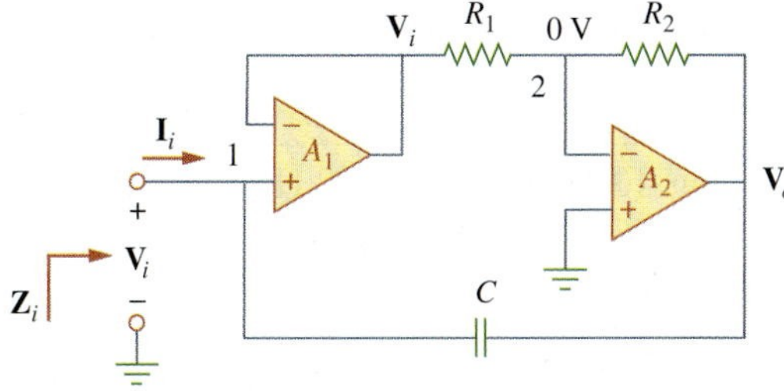

Figure 10.41
Capacitance multiplier.

In Fig. 10.41, the first op amp operates as a voltage follower, while the second one is an inverting amplifier. The voltage follower isolates the capacitance formed by the circuit from the loading imposed by the inverting amplifier. Since no current enters the input terminals of the op amp, the input current $\mathbf{I}_i$ flows through the feedback capacitor. Hence, at node 1,

$$\mathbf{I}_i = \frac{\mathbf{V}_i - \mathbf{V}_o}{1/j\omega C} = j\omega C(\mathbf{V}_i - \mathbf{V}_o) \tag{10.3}$$

Applying KCL at node 2 gives

$$\frac{\mathbf{V}_i - 0}{R_1} = \frac{0 - \mathbf{V}_o}{R_2}$$

or

$$\mathbf{V}_o = -\frac{R_2}{R_1}\mathbf{V}_i \tag{10.4}$$

Substituting Eq. (10.4) into (10.3) gives

$$\mathbf{I}_i = j\omega C\left(1 + \frac{R_2}{R_1}\right)\mathbf{V}_i$$

or

$$\frac{\mathbf{I}_i}{\mathbf{V}_i} = j\omega\left(1 + \frac{R_2}{R_1}\right)C \tag{10.5}$$

The input impedance is

$$\mathbf{Z}_i = \frac{\mathbf{V}_i}{\mathbf{I}_i} = \frac{1}{j\omega C_{\text{eq}}} \tag{10.6}$$

where

$$C_{\text{eq}} = \left(1 + \frac{R_2}{R_1}\right)C \tag{10.7}$$

Thus, by a proper selection of the values of R_1 and R_2, the op amp circuit in Fig. 10.41 can be made to produce an effective capacitance between the input terminal and ground, which is a multiple of the physical capacitance C. The size of the effective capacitance is practically limited by the inverted output voltage limitation. Thus, the larger the capacitance multiplication, the smaller is the allowable input voltage to prevent the op amps from reaching saturation.

A similar op amp circuit can be designed to simulate inductance. (See Prob. 10.89.) There is also an op amp circuit configuration to create a resistance multiplier.

Example 10.15

Calculate C_{eq} in Fig. 10.41 when $R_1 = 10\text{ k}\Omega$, $R_2 = 1\text{ M}\Omega$, and $C = 1\text{ nF}$.

Solution:

From Eq. (10.7)

$$C_{\text{eq}} = \left(1 + \frac{R_2}{R_1}\right)C = \left(1 + \frac{1 \times 10^6}{10 \times 10^3}\right)1\text{ nF} = 101\text{ nF}$$

Practice Problem 10.15

Determine the equivalent capacitance of the op amp circuit in Fig. 10.41 if $R_1 = 10\text{ k}\Omega$, $R_2 = 10\text{ M}\Omega$, and $C = 10\text{ nF}$.

Answer: $10\ \mu\text{F}$.

10.9.2 Oscillators

We know that dc is produced by batteries. But how do we produce ac? One way is using *oscillators*, which are circuits that convert dc to ac.

> An **oscillator** is a circuit that produces an ac waveform as output when powered by a dc input.

The only external source an oscillator needs is the dc power supply. Ironically, the dc power supply is usually obtained by converting the ac supplied by the electric utility company to dc. Having gone through the trouble of conversion, one may wonder why we need to use the oscillator to convert the dc to ac again. The problem is that the ac supplied by the utility company operates at a preset frequency of 60 Hz in the United States (50 Hz in some other nations), whereas many applications such as electronic circuits, communication systems, and microwave devices require internally generated frequencies that range from 0 to 10 GHz or higher. Oscillators are used for generating these frequencies.

This corresponds to $\omega = 2\pi f = 377$ rad/s.

In order for sine wave oscillators to sustain oscillations, they must meet the *Barkhausen criteria:*

1. The overall gain of the oscillator must be unity or greater. Therefore, losses must be compensated for by an amplifying device.
2. The overall phase shift (from input to output and back to the input) must be zero.

Three common types of sine wave oscillators are phase-shift, twin T, and Wien-bridge oscillators. Here we consider only the Wien-bridge oscillator.

The *Wien-bridge oscillator* is widely used for generating sinusoids in the frequency range below 1 MHz. It is an RC op amp circuit with only a few components, easily tunable and easy to design. As shown in Fig. 10.42, the oscillator essentially consists of a noninverting amplifier with two feedback paths: the positive feedback path to the noninverting input creates oscillations, while the negative feedback path to the inverting input controls the gain. If we define the impedances of the RC series and parallel combinations as $\mathbf{Z}_s$ and $\mathbf{Z}_p$, then

$$\mathbf{Z}_s = R_1 + \frac{1}{j\omega C_1} = R_1 - \frac{j}{\omega C_1} \tag{10.8}$$

$$\mathbf{Z}_p = R_2 \parallel \frac{1}{j\omega C_2} = \frac{R_2}{1 + j\omega R_2 C_2} \tag{10.9}$$

The feedback ratio is

$$\frac{\mathbf{V}_2}{\mathbf{V}_o} = \frac{\mathbf{Z}_p}{\mathbf{Z}_s + \mathbf{Z}_p} \tag{10.10}$$

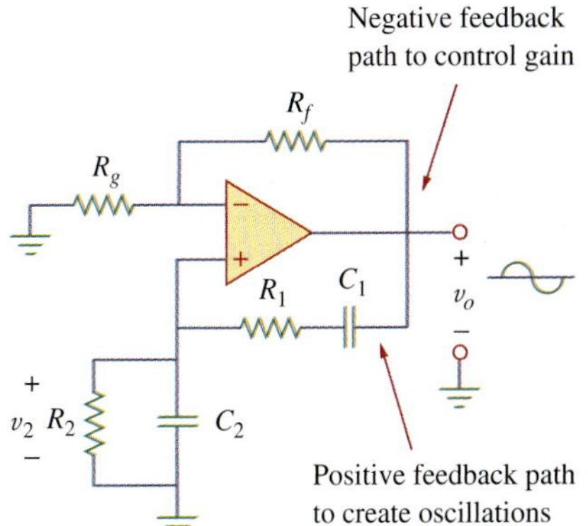

Figure 10.42 Wien-bridge oscillator.

Substituting Eqs. (10.8) and (10.9) into Eq. (10.10) gives

$$\frac{\mathbf{V}_2}{\mathbf{V}_o} = \frac{R_2}{R_2 + \left(R_1 - \dfrac{j}{\omega C_1}\right)(1 + j\omega R_2 C_2)} = \frac{\omega R_2 C_1}{\omega(R_2C_1 + R_1C_1 + R_2C_2) + j(\omega^2 R_1 C_1 R_2 C_2 - 1)} \tag{10.11}$$

To satisfy the second Barkhausen criterion, $\mathbf{V}_2$ must be in phase with $\mathbf{V}_o$, which implies that the ratio in Eq. (10.11) must be purely real. Hence, the imaginary part must be zero. Setting the imaginary part equal to zero gives the oscillation frequency ω_o as

$$\omega_o^2 R_1 C_1 R_2 C_2 - 1 = 0$$

or

$$\omega_o = \frac{1}{\sqrt{R_1 R_2 C_1 C_2}} \tag{10.12}$$

In most practical applications, $R_1 = R_2 = R$ and $C_1 = C_2 = C$, so that

$$\omega_o = \frac{1}{RC} = 2\pi f_o \tag{10.13}$$

or

$$\boxed{f_o = \frac{1}{2\pi RC}} \tag{10.14}$$

Substituting Eq. (10.13) and $R_1 = R_2 = R$, $C_1 = C_2 = C$ into Eq. (10.11) yields

$$\frac{\mathbf{V}_2}{\mathbf{V}_o} = \frac{1}{3} \tag{10.15}$$

Thus, in order to satisfy the first Barkhausen criterion, the op amp must compensate by providing a gain of 3 or greater so that the overall gain is at least 1 or unity. We recall that for a noninverting amplifier,

$$\frac{\mathbf{V}_o}{\mathbf{V}_2} = 1 + \frac{R_f}{R_g} = 3 \tag{10.16}$$

or

$$R_f = 2R_g \tag{10.17}$$

Due to the inherent delay caused by the op amp, Wien-bridge oscillators are limited to operating in the frequency range of 1 MHz or less.

Example 10.16

Design a Wien-bridge circuit to oscillate at 100 kHz.

Solution:

Using Eq. (10.14), we obtain the time constant of the circuit as

$$RC = \frac{1}{2\pi f_o} = \frac{1}{2\pi \times 100 \times 10^3} = 1.59 \times 10^{-6} \tag{10.16.1}$$

If we select $R = 10\ \text{k}\Omega$, then we can select $C = 159$ pF to satisfy Eq. (10.16.1). Since the gain must be 3, $R_f/R_g = 2$. We could select $R_f = 20\ \text{k}\Omega$ while $R_g = 10\ \text{k}\Omega$.

Practice Problem 10.16

In the Wien-bridge oscillator circuit in Fig. 10.42, let $R_1 = R_2 = 2.5\ \text{k}\Omega$, $C_1 = C_2 = 1$ nF. Determine the frequency f_o of the oscillator.

Answer: 63.66 kHz.

10.10 Summary

1. We apply nodal and mesh analysis to ac circuits by applying KCL and KVL to the phasor form of the circuits.
2. In solving for the steady-state response of a circuit that has independent sources with different frequencies, each independent source *must* be considered separately. The most natural approach to analyzing such circuits is to apply the superposition theorem. A separate phasor circuit for each frequency *must* be solved independently, and the corresponding response should be obtained in the time domain. The overall response is the sum of the time domain responses of all the individual phasor circuits.
3. The concept of source transformation is also applicable in the frequency domain.
4. The Thevenin equivalent of an ac circuit consists of a voltage source $\mathbf{V}_{\text{Th}}$ in series with the Thevenin impedance $\mathbf{Z}_{\text{Th}}$.
5. The Norton equivalent of an ac circuit consists of a current source $\mathbf{I}_N$ in parallel with the Norton impedance $\mathbf{Z}_N\ (=\mathbf{Z}_{\text{Th}})$.
6. *PSpice* is a simple and powerful tool for solving ac circuit problems. It relieves us of the tedious task of working with the complex numbers involved in steady-state analysis.
7. The capacitance multiplier and the ac oscillator provide two typical applications for the concepts presented in this chapter. A capacitance multiplier is an op amp circuit used in producing a multiple of a physical capacitance. An oscillator is a device that uses a dc input to generate an ac output.

Review Questions

10.1 The voltage $\mathbf{V}_o$ across the capacitor in Fig. 10.43 is:

(a) $5\angle 0^\circ$ V (b) $7.071\angle 45^\circ$ V
(c) $7.071\angle -45^\circ$ V (d) $5\angle -45^\circ$ V

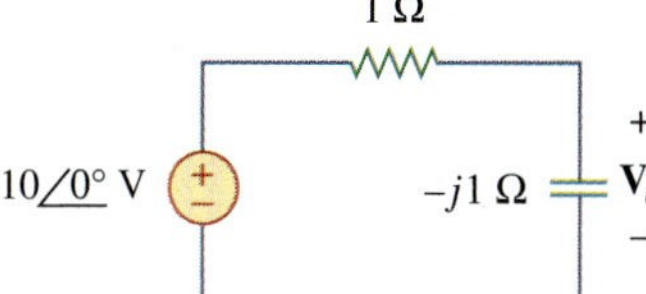

Figure 10.43
For Review Question 10.1.

10.2 The value of the current $\mathbf{I}_o$ in the circuit of Fig. 10.44 is:

(a) $4\angle 0^\circ$ A (b) $2.4\angle -90^\circ$ A
(c) $0.6\angle 0^\circ$ A (d) -1 A

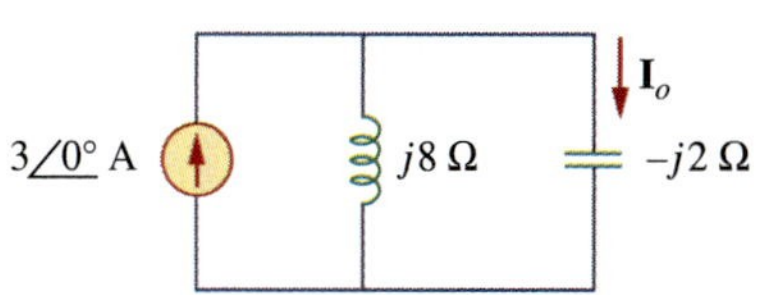

Figure 10.44
For Review Question 10.2.

10.3 Using nodal analysis, the value of $\mathbf{V}_o$ in the circuit of Fig. 10.45 is:

(a) −24 V (b) −8 V
(c) 8 V (d) 24 V

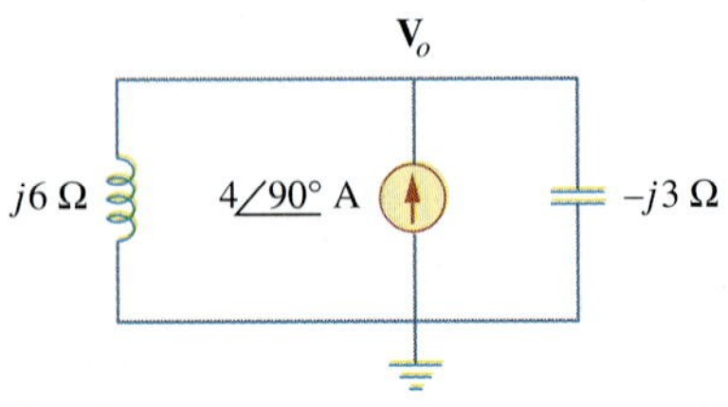

Figure 10.45
For Review Question 10.3.

10.4 In the circuit of Fig. 10.46, current $i(t)$ is:

(a) 10 cos t A (b) 10 sin t A (c) 5 cos t A
(d) 5 sin t A (e) 4.472 cos(t − 63.43°) A

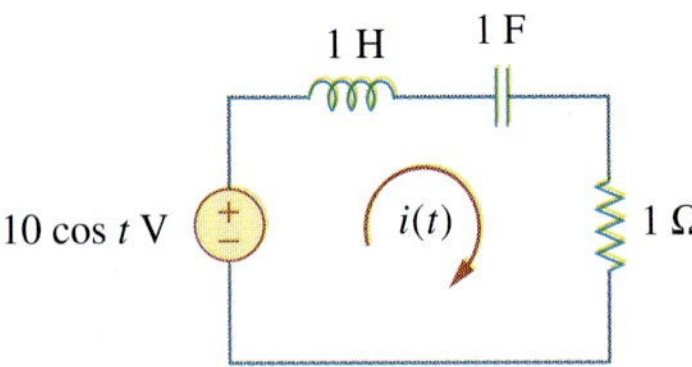

Figure 10.46
For Review Question 10.4.

10.5 Refer to the circuit in Fig. 10.47 and observe that the two sources do not have the same frequency. The current $i_x(t)$ can be obtained by:

(a) source transformation
(b) the superposition theorem
(c) *PSpice*

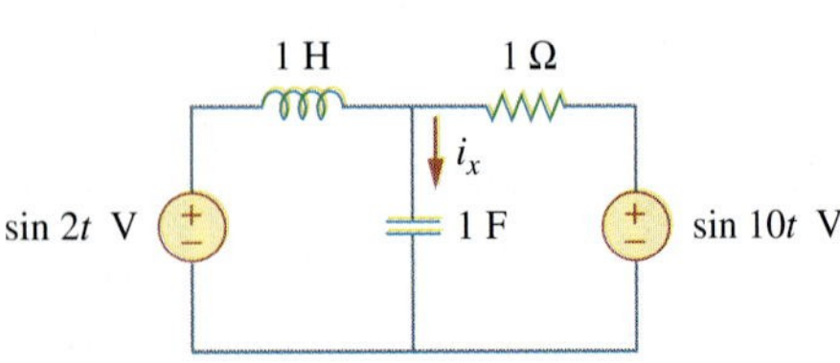

Figure 10.47
For Review Question 10.5.

10.6 For the circuit in Fig. 10.48, the Thevenin impedance at terminals a-b is:

(a) 1 Ω (b) 0.5 − j0.5 Ω
(c) 0.5 + j0.5 Ω (d) 1 + j2 Ω
(e) 1 − j2 Ω

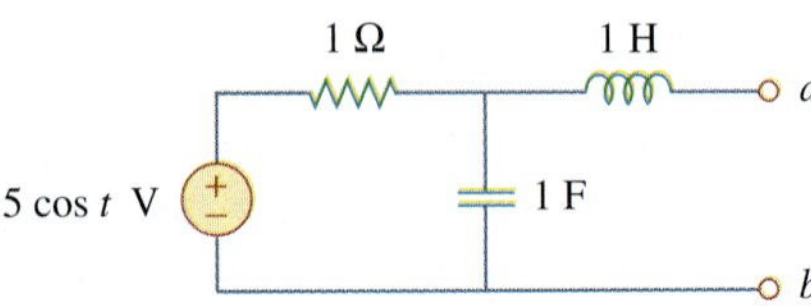

Figure 10.48
For Review Questions 10.6 and 10.7.

10.7 In the circuit of Fig. 10.48, the Thevenin voltage at terminals a-b is:

(a) $3.535\angle -45^\circ$ V (b) $3.535\angle 45^\circ$ V
(c) $7.071\angle -45^\circ$ V (d) $7.071\angle 45^\circ$ V

10.8 Refer to the circuit in Fig. 10.49. The Norton equivalent impedance at terminals a-b is:

(a) −j4 Ω (b) −j2 Ω
(c) j2 Ω (d) j4 Ω

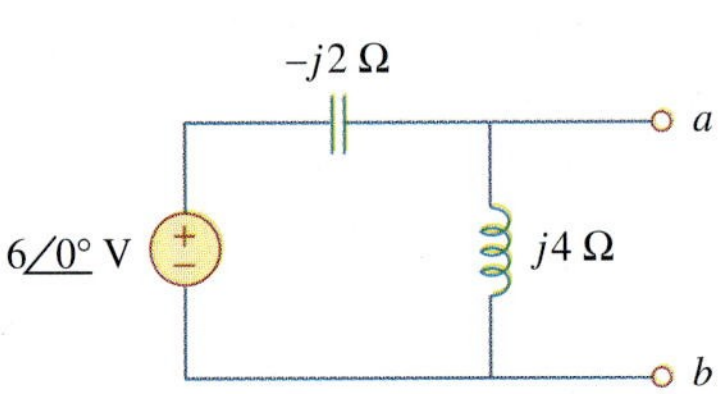

Figure 10.49
For Review Questions 10.8 and 10.9.

10.9 The Norton current at terminals a-b in the circuit of Fig. 10.49 is:

(a) $1\angle 0^\circ$ A (b) $1.5\angle -90^\circ$ A
(c) $1.5\angle 90^\circ$ A (d) $3\angle 90^\circ$ A

10.10 *PSpice* can handle a circuit with two independent sources of different frequencies.

(a) True (b) False

Answers: 10.1c, 10.2a, 10.3d, 10.4a, 10.5b, 10.6c, 10.7a, 10.8a, 10.9d, 10.10b.

Problems

Section 10.2 Nodal Analysis

10.1 Determine i in the circuit of Fig. 10.50.

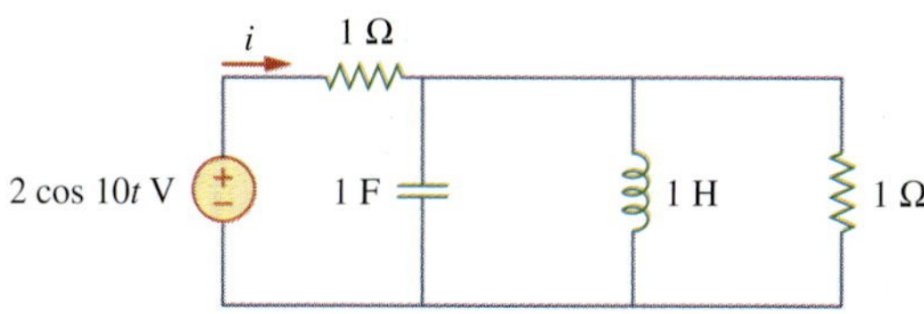

Figure 10.50
For Prob. 10.1.

10.2 Using Fig. 10.51, design a problem to help other students better understand nodal analysis.

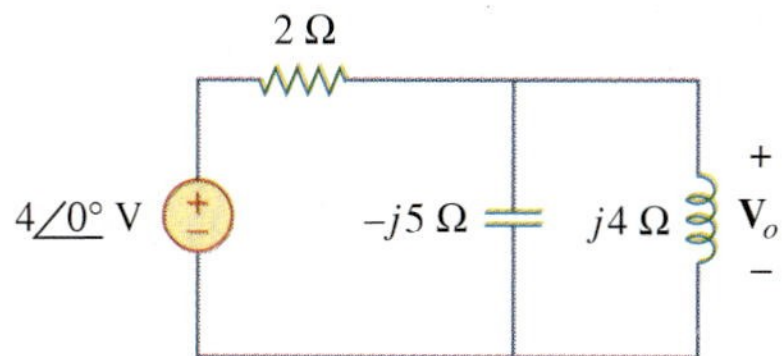

Figure 10.51
For Prob. 10.2.

10.3 Determine v_o in the circuit of Fig. 10.52.

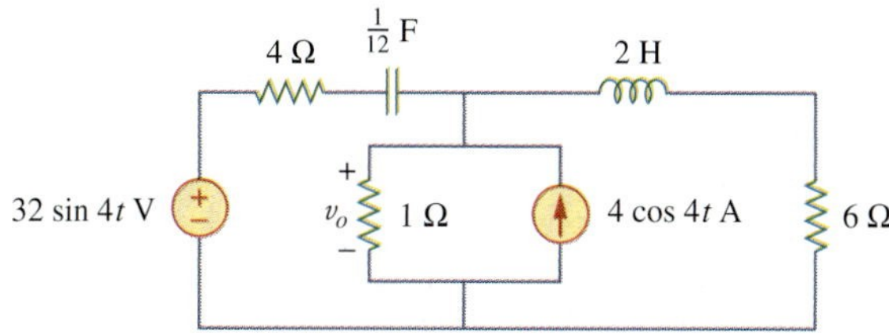

Figure 10.52
For Prob. 10.3.

10.4 Determine i_1 in the circuit of Fig. 10.53.

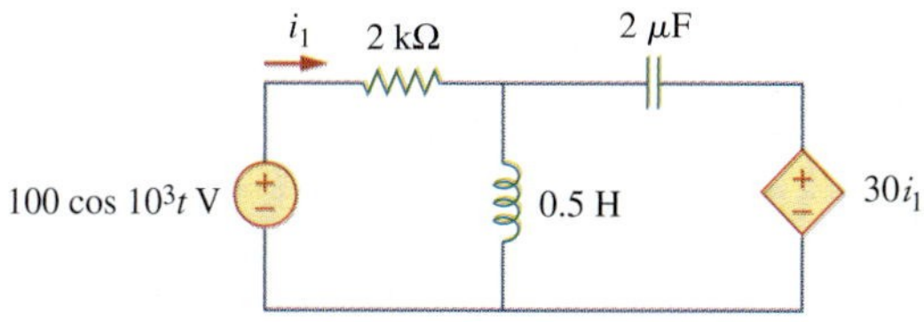

Figure 10.53
For Prob. 10.4.

10.5 Find i_o in the circuit of Fig. 10.54.

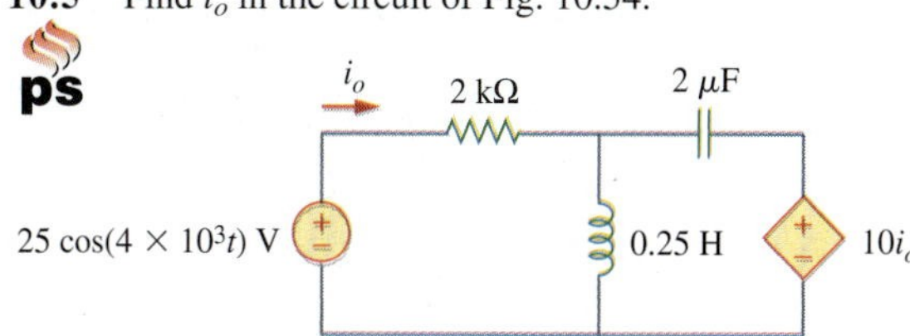

Figure 10.54
For Prob. 10.5.

10.6 Determine $\mathbf{V}_x$ in Fig. 10.55.

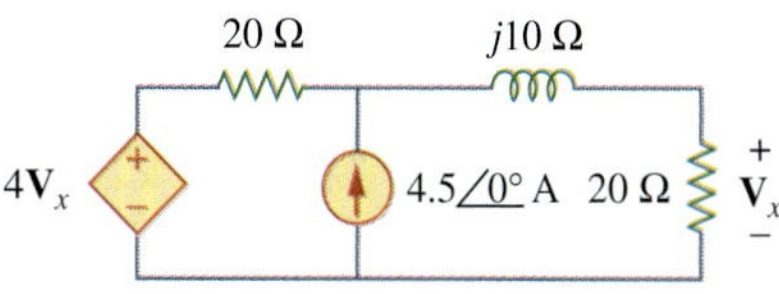

Figure 10.55
For Prob. 10.6.

10.7 Use nodal analysis to find $\mathbf{V}$ in the circuit of Fig. 10.56.

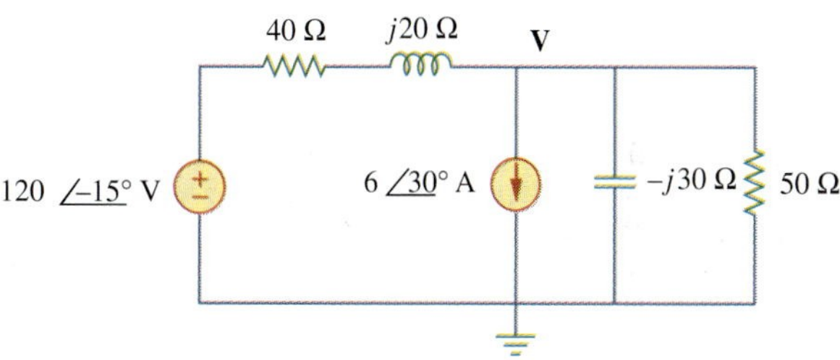

Figure 10.56
For Prob. 10.7.

10.8 Use nodal analysis to find current i_o in the circuit of Fig. 10.57. Let $i_s = 6\cos(200t + 15°)$ A.

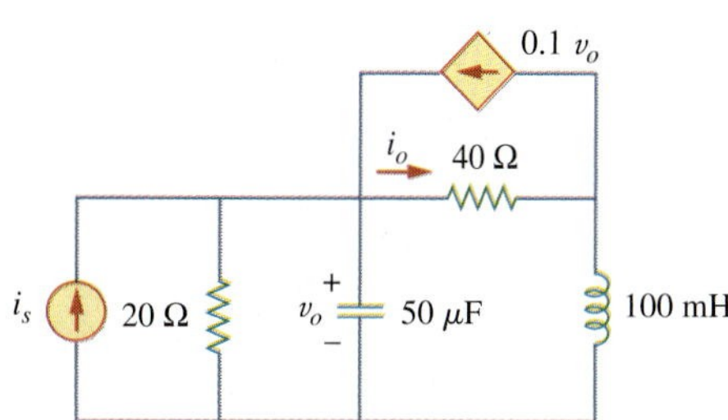

Figure 10.57
For Prob. 10.8.

10.9 Use nodal analysis to find v_o in the circuit of Fig. 10.58.

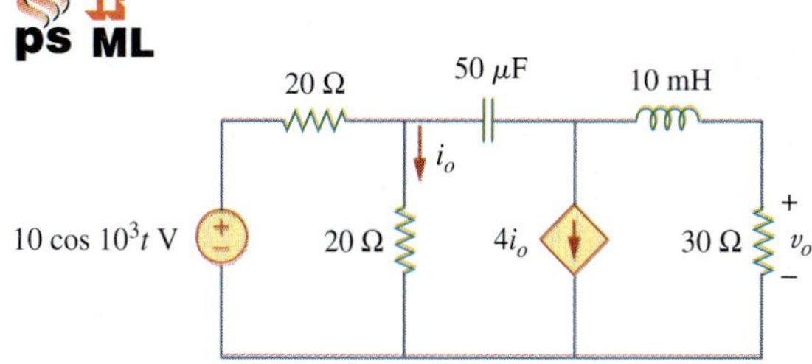

Figure 10.58
For Prob. 10.9.

10.10 Use nodal analysis to find v_o in the circuit of Fig. 10.59. Let $\omega = 2$ krad/s.

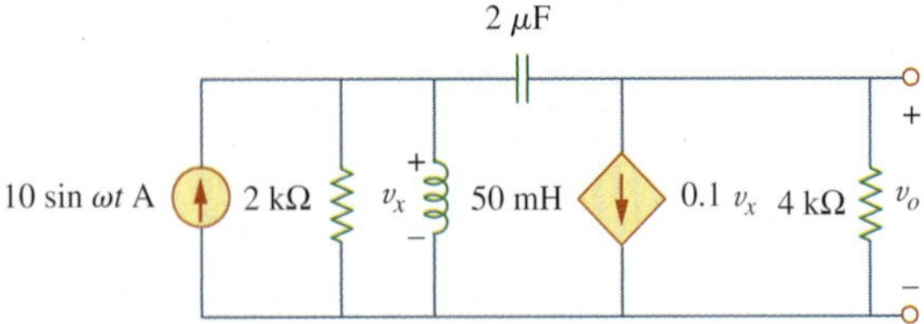

Figure 10.59
For Prob. 10.10.

10.11 Apply nodal analysis to the circuit in Fig. 10.60 and determine $\mathbf{I}_o$.

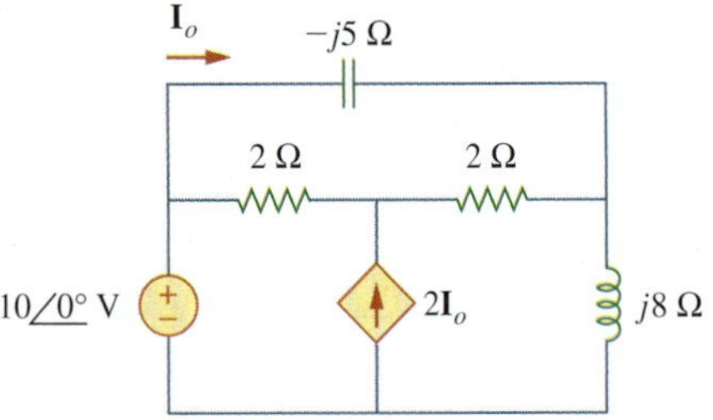

Figure 10.60
For Prob. 10.11.

10.12 Using Fig. 10.61, design a problem to help other students better understand nodal analysis.

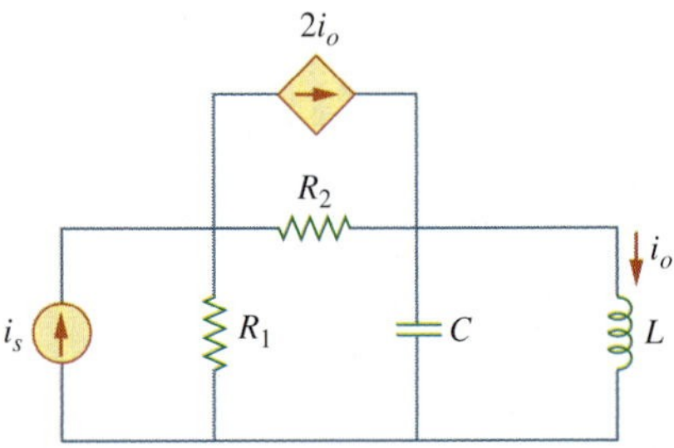

Figure 10.61
For Prob. 10.12.

10.13 Determine $\mathbf{V}_x$ in the circuit of Fig. 10.62 using any method of your choice.

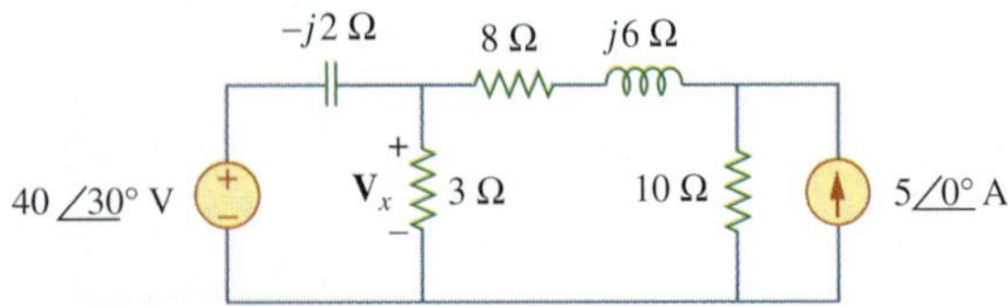

Figure 10.62
For Prob. 10.13.

10.14 Calculate the voltage at nodes 1 and 2 in the circuit of Fig. 10.63 using nodal analysis.

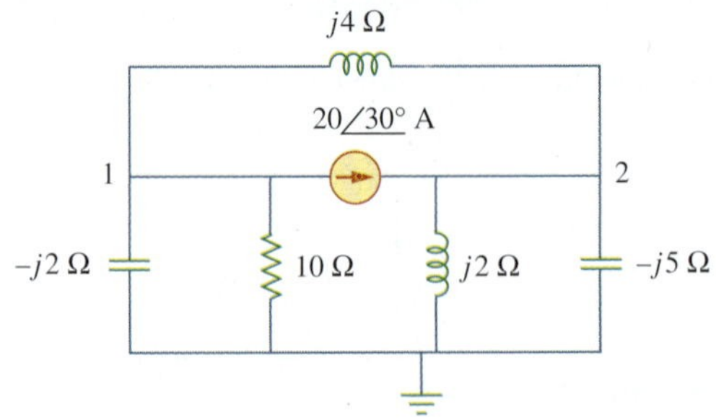

Figure 10.63
For Prob. 10.14.

10.15 Solve for the current **I** in the circuit of Fig. 10.64 using nodal analysis.

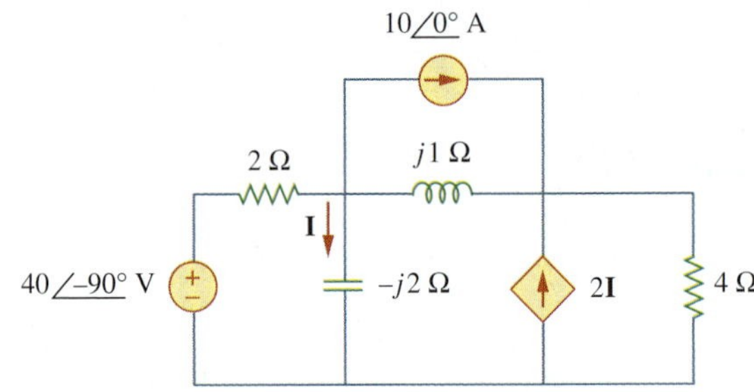

Figure 10.64
For Prob. 10.15.

10.16 Use nodal analysis to find $\mathbf{V}_x$ in the circuit shown in Fig. 10.65.

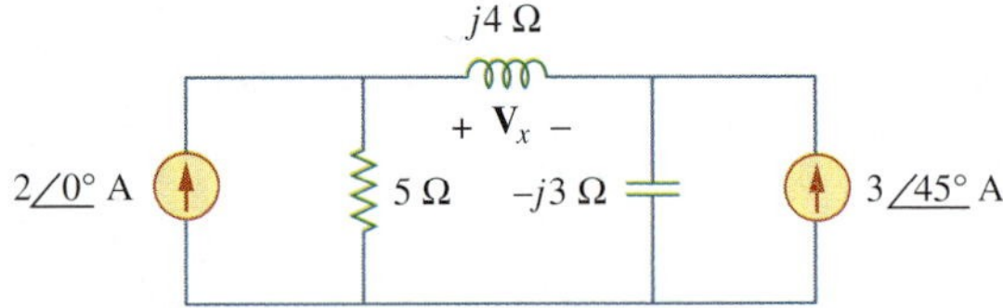

Figure 10.65
For Prob. 10.16.

10.17 By nodal analysis, obtain current $\mathbf{I}_o$ in the circuit of Fig. 10.66.

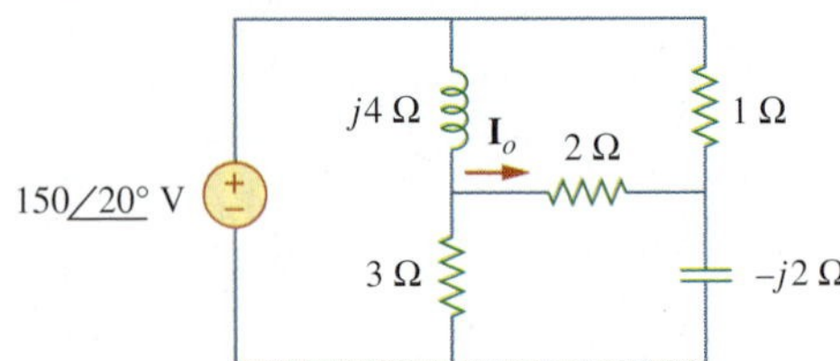

Figure 10.66
For Prob. 10.17.

10.18 Use nodal analysis to obtain $\mathbf{V}_o$ in the circuit of Fig. 10.67 below.

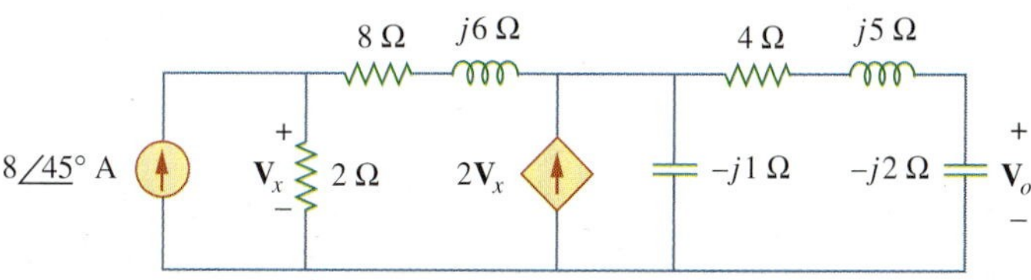

Figure 10.67
For Prob. 10.18.

10.19 Obtain $\mathbf{V}_o$ in Fig. 10.68 using nodal analysis.

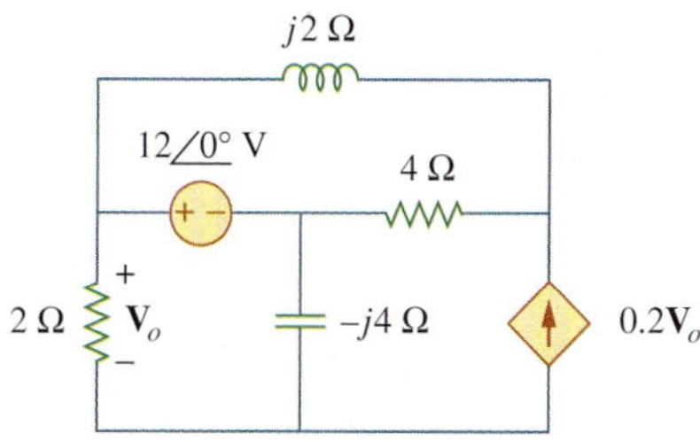

Figure 10.68
For Prob. 10.19.

10.20 Refer to Fig. 10.69. If $v_s(t) = V_m \sin\omega t$ and $v_o(t) = A\sin(\omega t + \phi)$, derive the expressions for A and ϕ.

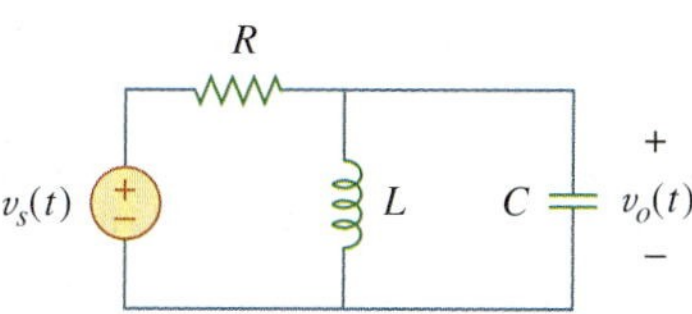

Figure 10.69
For Prob. 10.20.

10.21 For each of the circuits in Fig. 10.70, find $\mathbf{V}_o/\mathbf{V}_i$ for $\omega = 0$, $\omega \to \infty$, and $\omega^2 = 1/LC$.

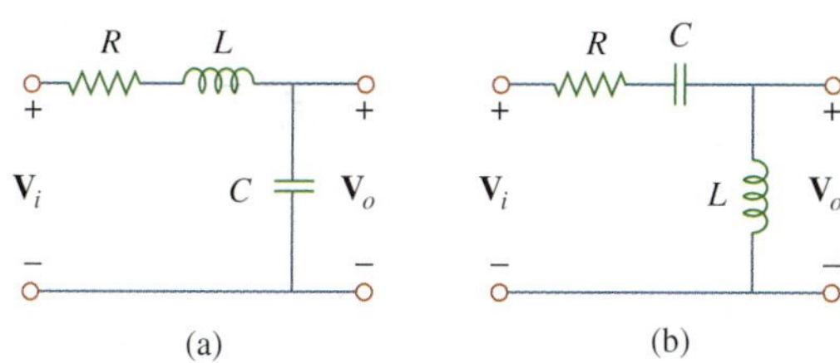

Figure 10.70
For Prob. 10.21.

10.22 For the circuit in Fig. 10.71, determine $\mathbf{V}_o/\mathbf{V}_s$.

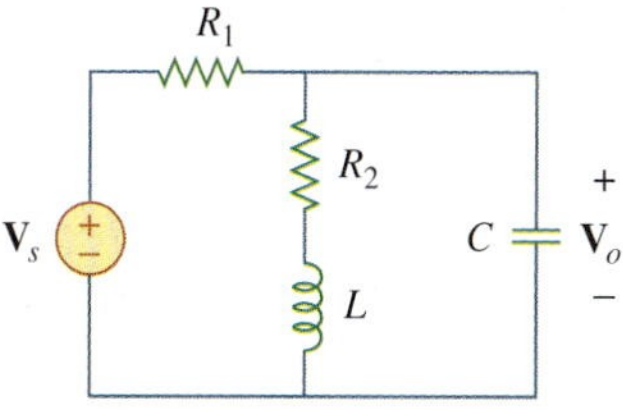

Figure 10.71
For Prob. 10.22.

10.23 Using nodal analysis obtain $\mathbf{V}$ in the circuit of Fig. 10.72.

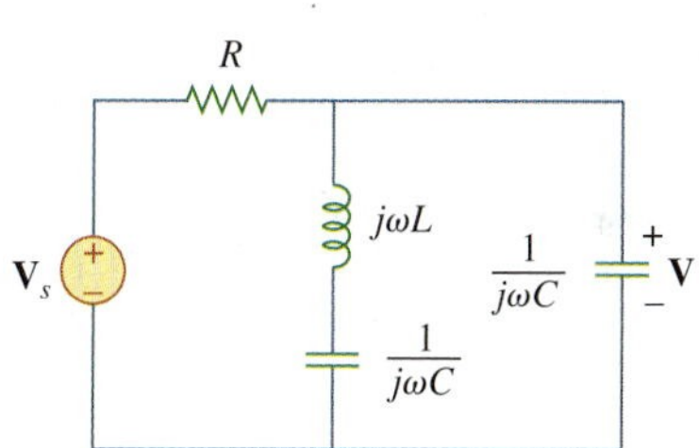

Figure 10.72
For Prob. 10.23.

Section 10.3 Mesh Analysis

10.24 Design a problem to help other students better understand mesh analysis.

10.25 Solve for i_o in Fig. 10.73 using mesh analysis.

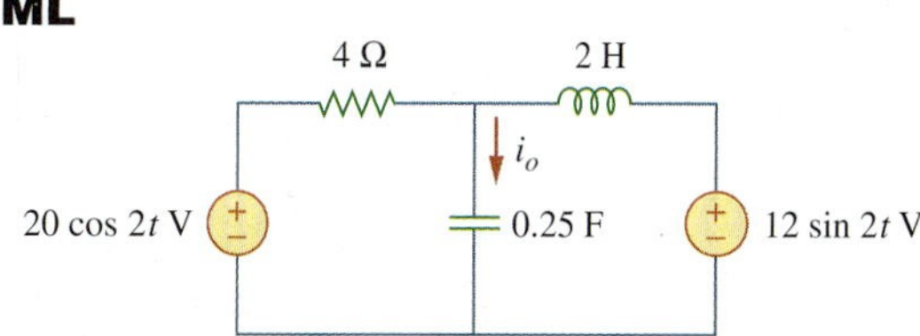

Figure 10.73
For Prob. 10.25.

10.26 Use mesh analysis to find current i_o in the circuit of Fig. 10.74.

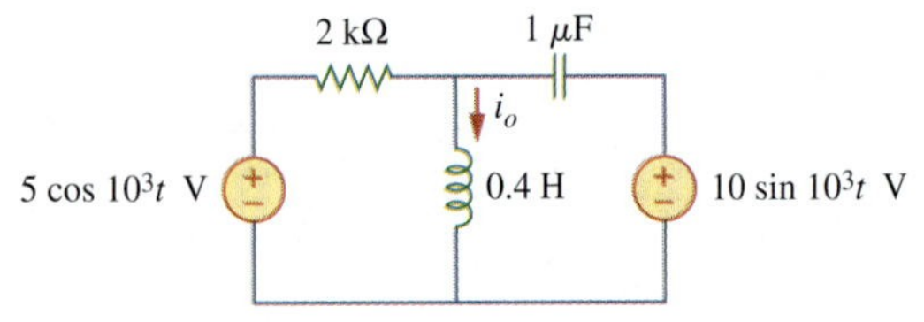

Figure 10.74
For Prob. 10.26.

10.27 Using mesh analysis, find $\mathbf{I}_1$ and $\mathbf{I}_2$ in the circuit of Fig. 10.75.

ML

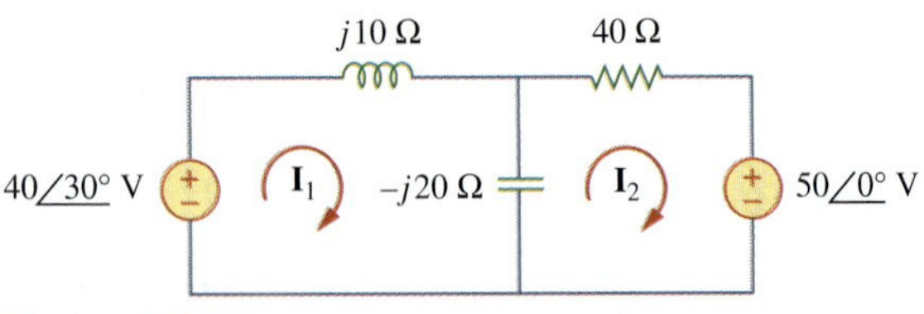

Figure 10.75
For Prob. 10.27.

10.28 In the circuit of Fig. 10.76, determine the mesh currents i_1 and i_2. Let $v_1 = 10 \cos 4t$ **V** and $v_2 = 20 \cos(4t - 30°)$ V.

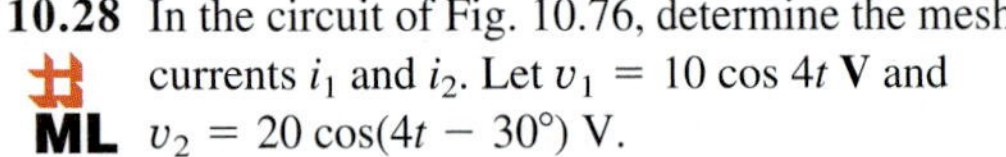

ML

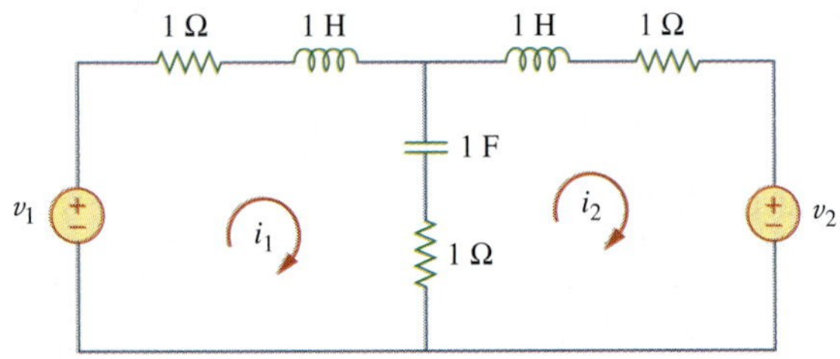

Figure 10.76
For Prob. 10.28.

10.29 Using Fig. 10.77, design a problem to help other students better understand mesh analysis.

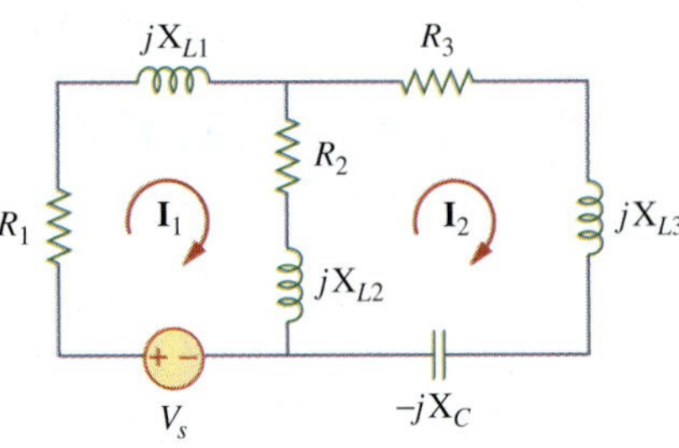

Figure 10.77
For Prob. 10.29.

10.30 Use mesh analysis to find v_o in the circuit of Fig. 10.78. Let $v_{s1} = 240 \cos(100t + 90°)$ V, $v_{s2} = 160 \cos 100t$ V.

ps ML

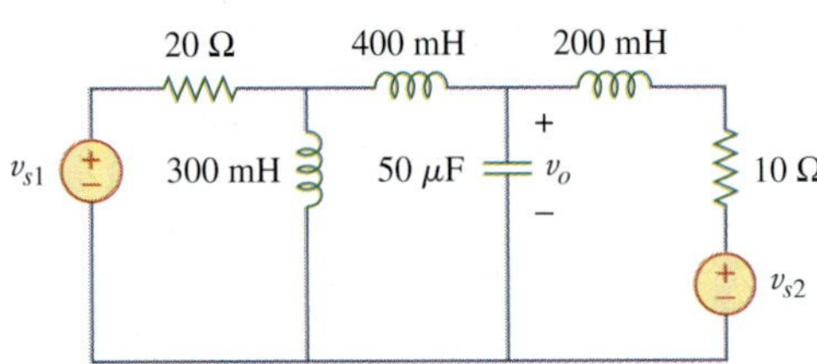

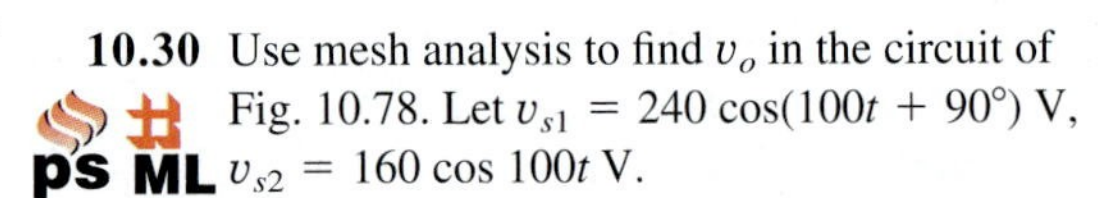

Figure 10.78
For Prob. 10.30.

10.31 Use mesh analysis to determine current $\mathbf{I}_o$ in the circuit of Fig. 10.79 below.

ps ML

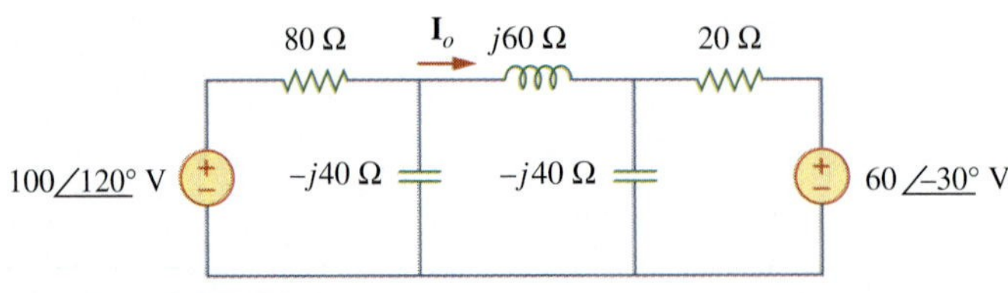

Figure 10.79
For Prob. 10.31.

10.32 Determine $\mathbf{V}_o$ and $\mathbf{I}_o$ in the circuit of Fig. 10.80 using mesh analysis.

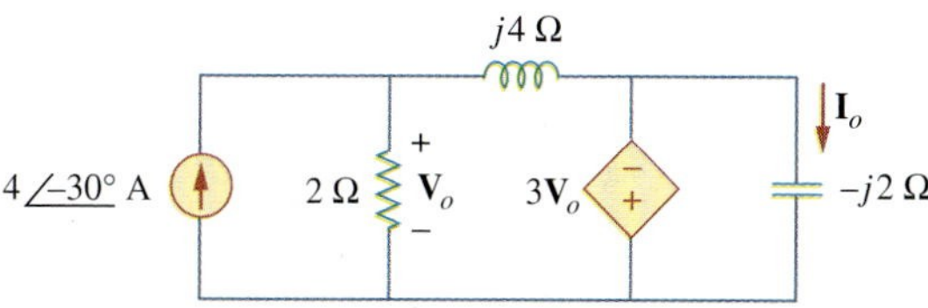

Figure 10.80
For Prob. 10.32.

10.33 Compute **I** in Prob. 10.15 using mesh analysis.

10.34 Use mesh analysis to find $\mathbf{I}_o$ in Fig. 10.28 (for Example 10.10).

10.35 Calculate $\mathbf{I}_o$ in Fig. 10.30 (for Practice Prob. 10.10) using mesh analysis.

10.36 Compute $\mathbf{V}_o$ in the circuit of Fig. 10.81 using mesh analysis.

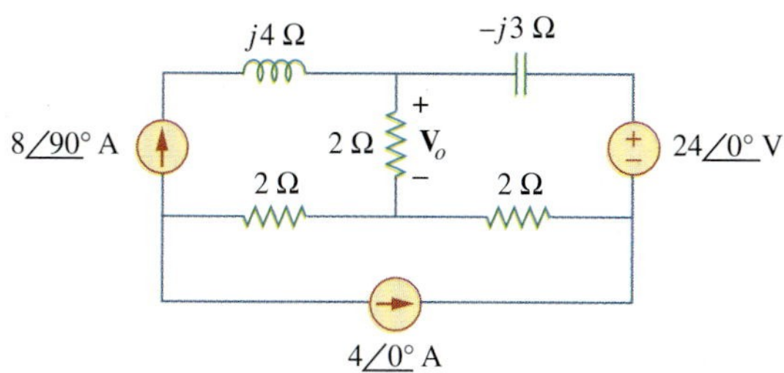

Figure 10.81
For Prob. 10.36.

10.37 Use mesh analysis to find currents $\mathbf{I}_1$, $\mathbf{I}_2$, and $\mathbf{I}_3$ in the circuit of Fig. 10.82.

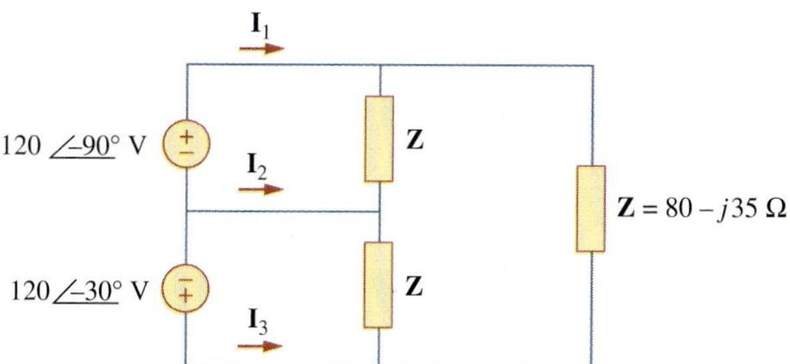

Figure 10.82
For Prob. 10.37.

10.38 Using mesh analysis, obtain $\mathbf{I}_o$ in the circuit shown in Fig. 10.83.

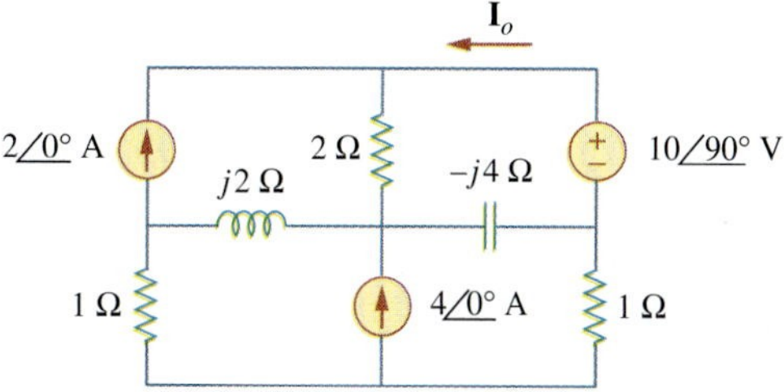

Figure 10.83
For Prob. 10.38.

10.39 Find $\mathbf{I}_1$, $\mathbf{I}_2$, $\mathbf{I}_3$, and $\mathbf{I}_x$ in the circuit of Fig. 10.84.

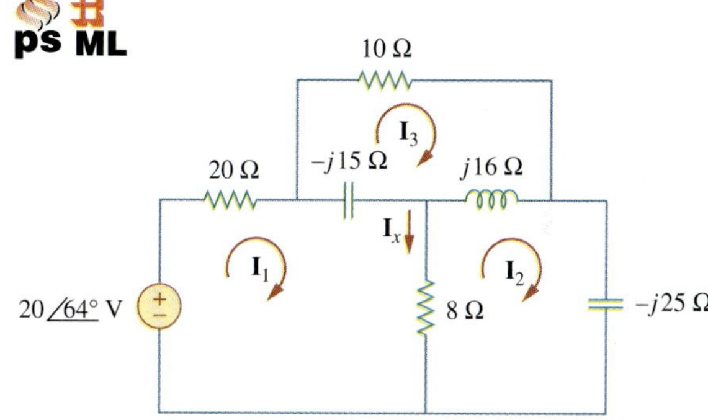

Figure 10.84
For Prob. 10.39.

Section 10.4 Superposition Theorem

10.40 Find i_o in the circuit shown in Fig. 10.85 using superposition.

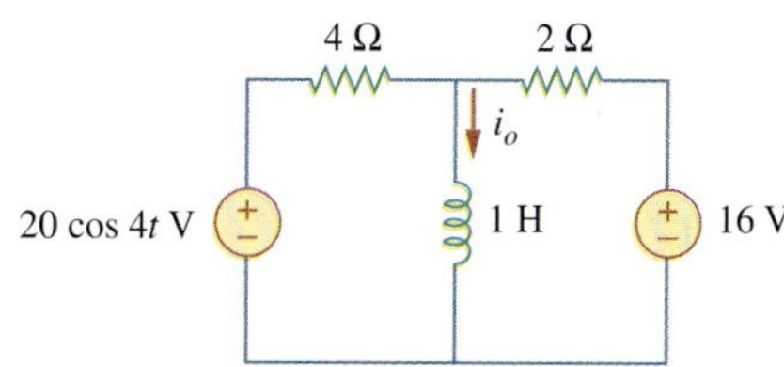

Figure 10.85
For Prob. 10.40.

10.41 Find v_o for the circuit in Fig. 10.86, assuming that $v_s = 3 \cos 2t + 8 \sin 4t$ V.

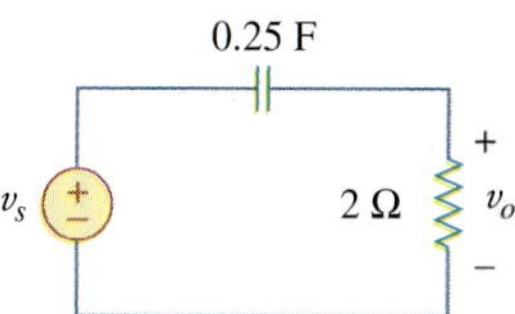

Figure 10.86
For Prob. 10.41.

10.42 Using Fig. 10.87, design a problem to help other students better understand the superposition theorem.

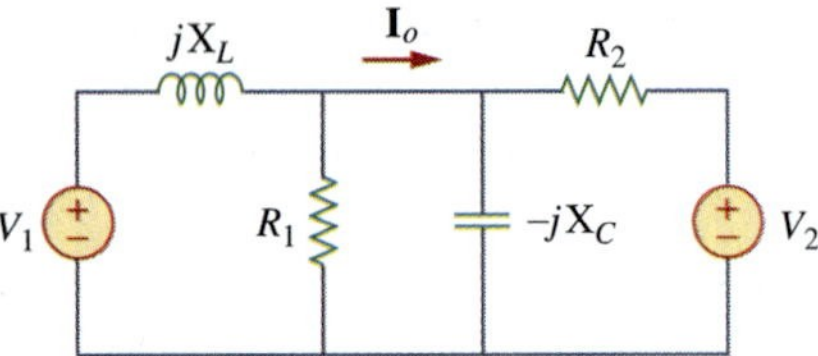

Figure 10.87
For Prob. 10.42.

10.43 Using the superposition principle, find i_x in the circuit of Fig. 10.88.

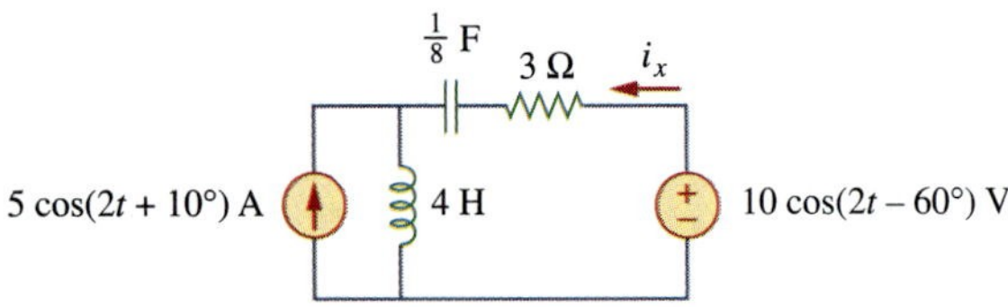

Figure 10.88
For Prob. 10.43.

10.44 Use the superposition principle to obtain v_x in the circuit of Fig. 10.89. Let $v_s = 25 \sin 2t$ V and $i_s = 6 \cos(6t + 10°)$ A.

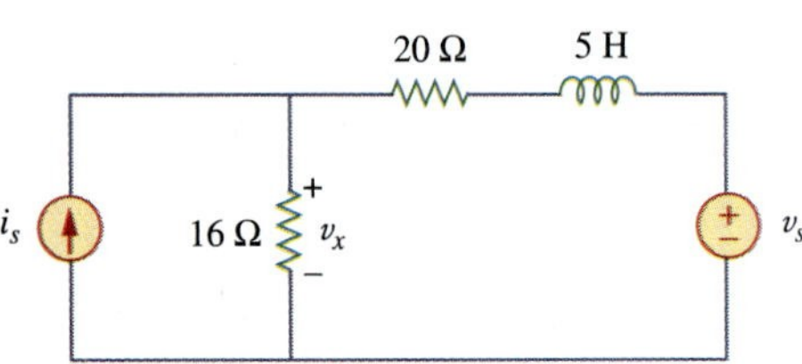

Figure 10.89
For Prob. 10.44.

10.45 Use superposition to find $i(t)$ in the circuit of Fig. 10.90.

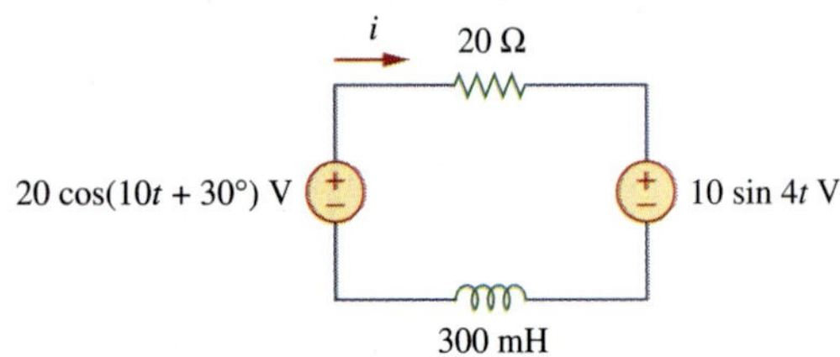

Figure 10.90
For Prob. 10.45.

10.46 Solve for $v_o(t)$ in the circuit of Fig. 10.91 using the superposition principle.

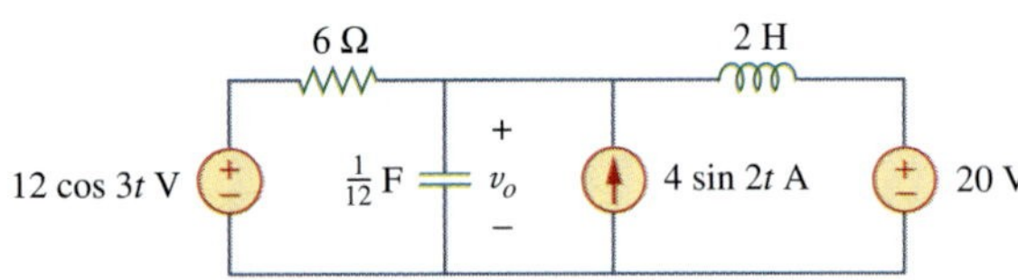

Figure 10.91
For Prob. 10.46.

10.47 Determine i_o in the circuit of Fig. 10.92, using the superposition principle.

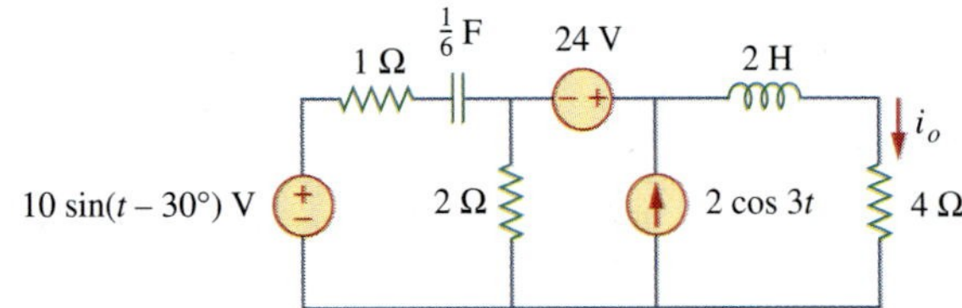

Figure 10.92
For Prob. 10.47.

10.48 Find i_o in the circuit of Fig. 10.93 using superposition.

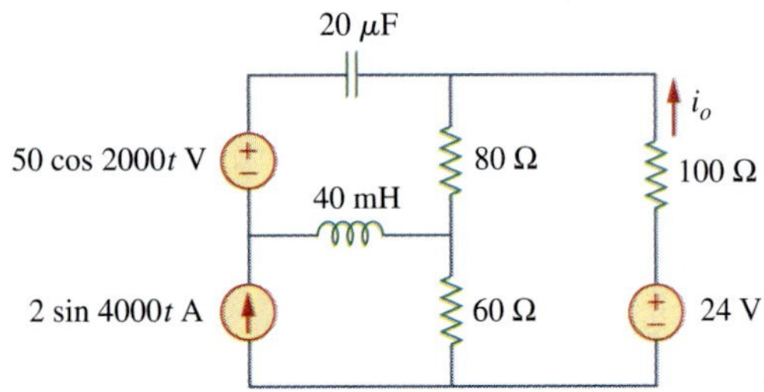

Figure 10.93
For Prob. 10.48.

Section 10.5 Source Transformation

10.49 Using source transformation, find i in the circuit of Fig. 10.94.

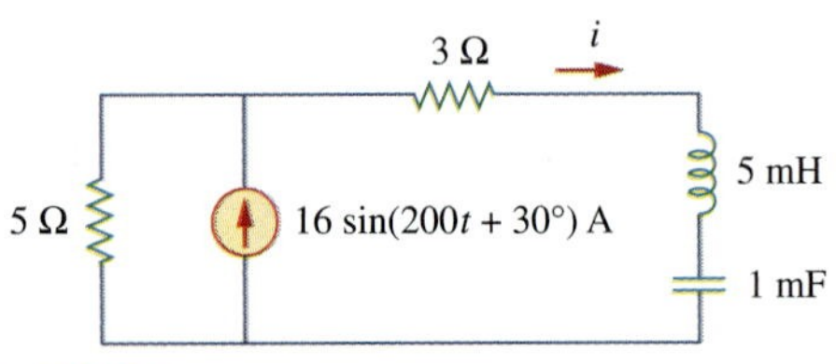

Figure 10.94
For Prob. 10.49.

10.50 Using Fig. 10.95, design a problem to help other students understand source transformation.

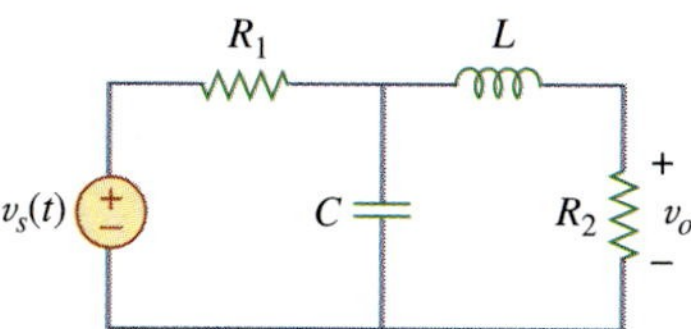

Figure 10.95
For Prob. 10.50.

10.51 Use source transformation to find $\mathbf{I}_o$ in the circuit of Prob. 10.42.

10.52 Use the method of source transformation to find $\mathbf{I}_x$ in the circuit of Fig. 10.96.

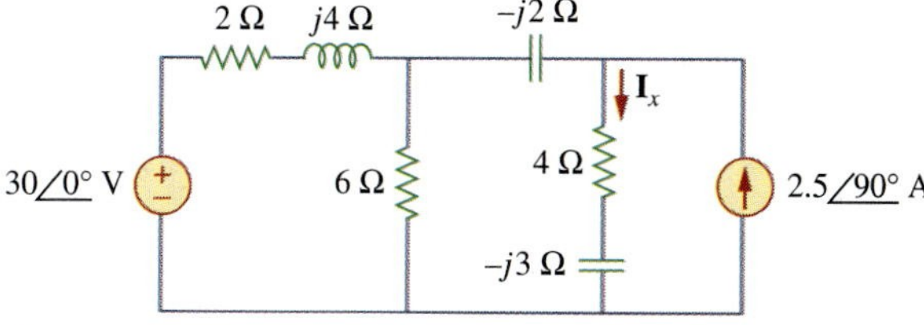

Figure 10.96
For Prob. 10.52.

10.53 Use the concept of source transformation to find $\mathbf{V}_o$ in the circuit of Fig. 10.97.

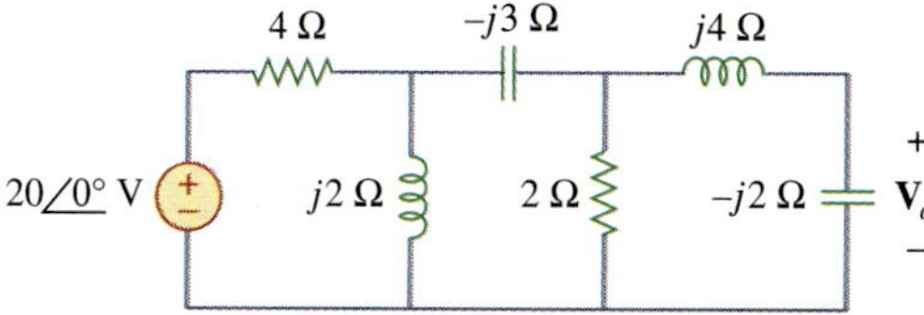

Figure 10.97
For Prob. 10.53.

10.54 Rework Prob. 10.7 using source transformation.

Section 10.6 Thevenin and Norton Equivalent Circuits

10.55 Find the Thevenin and Norton equivalent circuits at terminals *a-b* for each of the circuits in Fig. 10.98.

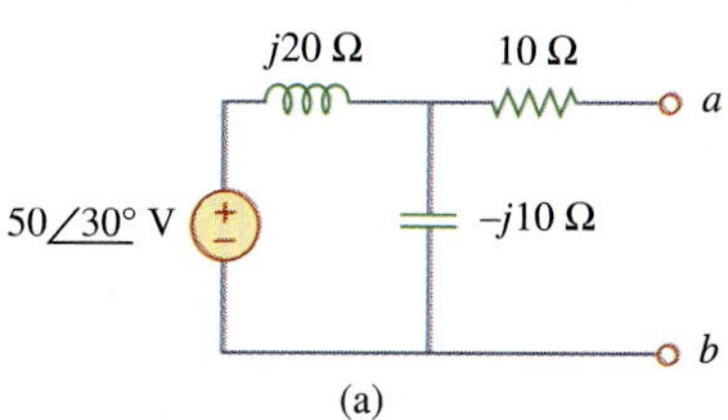

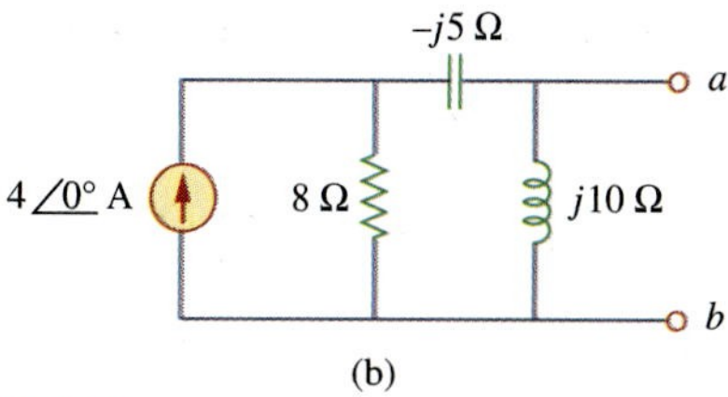

Figure 10.98
For Prob. 10.55.

10.56 For each of the circuits in Fig. 10.99, obtain Thevenin and Norton equivalent circuits at terminals *a-b*.

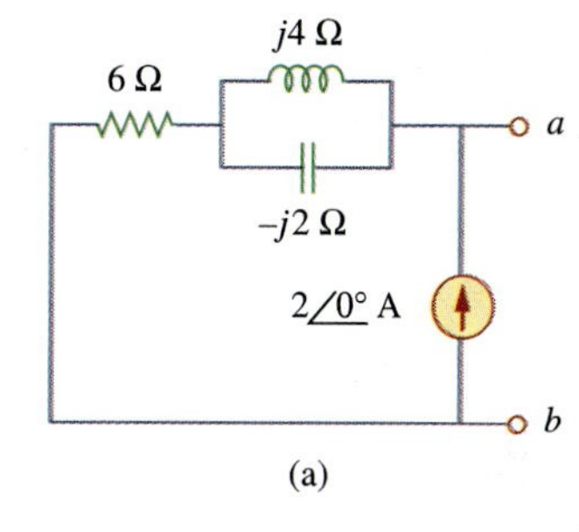

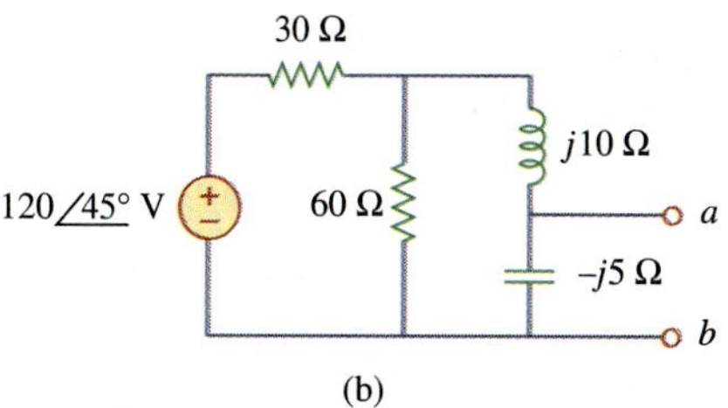

Figure 10.99
For Prob. 10.56.

10.57 Using Fig. 10.100, design a problem to help other students better understand Thevenin and Norton equivalent circuits.

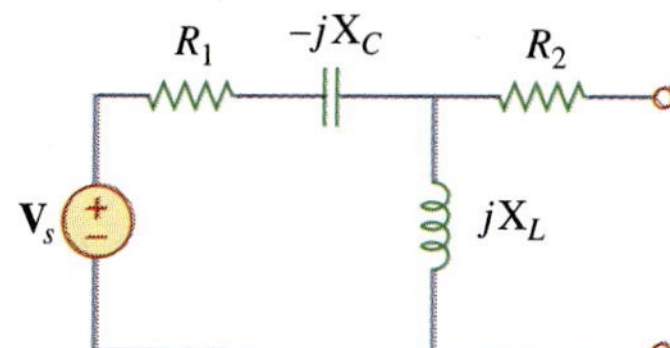

Figure 10.100
For Prob. 10.57.

10.58 For the circuit depicted in Fig. 10.101, find the Thevenin equivalent circuit at terminals *a-b*.

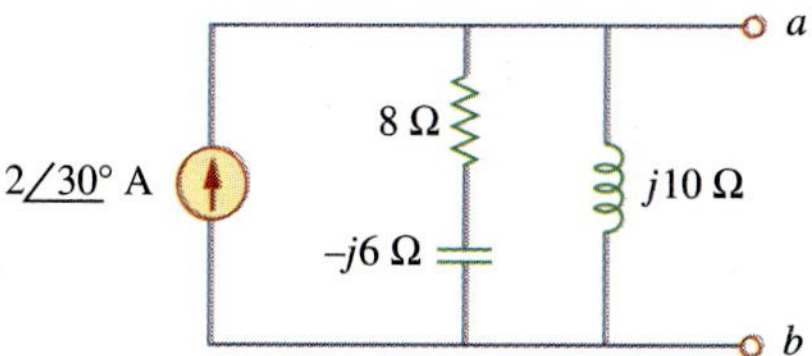

Figure 10.101
For Prob. 10.58.

10.59 Calculate the output impedance of the circuit shown in Fig. 10.102.

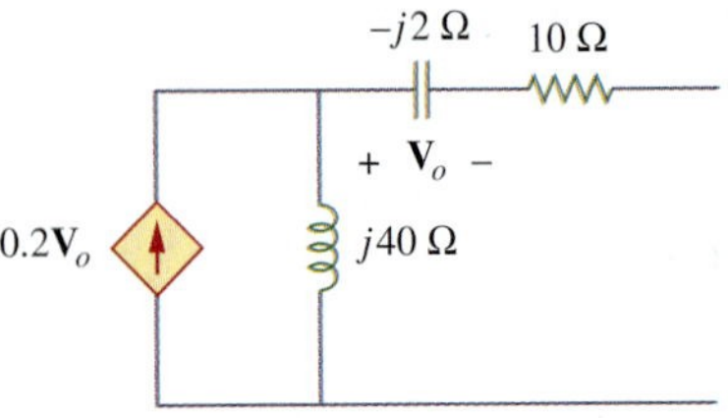

Figure 10.102
For Prob. 10.59.

10.60 Find the Thevenin equivalent of the circuit in Fig. 10.103 as seen from:

(a) terminals *a-b* (b) terminals *c-d*

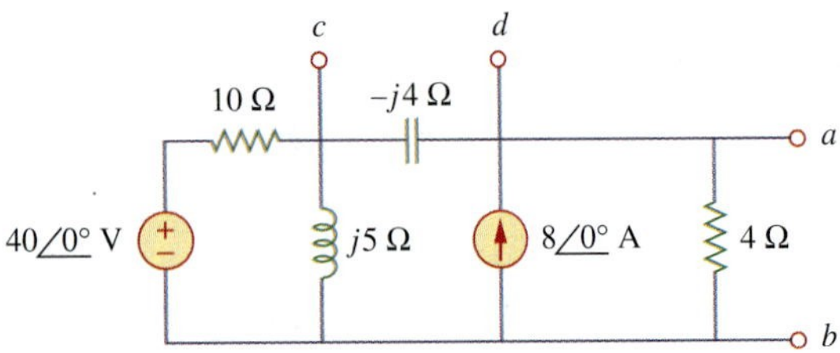

Figure 10.103
For Prob. 10.60.

10.61 Find the Thevenin equivalent at terminals *a-b* of the circuit in Fig. 10.104.

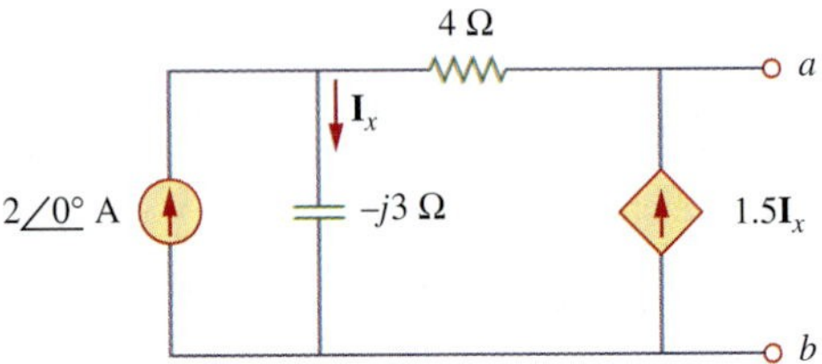

Figure 10.104
For Prob. 10.61.

10.62 Using Thevenin's theorem, find v_o in the circuit of Fig. 10.105.

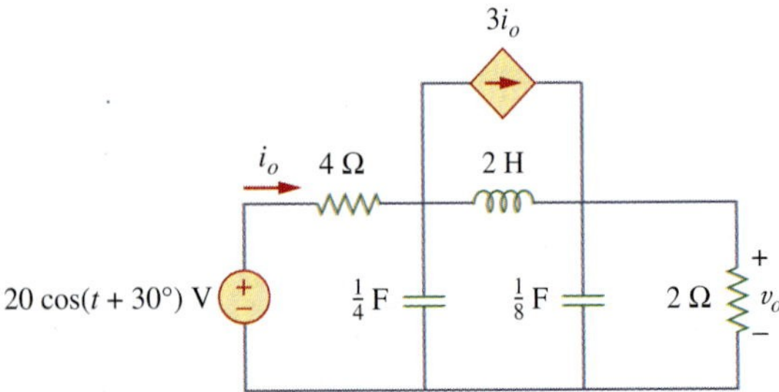

Figure 10.105
For Prob. 10.62.

10.63 Obtain the Norton equivalent of the circuit depicted in Fig. 10.106 at terminals *a-b*.

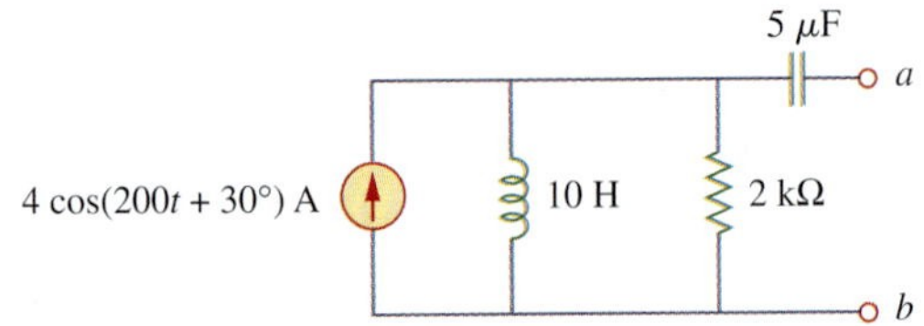

Figure 10.106
For Prob. 10.63.

10.64 For the circuit shown in Fig. 10.107, find the Norton equivalent circuit at terminals *a-b*.

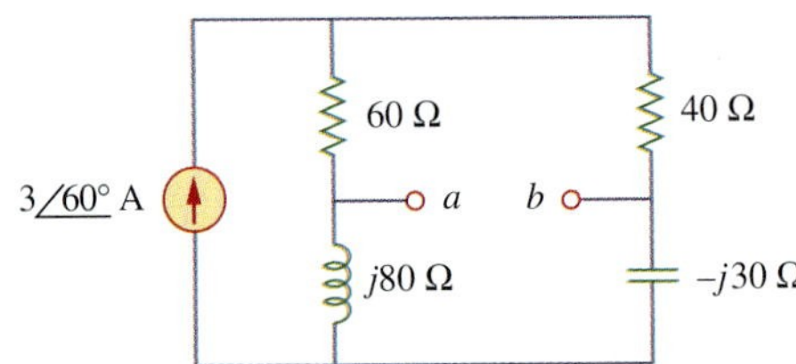

Figure 10.107
For Prob. 10.64.

10.65 Using Fig. 10.108, design a problem to help other students better understand Norton's theorem.

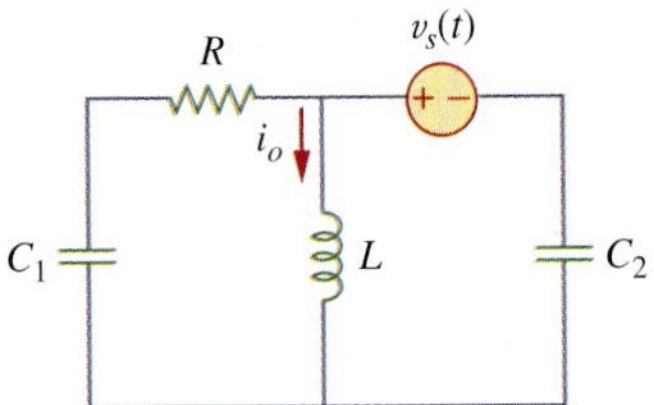

Figure 10.108
For Prob. 10.65.

10.66 At terminals *a-b*, obtain Thevenin and Norton equivalent circuits for the network depicted in Fig. 10.109. Take $\omega = 10$ rad/s.

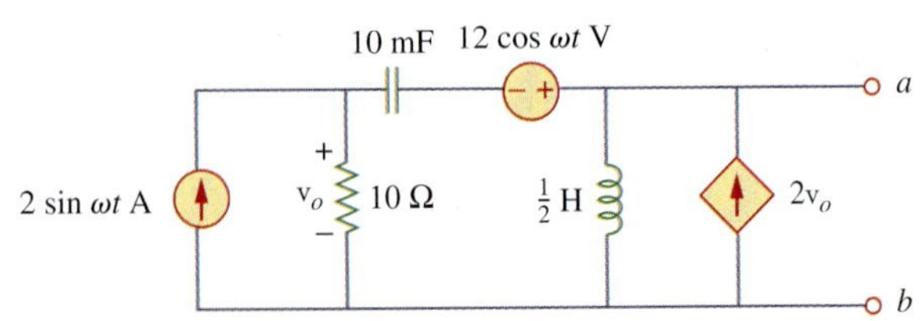

Figure 10.109
For Prob. 10.66.

10.67 Find the Thevenin and Norton equivalent circuits at terminals *a-b* in the circuit of Fig. 10.110.
ps ML

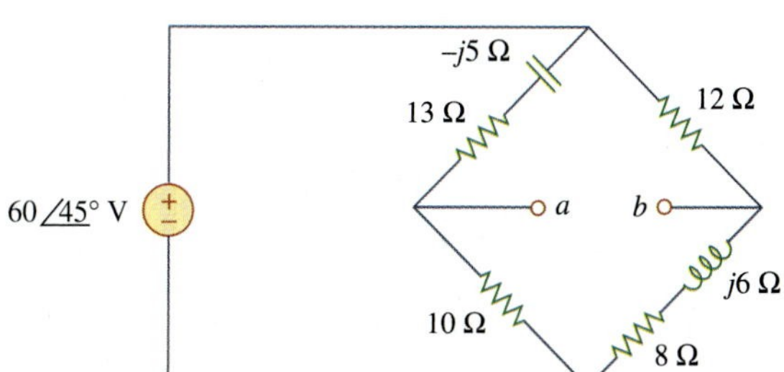

Figure 10.110
For Prob. 10.67.

10.68 Find the Thevenin equivalent at terminals *a-b* in the circuit of Fig. 10.111.
ps ML

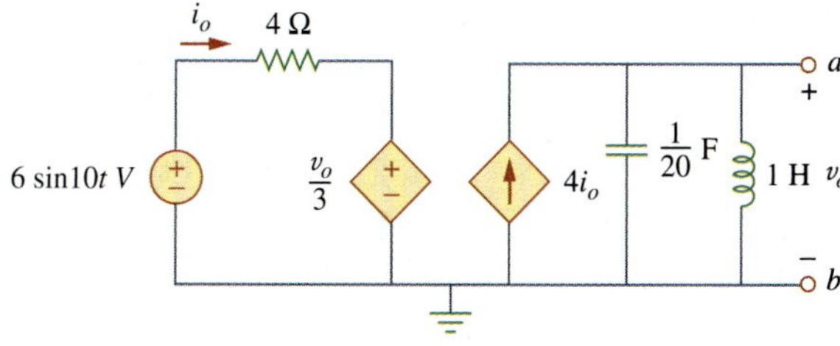

Figure 10.111
For Prob. 10.68.

Section 10.7 Op Amp AC Circuits

10.69 For the differentiator shown in Fig. 10.112, obtain $\mathbf{V}_o/\mathbf{V}_s$. Find $v_o(t)$ when $v_s(t) = \mathbf{V}_m \sin \omega t$ and $\omega = 1/RC$.

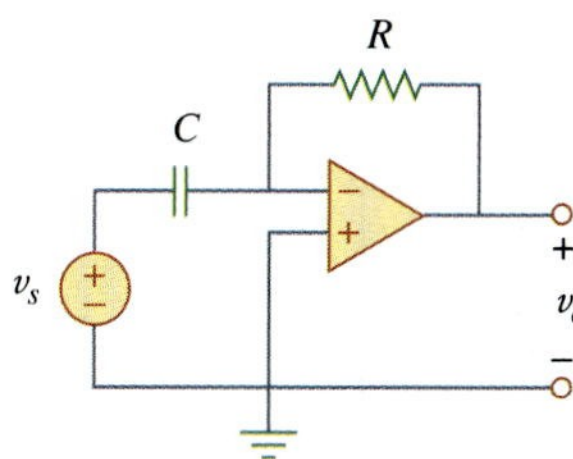

Figure 10.112
For Prob. 10.69.

10.70 Using Fig. 10.113, design a problem to help other students better understand op amps in AC circuits.
e⊘d

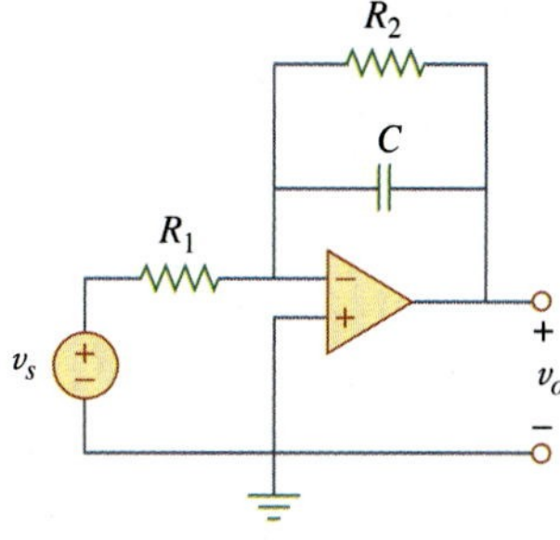

Figure 10.113
For Prob. 10.70.

10.71 Find v_o in the op amp circuit of Fig. 10.114.

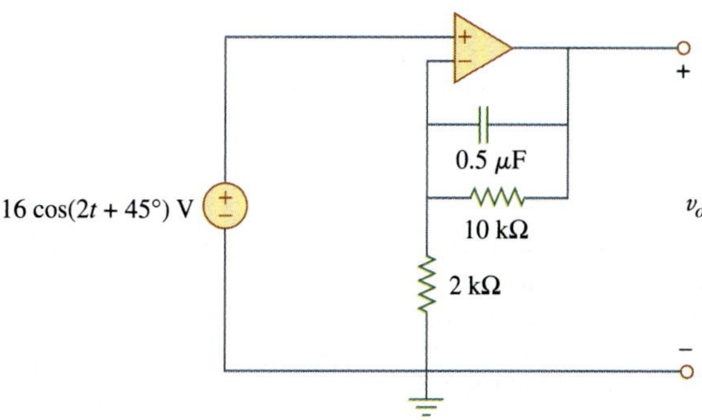

Figure 10.114
For Prob. 10.71.

10.72 Compute $i_o(t)$ in the op amp circuit in Fig. 10.115 if $v_s = 10 \cos(10^4 t + 30°)$ V.

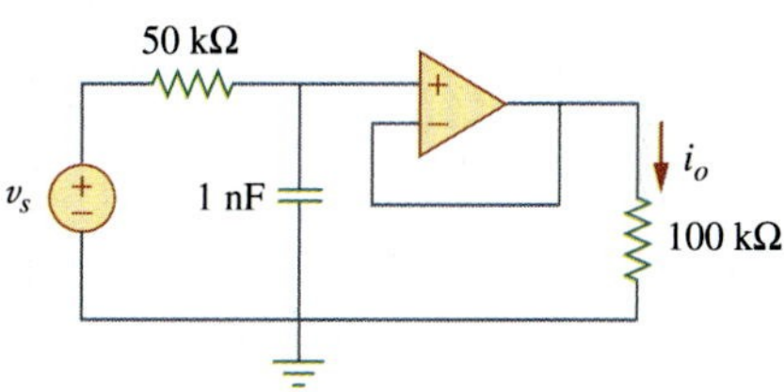

Figure 10.115
For Prob. 10.72.

10.73 If the input impedance is defined as $\mathbf{Z}_{in} = \mathbf{V}_s/\mathbf{I}_s$, find the input impedance of the op amp circuit in Fig. 10.116 when $R_1 = 10\ \text{k}\Omega$, $R_2 = 20\ \text{k}\Omega$, $C_1 = 10$ nF, $C_2 = 20$ nF, and $\omega = 5000$ rad/s.

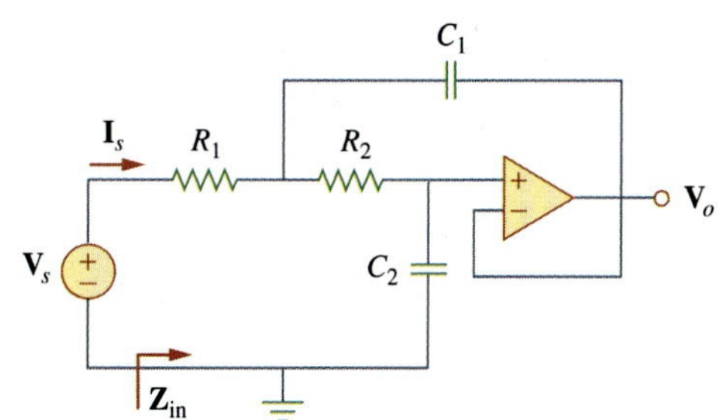

Figure 10.116
For Prob. 10.73.

10.74 Evaluate the voltage gain $\mathbf{A}_v = \mathbf{V}_o/\mathbf{V}_s$ in the op amp circuit of Fig. 10.117. Find $\mathbf{A}_v$ at $\omega = 0$, $\omega \to \infty$, $\omega = 1/R_1C_1$, and $\omega = 1/R_2C_2$.

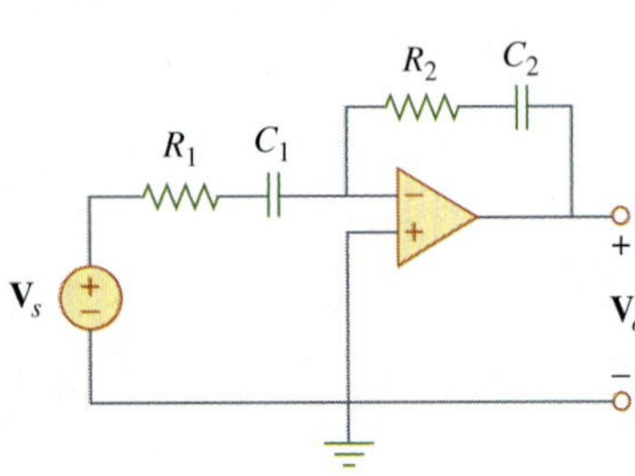

Figure 10.117
For Prob. 10.74.

10.75 In the op amp circuit of Fig. 10.118, find the closed-loop gain and phase shift of the output voltage with respect to the input voltage if $C_1 = C_2 = 1$ nF, $R_1 = R_2 = 100$ kΩ, $R_3 = 20$ kΩ, $R_4 = 40$ kΩ, and $\omega = 2000$ rad/s.

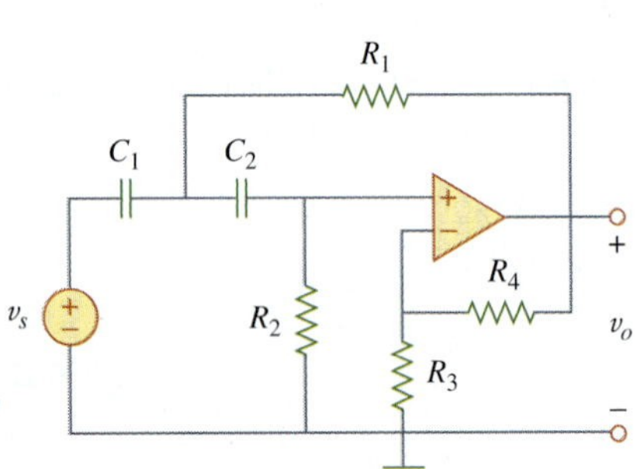

Figure 10.118
For Prob. 10.75.

10.76 Determine $\mathbf{V}_o$ and $\mathbf{I}_o$ in the op amp circuit of Fig. 10.119.

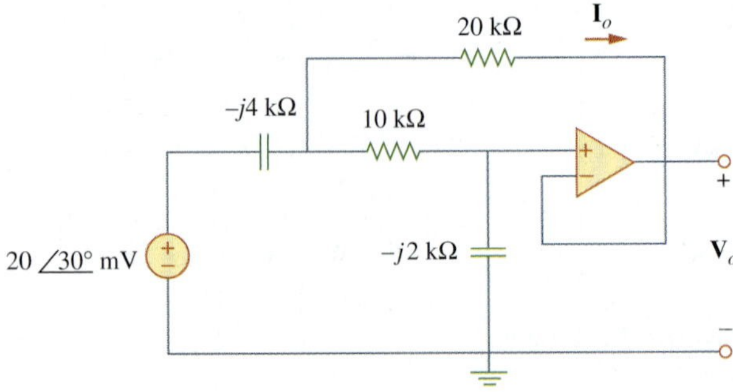

Figure 10.119
For Prob. 10.76.

10.77 Compute the closed-loop gain $\mathbf{V}_o/\mathbf{V}_s$ for the op amp circuit of Fig. 10.120.

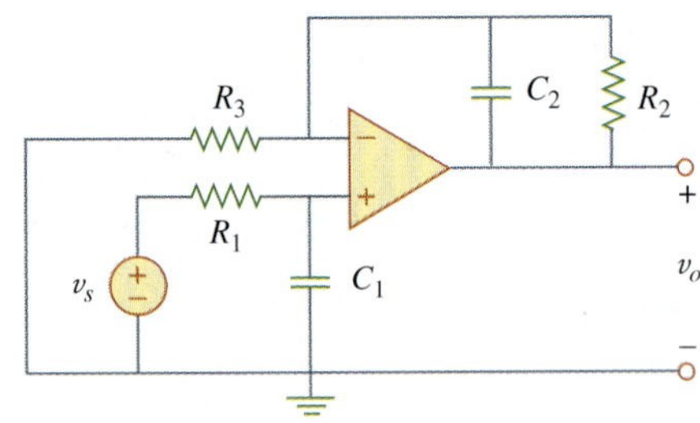

Figure 10.120
For Prob. 10.77.

10.78 Determine $v_o(t)$ in the op amp circuit in Fig. 10.121 below.

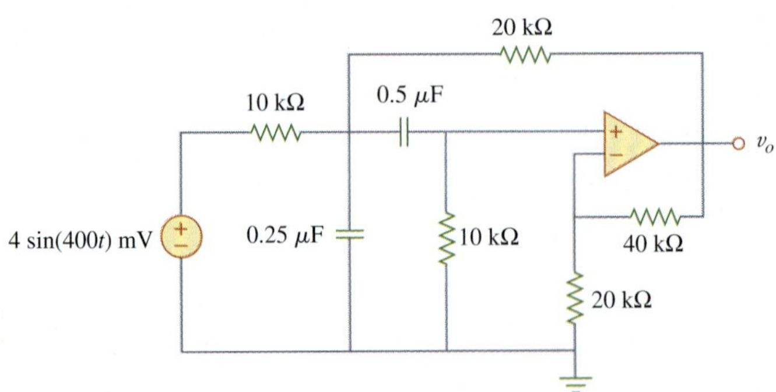

Figure 10.121
For Prob. 10.78.

10.79 For the op amp circuit in Fig. 10.122, obtain $v_o(t)$.

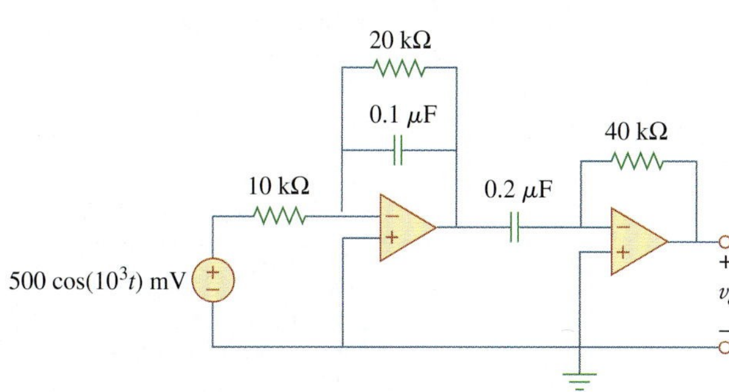

Figure 10.122
For Prob. 10.79.

10.80 Obtain $v_o(t)$ for the op amp circuit in Fig. 10.123 if $v_s = 4\cos(1000t - 60°)$ V.

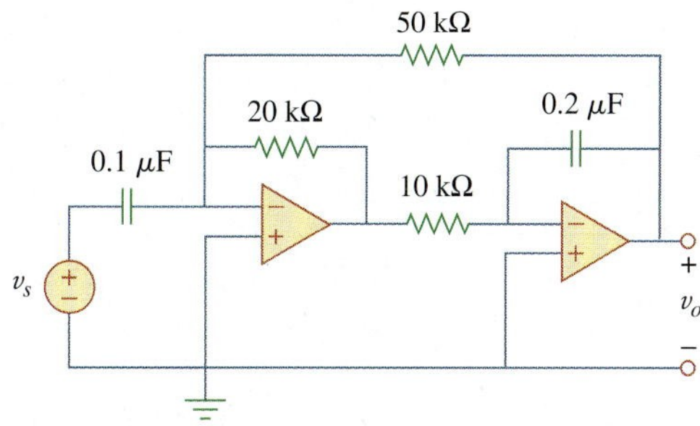

Figure 10.123
For Prob. 10.80.

Section 10.8 AC Analysis Using *PSpice*

10.81 Use *PSpice* to determine $\mathbf{V}_o$ in the circuit of Fig. 10.124. Assume $\omega = 1$ rad/s.

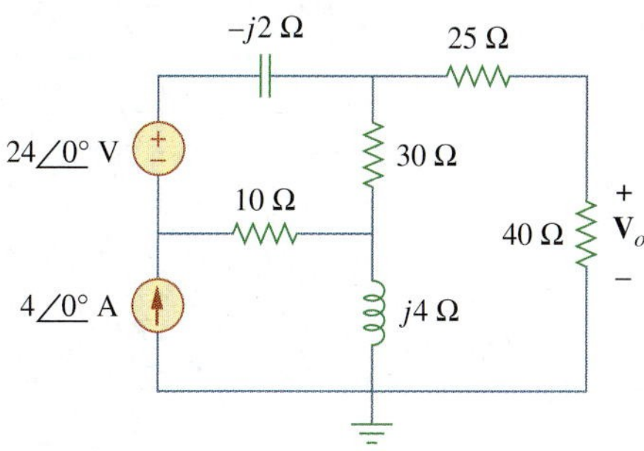

Figure 10.124
For Prob. 10.81.

10.82 Solve Prob. 10.19 using *PSpice*.

10.83 Use *PSpice* to find $v_o(t)$ in the circuit of Fig. 10.125. Let $i_s = 2\cos(10^3t)$ A.

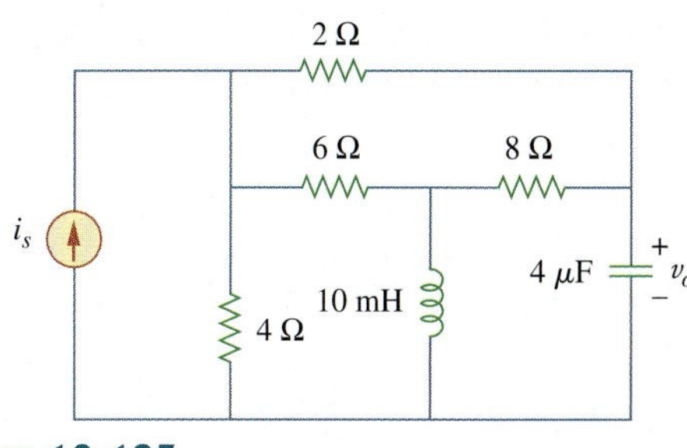

Figure 10.125
For Prob. 10.83.

10.84 Obtain $\mathbf{V}_o$ in the circuit of Fig. 10.126 using *PSpice*.

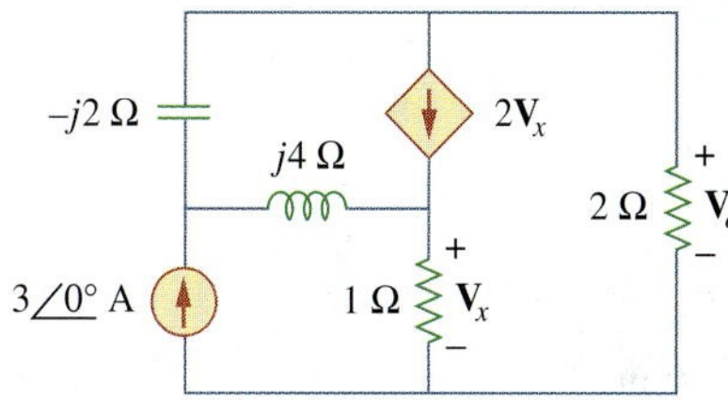

Figure 10.126
For Prob. 10.84.

10.85 Using Fig. 10.127, design a problem to help other students better understand performing AC analysis with *PSpice*.

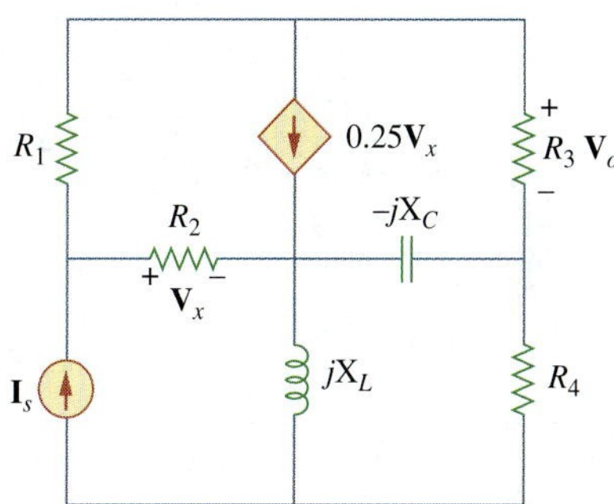

Figure 10.127
For Prob. 10.85.

10.86 Use *PSpice* to find $\mathbf{V}_1$, $\mathbf{V}_2$, and $\mathbf{V}_3$ in the network of Fig. 10.128.

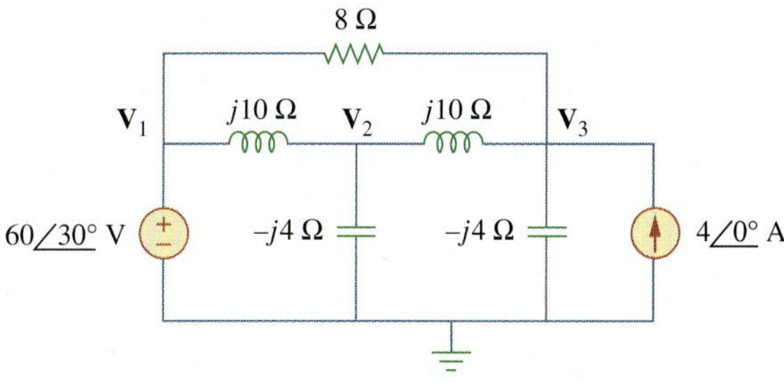

Figure 10.128
For Prob. 10.86.

10.87 Determine $\mathbf{V}_1$, $\mathbf{V}_2$, and $\mathbf{V}_3$ in the circuit of Fig. 10.129 using *PSpice*.

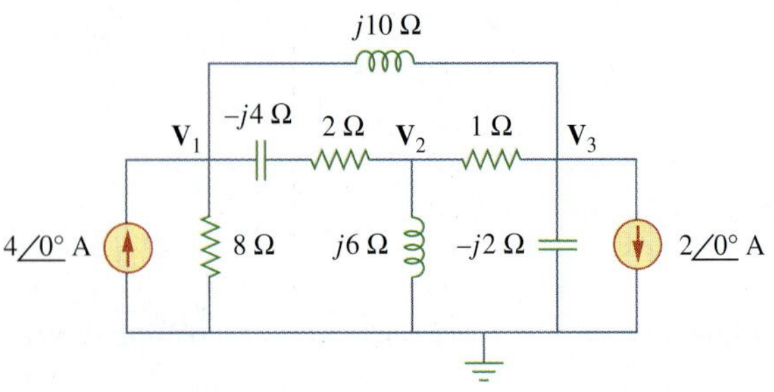

Figure 10.129
For Prob. 10.87.

10.88 Use *PSpice* to find v_o and i_o in the circuit of Fig. 10.130 below.

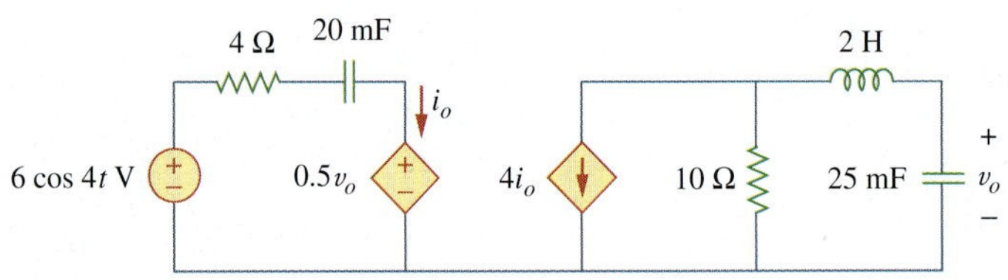

Figure 10.130
For Prob. 10.88.

Section 10.9 Applications

10.89 The op amp circuit in Fig. 10.131 is called an *inductance simulator*. Show that the input impedance is given by

$$\mathbf{Z}_{in} = \frac{\mathbf{V}_{in}}{\mathbf{I}_{in}} = j\omega L_{eq}$$

where

$$L_{eq} = \frac{R_1 R_3 R_4}{R_2} C$$

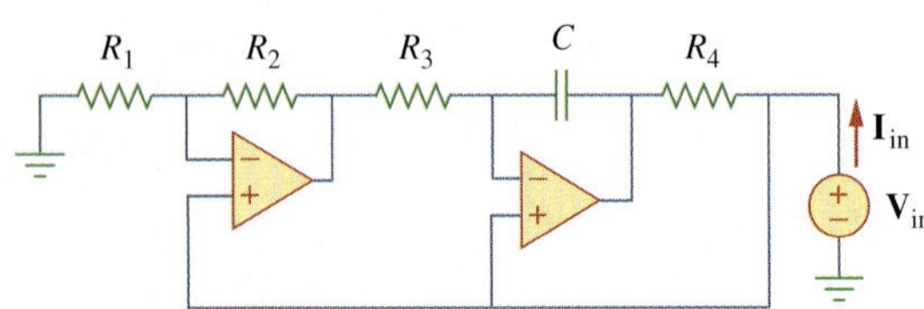

Figure 10.131
For Prob. 10.89.

10.90 Figure 10.132 shows a Wien-bridge network. Show that the frequency at which the phase shift between the input and output signals is zero is $f = \frac{1}{2}\pi RC$, and that the necessary gain is $\mathbf{A}_v = \mathbf{V}_o/\mathbf{V}_i = 3$ at that frequency.

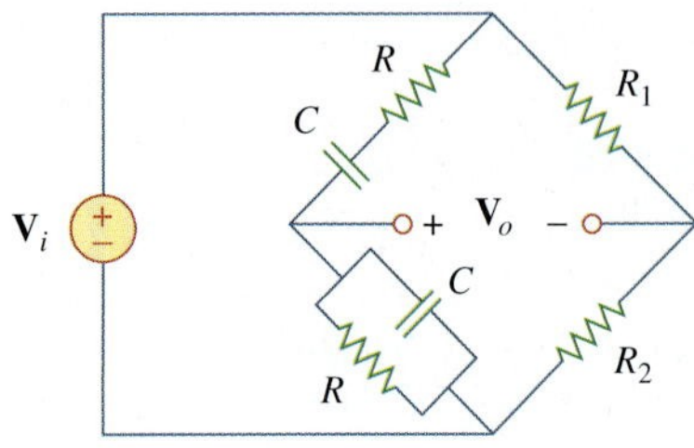

Figure 10.132
For Prob. 10.90.

10.91 Consider the oscillator in Fig. 10.133.

(a) Determine the oscillation frequency.

(b) Obtain the minimum value of R for which oscillation takes place.

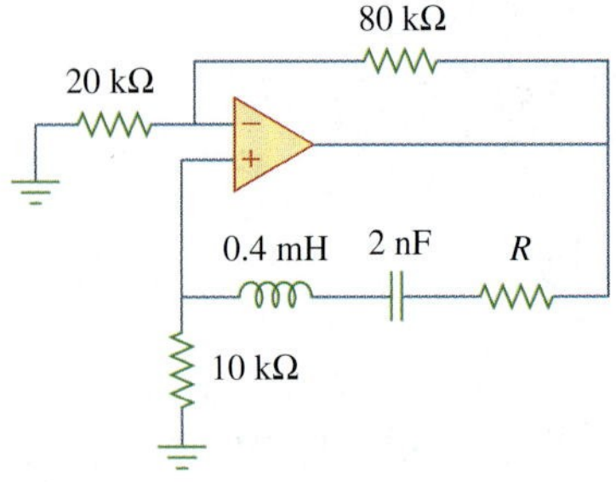

Figure 10.133
For Prob. 10.91.

10.92 The oscillator circuit in Fig. 10.134 uses an ideal op amp.

(a) Calculate the minimum value of R_o that will cause oscillation to occur.

(b) Find the frequency of oscillation.

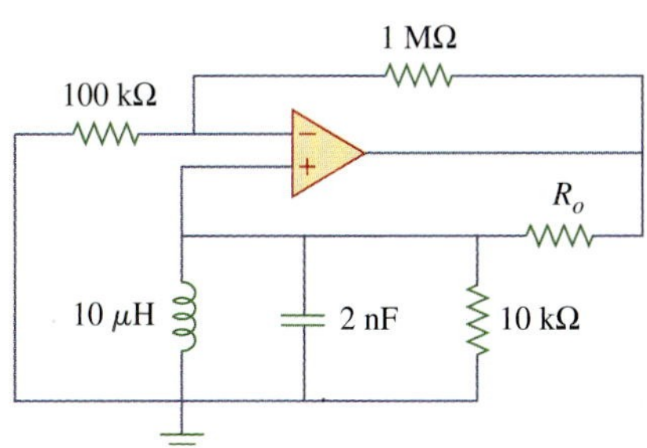

Figure 10.134
For Prob. 10.92.

10.93 Figure 10.135 shows a *Colpitts oscillator*. Show that the oscillation frequency is

$$f_o = \frac{1}{2\pi\sqrt{LC_T}}$$

where $C_T = C_1C_2/(C_1 + C_2)$. Assume $R_i \gg X_{C_2}$.

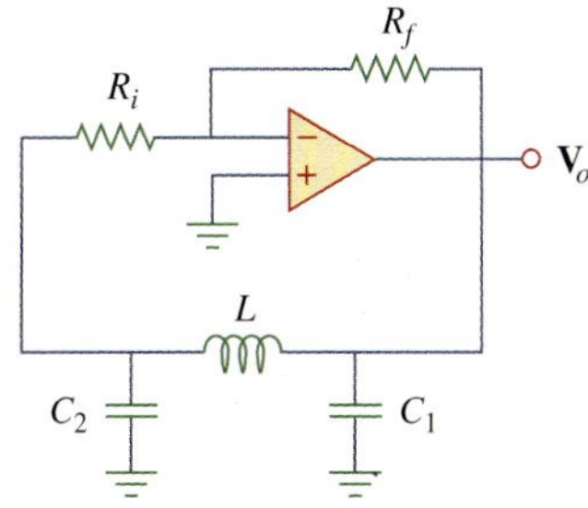

Figure 10.135
A Colpitts oscillator; for Prob. 10.93.

(*Hint:* Set the imaginary part of the impedance in the feedback circuit equal to zero.)

10.94 Design a Colpitts oscillator that will operate at 50 kHz.

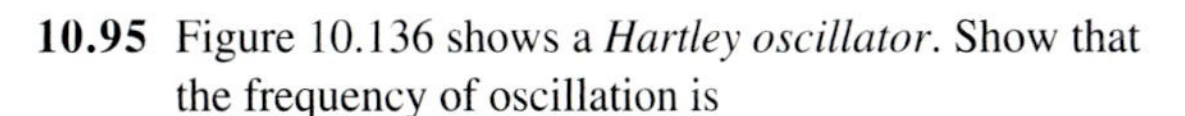

10.95 Figure 10.136 shows a *Hartley oscillator*. Show that the frequency of oscillation is

$$f_o = \frac{1}{2\pi\sqrt{C(L_1 + L_2)}}$$

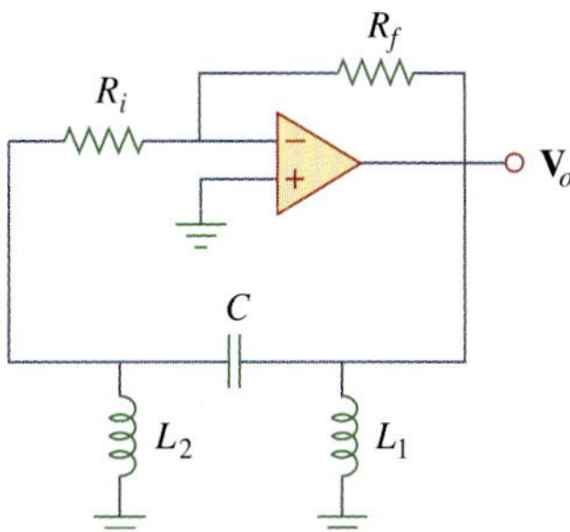

Figure 10.136
A Hartley oscillator; for Prob. 10.95.

10.96 Refer to the oscillator in Fig. 10.137.

(a) Show that

$$\frac{\mathbf{V}_2}{\mathbf{V}_o} = \frac{1}{3 + j(\omega L/R - R/\omega L)}$$

(b) Determine the oscillation frequency f_o.

(c) Obtain the relationship between R_1 and R_2 in order for oscillation to occur.

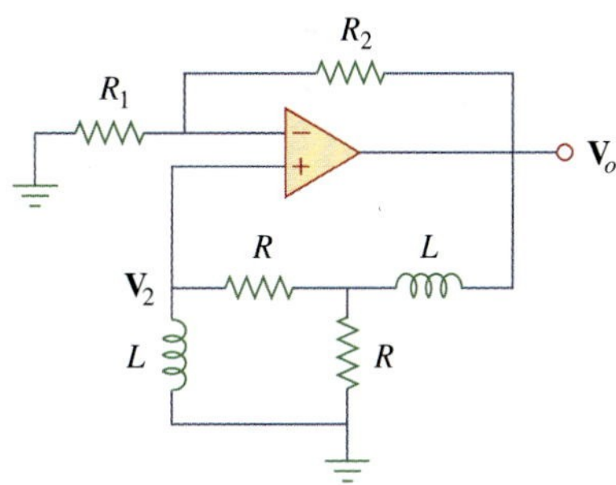

Figure 10.137
For Prob. 10.96.

chapter

11

AC Power Analysis

Four things come not back: the spoken word; the sped arrow; time past; the neglected opportunity.

—Al Halif Omar Ibn

Enhancing Your Career

Career in Power Systems

The discovery of the principle of an ac generator by Michael Faraday in 1831 was a major breakthrough in engineering; it provided a convenient way of generating the electric power that is needed in every electronic, electrical, or electromechanical device we use now.

A pole-type transformer with a low-voltage, three-wire distribution system. © Vol. 129 PhotoDisc/Getty

Electric power is obtained by converting energy from sources such as fossil fuels (gas, oil, and coal), nuclear fuel (uranium), hydro energy (water falling through a head), geothermal energy (hot water, steam), wind energy, tidal energy, and biomass energy (wastes). These various ways of generating electric power are studied in detail in the field of power engineering, which has become an indispensable subdiscipline of electrical engineering. An electrical engineer should be familiar with the analysis, generation, transmission, distribution, and cost of electric power.

The electric power industry is a very large employer of electrical engineers. The industry includes thousands of electric utility systems ranging from large, interconnected systems serving large regional areas to small power companies serving individual communities or factories. Due to the complexity of the power industry, there are numerous electrical engineering jobs in different areas of the industry: power plant (generation), transmission and distribution, maintenance, research, data acquisition and flow control, and management. Since electric power is used everywhere, electric utility companies are everywhere, offering exciting training and steady employment for men and women in thousands of communities throughout the world.

11.1 Introduction

Our effort in ac circuit analysis so far has been focused mainly on calculating voltage and current. Our major concern in this chapter is power analysis.

Power analysis is of paramount importance. Power is the most important quantity in electric utilities, electronic, and communication systems, because such systems involve transmission of power from one point to another. Also, every industrial and household electrical device—every fan, motor, lamp, pressing iron, TV, personal computer—has a power rating that indicates how much power the equipment requires; exceeding the power rating can do permanent damage to an appliance. The most common form of electric power is 50- or 60-Hz ac power. The choice of ac over dc allowed high-voltage power transmission from the power generating plant to the consumer.

We will begin by defining and deriving *instantaneous power* and *average power*. We will then introduce other power concepts. As practical applications of these concepts, we will discuss how power is measured and reconsider how electric utility companies charge their customers.

11.2 Instantaneous and Average Power

As mentioned in Chapter 2, the *instantaneous power* $p(t)$ absorbed by an element is the product of the instantaneous voltage $v(t)$ across the element and the instantaneous current $i(t)$ through it. Assuming the passive sign convention,

$$p(t) = v(t)i(t) \tag{11.1}$$

We can also think of the instantaneous power as the power absorbed by the element at a specific instant of time. Instantaneous quantities are denoted by lowercase letters.

The **instantaneous power** (in watts) is the power at any instant of time.

It is the rate at which an element absorbs energy.

Consider the general case of instantaneous power absorbed by an arbitrary combination of circuit elements under sinusoidal excitation, as shown in Fig. 11.1. Let the voltage and current at the terminals of the circuit be

$$v(t) = V_m \cos(\omega t + \theta_v) \tag{11.2a}$$

$$i(t) = I_m \cos(\omega t + \theta_i) \tag{11.2b}$$

where V_m and I_m are the amplitudes (or peak values), and θ_v and θ_i are the phase angles of the voltage and current, respectively. The instantaneous power absorbed by the circuit is

$$p(t) = v(t)i(t) = V_m I_m \cos(\omega t + \theta_v)\cos(\omega t + \theta_i) \tag{11.3}$$

We apply the trigonometric identity

$$\cos A \cos B = \frac{1}{2}[\cos(A - B) + \cos(A + B)] \tag{11.4}$$

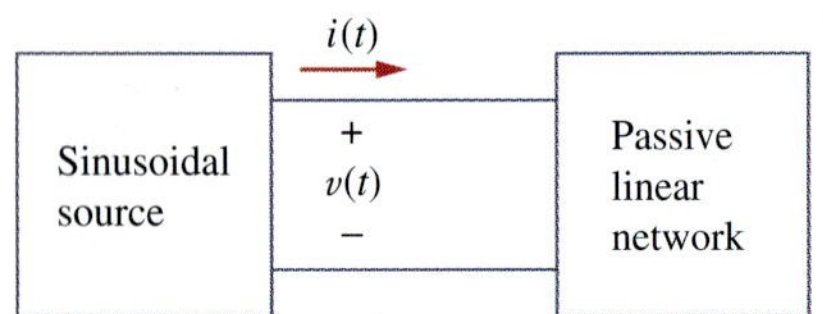

Figure 11.1
Sinusoidal source and passive linear circuit.

and express Eq. (11.3) as

$$p(t) = \frac{1}{2}V_m I_m \cos(\theta_v - \theta_i) + \frac{1}{2}V_m I_m \cos(2\omega t + \theta_v + \theta_i) \quad \textbf{(11.5)}$$

This shows us that the instantaneous power has two parts. The first part is constant or time independent. Its value depends on the phase difference between the voltage and the current. The second part is a sinusoidal function whose frequency is 2ω, which is twice the angular frequency of the voltage or current.

A sketch of $p(t)$ in Eq. (11.5) is shown in Fig. 11.2, where $T = 2\pi/\omega$ is the period of voltage or current. We observe that $p(t)$ is periodic, $p(t) = p(t + T_0)$, and has a period of $T_0 = T/2$, since its frequency is twice that of voltage or current. We also observe that $p(t)$ is positive for some part of each cycle and negative for the rest of the cycle. When $p(t)$ is positive, power is absorbed by the circuit. When $p(t)$ is negative, power is absorbed by the source; that is, power is transferred from the circuit to the source. This is possible because of the storage elements (capacitors and inductors) in the circuit.

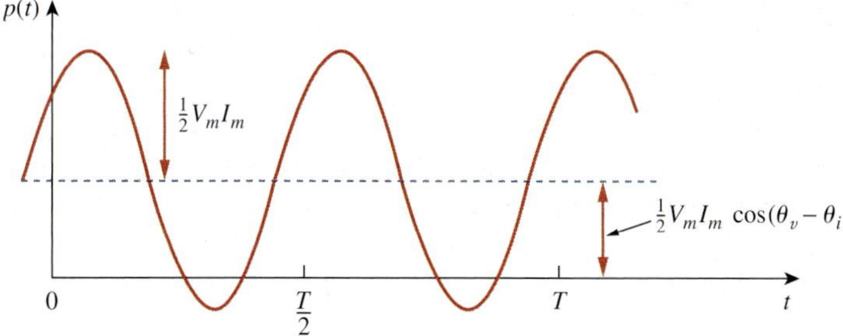

Figure 11.2
The instantaneous power $p(t)$ entering a circuit.

The instantaneous power changes with time and is therefore difficult to measure. The *average* power is more convenient to measure. In fact, the wattmeter, the instrument for measuring power, responds to average power.

> The **average power**, in watts, is the average of the instantaneous power over one period.

Thus, the average power is given by

$$P = \frac{1}{T}\int_0^T p(t)\, dt \quad \textbf{(11.6)}$$

Although Eq. (11.6) shows the averaging done over T, we would get the same result if we performed the integration over the actual period of $p(t)$ which is $T_0 = T/2$.

Substituting $p(t)$ in Eq. (11.5) into Eq. (11.6) gives

$$\begin{aligned} P &= \frac{1}{T}\int_0^T \frac{1}{2}V_m I_m \cos(\theta_v - \theta_i)\,dt \\ &\quad + \frac{1}{T}\int_0^T \frac{1}{2}V_m I_m \cos(2\omega t + \theta_v + \theta_i)\,dt \\ &= \frac{1}{2}V_m I_m \cos(\theta_v - \theta_i)\frac{1}{T}\int_0^T dt \\ &\quad + \frac{1}{2}V_m I_m \frac{1}{T}\int_0^T \cos(2\omega t + \theta_v + \theta_i)\,dt \end{aligned} \tag{11.7}$$

The first integrand is constant, and the average of a constant is the same constant. The second integrand is a sinusoid. We know that the average of a sinusoid over its period is zero because the area under the sinusoid during a positive half-cycle is canceled by the area under it during the following negative half-cycle. Thus, the second term in Eq. (11.7) vanishes and the average power becomes

$$P = \frac{1}{2}V_m I_m \cos(\theta_v - \theta_i) \tag{11.8}$$

Since $\cos(\theta_v - \theta_i) = \cos(\theta_i - \theta_v)$, what is important is the difference in the phases of the voltage and current.

Note that $p(t)$ is time-varying while P does not depend on time. To find the instantaneous power, we must necessarily have $v(t)$ and $i(t)$ in the time domain. But we can find the average power when voltage and current are expressed in the time domain, as in Eq. (11.8), or when they are expressed in the frequency domain. The phasor forms of $v(t)$ and $i(t)$ in Eq. (11.2) are $\mathbf{V} = V_m\underline{/\theta_v}$ and $\mathbf{I} = I_m\underline{/\theta_i}$, respectively. P is calculated using Eq. (11.8) or using phasors $\mathbf{V}$ and $\mathbf{I}$. To use phasors, we notice that

$$\begin{aligned} \frac{1}{2}\mathbf{V}\mathbf{I}^* &= \frac{1}{2}V_m I_m \underline{/\theta_v - \theta_i} \\ &= \frac{1}{2}V_m I_m[\cos(\theta_v - \theta_i) + j\sin(\theta_v - \theta_i)] \end{aligned} \tag{11.9}$$

We recognize the real part of this expression as the average power P according to Eq. (11.8). Thus,

$$P = \frac{1}{2}\text{Re}[\mathbf{V}\mathbf{I}^*] = \frac{1}{2}V_m I_m \cos(\theta_v - \theta_i) \tag{11.10}$$

Consider two special cases of Eq. (11.10). When $\theta_v = \theta_i$, the voltage and current are in phase. This implies a purely resistive circuit or resistive load R, and

$$P = \frac{1}{2}V_m I_m = \frac{1}{2}I_m^2 R = \frac{1}{2}|\mathbf{I}|^2 R \tag{11.11}$$

where $|\mathbf{I}|^2 = \mathbf{I} \times \mathbf{I}^*$. Equation (11.11) shows that a purely resistive circuit absorbs power at all times. When $\theta_v - \theta_i = \pm 90°$, we have a purely reactive circuit, and

$$P = \frac{1}{2}V_m I_m \cos 90° = 0 \tag{11.12}$$

showing that a purely reactive circuit absorbs no average power. In summary,

> A resistive load (R) absorbs power at all times, while a reactive load (L or C) absorbs zero average power.

Example 11.1

Given that

$$v(t) = 120\cos(377t + 45^\circ)\text{ V} \quad \text{and} \quad i(t) = 10\cos(377t - 10^\circ)\text{ A}$$

find the instantaneous power and the average power absorbed by the passive linear network of Fig. 11.1.

Solution:
The instantaneous power is given by

$$p = vi = 1200\cos(377t + 45^\circ)\cos(377t - 10^\circ)$$

Applying the trigonometric identity

$$\cos A \cos B = \frac{1}{2}[\cos(A + B) + \cos(A - B)]$$

gives

$$p = 600[\cos(754t + 35^\circ) + \cos 55^\circ]$$

or

$$p(t) = 344.2 + 600\cos(754t + 35^\circ)\text{ W}$$

The average power is

$$P = \frac{1}{2}V_m I_m \cos(\theta_v - \theta_i) = \frac{1}{2}120(10)\cos[45^\circ - (-10^\circ)]$$

$$= 600\cos 55^\circ = 344.2\text{ W}$$

which is the constant part of $p(t)$ above.

Practice Problem 11.1

Calculate the instantaneous power and average power absorbed by the passive linear network of Fig. 11.1 if

$$v(t) = 165\cos(10t + 20^\circ)\text{ V} \quad \text{and} \quad i(t) = 20\sin(10t + 60^\circ)\text{ A}$$

Answer: $1.0606 + 1.65\cos(20t - 10^\circ)$ kW, 1.0606 kW.

Example 11.2

Calculate the average power absorbed by an impedance $\mathbf{Z} = 30 - j70\ \Omega$ when a voltage $\mathbf{V} = 120\underline{/0^\circ}$ is applied across it.

Solution:
The current through the impedance is

$$\mathbf{I} = \frac{\mathbf{V}}{\mathbf{Z}} = \frac{120\underline{/0^\circ}}{30 - j70} = \frac{120\underline{/0^\circ}}{76.16\underline{/-66.8^\circ}} = 1.576\underline{/66.8^\circ}\text{ A}$$

The average power is

$$P = \frac{1}{2}V_m I_m \cos(\theta_v - \theta_i) = \frac{1}{2}(120)(1.576)\cos(0 - 66.8^\circ) = 37.24 \text{ W}$$

Practice Problem 11.2

A current $\mathbf{I} = 20\underline{/30^\circ}$ A flows through an impedance $\mathbf{Z} = 40\underline{/-22^\circ}\ \Omega$. Find the average power delivered to the impedance.

Answer: 3.709 kW.

Example 11.3

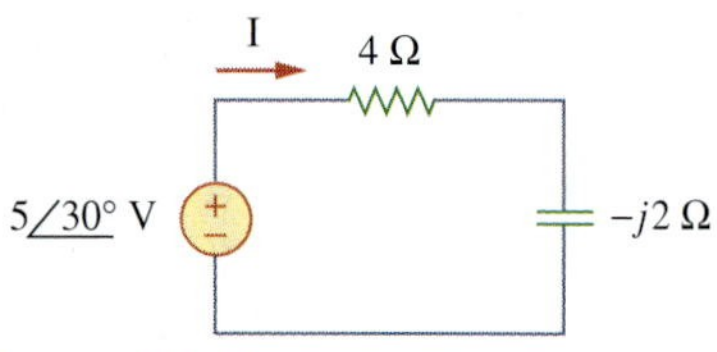

Figure 11.3
For Example 11.3.

For the circuit shown in Fig. 11.3, find the average power supplied by the source and the average power absorbed by the resistor.

Solution:
The current $\mathbf{I}$ is given by

$$\mathbf{I} = \frac{5\underline{/30^\circ}}{4 - j2} = \frac{5\underline{/30^\circ}}{4.472\underline{/-26.57^\circ}} = 1.118\underline{/56.57^\circ} \text{ A}$$

The average power supplied by the voltage source is

$$P = \frac{1}{2}(5)(1.118)\cos(30^\circ - 56.57^\circ) = 2.5 \text{ W}$$

The current through the resistor is

$$\mathbf{I}_R = \mathbf{I} = 1.118\underline{/56.57^\circ} \text{ A}$$

and the voltage across it is

$$\mathbf{V}_R = 4\mathbf{I}_R = 4.472\underline{/56.57^\circ} \text{ V}$$

The average power absorbed by the resistor is

$$P = \frac{1}{2}(4.472)(1.118) = 2.5 \text{ W}$$

which is the same as the average power supplied. Zero average power is absorbed by the capacitor.

Practice Problem 11.3

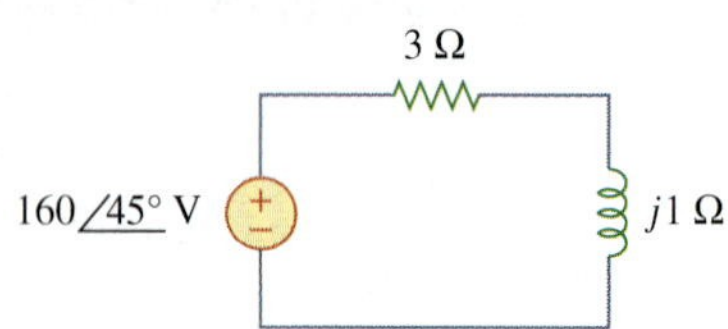

Figure 11.4
For Practice Prob. 11.3.

In the circuit of Fig. 11.4, calculate the average power absorbed by the resistor and inductor. Find the average power supplied by the voltage source.

Answer: 3.84 kW, 0 W, 3.84 kW.

Example 11.4

Determine the average power generated by each source and the average power absorbed by each passive element in the circuit of Fig. 11.5(a).

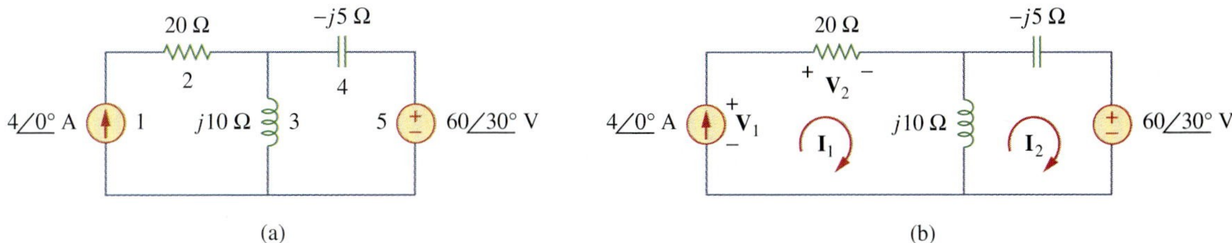

Figure 11.5
For Example 11.4.

Solution:

We apply mesh analysis as shown in Fig. 11.5(b). For mesh 1,

$$\mathbf{I}_1 = 4 \text{ A}$$

For mesh 2,

$$(j10 - j5)\mathbf{I}_2 - j10\mathbf{I}_1 + 60/\underline{30^\circ} = 0, \qquad \mathbf{I}_1 = 4 \text{ A}$$

or

$$j5\mathbf{I}_2 = -60/\underline{30^\circ} + j40 \qquad \Rightarrow \qquad \mathbf{I}_2 = -12/\underline{-60^\circ} + 8$$
$$= 10.58/\underline{79.1^\circ} \text{ A}$$

For the voltage source, the current flowing from it is $\mathbf{I}_2 = 10.58/\underline{79.1^\circ}$ A and the voltage across it is $60/\underline{30^\circ}$ V, so that the average power is

$$P_5 = \frac{1}{2}(60)(10.58)\cos(30^\circ - 79.1^\circ) = 207.8 \text{ W}$$

Following the passive sign convention (see Fig. 1.8), this average power is absorbed by the source, in view of the direction of $\mathbf{I}_2$ and the polarity of the voltage source. That is, the circuit is delivering average power to the voltage source.

For the current source, the current through it is $\mathbf{I}_1 = 4/\underline{0^\circ}$ and the voltage across it is

$$\mathbf{V}_1 = 20\mathbf{I}_1 + j10(\mathbf{I}_1 - \mathbf{I}_2) = 80 + j10(4 - 2 - j10.39)$$
$$= 183.9 + j20 = 184.984/\underline{6.21^\circ} \text{ V}$$

The average power supplied by the current source is

$$P_1 = -\frac{1}{2}(184.984)(4)\cos(6.21^\circ - 0) = -367.8 \text{ W}$$

It is negative according to the passive sign convention, meaning that the current source is supplying power to the circuit.

For the resistor, the current through it is $\mathbf{I}_1 = 4/\underline{0^\circ}$ and the voltage across it is $20\mathbf{I}_1 = 80/\underline{0^\circ}$, so that the power absorbed by the resistor is

$$P_2 = \frac{1}{2}(80)(4) = 160 \text{ W}$$

For the capacitor, the current through it is $\mathbf{I}_2 = 10.58\underline{/79.1^\circ}$ and the voltage across it is $-j5\mathbf{I}_2 = (5\underline{/-90^\circ})(10.58\underline{/79.1^\circ}) = 52.9\underline{/79.1^\circ - 90^\circ}$. The average power absorbed by the capacitor is

$$P_4 = \frac{1}{2}(52.9)(10.58)\cos(-90^\circ) = 0$$

For the inductor, the current through it is $\mathbf{I}_1 - \mathbf{I}_2 = 2 - j10.39 = 10.58\underline{/-79.1^\circ}$. The voltage across it is $j10(\mathbf{I}_1 - \mathbf{I}_2) = 10.58\underline{/-79.1^\circ + 90^\circ}$. Hence, the average power absorbed by the inductor is

$$P_3 = \frac{1}{2}(105.8)(10.58)\cos 90^\circ = 0$$

Notice that the inductor and the capacitor absorb zero average power and that the total power supplied by the current source equals the power absorbed by the resistor and the voltage source, or

$$P_1 + P_2 + P_3 + P_4 + P_5 = -367.8 + 160 + 0 + 0 + 207.8 = 0$$

indicating that power is conserved.

Practice Problem 11.4

Calculate the average power absorbed by each of the five elements in the circuit of Fig. 11.6.

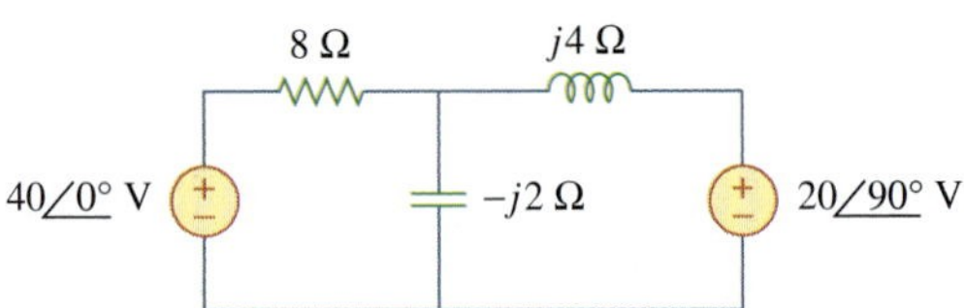

Figure 11.6
For Practice Prob. 11.4.

Answer: 40-V Voltage source: −60 W; j20-V Voltage source: −40 W; resistor: 100 W; others: 0 W.

11.3 Maximum Average Power Transfer

In Section 4.8 we solved the problem of maximizing the power delivered by a power-supplying resistive network to a load R_L. Representing the circuit by its Thevenin equivalent, we proved that the maximum power would be delivered to the load if the load resistance is equal to the Thevenin resistance $R_L = R_{\text{Th}}$. We now extend that result to ac circuits.

Consider the circuit in Fig. 11.7, where an ac circuit is connected to a load $\mathbf{Z}_L$ and is represented by its Thevenin equivalent. The load is usually represented by an impedance, which may model an electric motor, an antenna, a TV, and so forth. In rectangular form, the Thevenin impedance $\mathbf{Z}_{Th}$ and the load impedance $\mathbf{Z}_L$ are

$$\mathbf{Z}_{Th} = R_{Th} + jX_{Th} \tag{11.13a}$$

$$\mathbf{Z}_L = R_L + jX_L \tag{11.13b}$$

The current through the load is

$$\mathbf{I} = \frac{\mathbf{V}_{Th}}{\mathbf{Z}_{Th} + \mathbf{Z}_L} = \frac{\mathbf{V}_{Th}}{(R_{Th} + jX_{Th}) + (R_L + jX_L)} \tag{11.14}$$

From Eq. (11.11), the average power delivered to the load is

$$P = \frac{1}{2}|\mathbf{I}|^2 R_L = \frac{|\mathbf{V}_{Th}|^2 R_L/2}{(R_{Th} + R_L)^2 + (X_{Th} + X_L)^2} \tag{11.15}$$

Our objective is to adjust the load parameters R_L and X_L so that P is maximum. To do this we set $\partial P/\partial R_L$ and $\partial P/\partial X_L$ equal to zero. From Eq. (11.15), we obtain

$$\frac{\partial P}{\partial X_L} = -\frac{|\mathbf{V}_{Th}|^2 R_L (X_{Th} + X_L)}{[(R_{Th} + R_L)^2 + (X_{Th} + X_L)^2]^2} \tag{11.16a}$$

$$\frac{\partial P}{\partial R_L} = \frac{|\mathbf{V}_{Th}|^2[(R_{Th} + R_L)^2 + (X_{Th} + X_L)^2 - 2R_L(R_{Th} + R_L)]}{2[(R_{Th} + R_L)^2 + (X_{Th} + X_L)^2]^2} \tag{11.16b}$$

Setting $\partial P/\partial X_L$ to zero gives

$$X_L = -X_{Th} \tag{11.17}$$

and setting $\partial P/\partial R_L$ to zero results in

$$R_L = \sqrt{R_{Th}^2 + (X_{Th} + X_L)^2} \tag{11.18}$$

Combining Eqs. (11.17) and (11.18) leads to the conclusion that for maximum average power transfer, $\mathbf{Z}_L$ must be selected so that $X_L = -X_{Th}$ and $R_L = R_{Th}$, i.e.,

$$\boxed{\mathbf{Z}_L = R_L + jX_L = R_{Th} - jX_{Th} = \mathbf{Z}_{Th}^*} \tag{11.19}$$

For **maximum average power transfer,** the load impedance $\mathbf{Z}_L$ must be equal to the complex conjugate of the Thevenin impedance $\mathbf{Z}_{Th}$.

This result is known as the *maximum average power transfer theorem* for the sinusoidal steady state. Setting $R_L = R_{Th}$ and $X_L = -X_{Th}$ in Eq. (11.15) gives us the maximum average power as

$$\boxed{P_{max} = \frac{|\mathbf{V}_{Th}|^2}{8R_{Th}}} \tag{11.20}$$

In a situation in which the load is purely real, the condition for maximum power transfer is obtained from Eq. (11.18) by setting $X_L = 0$; that is,

$$R_L = \sqrt{R_{Th}^2 + X_{Th}^2} = |\mathbf{Z}_{Th}| \tag{11.21}$$

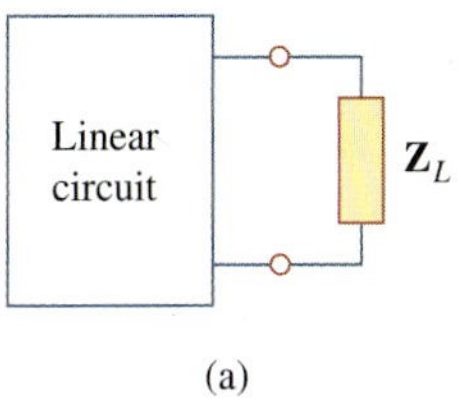

(a)

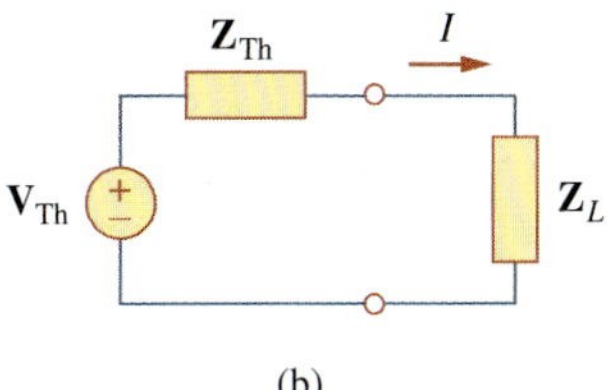

(b)

Figure 11.7
Finding the maximum average power transfer: (a) circuit with a load, (b) the Thevenin equivalent.

When $\mathbf{Z}_L = \mathbf{Z}_{Th}^*$, we say that the load is matched to the source.

This means that for maximum average power transfer to a purely resistive load, the load impedance (or resistance) is equal to the magnitude of the Thevenin impedance.

Example 11.5

Determine the load impedance $\mathbf{Z}_L$ that maximizes the average power drawn from the circuit of Fig. 11.8. What is the maximum average power?

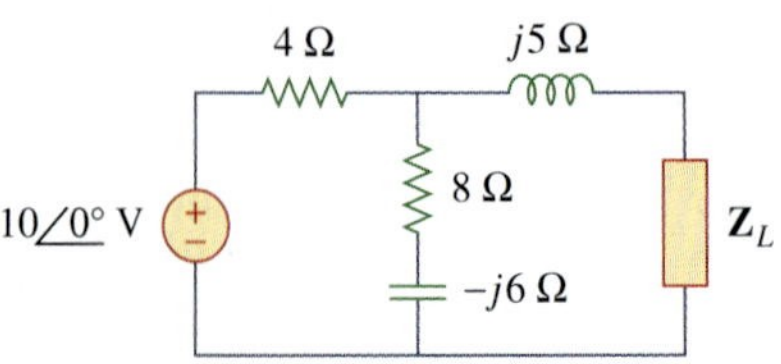

Figure 11.8
For Example 11.5.

Solution:
First we obtain the Thevenin equivalent at the load terminals. To get $\mathbf{Z}_{\text{Th}}$, consider the circuit shown in Fig. 11.9(a). We find

$$\mathbf{Z}_{\text{Th}} = j5 + 4 \parallel (8 - j6) = j5 + \frac{4(8 - j6)}{4 + 8 - j6} = 2.933 + j4.467\ \Omega$$

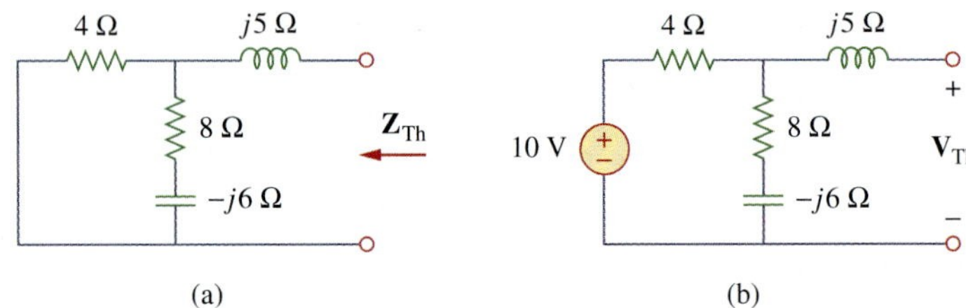

Figure 11.9
Finding the Thevenin equivalent of the circuit in Fig. 11.8.

To find $\mathbf{V}_{\text{Th}}$, consider the circuit in Fig. 11.8(b). By voltage division,

$$\mathbf{V}_{\text{Th}} = \frac{8 - j6}{4 + 8 - j6}(10) = 7.454\underline{/-10.3^\circ}\ \text{V}$$

The load impedance draws the maximum power from the circuit when

$$\mathbf{Z}_L = \mathbf{Z}_{\text{Th}}^* = 2.933 - j4.467\ \Omega$$

According to Eq. (11.20), the maximum average power is

$$P_{\text{max}} = \frac{|\mathbf{V}_{\text{Th}}|^2}{8R_{\text{Th}}} = \frac{(7.454)^2}{8(2.933)} = 2.368\ \text{W}$$

Practice Problem 11.5

For the circuit shown in Fig. 11.10, find the load impedance $\mathbf{Z}_L$ that absorbs the maximum average power. Calculate that maximum average power.

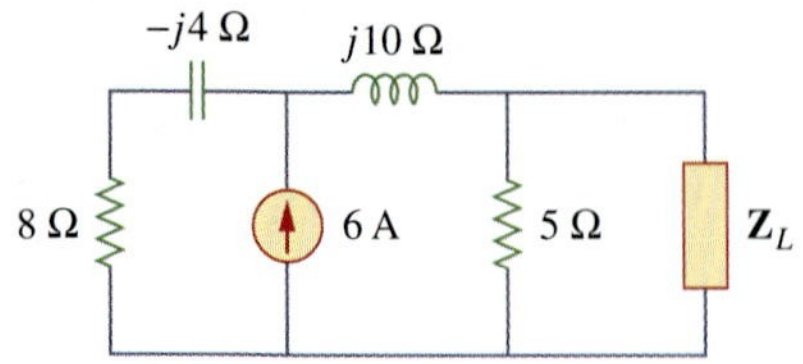

Figure 11.10
For Practice Prob. 11.5.

Answer: $3.415 - j0.7317\ \Omega$, 12.861 W.

Example 11.6

In the circuit in Fig. 11.11, find the value of R_L that will absorb the maximum average power. Calculate that power.

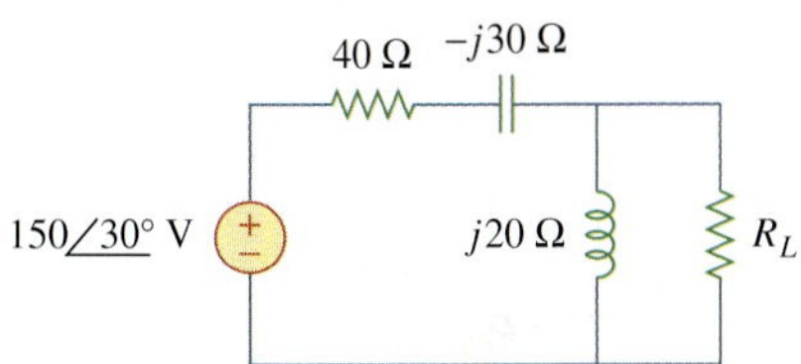

Figure 11.11
For Example 11.6.

Solution:
We first find the Thevenin equivalent at the terminals of R_L.

$$\mathbf{Z}_{\text{Th}} = (40 - j30) \parallel j20 = \frac{j20(40 - j30)}{j20 + 40 - j30} = 9.412 + j22.35\ \Omega$$

By voltage division,

$$\mathbf{V}_{\text{Th}} = \frac{j20}{j20 + 40 - j30}(150\underline{/30^\circ}) = 72.76\underline{/134^\circ}\text{ V}$$

The value of R_L that will absorb the maximum average power is

$$R_L = |\mathbf{Z}_{\text{Th}}| = \sqrt{9.412^2 + 22.35^2} = 24.25\ \Omega$$

The current through the load is

$$\mathbf{I} = \frac{\mathbf{V}_{\text{Th}}}{\mathbf{Z}_{\text{Th}} + R_L} = \frac{72.76\underline{/134^\circ}}{33.66 + j22.35} = 1.8\underline{/100.42^\circ}\text{ A}$$

The maximum average power absorbed by R_L is

$$P_{\max} = \frac{1}{2}|\mathbf{I}|^2 R_L = \frac{1}{2}(1.8)^2(24.25) = 39.29\text{ W}$$

Practice Problem 11.6

In Fig. 11.12, the resistor R_L is adjusted until it absorbs the maximum average power. Calculate R_L and the maximum average power absorbed by it.

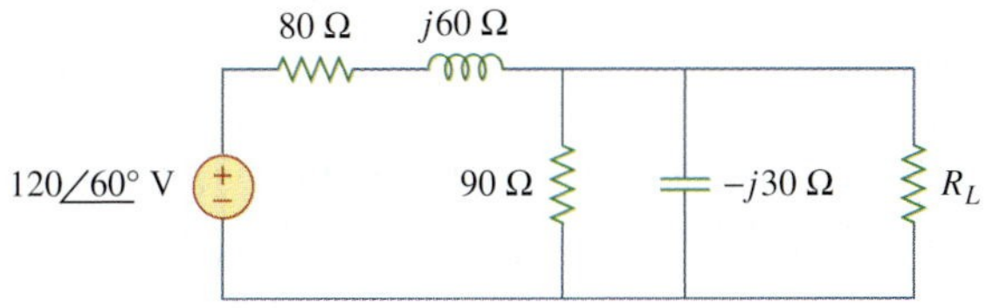

Figure 11.12
For Practice Prob. 11.6.

Answer: 30 Ω, 6.863 W.

11.4 Effective or RMS Value

The idea of *effective value* arises from the need to measure the effectiveness of a voltage or current source in delivering power to a resistive load.

> The **effective value** of a periodic current is the dc current that delivers the same average power to a resistor as the periodic current.

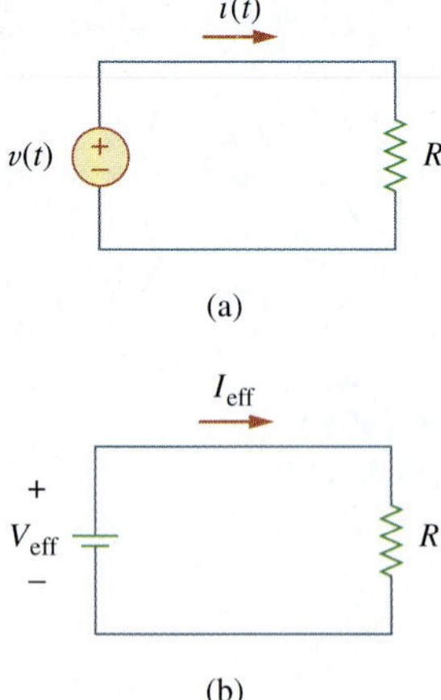

Figure 11.13
Finding the effective current: (a) ac circuit, (b) dc circuit.

In Fig. 11.13, the circuit in (a) is ac while that of (b) is dc. Our objective is to find I_{eff} that will transfer the same power to resistor R as the sinusoid i. The average power absorbed by the resistor in the ac circuit is

$$P = \frac{1}{T}\int_0^T i^2 R\, dt = \frac{R}{T}\int_0^T i^2\, dt \tag{11.22}$$

while the power absorbed by the resistor in the dc circuit is

$$P = I_{\text{eff}}^2 R \tag{11.23}$$

Equating the expressions in Eqs. (11.22) and (11.23) and solving for I_{eff}, we obtain

$$I_{\text{eff}} = \sqrt{\frac{1}{T}\int_0^T i^2\, dt} \tag{11.24}$$

The effective value of the voltage is found in the same way as current; that is,

$$V_{\text{eff}} = \sqrt{\frac{1}{T}\int_0^T v^2\, dt} \tag{11.25}$$

This indicates that the effective value is the (square) *root* of the *mean* (or average) of the *square* of the periodic signal. Thus, the effective value is often known as the *root-mean-square* value, or *rms* value for short; and we write

$$I_{\text{eff}} = I_{\text{rms}}, \qquad V_{\text{eff}} = V_{\text{rms}} \tag{11.26}$$

For any periodic function $x(t)$ in general, the rms value is given by

$$X_{\text{rms}} = \sqrt{\frac{1}{T}\int_0^T x^2\, dt} \tag{11.27}$$

The **effective value** of a periodic signal is its root mean square (rms) value.

Equation 11.27 states that to find the rms value of $x(t)$, we first find its *square* x^2 and then find the *mean* of that, or

$$\frac{1}{T}\int_0^T x^2\, dt$$

and then the square *root* ($\sqrt{\quad}$) of that mean. The rms value of a constant is the constant itself. For the sinusoid $i(t) = I_m \cos \omega t$, the effective or rms value is

$$\begin{aligned} I_{\text{rms}} &= \sqrt{\frac{1}{T}\int_0^T I_m^2 \cos^2 \omega t\, dt} \\ &= \sqrt{\frac{I_m^2}{T}\int_0^T \frac{1}{2}(1 + \cos 2\omega t)\, dt} = \frac{I_m}{\sqrt{2}} \end{aligned} \tag{11.28}$$

Similarly, for $v(t) = V_m \cos \omega t$,

$$V_{\text{rms}} = \frac{V_m}{\sqrt{2}} \tag{11.29}$$

Keep in mind that Eqs. (11.28) and (11.29) are only valid for sinusoidal signals.

The average power in Eq. (11.8) can be written in terms of the rms values.

$$P = \frac{1}{2}V_m I_m \cos(\theta_v - \theta_i) = \frac{V_m}{\sqrt{2}}\frac{I_m}{\sqrt{2}}\cos(\theta_v - \theta_i)$$
$$= V_{\text{rms}} I_{\text{rms}} \cos(\theta_v - \theta_i) \tag{11.30}$$

Similarly, the average power absorbed by a resistor R in Eq. (11.11) can be written as

$$P = I_{\text{rms}}^2 R = \frac{V_{\text{rms}}^2}{R} \tag{11.31}$$

When a sinusoidal voltage or current is specified, it is often in terms of its maximum (or peak) value or its rms value, since its average value is zero. The power industries specify phasor magnitudes in terms of their rms values rather than peak values. For instance, the 110 V available at every household is the rms value of the voltage from the power company. It is convenient in power analysis to express voltage and current in their rms values. Also, analog voltmeters and ammeters are designed to read directly the rms value of voltage and current, respectively.

Example 11.7

Determine the rms value of the current waveform in Fig. 11.14. If the current is passed through a 2-Ω resistor, find the average power absorbed by the resistor.

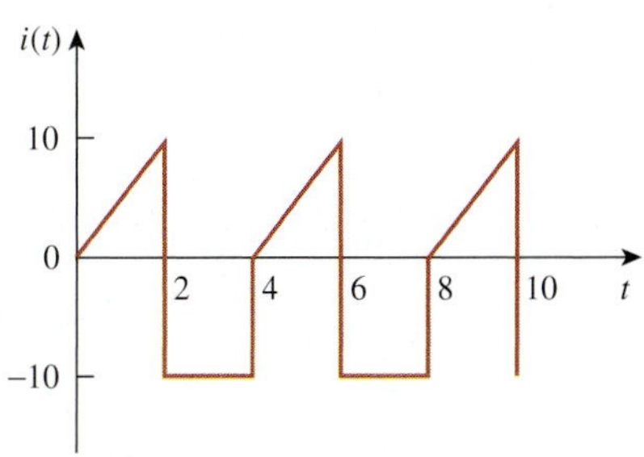

Figure 11.14
For Example 11.7.

Solution:
The period of the waveform is $T = 4$. Over a period, we can write the current waveform as

$$i(t) = \begin{cases} 5t, & 0 < t < 2 \\ -10, & 2 < t < 4 \end{cases}$$

The rms value is

$$I_{\text{rms}} = \sqrt{\frac{1}{T}\int_0^T i^2\,dt} = \sqrt{\frac{1}{4}\left[\int_0^2 (5t)^2\,dt + \int_2^4 (-10)^2\,dt\right]}$$
$$= \sqrt{\frac{1}{4}\left[25\frac{t^3}{3}\bigg|_0^2 + 100t\bigg|_2^4\right]} = \sqrt{\frac{1}{4}\left(\frac{200}{3} + 200\right)} = 8.165 \text{ A}$$

The power absorbed by a 2-Ω resistor is

$$P = I_{\text{rms}}^2 R = (8.165)^2(2) = 133.3 \text{ W}$$

Practice Problem 11.7

Find the rms value of the current waveform of Fig. 11.15. If the current flows through a 9-Ω resistor, calculate the average power absorbed by the resistor.

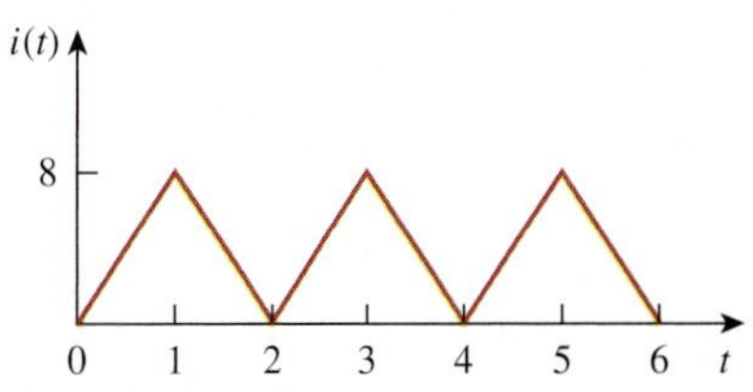

Figure 11.15
For Practice Prob. 11.7.

Answer: 4.318 A, 192 W.

Example 11.8

The waveform shown in Fig. 11.16 is a half-wave rectified sine wave. Find the rms value and the amount of average power dissipated in a 10-Ω resistor.

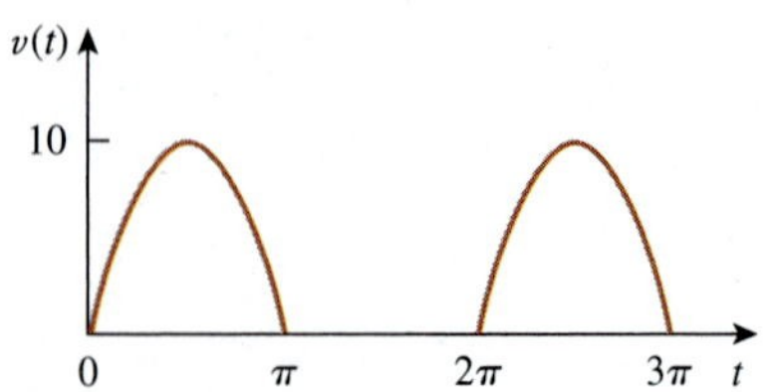

Figure 11.16
For Example 11.8.

Solution:
The period of the voltage waveform is $T = 2\pi$, and

$$v(t) = \begin{cases} 10\sin t, & 0 < t < \pi \\ 0, & \pi < t < 2\pi \end{cases}$$

The rms value is obtained as

$$V_{\text{rms}}^2 = \frac{1}{T}\int_0^T v^2(t)\,dt = \frac{1}{2\pi}\left[\int_0^{\pi}(10\sin t)^2\,dt + \int_{\pi}^{2\pi} 0^2\,dt\right]$$

But $\sin^2 t = \frac{1}{2}(1 - \cos 2t)$. Hence,

$$V_{\text{rms}}^2 = \frac{1}{2\pi}\int_0^{\pi}\frac{100}{2}(1 - \cos 2t)\,dt = \frac{50}{2\pi}\left(t - \frac{\sin 2t}{2}\right)\Bigg|_0^{\pi}$$

$$= \frac{50}{2\pi}\left(\pi - \frac{1}{2}\sin 2\pi - 0\right) = 25, \qquad V_{\text{rms}} = 5\text{ V}$$

The average power absorbed is

$$P = \frac{V_{\text{rms}}^2}{R} = \frac{5^2}{10} = 2.5\text{ W}$$

Practice Problem 11.8

Find the rms value of the full-wave rectified sine wave in Fig. 11.17. Calculate the average power dissipated in a 6-Ω resistor.

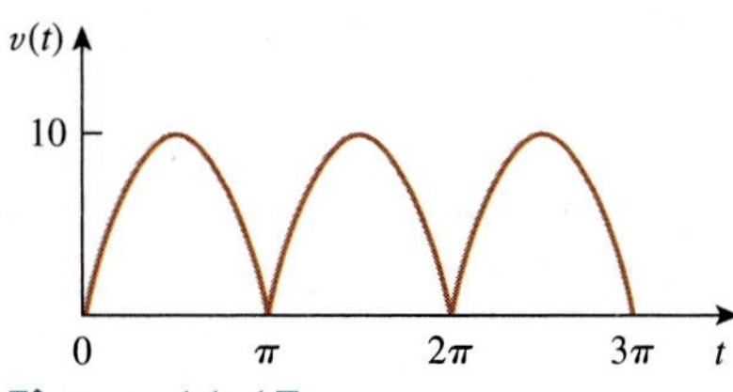

Figure 11.17
For Practice Prob. 11.8.

Answer: 7.071 V, 8.333 W.

11.5 Apparent Power and Power Factor

In Section 11.2 we saw that if the voltage and current at the terminals of a circuit are

$$v(t) = V_m\cos(\omega t + \theta_v) \quad \text{and} \quad i(t) = I_m\cos(\omega t + \theta_i) \tag{11.32}$$

or, in phasor form, $\mathbf{V} = V_m\underline{/\theta_v}$ and $\mathbf{I} = I_m\underline{/\theta_i}$, the average power is

$$P = \frac{1}{2}V_m I_m\cos(\theta_v - \theta_i) \tag{11.33}$$

In Section 11.4, we saw that

$$P = V_{\text{rms}}I_{\text{rms}}\cos(\theta_v - \theta_i) = S\cos(\theta_v - \theta_i) \tag{11.34}$$

We have added a new term to the equation:

$$\boxed{S = V_{\text{rms}}I_{\text{rms}}} \tag{11.35}$$

The average power is a product of two terms. The product $V_{rms} I_{rms}$ is known as the *apparent power S*. The factor $\cos(\theta_v - \theta_i)$ is called the *power factor* (pf).

The **apparent power** (in VA) is the product of the rms values of voltage and current.

The apparent power is so called because it seems apparent that the power should be the voltage-current product, by analogy with dc resistive circuits. It is measured in volt-amperes or VA to distinguish it from the average or real power, which is measured in watts. The power factor is dimensionless, since it is the ratio of the average power to the apparent power,

$$\text{pf} = \frac{P}{S} = \cos(\theta_v - \theta_i) \tag{11.36}$$

The angle $\theta_v - \theta_i$ is called the *power factor angle*, since it is the angle whose cosine is the power factor. The power factor angle is equal to the angle of the load impedance if **V** is the voltage across the load and **I** is the current through it. This is evident from the fact that

$$\mathbf{Z} = \frac{\mathbf{V}}{\mathbf{I}} = \frac{V_m\underline{/\theta_v}}{I_m\underline{/\theta_i}} = \frac{V_m}{I_m}\underline{/\theta_v - \theta_i} \tag{11.37}$$

Alternatively, since

$$\mathbf{V}_{rms} = \frac{\mathbf{V}}{\sqrt{2}} = V_{rms}\underline{/\theta_v} \tag{11.38a}$$

and

$$\mathbf{I}_{rms} = \frac{\mathbf{I}}{\sqrt{2}} = I_{rms}\underline{/\theta_i} \tag{11.38b}$$

the impedance is

$$\mathbf{Z} = \frac{\mathbf{V}}{\mathbf{I}} = \frac{\mathbf{V}_{rms}}{\mathbf{I}_{rms}} = \frac{V_{rms}}{I_{rms}}\underline{/\theta_v - \theta_i} \tag{11.39}$$

The **power factor** is the cosine of the phase difference between voltage and current. It is also the cosine of the angle of the load impedance.

From Eq. (11.36), the power factor may be seen as that factor by which the apparent power must be multiplied to obtain the real or average power. The value of pf ranges between zero and unity. For a purely resistive load, the voltage and current are in phase, so that $\theta_v - \theta_i = 0$ and pf $= 1$. This implies that the apparent power is equal to the average power. For a purely reactive load, $\theta_v - \theta_i = \pm 90°$ and pf $= 0$. In this case the average power is zero. In between these two extreme cases, pf is said to be *leading* or *lagging*. Leading power factor means that current leads voltage, which implies a capacitive load. Lagging power factor means that current lags voltage, implying an inductive

From Eq. (11.36), the power factor may also be regarded as the ratio of the real power dissipated in the load to the apparent power of the load.

load. Power factor affects the electric bills consumers pay the electric utility companies, as we will see in Section 11.9.2.

Example 11.9

A series-connected load draws a current $i(t) = 4\cos(100\pi t + 10^\circ)$ A when the applied voltage is $v(t) = 120\cos(100\pi t - 20^\circ)$ V. Find the apparent power and the power factor of the load. Determine the element values that form the series-connected load.

Solution:

The apparent power is

$$S = V_{\text{rms}} I_{\text{rms}} = \frac{120}{\sqrt{2}} \frac{4}{\sqrt{2}} = 240 \text{ VA}$$

The power factor is

$$\text{pf} = \cos(\theta_v - \theta_i) = \cos(-20^\circ - 10^\circ) = 0.866 \qquad \text{(leading)}$$

The pf is leading because the current leads the voltage. The pf may also be obtained from the load impedance.

$$\mathbf{Z} = \frac{\mathbf{V}}{\mathbf{I}} = \frac{120\underline{/-20^\circ}}{4\underline{/10^\circ}} = 30\underline{/-30^\circ} = 25.98 - j15\ \Omega$$

$$\text{pf} = \cos(-30^\circ) = 0.866 \qquad \text{(leading)}$$

The load impedance $\mathbf{Z}$ can be modeled by a 25.98-Ω resistor in series with a capacitor with

$$X_C = -15 = -\frac{1}{\omega C}$$

or

$$C = \frac{1}{15\omega} = \frac{1}{15 \times 100\pi} = 212.2\ \mu\text{F}$$

Practice Problem 11.9

Obtain the power factor and the apparent power of a load whose impedance is $\mathbf{Z} = 60 + j40\ \Omega$ when the applied voltage is $v(t) = 160\cos(377t + 10^\circ)$ V.

Answer: 0.8321 lagging, $177.5\underline{/33.69^\circ}$ VA.

Example 11.10

Determine the power factor of the entire circuit of Fig. 11.18 as seen by the source. Calculate the average power delivered by the source.

Solution:

The total impedance is

$$\mathbf{Z} = 6 + 4 \parallel (-j2) = 6 + \frac{-j2 \times 4}{4 - j2} = 6.8 - j1.6 = 7\underline{/-13.24^\circ}\ \Omega$$

The power factor is

$$\text{pf} = \cos(-13.24) = 0.9734 \qquad \text{(leading)}$$

since the impedance is capacitive. The rms value of the current is

$$\mathbf{I}_{\text{rms}} = \frac{\mathbf{V}_{\text{rms}}}{\mathbf{Z}} = \frac{30\underline{/0^\circ}}{7\underline{/-13.24^\circ}} = 4.286\underline{/13.24^\circ} \text{ A}$$

The average power supplied by the source is

$$P = V_{\text{rms}} I_{\text{rms}} \text{pf} = (30)(4.286)0.9734 = 125 \text{ W}$$

or

$$P = I_{\text{rms}}^2 R = (4.286)^2(6.8) = 125 \text{ W}$$

where R is the resistive part of $\mathbf{Z}$.

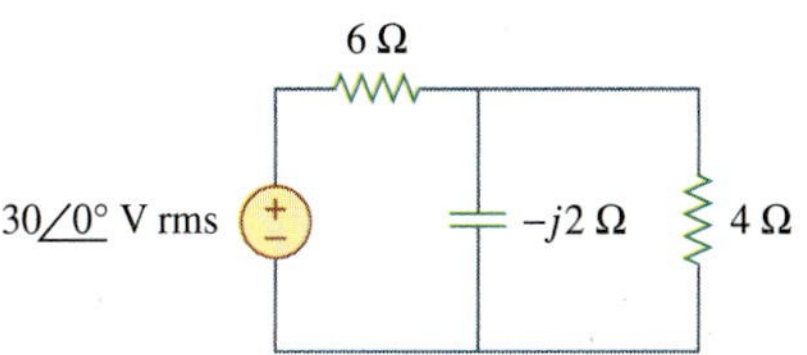

Figure 11.18
For Example 11.10.

Practice Problem 11.10

Calculate the power factor of the entire circuit of Fig. 11.19 as seen by the source. What is the average power supplied by the source?

Answer: 0.936 lagging, 1.062 kW.

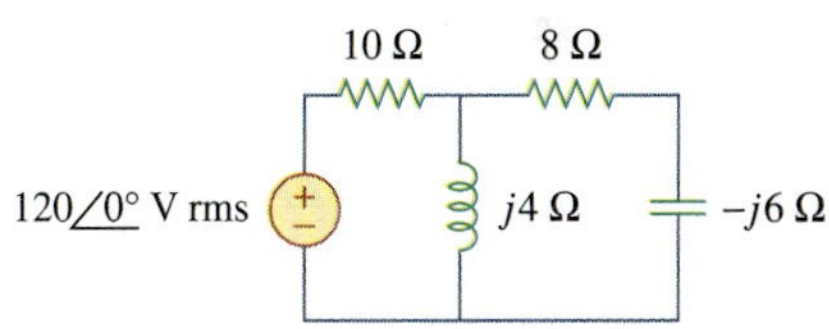

Figure 11.19
For Practice Prob. 11.10.

11.6 Complex Power

Considerable effort has been expended over the years to express power relations as simply as possible. Power engineers have coined the term *complex power*, which they use to find the total effect of parallel loads. Complex power is important in power analysis because it contains *all* the information pertaining to the power absorbed by a given load.

Consider the ac load in Fig. 11.20. Given the phasor form $\mathbf{V} = V_m\underline{/\theta_v}$ and $\mathbf{I} = I_m\underline{/\theta_i}$ of voltage $v(t)$ and current $i(t)$, the *complex power* $\mathbf{S}$ absorbed by the ac load is the product of the voltage and the complex conjugate of the current, or

$$\mathbf{S} = \frac{1}{2}\mathbf{V}\mathbf{I}^* \tag{11.40}$$

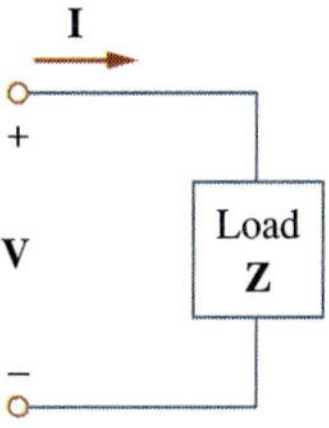

Figure 11.20
The voltage and current phasors associated with a load.

assuming the passive sign convention (see Fig. 11.20). In terms of the rms values,

$$\mathbf{S} = \mathbf{V}_{\text{rms}}\mathbf{I}_{\text{rms}}^* \tag{11.41}$$

where

$$\mathbf{V}_{\text{rms}} = \frac{\mathbf{V}}{\sqrt{2}} = V_{\text{rms}}\underline{/\theta_v} \tag{11.42}$$

and

$$\mathbf{I}_{\text{rms}} = \frac{\mathbf{I}}{\sqrt{2}} = I_{\text{rms}}\underline{/\theta_i} \tag{11.43}$$

When working with the rms values of currents or voltages, we may drop the subscript rms if no confusion will be caused by doing so.

Thus we may write Eq. (11.41) as

$$\begin{aligned}\mathbf{S} &= V_{\text{rms}} I_{\text{rms}} \underline{/\theta_v - \theta_i} \\ &= V_{\text{rms}} I_{\text{rms}} \cos(\theta_v - \theta_i) + jV_{\text{rms}} I_{\text{rms}} \sin(\theta_v - \theta_i)\end{aligned} \quad \textbf{(11.44)}$$

This equation can also be obtained from Eq. (11.9). We notice from Eq. (11.44) that the magnitude of the complex power is the apparent power; hence, the complex power is measured in volt-amperes (VA). Also, we notice that the angle of the complex power is the power factor angle.

The complex power may be expressed in terms of the load impedance **Z**. From Eq. (11.37), the load impedance **Z** may be written as

$$\mathbf{Z} = \frac{\mathbf{V}}{\mathbf{I}} = \frac{\mathbf{V}_{\text{rms}}}{\mathbf{I}_{\text{rms}}} = \frac{V_{\text{rms}}}{I_{\text{rms}}} \underline{/\theta_v - \theta_i} \quad \textbf{(11.45)}$$

Thus, $\mathbf{V}_{\text{rms}} = \mathbf{Z}\mathbf{I}_{\text{rms}}$. Substituting this into Eq. (11.41) gives

$$\mathbf{S} = I_{\text{rms}}^2 \mathbf{Z} = \frac{V_{\text{rms}}^2}{\mathbf{Z}^*} = \mathbf{V}_{\text{rms}} \mathbf{I}_{\text{rms}}^* \quad \textbf{(11.46)}$$

Since $\mathbf{Z} = R + jX$, Eq. (11.46) becomes

$$\mathbf{S} = I_{\text{rms}}^2 (R + jX) = P + jQ \quad \textbf{(11.47)}$$

where P and Q are the real and imaginary parts of the complex power; that is,

$$P = \text{Re}(\mathbf{S}) = I_{\text{rms}}^2 R \quad \textbf{(11.48)}$$

$$Q = \text{Im}(\mathbf{S}) = I_{\text{rms}}^2 X \quad \textbf{(11.49)}$$

P is the average or real power and it depends on the load's resistance R. Q depends on the load's reactance X and is called the *reactive* (or quadrature) power.

Comparing Eq. (11.44) with Eq. (11.47), we notice that

$$P = V_{\text{rms}} I_{\text{rms}} \cos(\theta_v - \theta_i), \qquad Q = V_{\text{rms}} I_{\text{rms}} \sin(\theta_v - \theta_i) \quad \textbf{(11.50)}$$

The real power P is the average power in watts delivered to a load; it is the only useful power. It is the actual power dissipated by the load. The reactive power Q is a measure of the energy exchange between the source and the reactive part of the load. The unit of Q is the *volt-ampere reactive* (VAR) to distinguish it from the real power, whose unit is the watt. We know from Chapter 6 that energy storage elements neither dissipate nor supply power, but exchange power back and forth with the rest of the network. In the same way, the reactive power is being transferred back and forth between the load and the source. It represents a lossless interchange between the load and the source. Notice that:

1. $Q = 0$ for resistive loads (unity pf).
2. $Q < 0$ for capacitive loads (leading pf).
3. $Q > 0$ for inductive loads (lagging pf).

Thus,

Complex power (in VA) is the product of the rms voltage phasor and the complex conjugate of the rms current phasor. As a complex quantity, its real part is real power P and its imaginary part is reactive power Q.

Introducing the complex power enables us to obtain the real and reactive powers directly from voltage and current phasors.

$$\begin{aligned}
\text{Complex Power} &= \mathbf{S} = P + jQ = \mathbf{V}_{\text{rms}}(\mathbf{I}_{\text{rms}})^* \\
&= V_{\text{rms}} I_{\text{rms}} \underline{/\theta_v - \theta_i} \\
\text{Apparent Power} &= S = |\mathbf{S}| = V_{\text{rms}} I_{\text{rms}} = \sqrt{P^2 + Q^2} \\
\text{Real Power} &= P = \text{Re}(\mathbf{S}) = S\cos(\theta_v - \theta_i) \\
\text{Reactive Power} &= Q = \text{Im}(\mathbf{S}) = S\sin(\theta_v - \theta_i) \\
\text{Power Factor} &= \frac{P}{S} = \cos(\theta_v - \theta_i)
\end{aligned} \tag{11.51}$$

This shows how the complex power contains *all* the relevant power information in a given load.

It is a standard practice to represent $\mathbf{S}$, P, and Q in the form of a triangle, known as the *power triangle*, shown in Fig. 11.21(a). This is similar to the impedance triangle showing the relationship between $\mathbf{Z}$, R, and X, illustrated in Fig. 11.21(b). The power triangle has four items—the apparent/complex power, real power, reactive power, and the power factor angle. Given two of these items, the other two can easily be obtained from the triangle. As shown in Fig. 11.22, when $\mathbf{S}$ lies in the first quadrant, we have an inductive load and a lagging pf. When $\mathbf{S}$ lies in the fourth quadrant, the load is capacitive and the pf is leading. It is also possible for the complex power to lie in the second or third quadrant. This requires that the load impedance have a negative resistance, which is possible with active circuits.

S contains *all* power information of a load. The real part of **S** is the real power P; its imaginary part is the reactive power Q; its magnitude is the apparent power S; and the cosine of its phase angle is the power factor pf.

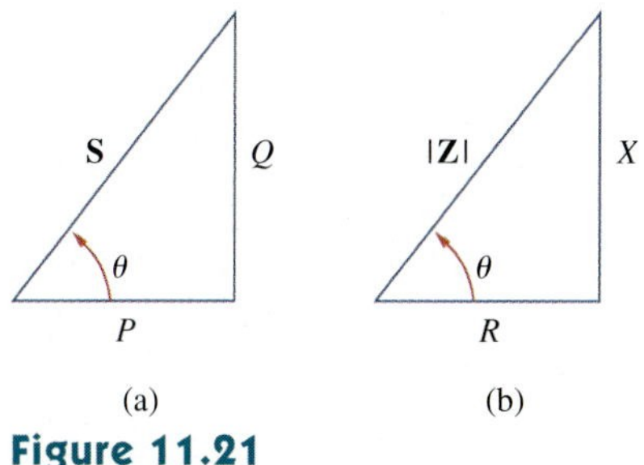

Figure 11.21
(a) Power triangle, (b) impedance triangle.

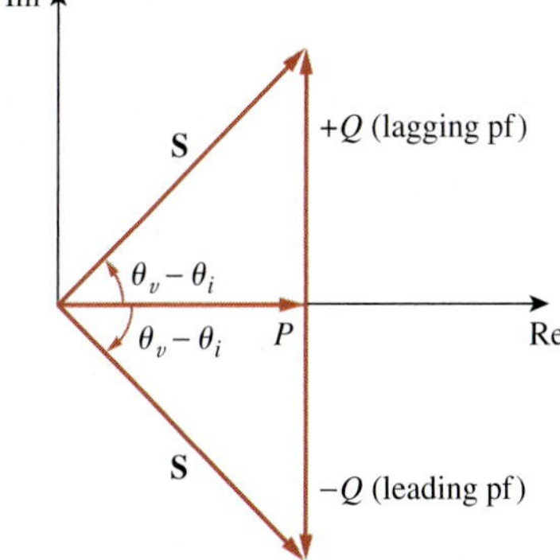

Figure 11.22
Power triangle.

Example 11.11

The voltage across a load is $v(t) = 60\cos(\omega t - 10°)$ V and the current through the element in the direction of the voltage drop is $i(t) = 1.5\cos(\omega t + 50°)$ A. Find: (a) the complex and apparent powers, (b) the real and reactive powers, and (c) the power factor and the load impedance.

Solution:

(a) For the rms values of the voltage and current, we write

$$\mathbf{V}_{rms} = \frac{60}{\sqrt{2}}\underline{/-10^\circ}, \qquad \mathbf{I}_{rms} = \frac{1.5}{\sqrt{2}}\underline{/+50^\circ}$$

The complex power is

$$\mathbf{S} = \mathbf{V}_{rms}\mathbf{I}^*_{rms} = \left(\frac{60}{\sqrt{2}}\underline{/-10^\circ}\right)\left(\frac{1.5}{\sqrt{2}}\underline{/-50^\circ}\right) = 45\underline{/-60^\circ} \text{ VA}$$

The apparent power is

$$S = |\mathbf{S}| = 45 \text{ VA}$$

(b) We can express the complex power in rectangular form as

$$\mathbf{S} = 45\underline{/-60^\circ} = 45[\cos(-60^\circ) + j\sin(-60^\circ)] = 22.5 - j38.97$$

Since $\mathbf{S} = P + jQ$, the real power is

$$P = 22.5 \text{ W}$$

while the reactive power is

$$Q = -38.97 \text{ VAR}$$

(c) The power factor is

$$\text{pf} = \cos(-60^\circ) = 0.5 \text{ (leading)}$$

It is leading, because the reactive power is negative. The load impedance is

$$\mathbf{Z} = \frac{\mathbf{V}}{\mathbf{I}} = \frac{60\underline{/-10^\circ}}{1.5\underline{/+50^\circ}} = 40\underline{/-60^\circ}\ \Omega$$

which is a capacitive impedance.

Practice Problem 11.11

For a load, $\mathbf{V}_{rms} = 110\underline{/85^\circ}$ V, $\mathbf{I}_{rms} = 0.4\underline{/15^\circ}$ A. Determine: (a) the complex and apparent powers, (b) the real and reactive powers, and (c) the power factor and the load impedance.

Answer: (a) $44\underline{/70^\circ}$ VA, 44 VA, (b) 15.05 W, 41.35 VAR, (c) 0.342 lagging, $94.06 + j258.4\ \Omega$.

Example 11.12

A load **Z** draws 12 kVA at a power factor of 0.856 lagging from a 120-V rms sinusoidal source. Calculate: (a) the average and reactive powers delivered to the load, (b) the peak current, and (c) the load impedance.

Solution:

(a) Given that pf $= \cos\theta = 0.856$, we obtain the power angle as $\theta = \cos^{-1} 0.856 = 31.13^\circ$. If the apparent power is $S = 12{,}000$ VA, then the average or real power is

$$P = S\cos\theta = 12{,}000 \times 0.856 = 10.272 \text{ kW}$$

while the reactive power is

$$Q = S \sin\theta = 12{,}000 \times 0.517 = 6.204 \text{ kVA}$$

(b) Since the pf is lagging, the complex power is

$$\mathbf{S} = P + jQ = 10.272 + j6.204 \text{ kVA}$$

From $\mathbf{S} = \mathbf{V}_{\text{rms}}\mathbf{I}^*_{\text{rms}}$, we obtain

$$\mathbf{I}^*_{\text{rms}} = \frac{\mathbf{S}}{\mathbf{V}_{\text{rms}}} = \frac{10{,}272 + j6204}{120\underline{/0^\circ}} = 85.6 + j51.7 \text{ A} = 100\underline{/31.13^\circ} \text{ A}$$

Thus $\mathbf{I}_{\text{rms}} = 100\underline{/-31.13^\circ}$ and the peak current is

$$I_m = \sqrt{2}I_{\text{rms}} = \sqrt{2}(100) = 141.4 \text{ A}$$

(c) The load impedance

$$\mathbf{Z} = \frac{\mathbf{V}_{\text{rms}}}{\mathbf{I}_{\text{rms}}} = \frac{120\underline{/0^\circ}}{100\underline{/-31.13^\circ}} = 1.2\underline{/31.13^\circ} \ \Omega$$

which is an inductive impedance.

Practice Problem 11.12

A sinusoidal source supplies 20 kVAR reactive power to load $\mathbf{Z} = 250\underline{/-75^\circ} \ \Omega$. Determine: (a) the power factor, (b) the apparent power delivered to the load, and (c) the rms voltage.

Answer: (a) 0.2588 leading, (b) 20.71 kVA, (c) 2.275 kV.

11.7 †Conservation of AC Power

The principle of conservation of power applies to ac circuits as well as to dc circuits (see Section 1.5).

In fact, we already saw in Examples 11.3 and 11.4 that average power is conserved in ac circuits.

To see this, consider the circuit in Fig. 11.23(a), where two load impedances $\mathbf{Z}_1$ and $\mathbf{Z}_2$ are connected in parallel across an ac source $\mathbf{V}$. KCL gives

$$\mathbf{I} = \mathbf{I}_1 + \mathbf{I}_2 \tag{11.52}$$

The complex power supplied by the source is (from now on, unless otherwise specified, all values of voltages and currents will be assumed to be rms values)

$$\mathbf{S} = \mathbf{V}\mathbf{I}^* = \mathbf{V}(\mathbf{I}_1^* + \mathbf{I}_2^*) = \mathbf{V}\mathbf{I}_1^* + \mathbf{V}\mathbf{I}_2^* = \mathbf{S}_1 + \mathbf{S}_2 \tag{11.53}$$

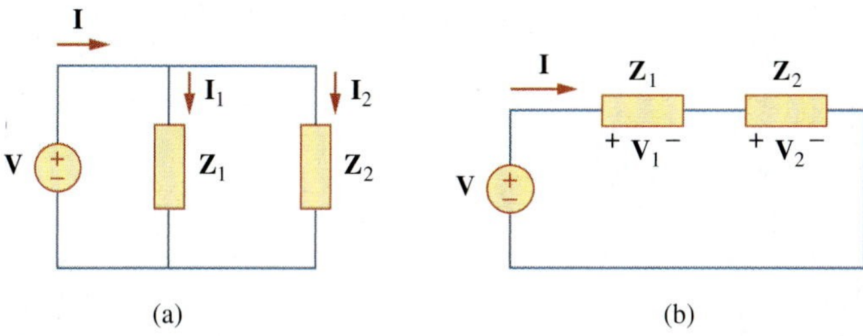

Figure 11.23
An ac voltage source supplied loads connected in: (a) parallel, (b) series.

where $\mathbf{S}_1$ and $\mathbf{S}_2$ denote the complex powers delivered to loads $\mathbf{Z}_1$ and $\mathbf{Z}_2$, respectively.

If the loads are connected in series with the voltage source, as shown in Fig. 11.23(b), KVL yields

$$\mathbf{V} = \mathbf{V}_1 + \mathbf{V}_2 \tag{11.54}$$

The complex power supplied by the source is

$$\mathbf{S} = \mathbf{V}\mathbf{I}^* = (\mathbf{V}_1 + \mathbf{V}_2)\mathbf{I}^* = \mathbf{V}_1\mathbf{I}^* + \mathbf{V}_2\mathbf{I}^* = \mathbf{S}_1 + \mathbf{S}_2 \tag{11.55}$$

where $\mathbf{S}_1$ and $\mathbf{S}_2$ denote the complex powers delivered to loads $\mathbf{Z}_1$ and $\mathbf{Z}_2$, respectively.

We conclude from Eqs. (11.53) and (11.55) that whether the loads are connected in series or in parallel (or in general), the total power *supplied* by the source equals the total power *delivered* to the load. Thus, in general, for a source connected to N loads,

$$\boxed{\mathbf{S} = \mathbf{S}_1 + \mathbf{S}_2 + \cdots + \mathbf{S}_N} \tag{11.56}$$

In fact, all forms of ac power are conserved: instantaneous, real, reactive, and complex.

This means that the total complex power in a network is the sum of the complex powers of the individual components. (This is also true of real power and reactive power, but not true of apparent power.) This expresses the principle of conservation of ac power:

> The complex, real, and reactive powers of the sources equal the respective sums of the complex, real, and reactive powers of the individual loads.

From this we imply that the real (or reactive) power flow from sources in a network equals the real (or reactive) power flow into the other elements in the network.

Example 11.13

Figure 11.24 shows a load being fed by a voltage source through a transmission line. The impedance of the line is represented by the $(4 + j2)\ \Omega$ impedance and a return path. Find the real power and reactive power absorbed by: (a) the source, (b) the line, and (c) the load.

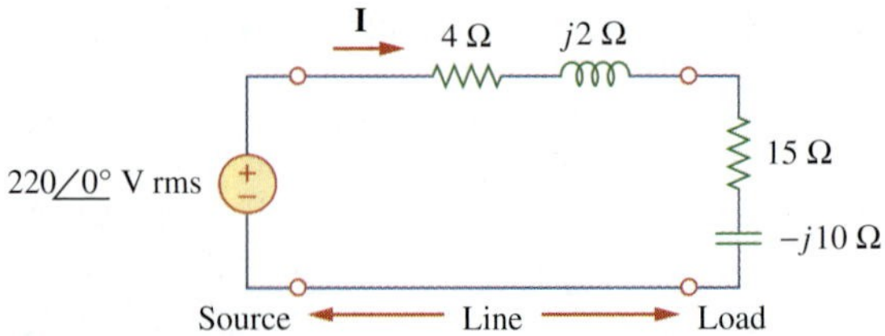

Figure 11.24
For Example 11.13.

Solution:
The total impedance is

$$\mathbf{Z} = (4 + j2) + (15 - j10) = 19 - j8 = 20.62\underline{/-22.83^\circ}\ \Omega$$

The current through the circuit is

$$\mathbf{I} = \frac{\mathbf{V}_s}{\mathbf{Z}} = \frac{220\underline{/0^\circ}}{20.62\underline{/-22.83^\circ}} = 10.67\underline{/22.83^\circ} \text{ A rms}$$

(a) For the source, the complex power is

$$\begin{aligned}\mathbf{S}_s = \mathbf{V}_s\mathbf{I}^* &= (220\underline{/0^\circ})(10.67\underline{/-22.83^\circ})\\ &= 2347.4\underline{/-22.83^\circ} = (2163.5 - j910.8) \text{ VA}\end{aligned}$$

From this, we obtain the real power as 2163.5 W and the reactive power as 910.8 VAR (leading).

(b) For the line, the voltage is

$$\begin{aligned}\mathbf{V}_{\text{line}} = (4 + j2)\mathbf{I} &= (4.472\underline{/26.57^\circ})(10.67\underline{/22.83^\circ})\\ &= 47.72\underline{/49.4^\circ} \text{ V rms}\end{aligned}$$

The complex power absorbed by the line is

$$\begin{aligned}\mathbf{S}_{\text{line}} = \mathbf{V}_{\text{line}}\mathbf{I}^* &= (47.72\underline{/49.4^\circ})(10.67\underline{/-22.83^\circ})\\ &= 509.2\underline{/26.57^\circ} = 455.4 + j227.7 \text{ VA}\end{aligned}$$

or

$$\mathbf{S}_{\text{line}} = |\mathbf{I}|^2\mathbf{Z}_{\text{line}} = (10.67)^2(4 + j2) = 455.4 + j227.7 \text{ VA}$$

That is, the real power is 455.4 W and the reactive power is 227.76 VAR (lagging).

(c) For the load, the voltage is

$$\begin{aligned}\mathbf{V}_L = (15 - j10)\mathbf{I} &= (18.03\underline{/-33.7^\circ})(10.67\underline{/22.83^\circ})\\ &= 192.38\underline{/-10.87^\circ} \text{ V rms}\end{aligned}$$

The complex power absorbed by the load is

$$\begin{aligned}\mathbf{S}_L = \mathbf{V}_L\mathbf{I}^* &= (192.38\underline{/-10.87^\circ})(10.67\underline{/-22.83^\circ})\\ &= 2053\underline{/-33.7^\circ} = (1708 - j1139) \text{ VA}\end{aligned}$$

The real power is 1708 W and the reactive power is 1139 VAR (leading). Note that $\mathbf{S}_s = \mathbf{S}_{\text{line}} + \mathbf{S}_L$, as expected. We have used the rms values of voltages and currents.

Practice Problem 11.13

In the circuit in Fig. 11.25, the 60-Ω resistor absorbs an average power of 240 W. Find **V** and the complex power of each branch of the circuit. What is the overall complex power of the circuit? (Assume the current through the 60-Ω resistor has no phase shift.)

Answer: $240.67\underline{/21.45^\circ}$ V (rms); the 20-Ω resistor: 656 VA; the $(30 - j10)$ Ω impedance: $480 - j160$ VA; the $(60 + j20)$ Ω impedance: $240 + j80$ VA; overall: $1376 - j80$ VA.

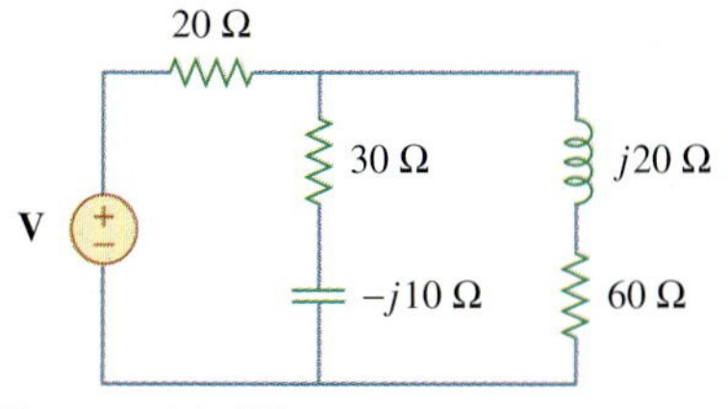

Figure 11.25
For Practice Prob. 11.13.

Example 11.14

In the circuit of Fig. 11.26, $\mathbf{Z}_1 = 60\angle -30^\circ\ \Omega$ and $\mathbf{Z}_2 = 40\angle 45^\circ\ \Omega$. Calculate the total: (a) apparent power, (b) real power, (c) reactive power, and (d) pf, supplied by the source and seen by the source.

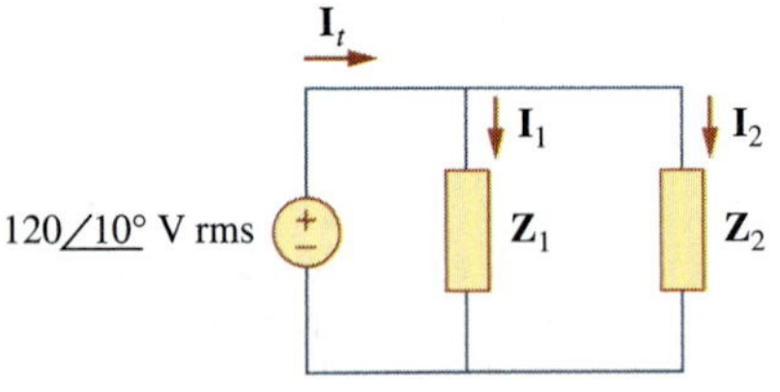

Figure 11.26
For Example 11.14.

Solution:

The current through $\mathbf{Z}_1$ is

$$\mathbf{I}_1 = \frac{\mathbf{V}}{\mathbf{Z}_1} = \frac{120\angle 10^\circ}{60\angle -30^\circ} = 2\angle 40^\circ \text{ A rms}$$

while the current through $\mathbf{Z}_2$ is

$$\mathbf{I}_2 = \frac{\mathbf{V}}{\mathbf{Z}_2} = \frac{120\angle 10^\circ}{40\angle 45^\circ} = 3\angle -35^\circ \text{ A rms}$$

The complex powers absorbed by the impedances are

$$\mathbf{S}_1 = \frac{V_{\text{rms}}^2}{\mathbf{Z}_1^*} = \frac{(120)^2}{60\angle 30^\circ} = 240\angle -30^\circ = 207.85 - j120 \text{ VA}$$

$$\mathbf{S}_2 = \frac{V_{\text{rms}}^2}{\mathbf{Z}_2^*} = \frac{(120)^2}{40\angle -45^\circ} = 360\angle 45^\circ = 254.6 + j254.6 \text{ VA}$$

The total complex power is

$$\mathbf{S}_t = \mathbf{S}_1 + \mathbf{S}_2 = 462.4 + j134.6 \text{ VA}$$

(a) The total apparent power is

$$|\mathbf{S}_t| = \sqrt{462.4^2 + 134.6^2} = 481.6 \text{ VA.}$$

(b) The total real power is

$$P_t = \text{Re}(\mathbf{S}_t) = 462.4 \text{ W or } P_t = P_1 + P_2.$$

(c) The total reactive power is

$$Q_t = \text{Im}(\mathbf{S}_t) = 134.6 \text{ VAR or } Q_t = Q_1 + Q_2.$$

(d) The pf $= P_t/|\mathbf{S}_t| = 462.4/481.6 = 0.96$ (lagging).

We may cross check the result by finding the complex power $\mathbf{S}_s$ supplied by the source.

$$\begin{aligned}\mathbf{I}_t = \mathbf{I}_1 + \mathbf{I}_2 &= (1.532 + j1.286) + (2.457 - j1.721)\\ &= 4 - j0.435 = 4.024\angle -6.21^\circ \text{ A rms}\\ \mathbf{S}_s = \mathbf{V}\mathbf{I}_t^* &= (120\angle 10^\circ)(4.024\angle 6.21^\circ)\\ &= 482.88\angle 16.21^\circ = 463 + j135 \text{ VA}\end{aligned}$$

which is the same as before.

Practice Problem 11.14

Two loads connected in parallel are respectively 2 kW at a pf of 0.75 leading and 4 kW at a pf of 0.95 lagging. Calculate the pf of the two loads. Find the complex power supplied by the source.

Answer: 0.9972 (leading), $6 - j0.4495$ kVA.

11.8 Power Factor Correction

Most domestic loads (such as washing machines, air conditioners, and refrigerators) and industrial loads (such as induction motors) are inductive and operate at a low lagging power factor. Although the inductive nature of the load cannot be changed, we can increase its power factor.

Alternatively, power factor correction may be viewed as the addition of a reactive element (usually a capacitor) in parallel with the load in order to make the power factor closer to unity.

> The process of increasing the power factor without altering the voltage or current to the original load is known as **power factor correction.**

An inductive load is modeled as a series combination of an inductor and a resistor.

Since most loads are inductive, as shown in Fig. 11.27(a), a load's power factor is improved or corrected by deliberately installing a capacitor in parallel with the load, as shown in Fig. 11.27(b). The effect of adding the capacitor can be illustrated using either the power triangle or the phasor diagram of the currents involved. Figure 11.28 shows the latter, where it is assumed that the circuit in Fig. 11.27(a) has a power factor of $\cos\theta_1$, while the one in Fig. 11.27(b) has a power factor of $\cos\theta_2$. It is evident from Fig. 11.28 that adding the capacitor has caused the phase angle between the supplied voltage and current to reduce from θ_1 to θ_2, thereby increasing the power factor. We also notice from the magnitudes of the vectors in Fig. 11.28 that with the same supplied voltage, the circuit in Fig. 11.27(a) draws larger current I_L than the current I drawn by the circuit in Fig. 11.27(b). Power companies charge more for larger currents, because they result in increased power losses (by a squared factor, since $P = I_L^2 R$). Therefore, it is beneficial to both the power company and the consumer that every effort is made to minimize current level or keep the power factor as close to unity as possible. By choosing a suitable size for the capacitor, the current can be made to be completely in phase with the voltage, implying unity power factor.

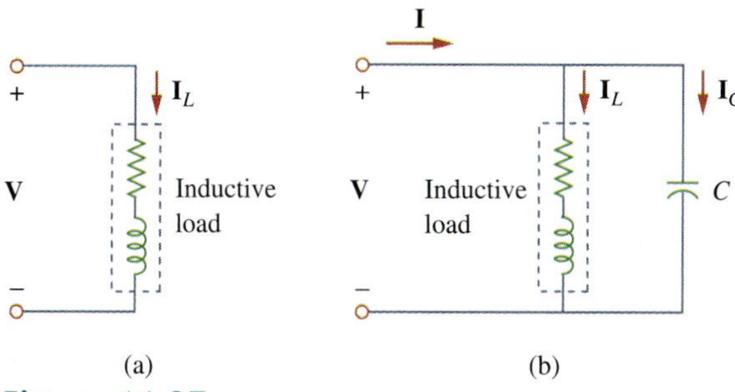

Figure 11.27
Power factor correction: (a) original inductive load, (b) inductive load with improved power factor.

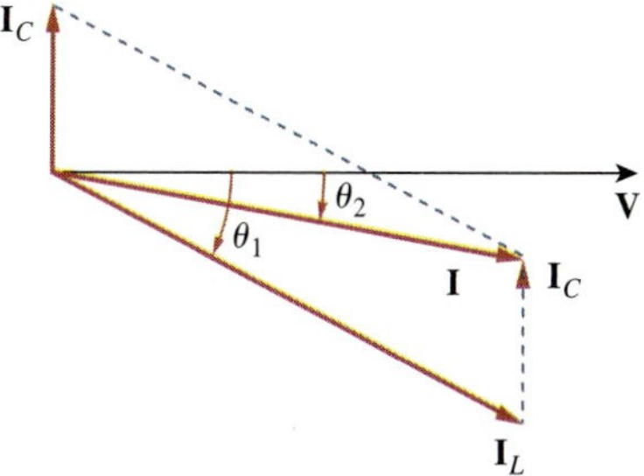

Figure 11.28
Phasor diagram showing the effect of adding a capacitor in parallel with the inductive load.

We can look at the power factor correction from another perspective. Consider the power triangle in Fig. 11.29. If the original inductive load has apparent power S_1, then

$$P = S_1 \cos\theta_1, \qquad Q_1 = S_1 \sin\theta_1 = P \tan\theta_1 \tag{11.57}$$

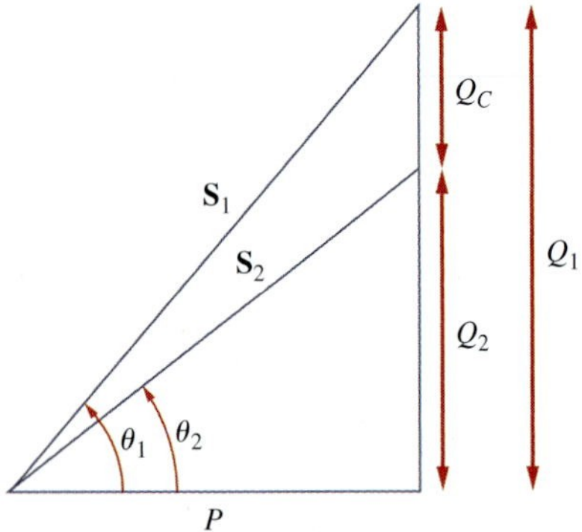

Figure 11.29
Power triangle illustrating power factor correction.

If we desire to increase the power factor from $\cos\theta_1$ to $\cos\theta_2$ without altering the real power (i.e., $P = S_2 \cos\theta_2$), then the new reactive power is

$$Q_2 = P \tan\theta_2 \tag{11.58}$$

The reduction in the reactive power is caused by the shunt capacitor; that is,

$$Q_C = Q_1 - Q_2 = P(\tan\theta_1 - \tan\theta_2) \tag{11.59}$$

But from Eq. (11.46), $Q_C = V_{\text{rms}}^2/X_C = \omega C V_{\text{rms}}^2$. The value of the required shunt capacitance C is determined as

$$C = \frac{Q_C}{\omega V_{\text{rms}}^2} = \frac{P(\tan\theta_1 - \tan\theta_2)}{\omega V_{\text{rms}}^2} \tag{11.60}$$

Note that the real power P dissipated by the load is not affected by the power factor correction because the average power due to the capacitance is zero.

Although the most common situation in practice is that of an inductive load, it is also possible that the load is capacitive; that is, the load is operating at a leading power factor. In this case, an inductor should be connected across the load for power factor correction. The required shunt inductance L can be calculated from

$$Q_L = \frac{V_{\text{rms}}^2}{X_L} = \frac{V_{\text{rms}}^2}{\omega L} \quad \Rightarrow \quad L = \frac{V_{\text{rms}}^2}{\omega Q_L} \tag{11.61}$$

where $Q_L = Q_1 - Q_2$, the difference between the new and old reactive powers.

Example 11.15

When connected to a 120-V (rms), 60-Hz power line, a load absorbs 4 kW at a lagging power factor of 0.8. Find the value of capacitance necessary to raise the pf to 0.95.

Solution:
If the pf = 0.8, then

$$\cos\theta_1 = 0.8 \quad \Rightarrow \quad \theta_1 = 36.87^\circ$$

where θ_1 is the phase difference between voltage and current. We obtain the apparent power from the real power and the pf as

$$S_1 = \frac{P}{\cos\theta_1} = \frac{4000}{0.8} = 5000 \text{ VA}$$

The reactive power is

$$Q_1 = S_1 \sin\theta = 5000 \sin 36.87 = 3000 \text{ VAR}$$

When the pf is raised to 0.95,

$$\cos\theta_2 = 0.95 \quad \Rightarrow \quad \theta_2 = 18.19^\circ$$

The real power P has not changed. But the apparent power has changed; its new value is

$$S_2 = \frac{P}{\cos\theta_2} = \frac{4000}{0.95} = 4210.5 \text{ VA}$$

The new reactive power is

$$Q_2 = S_2 \sin\theta_2 = 1314.4 \text{ VAR}$$

The difference between the new and old reactive powers is due to the parallel addition of the capacitor to the load. The reactive power due to the capacitor is

$$Q_C = Q_1 - Q_2 = 3000 - 1314.4 = 1685.6 \text{ VAR}$$

and

$$C = \frac{Q_C}{\omega V_{\text{rms}}^2} = \frac{1685.6}{2\pi \times 60 \times 120^2} = 310.5\ \mu\text{F}$$

Note: Capacitors are normally purchased for voltages they expect to see. In this case, the maximum voltage this capacitor will see is about 170 V peak. We would suggest purchasing a capacitor with a voltage rating equal to, say, 200 V.

Practice Problem 11.15

Find the value of parallel capacitance needed to correct a load of 140 kVAR at 0.85 lagging pf to unity pf. Assume that the load is supplied by a 110-V (rms), 60-Hz line.

Answer: 30.69 mF.

11.9 †Applications

In this section, we consider two important application areas: how power is measured and how electric utility companies determine the cost of electricity consumption.

11.9.1 Power Measurement

The average power absorbed by a load is measured by an instrument called the *wattmeter.*

Reactive power is measured by an instrument called the *varmeter.* The varmeter is often connected to the load in the same way as the wattmeter.

The **wattmeter** is the instrument for measuring the average power.

Figure 11.30 shows a wattmeter that consists essentially of two coils: the current coil and the voltage coil. A current coil with very low impedance (ideally zero) is connected in series with the load (Fig. 11.31) and responds to the load current. The voltage coil with very high impedance (ideally infinite) is connected in parallel with the load as shown in Fig. 11.31 and responds to the load voltage. The current coil acts like a short circuit because of its low impedance; the voltage coil

Some wattmeters do not have coils; the wattmeter considered here is the electromagnetic type.

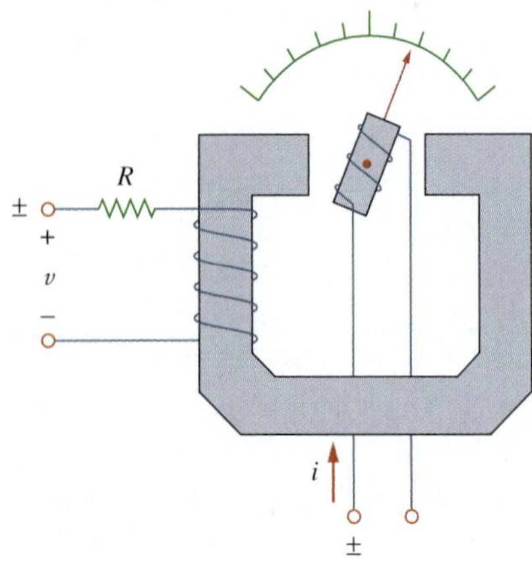

Figure 11.30
A wattmeter.

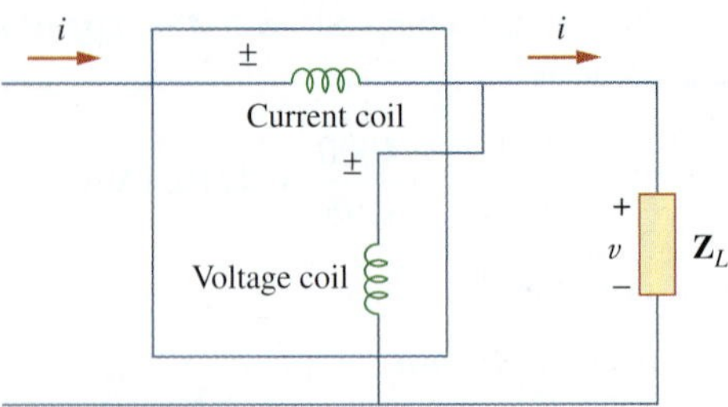

Figure 11.31
The wattmeter connected to the load.

behaves like an open circuit because of its high impedance. As a result, the presence of the wattmeter does not disturb the circuit or have an effect on the power measurement.

When the two coils are energized, the mechanical inertia of the moving system produces a deflection angle that is proportional to the average value of the product $v(t)i(t)$. If the current and voltage of the load are $v(t) = V_m \cos(\omega t + \theta_v)$ and $i(t) = I_m \cos(\omega t + \theta_i)$, their corresponding rms phasors are

$$\mathbf{V}_{\text{rms}} = \frac{V_m}{\sqrt{2}}\underline{/\theta_v} \quad \text{and} \quad \mathbf{I}_{\text{rms}} = \frac{I_m}{\sqrt{2}}\underline{/\theta_i} \tag{11.62}$$

and the wattmeter measures the average power given by

$$P = |\mathbf{V}_{\text{rms}}||\mathbf{I}_{\text{rms}}| \cos(\theta_v - \theta_i) = V_{\text{rms}} I_{\text{rms}} \cos(\theta_v - \theta_i) \tag{11.63}$$

As shown in Fig. 11.31, each wattmeter coil has two terminals with one marked ±. To ensure upscale deflection, the ± terminal of the current coil is toward the source, while the ± terminal of the voltage coil is connected to the same line as the current coil. Reversing both coil connections still results in upscale deflection. However, reversing one coil and not the other results in downscale deflection and no wattmeter reading.

Example 11.16

Find the wattmeter reading of the circuit in Fig. 11.32.

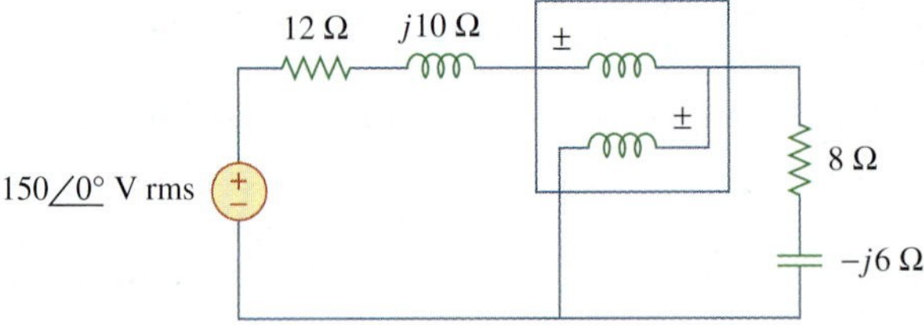

Figure 11.32
For Example 11.16.

Solution:

1. **Define.** The problem is clearly defined. Interestingly, this is a problem where the student could actually validate the results by doing the problem in the laboratory with a real wattmeter.

2. **Present.** This problem consists of finding the average power delivered to a load by an external source with a series impedance.
3. **Alternative.** This is a straightforward circuit problem where all we need to do is find the magnitude and phase of the current through the load and the magnitude and the phase of the voltage across the load. These quantities could also be found by using *PSpice*, which we will use as a check.
4. **Attempt.** In Fig. 11.32, the wattmeter reads the average power absorbed by the $(8 - j6)\ \Omega$ impedance because the current coil is in series with the impedance while the voltage coil is in parallel with it. The current through the circuit is

$$\mathbf{I}_{\text{rms}} = \frac{150\underline{/0^\circ}}{(12 + j10) + (8 - j6)} = \frac{150}{20 + j4}\ \text{A}$$

The voltage across the $(8 - j6)\ \Omega$ impedance is

$$\mathbf{V}_{\text{rms}} = \mathbf{I}_{\text{rms}}(8 - j6) = \frac{150(8 - j6)}{20 + j4}\ \text{V}$$

The complex power is

$$\mathbf{S} = \mathbf{V}_{\text{rms}}\mathbf{I}^*_{\text{rms}} = \frac{150(8 - j6)}{20 + j4} \cdot \frac{150}{20 - j4} = \frac{150^2(8 - j6)}{20^2 + 4^2}$$
$$= 423.7 - j324.6\ \text{VA}$$

The wattmeter reads

$$P = \text{Re}(\mathbf{S}) = \mathbf{432.7\ W}$$

5. **Evaluate.** We can check our results by using *PSpice.*

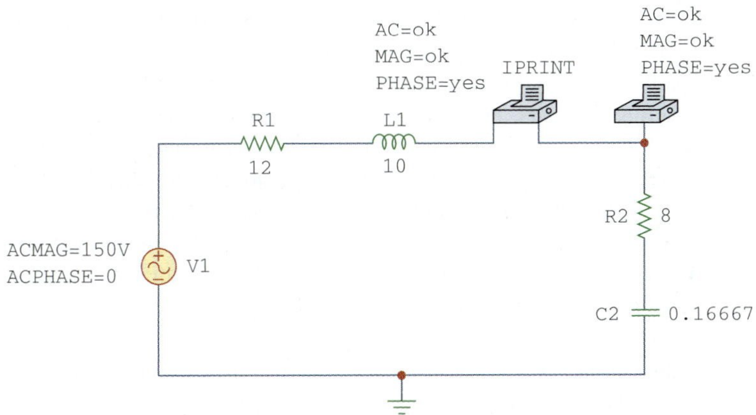

Simulation yields:

FREQ	IM(V_PRINT2)	IP(V_PRINT2)
1.592E-01	7.354E+00	-1.131E+01

and

FREQ	VM($N_0004)	VP($N_0004)
1.592E-01	7.354E+01	-4.818E+01

To check our answer, all we need is the magnitude of the current (7.354 A) flowing through the load resistor:

$$P = (I_L)^2R = (7.354)^2 8 = \mathbf{432.7\ W}$$

As expected, the answer does check!

6. **Satisfactory?** We have satisfactorily solved the problem and the results can now be presented as a solution to the problem.

Practice Problem 11.16

For the circuit in Fig. 11.33, find the wattmeter reading.

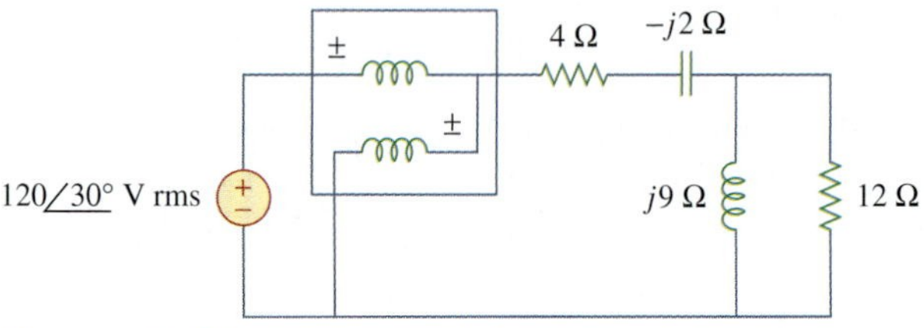

Figure 11.33
For Practice Prob. 11.16.

Answer: 1437 W.

11.9.2 Electricity Consumption Cost

In Section 1.7, we considered a simplified model of the way the cost of electricity consumption is determined. But the concept of power factor was not included in the calculations. Now we consider the importance of power factor in electricity consumption cost.

Loads with low power factors are costly to serve because they require large currents, as explained in Section 11.8. The ideal situation would be to draw minimum current from a supply so that $S = P$, $Q = 0$, and pf $= 1$. A load with nonzero Q means that energy flows back and forth between the load and the source, giving rise to additional power losses. In view of this, power companies often encourage their customers to have power factors as close to unity as possible and penalize some customers who do not improve their load power factors.

Utility companies divide their customers into categories: as residential (domestic), commercial, and industrial, or as small power, medium power, and large power. They have different rate structures for each category. The amount of energy consumed in units of kilowatt-hours (kWh) is measured using a kilowatt-hour meter installed at the customer's premises.

Although utility companies use different methods for charging customers, the tariff or charge to a consumer is often two-part. The first part is fixed and corresponds to the cost of generation, transmission, and distribution of electricity to meet the load requirements of the consumers. This part of the tariff is generally expressed as a certain price per kW of maximum demand. Or it may be based on kVA of maximum demand, to account for the power factor (pf) of the consumer. A pf penalty charge may be imposed on the consumer whereby a certain percentage of kW or kVA maximum demand is charged for every 0.01 fall in pf below a prescribed value, say 0.85 or 0.9. On the other hand, a pf credit may be given for every 0.01 that the pf exceeds the prescribed value.

The second part is proportional to the energy consumed in kWh; it may be in graded form, for example, the first 100 kWh at 16 cents/kWh, the next 200 kWh at 10 cents/kWh and so forth. Thus, the bill is determined based on the following equation:

$$\text{Total Cost} = \text{Fixed Cost} + \text{Cost of Energy} \qquad \textbf{(11.64)}$$

Example 11.17

A manufacturing industry consumes 200 MWh in one month. If the maximum demand is 1600 kW, calculate the electricity bill based on the following two-part rate:

Demand charge: \$5.00 per month per kW of billing demand.
Energy charge: 8 cents per kWh for the first 50,000 kWh,
5 cents per kWh for the remaining energy.

Solution:
The demand charge is

$$\$5.00 \times 1600 = \$8000 \qquad \textbf{(11.17.1)}$$

The energy charge for the first 50,000 kWh is

$$\$0.08 \times 50{,}000 = \$4000 \qquad \textbf{(11.17.2)}$$

The remaining energy is 200,000 kWh − 50,000 kWh = 150,000 kWh, and the corresponding energy charge is

$$\$0.05 \times 150{,}000 = \$7500 \qquad \textbf{(11.17.3)}$$

Adding the results of Eqs. (11.17.1) to (11.17.3) gives

$$\text{Total bill for the month} = \$8000 + \$4000 + \$7500 = \$19{,}500$$

It may appear that the cost of electricity is too high. But this is often a small fraction of the overall cost of production of the goods manufactured or the selling price of the finished product.

Practice Problem 11.17

The monthly reading of a paper mill's meter is as follows:

Maximum demand: 32,000 kW
Energy consumed: 500 MWh

Using the two-part rate in Example 11.17, calculate the monthly bill for the paper mill.

Answer: \$186,500.

Example 11.18

A 300-kW load supplied at 13 kV (rms) operates 520 hours a month at 80 percent power factor. Calculate the average cost per month based on this simplified tariff:

Energy charge: 6 cents per kWh
Power-factor penalty: 0.1 percent of energy charge for every 0.01 that pf falls below 0.85.
Power-factor credit: 0.1 percent of energy charge for every 0.01 that pf exceeds 0.85.

Solution:
The energy consumed is

$$W = 300 \text{ kW} \times 520 \text{ h} = 156{,}000 \text{ kWh}$$

The operating power factor pf = 80% = 0.8 is 5 × 0.01 below the prescribed power factor of 0.85. Since there is 0.1 percent energy charge for every 0.01, there is a power-factor penalty charge of 0.5 percent. This amounts to an energy charge of

$$\Delta W = 156{,}000 \times \frac{5 \times 0.1}{100} = 780 \text{ kWh}$$

The total energy is

$$W_t = W + \Delta W = 156{,}000 + 780 = 156{,}780 \text{ kWh}$$

The cost per month is given by

$$\text{Cost} = 6 \text{ cents} \times W_t = \$0.06 \times 156{,}780 = \$9{,}406.80$$

Practice Problem 11.18

An 800-kW induction furnace at 0.88 power factor operates 20 hours per day for 26 days in a month. Determine the electricity bill per month based on the tariff in Example 11.18.

Answer: $24,885.12.

11.10 Summary

1. The instantaneous power absorbed by an element is the product of the element's terminal voltage and the current through the element:

$$p = vi.$$

2. Average or real power P (in watts) is the average of instantaneous power p:

$$P = \frac{1}{T}\int_0^T p\, dt$$

If $v(t) = V_m \cos(\omega t + \theta_v)$ and $i(t) = I_m \cos(\omega t + \theta_i)$, then $V_{\text{rms}} = V_m/\sqrt{2}$, $I_{\text{rms}} = I_m/\sqrt{2}$, and

$$P = \frac{1}{2}V_m I_m \cos(\theta_v - \theta_i) = V_{\text{rms}} I_{\text{rms}} \cos(\theta_v - \theta_i)$$

Inductors and capacitors absorb no average power, while the average power absorbed by a resistor is $(1/2)I_m^2R = I_{\text{rms}}^2R$.

3. Maximum average power is transferred to a load when the load impedance is the complex conjugate of the Thevenin impedance as seen from the load terminals, $\mathbf{Z}_L = \mathbf{Z}_{\text{Th}}^*$.
4. The effective value of a periodic signal $x(t)$ is its root-mean-square (rms) value.

$$X_{\text{eff}} = X_{\text{rms}} = \sqrt{\frac{1}{T}\int_0^T x^2\,dt}$$

For a sinusoid, the effective or rms value is its amplitude divided by $\sqrt{2}$.

5. The power factor is the cosine of the phase difference between voltage and current:

$$\text{pf} = \cos(\theta_v - \theta_i)$$

It is also the cosine of the angle of the load impedance or the ratio of real power to apparent power. The pf is lagging if the current lags voltage (inductive load) and is leading when the current leads voltage (capacitive load).

6. Apparent power S (in VA) is the product of the rms values of voltage and current:

$$S = V_{\text{rms}}I_{\text{rms}}$$

It is also given by $S = |\mathbf{S}| = \sqrt{P^2 + Q^2}$, where P is the real power and Q is reactive power.

7. Reactive power (in VAR) is:

$$Q = \frac{1}{2}V_mI_m\sin(\theta_v - \theta_i) = V_{\text{rms}}I_{\text{rms}}\sin(\theta_v - \theta_i)$$

8. Complex power $\mathbf{S}$ (in VA) is the product of the rms voltage phasor and the complex conjugate of the rms current phasor. It is also the complex sum of real power P and reactive power Q.

$$\mathbf{S} = \mathbf{V}_{\text{rms}}\mathbf{I}_{\text{rms}}^* = V_{\text{rms}}I_{\text{rms}}\underline{/\theta_v - \theta_i} = P + jQ$$

Also,

$$\mathbf{S} = I_{\text{rms}}^2\mathbf{Z} = \frac{V_{\text{rms}}^2}{\mathbf{Z}^*}$$

9. The total complex power in a network is the sum of the complex powers of the individual components. Total real power and reactive power are also, respectively, the sums of the individual real powers and the reactive powers, but the total apparent power is not calculated by the process.
10. Power factor correction is necessary for economic reasons; it is the process of improving the power factor of a load by reducing the overall reactive power.
11. The wattmeter is the instrument for measuring the average power. Energy consumed is measured with a kilowatt-hour meter.

Review Questions

11.1 The average power absorbed by an inductor is zero.

(a) True (b) False

11.2 The Thevenin impedance of a network seen from the load terminals is $80 + j55\ \Omega$. For maximum power transfer, the load impedance must be:

(a) $-80 + j55\ \Omega$ (b) $-80 - j55\ \Omega$
(c) $80 - j55\ \Omega$ (d) $80 + j55\ \Omega$

11.3 The amplitude of the voltage available in the 60-Hz, 120-V power outlet in your home is:

(a) 110 V (b) 120 V
(c) 170 V (d) 210 V

11.4 If the load impedance is $20 - j20$, the power factor is

(a) $\angle -45^\circ$ (b) 0 (c) 1
(d) 0.7071 (e) none of these

11.5 A quantity that contains all the power information in a given load is the

(a) power factor (b) apparent power
(c) average power (d) reactive power
(e) complex power

11.6 Reactive power is measured in:

(a) watts (b) VA
(c) VAR (d) none of these

11.7 In the power triangle shown in Fig. 11.34(a), the reactive power is:

(a) 1000 VAR leading (b) 1000 VAR lagging
(c) 866 VAR leading (d) 866 VAR lagging

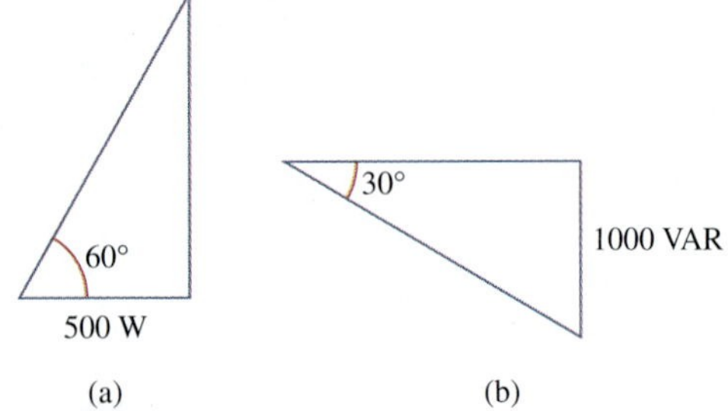

Figure 11.34
For Review Questions 11.7 and 11.8.

11.8 For the power triangle in Fig. 11.34(b), the apparent power is:

(a) 2000 VA (b) 1000 VAR
(c) 866 VAR (d) 500 VAR

11.9 A source is connected to three loads $\mathbf{Z}_1$, $\mathbf{Z}_2$, and $\mathbf{Z}_3$ in parallel. Which of these is not true?

(a) $P = P_1 + P_2 + P_3$ (b) $Q = Q_1 + Q_2 + Q_3$
(c) $S = S_1 + S_2 + S_3$ (d) $\mathbf{S} = \mathbf{S}_1 + \mathbf{S}_2 + \mathbf{S}_3$

11.10 The instrument for measuring average power is the:

(a) voltmeter (b) ammeter
(c) wattmeter (d) varmeter
(e) kilowatt-hour meter

Answers: 11.1a, 11.2c, 11.3c, 11.4d, 11.5e, 11.6c, 11.7d, 11.8a, 11.9c, 11.10c.

Problems[1]

Section 11.2 Instantaneous and Average Power

11.1 If $v(t) = 160 \cos 50t$ V and $i(t) = -20 \sin(50t - 30^\circ)$ A, calculate the instantaneous power and the average power.

11.2 Given the circuit in Fig. 11.35, find the average power supplied or absorbed by each element.

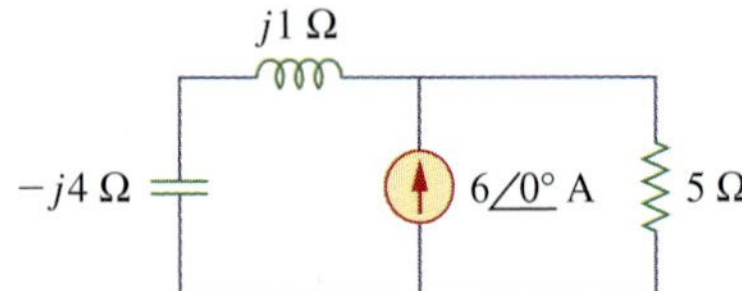

Figure 11.35
For Prob. 11.2.

11.3 A load consists of a 60-Ω resistor in parallel with a 90-μF capacitor. If the load is connected to a voltage source $v_s(t) = 40 \cos 2000t$, find the average power delivered to the load.

11.4 Using Fig. 11.36, design a problem to help other students better understand instantaneous and average power.

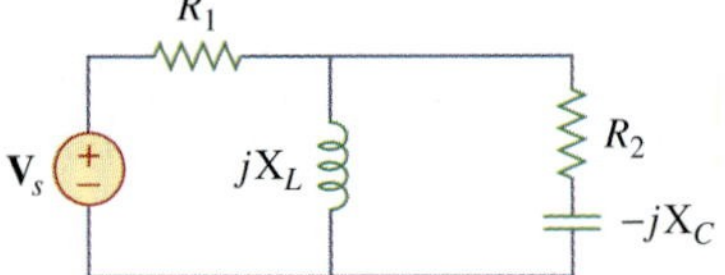

Figure 11.36
For Prob. 11.4.

[1] Starting with problem 11.22, unless otherwise specified, assume that all values of currents and voltages are rms.

11.5 Assuming that $v_s = 16 \cos(2t - 40°)$ V in the circuit of Fig. 11.37, find the average power delivered to each of the passive elements.

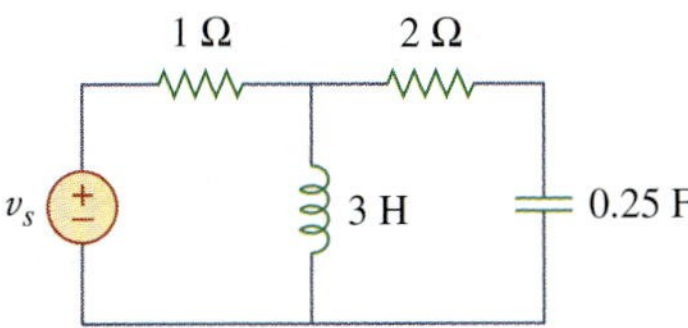

Figure 11.37
For Prob. 11.5.

11.6 For the circuit in Fig. 11.38, $i_s = 3 \cos 10^3 t$ A. Find the average power absorbed by the 50-Ω resistor.

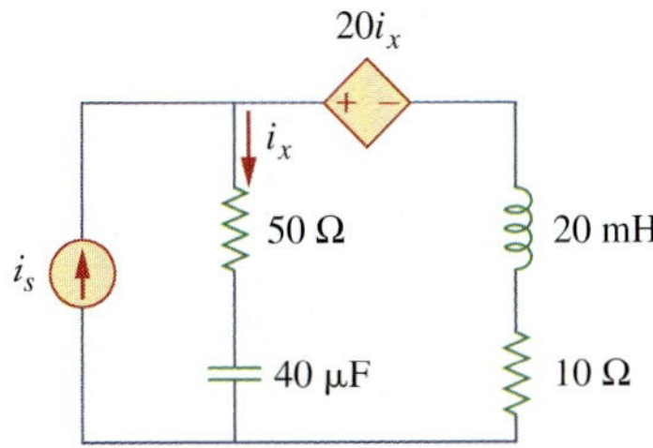

Figure 11.38
For Prob. 11.6.

11.7 Given the circuit of Fig. 11.39, find the average power absorbed by the 10-Ω resistor.

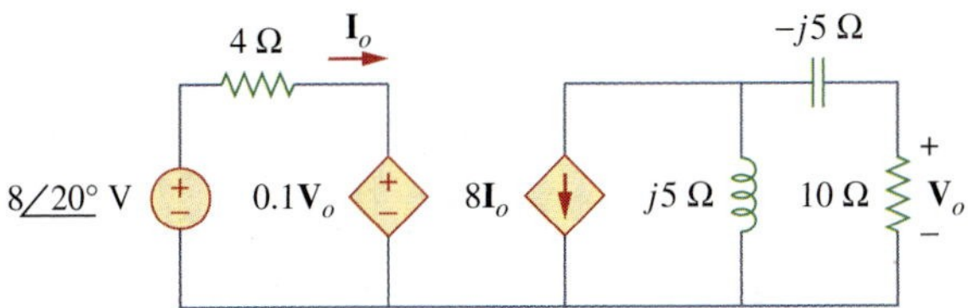

Figure 11.39
For Prob. 11.7.

11.8 In the circuit of Fig. 11.40, determine the average power absorbed by the 40-Ω resistor.

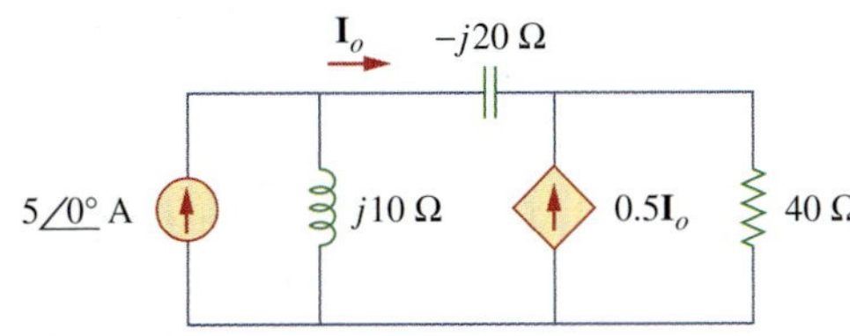

Figure 11.40
For Prob. 11.8.

11.9 For the op amp circuit in Fig. 11.41, $\mathbf{V}_s = 2\angle 30°$ V. Find the average power absorbed by the 20-kΩ resistor.

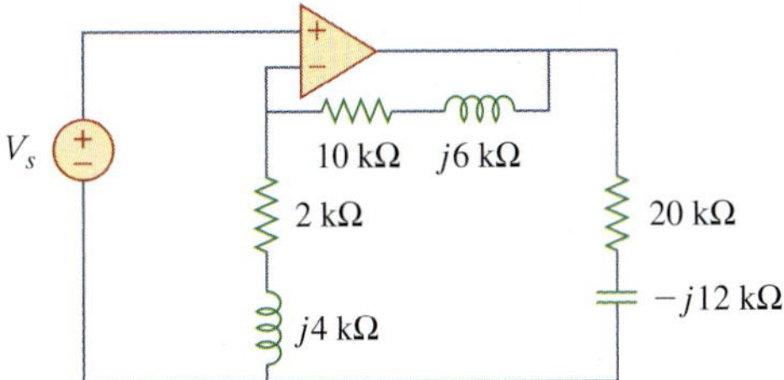

Figure 11.41
For Prob. 11.9.

11.10 In the op amp circuit in Fig. 11.42, find the total average power absorbed by the resistors.

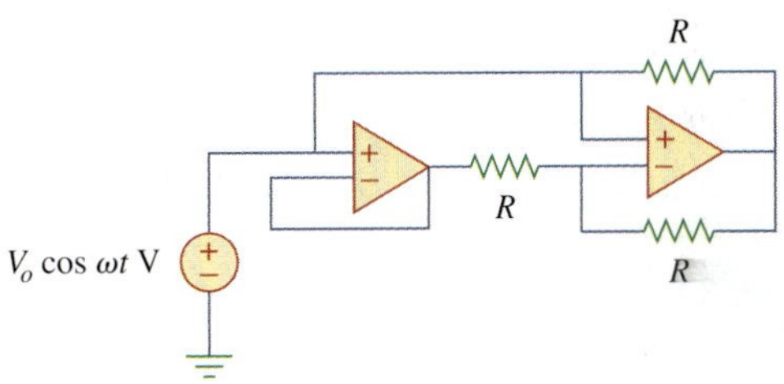

Figure 11.42
For Prob. 11.10.

11.11 For the network in Fig. 11.43, assume that the port impedance is

$$\mathbf{Z}_{ab} = \frac{R}{\sqrt{1 + \omega^2 R^2 C^2}} \angle -\tan^{-1} \omega RC$$

Find the average power consumed by the network when $R = 10$ kΩ, $C = 200$ nF, and $i = 2 \sin(377t + 22°)$ mA.

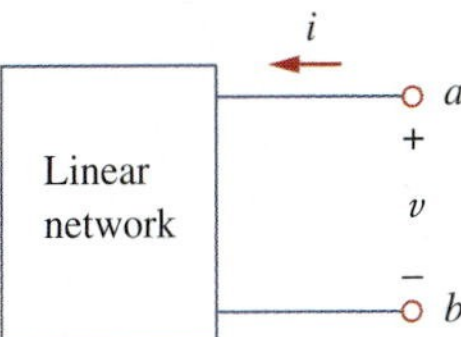

Figure 11.43
For Prob. 11.11.

Section 11.3 Maximum Average Power Transfer

11.12 For the circuit shown in Fig. 11.44, determine the load impedance **Z** for maximum power transfer (to **Z**). Calculate the maximum power absorbed by the load.

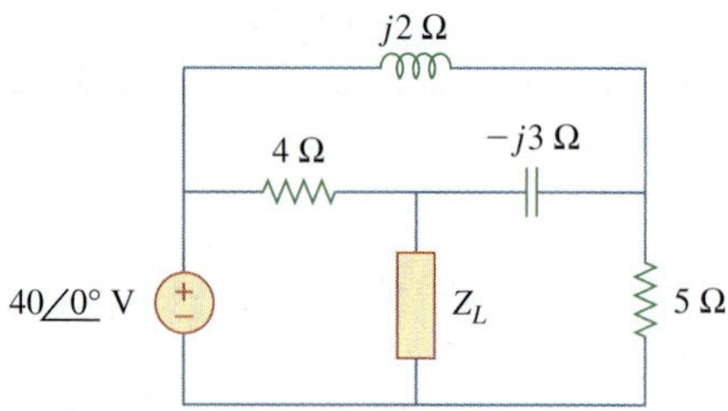

Figure 11.44
For Prob. 11.12.

11.13 The Thevenin impedance of a source is $\mathbf{Z}_{\text{Th}} = 120 + j60\ \Omega$, while the peak Thevenin voltage is $\mathbf{V}_{\text{Th}} = 110 + j0$ V. Determine the maximum available average power from the source.

11.14 Using Fig. 11.45, design a problem to help other students better understand maximum average power transfer.

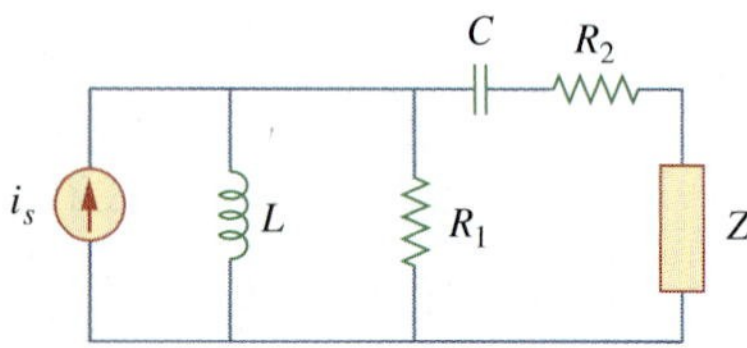

Figure 11.45
For Prob. 11.14.

11.15 In the circuit of Fig. 11.46, find the value of $\mathbf{Z}_L$ that will absorb the maximum power and the value of the maximum power.

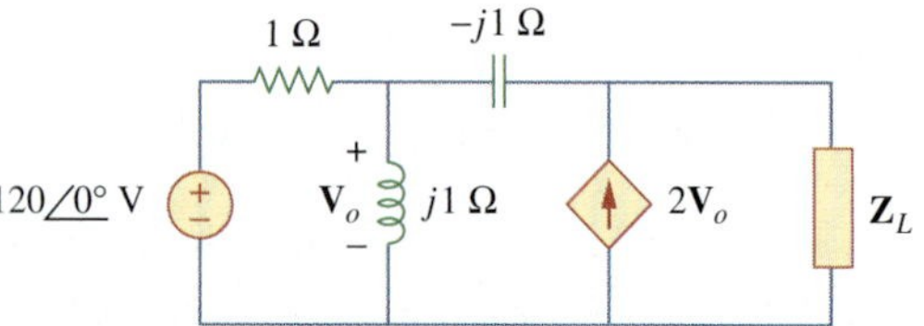

Figure 11.46
For Prob. 11.15.

11.16 For the circuit of Fig. 11.47, find the maximum power delivered to the load $\mathbf{Z}_L$.

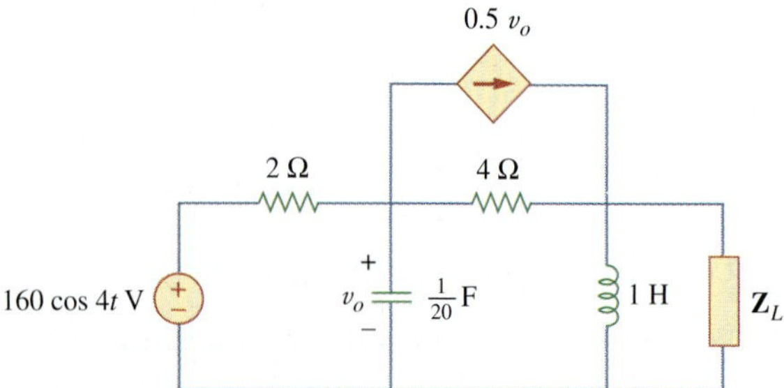

Figure 11.47
For Prob. 11.16.

11.17 Calculate the value of $\mathbf{Z}_L$ in the circuit of Fig. 11.48 in order for $\mathbf{Z}_L$ to receive maximum average power. What is the maximum average power received by $\mathbf{Z}_L$?

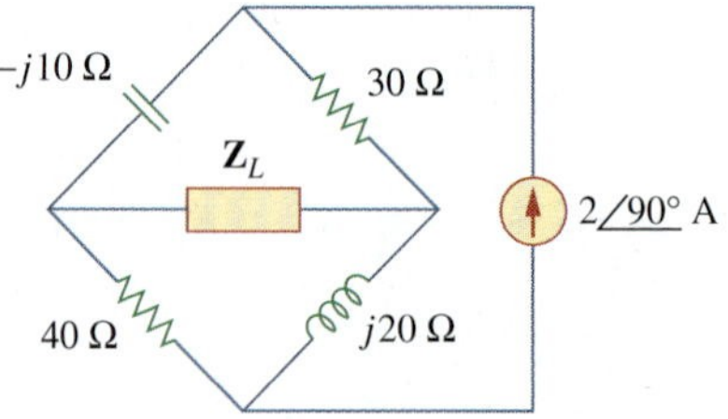

Figure 11.48
For Prob. 11.17.

11.18 Find the value of $\mathbf{Z}_L$ in the circuit of Fig. 11.49 for maximum power transfer.

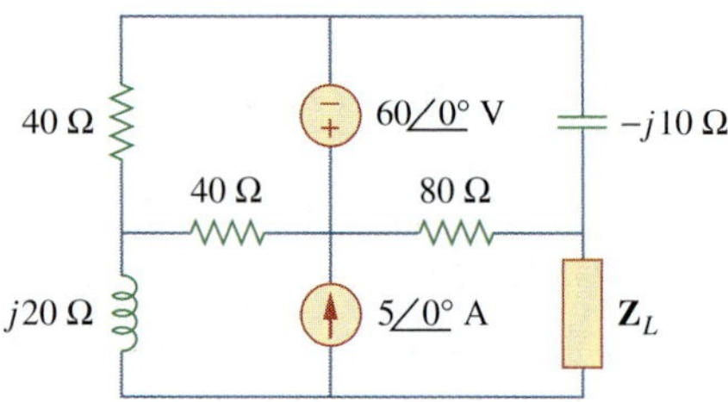

Figure 11.49
For Prob. 11.18.

11.19 The variable resistor R in the circuit of Fig. 11.50 is adjusted until it absorbs the maximum average power. Find R and the maximum average power absorbed.

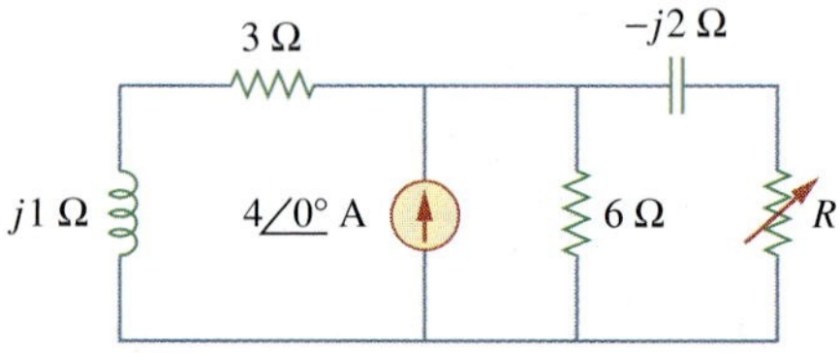

Figure 11.50
For Prob. 11.19.

11.20 The load resistance R_L in Fig. 11.51 is adjusted until it absorbs the maximum average power. Calculate the value of R_L and the maximum average power.

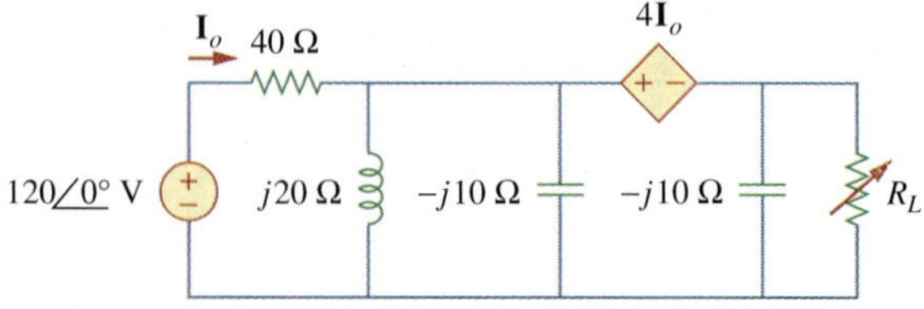

Figure 11.51
For Prob. 11.20.

11.21 Assuming that the load impedance is to be purely resistive, what load should be connected to terminals a-b of the circuits in Fig. 11.52 so that the maximum power is transferred to the load?

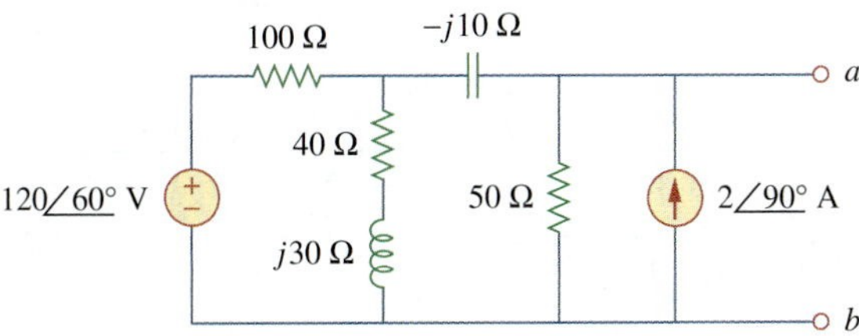

Figure 11.52
For Prob. 11.21.

Section 11.4 Effective or RMS Value

11.22 Find the rms value of the offset sine wave shown in Fig. 11.53.

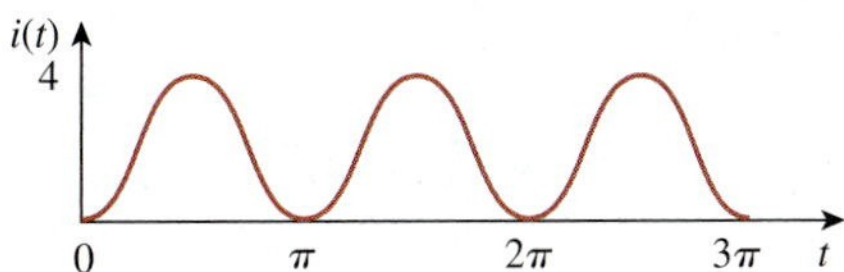

Figure 11.53
For Prob. 11.22.

11.23 Using Fig. 11.54, design a problem to help other **e⊘d** students better understand how to find the rms value of a waveshape.

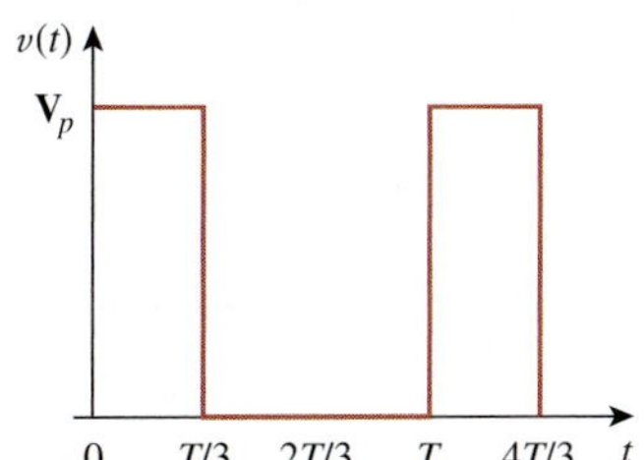

Figure 11.54
For Prob. 11.23.

11.24 Determine the rms value of the waveform in Fig. 11.55.

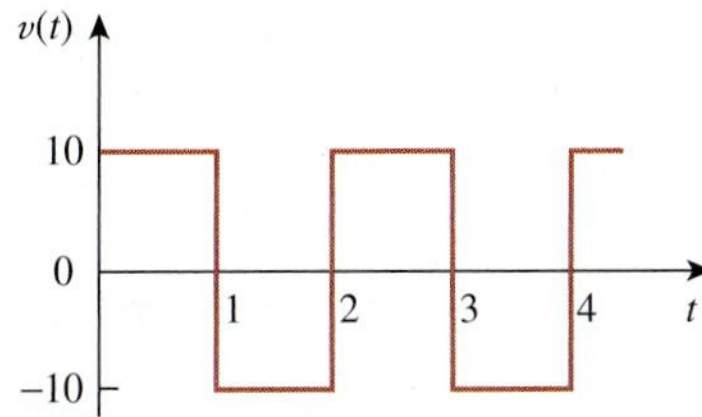

Figure 11.55
For Prob. 11.24.

11.25 Find the rms value of the signal shown in Fig. 11.56.

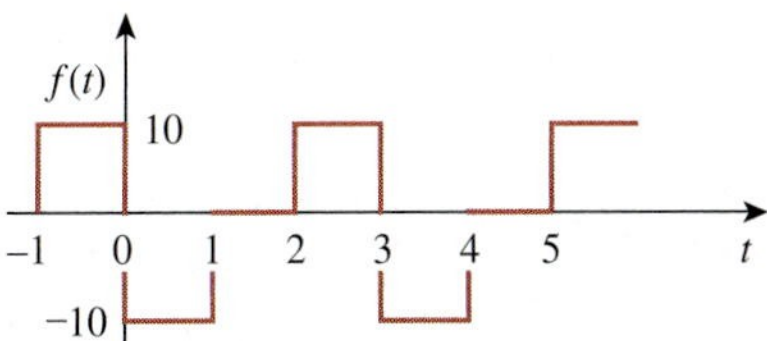

Figure 11.56
For Prob. 11.25.

11.26 Find the effective value of the voltage waveform in Fig. 11.57.

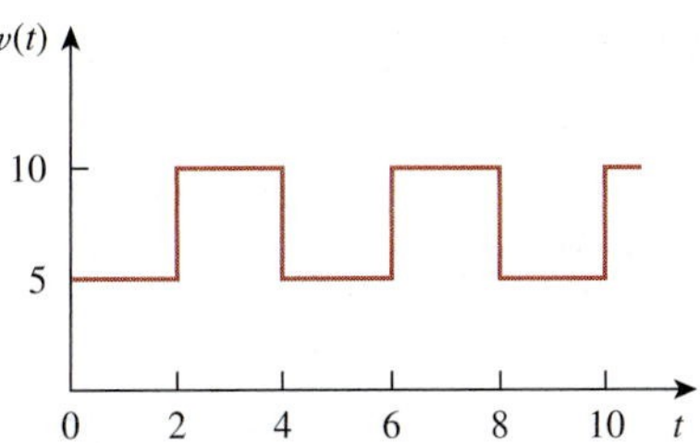

Figure 11.57
For Prob. 11.26.

11.27 Calculate the rms value of the current waveform of Fig. 11.58.

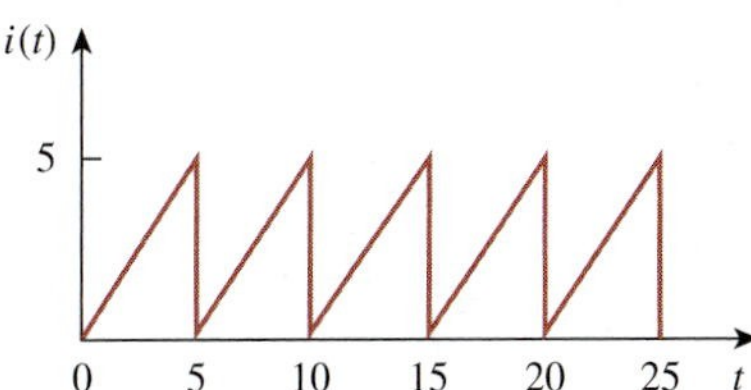

Figure 11.58
For Prob. 11.27.

11.28 Find the rms value of the voltage waveform of Fig. 11.59 as well as the average power absorbed by a 2-Ω resistor when the voltage is applied across the resistor.

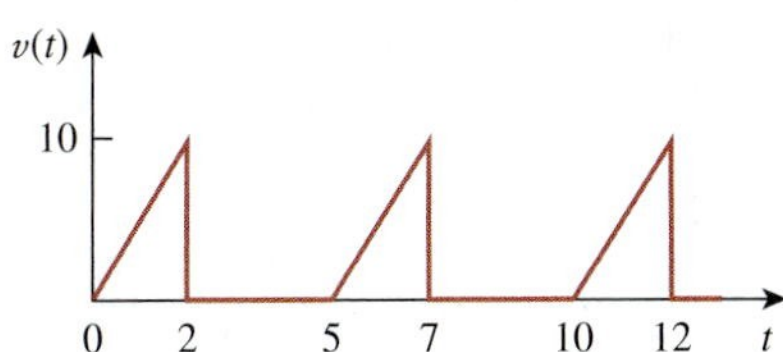

Figure 11.59
For Prob. 11.28.

11.29 Calculate the effective value of the current waveform in Fig. 11.60 and the average power delivered to a 12-Ω resistor when the current runs through the resistor.

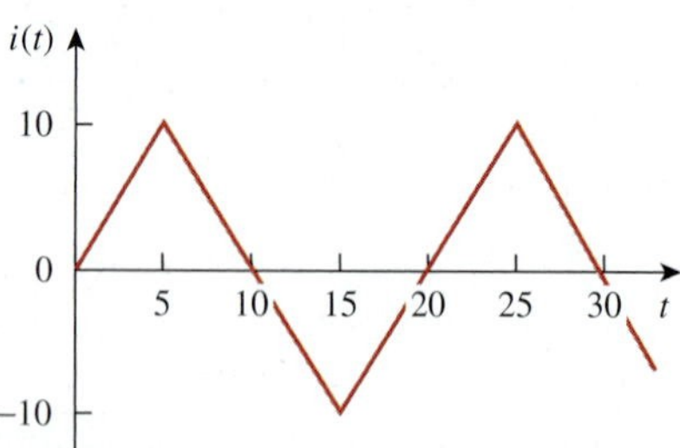

Figure 11.60
For Prob. 11.29.

11.30 Compute the rms value of the waveform depicted in Fig. 11.61.

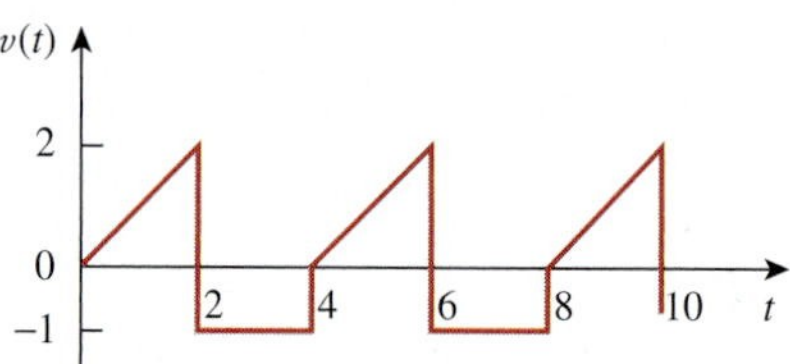

Figure 11.61
For Prob. 11.30.

11.31 Find the rms value of the signal shown in Fig. 11.62.

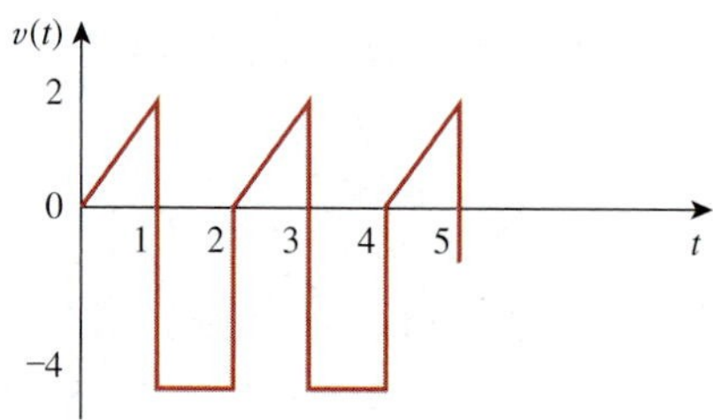

Figure 11.62
For Prob. 11.31.

11.32 Obtain the rms value of the current waveform shown in Fig. 11.63.

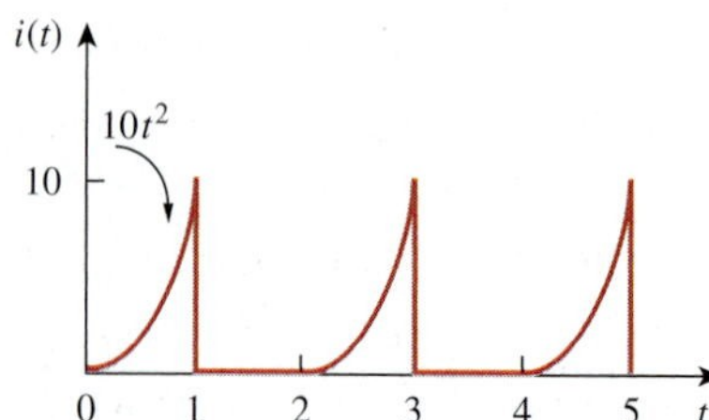

Figure 11.63
For Prob. 11.32.

11.33 Determine the rms value for the waveform in Fig. 11.64.

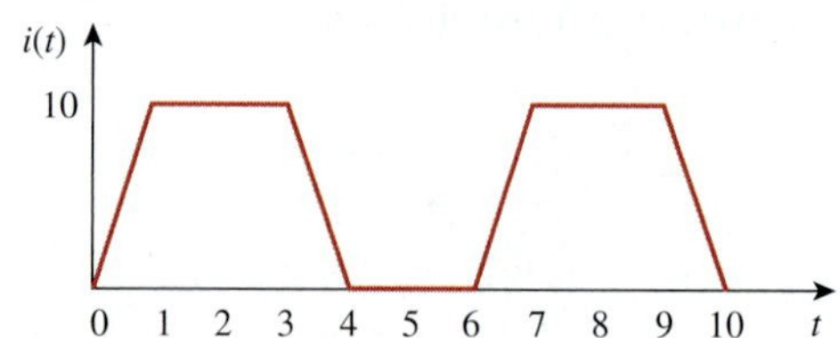

Figure 11.64
For Prob. 11.33.

11.34 Find the effective value of $f(t)$ defined in Fig. 11.65.

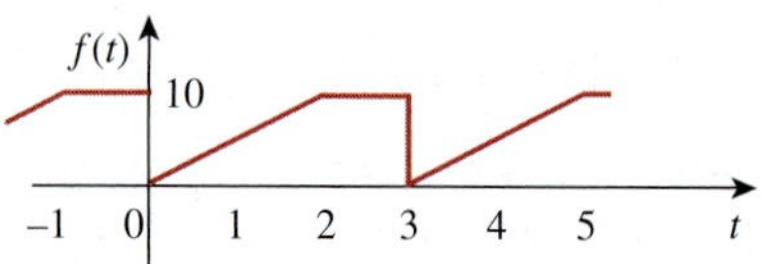

Figure 11.65
For Prob. 11.34.

11.35 One cycle of a periodic voltage waveform is depicted in Fig. 11.66. Find the effective value of the voltage. Note that the cycle starts at $t = 0$ and ends at $t = 6$ s.

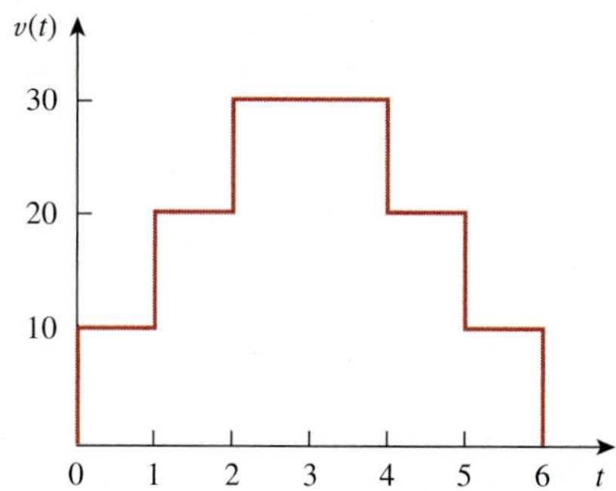

Figure 11.66
For Prob. 11.35.

11.36 Calculate the rms value for each of the following functions:

(a) $i(t) = 10$ A (b) $v(t) = 4 + 3 \cos 5t$ V

(c) $i(t) = 8 - 6 \sin 2t$ A (d) $v(t) = 5 \sin t + 4 \cos t$ V

11.37 Design a problem to help other students better understand how to determine the rms value of the sum of multiple currents.

Section 11.5 Apparent Power and Power Factor

11.38 For the power system in Fig. 11.67, find: (a) the average power, (b) the reactive power, (c) the power factor. Note that 220 V is an rms value.

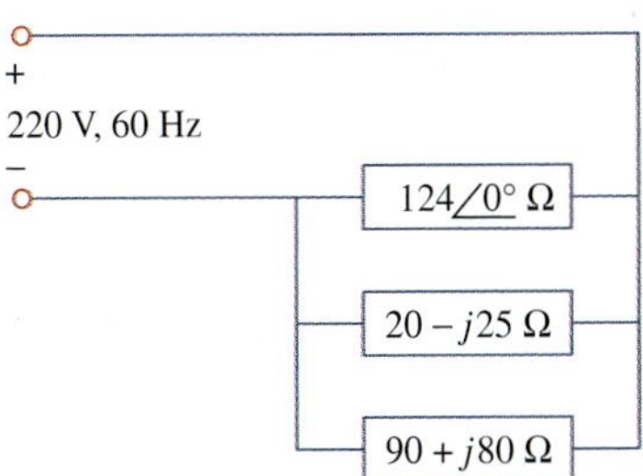

Figure 11.67
For Prob. 11.38.

11.39 An ac motor with impedance $\mathbf{Z}_L = 4.2 + j3.6\ \Omega$ is supplied by a 220-V, 60-Hz source. (a) Find pf, P, and Q. (b) Determine the capacitor required to be connected in parallel with the motor so that the power factor is corrected to unity.

11.40 e⇗d Design a problem to help other students better understand apparent power and power factor.

11.41 Obtain the power factor for each of the circuits in Fig. 11.68. Specify each power factor as leading or lagging.

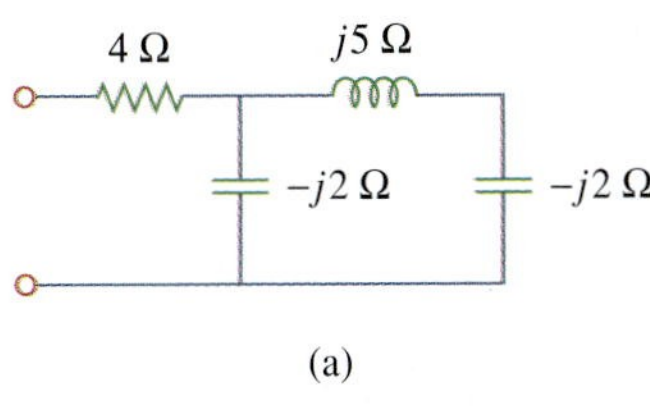

(a)

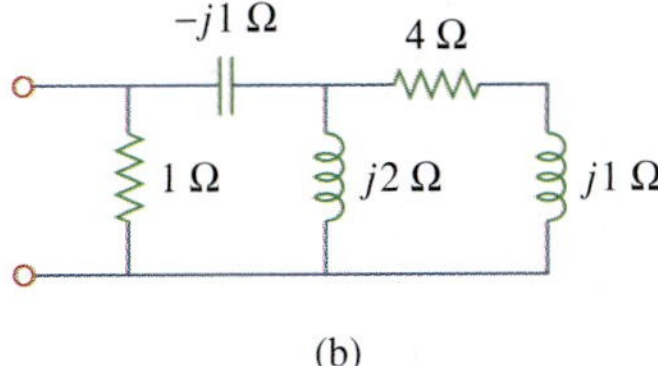

(b)

Figure 11.68
For Prob. 11.41.

Section 11.6 Complex Power

11.42 A 110-V rms, 60-Hz source is applied to a load impedance $\mathbf{Z}$. The apparent power entering the load is 120 VA at a power factor of 0.707 lagging.

(a) Calculate the complex power.

(b) Find the rms current supplied to the load.

(c) Determine $\mathbf{Z}$.

(d) Assuming that $\mathbf{Z} = R + j\omega L$, find the values of R and L.

11.43 e⇗d Design a problem to help other students understand complex power.

11.44 Find the complex power delivered by v_s to the network in Fig. 11.69. Let $v_s = 160 \cos 2000t$ V.

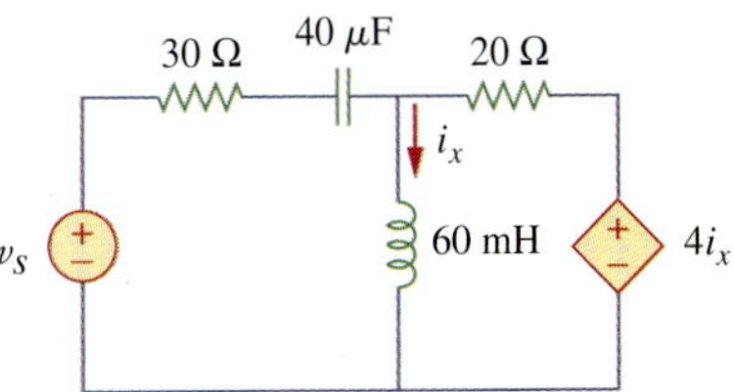

Figure 11.69
For Prob. 11.44.

11.45 The voltage across a load and the current through it are given by

$$v(t) = 20 + 60 \cos 100t \text{ V}$$
$$i(t) = 1 - 0.5 \sin 100t \text{ A}$$

Find:

(a) the rms values of the voltage and of the current

(b) the average power dissipated in the load

11.46 For the following voltage and current phasors, calculate the complex power, apparent power, real power, and reactive power. Specify whether the pf is leading or lagging.

(a) $\mathbf{V} = 220\angle 30^\circ$ V rms, $\mathbf{I} = 0.5\angle 60^\circ$ A rms

(b) $\mathbf{V} = 250\angle -10^\circ$ V rms,
$\mathbf{I} = 6.2\angle -25^\circ$ A rms

(c) $\mathbf{V} = 120\angle 0^\circ$ V rms, $\mathbf{I} = 2.4\angle -15^\circ$ A rms

(d) $\mathbf{V} = 160\angle 45^\circ$ V rms, $\mathbf{I} = 8.5\angle 90^\circ$ A rms

11.47 For each of the following cases, find the complex power, the average power, and the reactive power:

(a) $v(t) = 112 \cos(\omega t + 10^\circ)$ V,
$i(t) = 4 \cos(\omega t - 50^\circ)$ A

(b) $v(t) = 160 \cos 377t$ V,
$i(t) = 4 \cos(377t + 45^\circ)$ A

(c) $\mathbf{V} = 80\angle 60^\circ$ V rms, $\mathbf{Z} = 50\angle 30^\circ\ \Omega$

(d) $\mathbf{I} = 10\angle 60^\circ$ A rms, $\mathbf{Z} = 100\angle 45^\circ\ \Omega$

11.48 Determine the complex power for the following cases:

(a) $P = 269$ W, $Q = 150$ VAR (capacitive)

(b) $Q = 2000$ VAR, pf $= 0.9$ (leading)

(c) $S = 600$ VA, $Q = 450$ VAR (inductive)

(d) $V_{\text{rms}} = 220$ V, $P = 1$ kW,
$|\mathbf{Z}| = 40\ \Omega$ (inductive)

11.49 Find the complex power for the following cases:

(a) $P = 4$ kW, pf $= 0.86$ (lagging)

(b) $S = 2$ kVA, $P = 1.6$ kW (capacitive)

(c) $\mathbf{V}_{\text{rms}} = 208\underline{/20^\circ}$ V, $\mathbf{I}_{\text{rms}} = 6.5\underline{/-50^\circ}$ A

(d) $\mathbf{V}_{\text{rms}} = 120\underline{/30^\circ}$ V, $\mathbf{Z} = 40 + j60\ \Omega$

11.50 Obtain the overall impedance for the following cases:

(a) $P = 1000$ W, pf $= 0.8$ (leading), $V_{\text{rms}} = 220$ V

(b) $P = 1500$ W, $Q = 2000$ VAR (inductive), $I_{\text{rms}} = 12$ A

(c) $\mathbf{S} = 4500\underline{/60^\circ}$ VA, $\mathbf{V} = 120\underline{/45^\circ}$ V

11.51 For the entire circuit in Fig. 11.70, calculate:

(a) the power factor

(b) the average power delivered by the source

(c) the reactive power

(d) the apparent power

(e) the complex power

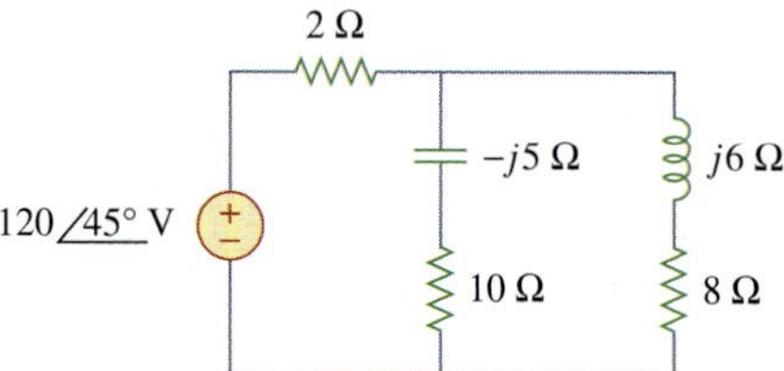

Figure 11.70
For Prob. 11.51.

11.52 In the circuit of Fig. 11.71, device A receives 2 kW at 0.8 pf lagging, device B receives 3 kVA at 0.4 pf leading, while device C is inductive and consumes 1 kW and receives 500 VAR.

(a) Determine the power factor of the entire system.

(b) Find **I** given that $\mathbf{V}_s = 240\underline{/45^\circ}$ V rms.

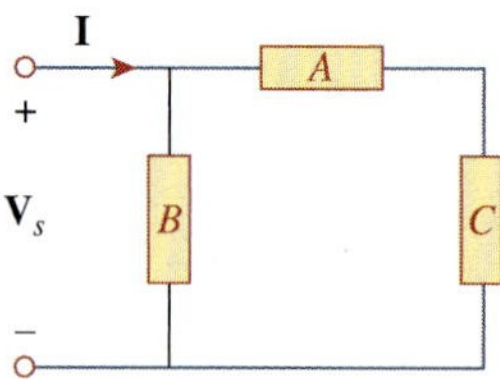

Figure 11.71
For Prob. 11.52.

11.53 In the circuit of Fig. 11.72, load A receives 4 kVA at 0.8 pf leading. Load B receives 2.4 kVA at 0.6 pf lagging. Box C is an inductive load that consumes 1 kW and receives 500 VAR.

(a) Determine **I**.

(b) Calculate the power factor of the combination.

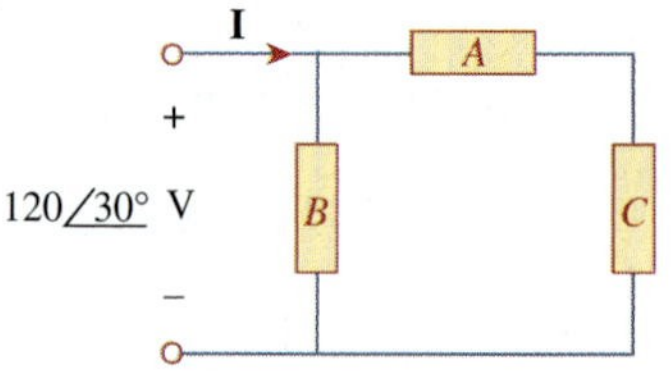

Figure 11.72
For Prob. 11.53.

Section 11.7 Conservation of AC Power

11.54 For the network in Fig. 11.73, find the complex power absorbed by each element.

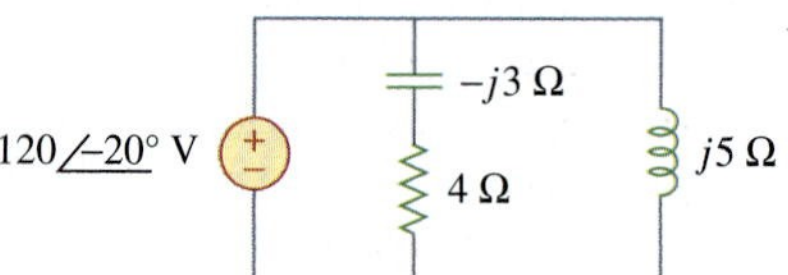

Figure 11.73
For Prob. 11.54.

11.55 Using Fig. 11.74, design a problem to help other students better understand the conservation of AC power.

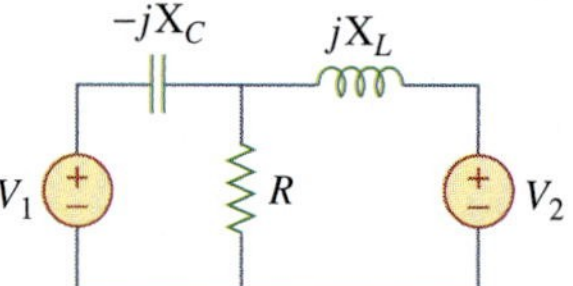

Figure 11.74
For Prob. 11.55.

11.56 Obtain the complex power delivered by the source in the circuit of Fig. 11.75.

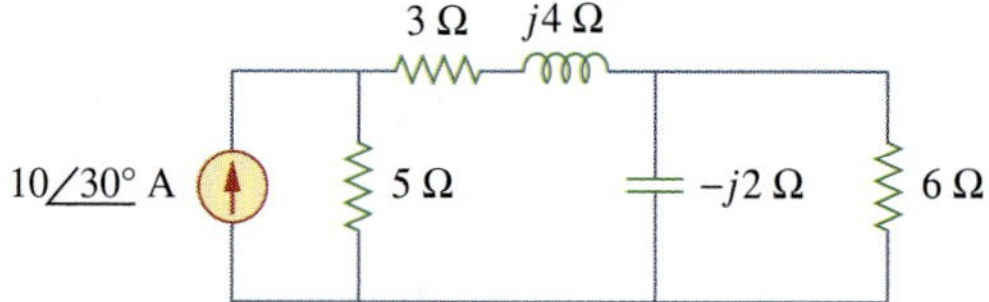

Figure 11.75
For Prob. 11.56.

11.57 For the circuit in Fig. 11.76, find the average, reactive, and complex power delivered by the dependent current source.

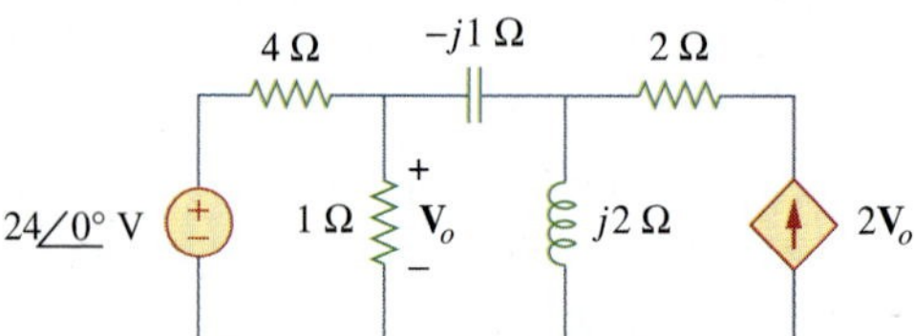

Figure 11.76
For Prob. 11.57.

11.58 Obtain the complex power delivered to the 10-kΩ resistor in Fig. 11.77 below.

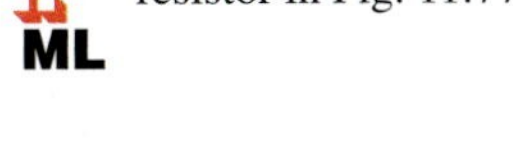

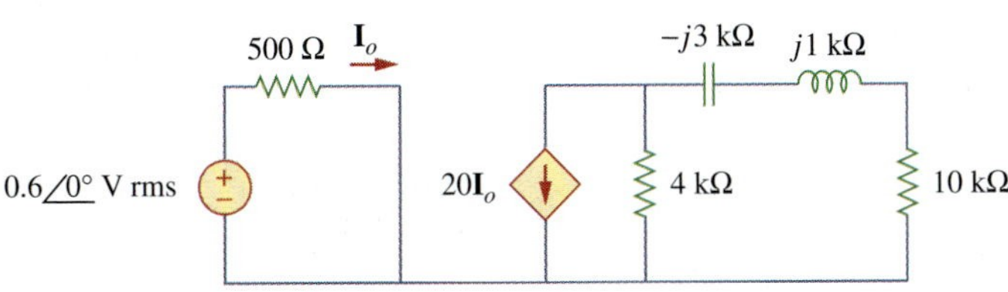

Figure 11.77
For Prob. 11.58.

11.59 Calculate the reactive power in the inductor and capacitor in the circuit of Fig. 11.78.

ML

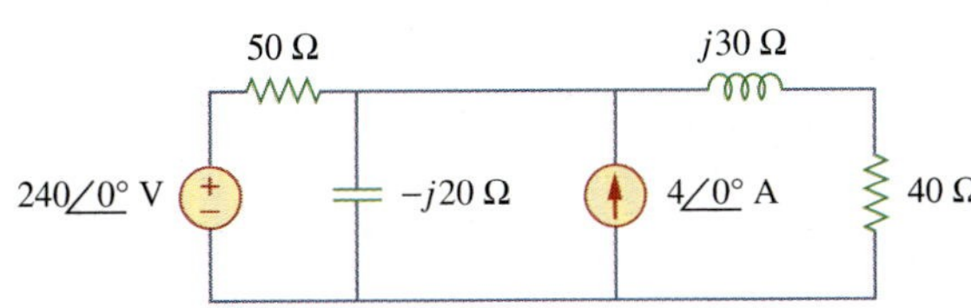

Figure 11.78
For Prob. 11.59.

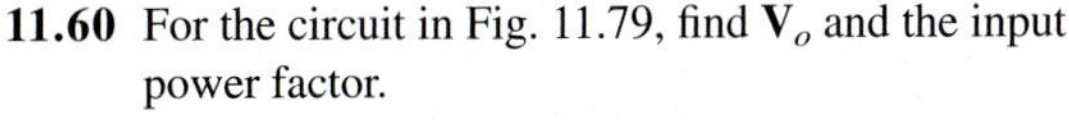

11.60 For the circuit in Fig. 11.79, find $\mathbf{V}_o$ and the input power factor.

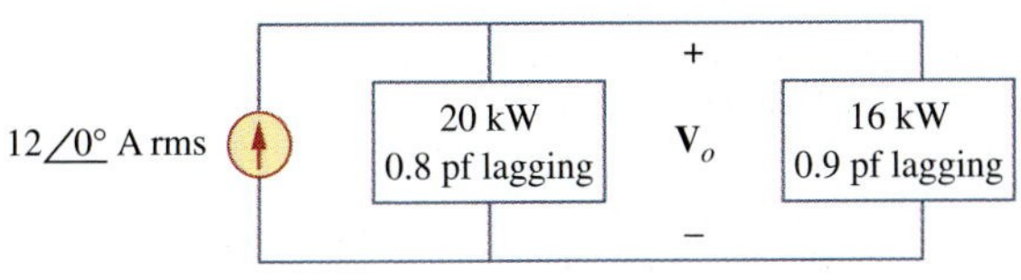

Figure 11.79
For Prob. 11.60.

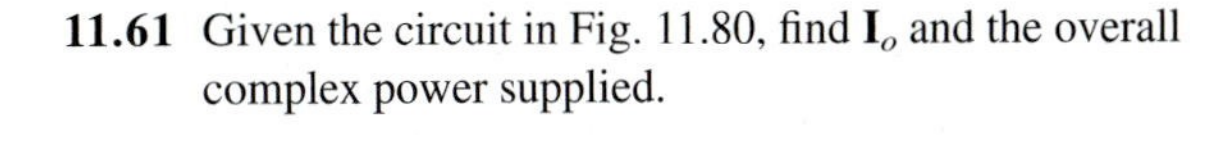

11.61 Given the circuit in Fig. 11.80, find $\mathbf{I}_o$ and the overall complex power supplied.

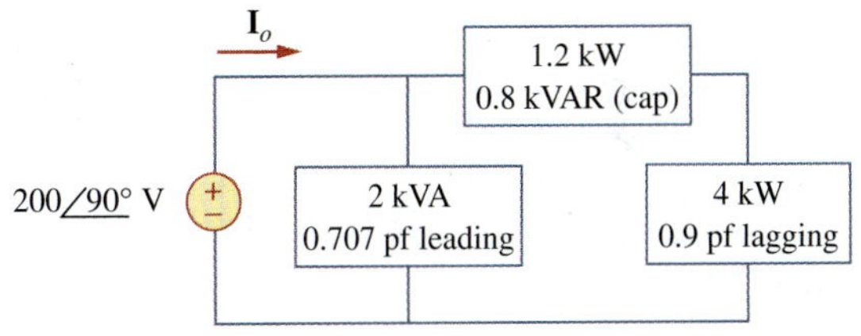

Figure 11.80
For Prob. 11.61.

11.62 For the circuit in Fig. 11.81, find $\mathbf{V}_s$.

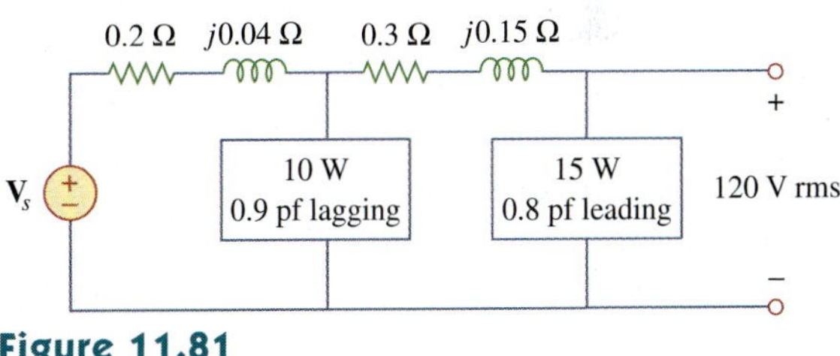

Figure 11.81
For Prob. 11.62.

11.63 Find $\mathbf{I}_o$ in the circuit of Fig. 11.82.

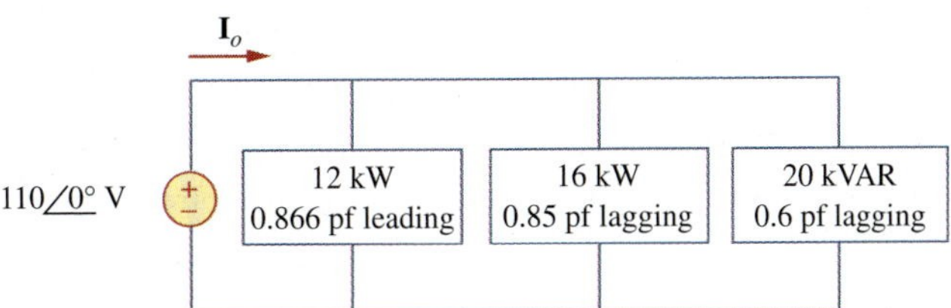

Figure 11.82
For Prob. 11.63.

11.64 Determine $\mathbf{I}_s$ in the circuit of Fig. 11.83, if the voltage source supplies 2.5 kW and 0.4 kVAR (leading).

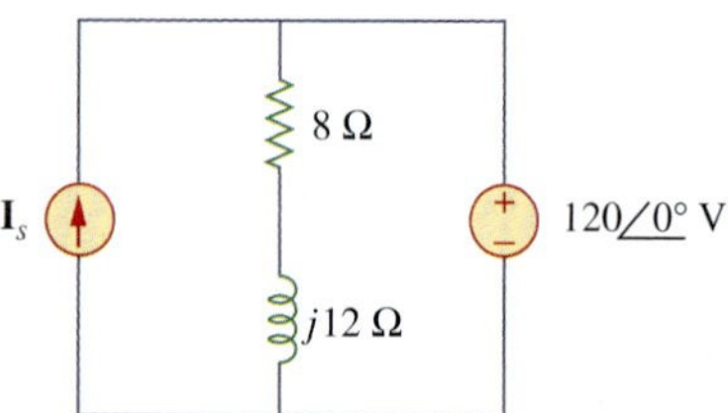

Figure 11.83
For Prob. 11.64.

11.65 In the op amp circuit of Fig. 11.84, $v_s = 4 \cos 10^4 t$ V. Find the average power delivered to the 50-kΩ resistor.

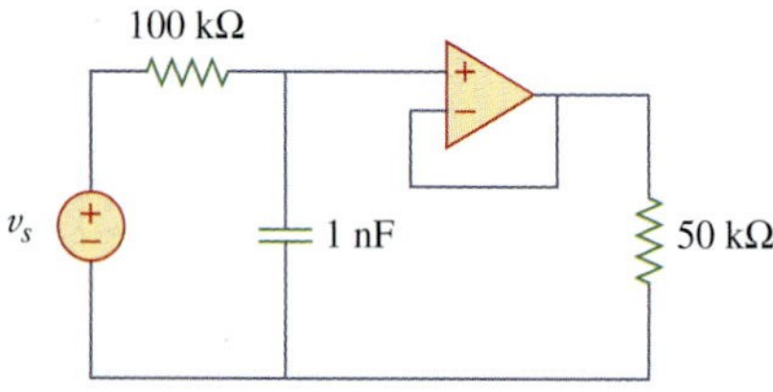

Figure 11.84
For Prob. 11.65.

11.66 Obtain the average power absorbed by the 6-kΩ resistor in the op amp circuit in Fig. 11.85.

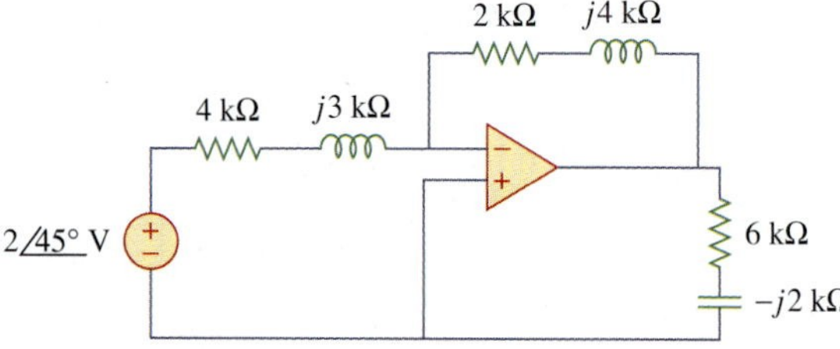

Figure 11.85
For Prob. 11.66.

11.67 For the op amp circuit in Fig. 11.86, calculate:

(a) the complex power delivered by the voltage source

(b) the average power dissipated in the 12-Ω resistor

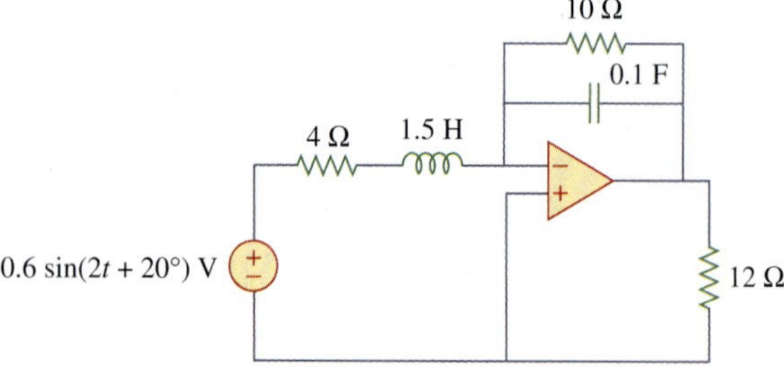

Figure 11.86
For Prob. 11.67.

11.68 Compute the complex power supplied by the current source in the series *RLC* circuit in Fig. 11.87.

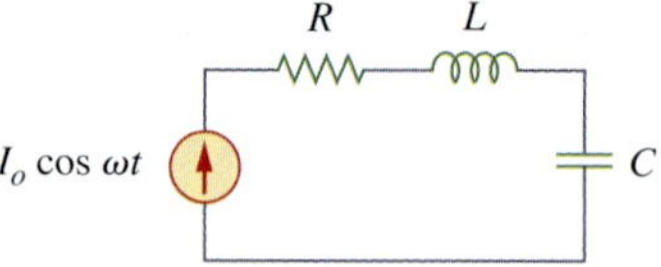

Figure 11.87
For Prob. 11.68.

Section 11.8 Power Factor Correction

11.69 Refer to the circuit shown in Fig. 11.88.

(a) What is the power factor?

(b) What is the average power dissipated?

(c) What is the value of the capacitance that will give a unity power factor when connected to the load?

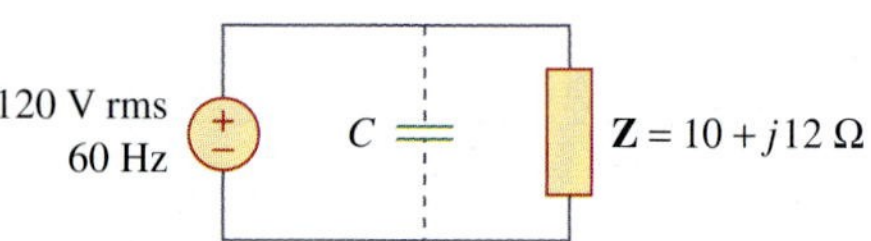

Figure 11.88
For Prob. 11.69.

11.70 Design a problem to help other students better understand power factor correction.

11.71 Three loads are connected in parallel to a 120/0° V rms source. Load 1 absorbs 60 kVAR at pf = 0.85 lagging, load 2 absorbs 90 kW and 50 kVAR leading, and load 3 absorbs 100 kW at pf = 1. (a) Find the equivalent impedance. (b) Calculate the power factor of the parallel combination. (c) Determine the current supplied by the source.

11.72 Two loads connected in parallel draw a total of 2.4 kW at 0.8 pf lagging from a 120-V rms, 60-Hz line. One load absorbs 1.5 kW at a 0.707 pf lagging. Determine: (a) the pf of the second load, (b) the parallel element required to correct the pf to 0.9 lagging for the two loads.

11.73 A 240-V rms 60-Hz supply serves a load that is 10 kW (resistive), 15 kVAR (capacitive), and 22 kVAR (inductive). Find:

(a) the apparent power

(b) the current drawn from the supply

(c) the kVAR rating and capacitance required to improve the power factor to 0.96 lagging

(d) the current drawn from the supply under the new power-factor conditions

11.74 A 120-V rms 60-Hz source supplies two loads connected in parallel, as shown in Fig. 11.89.

(a) Find the power factor of the parallel combination.

(b) Calculate the value of the capacitance connected in parallel that will raise the power factor to unity.

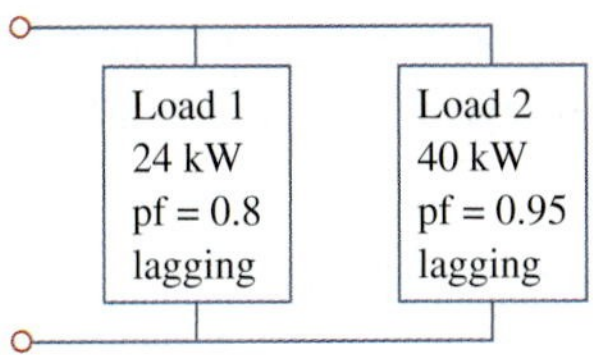

Figure 11.89
For Prob. 11.74.

11.75 Consider the power system shown in Fig. 11.90. Calculate:

(a) the total complex power

(b) the power factor

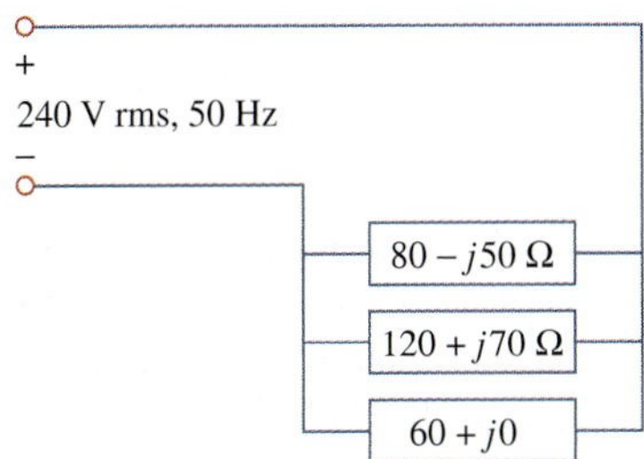

Figure 11.90
For Prob. 11.75.

Section 11.9 Applications

11.76 Obtain the wattmeter reading of the circuit in Fig. 11.91.

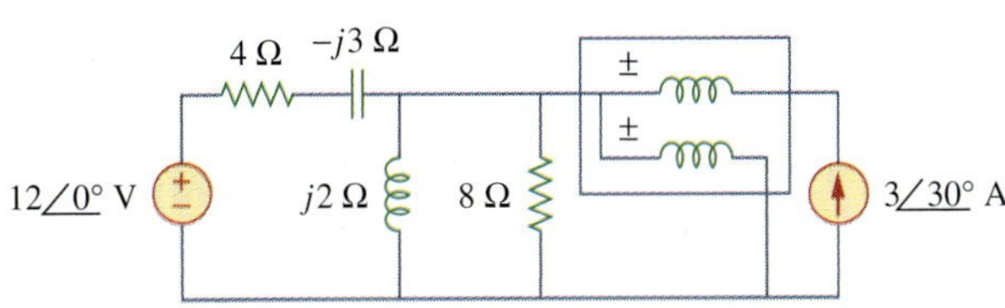

Figure 11.91
For Prob. 11.76.

11.77 What is the reading of the wattmeter in the network of Fig. 11.92?

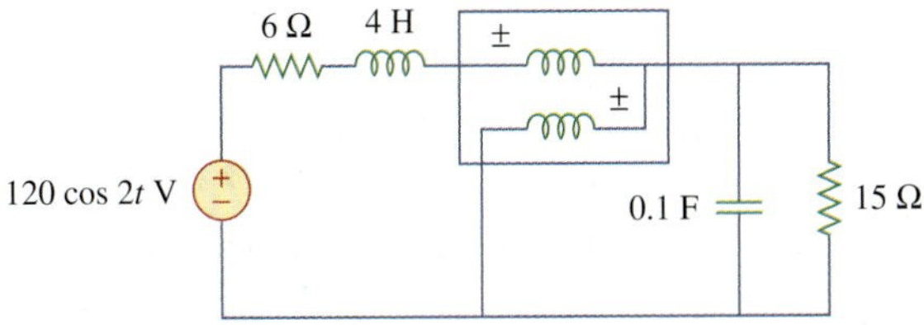

Figure 11.92
For Prob. 11.77.

11.78 Find the wattmeter reading of the circuit shown in Fig. 11.93.

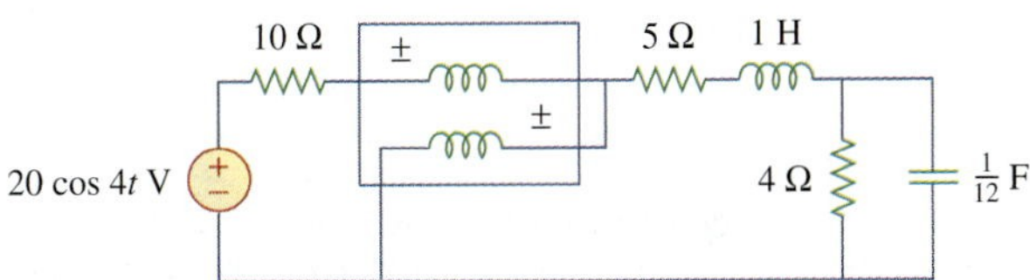

Figure 11.93
For Prob. 11.78.

11.79 Determine the wattmeter reading of the circuit in Fig. 11.94.

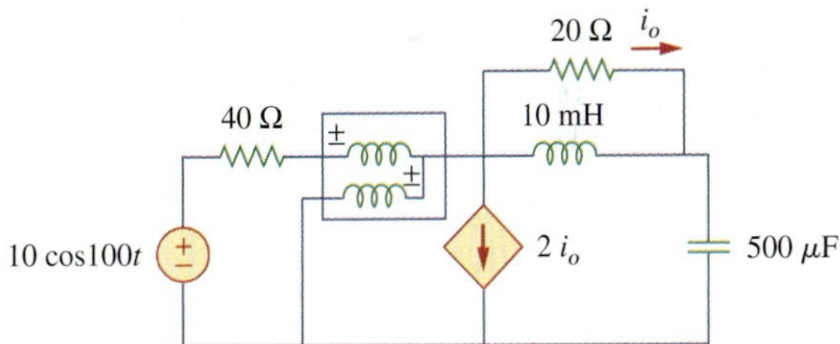

Figure 11.94
For Prob. 11.79.

11.80 The circuit of Fig. 11.95 portrays a wattmeter connected into an ac network.

(a) Find the load current.

(b) Calculate the wattmeter reading.

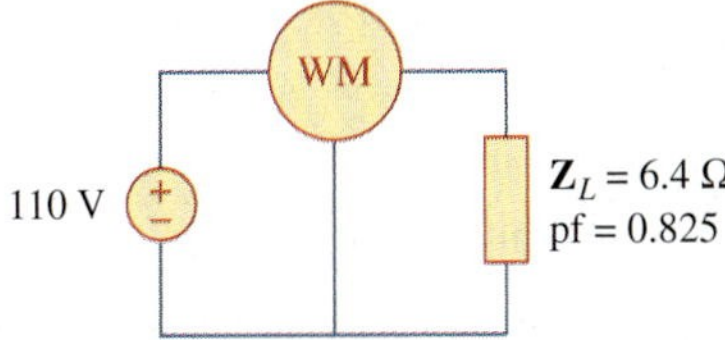

Figure 11.95
For Prob. 11.80.

11.81 Design a problem to help other students better understand **e⟋d** how to correct power factor to values other than unity.

11.82 A 240-V rms 60-Hz source supplies a parallel combination of a 5-kW heater and a 30-kVA induction motor whose power factor is 0.82. Determine:

(a) the system apparent power

(b) the system reactive power

(c) the kVA rating of a capacitor required to adjust the system power factor to 0.9 lagging

(d) the value of the capacitor required

11.83 Oscilloscope measurements indicate that the peak voltage across a load and the peak current through it are, respectively, $210\underline{/60^\circ}$ V and $8\underline{/25^\circ}$ A. Determine:

(a) the real power

(b) the apparent power

(c) the reactive power

(d) the power factor

11.84 A consumer has an annual consumption of 1200 **e⟋d** MWh with a maximum demand of 2.4 MVA. The maximum demand charge is \$30 per kVA per annum, and the energy charge per kWh is 4 cents.

(a) Determine the annual cost of energy.

(b) Calculate the charge per kWh with a flat-rate tariff if the revenue to the utility company is to remain the same as for the two-part tariff.

11.85 A regular household system of a single-phase three-wire circuit allows the operation of both 120-V and 240-V, 60-Hz appliances. The household circuit is modeled as shown in Fig. 11.96. Calculate:

(a) the currents $\mathbf{I}_1$, $\mathbf{I}_2$, and $\mathbf{I}_n$

(b) the total complex power supplied

(c) the overall power factor of the circuit

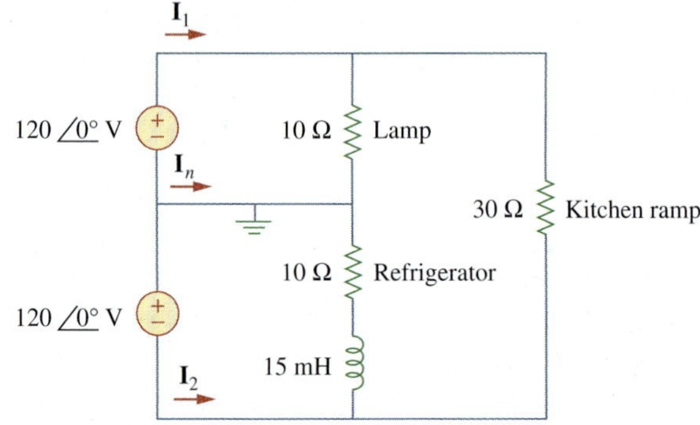

Figure 11.96
For Prob. 11.85.

Comprehensive Problems

11.86 A transmitter delivers maximum power to an antenna **e⟋d** when the antenna is adjusted to represent a load of 75-Ω resistance in series with an inductance of 4 μH. If the transmitter operates at 4.12 MHz, find its internal impedance.

11.87 In a TV transmitter, a series circuit has an impedance of 3 kΩ and a total current of 50 mA. If the voltage across the resistor is 80 V, what is the power factor of the circuit?

11.88 A certain electronic circuit is connected to a 110-V ac line. The root-mean-square value of the current drawn is 2 A, with a phase angle of 55°.

(a) Find the true power drawn by the circuit.

(b) Calculate the apparent power.

11.89 An industrial heater has a nameplate that reads: **e⟋d** 210 V 60 Hz 12 kVA 0.78 pf lagging
Determine:

(a) the apparent and the complex power

(b) the impedance of the heater

***11.90** A 2000-kW turbine-generator of 0.85 power factor **e⟋d** operates at the rated load. An additional load of 300 kW at 0.8 power factor is added. What kVAR of capacitors is required to operate the turbine-generator but keep it from being overloaded?

11.91 The nameplate of an electric motor has the following **e⟋d** information:

Line voltage: 220 V rms
Line current: 15 A rms
Line frequency: 60 Hz
Power: 2700 W

Determine the power factor (lagging) of the motor. Find the value of the capacitance C that must be connected across the motor to raise the pf to unity.

11.92 As shown in Fig. 11.97, a 550-V feeder line supplies an industrial plant consisting of a motor drawing 60 kW at 0.75 pf (inductive), a capacitor with a rating of 20 kVAR, and lighting drawing 20 kW.

(a) Calculate the total reactive power and apparent power absorbed by the plant.

(b) Determine the overall pf.

(c) Find the current in the feeder line.

* An asterisk indicates a challenging problem.

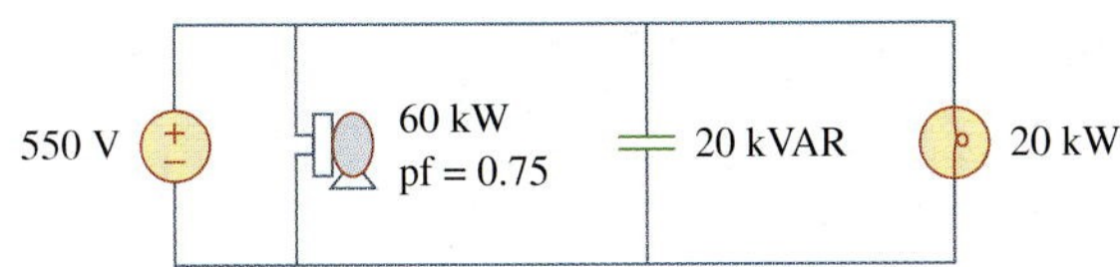

Figure 11.97
For Prob. 11.92.

11.93 A factory has the following four major loads:

- A motor rated at 5 hp, 0.8 pf lagging (1 hp = 0.7457 kW).
- A heater rated at 1.2 kW, 1.0 pf.
- Ten 120-W lightbulbs.
- A synchronous motor rated at 1.6 kVAR, 0.6 pf leading.

(a) Calculate the total real and reactive power.

(b) Find the overall power factor.

11.94 A 1-MVA substation operates at full load at 0.7 power factor. It is desired to improve the power factor to 0.95 by installing capacitors. Assume that new substation and distribution facilities cost \$120 per kVA installed, and capacitors cost \$30 per kVA installed.

(a) Calculate the cost of capacitors needed.

(b) Find the savings in substation capacity released.

(c) Are capacitors economical for releasing the amount of substation capacity?

11.95 A coupling capacitor is used to block dc current from an amplifier as shown in Fig. 11.98(a). The amplifier and the capacitor act as the source, while the speaker is the load as in Fig. 11.98(b).

(a) At what frequency is maximum power transferred to the speaker?

(b) If V_s = 4.6 V rms, how much power is delivered to the speaker at that frequency?

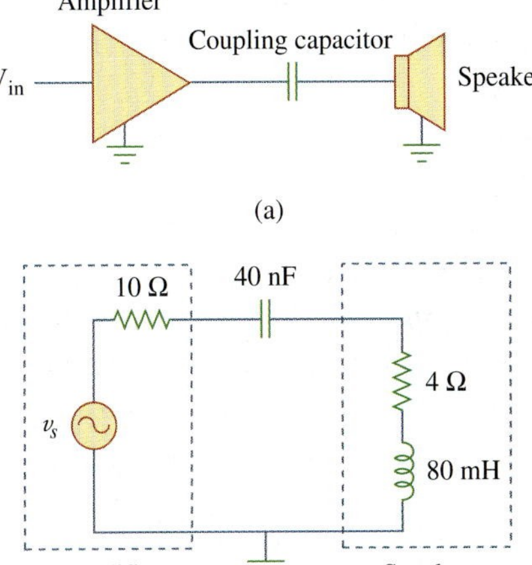

Figure 11.98
For Prob. 11.95.

11.96 A power amplifier has an output impedance of $40 + j8\ \Omega$. It produces a no-load output voltage of 146 V at 300 Hz.

(a) Determine the impedance of the load that achieves maximum power transfer.

(b) Calculate the load power under this matching condition.

11.97 A power transmission system is modeled as shown in Fig. 11.99. If $\mathbf{V}_s = 240\underline{/0^\circ}$ rms, find the average power absorbed by the load.

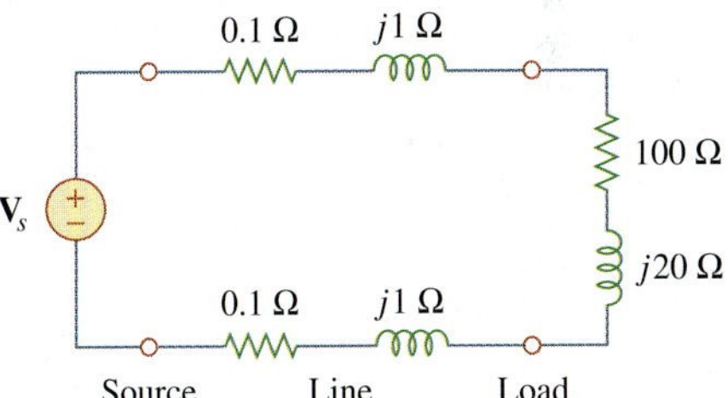

Figure 11.99
For Prob. 11.97.

chapter

12

Three-Phase Circuits

He who cannot forgive others breaks the bridge over which he must pass himself.

—G. Herbert

Enhancing Your Skills and Your Career

ABET EC 2000 criteria (3.e), *"an ability to identify, formulate, and solve engineering problems."*

Developing and enhancing your "ability to identify, formulate, and solve engineering problems" is a primary focus of textbook. Following our six step problem-solving process is the best way to practice this skill. Our recommendation is that you use this process whenever possible. You may be pleased to learn that this process works well for nonengineering courses.

ABET EC 2000 criteria (f), *"an understanding of professional and ethical responsibility."*

"An understanding of professional and ethical responsibility" is required of every engineer. To some extent, this understanding is very personal for each of us. Let us identify some pointers to help you develop this understanding. One of my favorite examples is that an engineer has the responsibility to answer what I call the "unasked question." For instance, assume that you own a car that has a problem with the transmission. In the process of selling that car, the prospective buyer asks you if there is a problem in the right-front wheel bearing. You answer no. However, as an engineer, you are required to inform the buyer that there is a problem with the transmission without being asked.

Your responsibility both professionally and ethically is to perform in a manner that does not harm those around you and to whom you are responsible. Clearly, developing this capability will take time and maturity on your part. I recommend practicing this by looking for professional and ethical components in your day-to-day activities.

Photo by Charles Alexander

12.1 Introduction

So far in this text, we have dealt with single-phase circuits. A single-phase ac power system consists of a generator connected through a pair of wires (a transmission line) to a load. Figure 12.1(a) depicts a single-phase two-wire system, where V_p is the rms magnitude of the source voltage and ϕ is the phase. What is more common in practice is a single-phase three-wire system, shown in Fig. 12.1(b). It contains two identical sources (equal magnitude and the same phase) that are connected to two loads by two outer wires and the neutral. For example, the normal household system is a single-phase three-wire system because the terminal voltages have the same magnitude and the same phase. Such a system allows the connection of both 120-V and 240-V appliances.

Historical note: Thomas Edison invented a *three-wire system,* using three wires instead of four.

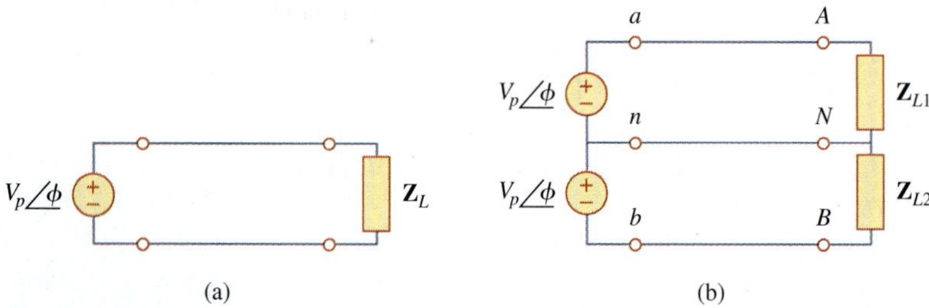

Figure 12.1
Single-phase systems: (a) two-wire type, (b) three-wire type.

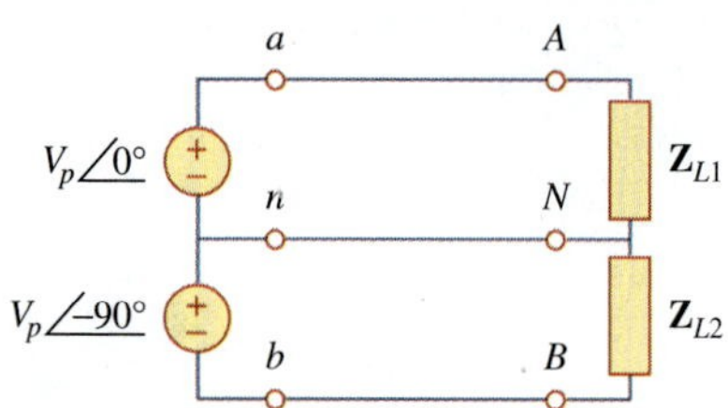

Figure 12.2
Two-phase three-wire system.

Circuits or systems in which the ac sources operate at the same frequency but different phases are known as *polyphase.* Figure 12.2 shows a two-phase three-wire system, and Fig. 12.3 shows a three-phase four-wire system. As distinct from a single-phase system, a two-phase system is produced by a generator consisting of two coils placed perpendicular to each other so that the voltage generated by one lags the other by 90°. By the same token, a three-phase system is produced by a generator consisting of three sources having the same amplitude and frequency but out of phase with each other by 120°. Since the three-phase system is by far the most prevalent and most economical polyphase system, discussion in this chapter is mainly on three-phase systems.

Three-phase systems are important for at least three reasons. First, nearly all electric power is generated and distributed in three-phase, at the operating frequency of 60 Hz (or $\omega = 377$ rad/s) in the United States or 50 Hz (or $\omega = 314$ rad/s) in some other parts of the world. When one-phase or two-phase inputs are required, they are taken from the three-phase system rather than generated independently. Even when more than three phases are needed—such as in the aluminum industry, where 48 phases are required for melting purposes—they can be provided by manipulating the three phases supplied. Second, the instantaneous power in a three-phase system can be constant (not pulsating), as we will see in Section 12.7. This results in uniform power transmission and less vibration of three-phase machines. Third, for the same amount of power, the three-phase system is more economical than the single-phase. The amount of wire required for a three-phase system is less than that required for an equivalent single-phase system.

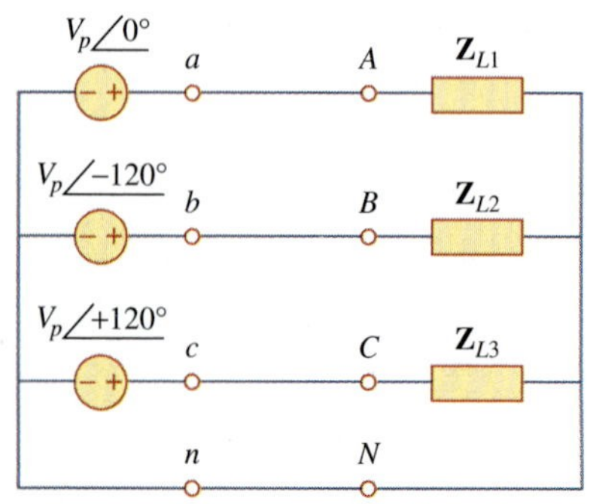

Figure 12.3
Three-phase four-wire system.

Historical

Courtesy Smithsonian Institution

Nikola Tesla (1856–1943) was a Croatian-American engineer whose inventions—among them the induction motor and the first polyphase ac power system—greatly influenced the settlement of the ac versus dc debate in favor of ac. He was also responsible for the adoption of 60 Hz as the standard for ac power systems in the United States.

Born in Austria-Hungary (now Croatia), to a clergyman, Tesla had an incredible memory and a keen affinity for mathematics. He moved to the United States in 1884 and first worked for Thomas Edison. At that time, the country was in the "battle of the currents" with George Westinghouse (1846–1914) promoting ac and Thomas Edison rigidly leading the dc forces. Tesla left Edison and joined Westinghouse because of his interest in ac. Through Westinghouse, Tesla gained the reputation and acceptance of his polyphase ac generation, transmission, and distribution system. He held 700 patents in his lifetime. His other inventions include high-voltage apparatus (the tesla coil) and a wireless transmission system. The unit of magnetic flux density, the tesla, was named in honor of him.

We begin with a discussion of balanced three-phase voltages. Then we analyze each of the four possible configurations of balanced three-phase systems. We also discuss the analysis of unbalanced three-phase systems. We learn how to use *PSpice for Windows* to analyze a balanced or unbalanced three-phase system. Finally, we apply the concepts developed in this chapter to three-phase power measurement and residential electrical wiring.

12.2 Balanced Three-Phase Voltages

Three-phase voltages are often produced with a three-phase ac generator (or alternator) whose cross-sectional view is shown in Fig. 12.4. The generator basically consists of a rotating magnet (called the *rotor*) surrounded by a stationary winding (called the *stator*). Three separate

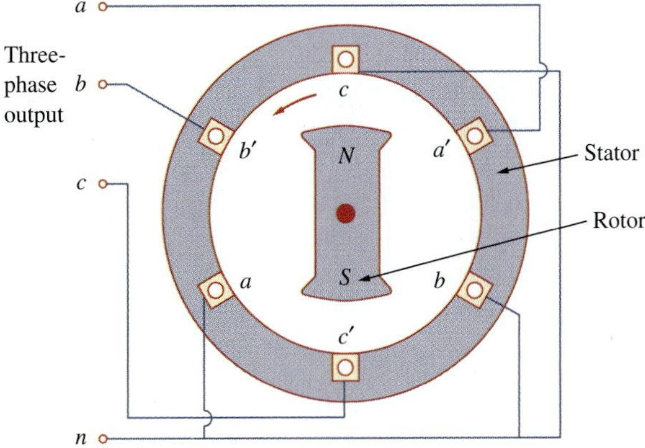

Figure 12.4
A three-phase generator.

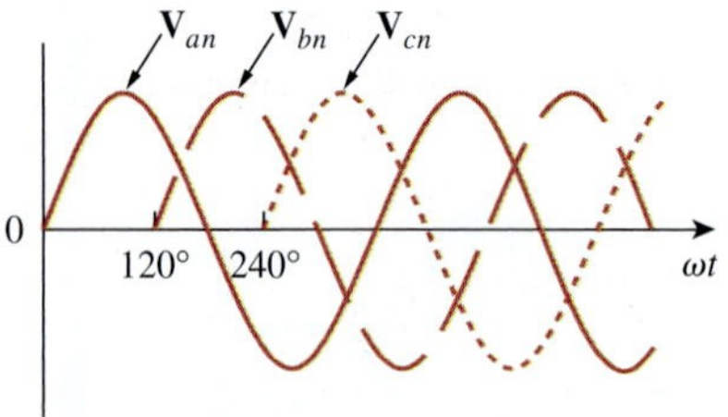

Figure 12.5
The generated voltages are 120° apart from each other.

windings or coils with terminals a-a', b-b', and c-c' are physically placed 120° apart around the stator. Terminals a and a', for example, stand for one of the ends of coils going into and the other end coming out of the page. As the rotor rotates, its magnetic field "cuts" the flux from the three coils and induces voltages in the coils. Because the coils are placed 120° apart, the induced voltages in the coils are equal in magnitude but out of phase by 120° (Fig. 12.5). Since each coil can be regarded as a single-phase generator by itself, the three-phase generator can supply power to both single-phase and three-phase loads.

A typical three-phase system consists of three voltage sources connected to loads by three or four wires (or transmission lines). (Three-phase current sources are very scarce.) A three-phase system is equivalent to three single-phase circuits. The voltage sources can be either wye-connected as shown in Fig. 12.6(a) or delta-connected as in Fig. 12.6(b).

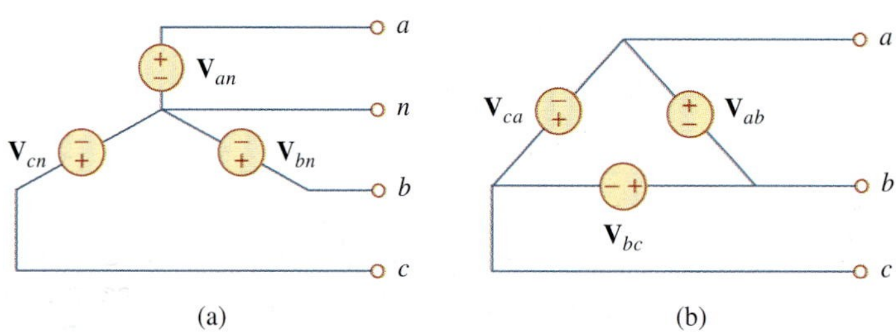

Figure 12.6
Three-phase voltage sources: (a) Y-connected source, (b) Δ-connected source.

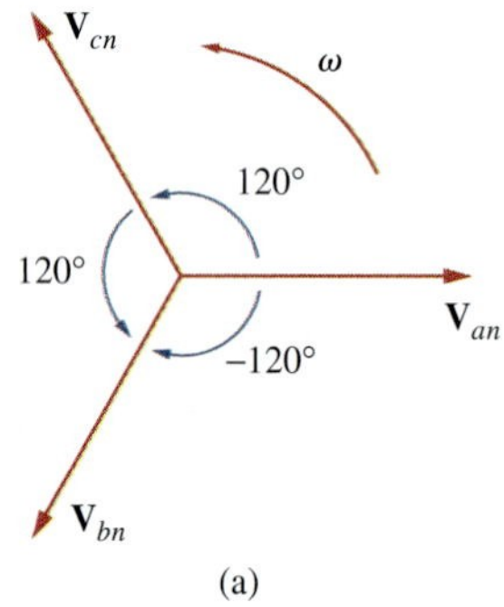

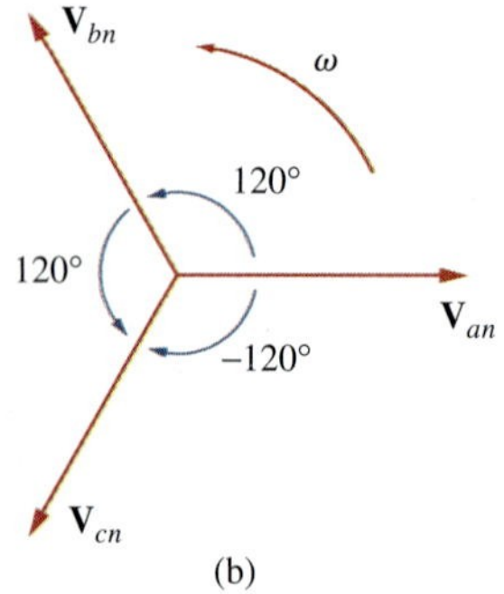

Figure 12.7
Phase sequences: (a) *abc* or positive sequence, (b) *acb* or negative sequence.

Let us consider the wye-connected voltages in Fig. 12.6(a) for now. The voltages $\mathbf{V}_{an}$, $\mathbf{V}_{bn}$, and $\mathbf{V}_{cn}$ are respectively between lines a, b, and c, and the neutral line n. These voltages are called *phase voltages*. If the voltage sources have the same amplitude and frequency ω and are out of phase with each other by 120°, the voltages are said to be *balanced*. This implies that

$$\mathbf{V}_{an} + \mathbf{V}_{bn} + \mathbf{V}_{cn} = 0 \tag{12.1}$$

$$|\mathbf{V}_{an}| = |\mathbf{V}_{bn}| = |\mathbf{V}_{cn}| \tag{12.2}$$

Thus,

> **Balanced phase voltages** are equal in magnitude and are out of phase with each other by 120°.

Since the three-phase voltages are 120° out of phase with each other, there are two possible combinations. One possibility is shown in Fig. 12.7(a) and expressed mathematically as

$$\begin{aligned} \mathbf{V}_{an} &= V_p\underline{/0^\circ} \\ \mathbf{V}_{bn} &= V_p\underline{/-120^\circ} \\ \mathbf{V}_{cn} &= V_p\underline{/-240^\circ} = V_p\underline{/+120^\circ} \end{aligned} \tag{12.3}$$

where V_p is the effective or rms value of the phase voltages. This is known as the *abc sequence* or *positive sequence*. In this phase sequence, $\mathbf{V}_{an}$ leads $\mathbf{V}_{bn}$, which in turn leads $\mathbf{V}_{cn}$. This sequence is produced when the rotor in Fig. 12.4 rotates counterclockwise. The other possibility is shown in Fig. 12.7(b) and is given by

As a common tradition in power systems, voltage and current in this chapter are in rms values unless otherwise stated.

$$\begin{aligned} \mathbf{V}_{an} &= V_p\underline{/0^\circ} \\ \mathbf{V}_{cn} &= V_p\underline{/-120^\circ} \\ \mathbf{V}_{bn} &= V_p\underline{/-240^\circ} = V_p\underline{/+120^\circ} \end{aligned} \tag{12.4}$$

This is called the *acb sequence* or *negative sequence*. For this phase sequence, $\mathbf{V}_{an}$ leads $\mathbf{V}_{cn}$, which in turn leads $\mathbf{V}_{bn}$. The *acb* sequence is produced when the rotor in Fig. 12.4 rotates in the clockwise direction. It is easy to show that the voltages in Eqs. (12.3) or (12.4) satisfy Eqs. (12.1) and (12.2). For example, from Eq. (12.3),

$$\begin{aligned} \mathbf{V}_{an} + \mathbf{V}_{bn} + \mathbf{V}_{cn} &= V_p\underline{/0^\circ} + V_p\underline{/-120^\circ} + V_p\underline{/+120^\circ} \\ &= V_p(1.0 - 0.5 - j0.866 - 0.5 + j0.866) \\ &= 0 \end{aligned} \tag{12.5}$$

The **phase sequence** is the time order in which the voltages pass through their respective maximum values.

The phase sequence may also be regarded as the order in which the phase voltages reach their peak (or maximum) values with respect to time.

The phase sequence is determined by the order in which the phasors pass through a fixed point in the phase diagram.

Reminder: As time increases, each phasor (or sinor) rotates at an angular velocity ω.

In Fig. 12.7(a), as the phasors rotate in the counterclockwise direction with frequency ω, they pass through the horizontal axis in a sequence *abcabca* . . . . Thus, the sequence is *abc* or *bca* or *cab*. Similarly, for the phasors in Fig. 12.7(b), as they rotate in the counterclockwise direction, they pass the horizontal axis in a sequence *acbacba* . . . . This describes the *acb* sequence. The phase sequence is important in three-phase power distribution. It determines the direction of the rotation of a motor connected to the power source, for example.

Like the generator connections, a three-phase load can be either wye-connected or delta-connected, depending on the end application. Figure 12.8(a) shows a wye-connected load, and Fig. 12.8(b) shows a delta-connected load. The neutral line in Fig. 12.8(a) may or may not be there, depending on whether the system is four- or three-wire. (And, of course, a neutral connection is topologically impossible for a delta connection.) A wye- or delta-connected load is said to be *unbalanced* if the phase impedances are not equal in magnitude or phase.

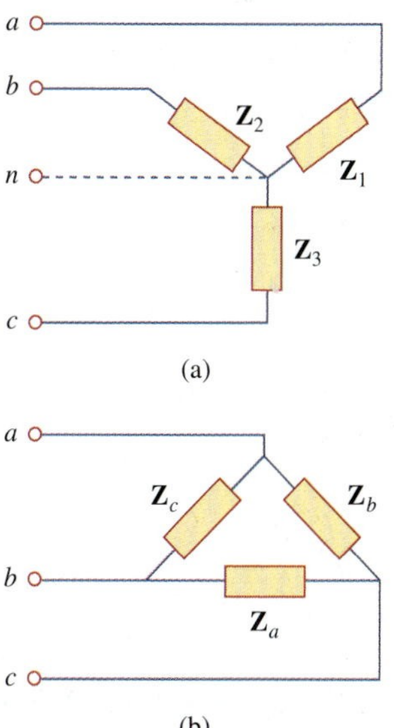

Figure 12.8
Two possible three-phase load configurations: (a) a Y-connected load, (b) a Δ-connected load.

A **balanced load** is one in which the phase impedances are equal in magnitude and in phase.

For a *balanced* wye-connected load,

$$\mathbf{Z}_1 = \mathbf{Z}_2 = \mathbf{Z}_3 = \mathbf{Z}_Y \tag{12.6}$$

Reminder: A Y-connected load consists of three impedances connected to a neutral node, while a Δ-connected load consists of three impedances connected around a loop. The load is balanced when the three impedances are equal in either case.

where $\mathbf{Z}_Y$ is the load impedance per phase. For a *balanced* delta-connected load,

$$\mathbf{Z}_a = \mathbf{Z}_b = \mathbf{Z}_c = \mathbf{Z}_\Delta \tag{12.7}$$

where $\mathbf{Z}_\Delta$ is the load impedance per phase in this case. We recall from Eq. (9.69) that

$$\mathbf{Z}_\Delta = 3\mathbf{Z}_Y \qquad \text{or} \qquad \mathbf{Z}_Y = \frac{1}{3}\mathbf{Z}_\Delta \tag{12.8}$$

so we know that a wye-connected load can be transformed into a delta-connected load, or vice versa, using Eq. (12.8).

Since both the three-phase source and the three-phase load can be either wye- or delta-connected, we have four possible connections:

- Y-Y connection (i.e., Y-connected source with a Y-connected load).
- Y-Δ connection.
- Δ-Δ connection.
- Δ-Y connection.

In subsequent sections, we will consider each of these possible configurations.

It is appropriate to mention here that a balanced delta-connected load is more common than a balanced wye-connected load. This is due to the ease with which loads may be added or removed from each phase of a delta-connected load. This is very difficult with a wye-connected load because the neutral may not be accessible. On the other hand, delta-connected sources are not common in practice because of the circulating current that will result in the delta-mesh if the three-phase voltages are slightly unbalanced.

Example 12.1

Determine the phase sequence of the set of voltages

$$v_{an} = 200\cos(\omega t + 10^\circ)$$

$$v_{bn} = 200\cos(\omega t - 230^\circ), \qquad v_{cn} = 200\cos(\omega t - 110^\circ)$$

Solution:

The voltages can be expressed in phasor form as

$$\mathbf{V}_{an} = 200\underline{/10^\circ}\text{ V}, \qquad \mathbf{V}_{bn} = 200\underline{/-230^\circ}\text{ V}, \qquad \mathbf{V}_{cn} = 200\underline{/-110^\circ}\text{ V}$$

We notice that $\mathbf{V}_{an}$ leads $\mathbf{V}_{cn}$ by 120° and $\mathbf{V}_{cn}$ in turn leads $\mathbf{V}_{bn}$ by 120°. Hence, we have an *acb* sequence.

Practice Problem 12.1

Given that $\mathbf{V}_{bn} = 110\underline{/30^\circ}$ V, find $\mathbf{V}_{an}$ and $\mathbf{V}_{cn}$, assuming a positive (*abc*) sequence.

Answer: $110\underline{/150^\circ}$ V, $110\underline{/-90^\circ}$ V.

12.3 Balanced Wye-Wye Connection

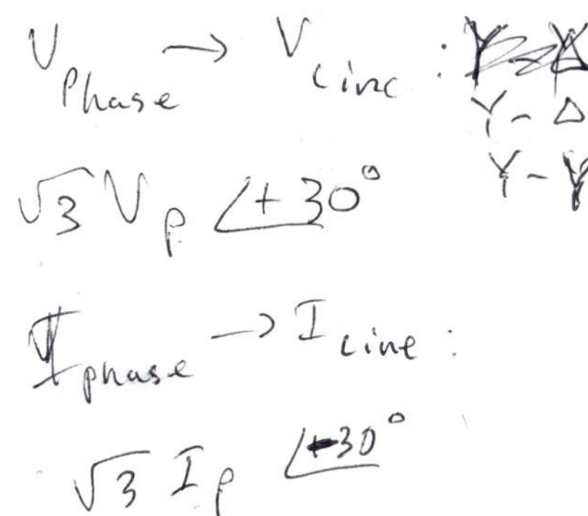

We begin with the Y-Y system, because any balanced three-phase system can be reduced to an equivalent Y-Y system. Therefore, analysis of this system should be regarded as the key to solving all balanced three-phase systems.

> A **balanced Y-Y system** is a three-phase system with a balanced Y-connected source and a balanced Y-connected load.

Consider the balanced four-wire Y-Y system of Fig. 12.9, where a Y-connected load is connected to a Y-connected source. We assume a balanced load so that load impedances are equal. Although the impedance $\mathbf{Z}_Y$ is the total load impedance per phase, it may also be regarded as the sum of the source impedance $\mathbf{Z}_s$, line impedance $\mathbf{Z}_\ell$, and load impedance $\mathbf{Z}_L$ for each phase, since these impedances are in series. As illustrated in Fig. 12.9, $\mathbf{Z}_s$ denotes the internal impedance of the phase winding of the generator; $\mathbf{Z}_\ell$ is the impedance of the line joining a phase of the source with a phase of the load; $\mathbf{Z}_L$ is the impedance of each phase of the load; and $\mathbf{Z}_n$ is the impedance of the neutral line. Thus, in general

$$\mathbf{Z}_Y = \mathbf{Z}_s + \mathbf{Z}_\ell + \mathbf{Z}_L \tag{12.9}$$

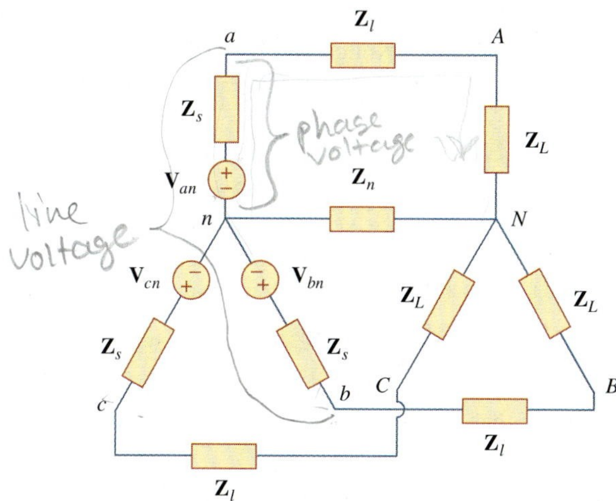

Figure 12.9
A balanced Y-Y system, showing the source, line, and load impedances.

$\mathbf{Z}_s$ and $\mathbf{Z}_\ell$ are often very small compared with $\mathbf{Z}_L$, so one can assume that $\mathbf{Z}_Y = \mathbf{Z}_L$ if no source or line impedance is given. In any event, by lumping the impedances together, the Y-Y system in Fig. 12.9 can be simplified to that shown in Fig. 12.10.

Assuming the positive sequence, the *phase* voltages (or line-to-neutral voltages) are

$$\mathbf{V}_{an} = V_p\underline{/0^\circ}$$
$$\mathbf{V}_{bn} = V_p\underline{/-120^\circ}, \qquad \mathbf{V}_{cn} = V_p\underline{/+120^\circ} \tag{12.10}$$

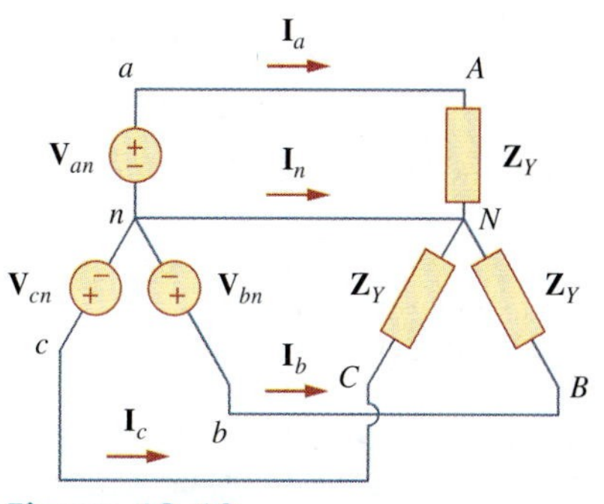

Figure 12.10
Balanced Y-Y connection.

The *line-to-line* voltages or simply *line* voltages $\mathbf{V}_{ab}$, $\mathbf{V}_{bc}$, and $\mathbf{V}_{ca}$ are related to the phase voltages. For example,

$$\begin{aligned}\mathbf{V}_{ab} &= \mathbf{V}_{an} + \mathbf{V}_{nb} = \mathbf{V}_{an} - \mathbf{V}_{bn} = V_p\underline{/0^\circ} - V_p\underline{/-120^\circ} \\ &= V_p\left(1 + \frac{1}{2} + j\frac{\sqrt{3}}{2}\right) = \sqrt{3}V_p\underline{/30^\circ}\end{aligned} \tag{12.11a}$$

Similarly, we can obtain

$$\mathbf{V}_{bc} = \mathbf{V}_{bn} - \mathbf{V}_{cn} = \sqrt{3}V_p\underline{/-90^\circ} \tag{12.11b}$$

$$\mathbf{V}_{ca} = \mathbf{V}_{cn} - \mathbf{V}_{an} = \sqrt{3}V_p\underline{/-210^\circ} \tag{12.11c}$$

Thus, the magnitude of the line voltages V_L is $\sqrt{3}$ times the magnitude of the phase voltages V_p, or

$$\boxed{V_L = \sqrt{3}V_p} \tag{12.12}$$

where

$$V_p = |\mathbf{V}_{an}| = |\mathbf{V}_{bn}| = |\mathbf{V}_{cn}| \tag{12.13}$$

and

$$V_L = |\mathbf{V}_{ab}| = |\mathbf{V}_{bc}| = |\mathbf{V}_{ca}| \tag{12.14}$$

Also the line voltages lead their corresponding phase voltages by 30°. Figure 12.11(a) illustrates this. Figure 12.11(a) also shows how to determine $\mathbf{V}_{ab}$ from the phase voltages, while Fig. 12.11(b) shows the same for the three line voltages. Notice that $\mathbf{V}_{ab}$ leads $\mathbf{V}_{bc}$ by 120°, and $\mathbf{V}_{bc}$ leads $\mathbf{V}_{ca}$ by 120°, so that the line voltages sum up to zero as do the phase voltages.

Applying KVL to each phase in Fig. 12.10, we obtain the line currents as

$$\begin{aligned}\mathbf{I}_a &= \frac{\mathbf{V}_{an}}{\mathbf{Z}_Y}, \qquad \mathbf{I}_b = \frac{\mathbf{V}_{bn}}{\mathbf{Z}_Y} = \frac{\mathbf{V}_{an}\underline{/-120^\circ}}{\mathbf{Z}_Y} = \mathbf{I}_a\underline{/-120^\circ} \\ \mathbf{I}_c &= \frac{\mathbf{V}_{cn}}{\mathbf{Z}_Y} = \frac{\mathbf{V}_{an}\underline{/-240^\circ}}{\mathbf{Z}_Y} = \mathbf{I}_a\underline{/-240^\circ}\end{aligned} \tag{12.15}$$

We can readily infer that the line currents add up to zero,

$$\mathbf{I}_a + \mathbf{I}_b + \mathbf{I}_c = 0 \tag{12.16}$$

so that

$$\mathbf{I}_n = -(\mathbf{I}_a + \mathbf{I}_b + \mathbf{I}_c) = 0 \tag{12.17a}$$

or

$$\mathbf{V}_{nN} = \mathbf{Z}_n\mathbf{I}_n = 0 \tag{12.17b}$$

that is, the voltage across the neutral wire is zero. The neutral line can thus be removed without affecting the system. In fact, in long distance power transmission, conductors in multiples of three are used with the earth itself acting as the neutral conductor. Power systems designed in this way are well grounded at all critical points to ensure safety.

While the *line* current is the current in each line, the *phase* current is the current in each phase of the source or load. In the Y-Y system, the line current is the same as the phase current. We will use single subscripts

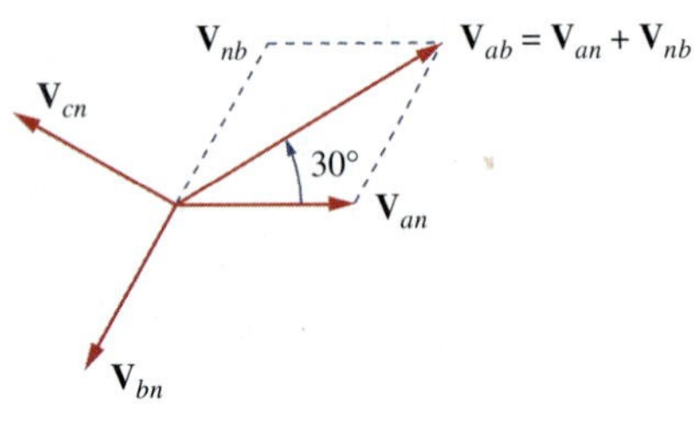

(a)

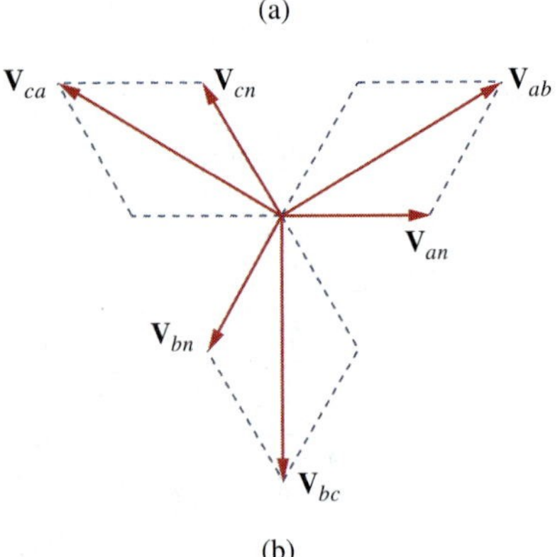

(b)

Figure 12.11
Phasor diagrams illustrating the relationship between line voltages and phase voltages.

for line currents because it is natural and conventional to assume that line currents flow from the source to the load.

An alternative way of analyzing a balanced Y-Y system is to do so on a "per phase" basis. We look at one phase, say phase a, and analyze the single-phase equivalent circuit in Fig. 12.12. The single-phase analysis yields the line current $\mathbf{I}_a$ as

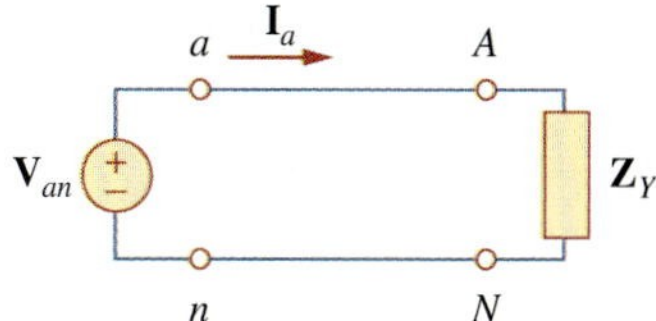

Figure 12.12
A single-phase equivalent circuit.

$$\mathbf{I}_a = \frac{\mathbf{V}_{an}}{\mathbf{Z}_Y} \tag{12.18}$$

From $\mathbf{I}_a$, we use the phase sequence to obtain other line currents. Thus, as long as the system is balanced, we need only analyze one phase. We may do this even if the neutral line is absent, as in the three-wire system.

Example 12.2

Calculate the line currents in the three-wire Y-Y system of Fig. 12.13.

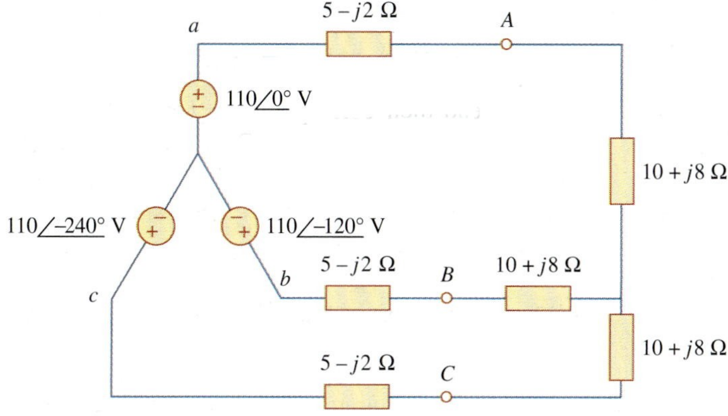

Figure 12.13
Three-wire Y-Y system; for Example 12.2.

Solution:

The three-phase circuit in Fig. 12.13 is balanced; we may replace it with its single-phase equivalent circuit such as in Fig. 12.12. We obtain $\mathbf{I}_a$ from the single-phase analysis as

$$\mathbf{I}_a = \frac{\mathbf{V}_{an}}{\mathbf{Z}_Y}$$

where $\mathbf{Z}_Y = (5 - j2) + (10 + j8) = 15 + j6 = 16.155\underline{/21.8^\circ}$. Hence,

$$\mathbf{I}_a = \frac{110\underline{/0^\circ}}{16.155\underline{/21.8^\circ}} = 6.81\underline{/-21.8^\circ} \text{ A}$$

Since the source voltages in Fig. 12.13 are in positive sequence, the line currents are also in positive sequence:

$$\mathbf{I}_b = \mathbf{I}_a\underline{/-120^\circ} = 6.81\underline{/-141.8^\circ} \text{ A}$$

$$\mathbf{I}_c = \mathbf{I}_a\underline{/-240^\circ} = 6.81\underline{/-261.8^\circ} \text{ A} = 6.81\underline{/98.2^\circ} \text{ A}$$

Practice Problem 12.2

A Y-connected balanced three-phase generator with an impedance of $0.4 + j0.3\ \Omega$ per phase is connected to a Y-connected balanced load with an impedance of $24 + j19\ \Omega$ per phase. The line joining the generator and the load has an impedance of $0.6 + j0.7\ \Omega$ per phase. Assuming a positive sequence for the source voltages and that $\mathbf{V}_{an} = 120\underline{/30^\circ}$ V, find: (a) the line voltages, (b) the line currents.

Answer: (a) $207.85\underline{/60^\circ}$ V, $207.85\underline{/-60^\circ}$ V, $207.85\underline{/-180^\circ}$ V, (b) $3.75\underline{/-8.66^\circ}$ A, $3.75\underline{/-128.66^\circ}$ A, $3.75\underline{/-111.34^\circ}$ A.

12.4 Balanced Wye-Delta Connection

A **balanced Y-Δ system** consists of a balanced Y-connected source feeding a balanced Δ-connected load.

This is perhaps the most practical three-phase system, as the three-phase sources are usually Y-connected while the three-phase loads are usually Δ-connected.

The balanced Y-delta system is shown in Fig. 12.14, where the source is Y-connected and the load is Δ-connected. There is, of course, no neutral connection from source to load for this case. Assuming the positive sequence, the phase voltages are again

$$\mathbf{V}_{an} = V_p\underline{/0^\circ}$$
$$\mathbf{V}_{bn} = V_p\underline{/-120^\circ}, \qquad \mathbf{V}_{cn} = V_p\underline{/+120^\circ} \tag{12.19}$$

As shown in Section 12.3, the line voltages are

$$\mathbf{V}_{ab} = \sqrt{3}V_p\underline{/30^\circ} = \mathbf{V}_{AB}, \qquad \mathbf{V}_{bc} = \sqrt{3}V_p\underline{/-90^\circ} = \mathbf{V}_{BC}$$
$$\mathbf{V}_{ca} = \sqrt{3}V_p\underline{/-150^\circ} = \mathbf{V}_{CA} \tag{12.20}$$

showing that the line voltages are equal to the voltages across the load impedances for this system configuration. From these voltages, we can obtain the phase currents as

$$\mathbf{I}_{AB} = \frac{\mathbf{V}_{AB}}{\mathbf{Z}_\Delta}, \qquad \mathbf{I}_{BC} = \frac{\mathbf{V}_{BC}}{\mathbf{Z}_\Delta}, \qquad \mathbf{I}_{CA} = \frac{\mathbf{V}_{CA}}{\mathbf{Z}_\Delta} \tag{12.21}$$

These currents have the same magnitude but are out of phase with each other by 120°.

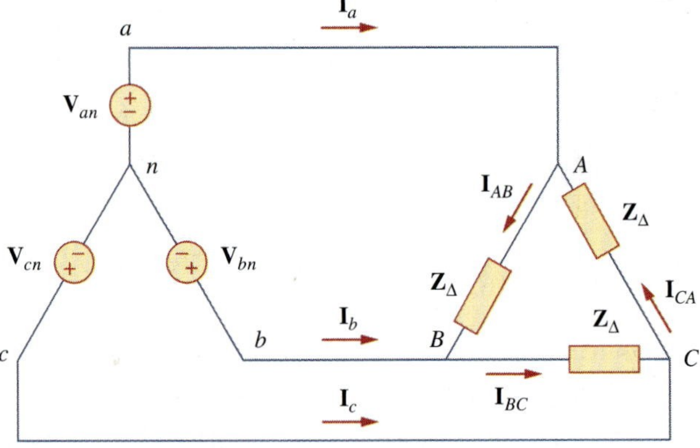

Figure 12.14
Balanced Y-Δ connection.

Another way to get these phase currents is to apply KVL. For example, applying KVL around loop $aABbna$ gives

$$-\mathbf{V}_{an} + \mathbf{Z}_\Delta \mathbf{I}_{AB} + \mathbf{V}_{bn} = 0$$

or

$$\mathbf{I}_{AB} = \frac{\mathbf{V}_{an} - \mathbf{V}_{bn}}{\mathbf{Z}_\Delta} = \frac{\mathbf{V}_{ab}}{\mathbf{Z}_\Delta} = \frac{\mathbf{V}_{AB}}{\mathbf{Z}_\Delta} \tag{12.22}$$

which is the same as Eq. (12.21). This is the more general way of finding the phase currents.

The line currents are obtained from the phase currents by applying KCL at nodes A, B, and C. Thus,

$$\mathbf{I}_a = \mathbf{I}_{AB} - \mathbf{I}_{CA}, \qquad \mathbf{I}_b = \mathbf{I}_{BC} - \mathbf{I}_{AB}, \qquad \mathbf{I}_c = \mathbf{I}_{CA} - \mathbf{I}_{BC} \tag{12.23}$$

Since $\mathbf{I}_{CA} = \mathbf{I}_{AB}\underline{/-240^\circ}$,

$$\begin{aligned}\mathbf{I}_a = \mathbf{I}_{AB} - \mathbf{I}_{CA} &= \mathbf{I}_{AB}(1 - 1\underline{/-240^\circ}) \\ &= \mathbf{I}_{AB}(1 + 0.5 - j0.866) = \mathbf{I}_{AB}\sqrt{3}\underline{/-30^\circ}\end{aligned} \tag{12.24}$$

showing that the magnitude I_L of the line current is $\sqrt{3}$ times the magnitude I_p of the phase current, or

$$\boxed{I_L = \sqrt{3} I_p} \tag{12.25}$$

where

$$I_L = |\mathbf{I}_a| = |\mathbf{I}_b| = |\mathbf{I}_c| \tag{12.26}$$

and

$$I_p = |\mathbf{I}_{AB}| = |\mathbf{I}_{BC}| = |\mathbf{I}_{CA}| \tag{12.27}$$

Also, the line currents lag the corresponding phase currents by 30°, assuming the positive sequence. Figure 12.15 is a phasor diagram illustrating the relationship between the phase and line currents.

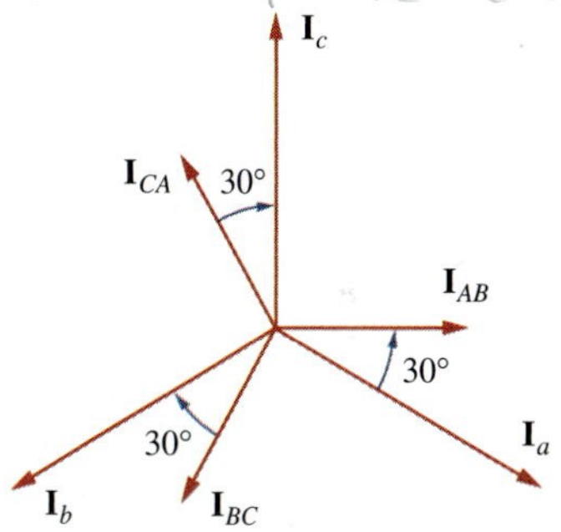

Figure 12.15 Phasor diagram illustrating the relationship between phase and line currents.

An alternative way of analyzing the Y-Δ circuit is to transform the Δ-connected load to an equivalent Y-connected load. Using the Δ-Y transformation formula in Eq. (12.8),

$$\boxed{\mathbf{Z}_Y = \frac{\mathbf{Z}_\Delta}{3}} \tag{12.28}$$

After this transformation, we now have a Y-Y system as in Fig. 12.10. The three-phase Y-Δ system in Fig. 12.14 can be replaced by the single-phase equivalent circuit in Fig. 12.16. This allows us to calculate only the line currents. The phase currents are obtained using Eq. (12.25) and utilizing the fact that each of the phase currents leads the corresponding line current by 30°.

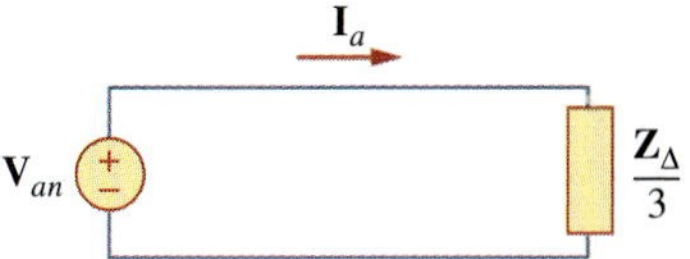

Figure 12.16 A single-phase equivalent circuit of a balanced Y-Δ circuit.

Example 12.3

A balanced abc-sequence Y-connected source with $\mathbf{V}_{an} = 100\underline{/10^\circ}$ V is connected to a Δ-connected balanced load $(8 + j4)\ \Omega$ per phase. Calculate the phase and line currents.

Solution:
This can be solved in two ways.

■ **METHOD 1** The load impedance is

$$\mathbf{Z}_\Delta = 8 + j4 = 8.944\underline{/26.57^\circ}\ \Omega$$

If the phase voltage $\mathbf{V}_{an} = 100\underline{/10^\circ}$, then the line voltage is

$$\mathbf{V}_{ab} = \mathbf{V}_{an}\sqrt{3}\underline{/30^\circ} = 100\sqrt{3}\underline{/10^\circ + 30^\circ} = \mathbf{V}_{AB}$$

or

$$\mathbf{V}_{AB} = 173.2\underline{/40^\circ}\text{ V}$$

The phase currents are

$$\mathbf{I}_{AB} = \frac{\mathbf{V}_{AB}}{\mathbf{Z}_\Delta} = \frac{173.2\underline{/40^\circ}}{8.944\underline{/26.57^\circ}} = 19.36\underline{/13.43^\circ}\text{ A}$$
$$\mathbf{I}_{BC} = \mathbf{I}_{AB}\underline{/-120^\circ} = 19.36\underline{/-106.57^\circ}\text{ A}$$
$$\mathbf{I}_{CA} = \mathbf{I}_{AB}\underline{/+120^\circ} = 19.36\underline{/133.43^\circ}\text{ A}$$

The line currents are

$$\mathbf{I}_a = \mathbf{I}_{AB}\sqrt{3}\underline{/-30^\circ} = \sqrt{3}(19.36)\underline{/13.43^\circ - 30^\circ}$$
$$= 33.53\underline{/-16.57^\circ}\text{ A}$$
$$\mathbf{I}_b = \mathbf{I}_a\underline{/-120^\circ} = 33.53\underline{/-136.57^\circ}\text{ A}$$
$$\mathbf{I}_c = \mathbf{I}_a\underline{/+120^\circ} = 33.53\underline{/103.43^\circ}\text{ A}$$

■ **METHOD 2** Alternatively, using single-phase analysis,

$$\mathbf{I}_a = \frac{\mathbf{V}_{an}}{\mathbf{Z}_\Delta/3} = \frac{100\underline{/10^\circ}}{2.981\underline{/26.57^\circ}} = 33.54\underline{/-16.57^\circ}\text{ A}$$

as above. Other line currents are obtained using the *abc* phase sequence.

Practice Problem 12.3

One line voltage of a balanced Y-connected source is $\mathbf{V}_{AB} = 240\underline{/-20^\circ}$ V. If the source is connected to a Δ-connected load of $20\underline{/40^\circ}\ \Omega$, find the phase and line currents. Assume the *abc* sequence.

Answer: $12\underline{/-60^\circ}$ A, $12\underline{/-180^\circ}$ A, $12\underline{/60^\circ}$ A, $20.79\underline{/-90^\circ}$ A, $20.79\underline{/-150^\circ}$ A, $20.79\underline{/30^\circ}$ A.

12.5 Balanced Delta-Delta Connection

A **balanced Δ-Δ system** is one in which both the balanced source and balanced load are Δ-connected.

The source as well as the load may be delta-connected as shown in Fig. 12.17. Our goal is to obtain the phase and line currents as usual.

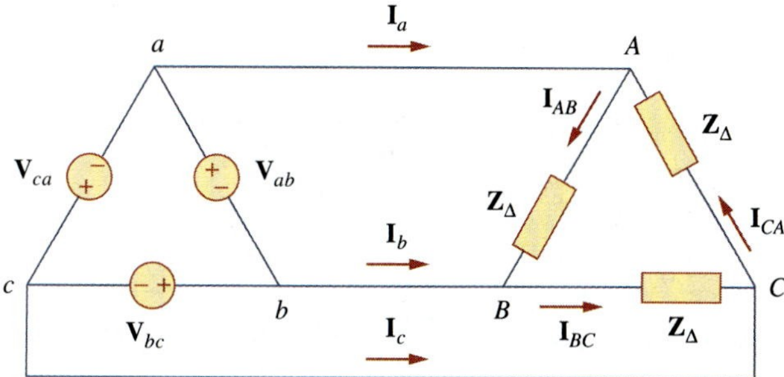

Figure 12.17
A balanced Δ-Δ connection.

Assuming a positive sequence, the phase voltages for a delta-connected source are

$$\begin{gathered} \mathbf{V}_{ab} = V_p\underline{/0^\circ} \\ \mathbf{V}_{bc} = V_p\underline{/-120^\circ}, \qquad \mathbf{V}_{ca} = V_p\underline{/+120^\circ} \end{gathered} \tag{12.29}$$

The line voltages are the same as the phase voltages. From Fig. 12.17, assuming there is no line impedances, the phase voltages of the delta-connected source are equal to the voltages across the impedances; that is,

$$\mathbf{V}_{ab} = \mathbf{V}_{AB}, \qquad \mathbf{V}_{bc} = \mathbf{V}_{BC}, \qquad \mathbf{V}_{ca} = \mathbf{V}_{CA} \tag{12.30}$$

Hence, the phase currents are

$$\begin{gathered} \mathbf{I}_{AB} = \frac{\mathbf{V}_{AB}}{\mathbf{Z}_\Delta} = \frac{\mathbf{V}_{ab}}{\mathbf{Z}_\Delta}, \qquad \mathbf{I}_{BC} = \frac{\mathbf{V}_{BC}}{\mathbf{Z}_\Delta} = \frac{\mathbf{V}_{bc}}{\mathbf{Z}_\Delta} \\ \mathbf{I}_{CA} = \frac{\mathbf{V}_{CA}}{\mathbf{Z}_\Delta} = \frac{\mathbf{V}_{ca}}{\mathbf{Z}_\Delta} \end{gathered} \tag{12.31}$$

Since the load is delta-connected just as in the previous section, some of the formulas derived there apply here. The line currents are obtained from the phase currents by applying KCL at nodes *A*, *B*, and *C*, as we did in the previous section:

$$\mathbf{I}_a = \mathbf{I}_{AB} - \mathbf{I}_{CA}, \qquad \mathbf{I}_b = \mathbf{I}_{BC} - \mathbf{I}_{AB}, \qquad \mathbf{I}_c = \mathbf{I}_{CA} - \mathbf{I}_{BC} \tag{12.32}$$

Also, as shown in the last section, each line current lags the corresponding phase current by 30°; the magnitude I_L of the line current is $\sqrt{3}$ times the magnitude I_p of the phase current,

$$I_L = \sqrt{3} I_p \tag{12.33}$$

An alternative way of analyzing the Δ-Δ circuit is to convert both the source and the load to their Y equivalents. We already know that $\mathbf{Z}_Y = \mathbf{Z}_\Delta/3$. To convert a Δ-connected source to a Y-connected source, see the next section.

Example 12.4

A balanced Δ-connected load having an impedance $20 - j15\ \Omega$ is connected to a Δ-connected, positive-sequence generator having $\mathbf{V}_{ab} = 330\underline{/0^\circ}$ V. Calculate the phase currents of the load and the line currents.

Solution:
The load impedance per phase is

$$\mathbf{Z}_\Delta = 20 - j15 = 25\underline{/-36.87^\circ}\ \Omega$$

Since $\mathbf{V}_{AB} = \mathbf{V}_{ab}$, the phase currents are

$$\mathbf{I}_{AB} = \frac{\mathbf{V}_{AB}}{\mathbf{Z}_\Delta} = \frac{330\underline{/0^\circ}}{25\underline{/-36.87}} = 13.2\underline{/36.87^\circ}\text{ A}$$
$$\mathbf{I}_{BC} = \mathbf{I}_{AB}\underline{/-120^\circ} = 13.2\underline{/-83.13^\circ}\text{ A}$$
$$\mathbf{I}_{CA} = \mathbf{I}_{AB}\underline{/+120^\circ} = 13.2\underline{/156.87^\circ}\text{ A}$$

For a delta load, the line current always lags the corresponding phase current by 30° and has a magnitude $\sqrt{3}$ times that of the phase current. Hence, the line currents are

$$\mathbf{I}_a = \mathbf{I}_{AB}\sqrt{3}\underline{/-30^\circ} = (13.2\underline{/36.87^\circ})(\sqrt{3}\underline{/-30^\circ}) = 22.86\underline{/6.87^\circ}\text{ A}$$
$$\mathbf{I}_b = \mathbf{I}_a\underline{/-120^\circ} = 22.86\underline{/-113.13^\circ}\text{ A}$$
$$\mathbf{I}_c = \mathbf{I}_a\underline{/+120^\circ} = 22.86\underline{/126.87^\circ}\text{ A}$$

Practice Problem 12.4

A positive-sequence, balanced Δ-connected source supplies a balanced Δ-connected load. If the impedance per phase of the load is $18 + j12\ \Omega$ and $\mathbf{I}_a = 19.202\underline{/35^\circ}$ A, find $\mathbf{I}_{AB}$ and $\mathbf{V}_{AB}$.

Answer: $11.094\underline{/65^\circ}$ A, $240\underline{/98.69^\circ}$ V.

12.6 Balanced Delta-Wye Connection

A **balanced Δ-Y system** consists of a balanced Δ-connected source feeding a balanced Y-connected load.

Consider the Δ-Y circuit in Fig. 12.18. Again, assuming the *abc* sequence, the phase voltages of a delta-connected source are

$$\mathbf{V}_{ab} = V_p\underline{/0^\circ}, \qquad \mathbf{V}_{bc} = V_p\underline{/-120^\circ}$$
$$\mathbf{V}_{ca} = V_p\underline{/+120^\circ} \tag{12.34}$$

These are also the line voltages as well as the phase voltages.

We can obtain the line currents in many ways. One way is to apply KVL to loop *aANBba* in Fig. 12.18, writing

$$-\mathbf{V}_{ab} + \mathbf{Z}_Y\mathbf{I}_a - \mathbf{Z}_Y\mathbf{I}_b = 0$$

or

$$\mathbf{Z}_Y(\mathbf{I}_a - \mathbf{I}_b) = \mathbf{V}_{ab} = V_p\underline{/0^\circ}$$

Thus,

$$\mathbf{I}_a - \mathbf{I}_b = \frac{V_p\underline{/0^\circ}}{\mathbf{Z}_Y} \tag{12.35}$$

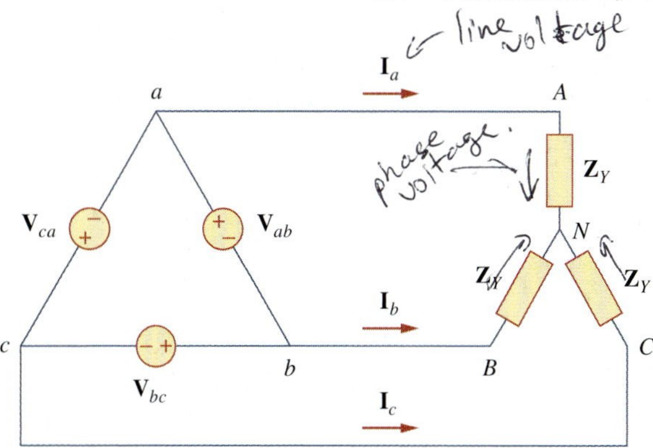

Figure 12.18
A balanced Δ-Y connection.

But $\mathbf{I}_b$ lags $\mathbf{I}_a$ by 120°, since we assumed the *abc* sequence; that is, $\mathbf{I}_b = \mathbf{I}_a\underline{/-120°}$. Hence,

$$\begin{aligned}\mathbf{I}_a - \mathbf{I}_b &= \mathbf{I}_a(1 - 1\underline{/-120°}) \\ &= \mathbf{I}_a\left(1 + \frac{1}{2} + j\frac{\sqrt{3}}{2}\right) = \mathbf{I}_a\sqrt{3}\underline{/30°}\end{aligned} \tag{12.36}$$

Substituting Eq. (12.36) into Eq. (12.35) gives

$$\mathbf{I}_a = \frac{V_p/\sqrt{3}\underline{/-30°}}{\mathbf{Z}_Y} \tag{12.37}$$

From this, we obtain the other line currents $\mathbf{I}_b$ and $\mathbf{I}_c$ using the positive phase sequence, i.e., $\mathbf{I}_b = \mathbf{I}_a\underline{/-120°}$, $\mathbf{I}_c = \mathbf{I}_a\underline{/+120°}$. The phase currents are equal to the line currents.

Another way to obtain the line currents is to replace the delta-connected source with its equivalent wye-connected source, as shown in Fig. 12.19. In Section 12.3, we found that the line-to-line voltages of a wye-connected source lead their corresponding phase voltages by 30°. Therefore, we obtain each phase voltage of the equivalent wye-connected source by dividing the corresponding line voltage of the delta-connected source by $\sqrt{3}$ and shifting its phase by −30°. Thus, the equivalent wye-connected source has the phase voltages

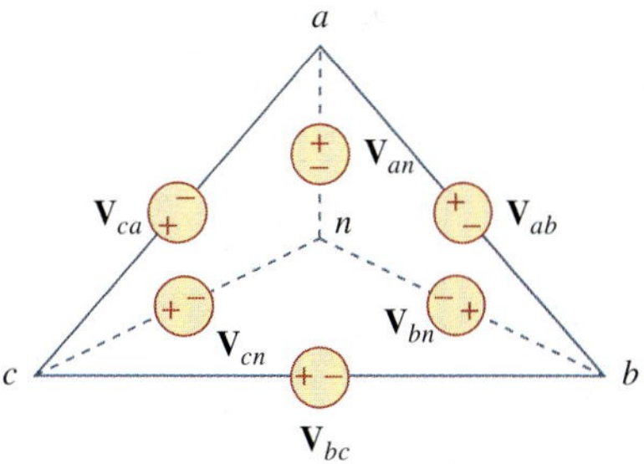

Figure 12.19
Transforming a Δ-connected source to an equivalent Y-connected source.

$$\mathbf{V}_{an} = \frac{V_p}{\sqrt{3}}\underline{/-30°}$$
$$\mathbf{V}_{bn} = \frac{V_p}{\sqrt{3}}\underline{/-150°}, \qquad \mathbf{V}_{cn} = \frac{V_p}{\sqrt{3}}\underline{/+90°} \tag{12.38}$$

If the delta-connected source has source impedance $\mathbf{Z}_s$ per phase, the equivalent wye-connected source will have a source impedance of $\mathbf{Z}_s/3$ per phase, according to Eq. (9.69).

Once the source is transformed to wye, the circuit becomes a wye-wye system. Therefore, we can use the equivalent single-phase circuit shown in Fig. 12.20, from which the line current for phase a is

$$\mathbf{I}_a = \frac{V_p/\sqrt{3}\underline{/-30°}}{\mathbf{Z}_Y} \tag{12.39}$$

which is the same as Eq. (12.37).

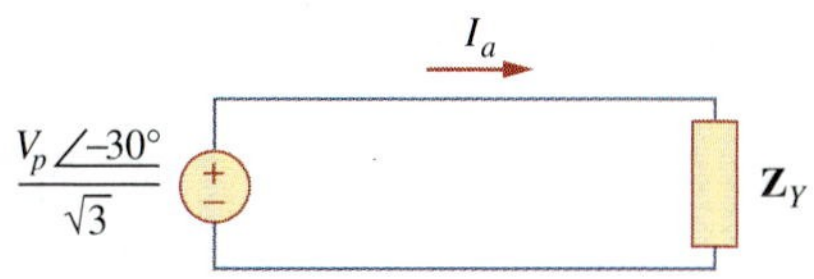

Figure 12.20
The single-phase equivalent circuit.

Alternatively, we may transform the wye-connected load to an equivalent delta-connected load. This results in a delta-delta system, which can be analyzed as in Section 12.5. Note that

$$\mathbf{V}_{AN} = \mathbf{I}_a\mathbf{Z}_Y = \frac{V_p}{\sqrt{3}}\angle{-30^\circ} \qquad \textbf{(12.40)}$$

$$\mathbf{V}_{BN} = \mathbf{V}_{AN}\angle{-120^\circ}, \qquad \mathbf{V}_{CN} = \mathbf{V}_{AN}\angle{+120^\circ}$$

As stated earlier, the delta-connected load is more desirable than the wye-connected load. It is easier to alter the loads in any one phase of the delta-connected loads, as the individual loads are connected directly across the lines. However, the delta-connected source is hardly used in practice, because any slight imbalance in the phase voltages will result in unwanted circulating currents.

Table 12.1 presents a summary of the formulas for phase currents and voltages and line currents and voltages for the four connections. Students are advised not to memorize the formulas but to understand how they are derived. The formulas can always be

TABLE 12.1

Summary of phase and line voltages/currents for balanced three-phase systems.[1]

Connection	Phase voltages/currents	Line voltages/currents
Y-Y	$\mathbf{V}_{an} = V_p\angle{0^\circ}$	$\mathbf{V}_{ab} = \sqrt{3}V_p\angle{30^\circ}$
	$\mathbf{V}_{bn} = V_p\angle{-120^\circ}$	$\mathbf{V}_{bc} = \mathbf{V}_{ab}\angle{-120^\circ}$
	$\mathbf{V}_{cn} = V_p\angle{+120^\circ}$	$\mathbf{V}_{ca} = \mathbf{V}_{ab}\angle{+120^\circ}$
	Same as line currents	$\mathbf{I}_a = \mathbf{V}_{an}/\mathbf{Z}_Y$
		$\mathbf{I}_b = \mathbf{I}_a\angle{-120^\circ}$
		$\mathbf{I}_c = \mathbf{I}_a\angle{+120^\circ}$
Y-Δ	$\mathbf{V}_{an} = V_p\angle{0^\circ}$	$\mathbf{V}_{ab} = \mathbf{V}_{AB} = \sqrt{3}V_p\angle{30^\circ}$
	$\mathbf{V}_{bn} = V_p\angle{-120^\circ}$	$\mathbf{V}_{bc} = \mathbf{V}_{BC} = \mathbf{V}_{ab}\angle{-120^\circ}$
	$\mathbf{V}_{cn} = V_p\angle{+120^\circ}$	$\mathbf{V}_{ca} = \mathbf{V}_{CA} = \mathbf{V}_{ab}\angle{+120^\circ}$
	$\mathbf{I}_{AB} = \mathbf{V}_{AB}/\mathbf{Z}_\Delta$	$\mathbf{I}_a = \mathbf{I}_{AB}\sqrt{3}\angle{-30^\circ}$
	$\mathbf{I}_{BC} = \mathbf{V}_{BC}/\mathbf{Z}_\Delta$	$\mathbf{I}_b = \mathbf{I}_a\angle{-120^\circ}$
	$\mathbf{I}_{CA} = \mathbf{V}_{CA}/\mathbf{Z}_\Delta$	$\mathbf{I}_c = \mathbf{I}_a\angle{+120^\circ}$
Δ-Δ	$\mathbf{V}_{ab} = V_p\angle{0^\circ}$	Same as phase voltages
	$\mathbf{V}_{bc} = V_p\angle{-120^\circ}$	
	$\mathbf{V}_{ca} = V_p\angle{+120^\circ}$	
	$\mathbf{I}_{AB} = \mathbf{V}_{ab}/\mathbf{Z}_\Delta$	$\mathbf{I}_a = \mathbf{I}_{AB}\sqrt{3}\angle{-30^\circ}$
	$\mathbf{I}_{BC} = \mathbf{V}_{bc}/\mathbf{Z}_\Delta$	$\mathbf{I}_b = \mathbf{I}_a\angle{-120^\circ}$
	$\mathbf{I}_{CA} = \mathbf{V}_{ca}/\mathbf{Z}_\Delta$	$\mathbf{I}_c = \mathbf{I}_a\angle{+120^\circ}$
Δ-Y	$\mathbf{V}_{ab} = V_p\angle{0^\circ}$	Same as phase voltages
	$\mathbf{V}_{bc} = V_p\angle{-120^\circ}$	
	$\mathbf{V}_{ca} = V_p\angle{+120^\circ}$	
	Same as line currents	$\mathbf{I}_a = \dfrac{V_p\angle{-30^\circ}}{\sqrt{3}\mathbf{Z}_Y}$
		$\mathbf{I}_b = \mathbf{I}_a\angle{-120^\circ}$
		$\mathbf{I}_c = \mathbf{I}_a\angle{+120^\circ}$

[1] Positive or abc sequence is assumed.

obtained by directly applying KCL and KVL to the appropriate three-phase circuits.

Example 12.5

A balanced Y-connected load with a phase impedance of $40 + j25\ \Omega$ is supplied by a balanced, positive sequence Δ-connected source with a line voltage of 210 V. Calculate the phase currents. Use $\mathbf{V}_{ab}$ as reference.

Solution:

The load impedance is

$$\mathbf{Z}_Y = 40 + j25 = 47.17\underline{/32^\circ}\ \Omega$$

and the source voltage is

$$\mathbf{V}_{ab} = 210\underline{/0^\circ}\ \text{V}$$

When the Δ-connected source is transformed to a Y-connected source,

$$\mathbf{V}_{an} = \frac{\mathbf{V}_{ab}}{\sqrt{3}}\underline{/-30^\circ} = 121.2\underline{/-30^\circ}\ \text{V}$$

The line currents are

$$\begin{aligned}\mathbf{I}_a &= \frac{\mathbf{V}_{an}}{\mathbf{Z}_Y} = \frac{121.2\underline{/-30^\circ}}{47.12\underline{/32^\circ}} = 2.57\underline{/-62^\circ}\ \text{A}\\ \mathbf{I}_b &= \mathbf{I}_a\underline{/-120^\circ} = 2.57\underline{/-178^\circ}\ \text{A}\\ \mathbf{I}_c &= \mathbf{I}_a\underline{/120^\circ} = 2.57\underline{/58^\circ}\ \text{A}\end{aligned}$$

which are the same as the phase currents.

Practice Problem 12.5

In a balanced Δ-Y circuit, $\mathbf{V}_{ab} = 240\underline{/15^\circ}$ and $\mathbf{Z}_Y = (12 + j15)\ \Omega$. Calculate the line currents.

Answer: $7.21\underline{/-66.34^\circ}$ A, $7.21\underline{/-173.66^\circ}$ A, $7.21\underline{/53.66^\circ}$ A.

12.7 Power in a Balanced System

Let us now consider the power in a balanced three-phase system. We begin by examining the instantaneous power absorbed by the load. This requires that the analysis be done in the time domain. For a Y-connected load, the phase voltages are

$$\begin{aligned}v_{AN} &= \sqrt{2}V_p \cos \omega t, \qquad v_{BN} = \sqrt{2}V_p \cos(\omega t - 120^\circ)\\ v_{CN} &= \sqrt{2}V_p \cos(\omega t + 120^\circ)\end{aligned} \tag{12.41}$$

where the factor $\sqrt{2}$ is necessary because V_p has been defined as the rms value of the phase voltage. If $\mathbf{Z}_Y = Z\underline{/\theta}$, the phase currents lag behind their corresponding phase voltages by θ. Thus,

$$\begin{aligned}i_a &= \sqrt{2}I_p \cos(\omega t - \theta), \qquad i_b = \sqrt{2}I_p \cos(\omega t - \theta - 120^\circ)\\ i_c &= \sqrt{2}I_p \cos(\omega t - \theta + 120^\circ)\end{aligned} \tag{12.42}$$

where I_p is the rms value of the phase current. The total instantaneous power in the load is the sum of the instantaneous powers in the three phases; that is,

$$\begin{aligned} p = p_a + p_b + p_c &= v_{AN}i_a + v_{BN}i_b + v_{CN}i_c \\ &= 2V_pI_p[\cos\omega t\cos(\omega t - \theta) \\ &\quad + \cos(\omega t - 120^\circ)\cos(\omega t - \theta - 120^\circ) \\ &\quad + \cos(\omega t + 120^\circ)\cos(\omega t - \theta + 120^\circ)] \end{aligned} \tag{12.43}$$

Applying the trigonometric identity

$$\cos A\cos B = \frac{1}{2}[\cos(A + B) + \cos(A - B)] \tag{12.44}$$

gives

$$\begin{aligned} p &= V_pI_p[3\cos\theta + \cos(2\omega t - \theta) + \cos(2\omega t - \theta - 240^\circ) \\ &\qquad + \cos(2\omega t - \theta + 240^\circ)] \\ &= V_pI_p[3\cos\theta + \cos\alpha + \cos\alpha\cos 240^\circ + \sin\alpha\sin 240^\circ \\ &\qquad + \cos\alpha\cos 240^\circ - \sin\alpha\sin 240^\circ] \\ &\qquad \text{where } \alpha = 2\omega t - \theta \\ &= V_pI_p\left[3\cos\theta + \cos\alpha + 2\left(-\frac{1}{2}\right)\cos\alpha\right] = 3V_pI_p\cos\theta \end{aligned} \tag{12.45}$$

Thus the total instantaneous power in a balanced three-phase system is constant—it does not change with time as the instantaneous power of each phase does. This result is true whether the load is Y- or Δ-connected. This is one important reason for using a three-phase system to generate and distribute power. We will look into another reason a little later.

Since the total instantaneous power is independent of time, the average power per phase P_p for either the Δ-connected load or the Y-connected load is $p/3$, or

$$P_p = V_pI_p\cos\theta \tag{12.46}$$

and the reactive power per phase is

$$Q_p = V_pI_p\sin\theta \tag{12.47}$$

The apparent power per phase is

$$S_p = V_pI_p \tag{12.48}$$

The complex power per phase is

$$\mathbf{S}_p = P_p + jQ_p = \mathbf{V}_p\mathbf{I}_p^* \tag{12.49}$$

where $\mathbf{V}_p$ and $\mathbf{I}_p$ are the phase voltage and phase current with magnitudes V_p and I_p, respectively. The total average power is the sum of the average powers in the phases:

$$P = P_a + P_b + P_c = 3P_p = 3V_pI_p\cos\theta = \sqrt{3}V_LI_L\cos\theta \tag{12.50}$$

For a Y-connected load, $I_L = I_p$ but $V_L = \sqrt{3}V_p$, whereas for a Δ-connected load, $I_L = \sqrt{3}I_p$ but $V_L = V_p$. Thus, Eq. (12.50) applies for both Y-connected and Δ-connected loads. Similarly, the total reactive power is

$$Q = 3V_pI_p\sin\theta = 3Q_p = \sqrt{3}V_LI_L\sin\theta \tag{12.51}$$

and the total complex power is

$$\mathbf{S} = 3\mathbf{S}_p = 3\mathbf{V}_p\mathbf{I}_p^* = 3I_p^2\mathbf{Z}_p = \frac{3V_p^2}{\mathbf{Z}_p^*} \quad \text{(12.52)}$$

where $\mathbf{Z}_p = Z_p\angle\theta$ is the load impedance per phase. ($\mathbf{Z}_p$ could be $\mathbf{Z}_Y$ or $\mathbf{Z}_\Delta$.) Alternatively, we may write Eq. (12.52) as

$$\mathbf{S} = P + jQ = \sqrt{3}V_L I_L \angle\theta \quad \text{(12.53)}$$

Remember that V_p, I_p, V_L, and I_L are all rms values and that θ is the angle of the load impedance or the angle between the phase voltage and the phase current.

A second major advantage of three-phase systems for power distribution is that the three-phase system uses a lesser amount of wire than the single-phase system for the same line voltage V_L and the same absorbed power P_L. We will compare these cases and assume in both that the wires are of the same material (e.g., copper with resistivity ρ), of the same length ℓ, and that the loads are resistive (i.e., unity power factor). For the two-wire single-phase system in Fig. 12.21(a), $I_L = P_L/V_L$, so the power loss in the two wires is

$$P_{\text{loss}} = 2I_L^2 R = 2R\frac{P_L^2}{V_L^2} \quad \text{(12.54)}$$

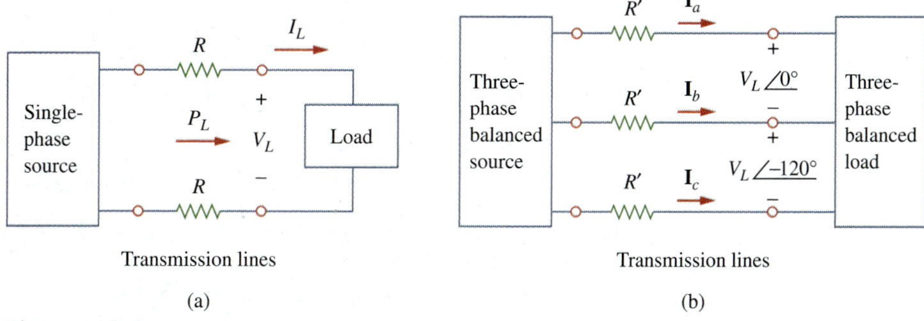

Figure 12.21
Comparing the power loss in (a) a single-phase system, and (b) a three-phase system.

For the three-wire three-phase system in Fig. 12.21(b), $I'_L = |\mathbf{I}_a| = |\mathbf{I}_b| = |\mathbf{I}_c| = P_L/\sqrt{3}V_L$ from Eq. (12.50). The power loss in the three wires is

$$P'_{\text{loss}} = 3(I'_L)^2 R' = 3R'\frac{P_L^2}{3V_L^2} = R'\frac{P_L^2}{V_L^2} \quad \text{(12.55)}$$

Equations (12.54) and (12.55) show that for the same total power delivered P_L and same line voltage V_L,

$$\frac{P_{\text{loss}}}{P'_{\text{loss}}} = \frac{2R}{R'} \quad \text{(12.56)}$$

But from Chapter 2, $R = \rho\ell/\pi r^2$ and $R' = \rho\ell/\pi r'^2$, where r and r' are the radii of the wires. Thus,

$$\frac{P_{\text{loss}}}{P'_{\text{loss}}} = \frac{2r'^2}{r^2} \tag{12.57}$$

If the same power loss is tolerated in both systems, then $r^2 = 2r'^2$. The ratio of material required is determined by the number of wires and their volumes, so

$$\begin{aligned}\frac{\text{Material for single-phase}}{\text{Material for three-phase}} &= \frac{2(\pi r^2\ell)}{3(\pi r'^2\ell)} = \frac{2r^2}{3r'^2} \\ &= \frac{2}{3}(2) = 1.333\end{aligned} \tag{12.58}$$

since $r^2 = 2r'^2$. Equation (12.58) shows that the single-phase system uses 33 percent more material than the three-phase system or that the three-phase system uses only 75 percent of the material used in the equivalent single-phase system. In other words, considerably less material is needed to deliver the same power with a three-phase system than is required for a single-phase system.

Example 12.6

Refer to the circuit in Fig. 12.13 (in Example 12.2). Determine the total average power, reactive power, and complex power at the source and at the load.

Solution:

It is sufficient to consider one phase, as the system is balanced. For phase *a*,

$$\mathbf{V}_p = 110\underline{/0^\circ} \text{ V} \qquad \text{and} \qquad \mathbf{I}_p = 6.81\underline{/-21.8^\circ} \text{ A}$$

Thus, at the source, the complex power absorbed is

$$\begin{aligned}\mathbf{S}_s = -3\mathbf{V}_p\mathbf{I}_p^* &= -3(110\underline{/0^\circ})(6.81\underline{/21.8^\circ}) \\ &= -2247\underline{/21.8^\circ} = -(2087 + j834.6) \text{ VA}\end{aligned}$$

The real or average power absorbed is -2087 W and the reactive power is -834.6 VAR.

At the load, the complex power absorbed is

$$\mathbf{S}_L = 3|\mathbf{I}_p|^2\mathbf{Z}_p$$

where $\mathbf{Z}_p = 10 + j8 = 12.81\underline{/38.66^\circ}$ and $\mathbf{I}_p = \mathbf{I}_a = 6.81\underline{/-21.8^\circ}$. Hence,

$$\begin{aligned}\mathbf{S}_L &= 3(6.81)^2 12.81\underline{/38.66^\circ} = 1782\underline{/38.66} \\ &= (1392 + j1113) \text{ VA}\end{aligned}$$

The real power absorbed is 1391.7 W and the reactive power absorbed is 1113.3 VAR. The difference between the two complex powers is absorbed by the line impedance $(5 - j2)\ \Omega$. To show that this is the case, we find the complex power absorbed by the line as

$$\mathbf{S}_\ell = 3|\mathbf{I}_p|^2\mathbf{Z}_\ell = 3(6.81)^2(5 - j2) = 695.6 - j278.3 \text{ VA}$$

which is the difference between $\mathbf{S}_s$ and $\mathbf{S}_L$; that is, $\mathbf{S}_s + \mathbf{S}_\ell + \mathbf{S}_L = 0$, as expected.

Practice Problem 12.6

For the Y-Y circuit in Practice Prob. 12.2, calculate the complex power at the source and at the load.

Answer: $-(1054 + j843.3)$ VA, $(1012 + j801.6)$ VA.

Example 12.7

A three-phase motor can be regarded as a balanced Y-load. A three-phase motor draws 5.6 kW when the line voltage is 220 V and the line current is 18.2 A. Determine the power factor of the motor.

Solution:

The apparent power is

$$S = \sqrt{3}V_L I_L = \sqrt{3}(220)(18.2) = 6935.13 \text{ VA}$$

Since the real power is

$$P = S\cos\theta = 5600 \text{ W}$$

the power factor is

$$\text{pf} = \cos\theta = \frac{P}{S} = \frac{5600}{6935.13} = 0.8075$$

Practice Problem 12.7

Calculate the line current required for a 30-kW three-phase motor having a power factor of 0.85 lagging if it is connected to a balanced source with a line voltage of 440 V.

Answer: 46.31 A.

Example 12.8

Two balanced loads are connected to a 240-kV rms 60-Hz line, as shown in Fig. 12.22(a). Load 1 draws 30 kW at a power factor of 0.6 lagging, while load 2 draws 45 kVAR at a power factor of 0.8 lagging. Assuming the *abc* sequence, determine: (a) the complex, real, and reactive powers absorbed by the combined load, (b) the line currents, and (c) the kVAR rating of the three capacitors Δ-connected in parallel with the load that will raise the power factor to 0.9 lagging and the capacitance of each capacitor.

Solution:

(a) For load 1, given that $P_1 = 30$ kW and $\cos\theta_1 = 0.6$, then $\sin\theta_1 = 0.8$. Hence,

$$S_1 = \frac{P_1}{\cos\theta_1} = \frac{30 \text{ kW}}{0.6} = 50 \text{ kVA}$$

and $Q_1 = S_1 \sin\theta_1 = 50(0.8) = 40$ kVAR. Thus, the complex power due to load 1 is

$$\mathbf{S}_1 = P_1 + jQ_1 = 30 + j40 \text{ kVA} \tag{12.8.1}$$

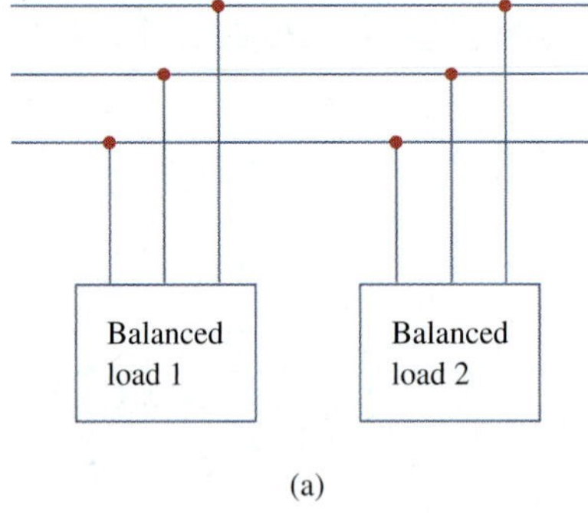

(a)

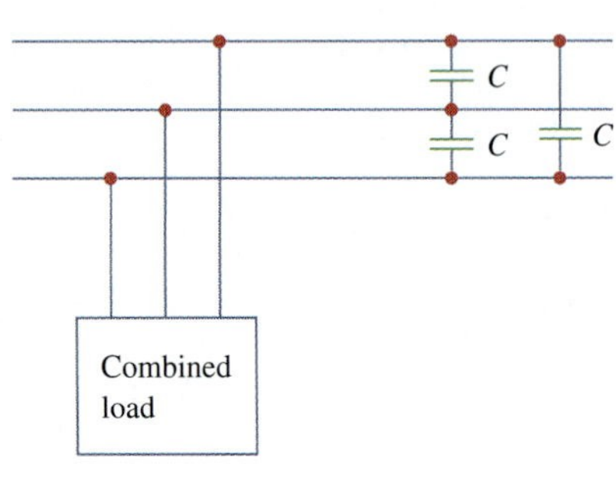

(b)

Figure 12.22
For Example 12.8: (a) The original balanced loads, (b) the combined load with improved power factor.

For load 2, if $Q_2 = 45$ kVAR and $\cos\theta_2 = 0.8$, then $\sin\theta_2 = 0.6$. We find

$$S_2 = \frac{Q_2}{\sin\theta_2} = \frac{45 \text{ kVA}}{0.6} = 75 \text{ kVA}$$

and $P_2 = S_2\cos\theta_2 = 75(0.8) = 60$ kW. Therefore the complex power due to load 2 is

$$\mathbf{S}_2 = P_2 + jQ_2 = 60 + j45 \text{ kVA} \tag{12.8.2}$$

From Eqs. (12.8.1) and (12.8.2), the total complex power absorbed by the load is

$$\mathbf{S} = \mathbf{S}_1 + \mathbf{S}_2 = 90 + j85 \text{ kVA} = 123.8\underline{/43.36^\circ} \text{ kVA} \tag{12.8.3}$$

which has a power factor of cos 43.36° = 0.727 lagging. The real power is then 90 kW, while the reactive power is 85 kVAR.

(b) Since $S = \sqrt{3}V_L I_L$, the line current is

$$I_L = \frac{S}{\sqrt{3}V_L} \tag{12.8.4}$$

We apply this to each load, keeping in mind that for both loads, $V_L =$ 240 kV. For load 1,

$$I_{L1} = \frac{50{,}000}{\sqrt{3}\, 240{,}000} = 120.28 \text{ mA}$$

Since the power factor is lagging, the line current lags the line voltage by $\theta_1 = \cos^{-1} 0.6 = 53.13^\circ$. Thus,

$$\mathbf{I}_{a1} = 120.28\underline{/-53.13^\circ}$$

For load 2,

$$I_{L2} = \frac{75{,}000}{\sqrt{3}\, 240{,}000} = 180.42 \text{ mA}$$

and the line current lags the line voltage by $\theta_2 = \cos^{-1} 0.8 = 36.87^\circ$. Hence,

$$\mathbf{I}_{a2} = 180.42\underline{/-36.87^\circ}$$

The total line current is

$$\begin{aligned}\mathbf{I}_a = \mathbf{I}_{a1} + \mathbf{I}_{a2} &= 120.28\underline{/-53.13^\circ} + 180.42\underline{/-36.87^\circ}\\ &= (72.168 - j96.224) + (144.336 - j108.252)\\ &= 216.5 - j204.472 = 297.8\underline{/-43.36^\circ} \text{ mA}\end{aligned}$$

Alternatively, we could obtain the current from the total complex power using Eq. (12.8.4) as

$$I_L = \frac{123{,}800}{\sqrt{3}\, 240{,}000} = 297.82 \text{ mA}$$

and

$$\mathbf{I}_a = 297.82\underline{/-43.36^\circ} \text{ mA}$$

which is the same as before. The other line currents, $\mathbf{I}_{b2}$ and $\mathbf{I}_{ca}$, can be obtained according to the *abc* sequence (i.e., $\mathbf{I}_b = 297.82\underline{/-163.36^\circ}$ mA and $\mathbf{I}_c = 297.82\underline{/76.64^\circ}$ mA).

(c) We can find the reactive power needed to bring the power factor to 0.9 lagging using Eq. (11.59),

$$Q_C = P(\tan\theta_{\text{old}} - \tan\theta_{\text{new}})$$

where $P = 90$ kW, $\theta_{old} = 43.36°$, and $\theta_{new} = \cos^{-1} 0.9 = 25.84°$. Hence,

$$Q_C = 90{,}000(\tan 43.36° - \tan 25.84°) = 41.4 \text{ kVAR}$$

This reactive power is for the three capacitors. For each capacitor, the rating $Q'_C = 13.8$ kVAR. From Eq. (11.60), the required capacitance is

$$C = \frac{Q'_C}{\omega V_{rms}^2}$$

Since the capacitors are Δ-connected as shown in Fig. 12.22(b), V_{rms} in the above formula is the line-to-line or line voltage, which is 240 kV. Thus,

$$C = \frac{13{,}800}{(2\pi 60)(240{,}000)^2} = 635.5 \text{ pF}$$

Practice Problem 12.8

Assume that the two balanced loads in Fig. 12.22(a) are supplied by an 840-V rms 60-Hz line. Load 1 is Y-connected with $30 + j40\ \Omega$ per phase, while load 2 is a balanced three-phase motor drawing 48 kW at a power factor of 0.8 lagging. Assuming the *abc* sequence, calculate: (a) the complex power absorbed by the combined load, (b) the kVAR rating of each of the three capacitors Δ-connected in parallel with the load to raise the power factor to unity, and (c) the current drawn from the supply at unity power factor condition.

Answer: (a) $56.47 + j47.29$ kVA, (b) 15.7 kVAR, (c) 38.813 A.

12.8 †Unbalanced Three-Phase Systems

This chapter would be incomplete without mentioning unbalanced three-phase systems. An unbalanced system is caused by two possible situations: (1) the source voltages are not equal in magnitude and/or differ in phase by angles that are unequal, or (2) load impedances are unequal. Thus,

> An **unbalanced system** is due to unbalanced voltage sources or an unbalanced load.

To simplify analysis, we will assume balanced source voltages, but an unbalanced load.

Unbalanced three-phase systems are solved by direct application of mesh and nodal analysis. Figure 12.23 shows an example of an unbalanced three-phase system that consists of balanced source voltages (not shown in the figure) and an unbalanced Y-connected load (shown in the figure). Since the load is unbalanced, $\mathbf{Z}_A$, $\mathbf{Z}_B$, and $\mathbf{Z}_C$ are not equal. The line currents are determined by Ohm's law as

$$\mathbf{I}_a = \frac{\mathbf{V}_{AN}}{\mathbf{Z}_A}, \qquad \mathbf{I}_b = \frac{\mathbf{V}_{BN}}{\mathbf{Z}_B}, \qquad \mathbf{I}_c = \frac{\mathbf{V}_{CN}}{\mathbf{Z}_C} \tag{12.59}$$

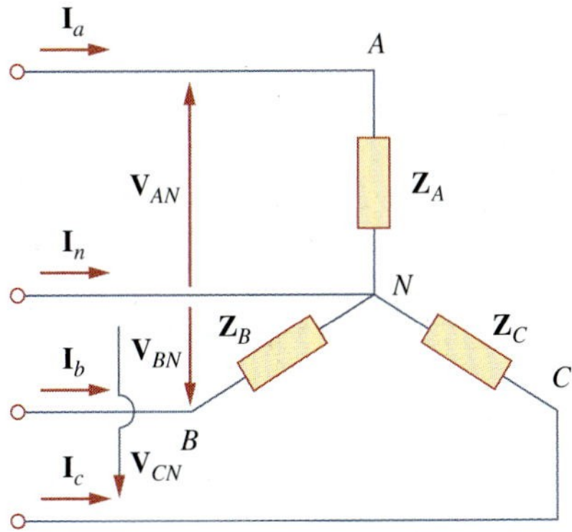

Figure 12.23
Unbalanced three-phase Y-connected load.

A special technique for handling unbalanced three-phase systems is the method of *symmetrical components*, which is beyond the scope of this text.

This set of unbalanced line currents produces current in the neutral line, which is not zero as in a balanced system. Applying KCL at node N gives the neutral line current as

$$\mathbf{I}_n = -(\mathbf{I}_a + \mathbf{I}_b + \mathbf{I}_c) \tag{12.60}$$

In a three-wire system where the neutral line is absent, we can still find the line currents $\mathbf{I}_a$, $\mathbf{I}_b$, and $\mathbf{I}_c$ using mesh analysis. At node N, KCL must be satisfied so that $\mathbf{I}_a + \mathbf{I}_b + \mathbf{I}_c = 0$ in this case. The same could be done for an unbalanced Δ-Y, Y-Δ, or Δ-Δ three-wire system. As mentioned earlier, in long distance power transmission, conductors in multiples of three (multiple three-wire systems) are used, with the earth itself acting as the neutral conductor.

To calculate power in an unbalanced three-phase system requires that we find the power in each phase using Eqs. (12.46) to (12.49). The total power is not simply three times the power in one phase but the sum of the powers in the three phases.

Example 12.9

The unbalanced Y-load of Fig. 12.23 has balanced voltages of 100 V and the *acb* sequence. Calculate the line currents and the neutral current. Take $\mathbf{Z}_A = 15\ \Omega$, $\mathbf{Z}_B = 10 + j5\ \Omega$, $\mathbf{Z}_C = 6 - j8\ \Omega$.

Solution:

Using Eq. (12.59), the line currents are

$$\mathbf{I}_a = \frac{100\angle 0^\circ}{15} = 6.67\angle 0^\circ \text{ A}$$

$$\mathbf{I}_b = \frac{100\angle 120^\circ}{10 + j5} = \frac{100\angle 120^\circ}{11.18\angle 26.56^\circ} = 8.94\angle 93.44^\circ \text{ A}$$

$$\mathbf{I}_c = \frac{100\angle -120^\circ}{6 - j8} = \frac{100\angle -120^\circ}{10\angle -53.13^\circ} = 10\angle -66.87^\circ \text{ A}$$

Using Eq. (12.60), the current in the neutral line is

$$\begin{aligned}\mathbf{I}_n &= -(\mathbf{I}_a + \mathbf{I}_b + \mathbf{I}_c) = -(6.67 - 0.54 + j8.92 + 3.93 - j9.2) \\ &= -10.06 + j0.28 = 10.06\angle 178.4^\circ \text{ A}\end{aligned}$$

Practice Problem 12.9

The unbalanced Δ-load of Fig. 12.24 is supplied by balanced line-to-line voltages of 240 V in the positive sequence. Find the line currents. Take $\mathbf{V}_{ab}$ as reference.

Answer: $21.66\angle -41.06^\circ$ A, $34.98\angle -139.8^\circ$ A, $38.24\angle 74.27^\circ$ A.

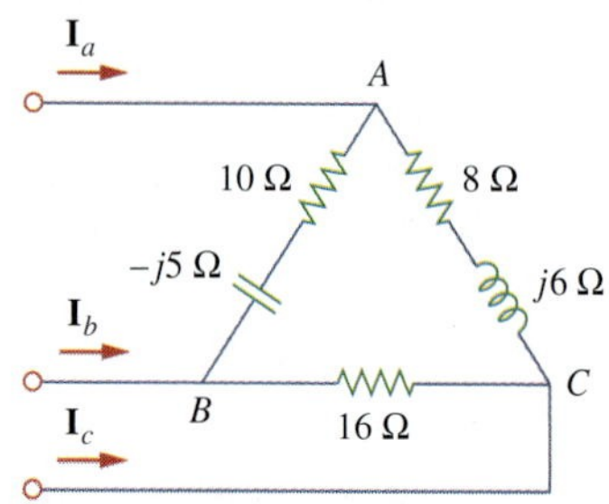

Figure 12.24
Unbalanced Δ-load; for Practice Prob. 12.9.

Example 12.10

For the unbalanced circuit in Fig. 12.25, find: (a) the line currents, (b) the total complex power absorbed by the load, and (c) the total complex power absorbed by the source.

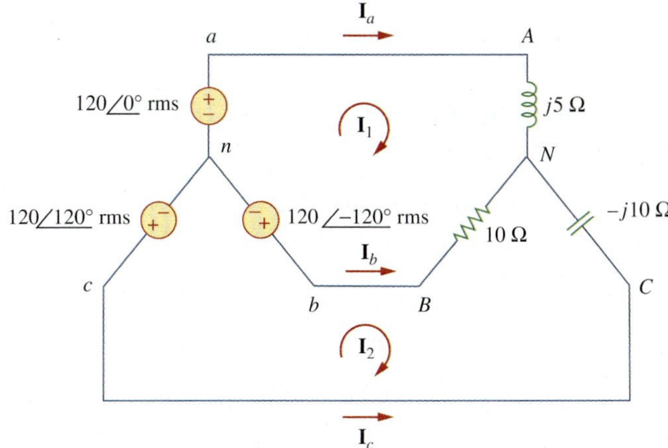

Figure 12.25
For Example 12.10.

Solution:

(a) We use mesh analysis to find the required currents. For mesh 1,

$$120\underline{/-120^\circ} - 120\underline{/0^\circ} + (10 + j5)\mathbf{I}_1 - 10\mathbf{I}_2 = 0$$

or

$$(10 + j5)\mathbf{I}_1 - 10\mathbf{I}_2 = 120\sqrt{3}\underline{/30^\circ} \qquad \textbf{(12.10.1)}$$

For mesh 2,

$$120\underline{/120^\circ} - 120\underline{/-120^\circ} + (10 - j10)\mathbf{I}_2 - 10\mathbf{I}_1 = 0$$

or

$$-10\mathbf{I}_1 + (10 - j10)\mathbf{I}_2 = 120\sqrt{3}\underline{/-90^\circ} \qquad \textbf{(12.10.2)}$$

Equations (12.10.1) and (12.10.2) form a matrix equation:

$$\begin{bmatrix} 10 + j5 & -10 \\ -10 & 10 - j10 \end{bmatrix} \begin{bmatrix} \mathbf{I}_1 \\ \mathbf{I}_2 \end{bmatrix} = \begin{bmatrix} 120\sqrt{3}\underline{/30^\circ} \\ 120\sqrt{3}\underline{/-90^\circ} \end{bmatrix}$$

The determinants are

$$\Delta = \begin{vmatrix} 10 + j5 & -10 \\ -10 & 10 - j10 \end{vmatrix} = 50 - j50 = 70.71\underline{/-45^\circ}$$

$$\Delta_1 = \begin{vmatrix} 120\sqrt{3}\underline{/30^\circ} & -10 \\ 120\sqrt{3}\underline{/-90^\circ} & 10 - j10 \end{vmatrix} = 207.85(13.66 - j13.66)$$

$$= 4015\underline{/-45^\circ}$$

$$\Delta_2 = \begin{vmatrix} 10 + j5 & 120\sqrt{3}\underline{/30^\circ} \\ -10 & 120\sqrt{3}\underline{/-90^\circ} \end{vmatrix} = 207.85(13.66 - j5)$$

$$= 3023.4\underline{/-20.1^\circ}$$

The mesh currents are

$$\mathbf{I}_1 = \frac{\Delta_1}{\Delta} = \frac{4015.23\underline{/-45^\circ}}{70.71\underline{/-45^\circ}} = 56.78 \text{ A}$$

$$\mathbf{I}_2 = \frac{\Delta_2}{\Delta} = \frac{3023.4\underline{/-20.1^\circ}}{70.71\underline{/-45^\circ}} = 42.75\underline{/24.9^\circ} \text{ A}$$

The line currents are

$$\mathbf{I}_a = \mathbf{I}_1 = 56.78 \text{ A}, \qquad \mathbf{I}_c = -\mathbf{I}_2 = 42.75\underline{/-155.1^\circ} \text{ A}$$
$$\mathbf{I}_b = \mathbf{I}_2 - \mathbf{I}_1 = 38.78 + j18 - 56.78 = 25.46\underline{/135^\circ} \text{ A}$$

(b) We can now calculate the complex power absorbed by the load. For phase A,

$$\mathbf{S}_A = |\mathbf{I}_a|^2\mathbf{Z}_A = (56.78)^2(j5) = j16{,}120 \text{ VA}$$

For phase B,

$$\mathbf{S}_B = |\mathbf{I}_b|^2\mathbf{Z}_B = (25.46)^2(10) = 6480 \text{ VA}$$

For phase C,

$$\mathbf{S}_C = |\mathbf{I}_c|^2\mathbf{Z}_C = (42.75)^2(-j10) = -j18{,}276 \text{ VA}$$

The total complex power absorbed by the load is

$$\mathbf{S}_L = \mathbf{S}_A + \mathbf{S}_B + \mathbf{S}_C = 6480 - j2156 \text{ VA}$$

(c) We check the result above by finding the power absorbed by the source. For the voltage source in phase a,

$$\mathbf{S}_a = -\mathbf{V}_{an}\mathbf{I}_a^* = -(120\underline{/0^\circ})(56.78) = -6813.6 \text{ VA}$$

For the source in phase b,

$$\begin{aligned}\mathbf{S}_b = -\mathbf{V}_{bn}\mathbf{I}_b^* &= -(120\underline{/-120^\circ})(25.46\underline{/-135^\circ}) \\ &= -3055.2\underline{/105^\circ} = 790 - j2951.1 \text{ VA}\end{aligned}$$

For the source in phase c,

$$\begin{aligned}\mathbf{S}_c = -\mathbf{V}_{bn}\mathbf{I}_c^* &= -(120\underline{/120^\circ})(42.75\underline{/155.1^\circ}) \\ &= -5130\underline{/275.1^\circ} = -456.03 + j5109.7 \text{ VA}\end{aligned}$$

The total complex power absorbed by the three-phase source is

$$\mathbf{S}_s = \mathbf{S}_a + \mathbf{S}_b + \mathbf{S}_c = -6480 + j2156 \text{ VA}$$

showing that $\mathbf{S}_s + \mathbf{S}_L = 0$ and confirming the conservation principle of ac power.

Practice Problem 12.10

Find the line currents in the unbalanced three-phase circuit of Fig. 12.26 and the real power absorbed by the load.

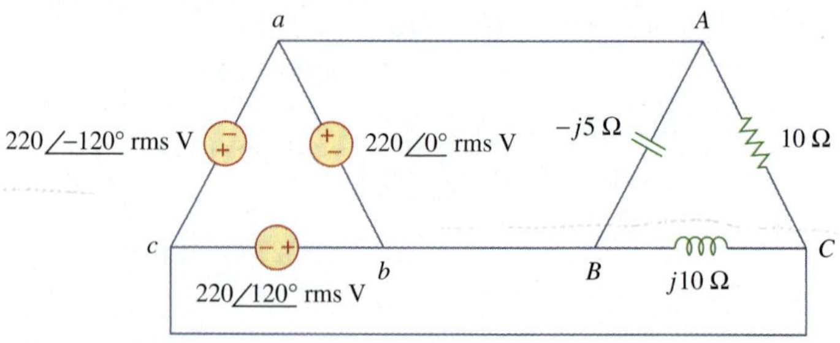

Figure 12.26
For Practice Prob. 12.10.

Answer: $64\underline{/80.1^\circ}$ A, $38.1\underline{/-60^\circ}$ A, $42.5\underline{/225^\circ}$ A, 4.84 kW.

12.9 *PSpice* for Three-Phase Circuits

PSpice can be used to analyze three-phase balanced or unbalanced circuits in the same way it is used to analyze single-phase ac circuits. However, a delta-connected source presents two major problems to *PSpice*. First, a delta-connected source is a loop of voltage sources—which *PSpice* does not like. To avoid this problem, we insert a resistor of negligible resistance (say, 1 $\mu\Omega$ per phase) into each phase of the delta-connected source. Second, the delta-connected source does not provide a convenient node for the ground node, which is necessary to run *PSpice*. This problem can be eliminated by inserting balanced wye-connected large resistors (say, 1 MΩ per phase) in the delta-connected source so that the neutral node of the wye-connected resistors serves as the ground node 0. Example 12.12 will illustrate this.

Example 12.11

For the balanced Y-Δ circuit in Fig. 12.27, use *PSpice* to find the line current $\mathbf{I}_{aA}$, the phase voltage $\mathbf{V}_{AB}$, and the phase current $\mathbf{I}_{AC}$. Assume that the source frequency is 60 Hz.

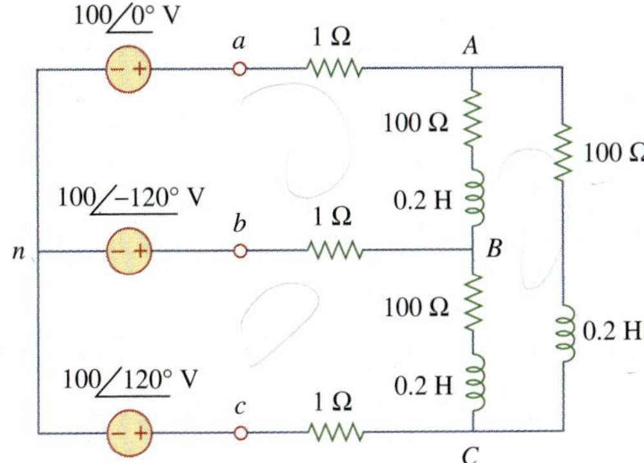

Figure 12.27
For Example 12.11.

Solution:

The schematic is shown in Fig. 12.28. The pseudocomponents IPRINT are inserted in the appropriate lines to obtain $\mathbf{I}_{aA}$ and $\mathbf{I}_{AC}$, while VPRINT2 is inserted between nodes A and B to print differential voltage $\mathbf{V}_{AB}$. We set the attributes of IPRINT and VPRINT2 each to *AC* = *yes*, *MAG* = *yes*, *PHASE* = *yes*, to print only the magnitude and phase of the currents and voltages. As a single-frequency analysis, we select **Analysis/Setup/AC Sweep** and enter *Total Pts* = 1, *Start Freq* = 60, and *Final Freq* = 60. Once the circuit is saved, it is simulated by selecting **Analysis/Simulate**. The output file includes the following:

```
FREQ          V(A,B)          VP(A,B)
6.000E+01     1.699E+02       3.081E+01

FREQ          IM(V_PRINT2)    IP(V_PRINT2)
6.000E+01     2.350E+00       -3.620E+01

FREQ          IM(V_PRINT3)    IP(V_PRINT3)
6.000E+01     1.357E+00       -6.620E+01
```

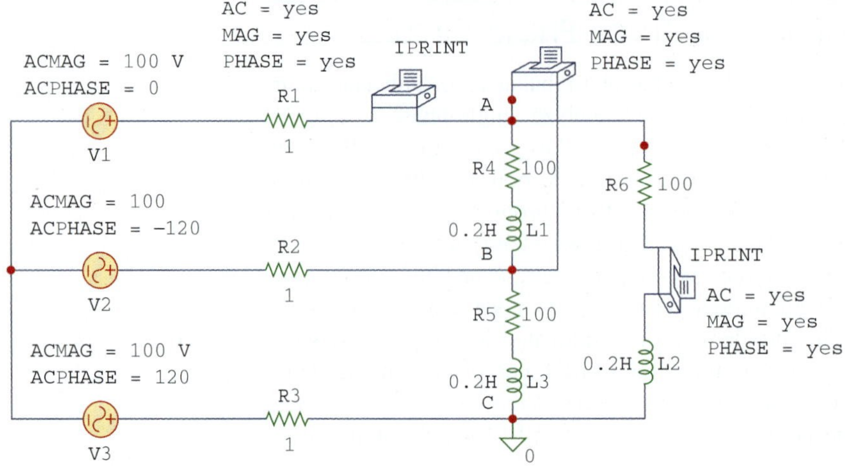

Figure 12.28
Schematic for the circuit in Fig. 12.27.

From this, we obtain

$$\mathbf{I}_{aA} = 2.35\underline{/-36.2^\circ} \text{ A}$$
$$\mathbf{V}_{AB} = 169.9\underline{/30.81^\circ} \text{ V}, \quad \mathbf{I}_{AC} = 1.357\underline{/-66.2^\circ} \text{ A}$$

Practice Problem 12.11

Refer to the balanced Y-Y circuit of Fig. 12.29. Use *PSpice* to find the line current $\mathbf{I}_{bB}$ and the phase voltage $\mathbf{V}_{AN}$. Take $f = 100$ Hz.

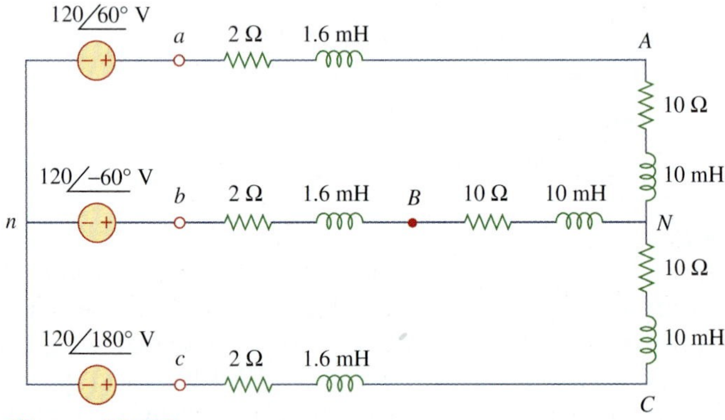

Figure 12.29
For Practice Prob. 12.11.

Answer: $100.9\underline{/60.87^\circ}$ V, $8.547\underline{/-91.27^\circ}$ A.

Example 12.12

Consider the unbalanced Δ-Δ circuit in Fig. 12.30. Use *PSpice* to find the generator current $\mathbf{I}_{ab}$, the line current $\mathbf{I}_{bB}$, and the phase current $\mathbf{I}_{BC}$.

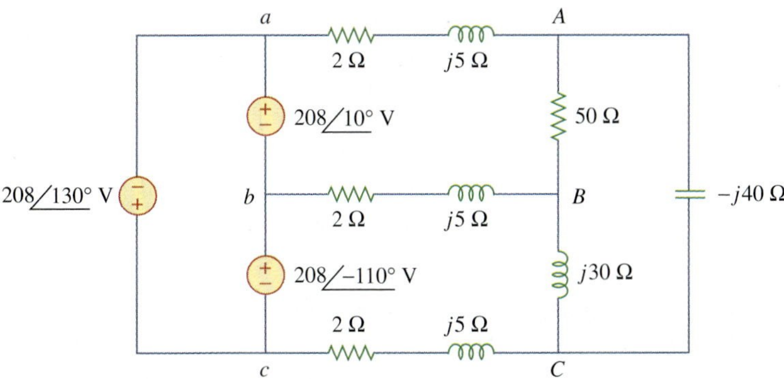

Figure 12.30
For Example 12.12.

Solution:

1. **Define.** The problem and solution process are clearly defined.
2. **Present.** We are to find the generator current flowing from a to b, the line current flowing from b to B, and the phase current flowing from B to C.
3. **Alternative.** Although, there are different approaches to solving this problem, the use of *PSpice* is mandated. Therefore, we will not use another approach.
4. **Attempt.** As mentioned above, we avoid the loop of voltage sources by inserting a 1-$\mu\Omega$ series resistor in the delta-connected source. To provide a ground node 0, we insert balanced wye-connected resistors (1 MΩ per phase) in the delta-connected source, as shown in the schematic in Fig. 12.31. Three IPRINT pseudocomponents with their

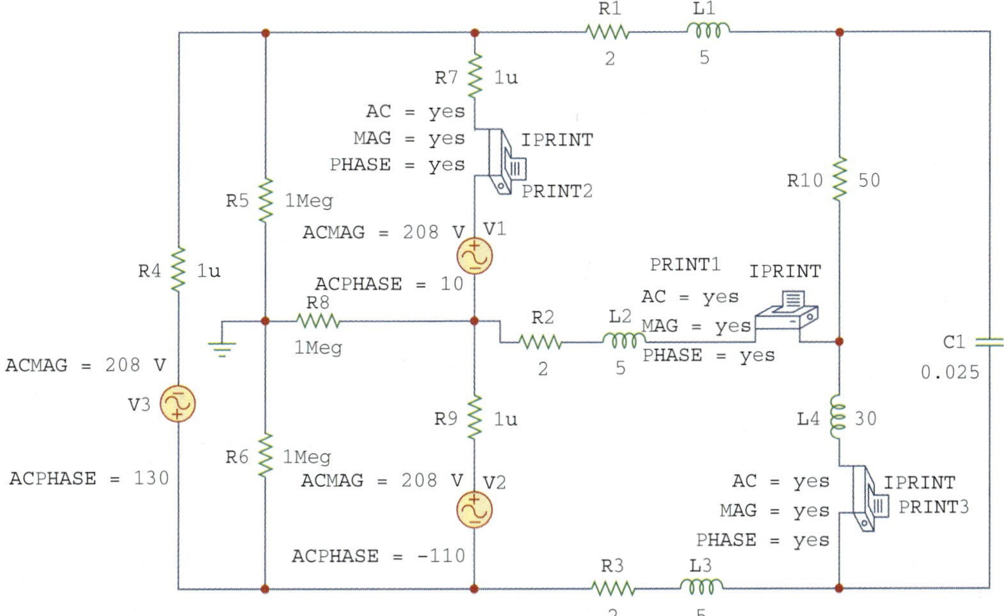

Figure 12.31
Schematic for the circuit in Fig. 12.30.

attributes are inserted to be able to get the required currents $\mathbf{I}_{ab}$, $\mathbf{I}_{bB}$, and $\mathbf{I}_{BC}$. Since the operating frequency is not given and the inductances and capacitances should be specified instead of impedances, we assume $\omega = 1$ rad/s so that $f = 1/2\pi = 0.159155$ Hz. Thus,

$$L = \frac{X_L}{\omega} \quad \text{and} \quad C = \frac{1}{\omega X_C}$$

We select **Analysis/Setup/AC Sweep** and enter *Total Pts* = 1, *Start Freq* = 0.159155, and *Final Freq* = 0.159155. Once the schematic is saved, we select **Analysis/Simulate** to simulate the circuit. The output file includes:

```
FREQ          IM(V_PRINT1)    IP(V_PRINT1)
1.592E-01     9.106E+00       1.685E+02

FREQ          IM(V_PRINT2)    IP(V_PRINT2)
1.592E-01     5.959E+00       -1.772E+02

FREQ          IM(V_PRINT3)    IP(V_PRINT3)
1.592E-01     5.500E+00       1.725E+02
```

which yields

$$I_{ab} = \mathbf{5.595\underline{/-177.2^\circ}\ A},\ I_{bB} = \mathbf{9.106\underline{/168.5^\circ}\ A}, \text{ and}$$
$$I_{BC} = \mathbf{5.5\underline{/172.5^\circ}\ A}$$

5. **Evaluate.** We can check our results by using mesh analysis. Let the loop *aABb* be loop 1, the loop *bBCc* be loop 2, and the loop *ACB* be loop 3, with the three loop currents all flowing in the clockwise direction. We then end up with the following loop equations:

Loop 1

$$(54 + j10)I_1 - (2 + j5)I_2 - (50)I_3 = 208\underline{/10^\circ} = 204.8 + j36.12$$

Loop 2

$$-(2 + j5)I_1 + (4 + j40)I_2 - (j30)I_3 = 208\underline{/-110^\circ} = -71.14 - j195.46$$

Loop 3

$$-(50)I_1 - (j30)I_2 + (50 - j10)I_3 = 0$$

Using MATLAB to solve this we get,

```
>>Z=[(54+10i),(-2-5i),-50;(-2-5i),(4+40i),
-30i;-50,-30i,(50-10i)]

Z=

54.0000+10.0000i-2.0000-5.0000i-50.0000
-2.0000-5.0000i 4.0000+40.0000i 0-30.0000i
-50.0000 0-30.0000i 50.0000-10.0000i
```

```
>>V=[(204.8+36.12i);(-71.14-195.46i);0]

V=

1.0e+002*
2.0480+0.3612i
-0.7114-1.9546i
    0
>>I=inv(Z)*V

I=

8.9317+2.6983i
0.0096+4.5175i
5.4619+3.7964i
```

$$I_{bB} = -I_1 + I_2 = -(8.932 + j2.698) + (0.0096 + j4.518)$$
$$= -8.922 + j1.82 = \mathbf{9.106\underline{/168.47^\circ}\ A} \qquad \text{Answer checks}$$
$$I_{BC} = I_2 - I_3 = (0.0096 + j4.518) - (5.462 + j3.796)$$
$$= -5.452 + j0.722 = \mathbf{5.5\underline{/172.46^\circ}\ A} \qquad \text{Answer checks}$$

Now to solve for I_{ab}. If we assume a small internal impedance for each source, we can obtain a reasonably good estimate for I_{ab}. Adding in internal resistors of 0.01Ω, and adding a fourth loop around the source circuit, we now get

Loop 1

$$(54.01 + j10)I_1 - (2 + j5)I_2 - (50)I_3 - 0.01I_4 = 208\underline{/10^\circ}$$
$$= 204.8 + j36.12$$

Loop 2

$$-(2 + j5)I_1 + (4.01 + j40)I_2 - (j30)I_3 - 0.01I_4$$
$$= 208\underline{/-110^\circ} = -71.14 - j195.46$$

Loop 3

$$-(50)I_1 - (j30)I_2 + (50 - j10)I_3 = 0$$

Loop 4

$$-(0.01)I_1 - (0.01)I_2 + (0.03)I_4 = 0$$

```
>>Z=[(54.01+10i),(-2-5i),-50,-0.01;(-2-5i),
(4.01+40i),-30i,-0.01;-50,-30i,(50-10i),
0;-0.01,-0.01,0,0.03]

Z=

54.0100+10.0000i  -2.0000-5.0000i,  -50.0000  -0.0100
-2.0000-5.0000i  4.0100-40.0000i  0-30.0000i  0.0100
-50.0000  0-30.0000i  50.0000-10.0000i  0
-0.0100  -0.0100  0  0.0300

>>V=[(204.8+36.12i);(-71.14-195.46i);0;0]
```

```
V=

1.0e+002*

2.0480+0.3612i
-0.7114-1.9546i
     0
     0
>>I=inv(Z)*V

I=

8.9309+2.6973i
0.0093+4.5159i
5.4623+3.7954i
2.9801+2.4044i
```

$$I_{ab} = -I_1 + I_4 = -(8.931 + j2.697) + (2.98 + j2.404)$$
$$= -5.951 - j0.293 = \mathbf{5.958\underline{/-177.18^\circ}\ A}. \qquad \text{Answer checks.}$$

6. **Satisfactory?** We have a satisfactory solution and an adequate check for the solution. We can now present the results as a solution to the problem.

Practice Problem 12.12

For the unbalanced circuit in Fig. 12.32, use *PSpice* to find the generator current $\mathbf{I}_{ca}$, the line current $\mathbf{I}_{cC}$, and the phase current $\mathbf{I}_{AB}$.

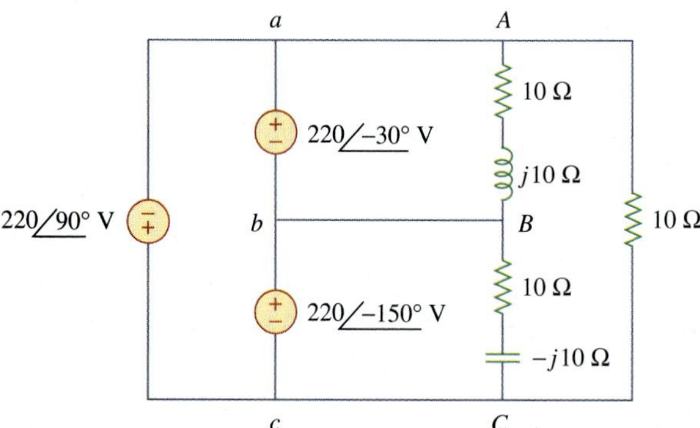

Figure 12.32
For Practice Prob. 12.12.

Answer: $24.68\underline{/-90^\circ}$ A, $37.25\underline{/83.8^\circ}$ A, $15.556\underline{/-75^\circ}$ A.

12.10 †Applications

Both wye and delta source connections have important practical applications. The wye source connection is used for long distance transmission of electric power, where resistive losses (I^2R) should be minimal. This is due to the fact that the wye connection gives a line voltage that is $\sqrt{3}$ greater than the delta connection; hence, for the same power, the line current is $\sqrt{3}$ smaller. The delta source connection is used when three single-phase circuits are desired from a three-phase source. This conversion from three-phase to single-phase is required in residential wiring, because household lighting and appliances use single-phase power. Three-phase power is used in industrial wiring where a large power is required. In some applications, it is immaterial whether the load is wye- or delta-connected. For example, both connections are satisfactory with induction motors. In fact, some manufacturers connect a motor in delta for 220 V and in wye for 440 V so that one line of motors can be readily adapted to two different voltages.

Here we consider two practical applications of those concepts covered in this chapter: power measurement in three-phase circuits and residential wiring.

12.10.1 Three-Phase Power Measurement

Section 11.9 presented the wattmeter as the instrument for measuring the average (or real) power in single-phase circuits. A single wattmeter can also measure the average power in a three-phase system that is balanced, so that $P_1 = P_2 = P_3$; the total power is three times the reading of that one wattmeter. However, two or three single-phase wattmeters are necessary to measure power if the system is unbalanced. The *three-wattmeter method* of power measurement, shown in Fig. 12.33, will work regardless of whether the load is balanced or unbalanced, wye- or delta-connected. The three-wattmeter method is well suited for power measurement in a three-phase system where the power factor is constantly changing. The total average power is the algebraic sum of the three wattmeter readings,

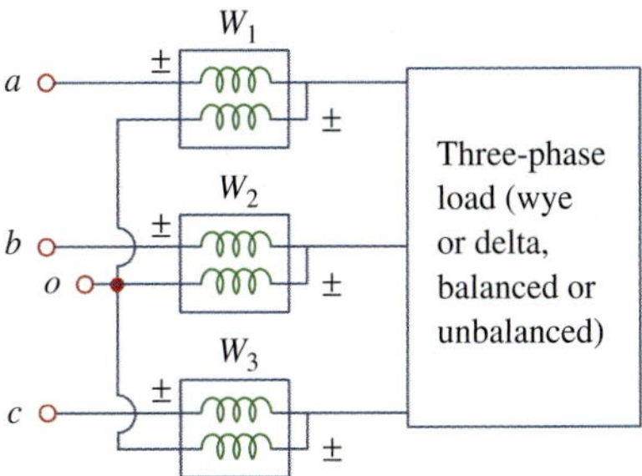

Figure 12.33
Three-wattmeter method for measuring three-phase power.

$$P_T = P_1 + P_2 + P_3 \tag{12.61}$$

where P_1, P_2, and P_3 correspond to the readings of wattmeters W_1, W_2, and W_3, respectively. Notice that the common or reference point o in Fig. 12.33 is selected arbitrarily. If the load is wye-connected, point o can be connected to the neutral point n. For a delta-connected load, point o can be connected to any point. If point o is connected to point b, for example, the voltage coil in wattmeter W_2 reads zero and $P_2 = 0$, indicating that wattmeter W_2 is not necessary. Thus, two wattmeters are sufficient to measure the total power.

The *two-wattmeter method* is the most commonly used method for three-phase power measurement. The two wattmeters must be properly connected to any two phases, as shown typically in Fig. 12.34. Notice that the current coil of each wattmeter measures the line current, while the respective voltage coil is connected between the line and the third line and measures the line voltage. Also notice that the ± terminal of the voltage coil is connected to the line to which the corresponding current coil is connected. Although the individual wattmeters no longer read the power taken by any particular phase, the algebraic sum of the two wattmeter readings equals the total average power absorbed by the load, regardless of whether it is wye- or delta-connected, balanced or

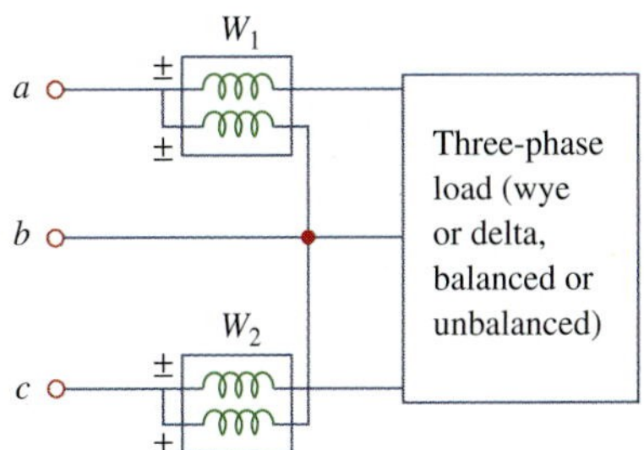

Figure 12.34
Two-wattmeter method for measuring three-phase power.

unbalanced. The total real power is equal to the algebraic sum of the two wattmeter readings,

$$P_T = P_1 + P_2 \tag{12.62}$$

We will show here that the method works for a balanced three-phase system.

Consider the balanced, wye-connected load in Fig. 12.35. Our objective is to apply the two-wattmeter method to find the average power absorbed by the load. Assume the source is in the *abc* sequence and the load impedance $\mathbf{Z}_Y = Z_Y\underline{/\theta}$. Due to the load impedance, each voltage coil leads its current coil by θ, so that the power factor is $\cos\theta$. We recall that each line voltage leads the corresponding phase voltage by 30°. Thus, the total phase difference between the phase current $\mathbf{I}_a$ and line voltage $\mathbf{V}_{ab}$ is $\theta + 30°$, and the average power read by wattmeter W_1 is

$$P_1 = \text{Re}[\mathbf{V}_{ab}\mathbf{I}_a^*] = V_{ab}I_a\cos(\theta + 30°) = V_L I_L \cos(\theta + 30°) \tag{12.63}$$

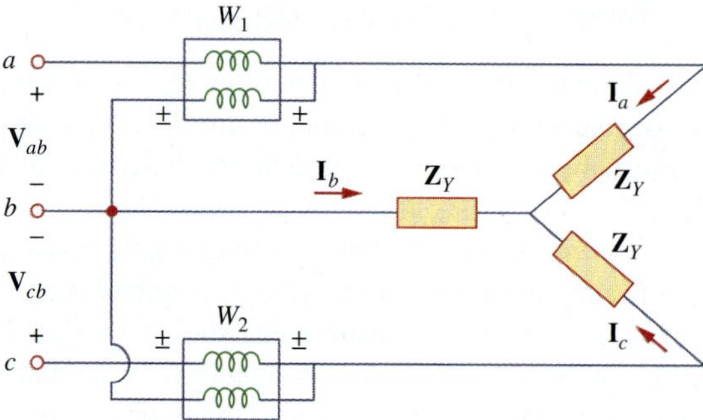

Figure 12.35
Two-wattmeter method applied to a balanced wye load.

Similarly, we can show that the average power read by wattmeter 2 is

$$P_2 = \text{Re}[\mathbf{V}_{cb}\mathbf{I}_c^*] = V_{cb}I_c\cos(\theta - 30°) = V_L I_L \cos(\theta - 30°) \tag{12.64}$$

We now use the trigonometric identities

$$\begin{aligned}\cos(A + B) &= \cos A \cos B - \sin A \sin B\\ \cos(A - B) &= \cos A \cos B + \sin A \sin B\end{aligned} \tag{12.65}$$

to find the sum and the difference of the two wattmeter readings in Eqs. (12.63) and (12.64):

$$\begin{aligned}P_1 + P_2 &= V_L I_L[\cos(\theta + 30°) + \cos(\theta - 30°)]\\ &= V_L I_L(\cos\theta\cos 30° - \sin\theta\sin 30°\\ &\qquad + \cos\theta\cos 30° + \sin\theta\sin 30°)\\ &= V_L I_L 2\cos 30°\cos\theta = \sqrt{3}V_L I_L\cos\theta\end{aligned} \tag{12.66}$$

since $2\cos 30° = \sqrt{3}$. Comparing Eq. (12.66) with Eq. (12.50) shows that the sum of the wattmeter readings gives the total average power,

$$\boxed{P_T = P_1 + P_2} \tag{12.67}$$

Similarly,

$$\begin{aligned}P_1 - P_2 &= V_L I_L[\cos(\theta + 30^\circ) - \cos(\theta - 30^\circ)] \\ &= V_l I_L(\cos\theta \cos 30^\circ - \sin\theta \sin 30^\circ \\ &\qquad - \cos\theta \cos 30^\circ - \sin\theta \sin 30^\circ) \\ &= -V_L I_L 2 \sin 30^\circ \sin\theta \\ P_2 - P_1 &= V_L I_L \sin\theta\end{aligned} \tag{12.68}$$

since $2 \sin 30^\circ = 1$. Comparing Eq. (12.68) with Eq. (12.51) shows that the difference of the wattmeter readings is proportional to the total reactive power, or

$$Q_T = \sqrt{3}(P_2 - P_1) \tag{12.69}$$

From Eqs. (12.67) and (12.69), the total apparent power can be obtained as

$$S_T = \sqrt{P_T^2 + Q_T^2} \tag{12.70}$$

Dividing Eq. (12.69) by Eq. (12.67) gives the tangent of the power factor angle as

$$\tan\theta = \frac{Q_T}{P_T} = \sqrt{3}\frac{P_2 - P_1}{P_2 + P_1} \tag{12.71}$$

from which we can obtain the power factor as pf $= \cos\theta$. Thus, the two-wattmeter method not only provides the total real and reactive powers, it can also be used to compute the power factor. From Eqs. (12.67), (12.69), and (12.71), we conclude that:

1. If $P_2 = P_1$, the load is resistive.
2. If $P_2 > P_1$, the load is inductive.
3. If $P_2 < P_1$, the load is capacitive.

Although these results are derived from a balanced wye-connected load, they are equally valid for a balanced delta-connected load. However, the two-wattmeter method cannot be used for power measurement in a three-phase four-wire system unless the current through the neutral line is zero. We use the three-wattmeter method to measure the real power in a three-phase four-wire system.

Example 12.13

Three wattmeters W_1, W_2, and W_3 are connected, respectively, to phases a, b, and c to measure the total power absorbed by the unbalanced wye-connected load in Example 12.9 (see Fig. 12.23). (a) Predict the wattmeter readings. (b) Find the total power absorbed.

Solution:

Part of the problem is already solved in Example 12.9. Assume that the wattmeters are properly connected as in Fig. 12.36.

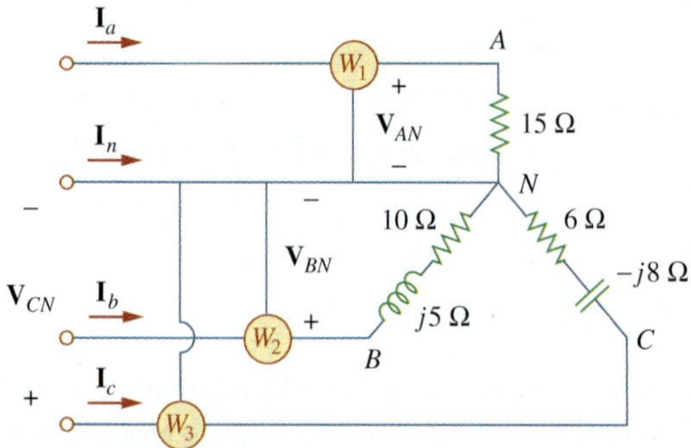

Figure 12.36
For Example 12.13.

(a) From Example 12.9,

$$\mathbf{V}_{AN} = 100\underline{/0^\circ}, \qquad \mathbf{V}_{BN} = 100\underline{/120^\circ}, \quad \mathbf{V}_{CN} = 100\underline{/-120^\circ} \text{ V}$$

while

$$\mathbf{I}_a = 6.67\underline{/0^\circ}, \qquad \mathbf{I}_b = 8.94\underline{/93.44^\circ}, \qquad \mathbf{I}_c = 10\underline{/-66.87^\circ} \text{ A}$$

We calculate the wattmeter readings as follows:

$$\begin{aligned}
P_1 &= \text{Re}(\mathbf{V}_{AN}\mathbf{I}_a^*) = V_{AN}I_a\cos(\theta_{\mathbf{V}_{AN}} - \theta_{\mathbf{I}_a}) \\
&= 100 \times 6.67 \times \cos(0^\circ - 0^\circ) = 667 \text{ W} \\
P_2 &= \text{Re}(\mathbf{V}_{BN}\mathbf{I}_b^*) = V_{BN}I_b\cos(\theta_{\mathbf{V}_{BN}} - \theta_{\mathbf{I}_b}) \\
&= 100 \times 8.94 \times \cos(120^\circ - 93.44^\circ) = 800 \text{ W} \\
P_3 &= \text{Re}(\mathbf{V}_{CN}\mathbf{I}_c^*) = V_{CN}I_c\cos(\theta_{\mathbf{V}_{CN}} - \theta_{\mathbf{I}_c}) \\
&= 100 \times 10 \times \cos(-120^\circ + 66.87^\circ) = 600 \text{ W}
\end{aligned}$$

(b) The total power absorbed is

$$P_T = P_1 + P_2 + P_3 = 667 + 800 + 600 = 2067 \text{ W}$$

We can find the power absorbed by the resistors in Fig. 12.36 and use that to check or confirm this result.

$$\begin{aligned}
P_T &= |I_a|^2(15) + |I_b|^2(10) + |I_c|^2(6) \\
&= 6.67^2(15) + 8.94^2(10) + 10^2(6) \\
&= 667 + 800 + 600 = 2067 \text{ W}
\end{aligned}$$

which is exactly the same thing.

Practice Problem 12.13

Repeat Example 12.13 for the network in Fig. 12.24 (see Practice Prob. 12.9). *Hint:* Connect the reference point *o* in Fig. 12.33 to point *B*.

Answer: (a) 3.92 kW, 0 W, 8.895 kW, (b) 12.815 kW.

Example 12.14

The two-wattmeter method produces wattmeter readings $P_1 = 1560$ W and $P_2 = 2100$ W when connected to a delta-connected load. If the line voltage is 220 V, calculate: (a) the per-phase average power, (b) the per-phase reactive power, (c) the power factor, and (d) the phase impedance.

Solution:

We can apply the given results to the delta-connected load. (a) The total real or average power is

$$P_T = P_1 + P_2 = 1560 + 2100 = 3660 \text{ W}$$

The per-phase average power is then

$$P_p = \frac{1}{3}P_T = 1220 \text{ W}$$

(b) The total reactive power is

$$Q_T = \sqrt{3}(P_2 - P_1) = \sqrt{3}(2100 - 1560) = 935.3 \text{ VAR}$$

so that the per-phase reactive power is

$$Q_p = \frac{1}{3}Q_T = 311.77 \text{ VAR}$$

(c) The power angle is

$$\theta = \tan^{-1}\frac{Q_T}{P_T} = \tan^{-1}\frac{935.3}{3660} = 14.33^\circ$$

Hence, the power factor is

$$\cos\theta = 0.9689 \text{ (lagging)}$$

It is a lagging pf because Q_T is positive or $P_2 > P_1$.
(c) The phase impedance is $\mathbf{Z}_p = Z_p\underline{/\theta}$. We know that θ is the same as the pf angle; that is, $\theta = 14.33^\circ$.

$$Z_p = \frac{V_p}{I_p}$$

We recall that for a delta-connected load, $V_p = V_L = 220$ V. From Eq. (12.46),

$$P_p = V_p I_p \cos\theta \qquad \Rightarrow \qquad I_p = \frac{1220}{220 \times 0.9689} = 5.723 \text{ A}$$

Hence,

$$Z_p = \frac{V_p}{I_p} = \frac{220}{5.723} = 38.44 \ \Omega$$

and

$$\mathbf{Z}_p = 38.44\underline{/14.33^\circ} \ \Omega$$

Practice Problem 12.14

Let the line voltage $V_L = 208$ V and the wattmeter readings of the balanced system in Fig. 12.35 be $P_1 = -560$ W and $P_2 = 800$ W. Determine:

(a) the total average power
(b) the total reactive power
(c) the power factor
(d) the phase impedance

Is the impedance inductive or capacitive?

Answer: (a) 240 W, (b) 2355.6 VAR, (c) 0.1014, (d) $18.25\underline{/84.18^\circ}$ Ω, inductive.

Example 12.15

The three-phase balanced load in Fig. 12.35 has impedance per phase of $\mathbf{Z}_Y = 8 + j6\ \Omega$. If the load is connected to 208-V lines, predict the readings of the wattmeters W_1 and W_2. Find P_T and Q_T.

Solution:

The impedance per phase is

$$\mathbf{Z}_Y = 8 + j6 = 10\underline{/36.87^\circ}\ \Omega$$

so that the pf angle is 36.87°. Since the line voltage $V_L = 208$ V, the line current is

$$I_L = \frac{V_p}{|\mathbf{Z}_Y|} = \frac{208/\sqrt{3}}{10} = 12\text{ A}$$

Then

$$\begin{aligned}P_1 &= V_L I_L \cos(\theta + 30^\circ) = 208 \times 12 \times \cos(36.87^\circ + 30^\circ)\\ &= 980.48\text{ W}\\ P_2 &= V_L I_L \cos(\theta - 30^\circ) = 208 \times 12 \times \cos(36.87^\circ - 30^\circ)\\ &= 2478.1\text{ W}\end{aligned}$$

Thus, wattmeter 1 reads 980.48 W, while wattmeter 2 reads 2478.1 W. Since $P_2 > P_1$, the load is inductive. This is evident from the load $\mathbf{Z}_Y$ itself. Next,

$$P_T = P_1 + P_2 = 3.459\text{ kW}$$

and

$$Q_T = \sqrt{3}(P_2 - P_1) = \sqrt{3}(1497.6)\text{ VAR} = 2.594\text{ kVAR}$$

Practice Problem 12.15

If the load in Fig. 12.35 is delta-connected with impedance per phase of $\mathbf{Z}_p = 30 - j40\ \Omega$ and $V_L = 440$ V, predict the readings of the wattmeters W_1 and W_2. Calculate P_T and Q_T.

Answer: 6.166 kW, 0.8021 kW, 6.968 kW, −9.291 kVAR.

12.10.2 Residential Wiring

In the United States, most household lighting and appliances operate on 120-V, 60-Hz, single-phase alternating current. (The electricity may also be supplied at 110, 115, or 117 V, depending on the location.) The local power company supplies the house with a three-wire ac system. Typically, as in Fig. 12.37, the line voltage of, say, 12,000 V is stepped down to 120/240 V with a transformer (more details on transformers in the next chapter). The three wires coming from the transformer are typically colored red (hot), black (hot), and white (neutral). As shown in Fig. 12.38, the two 120-V voltages are opposite in phase and hence add up to zero. That is, $\mathbf{V}_W = 0\underline{/0^\circ}$, $\mathbf{V}_B = 120\underline{/0^\circ}$, $\mathbf{V}_R = 120\underline{/180^\circ} = -\mathbf{V}_B$.

$$\mathbf{V}_{BR} = \mathbf{V}_B - \mathbf{V}_R = \mathbf{V}_B - (-\mathbf{V}_B) = 2\mathbf{V}_B = 240\underline{/0^\circ} \qquad \textbf{(12.72)}$$

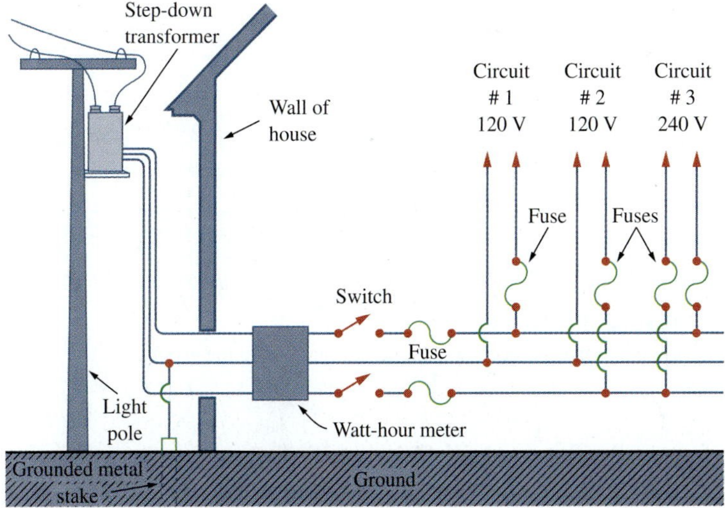

Figure 12.37
A 120/240 household power system.
A. Marcus and C. M. Thomson, *Electricity for Technicians,* 2nd ed. [Englewood Cliffs, NJ: Prentice Hall, 1975], p. 324.

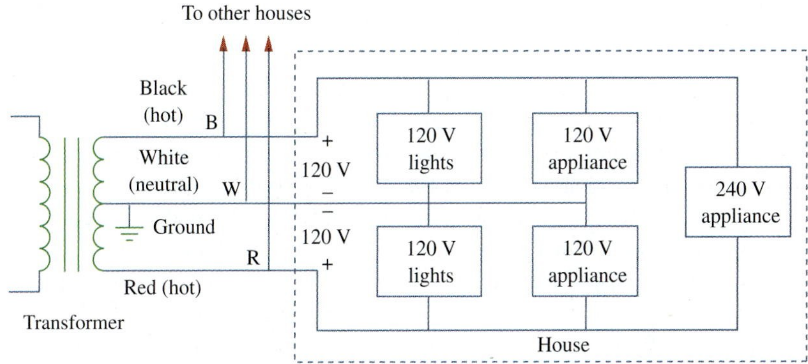

Figure 12.38
Single-phase three-wire residential wiring.

Since most appliances are designed to operate with 120 V, the lighting and appliances are connected to the 120-V lines, as illustrated in Fig. 12.39 for a room. Notice in Fig. 12.37 that all appliances are connected in parallel. Heavy appliances that consume large currents, such as air conditioners, dishwashers, ovens, and laundry machines, are connected to the 240-V power line.

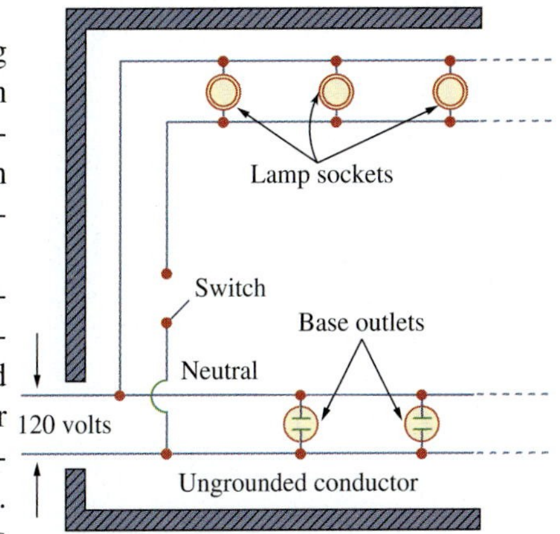

Figure 12.39
A typical wiring diagram of a room.
A. Marcus and C. M. Thomson, *Electricity for Technicians,* 2nd ed. [Englewood Cliffs, NJ: Prentice Hall, 1975], p. 325.

Because of the dangers of electricity, house wiring is carefully regulated by a code drawn by local ordinances and by the National Electrical Code (NEC). To avoid trouble, insulation, grounding, fuses, and circuit breakers are used. Modern wiring codes require a third wire for a separate ground. The ground wire does not carry power like the neutral wire but enables appliances to have a separate ground connection. Figure 12.40 shows the connection of the receptacle to a 120-V rms line and to the ground. As shown in the figure, the neutral line is connected to the ground (the earth) at many critical locations. Although the ground line seems redundant, grounding is important for many reasons. First, it is required by NEC. Second, grounding provides a

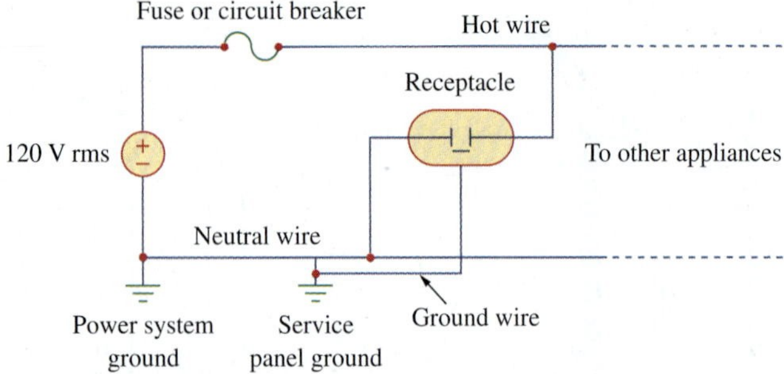

Figure 12.40
Connection of a receptacle to the hot line and to the ground.

convenient path to ground for lightning that strikes the power line. Third, grounds minimize the risk of electric shock. What causes shock is the passage of current from one part of the body to another. The human body is like a big resistor R. If V is the potential difference between the body and the ground, the current through the body is determined by Ohm's law as

$$I = \frac{V}{R} \tag{12.73}$$

The value of R varies from person to person and depends on whether the body is wet or dry. How great or how deadly the shock is depends on the amount of current, the pathway of the current through the body, and the length of time the body is exposed to the current. Currents less than 1 mA may not be harmful to the body, but currents greater than 10 mA can cause severe shock. A modern safety device is the *ground-fault circuit interrupter* (GFCI), used in outdoor circuits and in bathrooms, where the risk of electric shock is greatest. It is essentially a circuit breaker that opens when the sum of the currents i_R, i_W, and i_B through the red, white, and the black lines is not equal to zero, or $i_R + i_W + i_B \neq 0$.

The best way to avoid electric shock is to follow safety guidelines concerning electrical systems and appliances. Here are some of them:

- Never assume that an electrical circuit is dead. Always check to be sure.
- Use safety devices when necessary, and wear suitable clothing (insulated shoes, gloves, etc.).
- Never use two hands when testing high-voltage circuits, since the current through one hand to the other hand has a direct path through your chest and heart.
- Do not touch an electrical appliance when you are wet. Remember that water conducts electricity.
- Be extremely careful when working with electronic appliances such as radio and TV because these appliances have large capacitors in them. The capacitors take time to discharge after the power is disconnected.
- Always have another person present when working on a wiring system, just in case of an accident.

12.11 Summary

1. The phase sequence is the order in which the phase voltages of a three-phase generator occur with respect to time. In an *abc* sequence of balanced source voltages, $\mathbf{V}_{an}$ leads $\mathbf{V}_{bn}$ by 120°, which in turn leads $\mathbf{V}_{cn}$ by 120°. In an *acb* sequence of balanced voltages, $\mathbf{V}_{an}$ leads $\mathbf{V}_{cn}$ by 120°, which in turn leads $\mathbf{V}_{bn}$ by 120°.
2. A balanced wye- or delta-connected load is one in which the three-phase impedances are equal.
3. The easiest way to analyze a balanced three-phase circuit is to transform both the source and the load to a Y-Y system and then analyze the single-phase equivalent circuit. Table 12.1 presents a summary of the formulas for phase currents and voltages and line currents and voltages for the four possible configurations.
4. The line current I_L is the current flowing from the generator to the load in each transmission line in a three-phase system. The line voltage V_L is the voltage between each pair of lines, excluding the neutral line if it exists. The phase current I_p is the current flowing through each phase in a three-phase load. The phase voltage V_p is the voltage of each phase. For a wye-connected load,

$$V_L = \sqrt{3}V_p \qquad \text{and} \qquad I_L = I_p$$

For a delta-connected load,

$$V_L = V_p \qquad \text{and} \qquad I_L = \sqrt{3}I_p$$

5. The total instantaneous power in a balanced three-phase system is constant and equal to the average power.
6. The total complex power absorbed by a balanced three-phase Y-connected or Δ-connected load is

$$\mathbf{S} = P + jQ = \sqrt{3}V_L I_L \angle\theta$$

where θ is the angle of the load impedances.

7. An unbalanced three-phase system can be analyzed using nodal or mesh analysis.
8. *PSpice* is used to analyze three-phase circuits in the same way as it is used for analyzing single-phase circuits.
9. The total real power is measured in three-phase systems using either the three-wattmeter method or the two-wattmeter method.
10. Residential wiring uses a 120/240-V, single-phase, three-wire system.

Review Questions

12.1 What is the phase sequence of a three-phase motor for which $\mathbf{V}_{AN} = 220\angle{-100^\circ}$ V and $\mathbf{V}_{BN} = 220\angle{140^\circ}$ V?

(a) *abc* (b) *acb*

12.2 If in an *acb* phase sequence, $V_{an} = 100\angle{-20^\circ}$, then $\mathbf{V}_{cn}$ is:

(a) $100\angle{-140^\circ}$ (b) $100\angle{100^\circ}$
(c) $100\angle{-50^\circ}$ (d) $100\angle{10^\circ}$

12.3 Which of these is not a required condition for a balanced system:

(a) $|\mathbf{V}_{an}| = |\mathbf{V}_{bn}| = |\mathbf{V}_{cn}|$
(b) $\mathbf{I}_a + \mathbf{I}_b + \mathbf{I}_c = 0$
(c) $V_{an} + V_{bn} + V_{cn} = 0$
(d) Source voltages are 120° out of phase with each other.
(e) Load impedances for the three phases are equal.

12.4 In a Y-connected load, the line current and phase current are equal.

(a) True (b) False

12.5 In a Δ-connected load, the line current and phase current are equal.

(a) True (b) False

12.6 In a Y-Y system, a line voltage of 220 V produces a phase voltage of:

(a) 381 V (b) 311 V (c) 220 V
(d) 156 V (e) 127 V

12.7 In a Δ-Δ system, a phase voltage of 100 V produces a line voltage of:

(a) 58 V (b) 71 V (c) 100 V
(d) 173 V (e) 141 V

12.8 When a Y-connected load is supplied by voltages in *abc* phase sequence, the line voltages lag the corresponding phase voltages by 30°.

(a) True (b) False

12.9 In a balanced three-phase circuit, the total instantaneous power is equal to the average power.

(a) True (b) False

12.10 The total power supplied to a balanced Δ-load is found in the same way as for a balanced Y-load.

(a) True (b) False

Answers: 12.1a, 12.2a, 12.3c, 12.4a, 12.5b, 12.6e, 12.7c, 12.8b, 12.9a, 12.10a.

Problems[1]

Section 12.2 Balanced Three-Phase Voltages

12.1 If $\mathbf{V}_{ab} = 400$ V in a balanced Y-connected three-phase generator, find the phase voltages, assuming the phase sequence is:

(a) *abc* (b) *acb*

12.2 What is the phase sequence of a balanced three-phase circuit for which $\mathbf{V}_{an} = 160\underline{/30^\circ}$ V and $\mathbf{V}_{cn} = 160\underline{/-90^\circ}$ V? Find $\mathbf{V}_{bn}$.

12.3 Determine the phase sequence of a balanced three-phase circuit in which $\mathbf{V}_{bn} = 208\underline{/130^\circ}$ V and $\mathbf{V}_{cn} = 208\underline{/10^\circ}$ V. Obtain $\mathbf{V}_{an}$.

12.4 A three-phase system with *abc* sequence and $\mathbf{V}_L = 200$ V feeds a Y-connected load with $Z_L = 40\underline{/30^\circ}$ Ω. Find the line currents.

12.5 For a Y-connected load, the time-domain expressions for three line-to-neutral voltages at the terminals are:

$$v_{AN} = 150\cos(\omega t + 32^\circ)\text{ V}$$
$$v_{BN} = 150\cos(\omega t - 88^\circ)\text{ V}$$
$$v_{CN} = 150\cos(\omega t + 152^\circ)\text{ V}$$

Write the time-domain expressions for the line-to-line voltages v_{AB}, v_{BC}, and v_{CA}.

Section 12.3 Balanced Wye-Wye Connection

12.6 Using Fig. 12.41, design a problem to help other students better understand balanced wye-wye connected circuits.

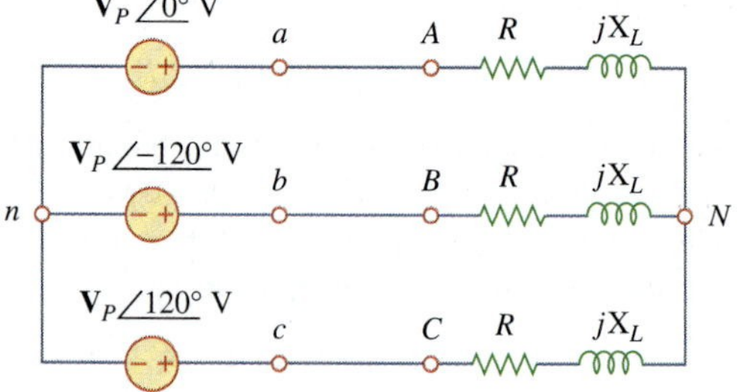

Figure 12.41
For Prob. 12.6.

12.7 Obtain the line currents in the three-phase circuit of Fig. 12.42 on the next page.

12.8 In a balanced three-phase Y-Y system, the source is an *abc* sequence of voltages and $\mathbf{V}_{an} = 220\underline{/20^\circ}$ V rms. The line impedance per phase is $0.6 + j1.2$ Ω, while the per-phase impedance of the load is $10 + j14$ Ω. Calculate the line currents and the load voltages.

12.9 A balanced Y-Y four-wire system has phase voltages

$$\mathbf{V}_{an} = 120\underline{/0^\circ}, \quad \mathbf{V}_{bn} = 120\underline{/-120^\circ}$$
$$\mathbf{V}_{cn} = 120\underline{/120^\circ}\text{ V}$$

The load impedance per phase is $19 + j13$ Ω, and the line impedance per phase is $1 + j2$ Ω. Solve for the line currents and neutral current.

[1] Remember that unless stated otherwise, all given voltages and currents are rms values.

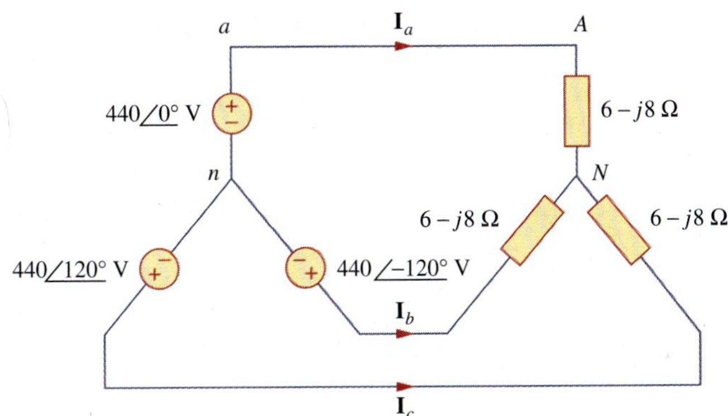

Figure 12.42
For Prob. 12.7.

12.10 For the circuit in Fig. 12.43, determine the current in the neutral line.

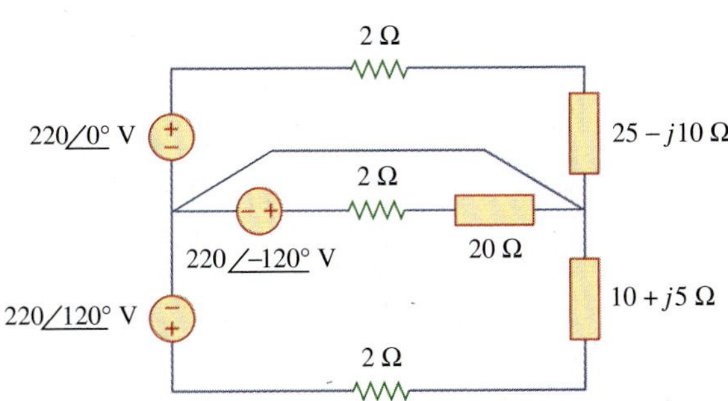

Figure 12.43
For Prob. 12.10.

Section 12.4 Balanced Wye-Delta Connection

12.11 In the Y-Δ system shown in Fig. 12.44, the source is a positive sequence with $\mathbf{V}_{an} = 120\underline{/0^\circ}$ V and phase impedance $\mathbf{Z}_p = 2 - j3\ \Omega$. Calculate the line voltage $\mathbf{V}_L$ and the line current $\mathbf{I}_L$.

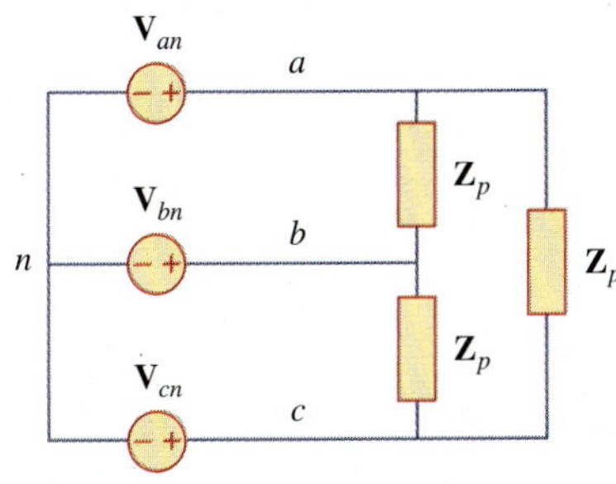

Figure 12.44
For Prob. 12.11.

12.12 Using Fig. 12.45, design a problem to help other students better understand wye-delta connected circuits.

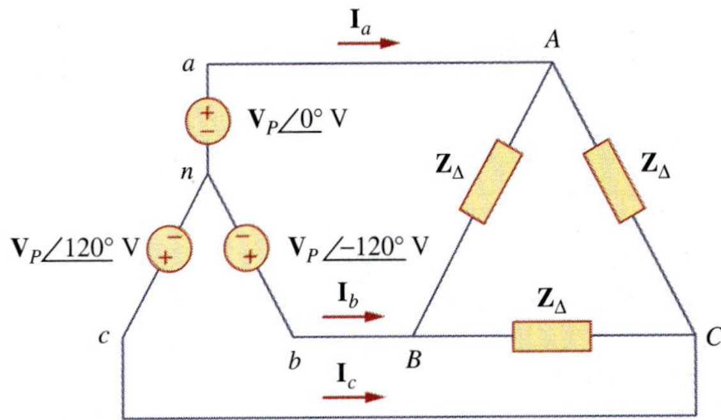

Figure 12.45
For Prob. 12.12.

12.13 In the balanced three-phase Y-Δ system in Fig. 12.46, find the line current I_L and the average power delivered to the load.

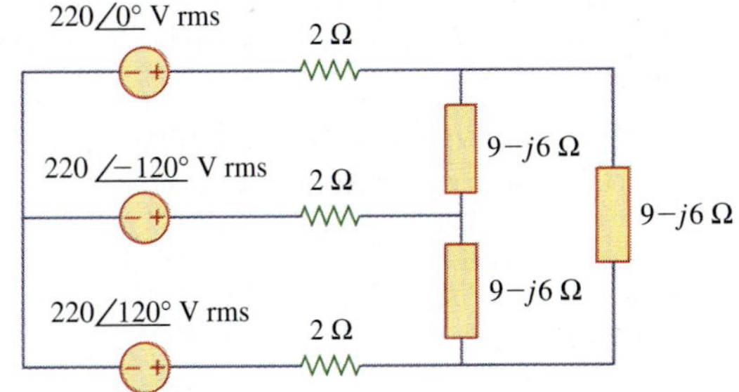

Figure 12.46
For Prob. 12.13.

12.14 Obtain the line currents in the three-phase circuit of Fig. 12.47 on the next page.

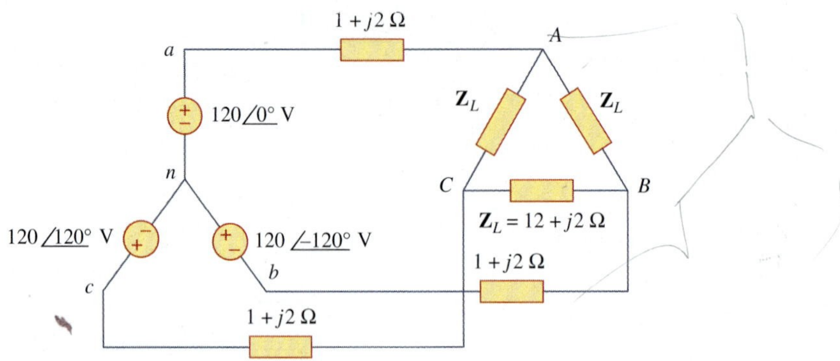

Figure 12.47
For Prob. 12.14.

12.15 The circuit in Fig. 12.48 is excited by a balanced three-phase source with a line voltage of 210 V. If $\mathbf{Z}_l = 1 + j1\ \Omega$, $\mathbf{Z}_\Delta = 24 - j30\ \Omega$, and $\mathbf{Z}_Y = 12 + j5\ \Omega$, determine the magnitude of the line current of the combined loads.

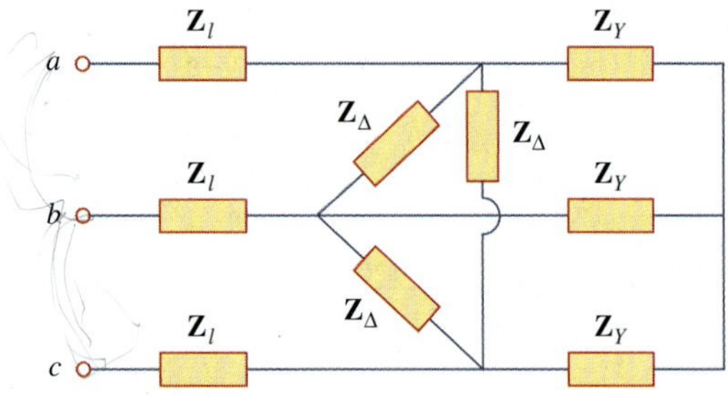

Figure 12.48
For Prob. 12.15.

12.16 A balanced delta-connected load has a phase current $\mathbf{I}_{AC} = 10\angle{-30^\circ}$ A.

(a) Determine the three line currents assuming that the circuit operates in the positive phase sequence.

(b) Calculate the load impedance if the line voltage is $\mathbf{V}_{AB} = 110\angle{0^\circ}$ V.

12.17 A balanced delta-connected load has line current $\mathbf{I}_a = 10\angle{-25^\circ}$ A. Find the phase currents $\mathbf{I}_{AB}$, $\mathbf{I}_{BC}$, and $\mathbf{I}_{CA}$.

12.18 If $\mathbf{V}_{an} = 440\angle{60^\circ}$ V in the network of Fig. 12.49, find the load phase currents $\mathbf{I}_{AB}$, $\mathbf{I}_{BC}$, and $\mathbf{I}_{CA}$.

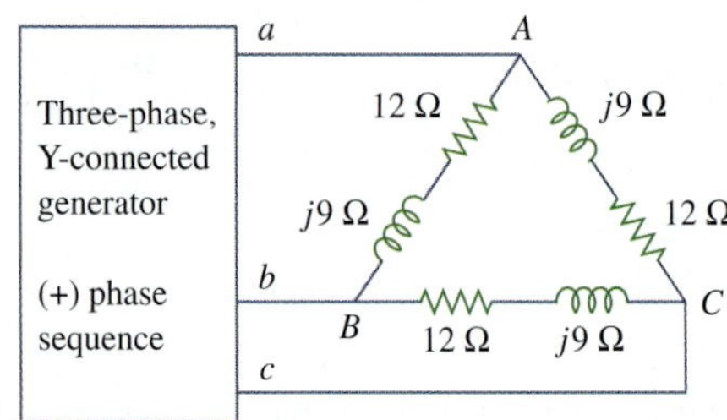

Figure 12.49
For Prob. 12.18.

Section 12.5 Balanced Delta-Delta Connection

12.19 For the Δ-Δ circuit of Fig. 12.50, calculate the phase and line currents.

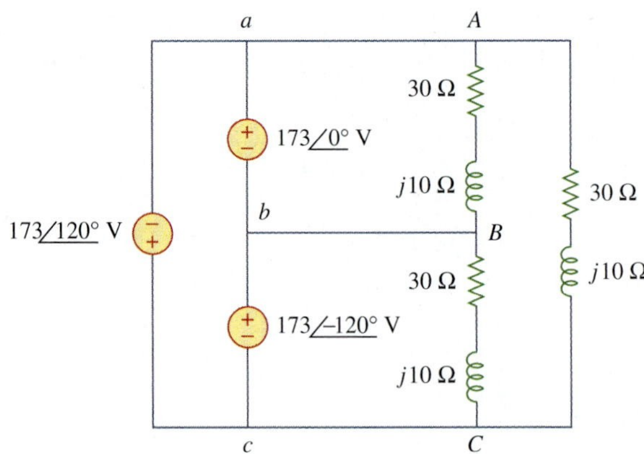

Figure 12.50
For Prob. 12.19.

12.20 Using Fig. 12.51, design a problem to help other students better understand balanced delta-delta connected circuits.

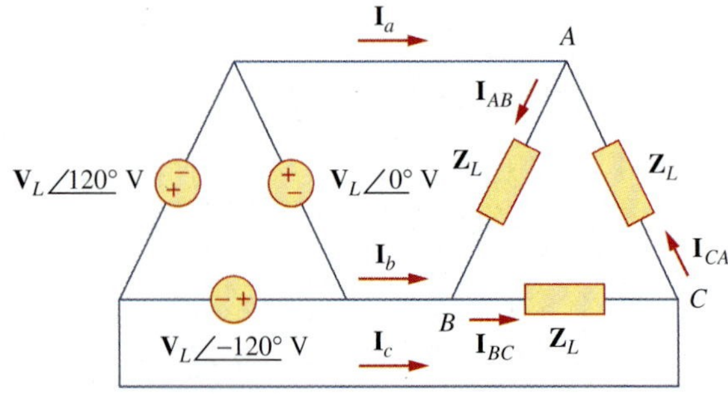

Figure 12.51
For Prob. 12.20.

12.21 Three 440-V generators form a delta-connected source that is connected to a balanced delta-connected load of $\mathbf{Z}_L = 10 + j8\ \Omega$ per phase as shown in Fig. 12.52.

(a) Determine the value of $\mathbf{I}_{AC}$.

(b) What is the value of $\mathbf{I}_{bB}$?

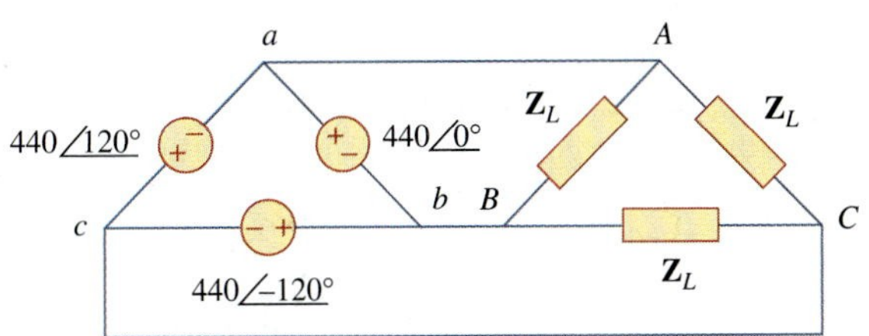

Figure 12.52
For Prob. 12.21.

12.22 Find the line currents $\mathbf{I}_a$, $\mathbf{I}_b$, and $\mathbf{I}_c$ in the three-phase network of Fig. 12.53 below. Take $\mathbf{Z}_\Delta = 12 - j15\ \Omega$, $\mathbf{Z}_Y = 4 + j6\ \Omega$, and $\mathbf{Z}_l = 2\ \Omega$.

12.23 A three-phase balanced system with a line voltage of 202 V rms feeds a delta-connected load with $\mathbf{Z}_p = 25\angle 60°\ \Omega$.

(a) Find the line current.

(b) Determine the total power supplied to the load using two wattmeters connected to the *A* and *C* lines.

12.24 A balanced delta-connected source has phase voltage $\mathbf{V}_{ab} = 440\angle 30°$ V and a positive phase sequence. If this is connected to a balanced delta-connected load, find the line and phase currents. Take the load impedance per phase as $60\angle 30°\ \Omega$ and line impedance per phase as $1 + j1\ \Omega$.

Section 12.6 Balanced Delta-Wye Connection

12.25 In the circuit of Fig. 12.54, if $\mathbf{V}_{ab} = 220\angle 10°$, $\mathbf{V}_{bc} = 220\angle -110°$, $\mathbf{V}_{ca} = 220\angle 130°$ V, find the line currents.

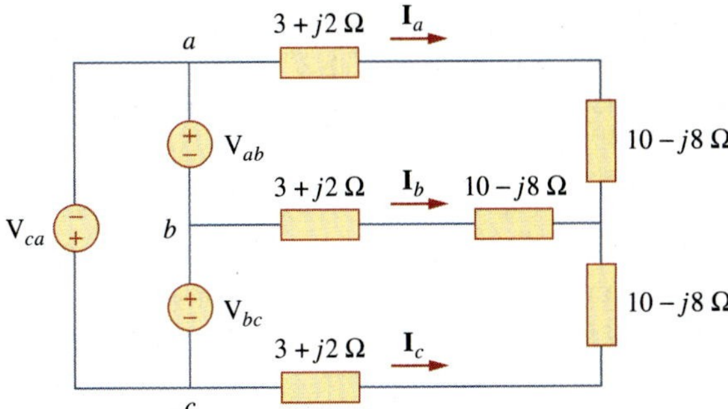

Figure 12.54
For Prob. 12.25.

12.26 Using Fig. 12.55, design a problem to help other students better understand balanced delta connected sources delivering power to balanced wye connected loads.

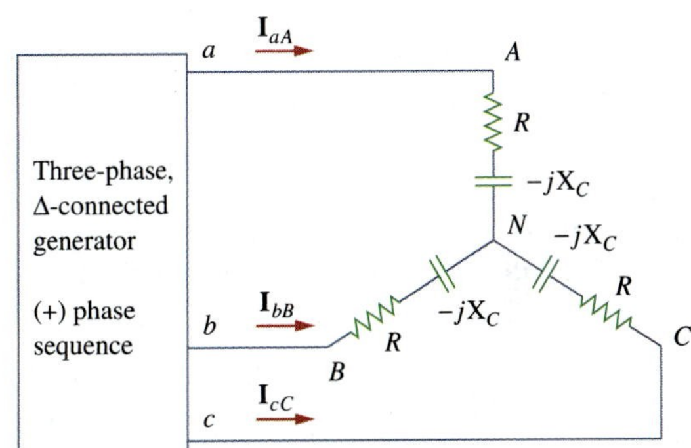

Figure 12.55
For Prob. 12.26.

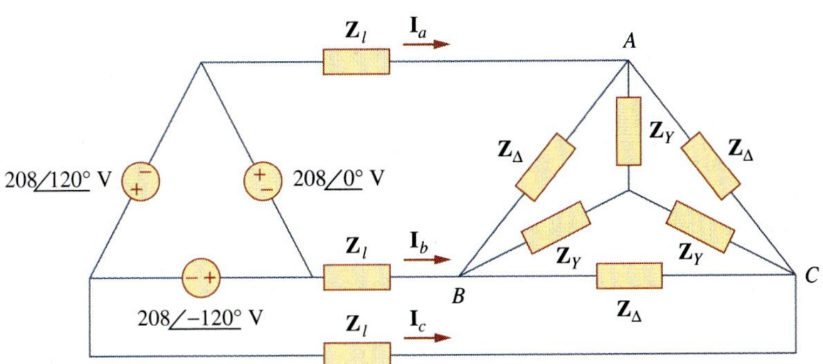

Figure 12.53
For Prob. 12.22.

12.27 A Δ-connected source supplies power to a Y-connected load in a three-phase balanced system. Given that the line impedance is $2 + j1\ \Omega$ per phase while the load impedance is $6 + j4\ \Omega$ per phase, find the magnitude of the line voltage at the load. Assume the source phase voltage $\mathbf{V}_{ab} = 208\underline{/0^\circ}$ V rms.

12.28 The line-to-line voltages in a Y-load have a magnitude of 220 V and are in the positive sequence at 60 Hz. If the loads are balanced with $Z_1 = Z_2 = Z_3 = 25\underline{/30^\circ}$, find all line currents and phase voltages.

Section 12.7 Power in a Balanced System

12.29 A balanced three-phase Y-Δ system has $\mathbf{V}_{an} = 120\underline{/0^\circ}$ V rms and $\mathbf{Z}_\Delta = 51 + j45\ \Omega$. If the line impedance per phase is $0.4 + j1.2\ \Omega$, find the total complex power delivered to the load.

12.30 In Fig. 12.56, the rms value of the line voltage is 208 V. Find the average power delivered to the load.

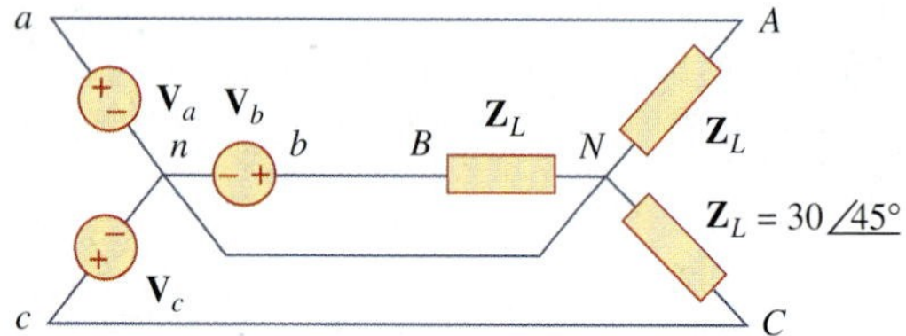

Figure 12.56
For Prob. 12.30.

12.31 A balanced delta-connected load is supplied by a 60-Hz three-phase source with a line voltage of 240 V. Each load phase draws 6 kW at a lagging power factor of 0.8. Find:

(a) the load impedance per phase

(b) the line current

(c) the value of capacitance needed to be connected in parallel with each load phase to minimize the current from the source

12.32 Design a problem to help other students better understand power in a balanced three-phase system.

12.33 A three-phase source delivers 9.6 kVA to a wye-connected load with a phase voltage of 208 V and a power factor of 0.9 lagging. Calculate the source line current and the source line voltage.

12.34 A balanced wye-connected load with a phase impedance of $10 - j16\ \Omega$ is connected to a balanced three-phase generator with a line voltage of 220 V. Determine the line current and the complex power absorbed by the load.

12.35 Three equal impedances, $60 + j30\ \Omega$ each, are delta-connected to a 230-V rms, three-phase circuit. Another three equal impedances, $40 + j10\ \Omega$ each, are wye-connected across the same circuit at the same points. Determine:

(a) the line current

(b) the total complex power supplied to the two loads

(c) the power factor of the two loads combined

12.36 A 4200-V, three-phase transmission line has an impedance of $4 + j\ \Omega$ per phase. If it supplies a load of 1 MVA at 0.75 power factor (lagging), find:

(a) the complex power

(b) the power loss in the line

(c) the voltage at the sending end

12.37 The total power measured in a three-phase system feeding a balanced wye-connected load is 12 kW at a power factor of 0.6 leading. If the line voltage is 208 V, calculate the line current I_L and the load impedance $\mathbf{Z}_Y$.

12.38 Given the circuit in Fig. 12.57 below, find the total complex power absorbed by the load.

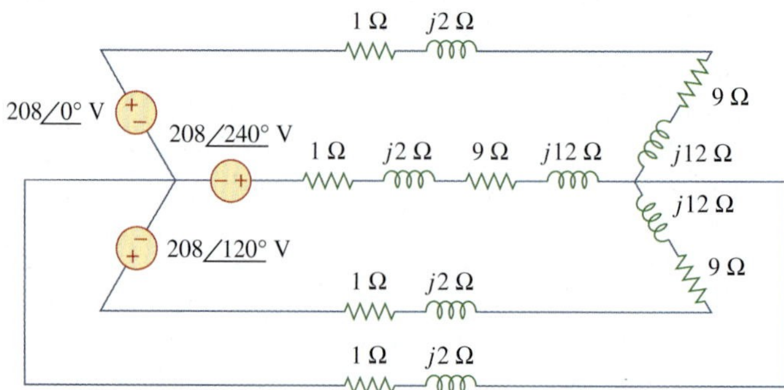

Figure 12.57
For Prob. 12.38.

12.39 Find the real power absorbed by the load in Fig. 12.58.

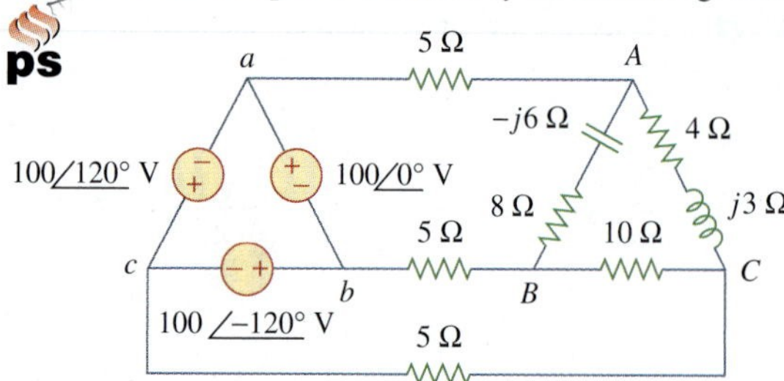

Figure 12.58
For Prob. 12.39.

12.40 For the three-phase circuit in Fig. 12.59, find the average power absorbed by the delta-connected load with $\mathbf{Z}_\Delta = 21 + j24\ \Omega$.

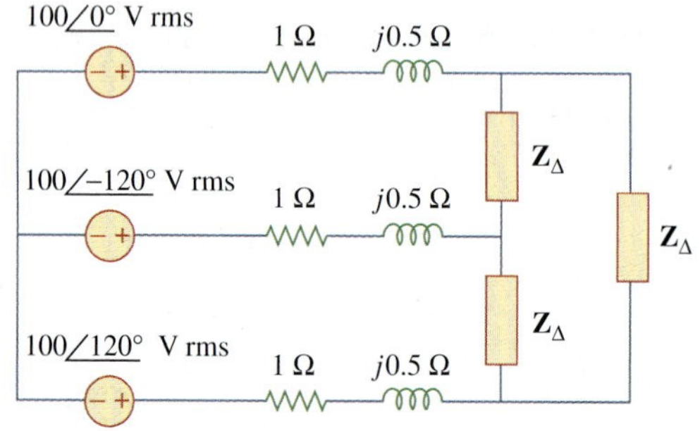

Figure 12.59
For Prob. 12.40.

12.41 A balanced delta-connected load draws 5 kW at a power factor of 0.8 lagging. If the three-phase system has an effective line voltage of 400 V, find the line current.

12.42 A balanced three-phase generator delivers 9.6 kW to a wye-connected load with impedance $30 - j40\ \Omega$ per phase. Find the line current I_L and the line voltage V_L.

12.43 Refer to Fig. 12.48. Obtain the complex power absorbed by the combined loads.

12.44 A three-phase line has an impedance of $1 + j3\ \Omega$ per phase. The line feeds a balanced delta-connected load, which absorbs a total complex power of $12 + j5$ kVA. If the line voltage at the load end has a magnitude of 240 V, calculate the magnitude of the line voltage at the source end and the source power factor.

12.45 A balanced wye-connected load is connected to the generator by a balanced transmission line with an impedance of $0.5 + j2\ \Omega$ per phase. If the load is rated at 450 kW, 0.708 power factor lagging, 440-V line voltage, find the line voltage at the generator.

12.46 A three-phase load consists of three 100-Ω resistors that can be wye- or delta-connected. Determine which connection will absorb the most average power from a three-phase source with a line voltage of 110 V. Assume zero line impedance.

12.47 The following three parallel-connected three-phase loads are fed by a balanced three-phase source:

Load 1: 250 kVA, 0.8 pf lagging
Load 2: 300 kVA, 0.95 pf leading
Load 3: 450 kVA, unity pf

If the line voltage is 13.8 kV, calculate the line current and the power factor of the source. Assume that the line impedance is zero.

12.48 A balanced, positive-sequence wye-connected source has $\mathbf{V}_{an} = 240\angle 0^\circ$ V rms and supplies an unbalanced delta-connected load via a transmission line with impedance $2 + j3\ \Omega$ per phase.

(a) Calculate the line currents if $\mathbf{Z}_{AB} = 40 + j15\ \Omega$, $\mathbf{Z}_{BC} = 60\ \Omega$, $\mathbf{Z}_{CA} = 18 - j12\ \Omega$.

(b) Find the complex power supplied by the source.

12.49 Each phase load consists of a 20-Ω resistor and a 10-Ω inductive reactance. With a line voltage of 220 V rms, calculate the average power taken by the load if:

(a) the three-phase loads are delta-connected

(b) the loads are wye-connected

12.50 A balanced three-phase source with $\mathbf{V}_L = 240$ V rms is supplying 8 kVA at 0.6 power factor lagging to two wye-connected parallel loads. If one load draws 3 kW at unity power factor, calculate the impedance per phase of the second load.

Section 12.8 Unbalanced Three-Phase Systems

12.51 Consider the Δ-Δ system shown in Fig. 12.60. Take $\mathbf{Z}_1 = 8 + j6\ \Omega$, $\mathbf{Z}_2 = 4.2 - j2.2\ \Omega$, $\mathbf{Z}_3 = 10 + j0\ \Omega$.

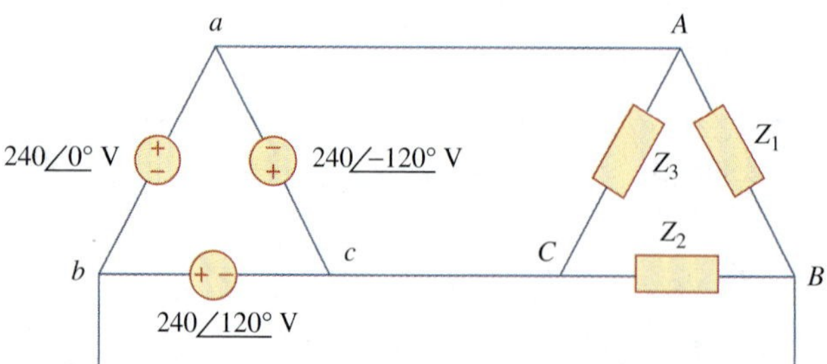

Figure 12.60
For Prob. 12.51.

(a) Find the phase current $\mathbf{I}_{AB}$, $\mathbf{I}_{BC}$, *and* $\mathbf{I}_{CA}$.

(b) Calculate line currents $\mathbf{I}_{aA}$, $\mathbf{I}_{bB}$, and $\mathbf{I}_{cC}$.

12.52 A four-wire wye-wye circuit has

$$\mathbf{V}_{an} = 120\angle 120°, \quad \mathbf{V}_{bn} = 120\angle 0°$$
$$\mathbf{V}_{cn} = 120\angle -120° \text{ V}$$

If the impedances are

$$\mathbf{Z}_{AN} = 20\angle 60°, \quad \mathbf{Z}_{BN} = 30\angle 0°$$
$$\mathbf{Z}_{cn} = 40\angle 30° \ \Omega$$

find the current in the neutral line.

12.53 e2d Using Fig. 12.61, design a problem that will help other students better understand unbalanced three-phase systems.

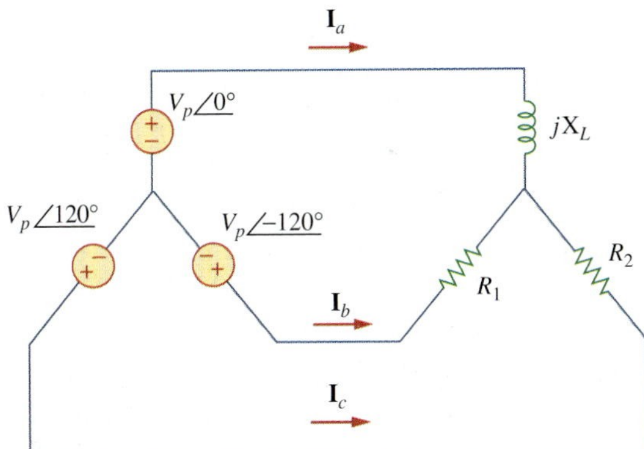

Figure 12.61
For Prob. 12.53.

12.54 A balanced three-phase Y-source with $V_P = 210$ V rms drives a Y-connected three-phase load with phase impedance $\mathbf{Z}_A = 80\ \Omega$, $\mathbf{Z}_B = 60 + j90\ \Omega$, and $\mathbf{Z}_C = j80\ \Omega$. Calculate the line currents and total complex power delivered to the load. Assume that the neutrals are connected.

12.55 A three-phase supply, with the line voltage 240 V rms positively phased, has an unbalanced delta-connected load as shown in Fig. 12.62. Find the phase currents and the total complex power.

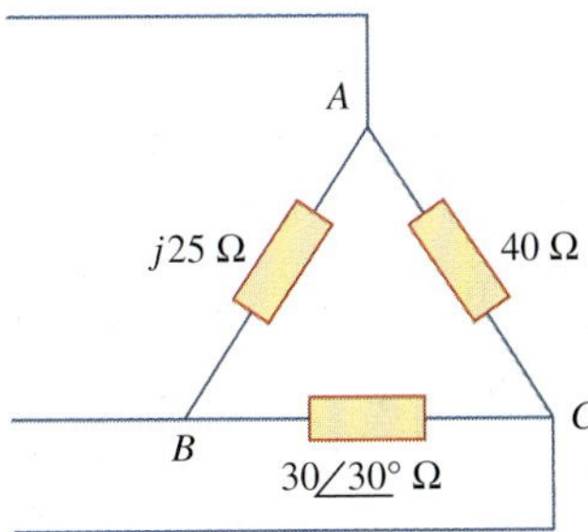

Figure 12.62
For Prob. 12.55.

12.56 e2d Using Fig. 12.63, design a problem to help other students to better understand unbalanced three-phase systems.

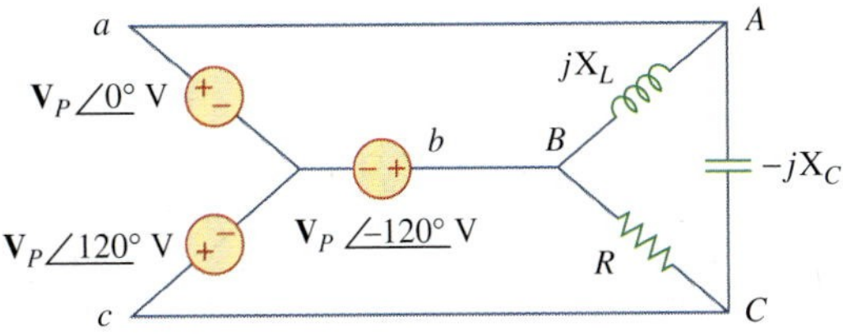

Figure 12.63
For Prob. 12.56.

12.57 Determine the line currents for the three-phase circuit of Fig. 12.64. Let $\mathbf{V}_a = 110\angle 0°$, $\mathbf{V}_b = 110\angle -120°$, $\mathbf{V}_c = 110\angle 120°$ V.

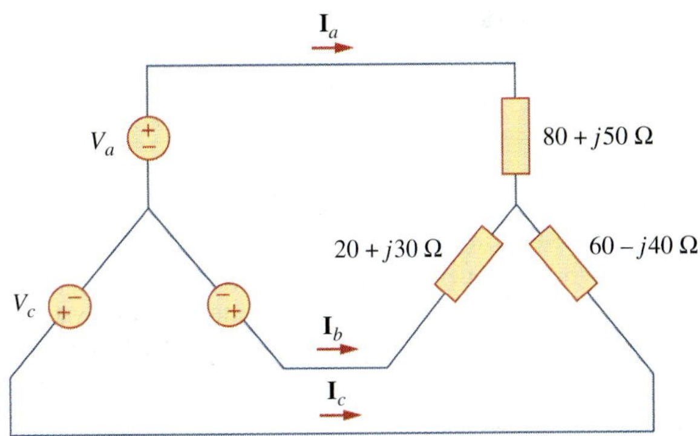

Figure 12.64
For Prob. 12.57.

Section 12.9 *PSpice* for Three-Phase Circuits

12.58 Solve Prob. 12.10 using *PSpice*.

12.59 The source in Fig. 12.65 is balanced and exhibits a positive phase sequence. If $f = 60$ Hz, use *PSpice* to find $\mathbf{V}_{AN}$, $\mathbf{V}_{BN}$, and $\mathbf{V}_{CN}$.

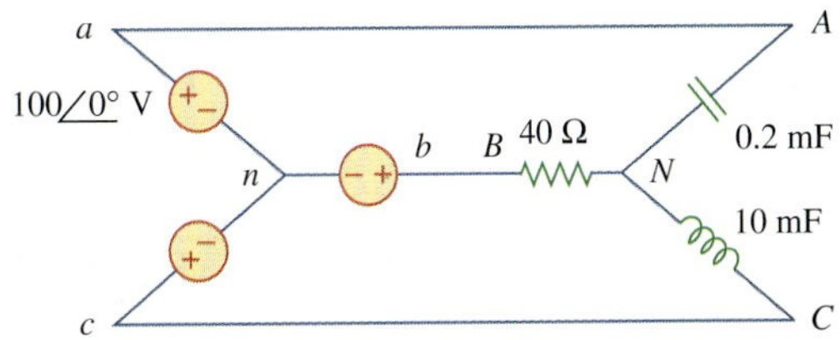

Figure 12.65
For Prob. 12.59.

12.60 Use *PSpice* to determine $\mathbf{I}_o$ in the single-phase, three-wire circuit of Fig. 12.66. Let $\mathbf{Z}_1 = 15 - j10\ \Omega$, $\mathbf{Z}_2 = 30 + j20\ \Omega$, and $\mathbf{Z}_3 = 12 + j5\ \Omega$.

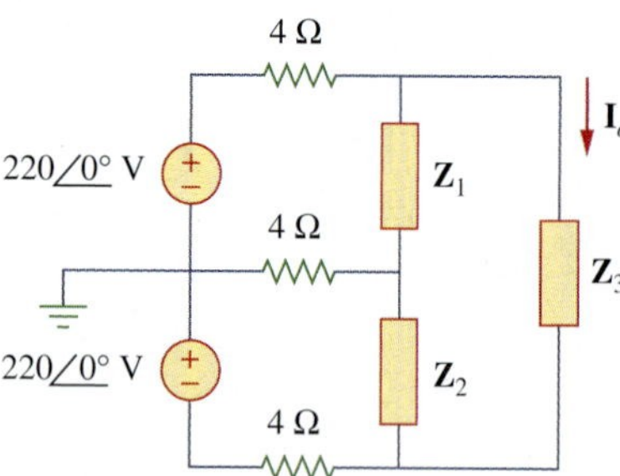

Figure 12.66
For Prob. 12.60.

12.61 Given the circuit in Fig. 12.67, use *PSpice* to determine currents $\mathbf{I}_{aA}$ and voltage $\mathbf{V}_{BN}$.

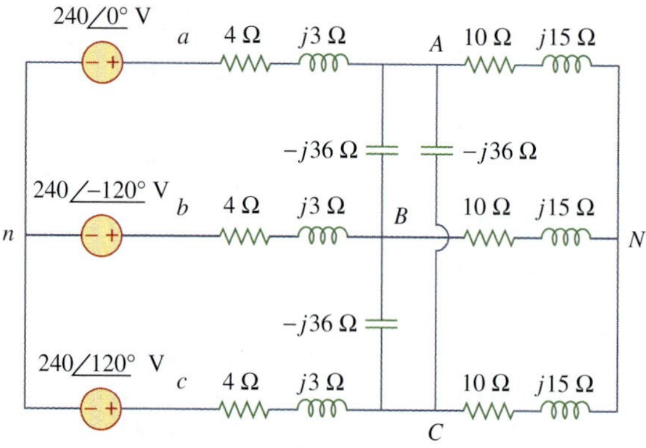

Figure 12.67
For Prob. 12.61.

12.62 Using Fig. 12.68, design a problem to help other **e⊿d** students better understand how to use *PSpice* to analyze three-phase circuits.

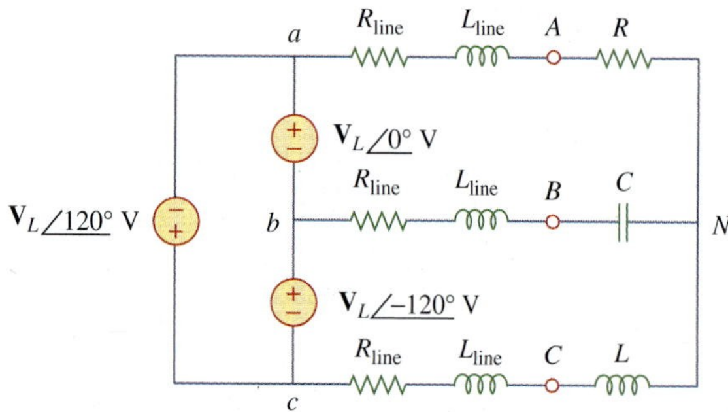

Figure 12.68
For Prob. 12.62.

12.63 Use *PSpice* to find currents $\mathbf{I}_{aA}$ and $\mathbf{I}_{AC}$ in the unbalanced three-phase system shown in Fig. 12.69. Let

$$\mathbf{Z}_l = 2 + j, \qquad \mathbf{Z}_1 = 40 + j20\ \Omega,$$
$$\mathbf{Z}_2 = 50 - j30\ \Omega, \qquad \mathbf{Z}_3 = 25\ \Omega$$

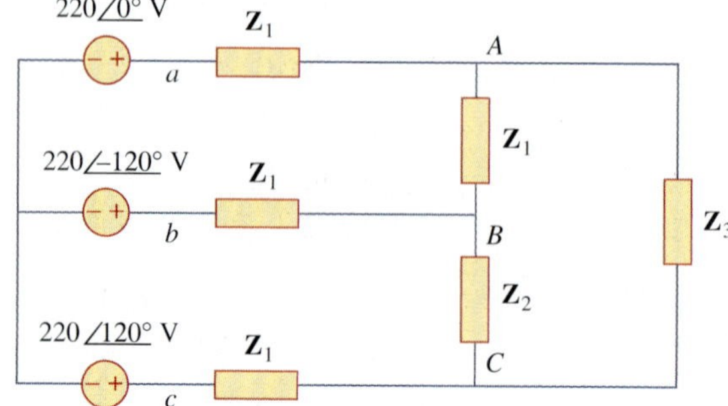

Figure 12.69
For Prob. 12.63.

12.64 For the circuit in Fig. 12.58, use *PSpice* to find the line currents and the phase currents.

12.65 A balanced three-phase circuit is shown in Fig. 12.70 on the next page. Use *PSpice* to find the line currents $\mathbf{I}_{aA}$, $\mathbf{I}_{bB}$, and $\mathbf{I}_{cC}$.

Section 12.10 Applications

12.66 A three-phase, four-wire system operating with a 208-V line voltage is shown in Fig. 12.71. The source voltages are balanced. The power absorbed by the resistive wye-connected load is measured by the three-wattmeter method. Calculate:

(a) the voltage to neutral

(b) the currents $\mathbf{I}_1$, $\mathbf{I}_2$, $\mathbf{I}_3$, and $\mathbf{I}_n$

(c) the readings of the wattmeters

(d) the total power absorbed by the load

***12.67** As shown in Fig. 12.72, a three-phase four-wire line **e⊿d** with a phase voltage of 120 V rms and positive phase sequence supplies a balanced motor load at 260 kVA at 0.85 pf lagging. The motor load is connected to the three main lines marked *a*, *b*, and *c*. In addition, incandescent lamps (unity pf) are connected as follows: 24 kW from line *c* to the neutral, 15 kW from line *b* to the neutral, and 9 kW from line *c* to the neutral.

(a) If three wattmeters are arranged to measure the power in each line, calculate the reading of each meter.

(b) Find the magnitude of the current in the neutral line.

12.68 Meter readings for a three-phase wye-connected alternator supplying power to a motor indicate that

* An asterisk indicates a challenging problem.

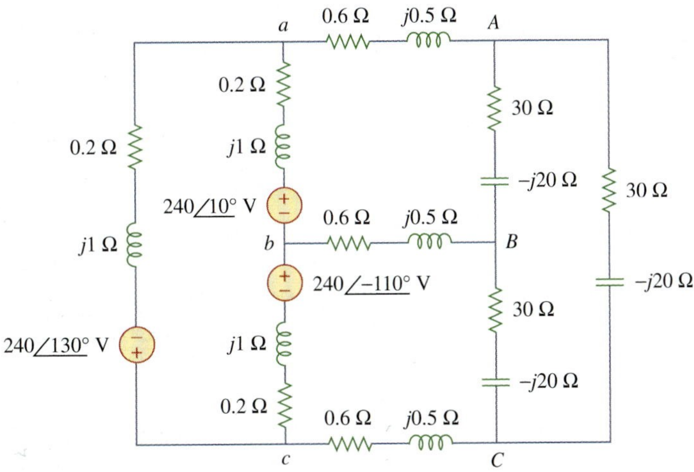

Figure 12.70
For Prob. 12.65.

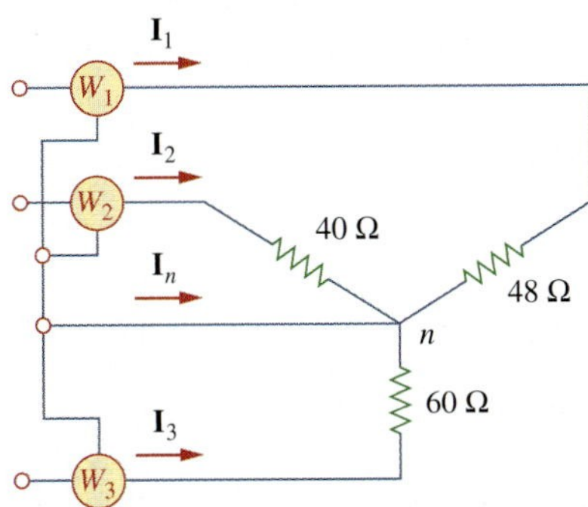

Figure 12.71
For Prob. 12.66.

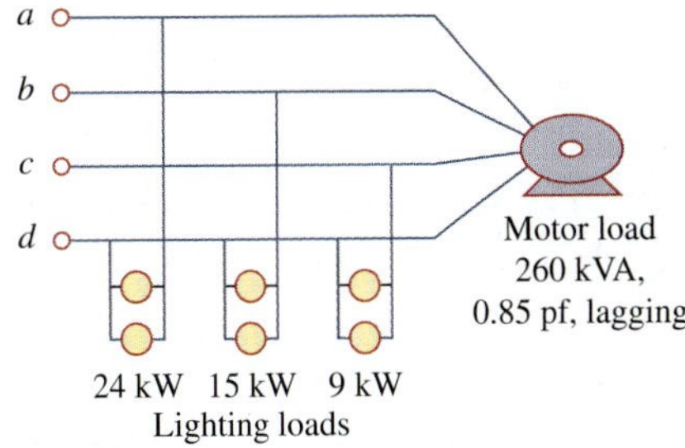

Figure 12.72
For Prob. 12.67.

the line voltages are 330 V, the line currents are 8.4 A, and the total line power is 4.5 kW. Find:

(a) the load in VA

(b) the load pf

(c) the phase current

(d) the phase voltage

12.69 A certain store contains three balanced three-phase loads. The three loads are:

Load 1: 16 kVA at 0.85 pf lagging
Load 2: 12 kVA at 0.6 pf lagging
Load 3: 8 kW at unity pf

The line voltage at the load is 208 V rms at 60 Hz, and the line impedance is $0.4 + j0.8\ \Omega$. Determine the line current and the complex power delivered to the loads.

12.70 The two-wattmeter method gives $P_1 = 1200$ W and $P_2 = -400$ W for a three-phase motor running on a 240-V line. Assume that the motor load is wye-connected and that it draws a line current of 6 A. Calculate the pf of the motor and its phase impedance.

12.71 In Fig. 12.73, two wattmeters are properly connected to the unbalanced load supplied by a balanced source such that $\mathbf{V}_{ab} = 208\underline{/0^\circ}$ V with positive phase sequence.

(a) Determine the reading of each wattmeter.

(b) Calculate the total apparent power absorbed by the load.

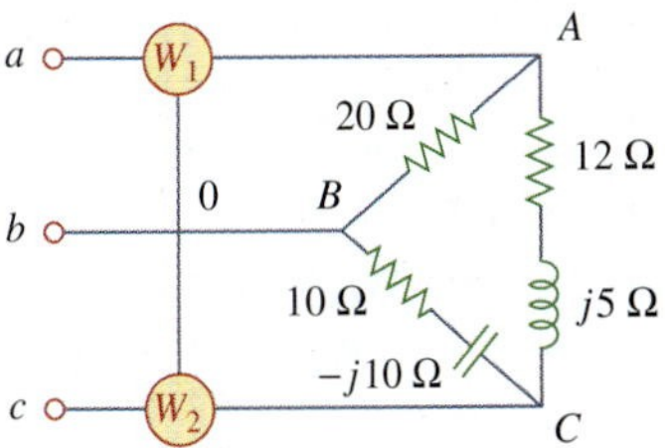

Figure 12.73
For Prob. 12.71.

12.72 If wattmeters W_1 and W_2 are properly connected respectively between lines *a* and *b* and lines *b* and *c* to measure the power absorbed by the delta-connected load in Fig. 12.44, predict their readings.

12.73 For the circuit displayed in Fig. 12.74, find the wattmeter readings.

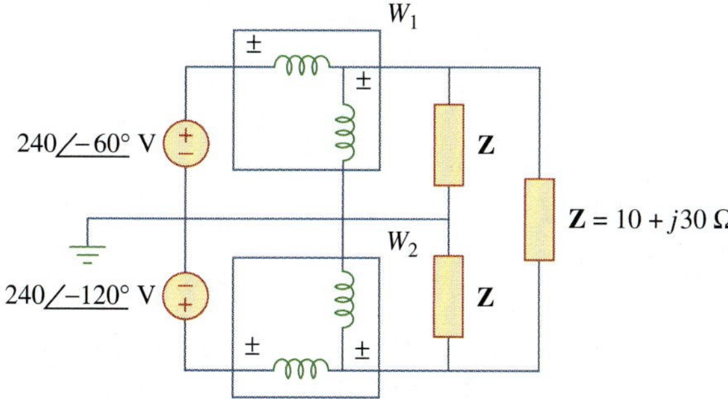

Figure 12.74
For Prob. 12.73.

12.74 Predict the wattmeter readings for the circuit in Fig. 12.75.

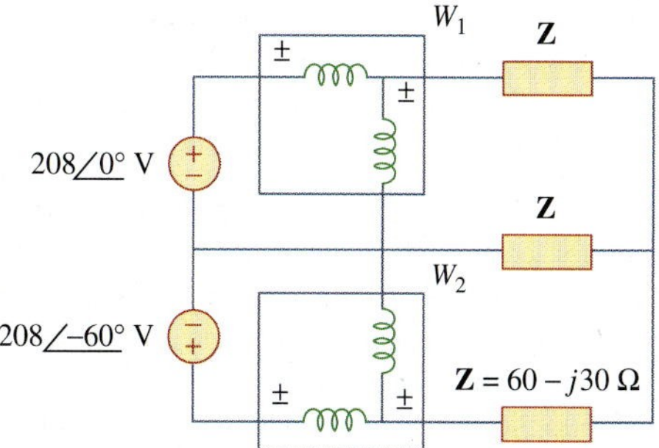

Figure 12.75
For Prob. 12.74.

12.75 A man has a body resistance of 600 Ω. How much current flows through his ungrounded body:

(a) when he touches the terminals of a 12-V autobattery?

(b) when he sticks his finger into a 120-V light socket?

12.76 Show that the I^2R losses will be higher for a 120-V appliance than for a 240-V appliance if both have the same power rating.

Comprehensive Problems

12.77 A three-phase generator supplied 3.6 kVA at a power factor of 0.85 lagging. If 2500 W are delivered to the load and line losses are 80 W per phase, what are the losses in the generator?

12.78 A three-phase 440-V, 51-kW, 60-kVA inductive load operates at 60 Hz and is wye-connected. It is desired to correct the power factor to 0.95 lagging. What value of capacitor should be placed in parallel with each load impedance?

12.79 A balanced three-phase generator has an *abc* phase sequence with phase voltage $\mathbf{V}_{an} = 255\underline{/0^\circ}$ V. The generator feeds an induction motor which may be represented by a balanced Y-connected load with an impedance of $12 + j5\ \Omega$ per phase. Find the line currents and the load voltages. Assume a line impedance of 2 Ω per phase.

12.80 A balanced three-phase source furnishes power to the following three loads:

Load 1: 6 kVA at 0.83 pf lagging

Load 2: unknown

Load 3: 8 kW at 0.7071 pf leading

If the line current is 84.6 A rms, the line voltage at the load is 208 V rms, and the combined load has a 0.8 pf lagging, determine the unknown load.

12.81 A professional center is supplied by a balanced three-phase source. The center has four balanced three-phase loads as follows:

Load 1: 150 kVA at 0.8 pf leading

Load 2: 100 kW at unity pf

Load 3: 200 kVA at 0.6 pf lagging

Load 4: 80 kW and 95 kVAR (inductive)

If the line impedance is $0.02 + j0.05\ \Omega$ per phase and the line voltage at the loads is 480 V, find the magnitude of the line voltage at the source.

12.82 A balanced three-phase system has a distribution wire with impedance $2 + j6\ \Omega$ per phase. The system supplies two three-phase loads that are connected in parallel. The first is a balanced wye-connected load that absorbs 400 kVA at a power factor of 0.8 lagging. The second load is a balanced delta-connected load with impedance of $10 + j8\ \Omega$ per phase. If the magnitude of the line voltage at the loads is 2400 V rms, calculate the magnitude of the line voltage at the source and the total complex power supplied to the two loads.

12.83 A commercially available three-phase inductive motor operates at a full load of 120 hp (1 hp = 746 W) at 95 percent efficiency at a lagging power factor of

0.707. The motor is connected in parallel to a 80-kW balanced three-phase heater at unity power factor. If the magnitude of the line voltage is 480 V rms, calculate the line current.

***12.84** Figure 12.76 displays a three-phase delta-connected motor load which is connected to a line voltage of 440 V and draws 4 kVA at a power factor of 72 percent lagging. In addition, a single 1.8 kVAR capacitor is connected between lines *a* and *b*, while a 800-W lighting load is connected between line *c* and neutral. Assuming the *abc* sequence and taking $\mathbf{V}_{an} = V_p\underline{/0^\circ}$, find the magnitude and phase angle of currents $\mathbf{I}_a$, $\mathbf{I}_b$, $\mathbf{I}_c$, and $\mathbf{I}_n$.

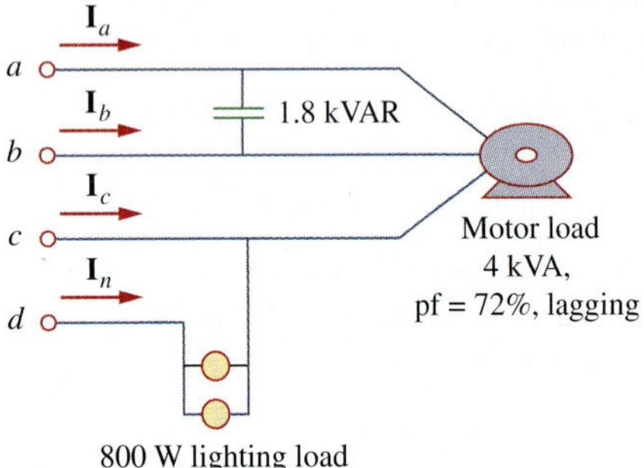

Figure 12.76
For Prob. 12.84.

12.85 Design a three-phase heater with suitable symmetric loads using wye-connected pure resistance. Assume that the heater is supplied by a 240-V line voltage and is to give 27 kW of heat.

12.86 For the single-phase three-wire system in Fig. 12.77, find currents $\mathbf{I}_{aA}$, $\mathbf{I}_{bB}$, and $\mathbf{I}_{nN}$.

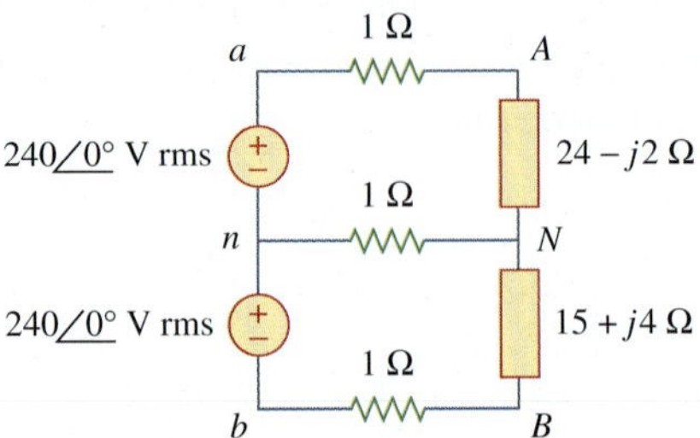

Figure 12.77
For Prob. 12.86.

12.87 Consider the single-phase three-wire system shown in Fig. 12.78. Find the current in the neutral wire and the complex power supplied by each source. Take $\mathbf{V}_s$ as a $115\underline{/0^\circ}$-V, 60-Hz source.

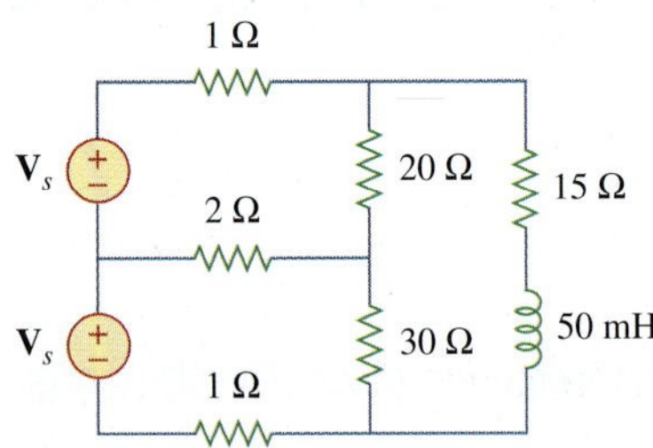

Figure 12.78
For Prob. 12.87.

chapter

13

Magnetically Coupled Circuits

If you would increase your happiness and prolong your life, forget your neighbor's faults . . . . Forget the peculiarities of your friends, and only remember the good points which make you fond of them . . . . Obliterate everything disagreeable from yesterday; write upon today's clean sheet those things lovely and lovable.

—Anonymous

Enhancing Your Career

Career in Electromagnetics

Electromagnetics is the branch of electrical engineering (or physics) that deals with the analysis and application of electric and magnetic fields. In electromagnetics, electric circuit analysis is applied at low frequencies.

The principles of electromagnetics (EM) are applied in various allied disciplines, such as electric machines, electromechanical energy conversion, radar meteorology, remote sensing, satellite communications, bioelectromagnetics, electromagnetic interference and compatibility, plasmas, and fiber optics. EM devices include electric motors and generators, transformers, electromagnets, magnetic levitation, antennas, radars, microwave ovens, microwave dishes, superconductors, and electrocardiograms. The design of these devices requires a thorough knowledge of the laws and principles of EM.

Telemetry receiving station for space satellites. © DV169/Getty Images

EM is regarded as one of the more difficult disciplines in electrical engineering. One reason is that EM phenomena are rather abstract. But if one enjoys working with mathematics and can visualize the invisible, one should consider being a specialist in EM, since few electrical engineers specialize in this area. Electrical engineers who specialize in EM are needed in microwave industries, radio/TV broadcasting stations, electromagnetic research laboratories, and several communications industries.

Historical

James Clerk Maxwell (1831–1879), a graduate in mathematics from Cambridge University, in 1865 wrote a most remarkable paper in which he mathematically unified the laws of Faraday and Ampere. This relationship between the electric field and magnetic field served as the basis for what was later called electromagnetic fields and waves, a major field of study in electrical engineering. The Institute of Electrical and Electronics Engineers (IEEE) uses a graphical representation of this principle in its logo, in which a straight arrow represents current and a curved arrow represents the electromagnetic field. This relationship is commonly known as *the right-hand rule.* Maxwell was a very active theoretician and scientist. He is best known for the "Maxwell equations." The maxwell, a unit of magnetic flux, was named after him.

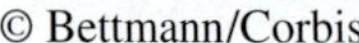

© Bettmann/Corbis

© Bettmann/Corbis

13.1 Introduction

The circuits we have considered so far may be regarded as *conductively coupled*, because one loop affects the neighboring loop through current conduction. When two loops with or without contacts between them affect each other through the magnetic field generated by one of them, they are said to be *magnetically coupled.*

The transformer is an electrical device designed on the basis of the concept of magnetic coupling. It uses magnetically coupled coils to transfer energy from one circuit to another. Transformers are key circuit elements. They are used in power systems for stepping up or stepping down ac voltages or currents. They are used in electronic circuits such as radio and television receivers for such purposes as impedance matching, isolating one part of a circuit from another, and again for stepping up or down ac voltages and currents.

We will begin with the concept of mutual inductance and introduce the dot convention used for determining the voltage polarities of inductively coupled components. Based on the notion of mutual inductance, we then introduce the circuit element known as the *transformer*. We will consider the linear transformer, the ideal transformer, the ideal autotransformer, and the three-phase transformer. Finally, among their important applications, we look at transformers as isolating and matching devices and their use in power distribution.

13.2 Mutual Inductance

When two inductors (or coils) are in a close proximity to each other, the magnetic flux caused by current in one coil links with the other coil, thereby inducing voltage in the latter. This phenomenon is known as *mutual inductance*.

Let us first consider a single inductor, a coil with N turns. When current i flows through the coil, a magnetic flux ϕ is produced around it (Fig. 13.1). According to Faraday's law, the voltage v induced in the coil is proportional to the number of turns N and the time rate of change of the magnetic flux ϕ; that is,

$$v = N\frac{d\phi}{dt} \tag{13.1}$$

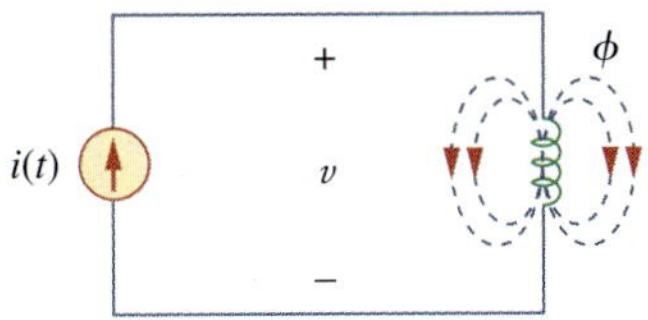

Figure 13.1
Magnetic flux produced by a single coil with N turns.

But the flux ϕ is produced by current i so that any change in ϕ is caused by a change in the current. Hence, Eq. (13.1) can be written as

$$v = N\frac{d\phi}{di}\frac{di}{dt} \tag{13.2}$$

or

$$v = L\frac{di}{dt} \tag{13.3}$$

which is the voltage-current relationship for the inductor. From Eqs. (13.2) and (13.3), the inductance L of the inductor is thus given by

$$L = N\frac{d\phi}{di} \tag{13.4}$$

This inductance is commonly called *self-inductance*, because it relates the voltage induced in a coil by a time-varying current in the same coil.

Now consider two coils with self-inductances L_1 and L_2 that are in close proximity with each other (Fig. 13.2). Coil 1 has N_1 turns, while coil 2 has N_2 turns. For the sake of simplicity, assume that the second inductor carries no current. The magnetic flux ϕ_1 emanating from coil 1 has two components: one component ϕ_{11} links only coil 1, and another component ϕ_{12} links both coils. Hence,

$$\phi_1 = \phi_{11} + \phi_{12} \tag{13.5}$$

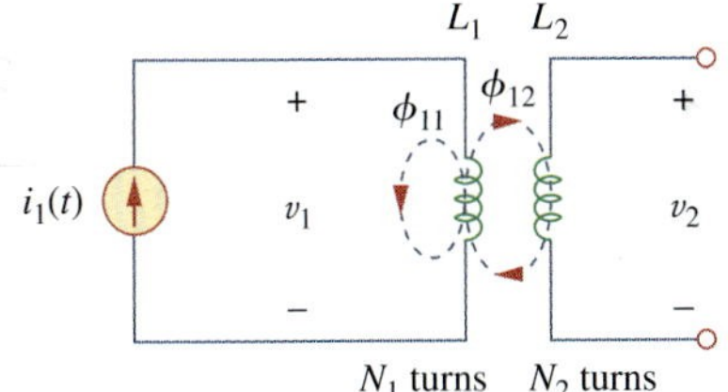

Figure 13.2
Mutual inductance M_{21} of coil 2 with respect to coil 1.

Although the two coils are physically separated, they are said to be *magnetically coupled*. Since the entire flux ϕ_1 links coil 1, the voltage induced in coil 1 is

$$v_1 = N_1\frac{d\phi_1}{dt} \tag{13.6}$$

Only flux ϕ_{12} links coil 2, so the voltage induced in coil 2 is

$$v_2 = N_2\frac{d\phi_{12}}{dt} \tag{13.7}$$

Again, as the fluxes are caused by the current i_1 flowing in coil 1, Eq. (13.6) can be written as

$$v_1 = N_1 \frac{d\phi_1}{di_1}\frac{di_1}{dt} = L_1 \frac{di_1}{dt} \tag{13.8}$$

where $L_1 = N_1\, d\phi_1/di_1$ is the self-inductance of coil 1. Similarly, Eq. (13.7) can be written as

$$v_2 = N_2 \frac{d\phi_{12}}{di_1}\frac{di_1}{dt} = M_{21} \frac{di_1}{dt} \tag{13.9}$$

where

$$M_{21} = N_2 \frac{d\phi_{12}}{di_1} \tag{13.10}$$

M_{21} is known as the *mutual inductance* of coil 2 with respect to coil 1. Subscript 21 indicates that the inductance M_{21} relates the voltage induced in coil 2 to the current in coil 1. Thus, the open-circuit *mutual voltage* (or induced voltage) across coil 2 is

$$\boxed{v_2 = M_{21} \frac{di_1}{dt}} \tag{13.11}$$

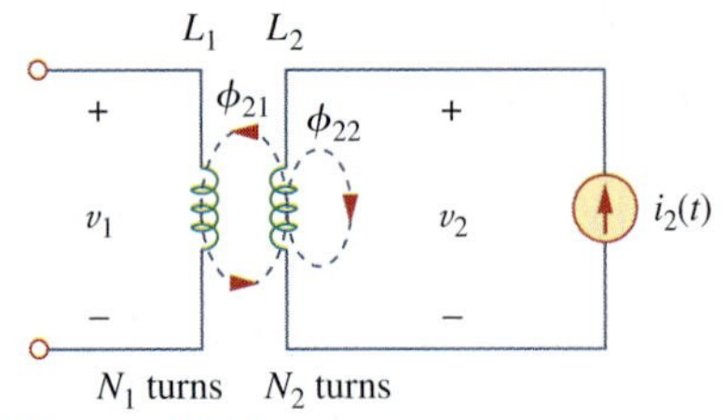

Figure 13.3
Mutual inductance M_{12} of coil 1 with respect to coil 2.

Suppose we now let current i_2 flow in coil 2, while coil 1 carries no current (Fig. 13.3). The magnetic flux ϕ_2 emanating from coil 2 comprises flux ϕ_{22} that links only coil 2 and flux ϕ_{21} that links both coils. Hence,

$$\phi_2 = \phi_{21} + \phi_{22} \tag{13.12}$$

The entire flux ϕ_2 links coil 2, so the voltage induced in coil 2 is

$$v_2 = N_2 \frac{d\phi_2}{dt} = N_2 \frac{d\phi_2}{di_2}\frac{di_2}{dt} = L_2 \frac{di_2}{dt} \tag{13.13}$$

where $L_2 = N_2\, d\phi_2/di_2$ is the self-inductance of coil 2. Since only flux ϕ_{21} links coil 1, the voltage induced in coil 1 is

$$v_1 = N_1 \frac{d\phi_{21}}{dt} = N_1 \frac{d\phi_{21}}{di_2}\frac{di_2}{dt} = M_{12} \frac{di_2}{dt} \tag{13.14}$$

where

$$M_{12} = N_1 \frac{d\phi_{21}}{di_2} \tag{13.15}$$

which is the *mutual inductance* of coil 1 with respect to coil 2. Thus, the open-circuit *mutual voltage* across coil 1 is

$$\boxed{v_1 = M_{12} \frac{di_2}{dt}} \tag{13.16}$$

We will see in the next section that M_{12} and M_{21} are equal; that is,

$$M_{12} = M_{21} = M \tag{13.17}$$

and we refer to M as the mutual inductance between the two coils. Like self-inductance L, mutual inductance M is measured in henrys (H). Keep in mind that mutual coupling only exists when the inductors or coils are in close proximity, and the circuits are driven by time-varying sources. We recall that inductors act like short circuits to dc.

From the two cases in Figs. 13.2 and 13.3, we conclude that mutual inductance results if a voltage is induced by a time-varying current in another circuit. It is the property of an inductor to produce a voltage in reaction to a time-varying current in another inductor near it. Thus,

> **Mutual inductance** is the ability of one inductor to induce a voltage across a neighboring inductor, measured in henrys (H).

Although mutual inductance M is always a positive quantity, the mutual voltage $M\,di/dt$ may be negative or positive, just like the self-induced voltage $L\,di/dt$. However, unlike the self-induced $L\,di/dt$, whose polarity is determined by the reference direction of the current and the reference polarity of the voltage (according to the passive sign convention), the polarity of mutual voltage $M\,di/dt$ is not easy to determine, because four terminals are involved. The choice of the correct polarity for $M\,di/dt$ is made by examining the orientation or particular way in which both coils are physically wound and applying Lenz's law in conjunction with the right-hand rule. Since it is inconvenient to show the construction details of coils on a circuit schematic, we apply the *dot convention* in circuit analysis. By this convention, a dot is placed in the circuit at one end of each of the two magnetically coupled coils to indicate the direction of the magnetic flux if current enters that dotted terminal of the coil. This is illustrated in Fig. 13.4. Given a circuit, the dots are already placed beside the coils so that we need not bother about how to place them. The dots are used along with the dot convention to determine the polarity of the mutual voltage. The dot convention is stated as follows:

> If a current **enters** the dotted terminal of one coil, the reference polarity of the mutual voltage in the second coil is **positive** at the dotted terminal of the second coil.

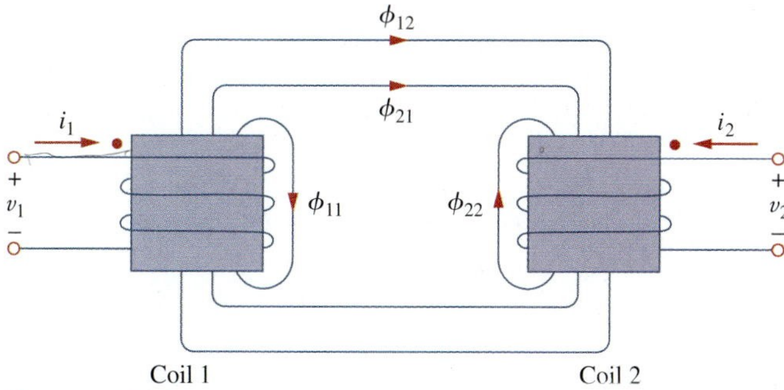

Figure 13.4
Illustration of the dot convention.

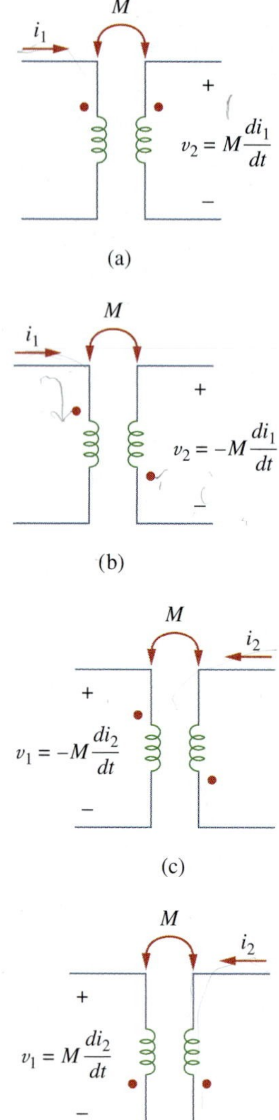

Figure 13.5
Examples illustrating how to apply the dot convention.

Alternatively,

> If a current **leaves** the dotted terminal of one coil, the reference polarity of the mutual voltage in the second coil is **negative** at the dotted terminal of the second coil.

Thus, the reference polarity of the mutual voltage depends on the reference direction of the inducing current and the dots on the coupled coils. Application of the dot convention is illustrated in the four pairs of mutually coupled coils in Fig. 13.5. For the coupled coils in Fig. 13.5(a), the sign of the mutual voltage v_2 is determined by the reference polarity for v_2 and the direction of i_1. Since i_1 enters the dotted terminal of coil 1 and v_2 is positive at the dotted terminal of coil 2, the mutual voltage is $+M\,di_1/dt$. For the coils in Fig. 13.5(b), the current i_1 enters the dotted terminal of coil 1 and v_2 is negative at the dotted terminal of coil 2. Hence, the mutual voltage is $-M\,di_1/dt$. The same reasoning applies to the coils in Fig. 13.5(c) and 13.5(d).

Figure 13.6 shows the dot convention for coupled coils in series. For the coils in Fig. 13.6(a), the total inductance is

$$L = L_1 + L_2 + 2M \qquad \text{(Series-aiding connection)} \tag{13.18}$$

For the coils in Fig. 13.6(b),

$$L = L_1 + L_2 - 2M \qquad \text{(Series-opposing connection)} \tag{13.19}$$

Now that we know how to determine the polarity of the mutual voltage, we are prepared to analyze circuits involving mutual inductance. As the first example, consider the circuit in Fig. 13.7. Applying KVL to coil 1 gives

$$v_1 = i_1R_1 + L_1\frac{di_1}{dt} + M\frac{di_2}{dt} \tag{13.20a}$$

For coil 2, KVL gives

$$v_2 = i_2R_2 + L_2\frac{di_2}{dt} + M\frac{di_1}{dt} \tag{13.20b}$$

We can write Eq. (13.20) in the frequency domain as

$$\mathbf{V}_1 = (R_1 + j\omega L_1)\mathbf{I}_1 + j\omega M\mathbf{I}_2 \tag{13.21a}$$

$$\mathbf{V}_2 = j\omega M\mathbf{I}_1 + (R_2 + j\omega L_2)\mathbf{I}_2 \tag{13.21b}$$

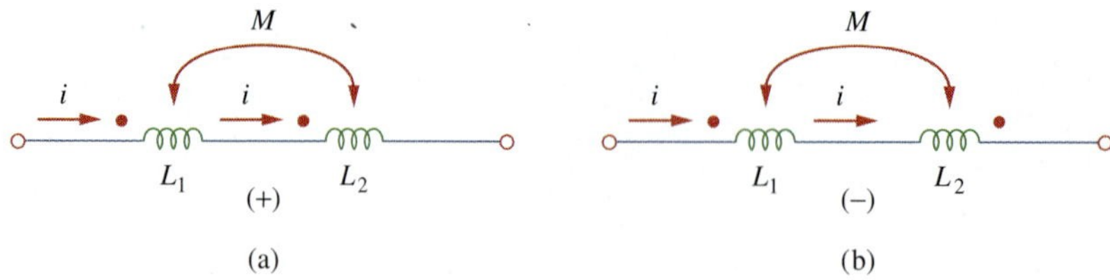

Figure 13.6
Dot convention for coils in series; the sign indicates the polarity of the mutual voltage: (a) series-aiding connection, (b) series-opposing connection.

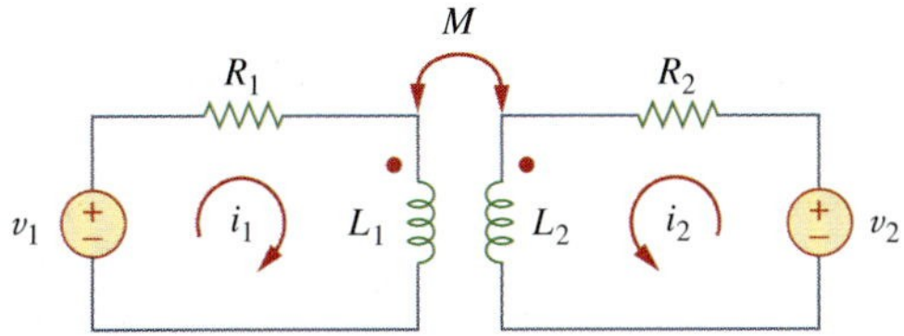

Figure 13.7
Time-domain analysis of a circuit containing coupled coils.

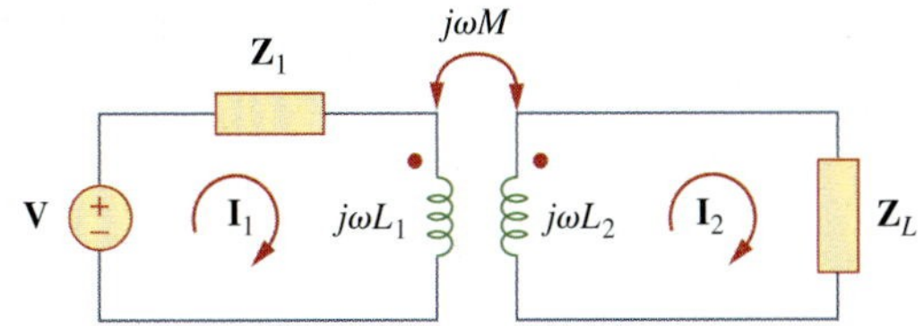

Figure 13.8
Frequency-domain analysis of a circuit containing coupled coils.

As a second example, consider the circuit in Fig. 13.8. We analyze this in the frequency domain. Applying KVL to coil 1, we get

$$\mathbf{V} = (\mathbf{Z}_1 + j\omega L_1)\mathbf{I}_1 - j\omega M\mathbf{I}_2 \tag{13.22a}$$

For coil 2, KVL yields

$$0 = -j\omega M\mathbf{I}_1 + (\mathbf{Z}_L + j\omega L_2)\mathbf{I}_2 \tag{13.22b}$$

Equations (13.21) and (13.22) are solved in the usual manner to determine the currents.

At this introductory level we are not concerned with the determination of the mutual inductances of the coils and their dot placements. Like R, L, and C, calculation of M would involve applying the theory of electromagnetics to the actual physical properties of the coils. In this text, we assume that the mutual inductance and the placement of the dots are the "givens" of the circuit problem, like the circuit components R, L, and C.

Example 13.1

Calculate the phasor currents $\mathbf{I}_1$ and $\mathbf{I}_2$ in the circuit of Fig. 13.9.

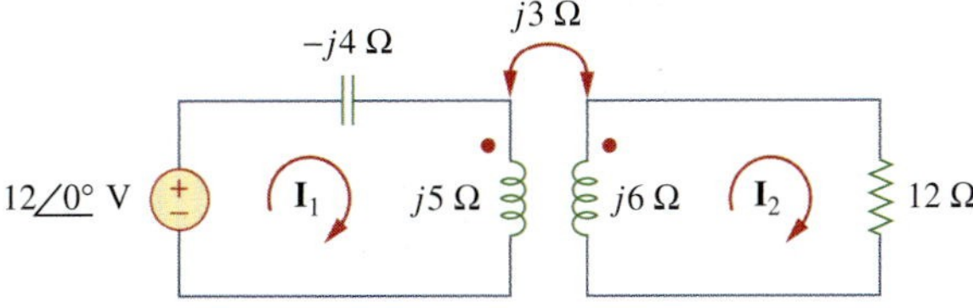

Figure 13.9
For Example 13.1.

Solution:

For coil 1, KVL gives

$$-12 + (-j4 + j5)\mathbf{I}_1 - j3\mathbf{I}_2 = 0$$

or

$$j\mathbf{I}_1 - j3\mathbf{I}_2 = 12 \tag{13.1.1}$$

For coil 2, KVL gives.

$$-j3\mathbf{I}_1 + (12 + j6)\mathbf{I}_2 = 0$$

or

$$\mathbf{I}_1 = \frac{(12 + j6)\mathbf{I}_2}{j3} = (2 - j4)\mathbf{I}_2 \tag{13.1.2}$$

Substituting this in Eq. (13.1.1), we get

$$(j2 + 4 - j3)\mathbf{I}_2 = (4 - j)\mathbf{I}_2 = 12$$

or

$$\mathbf{I}_2 = \frac{12}{4 - j} = 2.91\underline{/14.04^\circ} \text{ A} \tag{13.1.3}$$

From Eqs. (13.1.2) and (13.1.3),

$$\begin{aligned}\mathbf{I}_1 = (2 - j4)\mathbf{I}_2 &= (4.472\underline{/-63.43^\circ})(2.91\underline{/14.04^\circ}) \\ &= 13.01\underline{/-49.39^\circ} \text{ A}\end{aligned}$$

Practice Problem 13.1

Determine the voltage $\mathbf{V}_o$ in the circuit of Fig. 13.10.

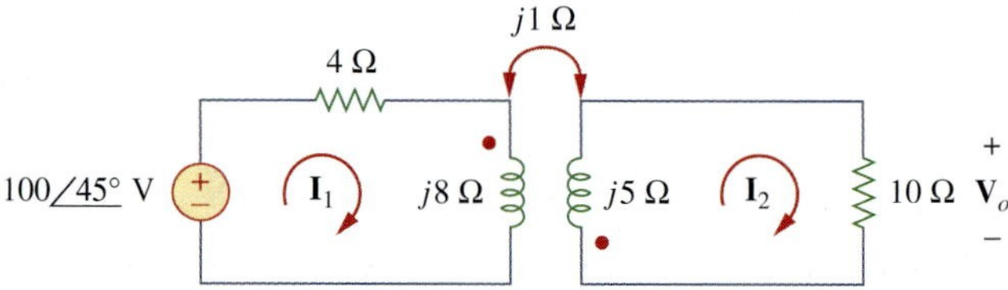

Figure 13.10
For Practice Prob. 13.1.

Answer: $10\underline{/-135^\circ}$ V.

Example 13.2

Calculate the mesh currents in the circuit of Fig. 13.11.

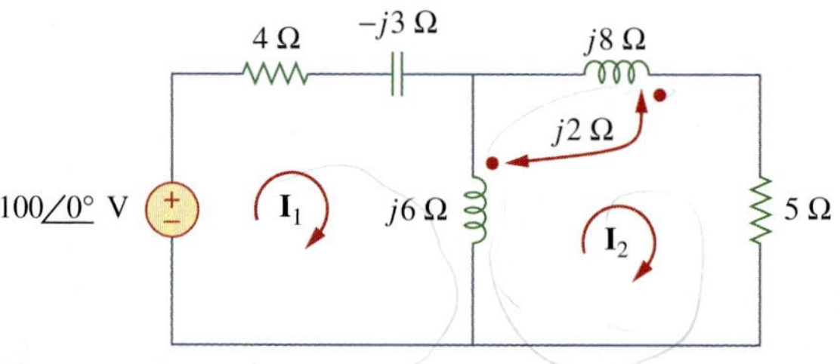

Figure 13.11
For Example 13.2.

Solution:
The key to analyzing a magnetically coupled circuit is knowing the polarity of the mutual voltage. We need to apply the dot rule. In Fig. 13.11, suppose coil 1 is the one whose reactance is 6 Ω, and coil 2 is the one whose reactance is 8 Ω. To figure out the polarity of the mutual voltage in coil 1 due to current $\mathbf{I}_2$, we observe that $\mathbf{I}_2$ leaves the dotted terminal of coil 2. Since we are applying KVL in the clockwise direction, it implies that the mutual voltage is negative, that is, $-j2\mathbf{I}_2$.

Alternatively, it might be best to figure out the mutual voltage by redrawing the relevant portion of the circuit, as shown in Fig. 13.12(a), where it becomes clear that the mutual voltage is $\mathbf{V}_1 = -2j\mathbf{I}_2$.

Thus, for mesh 1 in Fig. 13.11, KVL gives

$$-100 + \mathbf{I}_1(4 - j3 + j6) - j6\mathbf{I}_2 - j2\mathbf{I}_2 = 0$$

or

$$100 = (4 + j3)\mathbf{I}_1 - j8\mathbf{I}_2 \qquad \textbf{(13.2.1)}$$

Similarly, to figure out the mutual voltage in coil 2 due to current $\mathbf{I}_1$, consider the relevant portion of the circuit, as shown in Fig. 13.12(b). Applying the dot convention gives the mutual voltage as $\mathbf{V}_2 = -2j\mathbf{I}_1$. Also, current $\mathbf{I}_2$ sees the two coupled coils in series in Fig. 13.11; since it leaves the dotted terminals in both coils, Eq. (13.18) applies. Therefore, for mesh 2 in Fig. 13.11, KVL gives

$$0 = -2j\mathbf{I}_1 - j6\mathbf{I}_1 + (j6 + j8 + j2 \times 2 + 5)\mathbf{I}_2$$

or

$$0 = -j8\mathbf{I}_1 + (5 + j18)\mathbf{I}_2 \qquad \textbf{(13.2.2)}$$

Putting Eqs. (13.2.1) and (13.2.2) in matrix form, we get

$$\begin{bmatrix} 100 \\ 0 \end{bmatrix} = \begin{bmatrix} 4 + j3 & -j8 \\ -j8 & 5 + j18 \end{bmatrix} \begin{bmatrix} \mathbf{I}_1 \\ \mathbf{I}_2 \end{bmatrix}$$

The determinants are

$$\Delta = \begin{vmatrix} 4 + j3 & -j8 \\ -j8 & 5 + j18 \end{vmatrix} = 30 + j87$$

$$\Delta_1 = \begin{vmatrix} 100 & -j8 \\ 0 & 5 + j18 \end{vmatrix} = 100(5 + j18)$$

$$\Delta_2 = \begin{vmatrix} 4 + j3 & 100 \\ -j8 & 0 \end{vmatrix} = j800$$

Thus, we obtain the mesh currents as

$$\mathbf{I}_1 = \frac{\Delta_1}{\Delta} = \frac{100(5 + j18)}{30 + j87} = \frac{1{,}868.2\underline{/74.5^\circ}}{92.03\underline{/71^\circ}} = 20.3\underline{/3.5^\circ}\text{ A}$$

$$\mathbf{I}_2 = \frac{\Delta_2}{\Delta} = \frac{j800}{30 + j87} = \frac{800\underline{/90^\circ}}{92.03\underline{/71^\circ}} = 8.693\underline{/19^\circ}\text{ A}$$

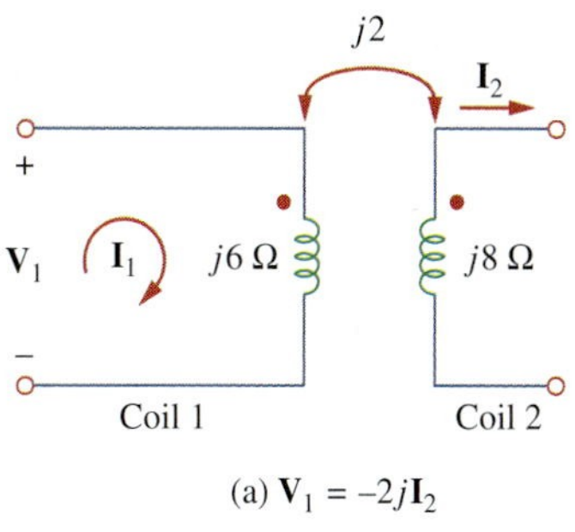

(a) $\mathbf{V}_1 = -2j\mathbf{I}_2$

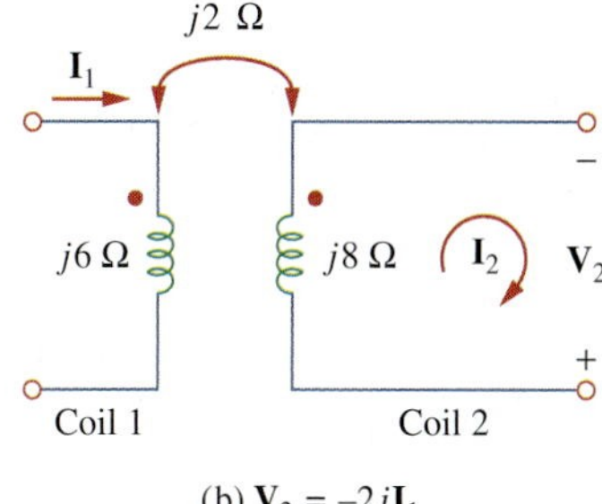

(b) $\mathbf{V}_2 = -2j\mathbf{I}_1$

Figure 13.12
For Example 13.2; redrawing the relevant portion of the circuit in Fig. 13.11 to find mutual voltages by the dot convention.

Practice Problem 13.2

Determine the phasor currents $\mathbf{I}_1$ and $\mathbf{I}_2$ in the circuit of Fig. 13.13.

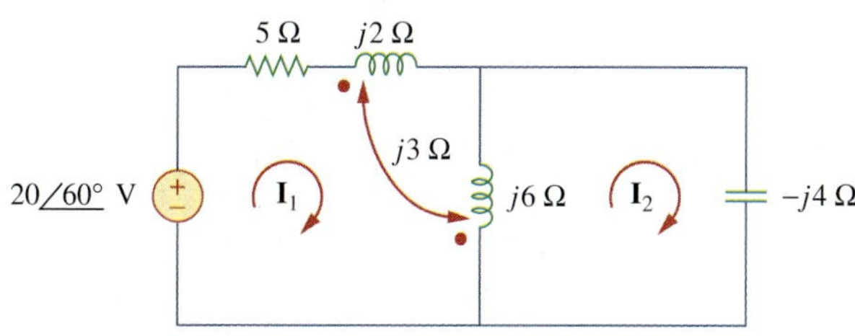

Figure 13.13
For Practice Prob. 13.2.

Answer: $3.583\underline{/86.56^\circ}$, $5.383\underline{/86.56^\circ}$ A.

13.3 Energy in a Coupled Circuit

In Chapter 6, we saw that the energy stored in an inductor is given by

$$w = \frac{1}{2}Li^2 \tag{13.23}$$

We now want to determine the energy stored in magnetically coupled coils.

Consider the circuit in Fig. 13.14. We assume that currents i_1 and i_2 are zero initially, so that the energy stored in the coils is zero. If we let i_1 increase from zero to I_1 while maintaining $i_2 = 0$, the power in coil 1 is

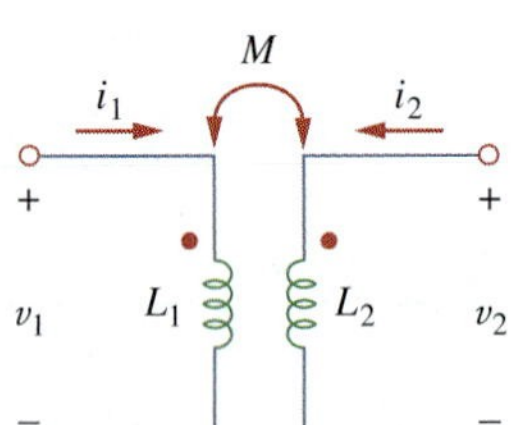

Figure 13.14
The circuit for deriving energy stored in a coupled circuit.

$$p_1(t) = v_1 i_1 = i_1 L_1 \frac{di_1}{dt} \tag{13.24}$$

and the energy stored in the circuit is

$$w_1 = \int p_1\, dt = L_1 \int_0^{I_1} i_1\, di_1 = \frac{1}{2}L_1 I_1^2 \tag{13.25}$$

If we now maintain $i_1 = I_1$ and increase i_2 from zero to I_2, the mutual voltage induced in coil 1 is $M_{12}\, di_2/dt$, while the mutual voltage induced in coil 2 is zero, since i_1 does not change. The power in the coils is now

$$p_2(t) = i_1 M_{12}\frac{di_2}{dt} + i_2 v_2 = I_1 M_{12}\frac{di_2}{dt} + i_2 L_2 \frac{di_2}{dt} \tag{13.26}$$

and the energy stored in the circuit is

$$\begin{aligned} w_2 &= \int p_2 dt = M_{12} I_1 \int_0^{I_2} di_2 + L_2 \int_0^{I_2} i_2 di_2 \\ &= M_{12} I_1 I_2 + \frac{1}{2} L_2 I_2^2 \end{aligned} \tag{13.27}$$

The total energy stored in the coils when both i_1 and i_2 have reached constant values is

$$w = w_1 + w_2 = \frac{1}{2}L_1 I_1^2 + \frac{1}{2}L_2 I_2^2 + M_{12} I_1 I_2 \tag{13.28}$$

If we reverse the order by which the currents reach their final values, that is, if we first increase i_2 from zero to I_2 and later increase i_1 from zero to I_1, the total energy stored in the coils is

$$w = \frac{1}{2}L_1 I_1^2 + \frac{1}{2}L_2 I_2^2 + M_{21} I_1 I_2 \tag{13.29}$$

Since the total energy stored should be the same regardless of how we reach the final conditions, comparing Eqs. (13.28) and (13.29) leads us to conclude that

$$M_{12} = M_{21} = M \tag{13.30a}$$

and

$$w = \frac{1}{2}L_1 I_1^2 + \frac{1}{2}L_2 I_2^2 + M I_1 I_2 \tag{13.30b}$$

This equation was derived based on the assumption that the coil currents both entered the dotted terminals. If one current enters one dotted terminal while the other current leaves the other dotted terminal, the mutual voltage is negative, so that the mutual energy MI_1I_2 is also negative. In that case,

$$w = \frac{1}{2}L_1I_1^2 + \frac{1}{2}L_2I_2^2 - MI_1I_2 \tag{13.31}$$

Also, since I_1 and I_2 are arbitrary values, they may be replaced by i_1 and i_2, which gives the instantaneous energy stored in the circuit the general expression

$$\boxed{w = \frac{1}{2}L_1i_1^2 + \frac{1}{2}L_2i_2^2 \pm Mi_1i_2} \tag{13.32}$$

The positive sign is selected for the mutual term if both currents enter or leave the dotted terminals of the coils; the negative sign is selected otherwise.

We will now establish an upper limit for the mutual inductance M. The energy stored in the circuit cannot be negative because the circuit is passive. This means that the quantity $1/2L_1i_1^2 + 1/2L_2i_2^2 - Mi_1i_2$ must be greater than or equal to zero:

$$\frac{1}{2}L_1i_1^2 + \frac{1}{2}L_2i_2^2 - Mi_1i_2 \geq 0 \tag{13.33}$$

To complete the square, we both add and subtract the term $i_1i_2\sqrt{L_1L_2}$ on the right-hand side of Eq. (13.33) and obtain

$$\frac{1}{2}(i_1\sqrt{L_1} - i_2\sqrt{L_2})^2 + i_1i_2(\sqrt{L_1L_2} - M) \geq 0 \tag{13.34}$$

The squared term is never negative; at its least it is zero. Therefore, the second term on the right-hand side of Eq. (13.34) must be greater than zero; that is,

$$\sqrt{L_1L_2} - M \geq 0$$

or

$$M \leq \sqrt{L_1L_2} \tag{13.35}$$

Thus, the mutual inductance cannot be greater than the geometric mean of the self-inductances of the coils. The extent to which the mutual inductance M approaches the upper limit is specified by the *coefficient of coupling* k, given by

$$k = \frac{M}{\sqrt{L_1L_2}} \tag{13.36}$$

or

$$\boxed{M = k\sqrt{L_1L_2}} \tag{13.37}$$

where $0 \leq k \leq 1$ or equivalently $0 \leq M \leq \sqrt{L_1L_2}$. The coupling coefficient is the fraction of the total flux emanating from one coil that links the other coil. For example, in Fig. 13.2,

$$k = \frac{\phi_{12}}{\phi_1} = \frac{\phi_{12}}{\phi_{11} + \phi_{12}} \tag{13.38}$$

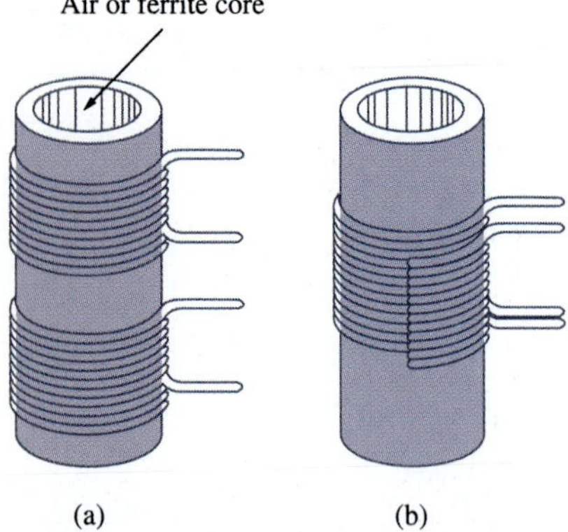

Figure 13.15
Windings: (a) loosely coupled, (b) tightly coupled; cutaway view demonstrates both windings.

and in Fig. 13.3,

$$k = \frac{\phi_{21}}{\phi_2} = \frac{\phi_{21}}{\phi_{21} + \phi_{22}} \tag{13.39}$$

If the entire flux produced by one coil links another coil, then $k = 1$ and we have 100 percent coupling, or the coils are said to be *perfectly coupled*. For $k < 0.5$, coils are said to be *loosely coupled*; and for $k > 0.5$, they are said to be *tightly coupled*. Thus,

> The **coupling coefficient** k is a measure of the magnetic coupling between two coils; $0 \le k \le 1$.

We expect k to depend on the closeness of the two coils, their core, their orientation, and their windings. Figure 13.15 shows loosely coupled windings and tightly coupled windings. The air-core transformers used in radio frequency circuits are loosely coupled, whereas iron-core transformers used in power systems are tightly coupled. The linear transformers discussed in Section 3.4 are mostly air-core; the ideal transformers discussed in Sections 13.5 and 13.6 are principally iron-core.

Example 13.3

Consider the circuit in Fig. 13.16. Determine the coupling coefficient. Calculate the energy stored in the coupled inductors at time $t = 1$ s if $v = 60\cos(4t + 30°)$ V.

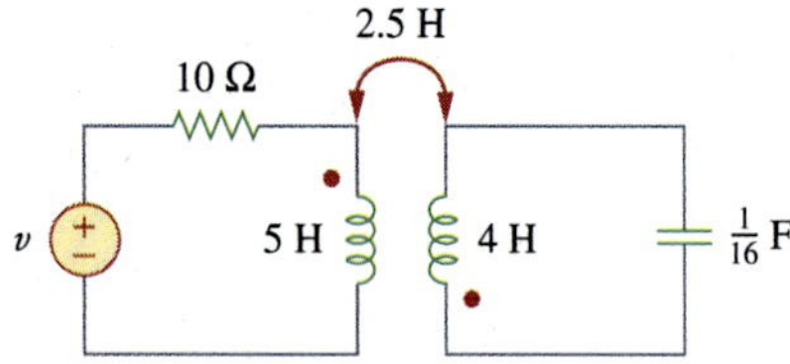

Figure 13.16
For Example 13.3.

Solution:
The coupling coefficient is

$$k = \frac{M}{\sqrt{L_1 L_2}} = \frac{2.5}{\sqrt{20}} = 0.56$$

indicating that the inductors are tightly coupled. To find the energy stored, we need to calculate the current. To find the current, we need to obtain the frequency-domain equivalent of the circuit.

$$\begin{aligned}
60\cos(4t + 30°) \quad &\Rightarrow \quad 60\angle 30°, \quad \omega = 4 \text{ rad/s} \\
5 \text{ H} \quad &\Rightarrow \quad j\omega L_1 = j20\ \Omega \\
2.5 \text{ H} \quad &\Rightarrow \quad j\omega M = j10\ \Omega \\
4 \text{ H} \quad &\Rightarrow \quad j\omega L_2 = j16\ \Omega \\
\frac{1}{16}\text{ F} \quad &\Rightarrow \quad \frac{1}{j\omega C} = -j4\ \Omega
\end{aligned}$$

The frequency-domain equivalent is shown in Fig. 13.17. We now apply mesh analysis. For mesh 1,

$$(10 + j20)\mathbf{I}_1 + j10\mathbf{I}_2 = 60\angle 30° \tag{13.3.1}$$

For mesh 2,

$$j10\mathbf{I}_1 + (j16 - j4)\mathbf{I}_2 = 0$$

or

$$\mathbf{I}_1 = -1.2\mathbf{I}_2 \tag{13.3.2}$$

Substituting this into Eq. (13.3.1) yields

$$\mathbf{I}_2(-12 - j14) = 60\underline{/30^\circ} \quad \Rightarrow \quad \mathbf{I}_2 = 3.254\underline{/160.6^\circ}\text{ A}$$

and

$$\mathbf{I}_1 = -1.2\mathbf{I}_2 = 3.905\underline{/-19.4^\circ}\text{ A}$$

In the time-domain,

$$i_1 = 3.905\cos(4t - 19.4^\circ), \qquad i_2 = 3.254\cos(4t + 160.6^\circ)$$

At time $t = 1$ s, $4t = 4$ rad $= 229.2^\circ$, and

$$i_1 = 3.905\cos(229.2^\circ - 19.4^\circ) = -3.389\text{ A}$$
$$i_2 = 3.254\cos(229.2^\circ + 160.6^\circ) = 2.824\text{ A}$$

The total energy stored in the coupled inductors is

$$\begin{aligned} w &= \frac{1}{2}L_1 i_1^2 + \frac{1}{2}L_2 i_2^2 + M i_1 i_2 \\ &= \frac{1}{2}(5)(-3.389)^2 + \frac{1}{2}(4)(2.824)^2 + 2.5(-3.389)(2.824) = 20.73\text{ J} \end{aligned}$$

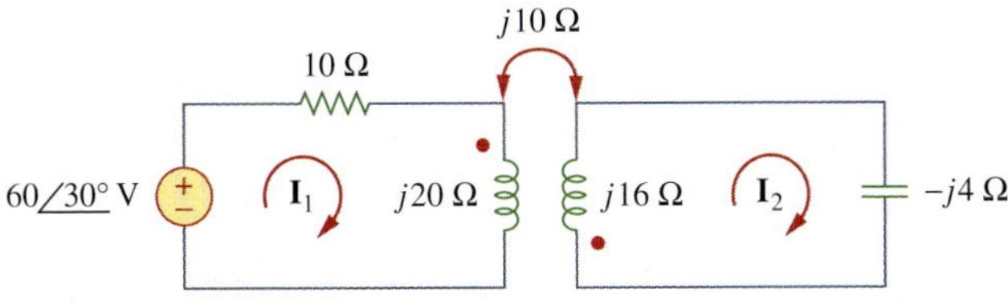

Figure 13.17
Frequency-domain equivalent of the circuit in Fig. 13.16.

Practice Problem 13.3

For the circuit in Fig. 13.18, determine the coupling coefficient and the energy stored in the coupled inductors at $t = 1.5$ s.

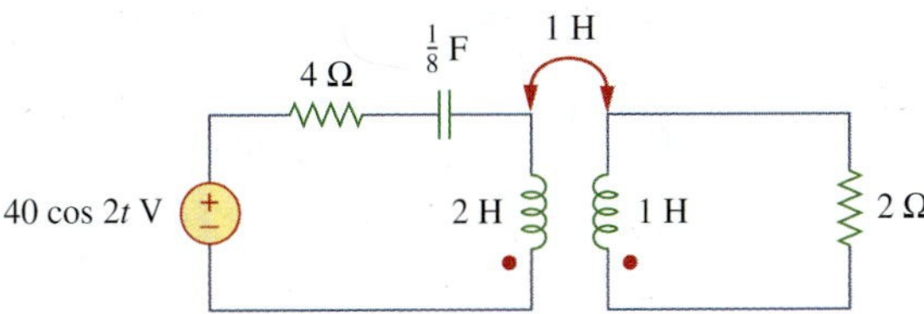

Figure 13.18
For Practice Prob. 13.3.

Answer: 0.7071, 39.4 J.

13.4 Linear Transformers

Here we introduce the transformer as a new circuit element. A transformer is a magnetic device that takes advantage of the phenomenon of mutual inductance.

> A **transformer** is generally a four-terminal device comprising two (or more) magnetically coupled coils.

As shown in Fig. 13.19, the coil that is directly connected to the voltage source is called the *primary winding*. The coil connected to the load is called the *secondary winding*. The resistances R_1 and R_2 are included to account for the losses (power dissipation) in the coils. The transformer is said to be *linear* if the coils are wound on a magnetically linear material—a material for which the magnetic permeability is constant. Such materials include air, plastic, Bakelite, and wood. In fact, most materials are magnetically linear. Linear transformers are sometimes called *air-core transformers*, although not all of them are necessarily air-core. They are used in radio and TV sets. Figure 13.20 portrays different types of transformers.

A linear transformer may also be regarded as one whose flux is proportional to the currents in its windings.

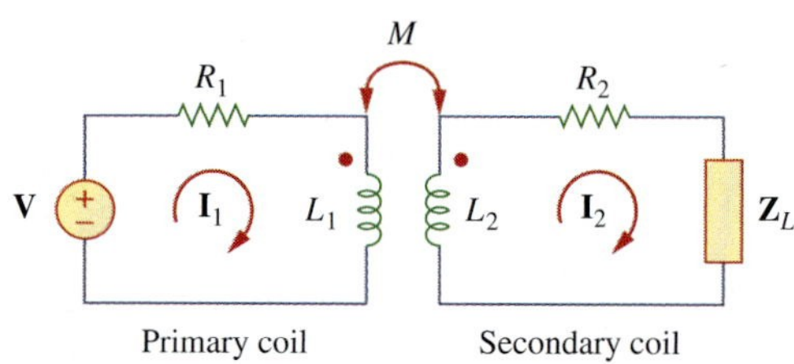

Figure 13.19
A linear transformer.

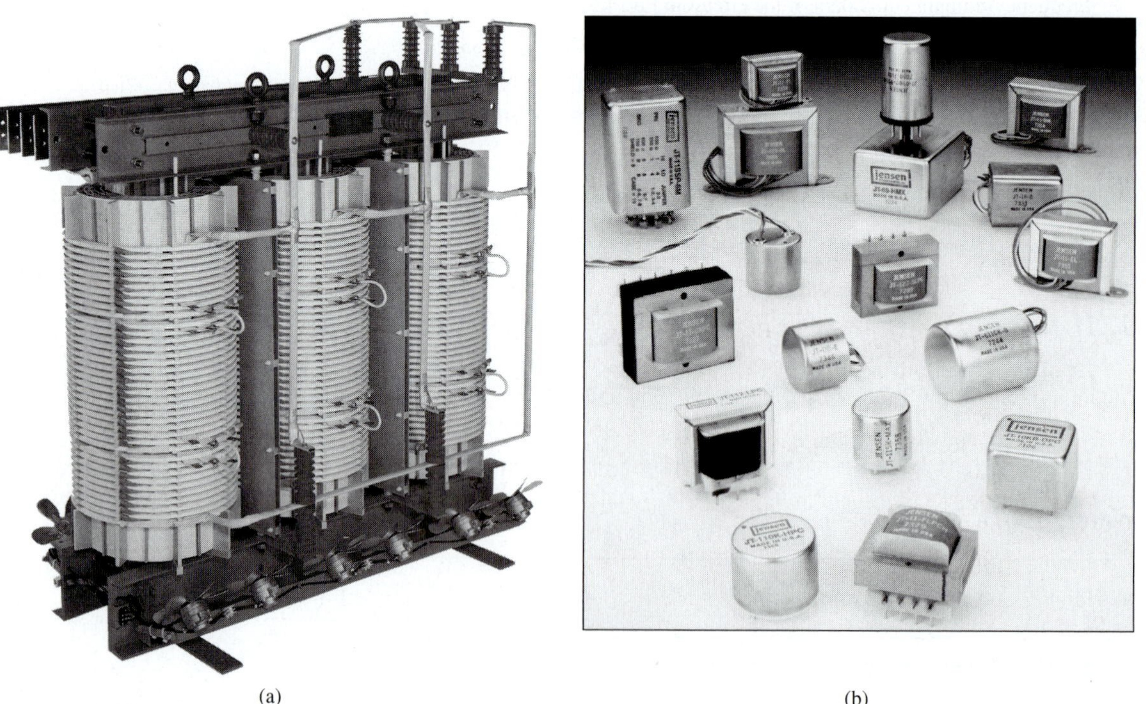

(a) (b)

Figure 13.20
Different types of transformers: (a) copper wound dry power transformer, (b) audio transformers.
Courtesy of: (a) Electric Service Co., (b) Jensen Transformers.

We would like to obtain the input impedance $\mathbf{Z}_{in}$ as seen from the source, because $\mathbf{Z}_{in}$ governs the behavior of the primary circuit. Applying KVL to the two meshes in Fig. 13.19 gives

$$\mathbf{V} = (R_1 + j\omega L_1)\mathbf{I}_1 - j\omega M\mathbf{I}_2 \tag{13.40a}$$

$$0 = -j\omega M\mathbf{I}_1 + (R_2 + j\omega L_2 + \mathbf{Z}_L)\mathbf{I}_2 \tag{13.40b}$$

In Eq. (13.40b), we express $\mathbf{I}_2$ in terms of $\mathbf{I}_1$ and substitute it into Eq. (13.40a). We get the input impedance as

$$\mathbf{Z}_{in} = \frac{\mathbf{V}}{\mathbf{I}_1} = R_1 + j\omega L_1 + \frac{\omega^2 M^2}{R_2 + j\omega L_2 + \mathbf{Z}_L} \tag{13.41}$$

Notice that the input impedance comprises two terms. The first term, $(R_1 + j\omega L_1)$, is the primary impedance. The second term is due to the coupling between the primary and secondary windings. It is as though this impedance is reflected to the primary. Thus, it is known as the *reflected impedance* $\mathbf{Z}_R$, and

Some authors call this the *coupled impedance.*

$$\mathbf{Z}_R = \frac{\omega^2 M^2}{R_2 + j\omega L_2 + \mathbf{Z}_L} \tag{13.42}$$

It should be noted that the result in Eq. (13.41) or (13.42) is not affected by the location of the dots on the transformer, because the same result is produced when M is replaced by $-M$.

The little bit of experience gained in Sections 13.2 and 13.3 in analyzing magnetically coupled circuits is enough to convince anyone that analyzing these circuits is not as easy as circuits in previous chapters. For this reason, it is sometimes convenient to replace a magnetically coupled circuit by an equivalent circuit with no magnetic coupling. We want to replace the linear transformer in Fig. 13.21 by an equivalent T or Π circuit, a circuit that would have no mutual inductance.

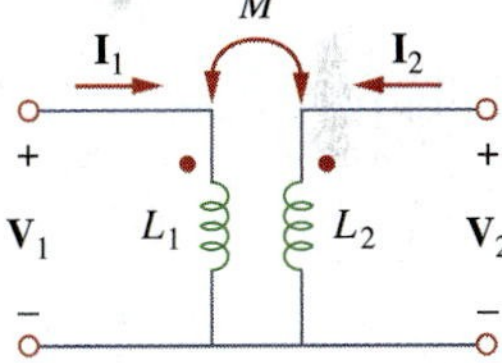

Figure 13.21
Determining the equivalent circuit of a linear transformer.

The voltage-current relationships for the primary and secondary coils give the matrix equation

$$\begin{bmatrix} \mathbf{V}_1 \\ \mathbf{V}_2 \end{bmatrix} = \begin{bmatrix} j\omega L_1 & j\omega M \\ j\omega M & j\omega L_2 \end{bmatrix} \begin{bmatrix} \mathbf{I}_1 \\ \mathbf{I}_2 \end{bmatrix} \tag{13.43}$$

By matrix inversion, this can be written as

$$\begin{bmatrix} \mathbf{I}_1 \\ \mathbf{I}_2 \end{bmatrix} = \begin{bmatrix} \dfrac{L_2}{j\omega(L_1L_2 - M^2)} & \dfrac{-M}{j\omega(L_1L_2 - M^2)} \\ \dfrac{-M}{j\omega(L_1L_2 - M^2)} & \dfrac{L_1}{j\omega(L_1L_2 - M^2)} \end{bmatrix} \begin{bmatrix} \mathbf{V}_1 \\ \mathbf{V}_2 \end{bmatrix} \tag{13.44}$$

Our goal is to match Eqs. (13.43) and (13.44) with the corresponding equations for the T and Π networks.

For the T (or Y) network of Fig. 13.22, mesh analysis provides the terminal equations as

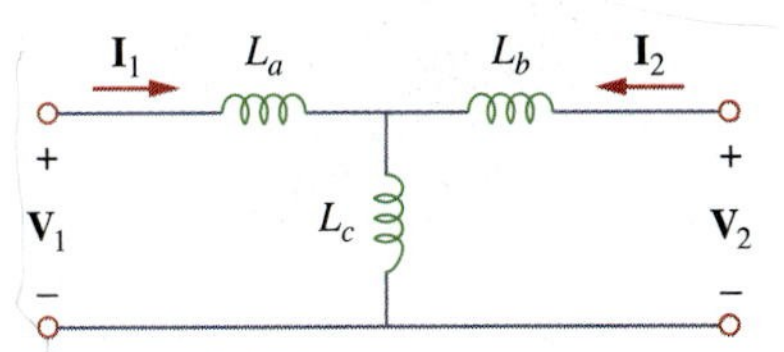

Figure 13.22
An equivalent T circuit.

$$\begin{bmatrix} \mathbf{V}_1 \\ \mathbf{V}_2 \end{bmatrix} = \begin{bmatrix} j\omega(L_a + L_c) & j\omega L_c \\ j\omega L_c & j\omega(L_b + L_c) \end{bmatrix} \begin{bmatrix} \mathbf{I}_1 \\ \mathbf{I}_2 \end{bmatrix} \tag{13.45}$$

If the circuits in Figs. 13.21 and 13.22 are equivalents, Eqs. (13.43) and (13.45) must be identical. Equating terms in the impedance matrices of Eqs. (13.43) and (13.45) leads to

$$L_a = L_1 - M, \qquad L_b = L_2 - M, \qquad L_c = M \tag{13.46}$$

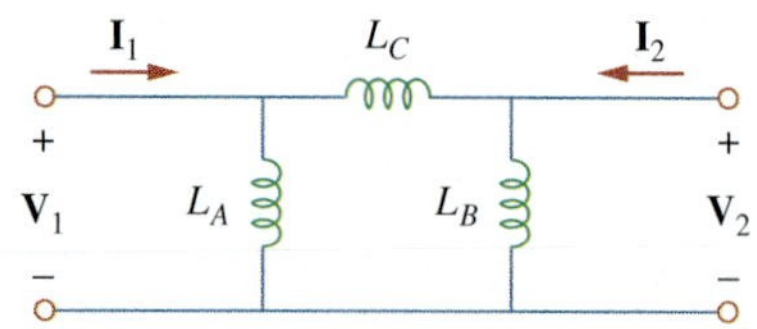

Figure 13.23
An equivalent Π circuit.

For the Π (or Δ) network in Fig. 13.23, nodal analysis gives the terminal equations as

$$\begin{bmatrix} \mathbf{I}_1 \\ \mathbf{I}_2 \end{bmatrix} = \begin{bmatrix} \dfrac{1}{j\omega L_A} + \dfrac{1}{j\omega L_C} & -\dfrac{1}{j\omega L_C} \\ -\dfrac{1}{j\omega L_C} & \dfrac{1}{j\omega L_B} + \dfrac{1}{j\omega L_C} \end{bmatrix} \begin{bmatrix} \mathbf{V}_1 \\ \mathbf{V}_2 \end{bmatrix} \tag{13.47}$$

Equating terms in admittance matrices of Eqs. (13.44) and (13.47), we obtain

$$L_A = \frac{L_1L_2 - M^2}{L_2 - M}, \qquad L_B = \frac{L_1L_2 - M^2}{L_1 - M}, \qquad L_C = \frac{L_1L_2 - M^2}{M} \tag{13.48}$$

Note that in Figs. 13.22 and 13.23, the inductors are not magnetically coupled. Also note that changing the locations of the dots in Fig. 13.21 can cause M to become $-M$. As Example 13.6 illustrates, a negative value of M is physically unrealizable but the equivalent model is still mathematically valid.

Example 13.4

In the circuit of Fig. 13.24, calculate the input impedance and current $\mathbf{I}_1$. Take $\mathbf{Z}_1 = 60 - j100\ \Omega$, $\mathbf{Z}_2 = 30 + j40\ \Omega$, and $\mathbf{Z}_L = 80 + j60\ \Omega$.

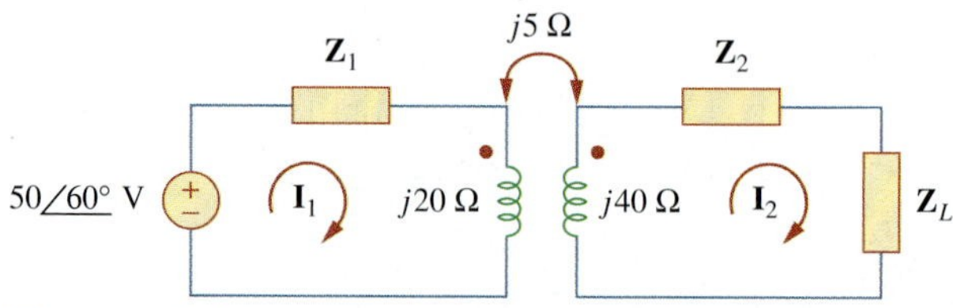

Figure 13.24
For Example 13.4.

Solution:
From Eq. (13.41),

$$\begin{aligned} \mathbf{Z}_{\text{in}} &= \mathbf{Z}_1 + j20 + \frac{(5)^2}{j40 + \mathbf{Z}_2 + \mathbf{Z}_L} \\ &= 60 - j100 + j20 + \frac{25}{110 + j140} \\ &= 60 - j80 + 0.14\underline{/-51.84^\circ} \\ &= 60.09 - j80.11 = 100.14\underline{/-53.1^\circ}\ \Omega \end{aligned}$$

Thus,

$$\mathbf{I}_1 = \frac{\mathbf{V}}{\mathbf{Z}_{in}} = \frac{50\angle 60^\circ}{100.14\angle -53.1^\circ} = 0.5\angle 113.1^\circ \text{ A}$$

Practice Problem 13.4

Find the input impedance of the circuit in Fig. 13.25 and the current from the voltage source.

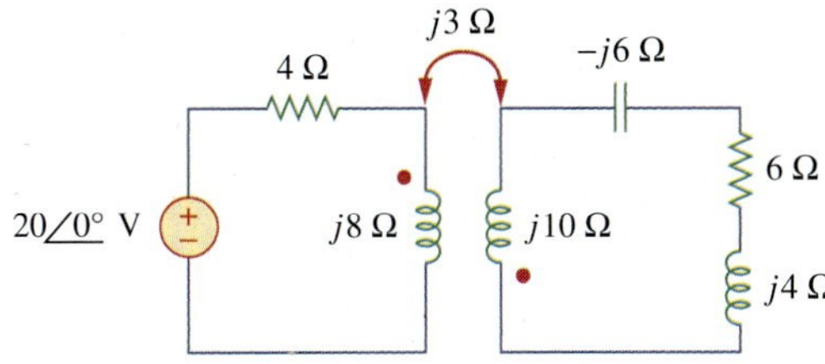

Figure 13.25
For Practice Prob. 13.4.

Answer: $8.58\angle 58.05^\circ\ \Omega$, $2.331\angle -58.05^\circ$ A.

Example 13.5

Determine the T-equivalent circuit of the linear transformer in Fig. 13.26(a).

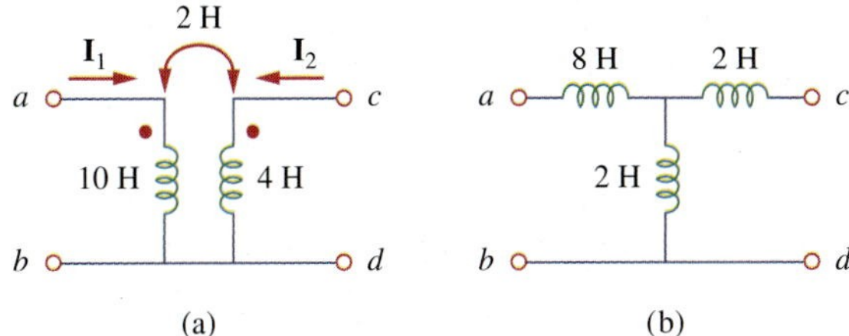

Figure 13.26
For Example 13.5: (a) a linear transformer, (b) its T-equivalent circuit.

Solution:
Given that $L_1 = 10$, $L_2 = 4$, and $M = 2$, the T-equivalent network has the following parameters:

$$L_a = L_1 - M = 10 - 2 = 8 \text{ H}$$

$$L_b = L_2 - M = 4 - 2 = 2 \text{ H}, \qquad L_c = M = 2 \text{ H}$$

The T-equivalent circuit is shown in Fig. 13.26(b). We have assumed that reference directions for currents and voltage polarities in the primary and secondary windings conform to those in Fig. 13.21. Otherwise, we may need to replace M with $-M$, as Example 13.6 illustrates.

Practice Problem 13.5

For the linear transformer in Fig. 13.26(a), find the Π equivalent network.

Answer: $L_A = 18$ H, $L_B = 4.5$ H, $L_C = 18$ H.

Example 13.6

Solve for $\mathbf{I}_1$, $\mathbf{I}_2$, and $\mathbf{V}_o$ in Fig. 13.27 (the same circuit as for Practice Prob. 13.1) using the T-equivalent circuit for the linear transformer.

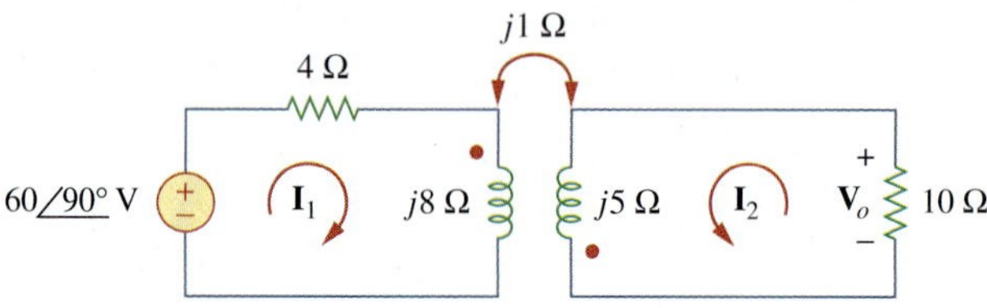

Figure 13.27
For Example 13.6.

Solution:
Notice that the circuit in Fig. 13.27 is the same as that in Fig. 13.10 except that the reference direction for current $\mathbf{I}_2$ has been reversed, just to make the reference directions for the currents for the magnetically coupled coils conform with those in Fig. 13.21.

We need to replace the magnetically coupled coils with the T-equivalent circuit. The relevant portion of the circuit in Fig. 13.27 is shown in Fig. 13.28(a). Comparing Fig. 13.28(a) with Fig. 13.21 shows that there are two differences. First, due to the current reference directions and voltage polarities, we need to replace M by $-M$ to make Fig. 13.28(a) conform with Fig. 13.21. Second, the circuit in Fig. 13.21 is in the time-domain, whereas the circuit in Fig. 13.28(a) is in the frequency-domain. The difference is the factor $j\omega$; that is, L in Fig. 13.21 has been replaced with $j\omega L$ and M with $j\omega M$. Since ω is not specified, we can assume $\omega = 1$ rad/s or any other value; it really does not matter. With these two differences in mind,

$$L_a = L_1 - (-M) = 8 + 1 = 9 \text{ H}$$

$$L_b = L_2 - (-M) = 5 + 1 = 6 \text{ H}, \qquad L_c = -M = -1 \text{ H}$$

Thus, the T-equivalent circuit for the coupled coils is as shown in Fig. 13.28(b).

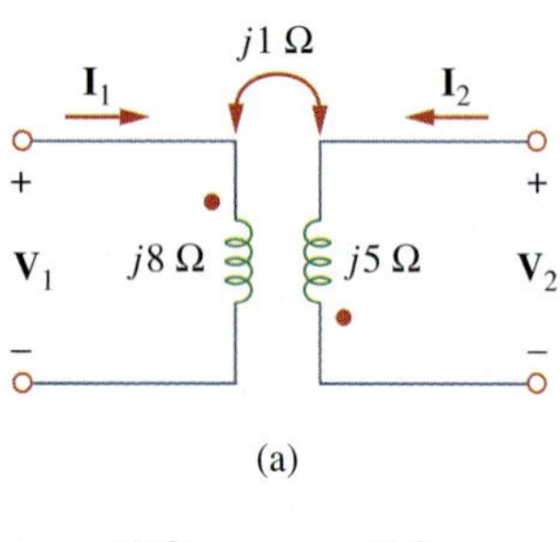

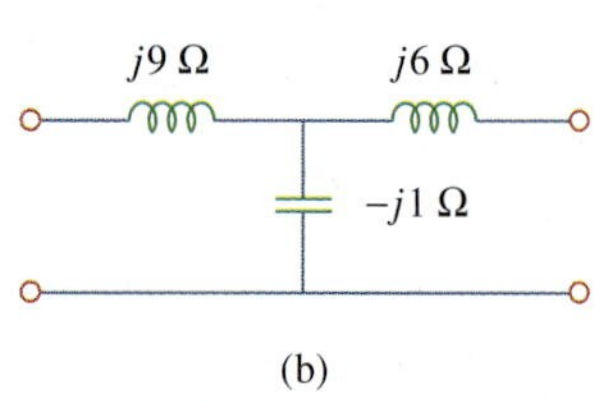

Figure 13.28
For Example 13.6: (a) circuit for coupled coils of Fig. 13.27, (b) T-equivalent circuit.

Inserting the T-equivalent circuit in Fig. 13.28(b) to replace the two coils in Fig. 13.27 gives the equivalent circuit in Fig. 13.29, which can be solved using nodal or mesh analysis. Applying mesh analysis, we obtain

$$j6 = \mathbf{I}_1(4 + j9 - j1) + \mathbf{I}_2(-j1) \tag{13.6.1}$$

and

$$0 = \mathbf{I}_1(-j1) + \mathbf{I}_2(10 + j6 - j1) \tag{13.6.2}$$

From Eq. (13.6.2),

$$\mathbf{I}_1 = \frac{(10 + j5)}{j}\mathbf{I}_2 = (5 - j10)\mathbf{I}_2 \tag{13.6.3}$$

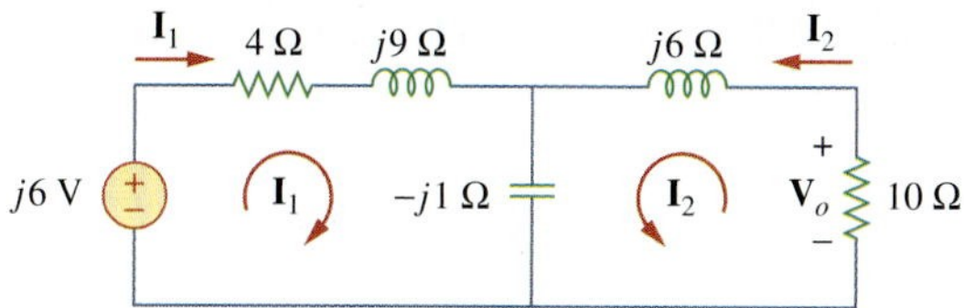

Figure 13.29
For Example 13.6.

Substituting Eq. (13.6.3) into Eq. (13.6.1) gives

$$j6 = (4 + j8)(5 - j10)\mathbf{I}_2 - j\mathbf{I}_2 = (100 - j)\mathbf{I}_2 \simeq 100\mathbf{I}_2$$

Since 100 is very large compared with 1, the imaginary part of $(100 - j)$ can be ignored so that $100 - j \simeq 100$. Hence,

$$\mathbf{I}_2 = \frac{j6}{100} = j0.06 = 0.06\underline{/90^\circ} \text{ A}$$

From Eq. (13.6.3),

$$\mathbf{I}_1 = (5 - j10)j0.06 = 0.6 + j0.3 \text{ A}$$

and

$$\mathbf{V}_o = -10\mathbf{I}_2 = -j0.6 = 0.6\underline{/-90^\circ} \text{ V}$$

This agrees with the answer to Practice Prob. 13.1. Of course, the direction of $\mathbf{I}_2$ in Fig. 13.10 is opposite to that in Fig. 13.27. This will not affect $\mathbf{V}_o$, but the value of $\mathbf{I}_2$ in this example is the negative of that of $\mathbf{I}_2$ in Practice Prob. 13.1. The advantage of using the T-equivalent model for the magnetically coupled coils is that in Fig. 13.29 we do not need to bother with the dot on the coupled coils.

Practice Problem 13.6

Solve the problem in Example 13.1 (see Fig. 13.9) using the T-equivalent model for the magnetically coupled coils.

Answer: $13\underline{/-49.4^\circ}$ A, $2.91\underline{/14.04^\circ}$ A.

13.5 Ideal Transformers

An ideal transformer is one with perfect coupling ($k = 1$). It consists of two (or more) coils with a large number of turns wound on a common core of high permeability. Because of this high permeability of the core, the flux links all the turns of both coils, thereby resulting in a perfect coupling.

To see how an ideal transformer is the limiting case of two coupled inductors where the inductances approach infinity and the coupling is perfect, let us reexamine the circuit in Fig. 13.14. In the frequency domain,

$$\mathbf{V}_1 = j\omega L_1\mathbf{I}_1 + j\omega M\mathbf{I}_2 \quad \textbf{(13.49a)}$$

$$\mathbf{V}_2 = j\omega M\mathbf{I}_1 + j\omega L_2\mathbf{I}_2 \quad \textbf{(13.49b)}$$

From Eq. (13.49a), $\mathbf{I}_1 = (\mathbf{V}_1 - j\omega M\mathbf{I}_2)/j\omega L_1$ (we could have also use this equation to develop the current ratios instead of using the conservation of power which we will do shortly). Substituting this in Eq. (13.49b) gives

$$\mathbf{V}_2 = j\omega L_2\mathbf{I}_2 + \frac{M\mathbf{V}_1}{L_1} - \frac{j\omega M^2\mathbf{I}_2}{L_1}$$

But $M = \sqrt{L_1L_2}$ for perfect coupling ($k = 1$). Hence,

$$\mathbf{V}_2 = j\omega L_2\mathbf{I}_2 + \frac{\sqrt{L_1L_2}\mathbf{V}_1}{L_1} - \frac{j\omega L_1L_2\mathbf{I}_2}{L_1} = \sqrt{\frac{L_2}{L_1}}\mathbf{V}_1 = n\mathbf{V}_1$$

where $n = \sqrt{L_2/L_1}$ and is called the *turns ratio*. As $L_1, L_2, M \to \infty$ such that n remains the same, the coupled coils become an ideal transformer. A transformer is said to be ideal if it has the following properties:

1. Coils have very large reactances ($L_1, L_2, M \to \infty$).
2. Coupling coefficient is equal to unity ($k = 1$).
3. Primary and secondary coils are lossless ($R_1 = 0 = R_2$).

> An **ideal transformer** is a unity-coupled, lossless transformer in which the primary and secondary coils have infinite self-inductances.

Iron-core transformers are close approximations to ideal transformers. These are used in power systems and electronics.

Figure 13.30(a) shows a typical ideal transformer; the circuit symbol is in Fig. 13.30(b). The vertical lines between the coils indicate an iron core as distinct from the air core used in linear transformers. The primary winding has N_1 turns; the secondary winding has N_2 turns.

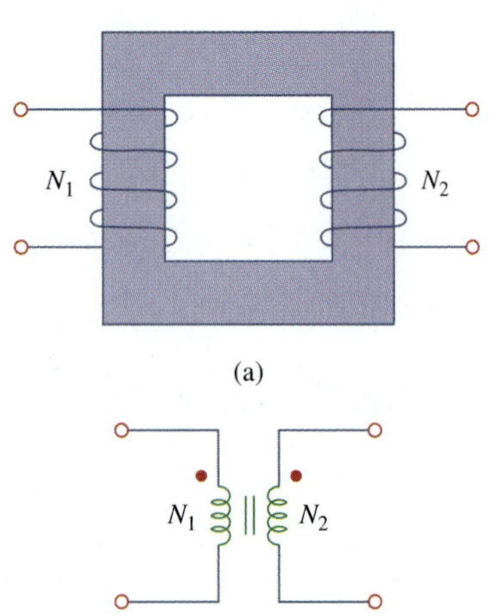

Figure 13.30
(a) Ideal transformer, (b) circuit symbol for ideal transformers.

When a sinusoidal voltage is applied to the primary winding as shown in Fig. 13.31, the same magnetic flux ϕ goes through both windings. According to Faraday's law, the voltage across the primary winding is

$$v_1 = N_1\frac{d\phi}{dt} \tag{13.50a}$$

while that across the secondary winding is

$$v_2 = N_2\frac{d\phi}{dt} \tag{13.50b}$$

Dividing Eq. (13.50b) by Eq. (13.50a), we get

$$\frac{v_2}{v_1} = \frac{N_2}{N_1} = n \tag{13.51}$$

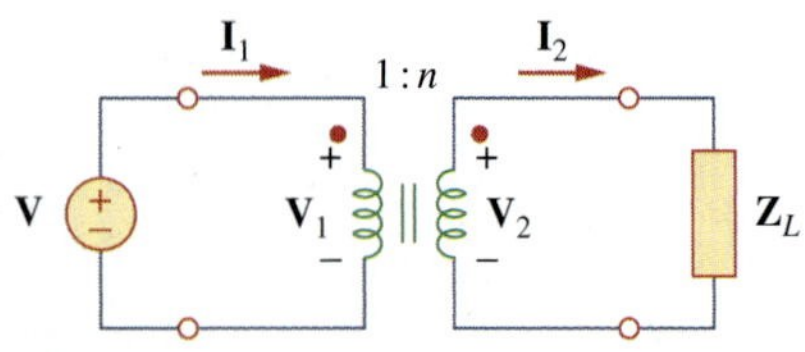

Figure 13.31
Relating primary and secondary quantities in an ideal transformer.

where n is, again, the *turns ratio* or *transformation ratio*. We can use the phasor voltages $\mathbf{V}_1$ and $\mathbf{V}_2$ rather than the instantaneous values v_1 and v_2. Thus, Eq. (13.51) may be written as

$$\boxed{\frac{\mathbf{V}_2}{\mathbf{V}_1} = \frac{N_2}{N_1} = n} \tag{13.52}$$

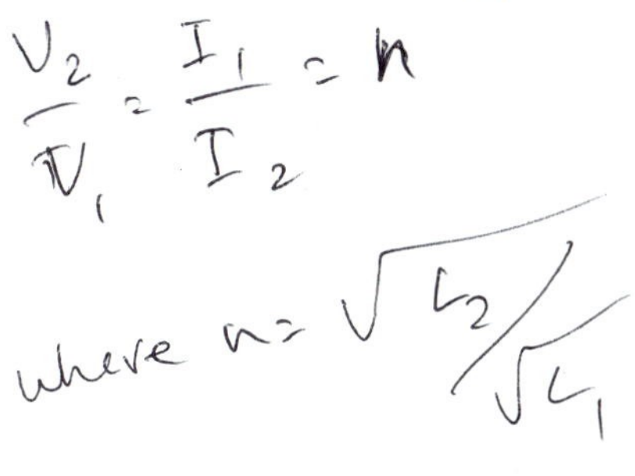

For the reason of power conservation, the energy supplied to the primary must equal the energy absorbed by the secondary, since there are no losses in an ideal transformer. This implies that

$$v_1 i_1 = v_2 i_2 \tag{13.53}$$

In phasor form, Eq. (13.53) in conjunction with Eq. (13.52) becomes

$$\frac{\mathbf{I}_1}{\mathbf{I}_2} = \frac{\mathbf{V}_2}{\mathbf{V}_1} = n \tag{13.54}$$

showing that the primary and secondary currents are related to the turns ratio in the inverse manner as the voltages. Thus,

$$\frac{\mathbf{I}_2}{\mathbf{I}_1} = \frac{N_1}{N_2} = \frac{1}{n} \tag{13.55}$$

When $n = 1$, we generally call the transformer an *isolation transformer*. The reason will become obvious in Section 13.9.1. If $n > 1$, we have a *step-up transformer*, as the voltage is increased from primary to secondary ($\mathbf{V}_2 > \mathbf{V}_1$). On the other hand, if $n < 1$, the transformer is a *step-down transformer*, since the voltage is decreased from primary to secondary ($\mathbf{V}_2 < \mathbf{V}_1$).

> A **step-down transformer** is one whose secondary voltage is less than its primary voltage.

> A **step-up transformer** is one whose secondary voltage is greater than its primary voltage.

The ratings of transformers are usually specified as V_1/V_2. A transformer with rating 2400/120 V should have 2400 V on the primary and 120 in the secondary (i.e., a step-down transformer). Keep in mind that the voltage ratings are in rms.

Power companies often generate at some convenient voltage and use a step-up transformer to increase the voltage so that the power can be transmitted at very high voltage and low current over transmission lines, resulting in significant cost savings. Near residential consumer premises, step-down transformers are used to bring the voltage down to 120 V. Section 13.9.3 will elaborate on this.

It is important that we know how to get the proper polarity of the voltages and the direction of the currents for the transformer in Fig. 13.31. If the polarity of $\mathbf{V}_1$ or $\mathbf{V}_2$ or the direction of $\mathbf{I}_1$ or $\mathbf{I}_2$ is changed, n in Eqs. (13.51) to (13.55) may need to be replaced by $-n$. The two simple rules to follow are:

1. If $\mathbf{V}_1$ and $\mathbf{V}_2$ are *both* positive or both negative at the dotted terminals, use $+n$ in Eq. (13.52). Otherwise, use $-n$.
2. If $\mathbf{I}_1$ and $\mathbf{I}_2$ *both* enter into or both leave the dotted terminals, use $-n$ in Eq. (13.55). Otherwise, use $+n$.

The rules are demonstrated with the four circuits in Fig. 13.32.

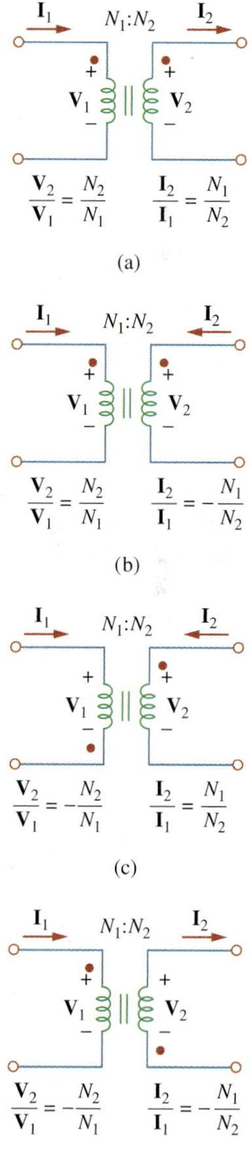

Figure 13.32
Typical circuits illustrating proper voltage polarities and current directions in an ideal transformer.

Using Eqs. (13.52) and (13.55), we can always express $\mathbf{V}_1$ in terms of $\mathbf{V}_2$ and $\mathbf{I}_1$ in terms of $\mathbf{I}_2$, or vice versa:

$$\mathbf{V}_1 = \frac{\mathbf{V}_2}{n} \quad \text{or} \quad \mathbf{V}_2 = n\mathbf{V}_1 \tag{13.56}$$

$$\mathbf{I}_1 = n\mathbf{I}_2 \quad \text{or} \quad \mathbf{I}_2 = \frac{\mathbf{I}_1}{n} \tag{13.57}$$

The complex power in the primary winding is

$$\mathbf{S}_1 = \mathbf{V}_1\mathbf{I}_1^* = \frac{\mathbf{V}_2}{n}(n\mathbf{I}_2)^* = \mathbf{V}_2\mathbf{I}_2^* = \mathbf{S}_2 \tag{13.58}$$

showing that the complex power supplied to the primary is delivered to the secondary without loss. The transformer absorbs no power. Of course, we should expect this, since the ideal transformer is lossless. The input impedance as seen by the source in Fig. 13.31 is found from Eqs. (13.56) and (13.57) as

$$\mathbf{Z}_{\text{in}} = \frac{\mathbf{V}_1}{\mathbf{I}_1} = \frac{1}{n^2}\frac{\mathbf{V}_2}{\mathbf{I}_2} \tag{13.59}$$

It is evident from Fig. 13.31 that $\mathbf{V}_2/\mathbf{I}_2 = \mathbf{Z}_L$, so that

$$\mathbf{Z}_{\text{in}} = \frac{\mathbf{Z}_L}{n^2} \tag{13.60}$$

Notice that an ideal transformer reflects an impedance as the square of the turns ratio.

The input impedance is also called the *reflected impedance*, since it appears as if the load impedance is reflected to the primary side. This ability of the transformer to transform a given impedance into another impedance provides us a means of *impedance matching* to ensure maximum power transfer. The idea of impedance matching is very useful in practice and will be discussed more in Section 13.9.2.

In analyzing a circuit containing an ideal transformer, it is common practice to eliminate the transformer by reflecting impedances and sources from one side of the transformer to the other. In the circuit of Fig. 13.33, suppose we want to reflect the secondary side of the circuit to the primary side. We find the Thevenin equivalent of the circuit to the right of the terminals *a-b*. We obtain $\mathbf{V}_{\text{Th}}$ as the open-circuit voltage at terminals *a-b*, as shown in Fig. 13.34(a).

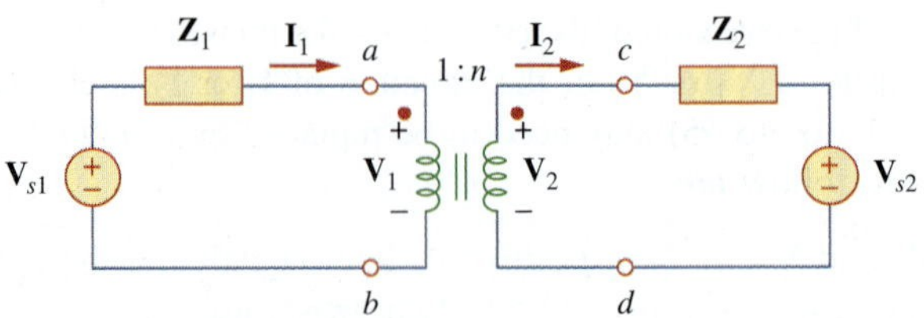

Figure 13.33
Ideal transformer circuit whose equivalent circuits are to be found.

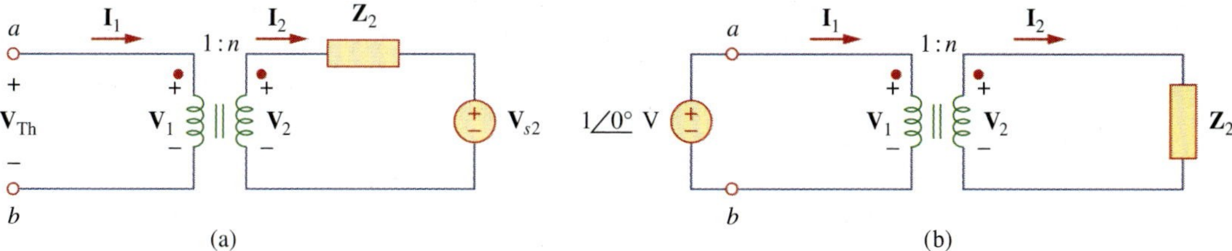

Figure 13.34
(a) Obtaining $\mathbf{V}_{\text{Th}}$ for the circuit in Fig. 13.33, (b) obtaining $\mathbf{Z}_{\text{Th}}$ for the circuit in Fig. 13.33.

Since terminals *a-b* are open, $\mathbf{I}_1 = 0 = \mathbf{I}_2$ so that $\mathbf{V}_2 = \mathbf{V}_{s2}$. Hence, From Eq. (13.56),

$$\mathbf{V}_{\text{Th}} = \mathbf{V}_1 = \frac{\mathbf{V}_2}{n} = \frac{\mathbf{V}_{s2}}{n} \tag{13.61}$$

To get $\mathbf{Z}_{\text{Th}}$, we remove the voltage source in the secondary winding and insert a unit source at terminals *a-b*, as in Fig. 13.34(b). From Eqs. (13.56) and (13.57), $\mathbf{I}_1 = n\mathbf{I}_2$ and $\mathbf{V}_1 = \mathbf{V}_2/n$, so that

$$\mathbf{Z}_{\text{Th}} = \frac{\mathbf{V}_1}{\mathbf{I}_1} = \frac{\mathbf{V}_2/n}{n\mathbf{I}_2} = \frac{\mathbf{Z}_2}{n^2}, \qquad \mathbf{V}_2 = \mathbf{Z}_2\mathbf{I}_2 \tag{13.62}$$

which is what we should have expected from Eq. (13.60). Once we have $\mathbf{V}_{\text{Th}}$ and $\mathbf{Z}_{\text{Th}}$, we add the Thevenin equivalent to the part of the circuit in Fig. 13.33 to the left of terminals *a-b*. Figure 13.35 shows the result.

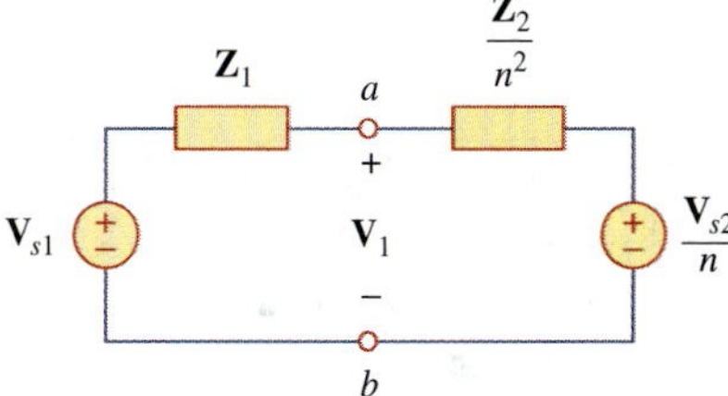

Figure 13.35
Equivalent circuit for Fig. 13.33 obtained by reflecting the secondary circuit to the primary side.

The general rule for eliminating the transformer and reflecting the secondary circuit to the primary side is: divide the secondary impedance by n^2, divide the secondary voltage by n, and multiply the secondary current by n.

We can also reflect the primary side of the circuit in Fig. 13.33 to the secondary side. Figure 13.36 shows the equivalent circuit.

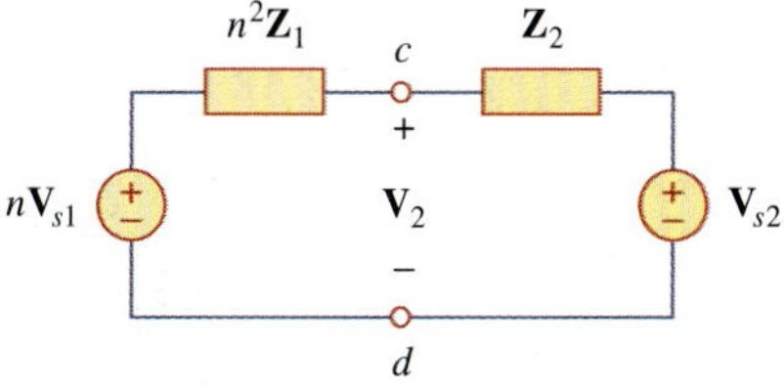

Figure 13.36
Equivalent circuit for Fig. 13.33 obtained by reflecting the primary circuit to the secondary side.

The rule for eliminating the transformer and reflecting the primary circuit to the secondary side is: multiply the primary impedance by n^2, multiply the primary voltage by n, and divide the primary current by n.

According to Eq. (13.58), the power remains the same, whether calculated on the primary or the secondary side. But realize that this reflection approach only applies if there are no external connections between the primary and secondary windings. When we have external connections between the primary and secondary windings, we simply use regular mesh and nodal analysis. Examples of circuits where there are external connections between the primary and secondary windings are in Figs. 13.39 and 13.40. Also note that if the locations of the dots in Fig. 13.33 are changed, we might have to replace n by $-n$ in order to obey the dot rule, illustrated in Fig. 13.32.

Example 13.7

An ideal transformer is rated at 2400/120 V, 9.6 kVA, and has 50 turns on the secondary side. Calculate: (a) the turns ratio, (b) the number of turns on the primary side, and (c) the current ratings for the primary and secondary windings.

Solution:

(a) This is a step-down transformer, since $V_1 = 2{,}400\text{ V} > V_2 = 120\text{ V}$.

$$n = \frac{V_2}{V_1} = \frac{120}{2{,}400} = 0.05$$

(b)

$$n = \frac{N_2}{N_1} \quad \Rightarrow \quad 0.05 = \frac{50}{N_1}$$

or

$$N_1 = \frac{50}{0.05} = 1{,}000 \text{ turns}$$

(c) $S = V_1 I_1 = V_2 I_2 = 9.6\text{ kVA}$. Hence,

$$I_1 = \frac{9{,}600}{V_1} = \frac{9{,}600}{2{,}400} = 4\text{ A}$$

$$I_2 = \frac{9{,}600}{V_2} = \frac{9{,}600}{120} = 80\text{ A} \quad \text{or} \quad I_2 = \frac{I_1}{n} = \frac{4}{0.05} = 80\text{ A}$$

Practice Problem 13.7

The primary current to an ideal transformer rated at 3300/110 V is 5 A. Calculate: (a) the turns ratio, (b) the kVA rating, (c) the secondary current.

Answer: (a) 1/30, (b) 16.5 kVA, (c) 150 A.

Example 13.8

For the ideal transformer circuit of Fig. 13.37, find: (a) the source current $\mathbf{I}_1$, (b) the output voltage $\mathbf{V}_o$, and (c) the complex power supplied by the source.

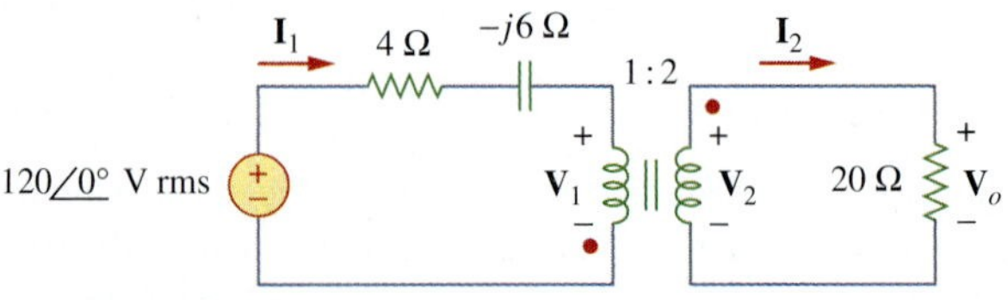

Figure 13.37
For Example 13.8.

Solution:

(a) The 20-Ω impedance can be reflected to the primary side and we get

$$\mathbf{Z}_R = \frac{20}{n^2} = \frac{20}{4} = 5\ \Omega$$

Thus,

$$\mathbf{Z}_{\text{in}} = 4 - j6 + \mathbf{Z}_R = 9 - j6 = 10.82\underline{/-33.69^\circ}\ \Omega$$

$$\mathbf{I}_1 = \frac{120\underline{/0^\circ}}{\mathbf{Z}_{\text{in}}} = \frac{120\underline{/0^\circ}}{10.82\underline{/-33.69^\circ}} = 11.09\underline{/33.69^\circ}\text{ A}$$

(b) Since both $\mathbf{I}_1$ and $\mathbf{I}_2$ leave the dotted terminals,

$$\mathbf{I}_2 = -\frac{1}{n}\mathbf{I}_1 = -5.545\underline{/33.69^\circ}\text{ A}$$

$$\mathbf{V}_o = 20\mathbf{I}_2 = 110.9\underline{/213.69^\circ}\text{ V}$$

(c) The complex power supplied is

$$\mathbf{S} = \mathbf{V}_s\mathbf{I}_1^* = (120\underline{/0^\circ})(11.09\underline{/-33.69^\circ}) = 1{,}330.8\underline{/-33.69^\circ}\text{ VA}$$

Practice Problem 13.8

In the ideal transformer circuit of Fig. 13.38, find $\mathbf{V}_o$ and the complex power supplied by the source.

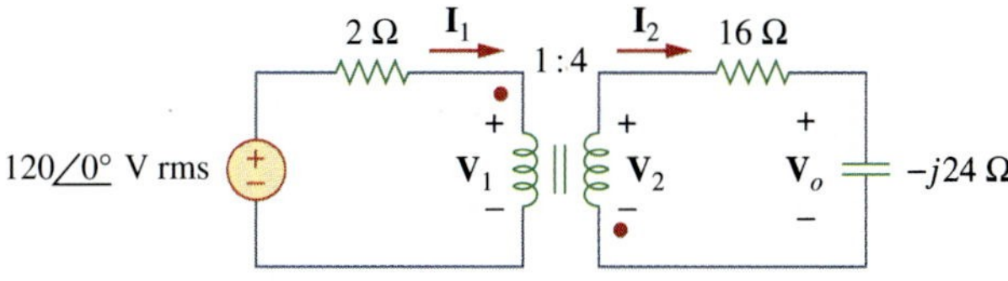

Figure 13.38
For Practice Prob. 13.8.

Answer: $214.7\underline{/116.56^\circ}$ V, $4.293\underline{/-26.56^\circ}$ kVA.

Example 13.9

Calculate the power supplied to the 10-Ω resistor in the ideal transformer circuit of Fig. 13.39.

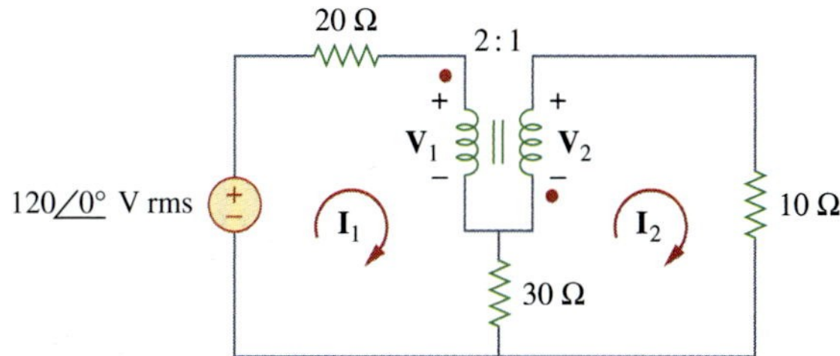

Figure 13.39
For Example 13.9.

Solution:

Reflection to the secondary or primary side cannot be done with this circuit: there is direct connection between the primary and secondary sides due to the 30-Ω resistor. We apply mesh analysis. For mesh 1,

$$-120 + (20 + 30)\mathbf{I}_1 - 30\mathbf{I}_2 + \mathbf{V}_1 = 0$$

or

$$50\mathbf{I}_1 - 30\mathbf{I}_2 + \mathbf{V}_1 = 120 \tag{13.9.1}$$

For mesh 2,

$$-\mathbf{V}_2 + (10 + 30)\mathbf{I}_2 - 30\mathbf{I}_1 = 0$$

or

$$-30\mathbf{I}_1 + 40\mathbf{I}_2 - \mathbf{V}_2 = 0 \tag{13.9.2}$$

At the transformer terminals,

$$\mathbf{V}_2 = -\frac{1}{2}\mathbf{V}_1 \tag{13.9.3}$$

$$\mathbf{I}_2 = -2\mathbf{I}_1 \tag{13.9.4}$$

(Note that $n = 1/2$.) We now have four equations and four unknowns, but our goal is to get $\mathbf{I}_2$. So we substitute for $\mathbf{V}_1$ and $\mathbf{I}_1$ in terms of $\mathbf{V}_2$ and $\mathbf{I}_2$ in Eqs. (13.9.1) and (13.9.2). Equation (13.9.1) becomes

$$-55\mathbf{I}_2 - 2\mathbf{V}_2 = 120 \tag{13.9.5}$$

and Eq. (13.9.2) becomes

$$15\mathbf{I}_2 + 40\mathbf{I}_2 - \mathbf{V}_2 = 0 \quad \Rightarrow \quad \mathbf{V}_2 = 55\mathbf{I}_2 \tag{13.9.6}$$

Substituting Eq. (13.9.6) in Eq. (13.9.5),

$$-165\mathbf{I}_2 = 120 \quad \Rightarrow \quad \mathbf{I}_2 = -\frac{120}{165} = -0.7272 \text{ A}$$

The power absorbed by the 10-Ω resistor is

$$P = (-0.7272)^2(10) = 5.3 \text{ W}$$

Practice Problem 13.9

Find $\mathbf{V}_o$ in the circuit of Fig. 13.40.

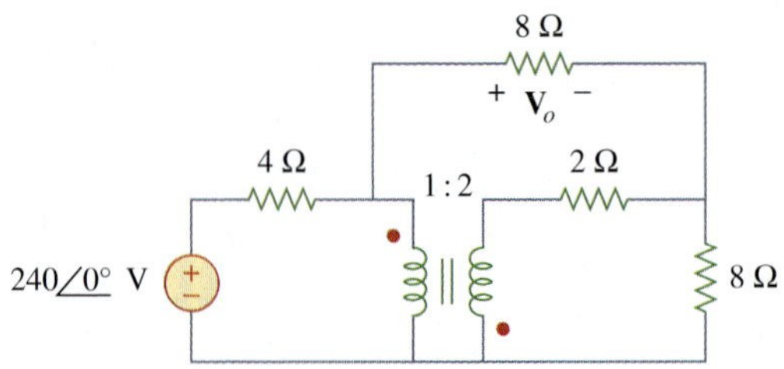

Figure 13.40
For Practice Prob. 13.9.

Answer: 96 V.

13.6 Ideal Autotransformers

Unlike the conventional two-winding transformer we have considered so far, an *autotransformer* has a single continuous winding with a connection point called a *tap* between the primary and secondary sides. The tap is often adjustable so as to provide the desired turns ratio for stepping up or stepping down the voltage. This way, a variable voltage is provided to the load connected to the autotransformer.

Figure 13.41
A typical autotransformer.
Courtesy of Todd Systems, Inc.

An **autotransformer** is a transformer in which both the primary and the secondary are in a single winding.

Figure 13.41 shows a typical autotransformer. As shown in Fig. 13.42, the autotransformer can operate in the step-down or step-up mode. The autotransformer is a type of power transformer. Its major advantage over the two-winding transformer is its ability to transfer larger apparent power. Example 13.10 will demonstrate this. Another advantage is that an autotransformer is smaller and lighter than an equivalent two-winding transformer. However, since both the primary and secondary windings are one winding, *electrical isolation* (no direct electrical connection) is lost. (We will see how the property of electrical isolation in the conventional transformer is practically employed in Section 13.9.1.) The lack of electrical isolation between the primary and secondary windings is a major disadvantage of the autotransformer.

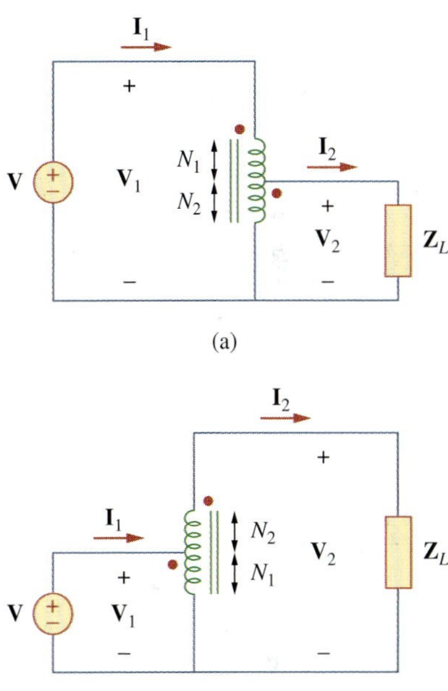

Figure 13.42
(a) Step-down autotransformer, (b) step-up autotransformer.

Some of the formulas we derived for ideal transformers apply to ideal autotransformers as well. For the step-down autotransformer circuit of Fig. 13.42(a), Eq. (13.52) gives

$$\frac{\mathbf{V}_1}{\mathbf{V}_2} = \frac{N_1 + N_2}{N_2} = 1 + \frac{N_1}{N_2} \quad \textbf{(13.63)}$$

As an ideal autotransformer, there are no losses, so the complex power remains the same in the primary and secondary windings:

$$\mathbf{S}_1 = \mathbf{V}_1\mathbf{I}_1^* = \mathbf{S}_2 = \mathbf{V}_2\mathbf{I}_2^* \quad \textbf{(13.64)}$$

Equation (13.64) can also be expressed as

$$V_1 I_1 = V_2 I_2$$

or

$$\frac{V_2}{V_1} = \frac{I_1}{I_2} \quad \textbf{(13.65)}$$

Thus, the current relationship is

$$\frac{\mathbf{I}_1}{\mathbf{I}_2} = \frac{N_2}{N_1 + N_2} \quad \textbf{(13.66)}$$

For the step-up autotransformer circuit of Fig. 13.42(b),

$$\frac{\mathbf{V}_1}{N_1} = \frac{\mathbf{V}_2}{N_1 + N_2}$$

or

$$\frac{\mathbf{V}_1}{\mathbf{V}_2} = \frac{N_1}{N_1 + N_2} \tag{13.67}$$

The complex power given by Eq. (13.64) also applies to the step-up autotransformer so that Eq. (13.65) again applies. Hence, the current relationship is

$$\frac{\mathbf{I}_1}{\mathbf{I}_2} = \frac{N_1 + N_2}{N_1} = 1 + \frac{N_2}{N_1} \tag{13.68}$$

A major difference between conventional transformers and autotransformers is that the primary and secondary sides of the autotransformer are not only coupled magnetically but also coupled conductively. The autotransformer can be used in place of a conventional transformer when electrical isolation is not required.

Example 13.10

Compare the power ratings of the two-winding transformer in Fig. 13.43(a) and the autotransformer in Fig. 13.43(b).

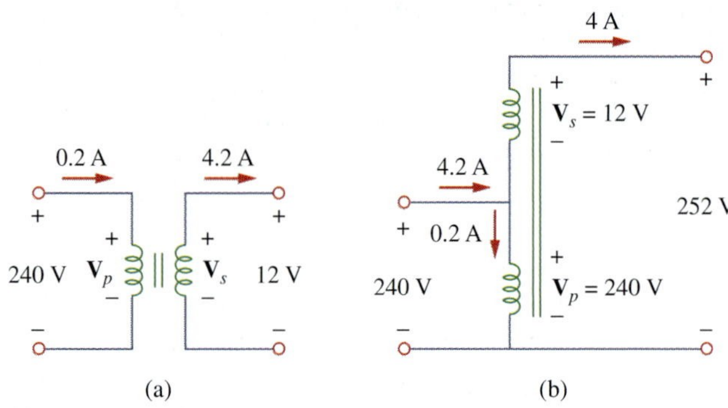

Figure 13.43
For Example 13.10.

Solution:

Although the primary and secondary windings of the autotransformer are together as a continuous winding, they are separated in Fig. 13.43(b) for clarity. We note that the current and voltage of each winding of the autotransformer in Fig. 13.43(b) are the same as those for the two-winding transformer in Fig. 13.43(a). This is the basis of comparing their power ratings.

For the two-winding transformer, the power rating is

$$S_1 = 0.2(240) = 48 \text{ VA} \qquad \text{or} \qquad S_2 = 4(12) = 48 \text{ VA}$$

For the autotransformer, the power rating is

$$S_1 = 4.2(240) = 1{,}008 \text{ VA} \qquad \text{or} \qquad S_2 = 4(252) = 1{,}008 \text{ VA}$$

which is 21 times the power rating of the two-winding transformer.

Practice Problem 13.10

Refer to Fig. 13.43. If the two-winding transformer is a 30-VA, 120 V/10 V transformer, what is the power rating of the autotransformer?

Answer: 390 VA.

Example 13.11

Refer to the autotransformer circuit in Fig. 13.44. Calculate: (a) $\mathbf{I}_1$, $\mathbf{I}_2$, and $\mathbf{I}_o$ if $\mathbf{Z}_L = 8 + j6\ \Omega$, and (b) the complex power supplied to the load.

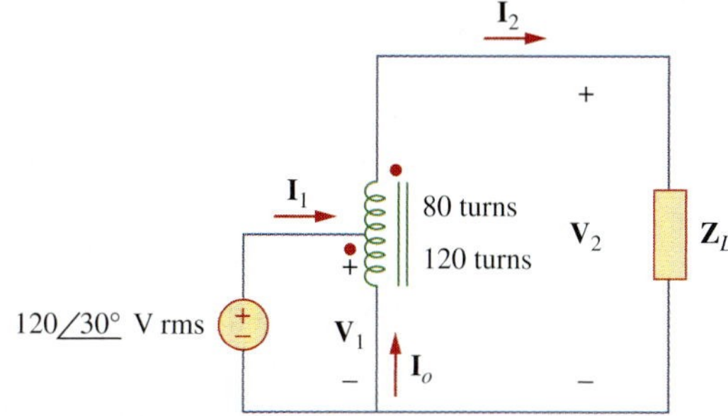

Figure 13.44
For Example 13.11.

Solution:

(a) This is a step-up autotransformer with $N_1 = 80$, $N_2 = 120$, $\mathbf{V}_1 = 120\underline{/30^\circ}$, so Eq. (13.67) can be used to find $\mathbf{V}_2$ by

$$\frac{\mathbf{V}_1}{\mathbf{V}_2} = \frac{N_1}{N_1 + N_2} = \frac{80}{200}$$

or

$$\mathbf{V}_2 = \frac{200}{80}\mathbf{V}_1 = \frac{200}{80}(120\underline{/30^\circ}) = 300\underline{/30^\circ}\text{ V}$$

$$\mathbf{I}_2 = \frac{\mathbf{V}_2}{\mathbf{Z}_L} = \frac{300\underline{/30^\circ}}{8 + j6} = \frac{300\underline{/30^\circ}}{10\underline{/36.87^\circ}} = 30\underline{/-6.87^\circ}\text{ A}$$

But

$$\frac{\mathbf{I}_1}{\mathbf{I}_2} = \frac{N_1 + N_2}{N_1} = \frac{200}{80}$$

or

$$\mathbf{I}_1 = \frac{200}{80}\mathbf{I}_2 = \frac{200}{80}(30\underline{/-6.87^\circ}) = 75\underline{/-6.87^\circ}\text{ A}$$

At the tap, KCL gives

$$\mathbf{I}_1 + \mathbf{I}_o = \mathbf{I}_2$$

or

$$\mathbf{I}_o = \mathbf{I}_2 - \mathbf{I}_1 = 30\underline{/-6.87^\circ} - 75\underline{/-6.87^\circ} = 45\underline{/173.13^\circ}\text{ A}$$

(b) The complex power supplied to the load is

$$\mathbf{S}_2 = \mathbf{V}_2\mathbf{I}_2^* = |\mathbf{I}_2|^2\mathbf{Z}_L = (30)^2(10\underline{/36.87^\circ}) = 9\underline{/36.87^\circ}\text{ kVA}$$

Practice Problem 13.11

In the autotransformer circuit of Fig. 13.45, find currents $\mathbf{I}_1$, $\mathbf{I}_2$, and $\mathbf{I}_o$. Take $\mathbf{V}_1 = 1{,}250$ V, $\mathbf{V}_2 = 500$ V.

Answer: 12.8 A, 32 A, 19.2 A.

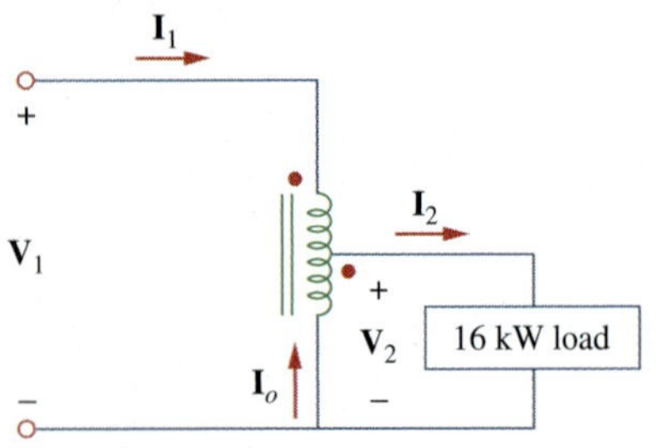

Figure 13.45
For Practice Prob. 13.11.

13.7 †Three-Phase Transformers

To meet the demand for three-phase power transmission, transformer connections compatible with three-phase operations are needed. We can achieve the transformer connections in two ways: by connecting three single-phase transformers, thereby forming a so-called *transformer bank*, or by using a special three-phase transformer. For the same kVA rating, a three-phase transformer is always smaller and cheaper than three single-phase transformers. When single-phase transformers are used, one must ensure that they have the same turns ratio n to achieve a balanced three-phase system. There are four standard ways of connecting three single-phase transformers or a three-phase transformer for three-phase operations: Y-Y, Δ-Δ, Y-Δ, and Δ-Y.

For any of the four connections, the total apparent power S_T, real power P_T, and reactive power Q_T are obtained as

$$S_T = \sqrt{3}V_L I_L \tag{13.69a}$$

$$P_T = S_T \cos\theta = \sqrt{3}V_L I_L \cos\theta \tag{13.69b}$$

$$Q_T = S_T \sin\theta = \sqrt{3}V_L I_L \sin\theta \tag{13.69c}$$

where V_L and I_L are, respectively, equal to the line voltage V_{Lp} and the line current I_{Lp} for the primary side, or the line voltage V_{Ls} and the line current I_{Ls} for the secondary side. Notice From Eq. (13.69) that for each of the four connections, $V_{Ls}I_{Ls} = V_{Lp}I_{Lp}$, since power must be conserved in an ideal transformer.

For the Y-Y connection (Fig. 13.46), the line voltage V_{Lp} at the primary side, the line voltage V_{Ls} on the secondary side, the line current I_{Lp} on the primary side, and the line current I_{Ls} on the secondary side are related to the transformer per phase turns ratio n according to Eqs. (13.52) and (13.55) as

$$V_{Ls} = nV_{Lp} \tag{13.70a}$$

$$I_{Ls} = \frac{I_{Lp}}{n} \tag{13.70b}$$

For the Δ-Δ connection (Fig. 13.47), Eq. (13.70) also applies for the line voltages and line currents. This connection is unique in the

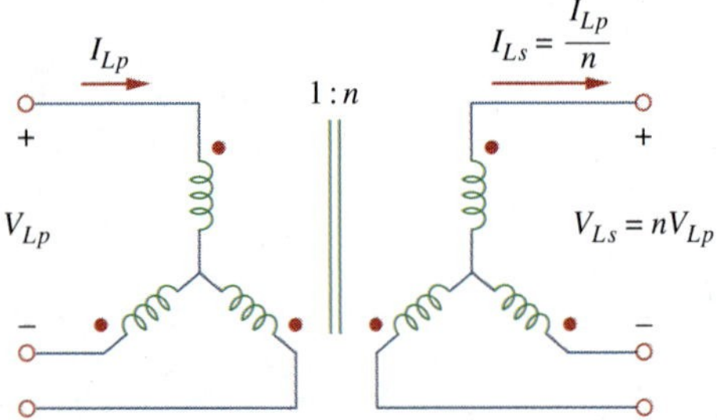

Figure 13.46
Y-Y three-phase transformer connection.

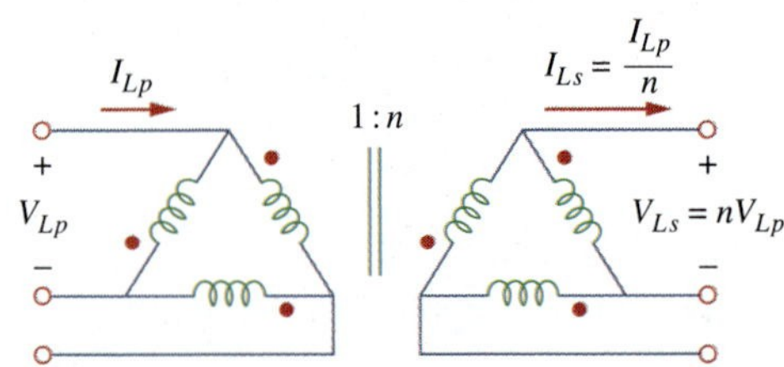

Figure 13.47
Δ-Δ three-phase transformer connection.

sense that if one of the transformers is removed for repair or maintenance, the other two form an *open delta*, which can provide three-phase voltages at a reduced level of the original three-phase transformer.

For the Y-Δ connection (Fig. 13.48), there is a factor of $\sqrt{3}$ arising from the line-phase values in addition to the transformer per phase turns ratio n. Thus,

$$V_{Ls} = \frac{nV_{Lp}}{\sqrt{3}} \tag{13.71a}$$

$$I_{Ls} = \frac{\sqrt{3}I_{Lp}}{n} \tag{13.71b}$$

Similarly, for the Δ-Y connection (Fig. 13.49),

$$V_{Ls} = n\sqrt{3}V_{Lp} \tag{13.72a}$$

$$I_{Ls} = \frac{I_{Lp}}{n\sqrt{3}} \tag{13.72b}$$

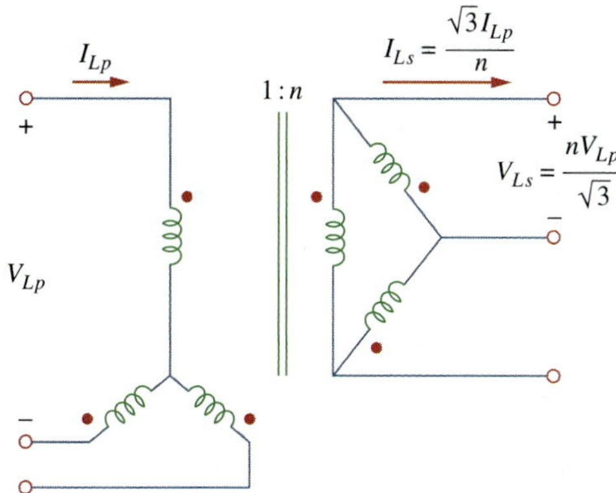

Figure 13.48
Y-Δ three-phase transformer connection.

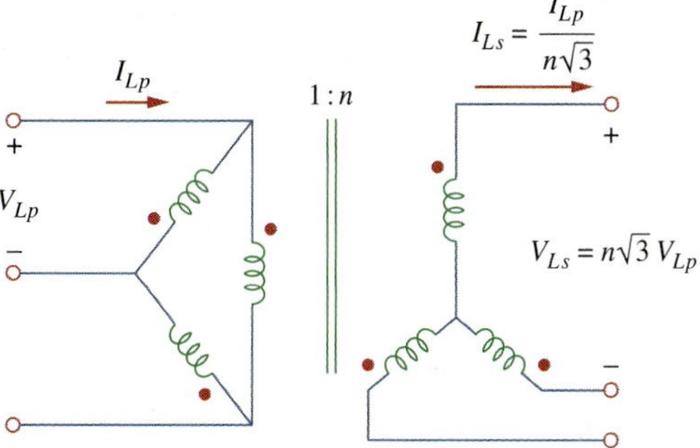

Figure 13.49
Δ-Y three-phase transformer connection.

Example 13.12

The 42-kVA balanced load depicted in Fig. 13.50 is supplied by a three-phase transformer. (a) Determine the type of transformer connections. (b) Find the line voltage and current on the primary side. (c) Determine the kVA rating of each transformer used in the transformer bank. Assume that the transformers are ideal.

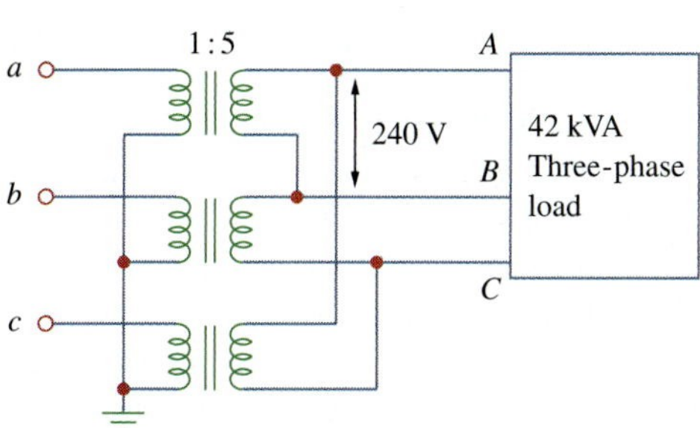

Figure 13.50
For Example 13.12.

Solution:

(a) A careful observation of Fig. 13.50 shows that the primary side is Y-connected, while the secondary side is Δ-connected. Thus, the three-phase transformer is Y-Δ, similar to the one shown in Fig. 13.48.
(b) Given a load with total apparent power $S_T = 42$ kVA, the turns ratio $n = 5$, and the secondary line voltage $V_{Ls} = 240$ V, we can find the secondary line current using Eq. (13.69a), by

$$I_{Ls} = \frac{S_T}{\sqrt{3}V_{Ls}} = \frac{42{,}000}{\sqrt{3}(240)} = 101 \text{ A}$$

From Eq. (13.71),

$$I_{Lp} = \frac{n}{\sqrt{3}} I_{Ls} = \frac{5 \times 101}{\sqrt{3}} = 292 \text{ A}$$

$$V_{Lp} = \frac{\sqrt{3}}{n} V_{Ls} = \frac{\sqrt{3} \times 240}{5} = 83.14 \text{ V}$$

(c) Because the load is balanced, each transformer equally shares the total load and since there are no losses (assuming ideal transformers), the kVA rating of each transformer is $S = S_T/3 = 14$ kVA. Alternatively, the transformer rating can be determined by the product of the phase current and phase voltage of the primary or secondary side. For the primary side, for example, we have a delta connection, so that the phase voltage is the same as the line voltage of 240 V, while the phase current is $I_{Lp}/\sqrt{3} = 58.34$ A. Hence, $S = 240 \times 58.34 = 14$ kVA.

Practice Problem 13.12

A three-phase Δ-Δ transformer is used to step down a line voltage of 625 kV, to supply a plant operating at a line voltage of 12.5 kV. The plant draws 40 MW with a lagging power factor of 85 percent. Find: (a) the current drawn by the plant, (b) the turns ratio, (c) the current on the primary side of the transformer, and (d) the load carried by each transformer.

Answer: (a) 2.1736 kA, (b) 0.02, (c) 43.47 A, (d) 15.69 MVA.

13.8 *PSpice* Analysis of Magnetically Coupled Circuits

PSpice analyzes magnetically coupled circuits just like inductor circuits except that the dot convention must be followed. In *PSpice* Schematic, the dot (not shown) is always next to pin 1, which is the left-hand terminal of the inductor when the inductor with part name L is placed (horizontally) without rotation on a schematic. Thus, the dot or pin 1 will be at the bottom after one 90° counterclockwise rotation, since rotation is always about pin 1. Once the magnetically coupled inductors are arranged with the dot convention in mind and their value attributes are set in henries, we use the coupling symbol K_LINEAR to define the coupling. For each pair of coupled inductors, take the following steps:

1. Select **Draw/Get New Part** and type K_LINEAR.
2. Hit <enter> or click **OK** and place the K_LINEAR symbol on the schematic, as shown in Fig. 13.51. (Notice that K_LINEAR is not a component and therefore has no pins.)
3. **DCLICKL** on COUPLING and set the value of the coupling coefficient *k*.
4. **DCLICKL** on the boxed K (the coupling symbol) and enter the reference designator names for the coupled inductors as values of Li, i = 1, 2, . . . , 6. For example, if inductors L20 and L23 are coupled, we set L1 = L20 and L2 = L23. L1 and at least one other Li must be assigned values; other Li's may be left blank.

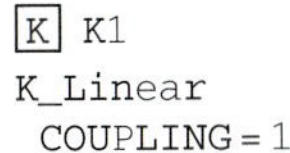

Figure 13.51
K_Linear for defining coupling.

In step 4, up to six coupled inductors with equal coupling can be specified.

For the air-core transformer, the partname is XFRM_LINEAR. It can be inserted in a circuit by selecting **Draw/Get Part Name** and then typing in the part name or by selecting the part name from the analog.slb library. As shown typically in Fig. 13.52(a), the main attributes of the linear transformer are the coupling coefficient *k* and the inductance values L1 and L2 in henries. If the mutual inductance *M* is specified, its value must be used along with L1 and L2 to calculate *k*. Keep in mind that the value of *k* should lie between 0 and 1.

For the ideal transformer, the part name is XFRM_NONLINEAR and is located in the breakout.slb library. Select it by clicking **Draw/Get Part Name** and then typing in the part name. Its attributes are the coupling coefficient and the numbers of turns associated with L1 and L2, as illustrated typically in Fig. 13.52(b). The value of the coefficient of mutual coupling $k = 1$.

PSpice has some additional transformer configurations that we will not discuss here.

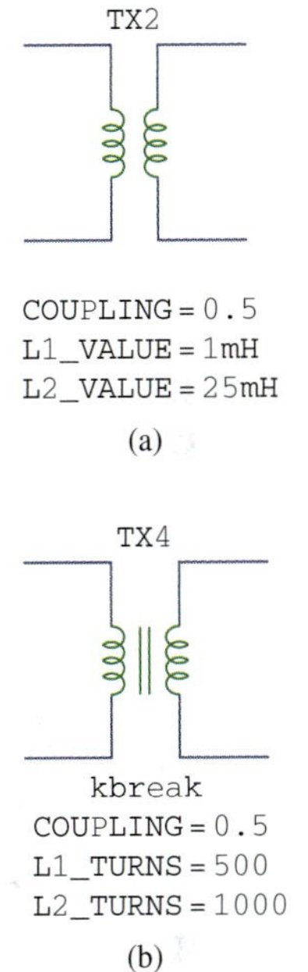

Figure 13.52
(a) Linear transformer XFRM_LINEAR, (b) ideal transformer XFRM_NONLINEAR.

Example 13.13

Use *PSpice* to find i_1, i_2, and i_3 in the circuit displayed in Fig. 13.53.

Figure 13.53
For Example 13.13.

Solution:

The coupling coefficients of the three coupled inductors are determined as follows:

$$k_{12} = \frac{M_{12}}{\sqrt{L_1 L_2}} = \frac{1}{\sqrt{3 \times 3}} = 0.3333$$

$$k_{13} = \frac{M_{13}}{\sqrt{L_1 L_3}} = \frac{1.5}{\sqrt{3 \times 4}} = 0.433$$

$$k_{23} = \frac{M_{23}}{\sqrt{L_2 L_3}} = \frac{2}{\sqrt{3 \times 4}} = 0.5774$$

The operating frequency f is obtained from Fig. 13.53 as $\omega = 12\pi = 2\pi f \rightarrow f = 6\text{Hz}$.

The schematic of the circuit is portrayed in Fig. 13.54. Notice how the dot convention is adhered to. For L2, the dot (not shown) is on pin 1 (the left-hand terminal) and is therefore placed without rotation. For L1, in order for the dot to be on the right-hand side of the inductor, the inductor must be rotated through 180°. For L3, the inductor must be rotated through 90° so that the dot will be at the bottom. Note that the 2-H inductor (L_4) is not coupled. To handle the three coupled inductors, we use three K_LINEAR parts provided in the analog library and set the following attributes (by double-clicking on the symbol K in the box):

The right-hand values are the reference designators of the inductors on the schematic.

```
K1 - K_LINEAR
L1 = L1
L2 = L2
COUPLING = 0.3333

K2 - K_LINEAR
L1 = L2
L2 = L3
COUPLING = 0.433

K3 - K_LINEAR
L1 = L1
L2 = L3
COUPLING = 0.5774
```

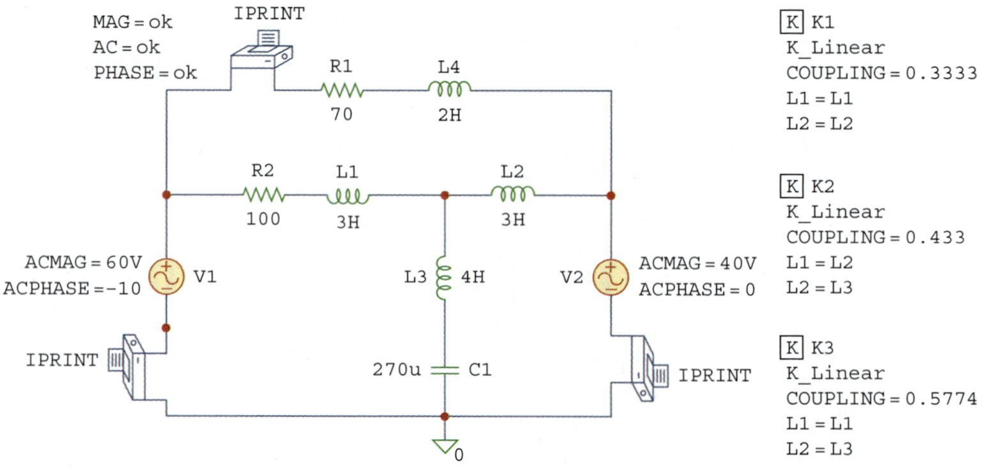

Figure 13.54
Schematic of the circuit of Fig. 13.53.

Three IPRINT pseudocomponents are inserted in the appropriate branches to obtain the required currents i_1, i_2, and i_3. As an AC single-frequency analysis, we select **Analysis/Setup/AC Sweep** and enter *Total Pts* = 1, *Start Freq* = 6, and *Final Freq* = 6. After saving the schematic, we select **Analysis/Simulate** to simulate it. The output file includes:

```
FREQ          IM(V_PRINT2)    IP(V_PRINT2)
6.000E+00     2.114E-01       -7.575E+01
FREQ          IM(V_PRINT1)    IP(V_PRINT1)
6.000E+00     4.654E-01       -7.025E+01
FREQ          IM(V_PRINT3)    IP(V_PRINT3)
6.000E+00     1.095E-01       1.715E+01
```

From this we obtain

$$\mathbf{I}_1 = 0.4654\underline{/-70.25^\circ}$$

$$\mathbf{I}_2 = 0.2114\underline{/-75.75^\circ}, \qquad \mathbf{I}_3 = 0.1095\underline{/17.15^\circ}$$

Thus,

$$i_1 = 0.4654\cos(12\pi t - 70.25^\circ)\text{ A}$$
$$i_2 = 0.2114\cos(12\pi t - 75.75^\circ)\text{ A}$$
$$i_3 = 0.1095\cos(12\pi t + 17.15^\circ)\text{ A}$$

Practice Problem 13.13

Find i_o in the circuit of Fig. 13.55, using *PSpice*.

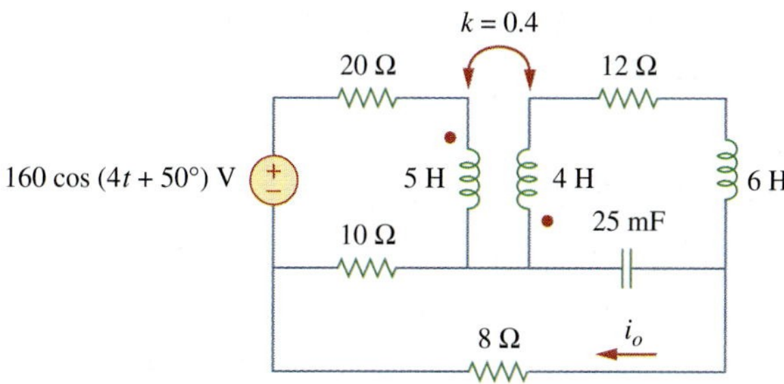

Figure 13.55
For Practice Prob. 13.13.

Answer: $2.012\cos(4t + 68.52^\circ)$ A.

Example 13.14

Find $\mathbf{V}_1$ and $\mathbf{V}_2$ in the ideal transformer circuit of Fig. 13.56 using *PSpice*.

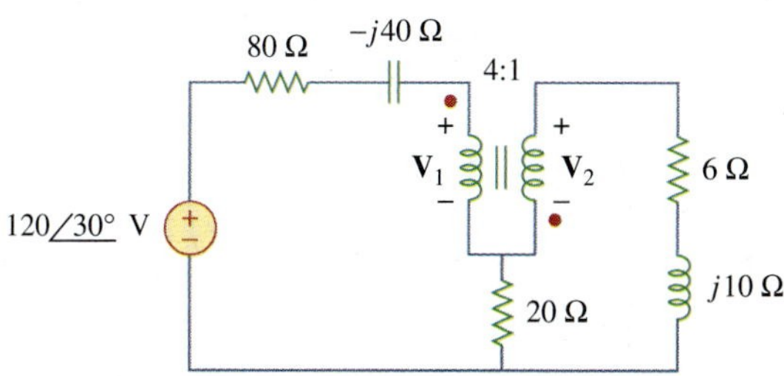

Figure 13.56
For Example 13.14.

Solution:

1. **Define.** The problem is clearly defined and we can proceed to the next step.
2. **Present.** We have an ideal transformer and we are to find the input and the output voltages for that transformer. In addition, we are to use *PSpice* to solve for the voltages.
3. **Alternative.** We are required to use *PSpice*. We can use mesh analysis to perform a check.
4. **Attempt.** As usual, we assume $\omega = 1$ and find the corresponding values of capacitance and inductance of the elements:

$$j10 = j\omega L \quad \Rightarrow \quad L = 10\text{ H}$$

$$-j40 = \frac{1}{j\omega C} \quad \Rightarrow \quad C = 25\text{ mF}$$

Reminder: For an ideal transformer, the inductances of both the primary and secondary windings are infinitely large.

Figure 13.57 shows the schematic. For the ideal transformer, we set the coupling factor to 0.99999 and the numbers of turns to 400,000 and 100,000. The two VPRINT2 pseudocomponents are connected across the transformer terminals to obtain $\mathbf{V}_1$ and $\mathbf{V}_2$. As a single-frequency analysis, we select **Analysis/Setup/AC Sweep** and enter *Total Pts* = 1, *Start Freq* = 0.1592, and *Final Freq* = 0.1592. After saving the schematic, we select **Analysis/Simulate** to simulate it. The output file includes:

```
FREQ         VM($N_0003,$N_0006) VP($N_0003,$N_0006)
1.592E-01    9.112E+01           3.792E+01

FREQ         VM($N_0006,$N_0005) VP($N_0006,$N_0005)
1.592E-01    2.278E+01           -1.421E+02
```

This can be written as

$$\mathbf{V}_1 = \mathbf{91.12\underline{/37.92^\circ}\text{ V}} \quad \text{and} \quad \mathbf{V}_2 = \mathbf{22.78\underline{/-142.1^\circ}\text{ V}}$$

5. **Evaluate.** We can check the answer by using mesh analysis as follows:

$$\text{Loop 1} \quad -120\underline{/30^\circ} + (80 - j40)I_1 + V_1 + 20(I_1 - I_2) = 0$$

$$\text{Loop 2} \quad 20(-I_1 + I_2) - V_2 + (6 + j10)I_2 = 0$$

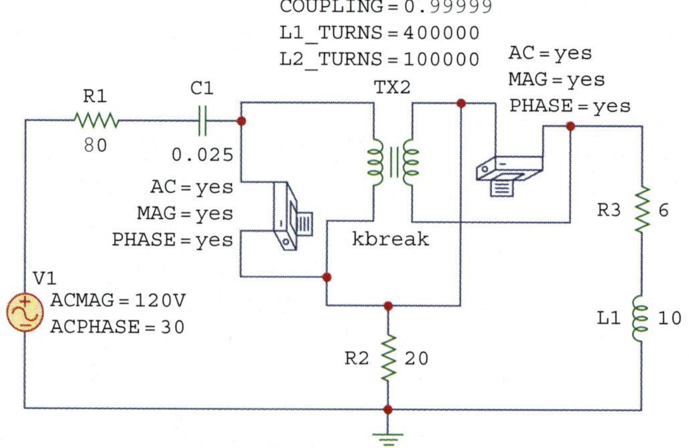

Figure 13.57
The schematic for the circuit in Fig. 13.56.

But $V_2 = -V_1/4$ and $I_2 = -4I_1$. This leads to

$$-120\underline{/30^\circ} + (80 - j40)I_1 + V_1 + 20(I_1 + 4I_1) = 0$$

$$(180 - j40)I_1 + V_1 = 120\underline{/30^\circ}$$

$$20(-I_1 - 4I_1) + V_1/4 + (6 + j10)(-4I_1) = 0$$

$$(-124 - j40)I_1 + 0.25V_1 = 0 \quad \text{or} \quad I_1 = V_1/(496 + j160)$$

Substituting this into the first equation yields

$$(180 - j40)V_1/(496 + j160) + V_1 = 120\underline{/30^\circ}$$

$$(184.39\underline{/-12.53^\circ}/521.2\underline{/17.88^\circ})V_1 + V_1$$

$$= (0.3538\underline{/-30.41^\circ} + 1)V_1 = (0.3051 + 1 - j0.17909)V_1 = 120\underline{/30^\circ}$$

$$V_1 = 120\underline{/30^\circ}/1.3173\underline{/-7.81^\circ} = \mathbf{91.1\underline{/37.81^\circ}\ V} \quad \text{and}$$

$$V_2 = \mathbf{22.78\underline{/-142.19^\circ}\ V}$$

Both answers check.

6. **Satisfactory?** We have satisfactorily answered the problem and checked the solution. We can now present the entire solution to the problem.

Practice Problem 13.14

Obtain $\mathbf{V}_1$ and $\mathbf{V}_2$ in the circuit of Fig. 13.58 using *PSpice*.

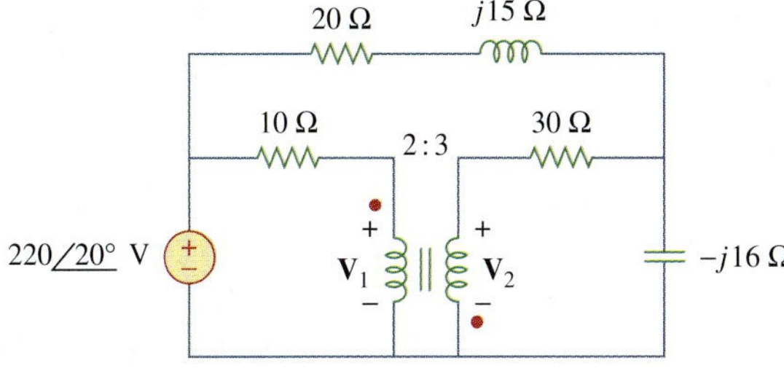

Figure 13.58
For Practice Prob. 13.14.

Answer: $138.82\underline{/28.65^\circ}$ V, $208.2\underline{/-151.4^\circ}$ V.

13.9 †Applications

Transformers are the largest, the heaviest, and often the costliest of circuit components. Nevertheless, they are indispensable passive devices in electric circuits. They are among the most efficient machines, 95 percent efficiency being common and 99 percent being achievable. They have numerous applications. For example, transformers are used:

- To step up or step down voltage and current, making them useful for power transmission and distribution.
- To isolate one portion of a circuit from another (i.e., to transfer power without any electrical connection).
- As an impedance-matching device for maximum power transfer.
- In frequency-selective circuits whose operation depends on the response of inductances.

Because of these diverse uses, there are many special designs for transformers (only some of which are discussed in this chapter): voltage transformers, current transformers, power transformers, distribution transformers, impedance-matching transformers, audio transformers, single-phase transformers, three-phase transformers, rectifier transformers, inverter transformers, and more. In this section, we consider three important applications: transformer as an isolation device, transformer as a matching device, and power distribution system.

For more information on the many kinds of transformers, a good text is W. M. Flanagan, *Handbook of Transformer Design and Applications*, 2nd ed. (New York: McGraw-Hill, 1993).

13.9.1 Transformer as an Isolation Device

Electrical isolation is said to exist between two devices when there is no physical connection between them. In a transformer, energy is transferred by magnetic coupling, without electrical connection between the primary circuit and secondary circuit. We now consider three simple practical examples of how we take advantage of this property.

First, consider the circuit in Fig. 13.59. A rectifier is an electronic circuit that converts an ac supply to a dc supply. A transformer is often used to couple the ac supply to the rectifier. The transformer serves two purposes. First, it steps up or steps down the voltage. Second, it provides electrical isolation between the ac power supply and the rectifier, thereby reducing the risk of shock hazard in handling the electronic device.

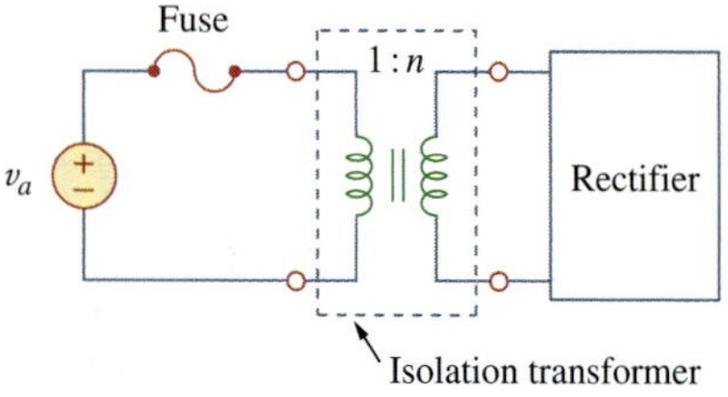

Figure 13.59
A transformer used to isolate an ac supply from a rectifier.

As a second example, a transformer is often used to couple two stages of an amplifier, to prevent any dc voltage in one stage from affecting the dc bias of the next stage. Biasing is the application of a dc voltage to a transistor amplifier or any other electronic device in order to produce a desired mode of operation. Each amplifier stage is biased separately to operate in a particular mode; the desired mode of operation will be compromised without a transformer providing dc isolation. As shown in Fig. 13.60, only the ac signal is coupled through the transformer from one stage to the next. We recall that magnetic coupling does not exist with a dc voltage source. Transformers are used in radio and TV receivers to couple stages of high-frequency amplifiers. When the sole purpose of a transformer is to provide isolation, its turns ratio n is made unity. Thus, an isolation transformer has $n = 1$.

As a third example, consider measuring the voltage across 13.2-kV lines. It is obviously not safe to connect a voltmeter directly to such high-voltage lines. A transformer can be used both to electrically isolate the line power from the voltmeter and to step down the voltage to a safe level, as shown in Fig. 13.61. Once the voltmeter is used to

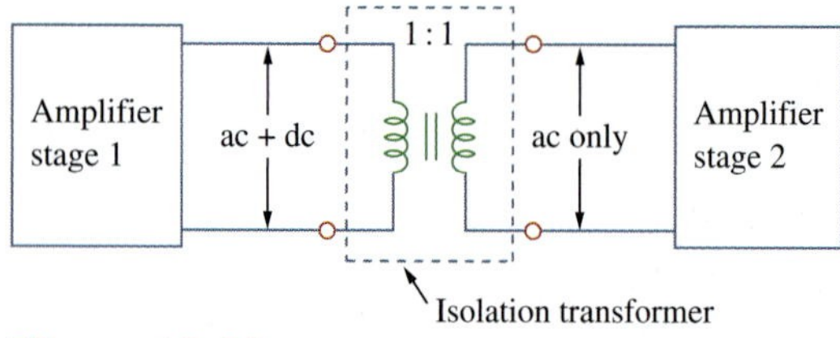

Figure 13.60
A transformer providing dc isolation between two amplifier stages.

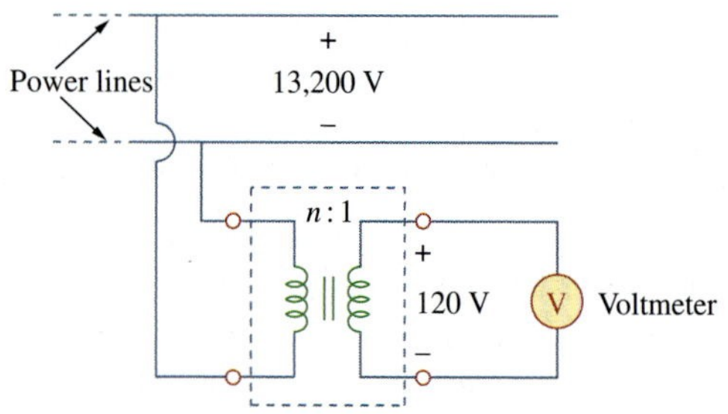

Figure 13.61
A transformer providing isolation between the power lines and the voltmeter.

measure the secondary voltage, the turns ratio is used to determine the line voltage on the primary side.

Example 13.15

Determine the voltage across the load in Fig. 13.62.

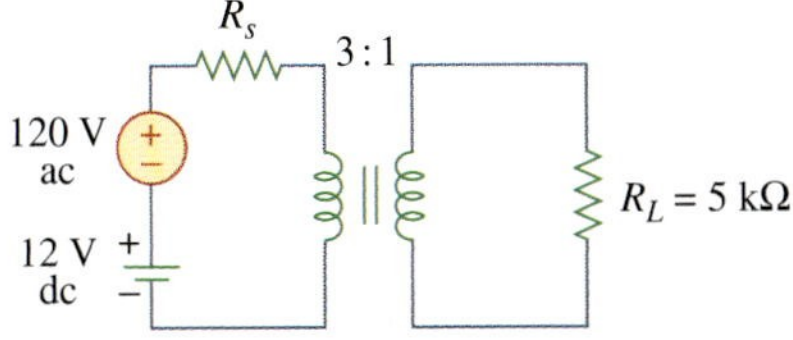

Figure 13.62
For Example 13.15.

Solution:
We can apply the superposition principle to find the load voltage. Let $v_L = v_{L1} + v_{L2}$, where v_{L1} is due to the dc source and v_{L2} is due to the ac source. We consider the dc and ac sources separately, as shown in Fig. 13.63. The load voltage due to the dc source is zero, because a time-varying voltage is necessary in the primary circuit to induce a voltage in the secondary circuit. Thus, $v_{L1} = 0$. For the ac source and a value of R_s so small it can be neglected,

$$\frac{\mathbf{V}_2}{\mathbf{V}_1} = \frac{\mathbf{V}_2}{120} = \frac{1}{3} \quad \text{or} \quad \mathbf{V}_2 = \frac{120}{3} = 40 \text{ V}$$

Hence, $\mathbf{V}_{L2} = 40$ V ac or $v_{L2} = 40 \cos \omega t$; that is, only the ac voltage is passed to the load by the transformer. This example shows how the transformer provides dc isolation.

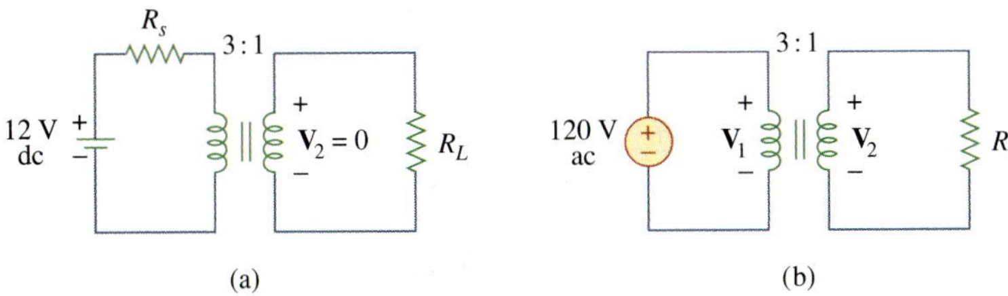

Figure 13.63
For Example 13.15: (a) dc source, (b) ac source.

Practice Problem 13.15

Refer to Fig. 13.61. Calculate the turns ratio required to step down the 13.2-kV line voltage to a safe level of 120 V.

Answer: 110.

13.9.2 Transformer as a Matching Device

We recall that for maximum power transfer, the load resistance R_L must be matched with the source resistance R_s. In most cases, the two resistances are not matched; both are fixed and cannot be altered. However, an iron-core transformer can be used to match the load resistance to the source resistance. This is called *impedance matching*. For example, to connect a loudspeaker to an audio power amplifier requires a transformer, because the speaker's resistance is only a few ohms while the internal resistance of the amplifier is several thousand ohms.

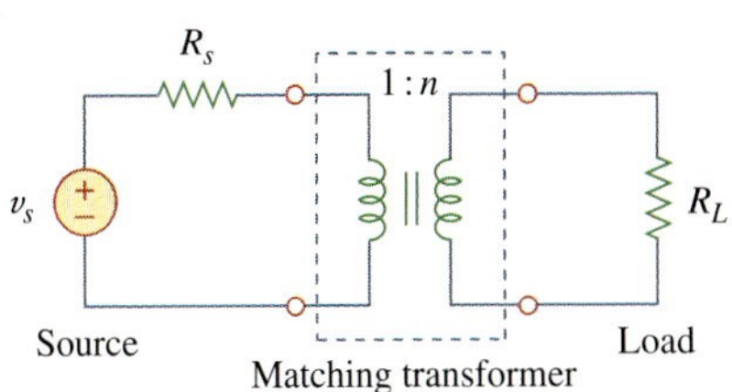

Figure 13.64
Transformer used as a matching device.

Consider the circuit shown in Fig. 13.64. We recall from Eq. (13.60) that the ideal transformer reflects its load back to the primary with a

13.10 Find v_o in the circuit of Fig. 13.79.

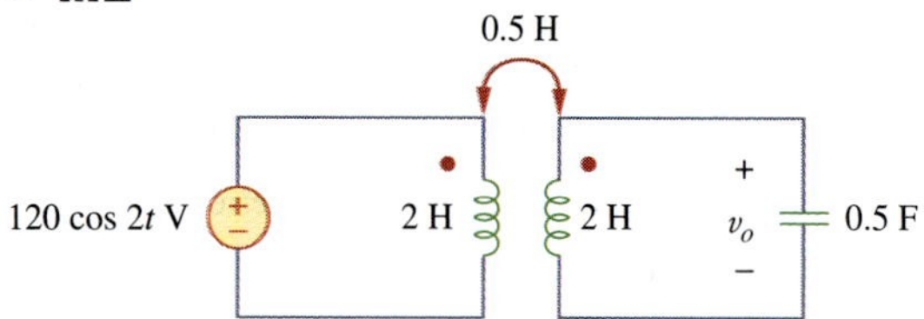

Figure 13.79
For Prob. 13.10.

13.11 Use mesh analysis to find i_x in Fig. 13.80, where

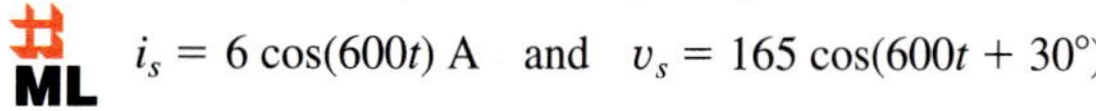

$$i_s = 6\cos(600t)\text{ A} \quad \text{and} \quad v_s = 165\cos(600t + 30°)$$

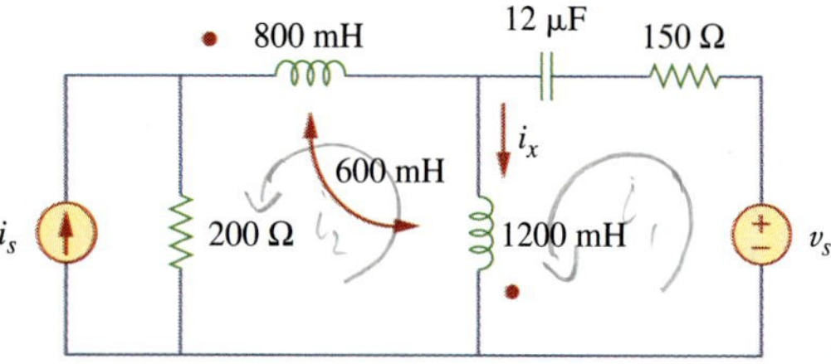

Figure 13.80
For Prob. 13.11.

13.12 Determine the equivalent L_{eq} in the circuit of Fig. 13.81.

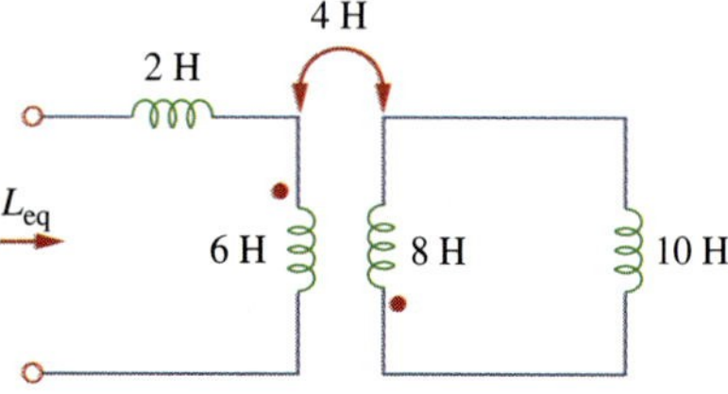

Figure 13.81
For Prob. 13.12.

13.13 For the circuit in Fig. 13.82, determine the impedance seen by the source.

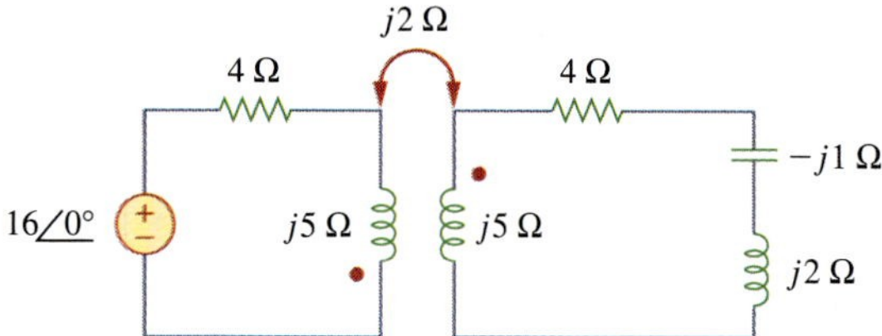

Figure 13.82
For Prob. 13.13.

13.14 Obtain the Thevenin equivalent circuit for the circuit in Fig. 13.83 at terminals *a-b*.

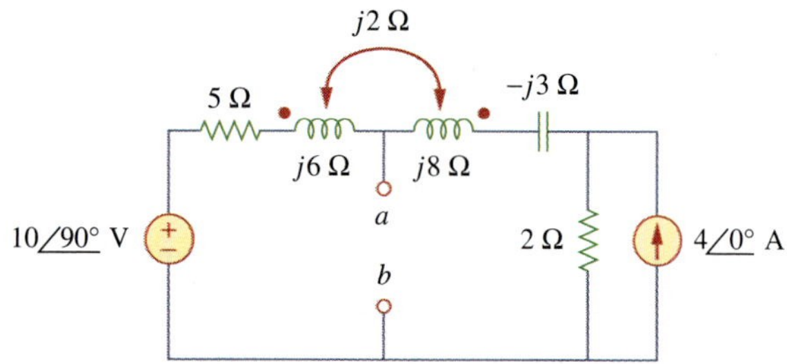

Figure 13.83
For Prob. 13.14.

13.15 Find the Norton equivalent for the circuit in Fig. 13.84 at terminals *a-b*.

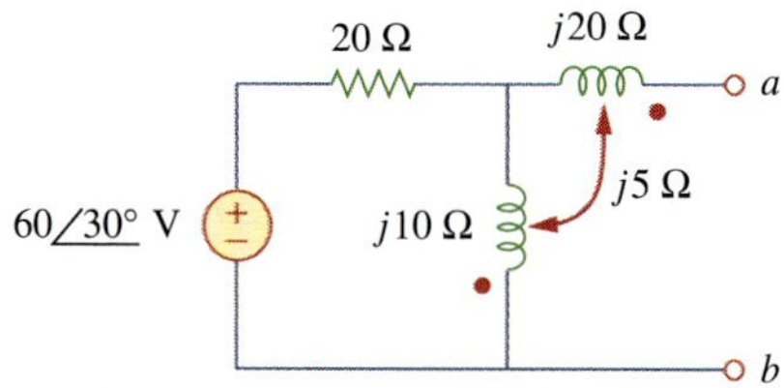

Figure 13.84
For Prob. 13.15.

13.16 Obtain the Norton equivalent at terminals *a-b* of the circuit in Fig. 13.85.

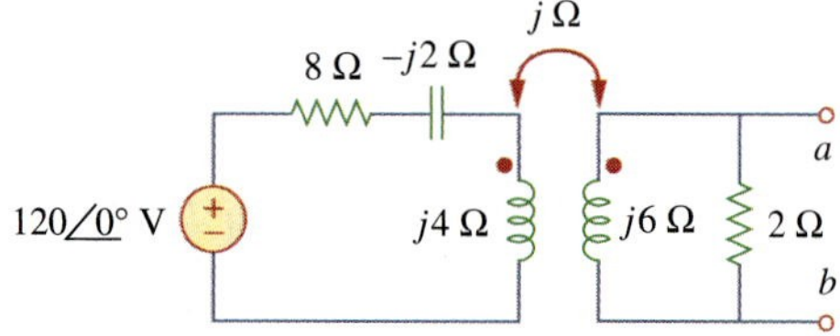

Figure 13.85
For Prob. 13.16.

13.17 In the circuit of Fig. 13.86, Z_L is a 15-mH inductor having an impedance of $j40\ \Omega$. Determine Z_{in} when $k = 0.6$.

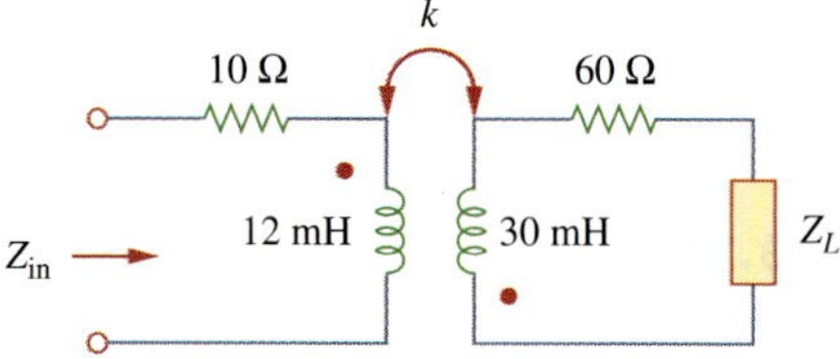

Figure 13.86
For Prob. 13.17.

13.18 Find the Thevenin equivalent to the left of the load **Z** in the circuit of Fig. 13.87.

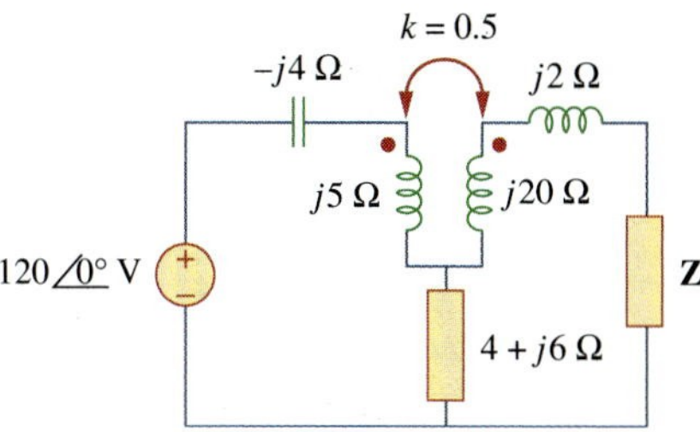

Figure 13.87
For Prob. 13.18.

13.19 Determine an equivalent T-section that can be used to replace the transformer in Fig. 13.88.

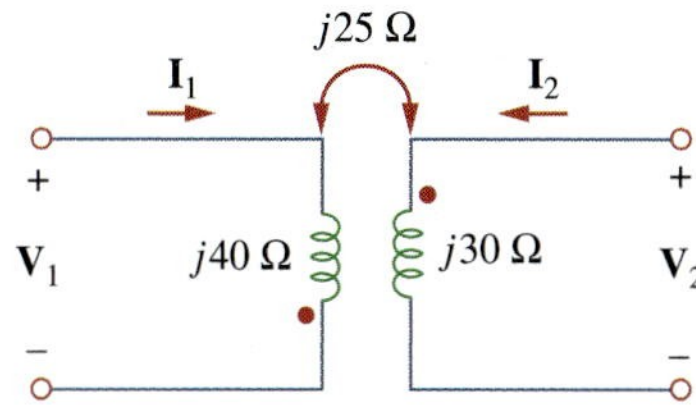

Figure 13.88
For Prob. 13.19.

Section 13.3 Energy in a Coupled Circuit

13.20 Determine currents $\mathbf{I}_1$, $\mathbf{I}_2$, and $\mathbf{I}_3$ in the circuit of Fig. 13.89. Find the energy stored in the coupled coils at $t = 2$ ms. Take $\omega = 1{,}000$ rad/s.

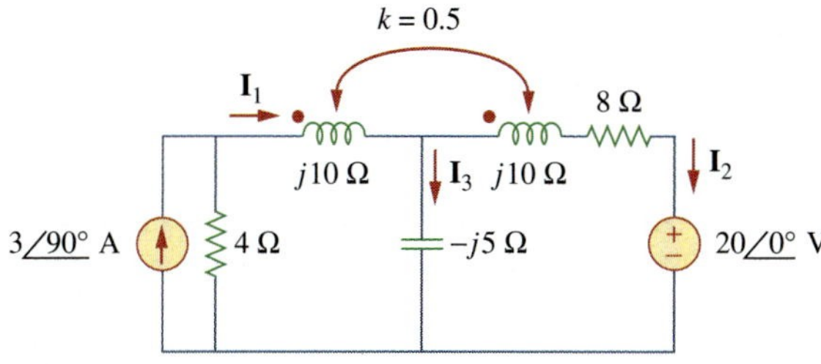

Figure 13.89
For Prob. 13.20.

13.21 Using Fig. 13.90, design a problem to help other students better understand energy in a coupled circuit.

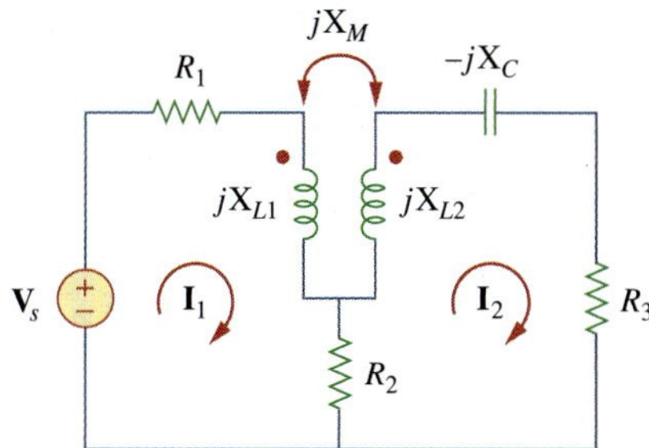

Figure 13.90
For Prob. 13.21.

***13.22** Find current $\mathbf{I}_o$ in the circuit of Fig. 13.91.

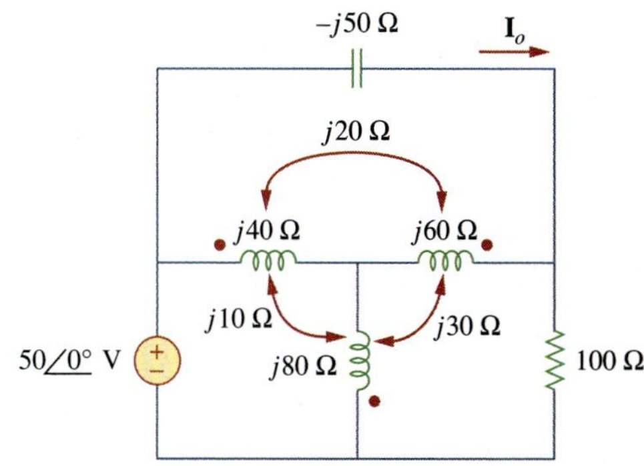

Figure 13.91
For Prob. 13.22.

13.23 If $M = 0.2$ H and $v_s = 120 \cos 10t$ V in the circuit of Fig. 13.92, find i_1 and i_2. Calculate the energy stored in the coupled coils at $t = 15$ ms.

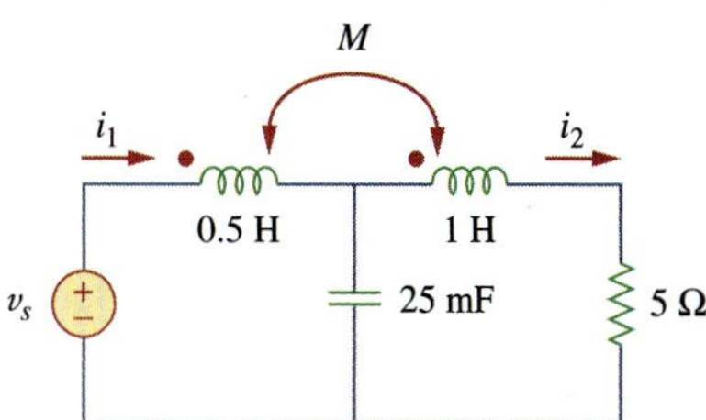

Figure 13.92
For Prob. 13.23.

13.24 In the circuit of Fig. 13.93,

(a) find the coupling coefficient,

(b) calculate v_o,

(c) determine the energy stored in the coupled inductors at $t = 2$ s.

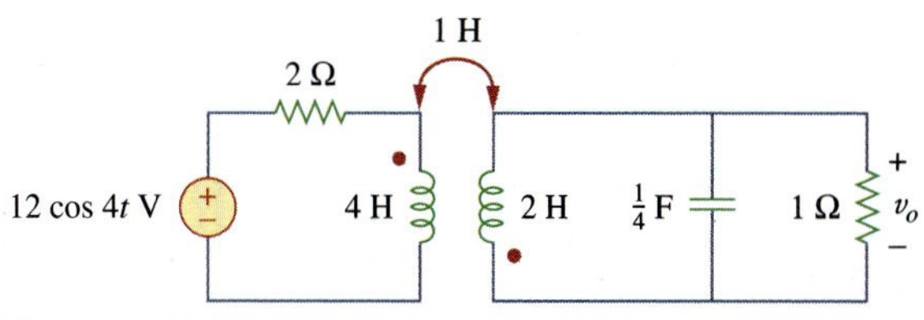

Figure 13.93
For Prob. 13.24.

* An asterisk indicates a challenging problem.

13.25 For the network in Fig. 13.94, find $\mathbf{Z}_{ab}$ and $\mathbf{I}_o$.

ps ML

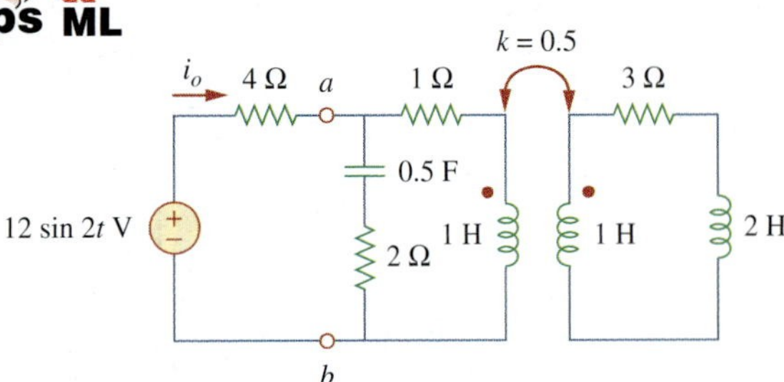

Figure 13.94
For Prob. 13.25.

13.26 Find $\mathbf{I}_o$ in the circuit of Fig. 13.95. Switch the dot on the winding on the right and calculate $\mathbf{I}_o$ again.

ps ML

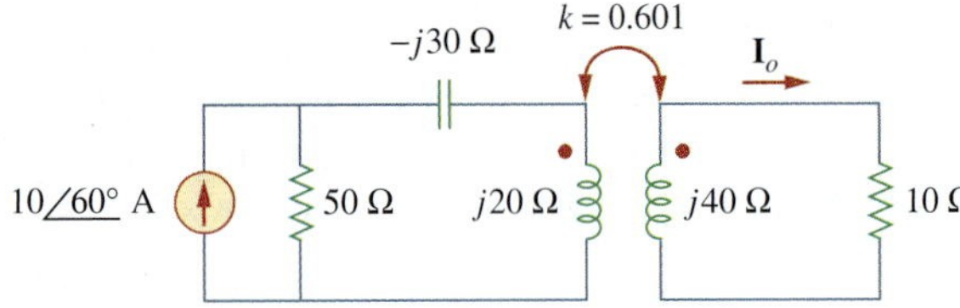

Figure 13.95
For Prob. 13.26.

13.27 Find the average power delivered to the 50-Ω resistor in the circuit of Fig. 13.96.

ps ML

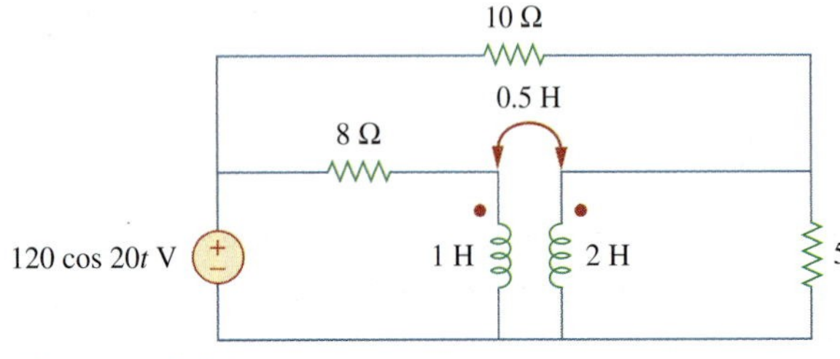

Figure 13.96
For Prob. 13.27.

***13.28** In the circuit of Fig. 13.97, find the value of X that will give maximum power transfer to the 20-Ω load.

ML

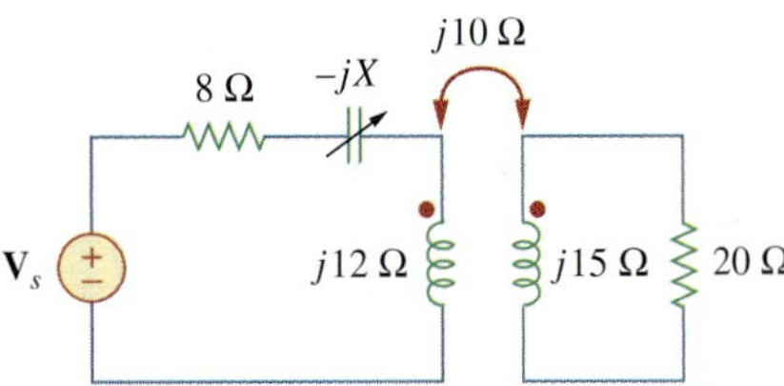

Figure 13.97
For Prob. 13.28.

Section 13.4 Linear Transformers

13.29 In the circuit of Fig. 13.98, find the value of the coupling coefficient k that will make the 10-Ω resistor dissipate 320 W. For this value of k, find the energy stored in the coupled coils at $t = 1.5$ s.

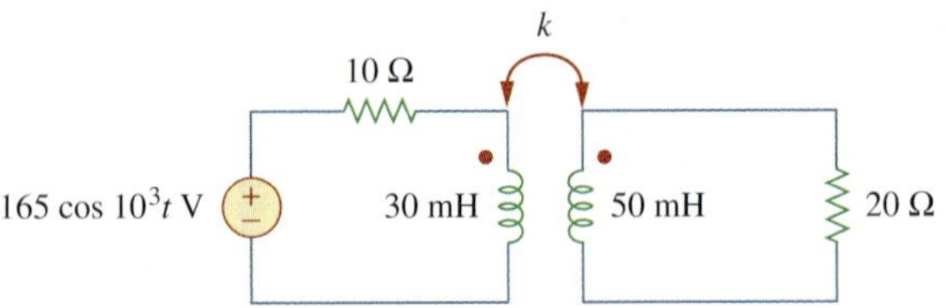

Figure 13.98
For Prob. 13.29.

13.30 (a) Find the input impedance of the circuit in Fig. 13.99 using the concept of reflected impedance.

(b) Obtain the input impedance by replacing the linear transformer by its T equivalent.

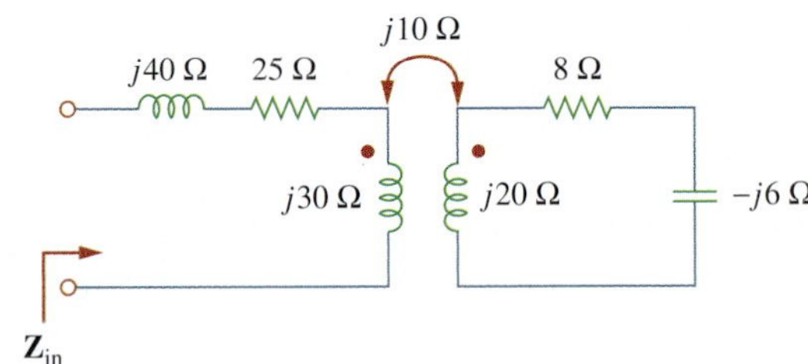

Figure 13.99
For Prob. 13.30.

13.31 Using Fig. 13.100, design a problem to help other students better understand linear transformers and how to find T-equivalent and Π-equivalent circuits.

e⊿d

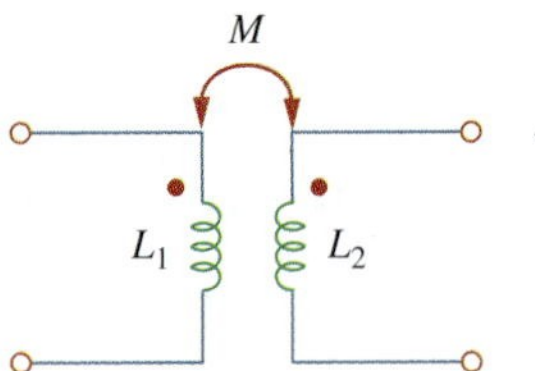

Figure 13.100
For Prob. 13.31.

***13.32** Two linear transformers are cascaded as shown in Fig. 13.101. Show that

$$\mathbf{Z}_{in} = \frac{\begin{array}{c}\omega^2 R(L_a^2 + L_aL_b - M_a^2) \\ +j\omega^3(L_a^2L_b + L_aL_b^2 - L_aM_b^2 - L_bM_a^2)\end{array}}{\omega^2(L_aL_b + L_b^2 - M_b^2) - j\omega R(L_a + L_b)}$$

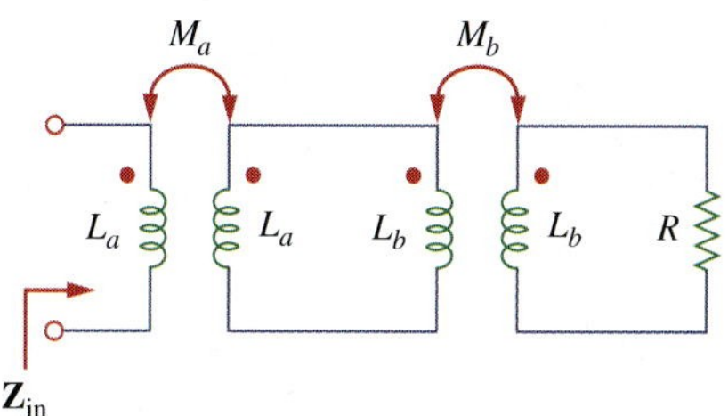

Figure 13.101
For Prob. 13.32.

13.33 Determine the input impedance of the air-core transformer circuit of Fig. 13.102.

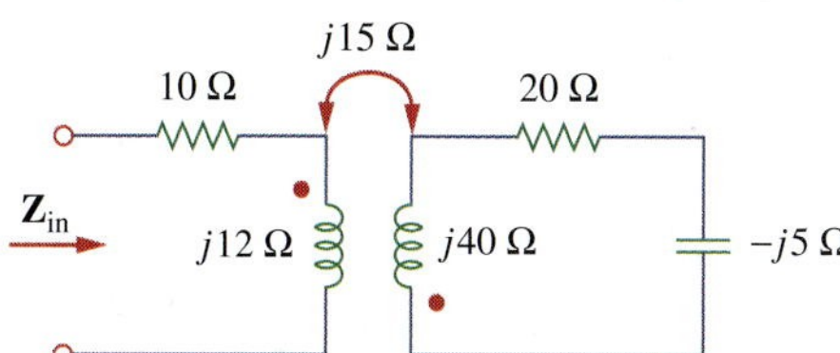

Figure 13.102
For Prob. 13.33.

13.34 Using Fig. 13.103, design a problem to help other students better understand how to find the input impedance of circuits with transformers.

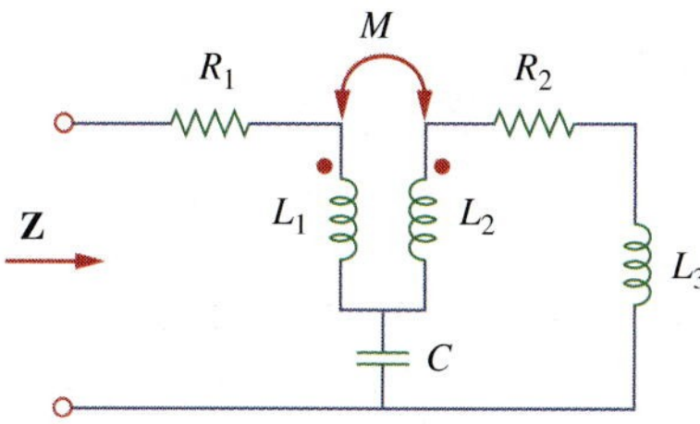

Figure 13.103
For Prob. 13.34.

***13.35** Find currents $\mathbf{I}_1$, $\mathbf{I}_2$, and $\mathbf{I}_3$ in the circuit of Fig. 13.104.

Section 13.5 Ideal Transformers

13.36 As done in Fig. 13.32, obtain the relationships between terminal voltages and currents for each of the ideal transformers in Fig. 13.105.

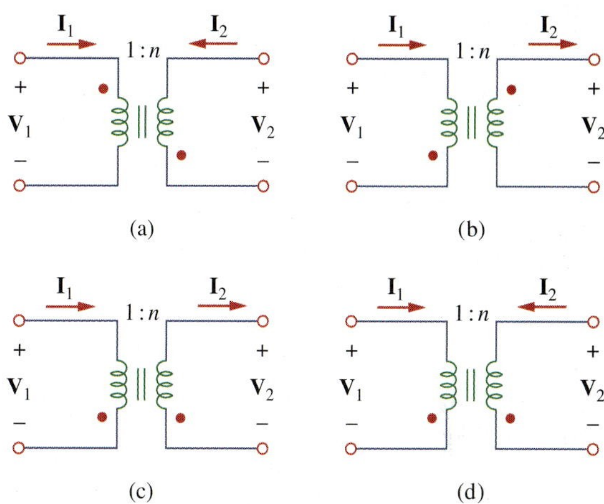

Figure 13.105
For Prob. 13.36.

13.37 A 480/2,400-V rms step-up ideal transformer delivers 50 kW to a resistive load. Calculate:

(a) the turns ratio

(b) the primary current

(c) the secondary current

13.38 Design a problem to help other students better understand ideal transformers.

13.39 A 1,200/240-V rms transformer has impedance $60\underline{/-30^\circ}$ Ω on the high-voltage side. If the transformer is connected to a $0.8\underline{/10^\circ}$-Ω load on the low-voltage side, determine the primary and secondary currents when the transformer is connected to 1200 V rms.

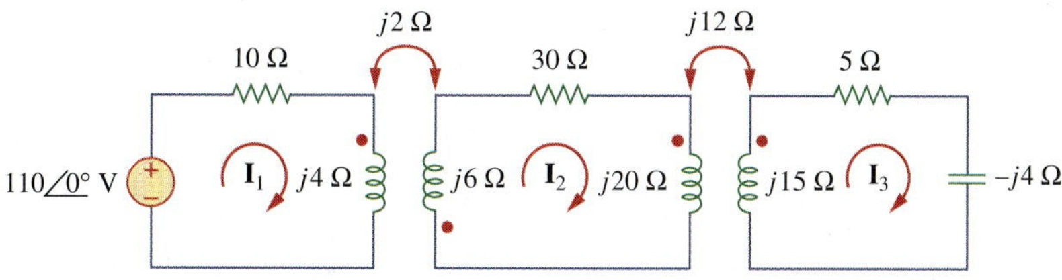

Figure 13.104
For Prob. 13.35.

13.40 The primary of an ideal transformer with a turns ratio of 5 is connected to a voltage source with Thevenin parameters $v_{Th} = 10 \cos 2000t$ V and $R_{Th} = 100\ \Omega$. Determine the average power delivered to a 200-Ω load connected across the secondary winding.

13.41 Determine $\mathbf{I}_1$ and $\mathbf{I}_2$ in the circuit of Fig. 13.106.

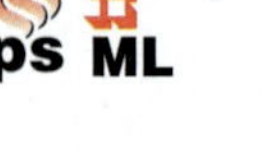

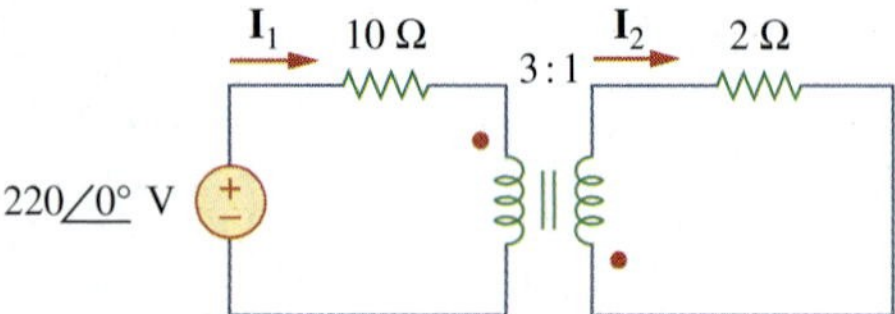

Figure 13.106
For Prob. 13.41.

13.42 For the circuit in Fig. 13.107, determine the power absorbed by the 2-Ω resistor. Assume the 80 V is an rms value.

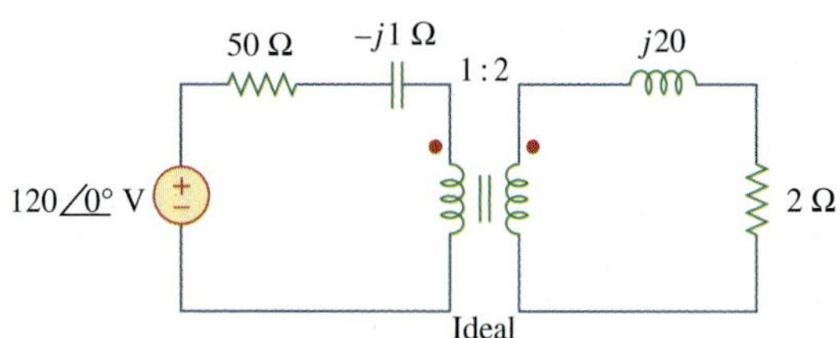

Figure 13.107
For Prob. 13.42.

13.43 Obtain $\mathbf{V}_1$ and $\mathbf{V}_2$ in the ideal transformer circuit of Fig. 13.108.

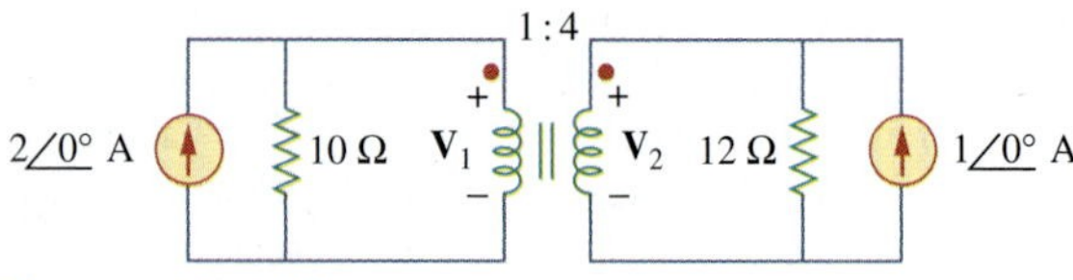

Figure 13.108
For Prob. 13.43.

***13.44** In the ideal transformer circuit of Fig. 13.109, find $i_1(t)$ and $i_2(t)$.

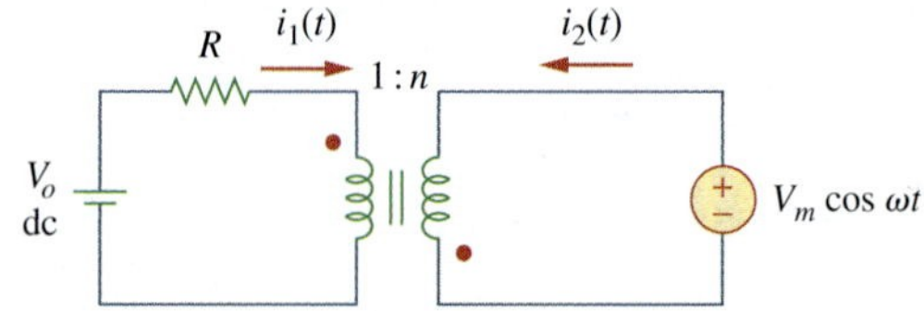

Figure 13.109
For Prob. 13.44.

13.45 For the circuit shown in Fig. 13.110, find the value of the average power absorbed by the 8-Ω resistor.

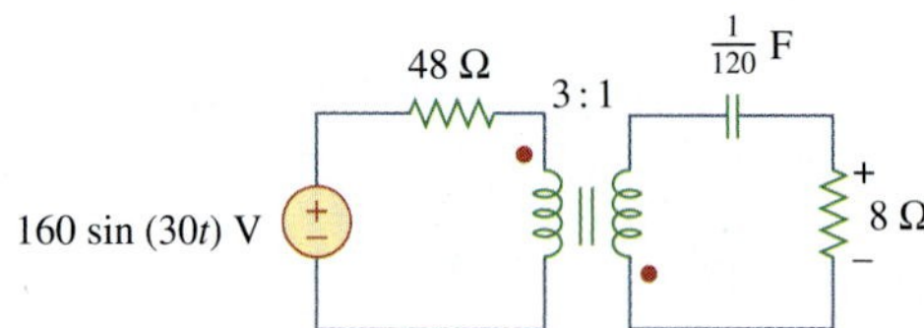

Figure 13.110
For Prob. 13.45.

13.46 (a) Find $\mathbf{I}_1$ and $\mathbf{I}_2$ in the circuit of Fig. 13.111 below.
(b) Switch the dot on one of the windings. Find $\mathbf{I}_1$ and $\mathbf{I}_2$ again.

13.47 Find $v(t)$ for the circuit in Fig. 13.112.

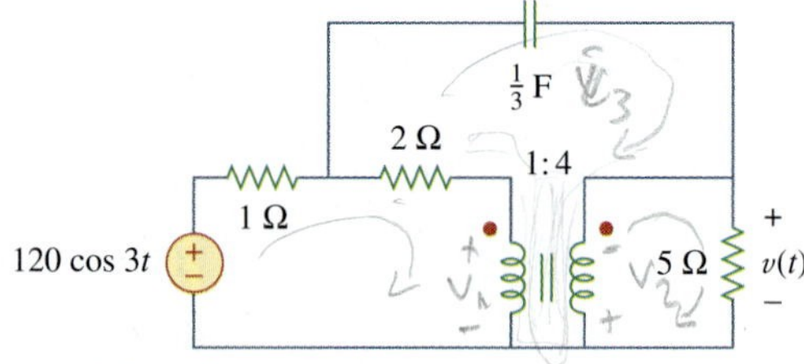

Figure 13.112
For Prob. 13.47.

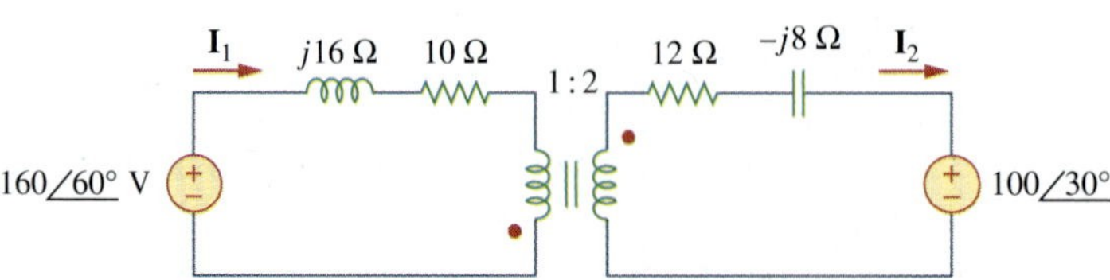

Figure 13.111
For Prob. 13.46.

13.48 Using Fig. 13.113, design a problem to help other students better understand how ideal transformers work.

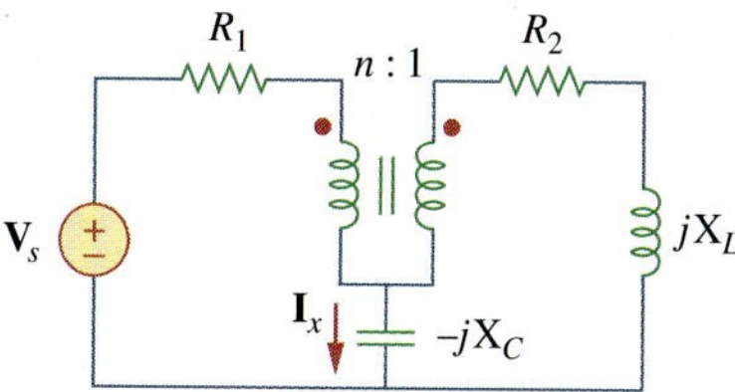

Figure 13.113
For Prob. 13.48.

13.49 Find current i_x in the ideal transformer circuit shown in Fig. 13.114.

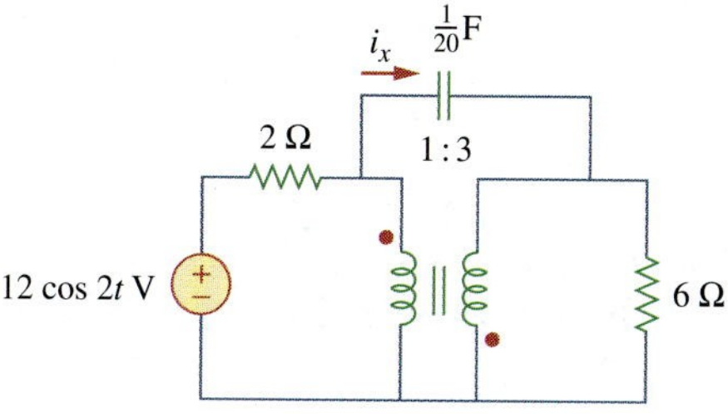

Figure 13.114
For Prob. 13.49.

13.50 Calculate the input impedance for the network in Fig. 13.115.

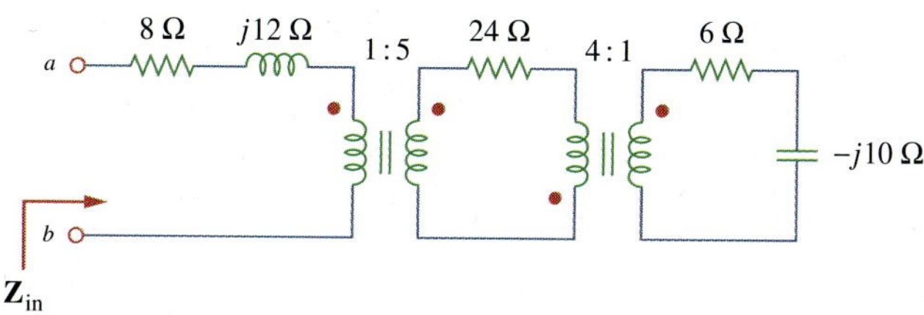

Figure 13.115
For Prob. 13.50.

13.51 Use the concept of reflected impedance to find the input impedance and current $\mathbf{I}_1$ in Fig. 13.116.

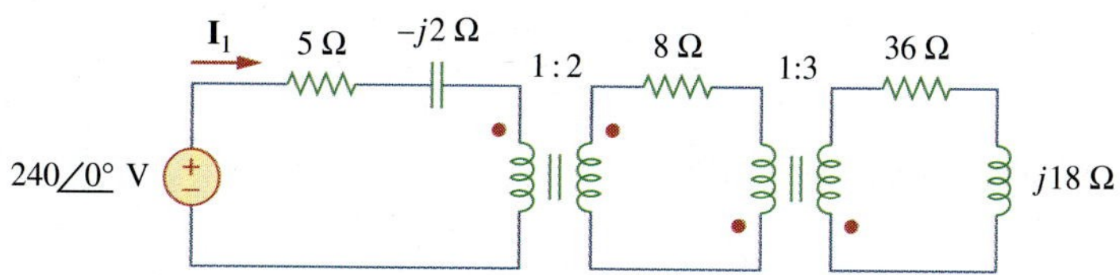

Figure 13.116
For Prob. 13.51.

13.52 For the circuit in Fig. 13.117, determine the turns ratio *n* that will cause maximum average power transfer to the load. Calculate that maximum average power.

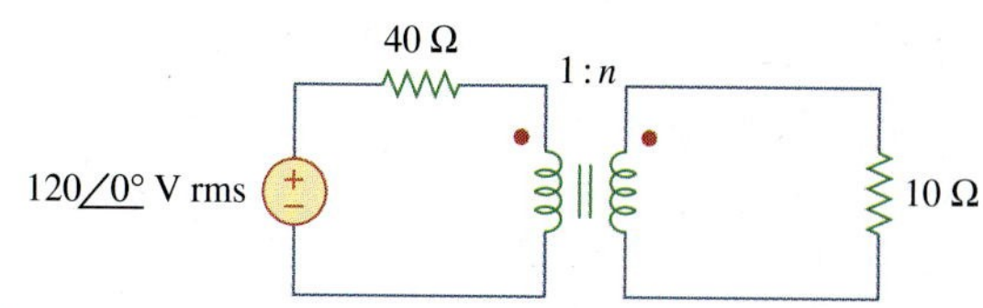

Figure 13.117
For Prob. 13.52.

13.53 Refer to the network in Fig. 13.118.

(a) Find *n* for maximum power supplied to the 200-Ω load.

(b) Determine the power in the 200-Ω load if $n = 10$.

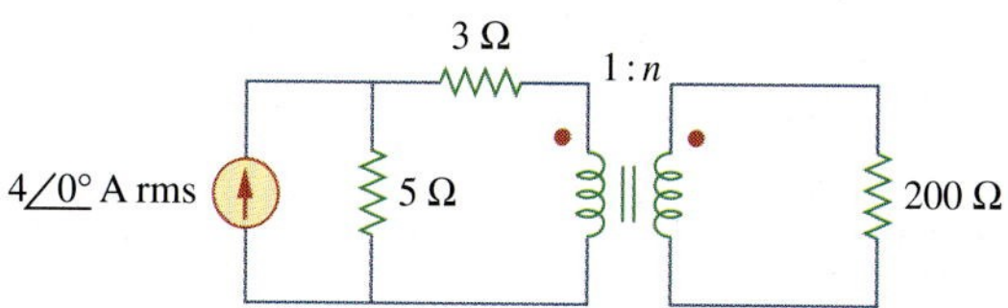

Figure 13.118
For Prob. 13.53.

13.54 A transformer is used to match an amplifier with an 8-Ω load as shown in Fig. 13.119. The Thevenin equivalent of the amplifier is: $V_{Th} = 10$ V, $Z_{Th} = 128\ \Omega$.

(a) Find the required turns ratio for maximum energy power transfer.

(b) Determine the primary and secondary currents.

(c) Calculate the primary and secondary voltages.

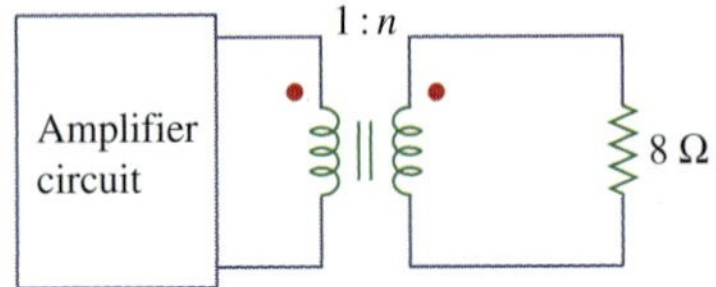

Figure 13.119
For Prob. 13.54.

13.55 For the circuit in Fig. 13.120, calculate the equivalent resistance.

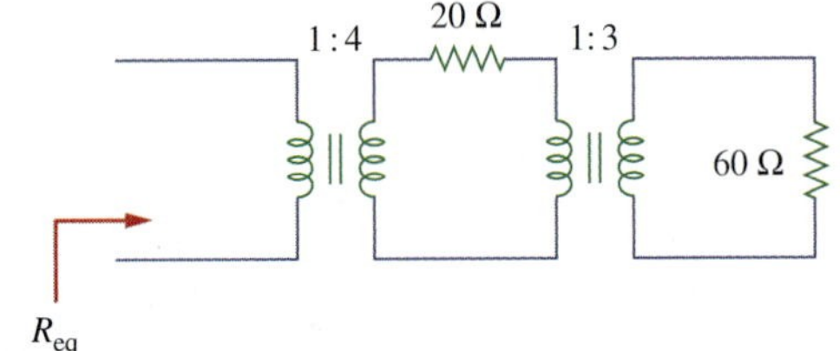

Figure 13.120
For Prob. 13.55.

13.56 Find the power absorbed by the 10-Ω resistor in the ideal transformer circuit of Fig. 13.121.

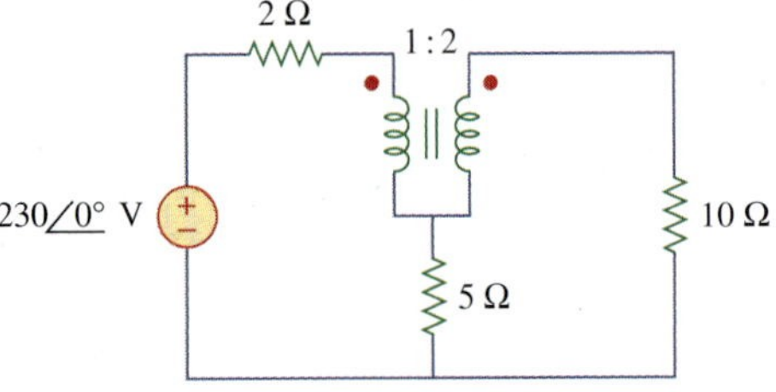

Figure 13.121
For Prob. 13.56.

13.57 For the ideal transformer circuit of Fig. 13.122 below, find:

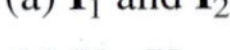

(a) $\mathbf{I}_1$ and $\mathbf{I}_2$,

(b) $\mathbf{V}_1$, $\mathbf{V}_2$, and $\mathbf{V}_o$,

(c) the complex power supplied by the source.

13.58 Determine the average power absorbed by each resistor in the circuit of Fig. 13.123.

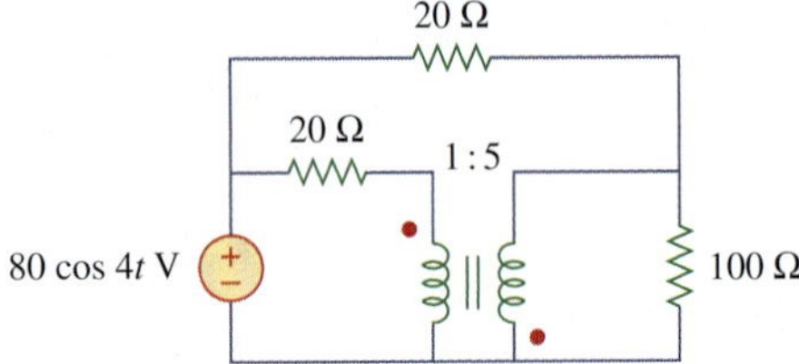

Figure 13.123
For Prob. 13.58.

13.59 In the circuit of Fig. 13.124, let $v_s = 160 \cos 1000t$. Find the average power delivered to each resistor.

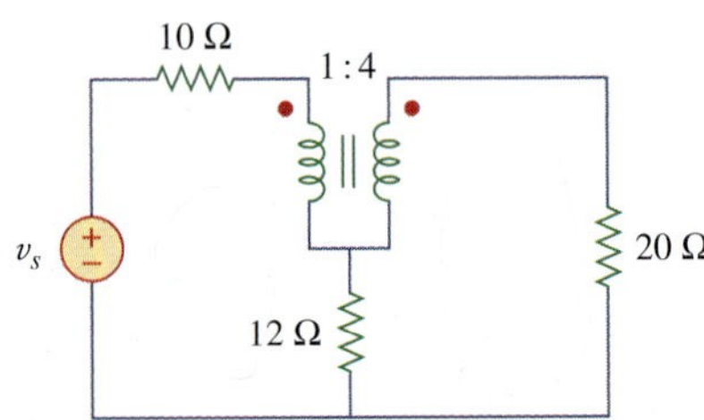

Figure 13.124
For Prob. 13.59.

13.60 Refer to the circuit in Fig. 13.125 on the following page.

(a) Find currents $\mathbf{I}_1$, $\mathbf{I}_2$, and $\mathbf{I}_3$.

(b) Find the power dissipated in the 40-Ω resistor.

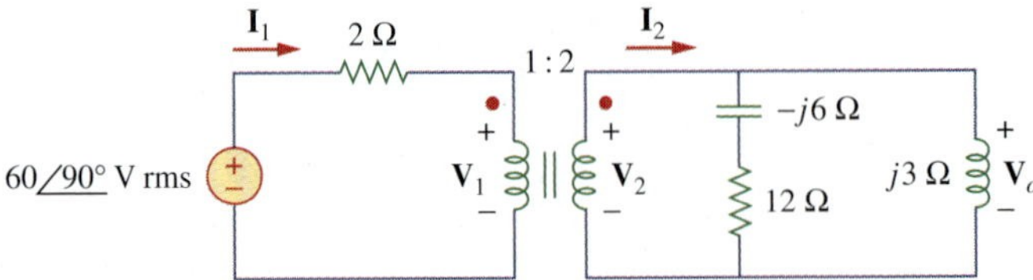

Figure 13.122
For Prob. 13.57.

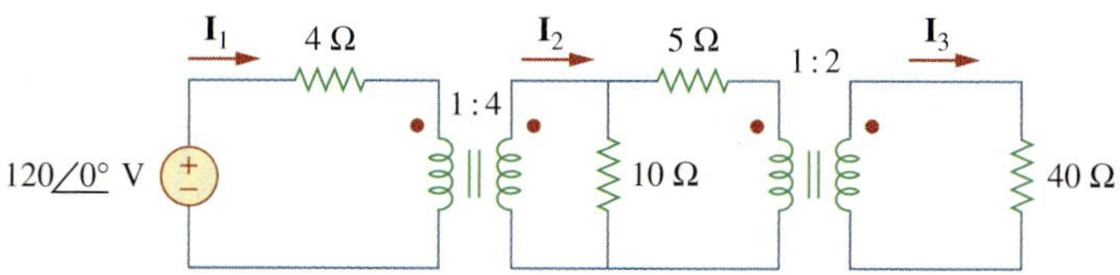

Figure 13.125
For Prob. 13.60.

*13.61 For the circuit in Fig. 13.126, find $\mathbf{I}_1$, $\mathbf{I}_2$, and $\mathbf{V}_o$.

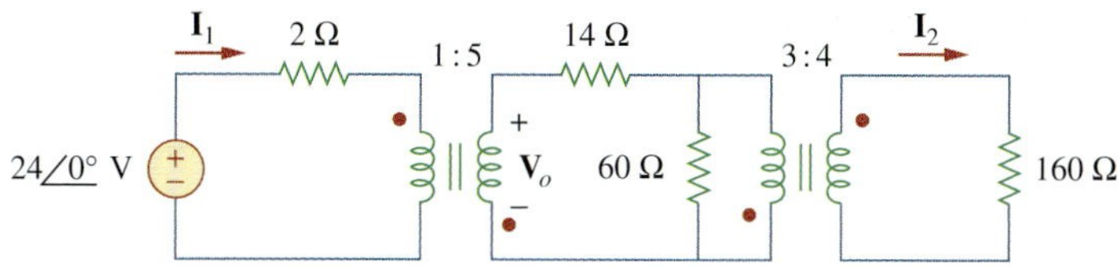

Figure 13.126
For Prob. 13.61.

13.62 For the network in Fig. 13.127, find

(a) the complex power supplied by the source,

(b) the average power delivered to the 18-Ω resistor.

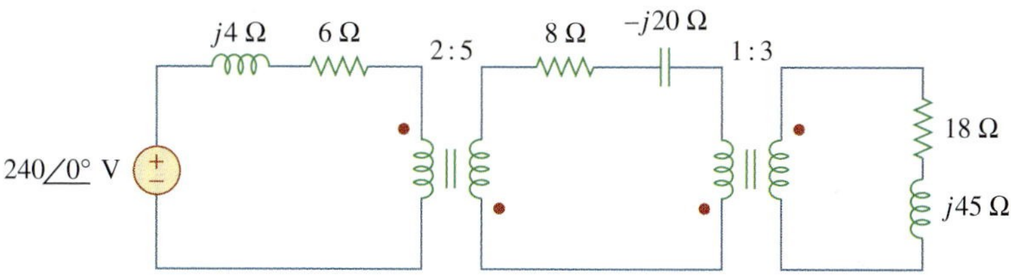

Figure 13.127
For Prob. 13.62.

13.63 Find the mesh currents in the circuit of Fig. 13.128

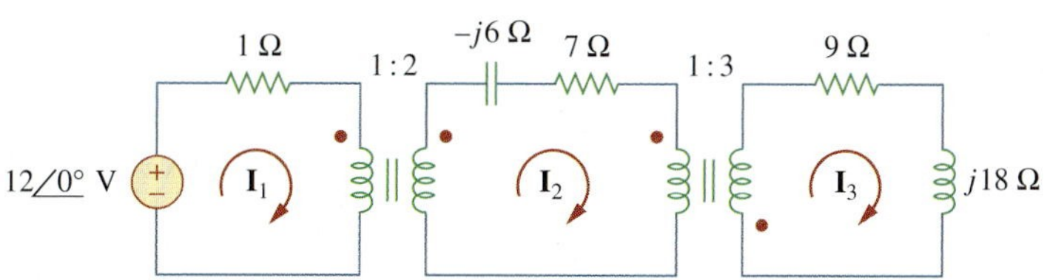

Figure 13.128
For Prob. 13.63.

13.64 For the circuit in Fig. 13.129, find the turns ratio so that the maximum power is delivered to the 30-kΩ resistor.

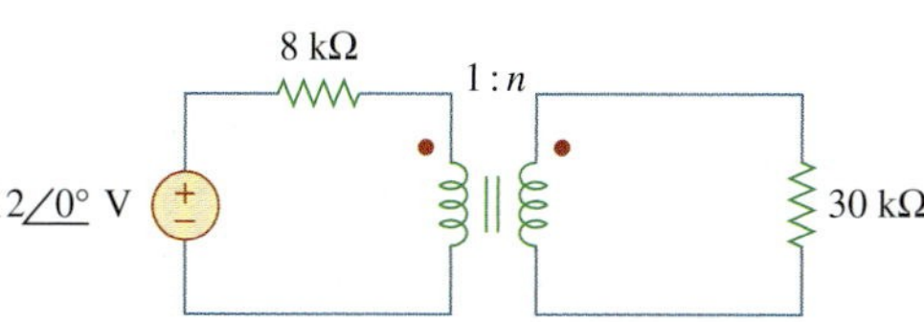

Figure 13.129
For Prob. 13.64.

*13.65 Calculate the average power dissipated by the 20-Ω resistor in Fig. 13.130.

ps ML

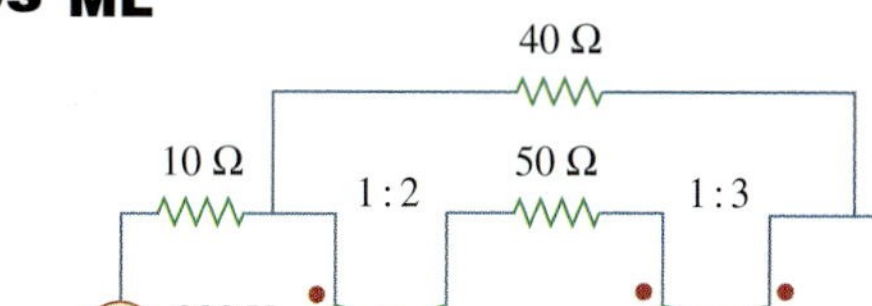

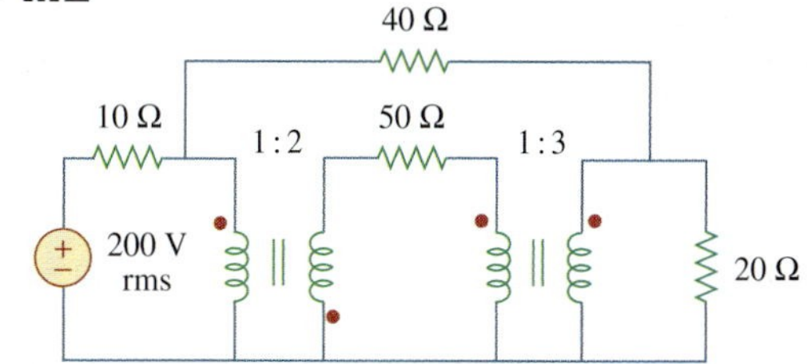

Figure 13.130
For Prob. 13.65.

Section 13.6 Ideal Autotransformers

13.66 Design a problem to help other students better understand how the ideal autotransformer works.

13.67 An autotransformer with a 40 percent tap is supplied by a 400-V, 60-Hz source and is used for step-down operation. A 5-kVA load operating at unity power factor is connected to the secondary terminals. Find:

(a) the secondary voltage

(b) the secondary current

(c) the primary current

13.68 In the ideal autotransformer of Fig. 13.131, calculate $\mathbf{I}_1$, $\mathbf{I}_2$, and $\mathbf{I}_o$. Find the average power delivered to the load.

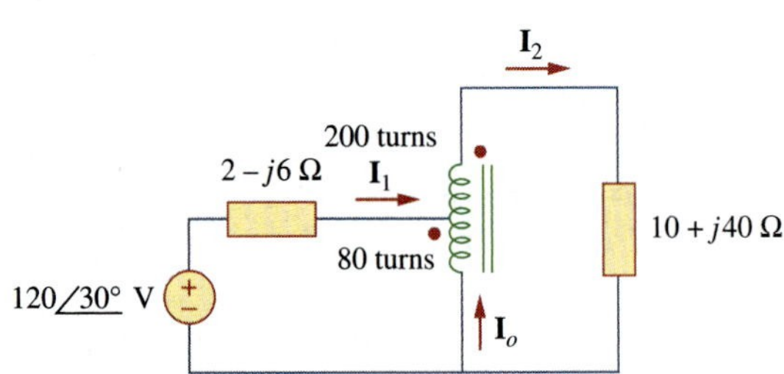

Figure 13.131
For Prob. 13.68.

***13.69** In the circuit of Fig. 13.132, $\mathbf{Z}_L$ is adjusted until maximum average power is delivered to $\mathbf{Z}_L$. Find $\mathbf{Z}_L$ and the maximum average power transferred to it. Take $N_1 = 600$ turns and $N_2 = 200$ turns.

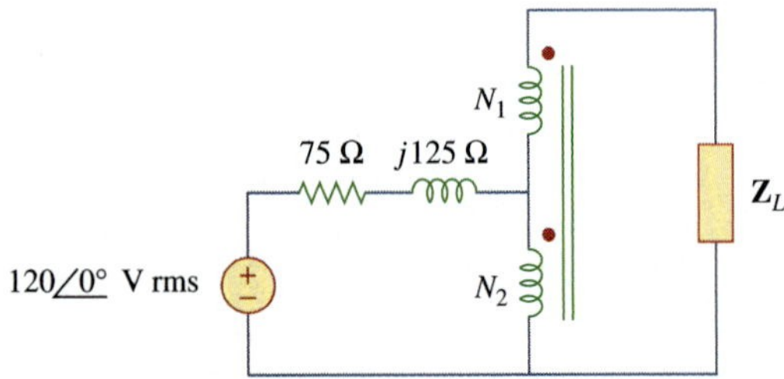

Figure 13.132
For Prob. 13.69.

13.70 In the ideal transformer circuit shown in Fig. 13.133, determine the average power delivered to the load.

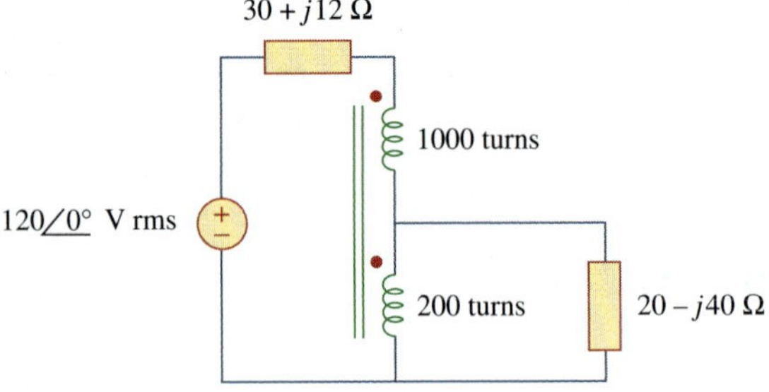

Figure 13.133
For Prob. 13.70.

13.71 In the autotransformer circuit in Fig. 13.134, show that

$$\mathbf{Z}_{\text{in}} = \left(1 + \frac{N_1}{N_2}\right)^2 \mathbf{Z}_L$$

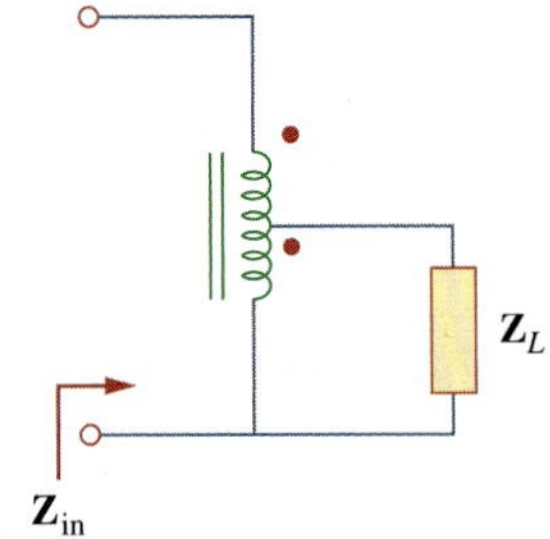

Figure 13.134
For Prob. 13.71.

Section 13.7 Three-Phase Transformers

13.72 In order to meet an emergency, three single-phase transformers with 12,470/7,200 V rms are connected in Δ-Y to form a three-phase transformer which is fed by a 12,470-V transmission line. If the transformer supplies 60 MVA to a load, find:

(a) the turns ratio for each transformer,

(b) the currents in the primary and secondary windings of the transformer,

(c) the incoming and outgoing transmission line currents.

13.73 Figure 13.135 on the following page shows a three-phase transformer that supplies a Y-connected load.

(a) Identify the transformer connection.

(b) Calculate currents $\mathbf{I}_2$ and $\mathbf{I}_c$.

(c) Find the average power absorbed by the load.

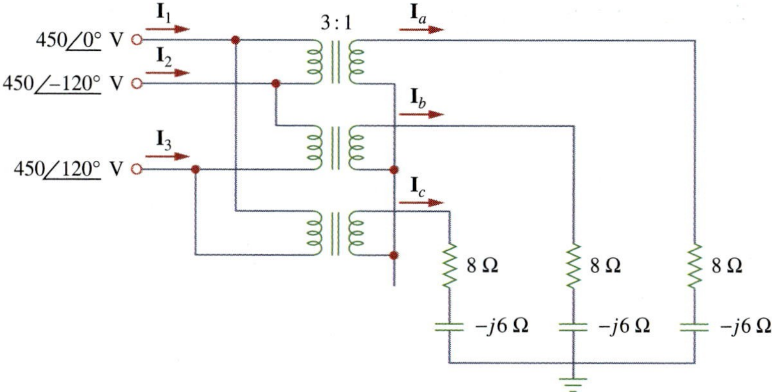

Figure 13.135
For Prob. 13.73.

13.74 Consider the three-phase transformer shown in Fig. 13.136. The primary is fed by a three-phase source with line voltage of 2.4 kV rms, while the secondary supplies a three-phase 120-kW balanced load at pf of 0.8. Determine:

(a) the type of transformer connections,

(b) the values of I_{LS} and I_{PS},

(c) the values of I_{LP} and I_{PP},

(d) the kVA rating of each phase of the transformer.

13.75 A balanced three-phase transformer bank with the Δ-Y connection depicted in Fig. 13.137 is used to step down line voltages from 4,500 V rms to 900 V rms. If the transformer feeds a 120-kVA load, find:

(a) the turns ratio for the transformer,

(b) the line currents at the primary and secondary sides.

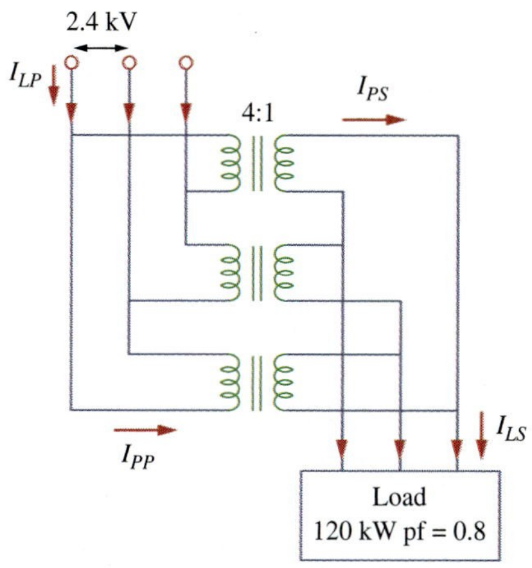

Figure 13.136
For Prob. 13.74.

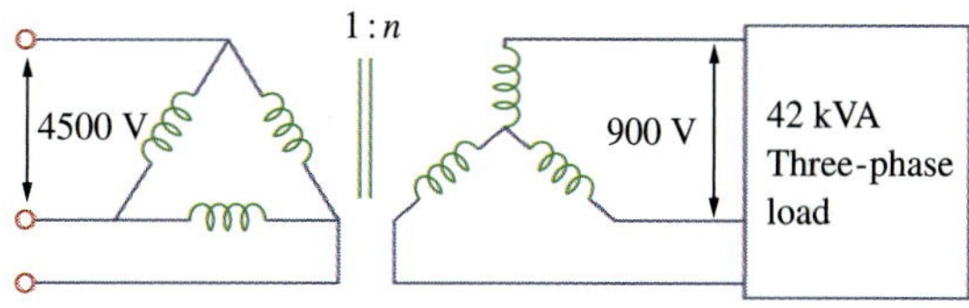

Figure 13.137
For Prob. 13.75.

13.76 **ed** Using Fig. 13.138, design a problem to help other students better understand a Y-Δ, three-phase transformer and how they work.

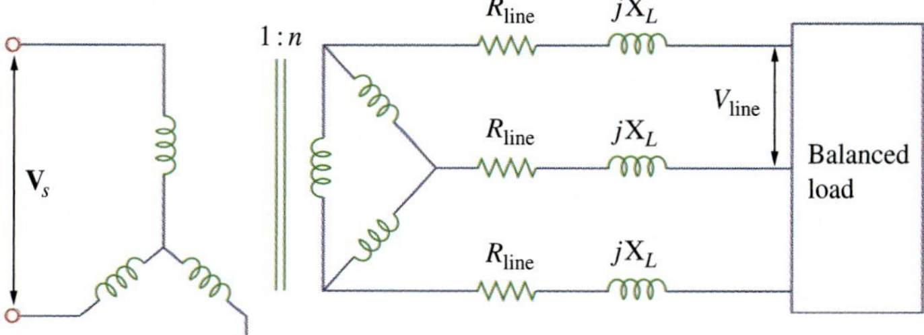

Figure 13.138
For Prob. 13.76.

13.77 The three-phase system of a town distributes power with a line voltage of 13.2 kV. A pole transformer connected to single wire and ground steps down the high-voltage wire to 120 V rms and serves a house as shown in Fig. 13.139.

(a) Calculate the turns ratio of the pole transformer to get 120 V.

(b) Determine how much current a 100-W lamp connected to the 120-V hot line draws from the high-voltage line.

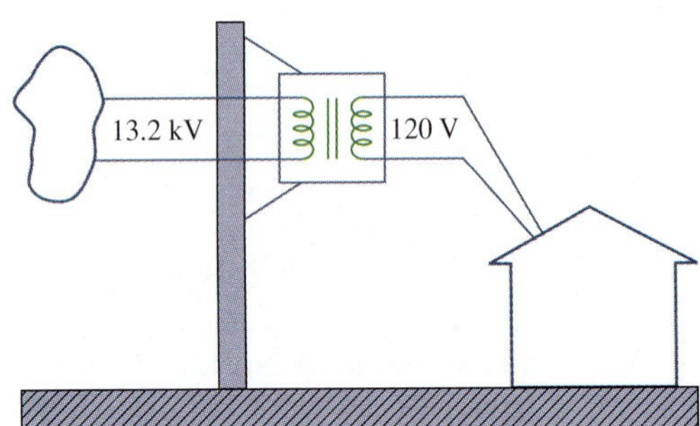

Figure 13.139
For Prob. 13.77.

Section 13.8 *PSpice* Analysis of Magnetically Coupled Circuits

13.78 Use *PSpice* to determine the mesh currents in the circuit of Fig. 13.140. Take $\omega = 1$ rad/s.

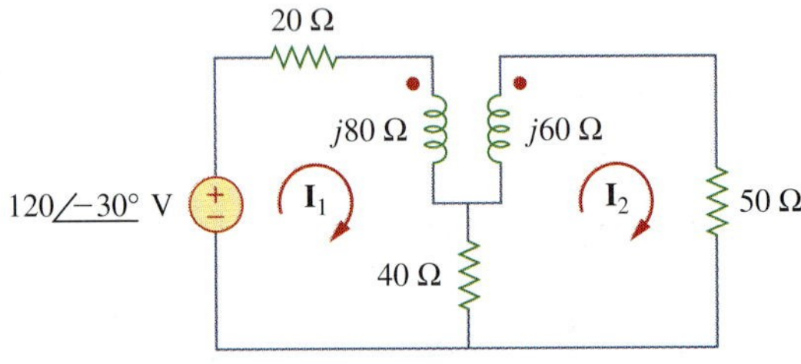

Figure 13.140
For Prob. 13.78.

13.79 Use *PSpice* to find $\mathbf{I}_1$, $\mathbf{I}_2$, and $\mathbf{I}_3$ in the circuit of Fig. 13.141.

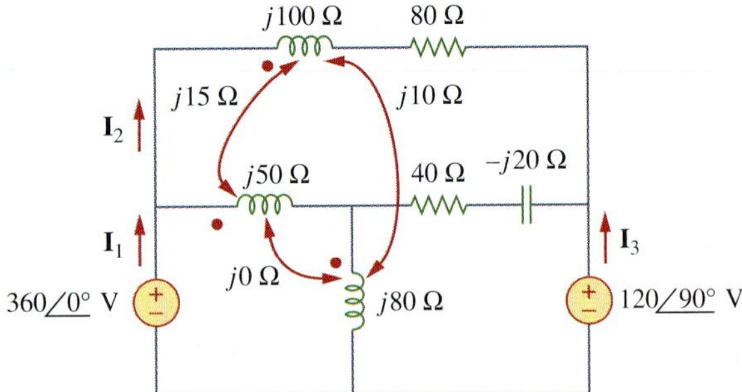

Figure 13.141
For Prob. 13.79.

13.80 Rework Prob. 13.22 using *PSpice*.

13.81 Use *PSpice* to find $\mathbf{I}_1$, $\mathbf{I}_2$, and $\mathbf{I}_3$ in the circuit of Fig. 13.142.

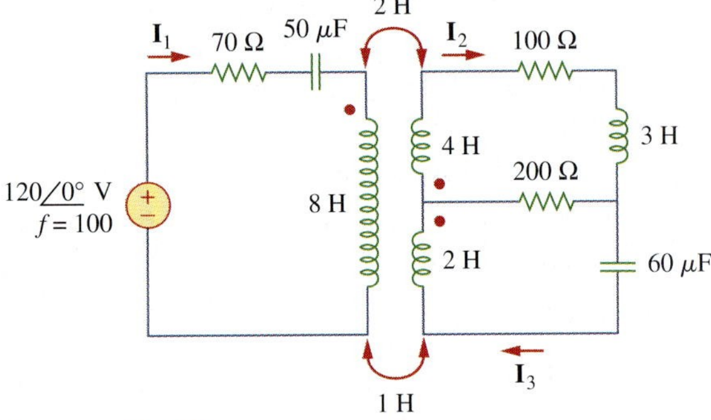

Figure 13.142
For Prob. 13.81.

13.82 Use *PSpice* to find $\mathbf{V}_1$, $\mathbf{V}_2$, and $\mathbf{I}_o$ in the circuit of Fig. 13.143.

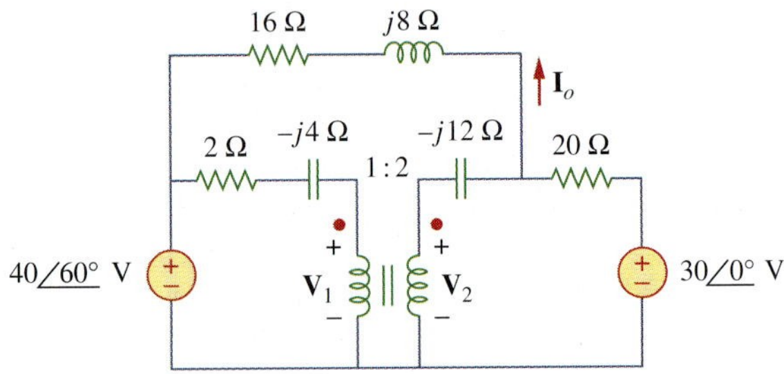

Figure 13.143
For Prob. 13.82.

13.83 Find $\mathbf{I}_x$ and $\mathbf{V}_x$ in the circuit of Fig. 13.144 using *PSpice*.

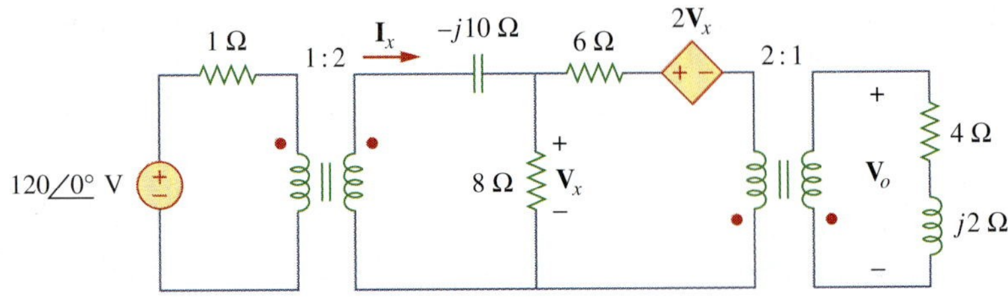

Figure 13.144
For Prob. 13.83.

13.84 Determine $\mathbf{I}_1$, $\mathbf{I}_2$, and $\mathbf{I}_3$ in the ideal transformer circuit of Fig. 13.145 using *PSpice*.

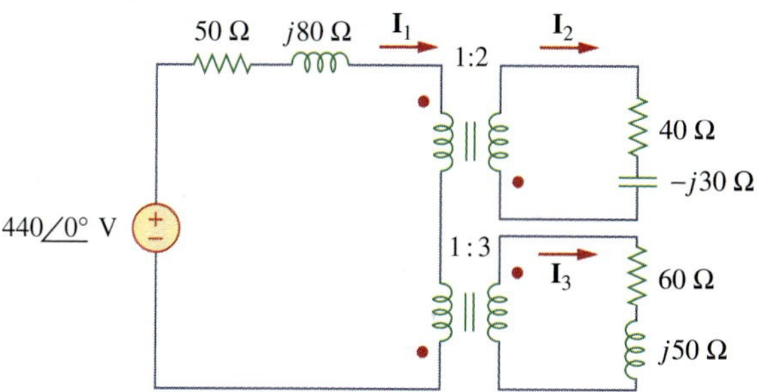

Figure 13.145
For Prob. 13.84.

Section 13.9 Applications

13.85 A stereo amplifier circuit with an output impedance of 7.2 kΩ is to be matched to a speaker with an input impedance of 8 Ω by a transformer whose primary side has 3,000 turns. Calculate the number of turns required on the secondary side.

13.86 A transformer having 2,400 turns on the primary and 48 turns on the secondary is used as an impedance-matching device. What is the reflected value of a 3-Ω load connected to the secondary?

13.87 A radio receiver has an input resistance of 300Ω. When it is connected directly to an antenna system with a characteristic impedance of 75 Ω, an impedance mismatch occurs. By inserting an impedance-matching transformer ahead of the receiver, maximum power can be realized. Calculate the required turns ratio.

13.88 A step-down power transformer with a turns ratio of $n = 0.1$ supplies 12.6 V rms to a resistive load. If the primary current is 2.5 A rms, how much power is delivered to the load?

13.89 A 240/120-V rms power transformer is rated at 10 kVA. Determine the turns ratio, the primary current, and the secondary current.

13.90 A 4-kVA, 2,400/240-V rms transformer has 250 turns on the primary side. Calculate:

(a) the turns ratio,

(b) the number of turns on the secondary side,

(c) the primary and secondary currents.

13.91 A 25,000/240-V rms distribution transformer has a primary current rating of 75 A.

(a) Find the transformer kVA rating.

(b) Calculate the secondary current.

13.92 A 4,800-V rms transmission line feeds a distribution transformer with 1,200 turns on the primary and 28 turns on the secondary. When a 10-Ω load is connected across the secondary, find:

(a) the secondary voltage,

(b) the primary and secondary currents,

(c) the power supplied to the load.

Comprehensive Problems

13.93 A four-winding transformer (Fig. 13.146) is often used in equipment (e.g., PCs, VCRs) that may be operated from either 110 V or 220 V. This makes the equipment suitable for both domestic and foreign use. Show which connections are necessary to provide:

(a) an output of 14 V with an input of 110 V,

(b) an output of 50 V with an input of 220 V.

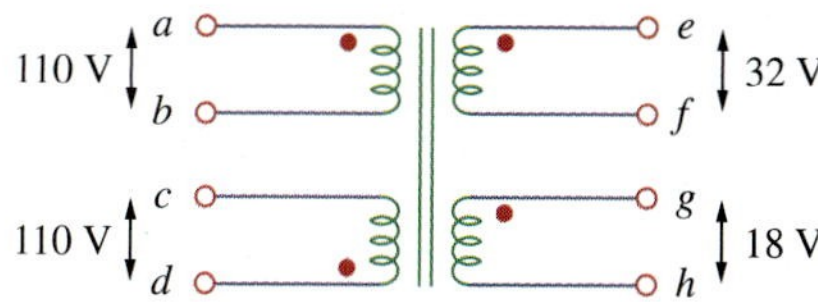

Figure 13.146
For Prob. 13.93.

***13.94** A 440/110-V ideal transformer can be connected to become a 550/440-V ideal autotransformer. There are four possible connections, two of which are wrong. Find the output voltage of:

(a) a wrong connection,

(b) the right connection.

13.95 Ten bulbs in parallel are supplied by a 7,200/120-V transformer as shown in Fig. 13.147, where the bulbs are modeled by the 144-Ω resistors. Find:

(a) the turns ratio n,

(b) the current through the primary winding.

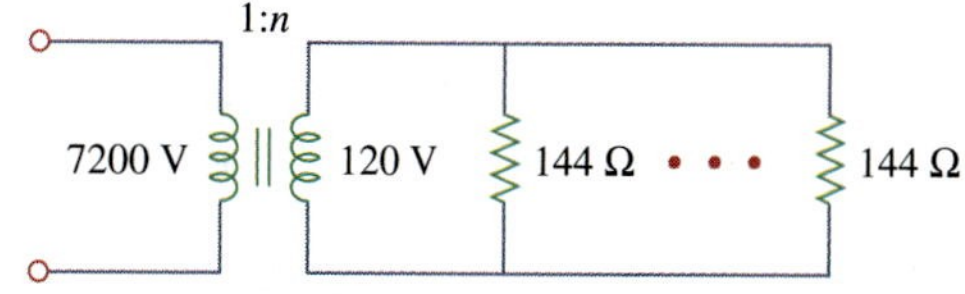

Figure 13.147
For Prob. 13.95.

***13.96** **e2d** Some modern power transmission systems now have major, high voltage DC transmission segments. There are a lot of good reasons for doing this but we will not go into them here. To go from the AC to DC, power electronics are used. We start with three-phase AC and then rectify it (using a full-wave rectifier). It was found that using a delta to wye and delta combination connected secondary would give us a much smaller ripple after the full-wave rectifier. How is this accomplished? Remember that these are real devices and are wound on common cores.

Hint: using Figs. 13.47 and 13.49, and the fact that each coil of the wye connected secondary and each coil of the delta connected secondary are wound around the same core of each coil of the delta connected primary so the voltage of each of the corresponding coils are in phase. When the output leads of both secondaries are connected through full-wave rectifiers with the same load, you will see that the ripple is now greatly reduced. Please consult the instructor for more help if necessary.

chapter

14

Frequency Response

Dost thou love Life? Then do not squander Time; for that is the stuff Life is made.

—Benjamin Franklin

Enhancing Your Career

Career in Control Systems

Control systems are another area of electrical engineering where circuit analysis is used. A control system is designed to regulate the behavior of one or more variables in some desired manner. Control systems play major roles in our everyday life. Household appliances such as heating and air-conditioning systems, switch-controlled thermostats, washers and dryers, cruise controllers in automobiles, elevators, traffic lights, manufacturing plants, navigation systems—all utilize control systems. In the aerospace field, precision guidance of space probes, the wide range of operational modes of the space shuttle, and the ability to maneuver space vehicles remotely from earth all require knowledge of control systems. In the manufacturing sector, repetitive production line operations are increasingly performed by robots, which are programmable control systems designed to operate for many hours without fatigue.

A welding robot. © Vol. 1 PhotoDisc/ Getty Images

Control engineering integrates circuit theory and communication theory. It is not limited to any specific engineering discipline but may involve environmental, chemical, aeronautical, mechanical, civil, and electrical engineering. For example, a typical task for a control system engineer might be to design a speed regulator for a disk drive head.

A thorough understanding of control systems techniques is essential to the electrical engineer and is of great value for designing control systems to perform the desired task.

14.1 Introduction

In our sinusoidal circuit analysis, we have learned how to find voltages and currents in a circuit with a constant frequency source. If we let the amplitude of the sinusoidal source remain constant and vary the frequency, we obtain the circuit's *frequency response*. The frequency response may be regarded as a complete description of the sinusoidal steady-state behavior of a circuit as a function of frequency.

> The **frequency response** of a circuit is the variation in its behavior with change in signal frequency.

The frequency response of a circuit may also be considered as the variation of the gain and phase with frequency.

The sinusoidal steady-state frequency responses of circuits are of significance in many applications, especially in communications and control systems. A specific application is in electric filters that block out or eliminate signals with unwanted frequencies and pass signals of the desired frequencies. Filters are used in radio, TV, and telephone systems to separate one broadcast frequency from another.

We begin this chapter by considering the frequency response of simple circuits using their transfer functions. We then consider Bode plots, which are the industry-standard way of presenting frequency response. We also consider series and parallel resonant circuits and encounter important concepts such as resonance, quality factor, cutoff frequency, and bandwidth. We discuss different kinds of filters and network scaling. In the last section, we consider one practical application of resonant circuits and two applications of filters.

14.2 Transfer Function

The transfer function $\mathbf{H}(\omega)$ (also called the *network function*) is a useful analytical tool for finding the frequency response of a circuit. In fact, the frequency response of a circuit is the plot of the circuit's transfer function $\mathbf{H}(\omega)$ versus ω, with ω varying from $\omega = 0$ to $\omega = \infty$.

A transfer function is the frequency-dependent ratio of a forced function to a forcing function (or of an output to an input). The idea of a transfer function was implicit when we used the concepts of impedance and admittance to relate voltage and current. In general, a linear network can be represented by the block diagram shown in Fig. 14.1.

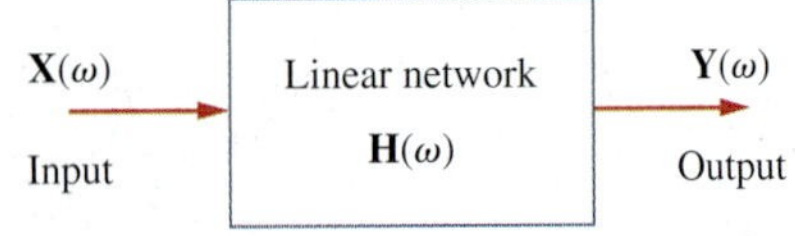

Figure 14.1
A block diagram representation of a linear network.

> The **transfer function** $\mathbf{H}(\omega)$ of a circuit is the frequency-dependent ratio of a phasor output $\mathbf{Y}(\omega)$ (an element voltage or current) to a phasor input $\mathbf{X}(\omega)$ (source voltage or current).

In this context, $\mathbf{X}(\omega)$ and $\mathbf{Y}(\omega)$ denote the input and output phasors of a network; they should not be confused with the same symbolism used for reactance and admittance. The multiple usage of symbols is conventionally permissible due to lack of enough letters in the English language to express all circuit variables distinctly.

Thus,

$$\mathbf{H}(\omega) = \frac{\mathbf{Y}(\omega)}{\mathbf{X}(\omega)} \tag{14.1}$$

assuming zero initial conditions. Since the input and output can be either voltage or current at any place in the circuit, there are four possible transfer functions:

Some authors use $\mathbf{H}(j\omega)$ for transfer instead of $\mathbf{H}(\omega)$, since ω and j are an inseparable pair.

$$\mathbf{H}(\omega) = \text{Voltage gain} = \frac{\mathbf{V}_o(\omega)}{\mathbf{V}_i(\omega)} \tag{14.2a}$$

$$\mathbf{H}(\omega) = \text{Current gain} = \frac{\mathbf{I}_o(\omega)}{\mathbf{I}_i(\omega)} \tag{14.2b}$$

$$\mathbf{H}(\omega) = \text{Transfer Impedance} = \frac{\mathbf{V}_o(\omega)}{\mathbf{I}_i(\omega)} \tag{14.2c}$$

$$\mathbf{H}(\omega) = \text{Transfer Admittance} = \frac{\mathbf{I}_o(\omega)}{\mathbf{V}_i(\omega)} \tag{14.2d}$$

where subscripts i and o denote input and output values. Being a complex quantity, $\mathbf{H}(\omega)$ has a magnitude $H(\omega)$ and a phase ϕ; that is, $\mathbf{H}(\omega) = H(\omega)\underline{/\phi}$.

To obtain the transfer function using Eq. (14.2), we first obtain the frequency-domain equivalent of the circuit by replacing resistors, inductors, and capacitors with their impedances R, $j\omega L$, and $1/j\omega C$. We then use any circuit technique(s) to obtain the appropriate quantity in Eq. (14.2). We can obtain the frequency response of the circuit by plotting the magnitude and phase of the transfer function as the frequency varies. A computer is a real time-saver for plotting the transfer function.

The transfer function $\mathbf{H}(\omega)$ can be expressed in terms of its numerator polynomial $\mathbf{N}(\omega)$ and denominator polynomial $\mathbf{D}(\omega)$ as

$$\boxed{\mathbf{H}(\omega) = \frac{\mathbf{N}(\omega)}{\mathbf{D}(\omega)}} \tag{14.3}$$

where $\mathbf{N}(\omega)$ and $\mathbf{D}(\omega)$ are not necessarily the same expressions for the input and output functions, respectively. The representation of $\mathbf{H}(\omega)$ in Eq. (14.3) assumes that common numerator and denominator factors in $\mathbf{H}(\omega)$ have canceled, reducing the ratio to lowest terms. The roots of $\mathbf{N}(\omega) = 0$ are called the *zeros* of $\mathbf{H}(\omega)$ and are usually represented as $j\omega = z_1, z_2, \ldots$. Similarly, the roots of $\mathbf{D}(\omega) = 0$ are the *poles* of $\mathbf{H}(\omega)$ and are represented as $j\omega = p_1, p_2, \ldots$.

A **zero,** as a *root* of the numerator polynomial, is a value that results in a zero value of the function. A **pole,** as a *root* of the denominator polynomial, is a value for which the function is infinite.

A zero may also be regarded as the value of $s = j\omega$ that makes $\mathbf{H}(s)$ zero, and a pole as the value of $s = j\omega$ that makes $\mathbf{H}(s)$ infinite.

To avoid complex algebra, it is expedient to replace $j\omega$ temporarily with s when working with $\mathbf{H}(\omega)$ and replace s with $j\omega$ at the end.

Example 14.1

For the RC circuit in Fig. 14.2(a), obtain the transfer function $\mathbf{V}_o/\mathbf{V}_s$ and its frequency response. Let $v_s = V_m \cos \omega t$.

Solution:

The frequency-domain equivalent of the circuit is in Fig. 14.2(b). By voltage division, the transfer function is given by

$$\mathbf{H}(\omega) = \frac{\mathbf{V}_o}{\mathbf{V}_s} = \frac{1/j\omega C}{R + 1/j\omega C} = \frac{1}{1 + j\omega RC}$$

Figure 14.2
For Example 14.1: (a) time-domain RC circuit, (b) frequency-domain RC circuit.

Comparing this with Eq. (9.18e), we obtain the magnitude and phase of $\mathbf{H}(\omega)$ as

$$H = \frac{1}{\sqrt{1 + (\omega/\omega_0)^2}}, \qquad \phi = -\tan^{-1}\frac{\omega}{\omega_0}$$

where $\omega_0 = 1/RC$. To plot H and ϕ for $0 < \omega < \infty$, we obtain their values at some critical points and then sketch.

At $\omega = 0$, $H = 1$ and $\phi = 0$. At $\omega = \infty$, $H = 0$ and $\phi = -90°$. Also, at $\omega = \omega_0$, $H = 1/\sqrt{2}$ and $\phi = -45°$. With these and a few more points as shown in Table 14.1, we find that the frequency response is as shown in Fig. 14.3. Additional features of the frequency response in Fig. 14.3 will be explained in Section 14.6.1 on lowpass filters.

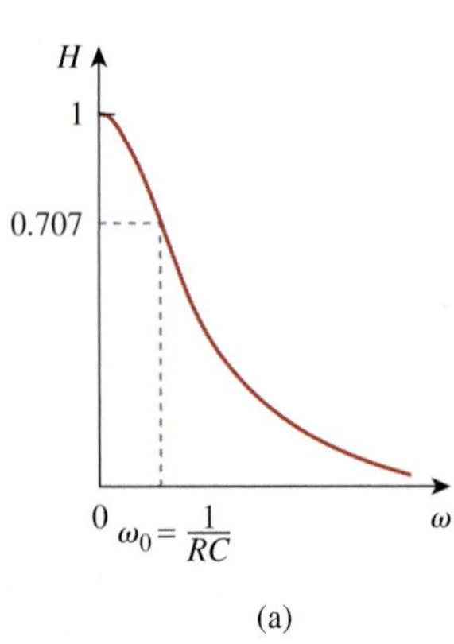

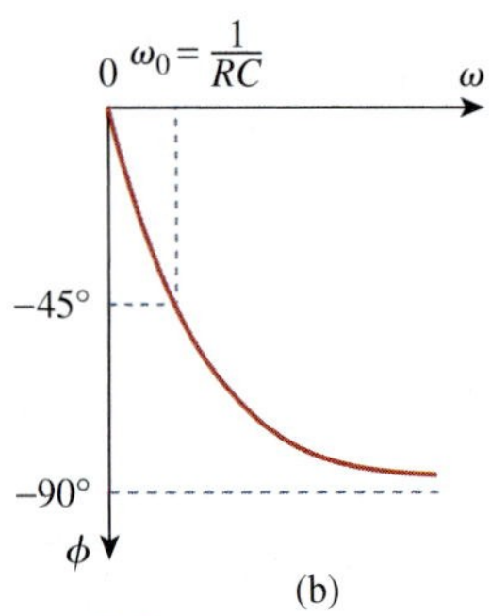

Figure 14.3
Frequency response of the RC circuit: (a) amplitude response, (b) phase response.

TABLE 14.1

For Example 14.1.

ω/ω_0	H	ϕ	ω/ω_0	H	ϕ
0	1	0	10	0.1	−84°
1	0.71	−45°	20	0.05	−87°
2	0.45	−63°	100	0.01	−89°
3	0.32	−72°	∞	0	−90°

Practice Problem 14.1

Obtain the transfer function $\mathbf{V}_o/\mathbf{V}_s$ of the RL circuit in Fig. 14.4, assuming $v_s = V_m \cos\omega t$. Sketch its frequency response.

Figure 14.4
RL circuit for Practice Prob. 14.1.

Answer: $j\omega L/(R + j\omega L)$; see Fig. 14.5 for the response.

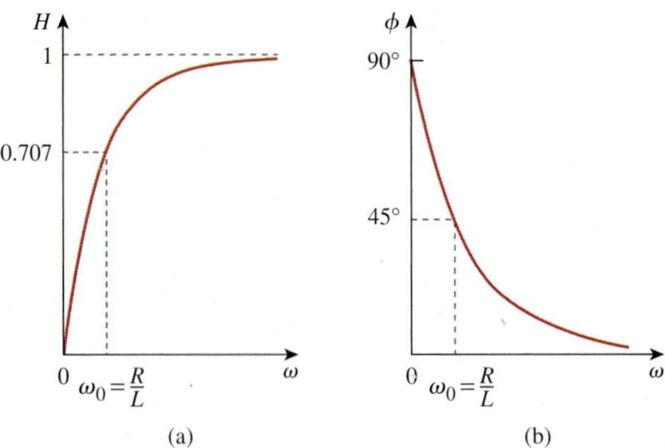

Figure 14.5
Frequency response of the RL circuit in Fig. 14.4.

Example 14.2

For the circuit in Fig. 14.6, calculate the gain $\mathbf{I}_o(\omega)/\mathbf{I}_i(\omega)$ and its poles and zeros.

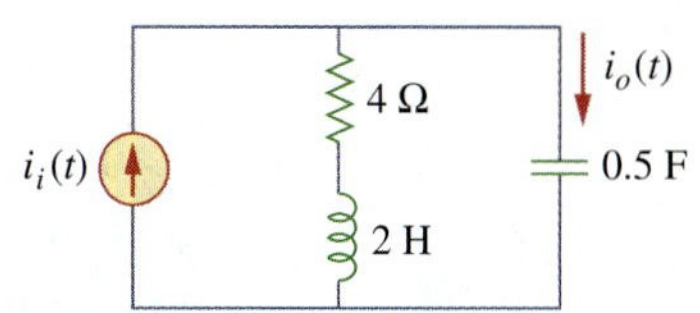

Figure 14.6
For Example 14.2.

Solution:
By current division,

$$\mathbf{I}_o(\omega) = \frac{4 + j2\omega}{4 + j2\omega + 1/j0.5\omega}\mathbf{I}_i(\omega)$$

or

$$\frac{\mathbf{I}_o(\omega)}{\mathbf{I}_i(\omega)} = \frac{j0.5\omega(4 + j2\omega)}{1 + j2\omega + (j\omega)^2} = \frac{s(s + 2)}{s^2 + 2s + 1}, \qquad s = j\omega$$

The zeros are at

$$s(s + 2) = 0 \qquad \Rightarrow \qquad z_1 = 0, z_2 = -2$$

The poles are at

$$s^2 + 2s + 1 = (s + 1)^2 = 0$$

Thus, there is a repeated pole (or double pole) at $p = -1$.

Practice Problem 14.2

Find the transfer function $\mathbf{V}_o(\omega)/\mathbf{I}_i(\omega)$ for the circuit in Fig. 14.7. Obtain its zeros and poles.

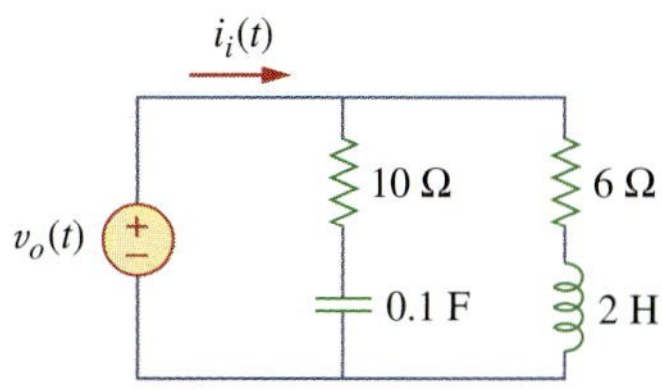

Figure 14.7
For Practice Prob. 14.2.

Answer: $\dfrac{10(s + 1)(s + 3)}{s^2 + 8s + 5}$, $s = j\omega$; zeros: $-1, -3$; poles: -0.683, -7.317.

14.3 †The Decibel Scale

It is not always easy to get a quick plot of the magnitude and phase of the transfer function as we did above. A more systematic way of obtaining the frequency response is to use Bode plots. Before we begin to construct Bode plots, we should take care of two important issues: the use of logarithms and decibels in expressing gain.

Since Bode plots are based on logarithms, it is important that we keep the following properties of logarithms in mind:

1. $\log P_1P_2 = \log P_1 + \log P_2$
2. $\log P_1/P_2 = \log P_1 - \log P_2$
3. $\log P^n = n \log P$
4. $\log 1 = 0$

In communications systems, gain is measured in *bels*. Historically, the bel is used to measure the ratio of two levels of power or power gain G; that is,

Historical note: The *bel* is named after Alexander Graham Bell, the inventor of the telephone.

$$G = \text{Number of bels} = \log_{10}\frac{P_2}{P_1} \qquad \textbf{(14.4)}$$

Historical

Alexander Graham Bell (1847–1922) inventor of the telephone, was a Scottish-American scientist.

Bell was born in Edinburgh, Scotland, a son of Alexander Melville Bell, a well-known speech teacher. Alexander the younger also became a speech teacher after graduating from the University of Edinburgh and the University of London. In 1866 he became interested in transmitting speech electrically. After his older brother died of tuberculosis, his father decided to move to Canada. Alexander was asked to come to Boston to work at the School for the Deaf. There he met Thomas A. Watson, who became his assistant in his electromagnetic transmitter experiment. On March 10, 1876, Alexander sent the famous first telephone message: "Watson, come here I want you." The bel, the logarithmic unit introduced in Chapter 14, is named in his honor.

The *decibel* (dB) provides us with a unit of less magnitude. It is 1/10th of a bel and is given by

$$G_{\text{dB}} = 10 \log_{10} \frac{P_2}{P_1} \tag{14.5}$$

When $P_1 = P_2$, there is no change in power and the gain is 0 dB. If $P_2 = 2P_1$, the gain is

$$G_{\text{dB}} = 10 \log_{10} 2 \simeq 3 \text{ dB} \tag{14.6}$$

and when $P_2 = 0.5P_1$, the gain is

$$G_{\text{dB}} = 10 \log_{10} 0.5 \simeq -3 \text{ dB} \tag{14.7}$$

Equations (14.6) and (14.7) show another reason why logarithms are greatly used: The logarithm of the reciprocal of a quantity is simply negative the logarithm of that quantity.

Alternatively, the gain G can be expressed in terms of voltage or current ratio. To do so, consider the network shown in Fig. 14.8. If P_1 is the input power, P_2 is the output (load) power, R_1 is the input resistance, and R_2 is the load resistance, then $P_1 = 0.5V_1^2/R_1$ and $P_2 = 0.5V_2^2/R_2$, and Eq. (14.5) becomes

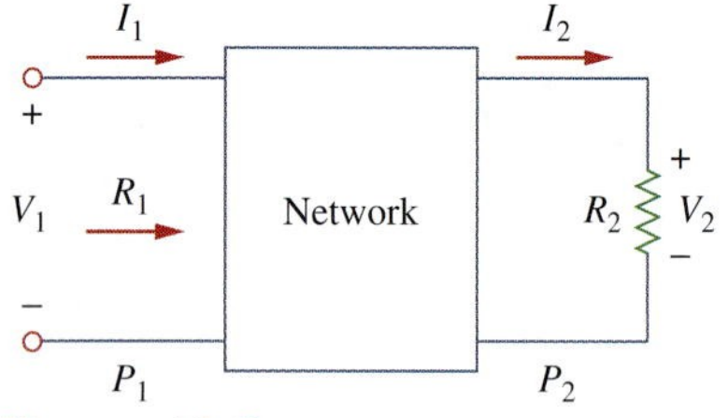

Figure 14.8
Voltage-current relationships for a four-terminal network.

$$\begin{aligned} G_{\text{dB}} &= 10 \log_{10} \frac{P_2}{P_1} = 10 \log_{10} \frac{V_2^2/R_2}{V_1^2/R_1} \\ &= 10 \log_{10}\left(\frac{V_2}{V_1}\right)^2 + 10 \log_{10} \frac{R_1}{R_2} \end{aligned} \tag{14.8}$$

$$G_{\text{dB}} = 20 \log_{10} \frac{V_2}{V_1} - 10 \log_{10} \frac{R_2}{R_1} \tag{14.9}$$

For the case when $R_2 = R_1$, a condition that is often assumed when comparing voltage levels, Eq. (14.9) becomes

$$G_{\text{dB}} = 20 \log_{10} \frac{V_2}{V_1} \tag{14.10}$$

Instead, if $P_1 = I_1^2 R_1$ and $P_2 = I_2^2 R_2$, for $R_1 = R_2$, we obtain

$$G_{\text{dB}} = 20 \log_{10} \frac{I_2}{I_1} \tag{14.11}$$

Three things are important to note from Eqs. (14.5), (14.10), and (14.11):

1. That $10 \log_{10}$ is used for power, while $20 \log_{10}$ is used for voltage or current, because of the square relationship between them ($P = V^2/R = I^2R$).
2. That the dB value is a logarithmic measurement of the *ratio* of one variable to another *of the same type*. Therefore, it applies in expressing the transfer function H in Eqs. (14.2a) and (14.2b), which are dimensionless quantities, but not in expressing H in Eqs. (14.2c) and (14.2d).
3. It is important to note that we only use voltage and current magnitudes in Eqs. (14.10) and (14.11). Negative signs and angles will be handled independently as we will see in Section 14.4.

With this in mind, we now apply the concepts of logarithms and decibels to construct Bode plots.

14.4 Bode Plots

Obtaining the frequency response from the transfer function as we did in Section 14.2 is an uphill task. The frequency range required in frequency response is often so wide that it is inconvenient to use a linear scale for the frequency axis. Also, there is a more systematic way of locating the important features of the magnitude and phase plots of the transfer function. For these reasons, it has become standard practice to plot the transfer function on a pair of semilogarithmic plots: the magnitude in decibels is plotted against the logarithm of the frequency; on a separate plot, the phase in degrees is plotted against the logarithm of the frequency. Such semilogarithmic plots of the transfer function—known as *Bode plots*—have become the industry standard.

Historical note: Named after Hendrik W. Bode (1905–1982), an engineer with the Bell Telephone Laboratories, for his pioneering work in the 1930s and 1940s.

> **Bode plots** are semilog plots of the magnitude (in decibels) and phase (in degrees) of a transfer function versus frequency.

Bode plots contain the same information as the nonlogarithmic plots discussed in the previous section, but they are much easier to construct, as we shall see shortly.

The transfer function can be written as

$$\mathbf{H} = H\underline{/\phi} = He^{j\phi} \tag{14.12}$$

Taking the natural logarithm of both sides,

$$\ln \mathbf{H} = \ln H + \ln e^{j\phi} = \ln H + j\phi \tag{14.13}$$

Thus, the real part of $\ln \mathbf{H}$ is a function of the magnitude while the imaginary part is the phase. In a Bode magnitude plot, the gain

$$\boxed{H_{\text{dB}} = 20 \log_{10} H} \tag{14.14}$$

TABLE 14.2

Specific gain and their decibel values.*

Magnitude H	$20 \log_{10} H$ (dB)
0.001	−60
0.01	−40
0.1	−20
0.5	−6
$1/\sqrt{2}$	−3
1	0
$\sqrt{2}$	3
2	6
10	20
20	26
100	40
1000	60

* Some of these values are approximate.

is plotted in decibels (dB) versus frequency. Table 14.2 provides a few values of H with the corresponding values in decibels. In a Bode phase plot, ϕ is plotted in degrees versus frequency. Both magnitude and phase plots are made on semilog graph paper.

A transfer function in the form of Eq. (14.3) may be written in terms of factors that have real and imaginary parts. One such representation might be

$$\mathbf{H}(\omega) = \frac{K(j\omega)^{\pm 1}(1 + j\omega/z_1)[1 + j2\zeta_1\omega/\omega_k + (j\omega/\omega_k)^2]\cdots}{(1 + j\omega/p_1)[1 + j2\zeta_2\omega/\omega_n + (j\omega/\omega_n)^2]\cdots} \quad \textbf{(14.15)}$$

which is obtained by dividing out the poles and zeros in $\mathbf{H}(\omega)$. The representation of $\mathbf{H}(\omega)$ as in Eq. (14.15) is called the *standard form*. $\mathbf{H}(\omega)$ may include up to seven types of different factors that can appear in various combinations in a transfer function. These are:

1. A gain K
2. A pole $(j\omega)^{-1}$ or zero $(j\omega)$ at the origin
3. A simple pole $1/(1 + j\omega/p_1)$ or zero $(1 + j\omega/z_1)$
4. A quadratic pole $1/[1 + j2\zeta_2\omega/\omega_n + (j\omega/\omega_n)^2]$ or zero $[1 + j2\zeta_1\omega/\omega_k + (j\omega/\omega_k)^2]$

The origin is where $\omega = 1$ or $\log \omega = 0$ and the gain is zero.

In constructing a Bode plot, we plot each factor separately and then add them graphically. The factors can be considered one at a time and then combined additively because of the logarithms involved. It is this mathematical convenience of the logarithm that makes Bode plots a powerful engineering tool.

We will now make straight-line plots of the factors listed above. We shall find that these straight-line plots known as Bode plots approximate the actual plots to a reasonable degree of accuracy.

Constant term: For the gain K, the magnitude is $20 \log_{10} K$ and the phase is 0°; both are constant with frequency. Thus, the magnitude and phase plots of the gain are shown in Fig. 14.9. If K is negative, the magnitude remains $20 \log_{10} |K|$ but the phase is ±180°.

A decade is an interval between two frequencies with a ratio of 10; e.g., between ω_0 and $10\omega_0$, or between 10 and 100 Hz. Thus, 20 dB/decade means that the magnitude changes 20 dB whenever the frequency changes tenfold or one decade.

Pole/zero at the origin: For the zero $(j\omega)$ at the origin, the magnitude is $20 \log_{10} \omega$ and the phase is 90°. These are plotted in Fig. 14.10, where we notice that the slope of the magnitude plot is 20 dB/decade, while the phase is constant with frequency.

The Bode plots for the pole $(j\omega)^{-1}$ are similar except that the slope of the magnitude plot is −20 dB/decade while the phase is −90°. In general, for $(j\omega)^N$, where N is an integer, the magnitude plot will have a slope of $20N$ dB/decade, while the phase is $90N$ degrees.

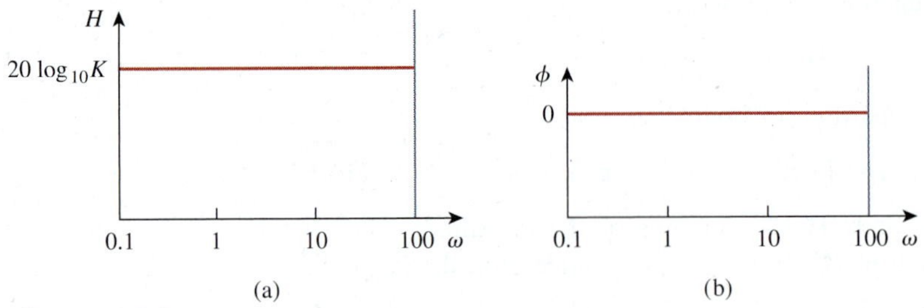

Figure 14.9
Bode plots for gain K: (a) magnitude plot, (b) phase plot.

Simple pole/zero: For the simple zero $(1 + j\omega/z_1)$, the magnitude is $20 \log_{10} |1 + j\omega/z_1|$ and the phase is $\tan^{-1} \omega/z_1$. We notice that

$$H_{\text{dB}} = 20 \log_{10} \left|1 + \frac{j\omega}{z_1}\right| \quad \Rightarrow \quad 20 \log_{10} 1 = 0 \quad \text{as} \quad \omega \to 0 \tag{14.16}$$

$$H_{\text{dB}} = 20 \log_{10} \left|1 + \frac{j\omega}{z_1}\right| \quad \Rightarrow \quad 20 \log_{10} \frac{\omega}{z_1} \quad \text{as} \quad \omega \to \infty \tag{14.17}$$

showing that we can approximate the magnitude as zero (a straight line with zero slope) for small values of ω and by a straight line with slope 20 dB/decade for large values of ω. The frequency $\omega = z_1$ where the two asymptotic lines meet is called the *corner frequency* or *break frequency*. Thus the approximate magnitude plot is shown in Fig. 14.11(a), where the actual plot is also shown. Notice that the approximate plot is close to the actual plot except at the break frequency, where $\omega = z_1$ and the deviation is $20 \log_{10} |(1 + j1)| = 20 \log_{10} \sqrt{2} \simeq 3$ dB.

The phase $\tan^{-1}(\omega/z_1)$ can be expressed as

$$\phi = \tan^{-1}\left(\frac{\omega}{z_1}\right) = \begin{cases} 0, & \omega = 0 \\ 45°, & \omega = z_1 \\ 90°, & \omega \to \infty \end{cases} \tag{14.18}$$

As a straight-line approximation, we let $\phi \simeq 0$ for $\omega \leq z_1/10$, $\phi \simeq 45°$ for $\omega = z_1$, and $\phi \simeq 90°$ for $\omega \geq 10z_1$. As shown in Fig. 14.11(b) along with the actual plot, the straight-line plot has a slope of 45° per decade.

The Bode plots for the pole $1/(1 + j\omega/p_1)$ are similar to those in Fig. 14.11 except that the corner frequency is at $\omega = p_1$, the magnitude has a slope of -20 dB/decade, and the phase has a slope of $-45°$ per decade.

Quadratic pole/zero: The magnitude of the quadratic pole $1/[1 + j2\zeta_2\omega/\omega_n + (j\omega/\omega_n)^2]$ is $-20 \log_{10} |1 + j2\zeta_2\omega/\omega_n + (j\omega/\omega_n)^2|$ and the phase is $-\tan^{-1}(2\zeta_2\omega/\omega_n)/(1 - \omega^2/\omega_n^2)$. But

$$H_{\text{dB}} = -20 \log_{10} \left|1 + \frac{j2\zeta_2\omega}{\omega_n} + \left(\frac{j\omega}{\omega_n}\right)^2\right| \quad \Rightarrow \quad 0 \quad \text{as} \quad \omega \to 0 \tag{14.19}$$

The special case of dc ($\omega = 0$) does not appear on Bode plots because log 0 = $-\infty$, implying that zero frequency is infinitely far to the left of the origin of Bode plots.

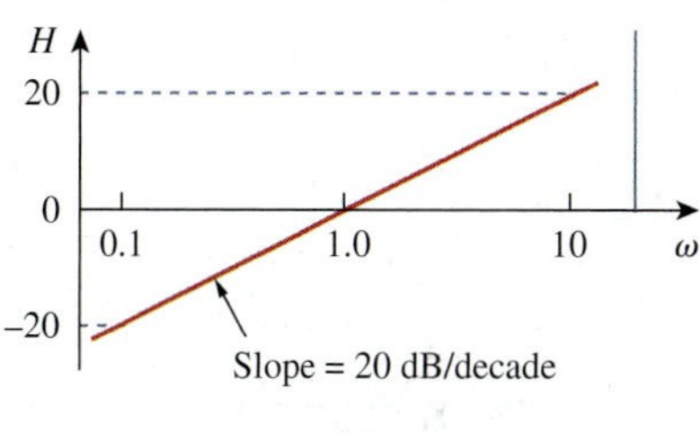

(a)

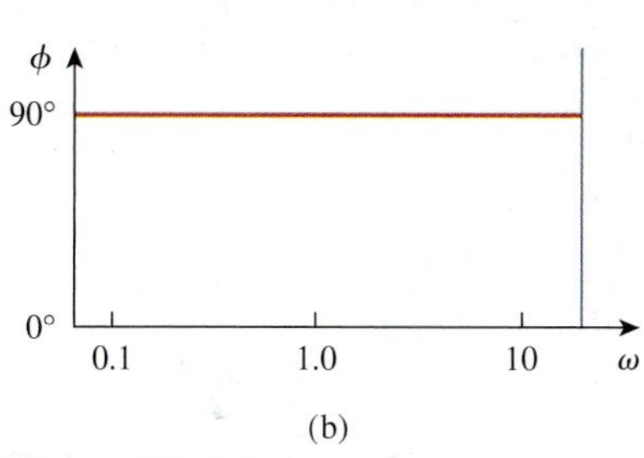

(b)

Figure 14.10
Bode plot for a zero ($j\omega$) at the origin: (a) magnitude plot, (b) phase plot.

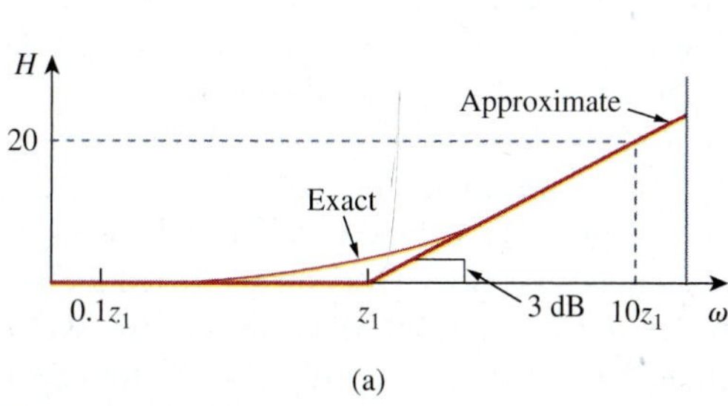

(a)

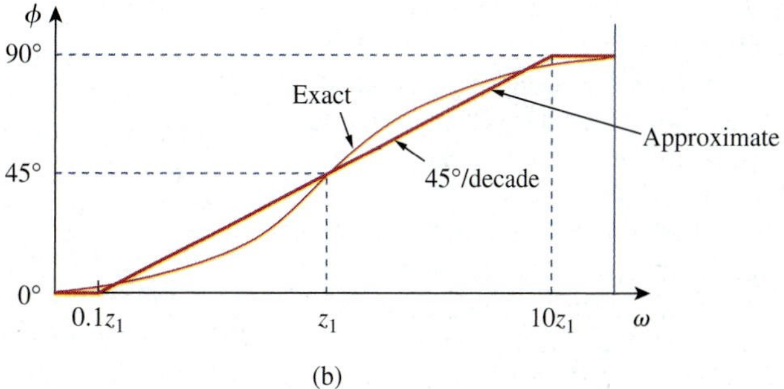

(b)

Figure 14.11
Bode plots of zero $(1 + j\omega/z_1)$: (a) magnitude plot, (b) phase plot.

and

$$H_{\text{dB}} = -20 \log_{10} \left| 1 + \frac{j2\zeta_2\omega}{\omega_n} + \left(\frac{j\omega}{\omega_n}\right)^2 \right| \quad \Rightarrow \quad -40 \log_{10} \frac{\omega}{\omega_n}$$

$$\text{as} \quad \omega \rightarrow \infty \qquad \textbf{(14.20)}$$

Thus, the amplitude plot consists of two straight asymptotic lines: one with zero slope for $\omega < \omega_n$ and the other with slope -40 dB/decade for $\omega > \omega_n$, with ω_n as the corner frequency. Figure 14.12(a) shows the approximate and actual amplitude plots. Note that the actual plot depends on the damping factor ζ_2 as well as the corner frequency ω_n. The significant peaking in the neighborhood of the corner frequency should be added to the straight-line approximation if a high level of accuracy is desired. However, we will use the straight-line approximation for the sake of simplicity.

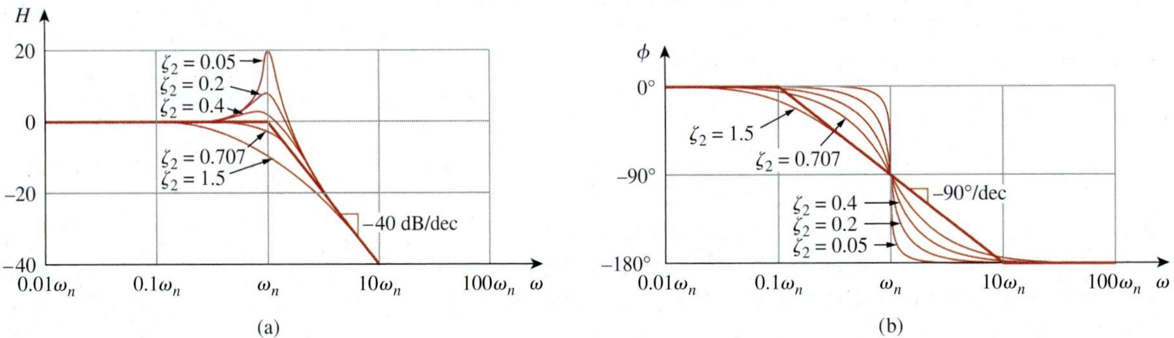

Figure 14.12
Bode plots of quadratic pole $[1 + j2\zeta\omega/\omega_n - \omega^2/\omega_n^2]^{-1}$: (a) magnitude plot, (b) phase plot.

There is another procedure for obtaining Bode plots that is faster and perhaps more efficient than the one we have just discussed. It consists in realizing that zeros cause an increase in slope, while poles cause a decrease. By starting with the low-frequency asymptote of the Bode plot, moving along the frequency axis, and increasing or decreasing the slope at each corner frequency, one can sketch the Bode plot immediately from the transfer function without the effort of making individual plots and adding them. This procedure can be used once you become proficient in the one discussed here.

Digital computers have rendered the procedure discussed here almost obsolete. Several software packages such as *PSpice*, *MATLAB*, Mathcad, and Micro-Cap can be used to generate frequency response plots. We will discuss *PSpice* later in the chapter.

The phase can be expressed as

$$\phi = -\tan^{-1} \frac{2\zeta_2\omega/\omega_n}{1 - \omega^2/\omega_n^2} = \begin{cases} 0, & \omega = 0 \\ -90^\circ, & \omega = \omega_n \\ -180^\circ, & \omega \rightarrow \infty \end{cases} \qquad \textbf{(14.21)}$$

The phase plot is a straight line with a slope of -90° per decade starting at $\omega_n/10$ and ending at $10\omega_n$, as shown in Fig. 14.12(b). We see again that the difference between the actual plot and the straight-line plot is due to the damping factor. Notice that the straight-line approximations for both magnitude and phase plots for the quadratic pole are the same as those for a double pole, i.e. $(1 + j\omega/\omega_n)^{-2}$. We should expect this because the double pole $(1 + j\omega/\omega_n)^{-2}$ equals the quadratic pole $1/[1 + j2\zeta_2\omega/\omega_n + (j\omega/\omega_n)^2]$ when $\zeta_2 = 1$. Thus, the quadratic pole can be treated as a double pole as far as straight-line approximation is concerned.

For the quadratic zero $[1 + j2\zeta_1\omega/\omega_k + (j\omega/\omega_k)^2]$, the plots in Fig. 14.12 are inverted because the magnitude plot has a slope of 40 dB/decade while the phase plot has a slope of 90° per decade.

Table 14.3 presents a summary of Bode plots for the seven factors. Of course, not every transfer function has all seven factors. To sketch the Bode plots for a function $\mathbf{H}(\omega)$ in the form of Eq. (14.15), for example, we first record the corner frequencies on the semilog graph paper, sketch the factors one at a time as discussed above, and then combine

TABLE 14.3

Summary of Bode straight-line magnitude and phase plots.

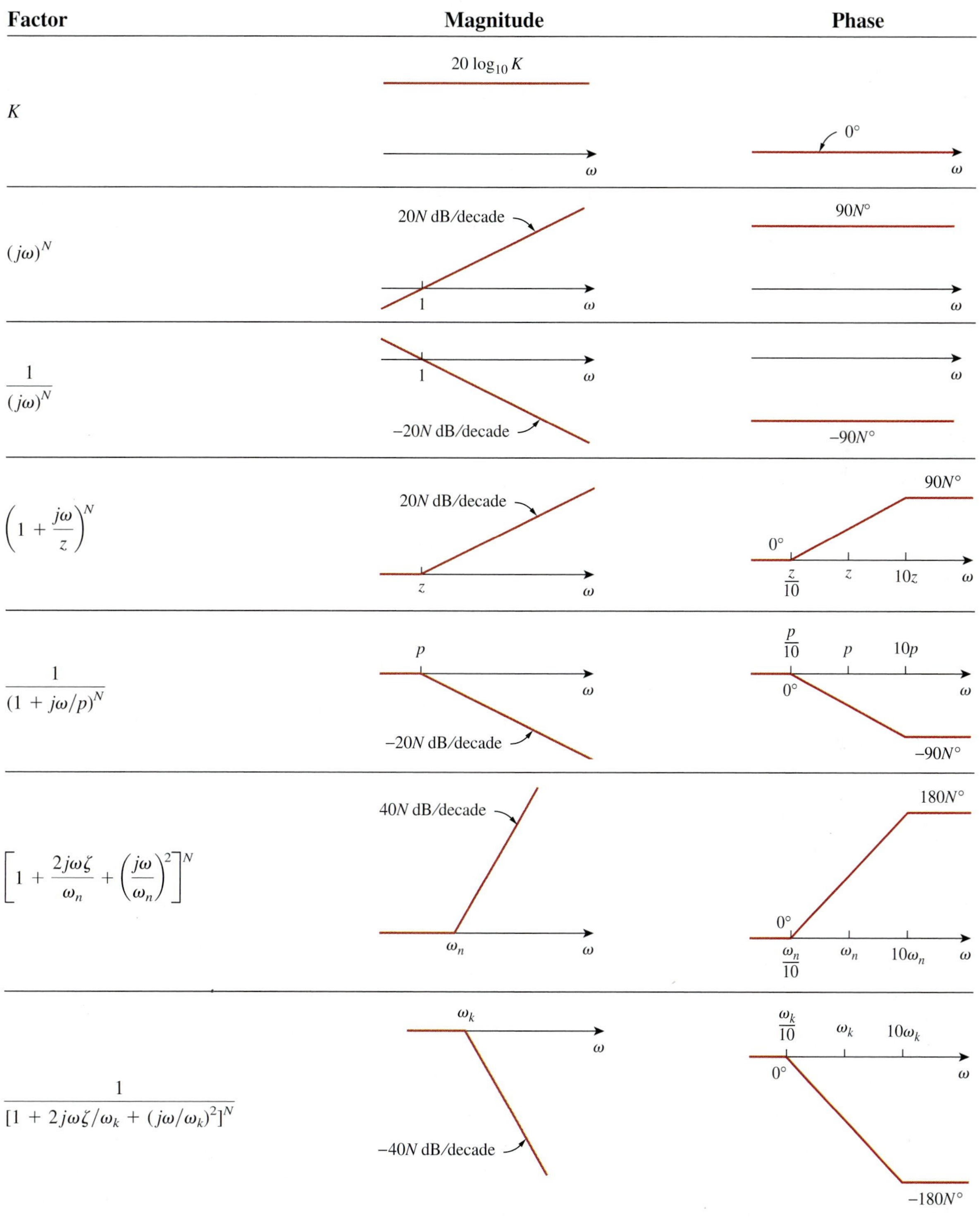

additively the graphs of the factors. The combined graph is often drawn from left to right, changing slopes appropriately each time a corner frequency is encountered. The following examples illustrate this procedure.

Example 14.3

Construct the Bode plots for the transfer function

$$\mathbf{H}(\omega) = \frac{200j\omega}{(j\omega + 2)(j\omega + 10)}$$

Solution:

We first put $\mathbf{H}(\omega)$ in the standard form by dividing out the poles and zeros. Thus,

$$\begin{aligned}\mathbf{H}(\omega) &= \frac{10j\omega}{(1 + j\omega/2)(1 + j\omega/10)}\\ &= \frac{10|j\omega|}{|1 + j\omega/2||1 + j\omega/10|}\angle 90^\circ - \tan^{-1}\omega/2 - \tan^{-1}\omega/10\end{aligned}$$

Hence, the magnitude and phase are

$$\begin{aligned}H_{\text{dB}} &= 20\log_{10}10 + 20\log_{10}|j\omega| - 20\log_{10}\left|1 + \frac{j\omega}{2}\right|\\ &\quad - 20\log_{10}\left|1 + \frac{j\omega}{10}\right|\\ \phi &= 90^\circ - \tan^{-1}\frac{\omega}{2} - \tan^{-1}\frac{\omega}{10}\end{aligned}$$

We notice that there are two corner frequencies at $\omega = 2, 10$. For both the magnitude and phase plots, we sketch each term as shown by the dotted lines in Fig. 14.13. We add them up graphically to obtain the overall plots shown by the solid curves.

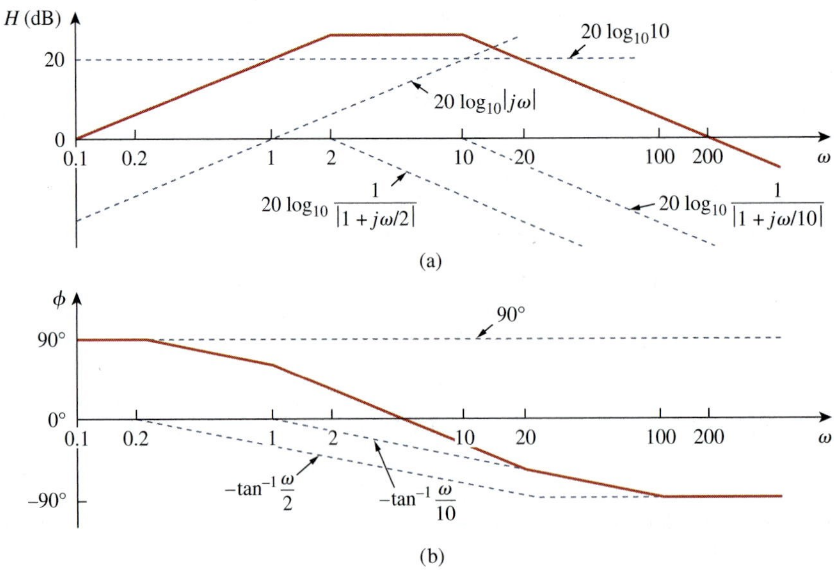

Figure 14.13
For Example 14.3: (a) magnitude plot, (b) phase plot.

Practice Problem 14.3

Draw the Bode plots for the transfer function

$$\mathbf{H}(\omega) = \frac{5(j\omega + 2)}{j\omega(j\omega + 10)}$$

Answer: See Fig. 14.14.

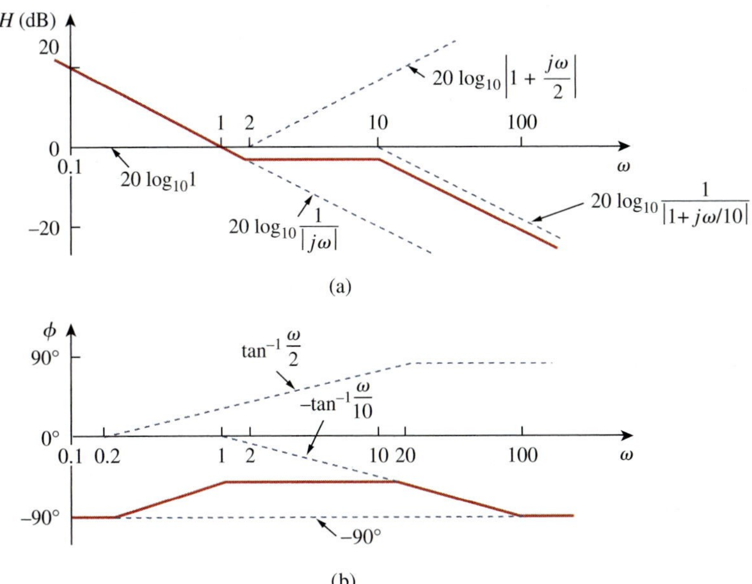

Figure 14.14
For Practice Prob. 14.3: (a) magnitude plot, (b) phase plot.

Example 14.4

Obtain the Bode plots for

$$\mathbf{H}(\omega) = \frac{j\omega + 10}{j\omega(j\omega + 5)^2}$$

Solution:
Putting $\mathbf{H}(\omega)$ in the standard form, we get

$$\mathbf{H}(\omega) = \frac{0.4(1 + j\omega/10)}{j\omega(1 + j\omega/5)^2}$$

From this, we obtain the magnitude and phase as

$$H_{\text{dB}} = 20 \log_{10} 0.4 + 20 \log_{10} \left|1 + \frac{j\omega}{10}\right| - 20 \log_{10} |j\omega| - 40 \log_{10} \left|1 + \frac{j\omega}{5}\right|$$

$$\phi = 0° + \tan^{-1}\frac{\omega}{10} - 90° - 2\tan^{-1}\frac{\omega}{5}$$

There are two corner frequencies at $\omega = 5$, 10 rad/s. For the pole with corner frequency at $\omega = 5$, the slope of the magnitude plot is -40 dB/decade and that of the phase plot is $-90°$ per decade due to the power of 2. The magnitude and the phase plots for the individual terms (in dotted lines) and the entire $\mathbf{H}(j\omega)$ (in solid lines) are in Fig. 14.15.

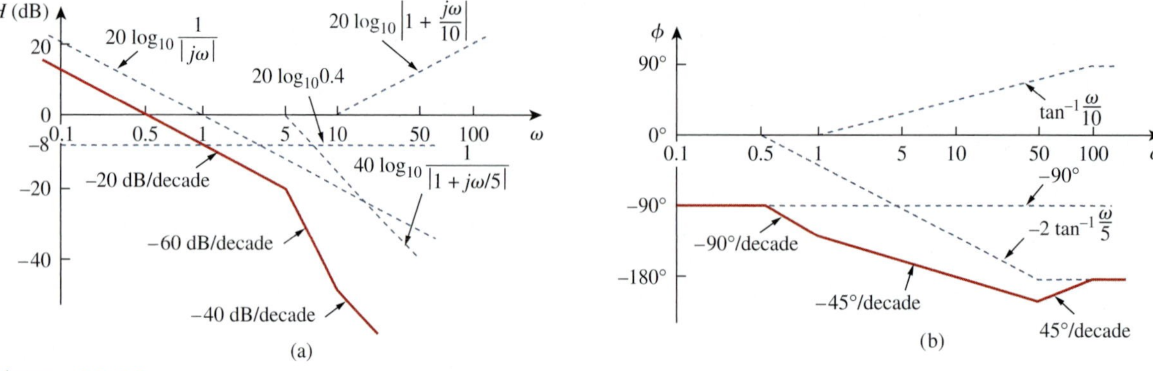

Figure 14.15
Bode plots for Example 14.4: (a) magnitude plot, (b) phase plot.

Practice Problem 14.4

Sketch the Bode plots for

$$\mathbf{H}(\omega) = \frac{50j\omega}{(j\omega + 4)(j\omega + 10)^2}$$

Answer: See Fig. 14.16.

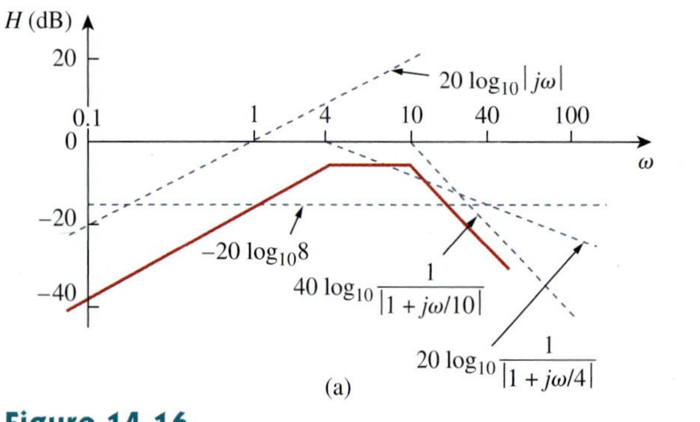

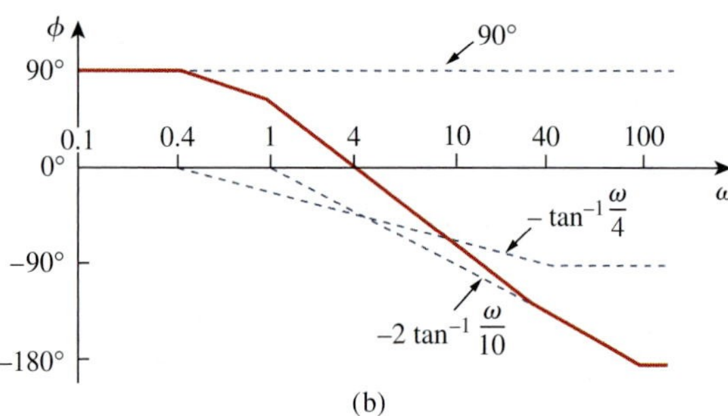

Figure 14.16
For Practice Prob. 14.4: (a) magnitude plot, (b) phase plot.

Example 14.5

Draw the Bode plots for

$$\mathbf{H}(s) = \frac{s + 1}{s^2 + 12s + 100}$$

Solution:

1. **Define.** The problem is clearly stated and we follow the technique outlined in the chapter.
2. **Present.** We are to develop the approximate bode plot for the given function, $\mathbf{H}(s)$.
3. **Alternative.** The two most effective choices would be the approximation technique outlined in the chapter, which we will

use here, and *MATLAB*, which can actually give us the exact Bode plots.

4. **Attempt.** We express $\mathbf{H}(s)$ as

$$\mathbf{H}(\omega) = \frac{1/100(1 + j\omega)}{1 + j\omega 1.2/10 + (j\omega/10)^2}$$

For the quadratic pole, $\omega_n = 10$ rad/s, which serves as the corner frequency. The magnitude and phase are

$$H_{\text{dB}} = -20 \log_{10} 100 + 20 \log_{10}|1 + j\omega| - 20 \log_{10}\left|1 + \frac{j\omega 1.2}{10} - \frac{\omega^2}{100}\right|$$

$$\phi = 0° + \tan^{-1}\omega - \tan^{-1}\left[\frac{\omega 1.2/10}{1 - \omega^2/100}\right]$$

Figure 14.17 shows the Bode plots. Notice that the quadratic pole is treated as a repeated pole at ω_k, that is, $(1 + j\omega/\omega_k)^2$, which is an approximation.

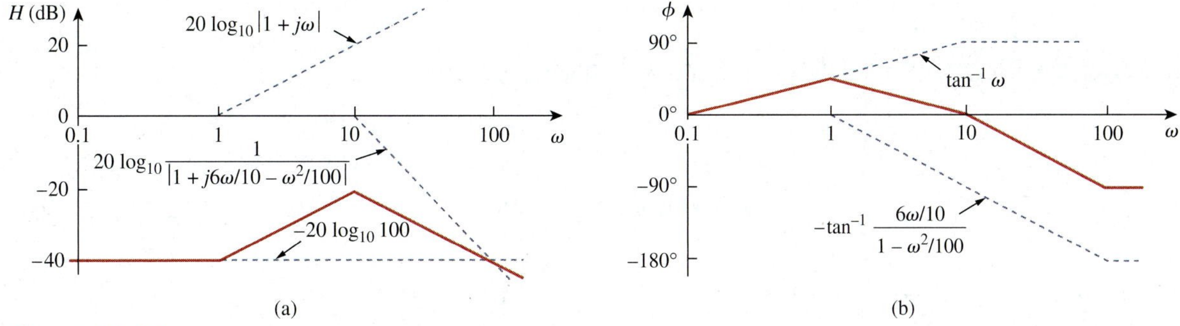

Figure 14.17
Bode plots for Example 14.5: (a) magnitude plot, (b) phase plot.

5. **Evaluate.** Although we could use *MATLAB* to validate the solution, we will use a more straightforward approach. First, we must realize that the denominator assumes that $\zeta = 0$ for the approximation, so we will use the following equation to check our answer:

$$\mathbf{H}(s) \simeq \frac{s + 1}{s^2 + 10^2}$$

We also note that we need to actually solve for H_{dB} and the corresponding phase angle ϕ. First, let $\omega = 0$.

$$H_{\text{dB}} = 20 \log_{10}(1/100) = -40 \qquad \text{and} \qquad \phi = 0°$$

Now try $\omega = 1$.

$$H_{\text{dB}} = 20 \log_{10}(1.4142/99) = -36.9 \text{ dB}$$

which is the expected 3 dB up from the corner frequency.

$$\phi = 45° \qquad \text{from} \qquad \mathbf{H}(j) = \frac{j + 1}{-1 + 100}$$

Now try $\omega = 100$.

$$H_{dB} = 20\log_{10}(100) - 20\log_{10}(9900) = 39.91 \text{ dB}$$

ϕ is 90° from the numerator minus 180°, which gives −90°. We now have checked three different points and got close agreement, and, since this is an approximation, we can feel confident that we have worked the problem successfully.

You can reasonably ask why did we not check at $\omega = 10$? If we just use the approximate value we used above, we end up with an infinite value, which is to be expected from $\zeta = 0$ (see Fig. 14.12a). If we used the actual value of $\mathbf{H}(j10)$ we will still end up being far from the approximate values, since $\zeta = 0.6$ and Fig. 14.12a shows a significant deviation from the approximation. We could have reworked the problem with $\zeta = 0.707$, which would have gotten us closer to the approximation. However, we really have enough points without doing this.

6. **Satisfactory?** We are satisfied the problem has been worked successfully and we can present the results as a solution to the problem.

Practice Problem 14.5

Construct the Bode plots for

$$H(s) = \frac{10}{s(s^2 + 80s + 400)}$$

Answer: See Fig. 14.18.

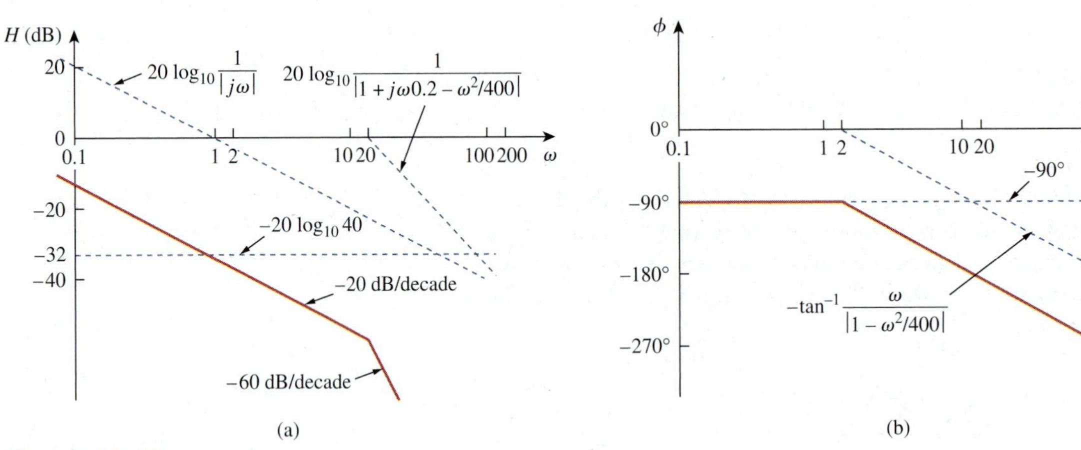

Figure 14.18
For Practice Prob. 14.5: (a) magnitude plot, (b) phase plot.

Example 14.6

Given the Bode plot in Fig. 14.19, obtain the transfer function $\mathbf{H}(\omega)$.

Solution:

To obtain $\mathbf{H}(\omega)$ from the Bode plot, we keep in mind that a zero always causes an upward turn at a corner frequency, while a pole causes a downward turn. We notice from Fig. 14.19 that there is a zero $j\omega$ at the origin which should have intersected the frequency axis at $\omega = 1$. This is indicated by the straight line with slope +20 dB/decade. The fact that this straight line is shifted by 40 dB indicates that there is a 40-dB gain; that is,

$$40 = 20 \log_{10} K \quad \Rightarrow \quad \log_{10} K = 2$$

or

$$K = 10^2 = 100$$

In addition to the zero $j\omega$ at the origin, we notice that there are three factors with corner frequencies at $\omega = 1, 5$, and 20 rad/s. Thus, we have:

1. A pole at $p = 1$ with slope −20 dB/decade to cause a downward turn and counteract the zero at the origin. The pole at $p = 1$ is determined as $1/(1 + j\omega/1)$.
2. Another pole at $p = 5$ with slope −20 dB/decade causing a downward turn. The pole is $1/(1 + j\omega/5)$.
3. A third pole at $p = 20$ with slope −20 dB/decade causing a further downward turn. The pole is $1/(1 + j\omega/20)$.

Putting all these together gives the corresponding transfer function as

$$\mathbf{H}(\omega) = \frac{100j\omega}{(1 + j\omega/1)(1 + j\omega/5)(1 + j\omega/20)} = \frac{j\omega 10^4}{(j\omega + 1)(j\omega + 5)(j\omega + 20)}$$

or

$$\mathbf{H}(s) = \frac{10^4 s}{(s + 1)(s + 5)(s + 20)}, \qquad s = j\omega$$

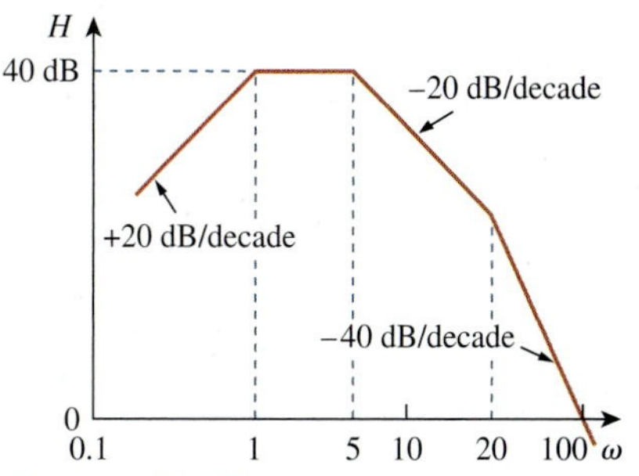

Figure 14.19
For Example 14.6.

Practice Problem 14.6

Obtain the transfer function $\mathbf{H}(\omega)$ corresponding to the Bode plot in Fig. 14.20.

Answer: $\mathbf{H}(\omega) = \dfrac{4{,}000(s + 5)}{(s + 10)(s + 100)^2}$.

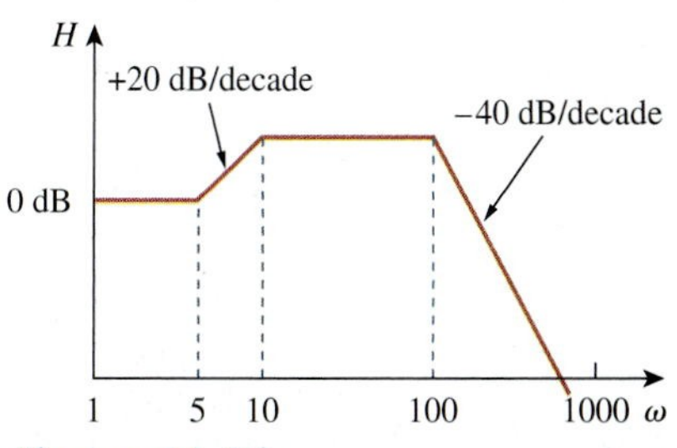

Figure 14.20
For Practice Prob. 14.6.

To see how to use *MATLAB* to produce Bode plots, refer to Section 14.11.

14.5 Series Resonance

The most prominent feature of the frequency response of a circuit may be the sharp peak (or *resonant peak*) exhibited in its amplitude characteristic. The concept of resonance applies in several areas of science and engineering. Resonance occurs in any system that has a complex conjugate pair of poles; it is the cause of oscillations of stored energy from one form to another. It is the phenomenon that allows frequency discrimination in communications networks. Resonance occurs in any circuit that has at least one inductor and one capacitor.

> **Resonance** is a condition in an *RLC* circuit in which the capacitive and inductive reactances are equal in magnitude, thereby resulting in a purely resistive impedance.

Resonant circuits (series or parallel) are useful for constructing filters, as their transfer functions can be highly frequency selective. They are used in many applications such as selecting the desired stations in radio and TV receivers.

Consider the series *RLC* circuit shown in Fig. 14.21 in the frequency domain. The input impedance is

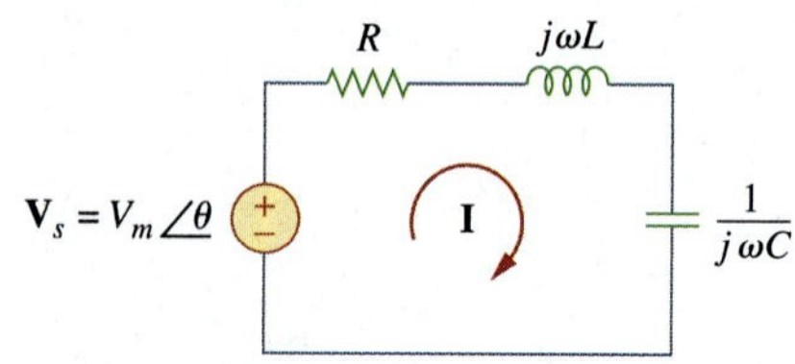

Figure 14.21
The series resonant circuit.

$$\mathbf{Z} = \mathbf{H}(\omega) = \frac{\mathbf{V}_s}{\mathbf{I}} = R + j\omega L + \frac{1}{j\omega C} \tag{14.22}$$

or

$$\mathbf{Z} = R + j\left(\omega L - \frac{1}{\omega C}\right) \tag{14.23}$$

Resonance results when the imaginary part of the transfer function is zero, or

$$\text{Im}(\mathbf{Z}) = \omega L - \frac{1}{\omega C} = 0 \tag{14.24}$$

The value of ω that satisfies this condition is called the *resonant frequency* ω_0. Thus, the resonance condition is

$$\omega_0 L = \frac{1}{\omega_0 C} \tag{14.25}$$

or

$$\boxed{\omega_0 = \frac{1}{\sqrt{LC}} \text{ rad/s}} \tag{14.26}$$

Since $\omega_0 = 2\pi f_0$,

$$f_0 = \frac{1}{2\pi\sqrt{LC}} \text{ Hz} \tag{14.27}$$

Note that at resonance:

1. The impedance is purely resistive, thus, $\mathbf{Z} = R$. In other words, the *LC* series combination acts like a short circuit, and the entire voltage is across *R*.
2. The voltage $\mathbf{V}_s$ and the current $\mathbf{I}$ are in phase, so that the power factor is unity.
3. The magnitude of the transfer function $\mathbf{H}(\omega) = \mathbf{Z}(\omega)$ is minimum.
4. The inductor voltage and capacitor voltage can be much more than the source voltage.

> Note No. 4 becomes evident from the fact that
>
> $$|\mathbf{V}_L| = \frac{V_m}{R}\omega_0 L = QV_m$$
>
> $$|\mathbf{V}_C| = \frac{V_m}{R}\frac{1}{\omega_0 C} = QV_m$$
>
> where Q is the quality factor, defined in Eq. (14.38).

The frequency response of the circuit's current magnitude

$$I = |\mathbf{I}| = \frac{V_m}{\sqrt{R^2 + (\omega L - 1/\omega C)^2}} \tag{14.28}$$

is shown in Fig. 14.22; the plot only shows the symmetry illustrated in this graph when the frequency axis is a logarithm. The average power dissipated by the *RLC* circuit is

$$P(\omega) = \frac{1}{2}I^2R \tag{14.29}$$

The highest power dissipated occurs at resonance, when $I = V_m/R$, so that

$$P(\omega_0) = \frac{1}{2}\frac{V_m^2}{R} \tag{14.30}$$

At certain frequencies $\omega = \omega_1, \omega_2$, the dissipated power is half the maximum value; that is,

$$P(\omega_1) = P(\omega_2) = \frac{(V_m/\sqrt{2})^2}{2R} = \frac{V_m^2}{4R} \tag{14.31}$$

Hence, ω_1 and ω_2 are called the *half-power frequencies.*

The half-power frequencies are obtained by setting Z equal to $\sqrt{2}R$, and writing

$$\sqrt{R^2 + \left(\omega L - \frac{1}{\omega C}\right)^2} = \sqrt{2}R \tag{14.32}$$

Solving for ω, we obtain

$$\begin{aligned} \omega_1 &= -\frac{R}{2L} + \sqrt{\left(\frac{R}{2L}\right)^2 + \frac{1}{LC}} \\ \omega_2 &= \frac{R}{2L} + \sqrt{\left(\frac{R}{2L}\right)^2 + \frac{1}{LC}} \end{aligned} \tag{14.33}$$

We can relate the half-power frequencies with the resonant frequency. From Eqs. (14.26) and (14.33),

$$\omega_0 = \sqrt{\omega_1\omega_2} \tag{14.34}$$

showing that the resonant frequency is the geometric mean of the half-power frequencies. Notice that ω_1 and ω_2 are in general not symmetrical around the resonant frequency ω_0, because the frequency response is not generally symmetrical. However, as will be explained shortly, symmetry of the half-power frequencies around the resonant frequency is often a reasonable approximation.

Although the height of the curve in Fig. 14.22 is determined by R, the width of the curve depends on other factors. The width of the response curve depends on the *bandwidth B*, which is defined as the difference between the two half-power frequencies,

$$B = \omega_2 - \omega_1 \tag{14.35}$$

This definition of bandwidth is just one of several that are commonly used. Strictly speaking, B in Eq. (14.35) is a half-power bandwidth, because it is the width of the frequency band between the half-power frequencies.

The "sharpness" of the resonance in a resonant circuit is measured quantitatively by the *quality factor Q*. At resonance, the reactive energy

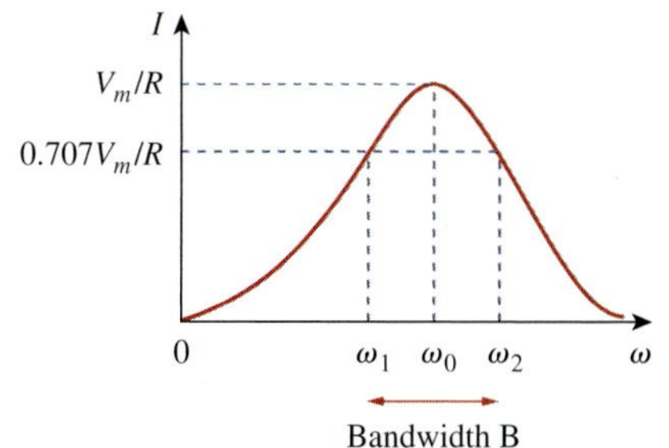

Figure 14.22
The current amplitude versus frequency for the series resonant circuit of Fig. 14.21.

Although the same symbol Q is used for the reactive power, the two are not equal and should not be confused. Q here is dimensionless, whereas reactive power Q is in VAR. This may help distinguish between the two.

in the circuit oscillates between the inductor and the capacitor. The quality factor relates the maximum or peak energy stored to the energy dissipated in the circuit per cycle of oscillation:

$$Q = 2\pi \frac{\text{Peak energy stored in the circuit}}{\text{Energy dissipated by the circuit in one period at resonance}} \tag{14.36}$$

It is also regarded as a measure of the energy storage property of a circuit in relation to its energy dissipation property. In the series RLC circuit, the peak energy stored is $\frac{1}{2}LI^2$, while the energy dissipated in one period is $\frac{1}{2}(I^2R)(1/f_0)$. Hence,

$$Q = 2\pi \frac{\frac{1}{2}LI^2}{\frac{1}{2}I^2R(1/f_0)} = \frac{2\pi f_0 L}{R} \tag{14.37}$$

or

$$Q = \frac{\omega_0 L}{R} = \frac{1}{\omega_0 CR} \tag{14.38}$$

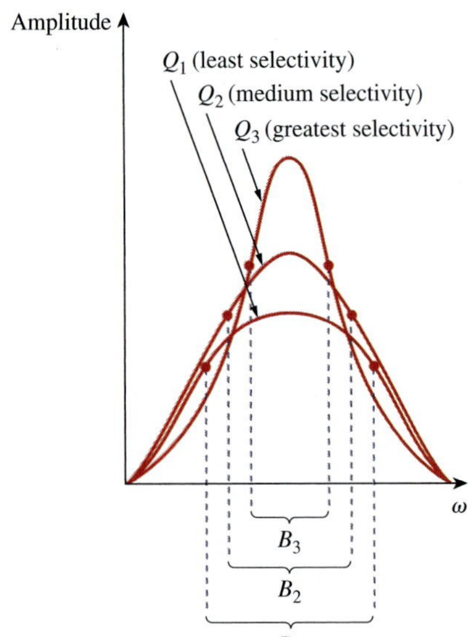

Figure 14.23
The higher the circuit Q, the smaller the bandwidth.

Notice that the quality factor is dimensionless. The relationship between the bandwidth B and the quality factor Q is obtained by substituting Eq. (14.33) into Eq. (14.35) and utilizing Eq. (14.38).

$$B = \frac{R}{L} = \frac{\omega_0}{Q} \tag{14.39}$$

or $B = \omega_0^2 CR$. Thus

The **quality factor** of a resonant circuit is the ratio of its resonant frequency to its bandwidth.

Keep in mind that Eqs. (14.33), (14.38), and (14.39) only apply to a series RLC circuit.

As illustrated in Fig. 14.23, the higher the value of Q, the more selective the circuit is but the smaller the bandwidth. The *selectivity* of an RLC circuit is the ability of the circuit to respond to a certain frequency and discriminate against all other frequencies. If the band of frequencies to be selected or rejected is narrow, the quality factor of the resonant circuit must be high. If the band of frequencies is wide, the quality factor must be low.

The quality factor is a measure of the selectivity (or "sharpness" of resonance) of the circuit.

A resonant circuit is designed to operate at or near its resonant frequency. It is said to be a *high-Q circuit* when its quality factor is equal to or greater than 10. For high-Q circuits ($Q \geq 10$), the half-power frequencies are, for all practical purposes, symmetrical around the resonant frequency and can be approximated as

$$\omega_1 \simeq \omega_0 - \frac{B}{2}, \qquad \omega_2 \simeq \omega_0 + \frac{B}{2} \tag{14.40}$$

High-Q circuits are used often in communications networks.

We see that a resonant circuit is characterized by five related parameters: the two half-power frequencies ω_1 and ω_2, the resonant frequency ω_0, the bandwidth B, and the quality factor Q.

Example 14.7

In the circuit of Fig. 14.24, $R = 2\ \Omega$, $L = 1$ mH, and $C = 0.4\ \mu$F. (a) Find the resonant frequency and the half-power frequencies. (b) Calculate the quality factor and bandwidth. (c) Determine the amplitude of the current at ω_0, ω_1, and ω_2.

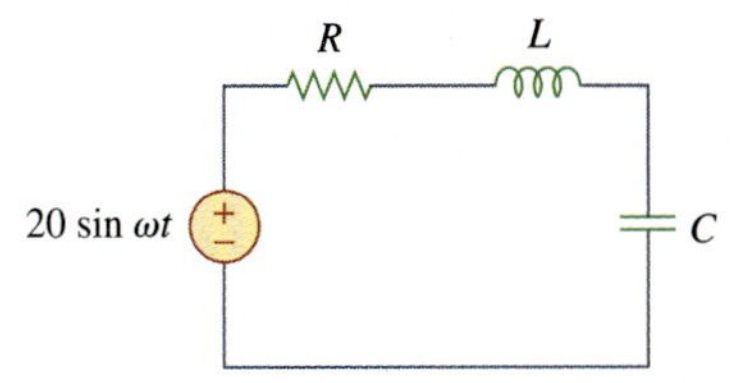

Figure 14.24
For Example 14.7.

Solution:

(a) The resonant frequency is

$$\omega_0 = \frac{1}{\sqrt{LC}} = \frac{1}{\sqrt{10^{-3} \times 0.4 \times 10^{-6}}} = 50 \text{ krad/s}$$

METHOD 1 The lower half-power frequency is

$$\begin{aligned}\omega_1 &= -\frac{R}{2L} + \sqrt{\left(\frac{R}{2L}\right)^2 + \frac{1}{LC}} \\ &= -\frac{2}{2 \times 10^{-3}} + \sqrt{(10^3)^2 + (50 \times 10^3)^2} \\ &= -1 + \sqrt{1 + 2500} \text{ krad/s} = 49 \text{ krad/s}\end{aligned}$$

Similarly, the upper half-power frequency is

$$\omega_2 = 1 + \sqrt{1 + 2500} \text{ krad/s} = 51 \text{ krad/s}$$

(b) The bandwidth is

$$B = \omega_2 - \omega_1 = 2 \text{ krad/s}$$

or

$$B = \frac{R}{L} = \frac{2}{10^{-3}} = 2 \text{ krad/s}$$

The quality factor is

$$Q = \frac{\omega_0}{B} = \frac{50}{2} = 25$$

METHOD 2 Alternatively, we could find

$$Q = \frac{\omega_0 L}{R} = \frac{50 \times 10^3 \times 10^{-3}}{2} = 25$$

From Q, we find

$$B = \frac{\omega_0}{Q} = \frac{50 \times 10^3}{25} = 2 \text{ krad/s}$$

Since $Q > 10$, this is a high-Q circuit and we can obtain the half-power frequencies as

$$\omega_1 = \omega_0 - \frac{B}{2} = 50 - 1 = 49 \text{ krad/s}$$

$$\omega_2 = \omega_0 + \frac{B}{2} = 50 + 1 = 51 \text{ krad/s}$$

as obtained earlier.

(c) At $\omega = \omega_0$,

$$I = \frac{V_m}{R} = \frac{20}{2} = 10 \text{ A}$$

At $\omega = \omega_1, \omega_2$,

$$I = \frac{V_m}{\sqrt{2}R} = \frac{10}{\sqrt{2}} = 7.071 \text{ A}$$

Practice Problem 14.7

A series-connected circuit has $R = 4\ \Omega$ and $L = 25$ mH. (a) Calculate the value of C that will produce a quality factor of 50. (b) Find ω_1, ω_2, and B. (c) Determine the average power dissipated at $\omega = \omega_0, \omega_1, \omega_2$. Take $V_m = 100$ V.

Answer: (a) 0.625 μF, (b) 7920 rad/s, 8080 rad/s, 160 rad/s, (c) 1.25 kW, 0.625 kW, 0.625 kW.

14.6 Parallel Resonance

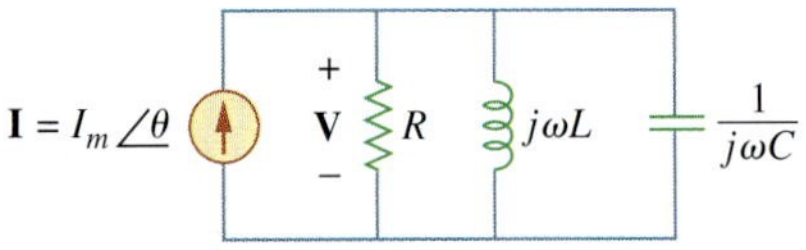

Figure 14.25
The parallel resonant circuit.

The parallel *RLC* circuit in Fig. 14.25 is the dual of the series *RLC* circuit. So we will avoid needless repetition. The admittance is

$$\mathbf{Y} = H(\omega) = \frac{\mathbf{I}}{\mathbf{V}} = \frac{1}{R} + j\omega C + \frac{1}{j\omega L} \tag{14.41}$$

or

$$\mathbf{Y} = \frac{1}{R} + j\left(\omega C - \frac{1}{\omega L}\right) \tag{14.42}$$

Resonance occurs when the imaginary part of $\mathbf{Y}$ is zero,

$$\omega C - \frac{1}{\omega L} = 0 \tag{14.43}$$

or

$$\boxed{\omega_0 = \frac{1}{\sqrt{LC}} \text{ rad/s}} \tag{14.44}$$

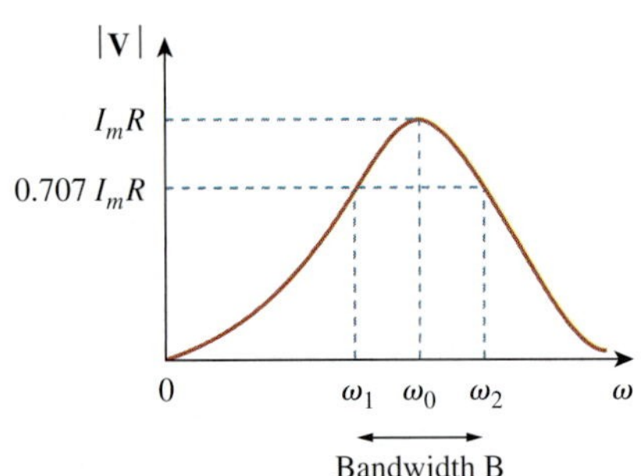

Figure 14.26
The current amplitude versus frequency for the series resonant circuit of Fig. 14.25.

which is the same as Eq. (14.26) for the series resonant circuit. The voltage |**V**| is sketched in Fig. 14.26 as a function of frequency. Notice that at resonance, the parallel *LC* combination acts like an open circuit, so that the entire current flows through *R*. Also, the inductor and capacitor current can be much more than the source current at resonance.

> We can see this from the fact that
>
> $$|\mathbf{I}_L| = \frac{I_m R}{\omega_0 L} = QI_m$$
>
> $$|\mathbf{I}_C| = \omega_0 C I_m R = QI_m$$
>
> where Q is the quality factor, defined in Eq. (14.47).

We exploit the duality between Figs. 14.21 and 14.25 by comparing Eq. (14.42) with Eq. (14.23). By replacing R, L, and C in the expressions for the series circuit with $1/R$, C, and L respectively, we obtain for the parallel circuit

$$\omega_1 = -\frac{1}{2RC} + \sqrt{\left(\frac{1}{2RC}\right)^2 + \frac{1}{LC}}$$
$$\omega_2 = \frac{1}{2RC} + \sqrt{\left(\frac{1}{2RC}\right)^2 + \frac{1}{LC}} \quad \textbf{(14.45)}$$

$$B = \omega_2 - \omega_1 = \frac{1}{RC} \quad \textbf{(14.46)}$$

$$Q = \frac{\omega_0}{B} = \omega_0 RC = \frac{R}{\omega_0 L} \quad \textbf{(14.47)}$$

It should be noted that Eqs. (14.45) to (14.47) apply only to a parallel *RLC* circuit. Using Eqs. (14.45) and (14.47), we can express the half-power frequencies in terms of the quality factor. The result is

$$\omega_1 = \omega_0\sqrt{1 + \left(\frac{1}{2Q}\right)^2} - \frac{\omega_0}{2Q}, \qquad \omega_2 = \omega_0\sqrt{1 + \left(\frac{1}{2Q}\right)^2} + \frac{\omega_0}{2Q} \quad \textbf{(14.48)}$$

Again, for high-Q circuits ($Q \geq 10$)

$$\omega_1 \simeq \omega_0 - \frac{B}{2}, \qquad \omega_2 \simeq \omega_0 + \frac{B}{2} \quad \textbf{(14.49)}$$

Table 14.4 presents a summary of the characteristics of the series and parallel resonant circuits. Besides the series and parallel *RLC* considered here, other resonant circuits exist. Example 14.9 treats a typical example.

TABLE 14.4

Summary of the characteristics of resonant *RLC* circuits.

Characteristic	Series circuit	Parallel circuit
Resonant frequency, ω_0	$\frac{1}{\sqrt{LC}}$	$\frac{1}{\sqrt{LC}}$
Quality factor, Q	$\frac{\omega_0 L}{R}$ or $\frac{1}{\omega_0 RC}$	$\frac{R}{\omega_0 L}$ or $\omega_0 RC$
Bandwidth, B	$\frac{\omega_0}{Q}$	$\frac{\omega_0}{Q}$
Half-power frequencies, ω_1, ω_2	$\omega_0\sqrt{1+\left(\frac{1}{2Q}\right)^2} \pm \frac{\omega_0}{2Q}$	$\omega_0\sqrt{1+\left(\frac{1}{2Q}\right)^2} \pm \frac{\omega_0}{2Q}$
For $Q \geq 10$, ω_1, ω_2	$\omega_0 \pm \frac{B}{2}$	$\omega_0 \pm \frac{B}{2}$

Example 14.8

In the parallel RLC circuit of Fig. 14.27, let $R = 8\ \text{k}\Omega$, $L = 0.2$ mH, and $C = 8\ \mu$F. (a) Calculate ω_0, Q, and B. (b) Find ω_1 and ω_2. (c) Determine the power dissipated at ω_0, ω_1, and ω_2.

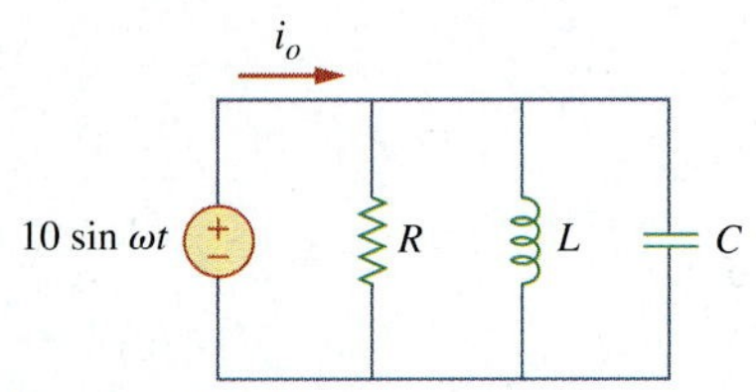

Figure 14.27
For Example 14.8.

Solution:

(a)

$$\omega_0 = \frac{1}{\sqrt{LC}} = \frac{1}{\sqrt{0.2 \times 10^{-3} \times 8 \times 10^{-6}}} = \frac{10^5}{4} = 25 \text{ krad/s}$$

$$Q = \frac{R}{\omega_0 L} = \frac{8 \times 10^3}{25 \times 10^3 \times 0.2 \times 10^{-3}} = 1{,}600$$

$$B = \frac{\omega_0}{Q} = 15.625 \text{ rad/s}$$

(b) Due to the high value of Q, we can regard this as a high-Q circuit, Hence,

$$\omega_1 = \omega_0 - \frac{B}{2} = 25{,}000 - 7.812 = 24{,}992 \text{ rad/s}$$

$$\omega_2 = \omega_0 + \frac{B}{2} = 25{,}000 + 7.812 = 25{,}008 \text{ rad/s}$$

(c) At $\omega = \omega_0$, $\mathbf{Y} = 1/R$ or $\mathbf{Z} = R = 8\ \text{k}\Omega$. Then

$$\mathbf{I}_o = \frac{\mathbf{V}}{\mathbf{Z}} = \frac{10\underline{/-90^\circ}}{8{,}000} = 1.25\underline{/-90^\circ} \text{ mA}$$

Since the entire current flows through R at resonance, the average power dissipated at $\omega = \omega_0$ is

$$P = \frac{1}{2}|\mathbf{I}_o|^2 R = \frac{1}{2}(1.25 \times 10^{-3})^2(8 \times 10^3) = 6.25 \text{ mW}$$

or

$$P = \frac{V_m^2}{2R} = \frac{100}{2 \times 8 \times 10^3} = 6.25 \text{ mW}$$

At $\omega = \omega_1, \omega_2$,

$$P = \frac{V_m^2}{4R} = 3.125 \text{ mW}$$

Practice Problem 14.8

A parallel resonant circuit has $R = 100\ \text{k}\Omega$, $L = 20$ mH, and $C = 5$ nF. Calculate ω_0, ω_1, ω_2, Q, and B.

Answer: 100 krad/s, 99 krad/s, 101 krad/s, 50, 2 krad/s.

Example 14.9

Determine the resonant frequency of the circuit in Fig. 14.28.

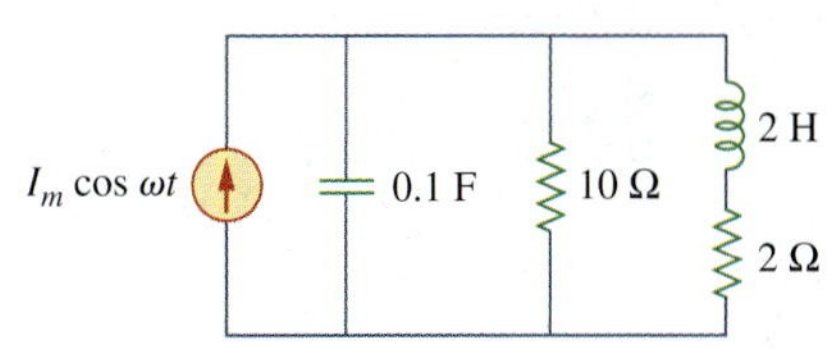

Figure 14.28
For Example 14.9.

Solution:
The input admittance is

$$\mathbf{Y} = j\omega 0.1 + \frac{1}{10} + \frac{1}{2 + j\omega 2} = 0.1 + j\omega 0.1 + \frac{2 - j\omega 2}{4 + 4\omega^2}$$

At resonance, $\text{Im}(\mathbf{Y}) = 0$ and

$$\omega_0 0.1 - \frac{2\omega_0}{4 + 4\omega_0^2} = 0 \quad \Rightarrow \quad \omega_0 = 2 \text{ rad/s}$$

Practice Problem 14.9

Calculate the resonant frequency of the circuit in Fig. 14.29.

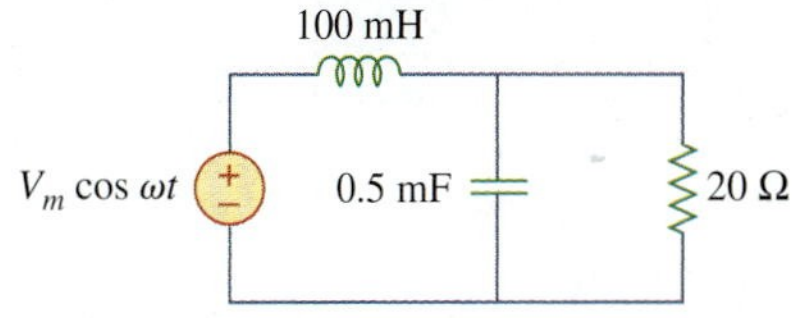

Figure 14.29
For Practice Prob. 14.9

Answer: 100 rad/s.

14.7 Passive Filters

The concept of filters has been an integral part of the evolution of electrical engineering from the beginning. Several technological achievements would not have been possible without electrical filters. Because of this prominent role of filters, much effort has been expended on the theory, design, and construction of filters and many articles and books have been written on them. Our discussion in this chapter should be considered introductory.

> A **filter** is a circuit that is designed to pass signals with desired frequencies and reject or attenuate others.

As a frequency-selective device, a filter can be used to limit the frequency spectrum of a signal to some specified band of frequencies. Filters are the circuits used in radio and TV receivers to allow us to select one desired signal out of a multitude of broadcast signals in the environment.

A filter is a *passive filter* if it consists of only passive elements R, L, and C. It is said to be an *active filter* if it consists of active elements (such as transistors and op amps) in addition to passive elements R, L, and C. We consider passive filters in this section and active filters in the next section. LC filters have been used in practical applications for more than eight decades. LC filter technology feeds related areas such as equalizers, impedance-matching networks, transformers, shaping networks, power dividers, attenuators, and directional couplers, and is continuously providing practicing engineers with oppurtunities to innovate and experiment. Besides the LC filters we study in these sections, there are other kinds of filters—such as digital filters, electromechanical filters, and microwave filters—which are beyond the level of this text.

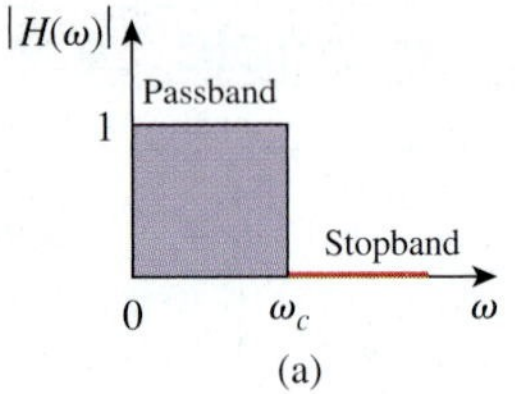

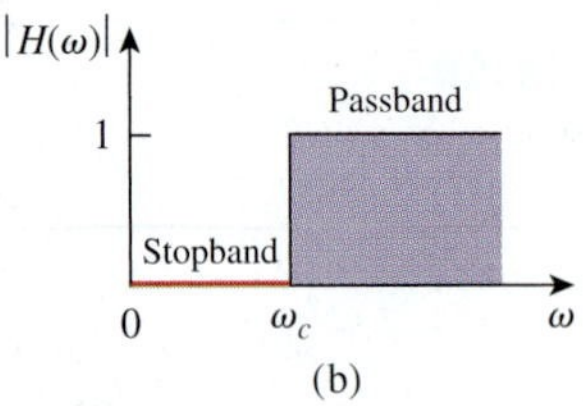

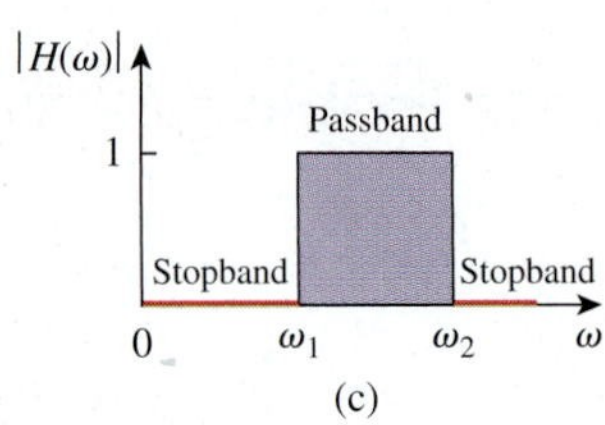

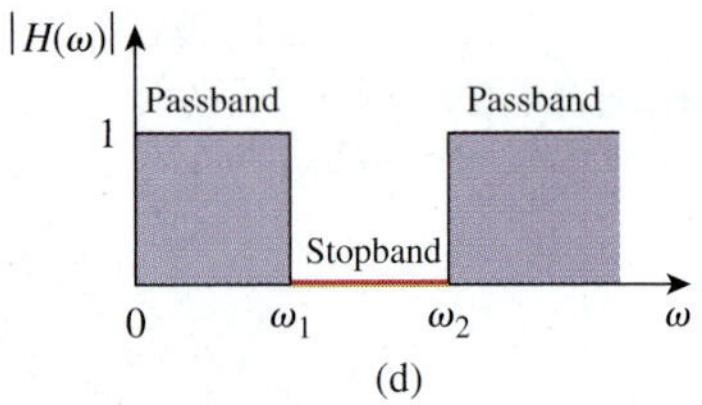

Figure 14.30
Ideal frequency response of four types of filter: (a) lowpass filter, (b) highpass filter, (c) bandpass filter, (d) bandstop filter.

As shown in Fig. 14.30, there are four types of filters whether passive or active:

1. A *lowpass filter* passes low frequencies and stops high frequencies, as shown ideally in Fig. 14.30(a).
2. A *highpass filter* passes high frequencies and rejects low frequencies, as shown ideally in Fig. 14.30(b).
3. A *bandpass filter* passes frequencies within a frequency band and blocks or attenuates frequencies outside the band, as shown ideally in Fig. 14.30(c).
4. A *bandstop filter* passes frequencies outside a frequency band and blocks or attenuates frequencies within the band, as shown ideally in Fig. 14.30(d).

Table 14.5 presents a summary of the characteristics of these filters. Be aware that the characteristics in Table 14.5 are only valid for first- or second-order filters—but one should not have the impression that only these kinds of filter exist. We now consider typical circuits for realizing the filters shown in Table 14.5.

TABLE 14.5

Summary of the characteristics of ideal filters.

Type of Filter	$H(0)$	$H(\infty)$	$H(\omega_c)$ or $H(\omega_0)$
Lowpass	1	0	$1/\sqrt{2}$
Highpass	0	1	$1/\sqrt{2}$
Bandpass	0	0	1
Bandstop	1	1	0

ω_c is the cutoff frequency for lowpass and highpass filters; ω_0 is the center frequency for bandpass and bandstop filters.

14.7.1 Lowpass Filter

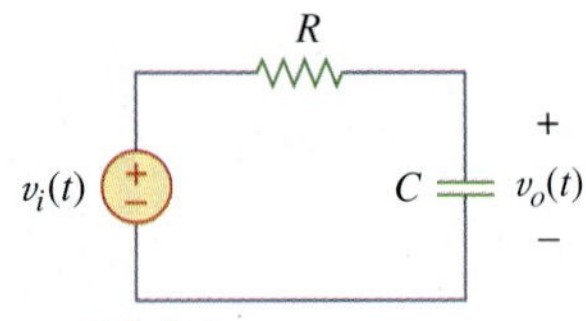

Figure 14.31
A lowpass filter.

A typical lowpass filter is formed when the output of an *RC* circuit is taken off the capacitor as shown in Fig. 14.31. The transfer function (see also Example 14.1) is

$$\mathbf{H}(\omega) = \frac{\mathbf{V}_o}{\mathbf{V}_i} = \frac{1/j\omega C}{R + 1/j\omega C}$$

$$\mathbf{H}(\omega) = \frac{1}{1 + j\omega RC} \tag{14.50}$$

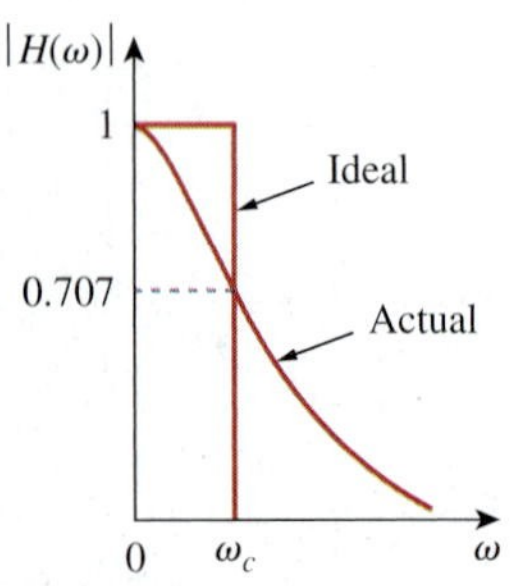

Figure 14.32
Ideal and actual frequency response of a lowpass filter.

Note that $\mathbf{H}(0) = 1$, $\mathbf{H}(\infty) = 0$. Figure 14.32 shows the plot of $|H(\omega)|$, along with the ideal characteristic. The half-power frequency, which is equivalent to the corner frequency on the Bode plots but in the context of filters is usually known as the *cutoff frequency* ω_c, is obtained by setting the magnitude of $\mathbf{H}(\omega)$ equal to $1/\sqrt{2}$, thus,

$$H(\omega_c) = \frac{1}{\sqrt{1 + \omega_c^2 R^2 C^2}} = \frac{1}{\sqrt{2}}$$

or

$$\omega_c = \frac{1}{RC} \tag{14.51}$$

The cutoff frequency is also called the *rolloff frequency.*

A **lowpass filter** is designed to pass only frequencies from dc up to the cutoff frequency ω_c.

A lowpass filter can also be formed when the output of an *RL* circuit is taken off the resistor. Of course, there are many other circuits for lowpass filters.

The cutoff frequency is the frequency at which the transfer function **H** drops in magnitude to 70.71% of its maximum value. It is also regarded as the frequency at which the power dissipated in a circuit is half of its maximum value.

14.7.2. Highpass Filter

A highpass filter is formed when the output of an *RC* circuit is taken off the resistor as shown in Fig. 14.33. The transfer function is

$$\mathbf{H}(\omega) = \frac{\mathbf{V}_o}{\mathbf{V}_i} = \frac{R}{R + 1/j\omega C}$$

$$\mathbf{H}(\omega) = \frac{j\omega RC}{1 + j\omega RC} \tag{14.52}$$

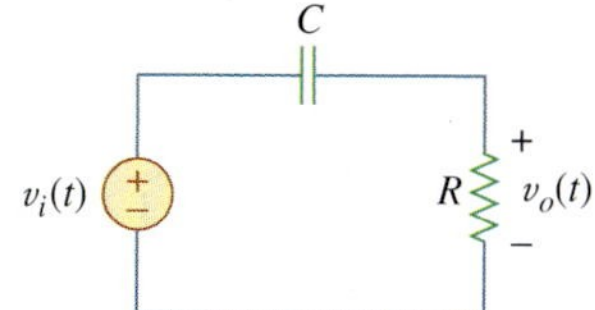

Figure 14.33
A highpass filter.

Note that $\mathbf{H}(0) = 0$, $\mathbf{H}(\infty) = 1$. Figure 14.34 shows the plot of $|H(\omega)|$. Again, the corner or cutoff frequency is

$$\omega_c = \frac{1}{RC} \tag{14.53}$$

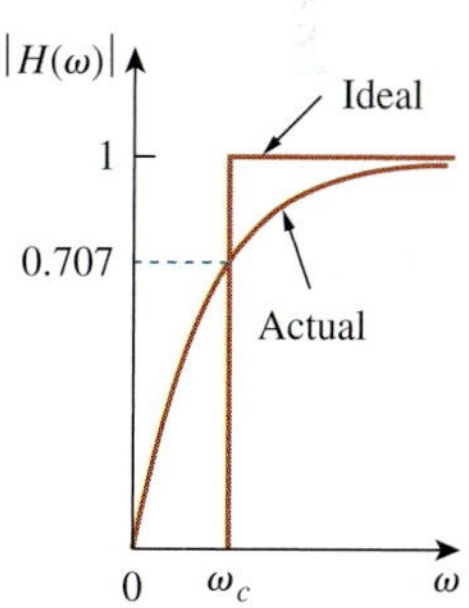

Figure 14.34
Ideal and actual frequency response of a highpass filter.

A **highpass filter** is designed to pass all frequencies above its cutoff frequency ω_c.

A highpass filter can also be formed when the output of an *RL* circuit is taken off the inductor.

14.7.3 Bandpass Filter

The *RLC* series resonant circuit provides a bandpass filter when the output is taken off the resistor as shown in Fig. 14.35. The transfer function is

$$\mathbf{H}(\omega) = \frac{\mathbf{V}_o}{\mathbf{V}_i} = \frac{R}{R + j(\omega L - 1/\omega C)} \tag{14.54}$$

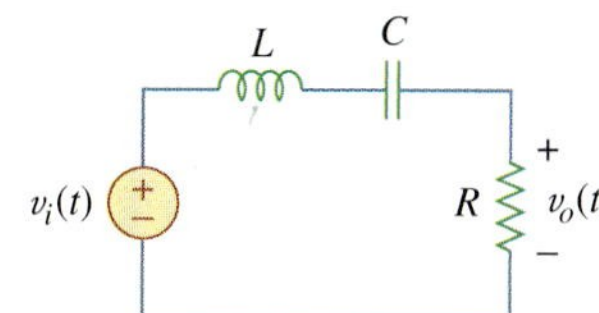

Figure 14.35
A bandpass filter.

We observe that $\mathbf{H}(0) = 0$, $\mathbf{H}(\infty) = 0$. Figure 14.36 shows the plot of $|H(\omega)|$. The bandpass filter passes a band of frequencies ($\omega_1 < \omega < \omega_2$) centered on ω_0, the center frequency, which is given by

$$\omega_0 = \frac{1}{\sqrt{LC}} \tag{14.55}$$

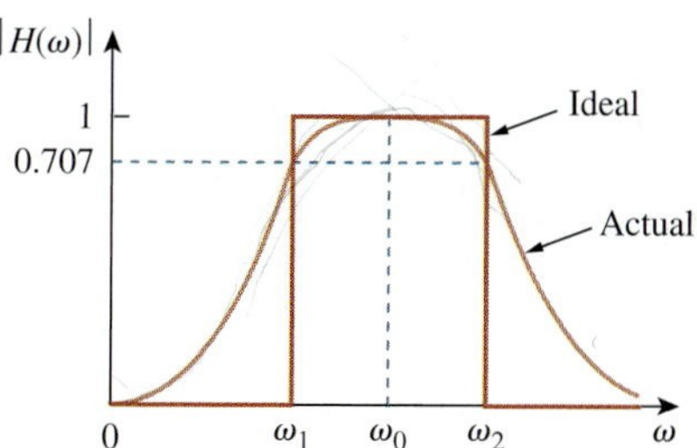

Figure 14.36
Ideal and actual frequency response of a bandpass filter.

A **bandpass filter** is designed to pass all frequencies within a band of frequencies, $\omega_1 < \omega < \omega_2$.

Since the bandpass filter in Fig. 14.35 is a series resonant circuit, the half-power frequencies, the bandwidth, and the quality factor are determined as in Section 14.5. A bandpass filter can also be formed by cascading the lowpass filter (where $\omega_2 = \omega_c$) in Fig. 14.31 with the highpass filter (where $\omega_1 = \omega_c$) in Fig. 14.33. However, the result would not be the same as just adding the output of the lowpass filter to the input of the highpass filter, because one circuit loads the other and alters the desired transfer function.

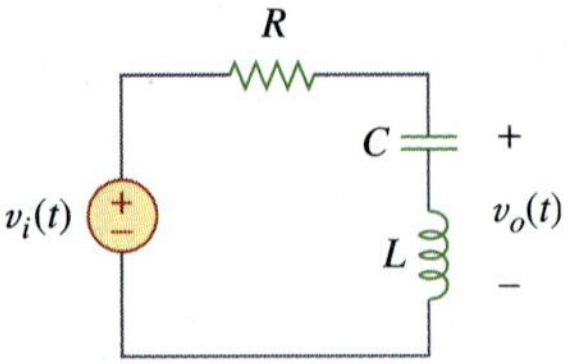

Figure 14.37
A bandstop filter.

14.7.4 Bandstop Filter

A filter that prevents a band of frequencies between two designated values (ω_1 and ω_2) from passing is variably known as a *bandstop, band-reject*, or *notch* filter. A bandstop filter is formed when the output *RLC* series resonant circuit is taken off the *LC* series combination as shown in Fig. 14.37. The transfer function is

$$\mathbf{H}(\omega) = \frac{\mathbf{V}_o}{\mathbf{V}_i} = \frac{j(\omega L - 1/\omega C)}{R + j(\omega L - 1/\omega C)} \tag{14.56}$$

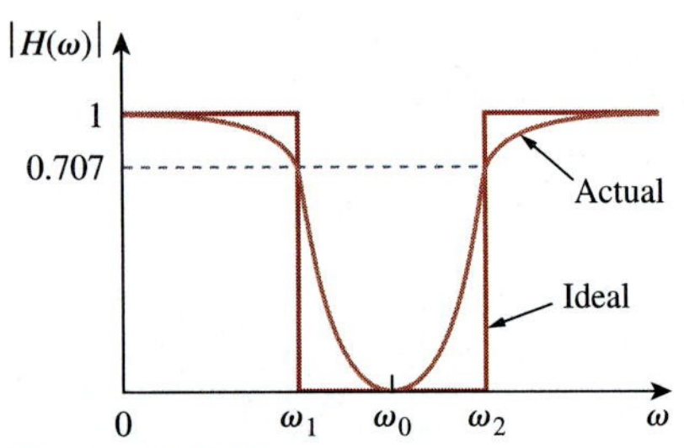

Figure 14.38
Ideal and actual frequency response of a bandstop filter.

Notice that $\mathbf{H}(0) = 1$, $\mathbf{H}(\infty) = 1$. Figure 14.38 shows the plot of $|H(\omega)|$. Again, the center frequency is given by

$$\omega_0 = \frac{1}{\sqrt{LC}} \tag{14.57}$$

while the half-power frequencies, the bandwidth, and the quality factor are calculated using the formulas in Section 14.5 for a series resonant circuit. Here, ω_0 is called the *frequency of rejection*, while the corresponding bandwidth ($B = \omega_2 - \omega_1$) is known as the *bandwidth of rejection*. Thus,

> A **bandstop filter** is designed to stop or eliminate all frequencies within a band of frequencies, $\omega_1 < \omega < \omega_2$.

Notice that adding the transfer functions of the bandpass and the bandstop gives unity at any frequency for the same values of *R*, *L*, and *C*. Of course, this is not true in general but true for the circuits treated here. This is due to the fact that the characteristic of one is the inverse of the other.

In concluding this section, we should note that:

1. From Eqs. (14.50), (14.52), (14.54), and (14.56), the maximum gain of a passive filter is unity. To generate a gain greater than unity, one should use an active filter as the next section shows.
2. There are other ways to get the types of filters treated in this section.
3. The filters treated here are the simple types. Many other filters have sharper and complex frequency responses.

Example 14.10

Determine what type of filter is shown in Fig. 14.39. Calculate the corner or cutoff frequency. Take $R = 2\ \text{k}\Omega$, $L = 2$ H, and $C = 2\ \mu$F.

Solution:
The transfer function is

$$\mathbf{H}(s) = \frac{\mathbf{V}_o}{\mathbf{V}_i} = \frac{R \parallel 1/sC}{sL + R \parallel 1/sC}, \qquad s = j\omega \tag{14.10.1}$$

But

$$R \,\Big\|\, \frac{1}{sC} = \frac{R/sC}{R + 1/sC} = \frac{R}{1 + sRC}$$

Substituting this into Eq. (14.10.1) gives

$$\mathbf{H}(s) = \frac{R/(1 + sRC)}{sL + R/(1 + sRC)} = \frac{R}{s^2RLC + sL + R}, \qquad s = j\omega$$

or

$$\mathbf{H}(\omega) = \frac{R}{-\omega^2RLC + j\omega L + R} \tag{14.10.2}$$

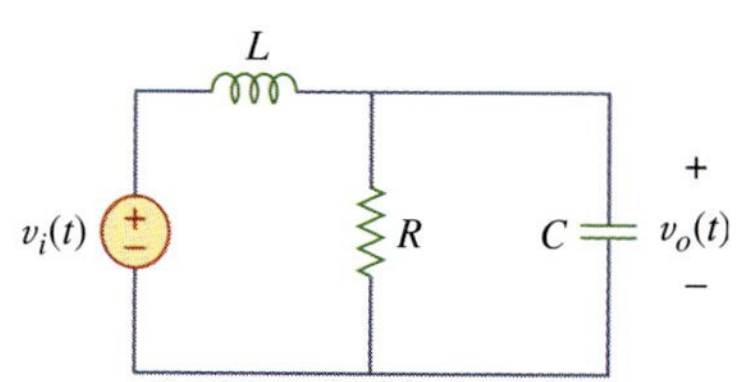

Figure 14.39
For Example 14.10.

Since $\mathbf{H}(0) = 1$ and $\mathbf{H}(\infty) = 0$, we conclude from Table 14.5 that the circuit in Fig. 14.39 is a second-order lowpass filter. The magnitude of $\mathbf{H}$ is

$$H = \frac{R}{\sqrt{(R - \omega^2RLC)^2 + \omega^2L^2}} \tag{14.10.3}$$

The corner frequency is the same as the half-power frequency, i.e., where $\mathbf{H}$ is reduced by a factor of $1/\sqrt{2}$. Since the dc value of $H(\omega)$ is 1, at the corner frequency, Eq. (14.10.3) becomes after squaring

$$H^2 = \frac{1}{2} = \frac{R^2}{(R - \omega_c^2RLC)^2 + \omega_c^2L^2}$$

or

$$2 = (1 - \omega_c^2LC)^2 + \left(\frac{\omega_cL}{R}\right)^2$$

Substituting the values of R, L, and C, we obtain

$$2 = (1 - \omega_c^2\, 4 \times 10^{-6})^2 + (\omega_c\, 10^{-3})^2$$

Assuming that ω_c is in krad/s,

$$2 = (1 - 4\omega_c^2)^2 + \omega_c^2 \qquad \text{or} \qquad 16\omega_c^4 - 7\omega_c^2 - 1 = 0$$

Solving the quadratic equation in ω_c^2, we get $\omega_c^2 = 0.5509$ and -0.1134. Since ω_c is real,

$$\omega_c = 0.742 \text{ krad/s} = 742 \text{ rad/s}$$

Practice Problem 14.10

For the circuit in Fig. 14.40, obtain the transfer function $\mathbf{V}_o(\omega)/\mathbf{V}_i(\omega)$. Identify the type of filter the circuit represents and determine the corner frequency. Take $R_1 = 100\ \Omega = R_2$, $L = 2$ mH.

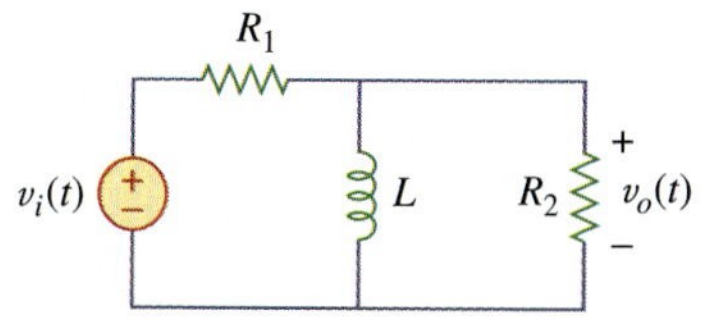

Figure 14.40
For Practice Prob. 14.10.

Answer: $\dfrac{R_2}{R_1 + R_2}\left(\dfrac{j\omega}{j\omega + \omega_c}\right)$, highpass filter

$\omega_c = \dfrac{R_1R_2}{(R_1 + R_2)L} = 25$ krad/s.

Example 14.11

If the bandstop filter in Fig. 14.37 is to reject a 200-Hz sinusoid while passing other frequencies, calculate the values of L and C. Take $R = 150\ \Omega$ and the bandwidth as 100 Hz.

Solution:

We use the formulas for a series resonant circuit in Section 14.5.

$$B = 2\pi(100) = 200\pi \text{ rad/s}$$

But

$$B = \frac{R}{L} \quad \Rightarrow \quad L = \frac{R}{B} = \frac{150}{200\pi} = 0.2387 \text{ H}$$

Rejection of the 200-Hz sinusoid means that f_0 is 200 Hz, so that ω_0 in Fig. 14.38 is

$$\omega_0 = 2\pi f_0 = 2\pi(200) = 400\pi$$

Since $\omega_0 = 1/\sqrt{LC}$,

$$C = \frac{1}{\omega_0^2 L} = \frac{1}{(400\pi)^2(0.2387)} = 2.653\ \mu\text{F}$$

Practice Problem 14.11

Design a bandpass filter of the form in Fig. 14.35 with a lower cutoff frequency of 20.1 kHz and an upper cutoff frequency of 20.3 kHz. Take $R = 20\ \text{k}\Omega$. Calculate L, C, and Q.

Answer: 15.92 H, 3.9 pF, 101.

14.8 Active Filters

There are three major limitations to the passive filters considered in the previous section. First, they cannot generate gain greater than 1; passive elements cannot add energy to the network. Second, they may require bulky and expensive inductors. Third, they perform poorly at frequencies below the audio frequency range (300 Hz $< f <$ 3,000 Hz). Nevertheless, passive filters are useful at high frequencies.

Active filters consist of combinations of resistors, capacitors, and op amps. They offer some advantages over passive *RLC* filters. First, they are often smaller and less expensive, because they do not require inductors. This makes feasible the integrated circuit realizations of filters. Second, they can provide amplifier gain in addition to providing the same frequency response as *RLC* filters. Third, active filters can be combined with buffer amplifiers (voltage followers) to isolate each stage of the filter from source and load impedance effects. This isolation allows designing the stages independently and then cascading them to realize the desired transfer function. (Bode plots, being logarithmic, may be added when transfer functions are cascaded.) However, active filters are less reliable and less stable. The practical limit of most active filters is about 100 kHz—most active filters operate well below that frequency.

Filters are often classified according to their order (or number of poles) or their specific design type.

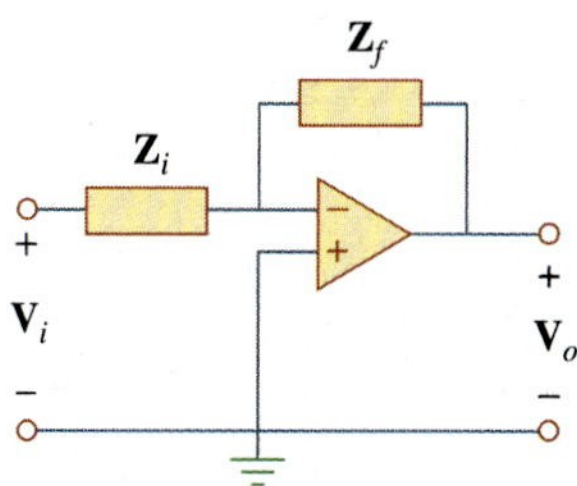

Figure 14.41
A general first-order active filter.

14.8.1 First-Order Lowpass Filter

One type of first-order filter is shown in Fig. 14.41. The components selected for Z_i and Z_f determine whether the filter is lowpass or highpass, but one of the components must be reactive.

Figure 14.42 shows a typical active lowpass filter. For this filter, the transfer function is

$$\mathbf{H}(\omega) = \frac{\mathbf{V}_o}{\mathbf{V}_i} = -\frac{\mathbf{Z}_f}{\mathbf{Z}_i} \tag{14.58}$$

where $\mathbf{Z}_i = R_i$ and

$$\mathbf{Z}_f = R_f \left\| \frac{1}{j\omega C_f} = \frac{R_f/j\omega C_f}{R_f + 1/j\omega C_f} = \frac{R_f}{1 + j\omega C_f R_f} \right. \tag{14.59}$$

Therefore,

$$\mathbf{H}(\omega) = -\frac{R_f}{R_i}\frac{1}{1 + j\omega C_f R_f} \tag{14.60}$$

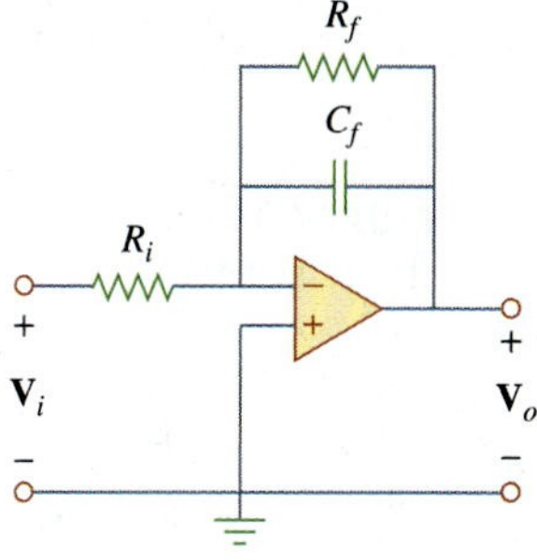

Figure 14.42
Active first-order lowpass filter.

We notice that Eq. (14.60) is similar to Eq. (14.50), except that there is a low frequency ($\omega \to 0$) gain or dc gain of $-R_f/R_i$. Also, the corner frequency is

$$\omega_c = \frac{1}{R_f C_f} \tag{14.61}$$

which does not depend on R_i. This means that several inputs with different R_i could be summed if required, and the corner frequency would remain the same for each input.

14.8.2 First-Order Highpass Filter

Figure 14.43 shows a typical highpass filter. As before,

$$\mathbf{H}(\omega) = \frac{\mathbf{V}_o}{\mathbf{V}_i} = -\frac{\mathbf{Z}_f}{\mathbf{Z}_i} \tag{14.62}$$

where $\mathbf{Z}_i = R_i + 1/j\omega C_i$ and $\mathbf{Z}_f = R_f$ so that

$$\mathbf{H}(\omega) = -\frac{R_f}{R_i + 1/j\omega C_i} = -\frac{j\omega C_i R_f}{1 + j\omega C_i R_i} \tag{14.63}$$

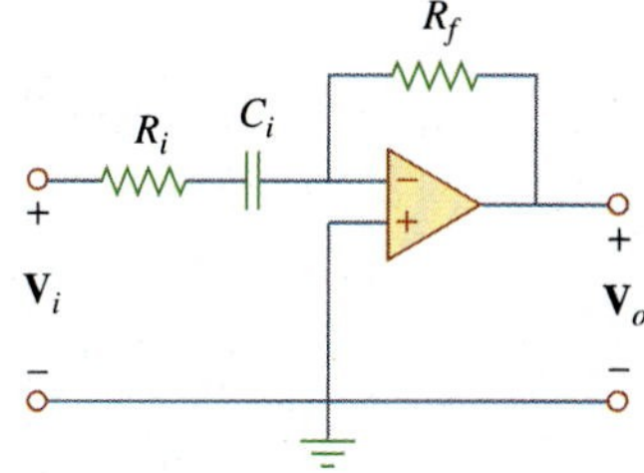

Figure 14.43
Active first-order highpass filter.

This is similar to Eq. (14.52), except that at very high frequencies ($\omega \to \infty$), the gain tends to $-R_f/R_i$. The corner frequency is

$$\omega_c = \frac{1}{R_i C_i} \tag{14.64}$$

14.8.3 Bandpass Filter

The circuit in Fig. 14.42 may be combined with that in Fig. 14.43 to form a bandpass filter that will have a gain K over the required range of frequencies. By cascading a unity-gain lowpass filter, a unity-gain

This way of creating a bandpass filter, not necessarily the best, is perhaps the easiest to understand.

highpass filter, and an inverter with gain $-R_f/R_i$, as shown in the block diagram of Fig. 14.44(a), we can construct a bandpass filter whose frequency response is that in Fig. 14.44(b). The actual construction of the bandpass filter is shown in Fig. 14.45.

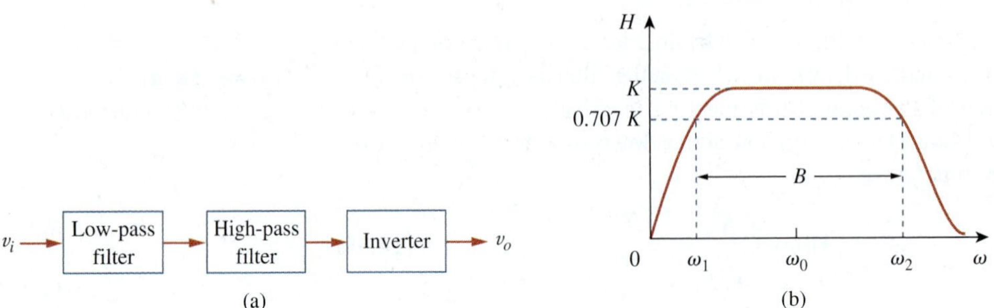

Figure 14.44
Active bandpass filter: (a) block diagram, (b) frequency response.

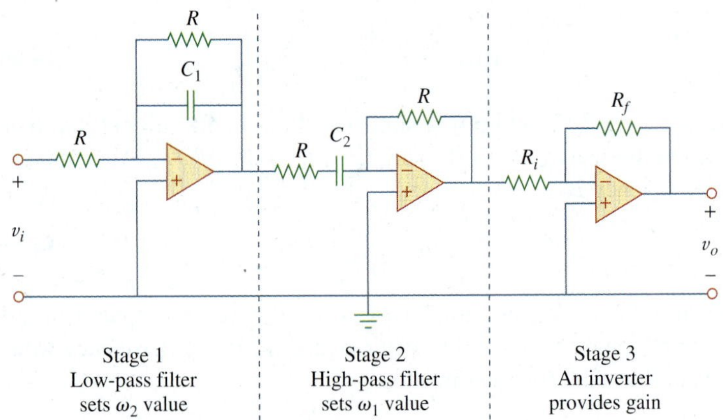

Figure 14.45
Active bandpass filter.

The analysis of the bandpass filter is relatively simple. Its transfer function is obtained by multiplying Eqs. (14.60) and (14.63) with the gain of the inverter; that is,

$$\begin{aligned}\mathbf{H}(\omega) = \frac{\mathbf{V}_o}{\mathbf{V}_i} &= \left(-\frac{1}{1 + j\omega C_1 R}\right)\left(-\frac{j\omega C_2 R}{1 + j\omega C_2 R}\right)\left(-\frac{R_f}{R_i}\right) \\ &= -\frac{R_f}{R_i}\frac{1}{1 + j\omega C_1 R}\frac{j\omega C_2 R}{1 + j\omega C_2 R}\end{aligned} \tag{14.65}$$

The lowpass section sets the upper corner frequency as

$$\omega_2 = \frac{1}{RC_1} \tag{14.66}$$

while the highpass section sets the lower corner frequency as

$$\omega_1 = \frac{1}{RC_2} \tag{14.67}$$

With these values of ω_1 and ω_2, the center frequency, bandwidth, and quality factor are found as follows:

$$\omega_0 = \sqrt{\omega_1 \omega_2} \tag{14.68}$$

$$B = \omega_2 - \omega_1 \tag{14.69}$$

$$Q = \frac{\omega_0}{B} \tag{14.70}$$

To find the passband gain K, we write Eq. (14.65) in the standard form of Eq. (14.15),

$$\mathbf{H}(\omega) = -\frac{R_f}{R_i} \frac{j\omega/\omega_1}{(1 + j\omega/\omega_1)(1 + j\omega/\omega_2)} = -\frac{R_f}{R_i} \frac{j\omega\omega_2}{(\omega_1 + j\omega)(\omega_2 + j\omega)} \tag{14.71}$$

At the center frequency $\omega_0 = \sqrt{\omega_1 \omega_2}$, the magnitude of the transfer function is

$$|\mathbf{H}(\omega_0)| = \left| \frac{R_f}{R_i} \frac{j\omega_0\omega_2}{(\omega_1 + j\omega_0)(\omega_2 + j\omega_0)} \right| = \frac{R_f}{R_i} \frac{\omega_2}{\omega_1 + \omega_2} \tag{14.72}$$

Thus, the passband gain is

$$K = \frac{R_f}{R_i} \frac{\omega_2}{\omega_1 + \omega_2} \tag{14.73}$$

14.8.4 Bandreject (or Notch) Filter

A bandreject filter may be constructed by parallel combination of a lowpass filter and a highpass filter and a summing amplifier, as shown in the block diagram of Fig. 14.46(a). The circuit is designed such that the lower cutoff frequency ω_1 is set by the lowpass filter while the upper cutoff frequency ω_2 is set by the highpass filter. The gap between ω_1 and ω_2 is the bandwidth of the filter. As shown in Fig. 14.46(b), the filter passes frequencies below ω_1 and above ω_2. The block diagram in Fig. 14.46(a) is actually constructed as shown in Fig. 14.47. The transfer function is

$$\mathbf{H}(\omega) = \frac{\mathbf{V}_o}{\mathbf{V}_i} = -\frac{R_f}{R_i}\left(-\frac{1}{1 + j\omega C_1 R} - \frac{j\omega C_2 R}{1 + j\omega C_2 R}\right) \tag{14.74}$$

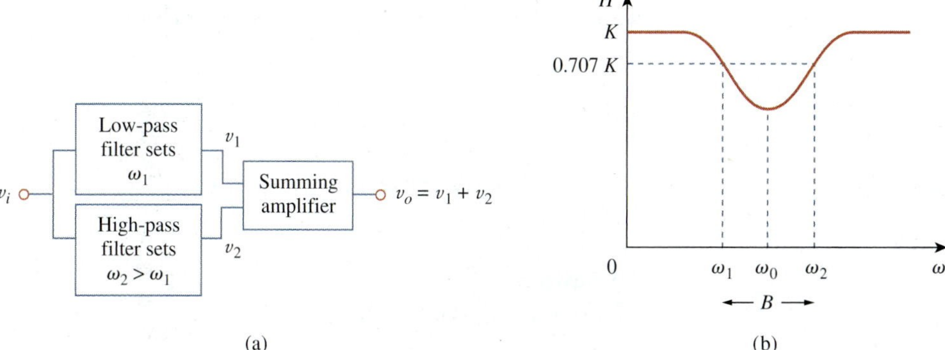

Figure 14.46
Active bandreject filter: (a) block diagram, (b) frequency response.

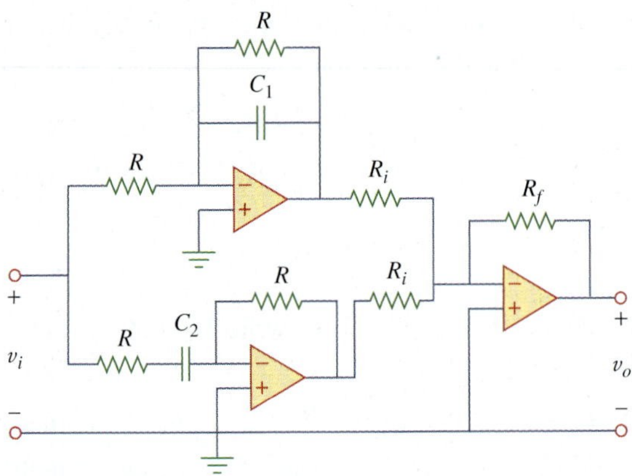

Figure 14.47
Active bandreject filter.

The formulas for calculating the values of ω_1, ω_2, the center frequency, bandwidth, and quality factor are the same as in Eqs. (14.66) to (14.70).

To determine the passband gain K of the filter, we can write Eq. (14.74) in terms of the upper and lower corner frequencies as

$$\begin{aligned}\mathbf{H}(\omega) &= \frac{R_f}{R_i}\left(\frac{1}{1 + j\omega/\omega_2} + \frac{j\omega/\omega_1}{1 + j\omega/\omega_1}\right) \\ &= \frac{R_f}{R_i}\frac{(1 + j2\omega/\omega_1 + (j\omega)^2/\omega_1\omega_1)}{(1 + j\omega/\omega_2)(1 + j\omega/\omega_1)}\end{aligned} \tag{14.75}$$

Comparing this with the standard form in Eq. (14.15) indicates that in the two passbands ($\omega \to 0$ and $\omega \to \infty$) the gain is

$$K = \frac{R_f}{R_i} \tag{14.76}$$

We can also find the gain at the center frequency by finding the magnitude of the transfer function at $\omega_0 = \sqrt{\omega_1\omega_2}$, writing

$$\begin{aligned}H(\omega_0) &= \left|\frac{R_f}{R_i}\frac{(1 + j2\omega_0/\omega_1 + (j\omega_0)^2/\omega_1\omega_1)}{(1 + j\omega_0/\omega_2)(1 + j\omega_0/\omega_1)}\right| \\ &= \frac{R_f}{R_i}\frac{2\omega_1}{\omega_1 + \omega_2}\end{aligned} \tag{14.77}$$

Again, the filters treated in this section are only typical. There are many other active filters that are more complex.

Example 14.12

Design a lowpass active filter with a dc gain of 4 and a corner frequency of 500 Hz.

Solution:
From Eq. (14.61), we find

$$\omega_c = 2\pi f_c = 2\pi(500) = \frac{1}{R_f C_f} \tag{14.12.1}$$

The dc gain is

$$H(0) = -\frac{R_f}{R_i} = -4 \tag{14.12.2}$$

We have two equations and three unknowns. If we select $C_f = 0.2\ \mu\text{F}$, then

$$R_f = \frac{1}{2\pi(500)0.2 \times 10^{-6}} = 1.59\ \text{k}\Omega$$

and

$$R_i = \frac{R_f}{4} = 397.5\ \Omega$$

We use a 1.6-kΩ resistor for R_f and a 400-Ω resistor for R_i. Figure 14.42 shows the filter.

Practice Problem 14.12

Design a highpass filter with a high-frequency gain of 5 and a corner frequency of 2 kHz. Use a 0.1-μF capacitor in your design.

Answer: $R_i = 800\ \Omega$ and $R_f = 4\ \text{k}\Omega$.

Example 14.13

Design a bandpass filter in the form of Fig. 14.45 to pass frequencies between 250 Hz and 3,000 Hz and with $K = 10$. Select $R = 20\ \text{k}\Omega$.

Solution:

1. **Define.** The problem is clearly stated and the circuit to be used in the design is specified.
2. **Present.** We are asked to use the op amp circuit specified in Fig. 14.45 to design a bandpass filter. We are given the value of R to use (20 kΩ). In addition, the frequency range of the signals to be passed is 250 Hz to 3 kHz.
3. **Alternative.** We will use the equations developed in Section 14.8.3 to obtain a solution. We will then use the resulting transfer function to validate the answer.
4. **Attempt.** Since $\omega_1 = 1/RC_2$, we obtain

$$C_2 = \frac{1}{R\omega_1} = \frac{1}{2\pi f_1 R} = \frac{1}{2\pi \times 250 \times 20 \times 10^3} = \mathbf{31.83\ nF}$$

Similarly, since $\omega_2 = 1/RC_1$,

$$C_1 = \frac{1}{R\omega_2} = \frac{1}{2\pi f_2 R} = \frac{1}{2\pi \times 3{,}000 \times 20 \times 10^3} = \mathbf{2.65\ nF}$$

From Eq. (14.73),

$$\frac{R_f}{R_i} = K\frac{\omega_1 + \omega_2}{\omega_2} = K\frac{f_1 + f_2}{f_2} = \frac{10(3{,}250)}{3{,}000} = 10.83$$

If we select $R_i = \mathbf{10\ k\Omega}$, then $R_f = 10.83R_i \simeq \mathbf{108.3\ k\Omega}$.

5. **Evaluate.** The output of the first op amp is given by

$$\frac{V_i - 0}{20\ \text{k}\Omega} + \frac{V_1 - 0}{20\ \text{k}\Omega} + \frac{s2.65 \times 10^{-9}(V_1 - 0)}{1}$$

$$= 0 \rightarrow V_1 = -\frac{V_i}{1 + 5.3 \times 10^{-5}s}$$

The output of the second op amp is given by

$$\frac{V_1 - 0}{20\ \text{k}\Omega + \dfrac{1}{s31.83\ \text{nF}}} + \frac{V_2 - 0}{20\ \text{k}\Omega} = 0 \rightarrow$$

$$V_2 = -\frac{6.366 \times 10^{-4}sV_1}{1 + 6.366 \times 10^{-4}s}$$

$$= \frac{6.366 \times 10^{-4}sV_i}{(1 + 6.366 \times 10^{-4}s)(1 + 5.3 \times 10^{-5}s)}$$

The output of the third op amp is given by

$$\frac{V_2 - 0}{10\ \text{k}\Omega} + \frac{V_o - 0}{108.3\ \text{k}\Omega} = 0 \rightarrow V_o = 10.83V_2 \rightarrow j2\pi \times 25°$$

$$V_o = -\frac{6.894 \times 10^{-3}sV_i}{(1 + 6.366 \times 10^{-4}s)(1 + 5.3 \times 10^{-5}s)}$$

Let $j2\pi \times 25°$ and solve for the magnitude of V_o/V_i.

$$\frac{V_o}{V_i} = \frac{-j10.829}{(1 + j1)(1)}$$

$|V_o/V_i| = \mathbf{(0.7071)10.829}$, which is the lower corner frequency point.

Let $s = j2\pi \times 3000 = j18.849\ \text{k}\Omega$. We then get

$$\frac{V_o}{V_i} = \frac{-j129.94}{(1 + j12)(1 + j1)}$$

$$= \frac{129.94\angle -90°}{(12.042\angle 85.24°)(1.4142\angle 45°)} = \mathbf{(0.7071)10.791\angle -18.61°}$$

Clearly this is the upper corner frequency and the answer checks.

6. **Satisfactory?** We have satisfactorily designed the circuit and can present the results as a solution to the problem.

Practice Problem 14.13

Design a notch filter based on Fig. 14.47 for $\omega_0 = 20$ krad/s, $K = 5$, and $Q = 10$. Use $R = R_i = 10\ \text{k}\Omega$.

Answer: $C_1 = 4.762$ nF, $C_2 = 5.263$ nF, and $R_f = 50\ \text{k}\Omega$.

14.9 Scaling

In designing and analyzing filters and resonant circuits or in circuit analysis in general, it is sometimes convenient to work with element values of 1 Ω, 1 H, or 1 F, and then transform the values to realistic values by *scaling*. We have taken advantage of this idea by not using realistic element values in most of our examples and problems; mastering circuit analysis is made easy by using convenient component values. We have thus eased calculations, knowing that we could use scaling to then make the values realistic.

There are two ways of scaling a circuit: *magnitude* or *impedance scaling*, and *frequency scaling*. Both are useful in scaling responses and circuit elements to values within the practical ranges. While magnitude scaling leaves the frequency response of a circuit unaltered, frequency scaling shifts the frequency response up or down the frequency spectrum.

14.9.1 Magnitude Scaling

> **Magnitude scaling** is the process of increasing all impedances in a network by a factor, the frequency response remaining unchanged.

Recall that impedances of individual elements R, L, and C are given by

$$\mathbf{Z}_R = R, \qquad \mathbf{Z}_L = j\omega L, \qquad \mathbf{Z}_C = \frac{1}{j\omega C} \tag{14.78}$$

In magnitude scaling, we multiply the impedance of each circuit element by a factor K_m and let the frequency remain constant. This gives the new impedances as

$$\begin{gathered}\mathbf{Z}'_R = K_m \mathbf{Z}_R = K_m R, \qquad \mathbf{Z}'_L = K_m \mathbf{Z}_L = j\omega K_m L \\ \mathbf{Z}'_C = K_m \mathbf{Z}_C = \frac{1}{j\omega C/K_m}\end{gathered} \tag{14.79}$$

Comparing Eq. (14.79) with Eq. (14.78), we notice the following changes in the element values: $R \rightarrow K_m R$, $L \rightarrow K_m L$, and $C \rightarrow C/K_m$. Thus, in magnitude scaling, the new values of the elements and frequency are

$$\boxed{\begin{aligned} R' &= K_m R, \qquad & L' &= K_m L \\ C' &= \frac{C}{K_m}, \qquad & \omega' &= \omega \end{aligned}} \tag{14.80}$$

The primed variables are the new values and the unprimed variables are the old values. Consider the series or parallel RLC circuit. We now have

$$\omega'_0 = \frac{1}{\sqrt{L'C'}} = \frac{1}{\sqrt{K_m L C/K_m}} = \frac{1}{\sqrt{LC}} = \omega_0 \tag{14.81}$$

showing that the resonant frequency, as expected, has not changed. Similarly, the quality factor and the bandwidth are not affected by magnitude scaling. Also, magnitude scaling does not affect transfer functions in the forms of Eqs. (14.2a) and (14.2b), which are dimensionless quantities.

14.9.2 Frequency Scaling

Frequency scaling is equivalent to relabeling the frequency axis of a frequency response plot. It is needed when translating frequencies such as a resonant frequency, a corner frequency, a bandwidth, etc., to a realistic level. It can be used to bring capacitance and inductance values into a range that is convenient to work with.

Frequency scaling is the process of shifting the frequency response of a network up or down the frequency axis while leaving the impedance the same.

We achieve frequency scaling by multiplying the frequency by a factor K_f while keeping the impedance the same.

From Eq. (14.78), we see that the impedances of L and C are frequency-dependent. If we apply frequency scaling to $\mathbf{Z}_L(\omega)$ and $\mathbf{Z}_C(\omega)$ in Eq. (14.78), we obtain

$$\mathbf{Z}_L = j(\omega K_f)L' = j\omega L \quad \Rightarrow \quad L' = \frac{L}{K_f} \tag{14.82a}$$

$$\mathbf{Z}_C = \frac{1}{j(\omega K_f)C'} = \frac{1}{j\omega C} \quad \Rightarrow \quad C' = \frac{C}{K_f} \tag{14.82b}$$

since the impedance of the inductor and capacitor must remain the same after frequency scaling. We notice the following changes in the element values: $L \rightarrow L/K_f$ and $C \rightarrow C/K_f$. The value of R is not affected, since its impedance does not depend on frequency. Thus, in frequency scaling, the new values of the elements and frequency are

$$\boxed{\begin{aligned} R' &= R, \qquad & L' &= \frac{L}{K_f} \\ C' &= \frac{C}{K_f}, \qquad & \omega' &= K_f \omega \end{aligned}} \tag{14.83}$$

Again, if we consider the series or parallel RLC circuit, for the resonant frequency

$$\omega_0' = \frac{1}{\sqrt{L'C'}} = \frac{1}{\sqrt{(L/K_f)(C/K_f)}} = \frac{K_f}{\sqrt{LC}} = K_f \omega_0 \tag{14.84}$$

and for the bandwidth

$$B' = K_f B \tag{14.85}$$

but the quality factor remains the same ($Q' = Q$).

14.9.3 Magnitude and Frequency Scaling

If a circuit is scaled in magnitude and frequency at the same time, then

$$\boxed{\begin{aligned} R' &= K_m R, \qquad & L' &= \frac{K_m}{K_f} L \\ C' &= \frac{1}{K_m K_f} C, \qquad & \omega' &= K_f \omega \end{aligned}} \tag{14.86}$$

These are more general formulas than those in Eqs. (14.80) and (14.83). We set $K_m = 1$ in Eq. (14.86) when there is no magnitude scaling or $K_f = 1$ when there is no frequency scaling.

Example 14.14

A fourth-order Butterworth lowpass filter is shown in Fig. 14.48(a). The filter is designed such that the cutoff frequency $\omega_c = 1$ rad/s. Scale the circuit for a cutoff frequency of 50 kHz using 10- kΩ resistors.

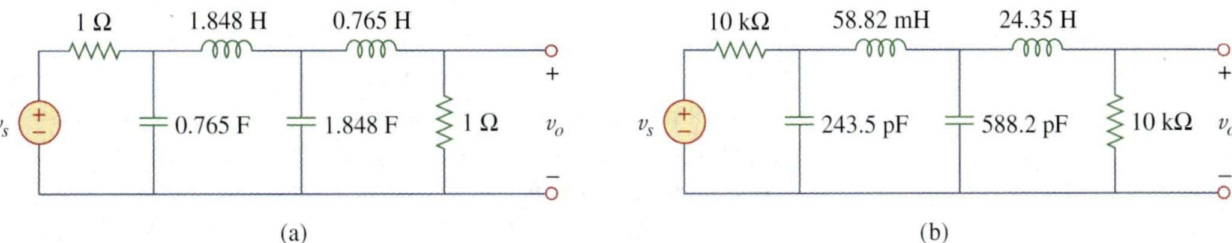

Figure 14.48
For Example 14.14: (a) Normalized Butterworth lowpass filter, (b) scaled version of the same lowpass filter.

Solution:
If the cutoff frequency is to shift from $\omega_c = 1$ rad/s to $\omega'_c = 2\pi(50)$ krad/s, then the frequency scale factor is

$$K_f = \frac{\omega'_c}{\omega_c} = \frac{100\pi \times 10^3}{1} = \pi \times 10^5$$

Also, if each 1-Ω resistor is to be replaced by a 10- kΩ resistor, then the magnitude scale factor must be

$$K_m = \frac{R'}{R} = \frac{10 \times 10^3}{1} = 10^4$$

Using Eq. (14.86),

$$L'_1 = \frac{K_m}{K_f}L_1 = \frac{10^4}{\pi \times 10^5}(1.848) = 58.82 \text{ mH}$$

$$L'_2 = \frac{K_m}{K_f}L_2 = \frac{10^4}{\pi \times 10^5}(0.765) = 24.35 \text{ mH}$$

$$C'_1 = \frac{C_1}{K_m K_f} = \frac{0.765}{\pi \times 10^9} = 243.5 \text{ pF}$$

$$C'_2 = \frac{C_2}{K_m K_f} = \frac{1.848}{\pi \times 10^9} = 588.2 \text{ pF}$$

The scaled circuit is shown in Fig. 14.48(b). This circuit uses practical values and will provide the same transfer function as the prototype in Fig. 14.48(a), but shifted in frequency.

Practice Problem 14.14

A third-order Butterworth filter normalized to $\omega_c = 1$ rad/s is shown in Fig. 14.49. Scale the circuit to a cutoff frequency of 10 kHz. Use 15-nF capacitors.

Answer: $R'_1 = R'_2 = 1.061$ kΩ, $C'_1 = C'_2 = 15$ nF, $L' = 33.77$ mH.

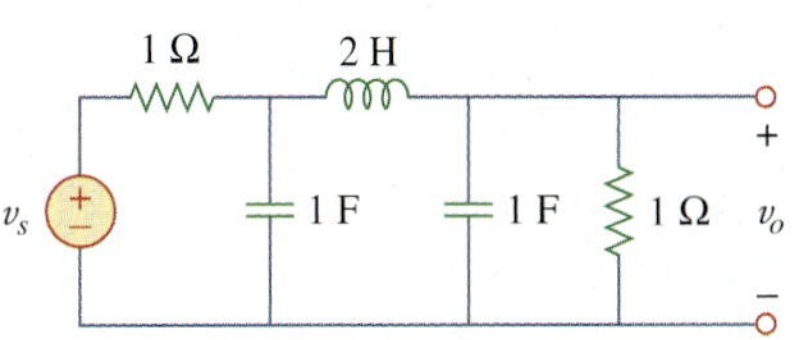

Figure 14.49
For Practice Prob. 14.14.

14.10 Frequency Response Using *PSpice*

PSpice is a useful tool in the hands of the modern circuit designer for obtaining the frequency response of circuits. The frequency response is obtained using the AC Sweep as discussed in Section D.5 (Appendix D). This requires that we specify in the AC Sweep dialog box *Total Pts, Start Freq, End Freq*, and the sweep type. *Total Pts* is the number of points in the frequency sweep, and *Start Freq* and *End Freq* are, respectively, the starting and final frequencies, in hertz. In order to know what frequencies to select for *Start Freq* and *End Freq*, one must have an idea of the frequency range of interest by making a rough sketch of the frequency response. In a complex circuit where this may not be possible, one may use a trial-and-error approach.

There are three types of sweeps:

Linear: The frequency is varied linearly from *Start Freq* to *End Freq* with *Total Pts* equally spaced points (or responses).

Octave: The frequency is swept logarithmically by octaves from *Start Freq* to *End Freq* with *Total Pts* per octave. An octave is a factor of 2 (e.g., 2 to 4, 4 to 8, 8 to 16).

Decade: The frequency is varied logarithmically by decades from *Start Freq* to *End Freq* with *Total Pts* per decade. A decade is a factor of 10 (e.g., from 2 Hz to 20 Hz, 20 Hz to 200 Hz, 200 Hz to 2 kHz).

It is best to use a linear sweep when displaying a narrow frequency range of interest, as a linear sweep displays the frequency range well in a narrow range. Conversely, it is best to use a logarithmic (octave or decade) sweep for displaying a wide frequency range of interest—if a linear sweep is used for a wide range, all the data will be crowded at the high- or low-frequency end and insufficient data at the other end.

With the above specifications, *PSpice* performs a steady-state sinusoidal analysis of the circuit as the frequency of all the independent sources is varied (or swept) from *Start Freq* to *End Freq*.

The *PSpice* A/D program produces a graphical output. The output data type may be specified in the *Trace Command Box* by adding one of the following suffixes to V or I:

M Amplitude of the sinusoid.

P Phase of the sinusoid.

dB Amplitude of the sinusoid in decibels, i.e., 20 $\log_{10}$ (amplitude).

Example 14.15

Determine the frequency response of the circuit shown in Fig. 14.50.

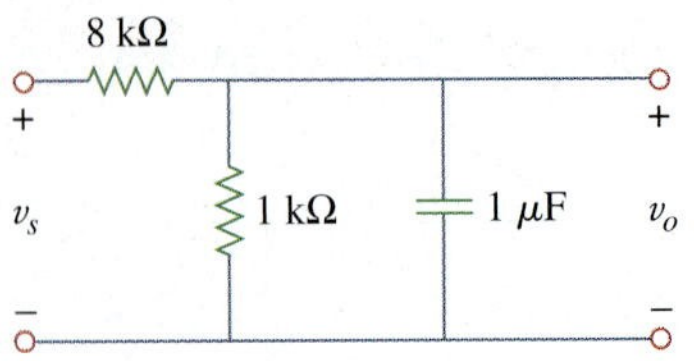

8 kΩ, 1 kΩ, 1 μF, v_s, v_o, +, −

Figure 14.50
For Example 14.15.

Solution:

We let the input voltage v_s be a sinusoid of amplitude 1 V and phase 0°. Figure 14.51 is the schematic for the circuit. The capacitor is rotated 270° counterclockwise to ensure that pin 1 (the positive terminal) is on top. The voltage marker is inserted to the output voltage across the capacitor. To perform a linear sweep for $1 < f < 1000$ Hz with 50 points, we select **Analysis/Setup/AC Sweep**, **DCLICK** *Linear*, type 50 in the *Total Pts* box, type 1 in the *Start Freq* box, and type 1000 in the *End Freq* box. After saving the file, we select **Analysis/Simulate** to simulate the circuit. If there are no errors, the *PSpice* A/D window will

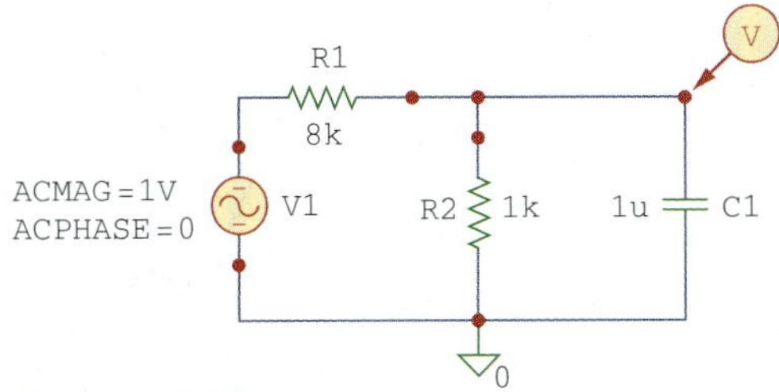

Figure 14.51
The schematic for the circuit in Fig. 14.50.

display the plot of V(C1:1), which is the same as V_o or $H(\omega) = V_o/1$, as shown in Fig. 14.52(a). This is the magnitude plot, since V(C1:1) is the same as VM(C1:1). To obtain the phase plot, select **Trace/Add** in the *PSpice* A/D menu and type VP(C1:1) in the **Trace Command** box. Figure 14.52(b) shows the result. By hand, the transfer function is

$$H(\omega) = \frac{V_o}{V_s} = \frac{1{,}000}{9{,}000 + j\omega 8}$$

or

$$H(\omega) = \frac{1}{9 + j16\pi \times 10^{-3}}$$

showing that the circuit is a lowpass filter as demonstrated in Fig. 14.52. Notice that the plots in Fig. 14.52 are similar to those in Fig. 14.3 (note that the horizontal axis in Fig. 14.52 is logrithic while the horizontal axis in Fig. 14.3 is linear.)

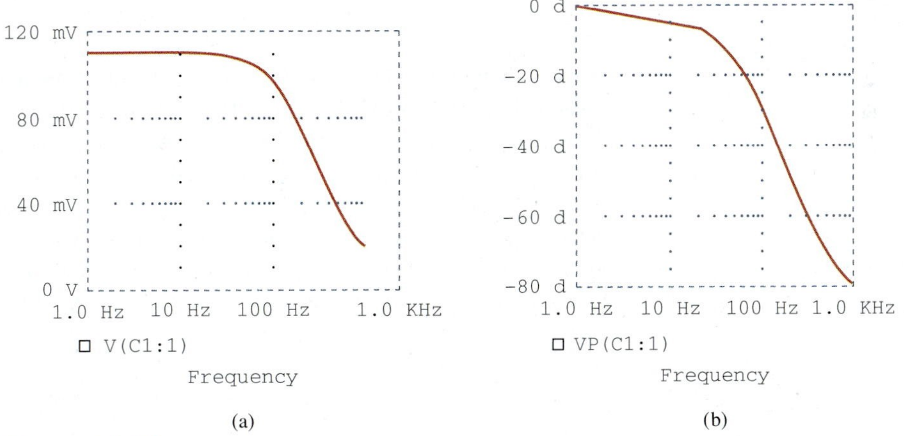

Figure 14.52
For Example 14.15: (a) magnitude plot, (b) phase plot of the frequency response.

Practice Problem 14.15

Obtain the frequency response of the circuit in Fig. 14.53 using *PSpice*. Use a linear frequency sweep and consider $1 < f < 1000$ Hz with 100 points.

Answer: See Fig. 14.54.

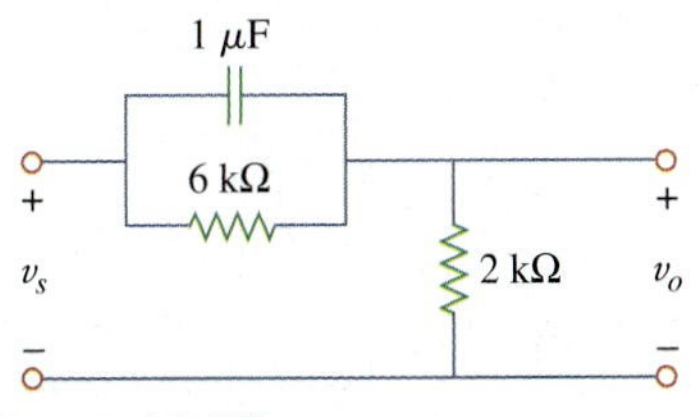

Figure 14.53
For Practice Prob. 14.15.

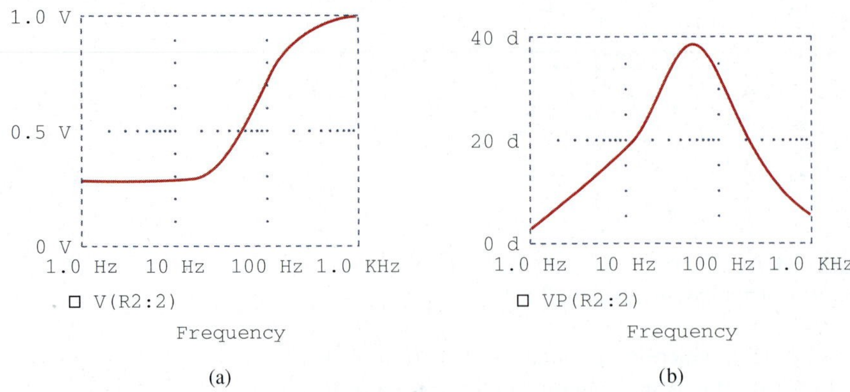

Figure 14.54
For Practice Problem 14.15: (a) magnitude plot, (b) phase plot of the frequency response.

Example 14.16

Use *PSpice* to generate the gain and phase Bode plots of V in the circuit of Fig. 14.55.

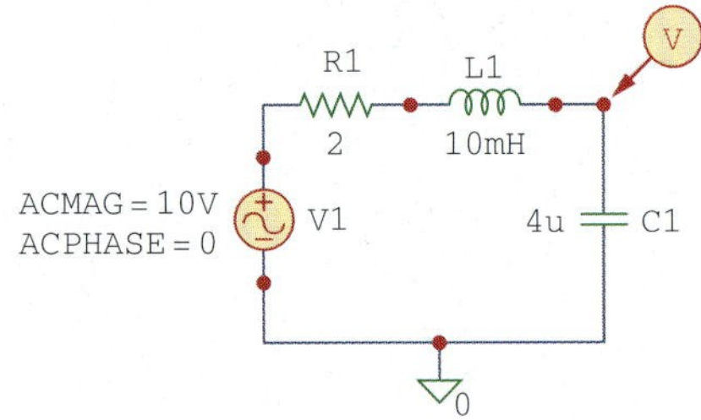

Figure 14.55
For Example 14.16.

Solution:

The circuit treated in Example 14.15 is first-order while the one in this example is second-order. Since we are interested in Bode plots, we use decade frequency sweep for $300 < f < 3{,}000$ Hz with 50 points per decade. We select this range because we know that the resonant frequency of the circuit is within the range. Recall that

$$\omega_0 = \frac{1}{\sqrt{LC}} = 5 \text{ krad/s} \qquad \text{or} \qquad f_0 = \frac{\omega}{2\pi} = 795.8 \text{ Hz}$$

After drawing the circuit as in Fig. 14.55, we select **Analysis/Setup/AC Sweep, DCLICK** *Decade*, enter 50 in the *Total Pts* box, 300 as the *Start Freq*, and 3,000 in the *End Freq* box. Upon saving the file, we simulate it by selecting **Analysis/Simulate**. This will automatically bring up the *PSpice* A/D window and display V(C1:1) if there are no errors. Since we are interested in the Bode plot, we select **Trace/Add** in the *PSpice* A/D menu and type dB(V(C1:1)) in the **Trace Command** box. The result is the Bode magnitude plot in Fig. 14.56(a). For the

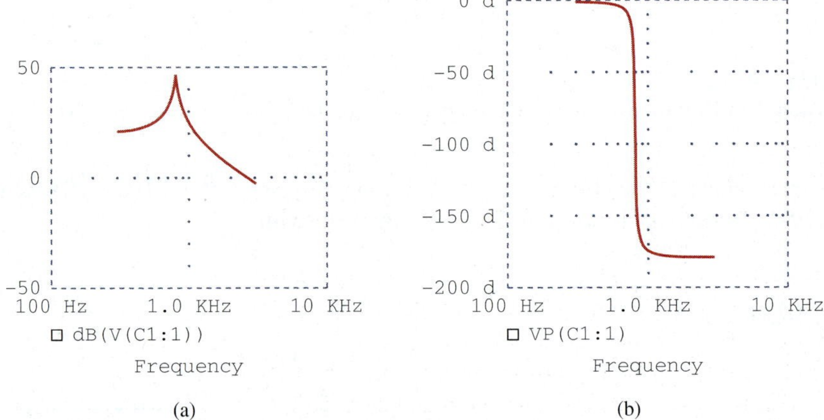

Figure 14.56
For Example 14.16: (a) Bode plot, (b) phase plot of the response.

phase plot, we select **Trace/Add** in the *PSpice* A/D menu and type VP(C1:1) in the **Trace Command** box. The result is the Bode phase plot of Fig. 14.56(b). Notice that the plots confirm the resonant frequency of 795.8 Hz.

Practice Problem 14.16

Consider the network in Fig. 14.57. Use *PSpice* to obtain the Bode plots for V_o over a frequency from 1 kHz to 100 kHz using 20 points per decade.

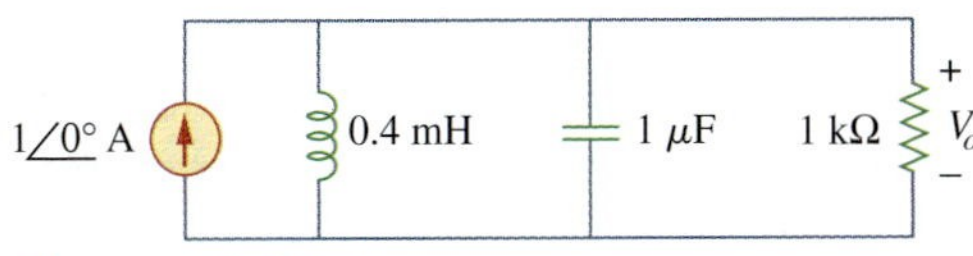

Figure 14.57
For Practice Prob. 14.16.

Answer: See Fig. 14.58.

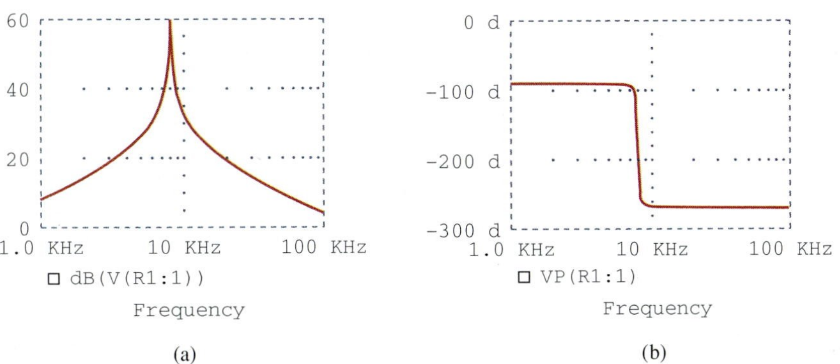

(a) (b)

Figure 14.58
For Practice Prob. 14.16: Bode (a) magnitude plot, (b) phase plot.

14.11 Computation Using *MATLAB*

MATLAB is a software package that is widely used for engineering computation and simulation. A review of *MATLAB* is provided in Appendix E for the beginner. This section shows how to use the software to numerically perform most of the operations presented in this chapter and Chapter 15. The key to describing a system in *MATLAB* is to specify the numerator (num) and denominator (den) of the transfer function of the system. Once this is done, we can use several *MATLAB* commands to obtain the system's Bode plots (frequency response) and the system's response to a given input.

The command **bode** produces the Bode plots (both magnitude and phase) of a given transfer function $H(s)$. The format of the command is **bode** (num, den), where num is the numerator of $H(s)$ and den is its denominator. The frequency range and number of points are automatically selected. For example, consider the transfer function in Example 14.3. It is better to first write the numerator and denominator in polynomial forms.

Thus,

$$H(s) = \frac{200j\omega}{(j\omega + 2)(j\omega + 10)} = \frac{200s}{s^2 + 12s + 20}, \qquad s = j\omega$$

Using the following commands, the Bode plots are generated as shown in Fig. 14.59. If necessary, the command **logspace** can be included to generate a logarithmically spaced frequency and the command **semilogx** can be used to produce a semilog scale.

```
>> num = [200 0];      % specify the numerator of H(s)
>> den = [1 12 20];    % specify the denominator of H(s)
>> bode(num, den);     % determine and draw Bode plots
```

The step response $y(t)$ of a system is the output when the input $x(t)$ is the unit step function. The command **step** plots the step response of a system given the numerator and denominator of the transfer function of that system. The time range and number of points are automatically selected. For example, consider a second-order system with the transfer function

$$H(s) = \frac{12}{s^2 + 3s + 12}$$

We obtain the step response of the system shown in Fig. 14.60 by using the following commands.

```
>> n = 12;
>> d = [1 3 12];
>> step(n,d);
```

We can verify the plot in Fig. 14.60 by obtaining $y(t) = x(t) * u(t)$ or $Y(s) = X(s)H(s)$.

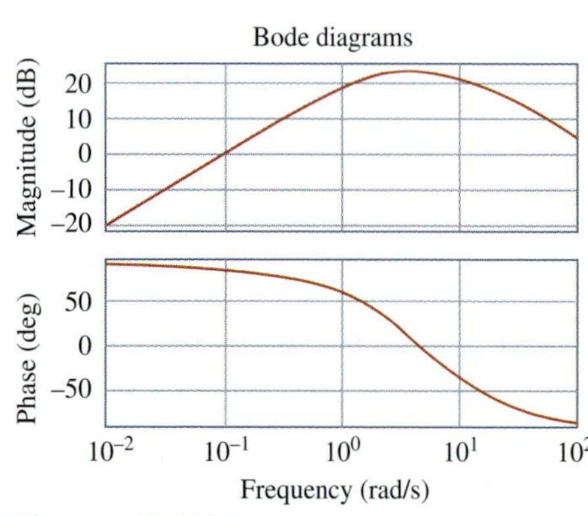

Figure 14.59
Magnitude and phase plots.

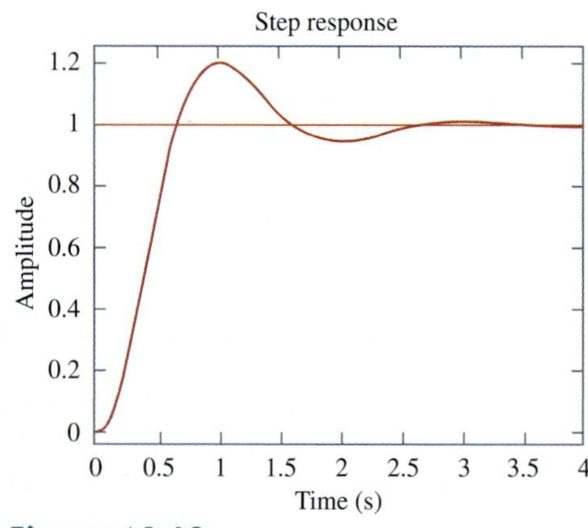

Figure 14.60
The Step response of $H(s) = 12/(s^2 + 3s + 12)$.

The command **lsim** is a more general command than **step**. It calculates the time response of a system to any arbitrary input signal. The format of the command is $y =$ **lsim** (num, den, x, t), where $x(t)$ is the input signal, t is the time vector, and $y(t)$ is the output generated. For example, assume a system is described by the transfer function

$$H(s) = \frac{s + 4}{s^3 + 2s^2 + 5s + 10}$$

To find the response $y(t)$ of the system to input $x(t) = 10e^{-t}u(t)$, we use the following *MATLAB* commands. Both the response $y(t)$ and the input $x(t)$ are plotted in Fig. 14.61.

```
>> t = 0:0.02:5; % time vector 0 < t < 5 with increment
      0.02
>> x = 10*exp(-t);
>> num = [1  4];
>> den = [1  2  5  10];
>> y = lsim(num,den,x,t);
>> plot(t,x,t,y)
```

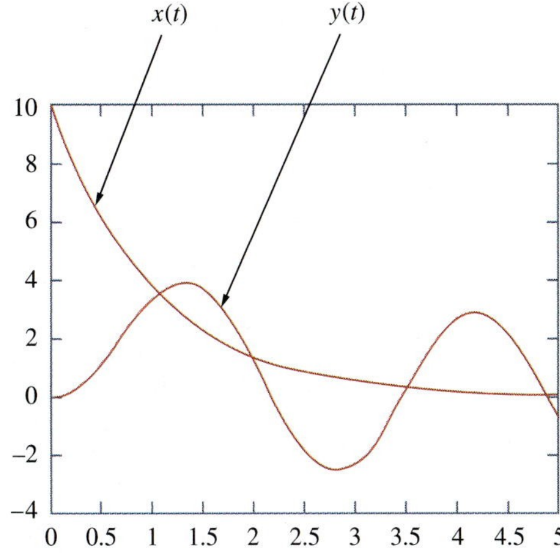

Figure 14.61
The response of the system described by $H(s) = (s + 4)/(s^2 + 2s^2 + 5s + 10)$ to an exponential input.

14.12 †Applications

Resonant circuits and filters are widely used, particularly in electronics, power systems, and communications systems. For example, a Notch filter with a cutoff frequency at 60 Hz may be used to eliminate the 60-Hz power line noise in various communications electronics. Filtering of signals in communications systems is necessary in order to select the desired signal from a host of others in the same range (as in the case of radio receivers discussed next) and also to minimize the effects of noise and interference on the desired signal. In this section, we consider one practical application of resonant circuits and two applications of filters. The focus of each application is not to understand the details of how each device works but to see how the circuits considered in this chapter are applied in the practical devices.

14.12.1 Radio Receiver

Series and parallel resonant circuits are commonly used in radio and TV receivers to tune in stations and to separate the audio signal from the radio-frequency carrier wave. As an example, consider the block diagram of an AM radio receiver shown in Fig. 14.62. Incoming amplitude-modulated radio waves (thousands of them at different frequencies from different broadcasting stations) are received by the antenna. A resonant circuit (or a bandpass filter) is needed to select just one of the incoming waves. The selected signal is very weak and is amplified in stages in order to generate an audible audio-frequency wave. Thus, we have the radio frequency (RF) amplifier to amplify the selected broadcast signal, the intermediate frequency (IF) amplifier to amplify an internally generated signal based on the RF signal, and the audio amplifier to amplify the audio signal just before it reaches the loudspeaker. It is much easier to amplify the signal at three stages than to build an amplifier to provide the same amplification for the entire band.

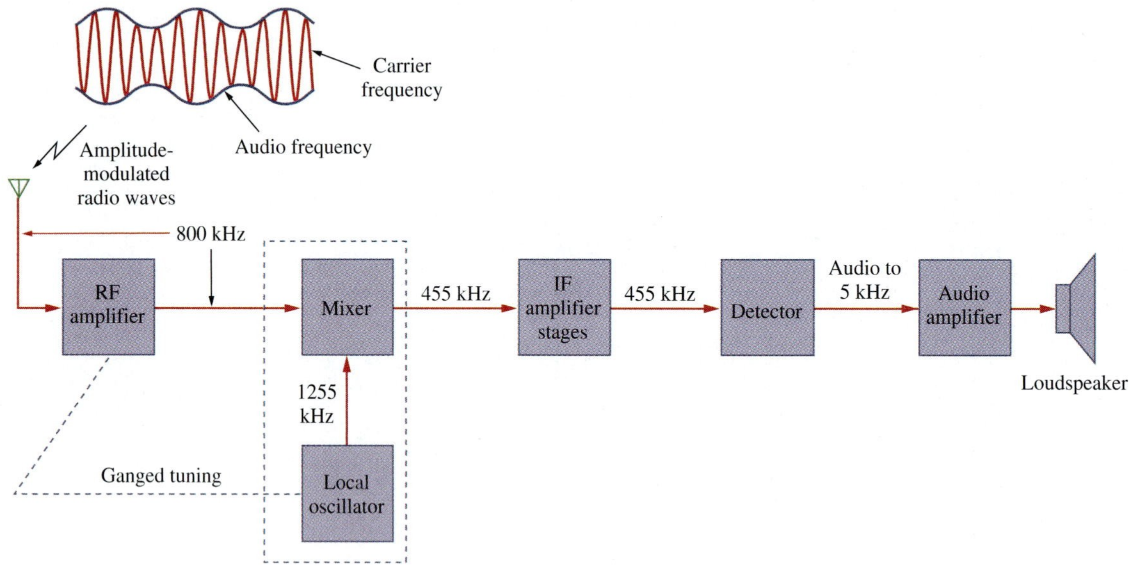

Figure 14.62
A simplified block diagram of a superheterodyne AM radio receiver.

The type of AM receiver shown in Fig. 14.62 is known as the *superheterodyne receiver*. In the early development of radio, each amplification stage had to be tuned to the frequency of the incoming signal. This way, each stage must have several tuned circuits to cover the entire AM band (540 to 1600 kHz). To avoid the problem of having several resonant circuits, modern receivers use a *frequency mixer* or *heterodyne* circuit, which always produces the same IF signal (445 kHz) but retains the audio frequencies carried on the incoming signal. To produce the constant IF frequency, the rotors of two separate variable capacitors are mechanically coupled with one another so that they can be rotated simultaneously with a single control; this is called *ganged tuning*. A *local oscillator* ganged with the RF amplifier produces an RF signal that is combined with the incoming wave by the frequency mixer to produce an output signal that contains the sum and the difference frequencies of the two signals. For example, if the resonant circuit is tuned to receive an 800-kHz incoming signal, the local oscillator must produce a 1,255-kHz signal, so that the sum (1,255 + 800 = 2,055 kHz) and the difference (1,255 − 800 = 455 kHz) of frequencies are available at the output of the mixer. However, only the difference, 455 kHz, is used in practice. This is the only frequency to which all the IF amplifier stages are tuned, regardless of the station dialed. The original audio signal (containing the "intelligence") is extracted in the detector stage. The detector basically removes the IF signal, leaving the audio signal. The audio signal is amplified to drive the loudspeaker, which acts as a transducer converting the electrical signal to sound.

Our major concern here is the tuning circuit for the AM radio receiver. The operation of the FM radio receiver is different from that of the AM receiver discussed here, and in a much different range of frequencies, but the tuning is similar.

Example 14.17

The resonant or tuner circuit of an AM radio is portrayed in Fig. 14.63. Given that $L = 1\ \mu$H, what must be the range of C to have the resonant frequency adjustable from one end of the AM band to another?

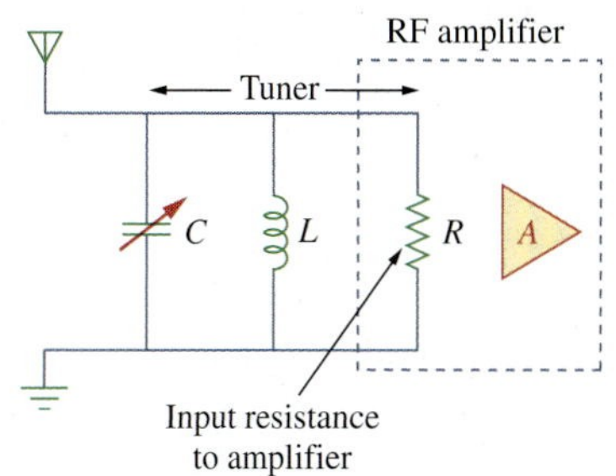

Figure 14.63
The tuner circuit for Example 14.17.

Solution:
The frequency range for AM broadcasting is 540 to 1,600 kHz. We consider the low and high ends of the band. Since the resonant circuit in Fig. 14.63 is a parallel type, we apply the ideas in Section 14.6. From Eq. (14.44),

$$\omega_0 = 2\pi f_0 = \frac{1}{\sqrt{LC}}$$

or

$$C = \frac{1}{4\pi^2 f_0^2 L}$$

For the high end of the AM band, $f_0 = 1{,}600$ kHz, and the corresponding C is

$$C_1 = \frac{1}{4\pi^2 \times 1{,}600^2 \times 10^6 \times 10^{-6}} = 9.9 \text{ nF}$$

For the low end of the AM band, $f_0 = 540$ kHz, and the corresponding C is

$$C_2 = \frac{1}{4\pi^2 \times 540^2 \times 10^6 \times 10^{-6}} = 86.9 \text{ nF}$$

Thus, C must be an adjustable (gang) capacitor varying from 9.9 nF to 86.9 nF.

Practice Problem 14.17

For an FM radio receiver, the incoming wave is in the frequency range from 88 to 108 MHz. The tuner circuit is a parallel *RLC* circuit with a 4-μH coil. Calculate the range of the variable capacitor necessary to cover the entire band.

Answer: From 0.543 pF to 0.818 pF.

14.12.2 Touch-Tone Telephone

A typical application of filtering is the touch-tone telephone set shown in Fig. 14.64. The keypad has 12 buttons arranged in four rows and three columns. The arrangement provides 12 distinct signals by using seven tones divided into two groups: the low-frequency group (697 to 941 Hz) and the high-frequency group (1,209 to 1,477 Hz). Pressing a button generates a sum of two sinusoids corresponding to its unique pair of frequencies. For example, pressing the number 6 button generates sinusoidal tones with frequencies 770 Hz and 1,477 Hz.

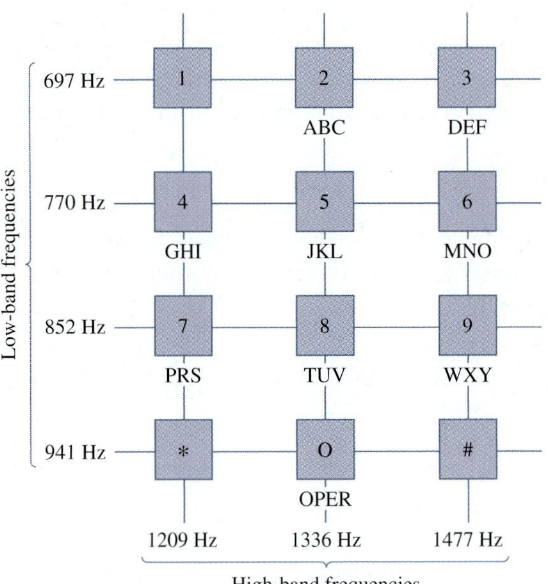

Figure 14.64
Frequency assignments for touch-tone dialing.
Adapted from G. Daryanani, *Principles of Active Network Synthesis and Design* [New York: John Wiley & Sons], 1976, p. 79.

When a caller dials a telephone number, a set of signals is transmitted to the telephone office, where the touch-tone signals are decoded by detecting the frequencies they contain. Figure 14.65 shows the block diagram for the detection scheme. The signals are first amplified and separated into their respective groups by the lowpass (LP) and highpass (HP) filters. The limiters (L) are used to convert the separated tones into square waves. The individual tones are identified using seven bandpass (BP) filters, each filter passing one tone and rejecting other tones. Each filter is followed by a detector (D), which is energized when its input voltage exceeds a certain level. The outputs of the detectors provide the required dc signals needed by the switching system to connect the caller to the party being called.

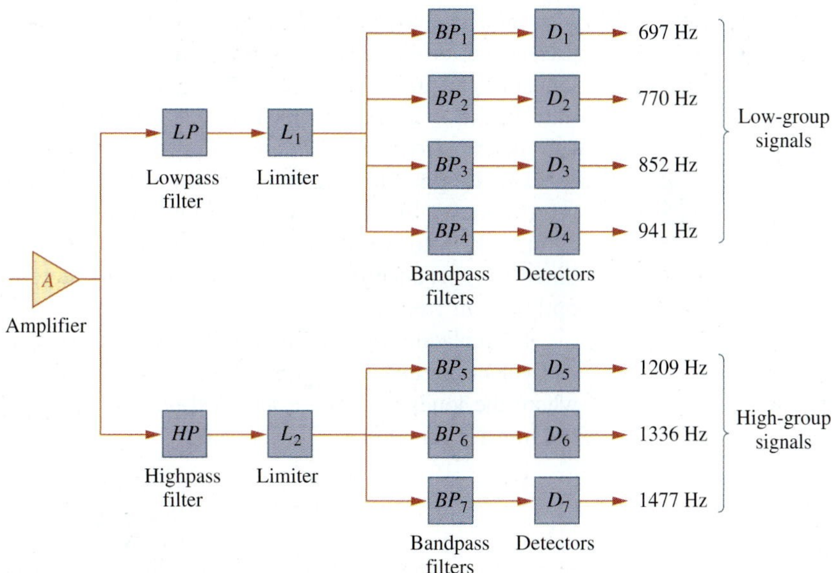

Figure 14.65
Block diagram of detection scheme.
G. Daryanani, *Principles of Active Network Synthesis and Design* [New York: John Wiley & Sons], 1976, p. 79.

Example 14.18

Using the standard 600-Ω resistor used in telephone circuits and a series *RLC* circuit, design the bandpass filter BP_2 in Fig. 14.65.

Solution:
The bandpass filter is the series *RLC* circuit in Fig. 14.35. Since BP_2 passes frequencies 697 Hz to 852 Hz and is centered at $f_0 = 770$ Hz, its bandwidth is

$$B = 2\pi(f_2 - f_1) = 2\pi(852 - 697) = 973.89 \text{ rad/s}$$

From Eq. (14.39),

$$L = \frac{R}{B} = \frac{600}{973.89} = 0.616 \text{ H}$$

From Eq. (14.27) or (14.55),

$$C = \frac{1}{\omega_0^2 L} = \frac{1}{4\pi^2 f_0^2 L} = \frac{1}{4\pi^2 \times 770^2 \times 0.616} = 69.36 \text{ nF}$$

Practice Problem 14.18

Repeat Example 14.18 for bandpass filter BP_6.

Answer: 0.356 H, 39.83 nF.

14.12.3 Crossover Network

Another typical application of filters is the *crossover network* that couples an audio amplifier to woofer and tweeter speakers, as shown in Fig. 14.66(a). The network basically consists of one highpass *RC*

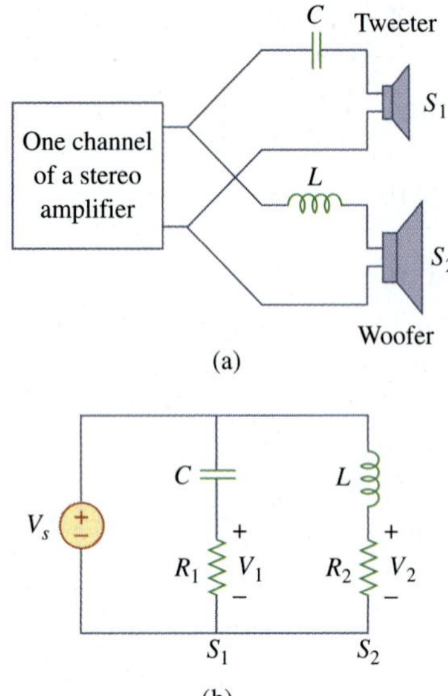

Figure 14.66
(a) A crossover network for two loudspeakers, (b) equivalent model.

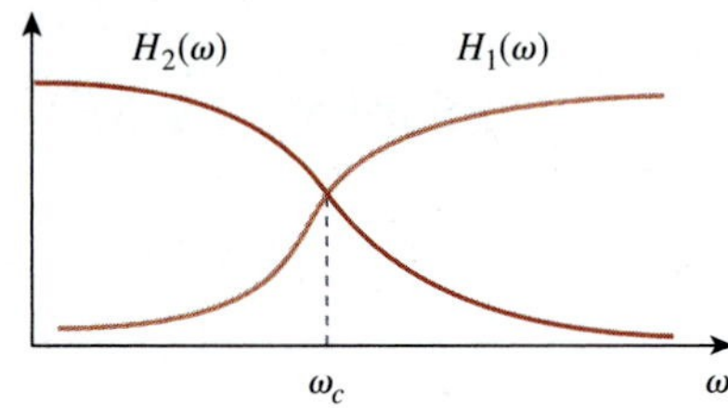

Figure 14.67
Frequency responses of the crossover network in Fig. 14.66.

filter and one lowpass *RL* filter. It routes frequencies higher than a prescribed crossover frequency f_c to the tweeter (high-frequency loudspeaker) and frequencies below f_c into the woofer (low-frequency loudspeaker). These loudspeakers have been designed to accommodate certain frequency responses. A woofer is a low-frequency loudspeaker designed to reproduce the lower part of the frequency range, up to about 3 kHz. A tweeter can reproduce audio frequencies from about 3 kHz to about 20 kHz. The two speaker types can be combined to reproduce the entire audio range of interest and provide the optimum in frequency response.

By replacing the amplifier with a voltage source, the approximate equivalent circuit of the crossover network is shown in Fig. 14.66(b), where the loudspeakers are modeled by resistors. As a highpass filter, the transfer function V_1/V_s is given by

$$H_1(\omega) = \frac{V_1}{V_s} = \frac{j\omega R_1 C}{1 + j\omega R_1 C} \tag{14.87}$$

Similarly, the transfer function of the lowpass filter is given by

$$H_2(\omega) = \frac{V_2}{V_s} = \frac{R_2}{R_2 + j\omega L} \tag{14.88}$$

The values of R_1, R_2, L, and C may be selected such that the two filters have the same cutoff frequency, known as the *crossover frequency*, as shown in Fig. 14.67.

The principle behind the crossover network is also used in the resonant circuit for a TV receiver, where it is necessary to separate the video and audio bands of RF carrier frequencies. The lower-frequency band (picture information in the range from about 30 Hz to about 4 MHz) is channeled into the receiver's video amplifier, while the high-frequency band (sound information around 4.5 MHz) is channeled to the receiver's sound amplifier.

Example 14.19

In the crossover network of Fig. 14.66, suppose each speaker acts as a 6-Ω resistance. Find C and L if the crossover frequency is 2.5 kHz.

Solution:

For the highpass filter,

$$\omega_c = 2\pi f_c = \frac{1}{R_1 C}$$

or

$$C = \frac{1}{2\pi f_c R_1} = \frac{1}{2\pi \times 2.5 \times 10^3 \times 6} = 10.61\ \mu\text{F}$$

For the lowpass filter,

$$\omega_c = 2\pi f_c = \frac{R_2}{L}$$

or

$$L = \frac{R_2}{2\pi f_c} = \frac{6}{2\pi \times 2.5 \times 10^3} = 382\ \mu\text{H}$$

Practice Problem 14.19

If each speaker in Fig. 14.66 has an 8-Ω resistance and $C = 10\ \mu$F, find L and the crossover frequency.

Answer: 0.64 mH, 1.989 kHz.

14.13 Summary

1. The transfer function $\mathbf{H}(\omega)$ is the ratio of the output response $\mathbf{Y}(\omega)$ to the input excitation $\mathbf{X}(\omega)$; that is, $\mathbf{H}(\omega) = \mathbf{Y}(\omega)/\mathbf{X}(\omega)$.
2. The frequency response is the variation of the transfer function with frequency.
3. Zeros of a transfer function $\mathbf{H}(s)$ are the values of $s = j\omega$ that make $H(s) = 0$, while poles are the values of s that make $H(s) \rightarrow \infty$.
4. The decibel is the unit of logarithmic gain. For a voltage or current gain G, its decibel equivalent is $G_{\text{dB}} = 20 \log_{10} G$.
5. Bode plots are semilog plots of the magnitude and phase of the transfer function as it varies with frequency. The straight-line approximations of H (in dB) and ϕ (in degrees) are constructed using the corner frequencies defined by the poles and zeros of $\mathbf{H}(\omega)$.
6. The resonant frequency is that frequency at which the imaginary part of a transfer function vanishes. For series and parallel RLC circuits.

$$\omega_0 = \frac{1}{\sqrt{LC}}$$

7. The half-power frequencies (ω_1, ω_2) are those frequencies at which the power dissipated is one-half of that dissipated at the resonant frequency. The geometric mean between the half-power frequencies is the resonant frequency, or

$$\omega_0 = \sqrt{\omega_1 \omega_2}$$

8. The bandwidth is the frequency band between half-power frequencies:

$$B = \omega_2 - \omega_1$$

9. The quality factor is a measure of the sharpness of the resonance peak. It is the ratio of the resonant (angular) frequency to the bandwidth,

$$Q = \frac{\omega_0}{B}$$

10. A filter is a circuit designed to pass a band of frequencies and reject others. Passive filters are constructed with resistors, capacitors, and inductors. Active filters are constructed with resistors, capacitors, and an active device, usually an op amp.
11. Four common types of filters are lowpass, highpass, bandpass, and bandstop. A lowpass filter passes only signals whose frequencies are below the cutoff frequency ω_c. A highpass filter passes only signals whose frequencies are above the cutoff frequency ω_c. A bandpass filter passes only signals whose frequencies are within a prescribed

range ($\omega_1 < \omega < \omega_2$). A bandstop filter passes only signals whose frequencies are outside a prescribed range ($\omega_1 > \omega > \omega_2$).

12. Scaling is the process whereby unrealistic element values are magnitude-scaled by a factor K_m and/or frequency-scaled by a factor K_f to produce realistic values.

$$R' = K_m R, \qquad L' = \frac{K_m}{K_f}L, \qquad C' = \frac{1}{K_m K_f}C$$

13. *PSpice* can be used to obtain the frequency response of a circuit if a frequency range for the response and the desired number of points within the range are specified in the AC Sweep.
14. The radio receiver—one practical application of resonant circuits—employs a bandpass resonant circuit to tune in one frequency among all the broadcast signals picked up by the antenna.
15. The touch-tone telephone and the crossover network are two typical applications of filters. The touch-tone telephone system employs filters to separate tones of different frequencies to activate electronic switches. The crossover network separates signals in different frequency ranges so that they can be delivered to different devices such as tweeters and woofers in a loudspeaker system.

Review Questions

14.1 A zero of the transfer function

$$H(s) = \frac{10(s+1)}{(s+2)(s+3)}$$

is at

(a) 10 (b) -1 (c) -2 (d) -3

14.2 On the Bode magnitude plot, the slope of $1/(5 + j\omega)^2$ for large values of ω is

(a) 20 dB/decade (b) 40 dB/decade
(c) -40 dB/decade (d) -20 dB/decade

14.3 On the Bode phase plot for $0.5 < \omega < 50$, the slope of $[1 + j10\omega - \omega^2/25]^2$ is

(a) 45°/decade (b) 90°/decade
(c) 135°/decade (d) 180°/decade

14.4 How much inductance is needed to resonate at 5 kHz with a capacitance of 12 nF?

(a) 2,652 H (b) 11.844 H
(c) 3.333 H (d) 84.43 mH

14.5 The difference between the half-power frequencies is called the:

(a) quality factor (b) resonant frequency
(c) bandwidth (d) cutoff frequency

14.6 In a series *RLC* circuit, which of these quality factors has the steepest magnitude response curve near resonance?

(a) $Q = 20$ (b) $Q = 12$
(c) $Q = 8$ (d) $Q = 4$

14.7 In a parallel *RLC* circuit, the bandwidth *B* is directly proportional to *R*.

(a) True (b) False

14.8 When the elements of an *RLC* circuit are both magnitude-scaled and frequency-scaled, which quality is unaffected?

(a) resistor (b) resonant frequency
(c) bandwidth (d) quality factor

14.9 What kind of filter can be used to select a signal of one particular radio station?

(a) lowpass (b) highpass
(c) bandpass (d) bandstop

14.10 A voltage source supplies a signal of constant amplitude, from 0 to 40 kHz, to an RC lowpass filter. A load resistor, connected in parallel across the capacitor, experiences the maximum voltage at:

(a) dc (b) 10 kHz
(c) 20 kHz (d) 40 kHz

Answers: 14.1b, 14.2c, 14.3d, 14.4d, 14.5c, 14.6a, 14.7b, 14.8d, 14.9c, 14.10a.

Problems

Section 14.2 Transfer Function

14.1 Find the transfer function $\mathbf{V}_o/\mathbf{V}_i$ of the *RC* circuit in Fig. 14.68. Express it using $\omega_0 = 1/RC$.

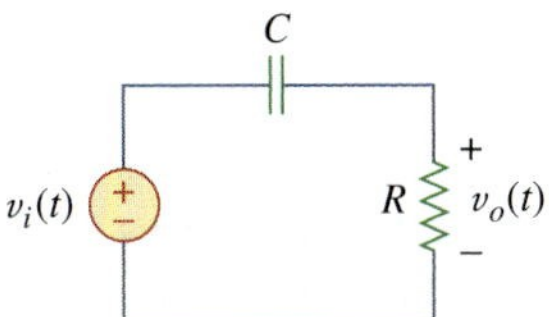

Figure 14.68
For Prob. 14.1.

14.2 **e⬈d** Using Fig. 14.69, design a problem to help other students better understand how to determine transfer functions.

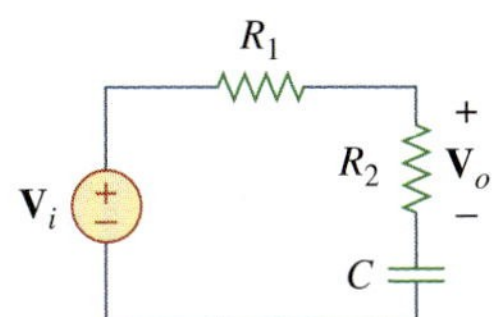

Figure 14.69
For Prob. 14.2.

14.3 Given the circuit in Fig. 14.70, $R_1 = 2\ \Omega$, $R_2 = 5\ \Omega$, $C_1 = 0.1$ F, and $C_2 = 0.2$ F, determine the transfer function $\mathbf{H}(s) = \mathbf{V}_o(s)/\mathbf{V}_i(s)$.

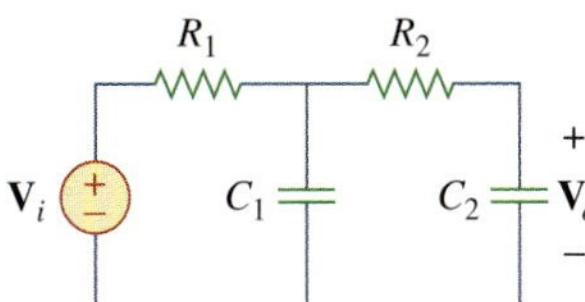

Figure 14.70
For Prob. 14.3.

14.4 Find the transfer function $\mathbf{H}(\omega) = \mathbf{V}_o/\mathbf{V}_i$ of the circuits shown in Fig. 14.71.

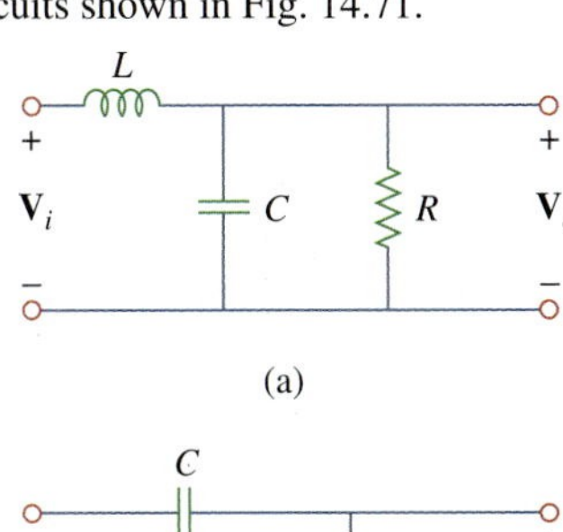

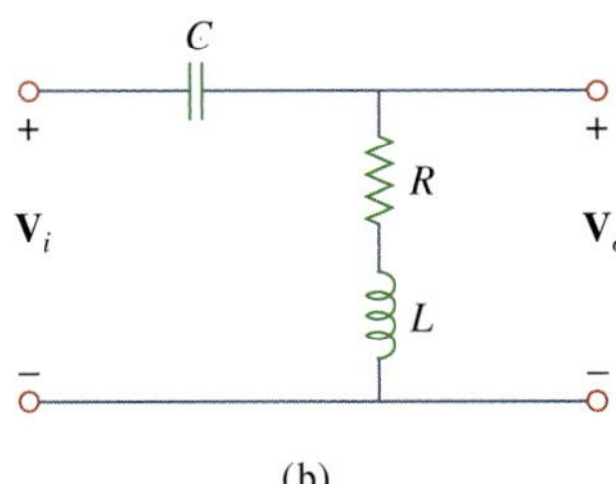

Figure 14.71
For Prob. 14.4.

14.5 For each of the circuits shown in Fig. 14.72, find $\mathbf{H}(s) = \mathbf{V}_o(s)/\mathbf{V}_s(s)$.

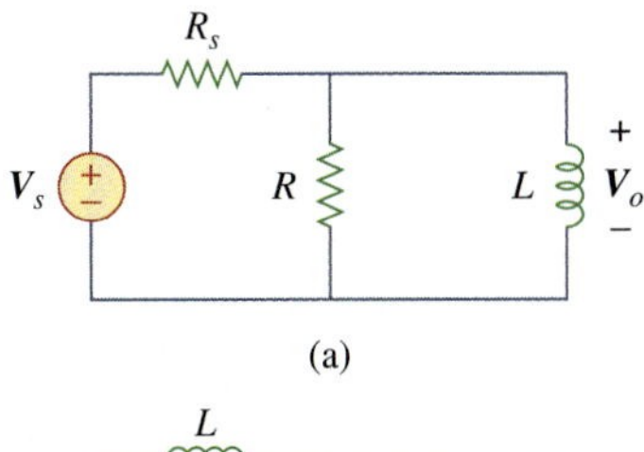

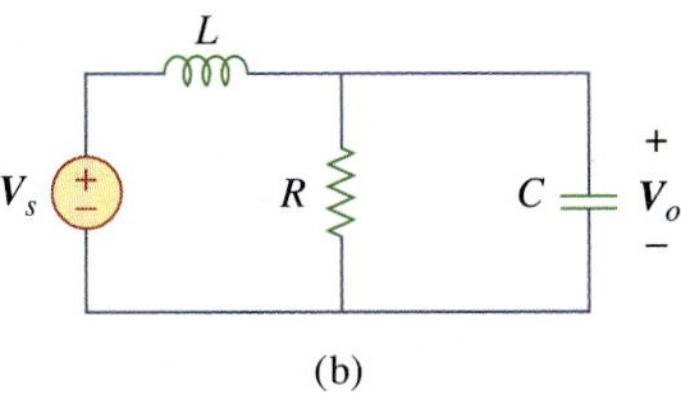

Figure 14.72
For Prob. 14.5.

14.6 For the circuit shown in Fig. 14.73, find $\mathbf{H}(s) = \mathbf{I}_o(s)/\mathbf{I}_s(s)$.

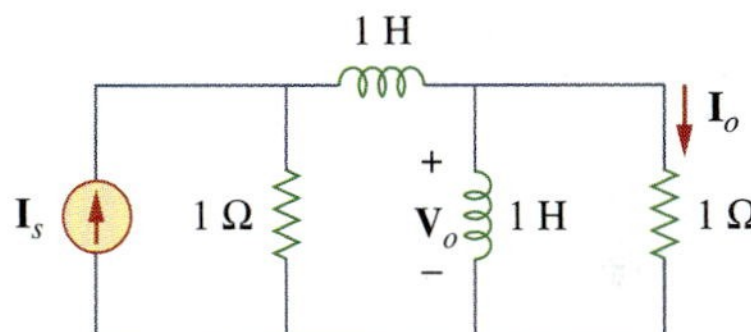

Figure 14.73
For Prob. 14.6.

Section 14.3 The Decibel Scale

14.7 Calculate $|\mathbf{H}(\omega)|$ if H_{dB} equals

(a) 0.05 dB (b) −6.2 dB (c) 104.7 dB

14.8 **e⬈d** Design a problem to help other students calculate the magnitude in dB and phase in degrees of a variety of transfer functions at a single value of ω.

Section 14.4 Bode Plots

14.9 A ladder network has a voltage gain of

$$\mathbf{H}(\omega) = \frac{10}{(1 + j\omega)(10 + j\omega)}$$

Sketch the Bode plots for the gain.

14.10 Design a problem to help other students better understand how to determine the Bode magnitude and phase plots of a given transfer function in terms of $j\omega$.

14.11 Sketch the Bode plots for

$$\mathbf{H}(\omega) = \frac{10 + j\omega}{j\omega(2 + j\omega)}$$

14.12 A transfer function is given by

$$T(s) = \frac{s + 1}{s(s + 10)}$$

Sketch the magnitude and phase Bode plots.

14.13 Construct the Bode plots for

$$G(s) = \frac{s + 1}{s^2(s + 10)}, \qquad s = j\omega$$

14.14 Draw the Bode plots for

$$\mathbf{H}(\omega) = \frac{50(j\omega + 1)}{j\omega(-\omega^2 + 10j\omega + 25)}$$

14.15 Construct the Bode magnitude and phase plots for

$$H(s) = \frac{40(s + 1)}{(s + 2)(s + 10)}, \qquad s = j\omega$$

14.16 Sketch Bode magnitude and phase plots for

$$H(s) = \frac{10}{s(s^2 + s + 16)}, \qquad s = j\omega$$

14.17 Sketch the Bode plots for

$$G(s) = \frac{s}{(s + 2)^2(s + 1)}, \qquad s = j\omega$$

14.18 A linear network has this transfer function

$$H(s) = \frac{7s^2 + s + 4}{s^3 + 8s^2 + 14s + 5}, \qquad s = j\omega$$

Use *MATLAB* or equivalent to plot the magnitude and phase (in degrees) of the transfer function. Take $0.1 < \omega < 10$ rad/s.

14.19 Sketch the asymptotic Bode plots of the magnitude and phase for

$$H(s) = \frac{100s}{(s + 10)(s + 20)(s + 40)}, \qquad s = j\omega$$

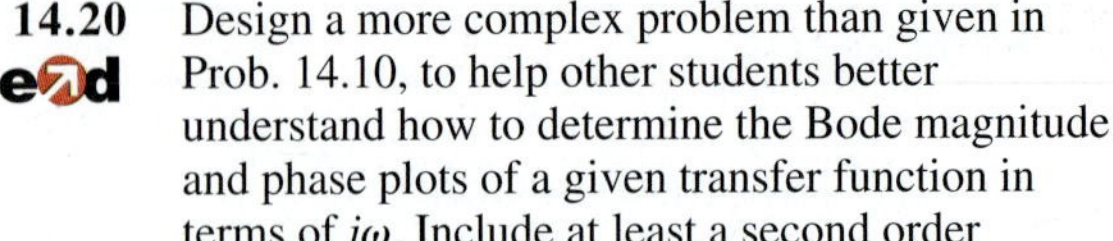

14.20 Design a more complex problem than given in Prob. 14.10, to help other students better understand how to determine the Bode magnitude and phase plots of a given transfer function in terms of $j\omega$. Include at least a second order repeated root.

14.21 Sketch the magnitude Bode plot for

$$H(s) = \frac{s(s + 20)}{(s + 1)(s^2 + 60s + 400)}, \qquad s = j\omega$$

14.22 Find the transfer function $\mathbf{H}(\omega)$ with the Bode magnitude plot shown in Fig. 14.74.

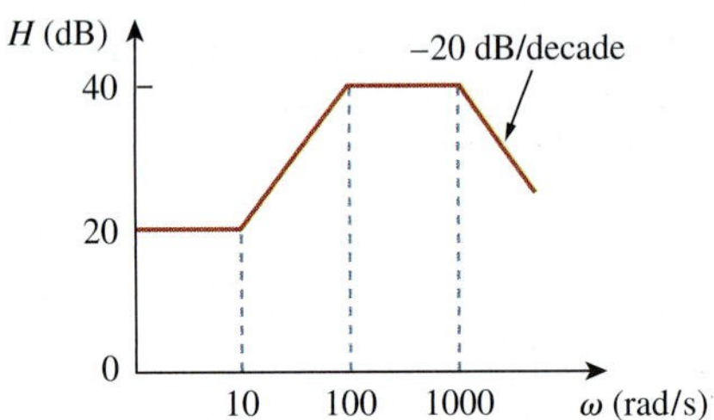

Figure 14.74
For Prob. 14.22.

14.23 The Bode magnitude plot of $\mathbf{H}(\omega)$ is shown in Fig. 14.75. Find $\mathbf{H}(\omega)$.

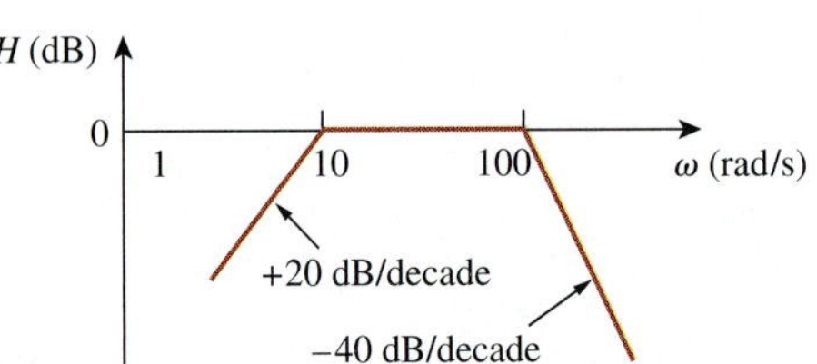

Figure 14.75
For Prob. 14.23.

14.24 The magnitude plot in Fig. 14.76 represents the transfer function of a preamplifier. Find $H(s)$.

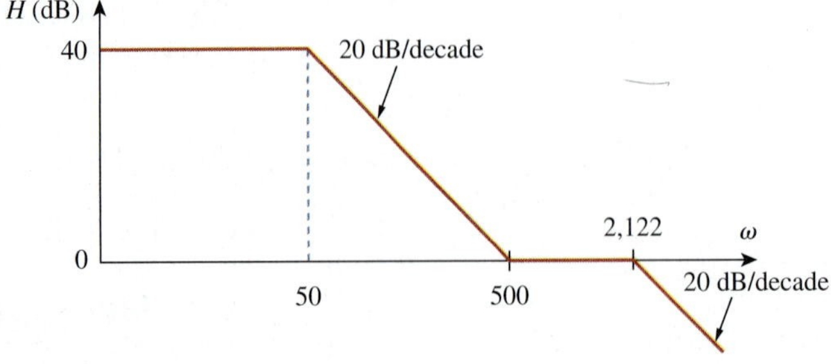

Figure 14.76
For Prob. 14.24.

Section 14.5 Series Resonance

14.25 A series *RLC* network has $R = 2\ \text{k}\Omega$, $L = 40$ mH, and $C = 1\ \mu$F. Calculate the impedance at resonance and at one-fourth, one-half, twice, and four times the resonant frequency.

14.26 **e⊘d** Design a problem to help other students better understand ω_0, Q, and B at resonance in series *RLC* circuits.

14.27 **e⊘d** Design a series *RLC* resonant circuit with $\omega_0 = 40$ rad/s and $B = 10$ rad/s.

14.28 Design a series *RLC* circuit with $B = 20$ rad/s and $\omega_0 = 1{,}000$ rad/s. Find the circuit's Q. Let $R = 10\ \Omega$.

14.29 Let $v_s = 120\cos(at)$ V in the circuit of Fig. 14.77. Find ω_0, Q, and B, as seen by the capacitor.

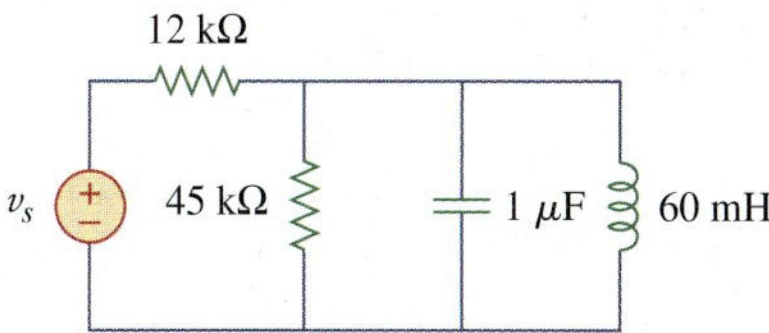

Figure 14.77
For Prob. 14.29.

14.30 A circuit consisting of a coil with inductance 10 mH and resistance 20 Ω is connected in series with a capacitor and a generator with an rms voltage of 120 V. Find:

(a) the value of the capacitance that will cause the circuit to be in resonance at 15 kHz

(b) the current through the coil at resonance

(c) the Q of the circuit

Section 14.6 Parallel Resonance

14.31 **e⊘d** Design a parallel resonant *RLC* circuit with $\omega_0 = 10$ rad/s and $Q = 20$. Calculate the bandwidth of the circuit. Let $R = 10\ \Omega$.

14.32 **e⊘d** Design a problem to help other students better understand the quality factor, the resonant frequency, and bandwidth of a parallel *RLC* circuit.

14.33 A parallel resonant circuit with quality factor 120 has a resonant frequency of 6×10^6 rad/s. Calculate the bandwidth and half-power frequencies.

14.34 A parallel *RLC* circuit is resonant at 5.6 MHz, has a Q of 80, and has a resistive branch of 40 kΩ. Determine the values of L and C in the other two branches.

14.35 A parallel *RLC* circuit has $R = 5\ \text{k}\Omega$, $L = 8$ mH, and $C = 60\ \mu$F. Determine:

(a) the resonant frequency

(b) the bandwidth

(c) the quality factor

14.36 It is expected that a parallel *RLC* resonant circuit has a midband admittance of 25×10^{-3} S, quality factor of 80, and a resonant frequency of 200 krad/s. Calculate the values of R, L, and C. Find the bandwidth and the half-power frequencies.

14.37 Rework Prob. 14.25 if the elements are connected in parallel.

14.38 Find the resonant frequency of the circuit in Fig. 14.78.

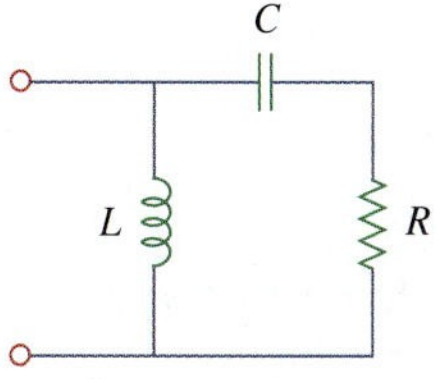

Figure 14.78
For Prob. 14.38.

14.39 For the "tank" circuit in Fig. 14.79, find the resonant frequency.

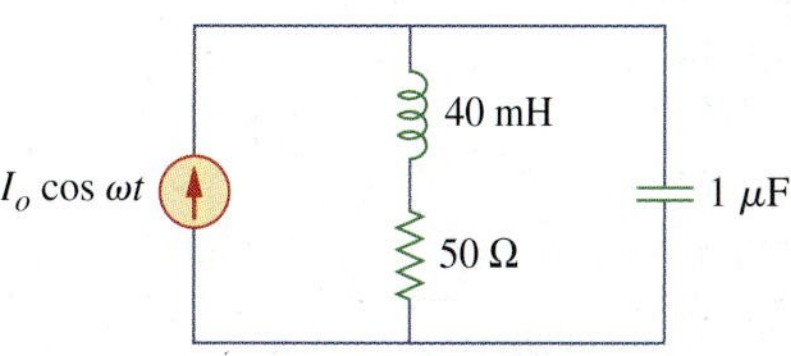

Figure 14.79
For Probs. 14.39 and 14.91.

14.40 A parallel resonance circuit has a resistance of 2 kΩ and half-power frequencies of 86 kHz and 90 kHz. Determine:

(a) the capacitance

(b) the inductance

(c) the resonant frequency

(d) the bandwidth

(e) the quality factor

14.41 **e⊘d** Using Fig. 14.80, design a problem to help other students better understand the quality factor, the resonant frequency, and bandwidth of *RLC* circuits.

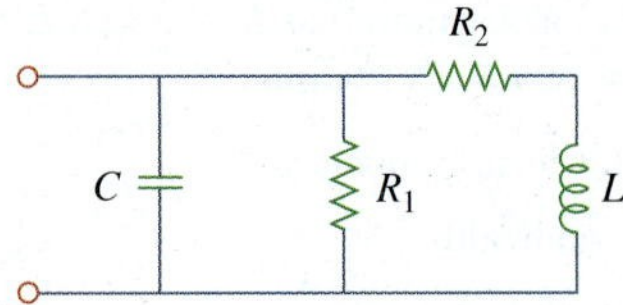

Figure 14.80
For Prob. 14.41.

14.42 For the circuits in Fig. 14.81, find the resonant frequency ω_0, the quality factor Q, and the bandwidth B.

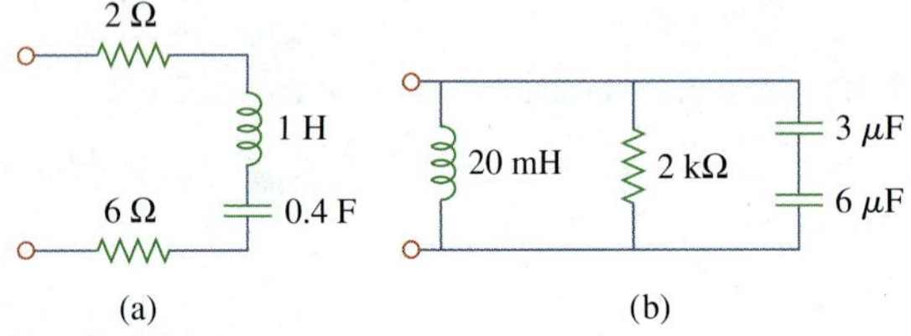

Figure 14.81
For Prob. 14.42.

14.43 Calculate the resonant frequency of each of the circuits in Fig. 14.82.

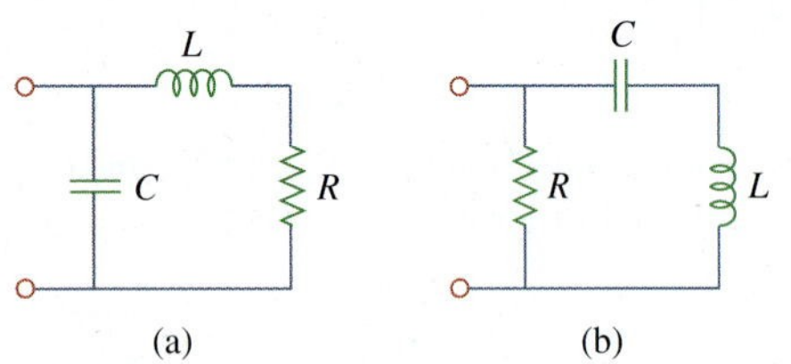

Figure 14.82
For Prob. 14.43.

***14.44** For the circuit in Fig. 14.83, find:

(a) the resonant frequency ω_0

(b) $\mathbf{Z}_{\text{in}}(\omega_0)$

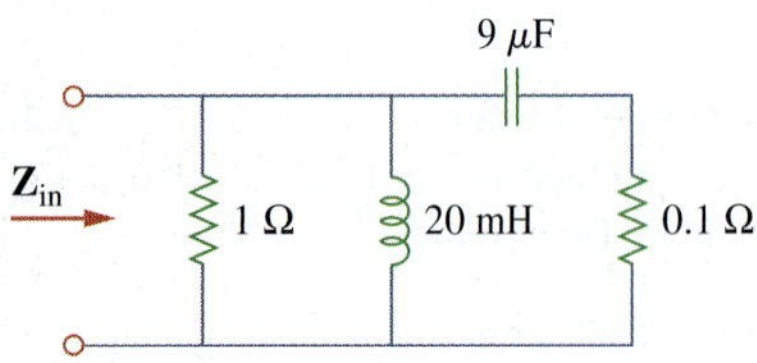

Figure 14.83
For Prob. 14.44.

14.45 For the circuit shown in Fig. 14.84, find ω_0, B, and Q, as seen by the voltage across the inductor.

* An asterisk indicates a challenging problem.

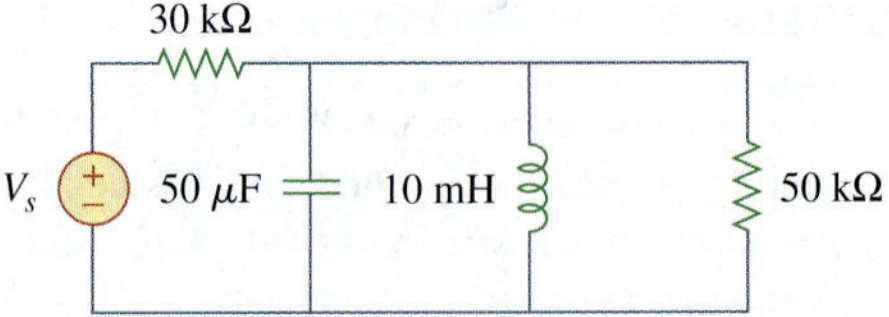

Figure 14.84
For Prob. 14.45.

14.46 For the network illustrated in Fig. 14.85, find

(a) the transfer function $\mathbf{H}(\omega) = \mathbf{V}_o(\omega)/\mathbf{I}(\omega)$,

(b) the magnitude of $\mathbf{H}$ at $\omega_0 = 1$ rad/s.

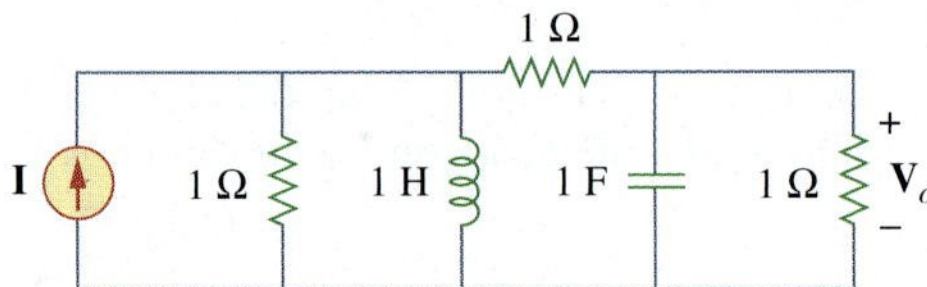

Figure 14.85
For Probs. 14.46, 14.78, and 14.92.

Section 14.7 Passive Filters

14.47 Show that a series LR circuit is a lowpass filter if the output is taken across the resistor. Calculate the corner frequency f_c if $L = 2$ mH and $R = 10$ kΩ.

14.48 Find the transfer function $\mathbf{V}_o/\mathbf{V}_s$ of the circuit in Fig. 14.86. Show that the circuit is a lowpass filter.

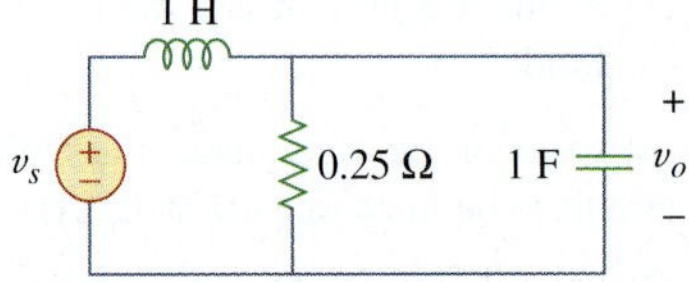

Figure 14.86
For Prob. 14.48.

14.49 **ed** Design a problem to help other students better understand lowpass filters described by transfer functions.

14.50 Determine what type of filter is in Fig. 14.87. Calculate the corner frequency f_c.

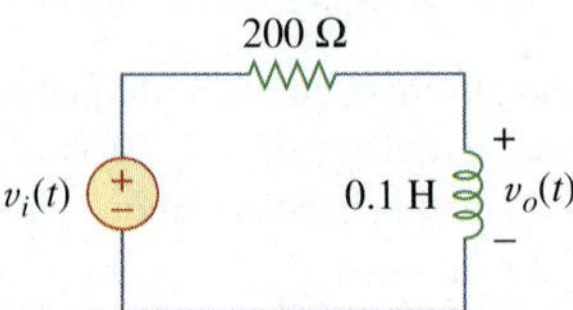

Figure 14.87
For Prob. 14.50.

14.51 Design an *RL* lowpass filter that uses a 40-mH coil and has a cutoff frequency of 5 kHz.

14.52 Design a problem to help other students better understand passive highpass filters.

14.53 Design a series *RLC* type bandpass filter with cutoff frequencies of 10 kHz and 11 kHz. Assuming $C = 80$ pF, find R, L, and Q.

14.54 Design a passive bandstop filter with $\omega_0 = 10$ rad/s and $Q = 20$.

14.55 Determine the range of frequencies that will be passed by a series *RLC* bandpass filter with $R = 10\ \Omega$, $L = 25$ mH, and $C = 0.4\ \mu$F. Find the quality factor.

14.56 (a) Show that for a bandpass filter,

$$\mathbf{H}(s) = \frac{sB}{s^2 + sB + \omega_0^2}, \qquad s = j\omega$$

where B = bandwidth of the filter and ω_0 is the center frequency.

(b) Similarly, show that for a bandstop filter,

$$\mathbf{H}(s) = \frac{s^2 + \omega_0^2}{s^2 + sB + \omega_0^2}, \qquad s = j\omega$$

14.57 Determine the center frequency and bandwidth of the bandpass filters in Fig. 14.88.

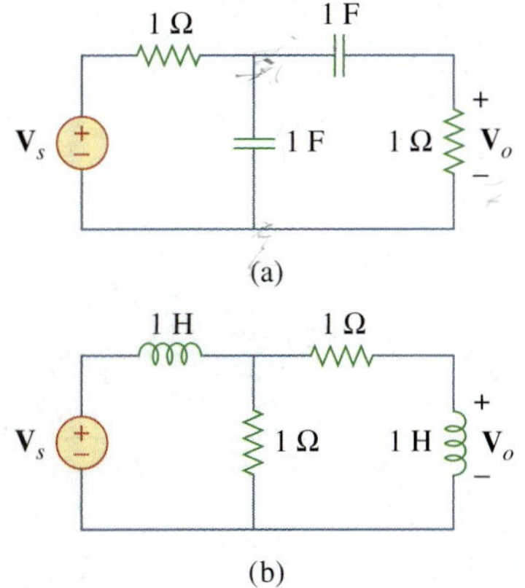

Figure 14.88
For Prob. 14.57.

14.58 The circuit parameters for a series *RLC* bandstop filter are $R = 2\ \text{k}\Omega$, $L = 0.1$ H, $C = 40$ pF. Calculate:

(a) the center frequency

(b) the half-power frequencies

(c) the quality factor

14.59 Find the bandwidth and center frequency of the bandstop filter of Fig. 14.89.

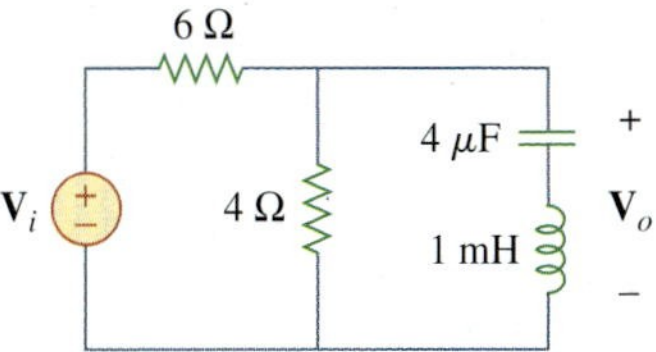

Figure 14.89
For Prob. 14.59.

Section 14.8 Active Filters

14.60 Obtain the transfer function of a highpass filter with a passband gain of 10 and a cutoff frequency of 50 rad/s.

14.61 Find the transfer function for each of the active filters in Fig. 14.90.

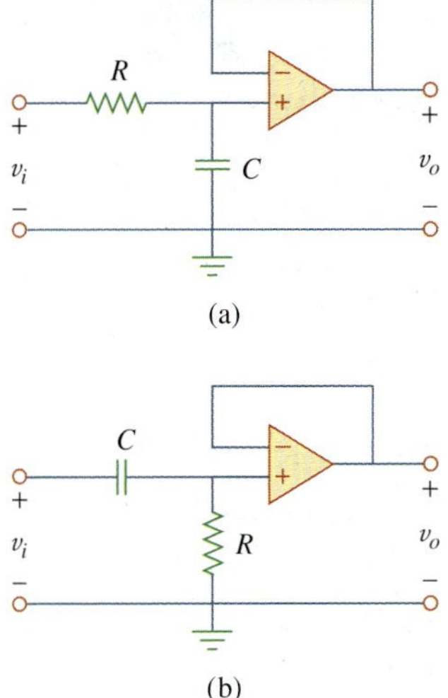

Figure 14.90
For Probs. 14.61 and 14.62.

14.62 The filter in Fig. 14.90(b) has a 3-dB cutoff frequency at 1 kHz. If its input is connected to a 120-mV variable frequency signal, find the output voltage at:

(a) 200 Hz (b) 2 kHz (c) 10 kHz

14.63 Design an active first-order highpass filter with

$$\mathbf{H}(s) = -\frac{100s}{s + 10}, \qquad s = j\omega$$

Use a 1-μF capacitor.

14.64 Obtain the transfer function of the active filter in Fig. 14.91 on the next page. What kind of filter is it?

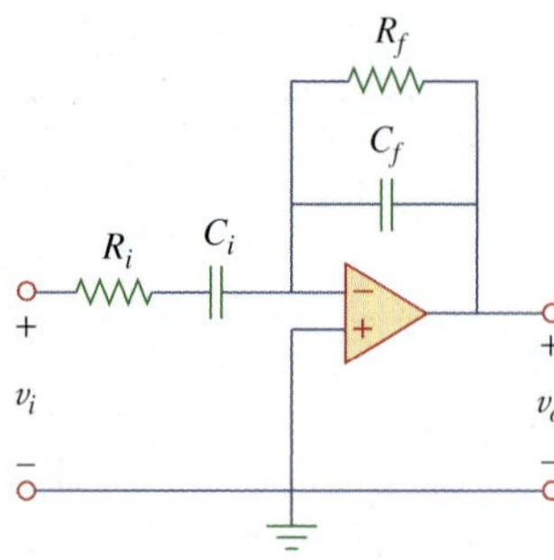

Figure 14.91
For Prob. 14.64.

14.65 A highpass filter is shown in Fig. 14.92. Show that the transfer function is

$$\mathbf{H}(\omega) = \left(1 + \frac{R_f}{R_i}\right) \frac{j\omega RC}{1 + j\omega RC}$$

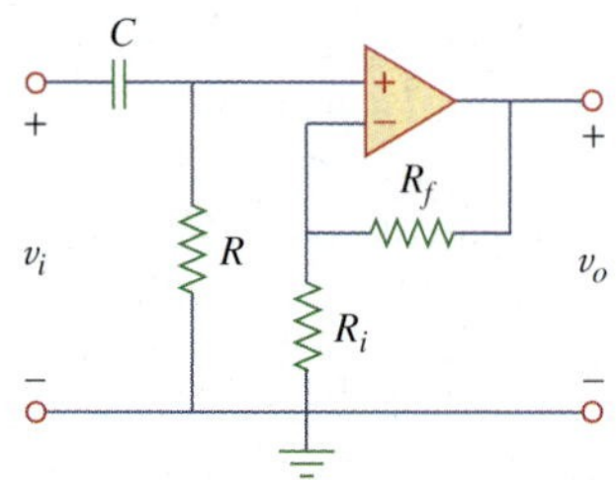

Figure 14.92
For Prob. 14.65.

14.66 A "general" first-order filter is shown in Fig. 14.93.

(a) Show that the transfer function is

$$\mathbf{H}(s) = \frac{R_4}{R_3 + R_4} \times \frac{s + (1/R_1C)[R_1/R_2 - R_3/R_4]}{s + 1/R_2C},$$

$s = j\omega$

(b) What condition must be satisfied for the circuit to operate as a highpass filter?

(c) What condition must be satisfied for the circuit to operate as a lowpass filter?

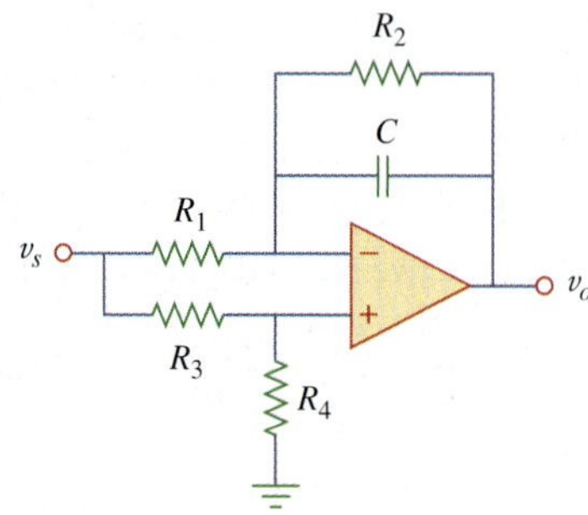

Figure 14.93
For Prob. 14.66.

14.67 **e⃗d** Design an active lowpass filter with dc gain of 0.25 and a corner frequency of 500 Hz.

14.68 **e⃗d** Design a problem to help other students better understand the design of active highpass filters when specifying a high-frequency gain and a corner frequency.

14.69 **e⃗d** Design the filter in Fig. 14.94 to meet the following requirements:

(a) It must attenuate a signal at 2 kHz by 3 dB compared with its value at 10 MHz.

(b) It must provide a steady-state output of $v_o(t) = 10 \sin(2\pi \times 10^8 t + 180°)$ V for an input $v_s(t) = 4 \sin(2\pi \times 10^8 t)$ V.

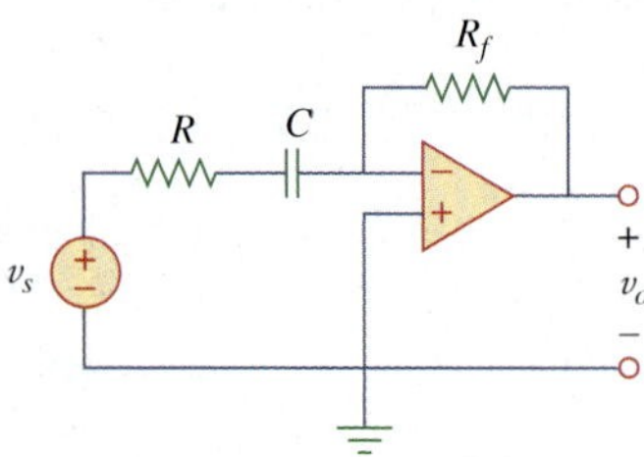

Figure 14.94
For Prob. 14.69.

***14.70** **e⃗d** A second-order active filter known as a Butterworth filter is shown in Fig. 14.95.

(a) Find the transfer function $\mathbf{V}_o/\mathbf{V}_i$.

(b) Show that it is a lowpass filter.

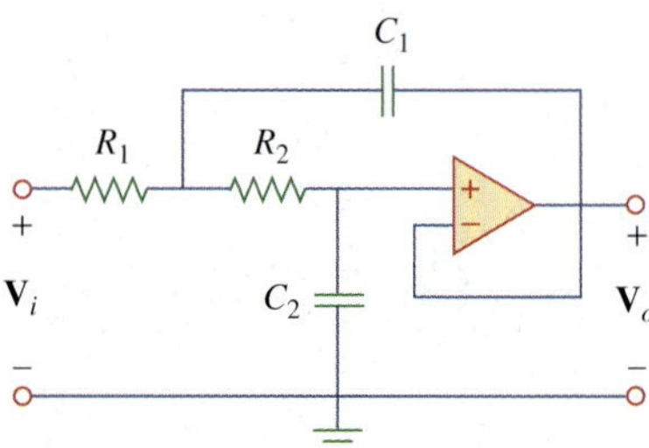

Figure 14.95
For Prob. 14.70.

Section 14.9 Scaling

14.71 Use magnitude and frequency scaling on the circuit of Fig. 14.76 to obtain an equivalent circuit in which the inductor and capacitor have magnitude 1 H and 1 F respectively.

14.72 **e⃗d** Design a problem to help other students better understand magnitude and frequency scaling.

14.73 Calculate the values of R, L, and C that will result in $R = 12$ kΩ, $L = 40$ μH, and $C = 300$ nF respectively when magnitude-scaled by 800 and frequency-scaled by 1000.

14.74 A circuit has $R_1 = 3\ \Omega$, $R_2 = 10\ \Omega$, $L = 2$H, and $C = 1/10$ F. After the circuit is magnitude-scaled by 100 and frequency-scaled by 10^6, find the new values of the circuit elements.

14.75 In an *RLC* circuit, $R = 20\ \Omega$, $L = 4$ H, and $C = 1$ F. The circuit is magnitude-scaled by 10 and frequency-scaled by 10^5. Calculate the new values of the elements.

14.76 Given a parallel *RLC* circuit with $R = 5$ kΩ, $L = 10$ mH, and $C = 20\ \mu$F, if the circuit is magnitude-scaled by $K_m = 500$ and frequency-scaled by $K_f = 10^5$, find the resulting values of *R*, *L*, and *C*.

14.77 A series *RLC* circuit has $R = 10\ \Omega$, $\omega_0 = 40$ rad/s, and $B = 5$ rad/s. Find *L* and *C* when the circuit is scaled:

(a) in magnitude by a factor of 600,

(b) in frequency by a factor of 1,000,

(c) in magnitude by a factor of 400 and in frequency by a factor of 10^5.

14.78 Redesign the circuit in Fig. 14.85 so that all resistive elements are scaled by a factor of 1,000 and all frequency-sensitive elements are frequency-scaled by a factor of 10^4.

***14.79** Refer to the network in Fig. 14.96.

(a) Find $\mathbf{Z}_{in}(s)$.

(b) Scale the elements by $K_m = 10$ and $K_f = 100$. Find $\mathbf{Z}_{in}(s)$ and ω_0.

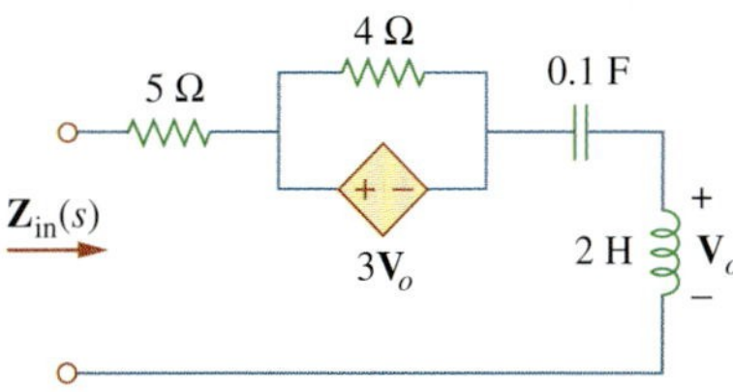

Figure 14.96
For Prob. 14.79.

14.80 (a) For the circuit in Fig. 14.97, draw the new circuit after it has been scaled by $K_m = 200$ and $K_f = 10^4$.

(b) Obtain the Thevenin equivalent impedance at terminals *a-b* of the scaled circuit at $\omega = 10^4$ rad/s.

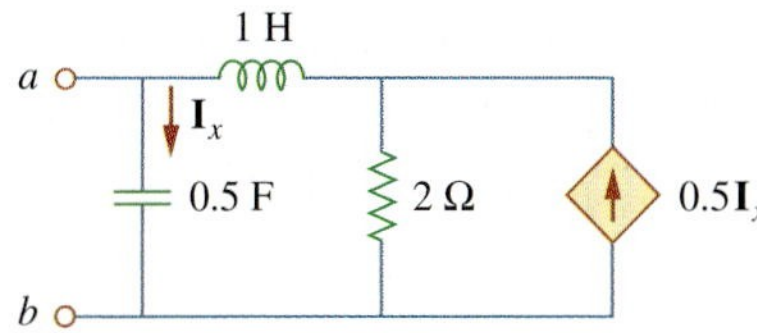

Figure 14.97
For Prob. 14.80.

14.81 The circuit shown in Fig. 14.98 has the impedance

$$Z(s) = \frac{1{,}000(s + 1)}{(s + 1 + j50)(s + 1 - j50)}, \qquad s = j\omega$$

Find:

(a) the values of *R*, *L*, *C*, and *G*

(b) the element values that will raise the resonant frequency by a factor of 10^3 by frequency scaling

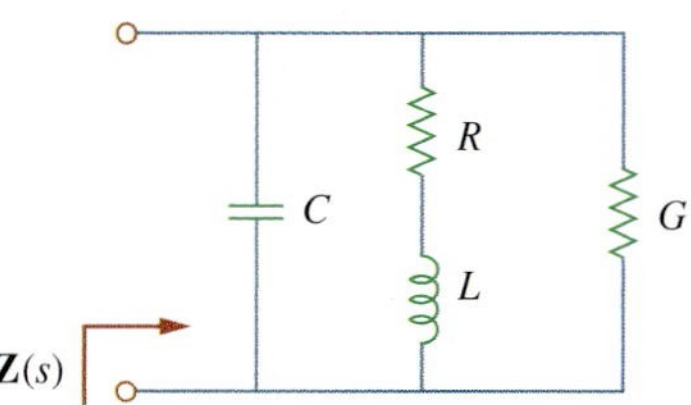

Figure 14.98
For Prob. 14.81.

14.82 Scale the lowpass active filter in Fig. 14.99 so that its corner frequency increases from 1 rad/s to 200 rad/s. Use a 1-μF capacitor.

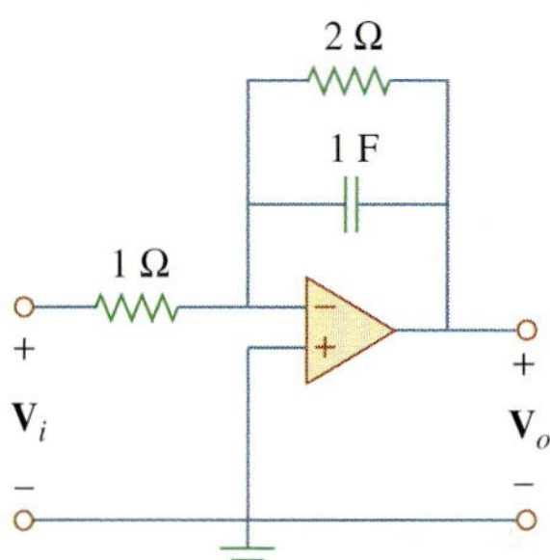

Figure 14.99
For Prob. 14.82.

14.83 The op amp circuit in Fig. 14.100 is to be magnitude-scaled by 100 and frequency-scaled by 10^5. Find the resulting element values.

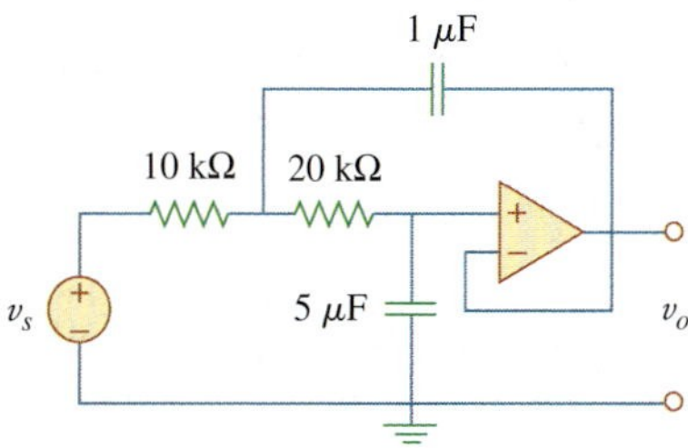

Figure 14.100
For Prob. 14.83.

Section 14.10 Frequency Response Using *PSpice*

14.84 Using *PSpice*, obtain the frequency response of the circuit in Fig. 14.101 on the next page.

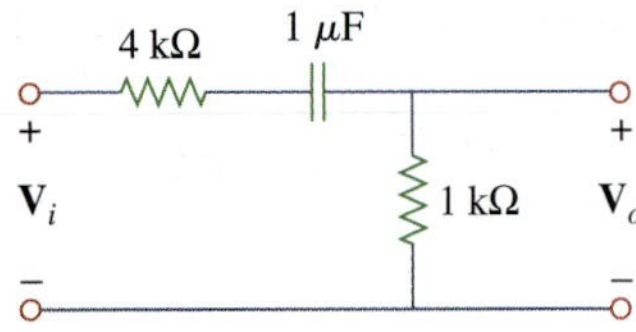

Figure 14.101
For Prob. 14.84.

14.85 Use *PSpice* to obtain the magnitude and phase plots of V_o/I_s of the circuit in Fig. 14.102.

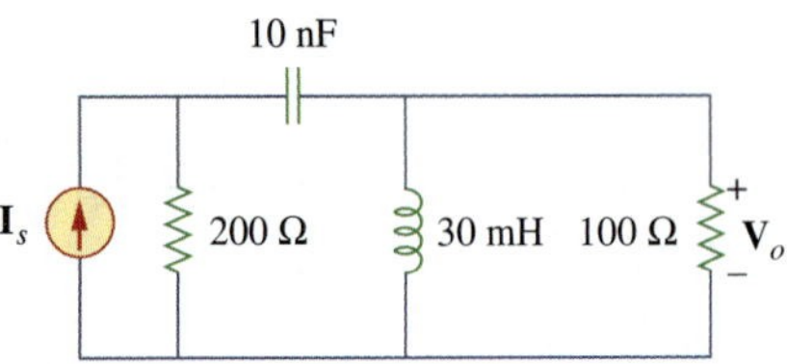

Figure 14.102
For Prob. 14.85.

14.86 Using Fig. 14.103, design a problem to help other students better understand how to use *PSpice* to obtain the frequency response (magnitude and phase of I) in electrical circuits.

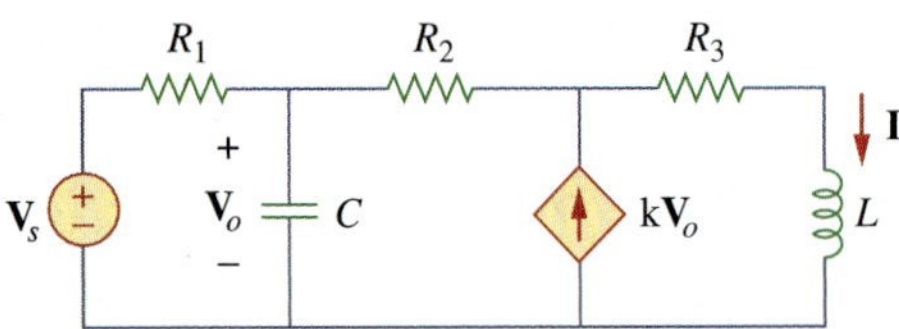

Figure 14.103
For Prob. 14.86.

14.87 In the interval $0.1 < f < 100$ Hz, plot the response of the network in Fig. 14.104. Classify this filter and obtain ω_0.

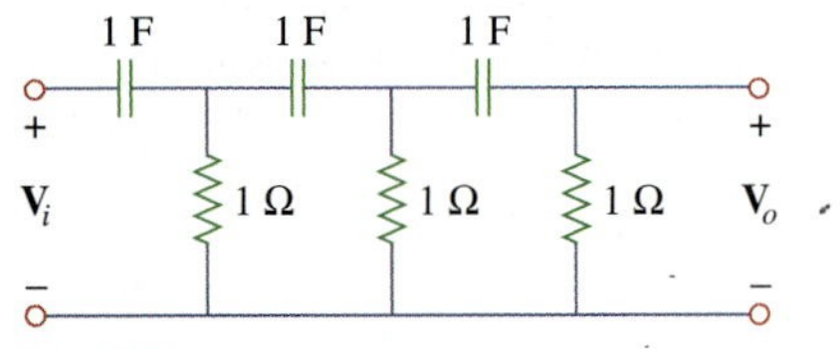

Figure 14.104
For Prob. 14.87.

14.88 Use *PSpice* to generate the magnitude and phase Bode plots of V_o in the circuit of Fig. 14.105.

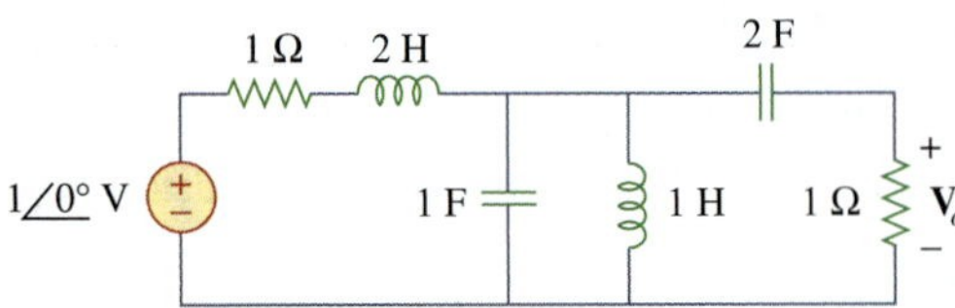

Figure 14.105
For Prob. 14.88.

14.89 Obtain the magnitude plot of the response V_o in the network of Fig. 14.106 for the frequency interval $100 < f < 1{,}000$ Hz.

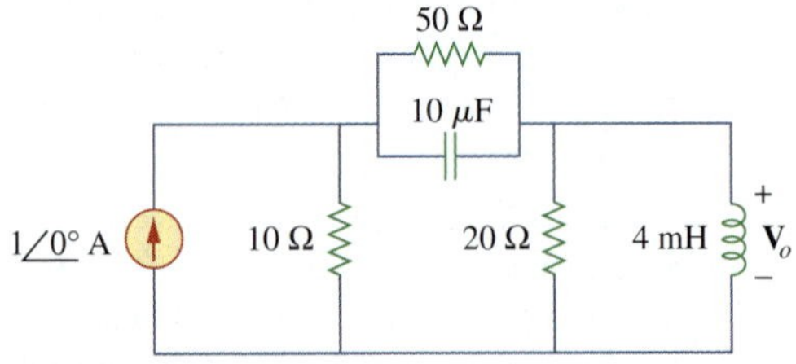

Figure 14.106
For Prob. 14.89.

14.90 Obtain the frequency response of the circuit in Fig. 14.40 (see Practice Problem 14.10). Take $R_1 = R_2 = 100\ \Omega$, $L = 2$ mH. Use $1 < f < 100{,}000$ Hz.

14.91 For the "tank" circuit of Fig. 14.79, obtain the frequency response (voltage across the capacitor) using *PSpice*. Determine the resonant frequency of the circuit.

14.92 Using *PSpice*, plot the magnitude of the frequency response of the circuit in Fig. 14.85.

Section 14.12 Applications

14.93 For the phase shifter circuit shown in Fig. 14.107, find $H = V_o/V_s$.

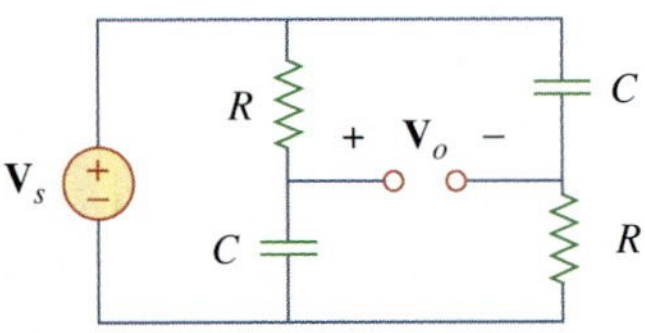

Figure 14.107
For Prob. 14.93.

14.94 For an emergency situation, an engineer needs to make an *RC* highpass filter. He has one 10-pF capacitor, one 30-pF capacitor, one 1.8-kΩ resistor, and one 3.3-kΩ resistor available. Find the greatest cutoff frequency possible using these elements.

14.95 A series-tuned antenna circuit consists of a variable capacitor (40 pF to 360 pF) and a 240-μH antenna coil that has a dc resistance of 12 Ω.

(a) Find the frequency range of radio signals to which the radio is tunable.

(b) Determine the value of *Q* at each end of the frequency range.

14.96 The crossover circuit in Fig. 14.108 is a lowpass filter that is connected to a woofer. Find the transfer function $\mathbf{H}(\omega) = \mathbf{V}_o(\omega)/\mathbf{V}_i(\omega)$.

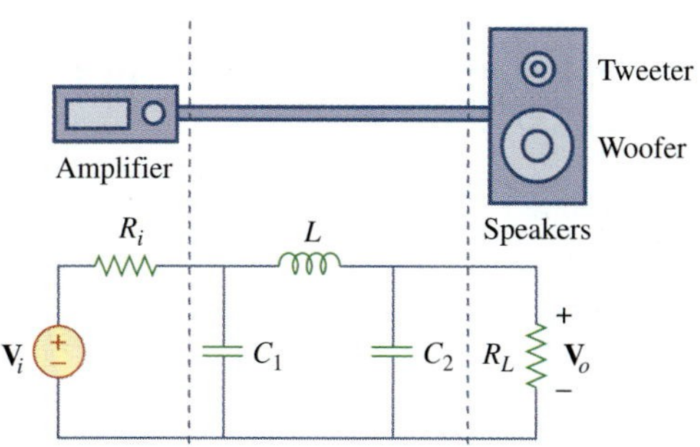

Figure 14.108
For Prob. 14.96.

14.97 The crossover circuit in Fig. 14.109 is a highpass filter that is connected to a tweeter. Determine the transfer function $\mathbf{H}(\omega) = \mathbf{V}_o(\omega)/\mathbf{V}_i(\omega)$.

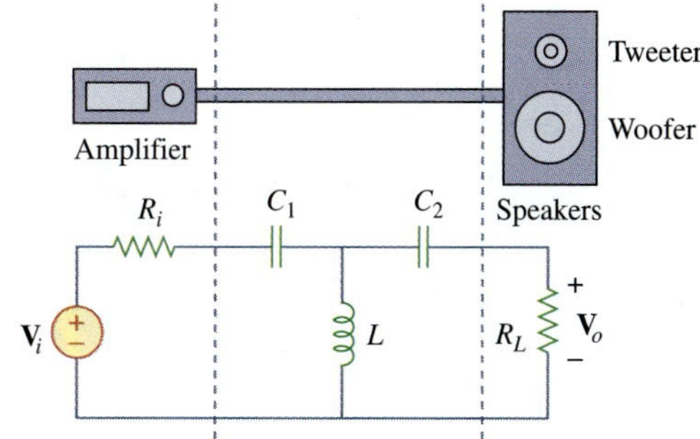

Figure 14.109
For Prob. 14.97.

Comprehensive Problems

14.98 A certain electronic test circuit produced a resonant curve with half-power points at 432 Hz and 454 Hz. If $Q = 20$, what is the resonant frequency of the circuit?

14.99 In an electronic device, a series circuit is employed that has a resistance of 100 Ω, a capacitive reactance of 5 kΩ, and an inductive reactance of 300 Ω when used at 2 MHz. Find the resonant frequency and bandwidth of the circuit.

14.100 In a certain application, a simple *RC* lowpass filter is designed to reduce high frequency noise. If the desired corner frequency is 20 kHz and $C = 0.5\ \mu$F, find the value of *R*.

14.101 In an amplifier circuit, a simple *RC* highpass filter is needed to block the dc component while passing the time-varying component. If the desired rolloff frequency is 15 Hz and $C = 10\ \mu$F, find the value of *R*.

14.102 Practical *RC* filter design should allow for source and load resistances as shown in Fig. 14.110. Let $R = 4$ kΩ and $C = 40$-nF. Obtain the cutoff frequency when:

(a) $R_s = 0$, $R_L = \infty$,

(b) $R_s = 1$ kΩ, $R_L = 5$ kΩ.

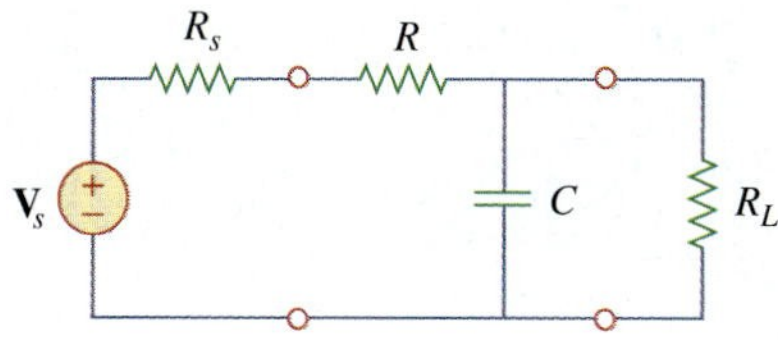

Figure 14.110
For Prob. 14.102.

14.103 The *RC* circuit in Fig. 14.111 is used for a lead compensator in a system design. Obtain the transfer function of the circuit.

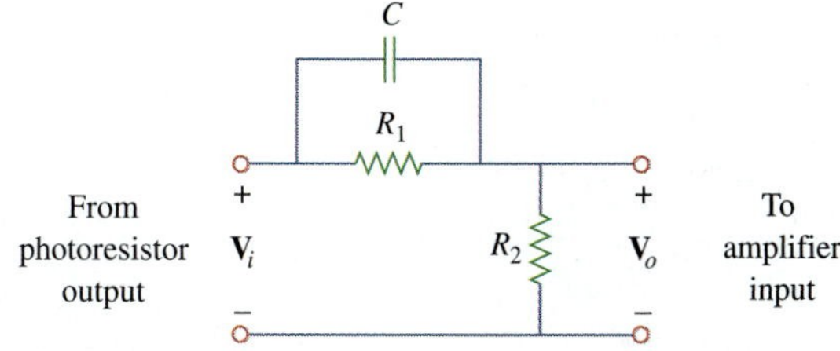

Figure 14.111
For Prob. 14.103.

14.104 A low-quality-factor, double-tuned bandpass filter is shown in Fig. 14.112. Use *PSpice* to generate the magnitude plot of $\mathbf{V}_o(\omega)$.

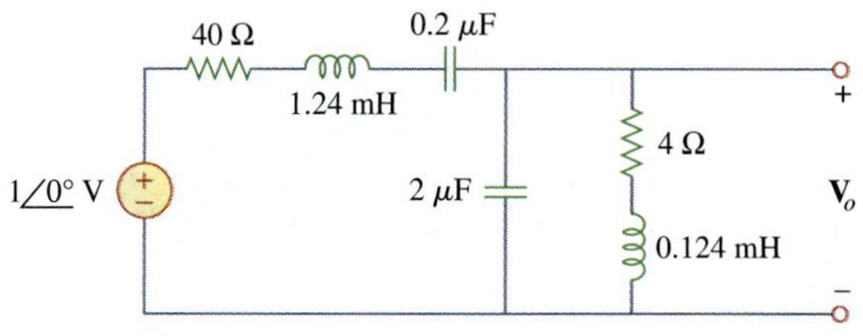

Figure 14.112
For Prob. 14.104.

Advanced Circuit Analysis

OUTLINE

chapter

15

Introduction to the Laplace Transform

The important thing about a problem is not its solution, but the strength we gain in finding the solution.

—Anonymous

Enhancing Your Skills and Your Career

ABET EC 2000 criteria (3.h), *"the broad education necessary to understand the impact of engineering solutions in a global and societal context."*

As a student, you must make sure you acquire "the broad education necessary to understand the impact of engineering solutions in a global and societal context." To some extent, if you are already enrolled in an ABET-accredited engineering program, then some of the courses you are required to take must meet this criteria. My recommendation is that even if you are in such a program, you look at all the elective courses you take to make sure that you expand your awareness of global issues and societal concerns. The engineers of the future must fully understand that they and their activities affect all of us in one way or another.

Photo by Charles Alexander

ABET EC 2000 criteria (3.i), *"need for, and an ability to engage in life-long learning."*

You must be fully aware of and recognize the "need for, and an ability to engage in life-long learning." It almost seems absurd that this need and ability must be stated. Yet, you would be surprised at how many engineers do not really understand this concept. The only way to be really able to keep up with the explosion in technology we are facing now and will be facing in the future is through constant learning. This learning must include nontechnical issues as well as the latest technology in your field.

The best way to keep up with the state of the art in your field is through your colleagues and association with individuals you meet through your technical organization or organizations (especially IEEE). Reading state-of-the-art technical articles is the next best way to stay current.

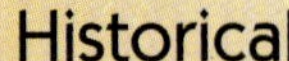

Pierre Simon Laplace (1749–1827), a French astronomer and mathematician, first presented the transform that bears his name and its applications to differential equations in 1779.

Born of humble origins in Beaumont-en-Auge, Normandy, France, Laplace became a professor of mathematics at the age of 20. His mathematical abilities inspired the famous mathematician Simeon Poisson, who called Laplace the Isaac Newton of France. He made important contributions in potential theory, probability theory, astronomy, and celestial mechanics. He was widely known for his work, *Traite de Mecanique Celeste (Celestial Mechanics)*, which supplemented the work of Newton on astronomy. The Laplace transform, the subject of this chapter, is named after him.

15.1 Introduction

Our goal in this and the following chapters is to develop techniques for analyzing circuits with a wide variety of inputs and responses. Such circuits are modeled by *differential equations* whose solutions describe the total response behavior of the circuits. Mathematical methods have been devised to systematically determine the solutions of differential equations. We now introduce the powerful method of *Laplace transformation*, which involves turning differential equations into *algebraic equations*, thus greatly facilitating the solution process.

The idea of transformation should be familiar by now. When using phasors for the analysis of circuits, we transform the circuit from the time domain to the frequency or phasor domain. Once we obtain the phasor result, we transform it back to the time domain. The Laplace transform method follows the same process: we use the Laplace transformation to transform the circuit from the time domain to the frequency domain, obtain the solution, and apply the inverse Laplace transform to the result to transform it back to the time domain.

The Laplace transform is significant for a number of reasons. First, it can be applied to a wider variety of inputs than phasor analysis. Second, it provides an easy way to solve circuit problems involving initial conditions, because it allows us to work with algebraic equations instead of differential equations. Third, the Laplace transform is capable of providing us, in one single operation, the total response of the circuit comprising both the natural and forced responses.

We begin with the definition of the Laplace transform which gives rise to its most essential properties. By examining these properties, we shall see how and why the method works. This also helps us to better appreciate the idea of mathematical transformations. We also consider some properties of the Laplace transform that are very helpful in circuit analysis. We then consider the inverse Laplace transform, transfer functions, and convolution. In this chapter, we will focus on the mechanics of the Laplace transformation. In Chapter 16 we will examine how the Laplace transform is applied in circuit analysis, network stability, and network synthesis.

15.2 Definition of the Laplace Transform

Given a function $f(t)$, its Laplace transform, denoted by $F(s)$ or $\mathcal{L}[f(t)]$, is defined by

$$\mathcal{L}[f(t)] = F(s) = \int_{0^-}^{\infty} f(t)e^{-st}\,dt \tag{15.1}$$

where s is a complex variable given by

$$s = \sigma + j\omega \tag{15.2}$$

Since the argument st of the exponent e in Eq. (15.1) must be dimensionless, it follows that s has the dimensions of frequency and units of inverse seconds (s^{-1}) or "frequency." In Eq. (15.1), the lower limit is specified as 0^- to indicate a time just before $t = 0$. We use 0^- as the lower limit to include the origin and capture any discontinuity of $f(t)$ at $t = 0$; this will accommodate functions—such as singularity functions—that may be discontinuous at $t = 0$.

For an ordinary function $f(t)$, the lower limit can be replaced by 0.

It should be noted that the integral in Eq. (15.1) is a definite integral with respect to time. Hence, the result of integration is independent of time and only involves the variable "s."

Equation (15.1) illustrates the general concept of transformation. The function $f(t)$ is transformed into the function $F(s)$. Whereas the former function involves t as its argument, the latter involves s. We say the transformation is from t-domain to s-domain. Given the interpretation of s as frequency, we arrive at the following description of the Laplace transform:

> The **Laplace transform** is an integral transformation of a function $f(t)$ from the time domain into the complex frequency domain, giving $F(s)$.

When the Laplace transform is applied to circuit analysis, the differential equations represent the circuit in the time domain. The terms in the differential equations take the place of $f(t)$. Their Laplace transform, which corresponds to $F(s)$, constitutes algebraic equations representing the circuit in the frequency domain.

We assume in Eq. (15.1) that $f(t)$ is ignored for $t < 0$. To ensure that this is the case, a function is often multiplied by the unit step. Thus, $f(t)$ is written as $f(t)u(t)$ or $f(t)$, $t \geq 0$.

The Laplace transform in Eq. (15.1) is known as the *one-sided* (or *unilateral*) Laplace transform. The *two-sided* (or *bilateral*) Laplace transform is given by

$$F(s) = \int_{-\infty}^{\infty} f(t)e^{-st}\,dt \tag{15.3}$$

The one-sided Laplace transform in Eq. (15.1), being adequate for our purposes, is the only type of Laplace transform that we will treat in this book.

$|e^{j\omega t}| = \sqrt{\cos^2 \omega t + \sin^2 \omega t} = 1$

A function $f(t)$ may not have a Laplace transform. In order for $f(t)$ to have a Laplace transform, the integral in Eq. (15.1) must converge to a finite value. Since $|e^{j\omega t}| = 1$ for any value of t, the integral converges when

$$\int_{0^-}^{\infty} e^{-\sigma t}|f(t)|\,dt < \infty \tag{15.4}$$

for some real value $\sigma = \sigma_c$. Thus, the region of convergence for the Laplace transform is $\text{Re}(s) = \sigma > \sigma_c$, as shown in Fig. 15.1. In this region, $|F(s)| < \infty$ and $F(s)$ exists. $F(s)$ is undefined outside the region of convergence. Fortunately, all functions of interest in circuit analysis satisfy the convergence criterion in Eq. (15.4) and have Laplace transforms. Therefore, it is not necessary to specify σ_c in what follows.

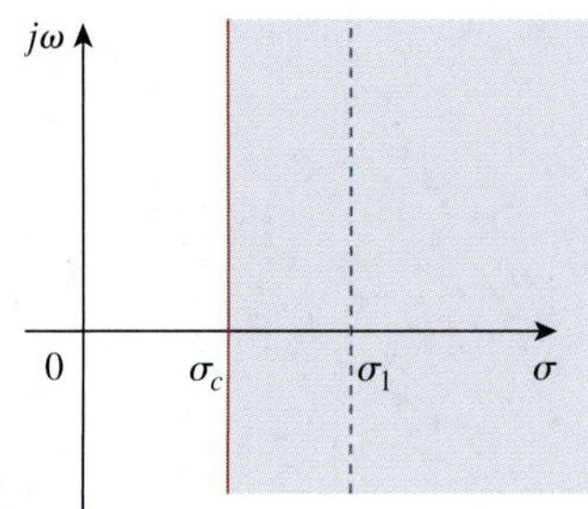

Figure 15.1
Region of convergence for the Laplace transform.

A companion to the direct Laplace transform in Eq. (15.1) is the *inverse* Laplace transform given by

$$\mathcal{L}^{-1}[F(s)] = f(t) = \frac{1}{2\pi j}\int_{\sigma_1 - j\infty}^{\sigma_1 + j\infty} F(s)e^{st}\,ds \tag{15.5}$$

where the integration is performed along a straight line $(\sigma_1 + j\omega, -\infty < \omega < \infty)$ in the region of convergence, $\sigma_1 > \sigma_c$. See Fig. 15.1. The direct application of Eq. (15.5) involves some knowledge about complex analysis beyond the scope of this book. For this reason, we will not use Eq. (15.5) to find the inverse Laplace transform. We will rather use a look-up table, to be developed in Section 15.3. The functions $f(t)$ and $F(s)$ are regarded as a Laplace transform pair where

$$f(t) \quad \Leftrightarrow \quad F(s) \tag{15.6}$$

meaning that there is one-to-one correspondence between $f(t)$ and $F(s)$. The following examples derive the Laplace transforms of some important functions.

Example 15.1

Determine the Laplace transform of each of the following functions: (a) $u(t)$, (b) $e^{-at}u(t)$, $a \geq 0$, and (c) $\delta(t)$.

Solution:

(a) For the unit step function $u(t)$, shown in Fig. 15.2(a), the Laplace transform is

$$\begin{aligned}\mathcal{L}[u(t)] &= \int_{0^-}^{\infty} 1e^{-st}\,dt = -\frac{1}{s}e^{-st}\Big|_0^{\infty} \\ &= -\frac{1}{s}(0) + \frac{1}{s}(1) = \frac{1}{s}\end{aligned} \tag{15.1.1}$$

(b) For the exponential function, shown in Fig. 15.2(b), the Laplace transform is

$$\begin{aligned}\mathcal{L}[e^{-at}u(t)] &= \int_{0^-}^{\infty} e^{-at}e^{-st}\,dt \\ &= -\frac{1}{s+a}e^{-(s+a)t}\Big|_0^{\infty} = \frac{1}{s+a}\end{aligned} \tag{15.1.2}$$

(c) For the unit impulse function, shown in Fig. 15.2(c),

$$\mathcal{L}[\delta(t)] = \int_{0^-}^{\infty} \delta(t)e^{-st}\,dt = e^{-0} = 1 \qquad \textbf{(15.1.3)}$$

since the impulse function $\delta(t)$ is zero everywhere except at $t = 0$. The sifting property in Eq. (7.33) has been applied in Eq. (15.1.3).

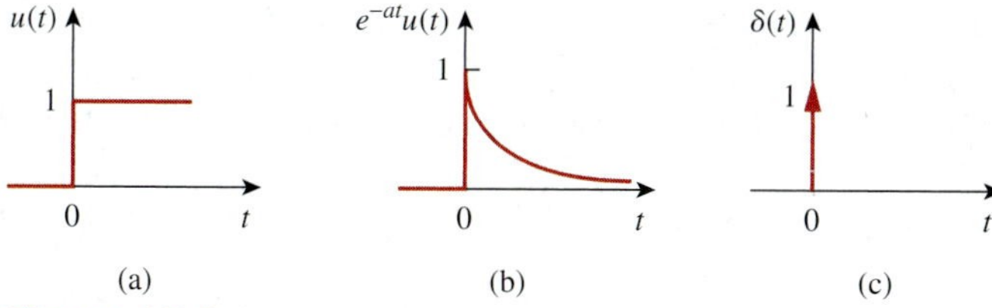

Figure 15.2
For Example 15.1: (a) unit step function, (b) exponential function, (c) unit impulse function.

Practice Problem 15.1

Find the Laplace transforms of these functions: $r(t) = tu(t)$, that is, the ramp function; $e^{-at}u(t)$; and $e^{-j\omega t}u(t)$.

Answer: $1/s^2$, $1/(s + a)$, $1/(s + j\omega)$.

Example 15.2

Determine the Laplace transform of $f(t) = \sin\omega t\, u(t)$.

Solution:
Using Eq. (B.27) in addition to Eq. (15.1), we obtain the Laplace transform of the sine function as

$$\begin{aligned} F(s) = \mathcal{L}[\sin\omega t] &= \int_0^{\infty} (\sin\omega t)e^{-st}\,dt = \int_0^{\infty} \left(\frac{e^{j\omega t} - e^{-j\omega t}}{2j}\right)e^{-st}\,dt \\ &= \frac{1}{2j}\int_0^{\infty} (e^{-(s-j\omega)t} - e^{-(s+j\omega)t})\,dt \\ &= \frac{1}{2j}\left(\frac{1}{s - j\omega} - \frac{1}{s + j\omega}\right) = \frac{\omega}{s^2 + \omega^2} \end{aligned}$$

Practice Problem 15.2

Find the Laplace transform of $f(t) = 10\cos\omega t\, u(t)$.

Answer: $10s/(s^2 + \omega^2)$.

15.3 Properties of the Laplace Transform

The properties of the Laplace transform help us to obtain transform pairs without directly using Eq. (15.1) as we did in Examples 15.1 and 15.2. As we derive each of these properties, we should keep in mind the definition of the Laplace transform in Eq. (15.1).

Linearity

If $F_1(s)$ and $F_2(s)$ are, respectively, the Laplace transforms of $f_1(t)$ and $f_2(t)$, then

$$\mathcal{L}[a_1 f_1(t) + a_2 f_2(t)] = a_1 F_1(s) + a_2 F_2(s) \tag{15.7}$$

where a_1 and a_2 are constants. Equation 15.7 expresses the linearity property of the Laplace transform. The proof of Eq. (15.7) follows readily from the definition of the Laplace transform in Eq. (15.1).

For example, by the linearity property in Eq. (15.7), we may write

$$\mathcal{L}[\cos\omega t\, u(t)] = \mathcal{L}\left[\frac{1}{2}(e^{j\omega t} + e^{-j\omega t})\right] = \frac{1}{2}\mathcal{L}[e^{j\omega t}] + \frac{1}{2}\mathcal{L}[e^{-j\omega t}] \tag{15.8}$$

But from Example 15.1(b), $\mathcal{L}[e^{-at}] = 1/(s + a)$. Hence,

$$\mathcal{L}[\cos\omega t\, u(t)] = \frac{1}{2}\left(\frac{1}{s - j\omega} + \frac{1}{s + j\omega}\right) = \frac{s}{s^2 + \omega^2} \tag{15.9}$$

Scaling

If $F(s)$ is the Laplace transform of $f(t)$, then

$$\mathcal{L}[f(at)] = \int_{0^-}^{\infty} f(at)e^{-st}\, dt \tag{15.10}$$

where a is a constant and $a > 0$. If we let $x = at$, $dx = a\, dt$, then

$$\mathcal{L}[f(at)] = \int_{0^-}^{\infty} f(x)e^{-x(s/a)}\frac{dx}{a} = \frac{1}{a}\int_{0^-}^{\infty} f(x)e^{-x(s/a)}\, dx \tag{15.11}$$

Comparing this integral with the definition of the Laplace transform in Eq. (15.1) shows that s in Eq. (15.1) must be replaced by s/a while the dummy variable t is replaced by x. Hence, we obtain the scaling property as

$$\mathcal{L}[f(at)] = \frac{1}{a}F\left(\frac{s}{a}\right) \tag{15.12}$$

For example, we know from Example 15.2 that

$$\mathcal{L}[\sin\omega t\, u(t)] = \frac{\omega}{s^2 + \omega^2} \tag{15.13}$$

Using the scaling property in Eq. (15.12),

$$\mathcal{L}[\sin 2\omega t\, u(t)] = \frac{1}{2}\frac{\omega}{(s/2)^2 + \omega^2} = \frac{2\omega}{s^2 + 4\omega^2} \tag{15.14}$$

which may also be obtained from Eq. (15.13) by replacing ω with 2ω.

Time Shift

If $F(s)$ is the Laplace transform of $f(t)$, then

$$\mathcal{L}[f(t - a)u(t - a)] = \int_{0^-}^{\infty} f(t - a)u(t - a)e^{-st}\, dt \qquad a \ge 0 \tag{15.15}$$

But $u(t-a) = 0$ for $t < a$ and $u(t-a) = 1$ for $t > a$. Hence,

$$\mathcal{L}[f(t-a)u(t-a)] = \int_a^\infty f(t-a)e^{-st}\,dt \tag{15.16}$$

If we let $x = t - a$, then $dx = dt$ and $t = x + a$. As $t \to a$, $x \to 0$ and as $t \to \infty$, $x \to \infty$. Thus,

$$\mathcal{L}[f(t-a)u(t-a)] = \int_{0^-}^\infty f(x)e^{-s(x+a)}\,dx$$

$$= e^{-as}\int_{0^-}^\infty f(x)e^{-sx}\,dx = e^{-as}F(s)$$

or

$$\mathcal{L}[f(t-a)u(t-a)] = e^{-as}F(s) \tag{15.17}$$

In other words, if a function is delayed in time by a, the result in the s-domain is found by multiplying the Laplace transform of the function (without the delay) by e^{-as}. This is called the *time-delay* or *time-shift property* of the Laplace transform.

As an example, we know from Eq. (15.9) that

$$\mathcal{L}[\cos\omega t\,u(t)] = \frac{s}{s^2+\omega^2}$$

Using the time-shift property in Eq. (15.17),

$$\mathcal{L}[\cos\omega(t-a)u(t-a)] = e^{-as}\frac{s}{s^2+\omega^2} \tag{15.18}$$

Frequency Shift

If $F(s)$ is the Laplace transform of $f(t)$, then

$$\mathcal{L}[e^{-at}f(t)u(t)] = \int_0^\infty e^{-at}f(t)e^{-st}\,dt$$

$$= \int_0^\infty f(t)e^{-(s+a)t}\,dt = F(s+a)$$

or

$$\mathcal{L}[e^{-at}f(t)u(t)] = F(s+a) \tag{15.19}$$

That is, the Laplace transform of $e^{-at}f(t)$ can be obtained from the Laplace transform of $f(t)$ by replacing every s with $s + a$. This is known as *frequency shift* or *frequency translation.*

As an example, we know that

$$\cos\omega t\,u(t) \quad \Leftrightarrow \quad \frac{s}{s^2+\omega^2}$$

and (15.20)

$$\sin\omega t\,u(t) \quad \Leftrightarrow \quad \frac{\omega}{s^2+\omega^2}$$

Using the shift property in Eq. (15.19), we obtain the Laplace transform of the damped sine and damped cosine functions as

$$\mathcal{L}[e^{-at}\cos\omega t u(t)] = \frac{s+a}{(s+a)^2+\omega^2} \tag{15.21a}$$

$$\mathcal{L}[e^{-at}\sin\omega t u(t)] = \frac{\omega}{(s+a)^2+\omega^2} \tag{15.21b}$$

Time Differentiation

Given that $F(s)$ is the Laplace transform of $f(t)$, the Laplace transform of its derivative is

$$\mathcal{L}\left[\frac{df}{dt}u(t)\right] = \int_{0^-}^{\infty} \frac{df}{dt}e^{-st}\,dt \tag{15.22}$$

To integrate this by parts, we let $u = e^{-st}$, $du = -se^{-st}\,dt$, and $dv = (df/dt)\,dt = df(t)$, $v = f(t)$. Then

$$\begin{aligned}\mathcal{L}\left[\frac{df}{dt}u(t)\right] &= f(t)e^{-st}\Big|_{0^-}^{\infty} - \int_{0^-}^{\infty} f(t)[-se^{-st}]\,dt \\ &= 0 - f(0^-) + s\int_{0^-}^{\infty} f(t)e^{-st}\,dt = sF(s) - f(0^-)\end{aligned}$$

or

$$\boxed{\mathcal{L}[f'(t)] = sF(s) - f(0^-)} \tag{15.23}$$

The Laplace transform of the second derivative of $f(t)$ is a repeated application of Eq. (15.23) as

$$\begin{aligned}\mathcal{L}\left[\frac{d^2f}{dt^2}\right] &= s\mathcal{L}[f'(t)] - f'(0^-) = s[sF(s) - f(0^-)] - f'(0^-) \\ &= s^2F(s) - sf(0^-) - f'(0^-)\end{aligned}$$

or

$$\boxed{\mathcal{L}[f''(t)] = s^2F(s) - sf(0^-) - f'(0^-)} \tag{15.24}$$

Continuing in this manner, we can obtain the Laplace transform of the nth derivative of $f(t)$ as

$$\boxed{\begin{aligned}\mathcal{L}\left[\frac{d^nf}{dt^n}\right] &= s^nF(s) - s^{n-1}f(0^-) \\ &\quad - s^{n-2}f'(0^-) - \cdots - s^0 f^{(n-1)}(0^-)\end{aligned}} \tag{15.25}$$

As an example, we can use Eq. (15.23) to obtain the Laplace transform of the sine from that of the cosine. If we let $f(t) = \cos\omega t u(t)$, then $f(0) = 1$ and $f'(t) = -\omega\sin\omega t u(t)$. Using Eq. (15.23) and the scaling property,

$$\begin{aligned}\mathcal{L}[\sin\omega t u(t)] &= -\frac{1}{\omega}\mathcal{L}[f'(t)] = -\frac{1}{\omega}[sF(s) - f(0^-)] \\ &= -\frac{1}{\omega}\left(s\frac{s}{s^2+\omega^2} - 1\right) = \frac{\omega}{s^2+\omega^2}\end{aligned} \tag{15.26}$$

as expected.

Time Integration

If $F(s)$ is the Laplace transform of $f(t)$, the Laplace transform of its integral is

$$\mathcal{L}\left[\int_0^t f(t)\,dt\right] = \int_{0^-}^{\infty}\left[\int_0^t f(x)\,dx\right]e^{-st}\,dt \tag{15.27}$$

To integrate this by parts, we let

$$u = \int_0^t f(x)\,dx, \qquad du = f(t)\,dt$$

and

$$dv = e^{-st}\,dt, \qquad v = -\frac{1}{s}e^{-st}$$

Then

$$\mathcal{L}\left[\int_0^t f(t)\,dt\right] = \left[\int_0^t f(x)\,dx\right]\left(-\frac{1}{s}e^{-st}\right)\Bigg|_{0^-}^{\infty} - \int_{0^-}^{\infty}\left(-\frac{1}{s}\right)e^{-st}f(t)\,dt$$

For the first term on the right-hand side of the equation, evaluating the term at $t = \infty$ yields zero due to $e^{-s\infty}$ and evaluating it at $t = 0$ gives $\frac{1}{s}\int_0^0 f(x)\,dx = 0$. Thus, the first term is zero, and

$$\mathcal{L}\left[\int_0^t f(t)\,dt\right] = \frac{1}{s}\int_{0^-}^{\infty} f(t)e^{-st}\,dt = \frac{1}{s}F(s)$$

or simply,

$$\boxed{\mathcal{L}\left[\int_0^t f(t)\,dt\right] = \frac{1}{s}F(s)} \tag{15.28}$$

As an example, if we let $f(t) = u(t)$, from Example 15.1(a), $F(s) = 1/s$. Using Eq. (15.28),

$$\mathcal{L}\left[\int_0^t f(t)\,dt\right] = \mathcal{L}[t] = \frac{1}{s}\left(\frac{1}{s}\right)$$

Thus, the Laplace transform of the ramp function is

$$\mathcal{L}[t] = \frac{1}{s^2} \tag{15.29}$$

Applying Eq. (15.28), this gives

$$\mathcal{L}\left[\int_0^t t\,dt\right] = \mathcal{L}\left[\frac{t^2}{2}\right] = \frac{1}{s}\frac{1}{s^2}$$

or

$$\mathcal{L}[t^2] = \frac{2}{s^3} \tag{15.30}$$

TABLE 15.1

Properties of the Laplace transform.

Property	$f(t)$	$F(s)$
Linearity	$a_1f_1(t) + a_2f_2(t)$	$a_1F_1(s) + a_2F_2(s)$
Scaling	$f(at)$	$\frac{1}{a}F\left(\frac{s}{a}\right)$
Time shift	$f(t-a)u(t-a)$	$e^{-as}F(s)$
Frequency shift	$e^{-at}f(t)$	$F(s+a)$
Time differentiation	$\frac{df}{dt}$	$sF(s) - f(0^-)$
	$\frac{d^2f}{dt^2}$	$s^2F(s) - sf(0^-) - f'(0^-)$
	$\frac{d^3f}{dt^3}$	$s^3F(s) - s^2f(0^-) - sf'(0^-) - f''(0^-)$
	$\frac{d^nf}{dt^n}$	$s^nF(s) - s^{n-1}f(0^-) - s^{n-2}f'(0^-) - \cdots - f^{(n-1)}(0^-)$
Time integration	$\int_0^t f(t)dt$	$\frac{1}{s}F(s)$
Frequency differentiation	$tf(t)$	$-\frac{d}{ds}F(s)$
Frequency integration	$\frac{f(t)}{t}$	$\int_s^\infty F(s)ds$
Time periodicity	$f(t) = f(t+nT)$	$\frac{F_1(s)}{1-e^{-sT}}$
Initial value	$f(0)$	$\lim_{s\to\infty} sF(s)$
Final value	$f(\infty)$	$\lim_{s\to 0} sF(s)$
Convolution	$f_1(t) * f_2(t)$	$F_1(s)F_2(s)$

TABLE 15.2

Laplace transform pairs.*

$f(t)$	$F(s)$
$\delta(t)$	1
$u(t)$	$\frac{1}{s}$
e^{-at}	$\frac{1}{s+a}$
t	$\frac{1}{s^2}$
t^n	$\frac{n!}{s^{n+1}}$
te^{-at}	$\frac{1}{(s+a)^2}$
t^ne^{-at}	$\frac{n!}{(s+a)^{n+1}}$
$\sin \omega t$	$\frac{\omega}{s^2+\omega^2}$
$\cos \omega t$	$\frac{s}{s^2+\omega^2}$
$\sin(\omega t + \theta)$	$\frac{s\sin\theta + \omega\cos\theta}{s^2+\omega^2}$
$\cos(\omega t + \theta)$	$\frac{s\cos\theta - \omega\sin\theta}{s^2+\omega^2}$
$e^{-at}\sin \omega t$	$\frac{\omega}{(s+a)^2+\omega^2}$
$e^{-at}\cos \omega t$	$\frac{s+a}{(s+a)^2+\omega^2}$

*Defined for $t \geq 0$; $f(t) = 0$, for $t < 0$.

Example 15.3

Obtain the Laplace transform of $f(t) = \delta(t) + 2u(t) - 3e^{-2t}u(t)$.

Solution:

By the linearity property,

$$F(s) = \mathcal{L}[\delta(t)] + 2\mathcal{L}[u(t)] - 3\mathcal{L}[e^{-2t}u(t)]$$
$$= 1 + 2\frac{1}{s} - 3\frac{1}{s+2} = \frac{s^2+s+4}{s(s+2)}$$

Practice Problem 15.3

Find the Laplace transform of $f(t) = (\cos(3t) + e^{-5t})u(t)$.

Answer: $\frac{2s^2+5s+9}{(s+5)(s^2+9)}$.

Example 15.4

Determine the Laplace transform of $f(t) = t^2 \sin 2t\, u(t)$.

Solution:

We know that

$$\mathcal{L}[\sin 2t] = \frac{2}{s^2 + 2^2}$$

Using frequency differentiation in Eq. (15.34),

$$F(s) = \mathcal{L}[t^2 \sin 2t] = (-1)^2 \frac{d^2}{ds^2}\left(\frac{2}{s^2 + 4}\right)$$

$$= \frac{d}{ds}\left(\frac{-4s}{(s^2 + 4)^2}\right) = \frac{12s^2 - 16}{(s^2 + 4)^3}$$

Practice Problem 15.4

Find the Laplace transform of $f(t) = t^2 \cos 3t\, u(t)$.

Answer: $\dfrac{2s(s^2 - 27)}{(s^2 + 9)^3}$.

Example 15.5

Find the Laplace transform of the gate function in Fig. 15.5.

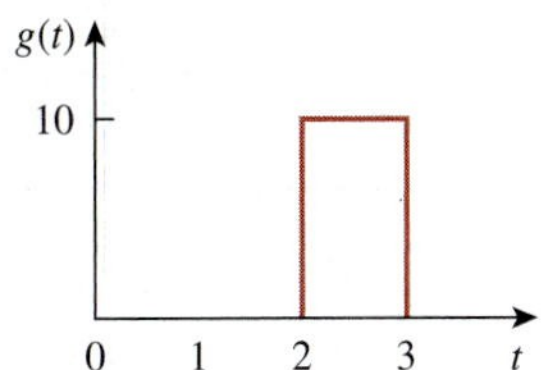

Figure 15.5
The gate function; for Example 15.5.

Solution:

We can express the gate function in Fig. 15.5 as

$$g(t) = 10[u(t - 2) - u(t - 3)]$$

Since we know the Laplace transform of $u(t)$, we apply the time-shift property and obtain

$$G(s) = 10\left(\frac{e^{-2s}}{s} - \frac{e^{-3s}}{s}\right) = \frac{10}{s}(e^{-2s} - e^{-3s})$$

Practice Problem 15.5

Find the Laplace transform of the function $h(t)$ in Fig. 15.6.

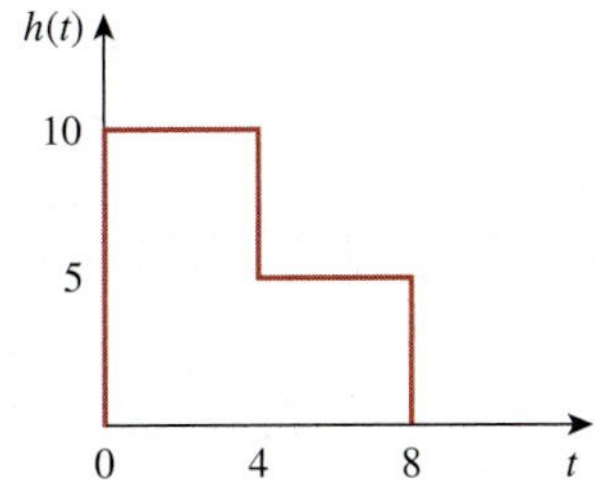

Figure 15.6
For Practice Prob. 15.5.

Answer: $\dfrac{5}{s}(2 - e^{-4s} - e^{-8s})$.

Example 15.6

Calculate the Laplace transform of the periodic function in Fig. 15.7.

Solution:
The period of the function is $T = 2$. To apply Eq. (15.40), we first obtain the transform of the first period of the function.

$$\begin{aligned} f_1(t) &= 2t[u(t) - u(t-1)] = 2tu(t) - 2tu(t-1) \\ &= 2tu(t) - 2(t-1+1)u(t-1) \\ &= 2tu(t) - 2(t-1)u(t-1) - 2u(t-1) \end{aligned}$$

Using the time-shift property,

$$F_1(s) = \frac{2}{s^2} - 2\frac{e^{-s}}{s^2} - \frac{2}{s}e^{-s} = \frac{2}{s^2}(1 - e^{-s} - se^{-s})$$

Thus, the transform of the periodic function in Fig. 15.7 is

$$F(s) = \frac{F_1(s)}{1 - e^{-Ts}} = \frac{2}{s^2(1 - e^{-2s})}(1 - e^{-s} - se^{-s})$$

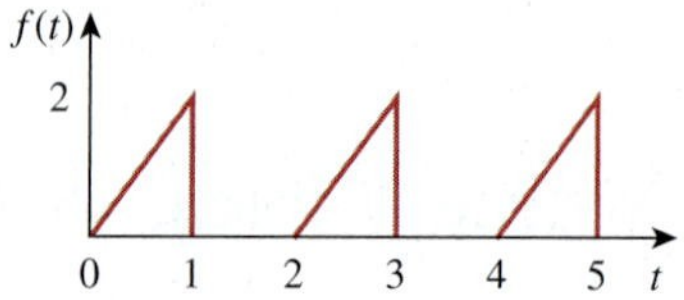

Figure 15.7
For Example 15.6.

Practice Problem 15.6

Determine the Laplace transform of the periodic function in Fig. 15.8.

Answer: $\dfrac{1 - e^{-2s}}{s(1 - e^{-5s})}$.

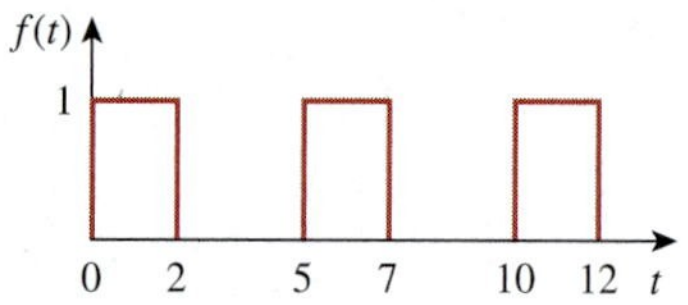

Figure 15.8
For Practice Prob. 15.6.

Example 15.7

Find the initial and final values of the function whose Laplace transform is

$$H(s) = \frac{20}{(s+3)(s^2 + 8s + 25)}$$

Solution:
Applying the initial-value theorem,

$$\begin{aligned} h(0) &= \lim_{s\to\infty} sH(s) = \lim_{s\to\infty} \frac{20s}{(s+3)(s^2+8s+25)} \\ &= \lim_{s\to\infty} \frac{20/s^2}{(1+3/s)(1+8/s+25/s^2)} = \frac{0}{(1+0)(1+0+0)} = 0 \end{aligned}$$

To be sure that the final-value theorem is applicable, we check where the poles of $H(s)$ are located. The poles of $H(s)$ are $s = -3, -4 \pm j3$, which all have negative real parts: they are all located on the left half of the s plane (Fig. 15.9). Hence, the final-value theorem applies and

$$\begin{aligned} h(\infty) &= \lim_{s\to 0} sH(s) = \lim_{s\to 0} \frac{20s}{(s+3)(s^2+8s+25)} \\ &= \frac{0}{(0+3)(0+0+25)} = 0 \end{aligned}$$

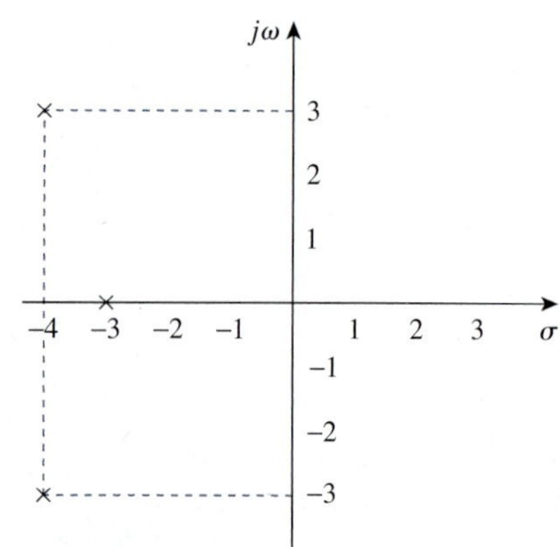

Figure 15.9
For Example 15.7: Poles of $H(s)$.

Both the initial and final values could be determined from $h(t)$ if we knew it. See Example 15.11, where $h(t)$ is given.

Practice Problem 15.7

Obtain the initial and the final values of

$$G(s) = \frac{3s^3 + 2s + 10}{s(s+2)^2(s+3)}$$

Answer: 3, 0.8333.

15.4 The Inverse Laplace Transform

Given $F(s)$, how do we transform it back to the time domain and obtain the corresponding $f(t)$? By matching entries in Table 15.2, we avoid using Eq. (15.5) to find $f(t)$.

Suppose $F(s)$ has the general form of

$$F(s) = \frac{N(s)}{D(s)} \tag{15.47}$$

where $N(s)$ is the numerator polynomial and $D(s)$ is the denominator polynomial. The roots of $N(s) = 0$ are called the *zeros* of $F(s)$, while the roots of $D(s) = 0$ are the *poles* of $F(s)$. Although Eq. (15.47) is similar in form to Eq. (14.3), here $F(s)$ is the Laplace transform of a function, which is not necessarily a transfer function. We use *partial fraction expansion* to break $F(s)$ down into simple terms whose inverse transform we obtain from Table 15.2. Thus, finding the inverse Laplace transform of $F(s)$ involves two steps.

Software packages such as *MATLAB*, Mathcad, and Maple are capable of finding partial fraction expansions quite easily.

Steps to Find the Inverse Laplace Transform:

1. Decompose $F(s)$ into simple terms using partial fraction expansion.
2. Find the inverse of each term by matching entries in Table 15.2.

Let us consider the three possible forms $F(s)$ may take and how to apply the two steps to each form.

15.4.1 Simple Poles

Recall from Chapter 14 that a simple pole is a first-order pole. If $F(s)$ has only simple poles, then $D(s)$ becomes a product of factors, so that

$$F(s) = \frac{N(s)}{(s+p_1)(s+p_2)\cdots(s+p_n)} \tag{15.48}$$

where $s = -p_1, -p_2, \ldots, -p_n$ are the simple poles, and $p_i \neq p_j$ for all $i \neq j$ (i.e., the poles are distinct). Assuming that the degree of $N(s)$ is

Otherwise, we must first apply long division so that $F(s) = N(s)/D(s) = Q(s) + R(s)/D(s)$, where the degree of $R(s)$, the remainder of the long division, is less than the degree of $D(s)$.

less than the degree of $D(s)$, we use partial fraction expansion to decompose $F(s)$ in Eq. (15.48) as

$$F(s) = \frac{k_1}{s + p_1} + \frac{k_2}{s + p_2} + \cdots + \frac{k_n}{s + p_n} \tag{15.49}$$

The expansion coefficients $k_1, k_2, \ldots, k_n$ are known as the *residues* of $F(s)$. There are many ways of finding the expansion coefficients. One way is using the *residue method.* If we multiply both sides of Eq. (15.49) by $(s + p_1)$, we obtain

$$(s + p_1)F(s) = k_1 + \frac{(s + p_1)k_2}{s + p_2} + \cdots + \frac{(s + p_1)k_n}{s + p_n} \tag{15.50}$$

Since $p_i \neq p_j$, setting $s = -p_1$ in Eq. (15.50) leaves only k_1 on the right-hand side of Eq. (15.50). Hence,

$$(s + p_1)F(s)\,|_{s=-p_1} = k_1 \tag{15.51}$$

Thus, in general,

$$\boxed{k_i = (s + p_i)F(s)\,|_{s=-p_i}} \tag{15.52}$$

This is known as *Heaviside's theorem*. Once the values of k_i are known, we proceed to find the inverse of $F(s)$ using Eq. (15.49). Since the inverse transform of each term in Eq. (15.49) is $\mathcal{L}^{-1}[k/(s + a)] = ke^{-at}\,u(t)$, then, from Table 15.2,

Historical note: Named after Oliver Heaviside (1850–1925), an English engineer, the pioneer of operational calculus.

$$\boxed{f(t) = (k_1e^{-p_1t} + k_2e^{-p_2t} + \cdots + k_ne^{-p_nt})u(t)} \tag{15.53}$$

15.4.2 Repeated Poles

Suppose $F(s)$ has n repeated poles at $s = -p$. Then we may represent $F(s)$ as

$$\begin{aligned} F(s) = {} & \frac{k_n}{(s + p)^n} + \frac{k_{n-1}}{(s + p)^{n-1}} + \cdots + \frac{k_2}{(s + p)^2} \\ & + \frac{k_1}{s + p} + F_1(s) \end{aligned} \tag{15.54}$$

where $F_1(s)$ is the remaining part of $F(s)$ that does not have a pole at $s = -p$. We determine the expansion coefficient k_n as

$$k_n = (s + p)^n F(s)\,|_{s=-p} \tag{15.55}$$

as we did above. To determine k_{n-1}, we multiply each term in Eq. (15.54) by $(s + p)^n$ and differentiate to get rid of k_n, then evaluate the result at $s = -p$ to get rid of the other coefficients except k_{n-1}. Thus, we obtain

$$k_{n-1} = \frac{d}{ds}[(s + p)^n F(s)]\,|_{s=-p} \tag{15.56}$$

Repeating this gives

$$k_{n-2} = \frac{1}{2!}\frac{d^2}{ds^2}[(s + p)^n F(s)]\,|_{s=-p} \tag{15.57}$$

The mth term becomes

$$k_{n-m} = \frac{1}{m!}\frac{d^m}{ds^m}[(s+p)^n F(s)]\,|_{s=-p} \tag{15.58}$$

where $m = 1, 2, \ldots, n-1$. One can expect the differentiation to be difficult to handle as m increases. Once we obtain the values of $k_1, k_2, \ldots, k_n$ by partial fraction expansion, we apply the inverse transform

$$\mathcal{L}^{-1}\left[\frac{1}{(s+a)^n}\right] = \frac{t^{n-1}e^{-at}}{(n-1)!}u(t) \tag{15.59}$$

to each term on the right-hand side of Eq. (15.54) and obtain

$$\begin{aligned} f(t) = \bigg(&k_1 e^{-pt} + k_2 t e^{-pt} + \frac{k_3}{2!}t^2 e^{-pt} \\ &+ \cdots + \frac{k_n}{(n-1)!}t^{n-1}e^{-pt}\bigg)u(t) + f_1(t) \end{aligned} \tag{15.60}$$

15.4.3 Complex Poles

A pair of complex poles is simple if it is not repeated; it is a double or multiple pole if repeated. Simple complex poles may be handled the same way as simple real poles, but because complex algebra is involved the result is always cumbersome. An easier approach is a method known as *completing the square*. The idea is to express each complex pole pair (or quadratic term) in $D(s)$ as a complete square such as $(s+\alpha)^2 + \beta^2$ and then use Table 15.2 to find the inverse of the term.

Since $N(s)$ and $D(s)$ always have real coefficients and we know that the complex roots of polynomials with real coefficients must occur in conjugate pairs, $F(s)$ may have the general form

$$F(s) = \frac{A_1 s + A_2}{s^2 + as + b} + F_1(s) \tag{15.61}$$

where $F_1(s)$ is the remaining part of $F(s)$ that does not have this pair of complex poles. If we complete the square by letting

$$s^2 + as + b = s^2 + 2\alpha s + \alpha^2 + \beta^2 = (s+\alpha)^2 + \beta^2 \tag{15.62}$$

and we also let

$$A_1 s + A_2 = A_1(s+\alpha) + B_1\beta \tag{15.63}$$

then Eq. (15.61) becomes

$$F(s) = \frac{A_1(s+\alpha)}{(s+\alpha)^2+\beta^2} + \frac{B_1\beta}{(s+\alpha)^2+\beta^2} + F_1(s) \tag{15.64}$$

From Table 15.2, the inverse transform is

$$f(t) = (A_1 e^{-\alpha t}\cos\beta t + B_1 e^{-\alpha t}\sin\beta t)u(t) + f_1(t) \tag{15.65}$$

The sine and cosine terms can be combined using Eq. (9.11).

Whether the pole is simple, repeated, or complex, a general approach that can always be used in finding the expansion coefficients

is the *method of algebra*, illustrated in Examples 15.9 to 15.11. To apply the method, we first set $F(s) = N(s)/D(s)$ equal to an expansion containing unknown constants. We multiply the result through by a common denominator. Then we determine the unknown constants by equating coefficients (i.e., by algebraically solving a set of simultaneous equations for these coefficients at like powers of s).

Another general approach is to substitute specific, convenient values of s to obtain as many simultaneous equations as the number of unknown coefficients, and then solve for the unknown coefficients. We must make sure that each selected value of s is not one of the poles of $F(s)$. Example 15.11 illustrates this idea.

Example 15.8

Find the inverse Laplace transform of

$$F(s) = \frac{3}{s} - \frac{5}{s+1} + \frac{6}{s^2+4}$$

Solution:

The inverse transform is given by

$$\begin{aligned} f(t) = \mathcal{L}^{-1}[F(s)] &= \mathcal{L}^{-1}\left(\frac{3}{s}\right) - \mathcal{L}^{-1}\left(\frac{5}{s+1}\right) + \mathcal{L}^{-1}\left(\frac{6}{s^2+4}\right) \\ &= (3 - 5e^{-t} + 3\sin 2t)u(t), \qquad t \geq 0 \end{aligned}$$

where Table 15.2 has been consulted for the inverse of each term.

Practice Problem 15.8

Determine the inverse Laplace transform of

$$F(s) = 1 + \frac{3}{s+4} - \frac{5s}{s^2+25}$$

Answer: $\delta(t) + (4e^{-4t} - 5\cos(5t))u(t)$.

Example 15.9

Find $f(t)$ given that

$$F(s) = \frac{s^2+12}{s(s+2)(s+3)}$$

Solution:

Unlike in the previous example where the partial fractions have been provided, we first need to determine the partial fractions. Since there are three poles, we let

$$\frac{s^2+12}{s(s+2)(s+3)} = \frac{A}{s} + \frac{B}{s+2} + \frac{C}{s+3} \tag{15.9.1}$$

where A, B, and C are the constants to be determined. We can find the constants using two approaches.

METHOD 1 **Residue method:**

$$A = sF(s)\big|_{s=0} = \frac{s^2 + 12}{(s + 2)(s + 3)}\bigg|_{s=0} = \frac{12}{(2)(3)} = 2$$

$$B = (s + 2)F(s)\big|_{s=-2} = \frac{s^2 + 12}{s(s + 3)}\bigg|_{s=-2} = \frac{4 + 12}{(-2)(1)} = -8$$

$$C = (s + 3)F(s)\big|_{s=-3} = \frac{s^2 + 12}{s(s + 2)}\bigg|_{s=-3} = \frac{9 + 12}{(-3)(-1)} = 7$$

METHOD 2 **Algebraic method:** Multiplying both sides of Eq. (15.9.1) by $s(s + 2)(s + 3)$ gives

$$s^2 + 12 = A(s + 2)(s + 3) + Bs(s + 3) + Cs(s + 2)$$

or

$$s^2 + 12 = A(s^2 + 5s + 6) + B(s^2 + 3s) + C(s^2 + 2s)$$

Equating the coefficients of like powers of s gives

Constant: $12 = 6A \quad \Rightarrow \quad A = 2$

s: $0 = 5A + 3B + 2C \quad \Rightarrow \quad 3B + 2C = -10$

s^2: $1 = A + B + C \quad \Rightarrow \quad B + C = -1$

Thus, $A = 2, B = -8, C = 7$, and Eq. (15.9.1) becomes

$$F(s) = \frac{2}{s} - \frac{8}{s + 2} + \frac{7}{s + 3}$$

By finding the inverse transform of each term, we obtain

$$f(t) = (2 - 8e^{-2t} + 7e^{-3t})u(t)$$

Practice Problem 15.9

Find $f(t)$ if

$$F(s) = \frac{6(s + 2)}{(s + 1)(s + 3)(s + 4)}$$

Answer: $f(t) = (e^{-t} + 3e^{-3t} - 4e^{-4t})u(t)$.

Example 15.10

Calculate $v(t)$ given that

$$V(s) = \frac{10s^2 + 4}{s(s + 1)(s + 2)^2}$$

Solution:

While the previous example is on simple roots, this example is on repeated roots. Let

$$V(s) = \frac{10s^2 + 4}{s(s + 1)(s + 2)^2} = \frac{A}{s} + \frac{B}{s + 1} + \frac{C}{(s + 2)^2} + \frac{D}{s + 2} \qquad \textbf{(15.10.1)}$$

METHOD 1 **Residue method:**

$$A = sV(s)\,|_{s=0} = \frac{10s^2 + 4}{(s+1)(s+2)^2}\bigg|_{s=0} = \frac{4}{(1)(2)^2} = 1$$

$$B = (s+1)V(s)\,|_{s=-1} = \frac{10s^2 + 4}{s(s+2)^2}\bigg|_{s=-1} = \frac{14}{(-1)(1)^2} = -14$$

$$C = (s+2)^2V(s)\,|_{s=-2} = \frac{10s^2 + 4}{s(s+1)}\bigg|_{s=-2} = \frac{44}{(-2)(-1)} = 22$$

$$D = \frac{d}{ds}[(s+2)^2V(s)]\bigg|_{s=-2} = \frac{d}{ds}\left(\frac{10s^2+4}{s^2+s}\right)\bigg|_{s=-2}$$

$$= \frac{(s^2+s)(20s) - (10s^2+4)(2s+1)}{(s^2+s)^2}\bigg|_{s=-2} = \frac{52}{4} = 13$$

METHOD 2 **Algebraic method:** Multiplying Eq. (15.10.1) by $s(s+1)(s+2)^2$, we obtain

$$\begin{aligned}10s^2 + 4 = {}& A(s+1)(s+2)^2 + Bs(s+2)^2 \\ &+ Cs(s+1) + Ds(s+1)(s+2)\end{aligned}$$

or

$$\begin{aligned}10s^2 + 4 = {}& A(s^3 + 5s^2 + 8s + 4) + B(s^3 + 4s^2 + 4s) \\ &+ C(s^2 + s) + D(s^3 + 3s^2 + 2s)\end{aligned}$$

Equating coefficients,

Constant:	$4 = 4A$	$\Rightarrow$	$A = 1$
s:	$0 = 8A + 4B + C + 2D$	$\Rightarrow$	$4B + C + 2D = -8$
s^2:	$10 = 5A + 4B + C + 3D$	$\Rightarrow$	$4B + C + 3D = 5$
s^3:	$0 = A + B + D$	$\Rightarrow$	$B + D = -1$

Solving these simultaneous equations gives $A = 1, B = -14, C = 22, D = 13$, so that

$$V(s) = \frac{1}{s} - \frac{14}{s+1} + \frac{13}{s+2} + \frac{22}{(s+2)^2}$$

Taking the inverse transform of each term, we get

$$v(t) = (1 - 14e^{-t} + 13e^{-2t} + 22te^{-2t})u(t)$$

Practice Problem 15.10

Obtain $g(t)$ if

$$G(s) = \frac{s^3 + 2s + 6}{s(s+1)^2(s+3)}$$

Answer: $(2 - 3.25e^{-t} - 1.5te^{-t} + 2.25e^{-3t})u(t)$.

Example 15.11

Find the inverse transform of the frequency-domain function in Example 15.7:

$$H(s) = \frac{20}{(s+3)(s^2 + 8s + 25)}$$

Solution:
In this example, $H(s)$ has a pair of complex poles at $s^2 + 8s + 25 = 0$ or $s = -4 \pm j3$. We let

$$H(s) = \frac{20}{(s+3)(s^2+8s+25)} = \frac{A}{s+3} + \frac{Bs+C}{(s^2+8s+25)} \tag{15.11.1}$$

We now determine the expansion coefficients in two ways.

■ **METHOD 1** **Combination of methods:** We can obtain A using the method of residue,

$$A = (s+3)H(s)\,|_{s=-3} = \frac{20}{s^2+8s+25}\bigg|_{s=-3} = \frac{20}{10} = 2$$

Although B and C can be obtained using the method of residue, we will not do so, to avoid complex algebra. Rather, we can substitute two specific values of s [say $s = 0, 1$, which are not poles of $F(s)$] into Eq. (15.11.1). This will give us two simultaneous equations from which to find B and C. If we let $s = 0$ in Eq. (15.11.1), we obtain

$$\frac{20}{75} = \frac{A}{3} + \frac{C}{25}$$

or

$$20 = 25A + 3C \tag{15.11.2}$$

Since $A = 2$, Eq. (15.11.2) gives $C = -10$. Substituting $s = 1$ into Eq. (15.11.1) gives

$$\frac{20}{(4)(34)} = \frac{A}{4} + \frac{B+C}{34}$$

or

$$20 = 34A + 4B + 4C \tag{15.11.3}$$

But $A = 2$, $C = -10$, so that Eq. (15.11.3) gives $B = -2$.

■ **METHOD 2** **Algebraic method:** Multiplying both sides of Eq. (15.11.1) by $(s+3)(s^2+8s+25)$ yields

$$\begin{aligned} 20 &= A(s^2+8s+25) + (Bs+C)(s+3) \\ &= A(s^2+8s+25) + B(s^2+3s) + C(s+3) \end{aligned} \tag{15.11.4}$$

Equating coefficients gives

$$\begin{aligned} &s^2: && 0 = A + B \quad\Rightarrow\quad A = -B \\ &s: && 0 = 8A + 3B + C = 5A + C \quad\Rightarrow\quad C = -5A \\ &\text{Constant:} && 20 = 25A + 3C = 25A - 15A \quad\Rightarrow\quad A = 2 \end{aligned}$$

That is, $B = -2$, $C = -10$. Thus,

$$\begin{aligned} H(s) &= \frac{2}{s+3} - \frac{2s+10}{(s^2+8s+25)} = \frac{2}{s+3} - \frac{2(s+4)+2}{(s+4)^2+9} \\ &= \frac{2}{s+3} - \frac{2(s+4)}{(s+4)^2+9} - \frac{2}{3}\frac{3}{(s+4)^2+9} \end{aligned}$$

Taking the inverse of each term, we obtain

$$h(t) = \left(2e^{-3t} - 2e^{-4t}\cos 3t - \frac{2}{3}e^{-4t}\sin 3t\right)u(t) \quad \textbf{(15.11.5)}$$

It is alright to leave the result this way. However, we can combine the cosine and sine terms as

$$h(t) = (2e^{-3t} - Re^{-4t}\cos(3t - \theta))u(t) \quad \textbf{(15.11.6)}$$

To obtain Eq. (15.11.6) from Eq. (15.11.5), we apply Eq. (9.11). Next, we determine the coefficient R and the phase angle θ:

$$R = \sqrt{2^2 + (\tfrac{2}{3})^2} = 2.108, \qquad \theta = \tan^{-1}\frac{\frac{2}{3}}{2} = 18.43^\circ$$

Thus,

$$h(t) = (2e^{-3t} - 2.108e^{-4t}\cos(3t - 18.43^\circ))u(t)$$

Practice Problem 15.11

Find $g(t)$ given that

$$G(s) = \frac{10}{(s+1)(s^2 + 4s + 13)}$$

Answer: $e^{-t} - e^{-2t}\cos 3t - \frac{1}{3}e^{-2t}\sin 3t,\ t \geq 0.$

15.5 The Convolution Integral

The term *convolution* means "folding." Convolution is an invaluable tool to the engineer because it provides a means of viewing and characterizing physical systems. For example, it is used in finding the response $y(t)$ of a system to an excitation $x(t)$, knowing the system impulse response $h(t)$. This is achieved through the *convolution integral*, defined as

$$y(t) = \int_{-\infty}^{\infty} x(\lambda)h(t - \lambda)\,d\lambda \quad \textbf{(15.66)}$$

or simply

$$y(t) = x(t) * h(t) \quad \textbf{(15.67)}$$

where λ is a dummy variable and the asterisk denotes convolution. Equation (15.66) or (15.67) states that the output is equal to the input convolved with the unit impulse response. The convolution process is commutative:

$$y(t) = x(t) * h(t) = h(t) * x(t) \quad \textbf{(15.68a)}$$

or

$$y(t) = \int_{-\infty}^{\infty} x(\lambda)h(t - \lambda)\,d\lambda = \int_{-\infty}^{\infty} h(\lambda)x(t - \lambda)\,d\lambda \quad \textbf{(15.68b)}$$

This implies that the order in which the two functions are convolved is immaterial. We will see shortly how to take advantage of this commutative property when performing graphical computation of the convolution integral.

The **convolution** of two signals consists of time-reversing one of the signals, shifting it, and multiplying it point by point with the second signal, and integrating the product.

The convolution integral in Eq. (15.66) is the general one; it applies to any linear system. However, the convolution integral can be simplified if we assume that a system has two properties. First, if $x(t) = 0$ for $t < 0$, then

$$y(t) = \int_{-\infty}^{\infty} x(\lambda)h(t - \lambda)\,d\lambda = \int_{0}^{\infty} x(\lambda)h(t - \lambda)\,d\lambda \tag{15.69}$$

Second, if the system's impulse response is *causal* (i.e., $h(t) = 0$ for $t < 0$), then $h(t - \lambda) = 0$ for $t - \lambda < 0$ or $\lambda > t$, so that Eq. (15.69) becomes

$$\boxed{y(t) = h(t) * x(t) = \int_{0}^{t} x(\lambda)h(t - \lambda)\,d\lambda} \tag{15.70}$$

Here are some properties of the convolution integral.

1. $x(t) * h(t) = h(t) * x(t)$ (Commutative)
2. $f(t) * [x(t) + y(t)] = f(t) * x(t) + f(t) * y(t)$ (Distributive)
3. $f(t) * [x(t) * y(t)] = [f(t) * x(t)] * y(t)$ (Associative)
4. $f(t) * \delta(t) = \int_{-\infty}^{\infty} f(\lambda)\delta(t - \lambda)\,d\lambda = f(t)$
5. $f(t) * \delta(t - t_o) = f(t - t_o)$
6. $f(t) * \delta'(t) = \int_{-\infty}^{\infty} f(\lambda)\delta'(t - \lambda)\,d\lambda = f'(t)$
7. $f(t) * u(t) = \int_{-\infty}^{\infty} f(\lambda)u(t - \lambda)\,d\lambda = \int_{-\infty}^{t} f(\lambda)\,d\lambda$

Before learning how to evaluate the convolution integral in Eq. (15.70), let us establish the link between the Laplace transform and the convolution integral. Given two functions $f_1(t)$ and $f_2(t)$ with Laplace transforms $F_1(s)$ and $F_2(s)$, respectively, their convolution is

$$f(t) = f_1(t) * f_2(t) = \int_{0}^{t} f_1(\lambda)f_2(t - \lambda)\,d\lambda \tag{15.71}$$

Taking the Laplace transform gives

$$F(s) = \mathcal{L}[f_1(t) * f_2(t)] = F_1(s)F_2(s) \tag{15.72}$$

To prove that Eq. (15.72) is true, we begin with the fact that $F_1(s)$ is defined as

$$F_1(s) = \int_{0^-}^{\infty} f_1(\lambda)e^{-s\lambda}\,d\lambda \tag{15.73}$$

Multiplying this with $F_2(s)$ gives

$$F_1(s)F_2(s) = \int_{0^-}^{\infty} f_1(\lambda)[F_2(s)e^{-s\lambda}]\,d\lambda \tag{15.74}$$

We recall from the time shift property in Eq. (15.17) that the term in brackets can be written as

$$F_2(s)e^{-s\lambda} = \mathcal{L}[f_2(t-\lambda)u(t-\lambda)] = \int_{0^-}^{\infty} f_2(t-\lambda)u(t-\lambda)e^{-s\lambda}\,dt \tag{15.75}$$

Substituting Eq. (15.75) into Eq. (15.74) gives

$$F_1(s)F_2(s) = \int_{0^-}^{\infty} f_1(\lambda)\left[\int_{0^-}^{\infty} f_2(t-\lambda)u(t-\lambda)e^{-s\lambda}\,dt\right]d\lambda \tag{15.76}$$

Interchanging the order of integration results in

$$F_1(s)F_2(s) = \int_{0^-}^{\infty}\left[\int_{0^-}^{t} f_1(\lambda)f_2(t-\lambda)\,d\lambda\right]e^{-s\lambda}\,d\lambda \tag{15.77}$$

The integral in brackets extends only from 0 to t because the delayed unit step $u(t-\lambda) = 1$ for $\lambda < t$ and $u(t-\lambda) = 0$ for $\lambda > t$. We notice that the integral is the convolution of $f_1(t)$ and $f_2(t)$ as in Eq. (15.71). Hence,

$$\boxed{F_1(s)F_2(s) = \mathcal{L}[f_1(t) * f_2(t)]} \tag{15.78}$$

as desired. This indicates that convolution in the time domain is equivalent to multiplication in the s-domain. For example, if $x(t) = 4e^{-t}$ and $h(t) = 5e^{-2t}$, applying the property in Eq. (15.78), we get

$$\begin{aligned} h(t) * x(t) &= \mathcal{L}^{-1}[H(s)X(s)] = \mathcal{L}^{-1}\left[\left(\frac{5}{s+2}\right)\left(\frac{4}{s+1}\right)\right] \\ &= \mathcal{L}^{-1}\left[\frac{20}{s+1} + \frac{-20}{s+2}\right] \\ &= 20(e^{-t} - e^{-2t}), \qquad t \geq 0 \end{aligned} \tag{15.79}$$

Although we can find the convolution of two signals using Eq. (15.78), as we have just done, if the product $F_1(s)F_2(s)$ is very complicated, finding the inverse may be tough. Also, there are situations in which $f_1(t)$ and $f_2(t)$ are available in the form of experimental data and there are no explicit Laplace transforms. In these cases, one must do the convolution in the time domain.

The process of convolving two signals in the time domain is better appreciated from a graphical point of view. The graphical procedure for evaluating the convolution integral in Eq. (15.70) usually involves four steps.

Steps to Evaluate the Convolution Integral:

1. Folding: Take the mirror image of $h(\lambda)$ about the ordinate axis to obtain $h(-\lambda)$.
2. Displacement: Shift or delay $h(-\lambda)$ by t to obtain $h(t-\lambda)$.
3. Multiplication: Find the product of $h(t-\lambda)$ and $x(\lambda)$.
4. Integration: For a given time t, calculate the area under the product $h(t-\lambda)x(\lambda)$ for $0 < \lambda < t$ to get $y(t)$ at t.

The folding operation in step 1 is the reason for the term *convolution.* The function $h(t - \lambda)$ scans or slides over $x(\lambda)$. In view of this superposition procedure, the convolution integral is also known as the *superposition integral.*

To apply the four steps, it is necessary to be able to sketch $x(\lambda)$ and $h(t - \lambda)$. To get $x(\lambda)$ from the original function $x(t)$ involves merely replacing every t with λ. Sketching $h(t - \lambda)$ is the key to the convolution process. It involves reflecting $h(\lambda)$ about the vertical axis and shifting it by t. Analytically, we obtain $h(t - \lambda)$ by replacing every t in $h(t)$ by $t - \lambda$. Since convolution is commutative, it may be more convenient to apply steps 1 and 2 to $x(t)$ instead of $h(t)$. The best way to illustrate the procedure is with some examples.

Example 15.12

Find the convolution of the two signals in Fig. 15.10.

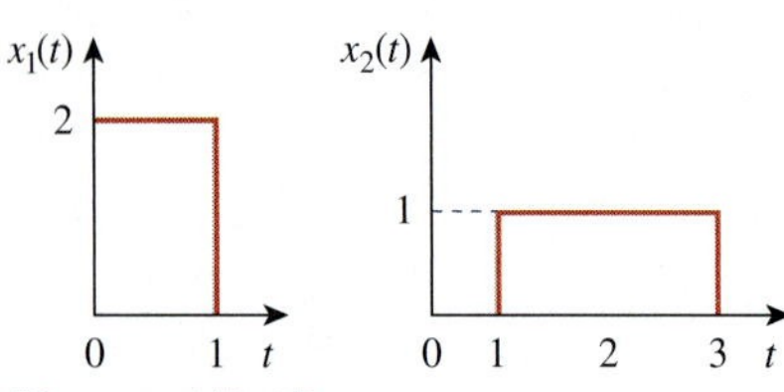

Figure 15.10
For Example 15.12.

Solution:

We follow the four steps to get $y(t) = x_1(t) * x_2(t)$. First, we fold $x_1(t)$ as shown in Fig. 15.11(a) and shift it by t as shown in Fig. 15.11(b). For different values of t, we now multiply the two functions and integrate to determine the area of the overlapping region.

For $0 < t < 1$, there is no overlap of the two functions, as shown in Fig. 15.12(a). Hence,

$$y(t) = x_1(t) * x_2(t) = 0, \qquad 0 < t < 1 \tag{15.12.1}$$

For $1 < t < 2$, the two signals overlap between 1 and t, as shown in Fig. 15.12(b).

$$y(t) = \int_1^t (2)(1)\, d\lambda = 2\lambda \Big|_1^t = 2(t - 1), \qquad 1 < t < 2 \tag{15.12.2}$$

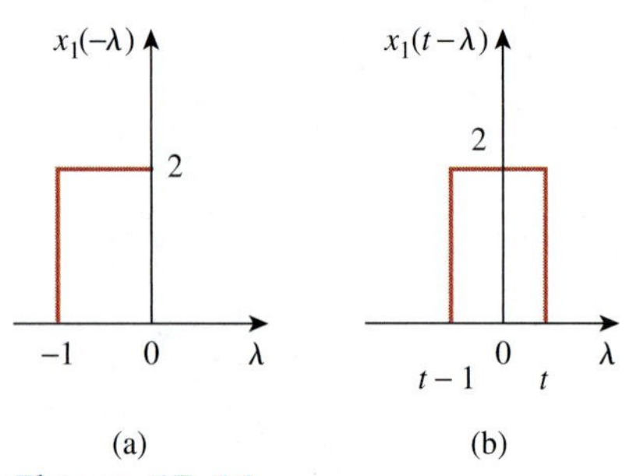

Figure 15.11
(a) Folding $x_1(\lambda)$, (b) shifting $x_1(-\lambda)$ by t.

For $2 < t < 3$, the two signals completely overlap between $(t - 1)$ and t, as shown in Fig. 15.12(c). It is easy to see that the area under the curve is 2. Or

$$y(t) = \int_{t-1}^t (2)(1)\, d\lambda = 2\lambda \Big|_{t-1}^t = 2, \qquad 2 < t < 3 \tag{15.12.3}$$

For $3 < t < 4$, the two signals overlap between $(t - 1)$ and 3, as shown in Fig. 15.12(d).

$$\begin{aligned} y(t) &= \int_{t-1}^3 (2)(1)\, d\lambda = 2\lambda \Big|_{t-1}^3 \\ &= 2(3 - t + 1) = 8 - 2t, \qquad 3 < t < 4 \end{aligned} \tag{15.12.4}$$

For $t > 4$, the two signals do not overlap [Fig. 15.12(e)], and

$$y(t) = 0, \qquad t > 4 \tag{15.12.5}$$

Combining Eqs. (15.12.1) to (15.12.5), we obtain

$$y(t) = \begin{cases} 0, & 0 \le t \le 1 \\ 2t - 2, & 1 \le t \le 2 \\ 2, & 2 \le t \le 3 \\ 8 - 2t, & 3 \le t \le 4 \\ 0, & t \ge 4 \end{cases} \tag{15.12.6}$$

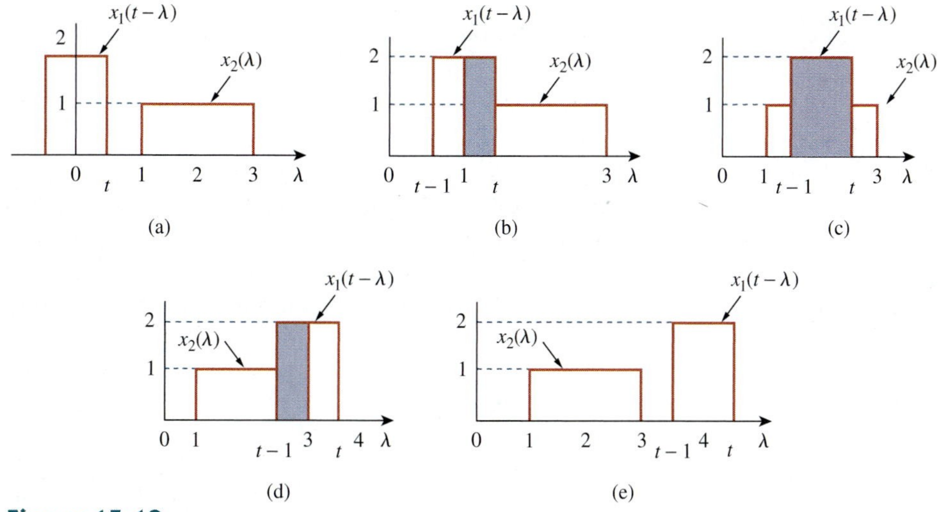

Figure 15.12
Overlapping of $x_1(t - \lambda)$ and $x_2(\lambda)$ for: (a) $0 < t < 1$, (b) $1 < t < 2$, (c) $2 < t < 3$, (d) $3 < t < 4$, (e) $t > 4$.

which is sketched in Fig. 15.13. Notice that $y(t)$ in this equation is continuous. This fact can be used to check the results as we move from one range of t to another. The result in Eq. (15.12.6) can be obtained without using the graphical procedure—by directly using Eq. (15.70) and the properties of step functions. This will be illustrated in Example 15.14.

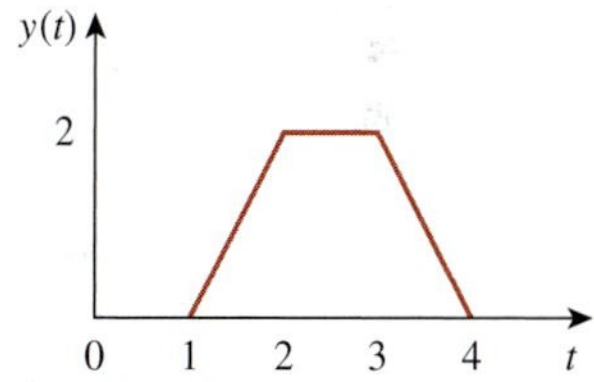

Figure 15.13
Convolution of signals $x_1(t)$ and $x_2(t)$ in Fig. 15.10.

Practice Problem 15.12

Graphically convolve the two functions in Fig. 15.14. To show how powerful working in the s-domain is, verify your answer by performing the equivalent operation in the s-domain.

Answer: The result of the convolution $y(t)$ is shown in Fig. 15.15, where

$$y(t) = \begin{cases} t, & 0 \le t \le 2 \\ 6 - 2t, & 2 \le t \le 3 \\ 0, & \text{otherwise.} \end{cases}$$

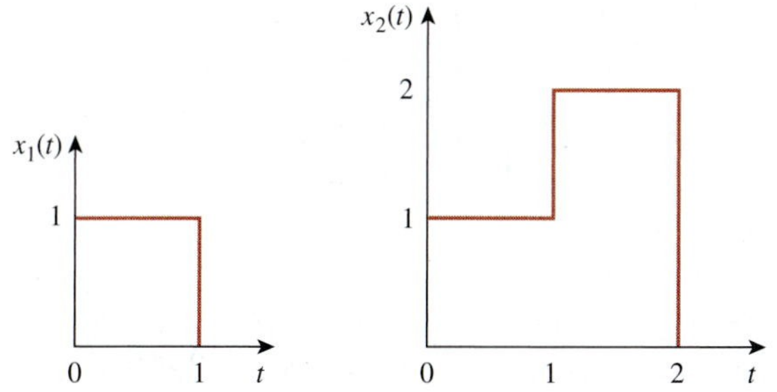

Figure 15.14
For Practice Prob. 15.12.

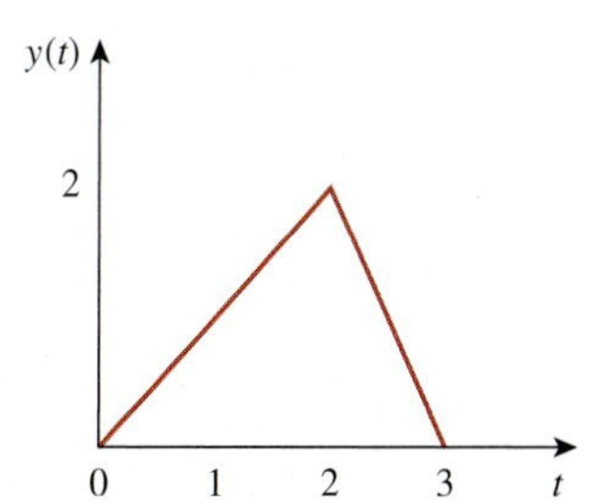

Figure 15.15
Convolution of the signals in Fig. 15.14.

Example 15.13

Graphically convolve $g(t)$ and $u(t)$ shown in Fig. 15.16.

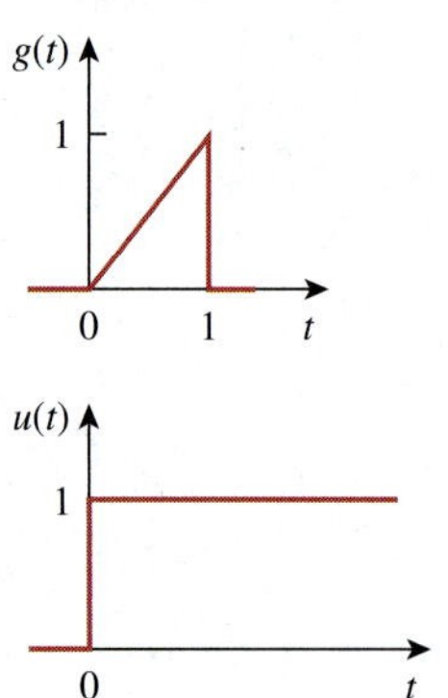

Figure 15.16
For Example 15.13.

Solution:
Let $y(t) = g(t) * u(t)$. We can find $y(t)$ in two ways.

METHOD 1 Suppose we fold $g(t)$, as in Fig. 15.17(a), and shift it by t, as in Fig. 15.17(b). Since $g(t) = t, 0 < t < 1$ originally, we expect that $g(t - \lambda) = t - \lambda, 0 < t - \lambda < 1$ or $t - 1 < \lambda < t$. There is no overlap of the two functions when $t < 0$ so that $y(0) = 0$ for this case.

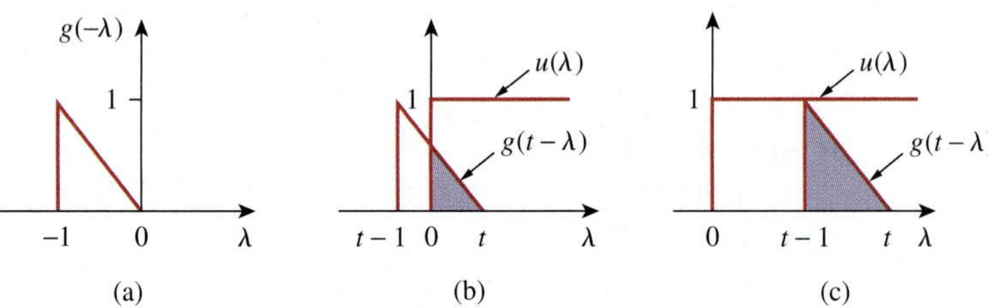

Figure 15.17
Convolution of $g(t)$ and $u(t)$ in Fig. 15.16 with $g(t)$ folded.

For $0 < t < 1$, $g(t - \lambda)$ and $u(\lambda)$ overlap from 0 to t, as evident in Fig. 15.17(b). Therefore,

$$\begin{aligned} y(t) &= \int_0^t (1)(t-\lambda)\,d\lambda = \left(t\lambda - \frac{1}{2}\lambda^2\right)\Bigg|_0^t \\ &= t^2 - \frac{t^2}{2} = \frac{t^2}{2}, \qquad 0 \le t \le 1 \end{aligned} \tag{15.13.1}$$

For $t > 1$, the two functions overlap completely between $(t - 1)$ and t [see Fig. 15.17(c)]. Hence,

$$\begin{aligned} y(t) &= \int_{t-1}^t (1)(t-\lambda)\,d\lambda \\ &= \left(t\lambda - \frac{1}{2}\lambda^2\right)\Bigg|_{t-1}^t = \frac{1}{2}, \qquad t \ge 1 \end{aligned} \tag{15.13.2}$$

Thus, from Eqs. (15.13.1) and (15.13.2),

$$y(t) = \begin{cases} \dfrac{1}{2}t^2, & 0 \le t \le 1 \\ \dfrac{1}{2}, & t \ge 1 \end{cases}$$

METHOD 2 Instead of folding g, suppose we fold the unit step function $u(t)$, as in Fig. 15.18(a), and then shift it by t, as in Fig. 15.18(b). Since $u(t) = 1$ for $t > 0$, $u(t - \lambda) = 1$ for $t - \lambda > 0$ or $\lambda < t$, the two functions overlap from 0 to t, so that

$$y(t) = \int_0^t (1)\lambda\,d\lambda = \frac{1}{2}\lambda^2\Bigg|_0^t = \frac{t^2}{2}, \qquad 0 \le t \le 1 \tag{15.13.3}$$

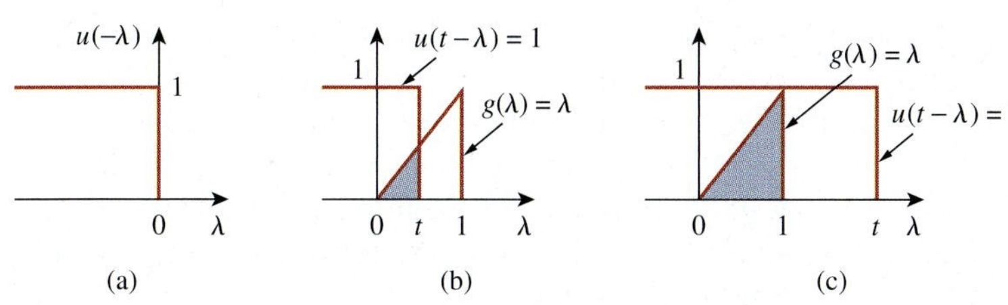

Figure 15.18
Convolution of $g(t)$ and $u(t)$ in Fig. 15.16 with $u(t)$ folded.

For $t > 1$, the two functions overlap between 0 and 1, as shown in Fig. 15.18(c). Hence,

$$y(t) = \int_0^1 (1)\lambda\, d\lambda = \frac{1}{2}\lambda^2 \Big|_0^1 = \frac{1}{2}, \qquad t \geq 1 \qquad \textbf{(15.13.4)}$$

And, from Eqs. (15.13.3) and (15.13.4),

$$y(t) = \begin{cases} \frac{1}{2}t^2, & 0 \leq t \leq 1 \\ \frac{1}{2}, & t \geq 1 \end{cases}$$

Although the two methods give the same result, as expected, notice that it is more convenient to fold the unit step function $u(t)$ than fold $g(t)$ in this example. Figure 15.19 shows $y(t)$.

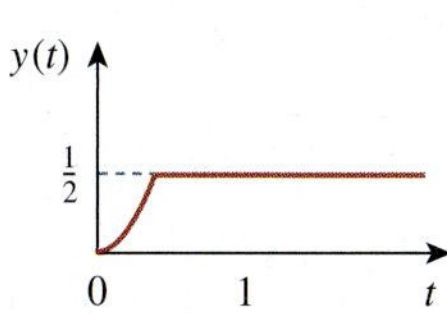

Figure 15.19
Result of Example 15.13.

Practice Problem 15.13

Given $g(t)$ and $f(t)$ in Fig. 15.20, graphically find $y(t) = g(t) * f(t)$.

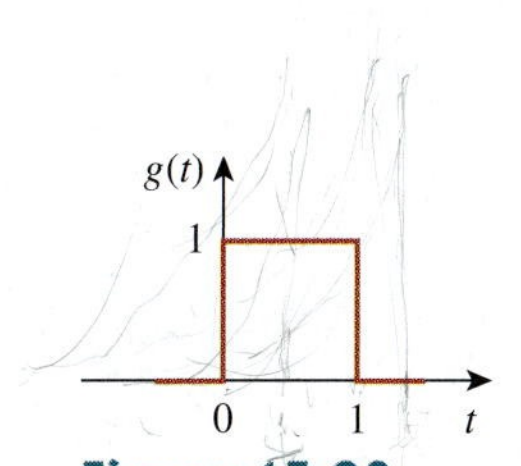

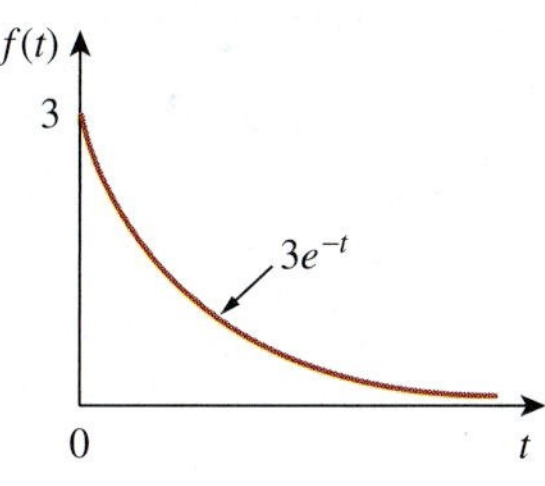

Figure 15.20
For Practice Prob. 15.13.

Answer: $y(t) = \begin{cases} 3(1 - e^{-t}), & 0 \leq t \leq 1 \\ 3(e - 1)e^{-t}, & t \geq 1 \\ 0, & \text{elsewhere.} \end{cases}$

Example 15.14

For the *RL* circuit in Fig. 15.21(a), use the convolution integral to find the response $i_o(t)$ due to the excitation shown in Fig. 15.21(b).

Solution:

1. **Define.** The problem is clearly stated and the method of solution is also specified.

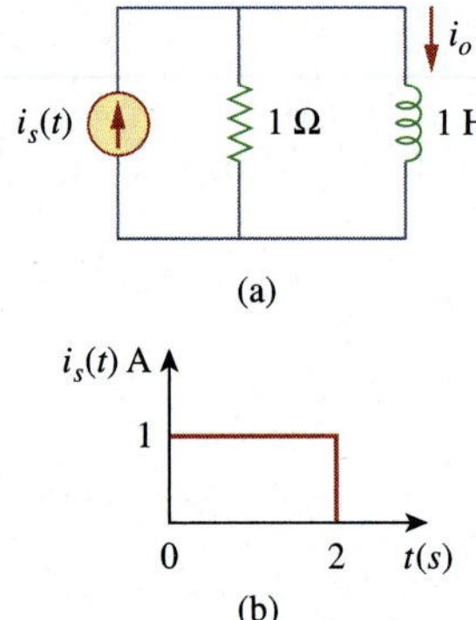

Figure 15.21
For Example 15.14.

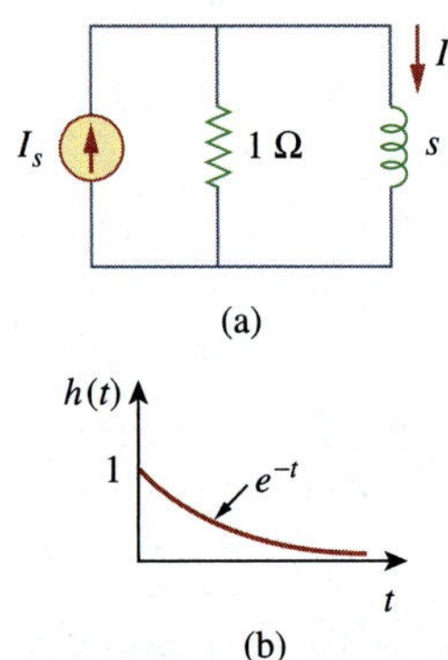

Figure 15.22
For the circuit in Fig. 15.21(a): (a) its s-domain equivalent, (b) its impulse response.

2. **Present.** We are to use the convolution integral to solve for the response $i_o(t)$ due to $i_s(t)$ shown in Fig. 15.21(b).
3. **Alternative.** We have learned to do convolution by using the convolution integral and how to do it graphically. In addition, we could always work in the s-domain to solve for the current. We will solve for the current using the convolution integral and then check it using the graphical approach.
4. **Attempt.** As we stated, this problem can be solved in two ways: directly using the convolution integral or using the graphical technique. To use either approach, we first need the unit impulse response $h(t)$ of the circuit. In the s-domain, applying the current division principle to the circuit in Fig. 15.22(a) gives

$$I_o = \frac{1}{s+1} I_s$$

Hence,

$$H(s) = \frac{I_o}{I_s} = \frac{1}{s+1} \tag{15.14.1}$$

and the inverse Laplace transform of this gives

$$h(t) = e^{-t}u(t) \tag{15.14.2}$$

Figure 15.22(b) shows the impulse response $h(t)$ of the circuit.

To use the convolution integral directly, recall that the response is given in the s-domain as

$$I_o(s) = H(s)I_s(s)$$

With the given $i_s(t)$ in Fig. 15.21(b),

$$i_s(t) = u(t) - u(t-2)$$

so that

$$\begin{aligned} i_o(t) = h(t) * i_s(t) &= \int_0^t i_s(\lambda)h(t-\lambda)\,d\lambda \\ &= \int_0^t [u(\lambda) - u(\lambda-2)]e^{-(t-\lambda)}\,d\lambda \end{aligned} \tag{15.14.3}$$

Since $u(\lambda - 2) = 0$ for $0 < \lambda < 2$, the integrand involving $u(\lambda)$ is nonzero for all $\lambda > 0$, whereas the integrand involving $u(\lambda - 2)$ is nonzero only for $\lambda > 2$. The best way to handle the integral is to do the two parts separately. For $0 < t < 2$,

$$\begin{aligned} i'_o(t) &= \int_0^t (1)e^{-(t-\lambda)}\,d\lambda = e^{-t}\int_0^t (1)e^{\lambda}\,d\lambda \\ &= e^{-t}(e^t - 1) = 1 - e^{-t}, \qquad 0 < t < 2 \end{aligned} \tag{15.14.4}$$

For $t > 2$,

$$\begin{aligned} i''_o(t) &= \int_2^t (1)e^{-(t-\lambda)}\,d\lambda = e^{-t}\int_2^t e^{\lambda}\,d\lambda \\ &= e^{-t}(e^t - e^2) = 1 - e^2e^{-t}, \qquad t > 2 \end{aligned} \tag{15.14.5}$$

Substituting Eqs. (15.14.4) and (15.14.5) into Eq. (15.14.3) gives

$$\begin{aligned} i_o(t) &= i_o'(t) - i_o''(t) \\ &= (1 - e^{-t})[u(t-2) - u(t)] - (1 - e^2 e^{-t})u(t-2) \\ &= \begin{cases} 1 - e^{-t}\text{ A}, & 0 < t < 2 \\ (e^2 - 1)e^{-t}\text{ A}, & t > 2 \end{cases} \end{aligned} \qquad \textbf{(15.14.6)}$$

5. **Evaluate.** To use the graphical technique, we may fold $i_s(t)$ in Fig. 15.21(b) and shift by t, as shown in Fig. 15.23(a). For $0 < t < 2$, the overlap between $i_s(t-\lambda)$ and $h(\lambda)$ is from 0 to t, so that

$$i_o(t) = \int_0^t (1)e^{-\lambda}\,d\lambda = -e^{-\lambda}\Big|_0^t = (\mathbf{1} - \boldsymbol{e}^{-t})\textbf{A}, \qquad 0 \le t \le 2 \qquad \textbf{(15.14.7)}$$

For $t > 2$, the two functions overlap between $(t-2)$ and t, as in Fig. 15.23(b). Hence,

$$\begin{aligned} i_o(t) &= \int_{t-2}^t (1)e^{-\lambda}\,d\lambda = -e^{-\lambda}\Big|_{t-2}^t = -e^{-t} + e^{-(t-2)} \\ &= (\boldsymbol{e}^2 - \mathbf{1})\boldsymbol{e}^{-t}\,\textbf{A}, \qquad t \ge 0 \end{aligned} \qquad \textbf{(15.14.8)}$$

From Eqs. (15.14.7) and (15.14.8), the response is

$$i_o(t) = \begin{cases} \mathbf{1} - \boldsymbol{e}^{-t}\,\textbf{A}, & 0 \le t \le 2 \\ (\boldsymbol{e}^2 - \mathbf{1})\boldsymbol{e}^{-t}\,\textbf{A}, & t \ge 2 \end{cases} \qquad \textbf{(15.14.9)}$$

which is the same as in Eq. (15.14.6). Thus, the response $i_o(t)$ along the excitation $i_s(t)$ is as shown in Fig. 15.24.

6. **Satisfactory?** We have satisfactorily solved the problem and can present the results as a solution to the problem.

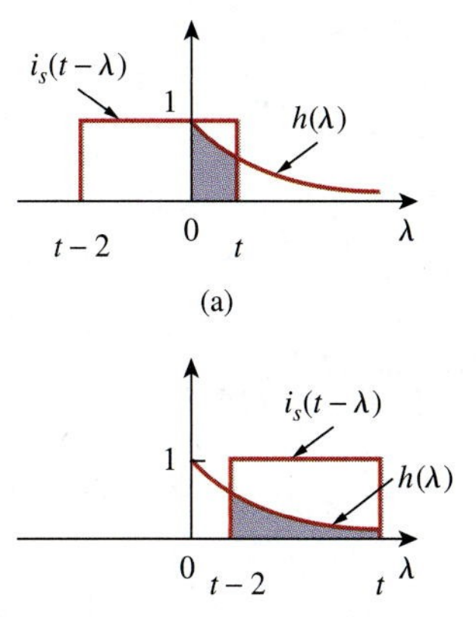

Figure 15.23
For Example 15.14.

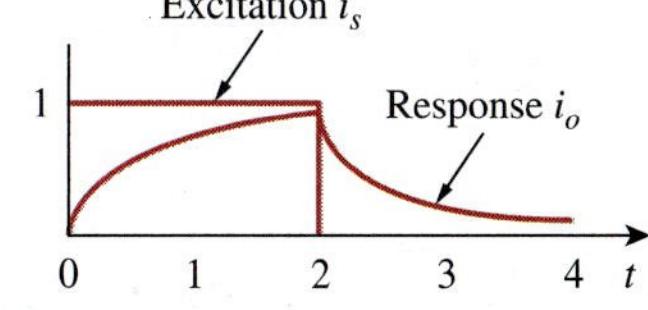

Figure 15.24
For Example 15.14; excitation and response.

Practice Problem 15.14

Use convolution to find $v_o(t)$ in the circuit of Fig. 15.25(a) when the excitation is the signal shown in Fig. 15.25(b). To show how powerful working in the s-domain is, verify your answer by performing the equivalent operation in the s-domain.

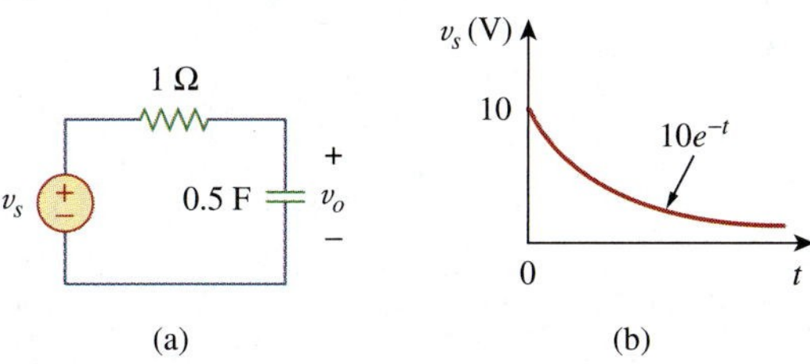

Figure 15.25
For Practice Prob. 15.14.

Answer: $20(e^{-t} - e^{-2t})$ V.

15.6 †Application to Integrodifferential Equations

The Laplace transform is useful in solving linear integrodifferential equations. Using the differentiation and integration properties of Laplace transforms, each term in the integrodifferential equation is transformed.

Initial conditions are automatically taken into account. We solve the resulting algebraic equation in the s-domain. We then convert the solution back to the time domain by using the inverse transform. The following examples illustrate the process.

Example 15.15

Use the Laplace transform to solve the differential equation

$$\frac{d^2v(t)}{dt^2} + 6\frac{dv(t)}{dt} + 8v(t) = 2u(t)$$

subject to $v(0) = 1, v'(0) = -2$.

Solution:

We take the Laplace transform of each term in the given differential equation and obtain

$$[s^2V(s) - sv(0) - v'(0)] + 6[sV(s) - v(0)] + 8V(s) = \frac{2}{s}$$

Substituting $v(0) = 1, v'(0) = -2$,

$$s^2V(s) - s + 2 + 6sV(s) - 6 + 8V(s) = \frac{2}{s}$$

or

$$(s^2 + 6s + 8)V(s) = s + 4 + \frac{2}{s} = \frac{s^2 + 4s + 2}{s}$$

Hence,

$$V(s) = \frac{s^2 + 4s + 2}{s(s + 2)(s + 4)} = \frac{A}{s} + \frac{B}{s + 2} + \frac{C}{s + 4}$$

where

$$A = sV(s)\,\big|_{s=0} = \left.\frac{s^2 + 4s + 2}{(s + 2)(s + 4)}\right|_{s=0} = \frac{2}{(2)(4)} = \frac{1}{4}$$

$$B = (s + 2)V(s)\,\big|_{s=-2} = \left.\frac{s^2 + 4s + 2}{s(s + 4)}\right|_{s=-2} = \frac{-2}{(-2)(2)} = \frac{1}{2}$$

$$C = (s + 4)V(s)\,\big|_{s=-4} = \left.\frac{s^2 + 4s + 2}{s(s + 2)}\right|_{s=-4} = \frac{2}{(-4)(-2)} = \frac{1}{4}$$

Hence,

$$V(s) = \frac{\frac{1}{4}}{s} + \frac{\frac{1}{2}}{s + 2} + \frac{\frac{1}{4}}{s + 4}$$

By the inverse Laplace transform,

$$v(t) = \frac{1}{4}(1 + 2e^{-2t} + e^{-4t})u(t)$$

Practice Problem 15.15

Solve the following differential equation using the Laplace transform method.

$$\frac{d^2v(t)}{dt^2} + 4\frac{dv(t)}{dt} + 4v(t) = e^{-t}$$

if $v(0) = v'(0) = 2$.

Answer: $(2e^{-t} + 4te^{-2t})u(t)$.

Example 15.16

Solve for the response $y(t)$ in the following integrodifferential equation.

$$\frac{dy}{dt} + 5y(t) + 6\int_0^t y(\tau)\,d\tau = u(t), \qquad y(0) = 2$$

Solution:

Taking the Laplace transform of each term, we get

$$[sY(s) - y(0)] + 5Y(s) + \frac{6}{s}Y(s) = \frac{1}{s}$$

Substituting $y(0) = 2$ and multiplying through by s,

$$Y(s)(s^2 + 5s + 6) = 1 + 2s$$

or

$$Y(s) = \frac{2s+1}{(s+2)(s+3)} = \frac{A}{s+2} + \frac{B}{s+3}$$

where

$$A = (s+2)Y(s)\,|_{s=-2} = \left.\frac{2s+1}{s+3}\right|_{s=-2} = \frac{-3}{1} = -3$$

$$B = (s+3)Y(s)\,|_{s=-3} = \left.\frac{2s+1}{s+2}\right|_{s=-3} = \frac{-5}{-1} = 5$$

Thus,

$$Y(s) = \frac{-3}{s+2} + \frac{5}{s+3}$$

Its inverse transform is

$$y(t) = (-3e^{-2t} + 5e^{-3t})u(t)$$

Practice Problem 15.16

Use the Laplace transform to solve the integrodifferential equation

$$\frac{dy}{dt} + 3y(t) + 2\int_0^t y(\tau)\,d\tau = 2e^{-3t}, \qquad y(0) = 0$$

Answer: $(-e^{-t} + 4e^{-2t} - 3e^{-3t})u(t)$.

15.7 Summary

1. The Laplace transform allows a signal represented by a function in the time domain to be analyzed in the s-domain (or complex frequency domain). It is defined as

$$\mathcal{L}[f(t)] = F(s) = \int_0^\infty f(t)e^{-st}\,dt$$

2. Properties of the Laplace transform are listed in Table 15.1, while the Laplace transforms of basic common functions are listed in Table 15.2.
3. The inverse Laplace transform can be found using partial fraction expansions and using the Laplace transform pairs in Table 15.2 as a look-up table. Real poles lead to exponential functions and complex poles to damped sinusoids.
4. The convolution of two signals consists of time-reversing one of the signals, shifting it, multiplying it point by point with the second signal, and integrating the product. The convolution integral relates the convolution of two signals in the time domain to the inverse of the product of their Laplace transforms:

$$\mathcal{L}^{-1}[F_1(s)F_2(s)] = f_1(t) * f_2(t) = \int_0^t f_1(\lambda)f_2(t-\lambda)\,d\lambda$$

5. In the time domain, the output $y(t)$ of the network is the convolution of the impulse response with the input $x(t)$,

$$y(t) = h(t) * x(t)$$

Convolution may be regarded as the flip-shift-multiply-time-area method.

6. The Laplace transform can be used to solve a linear integrodifferential equation.

Review Questions

15.1 Every function $f(t)$ has a Laplace transform.

(a) True (b) False

15.2 The variable s in the Laplace transform $H(s)$ is called

(a) complex frequency (b) transfer function
(c) zero (d) pole

15.3 The Laplace transform of $u(t-2)$ is:

(a) $\dfrac{1}{s+2}$ (b) $\dfrac{1}{s-2}$
(c) $\dfrac{e^{2s}}{s}$ (d) $\dfrac{e^{-2s}}{s}$

15.4 The zero of the function

$$F(s) = \frac{s+1}{(s+2)(s+3)(s+4)}$$

is at

(a) -4 (b) -3
(c) -2 (d) -1

15.5 The poles of the function

$$F(s) = \frac{s+1}{(s+2)(s+3)(s+4)}$$

are at

(a) -4 (b) -3
(c) -2 (d) -1

15.6 If $F(s) = 1/(s+2)$, then $f(t)$ is

(a) $e^{2t}u(t)$ (b) $e^{-2t}u(t)$
(c) $u(t-2)$ (d) $u(t+2)$

15.7 Given that $F(s) = e^{-2s}/(s + 1)$, then $f(t)$ is

(a) $e^{-2(t-1)}u(t - 1)$ (b) $e^{-(t-2)}u(t - 2)$
(c) $e^{-t}u(t - 2)$ (d) $e^{-t}u(t + 1)$
(e) $e^{-(t-2)}u(t)$

15.8 The initial value of $f(t)$ with transform

$$F(s) = \frac{s + 1}{(s + 2)(s + 3)}$$

is:

(a) nonexistent (b) ∞ (c) 0
(d) 1 (e) $\frac{1}{6}$

15.9 The inverse Laplace transform of

$$\frac{s + 2}{(s + 2)^2 + 1}$$

is:

(a) $e^{-t}\cos 2t$ (b) $e^{-t}\sin 2t$
(c) $e^{-2t}\cos t$ (d) $e^{-2t}\sin 2t$
(e) none of the above

15.10 The result of $u(t) * u(t)$ is:

(a) $u^2(t)$ (b) $tu(t)$
(c) $t^2u(t)$ (d) $\delta(t)$

Answers: 15.1b, 15.2a, 15.3d, 15.4d, 15.5a,b,c, 15.6b, 15.7b, 15.8d, 15.9c, 15.10b.

Problems

Sections 15.2 and 15.3 Definition and Properties of the Laplace Transform

15.1 Find the Laplace transform of:

(a) cosh at (b) sinh at

[*Hint:* $\cosh x = \frac{1}{2}(e^x + e^{-x})$,

$\sinh x = \frac{1}{2}(e^x - e^{-x})$.]

15.2 Determine the Laplace transform of:

(a) $\cos(\omega t + \theta)$ (b) $\sin(\omega t + \theta)$

15.3 Obtain the Laplace transform of each of the following functions:

(a) $e^{-2t}\cos 3tu(t)$ (b) $e^{-2t}\sin 4tu(t)$
(c) $e^{-3t}\cosh 2tu(t)$ (d) $e^{-4t}\sinh tu(t)$
(e) $te^{-t}\sin 2tu(t)$

15.4 **e⮵d** Design a problem to help other students better understand how to find the Laplace transform of different time varying functions.

15.5 Find the Laplace transform of each of the following functions:

(a) $t^2\cos(2t + 30^\circ)u(t)$ (b) $3t^4e^{-2t}u(t)$
(c) $2tu(t) - 4\frac{d}{dt}\delta(t)$ (d) $2e^{-(t-1)}u(t)$
(e) $5u(t/2)$ (f) $6e^{-t/3}u(t)$
(g) $\frac{d^n}{dt^n}\delta(t)$

15.6 Find $F(s)$ given that

$$f(t) = \begin{cases} 2t, & 0 < t < 1 \\ t, & 1 < t < 2 \\ 0, & \text{otherwise} \end{cases}$$

15.7 Find the Laplace transform of the following signals:

(a) $f(t) = (2t + 4)u(t)$
(b) $g(t) = (4 + 3e^{-2t})u(t)$
(c) $h(t) = (6\sin(3t) + 8\cos(3t))u(t)$
(d) $x(t) = (e^{-2t}\cosh(4t))u(t)$

15.8 Find the Laplace transform $F(s)$, given that $f(t)$ is:

(a) $2tu(t - 4)$
(b) $5\cos(t)\,\delta(t - 2)$
(c) $e^{-t}u(t - t)$
(d) $\sin(2t)u(t - \tau)$

15.9 Determine the Laplace transforms of these functions:

(a) $f(t) = (t - 4)u(t - 2)$
(b) $g(t) = 2e^{-4t}u(t - 1)$
(c) $h(t) = 5\cos(2t - 1)u(t)$
(d) $p(t) = 6[u(t - 2) - u(t - 4)]$

15.10 In two different ways, find the Laplace transform of

$$g(t) = \frac{d}{dt}(2e^{-4t}\cos(2t))$$

15.11 Find $F(s)$ if:

(a) $f(t) = 6e^{-t}\cosh 2t$ (b) $f(t) = 3te^{-2t}\sinh 4t$

(c) $f(t) = 8e^{-3t}\cosh tu(t-2)$

15.12 If $g(t) = e^{-2t}\cos 4t$, find $G(s)$.

15.13 Find the Laplace transform of the following functions:

(a) $t\cos tu(t)$ (b) $e^{-t}t\sin tu(t)$

(c) $\dfrac{\sin\beta t}{t}u(t)$

15.14 Find the Laplace transform of the signal in Fig. 15.26.

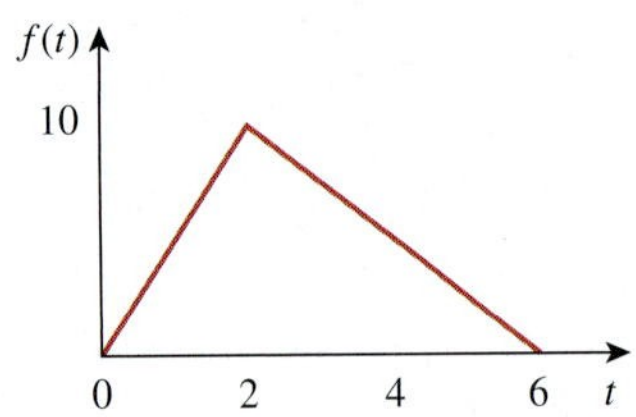

Figure 15.26
For Prob. 15.14.

15.15 Determine the Laplace transform of the function in Fig. 15.27.

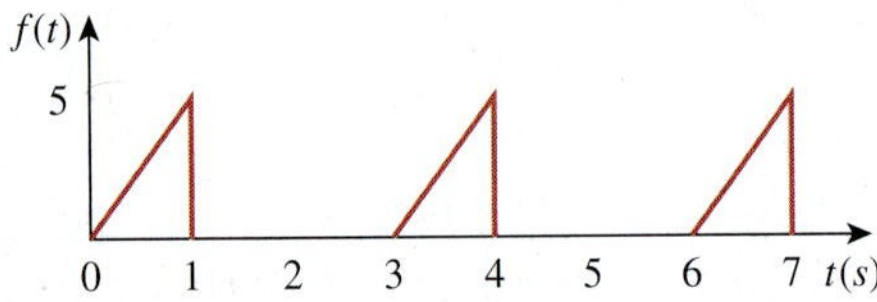

Figure 15.27
For Prob. 15.15.

15.16 Obtain the Laplace transform of $f(t)$ in Fig. 15.28.

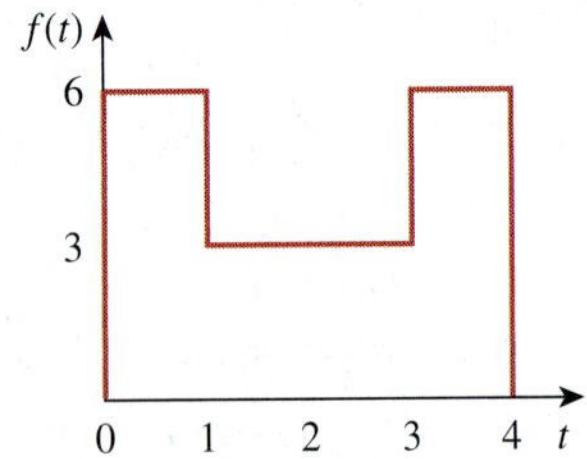

Figure 15.28
For Prob. 15.16.

15.17 Using Fig. 15.29, design a problem to help other **ed** students better understand the Laplace transform of a simple, non-periodic waveshape.

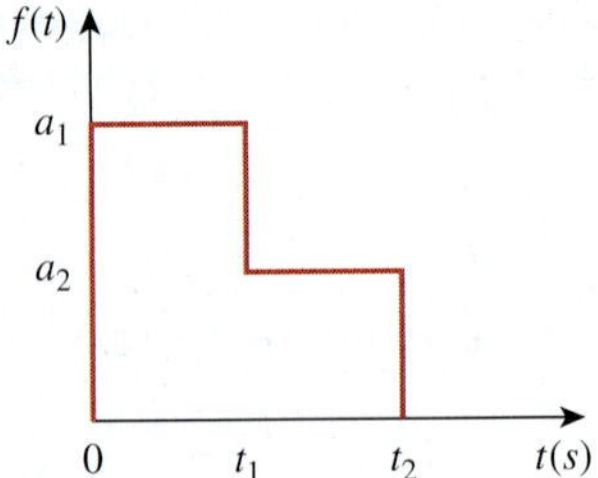

Figure 15.29
For Prob. 15.17.

15.18 Obtain the Laplace transforms of the functions in Fig. 15.30.

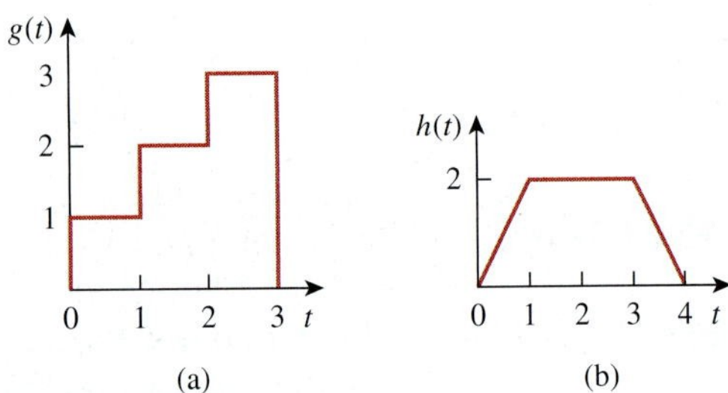

Figure 15.30
For Prob. 15.18.

15.19 Calculate the Laplace transform of the train of unit impulses in Fig. 15.31.

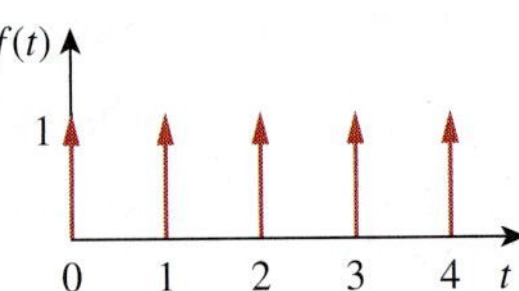

Figure 15.31
For Prob. 15.19.

15.20 Using Fig. 15.32, design a problem to help other **ed** students better understand the Laplace transform of a simple, periodic waveshape.

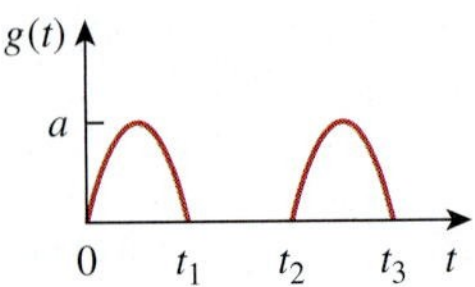

Figure 15.32
For Prob. 15.20.

15.21 Obtain the Laplace transform of the periodic waveform in Fig. 15.33.

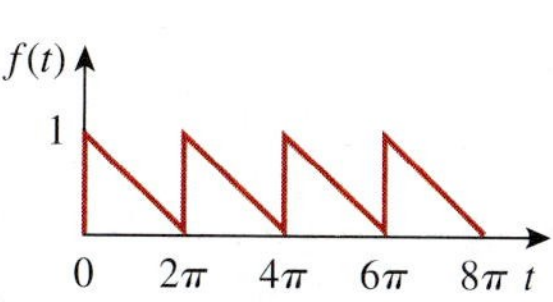

Figure 15.33
For Prob. 15.21.

15.22 Find the Laplace transforms of the functions in Fig. 15.34.

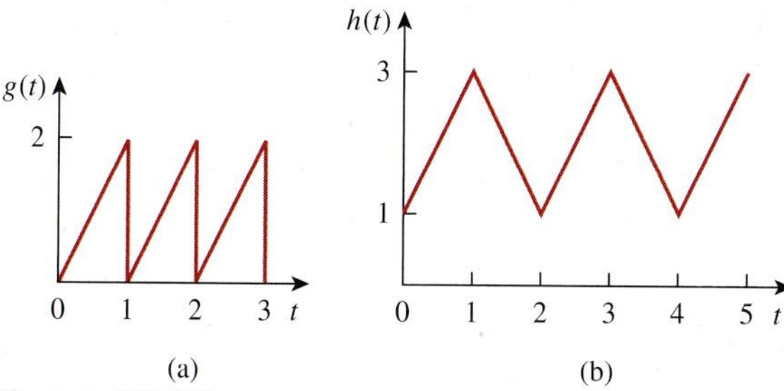

Figure 15.34
For Prob. 15.22.

15.23 Determine the Laplace transforms of the periodic functions in Fig. 15.35.

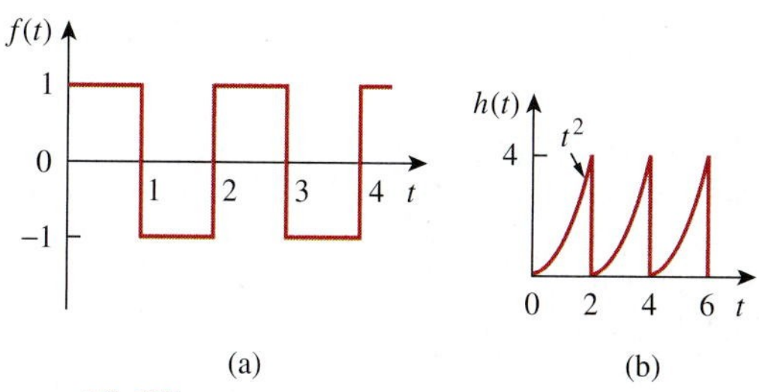

Figure 15.35
For Prob. 15.23.

15.24 Design a problem to help other students better understand how to find the initial and final value of a transfer function.

15.25 Let

$$F(s) = \frac{5(s+1)}{(s+2)(s+3)}$$

(a) Use the initial and final value theorems to find $f(0)$ and $f(\infty)$.

(b) Verify your answer in part (a) by finding $f(t)$, using partial fractions.

15.26 Determine the initial and final values of $f(t)$, if they exist, given that:

(a) $F(s) = \dfrac{s^2+3}{s^3+4s^2+6}$

(b) $F(s) = \dfrac{s^2-2s+1}{(s-2)(s^2+2s+4)}$

Section 15.4 The Inverse Laplace Transform

15.27 Determine the inverse Laplace transform of each of the following functions:

(a) $F(s) = \dfrac{1}{s} + \dfrac{2}{s+1}$

(b) $G(s) = \dfrac{3s+1}{s+4}$

(c) $H(s) = \dfrac{4}{(s+1)(s+3)}$

(d) $J(s) = \dfrac{12}{(s+2)^2(s+4)}$

15.28 Design a problem to help other students better understand how to find the inverse Laplace transform.

15.29 Find the inverse Laplace transform of:

$$V(s) = \frac{s+13}{s(s^2+4s+13)}$$

15.30 Find the inverse Laplace transform of:

(a) $F_1(s) = \dfrac{6s^2+8s+3}{s(s^2+2s+5)}$

(b) $F_2(s) = \dfrac{s^2+5s+6}{(s+1)^2(s+4)}$

(c) $F_3(s) = \dfrac{10}{(s+1)(s^2+4s+8)}$

15.31 Find $f(t)$ for each $F(s)$:

(a) $\dfrac{10s}{(s+1)(s+2)(s+3)}$

(b) $\dfrac{2s^2+4s+1}{(s+1)(s+2)^3}$

(c) $\dfrac{s+1}{(s+2)(s^2+2s+5)}$

15.32 Determine the inverse Laplace transform of each of the following functions:

(a) $\dfrac{8(s+1)(s+3)}{s(s+2)(s+4)}$

(b) $\dfrac{s^2-2s+4}{(s+1)(s+2)^2}$

(c) $\dfrac{s^2+1}{(s+3)(s^2+4s+5)}$

15.33 Calculate the inverse Laplace transform of:

(a) $\dfrac{6(s-1)}{s^4-1}$ (b) $\dfrac{e^{-\pi s}}{s^2+1}$

(c) $\dfrac{3}{s(s+1)^3}$

15.34 Find the time functions that have the following Laplace transforms:

(a) $F(s) = 10 + \dfrac{s^2+1}{s^2+4}$

(b) $G(s) = \dfrac{e^{-s} + 4e^{-2s}}{s^2+6s+8}$

(c) $H(s) = \dfrac{(s+1)e^{-2s}}{s(s+3)(s+4)}$

15.35 Obtain $f(t)$ for the following transforms:

(a) $F(s) = \dfrac{(s+3)e^{-6s}}{(s+1)(s+2)}$

(b) $F(s) = \dfrac{4 - e^{-2s}}{s^2+5s+4}$

(c) $F(s) = \dfrac{se^{-s}}{(s+3)(s^2+4)}$

15.36 Obtain the inverse Laplace transforms of the following functions:

(a) $X(s) = \dfrac{1}{s^2(s+2)(s+3)}$

(b) $Y(s) = \dfrac{1}{s(s+1)^2}$

(c) $Z(s) = \dfrac{1}{s(s+1)(s^2+6s+10)}$

15.37 Find the inverse Laplace transform of:

(a) $H(s) = \dfrac{s+4}{s(s+2)}$

(b) $G(s) = \dfrac{s^2+4s+5}{(s+3)(s^2+2s+2)}$

(c) $F(s) = \dfrac{e^{-4s}}{s+2}$

(d) $D(s) = \dfrac{10s}{(s^2+1)(s^2+4)}$

15.38 Find $f(t)$ given that:

(a) $F(s) = \dfrac{s^2+4s}{s^2+10s+26}$

(b) $F(s) = \dfrac{5s^2+7s+29}{s(s^2+4s+29)}$

***15.39** Determine $f(t)$ if:

(a) $F(s) = \dfrac{2s^3+4s^2+1}{(s^2+2s+17)(s^2+4s+20)}$

(b) $F(s) = \dfrac{s^2+4}{(s^2+9)(s^2+6s+3)}$

15.40 Show that

$$\mathcal{L}^{-1}\left[\frac{4s^2+7s+13}{(s+2)(s^2+2s+5)}\right] = \left[\sqrt{2}e^{-t}\cos(2t+45^\circ) + 3e^{-2t}\right]u(t)$$

Section 15.5 The Convolution Integral

***15.41** Let $x(t)$ and $y(t)$ be as shown in Fig. 15.36. Find $z(t) = x(t) * y(t)$.

15.42 Design a problem to help other students better understand how to convolve two functions together.

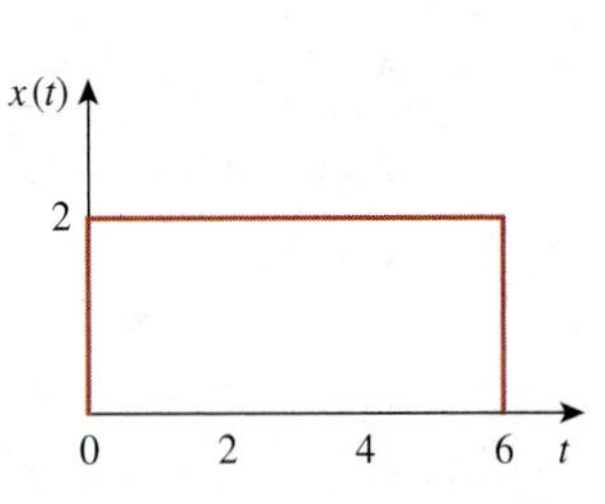

Figure 15.36
For Prob. 15.41.

* An asterisk indicates a challenging problem.

15.43 Find $y(t) = x(t) * h(t)$ for each paired $x(t)$ and $h(t)$ in Fig. 15.37.

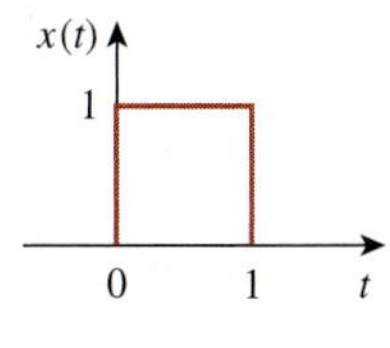

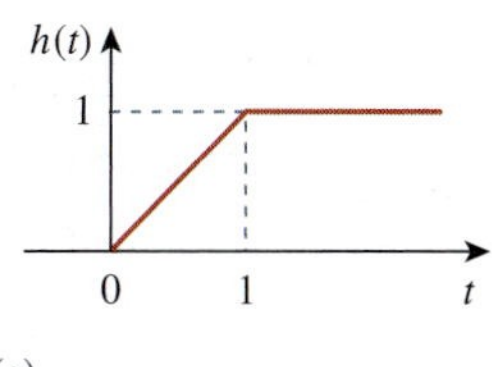

(a)

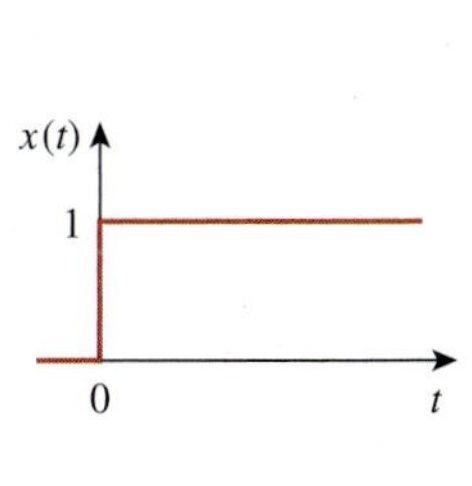

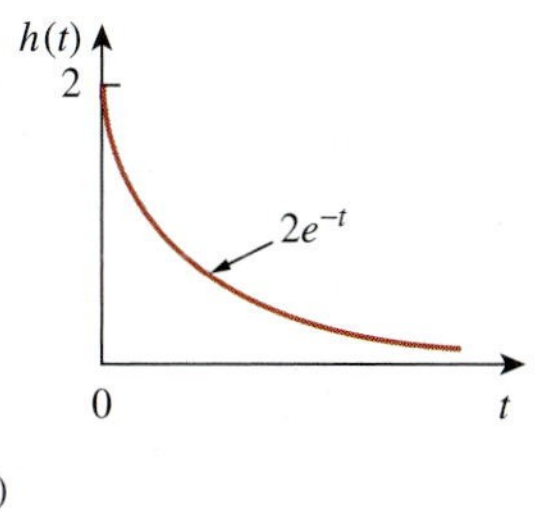

(b)

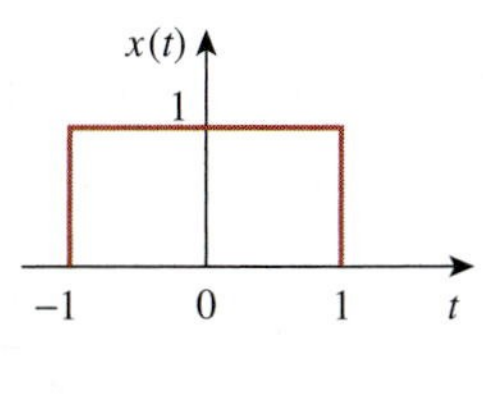

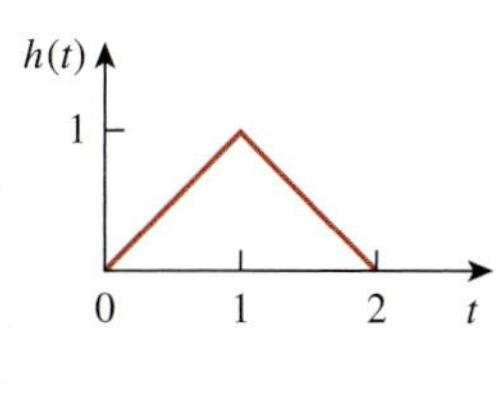

(c)

Figure 15.37
For Prob. 15.43.

15.44 Obtain the convolution of the pairs of signals in Fig. 15.38.

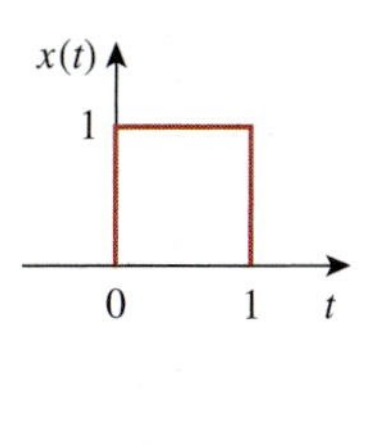

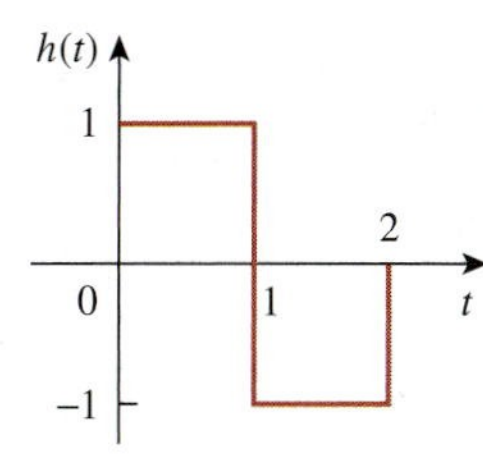

(a)

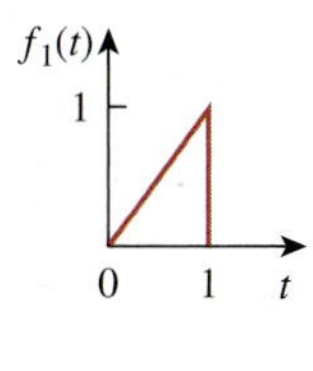

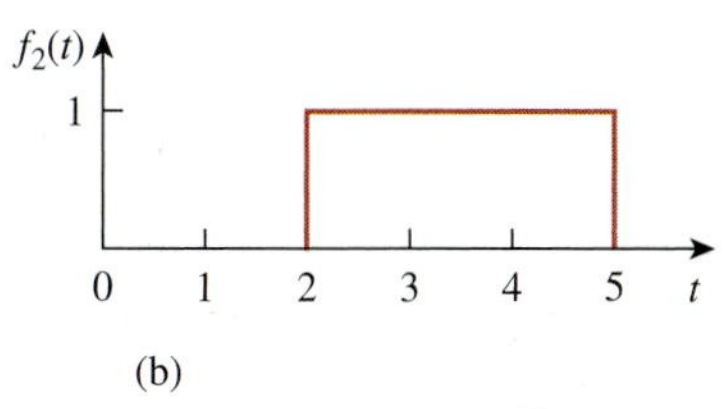

(b)

Figure 15.38
For Prob. 15.44.

15.45 Given $h(t) = 4e^{-2t}u(t)$ and $x(t) = \delta(t) - 2e^{-2t}u(t)$, find $y(t) = x(t) * h(t)$.

15.46 Given the following functions

$$x(t) = 2\delta(t), \qquad y(t) = 4u(t), \qquad z(t) = e^{-2t}u(t),$$

evaluate the following convolution operations.

(a) $x(t) * y(t)$

(b) $x(t) * z(t)$

(c) $y(t) * z(t)$

(d) $y(t) * [\,y(t) + z(t)]$

15.47 A system has the transfer function

$$H(s) = \frac{s}{(s+1)(s+2)}$$

(a) Find the impulse response of the system.

(b) Determine the output $y(t)$, given that the input is $x(t) = u(t)$.

15.48 Find $f(t)$ using convolution given that:

(a) $F(s) = \dfrac{4}{(s^2 + 2s + 5)^2}$

(b) $F(s) = \dfrac{2s}{(s+1)(s^2+4)}$

***15.49** Use the convolution integral to find:

(a) $t * e^{at}u(t)$

(b) $\cos(t) * \cos(t)u(t)$

Section 15.6 Application to Integrodifferential Equations

15.50 Use the Laplace transform to solve the differential equation

$$\frac{d^2v(t)}{dt^2} + 2\frac{dv(t)}{dt} + 10v(t) = 3\cos 2t$$

subject to $v(0) = 1$, $dv(0)/dt = -2$.

15.51 Given that $v(0) = 2$ and $dv(0)/dt = 4$, solve

$$\frac{d^2v}{dt^2} + 5\frac{dv}{dt} + 6v = 6e^{-t}u(t)$$

15.52 Use the Laplace transform to find $i(t)$ for $t > 0$ if

$$\frac{d^2i}{dt^2} + 3\frac{di}{dt} + 2i + \delta(t) = 0,$$

$$i(0) = 0, \qquad i'(0) = 3$$

***15.53** Use Laplace transforms to solve for $x(t)$ in

$$x(t) = \cos t + \int_0^t e^{\lambda - t}x(\lambda)\,d\lambda$$

15.54 Design a problem to help other students better understand solving second order differential equations with a time varying input.

15.55 Solve for $y(t)$ in the following differential equation if the initial conditions are zero.

$$\frac{d^3y}{dt^3} + 6\frac{d^2y}{dt^2} + 8\frac{dy}{dt} = e^{-t}\cos 2t$$

15.56 Solve for $v(t)$ in the integrodifferential equation

$$4\frac{dv}{dt} + 12\int_{-\infty}^{t} v\,dt = 0$$

given that $v(0) = 2$.

15.57 Design a problem to help other students better understand solving integrodifferential equations with a periodic input, using Laplace transforms.

e⊘d

15.58 Given that

$$\frac{dv}{dt} + 2v + 5\int_0^t v(\lambda)\,d\lambda = 4u(t)$$

with $v(0) = -1$, determine $v(t)$ for $t > 0$.

15.59 Solve the integrodifferential equation

$$\frac{dy}{dt} + 4y + 3\int_0^t y\,dt = 6e^{-2t}, \qquad y(0) = -1$$

15.60 Solve the following integrodifferential equation

$$2\frac{dx}{dt} + 5x + 3\int_0^t x\,dt + 4 = \sin 4t, \qquad x(0) = 1$$

chapter

16

Applications of the Laplace Transform

Communication skills are the most important skills any engineer can have. A very critical element in this tool set is the ability to ask a question and understand the answer, a very simple thing and yet it may make the difference between success and failure!

—James A. Watson

Enhancing Your Skills and Your Career

Asking Questions

In over 30 years of teaching, I have struggled with determining how best to help students learn. Regardless of how much time students spend in studying for a course, the most helpful activity for students is learning how to ask questions in class and then asking those questions. The student, by asking questions, becomes actively involved in the learning process and no longer is merely a passive receptor of information. I think this active involvement contributes so much to the learning process that it is probably the single most important aspect to the development of a modern engineer. In fact, asking questions is the basis of science. As Charles P. Steinmetz rightly said, "No man really becomes a fool until he stops asking questions."

Photo by Charles Alexander

It seems very straightforward and quite easy to ask questions. Have we not been doing that all our lives? The truth is to ask questions in an appropriate manner and to maximize the learning process takes some thought and preparation.

I am sure that there are several models one could effectively use. Let me share what has worked for me. The most important thing to keep in mind is that you do not have to form a perfect question. Since the question-and-answer format allows the question to be developed in an iterative manner, the original question can easily be refined as you go. I frequently tell students that they are most welcome to read their questions in class.

Here are three things you should keep in mind when asking questions. First, prepare your question. If you are like many students who are either shy or have not learned to ask questions in class, you may wish to start with a question you have written down outside of class. Second, wait for an appropriate time to ask the question. Simply use your judgment on that. Third, be prepared to clarify your question by paraphrasing it or saying it in a different way in case you are asked to repeat the question.

One last comment: not all professors like students to ask questions in class even though they may say they do. You need to find out which professors like classroom questions. Good luck in enhancing one of your most important skills as an engineer.

16.1 Introduction

Now that we have introduced the Laplace transform, let us see what we can do with it. Please keep in mind that with the Laplace transform we actually have one of the most powerful mathematical tools for analysis, synthesis, and design. Being able to look at circuits and systems in the s-domain can help us to understand how our circuits and systems really function. In this chapter we will take an in-depth look at how easy it is to work with circuits in the s-domain. In addition, we will briefly look at physical systems. We are sure you have studied some mechanical systems and may have used the same differential equations to describe them as we use to describe our electric circuits. Actually that is a wonderful thing about the physical universe in which we live; the same differential equations can be used to describe any linear circuit, system, or process. The key is the term *linear*.

A **system** is a mathematical model of a physical process relating the input to the output.

It is entirely appropriate to consider circuits as systems. Historically, circuits have been discussed as a separate topic from systems, so we will actually talk about circuits and systems in this chapter realizing that circuits are nothing more than a class of electrical systems.

The most important thing to remember is that everything we discussed in the last chapter and in this chapter applies to any linear system. In the last chapter, we saw how we can use Laplace transforms to solve linear differential equations and integral equations. In this chapter, we introduce the concept of modeling circuits in the s-domain. We can use that principle to help us solve just about any kind of linear circuit. We will take a quick look at how state variables can be used to analyze systems with multiple inputs and multiple outputs. Finally, we examine how the Laplace transform is used in network stability analysis and in network synthesis.

16.2 Circuit Element Models

Having mastered how to obtain the Laplace transform and its inverse, we are now prepared to employ the Laplace transform to analyze circuits. This usually involves three steps.

Steps in Applying the Laplace Transform:

1. Transform the circuit from the time domain to the s-domain.
2. Solve the circuit using nodal analysis, mesh analysis, source transformation, superposition, or any circuit analysis technique with which we are familiar.
3. Take the inverse transform of the solution and thus obtain the solution in the time domain.

Only the first step is new and will be discussed here. As we did in phasor analysis, we transform a circuit in the time domain to the frequency or s-domain by Laplace transforming each term in the circuit.

As one can infer from step 2, all the circuit analysis techniques applied for dc circuits are applicable to the s-domain.

For a resistor, the voltage-current relationship in the time domain is

$$v(t) = Ri(t) \tag{16.1}$$

Taking the Laplace transform, we get

$$V(s) = RI(s) \tag{16.2}$$

For an inductor,

$$v(t) = L\frac{di(t)}{dt} \tag{16.3}$$

Taking the Laplace transform of both sides gives

$$V(s) = L[sI(s) - i(0^-)] = sLI(s) - Li(0^-) \tag{16.4}$$

or

$$I(s) = \frac{1}{sL}V(s) + \frac{i(0^-)}{s} \tag{16.5}$$

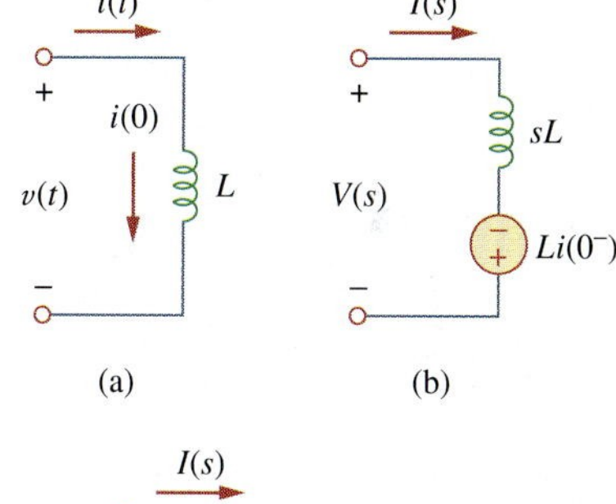

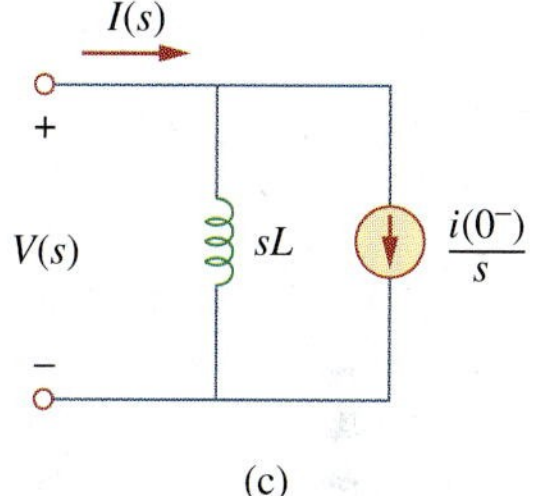

Figure 16.1
Representation of an inductor: (a) time-domain, (b,c) s-domain equivalents.

The s-domain equivalents are shown in Fig. 16.1, where the initial condition is modeled as a voltage or current source.

For a capacitor,

$$i(t) = C\frac{dv(t)}{dt} \tag{16.6}$$

which transforms into the s-domain as

$$I(s) = C[sV(s) - v(0^-)] = sCV(s) - Cv(0^-) \tag{16.7}$$

or

$$V(s) = \frac{1}{sC}I(s) + \frac{v(0^-)}{s} \tag{16.8}$$

The s-domain equivalents are shown in Fig. 16.2. With the s-domain equivalents, the Laplace transform can be used readily to solve first- and

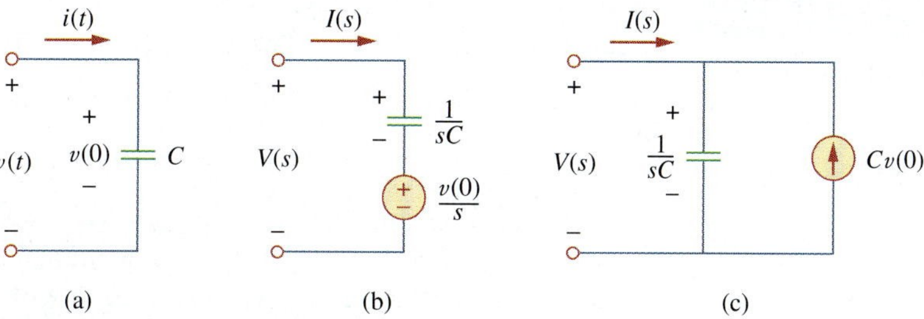

Figure 16.2
Representation of a capacitor: (a) time-domain, (b,c) s-domain equivalents.

The elegance of using the Laplace transform in circuit analysis lies in the automatic inclusion of the initial conditions in the transformation process, thus providing a complete (transient and steady-state) solution.

second-order circuits such as those we considered in Chapters 7 and 8. We should observe from Eqs. (16.3) to (16.8) that the initial conditions are part of the transformation. This is one advantage of using the Laplace transform in circuit analysis. Another advantage is that a complete response—transient and steady state—of a network is obtained. We will illustrate this with Examples 16.2 and 16.3. Also, observe the duality of Eqs. (16.5) and (16.8), confirming what we already know from Chapter 8 (see Table 8.1), namely, that L and C, $I(s)$ and $V(s)$, and $v(0)$ and $i(0)$ are dual pairs.

If we assume zero initial conditions for the inductor and the capacitor, the above equations reduce to:

$$\begin{aligned} &\text{Resistor:} && V(s) = RI(s) \\ &\text{Inductor:} && V(s) = sLI(s) \\ &\text{Capacitor:} && V(s) = \frac{1}{sC}I(s) \end{aligned} \tag{16.9}$$

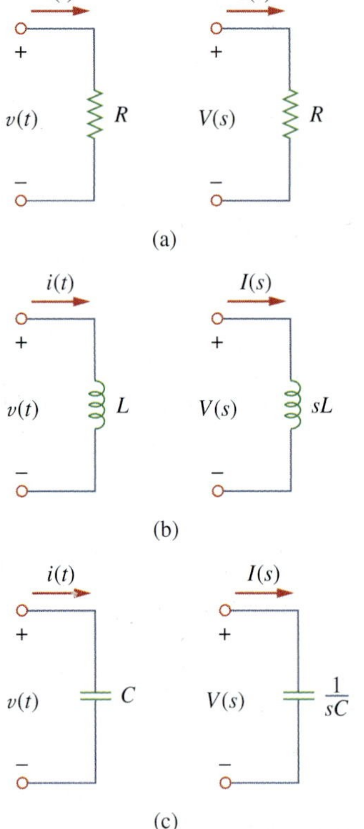

Figure 16.3
Time-domain and s-domain representations of passive elements under zero initial conditions.

The s-domain equivalents are shown in Fig. 16.3.

We define the impedance in the s-domain as the ratio of the voltage transform to the current transform under zero initial conditions; that is,

$$Z(s) = \frac{V(s)}{I(s)} \tag{16.10}$$

Thus, the impedances of the three circuit elements are

$$\begin{aligned} &\text{Resistor:} && Z(s) = R \\ &\text{Inductor:} && Z(s) = sL \\ &\text{Capacitor:} && Z(s) = \frac{1}{sC} \end{aligned} \tag{16.11}$$

Table 16.1 summarizes these. The admittance in the s-domain is the reciprocal of the impedance, or

$$Y(s) = \frac{1}{Z(s)} = \frac{I(s)}{V(s)} \tag{16.12}$$

The use of the Laplace transform in circuit analysis facilitates the use of various signal sources such as impulse, step, ramp, exponential, and sinusoidal.

The models for dependent sources and op amps are easy to develop drawing from the simple fact that if the Laplace transform of $f(t)$ is $F(s)$, then the Laplace transform of $af(t)$ is $aF(s)$—the linearity property. The dependent source model is a little easier in that we deal with a single value. The dependent source can have only two controlling values, a constant times either a voltage or a current. Thus,

$$\mathcal{L}[av(t)] = aV(s) \tag{16.13}$$

$$\mathcal{L}[ai(t)] = aI(s) \tag{16.14}$$

The ideal op amp can be treated just like a resistor. Nothing within an op amp, either real or ideal, does anything more than multiply a voltage by a constant. Thus, we only need to write the equations as we always do using the constraint that the input voltage to the op amp has to be zero and the input current has to be zero.

TABLE 16.1

Impedance of an element in the s-domain.*

Element	$Z(s) = V(s)/I(s)$
Resistor	R
Inductor	sL
Capacitor	$1/sC$

* Assuming zero initial conditions

Example 16.1

Find $v_o(t)$ in the circuit of Fig. 16.4, assuming zero initial conditions.

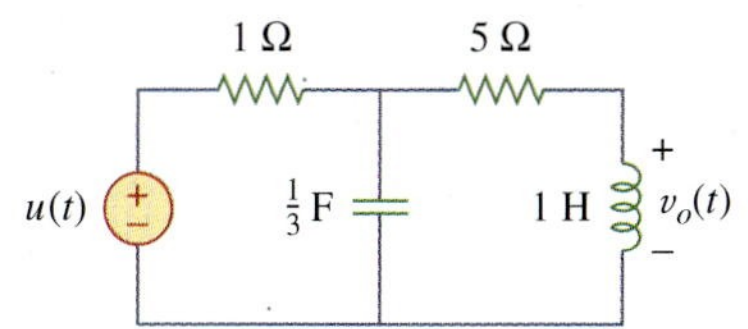

Figure 16.4
For Example 16.1.

Solution:
We first transform the circuit from the time domain to the s-domain.

$$u(t) \quad \Rightarrow \quad \frac{1}{s}$$

$$1\text{ H} \quad \Rightarrow \quad sL = s$$

$$\frac{1}{3}\text{F} \quad \Rightarrow \quad \frac{1}{sC} = \frac{3}{s}$$

The resulting s-domain circuit is in Fig. 16.5. We now apply mesh analysis. For mesh 1,

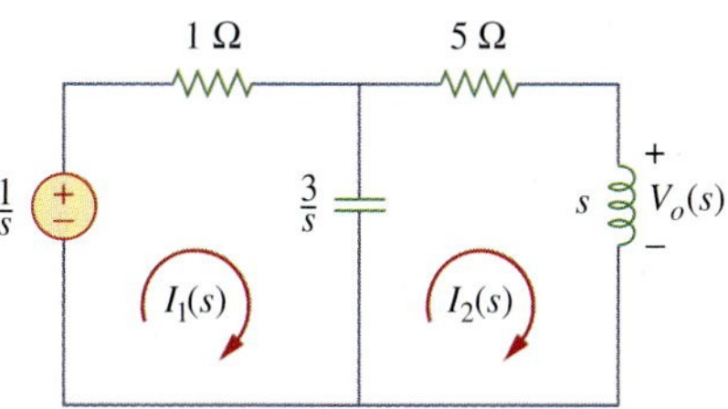

Figure 16.5
Mesh analysis of the frequency-domain equivalent of the same circuit.

$$\frac{1}{s} = \left(1 + \frac{3}{s}\right)I_1 - \frac{3}{s}I_2 \tag{16.1.1}$$

For mesh 2,

$$0 = -\frac{3}{s}I_1 + \left(s + 5 + \frac{3}{s}\right)I_2$$

or

$$I_1 = \frac{1}{3}(s^2 + 5s + 3)I_2 \tag{16.1.2}$$

Substituting this into Eq. (16.1.1),

$$\frac{1}{s} = \left(1 + \frac{3}{s}\right)\frac{1}{3}(s^2 + 5s + 3)I_2 - \frac{3}{s}I_2$$

Multiplying through by $3s$ gives

$$3 = (s^3 + 8s^2 + 18s)I_2 \quad \Rightarrow \quad I_2 = \frac{3}{s^3 + 8s^2 + 18s}$$

$$V_o(s) = sI_2 = \frac{3}{s^2 + 8s + 18} = \frac{3}{\sqrt{2}}\frac{\sqrt{2}}{(s+4)^2 + (\sqrt{2})^2}$$

Taking the inverse transform yields

$$v_o(t) = \frac{3}{\sqrt{2}}e^{-4t}\sin\sqrt{2}t\text{ V}, \qquad t \geq 0$$

Practice Problem 16.1

Determine $v_o(t)$ in the circuit of Fig. 16.6, assuming zero initial conditions.

Answer: $20(1 - e^{-2t} - 2te^{-2t})u(t)$ V.

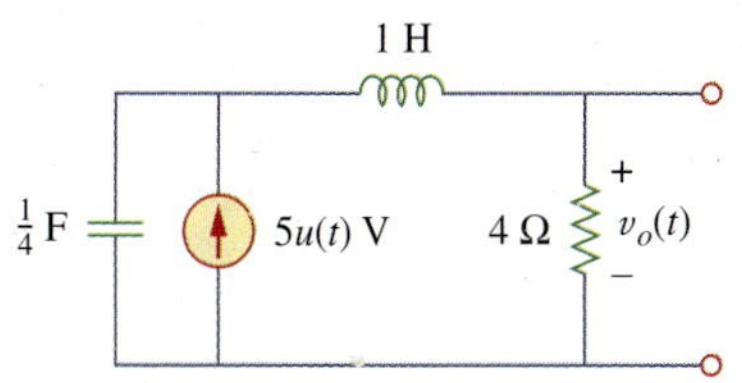

Figure 16.6
For Practice Prob. 16.1.

Example 16.2

Find $v_o(t)$ in the circuit of Fig. 16.7. Assume $v_o(0) = 5$ V.

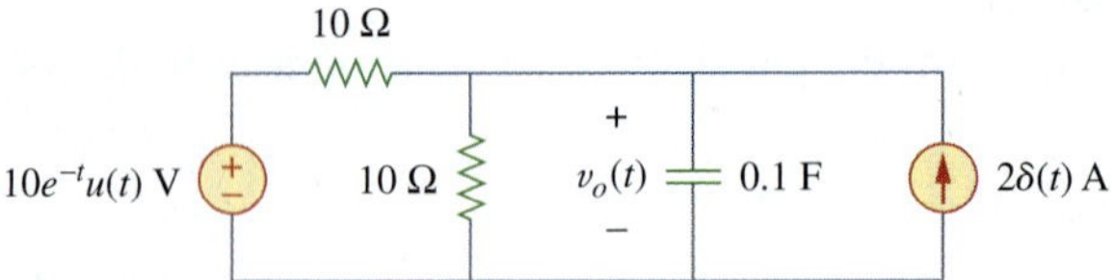

Figure 16.7
For Example 16.2.

Solution:

We transform the circuit to the s-domain as shown in Fig. 16.8. The initial condition is included in the form of the current source $Cv_o(0) = 0.1(5) = 0.5$ A. [See Fig. 16.2(c).] We apply nodal analysis. At the top node,

$$\frac{10/(s+1) - V_o}{10} + 2 + 0.5 = \frac{V_o}{10} + \frac{V_o}{10/s}$$

or

$$\frac{1}{s+1} + 2.5 = \frac{2V_o}{10} + \frac{sV_o}{10} = \frac{1}{10}V_o(s+2)$$

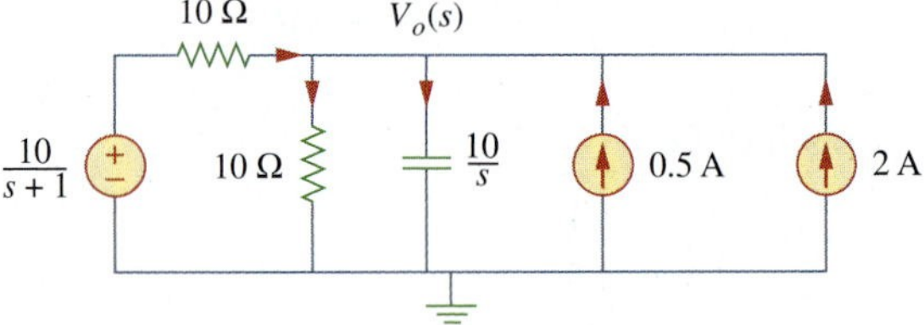

Figure 16.8
Nodal analysis of the equivalent of the circuit in Fig. 16.7.

Multiplying through by 10,

$$\frac{10}{s+1} + 25 = V_o(s+2)$$

or

$$V_o = \frac{25s + 35}{(s+1)(s+2)} = \frac{A}{s+1} + \frac{B}{s+2}$$

where

$$A = (s+1)V_o(s)\big|_{s=-1} = \frac{25s+35}{(s+2)}\bigg|_{s=-1} = \frac{10}{1} = 10$$

$$B = (s+2)V_o(s)\big|_{s=-2} = \frac{25s+35}{(s+1)}\bigg|_{s=-2} = \frac{-15}{-1} = 15$$

Thus,

$$V_o(s) = \frac{10}{s+1} + \frac{15}{s+2}$$

Taking the inverse Laplace transform, we obtain

$$v_o(t) = (10e^{-t} + 15e^{-2t})u(t) \text{ V}$$

Practice Problem 16.2

Find $v_o(t)$ in the circuit shown in Fig. 16.9. Note that, since the voltage input is multiplied by $u(t)$, the voltage source is a short for all $t < 0$ and $i_L(0) = 0$.

Answer: $(24e^{-2t} - 4e^{-t/3})u(t)$ V.

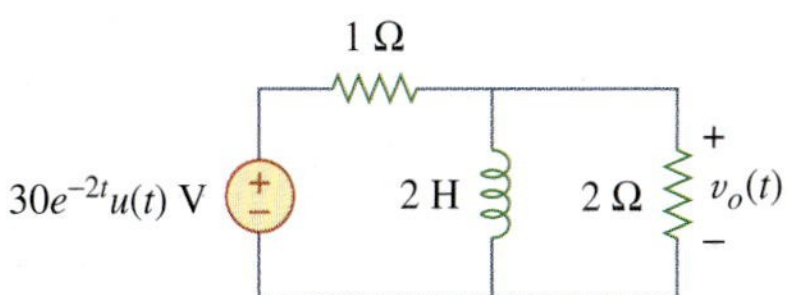

Figure 16.9
For Practice Prob. 16.2.

Example 16.3

In the circuit of Fig. 16.10(a), the switch moves from position a to position b at $t = 0$. Find $i(t)$ for $t > 0$.

Solution:
The initial current through the inductor is $i(0) = I_o$. For $t > 0$, Fig. 16.10(b) shows the circuit transformed to the s-domain. The initial condition is incorporated in the form of a voltage source as $Li(0) = LI_o$. Using mesh analysis,

$$I(s)(R + sL) - LI_o - \frac{V_o}{s} = 0 \tag{16.3.1}$$

or

$$I(s) = \frac{LI_o}{R + sL} + \frac{V_o}{s(R + sL)} = \frac{I_o}{s + R/L} + \frac{V_o/L}{s(s + R/L)} \tag{16.3.2}$$

Applying partial fraction expansion on the second term on the right-hand side of Eq. (16.3.2) yields

$$I(s) = \frac{I_o}{s + R/L} + \frac{V_o/R}{s} - \frac{V_o/R}{(s + R/L)} \tag{16.3.3}$$

The inverse Laplace transform of this gives

$$i(t) = \left(I_o - \frac{V_o}{R}\right)e^{-t/\tau} + \frac{V_o}{R}, \qquad t \geq 0 \tag{16.3.4}$$

where $\tau = R/L$. The term in parentheses is the transient response, while the second term is the steady-state response. In other words, the final value is $i(\infty) = V_o/R$, which we could have predicted by applying the final-value theorem on Eq. (16.3.2) or (16.3.3); that is,

$$\lim_{s\to 0} sI(s) = \lim_{s\to 0}\left(\frac{sI_o}{s + R/L} + \frac{V_o/L}{s + R/L}\right) = \frac{V_o}{R} \tag{16.3.5}$$

Equation (16.3.4) may also be written as

$$i(t) = I_o e^{-t/\tau} + \frac{V_o}{R}(1 - e^{-t/\tau}), \qquad t \geq 0 \tag{16.3.6}$$

The first term is the natural response, while the second term is the forced response. If the initial condition $I_o = 0$, Eq. (16.3.6) becomes

$$i(t) = \frac{V_o}{R}(1 - e^{-t/\tau}), \qquad t \geq 0 \tag{16.3.7}$$

which is the step response, since it is due to the step input V_o with no initial energy.

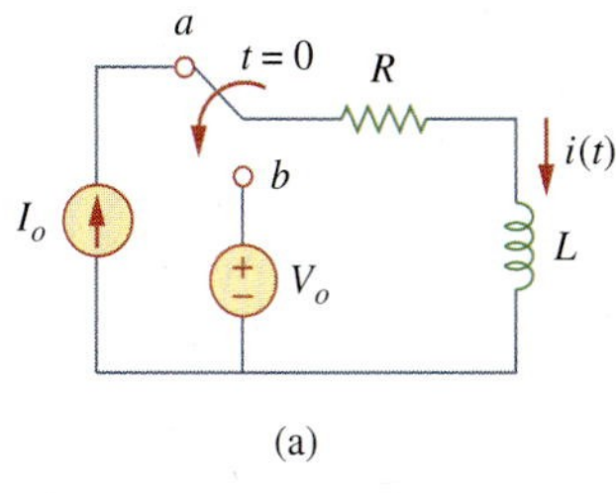

(a)

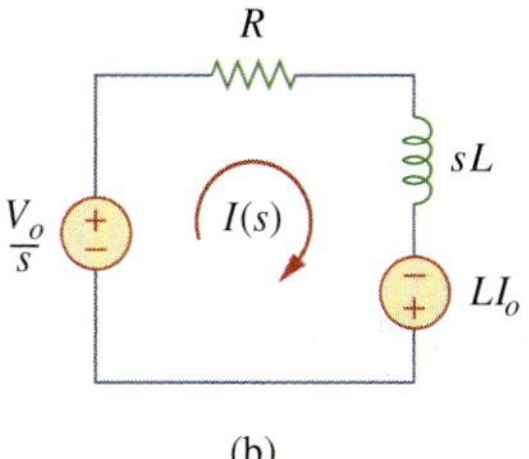

(b)

Figure 16.10
For Example 16.3.

Practice Problem 16.3

The switch in Fig. 16.11 has been in position b for a long time. It is moved to position a at $t = 0$. Determine $v(t)$ for $t > 0$.

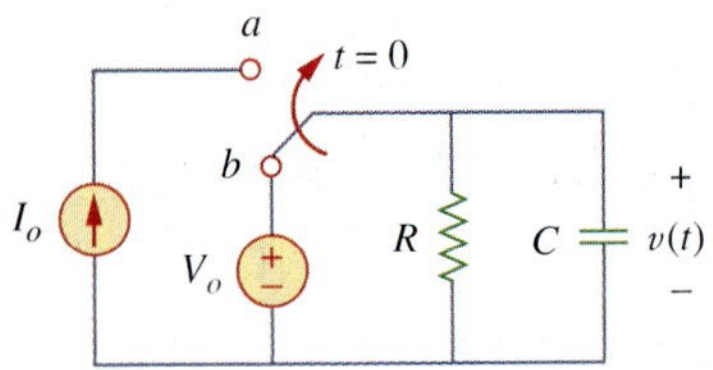

Figure 16.11
For Practice Prob. 16.3.

Answer: $v(t) = (V_o - I_oR)e^{-t/\tau} + I_oR,\ t > 0$, where $\tau = RC$.

16.3 Circuit Analysis

Circuit analysis is again relatively easy to do when we are in the s-domain. We merely need to transform a complicated set of mathematical relationships in the time domain into the s-domain where we convert operators (derivatives and integrals) into simple multipliers of s and $1/s$. This now allows us to use algebra to set up and solve our circuit equations. The exciting thing about this is that *all* of the circuit theorems and relationships we developed for dc circuits are perfectly valid in the s-domain.

Remember, **equivalent circuits,** with capacitors and inductors, only exist in the s-domain; they cannot be transformed back into the time domain.

Example 16.4

Consider the circuit in Fig. 16.12(a). Find the value of the voltage across the capacitor assuming that the value of $v_s(t) = 10u(t)$ V and assume that at $t = 0$, -1 A flows through the inductor and $+5$ V is across the capacitor.

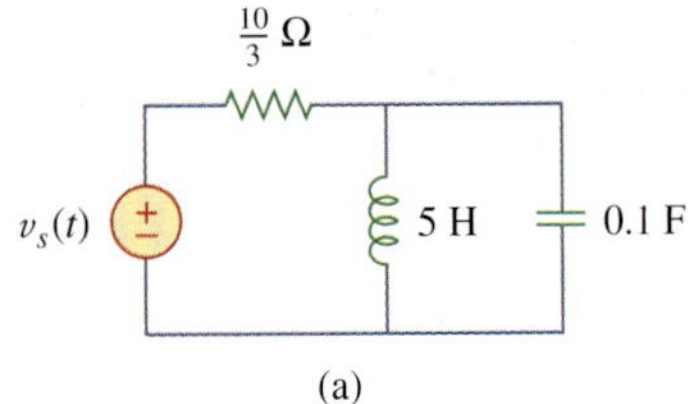

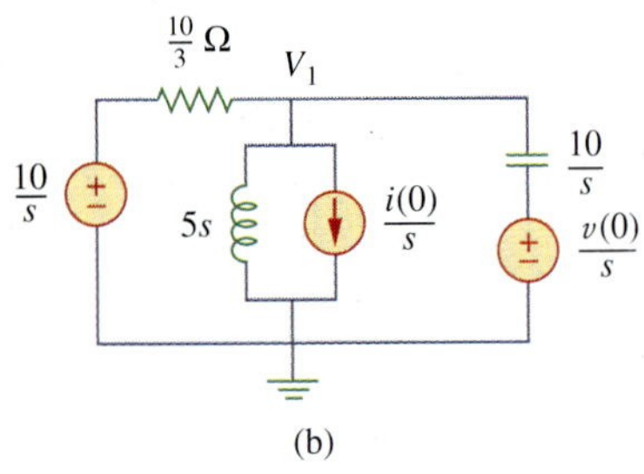

Figure 16.12
For Example 16.4.

Solution:

Figure 16.12(b) represents the entire circuit in the s-domain with the initial conditions incorporated. We now have a straightforward nodal analysis problem. Since the value of V_1 is also the value of the capacitor voltage in the time domain and is the only unknown node voltage, we only need to write one equation.

$$\frac{V_1 - 10/s}{10/3} + \frac{V_1 - 0}{5s} + \frac{i(0)}{s} + \frac{V_1 - [v(0)/s]}{1/(0.1s)} = 0 \qquad (16.4.1)$$

or

$$0.1\left(s + 3 + \frac{2}{s}\right)V_1 = \frac{3}{s} + \frac{1}{s} + 0.5 \qquad (16.4.2)$$

where $v(0) = 5$ V and $i(0) = -1$ A. Simplifying we get

$$(s^2 + 3s + 2)\,V_1 = 40 + 5s$$

or

$$V_1 = \frac{40 + 5s}{(s + 1)(s + 2)} = \frac{35}{s + 1} - \frac{30}{s + 2} \qquad (16.4.3)$$

Taking the inverse Laplace transform yields

$$v_1(t) = (35e^{-t} - 30e^{-2t})u(t)\ \text{V} \qquad (16.4.4)$$

Practice Problem 16.4

For the circuit shown in Fig. 16.12 with the same initial conditions, find the current through the inductor for all time $t > 0$.

Answer: $i(t) = (3 - 7e^{-t} + 3e^{-2t})u(t)$ A.

Example 16.5

For the circuit shown in Fig. 16.12, and the initial conditions used in Example 16.4, use superposition to find the value of the capacitor voltage.

Solution:

Since the circuit in the s-domain actually has three independent sources, we can look at the solution one source at a time. Figure 16.13 presents the circuits in the s-domain considering one source at a time. We now have three nodal analysis problems. First, let us solve for the capacitor voltage in the circuit shown in Fig. 16.13(a).

$$\frac{V_1 - 10/s}{10/3} + \frac{V_1 - 0}{5s} + 0 + \frac{V_1 - 0}{1/(0.1s)} = 0$$

or

$$0.1\left(s + 3 + \frac{2}{s}\right)V_1 = \frac{3}{s}$$

Simplifying we get

$$(s^2 + 3s + 2)V_1 = 30$$

$$V_1 = \frac{30}{(s+1)(s+2)} = \frac{30}{s+1} - \frac{30}{s+2}$$

or

$$v_1(t) = (30e^{-t} - 30e^{-2t})u(t) \text{ V} \qquad \textbf{(16.5.1)}$$

For Fig. 16.13(b) we get,

$$\frac{V_2 - 0}{10/3} + \frac{V_2 - 0}{5s} - \frac{1}{s} + \frac{V_2 - 0}{1/(0.1s)} = 0$$

or

$$0.1\left(s + 3 + \frac{2}{s}\right)V_2 = \frac{1}{s}$$

This leads to

$$V_2 = \frac{10}{(s+1)(s+2)} = \frac{10}{s+1} - \frac{10}{s+2}$$

Taking the inverse Laplace transform, we get

$$v_2(t) = (10e^{-t} - 10e^{-2t})u(t) \text{ V} \qquad \textbf{(16.5.2)}$$

For Fig. 16.13(c),

$$\frac{V_3 - 0}{10/3} + \frac{V_3 - 0}{5s} - 0 + \frac{V_3 - 5/s}{1/(0.1s)} = 0$$

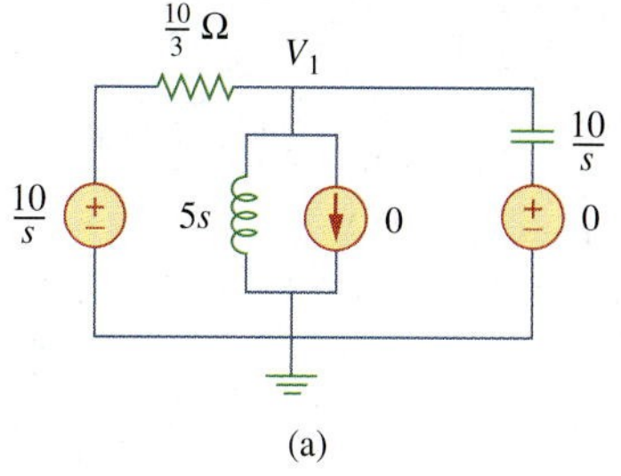

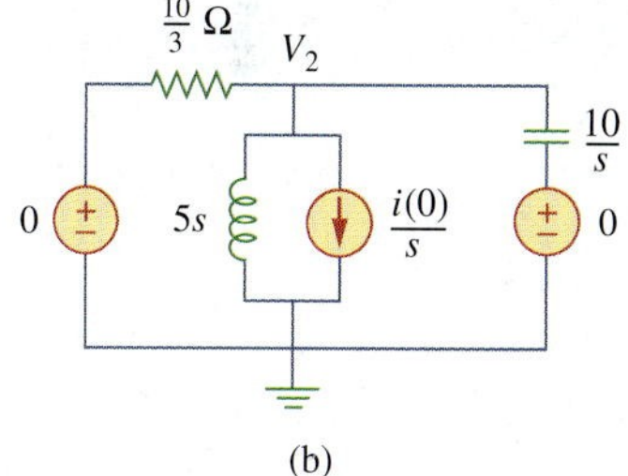

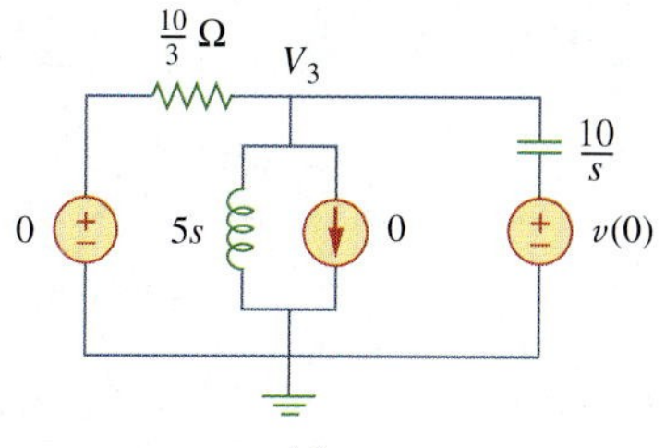

Figure 16.13
For Example 16.5.

or

$$0.1\left(s + 3 + \frac{2}{s}\right)V_3 = 0.5$$

$$V_3 = \frac{5s}{(s + 1)(s + 2)} = \frac{-5}{s + 1} + \frac{10}{s + 2}$$

This leads to

$$v_3(t) = (-5e^{-t} + 10e^{-2t})u(t)\text{ V} \qquad \textbf{(16.5.3)}$$

Now all we need to do is to add Eqs. (16.5.1), (16.5.2), and (16.5.3):

$$\begin{aligned} v(t) &= v_1(t) + v_2(t) + v_3(t) \\ &= \{(30 + 10 - 5)e^{-t} + (-30 + 10 - 10)e^{-2t}\}u(t)\text{ V} \end{aligned}$$

or

$$v(t) = (35e^{-t} - 30e^{-2t})u(t)\text{ V}$$

which agrees with our answer in Example 16.4.

Practice Problem 16.5

For the circuit shown in Fig. 16.12, and the same initial conditions in Example 16.4, find the current through the inductor for all time $t > 0$ using superposition.

Answer: $i(t) = (3 - 7e^{-t} + 3e^{-2t})u(t)$ A.

Example 16.6

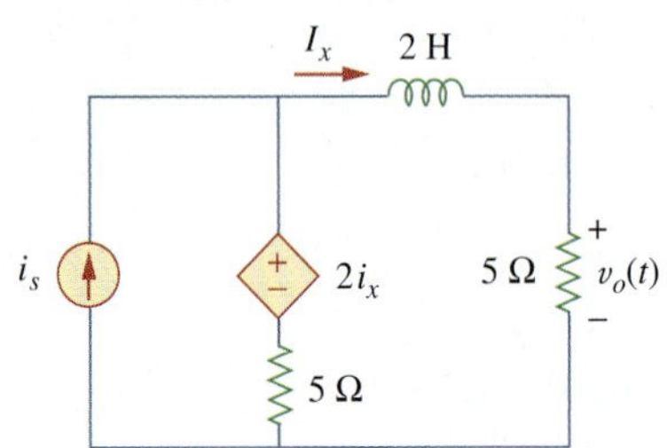

Figure 16.14
For Example 16.6.

Assume that there is no initial energy stored in the circuit of Fig. 16.14 at $t = 0$ and that $i_s = 10\,u(t)$ A. (a) Find $V_o(s)$ using Thevenin's theorem. (b) Apply the initial- and final-value theorems to find $v_o(0^+)$ and $v_o(\infty)$. (c) Determine $v_o(t)$.

Solution:

Since there is no initial energy stored in the circuit, we assume that the initial inductor current and initial capacitor voltage are zero at $t = 0$.

(a) To find the Thevenin equivalent circuit, we remove the 5-Ω resistor and then find V_{oc} (V_{Th}) and I_{sc}. To find V_{Th}, we use the Laplace-transformed circuit in Fig. 16.15(a). Since $I_x = 0$, the dependent voltage source contributes nothing, so

$$V_{oc} = V_{Th} = 5\left(\frac{10}{s}\right) = \frac{50}{s}$$

To find Z_{Th}, we consider the circuit in Fig. 16.15(b), where we first find I_{sc}. We can use nodal analysis to solve for V_1 which then leads to I_{sc} ($I_{sc} = I_x = V_1/2s$).

$$-\frac{10}{s} + \frac{(V_1 - 2I_x) - 0}{5} + \frac{V_1 - 0}{2s} = 0$$

along with

$$I_x = \frac{V_1}{2s}$$

leads to

$$V_1 = \frac{100}{2s + 3}$$

Hence,

$$I_{sc} = \frac{V_1}{2s} = \frac{100/(2s + 3)}{2s} = \frac{50}{s(2s + 3)}$$

and

$$Z_{Th} = \frac{V_{oc}}{I_{sc}} = \frac{50/s}{50/[s(2s + 3)]} = 2s + 3$$

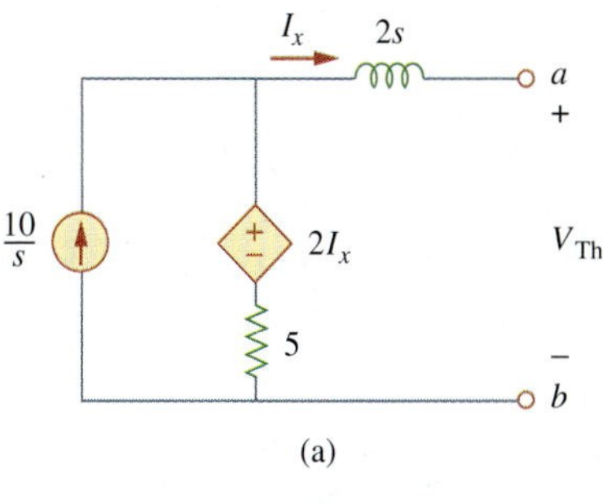

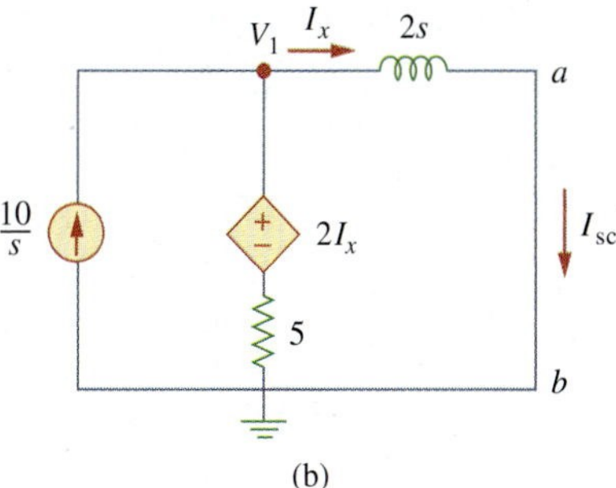

Figure 16.15
For Example 16.6: (a) finding V_{Th}, (b) determining Z_{Th}.

The given circuit is replaced by its Thevenin equivalent at terminals *a-b* as shown in Fig. 16.16. From Fig. 16.16,

$$V_o = \frac{5}{5 + Z_{Th}}V_{Th} = \frac{5}{5 + 2s + 3}\left(\frac{50}{s}\right) = \frac{250}{s(2s + 8)} = \frac{125}{s(s + 4)}$$

(b) Using the initial-value theorem we find

$$v_o(0) = \lim_{s\to\infty} sV_o(s) = \lim_{s\to\infty}\frac{125}{s + 4} = \lim_{s\to\infty}\frac{125/s}{1 + 4/s} = \frac{0}{1} = 0$$

Using the final-value theorem we find

$$v_o(\infty) = \lim_{s\to 0} sV_o(s) = \lim_{s\to 0}\frac{125}{s + 4} = \frac{125}{4} = 31.25 \text{ V}$$

(c) By partial fraction,

$$V_o = \frac{125}{s(s + 4)} = \frac{A}{s} + \frac{B}{s + 4}$$

$$A = sV_o(s)\bigg|_{s=0} = \frac{125}{s + 4}\bigg|_{s=0} = 31.25$$

$$B = (s + 4)V_o(s)\bigg|_{s=-4} = \frac{125}{s}\bigg|_{s=-4} = -31.25$$

$$V_o = \frac{31.25}{s} - \frac{31.25}{s + 4}$$

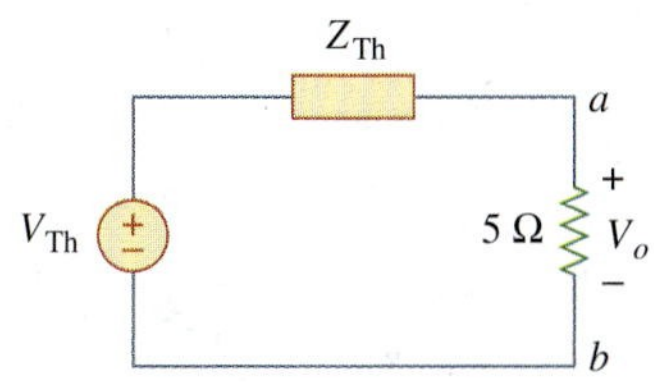

Figure 16.16
The Thevenin equivalent of the circuit in Fig. 16.14 in the *s*-domain.

Taking the inverse Laplace transform gives

$$v_o(t) = 31.25(1 - e^{-4t})u(t) \text{ V}$$

Notice that the values of $v_o(0)$ and $v_o(\infty)$ obtained in part (b) are confirmed.

Practice Problem 16.6

The initial energy in the circuit of Fig. 16.17 is zero at $t = 0$. Assume that $v_s = 15u(t)$ V. (a) Find $V_o(s)$ using the Thevenin theorem. (b) Apply the initial- and final-value theorems to find $v_o(0)$ and $v_o(\infty)$. (c) Obtain $v_o(t)$.

Answer: (a) $V_o(s) = \frac{4(s+0.25)}{s(s+0.3)}$, (b) 4, 3.333 V, (c) $(10 + 2e^{-0.3t})u(t)$ V.

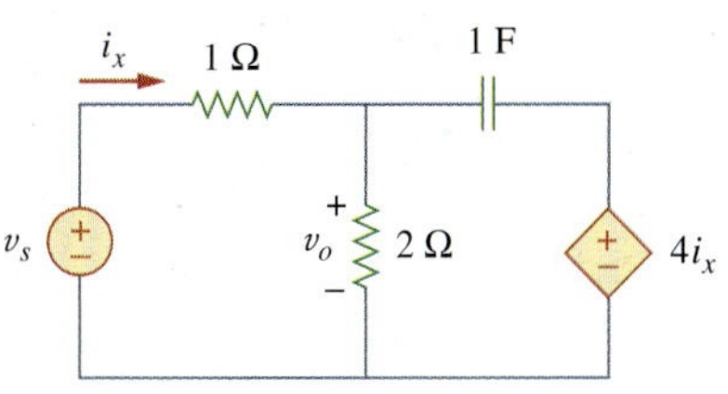

Figure 16.17
For Practice Prob. 16.6.

16.4 Transfer Functions

For electrical networks, the transfer function is also known as the *network function.*

The *transfer function* is a key concept in signal processing because it indicates how a signal is processed as it passes through a network. It is a fitting tool for finding the network response, determining (or designing for) network stability, and network synthesis. The transfer function of a network describes how the output behaves with respect to the input. It specifies the transfer from the input to the output in the s-domain, assuming no initial energy.

The **transfer function** $H(s)$ is the ratio of the output response $Y(s)$ to the input excitation $X(s)$, assuming all initial conditions are zero.

Thus,

$$H(s) = \frac{Y(s)}{X(s)} \tag{16.15}$$

The transfer function depends on what we define as input and output. Since the input and output can be either current or voltage at any place in the circuit, there are four possible transfer functions:

Some authors would not consider Eqs. (16.16c) and (16.16d) transfer functions.

$$H(s) = \text{Voltage gain} = \frac{V_o(s)}{V_i(s)} \tag{16.16a}$$

$$H(s) = \text{Current gain} = \frac{I_o(s)}{I_i(s)} \tag{16.16b}$$

$$H(s) = \text{Impedance} = \frac{V(s)}{I(s)} \tag{16.16c}$$

$$H(s) = \text{Admittance} = \frac{I(s)}{V(s)} \tag{16.16d}$$

Thus, a circuit can have many transfer functions. Note that $H(s)$ is dimensionless in Eqs. (16.16a) and (16.16b).

Each of the transfer functions in Eq. (16.16) can be found in two ways. One way is to assume any convenient input $X(s)$, use any circuit analysis technique (such as current or voltage division, nodal or mesh analysis) to find the output $Y(s)$, and then obtain the ratio of the two. The other approach is to apply the *ladder method*, which involves walking our way through the circuit. By this approach, we assume that the output is 1 V or 1 A as appropriate and use the basic laws of Ohm and Kirchhoff (KCL only) to obtain the input. The transfer function becomes unity divided by the input. This approach may be more convenient to use when the circuit has many loops or nodes so that applying nodal or mesh analysis becomes cumbersome. In the first method, we assume an input and find the output; in the second method, we assume the output and find the input. In both methods, we calculate $H(s)$ as the ratio of output to input transforms. The two methods rely on the linearity property, since we only deal with linear circuits in this book. Example 16.8 illustrates these methods.

Equation (16.15) assumes that both $X(s)$ and $Y(s)$ are known. Sometimes, we know the input $X(s)$ and the transfer function $H(s)$. We find the output $Y(s)$ as

$$Y(s) = H(s)X(s) \tag{16.17}$$

and take the inverse transform to get $y(t)$. A special case is when the input is the unit impulse function, $x(t) = \delta(t)$, so that $X(s) = 1$. For this case,

$$Y(s) = H(s) \quad \text{or} \quad y(t) = h(t) \tag{16.18}$$

where

$$h(t) = \mathcal{L}^{-1}[H(s)] \tag{16.19}$$

The unit impulse response is the output response of a circuit when the input is a unit impulse.

The term $h(t)$ represents the *unit impulse response*—it is the time-domain response of the network to a unit impulse. Thus, Eq. (16.19) provides a new interpretation for the transfer function: $H(s)$ is the Laplace transform of the unit impulse response of the network. Once we know the impulse response $h(t)$ of a network, we can obtain the response of the network to *any* input signal using Eq. (16.17) in the s-domain or using the convolution integral (section 15.5) in the time domain.

Example 16.7

The output of a linear system is $y(t) = 10e^{-t}\cos 4t\,u(t)$ when the input is $x(t) = e^{-t}u(t)$. Find the transfer function of the system and its impulse response.

Solution:

If $x(t) = e^{-t}u(t)$ and $y(t) = 10e^{-t}\cos 4t\,u(t)$, then

$$X(s) = \frac{1}{s+1} \quad \text{and} \quad Y(s) = \frac{10(s+1)}{(s+1)^2 + 4^2}$$

Hence,

$$H(s) = \frac{Y(s)}{X(s)} = \frac{10(s+1)^2}{(s+1)^2 + 16} = \frac{10(s^2 + 2s + 1)}{s^2 + 2s + 17}$$

To find $h(t)$, we write $H(s)$ as

$$H(s) = 10 - 40\frac{4}{(s+1)^2 + 4^2}$$

From Table 15.2, we obtain

$$h(t) = 10\delta(t) - 40e^{-t}\sin 4t\,u(t)$$

Practice Problem 16.7

The transfer function of a linear system is

$$H(s) = \frac{2s}{s+6}$$

Find the output $y(t)$ due to the input $5e^{-3t}u(t)$ and its impulse response.

Answer: $-10e^{-3t} + 20e^{-6t}$, $t \geq 0$, $2\delta(t) - 12e^{-6t}u(t)$.

Example 16.8

Determine the transfer function $H(s) = V_o(s)/I_o(s)$ of the circuit in Fig. 16.18.

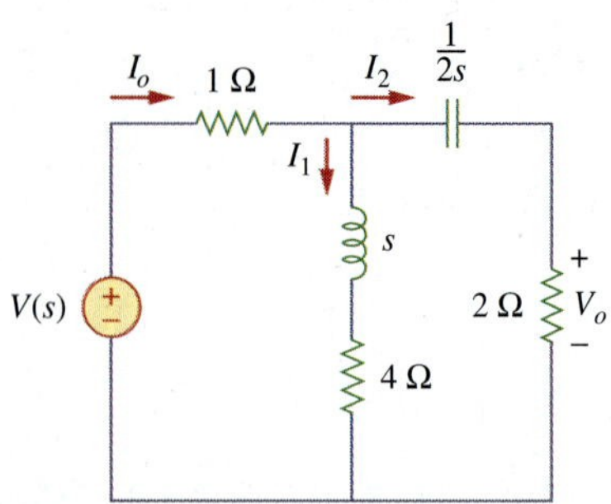

Figure 16.18
For Example 16.8.

Solution:

■ METHOD 1 By current division,

$$I_2 = \frac{(s+4)I_o}{s+4+2+1/2s}$$

But

$$V_o = 2I_2 = \frac{2(s+4)I_o}{s+6+1/2s}$$

Hence,

$$H(s) = \frac{V_o(s)}{I_o(s)} = \frac{4s(s+4)}{2s^2+12s+1}$$

■ METHOD 2 We can apply the ladder method. We let $V_o = 1$ V. By Ohm's law, $I_2 = V_o/2 = 1/2$ A. The voltage across the $(2 + 1/2s)$ impedance is

$$V_1 = I_2\left(2 + \frac{1}{2s}\right) = 1 + \frac{1}{4s} = \frac{4s+1}{4s}$$

This is the same as the voltage across the $(s + 4)$ impedance. Hence,

$$I_1 = \frac{V_1}{s+4} = \frac{4s+1}{4s(s+4)}$$

Applying KCL at the top node yields

$$I_o = I_1 + I_2 = \frac{4s+1}{4s(s+4)} + \frac{1}{2} = \frac{2s^2+12s+1}{4s(s+4)}$$

Hence,

$$H(s) = \frac{V_o}{I_o} = \frac{1}{I_o} = \frac{4s(s+4)}{2s^2+12s+1}$$

as before.

Practice Problem 16.8

Find the transfer function $H(s) = I_1(s)/I_o(s)$ in the circuit of Fig. 16.18.

Answer: $\dfrac{4s+1}{2s^2+12s+1}$.

Example 16.9

For the s-domain circuit in Fig. 16.19, find: (a) the transfer function $H(s) = V_o/V_i$, (b) the impulse response, (c) the response when $v_i(t) = u(t)$ V, (d) the response when $v_i(t) = 8 \cos 2t$ V.

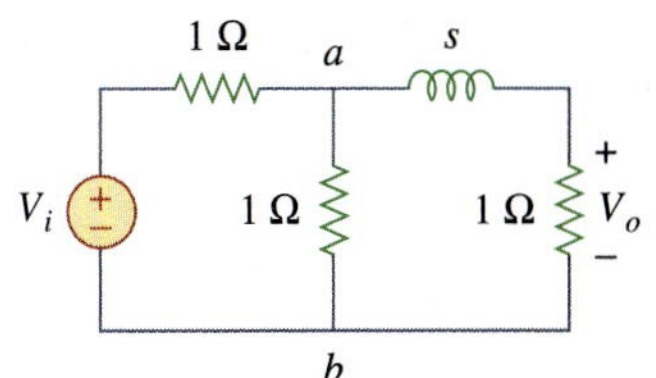

Figure 16.19
For Example 16.9.

Solution:

(a) Using voltage division,

$$V_o = \frac{1}{s+1} V_{ab} \tag{16.9.1}$$

But

$$V_{ab} = \frac{1 \| (s+1)}{1 + 1\|(s+1)} V_i = \frac{(s+1)/(s+2)}{1 + (s+1)/(s+2)} V_i$$

or

$$V_{ab} = \frac{s+1}{2s+3} V_i \tag{16.9.2}$$

Substituting Eq. (16.9.2) into Eq. (16.9.1) results in

$$V_o = \frac{V_i}{2s+3}$$

Thus, the transfer function is

$$H(s) = \frac{V_o}{V_i} = \frac{1}{2s+3}$$

(b) We may write $H(s)$ as

$$H(s) = \frac{1}{2} \frac{1}{s + \frac{3}{2}}$$

Its inverse Laplace transform is the required impulse response:

$$h(t) = \frac{1}{2} e^{-3t/2} u(t)$$

(c) When $v_i(t) = u(t)$, $V_i(s) = 1/s$, and

$$V_o(s) = H(s)V_i(s) = \frac{1}{2s(s + \frac{3}{2})} = \frac{A}{s} + \frac{B}{s + \frac{3}{2}}$$

where

$$A = sV_o(s)\big|_{s=0} = \frac{1}{2(s + \frac{3}{2})}\bigg|_{s=0} = \frac{1}{3}$$

$$B = \left(s + \frac{3}{2}\right) V_o(s)\bigg|_{s=-3/2} = \frac{1}{2s}\bigg|_{s=-3/2} = -\frac{1}{3}$$

Hence, for $v_i(t) = u(t)$,

$$V_o(s) = \frac{1}{3}\left(\frac{1}{s} - \frac{1}{s + \frac{3}{2}}\right)$$

and its inverse Laplace transform is

$$v_o(t) = \frac{1}{3}(1 - e^{-3t/2})u(t) \text{ V}$$

(d) When $v_i(t) = 8\cos 2t$, then $V_i(s) = \dfrac{8s}{s^2 + 4}$, and

$$\begin{aligned} V_o(s) = H(s)V_i(s) &= \frac{4s}{(s + \frac{3}{2})(s^2 + 4)} \\ &= \frac{A}{s + \frac{3}{2}} + \frac{Bs + C}{s^2 + 4} \end{aligned} \qquad \textbf{(16.9.3)}$$

where

$$A = \left(s + \frac{3}{2}\right)V_o(s)\bigg|_{s=-3/2} = \frac{4s}{s^2 + 4}\bigg|_{s=-3/2} = -\frac{24}{25}$$

To get B and C, we multiply Eq. (16.9.3) by $(s + 3/2)(s^2 + 4)$. We get

$$4s = A(s^2 + 4) + B\left(s^2 + \frac{3}{2}s\right) + C\left(s + \frac{3}{2}\right)$$

Equating coefficients,

$$\begin{aligned} &\text{Constant:} && 0 = 4A + \frac{3}{2}C && \Rightarrow && C = -\frac{8}{3}A \\ &s: && 4 = \frac{3}{2}B + C \\ &s^2: && 0 = A + B && \Rightarrow && B = -A \end{aligned}$$

Solving these gives $A = -24/25$, $B = 24/25$, $C = 64/25$. Hence, for $v_i(t) = 8\cos 2t$ V,

$$V_o(s) = \frac{-\frac{24}{25}}{s + \frac{3}{2}} + \frac{24}{25}\frac{s}{s^2 + 4} + \frac{32}{25}\frac{2}{s^2 + 4}$$

and its inverse is

$$v_o(t) = \frac{24}{25}\left(-e^{-3t/2} + \cos 2t + \frac{4}{3}\sin 2t\right)u(t) \text{ V}$$

Practice Problem 16.9

Rework Example 16.9 for the circuit shown in Fig. 16.20.

Answer: (a) $2/(s + 4)$, (b) $2e^{-4t}u(t)$, (c) $\frac{1}{2}(1 - e^{-4t})u(t)$ V, (d) $3.2(-e^{-4t} + \cos 2t + \frac{1}{2}\sin 2t)u(t)$ V.

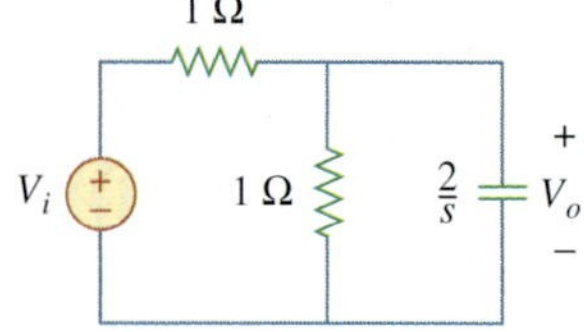

Figure 16.20
For Practice Prob. 16.9.

16.5 State Variables

Thus far in this book we have considered techniques for analyzing systems with only one input and only one output. Many engineering systems have many inputs and many outputs, as shown in Fig. 16.21. The state variable method is a very important tool in analyzing systems and understanding such highly complex systems. Thus, the state variable model is more general than the single-input, single-output model, such as a transfer function. Although the topic cannot be adequately covered in one chapter, let alone one section of a chapter, we will cover it briefly at this point.

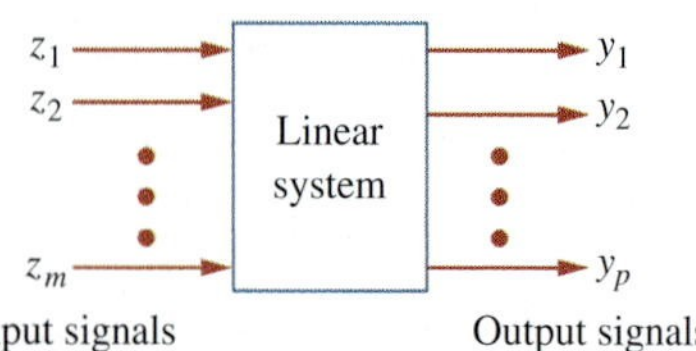

Figure 16.21
A linear system with m inputs and p outputs.

In the state variable model, we specify a collection of variables that describe the internal behavior of the system. These variables are known as the *state variables* of the system. They are the variables that determine the future behavior of a system when the present state of the system and the input signals are known. In other words, they are those variables which, if known, allow all other system parameters to be determined by using only algebraic equations.

A **state variable** is a physical property that characterizes the state of a system, regardless of how the system got to that state.

Common examples of state variables are the pressure, volume, and temperature. In an electric circuit, the state variables are the inductor current and capacitor voltage since they collectively describe the energy state of the system.

The standard way to represent the state equations is to arrange them as a set of first-order differential equations:

$$\dot{\mathbf{x}} = \mathbf{A}\mathbf{x} + \mathbf{B}\mathbf{z} \tag{16.20}$$

where

$$\dot{\mathbf{x}}(t) = \begin{bmatrix} x_1(t) \\ x_2(t) \\ \vdots \\ x_n(t) \end{bmatrix} = \text{state vector representing } n \text{ state vectors}$$

and the dot represents the first derivative with respect to time, i.e.,

$$\dot{\mathbf{x}}(t) = \begin{bmatrix} \dot{x}_1(t) \\ \dot{x}_2(t) \\ \vdots \\ \dot{x}_n(t) \end{bmatrix}$$

and

$$\mathbf{z}(t) = \begin{bmatrix} z_1(t) \\ z_2(t) \\ \vdots \\ z_m(t) \end{bmatrix} = \text{input vector representing } m \text{ inputs}$$

A and **B** are respectively $n \times n$ and $n \times m$ matrices. In addition to the state equation in Eq. (16.20), we need the output equation. The complete state model or state space is

$$\dot{\mathbf{x}} = \mathbf{A}\mathbf{x} + \mathbf{B}\mathbf{z} \tag{16.21a}$$

$$\mathbf{y} = \mathbf{C}\mathbf{x} + \mathbf{D}\mathbf{z} \tag{16.21b}$$

where

$$\mathbf{y}(t) = \begin{bmatrix} y_1(t) \\ y_2(t) \\ \vdots \\ y_p(t) \end{bmatrix} = \text{the output vector representing } p \text{ outputs}$$

C and **D** are, respectively, $p \times n$ and $p \times m$ matrices. For the special case of single-input single-output, $n = m = p = 1$.

Assuming zero initial conditions, the transfer function of the system is found by taking the Laplace transform of Eq. (16.21a); we obtain

$$s\mathbf{X}(s) = \mathbf{AX}(s) + \mathbf{BZ}(s) \quad \rightarrow \quad (s\mathbf{I} - \mathbf{A})\mathbf{X}(s) = \mathbf{BZ}(s)$$

or

$$\mathbf{X}(s) = (s\mathbf{I} - \mathbf{A})^{-1}\mathbf{BZ}(s) \tag{16.22}$$

where **I** is the identity matrix. Taking the Laplace transform of Eq. (16.21b) yields

$$\mathbf{Y}(s) = \mathbf{CX}(s) + \mathbf{DZ}(s) \tag{16.23}$$

Substituting Eq. (16.22) into Eq. (16.23) and dividing by $\mathbf{Z}(s)$ gives the transfer function as

$$\mathbf{H}(s) = \frac{\mathbf{Y}(s)}{\mathbf{Z}(s)} = \mathbf{C}(s\mathbf{I} - \mathbf{A})^{-1}\mathbf{B} + \mathbf{D} \tag{16.24}$$

where

$$\begin{aligned} \mathbf{A} &= \text{system matrix} \\ \mathbf{B} &= \text{input coupling matrix} \\ \mathbf{C} &= \text{output matrix} \\ \mathbf{D} &= \text{feedforward matrix} \end{aligned}$$

In most cases, $\mathbf{D} = \mathbf{0}$, so the degree of the numerator of $H(s)$ in Eq. (16.24) is less than that of the denominator. Thus,

$$\boxed{\mathbf{H}(s) = \mathbf{C}(s\mathbf{I} - \mathbf{A})^{-1}\mathbf{B}} \tag{16.25}$$

Because of the matrix computation involved, *MATLAB* can be used to find the transfer function.

To apply state variable analysis to a circuit, we follow the following three steps.

Steps to Apply the State Variable Method to Circuit Analysis:

1. Select the inductor current i and capacitor voltage v as the state variables, making sure they are consistent with the passive sign convention.
2. Apply KCL and KVL to the circuit and obtain circuit variables (voltages and currents) in terms of the state variables. This should lead to a set of first-order differential equations necessary and sufficient to determine all state variables.
3. Obtain the output equation and put the final result in state-space representation.

Steps 1 and 3 are usually straightforward; the major task is in step 2. We will illustrate this with examples.

Example 16.10

Find the state-space representation of the circuit in Fig. 16.22. Determine the transfer function of the circuit when v_s is the input and i_x is the output. Take $R = 1\Omega$, $C = 0.25$ F, and $L = 0.5$ H.

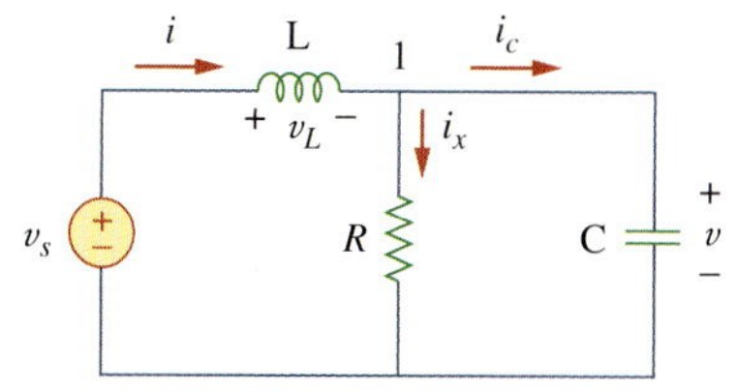

Figure 16.22
For Example 16.10.

Solution:
We select the inductor current i and capacitor voltage v as the state variables.

$$v_L = L\frac{di}{dt} \tag{16.10.1}$$

$$i_C = C\frac{dv}{dt} \tag{16.10.2}$$

Applying KCL at node 1 gives

$$i = i_x + i_C \quad \rightarrow \quad C\frac{dv}{dt} = i - \frac{v}{R}$$

or

$$\dot{v} = -\frac{v}{RC} + \frac{i}{C} \tag{16.10.3}$$

since the same voltage v is across both R and C. Applying KVL around the outer loop yields

$$v_s = v_L + v \quad \rightarrow \quad L\frac{di}{dt} = -v + v_s$$

$$\dot{i} = -\frac{v}{L} + \frac{v_s}{L} \tag{16.10.4}$$

Equations (16.10.3) and (16.10.4) constitute the state equations. If we regard i_x as the output,

$$i_x = \frac{v}{R} \tag{16.10.5}$$

Putting Eqs. (16.10.3), (16.10.4), and (16.10.5) in the standard form leads to

$$\begin{bmatrix} \dot{v} \\ \dot{i} \end{bmatrix} = \begin{bmatrix} \frac{-1}{RC} & \frac{1}{C} \\ \frac{-1}{L} & 0 \end{bmatrix} \begin{bmatrix} v \\ i \end{bmatrix} + \begin{bmatrix} 0 \\ \frac{1}{L} \end{bmatrix} v_s \tag{16.10.6a}$$

$$i_x = \begin{bmatrix} \frac{1}{R} & 0 \end{bmatrix} \begin{bmatrix} v \\ i \end{bmatrix} \tag{16.10.6b}$$

If $R = 1$, $C = \frac{1}{4}$, and $L = \frac{1}{2}$, we obtain from Eq. (16.10.6) matrices

$$\mathbf{A} = \begin{bmatrix} \frac{-1}{RC} & \frac{1}{C} \\ \frac{-1}{L} & 0 \end{bmatrix} = \begin{bmatrix} -4 & 4 \\ -2 & 0 \end{bmatrix}, \qquad \mathbf{B} = \begin{bmatrix} 0 \\ \frac{1}{L} \end{bmatrix} = \begin{bmatrix} 0 \\ 2 \end{bmatrix},$$

$$\mathbf{C} = \begin{bmatrix} \frac{1}{R} & 0 \end{bmatrix} = [1 \quad 0]$$

$$s\mathbf{I} - \mathbf{A} = \begin{bmatrix} s & 0 \\ 0 & s \end{bmatrix} - \begin{bmatrix} -4 & 4 \\ -2 & 0 \end{bmatrix} = \begin{bmatrix} s+4 & -4 \\ 2 & s \end{bmatrix}$$

Taking the inverse of this gives

$$(s\mathbf{I} - \mathbf{A})^{-1} = \frac{\text{adjoint of } \mathbf{A}}{\text{determinant of } \mathbf{A}} = \frac{\begin{bmatrix} s & 4 \\ -2 & s+4 \end{bmatrix}}{s^2 + 4s + 8}$$

Example 16.12

Assume we have a system where the output is $y(t)$ and the input is $z(t)$. Let the following differential equation describe the relationship between the input and the output.

$$\frac{d^2y(t)}{dt^2} + 3\frac{dy(t)}{dt} + 2y(t) = 5z(t) \qquad \textbf{(16.12.1)}$$

Obtain the state model and the transfer function of the system.

Solution:
First, we select the state variables. Let $x_1 = y(t)$, therefore

$$\dot{x}_1 = \dot{y}(t) \qquad \textbf{(16.12.2)}$$

Now let

$$x_2 = \dot{x}_1 = \dot{y}(t) \qquad \textbf{(16.12.3)}$$

Note that at this time we are looking at a second-order system that would normally have two first-order terms in the solution.

Now we have $\dot{x}_2 = \ddot{y}(t)$, where we can find the value $\dot{x}_2$ from Eq. (16.12.1), i.e.,

$$\dot{x}_2 = \ddot{y}(t) = -2y(t) - 3\dot{y}(t) + 5z(t) = -2x_1 - 3x_2 + 5z(t) \qquad \textbf{(16.12.4)}$$

From Eqs. (16.12.2) to (16.12.4), we can now write the following matrix equations:

$$\begin{bmatrix} \dot{x}_1 \\ \dot{x}_2 \end{bmatrix} = \begin{bmatrix} 0 & 1 \\ -2 & -3 \end{bmatrix} \begin{bmatrix} x_1 \\ x_2 \end{bmatrix} + \begin{bmatrix} 0 \\ 5 \end{bmatrix} z(t) \qquad \textbf{(16.12.5)}$$

$$\mathbf{y}(t) = [1 \quad 0] \begin{bmatrix} x_1 \\ x_2 \end{bmatrix} \qquad \textbf{(16.12.6)}$$

We now obtain the transfer function.

$$s\mathbf{I} - \mathbf{A} = s\begin{bmatrix} 1 & 0 \\ 0 & 1 \end{bmatrix} - \begin{bmatrix} 0 & 1 \\ -2 & -3 \end{bmatrix} = \begin{bmatrix} s & -1 \\ 2 & s+3 \end{bmatrix}$$

The inverse is

$$(s\mathbf{I} - \mathbf{A})^{-1} = \frac{\begin{bmatrix} s+3 & 1 \\ -2 & s \end{bmatrix}}{s(s+3)+2}$$

The transfer function is

$$\mathbf{H}(s) = \mathbf{C}(s\mathbf{I} - \mathbf{A})^{-1}\mathbf{B} = \frac{(1 \quad 0)\begin{bmatrix} s+3 & 1 \\ -2 & s \end{bmatrix}\begin{pmatrix} 0 \\ 5 \end{pmatrix}}{s(s+3)+2} = \frac{(1 \quad 0)\begin{pmatrix} 5 \\ 5s \end{pmatrix}}{s(s+3)+2}$$

$$= \frac{5}{(s+1)(s+2)}$$

To check this, we directly apply the Laplace transfer to each term in Eq. (16.12.1). Since initial conditions are zero, we get

$$[s^2 + 3s + 2]Y(s) = 5Z(s) \quad \rightarrow \quad H(s) = \frac{Y(s)}{Z(s)} = \frac{5}{s^2 + 3s + 2}$$

which is in agreement with what we got previously.

Practice Problem 16.12

Develop a set of state variable equations that represent the following differential equation.

$$\frac{d^3y}{dt^3} + 18\frac{d^2y}{dt^2} + 20\frac{dy}{dt} + 5y = z(t)$$

Answer:

$$\mathbf{A} = \begin{bmatrix} 0 & 1 & 0 \\ 0 & 0 & 1 \\ -5 & -20 & -18 \end{bmatrix}, \qquad \mathbf{B} = \begin{bmatrix} 0 \\ 0 \\ 1 \end{bmatrix}, \qquad \mathbf{C} = [1 \quad 0 \quad 0].$$

16.6 †Applications

So far we have considered three applications of Laplace's transform: circuit analysis in general, obtaining transfer functions, and solving linear integrodifferential equations. The Laplace transform also finds application in other areas in circuit analysis, signal processing, and control systems. Here we will consider two more important applications: network stability and network synthesis.

16.6.1 Network Stability

A circuit is *stable* if its impulse response $h(t)$ is bounded (i.e., $h(t)$ converges to a finite value) as $t \rightarrow \infty$; it is *unstable* if $h(t)$ grows without bound as $t \rightarrow \infty$. In mathematical terms, a circuit is stable when

$$\lim_{t\rightarrow\infty} |h(t)| = \text{finite} \tag{16.26}$$

Since the transfer function $H(s)$ is the Laplace transform of the impulse response $h(t)$, $H(s)$ must meet certain requirements in order for Eq. (16.26) to hold. Recall that $H(s)$ may be written as

$$H(s) = \frac{N(s)}{D(s)} \tag{16.27}$$

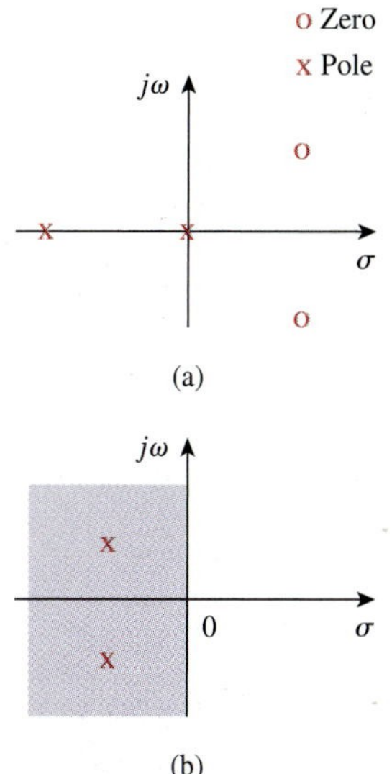

Figure 16.26
The complex s plane: (a) poles and zeros plotted, (b) left-half plane.

where the roots of $N(s) = 0$ are called the *zeros* of $H(s)$ because they make $H(s) = 0$, while the roots of $D(s) = 0$ are called the *poles* of $H(s)$ since they cause $H(s) \rightarrow \infty$. The zeros and poles of $H(s)$ are often located in the s plane as shown in Fig. 16.26(a). Recall from Eqs. (15.47) and (15.48) that $H(s)$ may also be written in terms of its poles as

$$H(s) = \frac{N(s)}{D(s)} = \frac{N(s)}{(s+p_1)(s+p_2)\cdots(s+p_n)} \tag{16.28}$$

$H(s)$ must meet two requirements for the circuit to be stable. First, the degree of $N(s)$ must be less than the degree of $D(s)$; otherwise, long division would produce

$$H(s) = k_n s^n + k_{n-1}s^{n-1} + \cdots + k_1 s + k_0 + \frac{R(s)}{D(s)} \tag{16.29}$$

where the degree of $R(s)$, the remainder of the long division, is less than the degree of $D(s)$. The inverse of $H(s)$ in Eq. (16.29) does not meet the condition in Eq. (16.26). Second, all the poles of $H(s)$ in

Eq. (16.27) (i.e., all the roots of $D(s) = 0$) must have negative real parts; in other words, all the poles must lie in the left half of the s plane, as shown typically in Fig. 16.26(b). The reason for this will be apparent if we take the inverse Laplace transform of $H(s)$ in Eq. (16.27). Since Eq. (16.27) is similar to Eq. (15.48), its partial fraction expansion is similar to the one in Eq. (15.49) so that the inverse of $H(s)$ is similar to that in Eq. (15.53). Hence,

$$h(t) = (k_1 e^{-p_1 t} + k_2 e^{-p_2 t} + \cdots + k_n e^{-p_n t})u(t) \tag{16.30}$$

We see from this equation that each pole p_i must be positive (i.e., pole $s = -p_i$ in the left-half plane) in order for $e^{-p_i t}$ to decrease with increasing t. Thus,

> A circuit is **stable** when all the poles of its transfer function $H(s)$ lie in the left half of the s plane.

An unstable circuit never reaches steady state because the transient response does not decay to zero. Consequently, steady-state analysis is only applicable to stable circuits.

A circuit made up exclusively of passive elements (R, L, and C) and independent sources cannot be unstable, because that would imply that some branch currents or voltages would grow indefinitely with sources set to zero. Passive elements cannot generate such indefinite growth. Passive circuits either are stable or have poles with zero real parts. To show that this is the case, consider the series RLC circuit in Fig. 16.27. The transfer function is given by

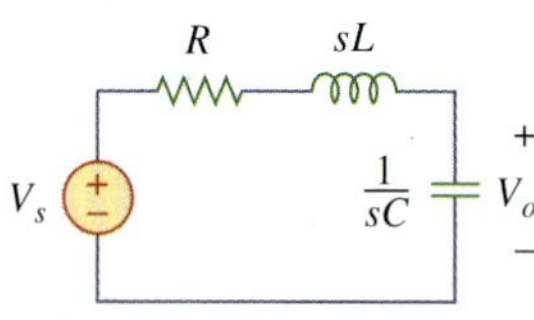

Figure 16.27
A typical RLC circuit.

$$H(s) = \frac{V_o}{V_s} = \frac{1/sC}{R + sL + 1/sC}$$

or

$$H(s) = \frac{1/L}{s^2 + sR/L + 1/LC} \tag{16.31}$$

Notice that $D(s) = s^2 + sR/L + 1/LC = 0$ is the same as the characteristic equation obtained for the series RLC circuit in Eq. (8.8). The circuit has poles at

$$p_{1,2} = -\alpha \pm \sqrt{\alpha^2 - \omega_0^2} \tag{16.32}$$

where

$$\alpha = \frac{R}{2L}, \qquad \omega_0 = \frac{1}{LC}$$

For $R, L, C > 0$, the two poles always lie in the left half of the s plane, implying that the circuit is always stable. However, when $R = 0$, $\alpha = 0$ and the circuit becomes unstable. Although ideally this is possible, it does not really happen, because R is never zero.

On the other hand, active circuits or passive circuits with controlled sources can supply energy, and they can be unstable. In fact, an oscillator is a typical example of a circuit designed to be unstable. An oscillator is designed such that its transfer function is of the form

$$H(s) = \frac{N(s)}{s^2 + \omega_0^2} = \frac{N(s)}{(s + j\omega_0)(s - j\omega_0)} \tag{16.33}$$

so that its output is sinusoidal.

Example 16.13

Determine the values of k for which the circuit in Fig. 16.28 is stable.

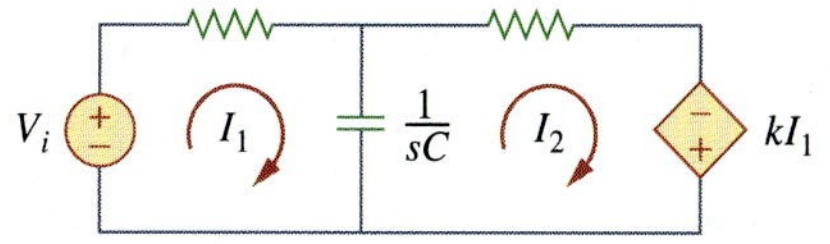

Solution:

Applying mesh analysis to the first-order circuit in Fig. 16.28 gives

$$V_i = \left(R + \frac{1}{sC}\right)I_1 - \frac{I_2}{sC} \tag{16.13.1}$$

and

$$0 = -kI_1 + \left(R + \frac{1}{sC}\right)I_2 - \frac{I_1}{sC}$$

or

$$0 = -\left(k + \frac{1}{sC}\right)I_1 + \left(R + \frac{1}{sC}\right)I_2 \tag{16.13.2}$$

Figure 16.28
For Example 16.13.

We can write Eqs. (16.13.1) and (16.13.2) in matrix form as

$$\begin{bmatrix} V_i \\ 0 \end{bmatrix} = \begin{bmatrix} \left(R + \dfrac{1}{sC}\right) & -\dfrac{1}{sC} \\ -\left(k + \dfrac{1}{sC}\right) & \left(R + \dfrac{1}{sC}\right) \end{bmatrix} \begin{bmatrix} I_1 \\ I_2 \end{bmatrix}$$

The determinant is

$$\Delta = \left(R + \frac{1}{sC}\right)^2 - \frac{k}{sC} - \frac{1}{s^2C^2} = \frac{sR^2C + 2R - k}{sC} \tag{16.13.3}$$

The characteristic equation ($\Delta = 0$) gives the single pole as

$$p = \frac{k - 2R}{R^2C}$$

which is negative when $k < 2R$. Thus, we conclude the circuit is stable when $k < 2R$ and unstable for $k > 2R$.

Practice Problem 16.13

For what value of β is the circuit in Fig. 16.29 stable?

Answer: $\beta > -1/R$.

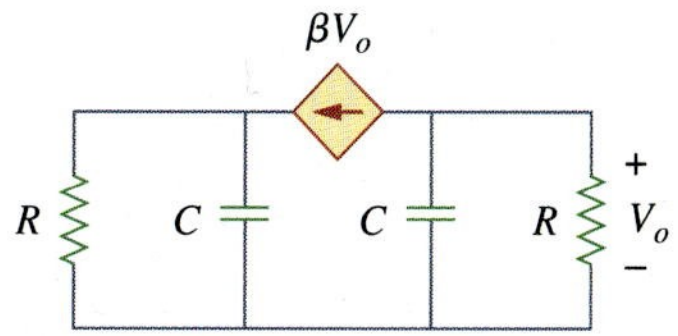

Figure 16.29
For Practice Prob. 16.13.

Example 16.14

An active filter has the transfer function

$$H(s) = \frac{k}{s^2 + s(4 - k) + 1}$$

For what values of k is the filter stable?

Solution:

As a second-order circuit, $H(s)$ may be written as

$$H(s) = \frac{N(s)}{s^2 + bs + c}$$

where $b = 4 - k$, $c = 1$, and $N(s) = k$. This has poles at $p^2 + bp + c = 0$; that is,

$$p_{1,2} = \frac{-b \pm \sqrt{b^2 - 4c}}{2}$$

For the circuit to be stable, the poles must be located in the left half of the s plane. This implies that $b > 0$.

Applying this to the given $H(s)$ means that for the circuit to be stable, $4 - k > 0$ or $k < 4$.

Practice Problem 16.14

A second-order active circuit has the transfer function

$$H(s) = \frac{1}{s^2 + s(25 + \alpha) + 25}$$

Find the range of the values of α for which the circuit is stable. What is the value of α that will cause oscillation?

Answer: $\alpha > -25$, $\alpha = -25$.

16.6.2 Network Synthesis

Network synthesis may be regarded as the process of obtaining an appropriate network to represent a given transfer function. Network synthesis is easier in the s-domain than in the time domain.

In network analysis, we find the transfer function of a given network. In network synthesis, we reverse the approach: given a transfer function, we are required to find a suitable network.

Network synthesis is finding a network that represents a given transfer function.

Keep in mind that in synthesis, there may be many different answers—or possibly no answers—because there are many circuits that can be used to represent the same transfer function; in network analysis, there is only one answer.

Network synthesis is an exciting field of prime engineering importance. Being able to look at a transfer function and come up with the type of circuit it represents is a great asset to a circuit designer. Although network synthesis constitutes a whole course by itself and requires some experience, the following examples are meant to whet your appetite.

Example 16.15

Given the transfer function

$$H(s) = \frac{V_o(s)}{V_i(s)} = \frac{10}{s^2 + 3s + 10}$$

realize the function using the circuit in Fig. 16.30(a). (a) Select $R = 5\ \Omega$, and find L and C. (b) Select $R = 1\ \Omega$, and find L and C.

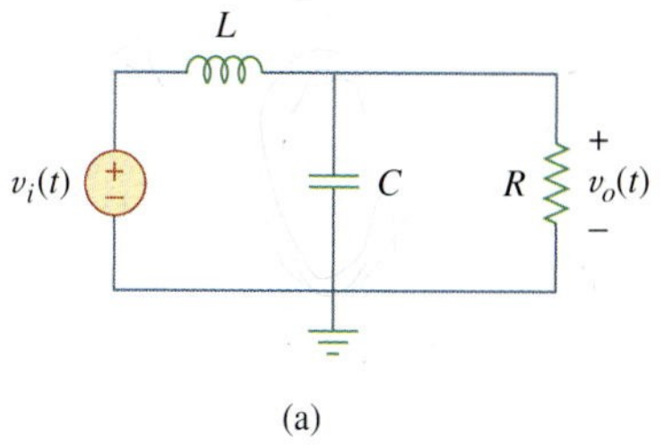

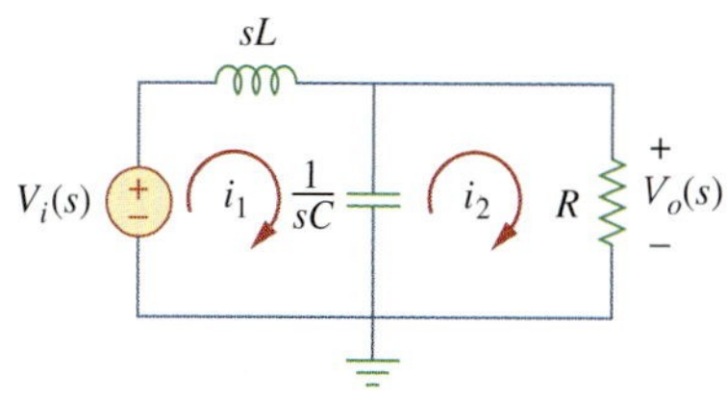

Figure 16.30
For Example 16.15.

Solution:

1. **Define.** The problem is clearly and completely defined. This problem is what we call a synthesis problem: given a transfer function, synthesize a circuit that produces the given transfer function. However, to keep the problem more manageable, we give a circuit that produces the desired transfer function.

   Had one of the variables, R in this case, not been given a value, then the problem would have had an infinite number of answers. An open-ended problem of this kind would require some additional assumptions that would have narrowed the set of solutions.
2. **Present.** A transfer function of the voltage out versus the voltage in is equal to $10/(s^2 + 3s + 10)$. A circuit, Fig. 16.30, is also given that should be able to produce the required transfer function. Two different values of R, $5\ \Omega$ and $1\ \Omega$, are to be used to calculate the values of L and C that produce the given transfer function.
3. **Alternative.** All solution paths involve determining the transfer function of Fig. 16.30 and then matching the various terms of the transfer function. Two approaches would be to use mesh analysis or nodal analysis. Since we are looking for a ratio of voltages, nodal analysis makes the most sense.
4. **Attempt.** Using nodal analysis leads to

$$\frac{V_o(s) - V_i(s)}{sL} + \frac{V_o(s) - 0}{1/(sC)} + \frac{V_o(s) - 0}{R} = 0$$

Now multiply through by sLR:

$$RV_o(s) - RV_i(s) + s^2RLCV_o(s) + sLV_o(s) = 0$$

Collecting terms we get

$$(s^2RLC + sL + R)V_o(s) = RV_i(s)$$

or

$$\frac{V_o(s)}{V_i(s)} = \frac{1/(LC)}{s^2 + [1/(RC)]s + 1/(LC)}$$

Matching the two transfer functions produces two equations with three unknowns.

$$LC = 0.1 \qquad \text{or} \qquad L = \frac{0.1}{C}$$

and

$$RC = \frac{1}{3} \qquad \text{or} \qquad C = \frac{1}{3R}$$

We have a constraint equation, $R = 5\ \Omega$ for (a) and $= 1\ \Omega$ for (b).

(a) $C = 1/(3 \times 5) =$ **66.67 mF** and $L =$ **1.5 H**

(b) $C = 1/(3 \times 1) =$ **333.3 mF** and $L =$ **300 mH**

5. **Evaluate.** There are different ways of checking the answer. Solving for the transfer function by using mesh analysis seems the most straightforward and the approach we could use here. However, it should be pointed out that this is mathematically more complex and will take longer than the original nodal analysis approach. Other approaches also exist. We can assume an input for $v_i(t)$, $v_i(t) = u(t)$ V and, using either nodal analysis or mesh analysis, see if we get the same answer we would get with just using the transfer function. That is the approach we will try using mesh analysis.

   Let $v_i(t) = u(t)\ V$ or $V_i(s) = 1/s$. This will produce

$$V_o(s) = 10/(s^3 + 3s^2 + 10s)$$

Based on Fig. 16.30, mesh analysis leads to

(a) For loop 1,

$$-(1/s) + 1.5sI_1 + [1/(0.06667s)](I_1 - I_2) = 0$$

or

$$(1.5s^2 + 15)I_1 - 15I_2 = 1$$

For loop 2,

$$(15/s)(I_2 - I_1) + 5I_2 = 0$$

or

$$-15I_1 + (5s + 15)I_2 = 0 \qquad \text{or} \qquad I_1 = (0.3333s + 1)I_2$$

Substituting into the first equation we get

$$(0.5s^3 + 1.5s^2 + 5s + 15)I_2 - 15I_2 = 1$$

or

$$I_2 = 2/(s^3 + 3s^2 + 10s)$$

but

$$V_o(s) = 5I_2 = 10/(s^3 + 3s^2 + 10s)$$

and the answer checks.

(b) For loop 1,

$$-(1/s) + 0.3sI_1 + [1/(0.3333s)](I_1 - I_2) = 0$$

or

$$(0.3s^2 + 3)I_1 - 3I_2 = 1$$

For loop 2,

$$(3/s)(I_2 - I_1) + I_2 = 0$$

or

$$-3I_1 + (s + 3)I_2 = 0 \qquad \text{or} \qquad I_1 = (0.3333s + 1)I_2$$

Substituting into the first equation we get

$$(0.09999s^3 + 0.3s^2 + s + 3)I_2 - 3I_2 = 1$$

or

$$I_2 = 10/(s^3 + 3s^2 + 10s)$$

but $V_o(s) = 1 \times I_2 = 10/(s^3 + 3s^2 + 10s)$
and the answer checks.

6. **Satisfactory?** We have clearly identified values of L and C for each of the conditions. In addition, we have carefully checked the answers to see if they are correct. The problem has been adequately solved. The results can now be presented as a solution to the problem.

Practice Problem 16.15

Realize the function

$$G(s) = \frac{V_o(s)}{V_i(s)} = \frac{4s}{s^2 + 4s + 20}$$

using the circuit in Fig. 16.31. Select $R = 2\ \Omega$, and determine L and C.

Answer: 0.5 H, 0.1 F.

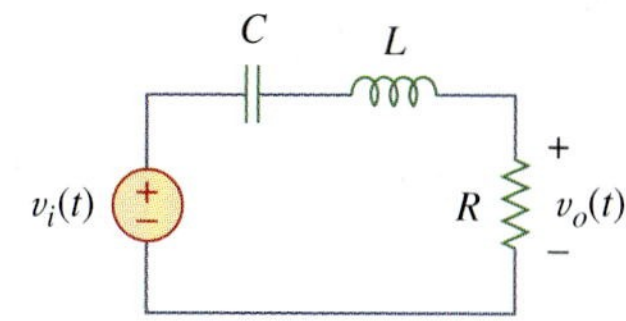

Figure 16.31
For Practice Prob. 16.15.

Example 16.16

Synthesize the function

$$T(s) = \frac{V_o(s)}{V_s(s)} = \frac{10^6}{s^2 + 100s + 10^6}$$

using the topology in Fig. 16.32.

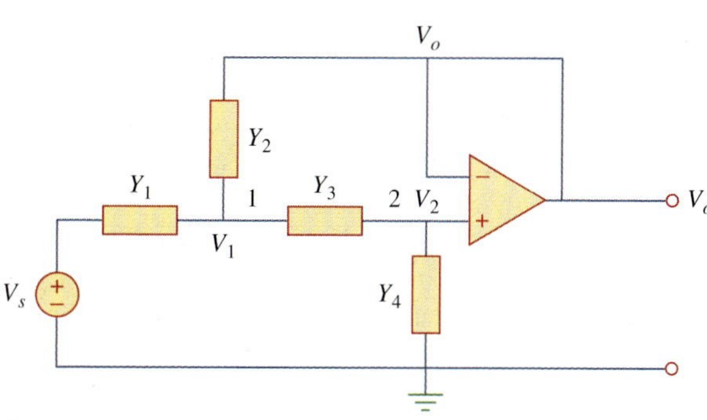

Figure 16.32
For Example 16.16.

Solution:
We apply nodal analysis to nodes 1 and 2. At node 1,

$$(V_s - V_1)Y_1 = (V_1 - V_o)Y_2 + (V_1 - V_2)Y_3 \qquad \textbf{(16.16.1)}$$

At node 2,

$$(V_1 - V_2)Y_3 = (V_2 - 0)Y_4 \qquad \textbf{(16.16.2)}$$

But $V_2 = V_o$, so Eq. (16.16.1) becomes

$$Y_1V_s = (Y_1 + Y_2 + Y_3)V_1 - (Y_2 + Y_3)V_o \tag{16.16.3}$$

and Eq. (16.16.2) becomes

$$V_1Y_3 = (Y_3 + Y_4)V_o$$

or

$$V_1 = \frac{1}{Y_3}(Y_3 + Y_4)V_o \tag{16.16.4}$$

Substituting Eq. (16.16.4) into Eq. (16.16.3) gives

$$Y_1V_s = (Y_1 + Y_2 + Y_3)\frac{1}{Y_3}(Y_3 + Y_4)V_o - (Y_2 + Y_3)V_o$$

or

$$Y_1Y_3V_s = [Y_1Y_3 + Y_4(Y_1 + Y_2 + Y_3)]V_o$$

Thus,

$$\frac{V_o}{V_s} = \frac{Y_1Y_3}{Y_1Y_3 + Y_4(Y_1 + Y_2 + Y_3)} \tag{16.16.5}$$

To synthesize the given transfer function $T(s)$, compare it with the one in Eq. (16.16.5). Notice two things: (1) Y_1Y_3 must not involve s because the numerator of $T(s)$ is constant; (2) the given transfer function is second-order, which implies that we must have two capacitors. Therefore, we must make Y_1 and Y_3 resistive, while Y_2 and Y_4 are capacitive. So we select

$$Y_1 = \frac{1}{R_1}, \qquad Y_2 = sC_1, \qquad Y_3 = \frac{1}{R_2}, \qquad Y_4 = sC_2 \tag{16.16.6}$$

Substituting Eq. (16.16.6) into Eq. (16.16.5) gives

$$\begin{aligned}\frac{V_o}{V_s} &= \frac{1/(R_1R_2)}{1/(R_1R_2) + sC_2(1/R_1 + 1/R_2 + sC_1)}\\ &= \frac{1/(R_1R_2C_1C_2)}{s^2 + s(R_1 + R_2)/(R_1R_2C_1) + 1/(R_1R_2C_1C_2)}\end{aligned}$$

Comparing this with the given transfer function $T(s)$, we notice that

$$\frac{1}{R_1R_2C_1C_2} = 10^6, \qquad \frac{R_1 + R_2}{R_1R_2\,C_1} = 100$$

If we select $R_1 = R_2 = 10\text{ k}\Omega$, then

$$C_1 = \frac{R_1 + R_2}{100R_1R_2} = \frac{20 \times 10^3}{100 \times 100 \times 10^6} = 2\ \mu\text{F}$$

$$C_2 = \frac{10^{-6}}{R_1R_2C_1} = \frac{10^{-6}}{100 \times 10^6 \times 2 \times 10^{-6}} = 5\text{ nF}$$

Thus, the given transfer function is realized using the circuit shown in Fig. 16.33.

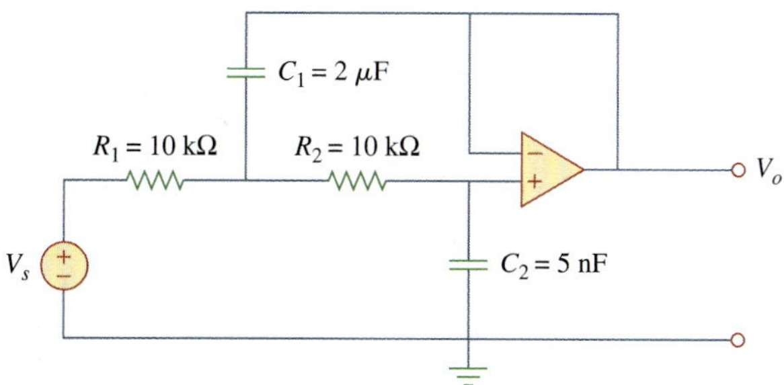

Figure 16.33
For Example 16.16.

Practice Problem 16.16

Synthesize the function

$$\frac{V_o(s)}{V_{in}} = \frac{-2s}{s^2 + 6s + 10}$$

using the op amp circuit shown in Fig. 16.34. Select

$$Y_1 = \frac{1}{R_1}, \qquad Y_2 = sC_1, \qquad Y_3 = sC_2, \qquad Y_4 = \frac{1}{R_2}$$

Let $R_1 = 1\ \text{k}\Omega$, and determine C_1, C_2, and R_2.

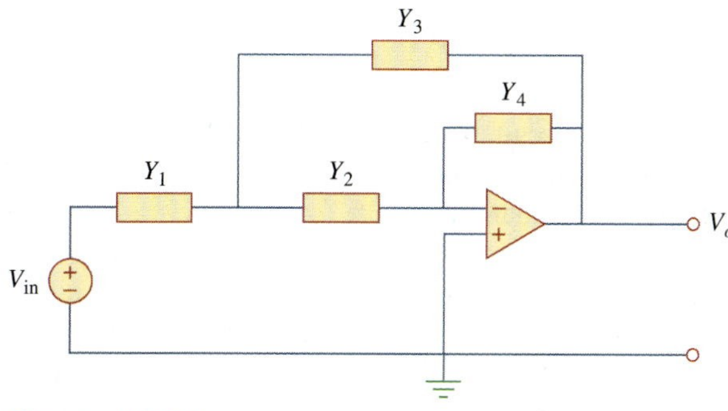

Figure 16.34
For Practice Prob. 16.16.

Answer: 0.1 mF, 0.5 mF, 2 kΩ.

16.7 Summary

1. The Laplace transform can be used to analyze a circuit. We convert each element from the time domain to the s-domain, solve the problem using any circuit analysis technique, and convert the result to the time domain using the inverse transform.
2. In the s-domain, the circuit elements are replaced with the initial condition at $t = 0$ as follows. (Please note, voltage models are

given below, but the corresponding current models work equally well.):

$$\text{Resistor:}\quad v_R = Ri \quad\rightarrow\quad V_R = RI$$

$$\text{Inductor:}\quad v_L = L\frac{di}{dt} \quad\rightarrow\quad V_L = sLI - Li(0^-)$$

$$\text{Capactior:}\quad v_C = \int i\,dt \quad\rightarrow\quad V_C = \frac{1}{sC} - \frac{v(0^-)}{s}$$

3. Using the Laplace transform to analyze a circuit results in a complete (both transient and steady state) response because the initial conditions are incorporated in the transformation process.
4. The transfer function $H(s)$ of a network is the Laplace transform of the impulse response $h(t)$.
5. In the s-domain, the transfer function $H(s)$ relates the output response $Y(s)$ and an input excitation $X(s)$; that is, $H(s) = Y(s)/X(s)$.
6. The state variable model is a useful tool for analyzing complex systems with several inputs and outputs. State variable analysis is a powerful technique that is most popularly used in circuit theory and control. The state of a system is the smallest set of quantities (known as state variables) that we must know in order to determine its future response to any given input. The state equation in state variable form is

$$\dot{\mathbf{x}} = \mathbf{A}x + \mathbf{B}z$$

while the output equation is

$$\mathbf{y} = \mathbf{C}x + \mathbf{D}z$$

7. For an electric circuit, we first select capacitor voltages and inductor current as state variables. We then apply KCL and KVL to obtain the state equations.
8. Two other areas of applications of the Laplace transform covered in this chapter are circuit stability and synthesis. A circuit is stable when all the poles of its transfer function lie in the left half of the s plane. Network synthesis is the process of obtaining an appropriate network to represent a given transfer function for which analysis in the s-domain is well suited.

Review Questions

16.1 The voltage through a resistor with current $i(t)$ in the s-domain is $sRI(s)$.

(a) True (b) False

16.2 The current through an RL series circuit with input voltage $v(t)$ is given in the s-domain as:

(a) $V(s)\left[R + \frac{1}{sL}\right]$ (b) $V(s)(R + sL)$

(c) $\frac{V(s)}{R + 1/sL}$ (d) $\frac{V(s)}{R + sL}$

16.3 The impedance of a 10-F capacitor is:

(a) $10/s$ (b) $s/10$ (c) $1/10s$ (d) $10s$

16.4 We can usually obtain the Thevenin equivalent in the time domain.

(a) True (b) False

16.5 A transfer function is defined only when all initial conditions are zero.

(a) True (b) False

16.6 If the input to a linear system is $\delta(t)$ and the output is $e^{-2t}u(t)$, the transfer function of the system is:

(a) $\dfrac{1}{s+2}$ (b) $\dfrac{1}{s-2}$ (c) $\dfrac{s}{s+2}$ (d) $\dfrac{s}{s-2}$

(e) None of the above

16.7 If the transfer function of a system is

$$H(s) = \frac{s^2 + s + 2}{s^3 + 4s^2 + 5s + 1}$$

it follows that the input is $X(s) = s^3 + 4s^2 + 5s + 1$, while the output is $Y(s) = s^2 + s + 2$.

(a) True (b) False

16.8 A network has its transfer function as

$$H(s) = \frac{s+1}{(s-2)(s+3)}$$

The network is stable.

(a) True (b) False

16.9 Which of the following equations is called the state equation?

(a) $\dot{\mathbf{x}} = \mathbf{Ax} + \mathbf{Bz}$

(b) $\mathbf{y} = \mathbf{Cx} + \mathbf{Dz}$

(c) $\mathbf{H}(s) = \mathbf{Y}(s)/\mathbf{Z}(s)$

(d) $\mathbf{H}(s) = \mathbf{C}(s\mathbf{I} - \mathbf{A})^{-1}\mathbf{B}$

16.10 A single-input, single-output system is described by the state model as:

$$\dot{x}_1 = 2x_1 - x_2 + 3z$$
$$\dot{x}_2 = -4x_2 - z$$
$$y = 3x_1 - 2x_2 + z$$

Which of the following matrices is incorrect?

(a) $\mathbf{A} = \begin{bmatrix} 2 & -1 \\ 0 & -4 \end{bmatrix}$ (b) $\mathbf{B} = \begin{bmatrix} 3 \\ -1 \end{bmatrix}$

(c) $\mathbf{C} = [3 \quad -2]$ (d) $\mathbf{D} = \mathbf{0}$

Answers: 16.1b, 16.2d, 16.3c, 16.4b, 16.5b, 16.6a, 16.7b, 16.8b, 16.9a, 16.10d.

Problems

Sections 16.2 and 16.3 Circuit Element Models and Circuit Analysis

16.1 Determine $i(t)$ in the circuit of Fig. 16.35 by means of the Laplace transform.

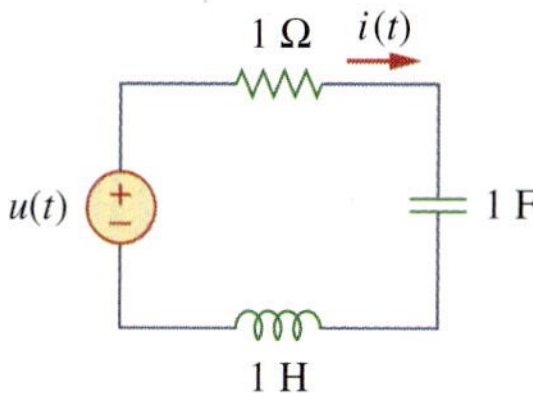

Figure 16.35
For Prob. 16.1.

16.2 **e2d** Using Fig. 16.36, design a problem to help other students better understand circuit analysis using Laplace transforms.

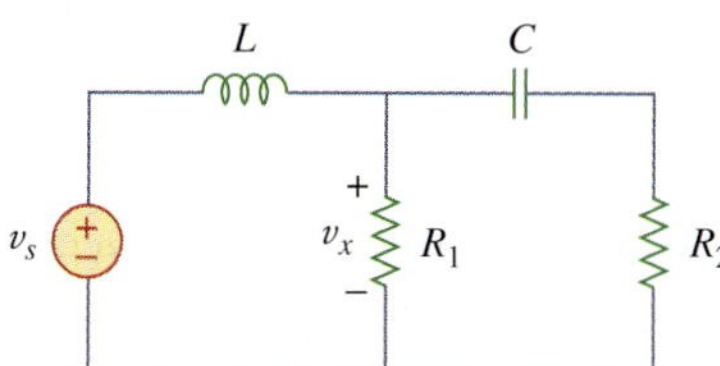

Figure 16.36
For Prob. 16.2.

16.3 Find $i(t)$ for $t > 0$ for the circuit in Fig. 16.37. Assume $i_s = 24u(t) + 12\delta(t)$ mA. (Hint: Can we use superposition to help solve this problem?)

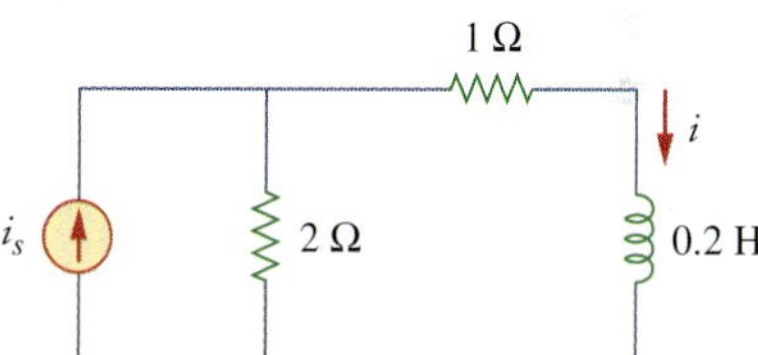

Figure 16.37
For Prob. 16.3.

16.4 The capacitor in the circuit of Fig. 16.38 is initially uncharged. Find $v_o(t)$ for $t > 0$.

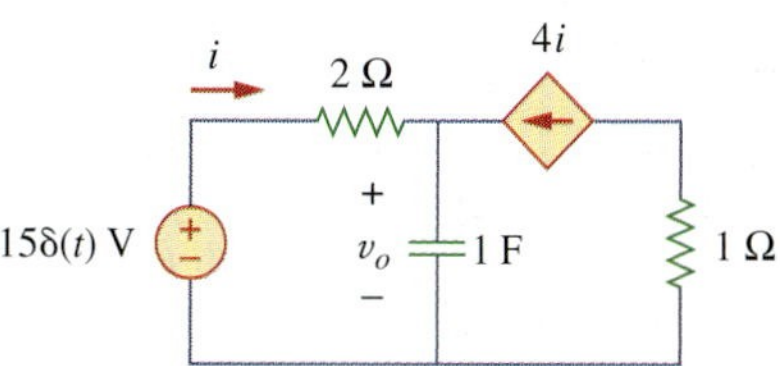

Figure 16.38
For Prob. 16.4.

16.5 If $i_s(t) = e^{-2t}u(t)$ A in the circuit shown in Fig. 16.39, find the value of $i_o(t)$.

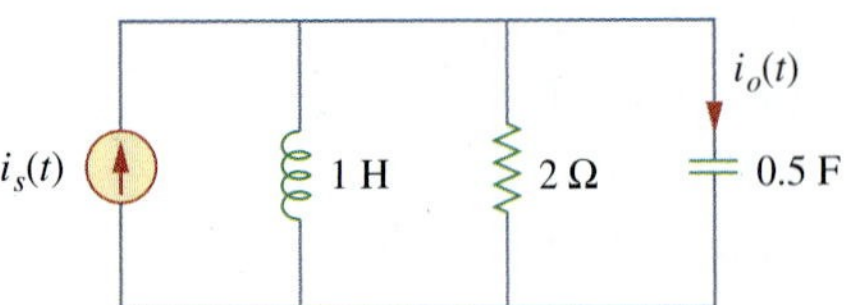

Figure 16.39
For Prob. 16.5.

16.6 Find $v(t)$, $t > 0$ in the circuit of Fig. 16.40. Let $v_s = 20$ V.

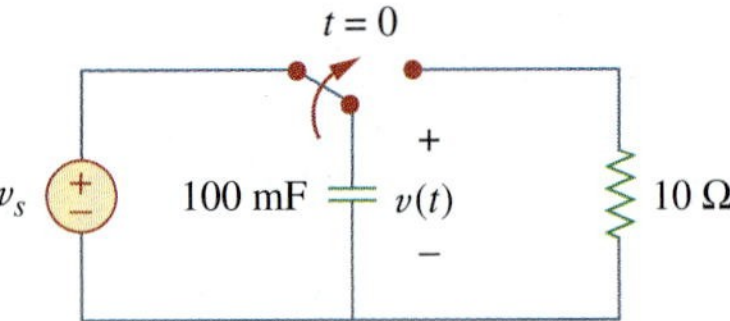

Figure 16.40
For Prob. 16.6.

16.7 Find $v_o(t)$, for all $t > 0$, in the circuit of Fig. 16.41.

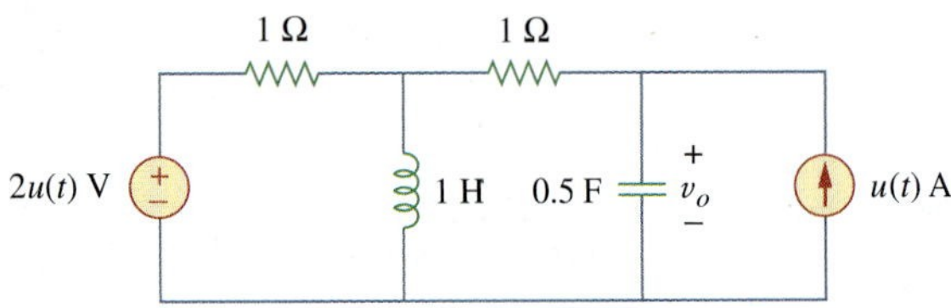

Figure 16.41
For Prob. 16.7.

16.8 If $v_o(0) = -1$ V, obtain $v_o(t)$ in the circuit of Fig. 16.42.

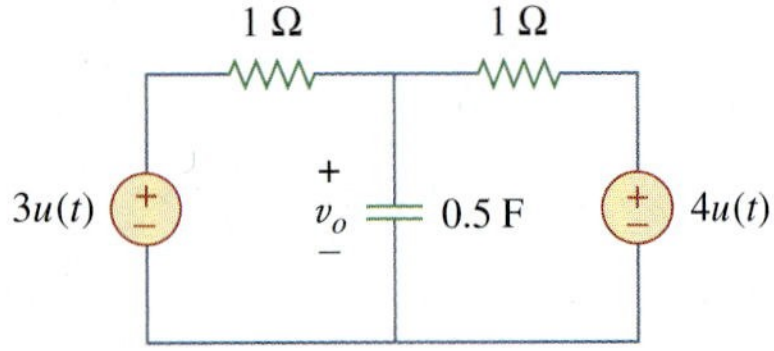

Figure 16.42
For Prob. 16.8.

16.9 Find the input impedance $Z_{in}(s)$ of each of the circuits in Fig. 16.43.

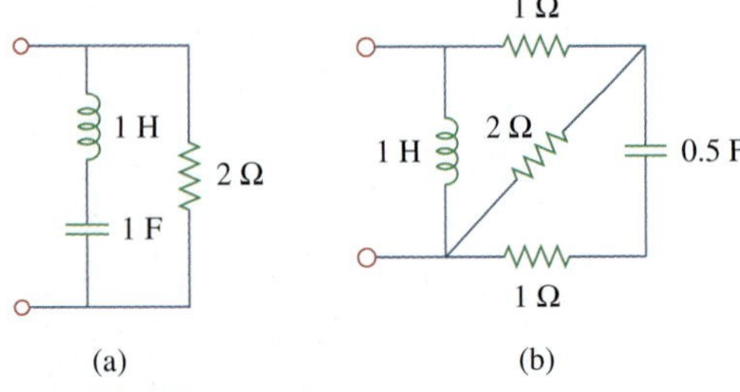

Figure 16.43
For Prob. 16.9.

16.10 Using Fig. 16.44, design a problem to help other students understand how to use Thevenin's theorem (in the s-domain) to aid in circuit analysis.

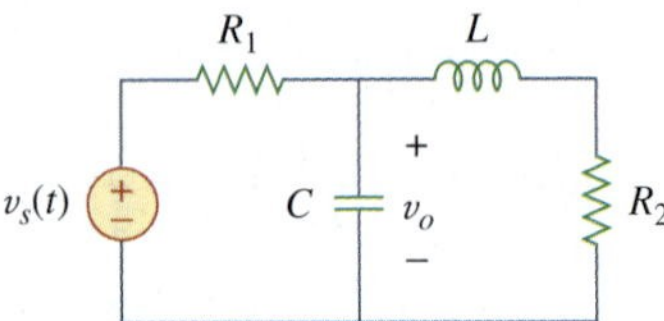

Figure 16.44
For Prob. 16.10.

16.11 Solve for the mesh currents in the circuit of Fig. 16.45. You may leave your results in the s-domain.

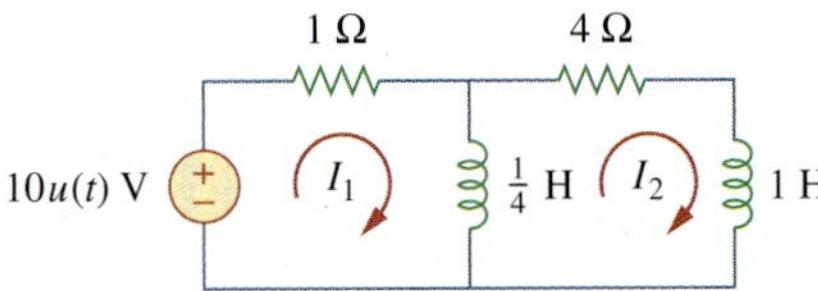

Figure 16.45
For Prob. 16.11.

16.12 Find $v_o(t)$ in the circuit of Fig. 16.46.

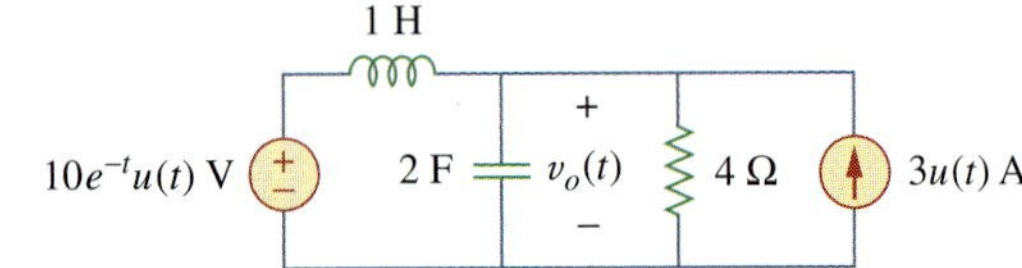

Figure 16.46
For Prob. 16.12.

16.13 Determine $i_o(t)$ in the circuit of Fig. 16.47.

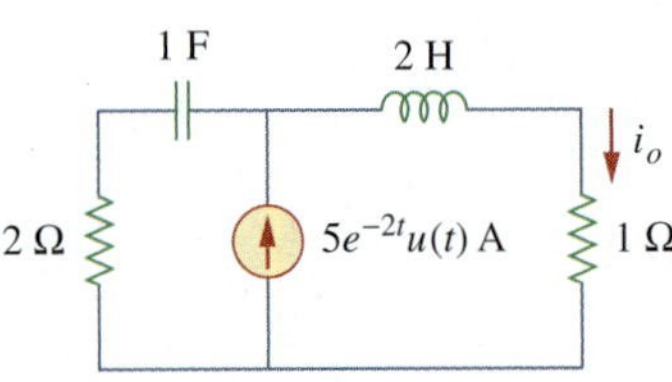

Figure 16.47
For Prob. 16.13.

***16.14** Determine $i_o(t)$ in the network shown in Fig. 16.48.

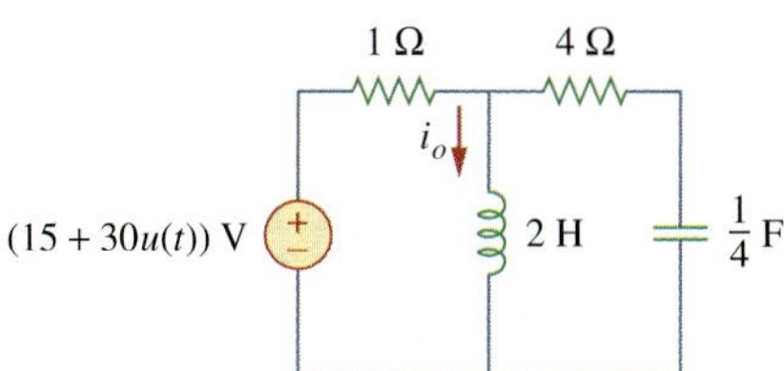

Figure 16.48
For Prob. 16.14.

16.15 Find $V_x(s)$ in the circuit shown in Fig. 16.49.

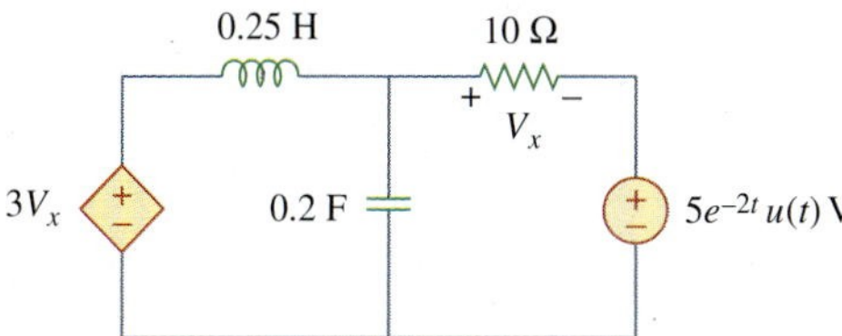

Figure 16.49
For Prob. 16.15.

***16.16** Find $i_o(t)$ for $t > 0$ in the circuit of Fig. 16.50.

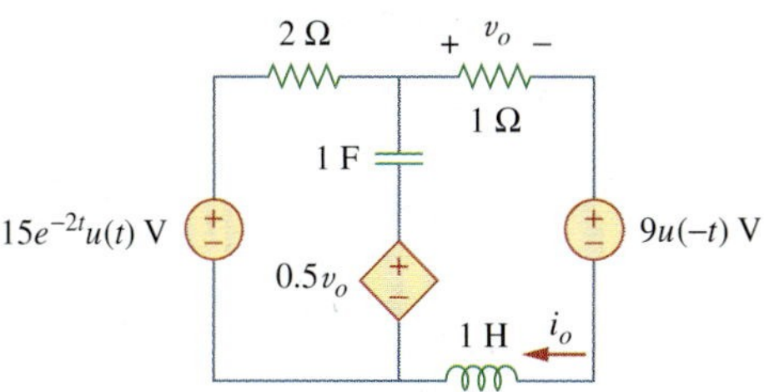

Figure 16.50
For Prob. 16.16.

16.17 Calculate $i_o(t)$ for $t > 0$ in the network of Fig. 16.51.

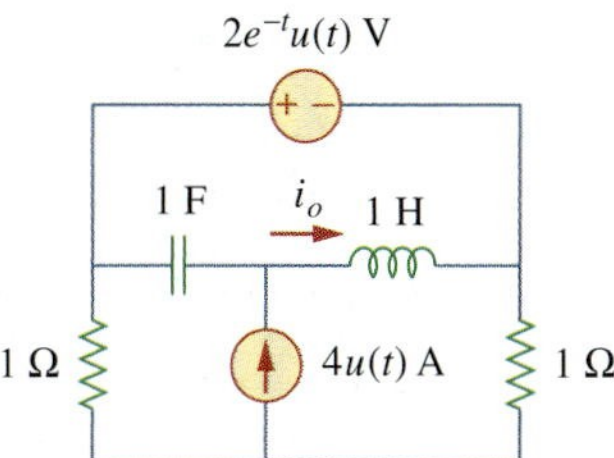

Figure 16.51
For Prob. 16.17.

16.18 (a) Find the Laplace transform of the voltage shown in Fig. 16.52(a). (b) Using that value of $v_s(t)$ in the circuit shown in Fig. 16.52(b), find the value of $v_o(t)$.

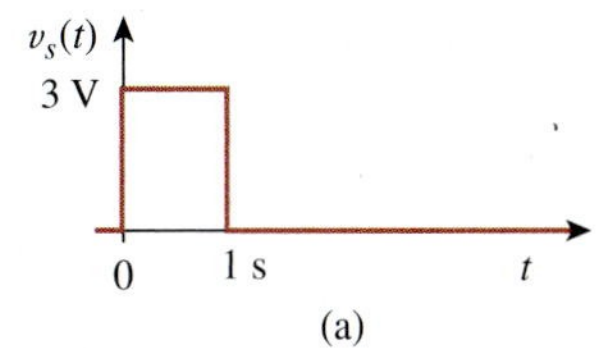

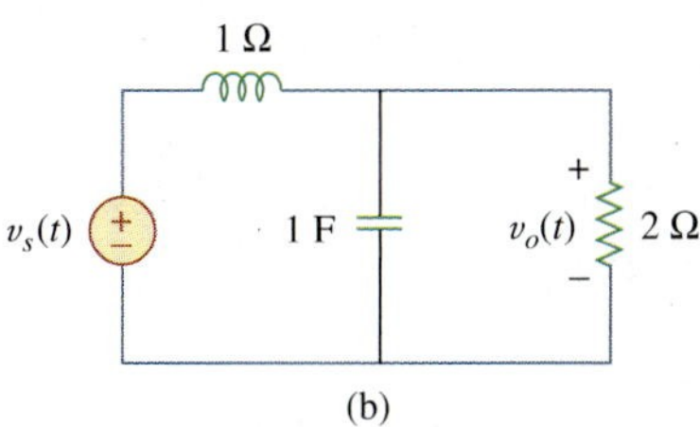

Figure 16.52
For Prob. 16.18.

16.19 Using Fig. 16.53, design a problem to help other students better understand circuit analysis in the *s*-domain with circuits that have dependent sources.

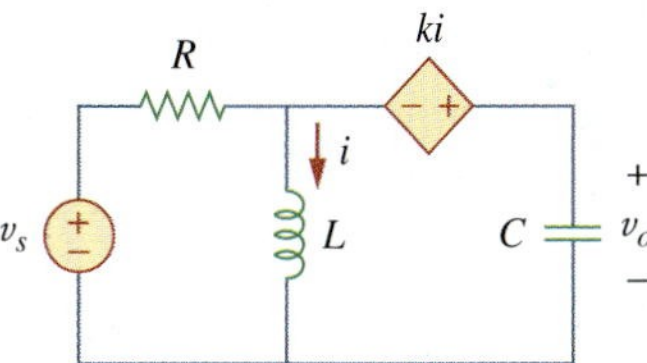

Figure 16.53
For Prob. 16.19.

16.20 Find $v_o(t)$ in the circuit of Fig. 16.54 if $v_x(0) = 2$ V and $i(0) = 1$ A.

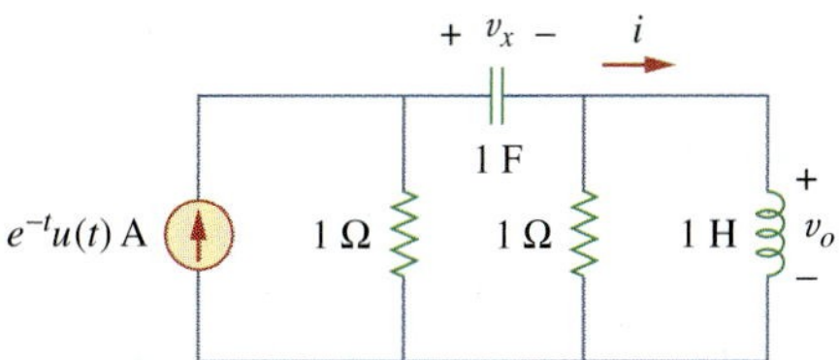

Figure 16.54
For Prob. 16.20.

16.21 Find the voltage $v_o(t)$ in the circuit of Fig. 16.55 by means of the Laplace transform.

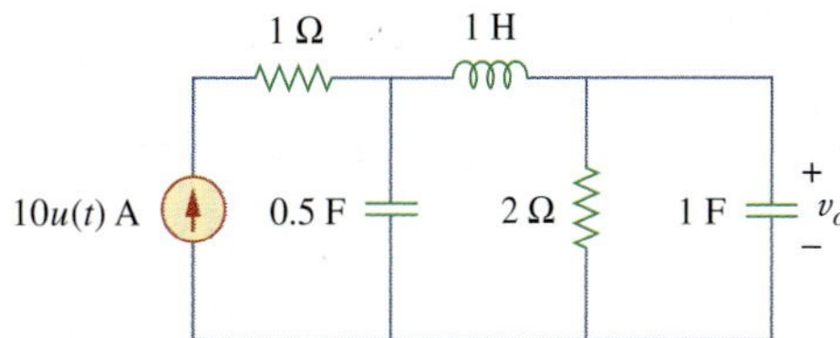

Figure 16.55
For Prob. 16.21.

* An asterisk indicates a challenging problem.

16.22 Using Fig. 16.56, design a problem to help other students better understand solving for node voltages by working in the s-domain.

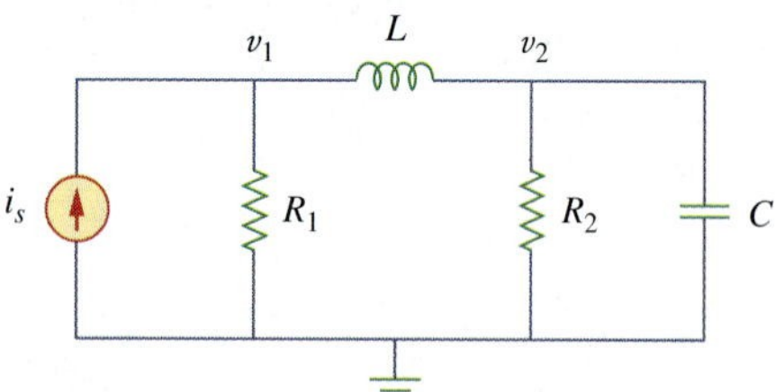

Figure 16.56
For Prob. 16.22.

16.23 Consider the parallel RLC circuit of Fig. 16.57. Find $v(t)$ and $i(t)$ given that $v(0) = 5$ and $i(0) = -2$ A.

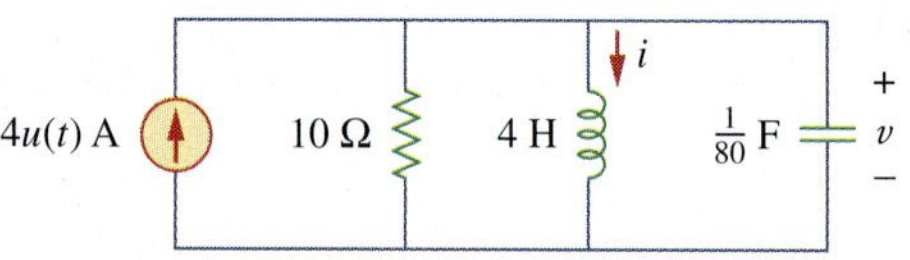

Figure 16.57
For Prob. 16.23.

16.24 The switch in Fig. 16.58 moves from position 1 to position 2 at $t = 0$. Find $v(t)$, for all $t > 0$.

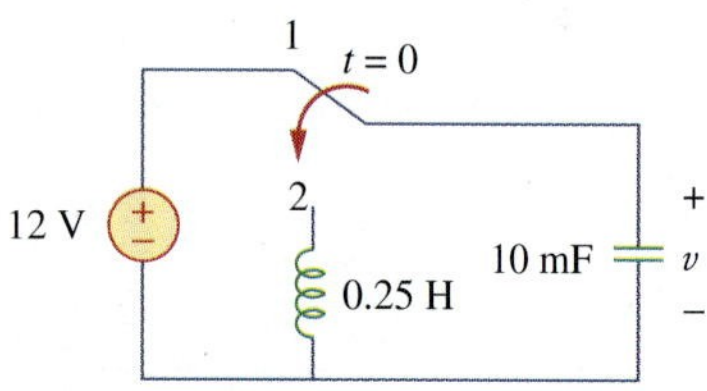

Figure 16.58
For Prob. 16.24.

16.25 For the RLC circuit shown in Fig. 16.59, find the complete response if $v(0) = 2$ V when the switch is closed.

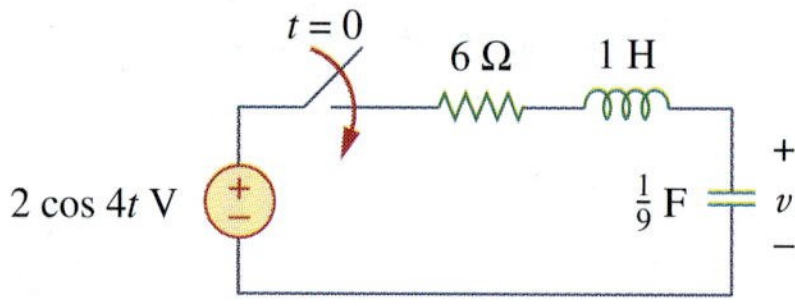

Figure 16.59
For Prob. 16.25.

16.26 For the op amp circuit in Fig. 16.60, find $v_o(t)$ for $t > 0$. Take $v_s = 10e^{-5t}u(t)$ V.

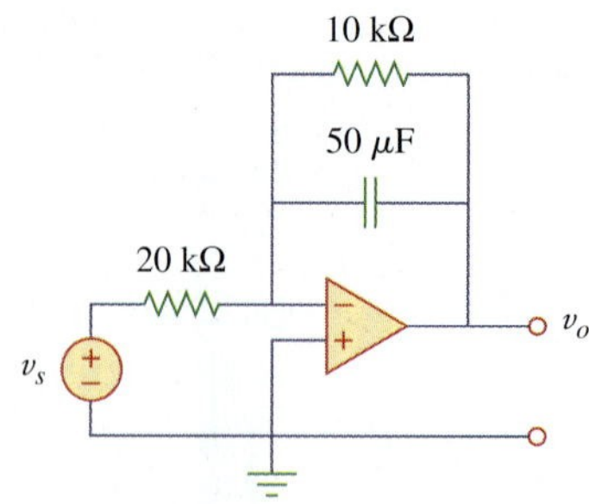

Figure 16.60
For Prob. 16.26.

16.27 Find $I_1(s)$ and $I_2(s)$ in the circuit of Fig. 16.61.

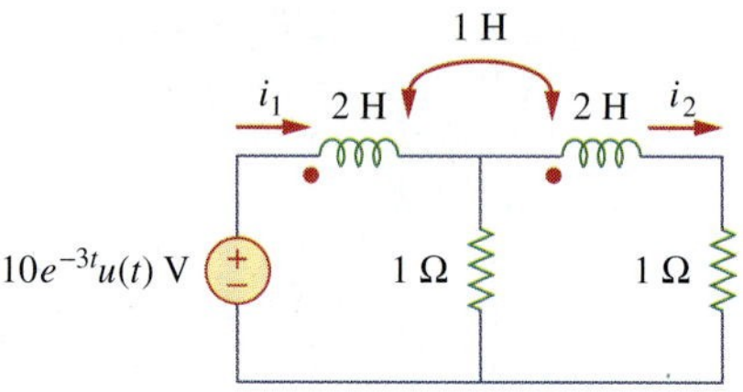

Figure 16.61
For Prob. 16.27.

16.28 Using Fig. 16.62, design a problem to help other students better understand how to do circuit analysis with circuits that have mutually coupled elements by working in the s-domain.

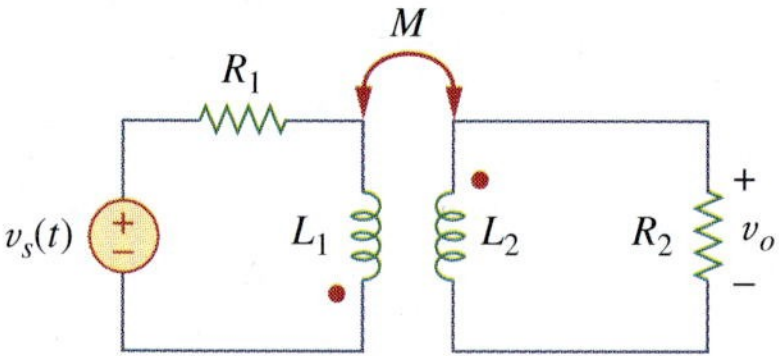

Figure 16.62
For Prob. 16.28.

16.29 For the ideal transformer circuit in Fig. 16.63, determine $i_o(t)$.

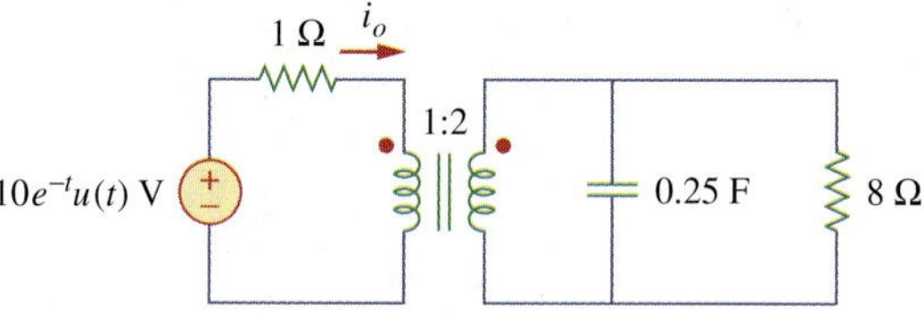

Figure 16.63
For Prob. 16.29.

Section 16.4 Transfer Functions

16.30 The transfer function of a system is

$$H(s) = \frac{s^2}{3s + 1}$$

Find the output when the system has an input of $4e^{-t/3}u(t)$.

16.31 When the input to a system is a unit step function, the response is $10 \cos 2tu(t)$. Obtain the transfer function of the system.

16.32 Design a problem to help other students better understand how to find outputs when given a transfer function and an input.

16.33 When a unit step is applied to a system at $t = 0$, its response is

$$y(t) = \left[4 + \frac{1}{2}e^{-3t} - e^{-2t}(2\cos 4t + 3\sin 4t)\right]u(t)$$

What is the transfer function of the system?

16.34 For the circuit in Fig. 16.64, find $H(s) = V_o(s)/V_s(s)$. Assume zero initial conditions.

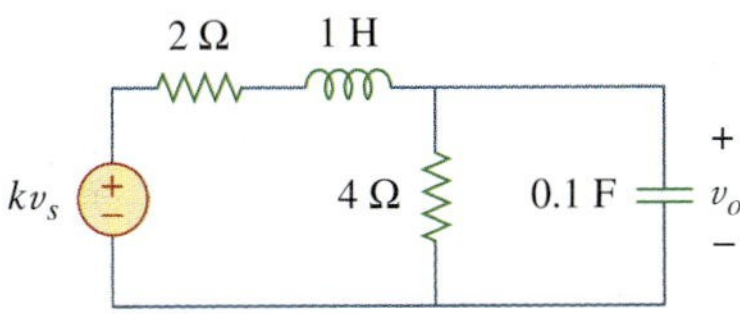

Figure 16.64
For Prob. 16.34.

16.35 Obtain the transfer function $H(s) = V_o/V_s$ for the circuit of Fig. 16.65.

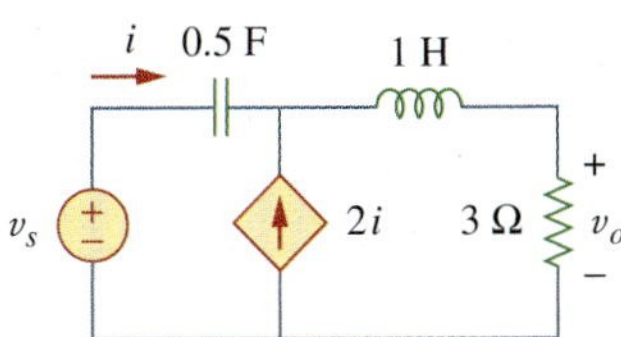

Figure 16.65
For Prob. 16.35.

16.36 The transfer function of a certain circuit is

$$H(s) = \frac{5}{s+1} - \frac{3}{s+2} + \frac{6}{s+4}$$

Find the impulse response of the circuit.

16.37 For the circuit in Fig. 16.66, find:

(a) I_1/V_s (b) I_2/V_x

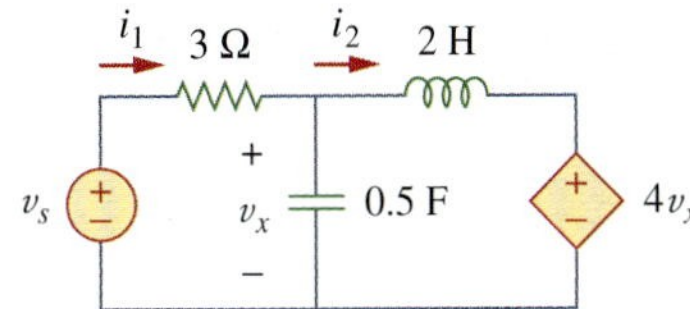

Figure 16.66
For Prob. 16.37.

16.38 Refer to the network in Fig. 16.67. Find the following transfer functions:

(a) $H_1(s) = V_o(s)/V_s(s)$
(b) $H_2(s) = V_o(s)/I_s(s)$
(c) $H_3(s) = I_o(s)/I_s(s)$
(d) $H_4(s) = I_o(s)/V_s(s)$

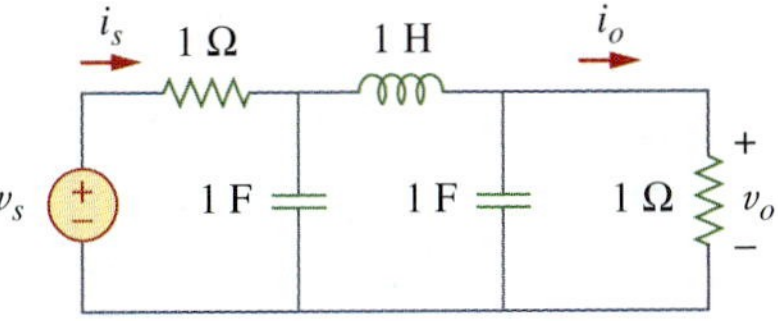

Figure 16.67
For Prob. 16.38.

16.39 Calculate the gain $H(s) = V_o/V_s$ in the op amp circuit of Fig. 16.68.

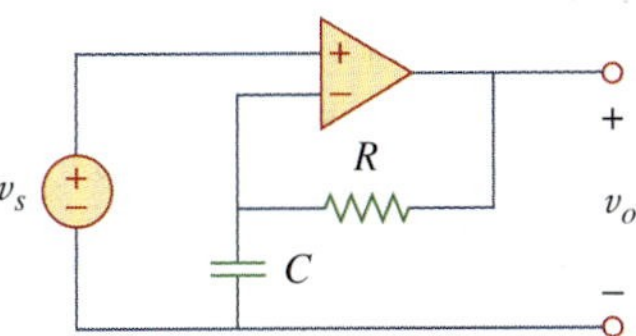

Figure 16.68
For Prob. 16.39.

16.40 Refer to the *RL* circuit in Fig. 16.69. Find:

(a) the impulse response $h(t)$ of the circuit.
(b) the unit step response of the circuit.

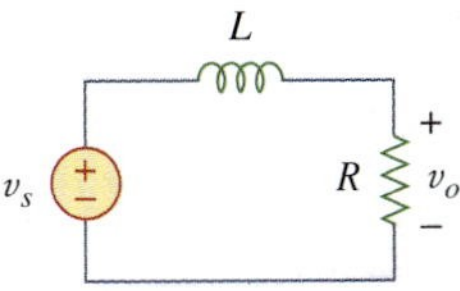

Figure 16.69
For Prob. 16.40.

16.41 A parallel *RL* circuit has $R = 4\ \Omega$ and $L = 1$ H. The input to the circuit is $i_s(t) = 2e^{-t}u(t)$ A. Find the inductor current $i_L(t)$ for all $t > 0$ and assume that $i_L(0) = -2$ A.

16.42 A circuit has a transfer function

$$H(s) = \frac{s+4}{(s+1)(s+2)^2}$$

Find the impulse response.

Section 16.5 State Variables

16.43 Develop the state equations for Prob. 16.1.

16.44 Develop the state equations for the problem you designed in Prob. 16.2.

16.45 Develop the state equations for the circuit shown in Fig. 16.70.

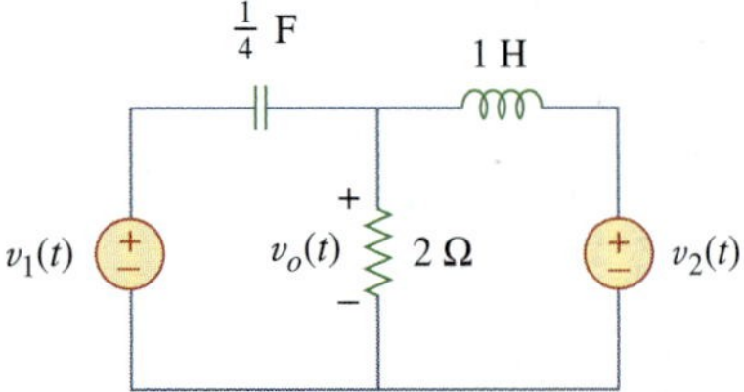

Figure 16.70
For Prob. 16.45.

16.46 Develop the state equations for the circuit shown in Fig. 16.71.

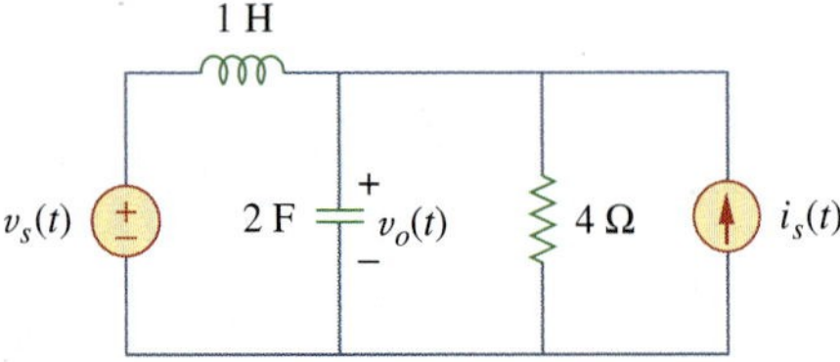

Figure 16.71
For Prob. 16.46.

16.47 Develop the state equations for the circuit shown in Fig. 16.72.

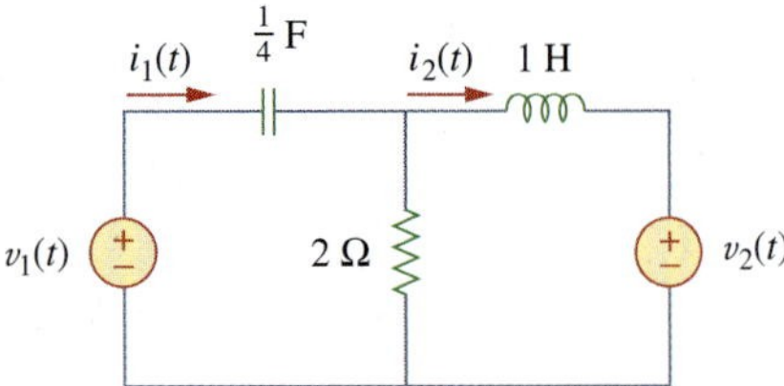

Figure 16.72
For Prob. 16.47.

16.48 Develop the state equations for the following differential equation.

$$\frac{d^2y(t)}{dt^2} + \frac{6\,dy(t)}{dt} + 7y(t) = z(t)$$

***16.49** Develop the state equations for the following differential equation.

$$\frac{d^2y(t)}{dt^2} + \frac{7\,dy(t)}{dt} + 9y(t) = \frac{dz(t)}{dt} + z(t)$$

***16.50** Develop the state equations for the following differential equation.

$$\frac{d^3y(t)}{dt^3} + \frac{6\,d^2y(t)}{dt^2} + \frac{11\,dy(t)}{dt} + 6y(t) = z(t)$$

***16.51** Given the following state equation, solve for $y(t)$:

$$\dot{\mathbf{x}} = \begin{bmatrix} -4 & 4 \\ -2 & 0 \end{bmatrix} x + \begin{bmatrix} 0 \\ 2 \end{bmatrix} u(t)$$

$$\mathbf{y}(t) = [1 \quad 0]x$$

***16.52** Given the following state equation, solve for $y_1(t)$ and $y_2(t)$.

$$\dot{\mathbf{x}} = \begin{bmatrix} -2 & -1 \\ 2 & -4 \end{bmatrix} x + \begin{bmatrix} 1 & 1 \\ 4 & 0 \end{bmatrix} \begin{bmatrix} u(t) \\ 2u(t) \end{bmatrix}$$

$$\mathbf{y} = \begin{bmatrix} -2 & -2 \\ 1 & -0 \end{bmatrix} x + \begin{bmatrix} 2 & 0 \\ 0 & -1 \end{bmatrix} \begin{bmatrix} u(t) \\ 2u(t) \end{bmatrix}$$

Section 16.6 Applications

16.53 Show that the parallel *RLC* circuit shown in Fig. 16.73 is stable.

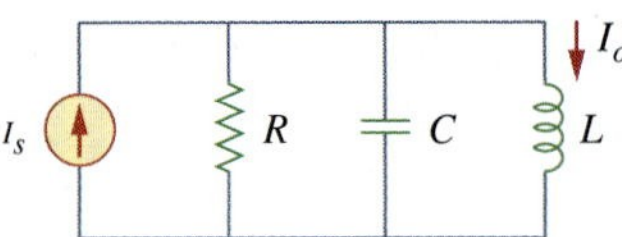

Figure 16.73
For Prob. 16.53.

16.54 A system is formed by cascading two systems as shown in Fig. 16.74. Given that the impulse response of the systems are

$$h_1(t) = 3e^{-t}u(t), \qquad h_2(t) = e^{-4t}u(t)$$

(a) Obtain the impulse response of the overall system.

(b) Check if the overall system is stable.

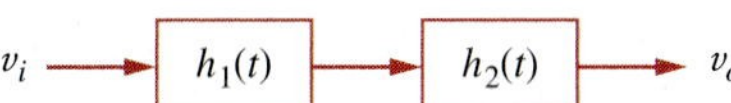

Figure 16.74
For Prob. 16.54.

16.55 Determine whether the op amp circuit in Fig. 16.75 is stable.

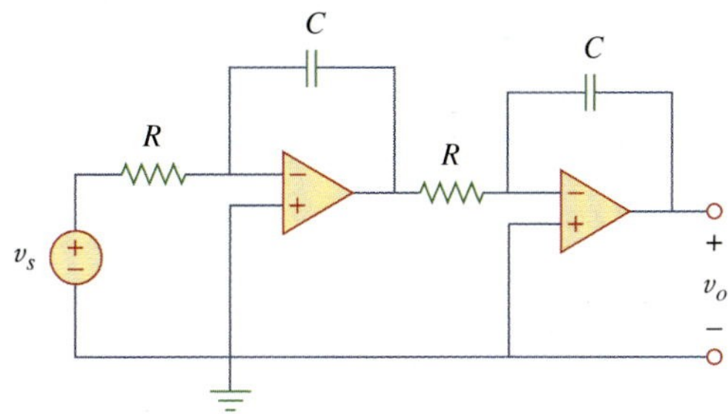

Figure 16.75
For Prob. 16.55.

16.56 It is desired to realize the transfer function

$$\frac{V_2(s)}{V_1(s)} = \frac{2s}{s^2 + 2s + 6}$$

using the circuit in Fig. 16.76. Choose $R = 1\ \text{k}\Omega$ and find L and C.

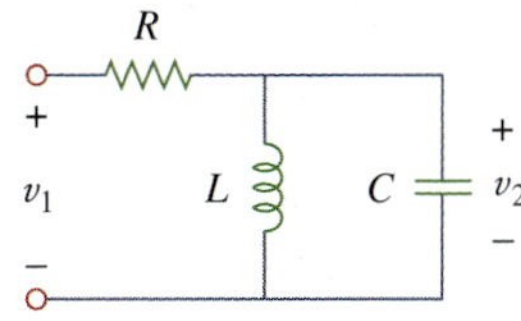

Figure 16.76
For Prob. 16.56.

16.57 Design an op amp circuit, using Fig. 16.77, that will realize the following transfer function:

$$\frac{V_o(s)}{V_i(s)} = -\frac{s + 1000}{2(s + 4000)}$$

Choose $C_1 = 10\ \mu\text{F}$; determine R_1, R_2, and C_2.

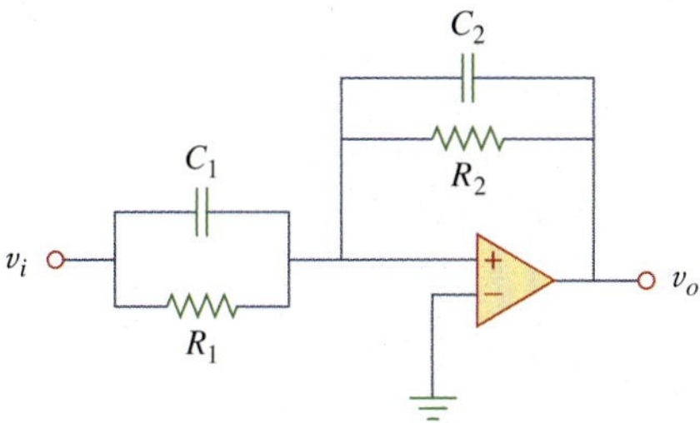

Figure 16.77
For Prob. 16.57.

16.58 Realize the transfer function

$$\frac{V_o(s)}{V_s(s)} = -\frac{s}{s + 10}$$

using the circuit in Fig. 16.78. Let $Y_1 = sC_1$, $Y_2 = 1/R_1$, $Y_3 = sC_2$. Choose $R_1 = 1\ \text{k}\Omega$ and determine C_1 and C_2.

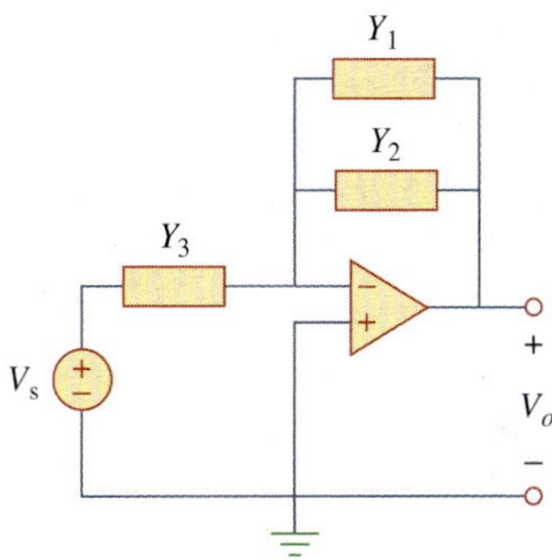

Figure 16.78
For Prob. 16.58.

16.59 Synthesize the transfer function

$$\frac{V_o(s)}{V_{in}(s)} = \frac{10^6}{s^2 + 100s + 10^6}$$

using the topology of Fig. 16.79. Let $Y_1 = 1/R_1$, $Y_2 = 1/R_2$, $Y_3 = sC_1$, $Y_4 = sC_2$. Choose $R_1 = 1\ \text{k}\Omega$ and determine C_1, C_2, and R_2.

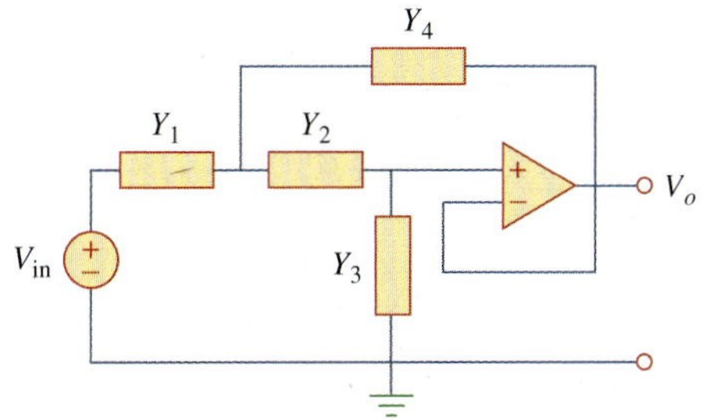

Figure 16.79
For Prob. 16.59.

Comprehensive Problems

16.60 Obtain the transfer function of the op amp circuit in Fig. 16.80 in the form of

$$\frac{V_o(s)}{V_i(s)} = \frac{as}{s^2 + bs + c}$$

where a, b, and c are constants. Determine the constants.

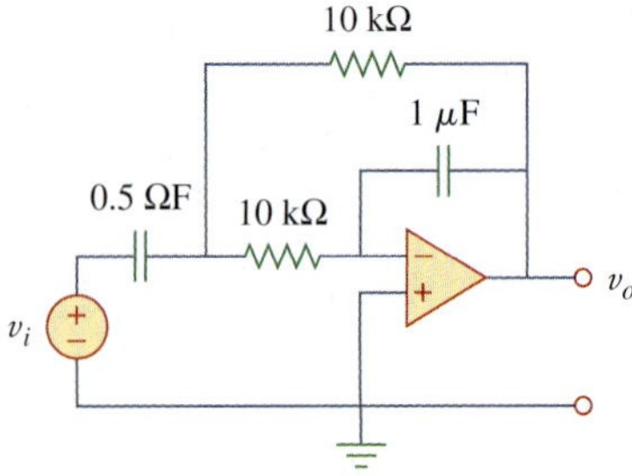

Figure 16.80
For Prob. 16.60.

16.61 A certain network has an input admittance $Y(s)$. The admittance has a pole at $s = -3$, a zero at $s = -1$, and $Y(\infty) = 0.25$ S.

(a) Find $Y(s)$.

(b) An 8-V battery is connected to the network via a switch. If the switch is closed at $t = 0$, find the current $i(t)$ through $Y(s)$ using the Laplace transform.

16.62 A gyrator is a device for simulating an inductor in a network. A basic gyrator circuit is shown in Fig. 16.81. By finding $V_i(s)/I_o(s)$, show that the inductance produced by the gyrator is $L = CR^2$.

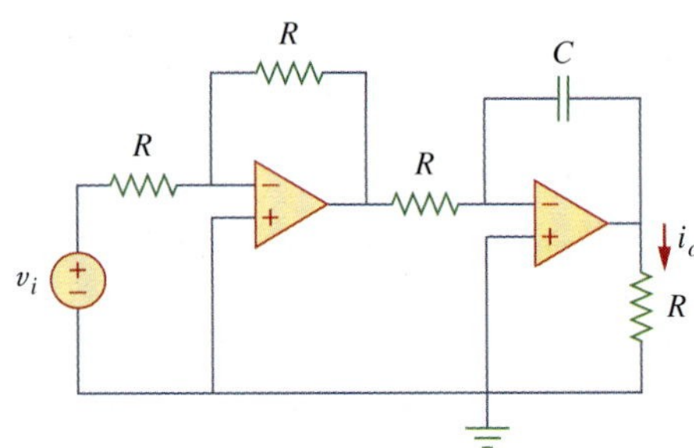

Figure 16.81
For Prob. 16.62.

chapter

17

The Fourier Series

Research is to see what everybody else has seen, and think what nobody has thought.

—Albert Szent Györgyi

Enhancing Your Skills and Your Career

Photo by Charles Alexander

ABET EC 2000 criteria (3.j), *"a knowledge of contemporary issues."*

Engineers must have knowledge of contemporary issues. To have a truly meaningful career in the twenty-first century, you must have knowledge of contemporary issues, especially those that may directly affect your job and/or work. One of the easiest ways to achieve this is to read a lot—newspapers, magazines, and contemporary books. As a student enrolled in an ABET-accredited program, some of the courses you take will be directed toward meeting this criteria.

ABET EC 2000 criteria (3.k), *"an ability to use the techniques, skills, and modern engineering tools necessary for engineering practice."*

The successful engineer must have the "ability to use the techniques, skills, and modern engineering tools necessary for engineering practice." Clearly, a major focus of this textbook is to do just that. Learning to use skillfully the tools that facilitate your working in a modern "knowledge capturing integrated design environment" (KCIDE) is fundamental to your performance as an engineer. The ability to work in a modern KCIDE environment requires a thorough understanding of the tools associated with that environment.

The successful engineer, therefore, must keep abreast of the new design, analysis, and simulation tools. That engineer must also use those tools until he or she is comfortable with using them. The engineer also must make sure software results are consistent with real-world actualities. It is probably in this area that most engineers have the greatest difficulty. Thus, successful use of these tools requires constant learning and relearning the fundamentals of the area in which the engineer is working.

Historical

Jean Baptiste Joseph Fourier (1768–1830), a French mathematician, first presented the series and transform that bear his name. Fourier's results were not enthusiastically received by the scientific world. He could not even get his work published as a paper.

Born in Auxerre, France, Fourier was orphaned at age 8. He attended a local military college run by Benedictine monks, where he demonstrated great proficiency in mathematics. Like most of his contemporaries, Fourier was swept into the politics of the French Revolution. He played an important role in Napoleon's expeditions to Egypt in the later 1790s. Due to his political involvement, he narrowly escaped death twice.

17.1 Introduction

We have spent a considerable amount of time on the analysis of circuits with sinusoidal sources. This chapter is concerned with a means of analyzing circuits with periodic, nonsinusoidal excitations. The notion of periodic functions was introduced in Chapter 9; it was mentioned there that the sinusoid is the most simple and useful periodic function. This chapter introduces the Fourier series, a technique for expressing a periodic function in terms of sinusoids. Once the source function is expressed in terms of sinusoids, we can apply the phasor method to analyze circuits.

The Fourier series is named after Jean Baptiste Joseph Fourier (1768–1830). In 1822, Fourier's genius came up with the insight that any practical periodic function can be represented as a sum of sinusoids. Such a representation, along with the superposition theorem, allows us to find the response of circuits to arbitrary periodic inputs using phasor techniques.

We begin with the trigonometric Fourier series. Later we consider the exponential Fourier series. We then apply Fourier series in circuit analysis. Finally, practical applications of Fourier series in spectrum analyzers and filters are demonstrated.

17.2 Trigonometric Fourier Series

While studying heat flow, Fourier discovered that a nonsinusoidal periodic function can be expressed as an infinite sum of sinusoidal functions. Recall that a periodic function is one that repeats every T seconds. In other words, a periodic function $f(t)$ satisfies

$$f(t) = f(t + nT) \tag{17.1}$$

where n is an integer and T is the period of the function.

According to the *Fourier theorem*, any practical periodic function of frequency ω_0 can be expressed as an infinite sum of sine or cosine functions that are integral multiples of ω_0. Thus, $f(t)$ can be expressed as

$$\begin{aligned} f(t) = a_0 &+ a_1 \cos\omega_0 t + b_1 \sin\omega_0 t + a_2 \cos 2\omega_0 t \\ &+ b_2 \sin 2\omega_0 t + a_3 \cos 3\omega_0 t + b_3 \sin 3\omega_0 t + \cdots \end{aligned} \tag{17.2}$$

or

$$f(t) = \underbrace{a_0}_{\text{dc}} + \underbrace{\sum_{n=1}^{\infty} (a_n \cos n\omega_0 t + b_n \sin n\omega_0 t)}_{\text{ac}} \tag{17.3}$$

where $\omega_0 = 2\pi/T$ is called the *fundamental frequency* in radians per second. The sinusoid $\sin n\omega_0 t$ or $\cos n\omega_0 t$ is called the nth harmonic of $f(t)$; it is an odd harmonic if n is odd and an even harmonic if n is even. Equation 17.3 is called the *trigonometric Fourier series* of $f(t)$. The constants a_n and b_n are the *Fourier coefficients.* The coefficient a_0 is the dc component or the average value of $f(t)$. (Recall that sinusoids have zero average values.) The coefficients a_n and b_n (for $n \neq 0$) are the amplitudes of the sinusoids in the ac component. Thus,

The harmonic frequency ω_n is an integral multiple of the fundamental frequency ω_0, i.e., $\omega_n = n\omega_0$.

The **Fourier series** of a periodic function $f(t)$ is a representation that resolves $f(t)$ into a dc component and an ac component comprising an infinite series of harmonic sinusoids.

A function that can be represented by a Fourier series as in Eq. (17.3) must meet certain requirements, because the infinite series in Eq. (17.3) may or may not converge. These conditions on $f(t)$ to yield a convergent Fourier series are as follows:

1. $f(t)$ is single-valued everywhere.
2. $f(t)$ has a finite number of finite discontinuities in any one period.
3. $f(t)$ has a finite number of maxima and minima in any one period.
4. The integral $\displaystyle\int_{t_0}^{t_0+T} |f(t)|\, dt < \infty$ for any t_0.

These conditions are called *Dirichlet conditions.* Although they are not necessary conditions, they are sufficient conditions for a Fourier series to exist.

A major task in Fourier series is the determination of the Fourier coefficients a_0, a_n, and b_n. The process of determining the

Historical note: Although Fourier published his theorem in 1822, it was P. G. L. Dirichlet (1805–1859) who later supplied an acceptable proof of the theorem.

A software package like *Mathcad* or *Maple* can be used to evaluate the Fourier coefficients.

Equating the coefficients of the series expansions in Eqs. (17.3) and (17.12) shows that

$$a_n = A_n \cos \phi_n, \qquad b_n = -A_n \sin \phi_n \tag{17.13a}$$

or

$$A_n = \sqrt{a_n^2 + b_n^2}, \qquad \phi_n = -\tan^{-1} \frac{b_n}{a_n} \tag{17.13b}$$

To avoid any confusion in determining ϕ_n, it may be better to relate the terms in complex form as

$$A_n \underline{/\phi_n} = a_n - jb_n \tag{17.14}$$

The convenience of this relationship will become evident in Section 17.6. The plot of the amplitude A_n of the harmonics versus $n\omega_0$ is called the *amplitude spectrum* of $f(t)$; the plot of the phase ϕ_n versus $n\omega_0$ is the *phase spectrum* of $f(t)$. Both the amplitude and phase spectra form the *frequency spectrum* of $f(t)$.

The frequency spectrum is also known as the *line spectrum* in view of the discrete frequency components.

The **frequency spectrum** of a signal consists of the plots of the amplitudes and phases of the harmonics versus frequency.

Thus, the Fourier analysis is also a mathematical tool for finding the spectrum of a periodic signal. Section 17.6 will elaborate more on the spectrum of a signal.

To evaluate the Fourier coefficients a_0, a_n, and b_n, we often need to apply the following integrals:

$$\int \cos at \, dt = \frac{1}{a} \sin at \tag{17.15a}$$

$$\int \sin at \, dt = -\frac{1}{a} \cos at \tag{17.15b}$$

$$\int t \cos at \, dt = \frac{1}{a^2} \cos at + \frac{1}{a} t \sin at \tag{17.15c}$$

$$\int t \sin at \, dt = \frac{1}{a^2} \sin at - \frac{1}{a} t \cos at \tag{17.15d}$$

It is also useful to know the values of the cosine, sine, and exponential functions for integral multiples of π. These are given in Table 17.1, where n is an integer.

TABLE 17.1

Values of cosine, sine, and exponential functions for integral multiples of π.

Function	Value
$\cos 2n\pi$	1
$\sin 2n\pi$	0
$\cos n\pi$	$(-1)^n$
$\sin n\pi$	0
$\cos \frac{n\pi}{2}$	$\begin{cases} (-1)^{n/2}, & n = \text{even} \\ 0, & n = \text{odd} \end{cases}$
$\sin \frac{n\pi}{2}$	$\begin{cases} (-1)^{(n-1)/2}, & n = \text{odd} \\ 0, & n = \text{even} \end{cases}$
$e^{j2n\pi}$	1
$e^{jn\pi}$	$(-1)^n$
$e^{jn\pi/2}$	$\begin{cases} (-1)^{n/2}, & n = \text{even} \\ j(-1)^{(n-1)/2}, & n = \text{odd} \end{cases}$

Example 17.1

Determine the Fourier series of the waveform shown in Fig. 17.1. Obtain the amplitude and phase spectra.

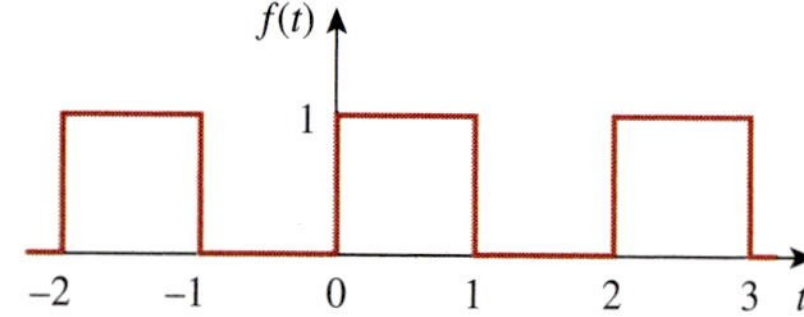

Figure 17.1
For Example 17.1; a square wave.

Solution:

The Fourier series is given by Eq. (17.3), namely,

$$f(t) = a_0 + \sum_{n=1}^{\infty} (a_n \cos n\omega_0 t + b_n \sin n\omega_0 t) \tag{17.1.1}$$

Equating the coefficients of the series expansions in Eqs. (17.3) and (17.12) shows that

$$a_n = A_n \cos \phi_n, \qquad b_n = -A_n \sin \phi_n \tag{17.13a}$$

or

$$A_n = \sqrt{a_n^2 + b_n^2}, \qquad \phi_n = -\tan^{-1} \frac{b_n}{a_n} \tag{17.13b}$$

To avoid any confusion in determining ϕ_n, it may be better to relate the terms in complex form as

$$A_n \underline{/\phi_n} = a_n - jb_n \tag{17.14}$$

The convenience of this relationship will become evident in Section 17.6. The plot of the amplitude A_n of the harmonics versus $n\omega_0$ is called the *amplitude spectrum* of $f(t)$; the plot of the phase ϕ_n versus $n\omega_0$ is the *phase spectrum* of $f(t)$. Both the amplitude and phase spectra form the *frequency spectrum* of $f(t)$.

The frequency spectrum is also known as the *line spectrum* in view of the discrete frequency components.

The **frequency spectrum** of a signal consists of the plots of the amplitudes and phases of the harmonics versus frequency.

Thus, the Fourier analysis is also a mathematical tool for finding the spectrum of a periodic signal. Section 17.6 will elaborate more on the spectrum of a signal.

To evaluate the Fourier coefficients a_0, a_n, and b_n, we often need to apply the following integrals:

$$\int \cos at\, dt = \frac{1}{a} \sin at \tag{17.15a}$$

$$\int \sin at\, dt = -\frac{1}{a} \cos at \tag{17.15b}$$

$$\int t \cos at\, dt = \frac{1}{a^2} \cos at + \frac{1}{a} t \sin at \tag{17.15c}$$

$$\int t \sin at\, dt = \frac{1}{a^2} \sin at - \frac{1}{a} t \cos at \tag{17.15d}$$

It is also useful to know the values of the cosine, sine, and exponential functions for integral multiples of π. These are given in Table 17.1, where n is an integer.

TABLE 17.1

Values of cosine, sine, and exponential functions for integral multiples of π.

Function	Value
$\cos 2n\pi$	1
$\sin 2n\pi$	0
$\cos n\pi$	$(-1)^n$
$\sin n\pi$	0
$\cos \frac{n\pi}{2}$	$\begin{cases} (-1)^{n/2}, & n = \text{even} \\ 0, & n = \text{odd} \end{cases}$
$\sin \frac{n\pi}{2}$	$\begin{cases} (-1)^{(n-1)/2}, & n = \text{odd} \\ 0, & n = \text{even} \end{cases}$
$e^{j2n\pi}$	1
$e^{jn\pi}$	$(-1)^n$
$e^{jn\pi/2}$	$\begin{cases} (-1)^{n/2}, & n = \text{even} \\ j(-1)^{(n-1)/2}, & n = \text{odd} \end{cases}$

Example 17.1

Determine the Fourier series of the waveform shown in Fig. 17.1. Obtain the amplitude and phase spectra.

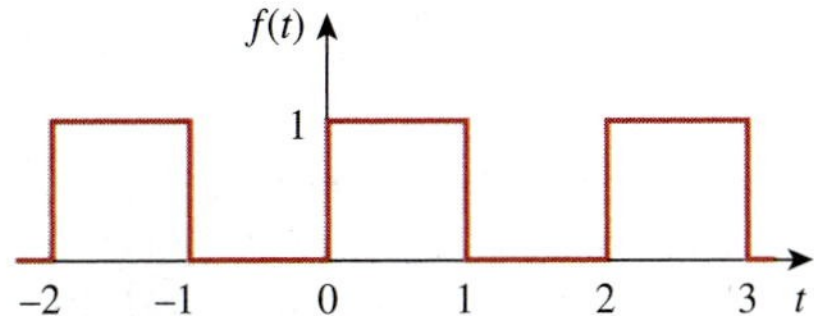

Figure 17.1
For Example 17.1; a square wave.

Solution:
The Fourier series is given by Eq. (17.3), namely,

$$f(t) = a_0 + \sum_{n=1}^{\infty} (a_n \cos n\omega_0 t + b_n \sin n\omega_0 t) \tag{17.1.1}$$

To evaluate a_n, we multiply both sides of Eq. (17.3) by $\cos m\omega_0 t$ and integrate over one period:

$$\begin{aligned}\int_0^T f(t)\cos m\omega_0 t\,dt &= \int_0^T \left[a_0 + \sum_{n=1}^{\infty}(a_n \cos n\omega_0 t + b_n \sin n\omega_0 t)\right]\cos m\omega_0 t\,dt \\ &= \int_0^T a_0 \cos m\omega_0 t\,dt + \sum_{n=1}^{\infty}\left[\int_0^T a_n \cos n\omega_0 t \cos m\omega_0 t\,dt \right. \\ &\quad \left. + \int_0^T b_n \sin n\omega_0 t \cos m\omega_0 t\,dt\right] dt \end{aligned} \tag{17.7}$$

The integral containing a_0 is zero in view of Eq. (17.4b), while the integral containing b_n vanishes according to Eq. (17.4c). The integral containing a_n will be zero except when $m = n$, in which case it is $T/2$, according to Eqs. (17.4e) and (17.4g). Thus,

$$\int_0^T f(t)\cos m\omega_0 t\,dt = a_n \frac{T}{2}, \qquad \text{for } m = n$$

or

$$a_n = \frac{2}{T}\int_0^T f(t)\cos n\omega_0 t\,dt \tag{17.8}$$

In a similar vein, we obtain b_n by multiplying both sides of Eq. (17.3) by $\sin m\omega_0 t$ and integrating over the period. The result is

$$b_n = \frac{2}{T}\int_0^T f(t)\sin n\omega_0 t\,dt \tag{17.9}$$

Be aware that since $f(t)$ is periodic, it may be more convenient to carry the integrations above from $-T/2$ to $T/2$ or generally from t_0 to $t_0 + T$ instead of 0 to T. The result will be the same.

An alternative form of Eq. (17.3) is the *amplitude-phase* form

$$f(t) = a_0 + \sum_{n=1}^{\infty} A_n \cos(n\omega_0 t + \phi_n) \tag{17.10}$$

We can use Eqs. (9.11) and (9.12) to relate Eq. (17.3) to Eq. (17.10), or we can apply the trigonometric identity

$$\cos(\alpha + \beta) = \cos\alpha\cos\beta - \sin\alpha\sin\beta \tag{17.11}$$

to the ac terms in Eq. (17.10) so that

$$\begin{aligned} a_0 + \sum_{n=1}^{\infty} A_n \cos(n\omega_0 t + \phi_n) = a_0 + \sum_{n=1}^{\infty} (A_n \cos\phi_n)\cos n\omega_0 t \\ - (A_n \sin\phi_n)\sin n\omega_0 t \end{aligned} \tag{17.12}$$

coefficients is called *Fourier analysis.* The following trigonometric integrals are very helpful in Fourier analysis. For any integers m and n,

$$\int_0^T \sin n\omega_0 t\, dt = 0 \tag{17.4a}$$

$$\int_0^T \cos n\omega_0 t\, dt = 0 \tag{17.4b}$$

$$\int_0^T \sin n\omega_0 t \cos m\omega_0 t\, dt = 0 \tag{17.4c}$$

$$\int_0^T \sin n\omega_0 t \sin m\omega_0 t\, dt = 0, \qquad (m \neq n) \tag{17.4d}$$

$$\int_0^T \cos n\omega_0 t \cos m\omega_0 t\, dt = 0, \qquad (m \neq n) \tag{17.4e}$$

$$\int_0^T \sin^2 n\omega_0 t\, dt = \frac{T}{2} \tag{17.4f}$$

$$\int_0^T \cos^2 n\omega_0 t\, dt = \frac{T}{2} \tag{17.4g}$$

Let us use these identities to evaluate the Fourier coefficients.

We begin by finding a_0. We integrate both sides of Eq. (17.3) over one period and obtain

$$\begin{aligned}\int_0^T f(t)\, dt &= \int_0^T \left[a_0 + \sum_{n=1}^{\infty} (a_n \cos n\omega_0 t + b_n \sin n\omega_0 t)\right] dt \\ &= \int_0^T a_0\, dt + \sum_{n=1}^{\infty} \left[\int_0^T a_n \cos n\omega_0 t\, dt \right. \\ &\qquad \left. + \int_0^T b_n \sin n\omega_0 t\, dt\right] dt\end{aligned} \tag{17.5}$$

Invoking the identities of Eqs. (17.4a) and (17.4b), the two integrals involving the ac terms vanish. Hence,

$$\int_0^T f(t)\, dt = \int_0^T a_0\, dt = a_0 T$$

or

$$\boxed{a_0 = \frac{1}{T}\int_0^T f(t)\, dt} \tag{17.6}$$

showing that a_0 is the average value of $f(t)$.

According to the *Fourier theorem*, any practical periodic function of frequency ω_0 can be expressed as an infinite sum of sine or cosine functions that are integral multiples of ω_0. Thus, $f(t)$ can be expressed as

$$f(t) = a_0 + a_1 \cos\omega_0 t + b_1 \sin\omega_0 t + a_2 \cos 2\omega_0 t + b_2 \sin 2\omega_0 t + a_3 \cos 3\omega_0 t + b_3 \sin 3\omega_0 t + \cdots \quad (17.2)$$

or

$$f(t) = \underbrace{a_0}_{\text{dc}} + \underbrace{\sum_{n=1}^{\infty} (a_n \cos n\omega_0 t + b_n \sin n\omega_0 t)}_{\text{ac}} \quad (17.3)$$

where $\omega_0 = 2\pi/T$ is called the *fundamental frequency* in radians per second. The sinusoid $\sin n\omega_0 t$ or $\cos n\omega_0 t$ is called the nth harmonic of $f(t)$; it is an odd harmonic if n is odd and an even harmonic if n is even. Equation 17.3 is called the *trigonometric Fourier series* of $f(t)$. The constants a_n and b_n are the *Fourier coefficients*. The coefficient a_0 is the dc component or the average value of $f(t)$. (Recall that sinusoids have zero average values.) The coefficients a_n and b_n (for $n \neq 0$) are the amplitudes of the sinusoids in the ac component. Thus,

The harmonic frequency ω_n is an integral multiple of the fundamental frequency ω_0, i.e., $\omega_n = n\omega_0$.

The **Fourier series** of a periodic function $f(t)$ is a representation that resolves $f(t)$ into a dc component and an ac component comprising an infinite series of harmonic sinusoids.

A function that can be represented by a Fourier series as in Eq. (17.3) must meet certain requirements, because the infinite series in Eq. (17.3) may or may not converge. These conditions on $f(t)$ to yield a convergent Fourier series are as follows:

1. $f(t)$ is single-valued everywhere.
2. $f(t)$ has a finite number of finite discontinuities in any one period.
3. $f(t)$ has a finite number of maxima and minima in any one period.
4. The integral $\displaystyle\int_{t_0}^{t_0+T} |f(t)|\, dt < \infty$ for any t_0.

These conditions are called *Dirichlet conditions.* Although they are not necessary conditions, they are sufficient conditions for a Fourier series to exist.

Historical note: Although Fourier published his theorem in 1822, it was P. G. L. Dirichlet (1805–1859) who later supplied an acceptable proof of the theorem.

A major task in Fourier series is the determination of the Fourier coefficients a_0, a_n, and b_n. The process of determining the

A software package like *Mathcad* or *Maple* can be used to evaluate the Fourier coefficients.

Our goal is to obtain the Fourier coefficients a_0, a_n, and b_n using Eqs. (17.6), (17.8), and (17.9). First, we describe the waveform as

$$f(t) = \begin{cases} 1, & 0 < t < 1 \\ 0, & 1 < t < 2 \end{cases} \tag{17.1.2}$$

and $f(t) = f(t + T)$. Since $T = 2$, $\omega_0 = 2\pi/T = \pi$. Thus,

$$a_0 = \frac{1}{T}\int_0^T f(t)\,dt = \frac{1}{2}\left[\int_0^1 1\,dt + \int_1^2 0\,dt\right] = \frac{1}{2}t\Big|_0^1 = \frac{1}{2} \tag{17.1.3}$$

Using Eq. (17.8) along with Eq. (17.15a),

$$\begin{aligned} a_n &= \frac{2}{T}\int_0^T f(t)\cos n\omega_0 t\,dt \\ &= \frac{2}{2}\left[\int_0^1 1\cos n\pi t\,dt + \int_1^2 0\cos n\pi t\,dt\right] \\ &= \frac{1}{n\pi}\sin n\pi t\Big|_0^1 = \frac{1}{n\pi}[\sin n\pi - \sin(0)] = 0 \end{aligned} \tag{17.1.4}$$

From Eq. (17.9) with the aid of Eq. (17.15b),

$$\begin{aligned} b_n &= \frac{2}{T}\int_0^T f(t)\sin n\omega_0 t\,dt \\ &= \frac{2}{2}\left[\int_0^1 1\sin n\pi t\,dt + \int_1^2 0\sin n\pi t\,dt\right] \\ &= -\frac{1}{n\pi}\cos n\pi t\Big|_0^1 \\ &= -\frac{1}{n\pi}(\cos n\pi - 1), \qquad \cos n\pi = (-1)^n \\ &= \frac{1}{n\pi}[1 - (-1)^n] = \begin{cases} \dfrac{2}{n\pi}, & n = \text{odd} \\ 0, & n = \text{even} \end{cases} \end{aligned} \tag{17.1.5}$$

Substituting the Fourier coefficients in Eqs. (17.1.3) to (17.1.5) into Eq. (17.1.1) gives the Fourier series as

$$f(t) = \frac{1}{2} + \frac{2}{\pi}\sin\pi t + \frac{2}{3\pi}\sin 3\pi t + \frac{2}{5\pi}\sin 5\pi t + \cdots \tag{17.1.6}$$

Since $f(t)$ contains only the dc component and the sine terms with the fundamental component and odd harmonics, it may be written as

$$f(t) = \frac{1}{2} + \frac{2}{\pi}\sum_{k=1}^{\infty}\frac{1}{n}\sin n\pi t, \qquad n = 2k - 1 \tag{17.1.7}$$

By summing the terms one by one as demonstrated in Fig. 17.2, we notice how superposition of the terms can evolve into the original square. As more and more Fourier components are added, the sum gets closer and closer to the square wave. However, it is not possible in practice to sum the series in Eq. (17.1.6) or (17.1.7) to infinity. Only a partial sum ($n = 1, 2, 3, \ldots, N$, where N is finite) is possible. If we plot the partial sum (or truncated series) over one period for a large N

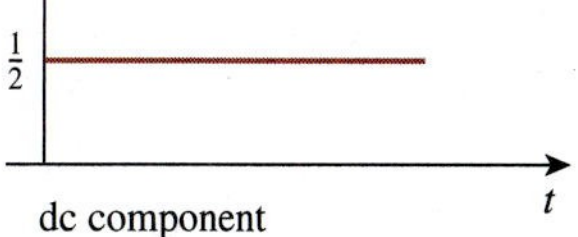

dc component

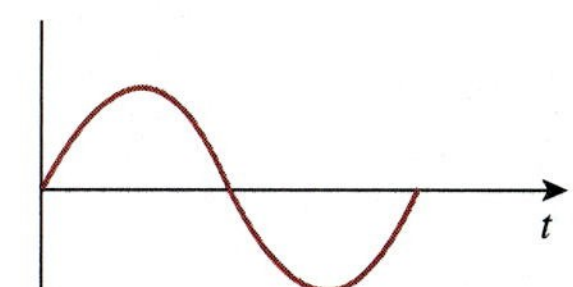

Fundamental ac component

(a)

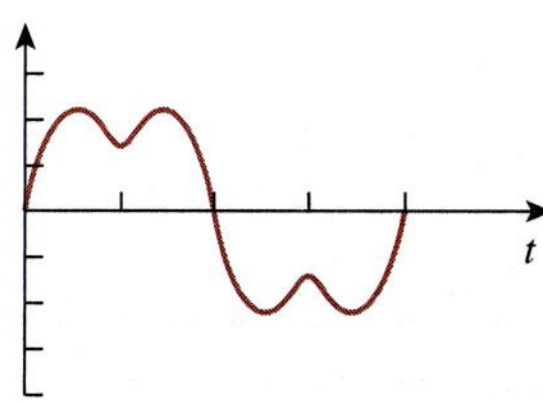

Sum of first two ac components

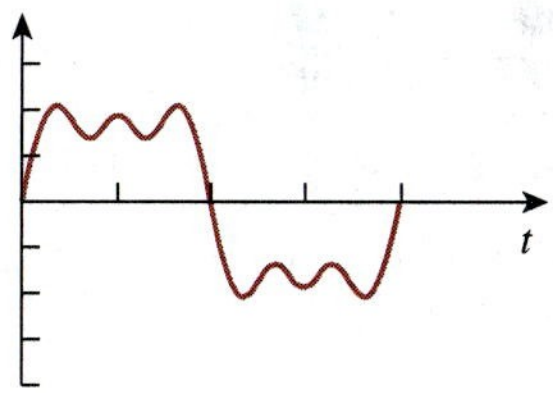

Sum of first three ac components

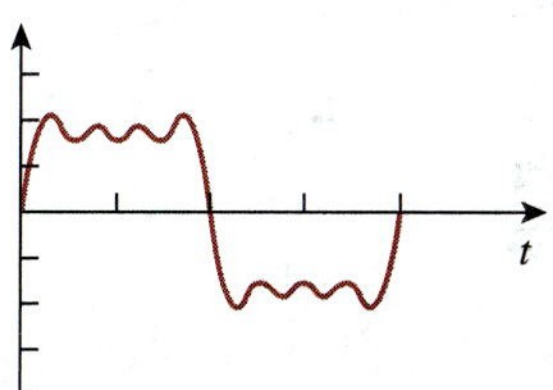

Sum of first four ac components

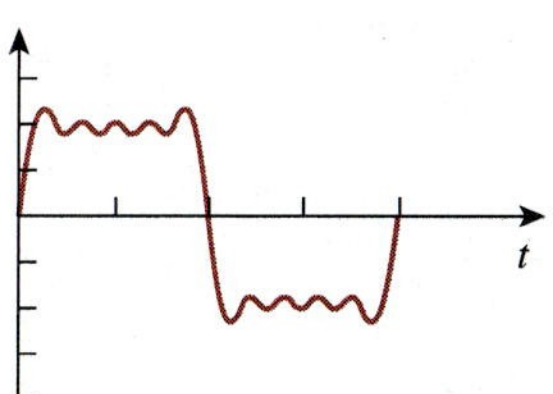

Sum of first five ac components

(b)

Figure 17.2
Evolution of a square wave from its Fourier components.

Summing the Fourier terms by hand calculation may be tedious. A computer is helpful to compute the terms and plot the sum like those shown in Fig. 17.2.

Historical note: Named after the mathematical physicist Josiah Willard Gibbs, who first observed it in 1899.

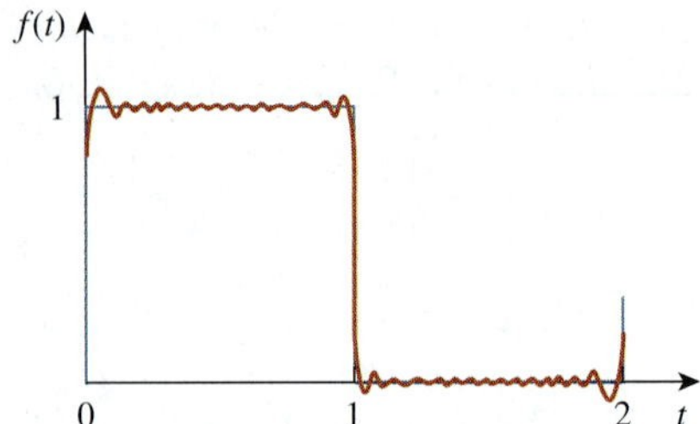

Figure 17.3
Truncating the Fourier series at $N = 11$; Gibbs phenomenon.

as in Fig. 17.3, we notice that the partial sum oscillates above and below the actual value of $f(t)$. At the neighborhood of the points of discontinuity $(x = 0, 1, 2, \dots)$, there is overshoot and damped oscillation. In fact, an overshoot of about 9 percent of the peak value is always present, regardless of the number of terms used to approximate $f(t)$. This is called the *Gibbs phenomenon.*

Finally, let us obtain the amplitude and phase spectra for the signal in Fig. 17.1. Since $a_n = 0$,

$$A_n = \sqrt{a_n^2 + b_n^2} = |b_n| = \begin{cases} \dfrac{2}{n\pi}, & n = \text{odd} \\ 0, & n = \text{even} \end{cases} \tag{17.1.8}$$

and

$$\phi_n = -\tan^{-1}\frac{b_n}{a_n} = \begin{cases} -90^\circ, & n = \text{odd} \\ 0, & n = \text{even} \end{cases} \tag{17.1.9}$$

The plots of A_n and ϕ_n for different values of $n\omega_0 = n\pi$ provide the amplitude and phase spectra in Fig. 17.4. Notice that the amplitudes of the harmonics decay very fast with frequency.

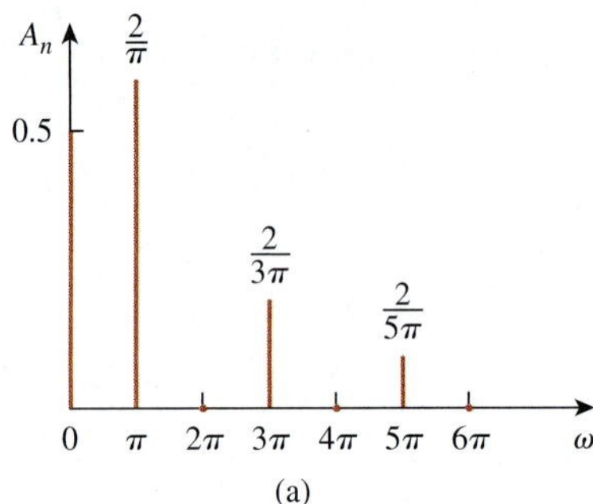

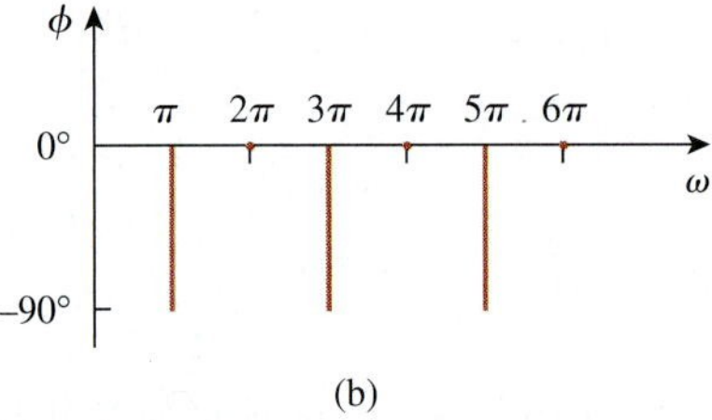

Figure 17.4
For Example 17.1: (a) amplitude and (b) phase spectrum of the function shown in Fig. 17.1.

Practice Problem 17.1

Find the Fourier series of the square wave in Fig. 17.5. Plot the amplitude and phase spectra.

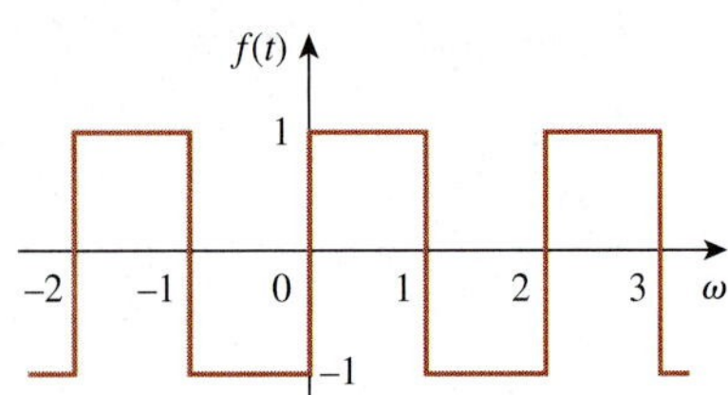

Figure 17.5
For Practice Prob. 17.1.

Answer: $f(t) = \dfrac{4}{\pi}\displaystyle\sum_{k=1}^{\infty}\dfrac{1}{n}\sin n\pi t,\ n = 2k - 1$. See Fig. 17.6 for the spectra.

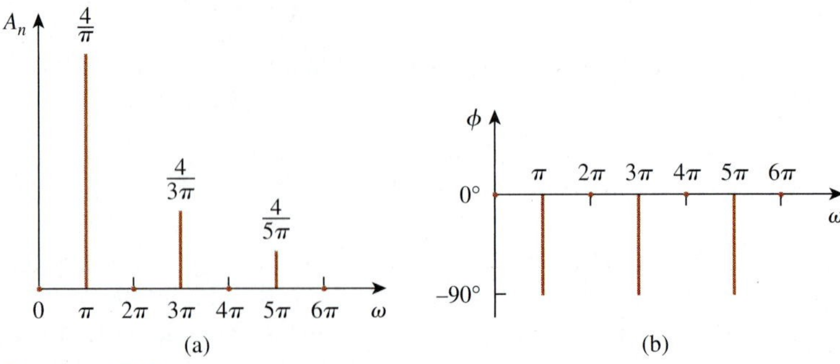

Figure 17.6
For Practice Prob. 17.1: amplitude and phase spectra for the function shown in Fig. 17.5.

Example 17.2

Obtain the Fourier series for the periodic function in Fig. 17.7 and plot the amplitude and phase spectra.

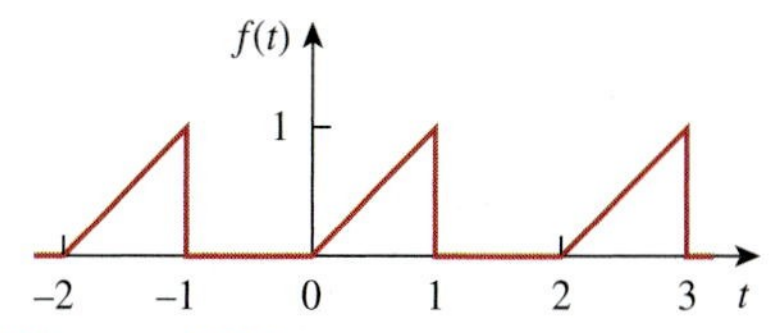

Figure 17.7
For Example 17.2.

Solution:
The function is described as

$$f(t) = \begin{cases} t, & 0 < t < 1 \\ 0, & 1 < t < 2 \end{cases}$$

Since $T = 2$, $\omega_0 = 2\pi/T = \pi$. Then

$$a_0 = \frac{1}{T}\int_0^T f(t)\,dt = \frac{1}{2}\left[\int_0^1 t\,dt + \int_1^2 0\,dt\right] = \frac{1}{2}\frac{t^2}{2}\Big|_0^1 = \frac{1}{4} \quad \textbf{(17.2.1)}$$

To evaluate a_n and b_n, we need the integrals in Eq. (17.15):

$$\begin{aligned} a_n &= \frac{2}{T}\int_0^T f(t)\cos n\omega_0 t\,dt \\ &= \frac{2}{2}\left[\int_0^1 t\cos n\pi t\,dt + \int_1^2 0\cos n\pi t\,dt\right] \\ &= \left[\frac{1}{n^2\pi^2}\cos n\pi t + \frac{t}{n\pi}\sin n\pi t\right]\Bigg|_0^1 \\ &= \frac{1}{n^2\pi^2}(\cos n\pi - 1) + 0 = \frac{(-1)^n - 1}{n^2\pi^2} \end{aligned} \quad \textbf{(17.2.2)}$$

since $\cos n\pi = (-1)^n$; and

$$\begin{aligned} b_n &= \frac{2}{T}\int_0^T f(t)\sin n\omega_0 t\,dt \\ &= \frac{2}{2}\left[\int_0^1 t\sin n\pi t\,dt + \int_1^2 0\sin n\pi t\,dt\right] \\ &= \left[\frac{1}{n^2\pi^2}\sin n\pi t - \frac{t}{n\pi}\cos n\pi t\right]\Bigg|_0^1 \\ &= 0 - \frac{\cos n\pi}{n\pi} = \frac{(-1)^{n+1}}{n\pi} \end{aligned} \quad \textbf{(17.2.3)}$$

Substituting the Fourier coefficients just found into Eq. (17.3) yields

$$f(t) = \frac{1}{4} + \sum_{n=1}^{\infty}\left[\frac{[(-1)^n - 1]}{(n\pi)^2}\cos n\pi t + \frac{(-1)^{n+1}}{n\pi}\sin n\pi t\right]$$

To obtain the amplitude and phase spectra, we notice that, for even harmonics, $a_n = 0$, $b_n = -1/n\pi$, so that

$$A_n\underline{/\phi_n} = a_n - jb_n = 0 + j\frac{1}{n\pi} \quad \textbf{(17.2.4)}$$

Hence,

$$\begin{aligned} A_n &= |b_n| = \frac{1}{n\pi}, \qquad n = 2, 4, \ldots \\ \phi_n &= 90^\circ, \qquad n = 2, 4, \ldots \end{aligned} \quad \textbf{(17.2.5)}$$

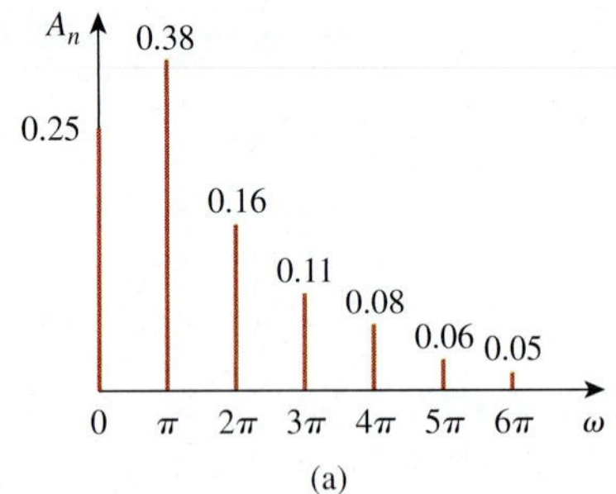

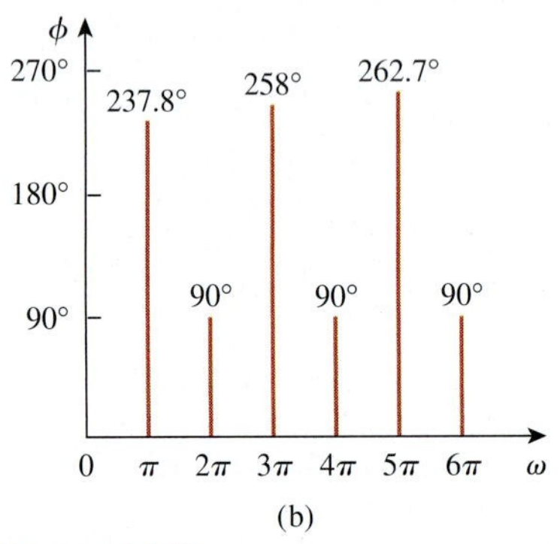

Figure 17.8
For Example 17.2: (a) amplitude spectrum, (b) phase spectrum.

For odd harmonics, $a_n = -2/(n^2\pi^2)$, $b_n = 1/(n\pi)$ so that

$$A_n\underline{/\phi_n} = a_n - jb_n = -\frac{2}{n^2\pi^2} - j\frac{1}{n\pi} \tag{17.2.6}$$

That is,

$$\begin{aligned} A_n &= \sqrt{a_n^2 + b_n^2} = \sqrt{\frac{4}{n^4\pi^4} + \frac{1}{n^2\pi^2}} \\ &= \frac{1}{n^2\pi^2}\sqrt{4 + n^2\pi^2}, \qquad n = 1, 3, \ldots \end{aligned} \tag{17.2.7}$$

From Eq. (17.2.6), we observe that ϕ lies in the third quadrant, so that

$$\phi_n = 180° + \tan^{-1}\frac{n\pi}{2}, \qquad n = 1, 3, \ldots \tag{17.2.8}$$

From Eqs. (17.2.5), (17.2.7), and (17.2.8), we plot A_n and ϕ_n for different values of $n\omega_0 = n\pi$ to obtain the amplitude spectrum and phase spectrum as shown in Fig. 17.8.

Practice Problem 17.2

Determine the Fourier series of the sawtooth waveform in Fig. 17.9.

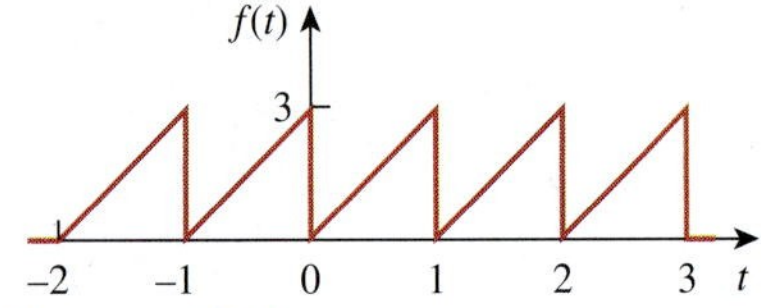

Figure 17.9
For Practice Prob. 17.2.

Answer: $f(t) = \dfrac{3}{2} - \dfrac{3}{\pi}\displaystyle\sum_{n=1}^{\infty}\frac{1}{n}\sin 2\pi nt.$

17.3 Symmetry Considerations

We noticed that the Fourier series of Example 17.1 consisted only of the sine terms. One may wonder if a method exists whereby one can know in advance that some Fourier coefficients would be zero and avoid the unnecessary work involved in the tedious process of calculating them. Such a method does exist; it is based on recognizing the existence of symmetry. Here we discuss three types of symmetry: (1) even symmetry, (2) odd symmetry, (3) half-wave symmetry.

17.3.1 Even Symmetry

A function $f(t)$ is *even* if its plot is symmetrical about the vertical axis; that is,

$$\boxed{f(t) = f(-t)} \tag{17.16}$$

Examples of even functions are t^2, t^4, and $\cos t$. Figure 17.10 shows more examples of periodic even functions. Note that each of these examples satisfies Eq. (17.16). A main property of an even function $f_e(t)$ is that:

$$\int_{-T/2}^{T/2} f_e(t)\,dt = 2\int_0^{T/2} f_e(t)\,dt \tag{17.17}$$

because integrating from $-T/2$ to 0 is the same as integrating from 0 to $T/2$. Utilizing this property, the Fourier coefficients for an even function become

$$\begin{aligned} a_0 &= \frac{2}{T}\int_0^{T/2} f(t)\,dt \\ a_n &= \frac{4}{T}\int_0^{T/2} f(t)\cos n\omega_0 t\,dt \\ b_n &= 0 \end{aligned} \tag{17.18}$$

Since $b_n = 0$, Eq. (17.3) becomes a *Fourier cosine series*. This makes sense because the cosine function is itself even. It also makes intuitive sense that an even function contains no sine terms since the sine function is odd.

To confirm Eq. (17.18) quantitatively, we apply the property of an even function in Eq. (17.17) in evaluating the Fourier coefficients in Eqs. (17.6), (17.8), and (17.9). It is convenient in each case to integrate over the interval $-T/2 < t < T/2$, which is symmetrical about the origin. Thus,

$$a_0 = \frac{1}{T}\int_{-T/2}^{T/2} f(t)\,dt = \frac{1}{T}\left[\int_{-T/2}^{0} f(t)\,dt + \int_0^{T/2} f(t)\,dt\right] \tag{17.19}$$

We change variables for the integral over the interval $-T/2 < t < 0$ by letting $t = -x$, so that $dt = -dx, f(t) = f(-t) = f(x)$, since $f(t)$ is an even function, and when $t = -T/2, x = T/2$. Then,

$$\begin{aligned} a_0 &= \frac{1}{T}\left[\int_{T/2}^{0} f(x)(-dx) + \int_0^{T/2} f(t)\,dt\right] \\ &= \frac{1}{T}\left[\int_0^{T/2} f(x)\,dx + \int_0^{T/2} f(t)\,dt\right] \end{aligned} \tag{17.20}$$

showing that the two integrals are identical. Hence,

$$a_0 = \frac{2}{T}\int_0^{T/2} f(t)\,dt \tag{17.21}$$

as expected. Similarly, from Eq. (17.8),

$$a_n = \frac{2}{T}\left[\int_{-T/2}^{0} f(t)\cos n\omega_0 t\,dt + \int_0^{T/2} f(t)\cos n\omega_0 t\,dt\right] \tag{17.22}$$

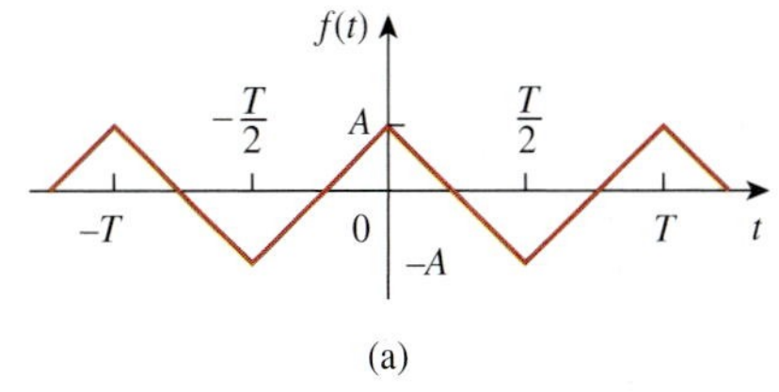

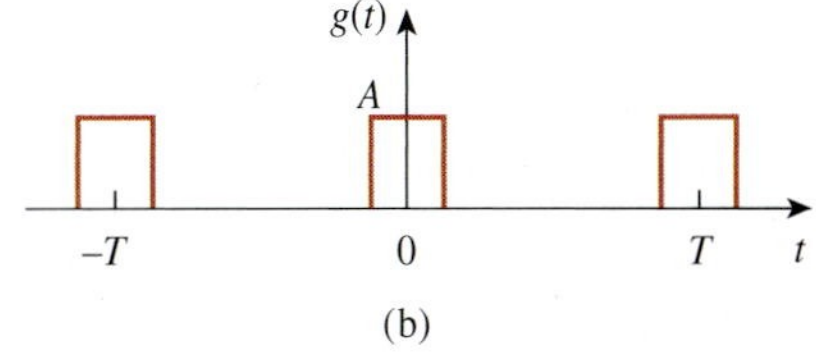

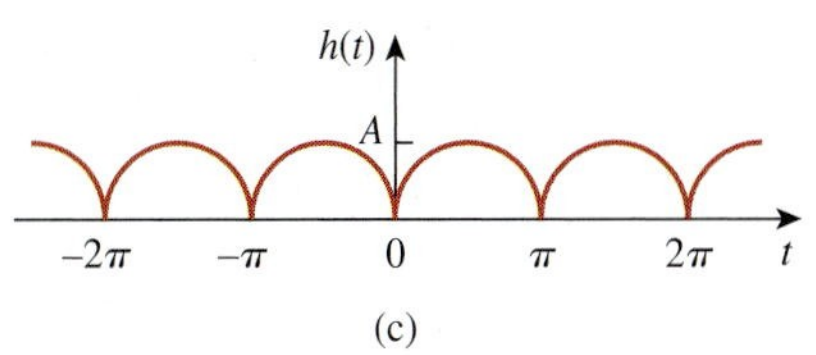

Figure 17.10
Typical examples of even periodic functions.

We make the same change of variables that led to Eq. (17.20) and note that both $f(t)$ and $\cos n\omega_0 t$ are even functions, implying that $f(-t) = f(t)$ and $\cos(-n\omega_0 t) = \cos n\omega_0 t$. Equation (17.22) becomes

$$\begin{aligned} a_n &= \frac{2}{T}\left[\int_{T/2}^{0} f(-x)\cos(-n\omega_0 x)(-dx) + \int_0^{T/2} f(t)\cos n\omega_0 t\, dt\right] \\ &= \frac{2}{T}\left[\int_{T/2}^{0} f(x)\cos(n\omega_0 x)(-dx) + \int_0^{T/2} f(t)\cos n\omega_0 t\, dt\right] \\ &= \frac{2}{T}\left[\int_0^{T/2} f(x)\cos(n\omega_0 x)\, dx + \int_0^{T/2} f(t)\cos n\omega_0 t\, dt\right] \end{aligned} \tag{17.23a}$$

or

$$a_n = \frac{4}{T}\int_0^{T/2} f(t)\cos n\omega_0 t\, dt \tag{17.23b}$$

as expected. For b_n, we apply Eq. (17.9),

$$b_n = \frac{2}{T}\left[\int_{-T/2}^{0} f(t)\sin n\omega_0 t\, dt + \int_0^{T/2} f(t)\sin n\omega_0 t\, dt\right] \tag{17.24}$$

We make the same change of variables but keep in mind that $f(-t) = f(t)$ but $\sin(-n\omega_0 t) = -\sin n\omega_0 t$. Equation (17.24) yields

$$\begin{aligned} b_n &= \frac{2}{T}\left[\int_{T/2}^{0} f(-x)\sin(-n\omega_0 x)(-dx) + \int_0^{T/2} f(t)\sin n\omega_0 t\, dt\right] \\ &= \frac{2}{T}\left[\int_{T/2}^{0} f(x)\sin n\omega_0 x dx + \int_0^{T/2} f(t)\sin n\omega_0 t dt\right] \\ &= \frac{2}{T}\left[-\int_0^{T/2} f(x)\sin(n\omega_0 x)\, dx + \int_0^{T/2} f(t)\sin n\omega_0 t\, dt\right] \\ &= 0 \end{aligned} \tag{17.25}$$

confirming Eq. (17.18).

17.3.2 Odd Symmetry

A function $f(t)$ is said to be *odd* if its plot is antisymmetrical about the vertical axis:

$$f(-t) = -f(t) \tag{17.26}$$

Examples of odd functions are t, t^3, and $\sin t$. Figure 17.11 shows more examples of periodic odd functions. All these examples satisfy Eq. (17.26). An odd function $f_o(t)$ has this major characteristic:

$$\int_{-T/2}^{T/2} f_o(t)\, dt = 0 \tag{17.27}$$

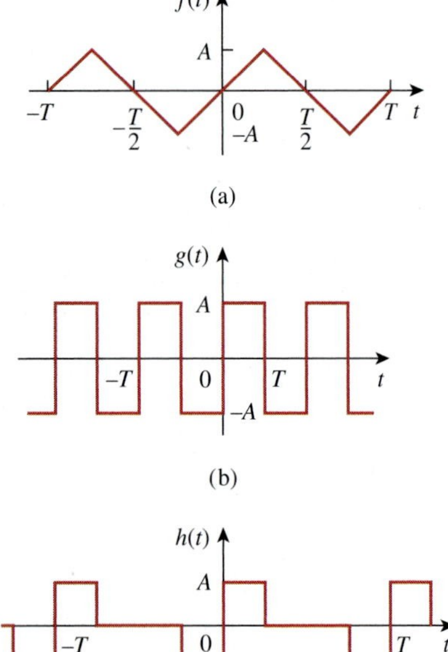

Figure 17.11
Typical examples of odd periodic functions.

because integration from $-T/2$ to 0 is the negative of that from 0 to $T/2$. With this property, the Fourier coefficients for an odd function become

$$a_0 = 0, \qquad a_n = 0$$
$$b_n = \frac{4}{T}\int_0^{T/2} f(t)\sin n\omega_0 t\, dt \tag{17.28}$$

which give us a *Fourier sine series*. Again, this makes sense because the sine function is itself an odd function. Also, note that there is no dc term for the Fourier series expansion of an odd function.

The quantitative proof of Eq. (17.28) follows the same procedure taken to prove Eq. (17.18) except that $f(t)$ is now odd, so that $f(t) = -f(t)$. With this fundamental but simple difference, it is easy to see that $a_0 = 0$ in Eq. (17.20), $a_n = 0$ in Eq. (17.23a), and b_n in Eq. (17.24) becomes

$$\begin{aligned} b_n &= \frac{2}{T}\left[\int_{T/2}^{0} f(-x)\sin(-n\omega_0 x)(-dx) + \int_0^{T/2} f(t)\sin n\omega_0 t\, dt\right] \\ &= \frac{2}{T}\left[-\int_{T/2}^{0} f(x)\sin n\omega_0 x\, dx + \int_0^{T/2} f(t)\sin n\omega_0 t\, dt\right] \\ &= \frac{2}{T}\left[\int_0^{T/2} f(x)\sin(n\omega_0 x)\, dx + \int_0^{T/2} f(t)\sin n\omega_0 t\, dt\right] \end{aligned}$$

$$b_n = \frac{4}{T}\int_0^{T/2} f(t)\sin n\omega_0 t\, dt \tag{17.29}$$

as expected.

It is interesting to note that any periodic function $f(t)$ with neither even nor odd symmetry may be decomposed into even and odd parts. Using the properties of even and odd functions from Eqs. (17.16) and (17.26), we can write

$$f(t) = \underbrace{\frac{1}{2}[f(t) + f(-t)]}_{\text{even}} + \underbrace{\frac{1}{2}[f(t) - f(-t)]}_{\text{odd}} = f_e(t) + f_o(t) \tag{17.30}$$

Notice that $f_e(t) = \frac{1}{2}[f(t) + f(-t)]$ satisfies the property of an even function in Eq. (17.16), while $f_o(t) = \frac{1}{2}[f(t) - f(-t)]$ satisfies the property of an odd function in Eq. (17.26). The fact that $f_e(t)$ contains only the dc term and the cosine terms, while $f_o(t)$ has only the sine terms, can be exploited in grouping the Fourier series expansion of $f(t)$ as

$$f(t) = \underbrace{a_0 + \sum_{n=1}^{\infty} a_n \cos n\omega_0 t}_{\text{even}} + \underbrace{\sum_{n=1}^{\infty} b_n \sin n\omega_0 t}_{\text{odd}} = f_e(t) + f_o(t) \tag{17.31}$$

It follows readily from Eq. (17.31) that when $f(t)$ is even, $b_n = 0$, and when $f(t)$ is odd, $a_0 = 0 = a_n$.

Also, note the following properties of odd and even functions:

1. The product of two even functions is also an even function.
2. The product of two odd functions is an even function.
3. The product of an even function and an odd function is an odd function.
4. The sum (or difference) of two even functions is also an even function.
5. The sum (or difference) of two odd functions is an odd function.
6. The sum (or difference) of an even function and an odd function is neither even nor odd.

Each of these properties can be proved using Eqs. (17.16) and (17.26).

17.3.3 Half-Wave Symmetry

A function is half-wave (odd) symmetric if

$$f\left(t - \frac{T}{2}\right) = -f(t) \tag{17.32}$$

which means that each half-cycle is the mirror image of the next half-cycle. Notice that functions $\cos n\omega_0 t$ and $\sin n\omega_0 t$ satisfy Eq. (17.32) for odd values of n and therefore possess half-wave symmetry when n is odd. Figure 17.12 shows other examples of half-wave symmetric functions. The functions in Figs. 17.11(a) and 17.11(b) are also half-wave symmetric. Notice that for each function, one half-cycle is the inverted version of the adjacent half-cycle. The Fourier coefficients become

$$\begin{aligned}
a_0 &= 0 \\
a_n &= \begin{cases} \dfrac{4}{T}\displaystyle\int_0^{T/2} f(t)\cos n\omega_0 t\, dt, & \text{for } n \text{ odd} \\ 0, & \text{for } n \text{ even} \end{cases} \\
b_n &= \begin{cases} \dfrac{4}{T}\displaystyle\int_0^{T/2} f(t)\sin n\omega_0 t\, dt, & \text{for } n \text{ odd} \\ 0, & \text{for } n \text{ even} \end{cases}
\end{aligned} \tag{17.33}$$

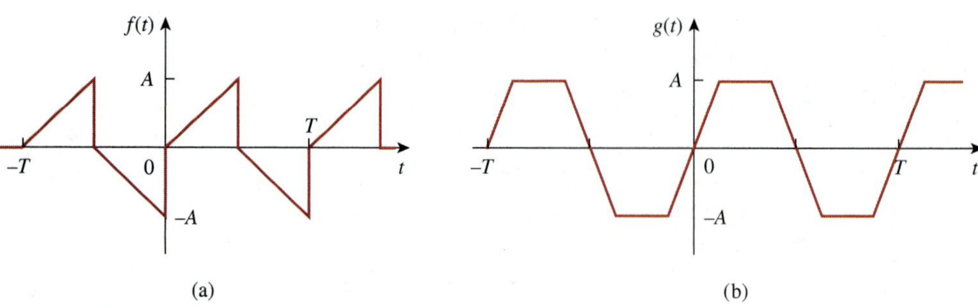

Figure 17.12
Typical examples of half-wave odd symmetric functions.

showing that the Fourier series of a half-wave symmetric function contains only odd harmonics.

To derive Eq. (17.33), we apply the property of half-wave symmetric functions in Eq. (17.32) in evaluating the Fourier coefficients in Eqs. (17.6), (17.8), and (17.9). Thus,

$$a_0 = \frac{1}{T}\int_{-T/2}^{T/2} f(t)\,dt = \frac{1}{T}\left[\int_{-T/2}^{0} f(t)\,dt + \int_{0}^{T/2} f(t)\,dt\right] \quad \textbf{(17.34)}$$

We change variables for the integral over the interval $-T/2 < t < 0$ by letting $x = t + T/2$, so that $dx = dt$; when $t = -T/2, x = 0$; and when $t = 0, x = T/2$. Also, we keep Eq. (17.32) in mind; that is, $f(x - T/2) = -f(x)$. Then,

$$\begin{aligned} a_0 &= \frac{1}{T}\left[\int_{0}^{T/2} f\left(x - \frac{T}{2}\right)dx + \int_{0}^{T/2} f(t)\,dt\right] \\ &= \frac{1}{T}\left[-\int_{0}^{T/2} f(x)\,dx + \int_{0}^{T/2} f(t)\,dt\right] = 0 \end{aligned} \quad \textbf{(17.35)}$$

confirming the expression for a_0 in Eq. (17.33). Similarly,

$$a_n = \frac{2}{T}\left[\int_{-T/2}^{0} f(t)\cos n\omega_0 t\,dt + \int_{0}^{T/2} f(t)\cos n\omega_0 t\,dt\right] \quad \textbf{(17.36)}$$

We make the same change of variables that led to Eq. (17.35) so that Eq. (17.36) becomes

$$\begin{aligned} a_n = \frac{2}{T}\bigg[&\int_{0}^{T/2} f\left(x - \frac{T}{2}\right)\cos n\omega_0\left(x - \frac{T}{2}\right)dx \\ &+ \int_{0}^{T/2} f(t)\cos n\omega_0 t\,dt\bigg] \end{aligned} \quad \textbf{(17.37)}$$

Since $f(x - T/2) = -f(x)$ and

$$\begin{aligned} \cos n\omega_0\left(x - \frac{T}{2}\right) &= \cos(n\omega_0 t - n\pi) \\ &= \cos n\omega_0 t \cos n\pi + \sin n\omega_0 t \sin n\pi \\ &= (-1)^n \cos n\omega_0 t \end{aligned} \quad \textbf{(17.38)}$$

substituting these in Eq. (17.37) leads to

$$\begin{aligned} a_n &= \frac{2}{T}[1 - (-1)^n]\int_{0}^{T/2} f(t)\cos n\omega_0 t\,dt \\ &= \begin{cases} \dfrac{4}{T}\displaystyle\int_{0}^{T/2} f(t)\cos n\omega_0 t\,dt, & \text{for } n \text{ odd} \\ 0, & \text{for } n \text{ even} \end{cases} \end{aligned} \quad \textbf{(17.39)}$$

confirming Eq. (17.33). By following a similar procedure, we can derive b_n as in Eq. (17.33).

Table 17.2 summarizes the effects of these symmetries on the Fourier coefficients. Table 17.3 provides the Fourier series of some common periodic functions.

TABLE 17.2

Effects of symmetry on Fourier coefficients.

Symmetry	a_0	a_n	b_n	Remarks
Even	$a_0 \neq 0$	$a_n \neq 0$	$b_n = 0$	Integrate over $T/2$ and multiply by 2 to get the coefficients.
Odd	$a_0 = 0$	$a_n = 0$	$b_n \neq 0$	Integrate over $T/2$ and multiply by 2 to get the coefficients.
Half-wave	$a_0 = 0$	$a_{2n} = 0$ $a_{2n+1} \neq 0$	$b_{2n} = 0$ $b_{2n+1} \neq 0$	Integrate over $T/2$ and multiply by 2 to get the coefficients.

TABLE 17.3

The Fourier series of common functions.

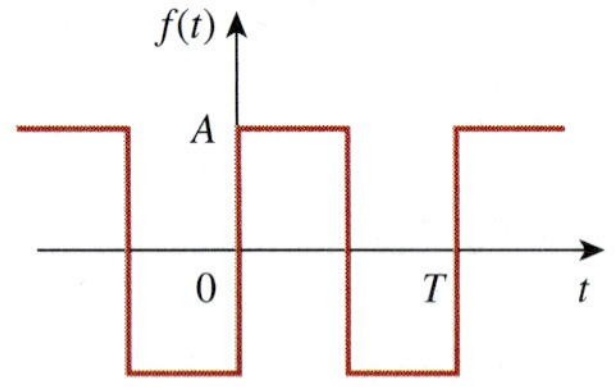

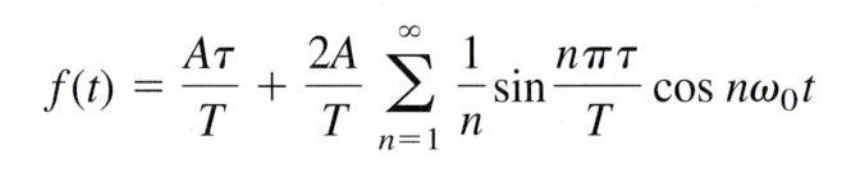

Function	Fourier series
1. Square wave 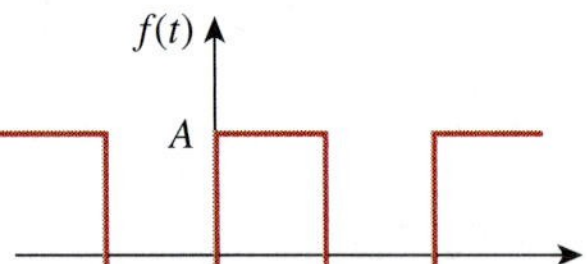 	$f(t) = \frac{4A}{\pi} \sum_{n=1}^{\infty} \frac{1}{2n-1} \sin(2n-1)\omega_0 t$
2. Rectangular pulse train 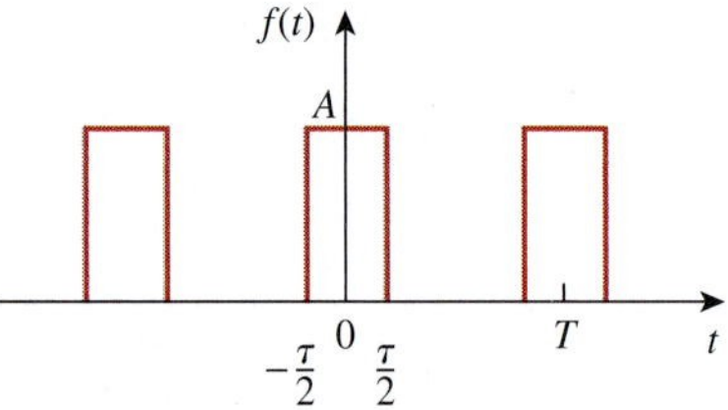	$f(t) = \frac{A\tau}{T} + \frac{2A}{T} \sum_{n=1}^{\infty} \frac{1}{n} \sin \frac{n\pi\tau}{T} \cos n\omega_0 t$
3. Sawtooth wave 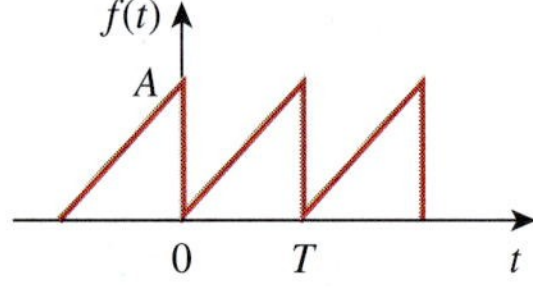	$f(t) = \frac{A}{2} - \frac{A}{\pi} \sum_{n=1}^{\infty} \frac{\sin n\omega_0 t}{n}$
4. Triangular wave 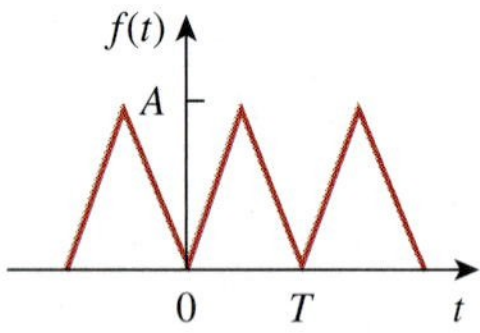	$f(t) = \frac{A}{2} - \frac{4A}{\pi^2} \sum_{n=1}^{\infty} \frac{1}{(2n-1)^2} \cos(2n-1)\omega_0 t$
5. Half-wave rectified sine 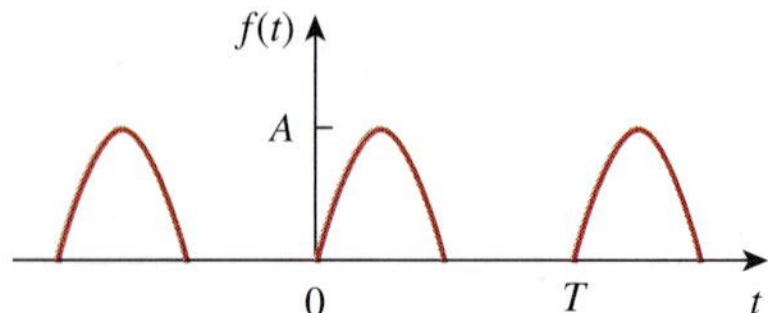	$f(t) = \frac{A}{\pi} + \frac{A}{2} \sin \omega_0 t - \frac{2A}{\pi} \sum_{n=1}^{\infty} \frac{1}{4n^2 - 1} \cos 2n\omega_0 t$
6. Full-wave rectified sine 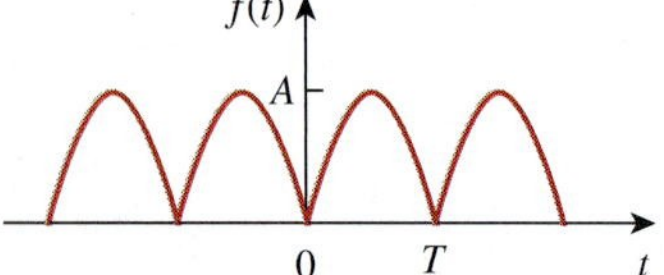	$f(t) = \frac{2A}{\pi} - \frac{4A}{\pi} \sum_{n=1}^{\infty} \frac{1}{4n^2 - 1} \cos n\omega_0 t$

Example 17.3

Find the Fourier series expansion of $f(t)$ given in Fig. 17.13.

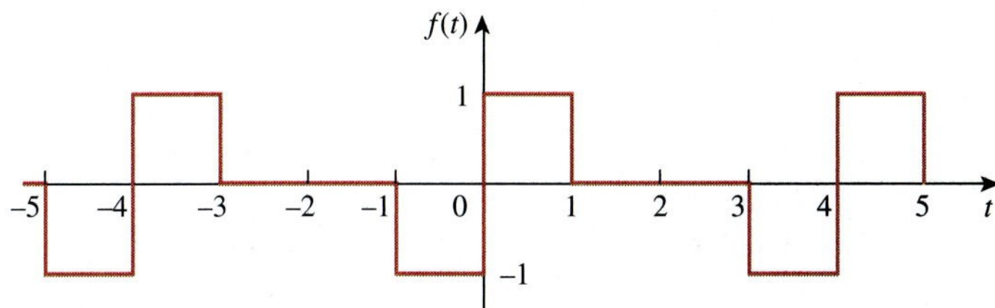

Figure 17.13
For Example 17.3.

Solution:

The function $f(t)$ is an odd function. Hence $a_0 = 0 = a_n$. The period is $T = 4$, and $\omega_0 = 2\pi/T = \pi/2$, so that

$$\begin{aligned} b_n &= \frac{4}{T}\int_0^{T/2} f(t)\sin n\omega_0 t\,dt \\ &= \frac{4}{4}\left[\int_0^1 1\sin\frac{n\pi}{2}t\,dt + \int_1^2 0\sin\frac{n\pi}{2}t\,dt\right] \\ &= -\frac{2}{n\pi}\cos\frac{n\pi t}{2}\bigg|_0^1 = \frac{2}{n\pi}\left(1-\cos\frac{n\pi}{2}\right) \end{aligned}$$

Hence,

$$f(t) = \frac{2}{\pi}\sum_{n=1}^{\infty}\frac{1}{n}\left(1-\cos\frac{n\pi}{2}\right)\sin\frac{n\pi}{2}t$$

which is a Fourier sine series.

Practice Problem 17.3

Find the Fourier series of the function $f(t)$ in Fig. 17.14.

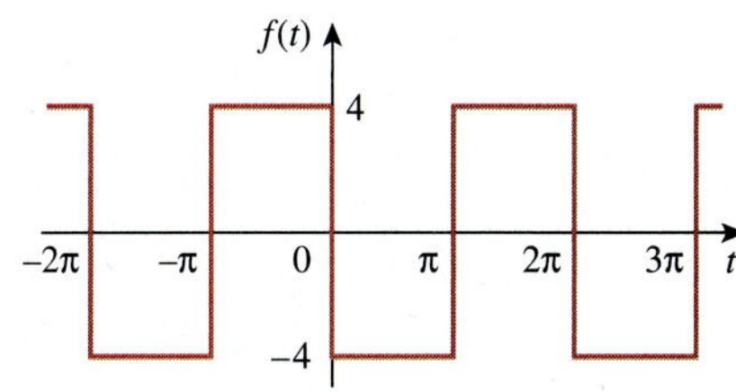

Figure 17.14
For Practice Prob. 17.3.

Answer: $f(t) = -\dfrac{16}{\pi}\displaystyle\sum_{k=1}^{\infty}\frac{1}{n}\sin nt,\ n = 2k-1.$

Example 17.4

Determine the Fourier series for the half-wave rectified cosine function shown in Fig. 17.15.

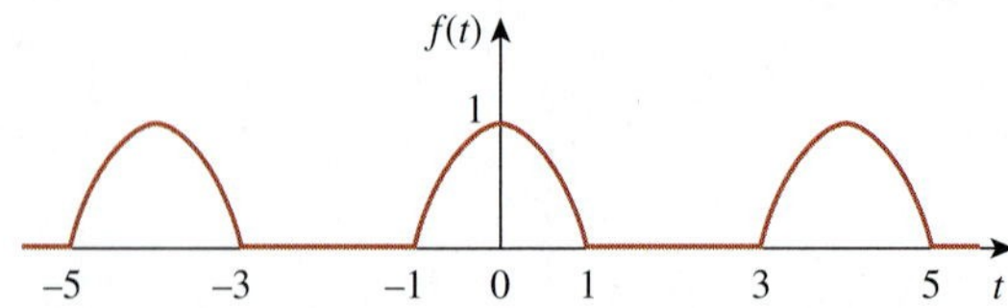

Figure 17.15
A half-wave rectified cosine function; for Example 17.4.

Solution:

This is an even function so that $b_n = 0$. Also, $T = 4$, $\omega_0 = 2\pi/T = \pi/2$. Over a period,

$$f(t) = \begin{cases} 0, & -2 < t < -1 \\ \cos\frac{\pi}{2}t, & -1 < t < 1 \\ 0, & 1 < t < 2 \end{cases}$$

$$a_0 = \frac{2}{T}\int_0^{T/2} f(t)\,dt = \frac{2}{4}\left[\int_0^1 \cos\frac{\pi}{2}t\,dt + \int_1^2 0\,dt\right]$$

$$= \frac{1}{2}\frac{2}{\pi}\sin\frac{\pi}{2}t\,\bigg|_0^1 = \frac{1}{\pi}$$

$$a_n = \frac{4}{T}\int_0^{T/2} f(t)\cos n\omega_0 t\,dt = \frac{4}{4}\left[\int_0^1 \cos\frac{\pi}{2}t\cos\frac{n\pi t}{2}\,dt + 0\right]$$

But $\cos A \cos B = \frac{1}{2}[\cos(A + B) + \cos(A - B)]$. Then

$$a_n = \frac{1}{2}\int_0^1 \left[\cos\frac{\pi}{2}(n + 1)t + \cos\frac{\pi}{2}(n - 1)t\right] dt$$

For $n = 1$,

$$a_1 = \frac{1}{2}\int_0^1 [\cos\pi t + 1]\,dt = \frac{1}{2}\left[\frac{\sin\pi t}{\pi} + t\right]\bigg|_0^1 = \frac{1}{2}$$

For $n > 1$,

$$a_n = \frac{1}{\pi(n + 1)}\sin\frac{\pi}{2}(n + 1) + \frac{1}{\pi(n - 1)}\sin\frac{\pi}{2}(n - 1)$$

For n = odd ($n = 1, 3, 5, \ldots$), $(n + 1)$ and $(n - 1)$ are both even, so

$$\sin\frac{\pi}{2}(n + 1) = 0 = \sin\frac{\pi}{2}(n - 1), \qquad n = \text{odd}$$

For n = even ($n = 2, 4, 6, \ldots$), $(n + 1)$ and $(n - 1)$ are both odd. Also,

$$\sin\frac{\pi}{2}(n + 1) = -\sin\frac{\pi}{2}(n - 1) = \cos\frac{n\pi}{2} = (-1)^{n/2}, \qquad n = \text{even}$$

Hence,

$$a_n = \frac{(-1)^{n/2}}{\pi(n + 1)} + \frac{-(-1)^{n/2}}{\pi(n - 1)} = \frac{-2(-1)^{n/2}}{\pi(n^2 - 1)}, \qquad n = \text{even}$$

Thus,

$$f(t) = \frac{1}{\pi} + \frac{1}{2}\cos\frac{\pi}{2}t - \frac{2}{\pi}\sum_{n=\text{even}}^{\infty}\frac{(-1)^{n/2}}{(n^2-1)}\cos\frac{n\pi}{2}t$$

To avoid using $n = 2, 4, 6, \ldots$ and also to ease computation, we can replace n by $2k$, where $k = 1, 2, 3, \ldots$ and obtain

$$f(t) = \frac{1}{\pi} + \frac{1}{2}\cos\frac{\pi}{2}t - \frac{2}{\pi}\sum_{k=1}^{\infty}\frac{(-1)^{k}}{(4k^2-1)}\cos k\pi t$$

which is a Fourier cosine series.

Practice Problem 17.4

Find the Fourier series expansion of the function in Fig. 17.16.

Answer: $f(t) = 2 - \frac{16}{\pi^2}\sum_{k=1}^{\infty}\frac{1}{n^2}\cos nt,\ n = 2k - 1.$

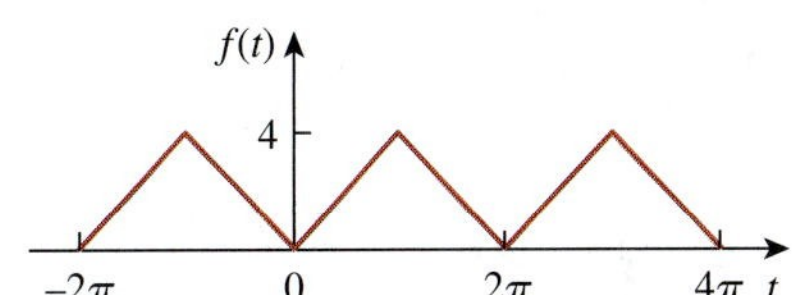

Figure 17.16
For Practice Prob. 17.4.

Example 17.5

Calculate the Fourier series for the function in Fig. 17.17.

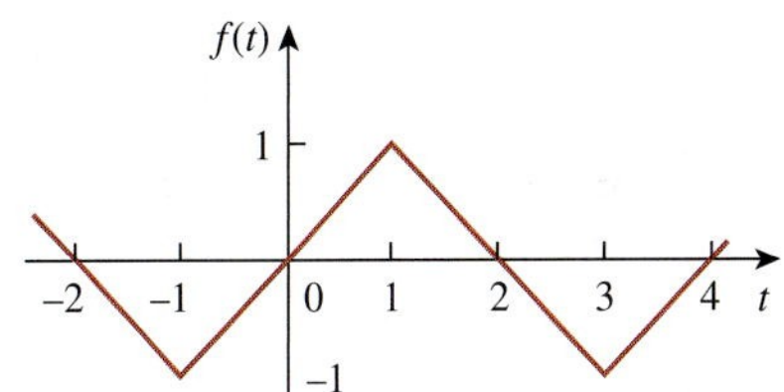

Figure 17.17
For Example 17.5.

Solution:

The function in Fig. 17.17 is half-wave odd symmetric, so that $a_0 = 0 = a_n$. It is described over half the period as

$$f(t) = t, \qquad -1 < t < 1$$

$T = 4$, $\omega_0 = 2\pi/T = \pi/2$. Hence,

$$b_n = \frac{4}{T}\int_0^{T/2} f(t)\sin n\omega_0 t\,dt$$

Instead of integrating $f(t)$ from 0 to 2, it is more convenient to integrate from -1 to 1. Applying Eq. (17.15d),

$$\begin{aligned} b_n &= \frac{4}{4}\int_{-1}^{1} t\sin\frac{n\pi t}{2}dt = \left[\frac{\sin n\pi t/2}{n^2\pi^2/4} - \frac{t\cos n\pi t/2}{n\pi/2}\right]\Bigg|_{-1}^{1} \\ &= \frac{4}{n^2\pi^2}\left[\sin\frac{n\pi}{2} - \sin\left(-\frac{n\pi}{2}\right)\right] - \frac{2}{n\pi}\left[\cos\frac{n\pi}{2} - \cos\left(-\frac{n\pi}{2}\right)\right] \\ &= \frac{8}{n^2\pi^2}\sin\frac{n\pi}{2} \end{aligned}$$

since $\sin(-x) = -\sin x$ is an odd function, while $\cos(-x) = \cos x$ is an even function. Using the identities for $\sin n\pi/2$ in Table 17.1,

$$b_n = \frac{8}{n^2\pi^2}(-1)^{(n-1)/2}, \quad n = \text{odd} = 1, 3, 5, \ldots$$

Thus,

$$f(t) = \sum_{n=1,3,5}^{\infty} b_n \sin\frac{n\pi}{2}t$$

where b_n is given above.

Practice Problem 17.5

Determine the Fourier series of the function in Fig. 17.12(a). Take $A = 2$ and $T = 2\pi$.

Answer: $f(t) = \frac{4}{\pi}\sum_{k=1}^{\infty}\left(\frac{-2}{n^2\pi}\cos nt + \frac{1}{n}\sin nt\right), n = 2k - 1.$

17.4 Circuit Applications

We find that in practice, many circuits are driven by nonsinusoidal periodic functions. To find the steady-state response of a circuit to a nonsinusoidal periodic excitation requires the application of a Fourier series, ac phasor analysis, and the superposition principle. The procedure usually involves four steps.

Steps for Applying Fourier Series:

1. Express the excitation as a Fourier series.
2. Transform the circuit from the time domain to the frequency domain.
3. Find the response of the dc and ac components in the Fourier series.
4. Add the individual dc and ac responses using the superposition principle.

The first step is to determine the Fourier series expansion of the excitation. For the periodic voltage source shown in Fig. 17.18(a), for example, the Fourier series is expressed as

$$v(t) = V_0 + \sum_{n=1}^{\infty} V_n \cos(n\omega_0 t + \theta_n) \quad \textbf{(17.40)}$$

(The same could be done for a periodic current source.) Equation (17.40) shows that $v(t)$ consists of two parts: the dc component V_0 and the ac component $\mathbf{V}_n = V_n\angle\theta_n$ with several harmonics. This Fourier series representation may be regarded as a set of series-connected sinusoidal sources, with each source having its own amplitude and frequency, as shown in Fig. 17.18(b).

The third step is finding the response to each term in the Fourier series. The response to the dc component can be determined in the

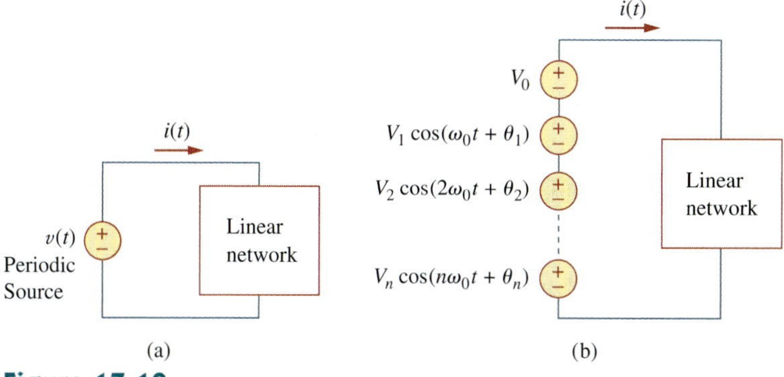

Figure 17.18
(a) Linear network excited by a periodic voltage source, (b) Fourier series representation (time-domain).

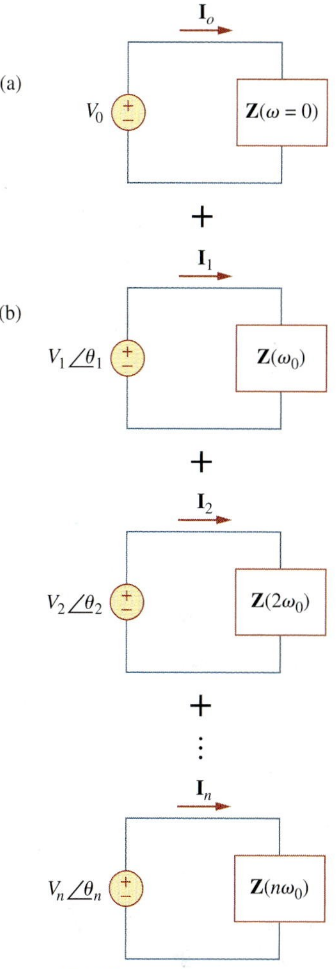

Figure 17.19
Steady-state responses: (a) dc component, (b) ac component (frequency domain).

frequency domain by setting $n = 0$ or $\omega = 0$ as in Fig. 17.19(a), or in the time domain by replacing all inductors with short circuits and all capacitors with open circuits. The response to the ac component is obtained by applying the phasor techniques covered in Chapter 9, as shown in Fig. 17.19(b). The network is represented by its impedance $\mathbf{Z}(n\omega_0)$ or admittance $\mathbf{Y}(n\omega_0)$. $\mathbf{Z}(n\omega_0)$ is the input impedance at the source when ω is everywhere replaced by $n\omega_0$, and $\mathbf{Y}(n\omega_0)$ is the reciprocal of $\mathbf{Z}(n\omega_0)$.

Finally, following the principle of superposition, we add all the individual responses. For the case shown in Fig. 17.19,

$$\begin{aligned} i(t) &= i_0(t) + i_1(t) + i_2(t) + \cdots \\ &= \mathbf{I}_0 + \sum_{n=1}^{\infty} |\mathbf{I}_n| \cos(n\omega_0 t + \psi_n) \end{aligned} \tag{17.41}$$

where each component $\mathbf{I}_n$ with frequency $n\omega_0$ has been transformed to the time domain to get $i_n(t)$, and ψ_n is the argument of $\mathbf{I}_n$.

Example 17.6

Let the function $f(t)$ in Example 17.1 be the voltage source $v_s(t)$ in the circuit of Fig. 17.20. Find the response $v_o(t)$ of the circuit.

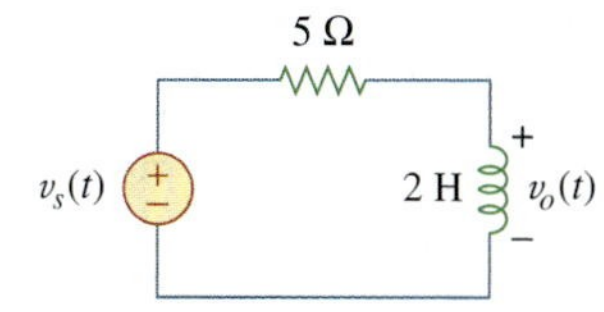

Figure 17.20
For Example 17.6.

Solution:
From Example 17.1,

$$v_s(t) = \frac{1}{2} + \frac{2}{\pi} \sum_{k=1}^{\infty} \frac{1}{n} \sin n\pi t, \qquad n = 2k - 1$$

where $\omega_n = n\omega_0 = n\pi$ rad/s. Using phasors, we obtain the response $\mathbf{V}_o$ in the circuit of Fig. 17.20 by voltage division:

$$\mathbf{V}_o = \frac{j\omega_n L}{R + j\omega_n L}\mathbf{V}_s = \frac{j2n\pi}{5 + j2n\pi}\mathbf{V}_s$$

For the dc component ($\omega_n = 0$ or $n = 0$)

$$\mathbf{V}_s = \frac{1}{2} \qquad \Rightarrow \qquad \mathbf{V}_o = 0$$

This is expected, since the inductor is a short circuit to dc. For the *n*th harmonic,

$$\mathbf{V}_s = \frac{2}{n\pi}\underline{/-90^\circ} \tag{17.6.1}$$

and the corresponding response is

$$\begin{aligned}\mathbf{V}_o &= \frac{2n\pi\underline{/90^\circ}}{\sqrt{25 + 4n^2\pi^2}\underline{/\tan^{-1} 2n\pi/5}}\left(\frac{2}{n\pi}\underline{/-90^\circ}\right) \\ &= \frac{4\underline{/-\tan^{-1} 2n\pi/5}}{\sqrt{25 + 4n^2\pi^2}}\end{aligned} \tag{17.6.2}$$

In the time domain,

$$v_o(t) = \sum_{k=1}^{\infty} \frac{4}{\sqrt{25 + 4n^2\pi^2}} \cos\left(n\pi t - \tan^{-1}\frac{2n\pi}{5}\right), \qquad n = 2k - 1$$

The first three terms ($k = 1, 2, 3$ or $n = 1, 3, 5$) of the odd harmonics in the summation give us

$$\begin{aligned}v_o(t) &= 0.4981 \cos(\pi t - 51.49^\circ) + 0.2051 \cos(3\pi t - 75.14^\circ) \\ &\quad + 0.1257 \cos(5\pi t - 80.96^\circ) + \cdots \text{ V}\end{aligned}$$

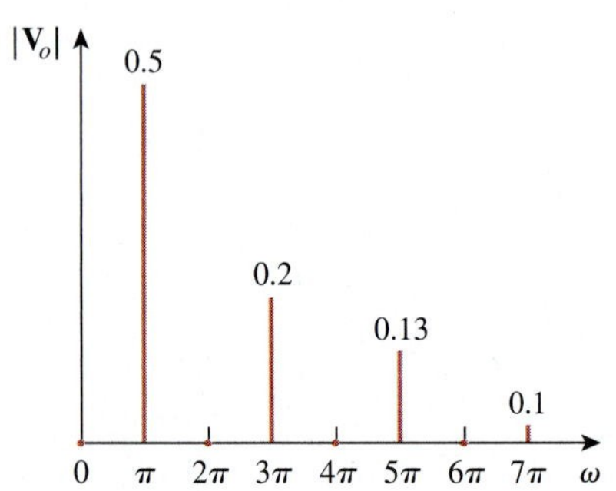

Figure 17.21
For Example 17.6: Amplitude spectrum of the output voltage.

Figure 17.21 shows the amplitude spectrum for output voltage $v_o(t)$, while that of the input voltage $v_s(t)$ is in Fig. 17.4(a). Notice that the two spectra are close. Why? We observe that the circuit in Fig. 17.20 is a highpass filter with the corner frequency $\omega_c = R/L = 2.5$ rad/s, which is less than the fundamental frequency $\omega_0 = \pi$ rad/s. The dc component is not passed and the first harmonic is slightly attenuated, but higher harmonics are passed. In fact, from Eqs. (17.6.1) and (17.6.2), $\mathbf{V}_o$ is identical to $\mathbf{V}_s$ for large *n*, which is characteristic of a highpass filter.

Practice Problem 17.6

If the sawtooth waveform in Fig. 17.9 (see Practice Prob. 17.2) is the voltage source $v_s(t)$ in the circuit of Fig. 17.22, find the response $v_o(t)$.

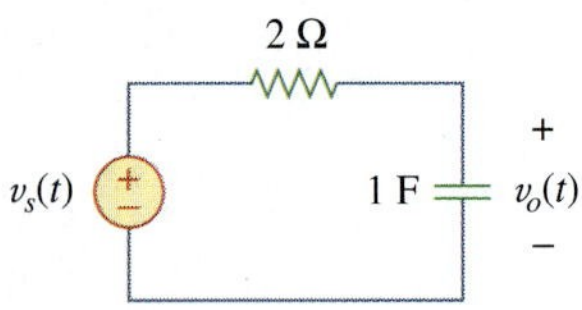

Figure 17.22
For Practice Prob. 17.6.

Answer: $v_o(t) = \dfrac{1}{2} - \dfrac{1}{\pi}\displaystyle\sum_{n=1}^{\infty} \frac{\sin(2\pi nt - \tan^{-1} 4n\pi)}{n\sqrt{1 + 16n^2\pi^2}}$ V.

Example 17.7

Find the response $i_o(t)$ of the circuit of Fig. 17.23 if the input voltage $v(t)$ has the Fourier series expansion

$$v(t) = 1 + \sum_{n=1}^{\infty} \frac{2(-1)^n}{1 + n^2}(\cos nt - n \sin nt)$$

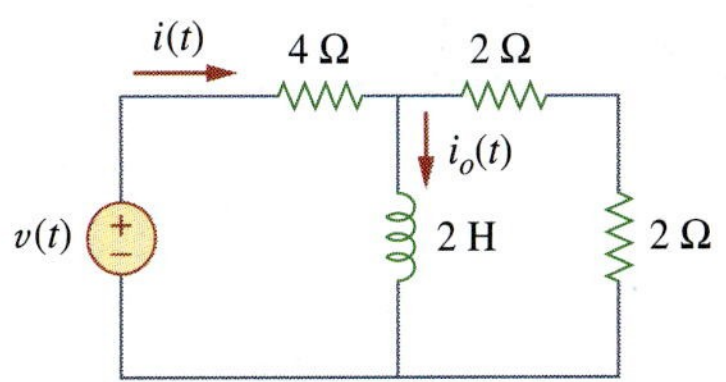

Figure 17.23
For Example 17.7.

Solution:
Using Eq. (17.13), we can express the input voltage as

$$\begin{aligned} v(t) &= 1 + \sum_{n=1}^{\infty} \frac{2(-1)^n}{\sqrt{1+n^2}} \cos(nt + \tan^{-1} n) \\ &= 1 - 1.414\cos(t + 45^\circ) + 0.8944\cos(2t + 63.45^\circ) \\ &\quad - 0.6345\cos(3t + 71.56^\circ) - 0.4851\cos(4t + 78.7^\circ) + \cdots \end{aligned}$$

We notice that $\omega_0 = 1$, $\omega_n = n$ rad/s. The impedance at the source is

$$\mathbf{Z} = 4 + j\omega_n 2 \,\|\, 4 = 4 + \frac{j\omega_n 8}{4 + j\omega_n 2} = \frac{8 + j\omega_n 8}{2 + j\omega_n}$$

The input current is

$$\mathbf{I} = \frac{\mathbf{V}}{\mathbf{Z}} = \frac{2 + j\omega_n}{8 + j\omega_n 8}\mathbf{V}$$

where $\mathbf{V}$ is the phasor form of the source voltage $v(t)$. By current division,

$$\mathbf{I}_o = \frac{4}{4 + j\omega_n 2}\mathbf{I} = \frac{\mathbf{V}}{4 + j\omega_n 4}$$

Since $\omega_n = n$, $\mathbf{I}_o$ can be expressed as

$$\mathbf{I}_o = \frac{\mathbf{V}}{4\sqrt{1+n^2}\,\underline{/\tan^{-1} n}}$$

For the dc component ($\omega_n = 0$ or $n = 0$)

$$\mathbf{V} = 1 \quad \Rightarrow \quad \mathbf{I}_o = \frac{\mathbf{V}}{4} = \frac{1}{4}$$

For the nth harmonic,

$$\mathbf{V} = \frac{2(-1)^n}{\sqrt{1+n^2}} \underline{/\tan^{-1} n}$$

so that

$$\mathbf{I}_o = \frac{1}{4\sqrt{1+n^2}\,\underline{/\tan^{-1} n}} \frac{2(-1)^n}{\sqrt{1+n^2}} \underline{/\tan^{-1} n} = \frac{(-1)^n}{2(1+n^2)}$$

In the time domain,

$$i_o(t) = \frac{1}{4} + \sum_{n=1}^{\infty} \frac{(-1)^n}{2(1+n^2)} \cos nt \text{ A}$$

Practice Problem 17.7

If the input voltage in the circuit of Fig. 17.24 is

$$v(t) = \frac{1}{3} + \frac{1}{\pi^2} \sum_{n=1}^{\infty} \left(\frac{1}{n^2} \cos nt - \frac{\pi}{n} \sin nt \right) \text{V}$$

determine the response $i_o(t)$.

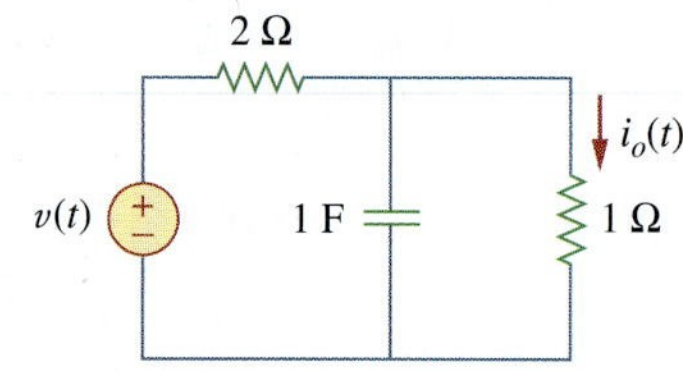

Figure 17.24
For Practice Prob. 17.7.

Answer: $\frac{1}{9} + \sum_{n=1}^{\infty} \frac{\sqrt{1+n^2\pi^2}}{n^2\pi^2\sqrt{9+4n^2}} \cos\left(nt - \tan^{-1}\frac{2n}{3} + \tan^{-1} n\pi\right)$ A.

17.5 Average Power and RMS Values

Recall the concepts of average power and rms value of a periodic signal that we discussed in Chapter 11. To find the average power absorbed by a circuit due to a periodic excitation, we write the voltage and current in amplitude-phase form [see Eq. (17.10)] as

$$v(t) = V_{dc} + \sum_{n=1}^{\infty} V_n \cos(n\omega_0 t - \theta_n) \tag{17.42}$$

$$i(t) = I_{dc} + \sum_{m=1}^{\infty} I_m \cos(m\omega_0 t - \phi_m) \tag{17.43}$$

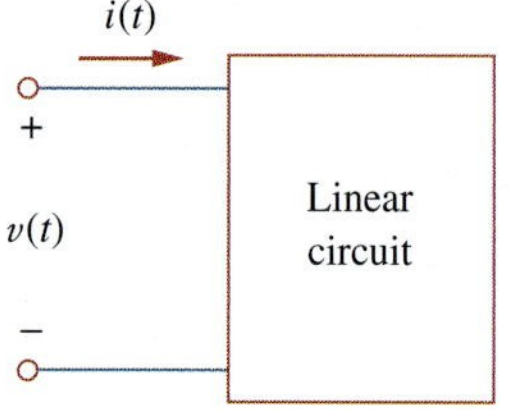

Figure 17.25
The voltage polarity reference and current reference direction.

Following the passive sign convention (Fig. 17.25), the average power is

$$P = \frac{1}{T}\int_0^T vi\,dt \tag{17.44}$$

Substituting Eqs. (17.42) and (17.43) into Eq. (17.44) gives

$$\begin{aligned} P = \frac{1}{T}\int_0^T V_{dc}I_{dc}\,dt &+ \sum_{m=1}^{\infty}\frac{I_m V_{dc}}{T}\int_0^T \cos(m\omega_0 t - \phi_m)\,dt \\ &+ \sum_{n=1}^{\infty}\frac{V_n I_{dc}}{T}\int_0^T \cos(n\omega_0 t - \theta_n)\,dt \\ &+ \sum_{m=1}^{\infty}\sum_{n=1}^{\infty}\frac{V_n I_m}{T}\int_0^T \cos(n\omega_0 t - \theta_n)\cos(m\omega_0 t - \phi_m)\,dt \end{aligned} \tag{17.45}$$

The second and third integrals vanish, since we are integrating the cosine over its period. According to Eq. (17.4e), all terms in the fourth integral are zero when $m \neq n$. By evaluating the first integral and applying Eq. (17.4g) to the fourth integral for the case $m = n$, we obtain

$$P = V_{dc}I_{dc} + \frac{1}{2}\sum_{n=1}^{\infty} V_n I_n \cos(\theta_n - \phi_n) \tag{17.46}$$

This shows that in average-power calculation involving periodic voltage and current, the total average power is the sum of the average powers in each harmonically related voltage and current.

Given a periodic function $f(t)$, its rms value (or the effective value) is given by

$$F_{rms} = \sqrt{\frac{1}{T}\int_0^T f^2(t)\,dt} \tag{17.47}$$

Substituting $f(t)$ in Eq. (17.10) into Eq. (17.47) and noting that $(a + b)^2 = a^2 + 2ab + b^2$, we obtain

$$\begin{aligned} F_{\text{rms}}^2 &= \frac{1}{T}\int_0^T \bigg[a_0^2 + 2\sum_{n=1}^{\infty} a_0 A_n \cos(n\omega_0 t + \phi_n) \\ &\quad + \sum_{n=1}^{\infty}\sum_{m=1}^{\infty} A_n A_m \cos(n\omega_0 t + \phi_n)\cos(m\omega_0 t + \phi_m) \bigg] dt \\ &= \frac{1}{T}\int_0^T a_0^2\, dt + 2\sum_{n=1}^{\infty} a_0 A_n \frac{1}{T}\int_0^T \cos(n\omega_0 t + \phi_n)\, dt \\ &\quad + \sum_{n=1}^{\infty}\sum_{m=1}^{\infty} A_n A_m \frac{1}{T}\int_0^T \cos(n\omega_0 t + \phi_n)\cos(m\omega_0 t + \phi_m)\, dt \end{aligned} \tag{17.48}$$

Distinct integers n and m have been introduced to handle the product of the two series summations. Using the same reasoning as above, we get

$$F_{\text{rms}}^2 = a_0^2 + \frac{1}{2}\sum_{n=1}^{\infty} A_n^2$$

or

$$F_{\text{rms}} = \sqrt{a_0^2 + \frac{1}{2}\sum_{n=1}^{\infty} A_n^2} \tag{17.49}$$

In terms of Fourier coefficients a_n and b_n, Eq. (17.49) may be written as

$$F_{\text{rms}} = \sqrt{a_0^2 + \frac{1}{2}\sum_{n=1}^{\infty} (a_n^2 + b_n^2)} \tag{17.50}$$

If $f(t)$ is the current through a resistor R, then the power dissipated in the resistor is

$$P = RF_{\text{rms}}^2 \tag{17.51}$$

Or if $f(t)$ is the voltage across a resistor R, the power dissipated in the resistor is

$$P = \frac{F_{\text{rms}}^2}{R} \tag{17.52}$$

One can avoid specifying the nature of the signal by choosing a 1-Ω resistance. The power dissipated by the 1-Ω resistance is

$$\boxed{P_{1\Omega} = F_{\text{rms}}^2 = a_0^2 + \frac{1}{2}\sum_{n=1}^{\infty} (a_n^2 + b_n^2)} \tag{17.53}$$

This result is known as *Parseval's theorem*. Notice that a_0^2 is the power in the dc component, while $\frac{1}{2}(a_n^2 + b_n^2)$ is the ac power in the nth harmonic. Thus, Parseval's theorem states that the average power in a periodic signal is the sum of the average power in its dc component and the average powers in its harmonics.

Historical note: Named after the French mathematician Marc-Antoine Parseval Deschemes (1755–1836).

Example 17.8

Determine the average power supplied to the circuit in Fig. 17.26 if $i(t) = 2 + 10\cos(t + 10°) + 6\cos(3t + 35°)$ A.

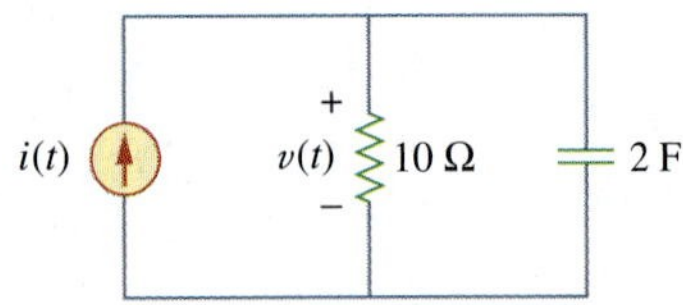

Figure 17.26
For Example 17.8.

Solution:

The input impedance of the network is

$$\mathbf{Z} = 10 \Bigg\| \frac{1}{j2\omega} = \frac{10(1/j2\omega)}{10 + 1/j2\omega} = \frac{10}{1 + j20\omega}$$

Hence,

$$\mathbf{V} = \mathbf{IZ} = \frac{10\mathbf{I}}{\sqrt{1 + 400\omega^2}\,\underline{/\tan^{-1} 20\omega}}$$

For the dc component, $\omega = 0$,

$$\mathbf{I} = 2 \text{ A} \quad \Rightarrow \quad \mathbf{V} = 10(2) = 20 \text{ V}$$

This is expected, because the capacitor is an open circuit to dc and the entire 2-A current flows through the resistor. For $\omega = 1$ rad/s,

$$\mathbf{I} = 10\underline{/10°} \quad \Rightarrow \quad \mathbf{V} = \frac{10(10\underline{/10°})}{\sqrt{1 + 400}\,\underline{/\tan^{-1} 20}} = 5\underline{/-77.14°}$$

For $\omega = 3$ rad/s,

$$\mathbf{I} = 6\underline{/35°} \quad \Rightarrow \quad \mathbf{V} = \frac{10(6\underline{/35°})}{\sqrt{1 + 3600}\,\underline{/\tan^{-1} 60}} = 1\underline{/-54.04°}$$

Thus, in the time domain,

$$v(t) = 20 + 5\cos(t - 77.14°) + 1\cos(3t - 54.04°) \text{ V}$$

We obtain the average power supplied to the circuit by applying Eq. (17.46), as

$$P = V_{\text{dc}} I_{\text{dc}} + \frac{1}{2} \sum_{n=1}^{\infty} V_n I_n \cos(\theta_n - \phi_n)$$

To get the proper signs of θ_n and ϕ_n, we have to compare v and i in this example with Eqs. (17.42) and (17.43). Thus,

$$P = 20(2) + \frac{1}{2}(5)(10)\cos[77.14° - (-10°)] + \frac{1}{2}(1)(6)\cos[54.04° - (-35°)]$$
$$= 40 + 1.247 + 0.05 = 41.5 \text{ W}$$

Alternatively, we can find the average power absorbed by the resistor as

$$P = \frac{V_{\text{dc}}^2}{R} + \frac{1}{2} \sum_{n=1}^{\infty} \frac{|V_n|^2}{R} = \frac{20^2}{10} + \frac{1}{2} \cdot \frac{5^2}{10} + \frac{1}{2} \cdot \frac{1^2}{10}$$
$$= 40 + 1.25 + 0.05 = 41.5 \text{ W}$$

which is the same as the power supplied, since the capacitor absorbs no average power.

Practice Problem 17.8

The voltage and current at the terminals of a circuit are

$$v(t) = 128 + 192\cos 120\pi t + 96\cos(360\pi t - 30^\circ)$$
$$i(t) = 8\cos(120\pi t - 10^\circ) + 3.2\cos(360\pi t - 60^\circ)$$

Find the average power absorbed by the circuit.

Answer: 22.23 kW.

Example 17.9

Find an estimate for the rms value of the voltage in Example 17.7.

Solution:
From Example 17.7, $v(t)$ is expressed as

$$v(t) = 1 - 1.414\cos(t + 45^\circ) + 0.8944\cos(2t + 63.45^\circ) - 0.6345\cos(3t + 71.56^\circ) - 0.4851\cos(4t + 78.7^\circ) + \cdots \text{ V}$$

Using Eq. (17.49), we find

$$V_{\text{rms}} = \sqrt{a_0^2 + \frac{1}{2}\sum_{n=1}^{\infty} A_n^2}$$
$$= \sqrt{1^2 + \frac{1}{2}\left[(-1.414)^2 + (0.8944)^2 + (-0.6345)^2 + (-0.4851)^2 + \cdots\right]}$$
$$= \sqrt{2.7186} = 1.649 \text{ V}$$

This is only an estimate, as we have not taken enough terms of the series. The actual function represented by the Fourier series is

$$v(t) = \frac{\pi e^t}{\sinh \pi}, \qquad -\pi < t < \pi$$

with $v(t) = v(t + T)$. The exact rms value of this is 1.776 V.

Practice Problem 17.9

Find the rms value of the periodic current

$$i(t) = 8 + 30\cos 2t - 20\sin 2t + 15\cos 4t - 10\sin 4t \text{ A}$$

Answer: 29.61 A.

17.6 Exponential Fourier Series

A compact way of expressing the Fourier series in Eq. (17.3) is to put it in exponential form. This requires that we represent the sine and cosine functions in the exponential form using Euler's identity:

$$\cos n\omega_0 t = \frac{1}{2}[e^{jn\omega_0 t} + e^{-jn\omega_0 t}] \tag{17.54a}$$

$$\sin n\omega_0 t = \frac{1}{2j}[e^{jn\omega_0 t} - e^{-jn\omega_0 t}] \tag{17.54b}$$

Substituting Eq. (17.54) into Eq. (17.3) and collecting terms, we obtain

$$f(t) = a_0 + \frac{1}{2}\sum_{n=1}^{\infty} [(a_n - jb_n)e^{jn\omega_0 t} + (a_n + jb_n)e^{-jn\omega_0 t}] \quad (17.55)$$

If we define a new coefficient c_n so that

$$c_0 = a_0, \qquad c_n = \frac{(a_n - jb_n)}{2}, \qquad c_{-n} = c_n^* = \frac{(a_n + jb_n)}{2} \quad (17.56)$$

then $f(t)$ becomes

$$f(t) = c_0 + \sum_{n=1}^{\infty} (c_n e^{jn\omega_0 t} + c_{-n}e^{-jn\omega_0 t}) \quad (17.57)$$

or

$$f(t) = \sum_{n=-\infty}^{\infty} c_n e^{jn\omega_0 t} \quad (17.58)$$

This is the *complex* or *exponential Fourier series* representation of $f(t)$. Note that this exponential form is more compact than the sine-cosine form in Eq. (17.3). Although the exponential Fourier series coefficients c_n can also be obtained from a_n and b_n using Eq. (17.56), they can also be obtained directly from $f(t)$ as

$$c_n = \frac{1}{T}\int_0^T f(t)e^{-jn\omega_0 t}\, dt \quad (17.59)$$

where $\omega_0 = 2\pi/T$, as usual. The plots of the magnitude and phase of c_n versus $n\omega_0$ are called the *complex amplitude spectrum* and *complex phase spectrum* of $f(t)$, respectively. The two spectra form the complex frequency spectrum of $f(t)$.

> The **exponential Fourier series** of a periodic function $f(t)$ describes the spectrum of $f(t)$ in terms of the amplitude and phase angle of ac components at positive and negative harmonic frequencies.

The coefficients of the three forms of Fourier series (sine-cosine form, amplitude-phase form, and exponential form) are related by

$$A_n\underline{/\phi_n} = a_n - jb_n = 2c_n \quad (17.60)$$

or

$$c_n = |c_n|\underline{/\theta_n} = \frac{\sqrt{a_n^2 + b_n^2}}{2}\underline{/-\tan^{-1} b_n/a_n} \quad (17.61)$$

if only $a_n > 0$. Note that the phase θ_n of c_n is equal to ϕ_n.

In terms of the Fourier complex coefficients c_n, the rms value of a periodic signal $f(t)$ can be found as

$$\begin{aligned} F_{\text{rms}}^2 &= \frac{1}{T}\int_0^T f^2(t)\,dt = \frac{1}{T}\int_0^T f(t)\left[\sum_{n=-\infty}^{\infty} c_n e^{jn\omega_0 t}\right]dt \\ &= \sum_{n=-\infty}^{\infty} c_n \left[\frac{1}{T}\int_0^T f(t)e^{jn\omega_0 t}\,dt\right] \\ &= \sum_{n=-\infty}^{\infty} c_n c_n^* = \sum_{n=-\infty}^{\infty} |c_n|^2 \end{aligned} \tag{17.62}$$

or

$$F_{\text{rms}} = \sqrt{\sum_{n=-\infty}^{\infty} |c_n|^2} \tag{17.63}$$

Equation (17.62) can be written as

$$F_{\text{rms}}^2 = |c_0|^2 + 2\sum_{n=1}^{\infty} |c_n|^2 \tag{17.64}$$

Again, the power dissipated by a 1-Ω resistance is

$$P_{1\Omega} = F_{\text{rms}}^2 = \sum_{n=-\infty}^{\infty} |c_n|^2 \tag{17.65}$$

which is a restatement of Parseval's theorem. The *power spectrum* of the signal $f(t)$ is the plot of $|c_n|^2$ versus $n\omega_0$. If $f(t)$ is the voltage across a resistor R, the average power absorbed by the resistor is F_{rms}^2/R; if $f(t)$ is the current through R, the power is $F_{\text{rms}}^2 R$.

As an illustration, consider the periodic pulse train of Fig. 17.27. Our goal is to obtain its amplitude and phase spectra. The period of the pulse train is $T = 10$, so that $\omega_0 = 2\pi/T = \pi/5$. Using Eq. (17.59),

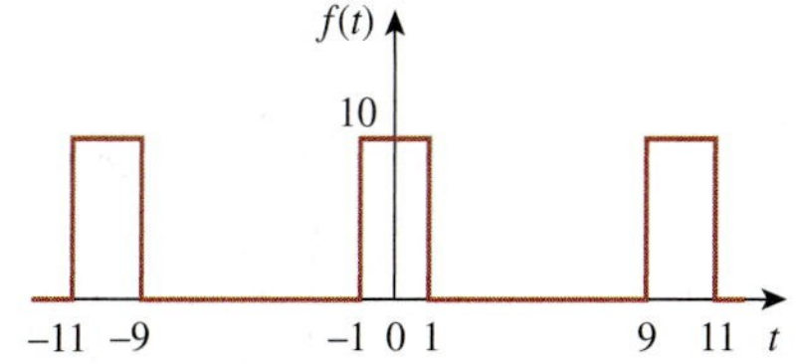

Figure 17.27
The periodic pulse train.

$$\begin{aligned} c_n &= \frac{1}{T}\int_{-T/2}^{T/2} f(t)e^{-jn\omega_0 t}\,dt = \frac{1}{10}\int_{-1}^{1} 10e^{-jn\omega_0 t}\,dt \\ &= \frac{1}{-jn\omega_0}e^{-jn\omega_0 t}\Big|_{-1}^{1} = \frac{1}{-jn\omega_0}(e^{-jn\omega_0} - e^{jn\omega_0}) \\ &= \frac{2}{n\omega_0}\,\frac{e^{jn\omega_0} - e^{-jn\omega_0}}{2j} = 2\frac{\sin n\omega_0}{n\omega_0}, \qquad \omega_0 = \frac{\pi}{5} \\ &= 2\frac{\sin n\pi/5}{n\pi/5} \end{aligned} \tag{17.66}$$

and

$$f(t) = 2\sum_{n=-\infty}^{\infty} \frac{\sin n\pi/5}{n\pi/5}e^{jn\pi t/5} \tag{17.67}$$

Notice from Eq. (17.66) that c_n is the product of 2 and a function of the form $\sin x/x$. This function is known as the *sinc function*; we write it as

The sinc function is called the *sampling function* in communication theory, where it is very useful.

$$\text{sinc}(x) = \frac{\sin x}{x} \tag{17.68}$$

Some properties of the sinc function are important here. For zero argument, the value of the sinc function is unity,

$$\text{sinc}(0) = 1 \tag{17.69}$$

This is obtained by applying L'Hopital's rule to Eq. (17.68). For an integral multiple of π, the value of the sinc function is zero,

$$\text{sinc}(n\pi) = 0, \qquad n = 1, 2, 3, \ldots \tag{17.70}$$

Also, the sinc function shows even symmetry. With all this in mind, we can obtain the amplitude and phase spectra of $f(t)$. From Eq. (17.66), the magnitude is

$$|c_n| = 2\left|\frac{\sin n\pi/5}{n\pi/5}\right| \tag{17.71}$$

while the phase is

$$\theta_n = \begin{cases} 0^\circ, & \sin\frac{n\pi}{5} > 0 \\ 180^\circ, & \sin\frac{n\pi}{5} < 0 \end{cases} \tag{17.72}$$

Examining the input and output spectra allows visualization of the effect of a circuit on a periodic signal.

Figure 17.28 shows the plot of $|c_n|$ versus n for n varying from -10 to 10, where $n = \omega/\omega_0$ is the normalized frequency. Figure 17.29 shows the plot of θ_n versus n. Both the amplitude spectrum and phase spectrum are called *line spectra*, because the values of $|c_n|$ and θ_n occur only at discrete values of frequencies. The spacing between the lines is ω_0. The power spectrum, which is the plot of $|c_n|^2$ versus $n\omega_0$, can also be plotted. Notice that the sinc function forms the envelope of the amplitude spectrum.

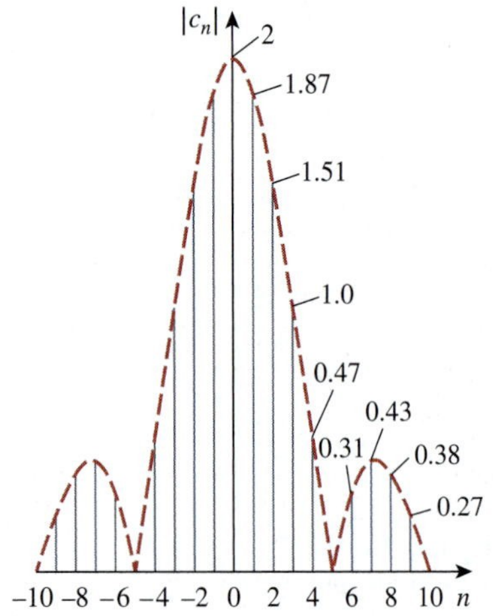

Figure 17.28
The amplitude of a periodic pulse train.

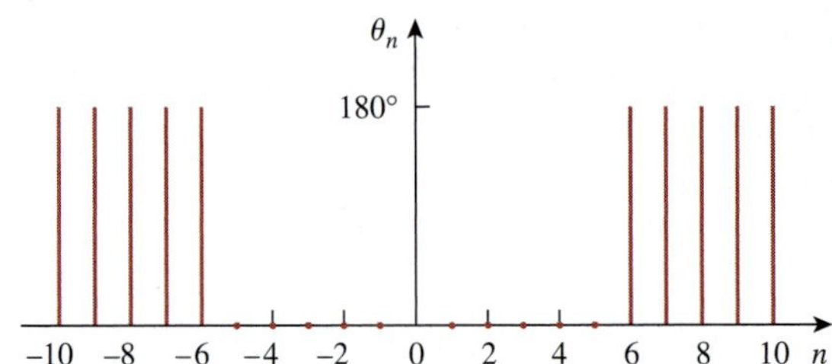

Figure 17.29
The phase spectrum of a periodic pulse train.

Example 17.10

Find the exponential Fourier series expansion of the periodic function $f(t) = e^t, 0 < t < 2\pi$ with $f(t + 2\pi) = f(t)$.

Solution:
Since $T = 2\pi$, $\omega_0 = 2\pi/T = 1$. Hence,

$$\begin{aligned} c_n &= \frac{1}{T}\int_0^T f(t)e^{-jn\omega_0 t}\,dt = \frac{1}{2\pi}\int_0^{2\pi} e^t e^{-jnt}\,dt \\ &= \frac{1}{2\pi}\frac{1}{1-jn}e^{(1-jn)t}\Big|_0^{2\pi} = \frac{1}{2\pi(1-jn)}[e^{2\pi}e^{-j2\pi n} - 1] \end{aligned}$$

But by Euler's identity,

$$e^{-j2\pi n} = \cos 2\pi n - j\sin 2\pi n = 1 - j0 = 1$$

Thus,

$$c_n = \frac{1}{2\pi(1 - jn)}[e^{2\pi} - 1] = \frac{85}{1 - jn}$$

The complex Fourier series is

$$f(t) = \sum_{n=-\infty}^{\infty} \frac{85}{1 - jn} e^{jnt}$$

We may want to plot the complex frequency spectrum of $f(t)$. If we let $c_n = |c_n|\underline{/\theta_n}$, then

$$|c_n| = \frac{85}{\sqrt{1 + n^2}}, \qquad \theta_n = \tan^{-1} n$$

By inserting in negative and positive values of n, we obtain the amplitude and the phase plots of c_n versus $n\omega_0 = n$, as in Fig. 17.30.

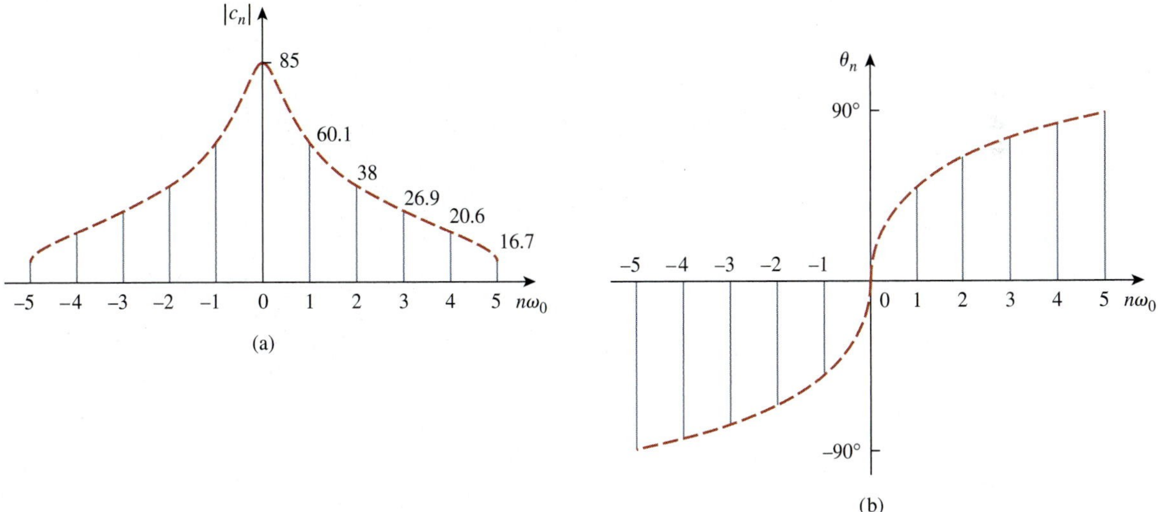

Figure 17.30
The complex frequency spectrum of the function in Example 17.10: (a) amplitude spectrum, (b) phase spectrum.

Practice Problem 17.10

Obtain the complex Fourier series of the function in Fig. 17.1.

Answer: $f(t) = \dfrac{1}{2} - \displaystyle\sum_{\substack{n=-\infty \\ n\neq 0 \\ n=\text{odd}}}^{\infty} \frac{j}{n\pi} e^{jn\pi t}.$

Example 17.11

Find the complex Fourier series of the sawtooth wave in Fig. 17.9. Plot the amplitude and the phase spectra.

Solution:

From Fig. 17.9, $f(t) = t, 0 < t < 1, T = 1$ so that $\omega_0 = 2\pi/T = 2\pi$. Hence,

$$c_n = \frac{1}{T}\int_0^T f(t)e^{-jn\omega_0 t}\, dt = \frac{1}{1}\int_0^1 te^{-j2n\pi t}\, dt \qquad \textbf{(17.11.1)}$$

But

$$\int te^{at}\,dt = \frac{e^{at}}{a^2}(ax - 1) + C$$

Applying this to Eq. (17.11.1) gives

$$\begin{aligned} c_n &= \frac{e^{-j2n\pi t}}{(-j2n\pi)^2}(-j2n\pi t - 1)\Big|_0^1 \\ &= \frac{e^{-j2n\pi}(-j2n\pi - 1) + 1}{-4n^2\pi^2} \end{aligned} \tag{17.11.2}$$

Again,

$$e^{-j2\pi n} = \cos 2\pi n - j\sin 2\pi n = 1 - j0 = 1$$

so that Eq. (17.11.2) becomes

$$c_n = \frac{-j2n\pi}{-4n^2\pi^2} = \frac{j}{2n\pi} \tag{17.11.3}$$

This does not include the case when $n = 0$. When $n = 0$,

$$c_0 = \frac{1}{T}\int_0^T f(t)\,dt = \frac{1}{1}\int_0^1 t\,dt = \frac{t^2}{2}\Big|_1^0 = 0.5 \tag{17.11.4}$$

Hence,

$$f(t) = 0.5 + \sum_{\substack{n=-\infty \\ n\neq 0}}^{\infty} \frac{j}{2n\pi}e^{j2n\pi t} \tag{17.11.5}$$

and

$$|c_n| = \begin{cases} \dfrac{1}{2|n|\pi}, & n \neq 0 \\ 0.5, & n = 0 \end{cases}, \qquad \theta_n = 90^\circ, \qquad n \neq 0 \tag{17.11.6}$$

By plotting $|c_n|$ and θ_n for different n, we obtain the amplitude spectrum and the phase spectrum shown in Fig. 17.31.

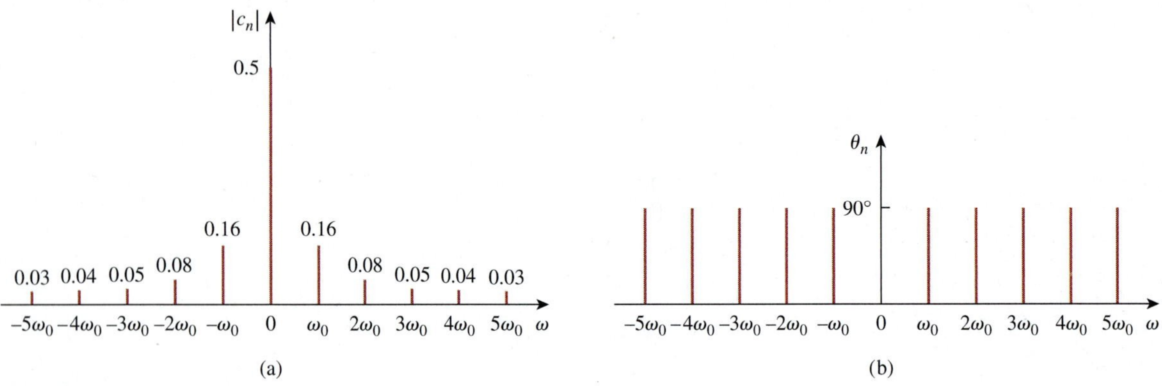

Figure 17.31
For Example 17.11: (a) amplitude spectrum, (b) phase spectrum.

Practice Problem 17.11

Obtain the complex Fourier series expansion of $f(t)$ in Fig. 17.17. Show the amplitude and phase spectra.

Answer: $f(t) = -\sum_{\substack{n=-\infty \\ n\neq 0}}^{\infty} \frac{j(-1)^n}{n\pi} e^{jn\pi t}$. See Fig. 17.32 for the spectra.

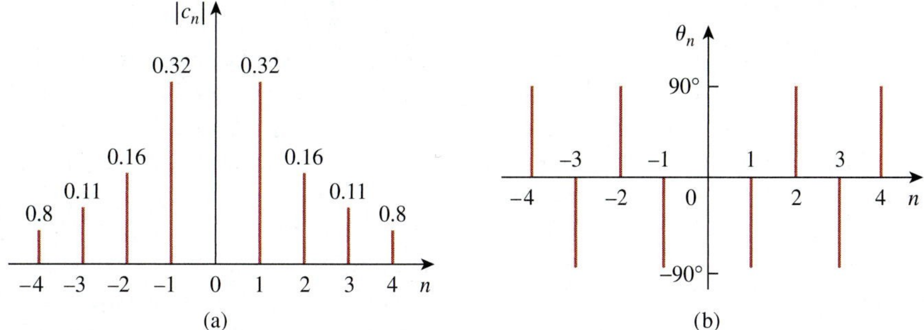

Figure 17.32
For Practice Prob. 17.11: (a) amplitude spectrum, (b) phase spectrum.

17.7 Fourier Analysis with *PSpice*

Fourier analysis is usually performed with *PSpice* in conjunction with transient analysis. Therefore, we must do a transient analysis in order to perform a Fourier analysis.

To perform the Fourier analysis of a waveform, we need a circuit whose input is the waveform and whose output is the Fourier decomposition. A suitable circuit is a current (or voltage) source in series with a 1-Ω resistor as shown in Fig. 17.33. The waveform is inputted as $v_s(t)$ using VPULSE for a pulse or VSIN for a sinusoid, and the attributes of the waveform are set over its period T. The output V(1) from node 1 is the dc level (a_0) and the first nine harmonics (A_n) with their corresponding phases ψ_n; that is,

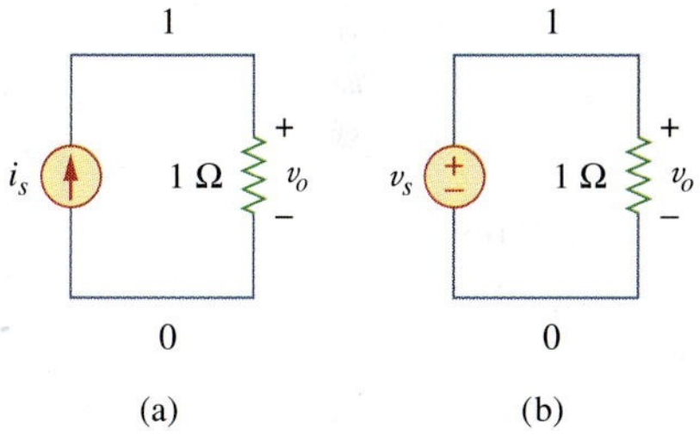

Figure 17.33
Fourier analysis with *PSpice* using: (a) a current source, (b) a voltage source.

$$v_o(t) = a_0 + \sum_{n=1}^{9} A_n \sin(n\omega_0 t + \psi_n) \tag{17.73}$$

where

$$A_n = \sqrt{a_n^2 + b_n^2}, \qquad \psi_n = \phi_n - \frac{\pi}{2}, \qquad \phi_n = \tan^{-1}\frac{b_n}{a_n} \tag{17.74}$$

Notice in Eq. (17.74) that the *PSpice* output is in the sine and angle form rather than the cosine and angle form in Eq. (17.10). The *PSpice* output also includes the normalized Fourier coefficients. Each coefficient a_n is normalized by dividing it by the magnitude of the fundamental a_1, so that the normalized component is a_n/a_1. The corresponding phase ψ_n is normalized by subtracting from it the phase ψ_1 of the fundamental, so that the normalized phase is $\psi_n - \psi_1$.

There are two types of Fourier analyses offered by *PSpice for Windows: Discrete Fourier Transform* (DFT) performed by the *PSpice* program and *Fast Fourier Transform* (FFT) performed by the *PSpice A/D* program. While DFT is an approximation of the exponential Fourier series, FTT is an algorithm for rapid efficient numerical computation of DFT. A full discussion of DFT and FTT is beyond the scope of this book.

17.7.1 Discrete Fourier Transform

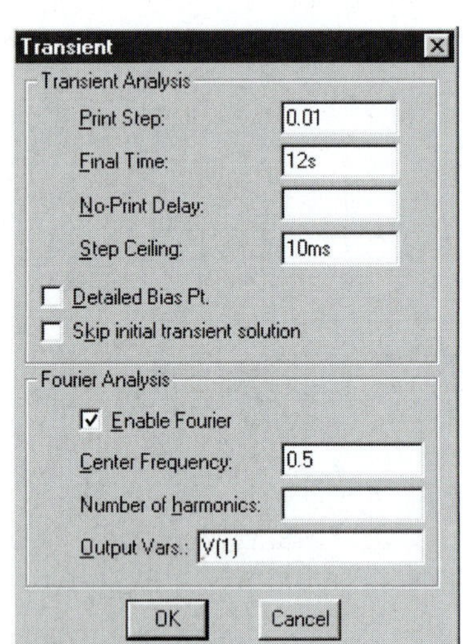

Figure 17.34
Transient dialog box.

A discrete Fourier transform (DFT) is performed by the *PSpice* program, which tabulates the harmonics in an output file. To enable a Fourier analysis, we select **Analysis/Setup/Transient** and bring up the Transient dialog box, shown in Fig. 17.34. The *Print Step* should be a small fraction of the period *T*, while the *Final Time* could be 6T. The *Center Frequency* is the fundamental frequency $f_0 = 1/T$. The particular variable whose DFT is desired, V(1) in Fig. 17.34, is entered in the **Output Vars** command box. In addition to filling in the Transient dialog box, **DCLICK** *Enable Fourier*. With the Fourier analysis enabled and the schematic saved, run *PSpice* by selecting **Analysis/Simulate** as usual. The program executes a harmonic decomposition into Fourier components of the result of the transient analysis. The results are sent to an output file which can be retrieved by selecting **Analysis/Examine Output**. The output file includes the dc value and the first nine harmonics by default, although you can specify more in the *Number of harmonics* box (see Fig. 17.34).

17.7.2 Fast Fourier Transform

A fast Fourier transform (FFT) is performed by the *PSpice A/D* program and displays as a *PSpice A/D* plot the complete spectrum of a transient expression. As explained above, we first construct the schematic in Fig. 17.33(b) and enter the attributes of the waveform. We also need to enter the *Print Step* and the *Final Time* in the Transient dialog box. Once this is done, we can obtain the FFT of the waveform in two ways.

One way is to insert a voltage marker at node 1 in the schematic of the circuit in Fig. 17.33(b). After saving the schematic and selecting **Analysis/Simulate**, the waveform V(1) will be displayed in the *PSpice A/D* window. Double clicking the FFT icon in the *PSpice A/D* menu will automatically replace the waveform with its FFT. From the FFT-generated graph, we can obtain the harmonics. In case the FFT-generated graph is crowded, we can use the *User Defined* data range (see Fig. 17.35) to specify a smaller range.

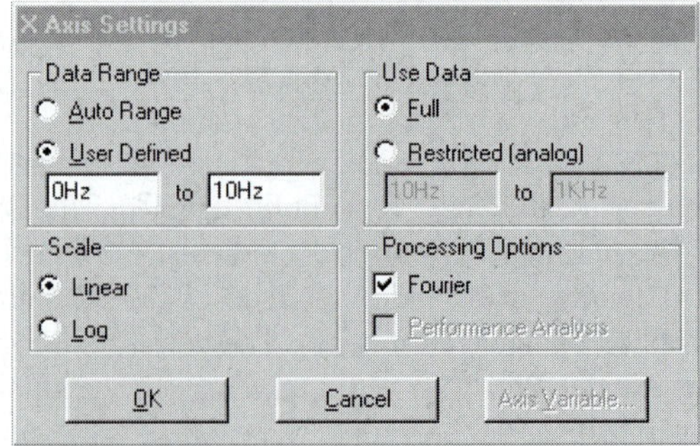

Figure 17.35
X axis settings dialog box.

Another way of obtaining the FFT of V(1) is to not insert a voltage marker at node 1 in the schematic. After selecting **Analysis/Simulate**, the *PSpice A/D* window will come up with no graph on it. We select **Trace/Add** and type V(1) in the **Trace Command** box and **DCLICKL OK**. We now select **Plot/X-Axis Settings** to bring up the *X-Axis Setting* dialog box shown in Fig. 17.35 and then select **Fourier/OK**. This will cause the FFT of the selected trace (or traces) to be displayed. This second approach is useful for obtaining the FFT of any trace associated with the circuit.

A major advantage of the FFT method is that it provides graphical output. But its major disadvantage is that some of the harmonics may be too small to see.

In both DFT and FFT, we should let the simulation run for a large number of cycles and use a small value of *Step Ceiling* (in the Transient dialog box) to ensure accurate results. The *Final Time* in the Transient dialog box should be at least five times the period of the signal to allow the simulation to reach steady state.

Example 17.12

Use *PSpice* to determine the Fourier coefficients of the signal in Fig. 17.1.

Solution:

Figure 17.36 shows the schematic for obtaining the Fourier coefficients. With the signal in Fig. 17.1 in mind, we enter the attributes of the voltage source VPULSE as shown in Fig. 17.36. We will solve this example using both the DFT and FFT approaches.

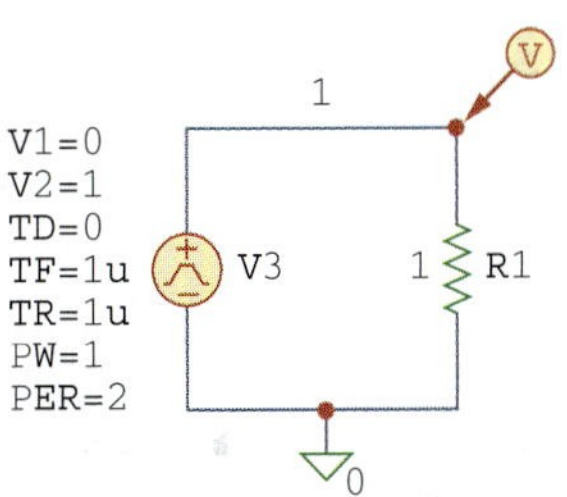

Figure 17.36
Schematic for Example 17.12.

■ **METHOD 1** **DFT Approach:** (The voltage marker in Fig. 17.36 is not needed for this method.) From Fig. 17.1, it is evident that $T = 2$ s,

$$f_0 = \frac{1}{T} = \frac{1}{2} = 0.5 \text{ Hz}$$

So, in the transient dialog box, we select the *Final Time* as $6T = 12$ s, the *Print Step* as 0.01 s, the *Step Ceiling* as 10 ms, the *Center Frequency* as 0.5 Hz, and the output variable as V(1). (In fact, Fig. 17.34 is for this particular example.) When *PSpice* is run, the output file contains the following result:

```
FOURIER COEFFICIENTS OF TRANSIENT RESPONSE V(1)

DC COMPONENT = 4.989950E-01

HARMONIC   FREQUENCY   FOURIER     NORMALIZED    PHASE       NORMALIZED
   NO        (HZ)      COMPONENT   COMPONENT     (DEG)       PHASE (DEG)

   1       5.000E-01   6.366E-01   1.000E+00   -1.809E-01    0.000E+00
   2       1.000E+00   2.012E-03   3.160E-03   -9.226E+01   -9.208E+01
   3       1.500E+00   2.122E-01   3.333E-01   -5.427E-01   -3.619E-01
   4       2.000E+00   2.016E-03   3.167E-03   -9.451E+01   -9.433E+01
   5       2.500E+00   1.273E-01   1.999E-01   -9.048E-01   -7.239E-01
   6       3.000E+00   2.024E-03   3.180E-03   -9.676E+01   -9.658E+01
   7       3.500E+00   9.088E-02   1.427E-01   -1.267E+00   -1.086E+00
   8       4.000E+00   2.035E-03   3.197E-03   -9.898E+01   -9.880E+01
   9       4.500E+00   7.065E-02   1.110E-01   -1.630E+00   -1.449E+00
```

Comparing the result with that in Eq. (17.1.7) (see Example 17.1) or with the spectra in Fig. 17.4 shows a close agreement. From Eq. (17.1.7), the dc component is 0.5 while *PSpice* gives 0.498995. Also, the signal has only odd harmonics with phase $\psi_n = -90^\circ$, whereas *PSpice* seems to indicate that the signal has even harmonics although the magnitudes of the even harmonics are small.

METHOD 2 **FFT Approach:** With voltage marker in Fig. 17.36 in place, we run *PSpice* and obtain the waveform V(1) shown in Fig. 17.37(a) on the *PSpice A/D* window. By double clicking the FFT icon in the *PSpice A/D* menu and changing the X-axis setting to 0 to 10 Hz, we obtain the FFT of V(1) as shown in Fig. 17.37(b). The FFT-generated graph contains the dc and harmonic components within the selected frequency range. Notice that the magnitudes and frequencies of the harmonics agree with the DFT-generated tabulated values.

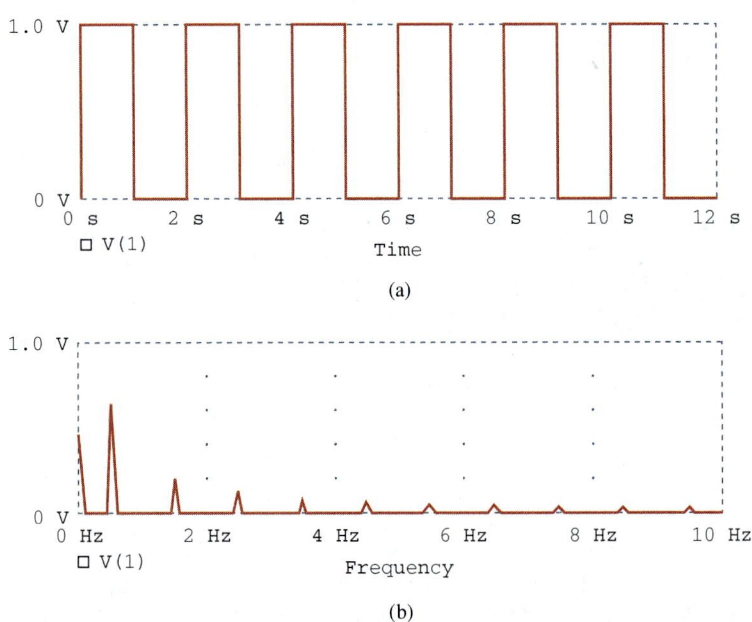

Figure 17.37
(a) Original waveform of Fig. 17.1, (b) FFT of the waveform.

Practice Problem 17.12

Obtain the Fourier coefficients of the function in Fig. 17.7 using *PSpice*.

Answer:

FOURIER COEFFICIENTS OF TRANSIENT RESPONSE V(1)

DC COMPONENT = 4.950000E-01

HARMONIC NO	FREQUENCY (HZ)	FOURIER COMPONENT	NORMALIZED COMPONENT	PHASE (DEG)	NORMALIZED PHASE (DEG)
1	1.000E+00	3.184E-01	1.000E+00	-1.782E+02	0.000E+00
2	2.000E+00	1.593E-01	5.002E-01	-1.764E+02	1.800E+00
3	3.000E+00	1.063E-01	3.338E-01	-1.746E+02	3.600E+00

(continued)

(continued)

4	4.000E+00	7.979E-02	2.506E-03	-1.728E+02	5.400E+00
5	5.000E+00	6.392E-01	2.008E-01	-1.710E+02	7.200E+00
6	6.000E+00	5.337E-02	1.676E-03	-1.692E+02	9.000E+00
7	7.000E+00	4.584E-02	1.440E-01	-1.674E+02	1.080E+01
8	8.000E+00	4.021E-02	1.263E-01	-1.656E+02	1.260E+01
9	9.000E+00	3.584E-02	1.126E-01	-1.638E+02	1.440E+01

Example 17.13

If $v_s = 12\sin(200\pi t)u(t)$ V in the circuit of Fig. 17.38, find $i(t)$.

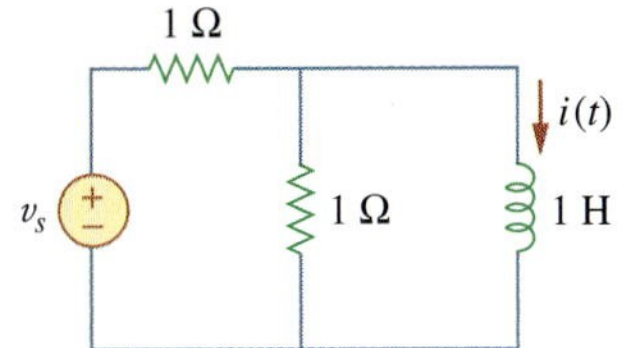

Figure 17.38
For Example 17.13.

Solution:

1. **Define.** Although the problem appears to be clearly stated, it might be advisable to check with the individual who assigned the problem to make sure he or she wants the transient response rather than the steady-state response; in the latter case the problem becomes trivial.
2. **Present.** We are to determine the response $i(t)$ given the input $v_s(t)$, using *PSpice* and Fourier analysis.
3. **Alternative.** We will use DFT to perform the initial analysis. We will then check using the FFT approach.
4. **Attempt.** The schematic is shown in Fig. 17.39. We may use the DFT approach to obtain the Fourier coefficents of $i(t)$. Since the period of the input waveform is $T = 1/100 = 10$ ms, in the Transient dialog box we select *Print Step:* 0.1 ms, *Final Time:* 100 ms, *Center Frequency:* 100 Hz, *Number of harmonics:* 4, and *Output Vars:* I(L1). When the circuit is simulated, the output file includes the following:

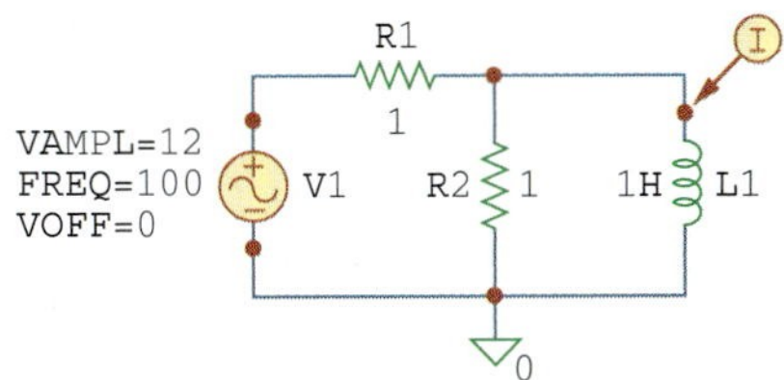

Figure 17.39
Schematic of the circuit in Fig. 17.38.

FOURIER COEFFICIENTS OF TRANSIENT RESPONSE I(VD)

DC COMPONENT = 8.583269E-03

HARMONIC NO	FREQUENCY (HZ)	FOURIER COMPONENT	NORMALIZED COMPONENT	PHASE (DEG)	NORMALIZED PHASE (DEG)
1	1.000E+02	8.730E-03	1.000E+00	-8.984E+01	0.000E+00
2	2.000E+02	1.017E-04	1.165E-02	-8.306E+01	6.783E+00
3	3.000E+02	6.811E-05	7.802E-03	-8.235E+01	7.490E+00
4	4.000E+02	4.403E-05	5.044E-03	-8.943E+01	4.054E+00

With the Fourier coefficients, the Fourier series describing the current $i(t)$ can be obtained using Eq. (17.73); that is,

$$\begin{aligned} i(t) &= 8.5833 + 8.73\sin(2\pi\cdot 100t - 89.84^\circ) \\ &\quad + 0.1017\sin(2\pi\cdot 200t - 83.06^\circ) \\ &\quad + 0.068\sin(2\pi\cdot 300t - 82.35^\circ) + \cdots \text{ mA} \end{aligned}$$

5. **Evaluate.** We can also use the FFT approach to cross-check our result. The current marker is inserted at pin 1 of the inductor as shown in Fig. 17.39. Running *PSpice* will automatically produce the plot of I(L1) in the *PSpice A/D* window, as shown in

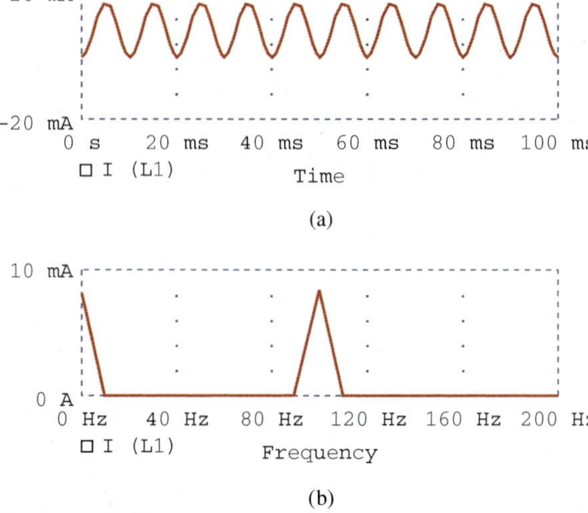

Figure 17.40
For Example 17.13: (a) plot of $i(t)$, (b) the FFT of $i(t)$.

Fig. 17.40(a). By double clicking the FFT icon and setting the range of the X-axis from 0 to 200 Hz, we generate the FFT of I(L1) shown in Fig. 17.40(b). It is clear from the FFT-generated plot that only the dc component and the first harmonic are visible. Higher harmonics are negligibly small.

One final observation, does the answer make sense? Let us look at the actual transient response, $i(t) = (9.549e^{-0.5t} - 9.549)\cos(200\pi t)u(t)$ mA. The period of the cosine wave is 10 ms while the time constant of the exponential is 2000 ms (2 seconds). So, the answer we obtained by Fourier techniques does agree.

6. **Satisfactory?** Clearly, we have solved the problem satisfactorily using the specified approach. We can now present our results as a solution to the problem.

Practice Problem 17.13

A sinusoidal current source of amplitude 4 A and frequency 2 kHz is applied to the circuit in Fig. 17.41. Use *PSpice* to find $v(t)$.

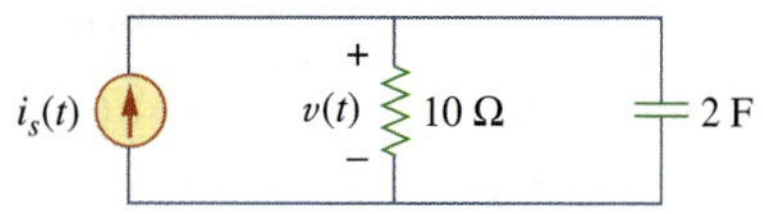

Figure 17.41
For Practice Prob. 17.13.

Answer: $v(t) = -150.72 + 145.5\sin(4\pi \cdot 10^3 t + 90°) + \cdots\ \mu$V. The Fourier components are shown below:

```
FOURIER COEFFICIENTS OF TRANSIENT RESPONSE V(R1:1)

DC COMPONENT = -1.507169E-04
```

HARMONIC NO	FREQUENCY (HZ)	FOURIER COMPONENT	NORMALIZED COMPONENT	PHASE (DEG)	NORMALIZED PHASE (DEG)
1	2.000E+03	1.455E-04	1.000E+00	9.006E+01	0.000E+00
2	4.000E+03	1.851E-06	1.273E-02	9.597E+01	5.910E+00
3	6.000E+03	1.406E-06	9.662E-03	9.323E+01	3.167E+00
4	8.000E+03	1.010E-06	6.946E-02	8.077E+01	-9.292E+00

17.8 †Applications

We demonstrated in Section 17.4 that the Fourier series expansion permits the application of the phasor techniques used in ac analysis to circuits containing nonsinusoidal periodic excitations. The Fourier series has many other practical applications, particularly in communications and signal processing. Typical applications include spectrum analysis, filtering, rectification, and harmonic distortion. We will consider two of these: spectrum analyzers and filters.

17.8.1 Spectrum Analyzers

The Fourier series provides the spectrum of a signal. As we have seen, the spectrum consists of the amplitudes and phases of the harmonics versus frequency. By providing the spectrum of a signal $f(t)$, the Fourier series helps us identify the pertinent features of the signal. It demonstrates which frequencies are playing an important role in the shape of the output and which ones are not. For example, audible sounds have significant components in the frequency range of 20 Hz to 15 kHz, while visible light signals range from 10^5 GHz to 10^6 GHz. Table 17.4 presents some other signals and the frequency ranges of their components. A periodic function is said to be *band-limited* if its amplitude spectrum contains only a finite number of coefficients A_n or c_n. In this case, the Fourier series becomes

$$f(t) = \sum_{n=-N}^{N} c_n e^{jn\omega_0 t} = a_0 + \sum_{n=1}^{N} A_n \cos(n\omega_0 t + \phi_n) \tag{17.75}$$

This shows that we need only $2N + 1$ terms (namely, $a_0, A_1, A_2, \ldots, A_N, \phi_1, \phi_2, \ldots, \phi_N$) to completely specify $f(t)$ if ω_0 is known. This leads to the *sampling theorem:* a band-limited periodic function whose Fourier series contains N harmonics is uniquely specified by its values at $2N + 1$ instants in one period.

A *spectrum analyzer* is an instrument that displays the amplitude of the components of a signal versus frequency. It shows the various frequency components (spectral lines) that indicate the amount of energy at each frequency.

It is unlike an oscilloscope, which displays the entire signal (all components) versus time. An oscilloscope shows the signal in the time domain, while the spectrum analyzer shows the signal in the frequency domain. There is perhaps no instrument more useful to a circuit analyst than the spectrum analyzer. An analyzer can conduct noise and spurious signal analysis, phase checks, electromagnetic interference and filter examinations, vibration measurements, radar measurements, and more. Spectrum analyzers are commercially available in various sizes and shapes. Figure 17.42 displays a typical one.

TABLE 17.4

Frequency ranges of typical signals.

Signal	Frequency Range
Audible sounds	20 Hz to 15 kHz
AM radio	540–1600 kHz
Short-wave radio	3–36 MHz
Video signals (U.S. standards)	dc to 4.2 MHz
VHF television, FM radio	54–216 MHz
UHF television	470–806 MHz
Cellular telephone	824–891.5 MHz
Microwaves	2.4–300 GHz
Visible light	10^5–10^6 GHz
X-rays	10^8–10^9 GHz

17.8.2 Filters

Filters are an important component of electronics and communications systems. Chapter 14 presented a full discussion on passive and active filters. Here, we investigate how to design filters to select the fundamental component (or any desired harmonic) of the input signal and reject other harmonics. This filtering process cannot be accomplished

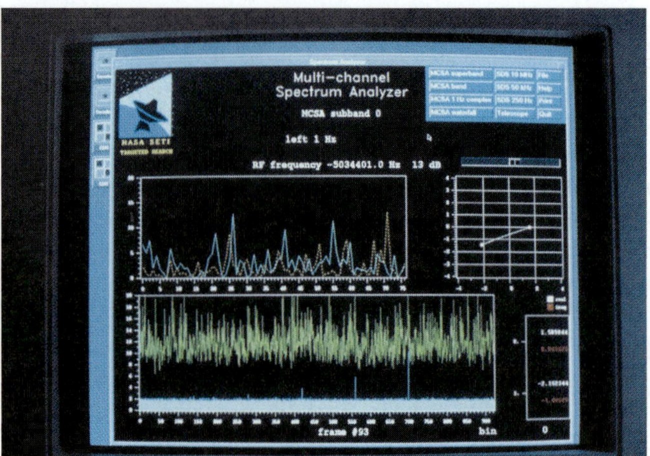

Figure 17.42
A typical spectrum analyzer. © SETI Institute/SPL/Photo Researchers, Inc.

without the Fourier series expansion of the input signal. For the purpose of illustration, we will consider two cases, a lowpass filter and a bandpass filter. In Example 17.6, we already looked at a highpass *RL* filter.

The output of a lowpass filter depends on the input signal, the transfer function $H(\omega)$ of the filter, and the corner or half-power frequency ω_c. We recall that $\omega_c = 1/RC$ for an *RC* passive filter. As shown in Fig. 17.43(a), the lowpass filter passes the dc and low-frequency components, while blocking the high-frequency components. By making ω_c sufficiently large ($\omega_c \gg \omega_0$, e.g., making C small), a large number of the harmonics can be passed. On the other hand, by making ω_c sufficiently small ($\omega_c \ll \omega_0$), we can block out all the ac components and pass only dc, as shown typically in Fig. 17.43(b). (See Fig. 17.2(a) for the Fourier series expansion of the square wave.)

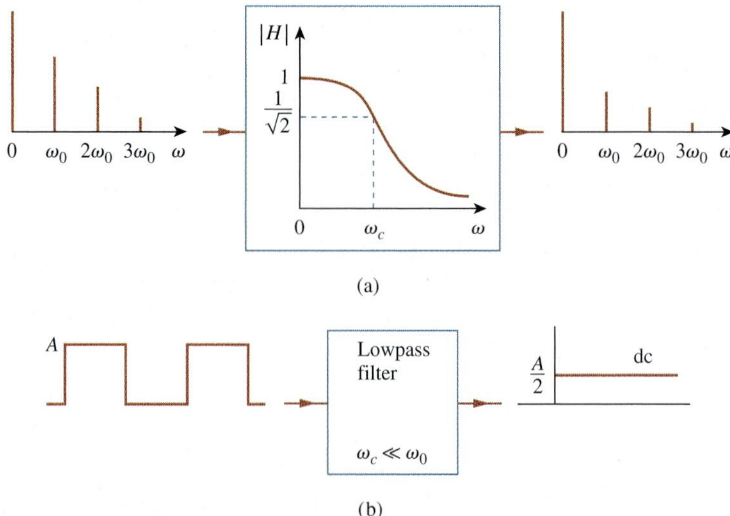

Figure 17.43
(a) Input and output spectra of a lowpass filter, (b) the lowpass filter passes only the dc component when $\omega_c \ll \omega_0$.

Similarly, the output of a bandpass filter depends on the input signal, the transfer function of the filter $H(\omega)$, its bandwidth B, and its center frequency ω_c. As illustrated in Fig. 17.44(a), the filter passes all the harmonics of the input signal within a band of frequencies ($\omega_1 < \omega < \omega_2$) centered around ω_c. We have assumed that ω_0, $2\omega_0$, and $3\omega_0$ are within that band. If the filter is made highly selective ($B \ll \omega_0$) and $\omega_c = \omega_0$, where ω_0 is the fundamental frequency of the input signal, the filter passes only the fundamental component ($n = 1$) of the input and blocks out all higher harmonics. As shown in Fig. 17.44(b), with a square wave as input, we obtain a sine wave of the same frequency as the output. (Again, refer to Fig. 17.2(a).)

In this section, we have used ω_c for the center frequency of the bandpass filter instead of ω_0 as in Chapter 14, to avoid confusing ω_0 with the fundamental frequency of the input signal.

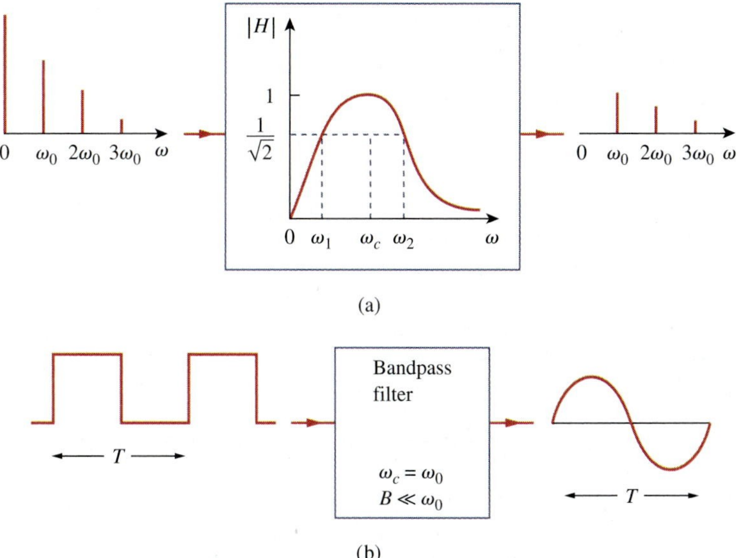

Figure 17.44
(a) Input and output spectra of a bandpass filter, (b) the bandpass filter passes only the fundamental component when $B \ll \omega_0$.

Example 17.14

If the sawtooth waveform in Fig. 17.45(a) is applied to an ideal lowpass filter with the transfer function shown in Fig. 17.45(b), determine the output.

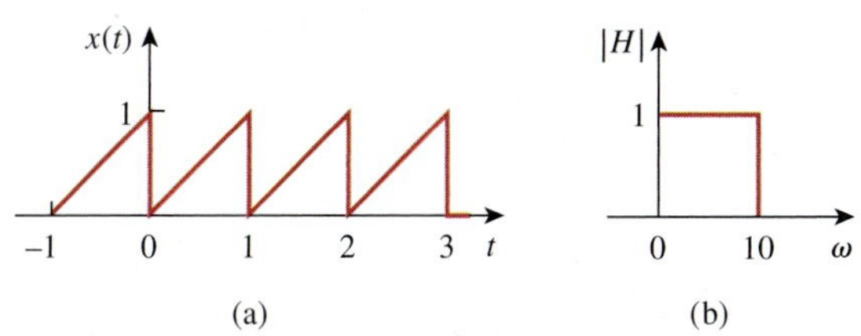

Figure 17.45
For example 17.14.

Solution:

The input signal in Fig. 17.45(a) is the same as the signal in Fig. 17.9. From Practice Prob. 17.2, we know that the Fourier series expansion is

$$x(t) = \frac{1}{2} - \frac{1}{\pi}\sin\omega_0 t - \frac{1}{2\pi}\sin 2\omega_0 t - \frac{1}{3\pi}\sin 3\omega_0 t - \cdots$$

where the period is $T = 1$ s and the fundamental frequency is $\omega_0 = 2\pi$ rad/s. Since the corner frequency of the filter is $\omega_c = 10$ rad/s, only the dc component and harmonics with $n\omega_0 < 10$ will be passed. For $n = 2$, $n\omega_0 = 4\pi = 12.566$ rad/s, which is higher than 10 rad/s, meaning that second and higher harmonics will be rejected. Thus, only the dc and fundamental components will be passed. Hence, the output of the filter is

$$y(t) = \frac{1}{2} - \frac{1}{\pi}\sin 2\pi t$$

Practice Problem 17.14

Rework Example 17.14 if the lowpass filter is replaced by the ideal bandpass filter shown in Fig. 17.46.

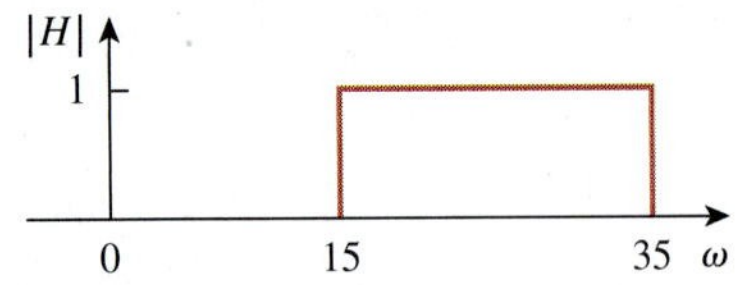

Figure 17.46
For Practice Prob. 17.14.

Answer: $y(t) = -\frac{1}{3\pi}\sin 3\omega_0 t - \frac{1}{4\pi}\sin 4\omega_0 t - \frac{1}{5\pi}\sin 5\omega_0 t.$

17.9 Summary

1. A periodic function is one that repeats itself every T seconds; that is, $f(t \pm nT) = f(t), n = 1, 2, 3, \ldots$.
2. Any nonsinusoidal periodic function $f(t)$ that we encounter in electrical engineering can be expressed in terms of sinusoids using Fourier series:

$$f(t) = \underbrace{a_0}_{\text{dc}} + \underbrace{\sum_{n=1}^{\infty}(a_n \cos n\omega_0 t + b_n \sin n\omega_0 t)}_{\text{ac}}$$

where $\omega_0 = 2\pi/T$ is the fundamental frequency. The Fourier series resolves the function into the dc component a_0 and an ac component containing infinitely many harmonically related sinusoids. The Fourier coefficients are determined as

$$a_0 = \frac{1}{T}\int_0^T f(t)\,dt, \qquad a_n = \frac{2}{T}\int_0^T f(t)\cos n\omega_0 t\,dt$$

$$b_n = \frac{2}{T}\int_0^T f(t)\sin n\omega_0 t\,dt$$

If $f(t)$ is an even function, $b_n = 0$, and when $f(t)$ is odd, $a_0 = 0$ and $a_n = 0$. If $f(t)$ is half-wave symmetric, $a_0 = a_n = b_n = 0$ for even values of n.
3. An alternative to the trigonometric (or sine-cosine) Fourier series is the amplitude-phase form

$$f(t) = a_0 + \sum_{n=1}^{\infty} A_n \cos(n\omega_0 t + \phi_n)$$

where

$$A_n = \sqrt{a_n^2 + b_n^2}, \qquad \phi_n = -\tan^{-1}\frac{b_n}{a_n}$$

4. Fourier series representation allows us to apply the phasor method in analyzing circuits when the source function is a nonsinusoidal periodic function. We use phasor technique to determine the response of each harmonic in the series, transform the responses to the time domain, and add them up.
5. The average-power of periodic voltage and current is

$$P = V_{\text{dc}}I_{\text{dc}} + \frac{1}{2}\sum_{n=1}^{\infty} V_n I_n \cos(\theta_n - \phi_n)$$

   In other words, the total average power is the sum of the average powers in each harmonically related voltage and current.
6. A periodic function can also be represented in terms of an exponential (or complex) Fourier series as

$$f(t) = \sum_{n=-\infty}^{\infty} c_n e^{jn\omega_0 t}$$

   where

$$c_n = \frac{1}{T}\int_0^T f(t)e^{-jn\omega_0 t}\,dt$$

   and $\omega_0 = 2\pi/T$. The exponential form describes the spectrum of $f(t)$ in terms of the amplitude and phase of ac components at positive and negative harmonic frequencies. Thus, there are three basic forms of Fourier series representation: the trigonometric form, the amplitude-phase form, and the exponential form.
7. The frequency (or line) spectrum is the plot of A_n and ϕ_n or $|c_n|$ and $\theta_n$versus frequency.
8. The rms value of a periodic function is given by

$$F_{\text{rms}} = \sqrt{a_0^2 + \frac{1}{2}\sum_{n=1}^{\infty} A_n^2}$$

   The power dissipated by a 1-Ω resistance is

$$P_{1\Omega} = F_{\text{rms}}^2 = a_0^2 + \frac{1}{2}\sum_{n=1}^{\infty}(a_n^2 + b_n^2) = \sum_{n=-\infty}^{\infty}|c_n|^2$$

   This relationship is known as *Parseval's theorem.*
9. Using *PSpice*, a Fourier analysis of a circuit can be performed in conjunction with the transient analysis.
10. Fourier series find application in spectrum analyzers and filters. The spectrum analyzer is an instrument that displays the discrete Fourier spectra of an input signal, so that an analyst can determine the frequencies and relative energies of the signal's components. Because the Fourier spectra are discrete spectra, filters can be designed for great effectiveness in blocking frequency components of a signal that are outside a desired range.

Review Questions

17.1 Which of the following cannot be a Fourier series?

(a) $t - \frac{t^2}{2} + \frac{t^3}{3} - \frac{t^4}{4} + \frac{t^5}{5}$

(b) $5 \sin t + 3 \sin 2t - 2 \sin 3t + \sin 4t$

(c) $\sin t - 2 \cos 3t + 4 \sin 4t + \cos 4t$

(d) $\sin t + 3 \sin 2.7t - \cos \pi t + 2 \tan \pi t$

(e) $1 + e^{-j\pi t} + \frac{e^{-j2\pi t}}{2} + \frac{e^{-j3\pi t}}{3}$

17.2 If $f(t) = t, 0 < t < \pi, f(t + n\pi) = f(t)$, the value of ω_0 is

(a) 1 (b) 2 (c) π (d) 2π

17.3 Which of the following are even functions?

(a) $t + t^2$ (b) $t^2 \cos t$ (c) e^{t^2}

(d) $t^2 + t^4$ (e) $\sinh t$

17.4 Which of the following are odd functions?

(a) $\sin t + \cos t$ (b) $t \sin t$

(c) $t \ln t$ (d) $t^3 \cos t$

(e) $\sinh t$

17.5 If $f(t) = 10 + 8 \cos t + 4 \cos 3t + 2 \cos 5t + \cdots$, the magnitude of the dc component is:

(a) 10 (b) 8 (c) 4

(d) 2 (e) 0

17.6 If $f(t) = 10 + 8 \cos t + 4 \cos 3t + 2 \cos 5t + \cdots$, the angular frequency of the 6th harmonic is

(a) 12 (b) 11 (c) 9

(d) 6 (e) 1

17.7 The function in Fig. 17.14 is half-wave symmetric.

(a) True (b) False

17.8 The plot of $|c_n|$ versus $n\omega_0$ is called:

(a) complex frequency spectrum

(b) complex amplitude spectrum

(c) complex phase spectrum

17.9 When the periodic voltage $2 + 6 \sin \omega_0 t$ is applied to a 1-Ω resistor, the integer closest to the power (in watts) dissipated in the resistor is:

(a) 5 (b) 8 (c) 20

(d) 22 (e) 40

17.10 The instrument for displaying the spectrum of a signal is known as:

(a) oscilloscope (b) spectrogram

(c) spectrum analyzer (d) Fourier spectrometer

Answers: 17.1a,d, 17.2b, 17.3b,c,d, 17.4d,e, 17.5a, 17.6d, 17.7a, 17.8b, 17.9d, 17.10c.

Problems

Section 17.2 Trigonometric Fourier Series

17.1 Evaluate each of the following functions and see if it is periodic. If periodic, find its period.

(a) $f(t) = \cos \pi t + 2 \cos 3\pi t + 3 \cos 5\pi t$

(b) $y(t) = \sin t + 4 \cos 2\pi t$

(c) $g(t) = \sin 3t \cos 4t$

(d) $h(t) = \cos^2 t$

(e) $z(t) = 4.2 \sin(0.4\pi t + 10°) + 0.8 \sin(0.6\pi t + 50°)$

(f) $p(t) = 10$

(g) $q(t) = e^{-\pi t}$

17.2 ML Using MATLAB, synthesize the periodic waveform for which the Fourier trigonometric Fourier series is

$$f(t) = \frac{1}{2} - \frac{4}{\pi^2}\left(\cos t + \frac{1}{9}\cos 3t + \frac{1}{25}\cos 5t + \cdots\right)$$

17.3 Give the Fourier coefficients a_0, a_n, and b_n of the waveform in Fig. 17.47. Plot the amplitude and phase spectra.

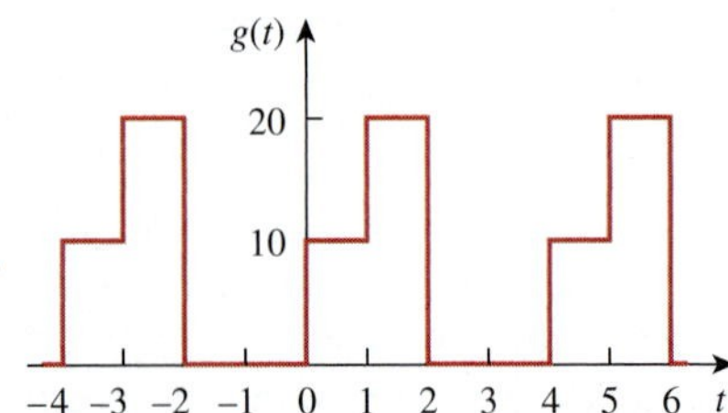

Figure 17.47
For Prob. 17.3.

17.4 Find the Fourier series expansion of the backward sawtooth waveform of Fig. 17.48. Obtain the amplitude and phase spectra.

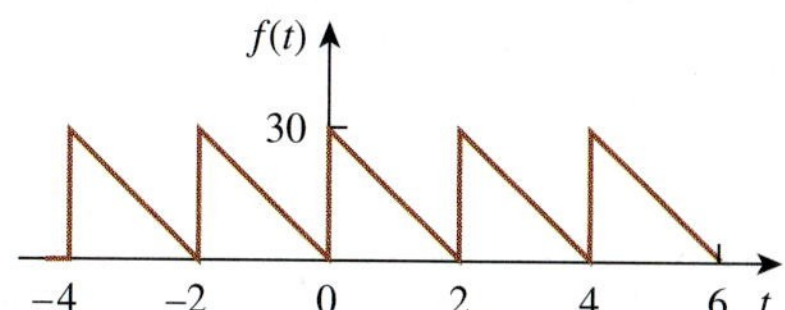

Figure 17.48
For Probs. 17.4 and 17.66.

17.5 Obtain the Fourier series expansion for the waveform shown in Fig. 17.49.

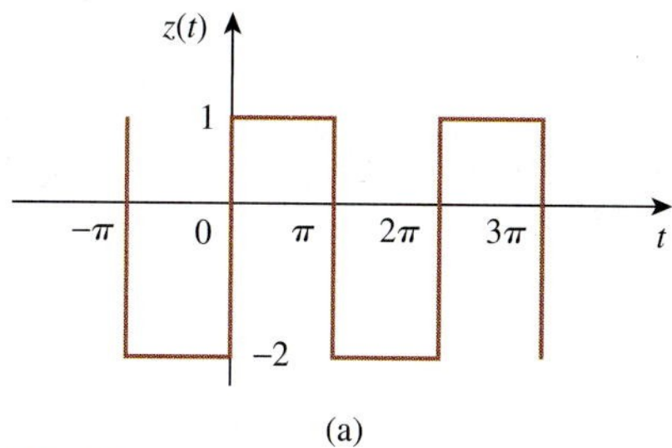

Figure 17.49
For Prob. 17.5.

17.6 Find the trigonometric Fourier series for

$$f(t) = \begin{cases} 20, & 0 < t < \pi \\ 40, & \pi < t < 2\pi \end{cases} \quad \text{and} \quad f(t + 2\pi) = f(t).$$

***17.7** Determine the Fourier series of the periodic function in Fig. 17.50.

ML

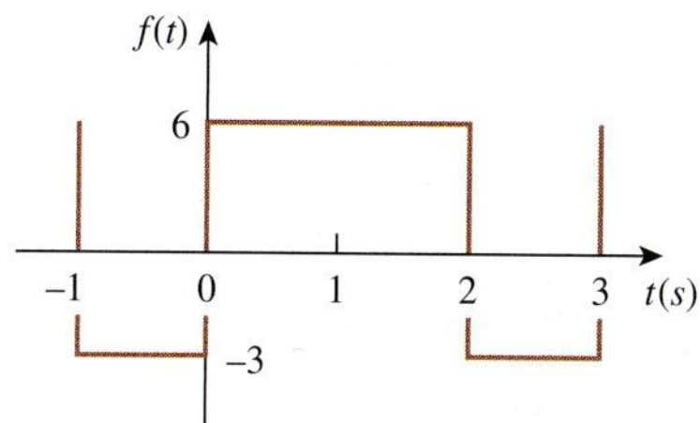

Figure 17.50
For Prob. 17.7.

17.8 Using Fig. 17.51, design a problem to help other students better understand how to determine the exponention Fourier series from a periodic wave shape.

ed

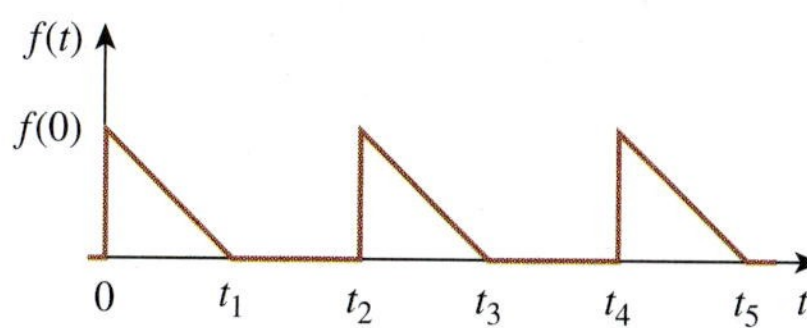

Figure 17.51
For Prob. 17.8.

* An asterisk indicates a challenging problem.

17.9 Determine the Fourier coefficients a_n and b_n of the first three harmonic terms of the rectified cosine wave in Fig. 17.52.

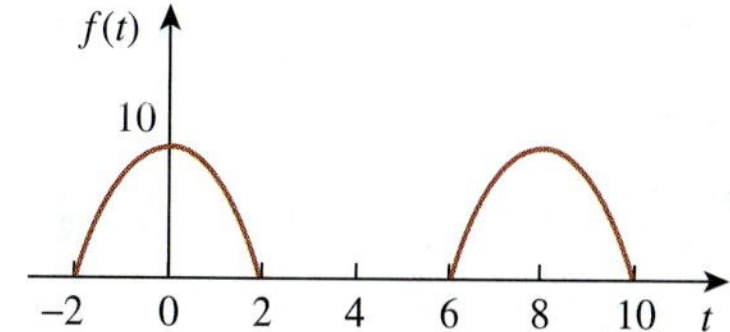

Figure 17.52
For Prob. 17.9.

17.10 Find the exponential Fourier series for the waveform in Fig. 17.53.

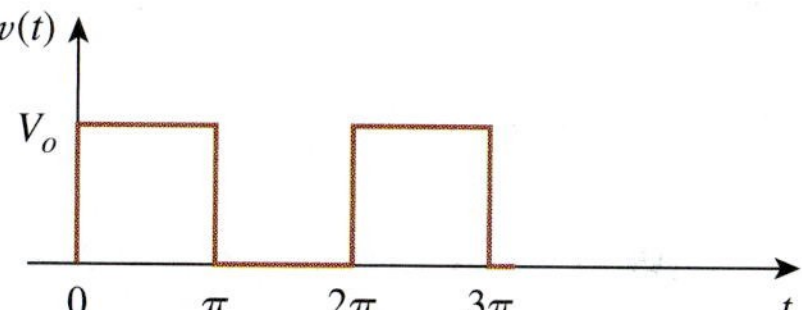

Figure 17.53
For Prob. 17.10.

17.11 Obtain the exponential Fourier series for the signal in Fig. 17.54.

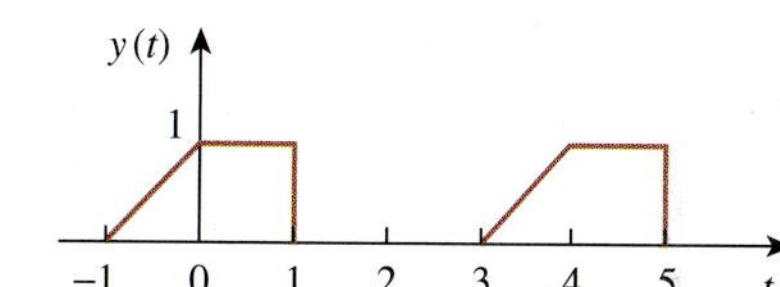

Figure 17.54
For Prob. 17.11.

***17.12** A voltage source has a periodic waveform defined over its period as

$$v(t) = t(2\pi - t) \text{ V}, \qquad 0 < t < 2\pi$$

Find the Fourier series for this voltage.

17.13 Design a problem to help other students better understand obtaining the Fourier series from a periodic function.

ed

17.14 Find the quadrature (cosine and sine) form of the Fourier series

$$f(t) = 2 + \sum_{n=1}^{\infty} \frac{10}{n^3 + 1} \cos\left(2nt + \frac{n\pi}{4}\right)$$

17.15 Express the Fourier series

$$f(t) = 10 + \sum_{n=1}^{\infty} \frac{4}{n^2 + 1} \cos 10nt + \frac{1}{n^3} \sin 10nt$$

(a) in a cosine and angle form,

(b) in a sine and angle form.

17.16 The waveform in Fig. 17.55(a) has the following Fourier series:

$$v_1(t) = \frac{1}{2} - \frac{4}{\pi^2}\left(\cos \pi t + \frac{1}{9}\cos 3\pi t + \frac{1}{25}\cos 5\pi t + \cdots\right) \text{V}$$

Obtain the Fourier series of $v_2(t)$ in Fig. 17.55(b).

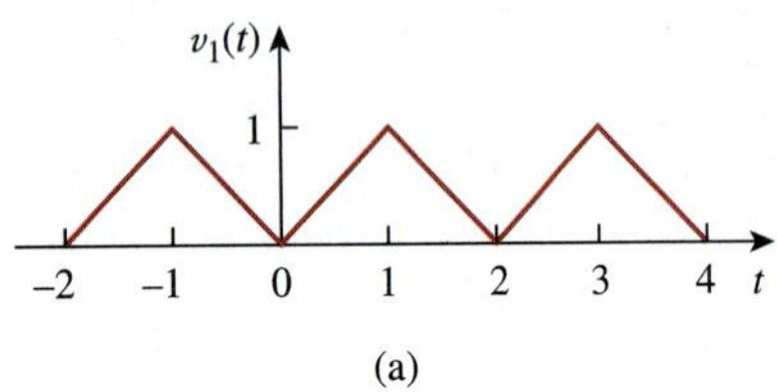

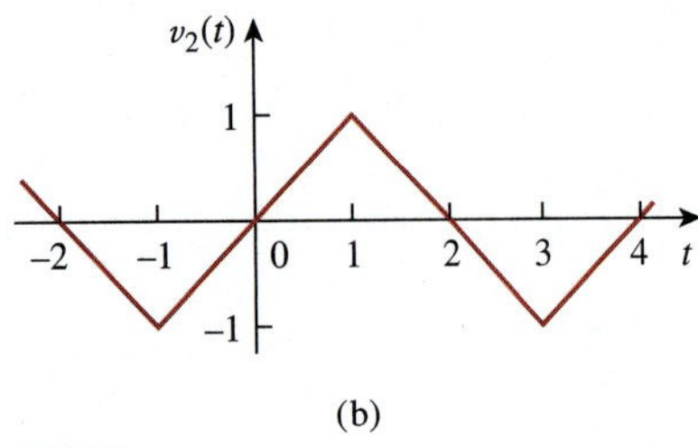

Figure 17.55
For Probs. 17.16 and 17.69.

Section 17.3 Symmetry Considerations

17.17 Determine if these functions are even, odd, or neither.

(a) $1 + t$ (b) $t^2 - 1$ (c) $\cos n\pi t \sin n\pi t$

(d) $\sin^2 \pi t$ (e) e^{-t}

17.18 Determine the fundamental frequency and specify the type of symmetry present in the functions in Fig. 17.56.

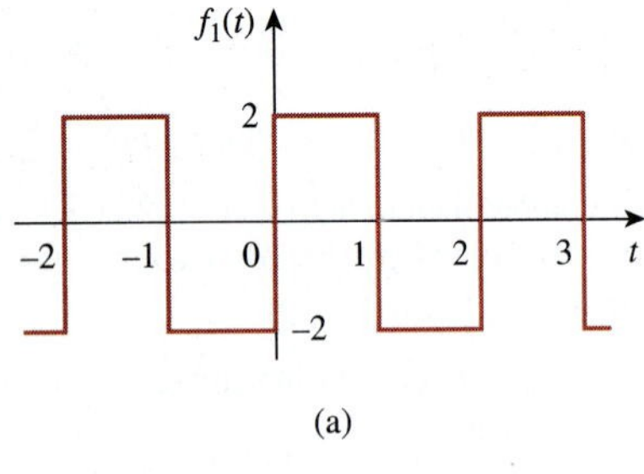

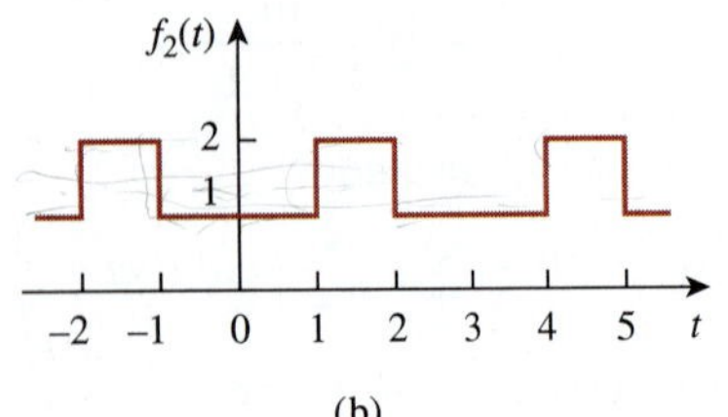

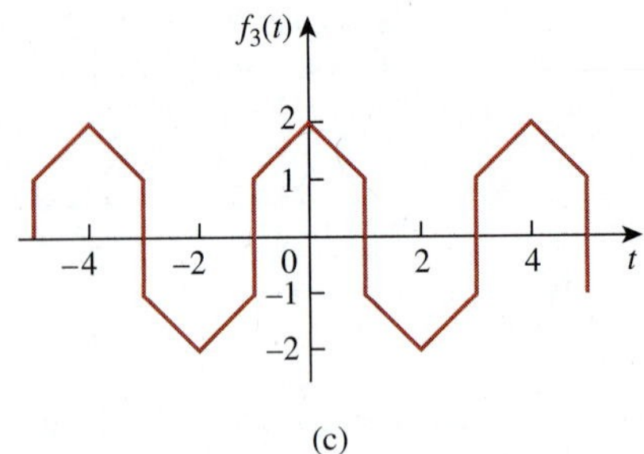

Figure 17.56
For Probs. 17.18 and 17.63.

17.19 Obtain the Fourier series for the periodic waveform in Fig. 17.57.

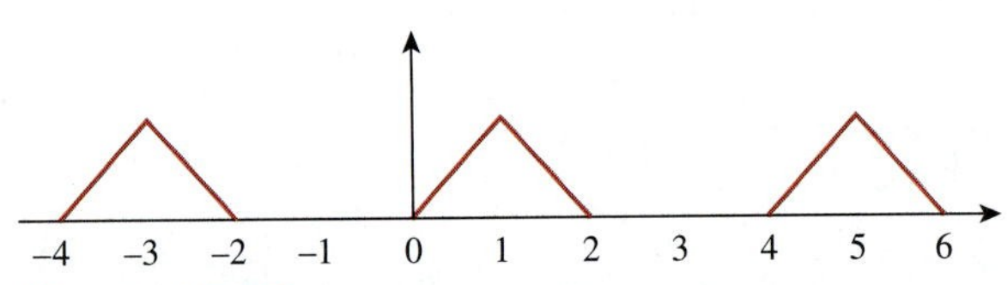

Figure 17.57
For Prob. 17.19.

17.20 Find the Fourier series for the signal in Fig. 17.58. Evaluate $f(t)$ at $t = 2$ using the first three nonzero harmonics.

ML

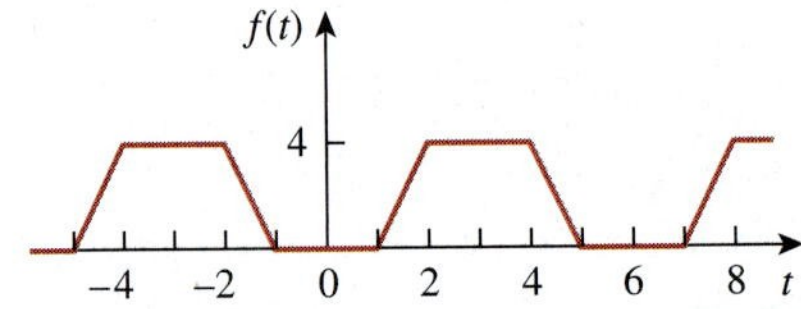

Figure 17.58
For Probs. 17.20 and 17.67.

17.21 Determine the trigonometric Fourier series of the signal in Fig. 17.59.

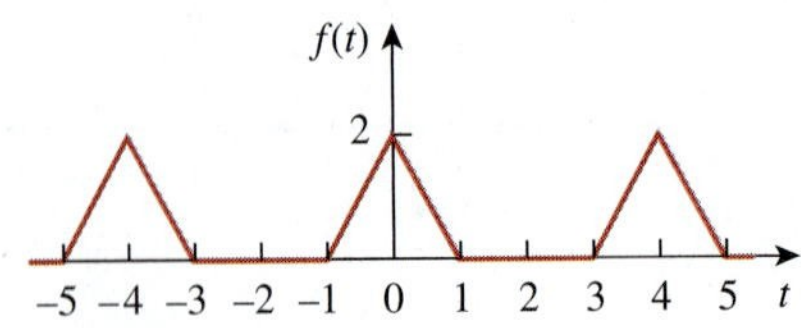

Figure 17.59
For Prob. 17.21.

17.22 Calculate the Fourier coefficients for the function in Fig. 17.60.

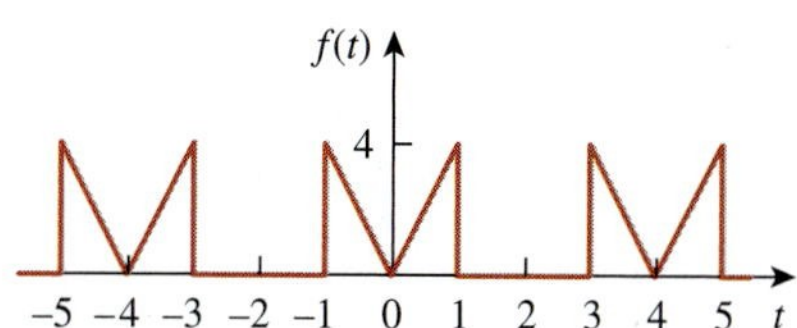

Figure 17.60
For Prob. 17.22.

17.23 Using Fig. 17.61, design a problem to help other students better understand finding the Fourier series of a periodic wave shape.

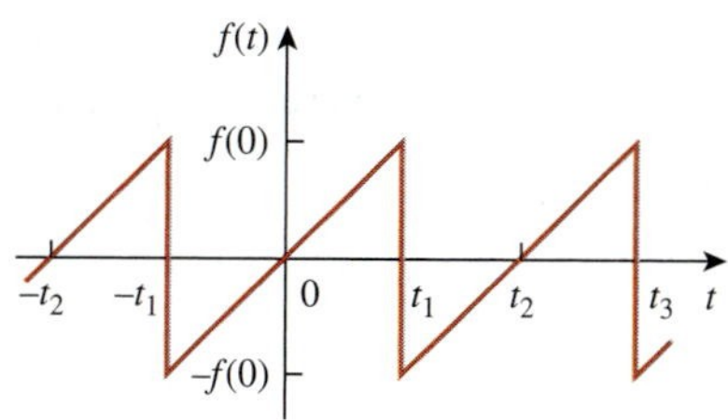

Figure 17.61
For Prob. 17.23.

17.24 In the periodic function of Fig. 17.62,

(a) find the trigonometric Fourier series coefficients a_2 and b_2,

(b) calculate the magnitude and phase of the component of $f(t)$ that has $\omega_n = 10$ rad/s,

(c) use the first four nonzero terms to estimate $f(\pi/2)$,

(d) show that

$$\frac{\pi}{4} = \frac{1}{1} - \frac{1}{3} + \frac{1}{5} - \frac{1}{7} + \frac{1}{9} - \frac{1}{11} + \cdots$$

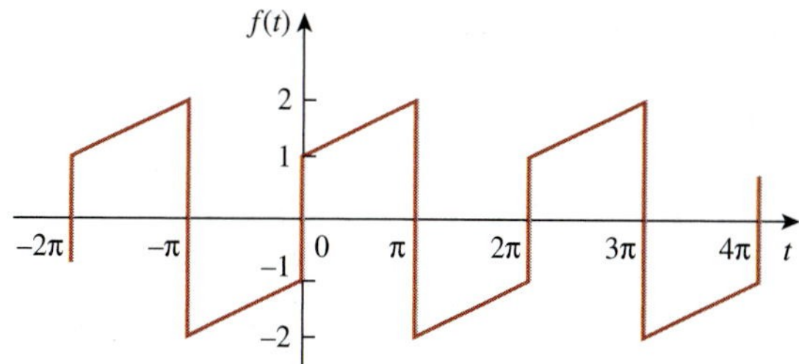

Figure 17.62
For Probs. 17.24 and 17.60.

17.25 Determine the Fourier series representation of the function in Fig. 17.63.

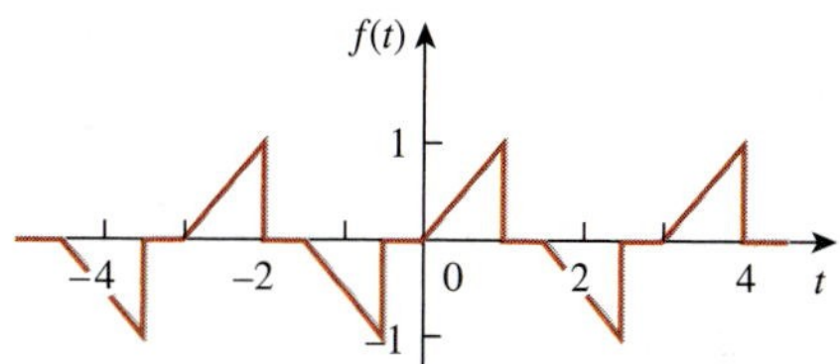

Figure 17.63
For Prob. 17.25.

17.26 Find the Fourier series representation of the signal shown in Fig. 17.64.

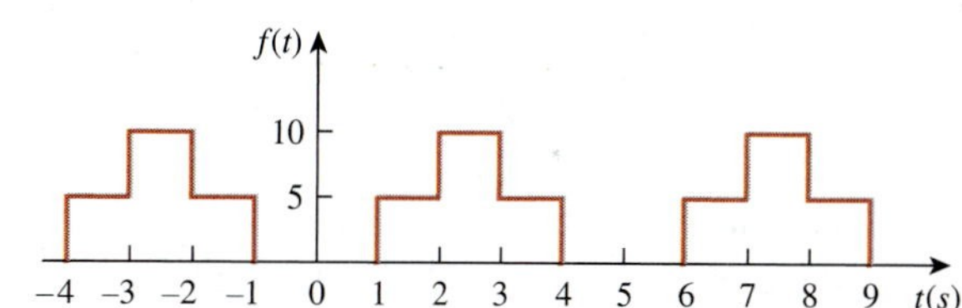

Figure 17.64
For Prob. 17.26.

17.27 For the waveform shown in Fig. 17.65 below,

(a) specify the type of symmetry it has,

(b) calculate a_3 and b_3,

(c) find the rms value using the first five nonzero harmonics.

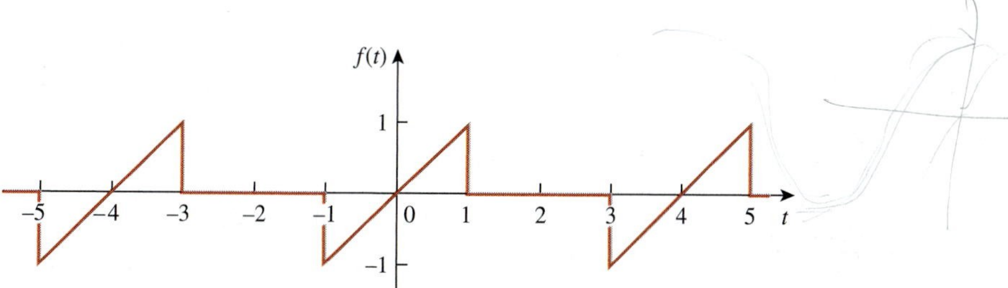

Figure 17.65
For Prob. 17.27.

17.28 Obtain the trigonometric Fourier series for the voltage waveform shown in Fig. 17.66.

ML

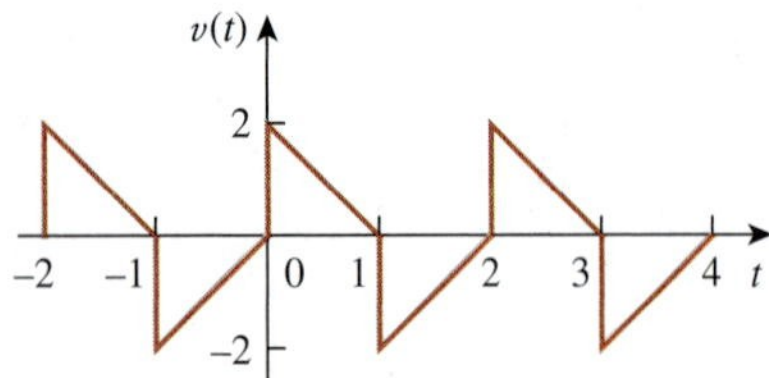

Figure 17.66
For Prob. 17.28.

17.29 Determine the Fourier series expansion of the sawtooth function in Fig. 17.67.

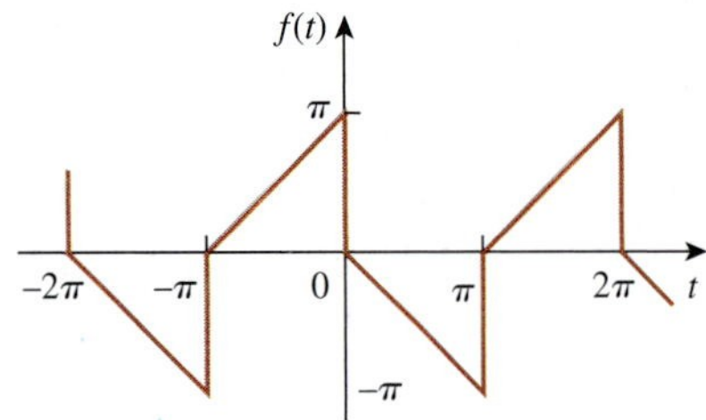

Figure 17.67
For Prob. 17.29.

17.30 (a) If $f(t)$ is an even function, show that

$$c_n = \frac{2}{T}\int_0^{T/2} f(t)\cos n\omega_o t\, dt$$

(b) If $f(t)$ is an odd function, show that

$$c_n = -\frac{j2}{T}\int_0^{T/2} f(t)\sin n\omega_o t\, dt$$

17.31 Let a_n and b_n be the Fourier series coefficients of $f(t)$ and let ω_o be its fundamental frequency. Suppose $f(t)$ is time-scaled to give $h(t) = f(\alpha t)$. Express the a_n' and b_n', and ω_o', of $h(t)$ in terms of a_n, b_n, and ω_o of $f(t)$.

Section 17.4 Circuit Applications

17.32 Find $i(t)$ in the circuit of Fig. 17.68 given that

$$i_s(t) = 1 + \sum_{n=1}^{\infty} \frac{1}{n^2}\cos 3nt \text{ A}$$

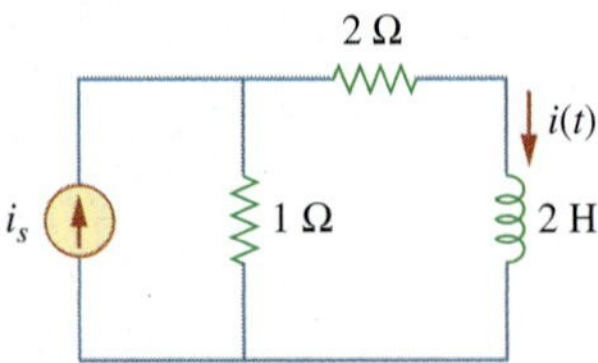

Figure 17.68
For Prob. 17.32.

17.33 In the circuit shown in Fig. 17.69, the Fourier series expansion of $v_s(t)$ is

$$v_s(t) = 3 + \frac{4}{\pi}\sum_{n=1}^{\infty} \frac{1}{n}\sin(n\pi t)$$

Find $v_o(t)$.

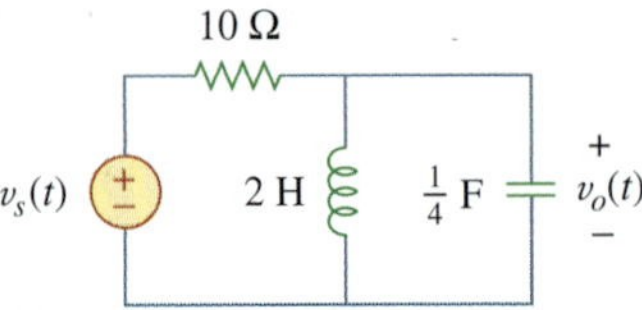

Figure 17.69
For Prob. 17.33.

17.34 Using Fig. 17.70, design a problem to help other students better understand circuit responses to a Fourier series.

e⁷d

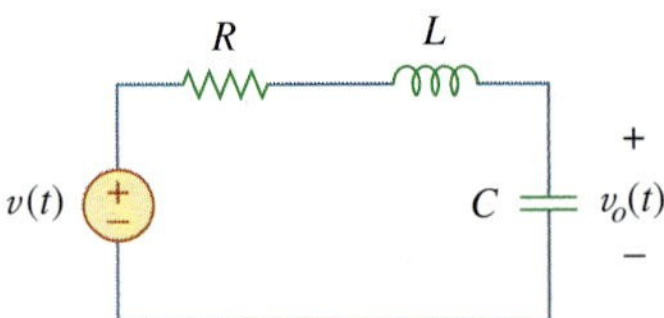

Figure 17.70
For Prob. 17.34.

17.35 If v_s in the circuit of Fig. 17.71 is the same as function $f_2(t)$ in Fig. 17.56(b), determine the dc component and the first three nonzero harmonics of $v_o(t)$.

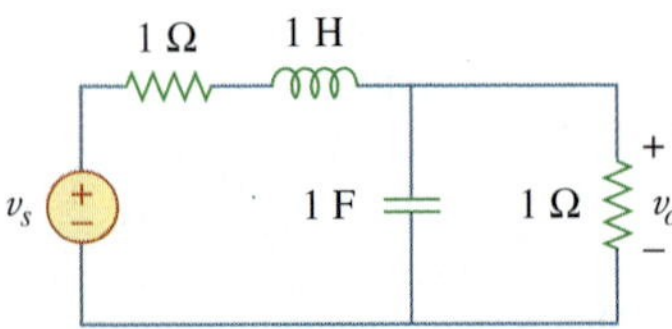

Figure 17.71
For Prob. 17.35.

***17.36** Find the response i_o for the circuit in Fig. 17.72(a), where $v_s(t)$ is shown in Fig. 17.72(b).

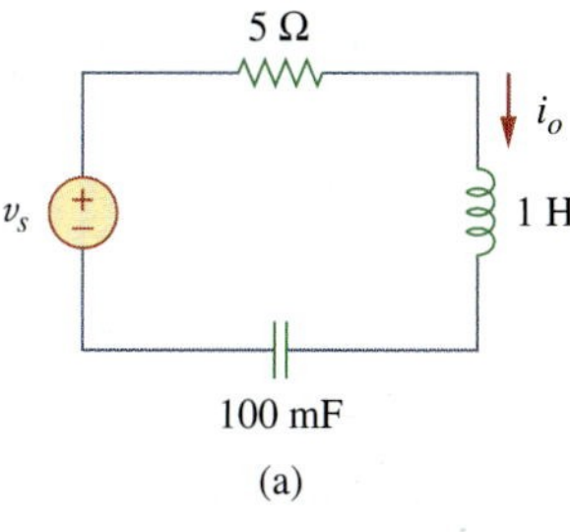

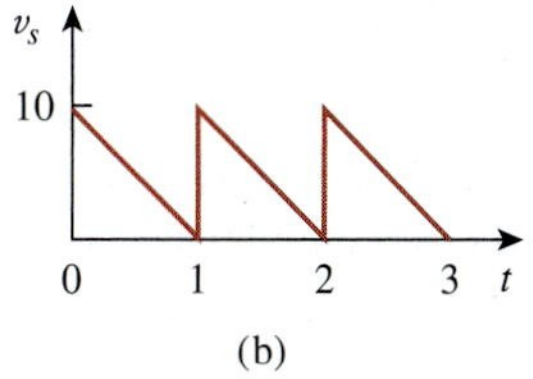

Figure 17.72
For Prob. 17.36.

17.37 If the periodic current waveform in Fig. 17.73(a) is applied to the circuit in Fig. 17.73(b), find v_o.

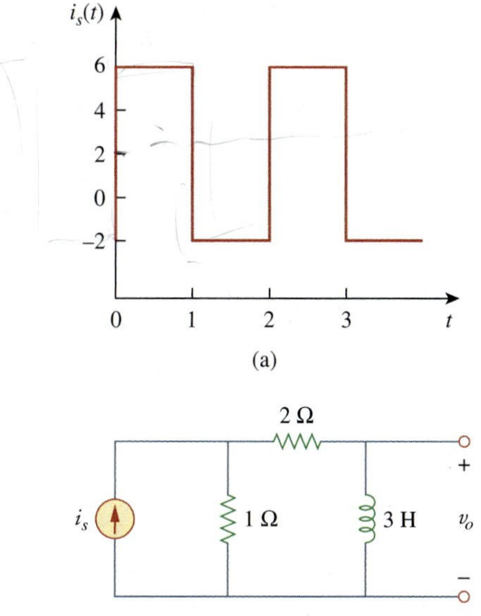

Figure 17.73
For Prob. 17.37.

17.38 If the square wave shown in Fig. 17.74(a) is applied to the circuit in Fig. 17.74(b), find the Fourier series for $v_o(t)$.

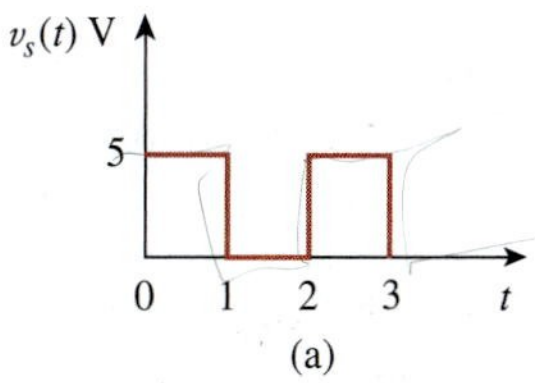

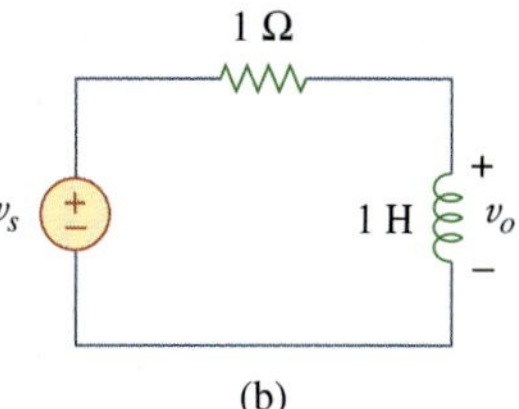

Figure 17.74
For Prob. 17.38.

17.39 If the periodic voltage in Fig. 17.75(a) is applied to the circuit in Fig. 17.75(b), find $i_o(t)$.

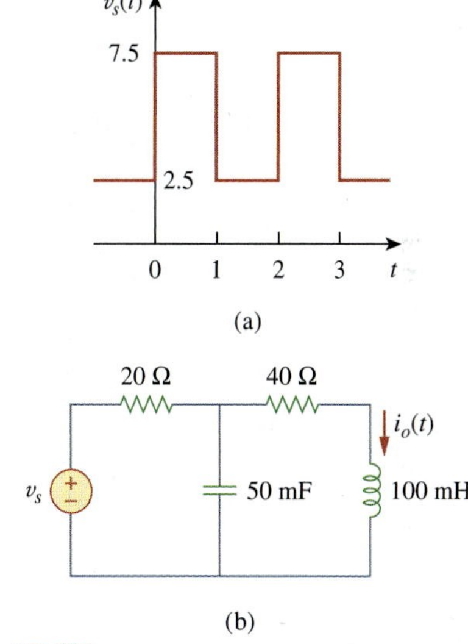

Figure 17.75
For Prob. 17.39.

***17.40** The signal in Fig. 17.76(a) is applied to the circuit in Fig. 17.76(b). Find $v_o(t)$.

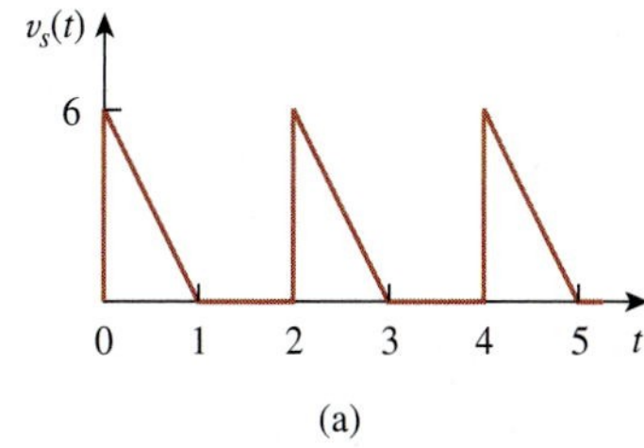

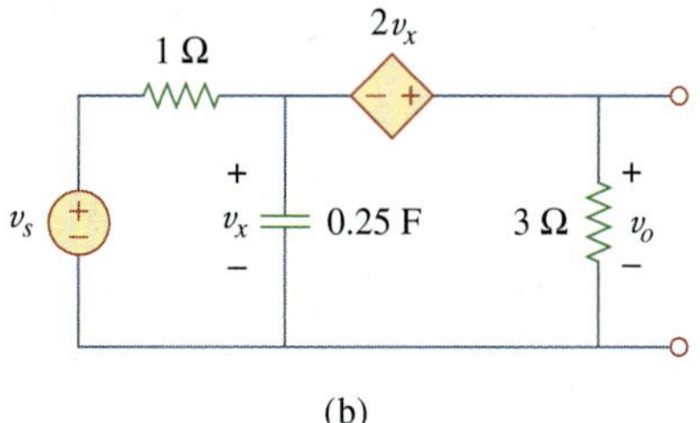

Figure 17.76
For Prob. 17.40.

17.41 The full-wave rectified sinusoidal voltage in Fig. 17.77(a) is applied to the lowpass filter in Fig. 17.77(b). Obtain the output voltage $v_o(t)$ of the filter.

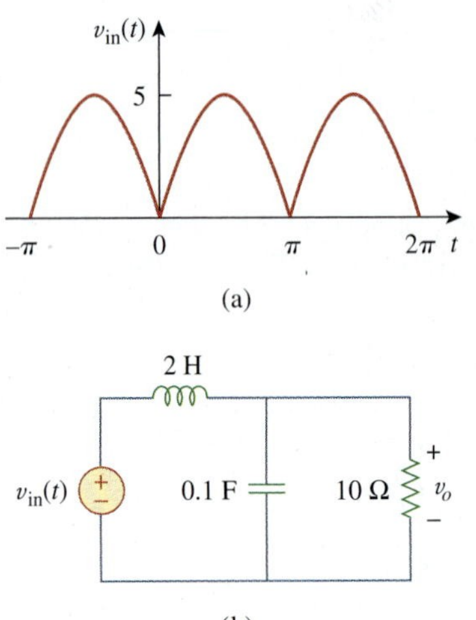

Figure 17.77
For Prob. 17.41.

17.42 The square wave in Fig. 17.78(a) is applied to the circuit in Fig. 17.78(b). Find the Fourier series of $v_o(t)$.

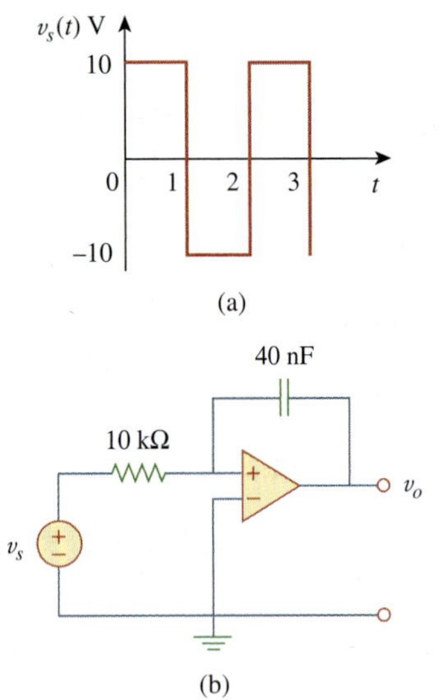

Figure 17.78
For Prob. 17.42.

Section 17.5 Average Power and RMS Values

17.43 The voltage across the terminals of a circuit is

$$v(t) = 30 + 20\cos(60\pi t + 45^\circ) + 10\cos(60\pi t - 45^\circ)\text{ V}$$

If the current entering the terminal at higher potential is

$$i(t) = 6 + 4\cos(60\pi t + 10^\circ) - 2\cos(120\pi t - 60^\circ)\text{ A}$$

find:

(a) the rms value of the voltage,

(b) the rms value of the current,

(c) the average power absorbed by the circuit.

***17.44** e⬈d Design a problem to help other students better understand how to find the rms voltage across and the rms current through an electrical element given a Fourier series for both the current and the voltage. In addition, have them calculate the average power delivered to the element and the power spectrum.

17.45 A series *RLC* circuit has $R = 10\ \Omega$, $L = 2$ mH, and $C = 40\ \mu$F. Determine the effective current and average power absorbed when the applied voltage is

$$v(t) = 100\cos 1000t + 50\cos 2000t + 25\cos 3000t\text{ V}$$

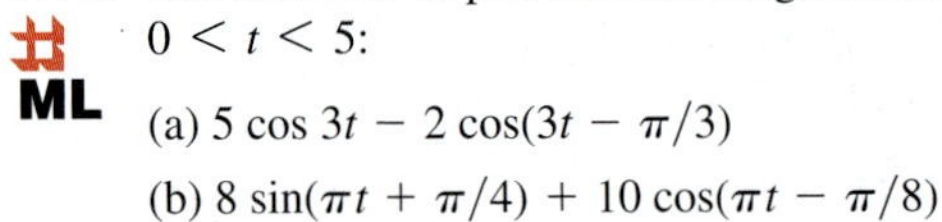

17.46 ML Use *MATLAB* to plot the following sinusoids for $0 < t < 5$:

(a) $5\cos 3t - 2\cos(3t - \pi/3)$

(b) $8\sin(\pi t + \pi/4) + 10\cos(\pi t - \pi/8)$

17.47 The periodic current waveform in Fig. 17.79 is applied across a 2-kΩ resistor. Find the percentage of the total average power dissipation caused by the dc component.

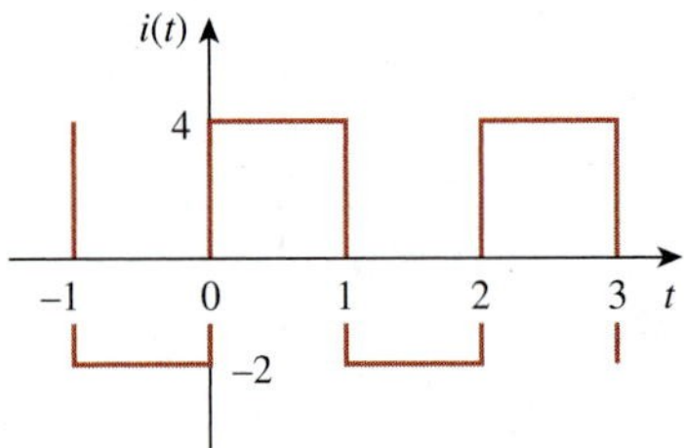

Figure 17.79
For Prob. 17.47.

17.48 For the circuit in Fig. 17.80,

$$i(t) = 20 + 16\cos(10t + 45^\circ) + 12\cos(20t - 60^\circ)\text{ mA}$$

(a) find $v(t)$, and

(b) calculate the average power dissipated in the resistor.

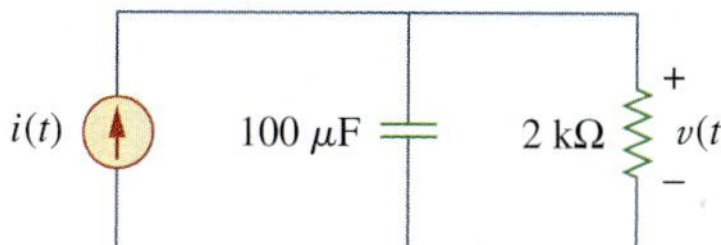

Figure 17.80
For Prob. 17.48.

17.49 (a) For the periodic waveform in Prob. 17.5, find the rms value.

(b) Use the first five harmonic terms of the Fourier series in Prob. 17.5 to determine the effective value of the signal.

(c) Calculate the percentage error in the estimated rms value of $z(t)$ if

$$\% \text{ error} = \left(\frac{\text{estimated value}}{\text{exact value}} - 1\right) \times 100$$

Section 17.6 Exponential Fourier Series

17.50 Obtain the exponential Fourier series for $f(t) = 2t$, $-1 < t < 1$, with $f(t + 2n) = f(t)$ for all integer values of n.

17.51 Design a problem to help other students better understand how to find the exponential Fourier series of a given periodic function.

17.52 Calculate the complex Fourier series for $f(t) = e^t$, $-\pi < t < \pi$, with $f(t + 2\pi n) = f(t)$ for all integer values of n.

17.53 Find the complex Fourier series for $f(t) = e^{-t}$, $0 < t < 1$, with $f(t + n) = f(t)$ for all integer values of n.

17.54 Find the exponential Fourier series for the function in Fig. 17.81.

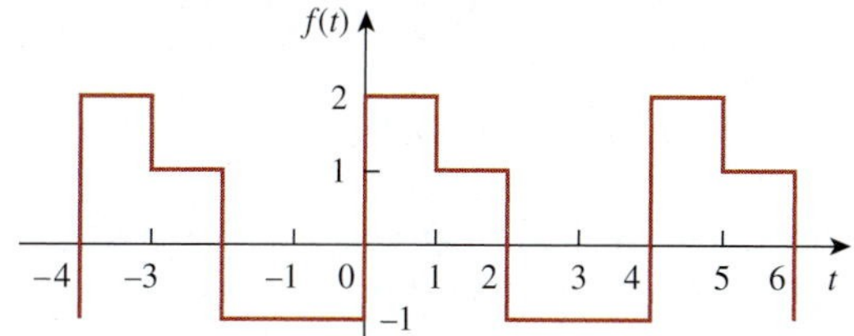

Figure 17.81
For Prob. 17.54.

17.55 Obtain the exponential Fourier series expansion of the half-wave rectified sinusoidal current of Fig. 17.82.

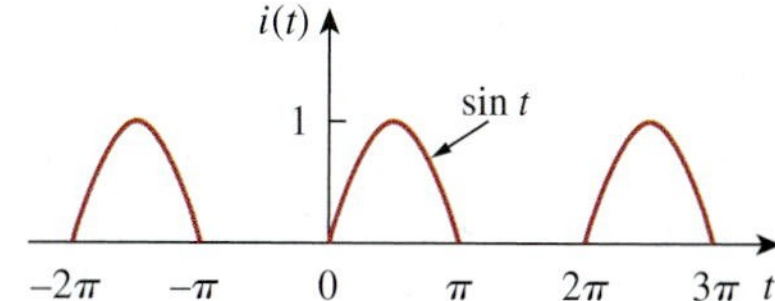

Figure 17.82
For Prob. 17.55.

17.56 The Fourier series trigonometric representation of a periodic function is

$$f(t) = 10 + \sum_{n=1}^{\infty}\left(\frac{1}{n^2 + 1}\cos n\pi t + \frac{n}{n^2 + 1}\sin n\pi t\right)$$

Find the exponential Fourier series representation of $f(t)$.

17.57 The coefficients of the trigonometric Fourier series representation of a function are:

$$b_n = 0, \qquad a_n = \frac{6}{n^3 - 2}, \qquad n = 0, 1, 2, \ldots$$

If $\omega_n = 50n$, find the exponential Fourier series for the function.

17.58 Find the exponential Fourier series of a function that has the following trigonometric Fourier series coefficients:

$$a_0 = \frac{\pi}{4}, \qquad b_n = \frac{(-1)^n}{n}, \qquad a_n = \frac{(-1)^n - 1}{\pi n^2}$$

Take $T = 2\pi$.

17.59 The complex Fourier series of the function in Fig. 17.83(a) is

$$f(t) = \frac{1}{2} - \sum_{n=-\infty}^{\infty}\frac{je^{-j(2n+1)t}}{(2n + 1)\pi}$$

Find the complex Fourier series of the function $h(t)$ in Fig. 17.83(b).

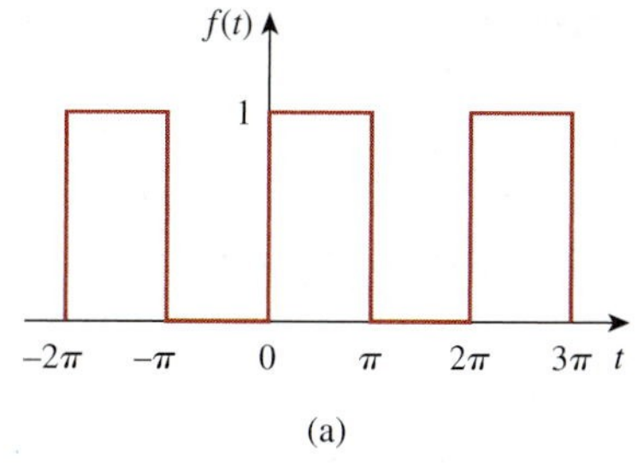

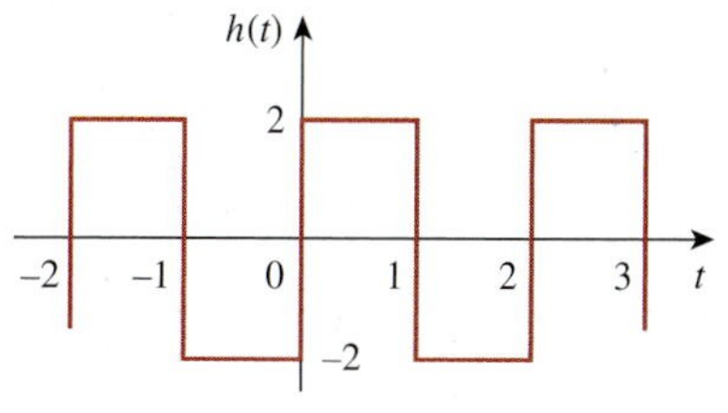

Figure 17.83
For Prob.17.59.

17.60 Obtain the complex Fourier coefficients of the signal in Fig. 17.62.

17.61 The spectra of the Fourier series of a function are shown in Fig. 17.84. (a) Obtain the trigonometric Fourier series. (b) Calculate the rms value of the function.

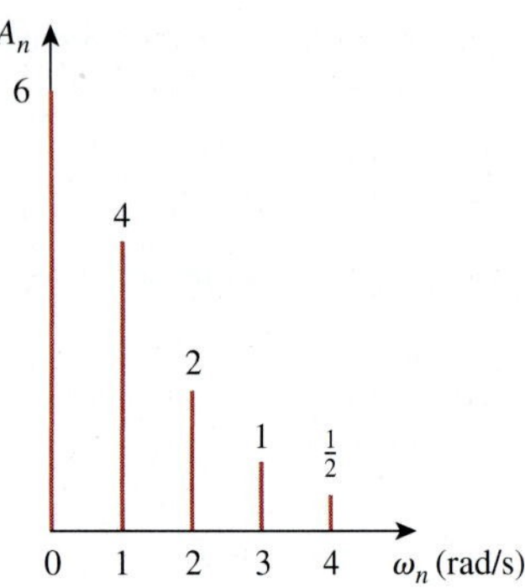

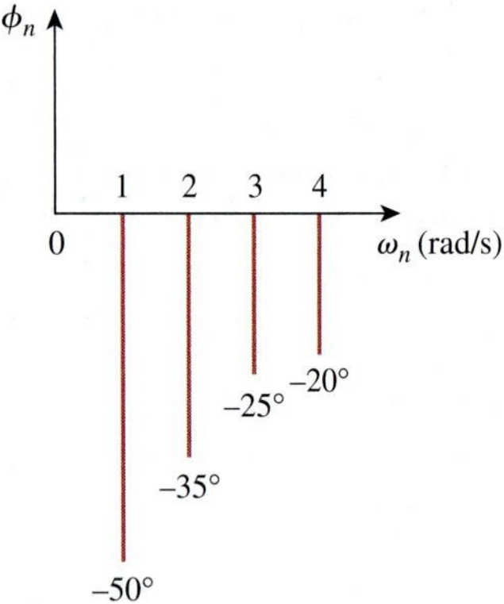

Figure 17.84
For Prob. 17.61.

17.62 The amplitude and phase spectra of a truncated Fourier series are shown in Fig. 17.85.

(a) Find an expression for the periodic voltage using the amplitude-phase form. See Eq. (17.10).

(b) Is the voltage an odd or even function of t?

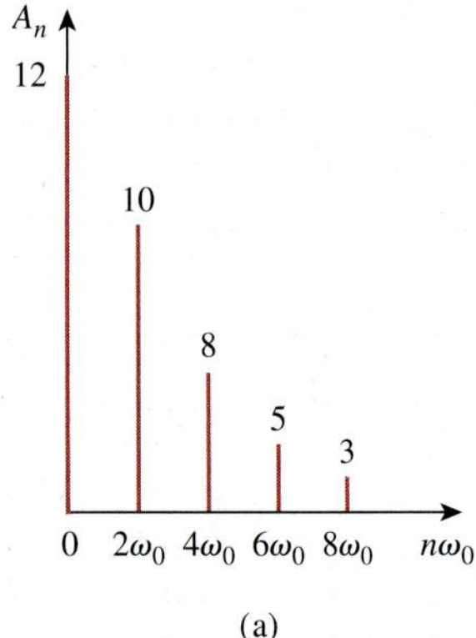

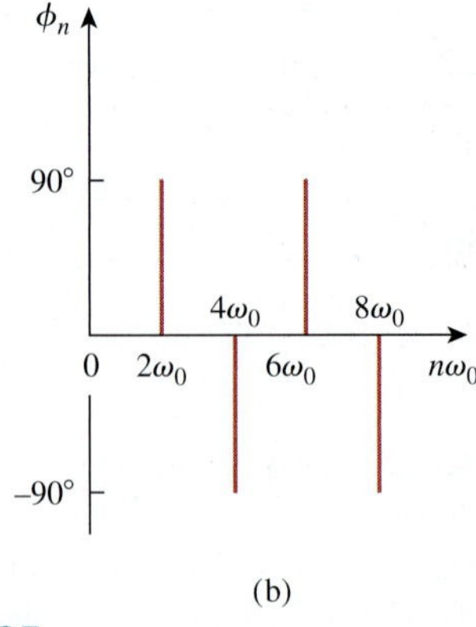

Figure 17.85
For Prob. 17.62.

17.63 Plot the amplitude spectrum for the signal $f_2(t)$ in Fig. 17.56(b). Consider the first five terms.

17.64  Design a problem to help other students better understand the amplitude and phase spectra of a given Fourier series.

17.65 Given that

$$f(t) = \sum_{\substack{n=1 \\ n=\text{odd}}}^{\infty} \left(\frac{20}{n^2\pi^2} \cos 2nt - \frac{3}{n\pi} \sin 2nt \right)$$

plot the first five terms of the amplitude and phase spectra for the function.

Section 17.7 Fourier Analysis with *PSpice*

ps

17.66 Determine the Fourier coefficients for the waveform in Fig. 17.48 using *PSpice*.

17.67 Calculate the Fourier coefficients of the signal in Fig. 17.58 using *PSpice*.

17.68 Use *PSpice* to find the Fourier components of the signal in Prob. 17.7.

17.69 Use *PSpice* to obtain the Fourier coefficients of the waveform in Fig. 17.55(a).

17.70 **e⬈d** Design a problem to help other students better understand how to use *PSpice* to solve circuit problems with periodic inputs.

17.71 Use *PSpice* to solve Prob. 17.39.

Section 17.8 Applications

17.72 The signal displayed by a medical device can be approximated by the waveform shown in Fig. 17.86. Find the Fourier series representation of the signal.

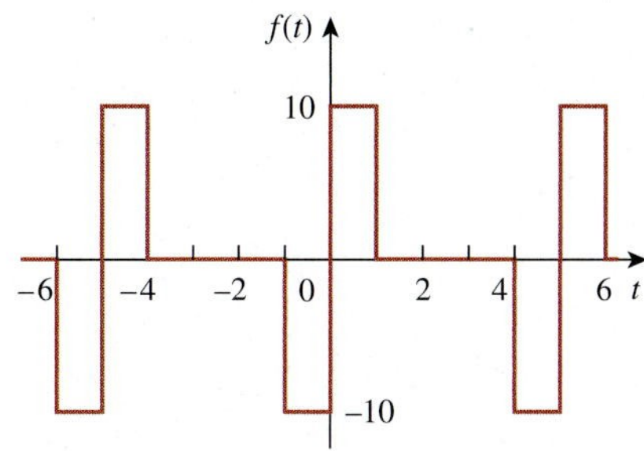

Figure 17.86
For Prob. 17.72.

17.73 A spectrum analyzer indicates that a signal is made up of three components only: 640 kHz at 2 V, 644 kHz at 1 V, 636 kHz at 1 V. If the signal is applied across a 10-Ω resistor, what is the average power absorbed by the resistor?

17.74 A certain band-limited periodic current has only three frequencies in its Fourier series representation: dc, 50 Hz, and 100 Hz. The current may be represented as

$$i(t) = 4 + 6 \sin 100\pi t + 8 \cos 100\pi t - 3 \sin 200\pi t - 4 \cos 200\pi t \text{ A}$$

(a) Express $i(t)$ in amplitude-phase form.

(b) If $i(t)$ flows through a 2-Ω resistor, how many watts of average power will be dissipated?

17.75 **e2d** Design a lowpass *RC* filter with a resistance $R = 2\ k\Omega$. The input to the filter is a periodic rectangular pulse train (see Table 17.3) with $A = 1$ V, $T = 10$ ms, and $\tau = 1$ ms. Select C such that the dc component of the output is 50 times greater than the fundamental component of the output.

17.76 A periodic signal given by $v_s(t) = 10$ V for $0 < t < 1$ and 0 V for $1 < t < 2$ is applied to the highpass filter in Fig. 17.87. Determine the value of R such that the output signal $v_o(t)$ has an average power of at least 70 percent of the average power of the input signal.

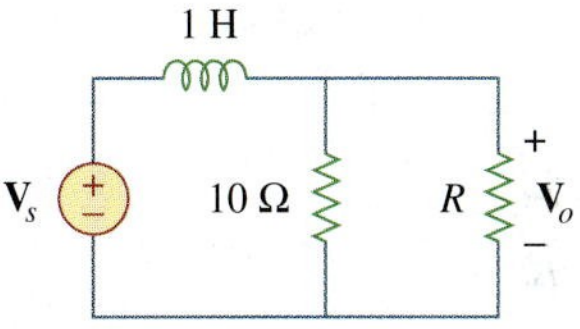

Figure 17.87
For Prob. 17.76.

Comprehensive Problems

17.77 The voltage across a device is given by

$$v(t) = -2 + 10 \cos 4t + 8 \cos 6t + 6 \cos 8t - 5 \sin 4t - 3 \sin 6t - \sin 8t \text{ V}$$

Find:

(a) the period of $v(t)$,

(b) the average value of $v(t)$,

(c) the effective value of $v(t)$.

17.78 A certain band-limited periodic voltage has only three harmonics in its Fourier series representation. The harmonics have the following rms values: fundamental 40 V, third harmonic 20 V, fifth harmonic 10 V.

(a) If the voltage is applied across a 5-Ω resistor, find the average power dissipated by the resistor.

(b) If a dc component is added to the periodic voltage and the measured power dissipated increases by 5 percent, determine the value of the dc component added.

17.79 Write a program to compute the Fourier coefficients (up to the 10th harmonic) of the square wave in Table 17.3 with $A = 10$ and $T = 2$.

17.80 Write a computer program to calculate the exponential Fourier series of the half-wave rectified sinusoidal current of Fig. 17.82. Consider terms up to the 10th harmonic.

17.81 Consider the full-wave rectified sinusoidal current in Table 17.3. Assume that the current is passed through a 1-Ω resistor.

(a) Find the average power absorbed by the resistor.

(b) Obtain c_n for $n = 1, 2, 3$, and 4.

(c) What fraction of the total power is carried by the dc component?

(d) What fraction of the total power is carried by the second harmonic ($n = 2$)?

17.82 A band-limited voltage signal is found to have the complex Fourier coefficients presented in the table below. Calculate the average power that the signal would supply a 4-Ω resistor.

$n\omega_0$	$\lvert c_n \rvert$	θ_n
0	10.0	0°
ω	8.5	15°
2ω	4.2	30°
3ω	2.1	45°
4ω	0.5	60°
5ω	0.2	75°

chapter

18

Fourier Transform

Planning is doing today to make us better tomorrow because the future belongs to those who make the hard decisions today.

—*Business Week*

Enhancing Your Skills and Your Career

Career in Communications Systems

Communications systems apply the principles of circuit analysis. A communication system is designed to convey information from a source (the transmitter) to a destination (the receiver) via a channel (the propagation medium). Communications engineers design systems for transmitting and receiving information. The information can be in the form of voice, data, or video.

We live in the information age—news, weather, sports, shopping, financial, business inventory, and other sources make information available to us almost instantly via communications systems. Some obvious examples of communications systems are the telephone network, mobile cellular telephones, radio, cable TV, satellite TV, fax, and radar. Mobile radio, used by police and fire departments, aircraft, and various businesses is another example.

Photo by Charles Alexander

The field of communications is perhaps the fastest growing area in electrical engineering. The merging of the communications field with computer technology in recent years has led to digital data communications networks such as local area networks, metropolitan area networks, and broadband integrated services digital networks. For example, the Internet (the "information superhighway") allows educators, business people, and others to send electronic mail from their computers worldwide, log onto remote databases, and transfer files. The Internet has hit the world like a tidal wave and is drastically changing the way people do business, communicate, and get information. This trend will continue.

A communications engineer designs systems that provide high-quality information services. The systems include hardware for generating, transmitting, and receiving information signals. Communications engineers are employed in numerous communications industries and places where communications systems are routinely used. More and more government agencies, academic departments, and businesses are demanding faster and more accurate transmission of information. To meet these needs, communications engineers are in high demand. Therefore, the future is in communications and every electrical engineer must prepare accordingly.

18.1 Introduction

Fourier series enable us to represent a periodic function as a sum of sinusoids and to obtain the frequency spectrum from the series. The Fourier transform allows us to extend the concept of a frequency spectrum to nonperiodic functions. The transform assumes that a nonperiodic function is a periodic function with an infinite period. Thus, the Fourier transform is an integral representation of a nonperiodic function that is analogous to a Fourier series representation of a periodic function.

The Fourier transform is an *integral transform* like the Laplace transform. It transforms a function in the time domain into the frequency domain. The Fourier transform is very useful in communications systems and digital signal processing, in situations where the Laplace transform does not apply. While the Laplace transform can only handle circuits with inputs for $t > 0$ with initial conditions, the Fourier transform can handle circuits with inputs for $t < 0$ as well as those for $t > 0$.

We begin by using a Fourier series as a stepping stone in defining the Fourier transform. Then we develop some of the properties of the Fourier transform. Next, we apply the Fourier transform in analyzing circuits. We discuss Parseval's theorem, compare the Laplace and Fourier transforms, and see how the Fourier transform is applied in amplitude modulation and sampling.

18.2 Definition of the Fourier Transform

We saw in the previous chapter that a nonsinusoidal periodic function can be represented by a Fourier series, provided that it satisfies the Dirichlet conditions. What happens if a function is not periodic? Unfortunately, there are many important nonperiodic functions—such as a unit step or an exponential function—that we cannot represent by a Fourier series. As we shall see, the Fourier transform allows a transformation from the time to the frequency domain, even if the function is not periodic.

Suppose we want to find the Fourier transform of a nonperiodic function $p(t)$, shown in Fig. 18.1(a). We consider a periodic function $f(t)$ whose shape over one period is the same as $p(t)$, as shown in Fig. 18.1(b). If we let the period $T \rightarrow \infty$, only a single pulse of width τ [the desired nonperiodic function in Fig. 18.1(a)] remains, because the adjacent pulses have been moved to infinity. Thus, the function $f(t)$ is no longer periodic. In other words, $f(t) = p(t)$ as $T \rightarrow \infty$. It is interesting to consider the spectrum of $f(t)$ for $A = 10$ and $\tau = 0.2$ (see Section 17.6). Figure 18.2 shows the effect of increasing T on the spectrum. First, we notice that the general shape of the spectrum remains the same, and the frequency at which the envelope first becomes zero remains the same. However, the amplitude of the spectrum and the spacing between adjacent components both decrease, while the number of harmonics increases. Thus, over a range of frequencies, the sum of the amplitudes of the harmonics remains almost constant. As the total "strength" or energy of the

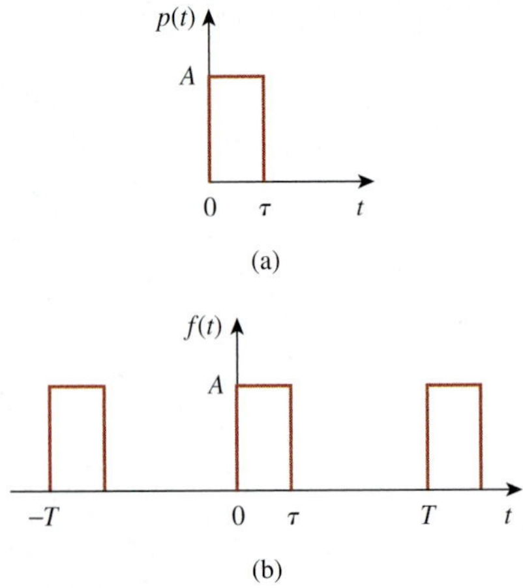

Figure 18.1
(a) A nonperiodic function, (b) increasing T to infinity makes $f(t)$ become the nonperiodic function in (a).

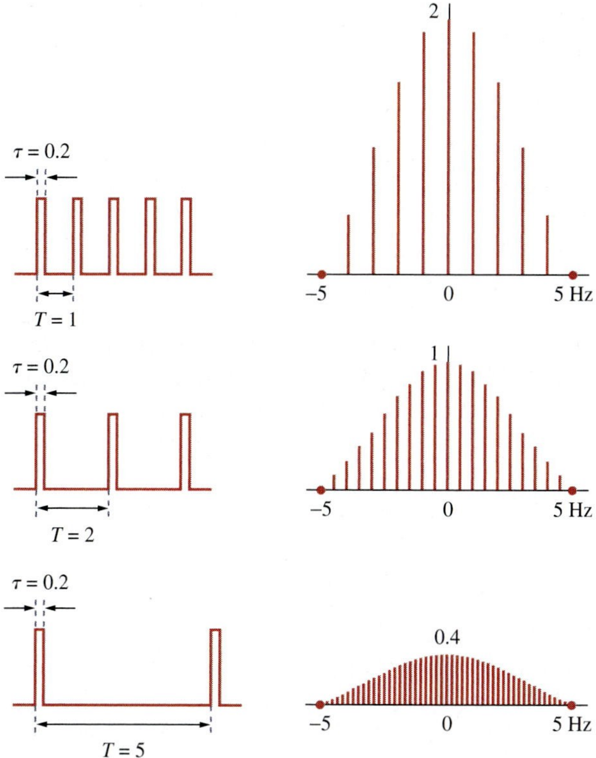

Figure 18.2
Effect of increasing T on the spectrum of the periodic pulse trains in Fig. 18.1(b).
L. Balmer, *Signals and Systems: An Introduction* [London: Prentice-Hall, 1991], p. 229.

components within a band must remain unchanged, the amplitudes of the harmonics must decrease as T increases. Since $f = 1/T$, as T increases, f or ω decreases, so that the discrete spectrum ultimately becomes continuous.

To further understand this connection between a nonperiodic function and its periodic counterpart, consider the exponential form of a Fourier series in Eq. (17.58), namely,

$$f(t) = \sum_{n=-\infty}^{\infty} c_n e^{jn\omega_0 t} \tag{18.1}$$

where

$$c_n = \frac{1}{T}\int_{-T/2}^{T/2} f(t)e^{-jn\omega_0 t}\,dt \tag{18.2}$$

The fundamental frequency is

$$\omega_0 = \frac{2\pi}{T} \tag{18.3}$$

and the spacing between adjacent harmonics is

$$\Delta\omega = (n+1)\omega_0 - n\omega_0 = \omega_0 = \frac{2\pi}{T} \tag{18.4}$$

Substituting Eq. (18.2) into Eq. (18.1) gives

$$\begin{aligned} f(t) &= \sum_{n=-\infty}^{\infty} \left[\frac{1}{T} \int_{-T/2}^{T/2} f(t) e^{-jn\omega_0 t}\, dt \right] e^{jn\omega_0 t} \\ &= \sum_{n=-\infty}^{\infty} \left[\frac{\Delta\omega}{2\pi} \int_{-T/2}^{T/2} f(t) e^{-jn\omega_0 t}\, dt \right] e^{jn\omega_0 t} \\ &= \frac{1}{2\pi} \sum_{n=-\infty}^{\infty} \left[\int_{-T/2}^{T/2} f(t) e^{-jn\omega_0 t}\, dt \right] \Delta\omega e^{jn\omega_0 t} \end{aligned} \tag{18.5}$$

If we let $T \to \infty$, the summation becomes integration, the incremental spacing $\Delta\omega$ becomes the differential separation $d\omega$, and the discrete harmonic frequency $n\omega_0$ becomes a continuous frequency ω. Thus, as $T \to \infty$,

$$\begin{aligned} \sum_{n=-\infty}^{\infty} \quad &\Rightarrow \quad \int_{-\infty}^{\infty} \\ \Delta\omega \quad &\Rightarrow \quad d\omega \\ n\omega_0 \quad &\Rightarrow \quad \omega \end{aligned} \tag{18.6}$$

so that Eq. (18.5) becomes

$$f(t) = \frac{1}{2\pi} \int_{-\infty}^{\infty} \left[\int_{-\infty}^{\infty} f(t) e^{-j\omega t}\, dt \right] e^{j\omega t}\, d\omega \tag{18.7}$$

Some authors use $F(j\omega)$ instead of $F(\omega)$ to represent the Fourier transform.

The term in the brackets is known as the *Fourier transform* of $f(t)$ and is represented by $F(\omega)$. Thus,

$$F(\omega) = \mathcal{F}[f(t)] = \int_{-\infty}^{\infty} f(t) e^{-j\omega t}\, dt \tag{18.8}$$

where $\mathcal{F}$ is the Fourier transform operator. It is evident from Eq. (18.8) that:

The **Fourier transform** is an integral transformation of $f(t)$ from the time domain to the frequency domain.

In general, $F(\omega)$ is a complex function; its magnitude is called the *amplitude spectrum*, while its phase is called the *phase spectrum*. Thus $F(\omega)$ is the *spectrum*.

Equation (18.7) can be written in terms of $F(\omega)$, and we obtain the *inverse Fourier transform* as

$$f(t) = \mathcal{F}^{-1}[F(\omega)] = \frac{1}{2\pi} \int_{-\infty}^{\infty} F(\omega) e^{j\omega t}\, d\omega \tag{18.9}$$

The function $f(t)$ and its transform $F(\omega)$ form the Fourier transform pairs:

$$f(t) \quad \Leftrightarrow \quad F(\omega) \tag{18.10}$$

since one can be derived from the other.

The Fourier transform $F(\omega)$ exists when the Fourier integral in Eq. (18.8) converges. A sufficient but not necessary condition that $f(t)$ has a Fourier transform is that it be completely integrable in the sense that

$$\int_{-\infty}^{\infty} |f(t)|\, dt < \infty \tag{18.11}$$

For example, the Fourier transform of the unit ramp function $tu(t)$ does not exist, because the function does not satisfy the condition above.

To avoid the complex algebra that explicitly appears in the Fourier transform, it is sometimes expedient to temporarily replace $j\omega$ with s and then replace s with $j\omega$ at the end.

Example 18.1

Find the Fourier transform of the following functions: (a) $\delta(t - t_0)$, (b) $e^{j\omega_0 t}$, (c) $\cos \omega_0 t$.

Solution:

(a) For the impulse function,

$$F(\omega) = \mathcal{F}[\delta(t - t_0)] = \int_{-\infty}^{\infty} \delta(t - t_0) e^{-j\omega t}\, dt = e^{-j\omega t_0} \tag{18.1.1}$$

where the sifting property of the impulse function in Eq. (7.32) has been applied. For the special case $t_0 = 0$, we obtain

$$\mathcal{F}[\delta(t)] = 1 \tag{18.1.2}$$

This shows that the magnitude of the spectrum of the impulse function is constant; that is, all frequencies are equally represented in the impulse function.

(b) We can find the Fourier transform of $e^{j\omega_0 t}$ in two ways. If we let

$$F(\omega) = \delta(\omega - \omega_0)$$

then we can find $f(t)$ using Eq. (18.9), writing

$$f(t) = \frac{1}{2\pi} \int_{-\infty}^{\infty} \delta(\omega - \omega_0) e^{j\omega t}\, d\omega$$

Using the sifting property of the impulse function gives

$$f(t) = \frac{1}{2\pi} e^{j\omega_0 t}$$

Since $F(\omega)$ and $f(t)$ constitute a Fourier transform pair, so too must $2\pi\delta(\omega - \omega_0)$ and $e^{j\omega_0 t}$,

$$\mathcal{F}[e^{j\omega_0 t}] = 2\pi\delta(\omega - \omega_0) \tag{18.1.3}$$

Alternatively, from Eq. (18.1.2),

$$\delta(t) = \mathcal{F}^{-1}[1]$$

Using the inverse Fourier transform formula in Eq. (18.9),

$$\delta(t) = \mathcal{F}^{-1}[1] = \frac{1}{2\pi} \int_{-\infty}^{\infty} 1 e^{j\omega t}\, d\omega$$

or

$$\int_{-\infty}^{\infty} e^{j\omega t}\, d\omega = 2\pi\delta(t) \tag{18.1.4}$$

Interchanging variables t and ω results in

$$\int_{-\infty}^{\infty} e^{j\omega t}\, dt = 2\pi\delta(\omega) \tag{18.1.5}$$

Using this result, the Fourier transform of the given function is

$$\mathcal{F}[e^{j\omega_0 t}] = \int_{-\infty}^{\infty} e^{j\omega_0 t} e^{-j\omega t}\, dt = \int_{-\infty}^{\infty} e^{j(\omega_0-\omega)}\, dt = 2\pi\delta(\omega_0 - \omega)$$

Since the impulse function is an even function, with $\delta(\omega_0 - \omega) = \delta(\omega - \omega_0)$,

$$\mathcal{F}[e^{j\omega_0 t}] = 2\pi\delta(\omega - \omega_0) \tag{18.1.6}$$

By simply changing the sign of ω_0, we readily obtain

$$\mathcal{F}[e^{-j\omega_0 t}] = 2\pi\delta(\omega + \omega_0) \tag{18.1.7}$$

Also, by setting $\omega_0 = 0$,

$$\mathcal{F}[1] = 2\pi\delta(\omega) \tag{18.1.8}$$

(c) By using the result in Eqs. (18.1.6) and (18.1.7), we get

$$\begin{aligned}\mathcal{F}[\cos\omega_0 t] &= \mathcal{F}\left[\frac{e^{j\omega_0 t} + e^{-j\omega_0 t}}{2}\right]\\ &= \frac{1}{2}\mathcal{F}[e^{j\omega_0 t}] + \frac{1}{2}\mathcal{F}[e^{-j\omega_0 t}]\\ &= \pi\delta(\omega - \omega_0) + \pi\delta(\omega + \omega_0)\end{aligned} \tag{18.1.9}$$

The Fourier transform of the cosine signal is shown in Fig. 18.3.

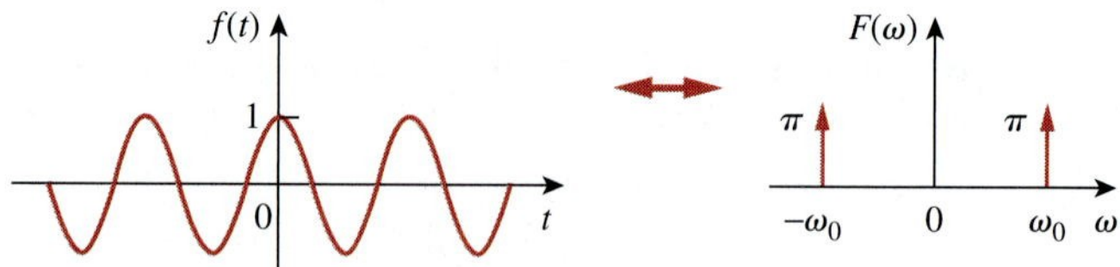

Figure 18.3
Fourier transform of $f(t) = \cos\omega_0 t$.

Practice Problem 18.1

Determine the Fourier transforms of the following functions: (a) gate function $g(t) = 4u(t-1) - 8u(t-2)$, (b) $4\delta(t+2)$, (c) $10\sin\omega_0 t$.

Answer: (a) $(4e^{-j\omega} - 8e^{-j2\omega})/j\omega$, (b) $4e^{j2\omega}$, (c) $j10\pi[\delta(\omega+\omega_0) - \delta(\omega-\omega_0)]$.

Example 18.2

Derive the Fourier transform of a single rectangular pulse of width τ and height A, shown in Fig. 18.4.

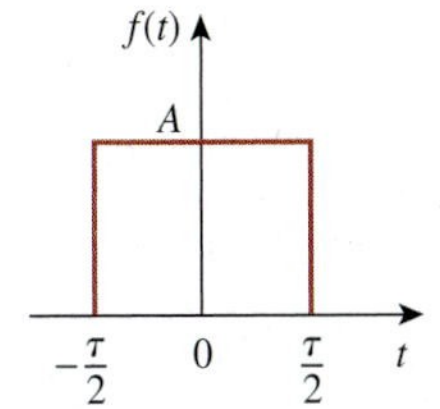

Figure 18.4
A rectangular pulse; for Example 18.2.

Solution:

$$F(\omega) = \int_{-\tau/2}^{\tau/2} Ae^{-j\omega t}\,dt = -\frac{A}{j\omega}e^{-j\omega t}\Big|_{-\tau/2}^{\tau/2}$$

$$= \frac{2A}{\omega}\left(\frac{e^{j\omega\tau/2} - e^{-j\omega\tau/2}}{2j}\right)$$

$$= A\tau\frac{\sin\omega\tau/2}{\omega\tau/2} = A\tau\,\mathrm{sinc}\,\frac{\omega\tau}{2}$$

If we make $A = 10$ and $\tau = 2$ as in Fig. 17.27 (like in Section 17.6), then

$$F(\omega) = 20\,\mathrm{sinc}\,\omega$$

whose amplitude spectrum is shown in Fig. 18.5. Comparing Fig. 18.4 with the frequency spectrum of the rectangular pulses in Fig. 17.28, we notice that the spectrum in Fig. 17.28 is discrete and its envelope has the same shape as the Fourier transform of a single rectangular pulse.

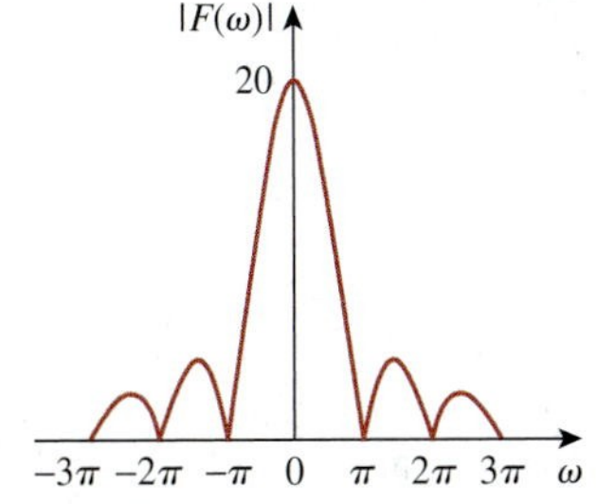

Figure 18.5
Amplitude spectrum of the rectangular pulse in Fig. 18.4: for Example 18.2.

Practice Problem 18.2

Obtain the Fourier transform of the function in Fig. 18.6.

Answer: $\dfrac{10(\cos\omega - 1)}{j\omega}$.

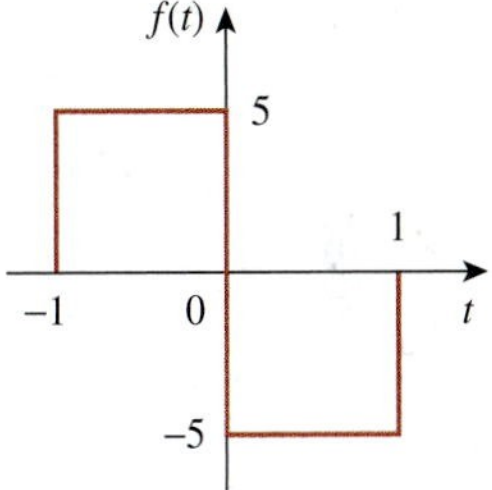

Figure 18.6
For Practice Prob. 18.2.

Example 18.3

Obtain the Fourier transform of the "switched-on" exponential function shown in Fig. 18.7.

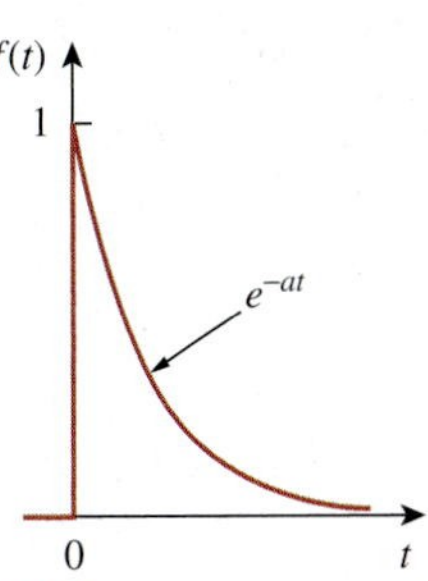

Figure 18.7
For Example 18.3.

Solution:
From Fig. 18.7,

$$f(t) = e^{-at}u(t) = \begin{cases} e^{-at}, & t > 0 \\ 0, & t < 0 \end{cases}$$

Hence,

$$F(\omega) = \int_{-\infty}^{\infty} f(t)e^{-j\omega t}\,dt = \int_{0}^{\infty} e^{-at}e^{-j\omega t}\,dt = \int_{0}^{\infty} e^{-(a+j\omega)t}\,dt$$

$$= \frac{-1}{a + j\omega}e^{-(a+j\omega)t}\Big|_0^{\infty} = \frac{1}{a + j\omega}$$

Practice Problem 18.3

Determine the Fourier transform of the "switched-off" exponential function in Fig. 18.8.

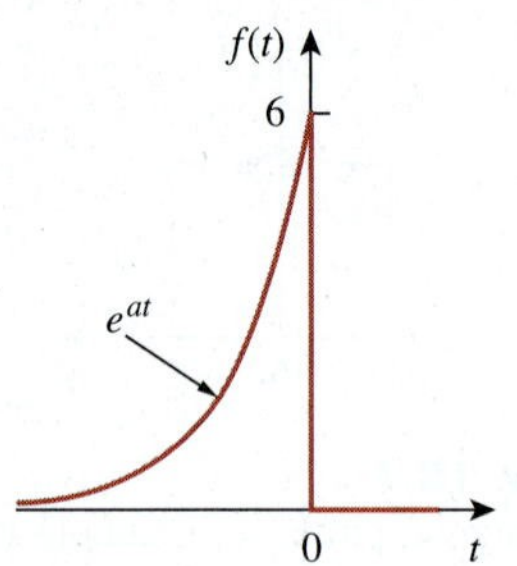

Figure 18.8
For Practice Prob. 18.3.

Answer: $\dfrac{6}{a - j\omega}$.

18.3 Properties of the Fourier Transform

We now develop some properties of the Fourier transform that are useful in finding the transforms of complicated functions from the transforms of simple functions. For each property, we will first state and derive it, and then illustrate it with some examples.

Linearity

If $F_1(\omega)$ and $F_2(\omega)$ are the Fourier transforms of $f_1(t)$ and $f_2(t)$, respectively, then

$$\mathcal{F}[a_1 f_1(t) + a_2 f_2(t)] = a_1 F_1(\omega) + a_2 F_2(\omega) \tag{18.12}$$

where a_1 and a_2 are constants. This property simply states that the Fourier transform of a linear combination of functions is the same as the linear combination of the transforms of the individual functions. The proof of the linearity property in Eq. (18.12) is straightforward. By definition,

$$\begin{aligned}\mathcal{F}[a_1 f_1(t) + a_2 f_2(t)] &= \int_{-\infty}^{\infty} [a_1 f_1(t) + a_2 f_2(t)] e^{-j\omega t}\, dt \\ &= \int_{-\infty}^{\infty} a_1 f_1(t) e^{-j\omega t}\, dt + \int_{-\infty}^{\infty} a_2 f_2(t) e^{-j\omega t}\, dt \\ &= a_1 F_1(\omega) + a_2 F_2(\omega)\end{aligned} \tag{18.13}$$

For example, $\sin \omega_0 t = \frac{1}{2j}(e^{j\omega_0 t} - e^{-j\omega_0 t})$. Using the linearity property,

$$\begin{aligned}F[\sin \omega_0 t] &= \frac{1}{2j}[\mathcal{F}(e^{j\omega_0 t}) - \mathcal{F}(e^{-j\omega_0 t})] \\ &= \frac{\pi}{j}[\delta(\omega - \omega_0) - \delta(\omega + \omega_0)] \\ &= j\pi[\delta(\omega + \omega_0) - \delta(\omega - \omega_0)\end{aligned} \tag{18.14}$$

Time Scaling

If $F(\omega) = \mathcal{F}[f(t)]$, then

$$\mathcal{F}[f(at)] = \frac{1}{|a|} F\left(\frac{\omega}{a}\right) \tag{18.15}$$

where a is a constant. Equation (18.15) shows that time expansion ($|a| > 1$) corresponds to frequency compression, or conversely, time compression ($|a| < 1$) implies frequency expansion. The proof of the time-scaling property proceeds as follows.

$$\mathcal{F}[f(at)] = \int_{-\infty}^{\infty} f(at)e^{-j\omega t}\,dt \tag{18.16}$$

If we let $x = at$, so that $dx = a\,dt$, then

$$\mathcal{F}[f(at)] = \int_{-\infty}^{\infty} f(x)e^{-j\omega x/a}\,\frac{dx}{a} = \frac{1}{a}F\left(\frac{\omega}{a}\right) \tag{18.17}$$

For example, for the rectangular pulse $p(t)$ in Example 18.2,

$$\mathcal{F}[p(t)] = A\tau \operatorname{sinc}\frac{\omega\tau}{2} \tag{18.18a}$$

Using Eq. (18.15),

$$\mathcal{F}[p(2t)] = \frac{A\tau}{2} \operatorname{sinc}\frac{\omega\tau}{4} \tag{18.18b}$$

It may be helpful to plot $p(t)$ and $p(2t)$ and their Fourier transforms. Since

$$p(t) = \begin{cases} A, & -\dfrac{\tau}{2} < t < \dfrac{\tau}{2} \\ 0, & \text{otherwise} \end{cases} \tag{18.19a}$$

then replacing every t with $2t$ gives

$$p(2t) = \begin{cases} A, & -\dfrac{\tau}{2} < 2t < \dfrac{\tau}{2} \\ 0, & \text{otherwise} \end{cases} = \begin{cases} A, & -\dfrac{\tau}{4} < t < \dfrac{\tau}{4} \\ 0, & \text{otherwise} \end{cases} \tag{18.19b}$$

showing that $p(2t)$ is time compressed, as shown in Fig. 18.9(b). To plot both Fourier transforms in Eq. (18.18), we recall that the sinc function has zeros when its argument is $n\pi$, where n is an integer. Hence, for the transform of $p(t)$ in Eq. (18.18a), $\omega\tau/2 = 2\pi f\tau/2 = n\pi \rightarrow f = n/\tau$, and for the transform of $p(2t)$ in Eq. (18.18b), $\omega\tau/4 = 2\pi f\tau/4 = n\pi \rightarrow f = 2n/\tau$. The plots of the Fourier transforms are shown in Fig. 18.9, which shows that time compression corresponds with frequency expansion. We should expect this intuitively, because when the signal is squashed in time, we expect it to change more rapidly, thereby causing higher-frequency components to exist.

Time Shifting

If $F(\omega) = \mathcal{F}[f(t)]$, then

$$\boxed{\mathcal{F}[f(t - t_0)] = e^{-j\omega t_0}F(\omega)} \tag{18.20}$$

that is, a delay in the time domain corresponds to a phase shift in the frequency domain. To derive the time shifting property, we note that

$$\mathcal{F}[f(t - t_0)] = \int_{-\infty}^{\infty} f(t - t_0)e^{-j\omega t}\,dt \tag{18.21}$$

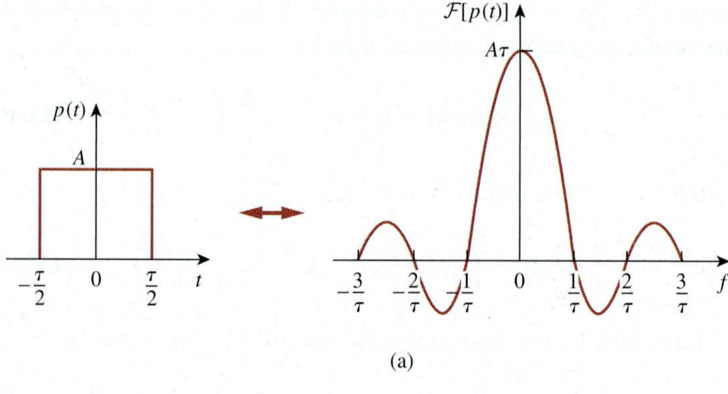

(a)

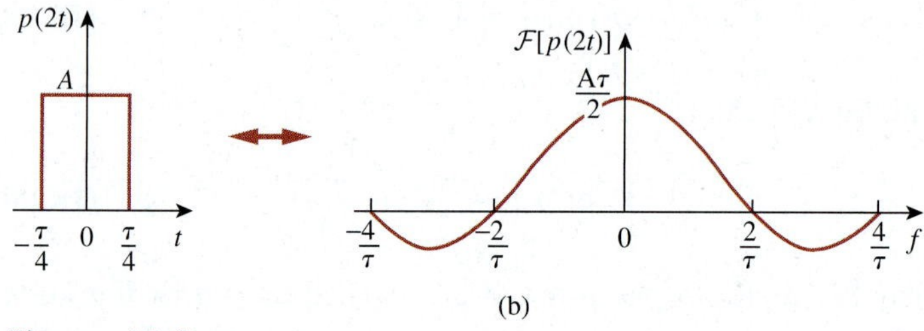

(b)

Figure 18.9
The effect of time scaling: (a) transform of the pulse, (b) time compression of the pulse causes frequency expansion.

If we let $x = t - t_0$ so that $dx = dt$ and $t = x + t_0$, then

$$\begin{aligned}\mathcal{F}[f(t - t_0)] &= \int_{-\infty}^{\infty} f(x)e^{-j\omega(x+t_0)}\,dx \\ &= e^{-j\omega t_0}\int_{-\infty}^{\infty} f(x)e^{-j\omega x}\,dx = e^{-j\omega t_0}F(\omega)\end{aligned} \tag{18.22}$$

Similarly, $\mathcal{F}[f(t + t_0)] = e^{j\omega t_0}F(\omega)$.

For example, from Example 18.3,

$$\mathcal{F}[e^{-at}u(t)] = \frac{1}{a + j\omega} \tag{18.23}$$

The transform of $f(t) = e^{-(t-2)}u(t - 2)$ is

$$F(\omega) = \mathcal{F}\left[e^{-(t-2)}u(t - 2)\right] = \frac{e^{-j2\omega}}{1 + j\omega} \tag{18.24}$$

Frequency Shifting (or Amplitude Modulation)

This property states that if $F(\omega) = \mathcal{F}[f(t)]$, then

$$\boxed{\mathcal{F}\left[f(t)e^{j\omega_0 t}\right] = F(\omega - \omega_0)} \tag{18.25}$$

meaning, a frequency shift in the frequency domain adds a phase shift to the time function. By definition,

$$\begin{aligned}\mathcal{F}[f(t)e^{j\omega_0 t}] &= \int_{-\infty}^{\infty} f(t)e^{j\omega_0 t}e^{-j\omega t}\,dt \\ &= \int_{-\infty}^{\infty} f(t)e^{-j(\omega-\omega_0)t}\,dt = F(\omega - \omega_0)\end{aligned} \tag{18.26}$$

For example, $\cos\omega_0 t = \frac{1}{2}(e^{j\omega_0 t} + e^{-j\omega_0 t})$. Using the property in Eq. (18.25),

$$\begin{aligned}\mathcal{F}[f(t)\cos\omega_0 t] &= \frac{1}{2}\mathcal{F}[f(t)e^{j\omega_0 t}] + \frac{1}{2}\mathcal{F}[f(t)e^{-j\omega_0 t}] \\ &= \frac{1}{2}F(\omega - \omega_0) + \frac{1}{2}F(\omega + \omega_0)\end{aligned} \tag{18.27}$$

This is an important result in modulation where frequency components of a signal are shifted. If, for example, the amplitude spectrum of $f(t)$ is as shown in Fig. 18.10(a), then the amplitude spectrum of $f(t)\cos\omega_0 t$ will be as shown in Fig. 18.10(b). We will elaborate on amplitude modulation in Section 18.7.1.

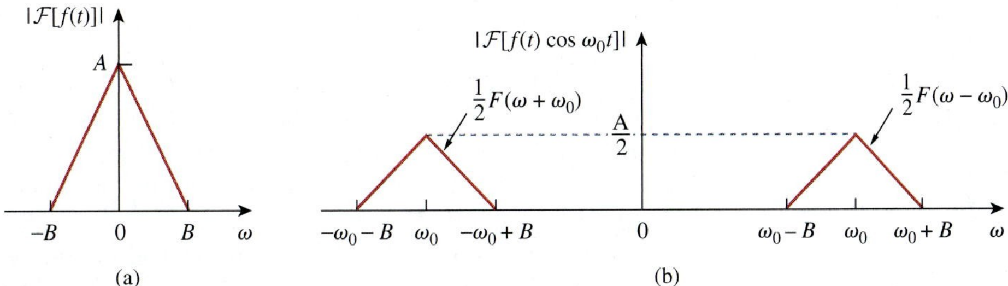

Figure 18.10
Amplitude spectra of: (a) signal $f(t)$, (b) modulated signal $f(t)\cos\omega_0 t$.

Time Differentiation

Given that $F(\omega) = \mathcal{F}[f(t)]$, then

$$\mathcal{F}[f'(t)] = j\omega F(\omega) \tag{18.28}$$

In other words, the transform of the derivative of $f(t)$ is obtained by multiplying the transform of $f(t)$ by $j\omega$. By definition,

$$f(t) = \mathcal{F}^{-1}[F(\omega)] = \frac{1}{2\pi}\int_{-\infty}^{\infty} F(\omega)e^{j\omega t}\,d\omega \tag{18.29}$$

Taking the derivative of both sides with respect to t gives

$$f'(t) = \frac{j\omega}{2\pi}\int_{-\infty}^{\infty} F(\omega)e^{j\omega t}\,d\omega = j\omega\mathcal{F}^{-1}[F(\omega)]$$

or

$$\mathcal{F}[f'(t)] = j\omega F(\omega) \tag{18.30}$$

Repeated applications of Eq. (18.30) give

$$\mathcal{F}\left[f^{(n)}(t)\right] = (j\omega)^n F(\omega) \tag{18.31}$$

For example, if $f(t) = e^{-at}u(t)$, then

$$f'(t) = -ae^{-at}u(t) + e^{-at}\delta(t) = -af(t) + e^{-at}\delta(t) \quad \textbf{(18.32)}$$

Taking the Fourier transforms of the first and last terms, we obtain

$$j\omega F(\omega) = -aF(\omega) + 1 \quad \Rightarrow \quad F(\omega) = \frac{1}{a + j\omega} \quad \textbf{(18.33)}$$

which agrees with the result in Example 18.3.

Time Integration

Given that $F(\omega) = \mathcal{F}[f(t)]$, then

$$\mathcal{F}\left[\int_{-\infty}^{t} f(t)\, dt\right] = \frac{F(\omega)}{j\omega} + \pi F(0)\delta(\omega) \quad \textbf{(18.34)}$$

that is, the transform of the integral of $f(t)$ is obtained by dividing the transform of $f(t)$ by $j\omega$ and adding the result to the impulse term that reflects the dc component $F(0)$. Someone might ask, "How do we know that when we take the Fourier transform for time integration, we should integrate over the interval $[-\infty, t]$ and not $[-\infty, \infty]$?" When we integrate over $[-\infty, \infty]$, the result does not depend on time anymore, and the Fourier transform of a constant is what we will eventually get. But when we integrate over $[-\infty, t]$, we get the integral of the function from the past to time t, so that the result depends on t and we can take the Fourier transform of that.

If ω is replaced by 0 in Eq. (18.8),

$$F(0) = \int_{-\infty}^{\infty} f(t)\, dt \quad \textbf{(18.35)}$$

indicating that the dc component is zero when the integral of $f(t)$ over all time vanishes. The proof of the time integration in Eq. (18.34) will be given later when we consider the convolution property.

For example, we know that $\mathcal{F}[\delta(t)] = 1$ and that integrating the impulse function gives the unit step function [see Eq. (7.39a)]. By applying the property in Eq. (18.34), we obtain the Fourier transform of the unit step function as

$$\mathcal{F}[u(t)] = \mathcal{F}\left[\int_{-\infty}^{t} \delta(t)\, dt\right] = \frac{1}{j\omega} + \pi\delta(\omega) \quad \textbf{(18.36)}$$

Reversal

If $F(\omega) = \mathcal{F}[f(t)]$, then

$$\mathcal{F}[f(-t)] = F(-\omega) = F^*(\omega) \quad \textbf{(18.37)}$$

where the asterisk denotes the complex conjugate. This property states that reversing $f(t)$ about the time axis reverses $F(\omega)$ about the frequency axis. This may be regarded as a special case of time scaling for which $a = -1$ in Eq. (18.15).

For example, $1 = u(t) + u(-t)$. Hence,

$$\begin{aligned}\mathcal{F}[1] &= \mathcal{F}[u(t)] + \mathcal{F}[u(-t)] \\ &= \frac{1}{j\omega} + \pi\delta(\omega) \\ &\quad - \frac{1}{j\omega} + \pi\delta(-\omega) \\ &= 2\pi\delta(\omega)\end{aligned}$$

Duality

This property states that if $F(\omega)$ is the Fourier transform of $f(t)$, then the Fourier transform of $F(t)$ is $2\pi f(-\omega)$; we write

$$\mathcal{F}[f(t)] = F(\omega) \quad \Rightarrow \quad \mathcal{F}[F(t)] = 2\pi f(-\omega) \tag{18.38}$$

This expresses the symmetry property of the Fourier transform. To derive this property, we recall that

$$f(t) = \mathcal{F}^{-1}[F(\omega)] = \frac{1}{2\pi}\int_{-\infty}^{\infty} F(\omega)e^{j\omega t}\, d\omega$$

or

$$2\pi f(t) = \int_{-\infty}^{\infty} F(\omega)e^{j\omega t}\, d\omega \tag{18.39}$$

Replacing t by $-t$ gives

$$2\pi f(-t) = \int_{-\infty}^{\infty} F(\omega)e^{-j\omega t}\, d\omega$$

If we interchange t and ω, we obtain

$$2\pi f(-\omega) = \int_{-\infty}^{\infty} F(t)e^{-j\omega t}\, dt = \mathcal{F}[F(t)] \tag{18.40}$$

Since $f(t)$ is the sum of the signals in Figs. 18.7 and 18.8, $F(\omega)$ is the sum of the results in Example 18.3 and Practice Prob. 18.3.

as expected.

For example, if $f(t) = e^{-|t|}$, then

$$F(\omega) = \frac{2}{\omega^2 + 1} \tag{18.41}$$

By the duality property, the Fourier transform of $F(t) = 2/(t^2 + 1)$ is

$$2\pi f(\omega) = 2\pi e^{-|\omega|} \tag{18.42}$$

Figure 18.11 shows another example of the duality property. It illustrates the fact that if $f(t) = \delta(t)$ so that $F(\omega) = 1$, as in Fig. 18.11(a), then the Fourier transform of $F(t) = 1$ is $2\pi f(\omega) = 2\pi\delta(\omega)$ as shown in Fig. 18.11(b).

Convolution

Recall from Chapter 15 that if $x(t)$ is the input excitation to a circuit with an impulse function of $h(t)$, then the output response $y(t)$ is given by the convolution integral

$$y(t) = h(t) * x(t) = \int_{-\infty}^{\infty} h(\lambda)x(t - \lambda)\, d\lambda \tag{18.43}$$

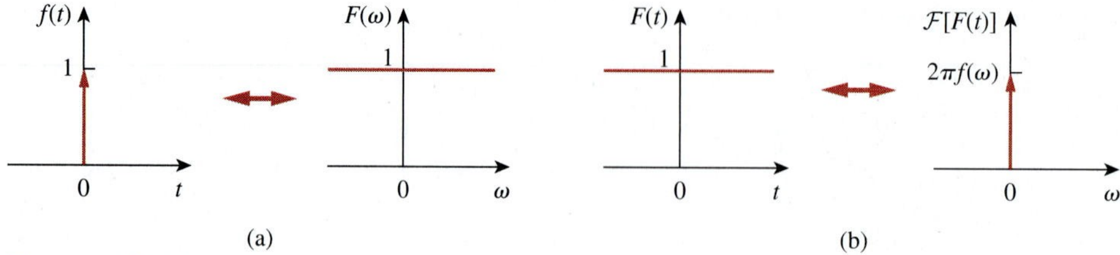

Figure 18.11
A typical illustration of the duality property of the Fourier transform: (a) transform of impulse, (b) transform of unit dc level.

If $X(\omega)$, $H(\omega)$, and $Y(\omega)$ are the Fourier transforms of $x(t)$, $h(t)$, and $y(t)$, respectively, then

$$Y(\omega) = \mathcal{F}[h(t) * x(t)] = H(\omega)X(\omega) \tag{18.44}$$

which indicates that convolution in the time domain corresponds with multiplication in the frequency domain.

To derive the convolution property, we take the Fourier transform of both sides of Eq. (18.43) to get

$$Y(\omega) = \int_{-\infty}^{\infty} \left[\int_{-\infty}^{\infty} h(\lambda)x(t-\lambda)\, d\lambda \right] e^{-j\omega t}\, dt \tag{18.45}$$

Exchanging the order of integration and factoring $h(\lambda)$, which does not depend on t, we have

$$Y(\omega) = \int_{-\infty}^{\infty} h(\lambda) \left[\int_{-\infty}^{\infty} x(t-\lambda)e^{-j\omega t}\, dt \right] d\lambda$$

For the integral within the brackets, let $\tau = t - \lambda$ so that $t = \tau + \lambda$ and $dt = d\tau$. Then,

$$\begin{aligned} Y(\omega) &= \int_{-\infty}^{\infty} h(\lambda) \left[\int_{-\infty}^{\infty} x(\tau)e^{-j\omega(\tau+\lambda)}\, d\tau \right] d\lambda \\ &= \int_{-\infty}^{\infty} h(\lambda)e^{-j\omega\lambda}\, d\lambda \int_{-\infty}^{\infty} x(\tau)e^{-j\omega\tau}\, d\tau = H(\omega)X(\omega) \end{aligned} \tag{18.46}$$

The important relationship in Eq. (18.46) is the key reason for using the Fourier transform in the analysis of linear systems.

as expected. This result expands the phasor method beyond what was done with the Fourier series in the previous chapter.

To illustrate the convolution property, suppose both $h(t)$ and $x(t)$ are identical rectangular pulses, as shown in Fig. 18.12(a) and 18.12(b). We recall from Example 18.2 and Fig. 18.5 that the Fourier transforms of the rectangular pulses are sinc functions, as shown in Fig. 18.12(c) and 18.12(d). According to the convolution property, the product of the sinc functions should give us the convolution of the rectangular pulses in the time domain. Thus, the convolution of the pulses in Fig. 18.12(e) and the product of the sinc functions in Fig. 18.12(f) form a Fourier pair.

In view of the duality property, we expect that if convolution in the time domain corresponds with multiplication in the frequency

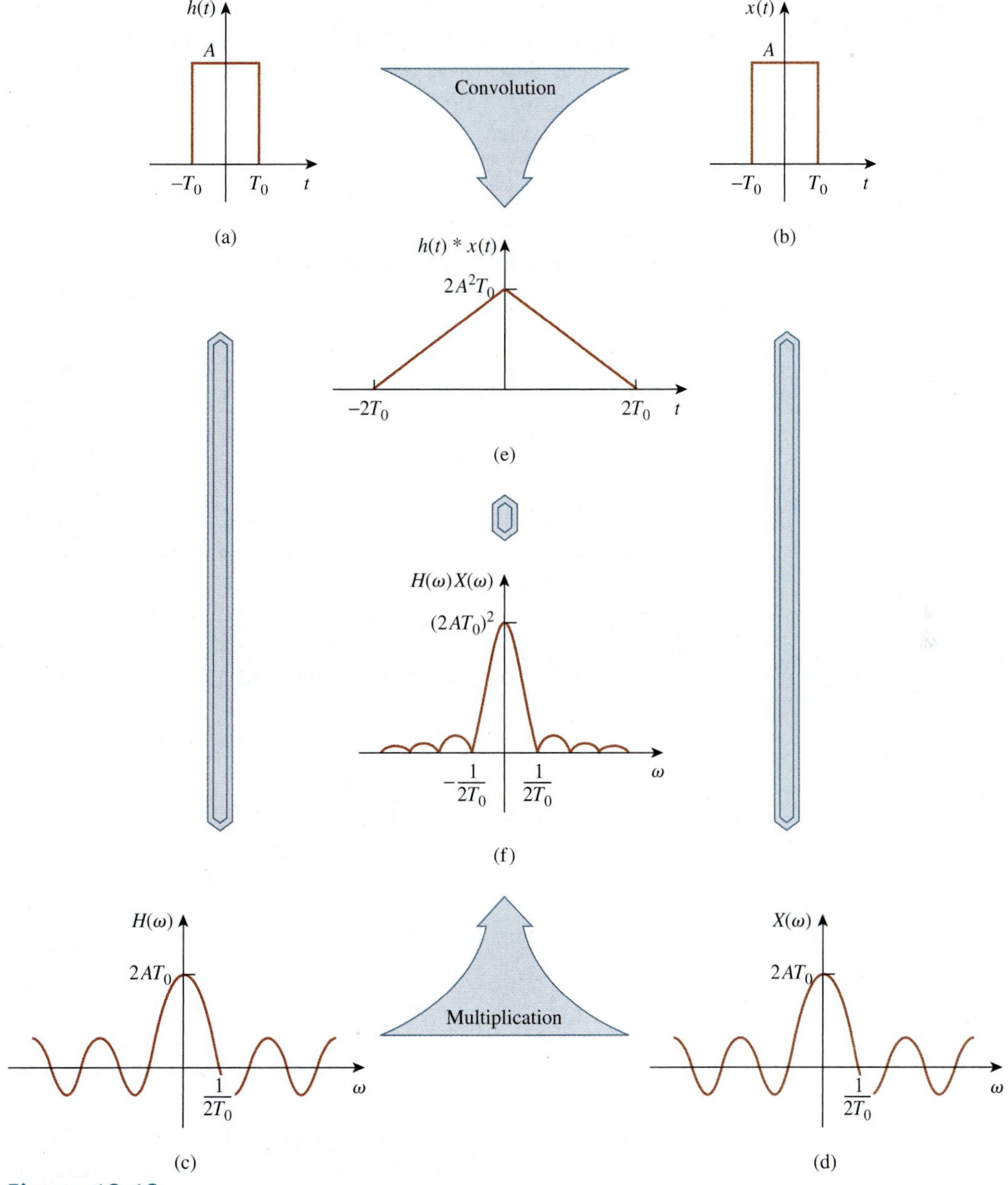

Figure 18.12
Graphical illustration of the convolution property.
E. O. Brigham, *The Fast Fourier Transform* [Englewood Cliffs, NJ: Prentice Hall, 1974], p. 60.

domain, then multiplication in the time domain should have a correspondence in the frequency domain. This happens to be the case. If $f(t) = f_1(t) f_2(t)$, then

$$F(\omega) = \mathcal{F}[f_1(t) f_2(t)] = \frac{1}{2\pi} F_1(\omega) * F_2(\omega) \tag{18.47}$$

or

$$F(\omega) = \frac{1}{2\pi} \int_{-\infty}^{\infty} F_1(\lambda) F_2(\omega - \lambda)\, d\lambda \tag{18.48}$$

which is convolution in the frequency domain. The proof of Eq. (18.48) readily follows from the duality property in Eq. (18.38).

Let us now derive the time integration property in Eq. (18.34). If we replace $x(t)$ with the unit step function $u(t)$ and $h(t)$ with $f(t)$ in Eq. (18.43), then

$$\int_{-\infty}^{\infty} f(\lambda)u(t-\lambda)\,d\lambda = f(t) * u(t) \tag{18.49}$$

But by the definition of the unit step function,

$$u(t-\lambda) = \begin{cases} 1, & t-\lambda > 0 \\ 0, & t-\lambda > 0 \end{cases}$$

We can write this as

$$u(t-\lambda) = \begin{cases} 1, & \lambda < t \\ 0, & \lambda > t \end{cases}$$

Substituting this into Eq. (18.49) makes the interval of integration change from $[-\infty, \infty]$ to $[-\infty, t]$, and thus Eq. (18.49) becomes

$$\int_{-\infty}^{t} f(\lambda)\,d\lambda = u(t) * f(t)$$

Taking the Fourier transform of both sides yields

$$\mathcal{F}\left[\int_{-\infty}^{t} f(\lambda)\,d\lambda\right] = U(\omega)F(\omega) \tag{18.50}$$

But from Eq. (18.36), the Fourier transform of the unit step function is

$$U(\omega) = \frac{1}{j\omega} + \pi\,\delta(\omega)$$

Substituting this into Eq. (18.50) gives

$$\begin{aligned} \mathcal{F}\left[\int_{-\infty}^{t} f(\lambda)\,d\lambda\right] &= \left(\frac{1}{j\omega} + \pi\,\delta(\omega)\right)F(\omega) \\ &= \frac{F(\omega)}{j\omega} + \pi F(0)\delta(\omega) \end{aligned} \tag{18.51}$$

which is the time integration property of Eq. (18.34). Note that in Eq. (18.51), $F(\omega)\delta(\omega) = F(0)\delta(\omega)$, since $\delta(\omega)$ is only nonzero at $\omega = 0$.

Table 18.1 lists these properties of the Fourier transform. Table 18.2 presents the transform pairs of some common functions. Note the similarities between these tables and Tables 15.1 and 15.2.

TABLE 18.1

Properties of the Fourier transform.

Property	$f(t)$	$F(\omega)$
Linearity	$a_1f_1(t) + a_2f_2(t)$	$a_1F_1(\omega) + a_2F_2(\omega)$
Scaling	$f(at)$	$\frac{1}{\lvert a\rvert}F\left(\frac{\omega}{a}\right)$
Time shift	$f(t-a)$	$e^{-j\omega a}F(\omega)$
Frequency shift	$e^{j\omega_0 t}f(t)$	$F(\omega-\omega_0)$

TABLE 18.1 *(continued)*

Property	$f(t)$	$F(\omega)$
Modulation	$\cos(\omega_0 t)f(t)$	$\frac{1}{2}[F(\omega + \omega_0) + F(\omega - \omega_0)]$
Time differentiation	$\frac{df}{dt}$	$j\omega F(\omega)$
	$\frac{d^n f}{dt^n}$	$(j\omega)^n F(\omega)$
Time integration	$\int_{-\infty}^{t} f(t)\,dt$	$\frac{F(\omega)}{j\omega} + \pi F(0)\,\delta(\omega)$
Frequency differentiation	$t^n f(t)$	$(j)^n \frac{d^n}{d\omega^n} F(\omega)$
Reversal	$f(-t)$	$F(-\omega)$ or $F^*(\omega)$
Duality	$F(t)$	$2\pi f(-\omega)$
Convolution in t	$f_1(t) * f_2(t)$	$F_1(\omega)F_2(\omega)$
Convolution in ω	$f_1(t)f_2(t)$	$\frac{1}{2\pi}F_1(\omega) * F_2(\omega)$

TABLE 18.2

Fourier transform pairs.

$f(t)$	$F(\omega)$
$\delta(t)$	1
1	$2\pi\,\delta(\omega)$
$u(t)$	$\pi\,\delta(\omega) + \frac{1}{j\omega}$
$u(t + \tau) - u(t - \tau)$	$2\frac{\sin \omega\tau}{\omega}$
$\lvert t \rvert$	$\frac{-2}{\omega^2}$
$\text{sgn}(t)$	$\frac{2}{j\omega}$
$e^{-at}u(t)$	$\frac{1}{a + j\omega}$
$e^{at}u(-t)$	$\frac{1}{a - j\omega}$
$t^n e^{-at}u(t)$	$\frac{n!}{(a + j\omega)^{n+1}}$
$e^{-a\lvert t \rvert}$	$\frac{2a}{a^2 + \omega^2}$
$e^{j\omega_0 t}$	$2\pi\,\delta(\omega - \omega_0)$
$\sin \omega_0 t$	$j\pi[\delta(\omega + \omega_0) - \delta(\omega - \omega_0)]$
$\cos \omega_0 t$	$\pi[\delta(\omega + \omega_0) + \delta(\omega - \omega_0)]$
$e^{-at} \sin \omega_0 t u(t)$	$\frac{\omega_0}{(a + j\omega)^2 + \omega_0^2}$
$e^{-at} \cos \omega_0 t u(t)$	$\frac{a + j\omega}{(a + j\omega)^2 + \omega_0^2}$

Example 18.4

Find the Fourier transforms of the following functions: (a) signum function sgn(t), shown in Fig. 18.13, (b) the double-sided exponential $e^{-a|t|}$, and (c) the sinc function $(\sin t)/t$.

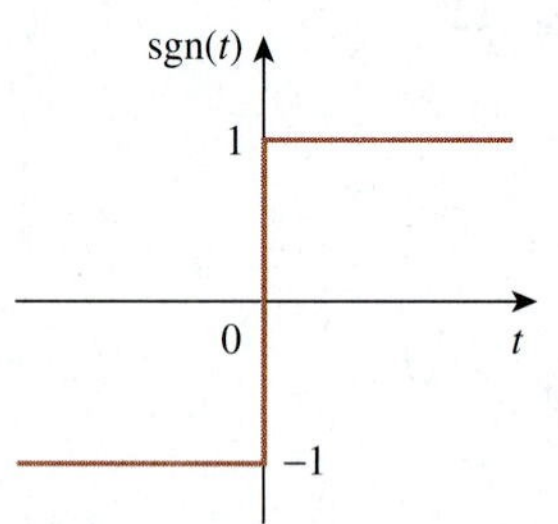

Figure 18.13
The signum function of Example 18.4.

Solution:

(a) We can obtain the Fourier transform of the *signum* function in three ways.

■ **METHOD 1** We can write the signum function in terms of the unit step function as

$$\text{sgn}(t) = f(t) = u(t) - u(-t)$$

But from Eq. (18.36),

$$U(\omega) = \mathcal{F}[u(t)] = \pi\delta(\omega) + \frac{1}{j\omega}$$

Applying this and the reversal property, we obtain

$$\begin{aligned}\mathcal{F}[\text{sgn}(t)] &= U(\omega) - U(-\omega) \\ &= \left(\pi\delta(\omega) + \frac{1}{j\omega}\right) - \left(\pi\delta(-\omega) + \frac{1}{-j\omega}\right) = \frac{2}{j\omega}\end{aligned}$$

■ **METHOD 2** since $\delta(\omega) = \delta(-\omega)$, another way of writing the signum function in terms of the unit step function is

$$f(t) = \text{sgn}(t) = -1 + 2u(t)$$

Taking the Fourier transform of each term gives

$$F(\omega) = -2\pi\delta(\omega) + 2\left(\pi\delta(\omega) + \frac{1}{j\omega}\right) = \frac{2}{j\omega}$$

■ **METHOD 3** We can take the derivative of the signum function in Fig. 18.13 and obtain

$$f'(t) = 2\delta(t)$$

Taking the transform of this,

$$j\omega F(\omega) = 2 \quad \Rightarrow \quad F(\omega) = \frac{2}{j\omega}$$

as obtained previously.
(b) The double-sided exponential can be expressed as

$$f(t) = e^{-a|t|} = e^{-at}u(t) + e^{at}u(-t) = y(t) + y(-t)$$

where $y(t) = e^{-at}u(t)$ so that $Y(\omega) = 1/(a + j\omega)$. Applying the reversal property,

$$\mathcal{F}\left[e^{-a|t|}\right] = Y(\omega) + Y(-\omega) = \left(\frac{1}{a + j\omega} + \frac{1}{a - j\omega}\right) = \frac{2a}{a^2 + \omega^2}$$

(c) From Example 18.2,

$$\mathcal{F}\left[u\left(t + \frac{\tau}{2}\right) - u\left(t - \frac{\tau}{2}\right)\right] = \tau\frac{\sin(\omega\tau/2)}{\omega\tau/2} = \tau \operatorname{sinc}\frac{\omega\tau}{2}$$

Setting $\tau/2 = 1$ gives

$$\mathcal{F}[u(t+1) - u(t-1)] = 2\frac{\sin\omega}{\omega}$$

Applying the duality property yields

$$\mathcal{F}\left[2\frac{\sin t}{t}\right] = 2\pi[U(\omega+1) - U(\omega-1)]$$

or

$$\mathcal{F}\left[\frac{\sin t}{t}\right] = \pi[U(\omega+1) - U(\omega-1)]$$

Practice Problem 18.4

Determine the Fourier transforms of these functions: (a) gate function $g(t) = u(t) - u(t-1)$, (b) $f(t) = te^{-2t}u(t)$, and (c) sawtooth pulse $p(t) = 10t[u(t) - u(t-2)]$.

Answer: (a) $(1 - e^{-j\omega})\left[\pi\delta(\omega) + \frac{1}{j\omega}\right]$, (b) $\frac{1}{(2+j\omega)^2}$, (c) $\frac{10(e^{-j2\omega} - 1)}{\omega^2} + \frac{20j}{\omega}e^{-j2\omega}$.

Example 18.5

Find the Fourier transform of the function in Fig. 18.14.

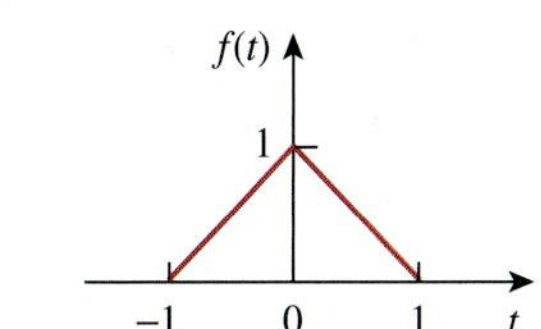

Figure 18.14
For Example 18.5.

Solution:
The Fourier transform can be found directly using Eq. (18.8), but it is much easier to find it using the derivative property. We can express the function as

$$f(t) = \begin{cases} 1+t, & -1 < t < 0 \\ 1-t, & 0 < t < 1 \end{cases}$$

Its first derivative is shown in Fig. 18.15(a) and is given by

$$f'(t) = \begin{cases} 1, & -1 < t < 0 \\ -1, & 0 < t < 1 \end{cases}$$

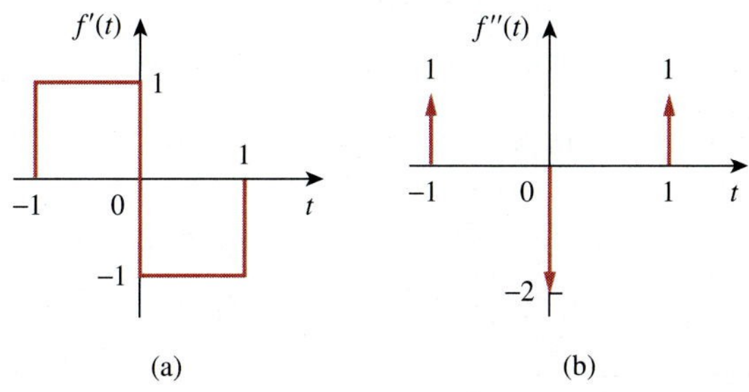

Figure 18.15
First and second derivatives of $f(t)$ in Fig. 18.14; for Example 18.5.

Its second derivative is in Fig. 18.15(b) and is given by

$$f''(t) = \delta(t+1) - 2\delta(t) + \delta(t-1)$$

Taking the Fourier transform of both sides,

$$(j\omega)^2 F(\omega) = e^{j\omega} - 2 + e^{-j\omega} = -2 + 2\cos\omega$$

or

$$F(\omega) = \frac{2(1-\cos\omega)}{\omega^2}$$

Practice Problem 18.5

Determine the Fourier transform of the function in Fig. 18.16.

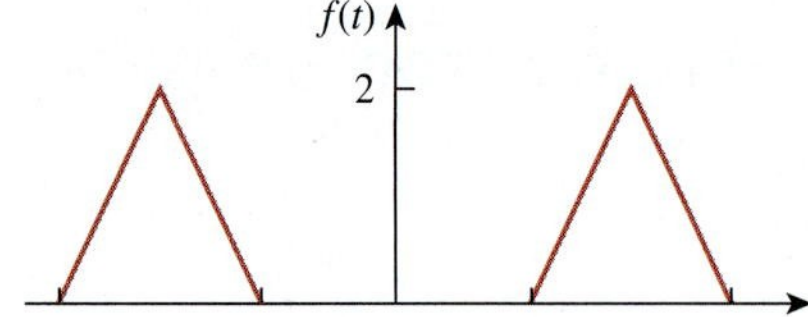

Figure 18.16
For Practice Prob. 18.5.

Answer: $(8\cos 3\omega - 4\cos 4\omega - 4\cos 2\omega)/\omega^2$.

Example 18.6

Obtain the inverse Fourier transform of:

(a) $F(\omega) = \dfrac{10j\omega + 4}{(j\omega)^2 + 6j\omega + 8}$ (b) $G(\omega) = \dfrac{\omega^2 + 21}{\omega^2 + 9}$

Solution:

(a) To avoid complex algebra, we can replace $j\omega$ with s for the moment. Using partial fraction expansion,

$$F(s) = \frac{10s+4}{s^2+6s+8} = \frac{10s+4}{(s+4)(s+2)} = \frac{A}{s+4} + \frac{B}{s+2}$$

where

$$A = (s+4)F(s)\big|_{s=-4} = \frac{10s+4}{(s+2)}\bigg|_{s=-4} = \frac{-36}{-2} = 18$$

$$B = (s+2)F(s)\big|_{s=-2} = \frac{10s+4}{(s+4)}\bigg|_{s=-2} = \frac{-16}{2} = -8$$

Substituting $A = 18$ and $B = -8$ in $F(s)$ and s with $j\omega$ gives

$$F(j\omega) = \frac{18}{j\omega+4} + \frac{-8}{j\omega+2}$$

With the aid of Table 18.2, we obtain the inverse transform as

$$f(t) = (18e^{-4t} - 8e^{-2t})u(t)$$

(b) We simplify $G(\omega)$ as

$$G(\omega) = \frac{\omega^2+21}{\omega^2+9} = 1 + \frac{12}{\omega^2+9}$$

With the aid of Table 18.2, the inverse transform is obtained as

$$g(t) = \delta(t) + 2e^{-3|t|}$$

Practice Problem 18.6

Find the inverse Fourier transform of:

(a) $H(\omega) = \dfrac{6(3 + j2\omega)}{(1 + j\omega)(4 + j\omega)(2 + j\omega)}$

(b) $Y(\omega) = \pi\delta(\omega) + \dfrac{1}{j\omega} + \dfrac{2(1 + j\omega)}{(1 + j\omega)^2 + 16}$

Answer: (a) $h(t) = (2e^{-t} + 3e^{-2t} - 5e^{-4t})u(t)$,
(b) $y(t) = (1 + 2e^{-t}\cos 4t)u(t)$.

18.4 Circuit Applications

The Fourier transform generalizes the phasor technique to nonperiodic functions. Therefore, we apply Fourier transforms to circuits with nonsinusoidal excitations in exactly the same way we apply phasor techniques to circuits with sinusoidal excitations. Thus, Ohm's law is still valid:

$$V(\omega) = Z(\omega)I(\omega) \tag{18.52}$$

where $V(\omega)$ and $I(\omega)$ are the Fourier transforms of the voltage and current and $Z(\omega)$ is the impedance. We get the same expressions for the impedances of resistors, inductors, and capacitors as in phasor analysis, namely,

$$\begin{aligned} R &\Rightarrow R \\ L &\Rightarrow j\omega L \\ C &\Rightarrow \frac{1}{j\omega C} \end{aligned} \tag{18.53}$$

Once we transform the functions for the circuit elements into the frequency domain and take the Fourier transforms of the excitations, we can use circuit techniques such as voltage division, source transformation, mesh analysis, node analysis, or Thevenin's theorem, to find the unknown response (current or voltage). Finally, we take the inverse Fourier transform to obtain the response in the time domain.

Although the Fourier transform method produces a response that exists for $-\infty < t < \infty$, Fourier analysis cannot handle circuits with initial conditions.

The transfer function is again defined as the ratio of the output response $Y(\omega)$ to the input excitation $X(\omega)$; that is,

$$H(\omega) = \frac{Y(\omega)}{X(\omega)} \tag{18.54}$$

or

$$Y(\omega) = H(\omega)X(\omega) \tag{18.55}$$

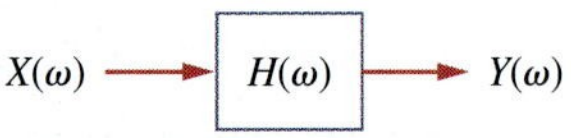

Figure 18.17
Input-output relationship of a circuit in the frequency domain.

The frequency domain input-output relationship is portrayed in Fig. 18.17. Equation (18.55) shows that if we know the transfer function and the input, we can readily find the output. The relationship in Eq. (18.54) is the principal reason for using the Fourier transform in circuit analysis. Notice that $H(\omega)$ is identical to $H(s)$ with $s = j\omega$. Also, if the input is an impulse function [i.e., $x(t) = \delta(t)$], then $X(\omega) = 1$, so that the response is

$$Y(\omega) = H(\omega) = \mathcal{F}[h(t)] \tag{18.56}$$

indicating that $H(\omega)$ is the Fourier transform of the impulse response $h(t)$.

Example 18.7

Find $v_o(t)$ in the circuit of Fig. 18.18 for $v_i(t) = 2e^{-3t}u(t)$.

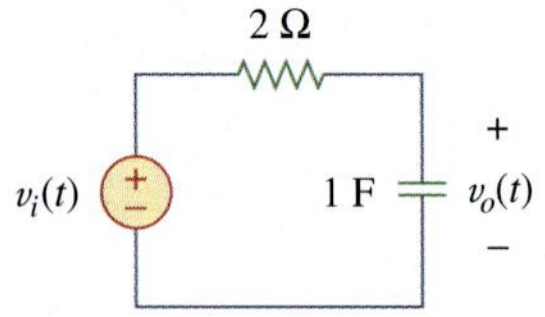

Figure 18.18
For Example 18.7.

Solution:
The Fourier transform of the input voltage is

$$V_i(\omega) = \frac{2}{3 + j\omega}$$

and the transfer function obtained by voltage division is

$$H(\omega) = \frac{V_o(\omega)}{V_i(\omega)} = \frac{1/j\omega}{2 + 1/j\omega} = \frac{1}{1 + j2\omega}$$

Hence,

$$V_o(\omega) = V_i(\omega)H(\omega) = \frac{2}{(3 + j\omega)(1 + j2\omega)}$$

or

$$V_o(\omega) = \frac{1}{(3 + j\omega)(0.5 + j\omega)}$$

By partial fractions,

$$V_o(\omega) = \frac{-0.4}{3 + j\omega} + \frac{0.4}{0.5 + j\omega}$$

Taking the inverse Fourier transform yields

$$v_o(t) = 0.4(e^{-0.5t} - e^{-3t})u(t)$$

Practice Problem 18.7

Determine $v_o(t)$ in Fig. 18.19 if $v_i(t) = 5\text{sgn}(t) = (-5 + 10u(t))$ V.

Answer: $-5 + 10(1 - e^{-4t})u(t)$ V.

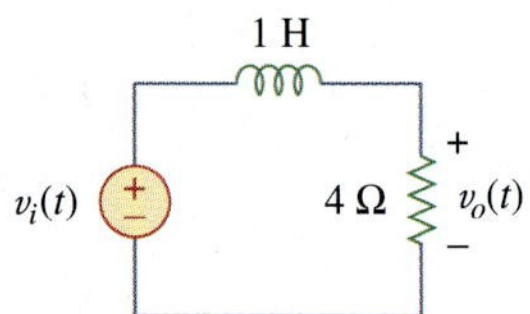

Figure 19.19
For Practice Prob. 18.7.

Example 18.8

Using the Fourier transform method, find $i_o(t)$ in Fig. 18.20 when $i_s(t) = 10 \sin 2t$ A.

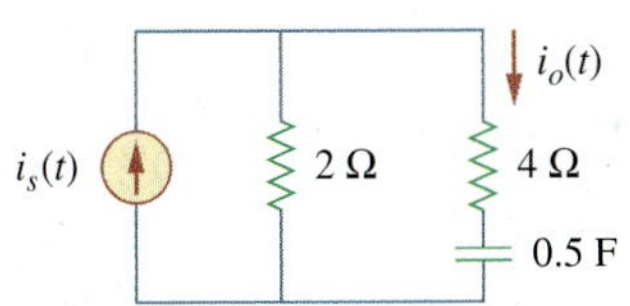

Figure 18.20
For Example 18.8.

Solution:
By current division,

$$H(\omega) = \frac{I_o(\omega)}{I_s(\omega)} = \frac{2}{2 + 4 + 2/j\omega} = \frac{j\omega}{1 + j\omega 3}$$

If $i_s(t) = 10 \sin 2t$, then

$$I_s(\omega) = j\pi 10[\delta(\omega + 2) - \delta(\omega - 2)]$$

Hence,

$$I_o(\omega) = H(\omega)I_s(\omega) = \frac{10\pi\omega[\delta(\omega - 2) - \delta(\omega + 2)]}{1 + j\omega 3}$$

The inverse Fourier transform of $I_o(\omega)$ cannot be found using Table 18.2. We resort to the inverse Fourier transform formula in Eq. (18.9) and write

$$i_o(t) = \mathcal{F}^{-1}[I_o(\omega)] = \frac{1}{2\pi}\int_{-\infty}^{\infty} \frac{10\pi\omega[\delta(\omega - 2) - \delta(\omega + 2)]}{1 + j\omega 3} e^{j\omega t}\, d\omega$$

We apply the sifting property of the impulse function, namely,

$$\delta(\omega - \omega_0) f(\omega) = f(\omega_0)$$

or

$$\int_{-\infty}^{\infty} \delta(\omega - \omega_0) f(\omega)\, d\omega = f(\omega_0)$$

and obtain

$$\begin{aligned} i_o(t) &= \frac{10\pi}{2\pi}\left[\frac{2}{1 + j6} e^{j2t} - \frac{-2}{1 - j6} e^{-j2t}\right] \\ &= 10\left[\frac{e^{j2t}}{6.082 e^{j80.54^\circ}} + \frac{e^{-j2t}}{6.082\, e^{-j80.54^\circ}}\right] \\ &= 1.644\left[e^{j(2t - 80.54^\circ)} + e^{-j(2t - 80.54^\circ)}\right] \\ &= 3.288 \cos(2t - 80.54^\circ) \text{ A} \end{aligned}$$

Practice Problem 18.8

Find the current $i_o(t)$ in the circuit in Fig. 18.21, given that $i_s(t) = 5 \cos 4t$ A.

Answer: $2.95 \cos(4t + 26.57^\circ)$ A.

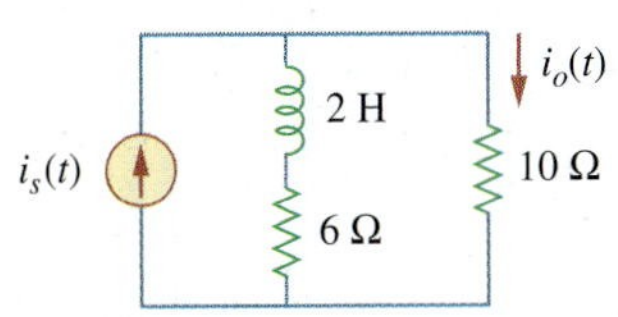

Figure 18.21
For Practice Prob. 18.8.

18.5 Parseval's Theorem

Parseval's theorem demonstrates one practical use of the Fourier transform. It relates the energy carried by a signal to the Fourier transform of the signal. If $p(t)$ is the power associated with the signal, the energy carried by the signal is

$$W = \int_{-\infty}^{\infty} p(t)\,dt \tag{18.57}$$

In order to be able to compare the energy content of current and voltage signals, it is convenient to use a 1-Ω resistor as the base for energy calculation. For a 1-Ω resistor, $p(t) = v^2(t) = i^2(t) = f^2(t)$, where $f(t)$ stands for either voltage or current. The energy delivered to the 1-Ω resistor is

$$W_{1\Omega} = \int_{-\infty}^{\infty} f^2(t)\,dt \tag{18.58}$$

Parseval's theorem states that this same energy can be calculated in the frequency domain as

$$W_{1\Omega} = \int_{-\infty}^{\infty} f^2(t)\,dt = \frac{1}{2\pi}\int_{-\infty}^{\infty} |F(\omega)|^2\,d\omega \tag{18.59}$$

> **Parseval's theorem** states that the total energy delivered to a 1-Ω resistor equals the total area under the square of $f(t)$ or $1/2\pi$ times the total area under the square of the magnitude of the Fourier transform of $f(t)$.

In fact, $|F(\omega)|^2$ is sometimes known as the energy spectral density of signal $f(t)$.

Parseval's theorem relates energy associated with a signal to its Fourier transform. It provides the physical significance of $F(\omega)$, namely, that $|F(\omega)|^2$ is a measure of the energy density (in joules per hertz) corresponding to $f(t)$.

To derive Eq. (18.59), we begin with Eq. (18.58) and substitute Eq. (18.9) for one of the $f(t)$'s. We obtain

$$W_{1\Omega} = \int_{-\infty}^{\infty} f^2(t)\,dt = \int_{-\infty}^{\infty} f(t)\left[\frac{1}{2\pi}\int_{-\infty}^{\infty} F(\omega)e^{j\omega t}\,d\omega\right]dt \tag{18.60}$$

The function $f(t)$ can be moved inside the integral within the brackets, since the integral does not involve time:

$$W_{1\Omega} = \frac{1}{2\pi}\int_{-\infty}^{\infty}\int_{-\infty}^{\infty} f(t)F(\omega)e^{j\omega t}\,d\omega\,dt \tag{18.61}$$

Reversing the order of integration,

$$\begin{aligned} W_{1\Omega} &= \frac{1}{2\pi}\int_{-\infty}^{\infty} F(\omega)\left[\int_{-\infty}^{\infty} f(t)e^{-j(-\omega)t}\,dt\right]d\omega \\ &= \frac{1}{2\pi}\int_{-\infty}^{\infty} F(\omega)F(-\omega)\,d\omega = \frac{1}{2\pi}\int_{-\infty}^{\infty} F(\omega)F^*(\omega)\,d\omega \end{aligned} \tag{18.62}$$

But if $z = x + jy$, $zz^* = (x + jy)(x - jy) = x^2 + y^2 = |z|^2$. Hence,

$$W_{1\Omega} = \int_{-\infty}^{\infty} f^2(t)\,dt = \frac{1}{2\pi}\int_{-\infty}^{\infty} |F(\omega)|^2\,d\omega \qquad \textbf{(18.63)}$$

as expected. Equation (18.63) indicates that the energy carried by a signal can be found by integrating either the square of $f(t)$ in the time domain or $1/2\pi$ times the square of $F(\omega)$ in the frequency domain.

Since $|F(\omega)|^2$ is an even function, we may integrate from 0 to ∞ and double the result; that is,

$$W_{1\Omega} = \int_{-\infty}^{\infty} f^2(t)\,dt = \frac{1}{\pi}\int_{0}^{\infty} |F(\omega)|^2\,d\omega \qquad \textbf{(18.64)}$$

We may also calculate the energy in any frequency band $\omega_1 < \omega < \omega_2$ as

$$W_{1\Omega} = \frac{1}{\pi}\int_{\omega_1}^{\omega_2} |F(\omega)|^2\,d\omega \qquad \textbf{(18.65)}$$

Notice that Parseval's theorem as stated here applies to nonperiodic functions. Parseval's theorem for periodic functions was presented in Sections 17.5 and 17.6. As evident in Eq. (18.63), Parseval's theorem shows that the energy associated with a nonperiodic signal is spread over the entire frequency spectrum, whereas the energy of a periodic signal is concentrated at the frequencies of its harmonic components.

Example 18.9

The voltage across a 10-Ω resistor is $v(t) = 5e^{-3t}u(t)$ V. Find the total energy dissipated in the resistor.

Solution:

1. **Define.** The problem is well defined and clearly stated.
2. **Present.** We are given the voltage across the resistor for all time and are asked to find the energy dissipated by the resistor. We note that the voltage is zero for all time less than zero. Thus, we only need to consider the time from zero to infinity.
3. **Alternative.** There are basically two ways to find this answer. The first would be to find the answer in the time domain. We will use the second approach to find the answer using Fourier analysis.
4. **Attempt.** In the time domain,

$$W_{10\Omega} = 0.1\int_{-\infty}^{\infty} f^2(t)\,dt = 0.1\int_{0}^{\infty} 25e^{-6t}\,dt$$

$$= 2.5\frac{e^{-6t}}{-6}\Bigg|_0^{\infty} = \frac{2.5}{6} = \mathbf{416.7\ mJ}$$

5. **Evaluate.** In the frequency domain,

$$F(\omega) = V(\omega) = \frac{5}{3 + j\omega}$$

so that

$$|F(\omega)|^2 = F(\omega)F(\omega)^* = \frac{25}{9 + \omega^2}$$

Hence, the energy dissipated is

$$W_{10\Omega} = \frac{0.1}{2\pi}\int_{-\infty}^{\infty} |F(\omega)|^2\, d\omega = \frac{0.1}{\pi}\int_0^{\infty} \frac{25}{9 + \omega^2}\, d\omega$$
$$= \frac{2.5}{\pi}\left(\frac{1}{3}\tan^{-1}\frac{\omega}{3}\right)\Big|_0^{\infty} = \frac{2.5}{\pi}\left(\frac{1}{3}\right)\left(\frac{\pi}{2}\right) = \frac{2.5}{6} = \mathbf{416.7\ mJ}$$

6. **Satisfactory?** We have satisfactorily solved the problem and can present the results as a solution to the problem.

Practice Problem 18.9

(a) Calculate the total energy absorbed by a 1-Ω resistor with $i(t) = 5e^{-2|t|}$ A in the time domain. (b) Repeat (a) in the frequency domain.

Answer: (a) 12.5 J, (b) 12.5 J.

Example 18.10

Calculate the fraction of the total energy dissipated by a 1-Ω resistor in the frequency band $-10 < \omega < 10$ rad/s when the voltage across it is $v(t) = e^{-2t}u(t)$.

Solution:

Given that $f(t) = v(t) = e^{-2t}u(t)$, then

$$F(\omega) = \frac{1}{2 + j\omega} \quad \Rightarrow \quad |F(\omega)|^2 = \frac{1}{4 + \omega^2}$$

The total energy dissipated by the resistor is

$$W_{1\Omega} = \frac{1}{\pi}\int_0^{\infty} |F(\omega)|^2\, d\omega = \frac{1}{\pi}\int_0^{\infty} \frac{d\omega}{4 + \omega^2}$$
$$= \frac{1}{\pi}\left(\frac{1}{2}\tan^{-1}\frac{\omega}{2}\Big|_0^{\infty}\right) = \frac{1}{\pi}\left(\frac{1}{2}\right)\frac{\pi}{2} = 0.25\ \text{J}$$

The energy in the frequencies $-10 < \omega < 10$ rad/s is

$$W = \frac{1}{\pi}\int_0^{10} |F(\omega)|^2\, d\omega = \frac{1}{\pi}\int_0^{10} \frac{d\omega}{4 + \omega^2} = \frac{1}{\pi}\left(\frac{1}{2}\tan^{1}\frac{\omega}{2}\Big|_0^{10}\right)$$
$$= \frac{1}{2\pi}\tan^{-1} 5 = \frac{1}{2\pi}\left(\frac{78.69^\circ}{180^\circ}\pi\right) = 0.218\ \text{J}$$

Its percentage of the total energy is

$$\frac{W}{W_{1\Omega}} = \frac{0.218}{0.25} = 87.4\ \%$$

Practice Problem 18.10

A 2-Ω resistor has $i(t) = 2e^{-t}u(t)$ A. What percentage of the total energy is in the frequency band $-4 < \omega < 4$ rad/s?

Answer: 84.4 percent.

18.6 Comparing the Fourier and Laplace Transforms

It is worthwhile to take some moments to compare the Laplace and Fourier transforms. The following similarities and differences should be noted:

1. The Laplace transform defined in Chapter 15 is one-sided in that the integral is over $0 < t < \infty$, making it only useful for positive-time functions, $f(t)$, $t > 0$. The Fourier transform is applicable to functions defined for all time.
2. For a function $f(t)$ that is nonzero for positive time only (i.e., $f(t) = 0, t < 0$) and $\int_0^\infty |f(t)|\, dt < \infty$, the two transforms are related by

$$F(\omega) = F(s)\big|_{s=j\omega} \tag{18.66}$$

   This equation also shows that the Fourier transform can be regarded as a special case of the Laplace transform with $s = j\omega$. Recall that $s = \sigma + j\omega$. Therefore, Eq. (18.66) shows that the Laplace transform is related to the entire s plane, whereas the Fourier transform is restricted to the $j\omega$ axis. See Fig. 15.1.
3. The Laplace transform is applicable to a wider range of functions than the Fourier transform. For example, the function $tu(t)$ has a Laplace transform but no Fourier transform. But Fourier transforms exist for signals that are not physically realizable and have no Laplace transforms.
4. The Laplace transform is better suited for the analysis of transient problems involving initial conditions, since it permits the inclusion of the initial conditions, whereas the Fourier transform does not. The Fourier transform is especially useful for problems in the steady state.
5. The Fourier transform provides greater insight into the frequency characteristics of signals than does the Laplace transform.

In other words, if all the poles of $F(s)$ lie in the left-hand side of the s plane, then one can obtain the Fourier transform $F(\omega)$ from the corresponding Laplace transform $F(s)$ by merely replacing s by $j\omega$. Note that this is not the case, for example, for $u(t)$ or $\cos atu(t)$.

Some of the similarities and differences can be observed by comparing Tables 15.1 and 15.2 with Tables 18.1 and 18.2.

18.7 †Applications

Besides its usefulness for circuit analysis, the Fourier transform is used extensively in a variety of fields such as optics, spectroscopy, acoustics, computer science, and electrical engineering. In electrical engineering, it is applied in communications systems and signal processing, where frequency response and frequency spectra are vital. Here we consider two simple applications: amplitude modulation (AM) and sampling.

18.7.1 Amplitude Modulation

Electromagnetic radiation or transmission of information through space has become an indispensable part of a modern technological society. However, transmission through space is only efficient and economical at high frequencies (above 20 kHz). To transmit intelligent signals—such as for speech and music—contained in the low-frequency range of 50 Hz to 20 kHz is expensive; it requires a huge amount of power and large antennas. A common method of transmitting low-frequency audio information is to transmit a high-frequency signal, called a *carrier*, which is controlled in some way to correspond to the audio information. Three characteristics (amplitude, frequency, or phase) of a carrier can be controlled so as to allow it to carry the intelligent signal, called the *modulating signal*. Here we will only consider the control of the carrier's amplitude. This is known as *amplitude modulation.*

> **Amplitude modulation (AM)** is a process whereby the amplitude of the carrier is controlled by the modulating signal.

AM is used in ordinary commercial radio bands and the video portion of commercial television.

Suppose the audio information, such as voice or music (or the modulating signal in general) to be transmitted is $m(t) = V_m \cos \omega_m t$, while the high-frequency carrier is $c(t) = V_c \cos \omega_c t$, where $\omega_c \gg \omega_m$. Then an AM signal $f(t)$ is given by

$$f(t) = V_c[1 + m(t)] \cos \omega_c t \tag{18.67}$$

Figure 18.22 illustrates the modulating signal $m(t)$, the carrier $c(t)$, and the AM signal $f(t)$. We can use the result in Eq. (18.27) together with the Fourier transform of the cosine function (see Example 18.1 or Table 18.1) to determine the spectrum of the AM signal:

$$\begin{aligned} F(\omega) &= \mathcal{F}[V_c \cos \omega_c t] + \mathcal{F}[V_c m(t) \cos \omega_c t] \\ &= V_c \pi[\delta(\omega - \omega_c) + \delta(\omega + \omega_c)] \\ &\quad + \frac{V_c}{2}[M(\omega - \omega_c) + M(\omega + \omega_c)] \end{aligned} \tag{18.68}$$

where $M(\omega)$ is the Fourier transform of the modulating signal $m(t)$. Shown in Fig. 18.23 is the frequency spectrum of the AM signal. Figure 18.23 indicates that the AM signal consists of the carrier and two other sinusoids. The sinusoid with frequency $\omega_c - \omega_m$ is known as the *lower sideband*, while the one with frequency $\omega_c + \omega_m$ is known as the *upper sideband.*

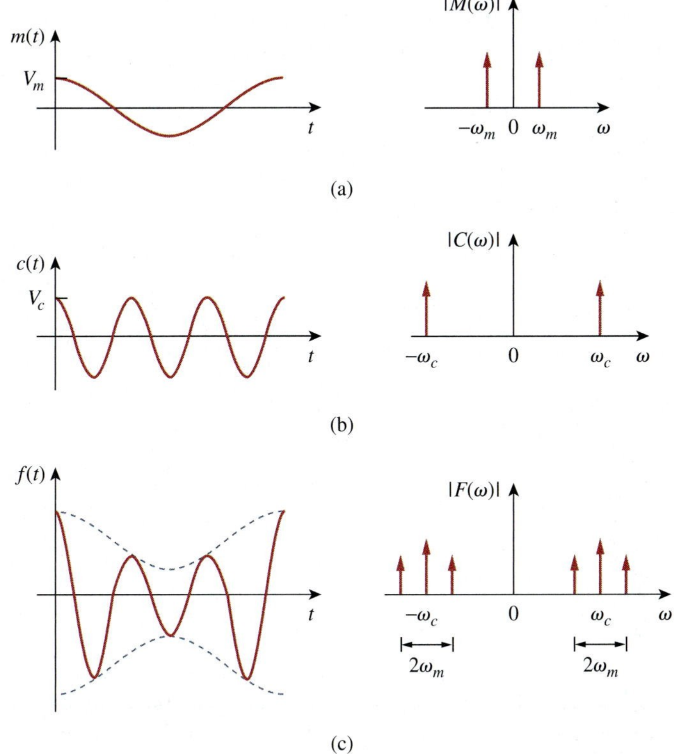

Figure 18.22
Time domain and frequency display of: (a) modulating signal, (b) carrier signal, (c) AM signal.

Figure 18.23
Frequency spectrum of AM signal.

Notice that we have assumed that the modulating signal is sinusoidal to make the analysis easy. In real life, $m(t)$ is a nonsinusoidal, band-limited signal—its frequency spectrum is within the range between 0 and $\omega_u = 2\pi f_u$ (i.e., the signal has an upper frequency limit). Typically, $f_u = 5$ kHz for AM radio. If the frequency spectrum of the modulating signal is as shown in Fig. 18.24(a), then the frequency spectrum of the AM signal is shown in Fig. 18.24(b). Thus, to avoid any interference, carriers for AM radio stations are spaced 10 kHz apart.

At the receiving end of the transmission, the audio information is recovered from the modulated carrier by a process known as *demodulation*.

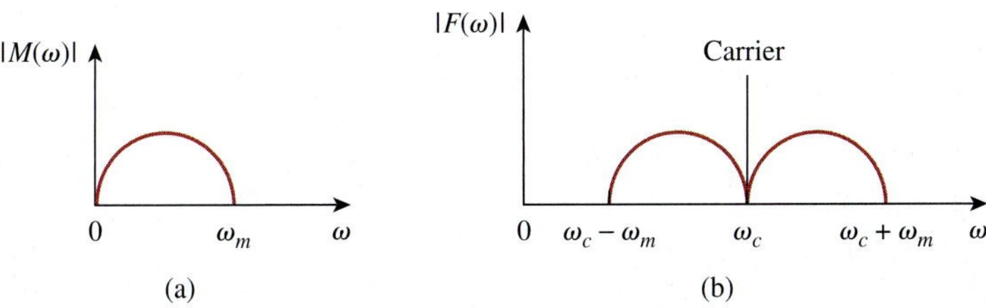

Figure 18.24
Frequency spectrum of: (a) modulating signal, (b) AM signal.

Example 18.11

A music signal has frequency components from 15 Hz to 30 kHz. If this signal could be used to amplitude modulate a 1.2-MHz carrier, find the range of frequencies for the lower and upper sidebands.

Solution:
The lower sideband is the difference of the carrier and modulating frequencies. It will include the frequencies from

$$1{,}200{,}000 - 30{,}000 \text{ Hz} = 1{,}170{,}000 \text{ Hz}$$

to

$$1{,}200{,}000 - 15 \text{ Hz} = 1{,}199{,}985 \text{ Hz}$$

The upper sideband is the sum of the carrier and modulating frequencies. It will include the frequencies from

$$1{,}200{,}000 + 15 \text{ Hz} = 1{,}200{,}015 \text{ Hz}$$

to

$$1{,}200{,}000 + 30{,}000 \text{ Hz} = 1{,}230{,}000 \text{ Hz}$$

Practice Problem 18.11

If a 2-MHz carrier is modulated by a 4-kHz intelligent signal, determine the frequencies of the three components of the AM signal that results.

Answer: 2,004,000 Hz, 2,000,000 Hz, 1,996,000 Hz.

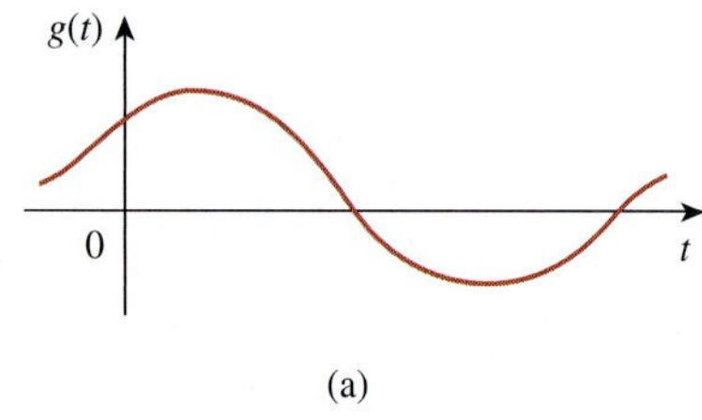

(a)

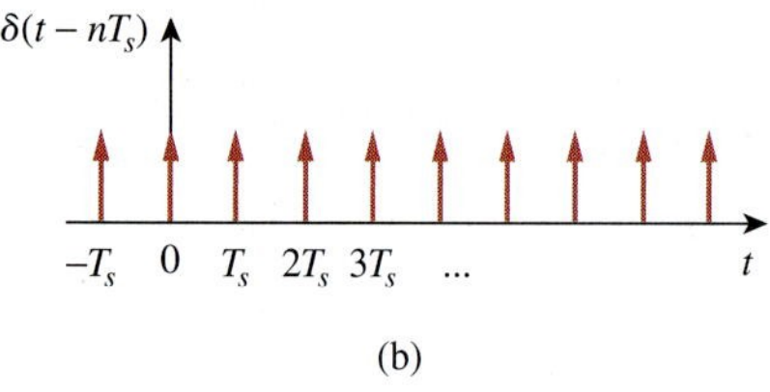

(b)

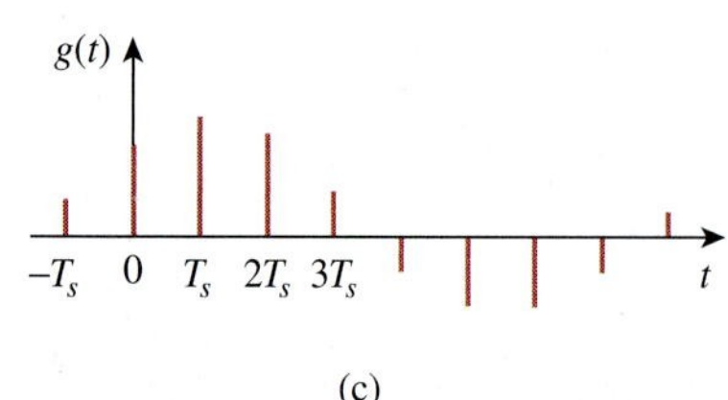

(c)

Figure 18.25
(a) Continuous (analog) signal to be sampled, (b) train of impulses, (c) sampled (digital) signal.

18.7.2 Sampling

In analog systems, signals are processed in their entirety. However, in modern digital systems, only samples of signals are required for processing. This is possible as a result of the sampling theorem given in Section 17.8.1. The sampling can be done by using a train of pulses or impulses. We will use impulse sampling here.

Consider the continuous signal $g(t)$ shown in Fig. 18.25(a). This can be multiplied by a train of impulses $\delta(t - nT_s)$ shown in Fig. 18.25(b), where T_s is the *sampling interval* and $f_s = 1/T_s$ is the *sampling frequency* or the *sampling rate*. The sampled signal $g_s(t)$ is therefore

$$g_s(t) = g(t) \sum_{n=-\infty}^{\infty} \delta(t - nT_s) = \sum_{n=-\infty}^{\infty} g(nT_s)\,\delta(t - nT_s) \tag{18.69}$$

The Fourier transform of this is

$$G_s(\omega) = \sum_{n=-\infty}^{\infty} g(nT_s)\mathcal{F}[\delta(t - nT_s)] = \sum_{n=-\infty}^{\infty} g(nT_s)e^{-jn\omega T_s} \tag{18.70}$$

It can be shown that

$$\sum_{n=-\infty}^{\infty} g(nT_s)e^{-jn\omega T_s} = \frac{1}{T_s} \sum_{n=-\infty}^{\infty} G(\omega + n\omega_s) \tag{18.71}$$

where $\omega_s = 2\pi/T_s$. Thus, Eq. (18.70) becomes

$$G_s(\omega) = \frac{1}{T_s} \sum_{n=-\infty}^{\infty} G(\omega + n\omega_s) \tag{18.72}$$

This shows that the Fourier transform $G_s(\omega)$ of the sampled signal is a sum of translates of the Fourier transform of the original signal at a rate of $1/T_s$.

In order to ensure optimum recovery of the original signal, what must be the sampling interval? This fundamental question in sampling is answered by an equivalent part of the sampling theorem:

> A band-limited signal, with no frequency component higher than W hertz, may be completely recovered from its samples taken at a frequency at least twice as high as $2W$ samples per second.

In other words, for a signal with bandwidth W hertz, there is no loss of information or overlapping if the sampling frequency is at least twice the highest frequency in the modulating signal. Thus,

$$\frac{1}{T_s} = f_s \geq 2W \tag{18.73}$$

The sampling frequency $f_s = 2W$ is known as the *Nyquist frequency* or rate, and $1/f_s$ is the *Nyquist interval.*

Example 18.12

A telephone signal with a cutoff frequency of 5 kHz is sampled at a rate 60 percent higher than the minimum allowed rate. Find the sampling rate.

Solution:
The minimum sample rate is the Nyquist rate = $2W = 2 \times 5 = 10$ kHz. Hence,

$$f_s = 1.60 \times 2W = 16 \text{ kHz}$$

Practice Problem 18.12

An audio signal that is band-limited to 12.5 kHz is digitized into 8-bit samples. What is the maximum sampling interval that must be used to ensure complete recovery?

Answer: 40 μs.

18.8 Summary

1. The Fourier transform converts a nonperiodic function $f(t)$ into a transform $F(\omega)$, where

$$F(\omega) = \mathcal{F}[f(t)] = \int_{-\infty}^{\infty} f(t)e^{-j\omega t}\,dt$$

2. The inverse Fourier transform of $F(\omega)$ is

$$f(t) = \mathcal{F}^{-1}[F(\omega)] = \frac{1}{2\pi}\int_{-\infty}^{\infty} F(\omega)e^{j\omega t}\,d\omega$$

3. Important Fourier transform properties and pairs are summarized in Tables 18.1 and 18.2, respectively.
4. Using the Fourier transform method to analyze a circuit involves finding the Fourier transform of the excitation, transforming the circuit element into the frequency domain, solving for the unknown response, and transforming the response to the time domain using the inverse Fourier transform.
5. If $H(\omega)$ is the transfer function of a network, then $H(\omega)$ is the Fourier transform of the network's impulse response; that is,

$$H(\omega) = \mathcal{F}[h(t)]$$

The output $V_o(\omega)$ of the network can be obtained from the input $V_i(\omega)$ using

$$V_o(\omega) = H(\omega)V_i(\omega)$$

6. Parseval's theorem gives the energy relationship between a function $f(t)$ and its Fourier transform $F(\omega)$. The 1-Ω energy is

$$W_{1\Omega} = \int_{-\infty}^{\infty} f^2(t)\,dt = \frac{1}{2\pi}\int_{-\infty}^{\infty} |F(\omega)|^2\,d\omega$$

The theorem is useful in calculating energy carried by a signal either in the time domain or in the frequency domain.
7. Typical applications of the Fourier transform are found in amplitude modulation (AM) and sampling. For AM application, a way of determining the sidebands in an amplitude-modulated wave is derived from the modulation property of the Fourier transform. For sampling application, we found that no information is lost in sampling (required for digital transmission) if the sampling frequency is equal to at least twice the Nyquist rate.

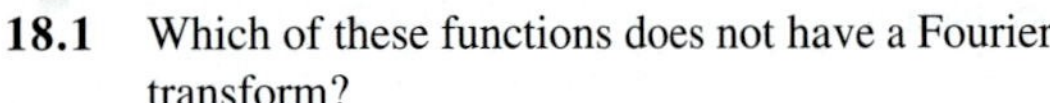

Review Questions

18.1 Which of these functions does not have a Fourier transform?

(a) $e^t u(-t)$ (b) $te^{-3t}u(t)$
(c) $1/t$ (d) $|t|u(t)$

18.2 The Fourier transform of e^{j2t} is:

(a) $\dfrac{1}{2 + j\omega}$ (b) $\dfrac{1}{-2 + j\omega}$
(c) $2\pi\delta(\omega - 2)$ (d) $2\pi\delta(\omega + 2)$

18.3 The inverse Fourier transform of $\dfrac{e^{-j\omega}}{2 + j\omega}$ is

(a) e^{-2t} (b) $e^{-2t}u(t - 1)$
(c) $e^{-2(t-1)}$ (d) $e^{-2(t-1)}u(t - 1)$

18.4 The inverse Fourier transform of $\delta(\omega)$ is:

(a) $\delta(t)$ (b) $u(t)$ (c) 1 (d) $1/2\pi$

18.5 The inverse Fourier transform of $j\omega$ is:

(a) $\delta'(t)$ (b) $u'(t)$
(c) $1/t$ (d) undefined

18.6 Evaluating the integral $\displaystyle\int_{-\infty}^{\infty} \frac{10\delta(\omega)}{4 + \omega^2}\,d\omega$ results in:

(a) 0 (b) 2 (c) 2.5 (d) ∞

18.7 The integral $\displaystyle\int_{-\infty}^{\infty} \frac{10\delta(\omega - 1)}{4 + \omega^2}\,d\omega$ gives:

(a) 0 (b) 2 (c) 2.5 (d) ∞

18.8 The current through an initially uncharged 1-F capacitor is $\delta(t)$ A. The voltage across the capacitor is:

(a) $u(t)$ V (b) $-1/2 + u(t)$ V
(c) $e^{-t}u(t)$ V (d) $\delta(t)$ V

18.9 A unit step current is applied through a 1-H inductor. The voltage across the inductor is:

(a) $u(t)$ V (b) sgn(t) V
(c) $e^{-t}u(t)$ V (d) $\delta(t)$ V

18.10 Parseval's theorem is only for nonperiodic functions.

(a) True (b) False

Answers: 18.1c, 18.2c, 18.3d, 18.4d, 18.5a, 18.6c, 18.7b, 18.8a, 18.9d, 18.10b

Problems

†Sections 18.2 and 18.3 Fourier Transform and its Properties

18.1 Obtain the Fourier transform of the function in Fig. 18.26.

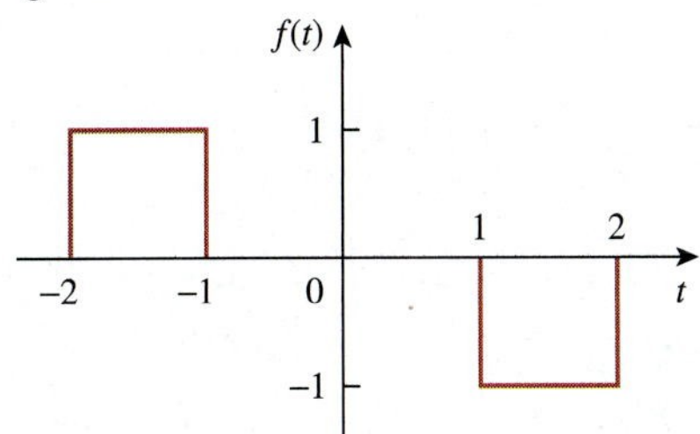

Figure 18.26
For Prob. 18.1.

18.2 Using Fig. 18.27, design a problem to help other students better understand the Fourier transform given a wave shape.

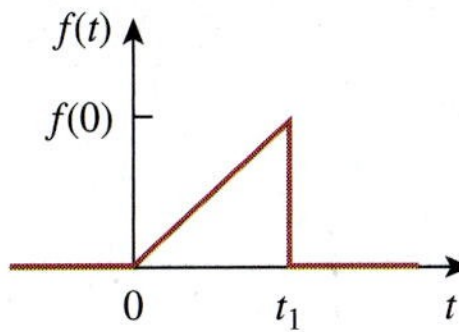

Figure 18.27
For Prob. 18.2.

18.3 Calculate the Fourier transform of the signal in Fig. 18.28.

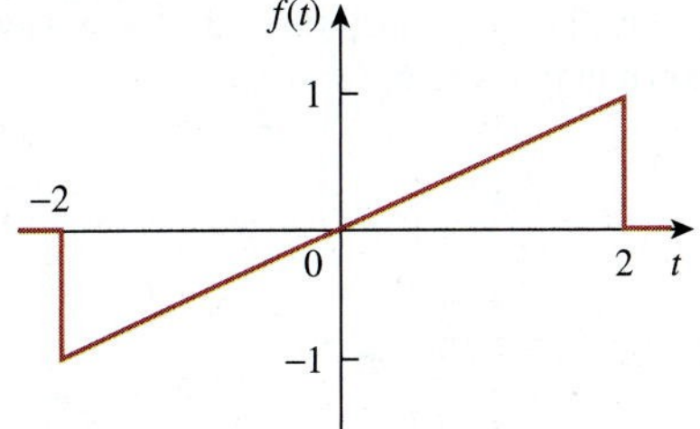

Figure 18.28
For Prob. 18.3.

18.4 Find the Fourier transform of the waveform shown in Fig. 18.29.

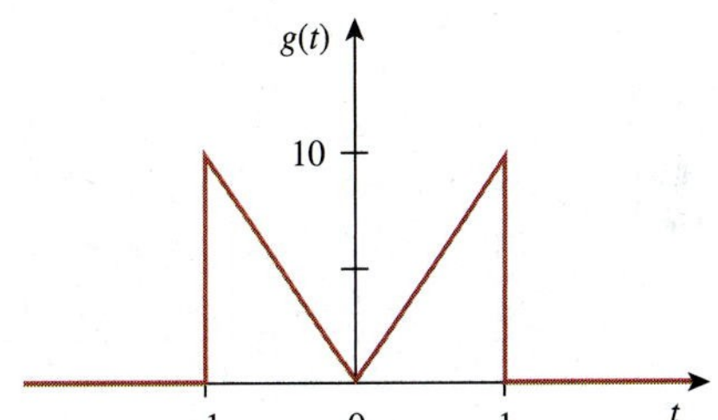

Figure 18.29
For Prob. 18.4.

18.5 Obtain the Fourier transform of the signal shown in Fig. 18.30.

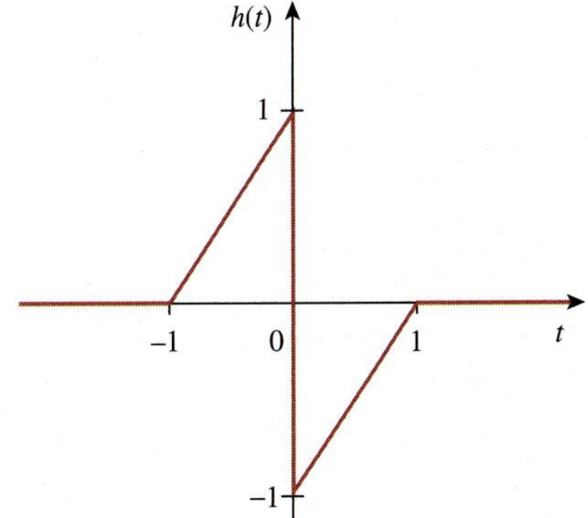

Figure 18.30
For Prob. 18.5.

18.6 Find the Fourier transforms of both functions in Fig. 18.31 on the following page.

† We have marked (with the *MATLAB* icon) the problems where we are asking the student to find the Fourier transform of a wave shape. We do this because you can use *MATLAB* to plot the results as a check.

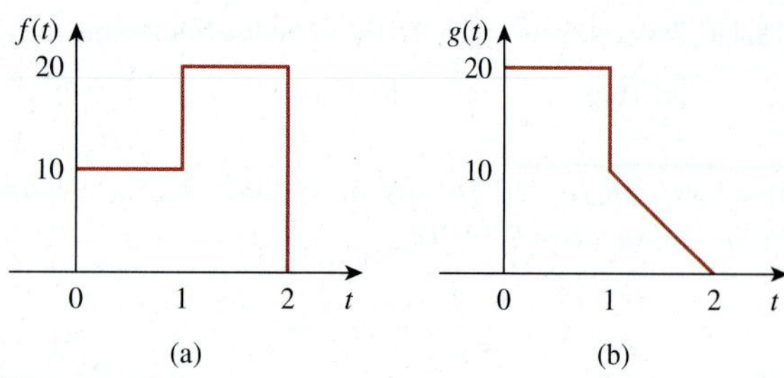

Figure 18.31
For Prob. 18.6.

18.7 Find the Fourier transforms of the signals in Fig. 18.32.

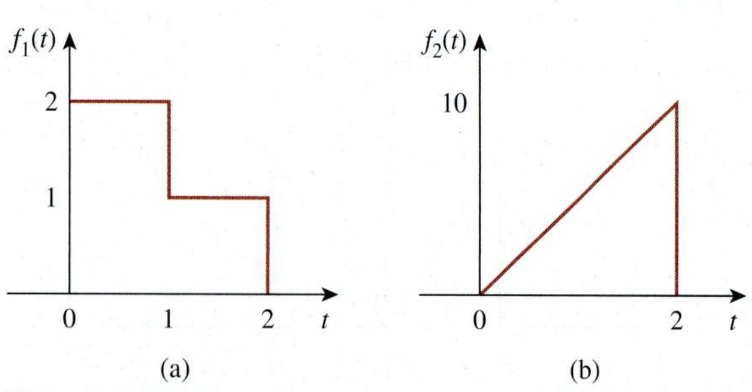

Figure 18.32
For Prob. 18.7.

18.8 Obtain the Fourier transforms of the signals shown in Fig. 18.33.

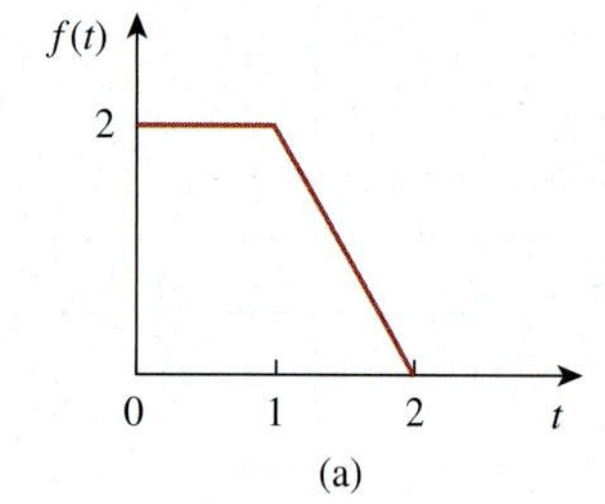

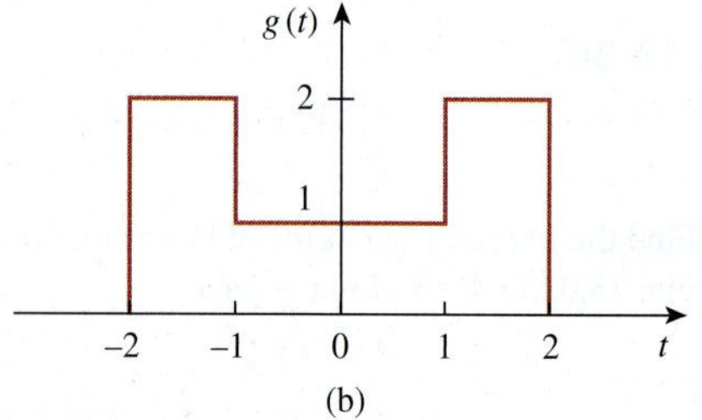

Figure 18.33
For Prob. 18.8.

18.9 Determine the Fourier transforms of the signals in Fig. 18.34.

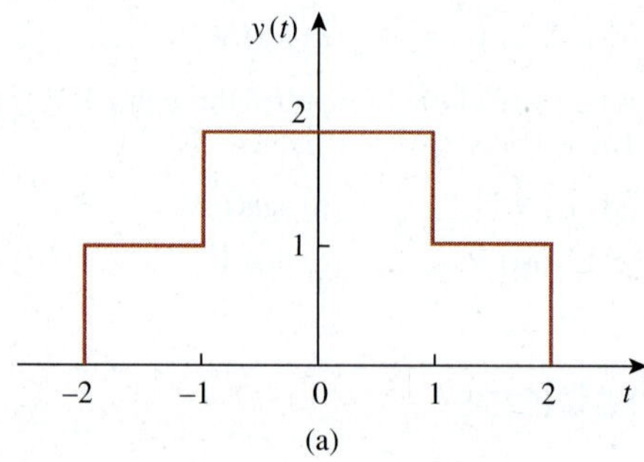

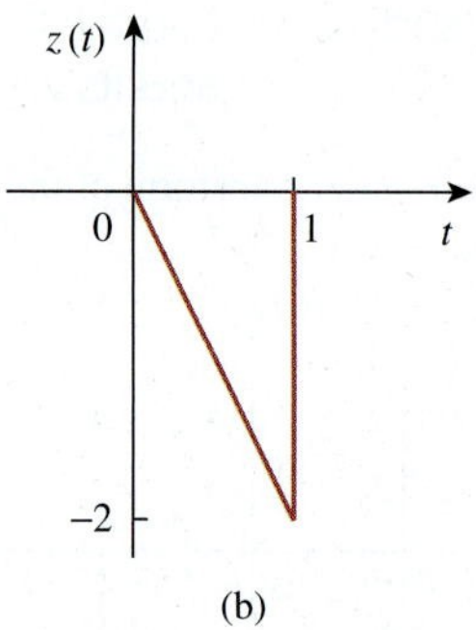

Figure 18.34
For Prob. 18.9.

18.10 Obtain the Fourier transforms of the signals shown in Fig. 18.35.

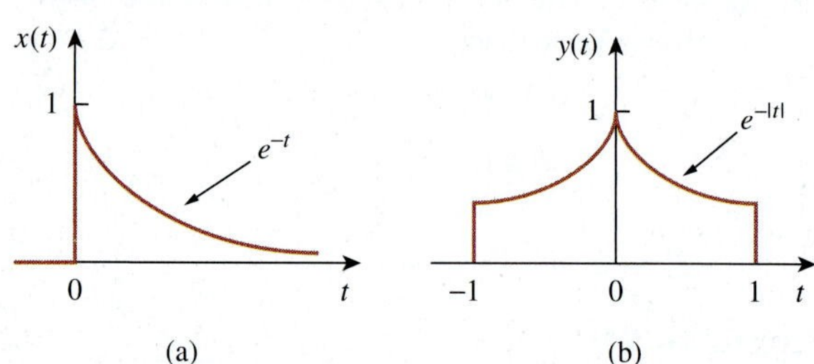

Figure 18.35
For Prob. 18.10.

18.11 Find the Fourier transform of the "sine-wave pulse" shown in Fig. 18.36.

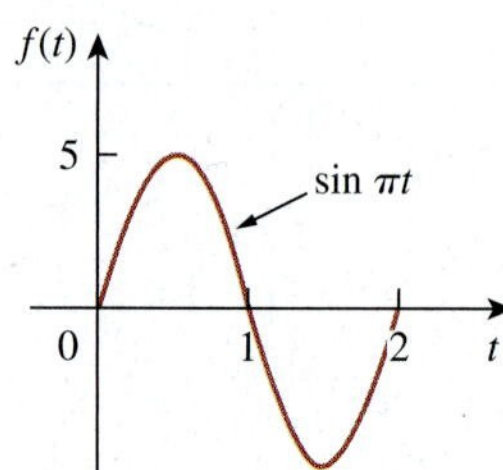

Figure 18.36
For Prob. 18.11.

18.12 Find the Fourier transform of the following signals.

(a) $f_1(t) = e^{-3t}\sin(10t)u(t)$

(b) $f_2(t) = e^{-4t}\cos(10t)u(t)$

18.13 Find the Fourier transform of the following signals:

(a) $f(t) = \cos(at - \pi/3), \quad -\infty < t < \infty$

(b) $g(t) = u(t + 1)\sin\pi t, \quad -\infty < t < \infty$

(c) $h(t) = (1 + A\sin at)\cos bt, \quad -\infty < t < \infty$, where A, a, and b are constants

(d) $i(t) = 1 - t, \quad 0 < t < 4$

18.14 ed Design a problem to help other students better understand finding the Fourier transform of a variety of time varying functions (do at least three).

18.15 Find the Fourier transforms of the following functions:

(a) $f(t) = \delta(t + 3) - \delta(t - 3)$

(b) $f(t) = \displaystyle\int_{-\infty}^{\infty} 2\delta(t - 1)\,dt$

(c) $f(t) = \delta(3t) - \delta'(2t)$

***18.16** Determine the Fourier transforms of these functions:

(a) $f(t) = 10/t^2$

(b) $g(t) = 20/(4 + t^2)$

18.17 Find the Fourier transforms of:

(a) $\cos 2tu(t)$

(b) $\sin 10tu(t)$

18.18 Given that $F(\omega) = \mathcal{F}[f(t)]$, prove the following results, using the definition of Fourier transform:

(a) $\mathcal{F}[f(t - t_0)] = e^{-j\omega t_0}F(\omega)$

(b) $\mathcal{F}\left[\dfrac{df(t)}{dt}\right] = j\omega F(\omega)$

(c) $\mathcal{F}[f(-t)] = F(-\omega)$

(d) $\mathcal{F}[tf(t)] = j\dfrac{d}{d\omega}F(\omega)$

18.19 Find the Fourier transform of

$$f(t) = \cos 2\pi t[u(t) - u(t - 1)]$$

18.20 ML (a) Show that a periodic signal with exponential Fourier series

$$f(t) = \sum_{n=-\infty}^{\infty} c_n e^{jn\omega_0 t}$$

has the Fourier transform

$$F(\omega) = \sum_{n=-\infty}^{\infty} c_n\delta(\omega - n\omega_0)$$

where $\omega_0 = 2\pi/T$.

(b) Find the Fourier transform of the signal in Fig. 18.37.

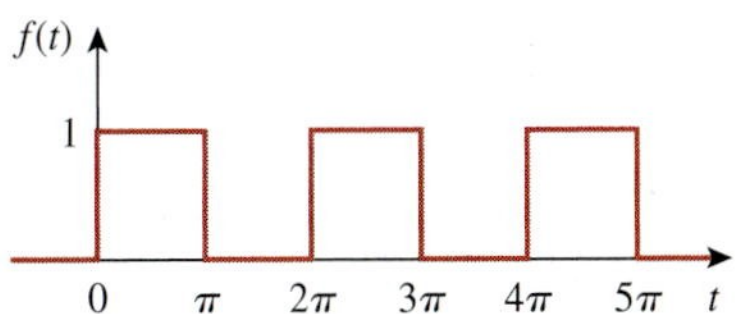

Figure 18.37
For Prob. 18.20(b).

18.21 Show that

$$\int_{-\infty}^{\infty}\left(\frac{\sin a\omega}{a\omega}\right)^2 d\omega = \frac{\pi}{a}$$

Hint: Use the fact that

$$\mathcal{F}[u(t + a) - u(t - a)] = 2a\left(\frac{\sin a\omega}{a\omega}\right).$$

18.22 Prove that if $F(\omega)$ is the Fourier transform of $f(t)$,

$$\mathcal{F}[f(t)\sin\omega_0 t] = \frac{j}{2}[F(\omega + \omega_0) - F(\omega - \omega_0)]$$

18.23 If the Fourier transform of $f(t)$ is

$$F(\omega) = \frac{10}{(2 + j\omega)(5 + j\omega)}$$

determine the transforms of the following:

(a) $f(-3t)$ (b) $f(2t - 1)$ (c) $f(t)\cos 2t$

(d) $\dfrac{d}{dt}f(t)$ (e) $\displaystyle\int_{-\infty}^{t} f(t)\,dt$

18.24 Given that $\mathcal{F}[f(t)] = (j/\omega)(e^{-j\omega} - 1)$, find the Fourier transforms of:

(a) $x(t) = f(t) + 3$ (b) $y(t) = f(t - 2)$

(c) $h(t) = f'(t)$

(d) $g(t) = 4f\left(\dfrac{2}{3}t\right) + 10f\left(\dfrac{5}{3}t\right)$

18.25 Obtain the inverse Fourier transform of the following signals.

(a) $G(\omega) = \dfrac{10}{j\omega - 2}$

(b) $H(\omega) = \dfrac{6}{\omega^2 + 4}$

(c) $X(\omega) = \dfrac{5}{(j\omega - 1)(j\omega - 2)}$

18.26 Determine the inverse Fourier transforms of the following:

(a) $F(\omega) = \dfrac{e^{-j2\omega}}{1 + j\omega}$

(b) $H(\omega) = \dfrac{1}{(j\omega + 4)^2}$

(c) $G(\omega) = 2u(\omega + 1) - 2u(\omega - 1)$

---

* An asterisk indicates a challenging problem.

18.27 Find the inverse Fourier transforms of the following functions:

(a) $F(\omega) = \dfrac{100}{j\omega(j\omega + 10)}$

(b) $G(\omega) = \dfrac{10j\omega}{(-j\omega + 2)(j\omega + 3)}$

(c) $H(\omega) = \dfrac{60}{-\omega^2 + j40\omega + 1300}$

(d) $Y(\omega) = \dfrac{\delta(\omega)}{(j\omega + 1)(j\omega + 2)}$

18.28 Find the inverse Fourier transforms of:

(a) $\dfrac{\pi\,\delta(\omega)}{(5 + j\omega)(2 + j\omega)}$

(b) $\dfrac{10\delta(\omega + 2)}{j\omega(j\omega + 1)}$

(c) $\dfrac{20\delta(\omega - 1)}{(2 + j\omega)(3 + j\omega)}$

(d) $\dfrac{5\pi\,\delta(\omega)}{5 + j\omega} + \dfrac{5}{j\omega(5 + j\omega)}$

***18.29** Determine the inverse Fourier transforms of:

(a) $F(\omega) = 4\delta(\omega + 3) + \delta(\omega) + 4\delta(\omega - 3)$

(b) $G(\omega) = 4u(\omega + 2) - 4u(\omega - 2)$

(c) $H(\omega) = 6\cos 2\omega$

18.30 For a linear system with input $x(t)$ and output $y(t)$, find the impulse response for the following cases:

(a) $x(t) = e^{-at}u(t), \quad y(t) = u(t) - u(-t)$

(b) $x(t) = e^{-t}u(t), \quad y(t) = e^{-2t}u(t)$

(c) $x(t) = \delta(t), \quad y(t) = e^{-at}\sin btu(t)$

18.31 Given a linear system with output $y(t)$ and impulse response $h(t)$, find the corresponding input $x(t)$ for the following cases:

(a) $y(t) = te^{-at}u(t), \quad h(t) = e^{-at}u(t)$

(b) $y(t) = u(t + 1) - u(t - 1), \quad h(t) = \delta(t)$

(c) $y(t) = e^{-at}u(t), \quad h(t) = \text{sgn}(t)$

***18.32** Determine the functions corresponding to the following Fourier transforms:

(a) $F_1(\omega) = \dfrac{e^{j\omega}}{-j\omega + 1}$ (b) $F_2(\omega) = 2e^{|\omega|}$

(c) $F_3(\omega) = \dfrac{1}{(1 + \omega^2)^2}$ (d) $F_4(\omega) = \dfrac{\delta(\omega)}{1 + j2\omega}$

***18.33** Find $f(t)$ if:

(a) $F(\omega) = 2\sin\pi\omega[u(\omega + 1) - u(\omega - 1)]$

(b) $F(\omega) = \dfrac{1}{\omega}(\sin 2\omega - \sin\omega) + \dfrac{j}{\omega}(\cos 2\omega - \cos\omega)$

18.34 Determine the signal $f(t)$ whose Fourier transform is shown in Fig. 18.38. (*Hint:* Use the duality property.)

ML

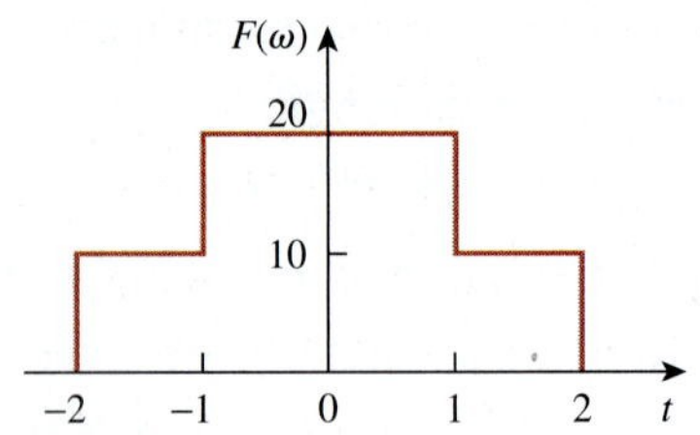

Figure 18.38
For Prob. 18.34.

18.35 A signal $f(t)$ has Fourier transform

$$F(\omega) = \frac{1}{2 + j\omega}$$

Determine the Fourier transform of the following signals:

(a) $x(t) = f(3t - 1)$

(b) $y(t) = f(t)\cos 5t$

(c) $z(t) = \dfrac{d}{dt}f(t)$

(d) $h(t) = f(t) * f(t)$

(e) $i(t) = tf(t)$

Section 18.4 Circuit Applications

18.36 The transfer function of a circuit is

$$H(\omega) = \frac{2}{j\omega + 2}$$

If the input signal to the circuit is $v_s(t) = e^{-4t}u(t)$ V, find the output signal. Assume all initial conditions are zero.

18.37 Find the transfer function $I_o(\omega)/I_s(\omega)$ for the circuit in Fig. 18.39.

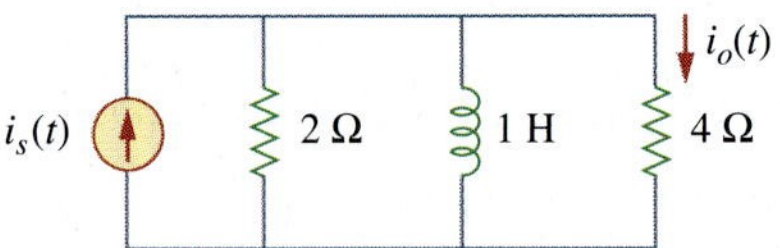

Figure 18.39
For Prob. 18.37.

18.38 Using Fig. 18.40, design a problem to help other students better understand using Fourier transforms to do circuit analysis.

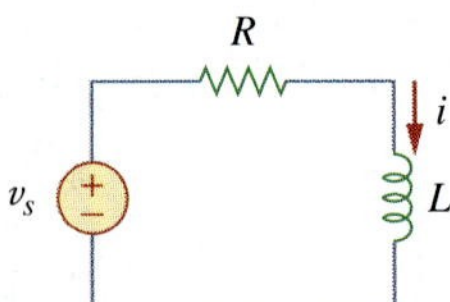

Figure 18.40
For Prob. 18.38.

18.39 Given the circuit in Fig. 18.41, with its excitation, determine the Fourier transform of $i(t)$.

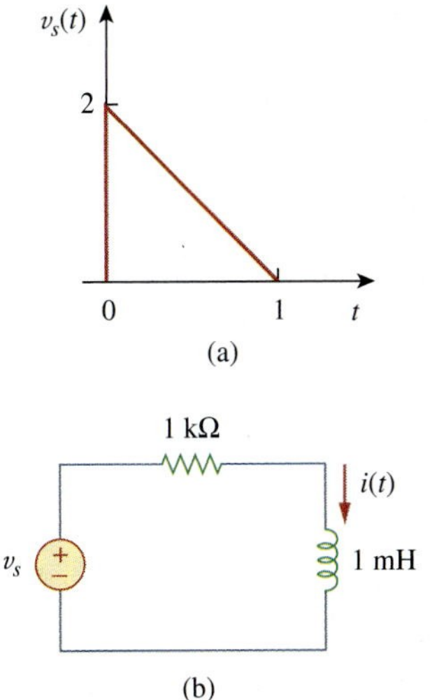

Figure 18.41
For Prob. 18.39.

18.40 Determine the current $i(t)$ in the circuit of Fig. 18.42(b), given the voltage source shown in Fig. 18.42(a).

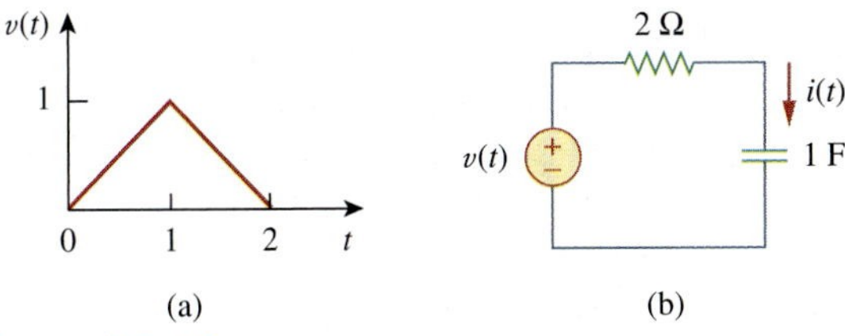

Figure 18.42
For Prob. 18.40.

18.41 Determine the Fourier transform of $v(t)$ in the circuit shown in Fig. 18.43.

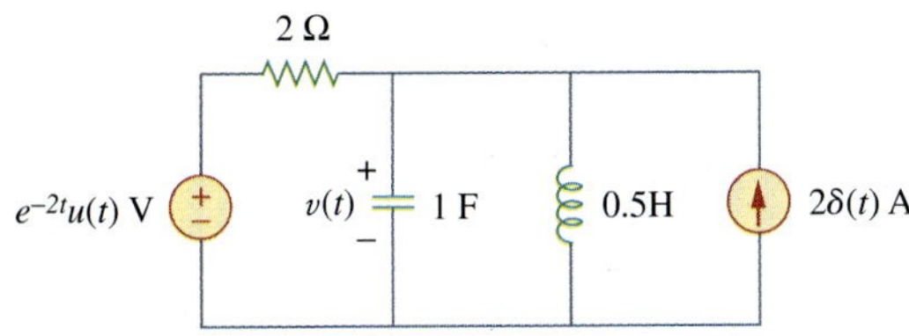

Figure 18.43
For Prob. 18.41.

18.42 Obtain the current $i_o(t)$ in the circuit of Fig. 18.44.

(a) Let $i(t) = \text{sgn}(t)$ A.

(b) Let $i(t) = 4[u(t) - u(t - 1)]$ A.

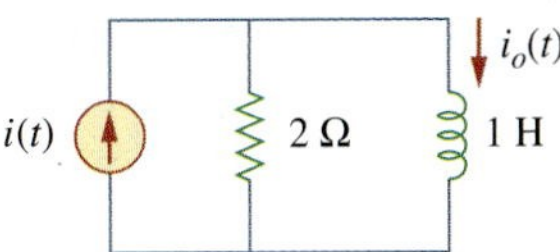

Figure 18.44
For Prob. 18.42.

18.43 Find $v_o(t)$ in the circuit of Fig. 18.45, where $i_s = 5e^{-t}u(t)$ A.

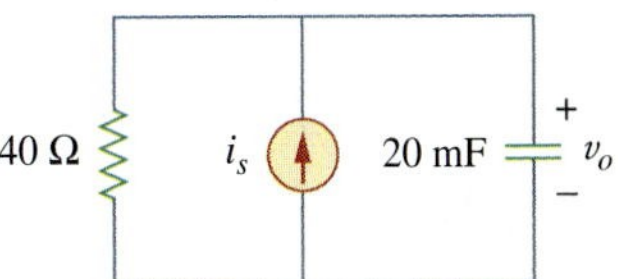

Figure 18.45
For Prob. 18.43.

18.44 If the rectangular pulse in Fig. 18.46(a) is applied to the circuit in Fig. 18.46(b), find v_o at $t = 1$ s.

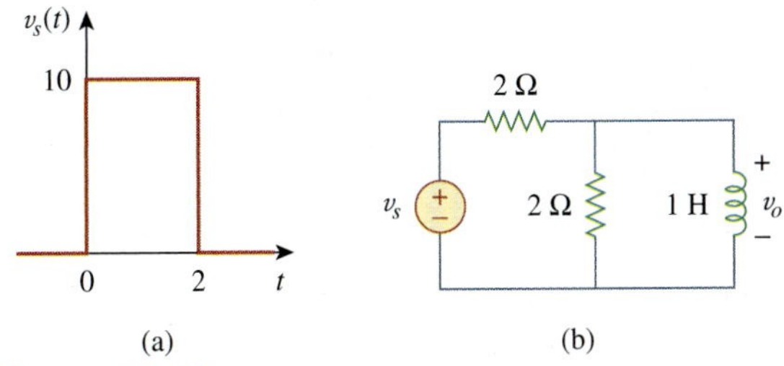

Figure 18.46
For Prob. 18.44.

18.45 Use the Fourier transform to find $i(t)$ in the circuit of Fig. 18.47 if $v_s(t) = 10e^{-2t}u(t)$.

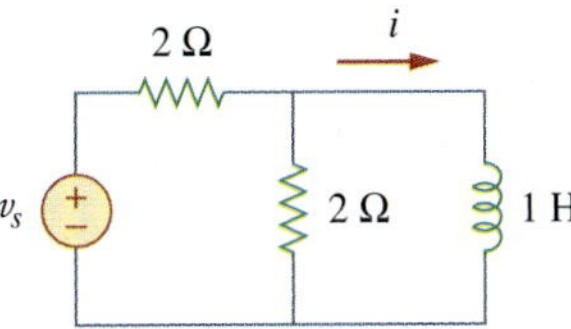

Figure 18.47
For Prob. 18.45.

18.46 Determine the Fourier transform of $i_o(t)$ in the circuit of Fig. 18.48.

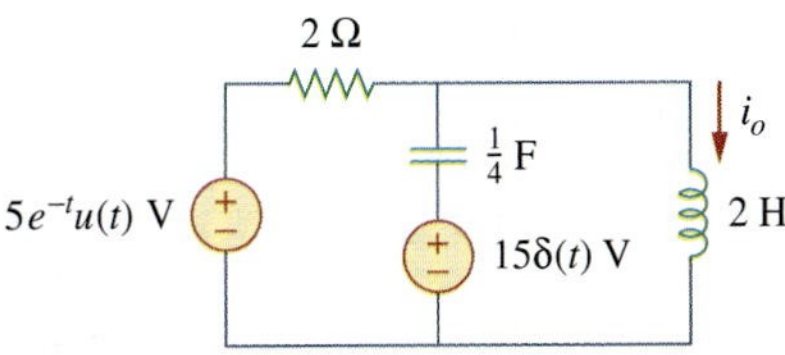

Figure 18.48
For Prob. 18.46.

18.47 Find the voltage $v_o(t)$ in the circuit of Fig. 18.49. Let $i_s(t) = 8e^{-t}u(t)$ A.

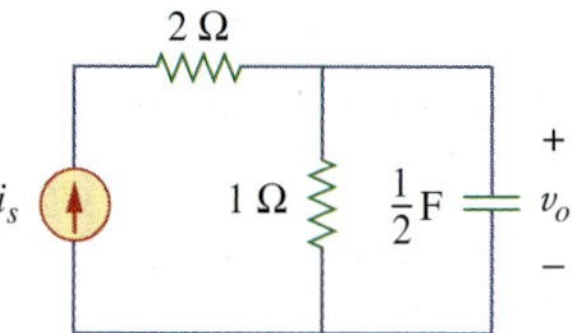

Figure 18.49
For Prob. 18.47.

18.48 Find $i_o(t)$ in the op amp circuit of Fig. 18.50.

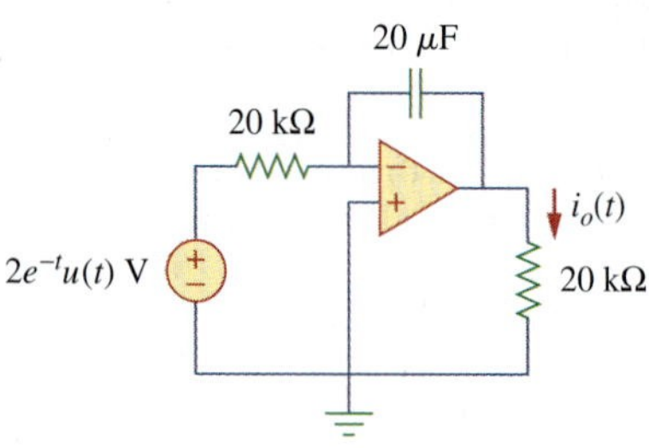

Figure 18.50
For Prob. 18.48.

18.49 Use the Fourier transform method to obtain $v_o(t)$ in the circuit of Fig. 18.51.

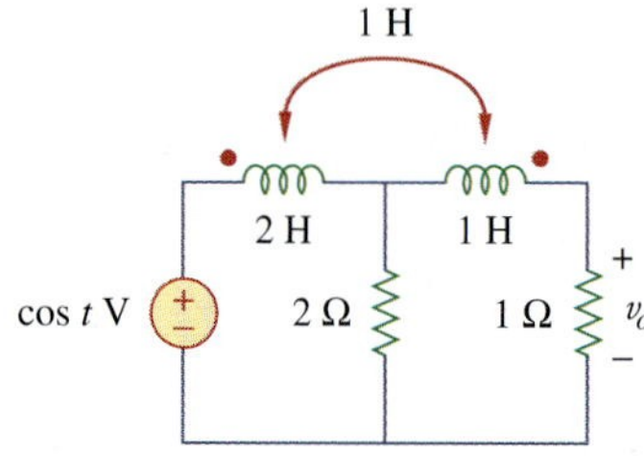

Figure 18.51
For Prob. 18.49.

18.50 Determine $v_o(t)$ in the transformer circuit of Fig. 18.52.

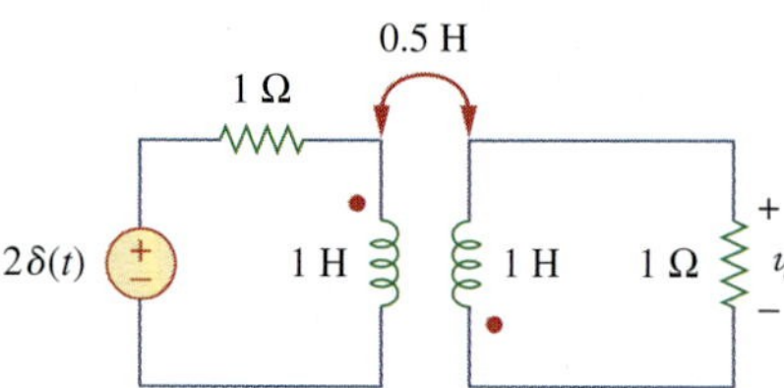

Figure 18.52
For Prob. 18.50.

18.51 Find the energy dissipated by the resistor in the circuit of Fig. 18.53.

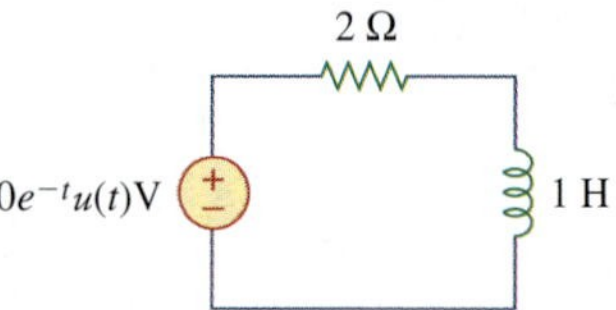

Figure 18.53
For Prob. 18.51.

Section 18.5 Parseval's Theorem

18.52 For $F(\omega) = \dfrac{1}{3 + j\omega}$, find $J = \displaystyle\int_{-\infty}^{\infty} f^2(t)\, dt$.

18.53 If $f(t) = e^{-2|t|}$, find $J = \displaystyle\int_{-\infty}^{\infty} |F(\omega)|^2\, d\omega$.

18.54 **e⃞d** Design a problem to help other students better understand finding the total energy in a given signal.

18.55 Let $f(t) = 5e^{-(t-2)}u(t)$. Find $F(\omega)$ and use it to find the total energy in $f(t)$.

18.56 The voltage across a 1-Ω resistor is $v(t) = te^{-2t}u(t)$ V. (a) What is the total energy absorbed by the resistor? (b) What fraction of this energy absorbed is in the frequency band $-2 \le \omega \le 2$?

18.57 Let $i(t) = 5e^{t}u(-t)$ A. Find the total energy carried by $i(t)$ and the percentage of the 1-Ω energy in the frequency range of $-5 < \omega < 5$ rad/s.

Section 18.6 Applications

18.58 **e⃞d** An AM signal is specified by

$$f(t) = 10(1 + 4\cos 200\pi t)\cos \pi \times 10^4 t$$

Determine the following:

(a) the carrier frequency,

(b) the lower sideband frequency,

(c) the upper sideband frequency.

18.59 For the linear system in Fig. 18.54, when the input voltage is $v_i(t) = 2\delta(t)$ V, the output is $v_o(t) = 10e^{-2t} - 6e^{-4t}$ V. Find the output when the input is $v_i(t) = 4e^{-t}u(t)$ V.

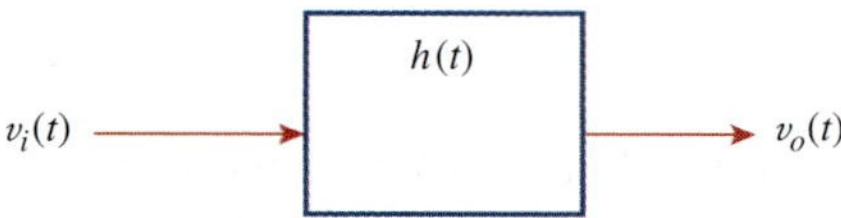

Figure 18.54
For Prob. 18.9.

18.60 A band-limited signal has the following Fourier series representation:

$$i_s(t) = 10 + 8\cos(2\pi t + 30^\circ) + 5\cos(4\pi t - 150^\circ)\text{mA}$$

If the signal is applied to the circuit in Fig. 18.55, find $v(t)$.

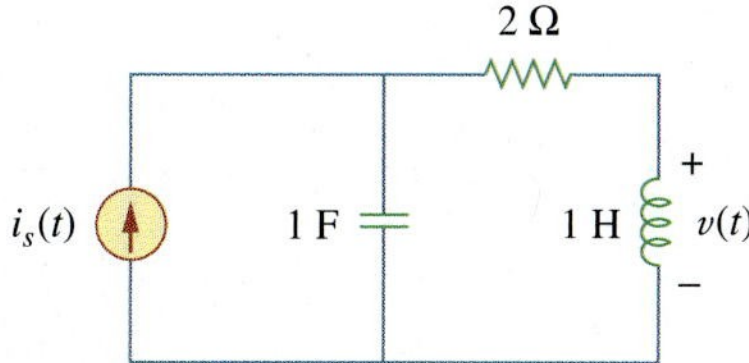

Figure 18.55
For Prob. 18.60.

18.61 In a system, the input signal $x(t)$ is amplitude-modulated by $m(t) = 2 + \cos\omega_0 t$. The response $y(t) = m(t)x(t)$. Find $Y(\omega)$ in terms of $X(\omega)$.

18.62 A voice signal occupying the frequency band of 0.4 to 3.5 kHz is used to amplitude-modulate a 10-MHz carrier. Determine the range of frequencies for the lower and upper sidebands.

18.63 For a given locality, calculate the number of stations allowable in the AM broadcasting band (540 to 1600 kHz) without interference with one another.

18.64 Repeat the previous problem for the FM broadcasting band (88 to 108 MHz), assuming that the carrier frequencies are spaced 200 kHz apart.

18.65 The highest-frequency component of a voice signal is 3.4 kHz. What is the Nyquist rate of the sampler of the voice signal?

18.66 A TV signal is band-limited to 4.5 MHz. If samples are to be reconstructed at a distant point, what is the maximum sampling interval allowable?

***18.67** Given a signal $g(t) = \text{sinc}(200\pi t)$, find the Nyquist rate and the Nyquist interval for the signal.

Comprehensive Problems

18.68 The voltage signal at the input of a filter is $v(t) = 50e^{-2|t|}$ V. What percentage of the total 1-Ω energy content lies in the frequency range of $1 < \omega < 5$ rad/s?

18.69 A signal with Fourier transform

$$F(\omega) = \frac{20}{4 + j\omega}$$

is passed through a filter whose cutoff frequency is 2 rad/s (i.e., $0 < \omega < 2$). What fraction of the energy in the input signal is contained in the output signal?

chapter

19

Two-Port Networks

Never put off till tomorrow what you can do today.
Never trouble another for what you can do yourself.
Never spend your money before you have it.
Never buy what you do not want because it is cheap.
Pride costs us more than hunger, thirst, and cold.
We seldom repent having eaten too little.
Nothing is troublesome that we do willingly.
How much pain the evils have cost us that have never happened!
Take things always by the smooth handle.
When angry, count ten before you speak; if very angry, a hundred.

—Thomas Jefferson

Enhancing Your Career

Career in Education

While two thirds of all engineers work in private industry, some work in academia and prepare students for engineering careers. The course on circuit analysis you are studying is an important part of the preparation process. If you enjoy teaching others, you may want to consider becoming an engineering educator.

Engineering professors work on state-of-the-art research projects, teach courses at graduate and undergraduate levels, and provide services to their professional societies and the community at large. They are expected to make original contributions in their areas of specialty. This requires a broad-based education in the fundamentals of electrical engineering and a mastery of the skills necessary for communicating their efforts to others.

Photo by James Watson

If you like to do research, to work at the frontiers of engineering, to make contributions to technological advancement, to invent, consult, and/ or teach, consider a career in engineering education. The best way to start is by talking with your professors and benefiting from their experience.

A solid understanding of mathematics and physics at the undergraduate level is vital to your success as an engineering professor. If you are having difficulty in solving your engineering textbook problems, start correcting any weaknesses you have in your mathematics and physics fundamentals.

Most universities these days require that engineering professors have a doctor's degree. In addition, some universities require that they be actively involved in research leading to publications in reputable journals.

To prepare yourself for a career in engineering education, get as broad an education as possible, because electrical engineering is changing rapidly and becoming interdisciplinary. Without doubt, engineering education is a rewarding career. Professors get a sense of satisfaction and fulfillment as they see their students graduate, become leaders in their profession, and contribute significantly to the betterment of humanity.

19.1 Introduction

A pair of terminals through which a current may enter or leave a network is known as a *port*. Two-terminal devices or elements (such as resistors, capacitors, and inductors) result in one-port networks. Most of the circuits we have dealt with so far are two-terminal or one-port circuits, represented in Fig. 19.1(a). We have considered the voltage across or current through a single pair of terminals—such as the two terminals of a resistor, a capacitor, or an inductor. We have also studied four-terminal or two-port circuits involving op amps, transistors, and transformers, as shown in Fig. 19.1(b). In general, a network may have n ports. A port is an access to the network and consists of a pair of terminals; the current entering one terminal leaves through the other terminal so that the net current entering the port equals zero.

In this chapter, we are mainly concerned with *two-port* networks (or, simply, *two-ports*).

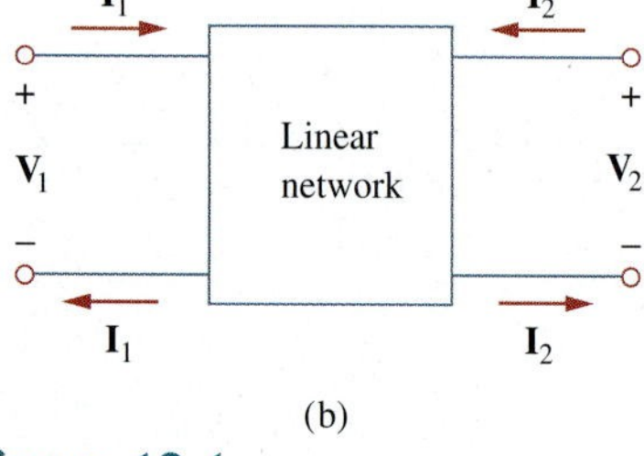

A **two-port network** is an electrical network with two separate ports for input and output.

Thus, a two-port network has two terminal pairs acting as access points. As shown in Fig. 19.1(b), the current entering one terminal of a pair leaves the other terminal in the pair. Three-terminal devices such as transistors can be configured into two-port networks.

Our study of two-port networks is for at least two reasons. First, such networks are useful in communications, control systems, power systems, and electronics. For example, they are used in electronics to model transistors and to facilitate cascaded design. Second, knowing the parameters of a two-port network enables us to treat it as a "black box" when embedded within a larger network.

To characterize a two-port network requires that we relate the terminal quantities $\mathbf{V}_1$, $\mathbf{V}_2$, $\mathbf{I}_1$, and $\mathbf{I}_2$ in Fig. 19.1(b), out of which two are independent. The various terms that relate these voltages and currents are called *parameters*. Our goal in this chapter is to derive six sets of these parameters. We will show the relationship between these parameters and how two-port networks can be connected in series, parallel, or cascade. As with op amps, we are only interested in the terminal behavior of the circuits. And we will assume that the two-port circuits contain no independent sources, although they can contain dependent sources. Finally, we will apply some of the concepts developed in this chapter to the analysis of transistor circuits and synthesis of ladder networks.

I
+
V
−
I
Linear network
(a)

I1 I2
+ +
V1 Linear network V2
− −
I1 I2
(b)

Figure 19.1
(a) One-port network, (b) two-port network.

19.2 Impedance Parameters

Impedance and admittance parameters are commonly used in the synthesis of filters. They are also useful in the design and analysis of impedance-matching networks and power distribution networks. We

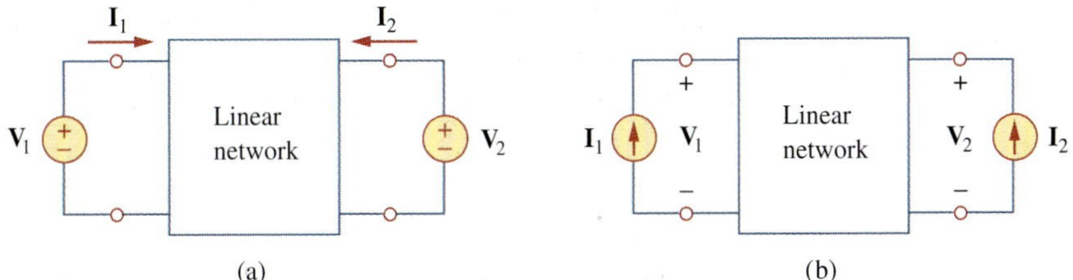

Figure 19.2
The linear two-port network: (a) driven by voltage sources, (b) driven by current sources.

discuss impedance parameters in this section and admittance parameters in the next section.

A two-port network may be voltage-driven as in Fig. 19.2(a) or current-driven as in Fig. 19.2(b). From either Fig. 19.2(a) or (b), the terminal voltages can be related to the terminal currents as

$$\begin{aligned} \mathbf{V}_1 &= \mathbf{z}_{11}\mathbf{I}_1 + \mathbf{z}_{12}\mathbf{I}_2 \\ \mathbf{V}_2 &= \mathbf{z}_{21}\mathbf{I}_1 + \mathbf{z}_{22}\mathbf{I}_2 \end{aligned} \tag{19.1}$$

Reminder: Only two of the four variables ($\mathbf{V}_1$, $\mathbf{V}_2$, $\mathbf{I}_1$, and $\mathbf{I}_2$) are independent. The other two can be found using Eq. (19.1).

or in matrix form as

$$\begin{bmatrix} \mathbf{V}_1 \\ \mathbf{V}_2 \end{bmatrix} = \begin{bmatrix} \mathbf{z}_{11} & \mathbf{z}_{12} \\ \mathbf{z}_{21} & \mathbf{z}_{22} \end{bmatrix} \begin{bmatrix} \mathbf{I}_1 \\ \mathbf{I}_2 \end{bmatrix} = [\mathbf{z}] \begin{bmatrix} \mathbf{I}_1 \\ \mathbf{I}_2 \end{bmatrix} \tag{19.2}$$

where the $\mathbf{z}$ terms are called the *impedance parameters*, or simply *z parameters*, and have units of ohms.

The values of the parameters can be evaluated by setting $\mathbf{I}_1 = 0$ (input port open-circuited) or $\mathbf{I}_2 = 0$ (output port open-circuited). Thus,

$$\begin{aligned} \mathbf{z}_{11} &= \frac{\mathbf{V}_1}{\mathbf{I}_1}\bigg|_{\mathbf{I}_2=0}, \qquad \mathbf{z}_{12} = \frac{\mathbf{V}_1}{\mathbf{I}_2}\bigg|_{\mathbf{I}_1=0} \\ \mathbf{z}_{21} &= \frac{\mathbf{V}_2}{\mathbf{I}_1}\bigg|_{\mathbf{I}_2=0}, \qquad \mathbf{z}_{22} = \frac{\mathbf{V}_2}{\mathbf{I}_2}\bigg|_{\mathbf{I}_1=0} \end{aligned} \tag{19.3}$$

Since the z parameters are obtained by open-circuiting the input or output port, they are also called the *open-circuit impedance parameters*. Specifically,

$$\begin{aligned} \mathbf{z}_{11} &= \text{Open-circuit input impedance} \\ \mathbf{z}_{12} &= \text{Open-circuit transfer impedance from port 1 to port 2} \\ \mathbf{z}_{21} &= \text{Open-circuit transfer impedance from port 2 to port 1} \\ \mathbf{z}_{22} &= \text{Open-circuit output impedance} \end{aligned} \tag{19.4}$$

According to Eq. (19.3), we obtain $\mathbf{z}_{11}$ and $\mathbf{z}_{21}$ by connecting a voltage $\mathbf{V}_1$ (or a current source $\mathbf{I}_1$) to port 1 with port 2 open-circuited as in Fig. 19.3(a) and finding $\mathbf{I}_1$ and $\mathbf{V}_2$; we then get

$$\mathbf{z}_{11} = \frac{\mathbf{V}_1}{\mathbf{I}_1}, \qquad \mathbf{z}_{21} = \frac{\mathbf{V}_2}{\mathbf{I}_1} \tag{19.5}$$

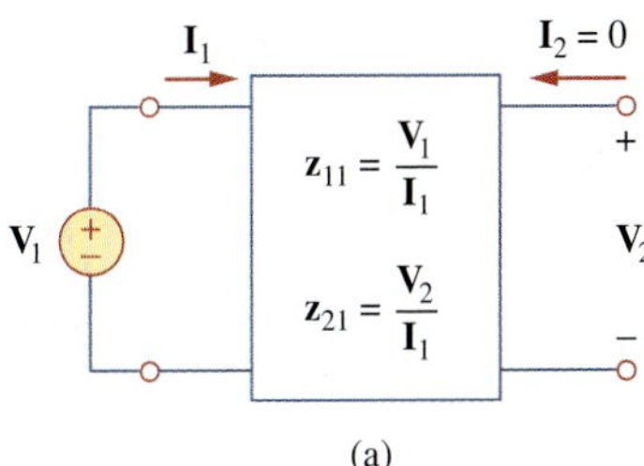

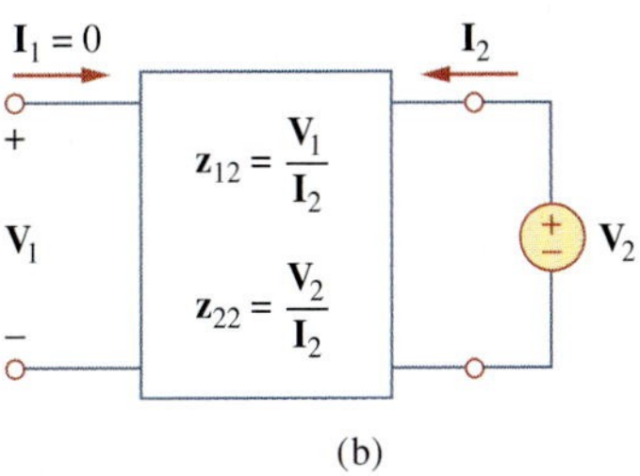

Figure 19.3
Determination of the z parameters: (a) finding $\mathbf{z}_{11}$ and $\mathbf{z}_{21}$, (b) finding $\mathbf{z}_{12}$ and $\mathbf{z}_{22}$.

Similarly, we obtain $\mathbf{z}_{12}$ and $\mathbf{z}_{22}$ by connecting a voltage $\mathbf{V}_2$ (or a current source $\mathbf{I}_2$) to port 2 with port 1 open-circuited as in Fig. 19.3(b) and finding $\mathbf{I}_2$ and $\mathbf{V}_1$; we then get

$$\mathbf{z}_{12} = \frac{\mathbf{V}_1}{\mathbf{I}_2}, \qquad \mathbf{z}_{22} = \frac{\mathbf{V}_2}{\mathbf{I}_2} \tag{19.6}$$

The above procedure provides us with a means of calculating or measuring the z parameters.

Sometimes $\mathbf{z}_{11}$ and $\mathbf{z}_{22}$ are called *driving-point impedances*, while $\mathbf{z}_{21}$ and $\mathbf{z}_{12}$ are called *transfer impedances*. A driving-point impedance is the input impedance of a two-terminal (one-port) device. Thus, $\mathbf{z}_{11}$ is the input driving-point impedance with the output port open-circuited, while $\mathbf{z}_{22}$ is the output driving-point impedance with the input port open-circuited.

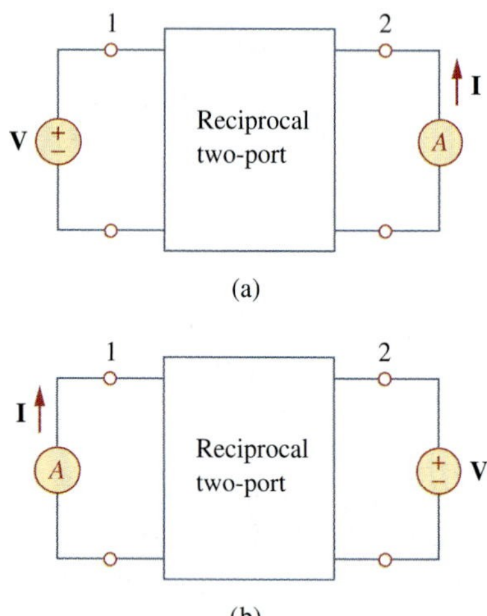

Figure 19.4
Interchanging a voltage source at one port with an ideal ammeter at the other port produces the same reading in a reciprocal two-port.

When $\mathbf{z}_{11} = \mathbf{z}_{22}$, the two-port network is said to be *symmetrical*. This implies that the network has mirrorlike symmetry about some center line; that is, a line can be found that divides the network into two similar halves.

When the two-port network is linear and has no dependent sources, the transfer impedances are equal ($\mathbf{z}_{12} = \mathbf{z}_{21}$), and the two-port is said to be *reciprocal*. This means that if the points of excitation and response are interchanged, the transfer impedances remain the same. As illustrated in Fig. 19.4, a two-port is reciprocal if interchanging an ideal voltage source at one port with an ideal ammeter at the other port gives the same ammeter reading. The reciprocal network yields $\mathbf{V} = \mathbf{z}_{12}\mathbf{I}$ according to Eq. (19.1) when connected as in Fig. 19.4(a), but yields $\mathbf{V} = \mathbf{z}_{21}\mathbf{I}$ when connected as in Fig. 19.4(b). This is possible only if $\mathbf{z}_{12} = \mathbf{z}_{21}$. Any two-port that is made entirely of resistors, capacitors, and inductors must be reciprocal. A reciprocal network can be replaced by the T-equivalent circuit in Fig. 19.5(a). If the network is not reciprocal, a more general equivalent network is shown in Fig. 19.5(b); notice that this figure follows directly from Eq. (19.1).

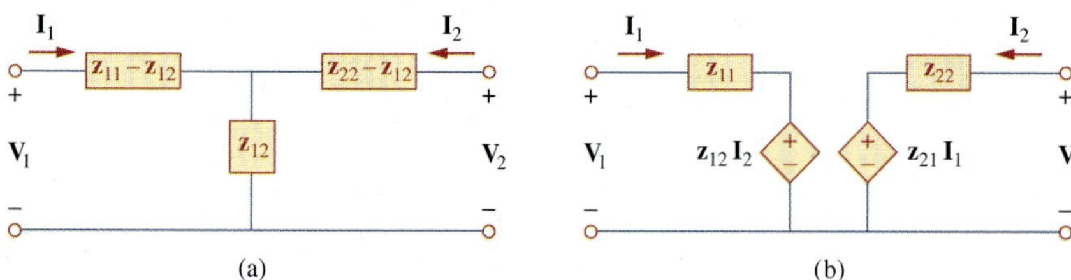

Figure 19.5
(a) T-equivalent circuit (for reciprocal case only), (b) general equivalent circuit.

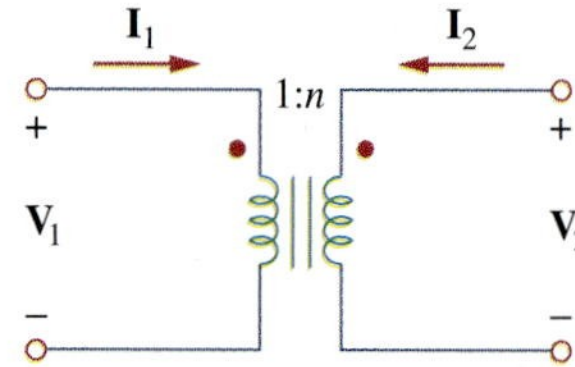

Figure 19.6
An ideal transformer has no z parameters.

It should be mentioned that for some two-port networks, the z parameters do not exist because they cannot be described by Eq. (19.1). As an example, consider the ideal transformer of Fig. 19.6. The defining equations for the two-port network are:

$$\mathbf{V}_1 = \frac{1}{n}\mathbf{V}_2, \qquad \mathbf{I}_1 = -n\mathbf{I}_2 \tag{19.7}$$

Observe that it is impossible to express the voltages in terms of the currents, and vice versa, as Eq. (19.1) requires. Thus, the ideal transformer has no z parameters. However, it does have hybrid parameters, as we shall see in Section 19.4.

Example 19.1

Determine the z parameters for the circuit in Fig. 19.7.

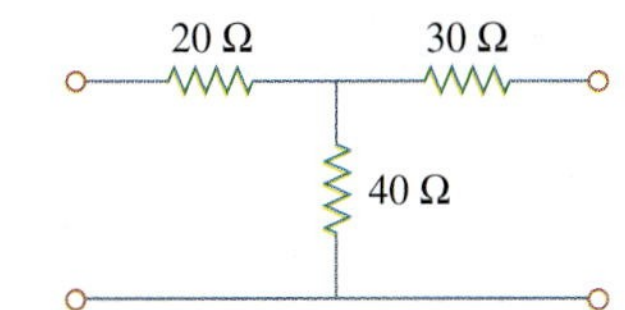

Figure 19.7
For Example 19.1.

Solution:

■ METHOD 1 To determine $\mathbf{z}_{11}$ and $\mathbf{z}_{21}$, we apply a voltage source $\mathbf{V}_1$ to the input port and leave the output port open as in Fig. 19.8(a). Then,

$$\mathbf{z}_{11} = \frac{\mathbf{V}_1}{\mathbf{I}_1} = \frac{(20 + 40)\mathbf{I}_1}{\mathbf{I}_1} = 60\ \Omega$$

that is, $\mathbf{z}_{11}$ is the input impedance at port 1.

$$\mathbf{z}_{21} = \frac{\mathbf{V}_2}{\mathbf{I}_1} = \frac{40\mathbf{I}_1}{\mathbf{I}_1} = 40\ \Omega$$

To find $\mathbf{z}_{12}$ and $\mathbf{z}_{22}$, we apply a voltage source $\mathbf{V}_2$ to the output port and leave the input port open as in Fig. 19.8(b). Then,

$$\mathbf{z}_{12} = \frac{\mathbf{V}_1}{\mathbf{I}_2} = \frac{40\mathbf{I}_2}{\mathbf{I}_2} = 40\ \Omega, \qquad \mathbf{z}_{22} = \frac{\mathbf{V}_2}{\mathbf{I}_2} = \frac{(30 + 40)\mathbf{I}_2}{\mathbf{I}_2} = 70\ \Omega$$

Thus,

$$[\mathbf{z}] = \begin{bmatrix} 60\ \Omega & 40\ \Omega \\ 40\ \Omega & 70\ \Omega \end{bmatrix}$$

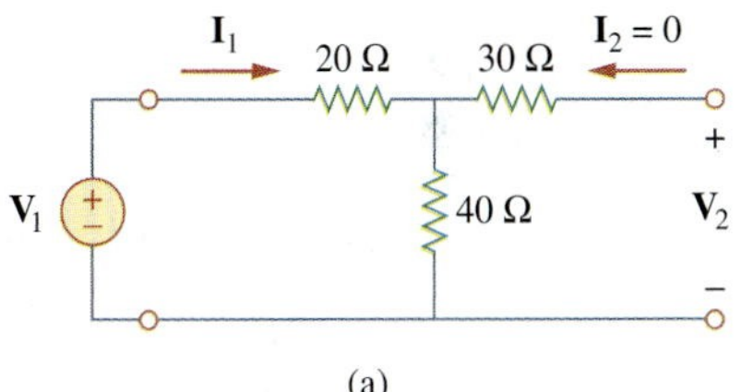

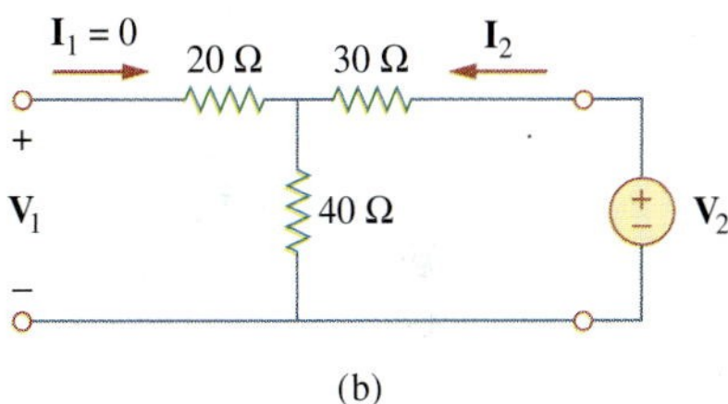

Figure 19.8
For Example 19.1: (a) finding $\mathbf{z}_{11}$ and $\mathbf{z}_{21}$, (b) finding $\mathbf{z}_{12}$ and $\mathbf{z}_{22}$.

■ METHOD 2 Alternatively, since there is no dependent source in the given circuit, $\mathbf{z}_{12} = \mathbf{z}_{21}$ and we can use Fig. 19.5(a). Comparing Fig. 19.7 with Fig. 19.5(a), we get

$$\mathbf{z}_{12} = 40\ \Omega = \mathbf{z}_{21}$$
$$\mathbf{z}_{11} - \mathbf{z}_{12} = 20 \quad \Rightarrow \quad \mathbf{z}_{11} = 20 + \mathbf{z}_{12} = 60\ \Omega$$
$$\mathbf{z}_{22} - \mathbf{z}_{12} = 30 \quad \Rightarrow \quad \mathbf{z}_{22} = 30 + \mathbf{z}_{12} = 70\ \Omega$$

Practice Problem 19.1

Find the z parameters of the two-port network in Fig. 19.9.

Answer: $\mathbf{z}_{11} = 28\ \Omega$, $\mathbf{z}_{12} = \mathbf{z}_{21} = \mathbf{z}_{22} = 12\ \Omega$.

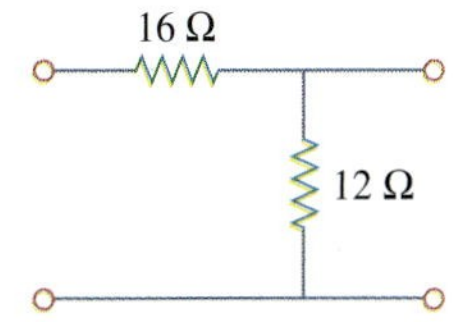

Figure 19.9
For Practice Prob. 19.1.

Example 19.2

Find $\mathbf{I}_1$ and $\mathbf{I}_2$ in the circuit in Fig. 19.10.

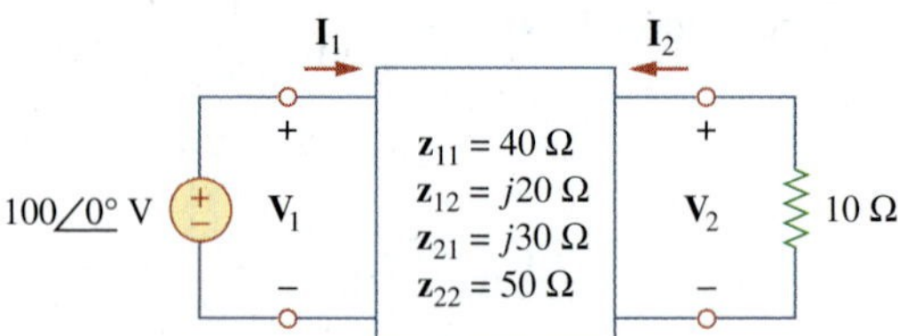

Figure 19.10
For Example 19.2.

Solution:
This is not a reciprocal network. We may use the equivalent circuit in Fig. 19.5(b) but we can also use Eq. (19.1) directly. Substituting the given z parameters into Eq. (19.1),

$$\mathbf{V}_1 = 40\mathbf{I}_1 + j20\mathbf{I}_2 \tag{19.2.1}$$

$$\mathbf{V}_2 = j30\mathbf{I}_1 + 50\mathbf{I}_2 \tag{19.2.2}$$

Since we are looking for $\mathbf{I}_1$ and $\mathbf{I}_2$, we substitute

$$\mathbf{V}_1 = 100\underline{/0°}, \qquad \mathbf{V}_2 = -10\mathbf{I}_2$$

into Eqs. (19.2.1) and (19.2.2), which become

$$100 = 40\mathbf{I}_1 + j20\mathbf{I}_2 \tag{19.2.3}$$

$$-10\mathbf{I}_2 = j30\mathbf{I}_1 + 50\mathbf{I}_2 \quad \Rightarrow \quad \mathbf{I}_1 = j2\mathbf{I}_2 \tag{19.2.4}$$

Substituting Eq. (19.2.4) into Eq. (19.2.3) gives

$$100 = j80\mathbf{I}_2 + j20\mathbf{I}_2 \quad \Rightarrow \quad \mathbf{I}_2 = \frac{100}{j100} = -j$$

From Eq. (19.2.4), $\mathbf{I}_1 = j2(-j) = 2$. Thus,

$$\mathbf{I}_1 = 2\underline{/0°}\text{ A}, \qquad \mathbf{I}_2 = 1\underline{/-90°}\text{ A}$$

Practice Problem 19.2

Calculate $\mathbf{I}_1$ and $\mathbf{I}_2$ in the two-port of Fig. 19.11.

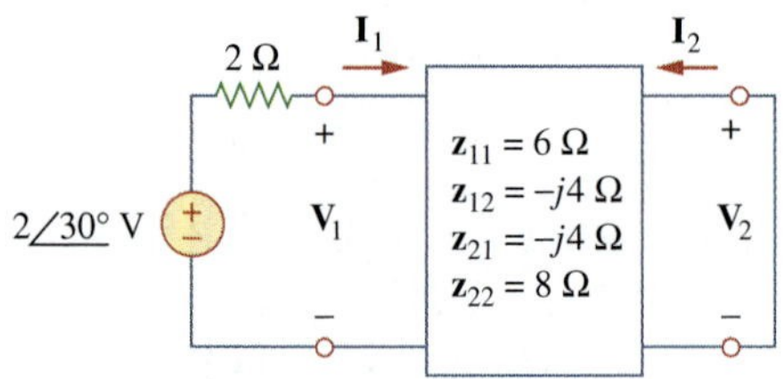

Figure 19.11
For Practice Prob. 19.2.

Answer: $200\underline{/30°}$ mA, $100\underline{/120°}$ mA.

19.3 Admittance Parameters

In the previous section we saw that impedance parameters may not exist for a two-port network. So there is a need for an alternative means of describing such a network. This need may be met by the second set of parameters, which we obtain by expressing the terminal currents in terms of the terminal voltages. In either Fig. 19.12(a) or (b), the terminal currents can be expressed in terms of the terminal voltages as

$$\begin{aligned} \mathbf{I}_1 &= \mathbf{y}_{11}\mathbf{V}_1 + \mathbf{y}_{12}\mathbf{V}_2 \\ \mathbf{I}_2 &= \mathbf{y}_{21}\mathbf{V}_1 + \mathbf{y}_{22}\mathbf{V}_2 \end{aligned} \tag{19.8}$$

or in matrix form as

$$\begin{bmatrix} \mathbf{I}_1 \\ \mathbf{I}_2 \end{bmatrix} = \begin{bmatrix} \mathbf{y}_{11} & \mathbf{y}_{12} \\ \mathbf{y}_{21} & \mathbf{y}_{22} \end{bmatrix} \begin{bmatrix} \mathbf{V}_1 \\ \mathbf{V}_2 \end{bmatrix} = [\mathbf{y}] \begin{bmatrix} \mathbf{V}_1 \\ \mathbf{V}_2 \end{bmatrix} \tag{19.9}$$

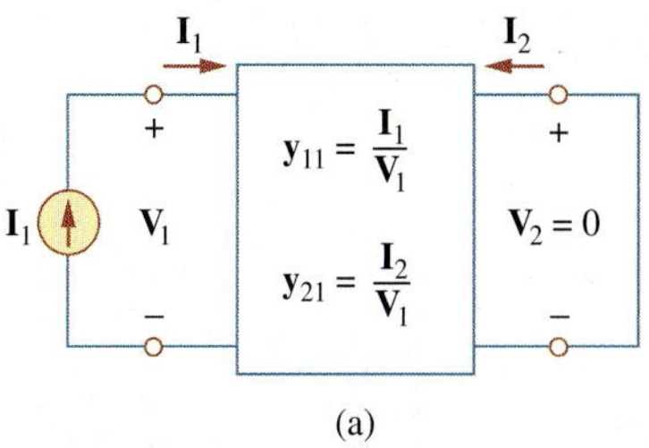

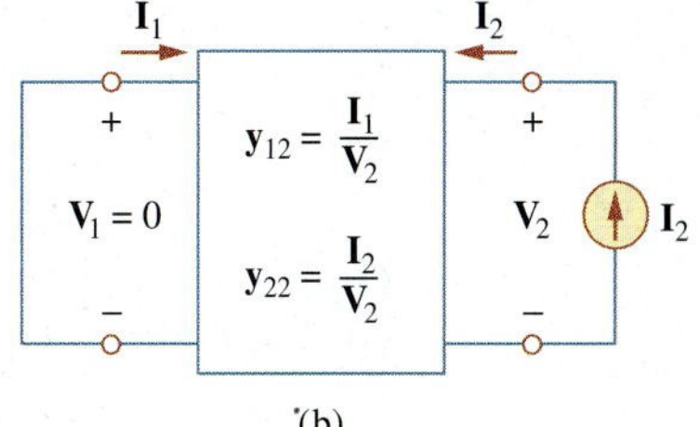

Figure 19.12
Determination of the y parameters: (a) finding y_{11} and y_{21}, (b) finding y_{12} and y_{22}.

The $\mathbf{y}$ terms are known as the *admittance parameters* (or, simply, *y parameters*) and have units of siemens.

The values of the parameters can be determined by setting $\mathbf{V}_1 = 0$ (input port short-circuited) or $\mathbf{V}_2 = 0$ (output port short-circuited). Thus,

$$\begin{aligned} \mathbf{y}_{11} &= \left.\frac{\mathbf{I}_1}{\mathbf{V}_1}\right|_{\mathbf{V}_2=0}, & \mathbf{y}_{12} &= \left.\frac{\mathbf{I}_1}{\mathbf{V}_2}\right|_{\mathbf{V}_1=0} \\ \mathbf{y}_{21} &= \left.\frac{\mathbf{I}_2}{\mathbf{V}_1}\right|_{\mathbf{V}_2=0}, & \mathbf{y}_{22} &= \left.\frac{\mathbf{I}_2}{\mathbf{V}_2}\right|_{\mathbf{V}_1=0} \end{aligned} \tag{19.10}$$

Since the y parameters are obtained by short-circuiting the input or output port, they are also called the *short-circuit admittance parameters*. Specifically,

$$\begin{aligned} \mathbf{y}_{11} &= \text{Short-circuit input admittance} \\ \mathbf{y}_{12} &= \text{Short-circuit transfer admittance from port 2 to port 1} \\ \mathbf{y}_{21} &= \text{Short-circuit transfer admittance from port 1 to port 2} \\ \mathbf{y}_{22} &= \text{Short-circuit output admittance} \end{aligned} \tag{19.11}$$

Following Eq. (19.10), we obtain $\mathbf{y}_{11}$ and $\mathbf{y}_{21}$ by connecting a current $\mathbf{I}_1$ to port 1 and short-circuiting port 2 as in Fig. 19.12(a), finding $\mathbf{V}_1$ and $\mathbf{I}_2$, and then calculating

$$\mathbf{y}_{11} = \frac{\mathbf{I}_1}{\mathbf{V}_1}, \qquad \mathbf{y}_{21} = \frac{\mathbf{I}_2}{\mathbf{V}_1} \tag{19.12}$$

Similarly, we obtain $\mathbf{y}_{12}$ and $\mathbf{y}_{22}$ by connecting a current source $\mathbf{I}_2$ to port 2 and short-circuiting port 1 as in Fig. 19.12(b), finding $\mathbf{I}_1$ and $\mathbf{V}_2$, and then getting

$$\mathbf{y}_{12} = \frac{\mathbf{I}_1}{\mathbf{V}_2}, \qquad \mathbf{y}_{22} = \frac{\mathbf{I}_2}{\mathbf{V}_2} \tag{19.13}$$

This procedure provides us with a means of calculating or measuring the y parameters. The impedance and admittance parameters are collectively referred to as *immittance* parameters.

For a two-port network that is linear and has no dependent sources, the transfer admittances are equal ($\mathbf{y}_{12} = \mathbf{y}_{21}$). This can be proved in the same way as for the z parameters. A reciprocal network ($\mathbf{y}_{12} = \mathbf{y}_{21}$) can be modeled by the Π-equivalent circuit in Fig. 19.13(a). If the network is not reciprocal, a more general equivalent network is shown in Fig. 19.13(b).

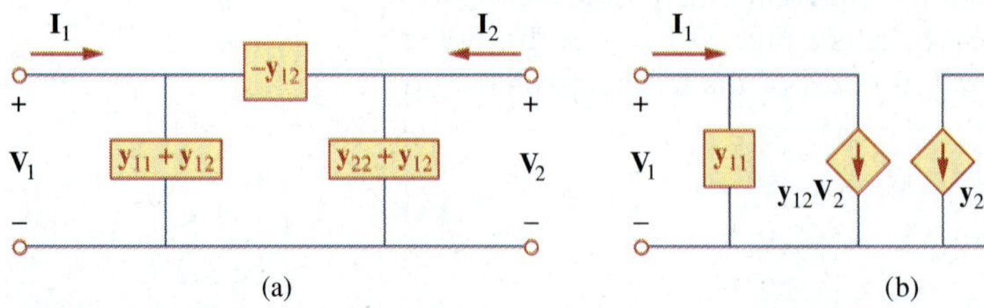

Figure 19.13
(a) Π-equivalent circuit (for reciprocal case only), (b) general equivalent circuit.

Example 19.3

Obtain the y parameters for the Π network shown in Fig. 19.14.

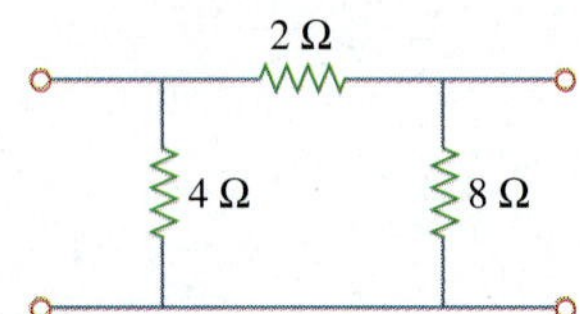

Figure 19.14
For Example 19.3.

Solution:

■ METHOD 1 To find $\mathbf{y}_{11}$ and $\mathbf{y}_{21}$, short-circuit the output port and connect a current source $\mathbf{I}_1$ to the input port as in Fig. 19.15(a). Since the 8-Ω resistor is short-circuited, the 2-Ω resistor is in parallel with the 4-Ω resistor. Hence,

$$\mathbf{V}_1 = \mathbf{I}_1(4 \parallel 2) = \frac{4}{3}\mathbf{I}_1, \qquad \mathbf{y}_{11} = \frac{\mathbf{I}_1}{\mathbf{V}_1} = \frac{\mathbf{I}_1}{\frac{4}{3}\mathbf{I}_1} = 0.75 \text{ S}$$

By current division,

$$-\mathbf{I}_2 = \frac{4}{4+2}\mathbf{I}_1 = \frac{2}{3}\mathbf{I}_1, \qquad \mathbf{y}_{21} = \frac{\mathbf{I}_2}{\mathbf{V}_1} = \frac{-\frac{2}{3}\mathbf{I}_1}{\frac{4}{3}\mathbf{I}_1} = -0.5 \text{ S}$$

To get $\mathbf{y}_{12}$ and $\mathbf{y}_{22}$, short-circuit the input port and connect a current source $\mathbf{I}_2$ to the output port as in Fig. 19.15(b). The 4-Ω resistor is short-circuited so that the 2-Ω and 8-Ω resistors are in parallel.

$$\mathbf{V}_2 = \mathbf{I}_2(8 \parallel 2) = \frac{8}{5}\mathbf{I}_2, \qquad \mathbf{y}_{22} = \frac{\mathbf{I}_2}{\mathbf{V}_2} = \frac{\mathbf{I}_2}{\frac{8}{5}\mathbf{I}_2} = \frac{5}{8} = 0.625 \text{ S}$$

By current division,

$$-\mathbf{I}_1 = \frac{8}{8+2}\mathbf{I}_2 = \frac{4}{5}\mathbf{I}_2, \qquad \mathbf{y}_{12} = \frac{\mathbf{I}_1}{\mathbf{V}_2} = \frac{-\frac{4}{5}\mathbf{I}_2}{\frac{8}{5}\mathbf{I}_2} = -0.5 \text{ S}$$

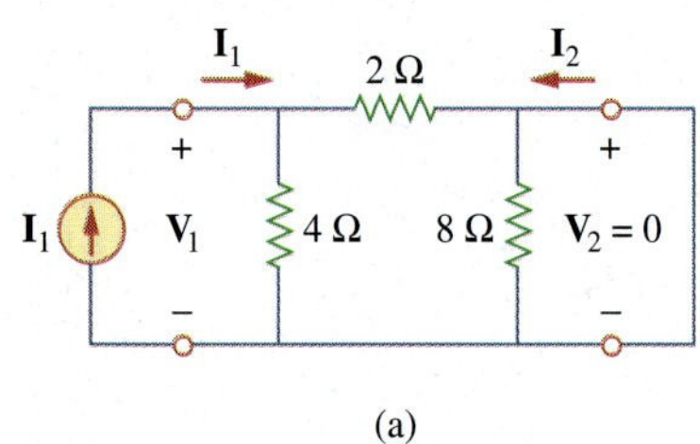

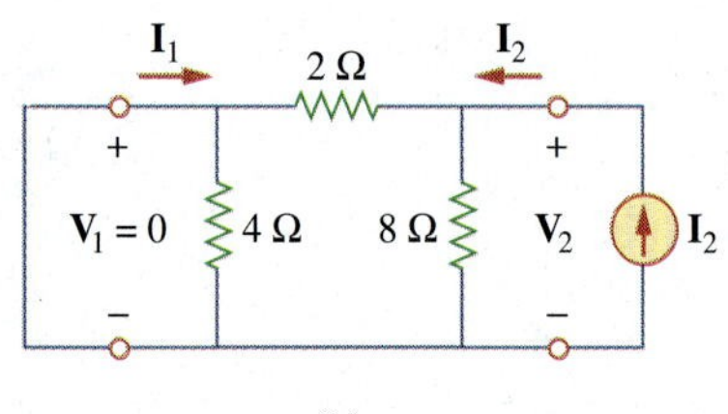

Figure 19.15
For Example 19.3: (a) finding $\mathbf{y}_{11}$ and $\mathbf{y}_{21}$, (b) finding $\mathbf{y}_{12}$ and $\mathbf{y}_{22}$.

■ METHOD 2 Alternatively, comparing Fig. 19.14 with Fig. 19.13(a),

$$\mathbf{y}_{12} = -\frac{1}{2}\text{ S} = \mathbf{y}_{21}$$

$$\mathbf{y}_{11} + \mathbf{y}_{12} = \frac{1}{4} \quad \Rightarrow \quad \mathbf{y}_{11} = \frac{1}{4} - \mathbf{y}_{12} = 0.75 \text{ S}$$

$$\mathbf{y}_{22} + \mathbf{y}_{12} = \frac{1}{8} \quad \Rightarrow \quad \mathbf{y}_{22} = \frac{1}{8} - \mathbf{y}_{12} = 0.625 \text{ S}$$

as obtained previously.

Practice Problem 19.3

Obtain the y parameters for the T network shown in Fig. 19.16.

Answer: $\mathbf{y}_{11} = 75.77$ mS, $\mathbf{y}_{12} = \mathbf{y}_{21} = -30.3$ mS, $\mathbf{y}_{22} = 45.47$ mS.

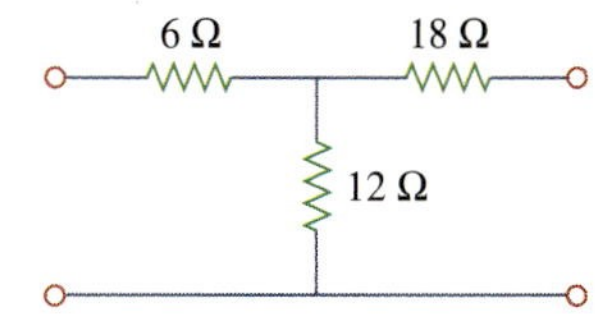

Figure 19.16
For Practice Prob. 19.3.

Example 19.4

Determine the y parameters for the two-port shown in Fig. 19.17.

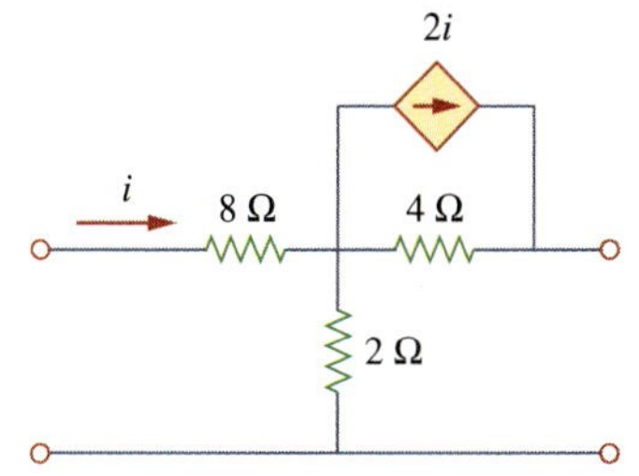

Figure 19.17
For Example 19.4.

Solution:
We follow the same procedure as in the previous example. To get $\mathbf{y}_{11}$ and $\mathbf{y}_{21}$, we use the circuit in Fig. 19.18(a), in which port 2 is short-circuited and a current source is applied to port 1. At node 1,

$$\frac{\mathbf{V}_1 - \mathbf{V}_o}{8} = 2\mathbf{I}_1 + \frac{\mathbf{V}_o}{2} + \frac{\mathbf{V}_o - 0}{4}$$

But $\mathbf{I}_1 = \dfrac{\mathbf{V}_1 - \mathbf{V}_o}{8}$; therefore,

$$0 = \frac{\mathbf{V}_1 - \mathbf{V}_o}{8} + \frac{3\mathbf{V}_o}{4}$$

$$0 = \mathbf{V}_1 - \mathbf{V}_o + 6\mathbf{V}_o \quad \Rightarrow \quad \mathbf{V}_1 = -5\mathbf{V}_o$$

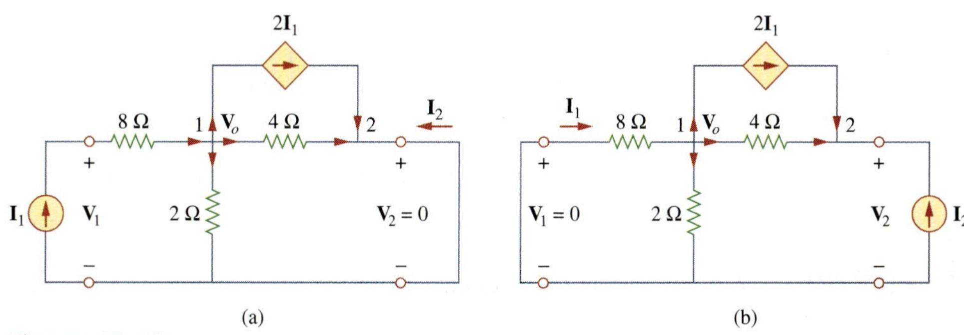

Figure 19.18
Solution of Example 19.4: (a) finding $\mathbf{y}_{11}$ and $\mathbf{y}_{21}$, (b) finding $\mathbf{y}_{12}$ and $\mathbf{y}_{22}$.

Hence,

$$\mathbf{I}_1 = \frac{-5\mathbf{V}_o - \mathbf{V}_o}{8} = -0.75\mathbf{V}_o$$

and

$$\mathbf{y}_{11} = \frac{\mathbf{I}_1}{\mathbf{V}_1} = \frac{-0.75\mathbf{V}_o}{-5\mathbf{V}_o} = 0.15 \text{ S}$$

At node 2,

$$\frac{\mathbf{V}_o - 0}{4} + 2\mathbf{I}_1 + \mathbf{I}_2 = 0$$

At the output,

$$\mathbf{I}_2 = 0 \tag{19.6.6}$$

Substituting Eqs. (19.6.5) and (19.6.6) into Eqs. (19.6.1) and (19.6.2), we obtain

$$60 - 40\mathbf{I}_1 = \mathbf{h}_{11}\mathbf{I}_1 + \mathbf{h}_{12}\mathbf{V}_2$$

or

$$60 = (\mathbf{h}_{11} + 40)\mathbf{I}_1 + \mathbf{h}_{12}\mathbf{V}_2 \tag{19.6.7}$$

and

$$0 = \mathbf{h}_{21}\mathbf{I}_1 + \mathbf{h}_{22}\mathbf{V}_2 \quad \Rightarrow \quad \mathbf{I}_1 = -\frac{\mathbf{h}_{22}}{\mathbf{h}_{21}}\mathbf{V}_2 \tag{19.6.8}$$

Now substituting Eq. (19.6.8) into Eq. (19.6.7) gives

$$60 = \left[-(\mathbf{h}_{11} + 40)\frac{\mathbf{h}_{22}}{\mathbf{h}_{21}} + \mathbf{h}_{12}\right]\mathbf{V}_2$$

or

$$\mathbf{V}_{\text{Th}} = \mathbf{V}_2 = \frac{60}{-(\mathbf{h}_{11} + 40)\mathbf{h}_{22}/\mathbf{h}_{21} + \mathbf{h}_{12}} = \frac{60\mathbf{h}_{21}}{\mathbf{h}_{12}\mathbf{h}_{21} - \mathbf{h}_{11}\mathbf{h}_{22} - 40\mathbf{h}_{22}}$$

Substituting the values of the h parameters,

$$\mathbf{V}_{\text{Th}} = \frac{60 \times 10}{-20.21} = -29.69 \text{ V}$$

Practice Problem 19.6

Find the impedance at the input port of the circuit in Fig. 19.27.

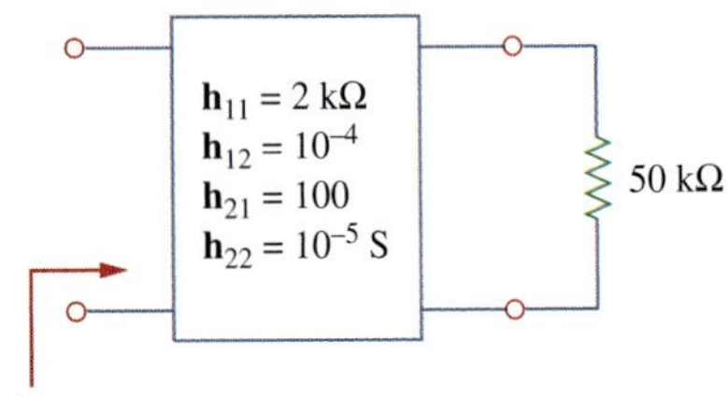

Figure 19.27
For Practice Prob. 19.6.

Answer: 1.6667 kΩ.

Example 19.7

Find the g parameters as functions of s for the circuit in Fig. 19.28.

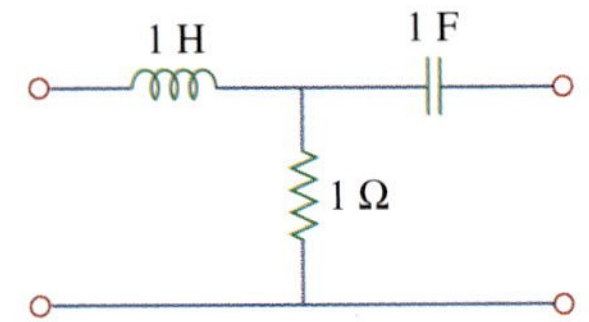

Figure 19.28
For Example 19.7.

Solution:

In the s domain,

$$1\text{ H} \quad \Rightarrow \quad sL = s, \qquad 1\text{ F} \quad \Rightarrow \quad \frac{1}{sC} = \frac{1}{s}$$

To get $\mathbf{g}_{11}$ and $\mathbf{g}_{21}$, we open-circuit the output port and connect a voltage source $\mathbf{V}_1$ to the input port as in Fig. 19.29(a). From the figure,

$$\mathbf{I}_1 = \frac{\mathbf{V}_1}{s + 1}$$

At the output,

$$\mathbf{I}_2 = 0 \tag{19.6.6}$$

Substituting Eqs. (19.6.5) and (19.6.6) into Eqs. (19.6.1) and (19.6.2), we obtain

$$60 - 40\mathbf{I}_1 = \mathbf{h}_{11}\mathbf{I}_1 + \mathbf{h}_{12}\mathbf{V}_2$$

or

$$60 = (\mathbf{h}_{11} + 40)\mathbf{I}_1 + \mathbf{h}_{12}\mathbf{V}_2 \tag{19.6.7}$$

and

$$0 = \mathbf{h}_{21}\mathbf{I}_1 + \mathbf{h}_{22}\mathbf{V}_2 \quad \Rightarrow \quad \mathbf{I}_1 = -\frac{\mathbf{h}_{22}}{\mathbf{h}_{21}}\mathbf{V}_2 \tag{19.6.8}$$

Now substituting Eq. (19.6.8) into Eq. (19.6.7) gives

$$60 = \left[-(\mathbf{h}_{11} + 40)\frac{\mathbf{h}_{22}}{\mathbf{h}_{21}} + \mathbf{h}_{12}\right]\mathbf{V}_2$$

or

$$\mathbf{V}_{\text{Th}} = \mathbf{V}_2 = \frac{60}{-(\mathbf{h}_{11} + 40)\mathbf{h}_{22}/\mathbf{h}_{21} + \mathbf{h}_{12}} = \frac{60\mathbf{h}_{21}}{\mathbf{h}_{12}\mathbf{h}_{21} - \mathbf{h}_{11}\mathbf{h}_{22} - 40\mathbf{h}_{22}}$$

Substituting the values of the h parameters,

$$\mathbf{V}_{\text{Th}} = \frac{60 \times 10}{-20.21} = -29.69 \text{ V}$$

Practice Problem 19.6

Find the impedance at the input port of the circuit in Fig. 19.27.

Answer: 1.6667 kΩ.

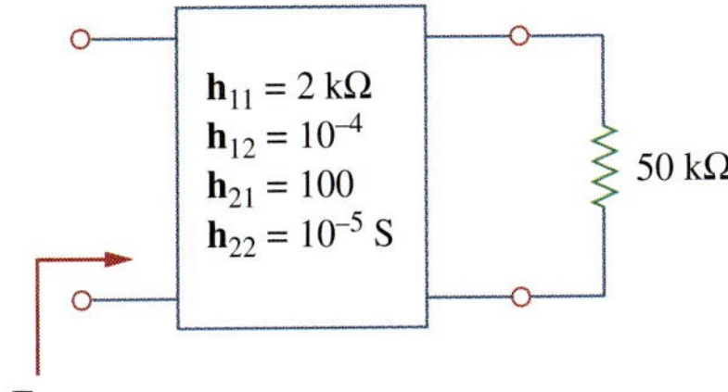

Figure 19.27
For Practice Prob. 19.6.

Example 19.7

Find the g parameters as functions of s for the circuit in Fig. 19.28.

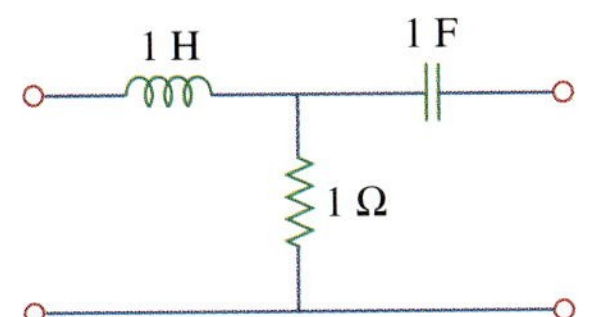

Figure 19.28
For Example 19.7.

Solution:
In the s domain,

$$1 \text{ H} \quad \Rightarrow \quad sL = s, \qquad 1 \text{ F} \quad \Rightarrow \quad \frac{1}{sC} = \frac{1}{s}$$

To get $\mathbf{g}_{11}$ and $\mathbf{g}_{21}$, we open-circuit the output port and connect a voltage source $\mathbf{V}_1$ to the input port as in Fig. 19.29(a). From the figure,

$$\mathbf{I}_1 = \frac{\mathbf{V}_1}{s + 1}$$

or in matrix form,

$$\begin{bmatrix} \mathbf{V}_1 \\ \mathbf{I}_2 \end{bmatrix} = \begin{bmatrix} \mathbf{h}_{11} & \mathbf{h}_{12} \\ \mathbf{h}_{21} & \mathbf{h}_{22} \end{bmatrix} \begin{bmatrix} \mathbf{I}_1 \\ \mathbf{V}_2 \end{bmatrix} = [\mathbf{h}] \begin{bmatrix} \mathbf{I}_1 \\ \mathbf{V}_2 \end{bmatrix} \tag{19.15}$$

The **h** terms are known as the *hybrid parameters* (or, simply, *h parameters*) because they are a hybrid combination of ratios. They are very useful for describing electronic devices such as transistors (see Section 19.9); it is much easier to measure experimentally the *h* parameters of such devices than to measure their *z* or *y* parameters. In fact, we have seen that the ideal transformer in Fig. 19.6, described by Eq. (19.7), does not have *z* parameters. The ideal transformer can be described by the hybrid parameters, because Eq. (19.7) conforms with Eq. (19.14).

The values of the parameters are determined as

$$\mathbf{h}_{11} = \left.\frac{\mathbf{V}_1}{\mathbf{I}_1}\right|_{\mathbf{V}_2=0}, \qquad \mathbf{h}_{12} = \left.\frac{\mathbf{V}_1}{\mathbf{V}_2}\right|_{\mathbf{I}_1=0}$$
$$\mathbf{h}_{21} = \left.\frac{\mathbf{I}_2}{\mathbf{I}_1}\right|_{\mathbf{V}_2=0}, \qquad \mathbf{h}_{22} = \left.\frac{\mathbf{I}_2}{\mathbf{V}_2}\right|_{\mathbf{I}_1=0} \tag{19.16}$$

It is evident from Eq. (19.16) that the parameters $\mathbf{h}_{11}$, $\mathbf{h}_{12}$, $\mathbf{h}_{21}$, and $\mathbf{h}_{22}$ represent an impedance, a voltage gain, a current gain, and an admittance, respectively. This is why they are called the hybrid parameters. To be specific,

$$\begin{aligned} \mathbf{h}_{11} &= \text{Short-circuit input impedance} \\ \mathbf{h}_{12} &= \text{Open-circuit reverse voltage gain} \\ \mathbf{h}_{21} &= \text{Short-circuit forward current gain} \\ \mathbf{h}_{22} &= \text{Open-circuit output admittance} \end{aligned} \tag{19.17}$$

The procedure for calculating the *h* parameters is similar to that used for the *z* or *y* parameters. We apply a voltage or current source to the appropriate port, short-circuit or open-circuit the other port, depending on the parameter of interest, and perform regular circuit analysis. For reciprocal networks, $\mathbf{h}_{12} = -\mathbf{h}_{21}$. This can be proved in the same way as we proved that $\mathbf{z}_{12} = \mathbf{z}_{21}$. Figure 19.20 shows the hybrid model of a two-port network.

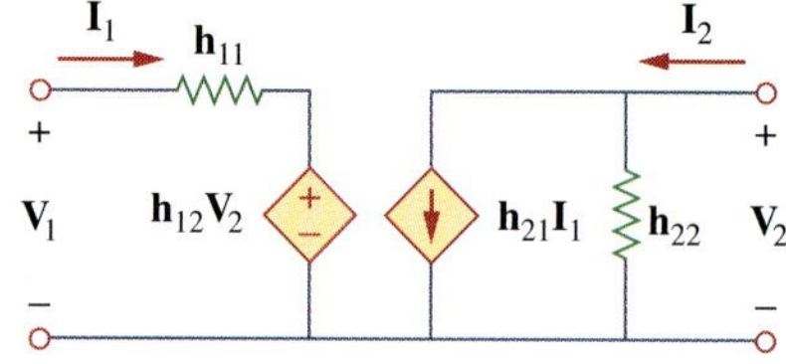

Figure 19.20
The *h*-parameter equivalent network of a two-port network.

A set of parameters closely related to the *h* parameters are the *g parameters* or *inverse hybrid parameters*. These are used to describe the terminal currents and voltages as

$$\begin{aligned} \mathbf{I}_1 &= \mathbf{g}_{11}\mathbf{V}_1 + \mathbf{g}_{12}\mathbf{I}_2 \\ \mathbf{V}_2 &= \mathbf{g}_{21}\mathbf{V}_1 + \mathbf{g}_{22}\mathbf{I}_2 \end{aligned} \tag{19.18}$$

or

$$\begin{bmatrix} \mathbf{I}_1 \\ \mathbf{V}_2 \end{bmatrix} = \begin{bmatrix} \mathbf{g}_{11} & \mathbf{g}_{12} \\ \mathbf{g}_{21} & \mathbf{g}_{22} \end{bmatrix} \begin{bmatrix} \mathbf{V}_1 \\ \mathbf{I}_2 \end{bmatrix} = [\mathbf{g}] \begin{bmatrix} \mathbf{V}_1 \\ \mathbf{I}_2 \end{bmatrix} \tag{19.19}$$

At the output,

$$\mathbf{I}_2 = 0 \tag{19.6.6}$$

Substituting Eqs. (19.6.5) and (19.6.6) into Eqs. (19.6.1) and (19.6.2), we obtain

$$60 - 40\mathbf{I}_1 = \mathbf{h}_{11}\mathbf{I}_1 + \mathbf{h}_{12}\mathbf{V}_2$$

or

$$60 = (\mathbf{h}_{11} + 40)\mathbf{I}_1 + \mathbf{h}_{12}\mathbf{V}_2 \tag{19.6.7}$$

and

$$0 = \mathbf{h}_{21}\mathbf{I}_1 + \mathbf{h}_{22}\mathbf{V}_2 \quad \Rightarrow \quad \mathbf{I}_1 = -\frac{\mathbf{h}_{22}}{\mathbf{h}_{21}}\mathbf{V}_2 \tag{19.6.8}$$

Now substituting Eq. (19.6.8) into Eq. (19.6.7) gives

$$60 = \left[-(\mathbf{h}_{11} + 40)\frac{\mathbf{h}_{22}}{\mathbf{h}_{21}} + \mathbf{h}_{12} \right] \mathbf{V}_2$$

or

$$\mathbf{V}_{\text{Th}} = \mathbf{V}_2 = \frac{60}{-(\mathbf{h}_{11} + 40)\mathbf{h}_{22}/\mathbf{h}_{21} + \mathbf{h}_{12}} = \frac{60\mathbf{h}_{21}}{\mathbf{h}_{12}\mathbf{h}_{21} - \mathbf{h}_{11}\mathbf{h}_{22} - 40\mathbf{h}_{22}}$$

Substituting the values of the h parameters,

$$\mathbf{V}_{\text{Th}} = \frac{60 \times 10}{-20.21} = -29.69 \text{ V}$$

Practice Problem 19.6

Find the impedance at the input port of the circuit in Fig. 19.27.

Answer: 1.6667 kΩ.

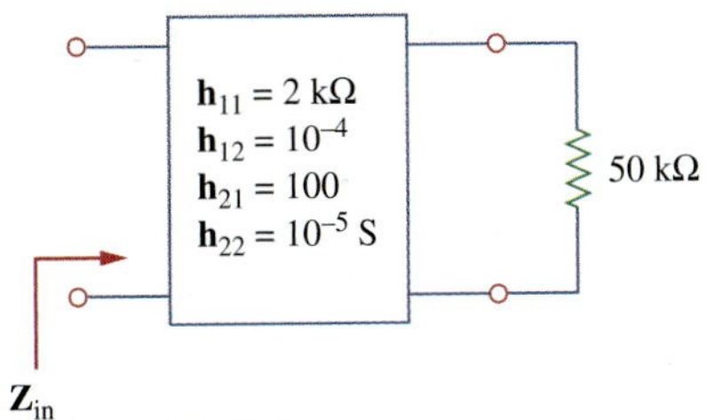

Figure 19.27
For Practice Prob. 19.6.

Example 19.7

Find the g parameters as functions of s for the circuit in Fig. 19.28.

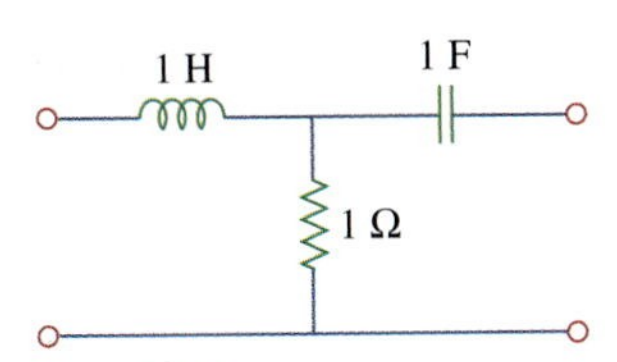

Figure 19.28
For Example 19.7.

Solution:
In the s domain,

$$1 \text{ H} \quad \Rightarrow \quad sL = s, \qquad 1 \text{ F} \quad \Rightarrow \quad \frac{1}{sC} = \frac{1}{s}$$

To get $\mathbf{g}_{11}$ and $\mathbf{g}_{21}$, we open-circuit the output port and connect a voltage source $\mathbf{V}_1$ to the input port as in Fig. 19.29(a). From the figure,

$$\mathbf{I}_1 = \frac{\mathbf{V}_1}{s + 1}$$

Thus,

$$\mathbf{h}_{22} = \frac{\mathbf{I}_2}{\mathbf{V}_2} = \frac{1}{9}\text{ S}$$

Practice Problem 19.5

Determine the h parameters for the circuit in Fig. 19.24.

Answer: $\mathbf{h}_{11} = 2.4\ \Omega$, $\mathbf{h}_{12} = 0.4$, $\mathbf{h}_{21} = -0.4$, $\mathbf{h}_{22} = 200$ mS.

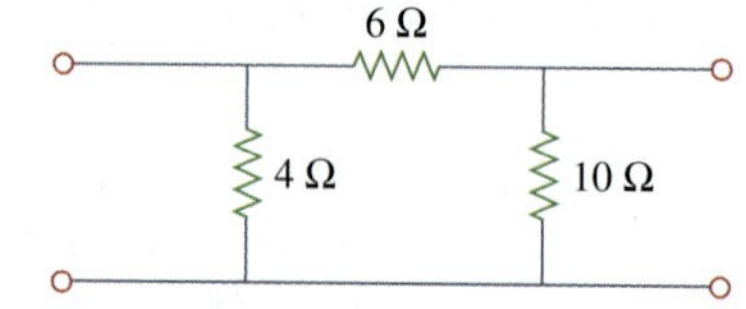

Figure 19.24
For Practice Prob. 19.5.

Example 19.6

Determine the Thevenin equivalent at the output port of the circuit in Fig. 19.25.

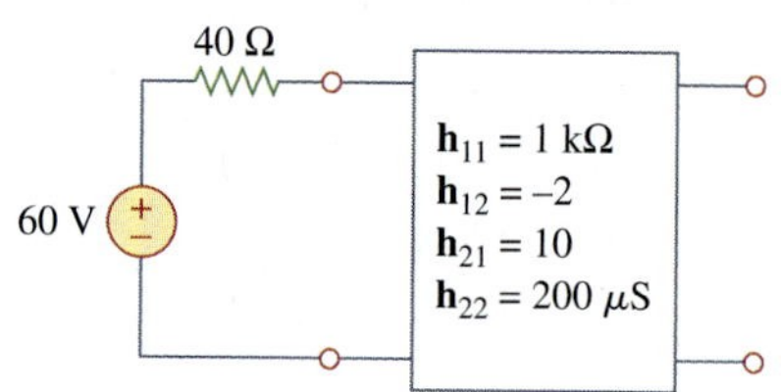

Figure 19.25
For Example 19.6.

Solution:
To find $\mathbf{Z}_{\text{Th}}$ and $\mathbf{V}_{\text{Th}}$, we apply the normal procedure, keeping in mind the formulas relating the input and output ports of the h model. To obtain $\mathbf{Z}_{\text{Th}}$, remove the 60-V voltage source at the input port and apply a 1-V voltage source at the output port, as shown in Fig. 19.26(a). From Eq. (19.14),

$$\mathbf{V}_1 = \mathbf{h}_{11}\mathbf{I}_1 + \mathbf{h}_{12}\mathbf{V}_2 \qquad \textbf{(19.6.1)}$$

$$\mathbf{I}_2 = \mathbf{h}_{21}\mathbf{I}_1 + \mathbf{h}_{22}\mathbf{V}_2 \qquad \textbf{(19.6.2)}$$

But $\mathbf{V}_2 = 1$, and $\mathbf{V}_1 = -40\mathbf{I}_1$. Substituting these into Eqs. (19.6.1) and (19.6.2), we get

$$-40\mathbf{I}_1 = \mathbf{h}_{11}\mathbf{I}_1 + \mathbf{h}_{12} \quad \Rightarrow \quad \mathbf{I}_1 = -\frac{\mathbf{h}_{12}}{40 + \mathbf{h}_{11}} \qquad \textbf{(19.6.3)}$$

$$\mathbf{I}_2 = \mathbf{h}_{21}\mathbf{I}_1 + \mathbf{h}_{22} \qquad \textbf{(19.6.4)}$$

Substituting Eq. (19.6.3) into Eq. (19.6.4) gives

$$\mathbf{I}_2 = \mathbf{h}_{22} - \frac{\mathbf{h}_{21}\mathbf{h}_{12}}{\mathbf{h}_{11} + 40} = \frac{\mathbf{h}_{11}\mathbf{h}_{22} - \mathbf{h}_{21}\mathbf{h}_{12} + \mathbf{h}_{22}40}{\mathbf{h}_{11} + 40}$$

Therefore,

$$\mathbf{Z}_{\text{Th}} = \frac{\mathbf{V}_2}{\mathbf{I}_2} = \frac{1}{\mathbf{I}_2} = \frac{\mathbf{h}_{11} + 40}{\mathbf{h}_{11}\mathbf{h}_{22} - \mathbf{h}_{21}\mathbf{h}_{12} + \mathbf{h}_{22}40}$$

Substituting the values of the h parameters,

$$\mathbf{Z}_{\text{Th}} = \frac{1000 + 40}{10^3 \times 200 \times 10^{-6} + 20 + 40 \times 200 \times 10^{-6}} = \frac{1040}{20.21} = 51.46\ \Omega$$

To get $\mathbf{V}_{\text{Th}}$, we find the open-circuit voltage $\mathbf{V}_2$ in Fig. 19.26(b). At the input port,

$$-60 + 40\mathbf{I}_1 + \mathbf{V}_1 = 0 \quad \Rightarrow \quad \mathbf{V}_1 = 60 - 40\mathbf{I}_1 \qquad \textbf{(19.6.5)}$$

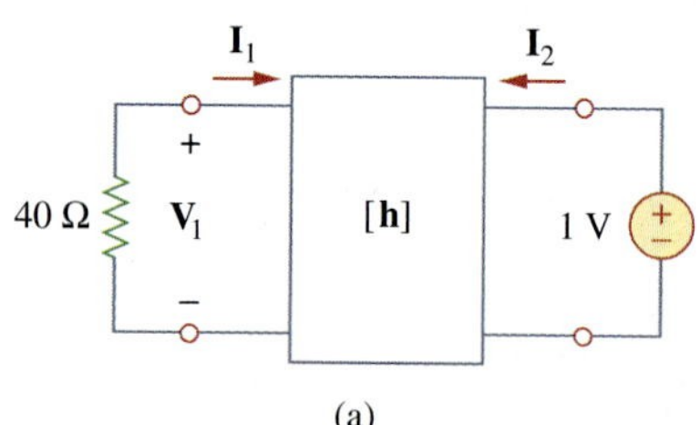

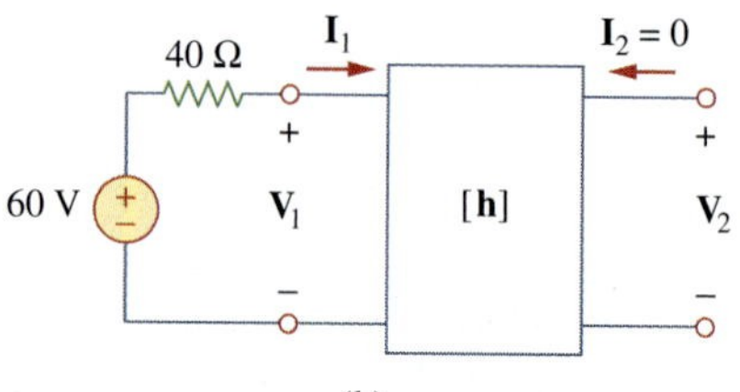

Figure 19.26
For Example 19.6: (a) finding $\mathbf{Z}_{\text{Th}}$, (b) finding $\mathbf{V}_{\text{Th}}$.

The values of the g parameters are determined as

$$\mathbf{g}_{11} = \frac{\mathbf{I}_1}{\mathbf{V}_1}\bigg|_{\mathbf{I}_2=0}, \qquad \mathbf{g}_{12} = \frac{\mathbf{I}_1}{\mathbf{I}_2}\bigg|_{\mathbf{V}_1=0}$$
$$\mathbf{g}_{21} = \frac{\mathbf{V}_2}{\mathbf{V}_1}\bigg|_{\mathbf{I}_2=0}, \qquad \mathbf{g}_{22} = \frac{\mathbf{V}_2}{\mathbf{I}_2}\bigg|_{\mathbf{V}_1=0} \tag{19.20}$$

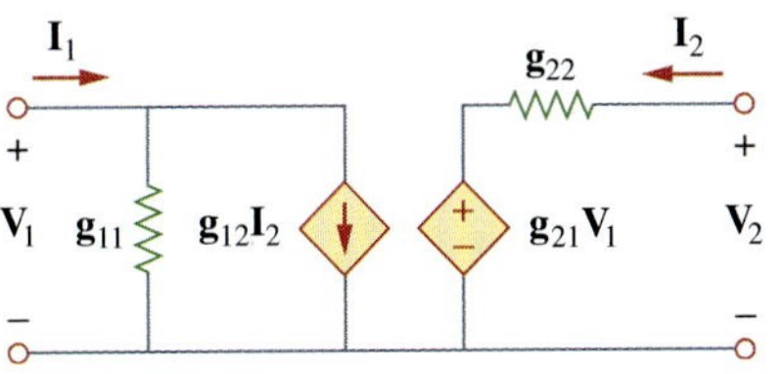

Figure 19.21
The g-parameter model of a two-port network.

Thus, the inverse hybrid parameters are specifically called

$$\begin{aligned} \mathbf{g}_{11} &= \text{Open-circuit input admittance} \\ \mathbf{g}_{12} &= \text{Short-circuit reverse current gain} \\ \mathbf{g}_{21} &= \text{Open-circuit forward voltage gain} \\ \mathbf{g}_{22} &= \text{Short-circuit output impedance} \end{aligned} \tag{19.21}$$

Figure 19.21 shows the inverse hybrid model of a two-port network. The g parameters are frequently used to model field-effect transistors.

Example 19.5

Find the hybrid parameters for the two-port network of Fig. 19.22.

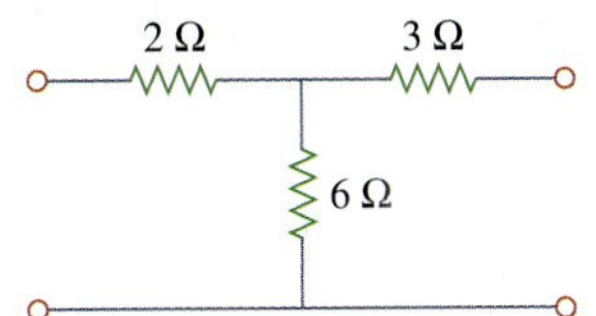

Figure 19.22
For Example 19.5.

Solution:
To find $\mathbf{h}_{11}$ and $\mathbf{h}_{21}$, we short-circuit the output port and connect a current source $\mathbf{I}_1$ to the input port as shown in Fig. 19.23(a). From Fig. 19.23(a),

$$\mathbf{V}_1 = \mathbf{I}_1(2 + 3 \parallel 6) = 4\mathbf{I}_1$$

Hence,

$$\mathbf{h}_{11} = \frac{\mathbf{V}_1}{\mathbf{I}_1} = 4\ \Omega$$

Also, from Fig. 19.23(a) we obtain, by current division,

$$-\mathbf{I}_2 = \frac{6}{6+3}\mathbf{I}_1 = \frac{2}{3}\mathbf{I}_1$$

Hence,

$$\mathbf{h}_{21} = \frac{\mathbf{I}_2}{\mathbf{I}_1} = -\frac{2}{3}$$

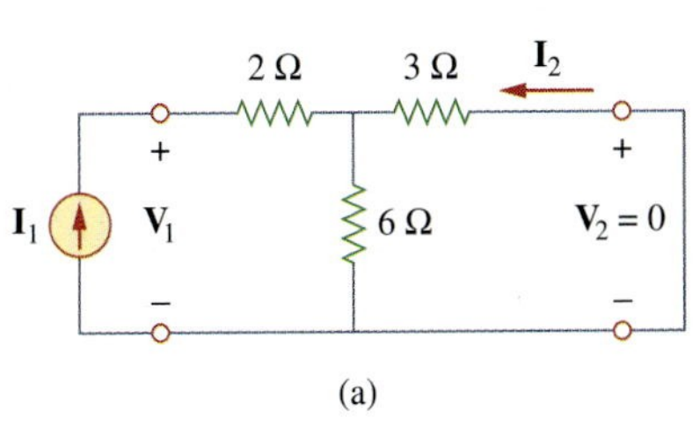

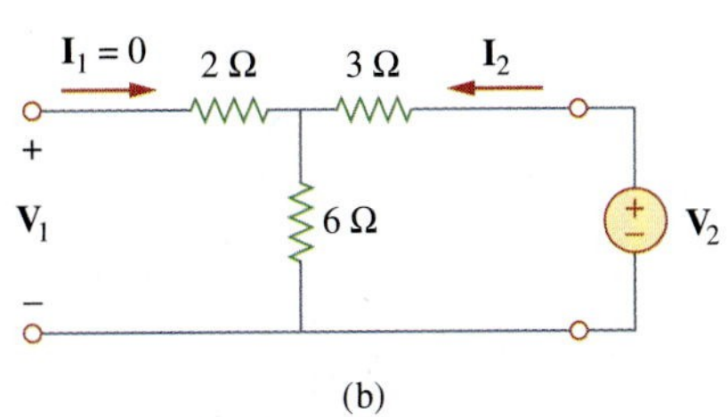

Figure 19.23
For Example 19.5: (a) computing $\mathbf{h}_{11}$ and $\mathbf{h}_{21}$, (b) computing $\mathbf{h}_{12}$ and $\mathbf{h}_{22}$.

To obtain $\mathbf{h}_{12}$ and $\mathbf{h}_{22}$, we open-circuit the input port and connect a voltage source $\mathbf{V}_2$ to the output port as in Fig. 19.23(b). By voltage division,

$$\mathbf{V}_1 = \frac{6}{6+3}\mathbf{V}_2 = \frac{2}{3}\mathbf{V}_2$$

Hence,

$$\mathbf{h}_{12} = \frac{\mathbf{V}_1}{\mathbf{V}_2} = \frac{2}{3}$$

Also,

$$\mathbf{V}_2 = (3 + 6)\mathbf{I}_2 = 9\mathbf{I}_2$$

or in matrix form,

$$\begin{bmatrix} \mathbf{V}_1 \\ \mathbf{I}_2 \end{bmatrix} = \begin{bmatrix} \mathbf{h}_{11} & \mathbf{h}_{12} \\ \mathbf{h}_{21} & \mathbf{h}_{22} \end{bmatrix} \begin{bmatrix} \mathbf{I}_1 \\ \mathbf{V}_2 \end{bmatrix} = [\mathbf{h}] \begin{bmatrix} \mathbf{I}_1 \\ \mathbf{V}_2 \end{bmatrix} \tag{19.15}$$

The **h** terms are known as the *hybrid parameters* (or, simply, *h parameters*) because they are a hybrid combination of ratios. They are very useful for describing electronic devices such as transistors (see Section 19.9); it is much easier to measure experimentally the *h* parameters of such devices than to measure their *z* or *y* parameters. In fact, we have seen that the ideal transformer in Fig. 19.6, described by Eq. (19.7), does not have *z* parameters. The ideal transformer can be described by the hybrid parameters, because Eq. (19.7) conforms with Eq. (19.14).

The values of the parameters are determined as

$$\begin{aligned} \mathbf{h}_{11} &= \left.\frac{\mathbf{V}_1}{\mathbf{I}_1}\right|_{\mathbf{V}_2=0}, \qquad \mathbf{h}_{12} = \left.\frac{\mathbf{V}_1}{\mathbf{V}_2}\right|_{\mathbf{I}_1=0} \\ \mathbf{h}_{21} &= \left.\frac{\mathbf{I}_2}{\mathbf{I}_1}\right|_{\mathbf{V}_2=0}, \qquad \mathbf{h}_{22} = \left.\frac{\mathbf{I}_2}{\mathbf{V}_2}\right|_{\mathbf{I}_1=0} \end{aligned} \tag{19.16}$$

It is evident from Eq. (19.16) that the parameters $\mathbf{h}_{11}$, $\mathbf{h}_{12}$, $\mathbf{h}_{21}$, and $\mathbf{h}_{22}$ represent an impedance, a voltage gain, a current gain, and an admittance, respectively. This is why they are called the hybrid parameters. To be specific,

$$\begin{aligned} \mathbf{h}_{11} &= \text{Short-circuit input impedance} \\ \mathbf{h}_{12} &= \text{Open-circuit reverse voltage gain} \\ \mathbf{h}_{21} &= \text{Short-circuit forward current gain} \\ \mathbf{h}_{22} &= \text{Open-circuit output admittance} \end{aligned} \tag{19.17}$$

The procedure for calculating the *h* parameters is similar to that used for the *z* or *y* parameters. We apply a voltage or current source to the appropriate port, short-circuit or open-circuit the other port, depending on the parameter of interest, and perform regular circuit analysis. For reciprocal networks, $\mathbf{h}_{12} = -\mathbf{h}_{21}$. This can be proved in the same way as we proved that $\mathbf{z}_{12} = \mathbf{z}_{21}$. Figure 19.20 shows the hybrid model of a two-port network.

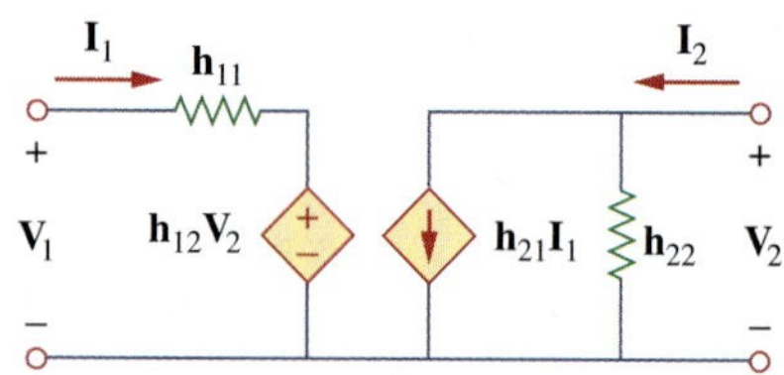

Figure 19.20
The *h*-parameter equivalent network of a two-port network.

A set of parameters closely related to the *h* parameters are the *g parameters* or *inverse hybrid parameters*. These are used to describe the terminal currents and voltages as

$$\begin{aligned} \mathbf{I}_1 &= \mathbf{g}_{11}\mathbf{V}_1 + \mathbf{g}_{12}\mathbf{I}_2 \\ \mathbf{V}_2 &= \mathbf{g}_{21}\mathbf{V}_1 + \mathbf{g}_{22}\mathbf{I}_2 \end{aligned} \tag{19.18}$$

or

$$\begin{bmatrix} \mathbf{I}_1 \\ \mathbf{V}_2 \end{bmatrix} = \begin{bmatrix} \mathbf{g}_{11} & \mathbf{g}_{12} \\ \mathbf{g}_{21} & \mathbf{g}_{22} \end{bmatrix} \begin{bmatrix} \mathbf{V}_1 \\ \mathbf{I}_2 \end{bmatrix} = [\mathbf{g}] \begin{bmatrix} \mathbf{V}_1 \\ \mathbf{I}_2 \end{bmatrix} \tag{19.19}$$

or

$$-\mathbf{I}_2 = 0.25\mathbf{V}_o - 1.5\mathbf{V}_o = -1.25\mathbf{V}_o$$

Hence,

$$\mathbf{y}_{21} = \frac{\mathbf{I}_2}{\mathbf{V}_1} = \frac{1.25\mathbf{V}_o}{-5\mathbf{V}_o} = -0.25 \text{ S}$$

Similarly, we get $\mathbf{y}_{12}$ and $\mathbf{y}_{22}$ using Fig. 19.18(b). At node 1,

$$\frac{0 - \mathbf{V}_o}{8} = 2\mathbf{I}_1 + \frac{\mathbf{V}_o}{2} + \frac{\mathbf{V}_o - \mathbf{V}_2}{4}$$

But $\mathbf{I}_1 = \dfrac{0 - \mathbf{V}_o}{8}$; therefore,

$$0 = -\frac{\mathbf{V}_o}{8} + \frac{\mathbf{V}_o}{2} + \frac{\mathbf{V}_o - \mathbf{V}_2}{4}$$

or

$$0 = -\mathbf{V}_o + 4\mathbf{V}_o + 2\mathbf{V}_o - 2\mathbf{V}_2 \quad \Rightarrow \quad \mathbf{V}_2 = 2.5\mathbf{V}_o$$

Hence,

$$\mathbf{y}_{12} = \frac{\mathbf{I}_1}{\mathbf{V}_2} = \frac{-\mathbf{V}_o/8}{2.5\mathbf{V}_o} = -0.05 \text{ S}$$

At node 2,

$$\frac{\mathbf{V}_o - \mathbf{V}_2}{4} + 2\mathbf{I}_1 + \mathbf{I}_2 = 0$$

or

$$-\mathbf{I}_2 = 0.25\mathbf{V}_o - \frac{1}{4}(2.5\mathbf{V}_o) - \frac{2\mathbf{V}_o}{8} = -0.625\mathbf{V}_o$$

Thus,

$$\mathbf{y}_{22} = \frac{\mathbf{I}_2}{\mathbf{V}_2} = \frac{0.625\mathbf{V}_o}{2.5\mathbf{V}_o} = 0.25 \text{ S}$$

Notice that $\mathbf{y}_{12} \neq \mathbf{y}_{21}$ in this case, since the network is not reciprocal.

Practice Problem 19.4

Obtain the y parameters for the circuit in Fig. 19.19.

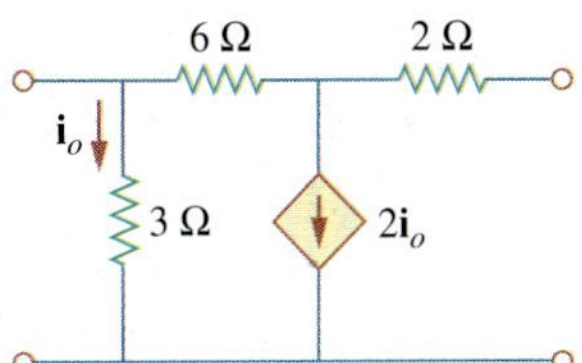

Figure 19.19
For Practice Prob. 19.4.

Answer: $\mathbf{y}_{11} = 0.625$ S, $\mathbf{y}_{12} = -0.125$ S, $\mathbf{y}_{21} = 0.375$ S, $\mathbf{y}_{22} = 0.125$ S.

19.4 Hybrid Parameters

The z and y parameters of a two-port network do not always exist. So there is a need for developing another set of parameters. This third set of parameters is based on making $\mathbf{V}_1$ and $\mathbf{I}_2$ the dependent variables. Thus, we obtain

$$\begin{aligned} \mathbf{V}_1 &= \mathbf{h}_{11}\mathbf{I}_1 + \mathbf{h}_{12}\mathbf{V}_2 \\ \mathbf{I}_2 &= \mathbf{h}_{21}\mathbf{I}_1 + \mathbf{h}_{22}\mathbf{V}_2 \end{aligned} \tag{19.14}$$

Practice Problem 19.3

Obtain the y parameters for the T network shown in Fig. 19.16.

Answer: $\mathbf{y}_{11} = 75.77$ mS, $\mathbf{y}_{12} = \mathbf{y}_{21} = -30.3$ mS, $\mathbf{y}_{22} = 45.47$ mS.

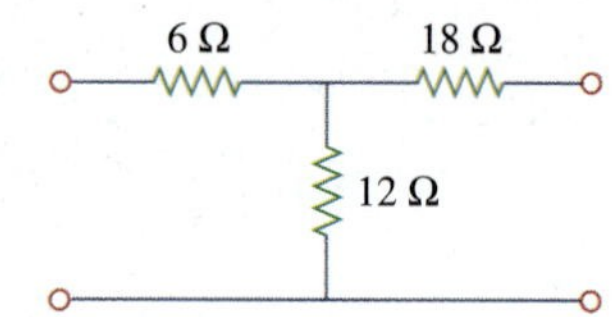

Figure 19.16
For Practice Prob. 19.3.

Example 19.4

Determine the y parameters for the two-port shown in Fig. 19.17.

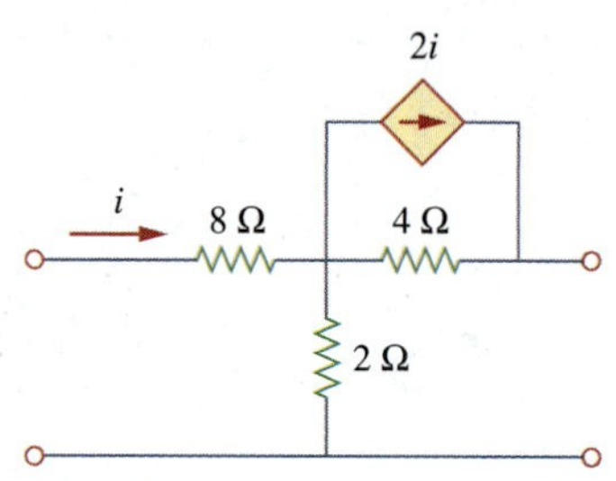

Figure 19.17
For Example 19.4.

Solution:
We follow the same procedure as in the previous example. To get $\mathbf{y}_{11}$ and $\mathbf{y}_{21}$, we use the circuit in Fig. 19.18(a), in which port 2 is short-circuited and a current source is applied to port 1. At node 1,

$$\frac{\mathbf{V}_1 - \mathbf{V}_o}{8} = 2\mathbf{I}_1 + \frac{\mathbf{V}_o}{2} + \frac{\mathbf{V}_o - 0}{4}$$

But $\mathbf{I}_1 = \dfrac{\mathbf{V}_1 - \mathbf{V}_o}{8}$; therefore,

$$0 = \frac{\mathbf{V}_1 - \mathbf{V}_o}{8} + \frac{3\mathbf{V}_o}{4}$$

$$0 = \mathbf{V}_1 - \mathbf{V}_o + 6\mathbf{V}_o \quad \Rightarrow \quad \mathbf{V}_1 = -5\mathbf{V}_o$$

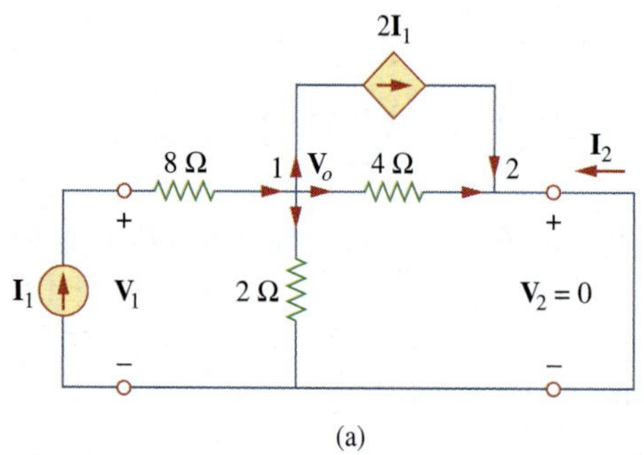

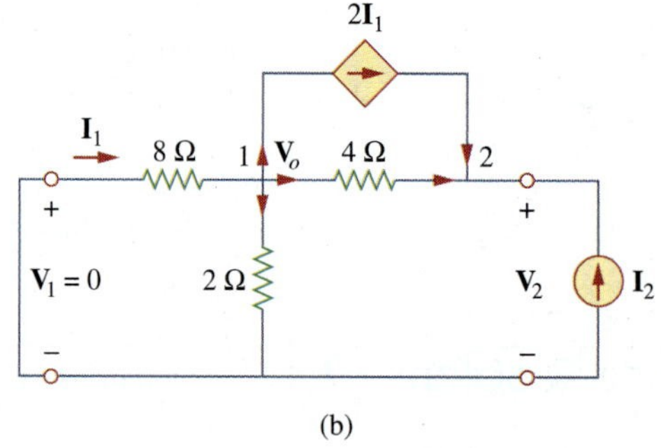

Figure 19.18
Solution of Example 19.4: (a) finding $\mathbf{y}_{11}$ and $\mathbf{y}_{21}$, (b) finding $\mathbf{y}_{12}$ and $\mathbf{y}_{22}$.

Hence,

$$\mathbf{I}_1 = \frac{-5\mathbf{V}_o - \mathbf{V}_o}{8} = -0.75\mathbf{V}_o$$

and

$$\mathbf{y}_{11} = \frac{\mathbf{I}_1}{\mathbf{V}_1} = \frac{-0.75\mathbf{V}_o}{-5\mathbf{V}_o} = 0.15 \text{ S}$$

At node 2,

$$\frac{\mathbf{V}_o - 0}{4} + 2\mathbf{I}_1 + \mathbf{I}_2 = 0$$

or

$$\mathbf{g}_{11} = \frac{\mathbf{I}_1}{\mathbf{V}_1} = \frac{1}{s + 1}$$

By voltage division,

$$\mathbf{V}_2 = \frac{1}{s + 1}\mathbf{V}_1$$

or

$$\mathbf{g}_{21} = \frac{\mathbf{V}_2}{\mathbf{V}_1} = \frac{1}{s + 1}$$

To obtain $\mathbf{g}_{12}$ and $\mathbf{g}_{22}$, we short-circuit the input port and connect a current source $\mathbf{I}_2$ to the output port as in Fig. 19.29(b). By current division,

$$\mathbf{I}_1 = -\frac{1}{s + 1}\mathbf{I}_2$$

or

$$\mathbf{g}_{12} = \frac{\mathbf{I}_1}{\mathbf{I}_2} = -\frac{1}{s + 1}$$

Also,

$$\mathbf{V}_2 = \mathbf{I}_2\left(\frac{1}{s} + s \parallel 1\right)$$

or

$$\mathbf{g}_{22} = \frac{\mathbf{V}_2}{\mathbf{I}_2} = \frac{1}{s} + \frac{s}{s + 1} = \frac{s^2 + s + 1}{s(s + 1)}$$

Thus,

$$[\mathbf{g}] = \begin{bmatrix} \dfrac{1}{s + 1} & -\dfrac{1}{s + 1} \\ \dfrac{1}{s + 1} & \dfrac{s^2 + s + 1}{s(s + 1)} \end{bmatrix}$$

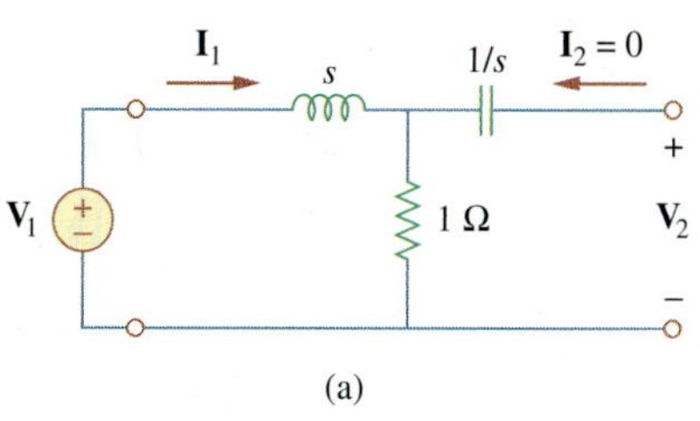

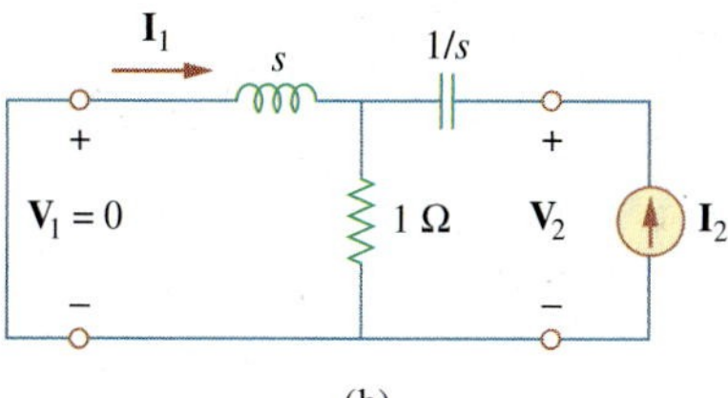

Figure 19.29
Determining the g parameters in the s domain for the circuit in Fig. 19.28.

Practice Problem 19.7

For the ladder network in Fig. 19.30, determine the g parameters in the s domain.

Answer: $[\mathbf{g}] = \begin{bmatrix} \dfrac{s + 2}{s^2 + 3s + 1} & -\dfrac{1}{s^2 + 3s + 1} \\ \dfrac{1}{s^2 + 3s + 1} & \dfrac{s(s + 2)}{s^2 + 3s + 1} \end{bmatrix}$.

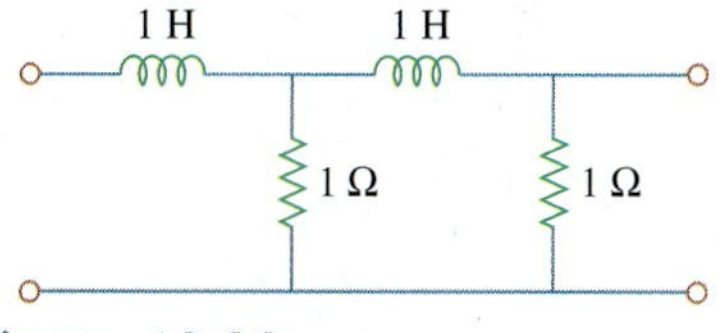

Figure 19.30
For Practice Prob. 19.7.

19.5 Transmission Parameters

Since there are no restrictions on which terminal voltages and currents should be considered independent and which should be dependent variables, we expect to be able to generate many sets of parameters.

Another set of parameters relates the variables at the input port to those at the output port. Thus,

$$\begin{aligned} \mathbf{V}_1 &= \mathbf{A}\mathbf{V}_2 - \mathbf{B}\mathbf{I}_2 \\ \mathbf{I}_1 &= \mathbf{C}\mathbf{V}_2 - \mathbf{D}\mathbf{I}_2 \end{aligned} \quad (19.22)$$

or

$$\begin{bmatrix} \mathbf{V}_1 \\ \mathbf{I}_1 \end{bmatrix} = \begin{bmatrix} \mathbf{A} & \mathbf{B} \\ \mathbf{C} & \mathbf{D} \end{bmatrix} \begin{bmatrix} \mathbf{V}_2 \\ -\mathbf{I}_2 \end{bmatrix} = [\mathbf{T}] \begin{bmatrix} \mathbf{V}_2 \\ -\mathbf{I}_2 \end{bmatrix} \quad (19.23)$$

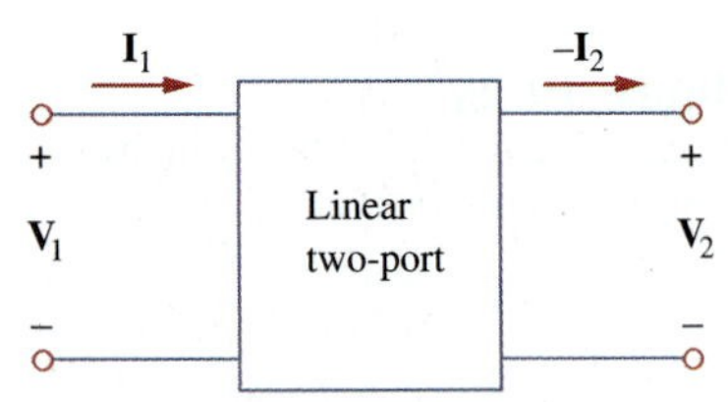

Figure 19.31
Terminal variables used to define the **ADCB** parameters.

Equations (19.22) and (19.23) relate the input variables ($\mathbf{V}_1$ and $\mathbf{I}_1$) to the output variables ($\mathbf{V}_2$ and $-\mathbf{I}_2$). Notice that in computing the transmission parameters, $-\mathbf{I}_2$ is used rather than $\mathbf{I}_2$, because the current is considered to be leaving the network, as shown in Fig. 19.31, as opposed to entering the network as in Fig. 19.1(b). This is done merely for conventional reasons; when you cascade two-ports (output to input), it is most logical to think of $\mathbf{I}_2$ as leaving the two-port. It is also customary in the power industry to consider $\mathbf{I}_2$ as leaving the two-port.

The two-port parameters in Eqs. (19.22) and (19.23) provide a measure of how a circuit transmits voltage and current from a source to a load. They are useful in the analysis of transmission lines (such as cable and fiber) because they express sending-end variables ($\mathbf{V}_1$ and $\mathbf{I}_1$) in terms of the receiving-end variables ($\mathbf{V}_2$ and $-\mathbf{I}_2$). For this reason, they are called *transmission parameters*. They are also known as **ABCD** parameters. They are used in the design of telephone systems, microwave networks, and radars.

The transmission parameters are determined as

$$\begin{aligned} \mathbf{A} &= \left.\frac{\mathbf{V}_1}{\mathbf{V}_2}\right|_{\mathbf{I}_2=0}, \qquad & \mathbf{B} &= -\left.\frac{\mathbf{V}_1}{\mathbf{I}_2}\right|_{\mathbf{V}_2=0} \\ \mathbf{C} &= \left.\frac{\mathbf{I}_1}{\mathbf{V}_2}\right|_{\mathbf{I}_2=0}, \qquad & \mathbf{D} &= -\left.\frac{\mathbf{I}_1}{\mathbf{I}_2}\right|_{\mathbf{V}_2=0} \end{aligned} \quad (19.24)$$

Thus, the transmission parameters are called, specifically,

$$\begin{aligned} \mathbf{A} &= \text{Open-circuit voltage ratio} \\ \mathbf{B} &= \text{Negative short-circuit transfer impedance} \\ \mathbf{C} &= \text{Open-circuit transfer admittance} \\ \mathbf{D} &= \text{Negative short-circuit current ratio} \end{aligned} \quad (19.25)$$

A and **D** are dimensionless, **B** is in ohms, and **C** is in siemens. Since the transmission parameters provide a direct relationship between input and output variables, they are very useful in cascaded networks.

Our last set of parameters may be defined by expressing the variables at the output port in terms of the variables at the input port. We obtain

$$\begin{aligned} \mathbf{V}_2 &= \mathbf{a}\mathbf{V}_1 - \mathbf{b}\mathbf{I}_1 \\ \mathbf{I}_2 &= \mathbf{c}\mathbf{V}_1 - \mathbf{d}\mathbf{I}_1 \end{aligned} \quad (19.26)$$

or

$$\begin{bmatrix} \mathbf{V}_2 \\ \mathbf{I}_2 \end{bmatrix} = \begin{bmatrix} \mathbf{a} & \mathbf{b} \\ \mathbf{c} & \mathbf{d} \end{bmatrix} \begin{bmatrix} \mathbf{V}_1 \\ -\mathbf{I}_1 \end{bmatrix} = [\mathbf{t}] \begin{bmatrix} \mathbf{V}_1 \\ -\mathbf{I}_1 \end{bmatrix} \tag{19.27}$$

The parameters **a**, **b**, **c**, and **d** are called the *inverse transmission*, or *t*, *parameters.* They are determined as follows:

$$\mathbf{a} = \frac{\mathbf{V}_2}{\mathbf{V}_1}\bigg|_{\mathbf{I}_1=0}, \qquad \mathbf{b} = -\frac{\mathbf{V}_2}{\mathbf{I}_1}\bigg|_{\mathbf{V}_1=0}$$
$$\mathbf{c} = \frac{\mathbf{I}_2}{\mathbf{V}_1}\bigg|_{\mathbf{I}_1=0}, \qquad \mathbf{d} = -\frac{\mathbf{I}_2}{\mathbf{I}_1}\bigg|_{\mathbf{V}_1=0} \tag{19.28}$$

From Eq. (19.28) and from our experience so far, it is evident that these parameters are known individually as

$$\begin{aligned} \mathbf{a} &= \text{Open-circuit voltage gain} \\ \mathbf{b} &= \text{Negative short-circuit transfer impedance} \\ \mathbf{c} &= \text{Open-circuit transfer admittance} \\ \mathbf{d} &= \text{Negative short-circuit current gain} \end{aligned} \tag{19.29}$$

While **a** and **d** are dimensionless, **b** and **c** are in ohms and siemens, respectively.

In terms of the transmission or inverse transmission parameters, a network is reciprocal if

$$\mathbf{AD} - \mathbf{BC} = 1, \qquad \mathbf{ad} - \mathbf{bc} = 1 \tag{19.30}$$

These relations can be proved in the same way as the transfer impedance relations for the *z* parameters. Alternatively, we will be able to use Table 19.1 a little later to derive Eq. (19.30) from the fact that $\mathbf{z}_{12} = \mathbf{z}_{21}$ for reciprocal networks.

Example 19.8

Find the transmission parameters for the two-port network in Fig. 19.32.

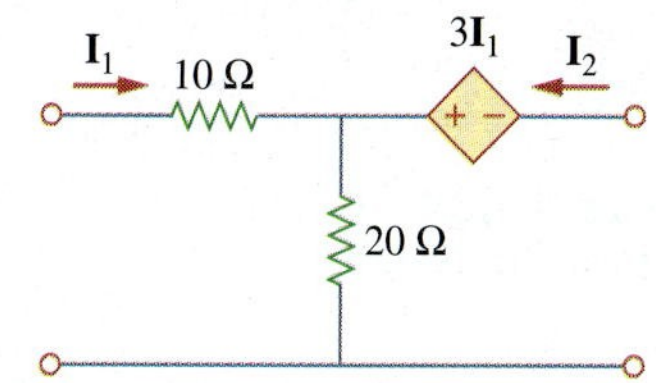

Figure 19.32
For Example 19.8.

Solution:

To determine **A** and **C**, we leave the output port open as in Fig. 19.33(a) so that $\mathbf{I}_2 = 0$ and place a voltage source $\mathbf{V}_1$ at the input port. We have

$$\mathbf{V}_1 = (10 + 20)\mathbf{I}_1 = 30\mathbf{I}_1 \qquad \text{and} \qquad \mathbf{V}_2 = 20\mathbf{I}_1 - 3\mathbf{I}_1 = 17\mathbf{I}_1$$

Thus,

$$\mathbf{A} = \frac{\mathbf{V}_1}{\mathbf{V}_2} = \frac{30\mathbf{I}_1}{17\mathbf{I}_1} = 1.765, \qquad \mathbf{C} = \frac{\mathbf{I}_1}{\mathbf{V}_2} = \frac{\mathbf{I}_1}{17\mathbf{I}_1} = 0.0588 \text{ S}$$

To obtain **B** and **D**, we short-circuit the output port so that $\mathbf{V}_2 = 0$ as shown in Fig. 19.33(b) and place a voltage source $\mathbf{V}_1$ at the input port. At node *a* in the circuit of Fig. 19.33(b), KCL gives

$$\frac{\mathbf{V}_1 - \mathbf{V}_a}{10} - \frac{\mathbf{V}_a}{20} + \mathbf{I}_2 = 0 \tag{19.8.1}$$

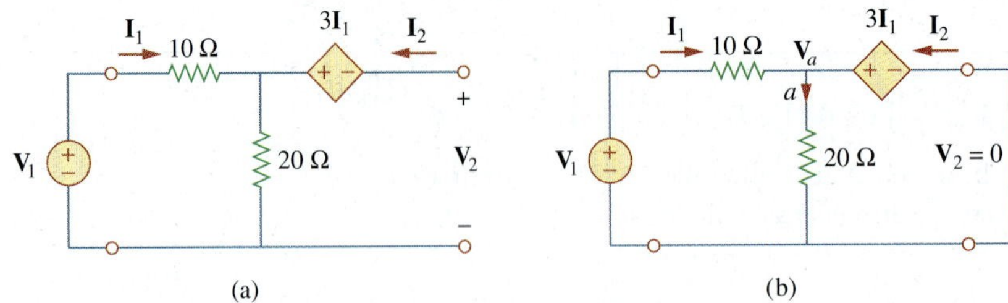

Figure 19.33
For Example 19.8: (a) finding **A** and **C**, (b) finding **B** and **D**.

But $\mathbf{V}_a = 3\mathbf{I}_1$ and $\mathbf{I}_1 = (\mathbf{V}_1 - \mathbf{V}_a)/10$. Combining these gives

$$\mathbf{V}_a = 3\mathbf{I}_1 \qquad \mathbf{V}_1 = 13\mathbf{I}_1 \tag{19.8.2}$$

Substituting $\mathbf{V}_a = 3\mathbf{I}_1$ into Eq. (19.8.1) and replacing the first term with $\mathbf{I}_1$,

$$\mathbf{I}_1 - \frac{3\mathbf{I}_1}{20} + \mathbf{I}_2 = 0 \qquad \Rightarrow \qquad \frac{17}{20}\mathbf{I}_1 = -\mathbf{I}_2$$

Therefore,

$$\mathbf{D} = -\frac{\mathbf{I}_1}{\mathbf{I}_2} = \frac{20}{17} = 1.176, \qquad \mathbf{B} = -\frac{\mathbf{V}_1}{\mathbf{I}_2} = \frac{-13\mathbf{I}_1}{(-17/20)\mathbf{I}_1} = 15.29\ \Omega$$

Practice Problem 19.8

Find the transmission parameters for the circuit in Fig. 19.16 (see Practice Prob. 19.3).

Answer: $\mathbf{A} = 1.5$, $\mathbf{B} = 22\ \Omega$, $\mathbf{C} = 125$ mS, $\mathbf{D} = 2.5$.

Example 19.9

The **ABCD** parameters of the two-port network in Fig. 19.34 are

$$\begin{bmatrix} 4 & 20\ \Omega \\ 0.1\ \text{S} & 2 \end{bmatrix}$$

The output port is connected to a variable load for maximum power transfer. Find R_L and the maximum power transferred.

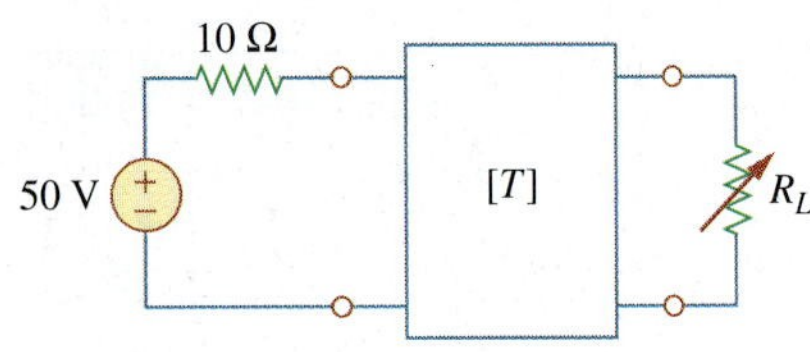

Figure 19.34
For Example 19.9.

Solution:
What we need is to find the Thevenin equivalent ($\mathbf{Z}_{\text{Th}}$ and $\mathbf{V}_{\text{Th}}$) at the load or output port. We find $\mathbf{Z}_{\text{Th}}$ using the circuit in Fig. 19.35(a). Our goal is to get $\mathbf{Z}_{\text{Th}} = \mathbf{V}_2/\mathbf{I}_2$. Substituting the given **ABCD** parameters into Eq. (19.22), we obtain

$$\mathbf{V}_1 = 4\mathbf{V}_2 - 20\mathbf{I}_2 \tag{19.9.1}$$

$$\mathbf{I}_1 = 0.1\mathbf{V}_2 - 2\mathbf{I}_2 \tag{19.9.2}$$

At the input port, $\mathbf{V}_1 = -10\mathbf{I}_1$. Substituting this into Eq. (19.9.1) gives

$$-10\mathbf{I}_1 = 4\mathbf{V}_2 - 20\mathbf{I}_2$$

or

$$\mathbf{I}_1 = -0.4\mathbf{V}_2 + 2\mathbf{I}_2 \tag{19.9.3}$$

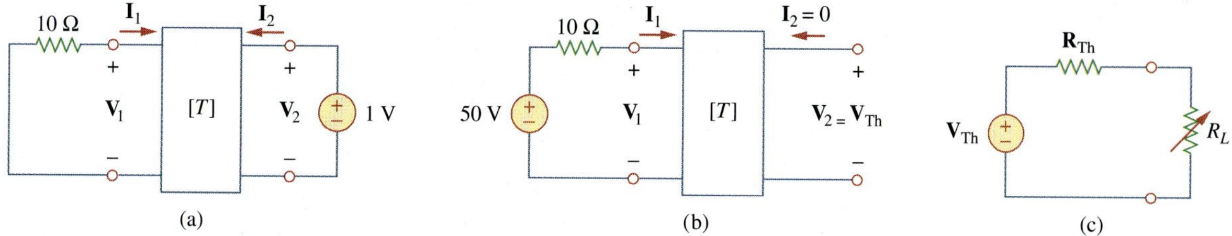

Figure 19.35
Solution of Example 19.9: (a) finding $\mathbf{Z}_{\text{Th}}$, (b) finding $\mathbf{V}_{\text{Th}}$, (c) finding R_L for maximum power transfer.

Setting the right-hand sides of Eqs. (19.9.2) and (19.9.3) equal,

$$0.1\mathbf{V}_2 - 2\mathbf{I}_2 = -0.4\mathbf{V}_2 + 2\mathbf{I}_2 \quad \Rightarrow \quad 0.5\mathbf{V}_2 = 4\mathbf{I}_2$$

Hence,

$$\mathbf{Z}_{\text{Th}} = \frac{\mathbf{V}_2}{\mathbf{I}_2} = \frac{4}{0.5} = 8\ \Omega$$

To find $\mathbf{V}_{\text{Th}}$, we use the circuit in Fig. 19.35(b). At the output port $\mathbf{I}_2 = 0$ and at the input port $\mathbf{V}_1 = 50 - 10\mathbf{I}_1$. Substituting these into Eqs. (19.9.1) and (19.9.2),

$$50 - 10\mathbf{I}_1 = 4\mathbf{V}_2 \tag{19.9.4}$$

$$\mathbf{I}_1 = 0.1\mathbf{V}_2 \tag{19.9.5}$$

Substituting Eq. (19.9.5) into Eq. (19.9.4),

$$50 - \mathbf{V}_2 = 4\mathbf{V}_2 \quad \Rightarrow \quad \mathbf{V}_2 = 10$$

Thus,

$$\mathbf{V}_{\text{Th}} = \mathbf{V}_2 = 10\ \text{V}$$

The equivalent circuit is shown in Fig. 19.35(c). For maximum power transfer,

$$R_L = \mathbf{Z}_{\text{Th}} = 8\ \Omega$$

From Eq. (4.24), the maximum power is

$$P = I^2 R_L = \left(\frac{\mathbf{V}_{\text{Th}}}{2R_L}\right)^2 R_L = \frac{\mathbf{V}_{\text{Th}}^2}{4R_L} = \frac{100}{4 \times 8} = 3.125\ \text{W}$$

Practice Problem 19.9

Find $\mathbf{I}_1$ and $\mathbf{I}_2$ if the transmission parameters for the two-port in Fig. 19.36 are

$$\begin{bmatrix} 5 & 10\ \Omega \\ 0.4\ \text{S} & 1 \end{bmatrix}$$

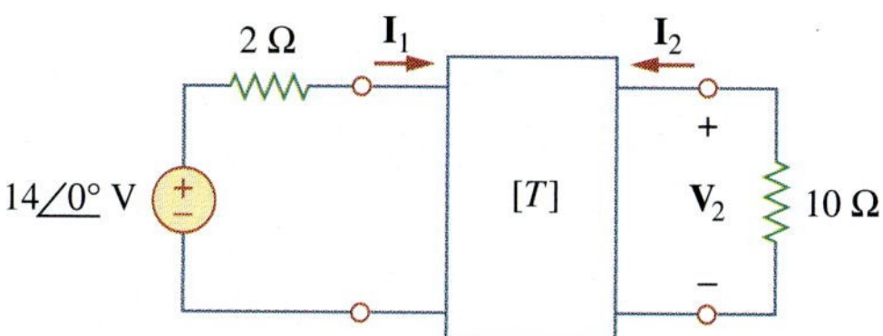

Figure 19.36
For Practice Prob. 19.9.

Answer: 1 A, −0.2 A.

19.6 †Relationships Between Parameters

Since the six sets of parameters relate the same input and output terminal variables of the same two-port network, they should be interrelated. If two sets of parameters exist, we can relate one set to the other set. Let us demonstrate the process with two examples.

Given the z parameters, let us obtain the y parameters. From Eq. (19.2),

$$\begin{bmatrix} \mathbf{V}_1 \\ \mathbf{V}_2 \end{bmatrix} = \begin{bmatrix} \mathbf{z}_{11} & \mathbf{z}_{12} \\ \mathbf{z}_{21} & \mathbf{z}_{22} \end{bmatrix} \begin{bmatrix} \mathbf{I}_1 \\ \mathbf{I}_2 \end{bmatrix} = [\mathbf{z}] \begin{bmatrix} \mathbf{I}_1 \\ \mathbf{I}_2 \end{bmatrix} \tag{19.31}$$

or

$$\begin{bmatrix} \mathbf{I}_1 \\ \mathbf{I}_2 \end{bmatrix} = [\mathbf{z}]^{-1} \begin{bmatrix} \mathbf{V}_1 \\ \mathbf{V}_2 \end{bmatrix} \tag{19.32}$$

Also, from Eq. (19.9),

$$\begin{bmatrix} \mathbf{I}_1 \\ \mathbf{I}_2 \end{bmatrix} = \begin{bmatrix} \mathbf{y}_{11} & \mathbf{y}_{12} \\ \mathbf{y}_{21} & \mathbf{y}_{22} \end{bmatrix} \begin{bmatrix} \mathbf{V}_1 \\ \mathbf{V}_2 \end{bmatrix} = [\mathbf{y}] \begin{bmatrix} \mathbf{V}_1 \\ \mathbf{V}_2 \end{bmatrix} \tag{19.33}$$

Comparing Eqs. (19.32) and (19.33), we see that

$$[\mathbf{y}] = [\mathbf{z}]^{-1} \tag{19.34}$$

The adjoint of the $[\mathbf{z}]$ matrix is

$$\begin{bmatrix} \mathbf{z}_{22} & -\mathbf{z}_{12} \\ -\mathbf{z}_{21} & \mathbf{z}_{11} \end{bmatrix}$$

and its determinant is

$$\Delta_z = \mathbf{z}_{11}\mathbf{z}_{22} - \mathbf{z}_{12}\mathbf{z}_{21}$$

Substituting these into Eq. (19.34), we get

$$\begin{bmatrix} \mathbf{y}_{11} & \mathbf{y}_{12} \\ \mathbf{y}_{21} & \mathbf{y}_{22} \end{bmatrix} = \frac{\begin{bmatrix} \mathbf{z}_{22} & -\mathbf{z}_{12} \\ -\mathbf{z}_{21} & \mathbf{z}_{11} \end{bmatrix}}{\Delta_z} \tag{19.35}$$

Equating terms yields

$$\mathbf{y}_{11} = \frac{\mathbf{z}_{22}}{\Delta_z}, \qquad \mathbf{y}_{12} = -\frac{\mathbf{z}_{12}}{\Delta_z}, \qquad \mathbf{y}_{21} = -\frac{\mathbf{z}_{21}}{\Delta_z}, \qquad \mathbf{y}_{22} = \frac{\mathbf{z}_{11}}{\Delta_z} \tag{19.36}$$

As a second example, let us determine the h parameters from the z parameters. From Eq. (19.1),

$$\mathbf{V}_1 = \mathbf{z}_{11}\mathbf{I}_1 + \mathbf{z}_{12}\mathbf{I}_2 \tag{19.37a}$$

$$\mathbf{V}_2 = \mathbf{z}_{21}\mathbf{I}_1 + \mathbf{z}_{22}\mathbf{I}_2 \tag{19.37b}$$

Making $\mathbf{I}_2$ the subject of Eq. (19.37b),

$$\mathbf{I}_2 = -\frac{\mathbf{z}_{21}}{\mathbf{z}_{22}}\mathbf{I}_1 + \frac{1}{\mathbf{z}_{22}}\mathbf{V}_2 \tag{19.38}$$

Substituting this into Eq. (19.37a),

$$\mathbf{V}_1 = \frac{\mathbf{z}_{11}\mathbf{z}_{22} - \mathbf{z}_{12}\mathbf{z}_{21}}{\mathbf{z}_{22}}\mathbf{I}_1 + \frac{\mathbf{z}_{12}}{\mathbf{z}_{22}}\mathbf{V}_2 \tag{19.39}$$

Putting Eqs. (19.38) and (19.39) in matrix form,

$$\begin{bmatrix} \mathbf{V}_1 \\ \mathbf{I}_2 \end{bmatrix} = \begin{bmatrix} \dfrac{\Delta_z}{\mathbf{z}_{22}} & \dfrac{\mathbf{z}_{12}}{\mathbf{z}_{22}} \\ -\dfrac{\mathbf{z}_{21}}{\mathbf{z}_{22}} & \dfrac{1}{\mathbf{z}_{22}} \end{bmatrix} \begin{bmatrix} \mathbf{I}_1 \\ \mathbf{V}_2 \end{bmatrix} \tag{19.40}$$

From Eq. (19.15),

$$\begin{bmatrix} \mathbf{V}_1 \\ \mathbf{I}_2 \end{bmatrix} = \begin{bmatrix} \mathbf{h}_{11} & \mathbf{h}_{12} \\ \mathbf{h}_{21} & \mathbf{h}_{22} \end{bmatrix} \begin{bmatrix} \mathbf{I}_1 \\ \mathbf{V}_2 \end{bmatrix}$$

Comparing this with Eq. (19.40), we obtain

$$\mathbf{h}_{11} = \frac{\Delta_z}{\mathbf{z}_{22}}, \qquad \mathbf{h}_{12} = \frac{\mathbf{z}_{12}}{\mathbf{z}_{22}}, \qquad \mathbf{h}_{21} = -\frac{\mathbf{z}_{21}}{\mathbf{z}_{22}}, \qquad \mathbf{h}_{22} = \frac{1}{\mathbf{z}_{22}} \tag{19.41}$$

Table 19.1 provides the conversion formulas for the six sets of two-port parameters. Given one set of parameters, Table 19.1 can be used to find other parameters. For example, given the *T* parameters, we find the corresponding *h* parameters in the fifth column of the third row.

TABLE 19.1

Conversion of two-port parameters.

	z		y		h		g		T		t	
z	$\mathbf{z}_{11}$	$\mathbf{z}_{12}$	$\frac{\mathbf{y}_{22}}{\Delta_y}$	$-\frac{\mathbf{y}_{12}}{\Delta_y}$	$\frac{\Delta_h}{\mathbf{h}_{22}}$	$\frac{\mathbf{h}_{12}}{\mathbf{h}_{22}}$	$\frac{1}{\mathbf{g}_{11}}$	$-\frac{\mathbf{g}_{12}}{\mathbf{g}_{11}}$	$\frac{\mathbf{A}}{\mathbf{C}}$	$\frac{\Delta_T}{\mathbf{C}}$	$\frac{\mathbf{d}}{\mathbf{c}}$	$\frac{1}{\mathbf{c}}$
	$\mathbf{z}_{21}$	$\mathbf{z}_{22}$	$-\frac{\mathbf{y}_{21}}{\Delta_y}$	$\frac{\mathbf{y}_{11}}{\Delta_y}$	$-\frac{\mathbf{h}_{21}}{\mathbf{h}_{22}}$	$\frac{1}{\mathbf{h}_{22}}$	$\frac{\mathbf{g}_{21}}{\mathbf{g}_{11}}$	$\frac{\Delta_g}{\mathbf{g}_{11}}$	$\frac{1}{\mathbf{C}}$	$\frac{\mathbf{D}}{\mathbf{C}}$	$\frac{\Delta_t}{\mathbf{c}}$	$\frac{\mathbf{a}}{\mathbf{c}}$
y	$\frac{\mathbf{z}_{22}}{\Delta_z}$	$-\frac{\mathbf{z}_{12}}{\Delta_z}$	$\mathbf{y}_{11}$	$\mathbf{y}_{12}$	$\frac{1}{\mathbf{h}_{11}}$	$-\frac{\mathbf{h}_{12}}{\mathbf{h}_{11}}$	$\frac{\Delta_g}{\mathbf{g}_{22}}$	$\frac{\mathbf{g}_{12}}{\mathbf{g}_{22}}$	$\frac{\mathbf{D}}{\mathbf{B}}$	$-\frac{\Delta_T}{\mathbf{B}}$	$\frac{\mathbf{a}}{\mathbf{b}}$	$-\frac{1}{\mathbf{b}}$
	$-\frac{\mathbf{z}_{21}}{\Delta_z}$	$\frac{\mathbf{z}_{11}}{\Delta_z}$	$\mathbf{y}_{21}$	$\mathbf{y}_{22}$	$\frac{\mathbf{h}_{21}}{\mathbf{h}_{11}}$	$\frac{\Delta_h}{\mathbf{h}_{11}}$	$-\frac{\mathbf{g}_{21}}{\mathbf{g}_{22}}$	$\frac{1}{\mathbf{g}_{22}}$	$-\frac{1}{\mathbf{B}}$	$\frac{\mathbf{A}}{\mathbf{B}}$	$-\frac{\Delta_t}{\mathbf{b}}$	$\frac{\mathbf{d}}{\mathbf{b}}$
h	$\frac{\Delta_z}{\mathbf{z}_{22}}$	$\frac{\mathbf{z}_{12}}{\mathbf{z}_{22}}$	$\frac{1}{\mathbf{y}_{11}}$	$-\frac{\mathbf{y}_{12}}{\mathbf{y}_{11}}$	$\mathbf{h}_{11}$	$\mathbf{h}_{12}$	$\frac{\mathbf{g}_{22}}{\Delta_g}$	$-\frac{\mathbf{g}_{12}}{\Delta_g}$	$\frac{\mathbf{B}}{\mathbf{D}}$	$\frac{\Delta_T}{\mathbf{D}}$	$\frac{\mathbf{b}}{\mathbf{a}}$	$\frac{1}{\mathbf{a}}$
	$-\frac{\mathbf{z}_{21}}{\mathbf{z}_{22}}$	$\frac{1}{\mathbf{z}_{22}}$	$\frac{\mathbf{y}_{21}}{\mathbf{y}_{11}}$	$\frac{\Delta_y}{\mathbf{y}_{11}}$	$\mathbf{h}_{21}$	$\mathbf{h}_{22}$	$-\frac{\mathbf{g}_{21}}{\Delta_g}$	$\frac{\mathbf{g}_{11}}{\Delta_g}$	$-\frac{1}{\mathbf{D}}$	$\frac{\mathbf{C}}{\mathbf{D}}$	$\frac{\Delta_t}{\mathbf{a}}$	$\frac{\mathbf{c}}{\mathbf{a}}$
g	$\frac{1}{\mathbf{z}_{11}}$	$-\frac{\mathbf{z}_{12}}{\mathbf{z}_{11}}$	$\frac{\Delta_y}{\mathbf{y}_{22}}$	$\frac{\mathbf{y}_{12}}{\mathbf{y}_{22}}$	$\frac{\mathbf{h}_{22}}{\Delta_h}$	$-\frac{\mathbf{h}_{12}}{\Delta_h}$	$\mathbf{g}_{11}$	$\mathbf{g}_{12}$	$\frac{\mathbf{C}}{\mathbf{A}}$	$-\frac{\Delta_T}{\mathbf{A}}$	$\frac{\mathbf{c}}{\mathbf{d}}$	$-\frac{1}{\mathbf{d}}$
	$\frac{\mathbf{z}_{21}}{\mathbf{z}_{11}}$	$\frac{\Delta_z}{\mathbf{z}_{11}}$	$-\frac{\mathbf{y}_{21}}{\mathbf{y}_{22}}$	$\frac{1}{\mathbf{y}_{22}}$	$-\frac{\mathbf{h}_{21}}{\Delta_h}$	$\frac{\mathbf{h}_{11}}{\Delta_h}$	$\mathbf{g}_{21}$	$\mathbf{g}_{22}$	$\frac{1}{\mathbf{A}}$	$\frac{\mathbf{B}}{\mathbf{A}}$	$\frac{\Delta_t}{\mathbf{d}}$	$-\frac{\mathbf{b}}{\mathbf{d}}$
T	$\frac{\mathbf{z}_{11}}{\mathbf{z}_{21}}$	$\frac{\Delta_z}{\mathbf{z}_{21}}$	$-\frac{\mathbf{y}_{22}}{\mathbf{y}_{21}}$	$-\frac{1}{\mathbf{y}_{21}}$	$-\frac{\Delta_h}{\mathbf{h}_{21}}$	$-\frac{\mathbf{h}_{11}}{\mathbf{h}_{21}}$	$\frac{1}{\mathbf{g}_{21}}$	$\frac{\mathbf{g}_{22}}{\mathbf{g}_{21}}$	$\mathbf{A}$	$\mathbf{B}$	$\frac{\mathbf{d}}{\Delta_t}$	$\frac{\mathbf{b}}{\Delta_t}$
	$\frac{1}{\mathbf{z}_{21}}$	$\frac{\mathbf{z}_{22}}{\mathbf{z}_{21}}$	$-\frac{\Delta_y}{\mathbf{y}_{21}}$	$-\frac{\mathbf{y}_{11}}{\mathbf{y}_{21}}$	$-\frac{\mathbf{h}_{22}}{\mathbf{h}_{21}}$	$-\frac{1}{\mathbf{h}_{21}}$	$\frac{\mathbf{g}_{11}}{\mathbf{g}_{21}}$	$\frac{\Delta_g}{\mathbf{g}_{21}}$	$\mathbf{C}$	$\mathbf{D}$	$\frac{\mathbf{c}}{\Delta_t}$	$\frac{\mathbf{a}}{\Delta_t}$
t	$\frac{\mathbf{z}_{22}}{\mathbf{z}_{12}}$	$\frac{\Delta_z}{\mathbf{z}_{12}}$	$-\frac{\mathbf{y}_{11}}{\mathbf{y}_{12}}$	$-\frac{1}{\mathbf{y}_{12}}$	$\frac{1}{\mathbf{h}_{12}}$	$\frac{\mathbf{h}_{11}}{\mathbf{h}_{12}}$	$-\frac{\Delta_g}{\mathbf{g}_{12}}$	$-\frac{\mathbf{g}_{22}}{\mathbf{g}_{12}}$	$\frac{\mathbf{D}}{\Delta_T}$	$\frac{\mathbf{B}}{\Delta_T}$	$\mathbf{a}$	$\mathbf{b}$
	$\frac{1}{\mathbf{z}_{12}}$	$\frac{\mathbf{z}_{11}}{\mathbf{z}_{12}}$	$-\frac{\Delta_y}{\mathbf{y}_{12}}$	$-\frac{\mathbf{y}_{22}}{\mathbf{y}_{12}}$	$\frac{\mathbf{h}_{22}}{\mathbf{h}_{12}}$	$\frac{\Delta_h}{\mathbf{h}_{12}}$	$-\frac{\mathbf{g}_{11}}{\mathbf{g}_{12}}$	$-\frac{1}{\mathbf{g}_{12}}$	$\frac{\mathbf{C}}{\Delta_T}$	$\frac{\mathbf{A}}{\Delta_T}$	$\mathbf{c}$	$\mathbf{d}$

$\Delta_z = \mathbf{z}_{11}\mathbf{z}_{22} - \mathbf{z}_{12}\mathbf{z}_{21}$, $\Delta_h = \mathbf{h}_{11}\mathbf{h}_{22} - \mathbf{h}_{12}\mathbf{h}_{21}$, $\Delta_T = \mathbf{AD} - \mathbf{BC}$

$\Delta_y = \mathbf{y}_{11}\mathbf{y}_{22} - \mathbf{y}_{12}\mathbf{y}_{21}$, $\Delta_g = \mathbf{g}_{11}\mathbf{g}_{22} - \mathbf{g}_{12}\mathbf{g}_{21}$, $\Delta_t = \mathbf{ad} - \mathbf{bc}$

Also, given that $\mathbf{z}_{21} = \mathbf{z}_{12}$ for a reciprocal network, we can use the table to express this condition in terms of other parameters. It can also be shown that

$$[\mathbf{g}] = [\mathbf{h}]^{-1} \tag{19.42}$$

but

$$[\mathbf{t}] \neq [\mathbf{T}]^{-1} \tag{19.43}$$

Example 19.10

Find [**z**] and [**g**] of a two-port network if

$$[\mathbf{T}] = \begin{bmatrix} 10 & 1.5\ \Omega \\ 2\ \text{S} & 4 \end{bmatrix}$$

Solution:

If $\mathbf{A} = 10$, $\mathbf{B} = 1.5$, $\mathbf{C} = 2$, $\mathbf{D} = 4$, the determinant of the matrix is

$$\Delta_T = \mathbf{AD} - \mathbf{BC} = 40 - 3 = 37$$

From Table 19.1,

$$\mathbf{z}_{11} = \frac{\mathbf{A}}{\mathbf{C}} = \frac{10}{2} = 5, \qquad \mathbf{z}_{12} = \frac{\Delta_T}{\mathbf{C}} = \frac{37}{2} = 18.5$$

$$\mathbf{z}_{21} = \frac{1}{\mathbf{C}} = \frac{1}{2} = 0.5, \qquad \mathbf{z}_{22} = \frac{\mathbf{D}}{\mathbf{C}} = \frac{4}{2} = 2$$

$$\mathbf{g}_{11} = \frac{\mathbf{C}}{\mathbf{A}} = \frac{2}{10} = 0.2, \qquad \mathbf{g}_{12} = -\frac{\Delta_T}{\mathbf{A}} = -\frac{37}{10} = -3.7$$

$$\mathbf{g}_{21} = \frac{1}{\mathbf{A}} = \frac{1}{10} = 0.1, \qquad \mathbf{g}_{22} = \frac{\mathbf{B}}{\mathbf{A}} = \frac{1.5}{10} = 0.15$$

Thus,

$$[\mathbf{z}] = \begin{bmatrix} 5 & 18.5 \\ 0.5 & 2 \end{bmatrix} \Omega, \qquad [\mathbf{g}] = \begin{bmatrix} 0.2\ \text{S} & -3.7 \\ 0.1 & 0.15\ \Omega \end{bmatrix}$$

Practice Problem 19.10

Determine [**y**] and [**T**] of a two-port network whose z parameters are

$$[\mathbf{z}] = \begin{bmatrix} 6 & 4 \\ 4 & 6 \end{bmatrix} \Omega$$

Answer: $[\mathbf{y}] = \begin{bmatrix} 0.3 & -0.2 \\ -0.2 & 0.3 \end{bmatrix} \text{S}, \quad [\mathbf{T}] = \begin{bmatrix} 1.5 & 5\ \Omega \\ 0.25\ \text{S} & 1.5 \end{bmatrix}.$

Example 19.11

Obtain the y parameters of the op amp circuit in Fig. 19.37. Show that the circuit has no z parameters.

Solution:

Since no current can enter the input terminals of the op amp, $\mathbf{I}_1 = 0$, which can be expressed in terms of $\mathbf{V}_1$ and $\mathbf{V}_2$ as

$$\mathbf{I}_1 = 0\mathbf{V}_1 + 0\mathbf{V}_2 \tag{19.11.1}$$

Comparing this with Eq. (19.8) gives

$$\mathbf{y}_{11} = 0 = \mathbf{y}_{12}$$

Also,

$$\mathbf{V}_2 = R_3\mathbf{I}_2 + \mathbf{I}_o(R_1 + R_2)$$

where $\mathbf{I}_o$ is the current through R_1 and R_2. But $\mathbf{I}_o = \mathbf{V}_1/R_1$. Hence,

$$\mathbf{V}_2 = R_3\mathbf{I}_2 + \frac{\mathbf{V}_1(R_1 + R_2)}{R_1}$$

which can be written as

$$\mathbf{I}_2 = -\frac{(R_1 + R_2)}{R_1R_3}\mathbf{V}_1 + \frac{\mathbf{V}_2}{R_3}$$

Comparing this with Eq. (19.8) shows that

$$\mathbf{y}_{21} = -\frac{(R_1 + R_2)}{R_1R_3}, \qquad \mathbf{y}_{22} = \frac{1}{R_3}$$

The determinant of the [**y**] matrix is

$$\Delta_y = \mathbf{y}_{11}\mathbf{y}_{22} - \mathbf{y}_{12}\mathbf{y}_{21} = 0$$

Since $\Delta_y = 0$, the [**y**] matrix has no inverse; therefore, the [**z**] matrix does not exist according to Eq. (19.34). Note that the circuit is not reciprocal because of the active element.

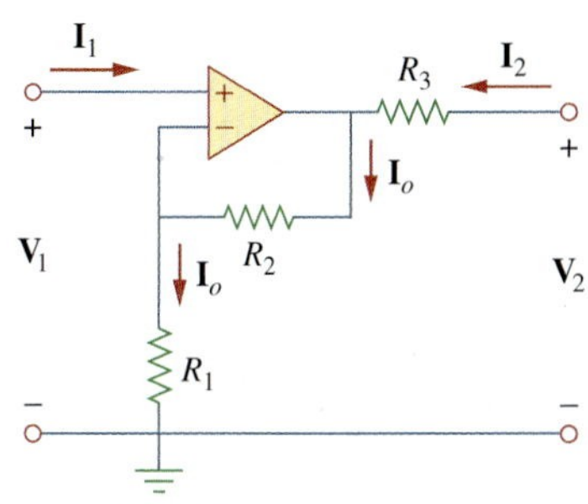

Figure 19.37
For Example 19.11.

Practice Problem 19.11

Find the z parameters of the op amp circuit in Fig. 19.38. Show that the circuit has no y parameters.

Answer: $[\mathbf{z}] = \begin{bmatrix} R_1 & 0 \\ -R_2 & 0 \end{bmatrix}$. Since $[\mathbf{z}]^{-1}$ does not exist, [**y**] does not exist.

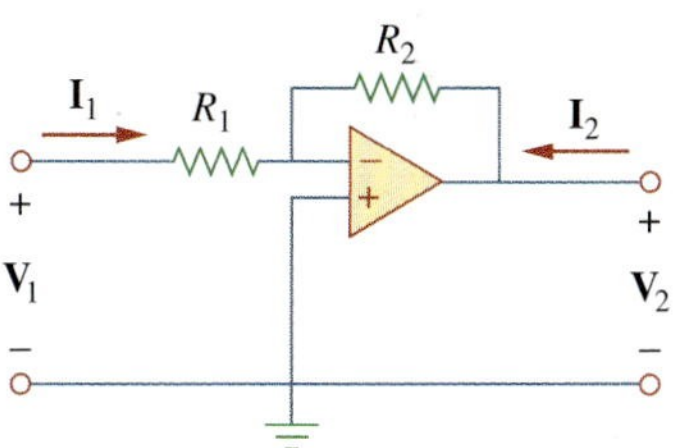

Figure 19.38
For Practice Prob. 19.11.

19.7 Interconnection of Networks

A large, complex network may be divided into subnetworks for the purposes of analysis and design. The subnetworks are modeled as two-port networks, interconnected to form the original network. The two-port networks may therefore be regarded as building blocks that can be interconnected to form a complex network. The interconnection can be in series, in parallel, or in cascade. Although the interconnected network can be described by any of the six parameter sets, a certain set of parameters may have a definite advantage. For example, when the networks are in series, their individual z parameters add up to give the z parameters of the larger network. When they are in parallel, their individual y parameters add up to give the y parameters of the larger network. When they are cascaded, their individual transmission parameters can be multiplied together to get the transmission parameters of the larger network.

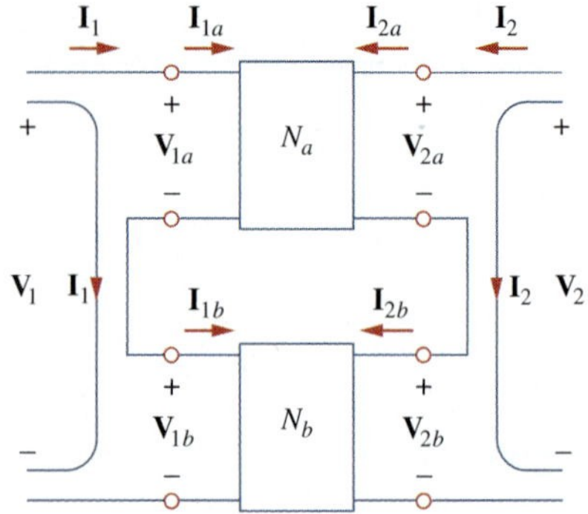

Figure 19.39
Series connection of two two-port networks.

Consider the series connection of two two-port networks shown in Fig. 19.39. The networks are regarded as being in series because their input currents are the same and their voltages add. In addition, each network has a common reference, and when the circuits are placed in series, the common reference points of each circuit are connected together. For network N_a,

$$\begin{aligned} \mathbf{V}_{1a} &= \mathbf{z}_{11a}\mathbf{I}_{1a} + \mathbf{z}_{12a}\mathbf{I}_{2a} \\ \mathbf{V}_{2a} &= \mathbf{z}_{21a}\mathbf{I}_{1a} + \mathbf{z}_{22a}\mathbf{I}_{2a} \end{aligned} \tag{19.44}$$

and for network N_b,

$$\begin{aligned} \mathbf{V}_{1b} &= \mathbf{z}_{11b}\mathbf{I}_{1b} + \mathbf{z}_{12b}\mathbf{I}_{2b} \\ \mathbf{V}_{2b} &= \mathbf{z}_{21b}\mathbf{I}_{1b} + \mathbf{z}_{22b}\mathbf{I}_{2b} \end{aligned} \tag{19.45}$$

We notice from Fig. 19.39 that

$$\mathbf{I}_1 = \mathbf{I}_{1a} = \mathbf{I}_{1b}, \qquad \mathbf{I}_2 = \mathbf{I}_{2a} = \mathbf{I}_{2b} \tag{19.46}$$

and that

$$\begin{aligned} \mathbf{V}_1 &= \mathbf{V}_{1a} + \mathbf{V}_{1b} = (\mathbf{z}_{11a} + \mathbf{z}_{11b})\mathbf{I}_1 + (\mathbf{z}_{12a} + \mathbf{z}_{12b})\mathbf{I}_2 \\ \mathbf{V}_2 &= \mathbf{V}_{2a} + \mathbf{V}_{2b} = (\mathbf{z}_{21a} + \mathbf{z}_{21b})\mathbf{I}_1 + (\mathbf{z}_{22a} + \mathbf{z}_{22b})\mathbf{I}_2 \end{aligned} \tag{19.47}$$

Thus, the z parameters for the overall network are

$$\begin{bmatrix} \mathbf{z}_{11} & \mathbf{z}_{12} \\ \mathbf{z}_{21} & \mathbf{z}_{22} \end{bmatrix} = \begin{bmatrix} \mathbf{z}_{11a} + \mathbf{z}_{11b} & \mathbf{z}_{12a} + \mathbf{z}_{12b} \\ \mathbf{z}_{21a} + \mathbf{z}_{21b} & \mathbf{z}_{22a} + \mathbf{z}_{22b} \end{bmatrix} \tag{19.48}$$

or

$$\boxed{[\mathbf{z}] = [\mathbf{z}_a] + [\mathbf{z}_b]} \tag{19.49}$$

showing that the z parameters for the overall network are the sum of the z parameters for the individual networks. This can be extended to n networks in series. If two two-port networks in the [**h**] model, for example, are connected in series, we use Table 19.1 to convert the **h** to **z** and then apply Eq. (19.49). We finally convert the result back to **h** using Table 19.1.

Two two-port networks are in parallel when their port voltages are equal and the port currents of the larger network are the sums of the individual port currents. In addition, each circuit must have a common reference and when the networks are connected together, they must all have their common references tied together. The parallel connection of two two-port networks is shown in Fig. 19.40. For the two networks,

$$\begin{aligned} \mathbf{I}_{1a} &= \mathbf{y}_{11a}\mathbf{V}_{1a} + \mathbf{y}_{12a}\mathbf{V}_{2a} \\ \mathbf{I}_{2a} &= \mathbf{y}_{21a}\mathbf{V}_{1a} + \mathbf{y}_{22a}\mathbf{V}_{2a} \end{aligned} \tag{19.50}$$

and

$$\begin{aligned} \mathbf{I}_{1b} &= \mathbf{y}_{11b}\mathbf{V}_{1b} + \mathbf{y}_{12b}\mathbf{V}_{2b} \\ \mathbf{I}_{2a} &= \mathbf{y}_{21b}\mathbf{V}_{1b} + \mathbf{y}_{22b}\mathbf{V}_{2b} \end{aligned} \tag{19.51}$$

But from Fig. 19.40,

$$\mathbf{V}_1 = \mathbf{V}_{1a} = \mathbf{V}_{1b}, \qquad \mathbf{V}_2 = \mathbf{V}_{2a} = \mathbf{V}_{2b} \tag{19.52a}$$

$$\mathbf{I}_1 = \mathbf{I}_{1a} + \mathbf{I}_{1b}, \qquad \mathbf{I}_2 = \mathbf{I}_{2a} + \mathbf{I}_{2b} \tag{19.52b}$$

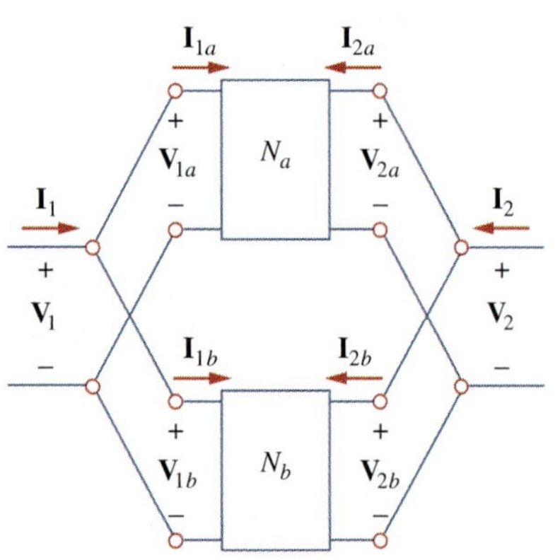

Figure 19.40
Parallel connection of two two-port networks.

Substituting Eqs. (19.50) and (19.51) into Eq. (19.52b) yields

$$\begin{aligned} \mathbf{I}_1 &= (\mathbf{y}_{11a} + \mathbf{y}_{11b})\mathbf{V}_1 + (\mathbf{y}_{12a} + \mathbf{y}_{12b})\mathbf{V}_2 \\ \mathbf{I}_2 &= (\mathbf{y}_{21a} + \mathbf{y}_{21b})\mathbf{V}_1 + (\mathbf{y}_{22a} + \mathbf{y}_{22b})\mathbf{V}_2 \end{aligned} \tag{19.53}$$

Thus, the y parameters for the overall network are

$$\begin{bmatrix} \mathbf{y}_{11} & \mathbf{y}_{12} \\ \mathbf{y}_{21} & \mathbf{y}_{22} \end{bmatrix} = \begin{bmatrix} \mathbf{y}_{11a} + \mathbf{y}_{11b} & \mathbf{y}_{12a} + \mathbf{y}_{12b} \\ \mathbf{y}_{21a} + \mathbf{y}_{21b} & \mathbf{y}_{22a} + \mathbf{y}_{22b} \end{bmatrix} \tag{19.54}$$

or

$$[\mathbf{y}] = [\mathbf{y}_a] + [\mathbf{y}_b] \tag{19.55}$$

showing that the y parameters of the overall network are the sum of the y parameters of the individual networks. The result can be extended to n two-port networks in parallel.

Two networks are said to be *cascaded* when the output of one is the input of the other. The connection of two two-port networks in cascade is shown in Fig. 19.41. For the two networks,

$$\begin{bmatrix} \mathbf{V}_{1a} \\ \mathbf{I}_{1a} \end{bmatrix} = \begin{bmatrix} \mathbf{A}_a & \mathbf{B}_a \\ \mathbf{C}_a & \mathbf{D}_a \end{bmatrix} \begin{bmatrix} \mathbf{V}_{2a} \\ -\mathbf{I}_{2a} \end{bmatrix} \tag{19.56}$$

$$\begin{bmatrix} \mathbf{V}_{1b} \\ \mathbf{I}_{1b} \end{bmatrix} = \begin{bmatrix} \mathbf{A}_b & \mathbf{B}_b \\ \mathbf{C}_b & \mathbf{D}_b \end{bmatrix} \begin{bmatrix} \mathbf{V}_{2b} \\ -\mathbf{I}_{2b} \end{bmatrix} \tag{19.57}$$

From Fig. 19.41,

$$\begin{bmatrix} \mathbf{V}_1 \\ \mathbf{I}_1 \end{bmatrix} = \begin{bmatrix} \mathbf{V}_{1a} \\ \mathbf{I}_{1a} \end{bmatrix}, \quad \begin{bmatrix} \mathbf{V}_{2a} \\ -\mathbf{I}_{2a} \end{bmatrix} = \begin{bmatrix} \mathbf{V}_{1b} \\ \mathbf{I}_{1b} \end{bmatrix}, \quad \begin{bmatrix} \mathbf{V}_{2b} \\ -\mathbf{I}_{2b} \end{bmatrix} = \begin{bmatrix} \mathbf{V}_2 \\ -\mathbf{I}_2 \end{bmatrix} \tag{19.58}$$

Substituting these into Eqs. (19.56) and (19.57),

$$\begin{bmatrix} \mathbf{V}_1 \\ \mathbf{I}_1 \end{bmatrix} = \begin{bmatrix} \mathbf{A}_a & \mathbf{B}_a \\ \mathbf{C}_a & \mathbf{D}_a \end{bmatrix} \begin{bmatrix} \mathbf{A}_b & \mathbf{B}_b \\ \mathbf{C}_b & \mathbf{D}_b \end{bmatrix} \begin{bmatrix} \mathbf{V}_2 \\ -\mathbf{I}_2 \end{bmatrix} \tag{19.59}$$

Thus, the transmission parameters for the overall network are the product of the transmission parameters for the individual transmission parameters:

$$\begin{bmatrix} \mathbf{A} & \mathbf{B} \\ \mathbf{C} & \mathbf{D} \end{bmatrix} = \begin{bmatrix} \mathbf{A}_a & \mathbf{B}_a \\ \mathbf{C}_a & \mathbf{D}_a \end{bmatrix} \begin{bmatrix} \mathbf{A}_b & \mathbf{B}_b \\ \mathbf{C}_b & \mathbf{D}_b \end{bmatrix} \tag{19.60}$$

or

$$[\mathbf{T}] = [\mathbf{T}_a][\mathbf{T}_b] \tag{19.61}$$

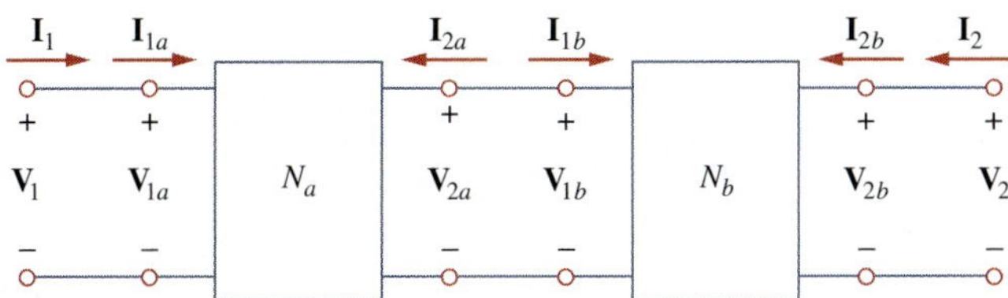

Figure 19.41
Cascade connection of two two-port networks.

It is this property that makes the transmission parameters so useful. Keep in mind that the multiplication of the matrices must be in the order in which the networks N_a and N_b are cascaded.

Example 19.12

Evaluate $\mathbf{V}_2/\mathbf{V}_s$ in the circuit in Fig. 19.42.

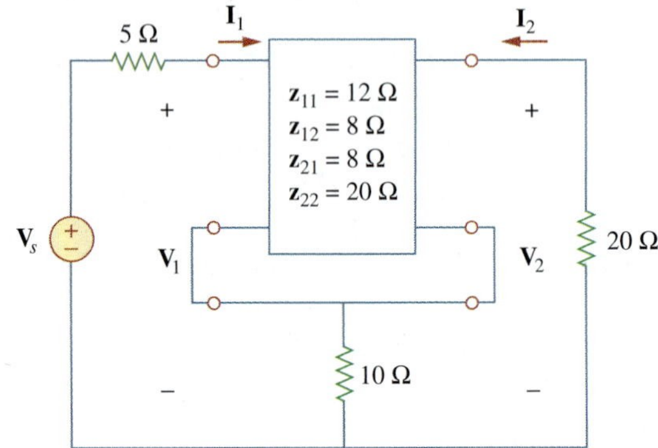

Figure 19.42
For Example 19.12.

Solution:

This may be regarded as two two-ports in series. For N_b,

$$\mathbf{z}_{12b} = \mathbf{z}_{21b} = 10 = \mathbf{z}_{11b} = \mathbf{z}_{22b}$$

Thus,

$$[\mathbf{z}] = [\mathbf{z}_a] + [\mathbf{z}_b] = \begin{bmatrix} 12 & 8 \\ 8 & 20 \end{bmatrix} + \begin{bmatrix} 10 & 10 \\ 10 & 10 \end{bmatrix} = \begin{bmatrix} 22 & 18 \\ 18 & 30 \end{bmatrix}$$

But

$$\mathbf{V}_1 = \mathbf{z}_{11}\mathbf{I}_1 + \mathbf{z}_{12}\mathbf{I}_2 = 22\mathbf{I}_1 + 18\mathbf{I}_2 \tag{19.12.1}$$

$$\mathbf{V}_2 = \mathbf{z}_{21}\mathbf{I}_1 + \mathbf{z}_{22}\mathbf{I}_2 = 18\mathbf{I}_1 + 30\mathbf{I}_2 \tag{19.12.2}$$

Also, at the input port

$$\mathbf{V}_1 = \mathbf{V}_s - 5\mathbf{I}_1 \tag{19.12.3}$$

and at the output port

$$\mathbf{V}_2 = -20\mathbf{I}_2 \quad \Rightarrow \quad \mathbf{I}_2 = -\frac{\mathbf{V}_2}{20} \tag{19.12.4}$$

Substituting Eqs. (19.12.3) and (19.12.4) into Eq. (19.12.1) gives

$$\mathbf{V}_s - 5\mathbf{I}_1 = 22\mathbf{I}_1 - \frac{18}{20}\mathbf{V}_2 \quad \Rightarrow \quad \mathbf{V}_s = 27\mathbf{I}_1 - 0.9\mathbf{V}_2 \tag{19.12.5}$$

while substituting Eq. (19.12.4) into Eq. (19.12.2) yields

$$\mathbf{V}_2 = 18\mathbf{I}_1 - \frac{30}{20}\mathbf{V}_2 \quad \Rightarrow \quad \mathbf{I}_1 = \frac{2.5}{18}\mathbf{V}_2 \tag{19.12.6}$$

Substituting Eq. (19.12.6) into Eq. (19.12.5), we get

$$\mathbf{V}_s = 27 \times \frac{2.5}{18}\mathbf{V}_2 - 0.9\mathbf{V}_2 = 2.85\mathbf{V}_2$$

And so,

$$\frac{\mathbf{V}_2}{\mathbf{V}_s} = \frac{1}{2.85} = 0.3509$$

Practice Problem 19.12

Find $\mathbf{V}_2/\mathbf{V}_s$ in the circuit in Fig. 19.43.

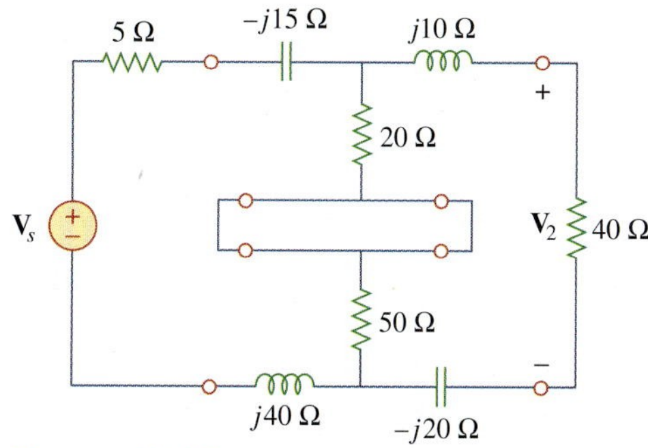

Figure 19.43
For Practice Prob. 19.12.

Answer: $0.6799\underline{/-29.05^\circ}$.

Example 19.13

Find the y parameters of the two-port in Fig. 19.44.

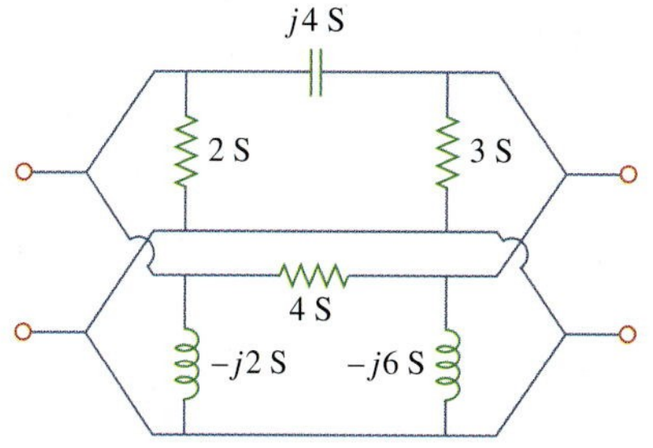

Figure 19.44
For Example 19.13.

Solution:
Let us refer to the upper network as N_a and the lower one as N_b. The two networks are connected in parallel. Comparing N_a and N_b with the circuit in Fig. 19.13(a), we obtain

$$\mathbf{y}_{12a} = -j4 = \mathbf{y}_{21a}, \qquad \mathbf{y}_{11a} = 2 + j4, \qquad \mathbf{y}_{22a} = 3 + j4$$

or

$$[\mathbf{y}_a] = \begin{bmatrix} 2 + j4 & -j4 \\ -j4 & 3 + j4 \end{bmatrix} \text{S}$$

and

$$\mathbf{y}_{12b} = -4 = \mathbf{y}_{21b}, \qquad \mathbf{y}_{11b} = 4 - j2, \qquad \mathbf{y}_{22b} = 4 - j6$$

or

$$[\mathbf{y}_b] = \begin{bmatrix} 4 - j2 & -4 \\ -4 & 4 - j6 \end{bmatrix} \text{S}$$

The overall y parameters are

$$[\mathbf{y}] = [\mathbf{y}_a] + [\mathbf{y}_b] = \begin{bmatrix} 6 + j2 & -4 - j4 \\ -4 - j4 & 7 - j2 \end{bmatrix} \text{S}$$

Practice Problem 19.13

Obtain the y parameters for the network in Fig. 19.45.

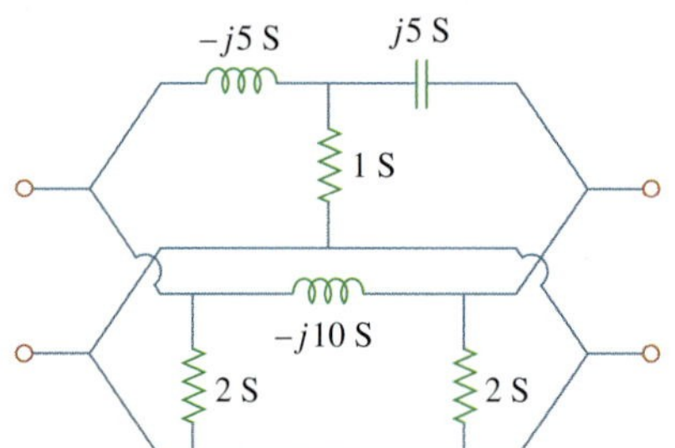

Figure 19.45
For Practice Prob. 19.13.

Answer: $\begin{bmatrix} 27 - j15 & -25 + j10 \\ -25 + j10 & 27 - j5 \end{bmatrix}$ S.

Example 19.14

Find the transmission parameters for the circuit in Fig. 19.46.

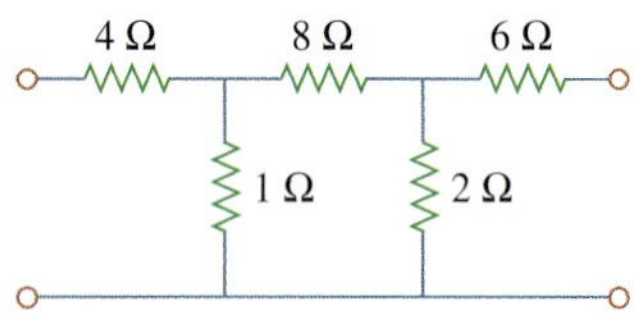

Figure 19.46
For Example 19.14.

Solution:

We can regard the given circuit in Fig. 19.46 as a cascade connection of two T networks as shown in Fig. 19.47(a). We can show that a T network, shown in Fig. 19.47(b), has the following transmission parameters [see Prob. 19.52(b)]:

$$\mathbf{A} = 1 + \frac{R_1}{R_2}, \qquad \mathbf{B} = R_3 + \frac{R_1(R_2 + R_3)}{R_2}$$

$$\mathbf{C} = \frac{1}{R_2}, \qquad \mathbf{D} = 1 + \frac{R_3}{R_2}$$

Applying this to the cascaded networks N_a and N_b in Fig. 19.47(a), we get

$$\mathbf{A}_a = 1 + 4 = 5, \qquad \mathbf{B}_a = 8 + 4 \times 9 = 44\ \Omega$$

$$\mathbf{C}_a = 1\text{ S}, \qquad \mathbf{D}_a = 1 + 8 = 9$$

or in matrix form,

$$[\mathbf{T}_a] = \begin{bmatrix} 5 & 44\ \Omega \\ 1\text{ S} & 9 \end{bmatrix}$$

and

$$\mathbf{A}_b = 1, \qquad \mathbf{B}_b = 6\ \Omega, \qquad \mathbf{C}_b = 0.5\text{ S}, \qquad \mathbf{D}_b = 1 + \frac{6}{2} = 4$$

i.e.,

$$[\mathbf{T}_b] = \begin{bmatrix} 1 & 6\ \Omega \\ 0.5\text{ S} & 4 \end{bmatrix}$$

Thus, for the total network in Fig. 19.46,

$$\begin{aligned} [\mathbf{T}] = [\mathbf{T}_a][\mathbf{T}_b] &= \begin{bmatrix} 5 & 44 \\ 1 & 9 \end{bmatrix} \begin{bmatrix} 1 & 6 \\ 0.5 & 4 \end{bmatrix} \\ &= \begin{bmatrix} 5 \times 1 + 44 \times 0.5 & 5 \times 6 + 44 \times 4 \\ 1 \times 1 + 9 \times 0.5 & 1 \times 6 + 9 \times 4 \end{bmatrix} \\ &= \begin{bmatrix} 27 & 206\ \Omega \\ 5.5\text{ S} & 42 \end{bmatrix} \end{aligned}$$

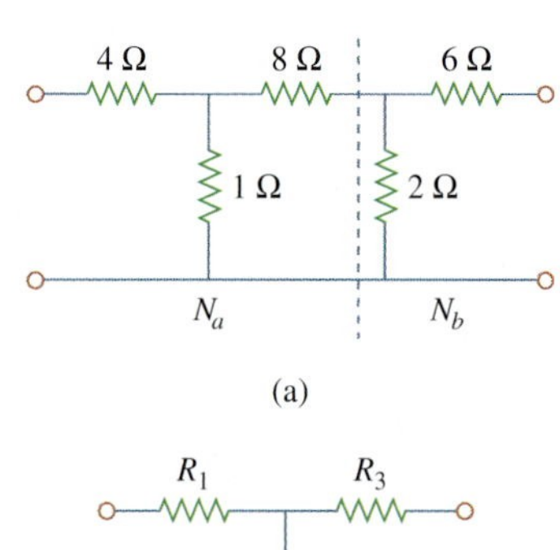

Figure 19.47
For Example 19.14: (a) Breaking the circuit in Fig. 19.46 into two two-ports, (b) a general T two-port.

Notice that

$$\Delta_{T_a} = \Delta_{T_b} = \Delta_T = 1$$

showing that the network is reciprocal.

Practice Problem 19.14

Obtain the **ABCD** parameter representation of the circuit in Fig. 19.48.

Answer: $[\mathbf{T}] = \begin{bmatrix} 29.25 & 2200\ \Omega \\ 0.425\ \text{S} & 32 \end{bmatrix}$.

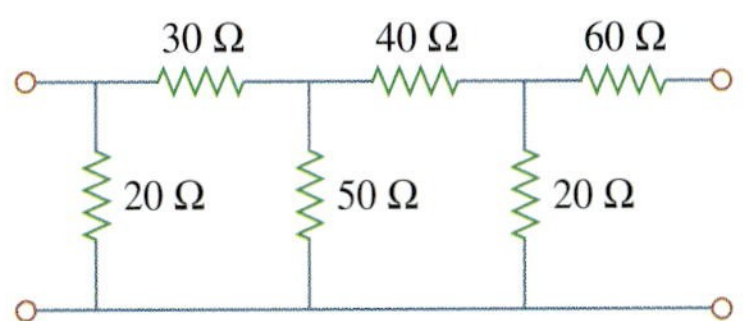

Figure 19.48
For Practice Prob. 19.14.

19.8 Computing Two-Port Parameters Using *PSpice*

Hand calculation of the two-port parameters may become difficult when the two-port is complicated. We resort to *PSpice* in such situations. If the circuit is purely resistive, *PSpice* dc analysis may be used; otherwise, *PSpice* ac analysis is required at a specific frequency. The key to using *PSpice* in computing a particular two-port parameter is to remember how that parameter is defined and to constrain the appropriate port variable with a 1-A or 1-V source while using an open or short circuit to impose the other necessary constraints. The following two examples illustrate the idea.

Example 19.15

Find the h parameters of the network in Fig. 19.49.

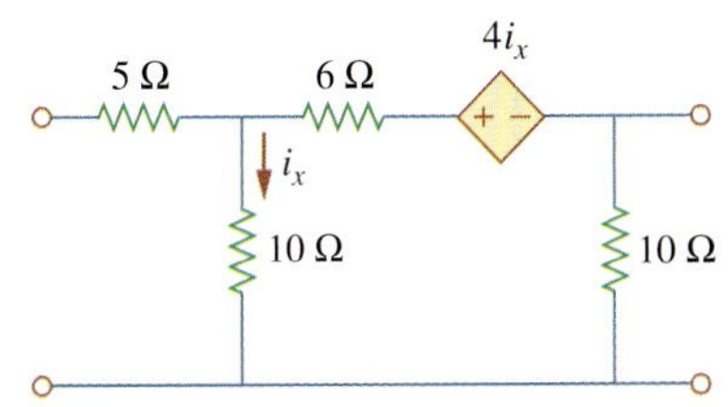

Figure 19.49
For Example 19.15.

Solution:
From Eq. (19.16),

$$\mathbf{h}_{11} = \frac{\mathbf{V}_1}{\mathbf{I}_1}\bigg|_{\mathbf{V}_2=0}, \qquad \mathbf{h}_{21} = \frac{\mathbf{I}_2}{\mathbf{I}_1}\bigg|_{\mathbf{V}_2=0}$$

showing that $\mathbf{h}_{11}$ and $\mathbf{h}_{21}$ can be found by setting $\mathbf{V}_2 = 0$. Also by setting $\mathbf{I}_1 = 1$ A, $\mathbf{h}_{11}$ becomes $\mathbf{V}_1/1$ while $\mathbf{h}_{21}$ becomes $\mathbf{I}_2/1$. With this in mind, we draw the schematic in Fig. 19.50(a). We insert a 1-A dc current source

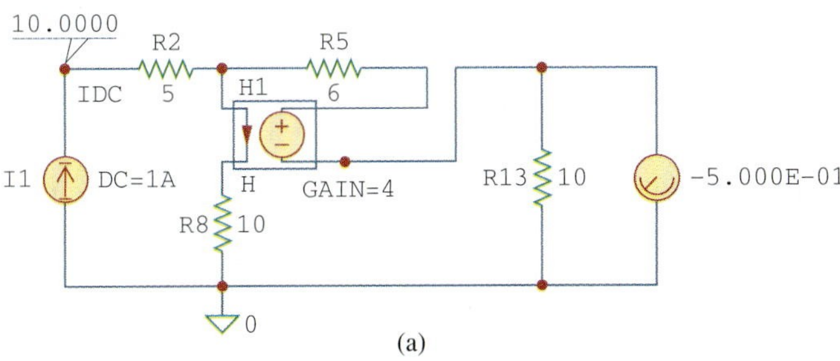

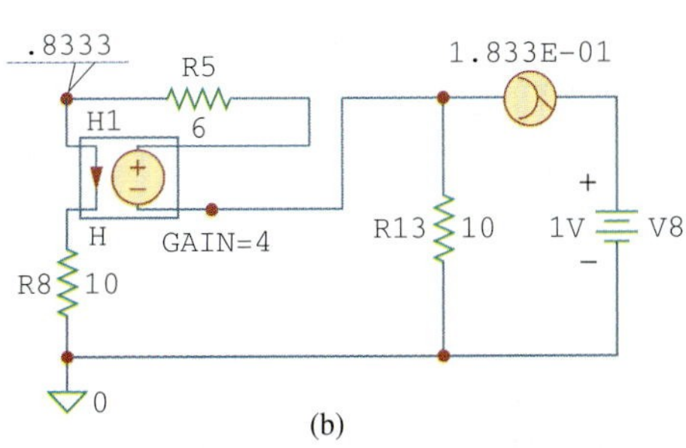

Figure 19.50
For Example 19.15: (a) computing $\mathbf{h}_{11}$ and $\mathbf{h}_{21}$, (b) computing $\mathbf{h}_{12}$ and $\mathbf{h}_{22}$.

IDC to take care of $\mathbf{I}_1 = 1$ A, the pseudocomponent VIEWPOINT to display $\mathbf{V}_1$ and pseudocomponent IPROBE to display $\mathbf{I}_2$. After saving the schematic, we run *PSpice* by selecting **Analysis/Simulate** and note the values displayed on the pseudocomponents. We obtain

$$\mathbf{h}_{11} = \frac{\mathbf{V}_1}{1} = 10\ \Omega, \qquad \mathbf{h}_{21} = \frac{\mathbf{I}_2}{1} = -0.5$$

Similarly, from Eq. (19.16),

$$\mathbf{h}_{12} = \left.\frac{\mathbf{V}_1}{\mathbf{V}_2}\right|_{\mathbf{I}_1=0}, \qquad \mathbf{h}_{22} = \left.\frac{\mathbf{I}_2}{\mathbf{V}_2}\right|_{\mathbf{I}_1=0}$$

indicating that we obtain $\mathbf{h}_{12}$ and $\mathbf{h}_{22}$ by open-circuiting the input port ($\mathbf{I}_1 = 0$). By making $\mathbf{V}_2 = 1$ V, $\mathbf{h}_{12}$ becomes $\mathbf{V}_1/1$ while $\mathbf{h}_{22}$ becomes $\mathbf{I}_2/1$. Thus, we use the schematic in Fig. 19.50(b) with a 1-V dc voltage source VDC inserted at the output terminal to take care of $\mathbf{V}_2 = 1$ V. The pseudocomponents VIEWPOINT and IPROBE are inserted to display the values of $\mathbf{V}_1$ and $\mathbf{I}_2$, respectively. (Notice that in Fig. 19.50(b), the 5-Ω resistor is ignored because the input port is open-circuited and *PSpice* will not allow such. We may include the 5-Ω resistor if we replace the open circuit with a very large resistor, say, 10 MΩ.) After simulating the schematic, we obtain the values displayed on the pseudocomponents as shown in Fig. 19.50(b). Thus,

$$\mathbf{h}_{12} = \frac{\mathbf{V}_1}{1} = 0.8333, \qquad \mathbf{h}_{22} = \frac{\mathbf{I}_2}{1} = 0.1833\ \text{S}$$

Practice Problem 19.15

Obtain the h parameters for the network in Fig. 19.51 using *PSpice*.

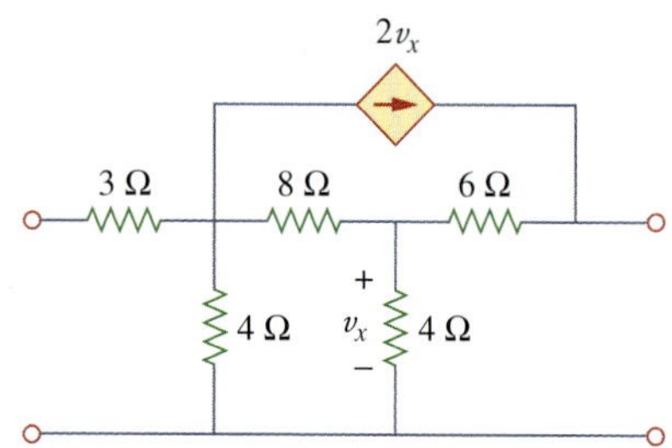

Figure 19.51
For Practice Prob. 19.15.

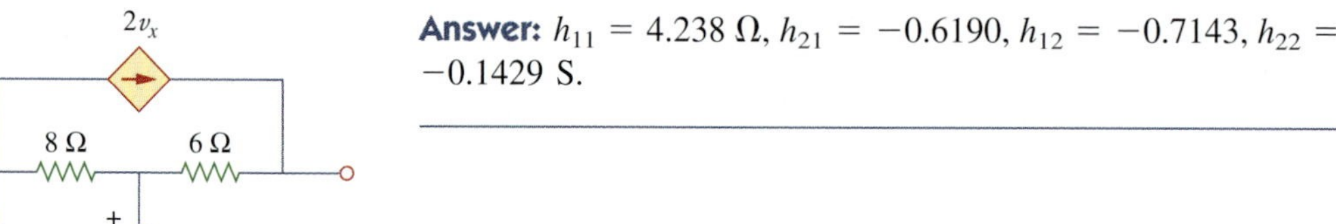

Answer: $h_{11} = 4.238\ \Omega$, $h_{21} = -0.6190$, $h_{12} = -0.7143$, $h_{22} = -0.1429$ S.

Example 19.16

Find the z parameters for the circuit in Fig. 19.52 at $\omega = 10^6$ rad/s.

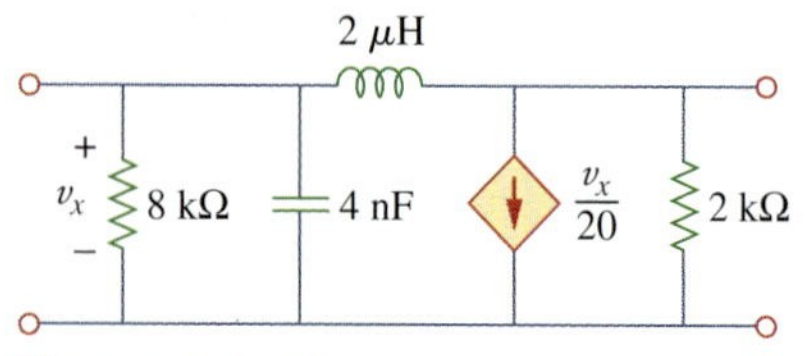

Figure 19.52
For Example 19.16.

Solution:

Notice that we used dc analysis in Example 19.15 because the circuit in Fig. 19.49 is purely resistive. Here, we use ac analysis at $f = \omega/2\pi = 0.15915$ MHz, because L and C are frequency dependent.

In Eq. (19.3), we defined the z parameters as

$$\mathbf{z}_{11} = \left.\frac{\mathbf{V}_1}{\mathbf{I}_1}\right|_{\mathbf{I}_2=0}, \qquad \mathbf{z}_{21} = \left.\frac{\mathbf{V}_2}{\mathbf{I}_1}\right|_{\mathbf{I}_2=0}$$

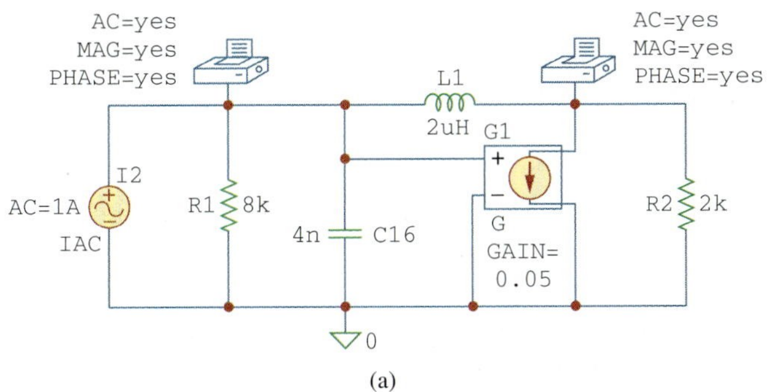

(a)

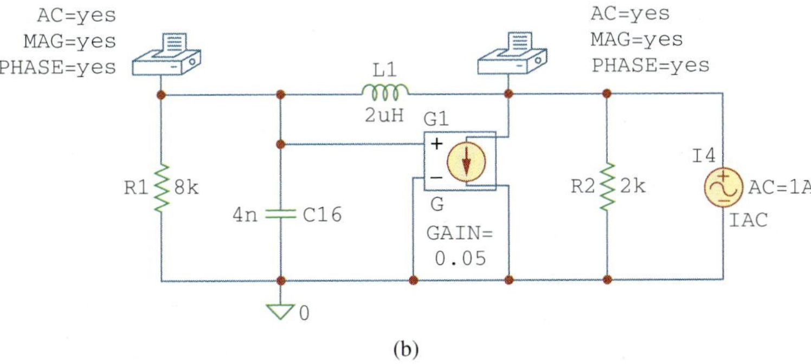

(b)

Figure 19.53
For Example 19.16: (a) circuit for determining $\mathbf{z}_{11}$ and $\mathbf{z}_{21}$, (b) circuit for determining $\mathbf{z}_{12}$ and $\mathbf{z}_{22}$.

This suggests that if we let $\mathbf{I}_1 = 1$ A and open-circuit the output port so that $\mathbf{I}_2 = 0$, then we obtain

$$\mathbf{z}_{11} = \frac{\mathbf{V}_1}{1} \quad \text{and} \quad \mathbf{z}_{21} = \frac{\mathbf{V}_2}{1}$$

We realize this with the schematic in Fig. 19.53(a). We insert a 1-A ac current source IAC at the input terminal of the circuit and two VPRINT1 pseudocomponents to obtain $\mathbf{V}_1$ and $\mathbf{V}_2$. The attributes of each VPRINT1 are set as *AC* = *yes*, *MAG* = *yes*, and *PHASE* = *yes* to print the magnitude and phase values of the voltages. We select **Analysis/Setup/AC Sweep** and enter 1 as *Total Pts*, 0.1519MEG as *Start Freq*, and 0.1519MEG as *Final Freq* in the **AC Sweep and Noise Analysis** dialog box. After saving the schematic, we select **Analysis/Simulate** to simulate it. We obtain $\mathbf{V}_1$ and $\mathbf{V}_2$ from the output file. Thus,

$$\mathbf{z}_{11} = \frac{\mathbf{V}_1}{1} = 19.70\underline{/175.7^\circ}\ \Omega, \qquad \mathbf{z}_{21} = \frac{\mathbf{V}_2}{1} = 19.79\underline{/170.2^\circ}\ \Omega$$

In a similar manner, from Eq. (19.3),

$$\mathbf{z}_{12} = \frac{\mathbf{V}_1}{\mathbf{I}_2}\bigg|_{\mathbf{I}_1=0}, \qquad \mathbf{z}_{22} = \frac{\mathbf{V}_2}{\mathbf{I}_2}\bigg|_{\mathbf{I}_1=0}$$

suggesting that if we let $\mathbf{I}_2 = 1$ A and open-circuit the input port,

$$\mathbf{z}_{12} = \frac{\mathbf{V}_1}{1} \quad \text{and} \quad \mathbf{z}_{22} = \frac{\mathbf{V}_2}{1}$$

This leads to the schematic in Fig. 19.53(b). The only difference between this schematic and the one in Fig. 19.53(a) is that the 1-A ac current source IAC is now at the output terminal. We run the schematic in Fig. 19.53(b) and obtain $\mathbf{V}_1$ and $\mathbf{V}_2$ from the output file. Thus,

$$\mathbf{z}_{12} = \frac{\mathbf{V}_1}{1} = 19.70\underline{/175.7^\circ}\ \Omega, \qquad \mathbf{z}_{22} = \frac{\mathbf{V}_2}{1} = 19.56\underline{/175.7^\circ}\ \Omega$$

Practice Problem 19.16

Obtain the z parameters of the circuit in Fig. 19.54 at $f = 60$ Hz.

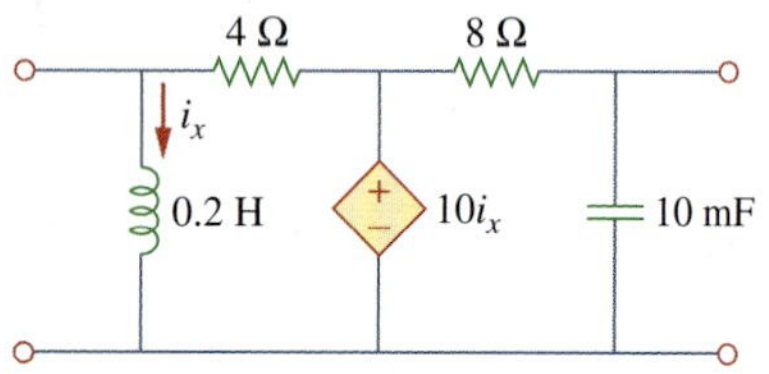

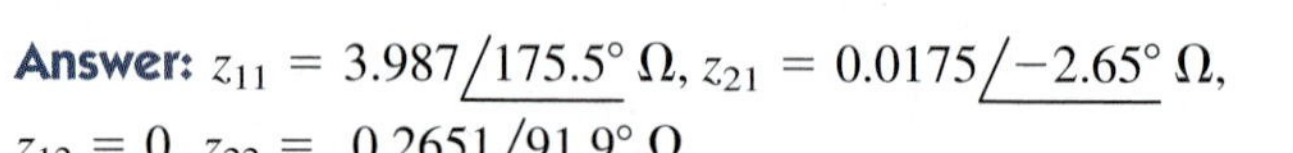

Answer: $z_{11} = 3.987\underline{/175.5^\circ}\ \Omega$, $z_{21} = 0.0175\underline{/-2.65^\circ}\ \Omega$, $z_{12} = 0$, $z_{22} = 0.2651\underline{/91.9^\circ}\ \Omega$.

Figure 19.54
For Practice Prob. 19.16.

19.9 †Applications

We have seen how the six sets of network parameters can be used to characterize a wide range of two-port networks. Depending on the way two-ports are interconnected to form a larger network, a particular set of parameters may have advantages over others, as we noticed in Section 19.7. In this section, we will consider two important application areas of two-port parameters: transistor circuits and synthesis of ladder networks.

19.9.1 Transistor Circuits

The two-port network is often used to isolate a load from the excitation of a circuit. For example, the two-port in Fig. 19.55 may represent an amplifier, a filter, or some other network. When the two-port represents an amplifier, expressions for the voltage gain A_v, the current gain A_i, the input impedance Z_{in}, and the output impedance Z_{out} can be derived with ease. They are defined as follows:

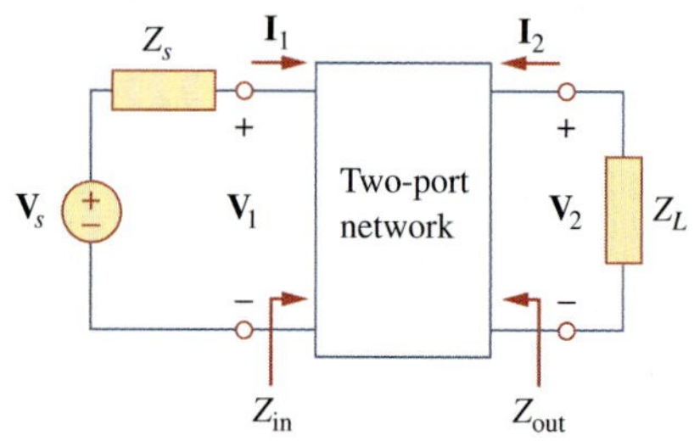

Figure 19.55
Two-port network isolating source and load.

$$A_v = \frac{V_2(s)}{V_1(s)} \tag{19.62}$$

$$A_i = \frac{I_2(s)}{I_1(s)} \tag{19.63}$$

$$Z_{\text{in}} = \frac{V_1(s)}{I_1(s)} \tag{19.64}$$

$$Z_{\text{out}} = \left.\frac{V_2(s)}{I_2(s)}\right|_{V_s=0} \tag{19.65}$$

Any of the six sets of two-port parameters can be used to derive the expressions in Eqs. (19.62) to (19.65). However, the hybrid (h) parameters are the most useful for transistors; they are easily measured and are often provided in the manufacturer's data or spec sheets for transistors. The h parameters provide a quick estimate of the performance of transistor circuits. They are used for finding the exact voltage gain, input impedance, and output impedance of a transistor.

The h parameters for transistors have specific meanings expressed by their subscripts. They are listed by the first subscript and related to the general h parameters as follows:

$$h_i = h_{11}, \qquad h_r = h_{12}, \qquad h_f = h_{21}, \qquad h_o = h_{22} \tag{19.66}$$

The subscripts i, r, f, and o stand for input, reverse, forward, and output. The second subscript specifies the type of connection used: e for common emitter (CE), c for common collector (CC), and b for common base (CB). Here we are mainly concerned with the common-emitter connection. Thus, the four h parameters for the common-emitter amplifier are:

$$\begin{aligned} h_{ie} &= \text{Base input impedance} \\ h_{re} &= \text{Reverse voltage feedback ratio} \\ h_{fe} &= \text{Base-collector current gain} \\ h_{oe} &= \text{Output admittance} \end{aligned} \tag{19.67}$$

These are calculated or measured in the same way as the general h parameters. Typical values are $h_{ie} = 6\ \text{k}\Omega$, $h_{re} = 1.5 \times 10^{-4}$, $h_{fe} = 200$, $h_{oe} = 8\ \mu\text{S}$. We must keep in mind that these values represent ac characteristics of the transistor, measured under specific circumstances.

Figure 19.56 shows the circuit schematic for the common-emitter amplifier and the equivalent hybrid model. From the figure, we see that

$$\mathbf{V}_b = h_{ie}\mathbf{I}_b + h_{re}\mathbf{V}_c \tag{19.68a}$$

$$\mathbf{I}_c = h_{fe}\mathbf{I}_b + h_{oe}\mathbf{V}_c \tag{19.68b}$$

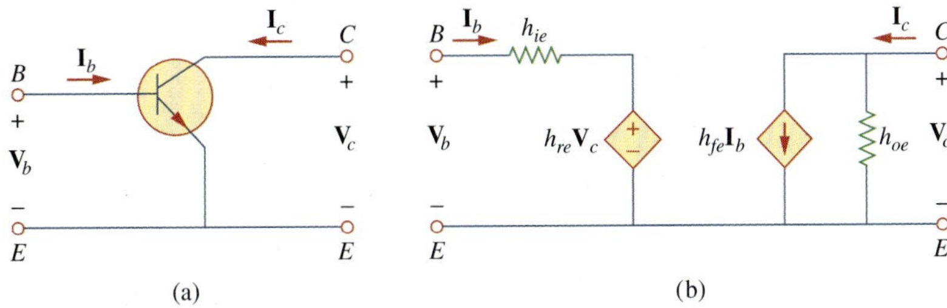

Figure 19.56
Common emitter amplifier: (a) circuit schematic, (b) hybrid model.

Consider the transistor amplifier connected to an ac source and a load as in Fig. 19.57. This is an example of a two-port network embedded within a larger network. We can analyze the hybrid equivalent circuit as usual with Eq. (19.68) in mind. (See Example 19.6.) Recognizing

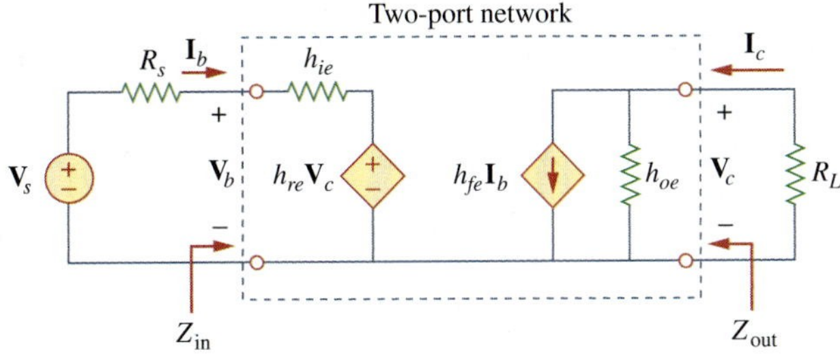

Figure 19.57
Transistor amplifier with source and load resistance.

from Fig. 19.57 that $\mathbf{V}_c = -R_L\mathbf{I}_c$ and substituting this into Eq. (19.68b) gives

$$\mathbf{I}_c = h_{fe}\mathbf{I}_b - h_{oe}R_L\mathbf{I}_c$$

or

$$(1 + h_{oe}R_L)\mathbf{I}_c = h_{fe}\mathbf{I}_b \tag{19.69}$$

From this, we obtain the current gain as

$$A_i = \frac{\mathbf{I}_c}{\mathbf{I}_b} = \frac{h_{fe}}{1 + h_{oe}R_L} \tag{19.70}$$

From Eqs. (19.68b) and (19.70), we can express $\mathbf{I}_b$ in terms of $\mathbf{V}_c$:

$$\mathbf{I}_c = \frac{h_{fe}}{1 + h_{oe}R_L}\mathbf{I}_b = h_{fe}\mathbf{I}_b + h_{oe}\mathbf{V}_c$$

or

$$\mathbf{I}_b = \frac{h_{oe}\mathbf{V}_c}{\dfrac{h_{fe}}{1 + h_{oe}R_L} - h_{fe}} \tag{19.71}$$

Substituting Eq. (19.71) into Eq. (19.68a) and dividing by $\mathbf{V}_c$ gives

$$\begin{aligned}\frac{\mathbf{V}_b}{\mathbf{V}_c} &= \frac{h_{oe}h_{ie}}{\dfrac{h_{fe}}{1 + h_{oe}R_L} - h_{fe}} + h_{re}\\ &= \frac{h_{ie} + h_{ie}h_{oe}R_L - h_{re}h_{fe}R_L}{-h_{fe}R_L}\end{aligned} \tag{19.72}$$

Thus, the voltage gain is

$$A_v = \frac{\mathbf{V}_c}{\mathbf{V}_b} = \frac{-h_{fe}R_L}{h_{ie} + (h_{ie}h_{oe} - h_{re}h_{fe})R_L} \tag{19.73}$$

Substituting $\mathbf{V}_c = -R_L\mathbf{I}_c$ into Eq. (19.68a) gives

$$\mathbf{V}_b = h_{ie}\mathbf{I}_b - h_{re}R_L\mathbf{I}_c$$

or

$$\frac{\mathbf{V}_b}{\mathbf{I}_b} = h_{ie} - h_{re}R_L\frac{\mathbf{I}_c}{\mathbf{I}_b} \tag{19.74}$$

Replacing $\mathbf{I}_c/\mathbf{I}_b$ by the current gain in Eq. (19.70) yields the input impedance as

$$Z_{\text{in}} = \frac{\mathbf{V}_b}{\mathbf{I}_b} = h_{ie} - \frac{h_{re}h_{fe}R_L}{1 + h_{oe}R_L} \tag{19.75}$$

The output impedance Z_{out} is the same as the Thevenin equivalent at the output terminals. As usual, by removing the voltage source and placing a

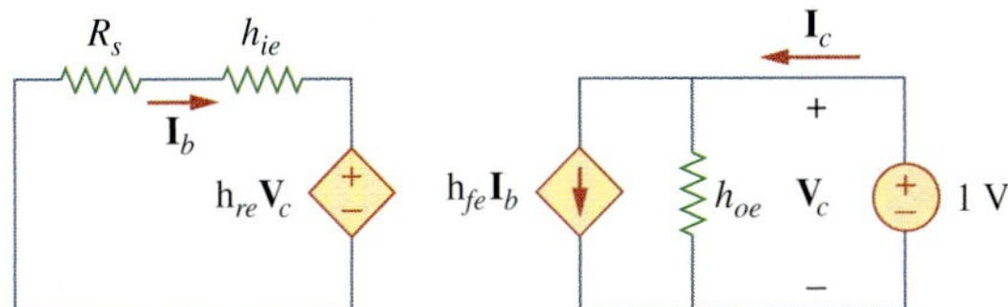

Figure 19.58
Finding the output impedance of the amplifier circuit in Fig. 19.57.

1-V source at the output terminals, we obtain the circuit in Fig. 19.58, from which Z_{out} is determined as $1/\mathbf{I}_c$. Since $\mathbf{V}_c = 1$ V, the input loop gives

$$h_{re}(1) = -\mathbf{I}_b(R_s + h_{ie}) \quad \Rightarrow \quad \mathbf{I}_b = -\frac{h_{re}}{R_s + h_{ie}} \tag{19.76}$$

For the output loop,

$$\mathbf{I}_c = h_{oe}(1) + h_{fe}\mathbf{I}_b \tag{19.77}$$

Substituting Eq. (19.76) into Eq. (19.77) gives

$$\mathbf{I}_c = \frac{(R_s + h_{ie})h_{oe} - h_{re}h_{fe}}{R_s + h_{ie}} \tag{19.78}$$

From this, we obtain the output impedance Z_{out} as $1/\mathbf{I}_c$; that is,

$$Z_{\text{out}} = \frac{R_s + h_{ie}}{(R_s + h_{ie})h_{oe} - h_{re}h_{fe}} \tag{19.79}$$

Example 19.17

Consider the common-emitter amplifier circuit of Fig. 19.59. Determine the voltage gain, current gain, input impedance, and output impedance using these h parameters:

$$h_{ie} = 1\text{ k}\Omega, \qquad h_{re} = 2.5 \times 10^{-4}, \qquad h_{fe} = 50, \qquad h_{oe} = 20\ \mu\text{S}$$

Find the output voltage $\mathbf{V}_o$.

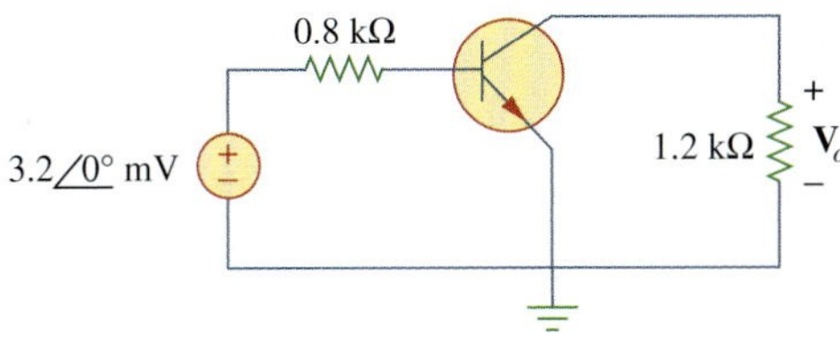

Figure 19.59
For Example 19.17.

Solution:

1. **Define.** In an initial look at this problem, it appears to be clearly stated. However, when we are asked to determine the input impedance and the voltage gain, do they refer to the transistor or the circuit? As far as the current gain and the output impedance are concerned, they are the same for both cases.

   We ask for clarification and are told that we should calculate the input impedance, the output impedance, and the voltage gain for the circuit and not the transistor. It is interesting to note that the problem can be restated so that it becomes a simple design problem: Given the h parameters, design a simple amplifier that has a gain of -60.

2. **Present.** Given a simple transistor circuit, an input voltage of 3.2 mV, and the h parameters of the transistor, calculate the output voltage.
3. **Alternative.** There are a couple of ways we can approach the problem, the most straightforward being to use the equivalent circuit shown in Fig. 19.57. Once you have the equivalent circuit you can use circuit analysis to determine the answer. Once you have a solution, you can check it by plugging in the answer into the circuit equations to see if they are correct. Another approach is to simplify the right-hand side of the equivalent circuit and work backward to see if you obtain approximately the same answer. We will use that approach here.
4. **Attempt.** We note that $R_s = 0.8\ \text{k}\Omega$ and $R_L = 1.2\ \text{k}\Omega$. We treat the transistor of Fig. 19.59 as a two-port network and apply Eqs. (19.70) to (19.79).

$$h_{ie}h_{oe} - h_{re}h_{fe} = 10^3 \times 20 \times 10^{-6} - 2.5 \times 10^{-4} \times 50$$
$$= 7.5 \times 10^{-3}$$

$$A_v = \frac{-h_{fe}R_L}{h_{ie} + (h_{ie}h_{oe} - h_{re}h_{fe})R_L} = \frac{-50 \times 1200}{1000 + 7.5 \times 10^{-3} \times 1200}$$
$$= -59.46$$

A_v is the voltage gain of the amplifier $= V_o/V_b$. To calculate the gain of the circuit we need to find V_o/V_s. We can do this by using the mesh equation for the circuit on the left and Eqs. (19.71) and (19.73).

$$-\text{V}_s + R_s\text{I}_b + \text{V}_b = 0$$

or

$$\text{V}_s = 800\frac{20 \times 10^{-6}}{\dfrac{50}{1 + 20 \times 10^{-6} \times 1.2 \times 10^3} - 50} - \frac{1}{59.46}\text{V}_o$$
$$= -0.03047\ \text{V}_o.$$

Thus, the circuit gain is equal to $-\mathbf{32.82}$. Now we can calculate the output voltage.

$$V_o = \text{gain} \times V_s = \mathbf{-105.09\underline{/0^\circ}\ mV}.$$

$$A_i = \frac{h_{fe}}{1 + h_{oe}R_L} = \frac{50}{1 + 20 \times 10^{-6} \times 1200} = 48.83$$

$$Z_{\text{in}} = h_{ie} - \frac{h_{re}h_{fe}R_L}{1 + h_{oe}R_L}$$
$$= 1000 - \frac{2.5 \times 10^{-4} \times 50 \times 1200}{1 + 20 \times 10^{-6} \times 1200}$$
$$= 985.4\ \Omega$$

You can modify Z_{in} to include the 800-ohm resistor so that

Circuit input impedance $= 800 + 985.4 = \mathbf{1785.4\ \Omega}$.

$$(R_s + h_{ie})h_{oe} - h_{re}h_{fe}$$
$$= (800 + 1000) \times 20 \times 10^{-6} - 2.5 \times 10^{-4} \times 50 = 23.5 \times 10^{-3}$$

$$Z_{\text{out}} = \frac{R_s + h_{ie}}{(R_s + h_{ie})h_{oe} - h_{re}h_{fe}} = \frac{800 + 1000}{23.5 \times 10^{-3}} = 76.6 \text{ k}\Omega$$

5. **Evaluate.** In the equivalent circuit, h_{oe} represents a resistor of 50,000 Ω. This is in parallel with a load resistor equal to 1.2 kΩ. The size of the load resistor is so small relative to the h_{oe} resistor that h_{oe} can be neglected. This then leads to

$$\text{I}_c = h_{fe}\text{I}_b = 50I_b, \quad \text{V}_c = -1200\text{I}_c,$$

and the following loop equation from the left-hand side of the circuit:

$$-0.0032 + (800 + 1000)\text{I}_b + (0.00025)(-1200)(50)\text{I}_b = 0$$
$$\text{I}_b = 0.0032/(1785) = 1.7927\ \mu\text{A}.$$
$$\text{I}_c = 50 \times 1.7927 = 89.64\ \mu\text{A and V}_c = -1200 \times 89.64 \times 10^{-6}$$
$$= -107.57 \text{ mV}$$

This is a good approximation to −105.09 mV.

$$\text{Voltage gain} = -107.57/3.2 = -33.62$$

Again, this is a good approximation to 32.82.

$$\text{Circuit input impedance} = 0.032/1.7927 \times 10^{-6} = \mathbf{1785\ \Omega}$$

which clearly compares well with the 1785.4 Ω we obtained before.

For these calculations, we assumed that $Z_{\text{out}} = \infty\ \Omega$. Our calculations produced 72.6 kΩ. We can test our assumption by calculating the equivalent resistance of this and the load resistance.

$$72{,}600 \times 1200/(72{,}600 + 1200) = 1{,}180.5 = 1.1805 \text{ k}\Omega$$

Again, we have a good approximation.

6. **Satisfactory?** We have satisfactorily solved the problem and checked the results. We can now present our results as a solution to the problem.

Practice Problem 19.17

For the transistor amplifier of Fig. 19.60, find the voltage gain, current gain, input impedance, and output impedance. Assume that

$$h_{ie} = 6 \text{ k}\Omega, \qquad h_{re} = 1.5 \times 10^{-4}, \qquad h_{fe} = 200, \qquad h_{oe} = 8\ \mu\text{S}$$

Answer: −123.61 for the transistor and −4.753 for the circuit, 194.17, 6 kΩ for the transistor and 156 kΩ for the circuit, 128.08 kΩ.

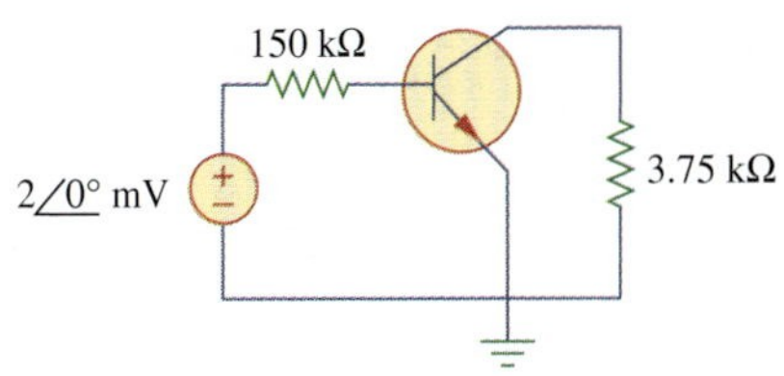

Figure 19.60
For Practice Prob. 19.17.

19.9.2 Ladder Network Synthesis

Another application of two-port parameters is the synthesis (or building) of ladder networks, which are found frequently in practice and

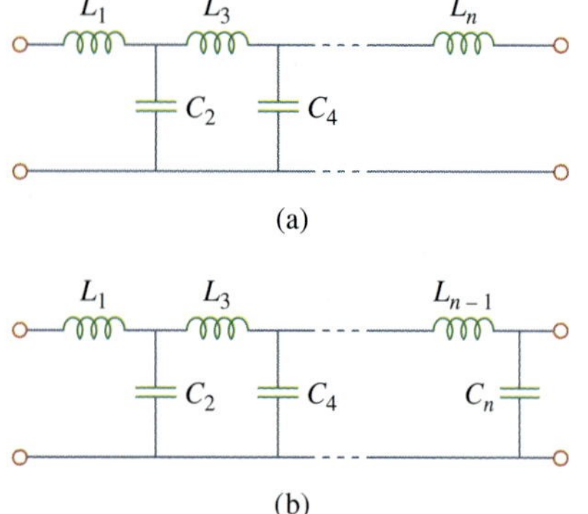

Figure 19.61
LC ladder networks for lowpass filters of: (a) odd order, (b) even order.

have particular use in designing passive lowpass filters. Based on our discussion of second-order circuits in Chapter 8, the order of the filter is the order of the characteristic equation describing the filter and is determined by the number of reactive elements that cannot be combined into single elements (e.g., through series or parallel combination). Figure 19.61(a) shows an *LC* ladder network with an odd number of elements (to realize an odd-order filter), while Fig. 19.61(b) shows one with an even number of elements (for realizing an even-order filter). When either network is terminated by the load impedance Z_L and the source impedance Z_s, we obtain the structure in Fig. 19.62. To make the design less complicated, we will assume that $Z_s = 0$. Our goal is to synthesize the transfer function of the *LC* ladder network. We begin by characterizing the ladder network by its admittance parameters, namely,

$$\mathbf{I}_1 = \mathbf{y}_{11}\mathbf{V}_1 + \mathbf{y}_{12}\mathbf{V}_2 \tag{19.80a}$$

$$\mathbf{I}_2 = \mathbf{y}_{21}\mathbf{V}_1 + \mathbf{y}_{22}\mathbf{V}_2 \tag{19.80b}$$

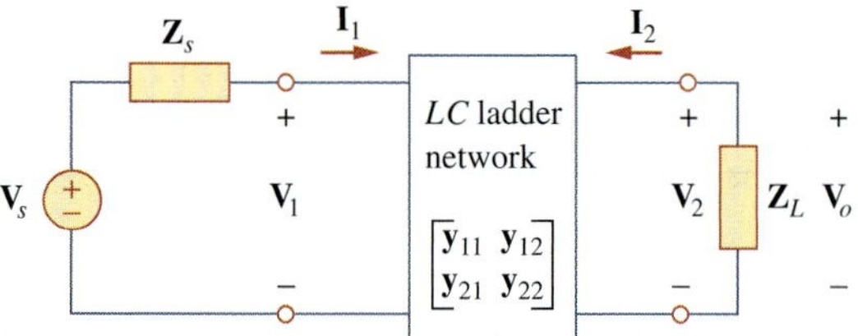

Figure 19.62
LC ladder network with terminating impedances.

(Of course, the impedance parameters could be used instead of the admittance parameters.) At the input port, $\mathbf{V}_1 = \mathbf{V}_s$ since $\mathbf{Z}_s = 0$. At the output port, $\mathbf{V}_2 = \mathbf{V}_o$ and $\mathbf{I}_2 = -\mathbf{V}_2/\mathbf{Z}_L = -\mathbf{V}_o\mathbf{Y}_L$. Thus, Eq. (19.80b) becomes

$$-\mathbf{V}_o\mathbf{Y}_L = \mathbf{y}_{21}\mathbf{V}_s + \mathbf{y}_{22}\mathbf{V}_o$$

or

$$\mathbf{H}(s) = \frac{\mathbf{V}_o}{\mathbf{V}_s} = \frac{-\mathbf{y}_{21}}{\mathbf{Y}_L + \mathbf{y}_{22}} \tag{19.81}$$

We can write this as

$$\boxed{\mathbf{H}(s) = -\frac{\mathbf{y}_{21}/\mathbf{Y}_L}{1 + \mathbf{y}_{22}/\mathbf{Y}_L}} \tag{19.82}$$

We may ignore the negative sign in Eq. (19.82) because filter requirements are often stated in terms of the magnitude of the transfer function. The main objective in filter design is to select capacitors and inductors so that the parameters $\mathbf{y}_{21}$ and $\mathbf{y}_{22}$ are synthesized, thereby realizing the desired transfer function. To achieve this, we take advantage of an important property of the *LC* ladder network: all *z* and *y* parameters are ratios of polynomials that contain only even powers of *s* or odd powers of s—that is, they are ratios of either $\text{Od}(s)/\text{Ev}(s)$ or $\text{Ev}(s)/\text{Od}(s)$, where Od and Ev are odd and even functions, respectively. Let

$$\mathbf{H}(s) = \frac{\mathbf{N}(s)}{\mathbf{D}(s)} = \frac{\mathbf{N}_o + \mathbf{N}_e}{\mathbf{D}_o + \mathbf{D}_e} \tag{19.83}$$

where $\mathbf{N}(s)$ and $\mathbf{D}(s)$ are the numerator and denominator of the transfer function $\mathbf{H}(s)$; $\mathbf{N}_o$ and $\mathbf{N}_e$ are the odd and even parts of $\mathbf{N}$; $\mathbf{D}_o$ and $\mathbf{D}_e$ are the odd and even parts of $\mathbf{D}$. Since $\mathbf{N}(s)$ must be either odd or even, we can write Eq. (19.83) as

$$\mathbf{H}(s) = \begin{cases} \dfrac{\mathbf{N}_o}{\mathbf{D}_o + \mathbf{D}_e}, & (\mathbf{N}_e = 0) \\ \dfrac{\mathbf{N}_e}{\mathbf{D}_o + \mathbf{D}_e}, & (\mathbf{N}_o = 0) \end{cases} \tag{19.84}$$

and can rewrite this as

$$\mathbf{H}(s) = \begin{cases} \dfrac{\mathbf{N}_o/\mathbf{D}_e}{1 + \mathbf{D}_o/\mathbf{D}_e}, & (\mathbf{N}_e = 0) \\ \dfrac{\mathbf{N}_e/\mathbf{D}_o}{1 + \mathbf{D}_e/\mathbf{D}_o}, & (\mathbf{N}_o = 0) \end{cases} \tag{19.85}$$

Comparing this with Eq. (19.82), we obtain the y parameters of the network as

$$\frac{\mathbf{y}_{21}}{\mathbf{Y}_L} = \begin{cases} \dfrac{\mathbf{N}_o}{\mathbf{D}_e}, & (\mathbf{N}_e = 0) \\ \dfrac{\mathbf{N}_e}{\mathbf{D}_o}, & (\mathbf{N}_o = 0) \end{cases} \tag{19.86}$$

and

$$\frac{\mathbf{y}_{22}}{\mathbf{Y}_L} = \begin{cases} \dfrac{\mathbf{D}_o}{\mathbf{D}_e}, & (\mathbf{N}_e = 0) \\ \dfrac{\mathbf{D}_e}{\mathbf{D}_o}, & (\mathbf{N}_o = 0) \end{cases} \tag{19.87}$$

The following example illustrates the procedure.

Example 19.18

Design the LC ladder network terminated with a 1-Ω resistor that has the normalized transfer function

$$\mathbf{H}(s) = \frac{1}{s^3 + 2s^2 + 2s + 1}$$

(This transfer function is for a Butterworth lowpass filter.)

Solution:

The denominator shows that this is a third-order network, so that the LC ladder network is shown in Fig. 19.63(a), with two inductors and one capacitor. Our goal is to determine the values of the inductors and

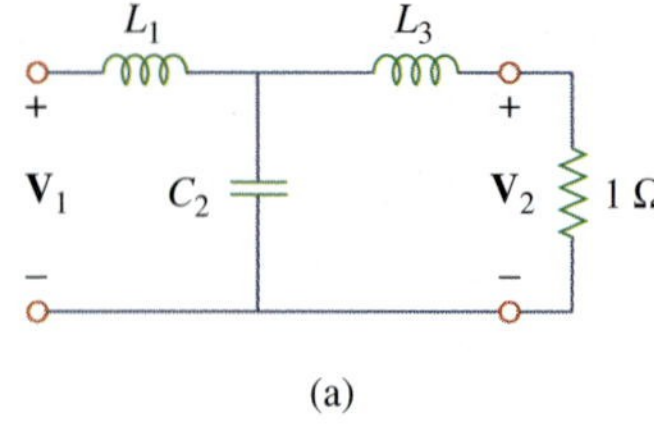

(a)

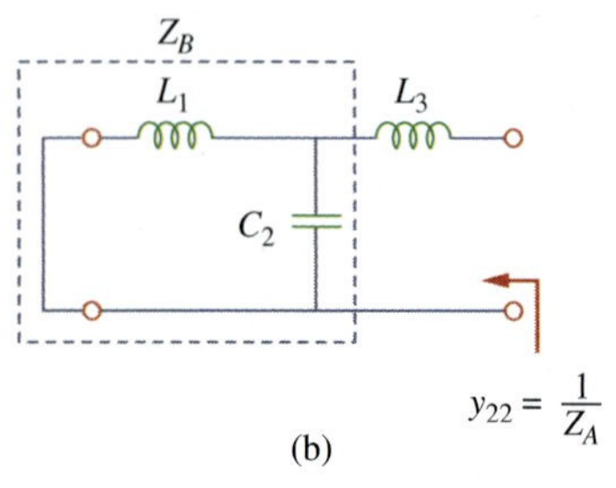

(b)

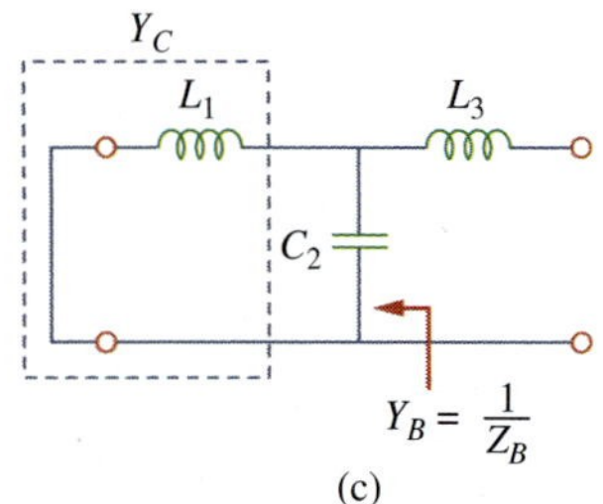

(c)

Figure 19.63
For Example 19.18.

capacitor. To achieve this, we group the terms in the denominator into odd or even parts:

$$\mathbf{D}(s) = (s^3 + 2s) + (2s^2 + 1\ )$$

so that

$$\mathbf{H}(s) = \frac{1}{(s^3 + 2s) + (2s^2 + 1)}$$

Divide the numerator and denominator by the odd part of the denominator to get

$$\mathbf{H}(s) = \frac{\dfrac{1}{s^3 + 2s}}{1 + \dfrac{2s^2 + 1}{s^3 + 2s}} \tag{19.18.1}$$

From Eq. (19.82), when $\mathbf{Y}_L = 1$,

$$\mathbf{H}(s) = \frac{-y_{21}}{1 + y_{22}} \tag{19.18.2}$$

Comparing Eqs. (19.19.1) and (19.19.2), we obtain

$$y_{21} = -\frac{1}{s^3 + 2s}, \qquad y_{22} = \frac{2s^2 + 1}{s^3 + 2s}$$

Any realization of y_{22} will automatically realize y_{21}, since y_{22} is the output driving-point admittance, that is, the output admittance of the network with the input port short-circuited. We determine the values of L and C in Fig. 19.63(a) that will give us y_{22}. Recall that y_{22} is the short-circuit output admittance. So we short-circuit the input port as shown in Fig. 19.63(b). First we get L_3 by letting

$$Z_A = \frac{1}{y_{22}} = \frac{s^3 + 2s}{2s^2 + 1} = sL_3 + Z_B \tag{19.18.3}$$

By long division,

$$Z_A = 0.5s + \frac{1.5s}{2s^2 + 1} \tag{19.18.4}$$

Comparing Eqs. (19.18.3) and (19.18.4) shows that

$$L_3 = 0.5\text{H}, \qquad Z_B = \frac{1.5s}{2s^2 + 1}$$

Next, we seek to get C_2 as in Fig. 19.63(c) and let

$$Y_B = \frac{1}{Z_B} = \frac{2s^2 + 1}{1.5s} = 1.333s + \frac{1}{1.5s} = sC_2 + Y_C$$

from which $C_2 = 1.33$ F and

$$Y_C = \frac{1}{1.5s} = \frac{1}{sL_1} \qquad \Rightarrow \qquad L_1 = 1.5\text{ H}$$

Thus, the LC ladder network in Fig. 19.63(a) with $L_1 = 1.5$ H, $C_2 = 1.333$ F, and $L_3 = 0.5$ H has been synthesized to provide the given transfer function $\mathbf{H}(s)$. This result can be confirmed by finding $\mathbf{H}(s) = \mathbf{V}_2/\mathbf{V}_1$ in Fig. 19.63(a) or by confirming the required y_{21}.

Practice Problem 19.18

Realize the following transfer function using an *LC* ladder network terminated in a 1-Ω resistor:

$$H(s) = \frac{2}{s^3 + s^2 + 4s + 2}$$

Answer: Ladder network in Fig. 19.63(a) with $L_1 = L_3 = 1.0$ H and $C_2 = 0.5$ F.

19.10 Summary

1. A two-port network is one with two ports (or two pairs of access terminals), known as input and output ports.
2. The six parameters used to model a two-port network are the impedance [**z**], admittance [**y**], hybrid [**h**], inverse hybrid [**g**], transmission [**T**], and inverse transmission [**t**] parameters.
3. The parameters relate the input and output port variables as

$$\begin{bmatrix} \mathbf{V}_1 \\ \mathbf{V}_2 \end{bmatrix} = [\mathbf{z}] \begin{bmatrix} \mathbf{I}_1 \\ \mathbf{I}_2 \end{bmatrix}, \quad \begin{bmatrix} \mathbf{I}_1 \\ \mathbf{I}_2 \end{bmatrix} = [\mathbf{y}] \begin{bmatrix} \mathbf{V}_1 \\ \mathbf{V}_2 \end{bmatrix}, \quad \begin{bmatrix} \mathbf{V}_1 \\ \mathbf{I}_2 \end{bmatrix} = [\mathbf{h}] \begin{bmatrix} \mathbf{I}_1 \\ \mathbf{V}_2 \end{bmatrix}$$

$$\begin{bmatrix} \mathbf{I}_1 \\ \mathbf{V}_2 \end{bmatrix} = [\mathbf{g}] \begin{bmatrix} \mathbf{V}_1 \\ \mathbf{I}_2 \end{bmatrix}, \quad \begin{bmatrix} \mathbf{V}_1 \\ \mathbf{I}_1 \end{bmatrix} = [\mathbf{T}] \begin{bmatrix} \mathbf{V}_2 \\ -\mathbf{I}_2 \end{bmatrix}, \quad \begin{bmatrix} \mathbf{V}_2 \\ \mathbf{I}_2 \end{bmatrix} = [\mathbf{t}] \begin{bmatrix} \mathbf{V}_1 \\ -\mathbf{I}_1 \end{bmatrix}$$

4. The parameters can be calculated or measured by short-circuiting or open-circuiting the appropriate input or output port.
5. A two-port network is reciprocal if $\mathbf{z}_{12} = \mathbf{z}_{21}$, $\mathbf{y}_{12} = \mathbf{y}_{21}$, $\mathbf{h}_{12} = -\mathbf{h}_{21}$, $\mathbf{g}_{12} = -\mathbf{g}_{21}$, $\Delta_T = 1$ or $\Delta_t = 1$. Networks that have dependent sources are not reciprocal.
6. Table 19.1 provides the relationships between the six sets of parameters. Three important relationships are

$$[\mathbf{y}] = [\mathbf{z}]^{-1}, \qquad [\mathbf{g}] = [\mathbf{h}]^{-1}, \qquad [\mathbf{t}] \neq [\mathbf{T}]^{-1}$$

7. Two-port networks may be connected in series, in parallel, or in cascade. In the series connection the *z* parameters are added, in the parallel connection the *y* parameters are added, and in the cascade connection the transmission parameters are multiplied in the correct order.
8. One can use *PSpice* to compute the two-port parameters by constraining the appropriate port variables with a 1-A or 1-V source while using an open or short circuit to impose the other necessary constraints.
9. The network parameters are specifically applied in the analysis of transistor circuits and the synthesis of ladder *LC* networks. Network parameters are especially useful in the analysis of transistor circuits because these circuits are easily modeled as two-port networks. *LC* ladder networks, important in the design of passive low-pass filters, resemble cascaded T networks and are therefore best analyzed as two-ports.

Review Questions

19.1 For the single-element two-port network in Fig. 19.64(a), $\mathbf{z}_{11}$ is:

(a) 0 (b) 5 (c) 10
(d) 20 (e) undefined

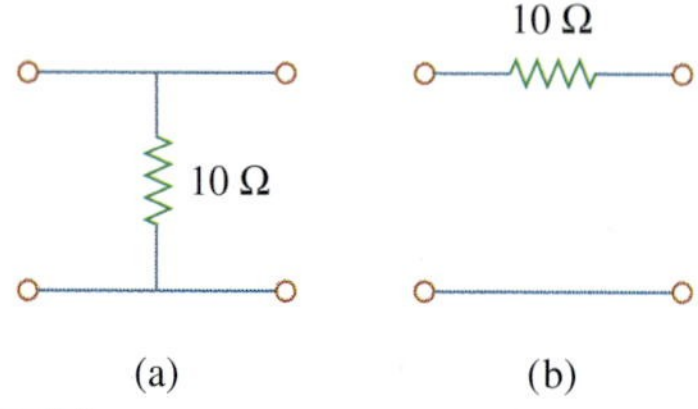

Figure 19.64
For Review Questions.

19.2 For the single-element two-port network in Fig. 19.64(b), $\mathbf{z}_{11}$ is:

(a) 0 (b) 5 (c) 10
(d) 20 (e) undefined

19.3 For the single-element two-port network in Fig. 19.64(a), $\mathbf{y}_{11}$ is:

(a) 0 (b) 5 (c) 10
(d) 20 (e) undefined

19.4 For the single-element two-port network in Fig. 19.64(b), $\mathbf{h}_{21}$ is:

(a) −0.1 (b) −1 (c) 0
(d) 10 (e) undefined

19.5 For the single-element two-port network in Fig. 19.64(a), **B** is:

(a) 0 (b) 5 (c) 10
(d) 20 (e) undefined

19.6 For the single-element two-port network in Fig. 19.64(b), **B** is:

(a) 0 (b) 5 (c) 10
(d) 20 (e) undefined

19.7 When port 1 of a two-port circuit is short-circuited, $\mathbf{I}_1 = 4\mathbf{I}_2$ and $\mathbf{V}_2 = 0.25\mathbf{I}_2$. Which of the following is true?

(a) $y_{11} = 4$ (b) $y_{12} = 16$
(c) $y_{21} = 16$ (d) $y_{22} = 0.25$

19.8 A two-port is described by the following equations:

$$\mathbf{V}_1 = 50\mathbf{I}_1 + 10\mathbf{I}_2$$
$$\mathbf{V}_2 = 30\mathbf{I}_1 + 20\mathbf{I}_2$$

Which of the following is *not* true?

(a) $\mathbf{z}_{12} = 10$ (b) $\mathbf{y}_{12} = -0.0143$
(c) $\mathbf{h}_{12} = 0.5$ (d) $\mathbf{A} = 50$

19.9 If a two-port is reciprocal, which of the following is *not* true?

(a) $\mathbf{z}_{21} = \mathbf{z}_{12}$ (b) $\mathbf{y}_{21} = \mathbf{y}_{12}$
(c) $\mathbf{h}_{21} = \mathbf{h}_{12}$ (d) $AD = BC + 1$

19.10 If the two single-element two-port networks in Fig. 19.64 are cascaded, then **D** is:

(a) 0 (b) 0.1 (c) 2
(d) 10 (e) undefined

Answers: 19.1c, 19.2e, 19.3e, 19.4b, 19.5a, 19.6c, 19.7b, 19.8d, 19.9c, 19.10c.

Problems

Section 19.2 Impedance Parameters

19.1 Obtain the z parameters for the network in Fig. 19.65.

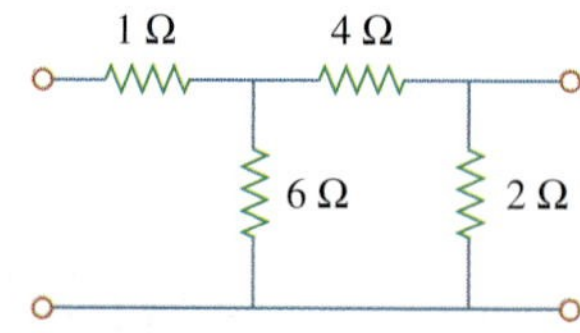

Figure 19.65
For Probs. 19.1 and 19.28

***19.2** Find the impedance parameter equivalent of the network in Fig. 19.66.

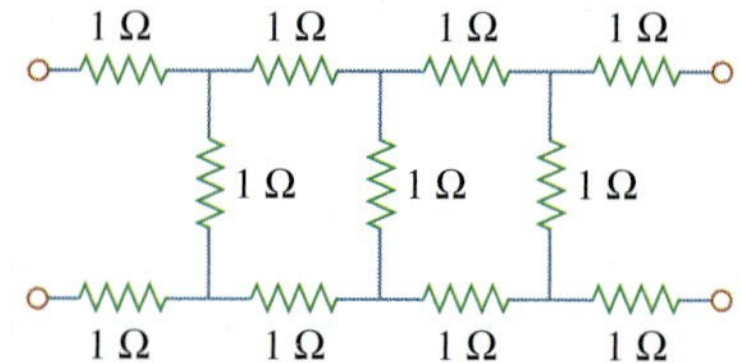

Figure 19.66
For Prob. 19.2.

* An asterisk indicates a challenging problem.

19.3 Find the z parameters of the circuit in Fig. 19.67.

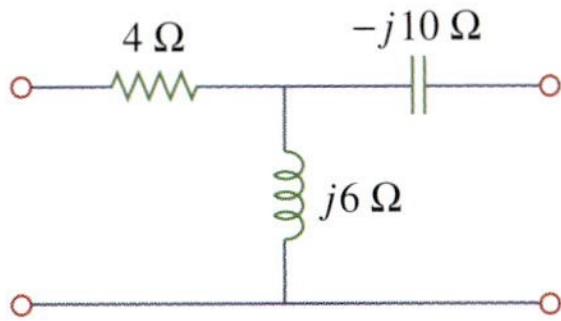

Figure 19.67
For Prob. 19.3.

19.4 Using Fig. 19.68, design a problem to help other students better understand how to determine z parameters from an electrical circuit.

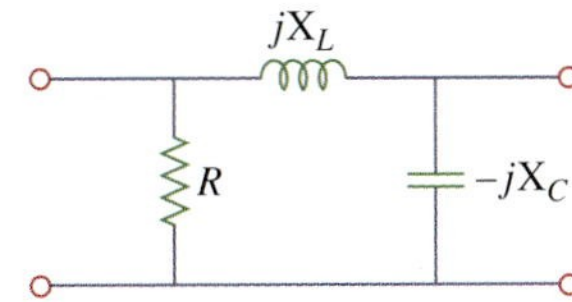

Figure 19.68
For Prob. 19.4.

19.5 Obtain the z parameters for the network in Fig. 19.69 as functions of s.

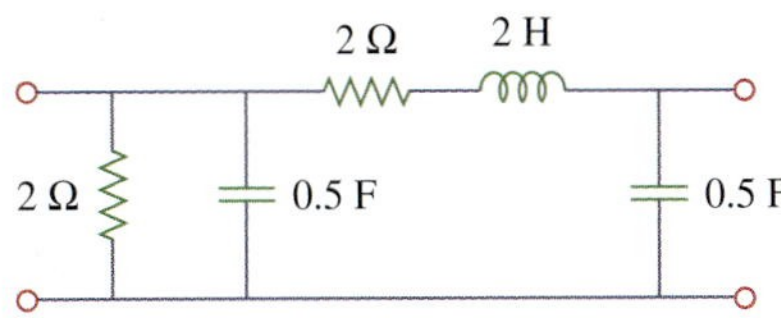

Figure 19.69
For Prob. 19.5.

19.6 Compute the z parameters of the circuit in Fig. 19.70.

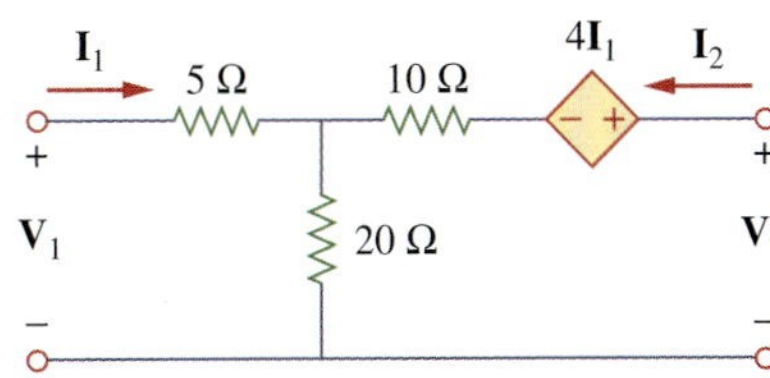

Figure 19.70
For Prob. 19.6 and 19.73.

19.7 Calculate the impedance-parameter equivalent of the circuit in Fig. 19.71.

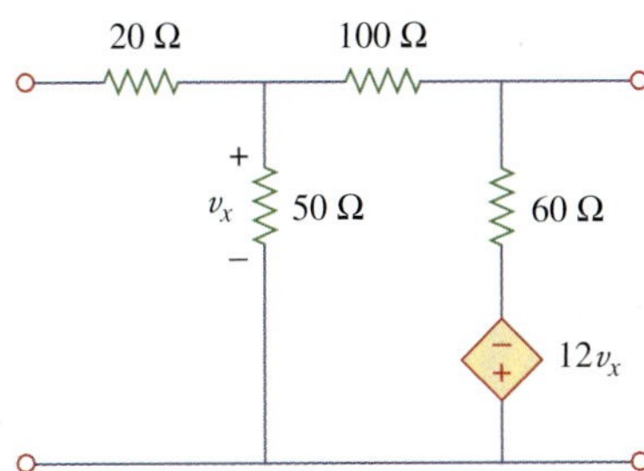

Figure 19.71
For Prob. 19.7 and 19.80.

19.8 Find the z parameters of the two-port in Fig. 19.72.

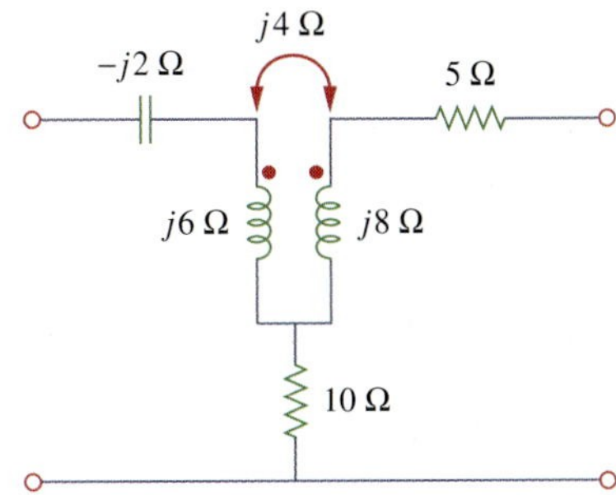

Figure 19.72
For Prob. 19.8.

19.9 The y parameters of a network are:

$$[\mathbf{y}] = \begin{bmatrix} 0.25 & -0.1 \\ -0.1 & 0.2 \end{bmatrix} S$$

Determine the z parameters for the network.

19.10 Construct a two-port that realizes each of the following z parameters.

(a) $[\mathbf{z}] = \begin{bmatrix} 25 & 20 \\ 5 & 10 \end{bmatrix} \Omega$

(b) $[\mathbf{z}] = \begin{bmatrix} 1 + \dfrac{3}{s} & \dfrac{1}{s} \\ \dfrac{1}{s} & 2s + \dfrac{1}{s} \end{bmatrix} \Omega$

19.11 Determine a two-port network that is represented by the following z parameters:

$$[\mathbf{z}] = \begin{bmatrix} 6 + j3 & 5 - j2 \\ 5 - j2 & 8 - j \end{bmatrix} \Omega$$

19.12 For the circuit shown in Fig. 19.73, let

$$[\mathbf{z}] = \begin{bmatrix} 10 & -6 \\ -4 & 12 \end{bmatrix} \Omega$$

Find I_1, I_2, V_1, and V_2.

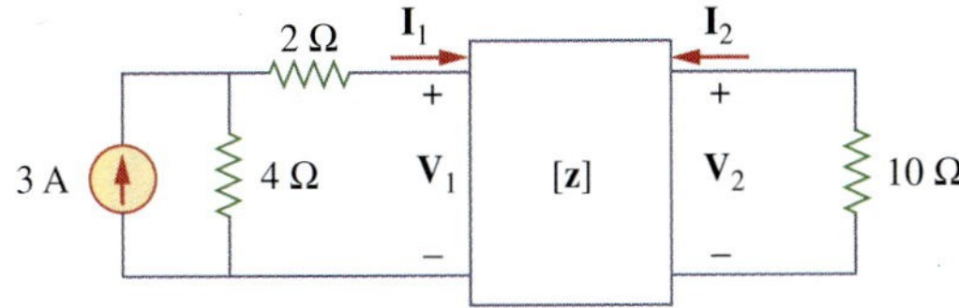

Figure 19.73
For Prob. 19.12.

19.13 Determine the average power delivered to $Z_L = 5 + j4$ in the network of Fig. 19.74. *Note:* The voltage is rms.

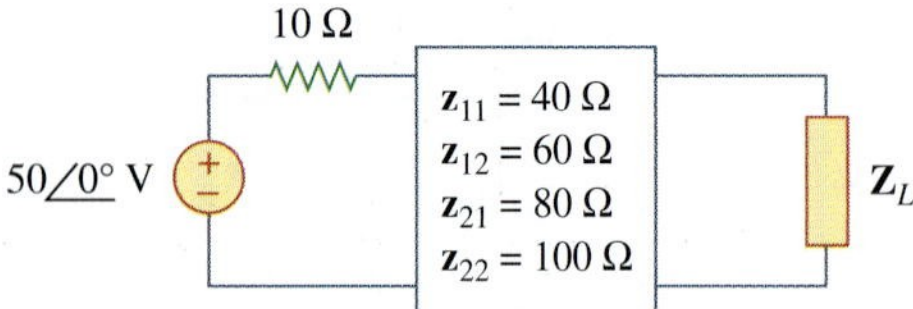

Figure 19.74
For Prob. 19.13.

19.14 For the two-port network shown in Fig. 19.75, show that at the output terminals,

$$\mathbf{Z}_{Th} = \mathbf{z}_{22} - \frac{\mathbf{z}_{12}\mathbf{z}_{21}}{\mathbf{z}_{11} + \mathbf{Z}_s}$$

and

$$\mathbf{V}_{Th} = \frac{\mathbf{z}_{21}}{\mathbf{z}_{11} + \mathbf{Z}_s}\mathbf{V}_s$$

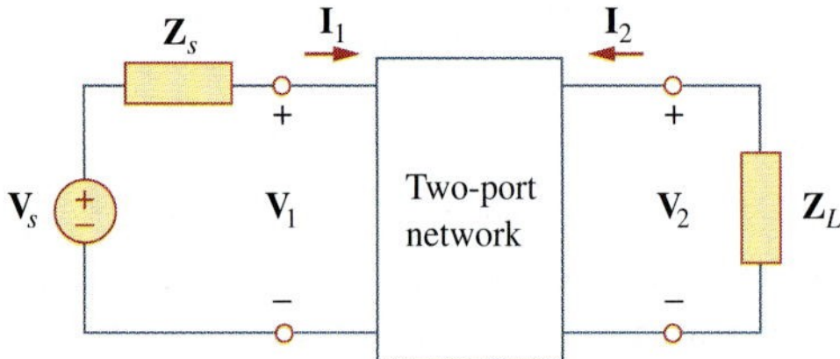

Figure 19.75
For Probs. 19.14 and 19.41.

19.15 For the two-port circuit in Fig. 19.76,

$$[\mathbf{z}] = \begin{bmatrix} 40 & 60 \\ 80 & 120 \end{bmatrix} \Omega$$

(a) Find $\mathbf{Z}_L$ for maximum power transfer to the load.

(b) Calculate the maximum power delivered to the load.

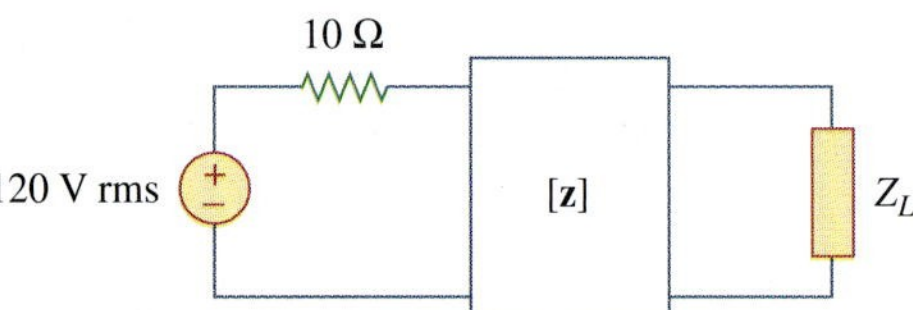

Figure 19.76
For Prob. 19.15.

19.16 For the circuit in Fig. 19.77, at $\omega = 2$ rad/s, $\mathbf{z}_{11} = 10\ \Omega$, $\mathbf{z}_{12} = \mathbf{z}_{21} = j6\ \Omega$, $\mathbf{z}_{22} = 4\ \Omega$. Obtain the Thevenin equivalent circuit at terminals *a-b* and calculate v_o.

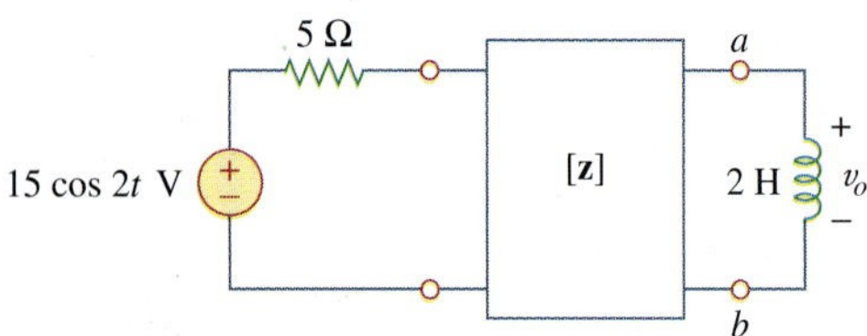

Figure 19.77
For Prob. 19.16.

Section 19.3 Admittance Parameters

***19.17** Determine the *z* and *y* parameters for the circuit in Fig. 19.78.

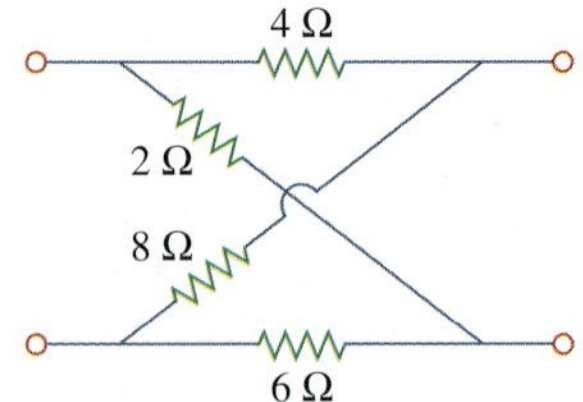

Figure 19.78
For Prob. 19.17.

19.18 Calculate the *y* parameters for the two-port in Fig. 19.79.

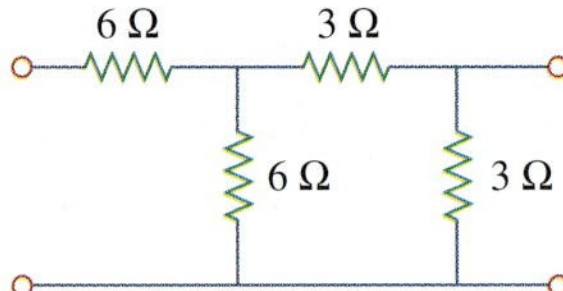

Figure 19.79
For Probs. 19.18 and 19.37.

19.19 **e2d** Using Fig. 19.80, design a problem to help other students better understand how to find *y* parameters in the *s*-domain.

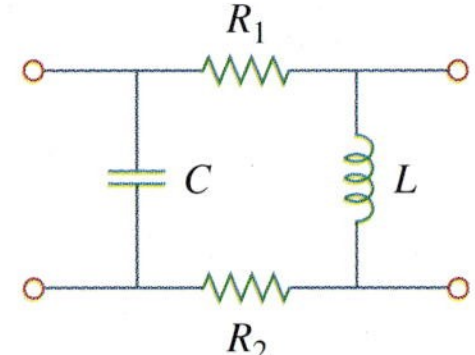

Figure 19.80
For Prob. 19.19.

19.20 Find the *y* parameters for the circuit in Fig. 19.81.

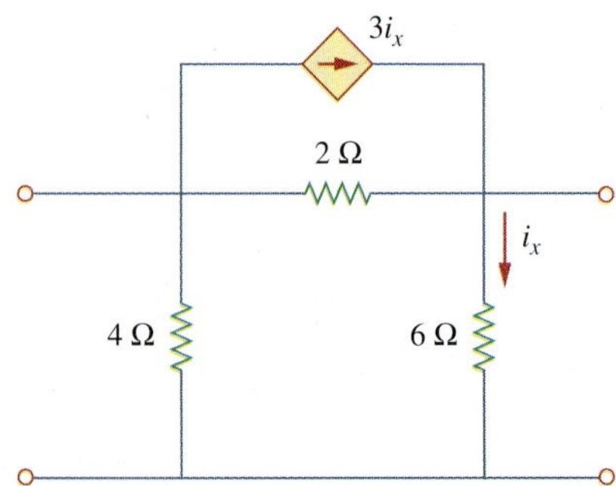

Figure 19.81
For Prob. 19.20.

19.21 Obtain the admittance parameter equivalent circuit of the two-port in Fig. 19.82.

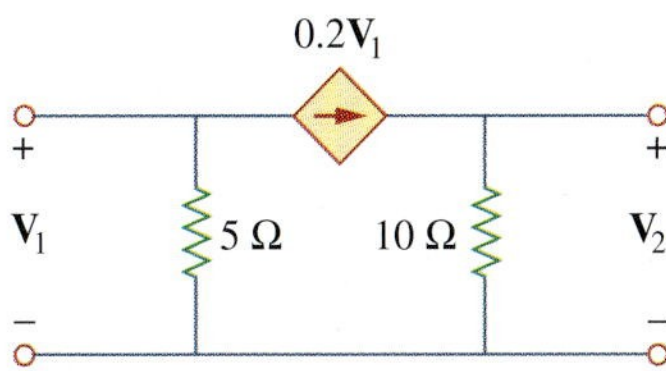

Figure 19.82
For Prob. 19.21.

19.22 Obtain the y parameters of the two-port network in Fig. 19.83.

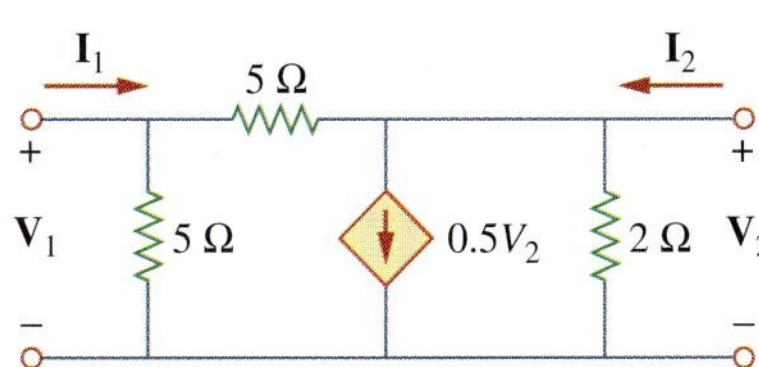

Figure 19.83
For Prob. 19.22.

19.23 (a) Find the y parameters of the two-port in Fig. 19.84.

(b) Determine $\mathbf{V}_2(s)$ for $v_s = 2u(t)$ V.

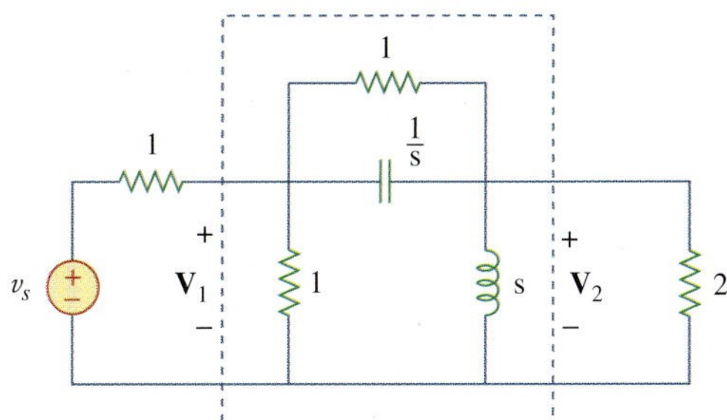

Figure 19.84
For Prob. 19.23.

19.24 Find the resistive circuit that represents these y parameters:

$$[\mathbf{y}] = \begin{bmatrix} \frac{1}{4} & -\frac{1}{8} \\ -\frac{1}{8} & \frac{3}{16} \end{bmatrix} \text{S}$$

19.25 Draw the two-port network that has the following y parameters:

$$[\mathbf{y}] = \begin{bmatrix} 1 & -0.5 \\ -0.5 & 1.5 \end{bmatrix} \text{S}$$

19.26 Calculate $[\mathbf{y}]$ for the two-port in Fig. 19.85.

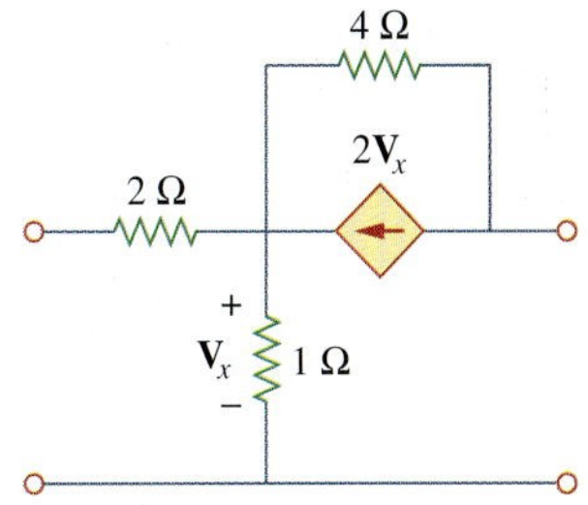

Figure 19.85
For Prob. 19.26.

19.27 Find the y parameters for the circuit in Fig. 19.86.

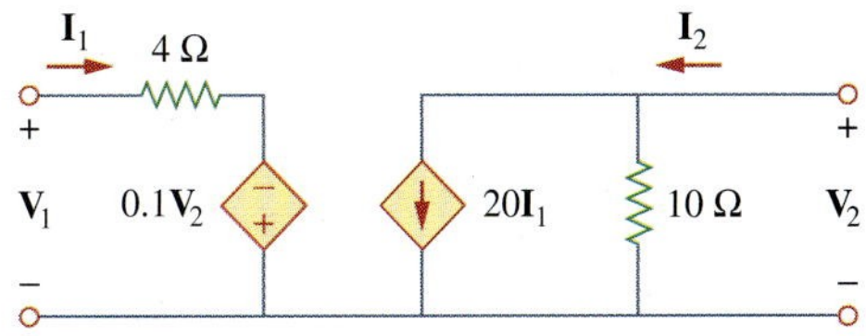

Figure 19.86
For Prob. 19.27.

19.28 In the circuit of Fig. 19.65, the input port is connected to a 1-A dc current source. Calculate the power dissipated by the 2-Ω resistor by using the y parameters. Confirm your result by direct circuit analysis.

19.29 In the bridge circuit of Fig. 19.87, $I_1 = 10$ A and $I_2 = -4$ A.

(a) Find V_1 and V_2 using y parameters.

(b) Confirm the results in part (a) by direct circuit analysis.

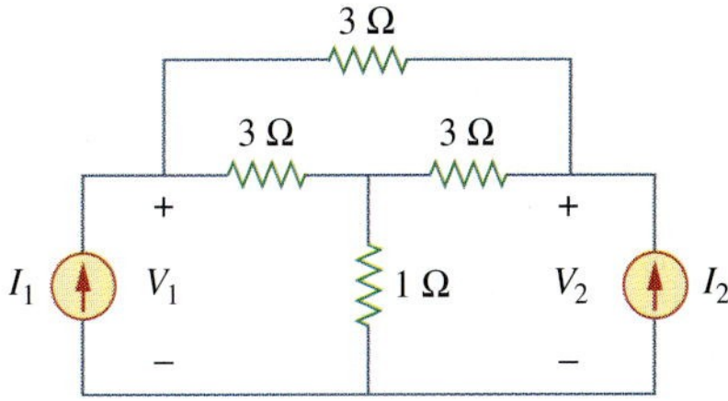

Figure 19.87
For Prob. 19.29.

Section 19.4 Hybrid Parameters

19.30 Find the h parameters for the networks in Fig. 19.88.

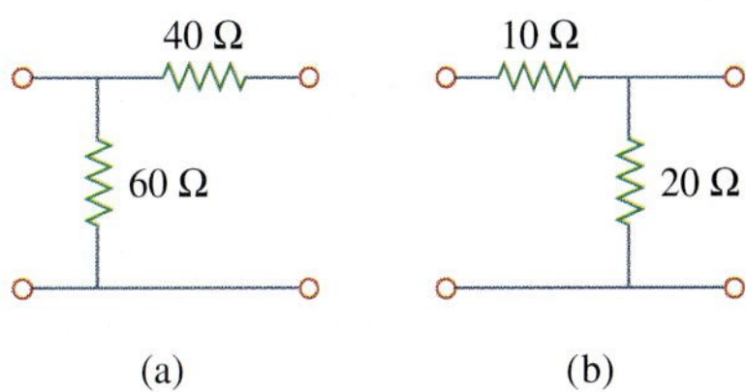

Figure 19.88
For Prob. 19.30.

19.31 Determine the hybrid parameters for the network in Fig. 19.89.

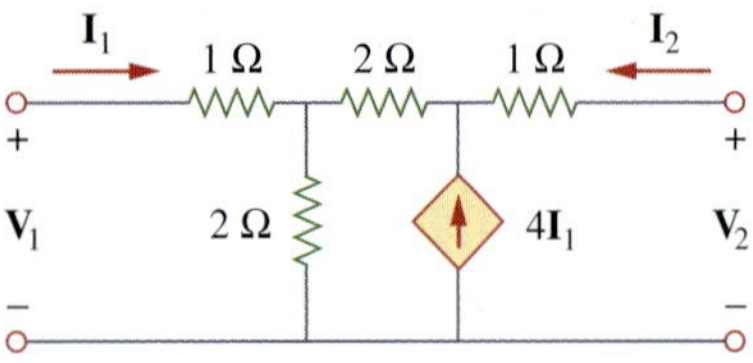

Figure 19.89
For Prob. 19.31.

19.32 Using Fig. 19.90, design a problem to help other students better understand how to find the h and g parameters for a circuit in the s-domain.

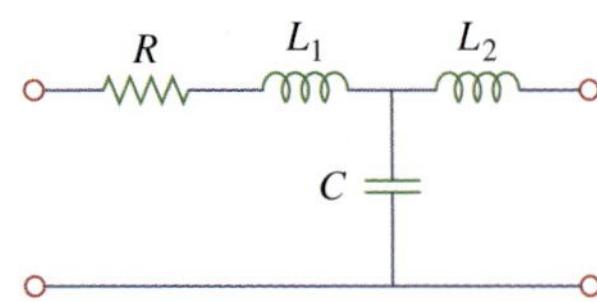

Figure 19.90
For Prob. 19.32.

19.33 Obtain the h parameters for the two-port of Fig. 19.91.

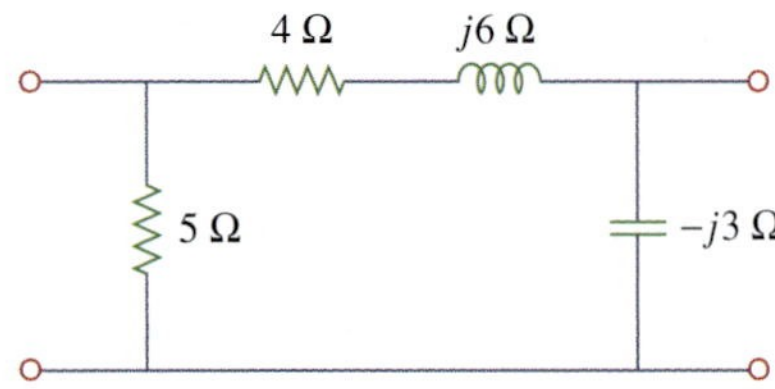

Figure 19.91
For Prob. 19.33.

19.34 Obtain the h and g parameters of the two-port in Fig. 19.92.

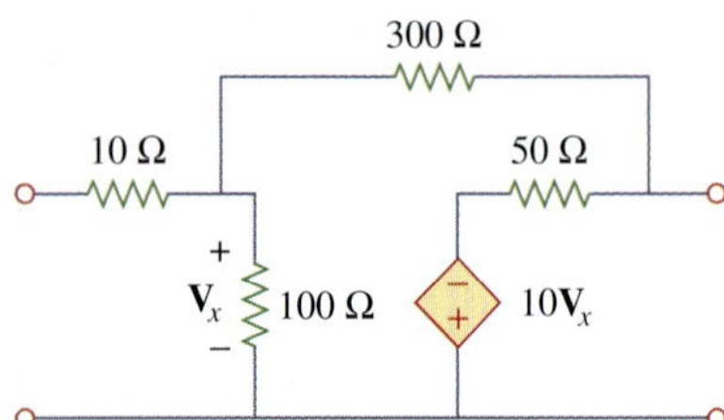

Figure 19.92
For Prob. 19.34.

19.35 Determine the h parameters for the network in Fig. 19.93.

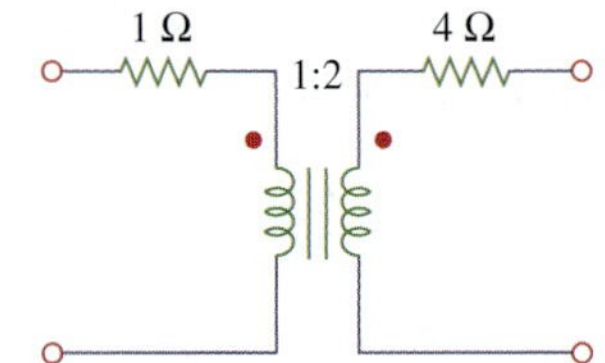

Figure 19.93
For Prob. 19.35.

19.36 For the two-port in Fig. 19.94,

$$[\mathbf{h}] = \begin{bmatrix} 16\ \Omega & 3 \\ -2 & 0.01\ \text{S} \end{bmatrix}$$

Find:

(a) V_2/V_1 (b) I_2/I_1
(c) I_1/V_1 (d) V_2/I_1

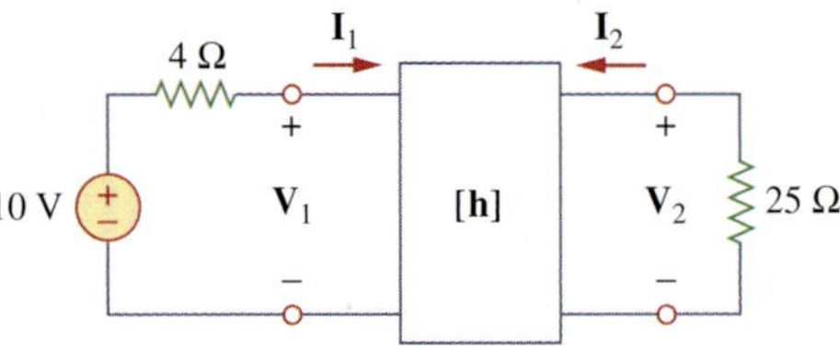

Figure 19.94
For Prob. 19.36.

19.37 The input port of the circuit in Fig. 19.79 is connected to a 10-V dc voltage source while the output port is terminated by a 5-Ω resistor. Find the voltage across the 5-Ω resistor by using h parameters of the circuit. Confirm your result by using direct circuit analysis.

19.38 The h parameters of the two-port of Fig. 19.95 are:

$$[\mathbf{h}] = \begin{bmatrix} 600\ \Omega & 0.04 \\ 30 & 2\ \text{mS} \end{bmatrix}$$

Given the $Z_s = 2$ kΩ and $Z_L = 400$ Ω, find Z_{in} and Z_{out}.

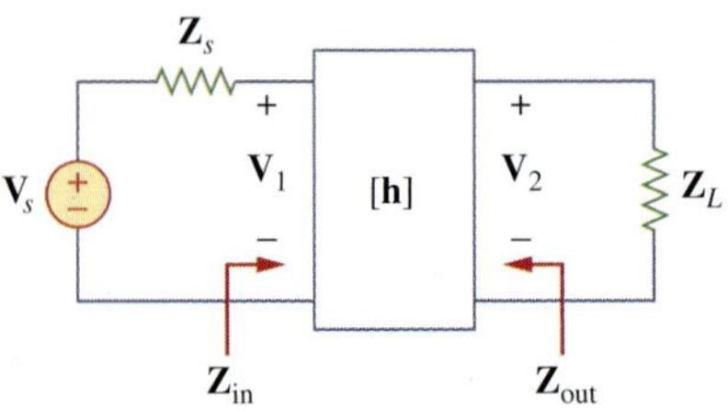

Figure 19.95
For Prob. 19.38.

19.39 Obtain the g parameters for the wye circuit of Fig. 19.96.

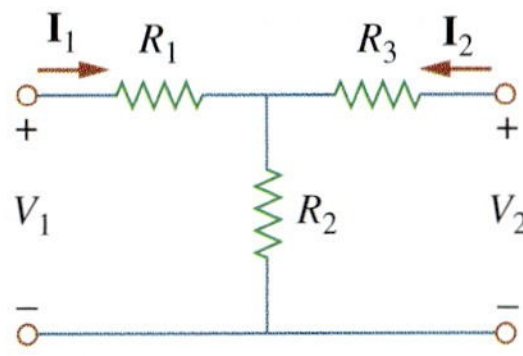

Figure 19.96
For Prob. 19.39.

19.40 Using Fig. 19.97, design a problem to help other students better understand how to find *g* parameters in an ac circuit.

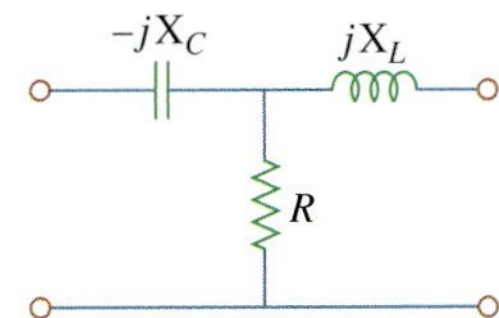

Figure 19.97
For Prob. 19.40.

19.41 For the two-port in Fig. 19.75, show that

$$\frac{\mathbf{I}_2}{\mathbf{I}_1} = \frac{-\mathbf{g}_{21}}{\mathbf{g}_{11}\mathbf{Z}_L + \Delta_g}$$

$$\frac{\mathbf{V}_2}{\mathbf{V}_s} = \frac{\mathbf{g}_{21}\mathbf{Z}_L}{(1 + \mathbf{g}_{11}\mathbf{Z}_s)(\mathbf{g}_{22} + \mathbf{Z}_L) - \mathbf{g}_{21}\mathbf{g}_{12}\mathbf{Z}_s}$$

where Δ_g is the determinant of [**g**] matrix.

19.42 The *h* parameters of a two-port device are given by

$$\mathbf{h}_{11} = 600\ \Omega, \quad \mathbf{h}_{12} = 10^{-3}, \quad \mathbf{h}_{21} = 120, \quad \mathbf{h}_{22} = 2 \times 10^{-6}\ \text{S}$$

Draw a circuit model of the device including the value of each element.

Section 19.5 Transmission Parameters

19.43 Find the transmission parameters for the single-element two-port networks in Fig. 19.98.

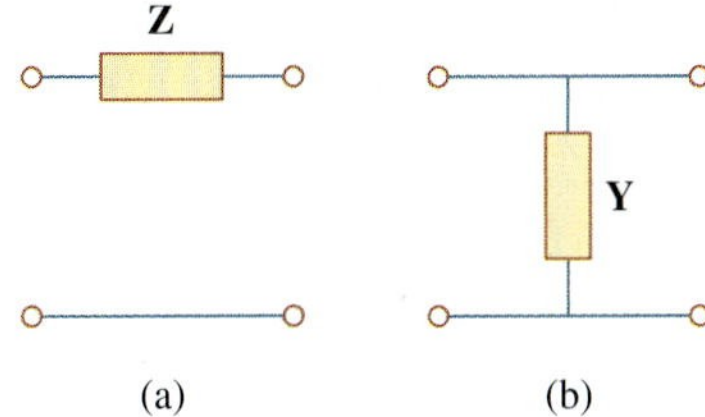

Figure 19.98
For Prob. 19.43.

19.44 Using Fig. 19.99, design a problem to help other students better understand how to find the transmission parameters of an ac circuit.

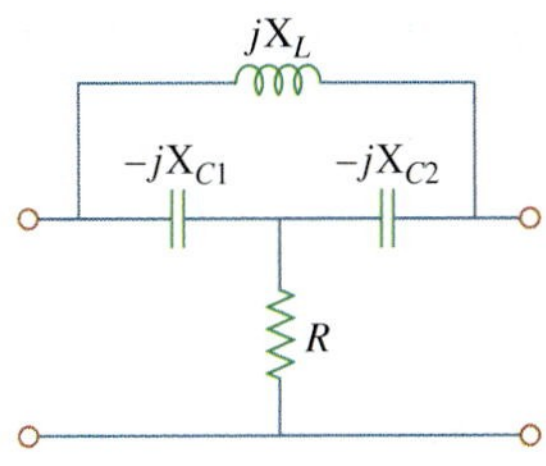

Figure 19.99
For Prob. 19.44.

19.45 Find the **ABCD** parameters for the circuit in Fig. 19.100.

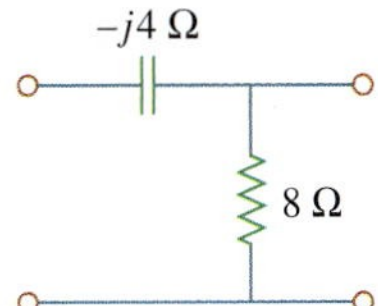

Figure 19.100
For Prob. 19.45.

19.46 Find the transmission parameters for the circuit in Fig. 19.101.

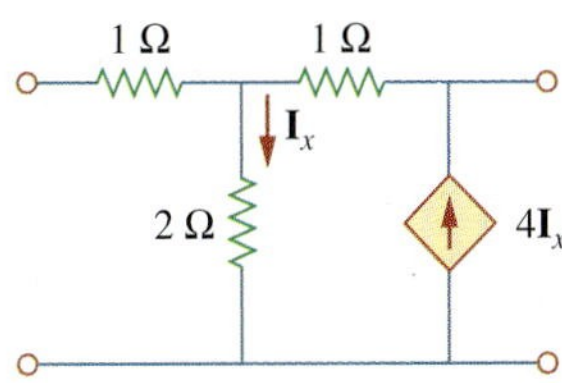

Figure 19.101
For Prob. 19.46.

19.47 Obtain the **ABCD** parameters for the network in Fig. 19.102.

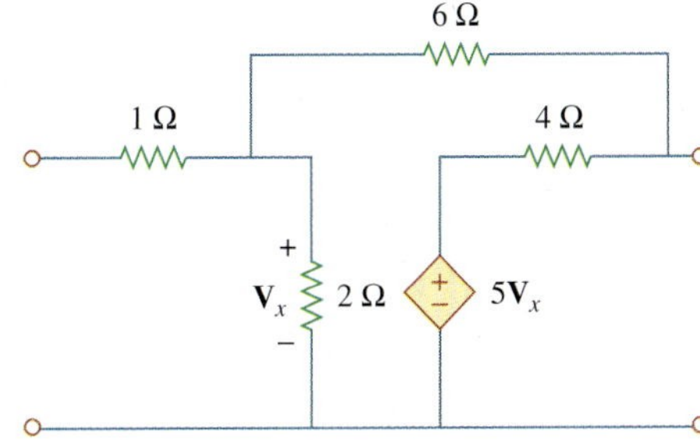

Figure 19.102
For Prob. 19.47

19.48 For a two-port, let **A** = 4, **B** = 30 Ω, **C** = 0.1 S, and **D** = 1.5. Calculate the input impedance $\mathbf{Z}_{in} = \mathbf{V}_1/\mathbf{I}_1$, when:

(a) the output terminals are short-circuited,

(b) the output port is open-circuited,

(c) the output port is terminated by a 10-Ω load.

19.49 Using impedances in the s-domain, obtain the transmission parameters for the circuit in Fig. 19.103.

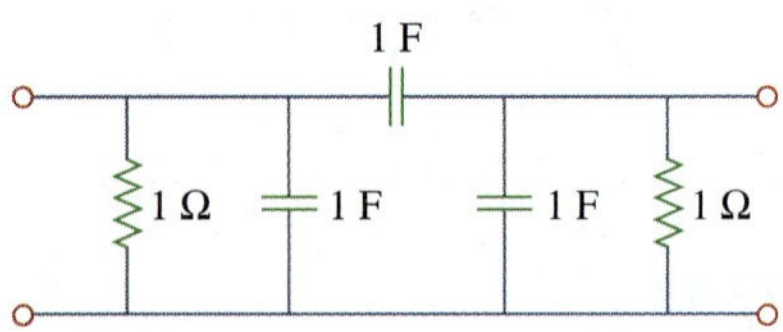

Figure 19.103
For Prob. 19.49.

19.50 Derive the s-domain expression for the t parameters of the circuit in Fig. 19.104.

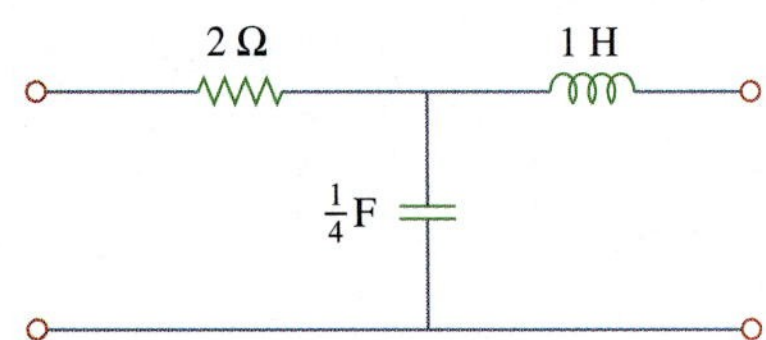

Figure 19.104
For Prob. 19.50.

19.51 Obtain the t parameters for the network in Fig. 19.105.

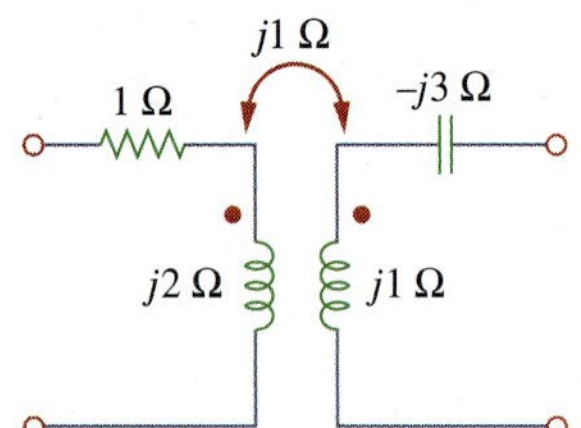

Figure 19.105
For Prob. 19.51.

Section 19.6 Relationships Between Parameters

19.52 (a) For the T network in Fig. 19.106, show that the h parameters are:

$$\mathbf{h}_{11} = R_1 + \frac{R_2R_3}{R_1 + R_3}, \qquad \mathbf{h}_{12} = \frac{R_2}{R_2 + R_3}$$

$$\mathbf{h}_{21} = -\frac{R_2}{R_2 + R_3}, \qquad \mathbf{h}_{22} = \frac{1}{R_2 + R_3}$$

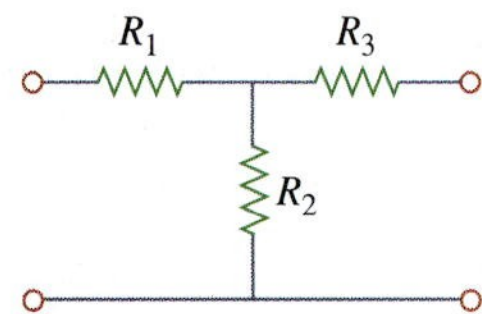

Figure 19.106
For Prob. 19.52.

(b) For the same network, show that the transmission parameters are:

$$\mathbf{A} = 1 + \frac{R_1}{R_2}, \qquad \mathbf{B} = R_3 + \frac{R_1}{R_2}(R_2 + R_3)$$

$$\mathbf{C} = \frac{1}{R_2}, \qquad \mathbf{D} = 1 + \frac{R_3}{R_2}$$

19.53 Through derivation, express the z parameters in terms of the **ABCD** parameters.

19.54 Show that the transmission parameters of a two-port may be obtained from the y parameters as:

$$\mathbf{A} = -\frac{\mathbf{y}_{22}}{\mathbf{y}_{21}}, \qquad \mathbf{B} = -\frac{1}{\mathbf{y}_{21}}$$

$$\mathbf{C} = -\frac{\Delta_y}{\mathbf{y}_{21}}, \qquad \mathbf{D} = -\frac{\mathbf{y}_{11}}{\mathbf{y}_{21}}$$

19.55 Prove that the g parameters can be obtained from the z parameters as

$$\mathbf{g}_{11} = \frac{1}{\mathbf{z}_{11}}, \qquad \mathbf{g}_{12} = -\frac{\mathbf{z}_{12}}{\mathbf{z}_{11}}$$

$$\mathbf{g}_{21} = \frac{\mathbf{z}_{21}}{\mathbf{z}_{11}}, \qquad \mathbf{g}_{22} = \frac{\Delta_z}{\mathbf{z}_{11}}$$

19.56 For the network of Fig. 19.107, obtain $\mathbf{V}_o/\mathbf{V}_s$.

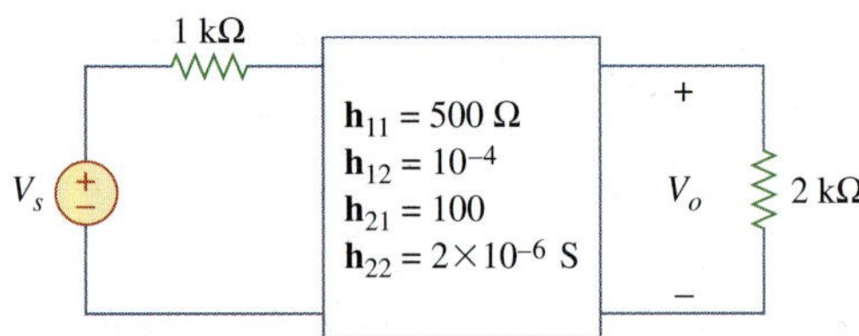

Figure 19.107
For Prob. 19.56.

19.57 Given the transmission parameters

$$[\mathbf{T}] = \begin{bmatrix} 3 & 20 \\ 1 & 7 \end{bmatrix}$$

obtain the other five two-port parameters.

19.58 Design a problem to help other students better understand how to develop the y parameters and transmission parameters, given equations in terms of the hybrid parameters.

19.59 Given that

$$[\mathbf{g}] = \begin{bmatrix} 0.06\text{ S} & -0.4 \\ 0.2 & 2\ \Omega \end{bmatrix}$$

determine:

(a) [**z**] (b) [**y**] (c) [**h**] (d) [**T**]

19.60 Design a T network necessary to realize the following z parameters at $\omega = 10^6$ rad/s.

ed

$$[\mathbf{z}] = \begin{bmatrix} 4 + j3 & 2 \\ 2 & 5 - j \end{bmatrix} \text{k}\Omega$$

19.61 For the bridge circuit in Fig. 19.108, obtain:

(a) the z parameters

(b) the h parameters

(c) the transmission parameters

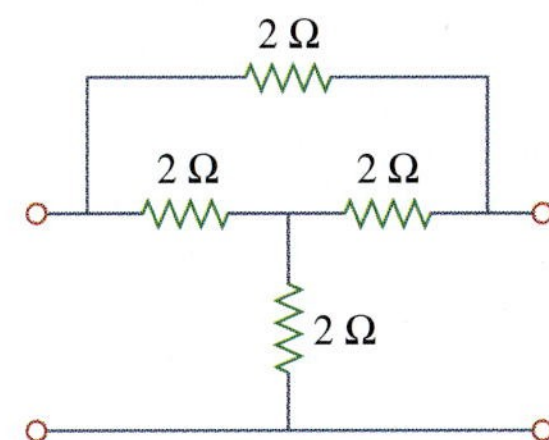

Figure 19.108
For Prob. 19.61.

19.62 Find the z parameters of the op amp circuit in Fig. 19.109. Obtain the transmission parameters.

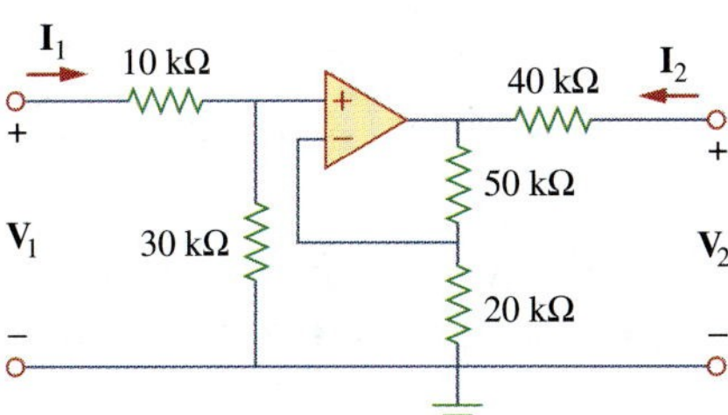

Figure 19.109
For Prob. 19.62.

19.63 Determine the z parameters of the two-port in Fig. 19.110.

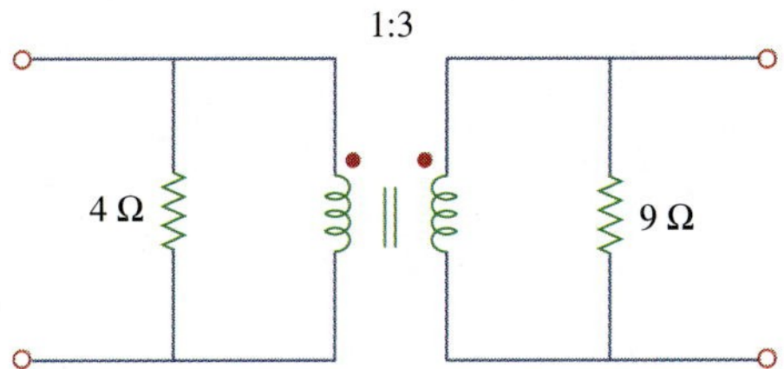

Figure 19.110
For Prob. 19.63.

19.64 Determine the y parameters at $\omega = 1{,}000$ rad/s for the op amp circuit in Fig. 19.111. Find the corresponding h parameters.

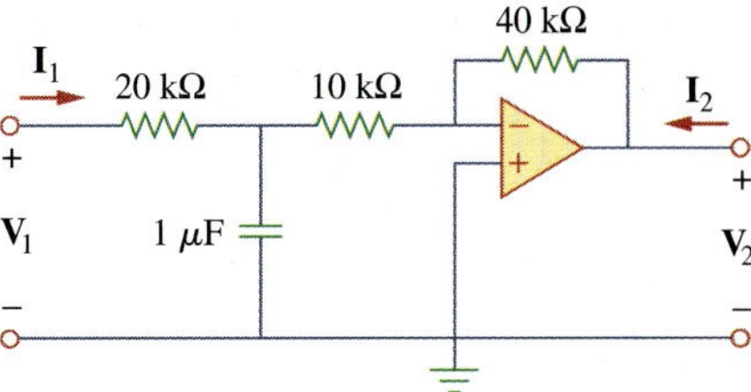

Figure 19.111
For Prob. 19.64.

Section 19.7 Interconnection of Networks

19.65 What is the y parameter presentation of the circuit in Fig. 19.112?

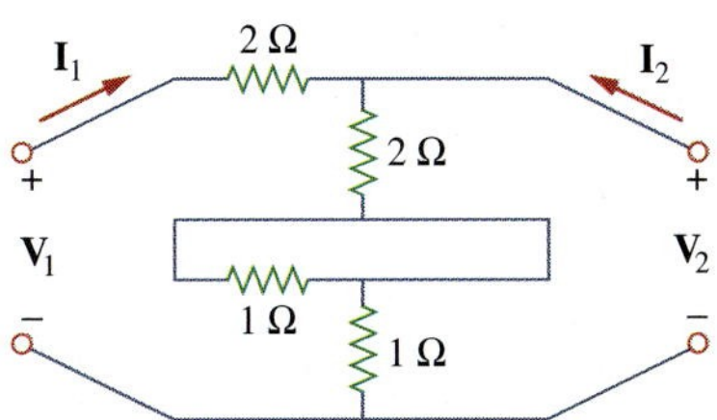

Figure 19.112
For Prob. 19.65.

19.66 In the two-port of Fig. 19.113, let $\mathbf{y}_{12} = \mathbf{y}_{21} = 0$, $\mathbf{y}_{11} = 2$ mS, and $\mathbf{y}_{22} = 10$ mS. Find $\mathbf{V}_o/\mathbf{V}_s$.

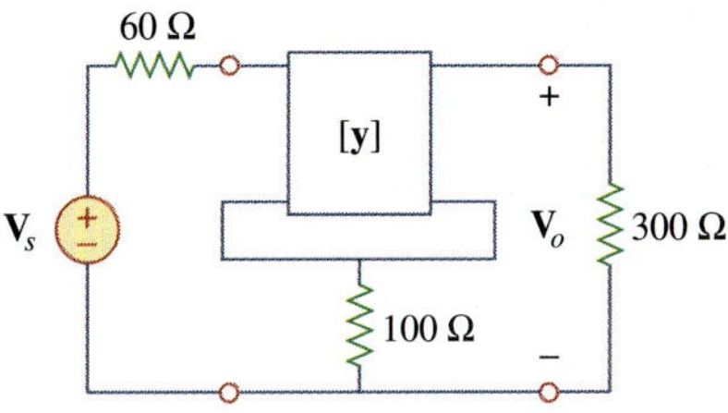

Figure 19.113
For Prob. 19.66.

19.67 If three copies of the circuit in Fig. 19.114 are connected in parallel, find the overall transmission parameters.

ML

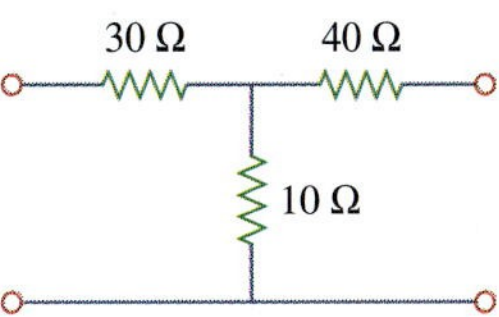

Figure 19.114
For Prob. 19.67.

19.68 Obtain the h parameters for the network in Fig. 19.115.

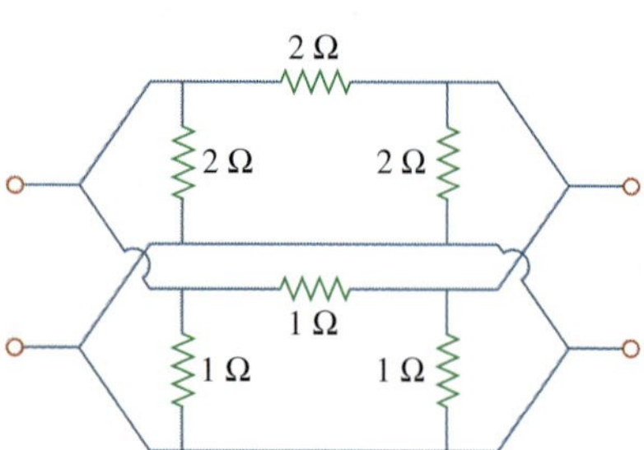

Figure 19.115
For Prob. 19.68.

***19.69** The circuit in Fig. 19.116 may be regarded as two two-ports connected in parallel. Obtain the y parameters as functions of s.

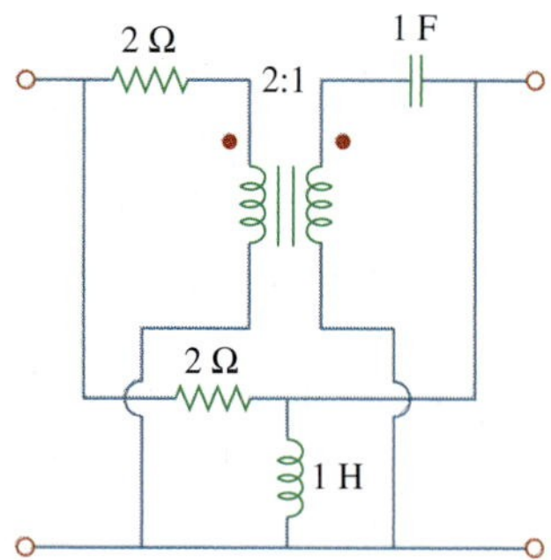

Figure 19.116
For Prob. 19.69.

***19.70** For the parallel-series connection of the two two-ports in Fig. 19.117, find the g parameters.

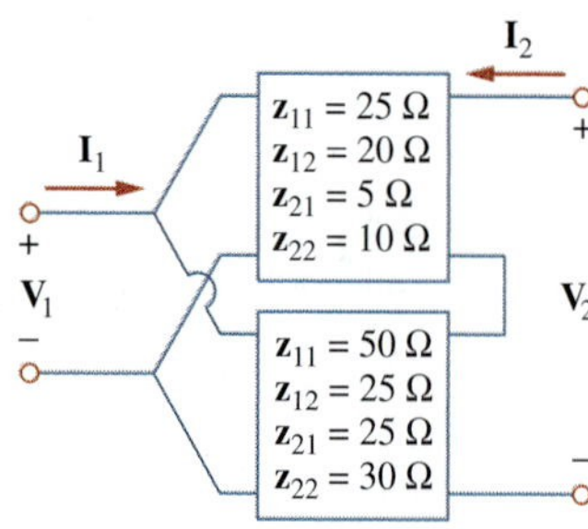

Figure 19.117
For Prob. 19.70.

***19.71** Determine the z parameters for the network in Fig. 19.118.

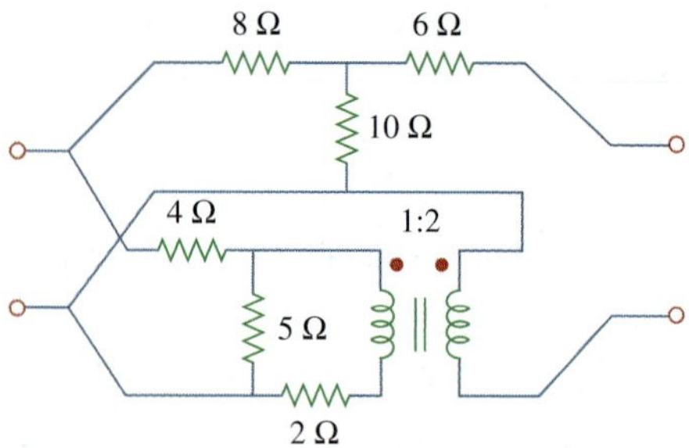

Figure 19.118
For Prob. 19.71.

***19.72** A series-parallel connection of two two-ports is shown in Fig. 19.119. Determine the z parameter representation of the network.

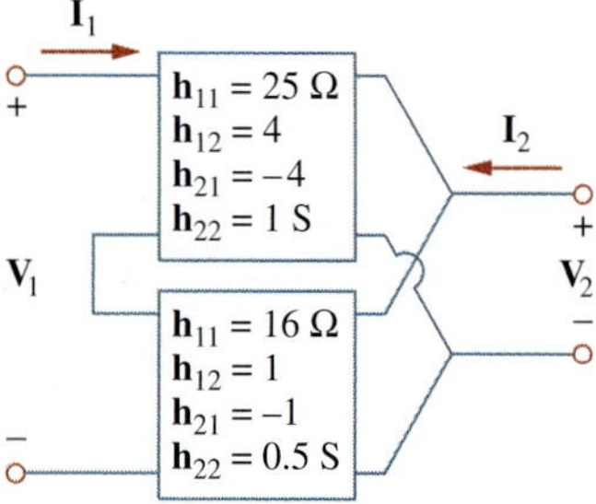

Figure 19.119
For Prob. 19.72.

19.73 Three copies of the circuit shown in Fig. 19.70 are connected in cascade. Determine the z parameters.

***19.74** Determine the **ABCD** parameters of the circuit in Fig. 19.120 as functions of s. (*Hint:* Partition the circuit into subcircuits and cascade them using the results of Prob. 19.43.)

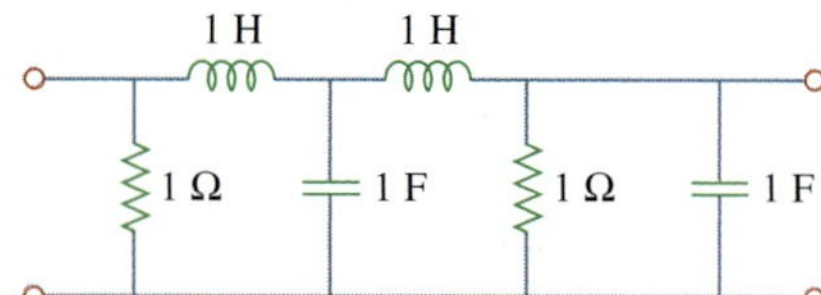

Figure 19.120
For Prob. 19.74.

***19.75** For the individual two-ports shown in Fig. 19.121 where,

$$[\mathbf{z}_a] = \begin{bmatrix} 8 & 6 \\ 4 & 5 \end{bmatrix} \Omega \quad [\mathbf{y}_b] = \begin{bmatrix} 8 & -4 \\ 2 & 10 \end{bmatrix} \text{S}$$

(a) Determine the y parameters of the overall two-port.

(b) Find the voltage ratio $\mathbf{V}_o/\mathbf{V}_i$ when $\mathbf{Z}_L = 2\ \Omega$.

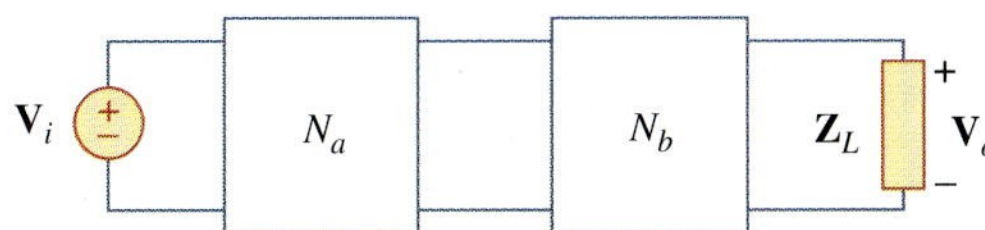

Figure 19.121
For Prob. 19.75.

Section 19.8 Computing Two-Port Parameters Using *PSpice*

19.76 Use *PSpice* to obtain the z parameters of the network in Fig. 19.122.

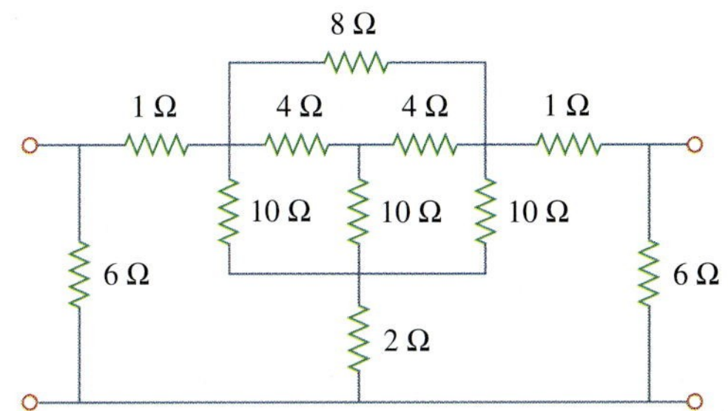

Figure 19.122
For Prob. 19.76.

19.77 Using *PSpice*, find the h parameters of the network in Fig. 19.123. Take $\omega = 1$ rad/s.

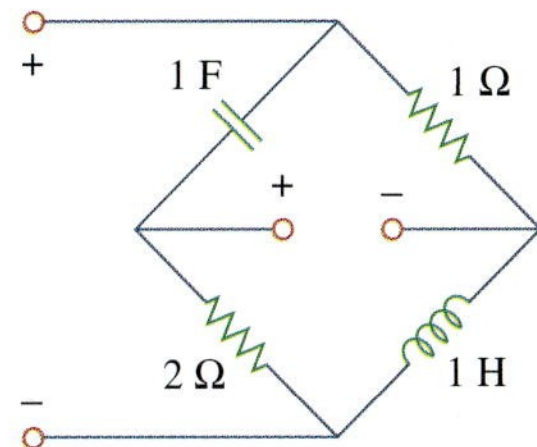

Figure 19.123
For Prob. 19.77.

19.78 Obtain the h parameters at $\omega = 4$ rad/s for the circuit in Fig. 19.124 using *PSpice*.

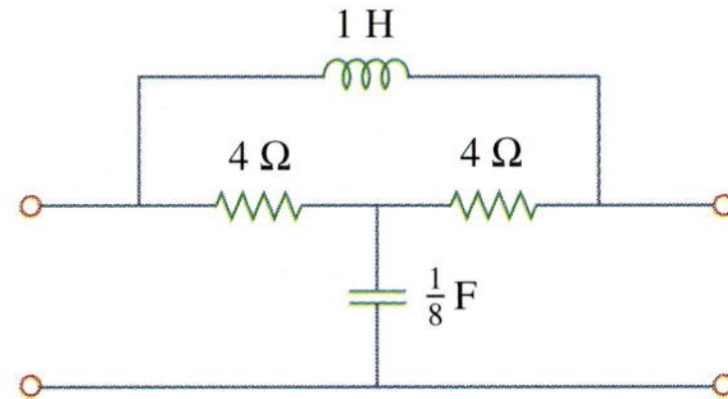

Figure 19.124
For Prob. 19.78.

19.79 Use *PSpice* to determine the z parameters of the circuit in Fig. 19.125. Take $\omega = 2$ rad/s.

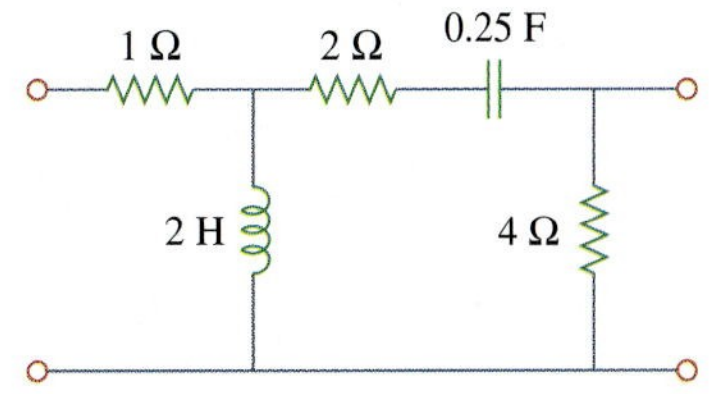

Figure 19.125
For Prob. 19.79.

19.80 Use *PSpice* to find the z parameters of the circuit in Fig. 19.71.

19.81 Repeat Prob. 19.26 using *PSpice*.

19.82 Use *PSpice* to rework Prob. 19.31.

19.83 Rework Prob. 19.47 using *PSpice*.

19.84 Using *PSpice*, find the transmission parameters for the network in Fig. 19.126.

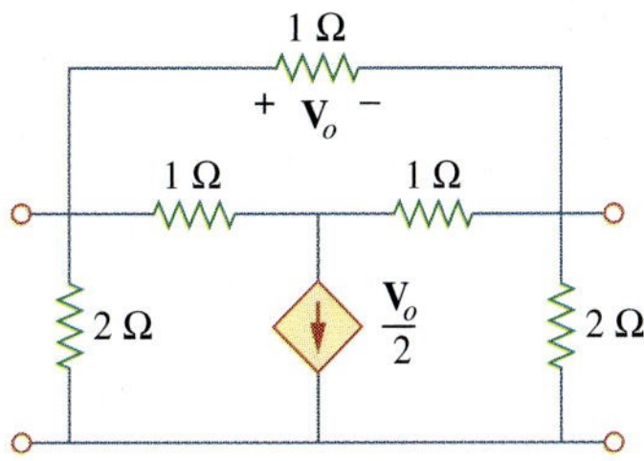

Figure 19.126
For Prob. 19.84.

19.85 At $\omega = 1$ rad/s, find the transmission parameters of the network in Fig. 19.127 using *PSpice*.

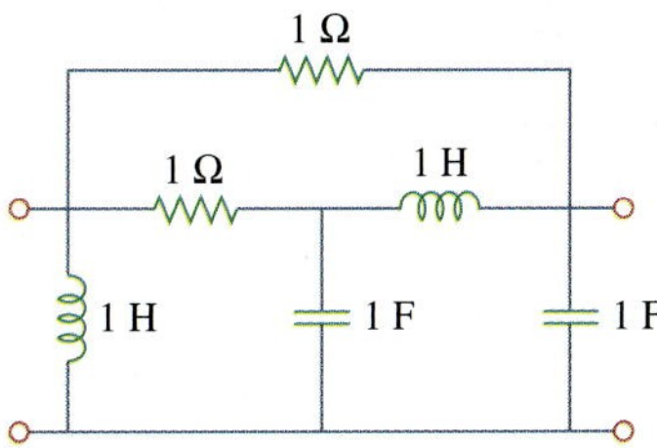

Figure 19.127
For Prob. 19.85.

19.86 Obtain the g parameters for the network in Fig. 19.128 using *PSpice*.

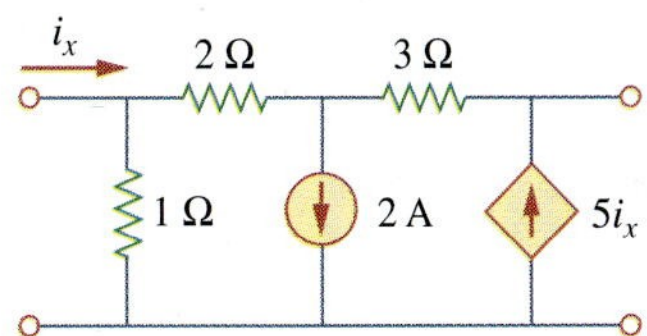

Figure 19.128
For Prob. 19.86.

19.87 For the circuit shown in Fig. 19.129, use *PSpice* to obtain the t parameters. Assume $\omega = 1$ rad/s.

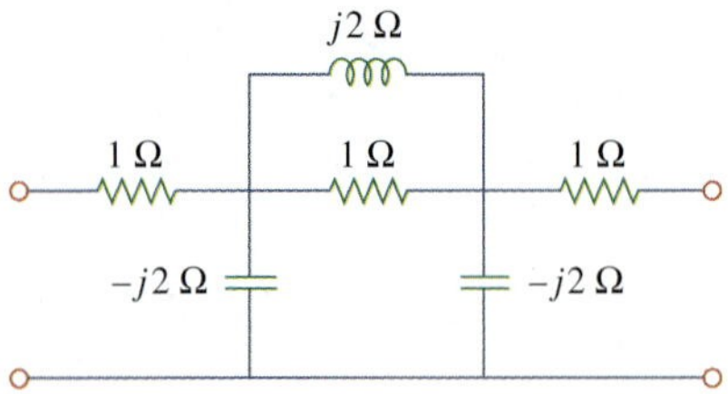

Figure 19.129
For Prob. 19.87.

Section 19.9 Applications

19.88 Using the y parameters, derive formulas for Z_{in}, Z_{out}, A_i, and A_v for the common-emitter transistor circuit.

19.89 A transistor has the following parameters in a common-emitter circuit:

$$h_{ie} = 2{,}640\ \Omega, \qquad h_{re} = 2.6 \times 10^{-4}$$
$$h_{fe} = 72, \qquad h_{oe} = 16\ \mu\text{S}, \qquad R_L = 100\ \text{k}\Omega$$

What is the voltage amplification of the transistor? How many decibels gain is this?

19.90 A transistor with

$$h_{fe} = 120, \qquad h_{ie} = 2\ \text{k}\Omega$$
$$h_{re} = 10^{-4}, \qquad h_{oe} = 20\ \mu\text{S}$$

is used for a CE amplifier to provide an input resistance of 1.5 kΩ.

(a) Determine the necessary load resistance R_L.

(b) Calculate A_v, A_i, and Z_{out} if the amplifier is driven by a 4-mV source having an internal resistance of 600 Ω.

(c) Find the voltage across the load.

19.91 For the transistor network of Fig. 19.130,

$$h_{fe} = 80, \qquad h_{ie} = 1.2\ \text{k}\Omega$$
$$h_{re} = 1.5 \times 10^{-4}, \qquad h_{oe} = 20\ \mu\text{S}$$

Determine the following:

(a) voltage gain $A_v = V_o/V_s$,

(b) current gain $A_i = I_o/I_i$,

(c) input impedance Z_{in},

(d) output impedance Z_{out}.

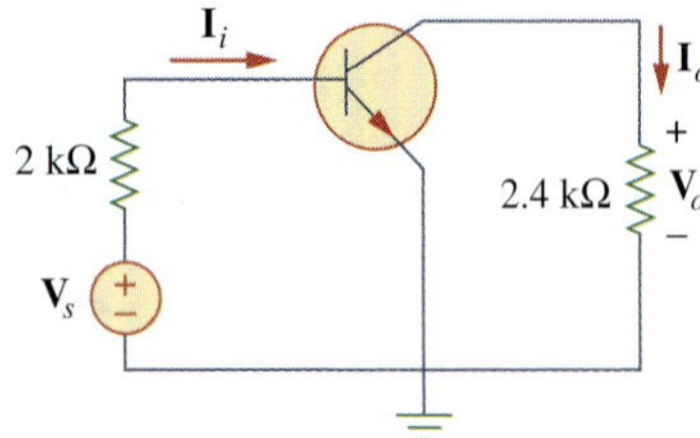

Figure 19.130
For Prob. 19.91.

***19.92** Determine A_v, A_i, Z_{in}, and Z_{out} for the amplifier shown in Fig. 19.131. Assume that

$$h_{ie} = 4\ \text{k}\Omega, \qquad h_{re} = 10^{-4}$$
$$h_{fe} = 100, \qquad h_{oe} = 30\ \mu\text{S}$$

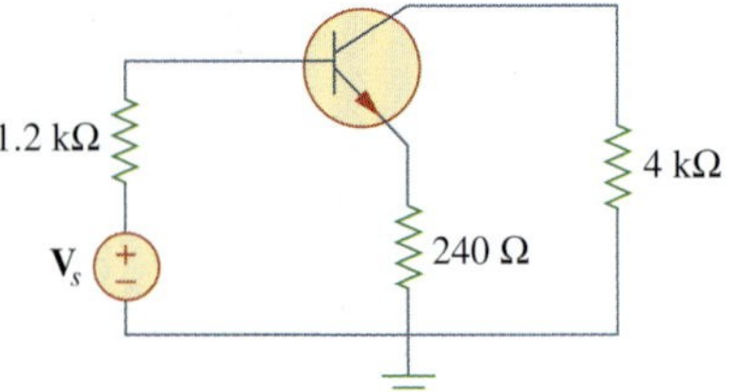

Figure 19.131
For Prob. 19.92.

***19.93** Calculate A_v, A_i, Z_{in}, and Z_{out} for the transistor network in Fig. 19.132. Assume that

$$h_{ie} = 2\ \text{k}\Omega, \qquad h_{re} = 2.5 \times 10^{-4}$$
$$h_{fe} = 150, \qquad h_{oe} = 10\ \mu\text{S}$$

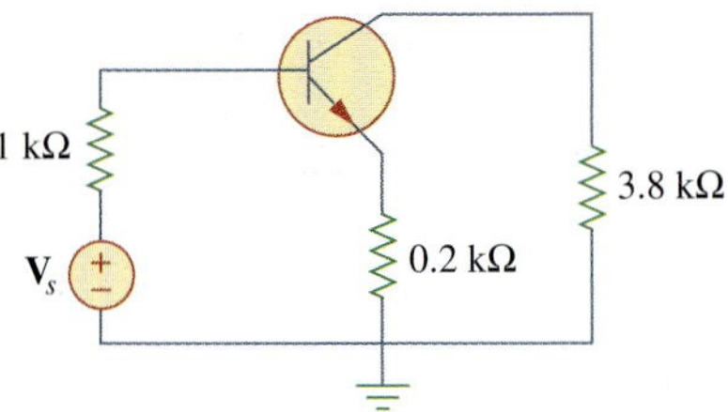

Figure 19.132
For Prob. 19.93.

19.94 A transistor in its common-emitter mode is specified by

$$[\mathbf{h}] = \begin{bmatrix} 200\ \Omega & 0 \\ 100 & 10^{-6}\ \text{S} \end{bmatrix}$$

Two such identical transistors are connected in cascade to form a two-stage amplifier used at audio frequencies. If the amplifier is terminated by a 4-kΩ resistor, calculate the overall A_v and Z_{in}.

19.95 Realize an LC ladder network such that

$$y_{22} = \frac{s^3 + 5s}{s^4 + 10s^2 + 8}$$

19.96 Design an LC ladder network to realize a lowpass filter with transfer function

$$H(s) = \frac{1}{s^4 + 2.613s^2 + 3.414s^2 + 2.613s + 1}$$

19.97 Synthesize the transfer function

$$H(s) = \frac{V_o}{V_s} = \frac{s^3}{s^3 + 6s + 12s + 24}$$

using the LC ladder network in Fig. 19.133.

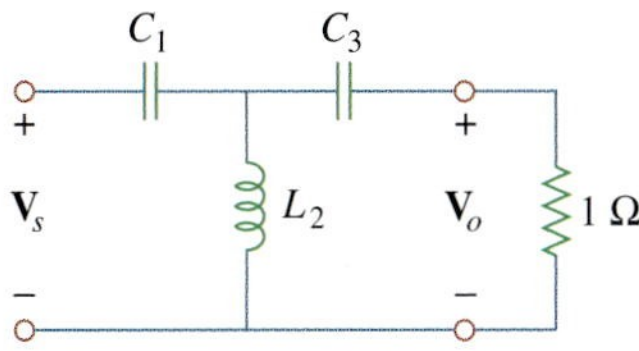

Figure 19.133
For Prob. 19.97.

19.98 A two-stage amplifier in Fig. 19.134 contains two identical stages with

$$[\mathbf{h}] = \begin{bmatrix} 2\text{ k}\Omega & 0.004 \\ 200 & 500\ \mu\text{S} \end{bmatrix}$$

If $\mathbf{Z}_L = 20$ kΩ, find the required value of $\mathbf{V}_s$ to produce $\mathbf{V}_o = 16$ V.

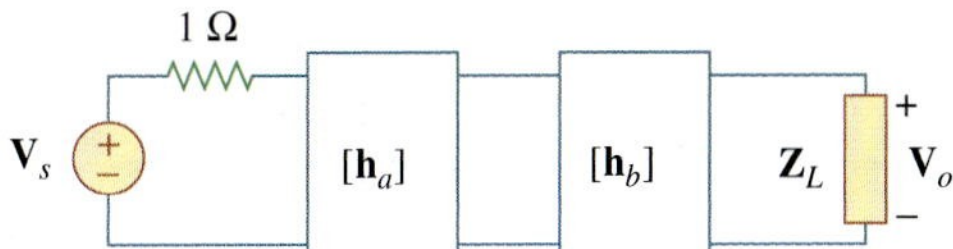

Figure 19.134
For Prob. 19.98.

Comprehensive Problem

19.99 Assume that the two circuits in Fig. 19.135 are equivalent. The parameters of the two circuits must be equal. Using this factor and the z parameters, derive Eqs. (9.67) and (9.68).

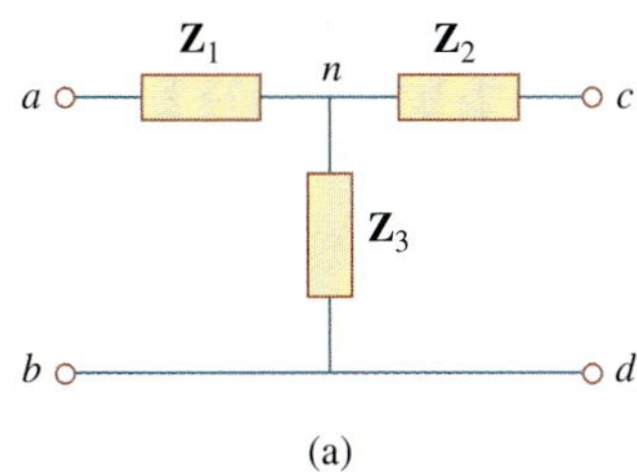

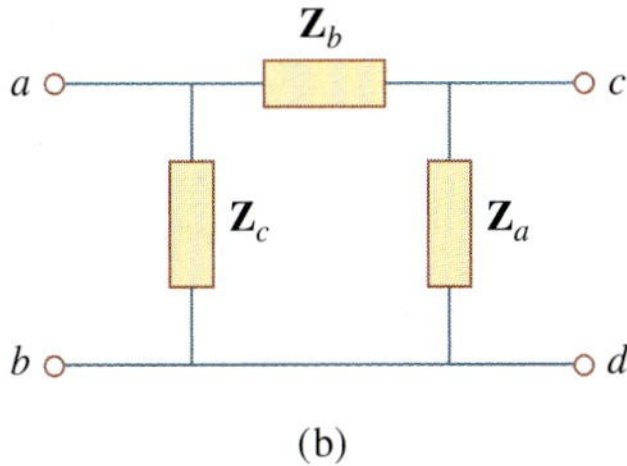

Figure 19.135
For Prob. 19.99.

Appendix A

Simultaneous Equations and Matrix Inversion

In circuit analysis, we often encounter a set of simultaneous equations having the form

$$\begin{aligned} a_{11}x_1 + a_{12}x_2 + \cdots + a_{1n}x_n &= b_1 \\ a_{21}x_1 + a_{22}x_2 + \cdots + a_{2n}x_n &= b_2 \\ \vdots \qquad\quad \vdots \qquad \vdots \\ a_{n1}x_1 + a_{n2}x_2 + \cdots + a_{nn}x_n &= b_n \end{aligned} \tag{A.1}$$

where there are n unknown $x_1, x_2, \ldots, x_n$ to be determined. Equation (A.1) can be written in matrix form as

$$\begin{bmatrix} a_{11} & a_{12} & \ldots & a_{1n} \\ a_{21} & a_{22} & \ldots & a_{2n} \\ \vdots & \vdots & \ldots & \vdots \\ a_{n1} & a_{n2} & \ldots & a_{nn} \end{bmatrix} \begin{bmatrix} x_1 \\ x_2 \\ \vdots \\ x_n \end{bmatrix} = \begin{bmatrix} b_2 \\ b_2 \\ \vdots \\ b_n \end{bmatrix} \tag{A.2}$$

This matrix equation can be put in a compact form as

$$\mathbf{AX} = \mathbf{B} \tag{A.3}$$

where

$$\mathbf{A} = \begin{bmatrix} a_{11} & a_{12} & \ldots & a_{1n} \\ a_{21} & a_{22} & \ldots & a_{2n} \\ \vdots & \vdots & \ldots & \vdots \\ a_{n1} & a_{n2} & \ldots & a_{nn} \end{bmatrix}, \qquad \mathbf{X} = \begin{bmatrix} x_1 \\ x_2 \\ \vdots \\ x_n \end{bmatrix}, \qquad \mathbf{B} = \begin{bmatrix} b_1 \\ b_2 \\ \vdots \\ b_n \end{bmatrix} \tag{A.4}$$

A is a square ($n \times n$) matrix while **X** and **B** are column matrices.

There are several methods for solving Eq. (A.1) or (A.3). These include substitution, Gaussian elimination, Cramer's rule, matrix inversion, and numerical analysis.

A.1 Cramer's Rule

In many cases, Cramer's rule can be used to solve the simultaneous equations we encounter in circuit analysis. Cramer's rule states that the solution to Eq. (A.1) or (A.3) is

$$\boxed{\begin{aligned} x_1 &= \frac{\Delta_1}{\Delta} \\ x_2 &= \frac{\Delta_2}{\Delta} \\ &\vdots \\ x_n &= \frac{\Delta_n}{\Delta} \end{aligned}} \tag{A.5}$$

where the Δ's are the determinants given by

$$\Delta = \begin{vmatrix} a_{11} & a_{12} & \cdots & a_{1n} \\ a_{21} & a_{22} & \cdots & a_{2n} \\ \vdots & \vdots & \cdots & \vdots \\ a_{n1} & a_{n2} & \cdots & a_{nn} \end{vmatrix}, \quad \Delta_1 = \begin{vmatrix} b_1 & a_{12} & \cdots & a_{1n} \\ b_2 & a_{22} & \cdots & a_{2n} \\ \vdots & \vdots & \cdots & \vdots \\ b_n & a_{n2} & \cdots & a_{nn} \end{vmatrix}$$

$$\vdots \qquad\qquad \vdots$$

$$\Delta_2 = \begin{vmatrix} a_{11} & b_1 & \cdots & a_{1n} \\ a_{21} & b_2 & \cdots & a_{2n} \\ \vdots & \vdots & \cdots & \vdots \\ a_{n1} & b_n & \cdots & a_{nn} \end{vmatrix}, \ldots, \Delta_n = \begin{vmatrix} a_{11} & a_{12} & \cdots & b_1 \\ a_{21} & a_{22} & \cdots & b_2 \\ \vdots & \vdots & \cdots & \vdots \\ a_{n1} & a_{n2} & \cdots & b_n \end{vmatrix} \tag{A.6}$$

Notice that Δ is the determinant of matrix **A** and Δ_k is the determinant of the matrix formed by replacing the kth column of **A** by **B**. It is evident from Eq. (A.5) that Cramer's rule applies only when $\Delta \neq 0$. When $\Delta = 0$, the set of equations has no unique solution, because the equations are linearly dependent.

The value of the determinant Δ, for example, can be obtained by expanding along the first row:

$$\Delta = \begin{vmatrix} a_{11} & a_{12} & a_{13} & \cdots & a_{1n} \\ a_{21} & a_{22} & a_{23} & \cdots & a_{2n} \\ a_{31} & a_{32} & a_{33} & \cdots & a_{3n} \\ \vdots & \vdots & \vdots & \cdots & \vdots \\ a_{n1} & a_{n2} & a_{n3} & \cdots & a_{nn} \end{vmatrix} \tag{A.7}$$

$$= a_{11}M_{11} - a_{12}M_{12} + a_{13}M_{13} + \cdots + (-1)^{1+n}a_{1n}M_{1n}$$

where the minor M_{ij} is an $(n-1) \times (n-1)$ determinant of the matrix formed by striking out the ith row and jth column. The value of Δ may also be obtained by expanding along the first column:

$$\Delta = a_{11}M_{11} - a_{21}M_{21} + a_{31}M_{31} + \cdots + (-1)^{n+1}a_{n1}M_{n1} \tag{A.8}$$

We now specifically develop the formulas for calculating the determinants of 2 × 2 and 3 × 3 matrices, because of their frequent occurrence in this text. For a 2 × 2 matrix,

$$\Delta = \begin{vmatrix} a_{11} & a_{12} \\ a_{21} & a_{22} \end{vmatrix} = a_{11}a_{22} - a_{12}a_{21} \tag{A.9}$$

For a 3 × 3 matrix,

$$\Delta = \begin{vmatrix} a_{11} & a_{12} & a_{13} \\ a_{21} & a_{22} & a_{23} \\ a_{31} & a_{32} & a_{33} \end{vmatrix} = a_{11}(-1)^2\begin{vmatrix} a_{22} & a_{23} \\ a_{32} & a_{33} \end{vmatrix} + a_{21}(-1)^3\begin{vmatrix} a_{12} & a_{13} \\ a_{32} & a_{33} \end{vmatrix}$$

$$+ a_{31}(-1)^4\begin{vmatrix} a_{12} & a_{13} \\ a_{22} & a_{23} \end{vmatrix}$$

$$= a_{11}(a_{22}a_{33} - a_{32}a_{23}) - a_{21}(a_{12}a_{33} - a_{32}a_{13})$$

$$+ a_{31}(a_{12}a_{23} - a_{22}a_{13}) \tag{A.10}$$

An alternative method of obtaining the determinant of a 3×3 matrix is by repeating the first two rows and multiplying the terms diagonally as follows.

$$\Delta = \begin{vmatrix} a_{11} & a_{12} & a_{13} \\ a_{21} & a_{22} & a_{23} \\ a_{31} & a_{32} & a_{33} \\ a_{11} & a_{12} & a_{13} \\ a_{21} & a_{22} & a_{23} \end{vmatrix}$$

$$= a_{11}a_{22}a_{33} + a_{21}a_{32}a_{13} + a_{31}a_{12}a_{23} - a_{13}a_{22}a_{31} - a_{23}a_{32}a_{11} - a_{33}a_{12}a_{21} \quad \textbf{(A.11)}$$

In summary:

The solution of linear simultaneous equations by Cramer's rule boils down to finding

$$x_k = \frac{\Delta_k}{\Delta}, \qquad k = 1, 2, \ldots, n \quad \textbf{(A.12)}$$

where Δ is the determinant of matrix **A** and Δ_k is the determinant of the matrix formed by replacing the kth column of **A** by **B**.

One may use other methods, such as matrix inversion and elimination. Only Cramer's method is covered here, because of its simplicity and also because of the availability of powerful calculators.

You may not find much need to use Cramer's method described in this appendix, in view of the availability of calculators, computers, and software packages such as *MATLAB*, which can be used easily to solve a set of linear equations. But in case you need to solve the equations by hand, the material covered in this appendix becomes useful. At any rate, it is important to know the mathematical basis of those calculators and software packages.

Example A.1

Solve the simultaneous equations

$$4x_1 - 3x_2 = 17, \qquad -3x_1 + 5x_2 = -21$$

Solution:
The given set of equations is cast in matrix form as

$$\begin{bmatrix} 4 & -3 \\ -3 & 5 \end{bmatrix} \begin{bmatrix} x_1 \\ x_2 \end{bmatrix} = \begin{bmatrix} 17 \\ -21 \end{bmatrix}$$

The determinants are evaluated as

$$\Delta = \begin{vmatrix} 4 & -3 \\ -3 & 5 \end{vmatrix} = 4 \times 5 - (-3)(-3) = 11$$

$$\Delta_1 = \begin{vmatrix} 17 & -3 \\ -21 & 5 \end{vmatrix} = 17 \times 5 - (-3)(-21) = 22$$

$$\Delta_2 = \begin{vmatrix} 4 & 17 \\ -3 & -21 \end{vmatrix} = 4 \times (-21) - 17 \times (-3) = -33$$

Hence,

$$x_1 = \frac{\Delta_1}{\Delta} = \frac{22}{11} = 2, \qquad x_2 = \frac{\Delta_2}{\Delta} = \frac{-33}{11} = -3$$

Practice Problem A.1

Find the solution to the following simultaneous equations:

$$3x_1 - x_2 = 4, \qquad -6x_1 + 18x_2 = 16$$

Answer: $x_1 = 1.833, x_2 = 1.5$.

Example A.2

Determine x_1, x_2, and x_3 for this set of simultaneous equations:

$$\begin{aligned} 25x_1 - 5x_2 - 20x_3 &= 50 \\ -5x_1 + 10x_2 - 4x_3 &= 0 \\ -5x_1 - 4x_2 + 9x_3 &= 0 \end{aligned}$$

Solution:
In matrix form, the given set of equations becomes

$$\begin{bmatrix} 25 & -5 & -20 \\ -5 & 10 & -4 \\ -5 & -4 & 9 \end{bmatrix} \begin{bmatrix} x_1 \\ x_2 \\ x_3 \end{bmatrix} = \begin{bmatrix} 50 \\ 0 \\ 0 \end{bmatrix}$$

We apply Eq. (A.11) to find the determinants. This requires that we repeat the first two rows of the matrix. Thus,

$$\Delta = \begin{vmatrix} 25 & -5 & -20 \\ -5 & 10 & -4 \\ -5 & -4 & 9 \end{vmatrix} = \begin{vmatrix} 25 & -5 & -20 \\ -5 & 10 & -4 \\ -5 & -4 & 9 \\ 25 & -5 & -20 \\ -5 & 10 & -4 \end{vmatrix}$$

$$\begin{aligned} &= 25(10)9 + (-5)(-4)(-20) + (-5)(-5)(-4) \\ &\quad - (-20)(10)(-5) - (-4)(-4)25 - 9(-5)(-5) \\ &= 2250 - 400 - 100 - 1000 - 400 - 225 = 125 \end{aligned}$$

Similarly,

$$\Delta_1 = \begin{vmatrix} 50 & -5 & -20 \\ 0 & 10 & -4 \\ 0 & -4 & 9 \end{vmatrix} = \begin{vmatrix} 50 & -5 & -20 \\ 0 & 10 & -4 \\ 0 & -4 & 9 \\ 50 & -5 & -20 \\ 0 & 10 & -4 \end{vmatrix}$$

$$= 4500 + 0 + 0 - 0 - 800 - 0 = 3700$$

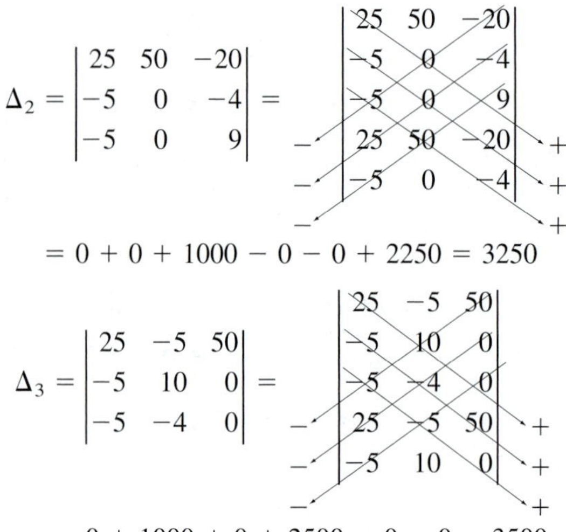

$$\Delta_2 = \begin{vmatrix} 25 & 50 & -20 \\ -5 & 0 & -4 \\ -5 & 0 & 9 \end{vmatrix} = \begin{vmatrix} 25 & 50 & -20 \\ -5 & 0 & -4 \\ -5 & 0 & 9 \\ 25 & 50 & -20 \\ -5 & 0 & -4 \end{vmatrix}$$

$$= 0 + 0 + 1000 - 0 - 0 + 2250 = 3250$$

$$\Delta_3 = \begin{vmatrix} 25 & -5 & 50 \\ -5 & 10 & 0 \\ -5 & -4 & 0 \end{vmatrix} = \begin{vmatrix} 25 & -5 & 50 \\ -5 & 10 & 0 \\ -5 & -4 & 0 \\ 25 & -5 & 50 \\ -5 & 10 & 0 \end{vmatrix}$$

$$= 0 + 1000 + 0 + 2500 - 0 - 0 = 3500$$

Hence, we now find

$$x_1 = \frac{\Delta_1}{\Delta} = \frac{3700}{125} = 29.6$$

$$x_2 = \frac{\Delta_2}{\Delta} = \frac{3250}{125} = 26$$

$$x_3 = \frac{\Delta_2}{\Delta} = \frac{3500}{125} = 28$$

Practice Problem A.2

Obtain the solution of this set of simultaneous equations:

$$\begin{aligned} 3x_1 - x_2 - 2x_3 &= 1 \\ -x_1 + 6x_2 - 3x_3 &= 0 \\ -2x_1 - 3x_2 + 6x_3 &= 6 \end{aligned}$$

Answer: $x_1 = 3 = x_3, x_2 = 2$.

A.2 Matrix Inversion

The linear system of equations in Eq. (A.3) can be solved by matrix inversion. In the matrix equation $\mathbf{AX} = \mathbf{B}$, we may invert $\mathbf{A}$ to get $\mathbf{X}$, i.e.,

$$\mathbf{X} = \mathbf{A}^{-1}\mathbf{B} \tag{A.13}$$

where $\mathbf{A}^{-1}$ is the inverse of $\mathbf{A}$. Matrix inversion is needed in other applications apart from using it to solve a set of equations.

By definition, the inverse of matrix $\mathbf{A}$ satisfies

$$\mathbf{A}^{-1}\mathbf{A} = \mathbf{A}\mathbf{A}^{-1} = \mathbf{I} \tag{A.14}$$

where **I** is an identity matrix. $\mathbf{A}^{-1}$ is given by

$$\mathbf{A}^{-1} = \frac{\text{adj } \mathbf{A}}{\det \mathbf{A}} \tag{A.15}$$

where adj **A** is the adjoint of **A** and det **A** = |**A**| is the determinant of **A**. The adjoint of **A** is the transpose of the cofactors of **A**. Suppose we are given an $n \times n$ matrix **A** as

$$\mathbf{A} = \begin{bmatrix} a_{11} & a_{12} & \cdots & a_{1n} \\ a_{21} & a_{22} & \cdots & a_{2n} \\ \vdots & & & \\ a_{n1} & a_{n2} & \cdots & a_{nn} \end{bmatrix} \tag{A.16}$$

The cofactors of **A** are defined as

$$\mathbf{C} = \text{cof}\,(\mathbf{A}) = \begin{bmatrix} c_{11} & c_{12} & \cdots & c_{1n} \\ c_{21} & c_{22} & \cdots & c_{2n} \\ \vdots & & & \\ c_{n1} & c_{n2} & \cdots & c_{nn} \end{bmatrix} \tag{A.17}$$

where the cofactor c_{ij} is the product of $(-1)^{i+j}$ and the determinant of the $(n-1) \times (n-1)$ submatrix is obtained by deleting the ith row and jth column from **A**. For example, by deleting the first row and the first column of **A** in Eq. (A.16), we obtain the cofactor c_{11} as

$$c_{11} = (-1)^2 = \begin{vmatrix} a_{22} & a_{23} & \cdots & a_{2n} \\ a_{32} & a_{33} & \cdots & a_{3n} \\ \vdots & & & \\ a_{n2} & a_{n3} & \cdots & a_{nn} \end{vmatrix} \tag{A.18}$$

Once the cofactors are found, the adjoint of **A** is obtained as

$$\text{adj}\,(\mathbf{A}) = \begin{bmatrix} c_{11} & c_{12} & \cdots & c_{1n} \\ c_{21} & c_{22} & \cdots & c_{2n} \\ \vdots & & & \\ c_{n1} & c_{n2} & \cdots & c_{nn} \end{bmatrix}^T = \mathbf{C}^T \tag{A.19}$$

where T denotes transpose.

In addition to using the cofactors to find the adjoint of **A**, they are also used in finding the determinant of **A** which is given by

$$|\mathbf{A}| = \sum_{j=1}^{n} a_{ij}c_{ij} \tag{A.20}$$

where i is any value from 1 to n. By substituting Eqs. (A.19) and (A.20) into Eq. (A.15), we obtain the inverse of **A** as

$$\boxed{\mathbf{A}^{-1} = \frac{\mathbf{C}^T}{|\mathbf{A}|}} \tag{A.21}$$

For a 2×2 matrix, if

$$\mathbf{A} = \begin{bmatrix} a & b \\ c & d \end{bmatrix} \tag{A.22}$$

its inverse is

$$\mathbf{A}^{-1} = \frac{1}{|\mathbf{A}|}\begin{bmatrix} d & -b \\ -c & a \end{bmatrix} = \frac{1}{ad - bc}\begin{bmatrix} d & -b \\ -c & a \end{bmatrix} \quad \textbf{(A.23)}$$

For a 3 × 3 matrix, if

$$\mathbf{A} = \begin{bmatrix} a_{11} & a_{12} & a_{13} \\ a_{21} & a_{22} & a_{23} \\ a_{31} & a_{32} & a_{33} \end{bmatrix} \quad \textbf{(A.24)}$$

we first obtain the cofactors as

$$\mathbf{C} = \begin{bmatrix} c_{11} & c_{12} & c_{13} \\ c_{21} & c_{22} & c_{23} \\ c_{31} & c_{32} & c_{33} \end{bmatrix} \quad \textbf{(A.25)}$$

where

$$\begin{aligned} c_{11} &= \begin{vmatrix} a_{22} & a_{23} \\ a_{32} & a_{33} \end{vmatrix}, & c_{12} &= -\begin{vmatrix} a_{21} & a_{23} \\ a_{31} & a_{33} \end{vmatrix}, & c_{13} &= \begin{vmatrix} a_{21} & a_{22} \\ a_{31} & a_{32} \end{vmatrix}, \\ c_{21} &= -\begin{vmatrix} a_{12} & a_{13} \\ a_{32} & a_{33} \end{vmatrix}, & c_{22} &= \begin{vmatrix} a_{11} & a_{13} \\ a_{31} & a_{33} \end{vmatrix}, & c_{23} &= -\begin{vmatrix} a_{11} & a_{12} \\ a_{31} & a_{32} \end{vmatrix}, \\ c_{31} &= \begin{vmatrix} a_{12} & a_{13} \\ a_{22} & a_{23} \end{vmatrix}, & c_{32} &= -\begin{vmatrix} a_{11} & a_{13} \\ a_{21} & a_{23} \end{vmatrix}, & c_{33} &= \begin{vmatrix} a_{11} & a_{12} \\ a_{21} & a_{22} \end{vmatrix} \end{aligned} \quad \textbf{(A.26)}$$

The determinant of the 3 × 3 matrix can be found using Eq. (A.11). Here, we want to use Eq. (A.20), i.e.,

$$|\mathbf{A}| = a_{11}c_{11} + a_{12}c_{12} + a_{13}c_{13} \quad \textbf{(A.27)}$$

The idea can be extended $n > 3$, but we deal mainly with 2 × 2 and 3 × 3 matrices in this book.

Example A.3

Use matrix inversion to solve the simultaneous equations

$$2x_1 + 10x_2 = 2, \qquad -x_1 + 3x_2 = 7$$

Solution:

We first express the two equations in matrix form as

$$\begin{bmatrix} 2 & 10 \\ -1 & 3 \end{bmatrix}\begin{bmatrix} x_1 \\ x_2 \end{bmatrix} = \begin{bmatrix} 2 \\ 7 \end{bmatrix}$$

or

$$\mathbf{AX} = \mathbf{B} \longrightarrow \mathbf{X} = \mathbf{A}^{-1}\mathbf{B}$$

where

$$\mathbf{A} = \begin{bmatrix} 2 & 10 \\ -1 & 3 \end{bmatrix}, \qquad \mathbf{X} = \begin{bmatrix} x_1 \\ x_2 \end{bmatrix}, \qquad \mathbf{B} = \begin{bmatrix} 2 \\ 7 \end{bmatrix}$$

The determinant of $\mathbf{A}$ is $|\mathbf{A}| = 2 \times 3 - 10(-1) = 16$, so the inverse of $\mathbf{A}$ is

$$\mathbf{A}^{-1} = \frac{1}{16}\begin{bmatrix} 3 & -10 \\ 1 & 2 \end{bmatrix}$$

Hence,

$$\mathbf{X} = \mathbf{A}^{-1}\mathbf{B} = \frac{1}{16}\begin{bmatrix} 3 & -10 \\ 1 & 2 \end{bmatrix}\begin{bmatrix} 2 \\ 7 \end{bmatrix} = \frac{1}{16}\begin{bmatrix} -64 \\ 16 \end{bmatrix} = \begin{bmatrix} -4 \\ 1 \end{bmatrix}$$

i.e., $x_1 = -4$ and $x_2 = 1$.

Practice Problem A.3

Solve the following two equations by matrix inversion.

$$2y_1 - y_2 = 4, \quad y_1 + 3y_2 = 9$$

Answer: $y_1 = 3, y_2 = 2$.

Example A.4

Determine x_1, x_2, and x_3 for the following simultaneous equations using matrix inversion.

$$\begin{aligned} x_1 + x_2 + x_3 &= 5 \\ -x_1 + 2x_2 &= 9 \\ 4x_1 + x_2 - x_3 &= -2 \end{aligned}$$

Solution:
In matrix form, the equations become

$$\begin{bmatrix} 1 & 1 & 1 \\ -1 & 2 & 0 \\ 4 & 1 & -1 \end{bmatrix}\begin{bmatrix} x_1 \\ x_2 \\ x_3 \end{bmatrix} = \begin{bmatrix} 5 \\ 9 \\ -2 \end{bmatrix}$$

or

$$\mathbf{AX} = \mathbf{B} \longrightarrow \mathbf{X} = \mathbf{A}^{-1}\mathbf{B}$$

where

$$\mathbf{A} = \begin{bmatrix} 1 & 1 & 1 \\ -1 & 2 & 0 \\ 4 & 1 & -1 \end{bmatrix}, \quad \mathbf{X} = \begin{bmatrix} x_1 \\ x_2 \\ x_3 \end{bmatrix}, \quad \mathbf{B} = \begin{bmatrix} 5 \\ 9 \\ -2 \end{bmatrix}$$

We now find the cofactors

$$c_{11} = \begin{vmatrix} 2 & 0 \\ 1 & -1 \end{vmatrix} = -2, \; c_{12} = -\begin{vmatrix} -1 & 0 \\ 4 & -1 \end{vmatrix} = -1, \; c_{13} = \begin{vmatrix} -1 & 2 \\ 4 & 1 \end{vmatrix} = -9$$

$$c_{21} = -\begin{vmatrix} 1 & 1 \\ 1 & -1 \end{vmatrix} = 2, \; c_{22} = \begin{vmatrix} 1 & 1 \\ 4 & -1 \end{vmatrix} = -5, \quad c_{23} = -\begin{vmatrix} 1 & 1 \\ 4 & 1 \end{vmatrix} = 3$$

$$c_{31} = \begin{vmatrix} 1 & 1 \\ 2 & 0 \end{vmatrix} = -2, \quad c_{32} = -\begin{vmatrix} 1 & 1 \\ -1 & 0 \end{vmatrix} = -1, \quad c_{33} = \begin{vmatrix} 1 & 1 \\ -1 & 2 \end{vmatrix} = 3$$

The adjoint of matrix **A** is

$$\text{adj}\,\mathbf{A} = \begin{bmatrix} -2 & -1 & -9 \\ 2 & -5 & 3 \\ -2 & -1 & 3 \end{bmatrix}^T = \begin{bmatrix} -2 & 2 & -2 \\ -1 & -5 & -1 \\ -9 & 3 & 3 \end{bmatrix}$$

We can find the determinant of **A** using any row or column of **A**. Since one element of the second row is 0, we can take advantage of this to find the determinant as

$$|\mathbf{A}| = -1c_{21} + 2c_{22} + (0)c_{23} = -1(2) + 2(-5) = -12$$

Hence, the inverse of **A** is

$$\mathbf{A}^{-1} = \frac{1}{-12}\begin{bmatrix} -2 & 2 & -2 \\ -1 & -5 & -1 \\ -9 & 3 & 3 \end{bmatrix}$$

$$\mathbf{X} = \mathbf{A}^{-1}\mathbf{B} = \frac{1}{-12}\begin{bmatrix} -2 & 2 & -2 \\ -1 & -5 & -1 \\ -9 & 3 & 3 \end{bmatrix}\begin{bmatrix} 5 \\ 9 \\ -2 \end{bmatrix} = \begin{bmatrix} -1 \\ 4 \\ 2 \end{bmatrix}$$

i.e., $x_1 = -1, x_2 = 4, x_3 = 2$.

Practice Problem A.4

Solve the following equations using matrix inversion.

$$\begin{aligned} y_1 - y_3 &= 1 \\ 2y_1 + 3y_2 - y_3 &= 1 \\ y_1 - y_2 - y_3 &= 3 \end{aligned}$$

Answer: $y_1 = 6, y_2 = -2, y_3 = 5$.

Appendix B

Complex Numbers

The ability to manipulate complex numbers is very handy in circuit analysis and in electrical engineering in general. Complex numbers are particularly useful in the analysis of ac circuits. Again, although calculators and computer software packages are now available to manipulate complex numbers, it is still advisable for a student to be familiar with how to handle them by hand.

B.1 Representations of Complex Numbers

A complex number z may be written in *rectangular form* as

$$z = x + jy \tag{B.1}$$

where $j = \sqrt{-1}$; x is the *real part* of z while y is the *imaginary part* of z; that is,

$$x = \text{Re}(z), \qquad y = \text{Im}(z) \tag{B.2}$$

The complex number z is shown plotted in the complex plane in Fig. B.1. Since $j = \sqrt{-1}$,

$$\begin{aligned} \frac{1}{j} &= -j \\ j^2 &= -1 \\ j^3 &= j \cdot j^2 = -j \\ j^4 &= j^2 \cdot j^2 = 1 \\ j^5 &= j \cdot j^4 = j \\ &\vdots \\ j^{n+4} &= j^n \end{aligned} \tag{B.3}$$

The complex plane looks like the two-dimensional curvilinear coordinate space, but it is not.

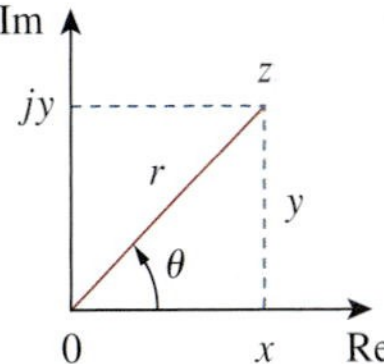

Figure B.1
Graphical representation of a complex number.

A second way of representing the complex number z is by specifying its magnitude r and the angle θ it makes with the real axis, as Fig. B.1 shows. This is known as the *polar form*. It is given by

$$z = |z|\underline{/\theta} = r\underline{/\theta} \tag{B.4}$$

where

$$r = \sqrt{x^2 + y^2}, \qquad \theta = \tan^{-1}\frac{y}{x} \tag{B.5a}$$

or

$$x = r\cos\theta, \qquad y = r\sin\theta \tag{B.5b}$$

that is,

$$z = x + jy = r\underline{/\theta} = r\cos\theta + jr\sin\theta \tag{B.6}$$

In converting from rectangular to polar form using Eq. (B.5), we must exercise care in determining the correct value of θ. These are the four possibilities:

$$\begin{aligned} z &= x + jy, & \theta &= \tan^{-1}\frac{y}{x} & &\text{(1st Quadrant)} \\ z &= -x + jy, & \theta &= 180^\circ - \tan^{-1}\frac{y}{x} & &\text{(2nd Quadrant)} \\ z &= -x - jy, & \theta &= 180^\circ + \tan^{-1}\frac{y}{x} & &\text{(3rd Quadrant)} \\ z &= x - jy, & \theta &= 360^\circ - \tan^{-1}\frac{y}{x} & &\text{(4th Quadrant)} \end{aligned} \quad \textbf{(B.7)}$$

assuming that x and y are positive.

In the exponential form, $z = re^{j\theta}$ so that $dz/d\theta = jre^{j\theta} = jz$.

The third way of representing the complex z is the *exponential form*:

$$z = re^{j\theta} \quad \textbf{(B.8)}$$

This is almost the same as the polar form, because we use the same magnitude r and the angle θ.

The three forms of representing a complex number are summarized as follows.

$$\begin{array}{lll} z = x + jy, & (x = r\cos\theta,\ y = r\sin\theta) & \text{Rectangular form} \\ z = r\underline{/\theta}, & \left(r = \sqrt{x^2 + y^2},\ \theta = \tan^{-1}\dfrac{y}{x}\right) & \text{Polar form} \\ z = re^{j\theta}, & \left(r = \sqrt{x^2 + y^2},\ \theta = \tan^{-1}\dfrac{y}{x}\right) & \text{Exponential form} \end{array} \quad \textbf{(B.9)}$$

The first two forms are related by Eqs. (B.5) and (B.6). In Section B.3 we will derive Euler's formula, which proves that the third form is also equivalent to the first two.

Example B.1

Express the following complex numbers in polar and exponential form: (a) $z_1 = 6 + j8$, (b) $z_2 = 6 - j8$, (c) $z_3 = -6 + j8$, (d) $z_4 = -6 - j8$.

Solution:

Notice that we have deliberately chosen these complex numbers to fall in the four quadrants, as shown in Fig. B.2.

(a) For $z_1 = 6 + j8$ (1st quadrant),

$$r_1 = \sqrt{6^2 + 8^2} = 10, \qquad \theta_1 = \tan^{-1}\frac{8}{6} = 53.13^\circ$$

Hence, the polar form is $10\underline{/53.13^\circ}$ and the exponential form is $10e^{j53.13^\circ}$.

(b) For $z_2 = 6 - j8$ (4th quadrant),

$$r_2 = \sqrt{6^2 + (-8)^2} = 10, \qquad \theta_2 = 360^\circ - \tan^{-1}\frac{8}{6} = 306.87^\circ$$

so that the polar form is $10\underline{/306.87^\circ}$ and the exponential form is $10e^{j306.87^\circ}$. The angle θ_2 may also be taken as -53.13°, as shown in Fig. B.2, so that the polar form becomes $10\underline{/-53.13^\circ}$ and the exponential form becomes $10e^{-j53.13^\circ}$.

(c) For $z_3 = -6 + j8$ (2nd quadrant),

$$r_3 = \sqrt{(-6)^2 + 8^2} = 10, \qquad \theta_3 = 180^\circ - \tan^{-1}\frac{8}{6} = 126.87^\circ$$

Hence, the polar form is $10\underline{/126.87^\circ}$ and the exponential form is $10e^{j126.87^\circ}$.

(d) For $z_4 = -6 - j8$ (3rd quadrant),

$$r_4 = \sqrt{(-6)^2 + (-8)^2} = 10, \qquad \theta_4 = 180^\circ + \tan^{-1}\frac{8}{6} = 233.13^\circ$$

so that the polar form is $10\underline{/233.13^\circ}$ and the exponential form is $10e^{j233.13^\circ}$.

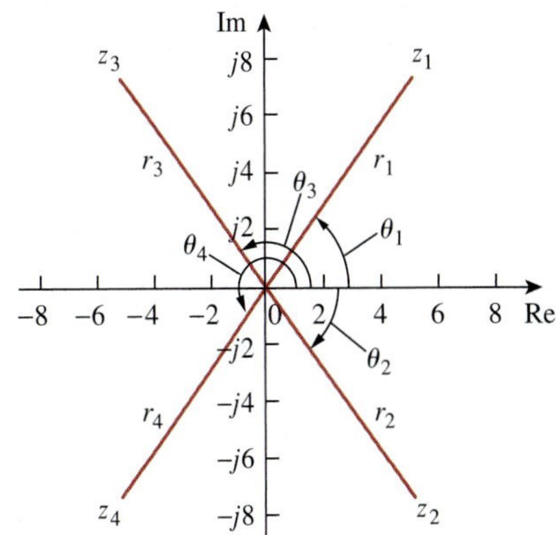

Figure B.2
For Example B.1.

Practice Problem B.1

Convert the following complex numbers to polar and exponential forms: (a) $z_1 = 3 - j4$, (b) $z_2 = 5 + j12$, (c) $z_3 = -3 - j9$, (d) $z_4 = -7 + j$.

Answer: (a) $5\underline{/306.9^\circ}$, $5e^{j306.9^\circ}$, (b) $13\underline{/67.38^\circ}$, $13e^{j67.38^\circ}$, (c) $9.487\underline{/251.6^\circ}$, $9.487e^{j251.6^\circ}$, (d) $7.071\underline{/171.9^\circ}$, $7.071e^{j171.9^\circ}$.

Example B.2

Convert the following complex numbers into rectangular form: (a) $12\underline{/-60^\circ}$, (b) $-50\underline{/285^\circ}$, (c) $8e^{j10^\circ}$, (d) $20e^{-j\pi/3}$.

Solution:

(a) Using Eq. (B.6),

$$12\underline{/-60^\circ} = 12\cos(-60^\circ) + j12\sin(-60^\circ) = 6 - j10.39$$

Note that $\theta = -60^\circ$ is the same as $\theta = 360^\circ - 60^\circ = 300^\circ$.

(b) We can write

$$-50\underline{/285^\circ} = -50\cos 285^\circ - j50\sin 285^\circ = -12.94 + j48.3$$

(c) Similarly,

$$8e^{j10^\circ} = 8\cos 10^\circ + j8\sin 10^\circ = 7.878 + j1.389$$

(d) Finally,

$$20e^{-j\pi/3} = 20\cos(-\pi/3) + j20\sin(-\pi/3) = 10 - j17.32$$

Practice Problem B.2

Find the rectangular form of the following complex numbers: (a) $-8\underline{/210^\circ}$, (b) $40\underline{/305^\circ}$, (c) $10e^{-j30^\circ}$, (d) $50e^{j\pi/2}$.

Answer: (a) $6.928 + j4$, (b) $22.94 - j32.77$, (c) $8.66 - j5$, (d) $j50$.

B.2 Mathematical Operations

We have used lightface notation for complex numbers—since they are not time- or frequency-dependent—whereas we use boldface notation for phasors.

Two complex numbers $z_1 = x_1 + jy_1$ and $z_2 = x_2 + jy_2$ are equal if and only if their real parts are equal and their imaginary parts are equal,

$$x_1 = x_2, \qquad y_1 = y_2 \tag{B.10}$$

The *complex conjugate* of the complex number $z = x + jy$ is

$$z^* = x - jy = r\underline{/-\theta} = re^{-j\theta} \tag{B.11}$$

Thus, the complex conjugate of a complex number is found by replacing every j by $-j$.

Given two complex numbers $z_1 = x_1 + jy_1 = r_1\underline{/\theta_1}$ and $z_2 = x_2 + jy_2 = r_2\underline{/\theta_2}$, their sum is

$$z_1 + z_2 = (x_1 + x_2) + j(y_1 + y_2) \tag{B.12}$$

and their difference is

$$z_1 - z_2 = (x_1 - x_2) + j(y_1 - y_2) \tag{B.13}$$

While it is more convenient to perform addition and subtraction of complex numbers in rectangular form, the product and quotient of the two complex numbers are best done in polar or exponential form. For their product,

$$z_1 z_2 = r_1 r_2\underline{/\theta_1 + \theta_2} \tag{B.14}$$

Alternatively, using the rectangular form,

$$\begin{aligned} z_1 z_2 &= (x_1 + jy_1)(x_2 + jy_2) \\ &= (x_1x_2 - y_1y_2) + j(x_1y_2 + x_2y_1) \end{aligned} \tag{B.15}$$

For their quotient,

$$\frac{z_1}{z_2} = \frac{r_1}{r_2}\underline{/\theta_1 - \theta_2} \tag{B.16}$$

Alternatively, using the rectangular form,

$$\frac{z_1}{z_2} = \frac{x_1 + jy_1}{x_2 + jy_2} \tag{B.17}$$

We rationalize the denominator by multiplying both the numerator and denominator by z_2^*.

$$\frac{z_1}{z_2} = \frac{(x_1 + jy_1)(x_2 - jy_2)}{(x_2 + jy_2)(x_2 - jy_2)} = \frac{x_1x_2 + y_1y_2}{x_2^2 + y_2^2} + j\frac{x_2y_1 - x_1y_2}{x_2^2 + y_2^2} \tag{B.18}$$

Example B.3

If $A = 2 + j5$, $B = 4 - j6$, find: (a) $A^*(A + B)$, (b) $(A + B)/(A - B)$.

Solution:

(a) If $A = 2 + j5$, then $A^* = 2 - j5$ and

$$A + B = (2 + 4) + j(5 - 6) = 6 - j$$

so that

$$A^*(A + B) = (2 - j5)(6 - j) = 12 - j2 - j30 - 5 = 7 - j32$$

(b) Similarly,

$$A - B = (2 - 4) + j(5 - -6) = -2 + j11$$

Hence,

$$\frac{A + B}{A - B} = \frac{6 - j}{-2 + j11} = \frac{(6 - j)(-2 - j11)}{(-2 + j11)(-2 - j11)}$$

$$= \frac{-12 - j66 + j2 - 11}{(-2)^2 + 11^2} = \frac{-23 - j64}{125} = -0.184 - j0.512$$

Practice Problem B.3

Given that $C = -3 + j7$ and $D = 8 + j$, calculate:
(a) $(C - D^*)(C + D^*)$, (b) D^2/C^*, (c) $2CD/(C + D)$.

Answer: (a) $-103 - j26$, (b) $-5.19 + j6.776$, (c) $6.045 + j11.53$.

Example B.4

Evaluate:

(a) $\dfrac{(2 + j5)(8e^{j10^\circ})}{2 + j4 + 2\underline{/-40^\circ}}$ (b) $\dfrac{j(3 - j4)^*}{(-1 + j6)(2 + j)^2}$

Solution:

(a) Since there are terms in polar and exponential forms, it may be best to express all terms in polar form:

$$2 + j5 = \sqrt{2^2 + 5^2}\underline{/\tan^{-1} 5/2} = 5.385\underline{/68.2^\circ}$$

$$(2 + j5)(8e^{j10^\circ}) = (5.385\underline{/68.2^\circ})(8\underline{/10^\circ}) = 43.08\underline{/78.2^\circ}$$

$$2 + j4 + 2\underline{/-40^\circ} = 2 + j4 + 2\cos(-40^\circ) + j2\sin(-40^\circ)$$

$$= 3.532 + j2.714 = 4.454\underline{/37.54^\circ}$$

Thus,

$$\frac{(2 + j5)(8e^{j10^\circ})}{2 + j4 + 2\underline{/-40^\circ}} = \frac{43.08\underline{/78.2^\circ}}{4.454\underline{/37.54^\circ}} = 9.672\underline{/40.66^\circ}$$

(b) We can evaluate this in rectangular form, since all terms are in that form. But

$$j(3 - j4)^* = j(3 + j4) = -4 + j3$$

$$(2 + j)^2 = 4 + j4 - 1 = 3 + j4$$

$$(-1 + j6)(2 + j)^2 = (-1 + j6)(3 + j4) = -3 - 4j + j18 - 24$$

$$= -27 + j14$$

Hence,

$$\frac{j(3 - j4)^*}{(-1 + j6)(2 + j)^2} = \frac{-4 + j3}{-27 + j14} = \frac{(-4 + j3)(-27 - j14)}{27^2 + 14^2}$$

$$= \frac{108 + j56 - j81 + 42}{925} = 0.1622 - j0.027$$

Practice Problem B.4

Evaluate these complex fractions:

(a) $\dfrac{6\angle 30^\circ + j5 - 3}{-1 + j + 2e^{j45^\circ}}$ (b) $\left[\dfrac{(15 - j7)(3 + j2)^*}{(4 + j6)^*(3\angle 70^\circ)}\right]^*$

Answer: (a) $3.387\angle -5.615^\circ$, (b) $2.759\angle -287.6^\circ$.

B.3 Euler's Formula

Euler's formula is an important result in complex variables. We derive it from the series expansion of e^x, $\cos\theta$, and $\sin\theta$. We know that

$$e^x = 1 + x + \frac{x^2}{2!} + \frac{x^3}{3!} + \frac{x^4}{4!} + \cdots \tag{B.19}$$

Replacing x by $j\theta$ gives

$$e^{j\theta} = 1 + j\theta - \frac{\theta^2}{2!} - j\frac{\theta^3}{3!} + \frac{\theta^4}{4!} + \cdots \tag{B.20}$$

Also,

$$\begin{aligned} \cos\theta &= 1 - \frac{\theta^2}{2!} + \frac{\theta^4}{4!} - \frac{\theta^6}{6!} + \cdots \\ \sin\theta &= \theta - \frac{\theta^3}{3!} + \frac{\theta^5}{5!} - \frac{\theta^7}{7!} + \cdots \end{aligned} \tag{B.21}$$

so that

$$\cos\theta + j\sin\theta = 1 + j\theta - \frac{\theta^2}{2!} - j\frac{\theta^3}{3!} + \frac{\theta^4}{4!} + j\frac{\theta^5}{5!} - \cdots \tag{B.22}$$

Comparing Eqs. (B.20) and (B.22), we conclude that

$$\boxed{e^{j\theta} = \cos\theta + j\sin\theta} \tag{B.23}$$

This is known as *Euler's formula.* The exponential form of representing a complex number as in Eq. (B.8) is based on Euler's formula. From Eq. (B.23), notice that

$$\boxed{\cos\theta = \text{Re}(e^{j\theta}), \qquad \sin\theta = \text{Im}(e^{j\theta})} \tag{B.24}$$

and that

$$|e^{j\theta}| = \sqrt{\cos^2\theta + \sin^2\theta} = 1$$

Replacing θ by $-\theta$ in Eq. (B.23) gives

$$e^{-j\theta} = \cos\theta - j\sin\theta \tag{B.25}$$

Adding Eqs. (B.23) and (B.25) yields

$$\boxed{\cos\theta = \frac{1}{2}(e^{j\theta} + e^{-j\theta})} \tag{B.26}$$

Subtracting Eq. (B.25) from Eq. (B.23) yields

$$\sin\theta = \frac{1}{2j}(e^{j\theta} - e^{-j\theta}) \tag{B.27}$$

Useful Identities

The following identities are useful in dealing with complex numbers. If $z = x + jy = r\underline{/\theta}$, then

$$zz^* = x^2 + y^2 = r^2 \tag{B.28}$$

$$\sqrt{z} = \sqrt{x + jy} = \sqrt{r}e^{j\theta/2} = \sqrt{r}\underline{/\theta/2} \tag{B.29}$$

$$z^n = (x + jy)^n = r^n\underline{/n\theta} = r^n e^{jn\theta} = r^n(\cos n\theta + j\sin n\theta) \tag{B.30}$$

$$z^{1/n} = (x + jy)^{1/n} = r^{1/n}\underline{/\theta/n + 2\pi k/n} \qquad k = 0, 1, 2, \dots, n-1 \tag{B.31}$$

$$\ln(re^{j\theta}) = \ln r + \ln e^{j\theta} = \ln r + j\theta + j2k\pi \qquad (k = \text{integer}) \tag{B.32}$$

$$\begin{aligned} \frac{1}{j} &= -j \\ e^{\pm j\pi} &= -1 \\ e^{\pm j2\pi} &= 1 \\ e^{j\pi/2} &= j \\ e^{-j\pi/2} &= -j \end{aligned} \tag{B.33}$$

$$\begin{aligned} \text{Re}(e^{(\alpha + j\omega)t}) &= \text{Re}(e^{\alpha t}e^{j\omega t}) = e^{\alpha t}\cos\omega t \\ \text{Im}(e^{(\alpha + j\omega)t}) &= \text{Im}(e^{\alpha t}e^{j\omega t}) = e^{\alpha t}\sin\omega t \end{aligned} \tag{B.34}$$

Example B.5

If $A = 6 + j8$, find: (a) $\sqrt{A}$, (b) A^4.

Solution:

(a) First, convert A to polar form:

$$r = \sqrt{6^2 + 8^2} = 10, \qquad \theta = \tan^{-1}\frac{8}{6} = 53.13^\circ, \qquad A = 10\underline{/53.13^\circ}$$

Then

$$\sqrt{A} = \sqrt{10}\underline{/53.13^\circ/2} = 3.162\underline{/26.56^\circ}$$

(b) Since $A = 10\underline{/53.13^\circ}$,

$$A^4 = r^4\underline{/4\theta} = 10^4\underline{/4 \times 53.13^\circ} = 10{,}000\underline{/212.52^\circ}$$

Practice Problem B.5

If $A = 3 - j4$, find: (a) $A^{1/3}$ (3 roots), and (b) $\ln A$.

Answer: (a) $1.71\underline{/102.3^\circ}$, $1.71\underline{/222.3^\circ}$, $1.71\underline{/342.3^\circ}$,
(b) $1.609 + j5.356 + j2n\pi$ $(n = 0, 1, 2, \dots)$.

Appendix C

Mathematical Formulas

This appendix—by no means exhaustive—serves as a handy reference. It does contain all the formulas needed to solve circuit problems in this book.

C.1 Quadratic Formula

The roots of the quadratic equation $ax^2 + bx + c = 0$ are

$$x_1, x_2 = \frac{-b \pm \sqrt{b^2 - 4ac}}{2a}$$

C.2 Trigonometric Identities

$$\sin(-x) = -\sin x$$

$$\cos(-x) = \cos x$$

$$\sec x = \frac{1}{\cos x}, \qquad \csc x = \frac{1}{\sin x}$$

$$\tan x = \frac{\sin x}{\cos x}, \qquad \cot x = \frac{1}{\tan x}$$

$$\sin(x \pm 90^\circ) = \pm\cos x$$

$$\cos(x \pm 90^\circ) = \mp\sin x$$

$$\sin(x \pm 180^\circ) = -\sin x$$

$$\cos(x \pm 180^\circ) = -\cos x$$

$$\cos^2 x + \sin^2 x = 1$$

$$\frac{a}{\sin A} = \frac{b}{\sin B} = \frac{c}{\sin C} \qquad \text{(law of sines)}$$

$$a^2 = b^2 + c^2 - 2bc\cos A \qquad \text{(law of cosines)}$$

$$\frac{\tan\frac{1}{2}(A - B)}{\tan\frac{1}{2}(A + B)} = \frac{a - b}{a + b} \qquad \text{(law of tangents)}$$

$$\sin(x \pm y) = \sin x \cos y \pm \cos x \sin y$$

$$\cos(x \pm y) = \cos x \cos y \mp \sin x \sin y$$

$$\tan(x \pm y) = \frac{\tan x \pm \tan y}{1 \mp \tan x \tan y}$$

$$2\sin x \sin y = \cos(x - y) - \cos(x + y)$$

$$2\sin x \cos y = \sin(x + y) + \sin(x - y)$$

$$2\cos x \cos y = \cos(x + y) + \cos(x - y)$$

$$\sin 2x = 2\sin x \cos x$$

$$\cos 2x = \cos^2 x - \sin^2 x = 2\cos^2 x - 1 = 1 - 2\sin^2 x$$

$$\tan 2x = \frac{2\tan x}{1 - \tan^2 x}$$

$$\sin^2 x = \frac{1}{2}(1 - \cos 2x)$$

$$\cos^2 x = \frac{1}{2}(1 + \cos 2x)$$

$$K_1 \cos x + K_2 \sin x = \sqrt{K_1^2 + K_2^2}\cos\left(x + \tan^{-1}\frac{-K_2}{K_1}\right)$$

$$e^{jx} = \cos x + j\sin x \qquad \text{(Euler's formula)}$$

$$\cos x = \frac{e^{jx} + e^{-jx}}{2}$$

$$\sin x = \frac{e^{jx} - e^{-jx}}{2j}$$

$$1 \text{ rad} = 57.296^\circ$$

C.3 Hyperbolic Functions

$$\sinh x = \frac{1}{2}(e^x - e^{-x})$$

$$\cosh x = \frac{1}{2}(e^x + e^{-x})$$

$$\tanh x = \frac{\sinh x}{\cosh x}$$

$$\coth x = \frac{1}{\tanh x}$$

$$\operatorname{csch} x = \frac{1}{\sinh x}$$

$$\operatorname{sech} x = \frac{1}{\cosh x}$$

$$\sinh(x \pm y) = \sinh x \cosh y \pm \cosh x \sinh y$$

$$\cosh(x \pm y) = \cosh x \cosh y \pm \sinh x \sinh y$$

C.4 Derivatives

If $U = U(x)$, $V = V(x)$, and a = constant,

$$\frac{d}{dx}(aU) = a\frac{dU}{dx}$$

$$\frac{d}{dx}(UV) = U\frac{dV}{dx} + V\frac{dU}{dx}$$

$$\frac{d}{dx}\left(\frac{U}{V}\right) = \frac{V\dfrac{dU}{dx} - U\dfrac{dV}{dx}}{V^2}$$

$$\frac{d}{dx}(aU^n) = naU^{n-1}$$

$$\frac{d}{dx}(a^U) = a^U \ln a \frac{dU}{dx}$$

$$\frac{d}{dx}(e^U) = e^U \frac{dU}{dx}$$

$$\frac{d}{dx}(\sin U) = \cos U \frac{dU}{dx}$$

$$\frac{d}{dx}(\cos U) = -\sin U \frac{dU}{dx}$$

C.5 Indefinite Integrals

If $U = U(x)$, $V = V(x)$, and $a =$ constant,

$$\int a\,dx = ax + C$$

$$\int U\,dV = UV - \int V\,dU \qquad \text{(integration by parts)}$$

$$\int U^n\,dU = \frac{U^{n+1}}{n+1} + C, \qquad n \neq 1$$

$$\int \frac{dU}{U} = \ln U + C$$

$$\int a^U\,dU = \frac{a^U}{\ln a} + C, \qquad a > 0, a \neq 1$$

$$\int e^{ax}\,dx = \frac{1}{a}e^{ax} + C$$

$$\int xe^{ax}\,dx = \frac{e^{ax}}{a^2}(ax - 1) + C$$

$$\int x^2e^{ax}\,dx = \frac{e^{ax}}{a^3}(a^2x^2 - 2ax + 2) + C$$

$$\int \ln x\,dx = x \ln x - x + C$$

$$\int \sin ax\,dx = -\frac{1}{a}\cos ax + C$$

$$\int \cos ax\,dx = \frac{1}{a}\sin ax + C$$

$$\int \sin^2 ax\,dx = \frac{x}{2} - \frac{\sin 2ax}{4a} + C$$

$$\int \cos^2 ax\,dx = \frac{x}{2} + \frac{\sin 2ax}{4a} + C$$

$$\int x \sin ax\,dx = \frac{1}{a^2}(\sin ax - ax\cos ax) + C$$

$$\int x \cos ax\,dx = \frac{1}{a^2}(\cos ax + ax\sin ax) + C$$

$$\int x^2 \sin ax\,dx = \frac{1}{a^3}(2ax\sin ax + 2\cos ax - a^2x^2\cos ax) + C$$

$$\int x^2 \cos ax\,dx = \frac{1}{a^3}(2ax\cos ax - 2\sin ax + a^2x^2\sin ax) + C$$

$$\int e^{ax}\sin bx\,dx = \frac{e^{ax}}{a^2+b^2}(a\sin bx - b\cos bx) + C$$

$$\int e^{ax}\cos bx\,dx = \frac{e^{ax}}{a^2+b^2}(a\cos bx + b\sin bx) + C$$

$$\int \sin ax\sin bx\,dx = \frac{\sin(a-b)x}{2(a-b)} - \frac{\sin(a+b)x}{2(a+b)} + C, \qquad a^2 \neq b^2$$

$$\int \sin ax\cos bx\,dx = -\frac{\cos(a-b)x}{2(a-b)} - \frac{\cos(a+b)x}{2(a+b)} + C, \qquad a^2 \neq b^2$$

$$\int \cos ax\cos bx\,dx = \frac{\sin(a-b)x}{2(a-b)} + \frac{\sin(a+b)x}{2(a+b)} + C, \qquad a^2 \neq b^2$$

$$\int \frac{dx}{a^2+x^2} = \frac{1}{a}\tan^{-1}\frac{x}{a} + C$$

$$\int \frac{x^2\,dx}{a^2+x^2} = x - a\tan^{-1}\frac{x}{a} + C$$

$$\int \frac{dx}{(a^2+x^2)^2} = \frac{1}{2a^2}\left(\frac{x}{x^2+a^2} + \frac{1}{a}\tan^{-1}\frac{x}{a}\right) + C$$

C.6 Definite Integrals

If m and n are integers,

$$\int_0^{2\pi} \sin ax\,dx = 0$$

$$\int_0^{2\pi} \cos ax\,dx = 0$$

$$\int_0^{\pi} \sin^2 ax\,dx = \int_0^{\pi} \cos^2 ax\,dx = \frac{\pi}{2}$$

$$\int_0^{\pi} \sin mx\sin nx\,dx = \int_0^{\pi} \cos mx\cos nx\,dx = 0, \qquad m \neq n$$

$$\int_0^{\pi} \sin mx\cos nx\,dx = \begin{cases} 0, & m+n = \text{even} \\ \dfrac{2m}{m^2-n^2}, & m+n = \text{odd} \end{cases}$$

$$\int_0^{2\pi} \sin mx\sin nx\,dx = \int_{-\pi}^{\pi} \sin mx\sin nx\,dx = \begin{cases} 0, & m \neq n \\ \pi, & m = n \end{cases}$$

$$\int_0^\infty \frac{\sin ax}{x}\,dx = \begin{cases} \dfrac{\pi}{2}, & a > 0 \\ 0, & a = 0 \\ -\dfrac{\pi}{2}, & a < 0 \end{cases}$$

C.7 L'Hopital's Rule

If $f(0) = 0 = h(0)$, then

$$\lim_{x \to 0} \frac{f(x)}{h(x)} = \lim_{x \to 0} \frac{f'(x)}{h'(x)}$$

where the prime indicates differentiation.

Appendix D

PSpice for Windows

There are several computer software packages, such as *Spice*, *Mathcad*, *Quattro*, *MATLAB*, and *Maple*, which can be used for circuit analysis. The most popular is *Spice*, which stands for *Simulation Program with Integrated-Circuit Emphasis*. *Spice* was developed at the Department of Electrical and Computer Engineering at the University of California at Berkeley in the 1970s for mainframe computers. Since then about 20 versions have been developed. *PSpice*, a version of *Spice* for personal computers, was developed by MicroSim Corporation in California and made available in 1984 and later by OrCAD and cadence. *PSpice* has been made available in different operating systems (DOS, Windows, Unix, etc.).

The student version of *PSpice* can be obtained free of charge.

If you do not have access to *PSpice*, you can find out how to obtain a free student copy by going to the text website (www.mhhe.com/alexander). The instructions and examples in this appendix were developed for version 9.1, but also work for later versions.

Assuming that you are using Windows and have the *PSpice* software installed in your computer, you can access *PSpice* by clicking the Start icon on the left-hand corner of your PC; drag the cursor to Programs, to *PSpice* students and to *Schematics*, and then click as shown in Fig. D.1.

The objective of this appendix is to provide a short tutorial on using the Windows-based *PSpice* on an IBM PC or equivalent.

PSpice can analyze up to roughly 130 elements and 100 nodes. It is capable of performing three major types of circuit analysis: dc analysis, transient analysis, and ac analysis. In addition, it can also perform transfer function analysis, Fourier analysis, and operating point analysis.The circuit can contain resistors, inductors, capacitors, independent and dependent voltage and current sources, op amps, transformers, transmission lines, and semiconductor devices.

We will assume that you are familiar with using the Microsoft Windows operating system and that *PSpice for Windows* is already installed in your computer. As with any standard Windows application, *PSpice* provides an on-line help system.

If you need help on any topic at any level, click Help, click Help Topics, click Search, and type in the topic.

D.1 Design Center for Windows

In earlier versions of *PSpice* prior to Windows 95, *PSpice for Windows* is formally known as the *MicroSim Design Center*, which is a computer environment for simulating electric circuits. The *Design Center for Windows* includes the following programs:

Schematics: This program is a graphical editor used to draw the circuit to be simulated on the screen. It allows the user to enter

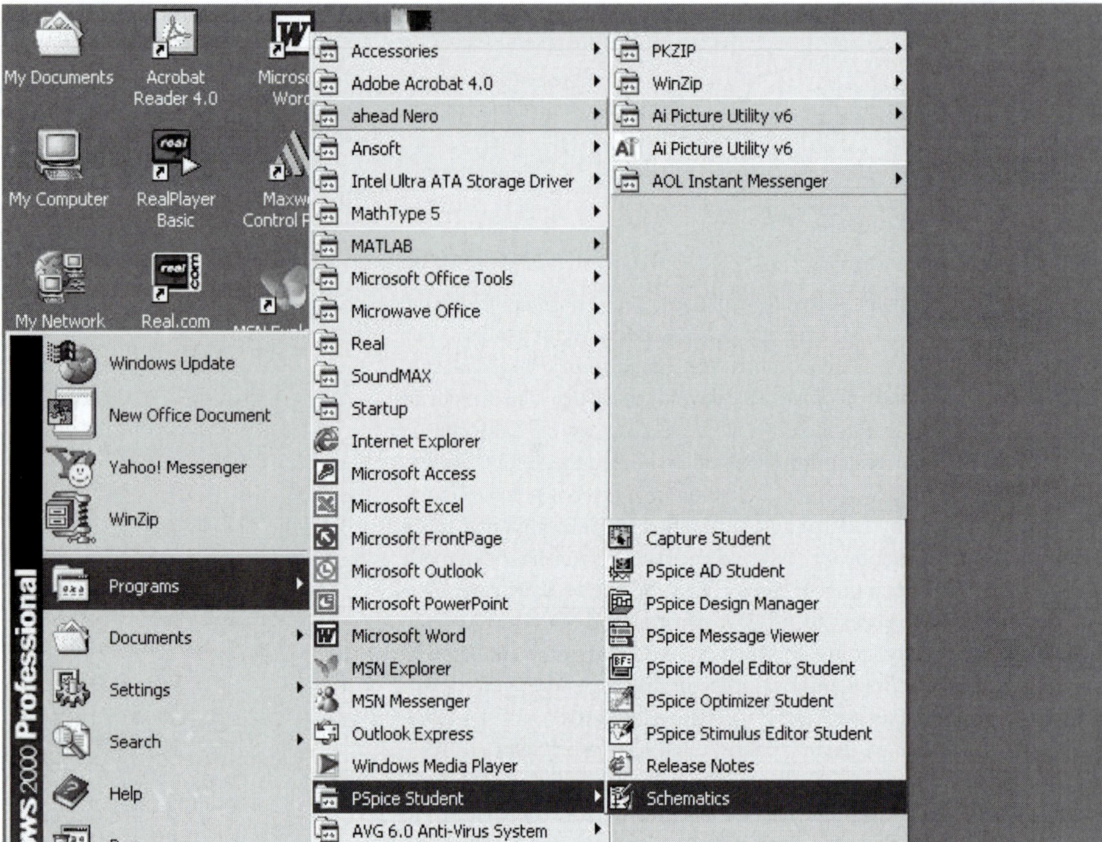

Figure D.1
Accessing *PSpice* on Windows.

the components, wire the components together to form the circuit, and specify the type of analysis to be performed.

Pspice: This program simulates the circuit created using *Schematics*. By simulation, we mean a method of analysis described in a program by which a circuit is represented by mathematical models of the components comprising the circuit.

Orcad PSpice: This program provides a graphic display of the output generated by the *PSpice* program. It can be used to observe any voltage or current in the circuit.

One may think of *Schematics* as the computer breadboard for setting up the circuit topology, *PSpice* as the simulator (performing the computation), and *Orcad PSpice* as the oscilloscope. Using the *Schematics* program is perhaps the hardest part of circuit simulation using *PSpice*. The next section covers the essential skills needed to operate the *Schematics*.

D.2 Creating a Circuit

For a circuit to be analyzed by *PSpice*, we must take three steps: (1) create the circuit, (2) simulate it, and (3) print or plot the results. In this section, we learn how to create the circuit using the Schematics program.

Before we discuss how to use the *Schematics* capture, we need to know how to use the mouse to select an object and perform an action. One uses the mouse in *Schematics* in conjunction with the keyboard to carry out various instructions. Throughout this text, we will use the following terms to represent actions to be performed by the mouse:

- **CLICKL** : click the left button once to select an item.
- **CLICKR** : click the right button once to abort a mode.
- **DCLICKL** : double-click the left button to edit a selection or end a mode.
- **DCLICKR** : double-click the right button to repeat an action.
- **CLICKLH** : click the left button, hold down, and move the mouse to drag a selected item. Release the left button after the item has been placed.
- **DRAG** : drag the mouse (without clicking) to move an item.

When the term "click" is used, it means that you quickly press and release the **left** mouse button. To select an item requires **CLICKL**, while to perform an action requires **DCLICKL**. Also, to avoid writing "click" several times, the menu to be clicked will be highlighted in bold. For example, "click **Draw**, click **Get New Part**" will be written as **Draw/Get New Part**. Of course, we can always press the <Esc> key to abort any action.

Assuming that you are using Windows, you can access *Pspice* by clicking the Start icon on the left-hand corner of your PC, drag the cursor to Programs, *PSpice* student; and to *Schematics*, as shown in Fig. D.1. Alternatively, you have the *PSpice* icon on your screen. **DCLICK** on it. Either way, a blank screen will appear as shown in Fig. D.2. The file

Figure D.2
Schematics window.

name [*Schematic1* **p.1**] next to *PSpice Schematics* is assigned to a circuit which is yet to be saved. You can change it by pulling down the *File* menu.

To create a circuit using *Schematics* requires three steps: (1) placing the parts or components of the circuit, (2) wiring the parts together to form the circuit, and (3) changing attributes of the parts.

Step 1: Placing the Parts

Each circuit part is retrieved by following this procedure:

- Select **Draw/Get New Part** to pull down the Draw menu (or type <Ctrl-G>).
- Use scroll bar to select the part (or type the part name, e.g., R for resistor, in the *PartName* box). Figures D.3 to D.5 show some part names and symbols for circuit elements and independent voltage and current sources.
- Click **Place & Close** (or press <Enter>).
- **DRAG** part to the desired location on the screen.
- **CLICKR** to terminate the placement mode.

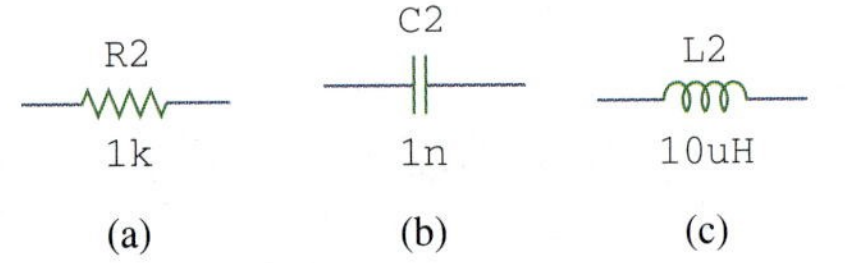

Figure D.3
Part symbols and attributes for circuit elements: (a) a resistor, (b) a capacitor, (c) an inductor.

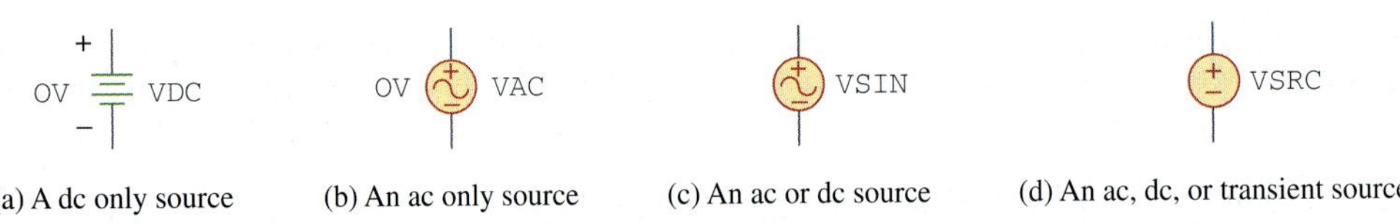

Figure D.4
Part symbols and attributes for independent voltage sources.

Sometimes, we want to rotate a part 90°. To rotate a resistor, for example, select the part R and click **Edit/Rotate** (or type <Ctrl R>). To delete a part, **CLICKL** to select (highlight red) the part, then click **Edit/Cut** (or press <Delete>).

Step 2: Wiring Parts Together

We complete the circuit by wiring the parts together. We first select **Draw/Wire** (or type <Ctrl-W>) to be in wiring mode. A pencil cursor will appear in place of an arrow cursor. **DRAG** the pencil cursor to the first point you want to connect and **CLICKL**. Next, **DRAG** the pencil cursor to the second point and **CLICKL** to change the dashed line to a solid line. (Only solid lines are wires.) **CLICKR** to end the wiring mode. To resume the wiring mode, press the <Space bar>. Repeat the above procedure for each connection in the circuit until all the parts are wired. The wiring is not complete without adding a ground connection (part AGND) to a schematic; *PSpice* will not operate without it. To verify that the parts are actually connected together, the *Junctions* option available in the **Options/Set Display Level** menu should be in the on position when wiring the parts. By default, the *Junctions* option is marked with a checksign (✓) in the dialog box, indicating that it is on.

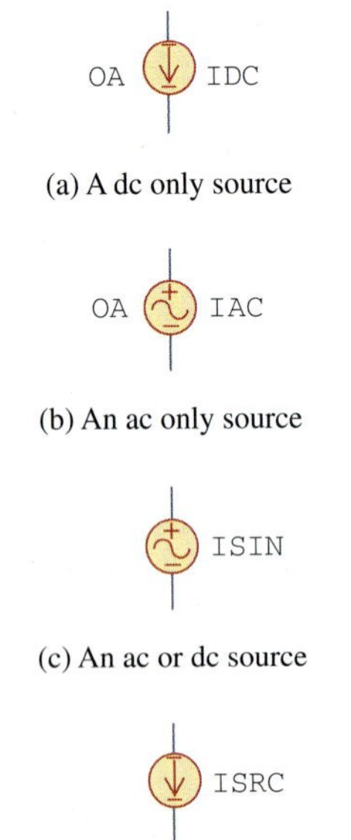

Figure D.5
Part symbols and attributes for independent current sources.

Some of the connections have a black dot indicating a connection. Although it is not necessary to have a dot where a wire joins a pin, having the dot shows the presence of a connection. To be sure a dot appears, make sure the wire overlaps the pin.

If you make a mistake, you can delete the part or wire by highlighting it (select **CLICKL**) and pressing the <Delete> key. Typing <Ctrl-L> will erase the fragments that are not really on the shematic.

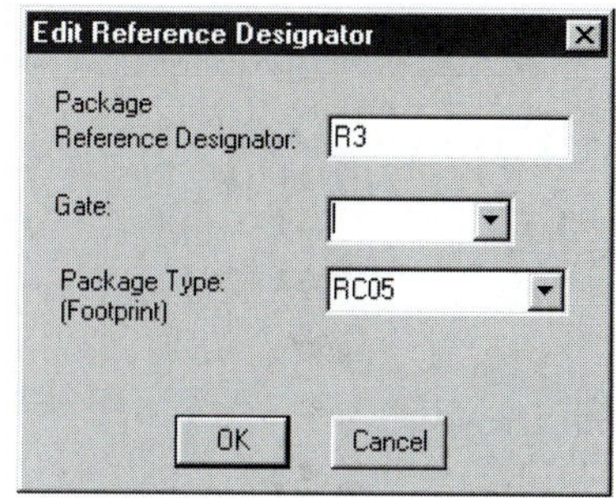

(a)

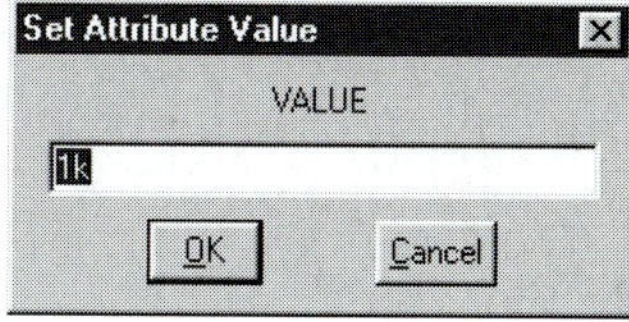

(b)

Figure D.6
(a) Changing name R3 to RX, (b) changing 1k to 10Meg.

Step 3: Changing Attributes of Parts

As shown in Figs. D.3 to D.5, each component has an attribute in addition to its symbol. Attributes are the labels for parts. Each attribute consists of a *name* and its designated *value*. For example, R and VSRC are the names of resistor and voltage source (dc, ac, or transient source), while 2k and DC = +10V are the designated values of the resistor and voltage source, respectively.

As parts are placed on the screen, they are automatically assigned names by successive numbers (R1, R2, R3, etc.). Also, some parts are assigned some predetermined values. For example, all resistors are placed horizontally and assigned a value of 1 kΩ. We may need to change the attributes (names and values) of a part. Although there are several ways of changing the attributes, the following is one simple way.

To change the name R3 to RX, for example, **DCLICKL** on the text R3 to bring up the *Edit Reference Designator* dialog box of Fig. D.6(a). Type the new name RX and click the **OK** button to accept the change. The same procedure can be used to change VDC to V1 or whatever.

To change the value 1k to 10Meg, for example, **DCLICKL** on the 1k attribute (not the symbol) to open up the *Set Attribute Value* dialog box of Fig. D.6(b). Type the new value 10Meg (no space between 10 and Meg) and click the **OK** button to accept the change. Similarly, to change the default value 0V to 15kV for voltage source VDC, **DCLICKL** the symbol for VDC to bring up the *PartName* dialog box. **DCLICKL** on the *DC* = attribute and type 15kV in the value box. For convenience, one can express numbers with the scale factors in Table D.1. For example, 6.6×10^{-8} can be written as 66N or 0.066U.

Except for the ground, which is automatically assigned node 0, every node is either given a name (or number) or is assigned one in the netlist. A node is labeled by giving a name to a wire connected to that node. **DCLICKL** the wire to open up the *Set Attribute Value* dialog box, and type the label.

To obtain a hard copy of the screen/schematic, click **File/Print/OK**. To save the schematic created, select **File/Save As** and type Filename. Click **OK** or press <Enter>. This creates a file named "filename" and saves it with extension .sch.

A component may have several attributes; some are displayed by default. If need be, we may add more attributes for display, but we should hide unimportant attributes to avoid clutter.

TABLE D.1

Scale factors.

Symbol	Value	Name of suffix
T	10^{12}	tera
G	10^{9}	giga
MEG	10^{6}	mega
K	10^{3}	kilo
M	10^{-3}	milli
U	10^{-6}	micro
N	10^{-9}	nano
P	10^{-12}	pico
F	10^{-15}	femto

It is always expedient to number the nodes by numbering the wires. Otherwise, *Schematics* will label the nodes its own way, and one may not understand which node is which in the output results.

Example D.1

Draw the circuit in Fig. D.7 using *Schematics*.

Solution:

We will follow the three steps mentioned above. We begin by double-clicking the *Schematics* icon. This provides us with a blank screen as

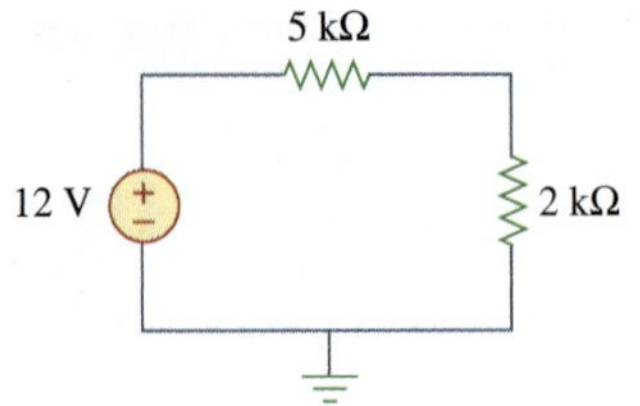

Figure D.7
For Example D.1.

a worksheet to draw the circuit on. We now take the following steps to create the circuit in Fig. D.7.
To place the voltage source, we need to:

1. Click **Draw/Get New Part** (or type <Ctrl-G>).
2. Type VSRC in the *Part Browser Basic* box.
3. Click **OK** (or type <Enter>).
4. **DRAG** the part to the desired location on the screen.
5. **CLICKL** to place VSRC and **CLICKR** to terminate placement mode.

At this point, only the voltage source V1 in Fig. D.8(a) is shown on the screen, highlighted red. To place the resistors, we need to:

1. Click **Draw/Get New Part.**
2. Type R in the *Part Browser Basic* box.
3. Click **OK.**
4. **DRAG** resistor to R1's location on the screen.
5. **CLICKL** to place R1.
6. **CLICKL** to place R2 and **CLICKR** to terminate placement mode.
7. **DRAG** R2 to its location.
8. **Edit/Rotate** (or type <Ctrl-R>) to rotate R2.

At this point, the three parts have been created as shown in Fig. D.8(a). The next step is to connect the parts by wiring. To do this:

1. Click **Draw/Wire** to be in wiring mode, indicated by the pencil cursor.
2. **DRAG** the pencil cursor to the top of V1.
3. **CLICKL** to join the wire to the top of V1.
4. **DRAG** the dotted wire to the top corner.
5. **CLICKL** to turn wire segment solid, and anchor at corner.
6. **DRAG** dotted wire to left of R1.
7. **CLICKL** to turn wire segment solid and anchor to left of R1.
8. **CLICKR** to end placement mode.

Follow the same steps to connect R1 with R2 and V1 with R2. (You can resume the wiring mode by pressing <Space bar>.) At this point, we have the circuit in Fig. D.8(b), except that the ground symbol is missing. We insert the ground by taking the following steps:

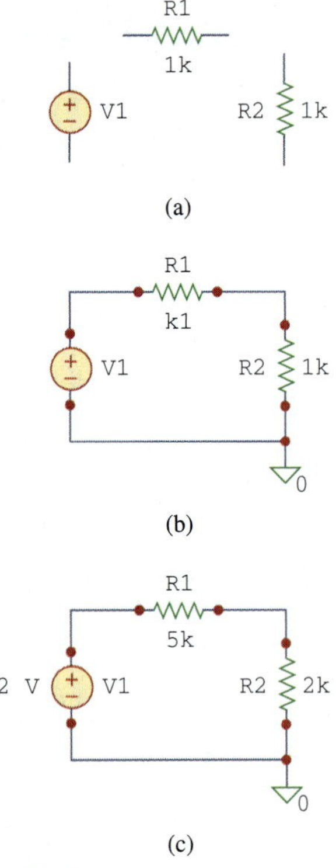

Figure D.8
Creating the circuit in Fig. D.7:
(a) placing the parts, (b) wiring the parts together, (c) changing the attributes.

1. Click **Draw/Get New Part.**
2. Type AGND in the *Part Browser Basic* box.
3. Click **OK.**
4. **DRAG** the part to the desired location on the screen.
5. **CLICKL** to place AGND and **CLICKR** to terminate placement mode.

The last thing to be done is to change or assign values to the attributes. To assign the attribute 12V to V1, we take these steps:

1. **DCLICKL** on the V1 symbol to open up the *PartName* dialog box.
2. **DCLICKL** on the *DC* = attribute.
3. Type +12V (or simply 12) in the *Value* box.
4. Click **Save Attr.**
5. Click **OK.**

To assign 5k to R1, we follow these steps:

1. **DCLICKL** on 1k attribute of R1 to bring up the *Set Attribute Value* dialog box.
2. Type 5k in the *Value* box.
3. Click **OK**.

Use the same procedure in assigning value 2k to R2. Figure D.8(c) shows the final circuit.

Practice Problem D.1

Construct the circuit in Fig. D.9 with *Schematics*.

Answer: See the schematic Fig. D.10.

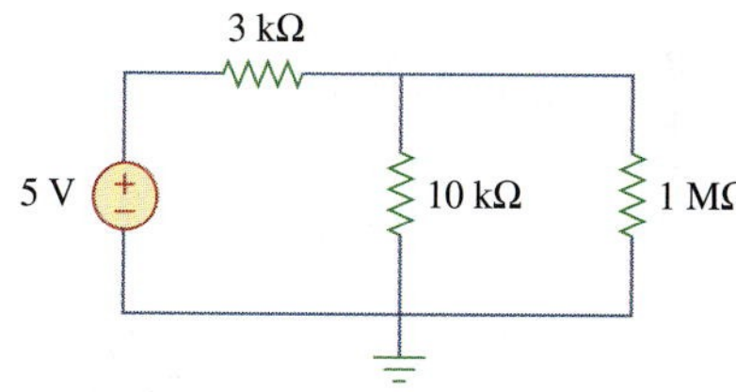

Figure D.9
For Practice Prob. D.1.

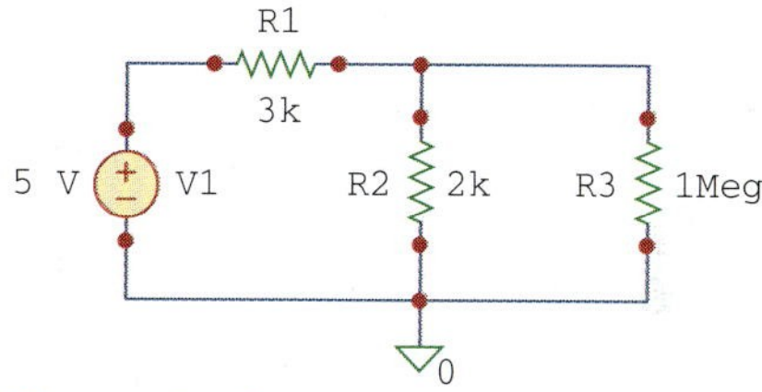

Figure D.10
For Practice Prob. D.1.

D.3 DC Analysis

DC analysis is one of the standard analyses that we can perform using *PSpice*. Other standard analyses include transient, AC, and Fourier. Under DC analysis, there are two kinds of simulation that *PSpice* can execute: DC nodal analysis and DC sweep.

1. DC Nodal Analysis

PSpice allows dc nodal analysis to be performed on sources with an attribute of the form DC = value and provides the dc voltage at each node of the circuit and dc branch currents if required. To view dc node voltages and branch currents requires adding two kinds of additional parts, shown in Fig. D.11. The symbol VIEWPOINT is connected to each node at which the voltage is to be viewed, while the symbol IPROBE is connected in the branch where the current is to be displayed. This necessitates modifying the schematic. For example, let us consider placing voltage VIEWPOINTS and current IPROBES to the schematic in Fig. D.8(c). To add VIEWPOINTS, we take the following steps:

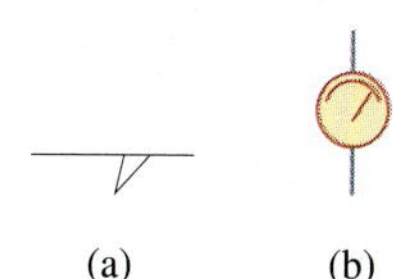

Figure D.11
Symbols for: (a) voltage VIEWPOINT, (b) current IPROBE.

1. Click **Draw/Get New Part** (or type <Ctrl-G>).
2. Type VIEWPOINT in the *Part Browser Basic* box.
3. Click **OK** (or type <Enter>).
4. **DRAG** to locate VIEWPOINT above V1 and **CLICKL**.
5. **DRAG** to locate VIEWPOINT above R2 and **CLICKL**.
6. **CLICKR** to end placement mode.

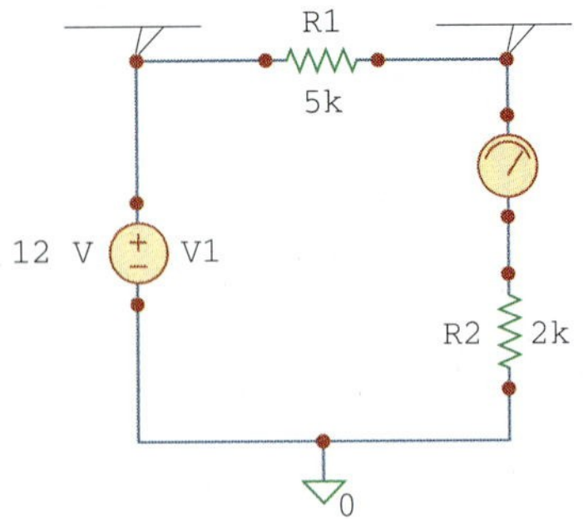

Figure D.12
Placing VIEWPOINTS and IPROBES.

Figure D.12 shows the two voltage VIEWPOINTS. Since the IPROBE symbol must be connected in series with a branch element, we need to move R2 down by clicking and dragging R2 and the wires. Once this is done, we add IPROBE as follows:

1. Click **Draw/Get New Part** (or type <Ctrl-G>).
2. Type IPROBE in the *Part Browser Basic* box.
3. Click **OK** (or type <Enter>).
4. **DRAG** to locate IPROBE above R2 and **CLICKL**.
5. **CLICKR** to end placement mode.
6. Use wiring to join all gaps.

The schematic becomes that shown in Fig. D.12. We are ready to simulate the circuit. At this point, we must save the schematic—*PSpice* will not run without first saving the schematic to be simulated. Before learning how to run *PSpice*, note the following points:

1. There must be a reference node or ground connection (part AGND) in the schematic. Any node can be used as ground, and the voltages at other nodes will be measured with respect to the selected ground.
2. Dependent sources are found in the Parts library. Obtain them by selecting **Draw/Get New Part** and typing the part name. Figure D.13 shows the part name for each type, with the gain. E is a voltage-controlled voltage source with gain e; F is a current-controlled current source with gain f; G is a voltage-controlled current source with a transconductance gain g; and H is a current-controlled voltage source with transresistance gain h.
3. By convention, we assume in dc analysis that all capacitors are open circuits and all inductors are short circuits.

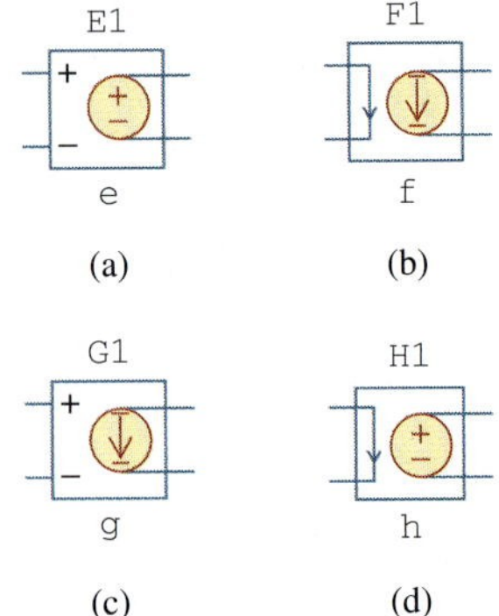

Figure D.13
Dependent sources:
(a) voltage-controlled voltage source (VCVS),
(b) current-controlled current source (CCCS),
(c) voltage-controlled current source (VCCS),
(d) current-controlled voltage source (CCVS).

A netlist can be generated manually or automatically by *Schematics*.

There are two kinds of common errors in *PSpice*: (1) errors involving wiring of the circuit, and (2) errors that occur during simulation.

We run *PSpice* by clicking **Analysis/Simulate**. This invokes the *electric rule check* (ERC), which generates the *netlist*. The ERC performs a connectivity check on the schematic before creating the netlist. The netlist is a list describing the operational behavior of each component in the circuit and its connections. Each line in the netlist represents a single component of the circuit. The netlist can be examined by clicking **Analysis/Examine Netlist** from the *Schematics* window. If there are errors in the schematic, an *error* window will appear. Click **OK** (or type <Enter>) to display the error list. After noting the errors, exit from the error list and go back to *Schematics* to correct the errors. If no errors are found, the system automatically enters *PSpice* and performs the simulation (nodal analysis). When the analysis is complete, the program displays **Bias point calculated**, and creates the result/output file with extension .out.

To examine the output file, click **Analysis/Examine Output** from the *Schematics* window (or click **File/Examine Output** from the *PSpice* window). To print the output file, click **File/Print**, and to exit the output file, click **File/Exit**.

We can also examine the results of the simulation by looking at the values displayed on the VIEWPOINTS and IPROBES parts of the *schematics* after the simulation is complete. The values displayed with VIEWPOINTS and IPROBES should be the same as those in the output file.

2. DC Sweep

DC nodal analysis allows simulation for DC sources with fixed voltages or currents. DC sweep provides more flexibility in that it allows the calculation of node voltages and branch currents of a circuit when a source is swept over a range of values. As in nodal analysis, we assume capacitors to be open circuits and inductors to be short circuits.

Suppose we desire to perform a DC sweep of voltage source V1 in Fig. D.12 from 0 to 20 volts in 1-volt increments. We proceed as follows:

1. Click **Analysis/Setup**.
2. **CLICKL DC Sweep** button.
3. Click *Name* box and type V1.
4. Click *Start Value* box and type 0.
5. Click *End Value* box and type 20.
6. Click *Increment* box and type 1.
7. Click **OK** to end the *DC Sweep* dialog box and save parameters.
8. Click **Close** to end the *Analysis Setup* menu.

Figure D.14 shows the *DC Sweep* dialog box. Notice that the default setting is *Voltage Source* for the *Swept Var. Type*, while it is *Linear* for *Sweep Type*. If needed, other options can be selected by clicking the appropriate buttons.

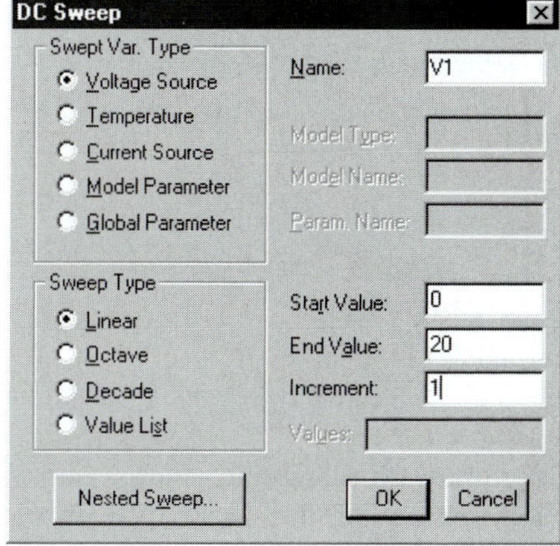

Figure D.14
DC sweep analysis dialog box.

To run DC sweep analysis, click **Analysis/Simulate**. *Schematics* will create a netlist and then run *PSpice* if no errors are found. If errors are found in the schematic, check for them in the *Error List* and correct them as usual. If no errors are found, the data generated by *PSpice* is passed to *Orcad PSpice*. The *Orcad PSpice* window will appear, displaying a graph in which the X axis is by default set to the DC sweep variable and range, and the Y axis is blank for now. To display some specific plots, click **Trace/Add** in the *Orcad PSpice* menu to open the *Add Traces* dialog box. The box contains traces, which are the output variables (node voltages and branch currents) in the data file available for display. Select the traces to be displayed by clicking or typing them, and click **OK**. The selected traces will be plotted and displayed on the screen. As many traces as you want may be added to the same plot or on different windows. Select a new window by clicking **Window/New**. To delete a trace, click the trace name in the legend of the plot to highlight it and click **Edit/Delete** (or press <Delete>).

It is important to understand how to interpret the traces. We must interpret the voltage and current variables according to the passive sign convention. As parts are initially placed horizontally in a schematic as shown typically in Fig. D.3, the left-hand terminal is named pin 1 while the right-hand terminal is pin 2. When a component (say R1) is rotated counterclockwise once, pin 2 would be on the top, since rotation is about pin 1. Therefore, if current enters through pin 2, the current I(R1) through R1 would be negative. In other words, positive current implies that the current enters through pin 1, and negative current means that the current enters through pin 2. As for voltage variables, they are always with respect to the ground. For example, V(R1:2) is the voltage (with respect to the ground) at pin 2 of resistor R1; V(V1:+) is the voltage (with respect to the ground) at the positive terminal of voltage source V1; and V(E2:1) is the voltage at pin 1 of component E2 with respect to ground, regardless of the polarity.

Example D.2

For the circuit in Fig. D.15, find the dc node voltages and the current i_o.

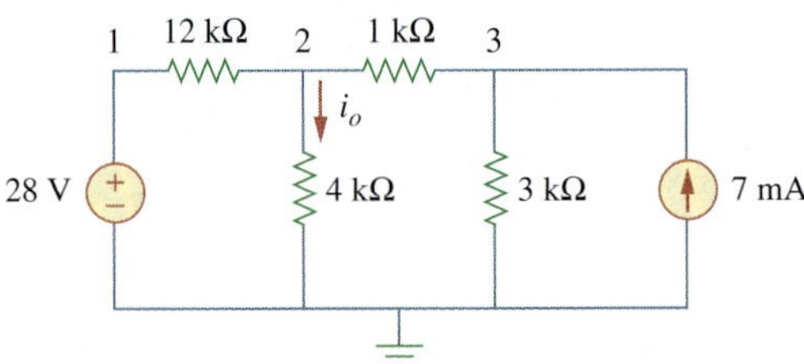

Figure D.15
For Example D.2.

Solution:
We use *Schematics* to create the circuit. After saving the circuit, click **Analysis/Simulate** to simulate the circuit. We obtain the results of the dc analysis from the output file or from the VIEWPOINT AND IPROBE parts, as shown in Fig. D.16. The netlist file is shown in Fig. D.17. Notice that the netlist contains the name, value, and

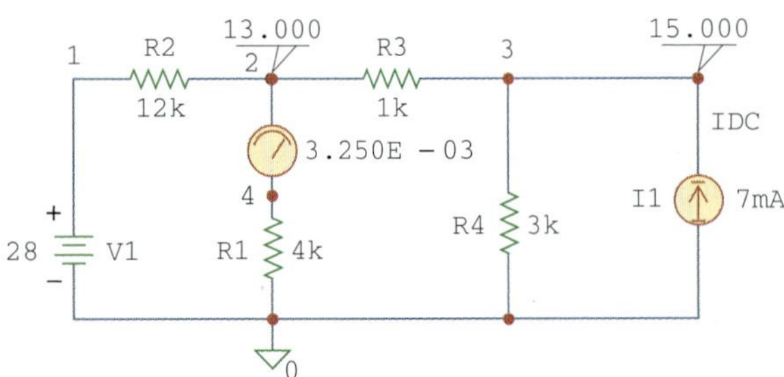

Figure D.16
For Example D.2; schematic for the circuit in Fig. D.15.

```
* Schematics Netlist *

V_V1     1 0 28
R_R1     0 4 4k
R_R2     1 2 12k
R_R3     2 3 1k
R_R4     0 3 3k
I_I1     0 3 DC 7mA
v_V2     2 4 0
```

Figure D.17
The Netlist file for Example D.2.

connection for each element in the circuit. First example, the first line shows that the voltage source V1 has a value of 28 V and is connected between nodes 0 and 1. Figure D.18 shows the edited version of the output file. The output file also contains the Netlist file, but this was removed from Fig. D.18. From IPROBE or the output file, we obtain i_o as 3.25 mA.

```
**** 11/26/99 20:56:05 ********* NT Evaluation PSpice (Nov. 1999) *********

* C:\ MSIMEV63\ examd2.sch

**** CIRCUIT DESCRIPTION

******************************************************************************

* Schematics Version 6.3 - April 1996
* Sat Jul 26 20:56:04 1997

**** INCLUDING examd2.als ****
* Schematics Aliases *

.ALIASES
V_V1          V1(+=1 -=0 )
R_R1          R1(1=0 2=4 )
R_R2          R2(1=1 2=2 )
R_R3          R3(1=2 2=3 )
R_R4          R4(1=0 2=3 )
I_I1          I1(+=0 -=3 )
v_V2          V2(+=2 -=4 )
_ _(1=1)
_ _(2=2)
_ _(3=3)
.ENDALIASES

.probe

.END

NODE VOLTAGE     NODE VOLTAGE      NODE VOLTAGE     NODE VOLTAGE

( 1) 28.0000     ( 2) 13.0000      ( 3) 15.0000     ( 4) 13.0000

VOLTAGE SOURCE CURRENTS
NAME     CURRENT

V_V1      -1.250E-03
v_V2       3.250E-03

TOTAL POWER DISSIPATION 3.50E-02 WATTS
```

Figure D.18
Output file (edited version) for Example D.2.

Practice Problem D.2

Use *PSpice* to determine the node voltages and the current i_x in the circuit of Fig. D.19.

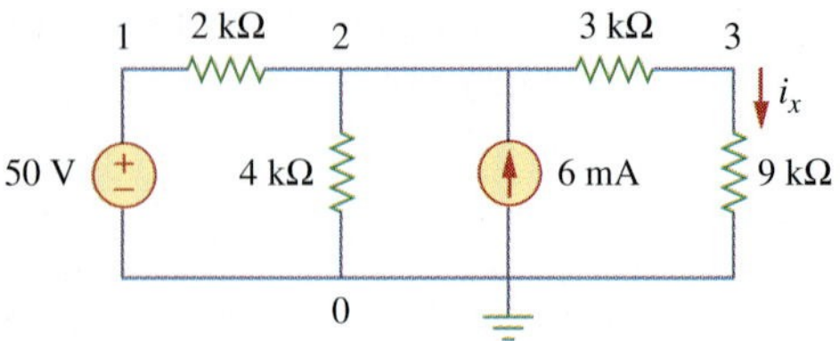

Figure D.19
For Practice Prob. D.2.

Answer: $V_1 = 50$, $V_2 = 37.2$, $V_3 = 27.9$, $i_x = 3.1$ mA.

Example D.3

Plot I_1 and I_2 if the dc voltage source in Fig. D.20 is swept from 2 V to 10 V.

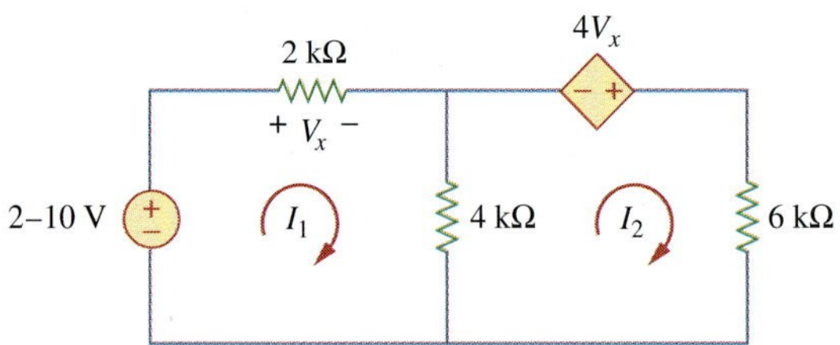

Figure D.20
For Example D.3.

Solution:
We draw the schematic of the circuit and set the attributes as shown in Fig. D.21. Notice how the voltage-controlled voltage source E1 is connected. After completing the schematic, we select **Analysis/Setup** and input the start, end, and increment values as 2, 10, and 0.5, respectively. By selecting **Analysis/Simulate**, we bring up the *Orcad PSpice* window. We select **Trace/Add** and click I(R1) and −I(R3) to

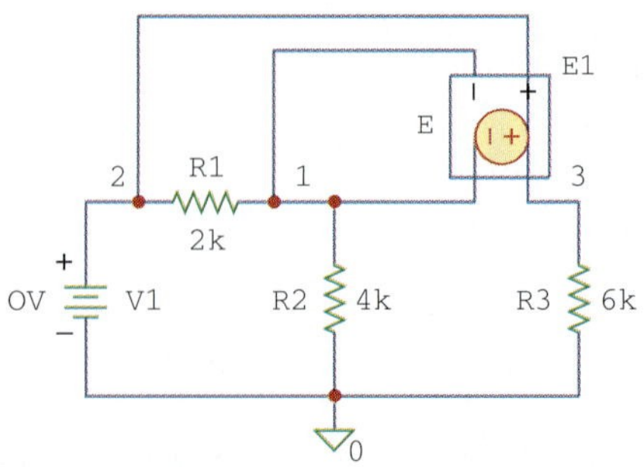

Figure D.21
The schematic for the circuit in Fig. D.20.

be displayed. (The negative sign is needed to make the current through R3 positive.) Figure D.22 shows the result.

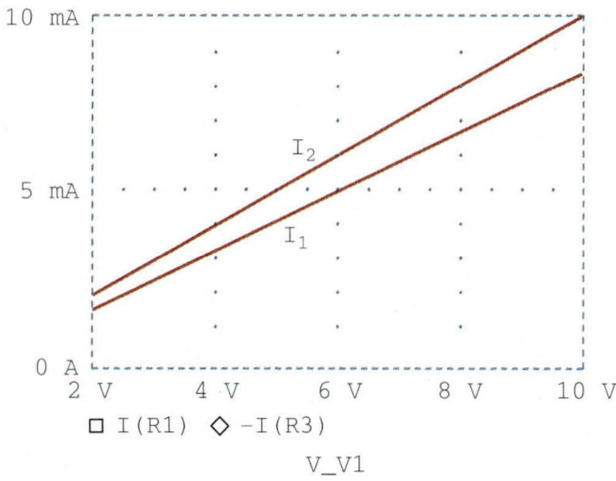

Figure D.22
Plots of I_1 and I_2 against V1.

Practice Problem D.3

Use *PSpice* to obtain the plots of i_x and i_o if the dc voltage source in Fig. D.23 is swept from 2 V to 10 V.

Answer: The plots of i_x and i_o are displayed in Fig. D.24.

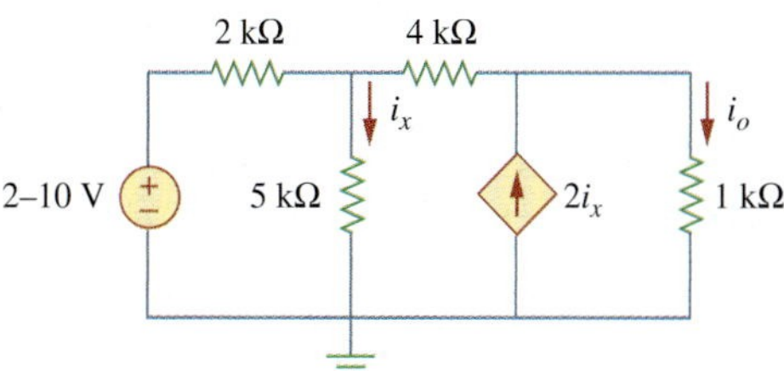

Figure D.23
For Practice Prob. D.3.

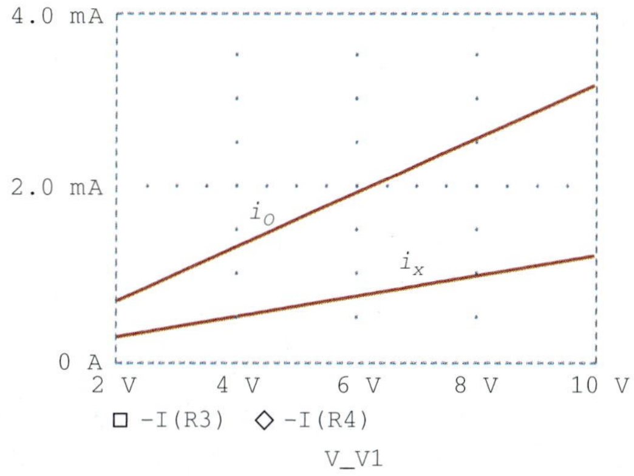

Figure D.24
Plots of i_x and i_o versus V1.

D.4 Transient Analysis

In *PSpice*, transient analysis is generally used to examine the behavior of a waveform (voltage or current) as time varies. Transient analysis solves some differential equations describing a circuit and obtains voltages and currents versus time. Transient analysis is also used to obtain Fourier analysis. To perform transient analysis on a circuit using *PSpice*

Transient analysis is used to view the transient response of inductors and capacitors.

usually involves these steps: (1) drawing the circuit, (2) providing specifications, and (3) simulating the circuit.

1. Drawing the Circuit

In order to run a transient analysis on a circuit, the circuit must first be created using *Schematics* and the source must be specified. *PSpice* has several time-varying functions or sources that enhance the performance of transient analysis. Sources used in the transient analysis include:

- VSIN, ISIN: damped sinusoidal voltage or current source, e.g., $v(t) = 10e^{-0.2t}\sin(120\pi t - 60°)$.
- VPULSE, IPULSE: voltage or current pulse.
- VEXP, IEXP: voltage or current exponential source, e.g., $i(t) = 6[1 - \exp(-0.5t)]$.
- VPWL, IPWL: piecewise linear voltage or current function, which can be used to create an arbitrary waveform.

It is expedient to take a close look at these functions.

VSIN is the exponentially damped sinusoidal voltage source, for example,

$$v(t) = V_o + V_m e^{-\alpha(t-t_d)} \sin[2\pi f(t - t_d) + \phi] \quad \textbf{(D.1)}$$

The VSIN source has the following attributes, which are illustrated in Fig. D.25 and compared with Eq. (D.1):

$$\begin{aligned} \text{VOFF} &= \text{Offset voltage, } V_o \\ \text{VAMPL} &= \text{Amplitude, } V_m \\ \text{TD} &= \text{Time delay in seconds, } t_d \\ \text{FREQ} &= \text{Frequency in Hz, } f \\ \text{DF} &= \text{Damping factor (dimensionless), } \alpha \\ \text{PHASE} &= \text{Phase in degrees, } \phi \end{aligned} \quad \textbf{(D.2)}$$

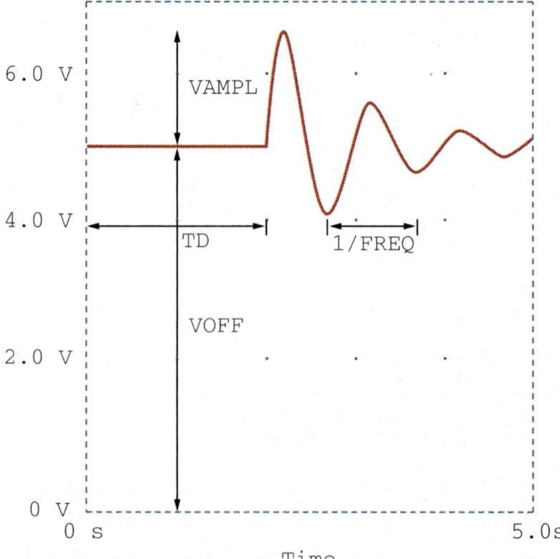

Figure D.25
Sinusoidal voltage source VSIN.

Attributes TD, DF, and PHASE are set to 0 by default but can be assigned other values if necessary. What has been said about VSIN is also true for ISIN.

The VPULSE source has the following attributes, which are portrayed in Fig. D.26:

$$
\begin{aligned}
\text{V1} &= \text{Low voltage} \\
\text{V2} &= \text{High voltage} \\
\text{TD} &= \text{Initial time delay in seconds} \\
\text{TR} &= \text{Rise time in seconds} \\
\text{TF} &= \text{Fall time in seconds} \\
\text{PW} &= \text{Pulse width in seconds} \\
\text{PER} &= \text{Period in seconds}
\end{aligned}
\tag{D.3}
$$

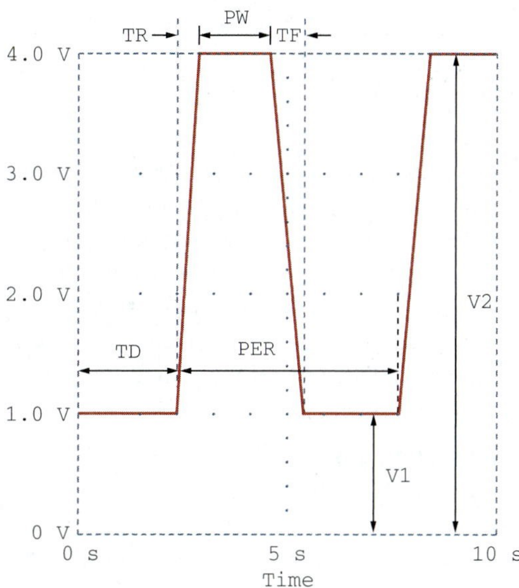

Figure D.26
Pulse voltage source VPULSE.

Attributes V1 and V2 must be assigned values. By default, attribute TD is assigned 0; TR and TF are assigned the *print step* value; and PW and PER are assigned the *final time* value. The values of the *print time* and *final time* are obtained as default values from the specifications provided by the user in the **Transient Analysis/Setup**, to be discussed a little later.

The exponential voltage source VEXP has the following attributes, typically illustrated in Fig. D.27:

$$
\begin{aligned}
\text{V1} &= \text{Initial voltage} \\
\text{V2} &= \text{Final voltage} \\
\text{TD1} &= \text{Rise delay in seconds} \\
\text{TC1} &= \text{Rise time constant in seconds} \\
\text{TD2} &= \text{Fall delay in seconds} \\
\text{TC2} &= \text{Fall time in seconds}
\end{aligned}
\tag{D.4}
$$

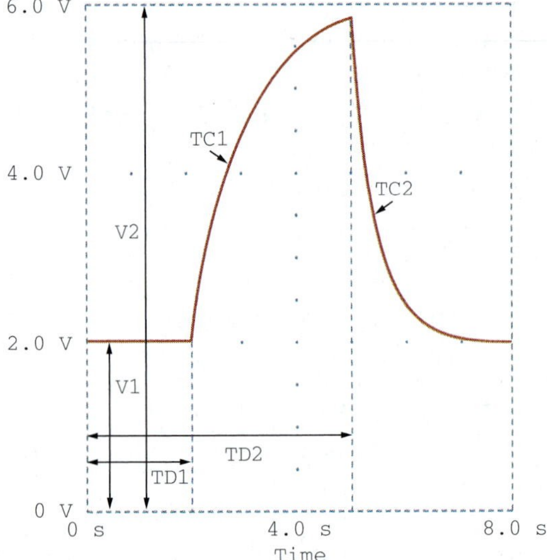

Figure D.27
Exponential voltage source VEXP.

The piecewise linear voltage source VPWL, such as shown in Fig. D.28, requires specifying pairs of TN, VN, where VN is the voltage at time TN for N = 1, 2, . . . , 10. For example, for the function in Fig. D.29, we will need to specify the attributes T1 = 0, V1 = 0, T2 = 2, V2 = 4, T3 = 6, V3 = 4, and T4 = 8, V4 = −2.

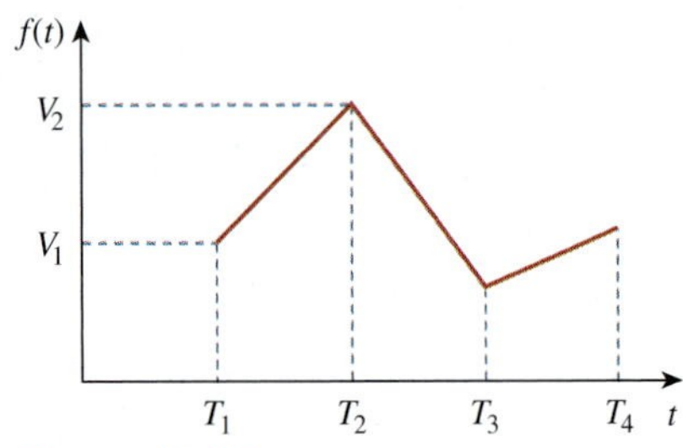

Figure D.28
Piecewise linear voltage source VPWL.

To obtain information about other sources, click **Help/Search for Help on** . . . and type in the name of the source. To add a source to the schematic, take the following steps:

1. Select **Draw/Get New Part**.
2. Type the name of the source.
3. Click **OK** and **DRAG** the symbol to the desired location.
4. **DCLICKL** the symbol of the source to open up the *PartName* dialog box.
5. For each attribute, **DCLICKL** on the attribute, enter the value, and click **Save Attr** to accept changes.
6. Click **OK** to accept new attributes.

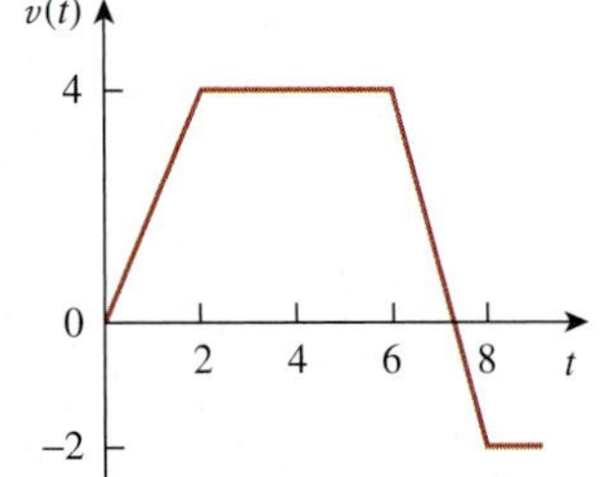

Figure D.29
An example of a piecewise linear voltage source VPWL.

In step 5, the attributes may not be shown on the schematic after entering their values. To display an attribute, select **Change Display/Both Name and Value** in the *PartName* dialog box.

In addition to specifying the source to be used in transient analysis, there may be need to set initial conditions on capacitors and inductors in the circuit. To do so, **DCLICKL** the part symbol to bring up the *PartName* dialog box, click **IC** = and type in the initial condition. The IC attribute allows for setting the initial conditions on a capacitor or inductor. The default value of IC is 0. The attributes of open/close switches (with part names Sw_tClose and Sw_tOpen) can be changed in a similar manner.

2. Providing Specifications

After the circuit is drawn and the source is specified with its attributes, we need to add some specifications for the transient analysis.

For example, suppose we want the analysis to run the simulation from 0 to 10 ms with a print interval of 2 ns; we enter these specifications as follows:

1. Select **Analysis/Setup/Transient** to open up the *Transient Analysis* dialog box.
2. **CLICKL** *Print Step* and type 2 ns.
3. **CLICKL** *Final Time* and type 10 ms.
4. **CLICKL** *Step Ceiling* and type 5 μs.
5. **CLICKL OK/Close** to accept specifications.

These specifications control the simulation and the display of output variables. *Final Time* specifies how long the simulation should run. In other words, the simulation runs from $t = 0$ to $t =$ Final Time. *Print Step* refers to the time interval the print part will print out; it controls how often simulation results are written to the output file. The value of *Print Step* can be any value less than the *Final Time*, but it cannot be zero. *Step Ceiling* is the maximum time between simulation points; specifying its value is optional. By selecting 10 ms as *Final Time* and 5 μs as *Step Ceiling*, the simulation will have a minimum of 10 ms/5 μs = 2000 points. When *Step Ceiling* is unspecified, *PSpice* selects its own internal time step—the time between simulation points. The time step is selected as large as possible to reduce simulation time. If the user has no idea of what the plot may look like, it is recommended that the value of *Step Ceiling* be unspecified. If the plot is jagged as a result of a large time step assumed by *PSpice*, the user may now specify a *Step Ceiling* that will smooth the plot. Keep in mind that a smaller value gives more points in the simulation but takes more time.

To obtain the Fourier component of a signal, we enable the Fourier option in the *Transient Analysis* dialog box. Chapter 17 has more on this.

3. Simulating the Circuit

After the circuit is drawn, the specifications for the transient analysis are given, and the circuit is saved, we are ready to simulate it. To perform transient analysis, we select **Analysis/Simulate**. If there are no errors, the *Orcad PSpice* window will automatically appear. As usual, the time axis (or X axis) is drawn but no curves are drawn yet. Select **Trace/Add** and click on the variables to be displayed.

An alternative way of displaying the results is to use *markers*. Although there are many types of markers, we will discuss only voltage and current markers. A voltage marker is used to display voltage at a node relative to ground; a current marker is for displaying current through a component pin. To place a voltage marker at a node, take the following steps while in the *Schematics* window:

1. Select **Markers/Mark Voltage/Level**.
2. **DRAG** the voltage marker to the desired node.
3. **CLICKL** to place the marker and **CLICKR** to end the placement mode.

This will cause two things to happen immediately. The voltage marker becomes part of the circuit and the appropriate node voltage is automatically displayed by the *Orcad PSpice* window when the *Schematics* is run. To place a current marker at a component pin, take the following steps in the *Schematics* window:

1. Select **Markers/Mark current into pin.**
2. **DRAG** the current marker to the desired pin.
3. **CLICKL** to place the marker and **CLICKR** to end the placement mode.

This will automatically add the current through the pin to your graph. It is important that the current marker be placed at the pin of the component; otherwise the system would reject the marker. You can place as many voltage and current markers as you want on a circuit. To remove the markers from the circuit as well as the plots from the *Orcad PSpice* window, select **Markers/Clear All** from *Schematics* window.

Example D.4

Assuming that $i(0) = 10\text{A}$, plot the zero-input response $i(t)$ in the circuit of Fig. D.30 for $0 < t < 4$ s using *PSpice.*

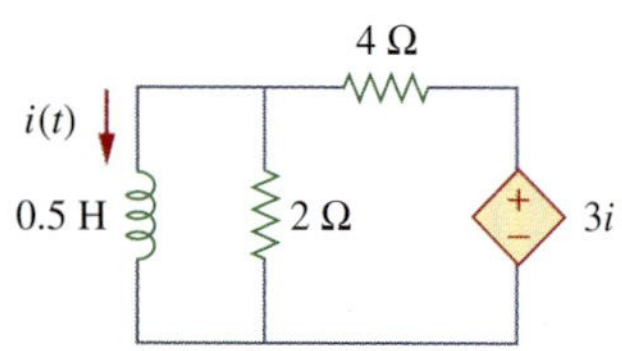

Figure D.30
For Example D.4.

Solution:
The circuit is the same as the one for Example 7.3, where we obtained the solution as

$$i(t) = 10e^{(-2/3)t}$$

For *PSpice* analysis, the schematic is in Fig. D.31, where the current-controlled source H1 has been wired to agree with the circuit in Fig. D.30. The voltage of H1 is 3 times the current through inductor L1. Therefore, for H1, we set GAIN = 3 and for the inductor L1, we set the initial condition IC = 10. Using the **Analysis/Setup/Transient** dialog box, we set *Print Step* = 0.25 s and *Final Time* = 4 s. After simulating the circuit, the output is taken as the inductor current $i(t)$, which is plotted in Fig. D.32.

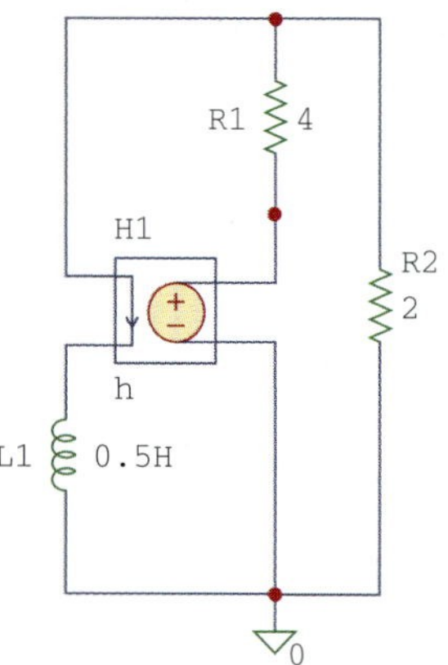

Figure D.31
The schematic of the circuit in Fig. D.30.

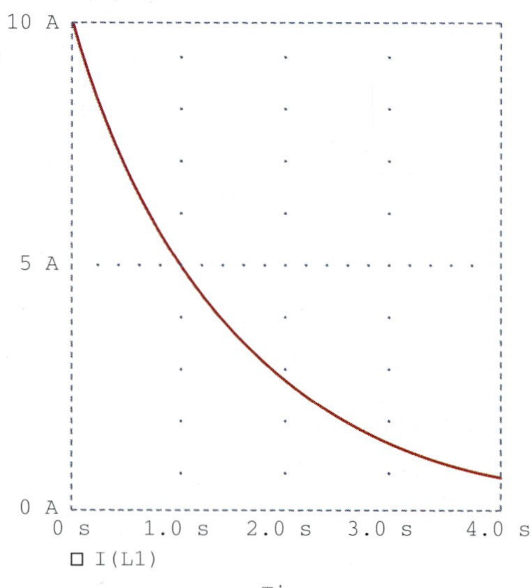

Figure D.32
Output plot for Example D.4.

Practice Problem D.4

Using *PSpice*, plot the source-free response $v(t)$ in the circuit of Fig. D.33, assuming that $v(0) = 10$ V.

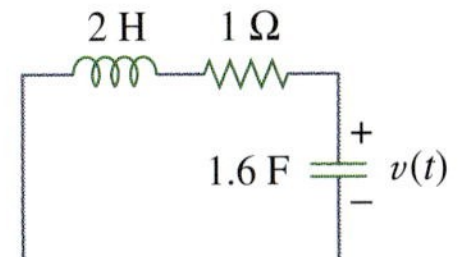

Figure D.33
For Practice Prob. D.4.

Answer: Figure D.34 shows the plot. Note that $v(t) = 10e^{-0.25t}\cos 0.5t + 5e^{-0.25t}\sin 0.5t$ V.

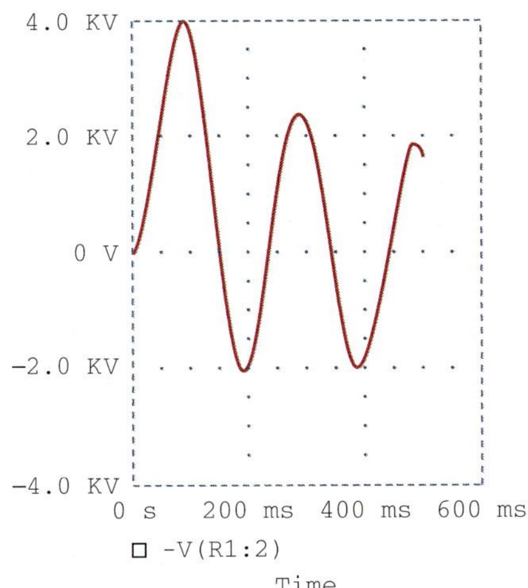

Figure D.34
Output plot for Practice Prob. D.4.

Example D.5

Plot the forced response $v_o(t)$ in the circuit of Fig. D.35(a) for $0 < t < 5$ s if the source voltage is as shown in Fig. D.35(b).

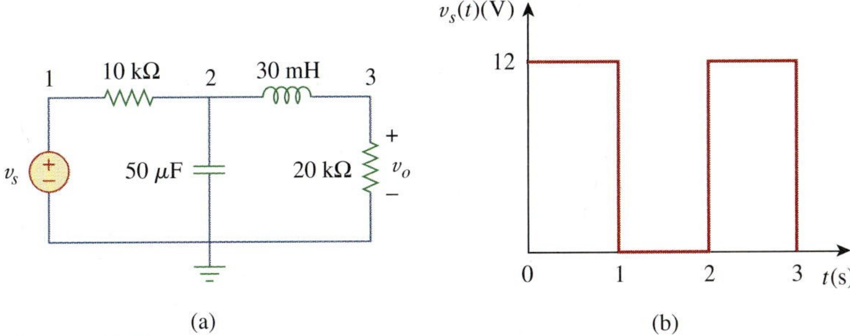

Figure D.35
For Example D.5.

Solution:

We draw the circuit and set the attributes as shown in Fig. D.36. We enter in the data in Fig. D.35(b) by double-clicking the symbol of the voltage source V1 and typing in T1 = 0, V1 = 0, T2 = 1 ns, V2 = 12, T3 = 1 s, V3 = 12, T4 = 1.001 s, V4 = 0, T5 = 2 s, V5 = 0, T6 = 2.001 s, V6 = 12, T7 = 3 s, V7 = 12, T8 = 3.001 s, V8 = 0. In the **Analysis/Setup/Transient** dialog box, we set *Print Step* = 0.2 s and *Final Time* = 5 s. When the circuit is simulated and we are in the *Orcad PSpice* window, we close or minimize the window to go back to the

Schematics window. We place two voltage markers as shown in Fig. D.36 to get the plots of input v_s and output v_o. We press <Alt-Esc> to get into the *Orcad PSpice* window and obtain the plots shown in Fig. D.37.

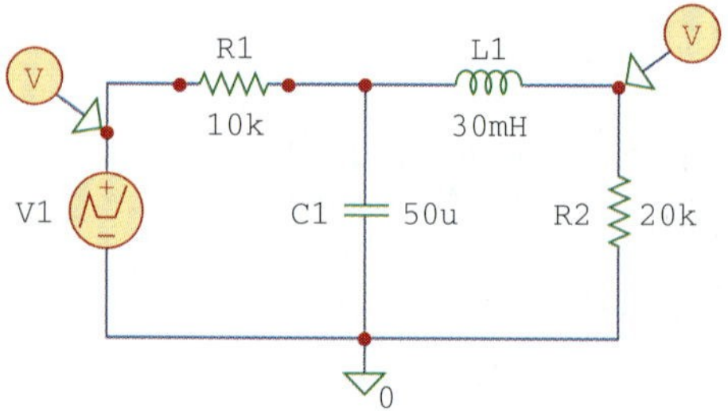

Figure D.36
The schematic of the circuit in Fig. D.35.

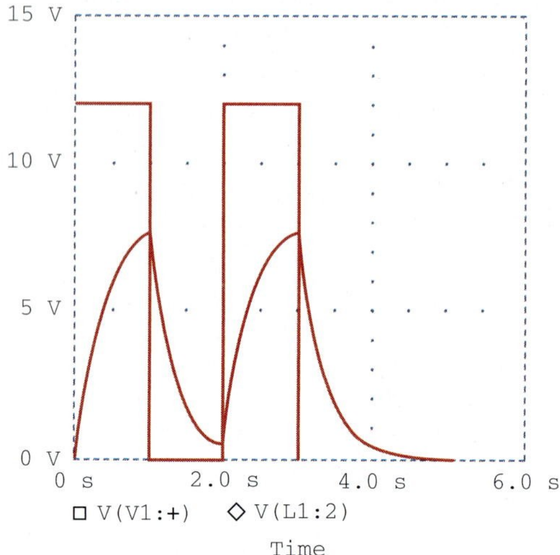

Figure D.37
Output plot for Example D.5.

Practice Problem D.5

Obtain the plot of $v(t)$ in the circuit in Fig. D.38 for $0 < t < 0.5$ s if $i_s = 2e^{-t} \sin 2\pi(5)t$ A.

Answer: See Fig. D.39.

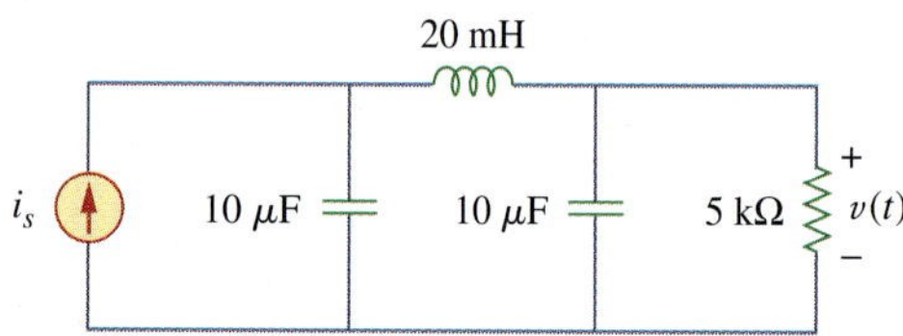

Figure D.38
For Practice Prob. D.5.

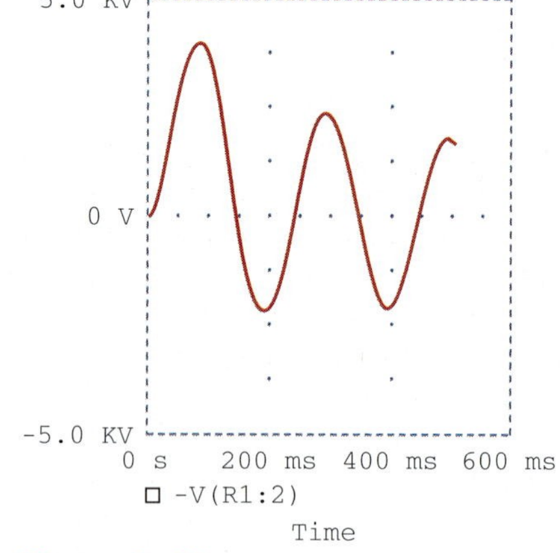

Figure D.39
Output plot for Practice Prob. D.5.

D.5 AC Analysis/Frequency Response

Using AC sweep, *PSpice* can perform AC analysis of a circuit for a single frequency or over a range of frequencies in increments that can vary linearly, by decade, or by octave. In AC sweep, one or more sources are swept over a range of frequencies while the voltages and currents of the circuit are calculated. Thus, we use AC sweep both for phasor analysis and for frequency response analysis: it will output Bode gain and phase plots. (Keep in mind that a phasor is a complex quantity with real and imaginary parts or with magnitude and phase.)

While transient analysis is done in the time domain, AC analysis is performed in the frequency domain. For example, if $v_s = 10\cos(377t + 40°)$, transient analysis can be used to display v_s as a function of time, whereas AC sweep will give the magnitude as 10 and phase as 40°. To perform AC sweep requires taking three steps similar to those for transient analysis: (1) drawing the circuit, (2) providing specifications, and (3) simulating the circuit.

1. Drawing the Circuit

We first draw the circuit using *Schematics* and specify the source(s). Sources used in AC sweep are AC sources VAC and IAC. The sources and attributes are entered into the *Schematics* as stated in the previous section. For each independent source, we must specify its magnitude and phase.

2. Providing Specifications

Before simulating the circuit, we need to add some specifications for AC sweep. For example, suppose we want a linear sweep at frequencies 50, 100, and 150 Hz. We enter these parameters as follows:

1. Select **Analysis/Setup/AC Sweep** to open up the dialog box for AC Sweep.
2. **CLICKL** *Linear* for the *X* axis to have a linear scale.
3. Type 3 in the *Total Pts* box.
4. Type 50 in the *Start Freq* box.
5. Type 150 in the *End Freq* box.
6. **CLICKL OK/Close** to accept specifications.

A linear sweep implies that simulation points are spread uniformly between the starting and ending frequencies. Note that the *Start Freq* cannot be zero because 0 Hz corresponds to DC analysis. If we want the simulation to be done at a single frequency, we enter 1 in step 3 and the same frequency in steps 4 and 5. If we want the AC sweep to simulate the circuit from 1 Hz to 10 MHz at 10 points per decade, we **CLICKL** on *Decade* in step 2 to make the *X* axis logarithmic, enter 10 in the *Total Pts* box in step 3, enter 1 in the *Start Freq* box, and enter 10Meg in the *End Freq* box. Keep in mind that a decade is a factor of 10. In this case, a decade is from 1 Hz to 10 Hz, from 10 Hz to 100, from 100 to 1 kHz, and so forth.

3. Simulating the Circuit

After providing the necessary specifications and saving the circuit, we perform the AC sweep by selecting **Analysis/Simulate**. If no errors are encountered, the circuit is simulated. At the end of the simulation, the system displays **AC analysis finished** and creates an output file with extension .out. Also, the *Orcad PSpice* program will automatically run if there are no errors. The frequency axis (or *X* axis) is drawn but no curves are shown yet. Select **Trace/Add** from the *Orcad PSpice* menu bar and click on the variables to be displayed. We may also use current or voltage markers to display the traces as explained in the previous section. To use advanced markers such as *vdb*, *idb*, *vphase*, *iphase*, *vreal*, and *ireal*, select **Markers/Mark Advanced**.

In case the resolution of the trace is not good enough, we may need to check the data points to see if they are enough. To do so, select **Tools/Options/Mark Data Points/OK** in the *Orcad PSpice* menu and the data points will be displayed. If necessary, we can improve the resolution by increasing the value of the entry in the **Total Pts** box in the **Analysis/Setup/AC Sweep and Noise Analysis** dialog box for AC sweep.

To generate Bode plots involves using the AC sweep and the dB command in *Orcad PSpice*.

Bode plots are separate plots of magnitude and phase versus frequency. To obtain Bode plots, it is common to use an AC source, say V1, with 1 volt magnitude and zero phase. After we have selected **Analysis/Simulate** and have the *Orcad PSpice* program running, we can display the magnitude and phase plots as mentioned above. Suppose we want to display a Bode magnitude plot of V(4). We select **Trace/Add** and type **dB(V(4))** in **Trace Command** box. dB(V(4)) is equivalent to 20log(V(4)), and because the magnitude of V1 or V(R1:1) is unity, dB(V(4)) actually corresponds to dB(V(4)/V(R1:1)), which is the gain. Adding the trace dB(V(4)) will give a Bode magnitude/gain plot with the *Y* axis in dB.

Once a plot is obtained in the *Orcad PSpice* window, we can add labels to it for documentation purposes. To add a title to the plot, select **Edit/Modify Title** in the *Orcad PSpice* menu and type the title in the dialog box. To add a label to the *Y* axis, select **Plot/Y Axis Settings**, type the label, and **CLICKL OK**. Add a label to the *X* axis in the same manner.

As an alternative approach, we can avoid running the *Orcad PSpice* program by using *pseudocomponents* to send results to the output file. Pseudocomponents are like parts that can be inserted into a schematic as if they were circuit elements, but they do not correspond to circuit elements. We can add them to the circuit for specifying initial conditions or for output control. In fact, we have already used two pseudocomponents, VIEWPOINT and IPROBE, for DC analysis. Other important pseudocomponents and their usage are shown in Fig. D.40 and listed in Table D.2. The pseudocomponents are added to the schematic. To add a pseudocomponent, select **Draw/Get New Parts** in the *Schematics* window, select the pseudocomponent, place it at the desired location, and add the appropriate attributes as usual. Once the pseudocomponents are added to the schematic, we select **Analysis/Setup/AC Sweep** and enter the specifications for the AC sweep, and finally, select **Analysis/Simulate** to perform AC sweep. If no errors are encountered, the voltages and currents specified in the pseudocomponents will be saved in the output file. We obtain the output file by selecting **Analysis/Examine Output** in the *Schematics* window.

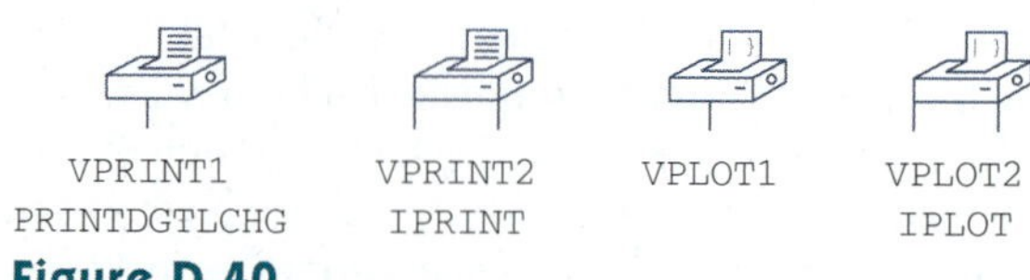

Figure D.40
Print and plot pseudocomponents.

TABLE D.2

Print and plot pseudocomponents.

Symbol	Description
IPLOT	Plot showing branch current; symbol must be placed in series
IPRINT	Table showing branch current; symbol must be placed in series
VPLOT1	Plot showing voltages at the node to which the symbol is connected
VPLOT2	Plot showing voltage differentials between two points to which the symbol is connected
VPRINT1	Table showing voltages at the node to which the symbol is connected
VPRINT2	Table showing voltage differentials between two points to which the symbol is connected

Example D.6

Find current i in the circuit in Fig. D.41.

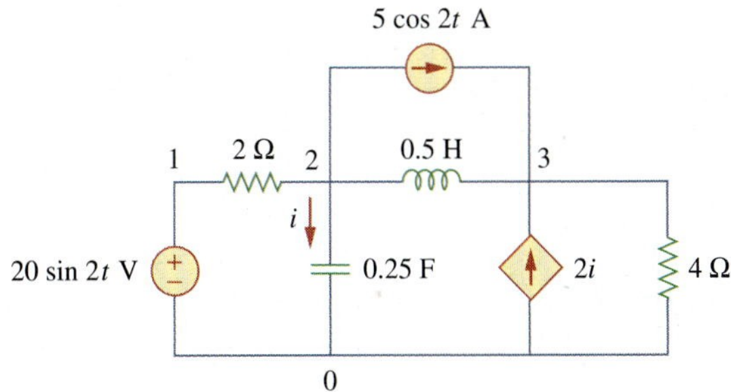

Figure D.41
For Example D.6.

Solution:

Recall that $20 \sin 2t = 20 \cos(2t - 90°)$ and that $f = \omega/2\pi = 2/2\pi = 0.31831$. The schematic is shown in Fig. D.42. The attributes of V1 are set as *ACMAG* = 20, *ACPHASE* = −90; while the attributes of IAC are set as AC = 5. The current-controlled current source is connected in such a way as to conform with the original circuit in Fig. D.41; its gain is set equal to 2. The attributes of the pseudocomponent IPRINT are set as *AC* = yes, *MAG* = yes, *PHASE* = ok, *REAL* =, and *IMAG* =. Since this is a single-frequency ac analysis, we select **Analysis/Setup/AC Sweep** and enter *Total Pts* = 1, *Start Freq* = 0.31831, and *Final Freq* = 0.31831. We save the circuit and select **Analysis/Simulate** for simulation. The output file includes

```
FREQ          IM(V_PRINT3)     IP(V_PRINT3)

3.183E-01     7.906E+00        4.349E+01
```

From the output file, we obtain $I = 7.906\underline{/43.49°}$ A or $i(t) = 7.906 \cos(2t + 43.49°)$ A. This example is for a single-frequency ac analysis; Example D.7 is for AC sweep over a range of frequencies.

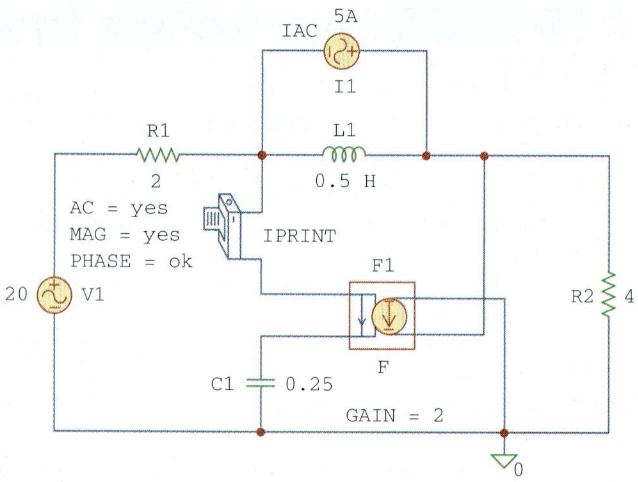

Figure D.42
The schematic of the circuit in Fig. D.41.

Practice Problem D.6

Find $i_x(t)$ in the circuit in Fig. D.43.

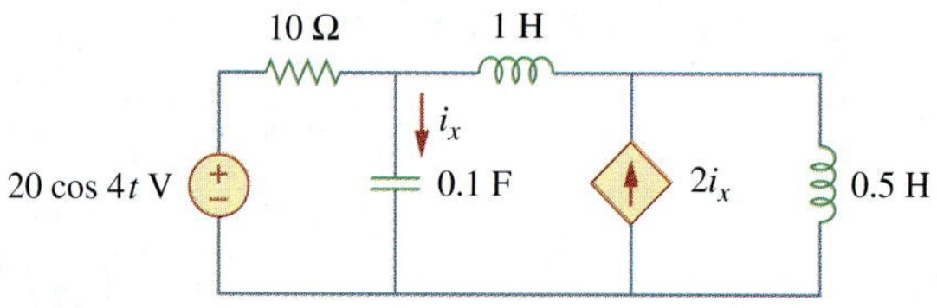

Figure D.43
For Practice Prob. D.6.

Answer: From the output file, $I_x = 7.59\underline{/108.43^\circ}$ or $i_x = 7.59\cos(4t + 108.43^\circ)$ A.

Example D.7

For the *RC* circuit shown in Fig. D.44, obtain the magnitude plot of the output voltage v_o for frequencies from 1 Hz to 10 kHz. Let $R = 1\ \text{k}\Omega$ and $C = 4\ \mu\text{F}$.

Solution:

The schematic is shown in Fig. D.45. We assume that the magnitude of V1 is 1 and its phase is zero; we enter these as the attributes of V1. We

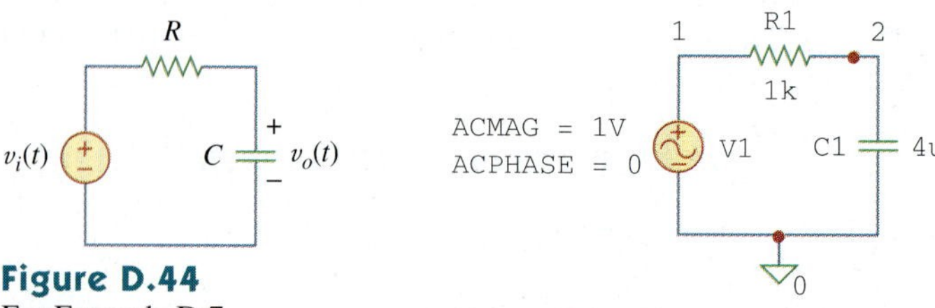

Figure D.44
For Example D.7.

Figure D.45
The schematic of the circuit in Fig. D.44.

also assume 10 points per decade. For the AC sweep specifications, we select **Analysis/Setup/AC Sweep** and enter 10 in the *Total Pts* box, 1 in the *Start Freq* box, and 10k in the *Final Freq* box. After saving the circuit, we select **Analysis/Simulate**. From the *Orcad PSpice* menu, we obtain the plot in Fig. D.46(a) by selecting **Traces/Add** and clicking V(2). Also, by selecting **Trace/Add** and typing dB(V(2)) in the *Trace Command* box, we obtain the Bode plot in Fig. D.46(b). The two plots in Fig. D.46 indicate that the circuit is a lowpass filter: low frequencies are passed while high frequencies are blocked by the circuit.

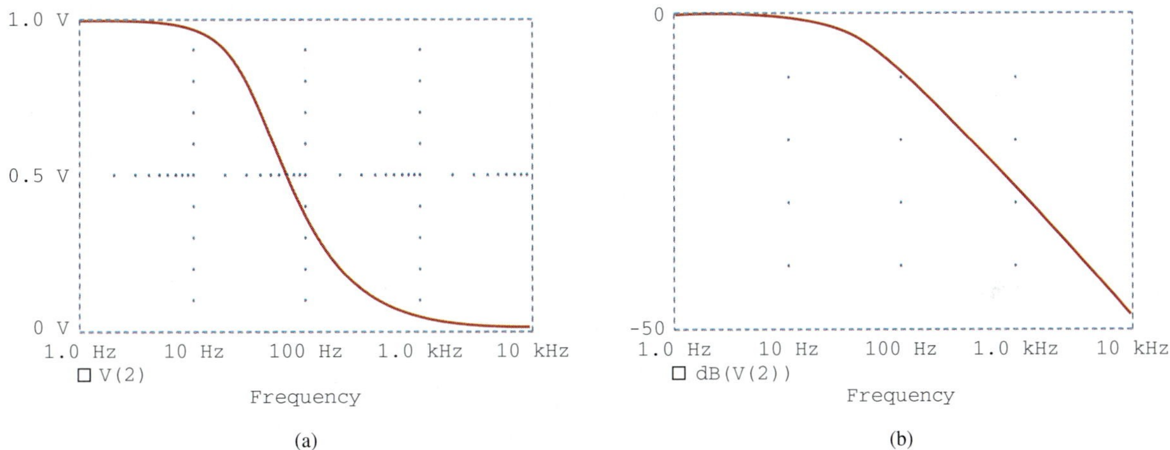

Figure D.46
Result of Example D.7: (a) linear, (b) Bode plot.

Practice Problem D.7

For the circuit in Fig. D.44, replace the capacitor C with an inductor $L = 4$ mH and obtain the magnitude plot (both linear and Bode) for v_o for $10 < f < 100$ MHz.

Answer: See the plots in Fig. D.47.

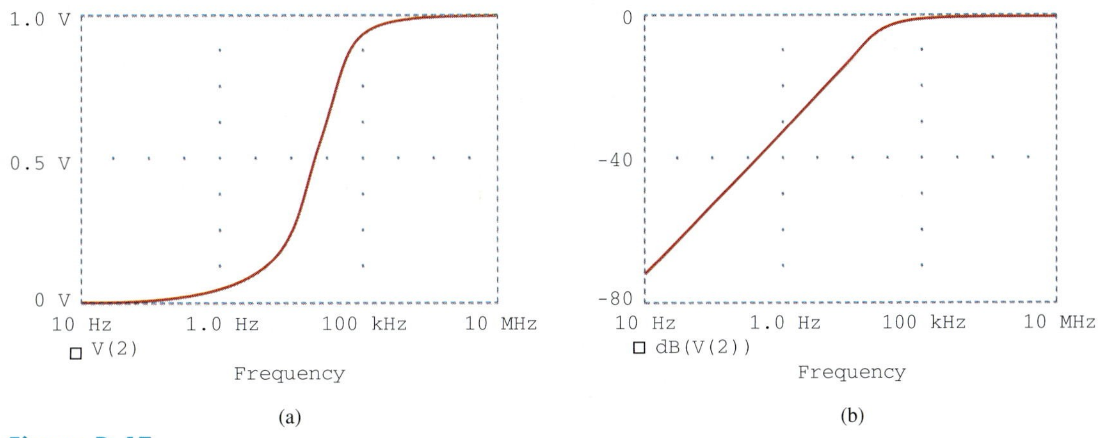

Figure D.47
Result of Practice Prob. D.7: (a) linear plot, (b) Bode plot.

Appendix E

MATLAB

MATLAB has become a powerful tool of technical professionals worldwide. The term *MATLAB* is an abbreviation for *MAT*rix *LAB*oratory, implying that *MATLAB* is a computational tool that uses matrices and vectors (or arrays) to carry out numerical analysis, signal processing, and scientific visualization tasks. Because *MATLAB* uses matrices as its fundamental building blocks, one can write mathematical expressions involving matrices just as easily as one would on paper. *MATLAB* is available for Macintosh, Unix, and Windows operating systems. A student version of *MATLAB* is available for personal computers (PCs). A copy of *MATLAB* can be obtained from

The Mathworks, Inc.
3 Apple Hill Drive
Natick, MA 01760-2098
Phone:(508) 647-7000
Website: http://www.mathworks.com

A brief introduction to *MATLAB* is presented in this appendix. What is presented is sufficient for solving problems in this book. More about *MATLAB* can be found in *MATLAB* books and from on-line help. The best way to learn *MATLAB* is to work with it after having learned the basics.

E.1 *MATLAB* Fundamentals

The Command window is the primary area where you interact with *MATLAB*. A little later, we will learn how to use the text editor to create M-files, which allow for execution of sequences of commands. For now, we focus on how to work in the Command window. We will first learn how to use *MATLAB* as a calculator.

Using *MATLAB* as a Calculator

The following are algebraic operators used in *MATLAB*:

- `+` Addition
- `-` Subtraction
- `*` Multiplication
- `^` Exponentiation
- `/` Right division (`a/b` means $a \div b$)
- `\` Left division (`a\b` means $b \div a$)

To begin to use *MATLAB*, we use these operators. Type commands to the *MATLAB* prompt ">>" in the Command window

TABLE E.3

Special matrices, variables, and constants.

Matrix, Variable, Constant	Remark
`eye`	Identity matrix
`ones`	An array of 1s
`zeros`	An array of 0s
`i or j`	Imaginary unit or `sqrt(-1)`
`pi`	3.142
`NaN`	Not a number
`inf`	Infinity
`eps`	A very small number, `2.2e -`
`rand`	Random element

```
>> H = eig(g)
H =
  -2.6861
   0.1861
```

Note that not all matrices can be inverted. A matrix can be invert and only if its determinant is nonzero. Special matrices, variables, constants are listed in Table E.3. For example, type

```
>> eye(3)
ans =
      1 0 0
      0 1 0
      0 0 1
```

to get a 3×3 identity matrix.

Plotting

To plot using *MATLAB* is easy. For a two-dimensional plot, us plot command with two arguments as follows:

```
>> plot(xdata,ydata)
```

where `xdata` and `ydata` are vectors of the same length contai the data to be plotted.

For example, suppose we want to plot `y = 10*sin(2*pi` from `0` to `5*pi`. We will proceed with the following commands:

```
%x is a vector, 0 <= x <= 5*pi, increments of pi/
%creates a vector y
%creates the plot
```

With this, *MATLAB* responds with the plot in Fig. E.1.

MATLAB will let you graph multiple plots together and di guish them with different colors. This is obtained with the fo `plot(xdata, ydata, 'color')`, where the color is indic by using a character string from the options listed in Table E.4.

For example,

```
>> plot(x1,y1,'r',x2,y2,'b',x3,y3,'--');
```

will graph data `(x1, y1)` in red, data `(x2, y2)` in blue, and `(x3, y3)` in dashed line all on the same plot.

(correct any mistakes by backspacing) and press the Enter key. For example,

```
>> a = 2; b = 4;c = -6;
>> dat = b^2 - 4*a*c
dat =
    64
>> e = sqrt(dat)/10
e =
    0.8000
```

The first command assigns the values 2, 4, and −6 to the variables `a`, `b`, and `c`, respectively. *MATLAB* does not respond because this line ends with a colon. The second command sets `dat` to $b^2 - 4ac$ and *MATLAB* returns the answer as 64. Finally, the third line sets `e` equal to the square root of `dat` and divides by 10. *MATLAB* prints the answer as 0.8. Other mathematical functions, listed in Table E.1, can be used similarly to how the function `sqrt` is used here. Table E.1 provides just a tiny sample of *MATLAB* functions. Others can be obtained from the on-line help. To get help, type

```
>> help
```

A long list of topics will come up. For a specific topic, type the command name. For example, to get help on "log to base 2," type

```
>> help log2
```

A help message on the log function will be displayed. Note that *MATLAB* is case sensitive, so `sin(a)` is not the same as `sin(A)`.

TABLE E.1

Typical elementary math functions.

Function	Remark
`abs(x)`	Absolute value or complex magnitude of x
`acos, acosh(x)`	Inverse cosine and inverse hyperbolic cosine of x in radians
`acot, acoth(x)`	Inverse cotangent and inverse hyperbolic cotangent of x in radians
`angle(x)`	Phase angle (in radian) of a complex number x
`asin, asinh(x)`	Inverse sine and inverse hyperbolic sine of x in radians
`atan, atanh(x)`	Inverse tangent and inverse hyperbolic tangent of x in radians
`conj(x)`	Complex conjugate of x
`cos, cosh(x)`	Cosine and hyperbolic cosine of x in radians
`cot, coth(x)`	Cotangent and hyperbolic cotangent of x in radians
`exp(x)`	Exponential of x
fix	Round toward zero
`imag(x)`	Imaginary part of a complex number x
`log(x)`	Natural logarithm of x
`log2(x)`	Logarithm of x to base 2
`log10(x)`	Common logarithms (base 10) of x
`real(x)`	Real part of a complex number x
`sin, sinh(x)`	Sine and hyperbolic sine of x in radians
`sqrt(x)`	Square root of x
`tan, tanh`	Tangent and hyperbolic tangent of x in radians

TABLE E.2

Matrix operations.

Operation	Remark
A′	Finds the transpose of matrix A
det(A)	Evaluates the determinant of matrix A
inv(A)	Calculates the inverse of matrix A
eig(A)	Determines the eigenvalues of matrix A
diag(A)	Finds the diagonal elements of matrix A

Try the following e

```
>> 3^(log10(25.6
>> y = 2* sin(p
>> exp(y+4-1)
```

In addition to ope allows one to work easil is a special matrix with

```
>> a = [1 -3 6
```

is a row vector. Definin example, a 3 × 3 matrix

```
>> A = [1 2 3;
```

or as

```
>> A = [ 1 2 3
         4 5 6
         7 8 9]
```

In addition to the arith matrix, the operations in

Using the operation follows:

```
>> B = A'
B =
      1   4   7
      2   5   8
      3   6   9
>> C = A + B
C =
      2   6  10
      6  10  14
     10  14  18
>> D = A^3 - B*
D =
    372   432   4
    948  1131  13
   1524  1830  21
>> e = [1 2; 3
e =
   1  2
   3  4
>> f = det(e)
f =
  -2
>> g = inv(e)
g =
  -2.0000   1.000
   1.5000  -0.500
```


TABLE E.4

Various color and line types.

y	Yellow	.	Point
m	Magenta	o	Circle
c	Cyan	x	x mark
r	Red	+	Plus
g	Green	-	Solid
b	Blue	*	Star
w	White	:	Dotted
k	Black	-.	Dashdot
		--	Dashed

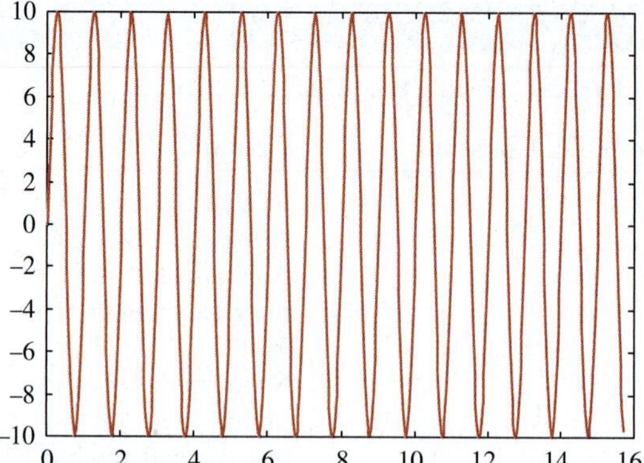

Figure E.1
MATLAB plot of `y = 10*sin(2*pi*x)`.

MATLAB also allows for logarithm scaling. Rather than using the plot command, we use

`loglog log(y)` versus `log(x)`

`semilogx y` versus `log(x)`

`semilogy log(y)` versus `x`

Three-dimensional plots are drawn using the functions mesh and meshdom (mesh domain). For example, to draw the graph of `z = x*exp( - x^2 - y^2)` over the domain `-1 < x, y < 1`, we type the following commands:

```
>> xx = -1:.1:1;
>> yy = xx;
>> [x,y] = meshgrid(xx,yy);
>> z = x.*exp(-x.^2 -y.^2);
>> mesh(z);
```

(The dot symbol used in `x.` and `y.` allows element-by-element multiplication.) The result is shown in Fig. E.2.

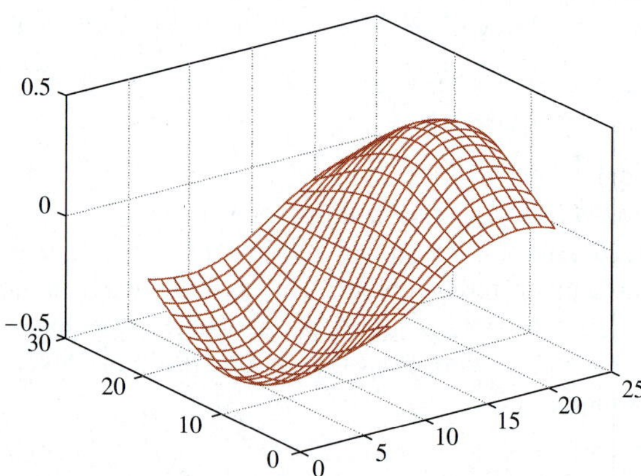

Figure E.2
A three-dimensional plot.

Programming *MATLAB*

So far we have used *MATLAB* as a calculator. You can also use *MATLAB* to create your own program. The command line editing in *MATLAB* can be inconvenient if one has several lines to execute. To avoid this problem, you can create a program that is a sequence of statements to be executed. If you are in the Command window, click **File/New/M**-files to open a new file in the *MATLAB* Editor/Debugger or simple text editor. Type the program and save it in a file with an extension `.m`, say `filename.m`; it is for this reason that it is called an M-file. Once the program is saved as an M-file, exit the Debugger window. You are now back in the Command window. Type the file without the extension `.m` to get results. For example, the plot that was made in Fig. E.2 can be improved by adding title and labels and being typed as an M-file called `example1.m`.

```
x = 0:pi/100:5*pi;                   % x is a vector, 0 <= x <= 5*pi, increments of pi/100
y = 10*sin(2*pi*x);                  % creates a vector y
plot(x,y);                           % create the plot
xlabel('x (in radians)');            % label the x axis
ylabel('10*sin(2*pi*x)');            % label the y axis
title('A sine functions');           % title the plot
grid                                 % add grid
```

Once the file is saved as `example1.m` and you exit the text editor, type

```
>> example1
```

in the Command window and hit **Enter** to obtain the result shown in Fig. E.3.

To allow flow control in a program, certain relational and logical operators are necessary. They are shown in Table E.5. Perhaps the most commonly used flow control statements are `for` and `if`. The `for`

TABLE E.5

Relational and logical operators.

Operator	Remark
`<`	less than
`<=`	less than or equal
`>`	greater than
`>=`	greater than or equal
`==`	equal
`~=`	not equal
`&`	and
`|`	or
`~`	not

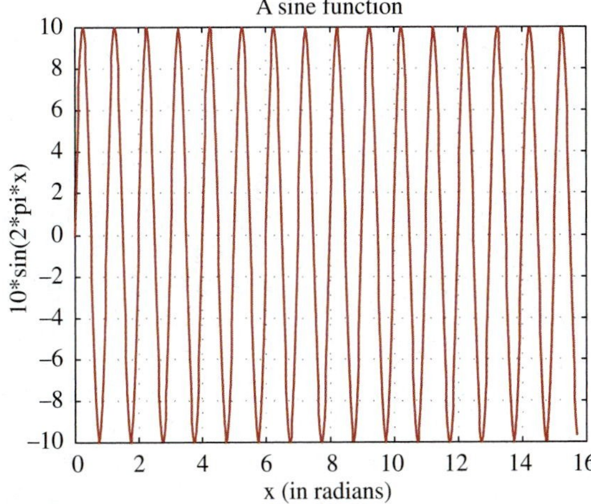

Figure E.3
MATLAB plot of `y = 10*sin(2*pi*x)` with title and labels.

statement is used to create a loop or a repetitive procedure and has the general form

```
for x = array
   [commands]
end
```

The `if` statement is used when certain conditions need to be met before an expression is executed. It has the general form

```
if expression
   [commands if expression is True]
else
   [commands if expression is False]
end
```

For example, suppose we have an array `y(x)` and we want to determine the minimum value of `y` and its corresponding index `x`. This can be done by creating an M-file as shown here.

```
% example2.m
% This program finds the minimum y value and
  its corresponding x index
x = [1 2 3 4 5 6 7 8 9 10]; %the nth term in y
y = [3 9 15 8 1 0 -2 4 12 5];
min1 = y(1); for k = 1:10
   min2 = y(k);
   if(min2 < min1)
      min1 = min2;
      xo = x(k);
   else
      min1 = min1;
   end
end
diary
min1, xo
diary off
```

Note the use of the `for` and `if` statements. When this program is saved as `example2.m`, we execute it in the Command window and obtain the minimum value of `y` as `-2` and the corresponding value of `x` as `7`, as expected.

```
>> example2
min1 =
  -2
xo =
   7
```

If we are not interested in the corresponding index, we could do the same thing using the command

```
>> min(y)
```

The following tips are helpful in working effectively with *MATLAB*:

- Comment your M-file by adding lines beginning with a % character.
- To suppress output, end each command with a semicolon (`;`); you may remove the semicolon when debugging the file.
- Press the up and down arrow keys to retrieve previously executed commands.
- If your expression does not fit on one line, use an ellipse (. . .) at the end of the line and continue on the next line. For example, *MATLAB* considers

  ```
  y = sin(x + log10(2x + 3)) + cos(x + ...
  log10(2x + 3));
  ```

  as one line of expression.
- Keep in mind that variable and function names are case sensitive.

Solving Equations

Consider the general system of n simultaneous equations:

$$\begin{aligned} a_{11}x_1 + a_{12}x_2 + \cdots + a_{1n}x_n &= b_1 \\ a_{21}x_1 + a_{22}x_2 + \cdots + a_{2n}x_n &= b_2 \\ &\vdots \\ a_{n1}x_1 + a_{n2}x_2 + \cdots + a_{nn}x_n &= b_n \end{aligned}$$

or in matrix form

$$\mathbf{AX} = \mathbf{B}$$

where

$$\mathbf{A} = \begin{bmatrix} a_{11} & a_{12} & \cdots & a_{1n} \\ a_{21} & a_{22} & \cdots & a_{2n} \\ \cdots & \cdots & \cdots & \cdots \\ a_{n1} & a_{n2} & a_{n3} & a_{n4} \end{bmatrix} \qquad \mathbf{X} = \begin{bmatrix} x_1 \\ x_2 \\ \cdots \\ x_n \end{bmatrix} \qquad \mathbf{B} = \begin{bmatrix} b_1 \\ b_2 \\ \cdots \\ b_n \end{bmatrix}$$

A is a square matrix and is known as the coefficient matrix, while **X** and **B** are vectors. **X** is the solution vector we are seeking to get. There are two ways to solve for **X** in *MATLAB*. First, we can use the backslash operator(\) so that

```
X = A\B
```

Second, we can solve for **X** as

$$\mathbf{X} = \mathbf{A}^{-1}\mathbf{B}$$

which in *MATLAB* is the same as

```
X = inv(A)*B
```

Example E.1

Use *MATLAB* to solve Example A.2.

Solution:
From Example A.2, we obtain matrix **A** and vector **B** and enter them in *MATLAB* as follows.

```
>> A = [25 -5 -20; -5 10 -4; -5 -4 9]
A =
    25  -5  -20
    -5  10   -4
    -5  -4    9
>> B = [50 0 0]'
B =
    50
     0
     0
>> X = inv(A)*B
X =
    29.6000
    26.0000
    28.0000
>> X = A\B
X =
    29.6000
    26.0000
    28.0000
```

Thus, `x1 = 29.6, x2 = 26`, and `x3 = 28.`

Practice Problem E.1

Solve the problem in Practice Prob. A.2 using *MATLAB*.

Answer: `x1 = 3 = x3, x2 = 2.`

E.2 DC Circuit Analysis

There is nothing special in applying *MATLAB* to resistive dc circuits. We apply mesh and nodal analysis as usual and solve the resulting simultaneous equations using *MATLAB* as is described in Section E.1. Examples E.2 to E.5 illustrate.

Example E.2

Use nodal analysis to solve for the nodal voltages in the circuit of Fig. E.4.

Solution:
At node 1,

$$2 = \frac{V_1 - V_2}{4} + \frac{V_1 - 0}{8} \rightarrow 16 = 3V_1 - 2V_2 \qquad \textbf{(E.2.1)}$$

At node 2,

$$3I_x = \frac{V_2 - V_1}{4} + \frac{V_2 - V_3}{2} + \frac{V_2 - V_4}{2}$$

But

$$I_x = \frac{V_4 - V_3}{4}$$

so that

$$3\left(\frac{V_4 - V_3}{4}\right) = \frac{V_2 - V_1}{4} + \frac{V_2 - V_3}{2} + \frac{V_2 - V_4}{2} \rightarrow \qquad \textbf{(E.2.2)}$$
$$0 = -V_1 + 5V_2 + V_3 - 5V_4$$

At node 3,

$$3 = \frac{V_3 - V_2}{2} + \frac{V_3 - V_4}{4} \rightarrow 12 = -2V_2 + 3V_3 - V_4 \qquad \textbf{(E.2.3)}$$

At node 4,

$$0 = 2 + \frac{V_4 - V_2}{2} + \frac{V_4 - V_3}{4} \rightarrow -8 = -2V_2 - V_3 + 3V_4 \qquad \textbf{(E.2.4)}$$

Combining Eqs. (E.2.1) to (E.2.4) gives

$$\begin{bmatrix} 3 & -2 & 0 & 0 \\ -1 & 5 & 1 & -5 \\ 0 & -2 & 3 & -1 \\ 0 & -2 & -1 & 3 \end{bmatrix} \begin{bmatrix} V_1 \\ V_2 \\ V_3 \\ V_4 \end{bmatrix} = \begin{bmatrix} 16 \\ 0 \\ 12 \\ -8 \end{bmatrix}$$

or

$$\mathbf{AV} = \mathbf{B}$$

We now use *MATLAB* to determine the nodal voltages contained in vector **V**.

```
>> A = [ 3 -2   0   0;
        -1   5   1 -5;
         0 -2   3 -1;
         0 -2 -1   3];
>> B = [16   0 12 -8]';
>> V = inv(A)*B
V =
    -6.0000
   -17.0000
   -13.5000
   -18.5000
```

Hence $V_1 = -6.0$, $V_2 = -17$, $V_3 = -13.5$, and $V_4 = -18.5$ V.

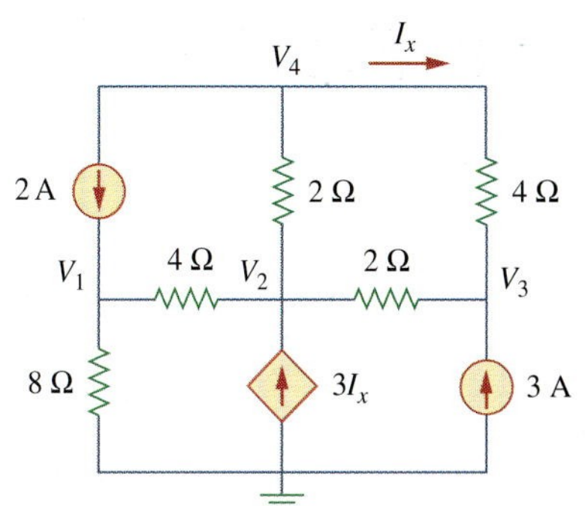

Figure E.4
For Example E.2.

Practice Problem E.2

Find the nodal voltages in the circuit in Fig. E.5 using *MATLAB*.

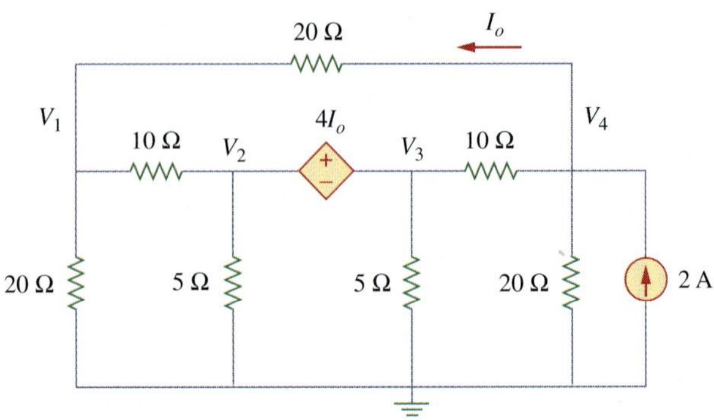

Figure E.5
For Practice Prob. E.2.

Answer:
$V_1 = 14.55$, $V_2 = 38.18$, $V_3 = -34.55$, and $V_4 = -3.636$ V.

Example E.3

Use *MATLAB* to solve for the mesh currents in the circuit in Fig. E.6.

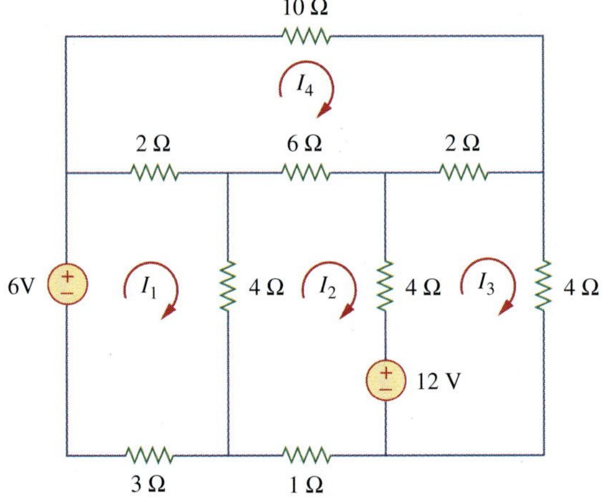

Figure E.6
For Example E.3.

Solution:
For the four meshes,

$$-6 + 9I_1 - 4I_2 - 2I_4 = 0 \longrightarrow 6 = 9I_1 - 4I_2 - 2I_4 \quad \textbf{(E.3.1)}$$

$$12 + 15I_2 - 4I_1 - 4I_3 - 6I_4 = 0 \longrightarrow$$

$$-12 = -4I_1 + 15I_2 - 4I_3 - 6I_4 \quad \textbf{(E.3.2)}$$

$$-12 + 10I_3 - 4I_2 - 2I_4 = 0 \longrightarrow 12 = -4I_2 + 10I_3 - 2I_4 \quad \textbf{(E.3.3)}$$

$$20I_4 - 2I_1 - 6I_2 - 2I_3 = 0 \longrightarrow 0 = -2I_1 - 6I_2 - 2I_3 + 20I_4 \quad \textbf{(E.3.4)}$$

Putting Eqs. (E.3.1) to (E.3.4) together in matrix form, we have

$$\begin{bmatrix} 9 & -4 & 0 & -2 \\ -4 & 15 & -4 & -6 \\ 0 & -4 & 10 & -2 \\ -2 & -6 & -2 & 20 \end{bmatrix} \begin{bmatrix} I_1 \\ I_2 \\ I_3 \\ I_4 \end{bmatrix} = \begin{bmatrix} 6 \\ -12 \\ 12 \\ 0 \end{bmatrix}$$

or $\mathbf{AI} = \mathbf{B}$, where the vector **I** contains the unknown mesh currents. We now use *MATLAB* to determine **I** as follows:

```
>> A = [9 -4  0 -2; -4 15 -4 -6;
        0 -4 10 -2;  -2 -6 -2 20]
A =
        9  -4   0  -2
       -4  15  -4  -6
        0  -4  10  -2
       -2  -6  -2  20
>> B = [6 -12 12 0]'
B =
        6
      -12
       12
        0
>> I = inv(A)*B
I =
        0.5203
       -0.3555
        1.0682
        0.0522
```

Thus, $I_1 = 0.5203$, $I_2 = -0.3555$, $I_3 = 1.0682$, and $I_4 = 0.0522$ A.

Practice Problem E.3

Find the mesh currents in the circuit in Fig. E.7 using *MATLAB*.

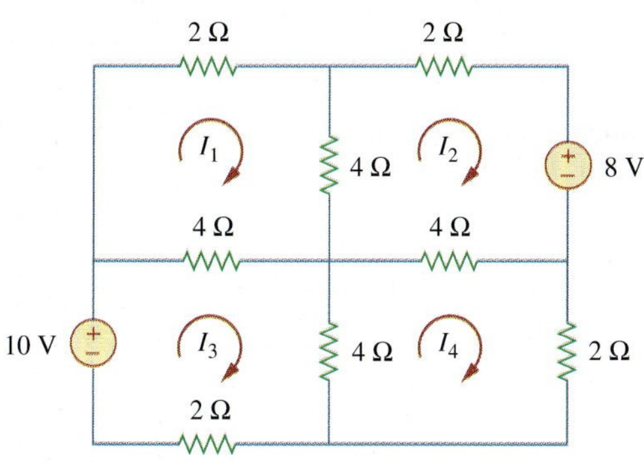

Figure E.7
For Practice Prob. E.3.

Answer: $I_1 = 0.2222$, $I_2 = -0.6222$, $I_3 = 1.1778$, and $I_4 = 0.2222$ A.

E.3 AC Circuit Analysis

Using *MATLAB* in ac circuit analysis is similar to how *MATLAB* is used for dc circuit analysis. We must first apply nodal or mesh analysis to the circuit and then use *MATLAB* to solve the resulting system of equations. However, the circuit is in the frequency domain, and we are dealing with phasors or complex numbers. So in addition to what we learned in Section E.2, we need to understand how *MATLAB* handles complex numbers.

MATLAB expresses complex numbers in the usual manner, except that the imaginary part can be either j or i representing $\sqrt{-1}$. Thus, $3 - j4$ can be written in *MATLAB* as `3 - j4`, `3 - j*4`, `3 - i4`, or `3 - I*4`. Here are the other complex functions:

`abs(A)`	Absolute value of magnitude of `A`
`angle(A)`	Angle of `A` in radians
`conj(A)`	Complex conjugate of `A`
`imag(A)`	Imaginary part of `A`
`real(A)`	Real part of `A`

Keep in mind that an angle in radians must be multiplied by $180/\pi$ to convert it to degrees, and vice versa. Also, the transpose operator (`'`) gives the complex conjugate transpose, whereas the dot-transpose (`.'`) transposes an array without conjugating it.

Example E. 4

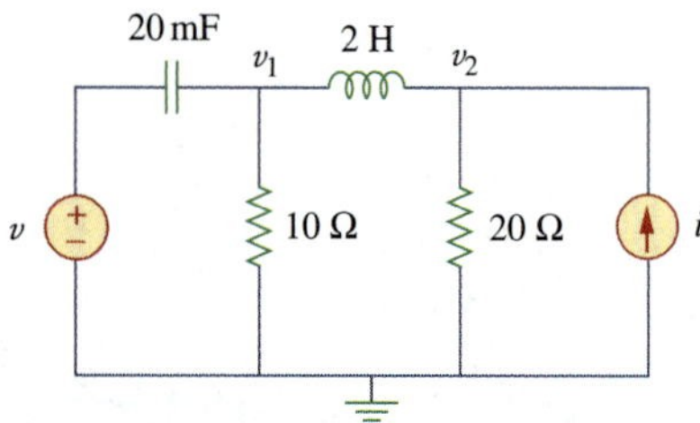

Figure E.8
For Example E.4.

In the circuit of Fig. E.8, let $v = 4\cos(5t - 30°)$ V and $i = 0.8\cos 5t$ A. Find v_1 and v_2.

Solution:

As usual, we convert the circuit in the time-domain to its frequency-domain equivalent.

$$v = 4\cos(5t - 30°) \longrightarrow \mathbf{V} = 4\underline{/-30°}, \quad \omega = 5$$

$$i = 0.8\cos 5t \longrightarrow \mathbf{I} = 8\underline{/0°}$$

$$2\text{ H} \longrightarrow j\omega L = j5 \times 2 = j10$$

$$20\text{ mF} \longrightarrow \frac{1}{j\omega C} = \frac{1}{j10\ \Omega \times 10^{-3}} = -j10$$

Thus, the frequency-domain equivalent circuit is shown in Fig. E.9. We now apply nodal analysis to this.

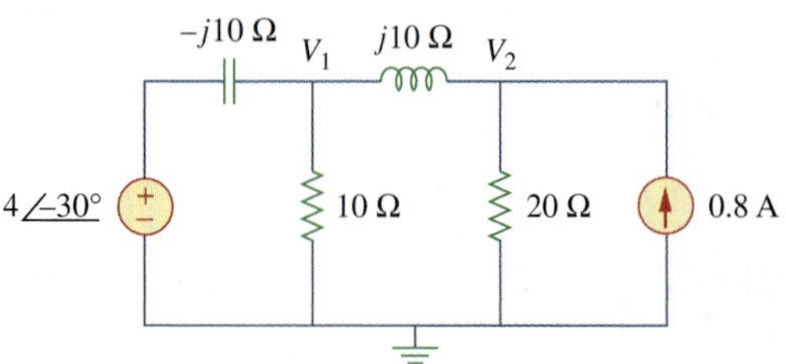

Figure E.9
The frequency-domain equivalent circuit of the circuit in Fig. E.8.

At node 1,

$$\frac{4\angle{-30^\circ} - V_1}{-j10} = \frac{V_1}{10} + \frac{V_1 - V_2}{j10} \longrightarrow 4\angle{-30^\circ} = 3.468 - j2$$

$$= -jV_1 + V_2 \qquad \textbf{(E.4.1)}$$

At node 2,

$$0.8 = \frac{V_2}{20} + \frac{V_2 - V_1}{j10} \longrightarrow j16 = -2V_1 + (2 + j)V_2 \quad \textbf{(E.4.2)}$$

Equations (E.4.1) and (E.4.2) can be cast in matrix form as

$$\begin{bmatrix} -j & 1 \\ -2 & (2 + j) \end{bmatrix} \begin{bmatrix} V_1 \\ V_2 \end{bmatrix} = \begin{bmatrix} 3.468 - j2 \\ j16 \end{bmatrix}$$

or **AV** = **B**. We use *MATLAB* to invert **A** and multiply the inverse by **B** to get **V**.

```
>> A = [-j 1; -2 (2+j)]
A =
    0 - 1.0000i 1.000
    -2.0000 2.0000 + 1.000 i
>> B = [(3.468 - 2j) 16j].' %note the dot-transpose
B =
    3.4680 - 2.0000i
    0 + 16.0000i
>> V = inv(A)*B
V =
    4.6055 - 2.4403i
    5.9083 + 2.6055i
>> abs(V(1))
ans =
    5.2121
>> angle(V(1))*180/pi %converts angle from
    radians to degrees
ans =
   -27.9175
>> abs(V(2))
ans =
   6.4573
>> angle(V(2))*180/pi
ans =
   23.7973
```

Thus,

$$V_1 = 4.6055 - j2.4403 = 5.212\angle{-27.92^\circ}$$
$$V_2 = 5.908 + j2.605 = 6.457\angle{23.8^\circ}$$

In the time domain,

$$v_1 = 4.605 \cos(5t - 27.92°) \text{ V}, \qquad v_2 = 6.457 \cos(5t + 23.8°) \text{ V}$$

Practice Problem E.4

Calculate v_1 and v_2 in the circuit in Fig. E.10 given $i = 4\cos(10t + 40°)$ A and $v = 12 \cos 10t$ V.

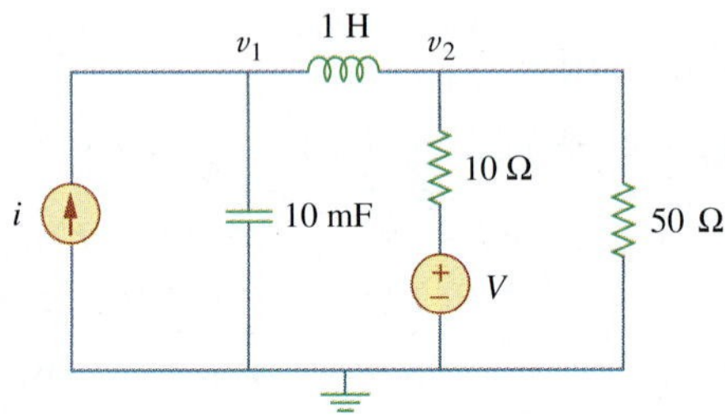

Figure E.10
For Practice Prob. E.4.

Answer: 63.58 cos(10t − 10.68°) V, 40 cos(10t − 50°) V.

Example E.5

In the unbalanced three-phase system shown in Fig. E.11, find currents I_1, I_2, I_3, and I_{Bb}. Let

$$Z_A = 12 + j10\ \Omega, \qquad Z_B = 10 - j8\ \Omega, \qquad Z_C = 15 + j6\ \Omega$$

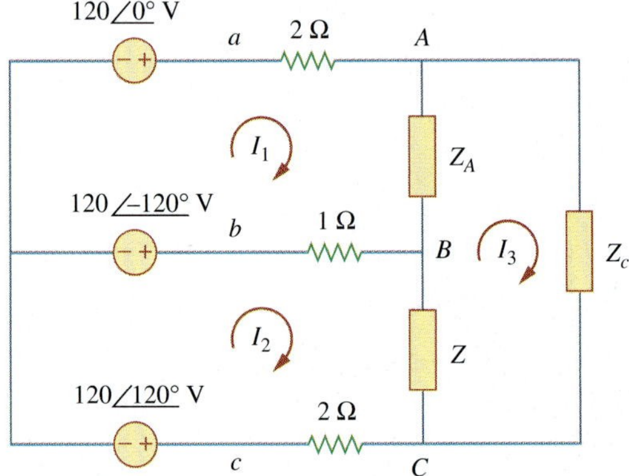

Figure E.11
For Example E.5.

Solution:
For mesh 1,

$$120\underline{/-120°} - 120\underline{/0°} + I_1(2 + 1 + 12 + j10) - I_2 - I_3(12 + j10) = 0$$

or

$$I_1(15 + j10) - I_2 - I_3(12 + j10) = 120\underline{/0°} - 120\underline{/-120°} \qquad \textbf{(E.5.1)}$$

For mesh 2,

$$120\underline{/120^\circ} - 120\underline{/-120^\circ} + I_2(2 + 1 + 10 - j8) - I_1 - I_3(10 - j8) = 0$$

or

$$-I_1 + I_2(13 - j8) - I_3(10 - j8) = 120\underline{/-120^\circ} - 120\underline{/120^\circ} \qquad \textbf{(E.5.2)}$$

For mesh 3,

$$I_3(12 + j10 + 10 - j8 + 15 + j6) - I_1(12 + j10) - I_2(10 - j8) = 0$$

or

$$-I_1(12 + j10) - I_2(10 - j8) - I_3(37 + j8) = 0 \qquad \textbf{(E.5.3)}$$

In matrix form, we can express Eqs. (E.5.1) to (E.5.3) as

$$\begin{bmatrix} 15 + j10 & -1 & -12 - j10 \\ -1 & 13 - j8 & -10 + j8 \\ -12 - j10 & -10 + j8 & 37 + j8 \end{bmatrix} \begin{bmatrix} I_1 \\ I_2 \\ I_3 \end{bmatrix} = \begin{bmatrix} 120\underline{/0^\circ} - 120\underline{/-120^\circ} \\ 120\underline{/-120^\circ} - 120\underline{/120^\circ} \\ 0 \end{bmatrix}$$

or

$$\mathbf{ZI} = \mathbf{V}$$

We input matrices **Z** and **V** into *MATLAB* to get I.

```
>> z = [(15 + 10j) -1 (-12 - 10j);
      -1 (13 - 8j) (-10 + 8j);
      (-12 - 10j) (-10 + 8j) (37 + 8j)];
>> c1=120*exp(j*pi*(-120)/180);
>> c2=120*exp(j*pi*(-120)/180);
>> a1=120 - c1; a2=c1 - c2;
>> V = [a1; a2; 0]
>> I = inv(z)*V
I=
   16.9910 - 6.5953i
   12.4023 - 16.9993i
    5.6621 - 6.0471i
>> IbB = I(2) - I(1)
IbB =
   -4.5887 - 10.4039i
>> abs(I(1))
ans =
    18.2261
>> angle(I(1))*180/pi
ans =
   -21.2146
```

```
>> abs  (I(2))
ans =
    21.0426
>> angle(I(2))*180/pi
ans =
  -53.8864
>> abs(I(3))
ans =
    8.2841
>> angle(I(3))*180/pi
ans =
  -46.8833
>> abs(IbB)
ans =
  11.3709
>> angle(IbB)*180/pi
ans =
  -113.8001
```

Thus, $I_1 = 18.23\underline{/-21.21^\circ}$, $I_2 = 21.04\underline{/-58.89^\circ}$, $I_3 = 8.284\underline{/-46.88^\circ}$, and $I_{bB} = 11.37\underline{/-113.8^\circ}$A.

Practice Problem E.5

In the unbalanced wye-wye three-phase system in Fig. E.12, find the line currents I_1, I_2, and I_3 and the phase voltage V_{CN}.

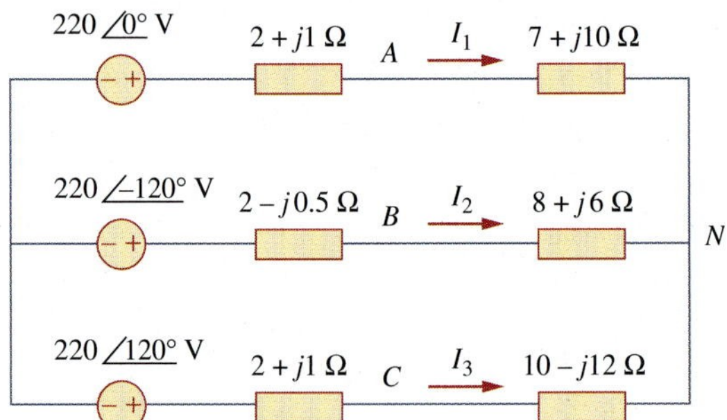

Figure E.12
For Practice Prob. E.5.

Answer: $22.66\underline{/-26.54^\circ}$ A, $6.036\underline{/-150.48^\circ}$ A, $19.93\underline{/138.9^\circ}$ A, $94.29\underline{/159.3^\circ}$ V.

E.4 Frequency Response

Frequency response involves plotting the magnitude and phase of the transfer function $H(s) = D(s)/N(s)$ or obtaining the Bode magnitude and phase plots of $H(s)$. One hard way to obtain the plots is to generate data using the `for` loop for each value of $s = j\omega$ for a given range of ω and then plot the data as we did in Section E.1. However, there is an easy way that allows us to use one of two *MATLAB* commands: `freqs` and `bode`. For each command, we must first specify `H(s)` as `num` and `den`, where `num` and `den` are the vectors of coefficients of the numerator `N(s)` and denominator `D(s)` in descending powers of `s`, i.e., from the highest power to the constant term. The general form of the `bode` function is

```
bode(num, den, range);
```

where `range` is a specified frequency interval for the plot. If `range` is omitted, *MATLAB* automatically selects the frequency range. The `range` could be linear or logarithmic. For example, for $1 < \omega < 1000$ rad/s with 50 plot points, we can specify a linear `range` as

```
range = linspace(1,1000,50);
```

For a logarithmic `range` with $10^{-2} < \omega < 10^4$ rad/s and 100 plot points in between, we specify `range` as

```
range = logspace(-2,4,100);
```

For the `freqs` function, the general form is

```
hs = freqs(num, den, range);
```

where `hs` is the frequency response (generally complex). We still need to calculate the magnitude in decibels as

```
mag = 20*log 10(abs(hs))
```

and phase in degrees as

```
phase = angle(hs)*180/pi
```

and plot them, whereas the `bode` function does it all at once. We illustrate with an example.

Example E.6

Use *MATLAB* to obtain the Bode plots of

$$G(s) = \frac{s^3}{s^3 + 14.8s^2 + 38.1s + 2554}$$

Solution:
With the explanation previously given, we develop the *MATLAB* code as shown here.

```
% for example e.6
num=[1 0 0 0];
den = [1 14.8 38.1 2554];
w = logspace(-1,3);
bode(num, den, w);
title('Bode plot for a highpass filter')
```

Running the program produces the Bode plots in Fig. E.13. It is evident from the magnitude plot that $G(s)$ represents a highpass filter.

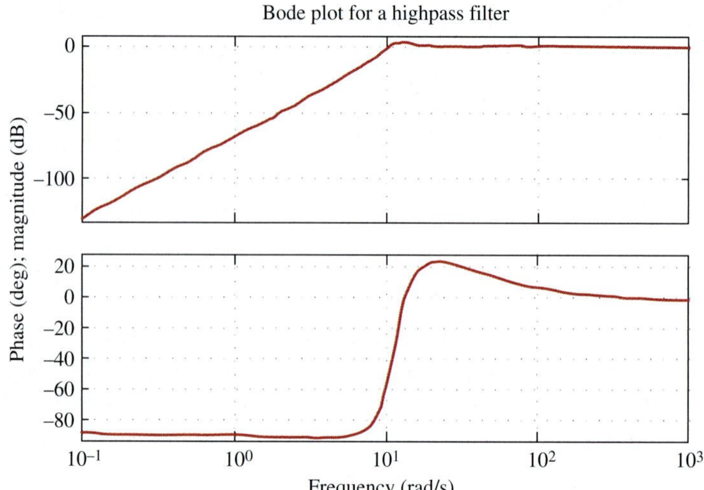

Figure E.13
For Example E.6.

Practice Problem E.6

Use *MATLAB* to determine the frequency response of

$$H(s) = \frac{10(s + 1)}{s^2 + 6s + 100}$$

Answer: See Fig. E.14.

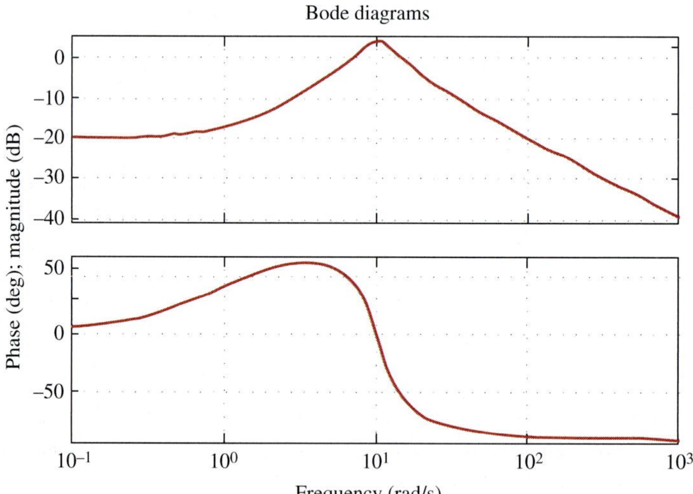

Figure E.14
For Practice Prob. E.6.

Appendix F

KCIDE for Circuits

Engineers of the twenty-first century will need to be able to work in a "knowledge capturing integrated design environment" (known as KCIDE). Essentially, engineers will go to their computer where they will do their work on a platform, much like Windows, where their various software packages, laboratory work, and other support software packages (such as *Word*) will all come together and interact with each other to help them with their work. These platforms will capture the work being done by the engineer and make the data available to be used in any manner the engineer chooses (such as preliminary design reports, user manuals, papers, books, proposals, or requests for proposals).

A detailed presentation of all the elements associated with learning how to work in such an environment is beyond the scope of this book. However, a platform to begin the process of training engineers to work in this environment is included in this textbook. *KCIDE for Circuits* was designed to assist the circuits student to learn how to work in a simplified KCIDE environment designed especially for electrical circuits students. The software that is used in the platform includes *PSpice*, *MATLAB*, *Excel*, *Word*, and *PowerPoint*.

In this appendix, we will help you to understand the *KCIDE for Circuits* platform and how to use it. The software can be obtained, free of any charges, from the website http://KCIDE.FennResearch.org. More details and examples are also included at the website. In addition, we will also have support services available at the website.

F.1 How to Work with *KCIDE for Circuits*

The structure of the platform and how it is effectively used follows the problem-solving process used throughout the text. This is essentially a systems approach to problem solving that uses a structured process to capture your work and present it in two different formats. It will be helpful to work through an example to see how to use the platform.

Example F.1

Use the *KCIDE for Circuits* platform to solve Example 3.2.

Solution:

Opening the software, we see the screen shown in Fig. F.1, where we define a new project. Although we can name the project anyway we

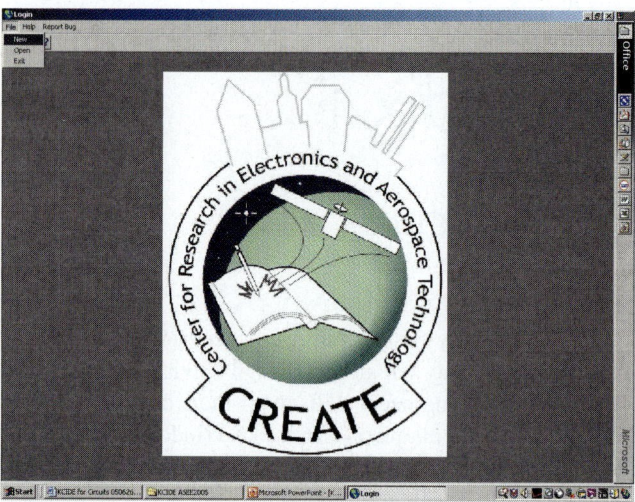

Figure F.1
Creating a new project in *KCIDE for Circuits.*

wish to, we name it KCIDE Example F-1 050626 (see Fig. F.2). Note that the last six digits are: year/month/day. The reason for this is that if we create different files corresponding to different dates, the files will always appear in chronological order.

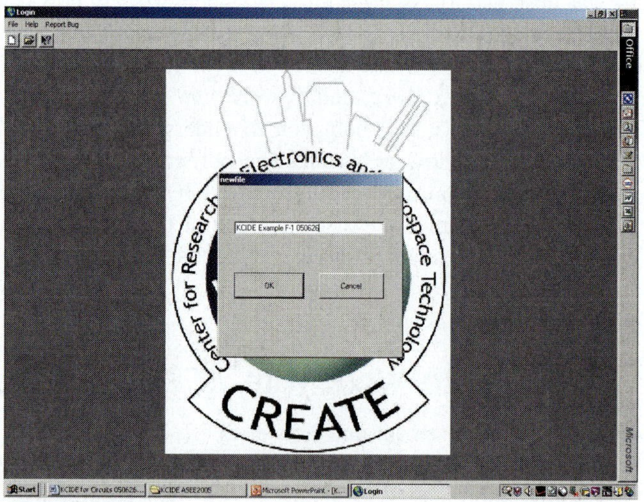

Figure F.2
Naming the project.

We now enter the problem statement into the screen shown in Fig. F.3. After we have entered the problem statement, we can click on the button to Open *PSpice*. The next screen, Fig. F.4, shows what is seen when the Open *PSpice* button is clicked. To open the *PSpice* schematic capture, we need to click on the page 1 icon. In Schematic, we create the circuit representing our problem. This is shown in Fig. F.5.

We now need to enter all we know about the problem by entering our problem analysis into the text box and then identifying the number

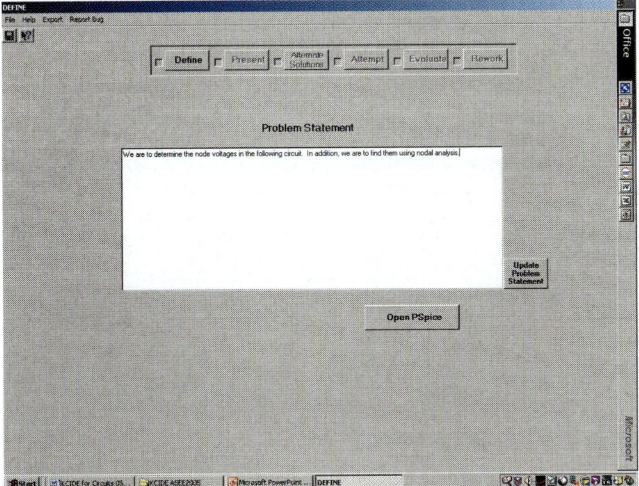

Figure F.3
Entering the problem statement.

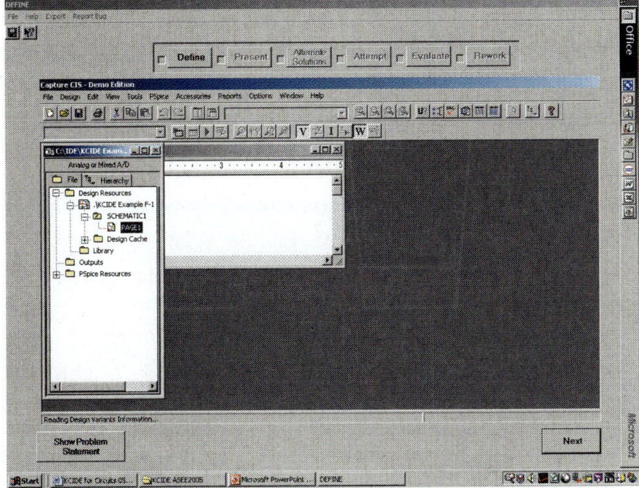

Figure F.4
How to open the schematic capture of *PSpice*.

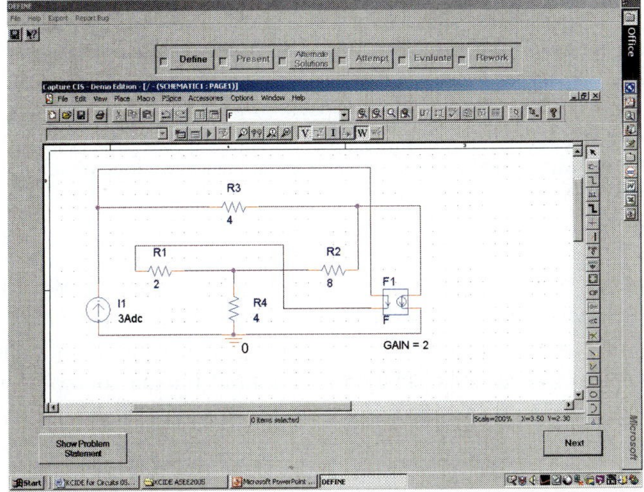

Figure F.5
Circuit for Example F.1.

of unknown nodes and loops for the circuit (see Fig. F.6). We continue this process by going to the next screen and entering the requested information (see Fig. F.7).

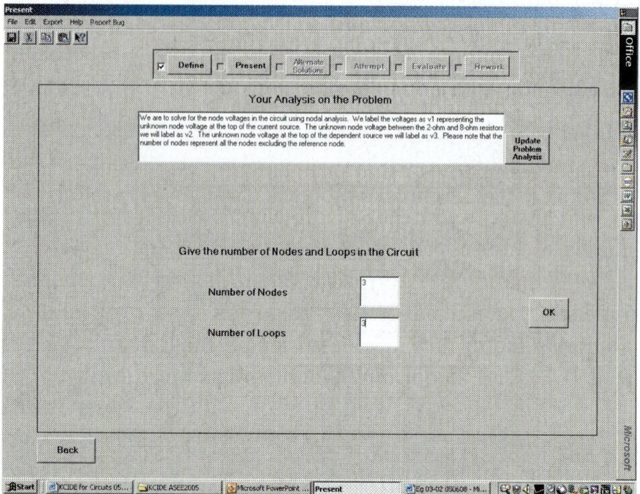

Figure F.6
Presenting what we know about the problem, part 1.

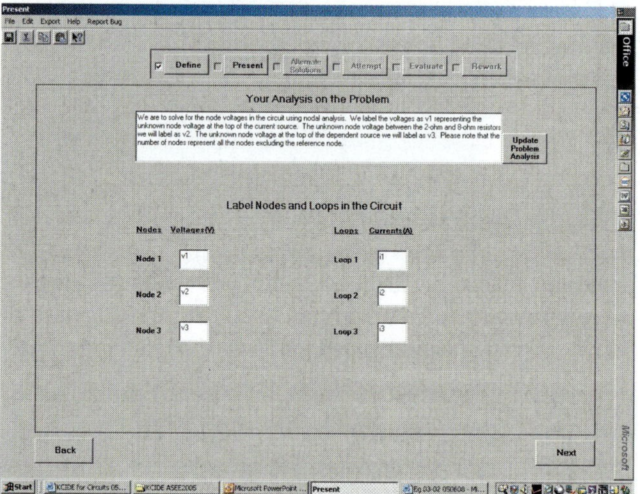

Figure F.7
Identifying the unknown node voltages and unknown loop currents, part 2.

We now proceed to selecting the method of solution. We do this by entering the requested information into the screen shown in Fig. F.8. Now we can develop the equations that will generate a solution for the problem. Since nodal analysis is required for the solution for the node voltages, all we need to do is to write the nodal equations. Once we have the appropriate equations, as shown in Fig. F.9, we can select a solution technique. In this case we chose Excel to solve our simultaneous equations.

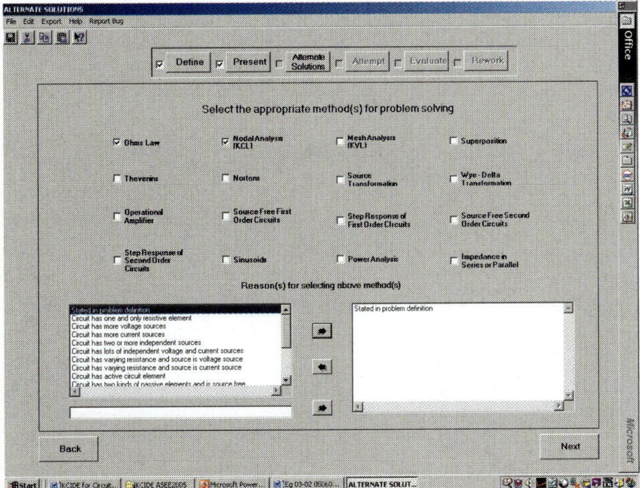

Figure F.8
Selecting the method of solving the problem.

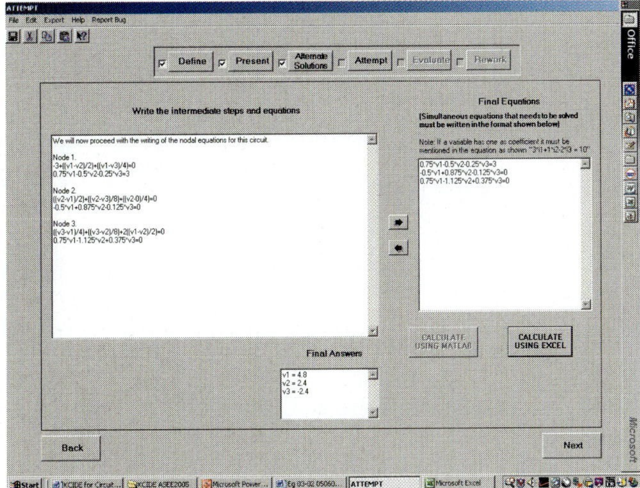

Figure F.9
Solving for the unknown node voltages.

Now, when we go to the next screen, the Evaluate portion of the solution, we actually open *PSpice* again. We need to open page 1 to retrieve our original circuit (see Fig. F.10). Once we have our original *PSpice* circuit, we need to prepare it for solving for our unknowns. The first step in this task is to go to the *PSpice* button and select New Simulation Profile (see Fig. F.11).

We next need to assign a name to the new simulation profile (Fig. F.12). Clicking on the Create button produces the screen shown in Fig. F.13. For this problem, we select Bias Point for the Analysis type.

Clicking on OK returns the screen to the original condition. Now we go to the *PSpice* button and select Run from the dropdown menu,

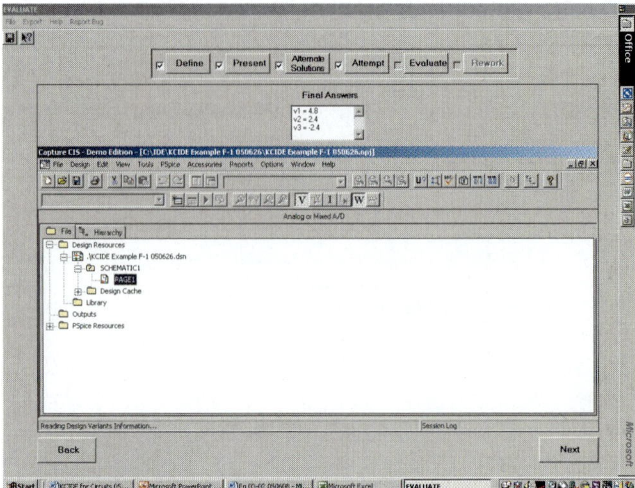

Figure F.10
Opening *PSpice* again.

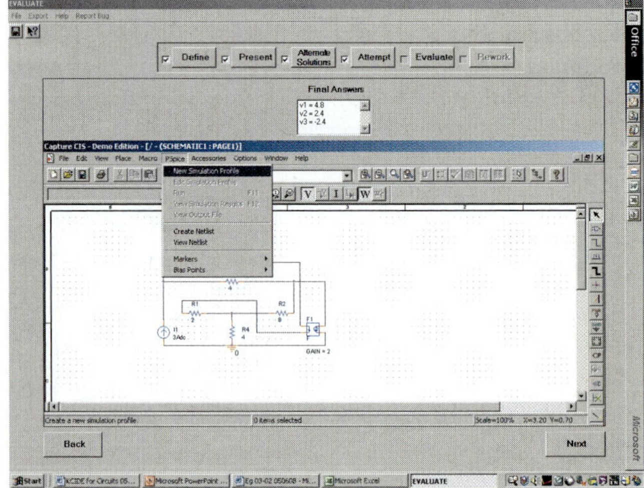

Figure F.11
Setting up our circuit for solution by *PSpice*.

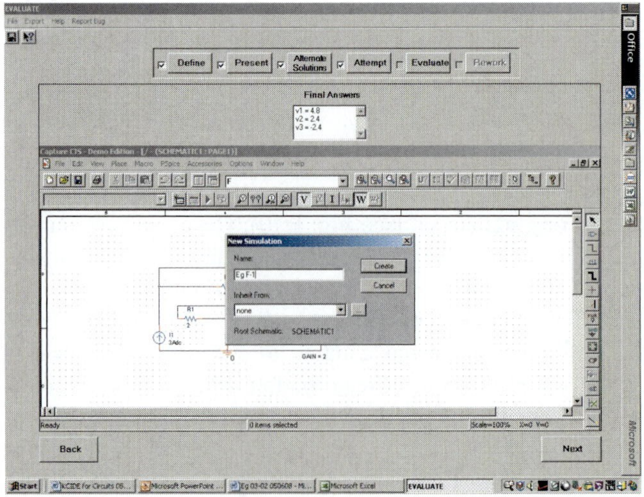

Figure F.12
Setting up our circuit for solution by *PSpice*.

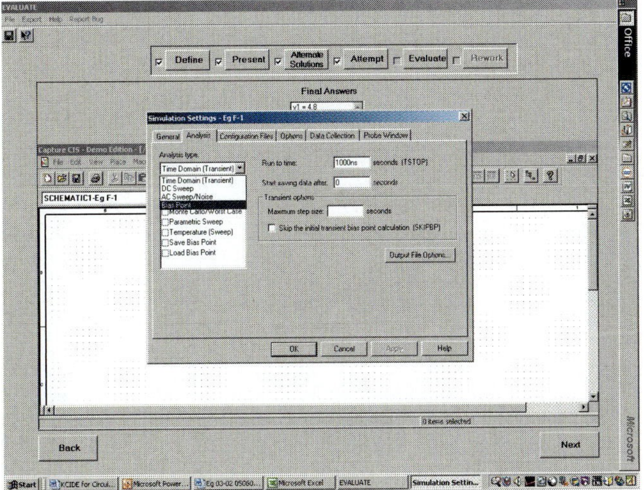

Figure F.13
Setting up our circuit for solution by *PSpice.*

Fig. F.14. Running *PSpice* produces the screen shown in Fig. F.15. We can immediately see that the voltages agree with the solution we obtained by using nodal analysis. Clicking on Next leads to our being asked, as in Fig. F.16, if we have any graphics to export. For this problem, we have no graphs.

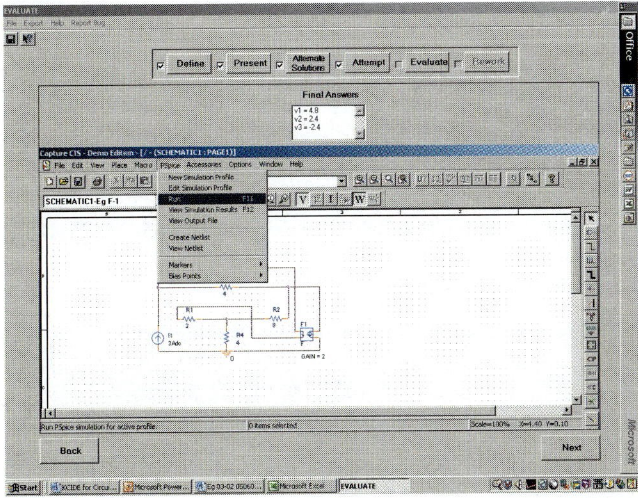

Figure F.14
Setting up our circuit for solution by *PSpice.*

We are now approaching the end of the process. We are asked to comment about the solution, Fig. F.17. And, we are asked if the answers agree with the *PSpice* solution. The answers do agree and we can proceed to determining what we want to export, Fig. F.18. We can generate *Word* and/or *PowerPoint* files, Fig. F.19. In this case, we select both but will only show the output of the *Word* file, Fig. F.20. Note: This output was modified so that it could be presented on two pages.

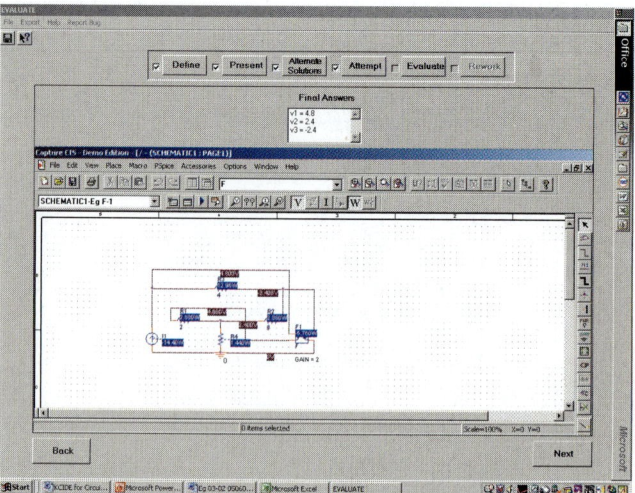

Figure F.15
Problem solution using *PSpice*.

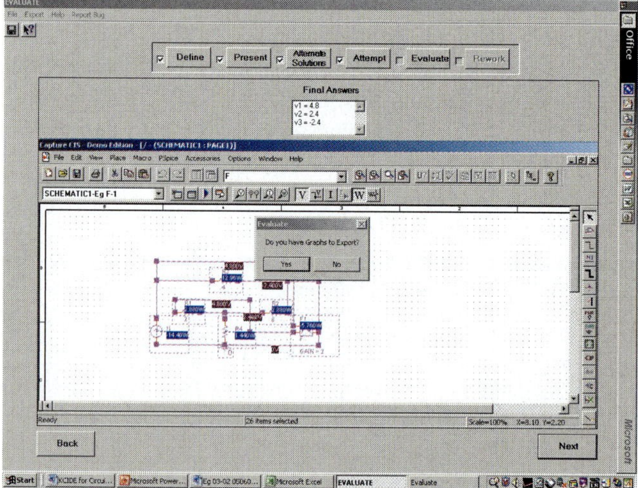

Figure F.16
Screen for exporting graphs.

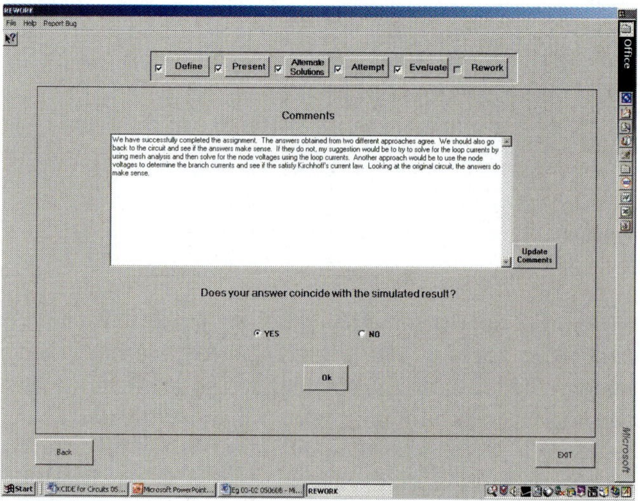

Figure F.17
Determining if the problem has been solved correctly.

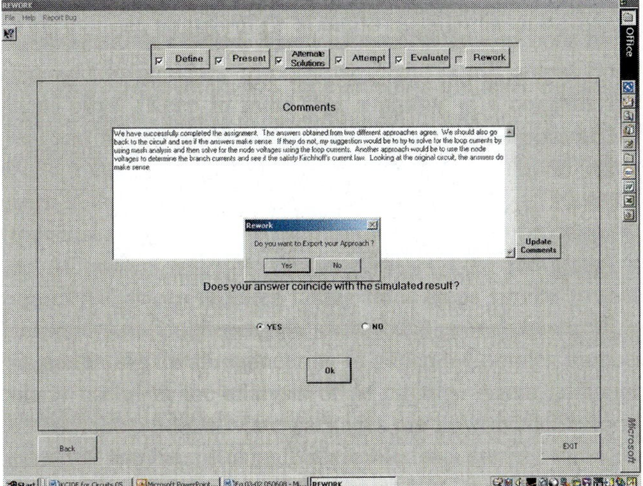

Figure F.18
Determining if you want to generate *Word* and/or *PowerPoint* documents.

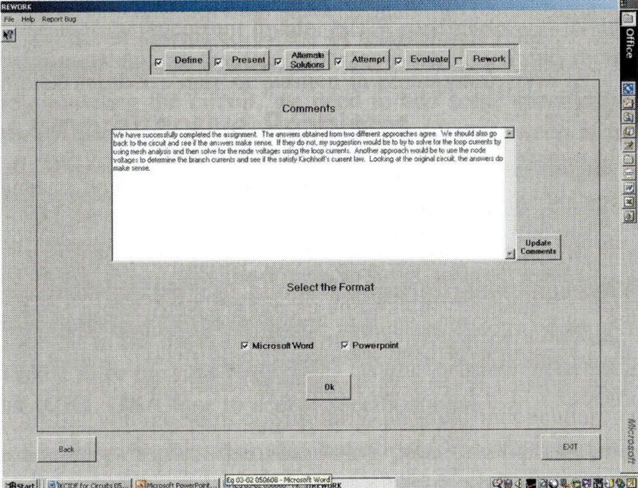

Figure F.19
Generating *Word* and *PowerPoint* files for Example F.1.

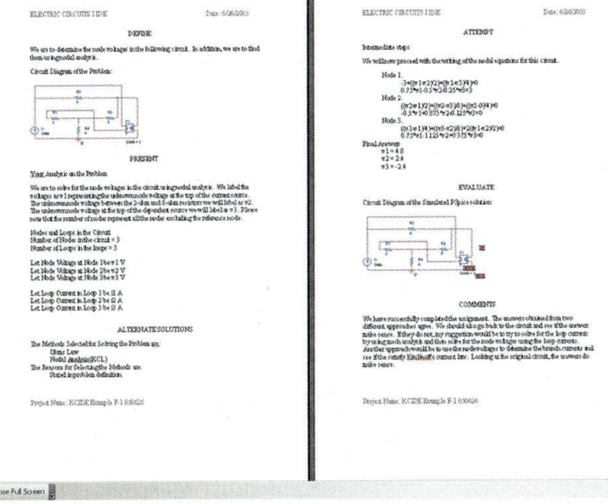

Figure F.20
Word file output.

We have now completed a detailed example. We suggest that you first try to do this with the platform and look at your output in both *Word* and *PowerPoint.* To help you to continue to develop your facility with the platform, try working the following practice problem, using mesh analysis. For more examples, please go to the website.

Practice Problem F.1

Use the *KCIDE for Circuits* platform to solve Practice Prob. 3.2.

Answer: $v_1 = 80$ V, $v_2 = -64$ V, and $v_3 = 156$ V.

Appendix G

Answers to Odd-Numbered Problems

Chapter 1

1.1 (a) -0.1038 C, (b) -0.19865 C, (c) -3.941 C, (d) -26.08 C

1.3 (a) $3t + 1$ C, (b) $t^2 + 5t$ mC, (c) $2\sin(10t + \pi/6) + 1\ \mu$C, (d) $-e^{-30t}[0.16\cos 40t + 0.12\sin 40t]$ C

1.5 25 C

1.7 $i = \begin{cases} 25\text{ A}, & 0 < t < 2 \\ -25\text{ A}, & 2 < t < 6 \\ 25\text{ A}, & 6 < t < 8 \end{cases}$

See the sketch in Fig. G.1.

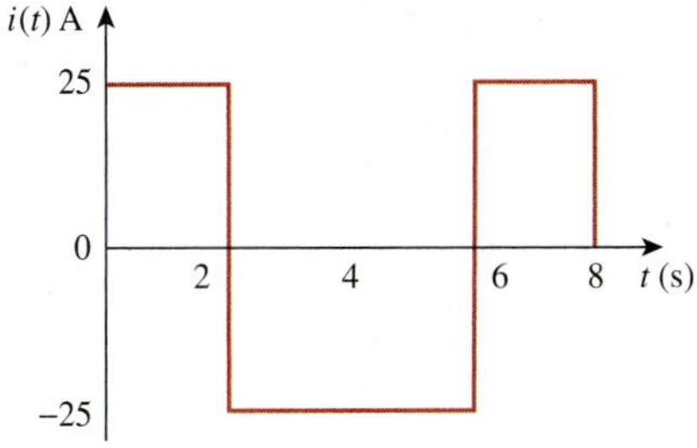

Figure G.1
For Prob. 1.7.

1.9 (a) 10 C, (b) 22.5 C, (c) 30 C

1.11 3.672 kC, 4.406 kJ

1.13 164.5 mW, 78.34 mJ

1.15 (a) 1.297 C, (b) $-90e^{-4t}$ W, (c) -22.5 J

1.17 70 W

1.19 3 A

1.21 2.696×10^{23} electrons, 43,200 C

1.23 $1.35

1.25 21.6 cents

1.27 (a) 43.2 kC, (b) 475.2 kJ, (c) 1.188 cents

1.29 39.6 cents

1.31 $42.05

1.33 6 C

1.35 2.333 MWh

1.37 29.84 kWh

1.39 24 cents

Chapter 2

2.1 This is a design problem with several answers.

2.3 184.3 mm

2.5 $n = 9, b = 15, l = 7$

2.7 (a) 6 branches and 5 nodes, and (b) 7 branches and 5 nodes.

2.9 14 A, -2 A, 10 A

2.11 6 V, 3 V

2.13 12 A, -10 A, 5 A, -2 A

2.15 10 V, -2 A

2.17 2 V, -22 V, 10 V

2.19 2 A, 12 W, 8 W, -40 W, 20 W

2.21 4.167 W

2.23 2 V, 1.92 W

2.25 0.1 A, 2 kV, 0.2 kW

2.27 6.4 V

2.29 1.625 Ω

2.31 11.2 A, 1.6 A, 9.6 A, 6.4 A, 3.2 A

2.33 3 V, 6 A

2.35 8 V, 0.2 A

2.37 2.5 Ω

2.39 (a) 727.3 Ω, (b) 3 kΩ

2.41 16 Ω

2.43 (a) 12 Ω, (b) 16 Ω

2.45 (a) 59.8 Ω, (b) 32.5 Ω

2.47 24 Ω

2.49 (a) 4 Ω, (b) $R_1 = 18\ \Omega$, $R_2 = 6\ \Omega$, $R_3 = 3\ \Omega$

2.51 (a) 9.231 Ω, (b) 36.25 Ω

2.53 (a) 142.32 Ω, (b) 33.33 Ω

2.55 997.4 mA

2.57 12.21 Ω, 1.64 A

2.59 1.2 A

2.61 Use R_1 and R_3 bulbs

2.63 0.4 Ω, ≅ 1 W

2.65 4 kΩ

2.67 (a) 4 V, (b) 2.857 V, (c) 28.57%, (d) 6.25%

2.69 (a) 1.278 V (with), 1.29 V (without)
(b) 9.30 V (with), 10 V (without)
(c) 25 V (with), 30.77 V (without)

2.71 10 Ω

2.73 45 Ω

2.75 (a) 19.9 kΩ, (b) 20 kΩ

2.77 (a) Four 20-Ω resistors in parallel
(b) One 300-Ω resistor in series with a 1.8-Ω resistor and a parallel combination of two 20-Ω resistors
(c) Two 24-kΩ resistors in parallel connected in series with two 56-kΩ resistors in parallel
(d) A series combination of a 20-Ω resistor, 300-Ω resistor, 24-kΩ resistor, and a parallel combination of two 56-kΩ resistors

2.79 75 Ω

2.81 38 kΩ, 3.333 kΩ

2.83 3 kΩ, ∞ Ω (best answer)

Chapter 3

3.1 This is a design problem with several answers.

3.3 4 A, 2 A, 1.3333 A, 0.667 A, 40 V

3.5 20 V

3.7 5.714 V

3.9 39.67 mA

3.11 293.9 W, 177.79 W, 238 W

3.13 8 V, 8 V

3.15 29.45 A, 144.6 W, 129.6 W, 12 W

3.17 1.73 A

3.19 10 V, 4.933 V, 12.267 V

3.21 1 V, 3 V

3.23 22.34 V

3.25 25.52 V, 22.05 V, 14.842 V, 15.055 V

3.27 625 mV, 375 mV, 1.625 V

3.29 −0.7708 V, 1.209 V, 2.309 V, 0.7076 V

3.31 4.97 V, 4.85 V, −0.12 V

3.33 (a) and (b) are both planar and can be redrawn as shown in Fig. G.2.

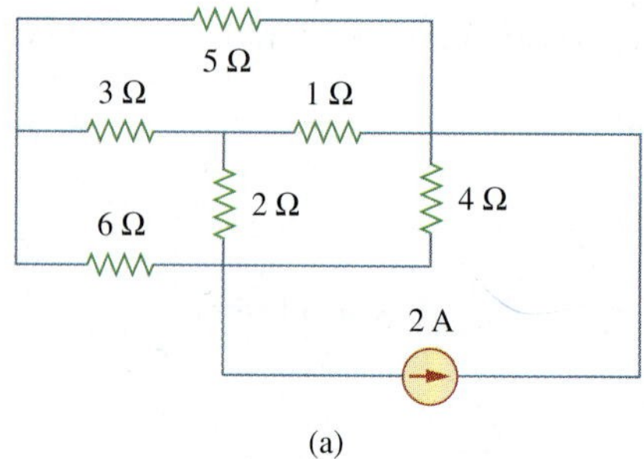

(a)

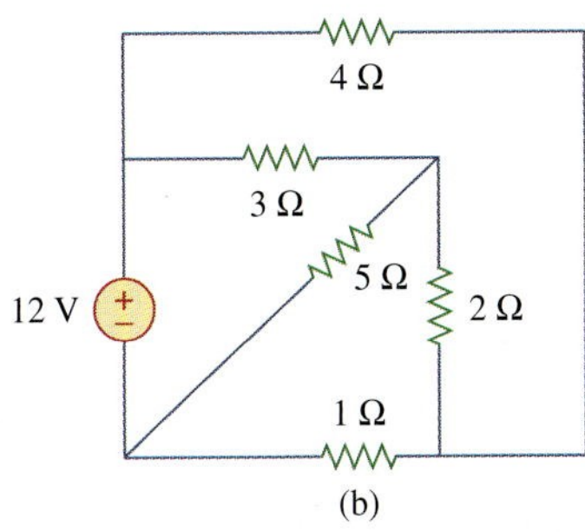

Figure G.2
For Prob. 3.33.

3.35 20 V

3.37 1.1111 V

3.39 0.8 A, −0.9 A

3.41 1.188 A

3.43 1.7778 A, 53.33 V

3.45 8.561 A

3.47 10 V, 4.933 V, 12.267 V

3.49 33.78 V, 10.67 A

3.51 20 V

3.53 1.6196 mA, −1.0202 mA, −2.461 mA, 3 mA, −2.423 mA

3.55 −1 A, 0 A, 2 A

3.57 3.23 kΩ, 28 V, 72 V

3.59 −1.344 kV, −5.6 A

3.61 −0.3

3.63 −4 V, 2.105 A

3.65 2.17 A, 1.9912 A, 1.8119 A, 2.094 A, 2.249 A

3.67 −12 V

3.69 $$\begin{bmatrix} 1.75 & -0.25 & -1 \\ -0.25 & 1 & -0.25 \\ -1 & -0.25 & 1.25 \end{bmatrix} \begin{bmatrix} V_1 \\ V_2 \\ V_3 \end{bmatrix} = \begin{bmatrix} 20 \\ 5 \\ 5 \end{bmatrix}$$

3.71 2.085 A, 653.3 mA, 1.2312 A

3.73 $$\begin{bmatrix} 9 & -3 & -4 & 0 \\ -3 & 8 & 0 & 0 \\ -4 & 0 & 6 & -1 \\ 0 & 0 & -1 & 2 \end{bmatrix} \begin{bmatrix} i_1 \\ i_2 \\ i_3 \\ i_4 \end{bmatrix} = \begin{bmatrix} 6 \\ 4 \\ 2 \\ -3 \end{bmatrix}$$

3.75 −3 A, 0 A, 3 A

3.77 3.111 V, 1.4444 V

3.79 −5.278 V, 10.28 V, 694.4 mV, −26.88 V

3.81 26.67 V, 6.667 V, 173.33 V, −46.67 V

3.83 See Fig. G.3; −12.5 V

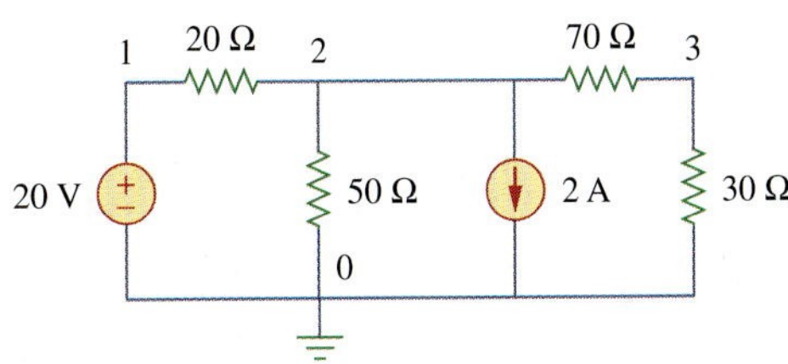

Figure G.3
For Prob. 3.83.

3.85 9 Ω

3.87 −8

3.89 30 μA, 12 V

3.91 0.61 μA, 8.641 V, 49 mV

Chapter 4

4.1 0.1 A, 1 A

4.3 (a) 0.5 V, 0.5 A, (b) 5 V, 5 A, (c) 5 V, 500 mA

4.5 4.5 V

4.7 888.9 mV

4.9 7 V

4.11 17.99 V, 1.799 A

4.13 8.696 V

4.15 1.875 A, 10.55 W

4.17 −8.571 V

4.19 −26.67 V

4.21 This is a design problem with multiple answers.

4.23 2 A, 32 W

4.25 −6.6 V

4.27 −48 V

4.29 3 V

4.31 3.652 V

4.33 (a) 8 Ω, 16 V, (b) 20 Ω, 50 V

4.35 −125 mV

4.37 10 Ω, 1 A

4.39 20 Ω, −16.4 V

4.41 4 Ω, −8 V, −2 A

4.43 10 Ω, 0 V

4.45 3 Ω, 3 A

4.47 476.2 mΩ, 1.9841 V, 4.176 A

4.49 28 Ω, 3.286 A

4.51 (a) 2 Ω, 7 A, (b) 1.5 Ω, 12.667 A

4.53 3 Ω, 1 A

4.55 100 kΩ, −20 mA

4.57 10 Ω, 166.67 V, 16.667 A

4.59 22.5 Ω, 40 V, 1.7778 A

4.61 1.2 Ω, 9.6 V, 8 A

4.63 −3.333 Ω, 0 A

4.65 $V_0 = (48 - 5I_0)$ V

4.67 25 Ω, 7.84 W

4.69 ∞ (theoretically)

4.71 8 kΩ, 1.152 W

4.73 20.77 W

4.75 $R_L = 10\ \Omega$, P_L tends toward infinity.

4.77 (a) 3.8 Ω, 4 V, (b) 3.2 Ω, 15 V

4.79 10 Ω, 167 V

4.81 3.3 Ω, 10 V (Note, values obtained graphically)

4.83 8 Ω, 12 V

4.85 (a) 24 V, 30 kΩ, (b) 9.6 V

4.87 (a) 10 mA, 8 kΩ, (b) 9.926 mA

4.89 (a) 99.99 μA, (b) 99.99 μA

4.91 (a) 100 Ω, 20 Ω, (b) 100 Ω, 200 Ω

4.93 $\dfrac{V_s}{R_s + (1 + \beta)R_o}$

4.95 5.333 V, 66.67 kΩ

4.97 2.4 kΩ, 4.8 V

Chapter 5

5.1 (a) 1.5 MΩ, (b) 60 Ω, (c) 98.06 dB

5.3 10 V

5.5 0.9999990

5.7 −100 nV, −10 mV

5.9 (a) 2 V, (b) 3 V

5.11 This is a design problem with multiple answers.

5.13 2.7 V, 288 μA

5.15 (a) $-\left(R_1 + R_3 + \dfrac{R_1R_3}{R_2}\right)$, (b) −92 kΩ

5.17 (a) −1.2, (b) −8, (c) −200

5.19 −0.375 mA

5.21 −4 V

5.23 $-\dfrac{R_f}{R_1}$

5.25 1.25 V

5.27 1.8 V

5.29 $\dfrac{R_2}{R_1}$

5.31 727.2 μA

5.33 −6 mA, 108 mW

5.35 If $R_1 = 10\ \text{k}\Omega$, then $R_f = 90\ \text{k}\Omega$.

5.37 -3 V

5.39 3 V

5.41 See Fig. G.4.

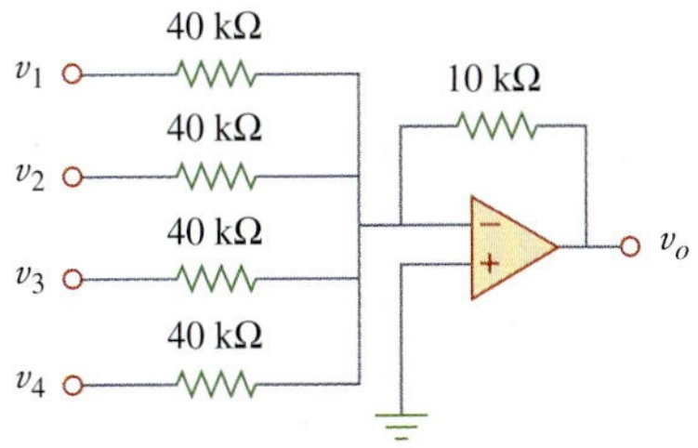

Figure G.4
For Prob. 5.41.

5.43 3 kΩ

5.45 See Fig. G.5, where $R \leq 100\ \text{k}\Omega$.

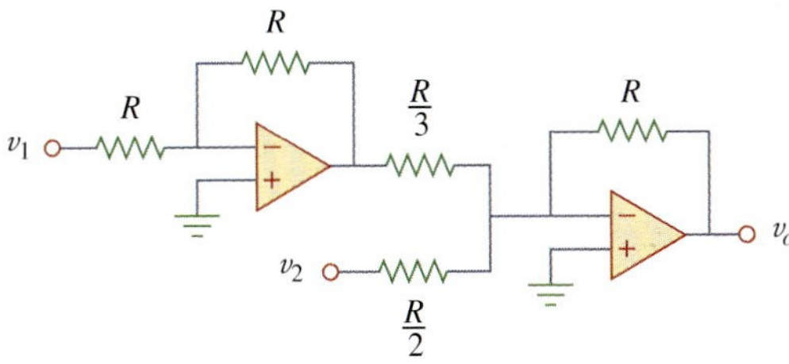

Figure G.5
For Prob. 5.45.

5.47 14.09 V

5.49 $R_1 = R_3 = 10\ \text{k}\Omega$, $R_2 = R_4 = 20\ \text{k}\Omega$

5.51 See Fig. G.6.

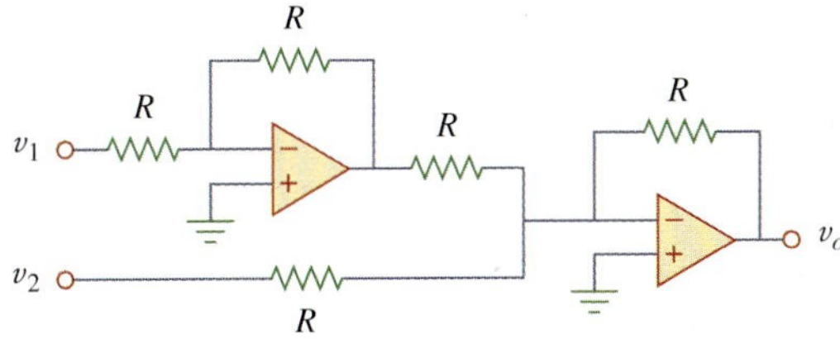

Figure G.6
For Prob. 5.51.

5.53 Proof.

5.55 7.956, 7.956, 1.989

5.57 $6v_{s1} - 6v_{s2}$

5.59 -16

5.61 -4.8 V

5.63 $\dfrac{R_2R_4/R_1R_5 - R_4/R_6}{1 - R_2R_4/R_3R_5}$

5.65 -21.6 mV

5.67 2.4 V

5.69 -17.143 mV

5.71 10 V

5.73 18 V

5.75 $-2,200\ \mu$A

5.77 -3.343 mV

5.79 -14.61 V

5.81 343.4 mV, 24.51 μA

5.83 The result depends on your design. Hence, let $R_G = 10$ k ohms, $R_1 = 10$ k ohms, $R_2 = 20$ k ohms, $R_3 = 40$ k ohms, $R_4 = 80$ k ohms, $R_5 = 160$ k ohms, $R_6 = 320$ k ohms, then,

$$-v_o = (R_f/R_1)v_1 + \text{———} + (R_f/R_6)v_6$$
$$= v_1 + 0.5v_2 + 0.25v_3 + 0.125v_4 + 0.0625v_5 + 0.03125v_6$$

(a) $|v_o| = 1.1875 = 1 + 0.125 + 0.0625 = 1 + (1/8) + (1/16)$, which implies, $[v_1\, v_2\, v_3\, v_4\, v_5\, v_6] = $ **[100110]**

(b) $|v_o| = 0 + (1/2) + (1/4) + 0 + (1/16) + (1/32) = (27/32) = $ **843.75 mV**

(c) This corresponds to [111111].
$|v_o| = 1 + (1/2) + (1/4) + (1/8) + (1/16) + (1/32)$
$= 63/32 = $ **1.96875 V**

5.85 160 kΩ

5.87 $\left(1 + \dfrac{R_4}{R_3}\right)v_2 - \left[\dfrac{R_4}{R_3} + \left(\dfrac{R_2R_4}{R_1R_3}\right)\right]v_1$

Let $R_4 = R_1$ and $R_3 = R_2$;

then $v_0 = \left(1 + \dfrac{R_4}{R_3}\right)(v_2 - v_1)$

a subtractor with a gain of $\left(1 + \dfrac{R_4}{R_3}\right)$.

5.89 A summer with $v_0 = -v_1 - (5/3)v_2$ where $v_2 = 6$ V battery and an inverting amplifier with $v_1 = -12\,v_2$.

5.91 9

5.93 $A = \dfrac{1}{(1 + \frac{R_1}{R_3})R_L - R_1(\frac{R_2 + R_L}{R_2R_3})(R_4 + \frac{R_2R_L}{R_2 + R_L})}$

Chapter 6

6.1 $10(1 - 3t)e^{-3t}$ A, $20t(1 - 3t)e^{-6t}$ W

6.3 This is a design problem with multiple answers.

6.5 $$v = \begin{cases} 20 \text{ mA}, & 0 < t < 2 \text{ ms} \\ -20 \text{ mA}, & 2 < t < 6 \text{ ms} \\ 20 \text{ mA}, & 6 < t < 8 \text{ ms} \end{cases}$$

6.7 $0.04t^2 + 10$ V

6.9 13.624 V, 70.66 W

6.11 $$v(t) = \begin{cases} 10 + 3.75t \text{ V}, & 0 < t < 2\text{s} \\ 22.5 - 2.5t \text{ V}, & 2 < t < 4\text{s} \\ 12.5 \text{ V}, & 4 < t < 6\text{s} \\ 2.5t - 2.5 \text{ V}, & 6 < t < 8\text{s} \end{cases}$$

6.13 30 V, 40 V

6.15 (a) 100 mJ, 150 mJ, (b) 36 mJ, 24 mJ

6.17 (a) 3 F, (b) 8 F, (c) 1 F

6.19 10 μF

6.21 2.5 μF

6.23 This is a design problem with multiple answers.

6.25 (a) For the capacitors in series,

$$Q_1 = Q_2 \rightarrow C_1 v_1 = C_2 v_2 \rightarrow \frac{v_1}{v_2} = \frac{C_2}{C_1}$$

$$v_s = v_1 + v_2 = \frac{C_2}{C_1} v_2 + v_2 = \frac{C_1 + C_2}{C_1} v_2$$

$$\rightarrow v_2 = \frac{C_1}{C_1 + C_2} v_s$$

Similarly, $v_1 = \dfrac{C_2}{C_1 + C_2} v_s$

(b) For capacitors in parallel,

$$v_1 = v_2 = \frac{Q_1}{C_1} = \frac{Q_2}{C_2}$$

$$Q_s = Q_1 + Q_2 = \frac{C_1}{C_2} Q_2 + Q_2 = \frac{C_1 + C_2}{C_2} Q_2$$

or

$$Q_2 = \frac{C_2}{C_1 + C_2}$$

$$Q_1 = \frac{C_1}{C_1 + C_2} Q_s$$

$$i = \frac{dQ}{dt} \rightarrow i_1 = \frac{C_1}{C_1 + C_2} i_{s,}$$

$$i_2 = \frac{C_2}{C_1 + C_2} i_s$$

6.27 1 μF, 16 μF

6.29 (a) 1.6 C, (b) 1 C

6.31 $$v(t) = \begin{cases} t^2 \text{ kV}, & 0 < t < 1\text{s} \\ 2t - 1 \text{ kV}, & 1 < t < 3\text{s} \\ 0.5t^2 - 5t + 15.5 \text{ kV}, & 3 < t < 5\text{s} \end{cases}$$

$$i_1(t) = \begin{cases} 12t \text{ mA}, & 0 < t < 1\text{s} \\ 12 \text{ mA}, & 1 < t < 3\text{s} \\ 6t - 30 \text{ mA}, & 3 < t < 5\text{s} \end{cases}$$

$$i_2(t) = \begin{cases} 8t \text{ mA}, & 0 < t < 1\text{s} \\ 8 \text{ mA}, & 1 < t < 3\text{s} \\ 4t - 20 \text{ mA}, & 3 < t < 5\text{s} \end{cases}$$

6.33 10 F, 7.5 V

6.35 6.4 mH

6.37 $4.8 \cos 100t$ V, 96 mJ

6.39 $(5t^3 + 5t^2 + 20t + 1)$ A

6.41 5.977 A, 35.72 J

6.43 144 μJ

6.45 $$i(t) = \begin{cases} 100t^2 \text{ A}, & 0 < t < 1\text{s} \\ [100t^2 - 400t + 400] \text{ A}, & 1 < t < 2\text{s} \end{cases}$$

6.47 5 Ω

6.49 3.75 mH

6.51 7.778 mH

6.53 20 mH

6.55 (a) 1.4 L, (b) 0.5 L

6.57 6.625 H

6.59 Proof.

6.61 (a) 6.667 mH, e^{-t} mA, $2e^{-t}$ mA
(b) $-20e^{-t}$ μV (c) 1.3534 nJ

6.63 See Fig. G.7.

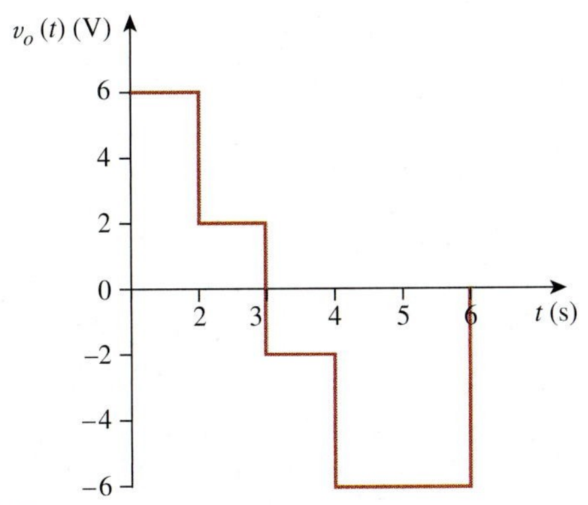

Figure G.7
For Prob. 6.63.

6.65 (a) 40 J, 40 J, (b) 80 J, (c) $5 \times 10^{-5}(e^{-200t} - 1) +$ 4 A, $1.25 \times 10^{-5}(e^{-200t} - 1) - 2$ A
(d) $6.25 \times 10^{-5}(e^{-200t} - 1) + 2$ A

6.67 $200 \cos(50t)$ mV

6.69 See Fig. G.8.

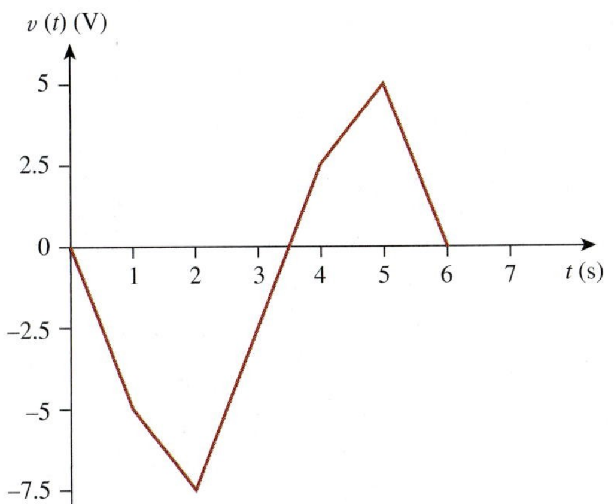

Figure G.8
For Prob. 6.69.

6.71 By combining a summer with an integrator, we have the circuit shown in Fig. G.9.

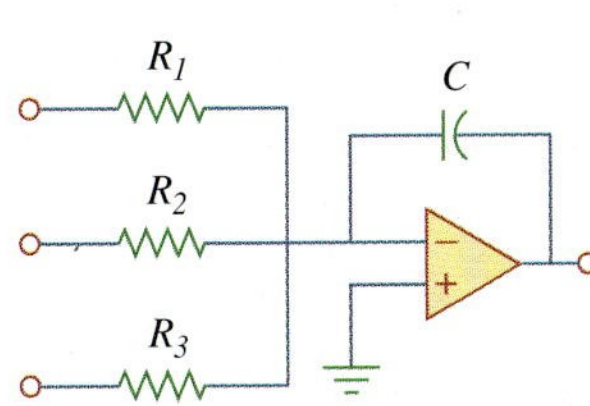

Figure G.9
For Prob. 6.71.

$$v_o = -\frac{1}{R_1C}\int v_1\,dt - \frac{1}{R_2C}\int v_2\,dt - \frac{1}{R_2C}\int v_2\,dt$$

For the given problem, $C = 2\mu\text{F} : R_1 = 500\text{ k}\Omega$, $R_2 = 125\text{ k}\Omega$, $R_3 = 50\text{ k}\Omega$.

6.73 Consider the op amp as shown in Fig. G.10.

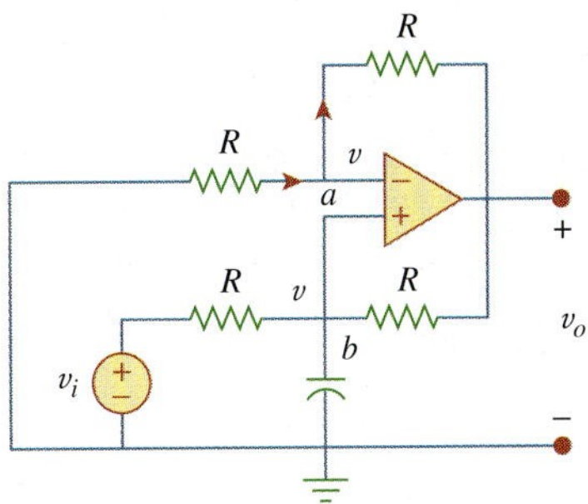

Figure G.10
For Prob. 6.73.

Let $v_a = v_b = v$. At node a,

$$\frac{0 - v}{R} = \frac{v - v_0}{R} \rightarrow 2v - v_0 = 0 \qquad (1)$$

At node b, $\dfrac{v_i - v}{R} = \dfrac{v - v_0}{R} + C\dfrac{dv}{dt}$

$$v_i = 2v - v_o + RC\frac{dv}{dt} \qquad (2)$$

Combining Eqs. (1) and (2),

$$v_i = v_o - v_o + \frac{RC}{2}\frac{dv_o}{dt} \quad \text{or} \quad v_o = \frac{2}{RC}\int v_i\,dt$$

showing that the circuit is a noninverting integrator.

6.75 −30 mV

6.77 See Fig. G.11.

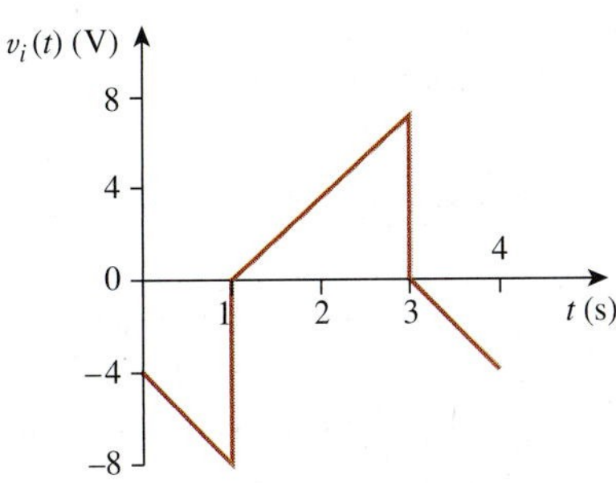

Figure G.11
For Prob. 6.77.

6.79 See Fig. G.12.

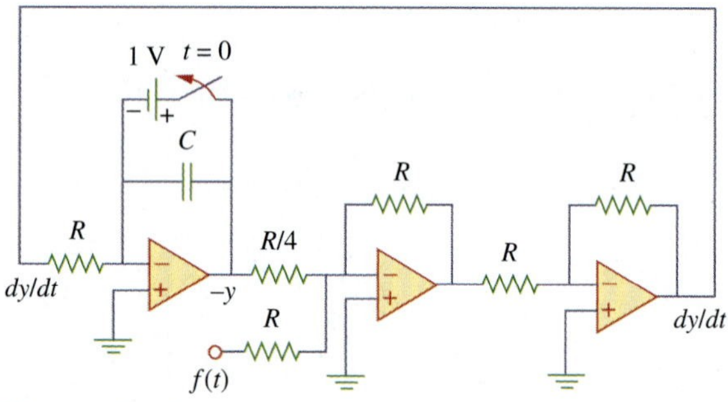

Figure G.12
For Prob. 6.79.

6.81 See Fig. G.13.

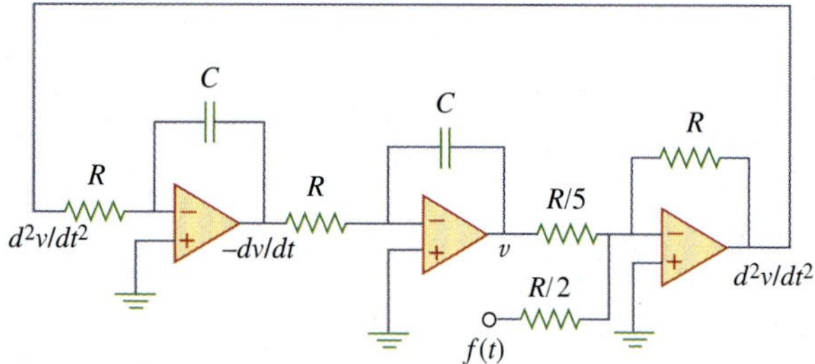

Figure G.13
For Prob. 6.81.

6.83 Eight groups in parallel with each group made up of two capacitors in series

6.85 1.25 mH inductor

Chapter 7

7.1 (a) 0.7143 μF, (b) 5 ms, (c) 3.466 ms

7.3 3.222 μs

7.5 This is a design problem with multiple answers.

7.7 $v_o(t) = [6 + 2e^{-t/20}]$ V for all $t > 0$.

7.9 $v_o(t) = 4e^{-5t}$ V for all $t > 0$.

7.11 $1.4118e^{-3t}$ A

7.13 (a) 5 kΩ, 5 H, 1 ms, (b) 25.28 μJ

7.15 (a) 0.25 s, (b) 0.5 ms

7.17 $-2e^{-16t}u(t)$ V

7.19 $2e^{-5t}u(t)$ A

7.21 13.333 Ω

7.23 $2e^{-4t}$ V, $t > 0$, $0.5e^{-4t}$ V, $t > 0$

7.25 This is a design problem with multiple answers.

7.27 $[10u(t + 1) + 20u(t) - 50u(t - 1) + 30u(t - 2)]$ V

7.29 (a) See Fig. G.14(a). (b) See Fig. G.14(b). (c) $z(t) = 5\cos 4t\,\delta(t - 1) = 5\cos 4\delta(t - 1) = -3.268\delta(t - 1)$, which is sketched in Fig. G.14(c).

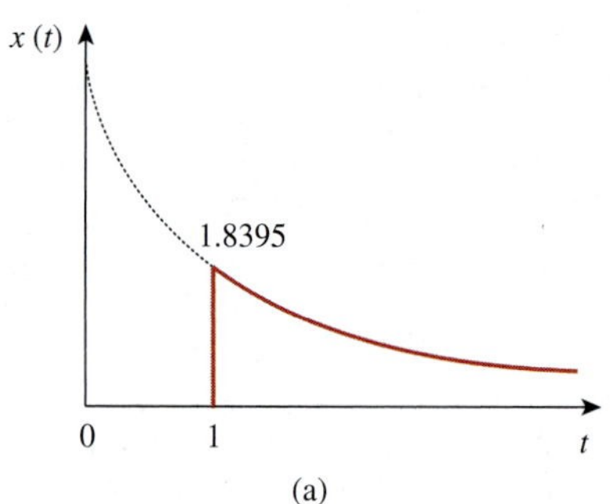

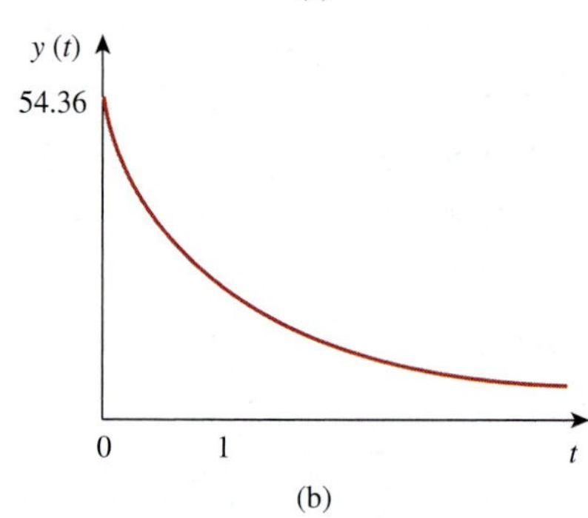

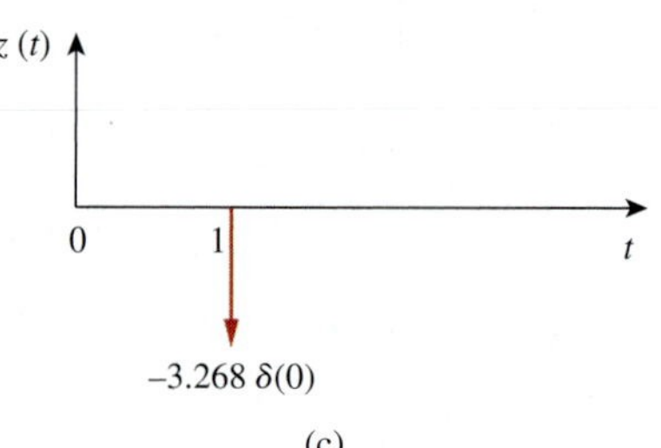

Figure G.14
For Prob. 7.29.

7.31 (a) 112×10^{-9}, (b) 7

7.33 $2u(t-2)$ A

7.35 (a) $-e^{-2t}u(t)$ V, (b) $2e^{1.5t}u(t)$ A

7.37 (a) 4 s, (b) 10 V, (c) $(10 - 8e^{-t/4})\,u(t)$ V

7.39 (a) 4 V, $t < 0$, $20 - 16e^{-t/8}$, $t > 0$,
(b) 4 V, $t < 0$, $12 - 8e^{-t/6}$ V, $t > 0$.

7.41 This is a design problem with multiple answers.

7.43 0.8 A, $0.8e^{-t/480}u(t)$ A

7.45 $(4 - 3e^{-14.286t})\,u(t)$ V

7.47 $\begin{cases} 48(1 - e^{-t})\text{ V}, & 0 < t < 1 \\ (60 - 29.66e^{-(t-1)})\text{ V}, & t > 1 \end{cases}$

7.49 $\begin{cases} 8(1 - e^{-t/5})\text{ V}, & 0 < t < 1 \\ 1.45e^{-(t-1)/5}\text{ V}, & t > 1 \end{cases}$

7.51 $V_S = Ri + L\dfrac{di}{dt}$

or $L\dfrac{di}{dt} = -R\left(i - \dfrac{V_S}{R}\right)$

$$\frac{di}{i - V_S/R} = \frac{-R}{L}dt$$

Integrating both sides,

$$\ln\left(i - \frac{V_S}{R}\right)\Bigg|_{I_0}^{i(t)} = \frac{-R}{L}t$$

$$\ln\left(\frac{i - V_S/R}{I_0 - V_S/R}\right) = \frac{-t}{\tau}$$

or $\dfrac{i - V_S/R}{I_0 - V_S/R} = e^{-t/\tau}$

$$i(t) = \frac{V_S}{R} + \left(I_0 - \frac{V_S}{R}\right)e^{-t/\tau}$$

which is the same as Eq. (7.60).

7.53 (a) 5 A, $5e^{-t/2}u(t)$ A, (b) 6 A, $6e^{-2t/3}u(t)$ A

7.55 96 V, $96e^{-4t}u(t)$ V

7.57 $4.8e^{-2t}u(t)$ A, $1.2e^{-5t}u(t)$ A

7.59 $3e^{-4t}u(t)$ V

7.61 $20e^{-80t}u(t)$ V, $(5 + (5 - 5e^{-80t})\,u(t))$ A

7.63 $-16e^{-8t}u(t)$ V, $(4 + (-4 + 4e^{-8t})u(t))$ A

7.65 $\begin{cases} 4(1 - e^{-2t})\text{ A} & 0 < t < 1 \\ 3.458e^{-2(t-1)}\text{ A} & t > 1 \end{cases}$

7.67 $10e^{-100t/3}u(t)$ A

7.69 $48(e^{-t/3000} - 1)\,u(t)$ V

7.71 $6(1 - e^{-5t})u(t)$ V

7.73 $-50u(t)$ mA

7.75 $(6 - 3e^{-50t})u(t)$ V, $-$ 0.2 mA

7.77 See Fig. G.15.

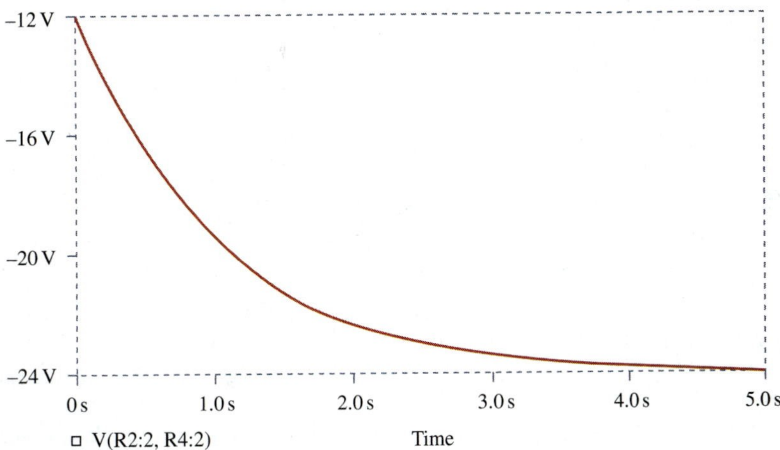

Figure G.15
For Prob. 7.77.

7.79 $(-0.5 + 4.5e^{-80t/3})u(t)$ A

7.81 See Fig. G.16.

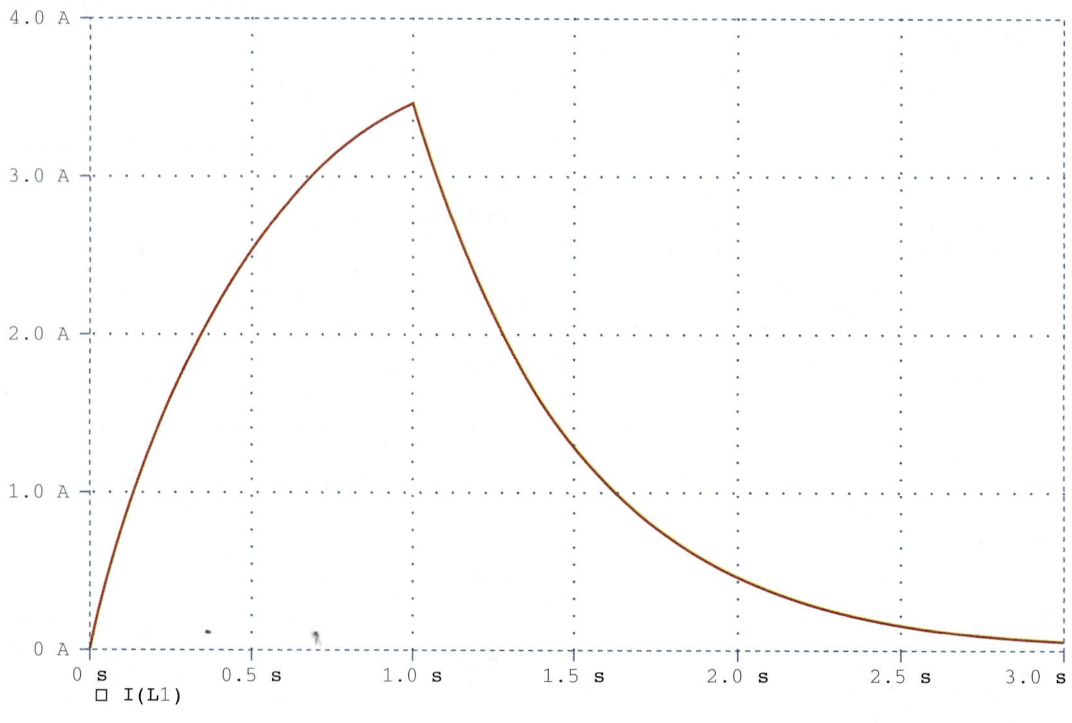

Figure G.16
For Prob. 7.81.

7.83 6.278 m/s

7.85 (a) 659.7 μs, (b) 16.636 s

7.87 441 mA

7.89 $L < 200$ mH

7.91 1.271 Ω

Chapter 8

8.1 (a) 2 A, 12 V, (b) −4 A/s, −5 V/s, (c) 0 A, 0 V

8.3 (a) 0 A, −10 V, 0 V, (b) 0 A/s, 8 V/s, 8 V/s, (c) 400 mA, 6 V, 16 V

8.5 (a) 0 A, 0 V, (b) 4 A/s, 0 V/s, (c) 2.4 A, 9.6 V

8.7 overdamped

8.9 $(2 + 10t)e^{-5t}u(t)$ A

8.11 $20(1 + t)e^{-t}$ V for $t > 0$.

8.13 120 Ω

8.15 750 Ω, 200 μF, 25 H

8.17 $(64.65e^{-2.679t} - 4.641e^{-37.32t})$ V

8.19 $18\sin(0.5t)$ V for $t > 0$.

8.21 $18e^{-t} - 2e^{-9t}$ V

8.23 40 mF

8.25 This is a design problem with multiple answers.

8.27 $(6 - 6(\cos(2t) + \sin(2t)e^{-2t})u(t))$ V

8.29 (a) $3 - 3\cos 2t + \sin 2t$ V,
(b) $2 - 4e^{-t} + e^{-4t}$ A,
(c) $3 + (2 + 3t)e^{-t}$ V,
(d) $2 + 2\cos 2te^{-t}$ A

8.31 80 V, 40 V

8.33 $[20 + 0.001125e^{-4.95t} - 10.001e^{-0.05t}]$ V

8.35 This is a design problem with multiple answers.

8.37 $5e^{-4t}$ A

8.39 $[30 + (0.021e^{-47.33t} - 6.021e^{-0.167t})]$ V

8.41 $(0.3638\sin(4.583t)e^{-2t})$ A for $t > 0$.

8.43 8 Ω, 2.392 mF

8.45 $[4 - [3\cos(1.3229t) + 1.1339\sin(1.3229t)]e^{-t/2}]$ A,
$[4.536\sin(1.3229t)e^{-t/2}]$ V

8.47 $(200te^{-10t})$ V

8.49 $[3 + (3 + 6t)e^{-2t}]$ A

8.51 $\left[-\dfrac{i_0}{\omega_o C}\sin(\omega_o t)\right]$ V where $\omega_o = 1/\sqrt{LC}$

8.53 $(d^2i/dt^2) + 0.125(di/dt) + 400i = 600$

8.55 $7.448 - 3.448e^{-7.25t}$ V, $t > 0$

8.57 (a) $s^2 + 20s + 36 = 0$,
(b) $-\dfrac{3}{4}e^{-2t} - \dfrac{5}{4}e^{-18t}$ A, $6e^{-2t} + 10e^{-18t}$ V

8.59 $-32te^{-t}$ V

8.61 $2.4 - 2.667e^{-2t} + 0.2667e^{-5t}$ A,
$9.6 - 16e^{-2t} + 6.4e^{-5t}$ V

8.63 $\dfrac{d^2i(t)}{dt^2} = -\dfrac{v_s}{RCL}$

8.65 $\dfrac{d^2v_o}{dt^2} - \dfrac{v_o}{R^2C^2} = 0$, $e^{10t} - e^{-10t}$ V
Note, circuit is unstable.

8.67 $-te^{-t}u(t)$ V

8.69 See Fig. G.17.

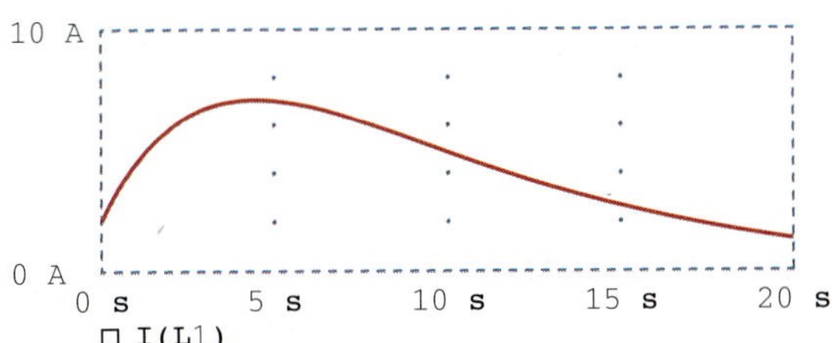

Figure G.17
For Prob. 8.69.

8.71 See Fig. G.18.

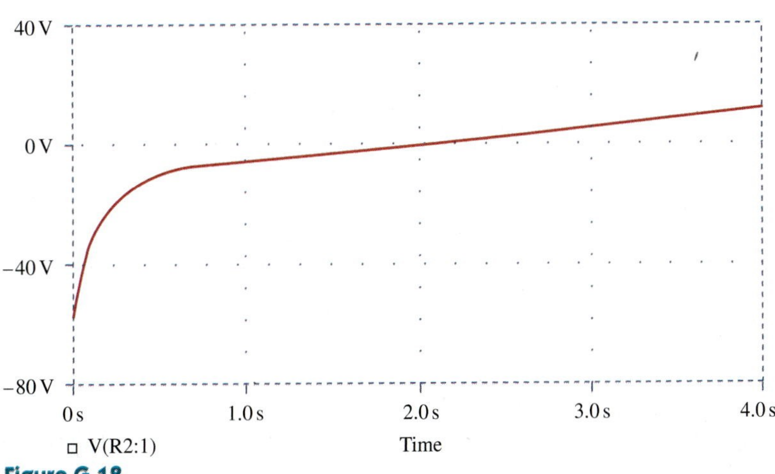

Figure G.18
For Prob. 8.71.

8.73 This is a design problem with multiple answers.

8.75 See Fig. G.19.

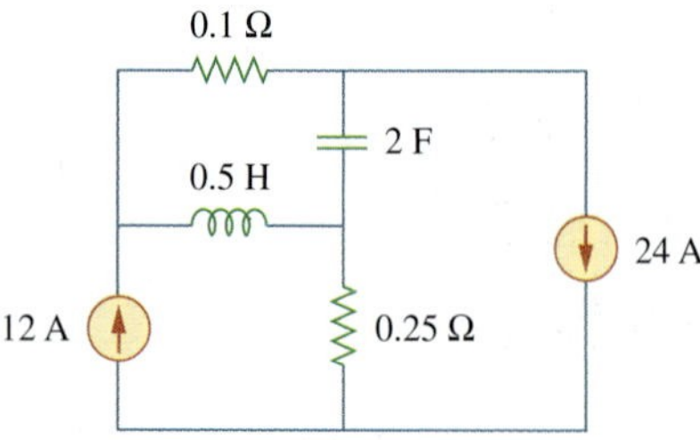

Figure G.19
For Prob. 8.75.

8.77 See Fig. G.20.

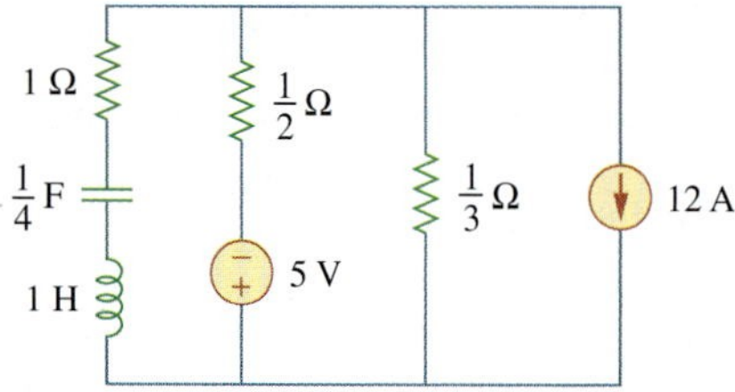

Figure G.20
For Prob. 8.77.

8.79 434 μF

8.81 2.533 μH, 625 μF

8.83 $\frac{d^2v}{dt^2} + \frac{R}{L}\frac{dv}{dt} + \frac{R}{LC}i_D + \frac{1}{C}\frac{di_D}{dt} = \frac{v_s}{LC}$

Chapter 9

9.1 (a) 50 V, (b) 209.4 ms, (c) 4.775 Hz, (d) 44.48 V, 0.3 rad

9.3 (a) $4\cos(\omega t - 120°)$, (b) $2\cos(6t + 90°)$, (c) $10\cos(\omega t + 110°)$

9.5 20°, v_1 lags v_2

9.7 Proof

9.9 (a) $50.88\angle{-15.52°}$, (b) $60.02\angle{-110.96°}$

9.11 (a) $21\angle{-15°}$ V, (b) $8\angle{160°}$ mA, (c) $120\angle{-140°}$ V, (d) $60\angle{190°}$ mA

9.13 (a) $-1.2749 + j0.1520$, (b) -2.083, (c) $35 + j14$

9.15 (a) $-6 - j11$, (b) $120.99 + j4.415$, (c) -1

9.17 $15.62\cos(50t - 9.8°)$ V

9.19 (a) $3.32\cos(20t + 114.49°)$, (b) $64.78\cos(50t - 70.89°)$, (c) $9.44\cos(400t - 44.7°)$

9.21 (a) $f(t) = 8.324\cos(30t + 34.86°)$, (b) $g(t) = 5.565\cos(t - 62.49°)$, (c) $h(t) = 1.2748\cos(40t - 168.69°)$

9.23 (a) $43.49\cos(\omega t - 6.59°)$ V, (b) $18.028\cos(\omega t + 78.69°)$ A

9.25 (a) $0.8\cos(2t - 98.13°)$ A, (b) $0.745\cos(5t - 4.56°)$ A

9.27 $0.289\cos(377t - 92.45°)$ V

9.29 $2\sin(10^6 t - 65°)$

9.31 $78.3\cos(2t + 51.21°)$ mA

9.33 69.82 V

9.35 $4.789\cos(200t - 16.7°)$ A

9.37 $(500 - j50)$ mS

9.39 $9.135 + j27.47\ \Omega$, $414.5\cos(10t - 71.6°)$ mA

9.41 $15.812\cos(t - 18.43°)$ V

9.43 $499.7\angle{-28.85°}$ mA

9.45 -5 A

9.47 $1.8428\cos(2{,}000t + 52.63°)$ A

9.49 $1.4142\sin(200t - 45°)$ V

9.51 $-12.5\cos(2t - 53.13°)$ A

9.53 $8.873\angle{-21.67°}$ A

9.55 $2.798 - j16.403\ \Omega$

9.57 $0.3171 - j0.1463$ S

9.59 $2.707 + j2.509$

9.61 $1 + j0.5\ \Omega$

9.63 $34.69 - j6.93\ \Omega$

9.65 $17.35\underline{/0.9^\circ}$ A, $6.83 + j1.094\ \Omega$

9.67 (a) $14.8\underline{/-20.22^\circ}$ mS, (b) $19.7\underline{/74.57^\circ}$ mS

9.69 $1.661 + j0.6647$ S

9.71 $1.058 - j2.235\ \Omega$

9.73 $0.3796 + j1.46\ \Omega$

9.75 Can be achieved by the RL circuit shown in Fig. G.21.

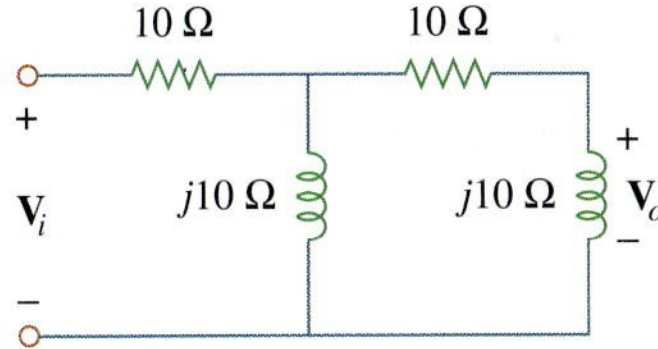

Figure G.21
For Prob. 9.75.

9.77 (a) 51.49° lagging, (b) 1.5915 MHz

9.79 (a) 140.2°, (b) leading, (c) 18.43 V

9.81 1.8 kΩ, 0.1 μF

9.83 104.17 mH

9.85 Proof

9.87 $38.21\underline{/-8.97^\circ}\ \Omega$

9.89 8.05 mH

9.91 235 pF

9.93 $1.7958\underline{/-38.66^\circ}$ A

Chapter 10

10.1 $1.9704\cos(10t + 5.65^\circ)$ A

10.3 $7.67\cos(4t - 35.02^\circ)$ V

10.5 $12.398\cos(4 \times 10^3 t + 4.06^\circ)$ mA

10.7 $124.08\underline{/-154^\circ}$ V

10.9 $6.154\cos(10^3 t + 70.26^\circ)$ V

10.11 $498.7\underline{/86.87^\circ}$ mA

10.13 $29.36\underline{/62.88^\circ}$ V

10.15 $15.812\underline{/43.49^\circ}$ A

10.17 $13.875\underline{/-162.12^\circ}$ A

10.19 $7.682\underline{/50.19^\circ}$ V

10.21 (a) $1, 0, -\dfrac{j}{R}\sqrt{\dfrac{L}{C}}$, (b) $0, 1, \dfrac{j}{R}\sqrt{\dfrac{L}{C}}$

10.23 $\dfrac{(1 - \omega^2 LC)V_s}{1 - \omega^2 LC + j\omega RC(2 - \omega^2 LC)}$

10.25 $2.828\cos(2t + 45^\circ)$ A

10.27 $4.698\underline{/95.24^\circ}$ A, $0.9928\underline{/37.71^\circ}$ A

10.29 This is a design problem with multiple answers.

10.31 $2.179\underline{/61.44^\circ}$ A

10.33 $15.92\underline{/43.49^\circ}$ A

10.35 $985.5\underline{/-2.1^\circ}$ mA

10.37 $2.38\underline{/-96.37^\circ}$ A, $2.38\underline{/143.63^\circ}$ A, $2.38\underline{/23.63^\circ}$ A

10.39 $0.6357\underline{/109.6^\circ}$ A, $0.5738\underline{/124.4^\circ}$ A, $0.2425\underline{/-60.42^\circ}$ A, $0.1675\underline{/48.5^\circ}$ A

10.41 $2.122\cos(2t + 45^\circ) + 7.156\sin(4t + 25.56^\circ)$ V

10.43 $9.902\cos(2t - 129.17^\circ)$ A

10.45 $[989.1\cos(10t + 21.47^\circ) + 499\sin(4t + 176.57^\circ)]$ mA

10.47 $[4 + 0.504\sin(t + 19.1^\circ) + 0.3352\cos(3t - 76.43^\circ)]$ A

10.49 $8.944\sin(200t + 56.56^\circ)$ A

10.51 $109.3\underline{/30^\circ}$ mA

10.53 $(3.529 - j5.883)$ V

10.55 (a) $\mathbf{Z}_N = \mathbf{Z}_{\text{Th}} = 22.63\underline{/-63.43^\circ}\ \Omega$, $\mathbf{V}_{\text{Th}} = -50\underline{/30^\circ}$ V, $\mathbf{I}_N = 2.236\underline{/273.4^\circ}$ A,

(b) $\mathbf{Z}_N = \mathbf{Z}_{\text{Th}} = 10\underline{/26^\circ}\ \Omega$, $\mathbf{V}_{\text{Th}} = 33.92\underline{/58^\circ}$ V, $\mathbf{I}_N = 3.392\underline{/32^\circ}$ A

10.57 This is a design problem with multiple answers.

10.59 $-6 + j38\ \Omega$

10.61 $-24 + j12$ V, $-8 + j6\ \Omega$

10.63 1 kΩ, 5.657 cos(200t + 75°) A

10.65 This is a design problem with multiple answers.

10.67 $4.945\angle{-69.76^\circ}$ V, $0.4378\angle{-75.24^\circ}$ A, $11.243 + j1.079\ \Omega$

10.69 $-j\omega RC$, $-V_m \cos \omega t$

10.71 96 cos(2t + 44.52°) V

10.73 $21.21\angle{-45^\circ}$ kΩ

10.75 $0.12499\angle{180^\circ}$

10.77 $\dfrac{R_2 + R_3 + j\omega C_2 R_2 R_3}{(1 + j\omega R_1 C_1)(R_3 + j\omega C_2 R_2 R_3)}$

10.79 3.578 cos(1,000t + 26.56°) V

10.81 $11.27\angle{128.1^\circ}$ V

10.83 6.611 cos(1,000t − 159.2°) V

10.85 This is a design problem with multiple answers.

10.87 $15.91\angle{169.6^\circ}$ V, $5.172\angle{-138.6^\circ}$ V, $2.27\angle{-152.4^\circ}$ V

10.89 Proof

10.91 (a) 180 kHz,
(b) 40 kΩ

10.93 Proof

10.95 Proof

Chapter 11

(Assume all values of currents and voltages are rms unless otherwise specified.)

11.1 800 + 1,600 cos(100t + 60°) W, 800 W

11.3 13.333 W

11.5 $P_{1\Omega} = 11.33$ W, $P_{2\Omega} = 40.79$ W,
$P_{3H} = P_{0.25F} = 0$

11.7 160 W

11.9 1.794 mW

11.11 12.751 mW

11.13 (a) $120 - j60\ \Omega$, (b) 12.605 W

11.15 $0.5 - j0.5\ \Omega$, 9 kW

11.17 20 Ω, 5 W

11.19 2.576 Ω, 3.798 W

11.21 19.58 Ω

11.23 This is a design problem with multiple answers.

11.25 8.165

11.27 2.887 A

11.29 5.773 A, 400 W

11.31 2.944 V

11.33 6.665

11.35 21.6 V

11.37 This is a design problem with multiple answers.

11.39 (a) 0.7592, 6.643 kW, 5.695 kVAR,
(b) 312 μF

11.41 (a) 0.5547 (leading), (b) 0.9304 (lagging)

11.43 This is a design problem with multiple answers.

11.45 (a) 46.9 V, 1.061 A, (b) 20 W

11.47 (a) $S = 112 + j194$ VA,
average power = 112 W,
reactive power = 194 VAR
(b) $S = 226.3 - j226.3$ VA,
average power = 226.3 W,
reactive power = −226.3 VAR
(c) $S = 110.85 + j64$ VA, average power = 110.85 W, reactive power = 64 VAR
(d) $S = 7.071 + j7.071$ kVA, average power = 7.071 kW, reactive power = 7.071 kVAR

11.49 (a) $4 + j2.373$ kVA,
(b) $1.6 - j1.2$ kVA,
(c) $0.4624 + j1.2705$ kVA,
(d) $110.77 + j166.16$ VA

11.51 (a) 0.9956 (lagging),
(b) 1.751 kW,
(c) 164.9 VAR,
(d) 1.7587 kVA,
(e) $(1{,}751 + j164.9)$ VA

11.53 (a) $93.97\angle{29.8^\circ}$ A, (b) 1.0 (lagging)

11.55 This is a design problem with multiple answers.

11.57 $(50.45 - j33.64)$ VA

11.59 $j339.3$ VAR, $-j1.4146$ kVAR

11.61 $33.1\angle 92.4^\circ$ A, $6.62\angle -2.4^\circ$ kVA

11.63 $443.3\angle -28.13^\circ$ A

11.65 80 μW

11.67 $36\angle 36.86^\circ$ mVA, 12.042 mW

11.69 (a) 0.6402 (lagging),
(b) 295.1 W,
(c) 130.4 μF

11.71 (a) $50.14 + j1.7509$ mΩ,
(b) 0.9994 lagging,
(c) $2.392\angle -2^\circ$ kA

11.73 (a) 12.21 kVA, (b) $50.86\angle -35^\circ$ A,
(c) 4.083 kVAR, 188.03 μF, (d) $43.4\angle -16.26^\circ$ A

11.75 (a) $1{,}835.9 - j114.68$ VA, (b) 0.998 (leading),
(c) no correction is necessary

11.77 157.69 W

11.79 50 mW

11.81 This is a design problem with multiple answers.

11.83 (a) 688.1 W, (b) 840 VA,
(c) 481.8 VAR, (d) 0.8191 (lagging)

11.85 (a) 20 A, $17.85\angle 163.26^\circ$ A, $5.907\angle -119.5^\circ$ A,
(b) $4{,}451 + j617$ VA, (c) 0.9904 (lagging)

11.87 0.5333

11.89 (a) 12 kVA, $9.36 + j7.51$ kVA,
(b) $2.866 + j2.3\ \Omega$

11.91 0.9775, 104 μF

11.93 (a) 7.328 kW, 1.196 kVAR, (b) 0.987

11.95 (a) 2.814 kHz,
(b) 431.8 mW

11.97 547.3 W

Chapter 12

(Assume all values of currents and voltages are rms unless otherwise specified.)

12.1 (a) $231\angle -30^\circ$, $231\angle -150^\circ$, $231\angle 90^\circ$ V,
(b) $231\angle 30^\circ$, $231\angle 150^\circ$, $231\angle -90^\circ$ V

12.3 *abc* sequence, $208\angle 250^\circ$ V

12.5 $260\cos(\omega t + 62^\circ)$ V, $260\cos(\omega t - 58^\circ)$ V,
$260\cos(\omega t + 182^\circ)$ V

12.7 $44\angle 53.13^\circ$ A, $44\angle -66.87^\circ$ A, $44\angle 173.13^\circ$ A

12.9 $4.8\angle -36.87^\circ$ A, $4.8\angle -156.87^\circ$ A, $4.8\angle 83.13^\circ$ A

12.11 207.8 V, 99.85 A

12.13 40.85 A, 15.02 kW

12.15 13.66 A

12.17 $5.773\angle 5^\circ$ A, $5.773\angle -115^\circ$ A,
$5.773\angle 125^\circ$ A

12.19 $5.47\angle -18.43^\circ$ A, $5.47\angle -138.43^\circ$ A, $5.47\angle 101.57^\circ$ A,
$9.474\angle -48.43^\circ$ A, $9.474\angle -168.43^\circ$ A,
$9.474\angle 71.57^\circ$ A

12.21 $34.36\angle -98.66^\circ$ A, $59.51\angle 171.34^\circ$ A

12.23 (a) 13.995 A,
(b) 2.448 kW

12.25 $8.87\angle 4.78^\circ$, $8.87\angle -115.22^\circ$, $8.87\angle 124.78^\circ$ A

12.27 91.79 V

12.29 $1.3 + j1.1465$ kVA

12.31 (a) $6.144 + j4.608\ \Omega$,
(b) 18.04 A, (c) 207.2 μF

12.33 15.385 A, 360.3 V

12.35 (a) $14.61 - j5.953$ A,
(b) $3.361 + j1.368$ kVA,
(c) 0.9261

12.37 55.51 A, $1.298 - j1.731\ \Omega$

12.39 431.1 W

12.41 9.021 A

12.43 $4.373 - j1.145$ kVA

12.45 $2.109\angle 24.83^\circ$ kV

12.47 39.19 A (rms), 0.9982 (lagging)

12.49 (a) 5.808 kW, (b) 1.9356 kW

12.51 (a) $19.2 - j14.4$ A, $-42.76 + j27.09$ A, $-12 - j20.78$ A,
(b) $31.2 + j6.38$ A, $-61.96 + j41.48$ A, $30.76 - j47.86$ A

12.53 This is a design problem with multiple answers.

12.55 $9.6\angle{-90°}$ A, $6\angle{120°}$ A, $8\angle{-150°}$ A, $3.103 + j3.264$ kVA

12.57 $I_a = 1.9585\angle{-18.1°}$ A, $I_b = 1.4656\angle{-130.55°}$ A, $I_c = 1.947\angle{117.8°}$ A

12.59 $220.6\angle{-34.56°}$, $214.1\angle{-81.49°}$, $49.91\angle{-50.59°}$ V, assuming that N is grounded.

12.61 $11.15\angle{37°}$ A, $230.8\angle{-133.4°}$ V, assuming N is grounded.

12.63 $18.67\angle{158.9°}$ A, $12.38\angle{144.1°}$ A

12.65 $11.02\angle{12°}$ A, $11.02\angle{-108°}$ A, $11.02\angle{132°}$ A

12.67 (a) 97.67 kW, 88.67 kW, 82.67 kW, (b) 108.97 A

12.69 $I_a = 94.32\angle{-62.05°}$ A, $I_b = 94.32\angle{177.95°}$ A, $I_c = 94.32\angle{57.95°}$ A, $28.8 + j18.03$ kVA

12.71 (a) 2,590 W, 4,808 W, (b) 8,335 VA

12.73 2,360 W, −632.8 W

12.75 (a) 20 mA, (b) 200 mA

12.77 320 W

12.79 $17.15\angle{-19.65°}$, $17.15\angle{-139.65°}$, $17.15\angle{100.35°}$ A, $223\angle{2.97°}$, $223\angle{-117.03°}$, $223\angle{122.97°}$ V

12.81 516 V

12.83 183.42 A

12.85 $Z_Y = 2.133\ \Omega$

12.87 $1.448\angle{-176.6°}$ A, $1{,}252 + j711.6$ VA, $1{,}085 + j721.2$ VA

Chapter 13

(Assume all values of currents and voltages are rms unless otherwise specified.)

13.1 10 H

13.3 150 mH, 50 mH, 25 mH, 0.2887

13.5 (a) 123.7 mH, (b) 24.31 mH

13.7 $540.5\angle{144.16°}$ mV

13.9 $4.148\angle{21.12°}$ V

13.11 $412.3 \cos(600t - 140.43°)$ mA

13.13 $4.308 + j6.538\ \Omega$

13.15 $1 + j19.5\ \Omega$, $1.404\angle{9.44°}$ A

13.17 $13.073 + j25.86\ \Omega$

13.19 See Fig. G.22.

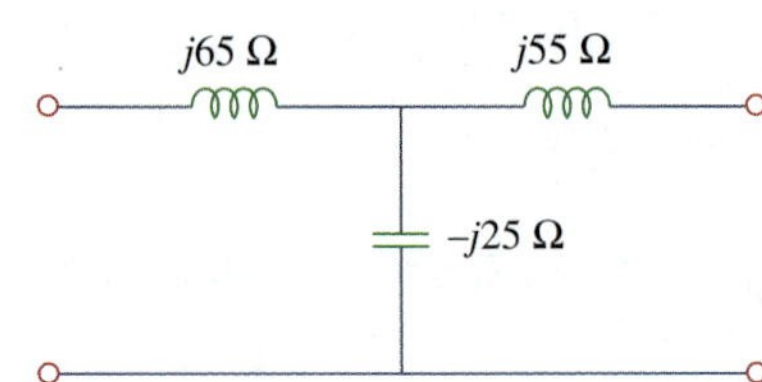

Figure G.22
For Prob. 13.19.

13.21 This is a design problem with multiple answers.

13.23 $50.68 \cos(10t + 52.54°)$ A, $27.19 \cos(10t - 100.89°)$ A, 1.5 kJ

13.25 $2.2 \sin(2t - 4.88°)$ A, $1.5085\angle{17.9°}\ \Omega$

13.27 1.567 W

13.29 0.984, 130.5 mJ

13.31 This is a design problem with multiple answers.

13.33 $12.769 + j7.154\ \Omega$

13.35 $10.143\angle{-21.4^\circ}$ A, $532.8\angle{-134.85^\circ}$ mA, $529.4\angle{-110.41^\circ}$ mA

13.37 (a) 5, (b) 104.17 A, (c) 20.83 A

13.39 $15.7\angle{20.31^\circ}$ A, $78.5\angle{20.31^\circ}$ A

13.41 7.857 A, −23.57 A

13.43 4.186 V, 16.744 V

13.45 58.72 W

13.47 $118.03 \cos(3t + 59.93^\circ)$ V

13.49 $0.937 \cos(2t + 51.34^\circ)$ A

13.51 $8 - j1.5\ \Omega$, $29.49\angle{10.62^\circ}$ A

13.53 (a) 5, (b) 8 W

13.55 1.6669 Ω

13.57 (a) $25.9\angle{69.96^\circ}$, $12.95\angle{69.96^\circ}$ A (rms), (b) $21.06\angle{147.4^\circ}$, $42.12\angle{147.4^\circ}$, $42.12\angle{147.4^\circ}$ V(rms), (c) $1554\angle{20.04^\circ}$ VA

13.59 $P_{10\Omega} = 395$ W, $P_{12\Omega} = 266.6$ W, $P_{20\Omega} = 49.39$ W

13.61 6 A, 0.36 A, −60 V

13.63 $3.795\angle{18.43^\circ}$ A, $1.8975\angle{18.43^\circ}$ A, $0.6325\angle{161.6^\circ}$ A

13.65 11.05 W

13.67 (a) 160 V, (b) 31.25 A, (c) 12.5 A

13.69 $(1.2 - j2)$ kΩ, 5.333 W

13.71 $[1 + (N_1/N_2)]^2 Z_L$

13.73 (a) three-phase Δ-Y transformer, (b) $8.66\angle{156.87^\circ}$ A, $5\angle{-83.13^\circ}$ A, (c) 1.8 kW

13.75 (a) 0.11547, (b) 76.98 A, 15.395 A

13.77 (a) a single-phase transformer, $1:n$, $n = 1/110$, (b) 7.576 mA

13.79 $7.836\angle{-68.01^\circ}$ A, $2.441\angle{-77.86^\circ}$ A, $8.016\angle{-54.92^\circ}$ A

13.81 $104.5\angle{13.96^\circ}$ mA, $29.54\angle{-143.8^\circ}$ mA, $208.8\angle{24.4^\circ}$ mA

13.83 $21.6\angle{33.91^\circ}$ A, $302.8\angle{-34.21^\circ}$ V

13.85 100 turns

13.87 0.5

13.89 0.5, 41.67 A, 83.33 A

13.91 (a) 1,875 kVA, (b) 7,812 A

13.93 (a) See Fig. G.23(a). (b) See Fig. G.23(b).

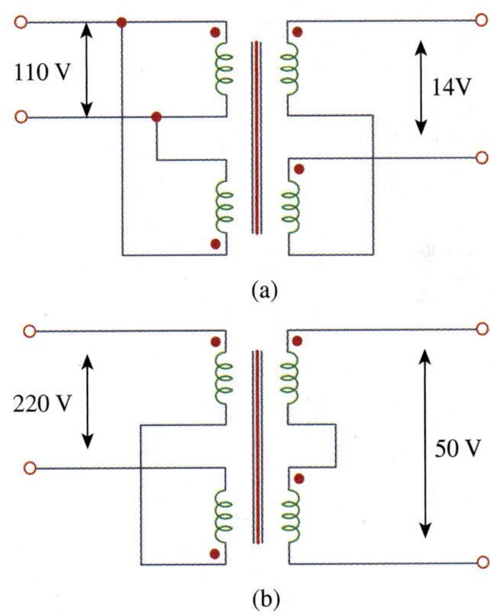

Figure G.23
For Prob. 13.93.

13.95 (a) 1/60, (b) 139 mA

Chapter 14

14.1 $\dfrac{j\omega/\omega_o}{1 + j\omega/\omega_o}$, $\omega_o = \dfrac{1}{RC}$

14.3 $\dfrac{5}{s^2 + 8s + 5}$

14.5 (a) $\dfrac{sRL}{(R + R_s)Ls + RR_s}$

(b) $\dfrac{R}{LRCs^2 + Ls + R}$

14.7 (a) 1.005773, (b) 0.4898, (c) 1.718×10^5

14.9 See Fig. G.24.

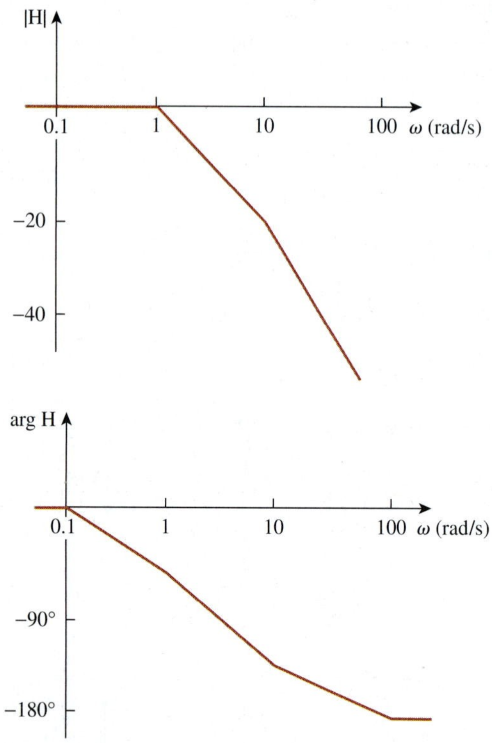

Figure G.24
For Prob. 14.9.

14.11 See Fig. G.25.

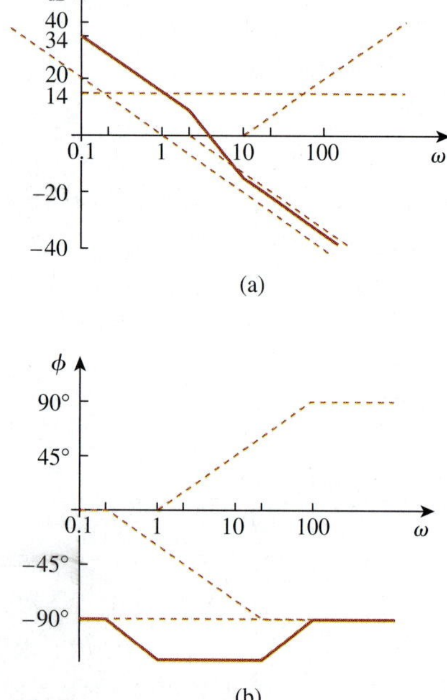

Figure G.25
For Prob. 14.11.

14.13 See Fig. G.26.

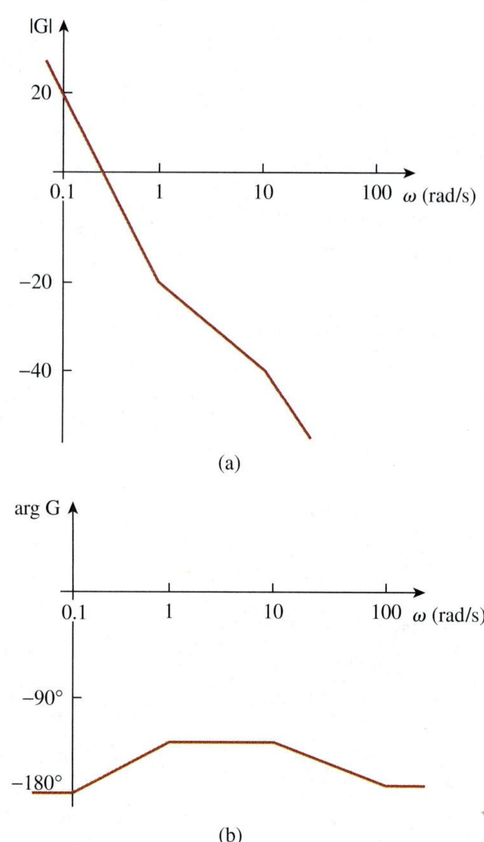

Figure G.26
For Prob. 14.13.

14.15 See Fig. G.27.

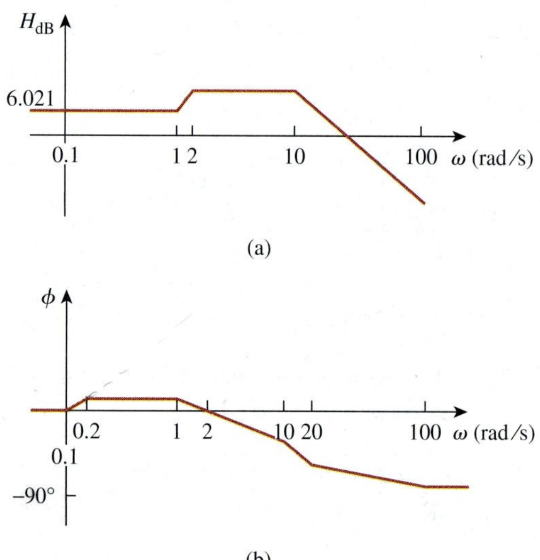

Figure G.27
For Prob. 14.15: (a) magnitude plot, (b) phase plot.

14.17 See Fig. G.28.

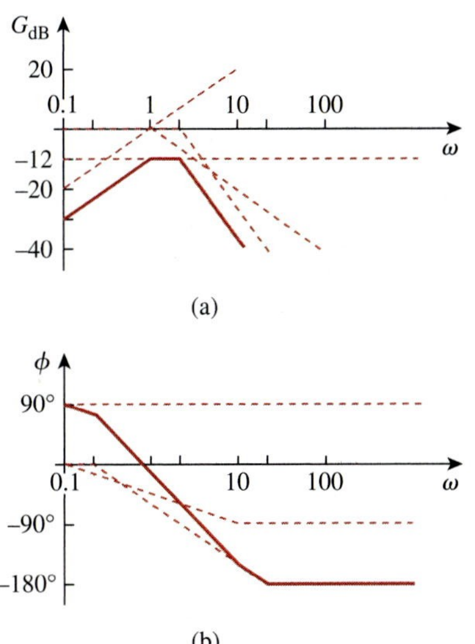

Figure G.28
For Prob. 14.17.

14.19 See Fig. G.29.

14.21 See Fig. G.30.

14.23 $\dfrac{10^4 j\omega}{(10 + j\omega)(100 + j\omega)^2}$

14.25 2 kΩ, 2 − j0.75 kΩ, 2 − j0.3 kΩ, 2 + j0.3 kΩ, 2 + j0.75 kΩ

14.27 $R = 1\ \Omega$, $L = 0.1$ H, $C = 25$ mF

14.29 4.082 krad/s, 38.67, 105.55 rad/s

14.31 8.796×10^6 rad/s,

14.33 14.21 μH, 56.84 pF

14.35 40 Ω, 2.5 μH, 10 μF, 2.5 krad/s, 198.75 krad/s, 202.25 krad/s

14.37 $\dfrac{1}{\sqrt{LC - R^2C^2}}$

14.39 (a) 19.89 nF (b) 164.45 μH, (c) 552.9 krad/s, (d) 25.13 krad/s, (e) 22

14.41 This is a design problem with multiple answers.

14.43 (a) 2.357 krad/s, (b) 1 Ω

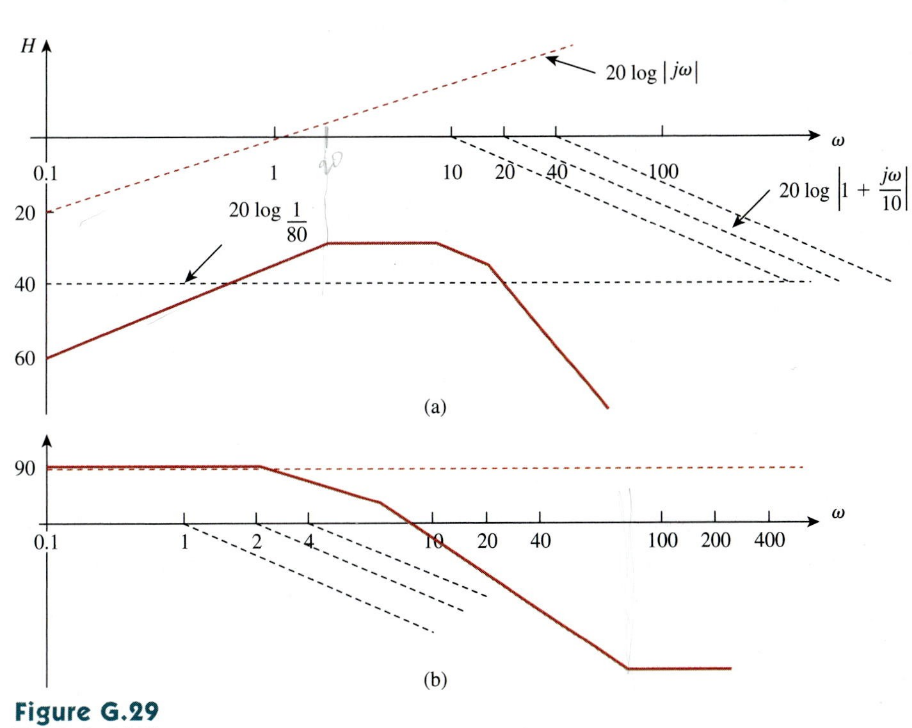

Figure G.29
For Prob. 14.19.

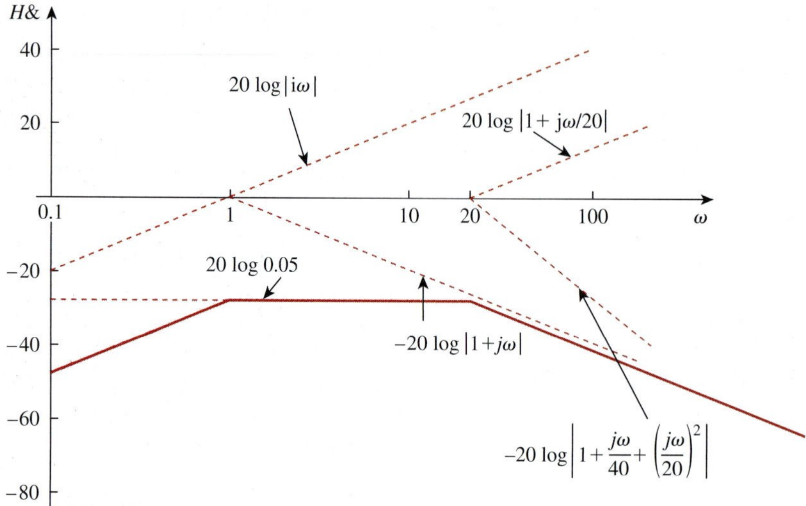

Figure G.30
For Prob. 14.21.

14.45 (a) $\frac{j\omega}{2(1 + j\omega)^2}$, (b) 0.25

14.47 796 kHz

14.49 This is a design problem with multiple answers.

14.51 1.256 kΩ

14.53 18.045 kΩ. 2.872 H, 10.5

14.55 1.56 kHz $< f <$ 1.62 kHz, 25

14.57 (a) 1 rad/s, 3 rad/s, (b) 1 rad/s, 3 rad/s

14.59 2.408 krad/s, 15.811 krad/s

14.61 (a) $\frac{1}{1 + j\omega RC}$, (b) $\frac{j\omega RC}{1 + j\omega RC}$

14.63 10 MΩ, 100 kΩ

14.65 Proof

14.67 If $R_f = 20$ kΩ, then $R_i = 80$ kΩ and $C = 15.915$ nF.

14.69 Let $R = 10$ kΩ, then $R_f = 25$ kΩ, $C = 7.96$ nF.

14.71 $K_f = 2 \times 10^{-4}$, $K_m = 5 \times 10^{-3}$

14.73 9.6 MΩ, 32 μH, 0.375 pF

14.75 200 Ω, 400 μH, 1 μF

14.77 (a) 1,200 H, 0.5208 μF, (b) 2 mH, 312.5 nF, (c) 8 mH, 7.81 pF

14.79 (a) $8s + 5 + \frac{10}{s}$, (b) $0.8s + 50 + \frac{10^4}{s}$, 111.8 rad/s

14.81 (a) 0.4 Ω, 0.4 H, 1 mF, 1 mS, (b) 0.4 Ω, 0.4 mH, 1 μF, 1 mS

14.83 0.1 pF, 0.5 pF, 1 MΩ, 2 MΩ

14.85 See Fig. G.31.

14.87 See Fig. G.32; highpass filter, $f_0 = 1.2$ Hz.

14.89 See Fig. G.33.

14.91 See Fig. G.34; $f_o = 800$ Hz.

14.93 $\frac{-RCs + 1}{RCs + 1}$

14.95 (a) 0.541 MHz $< f_o <$ 1.624 MHz, (b) 67.98, 204.1

14.97 $\frac{s^3LR_LC_1C_2}{(sR_iC_1+1)(s^2LC_2+sR_LC_2+1)+s^2LC_1(sR_LC_2+1)}$

14.99 8.165 MHz, 4.188×10^6 rad/s

14.101 1.061 kΩ

14.103 $\frac{R_2(1 + sCR_1)}{R_1 + R_2 + sCR_1R_2}$

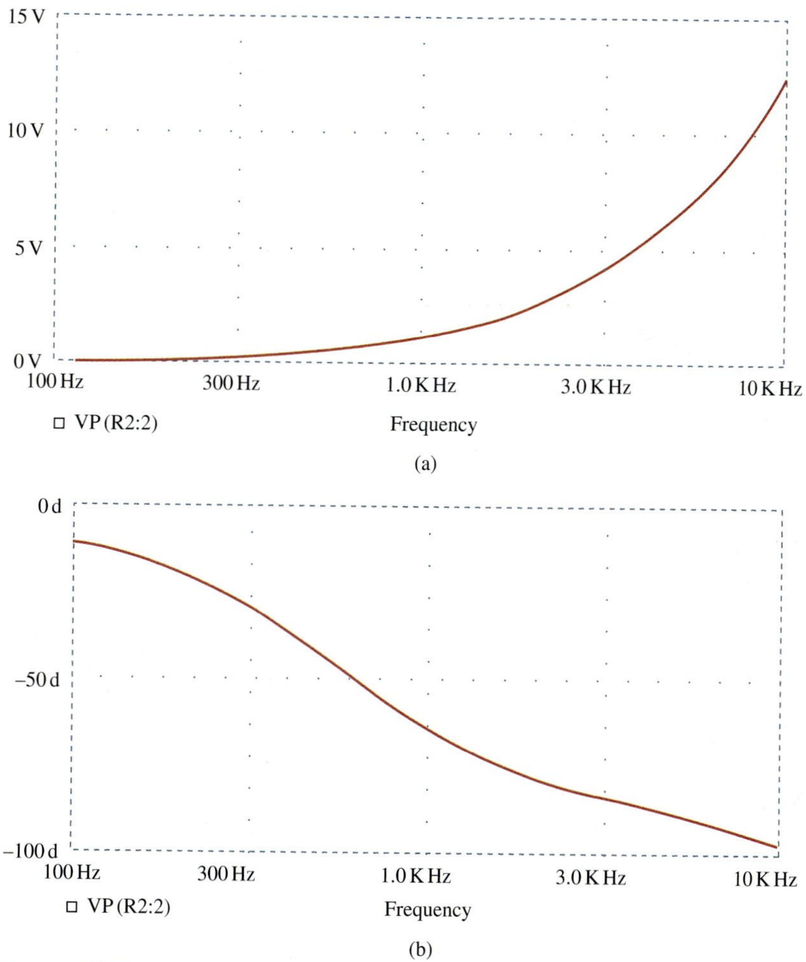

Figure G.31
For Prob. 14.85.

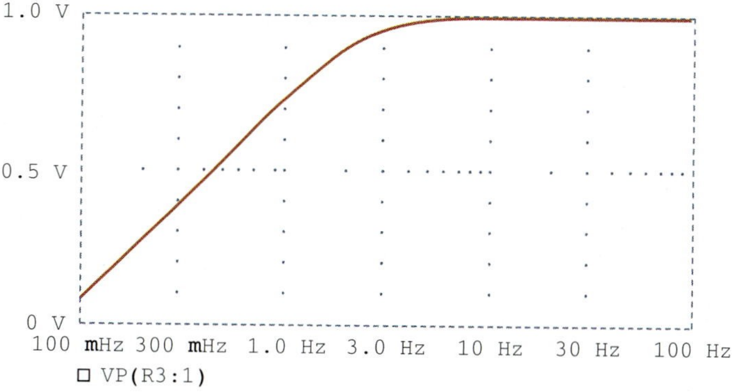

Figure G.32
For Prob. 14.87.

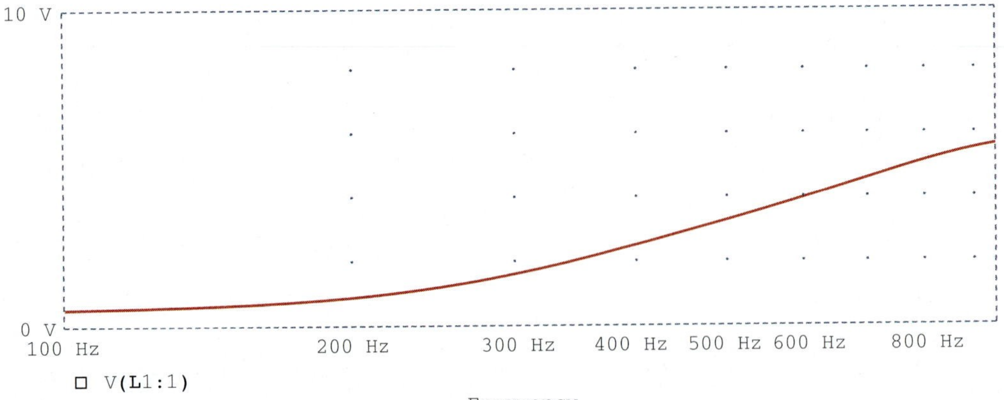

Figure G.33
For Prob. 14.89.

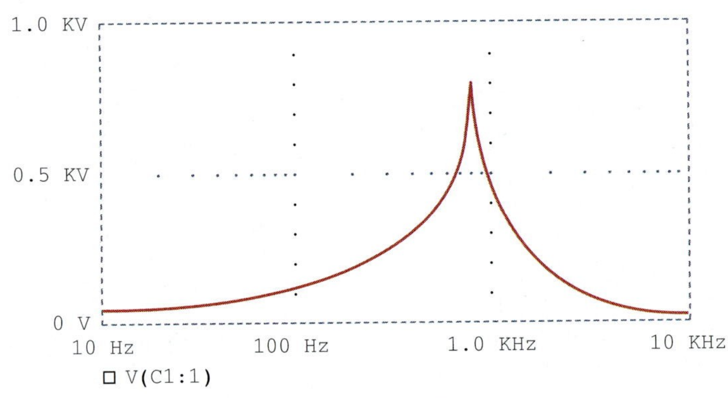

Figure G.34
For Prob. 14.91.

Chapter 15

15.1 (a) $\dfrac{s}{s^2 - a^2}$, (b) $\dfrac{a}{s^2 - a^2}$

15.3 (a) $\dfrac{s+2}{(s+2)^2+9}$, (b) $\dfrac{4}{(s+2)^2+16}$, (c) $\dfrac{s+3}{(s+3)^2-4}$ (d) $\dfrac{1}{(s+4)^2-1}$, (e) $\dfrac{4(s+1)}{[(s+1)^2+4]^2}$

15.5 (a) $\dfrac{8 - 12\sqrt{3}s - 6s^2 + \sqrt{3}s^3}{(s^2+4)^3}$, (b) $\dfrac{72}{(s+2)^5}$, (c) $\dfrac{2}{s^2} - 4s$, (d) $\dfrac{2e}{s+1}$, (e) $\dfrac{5}{s}$, (f) $\dfrac{18}{3s+1}$, (g) s^n

15.7 (a) $\dfrac{2}{s^2} + \dfrac{4}{s}$, (b) $\dfrac{4}{s} + \dfrac{3}{s+2}$, (c) $\dfrac{8s+18}{s^2+9}$, (d) $\dfrac{s+2}{s^2+4s-12}$

15.9 (a) $\dfrac{e^{-2s}}{s^2} - \dfrac{2e^{-2s}}{s^2}$, (b) $\dfrac{2e^{-s}}{e^4(s+4)}$, (c) $\dfrac{2.702s}{s^2+4} + \dfrac{8.415}{s^2+4}$, (d) $\dfrac{6}{s}e^{-2s} - \dfrac{6}{s}e^{-4s}$

15.11 (a) $\dfrac{6(s+1)}{s^2+2s-3}$, (b) $\dfrac{24(s+2)}{(s^2+4s-12)^2}$, (c) $\dfrac{e^{-(2s+6)}[(4e^2+4e^{-2})s + (16e^2+8e^{-2})]}{s^2+6s+8}$

15.13 (a) $\dfrac{s^2-1}{(s^2+1)^2}$,
(b) $\dfrac{2(s+1)}{(s^2+2s+2)^2}$,
(c) $\tan^{-1}\left(\dfrac{\beta}{s}\right)$

15.15 $5\dfrac{1-e^{-s}-se^{-s}}{s^2(1-e^{-3s})}$

15.17 This is a design problem with multiple answers.

15.19 $\dfrac{1}{1-e^{-s}}$

15.21 $\dfrac{(2\pi s-1+e^{-2\pi s})}{2\pi s^2(1-e^{-2\pi s})}$

15.23 (a) $\dfrac{(1-e^{-s})^2}{s(1-e^{-2s})}$,
(b) $\dfrac{2(1-e^{-2s})-4se^{-2s}(s+s^2)}{s^3(1-e^{-2s})}$

15.25 (a) 5 and 0, (b) 5 and 0

15.27 (a) $u(t)+2e^{-t}u(t)$, (b) $3\delta(t)-11e^{-4t}u(t)$,
(c) $(2e^{-t}-2e^{-3t})u(t)$,
(d) $(3e^{-4t}-3e^{-2t}+6te^{-2t})u(t)$

15.29 $(1-e^{-2t}\cos(3t)-(1/3)e^{-2t}\sin(3t))u(t)$

15.31 (a) $(-5e^{-t}+20e^{-2t}-15e^{-3t})u(t)$,
(b) $\left(-e^{-t}+\left(1+3t-\dfrac{t^2}{2}\right)e^{-2t}\right)u(t)$,
(c) $(-0.2e^{-2t}+0.2e^{-t}\cos(2t)$
$+0.4e^{-t}\sin(2t))u(t)$

15.33 (a) $(3e^{-t}-3\cos(t)+3\sin(t))u(t)$,
(b) $\sin(t-\pi)u(t-\pi)$,
(c) $3u(t)[1-e^{-t}-te^{-t}-0.5t^2e^{-t}]$

15.35 (a) $[2e^{-(t-6)}-e^{-2(t-6)}]u(t-6)$,
(b) $\dfrac{4}{3}u(t)[e^{-t}-e^{-4t}]-\dfrac{1}{3}u(t-2)[e^{-(t-2)}-e^{-4(t-2)}]$,
(c) $\dfrac{1}{13}u(t-1)[-3e^{-3(t-1)}+3\cos 2(t-1)$
$+2\sin 2(t-1)]$

15.37 (a) $(2-e^{-2t})u(t)$,
(b) $[0.4e^{-3t}+0.6e^{-t}\cos t+0.8e^{-t}\sin t]u(t)$,
(c) $e^{-2(t-4)}u(t-4)$,
(d) $\left(\dfrac{10}{3}\cos t-\dfrac{10}{3}\cos 2t\right)u(t)$

15.39 (a) $(-1.6e^{-t}\cos 4t-4.05e^{-t}\sin 4t$
$+3.6e^{-2t}\cos 4t+(3.45e^{-2t}\sin 4t)u(t)$,
(b) $[0.08333\cos 3t+0.02778\sin 3t$
$+0.0944e^{-0.551t}-0.1778e^{-5.449t}]u(t)$

15.41 $z(t)=\begin{cases}8t, & 0<t<2\\ 16-8t, & 2<t<6\\ -16, & 6<t<8\\ 8t-80, & 8<t<12\\ 112-8t, & 12<t<14\\ 0, & \text{otherwise}\end{cases}$

15.43 (a) $y(t)=\begin{cases}\dfrac{1}{2}t^2, & 0<t<1\\ -\dfrac{1}{2}t^2+2t-1, & 1<t<2\\ 1, & t>2\\ 0, & \text{otherwise}\end{cases}$
(b) $y(t)=2(1-e^{-t}),\ t>0$,
(c) $y(t)=\begin{cases}\dfrac{1}{2}t^2+t+\dfrac{1}{2}, & -1<t<0\\ -\dfrac{1}{2}t^2+t+\dfrac{1}{2}, & 0<t<2\\ \dfrac{1}{2}t^2-3t+\dfrac{9}{2}, & 2<t<3\\ 0, & \text{otherwise}\end{cases}$

15.45 $(4e^{-2t}-8te^{-2t})u(t)$

15.47 (a) $(-e^{-t}+2e^{-2t})u(t)$, (b) $(e^{-t}-e^{-2t})u(t)$

15.49 (a) $\left(\dfrac{t}{a}(e^{at}-1)-\dfrac{1}{a^2}-\dfrac{e^{at}}{a^2}(at-1)\right)u(t)$,
(b) $[0.5\cos(t)(t+0.5\sin(2t))$
$-0.5\sin(t)(\cos(t)-1)]u(t)$

15.51 $(3e^{-t}+4e^{-2t}-5e^{-3t})u(t)V$

15.53 $\cos(t)+\sin(t)$ or $1.4142\cos(t-45°)$

15.55 $\left(\dfrac{1}{40}+\dfrac{1}{20}e^{-2t}-\dfrac{3}{104}e^{-4t}-\dfrac{3}{65}e^{-t}\cos(2t)\right.$
$\left.-\dfrac{2}{65}e^{-t}\sin(2t)\right)u(t)$

15.57 This is a design problem with multiple answers.

15.59 $[-2.5e^{-t}+12e^{-2t}-10.5e^{-3t}]u(t)$

Chapter 16

16.1 $1.155e^{-0.5t}\sin(0.866t)u(t)$ A

16.3 $[16 + 104e^{-15t}]u(t)$ mA

16.5 $\left(e^{-2t} - \frac{2}{\sqrt{7}}\sin\left(\frac{\sqrt{7}}{2}t\right)\right)u(t)$ A

16.7 $\left(1 - e^{-3t/4}\cos\frac{\sqrt{7}}{4}t + 4.9135e^{-3t/4}\sin\frac{\sqrt{7}}{4}t\right)u(t)$ V

16.9 (a) $\frac{2(s^2+1)}{s^2+2s+1}$,
(b) $\frac{s(5s+6)}{3s^2+7s+6}$

16.11 $\frac{50s+160}{s(s^2+9s+16)}$

16.13 $5(e^{-t} - e^{-2t})\,u(t)$ A

16.15 $-\frac{5s(s^2+20)}{s(s+2)(s^2+0.5s+40)}$

16.17 $[4 - e^{-t} + 1.5811e^{-jt+161.57°}$
$+ 1.5811e^{jt-161.57°}]u(t)$ A

16.19 This is a design problem with multiple answers.

16.21 $v_o(t) = \frac{20}{3}[1 - e^{-t}\cos 0.7071t$
$- 1.414e^{-t}\sin 0.7071t]u(t)$V

16.23 $(5e^{-4t}\cos 2t + 230e^{-4t}\sin 2t)u(t)$ V,
$(6 - 6e^{-4t}\cos 2t - 11.37e^{-4t}\sin 2t)u(t)$ A

16.25 $[2.202e^{-3t} + 3.84te^{-3t} - 0.202\cos(4t)$
$+ 0.6915\sin(4t)]u(t)$ V

16.27 $\frac{20(s+1)}{(s+3)(3s^2+4s+1)}, \frac{10(s+1)}{(s+3)(3s^2+4s+1)}$

16.29 $10[2e^{-1.5t} - e^{-t}]u(t)$ A

16.31 $\frac{10s^2}{s^2+4}$

16.33 $4 + \frac{s}{2(s+3)} - \frac{2s(s+2)}{s^2+4s+20} - \frac{12s}{s^2+4s+20}$

16.35 $\frac{9s}{3s^2+9s+2}$

16.37 (a) $\frac{s^2-3}{3s^2+2s-9}$, (b) $\frac{-3}{2s}$

16.39 $sRC + 1$

16.41 $\left(2 + \frac{8}{3}e^{-t} - \frac{14}{3}e^{-4t}\right)u(t)$ A

16.43 $\begin{bmatrix} v_C' \\ i' \end{bmatrix} = \begin{bmatrix} 0 & 1 \\ -1 & -1 \end{bmatrix}\begin{bmatrix} v_C \\ i \end{bmatrix} + \begin{bmatrix} 0 \\ 1 \end{bmatrix}u(t);$

$i(t) = [0 \quad 1]\begin{bmatrix} v_C \\ i \end{bmatrix} + [0]u(t)$

16.45 $\begin{bmatrix} i_L' \\ v_C' \end{bmatrix} = \begin{bmatrix} 0 & -1 \\ 4 & -2 \end{bmatrix}\begin{bmatrix} i_L \\ v_C \end{bmatrix} + \begin{bmatrix} 1 & -1 \\ 2 & 0 \end{bmatrix}\begin{bmatrix} v_1(t) \\ v_2(t) \end{bmatrix};$

$v_o(t) = [0 \quad -1]\begin{bmatrix} i_L \\ v_C \end{bmatrix} + [1 \quad 0]\begin{bmatrix} v_1(t) \\ v_2(t) \end{bmatrix}$

16.47 $\begin{bmatrix} i_L' \\ v_C' \end{bmatrix} = \begin{bmatrix} 0 & -1 \\ 4 & -2 \end{bmatrix}\begin{bmatrix} i_L \\ v_C \end{bmatrix} + \begin{bmatrix} 1 & -1 \\ 2 & 0 \end{bmatrix}\begin{bmatrix} v_1(t) \\ v_2(t) \end{bmatrix};$

$\begin{bmatrix} i_1(t) \\ i_2(t) \end{bmatrix} = \begin{bmatrix} 1 & -0.5 \\ 1 & 0 \end{bmatrix}\begin{bmatrix} i_L \\ v_C \end{bmatrix} + [0.5 \quad 0]\begin{bmatrix} v_1(t) \\ v_2(t) \end{bmatrix}$

16.49 $\begin{bmatrix} 0 & 1 \\ -9 & -7 \end{bmatrix}, \begin{bmatrix} 1 \\ -5 \end{bmatrix}, \quad [1 \quad 0], [0]$

16.51 $[1 - e^{-2t}(\cos 2t + \sin 2t)]u(t)$

16.53 $s_{1,2} = \frac{-1}{2RC} \pm \sqrt{\frac{1}{(2RC)^2} - \frac{1}{LC}}$. Note that both roots lie in the left half-plane since R, L, and C are positive quantities; thus the circuit is stable.

16.55 The circuit is unstable.

16.57 100 Ω, 12.8 Ω, 20 μF

16.59 We have three equations and four unknowns. Thus, there is a family of solutions. One such solution is

$R_2 = 1\text{ k}\Omega,\ C_1 = 50\text{ nF},\ C_2 = 20\ \mu\text{F}$

16.61 See Fig. G.35.

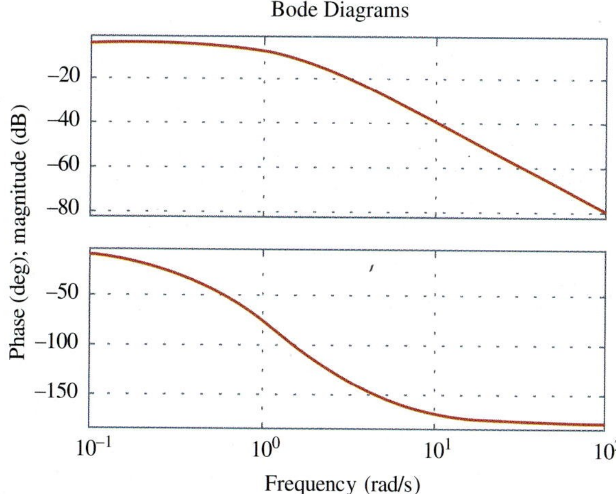

Figure G.35
For Prob. 16.61.

16.63 See Fig. G.36.

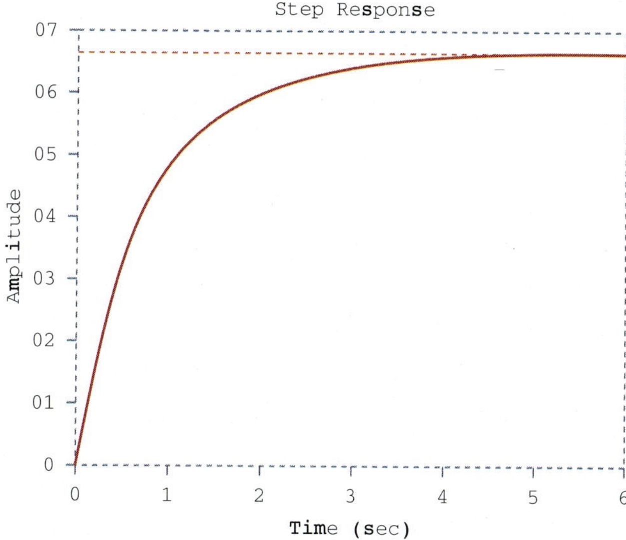

Figure G.36
For Prob. 16.63.

16.65 See Fig. G.37.

16.67 $a = -100$, $b = 400$, $c = 20{,}000$

16.69 Proof

Chapter 17

17.1 (a) periodic, 2, (b) not periodic, (c) periodic, 2π, (d) periodic, π, (e) periodic, 10, (f) not periodic, (g) not periodic

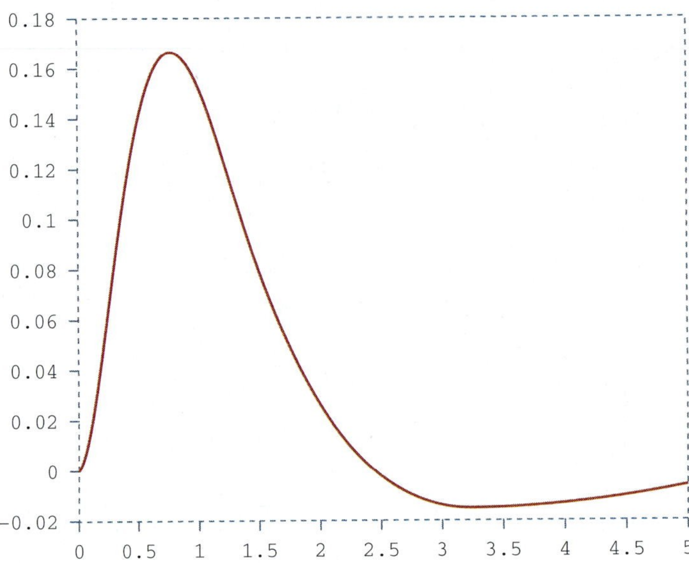

Figure G.37
For Prob. 16.65.

17.3 $a_0 = 7.5$

$$a_n = \begin{cases} \dfrac{10}{n\pi}(-1)^{(n+1)/2}, & n = \text{odd}, \\ & n = \text{even} \end{cases}$$

$$b_n = \frac{10}{n\pi}\left[3 - 2\cos n\pi - \cos\frac{n\pi}{2}\right]$$

17.5 $-0.5 + \sum_{\substack{n=1 \\ n=\text{odd}}}^{\infty} \frac{6}{n\pi} \sin nt$

17.7 $3 + \sum_{n=0}^{\infty}\left[\frac{9}{n\pi}\sin\frac{4n\pi}{3}\cos\frac{2n\pi t}{3} + \frac{9}{n\pi}\left(1 - \cos\frac{4n\pi}{3}\right)\sin\frac{2n\pi t}{3}\right]$

17.9 $a_0 = 3.183$, $a_1 = 10$, $a_2 = 6.362$, $a_3 = 0$, $b_1 = 0 = b_2 = b_3$

17.11 $\sum_{n=-\infty}^{\infty} \frac{1}{n^2\pi^2}[1 + j(jn\pi/2 - 1)\sin n\pi/2 + n\pi \sin n\pi/2]e^{jn\pi t/2}$

17.13 This is a design problem with multiple answers.

17.15 (a) $10 + \sum_{n=1}^{\infty}\sqrt{\frac{16}{(n^2+1)^2} + \frac{1}{n^6}}\cos\left(10nt - \tan^{-1}\frac{n^2+1}{4\pi^3}\right)$,

(b) $10 + \sum_{n=1}^{\infty}\sqrt{\frac{16}{(n^2+1)} + \frac{1}{n^6}}\sin\left(10nt + \tan^{-1}\frac{4n^3}{n^2+1}\right)$

17.17 (a) neither odd nor even, (b) even, (c) odd, (d) even, (e) neither odd nor even

17.19 $\frac{5}{n^2\omega_o^2}\sin n\pi/2 - \frac{10}{n\omega_o}(\cos \pi n - \cos n\pi/2) - \frac{5}{n^2\omega_o^2}(\sin \pi n - \sin n\pi/2) - \frac{2}{n\omega_o}\cos n\pi - \frac{\cos \pi n/2}{n\omega_o}$

17.21 $\frac{1}{2} + \sum_{n=1}^{\infty}\frac{8}{n^2\pi^2}\left[1 - \cos\left(\frac{n\pi}{2}\right)\right]\cos\left(\frac{n\pi t}{2}\right)$

17.23 This is a design problem with multiple answers.

17.25

$$\sum_{\substack{n=1\\n=\text{odd}}}^{\infty}\left\{\begin{aligned}&\left[\frac{3}{\pi^2n^2}\left(\cos\left(\frac{2\pi n}{3}\right)-1\right)+\frac{2}{\pi n}\sin\left(\frac{2\pi n}{3}\right)\right]\cos\left(\frac{2\pi n}{3}\right)\\&+\left[\frac{3}{\pi^2n^2}\sin\left(\frac{2\pi n}{3}\right)-\frac{2}{n\pi}\cos\left(\frac{2\pi n}{3}\right)\right]\sin\left(\frac{2\pi n}{3}\right)\end{aligned}\right\}$$

17.27 (a) odd, (b) -0.045, (c) 0.3829

17.29 $2\sum_{k=1}^{\infty}\left[\frac{2}{n^2\pi}\cos(nt)-\frac{1}{n}\sin(nt)\right],\ n=2k-1$

17.31 $\omega_o'=\frac{2\pi}{T'}=\frac{2\pi}{T/\alpha}=\alpha\omega_o$

$$a_n'=\frac{2}{T'}\int_0^{T'}f(\alpha t)\cos n\omega_o't\,dt$$

Let $\alpha t=\lambda$, $dt=d\lambda/\alpha$, and $\alpha T'=T$. Then

$$a_n'=\frac{2\alpha}{T}\int_0^{T}f(\lambda)\cos n\omega_o\lambda\,d\lambda/\alpha=a_n$$

Similarly, $b_n'=b_n$

17.33 $v_o(t)=\sum_{n=1}^{\infty}A_n\sin(n\pi t-\theta_n)$ V,

$$A_n=\frac{8(4-2n^2\pi^2)}{\sqrt{(20-10n^2\pi^2)^2-64n^2\pi^2}},$$

$$\theta_n=90^\circ-\tan^{-1}\left(\frac{8n\pi}{20-10n^2\pi^2}\right)$$

17.35 $\frac{3}{8}+\sum_{n=\text{odd}}^{\infty}A_n\cos\left(\frac{2\pi n}{3}+\theta_n\right)$, where

$$A_n=\frac{\frac{6}{n\pi}\sin\frac{2n\pi}{3}}{\sqrt{9\pi^2n^2+(2\pi^2n^2/3-3)^2}},$$

$$\theta_n=\frac{\pi}{2}-\tan^{-1}\left(\frac{2n\pi}{9}-\frac{1}{n\pi}\right)$$

17.37 $\sum_{n=1}^{\infty}\frac{8(1-\cos\pi n)}{\sqrt{1+n^2\pi^2}}\cos(n\pi t-\tan^{-1}n\pi)$ V

17.39 $\frac{1}{20}+\frac{200}{\pi}\sum_{k=1}^{\infty}I_n\sin(n\pi t-\theta_n),\ n=2k-1,$

$$\theta_n=90^\circ+\tan^{-1}\frac{2n^2\pi^2-1{,}200}{802n\pi},$$

$$I_n=\frac{1}{n\sqrt{(804n\pi)^2+(2n^2\pi^2-1{,}200)}}$$

17.41 $\frac{10}{\pi}+\sum_{n=1}^{\infty}A_n\cos(2nt+\theta_n)$,

$$A_n=\frac{100}{\pi(4n^2-1)\sqrt{16n^2-40n+29}},$$

$$\theta_n=90^\circ-\tan^{-1}(2n-2.5)$$

17.43 (a) 33.91 V,
(b) 6.782 A,
(c) 203.1 W

17.45 4.263 A, 181.7 W

17.47 10%

17.49 (a) 1.5326,
(b) 1.7086,
(c) 3.061%

17.51 This is a design problem with multiple answers.

17.53 $\sum_{n=-\infty}^{\infty}\frac{0.6321e^{j2n\pi t}}{1+j2n\pi}$

17.55 $\sum_{n=-\infty}^{\infty}\frac{1+e^{-jn\pi}}{2\pi(1-n^2)}e^{jnt}$

17.57 $-3+\sum_{n=\infty,\,n\neq0}^{\infty}\frac{3}{n^3-2}e^{j50nt}$

17.59 $-\sum_{\substack{n=-\infty\\n\neq0}}^{\infty}\frac{j4e^{-j(2n+1)\pi t}}{(2n+1)\pi}$

17.61 (a) $6+2.571\cos t-3.83\sin t+1.638\cos 2t-1.147\sin 2t+0.906\cos 3t-0.423\sin 3t+0.47\cos 4t-0.171\sin 4t$, (b) 6.828

17.63 See Fig. G.38.

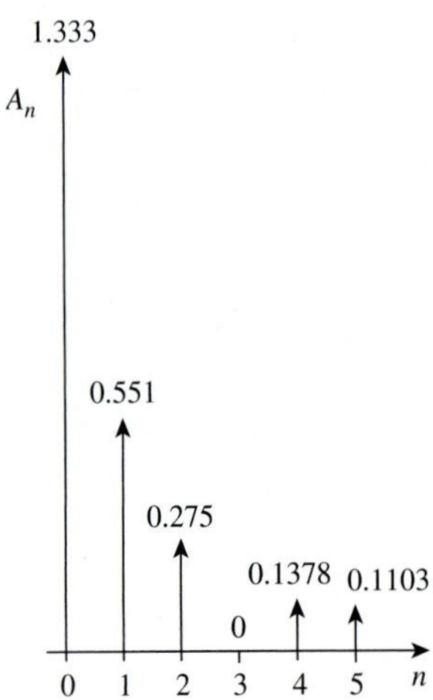

Figure G.38
For Prob. 17.63.

17.65 See Fig. G.39.

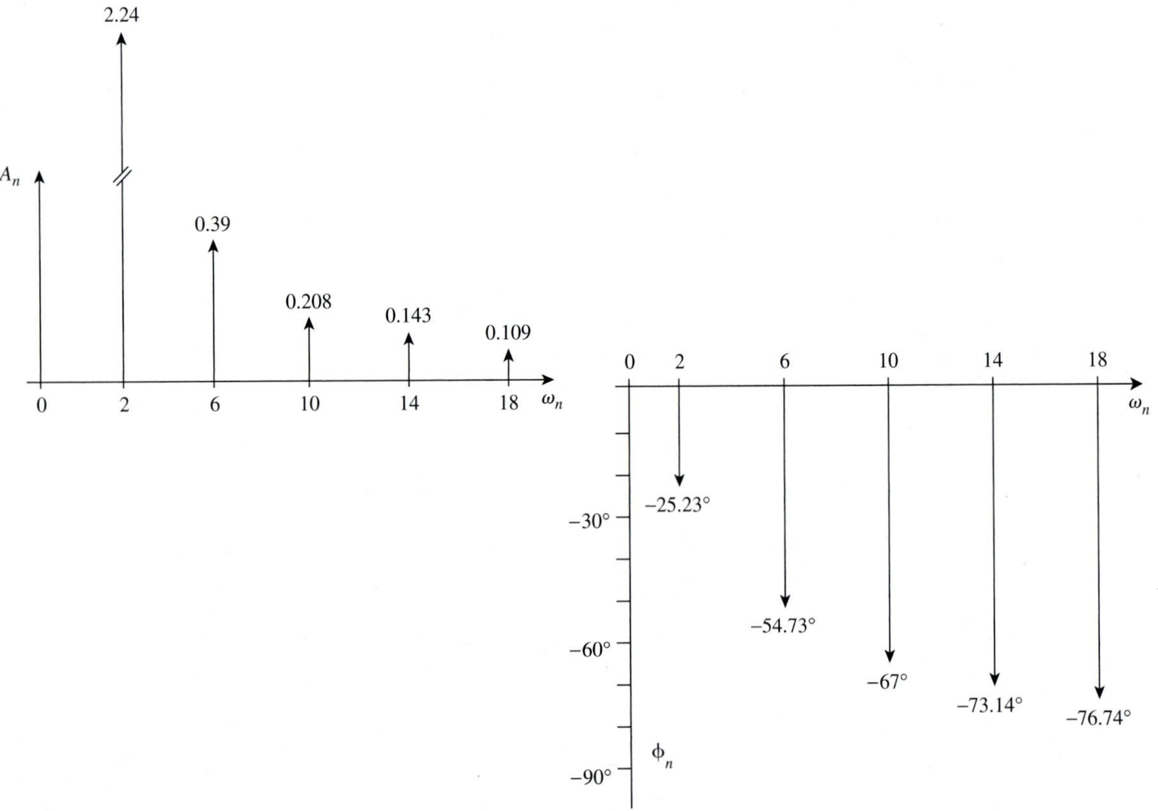

Figure G.39
For Prob. 17.65.

17.67 DC COMPONENT = 4.950000E-01

HARMONIC NO	FREQUENCY (HZ)	FOURIER COMPONENT	NORMALIZED COMPONENT	PHASE (DEG)	NORMALIZED PHASE (DEG)
1	1.667E-01	2.432E+00	1.000E+00	-8.996E+01	0.000E+00
2	3.334E-01	6.576E-04	2.705E-04	-8.932E+01	6.467E-01
3	5.001E-01	5.403E-01	2.222E-01	9.011E+01	1.801E+02
4	6.668E+00	3.343E-04	1.375E-04	9.134E+01	1.813E+02
5	8.335E-01	9.716E-02	3.996E-02	-8.982E+01	1.433E-01
6	1.000E+00	7.481E-06	3.076E-06	-9.000E+01	-3.581E-02
7	1.167E+00	4.968E-02	2.043E-01	-8.975E+01	2.173E-01
8	1.334E+00	1.613E-04	6.634E-05	-8.722E+01	2.748E+00
9	1.500E+00	6.002E-02	2.468E-02	-9.032E+01	1.803E+02

17.69

HARMONIC NO	FREQUENCY (HZ)	FOURIER COMPONENT	NORMALIZED COMPONENT	PHASE (DEG)	NORMALIZED PHASE (DEG)
1	5.000E-01	4.056E-01	1.000E+00	-9.090E+01	0.000E+00
2	1.000E+00	2.977E-04	7.341E-04	-8.707E+01	3.833E+00
3	1.500E+00	4.531E-02	1.117E-01	-9.266E+01	-1.761E+00
4	2.000E+00	2.969E-04	7.320E-04	-8.414E+01	6.757E+00
5	2.500E+00	1.648E-02	4.064E-02	-9.432E+01	-3.417E+00
6	3.000E+00	2.955E-04	7.285E-04	-8.124E+01	9.659E+00
7	3.500E+00	8.535E-03	2.104E-02	-9.581E+01	-4.911E+00
8	4.000E+00	2.935E-04	7.238E-04	-7.836E+01	1.254E+01
9	4.500E+00	5.258E-03	1.296E-02	-9.710E+01	-6.197E+00

TOTAL HARMONIC DISTORTION = 1.214285+01 PERCENT

17.71 See Fig. G.40.

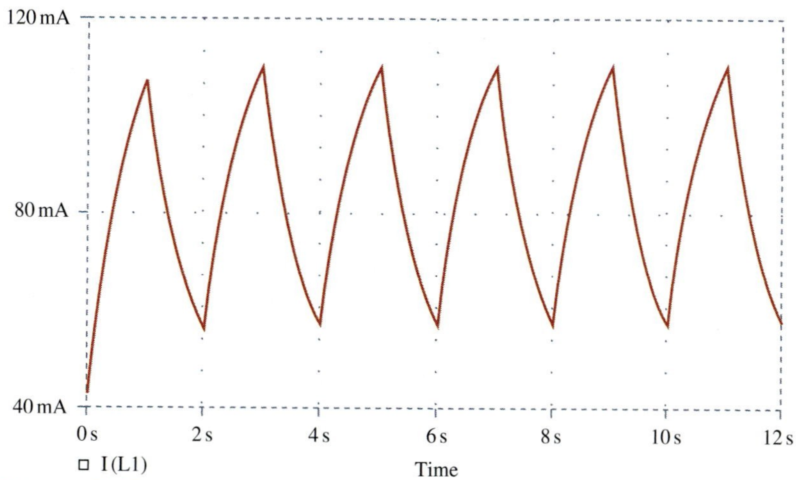

Figure G.40
For Prob. 17.71.

17.73 300 mW

17.75 24.59 mF

17.77 (a) π, (b) -2 V, (c) 11.02 V

17.79 See below for the program in *MATLAB* and the results.

```
% for problem 17.79
a = 10;
c = 4.*a/pi
for n = 1:10
  b(n)=c/(2*n-1);
end
diary
n, b
diary off
```

n	b_n
1	12.7307
2	4.2430
3	2.5461
4	1.8187
5	1.414
6	1.1573
7	0.9793
8	0.8487
9	0.7488
10	0.6700

17.81 (a) $\frac{A^2}{2}$, (b) $|c_1| = 2A/(3\pi)$, $|c_2| = 2A/(15\pi)$, $|c_3| = 2A/(35\pi)$, $|c_4| = 2A/(63\pi)$ (c) 81.1% (d) 0.72%

Chapter 18

18.1 $\frac{2(\cos 2\omega - \cos\omega)}{j\omega}$

18.3 $\frac{j}{\omega^2}(2\omega\cos 2\omega - \sin 2\omega)$

18.5 $\frac{2j}{\omega} - \frac{2j}{\omega^2}\sin\omega$

18.7 (a) $\frac{2 - e^{-j\omega} - e^{-j2\omega}}{j\omega}$, (b) $\frac{5e^{-j2\omega}}{\omega^2}(1 + j\omega 2) - \frac{5}{\omega^2}$

18.9 (a) $\frac{2}{\omega}\sin 2\omega + \frac{4}{\omega}\sin\omega$, (b) $\frac{2}{\omega^2} - \frac{2e^{-j\omega}}{\omega^2}(1 + j\omega)$

18.11 $\frac{5\pi}{\omega^2 - \pi^2}(e^{-j\omega 2} - 1)$

18.13 (a) $\pi e^{-j\pi/3}\delta(\omega - a) + \pi e^{j\pi/3}\delta(\omega + a)$, (b) $\frac{e^{j\omega}}{\omega^2 - 1}$, (c) $\pi[\delta(\omega + b) + \delta(\omega - b)] + \frac{j\pi A}{2}[\delta(\omega + a + b) - \delta(\omega - a + b) + \delta(\omega + a - b) - \delta(\omega - a - b)]$, (d) $\frac{1}{\omega^2} - \frac{e^{-j4\omega}}{j\omega} - \frac{e^{-j4\omega}}{\omega^2}(j4\omega + 1)$

18.15 (a) $2j\sin 3\omega$, (b) $\frac{2e^{-j\omega}}{j\omega}$, (c) $\frac{1}{3} - \frac{j\omega}{2}$

18.17 (a) $\frac{\pi}{2}[\delta(\omega + 2) + \delta(\omega - 2)] - \frac{j\omega}{\omega^2 - 4}$, (b) $\frac{j\pi}{2}[\delta(\omega + 10) - \delta(\omega - 10)] - \frac{10}{\omega^2 - 100}$

18.19 $\frac{j\omega}{\omega^2 - 4\pi^2}(e^{-j\omega} - 1)$

18.21 Proof

18.23 (a) $\frac{30}{(6 - j\omega)(15 - j\omega)}$, (b) $\frac{20e^{-j\omega/2}}{(4 + j\omega)(10 + j\omega)}$, (c) $\frac{5}{[2 + j(\omega + 2)][5 + j(\omega + 2)]} + \frac{5}{[2 + j(\omega - 2)][5 + j(\omega - 2)]}$, (d) $\frac{j\omega 10}{(2 + j\omega)(5 + j\omega)}$, (e) $\frac{10}{j\omega(2 + j\omega)(5 + j\omega)} + \pi\delta(\omega)$

18.25 (a) $10e^{2t}u(t)$, (b) $1.5e^{-2|t|}$, (c) $-5e^{t}u(t) + 5e^{2t}u(t)$

18.27 (a) $5\,\text{sgn}(t) - 10e^{-10t}u(t)$, (b) $4e^{2t}u(-t) - 6e^{-3t}u(t)$, (c) $2e^{-20t}\sin(30t)u(t)$, (d) $\frac{1}{4}\pi$

18.29 (a) $\frac{1}{2\pi}(1 + 8\cos 3t)$, (b) $\frac{4\sin 2t}{\pi t}$,
(c) $3\delta(t + 2) + 3\delta(t - 2)$

18.31 (a) $x(t) = e^{-at}u(t)$,
(b) $x(t) = u(t + 1) - u(t - 1)$,
(c) $x(t) = \frac{1}{2}\delta(t) - \frac{a}{2}e^{-at}\,u(t)$

18.33 (a) $\frac{2j\sin t}{t^2 - \pi^2}$, (b) $u(t - 1) - u(t - 2)$

18.35 (a) $\frac{e^{-j\omega/3}}{6 + j\omega}$, (b) $\frac{1}{2}\left[\frac{1}{2 + j(\omega + 5)} + \frac{1}{2 + j(\omega - 5)}\right]$,
(c) $\frac{j\omega}{2 + j\omega}$, (d) $\frac{1}{(2 + j\omega)^2}$, (e) $\frac{1}{(2 + j\omega)^2}$

18.37 $\frac{j\omega}{4 + j3\omega}$

18.39 $\frac{2x}{10^6 + j\omega}\left(\frac{1}{j\omega} + \frac{1}{\omega^2} - \frac{1}{\omega^2}e^{-j\omega}\right)$

18.41 $\frac{2j\omega(4.5 + j2\omega)}{(2 + j\omega)(4 - 2\omega^2 + j\omega)}$

18.43 $1000(e^{-1t} - e^{-1.25t})u(t)$ V

18.45 $5(e^{-t} - e^{-2t})u(t)$ A

18.47 $16(e^{-t} - e^{-2t})u(t)$ V

18.49 $0.542\cos(t + 13.64°)$ V

18.51 16.667 J

18.53 π

18.55 682.5 J

18.57 12.5 J, 87.41%

18.59 $(16e^{-t} - 20e^{-2t} + 4e^{-4t})u(t)$ V

18.61 $2X(\omega) + 0.5X(\omega + \omega_0) + 0.5X(\omega - \omega_0)$

18.63 106 stations

18.65 6.8 kHz

18.67 200 Hz, 5 ms

18.69 35.24%

Chapter 19

19.1 $\begin{bmatrix} 4 & 1 \\ 1 & 1.667 \end{bmatrix}\Omega$

19.3 $\begin{bmatrix} 4 + j6 & j6 \\ j6 & -j4 \end{bmatrix}\Omega$

19.5 $\begin{bmatrix} \frac{2(s^2 + s + 1)}{s^3 + 2s^2 + 3s + 1} & \frac{2}{s^3 + 2s^2 + 3s + 1} \\ \frac{2}{s^3 + 2s^2 + 3s + 1} & \frac{2(s^2 + 2s + 2)}{s^3 + 2s^2 + 3s + 1} \end{bmatrix}\Omega$

19.7 $\begin{bmatrix} 29.88 & 3.704 \\ -70.37 & 11.11 \end{bmatrix}\Omega$

19.9 $\begin{bmatrix} 5 & 2.5 \\ 2.5 & 6.25 \end{bmatrix}\Omega$

19.11 See Fig. G.41.

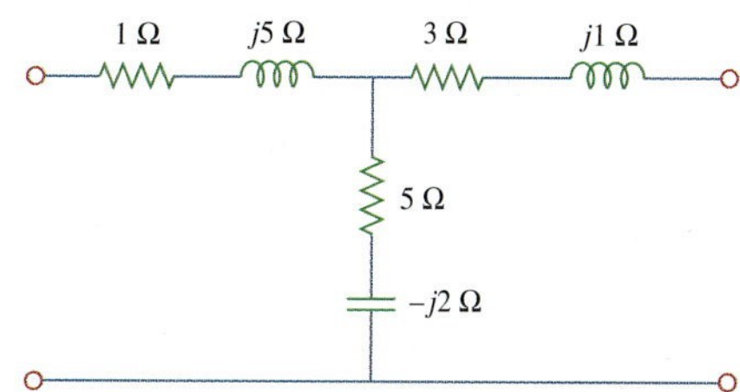

Figure G.41
For Prob. 19.11.

19.13 329.9 W

19.15 24 Ω, 384 W

19.17 $\begin{bmatrix} 4.8 & -0.4 \\ -0.4 & 4.2 \end{bmatrix}\Omega$, $\begin{bmatrix} 0.21 & 0.02 \\ 0.02 & 0.24 \end{bmatrix}$ S

19.19 This is a design problem with multiple answers.

19.21 See Fig. G.42.

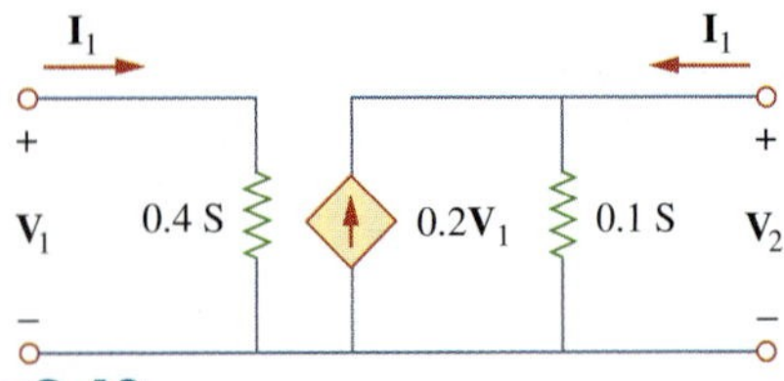

Figure G.42
For Prob. 19.21.

19.23 $\begin{bmatrix} s+2 & \dfrac{-(s+1)}{\dfrac{s^2+s+1}{s}} \\ -(s+1) & \end{bmatrix}, \dfrac{0.8(s+1)}{s^2+1.8s+1.2}$

19.25 See Fig. G.43.

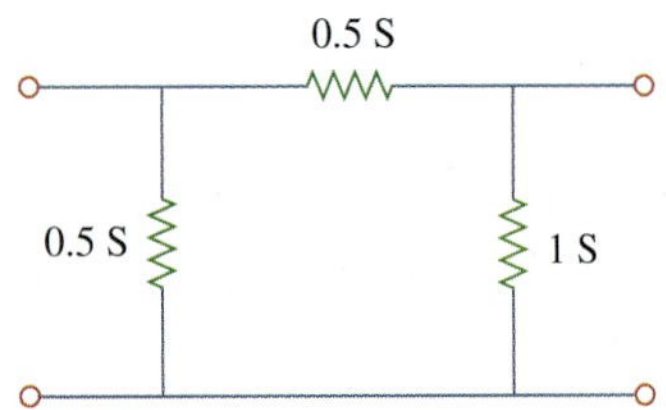

Figure G.43
For Prob. 19.25.

19.27 $\begin{bmatrix} 0.25 & 0.025 \\ 5 & 0.6 \end{bmatrix}$ S

19.29 (a) 22 V, 8 V, (b) same

19.31 $\begin{bmatrix} 3.8\ \Omega & 0.4 \\ -3.6 & 0.2\ \text{S} \end{bmatrix}$

19.33 $\begin{bmatrix} 3.077 + j1.2821 & 0.3846 - j0.2564 \\ -0.3846 + j0.2564 & 0.0769 + j0.2821 \end{bmatrix}$

19.35 $\begin{bmatrix} 2\ \Omega & 0.5 \\ -0.5 & 0 \end{bmatrix}$

19.37 1.1905 V

19.39 $g_{11} = \dfrac{1}{R_1 + R_2},\ g_{12} = -\dfrac{R_2}{R_1 + R_2}$
$g_{21} = \dfrac{R_2}{R_1 + R_2},\ g_{22} = R_3 + \dfrac{R_1 R_2}{R_1 + R_2}$

19.41 Proof

19.43 (a) $\begin{bmatrix} 1 & \mathbf{Z} \\ 0 & 1 \end{bmatrix}$, (b) $\begin{bmatrix} 1 & 0 \\ \mathbf{Y} & 1 \end{bmatrix}$

19.45 $\begin{bmatrix} 1 - j0.5 & -j4\ \Omega \\ 0.125\ \text{S} & 1 \end{bmatrix}$

19.47 $\begin{bmatrix} 0.3235 & 1.176 \\ 0.02941 & 0.4706 \end{bmatrix}$

19.49 $\begin{bmatrix} \dfrac{2s+1}{s} & \dfrac{1}{s}\ \Omega \\ \dfrac{(s+1)(3s+1)}{s}\ \text{S} & 2 + \dfrac{1}{s} \end{bmatrix}$

19.51 $\begin{bmatrix} 2 & 2 + j5 \\ j & -2 + j \end{bmatrix}$

19.53 $z_{11} = \dfrac{A}{C},\ z_{12} = \dfrac{AD - BC}{C},\ z_{21} = \dfrac{1}{C},\ z_{22} = \dfrac{D}{C}$

19.55 Proof

19.57 $\begin{bmatrix} 3 & 1 \\ 1 & 7 \end{bmatrix} \Omega,\ \begin{bmatrix} \dfrac{7}{20} & \dfrac{-1}{20} \\ \dfrac{-1}{20} & \dfrac{3}{20} \end{bmatrix} \text{S},\ \begin{bmatrix} \dfrac{20}{7}\ \Omega & \dfrac{1}{7} \\ \dfrac{-1}{7} & \dfrac{1}{7}\ \text{S} \end{bmatrix},$
$\begin{bmatrix} \dfrac{1}{3}\ \text{S} & \dfrac{-1}{3} \\ \dfrac{1}{3} & \dfrac{20}{3}\ \Omega \end{bmatrix},\ \begin{bmatrix} 7 & 20\ \Omega \\ 1\ \text{S} & 3 \end{bmatrix}$

19.59 $\begin{bmatrix} 16.667 & 6.667 \\ 3.333 & 3.333 \end{bmatrix} \Omega,\ \begin{bmatrix} 0.1 & -0.2 \\ -0.1 & 0.5 \end{bmatrix} \text{S},$
$\begin{bmatrix} 10\ \Omega & 2 \\ -1 & 0.3\ \text{S} \end{bmatrix},\ \begin{bmatrix} 5 & 10\ \Omega \\ 0.3\ \text{S} & 1 \end{bmatrix}$

19.61 $\begin{bmatrix} \dfrac{10}{3} & \dfrac{8}{3} \\ \dfrac{8}{3} & \dfrac{10}{3} \end{bmatrix} \Omega,\ \begin{bmatrix} \dfrac{6}{5}\ \Omega & \dfrac{4}{5} \\ \dfrac{-4}{5} & \dfrac{3}{10}\ \text{S} \end{bmatrix},\ \begin{bmatrix} \dfrac{5}{4} & \dfrac{3}{2}\ \Omega \\ \dfrac{3}{8}\ \text{S} & \dfrac{5}{4} \end{bmatrix}$

19.63 $\begin{bmatrix} 0.8 & 2.4 \\ 2.4 & 7.2 \end{bmatrix} \Omega$

19.65 $\begin{bmatrix} \dfrac{1}{3} & -\dfrac{1}{3} \\ -\dfrac{1}{3} & \dfrac{2}{3} \end{bmatrix}$ S

19.67 $\begin{bmatrix} 4 & 63.29 \\ 0.1576 & 4.994 \end{bmatrix}$

19.69 $\begin{bmatrix} \dfrac{s+1}{s+2} & \dfrac{-(3s+2)}{2(s+2)} \\ \dfrac{-(3s+2)}{2(s+2)} & \dfrac{5s^2+4s+4}{2s(s+2)} \end{bmatrix}$

19.71 $\begin{bmatrix} 2 & -3.334 \\ 3.334 & 20.22 \end{bmatrix}\ \Omega$

19.73 $\begin{bmatrix} 14.628 & 3.141 \\ 5.432 & 19.625 \end{bmatrix}$

19.75 (a) $\begin{bmatrix} 0.3015 & -0.1765 \\ 0.0588 & 10.94 \end{bmatrix}$, (b) -0.0051

19.77 $\begin{bmatrix} 0.9488\angle -161.6^\circ & 0.3163\angle 18.42^\circ \\ 0.3163\angle -161.6^\circ & 0.9488\angle -161.6^\circ \end{bmatrix}$

19.79 $\begin{bmatrix} 4.669\angle -136.7^\circ & 2.53\angle -108.4^\circ \\ 2.53\angle -108.4^\circ & 1.789\angle -153.4^\circ \end{bmatrix}\ \Omega$

19.81 $\begin{bmatrix} 1.5 & -0.5 \\ 3.5 & 1.5 \end{bmatrix}$ S

19.83 $\begin{bmatrix} 0.3235 & 1.1765 \\ 0.02941 & 0.4706 \end{bmatrix}$

19.85 $\begin{bmatrix} 1.581\angle 71.59^\circ & -j \\ j & 5.661 \times 10^{-4} \end{bmatrix}$

19.87 $\begin{bmatrix} -j1{,}765 & -j1{,}765 \\ j888.2 & j888.2 \end{bmatrix}$

19.89 $-1{,}613$, 64.15 dB

19.91 (a) -25.64 for the transistor and -9.615 for the circuit.

19.93 -17.74, 144.5, 31.17 Ω, $-$ 6.148 MΩ

19.95 See Fig. G.44.

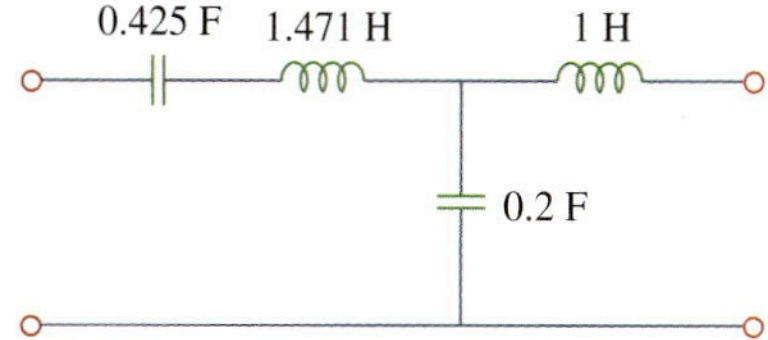

Figure G.44
For Prob. 19.95.

19.97 0.25 F, 0.3333 H, 0.5 F

19.99 Proof

Poularikas, A. D., (ed.). *The Transforms and Applications Handbook*. Boca Raton, FL: CRC Press, 2nd ed., 1999.

Ridsdale, R. E. *Electric Circuits*. 2nd ed. New York: McGraw-Hill, 1984.

Sander, K. F. *Electric Circuit Analysis: Principles and Applications*. Reading, MA: Addison-Wesley, 1992.

Scott, D. *Introduction to Circuit Analysis: A Systems Approach*. New York: McGraw-Hill, 1987.

Smith, K. C., and R. E. Alley. *Electrical Circuits: An Introduction*. New York: Cambridge Univ. Press, 1992.

Stanley, W. D. *Transform Circuit Analysis for Engineering and Technology*. 3rd ed. Upper Saddle River, NJ: Prentice Hall, 1997.

Strum, R. D., and J. R. Ward. *Electric Circuits and Networks*. 2nd ed. Englewood Cliffs, NJ: Prentice Hall, 1985.

Su, K. L. *Fundamentals of Circuit Analysis*. Prospect Heights, IL: Waveland Press, 1993.

Thomas, R. E., and A. J. Rosa. *The Analysis and Design of Linear Circuits*. 3rd ed. New York: John Wiley & Sons, 2000.

Tocci, R. J. *Introduction to Electric Circuit Analysis*. 2nd ed. Englewood Cliffs, NJ: Prentice Hall, 1990.

Tuinenga, P. W. *SPICE: A Guide to Circuit Simulation*. Englewood Cliffs, NJ: Prentice Hall, 1992.

Whitehouse, J. E. *Principles of Network Analysis*. Chichester, U.K.: Ellis Horwood, 1991.

Yorke, R. *Electric Circuit Theory*. 2nd ed. Oxford, U.K.: Pergamon Press, 1986.

Index

A

B

C

D

E

F

G

H

I

J

K

L

M

N

O

P

Q

R

S

T

U

V

W

Y

Z